DARWIN
Katherine
NORTHERN
TERRITORY
Tennant Creek
Alice Springs
GULF OF CARPENTARIA
CAPE YORK PENINSULA
Cairns
Townsville
Mount Isa
QUEENSLAND
Winton
Clermont
Rockhampton
Roma
BRISBANE
PACIFIC
OCEAN
Birdsville
Lake Eyre
Coober Pedy
SOUTH AUSTRALIA
Ceduna
Port Augusta
Broken Hill
Bourke
Darling R.
Moree
Grafton
NEW SOUTH WALES
Newcastle
SYDNEY
Mildura
Murrumbidgee R.
Murray R.
CANBERRA
Mt. Kosciusko
VICTORIA
MELBOURNE
ADELAIDE
Port Lincoln
Kangaroo Is.
Mount Gambier
LIAN BIGHT
TASMANIA
HOBART
YETS (mm)
N ANNUAL TEMPERATURE (°C)
0 300 400 500 600 700 800 900 1000 Kilometres
elaide, 1980

JOHN DIXON
1984
$92.92

SOILS

an Australian viewpoint

SOILS
an Australian viewpoint

Sponsored by
The Division of Soils
Commonwealth Scientific and Industrial
Research Organization, Australia

CSIRO/Academic Press

Citation of this book should take the following form:

Division of Soils, CSIRO (1983). 'Soils: an Australian viewpoint', xx + 928 pp. CSIRO: Melbourne/Academic Press: London.

Individual chapters should be cited as follows:

Author's name. (1983). Title of chapter. *In* 'Soils: an Australian viewpoint', Division of Soils, CSIRO, pp. 0-00. (CSIRO: Melbourne/Academic Press: London.)

Published jointly by CSIRO, 314 Albert Street, East Melbourne, Australia, 3002, and Academic Press Inc. (London) Ltd, 24-28 Oval Road, London NW1, U.K.

National Library of Australia Cataloguing-in-Publication Entry

Soils, an Australian viewpoint.

Bibliography.
Includes index.
ISBN 0 643 00400 9 CSIRO.
ISBN 0 12 654240 6 Academic Press.

1. Soils – Australia. I. Commonwealth Scientific and Industrial Research Organization (Australia). Division of Soils.

631.4'994

Production editors: John J. Lenaghan and Georgina Katsantoni. Typographic design: John Best

Typeset in Times and Helvetica by Jacobson Typesetters Ltd, Auckland, New Zealand, and printed at Griffin Press Limited, Adelaide, South Australia for CSIRO, Australia, and Academic Press, London, New York, Paris, San Diego, San Francisco, São Paulo, Tokyo, Toronto.

They are ill discoverers that think there is no land, when they can see nothing but sea.

FRANCIS BACON

Contents

Classification and mapping of Australian soils

Physical and chemical characteristics of Australian soils

Mineralogy

Chemistry

Physics and hydrology

Organic matter and its decomposition

Soil fauna

Soil/microbial/plant relationships

Soils and land management

List of contributors

Most of those who have contributed to this book are staff members of the Division of Soils, CSIRO, and their addresses can be found by reference to the list below of addresses of the Division's laboratories. All other authors' addresses are listed with their names.

Addresses of laboratories of the Division of Soils, CSIRO.

(1) Division of Soils, CSIRO,
Private Mail Bag No. 2, Glen Osmond, S.A. 5064.

(2) Division of Soils, CSIRO,
P.O. Box 639, Canberra City, A.C.T. 2601.

(3) Division of Soils, CSIRO,
Stowell Avenue, Hobart, Tas. 7000.

(4) Division of Soils, CSIRO,
Cunningham Laboratory, Mill Road, St Lucia, Qld 4067.

(5) Division of Soils, CSIRO,
Private Mail Bag, P.O. Aitkenvale, Qld 4814.

G. D. Aitchison
c/- Faculty of Science,
Monash University,
Clayton, Vic. 3168

G. B. Allison (1)

G. G. Beckmann (4)

R. S. Beckwith (1)

E. Bettenay
Division of Land
Resources Management, CSIRO,
Private Mail Bag,
Wembley, W.A. 6014

G. Blackburn (1)

G. D. Bowen (1)

R. Brewer (2)

B. J. Bridge (4)

B. E. Butler (2)

J. H. A. Butler (1)

H. M. Churchward
Division of Land
Resources Management, CSIRO,
Private Mail Bag,
Wembley, W.A. 6014

J. S. Colville (1)

J. D. Colwell (2)

R. A. Date
Division of Tropical
Crops and Pastures, CSIRO,
Cunningham Laboratory,
Mill Road,
St Lucia, Qld 4067

W. W. Emerson (1)

A. W. Fordham (1)

R. C. Foster (1)

F. R. Gibbons
Soil Conservation Authority,
378 Cotham Road,
Kew, Vic. 3101

G. P. Gillman (5)

E. L. Greacen (1)

P. Greenslade
c/-South Australian Museum,
North Terrace,
Adelaide, S.A. 5000

P. J. M. Greenslade (1)

R. H. Gunn
Division of Land
Use Research, CSIRO,
P.O. Box 1666,
Canberra City, A.C.T. 2601

J. R. Harris (1)
G. D. Hubble (4)
B. R. Hutson (1)
J. T. Hutton (1)
R. F. Isbell (5)
R. B. Jackson (1)
R. W. L. Kimber (1)
J. N. Ladd (1)
K. E. Lee (1)
J. Loveday (2)
A. E. Martin (1)
J. K. Martin (1)

W. M. McArthur
Division of Land
Resources Management, CSIRO,
Private Mail Bag,
Wembley, W.A. 6014

R. M. McKenzie (1)
A. R. Milnes (1)
A. W. Moore (4)
K. D. Nicolls (3)
K. Norrish (1)
K. H. Northcote (1)
J. G. Pickering (1)

J. A. Prescott
2 Netherby Avenue,
Netherby, S.A. 5062

M. E. Probert (5)
M. Raupach (1)
R. Reeve (4)
E. H. Ridge (1)
H. Rosser (1)
A. D. Rovira (1)

J. S. Russell
Division of Tropical
Crops and Pastures, CSIRO,
Cuningham Laboratory,
Mill Road,
St Lucia, Qld 4067

J. R. Sleeman (2)
A. V. Spain (5)
R. J. Swaby (1)
J. K. Taylor
(Died: 11 January 1982)

R. M. Taylor (1)
C. H. Thompson (4)
K. G. Tiller (1)
B. M. Tucker (1)
P. H. Walker (2)

J. H. Warcup
Waite Agricultural
Research Institute,
Department of Plant Pathology,
Private Mail Bag No. 1,
Glen Osmond, S.A. 5064

W. T. Ward (4)
C. B. Wells (1)
I. R. Willett (2)

C. H. Williams
Division of Plant
Industry, CSIRO,
P.O. Box 1600,
Canberra City, A.C.T. 2601

J. Williams (5)
M. J. Wright (1)

Colour plates for Chapter 3

Plate 3.1. Shallow stony soils (loamy *lithosols*) formed on steeply dipping Proterozoic shales near Mt Isa, north-western Queensland.

Plate 3.2. Ironstone gravel soils (Yalanbee series, KS-Uc 4.21) developed on laterite in the Darling Ranges, south-west Western Australia. All the sandy horizons contain more than 60% of ironstone gravel up to 2 cm diameter.

Plate 3.3. A deep sand profile (*siliceous sand*) formed on granitic colluvium near Coen, Cape York Peninsula, Queensland.

Plate 3.4. A cracking clay (*black earth*) showing prominent slickensides. The soil is formed from basalt, near Collinsville in northern Queensland.

Plate 3.5. A *solonized brown soil* from Coomealla, New South Wales, formed in sandy calcareous materials following aeolian activity. The parent material has had a complicated history.

Plate 3.6. A massive sesquioxidic soil (*yellow earth*) from northern Cape York Peninsula, Queensland. The soil is strongly red-mottled at depth.

Plate 3.7. A structured sesquioxidic soil (*krasnozem*) derived from basalt on the Atherton Tableland, northern Queensland. Note the very strong polyhedral structure in the B horizon, the detached units of which have formed a scree slope at the bottom of the photo, a characteristic feature of such soils.

Plate 3.8. A texture-contrast soil with a weak A_2 horizon. This is typical non-sodic *red-brown earth* (the Urrbrae loam), as originally defined by Prescott.

Plate 3.9. A texture-contrast soil with a strongly developed bleached A_2 horizon. This is a *solodic soil* developed on alluvium in central Queensland. Note the prominent prismatic structure in the heavy clay B horizon.

Plate 3.10. A texture-contrast soil with nodular laterite (ferruginous zone) and prominent deep subsoil development (mottled zone). This is a *lateritic podozolic soil* formed in sandstone, north of Brisbane, south-east Queensland.

Plate 3.11. Soils with sesquioxide-humus B horizons. This is a typical *podzol* formed on siliceous sand dunes at Cooloola, south-east Queensland.

Plate 3.12. An *alpine humus soil* formed on granitic rock in the Mt Kosciusko area of south-eastern Australia. The black organic loam surface soil often contains 15-20% organic matter.

Plate 3.13. A highly saline soil (*solonchak*) from tidal flats near Townsville, northern Queensland.

Plate 3.14. The mottled zone of a truncated laterite profile exposed in the Sutton River, central Queensland.

(*a*)

Plate 3.15. (*a*) Laterite profile showing red structured *krasnozem* soil grading into the ferruginous zone of vesicular laterite (at top of scale) and then into the mottled zone of white and red clays (at the bottom of the scale). Near Brisbane, south-eastern Queensland. (*b*) Close-up of the vesicular laterite and the top of the mottled zone.

(*b*)

Preface

MUCH HAS BEEN WRITTEN ABOUT AUSTRALIAN SOILS IN scientific journals, review articles and, occasionally, in the popular press. Two books that have dealt effectively with Australian soils are G. W. Leeper's 'Introduction to Soil Science', published in 1948, and Janice Corbett's 'The Living Soil', published in 1969, but even though these volumes serve a useful purpose, there is need for a more modern and comprehensive text on our soils and their relationships with the soils of the world; it was to meet this need, as well as to mark the first 50 years of soil research in the Division of Soils, CSIRO, that the present volume was planned.

This book is addressed mainly to professional soil scientists and to postgraduate research workers in soil science and related fields, but we hope that its contents will be useful to all students of soils in tertiary institutions or in industry.

You will see that the book is divided into a number of sections that deal with the occurrence and genesis of Australian soils, their chemistry, mineralogy, physics and biology; some final chapters concern soil and land management. Each section is relatively self contained, so we also hope that some of the material will be of interest to workers outside the field of soil science. At the same time, and in keeping with the view that pedology constitutes an important frame of reference for soil science and its application to the real world, we have attempted throughout to relate soil properties and processes to particular soils or soil groups, based on the Australian Great Soil Group classification. This was the preferred scheme; it is still in common use in Australia and was employed in the 'Handbook of Australian Soils' published in 1968. Soil taxonomists will find, in Chapter 22, approximate correlations between the Great Soil Group nomenclature and those used in the Northcote Factual Key, the U.S. Soil Taxonomy and the FAO-UNESCO classification.

To some, soil science is a discipline in its own right; to others, it is a meeting ground where many disciplines of the natural sciences can collaborate. Each approach has two aims: firstly to provide a scientific basis for land use and land management; secondly to add to the store of knowledge about world soils on which our survival depends. This book is a record of our experience, so far, in documenting, characterizing and managing the fragile soil resources of an ancient continent that is often bedevilled by a harsh and capricious climate.

The production of a multi-author and multidisciplinary volume such as this is an enormous task. A number of colleagues were concerned at various stages of the project, but special thanks are due to K. E. Lee, who was closely involved in the production, from the initial planning and organization of the chapters to the finalization of the typescript before despatch to the publishers: his systematic dedication to this task enabled this book to appear only 4 years after its conception. In this he was ably assisted by R. F. Isbell, who not only contributed substantially as an author, but undertook a significant overview of many phases of the book during its preparation and ensured as far as possible an integrated format. The task of editing all drafts of each chapter was shared between K. E. Lee, R. F. Isbell, K. A. Handreck, C. B. Wells and M. Raupach, and their cooperation in this important and often frustrating assignment is gratefully acknowledged. K. G. Tiller, J. Loveday, A. D. Rovira and R. F. Isbell undertook the management of major sections, while David Wright and his colleagues in the Division's Publications Section were responsible for the skilled preparation of the final drafts, figures and map. Finally, I should like to thank the 67 authors for their cooperation in meeting strict deadlines while being distracted from their research.

A. E. Martin
Chief, Division of Soils,
Adelaide

Chapter 1
The Australian environment

J. K. Taylor

The soils of Australia mirror the continent's great age, and are in many cases unique in the world's pedology.

Australia's latitudinal position dictates much of its climatic character (40% lies within the tropics), and its modest relief and insular nature produce broadly transitional climatic features (see Gentilli (1971) and the Bureau of Meteorology (1974) for detailed reviews).

It is a relatively dry continent, with 50% of its area having a median rainfall of < 300 mm per year, and 80% < 600 mm; the wetter areas are restricted to the north-eastern tropical coast and western Tasmania. Rainfall is the dominant component of precipitation over the major part of the continent and snowfall is confined to some highland areas in south-eastern Australia and Tasmania. Over much of the continent the rainfall pattern is strongly seasonal in character, with a winter rainfall regime in the south and a summer regime in the north. Variability of rainfall is generally high, particularly over the more arid interior and in parts of the north-east and north-west coasts, where the irregular incidence of tropical cyclones may cause great variation in yearly totals.

The temperature regime is essentially continental, with the high temperatures of the inland reaching almost to the coastline in some areas. Average annual air temperatures range from 29°C along the north-west coast area, to 5°C in the alpine areas of the south east. In January, average maximum temperatures exceed 36°C over vast areas of the interior and the north-west. In July, average minima fall below 6°C in areas south of the tropic and away from the coasts. Winter frosts may occur as far north as approximately 17°S latitutde.

Average annual evaporation is high over much of the continent, exceeding 2000 mm over about 75% of the total area. Over a similar area, rainfall does not exceed evaporation loss from a free water surface in any month of the year.

The present climate of much of Australia has little relevance dynamically to the character of the soils; much of the extensive lateritic remnants over large parts of the continent date from the Early to Mid-Tertiary, the great alluvial plains extending from central Queensland to Victoria were laid down in the Late Cainozoic, and much of the mallee landscape of South Australia accumulated by aeolian deposition in the Quaternary. This is in great contrast to conditions in the northern hemisphere, where much of the landscape familiar to soil scientists has been the outcome of relatively recent glacial periods.

The Australian inland has been dominated by arid and desert conditions since the end of the Tertiary. Since much of Australia has an unreliable annual rainfall, the current changes occurring in most soils are much more anthropogenic than climatic, especially considering the prevailing levels of evaporation.

The native vegetation of Australia is dictated, both in form and distribution, more by rainfall than by any other physical factor, so that there is a general sequence of plant formations related to broad climatic regions (Specht 1970; Carnahan 1976). These range from hummock grassland of the extreme arid regions through low shrubland or tussock grassland, to tall shrubland or scrub of the semi-arid regions, to woodland or low open-forest in the subhumid regions, to open forest in the humid regions, and eventually to closed forest in the wettest sites. However, much divergence, both floristically and structurally, from this climatic zonation arises from changes in soil and topography.

Australian trees and shrubs are predominantly evergreen, and their leaves are in general hard and tough (sclerophyllous). A few distinctively Australian genera, notably *Eucalyptus*, *Acacia* and *Triodia* (with the closely related *Plectrachne*) dominate, and give a characteristic appearance to much of the vegetation. In contrast with many other similarly arid regions of the world, the lower-rainfall areas of Australia support a high proportion of woody species, and have no large indigenous succulents.

Over much of the better-watered areas of southern and eastern Australia, the natural vegetation has either been modified, or cleared and replaced by exotic pastures, crops or forest plantations. Nevertheless, considerable areas of various types of vegetation have been preserved in a relatively natural condition.

Australia is a land of low relief. The only major highland is the mountain and tableland chain following the east coast from central Victoria to Cape York Peninsula. Although shown on maps as the Great Dividing Range, it is actually more a dissected tableland up to 300 km wide, with a rugged mountainous strip on parts of its eastern side from which flow short quick-running streams to the coast; on its western flank, the tableland descends in smooth dissected slopes to the belt of great plains almost continuous in extent from the Gulf of Carpentaria to the Southern Ocean. Westward of this, most of South Australia, the Northern Territory and Western Australia is occupied by a surface of ancient planation, less than 450 m in elevation; this is the region of deserts. Elsewhere, in parts of the coastal belt, low ranges provide catchment for a number of mostly short streams of variable flow.

The physiography of inland Australia, combined with the associated climatic features and immense areas of arid and semi-arid lands, is allied with a very poor external drainage system. Only two drainage basins are large, and only one of them reaches the sea. The Murray Basin, as one key to the agricultural wealth of south-eastern Australia, draws its tributaries from southern Queensland, eastern New South Wales and the Victorian mountains, in all occupying 1 million km^2. This system of normally slow-flowing streams finds its single outfall to the sea in South Australia; it drains vast plains of almost unrelieved flatness and low gradient.

The second great basin, one of the biggest centres of internal drainage in the world, leads from a catchment of about 1 million km^2 to Lake Eyre. A good illustration of the deficiency in external drainage is given in these figures for Australia (Australian Water Resources Council 1976).

Area (km^2)	Average (rainfall mm)	Evapotranspiration (% of rainfall)	Run-off (% of rainfall)
7 690 000	420	87	13

No other continent approaches these levels, and the data also make clear why so much of the southern half of the country has salinity problems.

Comprehensive accounts of the landforms of Australia are given by Mabbutt (1970, 1973), and Mabbutt and Sullivan (1970).

Many of the soils of Australia are pedologically very old, the key geological period being the Tertiary. This period covers the great development of laterite and silcrete and other formations caused by deep weathering, and the planation of the land to low relief. It was also the time of several major invasions of the sea, by extensive subsidence, and of a subsequent slow uplift with deposition of new materials; as an example of the latter, immense amounts of detritus were poured out during the Late Cainozoic by ancestral streams to form the Riverine Plain in southern New South Wales and Victoria.

Two features of marked significance, both pedologically and agriculturally, and which have provided the basis for a large part of soil research, are the effects of salinity, and the widespread problem of low soil fertility. A recent compilation by Northcote and Skene (1972) has shown the occurrence of saline soils, dominated by chlorides, over great areas of southern and central Australia with a low rainfall. Even greater areas of sodic soils occur over many parts of the continent, and include a considerable proportion of agriculturally useful land in regions with favourable rainfall and under irrigation development. The effects of salinity are an issue greater in Australia than in other continental areas of the world.

The widespread poverty of Australian soils has long been recognized, for example the low level of phosphorus was realized late in the nineteenth century. Perhaps no other large country has quite the array of deficiencies of trace elements in such a diversity of soils, so excitingly investigated between 1927 and 1947 especially.

The general level of soil fertility is low; it is emphasized in the ecology, both plant and animal, and it may be associated with the possibility that the geological raw material may have been through many cycles of weathering. Development of profitable agriculture in extensive areas of previously unproductive land is one of the highest achievements in Australian agricultural research. It has been the basis for the bulk of research on soils over the last 50 years.

References

Australian Water Resources Council (1976). Review of Australia's water resources, 1975. (Aust. Gov. Print. Serv.: Canberra.)

Bureau of Meteorology (1974). Climate and Physical Geography of Australia. *In* 'Year Book' (Aust. Bur. Stat.: Canberra.)

Carnahan, J. A. (1976). 'Natural Vegetation'. Atlas of Australian Resources, Second Series (Dep. Nat. Resour.: Canberra.)

Gentilli, J. (1971). 'Climates of Australia and New Zealand'. World Survey of Climatology, Vol. 13. (Elsevier: Amsterdam.)

Mabbutt, J. A. (1970). Landforms. *In* 'The Australian Environment'. 4th edn. (Ed. G. W. Leeper.) (CSIRO–Melbourne Univ. Press: Melbourne.)

Mabbutt, J. A. (1973). 'Landforms'. Atlas of Australian Resources, Second Series. (Dep. Min. Energy: Canberra.)

Mabbutt, J. A. and Sullivan, M. E. (1970). Landforms and Structure. *In* 'Australian Grasslands'. (Ed. R. Milton Moore.) (Aust. Nat. Univ. Press: Canberra.)

Northcote, K. H. and Skene, J. K. M. (1972). Australian soils with saline and sodic properties. CSIRO Aust., *Soil Publ.* No. 27.

Specht, R. L. (1970). Vegetation. *In* 'The Australian Environment'. 4th edn. (Ed. G. W. Leeper.) (CSIRO–Melbourne Univ. Press: Melbourne.)

Chapter 2

The origins and early development of soil science in Australia

C. B. Wells and J. A. Prescott

PERMANENT EUROPEAN SETTLEMENT AND THE introduction of agriculture came to Australia in 1788 (Australia 1979); concern for, and interest in, its soils came rather later.

The only interest shown in soils by the earliest settlers was in harvesting the crops that could be grown on them. This exploitation continued until all the conveniently situated land had been exhausted. As crops inevitably failed more and more often, settlers began looking for ways, particularly manures, to restore the soils' fertility. To help each other in this search, close neighbours formed Settlers Societies or Agricultural and Horticultural Societies, as had been very much in vogue in England from about 1760 (Prescott 1977; Orwin and Whetham 1964). There are records of such societies appearing in New South Wales as early as 1822 (Anon 1977), and many more appeared there and in the other colonies later on.

In the course of time, parliaments were persuaded to make grants in aid to the societies through central honorary boards that were established to distribute the money and to collect and disseminate information gained by local or foreign experience and experiment in agricultural practices (Victoria 1859; South Australia 1889; Tasmania 1891; Western Australia 1894*a*). The boards in many cases decided to collect independent information on manures and local soils, and engaged men trained to study and analyse them. This can be regarded as the beginning of soil science in Australia.

Agricultural chemistry and soil fertility

Some of the earliest work in soil science was published by Strzelecki (1845). From samples collected from farms in New South Wales and Van Diemen's Land (now Tasmania) in the early 1840s, he appended to his section on agriculture "a detailed examination of forty-one different soils . . . with the view . . . of contributing some fresh data to the important question relating to the causes of the fertility or sterility of soils in general" (Strzelecki 1845, pp. 386–7). Being a reputable scientist, he took some care to describe his sampling technique and the measurements he made (Strzelecki 1845, pp. 387–8), he assembled the data (Figs 2.1–3), and discussed (Strzelecki 1845, pp. 431–41) the conclusions to be drawn from them, adverting particularly to the

NEW SOUTH WALES.

SOIL No. 1.

The locality of this soil is an old basin in the valley of the Karua; it is taken from Strout (Garden of Eden), farm of the Australian Agricultural Company. It is an alluvial soil, of a clayey nature, formed from the adjacent hills, which are composed of greenstones and porphyries. It is light, soft to the touch, easily crushed between the fingers, of sufficient cohesion to resist drought, and of sufficient permeability to allow the quick escape of the superfluous rain-water. Its colour is dark brown; it is very easily worked, has not been manured, and is every second year under a crop of wheat. Its subsoil is a clayey diluvium. The acre produces 20 bushels of wheat. Specific gravity, 1·9.

Physical Character.

Power of absorption of solar heat	+ 14
Power of terrestrial radiation	— 4
Capacity for moisture	+ 6

Chemical Character.

Solubility of 100 parts:—	
Soluble in acids	19·74
Insoluble in acids	80·26
	100·00
Constituents in 100 parts:—	
Animal and vegetable matter	18·44
Water	6·53
Silica	56·63
Alumina	10·15
Peroxide of iron	4·00
Carbonate of lime	2·70
Sulphate of lime	1·20
Magnesia (traces).	
Loss	0·35
	100·00

Fig. 2.1. Description and analysis of a farmed soil, one of 41 such descriptions published by Strzelecki (1845, p. 389) for New South Wales and Van Diemen's Land.

SOIL No. 4.

This soil is the subsoil of the soil No. 3., and identical with it as regards its external character. It is taken at a foot deeper from the surface.

Chemical Character.

Solubility of 100 parts:—	
Soluble in acids - - - -	23·26
Insoluble in acids - - - -	76·74
	100·00
Constituents in 100 parts:—	
Animal and vegetable matter - -	8·10
Water - - - - -	3·00
Silica - - - - -	65·40
Alumina - - - - -	8·90
Peroxide of iron - - - -	4·50
Carbonate of lime - - - -	6·50
Sulphate of lime - - - -	2·50
Potash and soda (traces).	
Magnesia - - - - -	1·10
	100·00

Fig. 2.2. Description and analysis of a farmed soil (Strzelecki 1845, p. 391–2). Note particularly that this is a subsoil sample.

writings of Leibig and Boussingault. He also made a foray into the classification of the soils (Strzelecki 1845, pp. 362–3), the variety of which he "traced not only to the rocks characteristic of each colony, but to their respective configuration, and to the greater or lesser *denudation* and *renovation* of the surface, consequent upon that configuration"; Australian pedologists returned to this theme a century later (Stephens 1946). Strzelecki's two major classes were "the *first*, impoverished by denudation . . . the *last*, enriched by the drift".

As early as 1859 in Victoria, a careful description of the soils and subsoils of the newly established Experimental Farm at Royal Park, and the fullest information as to the chemical constituents of the soils, were called for by the Farm Committee of the Council of Agriculture (Victoria 1860–61). The Director of the Experimental Farm supplied the information, in part marked on a soil map, which received the Council's attention on 21 March 1860. The Council also had analyses made of various guanos of the Colony (Fig. 2.4).

This interest in soils was sustained and actively pursued. The Victorian Government Analyst reported in a 10-page document on 14 August 1863 the results of his analyses of 20 samples of soils, with a further document in 1864 of 10 pages on 20 virgin soils of Victoria and a 25-page general treatise on chemical analyses and constituents of crops and soils. The report includes such statements as "Fertility of a soil depends mainly upon the amount of phosphates, alkalies, and alkaline earths . . .", and "Of the former of these, viz. the phosphates, all the samples of soil examined contained very small proportions, varying from 1½ grains in 1 lb. . . . down to the barest traces. In the samples of subsoils . . . no essential difference . . . from the former". On 12 June 1866 the Government Analyst reported on a phosphoric-potash manure, and proposed experiments comparing it with guano, farm yard manure, and earth-closet manure (Victoria 1862–63, 1864–65, 1867).

There was no intention that this should be mere routine work; and, there being no satisfactory liaison then with the University of Melbourne, one parliamentarian in 1876 wished to attach a professor of chemistry to the newly formed Victorian Department of Agriculture who could analyse soils and determine the crops for which they had affinity (Victoria 1875–76). Three years earlier, the Secretary for Agriculture in his first report in 1873 had sought the appointment of an agricultural chemist "with a view to discover of what elements of fertility they (soils) have been exhausted.". He continued, "There is a wide difference between chemical research and routine analysis . . . It is the former which we require in agriculture, the latter is of little avail." (Victoria 1873). Then, as now, the gulf between professional opinion and political provision intervened; ten years later in 1884 the Secretary for Agriculture was still begging for his chemist: "The usefulness of the Department would be greatly increased by the appointment of a chemist . . . By analysis of soils, they (agriculturalists) would be guided as to the best uses to which they should place their land, and how to rectify any deficiencies. The same as regards manures; and now that irrigation is being adopted, it will be necessary to know whether the available water supply may or may not be injurious to vegetation." (Victoria 1885).

The South Australian parliament, wishing to provide farmers with a knowledge of sound and up to date farming techniques, appointed a Commission on 6 July 1875 to "inquire into the best means of providing agricultural and technical education". Among the recommendations in its report of 13 October 1875 was: "That for a moderate fee, any person engaged in cultivation should be entitled to apply to the (proposed) laboratory for an analysis of soils and manures". Another was that there should be a Model Farm near Adelaide . . . at which a Government Professor of Agriculture acquainted with analytical chemistry should give lectures and descriptive lessons on matters affecting the nature of the soils (South Australia 1875). As in Victoria, the parliamentarians took up the cry, suggesting that a University Professor of Geology could assist in analysis of soils, and in advising what manures to use (South Australia 1879).

A Government Professor of Agriculture was appointed to South Australia in 1881; part of his brief was to study the climate of South Australia, and to study the various soils. That he took the brief seriously is shown by

I.—TABLE OF AUSTRALIAN SOILS OF THE HIGHEST PRODUCTIVE POWER.

Number to which the Soil refers.	PHYSICAL CHARACTER.				CHEMICAL CHARACTER OF THE SOILS.												
						Proximate Constituents in 100 Parts.											
											LIME.						
	Absorption of Solar Rays.	Emission of Heat.	Capacity for Moisture.	Specific Gravity.	Soluble Portion of 100 Parts.	Vegetable and Animal Matter.	Water.	Silica.	Alumina.	Peroxide of Iron.	Carbonate.	Sulphate.	Potash and Soda.	Chlorides.	Magnesia.	Metallic Sulphurets and Oxides.	Loss.
No. 1	+ 14·0	− 1·0	+ 6·0	1·9	19·74	18·44	6·53	56·63	10·15	4·00	2·70	1·20	—	—	Traces.	—	0·35
No. 2	+ 11·0	− 2·5	+ 7·0	2·2	33·07	15·60	5·50	59·80	9·00	1·90	2·00	4·70	1·50	—	Traces.	—	—
No. 3	+ 7·0	− 2·0	+ 8·0	2·0	26·51	7·30	4·00	65·73	12·42	3·75	6·20	0·60	Traces.	—	—	—	—
No. 4	Subsoil not taken.				23·26	8·10	3·00	65·40	8·90	4·50	6·50	2·25	Traces.	—	1·10	—	—
No. 5	+ 12·0	− 4·0	+ 5·0	3·0	25·75	9·00	4·00	63·00	15·00	4·00	3·20	0·80	—	—	1·00	—	—
No. 6	+ 11·7	− 3·0	+ 7·0	2·2	30·30	10·00	5·20	53·70	19·20	3·20	2·50	4·20	0·80	—	1·20	—	—
No. 8	+ 14·0	− 2·5	+ 6·0	1·8	25·30	7·50	3·00	72·30	10·00	3·50	1·20	2·50	Traces.	—	Traces.	—	—
No. 10	+ 12·6	− 2·0	+ 18·0	1·1	40·30	37·75	24·40	14·86	2·75	Traces.	10·30	5·70	2·00	Traces.	Traces.	—	2·19
No. 14	+ 14·2	− 0·5	+ 17·0	1·0	38·10	34·05	22·80	19·54	8·32	3·50	1·50	2·80	Traces.	Traces.	Traces.	5·70	1·79
No. 15	+ 12·0	− 3·0	+ 4·0	2·0	32·80	11·00	6·00	56·50	10·10	4·50	2·10	4·80	1·50	Traces.	1·50	Traces.	2·00
No. 18	+ 14·0	− 3·5	+ 9·0	1·9	35·10	9·28	10·00	51·52	9·00	5·10	5·40	5·50	1·10	Traces.	2·00	—	1·10
No. 21	+ 24·7	− 7·0	+ 13·0	2·7	30·00	18·93	8·50	49·17	5·90	3·10	4·20	2·80	1·80	Traces.	3·10	Traces.	2·50
No. 22	+ 21·0	− 6·0	+ 3·0	2·2	25·20	8·50	6·60	54·50	15·10	3·10	8·10	1·40	—	—	1·50	1·20	—
No. 24	+ 10·5	− 0·4	+ 6·1	1·7	26·50	10·60	5·50	61·80	10·20	2·50	5·50	1·50	—	—	1·30	—	1·10
No. 27	+ 14·0	− 1·5	+ 6·0	1·9	22·00	11·50	5·20	66·65	5·30	3·20	3·80	0·80	Traces.	—	Traces.	2·80	0·75
No. 31	+ 19·0	− 2·0	+ 8·0	2·0	25·40	9·50	4·00	70·90	6·90	2·10	2·90	2·50	Traces.	Traces.	Traces.	1·20	—
No. 32	+ 15·0	− 3·0	+ 12·0	1·0	38·50	20·10	10·10	37·00	6·50	3·30	6·20	Traces.	1·90	Traces.	1·50	1·20	2·20
No. 39	+ 16·0	− 2·0	+ 9·0	1·3	41·20	18·00	7·50	52·60	10·00	2·10	7·10	Traces.	1·20	Traces.	1·50	Traces.	—
No. 40	+ 13·0	− 3·0	+ 10·0	1·6	37·50	15·00	8 00	50·50	12·00	3·20	8·00	Traces.	2·30	Traces.	Traces.	Traces.	1·00
Mean.	+ 13·4	− 2·5	+ 8·0	1·8	30·23	14·70	7·88	54·32	9·82	3·18	4·74	2·33	0·74	—	0·82	0·63	0·84

II.—TABLE OF THE AUSTRALIAN SOILS OF THE LOWEST PRODUCTIVE POWER.

Number to which the Soil refers.	PHYSICAL CHARACTER.				CHEMICAL CHARACTER OF THE SOILS.												
						Constituents in 100 Parts.											
											LIME.						
	Absorption of Solar Rays.	Emission of Terrestrial Heat.	Capacity for Moisture.	Specific Gravity.	Soluble Portion of 100 Parts.	Vegetable and Animal Matter.	Water.	Silica.	Alumina.	Peroxide of Iron.	Carbonate.	Sulphate.	Potash and Soda.	Chlorides.	Magnesia.	Sulphurets and Metallic Oxides.	Loss.
No. 7	+ 6·00	− 3·5	+ 2·0	1·40	10·00	6·80	2·00	70·10	16·20	0·20	2·50	—	—	—	0·20	—	2·00
No. 9	+ 15·00	− 6·0	+ 4·0	2·50	5·00	2·20	2·00	77·50	5·00	5·50	6·00	—	—	—	1·80	—	—
No. 11	+ 6·00	− 3·5	+ 3·0	2·10	9·00	5·00	1·00	70·70	8·50	4·50	7·20	—	2·00	Traces.	Traces.	—	1·10
No. 12	+ 10·00	− 4·0	+ 2·0	1·90	10·00	3·50	2·50	70·00	10·80	3·50	6·20	Traces.	—	—	2·20	—	1·30
No. 13	+ 11·80	− 6·5	+ 2·0	3·50	8·00	4·00	0·50	87·80	6·20	—	1·50	—	—	—	—	—	—
No. 16	+ 9·00	− 4·0	+ 8·0	1·70	11·00	9·70	8·00	44·91	6·70	27·69	—	1·70	1·30	Traces.	Traces.	—	—
No. 17	+ 9·80	− 4·5	+ 6·6	1·30	9·50	10·60	12·80	46·47	15·37	14·76	—	—	—	Traces.	Traces.	—	—
No. 19	+ 15·00	− 8·0	+ 3·5	2·50	7·50	8·00	5·00	69·00	10·20	1·00	2·00	—	1·30	—	2·00	—	1·50
No. 23	+ 27·50	− 12·0	+ 4·0	6·09	5·50	4·33	8·00	82·77	4·90	—	—	—	Traces.	—	—	—	—
No. 26	+ 26·00	− 14·0	+ 1·0	2·80	8·50	5·40	4·00	65·66	16·00	1·00	4·54	—	1·20	—	2·20	—	—
No. 28	+ 14·90	− 6·0	+ 7·0	2·50	10·50	4·00	3·00	80·29	5·70	0·21	4·80	—	—	—	2·00	—	—
No. 30	+ 15·40	− 8·0	+ 2·0	2·00	7·50	3·65	1·00	74·25	17·10	1·20	2·50	—	—	—	0·30	—	—
No. 41	+ 18·70	− 7·0	+ 2·0	3·20	7·00	6·50	2·30	68·10	10·00	4·20	6·10	—	0·50	—	1·20	—	1·10
No. 33	+ 18·00	− 4·5	+ 7·0	2·00	18·00	5·20	3·10	57·90	12·00	9·00	10·00	—	1·00	—	—	—	1·80
No. 34	+ 20·00	− 8·0	+ 4·0	1·90	10·20	7·20	3·10	70·00	11·20	3·50	2·10	—	1·80	—	1·10	—	—
No. 35	+ 10·00	− 4·0	+ 3·0	3·20	5·00	8·00	4·00	72·40	9·10	1·20	3·50	—	—	—	1·80	—	—
No. 36	+ 15·00	− 5·0	+ 3·0	2·50	4·50	4·00	3·00	68·00	10·20	2·10	10·90	—	—	—	1·80	—	—
No. 37	+ 10·00	− 4·0	+ 3·0	2·50	10·00	2·50	1·50	71·50	8·10	4·50	9·00	—	1·70	—	—	—	1·20
No. 38	+ 12·00	− 5·0	+ 3·0	2·00	[illegible]·00	4·00	3·05	82·70	7·20	1·10	1·50	—	—	—	—	—	0·45
Mean.	+ 14·21	− 6·1	+ 3·6	2·04	8·53	5·50	3·71	69·99	10·02	4·48	4·12	0·08	0·56	—	0·87	—	0·67

Fig. 2.3. Tables published by Strzelecki (1845, pp. 429–30; see also pp. 431–8) in his search to define the soil properties respectively characteristic of soils of the highest and lowest productive power for agriculture.

Sample of Guano, No. 2.

I have examined with care a sample of guano received from the Board of Agriculture, and marked No! 2, with the following results :—

Characters.—The sample is of a light greyish brown colour, presenting a mottled appearance, white masses being interspersed with dark particles and filaments. The former consist mainly of lime compounds, chiefly of the phosphate; and the latter are woody fibre and sand. When only pulverized and mixed, it assumes a drab color, like the tint of compound jalap powder. The white masses are smaller and fewer than in the sample No. 1, and the vegetable matter rather more abundant; it is consequently of a darker color.

It is almost devoid of smell, the only odour being a slight mustiness, proceeding from the decomposition of vegetable matter. In the natural state it has not any smell of actual ammonia, and the indications obtained by the qualitative tests are extremely faint.

The reaction to test-paper is neutral.

The specific gravity is 2·3.

When dried at 212 Fahr. it loses 11·5 per cent. of moisture. This is, of course, exclusive of the small quantity which resists separation below the temperature at which the organic matter suffers change. That portion is classed as combined water, and is estimated with the soluble organic matter. The constituents are stated as of the dried substance, a plan which I consider preferable to making the analysis directly of the normal guano, because the moisture is likely to vary in every sample; whereas, when that is eliminated, the constitution becomes definite, and the proportions may be readily calculated for a drier or wetter sample.

Constitution.—Its chemical composition is as follows :—

In 100 parts of the dried guano, 11·5 per cent. of water having been previously dispelled—

Soluble organic matter and combined water	4·12
Humic acid	1·00
Other insoluble organic matter	17·28
Soluable salts	1·87
Alkaline sulphates and chlorides—	
Phosphates of lime and magnesia	43·32
Sulphate of lime	4 13
Silica	28·28
	100·00
Available ammonia...	0·20

Remarks.—The silica, which is chiefly sea sand, is here in large proportion, and is valueless. Its presence in so great quantity may be partly attributed to want of care in collection, and may be obviated, much being capable of mechanical separation. If, as may be presumed, the sample is taken from the surface, there is additional ground for expecting the proportion of sand to decrease as the digging proceeds.

Fig. 2.4. Report of guano analysis made for the Council of Agriculture in 1860 (Victoria 1860–61).

his report of four foolscap pages on the composition of the soils of the Roseworthy Farm, and at Yongala (Fig. 2.5) and, for comparison, at Cirencester in the United Kingdom (South Australia 1883–84).

After 5 years of experimental work at the Roseworthy farm it was reported that "the great deficiency appears to be phosphates and potash. The result of application of phosphatic manures has always been increased yield." (South Australia 1887*a*).

As in Victoria, the Bureau of Agriculture in South Australia in 1888 included in its charter the "analysis of soils and manures"; likewise, the Council of Agriculture in Tasmania in 1891 (South Australia 1889; Tasmania 1891).

The idea that unhealthiness in animals may result from soil characteristics was evident in the South Australian Department of Agriculture's report of 1887, which stated that: "Among the soils presented for analysis have been several from districts where the stock suffer from 'Coast Disease', the most notable feature in all of them being the large amount of carbonate of lime present, and the absence of iron, while on the other hand the soils on which the diseased stock recover are . . . deficient in carbonate of lime, and contain a large percentage of silica, with a fair amount of oxide of iron . . ."; two sets of analyses are appended to the report (South Australia 1887*b*). It was almost 50 years before the true explanation (Lines 1935) for coast disease was discovered; and it did indeed arise from the influence of the 'carbonate of lime' upon the availability of nutrient elements (Marston 1935).

SECTION 10.

Importance of accurate information with regard to nature of soil.

The importance of accurate information is clearly shown by the following analysis of a soil near Yongala, which contained barren patches obstinately refusing to produce a crop, the cause being the presence of magnetic iron sand, and the want of phosphoric acid, potash, and nitrogen.

Mechanical Analysis.

Vegetable detritus (undecomposed)	·8
" fine	1·1
Sand, as air-dried	34·2
Fine sand, as air-dried	33·0
Fine siliceous sediment	28.0
Magnetic iron ore sand	1·8
Matter soluble in rain water	1·1
	100·

Chemical Composition.

Silica, as sand	89·12
Magnetic iron sand	1·8
Alumina and iron oxide, soluble in acid	·18
Vegetable detritus	1·9
Soluble salts	1·1
Hygroscopic moisture	1·4
Water of combination	4·5
Phosphoric acid and potash	none
Ammonia	trace
Sulphuric acid	none
	100·

Fig. 2.5. Report of analysis of soil near Yongala in South Australia for comparison with analysis of soils at Roseworthy Agricultural College and at Cirencester in England (South Australia 1883–84).

An unsuccessful attempt to initiate Australian university studies of soils appeared in the University of Sydney as the Belmore Medal, first awarded in 1873. A sum of £300 was given by the Right Honourable the Earl of Belmore in 1870 for the founding of a medal, to be awarded annually to a member of the University, under the standard of Master of Arts, for proficiency in Geology and Practical Chemistry, with special reference to Agriculture (Sydney 1874–75). There is no obvious evidence that university work on soils was promoted by this award; however the first bibliographical record in the University of Sydney is not until the 1910 calendar. The first professional staff, appointed in 1890, of the New South Wales Department of Agriculture included a consultant chemist; his colleagues were a vegetable pathologist, an entomologist and an economic botanist (New South Wales 1890). It was announced on 1 July 1890 that: "Analyses of soils and the ashes of plants will be undertaken on behalf of farmers, who are now in the dark as to the needs of their crops, the deficiencies of their soils, and the relative values of manures offered to them" (New South Wales 1892–93). Successive annual reports of the Department of Agriculture of N.S.W. from then onwards refer to soil analyses made during the past year, especially of soils typical of considerable areas (New South Wales 1893).

By July 1895, with 300 soils analysed from many parts of the Colony, the agricultural chemist proposed to present a report on the characteristics of the soils from different districts and from different geological formations – one of the early moves towards pedological studies. He went on to suggest preparing from these data a surface map of New South Wales as a valuable companion to the geological map of the Colony – an early move towards broad-scale pedological soil survey and mapping (New South Wales 1895). He was still churning out these analyses at the turn of the century. The mechanical analyses *(sic)* he reported were: reaction, capacity for water, capillary power, and granulometry; the chemical analyses on the fine-earth fraction were: moisture, organic matter, total nitrogen, hydrochloric acid soluble lime and potash and phosphoric acid. He added to this a bacterial examination to determine the nitrifying power, and to recommend methods of increasing it. To support these examinations, he was studying nitrification, and isolating the various organisms concerned – a remarkable range of soil data to be available to farmers in 1900 (New South Wales 1901).

Twenty years before the CSIR (see p. 10) Soil Investigation Unit began its work on soil survey for irrigation along the River Murray, the New South Wales Department of Agriculture was collecting and examining samples of soils typical of areas to be thrown open for closer settlement; in 1906–07 the Department presented to Parliament a report on the chemical and physical nature of the soil of the area proposed for irrigation from the Barren Jack, now called Burrenjuck, reservoir and on their fitness for irrigation. This was all that the Department had resources for, although its expressed wish for some years had been to have staff to undertake a systematic soil survey of the whole state, as being much more profitable than the analyses and reports on numerous isolated soils sent by farmers (New South Wales 1907).

A plea for systematic soil survey, after the manner of the United States Department of Agriculture, to replace haphazard soil testing was made by the Western Australian Department of Agriculture in 1919; this accompanied a move to devise a soil classification based on pedology and pedogenesis, to relate soil variation to timber growth variation (Western Australia 1915, 1919).

In the mid 1890s, Western Australia followed the lead of the other Colonies and offered settlers the opportunity of having their soils analysed. The principal reason for this was the great diversity of the soils in the Colony (Western Australia 1894*b*), another was the recognition by 1896 that a large portion of "the arenaceous or light sandy soils are saltiferous" (Western Australia 1896). Despite the report (Western Australia 1898) that an agricultural chemist had written a handbook of soil and manures for Western Australian farmers in 1898, there is a plea six years later (Western Australia 1904) of urgent need for an agricultural chemist because few soils had been analysed up to 1904. As in New South Wales, by 1905 soil inoculation was being practised (Western Australia 1905) and the knowledge on symbiotic nitrogen fixation was being spread and used. Progress in soil analysis cannot have been very rapid in Western Australia; the Royal Commission on Agricultural Industries report dated 22 November 1917 still recognized soil analysis, especially of 2nd and 3rd class land, to be one of the most urgent fields for investigation by the Department of Agriculture (Western Australia 1917–18).

The situation was less tardy in Tasmania. Once the agricultural laboratory had been built in 1900 (Tasmania 1901), the previous academic interest in soil analysis, generated by such talks as 'On the chemical analysis of soils as a guide to manuring' given to Branch Boards of Agriculture and published by the Council of Agriculture (Tasmania 1899), became a practical interest as more farmers submitted increasing numbers of samples for analysis and for recommendations for manuring. From these, by 1904, the Agricultural Expert found that most of the soils were deficient in organic matter, and stated that he had shown tests for citrate solubility of phosphates in the soils to be most important, "especially in those containing much iron, which reveal in some cases a very considerable difference between the total amount and the citrate-soluble . . . The soils of Tasmania, as well as those upon the mainland appear to be not so much lacking in phosphoric acid as possessing it in an unavailable form." (Tasmania 1905).

Some measure of the general acceptance of the importance of soil analysis, and understanding and interpretation, can be read into moves in 1904 to bring all the states' agricultural chemists into closer touch by forming the State Chemists Association (Tasmania 1905).

It seems likely that knowledge of the New South Wales thinking on soil survey as a replacement for haphazard soil testing reached Tasmania through this Association, so that by 1911/1912 a start had been made to analyse systematically the soils of the principal fruit-growing districts as a guide to correct management (Tasmania 1912). As a result, the Agricultural Chemist (formerly Agricultural Expert since 1909) concluded that many Tasmanian soils were deficient in potash, or that the potash was in an unavailable form. He also reported the discovery of a fertilizer interaction by a farmer who, acting on the Department's advice based on an analysis of his soil, found that crops gave greater yields after treatment with a mixed potash and superphosphate manure than only one or the other, alone or with lime (Tasmania 1914).

The nature and distribution of soils

Most of the earlier reports on soils were concerned with their qualities for crop production, as, for example, in the long series of publications in New South Wales, between 1898 and 1922, resulting from the association of F. B. Guthrie* and H. I. Jensen of the Department of Agriculture, which included one by Guthrie (1910*a*) on the soils of the proposed Murrumbidgee Irrigation Area prior to settlement. In this, and in other papers in the series (1910*b*), he discussed chemical analyses, and mechanical analyses and estimates of water-holding capacities as physical criteria. Agriculturalists in New South Wales were told of the "potential value of soil analysis" by Department of Agriculture officers who

* Guthrie, F. B. (1898–1922). A list of some of the many publications by F. B. Guthrie during the period 1890–1913 is in the University of Sydney Calendar for 1917, from p. 655. Later publications are listed in later volumes of the Calendar.

took upon themselves "the duty of inciting the farmer in season and out of season to adapt the teachings of science to his daily calling" (Anon. 1890). Masses of analytical data, largely from samples sent by farmers, were put on record by the Departments of Agriculture in New South Wales and Queensland; in Queensland, this work was pioneered by J. C. Brunnich.

A few papers, however, had been published on other aspects of the nature and distribution of Australian soils: papers such as Tepper's (1880) descriptive studies and hypotheses for the genesis of the deeply cracking clays near Adelaide and their, subsequently named, gilgai micro-relief; his description (1888) of other soils at Nuriootpa in the Barossa Valley of South Australia; Henson's (1887) paper read to the Sanitary Section of the Royal Society of New South Wales on 13 September 1887 makes mention of subsurface waterlogging and draining in the context of the origin and decomposition of natural and domestic accumulations of organic matter in the subsurface and subsoils around Sydney; David (1888/9) first described* and theorized on laterite as a derivative of volcanic rocks in the New England district of New South Wales; botanists such as Maiden (1903), concerned with soil drift and soil conservation, dealt with ecology and soils.

There was some soil survey work undertaken, but, although the United States Soil Survey was known (Guthrie discussed it favourably in 1903), there was no move to follow the principles that Milton Whitney and his co-workers were establishing by laborious effort in that country as early as 1892. Even though transport facilities were limited, the surveys by Jensen, covering widely spread parts of New South Wales, were conducted with only very sketchy field work; soils were described in the field with specific regard to their assumed parent material, and characterized by laboratory analyses according to the then standard procedures. The investigations culminated in the publication by Jensen (1914) entitled 'The Soils of New South Wales'.

Agricultural science in Australian universities

Universities were founded in most of the Australian colonies during the 19th century – Sydney in 1850, Melbourne in 1853, Adelaide in 1874, Tasmania in 1890 (New South Wales 1861; Victoria 1866; South Australia 1874; Tasmania 1889) – but the principal faculties were Arts, Law, Medicine, and later, Science. Reference has already been made to the Belmore Medal of the 1870s in the University of Sydney. Such bibliographies as are available in the university calendars up to 1900 reveal no evidence of an important interest in soil science in any faculty of arts, medicine or science.

A significant advance towards modern studies in soil science occurred at the turn of the century, when university teaching in agriculture started in 1901 with the University of Adelaide offering a four-year course leading to the Degree of Bachelor of Science with Agriculture (Adelaide 1901). The University of Melbourne instituted a Faculty of Agriculture in 1905, and offered courses leading to a Diploma in Agriculture or to the Degree of Bachelor of Agricultural Science (Melbourne 1906). The University of Sydney in 1910 was the first Australian university to create and fill a Chair of Agriculture, and offered at the same time courses leading to the Degree of Bachelor of Science in Agriculture (Sydney 1910). Western Australia followed in 1913, Queensland in 1927, and Tasmania founded a Chair in 1958 and first filled it in 1962, but had offered the first year of a degree course in Agriculture since 1954 (Western Australia 1915; Queensland 1927, 1928; Tasmania 1954, 1960).

Soil science in the first Australian universities

For the 25 years between the first offer of a university course in agriculture in 1901 in Adelaide, until the foundation of the Waite Agricultural Research Institute, also by the University of Adelaide, there was not a great deal of work published involving soils, and even less of it was primarily concerned with soils. Records of the study of soil animals can be found for the Biology Department of the University of Melbourne in 1910, on the subject of earthworms; the Botany Department was studying 'root tubercle bacteria' at the same time (Melbourne 1911). In Sydney a decade later the central theme of symbiosis in mycorhiza (as was the customary spelling) and nodules on tree roots (McLuckie 1922, 1923) was an expression of an indirect interest in soil; Watt (1912, 1926) and Waterhouse (1913), also from the University of Sydney, added further publications on phosphatic fertilizers and nitrogen; accompanying studies having the soil as a prime interest were being undertaken by Taylor (1922).

From the Geology and Mineralogy Department of the University of Adelaide, Teale (1918) published a soil survey of a site for an experimental forest at Kuitpo, in South Australia. This, and Jacobs' (1926/28) recognition that for his proposed classification of soils by mechanical analysis it was esential for "analyses to be made of both soil and subsoil" for surveys of forest soils, constituted an important landmark, because it required an examination of the soil profile to depth.

The Waite Institute and the CSIR

In the meantime, two events of the greatest significance to the development of soil science in Australia

* A list of Dr T. W. Edgeworth David's publications during the years 1890–1914 is on pp. 669–73 of the University of Sydney Calendar for 1916.

were first, Mr Peter Waite's offer to the University of Adelaide from 1913 (Adelaide 1914), that resulted in the foundation of the Waite Agricultural Research Institute in 1924 (Adelaide 1915, 1916, 1919, 1923, 1924), and second, the Federal Government's initiative in setting up an Advisory Council (later Institute) of Science and Industry, which eventually resulted in the formation of the Council for Scientific and Industrial Research (CSIR 1927), and of its Soils Investigation Section, formed at, and in co-operation with, the Waite Institute on 1 May 1927 (CSIR 1927; Adelaide 1928).

The charter for the Waite Institute extended to many fields, (Adelaide 1926, 1928, 1930; Waite Institute 1925–1932), and included, for agricultural chemistry and soil research:– the survey and classification of Australian soil types; – the investigation of fertility problems, particularly in the irrigation settlements; – the chemical study of the nutritive value of pastures, with special reference to mineral constituents; and – the adaptation of the newer standard laboratory methods to Australian soil problems.

The purpose and the first tasks of the Soils Investigation Section were to follow up, by soil survey and profile descriptions and by investigation of the behaviour of the soils under irrigation, the problems of waterlogging and salinity in the Murray Valley as then recently reported to CSIR as a result of a survey undertaken in 1926 by its predecessor, the Institute of Science and Industry.

In June 1929 the University of Adelaide received a private bequest for the building of a laboratory for soils research. The CSIR agreed to finance the equipment and maintenance of this laboratory, and jointly with the University to establish a Division of Soils Research with its headquarters at the Waite Institute (CSIR 1929; Adelaide 1930). These events initiated the development of modern soil science in Australia, and were the catalyst for the spread of soil research to other institutions.

A need for soil surveys along modern lines was submitted to the new Council for Science and Industry in 1916, but nothing further was done during the remainder of World War I; the subject was revived by the Australian National Research Council in 1922. Eventually, in 1927, the first detailed soil surveys (CSIR 1927; Adelaide 1928) along modern lines were made in the established irrigated areas served by the Murrumbidgee and Murray rivers. The need to understand the soils and their reactions to irrigation water required the systematic study of the soil profile. Thus a standard that facilitated critical international comparisons was established for all subsequent field work in Australia. The work (Taylor and England 1929) was begun at Renmark in South Australia; Prescott (1931) was the first to make such studies in Australia on a continental scale.

Simultaneously with the field observations, it became important to develop appropriate laboratory methods for the examination of soil samples. The examination and elaboration and standardization of methods to collect the soils in the field and to examine and analyse them there and in the laboratory (Prescott and Piper 1928) was undertaken with great care, and the techniques finally recommended (Piper 1942) received international recognition. Since the early work was concerned with irrigated soils, there was a stimulus for studies in soil physics, particularly for understanding water movement in soils (Marshall 1959); this work likewise received international acclaim.

The last fifty years, and the future

By the late 1920s the foundations had been laid for the development of soil science on a broad basis in universities, CSIR (and its successor, the Commonwealth Scientific and Industrial Research Organization, CSIRO), Departments of Agriculture or Primary Industries, and, later, State soil conservation organizations. During the past 50 years soil research, coupled with the research of agronomists, plant pathologists, entomologists and others, has greatly contributed to the prosperity of Australian agricultural, pastoral and forestry industries.

Australia's continuing prosperity depends in no small part on the continuing health of its primary industries and this requires the understanding and conservation of the soils and their fertility. The country's soils face increasing pressures as sources of food, fibre, and energy, and their ability to sustain these depends on the application of the results of past research and on a continuing research effort.

References

Adelaide (1901). Calendar of the University of Adelaide for the year 1901. pp. 80, 82, 88, 490.
Adelaide (1914). Annual report for the year 1913. Calendar of the University of Adelaide for the year 1914, p. 376.
Adelaide (1915). Annual report for the year 1914. Calendar of the University of Adelaide for the year 1915, p. 371.
Adelaide (1916). Annual report for the year 1915. Calendar of the University of Adelaide for the year 1916, p. 367.
Adelaide (1919). Annual report for the year 1918. Calendar of the University of Adelaide for the year 1919, p. 286.
Adelaide (1923). Annual report for the year 1922. Calendar of the University of Adelaide for the year 1923, p. 290.
Adelaide (1924). Annual report for the year 1923. Calendar of the University of Adelaide for the year 1924, pp. 296-7.
Adelaide (1926). Annual report for the year 1925; The Waite Agricultural Research Institute. Calendar of the University of Adelaide for the year 1926, pp. 246-7, 321.
Adelaide (1928). Annual report for the year 1927. Calendar of the University of Adelaide for the year 1928, pp. 257-9, 340-1.
Adelaide (1930). Annual report for the year 1929. Calendar of the University of Adelaide for the year 1930, pp. 278, 368, 370.
Anon. (1890). Soil analyses and their value. *Agric. Gaz. N.S.W.* 1 320-8.
Anon. (1977). 'The Australian Encyclopedia'. 3rd edn., Vol. 1, p. 90. (Grolier Society of Australia: Sydney.)
Australia (1979). 'Year Book Australia' Vol. 63, p. 6.

C.S.I.R. (1927). Council for Scientific and Industrial Research. First annual report for the period from 13th April 1926 to 30th June 1927, pp. 1, 23.

C.S.I.R. (1929). Third annual report of the Council for Scientific and Industrial Research for the year ended 30th June 1929, p. 16.

David, T. W. E. (1888/89). The origin of laterite in the New England district, New South Wales. Australas. Assoc. Adv. Sci. Rep. of 1st meeting 1888, Sydney, Vol. 1, pp. 233–41.

Guthrie, F. B. (1903). The chemical nature of the soils of New South Wales with special reference to irrigation. *R. Soc. N.S.W. J. Proc. Eng. Sect. Proc.*, **37**, LI-LXV.

Guthrie, F. B. (1910*a*). Notes on the soil occurring in the area served by the Barren Jack irrigation scheme. *Agric. Gaz. N.S.W.*, **21**, 663–6.

Guthrie, F. B. (1910*b*). The chemical nature of the black soil plains: with notes on the geological aspects of the question. *Agric. Gaz. N.S.W.* **21**, 855–60.

Henson, J. B. (1887). Soils and subsoils of Sydney and suburbs. *R. Soc. N.S.W. J. Proc. 1887* **21**, 220–6.

Jacobs, M. R. (1926/28). Soil survey of forest areas in South Australia. (Author's abstract). Australas. Assoc. Adv. Sci. Rep. of 18th meeting Perth 1926, Vol. 18, p. 787.

Jensen, H. I. (1914). 'The Soils of New South Wales.' (Dep. Agric. N.S.W.: Sydney).

Lines, E. W. L. (1935). The effect of the ingestion of minute quantities of cobalt by sheep affected with 'Coast Disease': Preliminary note. *J. Counc. Sci. Ind. Res.* **8**, 117–9.

Maiden, J. H. (1903). The sand drift problem in New South Wales. *R. Soc. N.S.W. J. Proc.* **37**, 82–106.

Marshall, T. J. (1959). 'Relations between water and soil.' Commonw. Bur. Soils Tech. Commun. No. 50. (C.A.B.: Harpenden.)

Marston, H. R. (1935). Problems associated with 'Coast Disease' in South Australia. *J. Counc. Sci. Ind. Res.* **8**, 111–16.

McLuckie, J. (1922). Studies in symbiosis. I. The mycorhiza of *Dipodium punctatum* R. Br. *Linn. Soc. N.S.W. Proc.* **47**, 293–310.

McLuckie, J. (1923). Studies in symbiosis. III. Contributions to the morphology and physiology of the root-nodules of *Podocarpus spinulosa* and *P. elata*. *Linn. Soc. N.S.W. Proc.* **48**, 82–93.

Melbourne (1906). Regulations. Annual report for 1904–05. The Melbourne University Calendar 1906, pp. 362, 366, 453–4.

Melbourne (1911). List of contributions to literature and science published by members of University staff and students working in the University laboratories, for the year ending 31st July, 1910. The Melbourne University Calendar 1911, p. 614.

New South Wales (1861). An Act to incorporate and endow the University of Sydney. 14° Vic. No. 31. The Public General Statutes of New South Wales from 11 Victoriae to 15 Victoriae inclusive (1847–1851). pp. 2179–84. (Govt. Printer: Sydney.)

New South Wales (1890). Votes and proceedings of the Legislative Assembly during the session of 1890. Vol. 5, p. 55.

New South Wales (1892-3). Department of Agriculture (Report for the year 1891). Votes and Proceedings of the Legislative Assembly during the session 1892-3. Vol. 8, p. 632 (p. 8 of the report).

New South Wales (1893). Department of Agriculture Annual Report. Journal of the Legislative Council of New South Wales. Session 1893. Vol. 51, p. 359 (p. 13 of the report).

New South Wales (1895). Agriculture and Forestry (Report on, from 1 January 1894 to 1 July 1895). Votes and Proceedings of the Legislative Assembly. Session 1895, Vol. 4, p. 1107 (p. 29 of report).

New South Wales (1901). Department of Agriculture (Report of, for 1900). Journal of the Legislative Council of New South Wales. Session 1901. Vol. 63 (1), p. 898, (p. 10 of report).

New South Wales (1907). Agriculture (Report of the Department for the year ended 30th June 1907). Joint volumes of papers presented to the Legislative Council and Legislative Assembly, 1907. Second Session, being the first session of the 21st Parliament. Vol. 1, p. 8.

Orwin, C. S. and Whetham, E. H. (1964). 'History of British Agriculture 1846–1914'. p. 49. (Longmans.)

Piper, C. S. (1942). 'Soil and Plant Analysis; Laboratory Manual of Methods for the Examination of Soils and the Determination of the Inorganic Constituents of Plants'. (Univ. of Adelaide: Adelaide).

Prescott, J. A. (1931). The soils of Australia in relation to vegetation and climate. Counc. Sci. Ind. Res. Aust. Bull. No. 52.

Prescott, J. A. (1977). The Russian Free Economic Society: foundation years. *Agric. Hist.* **51**, 503–12.

Prescott, J. A. and Piper, C. S. (1928). Methods for the examination of soils. Counc. Sci. Ind. Res. Aust. Pamphlet No. 8.

Queensland (1927). Statute relating to the faculties; Statute relating to the Degree of Bachelor of Science in Agriculture. Calendar of the University of Queensland for the year 1927, pp. 54, 98, 100.

Queensland (1928). Report of the Senate of the University of Queensland for the year ended 31st December 1926. Calendar of the University of Queensland for the year 1928, pp. 314-5.

South Australia (1874). An Act to incorporate and endow the University of Adelaide. Acts of the parliament of South Australia 1874 V.R. 37 and 38. No. 20. (Govt. Printer: Adelaide.)

South Australia (1875). Report of Commission appointed to inquire into the best means of providing agricultural and technical education; together with minutes of evidence and appendix. Proceedings of the Parliament of South Australia, 1875, Vol. 3. Paper No. 77, p. iii.

South Australia (1879). Debates in the Houses of Legislature during the 2nd session of the 9th Parliament of South Australia from May 29, 1879 to October 25, 1879. 8th October 1879, column 1320.

South Australia (1883–84). Department of Agriculture. First Annual Report, 31st December 1882. Proceedings of the Parliament of South Australia. Vol. 3. Paper No. 33, pp. 7–11.

South Australia (1887*a*). Department of Agriculture. Annual Progress Report, December 31st, 1886. Proceedings of the Parliament of South Australia 1887 First Session. Vol. 3. Paper No. 41, p. 13.

South Australia (1887*b*). Department of Agriculture. Progress Report for the first session of 1887. Proceedings of the Parliament of South Australia 1887, First Session Vol. 3. Paper No. 41A, p. 12.

South Australia (1889). Report of the Central Bureau of Agriculture – from 10th April 1888 to 30th June 1889. Blue Book 1889. Proceedings of the Parliament of South Australia 1889 Vol. 3. Paper No. 96, p. 1–2.

Stephens, C. G. (1946). Pedogenesis following the dissection of lateritic regions in southern Australia. Counc. Sci. Ind. Res. Aust. Bull. No. 206.

Strzelecki, Sir P. E. de (1845). 'Physical Description of New South Wales and Van Diemen's Land' (Longman, Brown, Green, and Longmans: London.)

Sydney (1874–75). The Sydney University Calendar 1874–75, p. 34.

Sydney (1910). By-laws of the University; report of the Senate of the University for the year ended 31st December 1909. Calendar of the University of Sydney for the year 1910, pp. 73, 496, 498.

Tasmania (1889). An Act to establish a university in Tasmania. 53 Vict. No. 41. Index to the Acts of Parliament passed session 53° Victoriae (1889) Vol. 9 (2), p. 627.

Tasmania (1891). An Act to provide for the establishment of a Council of Agriculture and for other purposes. 55 Vict. No. 43. Index to the Acts of Parliament passed Session 55° Victoriae (1891) Vol. 10 (2), p. 705.

Tasmania (1899). Council of Agriculture, Tasmania. Report for 1898. Journals and printed papers of the parliament of Tasmania, 1899 Vol. 41. Paper No. 12, p. 5.

Tasmania (1901). Council of Agriculture. Report for 1900. Journals and printed papers of the parliament of Tasmania, 1901 Vol. 45. Paper No. 53, p. 3.

Tasmania (1905). Council of Agriculture. Report for 1904. Journals and printed papers of the parliament of Tasmania 1905 Vol. 53. Paper No. 17, p. 4.

Tasmania (1912). Agricultural and Stock Department: report for 1911–12. Journals and printed papers of the parliament of Tasmania, 1912. Vol. 67. Paper No. 33, pp. 8, 18–23.

Tasmania (1914). Agricultural and Stock Department: report for 1913-14. Journals and printed papers of the parliament of Tasmania, 1914. Vol. 71. Paper No. 16, pp. 12–14.
Tasmania (1954). Calendar of the University of Tasmania for the year 1954, p. 76.
Tasmania (1960). The University of Tasmania. Annual report for the academic year 1958 and financial year 1957. Calendar of the University of Tasmania for the year 1960, p. A2.
Taylor, J. K. (1922). A chemical and bacteriological study of a typical wheat soil of N.S.Wales. *R. Soc. N.S.W. J. Proc.* **56**, 90–108.
Taylor, J. K. and England, H. N. (1929). A soil survey of Block E (Renmark) and Ral Ral (Chaffey) irrigation areas. Counc. Sci. Ind. Res. Aust. Bull. No. 42.
Teale, E. O. (1918). Soil survey and forest physiography of Kuitpo, S. A. Dept. For. Univ. Adelaide Bull. No. 6, 1–19.
Tepper, O. (1880). The 'Bay of Biscay' soil of South Australia and its formation. *R. Soc. S. Aust. Trans. Proc. Rep.* **3**, 91-8.
Tepper, O. (1887). Notes on the surface features and rocks of Nuriootpa and its neighbourhood. *R. Soc. S. Aust. Trans. Proc. Rep.* **11**, 25–9.
Victoria (1859). An Act for the establishment of a Board of Agriculture. 22 Vict. No. 83. Acts of Parliament of Victoria. Session 22 Victoriae 1858-9, pp. 135–7.
Victoria (1860-61). First Annual Report presented by the Council to the Board of Agriculture. Votes and proceedings of the Legislative Assembly, Session 1860–61 Vol. 3. Papers presented to both Houses of Parliament. Paper No. 1, p. 14.
Victoria (1862–63). Fourth annual report presented by the Council to the Board of Agriculture. Votes and Proceedings of the Legislative Assembly, Session 1862–3. Vol. 4. Papers presented to both Houses of Parliament. Paper No. 75, pp. 21-30.
Victoria (1864–65). Sixth annual report presented by the Council to the Board of Agriculture. Votes and proceedings of the Legislative Assembly, Session 1864–65 Vol. 4. Papers presented to both Houses of Parliament, Legislative Assembly. Paper No. 76, pp. 32–68.
Victoria (1866). An Act to incorporate and endow the University of Melbourne. 16 Vict. No. 34. The Victorian Statutes in 4 volumes. Vol. 4, p. 252. (Govt. Printer: Melbourne).
Victoria (1867). Seventh annual report presented by the Council to the Board of Agriculture 1866. Votes and proceedings of the Legislative Assembly; first session, 1867 Vol. 3. Papers presented to both Houses of Parliament. Legislative Assembly. Paper No. 10, p. 14.
Victoria (1873). Department of Lands and Agriculture. Report of the Secretary for Agriculture. Votes and proceedings of the Legislative Assembly. Session 1873. Vol. 3. Papers presented to both Houses of Parliament. Legislative Assembly. Paper No. 59, p. 10.
Victoria (1875–76). Parliamentary debates. Session 1875–6. Legislative Council and Legislative Assembly. Vol. 23, p. 2369.
Victoria (1885). Report of the Secretary for the Department of Agriculture for the year 1884. Votes and proceedings of the Legislative Assembly, Session 1885. Vol. 3. Papers presented to Parliament. Legislative Assembly. Paper No. 34, p. 5.
Waite Institute (1925–1932). Report of the Waite Agricultural Research Institute, Glen Osmond, South Australia 1925–1932, p. 9. (Published 1934) (Hassel Press: Adelaide.)
Waterhouse, W. L. (1913). The effect of superphosphate on the wheat yield in N.S.W. Dep. Agric. N.S.W. Sci. Bull. No. 10.
Watt, R. D. (1912). Some aspects of the nitrogen question. Australas. Assoc. Adv. Sci. Rep. 13th Meeting 1911 Sydney. Vol. 13, pp. 559-62.
Watt ,R. D. (1926). Note on the response of Australian wheat soils to small quantities of soluble phosphates. Pan Pacific Sci. Cong. 1923. Proc. Vol. 1, pp. 108–10.
Western Australia (1894*a*). Government Gazette of Western Australia, No. 3. Perth, Friday January 19th 1894.
Western Australia (1894*b*). Report of the Bureau of Agriculture of Western Australia for the six months ending July 30th, 1894. Minutes and votes and proceedings of the parliament during 1st session, 2nd Parliament 1894, Vol. 2. Paper No. A9, p.6.
Western Australia (1896). Third annual report of the Bureau of Agriculture of Western Australia for the year ending 30th June 1896. Minutes and votes and proceedings of the parliament during the 3rd session of the 2nd Parliament 1896 Vol. 1 (2). Paper No. A13, p.5.
Western Australia (1898). Department of Agriculture of Western Australia. Annual report for the financial year ending June 30th, 1898. Minutes and votes and proceedings of the parliament during the 3rd session of the 3rd Parliament 1898 Vol. 2. Paper No. 36, p. 30.
Western Australia (1904). Department of Agriculture of Western Australia. Annual report for the financial year ending June 30th, 1904. Minutes and votes and proceedings of the parliament during the 1st session of the 5th Parliament 1904. Vol. 2. Paper No. 25, p. 6.
Western Australia (1905). Department of Agriculture of Western Australia. Annual report for the financial year ended June 30th, 1905. Minutes and votes and proceedings of the parliament during the 1st session of the 6th Parliament 1905. Paper No. 1, pp. 15–16.
Western Australia (1915). Statutes of the University. Regulations. Summary of the report of the Professor of Agriculture for the year 1913. Report of the Senate of the University for the year ending 31st December 1913. Calendar of the University of Western Australia 1915, pp. 41, 71–80, 101, 209, 218. (Published 1914). (Govt. Printer: Perth.)
Western Australia (1917-18). Progress report of the Royal Commission on the agricultural industries of Western Australia, on the wheat-growing portion of the south-west Division of the State, together with minutes of evidence, indices, and appendices. Minutes and votes and proceedings of the parliament during the first session of the 10th Parliament 20/11/17 to 30/5/1918 Vol. 2, Paper No. 7, p. xxii.
Western Australia (1919). Department of Agriculture, Western Australia. Annual report for the financial year ended 30th June 1919. Minutes and votes and proceedings of the parliament during the 3rd session of the 10th Parliament Vol. 2. Paper No. 14, p. 25.

Soils and landscapes

Part 1
Features of Australian soils

Chapter 3
Features of Australian soils

G. D. Hubble, R. F. Isbell and K. H. Northcote

THIS CHAPTER IS CONCERNED WITH THE MORPHOLOGY OF Australian soils. It discusses some of their prominent features, gives a short account of the major soil groups recognized, and briefly compares the Australian soil pattern with that of some other countries. Because slightly different terminology has developed in different countries to describe the morphological features of soils, some of the more widely accept concepts about soils will be summarized.

Soils, soil horizons and the soil profile

There is no universally accepted definition of soil. This is largely due to the wide divergence of interests concerned with the use of soil, and consequently with studies about soils and soil materials. Traditionally, soil has been regarded as the natural medium for plant growth, but plants will also grow in other media that are not normally regarded as soil. Another problem concerns the lower limit of soil – when does weathered rock become soil?

Soils are independent natural bodies consisting of weathered mineral and organic matter often occurring in genetically related horizons formed in response to subaerial processes. They are anisotropic bodies which vary both vertically and horizontally. In a vertical section, soils usually consist of a series of layers more or less parallel to the earth's surface; these layers vary in thickness and are called horizons. The vertical section, from the surface to the underlying material from which the soil formed, is known as a soil profile and contains varying numbers of horizons. Because soils form a three-dimensional continuum it is not possible to study them completely. Most soil descriptions are based on the examination of a number of soil profiles, augmented where possible by information gained from more continuous exposures resulting from erosion or mechanical excavation.

Nomenclature of soil horizons

There is a fairly widely accepted schema and nomenclature for the designation of soil horizons. The common usage in Australia, which is based on that of the United States Department of Agriculture (Soil Survey Staff 1975), is shown in Fig. 3.1. Not all soils have this number of horizons, and both the relative and absolute thickness of horizons may vary widely. In addition, the boundaries between horizons may range from very distinct and abrupt to very diffuse, and may be even or highly irregular.

Opinions on the nomenclature and nature of the lower horizons vary. In the USA (Soil Survey Staff 1975; Soil Science Society of America 1975) the C horizon is designated as a layer of unconsolidated weathered material, relatively little affected by pedogenic processes and either like or unlike the material from which the overlying solum is presumed to have formed. The horizon designation R is given to underlying consolidated bedrock. The more common Australian usage is that proposed by Brewer (1964), who suggested that the soil profile should be regarded as the complete vertical succession of genetic horizons down to the unaltered parent material or parent rock. This accepts the C horizon concept but avoids the problems which arise in soils formed in unconsolidated alluvial, colluvial or aeolian deposits, where no 'consolidated bedrock' (R horizon in the USA sense) exists. Northcote (1971) has used the term D horizon for any soil material below the solum that is unlike the solum in its general character, and is not a C horizon as he defines it. It is usually applied to buried soils for which no reliable horizon designations can be given.

The horizons shown in Fig. 3.1 are the master horizons that may be present in soils. Many others, e.g. accumulations of carbonate or gypsum, gley horizons, are also recognized and have been adequately defined elsewhere (e.g. Soil Survey Staff 1975).

Soil morphology

Soil morphology is the form of the soil profile as expressed by the succession, transition, thickness and properties of the various horizons. Generally accepted standards for describing soil profile morphology are given in a number of publications (e.g. Soil Survey Staff 1951; Northcote 1971; Hodgson 1974), but one particular feature needs to be mentioned here. Textural classes (= texture grades of Northcote 1971), as commonly used in Australian soil descriptions and throughout this text, are not based on laboratory determinations of particle size as elsewhere, e.g. USA (Soil Survey Staff 1975). In

		Horizon	Description
	Organic horizons of mineral soils	O_1	Organic horizons in which the original form of most vegetative matter is visible to the naked eye.
		O_2	Organic horizons in which the organic debris is partly decomposed and the original form of most of it cannot be recognized by the naked eye.
The Solum	Horizons of maximum biological activity or eluviation, or both	A_1	Mineral horizons at or near the soil surface and having some accumulation of humified organic matter intimately associated with the mineral fraction.
		A_2	Mineral horizons, usually lighter in colour and having less organic matter, iron, aluminium and/or silicate clay then horizons above and/or below, with relative concentration of quartz or other resistant minerals.
		A_3	A transitional horizon between A and B, but dominated by properties characteristic of an overlying A_1 or A_2.
	Horizons of accumulation or maximum development of pedological organization	B_1	A transitional horizon between A and B which is dominated by properties characteristic of an underlying B_2.
		B_2	That part of the B horizon in which the dominant feature is one or more of the following (i) an illuvial, residual or other concentration of silicate clay, or iron, aluminium, or humus either alone or in combination; (ii) maximum of development of pedological organization as evidenced by having a different structure and/or consistence, and/or having stronger colours than the A horizons above or of any horizon immediately below.
		B_3	A transitional horizon between B and C or other sub-solum material in which the properties characteristic of an overlying B_2 are clearly expressed but are intergrading to those of the underlying material.
		C	A layer of consolidated or unconsolidated mineral material, usually partially weathered and either like or unlike the material from which the soil is considered to have formed. This horizon is relatively little affected by pedogenic processes and lacks properties characteristic of A or B horizons. Remnants of the original fabric of the parent rock or sediment can usually be recognized.

Fig. 3.1. Schema for the recognition and designation of the principal horizons. (Based on Soil Survey Staff (1975) and Soil Science Society of America (1975)).

Australia, texture (field texture) is assessed from the response of moist soil to hand moulding and thus reflects also the influence of organic matter, clay minerals, absorbed cations, sesquioxides, etc. (Northcote 1971).

Soil-forming factors and processes

Soil as an organized natural body is the product of its environment throughout its history. This proposition, first advanced by Dokuchaev (1899, referred to in Nikiforoff 1943), has long been accepted as a basic principle of soil science. Soil formation may be regarded as a function of *climate, organisms, topography, parent material and time* (Jenny 1941; Stephens 1947). The processes involved are both numerous and complex and many of them are still little understood.

The role of time differs from the roles of the other factors. It is the dimension within which the other factors operate and as such has dual significance. Firstly, many soil-forming processes are so slow that considerable time* is necessary for the processes associated with any set of environmental conditions to reach their maximum expression. Secondly, environmental conditions frequently change and thus the soils of any part of an old land surface may have been influenced by more than one combination of soil-forming factors. The resulting soils may bear the imprint of different soil-forming processes during their development. Most of these changes are brought about by changes in climate, which in turn bring further changes such as an alteration in the kind and activity of organisms affecting soil development or in the stability of the land surface. As a consequence further changes result, e.g. plant cover may be reduced, perhaps leading to varying truncation or stripping of soils and thus exposing fresh parent material to soil formation.

The distinction between factors and processes deserves emphasis. Reduced to their simplest form the latter are the basic physical, chemical and biological processes as detailed in most texts on soil formation. They include physical comminution of mineral particles and organic materials, oxidation and reduction, hydration and solution, leaching and precipitation, and mixing either by physical and chemical agencies, such as geomorphic processes, or biological activity. These processes

*As an example, Walker and Coventry (1976) have shown that in south-eastern Australia, leached soils with strong texture differentiation between A and B horizons are up to 30 000 years old.

vary in relative magnitude and are involved in the development of most soils from their parent materials.

Soil-forming processes can be viewed in many other ways. In the past, particular combinations of these basic processes have been identified with certain prominent and strongly contrasted soil groups. Thus we have podzolization, lateritization, solonization, gleization, etc. (Anon. 1938), emphasizing the dominant processes which in particular combination are considered to account for the distinctive features of the soils concerned. As thus conceived, soil-forming processes are few relative to the wide range of possible combinations of factors and basic processes.

Another way of viewing the soil-forming processes that can account for the main features of all soils is to consider them as components of three broader processes generalized as *soil wasting, organic cycle* and *inorganic cycle* (Taylor and Cox 1956). *Soil wasting* is the consequence of the percolation of rainwater through the developing soil. Labile primary minerals are decomposed, e.g. by oxidation, and some of the products are hydrated and dissolved; new secondary minerals, mainly clay minerals, hydrated oxides of iron and aluminium, are formed; bases and other soluble materials are redistributed within the developing soil or removed by leaching; and there is a relative accumulation of resistant minerals. Where rainfall exceeds evapotranspiration, leaching is particularly active and leads ultimately to soil impoverishment with the removal of nutrient elements and bases. The *organic cycle* involves the growth and decay of plants and other organisms, and both partly offset and contribute to soil wasting. Nutrients and salts, taken up by vegetation and fauna, are returned to the soil by excretion, death and decay. Organic matter accumulates on and in the surface soil forming O and A_1 horizons; organic acids, and other soluble organic compounds released on decomposition, contribute to mineral weathering and the mobilization and transport of ions, clay minerals, iron, and aluminium probably to B horizons at lower depth. The *inorganic cycle* embraces additions, mixing, and removal of mineral materials by physical agencies. Sediments are deposited on the soil surface; salts are brought in by rain; mixing takes place by physical and biological processes; and at most sites losses occur from erosion by wind or water. Organisms are usually the most active mixing agents in soils, e.g. by the burrowing of such mesofauna as ants, worms and dung beetles, although the cyclic growth and decay of plant roots, and the windthrowing of trees also cause significant mixing and oppose horizon differentiation. In practice, the features of the resulting soils arise from a combination of these three cycles.

The evaluation of processes contributing to soil formation is further advanced by Simonson's (1959) general theory of soil formation. This theory ascribed pedogenesis to four groups of processes – additions, removals, transformations and translocations. In these terms the processes going on in the development of all soils could be the same but the relative importance of each varies for different soils. Emphasis is placed on the combined operation of the processes, some processes promoting, and others offsetting or retarding soil development and profile differentiation. Thus it is the balance among individual processes and groups of processes occurring in a given combination that determines the nature of a soil. As an example, additions, transformations and translocation of organic matter are of little importance in the combination of processes that form the soils of desert areas, but are of major importance in the formation of *podzols**.

This theory and reasoning are applicable to each of the three cycles of Taylor and Cox (1956). Although presented as a pedological rationalization of soil-forming processes operating in the development of soil profiles, the cycles are equally applicable to the lateral differentiation of soils in catenary or toposequences and embrace such slope-dependent processes as sloughing and accretion (Butler 1959), and removal, downslope transport, and accumulation of soluble materials (Hallsworth *et al.* 1952).

Soil evolution

The various processes of soil formation proceed at different rates and vary greatly in different environments. Usually soil-profile development is very slow, with a time scale closer to that of landscapes and some geological phenomena than that of living things. There is, however, some support for the view that soil differentiation is fastest in the early stages, slowing as processes approach their maximum expression (in terms of solum differentiation and depth) for the parent material and environment concerned.

Many concepts of soil evolution have been advanced for defined situations. Nikiforoff (1949) held that the theory of soil evolution was based on two assumptions – one that soil development is continuous and progressive, and the other that pedogenic processes do not operate steadily. In this latter case it is postulated that the development of each process inevitably leads to its own decline and to its eventual replacement by another process, irrespective of any changes in the environment. The nature of the soil at any time would then be determined by the dominant process although the soil would also have other characteristics whose expression may be waxing or waning. Two extreme situations can be envisaged. At one extreme, soils developing in an unchanging environment on near-level land surfaces with no significant erosion or deposition would be expected to develop progressively. Where the depth of effective weathering

*Australian great soil group names (Stace *et al.* 1968) mentioned in the text are printed in italics.

and leaching exceeds that of penetration of plant root systems resulting in a decline of soil maintenance through biological cycling, these soils approach senility. The occurrence of very deep impoverished soils on old land remnants lends support to this proposition. At the other extreme, periodic partial stripping or burial of soils on hillslopes initiates new cycles of soil formation in the parent materials thus exposed (Butler 1959).

Surface soil features

Weak development of O horizons

Most Australian soils have only a thin surface scatter of plant debris, mainly in the form of dry leaves, grass and twigs (Northcote *et al.* 1975) and are generally too thin and discontinuous to be regarded as an O_1 horizon. Development of an O_2 is negligible in such situations because of this sparse supply of litter which is often woody and lying on a dominantly dry soil surface. The reasons for this widespread lack of O horizons are probably many. Large areas of the continent are arid or semi-arid with a sparse and often ephemeral vegetative cover, the grass component of which is generally of tussock form. In many better-watered areas the soil biomass is small because of low soil-nutrient contents, low soil-water storage capacities, high soil temperatures in summer, and the accidental or deliberate annual or biennial burning that occurs over large areas with the consequent destruction of accumulated litter.

Notwithstanding, soils do occur in Australia with prominent O_1 and/or O_2 horizons. These are mostly in higher rainfall zones under particular vegetation communities such as vine forests, wet sclerophyll forests, some coniferous forests, vine thickets and some heaths. The O horizons of Australian soils, particularly the biological aspects, are discussed further in Chapter 38.

Low organic matter contents of A_1 horizons

Australian soils generally have low organic matter contents in the A_1 horizons. This is discussed in more detail in Chapter 34, but from the viewpoint of soil morphology, dark, well-developed organic A_1 horizons are not common in Australian soils. Soils with mollic or umbric A horizons (both characterized by dark colours and relatively high organic contents) are largely confined to relatively small areas in the wetter, temperate south-east of the continent and Tasmania (FAO–UNESCO 1976). Notable exceptions are *krasnozems* and related structured sesquioxidic soils. These soils occur intermittently in the higher-rainfall areas from north Queensland to Tasmania and have high organic matter contents in the A_1 horizons – range 6–16%. However, these A_1 horizons are not usually very dark or noticeably organic because of the masking effect of the red colour of the soil. By contrast, many clay soils with very dark surface horizons, e.g. *black earths* and *rendzinas*, do not contain high amounts of organic matter. The A_1 horizons of some high-clay, high-sesquioxide soils with moderate to high organic matter contents (e.g. the *krasnozems*) are commonly weakly subplastic (Butler 1955); this means that the field texture-consistence properties of these horizons suggest less clay-sized particles than are actually present, an effect which is probably a consequence of both the organic matter and sesquioxide contents (Norrish and Tiller 1976).

Weak structure-development and hard-setting character

Large areas of Australian soils have weak structure-development in the A_1 horizon, i.e. massive, single-grain or a weak grade of pedality. This is true for a wide range of soils and is not surprising in view of the above generalizations on low organic matter contents in the surface horizons, and particularly in view of the widespread low earthworm populations (Chapter 39). Many of the massive to weakly structured A_1 horizons in a number of soil groups are also hard-setting (Northcote 1971), particularly the loamy surfaced *solodic soils.* This character, which develops on drying of compact surface soil of low macroporosity and is largely the consequence of low mesofaunal activity, is very common in areas where wet and dry seasons are pronounced, but is less common in wet soils with very high sand contents or soils with high contents of swelling clays.

Many of the soils of the Australian arid region (in particular the *desert loams*) possess a thin (usually less than 1 cm) surface crust which, although rather fragile, may be readily separated from the soil below. The lower part of this crust often has a strongly vesicular appearance.

Water repellence

The water repellence of some sandy soils (Bond 1968, 1969*a*, 1969*b*) causes difficulty and unevenness in water penetration with consequent uneven germination of annuals and poor growth of pasture plants. Thus well-grassed patches alternate with bare areas over very small distances. The water repellence has been found to be caused by hydrophobic organic films produced by fungi growing in the soil and forming thin coatings on the sand grains. The intensity of water repellence varies with the species of fungus and its periods of active growth. Strongly repellent soils contain less than 7% clay ($<2\mu$); the depth of sand has little effect. Water repellence occurs on calcareous and acid sands alike, and no simple relation is apparent between the total organic matter content and the degree of repellence.

Water-repellent sandy soils are widespread on the large sand plain extending from Lake Alexandrina in South Australia across into Victoria, and on the coastal sand plain near Perth, and also occur on other smaller areas in South Australia, coastal New South Wales and south-east coastal Queensland. The phenomenon is not obvious in tropical Australia, although there have been no specific investigations, nor does it occur on the sandy soils in inland Australia where the rainfall is low and the vegetation is sparse.

Subsoil features

Strong texture contrast

This term describes the pronounced texture difference, so widespread in Australian soils, between the less clayey A and more clayey B horizons. These changes in field texture are usually, but not always, reflected by the analysis of particle size and A/B horizon clay ratios. Although values of about 1:2 are very common, the ratios range up to 1:10. Strong texture contrasts occur in many Australian soils (Stace *et al.* 1968; Northcote *et al.* 1975) but are particularly characteristic of soils with sodic B horizons. Northcote (1971) used texture contrast as a criterion to separate duplex soils as a class in his classification system.

The origin of strong texture contrast is one of the unsolved problems of soil genesis. Traditionally, clay eluviation has been considered important, although evidence from Australian studies suggests that illuviated clay, identified as void argillans and some papules (Brewer 1964), may constitute only a small proportion (up to 15%) of the total clay in the B horizon (Brewer 1955; Oertel 1961; Green 1966; Brewer 1968; Oertel 1968). However, the clay identified as argillans or papules may not be a true measure of the total amount of illuviated clay which could, in time, become rearranged and no longer identifiable.

A second hypothesis involves both differential weathering in the A and B horizons and some resynthesis of clay in the B horizon from weathering products leached from the A horizon. This is given support by the fact that a large proportion of Australian texture-contrast soils occurs in regions with strongly seasonal rainfall regimes which favour prolonged wetting and drying cycles in at least the upper part of the soil, thus favouring clay loss from the A horizon.

The third major possibility is that the A and B horizons represent different depositional bodies modified by weathering and pedogenesis. Some studies (Sleeman 1964, 1975; Oertel 1974; Oertel and Blackburn 1970) support this but at present the evidence is not conclusive. Chittleborough and Oades (1979) re-examined Oertel's (1974) data and concluded that they did not contain clear evidence for separate sedimentary origins of the A and B horizons.

It should be noted that soil fauna, particularly termites, may also be involved in texture differentiation. Some evidence for this is discussed in Chapter 39.

In Australia there is considerable evidence to show that the degree of texture contrast often increases with soil age (e.g. Walker and Hawkins 1958; Walker 1962; Brewer and Walker 1969; Walker and Green 1976; Walker and Coventry 1976). These studies, which have been made on dated river terrace sequences, indicate that soils with strong texture contrast (e.g. *red* and *yellow podzolic soils*) are 25 000–30 000 years old. However, clay illuviation does not appear to have made a significant contribution to the build-up of clay in the B horizon of a number of these profiles, because Brewer's study (1972) has indicated that *in situ* weathering has strongly influenced the development of their B horizons.

Bleached A_2 horizons

Very pale A_2 horizons are prominent in many Australian soil groups, notably *podzols, solodized solonetz* and *solodic soils, soloths*, and some *desert loams, red-brown earths* and *podzolic soils*. These bleached horizons are best seen in dry soils and range from < 3 mm to > 1 m in thickness. Commonly they are about 5–25 cm thick and usually overlie much less permeable subsoils, pans, or hard rock. The bleached A_2 horizons often invade the underlying materials along structural planes or as irregular pockets and have been used as a differentiating feature in the classification scheme of Northcote (1971), who defined them in Munsell colour codes for dry soil. The thicker, more continuous bleaches are termed conspicuous, and the thinner, discontinuous ones, sporadic bleaches.

Few definitive studies of bleached horizons have been made although it is generally accepted that they constitute the most leached part of the soil (Jenny 1941; Leeper 1948); organic matter, clay, iron, aluminium and nutrient elements have been removed, giving a relative accumulation of silica which, if finely divided, contributes to the whitish colour of the bleached horizon. Field observations have established that bleached horizons are often saturated with water. Their occurrence can usually be accepted as clear evidence of periodic water-logging and they frequently indicate sodic subsoils where there is a strong texture-contrast between A and B horizons.

Structure of B horizons

Lack of structure or a weak grade of pedality in B horizons is a prominent feature of many Australian soils. Where structure is weakly developed it tends to be coarse blocky, generally coarser than that of similar soils in other countries, and the peds, visible in undisturbed soil or separating when the mass is broken, are mainly compound units with strong inter-ped coherence whether moist or dry. Coarse structure reaches the extreme in some of the *solodized-solonetz* soils where the round-topped columns characteristic of these soils may be so large that the term pedality seems quite inappropriate.

Both the grade and type of structure are closely related

Fig. 3.2. (*a*) *Solodized-solonetz* soil showing large rounded columns (scale in feet). Yuleba Plateau, southern Queensland. (*b*) Multiple-topped columns from a *solodized-solonetz* soil. From terrace of Macquarie River, Launceston Basin, Tasmania.

to soil texture. At one extreme are the *siliceous sands* and *calcareous sands*, which are single grain, structureless soils; at the other, the heavy clays have strong structure and a form which varies with the dominant minerals of the clay fraction. The kaolin–sesquioxide clay soils, typified by the *krasnozems*, generally have very strong, medium to fine (> 15 mm separating to < 5 mm) polyhedral peds with smooth faces and low coherence; montmorillonite-dominant heavy clays, e.g. the *black earths*, have moderate to strong, medium blocky subsoils grading into deep subsoils of very coarse to fine peds (> 50 mm separating down to < 10 mm units), which are lenticular or parallelepiped (Butler 1955) with highly glazed surfaces.

The massive, structureless soils are widespread. They have high coherence when dry, and are mostly associated with sandy-textured materials ranging from sandy loam or sandy clay loam to sandy medium clay. The subsoils of the *solonized brown soils, red earths*, and *yellow earths* are examples. In the *solonized brown soils* the subsoils tend to be compact whereas in the *red earths* and *yellow earths* the subsoils, although massive, are porous, have an earthy fabric (Northcote 1971), and break easily when moist to crumb-sized fragments. Many of the *earthy sands* are also sufficiently coherent for the structure of the B horizon to be massive and stable when dry, even in a vertical face.

Weak coarse angular blocky structure is characteristic of a range of texture-contrast soils, mostly those with significant subsoil solonization (the cation exchange complex saturated > 6 to 8% by sodium ions). The peds are typically dense, have smooth faces, and their size commonly ranges from 25 to 50 mm. Well-formed medium-blocky peds are characteristic of the typical *red brown earths* and *non-calcic brown soils*, and often tend in these and some related soils to be aligned vertically into compound prism-like forms. Among the texture-contrast soils, fine to medium structure is most strongly developed in some *red podzolic* and *yellow podzolic* soils. In these, the peds tend to be irregular polyhedral ranging from about 20 to 5 mm or less, and have smooth faces and low coherence.

The columnar forms of the *solodized solonetz* and some *soloths* are the most striking of all soil structures. In Australia these features are developed on a scale exceeding that found in all other countries except perhaps southern Africa. The columns are generally polygonal in cross section, with diameters typically 15–30 cm but as large as 45 cm. The columns have either simple rounded tops like loaves of bread (Fig. 3.2*a*) or multiple ones like cauliflowers (Fig. 3.2*b*). Where they lie beneath thick sandy A horizons they are often separated to depths of 25 cm or more by tongues of A_2 horizon (Fig. 3.2*a*). The soil within the columns is characteristically very dense with bulk densities up to 2.2 g ml^{-1} and there are few coarse pores or voids. Although planar voids may extend inward from the column sides they seldom join up to isolate smaller blocky peds so that most columns are essentially massive within. In some soils the columns are compound units composed of peds of moderate, medium, blocky structure.

There are many complex reasons for the absence or poor development of structure in so many Australian B horizons. One important reason is low biological activity, both plant and faunal. Others relate to the long history of weathering and leaching of parent materials and soils, to extensive and intense aridity, and to widespread solonization. Some of these factors also explain why many Australian B horizons have a very narrow range of moisture within which they are friable. As a result, consistence as seen in the field is usually either very hard (dry), very firm (moist), or very sticky (wet). These characteristics are most common in soils with > 6 to 8% sodium saturation and/or relatively high magnesium contents.

Subplasticity

A moist soil which appears to become more clayey with continued kneading is referred to as subplastic (Butler 1955). In extreme cases materials that initially behave as sands finally develop sandy clay loam or clay textures. Subsoils with this property have been reported mainly in the drier regions of south-eastern Australia. Commonly subplasticity is a feature of a relatively deep subsoil. It was first recognized in the Murrumbidgee Irrigation Area as a textural anomaly of some clay materials (Taylor and Hooper 1938) and was related to soil behaviour in irrigated horticulture. Degrees of subplasticity have been proposed (van Dijk 1958) according to the number of classes of texture change over several minutes of kneading.

A series of five papers has discussed various aspects of subplasticity (see Butler 1976). The most outstanding property of subplastic materials is the high resistance of their micro-aggregates to dispersion during field texturing and laboratory treatments. For materials below the A and B horizons of *red-brown earths* of the Riverine Plain of New South Wales, there are positive correlations between the subplasticity rating and content of strongly oriented clay, dithionite-soluble iron oxides, and stability of aggregates. No correlation was obtained with carbonate content. Ultrasonic treatment is successful in disaggregating subplastic soils. Their subplasticity is closely associated with the presence of a fine clay fraction, a high proportion of which is strongly bonded into silt-size aggregates. The reasons for clay aggregate stability differ for various subplastic soils.

Red-brown hardpans

Although they have a porous appearance these pans are massive and very hard, with an irregular laminar cleavage and some vertical fissures. Other more variable features are bedded and unsorted sand and gravel lenses, wavy black veining that is probably manganiferous and, less commonly, veins of calcium carbonate. This name does not apply only to hardpans of the *red* and *brown hardpan soils* (Teakle 1936) but also to pans below a variety of other soils (Litchfield and Mabbutt 1962; Northcote *et al.* 1975) in other areas of Australia.

Although red-brown hardpans are ferruginous, treatment with concentrated hydrochloric acid shows that iron is a deposit in a matrix cemented by silica. The pans are impermeable to water, and so resistant to weathering, but they may break into irregular pieces along cleavages and fissures. Pans identical with these do not seem to be recorded in the soils literature, although some similarity with other silica pans such as 'duripans of arid climates' may be suspected. According to Soil Survey Staff (1975) such duripans are, however, restricted largely to volcanic areas whereas most Australian occurrences are far removed from any such region.

Red-brown hardpans which vary from < 30 cm to > 30 m thick are found below a variety of soils such as *earthy sands, red* and *yellow earths* as well as *red* and *brown hardpan soils* in the arid to semi-arid parts of Western Australia, South Australia, New South Wales and Queensland. By far the most extensive occurrences cover some 300 000 km^2 in the central west of Western Australia. Here the *red* and *brown hardpan soils* predominate and the hardpan itself has been formally named the Wiluna Hardpan (Bettenay and Churchward 1974). The soils and hardpans have formed mainly in strongly weathered alluvial and colluvial materials, in some places eroded from an older lateritized surface (Stace *et al.* 1968). Brewer *et al.* (1972) found that the shallow *red earths* and their hardpans studied in Western Australia were formed in colluvium derived largely from older lateritic profiles which were to some extent silicified. The hardpans were formed by silicification interspersed with episodes of clay illuviation. One explanation (Teakle 1936) for the formation of the hardpan is as follows. Episodic flooding and saturation of the soil on smooth slopes and plains cause leaching and acidification of the upper soil horizons, the leached silica being deposited as a hardpan during subsequent drying. Evidence presented by Litchfield and Mabbutt (1962) and Bettenay and Churchward (1974) tends to support this view.

Accumulation of salts

Significant amounts of carbonates, sulphates and chlorides are found particularly, but not solely, in soils of the arid to sub-humid regions. The most extensive occurrences are in the southern third of the continent where some parent materials contain abundant salt. Salts accumulate in subsoils because annual precipitation is insufficient to leach the soils beyond depths of about 0.7–1 m, or at the soil surface because the water-table was, or is, at a depth shallow enough to maintain an upward flow of soil solution to the surface where evaporation occurs. In humid climates, sulphides may be found in subsoils of coastal swamplands. High salt content is the dominant characteristic of *solonchaks* but significant concentrations are also found in *grey-brown* and *red calcareous soils, desert loams, grey, brown* and *red clays, solonetz, solodized-solonetz, soloths, solonized brown soils, red-brown earths* and *brown earths.*

Calcium and magnesium carbonates

Calcium and sometimes magnesium carbonates occur in particular B horizons (B_{Ca}) of many soils over a wide range of climates from arid through to sub-humid. Carbonates are often present throughout the profile in the *grey, brown* and *red clays* and *solonized brown soils,* as well as in a wide range of shallow soils either formed from calcareous parent materials, or underlain by hard, secondary carbonates in the form of calcretes. Carbonates are discussed further in another chapter.

Sulfates

Calcium sulphate (gypsum) commonly occurs in subsoils of a wide range of soils in the arid to semi-arid areas and is usually found below a zone of concentration of calcium carbonate in the soil profile. In many cases the gypsum is probably pedogenetic, but in soils developed on parent materials rich in gypsum, e.g. Cretaceous sedimentary rocks in western Queensland and South Australia, it may be inherited from them. Again, where there is a thick accumulation of crystalline gypsum below the soil, it may have been formed in playa lakes, e.g. the gypsum layer beneath some *desert loams* (Wopfner and Twidale 1967).

Gypsum may also be a major component of the material that forms lunettes (Hills 1940) – crescentic dune ridges that lie on the lee shore of many shallow ephemeral lakes in southern Australia. These landforms range widely in composition from well-sorted quartz sand, through clayey sand and gypseous sandy clay, to almost pure gypsum dunes. Their composition in relation to past climate has been reviewed by Bowler (1976).

Chlorides

Large areas of Australian soils in subhumid or drier areas have accumulations of water-soluble salts of sodium, calcium, magnesium and potassium in their subsoils. Occasionally the salt shows as efflorescences at the surface. These salts are usually dominated by sodium chloride. The distribution of saline soils has been mapped by Northcote and Skene (1972), who have also shown some relationships with profile morphology. Salt accumulation is further discussed in Chapter 9.

Sulfides

In coastal swamplands, some grey estuarine clays contain sulfides close enough to the surface to oxidize to form complex sulfates and free sulfuric acid, e.g. 'catclays', which often have very low pH values, 2.0–2.5, in the swamplands of the Macleay River floodplain, Kempsey, N.S.W. (Walker 1963). Strong acidity from the oxidation of sulfides has also developed after draining coastal peat swamps, e.g. in the south-west of Western Australia (Teakle and Southern 1937) and in Tasmania (G. D. Hubble, unpublished data).

Deep profile developments

Solum depths of about 1 m are so common in many world soils that this tends to be accepted as a norm. In Australia, a considerable part of the soil mantle ranging over many great soil groups is much deeper than this. In some soils, the solum is of giant dimensions with little B_3 horizon development; in others the main part of the solum is extended below by thick B_3 horizons – ranging up to several metres. The deepest developments are seen in many laterite profiles. Here, unusually thick sola grade below into deep zones of alteration that show varying degrees of pedological organization in the upper part, finally giving way to C horizons.

In Australia's rather restricted humid tropical and sub-tropical areas, high rainfalls and temperatures give deep weathering and leaching. As a consequence, deep soils are found in all but the very young soils and recently active erosional areas. Elsewhere, and especially in the semi-arid to arid regions, very deep soils are unlikely to be products of current climates except for the *siliceous sands* of desert dunes and sand sheets. Most of the deep soils are likely to be relicts.

Soils with very deep sola

Soils exceeding 2 m depth and lacking extended B_3 horizons are found mainly in the *podzol, krasnozem, red earth,* and *grey, brown* and *red clay* great soil groups, and in minor areas of *black earths* and some texture-contrast soils, notably *soloths.* Their distribution spans the range of present-day climate from per-humid to arid.

Deep profile development is generally a result of the interplay of several soil-forming factors. The prerequisites are a combination of parent material and climate to ensure a deep penetration of water, no significant decrease in permeability within or immediately below the solum, and sufficient time. In some instances the developing profile gradually grows upward as a result of continuing slow accretion of parent materials in a depositional environment. Most deep soils are associated with level to gently undulating old land surfaces or with their remnants following dissection. The age of these formations implies that the soil has had a long time to form beneath a stable surface, and allows the possibility that past climatic conditions may have favoured deep development more than the present climate.

Because deep wetting is a prerequisite for deep pedological development, it is tempting to conclude that, in the drier regions at least, very deep soils are unlikely to form under present climatic conditions. This conclusion may not always hold since periods of heavy rain are not well reflected in the annual means of rainfall upon which interpretations are usually based. Periods of rain which are sufficient to wet the soil deeply may be important for deep pedological development if they occur only once in a decade.

The important factors determining the giant *podzols* of the high dunes of the eastern littoral (Thompson 1975; Thompson and Hubble 1980) are: great depth of highly permeable silica sands as parent material, high rainfall and time. In this region there is an age sequence from minimal *podzols* less than a metre deep on the younger beach ridges to the strongly differentiated giant forms with depths up to 15 m or more on the older dunes where the A_1 horizons are 1 m, the bleached A_2 up to 12 m, and the B_2 several metres thick.

Deep *red earths* have also formed mainly on permeable sandy parent materials and on old land surfaces. In the semi-arid regions these soils are likely to have developed most of their characteristics under different past climates. Except in the humid tropics where parent materials include basic metamorphic rocks, deep *krasnozems* are formed mainly from basalt under moderate to high rainfall, but with free drainage to the base of the profile. Occurrences in sub-humid or drier regions are on modified remnants of old land surfaces and are likely to have formed under different past climates.

Different combinations of factors probably account for the very deep development of various cracking clays. These soils, which have slowly permeable clays, are often associated with semi-arid to arid climates that would not seem to favour deep profile development. However, progressive deposition of clay alluvium and deep wetting from intermittent flooding could account for the development of the very deep clays of the alluvial plains and fans of inland streams, even those of the arid channel country in south-western Queensland. Past wetter conditions might also be involved.

The deep (3–7 m) *grey* and *brown clays* of gently undulating plains in the sub-humid to semi-arid region of eastern Queensland and New South Wales – many originally having carried brigalow (*Acacia harpophylla*) forests – are the product of a very long period of soil formation beneath a stable surface on the remnants of old deep-weathering profiles. The latter formed in labile sedimentary rocks beneath an earlier stable surface and different climates of the Tertiary. These clays are montmorillonite–kaolinite mixtures, saline, and often have a pronounced gilgai microrelief. Although the upper horizons of most are neutral to alkaline and frequently calcareous, at relatively shallow depths (50–120 cm) the clays are strongly acid (pH values approaching 4.0, and occasionally lower). Similar soils that are acid throughout the profile are also found. Details of the occurrence, characteristics and possible origin of these acid clays have been given by Hubble and Isbell (1958), Isbell (1957, 1962) and Veen (1973).

Protection afforded by a close cover of lag silcrete gravel (gibbers) is undoubtedly a most important factor in the occurrence of the deep *red* and *brown clays* of Sturt's Stony Desert and other gently undulating uplands in the arid region. Without this protection these highly dispersible clays with their scanty plant cover would have been reduced by erosion. The relative stability of these landscapes has ensured long periods for soil formation on parent materials weathered from labile fine-grained sedimentary rocks which have, in places, been chemically altered under a much earlier weathering regime. Given a long stable period, the soils could possibly have formed under the present climatic conditions but it is likely that much of their development was influenced by different past climates.

Unusually deep texture-contrast soils are also common. Most are associated with deep bodies of colluvial or alluvial parent materials largely derived from preexisting soils. However, a great depth of B horizon development beneath A horizons of normal depth in some soils is puzzling, e.g. the *soloth* profile in Stace *et al.* (1968, p. 177). Although age and past wetter conditions could account for the depth of pedological development, the A horizons are disproportionately shallow. Possible explanations are that the development of the texture-contrast postdates that of the deeper profile and was associated with a climatic change or, alternatively, that there has been sufficient erosion to keep the A horizon thin even on the near-level surfaces where they are found.

Deep soils with extended B_3 horizons

In most freely drained soils, developments of B_3 horizons are thin relative to the solum depth. However, in Australia, there are widespread occurrences of soils in which deep pedological differentiation below the B_2 horizon is prominent and comprises much of the solum. The depth of the extended B_3 horizons generally exceeds 50 cm and ranges up to more than 3 m thick, making up more than 25% of total solum depth. These horizons include deep pedological differentiations described as subsola (van Dijk 1969), and are considered to have developed as separate features below sola of normal depth.

Extended B_3 horizons are found in many great soil groups, including some freely drained ones, e.g. some giant *podzols*, but typically in those associated with moist environments; *red, yellow* and *gleyed podzolics, red* and *yellow earths, krasnozems, xanthozems, prairie soils* and some *solodic soils*. These horizons lie below the zone of main biological activity. Commonly there are features associated with gleying; distinct to prominent mottling, usually coarse and commonly light grey, pale brownish yellow and red. The patterns of mottling, and associated textural and structural differentiations are strongly influenced by the structure, texture, and void patterns of the parent materials.

A variety of features and patterns of extended B_3 horizons from the Southern Tablelands of N.S.W. have been described (van Dijk 1969). There they form part of a soil mantle of considerable age, occur on low catenary sites, and are believed to represent 'descendant' pseudogley related to an earlier climatic regime.

Extended B_3 horizons of this kind are common in the moister environments of eastern Australia, usually on mid- to lower-slope sites and sometimes beneath terraces. While many of these horizons could have formed under present climates, others are more likely to be the product of earlier different climates. The close similarity of their features to those of gleyed soil in the zone of water-table fluctuations, and the absence of underlying water tables, suggest that they are formed under conditions of intermittent partial saturation of the main conducting voids and bordering soil by descending water (or in some cases water moving laterally). Excess water draining freely from overlying soil is locally concentrated below the zone of pedoturbation into fewer structurally and/or texturally controlled voids. These voids drain much more slowly than those of the overlying soil and maintain partial saturations for significant periods. Underlying materials of lower permeability may also contribute to the slow drainage and on slopes may deflect water laterally, so thickening the partially saturated zone down slope.

The laterite profile

The occurrence in Australia of laterite, "a massive, vesicular, or concretionary ironstone formation nearly always associated with uplifted peneplains. . ." (Prescott and Pendleton 1952) was recognized by early geologists (see Prescott 1931). Although the laterite which is commonly recognized is dominantly ferruginous, alumina, as in bauxite, is often a significant component. The economic importance of bauxite has focussed attention on the aluminous form; one of the best known such deposits is at Weipa on Cape York Peninsula. This deposit consists of a mixture of gibbsite and boehmite and contains up to 55% Al_2O_3. The main ore body is the lower part of a nodular *red earth* profile and has pisolites 10–200 mm in diameter and generally loosely arranged in a red to reddish brown friable kaolinitic matrix.

Prescott directed attention to the importance of laterite in relation to the classification of Australian soils, pointing out that typical laterite is the fossil illuvial horizon of an ancient soil. This view has been generally accepted by Australian pedologists (Stephens 1946), despite the fact that the associated profile is commonly many times deeper than normal for soils.

Laterite *in situ* is part of a deeply weathered formation which, although varying significantly in form and components across the continent (Fig. 3.3), is distinctive enough to be recognized as the 'laterite profile' (Colour Plate 3.15). The complete laterite profile has three more or less distinct zones (Whitehouse 1940; Prescott and Pendleton 1952), beneath a cover of relict or derived soil. From top to bottom of the profile these are:

'ferruginous' zone – the main component is laterite which can be dominantly aluminous.

mottled zone – a zone of whitish kaolinitic clay with coarse red and ochreous mottles usually becoming coarser and decreasing in amount with depth.

pallid zone – near-white kaolinitic clay with few mottles, but often showing some fabric of the parent rock into which it merges through a transition below.

Silcrete (or 'grey billy'), a hard dense silica rock, may occur in the lower part of the profile, sometimes as discrete masses or as a distinct zone; in other profiles there is varying silica impregnation of the clay in parts of the mottled and pallid zones so that exposures of brittle to hard clayey or 'porcelainized' rock are formed. Research on silcrete in Australia was reviewed by Langford-Smith (1978); it is evident from this review that many

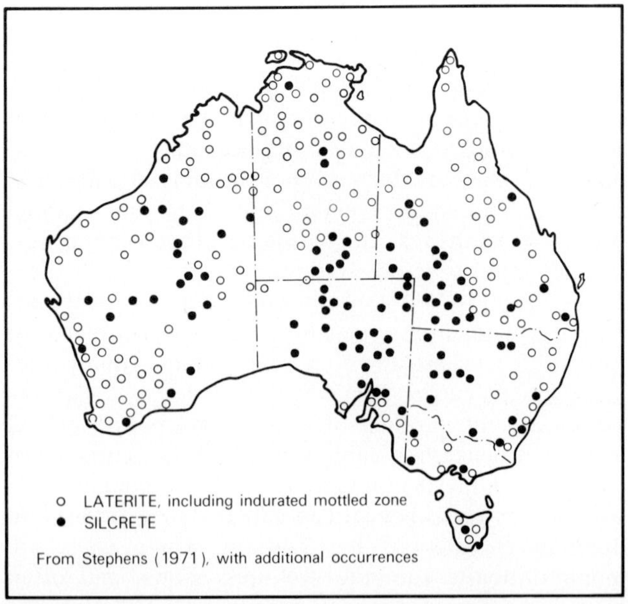

Fig. 3.3. Continental distribution of laterities, including indurated mottled zones, and silcrete. (From Stephens (1971), with additional occurrences).

Australian silcrete occurrences (see Fig. 3.3) are not directly associated with a 'laterite profile' as described above. Further discussion of silcrete will be found in Chapter 4. Except for the top of the laterite, all zone boundaries are merging ones and in some profiles there is a continuous gradual change from the laterite down to the parent rock. The laterite, together with either the mottled or pallid zone, is the minimum expression of the laterite profile, e.g. in the York district of Western Australia (Mulcahy 1960). The coarse red segregations of mottled zone exposures are hardened to some degree like the laterite itself and there is some evidence that even in these relict profiles hardening continues after exposure.

The above discussion indicates that the form of the laterite, associated relict soils, features of the mottled and pallid zones, and the overall dimension of profiles all vary greatly. The climate under which the profile formed, parent material, topographic position, and drainage have all contributed to this variety. At one extreme a fossil laterite profile on Kangaroo Island, South Australia (Northcote 1946) has a grey, sandy *lateritic podzolic* soil with a nodular laterite layer 50 cm thick and the whole profile is no more than 180 cm deep. At the other extreme a profile formed from basalt and exposed in the eastern scarp of the Binjour Plateau, Queensland (Teakle 1950) has several metres of *krasnozem* soil over several metres of massive, brick-like laterite, becoming cellular in the lower part and grading into a thick mottled zone. The whole profile is 65 m deep. Relict *lateritic podzolic soils* are common in southern and western Australia (Stephens 1946) and the more sesquioxidic forms (*red earths* and *krasnozems*) in northern and eastern Australia; differences in parent material rather than in climate, i.e. associated with latitude, seem to cause this. In Southern Queensland deeply weathered profiles similar to the laterite profile are found. These are up to 95 m thick and have formed in labile fine-grained sediments (Indurm and Senior 1978).

The term plinthite (Soil Survey Staff 1975) is used for laterite and for other sesquioxide-rich parts of the laterite profile with the potential to harden on exposure. It is defined as an iron-rich, humus-poor mixture of clay, quartz and other dilutents which changes irreversibly to an ironstone hardpan or to irregular aggregates on exposure to repeated wetting and drying. The word was coined to avoid the confusion arising from the earlier use in the U.S. of laterite as the name of a great soil group with deep, red B horizons over red or reticulately mottled, very deeply weathered parent material.

The term ferricrete has also been widely used in Australia, particularly by geologists and engineers, e.g. Grant and Aitchison (1970). In most, but not all, instances it has been used synonomously with laterite.

The intense and prolonged weathering and leaching during the formation of laterite profiles, together with the immobilization of some chemical elements in laterite, have resulted in severe nutritional impoverishment of the profiles and gross plant deficiencies have developed on the relict soils. This infertility is inherited by young soils formed in lateritic detritus and exposed remnants of laterite profile following dissection. Soil studies since the 1930s (e.g. Stephens 1946; Northcote 1946; Northcote and Tucker 1948; Stephens and Donald 1958; Mulcahy 1960; Gunn 1967) have gradually extended our knowledge of the important role of laterite formations and their dissection history in explaining the characteristics, occurrence, and distribution of many Australian soils. Some of these studies are discussed in Chapters 4 and 5.

Some features of Australian soil landscapes

Exposures of hardened materials of old, deeply weathered profiles

Exposures of laterite, weakly silicified, mottled and pallid zone materials, and silcrete are often important in controlling the shape of Australian landscapes. Erosion scarps cut in mottled and pallid zone materials or deeply weathered (kaolinized and ferruginized) sediments, and ranging from a few to more than 100 m high, extend from the humid coastal zone to the interior deserts and form features known as breakaways, jumpups, tenthills and tablelands. The continental distribution of these features follows that of the materials from which they have formed (see Fig. 3.3) and they appear as plateaux, tablelands, mesas, buttes, cuestas and lower rounded remnants thereof (Fig. 3.4). Most are erosion features produced by the dissection of extensive old plains initiated by uplift and tilting; a few may be the result of faulting (Whitehouse 1940; Stephens 1971). Climatic changes have also assisted this erosion. The dissection processes are discussed in Chapter 5.

Many of the landforms are protected by hard cappings

Fig. 3.4. Laterite-capped mesa forming part of the Kennedy Range, Western Australia.

Fig 3.5. *Desert loams* near the western Queensland–Northern Territory border with a lag gravel cover of silcrete.

Fig. 3.6. A shallow saline playa lake in arid inland Western Australia

of laterite, silcrete, and silicified altered sediments – the duricrust of Woolnough (1927) – which characteristically form jagged cliffs at the edge of the retreating scarp. The slopes below are partly mantled by fragments of the capping. Where the cappings are thin and the underlying materials are uniformly soft, straight bevelled scarps border the larger landforms and produce the tenthill forms of small remnants. In southern and south-western Queensland, variably hardened materials of the deeply weathered zones form prominent cliffed scarps. In this region as well as elsewhere in eastern Queensland, the mottled zone (usually coarsely mottled light grey and red and somewhat hardened by silification) is exposed in many positions in the landscape. These positions may vary from the beds of streams (Colour Plate 3.14) to the shoulders of low rounded landforms capped by modified lateritic *red earths*. Where the stripping of the laterite profile is more advanced, many of the rounded uplands forming the 'hard' mulga (*Acacia aneura*) country of this region are mantled by relatively shallow loamy *red earths* formed from the underlying mottled zone. At lower elevations the *red earths* of the 'soft' mulga country are formed in the partly sorted detritus from the *red earths* that previously formed the upper part of the laterite profile (see Chapter 18). In parts of Western Australia, e.g. the Mt Keith area, the low cliffed scarps of the dissected country are formed in the pallid zone, the pediment below being mantled by materials from this source.

The soils of the old laterite profiles are best preserved on extensive tableland remnants and, where erosion has been limited, on the gentle back slopes of cuestas. In southern Australia these sites preserve the least modified remnants of the *lateritic podzolic soils* but in the north *red* and *yellow earths* are found on them. Elsewhere dissection and differential erosion have provided a wide range of parent materials for younger soils, from transported and variously sorted detritus to *in situ* exposures of portions of the laterite profile (Stephens 1946; Northcote and Tucker 1948; Mulcahy and Hingston 1961; Gunn 1967).

Lag gravels

Lag gravels of fragments of laterite and silcrete are widespread on many landscapes, particularly in semi-arid and arid areas (Fig. 3.5). These residual surface concentrations resulting from the removal of finer materials by erosion may also be derived from several sources: from the underlying country rock, from a previously overlying formation that contained coarse resistant material, and from gravelly piedmont deposits, e.g. the outwash fans of the northern Flinders Ranges, South Australia. Many lag gravels have darkened and polished surfaces, desert varnish, and have to be broken to be identified.

Many areas in arid Australia would literally be blown or washed away were it not for their protective cover of gravels or stones, e.g. the soils of Sturt's Stony Desert. In arid regions lag gravels, commonly 1–20 cm diameter, are usually fairly tightly packed on the soil surface; typically they are only one layer thick. Elsewhere they are more scattered, e.g. on the lateritic tablelands of Kangaroo Island, and in the south-west of Western Australia. Gravel on the cracking clay soils of the rolling downs of central western Queensland may be the last resistant remains of once overlying lateritic formations which have been completely stripped as the present-day landscape has been shaped.

High salinity and salt scalds

Areas of high natural salinity as evidenced by bare salt-encrusted ground are widespread throughout arid and semi-arid parts of Australia. Playa floors and the ephemeral shallow salt lakes abundant in central Australia are examples (Fig. 3.6). These saline depressions have bare salt-encrusted pans in the lowest part, and often varying calcrete platforms and fringing gypsum and calcareous dunes along their margins (Bettenay 1962). Gypsum and sodium chloride are common in the pans which may have brine beneath them. Alunite (hydrated potassium and aluminium sulphate) may occur. Most saline lakes and valleys are the dismembered remnants of ancient drainageways that date back to Tertiary times (van de Graaff *et al.* 1977). At present they are often the only drainage systems in a very large endoreic drainage area. The *earthy sands, siliceous sands* and *red earths* situated above the saline valleys have been leached and contain only very small amounts of soluble salts.

Other highly saline natural features are the salt pans, marine forelands and delta fringes associated with many low-lying sections of the north Australian coastline. These formations are periodically flooded by saline or brackish waters and have saline water tables at shallow depths; salt concentrates on the surface owing to the high evaporation. Their soils are *solonchaks.*

Salt scalds appear as bare crusted patches in semi-arid areas, especially associated with sodic texture-contrast soils which are alkaline. The scalds are formed following erosional exposure of clay subsoils which disperse when wet and seal, thus preventing further water entry. High salinity, high sodicity and very low permeability of the affected soil surface are characteristic of this form of secondary dry-land salinization. The reclamation of salt-affected land is more fully discussed in Chapter 53.

Gilgai

Gilgai forms

Australia is unique in the variety and extent of gilgai associated with its soils. These micro-topographic features – small hummocks and hollows in patterns showing varying degrees of order – occur on a range of swelling clay and texture-contrast soils that have thick subsoil clay horizons. Early settlers gave gilgai such names as Bay of Biscay, crabhole and melon hole. Their continental distribution (Fig. 3.7) follows the two broad kinds of soils mentioned above, but not all such soils are gilgaied. Occurrences are on level or very gently sloping plains in regimes with pronounced wet–dry seasonal climates or subjected to more frequent intense drying, and are generally limited to soils at least 1 m deep except in arid areas where 60 cm seems to be a minimum depth (Jessup 1951).

Gilgai is an Australian aboriginal word meaning small waterhole. Its adoption for soil microrelief patterns recognizes the intermittent ponding of water in closed depressions of the larger forms. It is now firmly established in world soil literature (Soil Survey Staff 1975; FAO-UNESCO 1974) as a generic term embracing the great variety of microrelief forms typical of clayey soils that have a high coefficient of expansion with seasonal changes in water content. The term gilgai does not apply to microrelief apparently resulting from repeated freezing and thawing, solifluxion, or faunal activity (Anon. 1976). Gilgai are accompanied by orderly soil profile variation but the term should be restricted to the micro-topography.

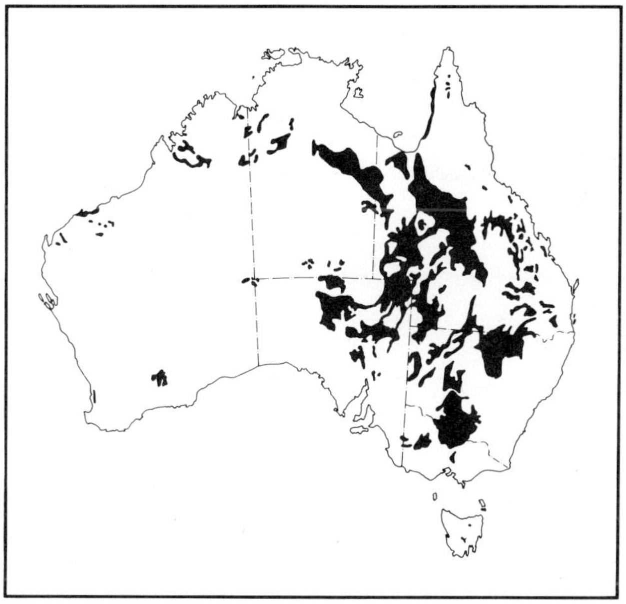

Fig. 3.7. Continental distribution of gilgai.

Three basic elements are recognized (Anon. 1976).

(i) Banks or mounds – respectively convex upward along one horizontal axis or two axes at right angles to each other. The term 'puff', previously used synonymously with 'mound' or 'banks' (Hallsworth *et al.* 1955) is more appropriate for the part of these forms that has a selfmulching surface of loose granular clay.

(ii) Depression – concave along one or both horizontal axes forming, respectively, an open-ended trough or a closed depression, subcircular to irregular in outline. A conspicuous sinkhole with vertical sides is sometimes present in the bottom of closed depressions.

(iii) Shelf – a nearly planar surface either intermediate in level between mound and depression or forming the general level in which they occur.

The various kinds of gilgai consist of two or more of these elements of different dimensions and spacing, arranged in random or ordered patterns. The dimensions commonly used in description are the vertical interval VI = the difference in height between adjacent low and high points, or amplitude ($VI/2$), and the wave length, horizontal interval, or repetition distance for mounds or depressions. Although distinct kinds of gilgai can be recognized visually, subjective descriptions supported by these measurements are usually inadequate to characterize them. The calculation of microrelief parameters such as irregularity (avoidance factor), expected range of heights (simple microrelief factor), and structural

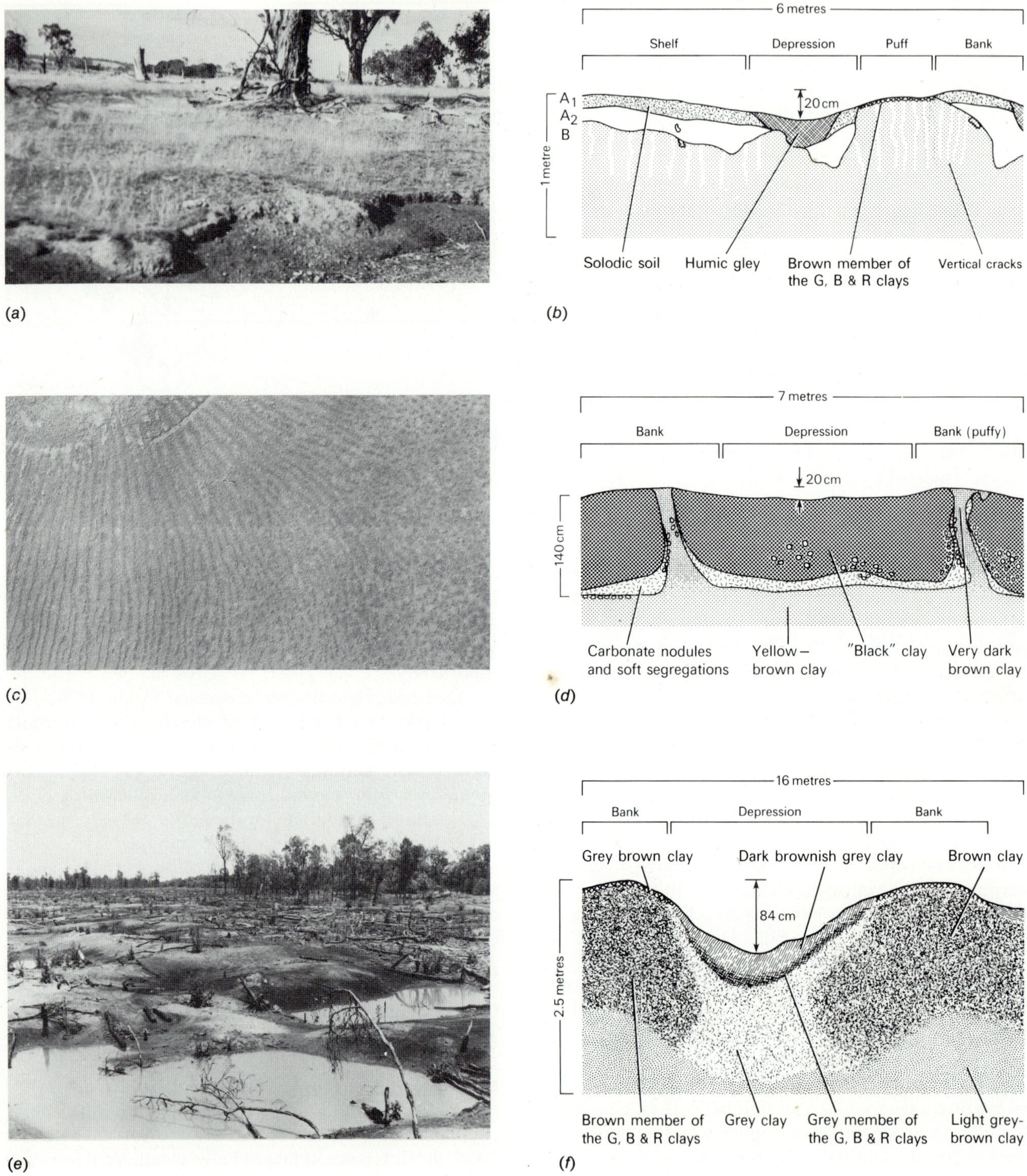

Fig. 3.8. (*a*) Crabhole gilgai, Tasmania. (*b*) Vertical section across crabhole gilgai on dominantly solodic soil in the Launceston Basin, Tasmania, showing microrelief features and associated soil profile variation. (*c*) Vertical aerial photograph showing normal and linear gilgai, Darling Downs, southern Queensland. (Original scale 1:6000). (*d*) Vertical section of linear gilgai on black earths on the Darling Downs, southern Queensland showing microrelief features and associated soil profile variation. (*e*) Melonhole gilgai, Tara district, southern Queensland. This is characteristic of much of the brigalow *(Acacia harpophylla)* country. (*f*) Vertical section of melonhole gilgai showing microrelief features and associated soil profile variation on grey and brown members of the grey, brown and red clays of brigalow forest lands, Chinchilla district, southern Queensland.

homogeneity has been suggested (Russell and Moore 1972) as a more adequate basis for their characterization and study.

Six gilgai types were recognised by Hallsworth *et al.* (1955); normal, lattice, wavy, tank, stony and melonhole. Of these, stony gilgai are stony phases of the first three types and lattice gilgai is a transitional form between wavy and normal gilgai. In order of increasing dimensions, types now commonly recognized (Anon. 1976) are as follows.

(i) Crabhole gilgai – randomly distributed small mounds and depressions separated by a more or less continuous shelf: *VI* < about 30 cm, wave length 3–20 m, surface almost level (Fig. 3.8 *a* and *b*).

(ii) Normal gilgai – small mounds and/or subcircular depressions varying in size and spacing and randomly distributed: *VI* < about 30 cm, wave length 3–10 m, surface near level (Fig. 3.8 *c* and *d*).

(iii) Linear gilgai – (Parberry 1933; Beckmann *et al.* 1973), also wavy gilgai – a pattern of long narrow banks and broader depressions more or less parallel to the maximum slope: *VI* < 30 cm, wave length 5–8 m, slope < 30' to 5° (Fig. 3.8 *c*).

(iv) Melonhole gilgai – large depressions usually 3 m diameter, subcircular or irregular and varying from closely spaced in a network of banks to isolated depressions set in an undulating shelf with occasional small mounds: *VI* 30 cm to 2 m, wavelength commonly 6–50 m, general surface almost level. (Fig. 3.8 *e* and *f*).

Probably the largest gilgai features, justifying the name tank, occur in scattered groups on the deep *grey and brown clays* of the brigalow (*Acacia harpophylla*) lands near Tara, Queensland (Fig. 3.8 *e*). These are closed linear depressions, irregular to arcuate ranging from 25–100 m long, 6–12 m wide, and up to 1.5 m deep.

Desert gilgai (Ollier 1966) is characterized by slightly depressed or slightly raised areas of cracked clay with little or no stone which are surrounded by a rim of gibbers and set in shelf areas covered by a stony pavement; it may be a variant of crabhole gilgai described by Jessup (1951) in the same region.

Soil variation

Profile variation associated with the components of gilgai may span from one to three great soil groups. *Black earths* associated with all elements of normal and linear gilgai on the Darling Downs, Queensland, exemplify the former; *brown clays, solodic soils*, and *humic gleys* on puff, shelf and depression respectively of crabhole gilgai in the Launceston basin, Tasmania, exemplify the latter. This profile variation can be explained as either rearrangement of horizons of the pre-gilgai soil by displacement during gilgai development (this gives antecedent horizon differentiation) or as profile differentiation resulting from gilgai development and the creation of new microenvironment (this gives consequent horizon differentiation). In the Tasmanian example given above, the pre-gilgai soil is the strongly differentiated *solodic* soil; the *brown clay* of the puff is the modified B horizon of this soil exposed on the slope of the mound following the stripping of the A horizon by erosion; and the *humic gley* is the result of post-gilgai gleying of the *solodic soil* along with accumulation of organic matter throughout its A horizons in the new microenvironment created by the development of the depression.

The origin of gilgai

There can be little doubt that the swelling and shrinking of clay subsoils are important in the development of gilgai. This is indicated by the association of gilgai with soils having thick subsoils of expansive clays in regions with annual or shorter cycles of thorough wetting and drying. Not all swelling soils are gilgaied, however, and the processes involved in gilgai formation are not fully understood. Consequently there is no consensus on how gilgai develop. The hypothesis most widely supported (Leeper 1948; Hallsworth *et al.* 1955) is that when the soil is dry, material from the surface and sides of the upper part of major cracks falls into or is washed into the deeper cracks, so reducing the volume available for expansion on rewetting of the subsoil. This creates pressures which are relieved by heaving of the soil between the major cracks which, once established, tend to be maintained on subsequent drying. This process is repeated, with the result that the subsoil is progressively displaced, a mound develops between the cracks, and the soil surface adjacent to the cracks is lowered to form depressions. Some observations support this hypothesis. In crabhole gilgai on some texture-contrast soils, wedge-shaped fills of loamy A horizon material are found in the cracks, and low mounds of self-mulching clay occur between them. Several observations also point to the movement of subsoil material to the surface of the mounds: carbonate nodules and gravel found on the mounds and banks of some gilgai formations are clearly lag concentrations of these materials which are otherwise confined to the subsoil between and beneath the mounds; the deep subsoil clay is similar to that exposed on puffs, and tongues of pale-coloured, calcareous clay rise to the surface beneath mounds of gilgai on black earths.

A different hypothesis for crabhole gilgai development on texture-contrast *solodic soils* (Howard 1939) emphasizes localized wetting of the impermeable clay subsoil at isolated points of higher permeability (e.g. those associated with old root channels) beneath a saturated A horizon. The resultant swelling of the subsoil initiates the development of a mound, shearing of the clay surrounding the wetted subsoil creates cracks which facilitate subsequent water entry, and lateral movement of water and salts into the more permeable clay under the developing mound support its growth. A combination of crack filling and differential wetting (via cracks) controlled by water movement down gentle slopes are thought to be the significant processes in the development of linear gilgai (Beckmann *et al.* 1971).

Whatever the mechanisms of gilgai development, observation and analysis of gilgai over wide areas lead to the conclusion that the following are significant factors: the characteristics of the soil including the thickness, swelling, size and spacing of the cracks; the frequency and intensity of wetting and drying; and the surface slope. In studies of gilgai soils in New South Wales, Hallsworth *et al.* (1955) found swelling to be significantly related to clay content, the exchange capacity of the clay, and the exchangeable sodium percentage (ESP); and that in the zone of 'optimum gilgai development' the vertical interval was significantly related to the swelling capacity of the clay and its ESP. They concluded that the distinguishing feature of gilgai soils was a characteristic swelling profile with a volume expansion of not less than 10% in the topsoil and increasing with depth, usually by 15–30%. However, similar measurements on gilgaied *grey clays* (Isbell 1957) and *black earths* (Beckmann, unpublished data) did not support this. Furthermore, well developed gilgai are associated with texture-contrast soils with sandy to loamy textured surface horizons that undoubtedly have volume expansion of less than 10%.

Land use effects

Gilgai are of considerable significance in land use. They give rise to differences in the distribution and growth of both native vegetation and crops, instability of foundations for roads, buildings and other structures, poor trafficability in virgin areas, and displacement of fence posts and poles buried to depths of less than 75 cm.

The effects on plants are due to differences in soil and water distribution associated with the microrelief and cracks. In native grasslands and volunteer pastures there are pronounced differences between mounds and depressions in the distribution of species, in population density, and in production. Often the mounds of crabhole gilgai are almost bare. In crops, differences in stand density and vigour are usually pronounced in the early years but decline with increasing periods under crop. On the *black earths* of the Darling Downs, low microrelief is commonly eliminated by a decade or so of cultivation and fairly uniform crops of wheat are then obtained (Thompson and Beckmann 1959); the soil differences remain below cultivation depth and may still affect crop growth in some seasons. On the *grey clays* of the cleared brigalow forest lands further west, melon hole gilgai with a vertical interval of 0.5 to 1.5 m interfere so seriously with all of the operations involved in grain cropping that the cultivation of many areas is precluded. Uneveness of crop growth is much greater than elsewhere and persists for decades. In wet years, the ponding of water in depressions may also prevent cultivation, seeding or harvesting of significant portions of fields. If the microrelief is reduced rapidly by shifting soil from crests to depressions, the uneveness of crop growth is increased because infertile subsoils on the crests are then exposed.

Engineering and foundation problems and the costs of overcoming them usually increase with the vertical interval and local soil variation associated with gilgai. Design and construction practices are aimed at preventing or minimizing changes in soil water content, hence restricting movement in the expansive clay on which the structures are founded, or at isolating buildings from movement of the clay. These objectives are achieved by constructing broad impermeable aprons around foundations, by raft foundations strengthened by beams, or by the use of pier and beam foundations anchored below the depth of significant changes in soil water content.

Generalized profile forms of Australian soils

Detailed descriptions and laboratory data for great soil groups recognized in Australia are given in the 'Handbook of Australian Soils' (Stace *et al.* 1968), and a comprehensive coverage of all known Australian soils as classes of the Factual Key (Northcote 1971) is presented in 'A Description of Australian Soils' (Northcote *et al.* 1975). In the Handbook the classes broader than great soil groups emphasize increasing degree of leaching and do not necessarily contain soils of similar morphology. In the case of 'A Description of Australian Soils', the broader categories are essentially based on the texture profile and contain soils of similar morphology.

To simplify an outline of the character of Australian soils, and to avoid repetition in the description of features common to several of them, the recognized great soil groups (Stace *et al.* 1968) and other important soils (Northcote *et al.* 1975) can be grouped into a number of loosely defined classes on the basis of the general appearance of the profile. This is determined by colour, texture, and structure; in some soils also visible segregations of ferro-manganiferous compounds, carbonate and salts.

The features of these 'generalized profile forms' and their component soils are outlined in this section and the continental distribution of each form is shown on the associated maps. Soils of minor importance in terms of aggregate area of occurrence or of use are generally not considered. Great soil group names of Stace *et al.* (1968) are italicized in the text and approximate equivalents given from the Factual Key. Unnamed soils are referred to by descriptive phrases (from Northcote *et al.* 1975) emphasizing key features at class level of the Factual Key. An approximate correlation between the groups of Stace *et al.* (1968), the Factual Key notation, and the units of Soil Taxonomy (Soil Survey Staff 1975) and FAO-UNESCO (1974) is given in Chapter 20.

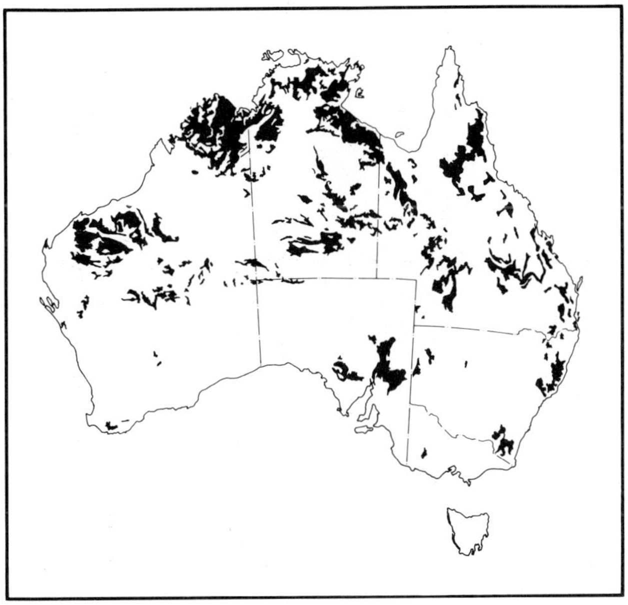

Fig. 3.9. Shallow stony soils.

Shallow stony soils

The essential features of this group of soils are the presence of gravelly or stony fragments of weathered and unweathered parent rock throughout the profile, usually the lack of marked horizons other than an A_1, and shallow depths (generally < 30–50 cm), see Fig. 3.9.

The component soils include *lithosols*, many of the *grey-brown* and *red calcareous soils*, and many of the shallow sands, loams and non-cracking clays of Northcote *et al.* (1975). The major occurrences are in the more arid regions; on steep slopes and ridge crests in the more humid zones, and on the more resistant rocks elsewhere. Parent rock outcrops are usually a feature of the soil occurrences.

The development of the A_1 horizon is variable, ranging from virtually none in many drier situations to prominent dark, moderately structured horizons in some humid areas. Many of the soils are not true lithosols in that they may possess thin A_2 horizons and weakly developed B horizons as evidenced by colour and/or structure.

The shallow stony soils are so diverse that it is not possible to give a generalized summary of their mineralogical and chemical properties, which are strongly influenced by both parent material and climate. The soils vary widely from those containing little other than quartz sand to those with moderate amounts of 2:1 clay minerals, unweathered rock and mineral fragments, and relatively high organic matter contents. Soil reaction, base status, and total nutrient element contents therefore vary widely. Most of the soils are little used other than for sparse grazing of native pastures.

The shallow stony soils are best discussed in terms of three broad texture groups.

(i) The sandy soils (shallow classes of Uc1.2, .4, Uc2.1, Uc4.1, .2, Uc5.1, Uc6.1) range from sands with minimal A_1 horizon development (Uc1.2) to soils that have a weak A_2 horizon and sometimes a thin, coloured B horizon (Uc4.1, .2). Such soils occur widely in tropical Australia on sandstone or quartzite hills, rocky ridges and dissected plateaux. Soils with a markedly bleached A_2 horizon are also found; in the Northern Territory these form on gently undulating terrain and overlie massive nodular ironstone or high amounts of ironstone nodules, whereas in eastern Queensland sandy soils with bleached A_2 horizons are formed on granite or acid volcanic rocks on hilly landscapes.

(ii) The loamy soils (shallow classes of Um1.4, Um2.1, .2, Um5.4, Um6.2) have a wider distribution with much variation in profile development. The simpler forms (Um1), with only weak A_1 horizon development, commonly occur on steep slopes and ridge crests in hilly to mountainous terrain and on dissected mesas. Parent rocks may include all but the more siliceous types. Those soils formed on limestone, dolomite and other calcareous rocks in more arid areas have been termed *grey-brown* and *red calcareous soils*. Other Um1 soils are commonly derived from the indurated remnants of dissected laterite profiles. Similar soils but with a weak A_2 horizon and/or a weakly-coloured B horizon (Um4.1, .2) are common in the more humid areas of eastern Australia.

Soils with weak profile development, which consists mainly of minor colour and texture changes, and with a markedly earthy fabric (Um5.5), are widespread in arid central western Australia where they are closely associated with, and intergrade into, the red earthy forms of the massive sesquioxidic soils. These soils which occur on a wide range of rocks, are often associated with dissected laterite profiles.

Loams with a bleached A_2 horizon (Um2.1) occur in subcoastal areas of north-eastern Australia, mainly on low hilly to mountainous terrain on argillaceous and sometimes on acid volcanic rocks. Loams with weak pedality throughout the profile (Um6.2) commonly occur on more basic rocks in the semi-arid and seasonally humid areas, mostly on stony ridges, hills, and low benched plateaux.

(iii) The non-cracking clays (Uf1.4) are much less common and are mainly restricted to profiles with little differentiation on the dissected argillaceous remnants of some lateritic mesas in western Queensland.

Ironstone gravel soils

The ironstone gravel soils contain more than 60% of ironstone gravel throughout the profile. The most widespread component soils (KS-Uc1.43, KS-Uc2.21, KS-Uc4.11, .12, .21, KS-Uc5.21, .22) are sandy and usually shallow (< 60 cm deep). In addition, there are smaller areas of *red* and *yellow earths* (KS-Gn2.11, .21) and *lateritic podzolic soils* (KS-Dy2.62). All these soils have

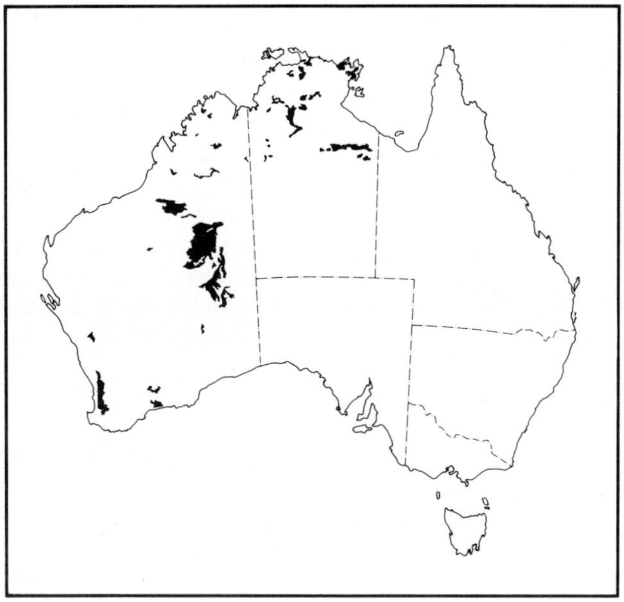

Fig. 3.10. Ironstone gravel soils.

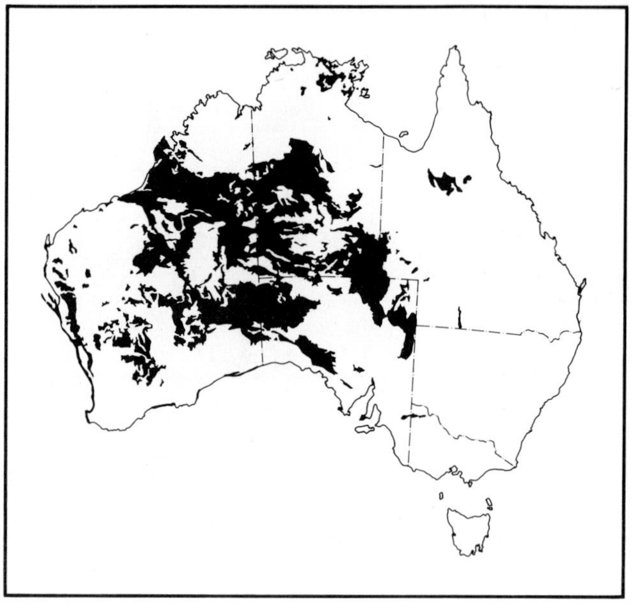

Fig. 3.11. Deep sands.

formed in colluvium composed largely of ironstone gravels, and are underlain by hard laterite or some other portion of a truncated laterite profile (Fig. 3.10).

By far the most extensive areas of ironstone gravel soils are on dissected lateritic plateaux in climates ranging from arid to humid. These soils are generally little used; in the Darling Range of Western Australia they support commercial hardwood forests of jarrah (*Eucalyptus marginata*). They are mildly acidic and, like all members of this generalized profile form, their inherent fertility is low.

Deep sands

These soils are sands throughout, have minimal profile development and are generally > 1 m deep. The group includes *calcareous sands, siliceous sands, earthy sands* and brownish sands. Major occurrences are on dunes, swales and plains of the arid and semi-arid regions and of the humid coastal areas. They occupy about 30% of the continent (Northcote *et al.* 1975). Their parent materials are calcareous and/or siliceous sands of aeolian, colluvial, alluvial or littoral origin (Fig. 3.11).

Deep sands are notable for excessive drainage and gross deficiences of both major and minor elements. Consequently many areas remain unused even in the humid regions. Some sheep and cattle grazing occurs on them and in specially favourable locations they also support limited cereal cropping, notably barley, and irrigated horticulture. Wind erosion is always a serious hazard when these soils are left with little vegetative cover.

The more important features of the component soils are given below.

(i) The *calcareous sands* (Uc1.1) are found on the coastal beach ridges, dunes and/or sand sheets that occur at intervals along the Australian coast, particularly on the southern half of the continent. They have developed in Quaternary sand formations containing many shell fragments. Carbonate contents vary from 10 to > 80% of the fine earth. The colour of the sands varies from off-white through yellows and reds to almost black. 'Coast disease' caused by copper and cobalt deficiency occurs in stock grazed on the natural grasslands growing on these soils on the Bass Strait islands and the southern mainland coast.

(ii) The *siliceous sands* (Uc1.2) cover about 10% of the continent. They are dominantly quartzose and vary in colour. The most extensive are the red forms (Uc1.23) of the Australian sandy deserts; yellow forms (Uc1.22) are common on Quaternary dunes and sand sheets adjacent to drainage lines of riverine plains west of the Great Dividing Range. Coastal dunes and beach ridges have both yellow and pale-coloured (Uc1.21) forms, and in these moister locations in south-west Western Australia and southern Victoria, softwood forests (*Pinus* spp.), vegetables, and grapevines are grown. Irrigated horticulture is practised on the Riverine Plain.

(iii) The *earthy sands* (Uc5.2) differ from the above soils in being generally more coherent and having an earthy fabric, especially in their subsoils. These sand to sandy loam soils may be red (Uc5.21) or yellow (Uc5.22), are always associated with highly siliceous parent materials, and often occur on remnants of deep-weathering profiles. They form extensive plains in the arid interior, especially in the western half of the continent, and sparse native herbage provides grazing for cattle and sheep. These soils are basically infertile but where they extend into semi-arid areas they are used for cereals and pastures with heavy applications of fertilizer.

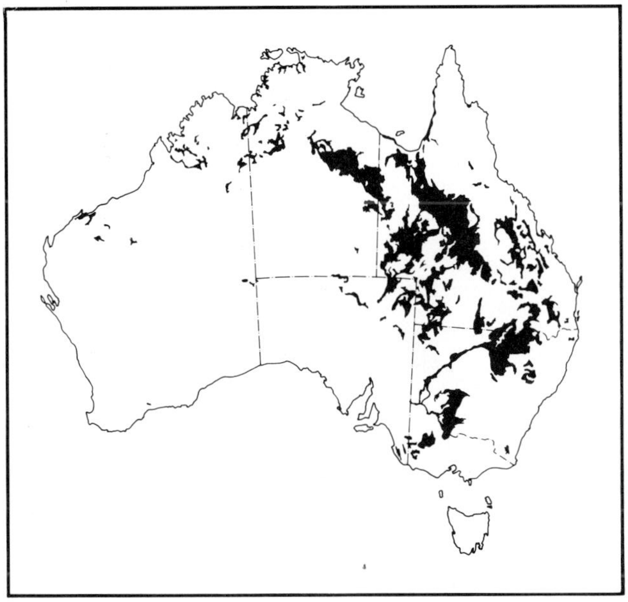

Fig 3.12. Cracking clays.

(iv) The brownish sands (Uc5.1) are chiefly associated with dunes and sand sheets in the semi-arid parts of southern Australia. They are generally loose sands, brownish to reddish in the upper part, and gradually becoming paler and yellower with depth where carbonates or earthy pans may occur. They are particularly liable to wind erosion where they are used for dry-farming as in the mallee lands of southern Australia. Some areas are used for irrigated horticulture which is mainly citrus.

Cracking clays

In Australia there are large areas of soils with high contents of clays which swell and shrink greatly on wetting and drying. These uniform clay profiles develop a characteristic structure and crack widely because of the swelling and shrinkage. Clear-cut colour and textural differences are lacking in the top metre or so of these profiles (Fig. 3.12).

Definitive structural features include (i) a subsurface horizon of moderate, medium to coarse, irregular blocky or polyhedral peds, grading below 30–50 cm depth into (ii) a subsoil of moderate, medium to coarse, lenticular or parallelepiped peds (Butler 1955) and with prominent diagonal shear planes or slickensides, both features continuing to the full depth of the soil. In the subsoil, ped surfaces are glazed and often striated. Slickensides are inclined 40–60° to the horizontal and may extend, uninterrupted, for several square metres. High coherence of peds is a feature of these soils for it is the compound aggregates that determine the visible structural profile.

The following may vary greatly: gilgai development (whether it is present or not, its kind and degree), the mineral suite of the clay fraction, the soil reaction profile, and salinity.

Fig. 3.13. Widely cracked surface of a *brown clay* (scale = 30 cm).

The main soils include the *black earths* and the *grey, brown*, and *red clays*, the latter having the largest area. Typically the cracking clays are soils of rolling plains, neutral to strongly alkaline, calcareous and moderately saline in the deeper subsoil, and generally fertile. Parent materials are either clay sediments or labile sandstones and basic rocks that produce high clay contents on weathering. Although the texture profiles are mostly uniform, a gradual increase in clay content with depth occurs in the upper part of some soils.

Characteristically the surface soil is self-mulching, forming a loose layer (2–5 cm thick) of fine to coarse granular and polyhedral peds (Fig. 3.13). In some soils, notably in the more arid regions, this layer is capped by a fragile laminated sandy clay crust 35 mm thick. When dry this forms polygonal plates up to 10 cm across, separated by fine cracks which quickly close with a light shower of rain to form a surface seal. Less common variants have a massive sandy clay surface up to 3 cm thick firmly attached to underlying blocky peds, or have weak, coarse blocky structure from the surface down.

Cation exchange capacities range from 20 to 80 m-equiv. per 100 g of soil, and vary with clay content (40–80%) and the clay mineral type. Calcium is usually the dominant exchangeable cation in the upper horizons and magnesium in the deep subsoils in which 10–25% sodium saturation is common. Many *grey, brown* and *red clays* of the semi-arid lands have appreciable exchangeable sodium contents in the surface soil or at shallow depth. Where the rainfall is adequate these soils are used for a range of summer and winter crops. Elsewhere they support good quality native pastures and considerable areas are irrigated for crops such as rice and cotton.

Black earths (Ug5.1, Ug5.4) are distinguished by dark moist colours — black to very dark grey or brown — to about 90 cm depth or to the bottom of the solum if this is shallower. Most occur on basaltic parent materials in areas of 500–700 mm strongly seasonal rainfall scattered through the near hinterland of eastern Australia. The clays are overwhelmingly montmorillonite-dominant (Norrish and Pickering 1977).

Typically they have a self-mulching granular surface. Soils deeper than about 60 cm have some free carbonates in the subsoil; those deeper than 80–100 cm have brown, yellowish brown, or light grey-brown deep subsoils and moderate gilgai, (15–20 cm vertical interval) which is linear on colluvial slopes and of the normal or crabhole type on the plains. Tongues of the paler deep subsoil with nodules and soft patches of carbonates come to the surface on gilgai crests.

Grey, brown and *red clays* (Ug5.2, Ug5.3, Ug5.5) are distinguished from *black earths* by greyer, browner or redder colours and very gradual colour changes to depths of as much as 2 m. The group is broad, accommodating wide ranges of some properties, particularly surface structure, reaction profiles and gilgai. In southern Australia the clay fractions are illite-dominant with 20–30% kaolin; in the east and north they are montmorillonite-kaolin mixtures with some illite (Norrish and Pickering 1977).

The *red clays* occur mainly in the drier semi-arid areas, the *grey* and *brown clays* range over annual rainfalls from 650 to less than 250 mm from the seasonally sub-humid regions to the desert margin. Most of the soils occur on plains underlain by clay alluvium or rocks other than the basic igneous types but of high clay-forming potential. The *grey clays* usually have the poorest drainage and include some soils with massive or coherent, coarse blocky surface structure. This poor structure which is generally associated with low-lying sites and occasional flooding, may be reflected in some rusty and ochreous spotting in the upper horizons and mottling below. The *brown clays* are commonly found on better-drained sites, e.g. broad banks on alluvial plains, and often occur with parent materials of higher calcium status. Thin fragile crusts that cap self-mulching clay surface soils are common in the drier regions where the soil cover is often sparse and there is an appreciable exchangeable sodium content near the surface. In these drier regions crystalline gypsum and carbonate nodules are features of the profile. Gilgai development ranges from none through crabhole and normal to melon-hole forms.

Calcareous soils

These soils all have calcareous material disseminated through the fine earth fraction of the soil profile, and soft and hard carbonate segregations, usually concentrated in the subsoil. Moderately deep profiles having a gradual texture increase with depth include the *solonized brown soils*; related soils without an increase in soil texture are the shallow *grey-brown* and *red calcareous soils, calcareous sands*, and some deep brownish sands (Fig. 3.14).

The *solonized brown soils* are widespread through the semi-arid to arid regions across southern Australia, with much smaller areas elsewhere. They cover undulating plains, dunes and low hills. Their commonest parent materials are unconsolidated coarse- to medium-textured, calcareous sediments modified by aeolian activity; other parent materials include limestones, calcretes, calcareous shales and lake parna (Bettenay 1962).

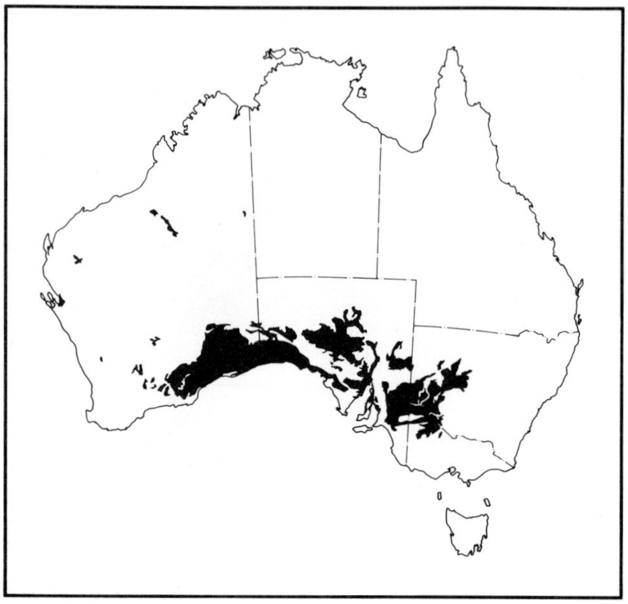

Fig. 3.14. Calcareous soils.

The calcareous materials include some magnesium carbonates (Northcote 1951) but calcium carbonate is dominant. The visible distribution of the carbonates down the profile, together with horizons of maximum carbonate concentration in subsoils, provide the most distinctive features of these soils, which otherwise have merging, indistinct horizons. The carbonate in the fine earth fraction contributes to the apparent clayiness of the field texture to such an extent that the subsoils seem more clayey than indicated by their analysis. The calcareous layers often have low density and bearing capacity with the result that they are likely to 'collapse' on wetting. They are strongly alkaline soils (pH > 9), often with moderate to high soluble-salt contents in their subsoils. Inherent fertility is low to moderate, and soil productivity is greatly improved by applications of superphosphate and the use of legumes to improve their nitrogen status. Zinc, manganese, iron and copper deficiencies have been recorded.

Three broad types of *solonized brown soils* have been recognized.

(i) The more calcareous and duller-coloured soils (Gc1.1). These are the most widespread. They contain the largest amount of carbonates (> 50% in the horizons of maximum accumulation), have the highest pH values and soluble-salt contents and have low permeabilities with subsoils often collapsing when wet. They tend to loamy

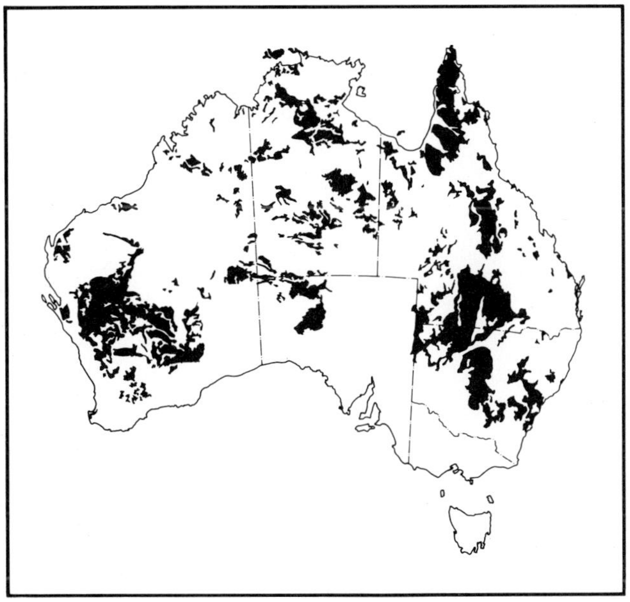

Fig. 3.15. Massive sesquioxidic soils.

textures in their thin surface soils and clay contents do not increase with depth, although fine earth carbonates in the clay fraction do.

(ii) Less calcareous and brighter-coloured soils (Gc1.2). These are commonly associated with (i) but are generally less extensive. They contain smaller amounts of carbonates, have slightly lower pH values and soluble-salt contents, tend to have sandy surface soils that are up to twice as thick as those of (i), and have clay contents that increase with depth to maxima in the subsoil. They are the most permeable of the *solonized brown soils.*

(iii) The structured soils (Gc2.2). These have loamy surface soils and moderate blocky structure throughout the profile; otherwise they are similar to (ii). Their occurrence is limited.

Massive sesquioxidic soils

These soils are massive throughout but have characteristic earthy fabrics and clay fractions dominated by kaolinite and iron oxides. Surface textures are sandy or loamy and most soils have a gradual increase in clay content with depth in the upper part of the profile. Most profiles are moderately deep to deep but some loamy forms are shallow and have a uniform texture throughout. Horizons are often only weakly developed and many of the soils are very permeable in their upper profiles (Fig. 3.15).

The component soils are *red earths, calcareous red earths, yellow earths*, grey massive earths, *red* and *brown hardpan soils*, and some earthy loams. They are very widespread and occupy about 20% of the continent (Northcote *et al.* 1975). The earths are by far the most extensive, and of these the *red earths* are dominant. The grey massive earths are mostly found in tropical Queensland; the *calcareous red earths, red* and *brown hardpan soils*, and the earthy loams are largely confined to the more arid areas of less than about 400 mm mean annual rainfall.

In the earths, B_2 horizons range from sandy clay loam to medium clay. The clay maximum usually occurs between 1 and 2 m in the *red earths* but is often shallower and may be more pronounced in the yellow and grey soils. Ironstone nodules are prominent in many of these massive sesquioxidic soils and may constitute more than 60% by weight of the lower B horizons; in other cases they may be cemented into a massive indurated layer.

The massive sesquioxidic soils have formed from a wide range of dominantly siliceous parent materials. Some are relict soils while others have formed on the truncated remnants of laterite profiles, or on their detritus. All have a major quartz component and usually the clay fractions are dominated by kaolinite with smaller amounts of illite or randomly-interstratified minerals; this is particularly so for *red earths* in which haematite and goethite are the dominant iron oxide minerals. In the *yellow earths* only goethite occurs (Isbell and Smith 1976). There is greater variation in chemical properties within *red* and *yellow earths* than between them (Dawson and Ahern 1974; Isbell and Smith 1976; Day and van Cuylenburg 1980). Generally the massive sesquioxidic soils have low contents of soluble salts, organic carbon, nitrogen, phosphorus, sulphur and exchangeable basic cations. Mostly they are moderately acid with pH values increasing slightly with depth, but the *calcareous red earths* are alkaline throughout. The base saturation in both surface soils and subsoils ranges from 10 to 80%, the higher values being more common in the lower rainfall soils. Cation exchange capacities of the B_2 horizon fall in the range 6–30 mequiv. per 100 g of clay. The higher values are more common on soils from the arid regions.

The majority of the soils are used only for the sparse grazing of native pastures by sheep and cattle.

The component soil groups are as follows:

(i) The *red earths* (Gn2.1, .4) are red to depths of 1 to > 6 m and usually occupy well-drained sites. In more humid areas, a darker A_1 and a paler A_2 horizon may occur. Although the B_2 horizons are seldom mottled, in many of the deeper profiles the B_3 horizons are prominently mottled with red, light grey, and yellow.

(ii) The *calcareous red earths* (Gn2.13) have weak A_1 and no A_2 horizons, and also the lower part of the profile has free carbonates, usually in the form of isolated soft segregations or fine nodules.

(iii) The *yellow earths* (Gn2.2, .3, .6, .7) vary considerably in profile form, horizon differentiation being frequently more pronounced than in the red earths. Although often closely associated with *red earths*, the *yellow earths* are mostly shallower and often have impeded internal drainage. Their A_1 horizons are usually grey-brown or brown, but the A_2 development varies, being prominent in some profiles. The B horizons range from yellow to yellowish brown or reddish yellow, and red mottling is common; many of the mottles are incipient ironstone nodules. The reticulate red and grey mottling present in the B_3 and C horizons is usually not as prominent as in many of the *red earth* deep subsoils. Variants with bleached A_2 horizons and a more pronounced increase in texture intergrade into the *yellow* and *lateritic podzolic soils.*

(iv) The grey massive earths (Gn2.8, .9) occur mostly in sites with seasonal saturation of the profile where the internal drainage is slow. The general profile form is very similar to the *yellow earths* but B horizons are grey to light grey or light grey-brown, and commonly mottled with red, reddish brown, or yellow. The deep subsoils are usually strongly mottled. The majority of the soils have a bleached A_2 horizon, and the ironstone nodules are often darker and more manganiferous than in the *yellow earths.* Many of these soils intergrade into the *lateritic* and *gleyed podzolic soils.*

(v) *The red and brown hardpan soils* (Um5.3) are typically red, lack appreciable horizon development, and generally have uniform texture profiles which vary from a sandy loam to a sandy clay in the shallower examples (< *c.* 60 cm). Deeper profiles, up to 150 cm, may show an increase in texture with depth. The hardpan is an essential feature and abruptly underlies the soil material as described earlier in this chapter.

(vi) The earthy loams (Um5.5) are commonly associated with remnants of old laterite profiles and have weakly-developed and mostly shallow profiles of uniform texture. The B horizon is normally red, red-brown or brown. Many of these soils intergrade into the loamy variants of the shallow stony soils described earlier.

Structured sesquioxidic soils

These soils are moderately to strongly structured and have a clay fraction dominated by kaolinite and iron oxides. Horizon development is weak to moderate and commonly the texture gradually increases with depth in the upper part of the profile. The main distinctive features within the group relate to the nature of the A_1 and A_2 horizons, to the colour, degree and kind of structure in the B horizon, and to the profile depth (Fig. 3.16).

The component soils are *krasnozems, xanthozems,* red or yellow smooth-ped earths, red or yellow pale loams, *euchrozems* and a group of related brown smooth-ped earths. With the exception of the *euchrozems,* some of which occur in the north of Western Australia, the Northern Territory and western Queensland, they are almost entirely confined to eastern Australia where they occur intermittently from Tasmania to far north Queensland and mostly in the higher rainfall areas. The *krasnozems, euchrozems* and the brown smooth-ped earths are formed from basalt or similar basic rocks; the other soils are mostly derived from acid igneous or metamorphic rocks.

Those soils which occur in areas of high rainfall (> 1000 mm) under vine forest or wet sclerophyll forest usually have well-developed A_1 horizons containing many roots and with high organic matter contents (6–16%) and a strong or moderate, fine granular or crumb structure. The A_1 horizons of the red soils, however, may not have a dark colour. Many of these soils also have seasonally prominent O_1 horizons of leaf litter, but marked O_2 horizons are less common. By contrast, the lower rainfall soils such as the *euchrozems* and related brown smooth-ped earths usually have thinner, less-well-defined A_1 horizons with lower organic matter contents and a weak to moderate, fine subangular blocky structure. The O horizons are virtually absent.

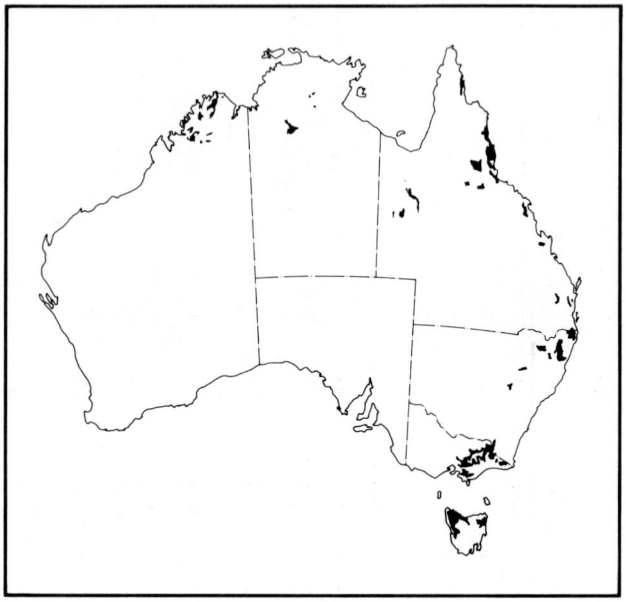

Fig. 3.16. Structured sesquioxidic soils.

The high-rainfall soils are very permeable and often have very deep profiles; this is especially so for the *krasnozems* which can have sola of 3–7 m and have a very gradual transition to weathered rock below. The other high-rainfall soils normally have thinner sola (1–3 m) often with considerable depths of saprolite which may be strongly mottled red, yellow and grey. In contrast, lower-rainfall soils such as the *euchrozems,* brown smooth-ped earths and some *krasnozems* are seldom deeper than 2 m and often < 1 m. The transition to weathered rock is usually over 20–30 cm.

Although all these soils are dominated by kaolinite (occasionally with some halloysite), there is some variation in their clay mineralogy. The red soils normally contain both haematite and goethite, gibbsite being present in some high-rainfall soils. The yellowish and brown soils have considerably lower iron oxide contents. Many of the lower-rainfall soils have appreciable amounts of randomly interstratified minerals.

The chemical properties of the structured sesquioxidic

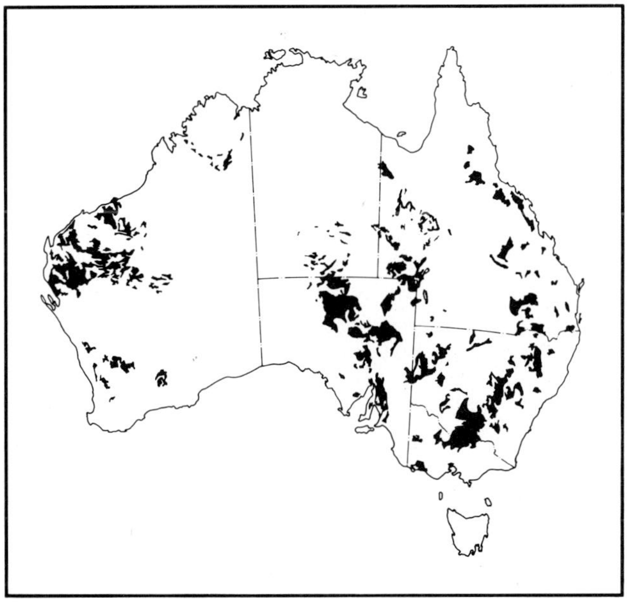

Fig. 3.17. Texture-contrast soils, with or without weak A_2 horizons.

soils are largely determined by the parent rock and the rainfall. Most red and brown soils derived from basalt and occurring in areas of lower rainfall are inherently very fertile, slightly acid to neutral in reaction, have moderate organic matter levels, a fairly high base status, and contain high amounts of potassium and phosphorus. The nutrient status of the high-rainfall soils is largely governed by their organic matter contents. Most of these soils are moderately to strongly acid, are extremely low in base status at depth (or with depletion of organic matter) and normally are strongly unsaturated. Their contents of phosphorus vary widely with many soils having very low contents of plant-available phosphorus and a high capacity to 'fix' fertilizer phosphorus.

The use of these soils varies widely. Many of the lower-rainfall soils are used only for cattle grazing of native pastures although some areas in eastern Australia are cropped. The high-rainfall soils, partly because of their excellent physical properties, are widely used for sugar cane, horticultural crops, peanuts, maize and oats. In the south they are widely used for dairying on improved pastures. They also occur in mountainous areas where they provide valuable timber.

The component soils are briefly described below.

(i) The *krasnozems* (Gn3.11, Gn4.11, .31) are mostly red throughout, less commonly brown. They are chiefly characterized by their structure profile and their uniform, very high clay contents (60–80% on an organic matter-free basis). However, textures range from loam or clay loam at the surface to medium clay at depth. The thick B_2 horizon is strongly structured with characteristic compound polyhedral units up to 1 cm in diameter which break progressively to smaller units. In the Queensland occurrences the peds are smooth and shiny faced, whereas further south the peds mostly have earthy porous surfaces. The Queensland soils usually contain many dense clay vermiform units 0.5–1.5 cm in diameter and up to 25 cm long (Isbell *et al.* 1976).

Many *krasnozems* contain low amounts of small ironstone nodules, but in a few soils there may be an horizon of massive or pisolitic laterite at a varying depth in the profile, grading below into a thick mottled zone of mainly light grey and red heavy clay.

(ii) Red smooth-ped earths (Gn3.14) have sometimes been regarded as a variant of the *krasnozems* (Stace *et al.* 1968) but they are probably more closely related to the *red podzolic soils*. The clay contents in the red smooth-ped earths increase gradually with depth but seldom exceed 60%. The A horizons of these soils have considerably lower clay contents than the *krasnozems*. They also have a weakly developed A_2 horizon and a B_2 horizon with a weak to moderate, medium subangular blocky structure and a consistence that is less friable than in the *krasnozems*. These soils occur in moderately high rainfall areas in northern New South Wales and north Queensland.

(iii) *Xanthozems* and yellow smooth-ped earths (Gn3.54, .71, .74, .84) have been regarded as yellow variants of the *krasnozems* (Stace *et al.* 1968), but they are more closely related to the *yellow podzolic soils*. They have a number of similarities with the red smooth-ped earths but the B_2 horizon is yellow, yellowish brown or reddish yellow and is often mottled with red, grey or brown. A_2 horizons are often present and are sometimes bleached.

(iv) Red or yellow pale loams (Um4.41, .43) are very similar to the red or yellow smooth-ped earths but are normally shallower and have a uniform texture profile of clay loam or silty clay loam. These loams are mostly confined to high-rainfall areas of north-east Queensland.

(v) *Euchrozems* (Gn3.12) differ from the *krasnozems* mainly in their chemistry, their shallower profiles, their less-well-developed A_1 horizons of lower clay content (40–50% on an organic matter-free basis), and their generally less-strongly-structured B horizons. Small ironstone nodules are normally present in low amounts.

(vi) Brown smooth-ped earths (Gn3.22) are closely related to the *euchrozems* and differ from them mainly in colour and in the nature of the B_2 horizon which is normally mottled with red or yellowish red in the lower part, has a subangular blocky rather than a polyhedral structure, and does not contain any vermiform units. This horizon usually contains at least moderate amounts of ironstone nodules, many of which are black and manganiferous. The main occurrence is in north-east Queensland (Isbell *et al.* 1977).

Texture-contrast soils with or without weak A_2 horizons

These soils are characterized by a marked difference in texture between the A and B horizons. If present the A_2 horizon is only weakly developed, in which case it is either thin (1–2 cm) and not strongly bleached, or thick and without a strong colour-contrast with the adjacent horizons. The textures of the A horizons usually range from sandy loam to clay loam and of the B horizons, from light to heavy clay. The A/B horizon boundary is normally clear and occasionally abrupt (Fig. 3.17).

The soils in this group are the *desert loams, red-brown earths* and *non-calcic brown soils*. The first group is confined to the more arid regions and the last is more common in sub-humid areas.

The soils which have formed from a wide variety of basic and intermediate parent materials and particularly from alluvium, range from shallow to moderately deep. The A horizons are 3–25 cm thick, reddish brown, brown or grey-brown, usually weak blocky to massive, and generally hard-setting when dry. The *desert loams* are characterized by a thin vesicular surface crust, and a distinct sporadic bleach in the lower part of the A horizon. This bleaching also occurs in many *red-brown earths*, and is more common when there is an abrupt texture change between the A and B horizons. The B_2 horizons are mostly uniform in colour, red, reddish brown or brown, and have at least a moderate grade of structural development.

The clay fractions of the *desert loams* are dominated by illite and randomly interstratified clay minerals, but those of most of the *red brown earths* have dominantly illite, whereas most of the *non-calcic brown soils*, and particularly those from northern Australia, have kaolinite.

The chemical properties vary widely both between and within these groupings. The *desert loams* are normally alkaline throughout, whereas the *red-brown earths* mostly have slightly acid to neutral A horizons but alkaline subsoils, and the *non-calcic brown soils* have slightly acid A horizons and neutral subsoils. Almost all the soils are low in both organic matter and phosphorus. Many *desert loams* and some *red-brown earths* are both saline and sodic (Northcote and Skene 1972), and hence intergrade with the *solodic soils*.

Although many of the soils, and in particular the *desert loams* and the *non-calcic brown soils*, are used for sparse grazing of native pastures, the *red-brown earths* are frequently used for wheat growing, improved pastures, and irrigated horticulture.

Descriptions of the component soils are as follows.

(i) *Desert loams* (Dr1.12, .13, .32, .33) have thin (3–10 cm) crusty A horizons and B_2 horizons with moderate, fine to medium, blocky or occasionally prismatic structure. The subsoils contain carbonate and/or gypsum. The soil surface is often strewn with rounded to subangular gravel and stones, and silcrete is often a major component.

(ii) *Red-brown earths* (Dr2.13, .23, .33; Dr3.23; Db1.13, .23, .33) have A horizons generally 15–25 cm thick. Normally an A_2 horizon is present, either as a paler relatively thick horizon or as a thin (1–2 cm) sporadically bleached layer immediately above the B horizon. The upper part of the B horizon often has prismatic structure, which breaks to medium blocky peds, and carbonate occurs as soft segregations or hard nodules in the B_3 horizon.

(iii) *Non-calcic brown soils* (Dr2.12, .22, Db1.12, .22) have A_1 horizons 10–20 cm thick and A_2 horizons are seldom present. Other features are similar to the *red-brown earths* but there is no carbonate in the subsoils.

Texture-contrast soils with strongly developed A_2 horizons

These soils have prominent A_2 horizons, which are either bleached and thin to moderately thick, or thick and pale, and also a strong texture-contrast between the A and B horizons (Fig. 3.18).

Other characteristics vary widely. The A horizons range from thick to thin and sandy to loamy, and the B horizons from sandy clay loams to heavy clays. The clay minerals are varying mixtures of kaolinite, illite, and montmorillonite, the frequency of dominance decreasing in that order. Structure in the A horizons is usually weakly developed or absent. Many A horizons are hard-setting. The B horizons range from massive, or weak coarse blocky, to moderate or strong angular blocky, and to coarse prismatic or columnar. Moist consistence of the B

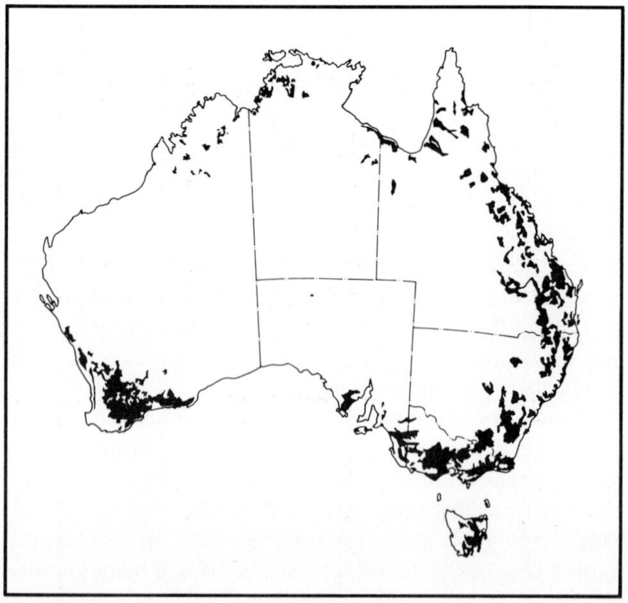

Fig. 3.18. Texture-contrast soils with strongly developed A_2 horizons.

horizons varies from friable to very firm or hard. Soil reaction profiles are acid throughout, or acid to neutral in the upper profile grading to neutral or strongly alkaline in the lower B horizons. The cation exchange capacity and base saturation vary widely.

These soils include the *solodized-solonetz and solodics*, the *soloths*, and most *red podzolics, yellow podzolics* and *gleyed podzolics*. They occur mostly in coastal and hinterland regions. The associated climates range from cool temperate to tropical and from semi-arid to humid. Parent rocks include all but extremes like quartzose sands, heavy clay sediments, basalts and other basic rocks, although some of these texture-contrast soils have formed on materials weathered from dolerite in Tasmania. Although the inherent fertility of these soils is generally low, their response to management varies considerably. Marked deficiencies of nitrogen, phosphorus, potassium, calcium and some other elements – notably molybdenum and sulphur – are common. Other general properties are briefly described below.

(i) *Solodized-solonetz and solodic soils.* (Dr2.32, .42, .43; Db1.4, Db2.4; Dy2.33, .42, .43; Dy3.32, .33, .42, .43, .82; Dy5.42, .43, .82, .83; Dd1.33, .43; Dd2.33) are identified by the combination of a generally abrupt A/B horizon boundary, B horizons with weak, coarse blocky to strong, coarse prismatic structure (columnar in the case of the *solodized-solonetz*), very hard consistence, and neutral to strongly alkaline soil reaction trends (Northcote 1971). The clay fractions are either dominated by kaolinite, with lesser amounts of illite and montmorillonite, or, in southern Australia, by illite. Magnesium is usually the dominant exchangeable cation; sodium is present in appreciable amounts (more than 6–8%) and is dominant in some soils, and calcium is usually low. Many of the B horizons are variously mottled yellow-brown, grey-brown and light grey, and have low macroporosity and high bulk density.

The soils have very poor water regimes owing to very slow permeability and wetting of the B horizons (McCown *et al.* 1976) and slow infiltration through some loamy A horizons. Saturation of the A_2 horizon following heavy or prolonged rains is common, and the soils are very difficult to manage for cropping over a range of environments from semi-arid to sub-humid. Hard setting A_1 horizons are a feature of all the sandy textured soils. Most soils in this group are highly susceptible to erosion (Fig. 3.19).

Fig. 3.19. Extreme gully erosion in a *solodic soil* in central Queensland.

(ii) *Soloths* (Dr2.41; Dy2.41; Dy3.41; Dy5.41, .81) are distinguished by an acid reaction throughout the solum and a tendency to clear rather than abrupt A/B horizon boundaries. Other characteristics are as for the *solodized-solonetz and solodics* except that the exchangeable cations are usually dominated by hydrogen and magnesium with considerable exchangeable sodium; even in the A_1 horizon the exchangeable calcium is often very low. The structure of the B horizon is typically moderate medium blocky, and water and root penetration are therefore better than in the *solodized-solonetz and solodic soils*. The A_2 horizon is usually strongly bleached.

(iii) *Red podzolic soils* (Dr2.21, .31, .41; Dr4.41) are identified by an acid reaction (they are often strongly acid in the deeper subsoil) and red clay B horizons whose structure is moderate, medium to fine, blocky and whose consistence is friable to firm when moist. The A/B boundary is commonly clear rather than abrupt and there is often some yellowish mottling in the B horizon.

The subsoils are typically strongly dominated by kaolinite but most contain a considerable proportion of illite. Usually in the deeper profile, some of the soils have small amounts of vermiculite and minor amounts of montmorillonite (Norrish and Pickering 1977). The A horizons are commonly loamy with a weak blocky structure and they usually set hard on drying. Cation exchange capacities are low: acidic ions (hydrogen and aluminium) are dominant, and exchangeable magnesium is commonly greater than calcium.

Red podzolic soils are found mostly on well-drained sites in humid to sub-humid regions of the coast and hinterland. They vary from shallow to deep and in the latter case often have an extended B_3 horizon which is coarsely mottled red, light grey, and yellow-brown.

(iv) *Yellow podzolic soils* (Dy2 and Dy3.21, .22; Dy5.61, .81) are distinguished from *red podzolics* by predominantly yellow-brown B horizons with some red and yellow-grey mottling. This mottling usually reflects restricted drainage and moister conditions often associated with mid- to lower-slope sites in sub-humid or humid regions. In all other respects the properties of these soils are much the same as for the *red podzolics* and intergrades with strongly mottled B horizons are common.

(v) *Gleyed podzolic soils* (Dy3.41, .42; Dg2.41, .42, .81, .82; Dg4.41, .42) are poorly drained soils distinguished by coarse ochreous, grey, and some reddish mottles in the B horizon; they also have some rusty to yellow-brown root tracings in the grey-brown loamy A_1 and pale A_2 horizons. These soils occur on poorly drained, lower slopes and on low, level sites. The A horizons usually contain more organic matter, have a better structure, and are less inclined to set hard on drying than the A horizons of the *red* and *yellow podzolics*. The A/B horizon boundary tends to be clear or gradual and the B horizons are rather lower in clay content than for the other podzolic soils and have a blocky to prismatic structure. The soil reaction, cation exchange properties, and fertility are not dissimilar to the *red podzolic* and *yellow podzolic soils*.

Texture-contrast soils with laterite and prominent deep subsoil development

The key features of this form are a texture-contrast solum, the presence of much ironstone gravel, and a thick lower B horizon of mottled whitish, red and ochreous kaolinitic clay. These soils, the *lateritic podzolic soils*, comprise one of the major forms associated with laterite profiles. (Fig. 3.20).

Lateritic podzolic soils (Dy3.61, .62, .81, .41, Dy5.84): vary in the texture and development of A horizons, texture of the B horizon, thickness of individual horizons, and the total depth of the profile. The soils are associated with land surfaces of low relief and characteristically occupy the higher levels of broad, subdued ridges or smaller gently convex interfluves under mean annual rainfalls of about 300 to 1500 mm. Parent materials have a high siliceous component and range from quartz sandstones to granite and gneiss. The soils have low inherent fertility with gross deficiencies of some of the major and trace elements.

The A_1 horizons (10–20 cm thick) vary from grey to brown and from sands to loams; they have low organic contents and weak crumb structure. Although they may contain some ironstone gravel, the main laterite horizon which is nodular, bouldery or conglomeratic, is characteristically in the lower A_2 and upper B horizon. The A_2 horizon is usually relatively thick, massive, and light grey to white, pale yellowish or pale brown. It has a clear to gradual boundary to the B horizon, which is commonly mottled strong brown, yellow-brown and red and varies from massive sandy clay loam to clay with medium blocky structure. Typically the deep subsoil is the whitish, red and ochreous clay of the mottled zone and/or the whitish clay of the pallid zone of the laterite profile.

The soil reaction is usually mildly to moderately acid in the A_1 horizon and becomes strongly acid in the B with pH values of 5.5 or less. Cation exchange capacities (10–20 mequiv. per 100 g of clay) and base saturation are usually low: exchangeable calcium and potassium contents are very low in both surface and subsoils. The clay fraction of the soil is strongly kaolinite-dominant with a little haematite and goethite; gibbsite may be present in the upper part of some profiles.

With added fertilizer these soils are used for legume-based sown pastures and *Pinus* spp. plantations.

Soils with sesquioxide–humus B horizons

These soils are characterized by illuvial B horizons dominated by sesquioxides and/or humus (usually with only low amounts of clay) underlying greyish A_1 and pale A_2 horizons. The nature of the B horizon, the degree of profile differentiation, and the depth of solum vary greatly. At one extreme are soils with a weak horizon development, brownish yellow B horizons, and sola as thin as 0.5 m; at the other are soils with sola more than 15 m deep, and having extreme horizon contrasts, bleached A_2 horizons more than 10 m thick, and dark humic B horizons up to 2.5 m thick (Fig. 3.21).

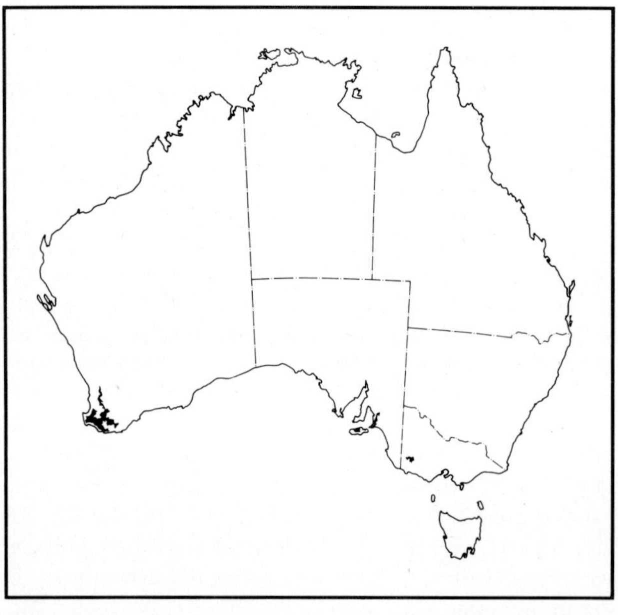

Fig. 3.20. Texture-contrast soils with laterite and prominent subsoil development.

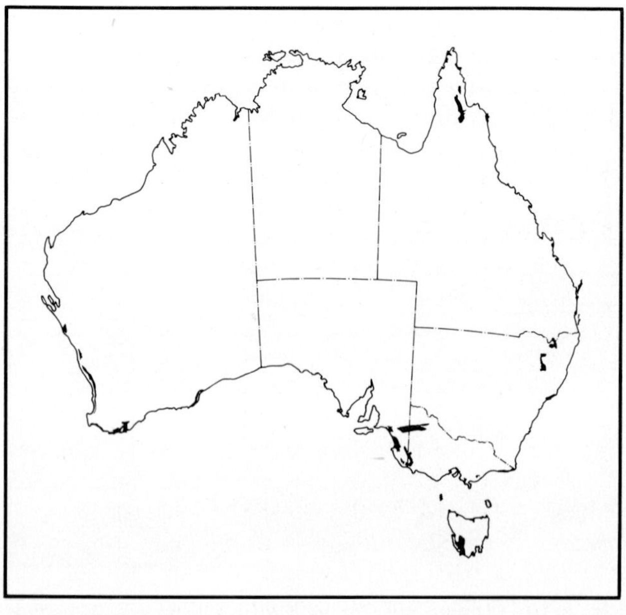

Fig. 3.21 Soils with sesquioxide — humus B horizons.

Two great soil groups are included, *podzols* and *humus podzols*. The soils are those of relatively humid environments with mean annual rainfalls ranging from 500 mm in the cool temperate south to 1800 mm or more in the tropics and sub-tropics of north-eastern Australia. They are all formed in quartz sands of dunes and plains along parts of the coastline or from sandy colluvial and alluvial materials derived from quartz sandstones and other quartz-dominant rocks in the subcoastal regions. Surface accumulations of litter or O horizons are rare. The A_1 horizons are greyish to dark grey sand or loamy sand ranging in depth from a few cm to more than 100 cm and speckled with particles of organic matter and clean sand grains near the surface but uniformly coloured below. The A_2 horizons vary from pale greyish brown in the weakly developed forms to light grey, white, or weakly mottled loose sands up to 12 m thick in strongly developed profiles on high dunes. The B horizons are loamy owing to the accumulation of translocated organic matter, sesquioxides, and a little clay, generally not more than 5 to 8% but rarely up to 15%. The clay mineralogy is characteristically kaolinite and quartz, sometimes with small amounts of gibbsite.

Soil reaction ranges from mildly to strongly acid. Both cation exchange capacity and base saturation of the A horizons are low, the former being determined by the low organic matter content of the A_1 horizon. Soil fertility is extremely low owing to gross deficiencies of both major and trace elements and to unsatisfactory soil water regimes. Land use is limited to some plantation forestry of *Pinus spp.* for softwood production and some sown pastures; both are dependent on correction of the nutrient deficiencies including that of cobalt in stock. Large areas are essentially unused.

(i) The *podzols* (Uc2.2, .3; Uc3.3; Uc4.2, .3) vary greatly in their characteristics (Thompson and Hubble 1980). Noteworthy features are irregular upper boundaries and uneven development of B horizons (as a consequence of deposition associated with very free drainage) which range from dominantly sesquioxidic (patchy brownish yellow to yellowish red with a little diffuse humic material) to dominantly humic (dark brown to very dark brown with some diffuse ochreous colours). Many of the B horizons are penetrated by 'pipes' of the A_2 horizon and most of them are dominantly friable with some patches of moderate cementation or thin ironpan bands.

(ii) The *humus podzols* (Uc2.2, .3; Uc3.3; Uc4.2, .3) have dominantly humic B horizons with an even upper boundary which overlie fluctuating water tables. The upper part of the B horizon has a more or less uniform colour which is very dark grey-brown to brownish black and often forms a hard pan; the lower part of the B horizon is usually mottled and has yellowish, rusty-brown, and pale greyish colours. Commonly sola are only moderately deep and drainage impedance by the B horizon causes some humic staining and mottling of the A_2 horizon and also development of a thicker, more organic A_1 horizon which tends to become peaty on very wet sites (Hubble 1951). Some of these soils have a dense but non-cemented pan in the lower part of the A_2 horizon (Hubble 1954) due to close packing of a graded mixture of sand, silt, and clay. Though previously called fragipans the features of these pans are the same as those of the recently defined densipans (Smith *et al.* 1975). Along with humus pans they severely restrict root development when they are present at shallow depths.

Humic soils

These soils are dominated by high organic matter contents in the surface horizons and this tends to obscure the effect of other soil components. They are all soils of humid environments, and include the *alpine humus soils* and *humic gleys* (Fig. 3.22).

(i) The *alpine humus soils* (Um7.1) are acid, organic loamy soils found in the alpine and sub-alpine areas of south-eastern Australia and Tasmania. The surface is often covered by a semi-decomposed litter layer. The A horizons are mostly thick and dark (near black), well-drained loams of granular to fine blocky structure. They pass gradually to a B horizon that is a brown and yellow-brown friable loam and then to C horizons that are usually grey, yellow-grey, or grey-brown, and coarser in texture. This is the most common form on the mainland. In Tasmania similar soils have a peat horizon below the litter layer, underlain by brown organic loam A horizons having about half the organic matter content of the mainland soils; their subsoils are similar but they may contain thin ironpans.

These soils occur on a wide range of rocks including granite, gneiss, basalt, dolerite and metasediments. Their

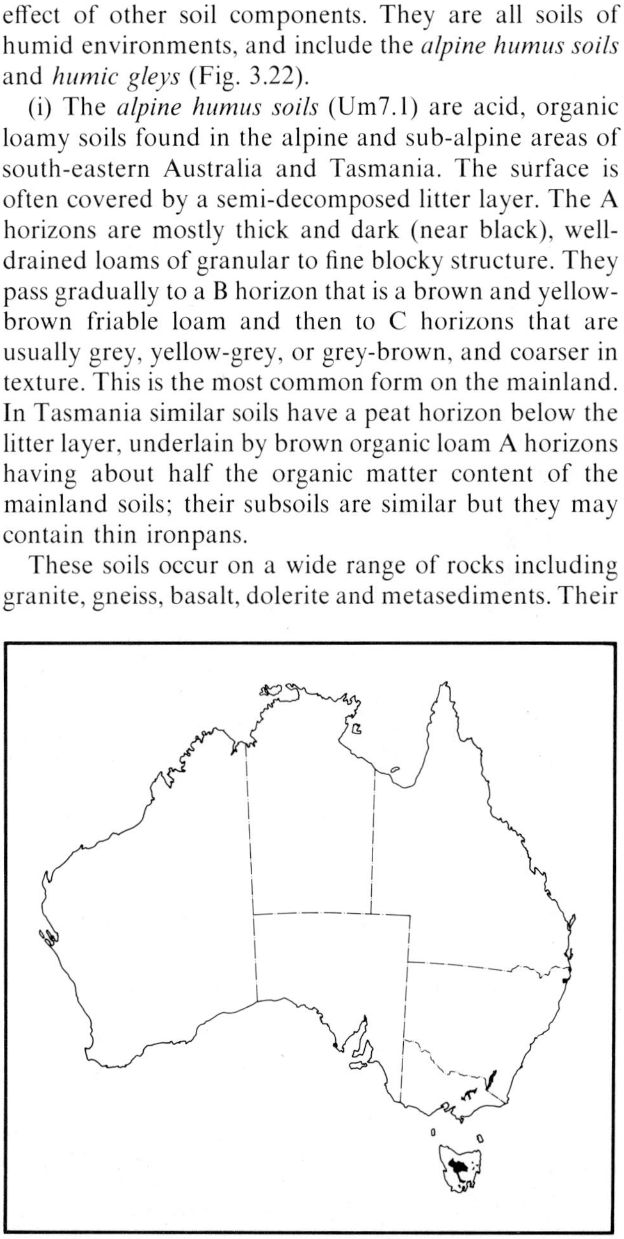

Fig. 3.22. Humic soils.

inherent fertility is generally low and large areas remain under native grassland, woodland, forest or heath. The former practice of summer grazing for sheep and cattle is now being restricted to preserve the alpine and subalpine tracts from over-grazing and subsequent erosion.

(ii) The *humic gleys* (Uf6.42, .61, .62; Gn3.9; Dy5.1; Dg4.11, .41, .81) include a range of acid to neutral, predominantly mineral soils varying from organic sands and loams to clays. They have significant but widely varying amounts of organic matter incorporated in their dark surface soils. These surface soils grade into predominantly clay subsoils marked by rusty and ochreous streaks and mottles in a pale grey matrix. The deeper soil is typically grey to bluish grey and permanently waterlogged, but the level of the watertable fluctuates, periodically rising almost to the surface; the soils may be flooded for short periods. They are formed in alluvium of low lying areas particularly on coastal and subcoastal lowlands, in interdune swales, and on estuarine plains extending inland along the lower reaches of river valleys. Where the topography allows them to be drained, they are used for a range of primary production from sugar cane to sown pastures and vegetables, especially potatoes.

Highly saline soils

The highly saline soils usually have bare surfaces often with a white salty crust, high salt contents in the upper part of the profile, and weak pedological development rarely exceeding 0.5 m deep. Their texture, water regimes, salt contents and sometimes their microrelief vary. Parent materials are mostly unconsolidated sediments ranging from loams to heavy clays. This group includes the *solonchaks* and the very salty soils with well-developed characteristics of other great soil groups. Owing to their high salinity the *solonchaks* are almost sterile soils; the few halophytes that grow on them commonly occupy scattered patches of less saline soil (Fig. 3.23).

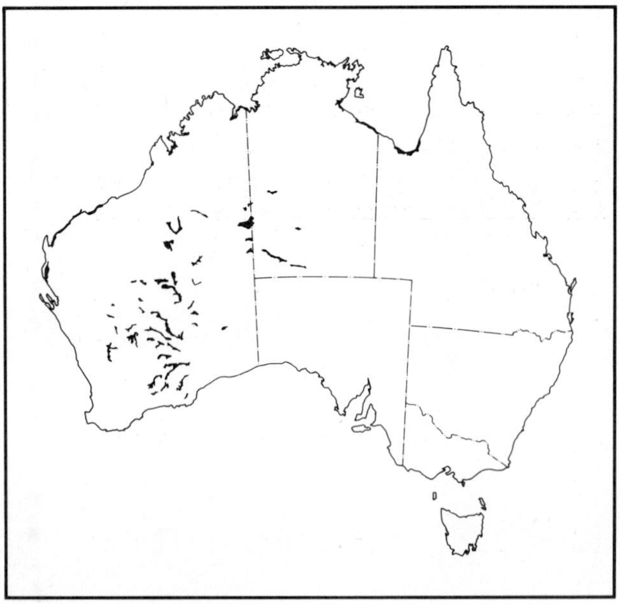

Fig. 3.23. Highly saline soils.

In addition to white salty crusts, characteristic surface features may include a thin layer of loose, powdery, fine crumb soil, massive crusts < 0.5–2 cm thick and vesicular in the lower part, and glazed, polygonally cracked, dense surface soils. When dry, fine cracks often separate the massive crusts into a mosaic of irregular polygons.

Solonchaks have been recognized in three different geomorphic environments:

(i) coastal saltpans, forelands and estuarine flats, occasionally flooded by sea or brackish waters. Most of these soils are olive brown, grey-brown, and grey clays (Uf1.4; Uf6.5, .6; Ug5.4, .5) becoming prominently mottled with ochreous, olive and red colours at shallow depth. Below the saline water table which may be little deeper than 0.5 m they become bluish grey. Some of these soils crack to 30 cm depth on prolonged drying.

(ii) playas and claypans of the arid inland. These soils (Dy1.43; Ug5.5) are seldom moistened below 30 cm depth and lack gley features in their subsoils. Generally they are grey-brown, brown, or reddish sandy clay loams to clays with pale, massive crusts overlying strong blocky clays which often have manganese staining on ped faces. The deeper subsoil below about 30 cm is paler, often yellow-brown, lighter-textured and massive.

(iii) floors of saline valleys of inland Western Australia. Loamy soils (Um1.1, .2) with little profile differentiation occur overlying sandy or clayey strata below 30 cm.

Total soluble salt contents (dominantly chlorides) commonly range from 1 to 5% but may be as high as 20–30% in subsoil layers containing gypsum; the surface crusts may contain > 50% salt. Generally the massive crusts have low salt contents but they and the underlying soil have high percentages of exchangeable sodium, often > 40%. Limited information suggests that the clay fractions are varying mixtures of kaolinite and montmorillonite with an important illite component in the arid lands. Soil reaction profiles vary from slightly acid to strongly alkaline throughout the profile. Carbonate contents are generally low and most soils contain gypsum.

Australian and world soils: some continental comparisons

Those familiar with the soils of other continents will recognize that few, if any, of the prominent features of Australian soils as outlined above, are unique to this country. Comparisons with other continents therefore rest on both similarities and differences in their soil cover. Two features stand out: firstly, the striking contrast between the overall character of the Australian soil cover and those of areas of the northern hemisphere affected by the Pleistocene glaciations; secondly, in those continents with similar soils or soil features, these are often relatively minor aspects of the landscape in contrast to their prominence in Australia.

The 1:5 000 000 soil map of Australia (Northcote *et al.* 1975) shows that a variety of sandy soils cover almost one third of the continent. Of these the deep *siliceous sands* and *earthy sands* of the arid interior are the most extensive, followed by other *siliceous sands* and sandy *lithosols*. The *red* and *yellow earths* and grey massive earths occupy 17% of the continent (the *red earths* are the most widespread), and the *black earths* and *grey, brown* and *red clays* account for another 11%. Other soils which occur extensively include the loamy *lithosols, grey, brown* and *red calcareous soils*, the *red* and *brown hardpan soils*, and a group dominated by the *red-brown earths* and *desert loams*. The *solodized-solonetz* and *solodic soils* and *soloths* also occupy important areas and are widely distributed.

Thus the continental soil pattern is dominated by little-differentiated sandy soils of varying depth. This and the associated large areas of (i) strongly leached and highly weathered kaolin–sesquioxide soils, (ii) cracking clay soils, (iii) little-differentiated, mostly shallow loamy soils, (iv) texture-contrast soils with neutral to alkaline red clay subsoils, and (v) soils with a strong texture-contrast and sodic, usually heavy clay subsoils, give Australia its distinctive soil character.

The strong development of certain soil features also imparts a uniqueness to the continent. Gilgai reach their maximum expression and extent on many of the widespread cracking clays. The remnants of old laterite profiles which occur throughout the arid and semi-arid zones are noteworthy and, depending on their degree of dissection, give rise to very different soils. The widespread presence of the siliceous red-brown hardpans and the extensive cover of lag gravels ('desert pavement') over large areas of the arid interior, often overlying stone-free clays, are further noteworthy features of this region. Contrary to what might be expected, in the Australian arid zone large areas have deep, acid to neutral soils with strong red colours, and although saline soils are common, they are often essentially confined to the playas and shallow valleys of the ancient river systems. If we look at the more humid eastern lands a striking feature is the presence of very deep soils, usually with very thick B_3 horizons. Here, too, the widespread occurrence is to be noted of texture-contrast soils with dense, sodic clay B horizons which often possess strikingly developed columnar or prismatic structure, underlying strongly bleached A_2 horizons.

The reasons for the distinctive soil pattern and the strong development of certain soil features are largely a consequence of soil parent material, climate and geomorphic history. Australia is an arid continent, and although many of the soils of the arid interior are obviously relict, the current aridity has had a marked effect on them. The consequences of the geomorphic history of many Australian soil landscapes, and particularly the role of different past climates, are evident in almost all parts of the continent. Most apparent discrepancies between present environment and the nature and distribution of soils in various localities can be explained on the basis of geomorphic history and the parent material factor. These matters are more fully discussed in Chapter 4.

If we look at soil patterns in other continents it is evident that there are some similarities but many differences. Parts of Africa come closest to the Australian scene, but even this comparison shows the widespread occurrence of strongly differentiated soils with sodic clay B horizons to be unique to Australia. The relative absence of such soils from South America, together with the minimal occurrence of cracking clay soils and vast areas of deep, highly leached, strongly acid, sesquioxidic soils uncommon in Australia, highlight any comparison with that continent. Further contrasts have been given by Sanchez and Isbell (1979).

Comparisons with the northern hemisphere show the differences to be even more striking. Although there are similarities between Australia and those parts of southern U.S.A. not affected by the Pleistocene glaciations, Australia has no frozen soils, lacks any great areas of podzols and podzolic soils, and certainly lacks the vast occurrences of various humic soils derived from loess that are a feature of much of Europe and North America. Certain other well-known world soils, such as those derived from volcanic ash, are unknown in Australia which has no active volcanism, and no widespread recent deposits of pyroclastic material. It may also be noted that anthropogenic soils are rare in Australia, as might be expected from its relatively short period of European settlement and its pre-history of indigenous people who were hunters and gatherers and seldom cultivated the soil. It also follows from Australia's sparse settlement that it is still possible in most areas to find soils that have been little, if at all, modified by man.

References

Anon. (1938). 'Soils and Men'. Yearbook Agric. U.S. Dep. Agric. (Gov. Printer: Washington D.C.)

Anon. (1976). Supplementary glossary of soil science terms used in Australia. Aust. Soc. Soil Sci. Inc. Pub. No. 6.

Beckmann, G. G., Hubble, G. D. and Thompson, C. H. (1971). Gilgai forms, distribution, and soil relationships in north-eastern Australia. Proc. Symp. 'Soils and earth structures in arid climates' Adelaide 1970. pp. 116-21. (Inst. Engr. Aust., Sydney.)

Beckmann, G. G., Thompson, C. H. and Hubble, G. D. (1973). Australian Landform example No. 22: Linear gilgai. *Aust. Geogr.* **12**, 363-6.

Bettenay, E. (1962). The salt lake systems and their associated aeolian features in the semi-arid regions of Western Australia. *J. Soil Sci.* **13**, 10-17.

Bettenay, E. and Churchward, H. M. (1974). Morphology and stratigraphic relationships of the Wiluna Hardpan in arid Western Australia. *J. Geol. Soc. Aust.* **21**, 73-80.

Bond, R. D. (1968). Water repellent sands. Trans. 9th Internat. Congr. Soil Sci. Adelaide. Vol. 1, pp. 339-47.

Bond, R. D. (1969*a*). The occurrence of water-repellent soils in Australia. *In* 'Symposium on Water Repellence of Soils'. Eds L. F. Debano and J. Letey. Vol. 16. (Univ. of California: Riverside.)

Bond, R. D. (1969*b*). Factors responsible for water repellence of soils. *In* 'Symposium on Water Repellence of Soils'. (Eds L. F. Debano and J. Letey), pp. 259-64. (Univ. of California: Riverside.)

Bowler, J. M. (1976). Aridity in Australia: age, origins and expression in aeolian landforms and sediments. *Earth-Sci. Rev.* **12**, 279-310.

Brewer, R. (1955). Mineralogical examination of a yellow podzolic soil formed on granodiorite. CSIRO (Aust.) Soil Publ. No. 5.

Brewer, R. (1964). 'Fabric and Mineral Analysis of Soils'. (John Wiley and Sons: New York.)

Brewer, R. (1968). Clay illuviation as a factor in particle size differentiation in soil profiles. Trans. 9th Int. Congr. Soil Sci. Adelaide. Vol. IV. pp. 489-99.

Brewer, R. (1972). Use of macro- and micro-morphological data in soil stratigraphy to educidate surficial geology and soil genesis. *J. Geol. Soc. Aust.* **19**, 331-44.

Brewer, R., Bettenay, E. and Churchward, H. M. (1972). Some aspects of the origin and development of the red and brown hardpan soils of Bulloo Downs, W.A. CSIRO (Aust.) Div. Soils Tech. Pap. No. 13.

Brewer, R. and Walker, P. H. (1969). Weathering and soil development on a sequence of river terraces. *Aust. J. Soil Res.* **20**, 293-305.

Butler, B. E. (1955). A system for the description of soil structure and consistence in the field. *J. Aust. Inst. Agric. Sci.* **24**, 239-49.

Butler, B. E. (1959). Periodic phenomena in landscapes as a basis for soil studies. CSIRO (Aust.) Soil Publ. No. 14.

Butler, B. E. (1976). Subplasticity in Australian soils. *Aust. J. Soil Res.* **14**, 225-89.

Chittleborough, D. J. and Oades, J. M. (1979). The development of a red-brown earth. 1. A reinterpretation of published data. *Aust. J. Soil Res.* **17**, 372-81.

Dawson, N. M. and Ahern, C. R. (1974). Soils. *In* 'Western Arid Region Land Use Study Part 1'. Div. Land Utilization Tech. Bull. No. 12. Qld Dep. Primary Industries, Brisbane.

Day, K. J. and van Cuylenburg, H. R. M. (1980). Characteristics of some red earth soils of the monsoonal region of the Northern Territory, Australia. Int. Soc. Soil Sci. Trans. Comm. IV and V. Kuala Lumpur, Malaysia 1977.

FAO-UNESCO (1974). Soil Map of the World, 1:5 000 000. Vol. 1. Legend. (UNESCO: Paris.)

FAO-UNESCO (1976). Soil Map of the World, 1:5 000 000. Vol. X. Australia. (UNESCO: Paris.)

Grant, K. and Aitchison, G. D. (1970). The engineering significance of silcretes and ferricretes in Australia. *Eng. Geol.* **4**, 93-120.

Green, P. (1966). Mineralogical and weathering study of a red-brown earth formed on granodiorite. *Aust. J. Soil Res.* 4, 181-97.

Gunn, R. H. (1967). A soil catena on denuded laterite profiles in Queensland. *Aust. J. Soil Res.* **5**, 117-32.

Hallsworth, E. G., Costin, A. B., Gibbons, F. R. and Robertson, G. K. (1952). Studies on pedogenesis in New South Wales. II. The chocolate soils. *J. Soil Sci.* **3**, 103-24.

Hallsworth, E. G., Robertson, G. K. and Gibbons, F. R. (1955). Studies in pedogenesis in New South Wales. VII. The gilgai soils. **6**, 1-31.

Hills, E. S. (1940). The lunette: a new landform of aeolian origin. *Aust. Geogr.* **3**, 15-21.

Hodgson, J. M. (1974). Soil survey field handbook. Soil Survey Technical Monograph No. 5. (Harpenden.)

Howard, A. (1939). Crab-hole, gilgai and self-mulching soils of the Murrumbidgee Irrigation Area. *Pedology* (Soviet Soil Science) **3**, 14-18.

Hubble, G. D. (1951). A reconnaissance soil survey of the coastal heath country, north-west Tasmania. CSIRO (Aust.) Div. Soils Div. Rep. No. 10/51.

Hubble, G. D. (1954). Some soils of the coastal lowlands north of Brisbane. CSIRO (Aust.) Div. Soils Div. Rep. No. 1/54.

Hubble, G. D. and Isbell, R. F. (1958). The occurrence of strongly acid clays beneath alkaline soils in Queensland. *Aust. J. Sci.* **20**, 186-7.

Idnurm, M. and Senior, B. R. (1978). Palaeomagnetic ages of Late Cretaceous and Tertiary weathered profiles in the Eromanga Basin, Queensland. *Palaeogeogr. Palaeoclimatol. Palaeoecol.* **24**, 263-77.

Isbell, R. F. (1957). The soils of the Inglewood-Talwood-Tara-Glenmorgan region, Queensland. Queensl. Bur. Invest. Tech. Bull. No. 5.

Isbell, R. F. (1962). Soils and vegetation of the brigalow lands, eastern Australia. CSIRO (Aust.) Div. Soils Soils Land Use Ser. No. 43.

Isbell, R. F., Gillman, G. P., Murtha, G. G. and Jones, P. N. (1977). Brown basaltic soils in north Queensland. CSIRO (Aust.) Div. Soils Tech. Pap. No. 34.

Isbell, R. F. and Smith, G. McL. (1976). Some properties of red, yellow, and grey massive earths in north Queensland. CSIRO (Aust.) Div. Soils Tech. Pap. No. 30.

Isbell, R. F., Stephenson, P. J. and Murtha, G. G. (1976). Red basaltic soils in north Queensland. 1. Environment, morphology, particle size characteristics and clay mineralogy. CSIRO (Aust.) Div. Soils Tech. Pap. No. 28.

Jenny, H. I. (1941). 'Factors of Soil Formation'. (McGraw Hill: New York.)

Jessup, R. W. (1951). The soils, geology and vegetation of north-western South Australia. *Trans. R. Soc. S. Aust.* **74**, 189-273.

Langford-Smith, T. (1978). A select review of silcrete research in Australia. *In* 'Silcrete in Australia'. (Ed. T. Langford-Smith). (Dep. Geogr., University of New England: Armidale.)

Leeper, G. W. (1948). 'Introduction to Soil Science'. (Melbourne Univ. Press: Melbourne).

Litchfield, W. H. and Mabbutt, J. A. (1962). Hardpan soils of semi-arid Western Australia. *J. Soil Sci.* **13**, 148-60.

McCown, R. L., Murtha, G. G. and Smith, G. D. (1976). Assessment of available water storage capacity of soils with restricted subsoil permeability. *Wat. Resour. Res.* **12**, 1255-59.

Mulcahy, M. J. (1960). Laterite and lateritic soils, south-west Australia. *J. Soil Sci.* 11, 205-25.

Mulcahy, H. J. and Hingston, F. J. (1961). The development and distribution of soils of the York-Quairading area, Western Australia, in relation to landscape evolution. CSIRO (Aust.) Soil Publ. No. 17.

Nikiforoff, C. C. (1943). Introduction of palaeopedology. *Amer. J. Sci.* **241**, 194-200.

Nikiforoff, C. C. (1949). Weathering and soil evolution. *Soil. Sci.* 67, 219-30.

Norrish, K. and Pickering, J. G. (1977). Clay mineralogic properties. *In* 'Soil Factors in Crop Production in a Semi-arid Environment'. (Eds J. S. Russell and E. L. Greacen). (Academy Press: Brisbane.)

Norrish, K. and Tiller, K. G. (1976). Subplasticity in Australian soils. V. Factors involved and techniques of dispersion. *Aust. J. Soil Res.* **14**, 273-89.

Northcote, K. H. (1946). A fossil soil from Kangaroo Island, South Australia. *Trans. R. Soc. S. Aust.* **70**, 294-6.

Northcote, K. H. (1951). A pedological study of the soils occurring at Comealla, New South Wales. CSIR (Aust.) Bull. No. 264.

Northcote, K. H. (1971). 'A Factual Key for the Recognition of Australian Soils'. 3rd edn. (Rellim: Glenside, S.A.)

Northcote, K. H. and Skene, J. K. M. (1972). Australian soils with saline and sodic properties. CSIRO (Aust.) Soil Publ. No. 27.

Northcote, K. H. and Tucker, B. M. (1948). A soil survey of the Hundred of Seddon and part of the Hundred of MacGillivray, Kangaroo Island, South Australia. CSIR (Aust.) Bull. No. 233.

Northcote, K. H., Hubble, G. D., Isbell, R. F., Thompson, C. H. and Bettenay, E. (1975). 'A Description of Australian Soils'. (CSIRO: Melbourne.)

Oertel, A. C. (1961). Pedogenesis of some red-brown earths based on trace element profiles. *J. Soil Sci.* **12**, 242-58.

Oertel, A. C. (1968). Some observations incompatible with clay illuvation. Trans. 9th Int. Congr. Soil Sci. Adelaide. Vol. IV. pp. 481-86.

Oertel, A. C. (1974). The development of a typical red-brown earth. *Aust. J. Soil Res.* **12**, 97-105.

Oertel, A. C. and Blackburn, G. (1970). Pedogenesis of a solodized-solonetz based on duplicate soil profiles. *Aust. J. Soil Res.* **8**, 59-70.

Ollier, C. D. (1966). Desert gilgai. *Nature* (London) **212**, 581-3.

Parberry, N. H. (1933). An unusual soil feature 'linear' gilgai. *Agric. Gaz. N.S.W.* **44**, 281-2.

Prescott, J. A. (1931). The soils of Australia in relation to vegetation and climate. CSIRO (Aust.) Bull. No. 52.

Prescott, J. A. and Pendleton, R. L. (1952). Laterite and lateritic soils. Commonw. Bur. Soil Sci. Tech. Comm. No. 47.

Russell, J. S., and Moore, A. W. (1972). Some parameters of gilgai micro-relief. *Soil Sci.* **114**, 82-7.

Sanchez, P. A. and Isbell, R. F. (1979). A comparison of the soils of tropical Latin America and tropical Australia. *In* 'Pasture Production in Acid Soils of the Tropics'. (Eds P. A. Sanchez and L. E. Tergas). Proc. Seminar held at CIAT, Cali, Colombia. pp. 25-52.

Simonson, R. W. (1959). Outline of a general theory of soil genesis. *Soil Sci. Soc. Am. Proc.* **23**, 152-6.

Sleeman, J. R. (1964). Structure variation within two red-brown earth profiles. *Aust. J. Soil Res.* **2**, 146-61.

Sleeman, J. R. (1975). Micromorphology and mineralogy of a layered red-brown earth profile. *Aust. J. Soil Res.* **13**, 101-17.

Smith, G. D., Arya, L. M. and Stark, J. (1975). The densipan, a diagnostic horizon of densiaquults for Soil Taxonomy. *Soil Sci. Soc. Am. Proc.* **39**, 369-70.

Soil Science Society of America (1975). Glossary of Soil Science Terms.

Soil Survey Staff (1951). Soil survey manual. Handb. U.S. Dep. Agric. No. 18 (Gov. Printer: Washington D.C.).

Soil Survey Staff (1975). Soil taxonomy; a basic system of soil classification for making and interpreting soil surveys. Handb. U.S. Dep. Agric. No. 436. (Gov. Printer: Washington D.C.).

Stace, H. C. T., Hubble, G. D., Brewer, R., Northcote, K. H., Sleeman, J. R., Mulcahy, M. J. and Hallsworth, E. G. (1968). 'A Handbook of Australian Soils'. (Rellim: Glenside, S.A.)

Stephens, C. G. (1946). Pedogenesis following dissection of lateritic regions in southern Australia. CSIR (Aust.) Bull. No. 206.

Stephens, C. G. (1947). Functional synthesis in pedogenesis. *Trans. R. Soc. S. Aust.* **71**, 168-81.

Stephens, C. G. (1971). Laterite and silcrete in Australia. *Geoderma*, **5**, 52.

Stephens, C. G. and Donald, C. M. (1958). Australian soils and their responses to fertiliser. *Adv. Agron.* 10, 167-256.

Taylor, N. H. and Cox, J. E. (1956). The soil pattern of New Zealand. N.Z. Soil Bur. Publ. No. 113.

Taylor, J. K. and Hooper, P. D. (1938). A soil survey of the horticultural soils in the Murrumbidgee Irrigation Areas N.S.W. CSIR (Aust.) Bull. No. 118.

Teakle, L. J. H. (1936). The red and brown hard-pan soils of the acacia semi-desert scrub of Western Australia. *J. Dep. Agric. West. Aust.* **25**, 480-93.

Teakle, L. J. H. (1950). A description of a laterite profile near Gayndah, Queensland. *Queensl. J. Agri. Sci.* **7**, 15-23.

Teakle, L. J. H. and Southern, B. L. (1937). The peat soils and related soils of Western Australia. *J. Dep. Agric. West. Aust.* **14**, 332-58, 404-24.

Thompson, C. H. (1975). Coastal areas of southern Queensland, some land use conflicts. *Proc. R. Soc. Queensl.* **86**, 109-20.

Thompson, C. H. and Beckmann, G. G. (1959). Soils of the Toowoomba area, Darling Downs, Queensland. CSIRO (Aust.) Div. Soils Soils Land Use Ser. No. 28.

Thompson, C. H. and Hubble, G. D. (1980). Sub-tropical podzols (Spodosols and related soils) of coastal eastern Australia: profile form, classification and use. Internat. Soc. Soil Sci. Trans. Comm. IV and V. Kuala Lumpur, Malaysia, 1977.

van Dijk, D. C. (1958). Soil distribution in the Griffith–Yenda District, N.S.W. CSIRO (Aust.) Soil Publ. No 11.

van Dijk, D. C. (1969). Pseudogley in Gundaroo subsola, Southern Tableland, New South Wales. *Aust. J. Soils Res.* **7**, 143-62.

van de Graaff, W. J. E., Crowe, R. W. A., Bunting, J. A. and Jackson, M. J. (1977). Relict early Cainozoic drainages in arid Western Australia. *Z. Geomorphol. N.F.* **21**, 379-400.

Veen, A. W. L. (1973). Evaluation of clay mineral equilibria in some clay soils (usterts) of the brigalow lands. *Aust. J. Soil Res.* **11**, 167-84.

Walker, P. H. (1962). Terrace chronology and soil formation on the south coast of N.S.W. *J. Soil Sci.* **13**, 178-86.

Walker, P. H. (1963). A reconnaissance of soils in the Kempsey district, N.S.W. CSIRO (Aust.) Div. Soils Soils Land Use Ser. No. 44.

Walker, P. H. and Coventry, R. J. (1976). Soil profile development in some alluvial deposits of eastern New South Wales. *Aust. J. Soil Res.* **14**, 305-17.

Walker, P. H. and Green, P. (1976). Soil trends in two valley fill sequences. *Aust. J. Soil Res.* **14**, 291-303.

Walker, P. H. and Hawkins, C. A. (1958). A study of river terraces and soil development on the Nepean River, N.S.W. *J. Proc. R. Soc. N.S.W.* **91**, 67-84.

Whitehouse, F. W. (1940). The lateritic soils of Western Queensland. Univ. Queensl. Pap. Dep. Geol. Vol. 2, No. 1.

Woolnough, W. G. (1927). The duricrust of Australia. *J. Proc. R. Soc. N.S.W.* **61**, 17-33.

Wopfner, H. and Twidale, C. R. (1967). Geomorphological history of the Lake Eyre Basin. *In* 'Landform Studies from Australia and New Guinea'. (Eds J. N. Jennings and J. A. Mabbutt). (A.N.U. Press: Canberra.)

Part 2
Evolution of Australian landscapes

Chapters 4-11

MOST OF AUSTRALIA HAS A LOW RELIEF AND THIS together with its location in mid to low latitudes (10–45°S), has restricted Pleistocene glaciation to higher areas in the south-east. There have been no orogenic movements for some 130 million years (since the end of the Jurassic Period) and all tectonic movements since then have consisted of warpings and vertical uplifts with some faulting. These have affected the type of land form and the manner of landscape development, while still allowing old landforms and weathered products to persist.

Past climates and their influence on parent material, as well as topography and the parent material itself, have been more important factors in determining Australia's soil pattern than modern climate.

Low relief and low rainfall in the immediate past have meant a low intensity of recent erosion over much of the continent. Under these conditions material eroded from hills and highlands has been transported only a comparatively short distance and has formed the extensive alluvial plains of the inland. Some of these earlier sediments have been blown out in more arid periods to form aeolian deposits on, or adjoining, alluvial plains.

This part of the book, comprising chapters 4–11, is concerned with the various events and processes, both past and present, that have been responsible for the development of modern Australian soil landscapes. The emphasis is on past landscape history and the major processes involved in the evolution of Australian soils. Descriptions of the regional soil patterns of the continent are given in Part 3.

Chapter 4
Development of old landscapes and soils

G. G. Beckmann

AUSTRALIA HAS A HIGHLY COMPLICATED SOIL PATTERN, as can be seen by even a cursory glance at a soil map of the continent. A closer perusal suggests that there is some order in this pattern, in that certain soil groups are restricted to certain areas and others have a broader, though scattered, distribution. This order is real, and the aim here is to demonstrate how the geological history of the continent and past and present climates have interacted on materials on land surfaces of different ages to produce the environments in which the modern soil pattern has developed.

It has often been said that Australia has an old land surface, the implication of this being that the soils and underlying materials have been so strongly weathered that they are now infertile. This idea has to be qualified. While many areas of Australia have been exposed to weathering for many millions of years, many of its soils and surfaces, particularly those in the eastern part, are quite young, in fact as young as many in other continents, including those which were widely affected by glaciation. Even in those parts of Australia where there has been severe and deep weathering, older surfaces have been dissected and progressively younger soils have been developing in the older weathered materials. Simultaneously new surfaces have been developing on the sediments eroded from the old land surfaces and these sediments have in turn been reworked by the action of wind and water.

This chapter is divided into three parts to illustrate how the soil pattern developed. The first deals with the period to the end of the Mesozoic era and summarizes the geological and structural background which have determined the form of the continent and the rock types from which soils have been derived. By the end of the Mesozoic era the geological base had been established and Australia was commencing to split off from other southern hemisphere continents and to behave as a separate entity.

The second part deals with Tertiary geology and geomorphology to the end of the Miocene Period. During this time most of Australia was stable, with low relief and a wet climate. Deep weathering occurred over wide areas. The effects of this can still be seen.

The third part is concerned with Pliocene and Quaternary events and their influences. Tectonic, eustatic, and climatic changes during this period modified the earlier landscapes and weathering products, and exposed relatively fresh rocks to soil formation.

Geological history to the end of the Mesozoic

The geological history of Australia can be effectively treated as the history of a number of distinct sedimentary basins. Some of these have been active over hundreds of millions of years. Some have become defunct with time and have become part of the stable continental mass, while others have been active for a period, became stabilized, and have then been reactivated hundreds of millions of years later. New basins have appeared in succeeding geological epochs and these in turn have become inactive and stabilized (Wellman 1976). Basins are important in the distribution of later features, e.g. the Precambrian Canning and Officer Basins are the loci of the Great Sandy and Great Victoria Deserts, and the Tertiary Lake Eyre Basin is the locus of the Simpson Desert.

The geological history commences in the Archaean period (about 2000 million years ago), at a time when the continent was part of a larger landmass Gondwanaland, which also included Antarctica, Africa and South America. At this time sediments were deposited in the area of the Yilgarn-Pilbara block of Western Australia. These were uplifted in the Middle Precambrian, (about 1500 million years ago) and, as far as can be ascertained, have never again been covered by the sea (Johnstone *et al.* 1973). Since that time there has been deposition in both subsiding troughs (geosynclines) which have later suffered orogenic (folding and uplift) movements, and in broad sedimentary basins which have been affected by epeirogenic (vertical uplift and warping) movements. The loci of orogenies moved progressively eastward in later periods and most later geosynclines have a roughly meridional axis. After each orogeny the sediments affected became welded to the earlier layers as part of a larger craton (stable block), and further movements within each craton were epeirogenic. From time to time overall uplifts of cratons, including local warpings, have

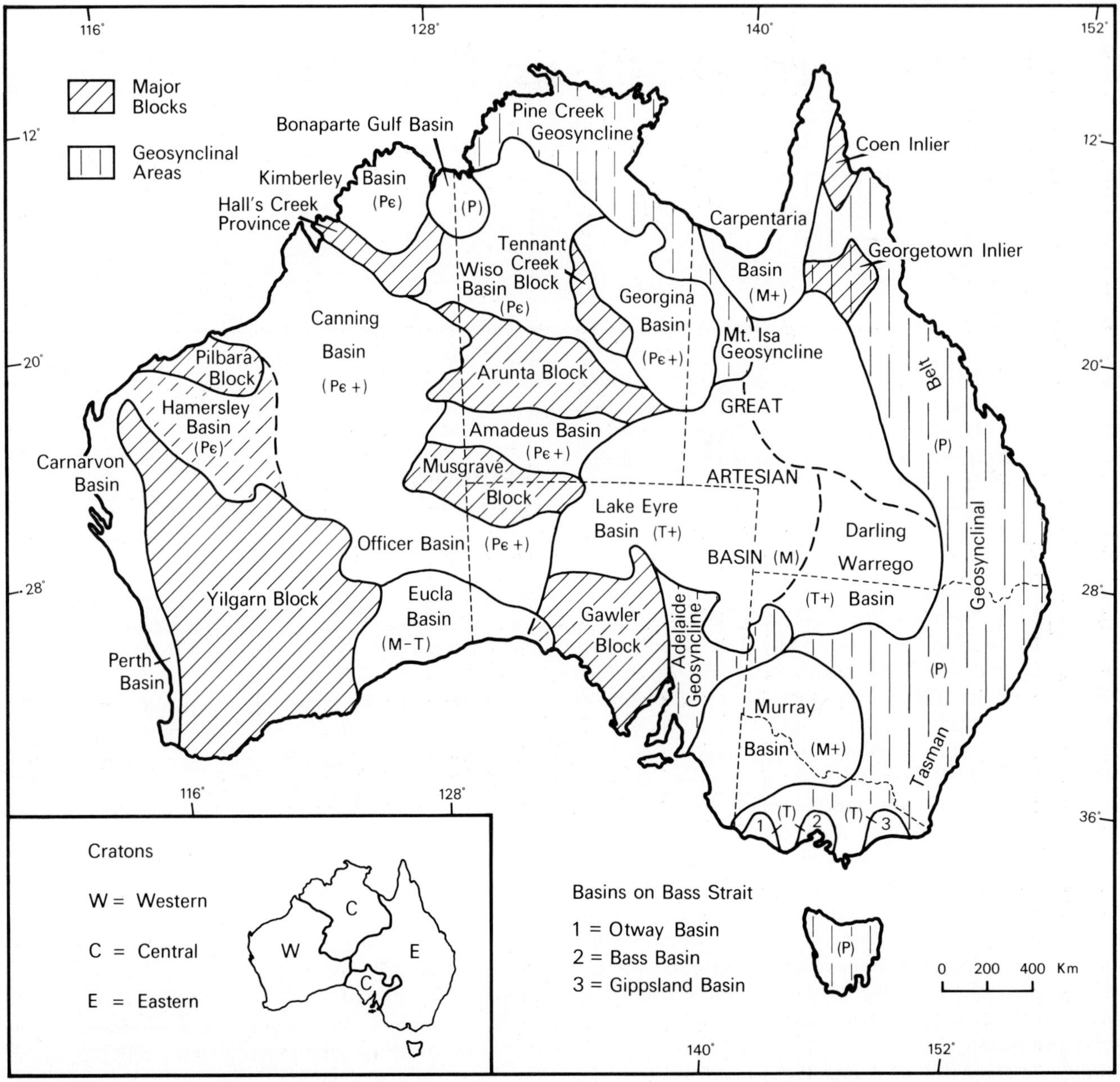

Fig. 4.1. Major structural elements of Australia (derived from Brown *et al.* 1968 and Wellman 1976), showing major blocks, basins and geosynclines. Age of basins shown in brackets: e.g. Pє = Precambrian, M = Mesozoic, T = Tertiary; + = continuing, sometimes with interruptions, through subsequent periods.

emphasized sags or depressions on the sites of older basins, as well as initiating new basins (Fig. 4.1).

Much of the following discussion has been derived from Brown *et al.* (1968) with additions from papers by other authors. Figs 4.2–4.5, showing the palaeogeography of Australia during the Phanerozoic, have been derived from maps in Brown *et al.* (1968).

Precambrian

The oldest rocks of Australia are high grade metamorphic schists and gneisses derived from both sedimentary and igneous rocks, of the Yilgarn and Pilbara blocks which form part of the Archaean Shield of Western Australia. Similar rocks from this period occur in the Arunta complex of the Macdonnell and Harts Ranges of the southern Northern Territory.

Sediments which were mainly laid down in shallow water accumulated during the Proterozoic in a number of basins and geosynclinal troughs, e.g. the Nullagine, Kimberley and McArthur Basins, and the Mt Isa Geosyncline in the earlier Proterozoic, and the Officer, Amadeus and Georgina Basins, and the Adelaide

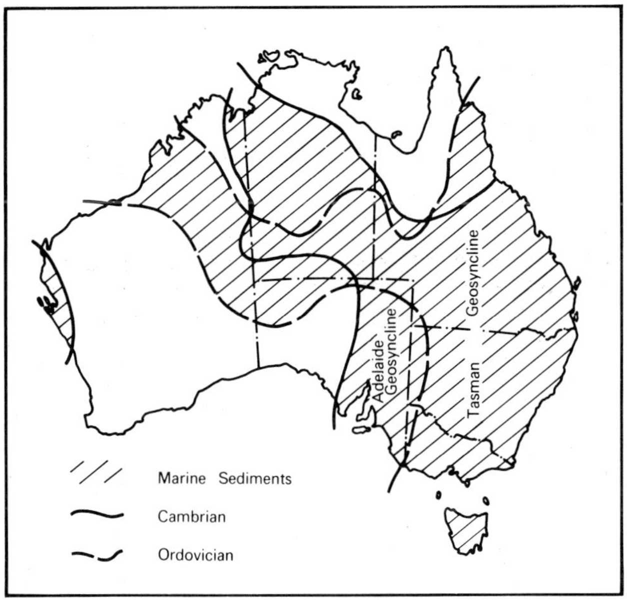

Fig. 4.2. Cambrian–Ordovician palaeogeography, showing distribution of land areas of Precambrian rocks and of early Paleozoic seas (derived from Brown *et al.* 1968).

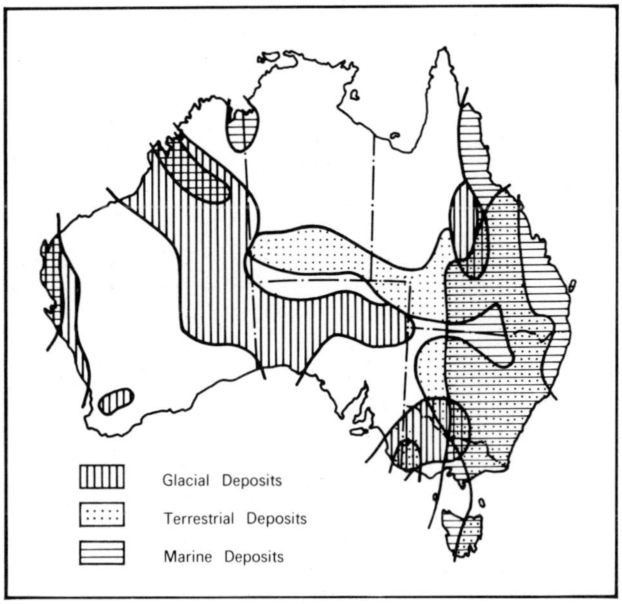

Fig. 4.3. Silurian to Permian palaeogeography, showing areas of terrestrial and glacial deposits (derived from Brown *et al.* 1968).

Geosyncline in the later Proterozoic. These latter basins became the sites for the deposition of sediments during many later periods.

Cambrian and Ordovician

Marine sedimentation continued in the Cambrian period (570–500 million years ago), some deposits being conformable on those of the Precambrian, e.g. in the Adelaide geosyncline. At this time sedimentation commenced in south-eastern Australia in the Tasman Geosynclinal Belt, a major geological feature which extended through much of Queensland, New South Wales and Western Victoria for the entire Paleozoic (Fig. 4.2).

Sedimentation continued within the confines of this eastern geosyncline for many millions of years, with minor structures developing at different times in different parts. While there were only epeirogenic movements over most of the Western Craton at this time, the Adelaide and Tasman Geosynclines were subjected to severe folding during the Middle to Late Cambrian.

Shallow water sediments were deposited during the Ordovician period, from 500 to 430 million years ago, within several of the basins extending from western Queensland through Northern Territory into north-western Western Australia. By this time the Adelaide Geosyncline had become land, but deposition continued through the Tasman Geosyncline into Tasmania. Only gentle epeiric movements occurred over most of the continent during this period.

Silurian to Permian

Essentially the same pattern continued in eastern Australia during the periods which followed (Fig. 4.3). In the Silurian and Early Devonian periods, from 430 to 370 million years ago, the Tasman Geosyncline was of more limited extent, most sedimentation being restricted to its eastern edge. Some shallow water sedimentation occurred within the major shield area of the west, but this was restricted mainly to the Amadeus, Carnarvon and Perth Basins.

Granitic rocks were emplaced in eastern New South Wales at the end of the Silurian.

The same sedimentary pattern continued in the area covered by the Tasman Geosyncline in eastern Australia until the middle of the Devonian (about 370 million years ago), when this pattern was changed completely with an orogeny which raised most of eastern Australia, producing °ide land areas but no mountains except possibly in Tasmania. The dominant types of sediments were then terrestrial deposits associated with major river systems. At this time, the Middle Devonian, the Amadeus Basin had become connected to the Drummond Basin in north-east Queensland.

Most of the continent was land also during the Carboniferous period (345 to 280 million years ago). Sedimentation was restricted to the marine basins in Western Australia and in northern and eastern Queensland, and to some restricted terrestrial basins, while glacial sediments were associated with terrestrial sediments in northern New South Wales. The continent was stable tectonically and free from igneous activity at this time except in the Tasman Geosyncline, where there were intermittent uplifts, emplacement of plutonic rocks, and outpouring of volcanics.

Most of the Permian rocks of Australia occur in the east, along the trend of the Tasman Geosyncline from central Queensland to the vicinity of Sydney. The greater part of the continent, especially in the south and west, was covered by land ice during at least part of the Permian, modifying earlier landscapes (Figs 4.3 and 4.9). There were both terrestrial and marine glacial deposits in connecting basins across it and shallow water marine sediments were deposited in the Canning, Carnarvon and Perth Basins after glacial deposits had accumulated within them. During this period the Canning Basin had become continuous with the Officer Basin and these with the Lake Eyre and Murray Basins contained largely terrestrial sediments. This continuity was important in later geomorphic and pedological developments.

The last major orogenic episode which affected any considerable extent of Australia, the Hunter-Bowen orogeny, occurred during the final stages of the Permian period, and was restricted to a zone paralleling the east coast. Many plutonic intrusions, both granites and granodiorites, were emplaced during these movements in north-eastern New South Wales and eastern and northern Queensland. The rest of the continent by this stage had become almost entirely a stable land mass, which was subsequently affected by epeiric movements only; some of these produced shallow seas within the basins.

Triassic and Jurassic

The continent was almost entirely land during the Triassic period which commenced 235 million years ago. In the Early Triassic, sedimentation continued in river systems within a mountain chain adjacent to the east coast. There were a few small basins in the eastern interior of the continent which also had fluviatile or lacustrine sediments. A distinct Mesozoic drainage system traversed the southern part of the shield in Western Australia in the Harvey area and to the east; the drainage was west towards the coast (Churchward and Bettenay 1973; van de Graaff *et al.* 1977). Marine and brackish water sediments were deposited in the Perth and Canning Basins.

The continent was then tectonically stable with only weak folding in any of the basins, whose areas extended only slightly during the Middle and Late Triassic, but rather more so in the Early Jurassic period. In inland north-eastern Australia, sedimentation commenced in the Great Artesian Basin, which had developed initially on an erosion surface cut into older rocks. This early form was continuous with basins in south-east Queensland which had been in existence during the Triassic and, like them, contained fluviatile sediment from an extensive river system (Fig. 4.4).

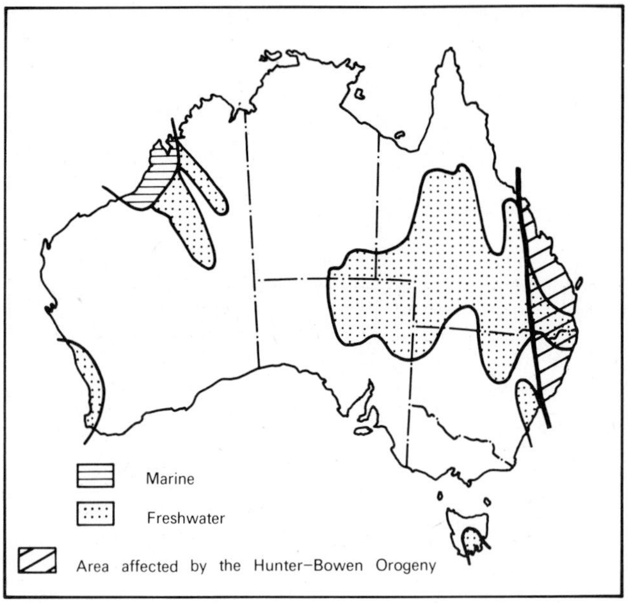

Fig. 4.4 Triassic–Jurassic palaeogeography, showing areas of terrestrial and of marine sedimentation, extent of land areas, and mountainous zone on east coast (derived from Brown *et al.* 1968).

The area of sedimentation increased considerably during the Middle Jurassic, so that the Great Artesian Basin spread over most of south-western Queensland into South Australia and New South Wales; this basin also extended even further west in the Middle and Late Jurassic, and sedimentation continued in smaller basins within the eastern mountain chain.

Most of Australia was land during the Jurassic and remnants of land surfaces of this age still persist not only in the west but also in the eastern highlands, e.g. in Victoria (Jenkin 1976*a*). Marine sediments were laid down only within the Canning and Perth Basins along the western fringe.

The only orogenic activity in the continent occurred during the Late Jurassic in the Clarence-Moreton and Maryborough Basins of south-eastern Queensland, where there was intense folding and uplift, followed by the intrusion of granites and extrusion of intermediate volcanics. These uplifts represented part of the last orogenic movements in Australia (the Maryborough Orogeny) and completed the formation in the east of the Great Divide, which had become clearly defined during the Triassic, following uplift of the Permian basins in the later Permian by the Hunter-Bowen Orogeny.

Doming, which preceded the separation of New Zealand from Australia in the first stages of the breakup of Gondwanaland, gave the first major uplift of the eastern highlands of Victoria (Wellman 1974). At the same time great thicknesses of dolerite were being emplaced as sills in central Tasmania, in association with similar sets of movements.

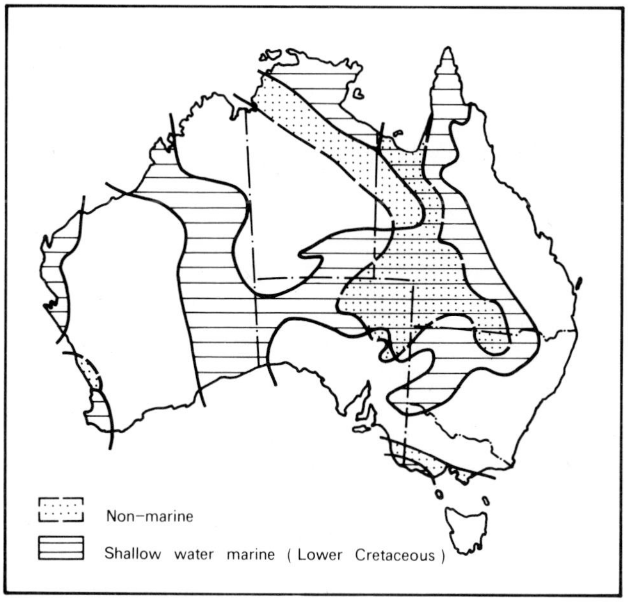

Fig. 4.5. Cretaceous palaeogeography, showing areas of Lower Cretaceous shallow water marine sediments and Upper Cretaceous terrestrial sediments (derived from Brown *et al.* 1968).

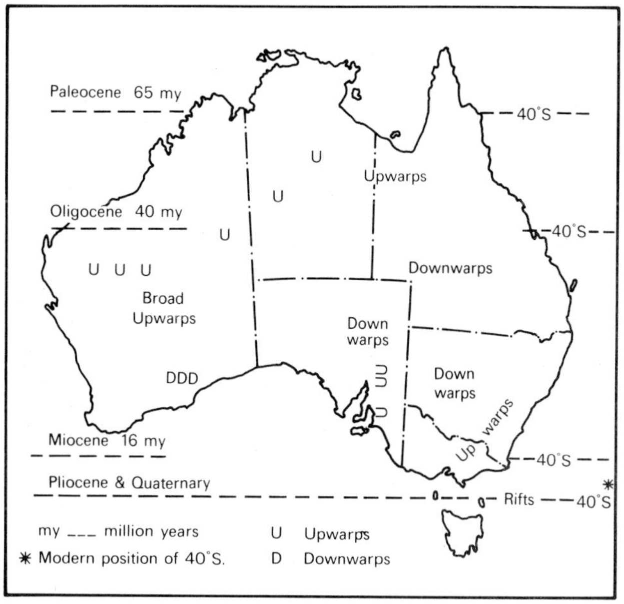

Fig. 4.6. Drift of Australia during the Tertiary (positions relative to 40°S at various stages shown) and location of major tectonic displacements.

Cretaceous

Deposition continued at the beginning of the Cretaceous Period, with fluviatile sediments in most of the basins which had been active in the Jurassic. There were marine transgressions early in the Cretaceous, however, and shallow seas inundated many of the early basins across the continent, entered the Murray Basin, and spread to the Eucla Basin in Western Australia (Fig. 4.5). According to Brown *et al.* (1968) the boundaries of the major basins extended until the seas met and the continent was separated into three major land masses –, an eastern land mass, forming an arc from Cairns in northern Queensland to Ceduna in South Australia, a north-western one, from the Kimberleys in north-western Western Australia to near Alice Springs, and a south-western one, occupying the south-western part of Western Australia. The area of the old basins covered by the seas in the west was very similar to that covered during the Permian. On the other hand Plumb (1979) recognizes the extension of the Basins, but does not indicate any connection between them.

Non-marine sediments accumulated in southern Victoria during the Early Cretaceous. These were lacustrine, deltaic and flood plain deposits, in troughs which developed at the time of east-west epeirogenic uplifts in Victoria in the Early Cretaceous (Jenkin 1976*a*, p. 329). In Western Australia the Carnarvon and Perth Basins also contain Early Cretaceous marine sediments.

At the end of the Early Cretaceous the sea withdrew from most of the submerged part of the continent but, as a result of movements associated with the rupture of Gondwanaland, a marine transgression entered the Otway Basin of Victoria from the west in the early Late Cretaceous. An extensive river system drained northward in the area of the Great Artesian Basin at this time.

Remnants of the extensive Cretaceous land surfaces remain in places throughout the continent. They have been recognized as the Ashburton surface of the Northern Territory (Hays 1967), the Aurukun surface of North Queensland (Doutch 1976), the summit plains of the Davenport-Murchison Ranges and Macdonnell Ranges (Mabbutt 1966, 1967) of the Northern Territory, and the upper surface of the Mount Lofty Ranges of South Australia (Maud 1972). These were modified during the Tertiary.

Tertiary geology and geomorphology, to the end of the Miocene

Australia attained its modern form during the Early Tertiary. It was tectonically stable and was not affected by any severe, widespread, folding movements. While different parts were subjected to gentle epeiric movements, long-continued weathering and denudation maintained the low relief of previous periods everywhere except in the low mountain ranges of the east and south east. Through much of the Tertiary a series of distinct drainage basins persisted in the western part of the Eastern Craton; here, and in places around the coast, there were periodic incursions of the sea.

Effects of continental drift

The southern super-continent, Gondwanaland, had

commenced to break apart in the Cretaceous, following fracturing, uplifts and igneous intrusions during the Jurassic. Australia, together with Antarctica to the south and New Guinea to the north, had begun to move away from eastern Africa during this time, while South America was moving to the west. The essentially northward drift continued during the whole of the Tertiary (Fig. 4.6). New Zealand, which had originally been part of the same land mass, had become separated early in the movement, during the Late Cretaceous, and continued to move further away to the east (Hayes and Ringis 1973). Similarly, India, which had lain between Western Australia and Africa for hundreds of millions of years, also became separated and moved northwards at a greater rate than Australia was moving (Veevers *et al.* 1971).

At the start of the Paleocene (65 million years ago) Australia was approximately 25° to the south of its present position, and at about 55 million years ago, at the start of the Eocene, was about 20° south of its present position. The northern margin of the continent at that time seems to have coincided with the mountainous spine of New Guinea. By the start of the Oligocene period, about 38 million years ago, Australia had moved another 10° to the north, and in the following 16 million years (by the start of the Miocene), it was within 5° of its present position (Veevers *et al.* 1971). This means that the continent had moved through about 20° of latitude in the first 40 million years of the Tertiary (Fig. 4.6).

The separation of Antarctica began about 55 million years ago, in the Late Paleocene (Weissel and Hayes 1972), and the main movement occurred during the Oligocene and Miocene, between 35 and about 10 million years ago (Jones 1971; Griffiths 1971, 1974; Gunn 1975). These movements began along fracture lines forming a rift valley which ran parallel to the southern coast of Western and South Australia, then turned through approximately 70° to the south-east to run parallel to the west coast of Tasmania. Other fracture lines penetrated into the coast of South Australia and into Bass Strait. By introducing a change of oceanic circulation near southern Australia, this separation led to far-reaching climatic changes.

Tectonic movements

Australia was relatively stable through the Early Tertiary. There was intense tectonic activity along the northern continental margin at the close of the Eocene, and in the Oligocene the eastern edge of the continental shelf emerged (Jones 1971) at the time when Antarctica was becoming separated in the south.

There were gentle epeiric movements within the Western and Central Cratons and over parts of the eastern highlands, with some major areas of subsidence which formed depositional basins such as the Lake Eyre, Darling-Warrego, and Murray Basins (Fig. 4.6). The total subsidence in the Murray Basin was about 500 m, while uplift due to faulting over this period was as much as 800 m. Tectonism in the eastern highlands was thought to have been uneven and intermittent, but to have continued throughout the early Cainozoic, culminating in the Kosciusko Uplifts of the Pliocene in the south-east (Wellman and McDougall 1974). By contrast Wellman (1979) has since proposed continuous uniform uplift in the south-eastern highlands during the Tertiary.

In Victoria the eastern highlands region had already been uplifted in the Mid-Mesozoic. A Late Mesozoic erosion surface was then raised up to 600 m before the Late Eocene; a second major uplift occurred in the Late Oligocene, and a third and last uplift in the Late Cainozoic (Wellman 1974). In parts of eastern New South Wales there had been a major uplift of the order of 300 m some time between the Mid-Cretaceous and the Late Oligocene (Wellman and McDougall 1974). The pattern of tectonism in Tasmania was different: horsts and grabens developed as a result of the uplift of the highlands (Solomon 1962), and the faulting which produced these continued during the Cainozoic.

There was intermittent and uneven uplift in eastern Queensland in this period, and broad warpings occurred in the centre of the eastern cratonic belt extending into New South Wales.

Differential movements occurred in the Eocene and Miocene in eastern central South Australia (Alley 1977). There was gentle warping near Lake Eyre during the Tertiary (Jessup and Norris 1971), and, according to Alley, faulting formed some basins of sedimentation. Uplifts in the Flinders Ranges area probably began in 'Late Medial Miocene times' (Callen 1977). The major period of uplift in the Adelaide region seems to have been Late Miocene also (Brown *et al.* 1968), although Early Tertiary uplifts have been recognized in the Fleurieu Peninsula by Brock (1971); according to his evidence sea levels due to these uplifts were up to 200 m above the present.

A similar set of uplifts occurred through the Northern Territory and north-western Queensland and there were broad warps on the shield in Western Australia in the later Tertiary, although Western Australia was basically stable during the Early Tertiary (Johnstone *et al.* 1973).

Major lineaments across the continent started to become prominent in the Early Tertiary. These have delimited the boundaries of some of the basins, e.g. the Murray Basin, and have determined major stream directions, e.g. of the Darling River (Hills 1956). Such lineaments have also affected the pattern of further denudation, through the influence of major structures of rock bodies on the courses of streams.

Marine deposits

These are of minor extent and are associated with a series of basins along the southern and western coasts of the continent. There were a series of distinct basins along

Bass Strait in the Early Tertiary. The first marine transgression of the Gippsland Basin occurred in the Late Eocene, related to the opening of the basin and subsidence at the time of separation of Australia from Antarctica (Gunn 1975). In Victoria the marine transgressions reached a maximum in the Early Miocene (Jenkin 1976*a*), when there were uplifts in the Victorian highlands, which preceded the Kosciusko Uplift. The seas began to retreat in the Middle Miocene.

The Murray and St Vincent Basins followed the trend of fracture lines associated with movements connected with the separation from Antarctica. The extent of the marine part of the Murray Basin shows that, during the Miocene, maritime conditions extended much further than they do today, with the result that parts of the present interior of the continent have previously had a climate influenced by the intrusion of the sea (Kemp 1978, pp. 174 and 195).

In the Tertiary, sedimentation recommenced in the Eucla Basin at the border of South and Western Australia, an area which had been part of the extensive Early Cretaceous sea; to the west an older basin, the Carnarvon Basin, was also active during this period. Subsequent uplifts were only small, so that the almost horizontal form of the surface of the sediments has been preserved in plains such as the Nullarbor Plain and those of the Murray-Darling area.

Terrestrial deposits

While marine deposits were of limited extent and marginal to the continent, terrestrial deposits were much more widespread in the east, particularly in major drainage basins within the Eastern Craton. There were no such extensive drainage basins within the Western Craton during the Tertiary.

The drainage basins are of two main types:

(1) Those associated with river systems and lakes within the eastern highlands belt. The modern drainage systems were established at more or less their present positions by the Early Tertiary (Wellman 1974), and the sedimentary sequences include quartzose gravels, shales, sandstones and clays with interbedded volcanics, e.g. as in south-east Queensland (Beckmann and Stevens 1978).

About 300 m of terrestrial sediments had accumulated in grabens in Tasmania, such as those of Launceston and Hobart, by the end of the Oligocene (Gill 1962, p. 253). These sediments have formed parent materials for a number of palaeosols which still persist on low erosion surfaces (Nicolls 1960).

(2) Those associated with inland drainage systems, both intracratonic basins such as the Lake Eyre Basin, and those older basins, such as the Georgina, Drummond, Amadeus and Upper Murray, in which renewed sagging in the Early Tertiary initiated the development of extensive stream systems.

The areas of deposition were originally somewhat restricted, the Eyre Formation (Wopfner *et al.* 1974) was laid down in parts of the Lake Eyre Basin in the Early Tertiary but the area of sedimentation extended later. There was only limited terrestrial sedimentation during the Early Tertiary outside the Lake Eyre Basin. This was the time of development of the extensive land surfaces of low relief in the inland, much of the Great Artesian Basin area being land at the start of the Oligocene.

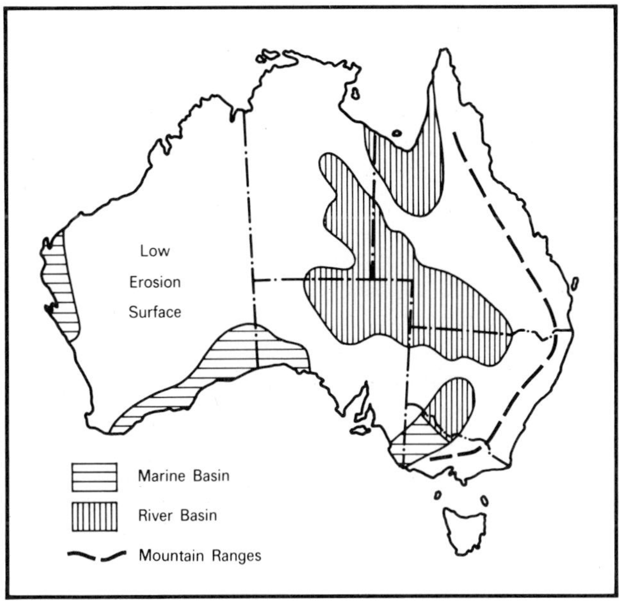

Fig. 4.7. Simplified representation of major elements of the early Tertiary landscapes of Australia.

Land surfaces and their relation to tectonics

At the start of the Tertiary the relief over most of the continent was low – even in the eastern highlands it was only moderate. Many pre-Tertiary sediments were flat-lying, and maintained this form following the slight amount of uplift they later underwent. There are still remnants of such pre-Tertiary surfaces through the continent. Hays (1967) has assigned a pre-Cretaceous age to the oldest surface he recognized in the Northern Territory, the 'Ashburton Surface'. Ayers Rock in the southern Northern Territory is fringed by Tertiary sediments and is a pre-Tertiary residual.

An extensive surface of very low relief, less than 15 m over many kilometres, developed over much of the Northern Territory outside the Lake Eyre Basin and over north-west Queensland in the Early Tertiary (Fig. 4.7). This was the 'Tennant Creek Surface' of Hays (1967), and its equivalents the 'Bradshaw Surface' of Wright (1963) and the 'Aurukun Surface' of Doutch (1976), in the Carpentaria Basin Area. It was formed partly on exposed Cretaceous sediments, and in part on older rocks and granites, during the Eocene and Oligocene.

The 'Mount Mee Tableland' of Marks (1933) in south-east Queensland is most likely a remnant of an Early Mesozoic surface (Beckmann and Stevens 1978), and Watkins (1967) has recognized a 'Gondwana surface' (Jurassic) near the Queensland–New South Wales border.

An Early Tertiary surface developed in Central Queensland mainly to the west of, but partly within, the eastern highlands (Galloway 1967*a* and *b*), and remnants of an equivalent surface continue west of The Divide through southern Queensland and into New South Wales. The Early Tertiary was the time of uplifts in the east (Wellmann and McDougall 1974), but remnants of this surface have been recognized in coastal New South Wales (King 1962). Distinct low erosion surfaces seem to have been absent from Victoria, where there has been warping and epeirogenic uplifts of the highlands through the Tertiary, with depressions in Gippsland and on the inland side of the range.

A series of distinct stepped erosion surfaces developed in Tasmania as a result of intermittent uplifts through the entire Tertiary. They range in height from 100 m to 1000 m, the stage of dissection reached by each depending upon the presence of resistant dolerite sheets and other hard rock layers.

The pre-Tertiary land surface in South Australia was already one of low relief, distorted at different times and in different places by earth movements which led to rejuvenations and the development of new land surfaces. The older surfaces in downwarped areas were buried by later Tertiary sediments (Alley 1977). The upwarped portions of the main surface in the mid-north, with a deep weathering profile (Alley 1977), are now seen at up to 600 m in the Adelaide Hills, as a result of a series of later uplifts associated with block faulting. Remnants of a Mesozoic to Early Tertiary surface, also deeply weathered, occur through the southern Mt Lofty Ranges to the Fleurieu Peninsula and are surrounded by fragments of younger surfaces which have developed below it. Pre-Tertiary land surfaces have been recognized also in the Flinders Ranges (Twidale 1966*a*); these were upwarped in the Early Tertiary and younger surfaces were cut out below them.

During the Tertiary, and subsequently, the extensive high surface of Western Australia, the Old Plateau of Jutson (1934) originating in the Mesozoic or earlier, was reduced with the development of a series of surfaces below it (Mulcahy 1960; Churchward and Bettenay 1973).

In short, it appears that, from the Cretaceous until the end of the Oligocene and into the Miocene, a long process of further reduction continued over most of the continent to produce what has been called the 'Australian surface' (King 1962). In the west, denudation may have been proceeding for hundreds of millions of years since the Precambrian, but in the east the surfaces developed in progressively uplifted Mesozoic and Early Tertiary sediments. This uplift produced multiple surfaces during the Mesozoic and Tertiary in the eastern half of the continent, but the relief was never high except in the eastern ranges.

Volcanism

There was continuous volcanism through the whole of the eastern highlands area during the Tertiary, and into the Quaternary (Ollier 1977). This continuous activity has been related to the spreading of the sea floor during the drift of the continent (Wellman and McDougall 1974) as tensional forces developed in the crust.

The flows occurred in shield volcanoes such as the Tweed Volcano of northern New South Wales, as fissure eruptions, and also as flows filling valleys which in places were interbedded with sediments. In addition, numerous volcanic plugs form conspicuous landscape features. The most common rock type is basalt. Some of the earliest flows (Paleocene to Eocene) were associated with Tertiary basins, where they are interbedded with sediments (Beckmann and Stevens 1978). Eruptions in Victorian and Tasmanian Tertiary basins have been thought to be associated with fault activity (Gunn 1975).

Basalts associated with deep weathering profiles in the Roma–Amby area of western Queensland have been dated radiometrically as Late Oligocene to Early Miocene (Exon *et al.* 1970), thus providing a means of putting an upper limit to the age of the local deep weathering profile.

Climates

Inferences on Tertiary climates have been pieced together from fragmentary data derived from the following.

(i) Theoretical considerations, e.g. pressure system changes, such as those following changes in continental position (Frakes and Kemp 1972; Kemp 1978).

(ii) Information on ocean temperature from marine fossils (such as forams) and also on oxygen isotope data from these.

(iii) Distribution of terrestrial plants, especially as inferred from microflora (Brown *et al.* 1968, p. 307; Alley 1977, p. 115).

This discussion aims to consider possible past climates during which weathering proceeded, on the basis of information from a number of independent sources. No inferences about climates based on weathering, laterite, or soil studies have therefore been incorporated. In this way the environment under which soils were formed can be deduced, and checked against conclusions based on an analysis of soil properties.

Theoretical considerations

As discussed above, Australia was 25° to the south of its present position at the beginning of the Tertiary. The continent then moved north through a series of climatic zones, which were themselves changing in position over the Tertiary. Higher latitudes initially did not mean substantially lower temperatures. The climatic gradient of temperature from equator to pole was lower than at present, and warm wet climates prevailed at high latitudes during the Paleocene (Frakes and Kemp 1972). At this time, the oceanic and atmospheric circulation patterns would have been quite different from the present because the continuity with Antarctica prevented any maritime influences along the southern shores even though rain-bearing winds penetrated far inland (Kemp 1978).

The circulation patterns were similar following the separation of Antarctica and Australia in the Late Paleocene, even with the extension of maritime influences further over the continent (Kemp 1978), as atmospheric circulation at the time was sluggish. Relief was lower than the present, at this time, even in the relatively high ranges in the east and south-east, and the altitudinal temperature gradient was also lower.

During the Oligocene the overall temperature became cooler and nearer to levels of the present (Frakes and Kemp 1972), and in the Miocene the sea penetrated well into the continent in the Murray Basin, bringing maritime conditions to the inland. Atmospheric circulation increased, but the northward drift of the continent intensified aridity in the north-west and west, and brought more rain to the south and south-east (Kemp 1978).

Ocean temperatures

These would have affected climate, especially in coastal areas, through their influence on air temperatures, and their possible effects on rainfall and cloud. Through the earlier Tertiary the ocean temperatures were warmer than the present. There was an increase from the Paleocene through the Eocene and earlier Oligocene (Frakes and Kemp 1972; Kemp 1978).

Abele (1976, p. 191) regards the very late Oligocene to Early Miocene as being a time of warm, though not tropical, climate in Victoria, on the basis of foraminifera species and oxygen isotope determination.

Thick ice sheets were already in existence in Antarctica during the Miocene and Pliocene, and Late Miocene cooling and the rapid expansion of the Antarctic ice sheet (Kemp 1978) gave eustatically lowered seas. While there were two distinct warmer periods during the Pliocene, the pattern of marine changes characteristic of the Quaternary were initiated at that time (Dorman 1966; Brown *et al.* 1968, p. 305).

Terrestrial climates

The presence of lakes, such as Lake Eyre, in the inland during the earlier Tertiary indicates either that there was sufficient overall moisture to maintain runoff, or that the evaporation was lower. Conditions were moist, but the amount of rain was not necessarily very high. Warmer conditions have been inferred for many places on the basis of the occurrence of particular plant species.

Palynological evidence, e.g. of the spread of rainforest in Australia from the east to the south-west during the Paleocene and Eocene (Kemp 1978), suggests high rainfall and moderate to high temperatures for the depositional environment of the Eyre Formation (Wopfner *et al.* 1974). General warming and increased precipitation occurred in this area throughout the Eocene. Alley (1977) also has inferred from local micro-flora a warm temperature, possibly subtropical, and a moist climate in the Eocene and Oligocene in the mid-north of South Australia. A warm climate with high rainfall has also been designated for the Miocene in the neighbouring Lake Frome area by Callen (1977). At this time the south-east and east of Australia were well watered and supported rainforest, whereas the north and north-west were becoming increasingly arid. Grassland developed in the centre.

Ollier (1977) has suggested that the Paleocene and Eocene were relatively cool and humid in Victoria, with a warming in the Late Eocene. Hot and wet conditions prevailed also in the Perth area during the Eocene (Quilty 1974). A pluvial climate over the area of the Yilgarn block also is inferred for the Early Tertiary from the existence of an extensive palaeodrainage system at that time (van de Graaff *et al.* 1977).

Weathering

As pointed out above, apart from some warping, which elevated upland areas and tilted depressions, the land surface of much of Australia was essentially stable through the Early Tertiary; under these conditions, surfaces of low relief continued to develop largely undisturbed, from the Cretaceous until the middle of the Tertiary.

Even though the continent was drifting northwards through 20° of latitude, warm and wet conditions seem to have prevailed at least over most of the southern part of the continent for much of the period from Paleocene to Miocene. Modern moisture barriers, such as the mountain ranges in the east and south-east, were less effective than at present, as their relief was lower. Maritime conditions, which had extended their influence following the separation from Antarctica, penetrated well inland with the expansion of the Murravian Gulf, suggesting the possibility of moist environments towards the centre of the continent.

Stable conditions for a long period on a surface of low relief allowed weathering to reach a uniformly advanced stage on a wide variety of rocks, with the removal by leaching of soluble constituents and the mobilization of ferrous iron, aluminium and silica. In certain areas the weathering front penetrated to depths of at least 60 m, the form and pattern of the weathering products depending on the type of parent rock (Senior 1976; Hays 1967).

Some remnants of weathered products which formed below the old land surface still exist, and these illustrate the uniformity of the surface over the continent. Many of the materials present can be identified with those characteristic of different layers of the laterite profile (see Chapter 3). Many surface materials are hardened remnants of the ferruginous zone of the former deep weathering profiles, and many residuals are composed of parts of the mottled and pallid zones. It is, however, rare for the complete profile or for the original surface soil layer to be found; nor are all deep weathering profiles part of laterite profiles and they do not necessarily resemble them. While such profiles can be considered as a weathering entity, the fabrics of some weathered materials are still largely those of the original rock, and little change to true soil fabric is evident in them.

Major changes of ground water levels, either seasonally or over periods of years, may be inferred to account for the movement and concentration of iron in the ferrous state within the profiles, and oxidation to the stable ferric state. The irreversible hardening, characteristic of laterite and evident in many lateritic materials here, may have occurred with local dehydration following regional lowering of the water table, or by exposure following uplift and progressive removal of parts of the profile, associated with scarp retreat away from the dissecting streams.

As a result of the dismembering of an old surface with an associated laterite or other deep-weathering profile, lower levels of the profile became exposed to surface weathering, and a variety of derivatives of the earlier weathered materials were formed; in some places these were covered by detritus transported from higher levels in the profile (Stephens 1946; Hays 1967).

Deeply weathered lateritic profiles have been dated on geological grounds, by using the age of the youngest sediments in which they formed, together with the age of the oldest sediments and volcanics burying them. Johnstone *et al.* (1973) established that laterite in the eastern arid area of Western Australia is Oligocene or Miocene, from the age of associated sediments. This technique is of limited application because many lateritic areas have developed in very old rocks on ancient land surfaces and have remained unburied since they were found. Upper limits to the age have also come from radiometrically determined ages of basalts which overlie them, e.g. the age of a deeply weathered profile determined by Exon *et al.* (1970) near Amby, in southern Queensland, can be no younger than Late Oligocene–Early Miocene, the age of the basalts above them.

More recently, deep-weathering profiles have been dated from measurements of the orientation of magnetic oxides. By this means two distinct profiles of different

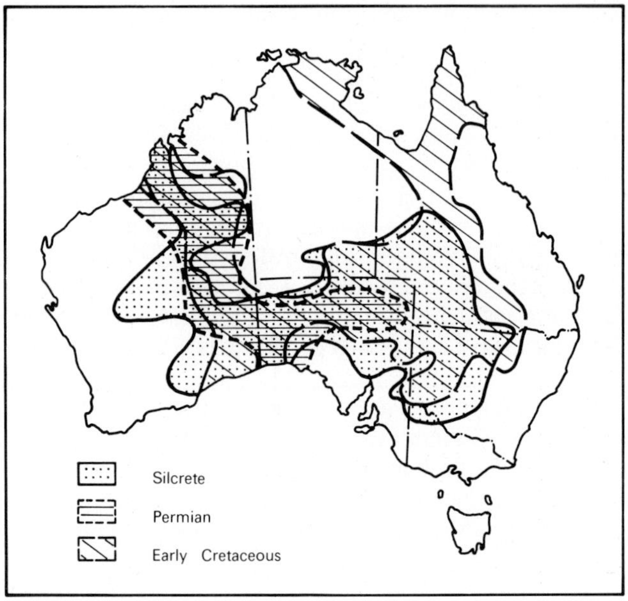

Fig. 4.8. Relationship of the distribution of silcrete (after Stephens 1971) to Permian and Early Cretaceous basins (after Brown *et al.* 1968).

ages have been recognized in south-western Queensland (Senior 1976; Idnurm and Senior 1978). The older, the Morney profile, is early Eocene and partly contemporaneous with depositions of the Eyre Formation. The younger, the Canaway profile, is Late Oligocene, a dating consistent with that obtained by Exon *et al.* (1970) on radiometric grounds.

Materials which can be related to a completely clay laterite profile (Thompson *et al.* 1954) developed on Late Oligocene to Early Miocene basalt at an elevation of about 600 m near Toowoomba, Queensland, i.e. on a rock of approximately the same age as that which covered the laterite profile at Amby. This indicates that conditions favourable to lateritization were present during the Miocene at this and similar sites, such as Kingaroy and Binjour Plateau to the north.

Similar, but much thinner, laterite profiles occur within coastal valleys of south-east Queensland (Bryan 1939). These may also be of Miocene age, but on physiographic grounds may be Pliocene or possibly younger (Beckmann and Stevens 1978). Apparently, conditions favourable to lateritization, both climatic and hydrologic, prevailed at different places during much of the Tertiary period in Queensland. Low level laterites in coastal valleys of southern New South Wales are regarded as post-uplift (Young 1978), i.e. of Late Tertiary age.

Laterite was forming during the pre-Eocene and Early Eocene in the mid-north of South Australia (Alley 1977), and was faulted and dissected later in the Eocene, while it continued to develop until the Miocene on more stable areas to the south. This laterite surface is contiguous with the main laterite surface in the Mount Lofty Ranges, which is of Late Mesozoic to Early Tertiary age (Maud 1972).

The main occurrences of silcrete in Australia are associated with a zone across the continent which is quite distinct from the adjoining areas where laterite profiles and deep-weathering materials are more common (Stephens 1971). This geographical distribution corresponds broadly to that of some of the basin areas extant in the Permian, and in the Early Cretaceous, viz. the Canning and Officers Basins in the west, and also the Great Artesian Basin and its extensions to the north and south (Fig. 4.8). Silcrete may have developed here because these areas were relatively low lying during the Tertiary, or because of favourable parent materials. There are small areas of silcrete in other parts of the continent, particularly in some river valleys of the eastern highlands and in some places associated with basalts.

The formation of silcrete appears to have been a two-stage process, which resulted in the absolute accumulation of silica in particular situations. In aluminosilicate systems, silica is dissolved and mobilized more readily below pH 6–7 and above pH 11 (see Chapter 26), the solubility increasing with increasing temperature. It is transported in solution before deposition as a silica gel, in a host rock which may also have been undergoing severe weathering.

Silcrete is commonly associated with silica-rich parent rocks such as quartz sandstones, where the indurating silica fills the pores, but occurs also in altered igneous rocks which have been depleted of most of their soluble constituents, and in which silica has formed pseudomorphs after the primary minerals (Langford-Smith 1978). Silcrete also appears in the lower layers of some laterite profiles, but may form a distinct cap over them as a separate body. Detrital silcretes may overlie deep weathered profiles, and these may show evidence of two periods of silcrete formation.

There is much evidence of warm moist conditions over the greater part of Australia during the Early Tertiary (Langford-Smith 1978; Wopfner 1978). These have favoured the development of silcrete in suitable environments. Laterite was forming under the same climatic conditions in the interior of Australia, the critical factor being interpreted by Alley (1977) for the mid-north of South Australia as the base level of erosion. This base level controlled local drainage and the leaching environment. Should the drainage situation change, e.g. by warping, silcrete could develop on downwarped laterite.

Silcrete was thus formed at many times during the Tertiary, if environmental conditions were favourable (Exon *et al.* 1970; Stephens 1971; Alley 1977), but the most widespread silcrete deposits are those of the Early Tertiary. These developed on a surface of low relief, the Cordillo surface of Wopfner (1978), on sediments of the Eyre Formation (Wopfner *et al.* 1974), and on some Cretaceous rocks. Late Cainozoic silcretes are more restricted in area than those of the earlier Cainozoic and typically have lower silica contents (Wopfner 1978).

Many silcretes in eastern Australia appear to be stratigraphically and topographically below basalts (the 'grey billy' of Browne 1972; Gunn and Galloway 1978). The latter authors have suggested that the silica necessary for the induration of the sandstone parent rock could have come from weathering of the overlying basalts. According to this theory silcrete in sedimentary rocks beyond the present limits of the basalt could have come from basalts previously overlying them, or from highly weatherable sediments in the area. Silcrete, however, occurs right across Australia in localities far removed from any known basalt areas, and where there is an adequate supply of local silica for the induration.

Early history of landscape development

When trying to unravel the problem of the history of weathering of parent materials and of soil development it is important to trace the history of land surface development through all geological periods (Table 4.1). By the end of the Miocene, Australia largely had its present form, and this was the summation of all the events it had undergone. This form, shown very broadly in Fig. 4.9, delineates areas exposed from varying geological periods up to the present, except for local sedimentation events. Portions of the old erosion surfaces of the southern and western parts of the continent, which were mainly land during the later Paleozoic, were planed off when mantled by continental ice sheets at various times during the Permian. Since the retreat of the ice the surfaces have again been subject to denudation.

The map gives the maximum age of landforms in any locality, but does not present a distribution of land surfaces of various ages. Each of the original post-uplift surfaces may have been partly or completely removed by subsequent denudation, so that in any area there may be several distinct erosional land surfaces with complementary depositional areas: the surfaces may occur by exhumation or as residuals which surmounted the depositional areas.

Points to note include the following:

(i) The extreme antiquity of erosion and weathering in parts of the Western and Central Cratons. It appears that parts of the Northern Territory have been exposed continuously since the Cambrian, while the Precambrian blocks were exposed until covered by ice in the Permian, and then continuously since that time.

(ii) The antiquity of parts of the eastern highlands, along uplifted geosyncline belts, in which there are remnants of pre-Mesozoic surfaces.

(iii) The extent of surfaces of Mesozoic age, which were affected by Early Tertiary weathering in Queensland

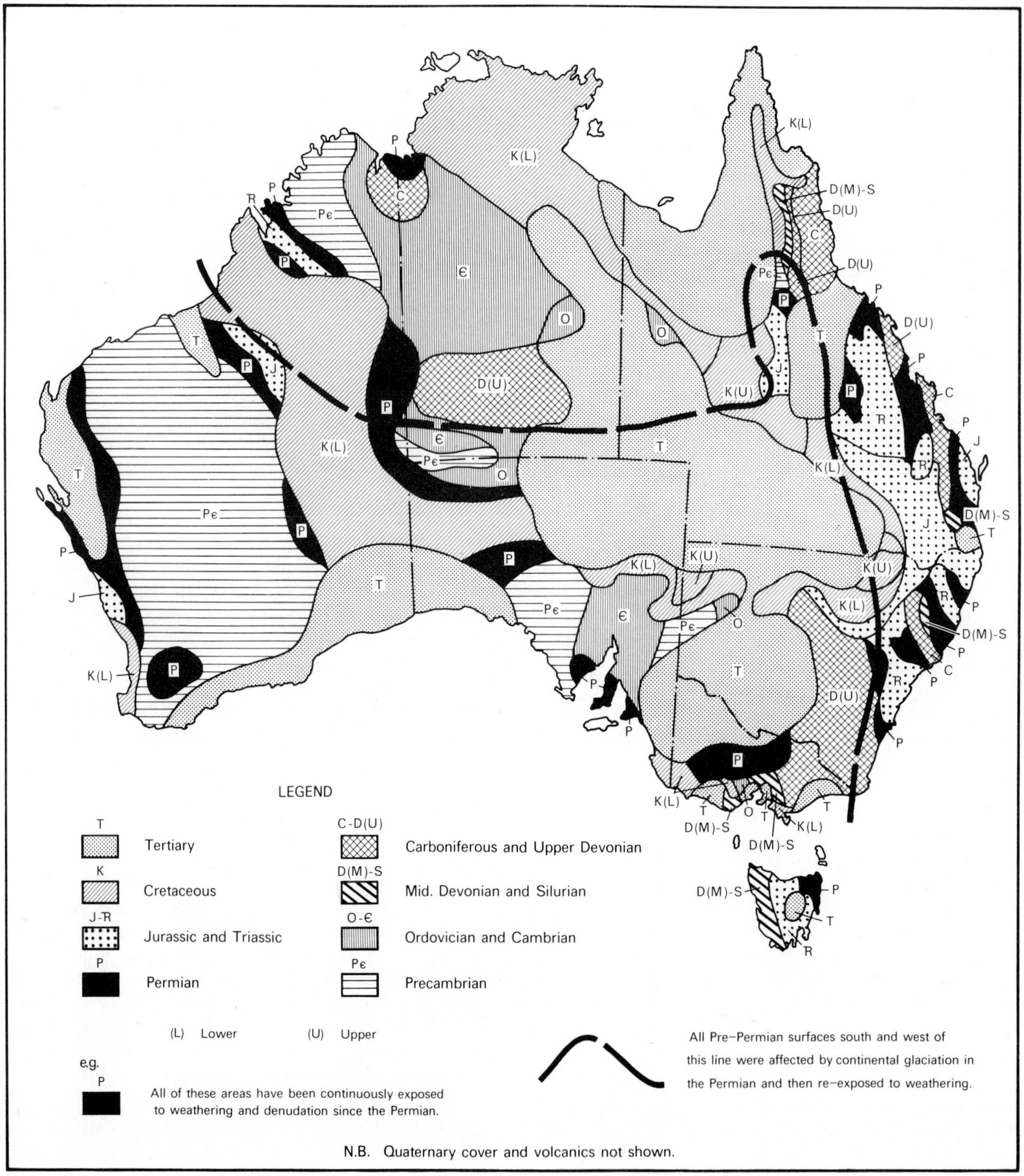

Fig. 4.9. Maximum age of exposure of various regions of Australia (derived from Maps in Brown *et al.* 1968).

the Northern Territory and parts of the Tertiary basin areas.

(iv) The extent of Tertiary lowlands in the south, representing raised, but largely undissected, basin floors.

Pliocene and Quaternary events modified this framework by dissection following further uplifts, by pedimentation, by infilling of depressions, and also by the consequences of numerous climatic changes.

Table 4.1. Outline of geologic and geomorphic history of Australia to end of Miocene

Ages (in millions of years) are on a log scale. P_ϵ = Precambrian; ϵ = Cambrian; O = Ordovician; S = Silurian; D = Devonian; C = Carboniferous; P = Permian; R = Triassic; J = Jurassic; K = Cretaceous; Pe = Paleocene; E = Eocene; Ol = Oligocene; M = Miocene.

Period & age		Form of continent and major Landscape events	Tectonic events	Type of sedimentation
		Dissection of old surface and production of new surfaces	Movements leading to Pliocene Kosciusko uplifts	Terrestrial sediments in W.A.
			Warping in W.A.; uplifts in Flinders Ra. in S.A.	
M		Deep weathering of volcanic rocks		
			Retreat of sea from basins in south	
	Miocene	Shield and fissure flows in Eastern Craton		
		Deep weathering, lateritisation, silcrete formation	Uplifts in Vic. raising marine basins	
Ol	to		Major uplift in N.S.W.	
		Development of low erosion surface over western half, and in east, west of Eastern Highlands	*Separation from Antarctica*	Fresh water sediments within eastern highlands and in grabens (E-Ol)
		Volcanism within valleys, as shields, and associated with faults in east	Uplift in Vic.; some upwarping in S.A. (Flinders Ra.)	Fluviatile sediments in inland basins e.g. Lake Eyre Basin and older basins, such as Amadeus Basin
E	Paleocene	Silcrete in some basin areas	Horsts and grabens in Tas.	
Pe		Most of continent had low relief, apart from highlands in east and south-east; deep weathering	*Separation from Africa*	Marine, in separate basins along the southern and western margins (Late Eocene–Late Miocene)
-65				
	Cretaceous	Most of continent was land with extensive terrestrial basins in north-east (Remnants of Cretaceous land surfaces in north west and through centre)	?Some folding in Maryborough Basin	Extensive fluviatile deposits in north; marine sediments in Otway Basin
K		Continent cut into three landmasses by shallow seas, south west W.A., N.T. and north west W.A., and an arc along east and south-east coasts	Mostly epeirogenic movements; i.e. broad warping	Shallow marine sedimentation within earlier basins and Great Artesian Basin and in Carnarvon and Perth Basins
		Divide continuous along east and south-east coasts		Fluviatile sedimentation in Jurassic basins
132				
			Maryborough orogeny	
J	Jurassic	Western and Central Cratons were land, as were southern N.S.W. and Vic. Basins in Qld extended during Middle and Late Jurassic		Marine sedimentation in W.A.
	and		Major uplifts in Vic. and Tas.	Terrestrial sedimentation in Great Artesian Basin
R	Triassic	Most of continent was land; mountain range along east coast; sedimentary basins within mountain range and to west	Weak folding in river basins	Fluviatile sedimentation in Qld and N.S.W. Marine sedimentation in W.A.
238				
P	Permian	Shallow seas across centre and west; southern half covered by land ice	Orogeny along east coast	Glacial sediments, especially in south overlain by marine (west) and terrestrial (east and centre)
		Most of continent was a stable land mass		
C	to	Seas in basins in north west and on east coast	Intermittent uplift in Tasman Geosyncline	Marine in north west and west, glacial with terrestrial in N.S.W.
D S	Silurian	Major river basins or troughs in east; Tasman Geosyncline, Amadeus and Carnarvon Basins active	Orogeny in (south-) east	Marine in Tasman Geosyncline, shallow water in western shield also
430				
O	Ordovician	Yilgarn–Pilbara blocks were land		Shallow water marine in north-west
	to	Adelaide Geosyncline, Gawler block, Northern territory were land	Only epeirogenic movements across continent	
ϵ	Cambrian			Marine in centre and east, Adelaide Geosyncline & Tasman Geosyncline
570				
P_ϵ	Pre-cambrian			

Pliocene and Quaternary events and influences

Situation at the end of the Miocene

In the course of its northward drift the continent had reached a point within 5° of latitude of its present position by the end of the Miocene (Veevers *et al.* 1971), so that the small amount of further movement which occurred subsequently would have had no more than a very slight influence on climatic patterns and on atmospheric circulations.

The shape of the continent and the landform pattern was broadly the same as it is today, even though the relief and the extent and types of landforms within each region differed. Most of the country consisted of low plateaux, with a number of extensive drainage basins spread down the eastern interior of the continent. These drained land west of the eastern and south-eastern mountain chains, and included the Murray, Lake Eyre, Darling-Warrego and Karumba Basins, i.e. the catchments of all the eastern inland streams.

Lateritic or deeply weathered profiles occurred on surfaces of low relief throughout the continent, including some areas within the eastern highlands. Some of the deeply weathered profiles were capped by silcrete, and this was extensive also in lower areas, particularly in the south–central part of the continent.

Tectonics

All the tectonic movements in the Late Cainozoic were vertical, with some broad warping and some local faulting. A series of uplifts, collectively known as the 'Kosciusko Uplifts', have been active intermittently since the Middle Miocene–Early Pliocene, particularly in the south-east of the continent (Abele 1976). There have been similar movements in other parts of the eastern highlands and corresponding uplifts in the more stable cratonic areas in the centre and west.

The uplifts were greatest along the eastern highlands, particularly in the north and south. In the area near the New South Wales–Victoria border, where the relief was already more than 1 000 m and the greatest height probably much more, the Kosciusko Uplifts were of the order of 300 m (Wellman 1974; Wellman and McDougall 1974); Wellman suggested in 1979 that such movements were part of a long-continued progressive uplift. Ollier (1977) has noted that late tectonic movements in Victoria have been important in producing physiographic effects in areas away from the highlands, in the form of major coastal inlets, in the outline of the Western Plains, and in supposed fault block topography in the south-east.

There are differences of opinion on the amount of late tectonic movements in Tasmania. Spry and Banks (1962) state that continued movement along faults during the Late Tertiary and Quaternary has lifted Tertiary sediments to as much as 510 m above sea level and disrupted previously formed land surfaces. On the other hand, Sutherland (1971) has pointed out that some of the supposed movements were probably earlier than Plio-Pleistocene; other explanations, such as eustasy, may explain some deep dissection and the persistence, in the undisturbed state, of some shallow-water Upper Pliocene sediments precludes any major uplifts after their deposition.

The main uplifts in central New South Wales have been pre-Pliocene; in southern Queensland Beckmann and Stevens (1978) have concluded that Late Pliocene–Pleistocene tectonic uplifts amounted to no more than a few tens of metres at most; in North Queensland, however, it has been inferred by Wyatt and Webb (1970) that there have been active uplifts along lineaments associated with the outpouring of basalts in Late Pliocene to Pleistocene time.

Uplift of the Flinders Ranges in South Australia probably commenced in 'late Medial Miocene' times (Callen 1977) and continued after this time. Alley (1977) noted that renewed tectonic activity in the mid-north of South Australia during the Late Cainozoic caused the dissection of earlier silcretes and faulting of Early Tertiary sediments. The main laterite surface of the Mt Lofty Range of South Australia (Late Mesozoic to Early Tertiary) was contiguous with that of the mid-north and was upwarped during a series of uplifts during the Tertiary; these uplifts reached their culmination at the end of the Pliocene, but continued during the Quaternary (Campana 1958). Brock (1971) has recognized two Late Cainozoic stages of uplift in the Fleurieu Peninsula, a southern extension of the Mt Lofty Ranges. The first of these was Late Miocene and the second Plio-Pleistocene. There was broad regional warping in the south-east of South Australia, with uplifts continuing during the Pliocene and Pleistocene (Cook *et al.* 1977).

In the Northern Territory a series of uplifts gave rise to the dissection of earlier surfaces and the development of younger surfaces below these. These surfaces show evidence of regional warping, which is also evident in corresponding surfaces in northern, north-western and central Queensland (Twidale 1966*b*; Galloway 1967*a,b*; Doutch 1976).

Downwarps in the area of the Gulf of Carpentaria provided environments for the deposition of sediments on the western side of Cape York Peninsula. Similar subsidences maintained the basins of the Lake Eyre–Lake Frome region, and also the Murray–Darling and Warrego Basins. These have persisted to the present day, but were limited by the position of major lineaments (Hills 1956), following dissection of the earlier Tertiary surfaces and exposure of the underlying less weathered rocks.

The shield area of Western Australia has been uplifted by about 300 m since the Eocene, and now has a relief of

300–600 m. Related to this uplift of the major part of the shield were downwarps of the southern edge so that equivalents of older surfaces pass below sea-level.

New land surfaces and materials within major sedimentary and structural basins

The modern drainage system and the pattern of valleys forming each drainage basin have been established by the tectonic pattern of uplifts and up- and downwarps during the whole of the Late Cainozoic. New surfaces were cut across the continent during backwearing of scarps, away from each of the major streams and their tributaries. At the same time any remnants of older surfaces above the widespread Early Tertiary surfaces would have become further reduced.

At the end of the Miocene period and into the Early Pliocene (Hays 1967), the Wave Hill Surface developed in the Northern Territory at the expense of the Tennant Creek Surface. Similarly, the Aurukun Surface of the Karumba Basin in north-west Queensland was reduced by pediplanation to produce the Late Miocene Strathgordon Surface (Doutch 1976). Later cycles of erosion during the later Pliocene and into the Quaternary resulted in the formation of the Koolpinyah Surface in the Northern Territory and the Campaspe Surface in the Karumba Basin (Doutch 1976).

Similar events occurred in Central Queensland in the Late Tertiary. The deeply weathered surface from the Early Tertiary was eroded and only parts of the original surface have survived (Galloway 1967*a*, 1967*b*; Wright 1968). During this period the landscape took on approximately its present form, of low hills in the east and a series of gently rolling plains in the west. Some of the clays of the 'brigalow' country of eastern Australia, which had been derived from Mesozoic sediments and later alluvia, were probably weathered during this period.

A succession of Late Tertiary events in south-eastern Queensland have been outlined by Watkins (1967) and by Beckmann and Stevens (1978). On the main divide the lateritized Miocene surface on basalt was cut down during the Late Miocene and Pliocene and a succession of younger surfaces were developed at various stages of the dissection. The influence of lineaments in the underlying rocks, and their structures, became evident as earlier surfaces were reduced, because the lineaments controlled the direction of stream courses which became the loci from which younger surfaces extended with time. Two major surfaces, each with a base level only a few tens of metres above present sea level, indicate some slight uplifts in the region in the Late Cainozoic, and these surfaces have been further dismembered by erosion during times of low sea level.

Large valleys were carved out below the level of the Miocene surface in New South Wales (Browne 1969), the dimensions of the new surfaces formed during the Pliocene depending on the original elevations of the older surfaces. Thus the Upper Macleay Valley on the east of The Divide and the valleys of most of the rivers west of The Divide are modified Pliocene valleys, some having cut back so far that the valley walls are hard to separate from the boundary with the Miocene surface. The Late Cainozoic uplift and erosion in eastern Victoria extended the down cutting from the Mid-Cainozoic cycle on the southern side of the highlands, but there has been insufficient time for distinct erosion surfaces to develop, especially on the northern side (Wellman 1974). Wellman (1979) has suggested that the presence of any such apparently distinct erosion surfaces in the east may be ascribed to differential erosion of rocks of varying hardness.

In Western Australia, the uplifts in the Late Tertiary and Quaternary have resulted in rejuvenation of drainage and the formation of new surfaces along the ancient river systems (Mulcahy 1960). Only remnants of older erosional surfaces are left, and older depositional surfaces have been buried. New soils have formed in the lower layers of former lateritic profiles exposed during the erosion cycles.

The shape of the younger landscapes in the Mt Lofty–Olary region of South Australia depends on both the amount of recent uplift and also on the local climate. In the northern part of the region uplift has been small and the climate comparatively arid, so that modern forms have a subdued relief. In the southern part, the Mt Lofty Ranges, uplift has been episodic and rainfall moderate and a series of 'young' valleys have been produced (Brock 1971).

There were outpourings of basaltic volcanic rocks during the Pliocene and Quaternary in both Western Victoria and south-eastern South Australia, in southern and more extensively in northern Queensland. Some of the earliest rocks which partly covered a dissected lateritic surface have been dissected. The youngest flows in Queensland, such as the Toomba basalt (Wyatt and Webb 1970), which is no more than 100 000 years old and may be as young as 13 000 years old (Stephenson *et al.*, 1980), have maintained their original form. From Pliocene to recent times the lava plains of south-western Victoria were poured out from a number of distinct centres over a Late Miocene to Early Pliocene erosion surface which had been cut in sediments of this age; the Otway Basin is a special example of this. These plains have also largely maintained many of their original characteristics, though the older areas are moderately deeply weathered (Ollier and Joyce 1976).

Colluvial and alluvial deposits occur through the major drainage basins between the main Eastern Divide and the Central Craton, and fill valleys up to 30 m or more deep. Fan deposits fringe the western slopes of the highlands above the Murray and Darling Basins, but the plains of these basins have thick fluviatile deposits,

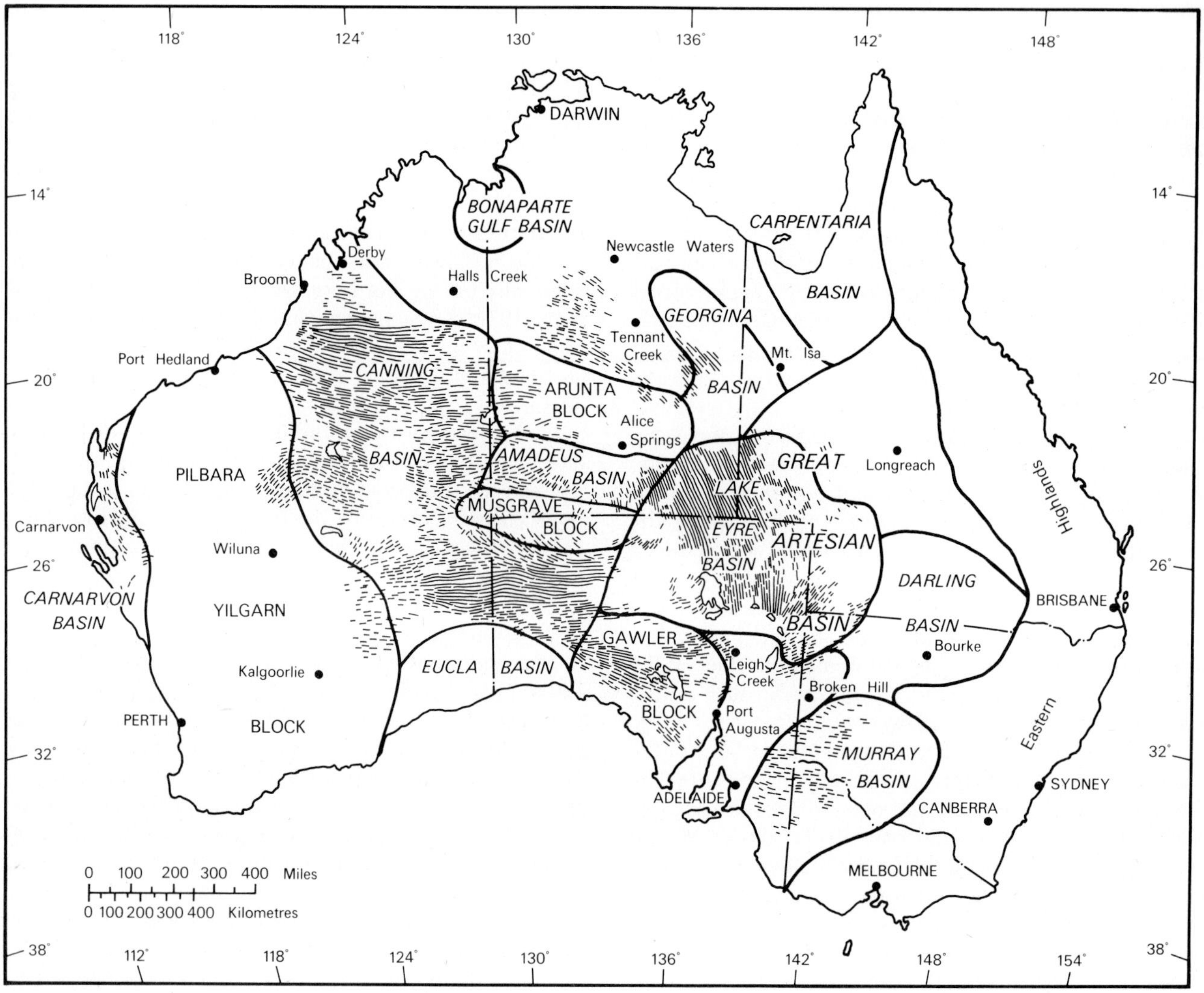

Fig. 4.10. Relationship of the distribution of desert dunes (after Jennings 1968) to the major structural basins and blocks (of Fig. 4.1.).

largely of sands and clays, the response to the influences of changing climatic conditions during the Quaternary, within the confines of major tectonic units. Both hillslope deposits and plain deposits have buried soils, prior streams and terraces which indicate a succession of changing conditions.

At intervals during the Quaternary, times of high rainfall or low evaporation caused extensive lakes to appear in south-western Queensland, western New South Wales, and north-eastern South Australia. Deposits within these lakes form the present localized plains. Some of the lake sediments have been blown out of the lake floors, giving fringing dunes and lunettes; such extensive low sand dune systems are found over the plains of the Lower Murray–Darling River area. Sets of low dunes, with their long axes aligned with the prevailing winds, occur over much of the major basin areas within the large craton occupying the western two-thirds of the continent (Jennings 1968 and Fig. 4.10; Brookfield 1970). These are areas which had been relative lowlands through much of Australia's history (Brown *et al.* 1968) and in part coincide with major areas of silcrete. They occur where rivers had deposited sandy alluvium in former humid periods (Mabbutt 1970).

Parna, a wind blown deposit of clay-sized particles, was interlayered with other aeolian deposits and alluvial bodies of south-eastern Australia during the later Quaternary (Butler 1956).

Eustasy and its effects

Cooling had commenced in the Late Miocene and continued during the Pliocene and Pleistocene with the development of thick ice sheets in Antarctica. The initial cooling in the Pliocene resulted in the first major eustatic lowering of the sea (Ollier 1977). A series of up to 17 rises and falls of sea level during the Quaternary were, however, controlled by the expansion and contraction of continental ice sheets in the Northern Hemisphere (Kukla 1977).

Lowered sea levels induced valley cutting on coastal streams, while the melting of the ice in the Pliocene caused inundation of the Lower Murray River Valley. The effects of all the Late Cainozoic sea level changes were superimposed on those of tectonic events in coastal areas in some parts of the continent, e.g. the Gippsland area in Victoria (Ward *et al.* 1971) and the Karumba Basin in North Queensland (Doutch 1976).

During times of low sea level the continental shelf was exposed to a maximum depth of more than 100 m below the present, and repeatedly entrenched. A lowering of the present sea level by as little as 18 m would close both Torres Strait and Bass Strait (Jennings 1971; Galloway and Löffler 1972) so that New Guinea and Tasmania would have been joined to the mainland on several occasions, producing temporary barriers to ocean circulation (Gill 1962). Coastal valleys were excavated, e.g. the Murray River at Murray Bridge, near its mouth, has a former bed at 36 m below the present, and greater depths have been recorded for the beds of buried valleys along the east coast (Johnson *et al.* 1978; Beckmann and Stevens 1978). In the south-west of Western Australia a succession of coastal sediments was dissected during times of low sea level in the Mid to Late Quaternary.

With each rise of the sea, coastal valleys were drowned and filled with sediment, commonly to a level below that of the previous infilling. A series of benches were cut along the coast, and there are Late Pleistocene and recent benches at low levels at many places, e.g. to 6 or 7.5 m on the New South Wales coast (Langford-Smith and Thom 1969). Islands such as New Guinea and Tasmania, and the smaller islands, were separated from the mainland.

With the successive rises of the sea producing infilling of valleys and falls of the sea rejuvenating streams, a series of terraces was formed near the mouths of the eastern and south-eastern coastal streams; these were continuous with terraces resulting from climatic change. The heights of the seas appear generally to have been less than 6 m above the present, suggesting that they may have been no earlier than Late Pleistocene. Only occasional remnants of alluvial fills and terraces exist above these heights. Alluvial plains of north Queensland coastal rivers contain material which may be Late Tertiary or Pleistocene (Lucas and de Keyser 1965); high level alluvia along the coast to the south may also be of this age.

Similar successions of alluvial bodies related to changing sea levels occur in the Swan Coastal Plain (McArthur and Bettenay 1960) of Western Australia, and in the Karumba Basin of the Gulf of Carpentaria (Grimes and Doutch 1978). These sets of fan deposits have been ascribed to eustatic changes of sea level, with the superimposed effects of climatic change. Correlations across the continent are difficult, however, even though heights are similar, as there may have been local variations in the amount of tectonic uplift.

Coastal sand dunes and beach ridges occur round much of the coast, and are both calcareous and siliceous. In southern and south-western Australia the individual dunes are widely spaced and appear to be parallel to the previous shore lines. McArthur and Bettenay (1960) related the sets of parallel dune systems on the Swan Coastal Plain, between the Darling Scarp and the sea, to higher sea levels from the present up to 7.5 m.

A series of beach dunes with intervening plains runs approximately parallel to the coast in the south-east of South Australia (Blackburn 1962). These span about 690 000 years of the Pleistocene and at least twenty sea level stands occurred during this time (Cook *et al.* 1977). In south-western Victoria there are similar lines of dunes produced during successive retreats of the sea (Gill 1976) and in Gippsland, eastern Victoria, dune sand sheets have accumulated on a sloping surface along the coast. These have been ascribed to eustatic changes of sea level combined with warping and tilting (Jenkin 1976*b*), or with continuous uplift (Ward 1977).

Whereas almost all of these sets of ridges and dunes are aligned approximately parallel to the present shore line, the longitudinal dunes of siliceous sands along the south-east Queensland and northern New South Wales coasts are oriented with the prevailing wind, and are commonly at an angle to the coast. Many of the dune ridges are up to 150 m above the present sea level (in some places 270 m) and they extend up to 60 m below the sea level. Such overlapping dune systems arise from several distinct episodes of aeolian deposition from the Mid-Pleistocene to the present. Low sea levels may expose a source of sand for a dune to move actively, but sand can also be mobilized by wind if the protecting vegetation should be removed, and it then encroaches on stabilized masses as blow outs.

Beach ridges in eastern Australia occur in sets with only narrow corridors between the adjoining ridges; these sets have varying ages; the older sets, away from the shore, are higher than the younger ones, and the component ridges are more widely spaced.

In New South Wales the older sequence of beach ridges, the Inner Barrier, has swales up to 6 m above the high water mark. The system is inferred to be of pre-Holocene age and may have formed during the last Interglacial stage. The Outer Barrier is of Holocene age, following the last major rise in sea level, and is probably 3000–6000 years old (Langford-Smith and Thompson 1969). Soil development is related to age of the beach ridge.

Older quartzose beach ridges (i.e. Late Pleistocene) and younger ridges (Holocene) of shelly quartzose sand have been recognized along the shores of the Gulf of Carpentaria (Smart 1976). The older ridges apparently formed when the sea level was at least 4–5 m above the present, and the younger ones when it was close to the present sea level. Further sets of these ridges occur along the eastern Queensland coast.

Climates

From a landscape and soils point of view, the most significant feature of the Quaternary was the repeated change of climate associated with the generally cooler conditions and with the intermittent warmer intervals. These climatic changes would have affected runoff and the behaviour of streams, as well as weathering and the rates of soil formation. This continued change has introduced much complexity into the landscape–soils pattern in many parts of the continent, particularly where there has been the greatest oscillation in climate.

Australia is a continent with low relief and it is in only the moderately high altitude and higher latitude areas, such as those of the south eastern highlands and in central Tasmania, that there is any direct evidence of glaciation. About 6 000 km^2 of Tasmania and about 50 km^2 in southern New South Wales around Mt Kosciusko were covered by ice (Peterson 1971), and the regional snowline was locally as much as 600 m lower than at present (Davies 1967).

While it appears that there was only one major period of glaciation in southern New South Wales (Derbyshire and Peterson 1971), three cold phases have been identified in Tasmania (Colhoun 1976). Cirques, roches moutonnées, and moraines developed in each glaciated area. The major cold period extended from about 30 000 to about 15 000 years ago (Bowler *et al.* 1976). During this period the higher parts of the Southern Highlands were as much as 6°C colder than at present (Coventry 1976) and mean annual precipitation was about half that of the present.

Periglacial effects were widespread along the high mountain areas, down to an elevation of 900 m in eastern New South Wales and Victoria, and in Tasmania; these effects included modification of landforms and the production of blockfields (Davies 1962; Galloway *et al.* 1973; Derbyshire 1973).

Beyond the areas affected by glacial and periglacial activity the effects of the changes are seen in the indirect effects on landscapes, soils, plants and animals, as well as on the interrelationships of these. Both coastal and inland areas showed effects of fluctuating environments related to changes in both precipitation and temperature. The pattern of changes across Australia was not uniform for several reasons: regional physiography, i.e. the distribution of mountains and plains, and the effects of varied relief on movement of pressure systems, continentality, i.e. distance of any area from the coast, and interruptions to oceanic circulation, when Torres and Bass Straits were closed at times of low sea level. Coaldrake (1967) pointed out that Australia is big enough to contain a number of distinct climatic regions, each with a different pattern of change – this has to be borne in mind in any extrapolations.

Indirect evidence on which to base climatic inferences can come from other sources such as stream patterns, slope deposits, the distribution and form of sand dunes, and the former distribution of animals and plants together with the acknowledged present relationship of such assemblages to the climate. Stratigraphic relationships of these can give a picture of changing conditions that when combined with radio-carbon datings can allocate inferred environments to particular stages in the Late Pleistocene and Holocene.

Conditions when the dune systems of inland Australia developed were drier or at least windier than at present. The dunes have been stabilized by subsequent wetter conditions which permitted protective vegetation to grow. By contrast, coastal sand dunes are less useful as indicators of past climatic conditions, as they can be rejuvenated at present with rapid changes of sediment supply, or if the modern vegetation is modified (Galloway 1971; Chappell 1978).

Changes in river systems have frequently been used to indicate changes in climatic conditions (e.g. van de Graaff *et al.* 1977). Butler (1950) recognized prior streams on the Riverine Plain of N.S.W. and the existence of ancient channels is incontrovertible. The climates which have controlled the development of these features are disputed, however (Bowler 1967). It has been proposed that stream erosion is increased with reduced rainfall through the effects on vegetative cover (Butler 1958), while the opposite interpretation, that erosion on streams has been due to wetter conditions to give the more vigorous cutting, has also been proposed by Langford-Smith (in Gill 1967).

Soils themselves have been used to infer climates of the past (e.g. Teakle 1952). As this discussion attempts to establish the climates under which soils formed, it is necessary to avoid circular arguments. Factors other than climate can affect gross profile features, chemistry, and clay mineralogy, and these are often hard to elucidate from the fragmentary evidence available. The presence of buried soils, which occur in slope deposits and on plains in south-eastern Australia has been related to cycles of stability, with soil formation, and of instability, which have permitted movement of surface material (Butler 1959). These cycles have become known as K-cycles, and the controlling influence producing these successive changes has been thought to be climatic change.

A more satisfactory and more reliable indication of climatic change can be gained from the study of the distribution of animals and plants, especially the latter. Animal species have the opportunity of moving rapidly in response to changing conditions, and the fossil record of plants is much richer than that of animals. Both macro- and micro-fossil evidence is valuable (Watts

1978; Davis 1978) and many studies have demonstrated the value of both marker species and assemblages of species in deducing past climates in many parts of Australia. Gentilli (1961) reviewed the distribution of species of animals and plant species in connection with climatic changes, particularly in southern Australia, on the basis of knowledge available at that time.

More recently Bowler *et al.* (1976) and Rognon and Williams (1977) have outlined the climatic changes over parts of the continent during the past 60 000 years, again largely on the basis of vegetation distribution. The following pattern has emerged.

From 60 000 to 40 000 years ago the climate as a whole was drier than at present. This aridity continued in tropical areas until less than 10 000 years ago, with the rainfall of areas which are humid today being one third to one half of that at present. In New South Wales and South Australia, on the other hand, cool to cold conditions prevailed from 40 000 to 30 000 years ago, with increased effective rainfall, higher lake levels and greater fluvial activity. The Snowy Mountains and Tasmania were glaciated about 25 000 years ago; at this time the rainfall became lower, lakes in South Australia contracted, and there was desiccation in semi-arid New South Wales. The lowered sea levels associated with the local glaciations at this time increased continentality, and this contributed to the extent of the aridity. The existence of barriers to oceanic circulation during times of low sea level, such as Bass Strait, but more particularly Torres Strait, also affected the climate in parts of the continent. Sea temperatures in the Timor Sea would have been lower with blocking of the warm current westward through Torres Strait. Most of north-western Australia would have been cooler as a consequence, and rainfall would have been reduced at such times (Webster and Streten 1978).

Ice commenced to retreat 20 000 years ago in the Snowy Mountains and about 14 000 years ago in Tasmania (Bowler *et al.* 1976). It was almost completely gone by 10 000 years ago. Lake levels were very high in both northern and southern Australia, absolute precipitation was higher, and the rainforest expanded in the tropics (Kershaw 1978).

The rainfall was marginally higher from 8 000 to 6 000 years ago, but there was a decrease in effective and absolute precipitation after this period. The climate has been essentially stable since then, with only minor oscillations.

Nevertheless, information on climatic conditions at precise times through the Quaternary is still very sparse. Most data refer to that part when carbon dating can be utilized, i.e. for only about 2% of Quaternary time, and this is available for only a few areas. Interpretations became less certain further back in time, and beyond the limits of the regions studied. We must perforce fall back on interpretations on the basis of known or inferred changes in climatic controls (Pittock *et al.* 1978). A problem in interpreting the effects of past climatic patterns is that intermittent, relatively severe events, such as marked drops of temperature, may have exerted a considerable influence on palaeoenvironments, especially in glacial times, when the general temperature was low (Webster and Streten 1978).

Modern pressure systems migrate a few degrees north and south with the seasons at present and there is evidence of such movements during the Quaternary with the onset and retreat of glacial conditions (Rognon and Williams 1977). Such movements intensified some winds and changed their direction, which affected elements of the landscape in some areas. Bowler (1976) notes that a movement of the summer anticyclones in full glacial time to a position 5° north of their present position would have strengthened westerly winds, and thus influenced dune building in the arid interior of the continent, so that the dunes extended beyond their present limits. On the other hand, south-easterly trades, which are important in the north and north-east today, would have maintained their direction.

Contrary to ideas which have previously been popular, the broad picture which is now developing for much of Australia during the Quaternary is one of dominant dryness. Lowering of sea temperatures in the Upper Miocene affected precipitation on adjoining land, as less warm and moist air was transmitted to the interior. By comparison with those of the Early Tertiary, the climates of Australia during the Quaternary have been dry, relatively cold and unstable (Bowler 1976). According to Galloway and Kemp (1979) "there is no reason to believe that Australia as a whole received high rainfall for significant periods". Indeed, what are now high rainfall areas in tropical north-eastern Australia experienced low rainfall conditions during glacials (Kershaw 1978), and slightly less than the present during warmer interglacial stages. The pattern, especially for southern Australia, is that of repeated oscillations from slightly wetter to drier conditions, but intermediate stages between full interglacial and glacial times could have been wetter (Rognon and Williams 1977).

Only in the eastern highlands, where some portion of the modern rainfall can be ascribed to orographic effects, would the total precipitation and possibly rainfall pattern in the past have been altered by changes in elevation due to lowering of the sea level in glacial times, as Ruhe (1965) inferred for Hawaii. While there may have been some response in this way here, the Australian continental shelf is wide and flat, in contrast to the submarine topography of Hawaii, so that a wide expanse of dry land would have been exposed along parts of the modern continental margins. Less water would have been available for evaporation and later precipitation, so that even mountainous areas now near the coast would have been in a continental situation during glacial times (Galloway and Kemp 1979), reducing any orographic effects. At the same time temperatures would have been affected by the increased continentality (Coventry 1976) thus lowering evaporation.

While some areas well within the limits of arid or wet regions may have had a climate which varied only slightly through much of the Quaternary, many other areas would show the effects of a succession of changing climates, in stream pattern and soil layering, or in the distribution of vegetation assemblages.

This then, is the geologic, geomorphological and palaeoclimatic setting, in which soil parent materials have weathered and profiles have developed in Australia, over periods varying from a few thousand years to as much as tens of millions of years. Some of the processes, mechanisms and many of the environments which have produced the variety of soils of the continent are discussed in the following Chapters. It will be seen that few of the soils are monogenetic, that many are derived from older weathered materials on progressively younger surfaces, and some have been on stable surfaces subjected to frequently changing climates. A number have developed from reworked older materials, which have been redistributed by water and/or wind. Apparent anomalies in the distribution of some soils are not yet completely understood, but Australia provides a natural laboratory for studying soil development through time, in mainly stable environments, which are representative of many low to moderate latitude non-glaciated areas.

Acknowledgments

The author would like to express his appreciation to Drs R. J. Coventry and R. W. Galloway for critical comments of drafts and useful suggestions for amendments.

References

Abele, C. (1976). Tertiary – Introduction. *In* 'Geology of Victoria'. (Eds J. G. Douglas and J. A. Ferguson). pp. 117-91. Geol. Soc. Aust. Spec. Publ. No. 5.

Alley, N. F. (1977). Age and origin of laterite and silcrete duricrusts and their relationship to episodic tectonism in the mid-north of South Australia. *J. Geol. Soc. Aust.* **24**, 107-16.

Beckmann, G. G. and Stevens, N. C. (1978). Geological history of the Brisbane River system. *Proc. R. Soc. Queensl.* **89**, 77-85.

Blackburn, G. (1962). Stranded coastal dunes of the Murray Basin in Victoria and South Australia. *Aust. J. Sci.* **24**, 388.

Bowler, J. M. (1967). Quaternary chronology of Goulburn Valley sediments and their correlation in south-eastern Australia. *J. Geol. Soc. Aust.* **14**, 287-92.

Bowler, J. M. (1976). Aridity in Australia: age, origins and expression in aeolian landforms and sediments. *Earth Sci. Rev.* **12**, 279-310.

Bowler, J. M., Hope, G. S., Jennings, J. N., Singh, G. and Walker, D. (1976). Late Quaternary climates of Australia and New Guinea. *Quat. Res.* **6**, 359-94.

Brock, E. J. (1971). The denudation chronology of the Fleurieu Peninsula, South Australia. *Trans. R. Soc. S. Aust.* **95**, 85-94.

Brookfield, M. (1970). Dune trends and wind regime in Central Australia. *Z. Geomorph. Supplbd.* **10**, 121-53.

Brown, D. A., Campbell, K. S. W. and Crook, K. A. W. (1968). 'The Geological Evolution of Australia and New Zealand'. (Pergamon Press: Oxford).

Browne, W. R. (1969). Geomorphology General Notes. *In* 'The Geology of New South Wales'. (Ed. G. H. Packham). *J. Geol. Soc. Aust.* **16**, 559-69.

Browne, W. R. (1972). Grey billy and its associates in Eastern Australia. *Proc. Linn. Soc. N.S.W.* **97**, 98-129.

Bryan, W. H. (1939). The Red-Earth Residuals and their significance in south-eastern Queensland. *Proc. R. Soc. Queensl.* **50**, 21-32.

Butler, B. E. (1950). Theory of prior streams as a causal factor in the distribution of soils in the Riverine plain of south-eastern Australia. *Aust. J. Agric. Res.* **1**, 231-52.

Butler, B. E. (1956). Parna, an aeolian clay. *Aust. J. Sci.* **18**, 145-51.

Butler, B. E. (1958). Depositional systems of the Riverine Plain of south-eastern Australia in relation to the soils. CSIRO Aust. Soil Publ. No. 10.

Butler, B. E. (1959). Periodic phenomena in landscapes as a basis for soil studies. CSIRO Aust. Soil Publ. No. 14.

Callen, R. A. (1977). Late Cainozoic environments of parts of north-eastern South Australia. *J. Geol. Soc. Aust.* **24**, 151-69.

Campana, B. (1958). The Mount Lofty-Olary region and Kangaroo Island *In* 'The Geology of South Australia'. (Eds M. F. Glaessner and L. W. Parkin) *J. Geol. Soc. Aust.* **5**, 3-27.

Chappell, J. (1978). Chronological methods and the ranges and rates of Quaternary physical changes. *In* 'Biology and Quaternary Environments'. (Eds D. Walker and J. C. Guppy) pp. 1-34. (Aust. Acad. Sci.: Canberra.)

Churchward, H. M. and Bettenay, E. (1973). The physiographic significance of conglomeratic sediments and associated laterites in valleys of the Darling Plateau, near Harvey, Western Australia. *J. Geol. Soc. Aust.* **20**, 309-17.

Coaldrake, J. E. (1967). Quaternary climates in north-eastern Australia and some of their effects. *Proc. R. Soc. Queensl.* **79**, v-xviii.

Colhoun, L. A. (1976). The glaciation of the lower Forth valley, north-western Tasmania. *Aust. Geogr. Studies* **14**, 83-102.

Cook, P. J., Colwell, J. B., Firman, J. B., Lindsay, J. M., Schwebel, D. A., von der Borch, C. C. (1977). The late Cainozoic sequence of south-east South Australia and Pleistocene sea-level changes. *B.M.R. J. Aust. Geol. Geophys.* **2**, 81-8.

Coventry, R. J. (1976). Abandoned shorelines and the Late Quaternary history of Lake George, New South Wales. *J. Geol. Soc. Aust.* **23**, 249-73.

Davies, J. L. (1962). Geomorphology and glaciation. *In* 'Geology of Tasmania'. (Eds A. Spry and M. R. Banks). *J. Geol. Soc. Aust.* **9**, 243-48.

Davies, J. L. (1967). Tasmanian landforms and Quaternary climates *In* 'Landform Studies from Australia and New Guinea'. (Eds J. N. Jennings and J. A. Mabbutt) pp. 1-25. (Aust. Nat. Univ. Press: Canberra.)

Davis, M. B. (1978). Climatic interpretation of pollen in Quaternary sediments. *In* 'Biology and Quaternary Environments' (Eds D. Walker and J. C. Guppy) pp. 33-51, (Aust. Acad. Sci.: Canberra.)

Derbyshire, E. (1973). Periglacial phenomena in Tasmania. *Biuletyn Peryglacjalny* **22**, 131-48.

Derbyshire, E. and Peterson, J. A. (1971). On the status and correlation of Pleistocene glacial episodes in south-eastern Australia. *Search* **2**, 285-88.

Dorman, F. H. (1966). Australian Tertiary palaeo-temperatures. *J. Geol.* **74**, 49-61.

Doutch, H. F. (1976). The Karumba Basin, north-eastern Australia and southern New Guinea. *B.M.R. J. Aust. Geol. Geophys.* **1**, 131-40.

Exon, N. F., Langford-Smith, T. and McDougall, I. (1970). The age and geomorphic correlations of deep weathering profiles, silcrete, and basalt in the Roma-Amby region, Qld. *J. Geol. Soc. Aust.* **17**, 21-30.

Frakes, L. A. and Kemp, E. M. (1972). Influence of continental positions on Early Tertiary climates. *Nature* (London) **240**, 97-100.

Galloway, R. W. (1967*a*). Geomorphology of the Nogoa-Belyando area. *In* 'Lands of the Nogoa-Belyando area Queensland'. CSIRO Aust. Land Res. Ser. No. 18, pp. 97-114.

Galloway, R. W. (1967*b*). Geomorphology of the Isaac-Comet area. *In* 'Lands of the Isaac-Comet Area'. CSIRO, Aust. Land Res. Ser. No. 19, pp. 77-88.

Galloway, R. W. (1971). Evidence for Late Quaternary Climates. *In* 'Aboriginal Man and Environment in Australia' (Eds D. J. Mulvaney and J. Golson) pp. 14-23. (Aust. Nat. Univ. Press: Canberra.)

Galloway, R. W., Hope, G. S., Loffler, E. and Peterson, J. A. (1973). Late Quaternary glaciation and periglacial phenomena in Australia and New Guinea. *In* 'Palaeoecology of Africa'. Vol. 8. (Ed. E. M. van Zinderen Bakker) pp. 125-36. (A. A. Balkema: Cape Town.)

Galloway, R. W. and Kemp, E. M. (1979). Late Cainozoic Environments in Australia. *In* 'Geological Biogeography in Australia' (Ed. Allen Keast). (W. Junk: The Hague.)

Galloway, R. W. and Loffler, E. (1972). Aspects of geomorphology and soils in the Torres Strait region. *In* 'Bridge and Barrier' (Ed. D. Walker) pp. 11-28. (Aust. Nat. Univ. Press: Canberra.)

Gentilli, J. (1961). Quaternary climates of the Australian region. *Ann. N.Y. Acad. Sci.* **95**, 465-501.

Gill, E. D. (1962). Cainozoic history. *In* 'Geology of Tasmania'. (Eds A. Spry and M. R. Banks) *J. Geol. Soc. Aust.* **9**, 248-53.

Gill, E. D. (1967). Further Australian research in Quaternary shorelines. (New South Wales by T. Langford Smith). *Aust. J. Sci.* **30**, 15-8.

Gill, E. D. (1976). Warrambool-Port Fairy District. *In* 'Geology of Victoria' (Eds J. G. Douglas and J. A. Ferguson) pp. 299-304. Geol. Soc. Aust. Spec. Publ. No. 5.

Griffiths, J. R. (1971). Continental margin tectonics and the evolution of South East Australia. *APEA J.* **11**, 75-9.

Griffiths, J. R. (1974). Revised continental fit of Australia and Antarctica. *Nature* (London) **249**, 336-8.

Grimes, K. G. and Doutch, H. F. (1978). The late Cainozoic evolution of the Carpentaria plains, North Queensland. *B.M.R. J. Aust. Geol. Geophys.* **3**, 101-12.

Gunn, P. J. (1975). Mesozoic-Cainozoic tectonics and igneous activity in south-eastern Australia. *J. Geol. Soc. Aust.* **22**, 215-21.

Gunn, R. H. and Galloway, R. W. (1978). Silcretes in south-central Queensland. *In* 'Silcrete in Australia'. (Ed. T. Langford-Smith) pp. 51-71. (Dep. Geogr. Univ. New England: Armidale.)

Hayes, D. E. and Ringis, J. (1973). Seafloor spreading in the Tasman Sea. *Nature* (London) **243**, 454-8.

Hays, J. (1967). Landsurfaces and laterites in the north of Northern Territory. *In* 'Landform Studies from Australia and New Guinea'. (Eds J. N. Jennings and J. A. Mabbutt) pp. 182-210 (Aust. Nat. Univ. Press: Canberra.)

Hills, E. S. (1956). A contribution to the morphotectonics of Australia. *J. Geol. Soc. Aust.* **3**, 1-15.

Idnurm, M. and Senior, B. R. (1978). Palaeomagnetic ages of Late Cretaceous and Tertiary weathered profiles in the Eromanga Basin, Queensland. *Palaeogeogr. Palaeoclimatol, Palaeoecol.* **24**, 263-77.

Jenkin, J. J. (1976*a*). Geomorphology. *In* 'Geology of Victoria' (Eds J. G. Douglas and J. A. Ferguson) pp. 329-42. Geol. Soc. Aust. Special Publ. No. 5.

Jenkin, J. J. (1976*b*). Westernport and Southern Gippsland. *In* 'Geology of Victoria'. (Eds J. G. Douglas and J. A. Ferguson) pp. 315-25. Geol. Soc. Aust. Special Publ. No. 5.

Jennings, J. N. (1968). A revised map of the desert dunes of Australia. *Aust. Geogr.* **10**, 408-9.

Jennings, J. N. (1971). Sea-level changes and land links. *In* 'Aboriginal Man and Environment in Australia' (Eds D. J. Mulvaney and J. Golson) pp. 11-3. (Aust. Nat. Univ. Press: Canberra.)

Jessup, R. W. and Norris, R. M. (1971). Cainozoic stratigraphy of the Lake Eyre basin and part of the arid region to the south. *J. Geol. Soc. Aust.* **18**, 303-31.

Johnson, B. D., Albani, A. D., Rickwood, P. C. and Tayton, J. W. (1978). The bedrock topography of the Botany Basin, New South Wales. *J. Geol. Soc. Aust.* **24**, 403-8.

Johnstone, M. H., Lowry, D. C. and Quilty, P. G. (1973). The geology of south-western Australia – a review. *J. R. Soc. W. Aust.* **56**, 5-15.

Jones, J. G. (1971). Australia's Cainozoic drift. *Nature* (London) **230**, 237-9.

Jutson, J. T. (1934). The physiography (geomorphology) of Western Australia. *Bull. Geol. Surv. West Aust.* No. 95.

Kemp, E. M. (1978). Tertiary climatic evolution and vegetation history in the south-east Indian Ocean region. *Palaeogeogr. Palaeoclimatol. Palaeoecol.* **24**, 169-208.

Kershaw, A. P. (1978). Record of last interglacial-glacial cycle from north-eastern Queensland. *Nature* (London) **272**, 159-61.

King, L. C. (1962). Morphology of the Earth. (Oliver and Boyd: Edinburgh.)

Kukla, G. J. (1977). Pleistocene land-sea correlations. I. Europe. *Earth-Sci. Rev.* **13**, 307-74.

Langford-Smith, T. (1978). A select review of silcrete research in Australia. *In* 'Silcrete in Australia'. (Ed. T. Langford-Smith) pp. 1-11. (Dep. Geogr. Univ. New England: Armidale.)

Langford-Smith, T. and Thom, B. G. (1969). New South Wales coastal morphology. *In* 'The Geology of New South Wales'. (Ed. G. H. Packham.) *J. Geol. Soc. Aust.* **16**, 572-80.

Lucas, K. G. and de Keyser, F. (1965). Cooktown, Qld 1:250,000 Geological Series. BMR Geol. Geophys. Aust., Explan. Notes, SD/55-13.

Mabbutt, J. A. (1966). Landforms of the Western Macdonnell Ranges. *In* 'Essays in Geomorphology'. (Ed. G. H. Dury) pp. 83-119. (Heinemann.)

Mabbutt, J. A. (1967). Denudation chronology in Central Australia: Structure, climate and landform inheritance in the Alice Springs area. *In* 'Landform Studies from Australia and New Guinea'. (Eds J. N. Jennings and J. A. Mabbutt) pp. 144-81. (Aust. Nat. Univ. Press: Canberra.)

Mabbutt, J. A. (1970). Landforms. *In* 'The Australian Environment'. (Ed. G. W. Leeper) pp. 1-11. (CSIRO and Melbourne Univ. Press: Melbourne.)

McArthur, W. and Bettenay, E. (1960). The development and distribution of soils on the Swan Coastal Plain, Western Australia. CSIRO Aust. Soil Publ. No. 16.

Marks, E. O. (1933). Some observations on the physiography of the Brisbane River and neighbouring watersheds. *Proc. R. Soc. Queensl.* **44**, 132-50.

Maud, R. R. (1972). Geology, geomorphology, and soils of Central County Hindmarsh (Mount Compass-Milang), South Australia. CSIRO Aust. Soil Publ. No. 29.

Mulcahy, M. J. (1960). Laterites and lateritic soils in south-western Australia. *J. Soil Sci.* **11**, 206-15.

Nicolls, K. D. (1960). Erosion surfaces, river terraces, and river capture in the Launceston Tertiary Basin. *Pap. Proc. R. Soc. Tas.* **94**, 1-12.

Ollier, C. D. (1977). Early landform evolution. *In* 'Australia: A Geography'. (Ed. D. N. Jeans) pp. 85-98. (Sydney Univ. Press. Sydney.)

Ollier, C. D. and Joyce, E. B. (1976). Newer volcanic landforms. *In* 'Geology of Victoria' (Eds J. G. Douglas and J. A. Ferguson) pp. 342-8. Geol. Soc. Aust. Special Publ. No. 5.

Peterson, J. A. (1971). The equivocal extent of glaciation in the south-eastern uplands of Australia. *Proc. R. Soc. Vic.* **84**, 207-12.

Pittock, A. B., Frakes, L. A., Jenssen, D., Peterson, J. A. and Zillman, J. W. (Eds) (1978). Climatic Change and Variability: A Southern Hemisphere Perspective. (Cambridge Univ. Press: London.)

Plumb, K. A. (1979). The tectonic evolution of Australia. *Earth Sci. Rev.* **14**, 205-49.

Quilty, P. G. (1974). Tertiary stratigraphy of Western Australia. *J. Geol. Soc. Aust.* **21**, 301-18.

Rognon, P. and Williams, M. A. J (1977). Late Quaternary climatic changes in Australia and North Africa: a preliminary interpretation. *Palaeogeogr. Palaeoclimatol. Palaeoecol.* **21**, 285-327.

Ruhe, R. V. (1964). An estimate of paleoclimate in Oahu, Hawaii. *Am. J. Sci.* **262**, 1098-115.

Senior, B. R. (1976). South-west Queensland Geomorphology. Summary of Activities, Geol. Branch, BMR, Rept. 196, pp. 25-8.

Smart, J. (1976). The nature and origin of beach ridges, western Cape York Peninsula, Queensland. *BMR J. Aust. Geol. Geophys.* **1**, 211-8.

Solomon, M. (1962). The tectonic history of Tasmania. *In* 'Geology of Tasmania' (Eds A. Spry and M. R. Banks) *J. Geol. Soc. Aust.* **9**, 311-39.

Spry, A. and Banks, M. R. (1967). The Geology of Tasmania. *J. Geol. Soc. Aust.* **9**, 362 pp.

Stephens, C. G. (1946). Pedogenesis following the dissection of lateritic regions in southern Australia. CSIR Aust. Bull. No. 206.

Stephens, C. G. (1971). Laterite and silcrete in Australia: a study of the genetic relationships of laterite and silcrete and their companion materials, and their collective significance in the formation of the weathered mantle, soils, relief and drainage of the Australian continent. *Geoderma* **5**, 5-52.

Stephenson, P. J., Griffin, T. J. and Sutherland, F. L. (1980). Cainozoic volcanism in north-eastern Australia. *In* 'The Geology and Geophysics of north-eastern Australia' (Eds R. A. Henderson and P. J. Stephenson). (Geol. Soc. Aust. (Qld. Div.): Brisbane.)

Sutherland, F. L. (1971). The question of Late Cainozoic uplifts in Tasmania. *Search* **2**, 430.

Teakle, L. J. H. (1952). An interpretation of the occurrences of diverse types of soils on basalt in northern New South Wales and Queensland. *Aust. J. Agric. Res.* **3**, 391-408.

Thompson, C. H., Hubble, G. D. and Beckmann, G. G. (1954). A soil survey of the Middle Ridge area, Parish of Drayton, Darling Downs, Queensland. CSIRO Aust. Div. Soils Div. Rep. No. 5/54.

Twidale, C. R. (1966*a*). Chronology and denudation in the Southern Flinders Ranges, South Australia. *Trans. R. Soc. S. Aust.* **90**, 3-32.

Twidale, C. R. (1966*b*). Late Cainozoic activity of the Selwyn upwarp north-west Queensland. *J. Geol. Soc. Aust.* **13**, 491-4.

Van de Graaff, W. J. E., Crowe, R. W. A., Bunting, J. A. and Jackson, M. J. (1977). Relict early Cainozoic drainages in arid Western Australia. *Z. Geomorph. N.F.* **21**, 379-400.

Veevers, J. J., Jones, J. G. and Talent, J. A. (1971). Indo-Australian stratigraphy and the configuration and dispersal of Gondwanaland. *Nature* (London) **229**, 383-8.

Ward, W. T. (1977). Geomorphology and soils of the Stratford-Bairnsdale area, East Gippsland, Victoria. CSIRO Aust., Soils and Land Use Ser. No. 57.

Ward, W. T., Ross, P. J. and Colquhoun, D. J. (1971). Interglacial high sea levels – an absolute chronology derived from shoreline elevations. *Palaeogeogr. Palaeoclimatol. Palaeoecol.* **9**, 77-99.

Watkins, J. R. (1967). The relationship between climate and the development of landforms in Cainozoic rocks of Queensland. *J. Geol. Soc. Aust.* **14**, 153-68.

Watts, W. A. (1978). Plant macro fossil and Quaternary paleoecology. *In* 'Biology and Quaternary Environments' (Eds D. Walker and J. C. Guppy) pp. 53-67. (Aust. Acad. Sci.: Canberra.)

Webster, P. J. and Streten, N. A. (1978). Later Quaternary Ice Age climates of tropical Australasia: interpretations and reconstructions. *Quat. Res.* **10**, 279-309.

Weissel, J. K. and Hayes, D. E. (1972). Magnetic anomalies in the southeast Indian Ocean. Antarctic Oceanology II: Australia New Zealand sector. *Aust. Res. Ser.* **19**, pp. 234-49.

Wellman, P. (1974). Potassium-argon ages on the Cainozoic volcanic rocks of eastern Victoria, Australia. *J. Geol. Soc. Aust.* **21**, 359-76.

Wellman, P. (1976). Gravity trends and the growth of Australia: a tentative correlation. *J. Geol. Soc. Aust.* **23**, 11-4.

Wellman , P. (1979). On the Cainozoic uplift of the south-eastern Australian highland. *J. Geol. Soc. Aust.* **26**, 1-9.

Wellman, P. and McDougall, I. (1974). Potassium-argon ages on the Cainozoic volcanic rocks of New South Wales. *J. Geol. Soc. Aust.* **21**, 247-72.

Wopfner, H. (1978). Silcretes in Northern South Australia and adjacent regions. *In* 'Silcrete in Australia'. (Ed. T. Langford-Smith) pp. 93-141. (Dep. Geogr. Univ. New England: Armidale).

Wopfner, H., Callen, R. and Harris, W. K. (1974). The lower Tertiary Eyre Formation of the South-western Great Artesian Basin. *J. Geol. Soc. Aust.* **21**, 17-51.

Wright, R. L. (1963). Deep weathering and erosion surfaces in the Daly River Basin, N. T. *J. Geol. Soc. Aust.* **10**, 151-63.

Wright, R. L. (1968). Geomorphology of the Dawson-Fitzroy area. *In* 'Lands of the Dawson-Fitzroy area'. CSIRO Aust. Land Research Series No. 21, pp. 117-34.

Wyatt, D. H. and Webb, A. W. (1970). Potassium-argon ages of some North Queensland basalts and an interpretation of late Cainozoic history. *J. Geol. Soc. Aust.* **17**, 39-51.

Young, R. W. (1978). Silcrete in a humid landscape: the Shoalhaven Valley and adjacent coastal plains of southern New South Wales. *In* 'Silcrete in Australia'. (Ed. T. Langford-Smith) pp. 195-207. (Dep. Geogr. Univ. New England: Armidale.)

Chapter 5
Stripping of deep weathered mantles and its significance to soil patterns

H. M. Churchward and R. H. Gunn

MESAS AND TABLELANDS CAPPED BY INDURATED materials (duricrusts) and flanked by slopes on which weathered rock is exposed are important features of large areas of Australia. These are interpreted as erosional remnants of an extensive, deep weathered mantle (e.g. Woolnough 1918; Connah and Hubble 1960; Mulcahy 1967; Stephens 1971) that formed in the Late Cretaceous to Late Tertiary (Gill 1961; Schmidt and Embleton 1976). This chapter considers the significance of old weathered landscapes on the broad pattern of soils in Australia, through its effect on relief and the development of soil parent materials. Emphasis is placed on south-west Australia and central Queensland since this reflects the more specific experience of the authors in these areas.

In stripped terrain there is a wide range of slope forms and degrees of stripping depending on the geomorphic history and on interactions between climate (past and present), relief and geology, and their effect on the slope budget. Residuals of the deep weathered mantle commonly have four main slope elements: an almost flat upper surface, a near-vertical free face or bluff, a steep concave or straight debris slope, and a gently sloping pediment in the lower situations (Fig. 5.1). The free face and debris slope together are often referred to as the breakaway. Where residuals are extensive the upper surface is gently undulating and has low concavo-convex slopes and a coarse drainage pattern. The more common ferruginous or siliceous capping materials are elements of a weathered mantle which frequently comprises variously mottled or pale coloured saprolite up to 100 m deep.

Regional pattern of stripping

There is considerable regional variation in the extent and degree of stripping. In the northern two-thirds and in the south-west of the continent the weathered mantle occurs extensively but in south-east Australia there are relatively small remnants on major divides such as in the Mt Lofty Ranges (Maud 1972; Alley 1973), the Dundas Tablelands (Gibbons and Downes 1964) and the Monaro (Hallsworth and Costin 1953), Shoalhaven (Young 1977) and New England regions (Voisey 1957; Francis and Walker 1978). On the gentle slopes which drain west and north from the south-eastern highlands, the mantle appears to have been stripped down to scattered pockets of saprolite and a few isolated occurrences of laterite (Hallsworth and Costin 1953; Langford-Smith and Dury 1965).

In Queensland large areas of the deep weathered mantle, dominated by *red* and *yellow earths*, are found

Fig. 5.1. Residuals of the weathered mantle in granite in (left) south-west Australia, and (right) central-west Australia. Slope elements indicated are: *a*, upper surfaces; *b*, free face or bluff; *c*, debris slope; *d*, pediment.

on regional divides such as the Alice Tableland (Connah and Hubble 1960; Isbell *et al.* 1967). Extensive stripping of the mantle along valleys draining east from the Alice Tableland has resulted in a complex scatter of small plateaux and residuals, separated by extensive pediments, alluvial tracts and hill belts, in which a wide array of duplex and grey clay soils occurs (Fig. 5.2*a*). West of the Alice Tableland in western Queensland widespread stripping in upper catchments has resulted in extensive, gently undulating plains, dominated by *brown clay* soils developed on exposed argillaceous Cretaceous rocks. Only small scattered residuals of the weathered mantle occur. To the west and south these clay plains give way to landscapes with frequent residuals and some large plateaux located on broad interfluves between the main drainages, such as the Diamantina and Barcoo Rivers.

In inland northern Australia, particularly north-west of the Isa Highlands, wide expanses of *red earths* with various textures are related to extensive preservation of the old weathered surface (Stewart *et al.* 1960; Wright 1964; Paterson 1970). Towards the coast, however, the old surface has been dissected, and plains, bevelled into the mantle, are veneered by cemented lateritic detritus and *red earths* (Wright 1963; Hays 1967), while some dissection and stripping has left laterite-capped plateaux (e.g. the Mitchell Plateau) isolated from the main areas to the south.

Southwards into central Australia the modification of the weathered surface is most notable on the flanks of emergent hill belts such as the Davenport Ranges (Mabbutt 1963). Local stripping of deeply weathered sedimentary rocks of the Canning Basin is evidenced by small lines of breakaways showing through extensive dune fields (Northcote *et al.* 1968). The many scattered plateaux which are often capped by duricrusts reflect a widespread dismemberment of the weathered mantle in south-central Australia (Jessup 1960; Mabbutt 1962; Langford-Smith and Dury 1965; Wopfner and Twidale 1967).

The western sector of the continent, dominated by the stable Archaean Shield, has been much less affected by stripping than in the east. In contrast to the eastern sector, there has been an increase in stripping from south to north (Churchward 1971) while trends along the valleys draining the Western Plateau, particularly in the south-western extremity (Mulcahy 1973), further highlight regional differences. These latter trends can be viewed as broad zones with different degrees of stripping (Fig. 5.2*b*), similar to those described by Mulcahy (1973). The upper catchments, associated with broad saline valleys, are dominated by coarse detritus derived from shallow stripping of the weathered mantle. Limited areas of more effective stripping vary from amphitheatre-like erosion scars with associated localized fans downslope, to pockets of exposed saprolite and granite with little topographic expression. Mid-catchments feature a second zone where the weathered mantle has been extensively stripped. However, saprolite still occurs widely on slopes between residuals and forms the subsoil of yellow duplex soils. Occasional laterite-capped spurs are interspersed.

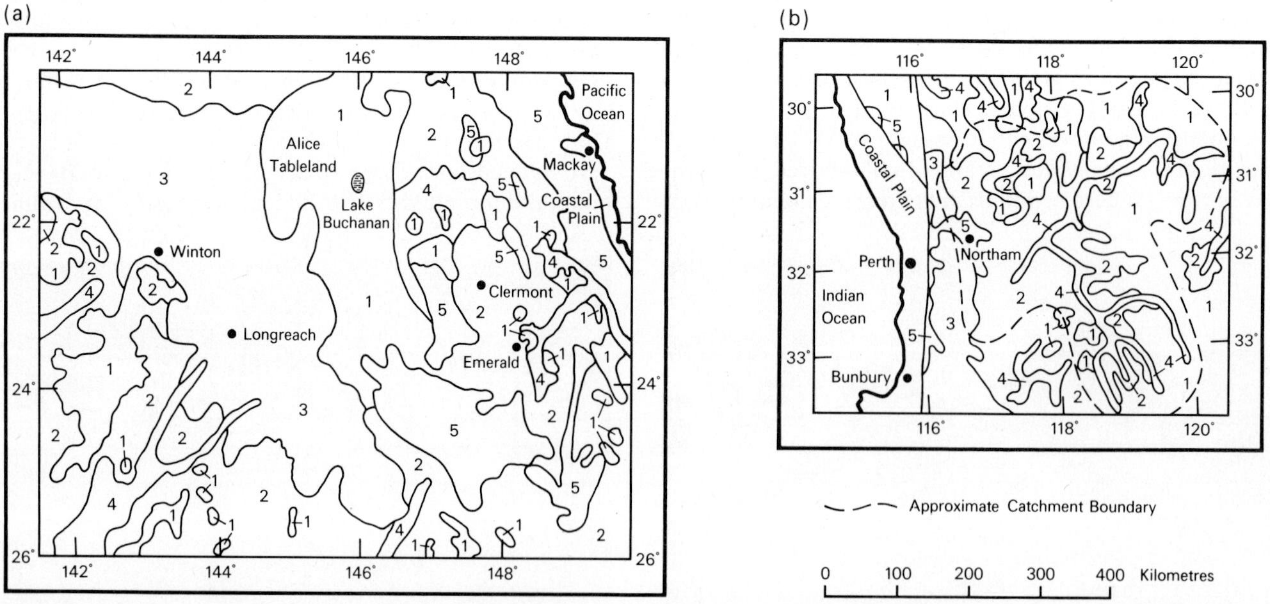

Fig. 5.2. Patterns of stripping in (*a*) western and east-central Queensland, and (*b*) in south-west Australia. (*a*) 1. Weathered landscapes, extensive plateaux. 2. Partially stripped tracts, residuals and pediments. 3. Clay plains, very few residuals. 4. Alluvial tracts. 5. Hill belts. (*b*) 1. Weathered landscapes, sand plains. 2. Partially stripped tracts, residuals and pediments. 3. Weathered landscapes, plateaux with bauxitic laterite. 4. Alluvial tracts and salinas. 5. Stripped steep valley slopes and scarps. (Avon catchment boundary indicated by the broken line.)

Towards the western fringes of the plateau is a third zone in which the weathered mantle is extensively preserved and often capped by a massive bauxitic duricrust. Mild stripping has yielded gravels and sands, which extend as colluvial sheets down the slopes of shallow valleys of the plateau surface. Although localized amphitheatre-like pockets of deep stripping occur along these shallow valleys the most extensive stripping in this zone has been on the steep slopes of the now deeply incised trunk valleys. Here a complex pattern of duplex soils and *red* and *yellow earths* occurs which is controlled largely by the nature of the colluvium and underlying remnants of saprolite (Churchward and Batini 1975; Finkl and Churchward 1976).

Northwards in the Murchison and Gascoyne River catchments of central-west Australia, stripping has been widespread and the second zone of widespread stripping is more extensive than in the catchments further south, such as that of the Avon River. As the regional divides are approached in these catchments stripping is progressively limited to discrete areas of localized erosion (Fig. 5.3) (Mabbutt 1963; Churchward 1977). Finally, in much of the Pilbara region of north-west Australia the weathered mantle has been widely removed and exposure of fresh rock is frequent.

Two contrasting regional trends therefore appear in the different patterns of stripping across the continent. The more common trend shows extensive residuals of the weathered mantle on regional divides, and increased stripping downstream; extensive areas of this are found in east-central Queensland. The second trend is obvious in south-west Australia where large areas of the weathered mantle are preserved adjacent to downstream sectors of catchments (as well as on the regional divides inland), whereas extensive stripping has occurred in mid catchments. This trend is somewhat similar to the pattern described above for western Queensland where the upper margins of catchments west of the Alice Tableland are extensively stripped, but where the degree of stripping decreases down-stream.

Stripping processes

Although a relation between slope elements and specific landform processes has often been suggested (e.g. King 1953; Schumm 1956), it appears to be more a matter of the degree of activity than the kind of process. In south-west Australia, for example, rain splash, splash creep and slope wash can be observed at present to move sediment down the concavo-convex slopes of weathered uplands. Nevertheless, occasional gravel lenses in the upland sandy deposits indicate localized channelled

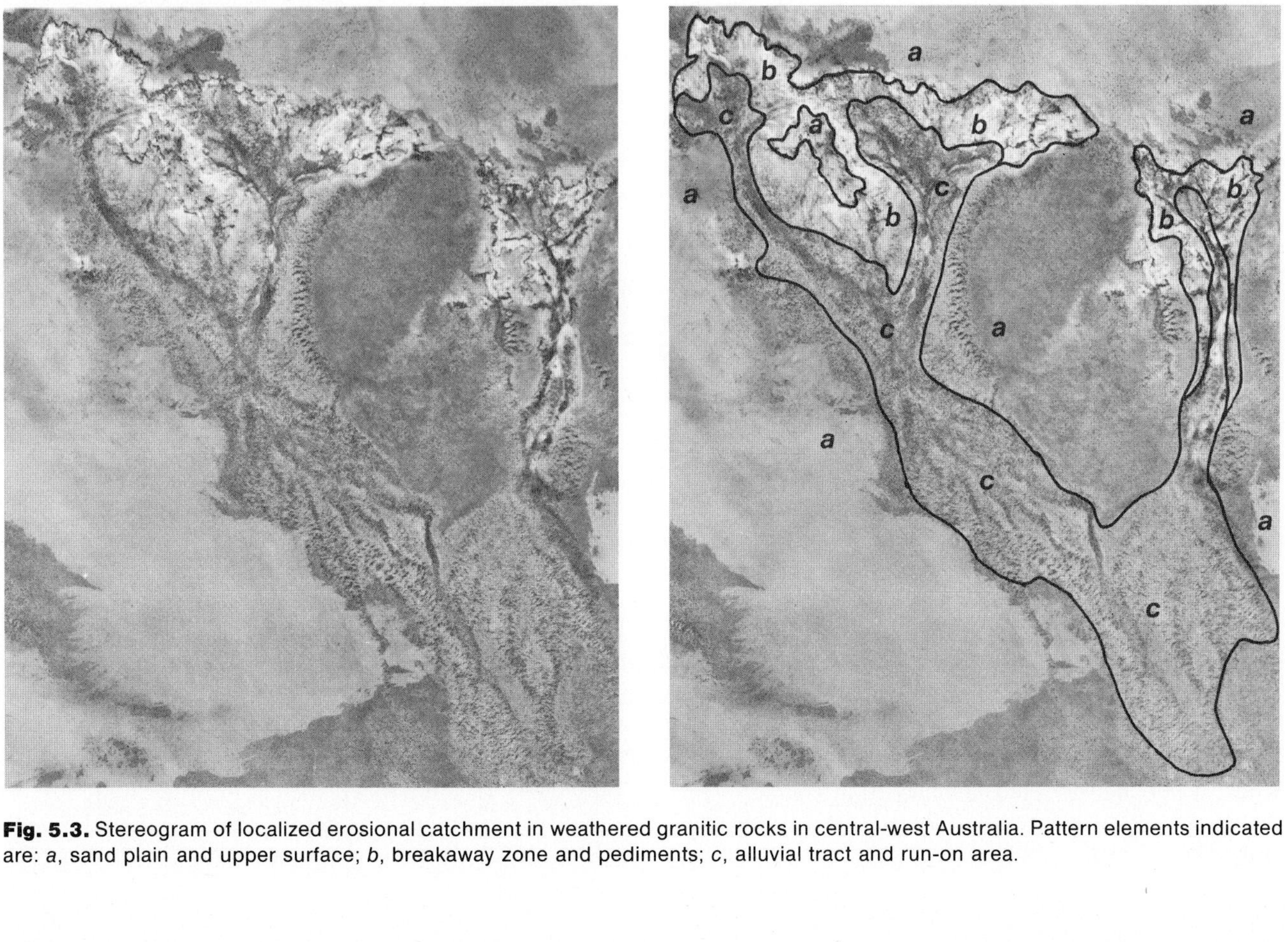

Fig. 5.3. Stereogram of localized erosional catchment in weathered granitic rocks in central-west Australia. Pattern elements indicated are: *a*, sand plain and upper surface; *b*, breakaway zone and pediments; *c*, alluvial tract and run-on area.

fluvial action (Mulcahy 1960). However, in these situations the overall erosional activity is not sufficient to effect large-scale removal of detritus.

In contrast, more effective erosional regimes, which naturally move a lot of detritus, give rise to striking physiographic features (such as vertical scarps and steep, concave or straight debris slopes), whose steep declivities are often maintained by the indurated ferruginous or siliceous cappings of the uplands. Where the cappings are fragmented by weathering and then removed, the steep slopes in non-resistant materials, now exposed, are rounded and lowered. The scarps often retreat as a result of cavernous 'weathering' and slumping, with their outline in plan modified by gullies eroding along joints in basement rock (Mabbutt 1961). Large caverns beneath the duricrust, at the junction of scarp and debris slope, are common in coarse-grained rocks, such as quartzose sandstones and granites (Fig. 5.4). In argillaceous rocks they are much smaller or absent. The cavernous 'weathering' is attributed to the granular disintegration and surface flaking of exposed saprolite, commonly aided by salt crystallization (Evans 1969). Appreciable quantities of soluble salts (mostly sodium chloride) are often present in mottled and pallid zones of the residual weathered mantle. Their origin has been attributed to atmospheric accessions derived from oceanic and/or terrestrial sources (Hubble and Isbell 1958; Dimmock *et al.* 1974; Hingston and Gailitis 1976), or to rock weathering (Gunn and Richardson 1979).

Fig. 5.4. Cavernous 'weathering' in pallid saprolite beneath duricrust near Lake Buchanan, Queensland.

As blocks of capping material, slumped on to the debris slopes, are further fragmented, finer material is distributed over the saprolite downslope, but coarse fragments remain as a surface lag. Granular fragmentation of granite material yields a gritty debris mantle and, in central-west Australia, this contrasts with the stony strew on debris slopes associated with adjacent metamorphic rocks. Unconfined slope wash and linear erosion are more obvious on the debris slopes than on the weathered uplands. Sometimes sufficient debris accumulates on these slopes to protect the soft saprolite beneath, resulting in relative slope stability. Selective breaching of this mantle allows for renewed linear erosion and ensures the resumption of more active scarp retreat. Relics of the more stable slope regime may remain as 'flat irons' isolated from the scarp by tracts of more active erosion (Mabbutt 1977; Churchward and McArthur 1980).

The debris slopes generally grade down to the pediments as shallow concavities. The pediments in the granitic terrain of central-west Australia are mainly bevelled in saprolite; alluvial fans and drainage floors are sometimes important associated features. The thickness and extent of alluvium usually increases with distance from the breakaway, and changes from coarse to fine textured material as a result of fluvial sorting. Low sinuous ridges of coarse alluvia commonly indicate greater fluviatile activity in the past (Galloway 1967; Churchward 1977).

Variations in the depth and pattern of stripping can be related to the depth of weathering, geology, the nature of the duricrust, relief and climate. The *depth of weathering*, and hence the amount of erodible material, is related in part to the *lithology of the country rock*. Deep chemical weathering followed by stripping of the friable saprolite is similar to the etch process of Wayland (1933). This process and the resultant development of etchplains is viewed by several authors as being significant in landform development (Wright 1963; Mabbutt 1965; Alley 1973). This is perhaps more so in areas of crystalline rocks, such as granite, where there is so often an abrupt change from weathered to unweathered rock, and where the depth of stripping is closely related to the depth of weathering (Mabbutt 1961, 1965). On the other hand, in 'soft' sedimentary rocks, there is a gradual change from weathered to unweathered rock, and in parts of central Australia for example, stripping has extended into the underlying shales (Ollier and Tuddenham 1962; Mabbutt 1963). In some rocks, such as granites, there is considerable variation in the depth of the weathered mantle (Ollier 1969). In south-west and central-west Australia the apparently random and often localized pockets of more effective stripping may in part relate to resistant elements of country rocks, extending up into the weathered mantle. Since there is less material to be removed at such sites, there could be earlier exposure of fresh rock than in adjacent areas. Groundwater impedance by these elements may have contributed to spring development, particularly when they are close to the surface. The springs act as foci for sapping and so facilitate stripping. Where the depth of weathered mantle is variable, rock exposure is sporadic. Such exposures form local base levels which limit deeper incision (Ollier 1965) and much saprolite is often retained in the landscape.

The development of seepages at interfaces of strata of contrasting lithology, have possibly influenced stripping in sedimentary terrain and areas with basalt flows in east-central Queensland.

The intricate pattern of stripping in east-central Queensland may relate, in part at least, to a complex geological pattern, particularly as it affects details of relief and drainage pattern. The geology of this region contrasts with the broad uniform areas of argillaceous sedimentary rocks in western Queensland, where the weathered mantle has been extensively stripped.

The *nature and incidence of the duricrust* has had a significant effect on the regional stripping pattern. In south-west Queensland extensive residuals of the weathered mantle, which formed in argillaceous sedimentary rocks, are capped by silicified Tertiary quartzose sandstones. In adjacent areas further north, (Fig. 5.2*a*) this silicified capping was absent or may have been much less extensive. Fresh rocks were thus readily exposed and on these, widespread clay plains developed.

On the western margins of the plateau in south-west Australia the massive bauxitic capping was probably a significant factor in preserving the broad area of minimal stripping which dominates this region, despite the deep incision of the trunk drainage. Inland, the thinner and less ferruginous or bauxitic capping was, apparently, less effective in limiting stripping in mid-catchment.

Clearly, local relief will affect the depth and pattern of stripping, particularly through its effect on stream incision and debris removal. In west Australia this pattern is in part related to the progressive increase of stripping downstream. Local pockets of erosion, on very gentle slopes in upper catchments, are often associated with short non-articulated streams. Downstream in mid-catchments local relief increases along with drainage density and drainage articulation, while stripped zones have extended and frequently merged. Finally, near the western margin of the plateau, the effect of relief on deep stripping appears to be restricted to the steeper slopes of the main valleys by other factors discussed above. Such an integrated trend may be in part the consequence of the frequent, though partial, preservation of the valleys of the ancient weathered surface of this sector of the continent (Mulcahy 1973). In contrast, the previous relief in east-central Queensland appears to have been extensively modified by erosion and uplift, and so its effect on the pattern of stripping cannot be readily assessed.

Twidale (1956), Voisey (1957) and Hays (1967) have related stripping to drainage rejuvenation following *tectonism* on the eastern and northern peripheries of the continent. Evidence indicates that stripping in central Australia is in part the result of downwarping of the Lake Eyre Basin (Mabbut 1962; Wopfner and Twidale 1967; Wopfner 1974). Exon and Senior (1976) relate the extensive stripping in western Queensland to possible uplift of the northern margin of the Eromanga Basin. Mulcahy (1967) concluded that partial drainage rejuvenation in south-west Australia occurred through several phases of uplift, while degrees of stripping have been related to phases of uplift in north Australia (Hays 1967) and in north-east Queensland (Coventry 1978). The model proposed by Woolnough (1918) of an ancient weathered surface of low relief, uplifted and eroded, is far too simple to adequately explain stripping patterns in Australia. Rather, landform studies such as those described by Grimes and Doutch (1978) for the late Cainozoic history of north Queensland, indicate the complex relationship between erosion, tectonism, and climatic change.

The *role of climate* in the stripping of the weathered mantle has been little discussed compared with that of tectonism. Climate is, nevertheless, widely accepted as having had major effects on landscapes in Australia (Galloway and Kemp 1979). The extensive stripping of midcatchments in south-west Australia and upper catchments in western Queensland is possibly related to past climatic conditions, which resulted in extensive erosion and sediment removal, as in the model proposed by Langbein and Schumm (1958). Many stripped tracts on the stable Archaean Shield are located in those parts of the catchments which have been interpreted as being beyond the upper limits of tectonic drainage rejuvenation. These are considered to be the result of erosion related to local base levels, during periods of slope instability, in response to climatic change (Mulcahy and Hingston 1961; Bettenay and Hingston 1964) and its effect on vegetative cover. Mabbutt (1977) has attributed the renewed linear erosion of debris slopes in central Australia to climatic change. The consequence of climatic change would be independent not only in tectonically stable areas but throughout the continent. It is clear that the relative importance of these two processes remains unresolved in many areas.

Soils in relation to stripping

The importance of the deeply weathered mantle to soil patterns in Australia received early recognition by Prescott (1931). Stephens (1946) indicated that the master horizons of the mantle formed distinctive soil parent materials, either exposed *in situ* or as transported sediments. The soil patterns depend partly on the degree of stripping of the weathered mantle, on the lithology and stratigraphy of the underlying rocks, and partly on the spread of weathering and erosional products across the various slope elements. The following soil sequences in granitic terrain in south-west Australia, and in areas underlain by sedimentary rocks and basalt in east-central Queensland, illustrate these relationships.

On the weathered plateau of the Archaean Shield in south-west Australia the weathered upland surfaces in *granitic terrain* are generally slightly stripped, and ferruginous duricrust is exposed on upper slopes and crests, while gravels and sands (the weathering products of the duricrust), have been transported downslope to form the so-called sand plains of this region (Mulcahy 1960). *Earthy sands* (Uc 5.22) are dominant. Indurated mottled zone commonly forms the scarp, while pallid saprolite, with a discontinuous colluvial mantle, is exposed extensively on the debris slope. On the pediments and other stripped surfaces below, *solodic soils* (Dy2.33; Dy3.42) predominate, but there are narrow belts of *non-calcic brown soils* (Dr2.22; Dr3.42) on basic dykes.

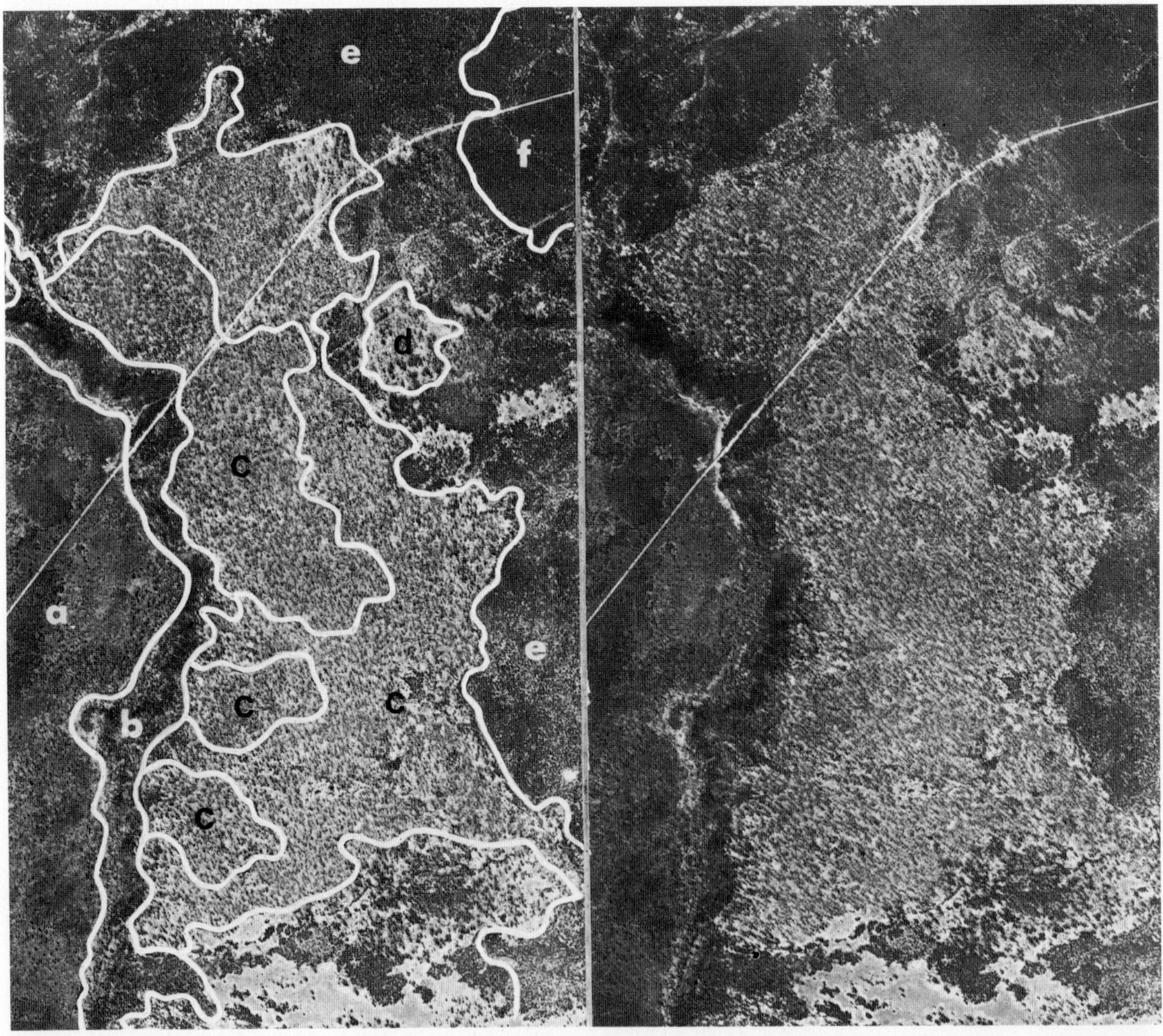

Fig. 5.5. Stereogram showing pattern of rocks and landforms developed by stripping of the weathered mantle in quartzose sandstone and underlying argillaceous sediments in east-central Queensland. The elements of the pattern indicated are: *a*, sandy red earths on slightly stripped upper surface; *b*, lithosols in breakaway zone; *c*, uniform sands on upper pediments and local fans; *d*, duplex soils with thick sandy surface horizons on mid-slopes of pediments; *e*, duplex soils with thin loamy surface horizons on lower pediments; *f*, cracking clay soils on exposed argillaceous sediments.

In upland areas underlain by weathered *sedimentary rocks* in east-central Queensland *red* and *yellow earths* (Gn2.12, 2.22) occur on the more-or-less intact weathered surface. In parts of this region, however, (e.g. on the Alice Tableland west of Lake Buchanan and in the Torrens Creek area), *yellow earths* occur on low ridges and the *red earths* downslope appear to have formed in transported materials, derived by stripping of a former more elevated surface (Gunn 1967; Coventry 1978).

Mottled or pallid saprolite is frequently exposed on flanking scarps and debris slopes, and shallow *solodic soils* (Dy2.43, Db1.43 and Dr2.13) occur on pediments below. These soils grade to a zone of similar but deeper soils with thicker surface horizons, which in turn grade to *solodized solonetz* (Db1.13 and Db1.33), with thin loamy surface horizons (Fig. 5.5). On the lower moist slopes are *grey clay* soils (Ug5.24), commonly with gilgai microrelief, which contain appreciable soluble salts and which are generally alkaline at or near the surface and strongly acid beneath (Hubble and Isbell 1958; Isbell 1962). In this region quartzose sandstones are commonly underlain by argillaceous strata, and variable exposure of these strata by stripping controls the occurrence and extent of the sandy and clayey elements in the sequence. The clay soils have formed either directly on the exposed weathered argillaceous sediments or in redistributed materials derived from them.

The *solodized-solonetz* and *solodic soils* on pediments underlain by weathered granitic or sedimentary rocks clearly reflect a strong halomorphic influence, particularly in the well-developed columnar-structured subsoils. This feature may relate in part to the release of soluble salts from the lower zones of the weathered mantle during stripping. The frequent occurrence of stone lines, particularly at the interface between the surface horizons and subsoil clays, as well as the progressive decrease in the thickness and the particle size of the surface horizons downslope from the scarp, indicate that sedimentary processes were important in the genesis of these profiles.

Small remnants of weathered *basalt* in this region have sequences of deep *red clay* soils (Uf6.53), overlying ferruginized basalt on crests and upper slopes, which grade to shallow, gravelly, acid clays (Uf6.12) on steeper margins, and reddish brown clays with carbonate enriched subsoils (Uf6.21) downslope. Deep *red clay* soils (Ug5.37) on lower slopes grade to shallower *black earths* (Ug5.12), formed on fresh rock in the lowermost zone of the sequence (Gunn 1974).

Kaolinite is the dominant clay mineral of soils in the granitic terrain of south-west Australia and reflects widespread incomplete removal of elements of the weathered mantle by the stripping process. Illite is significant, though not dominant, in soils on stripped surfaces, particularly where fresh rocks crop out, and small amounts of montmorillonite occur in soils derived from basic dykes (Mulcahy and Hingston 1961). The clay mineralogy of soils developed on weathered sedimentary rocks in Queensland varies considerably. Kaolinite predominates in the *red* and *yellow earths* and is generally dominant in the *solodized–solonetz* and *solodic soils* but appreciable illite and/or montmorillonite occur in some profiles. In the *grey clay* soils, kaolinite and montmorillonite are generally co-dominant. In basaltic terrain, the sequence of clay minerals ranges from kaolinite plus randomly interstratified material plus montmorillonite on weathered rock, to montmorillonite in soils formed on fresh rock exposed by complete stripping of the weathered mantle (Gunn 1967, 1974).

The large, relatively uniform areas of soil with low contrast sola, common on weathered landscapes, appear to be the product of both weathering and of sedimentary processes. Subtle lithological variations in the underlying rocks are masked by deep weathering and by local downslope accumulation of debris derived from the duricrust. The mild sedimentary processes on the low gradient slopes possibly allowed homogenization of fresh increments of sediment to a particular site to keep pace with deposition. Furthermore, at least on the weathered granitic terrain of south-west Australia, the generally sandy detritus shows little change in field texture downslope, so that fresh accessions to a particular site vary little in texture from the material already in place. Under these conditions, sharp sedimentary breaks are not common and low contrast sola occur frequently.

On the other hand, more complex soil patterns are common where slope processes are more active. Stripping is more effective and underlying rocks with contrasting lithology are often exposed, resulting in a wider variety of parent materials. In areas underlain by sandstones and granitic rocks, sandy detritus has been transported across the pediment and deposited on the exposed saprolite clays, resulting in high contrast sola. This contrast is further enhanced by the effect of saline ground waters on the physical condition of the clay subsoils. In finer-textured detritus such as that cited above for east-central Queensland, sedimentary sorting trends are more clearly expressed and soil patterns are more complex.

In mapping the soils and associated landscapes on the weathered plateau of south-west Australia, Mulcahy (1960) and Bettenay and Hingston (1964) related soil patterns to different degrees of stripping, through the recognition of several paired erosional and depositional surfaces. The various degrees of stripping of the weathered mantle are used as a prime basis for mapping land systems in Queensland, where several landform categories have been related to incipient, partial or complete removal of 'the Tertiary sediments and land surface' (Galloway 1967). Finkl and Churchward (1973) have applied the etchplain concept (Wayland 1933; Thomas 1965) to describe landscapes in south-west Australia.

The term etchplain has been used to describe landforms elsewhere in Australia by Mabbutt (1961, 1965), Wright (1963), Alley (1973) and by Twidale and Bourne (1975). With its central themes of deep chemical weathering and differential stripping, this concept could provide the basis of a continent-wide scheme for classifying these landscapes.

The fact that stripping has taken place in deeply weathered land surfaces has affected both soil development and the pattern of denudation in three ways. Firstly, the intensely weathered materials, having a range of physical characteristics, influence the rate and pattern of stripping. Secondly, stripping of the mantle exposes the various zones of the weathered profile and provides a range of pre-weathered soil parent materials in both erosional and depositional situations. Thirdly, the lower zones of the mantle commonly form stores of soluble salts to be released as erosion progresses with the development of halomorphic soils, as well as influencing stripping by including cavernous 'weathering'.

Variable stripping of deeply weathered landscapes is clearly the dominant process affecting the broad pattern of soil distribution over wide areas of Australia. Even in severely stripped terrain, where fresh rocks were widely exposed, the process has not been complete and remnants of the weathered mantle have continued to influence the soil pattern. Their relative importance in either erosional or depositional situations depends greatly on the generally low relief of these areas, as well as on geology and on the climatic and geomorphic history. Compared with continents in the northern hemisphere, where Pleistocene glaciation has exposed unweathered rocks and deposited materials derived from them, elements of the deeply weathered mantle that covered most of Australia in the Tertiary have been extensively preserved, on account of the generally low relief and tectonic stability of the continent and its position in mid latitudes. This has been a dominant influence on the soil pattern and to this may also be related the great extent of soils with inherently low levels of fertility.

References

Alley, N. F. (1973). Landsurface development in the Mid North of South Australia. *Trans. R. Soc. S. Aust.* **97**, 1-17.

Bettenay, E. and Hingston, F. J. (1964). Development and distribution of soils in the Merredin area, Western Australia. *Aust. J. Soil Res.* **2**, 173-86.

Churchward, H. M. (1971). Some soil and landform trends in Western Australia in relation to certain engineering problems. *In* 'Proceedings of a symposium on soils and earth structures in arid climates'. May 1970. Aust. Geomech. Soc. Inst. Eng. Aust.

Churchward, H. M. (1977). Landforms, regoliths and soils of the Sandstone – Mt Keith area, Western Australia. CSIRO Aust. Div. Land Resour. Management, Land Resour. Management Ser. No. 2.

Churchward, H. M. and Batini, F. (1975). Soil pattern and resources utilization in the Wongong Brook catchment, Western Australia. CSIRO Aust. Div. Land Resour. Management, Land Resour. Management Ser. No. 1.

Churchward, H. M. and McArthur, W. M. (1980). The soil pattern in relation to physiography and geology in the Northern Pilbara Region, Western Australia. CSIRO Aust. Div. Land Resour. Management Tech. Pap. No. 6.

Connah, R. T. and Hubble, G. D. (1960). Laterites. J. Geol. Soc. Aust. **7**, 373-86.

Coventry, R. J. (1978). Late Cainozoic geology, soils and landscape evolution of the Torrens Creek area, North Queensland. *J. Geol. Soc. Aust.* **25**, 415-27.

Dimmock, G. M., Bettenay, E. and Mulcahy, M. J. (1974). Salt content of lateritic profiles in the Darling Range, Western Australia. *Aust. J. Soil Res.* **12**, 63-9.

Evans, I.S. (1969). Salt crystallization and rock weathering: a review. *Rev. Geomorph. Dynam.* **19**, 153-77.

Exon, N. F. and Senior, B. R. (1976). The Cretaceous of the Eromanga and Surat Basins. Bur. Miner. Resour. *J. Aust. Geol. Geophys.* **1**, 33-50.

Finkl, C. W. and Churchward, H. M. (1973). The etched landsurfaces of south-western Australia. *J. Geol. Soc. Aust.* **20**, 295-307.

Finkl, C. W. and Churchward, H. M. (1976). Soil stratigraphy in a deeply weathered shield landscape in south-western Australia. *Aust. J. Soil Res.* **14**, 109-20.

Francis, G. and Walker, G. T. (1978). Silcretes of subaerial origin in southern New England. *Search* **9**, 321-23.

Galloway, R. W. (1967). Geomorphology of the Nogoa – Belyando area. CSIRO Aust. Land Res. Ser. No. 18.

Galloway, R. W. and Kemp, E. M. (1979). Late Cainozoic environments in Australia. *In* 'Biogeology and Ecology in Australia' (Ed. W. A. Keast). (W. Junk: The Hague.)

Gibbons, F. R. and Downes, R. G. (1964). A study of the land in south-western Victoria. Soil Cons. Auth., Vic., Aust.

Gill, E. D. (1961). The climates of Gondwanaland in Cainozoic times. *In* 'Descriptive Palaeoclimatology'. (Ed. A. E. M. Nairn). (Interscience Publishers Inc.: New York.)

Grimes, K. G. and Doutch, H. F. (1978). The late Cainozoic evolution of the Carpentaria Plains, North Queensland. Bur. Min. Resour. *J. Aust. Geol. Geophys.* **3**, 113-24.

Gunn, R. H. (1967). A soil catena on denuded laterite profiles in Queensland. *Aust. J. Soil Res.* **5**, 117-32.

Gunn, R. H. (1974). A soil catena on weathered basalt in Queensland. *Aust. J. Soil Res.* **12**, 1-14.

Gunn, R. H. and Richardson, D. P. (1979). The nature and possible origins of soluble salts in deeply weathered landscapes in eastern Australia. *Aust. J. Soil Res.* **17**, 197-215.

Hallsworth, E. G. and Costin, A. B. (1953). Studies in pedogenesis in New South Wales. IV. The iron stone soils. *J. Soil Sci.* **4**, 24–46.

Hays, J. (1967). Land surfaces and laterites in the north of the Northern Territory. *In* 'Landform studies from Australia and New Guinea'. (Eds J. N. Jennings and J. A. Mabbutt). (Aust. Nat. Univ. Press: Canberra.)

Hingston, F. J. and Gailitis, V. (1976). The geographic variation of salt precipitated over Western Australia. *Aust. J. Soil Res.* **14**, 319-35.

Hubble, G. D. and Isbell, R. F. (1958). The occurrence of strongly acid clays beneath alkaline soils in Queensland. *Aust. J. Sci.* **20**, 186-7.

Isbell, R. F. (1962). Soils and vegetation of the brigalow lands, eastern Australia. CSIRO Aust. Soils Land Use Ser. No. 43.

Isbell, R., Thompson, C. H., Hubble, G. D., Beckmann, G. G. and Paton, T. R. (1967). Atlas of Australian Soils. Explanatory Data for Sheet 4. Brisbane–Charleville–Rockhampton–Clermont area (CSIRO Aust. and Melb. Univ. Press: Melbourne.)

Jessup, R. W. (1960). The lateritic soils of the south-eastern portion of the Australian arid zone. *J. Soil Sci.* **11**, 106-13.

King, L. C. (1953). Canons of landscape evolution. *Bull. Geol. Soc. Am.* **64**, 721-51.

Langbein, W. B. and Schumm, S. A. (1958). Yield of sediment in relation to mean annual precipitation. *Trans. Am. Geophys. Un.* **39**, 1076-84.

Langford-Smith, T. and Dury, G. H. (1965). Distribution, character and attitude of the duricrust in the northwest of New South Wales and the adjacent areas of Queensland. *Am. J. Sci.* **263**, 170-90.
Mabbutt, J. A. (1961). A stripped land surface in Western Australia. *Trans. Inst. Br. Geogr.* **29**, 101-14.
Mabbutt, J. A. (1962). Geomorphology of the Alice Springs Area. CSIRO Aust. Land Res. Ser. No. 6.
Mabbutt, J. A. (1963). The weathered land surface in Central Australia. *Z. Geomorph.* **9**, 82-114.
Mabbutt, J. A. (1965). Landforms of the western MacDonnell Ranges. *In* 'Essays in Geomorphology'. (Ed. G. H. Dury). (Heinemann: London.)
Mabbutt, J. A. (1977). 'Desert Landforms'. (Aust. Nat. Univ. Canberra.)
Maud, R. R. (1972). Geology, geomorphology, and soils of central County Hindmarsh (Mount Compass–Milang) South Australia. CSIRO Aust. Soil Publ. No. 29.
Mulcahy, M. J. (1960). Laterites and lateritic soils in south-western Australia. *J. Soil Sci.* **11**, 206-25.
Mulcahy, M. J. (1967). Landscapes, laterites and soils in southwestern Australia. *In* 'Landform studies from Australia and New Guinea'. (Eds J. N. Jennings and J. A. Mabbutt). (Aust. Nat. Univ. Press: Canberra.)
Mulcahy, M. J. (1973). Landforms and soils of southwestern Australia. *J. R. Soc. W. Aust.* **56**, 16-22.
Mulcahy, M. J. and Hingston, F. J. (1961). Soils of the York-Quairading area, Western Australia, in relation to landscape evolution. CSIRO Aust. Soil Publ. No. 17.
Northcote, K. H., Isbell, R. F., Webb, A. A., Murtha, G. G., Churchward, H. M. and Bettenay, E. (1968). Atlas of Australian Soils. Explanatory Data for Sheet 10. Central Australia. (CSIRO Aust. and Melb. Univ. Press: Melbourne.)
Ollier, C. D. and Tuddenham, W. G. (1962). Slope development at Coober Pedy, South Australia. *J. Geol. Soc. Aust.* **9**, 91-105.
Ollier, C. D. (1965). Some features of granite weathering in Australia. *Z. Geomorphol. N.F.* **9**, 285-304.
Ollier, C. D. (1969). Weathering. (Elsevier: New York.)
Paterson, S. J. (1970). Geomorphology of the Ord-Victoria area. CSIRO Aust. Land Res. Ser. No. 28.
Prescott, J. A. (1931). The soils of Australia in relation to vegetation and climate. CSIRO Aust. Bull. No. 52.
Schmidt, P. W. and Embleton, B. J. J. (1976). Paleomagnetic results from sediments of the Perth Basin, Western Australia, and their bearing on the timing of regional lateritisation. *Palaeogeog. Paleoclimatol. Palaeoecol.* **19**, 257-73.
Schumm, S. A. (1956). The role of creep and rainwash on the retreat of badland slopes. *Am. J. Sci.* **254**, 693-706.
Stephens, C. G. (1946). Pedogenesis following dissection of lateritic regions in southern Australia. *CSIRO Aust. Bull.* No. 206.
Stephens, C. G. (1971). Laterite and silcrete in Australia. *Geoderma* **5**, 5-52
Stewart, G. A., Twidale, C. R. and Bradley, J. (1960). Geomorphology of the North Kimberley Area, W.A. CSIRO Aust. Land Res. Ser. No. 4.
Thomas, M. F. (1965). An approach to some problems of landform analysis in tropical environments. *In* 'Essays on geography for Austin Miller'. (Eds J. P. Whittow and A. D. Wood). (Univ. Reading: Reading, U.K.)
Twidale, C. R. (1956). Chronology of denudation in northwest Queensland. *Geol. Soc. Am. Bull.* **67**, 867-82.
Twidale, C. R. and Bourne, A. (1975). Geomorphological evolution of the eastern Mt. Lofty Ranges, South Australia. *Trans. R. Soc. S. Aust.* **99**, 197-210.
Voisey, A. H. (1957). Erosion surfaces around Armidale, N.S.W. *J. R. Soc. N.S.W.* **90**, 128-33.
Wayland, E. S. (1933). Peneplains and some erosional platforms. Annu. Rep. Bull. Protectorate Uganda Geol. Surv. Dep. Notes 1, pp.77-79.
Woolnough, W. G. (1918). The Darling peneplain of Western Australia. *J. Proc. R. Soc. N.S.W.* **52**, 385-96.
Wopfner, H. (1974). Post-Eocene history and stratigraphy of north-eastern South Australia. *Trans. R. Soc. S. Aust.* **98**, 1-12.
Wopfner, H. and Twidale, C. R. (1967). Geomorphological history of the Lake Eyre Basin. *In* 'Landform studies from Australia and New Guinea'. (Eds J. N. Jennings and J. A. Mabbutt). (Aust. Nat. Univ. Press: Canberra.)
Wright, R. L. (1963). Deep weathering and erosion surfaces in the Daly River basin, N. T. *J. Geol. Soc. Aust.* **10**, 151-64.
Wright, R. L. (1964). Geomorphology of the Kimberley area. CSIRO Aust. Land Res. Ser. No. 9.
Young, R. W. (1977). Landscape development in the Shoalhaven River catchment of southeastern New South Wales. *Z. Geomorph. N. F.* **21**, 262-83.

Chapter 6
Fluvial processes

P. H. Walker and B. E. Butler

THE FLUVIAL PROCESSES DESCRIBED IN THIS CHAPTER refer to the action of flowing water in a wide range of continental settings. At one end of the spectrum are hillslope processes, including overland flows under the influence of rainfall, and at the other end are deep water flows of rivers, downstream to where they enter the ocean or lacustrine environments. Fluvial processes involve the detachment, transport, sorting and deposition of granular solids in these water flows, and the associated sedimentary differentiations that occur. The transport of solutes by flowing water is also considered to be part of fluvial processes.

In Australian landscapes, fluvial processes are ubiquitous; the action of ephemeral water flows is evident even in arid regions. In order to understand the environmental setting of Australian soils it is therefore necessary to understand the role of fluvial processes in modifying surfaces and developing parent materials. We present here a brief outline of fluvial processes and attempt to integrate it into what is known of fluvial landscapes and related soils of Australia.

Fluvial processes and landscapes

Fig. 6.1 shows a four-way subdivision of fluvial processes representing, from left to right, a decrease in the ratio of water to solids in transport. Solutes, suspended load and bed load fractions are differentiated in both deep and shallow water flows (Moss and Walker 1978). By comparison with these processes, mass movements involve minimal differentiation into transport modes and are generally not considered to be fluvial in character. They are included here however, because mass movement landscapes overlap fluvial landscapes in

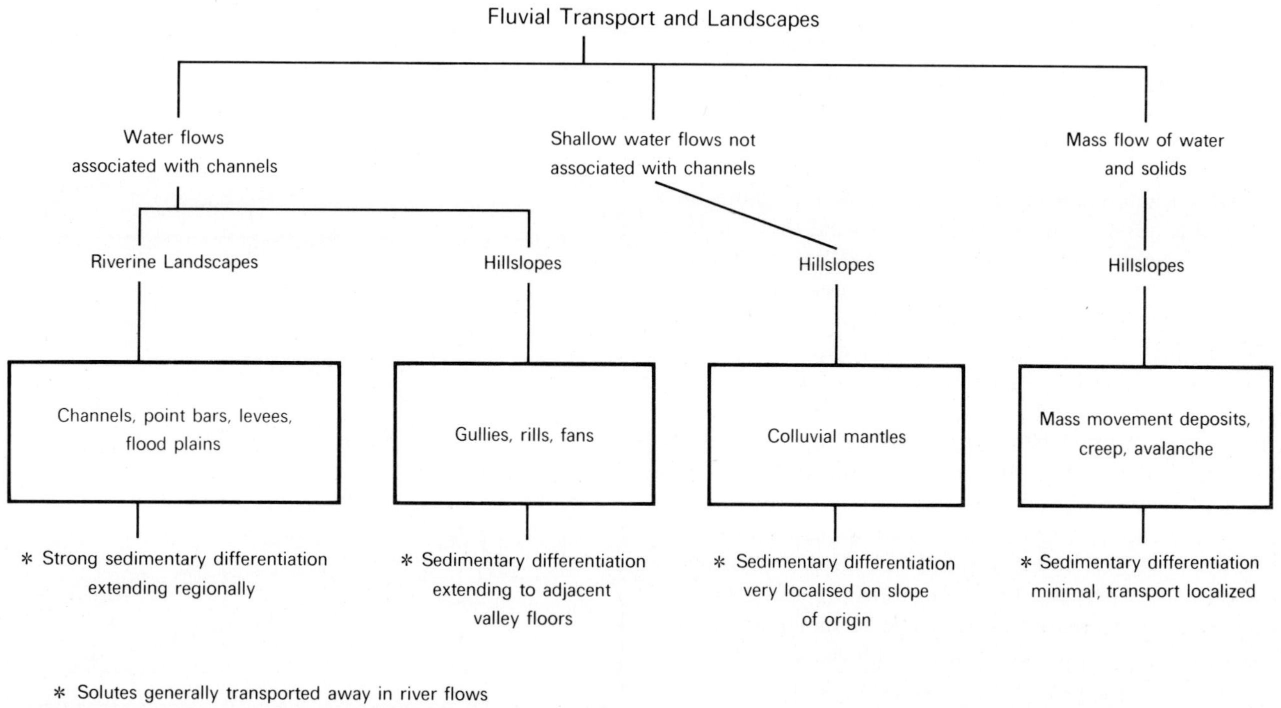

Fig. 6.1. Schematic representation of fluvial transport processes and associated landscapes. The ratio of water to solids decreases from left to right.

nature and some forms of intense bed-load transport in water flows are essentially rheological, and thus appear transitional towards mass movement (Moss *et al.* 1980).

Although Fig. 6.1 shows discrete transporting environments, the three water flows indicated form a sequence of continuously varying transporting conditions. In shallow water flows which are not associated with channels, transported materials are supplied by the action of raindrop impacts on bare soil surfaces and in the flow, to give the rain-flow transportation of Moss *et al.* (1979). Differentiation of bed-load and suspended load (fine-grained particles lifted from the bed and sustained in the flow) occurs, but it is spatially constrained due to the shallowness of the water layer. Nevertheless, saltation (the transport of grains by a series of leaping movements) and rolling of individual bed-load particles occur, even where flows are only of the same dimensions as the diameters of the largest grains being transported (Moss and Walker, 1978). In such flows, bed roughnesses are large relative to water depths so that hydraulic conditions vary greatly over short distances. Vegetation, litter and surface stones are elements of roughness that are major causes of hydraulic variability. In response to these features, hillslope surfaces commonly display very localized stepwise sequences of pools and scours (Moss and Walker 1978). Correspondingly, sedimentary differentiation and deposition tend to be very localized depending on variations in transporting energy downslope. Residual coarse fractions of the soil which are not moved by shallow water flows concentrate at the surface as a lag gravel and resist the ultimate progress of erosion. A characteristic of shallow water flows on hillslopes is the relatively high proportion of soil aggregates transported as bed load. These make the sedimentological identification of hillslope deposits difficult, particularly if the soil aggregates have undergone post-depositional breakdown due to pedogenesis. Bed-load aggregates do not readily survive fluvial transport beyond their slopes of origin because of the powerful effect of fragmentation due to grain collisions in water flows associated with channels (Moss 1972).

Water flows that are concentrated in hillslope gullies receive solutes and solids from shallow unconcentrated flows upslope, and detach further soil by scouring and channel expansion. These transported solids are readily differentiated into bed load and suspended load. The bed load components are deposited intermittently on the floors of channels; suspended load components and solutes are removed to adjacent rivers. The high transporting energy possessed by water flows in gullies is usually suddenly decreased as footslope elements of lower gradient are encountered down a hillside. Fan-shaped deposits are frequently built in response to these lower gradients, so that a relatively high proportion of the medium-to-coarse transported load is deposited locally (Bull 1968). The sedimentological nature of fan deposits is similar to that of riverine deposits, but the contrast in grain size between adjacent laminae in the fan deposits is usually greater, indicating more episodic transporting conditions.

The transport processes described here are related to the hillslope landscapes and soils depicted in Fig. 6.2*a*. The hillslope consists of an upper, steeper zone of net erosion and a lower zone of net deposition (after Moss and Walker 1978). Overland flows generated in the zone of net erosion are sheet flows and flows concentrated in rills and gullies. Soils in this hillslope zone are thus developed on coarse lag materials overlying weathered bedrock or other geological substrata and on dominantly bed-load deposits overlying rock. In the zone of net deposition, soils are developed in fluvial sediments which become progressively finer downslope. These sediments range from coarse bed load to fine suspended load, the latter usually being associated with valley floors. Sequences of sediments and soils in the zone of net deposition are essentially the lower parts of catenas as described by Milne(1936).

Water flows in rivers receive solutes from overland flows and valley slope seepage, solids from overland flows and gullies, and solids from erosion of the river bed

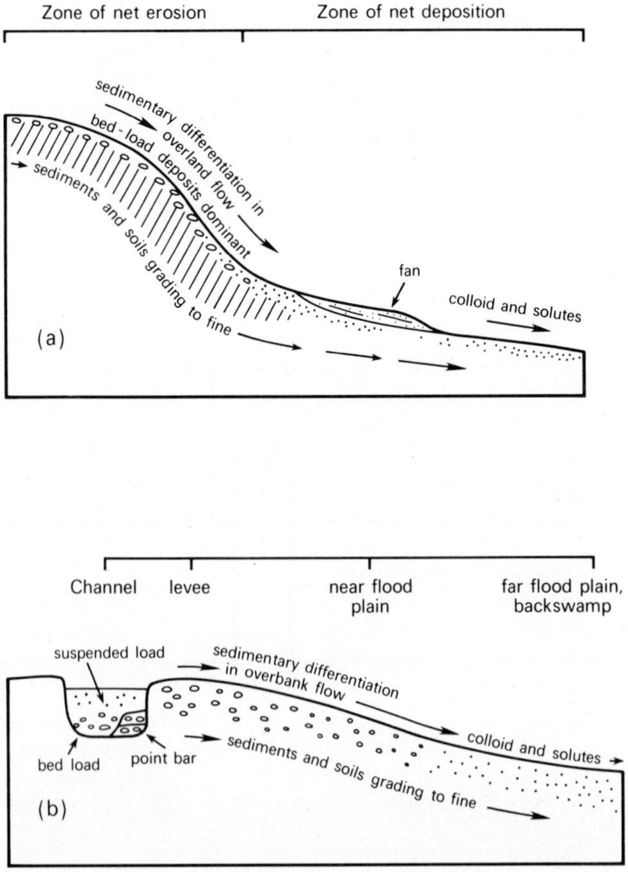

Fig. 6.2. Fluvial sediment differentiation in hillslope and riverine landscapes.

and banks. Solids transported in channel flows are differentiated into bed load and suspended load. The coarsest bed-load deposits form a more or less continuous layer on the river bed and are transported at times of major flooding. River gravel and sand fractions of the bed load are transported by saltation, rolling and sliding. Corresponding bed forms vary with the energy of river flows, e.g. fine ripples are associated with low energy flows and antidunes are associated with high energy flows. Suspended load, usually silt and clay, is distributed throughout the flow and is transported more rapidly downstream.

During high flood stages, river flows overtop their banks and sediment-laden waters spread laterally across flood plains (Leopold *et al.* 1964). Significant decreases in transportation energy are associated with the transition from channel to overbank flow so that the coarsest fractions held in suspension within the channel tend to be deposited on nearby levee banks. Fine suspended load and solutes are transported further laterally to low-energy transporting environments such as backswamps. The particle-size properties and internal drainage characteristics of soils developed in alluvium are thus dominated by the particle size of overbank deposits and the lateral differentiation of sediments in the near flood plain and backswamp environments (Fig. 6.2*b*). Sequences from coarse levee bank soils to clayey soils of flood plains have been described in many parts of Australia (e.g. Butler 1950; Bettenay *et al.* 1971). In the distal parts of large, inland river systems, flood plains tend to dominate the landscape. The Murray–Darling River system, for example, has developed extensive flood plains to the west in which a high proportion of the sediments and soils are of clay texture, often with >50% clay-size material (Butler 1950, 1958).

Because many river systems in Australia originate in humid coastal areas and flow through arid inland landscapes, their fluvial sediments become the source of aeolian sediments and landforms. Butler and Chuchward in the next chapter refer to stream-bordering dunes, i.e. those whose sediment is brought in by fluvial processes and therefore already sorted. Sandy fluvial sediments are relatively free of silt and clay and are readily blown into dune systems. Twidale (1972) and Butler and Churchward (Chapter 7) illustrate the dependency of dune systems adjacent to the Diamantina and Finke Rivers on the input of fluvial sands for dune development.

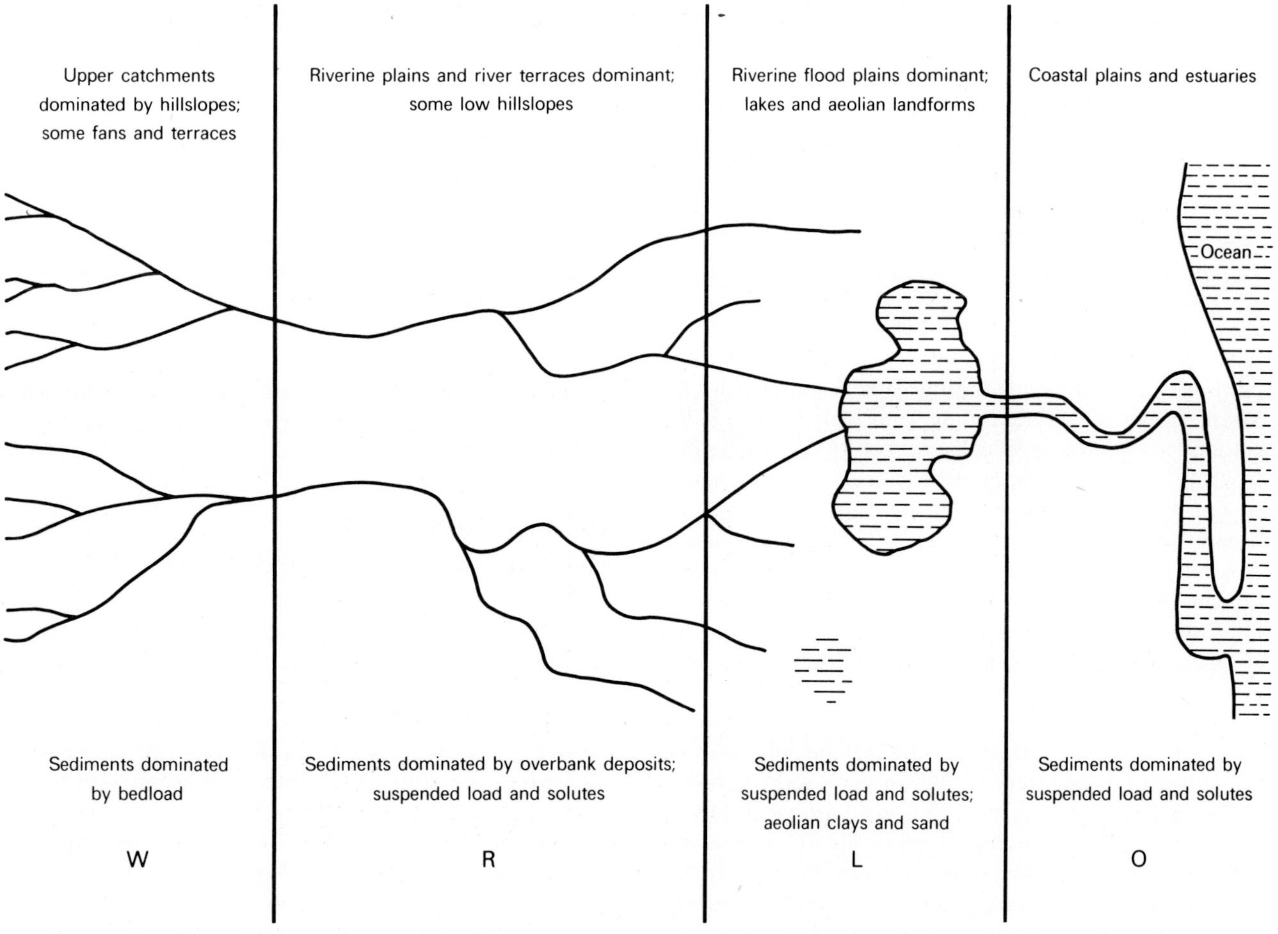

Fig. 6.3. General fluvial systems model showing the four major landscape components W, R, L, O and associated features.

Fluvial systems

The hillslope and riverine landscapes depicted in Fig. 6.2*a* and *b* are the basis of a more general model of fluvial systems presented in Fig. 6.3. Here fluvial systems are regarded as consisting of landscape components, each with particular sedimentological and geomorphological characteristics. River systems have an upper catchment component of essentially hillslope terrain (W) in which slope and riverine environments are dominated by overland flows and localized sedimentary differentiation where bed-load deposits are prominent. Down catchment, where gradients are lower, river systems enter a zone in which flood plains are extensive (R) and in which sediments are predominantly river channel and overbank deposits. Some river systems change their character as they pass through arid inland environments and either lose their discharge in aquifers, or by evaporation, or terminate in lakes (L). Associated with the termination of such systems are fine suspended-load sediments and solutes. Most river systems, however, flow to the ocean (O) into which they discharge their bed load, suspended load and solutes.

The terminus of a river system in an ocean or inland lake may occur at any given stage downstream. Some rivers enter the ocean after passing only through hillslope terrain and are designated here as WO. Other rivers terminate in the ocean after passing through hillslope and riverine plain landscapes (WRO); others either terminate in inland lakes (WL or WRL) or traverse lacustrine landscapes and then enter the ocean (WRLO).

Fig. 6.4. Fluvial systems of Australia in terms of the model presented in Fig. 6.3. WO systems consist of hillslope catchment and coastal fringe; WRO systems consist of hillslope catchment, riverine plains and coastal fringe; WRL systems consist of hillslope catchment, riverine plains and inland lakes; WRLO systems have the same components as WRL above but terminate in a coastal fringe; WL systems have hillslope and lake components characteristic of arid, aeolian landscapes.

All these fluvial systems are found in Australia (Fig. 6.4). WO systems characteristically occur in watersheds around much of the Australian mainland coast and in Tasmania. They usually have steep watersheds, originating in uplifted highlands, and flow to the ocean through hillslope terrain with varying degrees of development of estuarine and deltaic plain landscape. Examples are the Swan, Snowy, Derwent, Hunter and Daly Rivers. WRO systems are characteristically longer rivers which flow from hillslope watersheds in the north, west and southeast of Australia through extensive alluvial plains to the ocean. Examples are the Gascoyne River in Western Australia and rivers of the gulf country in northern Queensland such as the Leichhardt. The Murray–Darling River system is even more extensive and has characteristics of a WRLO system. It drains large areas of the highlands in eastern Australia, flows west through extensive alluvial plains and then through arid lands where its flow is at times intermittent and salt lakes occur. Eventually it enters the ocean via South Australia. WRL river systems occur in semi-arid and arid Australia. They are characterized by relatively low watersheds and extensive riverine plains and terminate in inland lakes which generally have associated aeolian features and salt deposits. Such rivers are the Diamantina and the Finke, both of which drain into Lake Eyre. A large area of arid Australia has poorly integrated drainage systems (WL) associated with aeolian landscapes in which runoff from occasional rainstorms is either dissipated very locally or discharged into inland lakes without significant development of riverine plain. Aeolian sand dunes and clays of the arid zone are in many places derived from the redistribution by wind of fluvial sediments. These sediments, however, relate to relict river systems which were active during the past when climatic regimes were different from the present.

Fluvial landscape development and reactivation

The development of fluvial landscapes is directly related to changes in the intensity of fluvial processes and to the varying degrees of sedimentary differentiation that result from them. Slope and discharge are the major variables controlling fluvial processes on hillslopes and in rivers and thus they control fluvial landscape development (Colby 1964; Wischmeier and Smith 1965; Moss and Walker 1978).

Increases in slope cause increases in the energy of water flows and increases in their effectiveness in detaching, transporting and differentiating erosional materials. Continental uplift causes regional increases in fluvial gradients and corresponding increases in rates of erosion and sediment accumulation in adjacent riverine plains, lakes and ocean floors. Although much of Australia was quiet tectonically during the Cainozoic (Rickard 1975), some regional upwarping occurred in both eastern and western margins of the continent. In Western Australia, limited uplift of the coastal range has led to some rejuvenation of hillslope erosion and river activity, as evidenced by partial stripping of relict weathered surfaces and the trenching of streams (Churchward 1970; Mulcahy and Bettenay 1972; Bettenay and Mulcahy 1972). Cainozoic uplift of the eastern highlands was more significant, in places about 600 m (Wellman and McDougall 1974), and most of it is now considered to have been of Early to Mid-Tertiary age rather than Late Tertiary (Ollier 1978). Evidence of this uplift is seen in the partial stripping of old weathered surfaces of the eastern highlands (Hallsworth and Costin 1953), in the canyons of some east-flowing rivers (Craft 1931), and in the large volume of Cainozoic alluvium deposited in extensive plains by west-flowing rivers in eastern Australia (Pels 1966; Lawrence 1976; Taylor 1977; Bowler 1978). A by-product of the uplift of the eastern highlands is that sufficient elevation existed in the south-east during the Pleistocene for localized glaciation to occur.

Regional variations in the quantity of overland flows and river flows result from variations in climate. Changes to an effectively moister climate tend to increase the volume of hillslope runoff and river flows, thus leading to an increase in transporting capacity. Because of the effect of climate on vegetation, stronger vegetative covers are associated with humid rather than arid climates; these stronger vegetative covers provide greater resistance to fluvial erosion and transport, particularly on hillslopes. Although maximum sediment yields have previously been thought to be associated with intermediate moisture regimes (Langbein and Schumm 1958), more recent evidence indicates a conflict about the relationship between the morphoclimatic zones and the mean annual sediment yields of rivers (Walling and Kleo 1979).

Extensive data reviewed by Walling and Kleo (1979) indicate that sediment concentrations in rivers draining arid lands are higher than those found in humid regions. For arid lands they argue that these higher levels result from lower runoff volumes and sparse vegetative covers, and in addition, the runoff has a relatively low ability to deliver sediment to through-flowing river systems. The alluvial fans and aggraded valley floors that are found in arid regions relate to this low sediment delivery capability.

Palaeohydrological and palaeobotanical evidence indicates that the Australian climate varied considerably during the Tertiary and the Quaternary; for long periods it was either effectively moister or drier than the present (Bowler *et al.* 1976; Kemp 1978). In south-eastern Australia, some later Quaternary climatic phases were colder and drier than the present and it is the change to cold conditions that is considered to have been most effective in modifying vegetation and increasing hillslope erosion (Costin 1971; Costin and Polach 1973; Coventry and Walker 1977).

Therefore, in the south-eastern highlands, the extensive fans and valley fills of Late Pleistocene age testify to the inability of runoff to deliver sediment through river channels, and this is consistent with the occurrence of effectively dry climatic phases. Holocene climates, being effectively moister, have produced runoff with an increased capacity to transport sediment. Late Pleistocene landscapes are prone to erode when cleared of vegetation, and this is attributable in part to their earlier adjustment to runoff of relatively low transporting capacity.

In a wide range of environments in Australia, fluvial landscape development has been episodic, e.g. arid central Australia (Litchfield 1969) and Western Australia (Bettenay *et al.* 1971), and humid south-eastern Australia (Butler 1967). These episodes represent minor phases of fluvial landscape instability relative to the gross effects of global climatic change and continental tectonics. Nevertheless, such episodes of landscape instability are highly significant to soils. Because of the need to define episodes of landscape development associated with soils, Butler (1959) proposed the term *K*-cycle. In fluvial landscapes, a *K*-cycle begins with an unstable phase (K_u), in which a previously stable surface is eroded and the erosion products are distributed downslope and downstream where sediments are deposited over existing stable depositional surfaces. Each *K*-cycle is completed by a phase of stability (K_s), characterized by soil development, which requires that erosion and deposition become sufficiently slow to allow significant pedogenesis. The sediments and soils of each *K*-cycle, the ground-surface, are related to the modern cycle (K_o) by a numbering system, the *n*th cycle before the present being K_n. Evidence of the *K*-cycle principle being involved in fluvial landscape development was deduced from the stratigraphic relations between prior stream deposits and their soils, and those of modern flood plains in the Riverine Plain of south-eastern Australia (Butler 1958, 1967) as well as in hillslope layering and river terraces of adjacent uplands (van Dijk 1959; Walker 1962; Beattie 1972).

The corollary of Butler's (1959) proposal, that *K*-cycles were associated with regional climatic changes and correlate regionally, has been questioned (Bowler 1967). However, more recent soil and landscape studies based on radiocarbon dating indicate that, in the highlands of south-eastern Australia, an intense episode of fluvial erosion and deposition was associated with the last major cold phase of the Late Quaternary, between 20 000 and 30 000 years ago (Costin and Polach 1973; Coventry and Walker 1977; Walker and Gillespie 1978). Dates from a range of sites in south-eastern Australia also led

Williams (1978) to suggest another, but less intense, regional episode of fluvial erosion and deposition associated with climatic change between about 1500 and 4000 years ago. It seems probable therefore, that the significant climatic changes of the Late Quaternary have resulted in some synchronous changes in fluvial landscapes and soils in parts of south-eastern Australia.

Pedogenesis in fluvial landscapes

In landscapes where fluvial erosion and deposition are active, pedogenesis interacts with the processes of sediment removal and sediment accession. To a large extent the rates of these fluvial processes and the amount and kind of sediment involved are controlling factors in pedogenesis.

Fig. 6.2*a* is a convenient model of fluvial processes in which to consider pedogenesis in non-cumulative (erosional) and cumulative landscapes (Nikiforoff 1949). In the zone of net erosion, surface soil is removed downslope and soil development involves a downward growth of the profile by weathering of the bedrock. If the rate of erosion exceeds the rate of weathering, the profile will be mostly rock outcrop. If the rate of erosion is less than the rate of weathering, a soil will develop in the bedrock or in a thin veneer of hillslope sediment overlying bedrock. In the zone of net deposition (Fig. 6.2*a*), soil development involves an upward growth of the profile through successive increments of new sediment. If the rate of deposition exceeds the rate of pedogenesis, the profile will consist of pedologically unmodified sediment. If the rate of deposition is less than the rate of pedogenesis, a soil will develop in the footslope fluvial sediments.

Pedogenesis in flood-plain landscapes (Fig. 6.2*b*) follows the same principles as in the hillslope zone of net deposition. In modern river channels and on point bars, rates of sediment removal and accession are so high that soils cannot develop. In overbank situations, however, rates of sediment accession are lower and soils may develop in levee-backswamp toposequences, the kind of profile being a function of the rate and kind of sediment deposited and the internal drainage at the site.

Rather than a series of steady fluvial processes, erosion and deposition in many parts of Australia were episodic during the Quaternary. These episodes were responsible for the truncation of soils in hillslope zones of net erosion, the burial of soils in fans of hillslope zones of net deposition, and the development of flights of river terraces in riverine landscapes. Evidence of alternating buried soils and fluvial sediments has been described in the alluvial plains of west-flowing river systems of the Murray–Darling Basin in south-eastern Australia (Pels 1966; Butler 1967; Lawrence 1976; Taylor 1977). Here, stratigraphic sequences have been described in which sediments and soils, associated with some older riverine systems (the prior streams of Butler 1950 and the ancestral rivers of Pels 1966), have been buried under sediments and soils of younger riverine systems. Thus pedogenesis has been interrupted by relatively rapid fluvial sedimentation and burial and then fluvial surfaces developed on which a new phase of pedogenesis took place.

Identification of soils and weathered zones in such stratigraphic sequences is basic to establishing landscape history and depends on the extent to which sediments, modified by pedogenesis, can be discriminated from unmodified sediments. Sleeman (1964) used the presence of such micromorphological features as papules, glaebules, plasma separations and cutans to discriminate soils from sedimentary layers in the riverine plain of south-eastern Australia. Green (1974) and Moss and Walker (1978) identified bed-load features from the grain elongation of fluvial soil parent materials, even where considerable soil particle-size differentiation had occurred. Such techniques offer the opportunity of resolving erosional, depositional and pedological episodes in stratigraphic sequences and of deciphering past environments.

In highland landscapes, narrow river flood plains and flights of terraces reflect the episodic nature of past fluvial processes. Alluvial terraces occur in valleys of all major rivers in Australia and many appear to relate to changes in river regime and sediment load associated with late Quaternary changes in climate (Mulcahy and Churchward 1973). Whereas sediments representing fluvial episodes in riverine plains are superimposed one on the other, in the highlands corresponding bodies of sediment tend to have been deposited as valley flood plains at separate levels. Terrace sequences usually range from highest and oldest to lowest and youngest, and soils on such sequences of surfaces vary accordingly (Walker and Green 1976). Soil development on high terraces proceeds undisturbed by modern fluvial processes whereas soil development on modern flood plains is frequently interrupted by sediment accession.

Terrace and soil sequences have been reported in several regions of Australia and some are summarized in Table 6.1. In subhumid to humid regions, sequences which span a significant part of the Quaternary broadly have the same succession of soils. They range from the relatively unmodified alluvium of point bars to biotic profiles with negligible or weak horizonation of low level alluvial surfaces, to soils of high terraces with weakly developed biotic profiles and strong texture differentiation (Walker and Coventry 1976). Soils from such alluvial landscapes have been considered to represent developmental sequences in which time is the dominant soil forming factor (Brewer and Walker 1969). The fact that such sequences can provide information about the rates of soil development has relevance to all landscapes affected by fluvial processes. For example, radiocarbon

Table 6.1. Summary of landscape and age data for soils developed in fluvial sediments in various regions of Australia

Location	Parent material	Dating[A]	Soils	Reference
New South Wales				
Coastal	Alluvium	0–120 years BP	Alluvial soils	Walker (1962)
		390 years BP	Prairie soils (minimal)	
		3 740 years BP	Podzolics (minimal)	
		29 000 years BP	Podzolics, Solodics	
Tablelands	Alluvium	255 years BP	Alluvial soils	Coventry and Walker (1977)
		855 years BP	Prairie soils (minimal)	Walker and Gillespie (1978)
		2 500 years BP	Prairie soils	Walker and Green (1976)
		2 750 years BP	Shallow earths	
		27 040 years BP	Podzolics	
	Fan	1 600–2 400 years BP	Prairie soils (minimal)	Coventry and Walker (1977)
		4 400–7 500 years BP	Podzolics (minimal)	
		26 800 years BP	Podzolics	
Australian Capital Territory				
Tablelands	Colluvium	17 170 years BP	Earth-podzolic intergrade	Walker and Gillespie (1978)
	Fan	27 800 years BP	Podzolics	Costin and Polach (1973)
		31 770 years BP	Podzolics	Walker and Gillespie (1978)
Riverine Plain	Alluvium	< 3 000 years BP	Alluvial soils	Butler (1958), Pels (1971), Lawrence (1976)
		4 000–13 000 years BP (ancestral rivers)	Minimal prairie, minimal red-brown earths	
		18 000 years BP	Red-brown earths	
		20 000 years BP (prior streams)	Red-brown earths	
Victoria				
Coastal	Alluvium	<3 000 years BP[B]	Alluvial soils	Ward (1977)
		42 000 years BP[B]	Gradational texture profiles	
		210 000 years BP[B]	Solodic soils	
South Australia	Alluvium	34 000 years BP	Red-brown earths	Williams (1969)
Northern Territory	Alluvium	undated	Alluvial soils	Litchfield (1969)
		50 000 – > 60 000 years BP[C]	Sandy red earths	
		100 000 – 130 000 years BP[C]	Red earths	
Western Australia				
Coastal	Alluvium	4 000 – 6 000 years BP[B]	Minimal prairie soils	McArthur and Bettenay (1960)
		10 000 years BP[B]	Minimal prairie soils	
		17 000 years BP[B]	Red podzolic soils	
		30 000 years BP[B]	Yellow podzolic soils	

[A]Based on ^{14}C dating unless otherwise indicated; BP, before present [B]Correlation by sea levels [C]Tentative correlation with glacial periods.

dates in Table 6.1 indicate that the biotic character of soils can be strongly developed in alluvium within 5000 years and that a texture B horizon can develop within 30 000 years.

References

Beattie, J. A. (1972). Groundsurfaces of the Wagga Wagga region, New South Wales. CSIRO Aust. Soil Publ. No. 28.

Bettenay, E. and Mulcahy, M. J. (1972). Soil and landscape studies in Western Australia. (2) Valley form and surface features of the south-west drainage division. *J. Geol. Soc. Aust.* **18**, 359-69.

Bettenay, E., Keay, J. and Churchward, H. M. (1971). Soils adjoining the Gascoyne River near Carnarvon, Western Australia. CSIRO Aust. Div. Soils, Soils Land Use Ser. No. 51.

Bowler, J. M. (1967). Quaternary chronology of Goulburn Valley sediments and their correlation in south-eastern Australia. *J. Geol. Soc. Aust.* **14**, 287-92.

Bowler, J. M. (1978). Quaternary climate and tectonics in the evolution of the Riverine Plain, south-eastern Australia. *In* 'Landform Evolution in Australasia'. (Eds J. L. Davies and M. A. J. Williams) pp 70-112. (Aust. Nat. Univ. Press: Canberra.)

Bowler, J. M., Hope, G. S. , Jennings, J. N., Singh, G. and Walker, D. (1976). Late Quaternary climates of Australia and New Guinea. *Quat. Res.* **6**, 359-94.

Brewer, R. and Walker, P. H. (1969). Weathering and soil development on a sequence of river terraces. Aust. J. Soil Res. **7**, 295-305.

Bull, W. B. (1968). Alluvial fan, cone. *In* 'Encyclopedia of Earth Science Series, 3. Encyclopedia of Geomorphology'. (Ed. R. W. Fairbridge) pp. 7-10. (Reinhold: New York.)

Butler, B. E. (1950). Theory of prior streams as a causal factor in the distribution of soils in the Riverine Plain of south-eastern Australia. *Aust. J. Agric. Res.* **1**, 231-52.

Butler, B. E. (1958). Depositional systems of the Riverine Plain of south-eastern Australia. CSIRO Aust. Soil Publ. No. 10.

Butler, B. E. (1959). Periodic phenomena in landscapes as a basis for soil studies. CSIRO Aust. Soil Publ. No. 14.

Butler, B. E. (1967). Soil periodicity in relation to land form development in south-eastern Australia. *In* 'Landform studies from Australia and New Guinea'. (Eds J. A. Mabbutt and J. N. Jennings) pp. 231-266. (Aust. Nat. Univ. Press: Canberra.)

Churchward, H. M. (1970). Erosional modification of a lateritized landscape over sedimentary rocks. Its effect on soil distribution. *Aust. J. Soil Res.* **8**, 1-20.

Colby, B. R. (1964). Discharge of sands and mean-velocity relationships in sand-bed streams. U.S. Geol. Surv. Prof. Paper 462-A. (U.S. Govt. Printing Office: Washington, D.C.).

Costin, A. B. (1971). Vegetation, soils and climate in late Quaternary south-eastern Australia. *In* 'Aboriginal Man and Environment in Australia'. (Eds D. J. Mulvaney and J. Golson) pp. 231-55. (Aust. Nat. Univ. Press: Canberra.)

Costin, A. B. and Polach, H. A. (1973). Age and significance of slope deposits, Black Mountain, Canberra. *Aust. J. Soil Res.* 11, 13-25.

Coventry, R. J. and Walker, P. H. (1977). Geomorphological significance of late Quaternary deposits of the Lake George area, N.S.W. *Aust. Geogr.* **13**, 369-76.

Craft, F. A. (1931). The physiography of the Shoalhaven River Valley. I. Tallong - Bungonia. *Proc. Linn. Soc. N.S.W.* **56**, 99-132.

Green, P. (1974). Recognition of sedimentary characteristics in soils by size-shape analysis. *Geoderma* **11**, 181-93.

Hallsworth, E. G. and Costin, A. B. (1953). Studies in pedogenesis in New South Wales. IV. The ironstone soils. *J. Soil Sci.* **4**, 24-47.

Kemp, E. M. (1978). Tertiary climatic evolution and vegetation history in the south-east Indian Ocean region. *Palaeogeogr. Palaeoclimatol. Palaeoecol.* **24**, 169-208.

Langbein, W. B. and Schumm, S. A. (1958). Yield of sediment in relation to mean annual precipitation. *Am. Geophys. Union Trans.* **39**, 1076-84.

Lawrence, C. R. (1976). Murray Basin. *In*'Geology of Victoria'. (Eds J. G. Douglas and J. A. Ferguson) Geol. Soc. Aust. Spec. Publ. 5, 276-88.

Leopold, L. B. , Wolman, M. G. and Miller, J. P. (1964). 'Fluvial Processes in Geomorphology'. (W. H. Freeman: San Francisco.)

Litchfield, W. H. (1969). Soil surfaces and sedimentary history near the MacDonnell Ranges, N.T. CSIRO Aust. Soil Publ. No. 25.

McArthur, W. H. and Bettenay, E. (1960). The development and distribution of the soils of the Swan Coastal Plain Western Australia. CSIRO Aust. Soil Publ. No. 16.

Milne, G. D. (1936). Normal erosion as a factor in soil profile development. *Nature* (London) **138**, 548-9.

Moss, A. J. (1972). Initial fluviatile fragmentation of granitic quartz. *J. Sediment. Petrol.* **42**, 905-16.

Moss, A. J. and Walker, P. H. (1978). Particle transport by continental water flows in relation to erosion, deposition, soils, and human activities. *Sediment. Geol.* **20**, 81-139.

Moss, A. J., Walker, P. H. and Hutka, J. (1979). Raindrop-stimulated transportation in shallow water flows: an experimental study. *Sediment. Geol.* **22**, 165-84.

Moss, A. J., Walker, P.H. and Hutka, J. (1980). Movement of loose, sandy detritus by shallow water flows: an experimental study. *Sediment. Geol.* **25**, 43-66.

Mulcahy, M. J. and Bettenay, E. (1972). Soil and landscape studies in Western Australia. (1) The major drainage divisions. *J. Geol. Soc Aust.* **18**, 349-57.

Mulcahy, M. J. and Churchward, H. M. (1973). Quaternary environments and soils in Australia. *Soil Sci.* **116**, 156-69.

Nikiforoff, C. C. (1949). Weathering and soil evolution. *Soil Sci.* **67**, 219-30.

Ollier, C. D. (1978). Tectonics and geomorphology of the eastern highlands. *In* 'Landform Evolution in Australasia'. (Eds J. L. Davies and M. A. J. Williams) pp. 5-47. (Aust. Nat. Univ. Press:Canberra.)

Pels, S. (1966). Late Quaternary chronology of the Riverine Plain of south-eastern Australia. *J. Geol. Soc. Aust.* **13**, 27-40.

Pels, S. (1971). River systems and climatic changes in south-eastern Australia. *In* 'Aboriginal Man and Environment in Australia'. (Eds D. J. Mulvaney and J. Golson) pp. 38-46. (Aust. Nat. Univ. Press: Canberra.)

Rickard, M. J. (1975). Australia — regional review. *In* 'The Encyclopedia of World Regional Geology, Part, 1: Western Hemisphere (including Antarctica and Australia).' (Ed. R. W. Fairbridge) pp. 21-28. (Dowden, Hutchison and Ross Inc.: Stroudsburg, U.S.A.)

Sleeman, J. R. (1964). Structure variation within two red-brown earth profiles. *Aust. J. Soil Res.* **2**, 146-61.

Taylor, G. (1977). The Barwon River, New South Wales — A study of basinfill by a low gradient stream in a semi-arid climate. Ph.D. Thesis in Geology, Aust. Nat. Univ.

Twidale, C. R. (1972). Evolution of sand dunes in the Simpson Desert, Central Australia. *Inst. Br. Geogr. Trans.* **56**, 77-109.

van Dijk, D. C. (1959). Soil features in relation to erosional history in the vicinity of Canberra. CSIRO Aust. Soil Publ. No. 13.

Walker, P. H. (1962). Terrace chronology and soil formation on the South Coast of N.S.W. *J. Soil Sci.* **13**, 178-86.

Walker, P. H. and Coventry, R. J. (1976). Soil profile development in some alluvial deposits of eastern New South Wales. *Aust. J. Soil Res.* **14**, 305-17.

Walker, P. H. and Gillespie, R. (1978). Notes on five radiocarbon dates from sites near Canberra. CSIRO Aust. Div. Soils Tech. Mem. No. 5/78.

Walker, P. H. and Green, P. (1976). Soil trends in two valley fill sequences. *Aust. J. Soil Res.* **14**, 291-303.

Walling, D. E. and Kleo, A. H. A. (1979). Sediment yields of rivers in areas of low precipitation: a global view. *In* 'The Hydrology of Areas of Low Precipitation'. Proc. Canberra Symp., Dec., 1979. Int. Assoc. Sci. Hydrol. Publ. No. 128, pp. 479-93.

Ward, W. T. (1977). Geomorphology and soils of the Stratford-Bairnsdale area, East Gippsland, Victoria. CSIRO Aust. Soils and Land Use Ser. No. 57.

Wellman, P. and McDougall, I. (1974). Potassium-argon ages on the Cainozoic volcanic rocks of New South Wales. *J. Geol. Soc. Aust.* **21**, 247-72.

Williams, G. E. (1969). Glacial age of piedmont alluvial deposits in the Adelaide area, South Australia. *Aust. J. Sci.* **32**, 257.

Williams, M. A. J. (1978). Late Holocene hillslope mantles and stream aggradation in the Southern Tablelands, N.S.W. *Search* **9**. 96-7.

Wischmeier, W. H. and Smith, D. D. (1965). Predicting rainfall erosion losses from cropland east of the Rocky Mountains. U.S. Dep. Agric. Handb. No. 282.

Chapter 7
Aeolian processes

B. E. Butler and H. M. Churchward

THE AEOLIAN PROCESSES REFERRED TO HERE ARE THOSE of the movement, sorting and shaping of earthy materials in the landscape by wind. The most familiar aeolian evidence is the form and material of the dunes which are widespread in Australia, but the dust component of the aeolian process is even more widespread and more subtle in its characterization. Several different kinds of dunes are recognized, but the dust materials, though they are probably even more variable, are not well recognized, neither by their generic name as 'dust mantle' nor in their specific forms as loess, aeolian clay, parna, lakulangite, loessial lime, etc.

In this chapter we propose a model for discussing the aeolian processes and make a preliminary inventory of the evidence for aeolian materials in Australia. We have also attempted to fit these materials with the aeolian model, and where necessary modify the model.

Elementary processes and concepts

'Wind work' model

As a basis for discussion, Fig. 7.1 is presented as the model of the effects of wind on earthy materials: 'wind work'. The model incorporates ideas on wind work by Chepil (1957), Churchward (1963*a*) and, in a general way, from Bagnold (1941). Wind erosion of soil is the initial process, and the entrainment of material, followed by its sorting and deposition of the components, are the subsequent activities.

In Sector A of Fig. 7.1 the land surface is pictured as being sufficiently exposed and loose for particles at the surface to become entrained in the moving air mass. Such entrained particles bounce or saltate down wind, and the impact energy of saltation abrades the soil surface further downwind, setting other particles into motion and breaking the massed material into its finer constituents, as shown in sector B. At this stage both the saltating particles and the finer particles released by abrasion are moving with the wind, but their sorting begins right away: the finer material is held in suspension by turbulence of the air, while the coarser saltating particles make intermittent contact with the ground. They need a hard target to maintain saltation. If they fall on a bed of incoherent particles (e.g. drift produced by earlier saltation) their energy is dissipated; they then stop and add to the accumulation of coarser particles. A stronger wind is needed to move these accumulations than serves to create them. However, as an interaction exists between movement and accretion at the site, the accumulations generally assume characteristic shapes in given localities. These accumulations of the coarser components are the dunes; they are made up of piled, weakly coherent or incoherent particulate materials, usually well sorted, with a modal size between 200 and 400 μm.

While a dune builds up nearer to the wind-eroded source the finer component moves on with the moving air mass, and slowly settles out more distantly as the air mass travels. In this way the finer component is deposited over an extended traverse of country, perhaps a hundred kilometres, with some sorting of finer material, and a decrease in its amount with distance down wind. A very small amount of the finest component enters into global circulations. The suspended or dust component is usually finer than 100 μm, but it is sometimes largely composed of clay aggregates, and samples of 50–70% clay are found in Australia.

The wind-worked area develops dunes and loses part

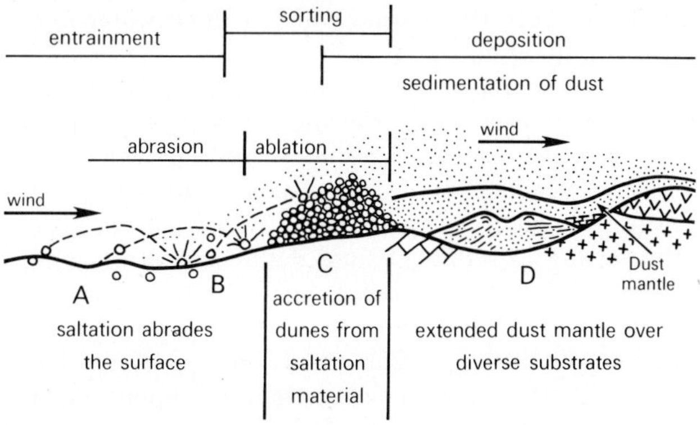

Fig. 7.1. A model illustrating aeolian processes and materials. Subsidiary processes: A, entrainment; B, abrasion; C, sorting ablation and dune accretion; D, sedimentation as dust mantle.

of its fine sedimenatry component, augmenting its apparent sandiness. Adjacent landscapes downwind acquire fine-textured mantles. Deposits of the fine aeolian sedimentary component will be called 'dust mantles' here, a broadly descriptive generic term which avoids distinctions in name on the basis of the composition of the material. A dust mantle has a locally uniform composition over a large area, although it lies over diverse substrates and in diverse topographic sites. Dust mantles from different sources may vary widely, and variations may also develop within a dust mantle after it is deposited.

Aeolian sediments and their characteristics

The various forms of dunes will not be enlarged upon in this chapter. While dunes are accepted as evidence of windwork, dust mantles are not always easily recognized. One looks for uniformity of the mantle over a diversity of substrates and land forms. The only likely confusion is with lacustrine sediments which can be distinguished by their associated beach formations. Wind erosion associated with aeolian sediments may lead to the development of residual lag gravels, and to certain residual landscape forms which, since they are well described in textbooks of geomorphology, will not be discussed further.

Dune fields and source-bordering dunes

Examination of dune distribution in Australia shows that the greater proportion of dunes are regularly spaced over great areas (to give dune fields), as distinct from dunes which occur linearly along some rivers and beaches (source-bordering dunes). The materials of source-bordering dunes are brought to the site of deposition by fluvial or marine agencies and are already sorted when wind takes effect to form dunes. There is no dust component in these dunes because that size fraction has already been removed by the marine agency, or has been drained off with the flood water in the case of a flooding river.

For the source-bordering dunes, exposed conditions need only prevail at the actual sources, but for dune fields to form, conditions for windwork must prevail over the whole area, and prior sedimentary sorting by other agencies is not necessarily involved. The sorting of sediments into dunes and dust mantles depends on the presence of the necessary size components at the source, and this may vary for reasons other than sedimentary sorting. Many landscape materials and environmental conditions do not lend themselves to windwork at all. In general terms windwork requires a loose granular, rather than a dense coherent soil surface, and a sparse cover of vegetation, both of which are favoured by low rainfall and by dry conditions at the time of the windwork.

The dust mantle and its variation

Dust enters global circulations (Jackson *et al.* 1973, and Mokma *et al.* 1972) and may be expected to occur everywhere, however it is reported in palpable form in only a few areas. Its persistence, quantity and character are liable to great variation; some of the main variations are as follows.

(i) Variation due to the lie of the receiving land surface. Dust that has settled out from air is too fine to be entrained again by the wind (Bagnold 1941), but it is nevertheless very susceptible to erosion by water. While dust mantles will be scanty or absent on steep, eroding hillsides, they should be sought on those parts of the landscape where erosion and other depositions are least: on terraces, abandoned levees, or inactive sectors of alluvial fans, on low spurs, plateaux, and the junction between the erosive and depositional sectors of a landscape. If dust is deposited on a surface such as a hillside or a riverine plain where sediments are being actively moved, the aeolian sediments become mixed with the other sediments and lose their identity. However, dust washed from hillsides may accumulate as valley fill in sufficient proportion to give it a 'foreign' character. Similarly the dust mantle falling on a plain of riverine deposition may impose a 'foreign' or unified character to the sediments of a variety of the deposits, as described by Butler and Hubble (1978) for the Murray River basin.

Similarly, settled dust may be entrained again by saltating sand (Bagnold 1941). Thus the part of a dust mantle which is deposited on active sand dunes will be put in motion again and moved into another distribution schedule. Provided there is no active saltation, however, dust that falls in swales or plains between dunes may be expected to persist and accumulate.

(ii) Variation of aeolian sediments due to variation in provenance. The composition of aeolian materials varies widely due to variation in the composition of source materials in different localities and in different environments. Some examples of this variation are presented in Table 7.1 and are referred to throughout the text, they include the following:

Typical dune sands: they have less than 10% clay, are well-sorted sands of 200–400 μm.

Clay dunes and loam lunettes have more than 50% clay, consisting largely of 200 μm aggregates, and may have up to 30% gypsum of comparable size: these deposits are often associated with dry salt lakes.

The classic aeolian sediment, loess, is characterized by low clay content (usually $>10\%$) but a high proportion of well-sorted silt (20–50μm) and is found adjacent to Pleistocene periglacial and glacial outwash plains of northern America and Europe: loess derives its special silty character from the physical weathering characteristics of the periglacial environment.

In Australia, dust mantles are often high in clay, as

Table 7.1. Some attributes of aeolian sediments

Entry	Selected attributes	References
North American loess	mean size silt + clay: 30 μm clay content 3% lime in sample	Smith (1942) sample 15524.28
Parna at Wagga Wagga	clay content 60–70% (sub-plastic, see Walker and Hutka, 1976) mode of non-clay fraction: 45 μm slightly calcareous	Beattie (1970)
Parna at Swan Hill	clay 50% mode of non-clay fraction: 90 μm 15–20% lime	Churchward (1963*a*)
Dunes at Swan Hill	10–15% clay mode of non-clay fraction: 400 μm no lime	Churchward (1963*a*)
Simpson Desert:		
(i) Dune sediment	modal size: 140 μm, <2% clay lime free	Crocker (1946*a*)
(ii) Swale sediment	modal size of non-clay fraction 80 μm 17% clay, 2–3% lime	Crocker (1946*a*)
Loam lunette Lake Albercutya	50% clay, mainly pelletted 180 μm 30% sand, 10% secondary gypsum 10% carbonate	Bowler (1970), p. 152–6
Lake parna Hines Hill (W.A.)	modal size of clay aggregates: 300 μm 20% clay, 10% carbonate, gypsum	Bettenay (1962)
South Australian marine aeolian clays		
(i) Ngaltinga Clay	'clay with kunkar'	Ward (1966)
(ii) Nurriga Clay	'highly calcareous clay'	Maud (1972)
Dust mantles at Whyalla (S.A.)		Jessup and Wright (1971)
(i) Whylie Member	30% clay, 25% lime, dust from the continental shelf	
(ii) Nonowie member	70% gypsum, 15% clay	Jessup and Wright (1971)
(iii) Koolee deposit (Yarraga pedoderm)	50% clay no lime, pH 4, 1% soluble salts	Jessup and Wright (1971)

indicated by Beattie's analysis (1970) of samples from Wagga Wagga: these had more than 50% clay, little silt, but a well-sorted mode for a fine sand at 45 μm.

In South Australia and New Zealand aeolian sediments which originated on the off-shore continental shelf are known to date from times of low sea level: in South Australia these are highly calcareous clays.

(iii) Variation of aeolian sediments due to soil development. Since the sorting and distribution of aeolian sediments was frequently an event of the past, there has been time since for marked pedogenetic development. Contrasting soil horizons have developed in the aeolian material, including concretions of lime, sesquioxides and hard pans of caliche, silicified layers and layers of sub-plastic clay. It is unusual for an aeolian sediment to be characterized by one kind of soil only; the range of sites over which dust has settled gives a potentially full catenary range of soils for each mantle, as in the Murray River basin (Butler and Hubble 1978).

Dust mantles may be eroded or truncated after soil profiles have developed. This has, as one result, the accentuation of any indurated horizons which have developed in the sediment. The widespread carbonate (calcrete) horizons of South Australia are examples.

Aeolian materials in Australia

Dunes

Three main types of dune occur in Australia, and these are described in some detail below.

Dunefields, which are very extensive in Australia, have a systematic pattern of orientation (see Fig. 4.10; also Jennings 1968; Anon. 1967). The most common form is of linear, regularly spaced dunes with the overall pattern for Australia of a whorl centred north west of Lake Eyre; to the south and west of this point the orientation is W–E, to the east it is SE–NW. There are three main concentrations of dunefields: the Simpson Desert, the Great Sandy Desert and the Great Victoria Desert. The last two are both broadly undulating, sandy, weathered plateaux developed over Permian or Cretaceous sediments. There are minor areas of alluvial plains. By contrast, the Simpson Desert has a large area of alluvial plains which are described below. While there seems to be justification for the view that the dunefields occur because the geologic sand source is there, the windwork that created the dunes augmented the sandiness by ablating the fine proportion from a more medium-textured sedimentary zone. Thus windwork, sandiness and desert are associated to a degree, but are not compatible with the occurrence of hard and stony ground, and geologic deposits of sand in these desert zones. Where fluvial sediments extend into the desert zone, windwork may create a dunefield, ablating the clay and giving an apparent increase in sandiness to the region.

Source-bordering dunes are abundant in certain localities, but have no general pattern on a continental scale and there has been no general study of them. They may be considered in two classes. (a) Dunes bordering streams and rivers. These dunes can be contemporary, but they are also found on 'prior streams' (Butler 1950) in the Riverine Plain of south-eastern Australia along stream courses now deserted. In the Canberra area it is claimed (Butler 1967) that several ages of dunes, marked by superposition and buried soils are to be found. (b) Dunes bordering beaches. These are common, and build up into dunefield-like structures where the shore has been advancing seawards for some time. Soil studies of such localities (Ward 1977) show progressive profile development within the dunes with distance back from the beach.

Lunettes are a special form of source-bordering dune. They occur mainly in semi-arid regions in southern Australia (Bowler 1973). The lunette is a crescent dune (or set of dunes) adjacent to a lake or claypan. This dune may be of sand, but mostly it is 'loam' (actually 'clay' with 30–60% clay-sized particles: Bowler 1970). The clay lunette has formed of saltation-sized aggregates, and particles in the 140–200 μm range, according to Bowler (1970) and Sleeman (1973). Bowler's material (see Table 7.1) contained gypsum crystals and quartz grains, as well as clay aggregates in this size, whereas Sleeman's material has no large grains but only clay and silt aggregates. The ability of clay aggregates to move about as saltating dune material in salt lake environments is shown also by Bettenay's study (1962) in the salt lake district of Western Australia. Here sheets of salty aeolian clay extended some miles beyond the dune at the lake's edge.

Dust mantles

The dust mantles produced by aeolian processes are not as simple to identify as the dune products. Principally one looks for fine material with a marked degree of continuity over large areas and diverse substrates, including contrasts in relief. Signs of these kinds are fairly common in southern Australia but have not been reported in northern Australia. Further discussion can be conveniently grouped on the basis of geography.

South-eastern evidence

A dust mantle under the name 'parna' was identified in the Riverine Plain of south-eastern Australia by Butler and Hutton (1956). This mantle was recognized by its uniformity, in terms of clay and lime content, and the fact that it overlies a diversity of substrates and topographic sites. It extends eastward from the Mallee region in Victoria and New South Wales. The material has 50–70% of clay and about 5% lime, and has strongly developed soil profiles, ranging from *red-brown earths** to *grey, brown* and *red clays* depending on the catenary position. The *red-brown* earths are leached of lime to 60 cm while the others are self-mulching and have lime to the surface.

Likewise, similar dust mantles have been identified by van Dijk (1958) at Griffith, New South Wales. He recognized three superimposed parnas, each with strongly developed soil profiles on them. The area studied included the foothills of Palaeozoic sediments which were mainly siliceous. The dust mantles extended from the alluvial plain onto the foothills, and the range of soils found, although similar, included *red earths* on the more elevated sites. The peculiar, sub-plastic nature of the *red earths* and better-drained *red brown earths* in the catenas formed on these parna mantles was emphasized by van Dijk. The three parna mantles have a total thickness of 6 m, but the uppermost one is only about 60 cm thick. The three are interlayered with unmodified riverine and hillside sediments.

Churchward (1961, 1963*a*) studied both dunes and dust mantles at Swan Hill. His work showed that parna originated in the wind erosion of soils, and by the separation of the calcareous, clayey dust component from the sandy, saltation component of the dunes. The dust mantle settled in the swales between the dunes: it contained 50% clay, 15–20% lime, and the modal size of the non-clay fraction was 80 μm. The dune crest sample however had a particle size of 400 μm, there was 10–15% clay and no lime. The dune and swale elements of the aeolian landscape are integrated into one system, with a gradual increase in the saltation fraction in traverses up the flanks of the dunes. Churchward (1961) found three main superimposed mantles; each had

*Great soil groups after Stace *et al.* (1968) are italicized.

strongly developed soil profiles ranging from *red earths* to *grey, brown* and *red clays*. The total thickness of the three aeolian mantles in the dune landscape was variable, a depth of 6 m being quoted by Churchward in one case.

Beattie (1970, 1972) also found three superimposed parna mantles in low, hilly terrain near Wagga Wagga with strongly developed soil profiles on them and which were similar to those found by van Dijk. Each mantle rested on current-bedded hillslope deposits, and in one case the basal layer was the mottled zone of a Tertiary laterite profile. Beattie (1970) (see Table 7.1) found that the non clay fraction of the dust mantle had a modal size of 45 μm. He examined the chemistry and mineralogy of these clays along with their highly developed sub-plasticity (McIntyre 1976); the clay contents were about 70%, when the soil was thoroughly dispersed (Walker and Hutka 1976). Zircon–tourmaline ratios (Beattie, unpublished data) showed that the dust mantle was unrelated to the substrate.

The mineralogy (Sleeman 1975) of soil profiles at Pyramid Hill (80 km south-east of Swan Hill), confirmed the field evidence for two dust mantles overlying weathering granite. These dust mantles were medium and heavy clays which were sub-plastic, 1.4 m thick, and had a modal size of 30–60 μm for the non-clay fraction.

Southern evidence

All of the dust mantles in this area are more highly calcareous than those in the south-east.

Ward (1966) mapped an aeolian clay south-west of Adelaide as Ngaltinga Clay and considered it to have been blown from the adjacent seabed at low sea level stages in the Pleistocene. This clay extends below the present sea level, contains characteristic marine microfossils, is grey in colour and has a high component of calcium carbonate. Horizons of the clay are often referred to (see Table 7.1) as marl, or, if indurated, as concretionary calcrete or kunkar. A variety of well-differentiated soil profiles have developed on this aeolian clay to give gradational red-brown soils with lime at 75–100 cm, and typical *red-brown earths* and calcareous *brown* and *grey clays* with gilgai. Sub-plastic properties in the soils of this mantle have not been reported. These calcareous aeolian deposits were previously recognized by Aitchison *et al.* (1954), who traced them throughout the Adelaide region and north to Gawler, and related them to the 'calcareous loess' of Crocker (1946*b*). Maud (1972) found aeolian clays east of the Mount Lofty Ranges, from Mt Compass to Lake Alexandrina, and correlated his Nurriga Clay with Ward's Ngaltinga Clay. It is widespread in his surveyed area on a terrace above the lake and is often found in a truncated form, as a cemented calcrete layer. Maud mentioned another highly calcareous mantle which overlies the Nurriga Clay in the eastern part of his area and which often occurs as cemented calcrete, or kunkar, 30–60 cm thick. This layer, which runs below sea level, was not given a formal name by Maud, but it is older than his 10-foot terrace and younger than his 26-foot terrace.

Firman (1969) in an extensive account of the Quaternary stratigraphy of South Australia found several Late Pleistocene aeolian formations, including his 'Loveday soil' and the 'Woorinen Formation'. He showed that there had been accession of calcareous loess at several times in the Mid-Pleistocene, of which he recognized and mapped the 'Bakara soil' and the 'Ripon calcrete'.

Like Ward, Firman (1967) associated his several 'calcretes' and 'loessial lime' members with low sea-level stages of the Pleistocene. Support for this is also given by Twidale *et al.* (1978).

Jessup and Wright (1971) found four soil layers above the Tertiary lateritic weathering zone at Whyalla, South Australia. Three of these were aeolian clays (see Table 7.1). The earliest (the Koolee deposit) is an acid, non-calcareous clay; the next (the Nonowie Member) is a gypsum–clay with 50–75% gypsum. The last one (the Whylie member) apparently came from the continental shelf in a low sea level stage; it is highly calcareous (40–50% lime) and has formed cemented caliche horizons during soil development.

Central Australian evidence

These mantles are geographically distinct and include saline clay mantles.

Jessup (1960, 1961*a* and *b*) found three mantles of soils of aeolian character in this region, the two uppermost being relatively similar to those at Swan Hill in the south-east. Each mantle rests on erosional discontinuities, and contains dunes in its sedimentary system. The lowest of these two mantles (the Parakylia layer) has well-differentiated soils varying from *red earths* in well-drained sites to calcareous, self-mulching *grey clays* in poorly drained sites. The source of these aeolian sediments was the Nullarbor Plain (Jessup 1961*a*).

The oldest soil mantle in Jessup's (1960) report is associated with stony tablelands. It is made up of a gypseous clay, varying in thickness from 50 cm to over 150 cm. The soil is a *brown clay*, free of lime in the A and B horizons, but containing a small amount below 30 cm. Fine crystalline gypsum appears at about 30 cm and increases in proportion with depth. The soil is normally in gilgai form, and often has pebbles on its surface. In poorly drained sites the soil is a *grey clay*. Because this soil mantle is widespread and is found over a diversity of substrates, including laterites and silcrete, and on diverse topography, including tablelands, Jessup inferred that it had been distributed as an aeolian dust mantle whose peculiar gypseous characteristics arose because of its source: the erosion of dry salt lakes. In support of this hypothesis Jessup pointed to the similarity of the aeolian sediments and the present sediments of the lake beds, and to the presence of erosion residuals, 12 m thick, of lake-bed clays described by King (1956). He gave the name 'lakulangite' to this aeolian gypseous clay mantle because of its origin in lake beds.

Wopfner and Twidale (1967) suggested a different hypothesis for the gypsum-rich layer in this region. They consider that it formed at the land-surface during a (Pleistocene) period when environmental conditions favoured high ground-water levels but low leaching potential: under these conditions salts accumulated and gypsite was deposited at the exposed ground surface. This hypothesis is not a valid alternative to Jessup's because it accounts for the gypsum and salts only and not for the clay that occurs with the gypsum.

Thus, evidence for the local entrainment and distribution of saline and gypseous clays by aeolian processes is not uncommon in southern Australia, and its occurrence on a broader scale in the arid zone is made plausible by the evidence presented here.

Fitting the model

The instances of aeolian sediments that have been described above are examined in this section in terms of the windwork model. Although the geographic zones are retained they cease to be clear cut.

South-eastern evidence

The distribution of aeolian sediments in the Murray Basin (Fig. 7.2) conforms reasonably well to the model. To the west of a meridian through Swan Hill there is an extensive dune field, and to the east a dust mantle has been identified at widely separated points as far away as Wagga Wagga. The orientation of the dunes is E–W. The dust mantle shows as relatively uniform fine clayey soils (both plastic and sub-plastic) over a variety of substrates. The soils and landscapes are described by Butler (1956) van Dijk (1958), Churchward (1963*b*) and Beattie (1972). Extension of the dust mantle to the north and north-east in Fig. 7.2 is based on recent field observations by Butler (unpublished data). Calcareous clay mantles, 75–100 cm thick, overlie the mottled and pallid zones of lateritic remnants in this area near Bourke, Cobar and Nyngan.

The source of the dust is inferred to be in the dune field region west of Swan Hill, where Churchward (1963*a*) has shown that wind erosion of soils involves the ablation of a fine sandy calcareous clay with similar composition to the dust mantle, and that the dune and swale soils both contain large amounts of clay and lime. Country like this could be the source of the dust mantle extending eastward from it, but there are some problems with this inference. Firstly, extensive areas in the Murray basin dunefield contain no clay or lime, and would not be a source for dust. Secondly, the high proportions of clay and lime remaining in other parts of this dunefield would be hard to explain if it were the source of an extensive dust mantle. This dunefield occupies the place of the Tertiary Murravian Gulf (Hills 1975) and surficial sediments would have originally been in littoral and riverine patterns of sorting. The reworking of these sediments to windwork patterns would not have produced a great amount of lime as dust because they are not particularly well supplied with it.

This problem can be met by proposing that the dunefield area has been *receiving* calcareous dust from marine sources further west at several intervals during the Pleistocene. It is feasible that the dunefields received dust material and also re-entrained it for another phase of aeolian transportation.

Southern evidence

The thesis in the south is that highly calcareous clays were blown inland from the continental shelf during low

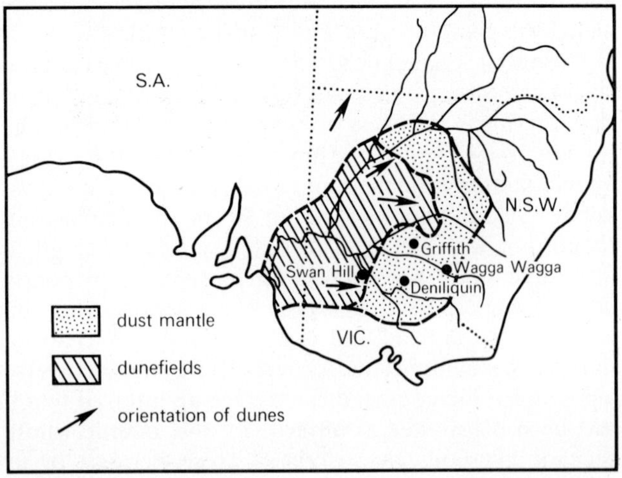

Fig. 7.2. Sketch showing the extent of the aeolian sedimentary systems in south-eastern Australia.

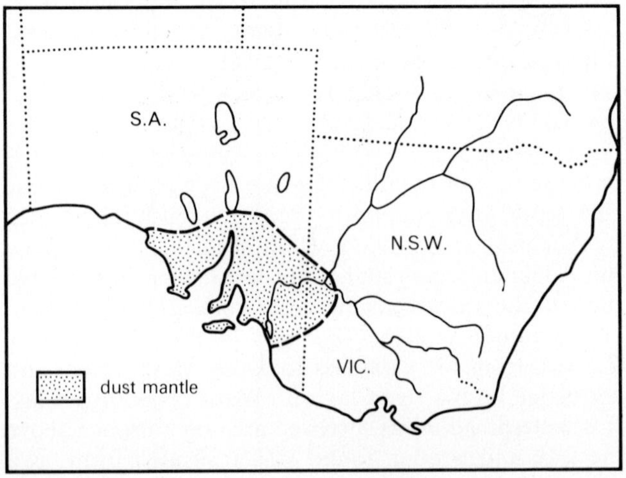

Fig. 7.3. Sketch showing the suggested extent of the aeolian dust mantle from the continental shelf in southern Australia.

sea-level phases of the Pleistocene. There were several episodes of this kind, separated by soil-forming phases, and the materials of each episode were similar. We make no attempt at correlation between the several authors here, but the general belief is that aeolian mantles occurred. Though many of the mantles have well-developed soils, others have suffered erosion, and cemented caliche benches now mark their geographic extent. Firman (1973) reports that several calcretes are widespread throughout the Murray Basin and the Central Basin in South Australia. Calcrete mantles extend further east into the Victorian mallee, and these, together with evidence from Eyre Peninsula from Crocker (1946*c*), have been used to compile Fig. 7.3. This dust mantle overlaps the dunefield section of Fig. 7.2

Jessup and Wright's study (1971), at Whyalla allows a connecting link between South Australian data and that for central Australia: there are several shared aeolian mantles at Whyalla, viz., the gypseous clay shared with central Australia and the marine clay shared with South Australia.

In the broad view it should be recognized that the propositions here of dust mantles from the continental shelf are a logical development of Crocker's (1946*b*) pioneering proposals.

Central Australia evidence

In spite of a degree of similarity between the younger soils and aeolian mantles in central, south and south-east Australia as mentioned above, the main interest here is in the evidence of extensive gypseous aeolian clay mantles, and in the apparent absence of any dust mantles associated with the extensive dunefields of the Simpson Desert.

Lake Eyre is the sink for a very extensive area of internal drainage with large river systems coming from the north-east and north-west. These streams are partly dispersed upon the plain which is the Simpson Desert before entering Lake Eyre, and in the present epoch floodwaters only occasionally reach the lake, bringing their load of silt, clay and salts which accumulate in the lake or otherwise disperse on the plain. Grey cracking

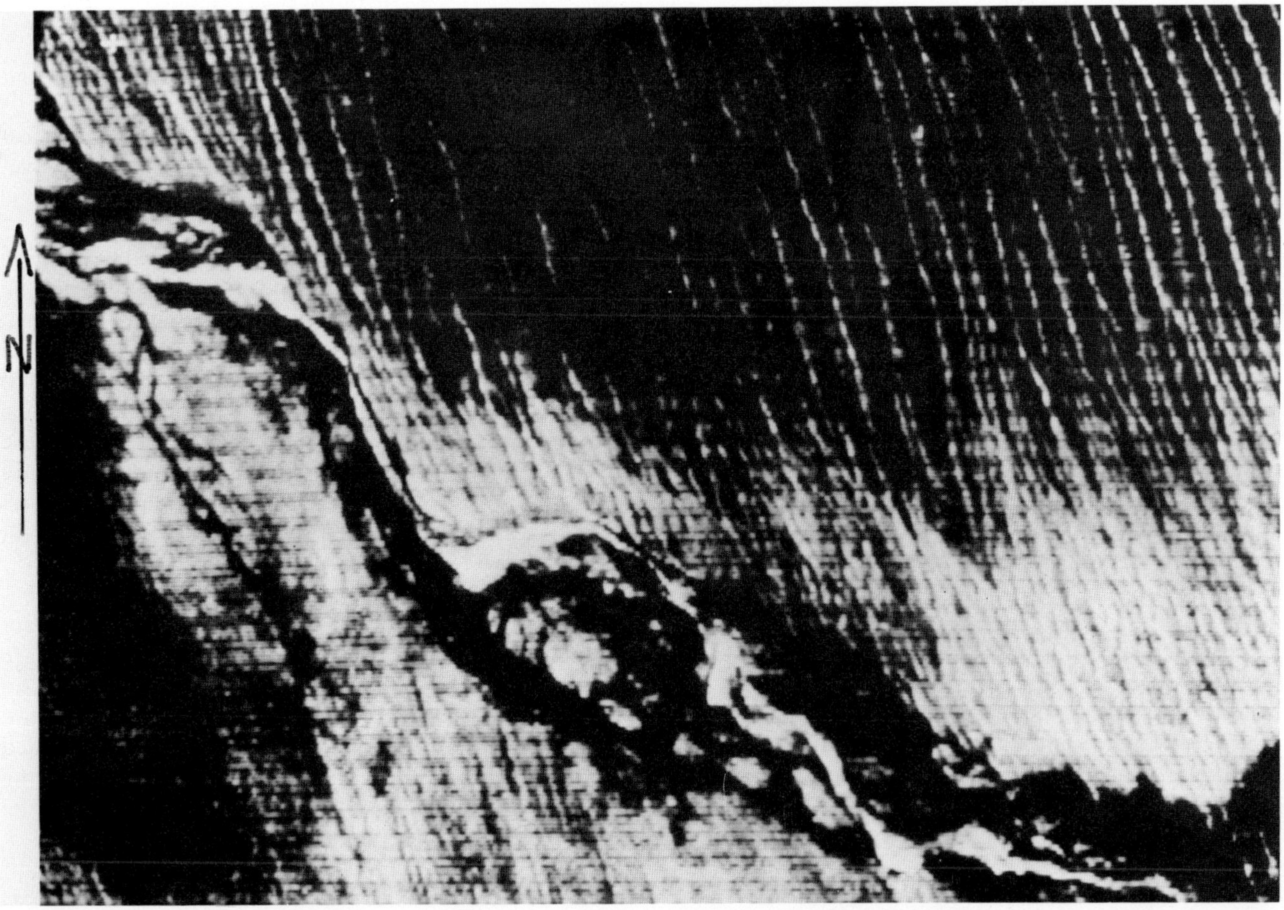

Fig. 7.4. Landsat image of the Finke River flowing to the south-east. Fluvial sediments (white) in the stream course, linear 'debris mounds' adjacent on the north, and these lead directly into the parallel linear dunes oriented west of north. The site is immediatly east of 'New Crown' station and the scale is 1:18 000.

clay (Ug5.2) occur along the main streams and in a few areas on the plain (Northcote *et al.* 1975), but the lake bed consists of clays with crystalline gypsum (Wopfner and Twidale 1967). Gypseous clays occur as extensive aeolian deposits in Australia but while those of southern and south-eastern Australia seem to be contemporary (since their orientation conforms to that of the linear dunes), those of central Australia are attributed to an earlier age. Jessup (1960) and Wopfner and Twidale (1967) refer them to the Pleistocene. It should be noted too that the central Australian aeolian gypseous clays are not oriented with the present linear dune pattern.

The main part of the Simpson Desert consists of a plain with red *siliceous sands* (Northcote *et al.* 1975) on linear parallel dunes which are oriented west of north; there are also some low tablelands of sandy laterite and silcrete. Though the dunefields are extensive, no associated dust mantles have yet been reported.

Twidale (1972) considered that an important source of the Simpson Desert sands were the contemporary sediments of the Diamantina and other streams on the eastern side of the desert. He demonstrated on aerial photos the river sediments first forming transverse 'debris mounds', and then these feeding into parallel linear dunes of the type common throughout the Desert. Twidale described the immediate source of the debris mounds as 'alluvial plains' and 'lake deposits' on riverine plains. A similar transformation of fluvial into aeolian sediments is shown in the Landsat image in Fig. 7.4 of the Finke River on the west side of the Simpson Desert. This conversion of riverine to aeolian sediments seems to be the contemporary sedimentary process, controlled probably by the existing environmental condition of the catchment and the desert. The grey cracking clays refer to an earlier period. The conversion of riverine to aeolian sediments in the Simpson Desert is an instance of the process inferred above for the Mallee region by Butler *et al.* (Chapter 18) for the Lower Murray and Darling Rivers in a prior period.

Crocker (1946*a*) crossed the Desert between Charlotte Waters and Birdsville and took a number of soil samples for particle size analysis. He found a well-sorted mode at 180 μm at the dune crests, with no clay, and a well sorted mode of 80 μm and 17% clay for samples from the swale. The dune crest sample in the present model would fit the description of a saltation sediment, and the swale sample would be a dust mantle sediment. Crocker remarked on the fineness of the Simpson Desert sands which is in agreement with their origin as river sediments at the end of an attenuated fluvial system, such as the Eyre river system.

Folk (1971) made a comprehensive study of the dunes of the Simpson Desert in the locality of Maryvale Station on the north-west fringe of the desert. He considered the source of the aeolian sediments to be the fluvial sediments of the alluvial plain of the desert, and that the crests of the linear dunes were well-sorted with a mode of 145 to 175 μm; the swale (or 'reg' sample) Folk found to be finer and less well-sorted. Both samples contained very little clay, but there was 5–20% of silt, of which 80% was coarser than 30 μm. Folk confirms Crocker's opinion that the sands are finer than the usual desert dune sands.

The evidence shows that fluvial sediments are the source of the Simpson Desert sands, and that they have been wind-sorted into dune and swale deposits. A dust sediment has been separated and its properties are indicated by Crocker's swale sediment (Table 7.1). This material has about 15% clay, is low in lime and has a modal size of its sand fraction of 80 μm. The *earthy sands* which Northcote *et al.* (1975) show occurring widely in the north-west of the Simpson Desert would fit this specification for a dust mantle sediment, and Northcote's description of its dominant soil, Uc5.21, as being about 1 m thick and overlying a variety of substrates including massive laterite, silicified hardpan and clay D horizons, confirms the identification of dust mantle.

Further support for the proposition of dust distribution towards the north-west comes from Kolla and Biscaye's study (1977) of sea-bed cores of the Indian Ocean.

Conclusions

Aeolian sediments are widespread in Australia. Their development has been episodic, and the recurrence of both like and contrasting materials are to be found. Correlations between episodes in different localities has yet to be considered, but some broad generalizations can be made.

Aeolian sediments of southern Australia have large proportions of clay and lime and some part of this came from low sea-level stages of the Pleistocene. To the north of this belt there is a proportion of gypseous clays as aeolian sediments, both dunes and dust mantles. In central Australia most of the aeolian sediments are sands. Dune sediments are widespread but it is proposed here that dust mantle sediments are also widespread: they are earthy sands, low in clay and lime, and of a fine texture. Loess of the kind familiar in Europe does not occur in Australia.

In Australia there seems to be an interrelation between fluvial and aeolian processes. Sediments are brought to a desert zone by fluvial processes and they are given an aeolian sorting, form and pattern of distribution by aeolian processes. This is happening in the Simpson Desert now, and there is evidence that it has happened in earlier times in the lower part of the Murray–Darling basin. Windwork is most active in a desert region but the dust mantle sediment may go well beyond its region of origin. Other environments for active windwork are indicated here as dry salt lakes and clay pans, and the exposed continental shelf.

The compilation of data for aeolian sediments in Table 7.1 indicates a very wide variation in clay, lime and gypsum. This precludes any definition of aeolian sediments on their composition, except on a local basis. Definition on the score of granulometry is not reliable because some sediments have a considerable component of clay aggregates. If this exception can be allowed for, the division ≤ 100 μm for the dust sediments, with most dune sediments having a well-sorted mode around 200 μm seems a satisfactory working basis.

References

Aitchison, G. D., Sprigg, R. C. and Cochrane, G. W. (1954). Soils and geology of Adelaide and suburbs. *Geol. Surv. S. Aust. Bull.* **32**.

Anon (1967). Geographic map of Australia 1:2.5 x 10^6 Div. National Mapping. (Dep. Nat. Dev.: Canberra.)

Bagnold, R. A. (1941). 'Physics of blown sand and desert dunes'. (Methuen: London.)

Beattie, J. A. (1970). Peculiar features of soil development in parna deposits in the eastern riverina, N.S.W. *Aust. J. Soil Res.* **8**, 145-56.

Beattie, J. A. (1972). Ground surfaces of the Wagga Wagga region, N.S.W. CSIRO Aust. Soil Pub. No. 28.

Bettenay, E. (1962). Salt Lake systems and their associated features in semi-arid regions of West. Australia. *J. Soil Sci.* **13**, 10-16.

Bowler, J. M. (1970). Late Quaternary environments: a study of lakes and associated sediments in south-eastern Australia. PhD Thesis, Australian National University, Canberra.

Bowler, J. M. (1973). Clay dunes: their occurrence, formation and environmental significance. *Earth Sci. Rev.* **9**, 315-38.

Butler, B. E. (1950). Theory of prior streams as a causal factor in the distribution of soils in the riverine plain of south-east Australia. *Aust. J. Agric. Res.* **1**, 231-52.

Butler, B. E. (1956). Parna, an aeolian clay. *Aust. J. Sci.* **18**, 145-51.

Butler, B. E. (1967). Soil periodicity in relation to landform development in south-eastern Australia. *In* 'Landform Studies from Australia and New Guinea' (Eds J. N. Jennings and J. A. Mabbutt). (Aust. Nat. Univ. Press: Canberra.)

Butler, B. E. and Hubble, G. D. (1978). The general distribution and character of soil in the Murray-Darling system. *Proc. R. Soc. Vic.* **90**, (1) 149-56.

Butler, B. E. and Hutton, J. T. (1956). Parna in the riverine plain of south-eastern Australia and the soils thereon. *Aust. J. Agric. Res.* **7**, 536-53.

Chepil, W. S. (1957). Sedimentary characteristics of dust storms: I. Sorting of wind eroded soil material. *Am. J. Sci.* **255**, 12-22.

Churchward, H. M. (1961). Soil studies at Swan Hill, Vic.: I. Soil Layering. *J. Soil Sci.* **12**, 73-86.

Churchward, H. M. (1963*a*). Soil studies at Swan Hill. II Dune moulding and parna formation. *Aust. J. Soil Res.* **1**, 103-16.

Churchward, H. M. (1963*b*). Soil studies at Swan Hill. IV Ground-surface history and its expression in the array of soils. *Aust. J. Soil Res.* **1**, 242-55.

Crocker, R. L. (1946*a*). Simpson desert expedition 1939. Sci. Report No. 8. The soils and vegetation of the Simpson desert and its borders. *Trans. R. Soc. S. Aust.* **70**, 235-58.

Crocker, R. L. (1946*b*). Post Miocene climatic and geologic history and its significance in relation to the genesis of the major soil types of South Australia. CSIR Aust. Bull. No. 193.

Crocker, R. L. (1946*c*). Introduction to soils and vegetation of Eyre Peninsula South Aust. *Trans. R. Soc. S. Aust.* **70**, 83-107.

van Dijk, D. C. (1958). Principles of soils distribution in the Griffith-Yenda district N.S.W. CSIRO Aust. Soil Pub. No. 11.

Firman, J. B. (1967). Stratigraphy of late Cainozoic deposits in South Australia. *Trans. R. Soc. S. Aust.* **91**, 165-80.

Firman J. B. (1969). The Quaternary period. *In* 'Handbook of South Australian Geology' (Ed. L. W. Parkin). (Gov. Printer: Adelaide.)

Firman, J. B. (1973). Regional stratigraphy of surficial deposits in the Murray Basin and Gambier Embayment. Rep. Invest. Geol. Surv. S. Aust. No. 39.

Folk, R. L. (1971). Longitudinal dunes of the north-western edge of the Simpson Desert, Northern Territory, Australia: I. Geomorphology and grain size relationships. *Sedimentology* **16**, 5-54.

Hills, E. S. (1975). 'Physiography of Victoria: an introduction to geomorphology'. Fifth edition. (Witcombe and Tombs: Melbourne.)

Jackson, M. L., Gillette, D. A., Danielsen, E. F., Blifford, I. H., Bryson, R. A. and Syers, J. K. (1973). Global dustfall during the Quaternary as related to environments. *Soil Sci.* **116**, 135-45.

Jennings, J. H. (1968). Revised map of the desert dunes of Australia. *Aust. Geogr.* **10**, 408-09.

Jessup, R. W. (1960). Stony tableland soils of the South-eastern portion of the Australian arid zone, and their evolutionary history. *J. Soil Sci.* **11**, 188-95.

Jessup, R. W. (1961*a*). Evolution of the two youngest Quaternary soil layers in the South-eastern portion of the Australian arid zone. I. The Parakylia layer. *J. Soil Sci.* **12**, 52-63.

Jessup, R. W. (1961*b*). Evolution of the two youngest Quaternary soil layers in the South-eastern portion of the Australian arid zone. II. The Bookaloo layer. *J. Soil Sci.* **12**, 64-72.

Jessup, R. W. and Wright, M. J. (1971). Cenozoic sediments, soils and climates at Whyalla, S. Aust. *Geoderma* **6**, 275-308.

King, D. (1956). Quaternary stratigraphic record at Lake Eyre north and the evolution of existing topographic forms. *Trans. R. Soc. S. Aust.* **79**, 93-103.

Kolla, V. and Biscaye, P. E. (1977). Distribution and origin of quartz in sediments of the Indian Ocean. *J. Sed. Petrol.* **47**, 647-49.

Maud, R. R. (1972). Geology, geomorphology and soils of central county, Hindmarsh, S. Aust. CSIRO Aust. Soil Publ. No. 29.

McIntyre, D. S. (1976). Subplasticity in Australian soils. I. Description, occurrence and some properties. *Aust. J. Soil Res.* **14**, 227-36.

Mokma, D. L., Syers, J. K., Jackson, M. L., Clayton, R. N. and Rex, R. W. (1972). Aeolian additions to soil and sediments in the south Pacific area. *J. Soil Sci.* **23**, 147-61.

Northcote, K. H., Hubble, G. D., Isbell, R. F., Thompson, C. H. and Bettenay, E. (1975). 'A Description of Australian soils'. (CSIRO: Melbourne.)

Sleeman, J. R. (1973). Microscopic examination of two parna deposits. CSIRO Aust. Div. Soils Tech. Mem. No. 76.

Sleeman, J. R. (1975). Micromorphology and mineralogy of a layered red-brown earth profile. *Aust. J. Soil Res.* **13**, 101-17.

Smith, G. D. (1942). Illinois loess, variations in its properties and distribution. *Univ. Illinois Ag. Exp. Stn Bull.* **490**, 139-84.

Stace, H. C. T., Hubble, G. D., Brewer, R., Northcote, K. H., Sleeman, J. R., Mulcahy, M. J. and Hallsworth, E. G. (1968). 'A Handbook of Australian Soils'. (Rellim Press: Glenside, S. Aust.)

Twidale, C.R. (1972). Evolution of sand dunes in the Simpson Desert, Central Australia. *Trans. Inst. Br. Geogr.* **56**, 77-109.

Twidale, C. R., Lindsay, J. M. and Bourne, J. H. (1978). Age and origin of the Murray River and gorge in South Australia. *Proc. R. Soc. Vic.* **90**, (1), 27-42.

Walker, P. H. and Hutka, J. (1976). Subplasticity in Australian soils. III. Particle-size properties of soil materials of ranging plasticity. *Aust. J. Soil Res.* **14**, 249-60.

Ward, W. T. (1966). Geology, geomorphology and soils of south-western part of county Adelaide, S. Aust. CSIRO Aust. Soil Pub. No. 23.

Ward, W. T. (1977). Geomorphology and soils of the Stratford-Bairnsdale area, East Gippsland, Victoria. CSIRO Aust. Div. Soils, Soils Land Use Ser. No. 57.

Wopfner, H. and Twidale, C. R. (1967). Geomorphological history of the Lake Eyre basin. *In* 'Landform Studies from Australia and New Guinea'. (Eds J. M. Jennings and J. A. Mabbutt) pp. 119-43. (Aust. Nat. Univ. Press: Canberra.)

Chapter 8
Soil formation on coastal lands and the effects of sea-level changes

W. T. Ward and W. M. McArthur

As a general rule the soils of young landscapes are less differentiated than those of older surfaces. This shows that a considerable time can be needed to form a soil, and it follows that one rational basis for soil classification is to sort the soils in the order of their formation, placing those which have developed on younger deposits before those formed at earlier dates. The different soils, arranged in this way, could represent a sequence in development. If past conditions of weathering should continue, the youngest soils might in time develop the characteristics now seen in the oldest. This is the most obvious interpretation, but others are possible. Apparent trends could be due, for instance, to secular changes in environmental factors like climate.

The methods developed by Butler (1959) for working out the age relationships of soils made full use of soil stratigraphic observations as a means of deciding the order of soil development. However, in many landscapes the fossil soil record is incomplete because the soils have not been preserved by burial but destroyed by erosion. In these circumstances the soil stratigraphic approach cannot be used. Fortunately, the relation between soil development and time can be studied by comparing the soils found on successively developed landforms. The two methods – stratigraphic study of former soils, and geomorphic study of the relations between soils and landscapes – are both necessary, and can complement one another. Because soil stratigraphic studies give emphasis to breaks in sequences, they tend to produce the impression that past changes in the environment were frequent and catastrophic. This defect is offset by studies which compare soils formed on sequentially developed and continuously exposed land surfaces. The signs of sudden disturbance are then balanced by an appreciation of the extended time available for weathering. Soil stratigraphy helps to identify the parent materials of the present-day soils and directs attention to past conditions. Soil – landscape studies based on the sequence of soil formation in time lead the observer to consider the effects of past environments on modern soil development.

Soil age

It is one thing to sort the soils in order of formation, but another to discover their true age. Radiocarbon dating has an evident value but it is limited by the availability of datable materials, and experience shows that it can be applied only to the youngest soils. The prospect of dating older soils is best at the coast, where sedimentation and subsequent soil development is closely associated with changes in sea level. If dates related to these changes in the level of the sea can be given to coastal landforms, the chronology can be extended inland by using local markers. At the coast abundant shells can supplement wood and peat as a basis for radiocarbon dating.

If there is unrecrystallized coral present, then dates back to 200 000 years ago can be obtained using uranium decay methods, and on volcanic shores useful information is also given by potassium/argon dating. However, for most coasts these methods cannot be used, so the principal means of dating then becomes the correlation of stranded shorelines. If the former shores in one locality cannot be dated directly, it may be possible to relate them to the shores of another locality, where dates exist, on the grounds that any changes in sea level are instantaneous and global.

Originally the signs of changed relationships between land and sea were attributed either to changes in the level of the ocean or to changes in the elevation of the land as a result of uplift or depression. In the first case, problems of dating were seen to be solved by identifying the relevant sea level for a particular soil-geomorphic unit, and adopting an age by reference to a supposed global standard. In the second case, it seemed to follow that no generally usable dates could be obtained. Close study of field evidence assessed in these different ways, geophysical considerations relating to changes in ocean volume, and coral datings of former shores have led to the realization nowadays that no coast can be regarded as stable, no matter how free it is of seismic activity. However, few coasts are so mobile that they show no part of the eustatic record. Movements of both the land and sea contribute to observed coastal landforms. The problem in reconstructing the past history of changing sea levels is to separate these two components.

The changes of sea level were mainly a consequence of changing global temperatures, and because past sea temperatures are now becoming well known from studies of deep-sea sediments, many workers today substitute the known temperature record for the as-yet unknown sea-level trace. This is obviously a provisional measure, and caution is needed with interpretations because the details of the record, the effects on the sea of short-term temperature excursions, and the effects on the earth of changing water and ice loads are still unknown.

Structure of coastal lands

The varied landforms of the coastal zone are not all the product of marine action. The outline of the shore and the profile of the coast can be attributed, in detail, to the actions of waves and currents but sediments deposited at the coast by rivers and winds are of prime importance, as are the effects of earth movement and changes of sea level. Differences in the vigour of marine action and differences in the availability and influx of sediment types largely control sediment texture and fabric, whereas the shape of the varied subaerial and submarine depositional landforms, and the thickness and relative position of the sedimentary bodies in a local sequence are frequently determined by long-term changes in the relative positions of land and sea.

The apparent simplicity of most coastal plains suggested by their general lack of relief disguises the fact that these plains are the terminal geomorphic surfaces of complex transgressive–regressive depositional units laid down during glacioeustatic fluctuations of sea level. The units, bounded at the surface by terrestrial and marine erosional, transportational and depositional landforms, are limited below by unconformities of terrestrial and marine origin. The varied sedimentary facies between these surfaces reflect coastal plain (alluvial, palustrine and aeolian), estuarine and marine open-shelf environments, and their full evaluation requires a complete stratigraphic examination.

Ideally, two main sedimentary sequences occur within a transgressive–regressive unit, a terrestrial and a marine sequence. The terrestrial sequence is developed as sea level is lowered, and results from the extension of alluvial erosion and sedimentation, and of soil formation, into the coastal zone. The marine sequence is formed during the subsequent rise of sea level. At this time marsh, estuarine, beach, dune and deltaic sediments are introduced (Fig. 8.1).

The relative thicknesses and significance of these sequences depend very largely on the vigour of coastal processes and on the volumes of sediment brought to the shore by rivers. So much alluvium is carried to the ocean by a major river that the marine contribution to the sedimentary pile at these localities can be severely limited; the coastal landforms are then predominantly the result of alluvial distribution.

Coastal strand plains, with no alluvial contribution, such as those forming the modern landscape in East Gippsland, Victoria (Ward 1977), lie at the other extreme of the scale. Similarly, the relative placement of the terrestrial and marine sediments is determined by the gradients at the coast, and by the history of sea-level movements relative to the land. If the sea should rise above an earlier high level, it is clear that the earlier deposits will be truncated or will occur as burials, or else as partially submerged landforms in quiet lagoon waters. On the other hand, a drop in sea level will leave the earlier deposits stranded as a terrace.

Although different conditions prevail in different localities, it is to be expected that general principles will emerge from regional comparisons. Present limitations to knowledge make it impossible to review the development of the entire Australian coast, but four cases with which we are familiar will be used for illustration. The first summarizes the development of the Swan Coastal Plain, an area produced partly by alluviation and partly

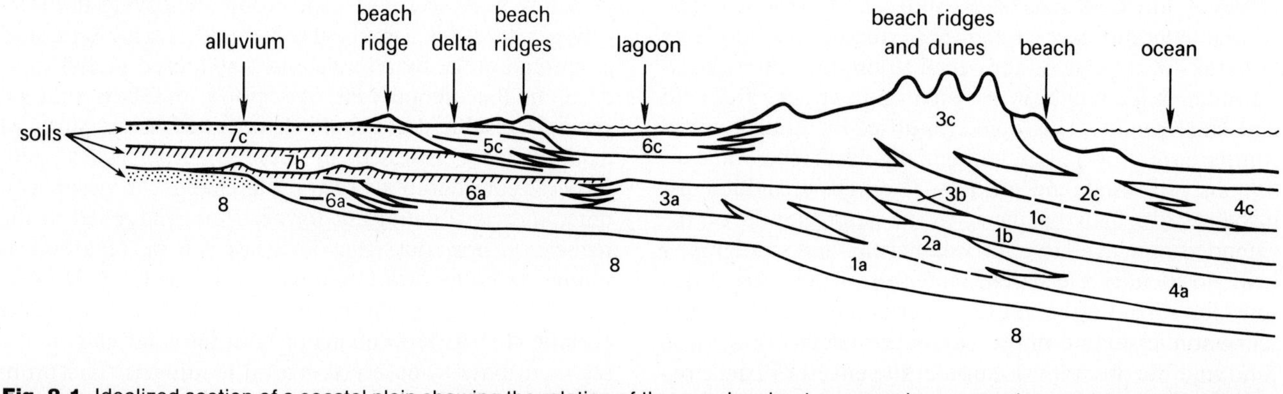

Fig. 8.1. Idealized section of a coastal plain showing the relation of three on-lapping transgressive–regressive sequences, identified by lower-case letters, a–c. The numbers identify depositional environments. 1, littoral transgressive gravels and sands; 2, littoral gravel and sand beach; 3, beach and foredune; 4, offshore sands and muds; 5, deltaic clays and peats; 6, estuarine sands, silts, clays and shell beds; 7, alluvial floodplain; 8, basement rocks. For simplicity, the section does not include redistributed aeolian sands, which are common in coastal lands, and may indeed dominate the landscape. In off-lapping sequences similar deposits occupy the terraces which are formed as the sea withdraws from the land.

by shoreline progradation. The coast south of Adelaide is cliffed and composed mostly of hard rocks, but there are several signs of changed sea level, and terrestrial deposits pass below the modern water line. In Gippsland, the coastal plain was produced by repeated marine transgressions during slow uplift of the land. The coast of southern Queensland is different. High dunes formed by the southeast trade winds are situated beside the present-day shore, with local areas of prograded beach and abandoned estuarine plains.

Swan Coastal Plain

In south-west Australia the coastline is associated with limestone, either as the main rock or as a veneer over older rocks, so that it is well suited to the recording and preservation of evidence of former sea levels above or below the datum level, even if they were of short duration. However, the record is limited because the limestone is formed of lithified dunes of Middle to Late Pleistocene age. Fairbridge (1961) has interpreted the field evidence as the result of movement of the sea against a stable land, and has incorporated it in a supposed global standard. The sequence of former sea levels was based on wave cut benches, submerged and stranded beach cobble deposits, karst, and marine bands associated with the Pleistocene limestone.

The Swan Coastal Plain (McArthur and Bettenay 1960) consists mainly of alluvium and dune sands. The fluviatile deposits, as alluvial fans, are of several different ages. Seven separate units, each with a characteristic soil, are recognized. With the passage of time, (i) the soil colour changes from dark brown through brown, red and yellow, to grey with yellow-brown mottles, due to changes in type and distribution of iron oxides; (ii) a texture profile develops and the colour becomes more marked; (iii) the proportion of kaolinite to illite increases, with gibbsite forming in the oldest soils; and (iv) the chemical properties change, with a general decline in nutrient levels.

The fluviatile deposits are limited to the west by sequences of coastal dunes. The oldest sequence is siliceous and *podzols* have developed on them. The next sequence has a core of aeolianite beneath weakly podzolised *siliceous sand.* At the coast there is a final sequence of calcareous dunes. The properties of the dune soils change in an orderly fashion with age. Assuming that the oldest was originally calcareous, there is progressive loss of carbonate, a lowerng of the pH, and formation of *podzols.* Between the two younger dune sequences is a zone that is marine in origin.

Thus there is well established evidence of former sea levels and clear evidence of periodicity in landscape events, but it is not possible, except in the most general way, to correlate the levels with the events. The fluviatile deposits all occur at a common level, so there is no opportunity to relate them to separate sea levels. Similarly the relationships of the aeolian materials to sea level is conjectural. McArthur and Bettenay (1960) have suggested that high sea levels were associated with warmer, drier conditions favouring land erosion, and they argued that during times of low sea level the weather was cooler and wetter, and soil formation could take place.

Adelaide area

South of Adelaide the coast is formed of ancient rocks, consolidated Tertiary sediments and Pleistocene alluvium. The major elements of the relief are tectonic: three faults downthrown to the north define three tilted fault blocks. In the fault-angle depressions the Tertiary rocks pass below young cover beds mostly produced by erosion of the high ground. Although alluvium is dominant, the cover includes fossiliferous marine sandstones, aeolian clays and wind-drifted sands. These beds are not dislocated by faulting except in the south.

The oldest alluvium formed ancient plains which are now represented by accordant summit levels with local remnants of alluvium. The youngest alluvium was spread in several distinct episodes as fans, and is built up at the coast to characteristic levels. It retains its original primary depositional form. The position of the marine base level for each surface is recorded by stranded coastal features; abandoned beach sediments, foredunes and sea cliffs. Considered along with the history of sedimentation, these elements enable Pliocene and Quaternary sea level movements to be reconstructed. Four major episodes of lowered sea level can be recognized, and were thought to indicate the four major European glaciations recognized at the time of survey (Ward 1966).

Sedimentation began in the Late Tertiary with Pliocene marine transgression. The sea rose relative to the land up to 180 m above present sea level. In the Pleistocene there were several later changes of level, and these led to altered drainage and to the development of superposed gorges along the coastal streams. The youngest deposits occur both above and below the gorges, showing that deposition is not to be attributed to the changes of sea level, but to the local environmental changes that were associated with the changing water levels. Aggradation occurred when the sea was low and continued while it was rising, and seems to have ceased when the highest levels were attained. The aeolian sediments (parna and wind-blown sand) drifted in from the west at times of low seas.

At first it would seem that the changes in climatic environment would interfere with soil development, but the soils of the depositional land surfaces form an orderly chronological sequence in terms of profile development. This suggests that the changes which induced large-scale erosion and deposition were nevertheless small enough not to affect weathering trends.

Gippsland area

The coastal zone east of Sale, as far as Lakes Entrance in East Gippsland, is part of an extensive lowland south of the Dividing Range in Victoria (Ward 1977). Large coastal lagoons are separated from the open sea by a narrow sand barrier. Broad low-lying sand plains and swamps surround the lagoons, whereas the ground elsewhere is formed by marine terraces which rise in shallow steps to the 130 m contour. Dune sands form low ridges on the marine terraces. They trend towards the north-east, parallel to several low marine cliffs now stranded on the plains. Alluvial terraces beside the main rivers terminate at former shorelines defined by the stranded marine terraces, except for the youngest one, which passes beneath present sea level. The detailed evidence of relative changes in level of land and sea provided by the large number of marine and alluvial terraces gives the area a special interest. Some of the fourteen marine terraces that are now known were attributed by earlier workers to supposed irregular movements of the land during the uplift of the Dividing Range, but the lowest terraces were all attributed to glacioeustatic changes in the level of the sea. The highest shorelines greatly exceed the level to which the sea would rise if the world's ice were to melt (50–55 m), and uplift is clearly implied. However, the stranded shores are level, and the subsurface structure shows that they were cut by transgressing seas. The apparent changes in sea level thus are not well explained by supposing that land movements were wholly responsible; there is no tilting, and too many reversals in direction would be required. It is more likely that steady uplift of the land has combined with glacioeustatic changes in ocean level. The ages of the shorelines cannot be found directly as there are no datable materials, but correlations with other localities with similar shoreline records are possible. Ward *et al.* (1971) have shown that the shoreline elevations in this area have the same ratios to one another as those in South Carolina, U.S.A., although there are differences in shoreline elevation. Sea-level change is therefore a major component of both records, and simple histories of uplift must apply to each locality. Ages based on this interpretation have been suggested for each shoreline.

In Gippsland, three groups of soils can be defined according to whether the parent sediment is aeolian sand, alluvium or stranded marine terrace. The soils on the aeolian sands show progressive development of podzol characteristics when viewed in order of age. In the other groups, the most striking feature is the progressive development, with increasing age, of a contrast in texture between topsoil and subsoil. Changes in clay mineral suites and chemical composition accompany these changes in morphology (Little and Ward 1981). The chemical studies indentify a level deep in the subsoil that marks a major change in the relationship between clay and the elements potassium, magnesium, zinc, cobalt, nickel, rubidium and lithium. Above this level the relationship is a systematic one whereas at greater depths there is no apparent relationship between clay and these elements. The depth of the level increases with estimated soil age in an approximately linear fashion reaching 165 cm in the soils of the oldest terrace. The level appears to mark the depth to which soil formation has extended.

South-east Queensland

In south-eastern Queensland the coast from Coolangatta to Sandy Cape is formed mostly by dunes, sandplains and swamps between a few headlands of hard rock. Beach ridges form narrow coastal strand plains in some places, and stranded estuarine sand flats are also well developed. Several abandoned shores mark changes in position of the sea but the elevations are not exceptional, the highest proven level being +3 m for the oldest known shoreline. Subsequent shores occupy positions both slightly above and slightly below present sea level. These small changes could be attributed wholly to variations in the level of the ocean. Nevertheless, the elevation of the youngest Pleistocene beach, −1 m, shown by the pumice deposits which are associated with it, is noteworthy in view of the now-common belief, founded on palaeotemperature measurements (Shackleton and Matthews 1977), that the ocean during the last interglacial was higher than now. At present, the evidence of coastal deposits is not enough to confirm that belief. If any subsidence has affected south-east Queensland it was remarkably uniform, for no stranded shore has been deformed. Moreover, the fact that in its changes from glacial to interglacial age the sea returned each time almost to its prior position is also important. A record like this must surely be the kind expected for a stable coast.

Ward and Little (1975) suggest that the highest level in south-east Queensland (+3 m) represents the shoreline at +49 m in Gippsland. On that basis a straightforward correlation of the raised shorelines in Gippsland with the shores identified in Queensland is possible. The fact that correlation can be made gives support to the ages which have been proposed for the Gippsland shores (Ward *et al.* 1971).

Although the stranded shores provide a chronology, the most obvious features of the Queensland coast are the great masses of sand dunes. Nine main periods of dune building are distinguished, and as the windblown sands pass below sea level, it is obvious that they were associated with glacial ages. The south-east trade winds which built them were stronger then.

The youngest dunes have soils classed as *siliceous sands*, but *podzols* are more generally developed. On the low ground *humus podzols* occur where groundwater levels are high. Depths of leaching increase with increasing age, to greater than 13 m on the oldest land surface. Parallel changes in the natural vegetation, and measured nutrient-element contents, show that there is an equivalent decline in soil fertility.

Conclusions

In the study of changes that occur in soils with the passage of time the soil stratigraphic approach gives too great an emphasis to breaks in weathering caused by erosion and sedimentation. The emphasis is on the discontinuities in a sequence, rather than on the sequence itself. One is persuaded that conditions in the past were markedly different to those at present. This argument is supported by the fact that soil stratigraphic reports frequently use the words 'inherited' and 'relict', when all that has been shown is that the soils under discussion are merely relatively old. Geological changes, especially those related to climate, are important for their effects on the landscape but it does not follow that changes in the landscape mean changes in the conditions of weathering. Proof is needed that the soils were affected.

Studies of soils arranged according to age are most likely to show whether environmental changes have had important effects. This question is best examined at the coast where sea-level changes continually generate new landforms and at the same time give a basis for dating.

The fluctuations in the sea which resulted from changes in the volume of the world's ice were uniform, all around the world, but every coast seems to have a different shoreline record. A need to reassess Fairbridge's classical work in Western Australia, fully referenced by Fairbridge (1961), is evident. Local earth movements affect the record everywhere. Nevertheless, shoreline correlation seems possible if realistic allowance is made for the movements, and the pairings made between Gippsland and south-east Queensland give confidence in shoreline ages that were based initially on some fairly extravagant assumptions.

References

Butler, B. E. (1959). Periodic phenomena in landscapes as a basis for soil studies. CSIRO Aust. Soil Publ. No. 14.

Fairbridge, R. W. (1961). Eustatic changes in sea level. *In* 'Physics and Chemistry of the Earth'. (Eds L. H. Ahrens *et al.*) Vol. 4. pp. 99-185.

Little, I. P. and Ward, W. T. (1981). Chemical and mineralogical trends in a chronosequence developed on alluvium in eastern Victoria, Australia. *Geoderma* **25**, 173-88.

McArthur, W. M. and Bettenay, E. (1960). The development and distribution of the soils of the Swan coastal plain, Western Australia. CSIRO Aust. Soil Publ. No. 16.

Shackleton, N. J. and Matthews, R. K. (1977). Oxygen isotope stratigraphy of Late Pleistocene coral terraces in Barbados. *Nature* (London) **268**, 618-20.

Ward, W. T. (1966). Geology, geomorphology, and soils of the south-western part of County Adelaide, South Australia. CSIRO Aust. Soil Publ. No. 23.

Ward, W. T. (1977). Geomorphology and soils of the Stratford-Bairnsdale area, East Gippsland, Victoria. CSIRO Aust. Div. Soils, Soil Land Use Ser. No. 57.

Ward, W. T. and Little, I. P. (1975). Times of coastal sand accumulation in south-east Queensland. *Proc. Ecol. Soc. Aust.* **9**, 313-17.

Ward, W. T., Ross, P. J. and Colquhoun, D. J. (1971). Interglacial high sea levels — an absolute chronology derived from shoreline elevations. *Palaeogeogr. Palaeoclimatol. Palaeoecol.* **9**, 77-99.

Chapter 9
Salt and sodicity

R. F. Isbell, R. Reeve and J. T. Hutton

MANY SOILS OF THE ARID TO SUBHUMID REGIONS OF Australia contain significant amounts of water-soluble salts – dominantly as sodium chloride – and their subsoils are frequently characterized by moderate to high amounts of exchangeable sodium and magnesium (see Chapter 3). The effects of such soil properties on land use are widely appreciated; this chapter will be mainly concerned with reasons for the distribution pattern of Australian saline and sodic soils, the origins of the salt and sodium, and their effects on soil genesis. There are large areas of secondary salinity in Australia but these will not be considered here since they are discussed in Chapter 53.

Northcote and Skene (1972) have shown that at least 5% of the surface area of Australia has saline soils and another 25–30% has sodic soils (Fig. 9.1). They defined saline soils as having more than 0.1% NaCl in surface sands and loams, more than 0.2% NaCl in surface clay loams and clays, or more than 0.3% NaCl in the B horizon. Sodic soils were defined as those having an exchangeable sodium percentage (ESP) of more than 6 within the top metre of the profile. However, this value commonly exceeds 15 (the figure widely adopted for the lower limit of sodic soils elsewhere in the world) and may frequently be as high as 30. It should be noted that whereas all the saline soils shown by Northcote and Skene (1972) are sodic, many of the sodic soils are non-saline. These definitions will be used here. Fig. 9.2 illustrates some characteristic profiles of Australian saline and sodic soils.

The occurrence of saline and sodic soils

Australia is the driest continent in the world. Of the average annual rainfall of 420 mm, 87% is lost to the atmosphere by evaporation and transpiration (Australian Water Resources Council 1975) and so it is hardly surprising that salts are found in a wide variety of soils. However, while the map of Northcote and Skene (1972) does show that the saline soils are most common in the more arid parts of the continent, the relationship is not a close one. The map also illustrates the fact that sodic soils, many of which have appreciably saline subsoils, are very common in the subhumid to humid parts of eastern and south-western Australia, but less so in the northern parts of the continent.

Saline soils

The saline soils of the arid interior and central west of the continent (Fig. 9.1) are mostly found in two major soil landscapes.

First, to the west of approximately 134°E they are mostly confined to saline depressions and playas that are the dismembered remnants of relict early Cainozoic drainageways (Bettenay 1962; Mulcahy and Bettenay 1972; Van de Graaff *et al.* 1977). The main soils are *solonchaks*. To the east of 134°E similar but frequently larger playa floors and ephemeral salt lakes are also common. These are the end of extensive internal drainage systems, some forming well-known features such as Lakes Eyre, Torrens, Frome and Gairdner. In south-west Queensland less well-defined salt pans occur as floodouts adjacent to the major rivers. Here again the soils are essentially *solonchaks*.

Secondly, the remaining extensive areas of saline soils in the arid interior, east of approximately 134°E, occupy very different soil landscapes. These are mostly undulating upland areas with a mosaic of *desert loams* and *red* or *brown clays*, usually with a prominent gravel pavement, that have mostly formed from Cretaceous and early Tertiary sediments of both marine and freshwater origin. In many cases the sediments have been subjected to deep weathering during the late Cretaceous to mid-Tertiary (e.g. Idnurm and Senior 1978). Clay soils in this region derived from alluvium also have high salt contents (Hubble and Reeve 1970; Dawson and Ahern 1974).

Permeable soils such as *siliceous sands, earthy sands, red earths* and *lithosols* occupy vast areas of the arid interior (Stace *et al.* 1968), and available data show that these soils are all low in salts. Hubble and Reeve (1970) have suggested that, even in this arid environment, the non-saline soils are sufficiently permeable to permit effective leaching under present rainfall conditions, particularly where individual monthly or even weekly rainfall totals may on occasion greatly exceed the long-term annual means.

Saline soils also frequently fringe low-lying sections of the northern Australian coastline (Chapter 3), which has areas periodically inundated by tidal waters. The soils are *solonchaks*. Few data are available for total soluble salt (TSS) contents of the interior playa-like depressions or the coastal tidal salt flats, but the salts in Lake Eyre have been examined by Bonython (1956). The surface crust is essentially pure halite (NaCl), but gypsum is the dominant salt below the halite crust. Some data are available for the saline *desert loams* and clay soils, but often exact figures for the relatively insoluble gypsum are not reported. For south-west Queensland Hubble and Reeve (1970) showed mean TSS contents to 1 m depth of about 4% for *desert loams*. For some clays TSS/NaCl ratios ranged from 8 to 15, although other clays in the area were considerably higher or lower.

Sodic soils

Following Northcote and Skene (1972), the remaining sodic soils excluding the *solonchaks* may be divided into

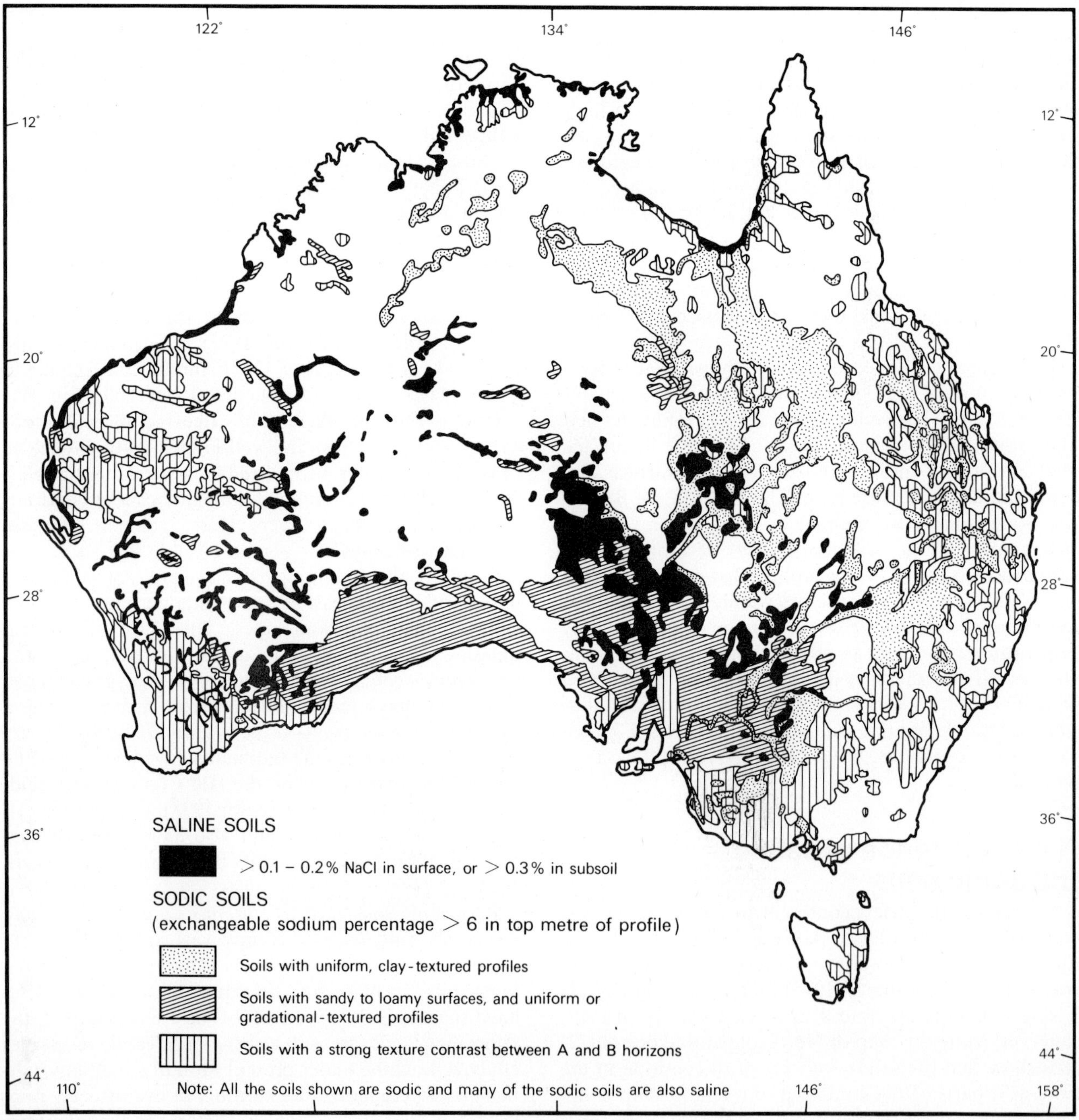

Fig. 9.1. A map of saline and sodic soils in Australia (based on Northcote and Skene 1972).

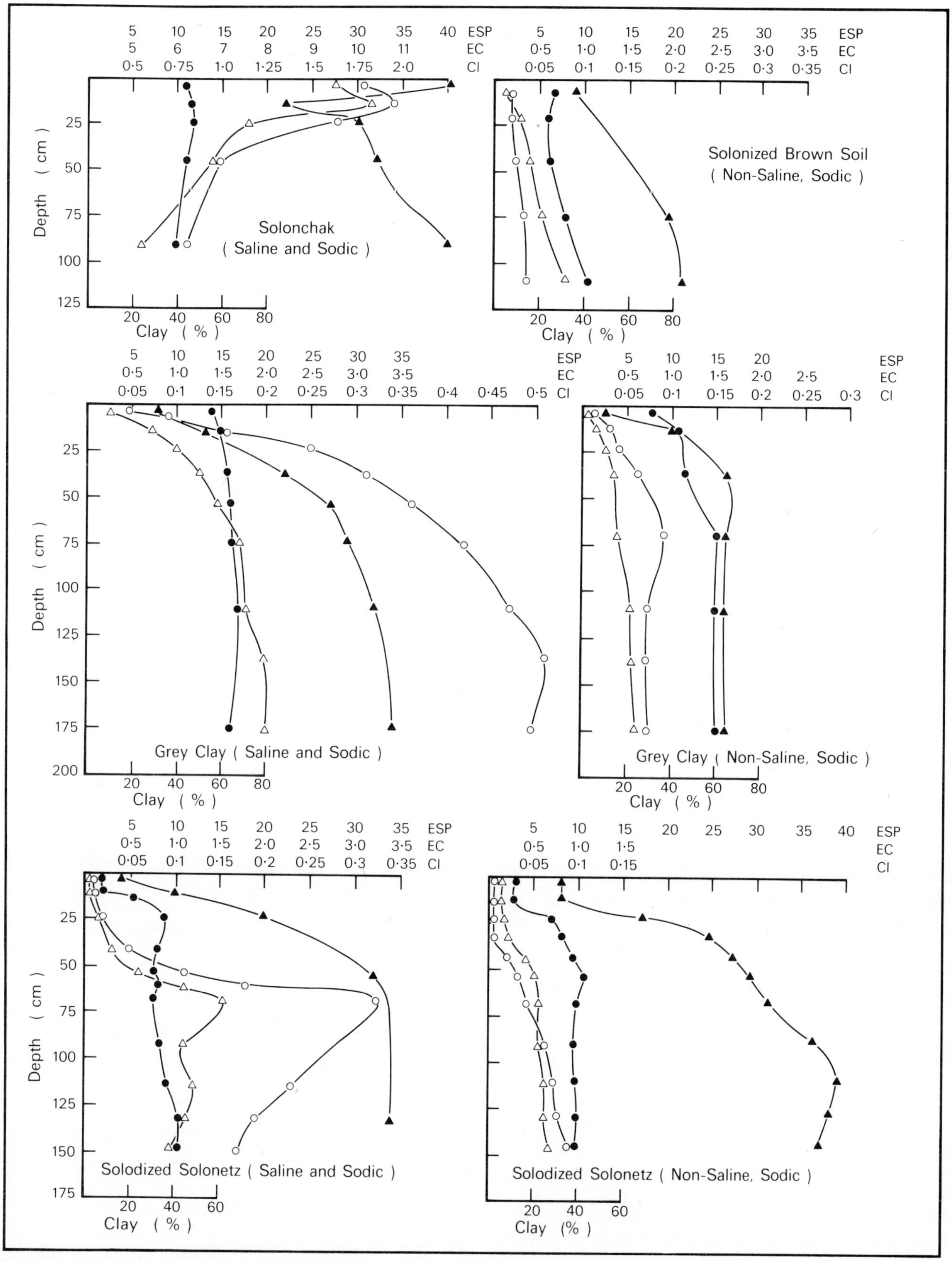

Fig. 9.2. Selected Australian soil profiles showing the relationship between clay content (●), chloride as percentage NaCl (○), electrical conductivity as mS cm^{-1} (△), and exchangeable sodium percentage (▲).

three major groups (Fig. 9.1). First, cracking clay soils (*grey, brown* and *red clays* and *black earths*) are an important component of the eastern third of Australia (see Fig. 3.12), and the great majority of these in the semi-arid to subhumid regions are sodic, at least in their subsoils (e.g. Fig. 9.2). Many also have appreciable subsoil salinity, e.g. about 30% of the 361 profiles examined by Northcote and Skene were saline according to their definition and 60% had an ESP of more than 15 within the top metre of the profile. The data also suggested that the *black earths* had generally lower ESP values than the *grey, brown* and *red clays*. A wide range of parent materials is involved: from basalt and other basic igneous rocks and derived alluvium for the *black earths* to a wide range of more labile sedimentary rocks and alluvium for the *grey, brown* and *red clays*. This variety means that it is seldom possible to relate degree of sodicity to particular parent materials.

Second, in the central southern parts of Australia (Fig. 9.1), Northcote and Skene show large areas of sodic soils that are generally medium-textured at the surface, and which may become finer textured at depth, usually because of increasing amounts of fine carbonate (<2μm). These sodic soils include the *solonized brown soils* (see Fig. 9.2) and the *grey-brown* and *red calcareous soils*, and are generally found in areas with < 350 mm mean annual rainfall. The data of Northcote and Skene show that of 197 profiles examined, only 9% of the soils were saline, but 64% had an ESP of 15 or more in the top metre. Soil parent materials, although variable, are usually calcareous – chiefly limestone, dolomite, calcareous shales, and calcareous aeolian and alluvial deposits.

Third, there are widespread areas of texture-contrast (duplex) soils with sodic subsoils. This group is slightly larger in area than either of the two previous groups of sodic soils and occurs widely in the central-east, south-east, and western parts of the continent (Fig. 9.1). Mean annual rainfall ranges from about 250 mm (mainly in Western Australia) to slightly over 1000 mm. The chief soils involved are *solodized-solonetz* (see Fig. 9.2), *solodic* and *soloth soils*, sodic *red-brown earths*, and the *desert loams* discussed earlier. Northcote and Skene considered 487 profiles in this duplex soil category (excluding the *desert loams*) and although only about 5% were saline according to their definition, 46% of profiles had an ESP of 15 or more in their subsoils. Slightly more than half of the profiles were alkaline (pH > 8.0) at depth, and of these 57% had an ESP of 15 or more. Of the acid subsoil forms (*soloths*), 22% had an ESP of 15 or higher. The duplex soils constitute a very diverse group, with a very wide range of parent materials. Whereas the other two broad groups of sodic soils commonly occur on landscapes of low relief, sodic duplex soils are also found on moderately steep slopes of hilly topography.

The origin of soluble salts and sodium

Present atmospheric accessions

Early views (e.g. Wilsmore and Wood 1929; Prescott 1931; Teakle 1937; Anderson 1941) were that much of the salt in soils originated from oceanic sources via rainfall, although Downes (1954) suggested that salt accumulation may also have occurred in the geological past (Holocene). More recent rainfall analyses in various States [Hutton and Leslie 1958 (Victoria); Wetselaar and Hutton 1963; Hutton 1962 (Northern Territory); Hingston and Gailitis 1976, 1977 (Western Australia); Probert 1976 (north Queensland); Reeve and Fergus, unpublished data (south Queensland); Hutton 1957, 1962 (South Australia); Hutton 1957, 1962 (New South Wales)] have given a better picture of the amount and ionic compositions of salts in rainfall, particularly in relation to distance from the sea. The main conclusions have been that the oceanic influence decreases rapidly with distance from the coast, and the increasing proportions of cations such as calcium and potassium suggest the increasing influence of terrestrial sources on rainwater composition.

Hutton (1976) proposed the formula $y = 0.99\ d^{-1/4} - 0.23$ to relate the decrease in chloride concentration in rainwater to the distance from the ocean, where y = weighted mean concentration (mequiv. l^{-1}) and d = km in the direction most likely to contribute maximum oceanic salt. This equation provided a good fit ($r^2 = 0.992$, $n = 42$) for the data of Hutton and Leslie (1958) for sites from Victoria receiving more than 500 mm of rain and located not more than 300 km from an oceanic

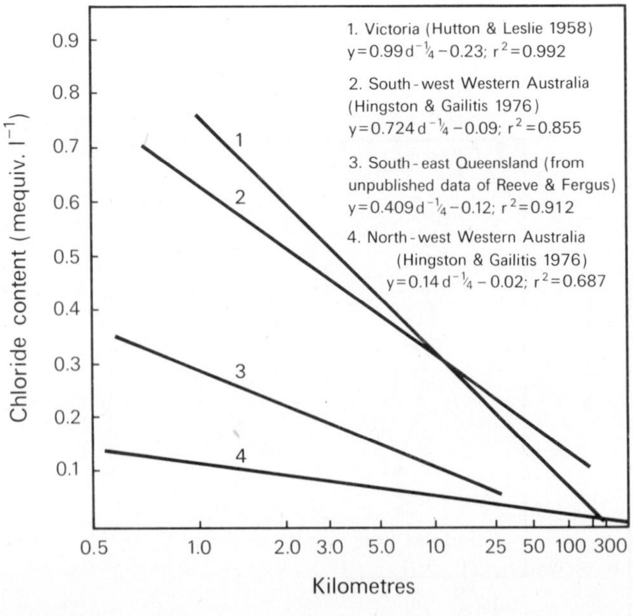

Fig. 9.3. Relationship between the chloride content of rain and distance from the sea for four widely separated areas of Australia.

Table 9.1. Annual deposition of chloride and some ionic ratios in rainwater at a selection of Australian sites

	Lat.	Long.	Distance (km) and direction to coast	Rainfall annual mean (mm)	Chloride (μequiv.l^{-1})	Chloride (kg ha^{-1})	Ionic ratios Na:Cl	Na:Ca	Na:Mg	Na:K	Cl:SO$_4$	Type of coastline	Ref.[A]
West coast													
Broome	17 58	122 04	3 (NW)	540	60	12	0.80	3.7	3.9	8.9	5.4	Exposed	1
Onslow	21 36	155 05	1 (NE)	267	100	10	0.77	0.88	3.0	12.0	6.9	Shoals	1
Carnarvon	24 53	113 38	2 (W)	230	670	55	0.85	4.7	4.0	22.8	5.6	Barrier Islands	1
Geraldton	28 45	114 36	1 (W)	475	440	74	0.86	9.1	4.7	31.8	7.8	Shoals	1
Perth (University)	31 56	115 47	4 (SW)	820	340	100	0.82	5.5	4.1	31.8	5.8	Exposed	1
South coast													
Albany	34 57	117 48	10 (SW)	815	530	153	0.85	11.4	4.3	38.3	8.4	Exposed	1
Esperance	33 52	121 54	3 (SW)	675	410	98	0.83	8.8	4.4	33.0	7.2	Shoals	1
Eucla	31 43	128 53	15 (SW)	255	370	33	0.83	0.67	3.1	22.0	6.8	Exposed	1
Mt Burr	37 33	140 24	20 (SW)	790	330	92	0.95	4.6	3.9	21.3	—	Exposed	2, 3
Cape Bridgewater	38 24	141 24	1.5 (SW)	800	700	200	0.89	8.0	4.1	40.0	—	Exposed	4
East coast													
Whian Whian	28 38	153 18	37 (SE)	2390	70	60	1.3	2.8	6.0	10.0	—	Exposed	2, 3
Cooloola	26 03	153 06	1.4 (ESE)	1430	220	112	0.83	16.0	4.1	43.7	9.8	Exposed	5
Toolara Forest	26 05	152 43	24 (ESE)	1430	115	58	0.78	12.0	4.4	29.7	7.4	Exposed	5
Townsville	19 19	146 47	9 (NE)	1120	110	44	0.67	1.8	3.6	15.0	3.8	Barrier Reef	6
Lansdown	19 37	146 48	40 (NE)	865	60	18	0.68	1.5	3.3	12.0	2.3	Barrier Reef	6
Inland stations													
Katherine	14 27	132 18	270 (NW)	870	5	1	0.9	2.0	3.0	6.0	—		8
Alice Springs	22 50	133 30	900 (NE)	245	25	2	1.1	0.9	1.2	3.1	—		3
Wittenoon Gorge	22 17	118 20	190 (NNW)	395	34	5	0.91	1.0	1.0	3.0	1.5		1
Meekatharra	26 35	118 30	445 (SW)	210	45	3	1.34	1.7	3.4	8.8	1.7		1
Kalgoorlie	30 47	121 28	335 (S)	250	70	6	0.79	0.67	1.6	6.3	1.3		1
Mt. Crawford	34 43	138 57	54 (SW)	785	120	33	1.0	3.6	3.5	11.3	—		2, 3
Coleraine	37 36	141 43	80 (SW)	605	140	30	0.86	6.0	4.0	24.0	—		4
Horsham	36 43	142 12	190 (SW)	450	60	10	1.0	3.0	3.0	10.0	—		4
Merbein	34 10	142 04	320 (SW)	250	65	6	1.15	0.39	1.4	6.6	—		2, 3
Mitta Mitta	36 32	147 22	290 (SW)	1050	8	3	1.0	1.0	1.0	5.0	—		4
Mundubbera[B]	25 40	150 50	230 (ESE)	705	28	7	1.2	2.2	3.8	3.4	1.7		7
Sea water ratio							0.86	22	4.3	47	10		

[A] 1. Hingston and Gailitis (1977); 2. Hutton (1957); 3. Hutton (1962); 4. Hutton and Leslie (1958); 5. Reeve and Fergus (unpublished data); 6. Probert (1976); 7. Reeve and Stirk (unpublished data); 8. Wetselaar and Hutton (1963).

[B] Narayen Research Station

coast. Data from Eire, New Zealand and the United Kingdom followed this relationship.

Data from other parts of Australia have followed regressions of the same form as Hutton's formula. Fig. 9.3 shows the straight lines obtained by plotting chloride concentration against the inverse of the fourth root of the distance from the ocean. The constants in the equations differ from those of Hutton.

Table 9.1 which complements Fig. 9.3, shows results for chloride accession and ionic ratios (on an equivalent basis) for a selection of coastal, near-coastal and inland Australian sites. There are indications in the table that the atmospheric salt load, besides being affected by distance from the sea, is also influenced by factors such as local wave energy, and wind speed and direction. Low concentrations might be expected in North Queensland coastal areas, where low wave energies prevail in the lee of the Great Barrier Reef. This is supported by the data of Douglas (1968) and Probert (1976). In the case of ionic ratios, values for a few near-coastal stations bear a close resemblance to those of sea water but the values at inland sites do not.

The ionic ratios of Table 9.1 also show the influence of local environment. Calcium enrichment at some coastal sites, as indicated by low Na:Ca ratios, may sometimes be explained by the presence of calcareous dunes, e.g. near Perth, Esperance, Eucla and Cape Bridgewater, although in areas of purely siliceous dunes, e.g. at Cooloola, the ratio is also significantly lower than that of seawater. In other areas reasons for calcium enrichment are not clear, e.g. Onslow, Lansdown. While it is difficult to define the true nature and extent of the contribution of ions from nearby areas (although Hutton (1968), Blackburn (1973) and Hingston and Gailitis (1976) have made some suggestions), both chloride and other ions are clearly being redistributed by wind action and some are being returned in the rain. Hence the input of salts as measured in rainfall may not represent a true gain.

While the data in Fig. 9.3 and Table 9.1 suggest that mean annual chloride fallout is low beyond about 200 km inland from the sea, it is of interest to consider the possible mechanisms of atmospheric salt transfer. In near coastal environments it is likely that simple open-funnel collectors would underestimate total meteoric input (Art *et al.* 1974). Thus in coastal southern Queensland the chloride concentration of drainage water leaving the siliceous sand dunes at Cooloola cannot be reconciled with that of the rainfall, unless allowance is made for an additional input which probably comes from the interception of aerosols by the vegetation (Reeve and Fergus, unpublished data). This effect has also been noted in the northern hemisphere (White *et al.* 1971; Potts 1978). While it is recognized that the oceanic aerosol particles generated by breaking waves tend to be of large diameter, apparently not all of them are so since it has been demonstrated that small aerosol particles are readily detectable many hundreds of kilometres from their source (King and Maher 1976; Anon 1977). Using such data as a basis, Gunn and Richardson (1979) have estimated that an annual salt fallout at Emerald, central Queensland (latitude 23°30′S, 280 km from the coast) would be of the order of 10 kg ha^{-1} $year^{-1}$. Consequently small net accessions of oceanic origin may well contribute significantly to the atmospheric salts measured in rainfall, but this is difficult to quantify against a background accession of terrestrial origin, particularly as Twomey (1960) has suggested that atmospheric particles may originate from drying soil. It is worth noting that limited analyses of some major inland Queensland rivers such as the Diamantina, Coopers Creek and Warrego (Reeve, unpublished data), show losses of chloride from their catchments to be so low (<0.4 kg ha^{-1}) that only small net annual accessions would be needed to allow accumulation of salt in the soils of their catchments.

Salt accession in the past

Although little direct evidence is available, the possible role of past salt accessions should be noted. Hubble and Isbell (1958) thought that salt accession from the *interior* of the continent in the Quaternary might explain the high salt and sodium contents of the *grey clays* of subhumid southern Queensland, a theory which takes into account the unexpectedly low sodium chloride tonnage found in Lake Eyre (Bonython 1956). Relict aeolian features in presently semi-arid to subhumid regions in eastern Australia provide evidence of more arid and windy conditions in the past (Bowler 1976), so there is good reason to believe that salt from both oceanic and terrestrial sources may have been transported considerable distances during such periods. Blackburn (1974) has suggested that aeolian movement of salt is significant in those parts of Australia where bare salty surfaces are exposed during hot windy weather, and that saline environments have been a feature of inland southern Australia for the last 20 000 years at least.

Primary and secondary rock sources

Although oceanic salt is an obvious direct source of both chloride and sodium, other sources have to be considered, even when allowance is made for aeolian recycling in inland areas. The average chloride content of crystal rocks is about 0.018% chlorine (Johns and Huang 1966), but some sediments may have higher contents than this, probably due to the inclusion of some sea water (with 2% chlorine) during their deposition. Although most of the common rock-forming minerals are low in chlorine, some trioctahedral micas, particularly biotites and phlogopites, and some amphiboles are reported to contain up to 2% chlorine. Some of the rarer minerals such as scapolite and sodalite contain up to 8%.

In the case of sodium an acceptable mean for crustal rocks is 2.5% of the element, with the principal sodium-bearing minerals being the plagioclases and the alkali feldspars, with 0.5–8.0% sodium. These minerals are abundant, particularly in granitic rocks. On weathering, the sodium that is released remains in soluble form, since practically all of the secondary sodium minerals are soluble in water.

Gunn and Richardson (1979) have recently presented data on the soluble chloride and sodium contents of various widespread rock types in eastern Australia. Their analyses were on saturation extracts of ground (< 1 mm) deeply weathered sedimentary and igneous rock samples, and unweathered sedimentary rocks from drill cores. Although saturation extracts should give a reliable estimate of the readily soluble ions, they may underestimate the amount of gypsum present. Additionally, the drill cores could have lost some of their more soluble constituents. However, the authors concluded that surface outcrops of deeply weathered samples had the highest soluble salt contents, probably due to evaporative concentrations of seepage water at exposed surfaces. The results from shallow bores are of the same order as those from deep drill samples and probably give accurate estimates of the salt contents in the upper part of deep-weathered zones. The results indicate that Na^+ and Cl^- are generally the dominant ions, although CA^{2+}, Mg^{2+} and $S^0{}_4{}^{2-}$ are often prominent.

Their data also provide some interesting comparisons between sediments of marine and fresh water origin. The unweathered drill core samples were from the Cretaceous Rolling Downs Group, which is a sequence of mudstones, siltstones and labile sandstones (Exon and Senior 1976) that underlies much of western Queensland, the north-east of South Australia and northern New South Wales. The results show that the contents of both Cl^- and Na^+ are appreciably higher in the unweathered freshwater rocks than in those of marine origin. This was confirmed by a number of X-ray fluorescence analyses for chlorine, and is also supported by available soil analyses for western Queensland (Hubble and Beckmann 1957; Dawson and Ahern 1974; Turner and Ahern 1978; Hubble and Reeve, unpublished data), which show *grey, brown* and *red clays* formed on freshwater Cretaceous sediments to be more often saline than those formed on marine sediments. On the other hand, weathered sedimentary rocks of marine origin contain appreciably more Cl^- and Na^+ than those of freshwater origin.

The data of Gunn and Richardson (1979) and unpublished data of Hutton suggest that, in the eastern half of the continent at least, sufficient Na^+ and Cl^- may occur in widespread parent rocks to provide suitable sources for the salt and sodium found in many soils, given sufficient time for rock weathering and salt accumulation in less permeable materials.

Salt derived from old deep-weathering profiles

As previously noted, salts are frequently present in the old deep-weathering profiles that are widespread in Australia. In some instances the entire profile may be relict, in others the deeply weathered profiles may have been variably stripped and soil formed on the eroded remnants.

There is considerable evidence to show that many deeply weathered profiles contain considerable salt at different depths, e.g. the laterite profiles in south-west Western Australia (Bettenay *et al.* 1964; Dimmock *et al.* 1974), where the ultimate source of most of the salt is thought to be atmospheric (oceanic). Peck and Hurle (1973) and Peck (1978) have reviewed the effect of these deep salt-rich zones on subsequent secondary salinization under various land use practices. By contrast, Gunn (1967) has claimed that, for central Queensland, the salts and sodium present at depth in *solodic* and *solonized-solonetz* soils that formed in mottled and pallid zone materials of truncated laterite profiles were due to weathering of the original primary minerals.

In southern Queensland, moderately saline *grey clays* of the brigalow (*Acacia harpophylla*) lands (Hubble and Isbell 1958; Isbell 1962) are very deep; they usually grade into strongly mottled materials with depth, and thereafter may eventually grade into presumed Cretaceous mudstones. Hubble and Reeve (unpublished data) have shown that in one profile 27 m deep (23 m of 'soil', 4 m of relatively unweathered mudstone), the weighted mean chloride content of the 'soil' was 0.286%; for the 4 m of mudstone it was 0.162%. Corresponding values for sulphate were 0.062% and 0.161%. Thus, although chloride values suggest enrichment for non-parent rock sources, the deficiency of sulphate is difficult to explain.

Local redistribution by drainage water, and the role of saline groundwater

It is commonly observed in many parts of Australia that salt and exchangeable sodium are more likely to occur where soils occupy lower sites in sloping situations. There seems little doubt that such concentrations of salt and sodium in solution occur irrespective of the ultimate origin of the salt. Unfortunately, few definitive studies have been made, but Conacher (1975) has discussed the likely role of throughflow in relation to salinization in Western Australia. It is widely believed that rising water tables, which are a common cause of secondary salinization, are produced by man's influence, e.g. irrigation, but such fluctuations can equally well result from long term climatic changes and short term (seasonal) events.

Saline and sodic soils are widespread in the Riverine Plain of south-eastern Australia (Northcote and Skene 1972), which consists of very gently sloping alluvial fans and the floodplains on to which they merge (Butler *et al.* 1973). These soils are frequently underlain by saline groundwaters which after long periods of irrigation may rise to the surface and produce extensive secondary salinization (Gutteridge *et al.* 1970). Butler (1950), in first postulating that there was a definite relationship between prior stream deposition on the Riverine Plain and the original salinity pattern, noted that the original salinity increased with distance down the prior stream channel. Talsma (1968) and Macumber (1969) have documented regional patterns of salinity in the soil mantle in various parts of the Plain. Pels (1978) concluded that, since the Plain was formed under conditions of an inland basin of occluded drainage, the water which accumulated in lakes and on flood plains towards the western margin was disposed of by evaporation; this explained the high salt status of both soils and groundwater in the region.

The primary origin of the salts is problematical. Pels (1978) thought that cyclic salt was mainly involved, Macumber (1969) was of the opinion that although cyclic salt had an obvious role, by far the largest source of the salt arose from rock weathering in the stream catchments. On the other hand, Lawrence (1975) postulated that much of the saline water of the region is of connate origin, derived from marine Tertiary sediments. More importantly, however, available data and irrigation experience indicate that in much of the Riverine Plain there is a delicate state of balance between hydrologic and climatic conditions. Macumber (1969) has suggested that, in the Kerang district of the Plain, past climatic cycles were coincident with cycles of salinization, hence the salt-affected soils of much of the Riverine Plain may have to be considered in terms of Quaternary events, rather than in terms of existing salt accessions.

Distribution and genesis of saline and sodic soils

Considering first some of the widespread saline and sodic soils of the arid to subhumid regions – the *desert loams* and the *grey, brown* and *red clays* – it is apparent that two sources of salt are available. First, atmospheric accessions (Table 9.1), although low in the interior, may still account for the estimated quantities of chloride in weathered sediments over, geologically speaking, relatively short periods. Thus Gunn and Richardson (1979) have estimated that the periods required for the chloride to accumulate are 0.642×10^6 and 0.725×10^6 years, assuming chloride accession rates of 2 and 1 kg ha^{-1} year^{-1} at locations 200 and 400 km from the ocean. This assumes no losses, but even if 50% drainage or other losses were assumed the resultant doubling of these periods would still give a feasible time estimate. Second, the data of Gunn and Richardson (1979) show that unweathered parent sediments could be adequate sources for salts in soils. However, it is difficult to understand the high quantities of salt present in the deeply weathered zones of the Cretaceous sediments if such weathering were the product of an intensive leaching environment, except by redistribution following cessation of leaching.

Gunn and Richardson have considered other data in central Queensland where there is evidence that unweathered basalt has been exposed at or near the surface for at least a million years. They claim that if atmospheric accessions were significant there, 250–300 km from the coast, about 2000 tonnes ha^{-1} of chloride would be deposited in 10^6 years. However, the relatively shallow (0.5–1 m) *black earths* which formed from basalt on the undulating terrain have very low salt contents. The unknown factor again is the rate of loss of soluble accessions. Although these are relatively impermeable cracking clay soils, the present rainfall environment (mean annual rainfall 650–700 mm), with its occasional more prolonged wet periods, may still be sufficient to leach soluble salts from the profile. This is supported by experimental results for water movement in these soils (Shaw and Yule 1978). Moreover, similar but generally deeper soils formed on deeply weathered basalt in the same area (McDonald 1970; Shaw and Yule 1978) contain appreciable chloride and sodium at depth, as do deeper soils in the area formed from basaltic alluvium and colluvium (McDonald and Baker in press).

A similar situation occurs in the equivalent-rainfall basaltic areas of the Darling Downs, south Queensland (150–180 km from the coast), where the shallow sedentary upland soils contain little or no chloride or sodium and yet appreciable quantities occur in the deeper soils derived from basaltic alluvium and colluvium (Reeve *et al.* 1960). As a result it is simply not possible to say whether the salts in these soils have originated from accessions, parent materials, or local lateral redistribution by drainage waters.

Gunn and Richardson (1979) have stressed the amounts of salt and sodium present in the lower zones of deeply weathered profiles in eastern Australia, and Gunn (1967) has illustrated the occurrence of *solodized-solonetz, solodic* and *grey* and *brown clay* soils where deeply weathered profiles have been eroded and truncated, so that they have acted as parent materials for the present soils. This may explain the salinity in many localities, particularly in western Queensland where available data indicate that *grey, brown* and *red clays* formed from deep-weathered Cretaceous sediments are usually more saline than similar soils formed on unweathered Cretaceous sediments. However, there are numerous occurrences of strongly developed sodic duplex soils in humid to subhumid areas of eastern Australia where there is little or no evidence for the presence of old deeply weathered, parent materials. In many instances the sodium may have been derived from the parent rocks, while in others (e.g. McGarity 1977 p. 65) accumulation through lateral drainage must be of considerable importance.

Two further factors may at least partly explain the sodic nature of many Australian soils. First, most plants and surface soils are much higher in calcium than sodium, and hence if, as postulated by Beadle (1962), the phosphate content of a soil is considerably reduced by the repeated removal by erosion of surface soil and plant debris over long periods of time, then more calcium would also be lost relative to sodium, which would tend to accumulate on the exchange complex in old soils. Second, in highly calcareous soils such as the *solonized brown soils* and the *grey-brown* and *red calcareous soils* with strongly alkaline reaction (pH 8.5–9.0), the concentration of calcium in the soil solution would be very low (Doner and Lynn 1977) compared with sodium and magnesium. This would favour preferential accumulation of sodium and magnesium as exchangeable ions (e.g. see profile A399 of de Mooy 1959).

There is a relative paucity of saline and sodic soils in the more northern parts of the continent. Three factors are probably involved. First, there is evidence which suggests that atmospheric accessions may be lower here than elsewhere. Second, the strong seasonality of the rainfall, with its relatively high incidence of regular, moderate to high totals, may ensure leaching of salts to considerable depths. Third, there are no large areas of parent materials which could provide suitable sources of salt or sodium. Much of the region consists of quartzose sandstones, which give rise to shallow sandy soils, and there are large areas of highly permeable, deeply weathered soils, such as *red earths* and *earthy sands* (see Region III, Chapter 14). It is worth noting that where suitable parent materials exist for the formation of deep cracking clay soils, these are usually sodic and may be saline (Fig. 9.1).

Finally, there is the question of the effect of salt and sodium on the genesis of soil profile morphology, which varies widely in salt-affected soils. These soils have been classified in a number of different ways (e.g. see Szabolcs 1974). The Australian classification (Stephens 1962; Stace *et al.* 1968) largely follows the classical ideas of Gedroiz (quoted in Joffe 1936). These authors have viewed the solonchaks, solonetz, solodized-solonetz and soloths or solods as idealized stages in a sequence of genesis involving salt accumulation (solonchak), followed by progressive leaching of salts, deflocculation of colloids, and downward translocation of clay to form the solonetz, and finally hydrolysis of the sodic clay subsoil, ending with the fully developed soloth. In keeping with similar soils elsewhere in the world, Australian salt-affected soils form a continuous array, including almost every form that could result from all possible combinations of these processes, as well as other genetic processes which are still poorly understood. However, it seems unlikely that many of the present diverse and widespread varieties of sodic soils ever went through the complete classical sequence mentioned above. In particular, much doubt has been cast on the role of clay translocation in the formation of the sodic clay B horizons in soils with a prominent texture contrast. This topic has been partly discussed in Chapter 3, but several points may be reemphasized.

First, the particle-size differentiation between A and B horizons in such soils can seldom, if ever, be explained by the amounts of illuviated clay present in the B horizons (e.g. Stace *et al.* 1968, pp. 162, 176; Oertel 1968). Second, there is evidence (e.g. Oertel and Blackburn 1970) that in at least some profiles the A and B horizons are no more than the end products of the weathering of distinct sedimentary layers. Churchward and Gunn (Chapter 5) have drawn attention to the importance of sedimentary processes in the genesis of some sodic soils with prominent texture-contrast profiles. Third, the A–B horizon differentiation may involve both differential weathering in each horizon and some resynthesis of the weathering products leached from the A horizon to give the clay in the B horizon. A slightly different hypothesis, invoked by Hallsworth and Waring (1964) for the formation of the B horizons of *solodized-solonetz*, involved a simple mechanical in-washing of clay by rainwater, relatively high in sodium compared to calcium, the parent sediments being coarse textured and low in calcium.

By no means do all soils with strong texture- contrast have impermeable saline and sodic B horizons (e.g. McCown *et al.* 1976), and additionally, the *solonized brown soils* and *grey-brown* and *red calcareous soils* so widespread in southern Australia are sodic, lack pronounced textural B horizons, and are permeable. It is thus difficult to resolve the cause/effect relationship in sodic soils. A self-perpetuating mechanism probably operates and low permeability is strongly indicated as an essential feature in most but not all saline and sodic soils. Other questions regarding the genesis and properties of sodic soils that need further investigations are (i) the role of cation composition, i.e. the relative proportions of sodium, magnesium and calcium both on the exchange complex and also in the parent material, and (ii) the nature of the clay minerals, particularly in relation to soil permeability.

Conclusion

It is evident that salts from a variety of sources have probably contributed to the present saline and sodic soils in Australia, but it is seldom possible to state unequivocally the ultimate source of the salts. There are several reasons for this difficulty. Salts, particularly NaCl, are mobile constituents in soils, and so we have to consider not only current fluxes, which are largely determined by climate and soil permeability, but additionally the effect of the very large time-scale involved in the history of many Australian soils. This problem is further complicated by the difficulty in distinguishing between the impact of present and past environments on the salt content of soils. The close relationship between deep-weathering profiles and salt-affected soils in various parts of the continent also raises the question of whether the presence of salts enhances the development of such profiles, or if deep profiles are simply a repository for salts to accumulate.

Although it is true that most of the less-permeable soils of the arid to subhumid regions of Australia are saline and/or sodic to varying degrees, in the case of the sodic soils it must be remembered that the source of sodium may not be NaCl alone. Sodium is relatively abundant in rock-forming minerals and hence sodic soils can form in the absence of NaCl. However, the question of causality arises: are the sodic soils relatively impermeable because of the effects of sodicity, or is sodium accumulation largely a result of low permeability? This question, together with the related one of the role of palaeoclimates, is still largely unresolved. With regard to soil genesis, it is still uncertain if salts and exchangeable sodium can be present without influencing the nature of soil development, or whether they do cause profile development to take a particular course.

References

Anderson, V. G. (1941). The origin of the dissolved inorganic salts in natural waters with special reference to the O'Shannassy River catchment. *J. Proc. Aust. Chem. Inst.* **8**, 130-50.

Anon. (1977). Air outback — clean but not virginal. *Ecos* **11**, 23-6.

Art, H. W., Borman, F. H., Voigt, G. K. and Woodwell, G. M. (1974). Barrier Island ecosystem: the role of meteorologic nutrient inputs. *Science* **184**, 60-2.

Australian Water Resources Council (1975). 'Review of Australia's Water Resources'. (Dep. Nat. Resour.: Canberra.)

Beadle, N. C. W. (1962). An alternative hypothesis to account for the generally low phosphate content of Australian soils. *Aust. J. Agric. Res.* **13**, 434-42.

Bettenay, E. (1962). The salt lake systems and their associated aeolian features in the semi-arid regions of Western Australia. *J. Soil Sci.* **13**, 10-17.

Bettenay, E., Blackmore, A. V. and Hingston, F. J. (1964). Aspects of the hydrological cycle and related salinity in the Belka Valley, Western Australia. *Aust. J. Soil Res.* **2**, 187-210.

Blackburn, G. (1973). Terrestrial salt dust as a probable contributor to River Murray salinity. *Search* **4**, 75-6.

Blackburn, G. (1974). Salinity cycles in Australia, with special reference to the Murray Basin. Trans. 10th Internat. Congr. Soil Sci. Soc. Moscow, Vol. *IV*, pp. 27-33.

Bonython, C. W. (1956). The salt of Lake Eyre — its occurrence in Madigan Gulf and its possible origin. *Trans. R. Soc. S. Aust.* **79**, 66-92.

Bowler, J. M. (1976). Aridity in Australia: Age, origins and expressions in aeolian landforms and sediments. *Earth Sci. Rev.* **12**, 279-310.

Butler, B. E. (1950). A theory of prior streams as a causal factor of soil occurrence in the Riverine Plain of south-eastern Australia. *Aust. J. Agric. Res.* **1**, 231-52.

Butler, B. E., Blackburn, G., Bowler, J. M., Lawrence, C. R., Newell, J. W. and Pels, S. (1973). 'A Geomorphic Map of the Riverine Plain of South-eastern Australia'. (Aust. Nat. Univ. Press: Canberra.)

Conacher, A. J. (1975). Throughflow as a mechanism responsible for excessive soil salinisation in non-irrigated, previously arable lands in the Western Australian wheatbelt: a field study. *Catena* **2**, 31-67.

Dawson, N. H. and Ahern, C. R. (1974). Soils. *In* 'Western Arid Region Land Use Study Part I'. Div. Land Utilization Tech. Bull. No. 12 (Queensl. Dep. Prim. Indust.: Brisbane.)

Dimmock, G. M., Bettenay, E. and Mulcahy, M. J. (1974). Salt content of laterite profiles in the Darling Range, Western Australia. *Aust. J. Soil Res.* **12**, 63-9.

Doner, H. E. and Lynn, W. C. (1977). Carbonate, halide, sulphate, and sulphide minerals. *In* 'Minerals in Soil Environments'. (Eds J. B. Dixon and S. B. Weed). (Soil Sci. Soc. Am.: Madison.)

Douglas, I. (1968). The effect of precipitation chemistry and catchment area lithology on the quality of river water in selected catchments in eastern Australia. *Earth Sci. J.* **2**, 126-44.

Downes, R. G. (1954). Cyclic salt as a factor in soil genesis. *Aust. J. Agric. Res.* **5**, 448-64.

Exon, N. F. and Senior, B. R. (1976). The Cretaceous of the Eromanga and Surat Basins. *BMR J. Aust. Geol. Geophys.* **1**, 35-50.

Gunn, R. H. (1967). A soil catena on denuded laterite profiles in Queensland. *Aust. J. Soil Res.* **5**, 117-32.

Gunn, R. H. and Richardson, D. P. (1979). The nature and possible origins of soluble salts in deeply weathered landscapes of eastern Australia. *Aust. J. Soil Res.* **17**, 197-215.

Gutteridge, Haskins, and Davey (1970). Murray Valley Salinity investigation. Vol. 1, The Report. (River Murray Commission: Canberra.)

Hallsworth, E. G. and Waring, H. D. (1964). Studies in pedogenesis in New South Wales. VIII. An alternative hypothesis for the formation of the solodized-solonetz of the Pilliga district. *J. Soil Sci.* **15**, 158-77.

Hingston, F. J. and Gailitis, V. (1976). The geographic variation of salt precipitation over Western Australia. *Aust. J. Soil Res.* **14**, 319-35.

Hingston, F. J. and Gailitis, V. (1977). Salts in rainfall in Western Australia (1973-74). CSIRO Aust. Div. Land Resour. Management Tech. Mem. 77/1.

Hubble, G. D. and Beckman, G. G. (1957). The soils of some western Queensland properties of Australian Estates Co. CSIRO (Aust.) Div. Soils Div. Rep. No. 6/56.

Hubble, G. D. and Isbell, R. F. (1958). The occurrence of strongly acid clays beneath alkaline soils in Queensland. *Aust. J. Sci.* **20**, 186-7.

Hubble, G. D. and Reeve, R. (1970). Soil salinity and stability in south-western Queensland. Proc. Symp. 'Soils and Earth Structures in Arid Climates, Adelaide, 1970', Sydney. Inst. Engr. Aust. pp. 111-15.

Hutton, J. T. (1957). Rainwater analyses. CSIRO (Aust.) Div. Soils Div. Rep. No. 8/57.

Hutton, J. T. (1962). Rainwater analysis July 1957 to March 1961. CSIRO (Aust.) Div. Soils Div. Rep. No. 7/62.

Hutton, J. T. (1968). The redistribution of the more soluble chemical elements associated with soils as indicated by analysis of rainwater, soils and plants. Trans. 9th Internat. Congr. Soil Sci., Adelaide, Vol. IV, pp. 313-322.

Hutton, J. T. (1976). Chloride in rainfall in relation to distance from the ocean. *Search* **7**, 207-8.

Hutton, J. T. and Leslie, T. I. (1958). Accession of non-nitrogenous ions dissolved in rainwater to soils in Victoria. *Aust. J. Agric. Res.* **9**, 492-507.

Idurm, A. and Senior, B. R. (1978). Palaeomagnetic ages of Late Cretaceous and Tertiary weathered profiles in the Eromanga Basin, Queensland. *Palaeogeogr. Palaeoclimatol. Palaeoecol.* **24**, 263-277.

Isbell, R. F. (1962). Soils and vegetation of the brigalow lands, eastern Australia. CSIRO (Aust.) Div. Soils, Soils and Land Use Ser. No. 43.

Joffe, J. S. (1936). 'Pedology'. (Rutgers University Press: New Brunswick.)

Johns, W. D. and Huang, W. H. (1966). Distribution of chlorine in terrestrial rocks. *Geochim. Cosmochim. Acta* **31**, 35-49.

King, W. D. and Maher, C. T. (1976). The spatial distribution of salt particles at cloud levels in central Queensland. *Tellus* **28**, 11-23.

Lawrence, C. R. (1975). Geology, hydrodynamics and hydrochemistry of the southern Murray Basin. Geol. Surv. Vic. Mem. No. 30.

McCown, R. L., Murtha, G. G. and Smith, G. D. (1976). Assessment of available water storage capacity of soils with restricted subsoil permeability. *Water Resour. Res.* **12**, 1255-59.

McDonald, R. C. (1970). Evaluation of soils for irrigation, Nogoa River, Emerald, Queensland. Agric. Chem. Branch Tech. Rep. No. 1, Queensl. Dep. Prim. Indust., Brisbane.

McDonald, R. C. and Baker, D. (in press). The soils of the left bank, Emerald irrigation area, Queensland. Agric. Chem. Branch Tech. Rep. No. 15, Queensl. Dep. Prim. Indust., Brisbane.

McGarity, J. W. (1977). Soils. *In* 'New England Atlas of Resources'. (Eds D. A. M. Lea, J. J. Pigram and L. Greenwood) Dep. Geog. Univ. New England: Armidale.)

Macumber, P. G. (1969). Inter-relationship between physiography, hydrology, sedimentation, and salinization of the Loddon River plains, Australia. *J. Hydrol.* **7**, 39-57.

de Mooy, C. J. (1959). Soils and potential land use of the area around Lake Alexandrina and Lake Albert, South Australia. CSIRO (Aust.) Div. Soils Land Use Ser. No. 29.

Mulcahy, M. J. and Bettenay, E. (1972). Soil and landcape studies in Western Australia (1) The major drainage divisions. *J. Geol. Soc. Aust.* **18**, 349-57.

Northcote, K. H. and Skene, J. K. M. (1972). Australian soils with saline and sodic properties. CSIRO (Aust.) Soil Publ. No. 27.

Oertel, A. C. (1968). Some observations incompatible with clay illuviation. Trans. 9th Internat. Congr. Soil Sci. Adelaide, Vol. IV, pp. 481-6.

Oertel, A. C. and Blackburn, G. (1970). Pedogenesis of a solodized solonetz, based on duplicate soil profiles. *Aust. J. Soil Res.* **8**, 59-70.

Peck, A. J. (1978). Salinization of non-irrigated soils and associated streams: a review. *Aust. J. Soil Res.* **16**, 157-68.

Peck, A. J. and Hurle, D. H. (1973). Chloride balance of some farmed and forested catchments in south-western Australia. *Water Resour. Res.* **9**, 648-57.

Pels, S. (1978). Disposal of saline drainage waters. Symp. Aust. Soc. Soil Sci. Inc. (Riverina Branch). 'The Hydrogeology of the Riverine Plain of South East Australia'. (Eds R. R. Storrier and I.D. Kelly) pp. 259-68.

Potts, M. J. (1978). The pattern of deposition of air-borne salt of marine origin under a forest canopy. *Plant Soil* **50**, 233-36.

Prescott, J. A. (1931). The soils of Australia in relation to vegetation and climate. CSIR (Aust.) Bull. No. 52.

Probert, M. E. (1976). The composition of rainwater at two sites near Townsville, Qld. *Aust. J. Soil Res.* **14**, 397-402.

Reeve, R., Thompson, C. H. and Beckmann, G. G. (1960). The laboratory examination of soils from the Toowoomba and Kurrawa areas, Darling Downs, Queensland. CSIRO (Aust.) Div. Soils, Div. Rep. No. 1/60.

Shaw, R. J. and Yule, D. F. (1978). The assessment of soils for irrigation, Queensland. Agric. Chem. Branch Tech. Rep. No. 13, Queensl. Dep. Prim. Indust., Brisbane.

Stace, H. C. T., Hubble, G. D., Brewer, R., Northcote, K. H., Sleeman, J. R., Mulcahy, M. J. and Hallsworth, E. G. (1968). 'A Handbook of Australian Soils'. (Rellim: Glenside, S.A.)

Stephens, C. G. (1962). 'A Manual of Australian Soils'. 3rd ed. (CSIRO: Melbourne.)

Szabolcs, I. (1974). 'Salt-affected Soils in Europe'. (Martinus Nijhoff: The Hague.)

Talsma, T. (1968). Environmental studies of the Coleambally Irrigation Area, and surrounding district. Part III. Soil salinity. Water Conserv. Irrig. Commn., N.S.W., Land Use Ser. No. 2: 35.

Teakle, L. J. H. (1937). Salt (sodium chloride) content of rain water. *J. Dep. Agric. W. Aust.* **14**, 115-23.

Turner, E. J. and Ahern, C. R. (1978). Soils. *In* 'Western Arid Region Land Use Study Part IV' Div. Land Utilization Tech. Bull. No. 23, Queensl. Dept. Prim. Indust., Brisbane.

Twomey, S. (1960). On the nature and origin of natural cloud nuclei. *Bull. Obs. Puy Dome,* 1-19.

Van de Graaff, W. J. E., Crowe, R. W. A., Bunting, J. A. and Jackson, M. J. (1977). Relict early Cainozoic drainages in arid Western Australia. *Z. Geomorph. N.F.* **21**, 379-400.

Wetselaar, R. and Hutton, J. T. (1963). The ionic composition of rainwater at Katherine, N. T., and its part in the cycling of plant nutrients. *Aust. J. Agric. Res.* **14**, 319-29.

White, E., Starkey, R. S. and Saunders, M. J. (1971). An assessment of the relative importance of several chemical sources to the waters of a small upland catchment. *J. Appl. Ecol.* **8**, 743-9.

Wilsmore, N. T. M. and Wood, W. E. (1929). Salinity of rain in Western Australia. *Proc. R. Soc. W. Aust.* **15**, XXII-XXX.

Chapter 10
Calcretes in Australia

A. R. Milnes and J. T. Hutton

In Australia in common with many semi-arid countries there are significant terrestrial accumulations of carbonate in soils and unconsolidated surficial sediments or on exhumed bedrock surfaces in the form of carbonate hardpans, nodular or pisolitic layers, mottled carbonate-rich layers, and calcareous fine earths. They are mainly distributed in southern Australia in a broad belt flanking the Great Australian Bight and the gulfs regions of South Australia, although some occur inland in the central and western parts of the continent and along the western coastline (Fig. 10.1). These materials have long been of interest to pedologists and geologists working on Quaternary sediments, but detailed investigations of their nature, relationships and origins are few. One of the problems has been the lack of an adequate terminology to encompass the wide variety of forms and relationships such carbonate accumulations display. Terrestrial carbonate accumulations are not confined to the Quaternary, but have been recognized in sedimentary sequences as old as Cambrian (Daily 1972; Daily *in* Thomson *et al.* 1976), where they are thought to mark unconformities due to subaerial exposure and reworking of calcareous sediments following emergence from a marine environment. However, in common with many terrestrial deposits, their preservation may be rarely achieved.

The Division of Soils undertook a program of investigation of carbonate accumulations in the regolith in southern Australia, principally in relation to the physical properties, chemical and mineralogical composition, nutritional status and origins of the agriculturally important calcareous soils. In this chapter, we have attempted to present a summary of our preliminary results and interpretations in the context of a review of the important Australian literature reporting observations and data on the origin and development of these surficial carbonate accumulations. The literature review is by no means exhaustive, but is intended to illustrate the sequence of development of ideas on the origins of carbonate accumulations in the regolith in southern Australia and the framework from which the current investigation has evolved. Much remains to be done. The most detailed and informative studies on the stratigraphic relationships and petrology of carbonate accumulations have been published by Western Australian geologists, particularly Logan *et al.* (1970) and Read (1974).

Terminology

One of the difficulties in a discussion of terminology in this field is the different meanings applied to the same term by geologists and pedologists. 'Horizon' is a conspicuous example of this ambivalence. We have attempted to use terms with a common meaning in both disciplines, but where this was not practicable the meaning of the term should be clear from its context.

Terrestrial carbonate accumulations in soils and unconsolidated surficial sediments and on exhumed bedrock surfaces in southern Australia display a wide variety of forms, as many workers have observed. Thick, multi-layered profiles overlying unconsolidated sediments or bedrock of varying composition and origin characteristically exhibit a surface layer of pisolitic pebbles, cobbles and sometimes boulders, and an underlying consolidated limestone sheet or hardpan which grades downwards through an unconsolidated nodular zone into a mottled to structureless calcareous fine earth. In many cases a thin soil mantle covers the profiles. There are also consolidated and often laminated carbonate-rich crusts on calcareous bedrock, carbonate-filled fractures and joints in the near-surface zones of non-calcareous bedrock, and nodular or mottled carbonate-rich horizons within soil profiles. The local tendency has been to group the multi-layered calcretes and consolidated carbonate crusts under the terms 'travertine' or 'kunkar', and to refer to the nodular or mottled horizons in soils as 'soil carbonate'. For example, many early geologists including Jack (1912, 1914), Ward (1913), Wade (1915) and Woolnough (1927) applied the term 'travertine' to widespread calcareous surface crusts over large areas of South Australia where strongly calcareous bedrock occurs.

Elsewhere, Lamplugh (1902) had introduced the term 'calcrete' for carbonate-cemented gravels, together with the parallel terms 'silcrete' and 'ferricrete' for materials cemented by silica and iron oxide. Lamplugh (1907) later used 'calcrete' to refer to carbonate-rich surface

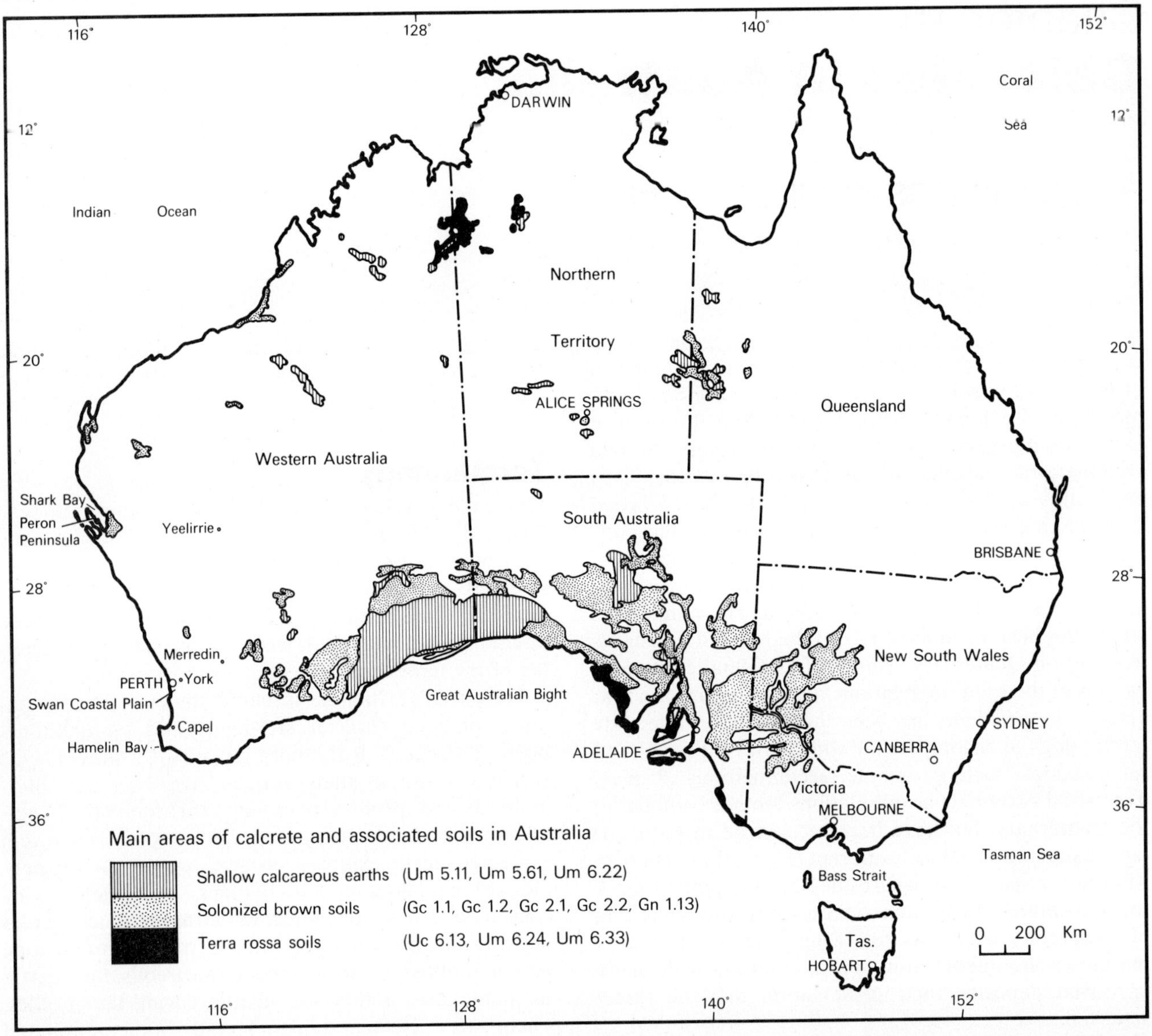

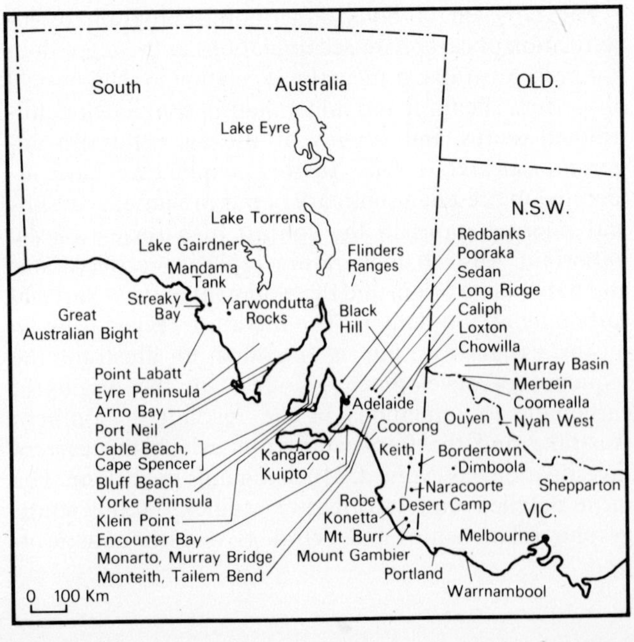

Fig. 10.1. Distribution of calcrete and associated soils in Australia (above) after Northcote *et al.* (1975), and Locality map (right).

materials deposited as a result of 'the evaporation of lime-charged waters' in the valley of the Zambesi River in central Africa and, in fact, coined the term 'valley-calcretes' for them. Heath (1966) considered the problem of nomenclature of carbonate nodules in soil profiles and came to the conclusion that the term 'kankar', apparently derived from a Sanskrit work meaning 'hard', was to be preferred for carbonate deposits formed by pedological processes. Goudie (1973) advocated use of 'calcrete' for carbonate-cemented crusts in preference to the many alternative names because it was consistent with the parallel terms formulated by Lamplugh (1902) and did not suffer the confusion inherent in the use of such names as 'caliche', 'kunkar' and 'travertine', which had been used at various times to describe quite different materials. In an attempt to formulate a statement embracing all impregnated regolith materials, Goudie selected the term 'duricrust', originally introduced by Woolnough (1930), and defined it as 'a product of terrestrial processes within the zone of weathering in which either iron and aluminium sesquioxides (in the case of ferricretes and alcretes) or silica (in the case of silcretes) or calcium carbonate (in the case of calcrete) or other compounds in the case of magnesicrete or the like have dominantly accumulated in and/or replaced a pre-existing soil, rock, or weathered material, to give a substance which may ultimately develop into an indurated mass'.

We have adopted the term 'calcrete' to refer to the range of carbonate accumulations in the regolith in southern Australia, although a great deal of flexibility has to be maintained when describing multi-layered carbonate profiles on the one hand and soil carbonate accumulations on the other to avoid ambiguity and confusion. Consequently, consistent with the approach of Netterberg (1978), we prefer to use 'calcrete' only in its broadest sense for carbonate accumulations, and more specific adjectival terms (for example pisolitic carbonate, vermicular carbonate, laminated carbonate layer, massive concretionary limestone or hardpan, calcareous fine earth) for detailed descriptions of their various forms. Read (1974) has conveniently escaped this potential confusion by using the term 'calcrete', contrary to Lamplugh's definition, to refer specifically to secondary cryptocrystalline carbonate exclusive of the inherited material in the host.

Logan *et al.* (1970) and Read (1974) regarded the terrestrial carbonate accumulations they studied as

Fig. 10.2. Cliff section (approximately 50 m high) at Point Labatt, Eyre Peninsula, showing sequence of Bridgewater Formation beach and back-shore dune sands with interbedded red palaeosols. Upper part of sequence is complex of thin sand-sheets with associated carbonate hardpans. Thick calcrete crust has developed on cliff top surface (right foreground). Unconsolidated sands cover this calcrete in places. Laterite profile in Precambrian granite preserved at base of calcarenite sequence. Shore platforms of granite are etch surfaces stripped of laterite profile.

Fig. 10.3. Breccia-like carbonate lithoclast deposit capped by thin crust of laminated carbonate and overlain by fretted fine-grained bioclastic sand with abundant small lithoclasts, Cable Beach, Yorke Peninsula. Note abundance of black carbonaceous clasts in both horizons. (Hammer 28 cm long.)

Fig. 10.4. Tangled mass of carbonate-replaced plant root channels in bioclastic sand layer within thick calcrete carapace on Bridgewater Formation beach and dune deposits, coastal cliff section, Cable Beach, Yorke Peninsula. Calcified root channels (rhizomorphs) stand out in relief due to cavernous weathering of relatively unconsolidated sand. (Hammer 28 cm long.)

variously consolidated calcareous scree and slope deposits and calcareous soils of Pleistocene to Recent age whose form and relationships were highly dependent on the local landscape and topography at the time of their formation. Although similar notions had been hinted at by many other authors of papers on calcretes, these appear to have been the first attempts to relate the various forms of calcrete to an environment of deposition and consequently a mode of origin. Our work in progress is more or less an extension of this approach, and is concerned with description of the field relationships and morphology of the various carbonate accumulations in southern Australia, together with a collection of geochemical and mineralogical data through selected profiles, in order to establish their origin. Ultimately it may be possible to devise a series of physical and chemical criteria, on which basis calcretes might be described objectively and allocated to various groups. Meanwhile, we consider it wise to maintain a flexible terminology.

Calcretes in southern Australia

Nature, occurrence and suggested origins

There are few published papers dealing exclusively with terrestrial carbonate accumulations in southern Australia. Early geologists rarely commented at length on calcretes because of the regional nature and broad scope of their investigations, but nevertheless often demonstrated their capacity for acute field observations of the characteristics of these materials as geological features. In general, the early geologists considered calcretes to be amongst the most recent of the superficial deposits of Pleistocene to Recent ages. On Kangaroo Island and Eyre Peninsula, calcrete was reported by Jack (1912, 1914) and Ward (1913) as a superficial capping over widespread aeolian sands (Fig. 10.2); in places it included a breccia-like form containing fragments of brown and black limestone (Fig. 10.3), and elsewhere exhibited structures of a root-like nature (Fig. 10.4). Red-coloured soil was commonly observed in hollows in the calcrete capping, and frequently sinkholes were recorded. Jack (1912) suggested that the calcrete was formed by evaporation of bicarbonate-rich waters brought to the surface from the underlying calcareous sands by capillary action.

Wade (1915) examined the calcrete crust in many parts of South Australia and described it as 'more or less massive limestone' varying from compact and dense or cellular material to a coarsely pisolitic and concretionary rock (Fig. 10.5). In places he observed ferruginous pisolites within it. Wade also commented on the vast areal extent of the calcrete, its comparative thinness, and its association with calcareous aeolian sands. He suggested that the calcrete formed as a result of 'solution and redistribution' of calcium carbonate from these sands. Moreover, he made the additional postulation that the

Fig. 10.5. Hard pisolitic and concretionary calcrete, Bluff Beach, Yorke Peninsula. (Hammer 28 cm long.)

Fig. 10.6. Bridgewater Formation exposed in cliff section at Cape Spencer, Yorke Peninsula. Red coloured palaeosols separate many cross-bedded dune deposits, which are only lightly consolidated. Some palaeosols appear to thin on flanks of dunes and thicken in swales. Thin sand sheets (some containing abundant rhizomorphs) with associated carbonate hardpans comprise upper part of section. Thickness of section approximately 80 m.

calcrete was laterally equivalent to the *Coxiella* limestones commonly found in old lagoonal or lacustrine environments in the region.

Woolnough (1927, 1930) examined the calcrete crusts near Adelaide, and regarded them as modern features of the landscape, suggesting that the calcium carbonate had originated from local calcareous rocks. Subsequently however, Segnit (*in* Segnit and Dridan 1938) reported an association between calcretes and former land surfaces identifiable in cliff sections along the coastline of western Eyre Peninsula.

Prescott and Piper (1932) referred to the abundance of calcium carbonate in various forms including fine earth, 'limestone rubble' and 'sheet limestone or travertine' as one of the characteristics of soils in the mallee regions in South Australia. Later workers, beginning with Crocker (1946), looked mainly at terrestrial carbonate accumulations from the pedogenic point of view and discussed the genesis of these materials in relation to soil-forming processes. In many cases, however, a genetic relationship between carbonate accumulations and soils could not be proved.

Crocker (1946) recognized two major types of calcrete (which he called 'travertine' in keeping with the geologists of the time) in particular soil zones in South Australia. One type, found in association with *terra rossa soils*, was cited as sandy limestone which capped Pleistocene calcareous beach and back-shore dune sand and sandrock (aeolianite) sequences of the coastal regions (Fig. 10.2) and ranged in thickness from 'a few inches to many feet'. This calcrete was regarded by Crocker as a fossil illuvial B horizon, formed by downward leaching of calcium carbonate in solution from surface materials and deposition at depth as a carbonate-rich layer. The loss of calcium carbonate in solution from the surface zones was thought to have concentrated the insoluble components (principally quartz) of the aeolianite, thus forming surficial siliceous sands. Exposure of the calcrete

Fig. 10.7. Multi-layered calcrete profile exposed in roadside quarry near Caliph, South Australia. Note shallow soil with abundant calcrete pisolites infilling hollows in thick massive limestone layer. Massive limestone gradational downwards through nodular zone into calcareous fine earth exposed in undercut at base of section. Marks on auger represent 10 cm intervals.

at the surface, for example over large areas of Eyre Peninsula, was reportedly achieved by the deflation of these surficial sand horizons, which may represent the parent materials for the terrestrial siliceous dunes and sand-spreads common inland from the aeolianite sequence. The *terra rossa soils* were considered by Crocker to be a modern veneer on the calcrete, although red-coloured, carbonate-impregnated palaeosols occur within the aeolianite complexes and evidently developed throughout the period of dune-building at times and in places of comparative stability. These palaeosols are clearly displayed in cliff-sections in many places along the coastline of southern Australia (Fig. 10.6).

The second calcrete type described by Crocker (1946) was the 'massive travertine and lime rubble' of the *solonized brown soil* ('mallee' soil) zone, constituting the largest single soil group in the agricultural regions of South Australia. The 'mallee' soils are particularly widespread on Eyre Peninsula, Yorke Peninsula and in the Murray Basin region, where surface and near-surface exposures of calcrete are common. The calcrete comprises a multi-layered profile described by Crocker as 'pisolitic rubble, or as massive travertine limestone'... 'normally immediately underlain by rubble and marly material, becoming less calcareous with depth', and then by a variety of clay layers, siliceous sands, or granitic and other bedrock (Figs 10.7, 10.8 and 10.9). Crocker (1946)

Fig. 10.8. Multi-layered calcrete profile in quarry exposure at Monarto, South Australia. Thin soil mantle with lenticular to platey calcrete boulders overlies massive to concretionary limestone, which grades down into nodular zone with crude horizontal laminations due to platey structure. (Hammer 28 cm long.)

Fig. 10.9. Multi-layered calcrete profile at Monteith, South Australia. Surface pisolitic layer (A) drapes irregular surface (outlined in part) in concretionary hardpan layer capped by thin laminated carbonate zone. Hardpan layer (B) significantly less well-developed than examples shown in Figs. 10.7 and 10.8. Grades down into nodular carbonate zone (C) with unconsolidated calcareous fine earth matrix. Note large boulders of concretionary calcrete near base of profile above calcareous fine earth. (Hammer 28 cm long.)

regarded the calcrete of the *solonized brown soil* zone as an illuvial B horizon. He suggested that the calcium carbonate was initially derived as fine material winnowed from the calcareous beach and back-shore dune complexes of the coastal regions during the Pleistocene, and was carried inland as loess 'under the south-western and southern components of the prevailing wind regime', possibly supplemented by additions of cyclic calcium salts. Leaching of the carbonate so accumulated on top of the existing soil and reprecipitation at depth within the profile under semi-arid conditions was thought to have formed the calcrete horizon. Many subsequent authors have supported Crocker's ideas on the formation of the calcrete. De Mooy (1959), for example, regarded the calcrete associated with the *solonized brown soils* of the south-western Murray Basin region as the calcareous accumulation layer of former soils, 'which were subsequently stripped, leading to consolidation of the calcareous sheet'. Other authors have questioned Crocker's hypothesis on the source of the calcium carbonate. Blackburn and Baker (1953) favoured a local source for the carbonate in the *solonized brown soils* of the southern Flinders Ranges, namely Precambrian limestone bedrock. They also described calcareous conglomerates that developed in creek beds as the result of fluviatile accumulation of calcium carbonate, presumably derived from the same source. These deposits might also be called 'groundwater calcretes' or 'valley-calcretes'.

Northcote (1951, 1979) cited the accumulation of large amounts of carbonate as the most striking morphological feature of the *solonized brown soils* developed in aeolian calcareous sands of the Coomealla region of the Murray Basin, and recognized soft, powdery carbonate, cemented hardpan and concretionary carbonate nodules in

Fig. 10.10. Gc1.22 soil profile (Matong sandy loam) from Coomealla showing accumulation horizon of soft carbonate nodules and patches. Scale intervals approximately 15 cm. (Photograph K. H. Northcote.)

the form of rubble. One or more of these forms of carbonate were described as illuvial B horizons within the soil profiles (Fig. 10.10 and colour Plate 3.5), but are rarely developed to the extent and thickness of the calcretes in the western portion of the Murray Basin, or elsewhere in South Australia. The immediate source of the carbonate in the soils of the Coomealla area was clearly the local aeolian calcareous sands, but Northcote found no evidence for an ultimate source for the carbonate and so proposed accession from an external source, favouring the loessal theory advanced by Crocker. More recently, Northcote (1979) suggested that the carbonate was ultimately derived from calcareous Tertiary and Pleistocene sediments in the region.

Crawford (1965) recognized Crocker's two-fold subdivision of calcrete on Yorke Peninsula in addition to the local development of calcrete crusts on calcareous bedrock, but he recorded difficulty in separately mapping the various types (which he referred to as 'kunkar'). Common evidence of reworking and recementation within calcrete profiles provided a further complication. On the one hand, he observed that aeolianite sequences exposed in coastal cliffs along the southern margin of the peninsula were capped by calcrete containing 'many angular and rounded dark patches rich in organic matter' (Fig. 10.3), which he suggested might have been derived from fossil soils. Along the east coast and in inland areas, Crawford described multi-layered calcrete profiles 'from less than 2 ft to over 20 ft' thick, typically comprising an *upper* nodular or boulder layer overlain by a thin soil, and a *middle* massive concretionary limestone sheet containing fragments of organic-rich calcareous rock, grading downwards into a *lower* zone of nodules in a matrix of calcareous fine earth. These profiles were reported to overlie red clays and older sediments assigned to the 'Ardrossan Clay and Sandrock' formation which Stuart (1970) mapped along the eastern coast of the peninsula and recognized as alluvial fan deposits with associated palaeosols. Crawford suggested that superimposition and coalescence of nodular layers by downward leaching of carbonate from repeated accumulations of Crocker's loess could account for the formation of the middle massive limestone sheet in the calcrete profiles. Abnormally thick limestone sheets observed in landscape depressions are likely to have formed from additional contributions of calcium carbonate from laterally moving groundwaters. However, the limited development of nodules and concretions in the lower calcareous fine earth zone in the profiles could be explained, according to Crawford, by restricted penetration of waters through the massive limestone sheet. Thus the calcareous fine earth zone represents the least-altered part of the calcrete profile, and may provide the greatest information about the nature of the original parent material. On the west coast of the peninsula, Crawford (1965) recorded a massive pisolitic calcrete layer (Fig. 10.5), which he correlated, at least in terms of origin, with the massive middle limestone sheet of the multi-layered calcrete profiles. He considered the karst upper surface of the pisolitic calcrete layer, which in places is overlain by stranded fossiliferous marine sediments, to mark a significant depositional time break. Thin layers of carbonate younger than the massive pisolitic calcrete were reported to occur in overlying longitudinal dunes.

In the south-east of South Australia, there is a series of uniquely preserved parallel ridges of Pleistocene aeolianite with associated calcrete cappings at successively greater distances inland from the coastline (Crocker 1946; Crocker and Cotton 1946; Hossfeld 1950; Sprigg 1952; Blackburn *et al.* 1965; Cook *et al.* 1977). The ridges overlie Tertiary marine limestones at shallow depths and represent strandline beach and back-shore dune deposits of calcareous sand formed during regression of the Pleistocene seas from the region. Marker (1975) described the lower South East as a 'karst province of considerable extent' in which the main karst development is hosted by Tertiary limestones assigned to the Oligo-Miocene Gambier Limestone and the Plio-Pleistocene Whalers Bluff Formation, and only minor karst development occurs on the Pleistocene calcarenite strandline ridges. Nevertheless, field studies by Blackburn *et al.* (1965) indicate that the strandline ridges display a well-developed suite of minor karst forms, including solution pipes and lapiés terrain infilled by *terra rossa* and *podzol soils* (Fig. 10.11 and Fig. 10.12). Calcrete carapaces on the calcarenite sand ridges appear to have developed largely by impregnation of the sands with calcium carbonate, but are by no means ubiquitous. In some places, laminated carbonate layers and blocky to nodular zones occur at the base of soil mantles. The variable development of these calcretes is well illustrated in the coastal cliffs south of Robe, where a thin calcrete crust is present on an exposed calcarenite dune crest but is absent from beneath the palaeosol exhibiting abundant solution pipes and rhizomorph structures in the adjacent swale (Fig. 10.13). Such calcretes are generally very thin by comparison with the multi-layered calcrete profiles and crusts in other parts of southern Australia.

Blackburn *et al.* (1965) regarded the calcretes in the South East as illuvial carbonate horizons and part of the soil profiles. Laboratory data given by Norrish and Rogers (1956) and Blackburn *et al.* indicate that the *terra rossa soils* were derived from calcareous sands similar to those forming the strandline ridges, but that many soil samples had an anomalous silt content. Accordingly, the mechanisms envisaged by Blackburn *et al.* for the formation of the *terra rossa soils* include leaching of the carbonate fraction of the calcareous sands and reprecipitation at depth to form calcrete, concomitant with

Fig. 10.11. Solution pipes filled with *terra rossa soil* in partly consolidated, cross-bedded Bridgewater Formation dune sands, Mount Burr, south-east Australia. Note absence of calcrete layer at interface between soil and calcareous sands. Soil in solution pipes contains scattered ferruginous pisolites.

reworking of the detrital lag and aeolian accession of silt from the adjacent swale areas. There are additional complications, however, because of the presence of ferruginous pisolites as scattered individuals or in gravel lenses within the *terra rossa soils*, and the absence in many places of a calcrete at the interface between the soil and the calcarenite. Near Mt Gambier, both *terra rossa* and *podzol soils* are covered by Recent volcanic ash deposits, for which radiocarbon ages of 4830 years and 1410 ± 90 years have been obtained (Blackburn 1966), so it is clear that these soils do not relate to the modern climatic and weathering regimes. At Naracoorte, on the

Fig. 10.12. Solution pipes (A) filled with *yellow podzolic soil* in partly consolidated cross-bedded Bridgewater Formation dune sands, Desert Camp, south-east South Australia. Light coloured calcrete crust occurs at top of quarry section as carapace over calcareous sands. Calcareous sands have developed grey surface colour since exposure.

Fig. 10.13. Thin white calcrete crust developed on exposed dune crest but absent beneath palaeosol with associated solution pipes and rhizomorphs in adjacent swale. Cliff section in Bridgewater Formation near Robe, South Australia.

other hand, extensive deposits of red sandy silts containing fossil marsupial bones occur as fill in caves in the Gambier Limestone, and radiocarbon data indicate that some of these bones are older than 40 000 years. As the red sediment is likely to be a residual weathering product of the strandline ridges (Marker 1975), it appears that soil formation was in progress at least 40 000 years ago.

Gill (1975) described various forms of carbonate accumulation, including hardpans and spectacular rhizomorph structures, associated with Pleistocene calcareous beach and dune deposits along the coastline of western Victoria. The calcretes form a carapace over stranded calcareous dunes (Dennington Sand) and include massive to crudely stratified limestone, consolidated carbonate breccias representing cemented slope deposits and confined to low-lying parts of the landscape, and laminated mamillary carbonate occurring as thin surface layers on the massive limestone. The surface of the calcretes displays karst weathering features, including lapiés terrain and solution pipes, and in places a relict *terra rossa soil* infills these structures. A bedded tuff, referred by Gill to the Tower Hill Tuff, overlies both the calcretes and the remnants of the *terra rossa soil.* A radiocarbon age of about 7300 years for the Tower Hill Tuff is therefore a minimum age for the formation of the *terra rossa soil*, consistent with similar relationships recorded by Blackburn *et al.* (1965) in the Mt Gambier district of South Australia. As in the case of Blackburn *et al.*, Gill (1975) regarded the calcretes as subsoil deposits of the *terra rossas* that were largely stripped from the landscape prior to deposition of the Tower Hill Tuff.

Stratigraphic relationships

The position of calcretes in Quaternary geological and pedological sequences in southern Australia has been considered by several authors in the context of their regional stratigraphic investigations. However, as mentioned previously, there are few studies related specifically to the field relationships and ages of calcretes.

Boutakoff (1963) described sections which located the Pleistocene aeolianites, for which he coined the name Bridgewater Formation, in the stratigraphic sequence of Tertiary and Quaternary sediments and volcanic rocks in the Portland area of western Victoria. The base of the Bridgewater Formation in some sections in south-western Victoria is represented by a basal Plio-Pleistocene sequence of calcarenites and calcareous silts with shell-beds (Whalers Bluff Formation) deposited during a gradual retreat of the sea. This sequence is reported by Boutakoff to pass upwards into Lower Pleistocene beach and dune calcarenites (with included soil horizons), which were truncated by erosion prior to deposition of four cycles of Upper Pleistocene calcarenite dunes separated by four palaeosols and topped by a partly eroded fifth soil. According to Boutakoff, cementation of each dune of the Bridgewater Formation occurred prior to soil formation and was achieved by downward percolation of calcium carbonate along bedding planes and solution channels.

Thus the Bridgewater Formation in its type area appears to span the whole of the Pleistocene. However, successive strandline ridges of beach and dune calcarenites, such as those particularly well developed in south-eastern South Australia, were apparently deposited during the Middle to Upper Pleistocene (Boutakoff 1963; Cook *et al.* 1977). According to Boutakoff, the siliceous sand lag produced by surface leaching of the calcarenite ridges, and the unconsolidated calcareous sand protected beneath the resulting calcrete carapace (until such time as it was perforated and eroded) were subsequently winnowed from the ridges to form a new sequence of aeolian structures. Many of these in fact developed on the truncated and eroded calcarenite dune ridges. Boutakoff referred these siliceous dunes and sand spreads to the Malanganee Formation, and reported their distribution along a wide belt on the inland margins of the Bridgewater Formation coastal belt from western Victoria to the western coast of Western Australia. These sands were subsequently named Malanganee Sands (Kenley 1976). The fine-grained sediments and alluvium deposited in linear estuaries, lakes, lagoons and swamps impounded between the calcarenite ridges during the Pleistocene marine regression were included by Boutakoff (1963) in his Malanganee Formation. Early estuarine sediments in this sequence include shell-beds containing *Anadara, Katelysia* and *Ostrea* (Crocker and Cotton 1946; Sprigg 1952; Blackburn *et al.* 1965) assigned to the Glanville Formation (Firman 1973). Younger estuarine or lagoonal sediments are mainly fine-grained carbonate-rich muds which contain the gastropod *Coxiella* (Blackburn *et al.* 1965; Cook *et al.* 1977). Boutakoff described the extension of sedimentation within the Malanganee Formation into modern times with the aeolian remobilization of the siliceous sands, and continued deposition in surviving lakes and lagoons between the Bridgewater Formation calcarenite ridges.

Aided by radiocarbon, uranium-thorium and potassium-argon determinations of age, Gill (1967, 1976) established a chrono-stratigraphic sequence of Pleistocene calcareous beach and dune deposits in the Warrnambool Embayment in southern Victoria. Stranded calcareous beach and dune sediments, including the Sunnyside Sand (approximate age 400 000 years from uranium-thorium data) and the Warrnambool Aeolianite were deposited over marine limestone throughout Middle Pleistocene times. Extensive palaeosols, karst features and the fossil remains of terrestrial gastropods were recorded within the aeolianite formations. The Woodbine Basalt, a Middle Pleistocene flow dated at 300 000 years from potassium-argon data, overlies the Warrnambool Aeolianite and older formations in a wide shallow valley within the Warrnambool Embayment, and is itself disconformably overlain by stranded shallow-water fossiliferous marine sediments assigned to the Late Pleistocene Port Fairy Calcarenite and containing *Anadara*. The latter formation, in which shell-beds have been dated at 125 000–110 000 years from uranium-thorium data, grades upwards into calcareous aeolian sands (Dennington Sand), in which the original dune morphology is preserved by an extensive carapace of calcrete hardpans (Gill 1975). Younger Holocene sediments include a sequence of tuff and lapilli horizons approximately 7 300 years old (Tower Hill Tuff), and stranded marine shell-beds (Pertobe Coquina).

Lawrence (1966, 1975, 1976) erected a comprehensive stratigraphic subdivision for the Quaternary deposits of the Victorian portion of the Murray Basin, based on a three-stage classification of the mode of deposition of the sediments, namely aeolian, lacustrine and alluvial. The main calcareous sediments are aeolian calcareous sandy clays and clayey sands in which at least five superimposed soils, characterized by illuviated carbonate accumulation layers, are present (Churchward 1960, 1963). These sediments, which Lawrence (1966) referred to the Woorinen Formation, account for a significant proportion of the aeolian landscape west of the Riverine Plain. Carbonate within the accumulation layers occurs in several forms, including friable material, pebble- to cobble-sized concretions often in a less indurated matrix, or hard indurated layers 'several feet thick'. The Woorinen Formation disconformably overlies either older Pleistocene sediments of the Blanchetown Clay or Bungunnia Limestone, or the Pliocene Parilla Sand. Lawrence (1979) considered that the calcium carbonate component of the Woorinen Formation was derived as cyclic salts or calcite dust from coastal zones, but that the clays had a local origin in either soil development processes or deflation of contiguous Quaternary lacustrine sediments. Associated and probably contemporaneous aeolian deposits in the region are siliceous sand dunes and spreads referred by Lawrence to the Lowan Sand. Crocker (1946) regarded these sands as the deflated siliceous A horizon of Pleistocene soils developed on the Bridgewater Formation aeolianites of the coastal regions, and Boutakoff (1963) and Kenley (1976) equated them with the Malanganee Sands of similar origin. According to Lawrence (1966, 1975, 1976, 1979), however, textural and mineralogical evidence indicate that the Lowan Sand originated by the erosion of the Pliocene Parilla Sand. The deposition of both the Lowan Sand and the Woorinen Formation is regarded by Lawrence to have spanned most of Quaternary time.

Since 1963, J. B. Firman has published many contributions on the regional stratigraphy of Late Cainozoic and Quaternary sequences of sediments and soils (including terrestrial carbonate accumulations) in South Australia. Unfortunately, most of these are short geological notes which present neither detailed descriptions of stated type sections and localities, nor adequate evidence for the interpolated local and regional stratigraphic correlations. In the inland regions of the South Australian portion of the Murray Basin, Firman (1963*a*, 1964) referred to the widespread multi-layered carbonate profile of Crocker's *solonized brown soil* zone as a fossil B horizon formed in an 'older brown soil' named the Bakara Soil, and as such considered it to be a soil stratigraphic unit. Together with reportedly similar calcretes in other areas, the Bakara Soil calcrete was tentatively correlated with the calcium carbonate crust developed on Glanville Formation fossiliferous marine sediments (regarded as Middle Pleistocene) in the Adelaide region by subaerial exposure following a major marine regression (Firman 1963*b*, 1964, 1965*a,b*, 1966*a*, 1967*a*, 1969*a*; Firman and Chugg 1963). Subsequently, Firman (1966*b*) separated the upper massive sheet limestone and pisolitic carbonate horizons from the underlying nodular to mottled carbonate and calcareous fine earth horizons of the Bakara Soil calcrete profile. He claimed that in the Chowilla area of the Murray Basin the unconsolidated horizons could be recognized as loess, and together with unspecified deposits which were reported to unconformably overlie the lacustrine Bungunnia Limestone in that area, predated the formation of the Bakara Soil. An alternative interpretation is given by Lawrence (1966, 1975, 1976), who regarded carbonate accumulations in this stratigraphic position in the Victorian portion of the Murray Basin as part of the Woorinen Formation.

Soft platey carbonate layers within aeolian calcareous silt and sand dunes overlying the Bakara Soil calcrete in the Murray Basin were regarded by Firman (1966*b*) as illuvial horizons developed within an Upper Pleistocene to Holocene soil named the Loveday Soil. Firman referred the calcareous aeolian sediments to the Woorinen Formation of Lawrence (1966). However, Churchward (1960, 1963) and Lawrence (1966, 1975, 1976, 1979) described several aeolian accretionary units within the Woorinen Formation, each (with the exception of the uppermost) displaying differentiated soil profiles and illuvial carbonate horizons. Later, Firman (1969*a*) mentioned 'successive horizons of soft nodular and earthy carbonate in the dunes of the Woorinen Formation' with which he identified the Loveday Soil. In the Adelaide region, layers of soft illuviated carbonate nodules and patches assigned to the Loveday Soil are developed in alluvial sediments near the top of the Pooraka Formation, and are overlain by thin deposits of red clay (Firman 1969*b*) which were subsequently claimed to have a widespread distribution in South Australia, and were named Callabonna Clay (Firman 1973, 1979; Firman and Lindsay *in* Daily *et al.* 1976). The upper part of the Pooraka Formation and the Loveday Soil carbonates at Pooraka (Fig. 10.14) were dated at 34 600 ± 2700 years and 18 900 ± 430 years by Williams (1969) using radiocarbon methods.

Ripon Calcrete was defined by Firman (1967*b*) as an 'off-white and pale reddish brown, thin bedded calcirudite' containing angular to rounded carbonate clasts of aeolianite and 'grey orthochemical' material, but detailed descriptions of the stated type sections were not presented. Strong cementation and the dark-coloured limestone fragments observed by many earlier workers were cited as characteristic of the calcrete (Fig. 10.15), regarded by Firman as a Middle Pleistocene unit occurring within the Bridgewater Formation (separating an upper and lower member) on the western coast of Eyre Peninsula. The process of formation of the Ripon Calcrete was thought to have involved development of soil carbonate horizons from 'a thin but extensive blanket of loess on the landscape', followed by repeated brecciation and cementation of the carbonate horizons with concomitant inmixing of clastic material (Firman 1967*a*). Subsequently, Firman (1969*a*) referred to the Ripon Calcrete as the earliest of the calcrete pans developed during periods of soil formation, and related its development to a major marine regression.

Elsewhere in South Australia, especially in the inland areas beyond the influence of the marine environment, Firman (1967*a*,*b*, 1969*a*) suggested that the Ripon Calcrete and the Bakara Soil calcrete coalesced 'to form a vertically continuous zone of soil carbonate accumulation, in many places with Ripon Calcrete at the base'. Firman (1967*a*, 1969*a*) also noted horizons of calcrete (called 'calcrete pans') within the sequence of calcareous beach and sand dunes of the Bridgewater Formation, and suggested that periods of calcrete formation alternated with aeolian sedimentation in coastal regions at this time.

Fig. 10.14. Cliff section in alluvial sands and gravels of Pooraka Formation, with carbonate stringers and mottles of relict calcareous palaeosol (Loveday Soil) near top, overlain by red Callabonna Clay and younger brown sandy loam. Mottled red and green-grey sandy clays of Hindmarsh Clay at base of section below gravels. Dry Creek, Pooraka, South Australia.

Fig. 10.15. Hard, pink-coloured, pisolitic and concretionary calcrete exposed in quarry at Mandama Tank west of Streaky Bay, type locality of Ripon Calcrete (Firman 1967*b*). Note black limestone clasts in cores of many pisolitic structures. Irregular white zones are due to surface alteration. (Hammer 28 cm long.)

Williams and Polach (1969, 1971) and Williams (1973) described three separate periods of soil formation, characterized by carbonate illuviation and concentration, within alluvial and aeolian deposits in the Lake Torrens piedmont plain in South Australia. Radiocarbon dates for carbonised detrital wood in the Pooraka Formation fan and valley-fill sediments, in which the oldest soil (Wilkatana palaeosol) is developed, indicate that deposition commenced in excess of 38 000 years ago and probably continued until about 30 000 years ago (Williams 1969, 1973; Williams and Polach 1969, 1971). This places the Pooraka Formation in the Late Pleistocene. The nodular calcareous horizon of the Wilkatana palaeosol, the possible equivalent of the Loveday Soil carbonate of Firman (1969*b*, 1979) and Firman and Lindsay (*in* Daily *et al.* 1976) at least in the Adelaide region, formed between 30 000 and 24 000 years ago. Two subsequent periods of soil formation, identified by carbonate horizons in aeolian and alluvial sediments younger than the Pooraka Formation and named the Motpena and Nacoona palaeosols, occurred between 16 000–12 000 years ago in the uppermost Pleistocene and 6 000–1 500 years ago in the Holocene respectively. The Wilkatana and Motpena palaeosols appear to have formed within the time range spanned by the three middle members of the Woorinen Formation (Churchward 1960, 1963; Lawrence 1966, 1975, 1976), according to the radiocarbon data published by Bowler and Polach (1971) for samples from Nyah West in Victoria.

Table 10.1 is a tentative summary of the chronology of the major sedimentary formations and terrestrial carbonate accumulations of the Quaternary in southern Australia. The essential correlations are based on radiocarbon ages for the Pooraka Formation and younger palaeosolic carbonate layers of the Adelaide and Lake Torrens regions of South Australia (Williams 1969, 1973; Williams and Polach 1969, 1971), and for members (again carbonate layers) within the Woorinen Formation in the Victorian portion of the Murray Basin (Bowler and Polach 1971). A combination of uranium-thorium, potassium-argon and radiocarbon ages for the formations in the Warrnambool Embayment (Gill 1967, 1975, 1976) and preliminary uranium-thorium data for the Glanville Formation and St Kilda Formation of the Adelaide region (G. W. O'Brien pers. comm. 1980) locate these formations in the chronologic succession. The age data, together with supporting palaeontological evidence, suggest probable correlations of the Port Fairy Calcarenite with the Glanville Formation, and the Pertobe Coquina with the St Kilda Formation. The positions of all other Formations in the chronologic succession are only approximate, based on a simplification of published stratigraphic sequences and chronologic data.

The location of Firman's Bakara Soil calcrete in the chronologic succession is awkward because of a lack of absolute ages and detailed stratigraphic information in the transitional zone of the Murray Basin between the marine–lacustrine sediments of South Australia and the aeolian sediments and landscapes of western Victoria. Calcretes in a similar stratigraphic position to the Bakara Soil calcrete have been assigned by Lawrence (1975) to the Woorinen Formation. Thus, if the radiocarbon ages and Lawrence's interpretation are correct, the Bakara Soil calcrete of the Murray Basin could be the chronologic equivalent of the oldest members of the Woorinen Formation in the Victorian mallee. There are similar difficulties with the stratigraphic and chronologic position of the Ripon Calcrete, which is considered to separate the upper and lower members of the Bridgewater Formation on the west coast of Eyre Peninsula (Firman 1967*b*) and near Naracoorte in the south-east of South Australia (Firman 1973), and to coalesce with the Bakara Soil calcrete in the multi-layered calcrete profiles of the Murray Basin (Firman 1967*a,b*, 1969*a*). In particular, the division between upper and lower members of the Bridgewater Formation in the South East appears to be marked by the Brunhes-Matuyama magnetic reversal 720 000 years ago (Cook *et al.* 1977), and so this ought to be the approximate age of the Ripon Calcrete. Yet in the Murray Basin, the multi-layered calcretes overlie Bungunnia Limestone and Blanchetown Clay, the upper part of which (according to Bowler 1979) was deposited during the Bruhnes Normal Polarity Epoch and might be as young as 400 000 years. Thus the multi-layered calcretes could be significantly younger than this.

The chrono-stratigraphic information summarized in Table 10.1 is confusing from the viewpoint of a regional assessment of the ages and relationships of calcretes; much critical stratigraphic and chronologic work needs to be done, particularly in the context of environments of formation and modes of origin of the various calcretes. Moreover, the significance of features such as the karst surfaces which may indicate time breaks within calcrete profiles have yet to be discovered. It appears, however, that the time interval established by Gill (1967, 1975) for the formation of calcrete hardpans on Dennington Sand in the Warrnambool Embayment, prior to deposition of the Tower Hill Tuff (Table 10.1), encompasses the main period of calcrete formation in southern Australia.

Calcretes in south-western Australia

Papers on terrestrial carbonate accumulations by Western Australian geologists are amongst the most informative published. Fairbridge (1954), and in other papers, described the association of calcrete with Pleistocene aeolianite in the coastal districts of Western Australia. As in southern Australia, the aeolianite occurs in parallel ridges up to 90 m high, with additional submerged ridges identifiable offshore, and was interpreted as consolidated beach and beach-ridge or littoral dune deposits. The aeolianite was consolidated by carbonate

Table 10.1. Tentative chronology for the major Quaternary sedimentary sequences and calcretes in southern Australia

	10^3 YEARS B.P.	1 LAKE TORRENS	2 ADELAIDE REGION	2 MURRAY BASIN	3 SOUTH EAST SOUTH AUSTRALIA	4 VICTORIAN MURRAY BASIN	5 SOUTH WESTERN VICTORIA	6 WARRNAMBOOL EMBAYMENT
HOLOCENE	1–2	fan dissection local alluviation and dune building	FULHAM SAND; SEMAPHORE SAND	COONAMBIDGAL FORMATION; BUNYIP SANDS; MOLINEAUX SAND				ARMSTRONG SAND; MOINE ALLUVIUM
	2–3	EYRE GRAVEL (member 2); THOMPSON CREEK FORMATION (member 2)	?			LOWAN SAND FORMATION; WOORINEN FORMATION (from c. 2)	MALANGANEE FORMATION (MALANGANEE SANDS) (from c. 2)	
	4–5	**NACOONA palaeosol**	**ST. KILDA FORMATION**		**ST. KILDA FORMATION**			**PERTOBE COQUINA**
	6–10	EYRE GRAVEL (member 1); THOMPSON CREEK FORMATION (member 1)	LIPSON FORMATION ? ; CALLABONNA CLAY	MONOMAN FORMATION ? ; WOORINEN FORMATION	PADTHAWAY FORMATION	? PIANGIL MEMBER	BRIDGEWATER FORMATION	**TOWER HILL TUFF**
PLEISTOCENE	10–20	**MOTPENA palaeosol**; LAKE TORRENS FORMATION	CALLABONNA CLAY; **LOVEDAY SOIL**	WOORINEN FORMATION	PADTHAWAY FORMATION	**KYALITE MEMBER**		terra rossa soils; #
	20–30	**WILKATANA palaeosol**	POORAKA FORMATION			**SPEEWA MEMBER**; **BYMUE MEMBER**		calcretes
	30–100	POORAKA FORMATION ?	POORAKA FORMATION	WOORINEN FORMATION	PADTHAWAY FORMATION	? MIRALIE MEMBER; ?		calcretes
	100–200		BAKARA SOIL CALCRETE; **GLANVILLE FORMATION**; RIPON CALCRETE	BAKARA SOIL CALCRETE; RIPON CALCRETE; BUNGUNNIA LIMESTONE	GLANVILLE FORMATION; BUNGUNNIA LIMESTONE equivalents	BUNGUNNIA LIMESTONE		DENNINGTON SAND; **PORT FAIRY CALCARENITE**
	200–1000		HINDMARSH CLAY	BLANCHETOWN CLAY	BRIDGEWATER FORMATION	BLANCHETOWN CLAY	BRIDGEWATER FORMATION	**WOODBINE BASALT**; WARRNAMBOOL AEOLIANITE; **SUNNYSIDE SAND**; early aeolianite
	1000–2000		**CARISBROOKE SAND**	BLANCHETOWN CLAY (unconformity)	**WHALERS BLUFF FORMATION**	(unconformity)	**WHALERS BLUFF FORMATION**	
PLIOCENE	2000–3000		**HALLETT COVE SANDSTONE, DRY CREEK SANDS** ?	**NORWEST BEND FORMATION**; **PARILLA SAND** ?	**PARILLA SAND** ?	**PARILLA SAND** ?		

* Bruhnes-Matuyama magnetic reversal at 720,000 years

\# rhizomorph ^{14}C age (Gill, 1976)

Dates are based on :

1 Williams (1969,1973); Williams and Polach (1969, 1971)
2 Firman (1969a, 1976); Lindsay (1969); Twidale *et al.* (1978)
3 Cook *et al.* (1977), Rogers (1980)
4 Lawrence (1966, 1975, 1976)
5 Boutakoff (1963); Kenley (1976)
6 Gill (1967, 1976)

cement, and the surface layers were leached leaving a siliceous sand blanket over a hard calcrete horizon capping the unleached material. The calcrete exhibits a karst surface topography, with prominent solution pipes that in some cases extend below sea level. According to Fairbridge (1954), the occurrence of such solution pipes indicates that karst weathering occurred during periods of low sea level. Within a given sequence, Fairbridge described individual dunes separated by soil horizons comprising 'grey-brown rendzinas in the south, red "terra rossas" in the west, and sometimes only travertine crusts farther north'. Fossiliferous marine sediments, in some cases containing *Anadara*, were reported in association with the aeolianite sequences.

Fairbridge and Teichert (1952) studied the palaeosols and marine horizons in the aeolianite sequences in south-western Western Australia and concluded that dune-building coincided with periods of high eustatic sea level, while during each succeeding eustatic low, the base level was lowered and rapid consolidation occurred with the formation of calcrete crusts and a karst topography. This seems to be a far more logical explanation than that expounded by Crocker (1946) and others in southern Australia in which vast areas of continental shelf were exposed during periods of low eustatic sea level to provide a source for the calcareous sands. In fact, Fairbridge and Teichert explained the coastal distribution of the aeolianite sequences by postulating dune-building on or adjacent to a beach of 'normal' width along a shoreline that was not very different in character from the modern shoreline. The palaeosols, identified as *rendzinas*, were reported to be 0.3–0.9 m thick on average (up to 2.4 m thick in hollows), carbonate-rich and with abundant organic matter, of soft earthy character where not cemented by carbonate, and without stratification. Many palaeosols were capped by a thin calcrete crust with abundant rhizomorph structures. Fossils included land snails (*Bothryembrion*) and the remains of insect pupal cases. The absence of thick calcrete crusts and karst features in the palaeosols suggested to Fairbridge and Teichert that periods of soil formation were comparatively short, and that dune-building stages occurred in rapid succession.

On the Swan Coastal Plain, McArthur and Bettenay (1960) described calcrete associated with stranded aeolianite ridges of the Spearwood Dune System, which they correlated with the aeolianite strandline ridges in south-eastern South Australia. Siliceous sand dunes on the inland side of the Spearwood Dunes were referred to the Bassendean Dune System, and were regarded as older calcareous dunes from which all the carbonate had been leached. Calcareous sand dunes of the Quindalup Dune System border the modern coastline to the west of the Spearwood Dunes. The calcrete was thought to have formed by leaching of carbonate from surface horizons and precipitation at depth, and its exposure at the surface was achieved by subsequent stripping of the leached surface horizons. McArthur and Bettenay (1960) reported a progressive increase in the degree of carbonate leaching from surface horizons of the Quindalup Dunes, and concomitant precipitation of carbonate in the lower horizons with incipient cementation, in an easterly direction away from the coastline. In the Spearwood Dunes on the other hand, there is a progressive increase in degree of erosion of the calcrete capping from west to east. These changes were interpreted as the result of the increasing age of the materials in an easterly direction away from the coastline. Based on the assumption that the Spearwood Dunes had an original composition similar to the modern Quindalup Dunes, Hingston (1957) and McArthur and Bettenay (1960) suggested that the significantly higher ratio of calcium to magnesium observed in samples of the Spearwood Dunes was due to selective loss of magnesium (as the result of the greater solubility of hydrated magnesium carbonates) during leaching and reprecipitation of the carbonates to form aeolianite.

Further north in the Shark Bay area, Logan *et al.* (1970) described Pleistocene and Holocene depositional environments that resemble in some respects the depositional environments of similar age in south-eastern South Australia and western Victoria. In particular, they referred to older Pleistocene dunes of red sandstone (Peron Sandstone) analogous in form and distribution pattern to the Bassendean Dune Sands of the Swan Coastal Plain (McArthur and Bettenay 1960), and suggested that they were residual sands after leaching and decalcification of original calcerenite dunes; younger dunes of bioclastic grainstone (Tamala Eolianite) formed subaerially during a marine transgression and exhibiting weathering phenomena such as solution pipes, rhizomorphs, and an eroded upper surface marked by a zone of calcrete nodules; and overlying marine and terrestrial limestones. The top of the Peron Sandstone is marked by a zone of calcrete nodules and pipes 0.3–4.5 m thick regarded by Logan *et al.* (1970) as a relict water table zone. Pelecypod limestones, foraminiferal limestones and lithoclast grainstones comprising the Dampier Formation were deposited in depressions in the calcrete zone at the top of the Peron Sandstone and on wave-planated flats and rock platforms cut in the Tamala Eolianite. The shell beds are commonly capped by a laminated calcrete crust, and are reported to be laterally continuous with the calcrete zone at the top of the Peron Sandstone. Fossils in the Dampier Formation include *Anadara, Pinctada* and *Marginopora*, which have also been reported from the Glanville Formation in South Australia (Howchin 1935; Cann 1978).

Distinctive limestones (Depuch Formation) conformably overlying the Dampier Formation beds were regarded by Logan *et al.* as alternating scree and calcareous soils deposited in a lowland terrestrial environment following emergence from the Dampier marine transgression. The lithology of the Depuch Formation is described as typically 'red-brown indurated, coarse to medium grainstone composed of pebble- to cobble-size fragments of laminar calcrete and calcrete pisolites'. Fossils, in addition to reworked material from the Tamala Eolianite or the Dampier Formation, include intact shells of the land snail *Bothryembrion*. Logan *et al.* (1970) suggested that the induration of the Depuch Formation was caused by the proximity of the water table to the surface of the interdune hollows, combined with splash and spray effects during later Pleistocene and Holocene marine transgressions.

A second Pleistocene marine transgression recognized by Logan *et al.* in the Shark Bay area deposited a 'thin unit of coral biostrome, pelecypod coquina and skeletal grainstone', referred to the Bibra Formation, on wave-bevelled surfaces cut in older Pleistocene units including the Depuch Formation. As yet, insufficient data exist for the southern Australian Quaternary sequences to know whether a similar second marine transgression occurred in addition to, or was encompassed by, the Glanville Formation.

Probably the most detailed and informative paper on calcretes in Australia to date was published by Read (1974), who described the nature and stratigraphic relationships of calcrete deposits in the Shark Bay area of Western Australia on the basis of the earlier work of Logan *et al.* (1970). Contrary to Lamplugh's definition, Read used 'calcrete' to refer to 'secondary cryptocrystalline calcite' exclusive of the detrital material of the host, but the term is not used in this sense here. According to Read, the presence of calcrete (as laminated carbonate, massive limestone or layers of carbonate-coated clasts) in a sedimentary section implies the presence of an unconformity with attendant subaerial exposure and diagenesis.

Read's descriptions refer to the Depuch Formation of Logan *et al.* (1970), a sequence of carbonate soils developed on the Tamala Eolianite and younger marine sediments of the Dampier Formation. The Depuch Formation is variable in thickness, but grades into thin laminated and massive carbonate zones in the upper parts of marine sedimentary horizons in offshore areas. Read subdivided calcrete profiles within the Depuch Formation developed on Tamala Eolianite into five distinct zones; namely, pisolitic soil at the land surface, underlain successively by laminated carbonate, massive limestone, a zone of secondary carbonate mottles, and unaltered parent sediment. Many variations on the development were reported, depending on the degree of weathering and the local topography and erosional history of the sites. It is of significance that this description of calcrete profiles applies equally well throughout southern Australia. The surface pisolitic soil zone was described by Read as a sandy soil composed of carbonate ooids, pisolites and lithoclasts of Tamala Eolianite. Fossils within the soils include insect pupal cases and shells of *Bothryembrion*. The zone is of variable thickness, but is considerably thinned over topographic highs as a result of downslope movement. Reference was made by Read to black lithoclast nuclei of unknown origin; as mentioned earlier, similar black clasts have been reported from calcretes in many parts of Australia. The laminated carbonate zone, underlying the pisolitic soil, is generally a thin irregular sheet-like deposit of finely laminated carbonate with sharply defined upper and lower contacts. The laminated carbonate drapes irregularities in the underlying surface, and tends to fill depressions. The upper surface of this zone has the form of low domes and small sinkholes, the latter being filled with pisolitic soil. Beneath the laminated calcrete, the massive limestone zone is up to 3 m thick. According to Read's descriptions, this material is dense and structureless and is transitional down into the carbonate mottled zone and the underlying sediment.

Read (1974) suggested that the development of the calcrete profiles was based on leaching of carbonate from sediments in the upper part of the profile and its precipitation as cryptocrystalline calcite, either in situ as grain coatings, or further down the profile as intergranular cements, laminated sheets, or grain replacements. Profile development was considered to begin with the formation of a thin soil with carbonate being deposited around sand grains to form ooids. With continuing precipitation, the carbonate coatings coalesced and the material lithified. Continued deposition of carbonate in the underlying sediment resulted ultimately in the filling of all intergranular spaces to form the massive limestone horizon in the zone of most frequent wetting. This horizon subsequently became impermeable to water percolation, and a layer of free water from which the carbonate precipitated as laminar sheets accumulated at its top. Soil creep downslope from topographic highs, according to Read, exposed cemented aeolianite in the dune crests, and subsequent erosion resulted in the shedding of lithoclasts onto dune flanks. The lithoclasts were admixed with ooids and became coated with carbonate to form pisolites. This movement of soil particles down slopes effectively prevented local carbonate accumulation on grain surfaces and was responsible for the concentric coatings around ooids and pisolites.

The stratigraphic relationships between the Depuch Formation soils and other sedimentary formations in Shark Bay were discovered by Read (1974) to be extremely varied and complex. Unconsolidated Depuch Formation soils have complex relationships with the Tamala Eolianite because of downslope movement away from topographically high dune crests, coupled with the development of multiple profiles due to shifts in groundwater zones. In some places, the Depuch Formation is locally underlain by Tamala Eolianite due to migrating dunes burying older soils. Downslope movement was also responsible for transportation of Depuch Formation lithofacies into the coastal zones of Shark Bay where induration occurred. In these areas, the Depuch Formation is interfingered with marine units of the Dampier and Bibra Formations and the Holocene sequence. It is highly probable that similarly complex relationships exist between calcretes and associated sedimentary units in southern Australia, but as yet insufficient critical field investigations have been undertaken.

Valley calcretes

Valley calcretes are crudely lenticular masses of alluvium and soil cemented by calcium and calcium-magnesium carbonates and are confined to the axial portions of broad fossil valleys and existing trunk drainage systems in the arid interior of Australia. Precipitation of carbonate (with associated silica) is believed to have occurred from saturated phreatic groundwaters, following lateral transport of at least calcium ions in vadose and phreatic waters in the regolith from zones of weathering rocks in the drainage divides (Carlisle 1978, 1980; Mann and Horwitz 1979). As an extensive modern literature on valley calcretes has developed following the discovery of carnotite uranium ore in calcrete at Yeelirrie in Western Australia (Carlisle 1978), and because they are assigned to a group of calcretes called non-pedogenic, they will not be discussed further here.

Mineralogy of calcretes

A discussion of the origin of calcretes hinges to a large degree on the mineralogical composition of both the detrital components (including clay minerals) and the carbonates. Unfortunately, such investigations have rarely been carried out in Australia: most interpretations of origin have hitherto been based on observations of field characteristics and relationships.

Carbonate minerals

Some of the earliest mineralogical data on calcretes were obtained by Norrish (reported in Northcote 1951), who identified both dolomite and calcite in the calcareous soils of the Coomealla region of New South Wales. Northcote noted that the zone of maximum dolomite accumulation coincided with the B_{Ca_2} horizons in the soil types he examined, and suggested that this feature was an important factor contributing to the distinction between the B_{Ca_1} and B_{Ca_2} horizons observed in the field.

Stace and Rogers (1954) and Norrish and Rogers (1956) published detailed mineralogical information on calcretes in the context of a study of *terra rossa soils* and *rendzinas* in the Adelaide and South East regions of South Australia. These calcretes, usually capped by a thin consolidated layer of laminated carbonate (Fig. 10.16), were reported by Norrish and Rogers (1956) at the base of practically all the soil profiles they examined, and were regarded as secondary carbonate and part of the soil profile. In all cases, calcite was the only carbonate mineral identified in the calcretes, although both calcite and dolomite were recorded in the calcareous bedrock beneath the soil profiles in the Adelaide region.

Fig. 10.16. Precambrian shales and siltstones blanketed by calcareous fine earth containing abundant rock clasts and small hard carbonate nodules. Surface of fine earth capped by thin layer of hard laminated carbonate. Overlying *terra rossa soil* mantle contains occasional clasts of calcrete near base, filling hollows in underlying carbonate deposit. Roadside cutting Section 79 Hd. Willunga. (Hammer 28 cm long.)

Bowler and Polach (1971) reported radiocarbon analyses of calcretes in Victoria with brief descriptions of their carbonate mineralogy. At Nyah West, superposed carbonate accumulation layers corresponding to the Kyalite, Speewa and Bymue Members of the Woorinen Formation (Churchward 1963; Lawrence 1966, 1975) were discovered to contain progressively more dolomite with age, with up to 40% dolomite in the carbonate fraction of the oldest Bymue Member. Samples of the carbonate accumulation layer in a *red-brown earth* south of Shepparton, on the other hand, have a carbonate fraction composed of almost pure dolomite. By way of contrast, the youngest of the Lake Torrens calcareous palaeosols described by Williams and Polach (1971) and Williams (1973) contains about equal proportions of calcite and dolomite, whereas the oldest palaeosols and the groundwater calcretes contain only calcite.

Lawrence (1975) recorded calcite with only minor dolomite in the carbonate accumulation layers within the Woorinen Formation in Victoria, contrasting with the dolomite-rich Bungunnia Limestone which underlies the Woorinen Formation. In the South Australian Murray Mallee, Wetherby and Oades (1975) identified calcite with traces of dolomite in both the unconsolidated and indurated portions of calcrete profiles representing Firman's Bakara Soil calcrete and Loveday Soil carbonates. They also recorded traces of ankerite. The tendency for dolomite to be concentrated in the lower portions of the profiles was regarded by Wetherby and Oades as a function of carbonate dissolution and subsequent disequilibrium precipitation of calcite in preference to dolomite in the surface zones, with magnesium being leached further and precipitated in the lower zones due to a progressive enrichment relative to calcium in the solution, or lost from the system altogether. Norrish and Pickering (1977) reported a similar disparity in the distribution of calcite and dolomite in calcareous soil profiles, but explained the effect as a 'soil chromatographic process' resulting from dolomite being more soluble in carbonic acid than calcite.

Read (1974) identified the secondary carbonate in calcretes in the Shark Bay area of Western Australia as dominantly small 'interlocking crystals of cryptocrystalline low-magnesium calcite'. However, other secondary minerals, including opaline silica, high-magnesium calcite and cryptocrystalline dolomite, were reported to be common.

Hutton and Dixon (1981) studied the mineralogy and geochemistry of multi-layered calcretes on the western edge of the Murray Basin and on Yorke Peninsula in South Australia. They identified calcite as the dominant carbonate mineral with minor dolomite in the consolidated limestone sheet, dolomite as the major carbonate mineral in the underlying calcareous fine earth, and approximately equal amounts of calcite and dolomite in nodules present in the transitional zone between them. The relative enrichment of dolomite with depth through the profiles was regarded as an effect of dolomitisation consequent upon in situ leaching of an original calcareous deposit of unspecified type.

The results of current mineralogical studies by Milnes on calcretes and calcareous sediments from several localities in South Australia are summarized in Table 10.2. Low-magnesium calcite (less than about 8 mol. %$MgCO_3$) is the dominant carbonate mineral in both the calcretes and the calcareous sediments, but minor dolomite occurs in some samples. Aragonite is present in those samples containing shell fragments.

Clay minerals

In a study of two of the calcareous Coomealla soils (Matong sandy loam and Boeil loam), Norrish (reported in Northcote 1951) found that illite and kaolinite in the ratio 7:3 were the dominant clay minerals and that montmorillonite was absent. Norrish and Rogers (1956) identified a similar clay mineralogy in *rendzinas* and *terra rossa soils* and the acid-insoluble residues of the underlying calcretes they examined in South Australia. Chlorite, montmorillonite and randomly interstratified mica were also reported in the clay fractions of some calcretes, together with ubiquitous quartz and in some cases, geothite. A progressive increase in clay content with depth was recorded in all soil profiles, with the acid residue of the calcretes being highest in clay. Norrish and Rogers suggested that the calcretes were part of the soil profiles, and that the presence of carbonate had played a significant part in effecting the characteristics of the soil types by inhibiting both clay movement within the profile and the weathering of the underlying bedrock.

Wetherby and Oades (1975) identified illite, kaolinite and randomly interstratified clays in the non-indurated portions of multi-layered calcrete profiles in the Murray Mallee of South Australia, but did not comment on the significance of these data.

Our preliminary mineralogical data on calcretes are consistent with the data of Norrish and Rogers (1956) and Wetherby and Oades (1975) in that illite, kaolinite and randomly interstratified clays are the dominant clay minerals. According to Norrish and Pickering (1977), a clay mineral assemblage such as this is characteristic of most soils in southern Australia. Hutton and Dixon (1981) recorded a similar clay mineral assemblage in samples of multi-layered calcrete profiles from Sedan, Long Ridge and Yorke Peninsula, but also identified unusual clay mineral assemblages including abundant sepiolite and palygorskite in samples from similar profiles at Tailem Bend, Monteith, Murray Bridge and Black Hill. Thus, the environment of formation of the calcretes in the latter region appears to have been distinctive in some geological and geochemical way. One possibility is that some of these profiles are genetically related to the lacustrine magnesium-rich Bungunnia Limestone inferred from Firman (1964, 1965*b*) to occur in a narrow zone adjacent to the Murray Valley. Alternatively, as the calcrete profiles in this area directly overlie a suite of basic igneous rocks, granites and metasedimentary rocks, the unusual magnesium-rich clay mineral assemblages could reflect the incorporation of weathered magnesium-rich bedrock. There is clear evidence of such inmixing of basic rock fragments in the calcrete profiles south-east of Tailem Bend and at Black Hill. The solution of this problem, however, requires more critical field and laboratory investigations.

Geochemical investigations

There is a significant amount of geochemical data on calcretes and associated materials in southern Australia. Although most of the data are for single specimens collected from various localities, some have been obtained for samples collected sequentially through calcrete profiles, and these data are potentially more valuable in assessing the origin and movement of carbonate in a given locality. To a large degree, the data are wet-chemical analyses of acid-soluble calcium and magnesium, but silicon, aluminium and iron have been determined in some samples. All of these are regarded as major elements in calcretes: calcium and magnesium characterize the carbonate fraction, whereas silicon, aluminium and iron typify the detrital fraction composed mainly of quartz and other relict primary minerals, together with secondary aluminosilicates and hydrated iron oxides. More recently, X-ray fluorescence techniques have been used for routine analysis of calcretes, and as a consequence, data for major elements plus many additional elements are now available. Although present in minor to trace amounts, some of these additional elements are potentially valuable genetic indicators.

Major elements

In very general terms, and in some ways following the broad subdivision adopted by Crocker (1946), geochemical information for the major elements is most easily presented in relation to the following.

(a) Calcretes in the calcareous Pleistocene beach and back-shore dune sand and aeolianite province (encompassed by the stratigraphic term Bridgewater Formation) in the coastal and near-coastal regions of southern Australia.

Table 10.2. Mineralogical analyses of some South Australian calcretes and calcareous sediments

	Bulk sample			Clay fraction ($<2\mu m$)		
	Carbonate[A](%)	Clay[B](%)			Illite/kaol	Carbonate
				Hindmarsh River mouth, Fleurieu Peninsula		
Glanville Formation sediments						
M195	17.8	45.1	–	illite, kaol, quartz, mixed-layer clays, mont.	1.0	aragonite
M196	68.0	23.1	–	illite, kaol, mixed-layer clays, quartz, mont.	1.3	calcite, dolomite
M197	13.2	24.1	–	illite, kaol, mixed-layer clays, quartz, mont.	1.8	calcite
M200	35.9	20.6	–	illite, kaol, mixed-layer clays, quartz, mont.	1.3	calcite
Indurated carbonate crust on Glanville Formation sediments						
M189a[C]	77.3	38.5	calcite, quartz, dolomite, feldspar	illite, kaol, quartz, mixed-layer clays, mont.	1.7	–
M189b[C]	74.2	19.2	calcite, quartz, feldspar	illite, kaol, mixed-layer clays, quartz, mont.	1.8	–
Carbonate-rich concretions in associated fluviatile sediments						
M205[C]	65.2	9.9	calcite, quartz, feldspar	illite, kaol, quartz, mixed–layer clays, feldspar, mont.	1.3	–
Lagoonal carbonate sediment, intermediate terrace						
M190[C]	50.8	6.1	calcite, quartz, feldspar	illite, kaol, quartz, feldspar	1.1	–
M191[C]	69.8	26.1	calcite, quartz, feldspar	illite, kaol, quartz, mixed-layer clays, feldspar	1.1	–
M202[C]	–	25.1	–	illite, kaol, quartz, feldspar	1.9	–
M203[C]	49.5	24.9	calcite, quartz, feldspar, dolomite	illite, kaol, quartz, mixed-layer clays	1.4	–
Holocene sands with shell fragments, lowest terrace						
M201	21.8	5.1	–	illite, kaol, quartz, mixed-layer clays	1.8	calcite, aragonite
				Encounter Bay		
Carbonate-cemented beach rock						
M171	66.2	–	calcite, quartz, aragonite, feldspar	–	–	–
M173	47.0	–	calcite, quartz, feldspar	–	–	–
M175	57.1	–	calcite, quartz, aragonite, feldspar	–	–	–
M176	44.5	–	quartz, calcite, aragonite, feldspar	–	–	–
M177	61.9	–	calcite, quartz, aragonite	–	–	–
M178	64.9	–	calcite, quartz	–	–	–
				Yarwondutta Rocks, Eyre Peninsula		
Calcretes						
M139[C]	75.8	5.3	calcite, quartz, feldspar	quartz, illite, kaol, mixed-layer clays	1.5	–
M140[C]	59.9	11.2	calcite, quartz, feldspar, dolomite	illite, kaol, quartz, mixed-layer clays	1.3	–
M165[C]	73.8	5.7	calcite, quartz, feldspar	illite, kaol, quartz, mixed-layer clays	1.5	–
M166[C]	83.6	4.9	calcite, quartz, feldspar	illite, kaol, quartz, mixed-layer clays	4.0	–
				Bremer River		
Calcrete						
M187[C]	68.6	6.8	calcite, quartz, dolomite, feldspar	illite, kaol, quartz, mixed-layer clays	3.4	–

[A] The percentage of carbonate is a direct conversion from total CaO and MgO concentrations determined by X-ray fluorescence analyses.
[B] Except for Glanville Formation sediments, clay concentrations are expressed as a percentage of the acid-insoluble residue.
[C] $<2\ \mu m$ clay fraction separated from acid-insoluble residue of original samples.

(b) Calcretes in the essentially non-calcareous continental province, which includes the Pleistocene to recent terrestrial deposits and soils of the inland regions of southern Australia.

These divisions appear to correspond to the extremes of geological environment in which calcretes are found. As yet we have very little information on the geological and pedological relationships at the boundary between the two provinces, and practically no understanding of the nature or influence of the critical processes

Table 10.3. Chemical analyses of Pleistocene and modern dune aeolianites and sands, calcretes and lacustrine carbonate sediments

	Acid-soluble			
	%CaO	%MgO	%(SiO_2+Al_2O_3+Fe_2O_3)	Ca/Mg
		Southern Australia		
Johns (1959)				
Eyre Peninsula				
Port Le Hunte dune sand	48.86	3.16	4.02	18.3
Blue Lake drift sand	48.11	3.19	5.44	17.9
Wanilla dune sand	53.34	1.05	1.68	60.2
Lake Wangary drift sand	51.27	2.72	1.22	22.3
Uley-Wanilla dune sand	52.11	1.91	1.06	32.3
Flinders dune sand	47.07	1.72	9.01	32.4
Yorke Peninsula				
Hardwicke Bay drift sand				
a.	39.95	1.63	24.04	29.0
b.	45.55	2.34	12.88	23.1
Stenhouse Bay aeolianite	48.58	3.06	4.02	18.8
South-east South Australia				
Woakwine dune aeolianite				
a.	43.37	1.91	17.27	26.9
b.	46.84	2.49	8.43	22.3
c.	45.10	1.67	14.57	32.0
d.	45.57	2.15	6.38	26.2
Woakwine dune drift sand	21.46	0.41	60.91	62.0
Robe dune aeolianite				
a.	45.66	1.63	13.32	33.2
b.	41.35	1.58	19.89	31.0
c.	41.74	1.58	16.35	31.3
d.	34.79	0.81	32.55	50.9
Rivoli Bay drift sand				
a.	47.40	1.96	9.72	28.7
b.	44.99	2.10	12.88	25.4
Johns (1963)				
Calcretes				
Prospect (Adelaide)	43.12	2.33	16.25	21.9
Tailem Bend				
a.	44.61	1.57	13.50	33.7
b.	39.32	1.48	22.36	31.5
c.	45.17	1.52	13.00	35.2
d.	48.30	0.90	8.46	63.6
Arno bay, Eyre Peninsula	29.80	21.71	0.88	1.63
Port Neill, Eyre Peninsula	29.63	21.67	1.49	1.62
Stansbury, Yorke Peninsula				
a.	49.83	0.90	7.13	65.6
b.	52.84	1.89	2.34	33.1
c.	47.23	1.92	11.92	29.2
Hd. Menzies, Kangaroo Is.				
a.	46.33	1.53	9.40	35.9
b.	46.54	2.67	7.32	20.7
c.	47.33	1.30	8.66	43.1
d.	45.78	1.86	11.74	29.2
e.	46.34	0.91	12.91	60.3
f.	50.43	0.48	4.62	—
Vivonne Bay, Kangaroo Is.	46.24	2.22	2.02	24.7
Stenhouse Bay, Yorke Peninsula aeolianite	47.63	1.15	10.88	49.0

responsible for the formation of calcretes in each province. Moreover, there is the complicating factor that calcareous materials in the form of bedrock limestones of various ages, and some Pleistocene aeolian deposits, have directly contributed to the formation of calcretes in the continental province. Nevertheless, this subdivision does at least give some potential for assessing the effect of parent materials and processes on the formation of calcretes.

Table 10.3. *cont.* Chemical analyses of Pleistocene and modern dune aeolianites and sands, calcretes and lacustrine carbonate sediments

		Acid-soluble				
		%CaO	%MgO	%(SiO + Al_2O_3 + Fe_2O_3)	Ca/Mg	
Modern bioclastic sands, east coast St Vincent Gulf						
Hd. Port Gawler						
	1a.	45.41	1.18	13.72	45.6	
	1e.	39.40	0.90	22.04	51.9	
	2.	50.11	2.01	2.9	29.5	
	4.	49.92	2.10	3.4	28.2	
	5.	44.82	1.77	13.1	30.0	
Hd. Dublin						
	13.	48.07	1.00	9.1	57.0	
Hd. Inkerman						
	14.	44.77	1.63	13.1	32.5	
	18.	48.86	2.39	5.4	24.2	
Lacustrine sediments						
Hd. Carribie, Yorke Peninsula						
	a.	24.37	21.70	6.82	1.33	
	b.	25.90	21.39	4.44	1.43	
Wanna, Eyre Peninsula		46.86	3.82	1.18^{A}	14.5	
Lake Pillie, Eyre Peninsula		36.57	12.47	3.92^{A}	3.47	
Coorong, South East		26.63	10.65	8.24^{A}	2.96	
Lake Hawdon north, South East		34.79	11.22	6.86^{A}	3.67	
Kingston Lake, South East		28.63	9.42	19.42^{A}	3.60	
Lake Eyre south shore						
	a.	27.42	17.40	2.50^{A}	1.87	
	b.	30.02	18.30	4.68^{A}	1.94	
Nixon (1959)						
Modern dune sands						
Hd. Moorowie, Yorke Peninsula						
	a.	39.95	1.63	24.04	29.0	
	b.	45.55	2.34	12.88	23.1	
			South-western Australia			
Fairbridge and Teichert (1952)						(Organic carbon)
Hamelin Bay palaeosols and aeolianite						
palaeosol A		49.25	2.75	2.95^{B}	21.2	1.45
palaeosol B		44.25	2.07	4.99^{B}	25.3	2.08
palaeosol C		51.31	2.33	0.49^{B}	26.1	1.73
palaeosol D		49.59	2.28	0.51^{B}	25.8	2.29
aeolianite		50.39	3.67	0.46^{B}	16.3	
Hingston (1957)						
Modern beach dunes, Capel						
P183.3		13.18	1.01	74.36^{C}	15.5	
P183.4		14.58	1.21	71.42^{C}	14.3	
P184.1		15.14	1.21	70.43^{C}	14.8	
P184.3		14.30	0.81	72.77^{C}	20.9	
P184.6		16.54	1.01	68.55^{C}	19.4	
Point Peron beach dune		46.55	3.02	10.51^{C}	18.3	
Spearwood aeolianite		46.27	0.60	16.07^{C}	91.4	
Stirling shelly layer, Capel		25.52	1.41	51.46^{C}	21.4	

^{A} SiO_2 only measured. ^{B} Acid-insoluble residue, assumed to be (SiO_2 + Al_2O_3 + Fe_2O_3). ^{C} 100%-($CaCO_3$% + $MgCO_3$%) assumed to be %(SiO_2 + Al_2O_3 + Fe_2O_3).

Bridgewater Formation province

Within the Bridgewater Formation province in the coastal areas of southern Australia, the composition of many calcretes can be compared with the compositions of calcareous Pleistocene sands, aeolianites and interbedded palaeosols, Pleistocene to Recent calcareous sediments of various types, and modern bioclastic sands. Published data on such materials from localities in South Australia (Johns 1959, 1963; Nixon 1959) and Western Australia (Fairbridge and Teichert 1952; Hingston 1957) are given in Table 10.3 and have been plotted in Fig. 10.17 in terms of the normalized end-member components CaO, MgO and $SiO_2 + Al_2O_3 + Fe_2O_3$, which may be directly correlated with the major mineral phases present in these materials, namely calcite, dolomite, magnesite and the minerals of the detrital fraction including quartz and aluminosilicates. In total, the samples exhibit a range of composition along the CaO-$(SiO_2 + Al_2O_3 + Fe_2O_3)$ side of the diagram in the calcite-dolomite-quartz/aluminosilicate field, but most contain more than 60% of end-member CaO. The Ca/Mg ratios of all samples are greater than 13.6, indicating that calcite is the dominant carbonate mineral, with dolomite (Ca/Mg = 1.4) forming less than 20% of the carbonate.

The calcretes, aeolianites and sands analysed by Johns (1959, 1963) and Nixon (1959) represent single specimens collected from many localities in South Australia (Table 10.3), and, as shown in Fig. 10.17, most contain more than 60% end-member CaO. With the exception of the Arno Bay and Port Neill samples, the calcretes on average contain about the same MgO concentration relative to CaO as the modern bioclastic sands, but less than the aeolianites.

Fairbridge and Teichert (1952) analysed four superposed palaeosol horizons and one aeolianite sample from a section at Hamelin Bay in Western Australia. The palaeosols contain similar carbonate concentrations to the aeolianite, but have higher Ca/Mg ratios (Table 10.3; Fig. 10.17). They also contain significant concentrations of organic carbon (Table 10.3).

In the case of the Western Australian samples studied by Hingston (1957), the modern calcareous dunes (Quindalup Dune System of McArthur and Bettenay 1960) from Capel and Point Peron are quite different in absolute composition, although their Ca/Mg ratios (17.2 ± 2.7) are similar. Moreover, the one sample of aeolianite from the Spearwood Dune System analysed by Hingston is significantly different in absolute composition and Ca/Mg ratio (91.4) both from the modern dunes and from the Hamelin Bay aeolianite sample (Ca/Mg = 16.3) and the suite of palaeosols (Ca/Mg = 24.6 ± 2.3) analysed by Fairbridge and Teichert (1952). Based on this variability in composition, further studies will have to be undertaken before the nature of the parent material for the aeolianites, as sought by Hingston

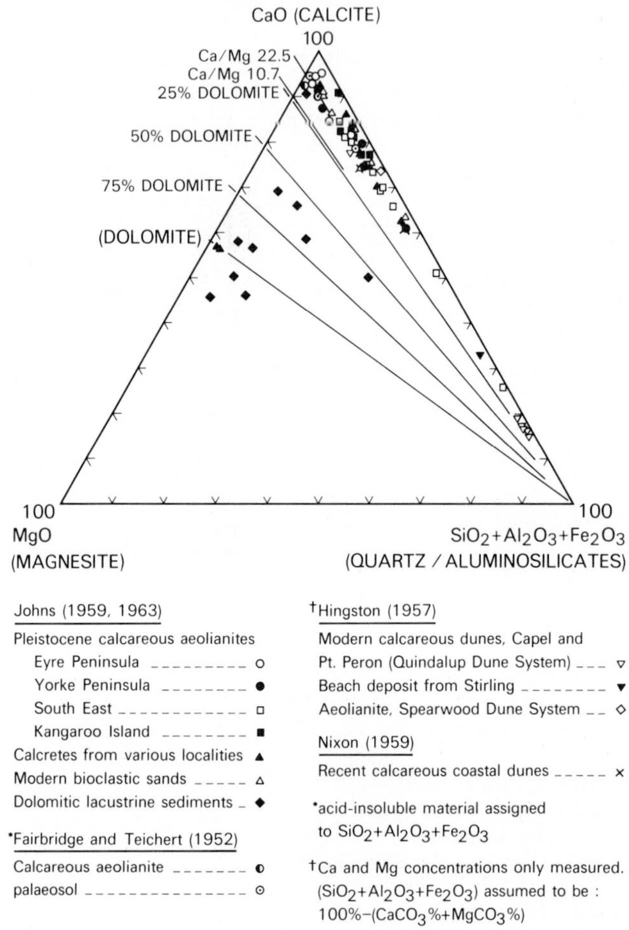

Fig. 10.17. Composition diagram for calcretes and calcareous sediments from data published by Johns (1959, 1963), Fairbridge and Teichert (1952), Hingston (1957) and Nixon (1959).

(1957) and McArthur and Bettenay (1960), can be adequately defined.

Johns (1963) published analyses of lacustrine carbonate sediments from many localities in South Australia (Table 10.3), and these have also been plotted in Fig. 10.17. In contrast to the aeolianites, sands and associated calcretes, the lacustrine sediments are all magnesium rich, and their compositional range spans the boundary between the calcite-dolomite-quartz/aluminosilicate and the dolomite-magnesite-quartz/aluminosilicate fields. Dolomite accounts for more than 50% of the carbonate in all samples, but in three samples magnesite would be expected to replace calcite as a stable phase (Fig. 10.17). Calcrete samples from Arno Bay and Port Neill (Johns 1963) also have compositions close to the dolomite-quartz/aluminosilicate tie-line (Fig. 10.17), but their field associations were not reported. It seems likely, however, that they may be recrystallised lacustrine dolomite sediments.

A. R. Milnes has analysed by X-ray fluorescence a

Table 10.4. Chemical analysis of carbonate-rich materials from southern Fleurieu Peninsula, South Australia

	%CaO	Total %MgO	%($SiO_2+Al_2O_3+Fe_2O_3$)	Ca/Mg
Hindmarsh River (intermediate terrace) stranded lagoonal carbonate sediment				
(Selected samples from various localities)				
M223D	24.0	1.98	45.84	14.4
M224D	9.76	0.71	74.45	16.3
M225A	11.01	0.65	73.17	20.1
M225B	7.70	0.62	78.87	14.7
M225C	15.19	1.03	65.31	17.5
M190	26.7	1.50	45.91	21.1
M203	25.6	1.84	44.75	16.5
M207	45.1	1.73	12.76	30.9
M214	20.3	1.12	60.11	21.5
(Profile through deposit: 10 cm sampling intervals)				
M238 = top	36.6	2.59	22.34	16.7
M239	36.9	3.24	22.0	13.5
M240	39.9	3.99	16.98	11.9
M241	24.6	3.69	42.75	7.9
M242	34.3	2.91	27.03	14.0
M243	35.4	2.50	25.84	16.8
M244	32.0	2.60	30.61	14.6
M245	21.1	2.53	50.43	9.9
M246	20.0	2.66	51.00	8.9
M247	7.77	2.40	71.71	3.8
M248	7.06	1.90	78.06	4.4
M249	6.95	2.02	75.59	4.1
(Carbonate-rich concretions in associated sediments)				
M205	34.8	1.46	29.99	28.2
M210	25.7	2.56	42.72	11.9
M216	28.5	1.45	41.89	23.3
Encounter Bay carbonate-cemented beach-rock				
M171	36.3	0.68	29.54	63.3
M173	25.5	0.73	46.82	40.8
M175	30.8	1.03	38.32	35.4
M176	24.2	0.60	51.29	47.8
M177	32.9	1.52	29.88	25.7
M178	34.9	1.25	29.29	33.1

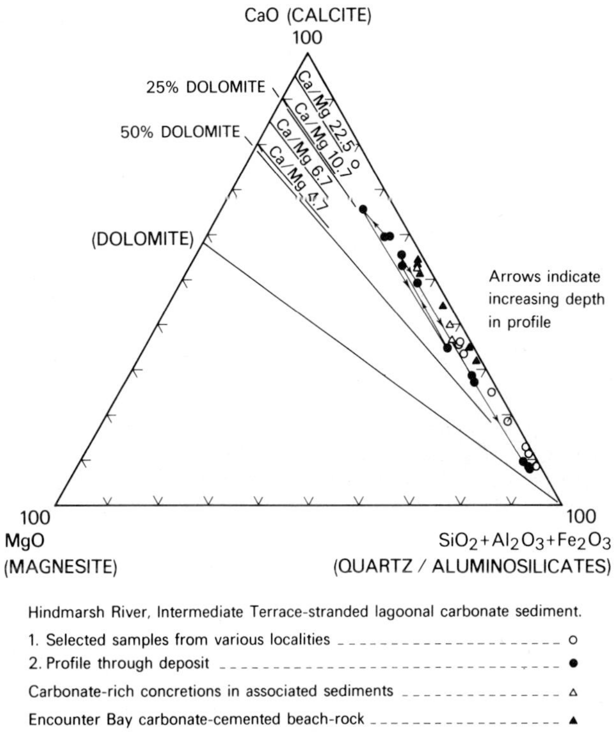

Fig. 10.18. Composition diagram for carbonate-rich materials from southern Fleurieu Peninsula, South Australia.

variety of calcareous materials of Pleistocene age from the Encounter Bay area of South Australia, including stranded and recrystallized lagoonal carbonate sediments, carbonate-rich concretions in estuarine and fluvial sediments, and beach-rock (Table 10.4). As shown in Fig. 10.18, these materials exhibit a range of compositions along the CaO-($SiO_2 + Al_2O_3 + Fe_2O_3$) join but compared with the aeolianites, sands and calcretes in Fig. 10.17, tend to be enriched in $SiO_2 + Al_2O_3 + Fe_2O_3$. Ca/Mg ratios for the samples are generally greater than about 10.7, indicating less than 25% dolomite in the carbonate fraction. In contrast to the calcareous concretions and beach-rock, the stranded lagoonal carbonate sediment has a wide range of composition between 8% and 76% end-member CaO (Fig. 10.18). Ca/Mg ratios are low near the base of the deposit where the carbonate concentrations are also low, and indicate up to 50% dolomite in the carbonate fraction. Trends in Ca/Mg ratios and absolute concentrations of CaO and MgO indicate discontinuities in the profile through the deposit (Table 10.4), and these are thought to record changes in carbonate chemistry during sedimentation.

In their study of *terra rossas* and *rendzinas* in the South East and Adelaide regions of South Australia, Stace and Rogers (1954) reported analyses of acid-soluble Ca and Mg in carbonate accumulation zones that occur beneath the soil horizons. These data (Table 10.5, Fig. 10.19) provide a comparison between calcretes in the Bridgewater Formation and continental provinces, where they overlie stranded Pleistocene calcareous beach dunes on the one hand, and calcareous bedrock ranging in age from Precambrian to Tertiary on the other. In total, the samples exhibit a composition range spanning practically the whole of the CaO-($SiO_2 + Al_2O_3 + Fe_2O_3$) join. However, the South East materials (including the shell sands) contain more than 50% end-member CaO, and this composition range agrees very well with the Johns (1959, 1963), Nixon (1959) and Fairbridge and Teichert (1952) calcretes, aeolianites and sands plotted in Fig. 10.17. On the other hand, samples of calcretes from the Adelaide region span the whole of the composition range. The thin, hard and usually laminated carbonate layers at the top of the carbonate zones generally contain more than 50% end-member CaO whereas the

Table 10.5. Chemical analyses of carbonate-rich materials beneath *terra rossa* and *rendzina* soils in the South East and near Adelaide, South Australia (after Stace and Rogers). Sample descriptions quoted from Stace and Rogers (1954)

			Acid-soluble		Acid-insoluble	
			%CaO	%MgO	residue[A]	Ca/Mg
		South East				
	Profile 11					
79 cm	16456	hard calcareous material	42.64	0.62	14.0	88.2
81-89 cm	16457	calcareous shell sand	39.48	0.76	27.5	60.9
91 cm	16459	calcareous shell sand	36.98	0.98	31.4	44.2
	Profile 18					
below 46 cm	16495	soft calcareous material	51.24	0.48	8.0	125
	Profile 14					
28-36 cm	16473	soft calcareous material	51.35	0.76	5.9	79.2
76 cm	16474	soft calcareous material	46.46	0.77	15.3	70.7
	Profile 16					
20-76 cm	16481	unconsolidated calcareous material	54.25	0.47	2.4	135
below 76 cm	16482	pink-brown mottled unconsolidated calcareous material	40.38	0.35	26.9	135
		Adelaide region				
	Profile 2					
31 cm	16057	pink-grey hard calcareous material	50.48	0.62	7.5	95.4
below 31 cm	16058	pink soft calcareous material	36.52	2.72	28.8	15.8
	16059		33.00	4.47	31.2	8.7
	Profile 3					
below 38 cm	16063	pink-yellow mottled soft calcareous material	34.93	1.97	31.2	20.8
	16064		27.80	1.32	46.7	24.7
	Profile 5					
25 cm	16071	grey hard calcareous material	43.68	0.92	19.8	55.7
below 25 cm	16072	yellow-pink-grey calcareous rock	25.36	0.80	52.3	37.2
	Profile 7					
below 51 cm	16120	pink broken stone with calcareous material	32.85	0.64	38.8	60.2
	16121	pink calcareous material	19.47	0.78	63.0	29.3
	Profile 10					
below 43 cm	16136	mottled soft calcareous material	20.14	6.24	50.7	3.8
	Profile 1					
below 28 cm	16054	pink-yellow soft calcareous material	42.91	0.84	20.4	59.9
	Profile 4					
27 cm	16067	grey-white hard calcareous material	50.51	0.90	6.3	65.8
below 27 cm	16068	pink-yellow mottled soft calcareous material	9.80	1.04	79.3	11.1
	Profile 8					
23 cm	16124	grey-white hard calcareous material	48.18	0.59	11.9	95.7
below 23 cm	16125	pink-yellow soft calcareous material	16.99	0.90	66.9	22.1
	Profile 9					
25-36 cm	16129	mottled white-brown soft calcarous material	28.41	0.76	46.8	43.8
below 36 cm	16130	mottled white-yellow soft calcareous material	19.70	0.60	61.1	38.5

[A]Assumed to be %($SiO_2 + Al_2O_3 + Fe_2O_3$).

underlying materials are relatively depleted in CaO (Fig. 10.19).

Ca/Mg ratios for most of the calcrete samples are greater than 22.5 indicating less than about 10% dolomite in the carbonate fraction, although in two samples from the Adelaide region, dolomite comprises more than 25% of the carbonate. Within all but two profiles in both the South East and Adelaide regions, there is a significant decrease in Ca/Mg ratio with depth (Table 10.5) signifying either a selective loss of magnesium from, or a selective enrichment of calcium in, the upper parts of the profiles during the formation of the carbonate-rich zones.

Unfortunately, Stace and Rogers (1954) did not publish sufficient geochemical data on the overlying soils or the underlying bedrock for us to define the relationship, if any, between these materials and the calcretes. Norrish and Rogers (1956) considered the carbonate to

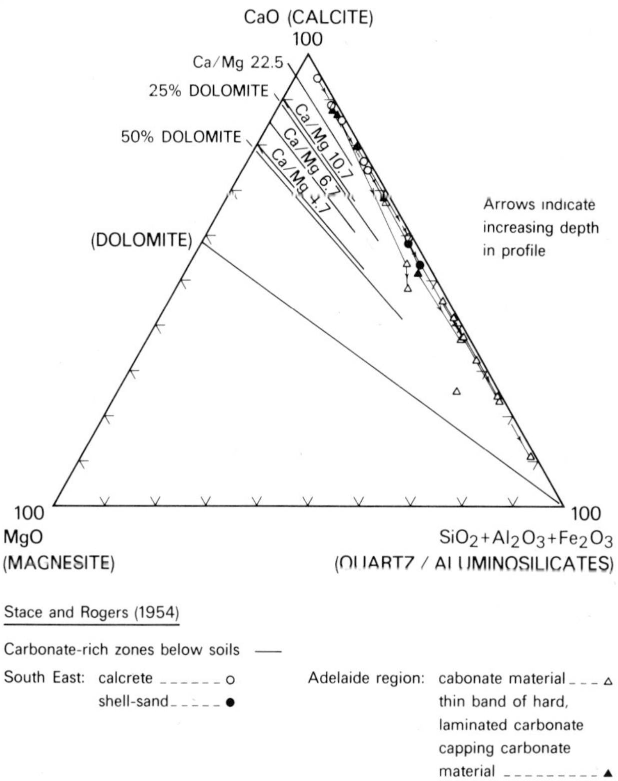

Fig. 10.19. Composition diagram for carbonate-rich materials beneath *terra rossa* and *rendzina soils* in the South East and near Adelaide, South Australia.

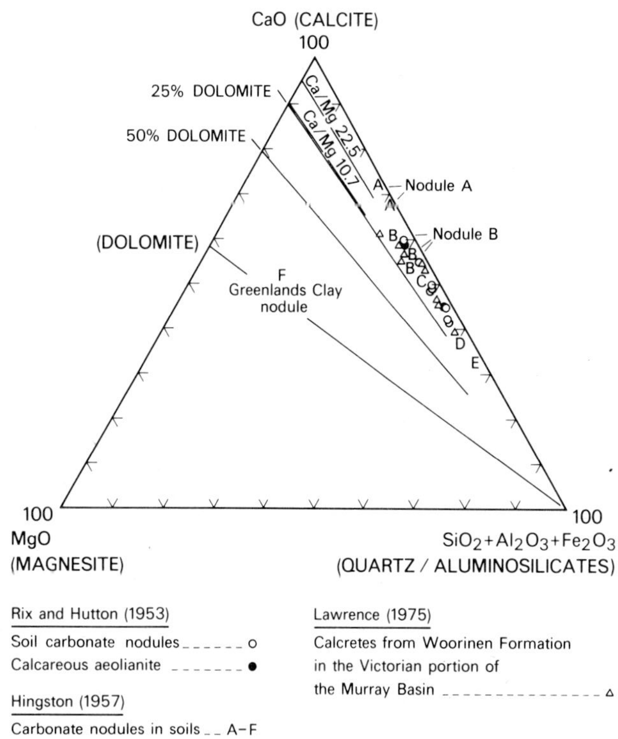

Fig. 10.20. Composition diagram for soil carbonate nodules and calcretes from data published by Rix and Hutton (1953), Hingston (1957) and Lawrence (1975).

have accumulated during soil-forming processes. However, Stace (1956) noted that in some localities in the Adelaide region, the soils directly overlie calcareous bedrock without the intervention of a carbonate-rich zone. Therefore it is possible that the calcretes are not related in a simple way to the overlying soils.

Continental province

Calcretes from the continental province of southern Australia include a much greater variety of materials and have a wider variety of associations than in the Bridgewater Formation province, due in part to the variable nature of the bedrock materials. Table 10.6 and Fig. 10.20 record the compositions of soil carbonate nodules from the Kuitpo area of South Australia (Rix and Hutton 1953) and from many areas of Western Australia (Hingston 1957), and the compositions of carbonate accumulation layers within the aeolian Woorinen Formation in the Victorian portion of the Murray Basin (Lawrence 1975). Rix and Hutton (1953) found that their samples of 'nodular lime' were similar in composition to calcareous dune sands collected from Kangaroo Island and Robe, and as shown in Fig. 10.20, occupy the same composition range (between 39% and 61% end-member CaO) as the calcrete samples collected by Lawrence. Thus both groups of samples are enriched in acid-insoluble detrital material compared with most of the calcretes, aeolianites and sands plotted in Fig. 10.17. Ca/Mg ratios for all samples are greater than 10.7 indicating that dolomite comprises less than 25% of the carbonate, as in the case of all Bridgewater Formation samples. With the exception of a carbonate nodule from the Greenlands Clay, which falls near the dolomite-quartz/aluminosilicate tie-line, Hingston's soil carbonate nodules occupy a similar but slightly enlarged composition range (between 32% and 72% end-member CaO) compared with the Rix and Hutton (1953) and the Lawrence (1975) samples (Fig. 10.20). Hingston (1957) also discovered that in some soil carbonate nodules the compositions of the outer layers and the inner layers or cores was different (samples A and B in Fig. 10.20), with the outer layers having a higher Ca/Mg ratio.

A variety of cobble- to boulder-size calcrete pisolites consisting of dark brown to black carbonaceous limestone cores encased by white rinds of concentric carbonate laminae has been collected from multi-layered carbonate accumulation profiles on Eyre Peninsula and analysed by Milnes using X-ray fluorescence techniques.

Table 10.6 Chemical analyses of soil carbonate nodules and calcretes

	Acid-soluble		Acid-insoluble residue	
	%CaO	%MgO		Ca/Mg
Rix and Hutton (1953)				
Kuitpo carbonate nodules				
1	30.37	1.72	41.1	20.9
2	30.82	1.77	41.0	20.6
3	34.29	1.58	35.1	25.7
4	37.37	1.05	29.9	42.2
5	34.91	1.24	34.0	33.4
6	34.35	1.43	34.7	28.5
7	39.56	1.58	25.1	29.7
8	32.89	1.20	39.1	32.5
Kangaroo Island dune sand	38.38	1.72	25.0	26.4
Robe dune sand	32.67	1.00	37.4	38.7
Hingston (1957)				
Greenlands Clay nodule (11)	29.72	17.53	10.26[A]	2.01
Tammin 'Morrell' soil nodule (12)				
outer shell	38.41	1.61	28.04[A]	28.3
middle shell (pink layers)	39.82	2.42	23.83[A]	19.5
centre (cemented soil)	36.73	2.42	29.35[A]	18.0
Merredin 'Morrell' soil nodule (13)	28.32	1.61	46.06[A]	20.8
Merredin sandy laterite nodule impregnated with carbonate (14)	25.85	0.56	52.66[A]	54.7
Kalgoorlie desert loam nodule (15)	36.17	1.21	32.88[A]	35.4
Madura 'Mallee' soil nodule (16)				
outer concentric layers	44.30	0.20	20.47[A]	–
core (travertine)	45.70	0.40	17.55[A]	–
Lawrence (1975)				
Calcretes from Woorinen Formation, Victorian portion of Murray Basin				
Boinka	31.42	1.72	36.29	21.6
Linga	35.33	3.63	25.70	11.5
Merbein a.	39.27	4.27	21.00	10.9
b.	38.92	1.95	25.75	23.7
Nyang	35.57	1.05	30.60	40.1
Ouyen	28.51	1.79	42.55	18.9
Piangil a.	37.60	0.91	30.30	49.0
b.	32.50	1.67	36.00	23.1
Tutye	36.49	2.49	25.60	17.4

[A] 100% − ($CaCO_3$% + $MgCO_3$%).

As shown in Table 10.7 and Fig. 10.21, the pisolites exhibit a range of composition between 50% and 92% end-member CaO. The carbonaceous material in all samples contains significantly less CaO than the surrounding rind, and in many cases has a lower Ca/Mg ratio. All but two samples have Ca/Mg ratios that are consistent with less than 25% dolomite in the carbonate fraction. Organic carbon concentrations in the carbonaceous pisolite cores are given in Table 10.7. Somewhat higher carbon concentrations were determined by Fairbridge and Teichert (1952) in palaeosols at Hamelin Bay in Western Australia (Table 10.3). The origin of the carbonaceous material is not known, but many samples contain microfossils. Coralline algae and foraminifera of undoubted marine origin show evidence of recrystallisation and reworking (Fig. 10.22). On the other hand, structures resembling algal and fungal remains are well preserved and are possibly of non-marine origin. Of three additional single specimens of calcrete plotted in Fig. 10.21 for comparison with the pisolites, one from the Bremer River contains about 35% dolomite in the carbonate phase.

Northcote (1951) published analyses of acid-soluble Ca and Mg in several soil profiles with carbonate accumulation (B_{Ca}) horizons from the Coomealla region of New South Wales (Table 10.8). Compared with previously described samples from both the continental and Bridgewater Formation provinces, these materials contain less than 25% end-member CaO, and so are significantly enriched in acid-soluble detrital minerals. Nevertheless, when plotted on a triangular composition diagram (Fig. 10.23), all soils display similar distinctive compositional trends with depth, with the exception of the Matong sandy loam. For example, there is a progressive enrichment in calcium and magnesium through the A and B horizons, but a sharp increase in

Table 10.7. Chemical analyses of carbonaceous calcrete pisolites and calcretes
A, carbonaceous calcrete core; B, laminated carbonate rind.

	Total %CaO	%MgO	%(SiO_2+Al_2O_3+ Fe_2O_3)	Ca/Mg	Organic carbon (%)
Carbonaceous pisolites with laminated carbonate rinds, Eyre Peninsula					
Yarwondutta Rocks					
M165A	40.2	0.99	20.06	48.1	–
M166A	45.5	1.12	13.36	48.1	–
Tcharkuldu Hill					
M221A	43.7	1.21	14.08	42.8	0.81
M221B	42.0	1.37	11.08	36.3	–
Mt Wudinna					
M228A	34.8	3.79	22.80	10.9	–
M229A	44.7	1.19	13.47	44.5	0.60
M230A	46.6	0.92	10.18	60.0	1.04
M231A	36.6	3.04	23.28	14.3	0.37
M231B	36.1	3.55	19.17	12.1	–
M232A	33.7	4.72	25.16	8.5	0.54
M232B	39.3	2.63	14.75	17.7	–
M233A	32.5	4.60	28.04	8.4	0.38
M233B	35.6	3.37	21.02	12.5	–
Pildappa Hill					
M234A	34.5	1.99	26.74	20.5	1.81
M234B	47.4	0.85	3.45	66.1	–
Kimba-Iron Knob road					
M237A	41.6	2.07	15.84	49.3	0.23
M237B	42.0	2.35	13.71	21.2	–
Calcrete single specimens					
M187 Bremer River	32.1	5.42	23.99	7.0	–
M153 Ucontitchie Hill	28.5	1.91	38.93	17.7	–
M237 Kimba-Iron Knob road	36.9	3.86	19.07	11.3	–

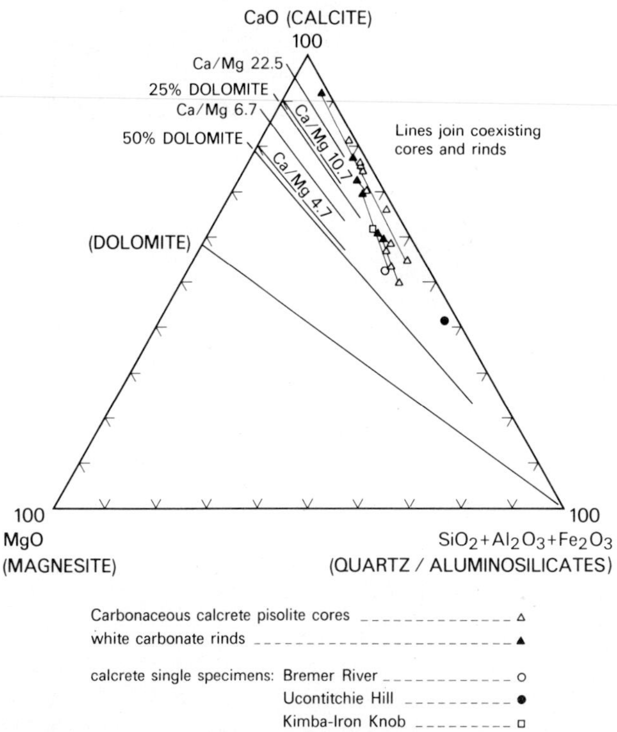

Fig. 10.21. Composition diagram for calcrete pisolites with carbonaceous cores and single specimens of calcrete from various localities.

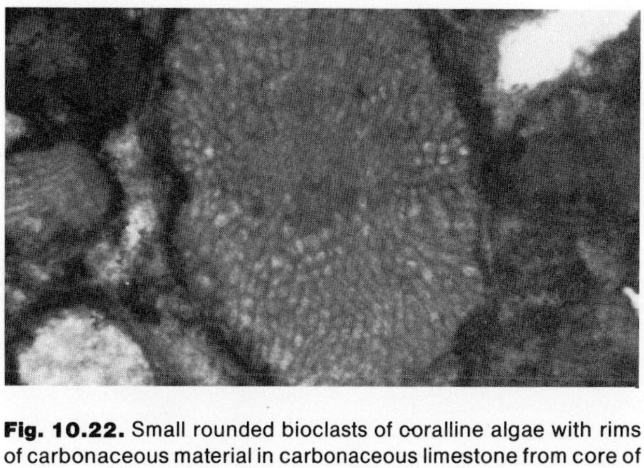

Fig. 10.22. Small rounded bioclasts of coralline algae with rims of carbonaceous material in carbonaceous limestone from core of calcrete pisolite M166, Yarwondutta Rocks, Eyre Peninsula. Bioclast in centre about 0.4 mm across.

concentration at the top of the B_{Ca} horizons. From this level in the profiles, there is a progressive increase in magnesium relative to calcium with depth to the base of the carbonate accumulation layers. In the surface horizons of most profiles and the C–D horizons at depth, the low concentrations of calcium and magnesium may include a significant proportion of the elements adsorbed on exchange complexes in the clay fraction, and so Ca/Mg ratios will not directly reflect carbonate chemistry. In those horizons containing abundant carbonate, however, the effect of exchangeable calcium and magnesium is less important, and Ca/Mg ratios may be used as an indicator of carbonate chemistry. Correction for exchangeable Ca and Mg on aluminosilicate clay complexes in four of the Coomealla soil profiles has been made by determining acid-soluble Ca and Mg after extraction of the sample with ammonium chloride at pH 8. Based on these data, Ca/Mg ratios (Table 10.8) are generally constant or increase through the A and B horizons, but decrease from the top of the carbonate zone in all profiles. The position of carbonate-rich A, B and B_{Ca1} horizon samples in the calcite-dolomite-quartz/aluminosilicate field of Fig. 10.23 indicates that dolomite

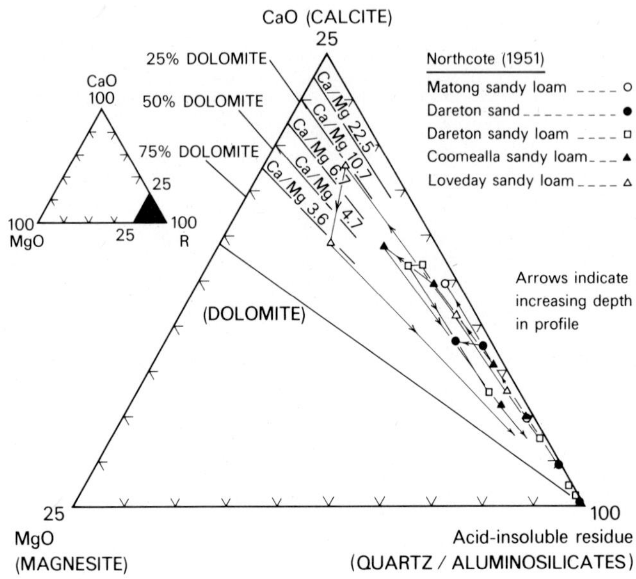

Fig. 10.23. Composition diagram for calcareous Coomealla soils from data published by Northcote (1951).

Table 10.8. Chemical analyses of calcareous soil profiles from Coomealla, N.S.W. (after Northcote 1951)

Soil type	Horizon	Acid-soluble %CaO	Acid-soluble %MgO	Acid-insoluble residue[A]	Ca/Mg	Ca/Mg[B]
Matong sandy loam	A	0.17	0.07	99.55	—	—
	A	0.21	0.09	99.44	—	2
	B_1	0.29	0.14	99.19	—	3
	B_2	4.56	0.36	91.11	15.0	17.8
	B_{Ca}	11.15	0.60	78.85	22.1	18.2
	C	0.76	0.46	97.68	1.6	2
Dareton sand	A	0.22	0.05	99.50	—	
	B	2.27	0.11	95.70	23.5	
	B_{Ca1}	8.29	0.42	84.32	23.4	
	B_{Ca2}	8.40	1.59	81.67	6.3	
Dareton sandy loam	A	0.29	0.07	99.33	—	—
	AB	1.17	0.13	97.64	10.3	5
	B	3.60	0.27	93.00	15.7	61
	B_{Ca1}	11.93	1.11	76.38	12.8	15.8
	B_{Ca2}	11.82	1.70	75.34	8.2	8.6
	C	5.94	1.36	86.56	5.2	4.9
	C-D	0.13	0.14	99.48	1.1	—
Coomealla sandy loam	A	4.74	0.34	90.84	16.7	29
	B	7.34	0.44	85.99	19.9	29
	B_{Ca1}	11.09	1.10	77.89	11.9	12.7
	B_{Ca2}	12.61	2.26	72.77	6.6	6.3
	$B_{Ca}D$	5.35	1.19	87.96	5.3	5.7
	D	0.53	0.31	98.39	2.0	2.4
	D	0.07	0.20	99.46	0.4	1.3
Loveday sandy loam	A	6.05	0.54	88.06	13.2	7.5
	B	9.64	0.87	80.98	13.1	7.9
	B_{Ca1}	16.08	1.86	67.40	10.2	5.8
	B_{Ca2}	12.44	4.30	68.80	3.4	1.5
	D	0.23	0.25	99.07	1.1	0.7

[A] 100% $-(CaCO_3\% + MgCO_3\%)$.

[B] Calculated from acid-soluble Ca and Mg concentrations determined after extraction of exchangeable Ca and Mg with ammonium chloride (repeat analyses, 1980).

Table 10.9. Chemical analyses of portions of calcareous profiles from the York and Merredin areas of Western Australia (after Hingston 1957)

	Acid-soluble		Acid-insoluble	
	%CaO	%MgO	residue[A]	Ca/Mg
Solonized brown soil on alluvial material, York				
P257.4	1.74	0.32	96.23	6.4
P257.5	8.16	1.07	83.19	8.9
P257.6	12.90	1.61	73.59	9.4
Brown soil of heavy texture on epidiorite, York				
P306.3	4.82	0.77	89.78	7.3
P306.4	8.19	2.18	80.81	4.4
P306.5	4.93	2.62	85.72	2.2
Grey soil on alluvial material, Merredin				
P292.4	7.18	3.55	79.75	2.4
P292.5	4.60	3.79	89.49	1.4
Brown soil of heavy texture on alluvial material, Merredin				
P289.1	4.04	0.85	91.01	5.6
P289.2	6.32	1.21	86.15	6.1
P289.3	6.00	2.01	85.08	3.5
P289.4	3.00	0.22	94.18	16.0
'Morrell' soil, Merredin				
P294.4	8.24	1.33	82.51	7.3
P294.5	10.54	2.10	76.79	5.9
'Morrell' soil on wind-transported material, Merredin				
P291.3	6.79	0.73	86.34	10.9
P291.4	8.92	0.68	82.65	15.4
P291.5	10.15	0.93	79.92	12.8

[A]100% − ($CaCO_3$% + $MgCO_3$%).

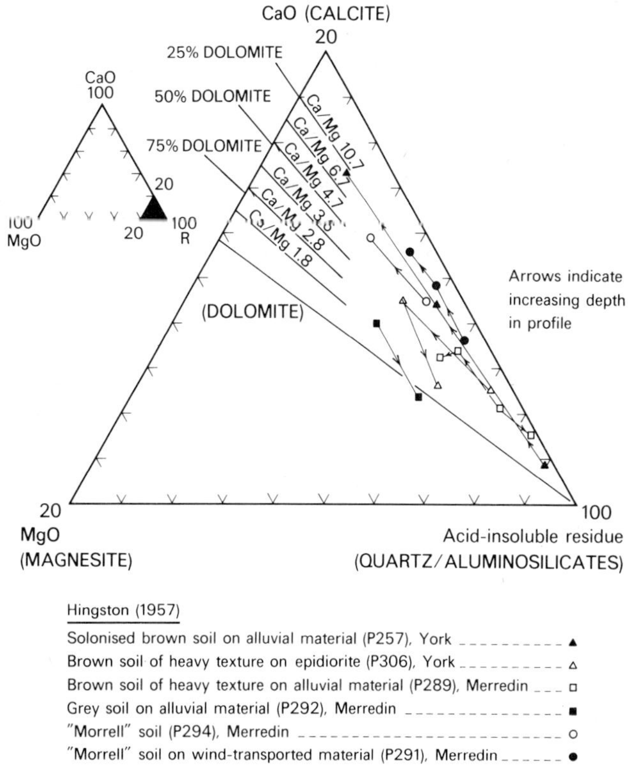

Fig. 10.24. Composition diagram for calcareous soils from York and Merredin areas of Western Australia from data published by Hingston (1957).

comprises less than 25% of the carbonate. In the B_{Ca2} horizons of most soil types, however, dolomite and calcite each account for about 50% of the carbonate. These soils are classified on morphological evidence as calcareous throughout the profile. The Matong sandy loam on the other hand, has A and B horizons that are essentially free of carbonate, and the carbonate accumulation layer displays an enrichment of calcium relative to magnesium with depth.

Based on these somewhat limited data, there appears to be a discontinuity in carbonate chemistry at the top of the carbonate accumulation horizons in the Coomealla soils, with the exception of the Matong sandy loam. In particular, there is a selective enrichment of magnesium with depth in the B_{Ca} horizons of all soils except the Matong sandy loam, but the suggestion of a selective enrichment of calcium with depth in the overlying A and B horizons. Clearly, further geochemical investigations are necessary. Nevertheless, there is the possibility that some of the profiles are polygenetic, as Northcote's other data suggest.

Hingston (1957) published partial analyses of several calcareous soil profiles of similar composition to Northcote's Coomealla soils from the York and Merredin areas of Western Australia (Table 10.9). As shown in Fig. 10.24, the samples cover a wide range of Ca/Mg values within the calcite-dolomite-quartz/aluminosilicate field, indicating dolomite concentrations between 15% and 100% in the carbonate fraction. Compositional trends with depth in the profiles, although few samples were analysed, appear similar to the two trends in the Coomealla soils. Discontinuities in Ca/Mg ratios are also evident, for example in the brown soil of heavy texture on alluvial material (P289) and the 'Morrell' soil on wind-transported material (P291), but there is insufficient information available upon which to judge their significance.

Comprehensive X-ray fluorescence analytical data for profiles through calcareous soils on various igneous rocks in the continental province have been obtained during recent studies of igneous rock weathering in South Australia. In the absence of direct measurements of the chemistry of the carbonate fraction using complementary wet-chemical techniques, the total element data can be used in conjunction with carbon dioxide analyses to provide an estimate of carbonate chemistry, although exchangeable calcium and magnesium and the presence of relict calcium and magnesium aluminosilicates can significantly influence the results. Hutton *et al.* (1977) described two weathered profiles on norite at Black Hill in which there appear to be distinctive changes in carbonate chemistry with

Table 10.10. Chemical analyses of calcareous soil profiles on weathered igneous rocks

		Total				
		%CaO	%MgO	%($SiO_2+Al_2O_3+Fe_2O_3$)	Ca/Mg	%CO_2
Hutton et al. (1977)						
Profile through soil on Black Hill Norite, Old Martin quarry (ignited basis)						
0- 8 cm	sandy loam; no carbonate	4.21	2.88	86.64	—	—
8-20 cm		4.37	2.67	88.44	—	—
20-30 cm	carbonate-rich; some clay	6.37	3.22	83.60	2.34	1.6
30-50 cm	but little structure	18.6	4.58	71.11	4.8	6.6
50-60 cm		29.5	5.19	61.99	5.9	14.1
60-100 cm	weathered norite	8.26	5.54	81.40	1.77	1.3
Profile through weathered Black Hill Norite, Council quarry (ignited basis)						
0-20 cm	carbonate-rich concretionary	35.8	10.7	51.08	3.96	29.6
20-40 cm	material	32.2	16.0	49.05	2.38	29.8
40-60 cm		23.1	15.4	58.16	1.78	25.5
60-75 cm	red clay with carbonate-rich patches	16.1	11.0	67.25	1.73	17.9
75-85 cm		9.31	6.21	79.41	1.78	7.9
85-100 cm		8.89	4.61	79.63	2.28	—
100-115 cm	weathered norite, no carbonate	8.43	3.90	80.97	2.56	—
rock (average of 4 samples)		9.49	5.64	77.11	1.99	
Monier quarry rock	I (7)	9.43	5.47	78.52	1.72	
	II (1)	9.23	4.64	78.74	1.99	
Old Martin quarry rock	(3)	8.83	5.54	80.39	1.59	
Unpublished data, A. R. Milnes and J. T. Hutton						
Profile through soil, calcretes and weathered granite at Yarwondutta Rocks						
0-10 cm	sand loam, calcareous	8.40	0.66	76.21	15.1	
10-20 cm	sandy loam with calcrete	14.92	0.89	65.00	19.9	
20-30 cm	sandy clay loam; calcrete	32.9	1.85	32.87	21.1	
30-40 cm	rubble	21.7	1.65	50.88	15.6	
40-50 cm		1.34	1.60	85.02	1.0	
50-60 cm	gritty medium clay, non-calcareous	0.74	0.27	89.01	3.3	
60-70 cm	weathered granite	0.52	0.34	89.18	1.5	
70-80 cm	weathered granite	0.56	0.41	88.57	1.4	
80-90 cm	weathered granite	0.56	0.51	87.03	1.1	
90-100 cm	weathered granite	0.67	0.58	86.38	1.2	
grus		0.66	0.87	84.45	0.8	
fresh granite		1.03	0.63	86.61	1.6	

depth (Table 10.10; Fig. 10.25). In the case of the soil over calcrete in the Old Martin quarry, there is an increase in both carbonate content and Ca/Mg ratio with depth below 20 cm in the profile to the base of the carbonate zone at about 0.6 m. Beyond this, the material encountered in the profile falls within the composition range of the norite (Fig. 10.25). However, the disparity between CaO and MgO concentrations and CO_2 content indicates that Ca/Mg ratios are not necessarily those of the carbonate fraction. It is interesting that the profile was considered by Hutton *et al.* (1977) to have formed in situ, and it is stated to have a composition that is consistent with alteration of the norite under present day climatic conditions. Calcium accumulated in the profile was thought by Hutton *et al.* to have been derived directly from weathering of the norite.

At the Council quarry at Black Hill, on the other hand, the profile on norite comprises surface carbonate-rich material down to 0.4 m containing between 50% and 75% dolomite. With increasing depth to 0.85 m, the carbonate content of the material decreases but is wholly composed of dolomite (Table 10.10; Fig. 10.25). At greater depths, the composition of the material is essentially that of fresh norite. In effect, the norite is progressively contaminated by dolomite towards the surface to 0.4 m depth: in shallower zones there is a progressive enrichment in calcite relative to dolomite to the surface. This is similar to geochemical trends observed through calcretes in other parts of the continental province. Hutton *et al.* (1977) regarded the dolomitic character of the carbonate in the Council quarry as inconsistent with the contemporary weathering environment, and suggested that it had been introduced into the profile at an earlier time, possibly during the influence of a Tertiary lacustrine environment in the area.

The calcareous soil profile on granite at Yarwondutta Rocks on Eyre Peninsula is characterised by an increase in both calcium and magnesium concentrations and Ca/Mg ratio with depth to the top of the carbonate-rich zone, beyond which the Ca/Mg ratio decreases with

Table 10.11. Chemical analyses of multi-layered calcrete profiles (Hutton and Dixon 1981)

		Total %CaO	Total %MgO	$\%(SiO_2+Al_2O_3+Fe_2O_3)$	Ca/Mg
		Murray Basin profiles			
Tailem Bend					
TB1	massive limestone	40.4	6.8	11.20	7.0
TB5	massive limestone	40.9	3.8	18.60	12.8
TB6	massive limestone	35.8	7.0	21.70	6.1
TB9	massive limestone	38.1	9.6	12.40	4.7
TB11	massive limestone	37.9	9.3	16.90	4.8
TB7	nodular calcrete	40.0	6.8	14.80	7.0
TB8	nodular calcrete	34.2	13.3	12.70	3.1
TB12	nodular calcrete	38.1	12.4	15.70	3.6
TBS1	calcareous fine earth	22.8	8.5	34.50	3.2
Monteith					
MQ001	massive limestone	33.9	3.1	29.00	12.8
MQ3	massive limestone	36.5	4.2	23.30	10.3
MQ4	massive limestone	29.0	5.0	35.90	6.9
MQ5	nodular calcrete	33.5	5.8	27.00	6.8
MQ6	nodular calcrete	31.2	8.0	27.60	4.6
MQ7	nodular calcrete	31.9	9.1	23.10	4.2
MQS 2	calcareous fine earth	12.6	8.3	53.60	1.8
OMQ 2	massive limestone	35.7	4.3	26.90	9.8
OMQ 3	massive limestone	28.8	13.2	20.60	2.6
OMQ 1	nodular calcrete	36.1	3.6	26.80	11.9
Murray Bridge					
MB1	massive limestone	39.4	4.6	20.30	10.2
MB2	massive limestone	36.3	2.5	30.20	17.2
Black Hill					
BHC 2	massive limestone	37.6	4.1	21.00	10.9
BHC 1	nodular calcrete	37.3	4.1	20.80	10.8
		Yorke Peninsula profiles			
Yorketown					
YT1	massive limestone	40.4	3.0	18.90	16.0
YT2	nodular calcrete	40.3	3.1	20.30	15.4
YP1	massive limestone	51.1	1.1	6.50	55.0
YPS	massive limestone	53.0	1.1	3.50	57.1
YP3	calcareous fine earth	22.8	8.5	34.50	3.2
Port Vincent					
PV1	massive limestone	31.5	2.4	34.80	15.6
PV2	nodular calcrete	31.4	2.5	36.30	14.9
Moonta Bay					
MB3	massive limestone	39.3	2.1	24.00	22.2
MB4	nodular calcrete	27.3	5.0	37.70	6.5
Port Julia					
PJ2	massive limestone	42.8	3.6	13.50	14.1
PJ1	nodular calcrete	43.5	2.0	15.50	25.8
Wool Bay					
WB1	massive limestone	42.0	3.5	18.10	14.2
WB2	nodular calcrete	44.9	4.8	11.40	11.1

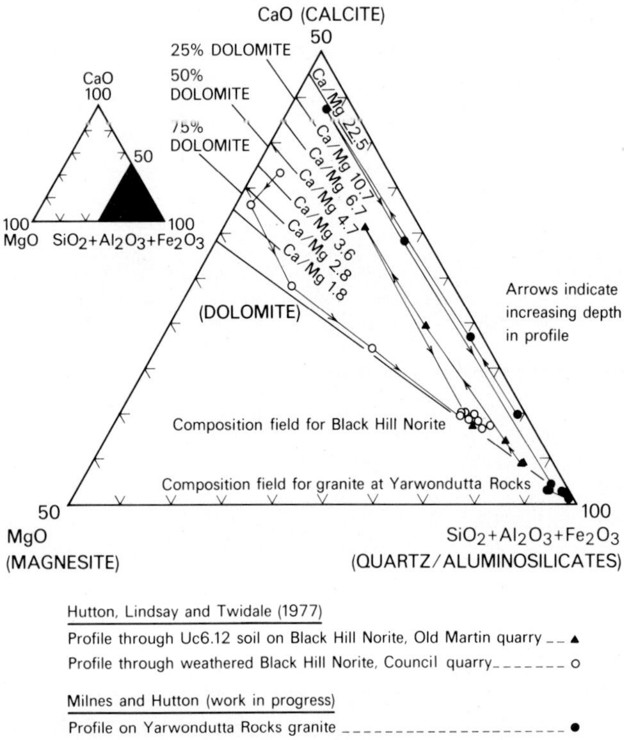

Fig. 10.25. Composition diagram for calcareous soil profiles on weathered igneous rocks.

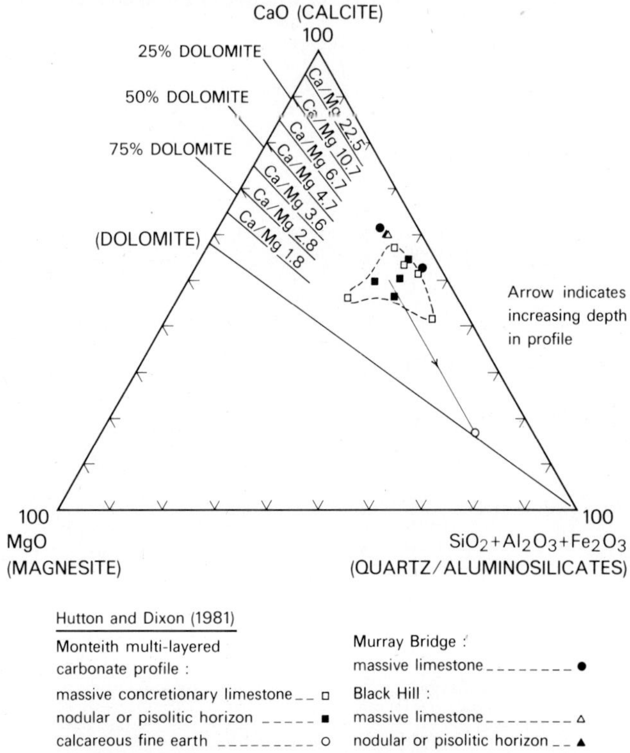

Fig. 10.27. Composition diagram for multi-layered calcretes in the Monteith area from data published by Hutton and Dixon (1981).

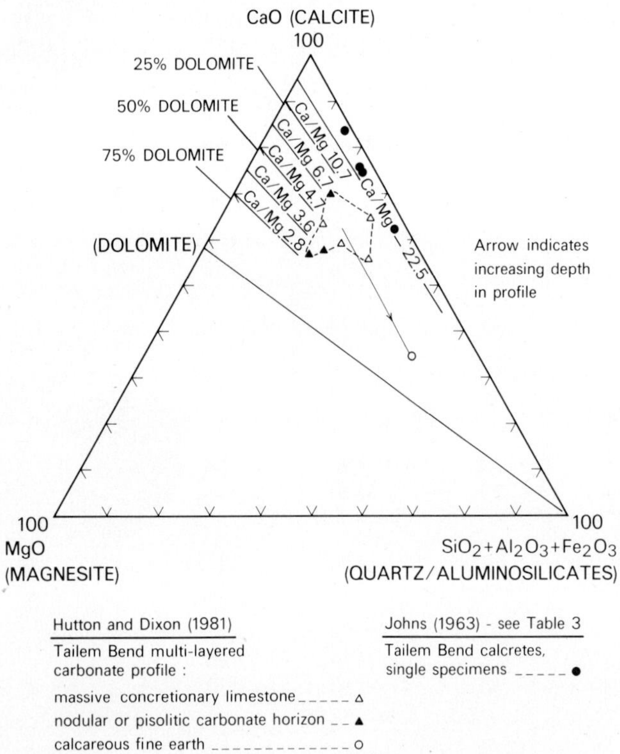

Fig. 10.26. Composition diagram for multi-layered calcretes in the Tailem Bend area from data published by Hutton and Dixon (1981).

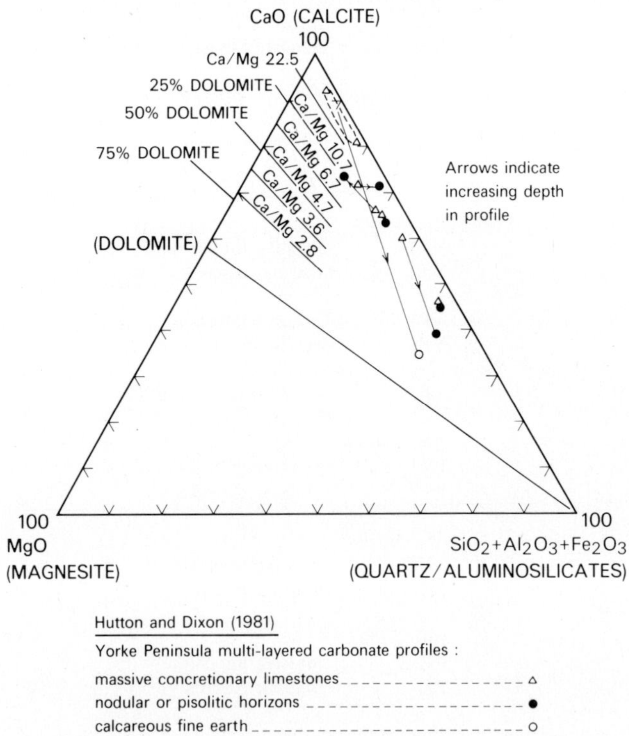

Fig. 10.28. Composition diagram for multi-layered calcretes on Yorke Peninsula from data published by Hutton and Dixon (1981)

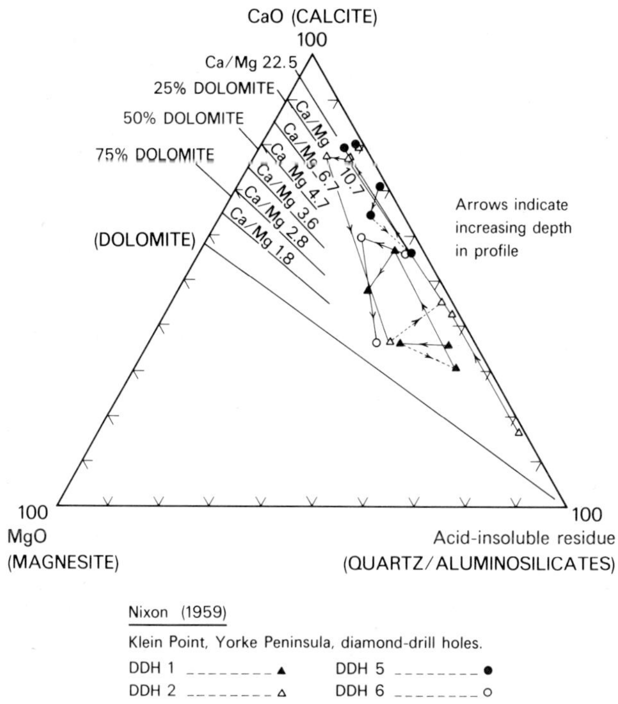

Fig. 10.29. Composition diagram for calcrete profiles from Klein Point quarry drill cores from data published by Nixon (1959).

depth until weathered granite is encountered at 0.5 m (Table 10.10; Fig. 10.25). In the absence of direct measurements, the influence on Ca/Mg ratios of magnesium from exchange complexes or relict biotite is difficult to assess, but the observed trends are similar to those in other soil profiles. The apparent discontinuity in carbonate chemistry at the top of the carbonate-rich zone is supported by thin-section studies of the samples, and may be interpreted as a disconformity within the profile: in other words, the soil mantle and the underlying carbonate-rich zone could be different deposits. The origin of the carbonate is not known, but the existence of microfossils (including foraminifera, coralline algae and some mollusc fragments) in carbonaceous and other lithoclasts within the zone suggest an influence of Pleistocene beach and dune sands.

In a recent study, Hutton and Dixon (1981) documented the progressive decrease in Ca/Mg ratio with depth in multi-layered calcrete profiles from the western Murray Basin and Yorke Peninsula (Table 10.11). They noted that the near-surface massive limestone layer had high Ca/Mg ratios compared with underlying nodular carbonate zones whilst the calcareous fine earth at depth had Ca/Mg ratios approaching that of dolomite. Such trends are common to many carbonate accumulations in the regolith, as described earlier. However, when compared in detail with geochemical trends for other calcretes (Figs 10.23–25), the multi-layered profiles display distinctive characteristics. In the case of theTailem Bend and Monteith profiles, for example, the samples of massive limestones and underlying nodular materials cluster in broad composition fields (Ca/Mg ratios ranging from about 20 to 3) which are grossly different in terms of carbonate content (but not necessarily Ca/Mg ratio) from the composition of single specimens of calcareous fine earths (Figs. 10.26 and 10.27). The same is true of a Yorke Peninsula profile (Fig. 10.28) where the distinction is even more pronounced. Within the composition field of the massive limestones and nodular zones there is no apparent consistent distinction between the two materials taken at random in a given locality, although, as Hutton and Dixon noted, there is such a consistent compositional relationship within particular profiles. It is interesting to note that four single specimen calcrete samples from the Tailem Bend district but west of the River Murray (Johns 1963) have quite different compositions from the Hutton and Dixon samples (Fig. 10.26).

Based on their geochemical data, Hutton and Dixon (1981) discussed possibilities that the calcrete profiles developed by leaching of an original carbonate deposit which could have been either calcite- or dolomite-rich, though in one part of their paper they suggested that the calcareous (dolomitic) fine earth was a sediment or soil buried by the carbonate deposit and subsequently impregnated by magnesium-rich carbonate leached down from above. The difficulty with this explanation is the unconsolidated nature of the calcareous fine earth, a condition considered unlikely to be maintained following impregnation by carbonates precipitated from solution. In fact, the geochemical data favour the existence of two different sedimentary layers comprising the massive and nodular zones on the one hand and the calcareous fine earth on the other. The presence of magnesium-rich clay minerals (sepiolite, palygorskite) in the calcretes in some localities, such as Monteith and Tailem Bend, but not in others, including Long Ridge and Yorke Peninsula, strongly supports the probability of a primary magnesium-rich environment at the time of deposition of the original materials, at least in some areas. The progressive decrease in Ca/Mg ratios from the massive to the nodular zones with depth in the profiles has been superimposed on these original materials as a result of subsequent leaching. Hutton and Dixon (1981) suggested that the carbonate in the fine earth was also leached from the overlying materials. However, when compared with the morphology of the massive and nodular zones nearer the surface, the unconsolidated fine earth at depth in the profile appears to be the least altered material and therefore is most likely to resemble the character and composition of an original calcareous deposit. This is a geologically reasonable hypothesis, and is consistent with the field interpretations of Crawford (1965) and

Table 10.12. Chemical analyses of samples from diamond drill cores through calcrete profiles at Klein Point, Yorke Peninsula (after Nixon 1959)

		Acid-soluble %CaO	Acid-soluble %MgO	%($SiO_2+Al_2O_3$ $+Fe_2O_3$)	Ca/Mg
Core 1					
0-1.50 m	calcrete	29.43	3.87	41.3	9.0
1.50-1.78 m	calcrete	22.75	9.37	30.9	2.9
1.78-2.13 m	calcrete	22.86	4.92	46.9	5.5
2.13-2.44 m	calcrete	36.81	3.54	24.3	12.3
2.44-5.03 m	calcrete	28.80	9.23	22.5	3.7
5.03-7,62 m	clay horizon	0.90	1.05	76.1	—
7.62-8.84 m	Port Vincent Ls.	52.33	1.48	3.8	41.9
Core 2					
0-0.99 m	surface soil	12.21	0.76	61.3	19.0
0.99-1.37 m	calcrete	30.20	0.91	39.7	39.3
1.37-2.34 m	calcrete	45.05	2.49	10.9	21.4
2.34-4.89 m	calcrete	46.06	2.96	11.0	18.4
4.89-7.90 m	calcrete	44.82	4.92	8.1	10.8
7.09-8.18 m	calcrete	23.64	10.76	30.5	2.6
8.18-9.14 m	calcrete	30.82	1.53	36.1	23.9
9.14-10.7 m	clay horizon	2.24	0.67	79.9	—
10.7-14.0 m	Port Vincent Ls.	47.40	1.00	11.3	56.2
Core 5					
0-0.76 m	surface soil	1.68	0.48	81.2	—
0.76-1.32 m	calcrete	43.09	1.00	16.6	51.1
1.32-2.44 m	calcrete	39.67	3.87	17.7	12.2
2.44-3.05 m	calcrete	35.69	1.72	25.9	24.6
3.05-3.96 m	calcrete	46.67	1.20	11.0	46.1
3.96-5.49 m	calcrete	47.01	0.96	10.8	58.0
5.49-6.71 m	calcrete	41.57	0.86	20.4	57.3
Core 6					
0-1.30 m	calcrete	35.86	2.47	25.8	17.2
1.30-3.05 m	calcrete	36.14	6.36	18.2	6.7
3.05-5.49 m	calcrete	22.75	12.09	28.1	2.2

Read (1974), who suggested that progressive leaching of surface carbonate deposits produced at first nodules and concretions in underlying layers, and then with continuing precipitation of carbonate in intergranular spaces, impermeable limestone sheets with thin surficial laminated carbonate layers and crusts which effectively preserved deeper materials from alteration by preventing further downward movement of waters.

By way of contrast, analytical data for drill-core samples through calcrete profiles at Klein Point on Yorke Peninsula (Nixon 1959) reveal what appear to be superposed carbonate deposits, each exhibiting a decrease in Ca/Mg ratio with depth until the underlying deposit is encountered and the pattern repeated (Table 10.12, Fig. 10.29). Unfortunately, the broad sampling interval adopted by Nixon precludes recognition of the complete pattern of carbonate chemistry with depth. The calcretes are underlain by alluvial fan deposits of sands, gravels and mottled sandy clays, and are overlain by shallow soils. Further field and laboratory studies are currently in progress at Klein Point.

Minor elements

There are few published data on the minor element content of calcretes and associated soil materials. Although our analyses are based on X-ray fluorescence techniques that routinely produce concentration data on many minor elements, most of our work to date has dealt with single specimen samples from different localities, and so only provides a reconnaissance view of minor element variability. Systematic geochemical studies of profiles through calcretes are required for an understanding of the origin of the carbonate deposits and the mechanism of carbonate movement, and such studies are currently in progress.

Together with the major elements, silicon (Si), aluminium (Al) and iron (Fe), minor elements such as titanium (Ti), zirconium (Zr), manganese (Mn), potassium (K) and phosphorus (P), are located in discrete minerals in the non-carbonate detrital fraction of calcretes. Data for these elements are sought as potential indicators of the provenance of the detrital materials. Sodium (Na), in addition to minor amounts of potassium, calcium (Ca) and magnesium (Mg), ought to be adsorbed on clay mineral exchange complexes and could be indicative of contemporary environmental conditions, although relict feldspars and other primary rock minerals including micas will account in part for these elements. Strontium (Sr), with its strong geochemical affinity for calcium, is likely to be associated with the carbonates, though a proportion may be present in plumbogummite minerals known to be common constituents of soils and detrital materials. Phosphorus is probably present as primary apatite or secondary plumbogummite, and sulphur as gypsum, but the specific mineralogical status of these and other nutrient elements in calcretes and their associated soils is not presently known.

X-ray fluoresence analyses by Milnes of samples in the profile through calcareous soil and calcrete on granite at Yarwondutta Rocks indicate a concentration of sulphur and a depletion of zinc (Zn) in the carbonate zone relative to the contiguous granite and soil mantle (Table 10.13). Unfortunately, there are no comparative data available for these elements in other profiles. Both the weathered granite (between 0.6 m and 1.0 m) and the fresh granite have Ca/Sr ratios of 60. On the other hand, the calcrete and its brown soil mantle are enriched in strontium relative to the granite, but have Ca/Sr ratios in excess of 280, indicating a significant concentration of allogenic calcium in addition to any that may have been derived from granite weathering. This may be taken as evidence that the calcrete and overlying soil are discrete deposits that are not simply related to the granite via a weathering process. Supporting evidence for this suggestion is provided by data for titanium and zirconium which are usually regarded as indicators of minerals that resist weathering (Hutton 1977), although there is evidence of titanium mobility in certain weathering environments (Hutton *et al.* 1972; Milnes and Hutton 1974). In particular, calculations on a carbonate-free basis show that the detrital fraction of the calcrete is depleted in titanium and zirconium relative to both the underlying granite and the detrital fraction of the soil mantle, but Si/Ti and Ti/Zr ratios reflect higher relative concentrations of silicon and titanium than in the contiguous materials, consistent with the carbonate layer and the overlying soil mantle being separate and discrete deposits. A similar conclusion was obtained from the data on Ca/Mg ratios for the calcrete and its soil mantle. In this case there was an apparent discontinuity in carbonate chemistry between the soil and the calcrete. Petrographic data indicate that the additional silicon in the calcrete and the overlying soil is present as rounded silt-size quartz grains of aeolian origin.

Si/Ti ratios for samples collected from profiles through carbonate-rich materials on norite at Black Hill (Hutton *et al.* 1977) are also given in Table 10.13. On average, the norite in the Council quarry profile has a Si/Ti ratio of 30, whereas the overlying dolomitic material has Si/Ti ratios ranging from 37 (between 1.0 m and 2.6 m) to 66 (between the surface and 0.4 m). Together with the lower titanium concentration in the detrital fraction of the dolomitic material (calculated on a carbonate-free basis) compared with the weathered norite, the Si/Ti ratios indicate that the detrital fraction is in part allogenic and is increasingly contaminated with silicon towards the surface. In the case of the calcareous soil on weathered norite in the Old Martin quarry, Si/Ti ratios in conjunction with titanium concentrations calculated on a carbonate-free basis indicate that minor allogenic titanium and silicon have been added to the profile, as noted by Hutton *et al.* (1977).

Hutton and Dixon (1981) found that Ca/Sr ratios follow the same trend as Ca/Mg ratios in the multi-layered calcrete profiles of the western Murray Basin and Yorke Peninsula, and decrease down the profile from the near-surface massive limestone through the nodular zone to the unconsolidated calcareous fine earth at depth. This concentration of strontium relative to calcium with depth was explained in terms of selective dissolution of strontium in the surface layers and precipitation lower in the profile, but as pointed out earlier the morphological evidence suggests that only the upper layers of the multi-layered profiles may have been affected by leaching and precipitation of carbonates. In this sense, the Ca/Sr ratio of the fine earth, in common with the Ca/Mg ratio, could be characteristic of an original calcareous deposit. The Ca/Sr ratios of the overlying nodular and near-surface massive limestone layers record a successively greater enrichment of calcium towards the surface consistent with the leaching hypothesis, but of a separate deposit. In support of this suggestion, Si/Ti ratios calculated from the Hutton and Dixon (1981) data show that the massive and nodular layers have similar values, but that these values are significantly lower than for the fine earth, except in the case of the Monteith samples (Table 10.13). In other words, the detrital fractions of the nodular and massive layers on the one hand, and the calcareous fine earth on the other, appear to have a different origin.

Carbon isotopes

Studies of stable carbon isotope ratios* in calcretes are usually undertaken in order to identify the carbon source and hence interpret the environment of deposition of the carbonate. Fractionation of the ^{13}C and ^{12}C isotopes is effected through two different reaction mechanisms, leading to a *photosynthetic* depletion of ^{13}C in CO_2 and enrichment of ^{12}C in the synthesized plant material on the one hand, and an *inorganic* enrichment of ^{13}C in the bicarbonate species in solution as a result of exchange in the system [atmospheric CO_2 / dissolved HCO_3^- / $CaCO_3$] on the other (Hoefs 1973). Bicarbonate is the dominant carbon species in ocean water, and so the isotopic composition of both ocean water in equilibrium with atmospheric CO_2 and inorganic marine carbonates is near zero. Some marine organisms, including molluscs

*$^{13}C/^{12}C$ expressed as the difference, δ "delta" $^{13}C‰$ (parts per thousand) of the sample $^{13}C/^{12}C$ ratio from that of a standard belemnite from the Peedee Formation.

Table 10.13. Minor element data for calcretes

	%SO_3	Zn(ppm)	%SiO_2	%TiO_2	Sr(ppm)	Zr(ppm)	Ca/Sr	Ca/Mg	Si/Ti	Ti/Zr
Yarwondutta profile										
0-10 cm	0.32	21	67.0	0.30	209	351	287	15.1	174	5.1
10-20 cm	0.36	17	57.3	0.24	241	196	442	19.9	186	7.3
20-30 cm	0.43	9	28.7	0.11	606	78	388	21.1	203	8.5
30-40 cm	0.35	10	42.7	0.15	382	132	406	15.6	222	6.8
40-50 cm	0.23	18	68.4	0.24	107	229	90	1.0	222	6.3
50-60 cm	0.23	15	72.1	0.26	86	282	62	3.3	216	5.5
60-70 cm	0.22	23	75.1	0.33	64	334	58	1.5	177	5.9
70-80 cm	0.23	29	71.9	0.31	79	298	51	1.4	179	6.2
80-90 cm	0.24	23	68.6	0.33	67	261	60	1.1	162	7.6
90-100 cm	0.27	23	69.7	0.35	68	285	70	1.2	155	7.4

Black Hill (Hutton et al. 1977)

Council quarry profile	%SiO_2	%TiO_2	Si/Ti
0-20 cm	43.3	0.52	64.9
20-40 cm	41.7	0.48	67.8
40-60 cm	46.4	0.66	54.8
60-75 cm	49.0	0.77	49.6
75-85 cm	54.8	0.98	43.6
85-100 cm	53.8	1.01	41.5
100-115 cm	55.3	0.92	46.9
115-135 cm	51.7	1.33	30.3
135-160 cm	54.6	1.14	37.4
160-200 cm	57.3	1.03	43.4
200-260 cm	52.7	1.57	26.2
260 cm	52.0	1.36	29.8
260-280 cm	57.9	1.27	35.6
fresh norite (4)	49.6	1.30	29.7
Old Martin quarry profile			
0-8 cm	64.2	1.28	39.0
8-20 cm	65.5	1.30	39.2
20-30 cm	60.5	1.22	38.8
30-50 cm	48.1	1.17	32.1
50-60 cm	41.1	1.05	30.5
60-100 cm	55.2	1.30	33.1
fresh norite (3)	53.3	1.12	37.2

Multi-layered calcrete profiles (Hutton and Dixon 1981)

Murray Basin		%SiO_2	%TiO_2	Si/Ti
Tailem Bend:	Massive	9.4, 15.7, 18.7, 10.4, 14.8,	0.08, 0.20, 0.20, 0.10, 0.20	72.9 ± 14.0
	Nodular	12.3, 10.8, 13.4	0.14, 0.12, 0.15	69.5 ± 0.9
	Fine earth	30.5	0.20	118.9
Monteith:	Massive	26.4, 20.7, 32.2, 24.3, 17.7	0.14, 0.13, 0.21, 0.15, 0.16	120.7 ± 21.9
	Nodular	24.0, 24.3, 20.3, 23.6	0.20, 0.20, 0.20, 0.17	93.9 ± 11.9
	Fine earth	48.0	0.31	120.8
M. Bridge:	Massive:	17.8, 27.3	0.12, 0.20	111.1 ± 6.6
Black Hill:	Massive:	17.0	0.22	60.3
	Nodular	16.8	0.22	59.6
Yorke Peninsula				
Yorketown 1:	Massive	15.1	0.14	84.1
	Nodular	16.0	0.16	78.0
Yorketown 2:	Massive	4.4, 2.5	0.06, 0.04	53.0 ± 6.0
	Fine earth	30.5	0.20	118.9
Port Vincent:	Massive	32.3	0.09	280
	Nodular	33.2	0.11	235
Moonta Bay:	Massive	20.9	0.13	125.4
	Nodular	36.0	0.06	468
Port Julia:	Massive	11.5	0.07	128.1
	Nodular	12.8	0.10	99.8
Wool Bay:	Massive	14.8	0.12	96.2
	Nodular	8.8	0.07	98.0

and brachiopods, deposit skeletal carbonate in equilibrium with seawater, but others such as corals and echinoderms do not, presumably because of local biological factors (Schwarcz 1969). However, freshwater (lacustrine and groundwater) carbonates are generally lighter and exhibit a broad range of carbon isotope compositions between -5 and -11 ‰ because a variable proportion of the bicarbonate in freshwaters is derived from organic CO_2 generated mainly in soils.

Determinations of ^{14}C concentrations in terrestrial carbonates have been popular in the recent past in attempts to discover the age of carbonate deposition, based on the known half-life of the isotope, but there are several difficulties inherent in these measurements. In terms of sampling problems, exchange with modern ^{14}C is a common phenomenon in carbonate accumulations in surficial environments where post-depositional pedogenic processes may have been, or are still, active. Alternatively, incorporation of old or radioactively dead carbon from groundwaters leaching ancient limestones may occur in certain depositional environments. Other problems are generated because of the comparatively short half-life of the isotope, in the sense that many calcretes are known on geological grounds or by means of alternative dating techniques to be older than the limit of ^{14}C dating, or of an age that spans the time interval corresponding to five or six half-lives of the isotope, thus making accurate dating difficult.

Bowler and Polach (1971) published both stable carbon isotope and radiocarbon analyses of superposed pedogenic calcrete layers in Victoria, but restricted their samples to soft and porous materials rather than hard concretions and crusts. They discovered that the radiocarbon age of the carbonate was variably younger than the age of the host sediment as a result of modern ^{14}C uptake during and after pedogenic carbonate segregation. $^{13}C/^{12}C$ ratios for all samples examined, although from widely differing environments and of different radiocarbon ages, were reported to be 'remarkably consistent' and 'apparently independent of the different carbon sources'. Variations in $\delta^{13}C$ between -8.9 ‰ and -2.3 ‰ were recorded and explained in terms of a fractionation process involving post-pedogenic dolomitisation.

Williams and Polach (1969, 1971) described calcareous palaeosols from the Lake Torrens area of South Australia where the carbonate was 450 to 6950 radiocarbon years older than coexisting organic carbon. They concluded therefore that the pedogenic carbonate was not equilibrated with ^{14}C in the soil atmosphere at the time of its segregation but incorporated dead carbon from circulating groundwaters. Post-pedogenic contamination with modern ^{14}C was evidently minimal. Williams and Polach regarded the carbonate in the accumulation horizons as largely illuvial: in the alluvial sediments, carbonate in the Wilkatana and Nacoona palaeosols was thought to have been derived from Precambrian limestones in the adjacent Flinders Ranges, whereas the carbonate in the Motpena palaeosols in dune sands probably originated as calcareous dust from the deflation of older soils.

Values of $\delta^{13}C$ from the calcareous palaeosols near Lake Torrens ranged from $+3.4$ to -5.0 ‰, with most values between -1 and -5 ‰. Williams and Polach suggested that the $^{13}C/^{12}C$ ratios did not appear to be significantly influenced by palaeosol age, sample type or parent material. On the other hand, groundwater calcretes in areas of present or former high water table exhibited anomalously young radiocarbon ages compared to equivalent calcareous palaeosols in well-drained areas, as well as significantly different $\delta^{13}C$ values between -9.6 and -6.9 ‰.

Values of $\delta^{13}C$ determined by Milnes for a range of calcretes and calcareous sediments are given in Table 10.14. Although based on limited data, $\delta^{13}C$ values for the estuarine Glanville Formation sediments sampled approximately 4.5 m below the surface of the highest terrace of the Hindmarsh River (M211, M212), and in outcrop near the river mouth (M189b), fall within the range of $\delta^{13}C$ values determined by Williams and Polach (1971) for groundwater calcretes near Lake Torrens. Based on published data for the distribution of $\delta^{13}C$ values (Schwarcz 1969; Hoefs 1973) and the above observation, together with evidence of significant ^{234}U contamination of Glanville Formation fossil (G. W. O'Brien pers. comm. 1980), it appears that the original carbon isotope composition of the sediments has been altered by exchange with groundwaters. The fact that $\delta^{13}C$ values for the Glanville Formation sediments (-7.8 ± 1.3 ‰) are not significantly different from values for carbonate-rich concretions within the formation and in overlying fluvial sediments (-6.8 ± 1.1 ‰) may be taken as additional evidence of this alteration. Values of $\delta^{13}C$ for estuarine marls gradational upwards into lagoonal carbonate sediments on the intermediate terrace of the Hindmarsh River are enriched in ^{13}C and average -5.0 ± 0.5 ‰. Younger Holocene shell sands from the lowermost terrace have a $\delta^{13}C$ value (-1.5 ‰) consistent with most marine limestones. Thus there is some indication of a direct relationship between $^{13}C/^{12}C$ ratios and environment of deposition which is masked in part by an effect due to post-depositional alteration of the oldest Glanville Formation sediments.

In this framework, $\delta^{13}C$ values for calcretes from Yarwondutta Rocks (Table 10.14) appear to be intermediate between those of marine and terrestrial carbonate sediments, and so the carbon may be of mixed origin. The presence of microfossils of marine origin in lithoclasts within the calcretes supports this contention. The $\delta^{13}C$ value for one sample of carbonate-cemented beach-rock from Encounter Bay is likewise inconsistent with values for marine limestones, although many shell fragments

Table 10.14. Stable carbon isotope analyses of calcretes and calcareous sediments

	Ca/Mg[A]	$\delta^{13}C$ ‰
Hindmarsh River, Fleurieu Peninsula		
Glanville formation fossiliferous estuarine sands and clays		
M189b Calcrete crust	22.1	−9.0
M211	6.3	−6.5
M212	21.9	−8.0
Carbonate-rich concretions in Glanville Formation and overlying fluvial sediments		
M205	28.2	−5.6
M206	16.4	−7.8
M208	16.2	−7.1
Stranded and recrystallized estuarine–lagoonal Coxiella marl, intermediate terrace		
M209	20.0	−5.9
Lagoonal carbonate sediment overlying Coxiella marl, intermediate terrace		
M223D	14.4	−4.6
M203	16.5	−5.0
M207	30.9	−4.8
M214	21.5	−5.2
Holocene sands with shell fragments, lowest terrace		
M201	9.1	−1.5
Eyre Peninsula		
Yarwondutta profile		
0-10 cm brown soil	15.1	−7.5
10-20 cm calcrete	19.9	−6.2
30-40 cm calcrete	15.6	−6.6
Contiguous calcrete samples		
M165	48.1	−4.8
M166	48.1	−5.9
M140	12.5	−7.0
Bremer River		
calcrete		
M187	7.0	−5.6
Encounter Bay		
carbonate-cemented beach-rock		
M171	63.3	−5.3

[A] See Tables 10.4 and 10.7 for CaO and MgO concentrations.

Table 10.15. Stable carbon isotope analyses of profiles at Caliph and Konetta, South Australia

		$\delta^{13}C$ ‰
	Caliph	
Profile CAQ1		
10 cm	Red-brown	−4.6
22 cm	massive limestone sheet	−4.8
50 cm	laminated calcrete	−5.6
149 cm	calcareous sand	−4.9
175 cm	calcareous sand	−5.2
204 cm	calcareous sand	−5.3
224 cm	calcareous sand	−5.3
252 cm	calcareous sand	−5.3
Profile CA20		
10 cm		−5.6
80 cm		−1.6
105 cm		−1.2
130 cm		0
152 cm		+0.6
177 cm		+0.7
187 cm		−3.5
	Konetta profile	
15 cm		−9.2
35 cm		−6.8
45 cm		−6.0
95 cm		−9.3
118 cm		−9.4
128 cm		−9.1
131 cm		8.5
138 cm		−6.9
145 cm		−6.6

are present in the sample. Consequently it appears that the carbonate cement may be of terrestrial origin, and that the beach-rock is not in equilibrium with the marine environment in which it presently exists.

From the available data there does not appear to be any relationship between $^{13}C/^{12}C$ and Ca/Mg ratios in the samples studied. Bowler and Polach (1971) correlated anomalous ^{14}C activity with dolomite content in their calcrete samples and implied that in addition to exchange with modern ^{14}C, dolomitization occurred during post-depositional alteration. Moreover, they suggested that the observed $\delta^{13}C$ values resulted from post-depositional fractionation processes in the calcretes and effectively mask any effects due to carbon source and primary pedogenic processes. Williams and Polach (1969, 1971) arrived at a similar conclusion based on $^{13}C/^{12}C$ ratios for calcareous palaeosols from Lake Torrens.

G. B. Allison (pers. comm. 1980) has determined $^{13}C/^{12}C$ ratios for samples through carbonate accumulations at Caliph in the Murray Mallee and at Konetta in the South East of South Australia (Table 10.15). At Caliph, profile CAQ1 from a roadside quarry is a typical multi-layered calcrete with a shallow red-brown soil mantle. $\delta^{13}C$ values are virtually constant with depth to about 2.5 m, and average −.1 ± 0.3 ‰ (Fig. 10.30). Profile CA20, on the other hand, was collected by hand auger from a secondary unconsolidated carbonate accumulation in the floor of a valley cut into the multi-layered calcrete, and exhibits a progressive enrichment in ^{13}C from a $\delta^{13}C$ value of −6.6 ‰ at the surface to +0.7 ‰ at 1.77 m depth, and then a sharp change to −3.5 ‰ at 1.87 m. The Konetta profile was collected by coring through a *rendzina soil* and underlying carbonate-rich material, and the $^{13}C/^{12}C$ ratios appear to show ^{13}C enrichment trends with depth independently in both the soil and the carbonate zone. On average, however,

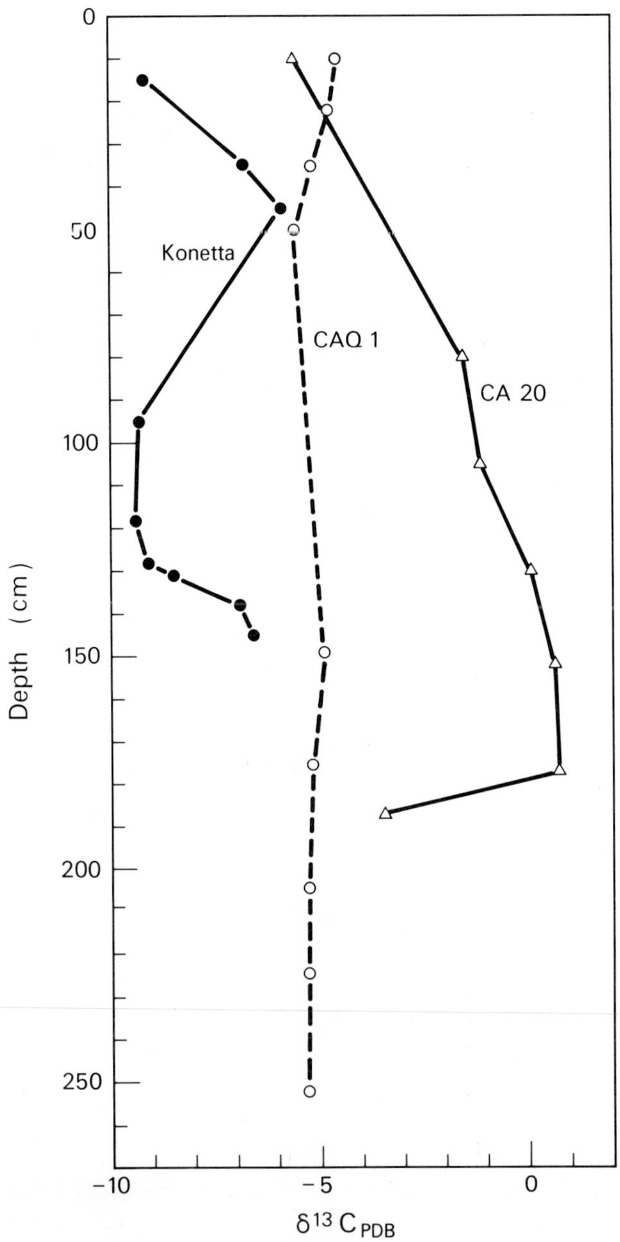

Fig. 10.30. Variations in $\delta^{13}C$ values through calcrete profiles at Caliph and Konetta in South Australia.

the isotope ratios for the Konetta samples (-8.1 ± 1.2 $^0/_{00}$) are depleted in ^{13}C compared with the Caliph samples.

The changes in carbon isotope ratios with depth in these calcrete profiles are not yet understood, but potentially have much to offer in aiding our understanding of carbonate origin and movement. As a consequence, further detailed studies of $^{13}C/^{12}C$ variations in profiles through calcretes are in progress in conjunction with investigations of field relationships and morphology as well as major and minor element geochemistry.

Discussion

Because of the large gaps in our knowledge and understanding, much work remains to be done on calcretes in southern Australia. Nevertheless, there are several points upon which pertinent comments can be made and some preliminary conclusions drawn.

Distribution of calcretes

Calcretes occur mainly in a broad zone flanking the concave southern coastline of the Australian continent (Fig. 10.1). This zone incorporates the Bridgewater Formation province of the coastal margins and the continental province of the inland areas, two informal subdivisions used herein to tentatively assign regional geological and pedological relationships to calcretes. In the Bridgewater Formation province, calcretes are associated with Pleistocene calcarenite dune and beach deposits and *terra rossa soils*, whereas in the continental province calcretes overlie a variety of calcareous and non-calcareous materials and are associated with an extensive blanket of calcareous earth soils. As yet, the nature of the boundary between the two provinces is not known. Outside the main area of distribution, calcretes occur in the Bridgewater Formation province of Pleistocene calcarenites along the coastlines of Western Australia and southern Victoria, and in the inland parts of the continent where they are often associated with zones of calcareous bedrock. This distribution of calcretes on the Australian continent can be explained broadly in terms of climatic regime (since carbonates are most likely to survive in the regolith in a semi-arid environment where leaching is not a dominant factor) and the proximity and availability of extensive and long-term carbonate sources.

Nature of calcretes

There are several calcrete types that can be easily recognized in the field and are comparatively well defined. Carbonate in soil profiles may occur as uniformly distributed fine earth material (as in many U soils), soft or hard patches and segregations (as in the lower A and the B horizons of many Gc soils, and the B horizons of some Gn and D soils), and distinct layers of concretionary nodules or pans in a more or less cemented state (Northcote *et al.* 1975). Hardened carbonate crusts on calcareous bedrock constitute another calcrete type. Examples of this are common in the Bridgewater Formation province where surficial layers of extensive calcareous sand deposits have been impregnated and cemented by secondary carbonate. In the continental province, there are consolidated calcrete crusts on calcareous bedrock surfaces as well as carbonate-filled fractures and joints in the near-surface zones of non-calcareous bedrock. Multi-layered calcrete profiles comprising mainly massive, nodular and fine earth layers with increasing depth were first described in some detail by Crocker (1946) as part of the extensive *solonized brown soil* ('mallee' soil) zone in southern Australia, and later by Crawford (1965) on Yorke Peninsula and Read (1974) in the Shark Bay area of Western Australia. In general, these profiles carry only a veneer of modern calcareous soil with reworked carbonate nodules and pisolites, and overlie sediments or bedrock of a wide variety of compositions.

Formation of calcretes

Based on established geological and pedological principles, studies of the detailed morphology of calcretes ought to give important clues to their mode and environment of formation. For example, certain forms of carbonate concentrated in the B horizons of soils, including vertically elongated stringers, mottles and vermicular patches are consistent with an illuvial origin, and appear to have formed by precipitation of carbonate leached down from the surface layers of the profile by soil water. Many of the Gc soils in the Coomealla region (Northcote 1951; Northcote *et al.* 1975) display carbonate of this type (Fig. 10.10), as do the calcareous palaeosols described and illustrated by Williams (1973) from the Lake Torrens area.

Hardened and/or cemented carbonate pans and zones or layers of hard carbonate nodules and pisolites are taken as evidence of disconformities that mark periods of subaerial exposure and diagenesis with attendant erosion and transport. Northcote's Coomealla sandy loam (colour Plate 5) has a zone of hard nodules at the top of the carbonate horizon that could be the result of local reworking prior to deposition of the overlying soil mantle. Thus, in view of the presence of well-developed vertical columns and mottles below the nodule zone, the carbonate horizon would be interpreted as the relict illuvial horizon of a former soil. Trends in Ca/Mg ratios down the profile, as discussed in the section on major element geochemistry, are consistent with this interpretation, but it is clear that more detailed geochemical data are warranted before a conclusive statement concerning the genesis of the profile can be made. A better developed carbonate nodule layer of probable colluvial origin occurs at the top of a carbonate-rich zone in Hindmarsh Clay at Redbanks in South Australia, and is overlain by a shallow Dr soil (Fig. 10.31). The boundaries at the top and bottom of this nodule layer are sharp, consistent with sedimentary layering. Examples of cemented carbonate pans marking the tops of superposed accretionary layers in a calcareous dune near Streaky Bay on Eyre Peninsula are illustrated in Fig. 10.32. Stripping of the leached soil mantle and cementation of the top of the carbonate accumulation zone appears to have taken place prior to deposition of the succeeding layer. Unfortunately, we have practically no detailed

Fig. 10.31. Layer of carbonate nodules at top of carbonate-rich zone in Hindmarsh Clay and below shallow duplex (Dr) soil, Redbanks, South Australia. (Hammer 33 cm long.)

Fig. 10.32. Partly cemented top of calcareous dune (A) with overlying calcrete pans marking tops of younger accretion layers in roadside quarry near Streaky Bay, Eyre Peninsula. Bottom carbonate pan shows evidence of disruption and reworking. (Section about 3.5 m high.)

micro-morphological or geochemical information about these materials.

Although calcretes of a wide variety of morphological types are evident in the Bridgewater Formation province, we have little detailed information about them either. Scree-slope lithoclast breccias (Fig. 10.3) and pisolitic soil layers provide evidence of erosion and reworking of both cemented calcrete pans and aeolianite bedrock on the flanks of the dune ridges, according to Read (1974) and Gill (1975). Layers of spectacular rhizomorph structures (Fig. 10.4) represent relict plant root zones in dunes and palaeosols, whereas the palaeosols themselves record the effects of soil-forming processes in stable non-accreting areas within the beach and backshore dune complex. Indurated sheet-like deposits of laminar calcrete on the surface of carbonate hardpans and in fractures and fissures within calcrete profiles (Fig. 10.3) are regarded by some to have formed by precipitation of carbonate from free water accumulations. However, micromorphological evidence for the presence in some laminae of algal filaments, fungal hyphae, and other fine structures including microborings and carbonate spherulites suggests an important biogenic influence on their formation (see for example Klappa 1979).

The typical multi-layered calcrete profiles of Crocker (1946), Crawford (1965), Read (1974) and Hutton and Dixon (1981) have distinct morphological layers including a surficial pisolitic soil and a massive limestone or carbonate hardpan, which usually grades downwards through a nodular zone to a mottled zone or calcareous fine earth (Figs 10.7–10.9). There are tremendous variations in the degree of development or degradation of the various layers in the profile, due in part to such factors as position in the landscape and duration of weathering and erosional processes. On the other hand, Read (1974) described multiple profiles formed by burial of old calcretes by mobile dunes or transported soils upon which the existing weathering environment was then imposed.

Leaching is an important illuviation process in pedogenesis, and is likely to constitute the major mechanism for moving carbonate through soil profiles to form accumulation layers. In ideal terms, it incorporates dissolution of carbonate in the surface layers of the soil profile by rainwater, and transportation in solution down the profile to a zone where precipitation occurs either due to evaporation of the soil solution or to degassing of CO_2 from the soil solution. The solubility of calcite depends essentially on the reversible reaction:

$$CO_2 + H_2O + CaCO_3 \rightleftharpoons Ca + 2HCO_3^-$$

although the kinetics of calcite dissolution are generally affected by the pH and carbonic acid concentration of the soil water, as discussed by Donner and Lynn (1977). As the solubility of carbonates increases with the partial pressure of CO_2 and the amount of H_2O present, supersaturation with respect to calcite could be achieved in soil water by loss of either component. Soil CO_2 is derived mainly from the plant root respiration zone, and so there could be a significant seasonal effect or even long term effect due to variations in the vigour of plant growth. CO_2 may also originate from groundwater charged with calcium and magnesium bicarbonates. With regard to H_2O loss, carbonate precipitation is expected to occur in the zone where high evapotranspiration causes large fluctuations in soil moisture. Unlike calcite, dolomite is *not* readily precipitated from solutions of Ca^{2+}, Mg^{2+} and HCO_3^- and the conditions for its formation are not fully understood (Hutton and Dixon 1981).

Some workers have attempted to assess climatic conditions favouring the formation of calcretes. With regard to the Lake Torrens palaeosols with well-defined illuvial carbonate horizons, Williams (1973) reviewed possible rainfall conditions under which calcretes were reported to form and concluded that the Pleistocene carbonate-rich Wilkatana and Motpena palaeosols developed under a significantly wetter climate than exists in that area today. The Holocene Nacoona palaeosol, on the other hand, contrasts with the other palaeosols in having a low carbonate content and a significant concentration of dolomite, and was supposed by Williams to have formed under drier and more saline conditions.

Mann and Horwitz (1979) suggested that low average rainfall and high potential evaporation were the dominant factors with respect to calcrete formation in Western Australia, but that the frequency and duration of rainfall were important contributing factors. Carlisle (1978) explained the distribution of calcretes in Australia and elsewhere using similar climatic parameters.

Hutton and Dixon (1981), on the other hand, proposed a wetter climate than present with cold wet winters and hot dry summers to account for the penetration of rainwater to depths they considered necessary to produce the multi-layered calcretes in South Australia. Thus, there are divergent opinions on the nature of the palaeoclimate during calcrete formations and no unambiguous solutions to the question of environmental conditions can be expected until more critical data are obtained for these materials.

Origin of the carbonate

Crocker (1946) proposed the interesting hypothesis that the calcium carbonate, from which the multi-layered calcretes in the *solonized brown soil* zone were developed by leaching, originated as fine material winnowed from the calcareous beach and back-shore dune complexes of the coastal regions during the Pleistocene and was carried inland as loess. As reported earlier, many workers have concurred with Crocker's hypothesis. In fact, Crawford (1965), Firman (1965*b*) and

Wetherby and Oades (1975) claimed to have recognized deposits of Crocker's loess in several localities. Others, including Blackburn *et al.* (1965) and Hutton and Dixon (1981) essentially rejected Crocker's hypothesis. In describing carbonate-rich sediments in north-western Victoria, Blackburn (1979) refuted the concept of aeolian accession of calcareous loess from the coastal zone because of the occurrence of extensive zones of siliceous Lowan Sand without trace of calcareous material surrounded by calcareous sediments of the Woorinen Formation. Accordingly, with respect to the localities described, Blackburn invoked local accession of calcareous material from adjacent lacustrine areas.

Hutton and Leslie (1958) demonstrated that the concentration of oceanic salts in rainwater decreases markedly with distance from the coast, and that in the inland regions calcium derived from locally produced aeolian dust is the dominant cation. Hutton (1979) presented data indicating significant amounts of calcium deposited per annum by rainfall at Merbein and in adjacent parts of the Murray Basin, and suggested that such atmospheric dust was obtained from the surrounding ground surface during seasons when plant cover was minimal. This could well constitute a 'piggy-back' method of distributing calcareous material, ultimately derived from the coastal zones, throughout inland areas under the influence of the prevailing wind regime over long periods of time. However, it is clear when looking in detail at Crocker's hypothesis that long distance aeolian transport from the coastal zone need not be envisaged. Calcareous strandlines marking Pleistocene coastal zones occur as far inland as an arc between Naracoorte and Bordertown, but older siliceous strandlines in north-western Victoria were recognized by Blackburn (1962) nd were described by Blackburn *et al.* (1967) as far eastwards as Dimboola. Sprigg (1979) mapped the furthest inland of these siliceous strandlines in an arc between Ouyen and Loxton. Similar materials in Western Australia were referred to the Bassendean Dune System of the Swan Coastal Plain (McArthur and Bettenay 1960) and the Peron Sandstone of the Shark Bay region (Logan *et al.* 1970), and were regarded as residual sands after leaching and decalcification of original calcareous dunes. There is no evidence reported for an original calcareous component in the Lowan Ridges and other strandlines. Nevertheless, they record former coastlines of the Pleistocene and possibly Late Pliocene seas well into those parts of the Murray Basin now covered by extensive calcrete deposits. At the time of development of these coastlines, calcareous material of a *local* origin was potentially available for transport short distances inland to zones where weathering could have initiated calcrete development. There would have been more than enough time to account for terrestrial reworking of existing calcareous deposits and accession of new marine carbonate during the protracted step-wise regression of the seas to the present coastline. During this time, local accession of carbonate via dissolution of atmospheric dust (Hutton 1979) and local reworking of lacustrine sediments (Blackburn 1979) could well have provided important contributions to the carbonate cycle.

Conclusion

Calcretes have generally been a neglected area of both geological and pedological study, and there are many questions concerning them that have yet to be answered. Problems, including the ultimate origin of the carbonate, the mechanism of hardening of calcretes, and the influence of climate and environment on their formation have not been seriously tackled. It would appear that unambiguous answers to many of these questions will only be forthcoming if comprehensive and systematic studies of calcretes, taking into account their stratigraphic relationships, geological and pedological characteristics, and mineralogical and geochemical composition are undertaken in selected areas of good exposure.

References

Blackburn, G. (1962). Stranded coastal dunes in north-western Victoria. *Aust. J. Sci.* **24**, 388.

Blackburn, G. (1966). Radiocarbon dates relating to soil development, coast-line changes, and volcanic ash deposition in south-east South Australia. *Aust. J. Sci.* **29**, 50-2.

Blackburn, G. (1979). Origin of aeolian carbonate-rich sediments considered by reference to sites near Walpeup and Tempy, N.W. Victoria. Aust. Soc. Soil Sci. Aeolian Landscapes Conf., Mildura, Vic. Oct. 1979.

Blackburn, G. and Baker, R. M. (1953). The soils of part of the southern Flinders Ranges, South Australia. CSIRO Aust. Soil Publ. No. 3.

Blackburn, G., Bond, R. D. and Clarke, A. R. P. (1965). Soil development associated with stranded beach ridges in south-east South Australia. CSIRO Aust. Soil Publ. No. 22.

Blackburn, G., Bond, R. D. and Clarke, A. R. P. (1967). Soil development in relation to stranded beach ridges of County Lowan, Victoria. CSIRO Aust. Soil Publ. No. 24.

Boutakoff, N. (1963). The geology and geomorphology of the Portland area. Geol. Surv. Victoria Memoir No. 22.

Bowler, J. M. (1979). Palaeoclimate and palaeohydrology with special reference to Mallee landscapes. Aust. Soc. Soil Sci. Aeolian Landscapes Conf., Mildura, Vic. Oct. 1979.

Bowler, J. M. and Polach, H. A. (1971). Radiocarbon analyses of soil carbonates: an evaluation from palaeosols in southeastern Australia. *In* 'Palaeopedology. Origin, Nature and Dating of Palaeosols'. (Ed. D. H. Yaalon). (International Society of Soil Science and Israel Universities Press.)

Cann, J. H. (1978). An exposed reference section for the Glanville Formation. *Q. Geol. Notes Geol. Surv. S. Aust.* 65, 1-4.

Carlisle, D. (1978). The distribution of calcretes and gypcretes in southwestern United States and their uranium favourability based on a study of deposits in Western Australia and South West Africa (Namibia). U. S. Department of Energy Open File Report.

Carlisle, D. (1980). Possible variations on the calcrete-gypsum uranium model. U.S. Department of Energy Open File report.

Churchward, H. M. (1960). Soils of the Woorinen Settlement, Victoria. CSIRO Aust. Div. Soils, Soils Land Use Ser. No. 36.

Churchward, H. M. (1963). Soil studies at Swan Hill, Victoria, Australia. III. Some aspects of soil development on aeolian materials. *Aust. J. Soil Res.* **1**, 117-28.

Cook, P. J., Colwell, J. B., Firman, J. B., Lindsay, J. M., Schwebel, D. A. and von der Borch, C. C. (1977). The late Cainozoic sequence of southeast South Australia and Pleistocene sea-level changes. *BMR J. Aust. Geol. Geophys.* **2**, 81-8.

Crawford, A. R. (1965). The geology of Yorke Peninsula. Geol. Surv. S. Aust. Bull. No. 39.

Crocker, R. L. (1946). Post-Miocene climatic and geologic history and its significance in relation to the genesis of the major soil types of South Australia. CSIR Bull. No. 193.

Crocker, R. L. and Cotton, B. C. (1946). Some raised beaches of the Lower South-East of South Australia and their significance. *Trans. R. Soc. S. Aust.* **70**, 64-82.

Daily, B. (1972). Aspects of carbonate sedimentation in the Cambrian of South Australia. Abstracts Joint Specialists Groups Meetings, Canberra. Geol. Soc. Aust., C10-14.

Daily, B., Firman, J. B., Forbes, B. G. and Lindsay, J. M. (1976). Geology. *In* 'Natural History of the Adelaide Region'. (Eds C. R. Twidale, M. J. Tyler and B. P. Webb). (Royal Society South Australia: Adelaide.)

de Mooy, C. J. (1959). Soils and potential land use of the area around Lake Alexandrina and Lake Albert, South Australia. CSIRO Aust. Div. Soils Soils Land Use Ser. No. 29.

Donner, H. E. and Lynn, W. C. (1977). Carbonate, halide, sulphate and sulfide minerals. *In* 'Minerals in Soil Environments'. (Eds J. B. Dixon and S. B. Weed). (Soil Science Society of America: Madison.)

Fairbridge, R. W. (1954). Quaternary eustatic data for Western Australia and adjacent states. Pan Indian Ocean Congr. Proc. Section C. Geology. Perth, W.A., Aug. 1954.

Fairbridge, R. W. and Teichert, C. (1952). Soil horizons and marine bands in the coastal limestones of Western Australia. *J. Proc. R. Soc. N.S.W.* 86, 68-87.

Firman, J. B. (1963*a*). Quaternary geological events near Swan Reach in the Murray Basin, South Australia. *Q. Geol. Notes Geol. Surv. S. Aust.* **5**.

Firman, J. B. (1963*b*). Quaternary geological events near Port Adelaide. *Q. Geol. Notes Geol. Surv. S. Aust.* **7**.

Firman, J. B. (1964). The Bakara Soil and other stratigraphic units of Late Cainozoic age in the Murray Basin, South Australia. *Q. Geol. Notes. Geol. Surv. S. Aust.* **10**, 2-5.

Firman, J. B. (1965*a*). Late Cainozoic sedimentation in northern Spencer Gulf, South Australia. *Trans. R. Soc. S. Aust.* **89**, 125-31.

Firman, J. B. (1965*b*). Late Cainozoic lacustrine deposits in the Murray Basin, South Australia. *Q. Geol. Notes Geol. Surv. S. Aust.* **16**, 1-2.

Firman, J. B. (1966*a*). Stratigraphic units of Late Cainozoic age in the Adelaide Plains Basin, South Australia. *Q. Geol. Notes Geol. Surv. S. Aust.* **17**, 6-9.

Firman, J. B. (1966*b*). Stratigraphy of the Chowilla area in the Murray Basin, *Q. Geol. Notes Geol. Surv. S. Aust.* **20**, 3-7.

Firman, J. B. (1967*a*). Stratigraphy of Late Cainozoic deposits in South Australia. *Trans. R. Soc. S. Aust.* **91**, 165-78.

Firman, J. B. (1967*b*). Late Cainozoic stratigraphic units in South Australia. *Q. Geol. Notes Geol. Surv. S. Aust.* **22**, 4-8.

Firman, J. B. (1969*a*). Palaeosols. A Stratigraphic Definition. *In* 'Etudes sur le Quaternaire dans le Monde'. Vol. I. Paris: Union Internat. pour l'etude du Quat. 8th INQUA, Paris, 1969.

Firman, J. B. (1969*b*). Stratigraphy and landscape relations of soil materials near Adelaide, South Australia. *Trans. R. Soc. S. Aust.* **93**, 39-54.

Firman, J. B. (1973). Regional stratigraphy of surficial deposits in the Murray Basin and Gambier Embayment. Geol. Surv. S. Aust. Rep. Invest. No. 39.

Firman, J. B. (1979). Palaeopedology applied to land use studies in southern Australia. *Geoderma* **22**, 105-17.

Firman, J. B. and Chugg, R. I. (1963). Marine erosion and stranded beach deposits near Port Wakefield. *Q. Geol. Notes Geol. Surv. S. Aust.* **8**, 4-5.

Gill, E. D. (1967). Evolution of the Warrnambool – Port Fairy coast and Tower Hill eruption, western Victoria. *In* 'Landform Studies in Australia and New Guinea'. (Eds J. N. Jennings and J. A. Mabbutt). (A.N.U. Press: Canberra.)

Gill, E. D. (1975). Calcrete hardpans and rhizomorphs in western Victoria, Australia. *Pac. Geol.* **9**, 1-16.

Gill, E. D. (1976). Warrnambool – Port Fairy district. *In* 'Geology of Victoria'. Chap. 9, Quaternary. (Eds J. G. Douglas and J. A. Ferguson). Geol. Soc. Aust. Spec. Pub. No. 5.

Goudie, A. (1973). 'Duricrusts in Tropical and Subtropical Landscapes'. (Clarendon Press: Oxford.)

Heath, G. R. (1966). Carbonate nodules formed in soil profiles: the nomenclature problem. *Aust. J. Sci.* **28**, 395-6.

Hingston, F. J. (1957). Calcium and magnesium contents of carbonate materials from some Western Australian soils. CSIRO Aust. Div. Soils Tech. Memo. No. 11/57.

Hoefs, J. (1973). 'Stable Isotope Geochemistry'. (Springer-Verlag: Berlin.)

Hossfeld, P. S. (1950). The late Cainozoic history of the southeast of South Australia. *Trans. R. Soc. S. Aust.* **73**, 232-79.

Howchin, W. (1935). Notes on the geological sections obtained by several borings situated on the plain between Adelaide and Gulf St. Vincent. Part I. *Trans. R. Soc. S. Aust.* **59**, 68-102.

Hutton, J. T. (1977). Titanium and zirconium minerals. *In* 'Minerals in Soil Environments'. (Eds J. B. Dixon and S. B. Weed). (Soil Science Society of America: Madison.)

Hutton, J. T. (1979). Influence of lack of rain on the inorganic ions in rain collected at Merbein, Victoria. Aust. Soc. Soil. Sci. Aeolian Landscapes Conf., Mildura, Vic. Oct. 1979.

Hutton, J. T. and Dixon, J. C. (1981). The chemistry and mineralogy of some South Australian calcretes and associated soft carbonates and their dolomitization. J. Geol. Soc. Aust. **28**, 71-9.

Hutton, J. T. and Leslie, T. I. (1958). Accession of non-nitrogenous ions dissolved in rainwater to soils in Victoria. *Aust. J. Agric. Res.* **9**, 492-507.

Hutton, J. T., Lindsay, D. S. and Twidale, C. R. (1977). The weathering of norite at Black Hill, South Australia. *J. Geol. Soc. Aust.* **24**, 37-50.

Hutton, J. T., Twidale, C. R., Milnes, A. R. and Rosser, H. (1972). Composition and genesis of silcretes and silcrete skins from the Beda Valley, southern Arcoona Plateau, South Australia. *J. Geol. Soc. Aust.* **19**, 31-9.

Jack, R. L. (1912). The geology of portions of the Counties of Le Hunte, Robinson, and Dufferin, with special reference to underground water supplies. Geol. Surv. S. Aust. Bull. No. 1.

Jack, R. L. (1914). The geology of the County of Jervois, and portions of the Counties of Buxton and York, with special reference to underground water supplies. Geol. Surv. S. Aust. Bull. No. 3.

Johns, R. K. (1959). Calcareous aeolianites of South Australia. S. Aust. Mines Dep. Mining Rev. **111**, 11-17.

Johns, R. K. (1963). Limestone, dolomite and magnesite resources of South Australia. Geol. Surv. S. Aust. Bull. No. 38.

Kenley, P. R. (1976). Southwestern Victoria. In 'Geology of Victoria'. Chap. 9, Quaternary. (Eds J. G. Douglas and J. A. Ferguson). Geol. Soc. Aust. Spec. Pub. No. 5.

Klappa, C. F. (1979). Lichen stromatolites: criteria for subaerial exposure and a mechanism for the formation of laminar calcretes (caliche). *J. sed. Petrol.* **49**, 387-400.

Lamplugh, G. W. (1902). Calcrete. *Geol. Mag.* **9**, 575.

Lamplugh, G. W. (1907). Geology of the Zambezi Basin around Batoka Gorge. *Q. J. Geol. Soc. Lond.* **63**, 162-216.

Lawrence, C. R. (1966). Cainozoic stratigraphy and structure of the Mallee region, Victoria. *Proc. R. Soc. Vic.* **79**, 520-53.

Lawrence, C. R. (1975). Geology, hydrodynamics and hydrochemistry of the southern Murray Basin. Geol. Surv. Victoria Memoir. No. 30.

Lawrence, C. R. (1976). Murray Basin. In 'Geology of Victoria'. (Eds

J. G. Lawrence and J. A. Ferguson). Chap. 9, Quaternary. Geol. Soc. Aust. Spec. Pub. No. 5.

Lawrence, C. R. (1979). Aeolian Landforms of the Lowan Sand and Woorinen Formation in north-western Victoria. Aust. Soc. Soil Sci. Aeolian Landscapes Conf., Mildura, Vic. Oct. 1979.

Lindsay, J. M. (1969). Cainozoic foraminifera and stratigraphy of the Adelaide Plains Sub-basin, South Australia. Geol. Surv. S. Aust. Bull. No. 42.

Logan, B. W., Read, J. F. and Davies, G. R. (1970). History of carbonate sedimentation, Quaternary Epoch, Shark Bay, Western Australia. Amer. Assoc. Petrol. Geol. Mem. **13**, 38-84.

McArthur, W. M. and Bettenay, E. (1960). The development and distribution of the soils of the Swan Coastal Plain, Western Australia. CSIRO Soil Pub. No. 16.

Mann, A. W. and Horwitz, R. C. (1979). Groundwater calcrete deposits in Australia: some observations from Western Australia. *J. Geol. Soc. Aust.* **26**, 293-303.

Marker, M. E. (1975). The lower southeast of South Australia: a karst province. Occasional Paper No. 13, Environmental Studies. Dep. Geography and Environmental Studies, Univ. Witwatersrand, Johannesburg.

Milnes, A. R. and Hutton, J. T. (1974). The nature of micro-cryptocrystalline titania in 'silcrete' skins from the Beda Hill area of South Australia. *Search* **5**, 153-4.

Netterberg, F. (1978). Dating and correlation of calcretes and other pedocretes. *Trans. Geol. Soc. S. Afr.* **81**, 379-91.

Nixon, L. G. B. (1959). Some Recent and Tertiary limestone deposits on Yorke Peninsula. *S. Aust. Mines Dep. Mining Rev.* **111**, 65-70.

Norrish, K. and Pickering, J. G. (1977). Clay mineralogic properties. *In* 'Soil factors in crop production in a semi-arid environment'. (Eds J. S. Russell and E. L. Greacen). (University Queensland Press, Australian Society Soil Science: Brisbane).

Norrish, K. and Rogers, L. E. R. (1956). The mineralogy of some terra rossas and rendzinas of South Australia. *J. Soil Sci.* **7**, 294-301.

Northcote, K. H. (1951). A pedological study of the soils occurring at Coomealla, New South Wales. CSIRO Aust. Bull. No. 264.

Northcote, K. H. (1979). Soils of aeolian landscapes in part of the Murray Basin of south-eastern Australia. Aust. Soil Sci. Soc. Aeolian Landscapes Conf., Mildura, Vic. Oct. 1979.

Northcote, K. H., Hubble, G. D., Isbell, R. F., Thompson, C. H. and Bettenay, E. (1975). 'A Description of Australian soils'. (CSIRO: Melbourne).

Prescott, J. A. and Piper, C. S. (1932). The soils of the South Australian mallee. *Trans. R. Soc. S. Aust.* **56**, 118-47.

Read, J. F. (1974). Calcrete deposits and Quaternary sediments, Edel Province, Shark Bay, Western Australia. *Amer. Assoc. Petrol. Geol. Mem.* **22**, 250-82.

Rix, C. E. and Hutton, J. T. (1953). A soil survey of the Hundred of Kuipto in the Mount Lofty Ranges of South Australia. S. Aust. Land Tax Dept. Bull. No. 1.

Rogers, P. A. (1980). Geology of the South East, South Australia. 1:500,000 map sheet. (Geol. Surv. S. Aust.: Adelaide.)

Schwarcz, H. P. (1969). The stable isotopes of carbon. *In* 'Handbook of Geochemistry'. (Ed. K. H. Wedepohl). Chap. 6, Carbon: Section B – Isotopes in Nature. (Springer-Verlag: Berlin.)

Segnit, R. W. and Dridan, J. R. (1938). Geology and development of ground water in the Robinson fresh water basin, Eyre's Peninsula. *Geol. Surv. S. Aust. Bull. No. 17.*

Sprigg, R. C. (1952). The geology of the South-East province, South Australia, with special reference to Quaternary coastline migrations and modern beach developments. Geol. Surv. S. Aust. Bull. No. 20.

Sprigg, R. C. (1979). Stranded and submerged sea-beach systems of southeast Australia and the aeolian desert cycle. *Sed. Geol.* **22**, 53-96.

Stace, H. C. T. (1956). Chemical characteristics of terra rossas and rendzinas of South Australia. *J. Soil Sci.* **7**, 280-93.

Stace, H. C. T. and Rogers, L. E. R. (1954). Morphological, chemical and mineralogical data for some South Australian terra rossa and rendzina soils. CSIRO Aust. Div. Soils Div. Rep. No. 10/54.

Stuart, W. J. (1970). The Cainozoic stratigraphy of the south-eastern coastal area of Yorke Peninsula, South Australia. *Trans. R. Soc. S. Aust.* **94**, 151-78.

Thomson, B. P., Daily, B., Coats, R. P. and Forbes, B. G. (1976). Late Precambrian and Cambrian geology of the Adelaide 'Geosyncline' and Stuart Shelf, South Australia. 25th International Geological Congress, Australia, Excursion 33A Guidebook.

Twidale, C. R., Lindsay, J. M. and Bourne, J. A. (1978). Age and origin of the Murray River and gorge in South Australia. *Proc. R. Soc. Vic.* **90**, 27-42.

Wade, A. (1915). The supposed oil-bearing areas of South Australia. Geol. Surv. S. Aust. Bull. No. 4.

Ward, L. K. (1913). The possibilities of the discovery of petroleum on Kangaroo Island and the western coast of Eyre's Peninsula. Geol. Surv. S. Aust. Bull. No. 2.

Wetherby, K. G. and Oades, J. M. (1975). Classification of carbonate layers in highland soils of the northern Murray Mallee, S.A., and their use in stratigraphic and land-use studies. *Aust. J. Soil Res.* **13**, 119-32.

Williams, G. E. (1969). Glacial age of piedmont alluvial deposits in the Adelaide area, South Australia. *Aust. J. Sci.* **32**, 257.

Williams, G. E. (1973). Late Quaternary piedmont sedimentation, soil formation and palaeoclimates in arid South Australia. *Z. Geomorphol. N.F.* **17**, 102-25.

Williams, G. E. and Polach, H. A. (1969). The evaluation of ^{14}C ages for soil carbonate from the arid zone. *Earth Planet. Sci. Lett.* **7**, 240-2.

Williams, G. E. and Polach, H. A. (1971). Radiocarbon dating of arid zone calcareous palaeosols. *Geol. Soc. Amer. Bull.* **82**, 3069-86.

Woolnough, W. G. (1927). The duricrust of Australia. *J. Proc. R. Soc. N.S.W.* **61**, 24-53.

Woolnough, W. G. (1930). Influence of climate and topography in the formation and distribution of products of weathering. *Geol. Mag.* **67**, 123-32.

Chapter 11
Development of soil landscapes

G. G. Beckmann

SOME AUSTRALIAN SOILS HAVE EXISTED FOR EXCEEDINGLY long periods of time, whereas others are as young as any on continents glaciated during the Quaternary. The previous chapters have summarized the geologic and geomorphic history of Australia and the influences of various processes during the later stages of the evolution of the continents. One of the aims of this chapter is to integrate the roles of the important soil-forming factors and processes in producing some of the major soil landscapes of Australia.

Parent materials and topography, and climates, both past and present, are the major controls on the distribution of Australian soils. The role of biological factors is little understood in the genesis of Australian soils. Some discussion of particular aspects, e.g. earthworms and termites (Chapter 39) is given later. Although the influence of vegetation on soil-forming processes is recognized, the interactions are highly complex, in that cause and effect are hard to separate, and have been little studied in Australia.

The present climate influences modern vegetation and both of these modify earlier soils and weathering environments. Many soils are polygenetic, in that soil material has been exposed on land surfaces through several climatic changes of the Quaternary, at least, but even many monogenetic soils are derived from preweathered materials. As a result, although the soil pattern of Australia is complex, an inherent order can be perceived through a systematic analysis of the relative importance of landscape history, parent material, and soil-forming process. Some important soil-landscape systems will be used to illustrate this.

Soils associated with Early Cainozoic, deeply weathered profiles

During the Early Tertiary most of Australia was tectonically stable for tens of millions of years, and relief and elevation were low; warm and moist to wet conditions prevailed almost continuously. Under these conditions rocks exposed at the surface were deeply weathered and, even where there had been uplifts and renewed denudation, the new surfaces exposed were similarly weathered, masking the initial variability of the rocks. The effects of this thorough 'homogenizing' were carried over into the parent materials exposed during later landscape changes.

Much of the eastern highlands projected above the general level of the landscape to the west and, because of its relief and the properties of the rocks, a range of weathering situations would have persisted. These situations would have disappeared with the higher relief and greater dissection which developed during the later Tertiary and Quaternary uplifts.

Soils on sedimentary rocks and granite

The dominant denudation process following moderate uplift was the backwearing of scarps, leaving a new low erosion surface, often of pediment form, below them. Weathering continued on this surface, and unless there had been a change in climate, the same weathering conditions also continued on the surfaces of low relief above the scarps. The soils which formed on the younger surfaces under these conditions, e.g. *red earths*, were so thoroughly weathered that they were little changed during the dominantly more arid conditions of the Quaternary.

There was, however, some local redistribution of highly weathered (often lateritic) material on old surfaces, giving deposits which formed the 'sand plains' which are characteristic of parts of south-western Western Australia (Mulcahy 1960). Relief was low at all times within most of the western craton, so that phases of uplift and dissection exposed only layers of the earlier weathered materials, and detritus was initially moved only short distances. The form that the weathering took has been described as 'etching' (Finkl and Churchward 1975), and the uneven amounts of stripping of the country rock, which had already been etched to varying degrees, produced a number of complex soil and topographic patterns during the Quaternary.

Features somewhat similar to those of the western part of the continent occur in the western parts of the eastern craton. Thus there are the *red earths* on deeply weathered sediments in the 'hard mulga' country of south-western Queensland and redistributed sandy *red earths* which make up the 'soft mulga' country (Chapter 18, Region VII).

Soils on basalt

Tertiary and Quaternary basalts are widely distributed in a belt throughout eastern Australia with the youngest rocks at the northern and southern extremities. The oldest of the more extensive rocks are Late Oligocene to Early Miocene and covered an uneven Oligocene landscape, which included deep-weathered materials. Some of these basalts have in turn been deeply weathered, as at Toowoomba, Kingaroy, and the Binjour Plateau in south-east Queensland. At Binjour, this weathering has penetrated to 60 m (Teakle 1950). Younger slopes, cut below the original slightly uneven surface and forming the fringing scarps of the plateaux, show examples of the different layers of the complete laterite profile. These layers have also been cut by dissection on the surfaces of the old plateau remnants and younger soils have been derived by re-organization of the materials of these layers, probably during the Late Miocene and Pliocene. The dominant soils in these situations are lateritic *krasnozems*, with *xanthozems* which contain lateritic materials just above drainage lines.

Soils associated with Late Cainozoic landscapes

Soils on dissected deeply weathered materials

With the dissection of former lateritic profiles in the southern part of the central craton, *lateritic podzolic soils* formed from the lower horizons of deeply weathered material in the Mount Lofty Ranges and on Kangaroo Island (Stephens 1946). With continuing uplift, as in the Mt Lofty Ranges, such soils were left as relicts, or were slightly modified, and modern soils, *red* and *yellow podzolic soils*, have formed on younger slopes below them, in much less weathered parent materials.

Grey clays, of the 'brigalow country' of eastern Queensland (Chapter 18, Region VII), partly within the western limits of the eastern highlands, and in north-eastern New South Wales, have been derived from fine-textured Mesozoic sediments. In places these rocks have been deeply weathered to as much as 30 m below surfaces younger and lower than those with the ferruginous deep weathering. These younger deeply weathered surfaces may date from the Later Miocene or Pliocene. Within this area, there are a number of progressively younger surfaces (possibly Late Pliocene and Quaternary) with distinctive soil and gilgai patterns (van Dijk, pers. commun.), but the reasons for the variety of soil features are not understood, nor is it feasible yet to date the younger surfaces.

Soils with the forms of *solodized solonetz, solodic soils*, and *soloths* occur in every State, and from tropical or subtropical summer rainfall areas in Queensland and northern New South Wales to winter rainfall areas, such as South Australia and south-western Western Australia. They occur on both alluvial plains and on slopes, generally on medium-textured parent materials, some of which were part of deeply weathered profiles.

Those on slopes which have developed since the dissection of the deeply weathered profiles grade into the lower levels of these profiles. Catenary relationships are evident at many places, as in eastern Queensland (e.g. Gunn 1967). The actual sequence may depend to a large degree on the textures of the underlying materials, e.g. Beckmann and Thompson (1960) have shown the following sequences:

(a) deep sandy *yellow podzolic* with A horizon to 1 m – sandy *yellow podzolic* – sand-surfaced *solodized solonetz* (just above the valley floors); and
(b) *red podzolic/soloth* – *solodic soil* with coarse textured surfaces – *solodic soil* with thinner, fine textured surface – *grey clays*.

Some profiles may be composite; i.e. the surface layer may be a separate depositional layer from the subsoil, as was concluded by Oertel and Blackburn (1970) from studies of a *solodic soil* on an alluvial plain in South Australia. It may also be the situation on pediment slopes where there is a progressive fining of texture downslope, in accord with the distribution of grain sizes downstream in alluvial systems. Later pedogenesis may have redistributed clays and ions within the soil horizons.

It is evident from a study of landscape relationships that many *solodic soils* are paleosols. Soil formation on slopes and on adjoining alluvial plains commenced before streams cut down below the plains, and continued after that. As Downes (1954) suggested, solodic soils may be polygenetic soils whose mode of formation is also related to climatic change or, as outlined above, they may be composite soils with a complex history. Their age is uncertain, but many of the surfaces on which they are found appear to be older than later Quaternary and younger than Miocene.

As discussed in Chapter 9, the sodium which is a feature of these soils may have come from groundwater, atmospheric accession of salts, weathering of sodium feldspars in parent rocks, or from combinations of these sources.

Soils on basalt

Krasnozems, deep red acid clay soils, have developed on younger, stable, moderately sloping to flat surfaces, probably of Pliocene and Pleistocene age, with present rainfall in excess of 900 mm year^{-1} and in some areas receiving over 2500 mm year^{-1}. They occur along the east coast from north Queensland to Tasmania, but, while many can be considered to be developing under modern climates, there has been disagreement about the importance of current climatic influences. Hallsworth (1951) claimed that an annual rainfall of at least 1250 mm was necessary for their formation in the Lismore district of northern New South Wales. Teakle (1952) suggested that these krasnozems were a relict from an earlier wetter period, while Nicolls and Tucker (1956) believed that they represented an advanced stage of weathering of the local black earths.

Euchrozems, moderately deep, neutral red soils (Hallsworth *et al.* 1953; Isbell *et al.* 1977), occur on flat to gently sloping stable surfaces on basalts and buried clays in areas whose present annual rainfall is of the order of 600 mm. Although these surfaces are now being reduced by denudation, many would have been in existence during the Quaternary, and so subject to the climatic changes of that period. Whereas *krasnozems* contain dominantly kaolinite and hematite in their clay fraction, *euchrozems* contain some 2:1 lattice clays as well, particularly in their lower horizons. They appear to represent stages in a trend of weathering made under subhumid environments in well-drained sites. The trend with increasing time is towards slightly deeper clay soils, with profile features like those of *krasnozems*.

Shallow *black earths* occur in shallow depressions, on concave slopes and on low rises in areas with rainfall the same as, or slightly higher than, the *euchrozems*. They are neutral to alkaline soils, dominantly montmorillonitic in the clay fraction, and their presence may be related to a weathering environment controlled largely by topography as well as by climate. They have been thought of as the first stages in a weathering sequence which leads with time to the formation of *krasnozems* (Nicholls and Tucker 1956), or as the end point of weathering in areas with restricted drainage (Hosking 1940). On the other hand, Beckmann *et al.* (1974) suggest that only montmorillonite clays could be produced on sloping sites where the runoff is high and little water is available for weathering. This trend continued during landscape development (backwearing of scarps and pediment production), so that a dominantly *black earth* soil landscape resulted. This could occur also in higher rainfall areas in which the topography was favourable. Black earths are thus commonly associated with surfaces which are younger than those for the *euchrozems*, but which have been stable during the later Quaternary.

The Newer Volcanics of Western Victoria, of Pliocene to Recent age, are largely basalts with some pyroclastics. The soil pattern is quite complex. Relief is low, most major slopes are moderate, and soil types depend on age of extrusion, structure of the lava, and local drainage conditions. Young flows, e.g. of ropy lava, and stony rises are often little weathered. Other flows have deeper soils, which locally may include *krasnozems, red-brown earths,* and *black earths* on the one flow (Ollier and Joyce 1976).

Sargeant and Skene (1970) observed that *krasnozems* on the older of these Newer Basalts had developed under higher rainfall than *krasnozems* on the younger basalts. They also noted the presence of quartz as the main component of the silt and fine sand fractions. Jackson *et al.* (1972) found that the amount of quartz increased with the age of the landscape. The quartz is most likely to be aeolian, as quartz sand dunes are found to windward (westward), and soils on flows nearer to the sources have higher quartz contents than soils on flows further away.

Soils on inland fans and plains

Throughout the Quaternary there were repeated deposition and incision in the upper reaches of streams of the Murray – Darling system. A series of terraces was developed along the streams. Each terrace, or set of terraces, had a characteristic soil pattern or sub-solum weathering pattern. Duplex (texture contrast) soils with mottled deep subsoils occur on higher terraces of the minor streams of the Albury–Wodonga area (van Dijk and Rowe 1980), and weakly differentiated soils on the younger terraces. *Red-brown earths* occupy fans derived from a variety of parent rocks along the western slopes of the highlands, with *grey clays* below and beyond them. Climatic change has been cited as the cause of the terrace development, and of changing conditions on hillslopes (Butler 1959). On such slopes, the soils which developed during a succession of stable phases within the Quaternary have been *red-brown earths.*

Red-brown earths and *grey clays* probably continued to develop through several phases of changing conditions, with rate of profile development being slower in colder, arid times, and more rapid in warmer and moister times. As soils in these environments are polygenetic to varying degrees it is difficult to use them in regional correlations on the basis of profile form and characteristics, or on position in sequences of K-cycles (*cf.* Walker 1962).

Soils on coastal alluvial systems

The repeated falls of sea level, to 100 m or more below the present, during the later Tertiary and Pleistocene, rejuvenated all the streams and induced erosion of adjoining surfaces, in places exposing fresh rock to weathering and hence permitted soil formation on the new surfaces. A succession of alluvial bodies was deposited during the later Quaternary along east coastal streams with each rise of sea-level, and these were left as terraces during later falls of sea-level (Chapter 6). A typical pattern of soil development on these terraces, from highest and oldest to lowest and youngest, has been established for eastern New South Wales (Walker and Coventry 1976). This sequence seems to represent the situation along the central coast of the continent at least.

This may be related to progressively improved drainage and age differences assuming that the initial alluvial parent materials had similar compositions. Climates may have varied from slightly drier during low sea level stages to slightly wetter when sea level was higher, but there has been a common pathway of soil development with time. The alluvium on the youngest terraces still shows alluvial layering (the stratic stage), but this has disappeared from soils of the next oldest level, in the minimal *prairie soils* of the cumulic stage. Soils on terraces above these exhibit low contrast to high contrast sola, i.e. they are weakly differentiated to strongly differentiated *red* and *yellow podzolic soils.* Older alluvial materials above these, in remnants on lower valley walls, may have *red earth* profiles, or texture contrast profiles, with much better developed subsoils.

Soils on aeolian materials

The parallel longitudinal dune systems in the old basin areas of the centre of the continent, such as the Simpson and Great Victoria Deserts, originated about 300 000 years ago (Bowler 1976), the time of active growth being correlated with world glacial maxima.

The soils of the dunes are largely red *siliceous sands* and *earthy sands* derived from deflation of laterite and silcrete material, and of redistributed weathered products from these. There has been minimal soil development on these dunes, even in intermittently somewhat moister intervals of the later Quaternary. The following factors have precluded anything but some transport of clay to the lower layers: the advanced stage of prior weathering of the source materials, their coarse texture, and the overall low rainfall of the area.

Crescentic dune systems, lunettes (Hills 1940), on the lee side of lakes in southern Australia are composed of material blown from exposed lake floors during dry periods of the Quaternary. A wide range of textures is represented in the materials of the dunes – from quartz sand to sandy clays. Some dunes are almost pure gypsum and many have an internal core of quartz sand overlain by clay and gypsum. A range of soils occurs, related to parent material and age. Some dunes show little profile development (viz. some of the gypsum dunes), while others, which in New South Wales and Victoria have strongly developed soils, have been related to the penultimate glacial episode of about 120 000 years ago (Bowler 1976).

Aeolianite dunes in southern Australia and siliceous dunes along the east coast both have *podzols* as the end product of soil development with time. Continuous leaching since the aeolianites of south-eastern South Australia accumulated about 700 000 years ago (Cook *et al.* 1977) has progressively altered these sands during several periods in which the climate changed.

The siliceous sand dunes of the east coast are in areas which now have annual rainfall of at least 1500 mm. While this may have been lower at times when sea-level was lower, there has been sufficient rainfall during interglacial periods for leaching to produce a series of *podzols* in well-drained sites on progressively younger dune systems. The oldest 'giant' *podzols* have A horizons 14 m thick (Thompson and Hubble 1980), while *podzols* on the youngest dunes have A horizons as thin as 20 cm.

A similar progression of soil development is evident on beach ridges on the coast. The older sets have *siliceous sands* and *podzols*, while the younger, which include those currently forming, have *siliceous* or *calcareous sands*, depending on the parent material.

Aeolian dusts are common in south-eastern Australia, and parna, an aeolian clay (Butler 1956 and Chapter 7), is believed to have mantled both plains and hillslopes during the Quaternary, but is most distinctive and best developed in association with alluvium on the plains (Churchward 1963). Such materials are calcareous and the well-developed soils on them include *red-brown earths* and *grey, brown* and *red clays*, depending on topographic sites. Younger mantles may have well-developed profiles in the upper layers, indicating either that soil formation may have proceeded more rapidly in later periods, or that the aeolian sediments were exposed for shorter periods earlier in the Quaternary.

Organic soils

These are all young soils and include *humic gleys, acid peats, neutral* to *alkaline peats*, all of which show the effects of ground water in their profiles, and the *alpine humus soils. Humic gleys* and *neutral* to *alkaline peats* are common in lowlands in coastal areas; the *acid peats* are more common in highlands to the south; and the *alpine humus soils* occur exclusively in these highlands, including areas which were glaciated in the Late Pleistocene.

In contrast to similar soils elsewhere, the *alpine humus soils* are found not only above the tree line in southern New South Wales, but also under woodlands; here they attain their best development (Costin *et al.* 1952) and at these altitudes they are deeper than their overseas equivalents. Their profiles are dominated by organo-mineral horizons which may penetrate decomposing rocks.

Physical weathering, which produces the coarser fractions of these soils, is the dominant weathering process at higher altitudes, while chemical weathering, producing the finer fractions, is more important at lower elevations.

Comparison of Australia with some other continents

For much of its geological history Australia was part of Gondwanaland, along with Africa, South America and India, prior to the disruption during the Cretaceous and the subsequent drifting apart through the earlier Tertiary. During this time it had a similar geomorphic history to the other continents – epeirogenic uplifts rather than orogenies (i.e. intense folding movements were absent), and broad warps resulted in areas of uplift and sagging, the latter producing large depositional basins. Distinct geomorphic surfaces which developed as a result of the periodic uplifts have been recognized in each of the continents, and a consistent sequence has been recognized (King 1962; 1976).

There are some similarities in the distribution of present climatic zones in these continents (e.g. Rognon and Williams 1977). While the continents moved during the Early Tertiary and when they became stationary in the Late Tertiary and Quaternary, they have always been in low to moderate latitudes. This indicates that throughout the recent periods of landscape and soil development the environments have been warm to cool with the only glaciations on high mountains, where elevation compensated for low latitude. One might therefore expect some similarities in soils and soil patterns between the Gondwanaland continents, and indeed this is so.

There are also some similarities with low latitude areas on the southern parts of the northern supercontinent, Laurasia, particularly in south-eastern United States and in parts of southern Europe (Chapter 3). The discussion here concerns some comparable soil–landscape relationships, and, where relevant, parallel climatic sequences. Some examples of common soil–landscape systems illustrate these relationships.

Surface materials of many of the distinct planation surfaces on the Gondwanaland continents, ranging in age from Cretaceous to Early Tertiary (the 'Moorland Planation' of King 1976), have been affected by lateritic weathering, and different forms of weathering have been recognized on the different surfaces (Vaidyanadhan 1965; Hays 1967). *Red* and *yellow earths*, which occupy a larger proportion of South America and Africa than they do of Australia, are associated with these surfaces and also with others less easy to distinguish but of later Tertiary and Quaternary age. These latter surfaces occur in North Queensland (Coventry 1978), in Latin America (Sanchez and Isbell 1979), and in southern Sudan (Fölster *et al.* 1971). The earth soils are more commonly developed on coarse-textured and more acid parent materials, sandstones, acid igneous rocks and young sediments derived from these rocks.

There are problems in assigning all such soils to previous weathering environments, as some, particularly in Africa and Latin America, are currently in humic tropical areas and may be forming at present, even though they are situated on old surfaces (Sanchez and Isbell 1979; van der Eyk *et al.* 1969).

Deep red clay soils on basic igneous rocks and similar clay-forming parent materials, under similar conditions, have weathered in a similar fashion to the red earths. Such soils occur in Hawaii, parts of Brazil, and in southern Africa as well as in eastern Australia (Isbell 1980). There are also similarities in the catenary relationships which exist on both old and modern surfaces. A toposequence of lateritic *krasnozems* and a *xanthozem*, occurring on a slightly dissected plateau surface in eastern Australia (Thompson *et al.* 1954), has elements in common with those of two catenas on a gneissic rock in the Sudan (Greene 1945).

A common catenary relationship of soils on comparatively young slopes on basic igneous rock includes *krasnozems* or *euchrozems* on better drained upper slopes and dark clay soils on less well-drained and/or more eroded lower hill slopes. This relationship is found in many parts of the world; e.g. Israel (Singer 1966), Kenya (Kantor and Schwertmann 1974), South Africa (Oberholster 1969), as well as in eastern Australia (Nicolls and Tucker 1956; Beckmann *et al.* 1974). While local variations in parent material and age of surface may have had some effects, differences in drainage controlled by local topography have been the dominant factor in the development of these catenas and soils.

Soils with solodic morphology also occur in other continents, e.g. in southern and western Africa, though nowhere as extensively as in Australia. In southern Africa they are associated with pedimented landscapes (van der Eyk *et al.* 1969), which can be assigned to King's (1976) 'Widespread' Pliocene planation or to the Quaternary 'Youngest' planation, a situation very similar to that in eastern Australia. Another example of close correspondence of soil landscapes in Africa and Australia is that of a toposequence in the Chad Basin (Bocquier 1968), in which there are sandy soils on upper slopes and *solodized solonetz* on footslopes of pediments on granite hills. This is very similar to the situation on such slopes in Australia. That the same controls seem to be operating in Australia and in Africa is seen from this study and that of van der Eyk *et al.* (1969, p.49).

It is thus apparent that similar soils, at the great group or suborder level, occur in Australia and the other post-Gondwana continents, South America, Africa and India, and there are also some similar soil landscapes within each of the continents. The continents contain broadly similar parent materials and have had corresponding climatic and geomorphic histories since the Cretaceous, and certain soil-forming environments appear to have occurred in them at the same time, or at least during corresponding periods. The relative importance of soils within the continents is not the same, however, e.g. dark clay soils are widespread in Africa, India, and Australia, but are relatively uncommon in South America (Isbell 1978).

There appear to be three main reasons for differences in soil pattern: differences in the proportions of the different parent materials; differences in the amount of uplift; and differences in the distribution of past or present climatic patterns across each continent.

(i) Large areas of southern Africa have a clearly defined sequence of erosion surfaces (King 1976), equivalents of which are also present in eastern Australia, but are less well-defined, i.e. the vertical separation is much less. In southern Africa these occur on large areas of Archaean nuclei, acid crystalline rocks whose only

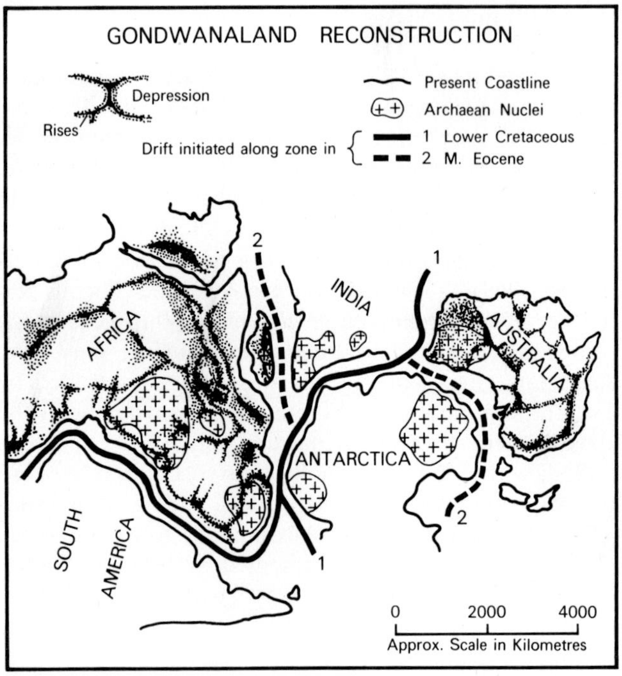

Fig. 11.1. Reconstruction of Gondwanaland continents (after Johnstone *et al.* 1973) showing distribution of Archean nuclei and the relation of these to later tectonic basins and rises in eastern Africa and in Australia. Tectonic basins, etc., of Africa, from A. Holmes 'Principles of Physical Geology', 2nd edn, (Nelson 1965), and those of Australia adapted from Fig. 4.1.

equivalents in Australia are in the western and central cratons (Johnstone *et al.* 1973, Fig. 2; White 1972, Fig. 1). These are areas which have been exposed to subaerial denudation and to weathering for hundreds of millions of years, apart from partial covering by continental ice during the Permian (see Chapter 4), and have experienced only minimal uplift. Areas of Australia with the greatest uplift, parts of the eastern highlands, have younger rocks, many of which escaped the deep weathering of the Early Tertiary in the west (Fig. 11.1).

(ii) Uplift in Africa, since the Miocene at least and possibly as far back as the Eocene, has been continually greater than in any other of the former Gondwanaland continents (Bond 1978), the differences being greatest when compared with uplifts in Australia. The zones of uplift which produced several stepped surfaces in Africa are matched by internal depositional basins filled with thick sediments eroded from the adjoining highlands. In the same periods the uplifts also induced vigorous denudation along coastal streams in both Africa and Australia.

India also experienced greater uplift than Australia as a whole, with Tertiary surfaces at about 1800 m (Lower Eocene), 900 m (Middle Miocene), 600 m (Lower Pliocene) and 360 m (Upper Pliocene) (Vaidyanadhan 1965). It is only in the eastern highlands of Australia that such a separation could exist, and in that region the differences are much less, and the distinctions between the surfaces are less clear. Broad warps, with little vertical separation between surfaces as a consequence, are features of Brazil and Western Australia.

(iii) Modern desert areas, such as the Kalahari Desert of south-west Africa and the much more extensive deserts of inland Australia, are associated with modern high pressure systems. Similarities of climatic pattern between southern Australia and parts of the Sahara region have been discussed by Rognon and Williams (1977), and there are some similarities in the soil patterns of the arid areas.

Little of tropical and subtropical Australia has the modern high rainfalls of central Africa and of parts of Brazil (Isbell 1978), although there are similarities in amount and annual distribution along corresponding latitudes on the east coast of each continent. The climatic patterns over Australia and Africa during the Quaternary differed also. The effects of eustatic sea-level changes would have been minor in Africa, where the relative relief was greater. Africa's continental outline was almost constant and there would have been very little effect on circulation patterns by such changes in level of the sea. Climates in Australia, on the other hand, were affected by intermittent closure of Torres and Bass Straits (Webster and Streten 1978), through the effects on oceanic and thence atmospheric circulations.

Australia's soil patterns can be attributed to the effects of changing environments on parent materials (often preweathered) which have been exposed by intermittent moderate tectonic uplifts, followed by accelerated erosion. Successive soil developments in more stable areas have been produced by climatic changes which have induced alternating phases of erosion and deposition, and of stability and weathering. In Africa, on the other hand, tectonics, especially epeirogenic uplifts, throughout the Tertiary and Quaternary, may have been more important than climatic change in producing the overall soil pattern, though not the soils themselves, whose presence depends on climate and local topography. Before the Late Tertiary and Quaternary histories can be understood, surfaces and soils will have to be correlated by many further studies.

References

Beckmann, G. G. and Thompson, C. H. (1960). Soils and Land Use in the Kurrawa Area, Queensland. CSIRO Aust. Soils Land Use Ser. No. 37.

Beckmann, G. G., Thompson, C. H. and Hubble, G. D. (1974). Genesis of red and black soils on basalt on the Darling Downs, Queensland, Australia. *J. Soil Sci.* **25,** 265-81.

Bocquier, G. (1968). Biogeocenoses et morphogenese actuelle de certain pediments du Bassin Tchadien. Trans. 9th Internat. Congr. Soil Sci., Vol. IV, pp. 605-14.

Bond, G. (1978). Evidence for late Tertiary uplift of Africa relative to North America, South America, Australia and Europe. *J. Geol.* **86,** 47-65.

Bowler, J. M. (1976). Aridity in Australia: age, origins and expression in aeolian landforms and sediments. *Earth Sci. Rev.* **12,** 279-310.

Butler, B. E. (1956). Parna, an aeolian clay. *Aust. J. Sci.* **18,** 145-51.

Butler, B. E. (1959). Periodic Phenomena in Landscapes as a Basis for Soil Studies. CSIRO (Aust.) Soil Publ. No. 4.

Churchward, H. M. (1963). Soil Studies at Swan Hill. II. Dune moulding and parna formation. *Aust. J. Soil Res.* **1,** 103-16.

Cook, P. J., Colwell, J. B., Firman, J. B., Lindsay, J. M., Schwebel, D. A. von der Borch, C. C. (1977). The late Cainozoic sequence of south-east South Australia and Pleistocene sea-level changes. *BMR J. Aust. Geol. Geophys.* **2,** 81-8.

Costin, A. B., Hallsworth, E. G. and Woof, M. (1952). Studies in pedogenesis in New South Wales. III. The Alpine humus soils. *J. Soil Sci.* **3,** 190-218.

Coventry, R. J. (1978). Late Cainozoic geology, soils and landscape evolution of the Torrens Creek area, north Queensland. *J. Geol. Soc. Aust.* **25,** 415-27.

Downes, R. G. (1954). Cyclic salt as a dominant factor in the genesis of soils in south-eastern Australia. *Aust. J. Agric. Res.* **5,** 448-64.

Finkl, C. W. and Churchward, H. M. (1975). The etched landsurfaces of southwestern Australia. *J. Geol. Soc. Aust.* **20,** 295-307.

Fölster, H., Kalk, E. and Moshrefi, N. (1971). Complex pedogenesis of ferallitic savanna soils in southern Sudan. *Geoderma* **6,** 135-49.

Greene, H. V. (1945). Classification and use of tropical soils. *Soil Sci. Soc. Amer. Proc.* **10,** 392-6.

Gunn, R. H. (1967). A soil catena on denuded laterite profiles in Queensland. *Aust. J. Soil Res.* **5,** 117-32.

Hallsworth, E G. (1951). An interpretation of the soil formations found on basalt in the Richmond-Tweed region of New South Wales. *Aust. J. Agric. Res.* **2,** 411-28.

Hallsworth, E. G., Colwell, J. D. and Gibbons, F. R. (1953). Studies in pedogenesis in New South Wales. V. The euchrozems. *Aust. J. Agric. Res.* **4,** 305-25.

Hays, J. (1967). Land surfaces and laterites in the north of Northern Territory. *In* 'Landform Studies from Australia and New Guinea'. (Eds J. N. Jennings and J. A. Mabbutt) pp. 182-210. (Aust. Nat. Univ. Press: Canberra.)

Hills, E. S. (1940). The lunette: a new landform of aeolian origin. *Aust. Geogr.* **3,** 13-21.

Holmes, A. (1965). 'Principles of Physical Geology.' 2nd edn (Nelson: London.)

Hosking, J. S. (1940). The clay mineralogy of some Australian soils developed from granite and basaltic parent material. *J. Counc. Sci. Ind. Res. Aust.* **13,** 206-16.

Isbell, R. F. (1978). Soils of the tropics and sub-tropics: genesis and characteristics. *In* 'Mineral Nutrition of Legumes in Tropical and Subtropical Pastures. (Eds C. S. Andrew and E. J. Kamprath) pp. 1-20. (CSIRO: Melbourne.)

Isbell, R. F. (1980). A comparison of the red basaltic soils of tropical North Queensland and those of Hawaii, Mauritius, Brazil and Natal. Proc. Conf. Class. Manage. Trop. Soils, 1977 (Ed. K. T. Joseph) pp. 214-22. Kuala Lumpur, Malaysia. Malays. Soc. Soil Sci.

Isbell, R. F., Gillman, G. P., Murtha, G. G. and Jones, P. N. (1977). Brown basaltic soils in north Queensland. CSIRO Aust., Div. Soils Tech. Pap. No. 34.

Jackson, M. L., Gibbons, F. R., Syers, J. K. and Mokma, D. L. (1972). Eolian influences on soils developed in a chronosequence of basalts of Victoria, Australia. *Geoderma* **8,** 147-63.

Johnstone, M. H., Lowry, D. C. and Quilty, P. G. (1973). The geology of south-western Australia – a review. *J. R. Soc. W. Aust.* **56,** 5-15.

Kantor, W. and Schwertmann, U. (1974). Mineralogy and genesis of clays in red-black soil sequences on basic igneous rocks in Kenya. *J. Soil Sci.* **25,** 67-78.

King, L. C. (1962). The Morphology of the Earth. (Oliver and Boyd: London.)

King, L. C. (1976). Planation remnants upon high lands. *Z. Geomorphol. N.F.* **20,** 133-48.

Mulcahy, M. J. (1960). Laterites and lateritic soils in south-western Australia. *J. Soil Sci.* **11,** 206-25.

Nicolls, K. D. and Tucker, B. M. (1956). Pedology and Chemistry of the Basaltic Soils of the Lismore District, N.S.W., CSIRO (Aust.) Soil Publ. No. 7.

Oberholster, R. E. (1969). Genesis of two different soils on basalt. 1. Mineralogical characteristics. *Agrochemophysica* **1,** 53-62.

Oertel, A. C. and Blackburn, G. (1970). Pedogenesis of a solodized solonetz, based on duplicate soil profiles. *Aust. J. Soil Res.* **8,** 59-70.

Ollier, C. D. and Joyce, E. G. (1976). Newer volcanic landforms. *In* 'Geology of Victoria' (Eds J. B. Douglas and J. A. Ferguson) pp. 342-51. Geol. Soc. Aust. Spec. Publ. 5.

Rognon, B. and Williams, M. A. J. (1977). Late Quaternary climatic changes in Australia and North Africa: a preliminary interpretation. *Paleogeogr. Paleoclimatol. Paleoecol.* **21,** 285-327.

Sanchez, P. A. and Isbell, R. F. (1979). A comparison of the soils of tropical Latin America and tropical Australia. *In* 'Pasture Production in Acid Soils of the Tropics'. (Eds P. A. Sanchez and L. E. Tergas). Proc. Seminar held at CIAT, Cali. Colombia, pp. 25-52.

Sargeant, L. J. and Skene, J. K. M. (1970). Some properties of the krasnozems of southern Victoria, Australia. *Aust. J. Soil Res.* **8,** 281-95.

Singer, A. (1966). The mneralogy of the clay fraction from basaltic soils in the Galilee, Israel. *J. Soil Sci.* **17,** 136-47.

Stephens, C. G. (1946). Pedogenesis following the Dissection of Lateritic Regions in Southern Australia. CSIR (Aust.) Bull. No. 206.

Teakle, L. J. H. (1950). Description of a laterite profile near Gayndah, Queensland. *Queensl. J. Agric. Sci.* **7,** 15-23.

Teakle, L. J. H. (1952). An interpretation of the occurrence of diverse types of soils on basalts in northern New South Wales and Queensland. *Aust. J. Agric. Res.* **3,** 391-408.

Thompson, C. H. and Hubble, G. D. (1980). Sub-tropical podzols (Spodosols and related soils) of coastal Eastern Australia: profile form, classification and use. Proc. Conf. Class. Manage. Trop. Soils, 1977 (Ed. K. T. Joseph) pp. 203-13. Kuala Lumpur, Malaysia. Malays. Soc. Soil Sci.

Thompson, C. H., Hubble, G. D. and Beckmann, G. G. (1954). A soil survey of the Middle Ridge area, Parish of Drayton, Darling Downs, Queensland. CSIRO (Aust.) Div. Soils Div. Rep. No. 5/54.

Vaidyanadhan, R. (1965). An outline of the geomorphic history of India south of No. latitude 18. Seminar Volume, Saugar, pp. 121-30.

van der Eyk et al. (1969). Soils of the Tugela Basin, Natal Town and Regional Planning Reports, Vol. 15. Town and Regional Planning Commission, Natal.

van Dijk, D. C. and Rowe, R. K. (1980). Soil stratigraphy in the Murray Valley at Albury–Wodonga: a new approach to surficial stratigraphy. *Proc. R. Soc. Vic.* **91,** 109-25.

Walker, P. H. (1962). Terrace chronology and soil formation on the south coast of N.S.W. *J. Soil Sci.* **13,** 178-86.

Walker, P. H. and Coventry, R. J. (1976). Soil profile development in some alluvial deposits of eastern New South Wales. *Aust. J. Soil Res.* **14,** 305-17.

Webster, P. J. and Streten, W. A. (1978). Late Quaternary Ice Age climates of tropical Australasia: interpretations and reconstructions. *Quat. Res.* **10,** 279-309.

White, A. (1972). Deep erosion by continental ice sheets. *Geol. Soc. Am. Bull.* **83,** 1037-56.

Part 3
Australian soil landscape regions

Chapters 12-19

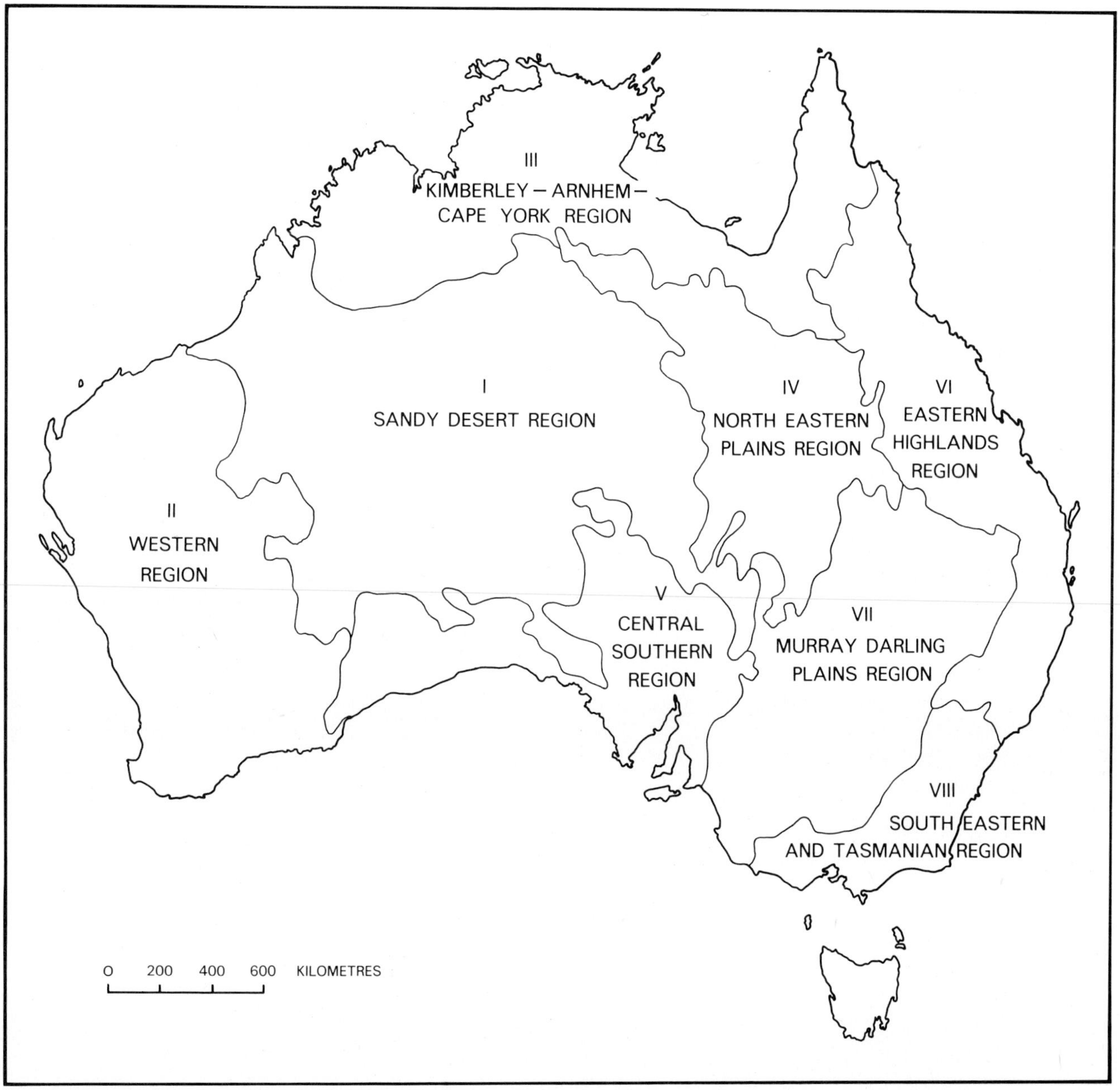

Soil landscape regions of Australia

PART TWO OF THIS SECTION HAS DEALT WITH THE various processes involved in the formation of Australian landscapes and soils, and hence has given an insight into the factors controlling their general distribution.

This section, Part 3, considers the regional soil patterns in more detail. In particular it considers the kinds of soil found in various parts of Australia, and attempts to offer an explanation of the more local factors controlling distribution. Brief reference is also made to relationships between the soils and their use; special attention is given to the way soil formation and history have influenced the use of soils.

A glance at the several more recent small-scale soil maps of Australia, e.g. Stace *et al.* (1968); Northcote *et al.* (1957)*, leaves the impression of great complexity and little order in soil distribution patterns, but closer study enables the delineation of a number of large regions, each of which is unique in terms of its soil landscapes. The unifying regional properties include major kinds of soils, their evolutionary history, and often their climate. It is also possible to further subdivide these major soil regions, which are shown in the map opposite, into a number of constituent soil landscape provinces. These may vary considerably in complexity, but again each is characterized by a particular array of soils that have formed either now or in the geological past because a certain set of processes have operated on a particular assemblage of parent materials to produce particular landscapes and soils. This subdivision, first of the continent and then of each region, makes possible more orderly treatment by which the continent-wide complexity may be understood.

*Northcote, K. H., Hubble, G. D., Isbell, R. F. Thompson, C. H. and Bettenay, E. (1975). 'A Description of Australian Soils'. (CSIRO: Melbourne.)

Stace, H. C. T., Hubble, G. D., Brewer, R., Northcote, K. H. Sleeman, J. R., Mulcahy, M. J. and Hallsworth, E. G. (1968). 'A Handbook of Australian Soils'. (Rellim: Glenside, S. Aust.)

Chapter 12
Sandy Desert Region (I)

K. H. Northcote and M. J. Wright

Region I is a vast, and largely interior, area extending from near the Queensland border to the 80-mile beach in Western Australia. It is composed of extensive erosional and depositional plains with salt lakes, and sandy dunefields, out of which mesas, buttes and ranges rise sharply. Although the greater part of the region is arid desert, there is very little that is not at least sparsely vegetated (Carnahan 1976).

A significant proportion of the Region is gazetted as Aboriginal reserves; much of the sandy dunefields outside these is unalienated, and land use in the remainder is confined almost exclusively to extensive beef cattle grazing. There are some mining ventures and the limited sheep grazing is virtually confined to the southern margin of the Region.

Average annual rainfall (Fig. 12.1) ranges from less than 100 mm in the driest part, the Simpson Desert, to more than 300 mm in the north where it is summer dominant; southwards the winter rainfall component increases and is dominant in the coastal strip of the adjoining Region V. Variability in annual rainfall ranges between 30 and 50%. Air temperatures range from average maxima of 33–36°C during January to 18–24°C during July with average minima of 18–24°C and 6–9°C during those same months.

The most significant aspect of the natural vegetation is that all plants are usually widely spaced, with a foliage cover of often less than 10% and rarely up to 30%. The northern part of the Region, from the Great Sandy Desert in the north-west to the Simpson Desert in the east, is largely open woodland, mainly *Acacia aneura*, with hummock grasses (spinifex), particularly *Triodia pungens* and *T. basedowii*. *Hakea*, *Grevillea* and mallee species of *Eucalyptus* are also prominent.

In the southern part this vegetation is replaced by low open woodlands of *Eucalyptus* spp. with hummock grasses (*T. basedowii*). The characteristic trees of the Great Victoria Desert are *E. gongylocarpa* and *Casuarina* spp. Features of the Region are extensive plains and basins underlain largely by sandstones and siltstones of Permian, Jurassic and Cretaceous age, with lesser areas of dolomites and limestones of similar age. These plains are part of a low plateau that begins near the coast of Western Australia and extends more than half-way across the continent; the plateau has an average elevation of about 600 m and is covered by a mantle of weathered sediments. It is in these surficial sediments, seldom adequately described on geological maps, that most soils of the Region have formed.

The ranges and hills of central Australia rise abruptly as much as 900 m above the general level of the plains. Proterozoic granites, gneisses, gabbros, basalts, acid volcanics, conglomerates and quartzites are widely exposed in these uplands (Macdonnell province, Fig. 12.1).

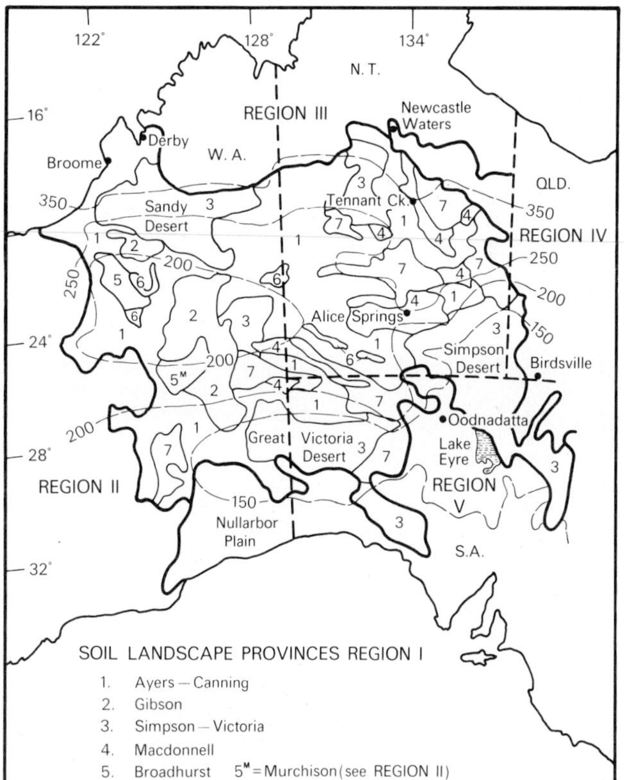

Fig. 12.1. The soil-landscape provinces of Region I.

Outline of landscape history

The landscapes of the Region are closely linked with basins that formed early in geological history and have persisted to the present day. Examples are the Precambrian Canning and Officer Basins which are the loci of the Great Sandy and Great Victoria Deserts (Simpson–Victoria Province, Fig. 12.1).

Of the uplands, the Macdonnell Ranges, for instance, have been subjected to several phases of downwearing and uplift since their initial formation in the Middle Precambrian, about 1500 m years ago. Perhaps the most spectacular uplift occurred following the Ordovician deposition when quartzites were folded and uplifted to form an immense mountain range 3000–4500 m high (Laseron 1972). The remnants of this range, subjected to less intense movements following deposition of Devonian conglomerates on its flanks, form the present Macdonnell Ranges.

In other parts, differential erosion of Cambrian sandstones and conglomerates has produced features such as Ayers Rock and Mt Olga. But apart from these major events, geological history for much of the Region has mainly involved a series of gentle upwarps followed by long periods of quiescence and erosion, with deposition in local basins, until the Cretaceous when a large part of the Region was inundated by marine incursions (Part 2).

Following the widespread Cretaceous sedimentation subsequent upwarping converted the western half of Australia into a vast plateau (Mabbutt 1969) probably never much more than 1500 m high. The upwarping was gradual, continued throughout the Tertiary, and allowed rivers to cut steep gorges in the central Australian mountains (Quinlan 1962). Elsewhere water was also the main erosive agent in the final fashioning of ridges, hills and ranges with their resultant shapes determined by their individual rock structures (Mabbutt 1962).

During the Quaternary, conditions oscillated between wetter and drier phases (Bowler 1976; Lamb 1977). The drier phases have been arid enough to destroy vegetation and allow wind to play a major part in transport, resulting in the formation of major sand deserts with their striking dunefields. Even the current climate oscillates between wetter and drier phases, although the difference between the two is probably not as pronounced as in former times (Lamb 1977).

Soil-landscape Provinces

Seven soil-landscape provinces may be identified within the Region (Fig. 12.1) and their main features are summarized in Table 12.1. In the discussion that follows

Table 12.1. Main features of the soil-landscape provinces of Region I

Province	Geological and geomorphological features	Major soil features
Ayers–Canning	Low tablelands of ferruginous and kaolinized materials with laterite and silcrete, resulting from deep weathering of Permian, Jurassic and Cretaceous sandstones and siltstones during the Tertiary	Most commonly red *earthy sands* on the plains, with *siliceous sands* on the sand dunes and *grey-brown and red calcareous soils* on areas of calcrete
Gibson	Low tablelands similar to Ayers–Canning but derived from Cretaceous sediments	Mostly *red earthy sands* but with much ironstone gravel. Otherwise as for Ayers–Canning
Simpson–Victoria	Large, long established basin plains with sands derived from erosion of deep-weathering profiles and re-sorted during Quaternary arid periods	*Siliceous sands* occur on the widespread dunes. These may be red, yellow or whitish, depending on source, but red are strongly dominant. Shallower red *siliceous sands* in swales
Macdonnell	High, rugged ranges of various Proterozoic to Devonian rocks. Localized plains, at foot of ranges, developed on alluvium derived from Proterozoic basalts	*Lithosols* are the common soils of the ranges but areas of flanking *red clays* are included
Broadhurst	Narrow terraces bordering drainageways and old watercourses, calcreted during late Tertiary.	Commonly *grey-brown and red calcareous soils* with areas of *siliceous sands*
Amadeus	Plains with playa lakes on alluvium periodically deposited in ancient valleys, deflated and resorted by wind	Mainly *solonchaks* with flanking *siliceous sands* and variable dune deposits such as granular gypsum
Stuart–Burt	Stepped plains of multicyclic origin on mixed sediments, some derived from Tertiary deep-weathering profile but intermixed with products of weathering of Proterozoic to Ordovician rocks	*Red earths* are the most extensive soils with significant areas of red *alluvial soils, desert loams* and *solonized brown soils;* minor areas of red *siliceous sands, earthy sands, red clays* and *red calcareous soils*

an attempt will be made to interpret the features of each province in relation to the formative elements. The main soil studies have been made by Jackson (1962) and Litchfield (1962). Litchfield (1969) also elucidated the stratigraphy of two floodplains near Alice Springs.

Ayers–Canning Province

This Province is the most extensive in the Region and consists of gently undulating and often stepped plains or low tablelands; in places, low but moderately steep marginal scarps are a feature. A light scatter of ironstone gravel mantles the ground surface except on sand dunes and in small valleys. Other landscape features are laterite-capped mesas and buttes, low silcrete-covered rises and hills, rock outcrops of sandstones, quartzites and granitic rock on mounds, ridges or low hills, and narrow benches or mounds of calcrete along the margins of small valleys. While these landscape features are universal for the unit, they vary locally in their relative proportions, for example sand dunes may be common or entirely absent (Fig. 12.2).

The soils of the Province are diverse, but *earthy sands* are by far the most common on the plains; red *siliceous sands* are found on the sand dunes (see Simpson–Victoria Province), powdery calcareous loams (Um5.11) or *grey-brown calcareous soils* occur on calcrete (see Broadhurst Province).

The red *earthy sands* of the plains are hard when dry and have clayey sand to sandy loam texture, earthy fabric, and ironstone gravel on the surface and in the profile. They are free of salts and carbonates, and are slightly to moderately acid. The soils are underlain at 1–2 m by ironstone gravel, laterite and/or mottled and pallid-zone material. They have formed in a residuum of colluvial sand derived from the erosion and dissection of pre-existing lateritic materials.

Gibson Province

This Province is very closely related to the Ayers–Canning Province; it is a slightly higher upland plain or tableland with more relief, more frequent breakaways and considerably more laterite and ironstone gravel. The soils are *earthy sands* with ironstone gravel content > 60% throughout the profile. Massive laterite is common, often at depths as shallow as 45 cm, as well as above the breakaways.

The distribution of this soil landscape closely parallels that of Cretaceous sandstones and siltstones shown on the Geological Map of Australia (Bureau of Mineral Resources 1976). This area was lateritized during the Tertiary but the residuum derived by subsequent dissection and erosion is largely redistributed ironstone gravels, rather than sand as in the previous Province.

Simpson–Victoria Province

This Province forms one of the most extensive systems of desert dunes in the world. It includes all the Australian sandy deserts: the Great Sandy Desert, the eastern part of the Gibson Desert, Great Victoria Desert and Simpson Desert, as well as several smaller unnamed sandy deserts. Dunes of red *siliceous sands* are their common feature, but these vary in height (10–30 m), length (up to 300 km) and shape (Fig. 12.3). The common dune form is longitudinal and their direction in each sand desert is shown in Fig. 4.10, p.66. There are other shapes, including crescents and circular forms. Dune convergences (King 1960), represented in ground plan by an asymmetric 'y', are an outstanding feature that contrasts with the general parallelism of the dune system.

In cross-section also, individual dunes are asymmetric, having one slope steep and the other longer and more gentle (Mabbutt 1968; Mabbutt *et al.* 1969). The dunes are generally 200 to 500 m apart with an intervening swale or corridor which may also be of red *siliceous sands*

Fig. 12.2. Ayers–Canning Province showing gully erosion of earthy sands along a former earth road near the edge of the Victoria Desert, South Australia.

Fig. 12.3. Simpson–Victoria Province. Red sand dunes of the Gibson Desert showing some 60% of shrub and tussock grass cover.

which are shallower than those of the dune, and which overlie laterite, ironstone gravels, sandy clays or other materials. King (1960) shows trends not only of the distribution and orientation of the sand dunes but also the direction of y-shaped dune convergence.

The red *siliceous sands* have the following properties: a soft, loose surface, 'fixed' where vegetated but otherwise easily blown away; quartzose sand throughout, coated with iron oxide to give a characteristic red colouring (Folk 1976); free of salts and carbonates and mildly acid to depths of 2 m. The sand often continues deeper than 2 m but in some cases compacted clayey sands or earthy pans are found at that depth; in other cases Pleistocene sediments have been recorded below 6 m (King 1956). As well as the dominant red *siliceous sands*, smaller areas of yellow and whitish sands also occur on dunes where the source of the sand (e.g. drainageways on the desert margins) is different.

The dune formations illustrate the major role of wind in erosion and transportation during Quaternary arid cycles. The dunes are probably windrifts resulting from relatively local deflation, including a channelling action, and shepherding of loose sand from the adjacent erosional plains and playa lakes (King 1960). The forms of the dunes, the soils, and the materials below the soils, suggest that sand migration has only taken place over relatively short distances. The properties of the red *siliceous sands* and their geographic relationships with the erosional plains suggest that the sands were derived from the surface portion of the Tertiary deep-weathering profile.

Macdonnell Province

Although these uplands are geologically diverse it is convenient to consider them within the one soil landscape unit. Their common features are widespread exposures of bare rock which are interspersed with shallow soils. They possess considerable relief, with the highest peak, Mt Liebig, rising 1524 m above sea level, and their structure indicates that erosion has reduced them to about a third of their original size (Fig. 12.4). There are many spectacular mountain gorges and valleys, where dense vegetation with vines, lianas, cycads and palm trees indicates the earlier existence of a flora which was very different to that of the present surrounding desert (Chippendale 1959). The gorges and valleys were cut by rivers that once joined the original Finke River as it flowed eastward to a Cretaceous sea.

Fig. 12.4. Macdonnell Province. Rock outcrops and lithosols on slopes above a valley of the Musgrave Ranges, Ernabella, South Australia.

The principal soils are shallow sandy *lithosols* which are free of salts and carbonates, and mildly acid to neutral. Included in the unit because of considerations of scale are the subjacent plains. Some of the more noteworthy associated soils on these are the *red clays* on outwash plains where Proterozoic basalts have been exposed and weathered in the adjoining ranges, e.g. those fronting the Blackstone and Tomkinson Ranges.

Broadhurst Province

Although only two areas of this Province are shown in Fig. 12.1 there are many smaller occurrences throughout the Region.

The main soils present in the Broadhurst Province are the powdery *grey-brown* and *red calcareous soils.* These occur throughout the entire region on narrow (< 10 km wide) calcrete terraces which formed (probably during the late Tertiary) mainly in drainage depressions and along old watercourses, some of which are now important aquifers. Many of the terraces are covered by more recent deposits, especially by aeolian sands. Where the calcretes are exposed, the calcareous, powdery loam soils (15–60 cm thick) are similar to those of the Nullarbor Plain (Region V, Nullarbor Province). These soils have developed from a parent material formed subsequent to the lateritic deep weathering phase of the Early Tertiary and their properties reflect processes of soil formation current during the more arid phases of the Cainozoic. These powdery *grey-brown* and *red calcareous soils* are thus closer to the accepted concepts of typical desert soils (Aubert 1960) with their lack of weathering and leaching.

The Province containing *red* and *brown hardpan soils*, shown as 5^M in Fig. 12.1, will not be considered further here, as these soils and landscapes are much more widespread in the adjoining Region II (Murchison Province). As well, soil landscapes of earthy loams occur but are too small to show in Fig. 12.1; they are erosional remnants of the *red earths* in Stuart–Burt Province discussed later.

Amadeus Province

Gently sloping, shallow basin plains typified by the presence of playa lakes and claypans which occur in all the lower-lying valleys constitute the Amadeus Province.

In the large playa lakes e.g. Lake Mackay and Lake Amadeus, sodium chloride, gypsum, calcium/magnesium carbonates and sometimes alunite have accumulated. These depressions contain surface water only periodically. They are most often fringed by dunes of sand, granular gypsum or calcareous materials, and remnants of calcrete terraces (Broadhurst Province). Their youthful appearance today is probably the result of wind erosion during Quaternary arid periods (Wopfner and Twidale 1967). In this sense the playas are hollows formed by deflation; the same is true of the great salt lakes like Lake Eyre and Lake Gairdner in Region V (Tarcoola–Quondong Province).

The soils are little known but are mainly saline, calcareous and siliceous loams (*solonchaks*) with very little profile development. They are juvenile desertic soils but the valleys in which they occur have long geologic histories, for these valleys are shrunken relicts of extensive exoreic drainage systems (van de Graaff *et al.* 1977), shaped in Early Tertiary times by structural lineaments and pluvial conditions. Since then they have been infilled by alluvium which reaches its maximum thickness in the lake basins, some 40–50 m of fill being recorded even in the smaller basins (Blake *et al.* 1977).

Stuart–Burt Province

This is a post-lateritic Province of stepped plains with variable drainageways and dunes generally flanking areas of Macdonnell Province (Fig. 12.1). Topography is mostly subdued but is nonetheless complex, especially in the vicinity of watercourses draining the plains, and around low, rocky hills and ridges of various basement rocks (remnants of Macdonnell Province). In a zone separating the landscapes of Macdonnell and Stuart–Burt there are sporadic remnants of the deep-weathering profile, laterite, silcrete and mottled and pallid zone materials, on residuals rather like terraces (Perry *et al.* 1962). The remnants are too small to be delineated at the present scale. These materials are also found sporadically beneath the sediments and soils of the stepped plain.

The soils of the Province are diverse (Jackson 1962; Litchfield 1969). *Red earths* are the most extensive and characteristic, with alluvial red sands (Uc1.43), crusty red duplex soils (Dr1.33), calcareous earths (Gc1.12, Gc2.12) and minor areas of *earthy sands*, red *siliceous sands*, *red clays* and powdery calcareous loams (Um5.11). The *red earths* occur on plains strewn with ironstone gravel and have hard or crusted surfaces when dry. They are sandy to loamy at the surface, gradually becoming more clayey with depth; ironstone and quartz gravels range up to 40–50% in some parts of their profiles and an earthy fabric is characteristic and pronounced, especially in their subsoils. The profiles are essentially free of salts, mildly acid to mildly alkaline, have dominantly kaolinitic clays and contain some unweathered minerals. They are underlain at 1–2 m by rounded gravels (quartz or ironstone) or blocky-structured clays; in some cases calcium carbonate occurs, either as calcrete cementing the gravels or clays, or in a finely divided state; palygorskite has sometimes been identified in these clays.

It seems likely that the alluvia in which the *red earths* formed were composed of lateritic materials mixed with weakly to moderately weathered detritus derived from the country rocks, which is in keeping with the close geographic association of this Province and the Macdonnell Province. This also explains the presence of 'fresh' minerals in an otherwise weathered soil matrix. The sporadic presence of carbonate and palygorskite below the solum relate to the occurrence of shallow groundwaters during calcrete formation.

Of the other three main soils, the alluvial red sands have formed in alluvium resulting from Late Quaternary erosion and deposition of *red earth* and red *siliceous sand* materials. Both the red duplex soil (Dr1.33) and the calcareous earths (Gc) are discussed in Region V where they are more widespread in the Oodnadatta–Murnpeowie and Tarcoola–Quondong Provinces respectively.

Main soil determinants

Parent material is obviously the principal factor determining the unusually large extent of leached soils (earthy sands and red earths) in this arid environment. The parent materials from which these soils formed are highly weathered and leached as a result of their geological and climatic history. It is only where fresh rocks have been exposed, or else on recent deposits, that lithosolic or regosolic soils occur.

Although there are a number of distinct relationships between soils and vegetation, e.g. dense groves of *Acacia aneura* on acid and neutral *red earths*, and grasslands of *Astrebla* spp. on red clays, the vegetation appears to have had little, if any, effect on soil development in the Region.

Land use aspects

The arid climate of the Region largely restricts land use to sparse grazing of native pastures. Soils influence the distribution of palatable plant species but there are significant areas of the Region, principally the Simpson–Victoria Province with its preponderance of *siliceous sands*, and non-availability of stock waters, where not even low intensity grazing of cattle is possible. Minor areas, particularly in the southern portion of the Region, are grazed by sheep rather than cattle. Where supplies of good quality underground water are available as in the Mereenie sandstone and Jay Creek limestone near Alice Springs, limited areas of many agricultural and horticultural crops are successfully grown under irrigation. The main soils used for irrigation are the deep sandy and loamy *alluvial soils*.

Most soils in the region are prone to damage if mis-managed. Those in the Ayers–Canning and Gibson Provinces have only sparse vegetative cover and are highly susceptible to any disturbance, either by natural causes (intense rainstorms, violent windstorms) or by man (wheel tracks, grazing animals). The sandy dune soils of the Simpson–Victoria Province are in a delicate state of balance even without external disturbance. Most dune crests are not 'fixed' by vegetation and therefore move readily when prevailing conditions of wind and moisture allow.

While the soils of the Macdonnell Province often occur on steep slopes, the high percentage of bare rock makes this perhaps the most stable of the Provinces considered. The powdery nature and high carbonate contents of the soils of Broadhurst Province make them very fragile, particularly with regard to bearing strength. Where they are exposed at the surface the calcretes are stable although they carry very little vegetation.

The floors of the lakes and playas of the Amadeus Province are normally stable although it is obvious that in previous dry periods they have been deflated to form fringing dunes. These must be regarded as equally susceptible to erosion, particularly by wind, as the dunes of Simpson–Victoria Province.

Apart from the Macdonnell Province, the Stuart–Burt Province is perhaps the most stable from the point of view of erodibility although, being so diverse, there are areas just as unstable as elsewhere. The *red earths*, while usually well vegetated, are subject to sheet erosion, and vehicle tracks often develop into localized gullies.

The fastest-growing and most valuable form of land use which has been extensively developed in recent years is tourism. The natural assets, such as the mountain gorges, stark ridges and ranges, monolithic rocks and vast open plains, have an aesthetic value but the natural fragility of many of the areas subjected to tourism has already caused concern about preserving these areas for the future. There is also a small industry based on the capture of feral animals such as donkeys, camels and goats, but the numbers coming out of the Region are likely to remain low.

References

Aubert, G. (1960). Arid zone soils. A study of their formation, characteristics, utilization and conservation. General Symposium on Arid Zone Problems, Paris, 11-18 May, 1960. Pap. No. 5.

Blake, D. H., Passmore, V. L. and Muhling, P. C. (1977). Billiluna W. A. 1:250,000 Geol. Series 2nd edn. Bur. Min. Resour. Aust. Explan. Notes SE/52-14.

Bowler, J. M. (1976). Aridity in Australia: Age, origins and expression in aeolian landforms and sediments. *Earth-Sci. Rev.* **12,** 279-310.

Bureau of Mineral Resources (1976). Map at 1:2,500,000. Geology and Geophysics, Australia. (Bulletin 181.)

Carnahan, J. A. (1976). Natural Vegetation. Atlas of Australian Resources, 2nd series, National Mapping, Canberra.

Chippendale, G. M. (1959). Plants of Palm Valley. *Vic. Nat.* **75,** 192-9.

Folk, R. L. (1976). Reddening of desert sands; Simpson Desert, N.T. Australia. *J. Sediment. Petrol.* **46,** 604-15.

van de Graaff, W. J. E., Crowe, R. W. A., Bunting, J. A. and Jackson, M. J. (1977). Relict Early Cainozoic drainages in arid Western Australia. *Z. Geomorphol. N.F.* **21,** 397-400.

Jackson, E. A. (1962). Soil studies in central Australia. CSIRO Aust. Soil Publ. No. 19.

King, D. (1956). The Quaternary stratigraphic record at Lake Eyre North and the evolution of existing topographic forms. *Trans. R. Soc. S. Aust.* **79,** 93-103.

King, D. (1960). The sand ridge deserts of South Australia and related aeolian landforms of the Quaternary arid cycles. *Trans. R. Soc. S. Aust.* **83,** 99-108.

Lamb, H. H. (1977). Climate: Present, Past and Future. Vol.2. Climatic History and the Future (Methuen and Co. Ltd.: London; Barnes and Noble Books: New York.)

Laseron, C. F. (1972). The Face of Australia: The Shaping of a Continent. 3rd edn. revised by J. N. Jennings. (Angus and Robertson: Sydney.)

Litchfield, W. H. (1962). Soils of the Alice Springs area. pp. 185-207. CSIRO Aust. Land Res. Ser. No. 6.

Litchfield, W. H. (1969). Soil surfaces and sedimentary history near the Macdonnell Ranges, N.T. CSIRO Aust. Soil Publ. No. 25.

Mabbutt, J. A. (1962). Geomorphology of the Alice Springs area. pp. 163-184. CSIRO Aust. Land Res. Ser. No. 6.

Mabbutt, J. A. (1968). Aeolian landforms in central Australia. *Aust. Geogr. Stud.* **6,** 139-50.

Mabbutt, J. A. (1969). Landforms of arid Australia. *In* 'Arid Lands of Australia'. (Eds R. O. Slatyer and R. A. Perry) (Aust. Nat. Univ. Press: Canberra.)

Mabbutt, J. A., Wooding, R. A. and Jennings, J. N. (1969). The asymmetry of Australian desert sandridges. *Aust. J. Sci.* **32,** 159-60.

Perry, R. A., Mabbutt, J. A., Litchfield, W. H. and Quinlan, T. (1962). Land systems of the Alice Springs area. pp. 20-108. CSIRO Aust. Land Res. Ser. No 6.

Quinlan, T. (1962). An outline of the geology of the Alice Springs area. pp. 129-46. CSIRO Aust. Land Res. Ser. No. 6.

Wopfner, H. and Twidale, C. R. (1967). Geomorphological history of the Lake Eyre basin. *In* 'Landform Studies from Australia and New Guinea'. (Eds J. N. Jennings and J. A. Mabbutt). (Aust. Nat. Univ. Press: Canberra).

Chapter 13

Western Region (II)

E. Bettenay

THE WESTERN REGION INCLUDES THE SOUTH-WESTERN Australian Agricultural Region, and the main areas of pastoral and mining activities in Western Australia exclusive of the Kimberley.

The Region is roughly divided by the tropic of Capricorn into zones of summer rainfall to the north and winter rainfall to the south. The annual rainfall, which varies from 1250 mm to less than 200 mm with high rainfall restricted to the south-west corner, decreases markedly away from the coast so that the central area receives less than 250 mm over both summer and winter periods. Perennial streams are restricted because of this and because of low relief. In the dry season most rivers are dry or are only a series of shallow pools.

The natural vegetation is closely related to the rainfall. In the southern part there are forests of karri (*Eucalyptus diversicolor*) and jarrah (*E. marginata*); further north these forests are replaced by woodlands which are at first eucalypts, and then mulga (*Acacia aneura* and spp.) in the zone where the rainfall is less reliable. Poor, sandy soils in the south have a heath vegetation of *Banksia*, *Casuarina* and *Acacia* species, whereas to the north hard grasslands of spinifex, largely *Triodia basedowii*, are dominant on stony and sandy areas. On some well watered inland areas forests of river gum (*E. camaldulensis*) and cajeput (*Melaleuca leucodendron*) are found but the inland drainage lines usually have a shrubland of halophytic vegetation.

Generally the broad physical features of the Region show very little variety. It consists of some narrow coastal plains and a plateau which is largely at an elevation of less than 600 m above sea level, and forms part of a vast gently undulating surface known as the Great Plateau of Western Australia (Jutson 1950). There are a few ranges above the general level of the plateau, the most notable being the Hamersley Range (Mt Meharry 1231 m) in the north, and the Stirling Range (Bluff Knoll 1092 m) in the south. The Great Plateau is bounded to the west by the Darling Fault Scarp which rises sharply from the coastal plain to between 240 and 300 m above sea level. This scarp can be traced northward, from near the south coast, a distance of 640 km to a point northeast of Geraldton.

Maps of Western Australia frequently show connected drainage in the north extending inland to the Great Northern Highway. In the south rivers are normally shown as confined to the western margin and only occasionally are they shown to connect with the salt lakes of the more arid interior. These salt lakes are remnants of ancient river systems which are now blocked and distorted in many ways, and which only function in very wet years. Details of their extent and modification have been given by many authors, including Gregory (1914) Bettenay (1962) and Morgan (1966).

The Region has a wide range of soils dominated by *lateritic podzolic soils*, *soloths* and *red-brown earths* in the south, *red* and *brown hardpan soils* and *red earths* in the central part, and *lithosols*, *red clays* and *red-brown earths* in the north. Throughout the Region *solonchaks* occur associated with coastal marshes and salt lakes, and *earthy sands* and *yellow earths* are common on divides. With few exceptions the profiles are deep and are underlain by an extensively weathered regolith.

Outline of landscape history

Regional geology has been mapped by the Geological Survey of Western Australia (Horwitz 1966). Most of the Region consists of a stable Precambrian shield. Rocks of the Pilbara area are largely Proterozoic and include some unmetamorphosed sediments, while those to the south are largely acid igneous rocks of Early Archaean age. Throughout the shield there are zones of Early Archaean volcanic and sedimentary rocks which have been largely metamorphosed; there are no extensive recent deposits of marine or continental origin.

Sediments were deposited on the downwarped margins of the shield in Palaeozoic times and are evident now in the Perth and Carnarvon Basins which form the western margin of the Region; at this stage also the Collie coal basins were formed. In Early Cretaceous times marine deposition began over the Palaeozoic rocks to both the east and west of the shield. This has been followed by marine deposition continuing until the Mid-Tertiary. More recently Quaternary terrestrial deposits have formed a thin mantle which covers most of the sedimentary rocks on the western margin of the shield.

In common with much of Australia, the Western Region has escaped the effects of glaciations, or orogenies, which have affected many other continental land masses since Palaeozoic times. For perhaps the past 100 million years it has been subject to the processes of sub-aerial weathering which has worn down most of its mountain ranges so reducing it largely to a low plateau with little relief. By the Late Tertiary, the landscape had been reduced practically to its present form, and had been extensively lateritized. The laterite profile consists of a surface layer of duricrust which is rich in iron and aluminium minerals, associated with mottled zones and pallid zones of deep weathering. The pallid zones are largely composed of chemically weathered saprolites of the basement rock to depths of up to 50 m.

Subsequent erosion, in part resulting from the establishment of new base levels near the coast, has removed parts of this lateritic mantle preferentially; erosion has proceeded back along the major drainage lines, reaching further inland along the rivers in the north than in the south. However, even further inland, stripping and dissection was independent of major base level changes, and apparently resulted from landscape instability following climatic fluctuations just before and during the Quaternary Era (Bettenay and Hingston 1964). As a result of this, latteritic remnants, of which red or yellow-sandplains are generally a feature, occur extensively on upland interfluval areas. Lateritic remnants in the form of mottled and pallid zones are also common on the valley sides and in the bottom of the valleys, where they lie buried under a shallow cover of recent sediments. The broad, gently sloping plains of the north have extensive colluvial deposits up to 30 m thick. These deposits have been silicified to form Wiluna Hardpan (Bettenay and Churchward 1974), which, in common with the laterite, is being destroyed by headward erosion of the major rivers.

As a result of climatic changes since the pluvial conditions during lateritization, the inland rivers have been reduced to wind-planed playa lakes and salt flats, with associated gypsum dunes, lunettes and saline lake deposits (Bettenay 1962). In the north the upper ends of the drainage lines are commonly occupied by deposits of calcrete which may be as much as 30 m thick.

Main soil determinants

Most of the Western Region is the oldest part of an ancient continent; this is reflected by the low relief of the landscape, the deep mantle of weathering, and the leached soils. Prescott (1931) recognized the pedological significance of Western Australia's old landscapes. Subsequently Stephens (1946) stressed the way in which the various horizons of old lateritic profiles may provide parent materials for present day soils, while Mabbutt (1961) emphasized the way in which the depth of the weathering profile may control the development of new etch plains at lower levels.

Fresh rocks are exposed to soil formation only in limited areas. These may contain ancient quartzose sediments which produce shallow skeletal sands, or base-rich doleritic or basaltic rocks in which case silt and clay predominate; in the main they are granitic and produce sand–clay mixtures on weathering.

Since patterns of soil distribution are often associated closely with changes of relief, an understanding of soil-geomorphological relationships provides a major tool for broad scale soil mapping. Upland areas are usually associated with relict sandy lateritic soils whereas the valley sides and floors mostly have duplex soils with sandy clay B horizons. The duplex soils have solodic and solonetzic characteristics which possibly reflect the salinity associated with the deeply weathered profiles on which they are developed (Dimmock *et al.*, 1974). The position of soils on slopes is also important in the broad mulga plains of the Murchison where increasing thickness and clay content of deposits leads to a change from uniform medium textured soils on the upper slopes, to gradational soils in the lower ones as the major drainage lines are approached: along with this the soil reaction changes from acid, to neutral, to alkaline.

Early climatic history has a particular bearing on the development of leached, lateritic, soil parent materials; later climatic changes, probably acting in concert with reduced vegetation cover, have led to landscape instability and episodic erosional phases. As a result there are large areas where both the physical and chemical properties of the soils, and the form of the landscape, are the results of previous cycles of weathering, even though the soils themselves may be young. In south-western Australia there is a particular example of the effect of current climate. As one proceeds inland leaching is reduced and duplex soils, which have acid subsoils close to the coast, become progressively neutral, and then alkaline on reaching the semi-arid interior.

Little work has been done on the effects of plants, animals and micro-organisms on the soils of the Western Region. Over large areas plant litter remains on the surface until it is destroyed by fire, and there is little build up of humus. Termites incorporate organic matter in the soil, and also accumulate fine materials in their nests, particularly in the sandy soils of the more arid interior. The influence of man in clearing native vegetation to develop cultivated agriculture is of particular importance; this has led to the extensive salinization of soils throughout the whole of south-western Australia (Peck 1978).

Soil-landscape provinces

The Western Region has been subdivided into eight soil-landscape provinces based on a consideration of

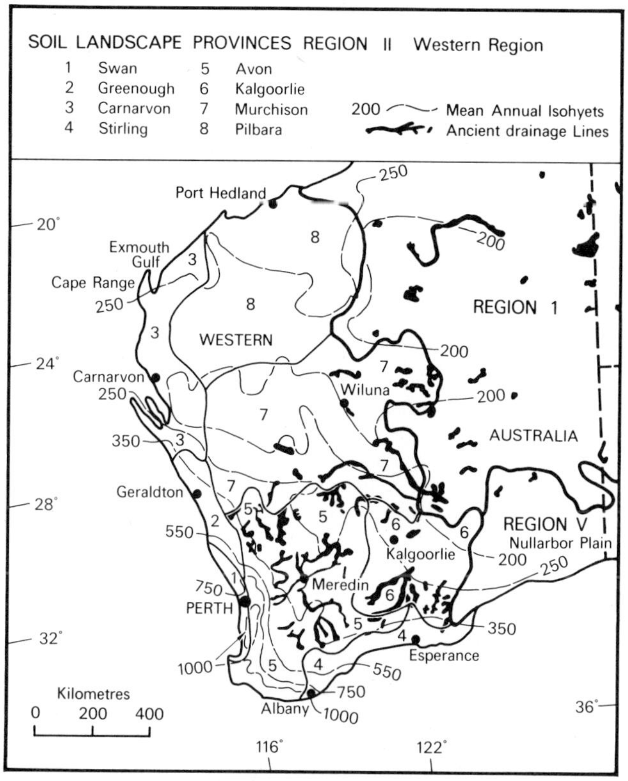

Fig. 13.1. The soil-landscape provinces of the Western Region.

landform, soils, superficial deposits and geology (Fig. 13.1). The main features of these provinces, which have been given local place names, are summarized in Table 13.1.

The Swan, Greenough, Carnarvon and Stirling Provinces include coastal plains with young soils developed on aeolian and alluvial deposits of Quaternary age; some older lateritic soils are also present. The Avon province contains the most extensively preserved lateritic landscapes, particularly on the western margin (which is wetter) and on the interior divide separating this Province from the Kalgoorlie and Murchison Provinces. In the Kalgoorlie and Murchison Provinces, belts of older metamorphic rocks form low hills with intervening plains on which Wiluna Hardpan is extensively developed. To the north the Pilbara Province has a soil and drainage pattern which is largely controlled by the basement geology.

Swan Province

This Province is a narrow coastal plain, seldom more than 40 km wide and having a surface cover of sedimentary material which is largely of Quaternary age. It has been discussed in detail by McArthur and Bettenay (1960). The most easterly feature is the Ridge Hill Shelf, which is dissected by streams and rivers issuing from the Darling Range to the east, to form a series of spurs on which *lateritic podzolic soils* have developed. To the west is a gently sloping plain, the Pinjarra Plain, built up of piedmont alluvium and riverine deposits of varying ages. The youngest of these deposits occurs along present river terraces and has undifferentiated *alluvial soils*; the oldest has *yellow podzolic soils* in which ferruginous gravels are common; deposits of intermediate age have soils ranging from *red podzolic soils* on levees, to *yellow earths* on plains, and *black earths* in swamps and depressions.

To the west of the Pinjarra Plain is a series of dunes aligned roughly parallel to the coast. The youngest of these, the Quindalup Dunes, flank the coast and are composed of siliceous and shelly detritus giving rise to *calcareous sands.*

Next inland from the Quindalup Dunes are the Spearwood Dunes which may attain an altitude of over 100 m. Their soils are typically *siliceous sands* overlying cemented aeolianite limestone at depth. These sands have brighter colours where the limestone is shallow (Uc4.22).

The most easterly dunes, the Bassendean Dunes, are more subdued with frequent swampy flats. The soils are largely *podzols*, with *humus podzols* (Uc2.33) occuring in wetter situations, and bleached sands (Uc2.22, .21) on the low dunes.

These three dune systems are commonly separated by shallow lakes and swamps in which a variety of organic soils and *podzols* are developed. Additional details of the soil stratigraphy of this Province have been given earlier (Chapter 8) by Ward and McArthur.

Greenough Province

The Greenough Province has a narrow coastal strip consisting largely of dunes comparable to those of the Swan Province to the south. Inland there are sediments of Cretaceous and Jurassic age which have been extensively lateritized on a landscape of broad valleys and undulating interfluves. The area has been mapped by Churchward (1970). The chief soils are *yellow podzolic soils* containing much ironstone gravel in the A horizons with a complex pattern of gravelly *lateritic podzolic soils* and sandy *podzols.*

Further to the north, where dissection has removed much of the lateritic profile, the terrain is gently undulating to rolling with some valley plains and terraces. The chief soils exhibit solonetzic characteristics and are *red-brown earths* with associated red duplex soils (Dr2.63, .73). Minor soils include sands (Uc1.43, .42) on the levees, gilgaied clay soils on the terraces, and *red earths* and *calcareous red earths* on the plains. Minor outcrops of granitic basement rocks also occur.

Carnarvon Province

This Province is largely composed of Cretaceous and Tertiary marine sediments, with an overlay of Quaternary deposits on the coastal fringe. The topography is generally subdued except for the steep Cape Range to the west of Exmouth Gulf, where the altitude exceeds

Table 13.1. Main features of the soil-landscape provinces of Region II

Province	Geological and geomorphological features	Major soil features
Swan	Coastal plain with piedmont and riverine alluvium, coastal marine and estuarine deposits and coastal dunes of Tertiary to Quaternary age	Soils range from recent *alluvial soils* through *red podzolic soils* and *yellow earths*, to *lateritic podzolic soils* on older alluvium. *Calcareous sands* on recent dunes, *siliceous sands* on dunes of intermediate age, and *humus podzols* and *bleached sands* on older dunes
Greenough	Lateritized plateau developed on Jurassic and Permian sediments and Proterozoic granites; dissected at fringes	Largely *yellow podzolic soils* with gravelly A horizons, gravelly *lateritic podzolic soils*, and sandy *podzols*. In more dissected areas to the north *red-brown earths* and solonetzic red duplex soils occur. Minor soils include *bleached sands*, gilgaied clay soils, *red earths*, and *calcareous red earths*. Dune soils as in Swan Province.
Carnarvon	Plains with limited seif dunes, developed on Tertiary, Mesozoic, and Palaeozoic sediments	*Calcareous sands* and *solonchaks* adjacent to the coast. Inland plain of red *earthy sands*, *calcareous red earths* and *solonized brown soils*. Some solonetzic *red-brown earths* on alluvium and pedimented areas. *Red earths* and *earthy sands* on plains, with *earthy sands* on dunes. *Red calcareous soils* with limestone outcrop in the Cape Range area
Stirling	Lateritized plateau on Tertiary sediments, dissected at fringes. Emergent quartzite ranges and coastal headlands of gneisses and migmatites. Deep lateritic weathering common	Mostly *lateritic podzolic*, *solodized solonetz* and *solodic soils*. Smaller areas of gravelly and sandy *podzols*, and minor cracking *red clays* near Ravensthorpe. *Lithosols* occur on the emergent hills and ranges, and there are *calcareous sands* and shallow sandy *podzols* near the coast
Avon	Lateritized plateau on Precambrian granites and gneisses, dissected at the margins. Saline drainage lines inland. Deep lateritic weathering common	*Lateritic podzolic* and *yellow podzolic soils* are widespread, with *red earths* and *red podzolic soils* occurring on the Darling Scarp, and *non-calcic brown soils* and *red-brown earths* in dissected valleys. Inland valley floors have *solodized-solonetz* and *solodic soils* with *solonchaks* in the main drainage lines. The main divides to the Murchison and Kalgoorlie Provinces to the north and east have *yellow earths* and *earthy sands*, while that to the Stirling Province to the south has extensive *solodized-solonetz* and *solodic soils*. Secondary salinisation is common
Kalgoorlie	Similar to Avon Province but with greater prevalence of greenstone belts. Much dissection near major salt lakes. Aeolian dust present	*Solonchaks* occur in the salt lakes and major valleys. *Solonized brown soils* associated with aeolian dust and greenstones in the south are largely replaced by *calcareous red earths* further north. Minor areas of *desert loams* and sandy and gravelly *lithosols* occur
Murchison	Extensive plains with residuals of laterite or Precambrian igneous rocks. Drainage lines have extensive saline or calcrete deposits. Wiluna Hardpan widespread	*Red* and *brown hardpan soils* and *red earths* with hardpan occupy most of the plains and gentle slopes. Major drainage lines have *solonchaks* or *grey-brown calcareous soils*, and there are solonetzic *red-brown earths* associated with the Murchison River. Red *earthy sands* occur on divides, and stony *lithosols* are found on the low ranges and hills
Pilbara	Rugged ranges and hills with well-defined river valleys developed in a terrain of Precambrian granites and metamorphics, and Palaeozoic sediments. Narrow coastal plains occur	The Chichester Range in the north is dominated by shallow brown loams with frequent basalt outcrops. Coastal plains further north have solonetzic *red-brown earths* and cracking *red clays*; coastal marshes have *solonchaks*. Red *lithosols* with frequent rock outcrop dominate the Hamersley Range, and to the south there are dissected plains with solonetzic *red-brown earths*. Non-cracking red clays, and *solonized brown soils* on calcrete occur in the valley of the Fortescue River. Limited areas of *red* and *brown hardpan soils* and *red earths* occur on plains marginal to other provinces

300 m. Inland *calcareous sands* which flank the coast, give way to red *earthy sands*, *calcareous red earths* and *solonized brown soils*; *solonchaks* occur along the coastal strips and as salt marshes. Alluvial tracts associated with the main rivers have non-calcareous loamy soils (Um5.2) on the levees. *Red-brown earths* with solonetzic characteristics are developed on the older alluvium, and on pedimented areas inland. Fig. 13.2, taken from the work of Bettenay *et al.* (1971), shows a typical cross section of the soil distribution and generalized layer properties on the Gascoyne River delta.

On Cape Range, the Province is dominated by bare limestone with pockets of *red calcareous soils*, and inland there are some plains with *red earths*, and parallel sand dunes with red *siliceous sands*, and *earthy sands*.

Stirling Province

The Stirling Province consists of a narrow strip less than 80 km wide and flanking the Southern Ocean; geologically it consists of Proterozoic gneisses and migmatites which are overlain sporadically by Tertiary marine and continental deposits. The Stirling and Mt Barren Ranges are composed of quartzites and phyllites which are younger than the surrounding Proterozoic rocks.

Generally the country has low relief and has been extensively lateritized; later dissection has exposed lateritic mottled and pallid zones to soil formation. This has given a very complex pattern of soils, with the greater part of the landscape occupied by *lateritic podzolic*, *solodized solonetz*, and *solodic soils*.

There are small areas of ironstone gravelly *podzols* and leached sandy *podzols*. *Lateritic podzolic soils* and *podzols* occur inland and to the west of Albany. Cracking *red clays* occupy small areas near Ravensthorpe. Because of their steep and rugged terrain, the Stirling and Mt Barren Ranges have soils which are largely uniform coarse-textured *lithosols*; these soils are also found on the emergent gneissic hills.

Much of the coast is flanked by *calcareous sands*, but there are also some sandy *podzols* near Israelite Bay. Elsewhere *yellow podzolic soils* extend to the coast, and there are headlands and offshore islands of bare granite with some shallow siliceous sandy *podzols* to the west of Bremmer Bay.

Avon Province

The Avon Province, which covers a large area of the southern Precambrian shield of south-western Australia, consists of a gently undulating plateau dissected at its margins and rising to some 600 m inland. Here the rocks are mainly granites and gneisses, with some dolerite

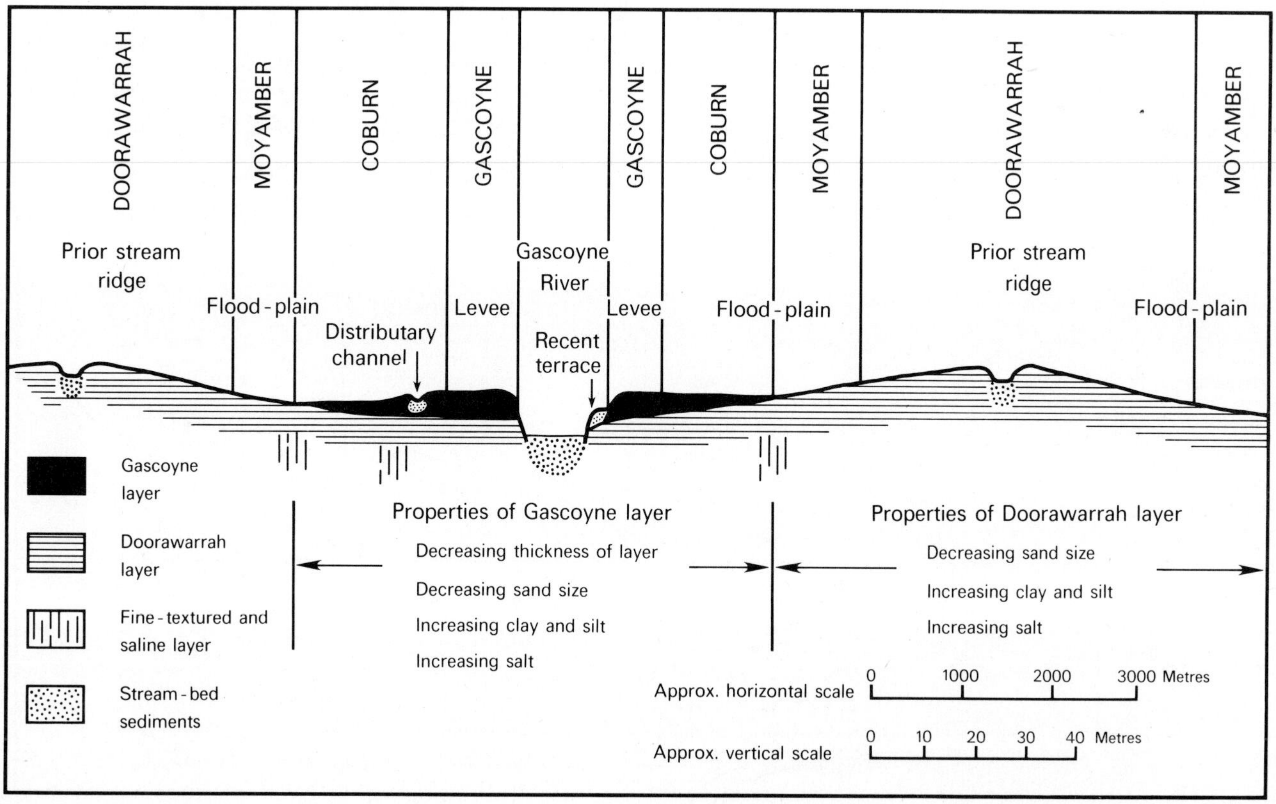

Fig. 13.2. Generalized section of the Gascoyne Delta showing major topographical features and general layer relationships. (From Bettenay *et al.* 1971.)

intrusives and limited areas of older metasedimentary and metavolcanic rocks. To the south-west is the Blackwood Plateau, a province of terrestrial sediments of Cretaceous age. Lateritization has been extensive over both igneous and sedimentary rocks, and many soils have older lateritic horizons as parent materials.

The rocky face of the Darling Scarp, and some steeply dissected valleys, are largely laterite-free with *red earths* and *red podzolic soils*. Close to the Darling Scarp *lateritic podzolic soils* are present, but inland they give way to extensive areas of *soloths* and *yellow podzolic soils*, with limited occurrences of *non-calcic brown soils* and *red-brown earths* in the more dissected areas.

Throughout the Province the uplands commonly have *yellow earths*, while on the divide between the Kalgoorlie and Murchison Provinces, to the east and north, *earthy sands* predominate; along the south coast there are areas of *podzols*.

The valley floors in the drier eastern part are of low gradient and have *solodized solonetz* and *solodic soils*, with *solonchaks* in the saline drainage lines and salt lakes. Some minor valleys have *red-brown earths* with solonetzic characteristics, and the divide to the Stirling Province to the south has extensive areas of *solodized solonetz* and *solodic soils*. The soils of both the valley floors and the uplands are extensively underlain by deep, lateritic, weathering zones, as shown in Fig. 13.3. This figure, which is taken from Bettenay and Hingston (1964), also shows characteristic erosional and depositional surfaces which are associated with the main landscape elements.

Secondary salinization resulting from clearing has resulted in the extensive development of *solonchak* soils (Fig. 13.4), which still retain a profile morphology characteristic of a prior soil condition, particularly on the low lying Belka surface.

Fig. 13.4. Dissected lateritic landscape with secondary salinity associated with dryland agriculture: Avon Province.

Kalgoorlie Province

This Province contains Precambrian rocks similar to those of the adjoining Avon Province, but having a higher proportion of north-north-westerly trending greenstone belts of metasedimentary and metavolcanic rocks. There are also extensive playa lakes, and aeolian dust, or parna, obscures the basement geology, particularly in the southern part of the Province. Generally the topography is subdued and falls gently to the Nullarbor Plain in the east.

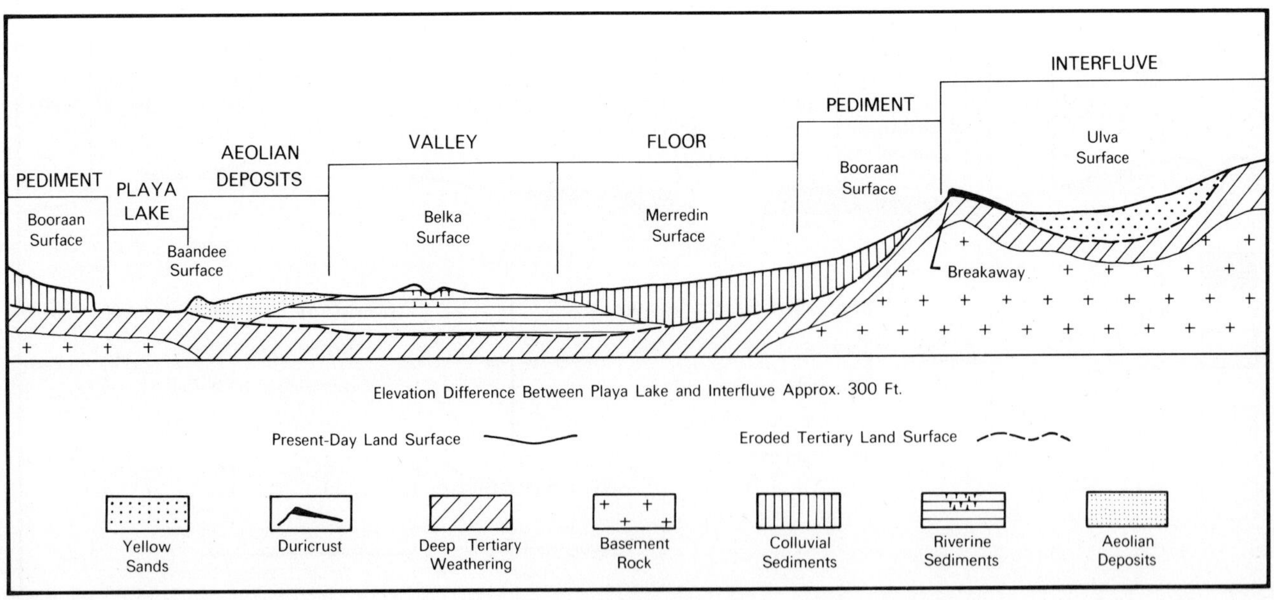

Fig. 13.3. Modification of the lateritized Tertiary landscape at Merredin, showing surfaces of erosion and deposition associated with the main landscape elements.

As with the Avon Province the soils in the major valleys and salt lakes are *solonchaks*. *Solonized brown soils* are associated with the parnas, and with the greenstone ridges; in the north they are replaced by *calcareous red earths*. There are also areas of saline *desert loams* which are probably associated with lateritic parent materials, and smaller areas of sandy *lithosols* on hills and ranges of granulitic rocks. Near the junction with the Avon and Stirling Provinces some lateritic, gravelly sandy *lithosols* occur on the uplands.

Murchison Province

This Province has a soil pattern which is distinctly different from the adjoining provinces. Geologically it is similar to the Kalgoorlie Province with subdued outcropping of Precambrian granites and greenstones. While some lateritic influence is evident from the presence of red *earthy sands* on the divides, the greater part of the area consists of flat and gently sloping plains that sometimes have a surface cover of gravels, and on which the Wiluna Hardpan frequently outcrops. The plains are covered by colluvial deposits, to 30 m in thickness, on which *red* and *brown hardpan soils* and *red earths* with hardpan are developed (Fig. 13.5), the latter soils being more common on the lower plains adjoining the drainage lines.

As with the Avon and Kalgoorlie Provinces the valley floors have low gradients and are frequently saline with *solonchaks*, but in the Murchison Province the upper valleys have *grey-brown calcareous soils* developed on calcrete. *Red-brown earths*, which are frequently saline, are associated with the main drainage of the Murchison River, and with some pedimented slopes. Areas of both acid and basic rocks, which form low ranges and stony hills throughout the Province, have shallow stony *lithosols*.

Pilbara Province

Geologically the Pilbara Province is very diverse. Precambrian granites are found in the northern part, but the main Pilbara block is composed of metamorphosed Proterozoic sediments, and acid and basic volcanic rocks. The drainage pattern is generally controlled by the basement configuration.

The topography is generally steep with altitudes rising to above 360 m in the Hamersley and Chichester Ranges. There are numerous steeply dissected gorges, and the valley of the Fortescue River, which flows to the west, divides the province into two distinct halves. To the north the Chichester Range is dominated by shallow brown loams (Um6.23) with frequent basalt outcrops. Yet further north are plains with cracking *red clays*, and with *red-brown earths* having solonetzic characteristics. The coast line is dominated by coastal marshes with *solonchaks*.

The Hamersley range has extensive upland areas of red *lithosols* with frequent rock outcrops and many bare cliff faces (Fig. 13.6). To the south the landscape is much more subdued and contains extensive dissected plains, largely dominated by solonetzic *red-brown earths* with occasional small steep stony hills with *red calcareous soils* and *lithosols*; less frequently cracking clays (Ug) occur on the more basic rocks.

The main valley of the Fortescue River has non cracking red clays (Uf6.71), and some *solonized brown soils* on calcrete. Generally laterite is not prominent but red *earthy sands* occur on the margins to the interior desert country. *Red* and *brown hardpan soils* and *red earths* are found on the plains particularly where the Province adjoins the Murchison Province to the south and the Carnarvon Province to the west.

Land use aspects

The present pattern of land use, which ranges from intensive horticulture to low intensity grazing, may be attributed to a number of factors.

Fig. 13.5. Gently sloping mulga (*Acacia aneura* and other species) plain with *red* and *brown hardpan soils*, and *red earths*: Murchison Province.

Fig. 13.6. Escarpment of the Hamersley Range viewed from the north. Rock outcrops are frequent and the main soils are shallow red *lithosols*: Pilbara Province.

The amount, reliability and effectiveness of the rainfall is a major determinant of land use. Permanent agriculture is confined to the Swan, Avon, Stirling and Greenough Provinces where the annual rainfall is winter dominant, relatively reliable and exceeds 260 mm. Agriculture ranges from intensive irrigated horticulture and dairying, carried on close to the major centres and particularly near Perth, to broad scale dryland farming with cereal crops in the drier areas. Permanent rivers are few, and confined to the high rainfall areas in the south-west.

Forestry too is largely confined to the south-west corner where the hardwoods jarrah and karri still form viable forests. There are also some forests inland, with the sandalwood industry, based on *Santalum spicatum*, still being of importance.

The pastoral industry, which is largely confined to areas with annual rainfall below 260 mm, is controlled by the presence of suitable native shrubs, trees and grasses. Grazing is extensive in the Kalgoorlie, Murchison, Pilbara and Carnarvon Provinces. Elsewhere the native vegetation is largely of low palatability and there are numerous poisonous plants, particularly the leguminous plants of the genus *Gastrolobium* and *Oxylobium*. The main areas of agriculture and forestry are thus separated from the grazing areas on the basis of both climate and vegetation.

Land settlement and the development of towns has also been greatly influenced by mining, firstly of gold, then nickel and iron ore. Sand mining and bauxite ventures are largely serviced by existing towns associated with agriculture.

Controls for agricultural development, apart from climate, have been largely a function of low soil fertility (Burvill 1965). Leached sandy *podzols* and *humus podzols* are common in the Swan, Greenough and Stirling Provinces, and lateritization and chemical depletion of their sedimentary rocks, and of the granites of the Avon Province, have had a further effect. Deficiencies of nitrogen, phosphorus, potassium, calcium, sulphur, copper, manganese, molybdenum, cobalt and selenium commonly have to be corrected by fertilizer application, licks, or bullets, before satisfactory crop and/or animal health can be maintained; nitrogen deficiencies have largely been overcome by the use of leguminous crops and pastures.

Low fertility however has played a part in the preservation of the jarrah and karri forests, and thus in maintaining water quality. Apart from agricultural development, the forests are still under threat from urbanization, roads and power lines, jarrah dieback – caused by *Phytophthora cinnamomi* – and from bauxite mining.

In addition to climate and nutrition, agricultural production is limited by other factors particularly by poor physical characterisitcs of soils, and by excess salinity. Northcote and Skene (1972) have documented the extent and nature of saline and sodic properties in Australian soils. In the Avon Province particularly, but extending into the other provinces where there is agriculture, both of these properties limit plant growth. The widespread occurrence of duplex soils in areas where the rainfall is concentrated into a few winter months, along with low relief and limited drainage, leads to extensive waterlogging; this results in poor aeration, limited root growth and the prevalence of root diseases.

Within the agricultural provinces excess salinity, largely as sodium chloride, is also common. It has been estimated (Burvill 1956; Malcolm and Stoneman 1976) that some 400 000 ha of salt land occurs naturally along the major drainage lines and rivers, and that a further 167 000 ha of land, which was once cultivated, is now unsuited to agriculture due to the excess salinity that has developed since clearing. Increased salinity has also led to the depletion of the already limited resource of potable water. Farm dams, streams and even major rivers such as the Swan-Avon, Murray, Collie and Blackwood have been affected in this way. Competition for water for agriculture, stock, domestic, and industrial use is strong and will increase as the region expands.

Within the grazing areas nutritional problems are not apparently widespread, but here again the physical properties of soils are important. In the Pilbara Province the prevalence of shallow *lithosols* leads to low water-holding capacity; the widespread occurrence of Wiluna Hardpan, particularly in the Murchison Province, has a similar effect. Overgrazing has led to serious depletion of the native plant cover, and to severe wind and water erosion which inhibits natural revegetation (Wilcox and McKinnon 1972).

References

Bettenay, E. (1962). The salt lake system and their associated aeolian features in the semi-arid regions of Western Australia. *J. Soil Sci.* **13**, 10-17.

Bettenay, E. and Churchward, H. M. (1974). Morphology and stratigraphic relationships of the Wiluna Hardpan in arid Western Australia. *J. Geol. Soc. Aust.* **21**, 73-80.

Bettenay, E. and Hingston, F. J. (1964). Development and distribution of soils in the Merredin area, Western Australia. *Aust. J. Soil Res.* **2**, 173-86.

Bettenay, E., Keay, J. and Churchward, H. M. (1971). Soils adjoining the Gascoyne River near Carnarvon, Western Australia. CSIRO Div. Soils, Soils Land Use Ser. No. 51.

Burvill, G. H. (1956). Salt land survey, 1955. *J. Dep. Agric. W. Aust.* **5** (3rd Ser.), 113–9.

Burvill, G. H. (1965). Plant nutrition in Western Australia. *J. Dep. Agric. W. Aust.* **6** (4th Ser.), 353-71.

Churchward, H. M. (1970). Erosional modification of a lateritized landscape over sedimentary rocks: its effect on soil distribution. *Aust. J. Soil Res.* **8**, 1–19.

Dimmock, G. M., Bettenay, E. and Mulcahy, M. J. (1974). Salt content of lateritic profiles in the Darling Range, Western Australia. *Aust. J. Soil Res.* **12**, 63-9.

Gregory, J. W. (1914). The lake systems of Westralia. *Geogr. J.* **43**, 656-64.

Horwitz, R. C. (1966). Geological map of Western Australia with explanatory notes. *Geol. Surv. West Aust.* Perth.

Jutson, J. T. (1950). The physiography of Western Australia. *Bull. Geol. Surv. W. Aust.* No. 95: (3rd edn.)
Mabbut, J. A. (1961). A stripped land surface in Western Australia. *Pub. Inst. Br. Geogr.* **29**, 101-14.
Malcolm, C. V. and Stoneman, T. C. (1976). Salt encroachment – the 1974 saltland survey. *J. Dept. Agric. W. Aust.* **17**, 42-9.
McArthur, W. M. and Bettenay, E. (1960). Development and distribution of soils of the Swan Coastal Plain, Western Australia. CSIRO Aust. Soil Pub. No. 16.
Morgan, K. H. (1966). Hydrogeology of the East Murchison and North Coolgardie Goldfields *Annu. Rep. Geol. Surv. W. Aust.* for 1965, pp. 14-19.
Northcote, K. H., and Skene, J. K. M. (1972). Australian soils with saline and sodic properties CSIRO Aust. Soil Pub. No. 27.
Peck, A. J. (1978). Salinization of non-irrigated soils and associated streams: a review. *Aust. J. Soil Res.* **16**, 157–68.
Prescott, J. A. (1931). The soils of Australia in relation to vegetation and climate. CSIR Aust. Bull. No. 52.
Stephens, C. G. (1946). Pedogenesis following the dissection of lateritic regions in southern Australia. CSIR Aust. Bull. No. 206.
Wilcox, D. G. and McKinnon, E. A. (1972). A report on conditions in the Gascoyne catchment. Dept. Agric., Western Australia.

Chapter 14

Kimberley—Arnhem—Cape York (III)

R. F. Isbell

THIS REGION ENCOMPASSES MUCH OF THE SPARSELY settled areas of northern Australia that have been little altered by man. Apart from several major mining ventures, land use is virtually confined to extensive grazing of beef cattle.

The Region has a monsoonal climate with little or no rain falling in the May–October period. The far northern extremities receive as much as 1600 mm mean annual rainfall, while over much of its extent the southern limit of the region approximately coincides with the 500 mm isohyet. In contrast to Region VI, rainfall gradients nearly everywhere decrease uniformly away from the coast. Summers are hot and humid, with the highest daily temperatures occurring in the north west; winters are mild and normally without frosts.

The native vegetation is dominated by eucalypt woodlands, with some open forests and small areas of rainforest in north-east Queensland. Lesser areas of tussock grassland are locally prominent. In general tree height and density decrease with rainfall, but exceptions due to edaphic factors are widespread.

Generally the relief of the Region is subdued; extensive, gently undulating, plains are common. There are, however, large areas in the north-west and in Arnhem Land which consist of rugged dissected plateaux and ranges. Land with an elevation of over 600 m occurs in the Kimberley area and in the eastern part of Cape York Peninsula. All drainage is to the coast except for a small endoreic basin in the southern-central part of the Northern Territory.

Although a wide range of soils occurs, a relatively small number of groups occupy most of the area. The most widespread are *red* and *yellow earths, siliceous* and *earthy sands, lithosols,* and *grey clays.* There is great variation in soil depth, ranging from the deep (> 3 m) profiles of many *red* and *yellow earths* to the widespread shallow (< 0.5 m) *lithosols* and *siliceous sands.*

Outline of landscape history

Because parts of the Region have been subaerially exposed since at least the Late Precambrian, the early geology plays an important role in later landscape history. The oldest established rocks (Archaean) are found only to the south of Darwin but Early Proterozoic low to high grade metasediments with dolerite and acid plutonic intrusives are more common to the east and south-east of Darwin and in the south of the Kimberley area. Much more widespread are the usually little-altered and only mildly deformed Lower to Middle Proterozoic sediments and associated basic volcanics of the Kimberley, Victoria River and McArthur Basins (Plumb and Derrick 1975). There are no equivalents of these sediments in Cape York Peninsula, where the oldest rocks are Proterozoic low and high grade metamorphics that are mostly highly deformed and intruded by later granitic bodies (Willmott *et al.* 1973.)

Other important early geologic elements in the central and western part of the Region are the widespread Lower Cambrian Antrim Basalt and the Cambrian limestones and associated sediments of the Daly River and Arafura Basins. The next event of major importance to landscape history took place in the late Mesozoic with the widespread deposition of both marine and terrestrial sediments over large areas of the eastern and southern central parts of the Region. Although much of these subhorizontal strata have subsequently been removed by erosion, their remnants are found over large areas.

The later geological history of the central and western part of the Region is not well documented although a map is available showing the Cainozoic geology of the Northern Territory (Bureau of Mineral Resources 1976). However, for north Queensland the Cainozoic geology is dominated by the development of the Karumba Basin which is known in some detail (e.g. Doutch 1976; Grimes 1980; Grimes and Doutch 1978). In its present form this basin coincides areally with the Gulf of Carpentaria and the river systems draining in to it. The earliest continental sedimentation began with the Bulimba Formation of probable Late Cretaceous–Early Palaeocene age. These arkoses, quartzose sandstones and claystones now outcrop only in the central and north western parts of Cape York Peninsula where they are usually extensively altered. The younger (probably Late Miocene–Pliocene), and generally similar Wyaaba Beds which outcrop widely in the central western parts of the Peninsula, are frequently covered by more recent widespread fan deposits of the Karumba Basin.

The most important factor in relation to landscape and soil history since the Late Cretaceous has been the formation and partial destruction of a number of widespread erosional and depositional surfaces, together with associated deep weathering which has produced thick laterite profiles or has otherwise extensively modified pre-existing landscapes. Although the chronology and nature of these events are not well known in many parts of the Region, sufficient evidence is available from some areas to indicate their scale and nature and hence their importance to the present soil landscapes.

In the Northern Territory, Hays (1967) has recognized four mature erosion surfaces which he named, from oldest to youngest, the Ashburton, Tennant Creek, Wave Hill and Koolpinyah Surfaces. The Ashburton occurs to the south of the Region, but the Tennant Creek Surface (probably equivalent to the Bradshaw Surface of Wright (1963)), although least disturbed in the south, is also widely distributed as remnants elsewhere. It usually has a thick (up to 50 m) laterite profile with the three-part development of ferruginous, mottled and pallid zones (Chapter 3). This profile is formed on a variety of substrates. The age of the surface and its associated laterite profile is thought to range from Late Cretaceous in the south to Early or Mid-Tertiary in the north.

The Wave Hill Surface (equated with the Maranboy Surface of Wright (1963)) exists as extensive erosional plains below the main lateritized surface. It also occurs on a wide variety of rocks and is thought to be of Miocene–Pliocene age. The Koolpinyah Surface is developing at present in most coastal areas on rocks of all ages after the Lower Proterozoic. The weathering profiles of the Wave Hill and Koolpinyah Surfaces vary according to the substrates. Inland, where erosion is still active, they have truncated unweathered rocks in many areas and there is little evidence of modern deep weathering. However, in the coastal areas truncated laterite profiles underlie these two surfaces and are overlain by detrital laterite resulting from erosion which has predominated in the formation and subsequent dissection of the two surfaces. However, it is possible that lateritization is still active on the near-coastal parts of the Koolpinyah Surface.

A number of erosion surfaces have been recognized in the Karumba Basin (Grimes 1979). The earliest of significance to the present study is the Early to Mid-Tertiary Aurukun Surface (Doutch 1976), which is best preserved around Weipa on the Bulimba Formation (Smart 1977). Deep weathering of this surface has resulted in a well-developed laterite profile which is highly aluminous in places. However, later deep-weathering events also appear to be superimposed on the original Mid-Tertiary profile (Grimes 1980; Smart 1977) and these are thought to be involved in the formation of the economic bauxites of the area. The Aurukun Surface has been correlated with the Tennant Creek Surface of the Northern Territory and the Mt Isa highlands (Grimes 1980).

The Kendall Surface is widespread around the upper reaches of many of the streams draining to the Gulf. It is developed as a depositional surface on the Wyaaba Beds, and as an erosional surface developed from the Aurukun Surface. It has been lateritized and is silicified in some places. The Kendall Surface is probably of Pliocene age, and may be a correlative of the Wave Hill Surface in the Northern Territory.

The current depositional cycle of the Karumba Basin is characterized by widespread sedimentation. Grimes and Doutch (1978) have recognized five phases of alluvial fan deposition, and Smart (1976) recognized two phases of Late Pleistocene and Holocene beach deposition along much of the coast of the Gulf.

Main soil determinants

The main factors determining soil formation and distribution in the Region are the lithology of soil parent materials and the geomorphic history of the present landscape. Present climate sets some limits to the kinds of soils occurring on young landscapes, but climate is not a particularly strong factor in soil differentiation on well-drained sites within the Region. The role of past climates, and this is intimately related to geomorphic history, is of obvious importance since relict soils are widespread. However, in many cases all that can be safely inferred is that certain soils have formed in a different environment to that of the present.

An interpretation of late Quaternary climates in northern Australia has been attempted by Webster and Streten (1972, 1978) and Nix and Kalma (1972), and these theories, together with the evidence of Kershaw (1978), were incorporated by Grimes and Doutch (1978) in their account of the Quaternary history of the Carpentaria Plains. This work suggests that marked climatic changes occurred during the Quaternary, and evidence from northwest Western Australia (Jennings 1975) suggests strongly that marked changes also occurred there. However, in most instances it is difficult to see the effect of these inferred climatic changes on present soil development and distribution.

The role of lithology, which together with geomorphic history largely determines present topography, is clear. One of the most striking and widespread examples is seen in those areas which consist of massive siliceous sandstones, another is the close association of particular soils and basic igneous rocks. Vegetation seems to depend on soil properties rather than itself influencing soil profile development.

Soil landscape provinces

Eleven soil landscape provinces (Fig. 14.1 and Table 14.1) have been delineated, although this represents a

Table 14.1 Main features of the soil-landscape provinces of Region III

Province	Geological and geomorphological features	Major soil features
Littoral	Depositional landscape of estuarine and riverine alluvial deposits with some beach ridge development	Highly saline soils (*solonchaks*) adjacent to coast; *black earths* and *grey clays* on floodplains; *calcareous* and *siliceous sands* on beach ridges
Carpentaria	Marginal Pliocene erosional suface; very extensive younger depositional surface of alluvial fan plains and floodplains; small remnants of Early to Mid-Tertiary erosional surface in south	Mainly *yellow earths* on Pliocene surface; grey massive earths and duplex soils on older fans; *alluvial soils*, *red earths*, *solodic soils* and *grey clays* on younger fans; southern floodplains have deep *grey clays*. Oldest surface in south has loamy *red* and *yellow earths* and *solodic soils*
Cape York–Arnhem	Mainly undulating low plateaux formed by erosional surfaces of several ages on Mesozoic or Tertiary sediments. Older surfaces have been extensively lateritized and have some economic bauxite deposits. Small depositional areas with aeolian deposits on the north-east Queensland coast	*Red earths* (often bauxitic and invariably nodular) on most older surfaces, some nodular *earthy sands*. On younger surfaces *yellow earths* are more common. Giant *podzols* on the aeolian deposits
Kimberley–Arnhem–Macarthur	Erosional landscape of rugged, dissected tablelands and plateaux formed from mainly siliceous Proterozoic sandstones. Some interbedded basic volcanics	Main soils are sandy *lithosols* and shallow stony *siliceous sands*. In local depositional sites are *earthy sands* with some *red* and *yellow earths*. Basic volcanics have loamy *lithosols* and shallow stony *euchrozems*
Borroloola	Gently undulating sloping plains broken in places by low hills of sandstone or laterite. Most of the province is probably part of an Early to Mid-Tertiary surface that was extensively lateritized	Dominant soils are shallow bleached sands overlying massive laterite or large amounts of ironstone nodules. Other nodular *siliceous sands* and *earthy sands* are associated, with smaller areas of nodular *red* or *yellow earths*
Daly	Southern area is part of a largely undissected, deeply lateritized Early or Mid-Tertiary surface formed on Cretaceous sediments. In the north is mainly a younger surface formed on Cambrian sedimentary rocks with small residuals of the older surface	In the south are mainly nodular *red* and *yellow earths* with *grey clays* in wide shallow drainageways. Similar soils occur in the north
Victoria River	A mainly erosional undulating landscape derived from calcareous, arenaceous and argillaceous Proterozoic sediments and Lower Cambrian volcanics. Some alluvial plains formed by major rivers	*Grey* and *brown clays* predominate with lesser *black earths*. Hilly volcanic areas have shallow stony *lithosols* and *euchrozems*. Alluvial plains have deep *grey clays* and marginal areas of *yellow earths*
Ord–Fitzroy	Mainly rugged topography of considerable relief derived from Lower Cambrian basalt and associated limestone, and Lower Proterozoic steeply folded metasediments, metavolcanics, and associated intrusions. Small areas of alluvial plains	Mainly *lithosols* and stony *euchrozems* on basalt and loamy, often calcareous *lithosols* on other rocks. Deep *grey clays* on the alluvial plains
Pine Creek	Strongly undulating foothills and strike ridges from erosion of Archaean metasediments, granites and gneisses, and from Lower Proterozoic metasediments and igneous intrusives. In the south west undulating plains on Palaeozoic sediments	On areas of greater relief *lithosols* are dominant with some *earthy sands* and *red* or *yellow earths*. Alluvial valleys and lower side slopes have *lateritic podzolic*, *yellow earths* and *solodic soils*. Palaeozoic area has mainly *lateritic podzolic soils* and *red earths*
Peninsula	Erosional strongly undulating to mountainous landscape formed from Proterozoic metamorphics and Palaeozoic granitic rocks and acid volcanics. Small areas of plains on Cainozoic sediments	In lower rainfall areas chief soils are *lithosols*, *red* and *yellow podzolics* and weakly developed *podzols*. Higher rainfall areas have *xanthozems*, *yellow* and *red podzolics*. Grey massive earths and *lateritic podzolics* occur on the Cainozoic sediments
Merluna	Erosional undulating plain formed on marine Cretaceous sediments. Isolated higher remnants of an older (Pliocene) lateritized surface	Main soils are *yellow earths*, with slightly gilgaied yellow-brown structured soils and *grey clays* in lower sites. *Red earths* occur on the older surface remnants

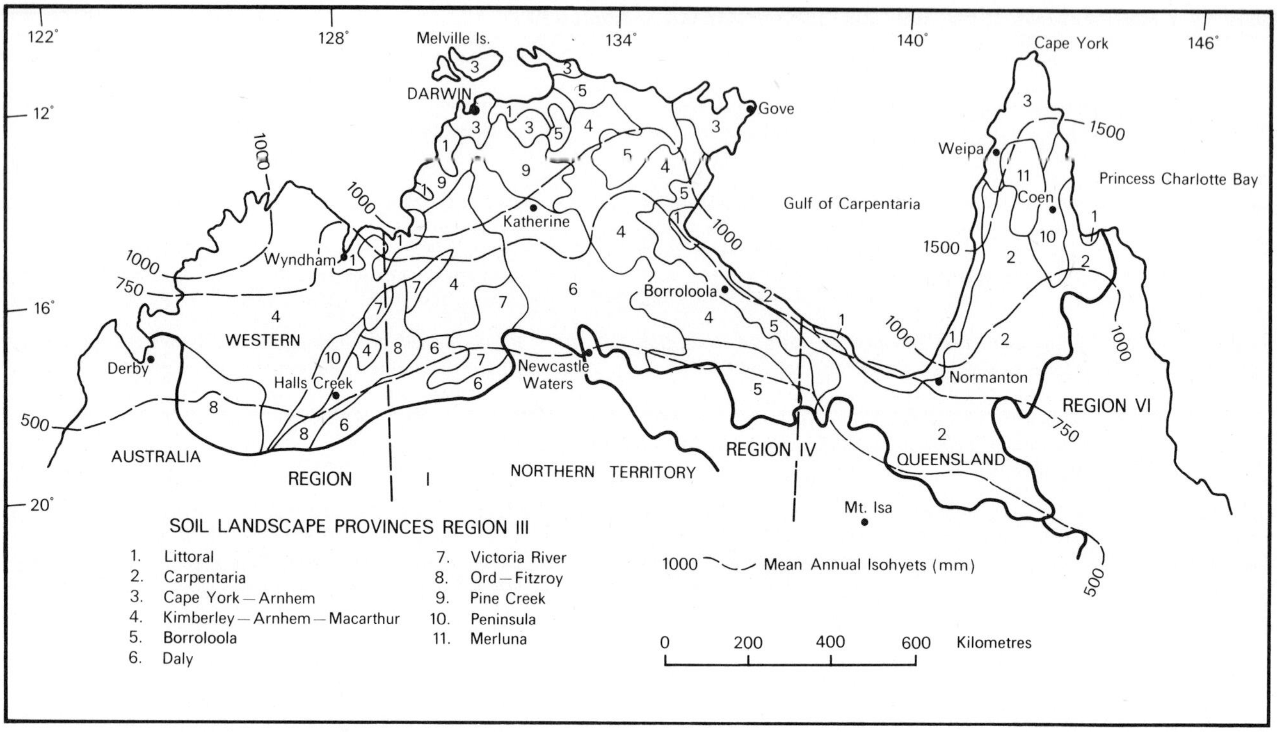

Fig. 14.1. Soil-landscape provinces of Region III.

gross simplification of the complex soil pattern present over much of the Region. Further, the somewhat unequal treatment of the various provinces below is largely a reflection of the extent of existing soil knowledge. Information on soil occurrence has been extensively drawn from McArthur and Wright (1967), Northcote (1968) and Isbell *et al.* (1968).

Littoral Province

This Province occurs as a narrow zone fringing many of the less-exposed coasts of northern Australia, particularly in the Gulf of Carpentaria. It consists mostly of fine-textured estuarine and riverine alluvial deposits, although coarser textured beach ridges are important at some localities.

On western Cape York Peninsula, Smart (1976) has recorded two sets of beach ridges, the older set are thought to be Late Pleistocene and the younger set range from about 6500–7000 years ago to the present. The youngest ridges consist of 50% shelly material and their soils are little-developed *calcareous sands.* The Holocene ridges show a landwards decline in carbonate content and an increasing depth of carbonate-free material; the older soils of these ridges are therefore mainly *siliceous sands.* The Pleistocene ridges are now essentially devoid of carbonate and their soils are either *siliceous sands,* usually with a colour B horizon, or more coherent yellowish *earthy sands.* No *podzols* have been reported, although these are common on similar beach ridges of the higher rainfall areas of the north-east Queensland coast.

A much more widespread component of this Province consists of clayey sediments of estuarine or riverine alluvial origin. These occur as near-shore or near-stream tidal salt flats and mangrove areas, often traversed by well-developed estuarine meanders and dendritic tidal channels (Whitehouse 1944; Twidale 1956; and Fig. 14.2). Most areas are inundated by high tides and floodwaters. The soils are *solonchaks.*

Another major component of the Province consists of low, poorly drained coastal floodplains that are seasonally inundated for 2–6 months. Although they occur in that part of the province which fringes the Gulf of Carpentaria, they are best developed in the higher-rainfall northern coastal areas, e.g. east of Darwin. The soils are all cracking clays: *black earths* and *grey clays.* Carbonate occurs in some profiles and yellow-grey, yellow-brown or red mottling is present in many below about 75 cm. A water table may occur in the deeper subsoil. These clays normally exhibit a variable gilgai microrelief, with a vertical interval ranging up to 20 cm. In more swampy sites there are soils with dark peaty clay loam or clay surface horizons which merge into mottled clay subsoils.

Although it has long been thought that the clay plains are currently aggrading, recent studies of the lower Carpentaria plains suggest that the present-day normal wet seasons and the more irregular major flooding have very little erosional or depositional effect on the landforms of this region (Simpson and Doutch 1977). These authors

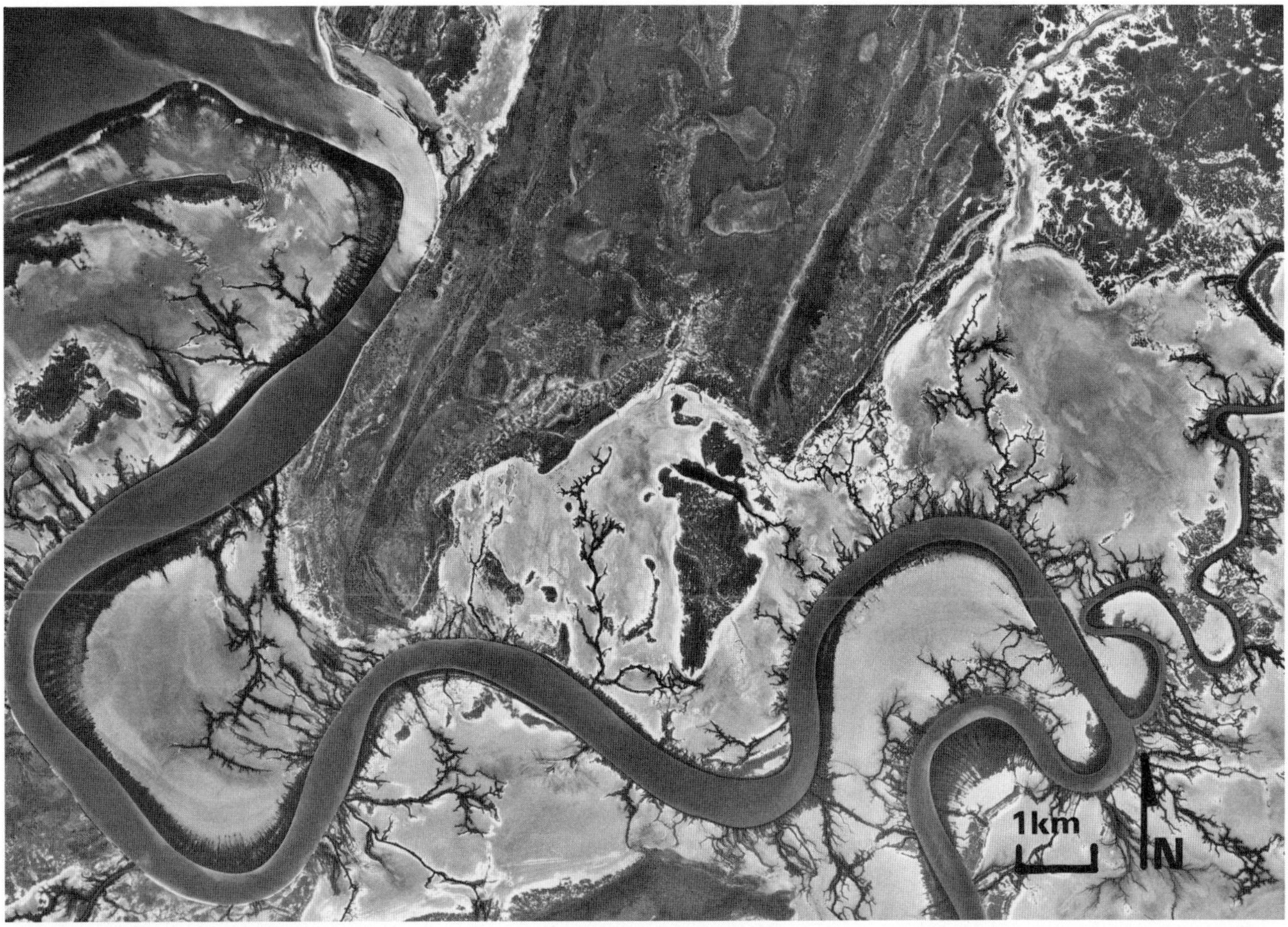

Fig. 14.2. Meandrine course of the Norman River near its mouth in the Gulf of Carpentaria at Karumba (N.W. corner of photo). The light-toned areas are saline tidal flats with *solonchaks*, the darker areas in the north centre are old beach ridges with *siliceous sands* and *earthy sands*. Photo courtesy of Division of National Mapping.

concluded that the major landforms probably developed in the Late Pleistocene under wetter conditions than those which currently prevail, and that they have not been substantially modified since their formation. This is supported by the configuration of the major river exits and beach ridges: the former have not substantially altered since the development of the Pleistocene ridges, inferred to be about 120 000 years old.

Carpentaria Province

The main occurrence of this large Province is contained within the Cainozoic Karumba Basin (Doutch 1976) and the soil pattern is controlled by the later history of this Basin (Grimes and Doutch 1978). In the central-western part of Cape York Peninsula the eastern margin of the Basin is a stable and more elevated old planation surface, whereas there is a widespread, complex series of fans to the west, which become progressively younger towards the Gulf coast. To the south of the Gulf there are extensive flood plains. The formation of the depositional plains began in Pliocene times with upwarping of the marginal parts of the Karumba Basin to create an erosional source area. The deposition was in five stages separated by periods of stability or incision, these stages were affected by continuing marginal source area movements, changes in Pleistocene sea levels, and climatic changes. The younger fans are still accumulating locally.

The eastern margin of the Province in Cape York Peninsula consists of remnants of the Pliocene Kendall Surface (Holroyd Plain of Grimes and Doutch (1978)). This is now a gently undulating plain consisting of low, very broad sandy rises and many sharply defined shallow drainage depressions together with prominent, small subcircular swampy areas (Fig. 14.3). The dominant soils of the rises are sandy *yellow earths* with some grey massive earths (Gn2.84, .94), and on higher ridge crests some sandy *red earths*. These soils may have ironstone nodule horizons at depth. The drainage depressions have sandy to loamy-surfaced *solodic soils* and bleached grey massive earths (Gn2.94, .95).

The older (more inland) of the Gilbert and Mitchell

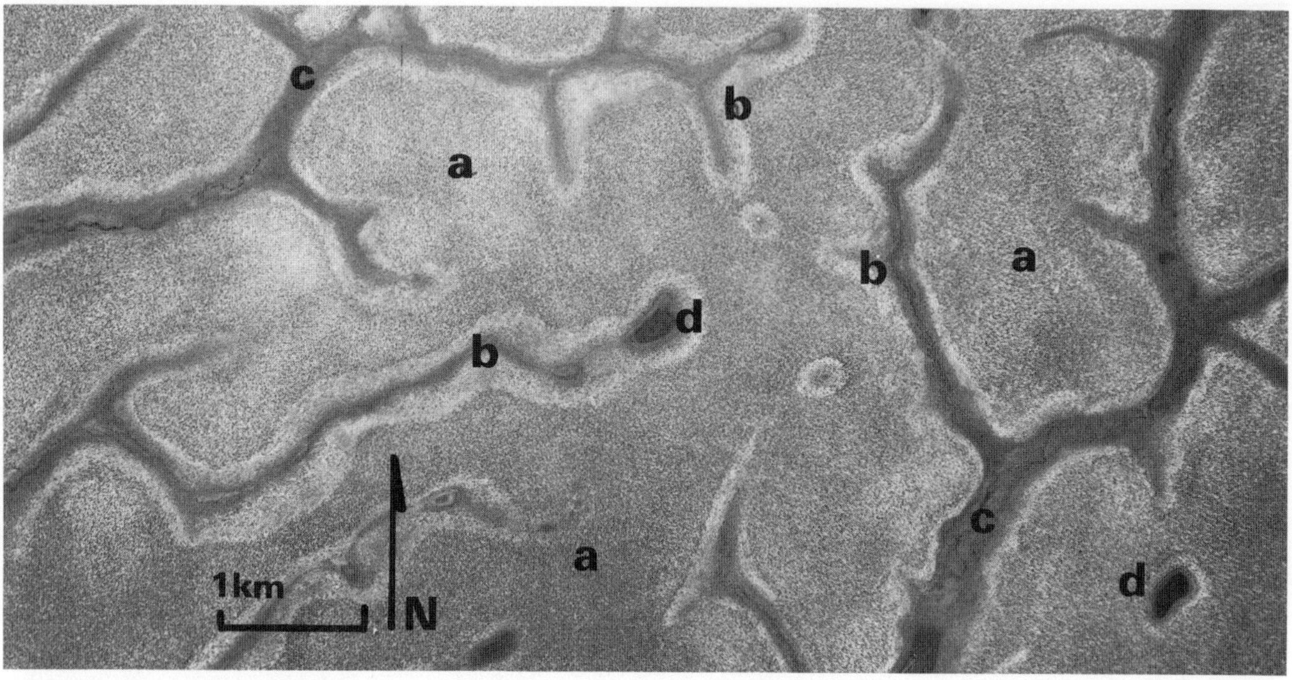

Fig. 14.3. Gently undulating plains of the Pliocene Kendall Surface forming the central eastern margin of the Carpentaria Province in Cape York Peninsula. This air photo shows (*a*) sandy *yellow* or *red earths* on the broad ridge crests; (*b*) bleached grey massive earths on the lower slopes; (*c*) *solodic soils* in the shallow drainage depressions; (*d*) small subcircular swampy areas. Photo courtesy Division of National Mapping.

Fans (Grimes and Doutch 1978) are near-level plains with loamy-surfaced bleached and mottled grey massive earths (Gn2.91, .94). Ironstone nodule horizons may occur at depth and in some soils a cemented hardpan may be present. Other earth soils are associated. Adjacent to small drainage lines *solodic soils* are usually dominant.

The younger fans of the Gilbert, Mitchell and smaller rivers to the north have a complex soil pattern. Adjacent to the present active streams there are usually low terraces or young levees with deep earthy loams (Um5.52) or sands (Uc5.22) that are usually stratified at depth. Associated on levee back slopes are brown massive earths (Gn2.41) and occasionally sandy *red earths.* Also common are slightly elevated old stream terraces, levees and infilled channels, these have sandy or loamy *red earths.* The greatest area of the fans is occupied by floodplain deposits with loamy-or silty-surfaced *solodic soils.* These plains are often cut by numerous old infilled channels. In slightly lower sites there are extensive areas of massive-surfaced *grey clays* or *black earths.* Some self-mulching *grey clays* also occur.

The depositional plains to the south of the Gulf fall into two distinctive groups: sandy plains and clay plains.

To the east of the Norman and Saxby Rivers the deposits are mostly sandy – the Claraville and Millungera Plains of Grimes and Doutch (1978). The Claraville Plain also shows a five-fold stage of evolution similar to that of the Gilbert and Mitchell Fans but the divisions are not as clear and it does not reach the coast. The fan-plains are very gently undulating with numerous sandy rises, relict channels, and levees adjacent to functional streams. The chief soils are deep, sandy or occasionally loamy, mottled grey massive earths (Gn2.84, .94) and duplex soils – *lateritic podzolics* – (Webb *et al.* 1974). Prominent ironstone nodule horizons usually occur in these soils. The associated old infilled channels are evident as low sandy rises with deep sandy *yellow earths*, grey massive earths (Gn2.94, .95, .81, .82) and *lateritic podzolic* or *soloth* soils. At the southern end of the Claraville Plain the soils are considerably sandier and show little increase in clay with depth. They are yellowish *earthy sands*, although sandy *yellow earths* are commonly associated. The smaller Millungera Plain has similar soils, but broad shallow depressions with *grey clays* are a more common feature.

Grimes and Doutch (1978) have shown the Millungera Plain to consist of a thin sandy cover overlying the clays that characterize the grassy treeless plains of the Wondoola and Armraynald Plains to the west. The striking lithofacies difference between these latter plains and their sandy counterparts to the east and north has been attributed to differences between the source areas (the labile Mesozoic mudstones to the south versus the siliceous sandstones and acid igneous rock areas to the south-east), and by the prior topography of the depositional area which guided streams from the different source areas and prevented mixing of the different types of sediment.

The clay plains are characterized by self-mulching *grey clays*. Associated with past or prior stream channels are small areas of loamy-surfaced (often scalded) *solodic soils*, while adjacent to some major streams and channels are deep loamy *calcareous red earths*.

The Wondoola and Armraynald clay plains are separated by an elevated area, the Donors Plateau, which together with the smaller Normanton Plateau (Grimes and Doutch 1978) represents a dissected remnant of the Early to Mid-Tertiary Aurukun Surface. The soils on these low dissected plateaux are shallow, gravelly earthy loams (Um5.51) that may contain massive nodular laterite, gravelly loamy *red earths* or *yellow earths*, and loamy *solodic soils* on gentle slopes. The soils appear to be derived from the dissected remnants of the eroded laterite profile which originally covered the Aurukun Surface.

The small occurrence of this Carpentaria Province in the Princess Charlotte Bay region has an inland margin of mainly dissected Kendall and Aurukun Surfaces with deep sandy *red earths*, and a depositional area nearer the coast with *yellow earths*, grey massive earths (Gn2.91), and *solodic soils*.

Cape York–Arnhem Province

The Cape York Peninsula occurrence of this Province is for the most part an undulating plateau. The oldest surface is the Weipa Plateau (Smart 1977), a dissected remnant of the Early to Mid-Tertiary Aurukun Surface. Where best preserved near the coast, as around Weipa, an aluminous (bauxitic) laterite is widely exposed. The soils forming the upper part of this sequence are bauxitic *red earths*. However, further inland where the surface has been stripped, the upper part is ferruginous and the soils are *red earths* with prominent ironstone nodule or laterite horizons at depth.

Elsewhere in Cape York Peninsula this Province consists mainly of much younger surfaces. In the central northern part a largely erosional, moderately to strongly undulating surface has developed on Late Jurassic–Early Cretaceous quartzose sandstones. However, in many areas these are deeply weathered with *yellow earth* and *red earth* soils which have prominent ironstone nodule horizons at depth. Where exposed by coastal erosion these horizons have hardened to massive nodular laterite. Depositional areas are also prominent, particularly fringing the Jardine River. Here on low-lying plains *podzols* and other bleached sandy soils (Uc2.21) are common. Fringing the east coast there is a zone of largely vegetated, elongated parabolic dunes with ridges up to 30 m high (Fig. 14.4). The soils are mostly giant *podzols* and deep *siliceous sands*.

The Northern Territory portions of the Province are less well known. The occurrence in east Arnhem Land and on the Cobourg Peninsula, Bathurst and Melville Islands is an undulating plateau up to 100 m that has been variably dissected. On the two islands (van-Cuylenburg and Dunlop 1973) the flat plateau surfaces – probably the Early or Mid-Tertiary Tennant Creek Surface of Hays (1967) – have been extensively lateritized and carry deep *red earths*. More shallow and gravelly soils dominate the plateau margins and dissected areas. The plateau areas are fringed by extensive colluvial plains with deep, very sandy, gravel-free *red earths*.

In some areas, the *red earths* are bauxitic, e.g. Gove and Cobourg Peninsula. Here it is thought that the higher grade bauxite deposits were formed by warping and partial truncation of the earlier surface, leading to detrital accumulation and further alumina enrichment of the earlier bauxite (Plumb and Gostin 1973; Hughes 1978). This younger surface is also dominated by *red earths* which contain variable amounts of ironstone and/or bauxite nodules (pisolites). Associated soils in all occurrences are chiefly *earthy sands* which also are often very gravelly.

The area to the east of Darwin is essentially an undulating terrain produced by variable stripping and dissection of the Cainozoic Koolpinyah Surface of Hays (1967). Much of it is lateritized but it has been suggested (Hooper 1969; Williams 1969) that many of the laterites are of secondary, detrital origin. The soils are mainly *red earths* and brown massive earths (Gn2.41) which often contain much ironstone gravel. With them are large areas of ironstone-gravelly red and yellow *earthy sands* and other sands with an ironstone gravel matrix.

Kimberley–Arnhem–Macarthur Province

This extensive province is largely underlain by little-deformed Proterozoic sedimentary rocks (often siliceous sandstones and quartzites) with some basic volcanics. Much of the area is extremely rugged, consisting of dissected sandstone tablelands and plateaux with many

Fig. 14.4. Semi-active, elongate parabolic dunes fringing the north east coast of Cape York Peninsula. The soils are mostly giant *podzols* and deep *siliceous sands*.

Fig. 14.5. Rugged quartzite escarpment forming the western edge of the Arnhem Land plateau. Much of the Kimberley–Arnhem–Macarthur Province consists of similar country, the soils being dominated by *lithosols*. Photo courtesy CSIRO Division of Land Use Research.

precipitous gorges (Fig. 14.5). As much of the area is remote and with little land use potential, it has not been studied in any detail. More specific information is available for the western part of the Arnhem Land plateau in Galloway (1976) and Aldrick (1976).

In the Northern Territory, Hays (1967) has considered the Province to form part of his Miocene–Pliocene Wave Hill Surface. Siliceous sedimentary rocks are widely exposed in the more rugged areas and the soils are dominantly *lithosols* and shallow *siliceous sands*. However, on areas of lesser relief, and on depositional sites, deeper *siliceous sands* and *earthy sands* are common; many of these contain ironstone gravel. Lesser areas of *red* and *yellow earths* occur on major stream levees or old terraces. Where the upland areas contain less-siliceous parent materials, loamy *lithosols* are more common.

In the Kimberley region there are extensive areas of basic volcanics with hilly topography where the soils are mostly shallow stony *euchrozems* on more moderate slopes and loamy *lithosols* on the steeper slopes and on hill crests. Very small areas of *black earths* and *grey clays* occur in some lower sites, together with loamy, structured yellow-brown soils (Gn3.22). In some areas (e.g. Mitchell Plateau), the volcanics have been lateritized to form bauxite; the soils are mainly pisolitic *red* or *yellow earths*.

Borroloola Province

This Province consists of gently undulating sloping plains, broken in places by low hills of sandstone or laterite. It appears that much of the province is underlain by laterite which may have been differentially eroded. Stewart (1954) referred to the south-eastern part as a low-level Tertiary plain which he thought represented the downwarped part of a much more extensive Tertiary lateritic landsurface to the west. This was subsequently identified as the Early or Mid-Tertiary Tennant Creek Surface by Hays (1967). However, it is also possible that the lowlands represent the younger Koolpinyah Surface of Hays.

The portion of the Province south-east from the Roper River and adjacent to the Gulf is characterized by shallow bleached sands (Uc2.12) overlying massive laterite or large amounts of ironstone nodules. Other shallow sands with ironstone gravels are associated with these bleached sands. The area to the east of Newcastle Waters has *red earths* and sands usually containing high amounts of ironstone nodules and frequently overlying massive laterite at shallow depths. In the shallow broad valleys at both localities the soils are mainly acid *yellow earths*, with some small areas of *grey clays*.

The Arnhem Land portion of the Province has slightly more relief in the form of residual hills of sandstone and laterite. Here the soils are generally deeper, mainly *earthy sands* and *siliceous sands* with variable amounts of ironstone nodules. Associated are ironstone–gravelly *red* and *yellow earths* and some *lateritic podzolic soils* in alluvial valleys.

Daly Province

This Province is mainly situated in the Daly River Basin to the north of about Mataranka, and extends south to Newcastle Waters. Its northern area, and also the portion across the State border, is part of the largely undissected Tennant Creek Surface of Hays (1967). The Province has a deeply lateritized surface formed on Cretaceous sediments in the north and on Lower Cambrian or Proterozoic volcanics and sediments across the border. The landscape is gently undulating and the soils are dominantly *red earths* with associated *yellow earths*; all soils often contain ironstone nodule horizons. The area is traversed by a number of wide, shallow drainageways with deep *grey clays*.

The part of the Province in the Daly River Basin (Wright 1963, 1965; van der Graaff 1965; Aldrick and Robinson 1972) may be divided into two sections. In the south are low plateaux of lateritized Cretaceous sediments which form a dissected part of the Tennant Creek Surface (Bradshaw Surface of Wright 1963). The dominant soils are ironstone–gravelly sands and *red earths*. In the north of the Basin, the Tennant Creek Surface has largely been removed to expose the underlying Cambrian sediments. This area now comprises extensive plains and undulating terrain and has been identified as the Koolpinyah Surface (Hays 1967). The soils are dominantly *red earths*, with associated *yellow earths*. Floodplains of major streams are characterized by massive-surfaced *grey clays*, with lesser self-mulching forms. Remnants of the Tennant Creek Surface also occur throughout the area as flat-topped laterite residuals forming plateau summits on which there is a cover of *red earths*.

Victoria River Province

This Province comprises parts of the Wave Hill and Koolpinyah Surfaces of Hays (1967) and consists largely of undulating plains that bevel calcareous, arenaceous and argillaceous Proterozoic sediments and Lower Cambrian volcanics.

The soil pattern is dominated by moderately deep *grey* and *brown clays* with lesser areas of *black earths*. Associated hilly areas of basic volcanics have mainly *lithosols* with relatively shallow, stony *euchrozems* on the more moderate slopes.

The small unit of this Province near the junction of the Baines and Victoria Rivers is an alluvial plain of *grey clays* with marginal areas of *yellow earths*.

Ord–Fitzroy Province

This Province occupies part of the Ord and Fitzroy River Basins. Most of the Province has a rugged topography of considerable relief. In the Ord Basin (Aldrick *et al.* 1978) there are two main components: the mesas, buttes, structural benches and cuestas of the Antrim Plateau Lower Cambrian basalts and associated limestones, and the steep ridges and hills of the eastern part of the Lower Proterozoic Hall's Creek Province with its steeply folded sediments and basic volcanics (Plumb and Derrick 1975).

The Antrim basalt area has loamy *lithosols* on ridge crests and steeper slopes, and shallow, usually stony *euchrozems* on moderate to gentle slopes. In lower, flatter situations there are small areas of *grey*, *brown* and *red clays*. Shallow loamy soils, some of which are calcareous throughout, occur on the dissected limestone cuestas. The Lower Proterozoic folded area of the upper Ord River Basin consists mainly of gravelly loamy *lithosols* with many rocky outcrops.

The upper Fitzroy River Basin part of the Province is an extremely complex, often rugged, dissected area of metasediments, metavolcanics and associated intrusives. The area consists broadly of a series of hills and ridges with NW–SE trending strikes and intervening narrow valleys. The geomorphic history of the area has been discussed by Wright (1964). The soil pattern is much dependent on the underlying lithology, but on the hilly areas loamy *lithosols*, often calcareous, are dominant. On granitic rocks, soils related to *non-calcic brown soils* are prominent, together with sandy *lithosols*.

Two other soil landscapes occur in this part of the Province. Extensive alluvial plains with *grey clays* are associated with the major rivers, and in the south there are *yellow earths*, shallow *red earths* developed on sedimentary rocks, and associated eroded *solodic soils*.

Pine Creek Province

To the south-east of Darwin this Province is underlain by the Lower Proterozoic metasediments and igneous intrusives of the Pine Creek Geosyncline (Plumb and Derrick 1975); to the south by Archaean metasediments, granites and gneisses; and to the south-west mainly by Palaeozoic sediments. The soils of the region were first described by Stewart (1956), and more recently in the south-eastern part by Hooper (1969) and Williams (1969), and in the lower Daly River area by Fogarty and Gibbs (1976).

The south and south-eastern area mostly consists of foothills cut into or below the plateau formed by remnants of the Tennant Creek Surface of Hays (1967). Typically these hills form a sequence of rocky strike ridges flanked by narrow pediments and separated from one another by alluvial flats of variable width. The soils on slopes steeper than about 5% are almost invariably sandy or loamy *lithosols*, but some red *earthy sands* or *red* or *yellow earths* are also associated. On some granitic areas the major soils are sandy-surfaced *yellow* and *lateritic podzolics*. The intervening alluvial valleys and lower side slopes have *lateritic podzolic* and *solodic soils*, and *yellow earths*. On the larger floodplains associated with major streams the soil pattern is generally one of deep silty loams or loamy *red earths* on recent and older levees, while on the stable floodplains loamy *yellow earths, lateritic podzolic* and *solodic soils* are predominant.

The Palaeozoic rock area to the south west of Darwin is mostly part of the Koolpinyah Surface of Hays (1967). It consists of undulating plains with *lateritic podzolic soils* and some *red earths*.

Peninsula Province

In the central–eastern part of Cape York Peninsula this Province consists of an erosional landscape containing Proterozoic metamorphics and Palaeozoic granitic rocks and acid volcanics. Small areas of Cainozoic continental sandstone occur in association with the older rocks. The metamorphics are most common in the south and mostly form areas of lower relief. They are partly covered by remnants of a lateritized surface with mainly *yellow earth* soils which have prominent ironstone nodule horizons and with massive nodular laterite often outcropping. More resistant remnants of metasediments occur as prominent strike ridges and have mostly *lithosols*, with shallow gravelly *yellow podzolic* and *red podzolic soils* on schistose rocks. Narrow, linear dolerite intrusions are common with deeper *euchrozems*. Stream levees throughout the area have *red earths*.

The granitic and acid volcanic rocks of this area have a high hilly to mountainous topography, and *lithosols* are the most common soil in the lower rainfall areas. They are mainly sandy, but shallow stony *red* and *yellow podzolic soils* also occur. Nearer to the coast in high rainfall zones the major soils are *xanthozems, yellow podzolics* and *red podzolics*. In the vicinity of Coen an extensive undulating plateau underlain by granitic rocks has deep bleached sands that might be regarded as weakly developed *podzols*.

The small areas of Cainozoic sediments form near-level plains, which may be elevated, with grey massive earths (Gn2.94) or *lateritic podzolic soils*.

In the south-east Kimberley area this Province has high hilly lands developed from lower Proterozoic granite in the north, and similar less hilly lands in the south. The major soils are *lithosols* with many rock outcrops on the steeper slopes, and *red* and *yellow podzolic soils* on areas of lesser relief.

Merluna Province

This small Province is an undulating erosional plain formed on marine Cretaceous sediments, with isolated higher remnants of the older lateritized Kendall Surface (Grimes 1979) on which there are *red earth* soils.

The dominant soils are loamy *yellow earths*, almost always with a prominent ironstone nodule horizon at depth and in many areas massive nodular laterite outcrops at the surface. There often seems to be a discontinuity at depth where there is a heavy textured, structured clay D horizon apparently derived *in situ* from the underlying Cretaceous argillaceous sediments. Throughout the area, mostly in lower sites or adjacent to drainage lines, minor occurrences of slightly gilgaied brown friable earths (Gn3.51, .71), *grey clays*, and occasionally *brown clays* are found.

Land use aspects

There are a number of factors that have been responsible for the present low level of land use in the Region. Some are historical and socio-economic and have been partly discussed elsewhere (e.g. Davidson 1965). The potential of the region for more intensive agriculture and pastoral production has also been discussed, by Davies and Eyles (1965), Henzell (1975) and t'Mannetje *et al.* (1976), but it is appropriate here to summarize some of the more important edaphic factors that are likely to influence future rural development. At least moderate annual rainfall totals are recorded with a reasonable degree of reliability over most of the Region. However, the rainfall is extremely seasonal and thus in the wet season extended waterlogging of a number of soils, e.g. *yellow earths, solodics, lateritic podzolics, grey clays* and *black earths*, presents serious problems for their utilization by plants, and their ability to support traffic.

Much of the land is unsuitable for intensive use because of limitations of topography and shallow soils. *Lithosols* are one of the major soil groups throughout the Region. However, there are large areas of soils that do have favourable physical properties. The most important and widespread of these are the *red earths*, which, although varying in moisture-holding capacity (Day 1977), can usually store considerable amounts of water that would be available to deep-rooted plants. Their disadvantages are the tendency of some forms to crust severely if plant cover is removed, their erodibility following cultivation, and the occurrence of massive or nodular laterite in many of their profiles.

Soil fertility is a major problem through the Region. Although data are relatively limited, the evidence suggests that nitrogen is almost invariably low throughout, and there is an almost universal phosphorus deficiency. Even soils formed from parent materials such as basalt, which elsewhere in Australia commonly have adequate to high phosphorus status, have very low levels of plant-available phosphorus in this Region (Isbell, Chapman and Winter, unpublished data). Jones *et al.* (1975) have drawn attention to the likelihood of widespread sulphur deficiency. Presently available data for trace element contents in the soils suggest copper and zinc may be limiting in some areas. Work on sandy soils in Cape York Peninsula (Jones 1973; Isbell *et al.* 1976; Winter and Jones 1977) has highlighted the occurrence of multiple deficiencies. However, soil pH values seldom fall below about 5.0 and are commonly higher.

It is difficult to attribute the overall low soil fertility levels entirely to soil parent materials and present climatic features. It seems likely that the long history of deep weathering of many parts of the Region, operating on parent materials that may have already undergone previous cycles of weathering, is the major reason for the present soil impoverishment. Fertilizers need to be added to most soils in the Region to ensure even a modest productivity.

References

Aldrick, J. M. (1976). Soils of the Alligator Rivers area, Northern Territory. CSIRO Aust. Land Res. Ser. No. 38.

Aldrick, J. M., Howe, D. F. and Dunlop, C. R. (1978). Report on the lands of the Ord River catchment, Northern Territory. Anim. Industry and Agric. Branch, Dep. Northern Territory Tech. Bull. No. 24.

Aldrick, J. M. and Robinson, C. S. (1972). Report on the land units of the Katherine–Douglas area, N.T. 1970. Land Conservation Ser. No. 1. (Dep. Interior: Canberra.)

Bureau of Mineral Resources (1976). Map of the Cainozoic geology of the Northern Territory. (Dep. Nat. Resour.: Canberra.)

Davidson, B. R. (1965). 'The Northern Myth'. (Melbourne University Press: Melbourne.)

Davies, J. G. and Eyles, A. G. (1965). Expansion of Australian pastoral production. *J. Aust. Inst. Agric. Sci.* **31**. 77-93.

Day, K. J. (1977). Fertility studies on three red earth soils of the Daly basin, Northern Territory. 1. Morphological, physical and chemical characteristics. Anim. Industry and Agric. Branch, Dep. Northern Territory Tech. Bull. No. 22.

Doutch, H. F. (1976). The Karumba Basin, north-eastern Australia and southern New Guinea. *BMR J. Aust. Geol. Geophys.* **1**, 131-40.

Fogarty, P. J. and Gibbs, C. R. (1976). Report on the land units of the Daly River agricultural area. Anim. Industry and Agric. Branch, Dep. Northern Territory Tech. Bull. No. 23.

Galloway, R. W. (1976). Geomorphology of the Alligator Rivers area, Northern Territory. CSIRO Aust. Land Res. Ser. No. 38.

Grimes, K. G. (1979). The stratigraphic sequence of old land surfaces in northern Queensland. *BMR J. Aust. Geol. Geophys.* **4**, 33-46.

Grimes, K. G. (1980). The Cainozoic geology of north Queensland. *In*

'The Geology and Geophysics of North-eastern Australia'. (Eds. R. A. Henderson and P. J. Stephenson). (Geol. Soc. Aust. (Queensl. Div.): Brisbane.)

Grimes, K. G. and Doutch, H. F. (1978). The late Cainozoic evolution of the Carpentaria Plains, North Queensland. *BMR J. Aust. Geol. Geophys.* **3**, 101-12.

Hays, J. (1967). Land surfaces and laterites in the north of the Northern Territory. *In* 'Landform Studies from Australia and New Guinea'. (Eds. J. N. Jennings and J. A. Mabbutt). (Aust. Nat. Univ. Press: Canberra.)

Henzell, E. F. (1975). Agricultural development in tropical Australia. *Etruscan* **24**, 18-23.

Hooper, A. D. L. (1969). Soils of the Adelaide-Alligator area, Northern Territory. CSIRO Aust. Land Res. Ser. No. 25.

Hughes, R. J. (1978). The geology and mineral occurrences of Bathurst Island, Melville Island, and Cobourg Peninsula, Northern Territory. Bur. Min. Resour. Aust. Bull. No. 177.

Isbell, R. F., Jones, R. K. and Gillman, G. P. (1976). Plant nutrition studies on some yellow and red earth soils in northern Cape York Peninsula. 1. Soils and their nutrient status. *Aust. J. Exp. Agric. Anim. Husb.* **16**, 532-41.

Isbell, R. F., Webb, A. A. and Murtha, G. G. (1968). 'Atlas of Australian Soils'. Sheet 7, North Queensland. With explanatory data. (CSIRO Aust. and Melbourne University Press: Melbourne.)

Jennings, J. N. (1975). Desert dunes and estuarine fill in the Fitzroy estuary (north-western Australia). *Catena* **2**, 215-62.

Jones, R. K. (1973). Studies on some deep sandy soils in Cape York Peninsula, North Queensland. 2. Plant nutrient status. *Aust. J. Exp. Agric. Anim. Husb.* **13**, 89-97.

Jones, R. K., Probert, M. E. and Crack, B. J. (1975). The occurrence of sulphur deficiency in the Australian tropics. Proc. Symp. on Sulphur in Australasian Agriculture, Canberra.

Kershaw, A. P. (1978). Record of last interglacial-glacial cycle from north-eastern Queensland. *Nature* (London) **272**, 159-61.

McArthur, W. M. and Wright, M. J. (1967). 'Atlas of Australian Soils'. Sheet 9, Kimberley Area. With explanatory data. (CSIRO Aust. and Melbourne University Press: Melbourne.)

't Mannetje, L., Durand, M. R. E., Isbell, R. F. and Sturtz, J. D. (1976). Options for improved efficiency in the beef industry of non-arid tropical Australia. *Trop. Grassl.* **10**, 151-64.

Nix, H. A. and Kalma, J. D. (1972). Climate as a dominant control in the biogeography of northern Australia and New Guinea. *In* 'Bridge and Barrier: the Natural and Cultural History of Torres Strait'. (Ed. D. Walker) (Aust. Nat. Univ. Press: Canberra.)

Northcote, K. H. (1968). 'Atlas of Australian Soils'. Sheet 8, Northern Part of Northern Territory. With explanatory data. (CSIRO Aust. and Melbourne University Press: Melbourne.)

Plumb, K. A. and Derrick, G. M. (1975). Geology of the Proterozoic rocks of the Kimberley to Mt. Isa region. *In* 'Economic Geology of Australia and Papua-New Guinea'. (Ed. C. J. Knight) (A.I.M.M.: Melbourne.)

Plumb, K. A. and Gostin, V. A. (1973). Origin of Australian bauxite deposits. Bur. Min. Resour. Aust. Rec. 1973/156 (unpubl.)

Simpson, C. J. and Doutch, H. F. (1977). The 1974 wet-season flooding of the southern Carpentaria Plains, northwest Queensland. *BMR J. Aust. Geol. Geophys.* **2**, 43-51.

Smart, J. (1976). The nature and origin of beach ridges, western Cape York Peninsula, Queensland. *BMR J. Aust. Geol. Geophys.* **1**, 211-18.

Smart, J. (1977). Weipa, Qld. – 1:250000 Geological series. Bur. Min. Resour. Aust. Explan. Notes. SD/54-3.

Stewart, G. A. (1954). Geomorphology of the Barkly region. CSIRO Aust. Land Res. Ser. No. 3.

Stewart, G. A. (1956). The soils of the Katherine-Darwin region, Northern Territory. CSIRO Aust. Soil Publ. No. 6.

Twidale, C. R. (1956). A reconnaissance survey of the coastline of the Leichardt-Gilbert area, north-west Queensland. *Aust. Geogr.* **6**, 14-20.

van-Cuylenburg, H. R. M. and Dunlop, C. R. (1973). Land units of the seventeen mile plain, Melville Island, N.T. Anim. Industry and Agric. Branch, Dept. of Northern Territory, Tech. Bull. No. 14.

van der Graff, R. H. M. (1965). Soils of the Tipperary area, Northern Territory. CSIRO Aust. Land Res. Ser. No. 13.

Webb, A. A., Beeston, G. R. and Hall, T. J. (1974). The soils and vegetation of part of the Mayvale land system in the Gulf of Carpentaria region. Agric. Chem. Branch Tech. Rep. No. 5, Queensl. Dept. Primary Industries, Brisbane.

Webster, P. J. and Streten, N. A. (1972). Aspects of late Quaternary climate in tropical Australasia. *In* 'Bridge and Barrier: the Natural and Cultural History of Torres Strait'. (Ed. D. Walker) (Aust. Nat. Univ. Press: Canberra.)

Webster, P. J. and Streten, N. A. (1978). Late Quaternary ice age climates of tropical Australasia: interpretations and reconstructions. *Quat. Res.* **10**, 279-309.

Whitehouse, F. W. (1944). The natural drainage of some very flat monsoonal lands. *Aust. Geogr.* **4**, 183-96.

Williams, M. A. J. (1969). Geomorphology of the Adelaide-Alligator area, Northern Territory. CSIRO Aust. Land Res. Ser. No. 25.

Willmott, W. F., Whitaker, W. G., Palfreyman, W. D. and Trail, D. S. (1973). Igneous and metamorphic rocks of Cape York Peninsula and Torres Strait. Bur. Min. Resour. Aust. Bull. No. 135.

Winter, W. H. and Jones, R. K. (1977). Nutrient responses on a yellow earth soil in northern Cape York Peninsula. *Trop. Grassl.* **11**, 247-55.

Wright, R. L. (1963). Deep weathering and erosion surfaces in the Daly River basin, Northern Territory. *J. Geol. Soc. Aust.* **10**, 151-64.

Wright, R. L. (1964). Geomorphology of the west Kimberley area, W.A. CSIRO Aust. Land Res. Ser. No. 9.

Wright, R. L. (1965). Geomorphology of the Tipperary area, Northern Territory. CSIRO Aust. Land Res. Ser. No. 13.

Chapter 15
North-eastern Plains (IV)

R. F. Isbell and G. D. Hubble

THIS SPARSELY SETTLED, ARID TO SEMI-ARID INLAND region occupies much of western Queensland, with small extensions into adjoining states. It is essentially a land of vast plains, although some areas, e.g. Cloncurry–Mt Isa, are characterized by rugged, dissected uplands in which local relief may be of the order of several hundred metres; the highest elevation is, however, only about 600 m. In contrast there are areas in the extreme south where elevation is less than 80 m. Drainage in all except the far north of the Region is into the playas of the Lake Eyre Basin, but all streams are intermittent and in many years fail to reach their ultimate outfalls.

The climate is extremely hot although winter frosts are experienced in all areas except the far north. Mean annual rainfall ranges from 500 mm in the north and east to less than 150 mm in the south-west, which borders the Simpson Desert (Fig. 15.1); it is highly variable but mostly summer dominant, and prolonged droughts are a feature.

Plant communities are generally open with extensive *Astrebla* spp. tussock grasslands a feature in the north-east. Areas of *Acacia*-dominant, low open woodlands (particularly mulga – *A. aneura*) are also prominent in parts of the Region, while the lowest rainfall areas are characterized by sparse open herbfields (Carnahan 1976) with species of Chenopodiaceae.

The soil pattern in most areas is relatively simple and the range of soils limited. The most extensive are *grey, brown* and *red clays, desert loams, lithosols*, solonized *red brown earths, red earths* which are often shallow, and *grey-brown* and *red calcareous soils*.

Outline of landscape history

The Region consists of three major geological elements: the Mt Isa Inlier in the north, the Georgina Basin in the north-west, and the greater part of the Eromanga Basin (Geological Survey of Queensland 1975). The Precambrian Mt Isa Inlier contains a wide variety of metamorphic, sedimentary and igneous rocks (Plumb *et al.* 1980) and has probably been a land area since Cambrian times. The Georgina Basin (Smith 1972; Shergold and Druce 1980) contains a sequence of Cambrian to Ordovician limestones, dolomites and other sediments which in many places are overlain by Tertiary limestones and Cainozoic surficial deposits (Paten 1964; Grimes 1980). The Eromanga Basin (Senior *et al.* 1978) is a large sedimentary basin that is part of the earlier-named 'Great Artesian Basin'. It contains both freshwater and marine sediments ranging from Jurassic to Cainozoic age and varying from quartzose and labile sandstone to siltstone and mudstone which may be calcareous.

Present landscapes in the two sedimentary basins owe much of their form to various Cainozoic events that have

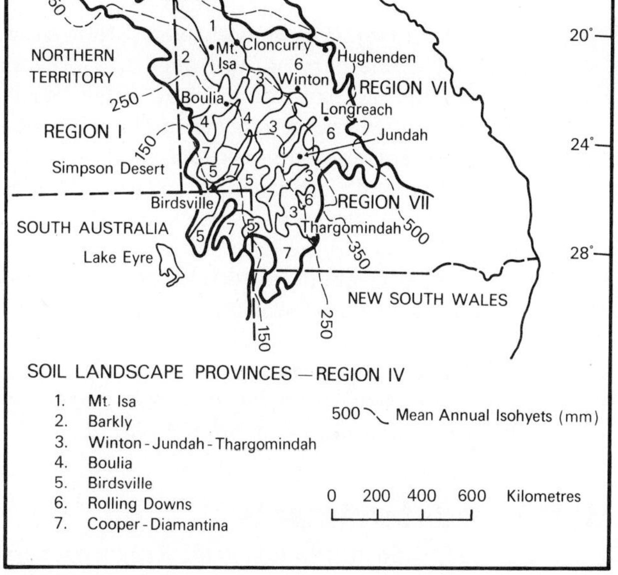

Fig. 15.1. Soil-landscape provinces of Region IV.

affected the earlier sedimentary deposits. The most important was the formation of various land surfaces during the Tertiary, associated with deep weathering under climatic regimes very different from those of the present. Subsequent widespread differential stripping of the old deeply weathered mantle has largely resulted in the present day landscape. These events have also been discussed in Part 2, so only an outline of the more relevant aspects is given here.

Much of the present day surface of the Eromanga Basin consists of the remnants of two deeply weathered (up to 90 m) and extensively silicified zones; the Morney and Canaway profiles of Idnurm and Senior (1978). The surface also consists of younger terrestrial deposits that are also extensively silicified, e.g. Wopfner *et al.* (1974), Wopfner (1978). Later Cainozoic erosion has truncated or completely removed these profiles, particularly along the axial zones of anticlinal folds and near faults. Elsewhere the profiles form low mesas and cuestas which are in part scarp-bounded or occur as low gravelly rises on an extensively pedimented surface.

The two deeply weathered profiles have been dated by palaeomagnetic means, the Morney as Late Upper Cretaceous to Early Eocene (60 million years), and the Canaway tentatively as Late Oligocene (30 million years) (Idnurm and Senior 1978). Although tentative correlations of the Cainozoic geological events in this Region have been made with those elsewhere in Queensland (Grimes 1980), the relationships between the western Queensland deep-weathering events, lateritization and silicification are far from fully elucidated.

The Cainozoic climatic history of the Region is also uncertain. It seems reasonable to relate the deep weathering to climates far more humid than those of the present, although doubt exists as to the environment which accompanied the widespread silicification (Langford-Smith 1978). The major rivers often show several levels of Quaternary alluvial deposits (Veevers and Rundle 1979), and sand dunes, now often partly fixed by vegetation, extend as relict formations well to the east of the present desert dune system into areas with a mean annual rainfall approaching 500 mm. Bowler (1976) concluded that although the development of aridity in Australia dates from the Late Tertiary, it probably reached its maximum about 18 000–16 000 years ago; since then, and particularly over the past 10 000 years, the climate has been relatively stable, probably fluctuating no more than at present. Further details on Cainozoic climates have been given in Part 2 and by Kemp (1978) and Galloway and Kemp (1981).

Main soil determinants

Soil occurrence and distribution in the Region are very much governed by the earlier landscape history and the lithology of the soil parent materials. The most striking examples of the latter are the *grey, brown* and *red clays* of the rolling downs, and in the uplands of the Mt Isa Inlier where *lithosols* are predominant on all of the more resistant rocks, particularly the siliceous forms, while soils related to *red-brown earths* and *non-calcic brown soils* are usually found on the more easily weathered parent materials. Another example is the usual occurrence of *grey-brown* and *red calcareous soils* on limestone.

Elsewhere, the soil pattern is largely determined by the degree of dissection of the Tertiary deeply-weathered profiles and by associated depositional events. Every stage of dissection exists from relatively intact remnants of the Tertiary profiles through to completely stripped exposures of the underlying fresh Cretaceous sedimentary rocks. The widespread silicification of the old weathered profiles has led to the extensive occurrence of lag gravels of silcrete, and silicified and kaolinized materials following their erosion. The soils thus include shallow *red earths* and *lithosols, desert loams* and *red-brown earth*-like soils, and very widespread *grey, brown* and *red clays*, all of which may be variably mantled with siliceous gravel.

The depositional sites are largely characterized by various forms of *grey, brown* and *red clays*, many of which have extensively scalded surfaces where inundation by flood waters has been infrequent. Other main soils found on depositional sites are the *siliceous sands* of the sand dunes.

Soil landscape provinces

Seven soil-landscape provinces have been delineated (Fig. 15.1) and their main features are summarized in Table 15.1. Some provinces are much more complex than others, but each has relative internal consistency in terms of geology, geomorphology and soil pattern. The soils information has been extensively drawn from Northcote (1968) and Northcote *et al.* (1968).

Mt Isa Province

The Mt Isa Inlier forms the greater part of this Province. A wide variety of Precambrian rocks occur, with granite, quartzite, gneiss, schist, dolerite, phyllite, limestone and shale being the most common (Plumb *et al.* 1980); many strata are highly contorted and steeply dipping. The area is a dissected plateau consisting mainly of predominantly N–S trending ridges and hills which average between 350 and 450 m above sea level (Twidale 1966). Small mesas of Mesozoic sandstone unconformably overlie the Precambrian rocks in several areas. This unconformity probably represents an Early Mesozoic erosion surface, whereas the deeply weathered and lateritized upper surface of the mesas has been equated by Grimes (1980) with the Tennant Creek surface of Early to Mid-Tertiary age which is widespread in the

Table 15.1. Main features of the soil-landscape provinces of Region IV

Province	Geological and geomorphological features	Major soil features
Mt Isa	Dissected plateau of Precambrian igneous, metamorphic and sedimentary rocks. Small remnants of Mesozoic sediments	*Lithosols* are predominant with much rock outcrop. On more gentle slopes are shallow soils related to *red-brown earths* and *non-calcic brown soils. Euchrozems* occur on more basic rocks, and *red earths* and *lithosols* on Mesozoic sandstone plateau remnants
Barkly	Extensive near-level plains underlain by Early Palaeozoic and Tertiary limestones. The south east has greater relief due to more resistant silicified limestones, and Early Palaeozoic sediments	The plains are dominated by deep *grey clays* and may be gravel-strewn with chert. In the south-east, shallow, gravelly *grey-brown* and *red calcareous soils* and calcareous earths occur on limestone and dolomite, and solonised *red-brown earths* on Early Palaeozoic sandstones
Winton–Jundah–Thargomindah	Strongly erosional terrain of dissected tablelands, plateaux, mesas and buttes, many with silicified caps. All are variably stripped remnants of deeply weathered and silicified Cretaceous and Tertiary sediments	*Red earths*, some shallow and gravelly, occur on more extensive residual surfaces. Shallow loams and loamy *lithosols* are dominant in areas of greater dissection with solonised *red-brown earths* common on undulating pediments. *Red and brown hardpan soils* occur in the south
Boulia	Extensive stripping of the altered Cretaceous and Tertiary sediments has produced residual plateaux, mesas and buttes, and rolling plains on unweathered Cretaceous sediments	*Lithosols* and solonized *red-brown earths* are characteristic of the higher dissected areas. *Brown and red clays* have formed on the rolling plains. Most soils are extensively strewn with siliceous gravel. *Brown clays* with a scalded surface dominate the strongly braided stream alluvia
Birdsville	Extensive undulating plains with low scarped plateaux, mesas and buttes, all mantled with a near-continuous cover of mostly siliceous gravel and stone. Lowland areas have alluvial flood plains, small dunefields and sandplains	The upland areas have *desert loams* and *brown* and *red clays* derived from the deep-weathered and silicified Cretaceous and Tertiary sediments. *Red, brown and grey clays* occur on the alluvial flood plains and red *siliceous sands* and *earthy sands* on the dunes and sandplains
Rolling Downs	Gently undulating to rolling plains formed by complete removal of the deep-weathered parts of the Cretaceous–Tertiary sedimentary sequence. Occasional residual mesas and buttes adjacent to province margins	*Brown and grey clays* derived from unweathered Cretaceous labile sediments are overwhelmingly dominant. Deeper forms occur on local stream alluvium. Very minor areas of solonized *red-brown earths* adjacent to western margin
Cooper–Diamantina	Flood plains of the major river systems draining into Lake Eyre. All are subject to extensive but irregular flooding. Streams are strongly braided in their upper sections with large-scale floodouts in their lower sections	Deep *red, brown and grey clays* are almost universal and many have a scalded or crusted surface. Red *siliceous sands* occur on small dunefields with active or semi-active dunes

Northern Territory (see Region III). Later Cainozoic deposits in the Province are of limited extent and mainly confined to small colluvial piedmont deposits and alluvial fans.

Soil occurrence in the Province is strongly controlled by parent rock lithology and the degree of jointing and cleavage present. Reddish to brown *lithosols* are the most extensive soils on upland sites; shallow gravelly loams are most common and rock outcrops are normally prominent; shallow gravelly sands are usual on more siliceous rocks. On gentler slopes there are shallow gravelly *red earths* and duplex (texture-contrast) soils related to *red brown earths* and *non-calcic brown soils*. These have a thin loamy A_1 horizon and a neutral to alkaline, red structured clay B horizon.

On areas of lesser relief throughout the Province the soils are deeper (0.5–1 m) and less gravelly. *Red earths* and red duplex soils (Dr2.13, .12) are widespread. Some of the latter have thin bleached A_2 horizons (Dr2.43, .33) and are probably solonized. On the more basic rocks, e.g. dolerite, some schists and some gneissic granites, *euchrozems* are characteristic.

The lateritized Mesozoic sandstone plateau remnants are mostly strongly dissected and the soils are loamy *lithosols* and shallow *red earths*. The latter are also predominant on gently sloping Cainozoic piedmont and fan deposits but here they are usually deeper and gravel-free.

Barkly Province

This Province is one of low relief, much of it characterized by extensive, flat or gently undulating and almost treeless plains that have long been inappropriately referred to as the Barkly Tableland. The Province coincides areally with most of the Georgina Basin whose geomorphology and soils have been discussed by Stewart (1954*a*,*b*).

The plains of the Northern Territory part of the Province in particular have deep *grey clays* which have a very slight gilgai microrelief (10–20 cm vertical interval) and in many areas a gravel strewn (mostly chert) surface. The soils have formed from limestone, or alluvium derived from it. Much of the drainage is uncoordinated and there are a number of shallow lake basins, some of which date from the Tertiary and contain silicified limestones. The soils in these are also *grey clays*, but most are characterised by a very soft self-mulching surface layer up to 40 cm thick and a sinkhole-type gilgai microrelief. There are also slightly higher, stony limestone rises with fine-textured, shallow, *grey-brown* and *red calcareous soils*.

The south-eastern part of the Province is more complex and has greater relief. Here both Early Palaeozoic limestones, dolomites and sandstones, and Tertiary silicified limestones, outcrop more widely. Whereas the older calcareous sediments form low rolling hills, the Tertiary limestones form low, dissected plateaux (Paten 1964). Some of the Early Palaeozoic sandstones occur as dissected low ranges of cuestas and strike ridges e.g. the Toko Range.

The most common soils on the limestones and dolomites are shallow gravelly *grey-brown* and *red calcareous soils* and deeper brown or red-brown calcareous earths (Gc2.2, Gc1.2) that resemble *solonized brown soils*. On other calcareous sediments the chief soils are gravel-strewn, solonized *red-brown earths* and occasionally *desert loams*; both are often strongly gypseous at depth. These soils also occur on the Early Palaeozoic sandstones, together with loamy *lithosols*.

Areas of deep *red earths* are also common in this south-eastern portion, particularly where it is marginal to the Mt Isa Province. These soils have formed on extensive, level, old alluvial fans.

Winton–Jundah–Thargomindah Province

Much of the Eromanga Basin was affected by deep weathering events after the Late Cretaceous (Senior *et al.* 1978). These events, together with associated widespread silicification and subsequent differential erosion, have largely shaped the present day landscapes. This Province represents the central part of the Region where there has been the least stripping of the deeply weathered and silicified mantle so that the underlying unweathered Cretaceous sediments of the Eromanga Basin are seldom exposed.

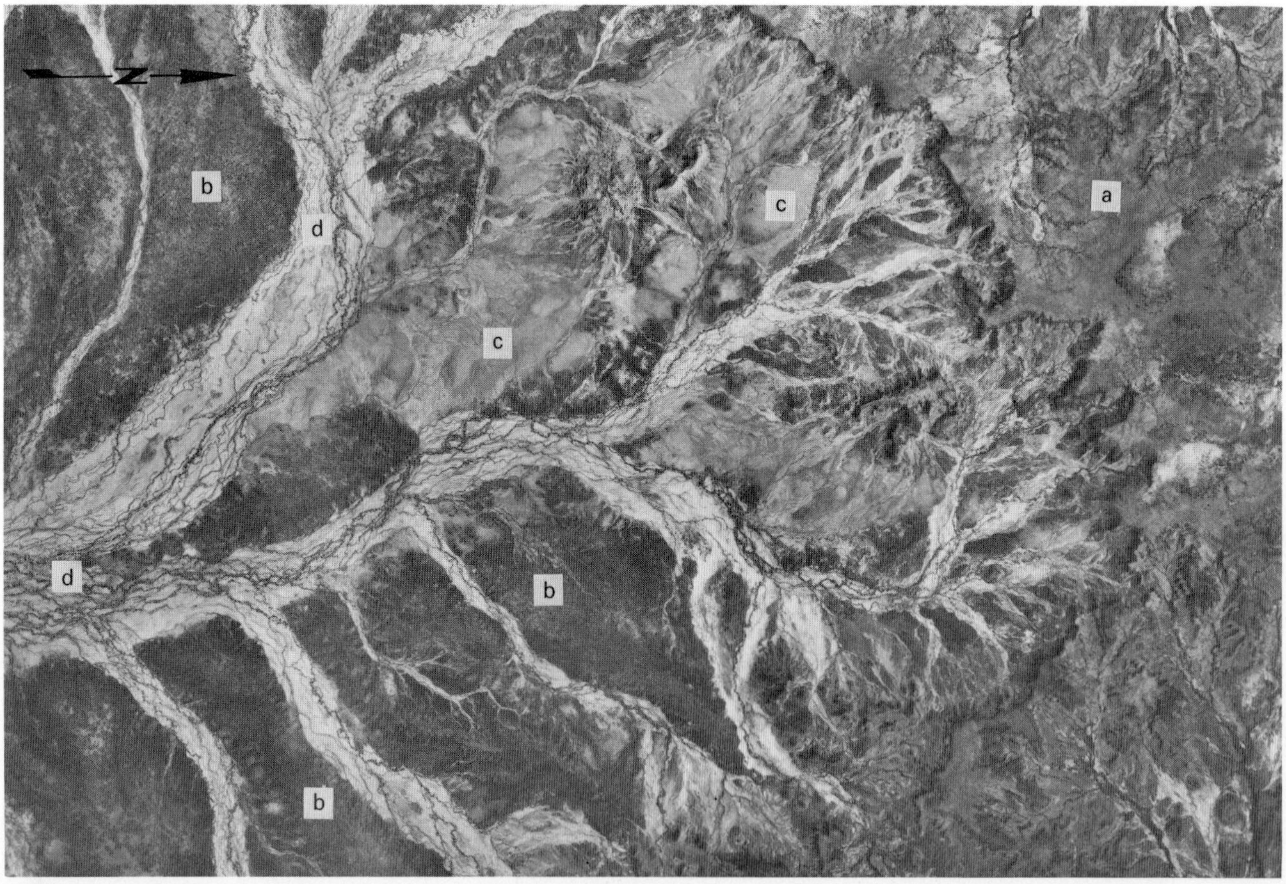

Fig. 15.2. Vertical aerial photograph (scale 1:50 000) showing dissection and stripping of deeply weathered and silicified Late Cretaceous and Tertiary sediments in western Queensland. The photo shows (*a*) a little-dissected surface with shallow, gravelly *red earths*; (*b*) undulating pediments with gravel-strewn, solonized *red-brown earths*; (*c*) small areas of *red clays*; and (*d*) braided stream alluvia with *brown clays* which have a bare scalded surface. Photo: Courtesy Division of National Mapping.

Extensive, low, dissected tablelands are common in the south but are progressively even more dissected towards the north where they form erosional residuals such as small plateaux, mesas and buttes, many of which have silcrete caps, bouldery scree slopes and gravelly pediments. Occasional scarps may be as high as 60 m.

The soils have been most studied in the south (Dawson and Ahern 1974) and a northern section has been covered by terrain classification and mapping (Grant 1968). Soil occurrence and distribution are closely related to the degree of erosion of the deeply weathered (kaolinized and ferruginized) and silicified Late Cretaceous and Tertiary sediments (Fig. 15.2) *Red earths*, some of them shallow and gravelly, are widespread on the more extensive residual surfaces in the central parts of the Province, whereas red, shallow porous loams (Um5.51) with some *red* and *brown hardpan soils* are more common on the less-dissected areas in the south. Over the remainder of the Province, i.e. in areas of greater dissection, loamy *lithosols* are the major soils, many with silcrete gravel and boulders both on the surface and throughout their shallow profiles. In the north particularly, greater stripping of the upper parts of the old deeply weathered profiles has exposed the less-altered sediments, so that the soils on the undulating pediments are gravel-strewn, thin-surfaced, solonized *red-brown earths* and similar brown friable clays (Uf6.31).

Throughout the Province the larger streams have braided alluvial tracts with *brown* and *red clays*, most of which have a bare scalded surface.

Boulia Province

More extensive stripping of the deep-weathering mantles of the Eromanga Basin has occurred in the northern central part of the Region. This has resulted in a characteristic pattern of land forms and soils that comprise the Boulia Province, which has features in common with the Winton–Jundah–Thargomindah Province, the Rolling Downs Province and the Birdsville Province. Topography ranges from residual plateaux, mesas, and buttes (with relief up to 50 m) of the old weathered mantle, to undulating or rolling plains formed on fresh or relatively unweathered Cretaceous sediments, and level alluvial plains.

The soil pattern is governed very largely by the history of landscape dissection. The soils include gravelly loamy *lithosols*, gravel-strewn, solonized *red brown earths*, and similar cracking and non-cracking *brown* and *red clays* where remnants of the deeply-weathered and silicified Cretaceous and Tertiary sediments remain. Much of the gravel is subrounded silcrete up to 10 cm in diameter. The clay soils are commonly slightly gilgaied and characteristically have stone-free depressions.

The undulating plains have a gravel-strewn surface (silcrete or ironstone) and the soils are *brown* and *red clays*, derived either from slightly weathered or fresh Cretaceous sediments where the old weathered profile has been stripped away. Small areas of gravel-strewn, solonized *red-brown earths* are commonly found along with these soils.

Alluvial plains of deep *brown clays* associated with major streams are also a feature of the Province, and in many instances they form the upper reaches of the strongly braided alluvial systems of the Cooper–Diamantina Province described below. The plains have severely scalded surfaces, and are subject to irregular inundation by floodwaters. The older alluvial floodplains of the major streams are not braided and are rarely flooded: they too are dominated by scalded *brown clays*. Associated with these floodplains are linear, semi-active, relict sand dunes of red *siliceous sands* with mobile crests.

Minor soils in the Province are shallow *grey-brown* and *red calcareous soils* formed on Tertiary limestones.

Birdsville Province

The main topographic features are the extensive treeless undulating plains mantled with an almost continuous cover of gravel of a size ranging from a few to 20 cm or more and of a composition ranging from silcrete and silicified, kaolinized sediments to ironstone. This landscape has been commonly referred to as 'stony desert', 'stony downs', or 'stony desert tableland'. The undulating plains are frequently interrupted by low scarped plateaux, mesas and buttes, which are extremely stony with many silcrete outcrops. The higher lands are mainly underlain by silicified Tertiary sandstones. More details of the surface features are given by Mabbutt (1977).

This Province has a number of features in common with the Boulia Province although the proportional distribution of soils is different. *Desert loams* and *red clays* are dominant. Widespread differential stripping of the deeply weathered and silicified Cretaceous and Tertiary sediments is again the main factor controlling soil distribution. However, in the Province, epeirogenic earth movements resulting in broad fold structures have largely directed erosional and depositional processes and hence have markedly influenced soil distribution patterns (Wopfner *et al.* 1974; Mabbutt 1977; Senior *et al.* 1978). Few soils seem to have formed from unweathered Cretaceous sediments, most parent materials of the sedentary soils apparently being part of the old deeply-weathered profiles – a conclusion supported by the fact that many of the soils and their C horizons are mildly acid to neutral in reaction. This Province is also similar to the Oodnadatta–Murnpeowie Province of Region V.

The two most widespread sedentary soils are the *desert loams* and the *red and brown clays* which occupy the undulating plains as well as many of the higher landscape features (Hubble and Reeve 1970; Dawson and Ahern 1974). They may occur in mosaic-like patterns with gilgai forms, e.g. the stony tableland soils of Jessup (1960), or they may occupy large discrete areas. Beneath the near-continuous cover of surface stone, the profiles of these soils are usually stone-free to depths of at least a metre.

In lowland sites adjacent to drainage lines there are *grey, brown* and *red clays* derived from alluvium, mostly with bare scalded or crusted surfaces and stone-free if they are associated with the larger streams. Aeolian dune fields and sandplains make up another important lowland soil unit. Red *siliceous sands* form elongate ridges up to 50 km long and 12–15 m high; these ridges are partly stablized by a sparse cover of spinifex but their crests are bare loose sand. The dune ridges mostly have the same north–north-west alignment as those of the Simpson Desert to the west, although they tend more north–north-easterly in the east. The interdune areas are variable and many contain clay corridors. The sandplains have fewer well-defined elongated dunes and the soils are red *siliceous sands* and *earthy sands*.

Rolling Downs Province

This Province of gently undulating to rolling, grassy clay plains forms the eastern part of the region (Fig. 15.3). It represents the final stage of landscape development in the Eromanga Basin in that virtually all of the deeply weathered and silicified portions of the Cretaceous and Tertiary sequences have been stripped away so that the soils on nearly all erosional sites have formed from fresh, labile Cretaceous sediments which are sometimes calcareous.

Fig. 15.3. Gently undulating grassy plains with *brown clays* derived from Cretaceous sediments in central western Queensland.

The soil pattern is relatively simple, and has been described in part by Hubble and Beckmann (1957). Vast clay plains are the most characteristic feature, often consisting of an intimate mixture of both erosional undulating plains formed on Cretaceous sediments and narrow alluvial plains of Late Cainozoic sediments. The whole province has been eroded to give low relief, particularly in the north where local relief is of the order of only tens of metres or less. Along its margins, there are a few small mesas or eroded remnants. The surface of the clay plains may have a gravel-strewn surface in these areas.

The sedentary clay soils are remarkably uniform overlarge areas: their average depth is about a metre or slightly less. Almost all these soils have slight amounts of carbonate and moderate amounts of gypsum. *Brown clays* are the most widespread of the sedentary clay soils, although red-brown forms are prominent in the western marginal parts of the Province and *grey clays* in the south-east. A slight gilgai microrelief (5–10 cm) is usually present. The soils are normally saline and sodic (Hubble and Reeve 1970; Northcote and Skene 1972) and have a relatively high phosphorus content (Hubble and Beckmann 1957).

The clay soils formed on alluvium are deeper and usually have much lower phosphorus contents (Hubble and Beckmann 1957). They are mostly associated with streams; along the major streams numerous braided distributary channels occur and many of the soils have a strongly scalded or crusted surface.

Near the western margin of the Province, thin-surfaced, gravel-strewn solonized *red-brown earths* are common. Adjacent to the outcrops of more calcareous sediments there are small areas of shallow *grey-brown* and *red calcareous soils*.

Cooper–Diamantina Province

Deep *grey, brown* and *red clays* characterize this entirely depositional Province which forms one of the more unusual drainage systems on the Australian continent. It comprises the wide flood plains of the major river systems draining into Lake Eyre, and the uncoordinated system of the Bulloo River. These extensive alluvial plains with their history of irregular flooding have long been referred to as the 'Channel Country' e.g. Whitehouse (1948). Further details of the morphology of these drainage systems have been given by Whitehouse (1941) and Mabbutt (1977) and some of their soils have been described by Dawson and Ahern (1974).

The alluvia of the major streams have a very low gradient (commonly less than 1:5000) and they are invariably multi-channelled (Fig. 15.4). In the upper and middle reaches well-defined braided channels occur with wider and usually marginal older alluvial plains; the latter characteristically have bare, scalded surfaces. In the lower reaches, although the braided stream pattern continues, it is less well defined and other features become prominent. Thus there are large-scale floodouts which take the form of temporary lakes during times of large floods, e.g. Lake Yamma Yamma adjacent to Cooper Creek, and other temporarily flooded swamps, playas and clay pans where the flood waters spread after being deflected or terminated by barriers such as sand dunes and sandplains.

Red and red-brown or *brown clays* are dominant in the upper reaches, and grey-brown and grey forms are more common in the lower sections. The surface is stone-free except for most marginal areas, although a gravel-strewn surface is more widespread in the upper reaches. Typically the clay soils are self-mulching where they are more regularly inundated; otherwise the surface is commonly crusted or scalded.

Minor but prominent associated soil landscapes are small dunefields with active or semi-active linear dunes of red *siliceous sands*, and clay pans and salt pans with massive-surfaced *grey clays*.

Land use aspects

The semi-arid and arid climates of this Region confine land use to sparse grazing of native pastures by beef cattle and sheep. Nevertheless soil landscapes have significant effects on both the kind of stock grazed and on carrying capacities. Thus the more fertile *grey* and *brown clays* of the Rolling Downs Province support productive *Astrebla* spp. grasslands which are utilized for wool production. By contrast the *lithosol*-dominant Mt Isa Province supports only sparse cattle grazing on its inferior *Triodia* sp. pastures. The irregularly flooded alluvial plains of the Cooper–Diamantina Province have been likened to a large-scale natural irrigation system and considerable numbers of cattle are fattened on the flush of ephemeral herbage following the larger floods

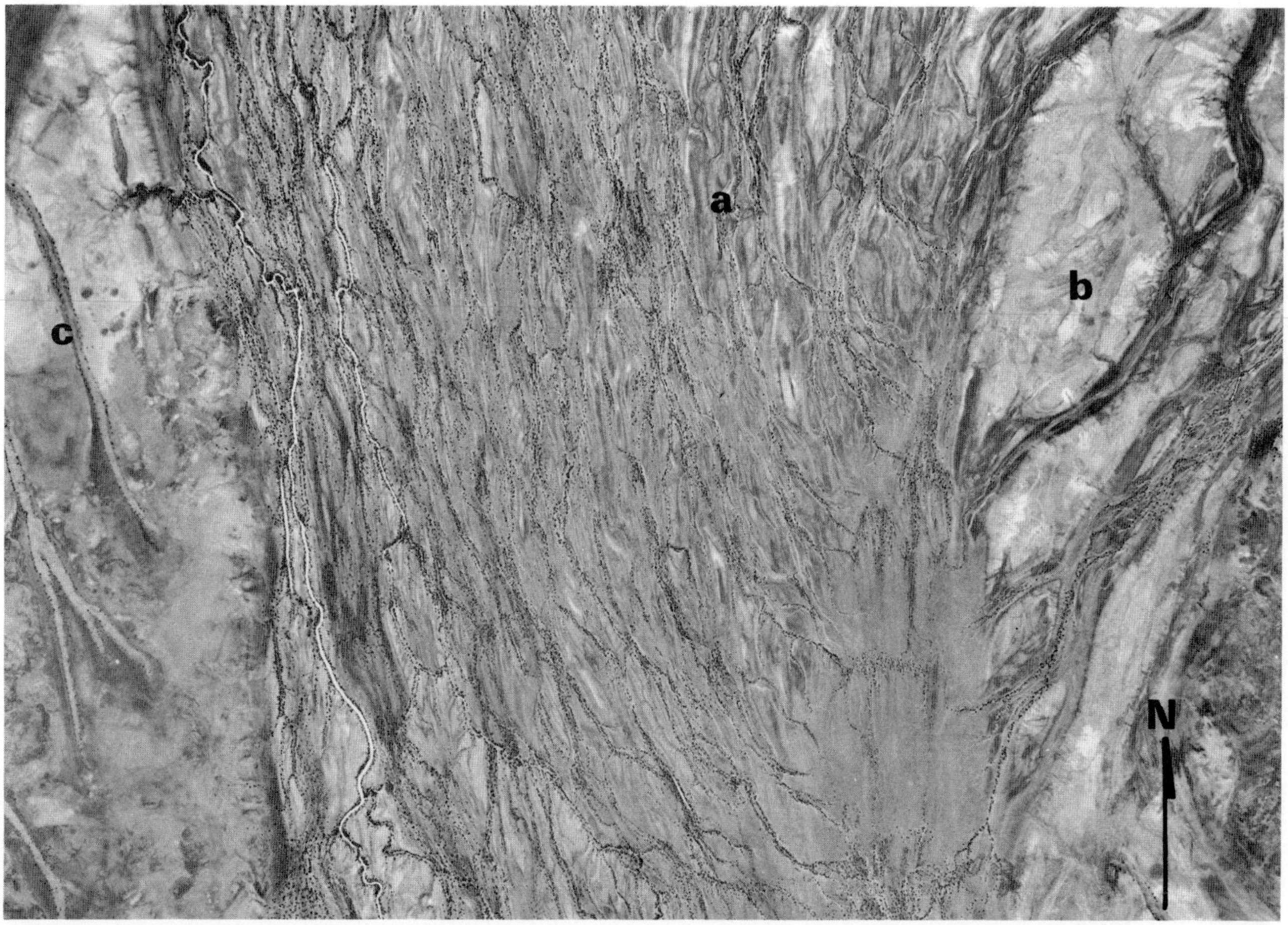

Fig. 15.4. Vertical aerial photograph (scale 1:86 000) showing (*a*) the strongly braided channels of the Diamantina River in south-west Queensland. Also shown are (*b*) older alluvial *grey* or *brown clay* plains with a bare, scalded surface, and (*c*) linear dunes of red *siliceous sands*. Photo: Courtesy Division of National Mapping.

which have occurred with a frequency of about one in three years.

Low rainfall and low soil nutrient status, particularly of phosphorus and nitrogen, are significant factors in the different character and low carrying capacities of herbage on a number of soils: the strongly weathered *red earths*, soils formed on materials of the old deep-weathering profiles, e.g. *desert loams* and some *grey, brown* and *red clays*, and the sandy soils of the dune and sand sheet areas. Soil salinity (Hubble and Reeve 1970; Dawson and Ahern 1974) interacting with soil texture and low rainfall has affected the vegetation and production on soils of the arid provinces e.g. the saltbush herbage on the clay soils and *desert loams* of the Birdsville Province. Salinity and concomitant soil sodicity are contributing factors to the ever-present erosion hazard in the drier parts of the Region where recurrent droughts leave the soil surface devoid of plant cover for long periods. Consequently bare scalded surfaces caused by deflation, water erosion, and high salinity are common features. The lack of vegetation due to aridity exposes the red *siliceous sands* on the dune crests to wind erosion; at best they may be partly stabilized for short periods by ephemeral vegetation. The cover of lag gravel (Fig. 3.5) on large areas of *grey, brown* and *red clays* and *desert loams* in the Birdsville Province effectively protect these soils from erosion even when plant cover is absent. However, removal of this protection when earth roads are formed may lead to rapid gullying during occasional periods of heavy rain.

References

Bowler, J. M. (1976). Aridity in Australia: age, origins and expression in aeolian landforms and sediments. *Earth-Sci. Rev.* **12**, 179-310.

Carnahan, J. A. (1976). 'Natural Vegetation'. Atlas of Australian Resources, Second Series. (Dep. Nat. Map. Canberra.)

Dawson, N. H. and Ahern, C. R. (1974). 'Soils'. In Western Arid Region Land Use Study Part 1. Div. Land Utilization Tech. Bull. No. 12. (Queensl. Dep. Prim. Indust.: Brisbane.)

Galloway, R. W. and Kemp, E. M. (1981). Late Cainozoic environments in Australia. *In* 'Ecological Biogeography in Australia'. (Ed. Allen Keast.) (W. Junk: The Hague.)

Geological Survey of Queensland (1975). Queensland Geology, Scale 1:2500000. (Dep. Mines: Brisbane.)

Grant, K. (1968). Terrain classification for engineering purposes of the Rolling Downs Province, Queensland. CSIRO Aust. Div. Soil Mechanics Tech. Pap. No. 3.

Grimes, K. G. (1980). The Cainozoic geology of north Queensland. *In* 'The Geology and Geophysics of North-eastern Australia'. (Eds. R. A. Henderson and P. J. Stephenson). (Geol. Soc. Aust. (Queensl. Div.): Brisbane.)

Hubble, G. D. and Beckmann, G. G. (1957). The soils of some western Queensland properties of Australian Estates Co. CSIRO Aust. Div. Soils Div. Rep. No. 6/56.

Hubble, G. D. and Reeve, R. (1970). Soil salinity and stability in south-western Queensland. Proc. Symp. 'Soils and Earth Structures in Arid Climates, Adelaide, 1970'. pp. 111-15. (Inst. Engr. Aust.: Sydney.)

Idnurm, M. and Senior, B. R. (1978). Palaeomagnetic ages of late Cretaceous and Tertiary weathered profiles in the Eromanga basin, Queensland. *Palaeogeogr. Palaeoclimatol. Palaeoecol.* **24**, 263-77.

Jessup, R. W. (1960). The stony tableland soils of the south-eastern portion of the Australian arid zone and their evolutionary history. *J. Soil Sci.* **11**, 188-96.

Kemp, E. M. (1978). Tertiary climatic evolution and vegetation history in the southeast Indian ocean region. *Palaeogeogr. Palaeoclimatol. Palaeoecol.* **24**, 169-208.

Langford-Smith, T. (1978). A select review of silcrete research in Australia. *In* 'Silcrete in Australia'. (Ed. T. Langford-Smith) (Dep. Geogr. University of New England: Armidale.)

Mabbutt, J. A. (1977). Desert lands. *In* 'Australia: a Geography' (Ed. D. N. Jeans). (Sydney University Press: Parramatta.)

Northcote, K. H. (1968). 'Atlas of Australian Soils'. Sheet 8, Northern Part of Northern Territory. With explanatory data. (CSIRO Aust. and Melbourne University Press: Melbourne.)

Northcote, K. H., Isbell, R. F., Webb, A. A., Murtha, G. G., Churchward, H. M. and Bettenay, E. (1968). 'Atlas of Australian Soils'. Sheet 10, Central Australia. With explanatory data (CSIRO Aust. and Melbourne University Press: Melbourne.)

Northcote, K. H. and Skene, J. K. M. (1972). Australian soils with saline and sodic properties. CSIRO Aust. Soil Publ. No. 27.

Paten, R. J. (1964). The Tertiary geology of the Boulia region, Western Queensland. Bur. Min. Resour. Aust. Rep. No. 77.

Plumb, K. A., Derrick, G. M. and Wilson, I. H. (1980). The Precambrian geology of the McArthur River–Cloncurry area. *In* 'The Geology and Geophysics of North-eastern Australia'. (Eds R. A. Henderson and P. J. Stephenson). (Geol. Soc. Aust. (Queensl. Div.): Brisbane.)

Senior, B. R., Mond, A. and Harrison, P. (1978). Geology of the Eromanga Basin. Bur. Min. Resour. Aust. Bull. No. 167.

Shergold, J. H. and Druce, E. C. (1980). Cambrian geology of the Georgina Basin. *In* 'The Geology and Geophysics of North-eastern Australia. (Eds R. A. Henderson and P. J. Stephenson.) (Geol. Soc. Aust. (Queensl. Div): Brisbane.)

Smith, K. G. (1972). Stratigraphy of the Georgina Basin. Bur. Min. Resour. Aust. Bull. No. 111.

Stewart, G. A. (1954*a*). Geomorphology of the Barkly region, Northern Territory and Queensland. CSIRO Aust. Land Res. Ser. No. 3.

Stewart, G. A. (1954*b*). Soils of the Barkly region, Northern Territory and Queensland. CSIRO Aust. Land Res. Ser. No. 3.

Twidale, C. R. (1966). Geomorphology of the Leichardt–Gilbert area, north-west Queensland. CSIRO Aust. Land Res. Ser. No. 16.

Veevers, J. J. and Rundle, A. S. (1979). Channel country fluvial sands and associated facies of central–eastern Australia: modern analogues of Mesozoic desert sands of South America. *Palaeogeogr. Palaeoclimatol. Palaeoecol.* **26**, 1-16.

Whitehouse, F. W. (1941). The surface of western Queensland. *Proc. R. Soc. Queensl.* **53**, 1-22.

Whitehouse, F. W. (1948). The geology of the Channel Country of south-western Queensland. Queensl. Bur. Investig. Tech. Bull. No. 1.

Wopfner, H. (1978). Silcretes of northern South Australia and adjacent regions. *In* 'Silcrete in Australia' (Ed. T. Langford-Smith). (Dep. Geogr. University of New England: Armidale.)

Wopfner, H., Callen, R. and Harris, W. (1974). The lower Tertiary Eyre Formation of the south-western Great Artesian Basin. *J. Geol. Soc. Aust.* **21**, 17-51.

Chapter 16

Central Southern Region (V)

K. H. Northcote

This Region borders the Great Australian Bight, Spencer Gulf and the Gulf St Vincent, and extends north to the Simpson and Great Victoria Deserts to meet Region I. Its annual rainfall, which ranges from < 100 mm in the north to > 1000 mm in the Mt Lofty Ranges (Fig. 16.1), is winter dominant towards the coast. An agriculturally significant part of this winter rainfall zone receives between 250 and 500 mm annually. Annual variability of rainfall ranges from 40% in the dry north to 20% along the south coast of Kangaroo Island. During January average air temperatures range from maxima of 24–30°C to minima of 12–18°C; during July the average maxima are 12–18°C and the minimum is about 6°C.

The natural vegetation is more varied than in Region I (Carnahan 1976). It is largely dominated by *Eucalyptus* species, ranging from tall shrublands with mallee species in the drier areas to dry sclerophyll forest with low shrubs in the Mt Lofty Ranges. There are many exceptions to this. The Nullarbor, Millers Creek–Waukaringa and Oodnadatta–Murnpeowie soil-landscape provinces have mainly low halophytic shrublands of Chenopodiaceae; the Gawler Ranges have low open woodlands dominated by *Acacia sowdenii*, low open woodlands of *Callitris glauca* are characteristic of the northern Flinders Ranges, and *Eucalyptus* woodlands with tussock grasses dominate the southern Flinders Ranges. Extensive areas of open tussock grasslands also occur, especially in the Adelaide–Belalie soil landscape province.

As in Region I, Proterozoic and Cambrian rocks form the basement of Region V and outcrop in the ranges of the Mt Lofty–Flinders horst and the Gawler Platform (Eyre Peninsula). In many places they are overlain by younger geological formations; both the younger and older rocks may have been deeply weathered through geological time, although some exposures of relatively fresh rock are found.

Outline of landscape history

The landscapes of the Region are closely linked with its geological history. Although the Mt Lofty, Flinders and Gawler Ranges are composed of Proterozoic and Cambrian sandstones, quartzites, shales, slates, limestones, dolomites, acid volcanics and granites, they were not elevated until comparatively recently. The subsidence of parts of southern Australia in the Cretaceous allowed much land to the west and north to be submerged beneath the sea and further subsidence in Early Tertiary times probably submerged all but the higher parts of the present ranges. Uplift began and continued through the Pleistocene. During this process the Mt Lofty and Flinders Ranges came into existence as a large horst. In the later stages of elevation, land to the west began to sink, resulting in great north–south faults that produced the rift valleys extending north to Lake Torrens. This also led to the formation of the present Spencer and St Vincent Gulfs. During this long process

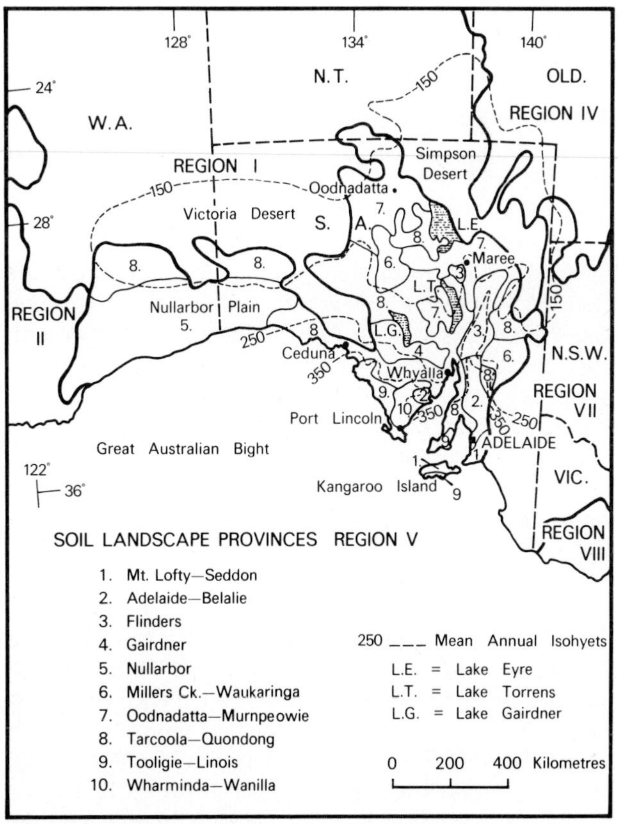

Fig. 16.1. Soil-landscape provinces of Region V.

of elevation, erosion moulded the ranges and stripped away most of the soft Tertiary marine sediments.

Surrounding the northern tips of the Flinders and Gawler Ranges are ephemeral salt lakes. The largest of these is Lake Eyre which is depressed nearly 16 m below sea level. It occupies the lowest portion of the Lake Eyre basin and was much more extensive during Tertiary times. Underlying this whole area are Cretaceous rocks, siltstones, sandstones, limestones, capped by remnants of a Tertiary deep weathering profile, including silcretes. Younger deposits of gypsite and clays occur on an extensive Pleistocene land surface (Wopfner and Twidale 1967). All these materials have been eroded by water and wind into a series of plateau remnants, plains and sand dunes. The last are more common in structural lows throughout this vast Region.

To the west of Eyre Peninsula, the Proterozoic rocks dip deeply below the Nullarbor Plain. This area was first submerged in the Cretaceous; later, further subsidence in the Miocene allowed the accumulation of thick (about 275 m), flat-lying beds of marine limestones. These form the remarkably flat Nullarbor Plain, which now provides a kind of causeway between the eastern and western remnants of the Proterozoic rocks.

Soil-landscape provinces

Ten soil-landscape provinces may be identified within the Region (Fig. 16.1). Their main features are summarized in Table 16.1.

Mt Lofty – Seddon Province

This small province forms the southern part of the Mt Lofty–Flinders horst, and is characterized by erosional plateaux and ranges with *podzolic soils* and *non-calcic brown soils*. It includes the Mt Lofty Ranges (up to 727 m elevation) with their prominent escarpments, the Fleurieu Peninsula and the Kangaroo Island plateau. The Province is narrow (30–35 km at its widest) with short rivers, e.g. the Torrens, that have cut steep-sided gullies or gorges < 150 m deep. The whole area consists of a series of fault blocks with flat-topped remnants of a Tertiary landsurface, commonly tilted to the south. These remnants range from small hill-top residuals in the north to extensive plateau surfaces such as those on Kangaroo Island in the south.

The soils of the Province are diverse; they exemplify all stages of development following dissection of the lateritic, deeply-weathered profiles of Tertiary age (Northcote and Tucker 1948; Rix and Hutton 1953; Jackson 1957; Maud 1972; Northcote 1976). Such dissection has resulted in a range of soils not only on variously truncated, residual sites, but also on depositional sites. The oldest soils are the relict forms on Kangaroo Island that contain a layer of massive laterite between their surface sands and their clay subsoils (Northcote 1946). The next oldest are the *lateritic podzolic soils* with sandy to loamy A horizons containing large amounts of ironstone (lateritic) gravel, notably just above the clay subsoil. Both of these soils are evidence of a relatively undissected Tertiary landsurface. Where dissection has stripped off the lateritic material, various *red* and *yellow podzolic soils* and *soloths* have developed on mottled and pallid zones of the deep weathering profile, or on newly exposed and weathered Cambrian and Precambrian rocks on hillslopes. *Non-calcic brown soils* occur on weathered shales, and *grey-brown podzolic soils* on mica schists (Northcote and de Mooy 1957). These are among the more fertile soils of the Province. Shallow *lithosols* are present on hillslopes, often forming a complex with the duplex soils. Small areas of other soils occur on specific parent materials, such as *terra rossas* and *rendzinas* on highly calcareous parent materials, and various cracking clay soils, including *black earths*, on clay sediments.

The soils developed on transported materials in valleys and other depositional sites include *podzols, soloths* and *solodized-solonetz* and *solodic soils.*

Adelaide–Belalie Province

Within this Province of sub-parallel, more or less north–south trending ridges and valleys, the northern Mt Lofty Ranges pass into the southern Flinders Ranges. This unit forms the middle section of the horst which continues north to include the rugged northern part of the Flinders Ranges (Flinders Province). There are no topographic remnants of the Tertiary surface here, only scattered blocks of laterite and silcrete, and some ironstone gravels. The distinctive ridge and valley topography of this mid-section is related to resistant beds of quartzites, and overlying softer tillites. The latter rocks have provided the sediments covering the floors of the valleys in the form of coalescing alluvial fans and colluvial aprons. The relief varies in elevation from 120 to 300 m but the regularity of the system is broken by higher and more distinct ranges, such as those near Clare and Wirrabara.

Red-brown earths are the characteristic soils of the valleys and hillslopes and cover > 60% of the total area (Stephens *et al.* 1945; Blackburn and Baker 1953; Northcote *et al.* 1954; French *et al.* 1968; Taylor *et al.* 1974). They have developed either *in situ* on Precambrian tillites, shales and impure limestones, or on colluvium and alluvium derived from them. On the Adelaide plains and in the Barossa Valley, the *red-brown earths* have friable non-sodic subsoils but further north, less friable sodic forms are dominant and salt patches may occur, e.g. in County Victoria. All these soils have developed from rocks exposed as a result of the stripping of the Tertiary deep-weathering profile and in this context are related to the *non-calcic brown soils* of the Mt Lofty–Seddon Province.

Grey, brown and *red clays*, and *black earths*, are often associated where clay sediments predominate in the valleys, in shallow basin plains (remnants of Pleistocene swamps and lakes such as those near Merilden) and on slopes in a mosaic with *red-brown earths* (Wells 1959). On the ridges *lithosols* and rock outcrops dominate, but *soloths* and *solodized-solonetz* and *solodic soils* are common on erosional and depositional sites in the ranges, e.g. near Clare.

Solonized brown soils, common in neighbouring provinces, become associated with the *red-brown earths* in the northern part of the unit.

Flinders Province

The northern Flinders Ranges, which comprise this Province, is an area that is most attractive to tourists. It has higher (up to 1165 m) peaks than are found in the south, and the ranges rise directly up from the flat surface of the great rift valley that parallels them to the west. Their summits usually consist of bare quartzite ridges; the valleys are narrow rocky gorges and the streams are intermittent. There are many unusual topographical features, e.g. Wilpena Pound, caused by the erosion of the folded Proterozoic and Cambrian rocks.

There are no remnants of the Tertiary deep-weathering profile left, and only small areas of well developed soils e.g. *red-brown earths* exist. Most of the area is bare rock, with associated *lithosols*. Shallow powdery calcareous loams (Um5.11) and dense loams (Um5.41) cover many lower slopes. Smaller areas of pediments and small basin plains have *red-brown earths*, developed either *in situ* on the more easily weathered parent rocks, e.g. limestones, tillites, shales, or on colluvium and alluvium derived from them.

Gairdner Province

This Province mainly comprises the Gawler Ranges in the northern part of Eyre Peninsula. Here the

Table 16.1. Main features of the soil-landscape provinces of Region V

Province	Geological and geomorphological features	Major soil features
Mt Lofty-Seddon	Dissected, stepped, block-faulted lateritic plateau developed on Cambrian and Proterozoic folded sedimentary rocks	Relict laterites and *lateritic podzolic soils* on residual sites with *red and yellow podzolic soils* and *soloths* on truncated sites, and *grey-brown podzolic soils* on mica schists. *Non-calcic brown soils* and *lithosols* on shales
Adelaide-Belalie	Sub-parallel ridges (mainly quartzites) and valleys (softer rocks e.g. tillites or sediments derived from them) formed in relatively recently exposed Proterozoic rocks	*Red-brown earths* in the valleys and on the slopes. In the north they are sodic with *solonized brown soils* associated. Small areas of many other soils including cracking clays (Ug5). *Lithosols* are common on the ridges
Flinders	Mountain ranges developed in unweathered Proterozoic sedimentary rocks	*Lithosols*, notably powdery calcareous loams (Um5.11) and dense loams (Um5.41). Small areas of *red-brown earths*
Gairdner	Rounded knolls and hills with small valleys developed in unweathered Proterozoic rocks (mainly acid volcanics)	*Lithosols*, mainly dense loams (Um5.41). Small areas of *desert loams* and *red-brown earths*
Nullarbor	Monotonously flat plain developed on Miocene limestones now with a calcreted surface	Powdery calcareous loams (Um5.11) (*lithosols* or *grey-brown and red calcareous soils*)
Millers Creek-Waukaringa	Dissected low tablelands with calcretes derived from Cretaceous limestones (west) or surficial calcareous sediments (east)	Powdery calcareous loams (Um5.11) (*lithosols* or *grey-brown and red calcareous soils*)
Oodnadatta-Murnpeowie	Stepped low tablelands with mesas and buttes developed in Cretaceous and early Tertiary sediments that were deeply weathered and silicified. Silcrete occurs on mesas and buttes. Later covered by Quaternary sediments characterized as the gypsum surface	*Desert loams* with red cracking clays
Tarcoola-Quondong	Plains and dunes of Cainozoic sandy calcareous sediments	*Solonized brown soils* that encompass (1) calcareous earths (Gc), (2) brownish sands (Uc5.1) of the dunes, (3) red sandy earths (Gn1.13), and (4) crusty red duplex soils (Dr1.33)
Tooligie-Linois	Uneven and irregular plains and hills developed on Cainozoic calcretes margined by Recent shelly coastal sands	*Terra rossa* and *rendzina* soils. *Calcareous sands* on coastal dunes and sand sheets
Wharminda-Wanilla	Hills developed on Cambrian and Proterozoic rocks with intervening plains and dunes of Cainozoic sediments	*Lateritic podzolic soils* in the southern hills and *red-brown earths* in the north. Various *solodized-solonetz and solodic soils* on the plains. Bleached sands (Uc2.2) on the dunes

Proterozoic crystalline basement is composed of acid volcanics which form rounded knolls and hills separated by small valleys and plains.

The soils are dominantly *lithosols*, mainly dense loams (Um5.41) with a surface stone cover and with frequent rock outcrops. Some powdery calcareous loams (Um5.11) occur on calcareous parent materials. Deeper soils common to nearby provinces occupy the small valleys and plains. These are mainly calcareous earths (Gc) and *desert loams*, but some *red-brown earths* are present in the southern portion of the unit.

Nullarbor Province

The monotonously level Nullarbor Plain extends about 240 km north of the present coast and east–west for about 750 km; it is underlain by Miocene limestones. At the coast there is a striking and continuous escarpment (45–90 m high) bordering most of the Great Australian Bight. Only between Eyre and Eucla is there a narrow coastal plain.

The Nullarbor Plain has no surface drainage although there are many caves in the limestone and some, perhaps most, have underground lakes in them. The surface of the Plain is covered by calcrete (Fig. 16.2) which was formed during the late Cainozoic after its emergence as a plain. Present day soils have developed on the calcrete and are the typical shallow, (15–60 cm) powdery, calcareous loams (Um5.11) of desertic regions in Australia (*grey-brown and red calcareous soils* or calcareous *lithosols*). On the plain they are associated with smaller areas of shallow forms of calcareous earths (Gc), with some *red earths* in the northern portions.

The narrow coastal plain between Eucla and Eyre has *calcareous sands* with some *siliceous sands* on the dunes fronting the sea.

Millers Creek–Waukaringa Province

Two widely separated soil landscapes comprise this Province. Both contain large areas of calcrete, and/or other calcareous rocks and sediments, which have given rise to *grey-brown* and *red calcareous soils* and less commonly to calcareous earths (Gc). Some dissimilar features are found in the two areas due to differences in their geology. In the west, the Millers Creek tableland with its mesas, buttes, claypans and small valleys, is underlain by flat-lying Cretaceous calcareous rocks and the subdominant calcareous earths (Gc) occur. These are discussed in the Tarcoola–Quondong Province (see also Jessup 1951). The eastern unit has a basement of Proterozoic rocks with slightly more relief and low rocky ridges separated by pediments and plains. The dominant soils have developed on the surficial calcareous sediments. *Desert loams* occur on the pediments.

Oodnadatta – Murnpeowie Province

The dissected low tablelands of this Province, with their broad rolling downs, mesas, buttes and scarps, all of which are often mantled with stones, extend from the limits of the northern Flinders Ranges to the Simpson Desert which they then adjoin to the north-west along the Finke River to the central Australian highlands (Fig. 16.3). The major ephemeral salt lakes, e.g. Lake Eyre and Lake Torrens, are closely associated with the tablelands. Similar and related soil landscapes also occur in south-western Queensland and western New South Wales. These low tablelands are remnants of Cretaceous and Early Tertiary sediments that were subsequently subjected to pronounced deep-weathering and silicification (Part 2). In the Late Cainozoic the whole area was uplifted so that erosion products were carried out of the area by rivers flowing south to the sea through the Lake Torrens depression. Subsequently, drainage to the sea became less effective, and gypsum crusts formed over the alluvial plains; artesian springs fed pools inwhich freshwater limestones accumulated. Bones of the marsupial *Diprotodon* have been found in Pleistocene sediments (Parkin 1969), allowing the gypsum surface (the gypsite of Wopfner and Twidale 1967) to be dated at between 80 000 and 40 000 years ago. This gypsum surface also includes clay sediments and sodium chloride.

Fig. 16.2. Nullarbor Province showing the flat, calcreted surface of the vast Nullarbor Plain with its shallow calcareous soil and typical shrub–steppe vegetation.

Fig. 16.3. Oodnadatta–Murnpeowie Province. A dissection scarp separating the stone-strewn upper and lower erosional plains of desert loams at Murnpeowie, S.A.

The common soils of this vast tableland area are *desert loams* (Jessup 1951; Jackson 1958), often with shallow gilgai depressions of red cracking clays (Ug5.38). Except where there are gilgai depressions, the soil surface is nearly always covered by stones that may vary from silcrete fragments and ironstone gravels to fragments of quartzites. The *desert loams* have loamy surfaces, 1–30 cm thick which sharply overlie red clays with a polyhedral, blocky or prismatic structure with gypsum and/or calcium carbonate in their deeper part. The whole profile varies from 0.5 to 3 m thick but is commonly > 1 m thick and overlies gypseous sediments, weathered rock, heavy stone-lines or a red-brown hardpan (Chapter 3). All the subsoils are strongly saline and sodic and the surface soils frequently contain > 0.1% of sodium chloride (Chapter 3). Although they contain appreciable amounts of salt, these soils have well developed and strongly differentiated profiles not usual in arid regions. Jessup (1960) considered that these soils had formed in deposits of aeolian clays but the micromorphological studies reported by Stace *et al.* (1968) on a number of profiles provided by Jessup do not support his suggestion.

Soils with the general properties mentioned above are particularly susceptible to gully erosion following heavy storms. This is most marked if their dispersible clay subsoils are exposed by a loss of surface soil due to (i) destruction of protective vegetation, (ii) disturbance of the gravel and stone mantle, and (iii) a break in the soil surface crust.

Tarcoola–Quondong Province

In southern central Australia large areas of carbonate-rich sandy soils on plains or low tablelands with variable tracts of dunes form the Tarcoola–Quondong Province. Somewhat similar areas in Regions II and VII give rise to westerly and easterly extensions of these soil landscapes. Generally, these areas, together with the Nullarbor Province, coincide with the area covered by an incursion of the sea into southern Australia in Miocene times. However, the soils do not appear to have developed from marine rocks but rather from re-sorted carbonate-rich sandy sediments, which at least in part are Pleistocene aeolian deposits.

All the soils of these areas have been called *solonized brown soils* but they include a rather diverse range of soils. Among these four main groups may be recognized.

(i) The most extensive soils are the calcareous earths (Gc). This name is preferred because of the large amounts of calcium carbonate always found in them. There is no build-up of silicate clay in their subsoils but simply clay-sized carbonate (Northcote 1951). The soils are sodic to strongly sodic and are of at least three ages corresponding to three maxima of aeolian activity (Northcote 1951).

(ii) Many of the dunes are characterized by deep (2–6 m), loose, calcareous sands. These brownish sands (Uc5.1) have slight to large amounts of carbonate through the profile and may have formed during each maximum of aeolian activity during arid phases of the Quaternary. Vegetation has fixed and stabilized the dunes; but where it has been subsequently lost, these sands are again moving eastwards.

(iii) Red sandy earths (Gn1.13) in which there is an increase in silicate clay in the subsoils and with carbonates below the depths of the clay maxima. These soils are also sodic and are probably to be equated with the first maximum of aeolian activity. They are the oldest of the aeolian soils, and are of limited occurrence.

(iv) Finally, there are the crusty red duplex soils (Dr1.33) of the lower lying depressions. These sodic soils, formed in the clay alluvium laid down before the sandy calcareous deposits, only occur where the alluvium is exposed.

Tooligie – Linois Province

This Province lies along the western and southern coastal sections of Eyre Peninsula where the low hilly to undulating terrain is characterized by outcropping calcrete, and also on similar areas in the lower part of Yorke Peninsula and along the southern coastal strip of Kangaroo Island. These areas have been largely stripped of the sandy calcareous sediments in which the calcareous earths formed, and there are extensive occurrences of very shallow to shallow, red and dark grey to black soils.

Red to red-brown, weakly structured, acid to neutral loamy sands (Uc6.13), usually about 10–20 cm thick, occupy hollows in the calcretes and may form a reddish sandy cover < 10 cm thick over calcrete ridges in the drier parts, notably on western Eyre Peninsula and southern Kangaroo Island. These very shallow soils are formed in part from the dissolution of the calcretes with loss of carbonates, and in part from small accessions of non-calcareous aeolian material. Such soils have been called *terra rossa*. In wetter sites, notably on southern Eyre Peninsula and on Yorke Peninsula, there are shallow, friable, structured loams that vary in colour and reaction. The near-black loams (Um6.21) are alkaline and fragments of calcrete may be present in the soil. They are often called *rendzinas*. The red (Um6.24) and brown loams (Um6.23) are mildly acid, and are often labelled *terra rossa*.

Calcareous sands, very young, undifferentiated soils, have formed on recent coastal shell sands, largely by solution of carbonates and accumulation of small amounts of organic matter in the top few centimetres.

Wharminda – Wanilla Province

Proterozoic rocks which are part of the ancient Gawler Block outcrop in the ranges and hills of this portion of Eyre Peninsula. Remnants of the Tertiary deep-weathering profile are associated with the most prominent features which are the Marble Range in the west and the Lincoln Range in the south-east. The most common soils are the *lateritic podzolic soils* on the flanks of ranges and there are similarities with the Mt Lofty–Seddon Province. In the north-east, however, the Cleve–Mangalo hills have similarities with the Adelaide–Belalie Province and the common soils are *red-brown earths*.

Between the Cleve–Mangalo hills and the southern ranges there are undulating plains and basins. They present a variable range of sodic to strongly sodic soils developed mainly on Quaternary colluvium derived from the older geological materials. Subdued dunes present in the northern part testify to limited Quaternary aeolian activity.

The sodic soils (*solodized-solonetz* and *solodic soils*) are variable but all have tough, slowly permeable clay subsoils on which subsurface waterlogging develops periodically in winter and spring. Some are hardsetting, but others have loose, sandy surface horizons that make them more liable to wind erosion when cultivated. Bleached sands (Uc2.2) are an important associated soil on dunes in the drier northern part of the unit. They are particularly liable to wind erosion when cultivated, as evidenced by the blow-outs on many dunes (see also French 1958).

Main soil determinants

It is clear from the previous discussion that the present soil landscapes of Region V and the variety of its landforms result largely from the basic regional geology and the weathering processes that have prevailed during more recent geological periods. In particular, the varying climates that have been operative from Early Tertiary through to the present have left such indelible marks as deep-weathering profiles, dune formations, calcretes and sodicity. Nevertheless, it is evident that parent material has largely influenced the present soil pattern and many of the soil properties. In the Mt Lofty–Seddon Province, for example, all soils formed from the Tertiary deep-weathering profile have a large range of nutrient deficiencies, whereas those formed from newly exposed rocks, even geologically ancient ones, do not.

Land use aspects

The present pattern of land use, ranging from low intensity grazing in the north to intensive horticulture and viticulture in the south, may be attributed largely to rainfall and soils. Economic factors also play an important role, especially in the region of the 250 mm annual rainfall isohyet (Fig. 16.1) which marks a somewhat fluctuating boundary between dominantly pastoral activity to the north and cropping-type agriculture to the south.

Extensive grazing in the Flinders, Gairdner, Nullarbor, Millers Creek–Waukaringa, Oodnadatta–Murnpeowie and Tarcoola–Quondong Provinces is based entirely on native pastures.

Soil erosion is always a constraint in areas of shallow soils and low, unreliable rainfall because of the danger of over-grazing the more palatable plants; all the pastoral areas of Region V are implicated here. Wind erosion provides an added hazard in the Tarcoola–Quondong Province. The results of water erosion are frequently particularly devastating in the Oodnadatta–Murnpeowie Province for reasons discussed above.

Although the Flinders Province has a broad-scale pastoral value, the natural features and scenic grandeur of these Ranges suggests that they could be deservedly given the status of a national park. The Nullarbor Province, which is a young stable landscape with large areas fronting the coast, deserves further study as an area of natural potential to cater for some of man's future urban and recreational activities; this potential is greater than is generally realized at present (Hallsworth 1976). Water is a limiting factor in this riverless tract.

Permanent agriculture is confined to the Mt Lofty–Seddon, Adelaide–Belalie, Wharminda–Wanilla, and the southern-most limits of the Tarcoola–Quondong Province. In all these areas, the rainfall is winter dominant and is in excess of 250 mm per year; its reliability is moderate and improves towards the coasts. Broad-scale dryland farming of cereal crops, together with a legume-dominant ley pasture phase of variable duration, is common in the drier areas. The *red-brown earths* and associated soils of the Adelaide–Belalie Province have excelled in cereal production since the early days of European settlement. Improved, sown pastures based on subterranean clovers have been develoepd extensively in the Mt Lofty–Seddon Province and in the southern parts of the Wharminda–Wanilla Province.

South Australia is pre-eminent in wine production in Australia. The main viticultural areas are near Adelaide, in the Barossa district and around Clare. All these areas are in the Adelaide–Belalie Province or close to its boundary with the Mt Lofty–Seddon Province where the rainfall exceeds about 500 mm per year. *Red-brown earths* are among the most significant soils used for viticulture but many other soils are also important. The vineyard with the highest altitude in Australia is at Pewsey Vale in the Mt Lofty–Seddon Province on *yellow podzolic soils*; this area was developed after soil studies in the Barossa district by Northcote and de Mooy (1957). The growing of fine grapes for the production of quality wine requires good soils combined with a suitable climate.

Horticultural crops and especially pome fruits and cherries are grown in the higher rainfall areas of the Mt Lofty–Seddon Province on *podzolic soils*. Almonds are a speciality crop grown south of Adelaide on *red-brown earths* and *solodized-solonetz* and *solodic soils*.

Forest plantations of *Pinus* spp. have been established in the higher rainfall areas, notably in the Mt Lofty–Seddon Province, but also to a limited extent in the higher ranges of the Adelaide–Belalie Province. The soils commonly used have been the various *podzolic soils* and *soloths*. These pine plantations do not generally have such rapid growth rates as those of the south-east of South Australia.

The low chemical fertility of many of the soils, particularly those developed on dissected and truncated laterite profiles, played an important part in delaying the early development of many areas in the Mt Lofty–Seddon Province (Northcote 1976). Nutrient problems are also evident in the cereal growing areas, notably on the calcareous earths of the Tarcoola–Quondong Province in northern Eyre Peninsula where high contents of calcium carbonate limit the availability of certain nutrients such as iron. In addition, many of the duplex soils, particularly those with sodic clay subsoils (Chapter 3), have poor subsoil moisture regimes for plant growth as well as low chemical soil fertility. Waterlogging of surface soils and relatively dry subsoils causes yields to be low even in cereal crops. Soil salinity is an ever present and generally low-key problem in most agricultural areas. The extent and nature of saline and sodic soils has been discussed by Northcote and Skene (1972), but see also Chapter 9.

References

Blackburn, G. and Baker, R. M. (1953). The soils of part of the southern Flinders Ranges, South Australia. CSIRO Aust. Soil Publ. No. 3.

Carnahan, J. A. (1976). 'Natural Vegetation'. Atlas of Australian Resources, Second Series, (Dep. Nat. Map.: Canberra.)

French, R. J. (1958). Soils of Eyre Peninsula, South Australia. Dep. Agric. S. Aust. Bull. No. 457.

French, R. J., Matheson, W. E. and Clarke, A. L. (1968). Soils and Agriculture of the northern and Yorke Peninsula regions in South Australia. Dept. Agric. S. Aust. Spec. Bull. No. 1.68.

Hallsworth, E. G. (1976). A forward look at a land policy for Australia. *Search* **8**, 58-67.

Jackson, E. A. (1957). A survey of the soils and their utilization in portion of the Mt Lofty Ranges, South Australia. CSIRO Aust. Div. Soils, Soils Land Use Ser. No. 21.

Jackson, E. A. (1958). A study of the soils and some aspects of the hydrology at Yudnapinna Station, South Australia. CSIRO Aust. Div. Soils, Soils Land Use Ser. No. 24.

Jessup, R. W. (1951). The soils, geology and vegetation of north-western South Australia. *Trans. R. Soc. S. Aust.* **74**, 189-273.

Jessup, R. W. (1960). The stony tableland soils of the south-eastern portion of the Australian arid zone and their evolutionary history. *J. Soil Sci.* **11**, 188 96.

Maud, R. R. (1972). Geology, geomorphology and soils of central County Hindmarsh (Mount Compass – Milang), South Australia. CSIRO Soil Publ. No. 29.

Northcote, K. H. (1946). A fossil soil from Kangaroo Island, South Australia. *Trans. R. Soc. S.Aust.* **70**, 294-6.

Northcote, K. H. (1951). A pedological study of the soils occurring at Coomealla, New South Wales. CSIRO Aust. Bull. No. 264.

Northcote, K. H. (1976). 'Soils'. *In* 'Natural History of the Adelaide Region' (Eds C. R. Twidale, M. J. Tyler and B. P. Webb). (R. Soc. S. Aust.: Adelaide.)

Northcote, K. H. and de Mooy, C. J. (1957). Soils and land use in the Barossa district, South Australia. Part A. The Angaston–Springton area. CSIRO Aust. Div. Soils, Soils Land Use Ser. No. 22.

Northcote, K. H. and Skene, J. K. M. (1972). Australian soils with saline and sodic properties. CSIRO Aust. Soil Publ. No. 27.

Northcote, K. H. and Tucker, B. M. (1948). A soil survey of the Hundred of Seddon and part of the Hundred of MacGillivray, Kangaroo Island, South Australia. CSIR Aust. Bull. No. 233.

Northcote, K. H., Russell, J. S. and Wells, C. B. (1954). Soils and land use in the Barossa district, South Australia. Zone 1. The Nuriootpa area. CSIRO Aust. Div. Soils, Soils Land Use Ser. No. 13.

Parkin, L. W. ed. (1969). Handbook of South Australian Geology. Geol. Surv. S. Aust. (Govt. Printer: S. Aust.)

Rix, C. E. and Hutton, J. T. (1953). A soil survey of the Hundred of Kuitpo in the Mt Lofty Ranges of South Australia. Land Tax Dep. S. Aust. Bull. No. 1.

Stace, H. C. T., Hubble, G. D., Brewer, R., Northcote, K. H., Sleeman, J. R., Mulcahy, M. J. and Hallsworth, E. G. (1968). A Handbook of Australian Soils. (Rellim: Glenside, S. Aust.)

Stephens, C. G., Herriot, R. I., Downes, R. G., Langford-Smith, T. and Acock, A. M. (1945). A soil, land-use and erosion survey of part of County Victoria, South Australia. CSIR Aust. Bull. No. 188.

Taylor, J. K., Thomson, B. P., Shepherd, R. G. (1974). The soils and geology of the Adelaide area. Bull. Geol. Surv. S. Aust. No. 46.

Wells, C. B. (1959). Soils and land use in the Barossa district, South Australia. The Greenock-Gomersal area. CSIRO Aust. Div. Soils, Soils Land Use Ser. No. 30.

Wopfner, H. and Twidale, C. R. (1967). Geomorphological history of the Lake Eyre basin. *In* 'Landform Studies from Australia and New Guinea'. (Eds J. N. Jennings and J. A. Mabbutt) (Aust. Nat. Univ. Press: Canberra.)

Chapter 17
Eastern Highlands (VI)

G. D. Hubble and R. F. Isbell

THIS REGION HAS A DOMINANTLY EROSIONAL TERRAIN and embraces all of the hilly to mountainous and undulating lands of eastern Australia between Cooktown and Newcastle (latitudes 15° to 33°S); it extends 400–500 km inland (Fig. 17.1). The Great Dividing Range traverses its length, reaching maximum elevations of 1600 m where it comes close to the coast in the extreme north and south. In central Queensland this range lies near the western margin of the Region and is reduced to a subdued feature, a broad upwarp, with a maximum height of 500 m. Overall, the relief is moderate and dominated by valleys and ranges approximately parallel to the coast with crest heights of up to 600–900 m, but extensive lowland areas fall within the boundaries. Most of the Region is drained eastwards by coastal streams although the headwaters of the Darling River and some Gulf of Carpentaria streams are also included.

The climate is generally subhumid, but varies from humid to semi-arid and from tropical in the north to temperate in the south. The rainfall is summer dominant; strongly so in the north where cyclones are an important influence, but less so in the south. Mean annual rainfalls are usually between 600 and 1400 mm, although some scattered areas along the coast receive more than 1600 mm. The upper limit is 4500 mm in Australia's highest rainfall area at Tully in North Queensland.

The vegetation is very varied (Carnahan 1976) and although eucalypt open forests and woodlands are most common, areas of tropical and sub-tropical rain forests, vine thickets, *Acacia*-dominant forests, grasslands and heaths occur. Strong climatic and edaphic relationships are evident in the distribution of plant communities.

The soils are characterized by their great variety and form complex patterns, texture-contrast soils of podzolic and solodic forms being the most extensive and widespread. Moderate soil depth and relatively thick clay B horizons are common features. The Region also includes important areas of some of Australia's most fertile soils –*black earths, red-brown earths, grey clays, krasnozems* and *prairie soils* – along with large areas of poorer soils such as *red* and *yellow earths, lithosols, soloths* and *podzols.*

Outline of landscape history

The oldest rocks are Proterozoic metamorphics which form a small shield area near Georgetown in the far north. Early in the Palaeozoic, the Tasman Geosyncline was formed and extensive sedimentation continued throughout the length of the Region until Late Permian (Part 2). These sediments, now mainly represented by greywackes, shales, slates, phyllites and schists, were folded and widely intruded by acidic magmas in the Mid to Late Palaeozoic to form most of the present structural

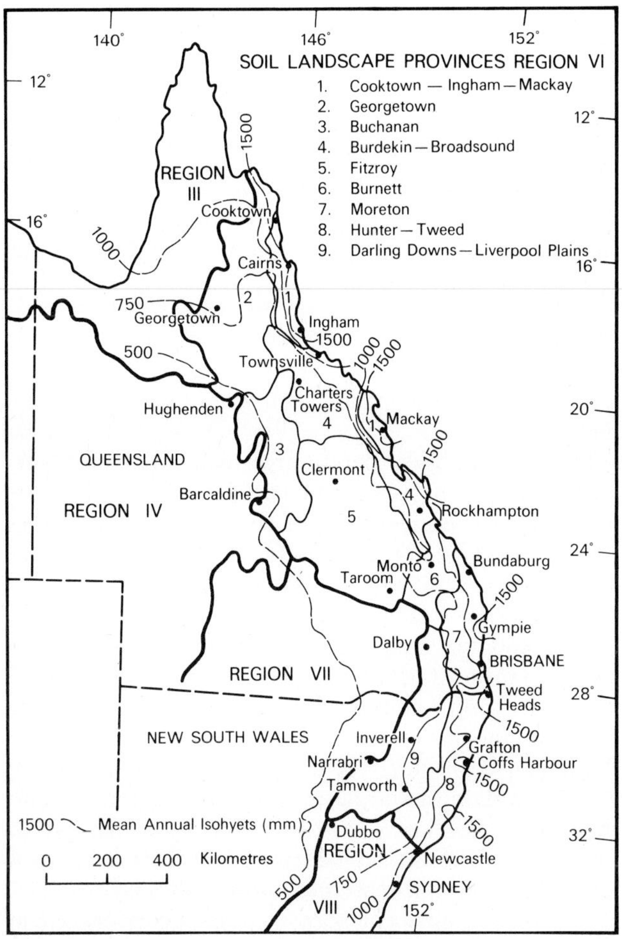

Fig. 17.1. Soil-landscape provinces of Region VI.

highs. The intervening lower areas represent basins in which younger sediments were deposited during Late Palaeozoic, Mesozoic and Cainozoic times. These younger rocks have been little affected by subsequent earth movements.

The highlands probably had their lowest relief in the Late Mesozoic when the oldest planation surface that can now be recognized (Wellman and McDougall 1974) was formed. This is equated with the Gondwana Surface of King (1950) and is only represented by isolated stripped remnants. The major uplifts, intervals of erosion, vulcanism (Wyatt and Webb 1970; Wellman and McDougall 1974; Wellman 1978) and sedimentation that have shaped the Region's landscapes date from this time. At least two distinct periods of uplift and warping, and two of downcutting are suggested (Wellman and McDougall 1974). Widespread remnants of well developed erosion surfaces are a feature of the Region.

Vulcanism became much more extensive in the Late Eocene and extended through to the Early Miocene. Warping at this time initiated a longer period of erosion, interrupted locally by episodic lava extrusions. By the Late Oligocene (see Part 2) this had produced an extensive planation surface (the Australian Surface of King 1950) with lateritic weathering up to 80 m in depth which mantled all kinds of rocks, including the earlier basaltic lavas; however, locally there may have been multiple surfaces with shallower lateritic weathering profiles (Francis and Walker 1978). A subsequent differential uplift (the Kosciusko Uplift of Browne 1969), which seems to have taken place in several stages, produced varying altitudes along the length of the highlands; this uplift arched across an axis parallel to the coast to form the gentler western slopes and the steeper eastern ones. Renewed erosion then dissected and stripped much of the older surface to form a much less extensive Pliocene level which was also mantled with deep lateritic weathering.

In southern Queensland, three major planation surfaces have been recognized (Watkins 1967). Each is deeply lateritized and more recent work (Murphy *et al.* 1976) has indicated the possibility of subsidiary surfaces at intermediate levels. The oldest and most extensive 'upper erosion surface' has an elevation of 690—420 m and is equated with King's Australian Surface; the 'middle erosion surface' at 210—120 m is very restricted in its extent and is considered to have formed by the Late Miocene; the 'lower erosion surface', thought to be Late Pliocene, occurs near the coast at 60—15 m elevation. Erosion and, in the far north, vulcanism continued into the Pleistocene; this erosion further reduced the earlier planation surfaces and dissected landscapes, so supplying material for deposition in the valleys.

The depositional landscapes are largely restricted to plains of sediments which were deposited in the lower valleys of streams and along the coast. Siliceous sands and other estuarine and deltaic sediments were deposited on a submerging coastline. Later emergence then exposed sections of low-lying sandy and clayey coastal plains, and wind action along the seaward margin and on the large islands piled up the sands into overlapping dune systems.

Main soil determinants

The great diversity in the Region's soils is primarily the consequence of its geological complexity. The soils have developed on parent materials derived from a wide range of rocks extending from loose silica sands and quartzose sandstones to clay alluvium and shales, from rhyolite to basalt, and from granite to gabbro. Because of its influence on water movement, erosion rates and soil water regimes, the Region's widely varying topography is the next most important soil determinant. Though the climatic range is considerable it ranks third in determining soil variety. Adding to this complexity are further effects e.g. materials exposed by differential stripping of the old deep-weathered profiles, differences in the length of time that soils have been forming and the survival of ancient soils on old land surfaces

Examples of differences due to parent materials exposed under relatively uniform climate and free drainage are the giant *podzols* on the coastal sand dunes, *red earths* on sandstones, *red podzolic soils* on parent materials such as phyllite that produce moderate to high amounts of clay on weathering, and *krasnozems* on basalt. Nearby in the same area, differences in water regimes and drainage related to topographic position result in shallow *humus podzols* on the silica sands of low lying plains adjacent to the giant *podzols*, in *yellow* and mottled *podzolic soils* on mid-to-lower slopes and *gleyed podzolic soils* on foot slopes and in depressions below the *red podzolic soils* on clay-forming parent materials. Minor differences in topography which cause variations in the drainage-water regime result in *black earths* (poorer drainage) and *euchrozems* (free drainage) on some of the younger basalts of similar age (Beckmann *et al.* 1974). However, under the same present-day climate, ancient lateritic *krasnozems* which are deep, strongly weathered and leached soils, survive as relicts formed on parent materials derived from basalt on the Mid-Miocene upper erosion surface. The occurrence of *krasnozems, euchrozems, chocolate soils* and *black earths* on similar and recently formed basaltic parent materials in different parts of the Region seems to result from broad climatic differences.

Materials of the old deep-weathering profile are characteristically high in clay content and sesquioxides, impovorished by strong leaching, and usually strongly acid. Thus these materials or their erosion products give rise to acid soils in semi-arid, non-leaching environments.

Soil-landscape provinces

The Region has been subdivided into nine soil-landscape provinces (Fig. 17.1 and Table 17.1). Many of these have an extremely complex soil pattern for the reasons given above; hence only a broad account of the dominant soil patterns will be given in the descriptions which follow.

Cooktown—Ingham—Mackay Province

This Province occupies Australia's only high-rainfall humid tropical environment where the mean annual rainfall is from 1500 to 4300 mm. Most of the Province consists of narrow coastal plains flanked by high mountains (up to 1600 m but more commonly at 600–900 m) with steep slopes (Fig. 17.2). The western margin is an elevated undulating to low hilly plateau and a considerable part of the Province was originally covered by rainforest. No detailed studies of the soil distribution have been published although reconnaissance maps cover part of the Province (Isbell and Murtha 1970).

The upland areas have four major soil parent materials. The oldest rocks (Silurian–Devonian) are greywacke, siltstone and slate and their low-grade metamorphic equivalents which occur only in the northern part of the Province. All these rocks are intruded by Upper Palaeozoic granites which are also widespread in the south. Upper Palaeozoic volcanics, acid in the north but generally more basic in the south, also occur widely. In addition, small but important areas of Cainozoic basalt are found, mostly in the north. These major parent rock groups, together with alluvium derived from them, are the main factors controlling the kind and distribution of soils, but a rainfall gradient (Fig. 17.1) also contributes markedly to soil differentiation.

The oldest sediments and metamorphic rocks have weathered to moderately deep, mostly medium-textured soils which are structured, red or yellow pale loams (Um4.41, .43). The granitic rocks and acid volcanics have somewhat similar but generally deeper soils whose texture gradually increases with depth. Two forms occur: red soils, related to *red podzolic soils*, and yellow soils that have been called *xanthozems* and *yellow podzolic soils*. With increasing rainfall there is a trend to greater acidity and lower cation exchange capacity. *Red earths* are the major soils of the upper piedmont slopes and fans derived from the acid rocks, while yellow *siliceous sands* and *yellow earths* are dominant on the lower, flatter parts of the fans. In the south (Mackay–Proserpine area) brown friable earths (Gn3.24, .21) are the chief soils associated with the intermediate volcanic rocks (e.g. andesites).

Fig. 17.2. The coastal plains near Cairns in north Queensland are intensively utilized for sugar cane. The mountainous areas are covered with dense rainforest.

The Cainozioc basalts have a characteristic cover of deep to very deep *krasnozems* (Isbell *et al.* 1976; Gillman 1976) whose chemical and mineralogical properties are closely related to rainfall and leaching intensity.

Most of the soils on the narrow coastal plains are derived from alluvium and, except where this is entirely basaltic, soil occurrences are largely determined by drainage. Thus on the better-drained levees yellowish friable loams (Um6.34) predominate, and there is a progression to *gleyed podzolic soils*, *humic gleys* and *acid peats* as the site drainage becomes progressively poorer. In some lower-rainfall situations *soloths* and *solodic soils* may occur. These are more common in the south where the *soloths* are also found associated with *yellow podzolic soils* on the foothills. Small areas of beach ridges occur adjacent to the coast, with *calcareous sands* or *siliceous sands* on the younger and *podzols* on the older ridges.

Georgetown Province

This Province is largely an upland erosional terrain which forms the upper catchments of streams draining to the Gulf of Carpentaria and to a lesser extent to the Burdekin River. The soils in the central and southern part of the Province have been described at reconnaissance level (Sleeman 1964) and in the south-east by Isbell and Murtha (1970).

The Province is geologically complex and the character of soils is very strongly influenced by parent material leading to a diverse range of soils in a complicated pattern. In the central-west, Proterozoic high-grade metamorphics and some acid volcanics, which are intruded by granites, are found in the Georgetown Inlier (Oversby *et al.* 1975). In the northern half of the Province there are extensive areas of low hilly to mountainous terrain formed on Middle Palaeozoic greywackes, siltstone and slate of the Hodgkinson Basin (de Keyser and Lucas 1968). In the central part of the region, extensive fault blocks of Carboniferous acid volcanics have produced resistant high hilly areas. Mesozoic siliceous sandstones form strongly dissected plateaux in the far north and south-western parts of the Province, and in the latter area they may be deeply lateritized. In the centre and south there are extensive undulating plateaux of Cainozoic basalt. Cainozoic sediments occupy locally

Table 17.1. Main features of the soil-landscape provinces of Region VI

Province	Geological and geomorphological features	Major soil features
Cooktown-Ingham-Mackay	High mountains of acid igneous and metamorphic rocks, hilly basalt plateaux and narrow coastal plains	*Red podzolics* and *xanthozems* or *yellow podzolics* on hilly to mountainous areas with *krasnozems* on basalt. On coastal plains friable loams, *gleyed podzolics, humic gleys* and *acid peats* form a sequence with progressively poorer drainage. *Siliceous sands* and *podzols* occur on beach ridges
Georgetown	Upland erosional terrain of Proterozoic metamorphic rocks and Palaeozoic sediments, both intruded by granite. Extensive fault blocks of acid volcanics, strongly dissected Mesozoic sandstone plateaux, and undulating basalt plateaux	*Lithosols, red* and *yellow podzolics* and *soloths* on the metamorphics and Palaeozoic sediments; sandy *lithosols, siliceous* and *earthy sands* on granite and sandstone *(red earths* where lateritized); *lithosols* on the acid volcanics and *euchrozems* on basalt
Buchanan	Mesozoic and Tertiary sediments which have been deeply weathered underlie most of the Province. Relief is low except for the few areas of exposed Mesozoic sandstone	*Red* and *yellow earths* dominate but seldom occur in a regular catenary sequence. The former are very deep soils while the latter are invariably shallower and more nodular. *Lithosols* are dominant on exposed Mesozoic sandstones
Burdekin-Broadsound	Mainly erosional, often hilly terrain with complex geology consisting chiefly of Palaeozoic sediments and volcanics and intrusive granitic rocks. Some small remnants of Cainozoic lateritized landsurfaces, and coastal alluvial plains	*Solodized solonetz* and *solodic soils* are widespread on Palaeozoic sediments and on the coastal plains, with *non-calcic brown soils* predominant on intermediate igneous rocks. *Black earths* and *grey* and *brown clays* occur on more lithic sediments. *Lithosols* are prominent on all resistant rocks and *red earths* on the old landsurface remnants
Fitzroy	Extensive undulating lowlands, low ranges and high dissected sandstone plateaux. Much of the Province consists of a variably stripped Tertiary landsurface which was originally deeply lateritized	*Solodized solonetz, solodic soils, grey* and *brown clays* and *black earths* are widespread on the partially stripped Tertiary landsurface and on depositional sites. More resistant freshly exposed rocks have *lithosols, siliceous sands* and shallow *solodic soils. Black earths* have formed from Tertiary basalts while the few remnants of the original Tertiary landsurface have *red* and *yellow earths*
Burnett	Mostly hilly lands of Palaeozoic metamorphic and granitic rocks, Mesozoic and Cainozoic sediments, and basalt. Modified plateau remnants of a mid-Tertiary landsurface are common	Texture contrast soils, *podzolics, soloths, solodized solonetz* and *solodics,* are predominant; the acid forms are most common in the east on metamorphic rocks. The basaltic soils include *krasnozems, euchrozems* and *black earths,* and on the plateau remnants over more siliceous rocks, *red earths* and *lateritic podzolic soils* are common
Moreton	Generally low to moderate relief but some elevated plateaux with steep scarps. Mostly geologically complex with some old erosion surface remnants. Large sand masses of giant compound dune systems border the coast	Complex soil pattern with various *podzolic soils* dominant nearer the coast and *solodized solonetz* and *solodic soils* on drier, more inland sites. *Lithosols* are also extensive, with basaltic *krasnozems* and *black earths* locally prominent. Various *podzols, gleyed podzolics* and *humic gleys* characterize the near-coastal deposits
Hunter-Tweed	Mountainous lands with steep escarpments on Palaeozoic sediments and granite, rolling to hilly basaltic plateaux, lower hilly areas of Mesozoic sediments, and narrow valley plains and coastal lowlands	*Lithosols* and various *podzolic soils* are dominant. Basaltic *krasnozems* and *prairie soils* are also common in high rainfall areas, with some *black earths, red-brown earths* and *solodized solonetz* in drier sites. On the coastal plains *alluvial soils,* minimal *prairie soils* and *humic gleys* are common
Darling Downs-Liverpool Plains	Erosional uplands of Palaeozoic metamorphic and granitic rocks, Mesozoic sediments and Tertiary basalts. Generally rounded hilly with some high basaltic tablelands and ranges, lower hilly plateaux and undulating colluvial-alluvial plains. Minor remnants of old landsurfaces	*Podzolic, solodized solonetz* and *solodic soils* are widespread on granite, siliceous metamorphic rocks and sediments; *black earths, prairie* and *chocolate soils* with minor *euchrozems* and *krasnozems* occur on basalt. *Non-calcic brown soils, red-brown earths* and *solodic soils* are common on the less siliceous sediments and in drier areas. Smaller areas of *lithosols, gleyed podzolics, yellow* and *red earths* and *lateritic podzolics*

important areas in the central part of the Province and are frequently lateritized.

On the more resistant metamorphic rocks *lithosols* are dominant, whereas on more basic schist and gneiss, gravelly-surfaced *red podzolic* and associated *yellow podzolic soils* are most common. Loamy *lithosols* are the most widespread soils on the Hodgkinson Basin sediments and rock outcrops are common, but shallow gravelly *soloths* and *red podzolic soils* usually occur on the gentler slopes with small areas of *solodic soils* on lower slopes and valley floors. The acid volcanic rocks are very resistant to weathering, resulting in very shallow and stony *lithosols* with much massive rock outcrop.

The granitic rocks of the region mostly have weathered to sandy soils, particularly *lithosols* although there are some deeper *siliceous sands* and *earthy sands*. Small areas of *yellow podzolic soils* and *soloths*, and less commonly *red podzolic soils* and *solodic* or *solodized solonetz* soils also occur, with the latter usually restricted to topographically low sites. In the higher rainfall areas in the east, *xanthozems* and *yellow podzolic soils* have formed on granitic rocks and acid volcanics.

The Mesozoic sandstones have dominantly *lithosols* on the steeper topography and deeper *siliceous sands* on gentler slopes. There are sandy *red earths* and *yellow earths* on the flatter plateaux with some extensive areas of loamy *red earths* on relict Middle or Late Tertiary surfaces that have been deeply lateritized.

The soils derived from basalt fall into three distinct groups. Shallow to moderately deep red soils, *euchrozems*, are the most widespread, and associated with them are lesser areas of similar brown soils (Isbell *et al.* 1976; Gillman 1976; Isbell *et al.* 1977). There are also small areas of *black earths* developed *in situ* on gently sloping terrain, on stream alluvium, and on marshy areas surrounding springs.

The largest area of Cainozoic sediments is in the east-central part of the Province. Here deep *red earths* with associated *yellow earths* are the major soils. Elsewhere in the Province *red earths* are the characteristic soils of the older river terraces and stream levees.

Buchanan Province

This relatively arid central Queensland Province is of low relief, and lies astride the broadly upwarped and little-dissected Great Divide whose elevation seldom exceeds 500 m. Extensive plains are common and there are two small internal drainage basins containing the usually saline Buchanan and Galilee Lakes. The soils in the northern part of the Province have been described by Isbell and Murtha (1970), and in the eastern part by Gunn (1967*a*).

The Province is underlain by Mesozoic and Tertiary sediments (Geological Survey of Queensland 1975). However, deep weathering and local erosional and depositional events during the Cainozoic have produced a subdued landscape which in many places largely obscures the underlying geology. The Mesozoic sandstones are only exposed where the greatest relief in the Province occurs. The soils on the more resistant rocks are *lithosols*, with sandy forms where the rocks are the more siliceous sandstones, and shallow gravelly loams where they are the widespread more labile sandstones. On the main occurrences of argillaceous rocks which are south of Lake Buchanan, the soils are brown and red *solodic soils* with small areas of *brown clays*.

Over the remainder of the Province *red earths* and *yellow earths* are dominant but the factors controlling the distribution of these soils are not well known. Since the Cretaceous, the area has been subject to alternating periods of stability with deep weathering, and of erosion and deposition forming Tertiary sediments and more recent extensive fans. Because much of the material was preweathered, and because Late Tertiary–Quaternary events have also produced deeply-weathered earths (Coventry 1978), it is difficult to assign even approximate ages to many of the *red* and *yellow earths* even though they are mostly relict soils. Gunn (1976*b*) proposed a fairly simple catenary sequence of *red* and *yellow earths* resulting from the denudation of a widespread Tertiary laterite profile, but more recent studies have established that this simple model does not apply widely to this Province.

In spite of uncertainties regarding the origin of the soils and doubts about the factors controlling their distribution, significant facts are known about them. All the soils may occupy a variety of sites on the gently undulating terrain where they occur, and nearly all are low in nutrients, particularly phosphorus. The *red earths* are very permeable and are deep to extremely deep soils (3–36 m or deeper); they may contain prominent ironstone nodule horizons or laterite at depth (or shallower when they have been truncated). The *yellow earths*, and some restricted occurrences of associated grey massive earths (Gn2.8, .9), are not only much shallower soils (1–2 m) but are frequently saturated during wet periods, and have varying and frequently high amounts of ironstone nodules in their subsoils. Most appear to overlie relatively impermeable substrates which may include truncated laterite profiles.

Minor soils include *solodized solonetz* and *solodic soils* in some of the broad valleys associated with the *yellow earth* landscapes, and small areas of deep *grey clays* on some of the old stream floodplains and in local internal drainage basins.

Burdekin–Broadsound Province

This subhumid Province is dominated by erosional terrain but includes some narrow coastal plains, and some old depositional areas further inland. The northern section is part of the Burdekin River catchment while the southern portion drains into the Fitzroy River. The relief is very variable, being greatest in the dissected near-coastal ranges which rise to 500 m or more. The soils have been widely studied (Isbell and Hubble 1967; Isbell and Murtha 1970; with more detail in some parts e.g. Hubble and Thompson 1953; Isbell 1966; Gunn 1972; Murtha 1975, and in press).

The Province, which is geologically complex (Geological Survey of Queensland 1975), consists mainly of the remnants of a series of sedimentary basins with both sediments and volcanics, and of large hilly areas of Late Palaeozoic granitic rocks. On most upland sites the soils have formed from the underlying hard rocks but there are also some small elevated remnants of old land surfaces underlain by Tertiary sediments (Grimes 1980) and on which there are mostly *red earths* and to a lesser extent *yellow earths.*

On Palaeozoic lithic and felspathic sandstone, siltstone and greywacke, common in the far north and central inland parts of the Province, the characteristic soils are usually shallow and often gravelly *solodized solonetz* and *solodic soils.* On similar rocks in the higher rainfall areas north of Rockhampton, *soloths* are dominant while on Permian lithic sandstones to the south of Collinsville there are deep *grey* and *brown clays* and *black earths* of moderate to high fertility. The more resistant rocks of these sedimentary sequences have produced *lithosols.*

Deep *solodized solonetz* and *solodic soils* are widespread in the Bowen–Townsville and Broadsound coastal areas where they have formed on alluvial deposits. Similar browner soils are common on old outwash plains to the south of Charters Towers, along with some associated gilgaied *grey clays.*

The soils derived from igneous rocks fall into a number of distinct groups. On acid volcanics and some acid granitic rocks, sandy and loamy *lithosols* are dominant. On other acid granites deeper soils occur – generally *siliceous sands* in the more inland areas and *soloths* or *yellow podzolic soils* on the near-coastal ranges. The intermediate igneous rocks, granodiorite, diorite and andesite, have a characteristic set of soils on topography varying from undulating to high hilly: shallow to moderately deep *non-calcic brown soils,* frequently with a cover of surface stone, are overwhelmingly dominant and are among the most widespread soils in the Province; some *solodic soils* also occur, particularly on lower slopes. The Permian and Tertiary basalts also have characteristic soils although their areas are limited; these are mostly *black earths,* in contrast to the dominance of *euchrozems* on the much younger basalts of the Georgetown Province to the north. Finally, shallow *euchrozems* have developed on the small areas of serpentinite to the north-west of Rockhampton.

Certain other soils in the Province warrant mention, although most are of limited extent. Deep *grey clays* are prominent on the lower alluvial plains of the Burdekin River, and non-self-mulching *black earths* occur on marine plains fringing Broad Sound. The lower Burdekin delta is dominated by stratified *alluvial soils* and weakly developed brown massive earths (Gn2.42, .44). Salt pans with *solonchaks* occur adjacent to many parts of the coast, while to the north of Rockhampton giant *podzols, humus podzols* and deep *siliceous sands* have formed in old dune systems.

Fitzroy Province

This relatively dry central Queensland Province occupies much of the Fitzroy River catchment. A wide variety of land forms is found, ranging from relatively narrow alluvial plains to very extensive undulating lowlands and high (500–900 m) dissected plateaux. Large areas of the lowlands were originally covered with *Acacia* forests, particularly brigalow (*A. harpophylla*). The many soil studies in the Province (Isbell and Hubble 1967; Gunn 1967*a*, 1967*c*, 1977; Sweeney 1968) have shown that there are large areas of *solodized solonetz* and *solodic soils, grey* and *brown clays, black earths,* and less extensive areas of *lithosols* and *red earths.*

Although rocks ranging in age from Early Palaeozoic to recent are present, the Province is not unduly complex lithologically (Geological Survey of Queensland 1975). Sedimentary rocks and unconsolidated sediments are predominant, although extensive areas of Tertiary basalt flows are also a feature. Studies of the geomorphic history (Galloway 1967*a*, 1967*b*; Wright 1968; and Grimes 1980) indicate that a mantle up to 90 m thick and mainly made up of Tertiary terrestrial deposits, once covered most of the Province. This mantle was subjected to intense weathering and developed deep laterite profiles, sometimes extending into the underlying pre-Tertiary rocks. Subsequent widespread differential erosion of this weathered mantle has resulted in the following geomorphic categories (Galloway 1967*a*; Gunn 1977) which largely determine the gross soil pattern in the Province: (i) intact to moderately stripped Tertiary land surfaces (ii) erosional-depositional surfaces within the Tertiary weathered zone (iii) post-Tertiary alluvial land surfaces. Within each of these categories, the character of the soil is largely controlled by the nature of the parent material.

Most of the original Tertiary surface with its three-part laterite profile has now been removed, and the few remaining portions are mostly level with undulating plains and low plateaux and have deep to moderately deep *red earths* and some associated *yellow earths.*

Soils formed on materials exposed by partial stripping and redeposition of the deeply weathered mantle occupy large areas and are varied (Gunn 1967*a*, 1974). They include a range of *solodized solonetz* and *solodic* *soils* with neutral to strongly alkaline subsoils, and deep and often strongly gilgaied *grey* and *brown clays*. Many of these soils are saline at relatively shallow depths and most of the clays have strongly acid subsoil beneath their neutral to alkaline surface soils.

The soils formed on rocks unaffected by deep weathering show a strong lithological control. Resistant quartz sandstones form high hilly to mountainous dissected plateaux where sandy *lithosols* are dominant, with associated *siliceous sands* and either *yellow podzolic* or *solodic soils*. *Non-calcic brown soils* are found on the small areas of granitic rocks in the Province, and are also prominent on the schistose metamorphic rocks which occur in the western part where loamy *lithosols* are usually dominant.

In the west, mixed sediments of the Middle to Late Palaeozoic and ranging from lithic and quartzose sandstones to siltstones and shale, have been strongly folded. These give rise to the mountainous to low hilly areas of the Drummond Range which have a complex pattern of narrow linear belts of differing lithology following strike lines; this leads to a closely related complex soil pattern. On the more argillaceous sediments, *grey* and *brown clays* and *black earths* occur but the greater part of the area is dominated by a very wide range of shallow, often stony, *solodized solonetz* and *solodic soils* with red, brown, dark grey or yellow-grey subsoils. The more resistant rocks form prominent strike ridges and structural benches which have loamy *lithosols*.

In the central and southern parts of the Province gently folded Permian and Mesozoic argillaceous sediments give rise to extensive rolling plains of *grey* and *brown clays* and *black earths* along with associated areas of brown or grey *solodic soils*. *Black earths* derived from Tertiary basalt flows are also widespread in the Province but are relatively shallow (< 1m) and subject to serious erosion when cultivated. Other basalt areas have steeper topography with loamy or clayey *lithosols* as well as shallow stony *black earths*.

Soils derived from more recent alluvium are also extensive and many are subject to flooding from adjacent rivers; *grey clays, black earths* and dark grey *solodic soils* are the most common.

Burnett Province

The catchments of the Burnett River and of some small coastal streams comprise this Province. These are hilly lands cut into both fresh rocks, Palaeozoic granitoids and metamorphics, Mesozoic and Cainozoic sediments and basalts, and into deep lateritic weathering mantles formed on them. Relief is greatest in the sub-coastal ranges. In the western half of the Province variously modified plateaux and mesa remnants of a Mid-Tertiary land surface are common. Although reconnaissance soil data are available for the whole area (Isbell *et al.* 1967) there is little detailed information (Teakle 1950; de Mooy *et al.* 1977).

Texture-contrast soils (*podzolic soils, soloths, solodized solonetz* and *solodic soils*) characterize the Province. The acid *podzolic soils* and *soloths*, commonly with mottled yellow clay subsoils and mainly on metamorphic rocks, are dominant in the east. These soils extend into the drier central and western areas as shallow soils on narrow crests and upper slopes over the more siliceous rocks, and as deeper soils on lower slopes where the parent materials are from the acid deep-weathering mantle. Soils with neutral to alkaline reaction trends (neutral pedal red (Dr2.22) and yellow (Dy2.22) duplex soils, *solodized solonetz* and *solodic soils*) are associated on the mid-to-lower slopes and on areas of lower relief with less-siliceous parent materials. *Lithosols* are found mainly in the eastern areas of stronger relief on narrow crests, steep slopes and scarps of the harder rocks, and on hardened mottled zone remnants of the deep-weathering mantle. Other soils on gentler topography over intermediate volcanic rocks include shallow stony *non-calcic brown soils, prairie soils* and alkaline, red smooth-ped earths (Gn3.13).

In the central and southern sections the soils on basalt conform to an idealized topographic–drainage sequence. Lateritic *kraznozems* lie on the plateau remnants of the Mid-Tertiary surfaces, normal *kraznozems* with minor areas of *xanthozems* are formed on detrital red clays and on *in situ* materials on slopes within the depth of the deep weathering, and *black earths* occur on fresh basalt exposed by deeper dissection and on colluvial–alluvial spreads from it. In the Coalstoun Lakes area on the flows of Pleistocene basalt, stony *euchrozems* are dominant but *black earths* are found on areas with slower drainage.

Both *red earths* and small areas of *lateritic podzolic soils* are found on remnants of old sufaces over granitic, and moderately quartzose sedimentary and metamorphic rocks. In the Mundubbera district, occurrences of *red earths* on plateaux at elevations of greater than 300 m, about 180 m and 120 m, suggest two or more old surfaces of different age. *Red podzolic soils* with lateritic gravels are common on dissected areas bordering the plateau, whereas a range of soils occurs on the plains of old alluvia, including various *solodic soils, grey clays*, and some *black earths* according to the nature of the patent materials.

Moreton Province

This Province embraces the higher-rainfall subtropical areas of south-eastern Queensland. The topography is dominantly erosional and the relief low to moderate, but elevated plateaux (sometimes basalt-capped), steep scarps and ranges, narrow valley floors and coastal plains are also included. Dune systems of siliceous sands border the coastal plain and make up the large islands rising to 270 m elevation off the south Queensland coast. Plateau remnants of old erosion surfaces with deep-weathering profiles are found and span the upper, middle and lower erosion surfaces of Watkins (1967). Soils data are available for the Province as a whole (e.g. Hubble *et al.* 1972) and for certain specific areas (Hubble 1954; Coaldrake 1961; Thompson 1965; Beckmann 1967; Paton 1971; Beckmann and Reeve 1972; Thompson 1975; Powell 1977).

Texture-contrast soils are the most extensive but the soil cover is characterized by great variety and complex distribution patterns which reflect the geological complexity of the Province; superimposed on this geological complexity are the effects of the variety of land forms, geomorphic and weathering histories, and climatic differences, past and present.

The *podzolic soils* are dominant and occupy low hilly to hilly landscapes in the moister parts of the Province, which are mainly near the coast. Forms with red, yellow, and mottled clay subsoils are common often occurring in toposequence with *red podzolic soils* on the higher sites (Fig. 17.3). They overlie such widely differing rocks as phyllites, greywackes and shales, granodiorite and granite, and sandstones of the Mesozoic sedimentary basins. Most are moderately deep soils with hard-setting loamy surfaces. On the more clayey or feldspathic Mesozoic sediments, or in drier areas and especially on lower slope sites, *solodized solonetz* and *solodic soils* or *soloths* replace the *podzolic soils.*

Shallow loamy *lithosols* are also extensive; they are found mainly on steep slopes, ridge crests, and stripped weakly dissected plateau surfaces over the more resistant rocks, particularly the metamorphics.

The more fertile soils of the Province have formed from basic and intermediate igneous rocks. Scattered areas of deep *krasnozems*, some of them lateritic, have formed in several situations: on basaltic remnants of the upper erosion surface as at Mt Tambourine and the Maleny plateau, on the lower erosion surface at Redland Bay and on young flow surfaces as at Bundaberg. Larger areas of shallow stony *prairie soils* are associated with freshly weathered basalt on the slopes of the south-western ranges and with some low hilly areas of andesitic rocks and fine sediments. Small areas of shallow *black earths* have formed on basalt near Brisbane. Much more significant are the larger areas of deep *black earths* and prairie soils formed in the alluvium of streams rising in basaltic country, e.g. Lockyer Creek. Where there is little or no basaltic component, the soils of the minor terraces along the streams are commonly *solodic soils* on the highest terrace and mixed *alluvial soils* and weakly developed *prairie soils* on the lower terraces.

In the coastal lowlands *lateritic podzolic soils* with thick sandy A horizons and deep weathering profiles (to 10 m) occur on the lateritized sandstone remnants of the lower erosion surface; associated with these soils are minor areas of deep *red earths* on well-drained rises and *gleyed podzolic soils* on wetter sites. The red earth residuals (Bryan 1939) on Tertiary sediments are also remnants of this surface: other *red earths* on hilly lands near Esk and Nambour are formed on materials derived from ferruginous quartzose sandstone.

The soils on the Quaternary sediments of the coastal plain range from *humus podzols* on quartz sands to *gleyed podzolic soils* and *humic gleys* on medium to fine

Fig. 17.3. Smooth hilly lands formed on phyllites near Gympie in south-east Queensland. The *red podzolic soils* are here used for bean crops.

textured sediments. *Podzols* of widely differing age and degree of development have formed on the several systems of silica sands forming the coastal dunes (Thompson 1975). The youngest of these show minimal profile differentiation with a weakly-developed ferruginous B horizon but the oldest are giant forms with iron-humus B horizons, A_2 horizons many metres thick, and sola up to 15 m deep (Thompson and Hubble 1979).

Hunter–Tweed Province

This mostly humid coastal Province embraces the eastern slopes of the Dividing Range from the Queensland border ranges southward to the Hunter Valley and has great complexity. Most of it consists of denudational lands cut in fresh rocks with the result that both landforms and soils are dominated by the influence of the underlying lithology. Areas of deep weathering are limited to some basaltic plateaux and re-exposed pre-basalt land surfaces. Mountains with summits rising 650 to 1600 m, steep escarpments and rugged hills are all associated with the harder Palaeozoic rocks – sandstones, siltstones, mudstones, greywackes, cherts, tuffs, and granites – whereas rolling to hilly plateaux and some steep ranges are associated with Tertiary basalt, and lower hilly to rolling topography with softer sedimentary rocks including the Jurassic-Cretaceous mudstones, siltstones and sandstones of the Clarence Basin. Constructional lands are restricted to narrow valley plains which are often terraced and to the swampy plains and dunes of the discontinuous coastal lowlands.

Reconnaissance soil maps cover the whole Province (van de Graaff 1963; Northcote 1966; McGarity 1977; Beckman and Thompson 1977) and more detailed soil studies have been made of several districts (Brewer and Butler 1953; Nicolls and Tucker 1956; Walker 1963; McArthur 1964; Jessup 1965). The soil patterns closely follow the topography, lithology and climate (mean annual rainfalls range from 500 to 1600 mm) in that order but soil toposequences on particular rocks vary considerably. The most extensive soils are the *lithosols* and *podzolic soils* of several kinds. The *lithosols* occupy all the steep rugged lands irrespective of rock type but minor variations are found; loamy forms on the harder sedimentary rocks, sandy on the granitic rocks and dark friable clay loams on basalt. Soils commonly associated in toposequence over the harder sediments and granite, from freely-drained crests and upper slopes down to poorly-drained foot slopes, are *grey-brown, red, yellow,* and *gleyed podzolic soils.* Reddish and yellowish *prairie soils* occur in some sequences on more basic rocks and fine sediments. On hilly country of lower relief and easier slopes mottled *yellow podzolic soils* dominate. In some areas *red earths* and *yellow earths* occur on gentle crests and slopes, while in the drier parts of the Hunter Valley, *red-brown earths, solodized solonetz* and degraded *black earths* are common on sedimentary rocks.

Throughout the Province *krasnozems* are characteristic of the hilly basaltic plateau remnants under high rainfall but they also occur on some steeper slopes. Other basaltic soils are *chocolate, prairie* and transitional *alpine humus soils,* with minor areas of *humic gleys* and *acid peats* on the higher colder plateaux. Dark rough ped earths (Gn4.41, .42) and *lithosols* occur on moderate to steep slopes and narrow crests, while *black earths* are generally dominant in the lower and drier areas.

The range of soils on alluvium and coastal sediments reflects the source and degree of weathering of the parent materials, drainage, and soil age. In the Lismore district the soils include *krasnozems, black earths* and *alluvial soils*; in the Hunter Valley, *chernozems, solonetz, alluvial soils* and some *grey clays.* On the wet flood plains of the lower valleys and merging into swampy estuarine and coastal plains *humic gleys* are found with minor areas of *acid peats* and *humus podzols.* On the coastal dunes there are *podzols, siliceous sands* and some *acid peats* in wet depressions.

Darling Downs–Liverpool Plains Province

This Province encompasses the western slopes of the Dividing Range from the Darling Downs southward to the Liverpool Plains. These are hilly erosional lands cut into mainly fresh rocks as in the Hunter-Tweed Province and the range of elevation is similar though the local relief is generally much lower and slopes are gentler. However rainfalls are lower and the range (800–500 mm) narrower producing a gentle decreasing gradient westward. In this Province, compared with the Hunter-Tweed Province, there is a lower proportion of Palaeozoic metasediments, and a higher proportion of granitic rocks, Jurassic sediments, intermediate to basic lavas, Tertiary basalts and Quaternary clayey alluvia. The associated landforms include high basaltic tablelands, intermediate-level erosional granitic plateaux, steep mountain ranges, rounded medium to low hills, and rolling terrain in the softer sedimentary rocks. The larger areas of depositional landforms are the plains of the western Darling Downs and broad valleys northward from the Liverpool Plains.

Northcote (1966) integrated data from all except the most recent of the many soil studies made within the Province (Hallsworth *et al.* 1952, 1953*a*, *b* and *c*; Thompson and Beckmann 1959; Beckmann and Thompson 1960; Hallsworth and Waring 1964; Jessup 1965; Powell 1976; McGarity 1977). The Province is characterized by a complex soil pattern which reflects differences in parent materials, present climate (associated with altitude and rainfall), and local differences in topography and drainage. The overall dominance of neutral to alkaline soil reaction trends is in keeping with the limited leaching potential of the rainfall on dominantly clayey soils.

On basalt, the characteristic soil sequence from the cool higher rainfall plateaux to the warmer lower-rainfall plains is; *kraznozems, chocolate soils, prairie soils, euchrozems* and *black earths* (Fig. 17.4). A common toposequence in the moister areas is; *chocolate soils, prairie soils and black earths.*

On granitic rocks, sandy *yellow podzolic soils* are dominant on the wetter tablelands. The toposequence from stony crest to drainage depression is; gritty *siliceous sands* (among outcropping rock), *yellow podzolic soils* and *gleyed podzolic soils*. Minor areas of *podzols* can also occur. In some areas of deeply-weathered rocks containing sodium-rich feldspars *solodic soils* occur on lower slopes (McGarity 1977).

The dominant soils on the metasediments change progressively to the west with decreasing rainfall and leaching; the change goes from a limited area of *red podzolic soils*, through *yellow podzolic* and *yellow solodic soils* to *non-calcic brown soils* and *red-brown earths* which are often solonized at the drier limit. In the 'trap rock' country of southern Queensland (Powell 1976) the characteristic toposequence from hillcrest to narrow valley floor is; shallow gravelly *lithosols*, shallow *soloths, solodized solonetz* and *solodic soils* with varying colours,and yellow-brown *solodic soils* in the lowest position.

The variety of soils on the Mesozoic sediments along the drier western margin of the Province is strongly influenced by lithology. On the sandstone of the rolling to hilly Pilliga area *yellow earths* and deep sandy *solodized solonetz* are usually dominant. The full toposequence from the stonier ridge crests to colluvial-alluvial footslopes is; sandy *lithosols*, acid sandy *yellow earths* (or acid to neutral *red earths* on the more ferruginous rocks), *soloths* and *solonized solonetz*. On argillaceous rocks loamy *solodized solonetz* are dominant with loamy *lithosols* on the stony crests.

Land use aspects

Land use differs widely through the Region, often over short distances, as might be expected from the range in climate and the wide diversity of landscapes and soils. In the high-rainfall northern coastal areas sugar cane is grown intensively and, although this crop is also grown in southern Queensland and northern New South Wales, the far northern areas account for the bulk of Australia's production. Both summer and winter grain crops are grown in the central Queensland Fitzroy Province, and wheat in particular, but also oats, barley, sorghum, sunflower and soybeans, become increasingly important to the south where some areas, e.g. Darling Downs, are noted for their productivity.

Dairying on improved pastures is important on the near coastal regions of New South Wales and southern Queensland, with a locally prominent occurrence on the Atherton Tableland in far north Queensland. Elsewhere, beef cattle grazing on both native and improved pastures is widespread, with sheep also on the southern inland areas. Timber is produced from the remaining rainforest and eucalypt forest areas and from plantations of both indigenous and exotic conifers. There are limited areas of tropical and subtropical horticulture, and vineyards are a feature of the Hunter Valley.

While climatic factors largely determine the kind of agriculture practised, the distribution of cropped land is very much landscape and soil dependent. This is also true, although perhaps to a lesser extent, in the case of dairying and beef production based on improved pastures. It has been noted earlier that the landscape history of the region is of great importance in determining the soil pattern, present climate often playing a subordinate

Fig. 17.4. Undulating uplands of *black earths* on basalt on the eastern Darling Downs, south Queensland. These fertile soils are widely used for grain crops. Photo courtesy of the Queensland Government Tourist Bureau.

role. It has been shown that relict soils are widespread, that the nature and distribution of the soils are often closely related to the degree of dissection of prior deeply-weathered mantles, and that where fresh rocks are now exposed, the underlying lithology exerts a dominant role in soil formation. All of these factors are of importance to current land use.

Grain cropping affords a good example of the influence of soils for it is largely associated with soils of higher inherent fertility, such as the basaltic *black earths* of central and southern Queensland. These soils together with *red brown earths, non-calcic brown soils* and *prairie soils* in inland New South Wales comprise most of the grain growing areas. In sharp contrast, the high-value sugar crops with their large fertilizer inputs, are grown on a very wide range of soils although it may be necessary to drain low-lying sites.

The horticultural areas also show a soil dependence, although here the emphasisis is more on soil physical conditions than on inherent fertility. The widespread improved grass pastures of central Queensland are almost entirely sown on the moderately fertile *grey* and *brown clays* of the cleared brigalow (*Acacia harpophylla*) lands, and the important paspalum (*P. dilatatum*)-white clover (*Trifolium repens*) dairy pastures of the southern coastal and subcoastal regions are mainly on the more fertile soils of the river valleys. With few exceptions, the infertile *lithosols, podzolic* and *solodic soils, soloths* and the *red* and *yellow earths* throughout the Region have largely remained under inferior native pastures and hence support only low-intensity beef cattle grazing.

Soil deficiencies of phosphorus and nitrogen of varying severity are almost ubiquitous. The main exceptions are some of the basalt-derived soils e.g. the *black earths* of the Darling Downs, and some of the *grey* and *brown clays* that previously supported brigalow in central Queensland. Severe phosphorus deficiencies are common from the coast through to the drier margin of the Region on such relict soils as *red* and *yellow earths* and a wide range of other soils, notably the *solodized solonetz* and *solodic soils* formed on the more quartzose rocks and materials of the old deep-weathering profiles. Multiple deficiencies, including several trace elements, are associated with the strongly leached soils of the higher-rainfall coastal areas e.g. the *lateritic podzolic soils, podzols* and *humus podzols* of southern Queensland.

Erosion is a hazard on many of the sloping to undulating lands of the Region. Even without cultivation, whenever the native vegetation is disturbed or removed from the *solodized solonetz* and *solodic soils*, gully erosion follows and is often spectacular (Fig.3.19, p.41). Gully and sheet erosion of varying severity is a problem of cultivated sloping lands on most of the Region's soils, even on the permeable *krasnozems*. It has been particularly serious on the upland *black earths* of the Darling Downs and central Queensland where cultivated fallows for both summer and winter crops annually expose large areas to the heavy storm rains common in the subtropics.

References

Beckmann, G. G. (1967). Soils and land use in the Beenleigh-Brisbane area, southern Queensland. CSIRO Aust. Div. Soils, Soils Land Use Ser. No. 50.

Beckmann, G. G. and Reeve, R. (1972). Classification and chemical features of soils of the Beenleigh-Brisbane area, south-east Queensland. CSIRO Aust. Div. Soils Tech. Pap. No. 11.

Beckmann, G. G. and Thompson, C. H. (1977). The Soils. *In* 'The Border Ranges, a land use conflict in regional perspective'. (R. Soc. Queensl: Brisbane.)

Beckmann, G. G. and Thompson, C. H. (1960). Soils and land use in the Kurrawa area, Queensland. CSIRO Aust. Div. Soils, Soils Land Use Ser. No. 37.

Beckmann, G. G., Thompson, C. H. and Hubble, G. D. (1974). The genesis of red and black clay soils on basalt on the Darling Downs, Queensland. *J. Soil Sci.* **25**, 265-81.

Brewer, R. and Butler, B. E. (1953). Pedology of part of the Upper Hunter Valley, New South Wales. CSIRO Aust. Soil Publ. No. 2.

Browne, W. R. (1969). Geomorphology of New South Wales. *J. Geol. Soc. Aust.* **16**, 559-69.

Bryan, W. H. (1939). The red earth residuals and their significance in south-eastern Queensland. *Proc. R. Soc. Queensl.* **50**, 21-32.

Carnahan, J. A. (1976). 'Natural Vegetation'. Atlas of Australian Resources. Second Series. (Div. Nat. Map.: Canberra.)

Coaldrake, J. E. (1961). The ecosystem of the coastal lowlands ('Wallum') of southern Queensland. CSIRO Aust. Bull No. 283.

Coventry, R. J. (1978). Late Cainozoic geology, soils, and landscape evolution of the Torrens Creek area, north Queensland. *J. Geol. Soc. Aust.* **25**, 415-27.

de Keyser, F. and Lucas, K. G. (1968). Geology of the Hodgkinson and Laura basins, north Queensland. Bur. Min. Resour. Aust. Bull. No. 84.

de Mooy, C. J., Hubble, G. D. and Beckmann, G. G. (1977). The soils of the central Burnett area, Queensland. CSIRO Aust. Div. Soils, Soils Land Use Ser. No. 56.

Francis, G. and Walker, G. T. (1978). Multiple laterite surfaces near Armidale. *Search* **9**, 191-2.

Galloway, R. W. (1967a). Geomorphology of the Nogoa-Belyando area, Queensland. CSIRO Aust. Land Res. Ser. No. 18.

Galloway, R. W. (1976b). Geomorphology of the Isaac-Comet area, Queensland. CSIRO Aust. Land Res. Ser. No. 19.

Geological Survey of Queensland (1975). Queensland Geology, Scale 1:2 500 000. (Dep. Mines: Brisbane.)

Gillman, G. P. (1976). Red basaltic soils in north Queensland. II. Chemistry. CSIRO Aust. Div. Soils, Tech. Pap. No. 28.

Grimes, K. G. (1980). The Cainozoic geology of north Queensland. *In* 'The Geology and Geophysics of north-eastern Australia'. (Eds R. A. Henderson and P. J. Stephenson). (Geol. Soc. Aust. (Queensl. Div.): Brisbane.)

Gunn, R. H. (1967*a*). Soils of the Nogoa-Belyando area, Queensland. CSIRO Aust. Land Res. Ser. No. 18.

Gunn, R. H. (1967*b*). A soil catena on denunded laterite profiles in Queensland. *Aust. J. Soil Res.* **5**, 117-32.

Gunn, R. H. (1967*c*). Soils of the Isaac-Comet area, Queensland. CSIRO Aust. Land Res. Ser. No. 19.

Gunn, R. H. (1972). Soils of the Shoalwater Bay area, Queensland. CSIRO Aust. Div. Land Res. Tech. Memo. 72/10.

Gunn, R. H. (1974). A soil catena on weathered basalt in Queensland. *Aust. J. Soil Res.* **12**, 1-14.

Gunn, R. H. (1977). Land classification criteria: geomorphology, soils and mode of occurrence. *In* 'Land Units of the Fitzroy Region, Queensland'. CSIRO Aust. Land Res. Ser. No. 39.

Hallsworth, E. G., Colwell, J. D. and Gibbons, F. R. (1953*a*). Studies in pedogenesis in New South Wales. V. The euchrozems. *Aust. J. Agric. Res.* **4**, 305-25.

Hallsworth, E. G. and Costin, A. B. (1953*b*). Studies in pedogenesis in New South Wales. IV. The ironstone soils. *J. Soil Sci.* **4**, 24-47.

Hallsworth, E. G., Costin, A. B. and Gibbons, F. R. (1953*c*). Studies in pedogenesis in New South Wales. VI. On the classification of soils showing features of podzol morphology. *J. Soil Sci.* **4**, 241-56.

Hallsworth, E. G., Costin, A. B., Gibbons, F. R. and Robertson, G. K. (1952). Studies in pedogenesis in New South Wales. II. The chocolate soils. *J. Soil Sci.* **3**, 103-24.

Hallsworth, E. G. and Waring, H. D. (1964). Studies in pedogenesis in New South Wales. VIII. An alternative hypothesis for the formation of the solodized solonetz of the Pilliga scrub. *J. Soil Sci.* **15**, 158-77.

Hubble, G. D. (1954). Some soils of the coastal lowlands north of Brisbane, Queensland. CSIRO Aust. Div. Soils Div. Rep. No. 1/54.

Hubble, G. D. and Thompson, C. H. (1953). The soils and land use potential of the lower Burdekin valley, north Queensland. CSIRO Aust. Div. Soils, Soils Land Use Ser. No. 10.

Hubble, G. D., Thompson, C. H. and Beckmann (1972). Soil Map. In 'Moreton Region-Natural Environment'. (Co-ordinator-General's Dep.: Brisbane.)

Isbell, R. F. (1966). Soils of the east Bald Hill area, Collinsville district, north Queensland. CSIRO Aust. Div. Soils, Soils Land Use Ser. No. 48.

Isbell, R. F., Gillman, G. P., Murtha, G. G. and Jones, P. N. (1977). Brown basaltic soils in north Queensland. CSIRO Aust. Div. Soils Tech. Pap. No. 34.

Isbell, R. F. and Hubble, G. D. (1967). 'Soils', Fitzroy Region (Queensland), Resources Series. (Dep. Nat. Dev. Geogr. Section: Canberra.)

Isbell, R. F. and Murtha, G. G. (1970). 'Soils', Burdekin-Townsville Region (Queensland), Resources Series (Dep. Nat. Dev. Geogr. Section: Canberra.)

Isbell, R. F., Stephenson, P. J. and Murtha, G. G. (1976). Red basaltic soils in north Queensland. I. Environment, morphology, particle size characteristics and clay mineralogy. CSIRO Aust. Div. Soils Tech. Pap. No. 28.

Isbell, R. F., Thompson, C. H., Hubble, G. D., Beckmann, G. G. and Paton, T. R. (1967). 'Atlas of Australian Soils'. Sheet 4, Brisbane-Charleville-Rockhampton-Clermont Area. With explanatory data. (CSIRO Aust. and Melbourne University Press: Melbourne.)

Jessup, R. W. (1965). Soils of the central portion of the New England region, New South Wales. CSIRO Aust. Soil Publ. No. 21.

King, L. C. (1950). The cyclic land surfaces of Australia. *Proc. R. Soc. Vic.* **62**, 79-95.

McArthur, W. M. (1964). Soils and land use in the Dorrigo-Ebor-Tyringham area, New South Wales. CSIRO Aust. Div. Soils, Soils Land Use Ser. No. 46.

McGarity, J. W. (1977). Soils. *In* 'New England Atlas of Resources'. (Eds D. A. M. Lea, J. J. Pigram and L. Greenwood). (Dep. Geogr. University of New England: Armidale.)

Murphy, P. R., Schwarzbock, H., Cranfield, L. C., Withnall, I. W. and Murray, G. G. (1976). Geology of the Gympie 1:250 000 sheet area. Rep. Geol. Surv. Queensl. No. 26.

Murtha, G. G. (1975). Soils and land use on the northern section of the Townsville coastal plain, north Queensland. CSIRO Aust. Div. Soils, Soils Land Use Ser. No. 55.

Murtha, G. G. (in press). Soils and land use on the southern section of the Townsville coastal plain, north Queensland. CSIRO Aust. Div. Soils, Soils Land Use Ser.

Nicolls, K. D. and Tucker, B. M. (1956). Pedology and chemistry of the basaltic soils of the Lismore district, New South Wales. CSIRO Aust. Soil Publ. No. 7.

Northcote, K. H. (1966). 'Atlas of Australian Soils'. Sheet 3, Sydney-Canberra-Bourke-Armidale Area. With explanatory data. (CSIRO Aust. and Melbourne University Press: Melbourne.)

Oversby, B. S., Palfreyman, W. D., Black, L. P., Cooper, J. A. and Bain, J. H. C. (1975). Georgetown, Yambo and Coen inliers – regional geology. *In* 'Economic Geology of Australia and Papua-New Guinea'. (Ed. C. J. Knight). (A.I.M.M.: Melbourne).

Paton, T. R. (1971). A reconnaissance survey of soils in the Boonah-Beaudesert district, Queensland. CSIRO Aust. Div. Soils, Soils Land Use Ser. No. 52.

Powell, B. (1976). Soils. *In* 'The Granite and Traprock Area of south-east Queensland. Part 1. – Land Inventory'. Div. Land Utilization Tech. Bull. No. 13 (Queensl. Dep. Prim. Indust.: Brisbane.)

Powell, B. (1977). Soils of the major pineapple growing districts of Queensland. Agric. Chem. Branch Tech. Rep. No. 7. (Queensl. Dep. Prim. Indust. Brisbane.)

Sleeman, J. R. (1964). Soils of the Leichardt-Gilbert area, Queensland. CSIRO Aust. Land Res. Ser. No. 11.

Sweeney, F. (1968). Soils of the Dawson-Fitzroy area, Queensland. CSIRO Aust. Land Res. Ser. No. 21.

Teakle, L. J. H. (1950). Description of a laterite profile near Gayndah, Queensland. *Queensl. J. Agric. Sci.* **7**, 15-23.

Thompson, C. H. (1965). Soils and soil problems of the Wide Bay district. *Trop. Grassl.* **6**, 7-12.

Thompson, C. H. (1975). Coastal areas of southern Queensland: some land-use conflicts. *Proc. R. Soc. Queensl.* **86**, 9-14.

Thompson, C. H. and Beckmann, G. G. (1959). Soils and land use in the Toowoomba area, Darling Downs, Queensland. CSIRO Aust. Div. Soils, Soils Land Use Ser. No. 28.

Thompson, C. H. and Hubble, G. D. (1979). Sub-tropical podzols (Spodsols and related soils) of coastal eastern Australia: profile form, classification and use. Int. Soc. Soil Sci. Trans. Comm. IV and V. Kuala Lumpur, Malaysia.

van de Graaff, R. H. M. (1963). Soils of the Hunter Valley. CSIRO Aust. Land Res. Ser. No. 8.

Walker, P. H. (1963). A reconnaissance of soils of the Kempsey district, New South Wales. CSIRO Aust. Div. Soils, Soils Land Use Ser. No. 44.

Watkins, J. R. (1967). The relationship between climate and the development of landforms in the Cainozoic rocks of Queensland. *J. Geol. Soc. Aust.* **14**, 153-68.

Wellman, P. (1978). Potassium-argon ages of Cainozoic volcanic rocks from the Bundaberg, Rockhampton and Clermont areas of eastern Queensland. *Proc. R. Soc. Queensl.* **89**, 59-64.

Wellman, P. and McDougall, I. (1974). Potassium-argon ages of Cainozoic volcanic rocks of New South Wales. *J. Geol. Soc. Aust.* **21**, 247-72.

Wright, R. L. (1968). Geomorphology of the Dawson-Fitzroy area, Queensland. CSIRO Aust. Land Res. Ser. No. 21.

Wyatt, D. H. and Webb, A. W. (1970). Potassium-argon ages of some northern Queensland basalts and an interpretation of late Cainozoic history. *J. Geol. Soc. Aust.* **17**, 39-51.

Chapter 18
Murray–Darling Plains (VII)

B. E. Butler, G. Blackburn and G. D. Hubble

THIS REGION COMPRISES ALL OF THE PLAIN COUNTRY OF the Murray–Darling River System and the enclosed hilly sections (Fig. 18.1). It specifically excludes the main hilly catchment of these rivers. Large sections are essentially areic (that is having the characteristics of very low rainfall), with no consistent pattern of streams, frequently comprised of dunelands, and where any streams which may be present are trunk streams of transit only. About one fifth of the Region has an annual rainfall of less than 250 mm and only in a small section of the Region's north-eastern and south-western extremities does the rainfall exceed 500 mm per annum. There, and in sections of the riverine plains with moderate rainfall, the native vegetation is open woodland while in the lower rainfall areas it is grassland or shrubland of saltbush (*Atriplex* spp., *Kochia* spp.), mallee (*Eucalyptus oleosa*) or mulga (*Acacia aneura*).

Large portions of the Region consist of fields of parallel linear dunes. These are generally closely spaced and thinly covered with vegetation: their orientation is consistent within a particular locality (Fig. 18.1). Several of the sandier areas are popularly designated 'deserts'.

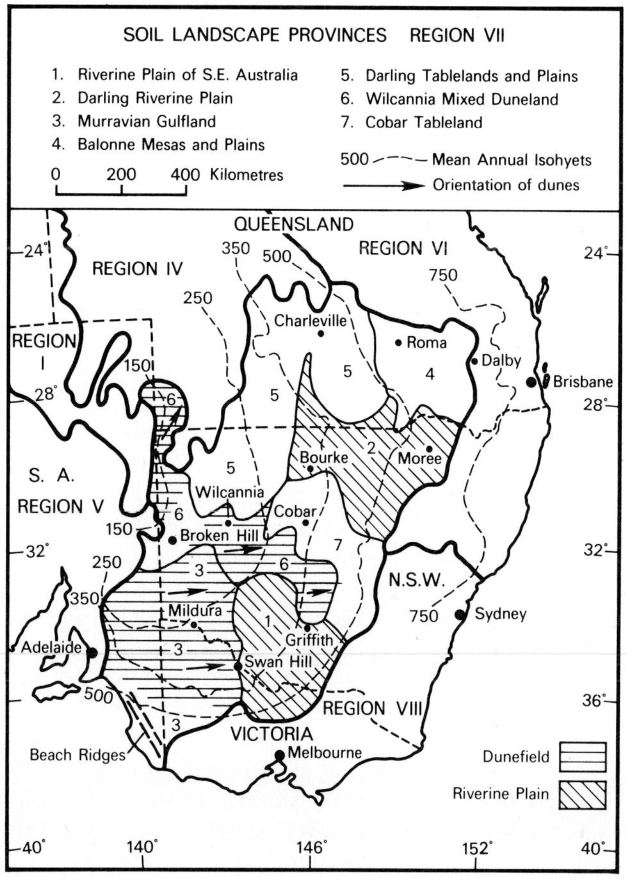

Fig. 18.1. Soil-landscape provinces of Region VII.

The provinces of the Murray–Darling Plains

The Region is subdivided for convenience of description into seven provinces. The criteria for subdivision are the relevant landscape processes and structures of each locality. These are illustrated in Fig. 18.2. As an

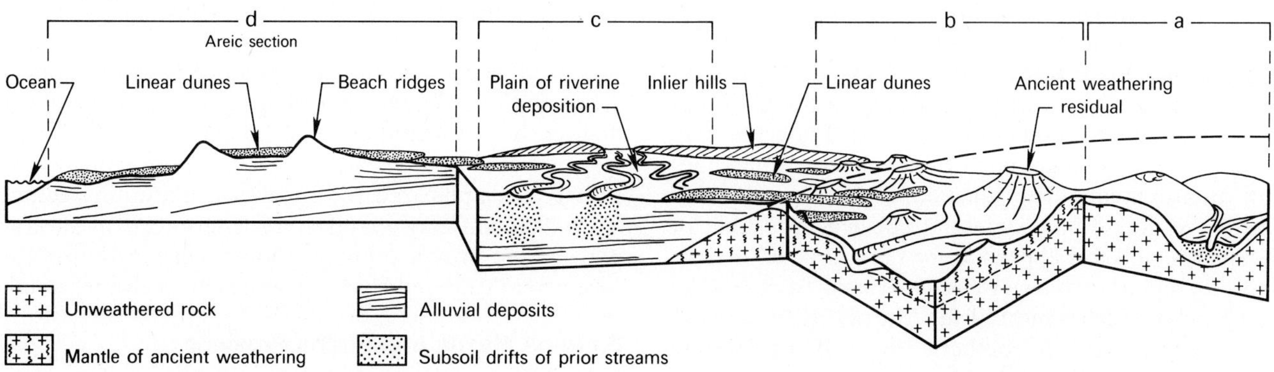

Fig. 18.2. Illustration of landscapes in the Murray–Darling region.

introduction to the description of the provinces, this figure may be discussed in terms of the four sections, *a*, *b*, *c* and *d*.

Section *a* in Fig. 18.2 illustrates the normal processes of denudation associated with current erosion and weathering in the landscape mantle. Most of the soils in this type of country are formed on a shallow colluvial mantle and only a small proportion are formed on alluvial deposits. However, alluvium is entrained in this section and most of it is carried beyond the province and deposited as terraces and alluvial plains lower down the course of the river.

Section *b* is similar to *a* in its denudational character and processes but here the processes are working into Tertiary weathered zones and palaeosols which occur extensively throughout this Region. Excision of portions of the older mantles presents variable exposure sequences of weathered and unweathered sediments for new soil development. Flat-topped hills and extensive pediments are the characteristic landforms of these landscapes, with plains of characteristically kaolinitic alluvium further downstream.

Section *c* illustrates the plains of riverine deposition, which in their most typical form are very extensive and are formed of coalesced alluvial fans. Less extensive alluvial deposits occur as valley plains upstream of the main riverine plains. The distal parts of the river systems consist characteristically of alluvial clays, and tend to show a build-up of soluble salts, both because they extend into low rainfall zones and because the rivers' flow is dissipated on the far flood plains.

Section *d* illustrates the complex pattern of riverine and littoral structures of beach, beach-ridge, and lagoonal components, which have been re-worked by wind and overlain by a pattern of linear and transverse dunes. The main beach-ridge structures remain, but the inferred riverine pattern in the surficial sediments is very largely obliterated and replaced at the surface by an aeolian one.

The Murray–Darling river system comprises two parts: The Darling subsystem, and the Murray subsystem. Each has the same sequence of landscapes: a catchment (*a* or *b* in Fig. 18.2), a riverine plain (*c*), and an areic (aeolian) plain (*d*), through which the rivers flow as trunk streams to their junction. The centres of the two riverine plains are 600 km apart. Their junction is in their areic section, a further 600 km from their debouchment into the ocean. Fig. 18.2 shows how the river system effects an integration of these diverse landscapes and environments. The junction of the Murray and Darling sectors of the system is effected in a zone of dunefields, and this appears to be the source of an aeolian clay mantle which extends eastward from it. The aeolian clay overlies most of the Murray riverine plain, but disappears on the southern fringe of the Darling riverine plain. It is described more fully below.

In many parts of the Region there are well-differentiated soil profiles on the main fluvial and aeolian sedimentary bodies. This proves that their deposition is to be associated with a *prior* regime and environment. Time is thus drawn in as an important consideration in the study of soils.

The provinces of the Region

Fig. 18.2 provides a general model for the description of the seven Provinces. These provinces are shown in Fig. 18.1, they are described now and their soils are described later.

Riverine Plain of south-eastern Australia Province

This plain (Butler 1950, 1958; Butler *et al.* 1973) comprises the coalescent alluvial fans of prior streams of the Murray, Murrumbidgee and Lachlan Rivers and their tributaries. Section *c* in Fig. 18.2 illustrates the situation where surficial sediments are flood-plain clays and the sandy loams of radial and anastomosing river ridges. A mantle of aeolian clay (parna) is widespread (Butler 1956; Butler and Hutton 1956). The parna is most conspicuous on foothills and hilly inliers, and it gives a unifying trend to the alluvia of the different catchments. The source of these aeolian deposits appears to be in parts of the Murravian Gulfland Province.

Darling Riverine Plain Province

This Province comprises the alluvial plains of the Macquarie, Namoi, MacIntyre, Balonne, Maranoa, Warrego and Paroo Rivers. Much of this Province, like the preceding one, is represented by Section *c* in Fig. 18.2, but it includes portions typified by both Sections *a* and *b*, a point of difference from the Murray riverine plain. Furthermore, mantling by parna is evident only along the Macquarie River and the Bogan River, and the unifying effect of the parna in the Murray riverine plain is lacking in this Province.

Murravian Gulfland Province

This Province comprises the area of the infilling materials of the Tertiary Murravian Gulf; the boundary of marine incursion is taken from Hills (1975). These structures are illustrated by Section *d* in Fig. 18.2. The two rivers which traverse the Province have little interaction with it, though the filling of the Gulf involved their earlier contribution of sediments; both these and the littoral sediments have been re-worked by wind, which activity is emphasized by the almost ubiquitous dunefields. Aeolian clay mantles are widespread in several parts of the Province but are not evident in the southwest where stranded coastal features are well preserved.

Balonne Mesas and Plains Province

This Province is part of the catchment of the northeastern tributaries of the Darling River. It is mainly a

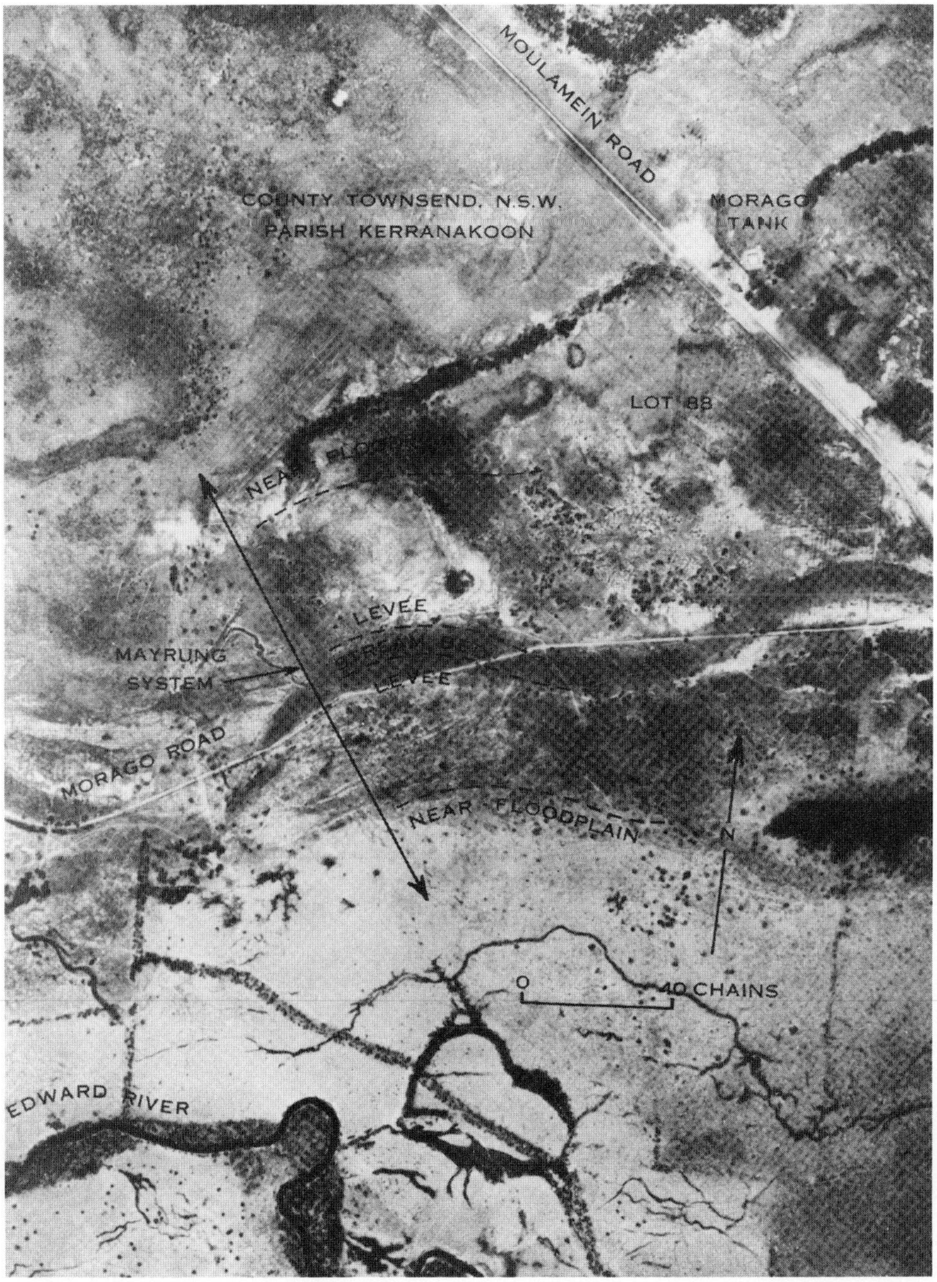

Fig. 18.3. Vertical air photo showing the pattern of soils due to a prior stream in the Murray basin: just north of Deniliquin, N.S.W.

denudational landscape with current erosion working into the Tertiary weathered mantle (Section *b* in Fig. 18.2).

Darling Tablelands and Plains Province

This Province comprises the greater part of the denudational section of the Darling catchment in the low rainfall zones. The Province is illustrated by Section *b* in Fig. 18.2, and it is characterized by plateaux or mesas and pediments cut into the Tertiary mantles (Idnurm and Senior 1978) with the occasional exposure of the unweathered rock below. Dunes are almost lacking, in fact the Province was mapped on this criterion. Some alluvial deposits occur along the main stream lines.

Wilcannia Mixed Dunefield Province

This Province shares the characteristics of adjacent provinces and contains the essential fluvial link between the Darling Riverine Plain and the Murravian Gulfland. Eastward of the Darling River there are many dunes, often arranged in parallel, closely spaced formations. West, near Broken Hill, plateaux and pediments of the denudational country, Section *b* of Fig. 18.2 are conspicuous, while dunes are rare.

Cobar Tableland Province

This is a low tableland which separates the Murray riverine plain from the Darling riverine plain. It also separates the two geological structures, the Great Artesian Basin and the Murray Basin. It comprises the denudational landscape, commonly with a Tertiary weathering mantle on it, but much of this is covered by a thin mantle of parna.

The soils of the provinces

The main soils in each province are introduced and briefly described as constituents of the *sequence* of both the soil profiles and the topography typifying the province. Where there is a second, subsidiary soil-landscape pattern or any major trends in soils across the province this is referred to also. The soils are classified in the classes of 'A Handbook of Australian Soils' (Stace *et al.* 1968) and these are *italicized* to distinguish them from use in the text of normal descriptive terms.

Riverine Plain of south-eastern Australia Province

The characteristic sequence of soils is associated with the levee-floodplain transect of prior streams (see Fig. 18.3). This is described in Butler (1958) and reproduced here as Fig. 18.4 from Butler and Hubble (1978). The soils on the levee are *red-brown earths*, sandy on the crest of the levee and loamy on the backslope. The soils in the far flood plain are *grey, brown* and *red clays*, often with gilgai formation and with soluble salts increasing downstream and with distance transverse from the levee.

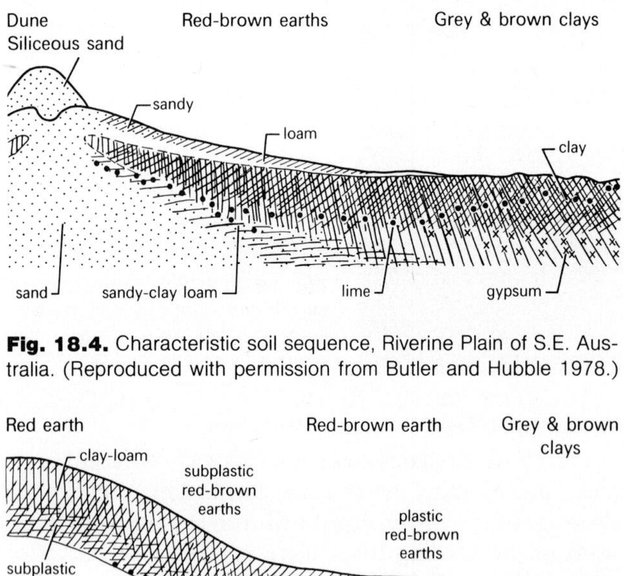

Fig. 18.4. Characteristic soil sequence, Riverine Plain of S.E. Australia. (Reproduced with permission from Butler and Hubble 1978.)

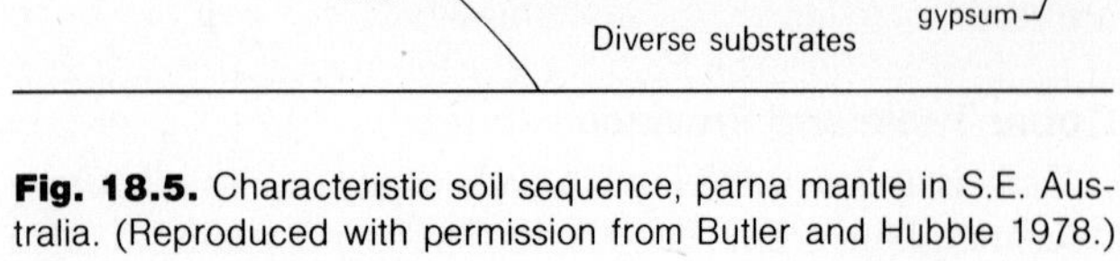

Fig. 18.5. Characteristic soil sequence, parna mantle in S.E. Australia. (Reproduced with permission from Butler and Hubble 1978.)

These clays occupy the greater part of the area of the plain.

A subsidiary sequence of soils is associated with the parna mantle of Butler and Hutton (1956). This is illustrated by Fig. 18.5 which is reproduced here from Butler and Hubble (1978). Its profile variation is associated with the varying drainage of different sites. Well-drained sites are found where the parna mantles foothills, inlier hills and river levees. Poorly drained sites are where the parna occurs on the flood plain. The soils of the well-drained sites are *red earths* and *red-brown earths* with significant subplastic characteristics (McIntyre 1976). Less-well-drained sites have normal plastic *red-brown earths* and on poorly drained sites the soils are *grey clays*, calcareous, self-mulching and gilgaied. The last-mentioned two types derived from parna are more calcareous than their counterpart soils on riverine sediments.

As the riverine pattern of soils in this province covers a range of isohyets, some soil features associated with rainfall can be distinguished. Soils in drier areas have higher contents of soluble salts and less depth to lime, but otherwise there is little change in soil morphology.

Specimen studies of the soils in this province are Skene and Harford (1964) for the Rochester district, Churchward (1958) near Deniliquin and van Dijk and Talsma (1964) near Griffith.

Darling Riverine Plain Province

The general pattern of soils in this Province compares with that of the previous province i.e. it is based on coalescent alluvial fans of prior-stream river ridges and clay flood-plains, and shows a general dominance of *grey, brown* and *red clays.* A parna mantle is lacking (except in the southernmost alluvial fan) and in consequence lithological differences between individual catchments are reflected in soil differences on the plains as described by Butler and Hubble (1978).

The tributary catchments may be divided into two sub-provinces. One includes the west-flowing (i.e. eastern) tributaries which rise in little-weathered rocks including basalt (as in Section *a* of Fig. 18.2). The other includes the south-flowing (i.e. northern and north-eastern) tributaries that drain the catchments of subdued relief cut in the Tertiary deep-weathering mantle of quartzose and labile sediments (as in Section *b* of Fig. 18.2).

Fig. 18.6 illustrates the character and relationship of soils on the west-flowing tributaries, generalized from the reconnaissance studies by Sleeman (1952), Downes and Sleeman (1955), Northcote (1966), McArthur* Stannard and Kelly (1968, 1977). In the southern-most tributaries of the west-flowing alluvial plains sandy and

*Information on the Boomi–Goondiwindi–Moree–Warialda Area, N.S.W., is available in CSIRO Aust. Div. Soils, Tech. Mem. No. 6, p59.

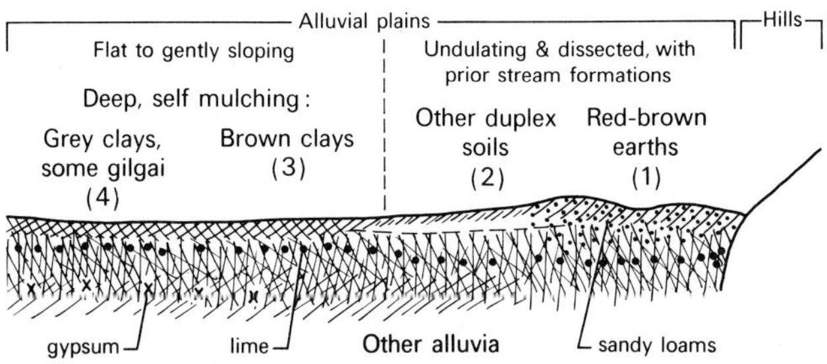

Fig. 18.6. Generalized soil relationships on the west-flowing tributaries of the Darling Riverine Plain Province.

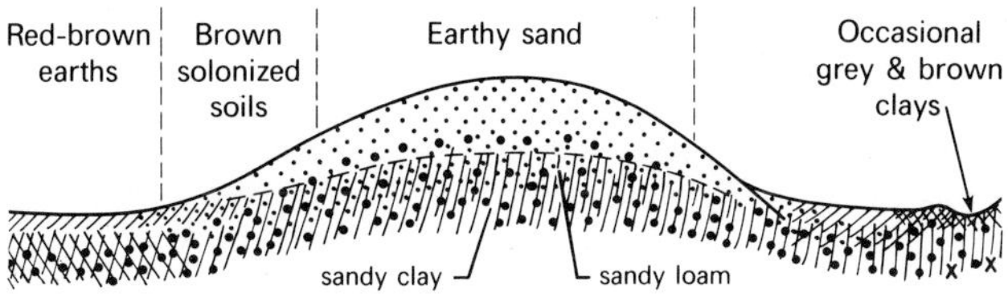

Fig. 18.7. Transect of east–west dunes showing characteristic soil sequence in Murravian Gulfland Province.

loamy *red-brown earths* ((1) in Fig. 18.6) occupy the better-drained sites on a broken belt of undulating older alluvium bordering the hills and on the levees of the prior streams extending westward onto the plains. Dense duplex soils form a transition to non-self-mulching clays on the main plain. Further north where the streams rise in catchments containing much basalt, the soil pattern on the plain is varied, with *black earths* on extensive level plains replacing (1) in Fig. 18.6. Self-mulching *grey clays* continue to occupy the plain, (4) in Fig. 18.6. In contrast to this, some occurrences of sandy *solodized solonetz* and deep *siliceous sands* further north are equivalent to (1) in Fig. 18.6, reflecting a dominance of quartzitic rocks in the catchment.

The soils of the south-flowing tributaries (Isbell 1957*a*, *b*; Northcote 1966; and Gunn 1974) differ from those of the west-flowing tributaries in both character and relative extent. They occur in more distinct valley plains, but though the dominant soils are *grey*, *brown* and *red clays* in both sub-provinces, those in the south-flowing sub-province tend to have a poorer structure, probably associated with their higher proportion of kaolinitic clays, their higher exchangeable magnesium and sodium, and their lower calcium contents; these properties are largely inherited from the Tertiary weathered materials of their catchment areas. Deserted stream structures are again evident in this sub-province, but sandy and loamy *solodic soils* and *solodized solonetz soils* are prominent, largely replacing *red-brown earths*. Deep *siliceous sands* occupy the levees and the associated dunes and sand sheets. Various texture-contrast, sandy and clayey *alluvial* and *prairie*-like soils are minor components with more restricted distribution, mainly in the valley.

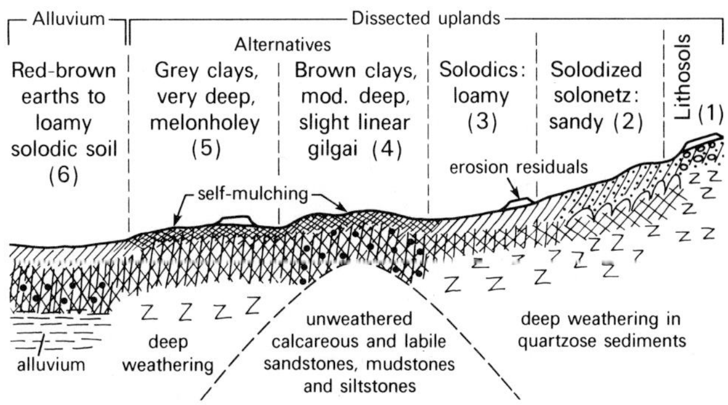

Fig. 18.8. Idealized relationships of important soils in the Balonne Mesas and Plains Province.

Murravian Gulfland Province

The sequence of soil profiles on a traverse across a linear sand-dune is characteristic for the Province, though soil variation is considerable. Fig. 18.7 illustrates this sequence. The sequence is generally associated with a series of parallel dune ridges such as are common in the central part of the Province. Reference may be made to Hubble and Crocker (1941), Churchward (1960) and Blackburn and Wright (1972) for particular soil studies in this Province.

The characteristic sequence of soils has *earthy sands* at the crest of the dune, *solonized brown soils* on their flanks with the upper slopes having sandier members, and *red-brown earths* on adjacent flats, which may also include areas of *grey, brown* and *red clays.* Variation in this sequence is expressed in two ways. In a given locality a greater depth of sand may occur at the east end of the ridge than at the west, and inclusions of clays are generally found on wider strips between ridges. The other kind of variation involves widely separated parts of the Province: abundant calcrete at shallow depth is a feature of the west central part of the Province, while elsewhere the subsoil of red sandy clay rarely contains this material.

Within about 100 km from the sea in the south-west of this Province, stranded coastal beach ridges are arranged sub-parallel to the coastline. The associated sequence of soils commonly has *podzols* developed in the residual quartz component of the coastal shelly sands in the dunes. *Terra rossa soils* are associated with the consolidated undifferentiated dune sands. Intervening flat areas have *solodized solonetz* or *rendzina* soils, the latter being restricted to areas of fine-textured estuarine or lacustrine sediments (Blackburn 1964).

A third important sequence is evident in inland situations where ancient coastal ridges aligned from north-west to south-east alternate with wide strips of clayey sediments bearing *grey, brown* and *red clays* (Blackburn *et al.* 1967). Here the high ground is covered by *podzols, solodized solonetz, red-brown earths* or the clays, apparently depending on the extent of terrestrial deposition on the ridges.

Balonne Mesas and Plains Province

The characteristic soils of this Province are cracking clays and a range of texture-contrast profiles dominated by *solodic soils*. Relationships between more extensive soils of the Province (Beckmann and Thompson 1960; Isbell 1957*b*, 1962; Isbell *et al.* 1967; Dawson 1972*a* and *b*; and Reeve *et al.* 1960, 1961) and landscapes are shown schematically in Fig. 18.8 which represents a generalised transect of the Province.

Major contrasts between soils can be attributed to differences in source rocks, in the survival of remnants of the Tertiary deep-weathering profile (or portion of it) and in more recent salinization and solonization.

The 'Tertiary deep-weathering profile' referred to here and above is frequently encountered throughout northern and western Australia: it is generally recognized by an upper horizon of sesquioxides and/or silica overlying a white and red reticulately mottled horizon of clay material, gradually passing in its turn to a white (pallid) horizon of highly weathered rock. The clay mineralogy of the profile is dominantly kaolinite with oxides. For a recent reference to these deep-weathering profiles in this part of Australia see Idnurm and Senior (1978).

Soils (1), (2) and (3) of Fig. 18.8 are associated with the quartzose lateritized rocks. The *lithosols* (1) are shallow and acid with much ironstone and lateritized sandstone gravel. Sandy *solodized solonetz* (2) and hard-setting loamy *solodic soils* (3) occupy the slopes and minor alluvial floors among the dissected residuals. Soils (4) and (5) occur as alternates to each other: the *grey clays* (5) of the brigalow (*Acacia harpophylla*) lands are typically strongly acid (pH 4–5), and they are strongly saline (0.5–1.0% total salts) below 30–80 cm. In contrast, the *brown clays* (4), which have formed in the freshly weathered materials of the same rocks, are alkaline, calcareous and low in salt. The texture-contrast soils (6) range from typical *red-brown earths* through hard pedal duplex soils (Db1.23–43) to *solodic soils* with thick bleached A_2 horizons.

Departures from this sequence include deep *siliceous sands* and sandy *red podzolics* and *yellow podzolics* among soils (1) and (2); low residuals with *lithosols* and modified lateritic *red earths* scattered through (5) along with minor *solodic soils* and small areas of texture-contrast soils and *black earths* associated with (4).

Darling Tablelands and Plains Province

Red earths are overwhelmingly dominant in the soils of this Province (James 1960; Northcote 1966; Isbell *et al.* 1967; Dawson and Ahern 1974; Dawson 1974 and Gunn 1974). The characteristic sequence of the soils is related to the topography, and is determined by exposure of parent materials after erosion into the Tertiary weathering mantle. Fig. 18.9 illustrates these relations.

The *lithosols* (1) and shallow stony *red earths* (2) on the stripped flat-topped residuals are acid sandy loams to clay loams with some silcrete gibbers on the surface (see Fig. 18.9). Where the hardened cappings have been removed, hard loamy *red earths* (3*a*) have formed in the underlying softer, deeply weathered materials. These are the soils of the 'hard mulga (*Acacia aneura*) country' (see Fig. 18.10), moderately deep, acid, and characteristically containing fine lateritic gravel. The softer loamy (3*b*) and sandy (3*c*) *red earths*, and *earthy sands* (4) constitute the 'soft mulga country', and have formed in redistributed and sorted detritus from the Tertiary soils. All are deep and essentially gravel-free; their soil reaction trends (Northcote 1971) range from acid to neutral to alkaline. The *earthy sands* and associated small areas of red *siliceous sands* occupy sand sheets and dunes respectively of material blown eastward from the floors of the main river valleys. Mixed alluvium from the deeply weathered and fresh country rock is the parent material of the deep, dominantly *grey clays* of the valley floors. These soils have been flooded frequently during their formation.

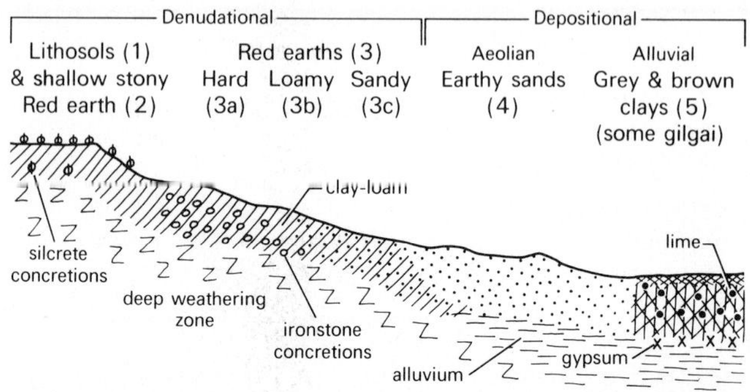

Fig. 18.9. Characteristic soil sequence for the Darling Tablelands Province.

Among soils of restricted area, the more important and widespread are various loamy, *solodic soils* with alkaline clay subsoils. These are found in detrital or deeply weathered material below the scarps of the residuals and in minor drainage depressions. Where landscapes have cut into little-weathered country rock (mainly in the west) shallow to deep *red* and *brown clays* with scattered lag silcrete gravels replace the hard loamy *red earths*.

Wilcannia Mixed Duneland Province

Information on this Province is scanty. On the extensive gently undulating plains separating the low rocky hills, the main elements of the soil sequence are *solonized brown soils* on convex surfaces, and *desert loams* on concave ones. *Grey, brown* and *red clays* occur on some river flats and dry lake floors. East–west sand ridges with *earthy sands* are common only near the Murravian Gulfland. Elsewhere the *earthy sands* occur sporadically on isolated low dunes without marked orientation. Toward the north of the Province are *red earths* on the Tertiary weathering mantle similar to those of the Darling Tablelands and Plains Province.

Fig. 18.10. Undulating 'hard mulga country' with silcrete (billy) exposed on low scarp of stripped tableland remnant: east of Thargomindah, Queensland.

This Province brings together the several different morphologic processes of its adjacent provinces (and therefore offers unique opportunity for studying their inter-relations).

Cobar Tablelands Province

Only general information is available on the soils of this Province (Northcote 1966; Downes and Sleeman 1955). Northcote describes the dominant soil as "red massive earths with neutral reaction trend". Downes and Sleeman, using a different classification system, call them "brown acid soils" and describe them as friable (sub-plastic) in the lower part of the profile: they also mention the occurrence of shallow buried soils. Recent examinations by one of the present authors (B.E.B.) in the western part of the Province revealed the occurrence of a shallow (about 50 cm) mantle of parna overlying a gently undulating Tertiary weathering zone and soil. This soil relationship seems to be widespread in the Province.

The topography is dominantly a low undulating tableland, but tending toward hilly terrain with residual soils towards the eastern boundary.

The characteristic sequence of soils is related to the topography and is similar to that shown for the parna mantle, Fig. 18.5, except that the soils are generally shallower and more gravelly in the upper part of the catena. The gravel is often silicified pebbles from the underlying Tertiary soil. The upper soil member in the characteristic sequence is a *red earth*, subplastic to a marked degree, followed down the catena by normal plastic *red-brown earths*, and *grey*, *brown* and *red clays* in that order proceeding to the more-poorly drained sites.

A summary of the soils and landscapes of the provinces is given in Table 18.1.

Regional land use problems in relation to soil features

The soil-related problems in this Region concern nutrient status, water availability for plant growth, saline deterioration under irrigation, and the variability of herbage under semi-arid conditions. These problems relate to soil parent material, landscape structures and soil texture.

Throughout the Darling Tablelands and Plains Province, and the Balonne Mesas and Plains Province, *red earths* and *solodic soils* formed on Tertiary weathering materials are very deficient in phosphorus and nitrogen, and some soils are deficient in copper and zinc for leguminous species. In these Provinces the *grey clays* of the brigalow lands (see Fig. 18.11), while low in phosphorus

Fig. 18.11. Grain crops on cleared brigalow land.

Table 18.1. Main features of the soil-landscape provinces of Region VII

Province	Geological and geomorphological features	Major soil features
Riverine Plain of south eastern Australia	The riverine plain of the Murray–Murrumbidgee–Lachlan Rivers, and few inlier and fringing hills. The plain is composed of 'prior stream' structures with a thin sheet of parna covering it	Main area is of *grey, brown* and *red clays* with soluble salts increasing toward N.W. *Red-brown earths* occur in the S. and E. and along prior stream levees. Subplastic *red earths* and *red-brown earths* occur on the hills
Darling Riverine Plain	Riverine plain of the Darling River and its tributaries. The northern tributaries rise in Tertiary weathering country: the eastern tributaries rise in recently weathered country including basalt. Parna cover is only in the S.E. catchment. Prior streams as above	Main area is *grey, brown* and *red clays* with soluble salts increasing toward S.W., *solodized solonetz* occur on the levees in the northern tributaries, and *red-brown earths* on those to the S.E. *Black earths* are associated with plains of basaltic alluvium
Murravian Gulfland Province	Comprising the area and material infilling the Tertiary Murravian gulf. Most of the area is a dunefield of reworked riverine and littoral deposits. Littoral structures dominate in the S.W. of the Province	In the area of dunefields *solonized brown soils* dominate with a varying proportion of *earthy sands* and minor *grey, brown* and *red clays*. The littoral deposits have *podzols, solodized solonetz* and *grey, brown* and *red clays*
Balonne Mesas and Plains Province	Undulating to rolling lands of the mid-course of the Balonne River and adjacent tributaries. Mainly a landscape of Tertiary materials, but partly cut into recently weathered rocks	In the north and east *solodized-solonetz* and loamy *solodics* are the major upland soils: elsewhere *grey, brown* and *red clays* are characteristic. The *grey clays* are 'brigalow clays', acid and saline; the *brown clays* are on recently weathered rocks and are alkaline and low in salt
Darling Tablelands and Plains Province	The portion of the Darling River catchment in the low rainfall zone. Most of this is in the Tertiary weathering mantle with little exposure of fresh mineral weathering	*Red earths* dominate the area, some being reworked as 'soft mulga soils' and red *siliceous sands*. In the broad valley floors there are *grey, brown* and *red clays*, alkaline and with gypsum: on the residuals *lithosols* and shallow *red earths*
Wilcannia Mixed Dunefield Province	The province joining the Darling valley and the Murray–Murrumbidgee valley. Partly dunefields and partly Tertiary residual landscape	*Solonized brown soils* are dominant in the dunefields with associated *earthy sands* and *desert loams*. Towards the north *red earths* are frequent
Cobar Tableland Province	The low tableland separating the Darling riverine plain and the riverine plain of the Murray–Murrumbidgee	*Red earths* are dominant overlying Tertiary weathering at shallow depth. Minor soils are *grey, brown* and *red clays* and *red-brown earths*

have high nitrogen contents, and by contrast the *brown clays* formed on freshly weathered country rock have moderate contents of nitrogen and available phosphorus.

Towards the centre of the Darling Riverine Plain Province the Tertiary landsurface dips below riverine and aeolian sediments which are relatively fertile, especially the aeolian clays. The littoral landscapes in the Murravian Gulf Province are noted for the low fertility of their soils, which are generally deficient in phosphorus, and also are known for their widespread deficiencies of copper, zinc and cobalt.

Throughout most of the Region and where irrigation is not practised, climatic conditions (chiefly limited rainfall) limit the potential of the land for dryland grazing. Local variation in grazing capacity is determined by the efficiency of the soils and the landscape to absorb, store and make the limited supply of water available to plants. Plains of fine sandy loam have these attributes to an optimum degree: clays and coarse sand soils are less desirable, as are hilly landscapes which shed much of their rainwater into the playas. Interactions between climatic factors and soil moisture parameters are significant, and are discussed by Butler and Hubble (1977) elsewhere. Earth-type and texture-contrast soils vary in their relative suitability according to their location in wetter or drier, hotter or cooler localities.

In semi-arid climates even under dry-land conditions soluble salts tend to accumulate in soils and to build up around the edge of run-on areas. A tendency to crust formation and scald at the soil surface is a common expression of this build-up which is one to which fine sandy loams are most prone. Contour furrowing is sometimes used to reclaim these scalded areas.

Irrigation schemes have been developed at a number of locations in the Murray–Darling Plains Region both on the riverine plains and on dunefield structures. Pastures, row crops and horticultural crops have been successfully grown; pastures are least fastidious and horticultural crops are most particular about their soil needs. Horticultural crops generally require medium to coarse textured soils. However, many of the irrigated soils are clays, and for satisfactory production an improvement of their structure is essential. In the

Murrumbidgee Irrigation Areas the clays of the horticultural areas are so subplastic that the general rule that clays are unsatisfactory for horticulture is overturned.

Several of the irrigation areas have been troubled by rising groundwater levels and by consequent increases in salinity. This has occurred on both the duneland and the riverine plains. Investigations have shown that reclamation can only be effective and satisfactory production can only be maintained if very precise water management is practised. Soil drainage is necessary to control groundwater and on the riverine plain it is often difficult to locate a drainage outfall without causing further drainage difficulties elsewhere.

If further development of the irrigation resources of the Region is required in the future, further exploration and investigation of the soil and landscape resources will be necessary. Inevitably the study and management of the system as a whole unit will have to be undertaken because the river serves both as a supply channel and as a drain.

References

Beckmann, G. G. and Thompson, C. H. (1960). Soils and land use in the Kurrawa area, Darling Downs, Queensland. CSIRO Aust. Soils Land Use Ser. No. 37.

Blackburn, G. (1964). The soils of Counties Macdonnell and Robe, South Australia. CSIRO Aust. Soils Land Use Ser. No. 45.

Blackburn, G. and Wright, D. A. (1972). Soil survey of the Loxton Irrigation Area, South Australia. CSIRO Aust. Soils Land Use Ser. No. 53.

Blackburn, G., Bond, R. D. amd Clarke, A. R. P. (1967). Soil development in relation to stranded beach ridges of County Lowan, Victoria. CSIRO Aust. Soil Publ. No. 24.

Butler, B. E. (1950). A theory of prior streams as a causal factor of soil occurrence in the Riverine Plain of south-eastern Australia. *Aust. J. Agric. Res.* **1,** 231-52.

Butler, B. E. (1956). Parna – an aeolian clay. *Aust. J. Sci.* **18,** 145-51.

Butler, B. E. (1958). Depositional systems of the Riverine Plain in relation to soils. CSIRO Aust. Soil Publ. No. 10.

Butler, B. E. and Hutton, J. T. (1956). Parna in the Riverine Plain of south-eastern Australia and the soils thereon. *Aust. J. Agric. Res.* **7,** 536-53.

Butler, B. E. and Hubble, G. D. (1977). Morphologic properties. *In* 'Soil Factors in Crop Production in a Semi-arid Environment'. (Eds J. S. Russell and E. L. Greacen) pp.9-32. (Queensl. Univ. Press: Brisbane.)

Butler, B. E. and Hubble, G. D. (1978). The general distribution and character of the soils in the Murray–Darling River System. *Proc. R. Soc. Vic.* **90,** 149–56.

Butler, B. E., Blackburn, G., Bowler, J. M., Lawrence, C. R., Newell, J. W. and Pels, S. (1973). 'A Geomorphic map of the Riverine Plain of South-Eastern Australia'. (Aust. Nat. Univ. Press: Canberra.)

Churchward, H. M. (1958). The soils and land use of the Denimein Irrigation District, N.S.W. CSIRO Aust. Soils Land Use Ser. No. 27.

Churchward, H. M. (1960). Soils of the Woorinen Settlement, Swan Hill Irrigation District, Victoria. CSIRO Aust. Div. Soils, Soil Land Use Ser. No. 36.

Dawson, N. M. (1972*a*). Land Inventory and Technical Guide, Jandowae Area, Queensland. Queensl. Dep. Prim. Indust. Div. Land Utilization Tech. Bull. No. 3.

Dawson, N. M. (1972*b*). Land Inventory and Technical Guide, Miles Area, Queensland. Queensl. Dep. Prim. Indust. Div. Land Utilization Tech. Bull. No. 5.

Dawson, N. M. (1974). Land Systems. *In* 'Western Arid Region Land Use Study – Part 1. Queensl. Dep. Prim. Indust., Div. Land Utilization Tech. Bull. No. 12.

Dawson, N. M. and Ahern, C. R. (1974). Soils. *In* 'Western Arid Region Land Use Study – Part 1. Queensl. Dept. Prim. Indust., Div. Land Utilization Tech. Bull. No. 12.

van Dijk, D. C. and Talsma, T. (1964). Soils of portion of the Coleambally Irrigation Area, N.S.W. CSIRO Aust. Soils Land Use Ser. No. 47.

Downes, R. G. and Sleeman, J. R. (1955). The soils of the Macquarie Region, New South Wales. CSIRO Aust. Soil Publ. No. 4.

Gunn, R. H. (1974). Soils of the Balonne–Maranoa Area. *In* 'Lands of the Balonne–Maranoa Area, Queensland'. CSIRO Aust. Land Res. Ser. No. 34.

Hills, E. S. (1975). 'Physiography of Victoria: an introduction to geomorphology'. Fifth Edition (Whitcombe and Tombs: Melbourne.)

Hubble, G. D. and Crocker, R. L. (1941). A soil survey of the Red Cliffs Irrigation District, Victoria. CSIRO Aust. Bull. No. 137.

Idnurm, M. and Senior, B. R. (1978). Palaeomagnetic ages of Late Cretaceous and Tertiary weathered profiles in the Eromanga Basin, Queensland. *Palaeogeogr. Palaeoclimatol. Paleoecol.* **24,** 263–77.

Isbell, R. F. (1957*a*). The soils of the Inglewood–Talwood–Tara–Glenmorgan Region, Queensland. Queensl. Bur. Investig. Tech. Bull. No. 5.

Isbell, R. F. (1957*b*). Soil survey of the proposed extension of the St George Irrigation Area, Queensland. Queensl. Bur. Investig. (unpublished report).

Isbell, R. F. (1962). Soils and vegetation of the brigalow lands, eastern Australia. CSIRO Aust. Soils Land Use Ser. No. 43.

Isbell, R. F., Thompson, C. H., Hubble, G. D., Beckmann, G. G. and Paton, T. R. (1967). 'Atlas of Australian Soils'. Sheet 4, Brisbane–Charleville–Rockhampton–Claremont area. With explanatory data. (CSIRO and Melbourne University Press: Melbourne.)

James, J. W. (1960). Erosion survey of the Paroo-Upper Darling Region. Part II. Land form and soils. *J. Soil Conserv. Serv. N.S.W.* **16,** 41–53.

Löffler, E. and Ruxton, B. P. (1969). 'Relief and landform map of Australia'. (CSIRO Div. Land Res.: Canberra.)

McIntyre, D. S. (1976). Subplasticity in Australian soils. I. Description, occurrence and some properties. *Aust. J. Soil Res.* **14,** 227-36.

Northcote, K. H. (1966). 'Atlas of Australian Soils'. Sydney–Bourke–Armidale area. With explanatory data. Sheet 3, (CSIRO and Melbourne University Press: Melbourne.)

Northcote, K. H. (1971). 'A Factual Key for the Recognition of Australian Soils'. 3rd edn. (Rellim: Glenside, S. Aust.)

Reeve, R., Isbell, R. F. and Hubble, G. D. (1961). Soil and climatic data for the brigalow lands, eastern Australia. CSIRO Aust. Div. Soils Div. Rep. No. 7/61.

Reeve, R., Thompson, C. H. and Beckmann, G. G. (1960). The laboratory examination of soils from the Toowoomba and Kurrawa areas, Darling Downs, Queensland. CSIRO Aust. Div. Soils Div. Rep. No. 1/60.

Skene, J. K. M. and Harford, L. B. (1964). Soils and land use in the Rochester and Echuca districts. Vic. Dep. Agric. Tech. Bull. No. 17.

Sleeman, J. R. (1952).The soils and irrigation potential of proposed irrigation area Mullengudgery N.S.W. CSIRO Aust. Div. Soils Div. Rep. No. 4.

Stace, H. C. T., Hubble, G. D., Brewer, R., Northcote, K. H., Sleeman, J. R., Mulcahy, M. J. and Hallsworth, E. G. (1968). 'A Handbook of Australian soils'. (Rellim Press: Glenside, S. Aust.)

Stannard, M. E. and Kelly, I. D. (1968). The irrigation potential of the lower Gwydir Valley. Water Conservation and Irrigation Commission, N.S.W.

Stannard, M. E. and Kelly, I. D. (1977). The irrigation potential of the lower Namoi Valley. Water Resources Commission, N.S.W.

Chapter 19
South-eastern Region and Tasmania (VIII)

P. H. Walker, K. D. Nicolls and F. R. Gibbons

THIS REGION INCLUDES SOUTH-EASTERN NEW SOUTH Wales, the Australian Capital Territory, eastern and southern Victoria and all of Tasmania (Fig. 19.1). In New South Wales and Victoria, it is characterized by the Dividing Range which is formed by the eastern highlands and alpine tract, and includes Mt Kosciusko, which at an elevation of 2230 m is the highest point in Australia. The topography of Tasmania is similarly dominated by western highlands and an extensive alpine tract. Coastal margins flank the highlands throughout the Region. In New South Wales and Victoria, coastal slopes and plains are up to 200 km wide. The Tasmanian highlands are circled by a narrower coastal belt varying from steeply sloping land in the south and west to relatively gently sloping land in the east and north. High population densities are associated with the coastal lands, particularly in the vicinity of Newcastle, Sydney, Wollongong, Melbourne, Geelong and Hobart. Apart from intensive urban use, these lands also have the longest history of forestry and agricultural use in Australia.

Most of the Region lies within Koeppen's climatic region Cfb, that is, a warm climate with uniformly distributed rainfall and temperate, long summers (Gentilli 1948). Mean annual rainfalls in the Region range from 350 to 3590 mm, the highest rainfalls being in Tasmania. The alpine tracts are designated Cfc, that is, similar to the above but with temperate short summers. Significant parts of the alpine tracts of the mainland and Tasmania are covered with snow during the winter months. Natural vegetation is dominated by eucalypt forests and woodlands with some relatively small areas of temperate and subtropical rainforest under high annual rainfall (> 1250 mm). Areas of sedgeland ('button grass plains') occur in the moist environments of south-western Tasmania. Alpine herbfields, grassland and heath are found at the highest altitudes in the Region.

The soils of the Region are very diverse. They range from *alluvial soils* of modern flood plains to podzolic and *solodic soils*, and soils developed on laterite and remnants of deeply weathered landscapes. In Tasmania, however, the proportion of alpine and highland landscapes is greater than on the mainland so that the soils tend to be more organic and less strongly differentiated. The diversity and intricate patterns of soils that result from complex landscape histories in the Region are illustrated by the occurrence, in Tasmania, of acid peats up to 50 cm thick and mildly leached soils with calcareous B horizons at virtually the same elevation and only 70 km apart.

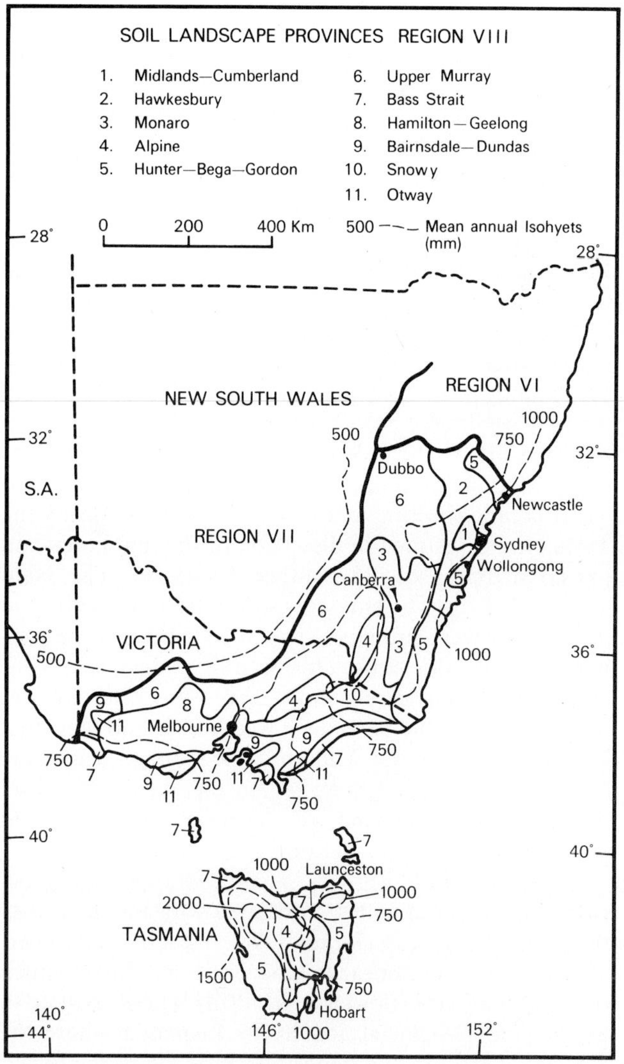

Fig. 19.1. Soil-landscape provinces of Region VIII.

Geology and landscape history

The oldest exposed rocks of the Region are the Precambrian schists of western Tasmania. Over most of the Region, Palaeozoic granitoids and metamorphosed sedimentary rocks of the Tasman Geosyncline are the predominant outcrop. Massive Late Palaeozoic and Mesozoic sedimentary rocks cover these older rocks in central New South Wales and have suffered only minor warping during the Cainozoic. Mesozoic sedimentary rocks are extensive in southern Victoria while in Tasmania mafic intrusive Mesozoic rocks occupy an important central area. Cainozoic volcanics and sedimentary rocks occur in large areas in Victoria and are also significant in Tasmania; their area in south-eastern New South Wales is limited.

Late Quaternary glaciation was extensive in Tasmania but affected only a small proportion of the mainland in the vicinity of Mt Kosciusko. But, as judged by extensive hillslope mantles and valley fills (Galloway 1965; Davies 1967), periglacial conditions appear to have been much more extensive than this both in Tasmania and in the highlands of the mainland. Alluvial terraces and modern flood plains occur along all rivers of the region and represent changing river regimes of the Quaternary. Along coastlines, extensive depositional plains have been built (Walker 1972) in response to the world-wide rises of sea level.

During the Cainozoic, rocks of the Palaeozoic and Mesozoic eras in the Region have been widely exposed to weatheEng, soil development and surficial transport processes. Weathering during the Early Tertiary took place under relatively warm and moist climates (Kemp 1978) and it is probable that some landforms and strongly weathered profiles of the Early Tertiary occur as relict features in the present highland landscapes (Pillans 1977). Significant uplift in eastern New South Wales and Victoria, and faulting in Tasmania, occurred during the Tertiary and resulted in dissection of the highlands and partial stripping of old weathered surfaces. The main outline and form of landscapes that are now evident in the Region were probably shaped by Mid-to Late Tertiary. During this time, erosion and pedogenesis continued under effectively moist and temperate climates which were conducive to deep weathering and leaching. Lateritic soils of the Midlands Graben of Tasmania (Nicolls 1960), the Sydney region (Walker 1960*a*) and the Dundas Tableland of Victoria (Jenkin 1976) probably relate to this period.

During the Late Tertiary, a considerable change occurred to cooler climates, associated with the growth of the Antarctic ice cap (Kemp 1978). Variable but generally cool climates appear to have prevailed during much of the Quaternary (Bowler *et al.* 1976). These resulted in glacial and periglacial activity in Tasmania where the associated erosion and deposition have led to the development of extensive, young geomorphic surfaces. On the mainland, the Quarternary was characterized by further erosional stripping of the uplifted highlands under the influence of marked climatic changes from temperate to cold and effectively moist conditions (Bowler *et al.* 1976). The episodic nature of these changes is indicated in sequences of hillslope mantles, alluvial fans and river terraces in the dissected landscapes of the Region (van Dijk 1959; Nicolls 1960; Walker 1962; Newell 1970; Lawrence 1976; Coventry and Walker 1977). On the mainland, erosion products from the highlands were differentiated and deposited in the extensive alluvial plains of the west-flowing rivers of the Murray–Darling system (Butler *et al.* 1973). Erosion products from the highlands were also deposited in coastal flood plains of the mainland and Tasmania where baselevels varied with rising and falling sea levels. These deposits have been extensively re-worked by wind into dunes and sand sheets along the coastal fringe. Quaternary glaciation had its greatest effect in the erosion of alpine landscapes (Costin 1954, 1971). Synchronous erosional instability of hillslopes, increased river discharges and high inland lake levels also occurred in the adjacent highlands during glacial phases (Coventry 1976). Soil parent materials of many hillslopes and valley bottoms relate to these phases of landscape instability.

Main soil determinants

The main factors in soil development in the Region are: the nature of the parent rocks or of parent materials derived from weathering and transport of weathering detritus; topography, including slope gradient and the relative position of soils on slopes; landscape stability, which includes the history of erosion and deposition; climate, including past climate and climatic changes; time, which in this Region may involve spans of up to 100 million years; and vegetation. Only in a very few cases can the differences between soils be said to reflect the variation of the one factor, e.g. climate rather than topography. Differences between soils usually involve several factors varying together. However, in many cases, the effects of variations in one factor appear to be dominant at a given location (Hallsworth and Gibbons 1951) and these instances provide the basis for the following paragraphs.

Major soil variations in the Region are associated with variations in parent material. Fine-grained, base-rich rocks such as basalt and limestone often weather to friable red clay soils such as *krasnozems, chocolate soils* and *terra rossa soils*. More siliceous rocks such as shales, acid volcanics and granites tend to weather to soils with moderately to strongly developed texture B horizons such as *red earths, red podzolic soils* and *red-brown earths*. Coastal aeolian sands and highly siliceous sandstones weather to soils with little texture differentiation such as *siliceous sands* and *podzols*.

Topographic effects due to slope gradient are evident in soils throughout the Region. On very steep slopes (> 40%), rock outcrop is common and thin soils such as *lithosols* occur. On gentler slopes, profiles are thicker and tend to have both A and B horizons. Gunn *et al.* (1969) noted in the southern highlands of New South Wales that the thickness of soil mantles was associated with slope convexity and concavity. Convex slopes (> 9%) on sedimentary rocks tended to have shallow *lithosols*, whereas thick colluvial mantles occurred on concave slopes up to 18%. Thick soil mantles also occurred on hillslopes up to 40% on granitic and volcanic rocks. Soil toposequences have been described in various parts of the Region. Loveday (1957) described toposequences on basalt near Hobart which ranged from *krasnozems* on upper slopes to hydromorphic soils on lower slopes. Walker (1960*a*) described sequences of soils on Triassic shale near Sydney which varied from *red podzolic soils* on upper slopes to *yellow podzolic soils* on lower slopes. In some landscapes, the toposequences are complicated by the superposition of soil layers which are not pedogenetically related but which originated from the erosional instability of the slopes. Van Dijk and Woodyer (1961) mapped extensive areas of hillslope landscape near Canberra where such complex surficial layering was believed to determine the soil properties in toposequences. Where such complexities arise, detailed stratigraphic studies are required to determine pedogenetic horizon relationships and the degree of association of soil horizons with underlying bedrock.

The present climate appears to relate to the soil patterns in the Region to a certain extent. For example, on the mainland, the 500 mm mean annual rainfall isohyet occurs approximately where large areas of leached soils, such as *red podzolic soils*, are delineated from soils with calcium carbonate or otherwise alkaline subsoil, such as *red-brown earths* and *solodic soils*. In Tasmania, on dolerite, the delineation of *black earths* and *non-calcic brown soils* from *grey-brown podzolic soils* occurs close to the 600 mm mean annual rainfall isohyet. The basis for relating soils to the present climate is often uncertain. Given the considerable age of many Australian soil landscapes, it is probable that soil distributions generally relate to climatic patterns of the past. By contrast, *alpine humus soils* seem to relate to the present climate. These soils occur in the alpine tract of the mainland and Tasmania under present-day high rainfalls and low temperatures. The weaker differentiation of these soils compared with podzolic soils on similar rocks at lower elevation has been attributed (Costin 1954) to the intense biotic activity and greater accumulation of organic matter that is found in cool, moist alpine environments.

The effects of Quaternary climatic changes on pedogenesis in the Region are not clearly understood. Great soil groups with strongly developed texture profile occur over a wide range of present environments (e.g. mean annual rainfalls 350 to 1000 mm). Examples are *red-brown earths* at the dry extreme and podzolic soils at the moist extreme of the range of environments. This suggests that the processes involved in particle-size differentiation are relatively insensitive to past climatic variations. Quaternary climatic changes have had the more important result of changing the vegetation in the highlands of the Region, thus changing landscape stability there. These effects have led to extensive erosional modification of soil landscapes (van Dijk 1959).

Differences in landscape age and associated soil variations have been noted frequently in the Region. Relict soil features, especially laterite which is widely attributed to pre-Quaternary weathering, are in marked contrast to the Quaternary soils (Hallsworth and Costin 1953; Jenkin 1976). Tertiary soils and weathering profiles must therefore represent the summed effects of many climates, some of which were probably very different from the present. Although the necessary association of lateritic materials with Tertiary weathering has been questioned at localities near Sydney (Hunt *et al.* 1977), the topographic association of some deep-weathering profiles displaced by uplift (Coventry 1967) and the palaeomagnetic dating of others (Pillans 1977) confirm that pre-Quaternary soil materials occur as relict features in modern landscapes.

Rowe *et al.* (1978) proposed a systematic relationship between soils, landscape forms and history (pedomorpholith) for the upper Murray Basin in southern New South Wales and Victoria. Although variations in soil drainage and particle size of parent materials were evident, the effect of age on soil development was very pronounced.

Flights of river terraces similarly provide surfaces of increasing age with increasing height above river level. Such sequences of Quaternary age have been described in Tasmania (Nicolls 1960), Victoria (Newell 1970) and New South Wales (Walker and Coventry 1976). In general, *alluvial soils* of modern flood plains have sedimentary features more or less modified by bioturbation. On successively higher surfaces above flood plains, soils have more strongly developed horizons. In many localities, high terraces above the level of extreme flooding have soils with strong texture profile development (e.g. *red podzolic soils, solodic soils, red-brown earths* and *lateritic podzolic soils*). Ward (1977) described a sequence of marine terraces in East Gippsland, Victoria, on which a comparable range of profile development occurred. In the Western District of Victoria, volcanic rocks of ages ranging from Holocene to Pliocene are weathered differentially according to age. The youngest volcanics are relatively fresh with shallow, stony *chocolate soils*; on the oldest volcanics, soils with relict ferruginous features and deep-weathering profiles are found (Gibbons and Gill 1964). The effect of time is also seen in the soils of coastal dunes. On the west coast of Tasmania (Nicolls 1955), the sequence ranges from *calcareous sands* of the youngest deposits to *podzols* and *humus podzols* on the oldest dunes.

Clear associations occur between vegetation and soils in parts of the Region and these can be related to factors affecting the ecosystem as a whole. For example, in Tasmania, the button grass and alpine herbfield communities are associated with *acid peats.* Both local drainage and orographic factors affect these ecosystems. Rainforests and wet sclerophyll forests are associated with *krasnozems* on basaltic rocks under high rainfalls throughout the Region. Here, climate and the base status of the parent rock are factors controlling the association of soil and vegetation. Steep mountainous terrain of the mainland and Tasmania is characterized by forest and deep soils. The effect of vegetative cover is to protect hillslopes from excessive erosion and to allow soils such as *brown earths* to develop. The high rainfall of these areas is an important factor in maintaining vegetative cover and the stability of the whole ecosystem.

As mentioned previously, vegetation played a key role in slope stability and valley sedimentation during the Quaternary when climatic changes caused changes in the vegetation. It appears (Bowler *et al.* 1976) that climatic changes to cold conditions did most to promote reduced vegetative cover, slope erosion and valley sedimentation. This was particularly so in the alpine tracts and adjacent highlands.

Soil-landscape provinces

Eleven soil-landscape provinces have been delineated in the Region (Fig. 19.1) and their main features are summarized in Table 19.1 The map and explanatory notes of Northcote *et al.* (1960–68) give the nature and distribution of soils in more detail. In most cases the soils in each province reflect the dominance of one or two of the soil-forming factors described in the previous section. Despite this simplification, the complex relationships between soils and landscapes throughout the Region have resulted in a wide range of soils being included in each province. In the following outline, many of these relationships are indicated.

Midlands–Cumberland Province

This Province, occupies the central part of the Tasmanian Midlands north of Hobart, and the Sydney Basin. In Tasmania the landscapes are of relatively low relief and constitute a corridor between higher landforms, and in the north, they extend into the Midlands Graben. Similarly, in the Sydney Basin, this soil landscape has been insulated from the full impact of Late Cainozoic erosion by the protection afforded by the surrounding resistant Mesozoic sandstone and by its low relief.

Landscapes of this Province consist of dissected relict landforms (Tertiary?) on bedrock and associated thick weathering profiles which may occur on the oldest river terrace remnants. *Lateritic podzolic soils* are associated with these relict surfaces on sedimentary rocks and Cainozoic sediments. However, thick and often ferruginous weathered zones occur below *krasnozems* together with massive laterite where these relict surfaces have developed on dolerite or basalt in Tasmania.

In the Sydney Basin, soil parent materials are developed mainly from Triassic shales, and the soils have strongly differentiated texture profiles and acid sola. The typical toposequence on the shale is from *red podzolic soils* of upper slopes to *yellow podzolic soils* of lower slopes and *gleyed podzolic soils* of bottom lands (Walker 1960*a*). *Yellow podzolic soils* occur in most slope situations on sandy Triassic shales, especially in Tasmania. Triassic sandstones in this Province carry *yellow podzolic soils* and some *humus podzols*; *yellow podzolic soils* and *gleyed podzolic soils* are also associated with Permian mudstones. In drier parts of the Province, *chocolate soils, non-calcic brown soils* and *black earths* occur on base-rich rocks such as dolerite and basalt; under higher rainfalls, *grey-brown podzolic soils* are found on dolerite. A range of podzolic soils is associated with high river terraces and *alluvial soils* occur on modern flood plains.

Hawkesbury Province

The Hawkesbury Province is associated with a large area of massive, siliceous, upper Palaeozoic and Mesozoic sandstones between Jervis Bay and Newcastle on the central coast of New South Wales. It is characterized by partially dissected and often rocky uplands, deep gorges and steep talus slopes (Fig. 19.2)

Lateritic podzolic soils occur in relict landscapes north and south of Sydney and profiles occasionally contain massive ferruginous crusts and thick weathering zones. For the most part, however, the soils tend to be sandy and have either uniform or gradational texture profiles. On dissected Triassic sandstone landscapes, the typical

Fig. 19.2. The Hawkesbury soil-landscape Province, near Fitzroy Falls in southern New South Wales, showing the characteristic upland, sandstone escarpment and talus slope topography.

Table 19.1. Main features of the soil-landscape provinces of Region VIII

Province	Geomorphological and geological features	Major soil features
Midlands–Cumberland	Landscapes of moderate to low relief, including relict surfaces. Rocks dominantly Mesozoic shales, sandstone and dolerite with Cainozoic fluvial and aeolian sediments and volcanics	*Lateritic podzolic soils* and thick weathered zones on relict surfaces; *red podzolic soils* in well drained sites on shales; *yellow podzolic soils* and *gleyed podzolic soils* in poorly drained sites; *grey-brown podzolic soils, non-calcic brown soils* and *black earths* in toposequences on dolerite; *podzols* and *humus podzols* on aeolian sands and sandstones; *alluvial soils* on modern flood plains
Hawkesbury	Partially dissected upland on Mesozoic sandstone caprock with bordering scarplands; some Mesozoic shales and Tertiary volcanics on upland	*Lateritic podzolic soils* and thick weathering zones on relict surfaces. Sandstone outcrop common; between outcrops, *siliceous sands, podzols, yellow earths* and *yellow podzolic soils. Red podzolic soils – yellow podzolic soils* toposequence on upland shales; *krasnozems* on basalt
Monaro	Partially dissected upland of moderate to low relief developed in dominantly Palaeozoic sedimentary and igneous rocks; Cainozoic volcanics important in the south; minor Cainozoic fluvial, lacustrine and aeolian sediments	*Lithosols* and shallow earths on upper slopes of steeper terrain; *red earths, red podzolic soils, yellow earths, yellow podzolic soils* and *solodic soils* in toposequences of rolling terrain on Palaeozoic rocks; thick weathered zones common on mid- and lower slopes. *Chocolate soils* and *chernozems* on basalt; *siliceous sands* and podzolic soils on aeolian sands; *alluvial soils* on modern flood plains
Alpine	Glacial and periglacial erosional and depositional landscapes of moderate to high relief on Palaeozoic and Mesozoic sedimentary and igneous rocks; minor Cainozoic volcanics	*Alpine humus soils* on well drained slopes on a wide range of parent materials; *acid peats* on poorly drained slope positions and in valley bottoms
Hunter–Bega–Gordon	Erosional landscapes of the coastal ranges on a wide range of bedrock; Precambrian metamorphics, Palaeozoic and Mesozoic sedimentary and igneous rocks dominant with Cainozoic volcanics and minor fluvial and aeolian sediments	*Lithosols* on steepest slopes; *red earths, red podzolic soils, grey-brown podzolic soils, yellow earths, yellow podzolic soils* and *solodic soils* in toposequences on a wide range of erosional slopes and bedrock. *Krasnozems* on basalt; *podzols* and *humus podzols* in toposequences on aeolian sands; *alluvial soils* in coastal flood plains
Upper Murray	Erosional landscapes of high to low relief in the upper Murray River catchment. Palaeozoic sedimentary and igneous rocks dominant; minor Cainozoic volcanics and fluvial and aeolian sediments	*Lithosols* on steepest slopes; *red earths, red podzolic soils, red-brown earths, yellow earths, yellow podzolic soils* and *solodic soils* on a wide range of erosional slopes and sediments. *Krasnozems* on Tertiary basalts; *siliceous sands* on aeolian sands and *alluvial soils* on modern river flood plains
Bass Strait	Coastal erosional and depositional landscapes mostly of low relief. Cainozoic aeolian and marine sediments with older sedimentary and igneous rocks	*Calcareous sands* and *siliceous sands* adjacent to coast; *podzols* on upper slopes of inland dunes; *humus podzols, acid peats* and *neutral to alkaline peats* in adjacent depressions. *Yellow podzolic soils* and *lithosols* on older sedimentary rocks and Cainozoic sediments; *krasnozems* on basalt
Hamilton–Geelong	Erosional landscapes of low relief in western Victoria derived from scoria and basalt of Holocene to Pliocene age	Stony, *chocolate soils* and *brown earths* on Holocene volcanics; *chocolate soils* and *prairie soils* in toposequences on Late Pleistocene volcanics; *yellow podzolic soils, solodic soils* and *prairie soils* in toposequences on Mid- to Late Pleistocene volcanics; *brown podzolic soils, soloths* and *prairie soils* with thick weathered zones in toposequences on Early Pleistocene to Late Pliocene volcanics
Bairnsdale–Dundas	Erosional and depositional landscapes of low to moderate relief developed mainly in sedimentary and acid igneous rocks of Palaeozoic and Cainozoic age; minor Cainozoic basalt and coastal dunes	*Solodic soils, red-brown earths, red podzolic soils* and *yellow podzolic soils* on upper Mesozoic and Cainozoic sedimentaries; *humus podzols* on coastal dunes; *krasnozems* on basalt
Snowy	Erosional landscapes of moderate to high relief developed mainly in Palaeozoic sedimentary and igneous rocks	*Lithosols, brown earths* and *red earths* on the steepest slopes; *brown podzolic soils* and *grey-brown podzolic soils* in toposequences on moderately sloping terrain
Otway	Erosional landscapes of moderate to high relief developed in upper Mesozoic calcareous feldspathic sedimentary rocks.	*Brown podzolic soils, yellow podzolic soils* and *prairie soils* under high, moderate and low rainfall respectively; *solodic soils* near the coast.

toposequence ranges from *yellow podzolic soils* on sloping divides to sandy *yellow earths* about rock outcrops and at breaks of slope (Walker 1960*a*). At the base of escarpments, coarse *siliceous sands* occur in bouldery talus, with small areas of sandy *alluvial soils* in narrow flood plains. *Red podzolic soil–yellow podzolic soil* toposequences occur intermittently on remnants of Triassic shale on the sandstone upland. To the south of Sydney, about Robertson, *krasnozems* have developed on Tertiary basalts which cap the sandstone upland (Walker 1960*b*). These soils provide a strong contrast because of their natural fertility compared with the adjacent infertile soils derived from siliceous sandstone.

Monaro Province

This Province occupies a large area of the southern highlands of New South Wales and the Australian Capital Territory, and consists of an old erosional upland of moderate to low relief developed in predominantly Palaeozoic sedimentary and igneous rocks. In a significant area near Cooma, Cainozoic basalts cap this ancient erosion surface. Prolonged exposure to weathering and surficial transport processes have resulted in complex soils. Thick weathering zones are common on hillslopes and paleosols are frequently found in depositional landscapes (van Dijk 1959). Generally, the soils have gradational or strongly differentiated texture profiles with slightly acid or neutral to alkaline sola.

Lithosols and shallow earths occur on steep slopes on Palaeozoic rocks. Large areas of moderate hilly to rolling terrain have toposequences ranging from *red earths* or *red podzolic soils* on upper slopes to *yellow earths, yellow podzolic soils* and *solodic soils* on lower slopes and valley bottoms (Gunn *et al.* 1969; Sleeman and Walker 1979). *Alluvial soils* occur on modern river flood plains. On Cainozoic basalts, in the southern part of the Province, toposequences consist of reddish *chocolate soils* on upper slopes and *chocolate soils* and *chernozems* on lower slopes (Costin 1954). Aeolian sands have been blown from river flood plains and lake shores during the Late Quaternary throughout the Province. *Siliceous sands* commonly occur on these deposits.

Alpine Province

Alpine lands occur in the highlands of southern New South Wales, Victoria and Tasmania. These landscapes

Fig. 19.3. Landscape in the Alpine Province. Lake St Clair (elevation 737 m), a deep piedmont lake mainly excavated by ice, on the Central Plateau of Tasmania. The surrounding mountains, capped by dolerite, show characteristic glacial features. In the foreground is part of the outwash plain, with recessional moraines. Photo: Department of Tourism, Tasmania.

are developed on a wide variety of bedrock including Palaeozoic and Mesozoic sedimentary and igneous rocks and Cainozoic basalt. In Tasmania, particularly, Quaternary glacial erosional and depositional landforms are prominent, with frequent lakes, small moraines and bare rock exposed as outcrop or talus (Fig. 19.3). As a result of pre-glacial weathering and of solifluction under late periglacial conditions, much of the remainder of the Province is mantled by deposits in which boulders and rock fragments occur in an earthy matrix.

Generally, soils of the Province have surface horizons of high organic content, profiles of weak to moderate texture differentiation and acid to neutral sola on a wide range of parent materials (Costin 1954). The most common soils are *alpine humus soils*, often with thin convoluted iron pans in relatively well drained hillslope positions and *acid peats* in poorly drained slope positions and valley bottoms.

Hunter–Bega–Gordon Province

Included in this Province are erosional landscapes of coastal ranges in southern New South Wales and Tasmania, where the rainfall is relatively high. Terrains vary from steep and mountainous to rolling (Fig. 19.4). A wide range of bedrock is associated with this Province. Palaeozoic and Mesozoic sedimentary and igneous rocks occur generally, and in Tasmania, together with Precambrian metamorphics. Minor areas of Cainozoic volcanics, and fluvial and aeolian sediments are also found. Along coastal margins, more or less extensive areas of flood plain have developed in response to the Holocene rise of the sea.

Generally, the soils have moderately to strongly developed texture profiles and acid to neutral sola (Walker 1960*a, b*). On siliceous rocks, *lithosols* are associated with the steepest terrain; elsewhere *red earths* and *red podzolic soils* occur on better drained slopes and *yellow earths* and *solodic soils* are associated with poorer drainage. *Grey-brown podzolics* and *yellow podzolic soils* are also common. *Krasnozems* occur on basalt and on dolerite under high rainfall (e.g. north-western Tasmania and the south coast of New South Wales). *Calcareous sands, siliceous sands* and *podzols* occur on Cainozoic sand deposits along coastal margins. *Alluvial soils* of coastal flood plains and *krasnozems* on basalt, although of relatively small area, are of considerable agricultural importance in the Province.

Fig. 19.4. Soil landscapes of the Hunter–Bega–Gordon Province sloping away belc w escarpments of the Hawkesbury Province near Nowra, southern New South Wales.

Upper Murray Province

This Province incorporates drainage basins of streams in southern New South Wales and Victoria which are tributaries of the Murray River system. It includes elements of the Central Tablelands of New South Wales and erosional landscapes of moderate relief in the main Dividing Range, and extends to landscapes of low relief where rivers enter the extensive alluvial plains of the Murray Basin. Relict depositional and weathering features of these landscapes (e.g. the lateritic soils of the Dundas Tableland, Victoria) reflect major episodes of uplift and climatic change that affected large areas of south-eastern Australia during the Cainozoic. Most landscapes of the Province are developed in Palaeozoic sedimentary and igneous rocks. Minor areas of Cainozoic basalt occur in the Central Tablelands of New South Wales, and Cainozoic fluvial sediments are extensive as streams enter the Riverine Plain of south-eastern Australia.

For the most part, the soils of the Province have gradational to strongly differentiated texture profiles and acid to neutral sola. *Lithosols* occur on the steepest terrain and *red earths, red podzolic soils, red-brown earths, yellow podzolic soils* and *solodic soils* on terrain of moderate to low relief developed in siliceous Palaeozoic rocks (Newell 1970; Rowe 1972; Crouch and Junor 1976). *Krasnozems* occur on Cainozoic basalts and *red-brown earths* and *red podzolic soils* are found on older Cainozoic sediments. Small areas of *alluvial soils* occur on modern river flood plains.

Bass Strait Province

Landscapes in Tasmania and Victoria which immediately border Bass Strait are included in this Province together with several islands in the Strait. Coastal aeolian dune landscapes are prominent. These consist of dune and sand sheet sequences and include interdune swamps. Areas of older bedrock are also found and these include sedimentary amd metamorphic rocks and granite.

Soils of the dune systems range from *calcareous sands* at the coastline to *siliceous sands* and *podzols* inland; *humus podzols, acid peats* and *neutral and alkaline peats* occur in the interdune corridors (Dimmock 1957). On the Tasmanian coast and Bass Strait islands, areas of *yellow podzolic soils* occur on siliceous bedrock and *krasnozems* on basalt.

Hamilton–Geelong Province

The volcanic plains of western Victoria which comprise this Province are the third largest system of this kind in the world. Basalt and scoria of six different ages ranging from Late Pliocene to Holocene have been identified (Gibbons and Gill 1964), each with a characteristic pattern of soils.

The youngest volcanic landscape consists of basalt and scoria of Holocene age and has a rough topography (Fig. 19.5). Weathering is minimal and the soils are correspondingly shallow, stony *chocolate soils* with *brown earths* occurring on cinder cones. Late Pleistocene volcanic features have produced a subdued landscape with stony rises surrounding cinder cones. Soils on the rises are stony *chocolate soils* with *prairie soils* in the adjacent swales. Basalt flows of Late to Mid-Pleistocene age also form a landscape of low relief on which *yellow podzolic soils* and *solodic soils* have developed; the latter are characterised by A horizons rich in ferruginous buckshot gravel. The two oldest basalt flows of this Province are of Early Pleistocene to Late Pliocene age. Relict kaolinized weathering zones are associated with the Late Pliocene flow. *Brown podzolic soils* occur on the Early Pleistocene basalt and *soloths* on the Late Pliocene basalt. *Prairie soils* occur along drainage lines developed in flows of Pleistocene and Pliocene age.

Bairnsdale–Dundas province

This Province occurs in southern and western Victoria and comprises erosional and depositional landscapes of moderate to low relief developed mainly on Palaeozoic and Cainozoic sedimentary and acid igneous rocks. Some of the Early and Mid-Pleistocene marine strandlines of eastern Gippsland described by Ward (1977) are included. Minor areas of Cainozoic basalt and coastal dunes also occur in the Province.

Fig. 19.5. Soil landscapes of the Hamilton–Geelong Province showing the rough topography associated with Holocene volcanic rocks in western Victoria.

Soils of the Province tend to have strongly differentiated texture profiles and acid to neutral sola, but some are alkaline. Dominant soils on the sedimentary rocks are *red podzolic soils, red-brown earths, yellow podzolic soils, solodic soils* and *soloths. Calcareous sands, siliceous sands, podzols* and *humus podzols* occur on coastal aeolian dune landscapes and *krasnozems* on basalt.

Snowy province

This province occurs in southern New South Wales and eastern Victoria and comprises the upper watersheds of river systems draining eastwards from the alpine tract such as the Snowy and Mitchell Rivers. It includes some sub-alpine and high, mountainous terrain mainly of Palaeozoic sedimentary and igneous rocks. Where slopes are steep and forested, soils tend to be friable and to have gradational texture profiles with acid to neutral sola (Northcote 1963). The dominant soils are *lithosols, brown earths* and *red earths* on the steepest slopes; *brown podzolic soils* and *grey-brown podzolic soils* occur on moderately sloping landscapes with *gleyed podzolic soils* and *acid peats* in poorly drained situations (Costin 1957).

Otway Province

In southern and south-western Victoria, distinctive soil landscapes are associated with mountainous to steep hilly terrain. Soils of the Province are developed on weathered calcareous feldspathic sediments of Late Mesozoic age in the Strzelecki and Otway Ranges and in the Casterton Hills. In the Strzelecki and Otway Ranges, *yellow podzolic soils* and *brown podzolic soils* are dominant in areas of moderate to high rainfall respectively. *Solodic soils* occur near the coast. In the Casterton Hills, where rainfall is lower, *prairie soils* predominate.

Land use

Land use varies widely in the Region from grazing on the dry rangelands of western Victoria and cash cropping in humid coastal areas to the industrial and urban use associated with major cities. The present broad patterns of land use outside the cities were established in the early days of settlement of the Region, over 100 years ago, and are controlled to a large extent by climate and topography with inherent soil fertility being subordinate. Much of the land of high relief in the Region, under humid climates, is used for forestry or is reserved as water supply catchment. Hillslope terrain of moderate to low relief under humid climates tends to be used as grazing land, with improved pasture, or as grain cropping land. On most of these soil landscapes, pastures and crops respond strongly to added nitrogenous and phosphatic fertilizers. The most productive agricultural lands are those used for vegetable and fruit growing, such as coastal flood plains and basaltic landscapes under high rainfalls. Both kinds of landscape have relatively high levels of available soil moisture and initial natural soil fertility; both have the added advantage of supplementary irrigation from rivers and springs.

This Region includes several large cities and associated areas with the highest population densities in Australia. In the vicinity of these cities, land use is devoted to construction, urban development, waste disposal, recreation reserves and city water supplies. In general, the requirements of these cities have been met despite the diversity of surrounding soils and landscapes. For example, much of the Sydney and Hobart city areas have been built on *red podzolic soils* and *yellow podzolic soils* (Midlands–Cumberland Province); large areas of *solodic soils* occur in the vicinity of Melbourne on Cainozoic sedimentary and volcanic rocks (Bairnsdale–Dundas and Hamilton–Geelong Provinces). Thick subsoils of low permeability characterize these soils and provide difficult conditions for housing foundations. The most productive agricultural lands near these cities are the areas of flood plain associated with the local rivers. Here, *alluvial soils* are intensively used for vegetable and fruit growing, with supplementary irrigation. It is important that encroachment of urban development on high quality land of this kind should be restricted as far as possible.

Near Sydney, soil landscapes of the Hawkesbury Province are used extensively for water catchment, parkland reserves and forestry. The shallow and highly permeable *siliceous sands,* sandy *yellow earths* and *yellow podzolic soils* of this Province are seldom used for agriculture. The underlying Triassic sandstone provides a firm base for city building, but is less desirable for suburban development because excavation costs are high. Precipitous slopes characterize large areas of the Hawkesbury Province and limit intensive use.

Extending between Sydney and Melbourne are highlands of the main divide which consist broadly of tablelands (Monaro Province) and alpine tract (Alpine Province). The earths, *chocolate soils*, podzolics and *solodic soils* of the Monaro Province support improved pasture and are used extensively for sheep and cattle grazing. Prolonged grazing of these soils has resulted in extensive sheet and gully erosion. On lower slopes and valley floors, the alkaline subsoils of *solodic soils* tend to be dispersed readily by subsurface water and gully heads often display severe tunnel erosion. In high rainfall areas of the Monaro Province, *krasnozems* on basalt are used for vegetable growing. Landscapes of the Alpine Province on the mainland and in Tasmania are intermittently covered by snow during most winters and most of these areas have been designated as reserves and water supply catchments. Selected areas of snowfield have been set aside for recreational purposes. To the east of the mainland alpine tract are steeply sloping landscapes of the Snowy Province (Fig. 19.1). The friable *brown earths* of this Province support forests under high rainfall conditions. Soil landscapes of the Alpine and Snowy Provinces are particularly susceptible to erosion following grazing and bush fires. Friable soils of other steeply sloping lands in the Strzelecki and Otway Ranges of southern Victoria (Otway Province) also occur under high rainfall but in a more temperate climate. These soils were formerly used for dairying and potato growing but are now being managed as forestry and water supply areas.

Some of the more critical areas for land use occur in the Hunter–Bega–Gordon Province. This Province includes some very intensively used agricultural soils under high rainfall, and also much highly productive forest both natural and planted. *Alluvial soils* of coastal flood plains are used for vegetable and orchard crops and for dairying; *krasnozems* and *prairie soils* on volcanic rocks are used for vegetable growing and as pasture land; earths and podzolic soils support economically useful eucalypt forests. However, these soils and landscapes occupy much of the coastal fringe which has optimal value as adjunct urban and recreational land, servicing cities such as Newcastle, Sydney, Wollongong, Burnie and Devonport. Future decisions about the use of these soils will need to take into account the value of particular soils for urban as against agricultural and forestry use. To some extent, similar considerations also apply to soils of the Upper Murray Province which are important grain crop and pasture soils, and to the foothill lands of the Dividing Range in Victoria (Bairnsdale–Dundas Province). In these two Provinces, extensive soil erosion testifies to the past overuse of many soils. In particular, gully erosion associated with the dispersible, alkaline subsoils of *solodic soils* and some *red-brown earths* poses significant management problems for agricultural land and has environmental implications in relation to regional water quality.

Soils and landscapes of the Bass Strait Province, comprising wind-blown sands and some residual terrain, support grazing and recreational activities. The *calcareous sands, siliceous sands* and *podzols* of this Province, however, are prone to wind erosion and require careful management of their vegetative covers.

The soils of the basalt lands of western Victoria (Hamilton–Geelong Province) are characterised by plastic clay subsoils and thus are in marked contrast to friable *krasnozems* on basalt in other parts of the Region. Soils of this Province are mainly used for grazing with some vegetable growing and dairying, especially on the *chocolate soils.*

References

Bowler, J. M., Hope, G. S., Jennings, J. N., Singh, G. and Walker, D. (1976). Late Quaternary climates of Australia and New Guinea. *Quat. Res. (N.Y.)* **6**, 359-94.

Butler, B. E., Blackburn, G., Bowler, J. M., Lawrence, C. R., Newell, J. W. and Pels, S. (1973). 'A Geomorphic Map of the Riverine Plain of south-eastern Australia'. (Aust. Nat. Univ. Press: Canberra.)

Costin, A. B. (1954). 'A Study of the Ecosystems of the Monaro Region of New South Wales with Special Reference to Soil Erosion'. (N.S.W. Govt. Printer: Sydney.)

Costin, A. B. (1957). 'High Mountain Catchments in Victoria in Relation to Land Use'. (Soil Cons. Authority Vic.: Melbourne.)

Costin, A. B. (1971). Vegetation, soils and climate in late Quaternary south-eastern Australia. *In* 'Aboriginal Man and Environment in Australia'. (Eds. D. J. Mulvaney and J. Golson) pp. 26-37. (Aust. Nat. Univ. Press: Canberra.)

Coventry, R. J. (1967). The geology of Shingle House Creek Valley with particular reference to its Cenozoic history. B.Sc. Hons. Thesis, Dep. Geol., Aust. Nat. Univ. Canberra.

Coventry, R. J. (1976). Abandoned shorelines and the late Quaternary history of Lake George, New South Wales. *J. Geol. Soc. Aust.* **23**, 249-73.

Coventry, R. J. and Walker, P. H. (1977). Geomorphological significance of late Quaternary deposits of the Lake George area, N.S.W. *Aust. Geogr.* **13**, 369-76.

Crouch, R. J. and Junor, R. S. (1976). Soils and geology of the Albury district, N.S.W. Soil Cons. Serv. N.S.W. Wagga Research Centre Tech. Bull. No. 14.

Davies, J. L. (1967). Tasmanian landforms and Quaternary climates. *In* 'Landform Studies from Australia and New Guinea'. (Eds J. N. Jennings and J. A. Mabbutt) pp. 1–25. (Aust. Nat. Univ. Press: Canberra.)

van Dijk, D. C. (1959). Soil features in relation to erosional history in the vicinity of Canberra. CSIRO Aust. Div. Soils, Soil Publ. 13.

van Dijk, D. C. and Woodyer, K. D. (1961). The soils of the Yass River Valley. Regional Research and Extension Study, Southern Tablelands, N.S.W., Joint Planning Committee, Rep. No. 6.

Dimmock, G. M. (1957). The soils of Flinders Island, Tasmania. CSIRO Aust., Div. Soils, Soils Land Use Ser. No. 23.

Galloway, R. W. (1965). Late Quaternary climates in Australia. *J. Geol.* **73**, 603–18.

Gentilli, J. (1948). Two climatic systems applied to Australia. *Aust. J. Sci.* **11**, 13–16.

Gibbons, F. R. and Gill, E. D. (1964). Terrains and soils of the basaltic plains of far western Victoria. *Proc. R. Soc. Vic. New Ser.* **77**, 387-95.

Gunn, R. H., Story, R., Galloway, R. W., Duffy, P. J. B., Yapp, G. A. and McAlpine, J. R. (1969). Lands of the Queanbeyan-Shoalhaven Area, A.C.T. and N.S.W. CSIRO Aust. Div. Land Use Res. Ser. No. 24.

Hallsworth, E. G. and Costin, A. B. (1953). Studies in pedogenesis in New South Wales IV. The ironstone soils. *J. Soil Sci.* **4**, 24-47.

Hallsworth, E. G. and Gibbons, F. R. (1951). Factors controlling soil formation in New South Wales. *Aust. J. Sci.* **14**, 9–13.

Hunt, P. A., Mitchell, P. B. and Paton, T. R. (1977). 'Laterite profiles' and 'lateritic ironstones' on the Hawkesbury sandstone, Australia. *Geoderma* **19**, 105-21.

Jenkin, J. J. (1976). Geomorphology. *In* 'Geology of Victoria'. Geol. Soc. Aust. Spec. Publ. 5. (Eds J. G. Douglas and J. A. Ferguson) pp.329–42. (Graphic Services: Adelaide.)

Kemp, E. M. (1978). Tertiary climatic evolution and vegetation history in the southeast Indian Ocean region. *Palaeogeogr. Palaeoclimatol. Palaeoecol.* **24**, 169-208.

Lawrence, C. R. (1976). Murray Basin. *In* 'Geology of Victoria'. Geol. Soc. Aust. Spec. Publ. 5. (Eds J. G. Douglas and J. A. Ferguson) pp. 276-288. (Graphic Services: Adelaide).

Loveday, J. (1957). The soils of the Sorell-Carlton-Copping area, south-east Tasmania, with special reference to the soils formed on basalt. CSIRO Aust. Div. Soils, Soil Publ. No. 8.

Newell, J. W. (1970). Soils and land use in the Ovens and Buffalo River valleys, Victoria. Vic. Dep. Agric. Tech. Bull. No. 21.

Nicolls, K. D. (1955). The soils, geomorphology and climate of an area between the Lagoon and Arthur Rivers, west coast of Tasmania. CSIRO Aust. Div. Soils Div. Rep. No. 7/55.

Nicolls, K. D. (1960). Erosion surfaces, river terraces, and river capture in the Launceston Tertiary Basin. *Pap. Proc. R. Soc. Tas.* **94**, 1-12.

Northcote, K. H. (1963). 'Soils of Victoria'. Extract from Victorian Year Book No. 78.

Northcote, K. H., with Beckmann, G. G., Bettenay, E., Churchward, H. M., van Dijk, D. C., Dimmock, G. M., Hubble, G. D., Isbell, R. F., McArthur, W. M., Murtha, G. G., Nicolls, K. D., Paton, T. R., Thompson, C. H., Webb, A. A. and Wright, M. J. (1960-68). 'Atlas of Australian Soils'. Sheets 1-10 with explanatory booklets. (CSIRO and Melbourne University Press: Melbourne.)

Pillans, B. J., (1977). An Early Tertiary age for deep weathering at Bredbo, southern N.S.W. *Search* **8**, 81–3.

Rowe, R. K. (1972). A study of the land in the catchment of the Kiewa River. Soil Cons. Authority Vic. Tech. Comm. No. 8.

Rowe, R. K., Crouch, R. J. and van Dijk, D. C. (1978). Soils in the upper valleys of the Murray River basin. Symposium on the Murray-Darling River System. *Proc. R. Soc. Vic.* **90**, 157-66.

Sleeman, J. R. and Walker, P. H. (1979). The soils of the Canberra District. CSIRO Aust. Div. Soils, Soils Land Use Ser. No. 59.

Walker, P. H. (1960*a*). A soil survey of the County of Cumberland, Sydney Region, New South Wales. N.S.W. Dep. Agric. Soil Surv. Unit. Bull. No. 2.

Walker, P. H. (1960*b*). A reconnaissance of soils and land use of part of the south coast of N.S.W. CSIRO Aust Div. of Soils, Soils Land Use Ser. No. 38.

Walker, P. H. (1962). Terrace chronology and soil formation on the south coast of N.S.W. *J. Soil Sci.* **13**, 178-86.

Walker, P. H. (1972). Seasonal and stratigraphic controls in coastal flood plain soils. *Aust. J. Soil Res.* **10**, 127-42.

Walker, P. H. and Coventry, R. J. (1976). Soil profile development in some alluvial deposits of eastern New South Wales. *Aust. J. Soil Res.* **14**, 305–17.

Ward, W. T. (1977). Geomorphology and soils of the Stratford-Bairnsdale area, East Gippsland, Victoria. CSIRO Aust. Div. Soils, Soils Land Use Ser. No. 57.

Classification and mapping of Australian soils

Chapter 20
Classification of Australian soils

A. W. Moore, R. F. Isbell and K. H. Northcote

THIS CHAPTER HAS TWO MAIN AIMS. FIRST IT IS INTENDED to trace briefly the history of two Australian and two overseas soil classifications, describe them, and consider their future. Secondly, a comparison of the current Australian classifications with the overseas classifications is attempted. Some discussion of the processes involved in soil systematics, and of the nature of classifications, is given as background information.

The classification aspects of any science should reflect its current state of knowledge, although there will always be some lag between the appearance of information in that science and its application to classification problems. Because knowledge is continually increasing, and because classification in turn may indicate new directions for scientific inquiry, the task of developing the ultimate classification is never complete. Classifications should be continuously changing and evolving, and this is as true in soil science as in any other scientific field. It follows, therefore, that all soil classifications are inadequate and imperfect, no matter what their purpose. It is also axiomatic that there are no right or wrong classifications, no true or false classifications, and no natural or artificial classifications, except that they are all artificial in the sense that they are man-made; there are simply classifications which have varying degrees of usefulness for any particular purpose. A classification is a product of the human mind, devised for one or more purposes that may or may not be explicitly enunciated.

Definitions

Several terms commonly used in soil classification are defined here to avoid confusion in the subsequent discussion. Where definitions are quoted, they are from Grove (1961).

Three general terms frequently encountered are classification, taxonomy and systematics.

To classify is defined as "to group or segregate in classes that have systematic relations usually founded on common properties or characters . . .". *Classification* may have one of two meanings: "the act or a method of classifying . . . assigning to a proper class . . ." or "the result of classifying: a system of classes or groups . . .". *Identification*, the referral of an unknown entity to an appropriate existing class, is synonymous with the second part of the first meaning above, and is used in this paper in this sense in preference to classification. Confusion between these two meanings of classification also leads to confusion in nomenclature. It is an almost universal practice in soil science to transfer the name of a class (an abstraction) to an individual soil (which is real). For example, we may look at a profile in the field and say "That is a podzol", i.e. we are identifying it and giving it as a proper name the name that is strictly the name of an abstract group concept. In the absence of common or other proper names, this practice is probably unavoidable; nevertheless, users should be aware of what they are doing.

Taxonomy also has two meanings: "the study of the general principles of scientific classification . . ." and "the systematic distinguishing, ordering, and naming of type groups within a subject field . . .". The second meaning is narrower than the first and is synonymous with classification as defined above (but excluding identification). This is the sense in which the United States Department of Agriculture (USDA) has chosen to use the term taxonomy; in fact, the title of its handbook on soil classification (Soil Survey Staff 1975) is 'Soil Taxonomy – a Basic System of Soil Classification . . .'. To simplify matters, taxonomy and classification are used synonymously in this paper, with the second meaning given above. On the other hand, its first meaning is synonymous with *systematics*, which may be defined as "the science of classification: classificatory method . . . a system of classification; usually the classification and study of organisms with regard to their natural relationships. . .". The idea of natural relationships is certainly invoked as an integral part of taxonomy in the biological sciences, perhaps not unreasonably since all living beings possess phylogenies, which it is hoped are reflected in taxonomies. Soils do not have phylogenies; nevertheless, the idea of relationships is carried over into the term taxonomy as used by the USDA Soil Survey Staff (1975). In this paper the terms taxonomy and systematics are also used interchangeably.

Purposes of soil classifications

Reduction of complexity. The marshalling and ordering

of data within a taxonomic framework enable the human mind to recall information and relate entities and attributes to each other. This is not possible if a large unstructured mass of information is presented to the mind, which is able to comprehend and work with a very limited amount of information at any one time.

Correlation of information. Data sets arising from different facets of soil science can be related to each other on the premise that the entities from which the data sets are derived belong to the same taxonomic group.

Information transfer. Soil classification provides a means of transferring information, e.g. extrapolation or interpolation from experimental results via the medium of publication, word of mouth, or other means. However, for this to occur it is necessary that soil scientists themselves, or pedologists with whom they consult, be familiar with and understand the classification or classifications that are used in the places between which it is desired to transfer information.

Basis of soil survey and field research. Although it is theoretically not essential, it is customary for soil classification to precede a soil survey. In fact, most soil classifications have as a major purpose the facilitation of soil mapping; for example, the Factual Key classification of Northcote (1979), with its emphasis on attributes observable in the field, was devised to provide a consistent Australia-wide classification that could be used in the production of the 'Atlas of Australian Soils' (Northcote *et al.* 1960–68). Similarly, it has been made quite clear that the U.S. Soil Taxonomy (Soil Survey Staff 1975) has been developed for making and interpreting soil surveys. Soil classification may also be a vital component in other classifications (e.g. land classification and terrain classification) and soil classification, at various categorical levels, helps provide a sound basis for the selection of sites for much agricultural and other environmental experimentation.

Psychological satisfaction to the classifier. Classification and related areas such as pattern recognition seem to be innate compulsive functions of the human brain (and perhaps those of other animals too). Few people these days, however, are able to justify their activities on this ground, although it is likely that it is, in fact, a component of many excursions into classification.

Soil systematics

Two major questions have to be answered before any study of soil systematics can proceed: what are we to classify, and how are we to classify?

What are we to classify?

First it is necessary to define the universe with which we wish to deal. Most soil scientists would conjure up a reasonably similar picture of what is meant when 'soil' is mentioned, and of how it is distributed, but each individual's perception of the universe of soil is conditioned by his experience of it. His perceived universe may be the result of chance encounters; often it is deliberately limited for a particular purpose, e.g. by considering only the soils of a geographical area, such as Australia, or of a conceptual group, such as cracking clays. In the latter case classification has already begun.

These universes are real, in the sense that we can walk on soil, plough it, and dig holes in it. Individual soils (whatever they may be) that constitute the universe of soil are also real, but we cannot classify them. What, then, is it that we classify? In the first place, the definition of a soil individual or entity type is difficult. The solution to this problem varies with the situation and with the classifier. The USDA has attempted to rationalize this by considering the soil entity type to be the pedon, or the polypedon (an aggregation of contiguous pedons), defined largely on the basis of dimensions, although usually a pedon is described in terms of a part of it, viz. the soil profile. The pedounit of Fitzpatrick (1971) is similar in concept to the pedon, in that it also is defined as a three-dimensional body.

Most soil scientists, most of the time, explicitly or implicitly, consider the soil profile as the soil entity type. The soil profile is likewise frequently defined as a column of soil of small cross-sectional area. In practice, however, the properties of this three-dimensional entity are usually inferred from observation of and measurement on a vertical section or small auger core. What is usually meant by soil profile is a vertical section through the soil body. This section ideally extends from the soil surface to the parent material, although in practice the latter may not always be identifiable. The lateral dimensions of this section may range from about 5 cm to a metre, depending upon the method of examination. Commonly, most emphasis is given to the solum, i.e. the A plus B horizons, although it is recognized that materials below this may interact with the solum.

It may be argued that soil classification should be concerned with entity types other than the soil profile, but in general the profile has been the entity type that soil taxonomists have dealt with up to the present time, and the practice in Australia has been no exception. In any account of soil classifications in Australia, and comparisons with other classifications overseas, the essential basis has been, and is, the soil profile. There is, however, no reason why different entity types, e.g pedons or pedounits, should not be used in the future, although this would present formidable problems to the taxonomist.

From the (real) universe, i.e. the aggregate of (real) entities that constitute that universe, we may select one entity. This entity is described in terms of a set of attributes selected from a suite of attributes, each of which has been defined. The resulting description model is an *ens type*. A set of entia (instances of a particular ens type)

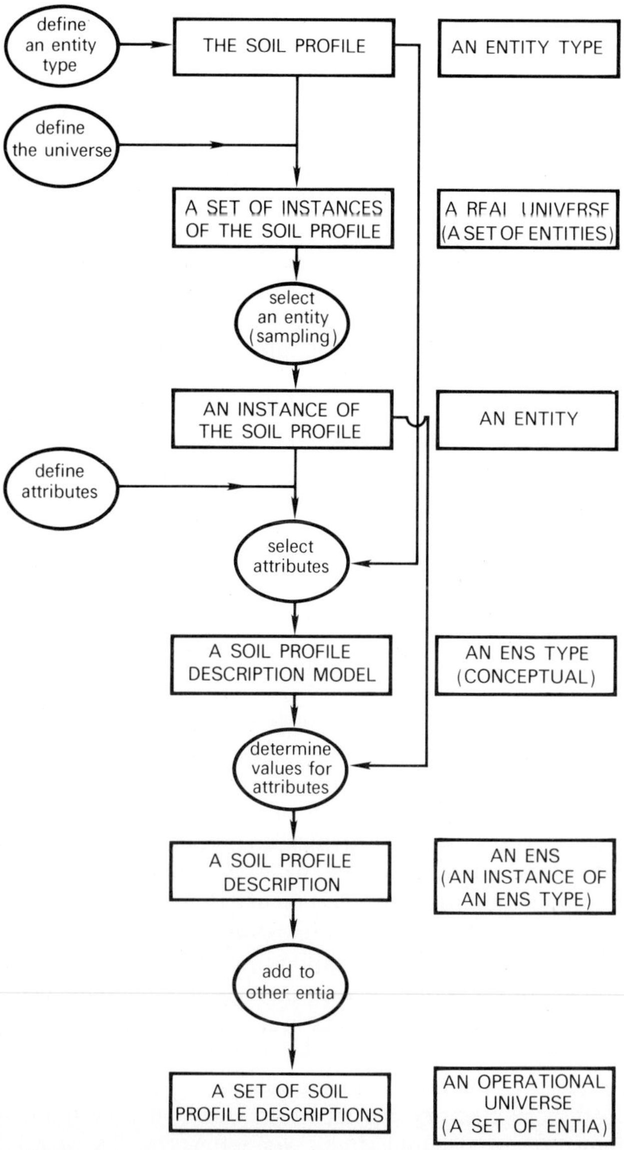

Fig. 20.1. What is classified.

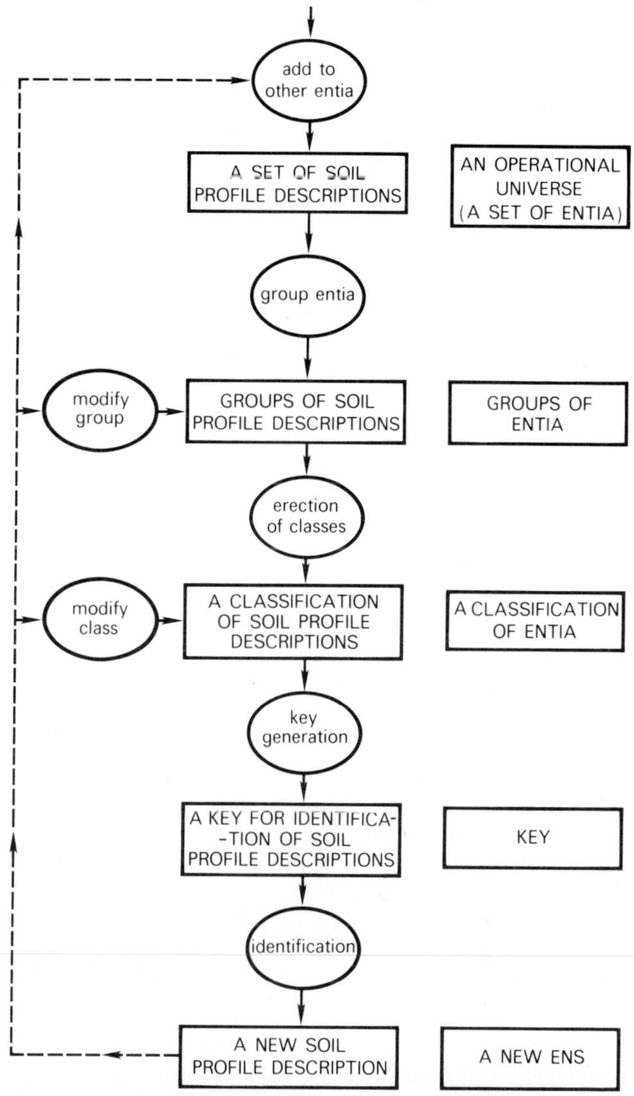

Fig. 20.2. The classification and identification process.

constitutes an operational universe to which classification procedures may be applied. There is always a nexus between the entity and the ens. Note, however, that one entity type may be the source of any number of different ens types and therefore different entia. For example, a soil profile description model that included measurements of free iron oxide and particle size according to the USDA system would be a different ens type from a description model of the same profile that did not include free iron oxide values and gave its particle-size measurements according to the international system. Many problems in soil taxonomy arise from an insufficient appreciation of these relationships. The procedure described above is shown diagramatically in Fig. 20.1, both in a generalized fashion and specifically in relation to soil.

Selection of attributes

The reason why many soil profile description models (ens types) may be derived from the soil profile (a particular entity type) is that any number of different sets of attributes may be used to describe the profile. The choice of a particular attribute for inclusion in such a set may be made because of one or more of the following reasons: because the attribute is considered to be a cardinal one, i.e. shows a high correlation with a number of others; because it is known or suspected to be relevant to the purpose of the classification being erected; because it is easily observed or measured; because it has been included traditionally; or because it is of special interest to the classifier.

A further complication is often introduced as a result of different soil scientists using the same name for different attributes.

How are we to classify?

Although a multitude of soil classifications exists, and they appear to have been arrived at by a variety of routes, it is probable that the procedure leading to each one is much the same in each case (Fig. 20.2).

As a first step, the selected operational universe (e.g. a particular set of soil profile descriptions) is partitioned in some way, either by dividing it into groups that are internally homogeneous in some sense or by aggregating similar soil profile descriptions. The ways in which this is done cover a spectrum from informal and intuitive on the one hand to formal and mathematical on the other. In all cases the procedure is subjective in the sense that the taxonomist has to decide on a certain procedure, implicit or explicit, for the partitioning. A large amount of trial-and-error is probably involved in most instances also, particularly in relation to selection of attributes.

Such a grouping procedure conforms to the dictionary definition of 'classify', particularly in the special case having one-to-one correspondence between the real and operational universes; then groups and classes show the same correspondence also. In taxonomy, generally, the operational universe represents only a sample of the real one, and it is necessary to devise a classification to which the whole of the real universe can be related. How successful this is depends to some extent at least on how representative that sample is. Classes are abstractions devised by the human mind and they may be derived directly from the groups mentioned above. However, this is not necessarily so; the groups may only provide starting points from which classes are derived intuitively or by some more formal means.

Identification

If we now select a new instance of a soil profile, and derive a new soil profile description from it, the latter can be referred to the classification that has been set up. This is frequently done by means of a key*. It may not be possible to identify this new soil profile description in relation to this classification if a suitable class has not yet been erected ; on the other hand, if identification is made it may or may not be possible to indicate with what degree of certainty this has been done.

The need for dynamism in taxonomy

Classification is normally a dynamic process, with strong feedback occurring when new entia are encountered. This may occur through a number of channels (broken line in Fig. 20.2): if the certainty of identification of a new ens is high, the class to which it has been referred may be modified directly; the new ens may be added to one of the groups of entia and the classification modified in light of this if it is felt desirable; or the new ens may be added to the original operational universe to give a new operational universe, which may in turn be partitioned and give rise to a modified classification.

If enough entia (soil profile descriptions) of a different nature are added to the operational universe a radically different classification may emerge. If, as a classification becomes older, various vested interests in it increase, there is an increasing reluctance to accept the feedback described above, and the classification is on the way to becoming fossilized.

Types of classifications

Classifications, like anything else, can be classified. It is perhaps not surprising that the people who have tried to do this explicitly are mostly those who are interested in the application of mathematics to classificatory problems (e.g. de Gruijter 1977, Williams 1976*b*). Williams, for example, presents a hierarchical classification of classifications.

Hierarchical and non-hierarchical classifications

Theoretically, a distinction may be made between these; but it can be argued that even a non-hierarchical classification is formally part of a three-level hierarchy because the universe provides an upper level and the individual entia a lower, with the non-hierarchical classification in between. In practice in soil science, hierarchical classifications are almost invariably used.

The reason why classifications are almost always constructed in a hierarchical fashion is probably largely psychological. The human mind is capable of comprehending and manipulating only a limited amount of information at any one time. For example, a person with a good short-term memory can recall the order of nine binary digits, seven decimal digits, six letters or five words (Hicks 1975). The number of 'bits' of information contained in each example ranges from 9 (9 x 1) in the binary digits to 50 (5 x 10) in the 'words' example. Thus by coding, i.e. packing more information into a symbol, man can recall more information, but there is still a finite, and small, limit. The same is true of other mental processes.

To accommodate these limitations, when moving to and fro between coarser and finer patterns of classification (corresponding to some degree with smaller and larger scales when the classification has geographical connotations), some link between categories is necessary. The most common and popular structure employed to achieve this is a hierarchy.

Disjoint, conjoint and fuzzy classes

In exclusive classification the operational universe is divided into mutually exclusive classes which nowhere

* A good definition of a key has been provided by Lawrence (1951): "Keys are devices useful in identifying an unknown. They represent one type of taxonomic literature. A key is an artificial analytical device or arrangement whereby a choice is provided between two contradictory propositions (a couplet), or it may be composed of an extensive series of these."

overlap; the classes are said to be disjoint; an ens can be a member of only one such class. Conversely, a non-exclusive classification possesses some classes which overlap; those classes that do overlap are said to be conjoint; an ens can be a member of more than one class. To take a trivial example, sands and non-sands are disjoint classes, while sandy soils and clayey soils are conjoint classes.

The above definitions imply clear and unambiguous boundaries around classes. In practice this is sometimes not the case, either by accident or by design; the classes are then said to be fuzzy; an ens can only have a degree of membership in any particular class.

Fuzzy classes have the advantage that they may allow the classification to reflect the real world more faithfully. Disjoint classes have the disadvantage that information that an ens is near a boundary (or far from a central concept) is lost; this is not so with fuzzy classes, and to a lesser extent with overlapping classes. However, because much soil classification has been associated with soil mapping (which requires the drawing of non-fuzzy geographical boundaries) most soil classifications have disjoint classes for practical reasons. To some extent the problem of fuzziness in the real world, transmitted through into the operational universe, is recognized in the terms intergrade and extragrade, but the practical solution to this problem has usually been the creation of extra disjoint classes to accommodate those parts of the universe which appear conspicuously fuzzy.

Australian and overseas classifications

There are two soil classifications currently in use in Australia, viz. the great soil group classification (Stace *et al.* 1968) and the classification associated with 'A Factual Key for the Recognition of Australian Soils' (Northcote 1979). Two overseas classifications are also of interest to Australian soil scientists because one, associated with the 'World Soil Map' (FAO 1974), purports to be an international classification, and the other, the U.S. Soil Taxonomy (Soil Survey Staff 1975), is rapidly becoming a *de facto* international classification.

There are pronounced differences between these four classifications in terms of the processes outlined in the previous section, and some of these will become obvious in the brief discussions of each of the classifications given below. Probably the only feature common to all four is that the entity type dealt with is the soil profile.

Australian Great Soil Group classification

This had its origin in the early Russian(Dokuchaev) system; its first published expression in Australia was in association with the soil map of the whole continent presented by Prescott (1931). This was expanded and developed over the next three decades (Stace 1961; Stephens 1962), with its most recent revision being in 'A Handbook of Australian Soils' (Stace *et al.* 1968).

The operational universe on which successive revisions of the classification have been based has increased steadily in size (i.e. in number of profile descriptions) and representativeness for the universe of Australian soils over the past half century; however, it probably contained no more than a few thousands at the time of its last revision. The ens type (soil profile description model) has probably not changed markedly during this period, since little of the micromorphological and laboratory data now available has been taken into consideration in developing the classification.

Is it nominally a many-level hierarchical classification (Stephens 1962) but the categories of class, order and suborder (superior to great soil group), and family (inferior) have been little used. The reasons for this are obscure but two may be: levels of soil survey have probably been such that the great soil group and series levels have been the most appropriate at which to operate (for reconnaissance and detailed surveys respectively), and the dearth of soil taxonomists in Australia has made filling-in at the family level unattainable.

It is essentially a fuzzy classification (the only one of the four) but, for the most part, central concepts have not been explicitly enunciated. Feedback operating in a fragmentary way (i.e. for individual scientists) at the class (rather than universe) level has resulted in differences in perception of classes and concepts by different individuals, e.g. the profiles recorded in Stace *et al.* (1968) for *grey-brown and red calcareous soils*, *solonized brown soils*, and *red-brown earths*. Classes become unwieldy and of decreasing usefulness under these conditions, leading eventually to breakdown in communication between soil scientists.

Advantages. Its main advantage is that it has had wide currency in Australia for many years, and consequently many people are familiar with the main properties of many of the classes and can readily visualize them in broad terms.

Disadvantages. No key has been constructed (and it is unlikely that one could be), and hence identification of new entia (soil profile descriptions) is highly subjective. A large number of entia, representing large areas of soil in Australia, either are not identifiable or are likely to be misidentified using this classification, i.e. feedback to the operational universe has not been occurring and the classification is becoming obsolete. A related problem is that individual scientists are independently modifying classes (limited feedback), leading to distortion and excessive heterogeneity in these classes and to an obvious need for some sort of national coordinating body for soil classification. Its nomenclature has been borrowed largely from older U.S. or European classifications; unfortunately the meanings of many of the names have been greatly changed from their original meanings, e.g. the *podzolic* group, *prairie soils* and *non-calcic brown soils*.

Northcote Factual Key classification

Although essentially general-purpose, the Factual Key classification (Northcote 1971) was devised in part to facilitate the production of a soil map consistent across the whole of the Australian continent. In that this use was clearly defined before development of the classification began, it resembles the two overseas classifications mentioned previously. It owes something, at least philosophically, to ideas put forward by Leeper (1954, 1956).

The operational universe on which this classification was originally based consisted of descriptions of about 500 soil profiles widely distributed across Australia, although most came from the south-east portion of the country. The ens type (soil profile description model) is probably reasonably similar to that used for the great soil group classification. It should be noted that the attributes used in the key (which are observable or measureable in the field) do not in the aggregate constitute the ens type used in the grouping procedure (see Fig. 20.1). However, the soil profile description model used has not been presented in published form.

A distinction is made between *morphological soil type* (lowest taxon), which is a set of soil profile descriptions with a narrow range in described features, and *generalized soil type*, which is a geographic unit usually synonymous with soil type (in the American, not Russian, sense) as used in soil surveys. It is stated explicitly that the Factual Key can be used for identification of a morphological soil type, but not of a generalized soil type.

The classification is largely a bifurcating hierarchical scheme with five categorical levels: divisions, subdivisions, sections, classes, and principal profile forms. It is strictly exclusive.

Advantages. It is simple and unambiguous to use. The keying attributes can be determined in the field (e.g. that used at the highest categorical level is the textural profile form) and therefore identification can be made there also, which is very convenient if the classification is being used in soil mapping. Most of these keying attributes are physical and a high degree of covariance with other physical attributes might be expected, e.g. texture is related to soil moisture characteristics.

Disadvantages. The non-use of laboratory data in defining class boundaries means that the classes may be more variable in such properties than in morphological ones unless there is a high covariance between them; this has not been generally established, although it is not unreasonable to expect such a covariance to exist on deductive grounds in some instances. Although non-identification or mis-identification are virtually impossible because of its exclusive nature and largely binary division, some new soil profile descriptions will have to have new classes created for them, the proportion depending on how representative of the universe of Australian soils the original operational universe was. Nomenclature is somewhat confusing. While the alphabetic characters in the code are mnemonic and have consistent meanings, the decimal digits do not. An English word nomenclature has also been devised for many of the classes (Northcote *et al.* 1975) but many of the names are lengthy, particularly at the lower categorical levels. The emphasis given to surface and near-surface (A_1 and A_2) horizons may lead to classification difficulties in disturbed soils because, as in the great soil group classification, no provision is made for soils with disturbed (A_p) horizons.

U. S. Soil Taxonomy

Although stated explicitly to be a classification 'for making and interpreting soil surveys', this is essentially a general-purpose classification. Prior to 1951, soil classification in the U.S. owed much to the early Russian system (USDA 1938) and consequently bore many resemblances to the Australian great soil group classification. In that year a start was made on an entirely new system of classification, which over the next quarter of a century proceeded through seven approximations, finally culminating in the publication of the U.S. 'Soil Taxonomy' (Soil Survey Staff 1975).

A description of the operational universe on which this classification was originally based has not appeared in published form, but as about 10 000 soil series were being used in the U.S. when the final draft of the classification was completed in 1971, it is reasonable to assume that the number of soil profile descriptions available would have been of that order. This was probably a representative sample of U.S. soils, but inadequate as far as the rest of the world is concerned. The ens type (soil profile description model) has not been published, but it differs from those used in Australian classifications, at least in some laboratory-determined attributes.

The classification is strictly exclusive and hierarchical, with six categorical levels: order, suborder, great group, subgroup, family, and series. There is considerable confusion in the use of the term series; it is used as the lowest taxon in the classification hierarchy as well as in its traditional geographical sense.

A completely new nomenclature has been devised from classical Greek and Latin roots. The class names are mnemonic and are intended to indicate the main properties of the class, the level of generalization, and relations with certain other classes.

A key is provided down to the great group level.

Advantages. It attempts to be more straightforward, explicit and comprehensive than any other existing system of soil classification. The idea of definition of terms and specification of classes is praiseworthy, even though there is an unfortunate tendency for the definitions to be mixed up with non-specific descriptive material. It has considerable value as a means of communication about the soils of large areas of the world. The development of a new nomenclature that avoids long-established, but often loosely or erroneously used, names for various groups of soils is a welcome feature.

Disadvantages. The problem of loss of information about similarity of entia in different disjoint classes but near their common boundary exists in this as in all exclusive classifications. An attempt to overcome this problem of intergrades and extragrades at the subgroup level results in further disjoint classes. Non-identification can occur because not all possible classes in categories below order level have been described. A particular disadvantage for Australian soil scientists at the moment is the requirement for laboratory data obtained by methods which are little or not used in Australia. Even so, it is possible that a large number of our soils could be identified with reasonable confidence using available data and some guesswork. The use of soil moisture and temperature data in many cases at high levels (suborder) in the hierarchy is questioned by many Australian pedologists. Although of obvious edaphic importance, this can be misleading in genetic studies. In addition, both are difficult to determine for individual entia.

World Soil Map classification

In many ways this is a compromise scheme which adopts many of the concepts and definitions of the U.S. Soil Taxonomy and some from European sources. It is the basis of the mapping units used for the 'FAO-Unesco Soil Map of the World' (FAO 1974). It is the only classification among the four considered here which was conceived from the start as a classification to encompass soils on a global scale.

The scheme is not a hierarchical classification in the sense of being a taxonomic system subdivided into categories at different levels of generalization. Rather it is a two-category system with the higher category varying quite widely in concept, being in some cases equivalent to the order in the U.S. Soil Taxonomy and in other cases being closer to the great group level. The lower category is composed of intergrades or soils with special horizons or features of note. Phases have been used to subdivide the secondary classes according to differences in characteristics important in use and management: e.g. stony, saline, etc. phases. It is an exclusive classification.

The nomenclature for many of the diagnostic horizons and soil properties, as well as names for several of the major classes, has been drawn from the U.S. Soil Taxonomy. The names for the remaining classes stem in part from classic soil names, particularly the original Russian soil type terminology (e.g. chernozem), and in part from some of the soil names recently coined in western Europe and Canada.

A key is provided to facilitate the identification of new soil profile descriptions.

Advantages. The scheme is much simpler than the U.S. Soil Taxonomy and involves much less information about soils. Many of the diagnostic horizons and soil properties are the same as those used in the U.S. Soil Taxonomy. Probably the main benefit of the scheme is that it is the basis of small-scale soil maps of all continents now being published, which will permit some inter-regional comparisons.

Disadvantages. Its simplicity means that it provides a coarser framework with many fewer classes than at equivalent levels of the U.S. Soil Taxonomy; as a consequence its lower classes have a much lower information content. It has the usual defects of an exclusive classification. Although non-identification is unlikely to occur, some new soil profile descriptions will be allocated to heterogeneous non-conformist classes simply because they are very dissimilar to most other profile descriptions. As with the U.S. Soil Taxonomy, the recognition of a number of diagnostic horizon and soil properties depends heavily on analytical data, much of it obtained by methods little used in Australia at present. Again, soil moisture criteria are used to separate several classes that may have similar morphology. Class names are a compromise selection of old and new terms borrowed from previous and existing soil classifications, with the addition of some new names. The result is a mixture that is a source of confusion, particularly where some of the older terms are now redefined to have rather different meanings from the original.

Current problems

One of the main roles of classification is the provision of means of communication among those interested in a particular field of endeavour. Within Australia, people concerned with soils presently use one or both of the existing Australian soil classifications for communication. At a local level a third system is also in use to some extent, viz. the collection of series or equivalent classes frequently presented in large-scale maps. Because they are essentially geographical land classes, however, we exclude them from consideration in this paper (see Chapter 21 for a discussion of soil survey in Australia).

The concurrent existence and use of two classifications is found confusing by some soil scientists and others involved with soils. Whilst there obviously can be as many special-purpose classifications as there are special purposes, in the case of general-purpose classifications there does not seem to be any reason why there should be more than one, provided everybody's perception of 'general-purpose' is the same. Almost certainly this is not the case, although we cannot be sure because only rarely has the purpose of any general-purpose classification been defined.

Table 20.1. Approximate correlations

Approximate correlations between: the great soil groups of 'A Handbook of Australian Soils' (Stace *et al.* 1968), the principal profile forms or classes of 'A Factual Key' (Northcote 1979), great groups of 'Soil Taxonomy' (Soil Survey Staff 1975), and the soil units of 'FAO-Unesco World Soil Map' (FAO 1974). The correlatives shown in *italics* are expected to be the most common, based on the actual profiles examined (numbers given in brackets) and wider Australian experience.

Handbook of Australian Soils	Factual Key	Soil Taxonomy	World Soil Map
Solonchaks (3)	Um1.1, .2; Um2.23; Uf1.41; Uf6.51, .61, .62	Haplargid, Salorthid, Natrargid	Solonchaks, Luvic Xerosol, Orthic Solonetz
Alluvial soils (9)	Um1.1, .2; Um5.22, .42	*Ustifluvent*, Xerofluvent, Torrifluvent	*Eutric Fluvisol*
Lithosols (10)	Uc1.4; Uc2.12; Uc6.11, .12; Um1.4; Um2.1; Um4.1, .2; Um5.5; Um6.2, .4; Uf1.4	*Ustorthent*, Xerorthent, Ustochrept, Camborthid, Haplargid	*Eutric Regosol*, Cambic Arenosol, Eutric and Dystric Cambisol, Haplic and Luvic Yermosol
Calcareous sands (6)	Uc1.1	*Xeropsamment*, Ustipsamment, Torripsamment	*Calcaric Regosol*
Siliceous sands (11)	Uc1.2; Uc4.2; Uc5.1	*Quartzipsamment, Torripsamment*, Torriorthent	*Cambic* and Albic *Arenosol*, Eutric Regosol
Earthy sands (7)	Uc5.2	*Quartzipsamment, Torripsamment*, Haplargid, Xerorthent	*Cambic, Ferralic* and *Albic Arenosol*, Luvic Yermosol, Dystric Regosol
Grey-brown and red calcareous soils (9)	Um1.3; Um5.1, .6	*Calciorthid*, Haplargid, Paleorthid, Haploxeralf	*Calcic* and Luvic *Xerosol*, Calcic Yermosol
Desert loams (11)	Dr1.12, .13, .32, .33, .42, .43	*Paleargid, Natrargid*, Haplargid, Natrixeralf	*Luvic Xerosol*, *Luvic Yermosol*, Orthic Solonetz
Red and brown hardpan soils (7)	Um5.31	*Durorthid, Durargid*	*Haplic* and Luvic *Yermosol*
Grey, brown and red clays (43)	Ug5.2,.3, .5	*Torrert, Chromustert*, Pellustert, Chromoxerert	*Chromic* and Pellic *Vertisol*
Black earths (18)	Ug5.1, .4	*Pellustert, Chromustert*, Torrert	*Pellic* and *Chromic Vertisol*
Rendzinas (19)	Um6.21; Uf6.11; Ug5.11	*Haploxeroll, Calcixeroll*, Calciustoll, Calciorthid,	*Rendzina*, Luvic Chernozem, Calcaric Phaeozem
Chernozems (6)	Um6.11; Gn3.43, .46	*Haplustoll*, Haploxeroll, Argiustoll	*Haplic* and Luvic *Phaeozem*, Chernozem
Prairie soils (27)	Um6.31, .41; Gn3.41, .42, .44, .45	*Haplustoll, Argiustoll, Haplustalf, Ustifluvent*, Palexeralf, Paleustalf	*Haplic* and *Luvic Phaeozem, Chromic* and Vertic *Luvisol*, Eutric Fluvisol
Wiesenboden (7)	Ug5.16, .4	*Pellustert*, Pelloxerert, Chromoxerert	*Pellic* and Chromic *Vertisol*
Solonetz (10)	Dy2.43; Dy3.43; Dd1.43	*Natrixeralf, Natrustalf, Natrargid*	*Orthic Solonetz*
Solodized solonetz and solodic soils (50)	Dr2; Dr3; Db1; Db2; Dy2; Dy3; Dy5; Dd1; :33, .43	*Natrustalf, Natrixeralf, Paleustalf*, Haplustalf	*Orthic Solonetz, Albic Luvisol*, Solodic Planosol
Soloths (19)	Dr2; Dr3; Db1; Db2; Dy2; Dy3; Dy5; Dd1; Dg2.31, .41, .81	*Natrustalf, Natrixeralf, Paleustalf*, Haplustalf	*Orthic Solonetz, Albic Luvisol, Solodic Planosol*

Table 20.1. *continued.* Approximate correlations

Handbook of Australian Soils	Factual Key	Soil Taxonomy	World Soil Map
Solonized brown soils (19)	Gc1.1, .2; Gc2.2; Dr2.13, .33; Dr4.63	*Calciorthid, Palexeralf,* Haplargid, Natrixeralf, Calcixeroll	*Calcic Luvisol, Calcic Xerosol, Calcic* and Luvic *Yermosol,* Orthic Solonetz
Red-brown earths (sodic) (33)	Dr2.33; Db1.33; Db2.33	*Natrixeralf, Palexeralf,* Paleustalf, Haploxeralf	*Orthic Solonetz, Chromic, Calcic* and *Albic Luvisol*
Red-brown earths (non-sodic) (16)	Dr2.13, .23; Dr3.23; Db1.23	*Rhodoxeralf, Haplustalf,* Palexeralf, Haploxeralf	*Calcic,* Chromic and *Albic Luvisol*
Non-calcic brown soils (18)	Dr2.12, .22; Db1.12, .22	*Paleustalf,* Rhodustalf, Rhodoxeralf, Haplustalf	*Chromic Luvisol*
Chocolate soils (9)	Um6.43; Uf6.32; Gn4.4; Dr4.1	Argixeroll, Haploxeroll, Haplustalf, Argiustoll	Luvic and Haplic Phaeozem, Chromic Luvisol
Brown earths (7)	Gn3.21, .22, .23, .24; Gn3.51, .52	*Dystrochrept,* Haplohumult, Haploxeroll, Haplustalf	*Dystric Cambisol* Humic Acrisol, Calcic Kastanozem
Calcareous red earths (8)	Gn2.13, .16	*Haplargid,* Paleorthid	*Luvic Yermosol,* Calcic Xerosol
Red earths (50)	Um5.5; Gn2.11, .12, .14, .15	*Paleustalf, Haplustalf,* Torrox, Haplargid, Haploxeralf, Paleustult, Paleudult	*Eutric Nitosol Ferric Luvisol,* Orthic Ferralsol, Luvic Yermosol, Dystric Nitosol
Yellow earths (22)	Gn2.21, .22, .24, .31, .32, .34, .61, .62, .64, .71, .72, .74	*Haplustalf,* Paleustult, Haplargid, Paleustalf, Plinthustult, Palexeralf	*Ferric,* Albic and Chromic *Luvisol,* Ferric Acrisol, Luvic Xerosol, Plinthic Luvisol
Terra rossa soils (18)	Uc6.13; Um6.24; Uf5.31; Gn4.13, .34	*Rhodoxeralf,* Xerochrept, Xerorthent, Paleustalf, Haploxeralf	*Chromic* and *Calcic Luvisol,* Chromic Cambisol, Eutric Regosol
Euchrozems (30)	Uf6.31; Gn3.12	*Rhodustalf, Ustochrept,* Haplustalf, Paleustalf	*Chromic Luvisol, Chromic Cambisol,* Eutric Nitosol
Xanthozems (8)	Gn3.71, .74, .84	Haplorthox, Haplustult, Acrohumox, Palehumult	Xanthic and Humic Ferralsol, Humic and Ferric Acrisol
Krasnozems (40)	Uf5.21; Gn3.11; Gn4.11, .31	*Paleustalf, Acrohumox, Palehumult, Haplohumox,* Acrorthox, Eutrustox	*Eutric* and Humic *Nitosol, Humic* and Rhodic *Ferralsol,* Humic Acrisol
Grey-brown podzolic soils (5)	Db1.42; Dr3.12; Dy3.22	Haplustalf, Palexeralf, Haploxeralf, Haploxerult	Orthic and Albic Luvisol, Orthic Acrisol
Red podzolic soils (32)	Gn3.14; Dr2.21, .31, .41; Dr4.21, .41; Dr2.61, .81; Dr3.21	*Paleustalf,* Palexeralf, Haplustalf, Paleustult, Paleudult, Haploxerult	*Albic* and *Chromic Luvisol,* Eutric and Dystric Nitosol, Ferric Acrisol

Table 20.1. *continued.* Approximate correlations

Handbook of Australian Soils	Factual Key	Soil Taxonomy	World Soil Map
Yellow podzolic soils (32)	Gn2.31, .71; Gn3.84; Dy2.21, .22, .31, .41; Dy3.21, .31, .41, .61; Dy5.41, .81	*Haplustalf, Haplustult,* Paleustalf, Haploxeralf, Paleustult	*Albic Luvisol, Orthic* and Ferric *Acrisol*
Brown podzolic soils (7)	Dy2.62; Dy3.61; Dy4.41	*Palexeralf,* Paleustalf, Paleudult	*Albic Luvisol,* Ferric Acrisol
Lateritic podzolic soils (30)	Dy2.61, .62; Dy3.61, .62, .81, .82; Dy5.61, .62	*Palexeralf,* Plinthoxeralf, Paleustalf, Paleustult, Plinthustult, Haploxerult, Albaqualf	*Albic* and *Plinthic Luvisol,* Ferric and Plinthic Acrisol, Eutric Planosol
Gleyed podzolic soils (19)	Dg2.41, .42, .81; Dg4.41, .42, .81	*Albaquult, Paleustalf,* Paleaquult, Ochraqualf	*Dystric, Eutric* and *Solodic Planosol,* Gleyic Acrisol
Podzols (10)	Uc2.2, .3; Uc4.2	Troporthod Haplorthod, Placorthod, Quartzipsamment	*Orthic* and *Placic* Podzol
Humus podzols (8)	Uc2.20, .33, .34, .35, .36	*Tropohumod,* Haplohumod, Placohumod, Quartzi-psamment	*Humic* and Placic *Podzol*
Peaty podzols (2)	Uc4.33	Sideraquod, Tropohumod	Gleyic and Humic Podzol
Alpine humus soils (6)	Um7.11, .12	*Haplumbrept,* Dystrochrept	*Humic* and Dystric *Cambisol*
Humic gleys (8)	Uf6.42, .61; Gn3.9; Dy5.81; Dg4.11, .41, .81	*Paleaquult, Albaquult,* Humaquult, Haplaquept	*Gleyic Acrisol,* Humic and Dystric Gleysol
Neutral to alkaline peats (7)	0	Medihemist, Medifibrist, Medisaprist	*Eutric Histosol*
Acid peats (4)	0	Medifibrist, Medisaprist	*Dystric Histosol*

Table 20.2. Soils described in Chapter 3 which cannot be referred to great soil groups in a 'Handbook of Australian Soils'

Group	Factual Key	Soil Taxonomy	World Soil Map
Brownish sands (8)	Uc5.11, .12, .13	*Torripsamment,* Xero-psamment, Haplargid	*Eutric Regosol,* Luvic Yermosol
Red or yellow pale loams (3)	Um4.41, .43	Dystropept, Haplorthox	Ferralic Cambisol, Orthic and Xanthic Ferralsol
Earthy loams (6)	Um5.51, .52	*Torrox,* Haplargid, Camborthid	*Orthic Ferralsol,* Luvic Xerosol, Haplic Yermosol
Grey massive earths (7)	Gn2.81, .82, .84, .85, .91, .92, .94, .95	*Paleustalf,* Tropaqualf, Haplustalf	*Gleyic Luvisol,* Albic Luvisol
Brown smooth-ped earths (10)	Gn3.22, .52	*Haplustalf,* Paleustalf, Ustochrept	*Ferric* and orthic *Luvisol,* Eutric Nitosol, Eutric and Chromic Cambisol

Explanatory Notes to Tables 20.1 and 20.2

In constructing these tables, approximately 750 profiles were examined for which appropriate morphological and analytical data were available. The profiles covered all Australian states but, because data are lacking for many areas and for particular soils, the sample cannot be considered completely representative of Australian soils. Some extrapolation was often necessary because analytical data was not always available by methods specified in 'Soil Taxonomy' (Soil Survey Staff 1975) and in the 'World Soil Map' legend (FAO 1974). Because of a general lack of micromorphological data, the presence of an argillic horizon was decided solely on the basis of particle-size specifications; however, data exist to show that argillic horizons identified in this way may not necessarily contain the required amounts of illuviated clay. Soil temperature and moisture data are mostly lacking for Australian soils, hence the identification of soil moisture regimes, in particular, is often very subjective; soil temperature data have been estimated from air temperatures.

In view of the above, it is stressed that many correlations are approximate only. The wide overlap that occurs for some soil groups is due to several factors: (1) the use of different soil attributes to differentiate between units of each classification must inevitably lead to some overlap; (2) the Australian great soil groups are not rigorously defined; (3) the correlation of Paleustalfs with a number of widely different Australian great soil groups is largely because the present definition of this great group accommodates both deep soils that show little clay decrease with depth *and* soils with an abrupt clay increase between A and B horizons.

Within Australia then, it is possible to communicate reasonably well using one or other (or both) of the two classifications. Would it be possible to meld them into a single classification in the future? While there may be some discrepancies between the ens types (soil profile description models) in the two systems, they are probably minor; and at least the operational universes are derived from samples of the same universe of soils (Australian soils). However, there are likely to be formidable problems arising from differences in the procedures used in grouping and classification (Fig. 20.2). In particular, the choice of different attribute sets for these activities and the different types of classes erected in the two classifications (disjoint in one, fuzzy in the other) would make any rapprochement between the two systems very difficult. For the present, it is necessary to live with the existence [f both classifications. A concordance, insofar as it is possible, between great soil groups and various classes of the Factual Key classification has been provided by Northcote *et al.* (1975); a revision of it is included in Tables 20.1 and 20.2.

Even greater difficulties are encountered if Australian soil scientists wish to communicate with those overseas. It is necessary to be pragmatic and accept that overseas soil scientists will not identify their soils in terms of either of the Australian classifications. While existing overseas classifications may not be particularly appropriate as vehicles of communication within Australia, they provide the only possible avenue for communication outside. However, limited experience in using both the World Soil Map classification and the U.S. Soil Taxonomy to identify Australian soils indicates that considerable problems exist.

For purely pragmatic reasons (greater financial and human resources) it is likely that the U.S. Soil Taxonomy will become a *de facto* international classification over the next decade or two. To develop lower categorical levels for the World Soil Map classification similar to those of the U.S. Soil Taxonomy would require resources which are unlikely to be made available to the United Nations Organization agencies concerned with it within the foreseeable future. This being the case, it is necessary to try to pinpoint the factors contributing to the difficulties encountered in trying to identify Australian soils in terms of the U.S. Soil Taxonomy. Two major areas of difficulty exist; the first concerns the identification process, the second the nature of the classification itself.

The first is a local problem. At present some of the laboratory data used as class criteria are not provided on a routine basis in the many Australian organizations concerned with soil, i.e. our ens type (soil profile description model) does not correspond to the U.S. ens type. This can only be overcome by placing an added burden on soil analytical services, or by establishing, where possible, empirical correlations between different methods, or (with the accumulation of experience) by substituting guesswork. The same problems exist to some extent in the acquisition of soil moisture and temperature data, which are also used as class criteria.

The second, and more serious, area of difficulty relates to the fact that the U.S. Soil Taxonomy is in some sense suboptimal as far as Australian soils (and probably those of other areas outside the U.S. and Europe) are concerned. This arises in the first instance from a restricted real universe, which in turn leads to an inadequate operational universe, even assuming that sampling of entities was representative. In short, a classification intended primarily for use within the U.S. is now being used widely outside that country, without adequate feedback to the operational universe and appropriate modification of the classification. Hence many Australian soils, particularly those that are highly weathered and those that possess strong texture contrast between A and B horizons, are very poorly catered for in the U.S. Soil Taxonomy.

If an updated operational universe were used now, it is very likely that it would be necessary to reconsider the choice of attributes in the grouping and class erection steps; for example, in the case of highly weathered soils many of the chemical criteria used for class differentiation are now being shown to rest on very uncertain grounds.

Again, however, we need to be pragmatic. If Australian soil scientists wish to communicate with many of those overseas, for the immediate future they will probably need to identify their soils in terms of the U.S. Soil Taxonomy as far as this is possible, at least for publication purposes.

On a broad scale there have been attempts to determine what relationships exist between the four classifications discussed above; for example, the legend of the 'Atlas of Australian Soils' (Northcote *et al.* 1960–68) has been translated to that of the World Soil Map sheet for the Australasian area by one of us (KHN), and the concordance between the two Australian classifications and the World Soil Map classification has been presented by Northcote *et al.* (1975). Unfortunately, Soil Survey Staff (1975) did not attempt to relate the U.S. Soil Taxonomy to any of the other three classifications of interest here, although they did for the U.S. 1938 classification and the current Canadian, French and Russian systems. Dregne (1976) has shown the approximate correlations between the U.S. Soil Taxonomy orders on the one hand and the Australian great soil groups and World Soil Map orders on the other.

Tables 20.1 and 20.2 attempt to show the concordance between all four classifications. This should help non-Australian readers of this book to relate Australian situations to their own reference framework.

Looking ahead

There is an eternal paradox in systematics. On one hand, there is a need for a reasonably stable classification that provides a continuous framework, both spatial and temporal, within which the host science can operate in an orderly fashion; on the other hand, there is a need for change in classification to reflect and make use of new knowledge acquired in that science. The taxonomist must try to tread the hazardous path between hyperactivity and rigor mortis.

The publication in recent years of 'A Description of Australian Soils', 'A Handbook of Australian Soils', the 'Atlas of Australian Soils', the 'World Soil Map', and the U.S. 'Soil Taxonomy', provides a platform from which Australian pedologists can survey and take stock of the present position of soil classification in their country. We believe that there are three broad objectives in this area that need to be kept in mind in the coming one or two decades; they cannot be achieved overnight, but ignoring them will result in deterioration in the overall quality of soil science in Australia.

Objective 1

The development of one general classification suitable for Australian conditions, in preference to the two classifications available and in use at present. This should be the best soil classification in the sense that, given the known universe of Australian soils at any particular time, it is the classification that, intrinsically minimizes variability within and maximizes variability between taxa, and extrinsically is most useful to the largest number of users.

It is difficult to sustain an argument to maintain and develop two general purpose classifications, given that we are dealing with Australian soils only and that the 'general purpose' is the same in both cases, viz. relating primarily to land use decisions, management of land, and to an understanding of soil genesis and soil-landscape relationships. The existence of two Australian classifications at present stems partly from differences in the philosophies of their developers, partly from differences in objectives, and partly from a tendency to cling to tradition.

Undoubtedly there will always be a need for special-purpose classifications. However, it should be noted that some special-purpose classifications can be derived from general-purpose classification(s).

Objective 2

Maintenance of a watching brief on international developments in soil classification, and where possible, making appropriate contributions to currently evolving world soil classification(s).

We consider it highly unlikely that any widely accepted international soil classification will also be the most suitable one for Australia, at least in the immediate future. Nevertheless, we must be able to communicate with the rest of the soil science world. Over the past decade Australian soil scientists have become increasingly isolated from other soil scientists overseas, partly because of their inability or unwillingness to become involved in soil classification and related aspects of pedology on a world-wide scale. It is worth emphasising that to make meaningful contributions to the development of either an Australian classification or an international classification, pedologists need to have a wide background experience of soils both in Australia and overseas.

Objective 3

Improved understanding of relationships between soil attributes, to provide a better basis for soil classification. We consider that research aimed at such understanding is essential for the evaluation of existing classifications and the formulation of future classifications.

Several measures must be taken in order to achieve this objective: Standardization of experimental procedures (by compiling handbooks for field observations and measurements and for chemical analysis methods) in order to achieve compatibility of soils data collected by the many different organizations concerned with soils in Australia; determination of minimum acceptable precision in measurements used for taxonomic purposes (e.g. particle-size analysis, exchangeable cations and organic carbon) and tailoring of methods to conform with this limit; increased use of computers in the processing and analysis of soil data by statistical and other mathematical techniques (see below); re-evaluation of the use of attributes of long standing as taxonomic criteria, e.g. soil colour and structure, and of such constructs as so-called diagnostic horizons; and evaluation of attributes not previously used as criteria in soil classification (including certain chemical and physical criteria at present not used, soil composition and fabric properties as ascertained by studies at a microscopic level, and engineering tests).

Potential uses of the computer in soil taxonomy

The central activity of taxonomy, the erection of the abstract constructs known as classes, is likely to remain the province of the human brain for the foreseeable future. Terms such as art, intuition, subjective, etc. used in reference to this activity simply indicate that we are a long way from understanding how a taxonomist's mind works.

In spite of this, the taxonomist can be, and often is, assisted by mathematics and logic in a number of activities associated with classification. In most cases the amount of data to be handled and the complicated calculations involved make the use of a computer imperative.

Data logging and data processing

It is now common practice to use the computer to carry out any calculations necessary to convert raw measurements to attribute values.

Data base management

The basis of any classification is one or more large data bases. When a data base contains more than a few hundred records, using computer data base management systems is probably more efficient than using manual systems (Moore and Bie 1977; Moore *et al.* 1981). For a data base of up to approximately 10^3–10^4 entia (e.g. soil profile descriptions) sequential accessing is reasonably efficient and has the advantage that any attribute can be used as a retrieval criterion; for a larger data base, direct access is necessary for efficiency. Activities in which taxonomists commonly engage, e.g. sorting, inversion of files, retrieval of data using sophisticated criteria etc., can be carried out rapidly by computer. At present, however, there is considerable uncertainty about the desirability of establishing computer-oriented data bases, on the grounds of data incompatibility and/or high costs associated with computing.

Pattern analysis

A wide range of multivariate techniques is now available (Williams 1976*a*) to assist in the elucidation of patterns in sets of data. Such techniques (e.g. ordination, clustering, correlation) are appropriate where no strong opinions on structure in the data are already held by the investigator. The use of pattern analysis was not feasible before the advent of computers, and even now the numbers of entia and attributes that can be handled are no more than several hundreds. Problems related to anisotrophy of soil profile properties (Russell and Moore 1968) have been made more obvious by attempts to use pattern analysis and statistical techniques to analyze soils data, but no general solution has been forthcoming.

Orthodox statistics

Where structure is already perceived in the data, the techniques of orthodox statistics are the appropriate tools to determine the relationships between entia and between attributes (e.g. analysis of variance, discriminant function analysis, regression). The computer has made the application of these techniques commonplace nowadays.

Key generation.

The traditional tool used to identify an unknown ens is a key. A number of keys may be constructed for any given classification, but the one usually favoured is that which uses easily measured attributes and/or is efficient in the sense that the endpoint taxon is reached via the smallest possible number of steps. Such a key can be derived more easily by using a computer than by an intuitive, heuristic approach (Pankhurst 1975).

Computer identification by key

Once a key has been devised it is a relatively simple matter to program a computer to identify an unknown when the appropriate attribute values are provided.

Direct computer identification

A key is not essential for identification of an unknown. By comparing the profile (in the general rather than the pedological sense) of an unknown ens with those of entia already allocated to appropriate classes, the unknown can be identified on the basis of degree of similarity (the mechanics are essentially the same as those used in pattern analysis).

From the above it is evident that the taxonomist can obtain considerable assistance from the computer. The soil taxonomists of the future must either be able to use the computer themselves or be sufficiently familiar with its potentialities in the areas discussed above to be able to cooperate with those who can use it.

References

de Gruijter, J. J. (1977). 'Numerical Classification of Soils and its Application in Survey.' (Pudoc: Wageningen.)

Dregne, H. E. (1976). 'Soils of Arid Regions.' Developments in Soil Science 6. (Elsevier: Amsterdam.)

FAO (1974). 'FAO-Unesco Soil Map of the World 1:5 000 000, Vol. 1: Legend.' (Unesco: Paris.)

Fitzpatrick, E. A. (1971). 'Pedology – a Systematic Approach to Soil Science.' (Oliver and Boyd: Edinburgh.)

Grove, P. B. (ed.) (1961). 'Webster's Third New International Dictionary of the English Language.' (G. Bell and Sons Ltd: London.)

Hicks, C. S. (1975). 'Man and Natural Resources – an Agricultural Perspective.' p. 94. (Croom Helm: London.)

Lawrence, G. M. H. (1951). 'Taxonomy of Vascular Plants.' (MacMillan: New York.)

Leeper, G. W. (1954). The classification of soils: an Australian approach. Trans. Fifth Internat. Congr. Soil Sci. Vol. 4, pp. 217-26.

Leeper, G. W. (1956). The classification of soils. *J. Soil Sci.* **7**, 59-64.

Moore, A. W. and Bie, S. W. (eds) (1977). 'Uses of Soil Information Systems.' Proceedings of the First Australian Meeting of the ISSS Working Group of Soil Information Systems, Canberra, Australia, 2-4 March 1976. (Pudoc: Wageningen.)

Moore, A. W. Cook, B. G. and Lynch L. G. (eds) (1981). 'Information Systems for Soil and Related Data.' Proceedings of the Second Australian Meeting of the ISSS Working Group on Soil Information Systems, Canberra, Australia, 19-20 February 1980. (Pudoc: Wageningen.)

Northcote, K. H. (1979). 'A Factual Key for the Recognition of Australian Soils.' 4th edn (Rellim Tech Publs: Glenside, S.Aust.)

Northcote, K. H., *et al.* (1960-68). 'Atlas of Australian Soils.' Sheets 1-10, with explanatory booklets. (CSIRO and Melb. Univ. Press: Melbourne.)

Northcote, K. H., Hubble, G. D., Isbell, R. F., Thompson, C. H. and Bettenay, E. (1975). 'A Description of Australian Soils.' (CSIRO: Australia.)

Pankhurst, R. J. (ed.) (1975). 'Biological Identification with Computers'. Proceedings of a Meeting held at King's College, Cambridge, 27-28 Sept. 1973. (Systematics Association and Academic Press: London.)

Prescott, J. A. (1931). The Soils of aAustralia in Relation to Vegetation and Climate. CSIR Aust. Bull. No. 52.

Russell, J. S. and Moore, A. W. (1968). Comparison of different depth weightings in numerical analysis of some anisotropic soil profile data. Trans. Ninth Internat. Congr. Soil Sci. Vol. 4, pp. 205-13.

Soil Survey Staff (1975). 'Soil Taxonomy – a Basic System of Soil Classification for Making and Interpreting Soil Surveys.' (USDA: Washington, D.C.)

Stace, H. C. T. (1961). 'The Morphological and Chemical Characteristics of Representative Profiles of the Great Soil Groups of Australia (A Supplement to 'A Manual of Australian Soils'. Second edition, 1956, by C. G. Stephens)'. (Division of Soils, CSIRO: Adelaide.)

Stace, H. C. T., Hubble, G. D., Brewer, R., Northcote, K. H., Sleeman, J. R., Mulcahy, M. J. and Hallsworth, E. G. (1968). 'A Handbook of Australian Soils.' (Rellim Tech. Publs: Glenside, S. Aust.)

Stephens, C. G. (1962). 'A Manual of Australian Soils.' 3rd edn. (CSIRO: Melbourne.)

USDA (1938). 'Soils and Men.' Yearbook of Agriculture 1938. (U.S. Govt. Printing Office: Washington, D.C.)

Williams, W. T. (ed.) (1976*a*). 'Pattern Analysis in Agricultural Research.' (CSIRO and Elsevier: Amsterdam.)

Williams, W. T. (1976*b*). Types of classifications. In 'Pattern Analysis in Agricultural Science', ed. W. T. Williams pp. 76-83. (CSIRO and Elsevier: Amsterdam.)

Chapter 21
Soil mapping in Australia

Frank R. Gibbons

In 1977, the Division of Soils celebrated the jubilee of its birth in 1927. But it was in 1827 that soil mapping began in the State of Victoria. William Dutton, a sealer who had called in at Portland Bay to take on fresh water, found a stretch of well-structured dark clay loam along the cliff tops there; he noted the fact, and later ploughed the first furrow in Victoria to plant his crops. For the next 150 years, people have been doing the same: locating areas of soil which appear to be suitable for their purpose, and then using them in that way.

There are the three essentials in soil mapping, identifying the soil, that is, both characterizing and classifying it on the basis of its features; assessing its suitability for use in one way or another; and locating and recording areas of the identified soil.

A number of issues is involved, as follows. Which, if any, soil uses have we in mind? Which soil characteristics do we choose to help identify the soil, and how do we measure them? Should we consider other environmental features, such as climate, which affect the performance of the soil? Does it matter what soil classification system we use, and what level of class? What are suitable units of mapping, and satisfactory techniques of locating? How best should we record and present the results, and to whom should they be presented?

It is useful to trace the development of soil mapping in Australia with the above issues in mind. The basic source of information is the nearly 300 separately published maps (with memoirs embracing a great range of approach) plus about 50 papers relevant to soil mapping. The chief immediate sources of information are a monograph by Taylor (1970) on the development of soil survey and field pedology over the period 1927–67 and the publication by Beckett and Bie (1978) on their examination of Australian soil and land-systems maps. A complete catalogue of Australian land-resource surveys is also available (Hallsworth and Archibald 1978).

Taylor distinguishes three phases in the 40-year period. Here, two earlier phases are added. The five are: up to 1890; 1890 to 1927; 1927 to 1940; 1940 to 1955; and 1955 to present.

During the first phase, soil was mapped by surveyors in the Departments of Lands, Agriculture, and Minerals, in the various Colonies. Features recorded were colour, depth and texture of soil, formation and chief species of vegetation, rock type and topography. The scale was normally 1:31 680 (40 chains to 1 inch). This was suitable for use by the Settlement Boards, although the land types were usually not delineated, only sketched in on the map. Two examples are the April 1855 parish plan of Portland, indicating the whereabouts of "limestone soil of fair quality, not timbered"; and the January 1876 parish plan of Kanawinka, showing the general locality of "alternate flats and ridges of poor sandy soil, free from stone, with clay subsoil, covered with heath and scrub and moderately timbered with stringybark".

By the 1860s, this type of information was being interpreted in rating systems with three or more classes, according to Colony, indicating the relative suitability of land for agricultural settlement.

In the mid-1800s, settlement was preceding the recording of such information, and eventually areas which had been tried but found wanting reverted to the Crown. With later settlement, for example the southern Victorian Mallee, the information was drawn upon to allow the selective release of the more suitable land. Indeed, the substantial proportion of Crown land still held by the States is partly a witness to what, on the whole, has been a far-sighted soil-conservationist attitude taken by the various Colonial and State Lands Departments. As Taylor remarks, "It is a sobering thought to all soil-scientists that the development of Australian virgin lands by pioneers was frequently successful through recognition of surface features of the soils and their associated vegetation long before any soil-surveyor came on the scene".

Similar stories can be told of attempts during the same period in other countries to indicate differences in soil/land types: for example Canada, where in 1857 Palliser correctly predicted the hazards of soil erosion by wind in parts of Alberta; and The Netherlands, which in 1856 was mapped by Staring (1856) at the scale of about 1:100 000 into areas each characterized in terms of geological material, relief, climate and soil features. We should not disparage the surveyors and administrators of this first phase and their achievements. Not blinded by science, they applied their powers of observation, perception and reasoning to the point where their empirical assessments were reliable and useful. They did this with dedication, often under conditions of great hardship.

Nevertheless, the interest "was almost entirely pragmatic" and the land classification elementary, as Stephens (1977) writes; "the approach was empirical, uncoordinated, and without the benefit of modern soil science".

In the second phase, Guthrie and Jensen in New South Wales attempted to improve the approach in two ways. First, they amassed information, having a modicum of scientific content, about soils submitted to them by farmers. This information included data on nutrient content, particle size and water-holding capacity. Secondly, in 'Soils of New South Wales' (Jensen 1914), they presented this information on a coordinated standardized basis for relatively large areas of eastern New South Wales, and later for parts of Queensland. It was at a very small scale, was not directly applicable by landholders, and not connected to the previous uncoordinated work at large scale, but based on what Taylor calls sketchy field work, and using a soil-classification system emphasizing topsoil and underlying rock (viz. the influence of Departments of Agriculture and of Minerals).

While this was happening in New South Wales (in 1899) in the United States the Soil Survey was under way, with a nationwide program of large-scale (1:63 360 or 1 mile to 1 inch) surveys at series and type level: that is based on using objective accurate data on colour and texture of topsoil and subsoil, and on drainage and relief. This was shortly to be linked to a comprehensive continent-wide classification based on the whole profile – the Great Soil Group system. In fact, the United States was already one phase ahead, looking at soils scientifically and in their own right, not just as a by-product of rocks and a raw material for agriculture. It was only after Prescott arrived on the scene in 1925, familiar with the American and Russian pedological approaches, and after the Council for Scientific and Industrial Research (CSIR) committee on which he served recommended 'the first organized soil-survey', that the opportunity came for the third phase to get under way in Australia.

But why were we behind? The reason is plain; then, as now, the resources available for soil mapping in the U.S. were greater by an order of magnitude. On a conservative estimate, the U.S. has committed to soil mapping and field pedology, for each acre of land, about twenty times the resources of trained people and of money as has Australia. It is easy to be critical of the Australian soils workers of the turn of the century for not building on the first phase, but they were so very few; and they did try to bring us into the modern world.

A suitable title for the third phase is 'The Science of Soils', with a subtitle 'Soils in Their Own Right'. When Prescott opened the door to the third phase, three courses could have been considered for soil mapping here. One was to copy the Americans by embarking on a nation-wide survey at large scale with standard criteria and categorical levels for the soil classes. Another was to examine the existing information in the States, to adopt a uniform approach with respect to scale and criteria, and taxa and procedures compatible with that information, and so to work towards a National Survey by coordinating the States and building on their information. Yet another was to ensure the independence of soil science and soil survey from particular factors of soil formation (e.g. rock type) and particular forms of land use (e.g. agriculture), and to start afresh, feeling the way, with one eye on American and Russian ideas and the other on Australian development problems (and funding).

The first course was impossible because of lack of resources, the second was what the Australian National Research Council of 1922 had wanted, but the third was what actually happened when CSIR started the first of its soil surveys at Renmark in South Australia (Taylor and England 1929). The glances overseas and at home had matched quite well: overseas was the large-scale mapping based on the detailed morphology of the soil profile, plus physical and chemical analyses; and at home were problems of water-logging and salinity in the intensively developed irrigation areas of the Murray Valley. Perhaps, in taking the third rather than the second course, the usefulness of a national mapping programme based on State co-ordination and data had been sacrificed on the altars of soil science and of the preoccupation of the southern States with their immediate problems. It is only now, 50 years later, that a National Soil and Land Resources Survey has (through state co-ordination) become a real possibility. Perhaps, however, the idea was to be a pathfinder for the States.

"The Renmark survey served as a pattern for the long succession of surveys in the irrigated lands in the Murray and Murrumbidgee valleys for the next fifteen years" writes Taylor. Traverses at close intervals with closer borings to six feet were made; the general concept of the U.S. soil series, type and phase was adopted. Practical differences were as follows. Australian soil series were based on a type profile, but the range of variation of profile characteristics was not usually defined; the textural variation was often wider than in the U.S.. The use of phases in Australia is different from that in the U.S. often being, by American definition, other series. In Australian surveys, many series and types are not named.

Taylor continues: "Soil types were defined on physical profile morphology using the Australian textural classes (Marshall 1947), parent material, crop-response, microrelief and approximate field assessments of the permeability of key horizons controlling internal drainage. They resulted from a combination of factual observations and intelligent judgements, from which the soil surveyor decided his mapping units ... Meticulous recording of soil and crop features gave the work a value still recognised today in the Irrigation Areas. The old soil types still stand".

Of course, they were not perfect, and it is clear from Talsma and Haskew (1959) that measurements of hydraulic conductivity are very variable, and that, for the information to be useful for tile-drainage schemes, it was necessary to consider soil features to greater depths. Nevertheless, the point is made by Beckett and Bie (1978) that "We encountered no map or survey of which the soil-classifications were more consistently praised than some of the early surveys of irrigation areas in soil types or phases".

This detailed approach was then adopted for dryland areas; adopted by CSIR for testing in representative areas such as the South Australian Ninety-Mile Desert (Taylor 1933), and in earnest by some States, particularly Western Australia, where, with enormous effort, 1.5 million acres (600 000 ha) were so examined for opening up 3500 proposed new wheat farms (Teakle 1939; Teakle *et al.* 1940).

But is was still too slow; there were too few surveyors for too many acres. A change had to be made from the American techniques. The point of departure and the two pathways leading from it are revealed in Taylor's comment "the Australian use of the soil series has always been more flexible than the American concept".

One apparent way of speeding things up was to reduce the precision. The broader use of the series – the wider textural variation – had already allowed a relatively high rate of mapping at 1:63 360, and it was broadened further in the King Island survey (Stephens and Hosking 1932) where series and their topographic phases were used at 2 miles to 1 inch (1:126 720). To broaden it further still, however, would take matters beyond the range of the American series-type system, so another and broader system was sought.

Now, having set up a Commonwealth-wide mapping organization, it was desirable that its output should be as widely applicable as possible, that is to be general purpose. The requirement for success in a general purpose classification is a high covariance of attributes, at least in each of the various categories of the system; that is, as distinct from a special purpose system, in which the requirement is a knowledge beforehand of the relevant attributes (Goodall 1966). The former requirement will be met in a classificatory system based on features with genetic significance, because in each category of the system a particular process will be operating, and therefore affecting a range of the properties of the various soils in that category.

A system of broad groupings of soil at higher level, purported to have genetic significance, was then developing overseas – the Great Soil Groups.

Prescott (1931) had used his own modification of these for the first map of The Soils of Australia at 1:19 million. Ten major soil groups were the basis for ten major soil zones (Prescott was careful to distinguish taxonomic and mapping units). Taylor and Hooper (1938) had acknowledged them in their very detailed (1:15 000) survey of the Murrumbidgee Irrigation Area. Stephens turned to them as a solution to the problem of logistics, by allowing quicker mapping at smaller scales, and also to the problems of ensuring that the information would be of general usefulness.

The possibility of this is seen first in the detailed 1 mile to 1 inch survey of County Grey in South Australia (Stephens *et al.* 1941), probably the quintessence of the third phase and in its application for a range of kinds of land use. Series and types were there colour coded into great soil groups. Then the approach was applied for the Soils of Tasmania (Stephens 1941), wherein the mapping units at 1:1 million were largely based on great soil groups. Stephens, through the soil maps of Australia at 1 to 5 million (Stephens, 1961, 1962) and through his writings to this day (Stephens 1977), has maintained his allegiance to great soil groups, and to mapping units derived from them, as appropriate vehicles for conveying generally useful information at small scale.

Clearly, however, the higher the taxonomic level, the less the precision and content of information. Also, because of the nature of soil distribution, a mapping unit at high taxonomic level usually in fact does not allow more rapid survey, except by ignoring the detail; thus, "rarely does grouping according to taxonomic similarity reduce the complexity of boundaries" (Soil Survey Staff 1951, p.302). The other departure path revealed in Taylor's comment was then tried; it ushered in the fourth phase, and the key word was "flexibility".

A willingness to experiment and to change is how Taylor describes this fourth phase. A start was made by looking not at the breadth of the taxonomic unit, but at the mapping-unit.

The repetitive sequence of soils with topography had been pointed out in East Africa by Milne (1935) who regarded it as a convenient mapping unit. Similar patterns in the Mount Gellibrand area of Western Victoria were recognized and described by Leeper *et al.* (1936) in the classic first of the modern soil studies in Australia. In County Moira in Victoria the patterns, as well as the series and types, were recognized and mapped as such by Butler *et al.* (1942). It was realized that an understanding of the patterns enabled the surveyor to predict the location of the component soil types from a knowledge of the topography or of some easily identifiable feature linked with it.

This is a mapping tool of great strength, because, for the same effort, other than the need to confirm the predictions, it allows smaller scales and more rapid coverage with no corresponding reduction in precision of information; it presumes an entirely consistent association of environmental variables, and its use relies on a knowledge of the nature of that association in the area concerned. The important part of the soil-mapper's work in this approach is to determine that nature: the environmental pattern, the local or regional model for soil distribution (if it exists), and the basis for it. Some at least of the soil mappers must become pedologists too.

The local pattern of soils was termed the soil association, and techniques were evolved to determine it by the detailed examination of small representative areas. The patterns themselves were mapped at various scales, and the component soils were identified at taxonomic levels varying from type (Skene and Walbran 1949), through series or families (Hubble and Thompson 1953), to great soil groups (Downes and Sleeman 1955), according to level of pattern.

The approach was adopted whole heartedly; it was not really new and must have been used informally in most previous surveys. Indeed, Taylor says of the early surveys . . . "there is no doubt that this early classification, while firmly based on the whole soil, made use of a synthesis of soil and environment in its application to mapping". It is almost impossible not to become aware of environmental relationships and patterns when mapping soils, especially in relatively undisturbed areas as in many parts of Australia. The Hallsworth school at Sydney, a little later but independently, happened on the approach in their Studies in Pedogenesis in New South Wales (Hallsworth *et al.* 1952), and applied it in regional surveys there. Overseas too, a similar approach was adopted, for example in the Netherlands (Steur 1961).

So, both the soil-association approach and the detailed soil-type-phase technique were available to meet the great demands for information during the period immediately following the Second World War. This demand had been stimulated by the reports of the Rural Reconstruction Commission (1944) which recommended soil surveys as a prerequisite in reconstruction schemes.

Requests involved assessments of Crown lands for settlement, or of occupied land for closer settlement, especially in areas deficient in trace elements, viz. South Australia, Tasmania and Western Australia. Here, respectively, during the fourth phase, nearly 2 million, 1 million and 0.5 million acres (800 000, 400 000 and 200 000 ha) were surveyed in detail, and 4000, 2000, and 5000 square miles (10 000, 5000 and 12 500 km^2) in soil association. In subtropical areas, too, the requests were met: 9000 square miles in Queensland and 1000 square miles in North-Western Australia were mapped by soil association, and 80 000 acres (32 000 ha) near the Ord River in detail. In addition, the detailed surveys of irrigation areas still proceeded apace in New South Wales and Victoria, totalling nearly 1.5 million acres (600 000 ha). It was the heyday of soil mapping in Australia – 'the golden age', says Taylor.

But change was in the air. First, with all the effort being put into soil mapping, people were questioning how useful it was going to be. Plainly, the surveyed soils were intended for many different kinds of use in different environments. Was the same information going to meet all the requirements? Butler (1958) thought not. Distinguishing the edaphic (crop response), geographic and pedological aspects of field soils studies, he asserted that the criteria of none of the taxonomic units from type to great soil group were adequate for the edaphic aspect. He proposed a programme of determining jointly with the mapping programme which soil features are relevant for particular kinds of production. This developed strongly in the fifth phase, but rarely, however, jointly with the mapping programme.

Secondly, Leeper (1952, 1954, 1956) was persistently questioning the basis of the great soil group system, "guesses about their origin", not proven genesis. The point is important. If the genetic significance given to the features that are the basis of the great soil groups is not valid, then the system cannot be expected to be successful as a general purpose system nor can the mapping units and programme derived from it. Leeper wanted to replace it with one or more systems based on a few selected soil attributes, and constructed on known values of those attributes. Butler agreed, but specified that those few soil attributes should be selected as relevant to the purpose. Northcote, with the responsibility for preparing the Atlas of Australian Soils, took the point.

Thirdly, the concept of land as distinct from soil was developing. Related to the ecosystem concept, it is the idea of the whole complex of environmental factors, environmental features and site conditions operating in inter-relationship. Land is an area of such a complex. An ecosystem is such a complex operating as a whole unit (Ovington 1962). A land system is an area of an ecosystem where some of the land features covary. Just after World War II, extensive land appraisals were made in northern Australia with a view to possible development, and the land system approach was first used there. A land system was defined as "an area, or groups of areas, throughout which there is a recurring pattern of topography, soils and vegetation. . . . the product of a common geomorphological phenomenon" (Christian and Stewart 1953).

It could be likened to a small-scale version (1:1 million to 1:0.25 million) of the soil association approach, with soil taxonomic units at the great soil group to high family level, but with the very important difference that other environmental features were included in the pattern because of their relevance to land performance, as well as their possibly being a diagnostic mapping tool. The reports which appeared in a long series beginning in 1953 (e.g. Christian and Stewart 1953; Perry *et al.* 1962) contain representations of patterns of the environment on which, usually on the basis of physiography, is hung the associated information about the other features of the land. So, as well as the relevant soil attributes being sought by Butler, other environmental features were now being added. It was at this stage, when large areas were being mapped with the use of diagnostic features, that heavy reliance was placed on the interpretation of aerial photography, which then became a standard technique.

In his monograph on the Ecosystems of the Monaro Region of New South Wales, Costin (1954) used a somewhat similar approach, but emphasized in his treatment of the soils and vegetation the ecological processes and the possibly harmful consequences of interfering with them. Costin's approach was adopted by the Soil Conservation Authority of Victoria (Rowan and Downes 1963). One aspect of soil mapping in Australia had come full circle. The importance put on vegetation by the surveyors of phase 1, but discouraged during the second and third phases, was again appearing in the guise of ecosystems. In fact, later still, it appeared in the form of an extensive survey, with considerable land use consequences, of soils of one particular vegetative community (Isbell 1962). It was inevitable, however, in Australia where there are large areas with relatively undisturbed vegetation that are highly diagnostic of site condition and processes, for such relationships to be used as a mapping tool, and also as an aid to understanding what was being done to the land. For if we seek out and use soils, should not we husband them too?

The stirrings of the wider awareness of conservation are seen in Costin's work. The concern with soil conservation is explicit in the survey by Stephens *et al.* (1945), Blackburn and Leslie (1949) and Downes (1949), wherein Australian soils were mapped primarily for the control of soil erosion. Emphasis was shifting from productivity to the avoidance of hazard. By the end of the fourth phase, the Soil Conservation Services of the various States had become established and were moving into soil mapping. This commitment was to become very significant some time later, because, towards the end of the fourth phase, the volume of requests for soil surveys dwindled sharply. 'The high tide rather quickly began to ebb' (Taylor); the heyday noon had passed, and become clouded over with confusion, doubt, and even disbelief.

The fifth phase, in response to that ebb, is the rationalizing of soil-land mapping so that it will be worthwhile. Let us keep in mind that the worth of soil mapping is determined by its usefulness compared to cost. In turn, usefulness depends on two things; the predictive ability of the mapping programme, including both the delineating and the classifying, and the relevance of the predicted information to the purpose.

There have been five distinct areas of rationalizing in the fifth phase.

One area is the predictive ability of the mapping programme. Part of this was the development of the regional soil-distribution models necessary for the successful use of the soil association, land system, and ecosystem approaches. Such work had been carried out during the earlier phases, notably on lateritic dissection (Stephens 1946), on pedogenesis in arid areas (Crocker 1946), and on pedogenesis in New South Wales by the Hallsworth school, but it developed vigorously during the earlier part of the fifth phase.

The key figure was Butler who, with his colleagues van Dijk, Churchward, Walker and Beattie, put out a series of publications on periodic phenomena in landscapes as a basis for soil-studies (Butler 1959; van Dijk 1959; Churchward 1963; Walker 1963*a*, *b*; Pels 1964; Beattie 1972). The studies involve soil-layering and former surfaces which may be buried or exposed, and relationships with age. The principles appear to be applicable in a wide range of geography and circumstances: for example, Alice Springs (Litchfield 1969) and a zone in north-eastern Victoria (Rowe *et al* 1978). The most recent version, called the pedomorpholith concept (van Dijk *et al.* 1968), is proving particularly useful along the fringes of the Highland zone (J. N. Rowan personal communication).

Other more localized geomorphic models are of the Lowan ridges of the Murray Basin (Blackburn *et al.* 1965, 1967), soils of the arid zone (Jessup 1960), lateritic areas in Western Australia (Mulcahy and Hingston 1961), basaltic areas (Nicolls and Tucker 1956; Loveday 1957) plus the pedogenetic bases for these soils (Oertel 1968; Brewer 1964). These studies provided much of the impetus for the soil association mapping (e.g. Jackson 1962; Paton 1971) and land-system mapping of the fifth phase.

Now, when these various studies are applied appropriately in a mapping programme, they should allow a high predictive ability for the component soils of the pattern, viz. a high purity of the mapping unit. The need then develops for a high predictive ability regarding the attributes of the soil so identified, viz. a high uniformity of the soil class. This requires either a soil classification system with known high covariance of attributes, or a system with selected attributes and constructed on known values of those attributes. The Great Soil Groups classification purported to be the former, but had been strongly challenged. The latter, however, became available in 1960 in the form of Northcote's Factual Key (Northcote 1960, 1971) which he had prepared for the Atlas of Australian Soils (Northcote *et al.* 1960–68). The use of the Factual Key, together with the models of soil distribution, may complete the requirements for high predictive ability in soil mapping.

Have they done so? As to purity, which is measured as the percentage of random sites correctly predicted for soil class (or proportion of soil classes for compound mapping units), a study by Chittleborough (1978) of land systems around Monarto in South Australia showed that the survey there was a success.

Uniformity is measured in terms of the statistical variance of individual soil properties in relation to soil class (Bie and Ulph 1972): for example, intra-class correlation coefficient, relative variance (Beckett and Burrough

1971), and Wilkes' Criterion (Webster 1971). The Monarto study showed success for texture and depth, but not for many other properties. Russell and Moore (1968) show that Northcote's Principal Profile Forms are as good as the great soil groups for a wide range of soil attributes and clearly must have high uniformity for those attributes on which Northcote's Key is based. However, Moore *et al.* (1972) show little strong correlation between attributes for south-eastern Queensland soils, and the most exhaustive study of uniformity in Australian soils (Webster and Butler 1976) is still less encouraging. There, a map, two systematic classifications and a numerical classification, of soils of an area said to be representative of the Southern Tablelands of NSW 'were unable to predict values of soil attributes sufficiently well to be useful'. General purpose soil classification and soil mapping were not worthwhile, in that area, for the attributes considered. Moreover, as much of the variation for some properties was over very short distances, it seems that (depending on the attributes considered) even special purpose mapping at large scale can show only the general trends in variation. Nevertheless, the variance in an area less than that to which separate management can be given is of little practical interest in soil mapping.

On the whole, the tests so far of the predictive ability of Australian soil mapping have not been encouraging, but the weakness lies more in the soil classes than in the mapping techniques. Encouragement does come, however, from the other four areas of rationalizing.

The second area of rationalizing is the knowledge of the preferences of the users of soil maps and of how to meet those preferences.

As a result of the study by Beckett and Bie (1978), there is now a better understanding in this area. They have shown that the intensity of land use and the function of user determine the preferred level of soil class, which in turn, has largely determined the kind of mapping unit and the scale. Nowhere is the preferred level of soil class higher than the family. The lowest level of Northcote's system, the Principal Profile Form, has a degree of detail which just comes within this range; happily so, because the Atlas of Australian Soils (Northcote *et al.* 1960–68), which is the largest scale (1:2 million), complete, uniform coverage of the whole continent, has mapping-units in which the listed soil classes are at that level. That the other influence on mapping unit and scale has been the availability of resources is borne out in the strong relationships between effort (cost) and both mapping-unit and scale. The understanding developed by these studies will lead to the choice of strategies and procedures of mapping, and forms of presentation of maps, in tune with the needs of the user.

The third area of rationalizing follows logically from the second. It is the consideration of the kind of use to be made of the soil or land – the special-purpose approach; the directing of attention to what is relevant (relevance being the other component of utility). This third area has developed from Butler's 1958 advocacy of edaphic studies, and these, too, were started early in the fifth phase. Loveday (1962), Cockcroft *et al.* (1962), Spencer and MacArthur (1962), Fergus (1962) and Colwell (1968) are some of the workers.

The formalizing of such an application of soil mapping has come late to Australia. The starting point for the formalizing was the land capability classification of the USDA Soil Conservation Service (Klingebiel and Montgomery 1961), an eight-class rating system of land for the hazard of soil erosion, based on eight value ranges of selected soil/land attributes. Local modifications for use in the various Australian States led to the recognition of areas to which the classes applied, for example the land-husbandry units of Downes (1949). They also led to the identification of further parameters relevant to productivity as well as to erosion hazard under agriculture (Dawson 1972). Ratings of mapping units for agricultural suitability then became an integral part of mapping programmes, particularly in the Victorian Irrigation Areas (Skene and Poutsma 1962), in proposed irrigation areas elsewhere (McDonald 1969) and also, at a later stage, in the CSIRO Soils and Land Use Series (Murtha 1975). That development is now going to its logical conclusion with the identification of relevant attributes, and the making of separate rating systems, for each of a comprehensive range of specific kinds of land use (Officers of Soil Cons. Bch. DPI 1974; Rowe *et al.* 1979). The rating systems are loosely modelled on the Soil Survey Interpretations of the USDA (United States Department of Agriculture SCS 1973), and the latter themselves are being widely applied in soil mapping for urban development (Quilty *et al.* 1978; Jeffery and Howe 1976). So, once again, Australia takes an American concept, now on the interpretation of mapping units, and uses it more flexibly.

This third area is already influencing soil mapping. Thus, because some of the specific land uses are intensive, for example urban development, there is an increasing proportion of surveys with mapping units at large scale and soil classes at low level, regardless of mapping technique. The corollary is that any system of classification which is intended to be widely applicable must be developed to low categorical levels. This is already happening in some States through the use of appropriate extensions of Northcote's Principal Profile Forms.

Also, because, as Leeper and Butler wanted, soil classes are becoming more based on selected attributes as these are being added from the rating systems to the existing classification systems, greater predictive ability is likely to result. As those attributes are relevant for the purpose, greater utility is to be expected.

The difficulty is that attributes which are relevant are

not necessarily mappable, just as those that are mapped are not always relevant. This makes mappers still hope for correlations between mappable and relevant attributes. It is of some significance then, that often the parameters relevant for particular kinds of land use are not the individual soil attributes (e.g. content of aluminium) but site conditions compounded of various soil and environmental attributes (e.g. availability of water, erodibility, and soil workability). The latter may have stronger correlations than the individual attributes, according to Webster and Butler (1976, p.17). Consequently, the general-purpose approach cannot yet be dismissed, a strange byproduct of special purpose mapping! It seems to be a characteristic of Australian soil mapping to have it both ways, a general-purpose and special-purpose, high level and low level of soil class, and all used side by side.

The third area includes the consequences of land use too. Outstanding examples are the assessment of the salinity status of Australian soils at the scale of 1:5 million by Northcote and Skene (1972), and the mapping at 1:375 000 of salt-affected soils in north-western Victoria by Rowan (1971).

The fourth area of rationalizing is the development of the land system approach.

Thus, the South Australian Department of Agriculture (Potter *et al.* 1973) makes simple clear presentations of the wanted information by the use of imaginative arrangements of annotated landscape diagrams, photographs and abbreviated texts.

A rationale of the land system approach of the Victorian Soil Conservation Authority has been attempted by Gibbons and Haans (1976). They claim that the use of patterns at three levels of resolution allow, at the one scale of publication, the identification of soil/land types down to quite low class level, plus the information required for their rapid delineation. With the participation of extension staff in applying the approach, it is seen as a partial solution to the problem of scale in relation to effort required. Attention is also drawn to the processes operating in the land system as a whole, as the proper basis for assessing hazard and the means of avoiding it, (Rowan and Aldrick 1980).

The PUCE (Pattern, Unit, Component, Evaluation) Programme (Aitchison and Grant 1968) is an evaluation of terrain for engineering purposes. It attempts to rationalise procedures for the gathering and processing of engineering information to be used in the preplanning, planning, and project stages of civil engineering works. Being quantitative, it draws attention to the use of objectively evaluated parameters, some detected by remote sensing.

In all these variants of the land system approach, the component soil class or detailed land-type is the vehicle for conveying the use information, such as ratings.

The fifth area of rationalizing is that of cost, the remaining component of worthwhileness.

Again, we turn to Beckett and Bie and their colleagues (Bie and Beckett 1971*a*, *b*; Burrough *et al.* 1971; Bie and Ulph 1972). On Australian data, they show that effort or cost is related unfavourably to length of mapped boundary, favourably to the use of a pattern approach, and favourably to aerial photography interpretation. They also point to the disproportionately high cost of professional staff. Their treatment reveals the important point that soil differences exert their effect through determining the required management in land use, rather than the potential productivity, and they offer a method for relating cost and usefulness of soil mapping in terms of management alternatives.

Their work indicates the direction of some of the desirable changes in soil mapping: for example, the development of profitable remote-sensing techniques, especially those used with computer pattern analysis of particular selected attributes.

The main value of their work, however, lies in showing both the need and the means for making presurvey appraisals of proposed mapping projects. These are now possible on the basis of data on the proposed land use, on management alternatives with their relative costs, on purity and uniformity, and on the relationships of all these to survey techniques, to mapping-units and to level of soil class. A reconnaissance of the survey area and discussions between soil mappers and map users are an essential part of this; the programme, procedures, staffing, information, and presentation can then be attuned to needs.

The fifth phase of soil mapping is now finished; not that rationalizing will cease, but radical changes in land use are going on in Australia and we must apply the results of the last twenty years of rationalizing.

I will not speculate on the future, I have only been digging into history. I did not even want to write this Chapter, but now I am glad that I have done so, even if others are not, because it has shown me things which I had not appreciated before.

First, to quote Beckett and Bie (1978) . . . 'Australia has produced an extraordinarily wide range of soil and land-system maps. Together these provide a major contribution to soil survey methodology'. The scope and variety of approach is remarkable. This can be attributed, as those authors suggest, to the unique combination of a whole continent to map, a similarly wide range of environments, contrasting levels of development, one political unit, but a number of soil-mapping organizations not all centralised into one national organization.

Secondly, I think that it is still people rather than standard techniques that are important in soil mapping. The unique combination has allowed personalities to contribute to the development of the various approaches, each appropriate to particular circumstances and needs.

Finally, like J. K. Taylor, I have found a great admiration for those people right back to William Dutton who have contributed to the mapping, and use, of the soils of Australia.

Acknowledgments

For help given in the preparation of the Chapter, I thank particularly my colleagues, Mr R. Carne, Dr R. H. M. van de Graaff, Mr J. N. Rowan, Mr T. I. Leslie, Dr J. J. Jenkin, Dr N. Alley, Mr G. D. Hubble and Mr R. F. Isbell, and my wife Pat. I also acknowledge the help of the Soil Conservation Authority of Victoria through its Chairman, Mr A. Mitchell, in allowing the time and facilities needed.

References

Aitchison, G. D. and Grant, K. (1968). The PUCE programme of terrain description, evaluation and interpretation for engineering purposes. Proc. 4th Reg. Conf. (Africa) Soil Mech. and Found. Engng. Vol. 1, pp. 1-8.

Beattie, J. A. (1972). Ground surfaces of the Wagga Wagga region New South Wales. CSIRO Aust. Soil Publ. No. 28.

Beckett, P. H. T. and Bie, S. W. (1978). Use of soil and land system maps to provide soil information in Australia. CSIRO Aust. Div. Soils Tech. Pap. No. 33.

Beckett, P. H. T. and Burrough, P. A. (1971). The relation between cost and utility in soil survey. V. The cost-effectiveness of different soil survey procedures. *J. Soil Sce.* **22**, 481-9.

Bie, S. W. and Beckett, P. H. T. (1971*a*). Quality control in soil survey. I. The choice of mapping unit. *J. Soil Sci.* **22**, 32-49.

Bie, S. W. and Beckett, P. H. T. (1971*b*). Quality control in soil survey. II. The costs of soil survey. *J. Soil Sci.* **22**, 453-65.

Bie, S. W. and Ulph, A. (1972). The economic value of soil survey information. *J. Agric. Econ.* **23**, 285-97.

Blackburn, G., Bond, R. D. and Clarke, A. R. P. (1965). Soil development associated with stranded beach ridges in south-east South Australia. CSIRO Aust. Soil Publ. No. 22.

Blackburn, G., Bond, R. D. and Clarke, A. R. P. (1967). Soil development in relation to stranded beach ridges of County Lowan, Victoria. CSIRO Aust. Soil Publ. No. 24.

Blackburn, G. and Leslie, T. I. (1949). Survey of soils, land use and soil erosion in the Coleraine district. CSIRO Aust. Div. of Soils, Soil and Land Use Ser. No. 3.

Brewer, R. (1964). 'Fabric and Mineral Analysis of Soils'. (John Wiley and Sons Inc.: New York.)

Burrough, P. A., Beckett, P. H. T. and Jarvis, M. G. (1971). The relation between cost and utility in soil survey. I to V. *J. Soil Sci.* **22**, 360-94, 466-89.

Butler, B. E. (1958). Diversity of concepts about soils. *J. Aust. Inst. Agric. Sci.* **24**, 14-20.

Butler, B. E. (1959). Periodic phenomena in landscape as a basis for soil studies. CSIRO Aust. Soil Publ. No. 14.

Butler, B. E., Baldwin J. G., Penman, F. and Downes, R. G. (1942). Soil survey of part of County Moira, Victoria. CSIR Aust. Bull. No. 152.

Chittleborough, D. J. (1978). Soil variability within land systems and land units at Monarto, South Australia. *Aust. J. Soil Res.* **16**, 137-56.

Christian, C. S. and Stewart, G. A. (1953). General report on survey of Katherine-Darwin Region 1946. CSIRO Aust. Div. Land Res. Land Res. Ser. No. 1.

Churchward, M. (1963). Soil studies at Swan Hill, Victoria. IV. Ground surface history and its expression in the array of soils. *Aust. J. Soil Res.* **1**, 242-55.

Cockcroft, B., Bakker, A. C. and Wallbrink, J. C. (1962). The relation between soil features and fruit tree growth. Proc. 3rd Aust. Conf. in Soil Sci., Canberra, Vol. 1, p. 26.

Colwell, J. D. (1968). Calibration and assessment of soil tests for estimating fertiliser requirements. 2. Fertiliser requirements and an evaluation of soil testing. *Aust. J. Soil Res.* **6**, 93-104.

Crocker, R. L. (1946). Post-Miocene climatic and geologic history and its significance in relation to the genesis of the major soil types of South Australia. CSIR Aust. Bull. No. 193.

Costin, A. B. (1954). A study of the ecosystems of the Monaro Region of New South Wales with special reference to soil erosion. Soil Cons. Serv. N.S.W. (Govt. Printer: Sydney.)

Dawson, N. M. (1972). Land inventory and technical guide, Miles area Queensland. Div. Land Utilization, Dep. Prim. Ind. Queensl. Tech. Bull. No. 5.

Downes, R. G. (1949). A soils, land use and erosion survey of parts of the Counties of Moira and Delatite, Victoria. CSIR Aust. Bull. No. 243.

Downes, R. G. and Sleeman, J. R. (1955). Soils of the Macquarie Region, New South Wales. CSIRO Aust. Div. Soils, Soils Publ. No. 4.

Fergus, I. (1962). Nutrient status of some soils of the brigalow lands. CSIRO Aust. Div. Soils Div. Rep. No. 1/62.

Gibbons, F. R. and Haans, J. C. F. M. (1976). Dutch and Victorian approaches to land appraisal. Soil Surv. Inst., Wageningen, Soil Surv. Pap. No. 11.

Goodall, D. (1966). Classification, probability and utility. *Nature* (London) **211**, 53-64.

Hallsworth, E. G. and Archibald, J. (1978). Catalogue of Australian land resource surveys; Rep. No. 4; Res. Directrte, Comm. Dept. Envnmt., Housg, and Commty. Devt. (AGPS: Canberra.)

Hallsworth, E. G., Costin, A. B., Gibbons, F. R. and Robertson, Gwen K. (1952). Studies in pedogenesis in New South Wales; II. The Chocolate Soils. *J. Soil Sci.* **3**, 103-24.

Hubble, G. D. and Thompson, C. H. (1953). The soils and land use potential of the Lower Burdekin Valley, North Queensland. CSIRO Aust. Div. Soils, Soils and Land Use Ser. No. 10.

Isbell, R. F. (1962). Soils and vegetation of the brigalow lands in eastern Australia. CSIRO Aust. Div. Soils, Soils and Land Use Series No. 48.

Jackson, E. A. (1962). Soils Studies in Central Australia. CSIRO Aust. Soils Publ. No. 19.

Jeffery, P. J. and Howe, D. F. (1976). A study of the land in the proposed Yarra Valley Metropolitan Park and interpretations for land based recreational and related activities. Foolscap Rep. (Soil Conserv. Auth., Vic.)

Jensen, H. I. (1914). 'The Soils of New South Wales'. (Dep. Agric. N.S.W.: Sydney.)

Jessup, R. W. (1960). Introduction to the soils of the south-eastern portion of the Australian arid Zone. *J. Soil Sci.* **11**, 92-105.

Klingebiel, A. A. and Montgomery, P. H. (1961). Land Capability Classification; Agricultural Handbook No. 210 (U.S. Gov. Print. Off.; Washington, D.C.)

Leeper, G. W. (1952). On classifying soils. *J. Aust. Inst. Agric. Sci.* **18**, 77-80.

Leeper, G. W. (1954). Classification of soils. *Proc. 5th Cong. Internat. Soc. Soil Sci.* **4**, 217-26.

Leeper, G. W. (1956). The classification of soils. *J. Soil Sci.* **7**, 59-64.

Leeper, G. W., Nicholls, A. and Wadham, S. M. (1936). Soil and pasture studies in the Mount Gellibrand area, Western District of Victoria. *Proc. R. Soc. Vic.* **49** (N.S.) Pt 1, 77-138.

Litchfield, W. H. (1969). Soil surfaces and sedimentary history near the MacDonnell Ranges, Northern Territory. CSIRO Aust. Soil Publ. No. 25.

Loveday, J. (1957). The soils of the Sorrel-Carlton-Copping Area, South East Tasmania; with special reference to the soils formed on basalt. CSIRO Aust. Soil Publ. No. 8.

Loveday, J. (1962). Study of the relationships between yield of irrigated lucerne and the properties of some grey and brown soils of heavy texture in south-western New South Wales. *Aust. J. Soil Res.* **2**, 96.

McDonald, R. C. (1969). Evaluation of soils for irrigation, Nogoa River, Emerald, Queensland. Chem. Branch. Queensl. Dep. Primary Ind. Tech. Rep. No. 1.

Marshall, T. J. (1947). Mechanical composition of soil in relation to field descriptions of texture. CSIR Aust. Bull. No. 224.

Milne, G. (1935). Some suggested units of classification and mapping, particularly for East African soils. *Soil Res.* **4**, 183-98.

Moore, A. W., Russell, J. S. and Ward, W. T. (1972). Numerical analysis of soils: a comparison of three soil profile models with field classification. *J. Soil Sci.* **23**, 193-209.

Mulcahy, M. J. and Hingston, F. J. (1961). The development and distribution of soils of the York-Quairading Area, Western Australia, in relation to landscape evolution. CSIRO Aust. Soil Publ. No. 17.

Murtha, G. G. (1975). Soils and land use on the northern section of the Townsville coastal plain, north Queensland. CSIRO Aust. Div. Soils, Soils and Land Use Ser. No. 55.

Nicolls, K. D. and Tucker, B. M. (1956). The pedology and chemistry of the basaltic soils of the Lismore District, N.S.W. CSIRO Aust. Soil Publ. No. 7.

Northcote, K.H. and Skene, J.K.M. (1972). Australian soils with saline and sodic properties. CSIRO Aust. Soil Publ. No 27.

Northcote, K. H. (1960). A factual key for the recognition of Australian soils. CSIRO Aust. Div. Soils Div. Rep. No. 4/60.

Northcote, K. H. (1971). 'A Factual Key for the Recognition of Australian Soils'. 3rd edn. (Rellim Tech. Pub.: S. Aust.)

Northcote, K. H. with Beckmann, G. G. , Bettenay, E., Churchward, H. M., Van Dijk, D. C., Dimmock, G. M., Hubble, G. D., Isbell, R. F., McArthur, W. M., Murtha, G. G., Nicholls, K. D., Paton, T. R., Thompson, C. H., Webb, A. A. and Wright, M. J. (1960-1968). 'Atlas of Australian Soils' Sheets 1 to 10 with Explanatory Booklets. (CSIRO and Melb. Univ. Press: Melbourne.)

Oertel, A. C. (1968). some observations incompatible with clay illuviation. *Ninth Internat. Congr. Soil Sci. Trans.* Vol. 4, pp. 481-8.

Officers of Soil Conservation Branch, Department of Primary Industry (1974). Moreton Region: Non-Urban Land Suitability Study; Part I; Tech. Bull. No. 11; Div. Land Utilis. Dep. Prim. Ind. Brisbane, Queensland.

Ovington, J. D. (1962). Quantitative ecology and the woodland ecosystem concept. *Adv. Ecol. Res.* **5**, 213-17.

Paton, T. R. (1971). A reconnaissance survey of soils in the Boonah-Beaudesert district, Queensland CSIRO Aust. Div. Soils, Soils and Land Use Ser. No. 52.

Pels, S. (1964). Quaternary sedimentation by prior streams in the riverine plain south-west of Griffith, N.S.W. *J. Proc. R. Soc. N.S.W.* **97**, 107-15.

Perry, R. A., Mabbutt, J. A., Litchfield, W. H., Quinlan, T., Lazarides, M., Jones, N. O., Slatyer, R. O., Stewart, G. A., Bateman, W. and Ryan, G. R. (1962). General report on land of the Alice Springs area, Northern Territory 1956-57. CSIRO Aust. Div. Land Res., Land Res. Ser. No. 6.

Potter, J. S., Wetherby, K. G. and Chittleborough, D. J. (1973). A description of land in County Albert, County Alfred and part of County Eyre, South Australia. Soil. Cons. Branch Dep. Agric. S. Aust. Rep. LD.1.

Prescott, J. A. (1931). The soils of Australia in relation to vegetation and climate. CSIR Aust. Bull. No. 52.

Quilty, J. A., Murphy, B. W. and Voler, A. (1978). Bathurst village studies; Pt III, Urban capability study, Raglan. (Soils Cons. Serv. N.S.W.: Sydney.)

Rowan, J.N. (1971). Salting on dryland farms in north-western Victoria. Soil Cons. Auth. Vic. Tech. Comm. No. 7.

Rowan, J.N. and Aldrick, J.M. (1980). Presenting data on land deterioration in land system surveys. Paper presented to National Soils Conference, Sydney, 1980.

Rowan, J. N. and Downes, R. G. (1963). A study of the land in north-western Victoria. Soil Cons. Auth. Vic. Tech. Comm. No. 2.

Rowe, R. K., Crouch, F. J. and van Dijk, D. C. (1978). Soils of the upper valleys of the river Murray basin. *Proc. R. Soc. Vic.* **90**, Pts 1 and 2, pp. 157-66.

Rowe, R. K., Howe, D. F. and Alley, N. (1979). Guidelines for land capability assessment. Foolscap Rep. (Soil Cons. Auth.: Vic.)

Rural Reconstruction Commission (1944). 'A general rural survey'. (The Rural Reconstr. Commsn AGPS.: Canberra.)

Russell, J. S. and Moore, A. W. (1968). Comparison of different depth weightings in the numerical analysis of anisotropic soil profile data. Ninth Internat. Congr. of Soil Sci. Trans. Vol. 4, pp. 205–13.

Skene, J. K. M. and Poutsma, T. J. (1962). Soils and land use in part of the Goulburn Valley, Victoria. Dep. Agric. Vic. Tech. Bull. No. 14.

Skene, J. K. M. and Walbran, W. I. (1949). Soil survey of parts of the parishes of Nuntin and Bundalaguah, County of Tanjil, Victoria. Dep. Agric. Vic. Tech. Bull. No. 8.

Soil Survey Staff (1951). 'Soil Survey Manual'. U.S. Dep. Agric. Hand. No. 18. (Govt. Printer: Washington, D. C.)

Spencer, K. A. and McArthur, W. M. (1962). Variability in nutrient status between apparently similar soils. Proc. 3rd Aust. Conf. Soil Sci., Canberra. Vol. 1, paper. No. 25.

Staring, W. C. H. (1856). 'De Bodem van Nederland. De samenstelling en het onstaan der gronden in Nederland ten behoeve van het algemeen beschreven'. (Kruseman: Haarlem).

Stephens, C. G. (1941). The soils of Tasmania. CSIR Aust. Bull. No. 139.

Stephens, C. G. (1946). Pedogenesis following the dissection of lateritic regions in southern Australia. CSIR Aust. Bull. No. 206.

Stephens, C. G. (1961). The soil landscapes of Australia. CSIRO Aust. Soil Publ. No. 18.

Stephens, C. G. (1962). 'A Manual of Australian Soils' (CSIRO: Melbourne.)

Stephens, C. G. (1977). Soils. *In* 'Australia – a Geography'. (Ed. D. N. Jeans). (Sydney Univ. Press: Sydney.)

Stephens, C. G. , Crocker, R. L., Butler, B. E. and Smith, R. (1941). A soil and land use survey of the Hundreds of Riddoch, Hindmarsh, Grey, Young and Nangwarry, County Grey, South Australia. CSIRO Aust. Bull. No. 142.

Stephens, C. G. and Hosking, J. S. (1932). A soil survey of King Island. CSIR Aust. Bull. No. 70.

Stephens, C. G., Herriot, R. I., Downes, R. G., Langford-Smith, T. and Acock, A. M. (1945). A soil, land use and erosion survey of part of County Victoria, South Australia. CSIR Aust. Bull. No. 188.

Steur, G. L. (1961). Methods of soil surveying in use at the Netherlands Soil Survey Institute. *Boor en Spade* **11**, 59-77.

Talsma, T. and Haskew, H. C. (1959). Investigation of water table response to tile drains in comparison with theory. *J. Geophys. Res.* **64**, 1933-44.

Taylor, J. K. (1933). A soil survey of the Hundreds of Laffer and Willalooka, South Australia. CSIR Aust. Bull. No. 76.

Taylor, J. K. (1970). 'The Development of Soil Survey and Field Pedology in Australia, 1927–1967.' (CSIRO: Melbourne.)

Taylor, J. K. and England, H. N. (1929). A soil survey of Block E (Renmark) and Ral Ral (Chaffey) Irrigation Areas. CSIR Aust. Bull. No. 42.

Taylor, J. K. and Hooper, P. D. (1938). A soil survey of the horticultural soils in the Murrumbidgee Irrigation Areas, New South Wales. CSIR Aust. Bull. No. 118.

Teakle, L. J. H. (1939). Soils of the 3500 Farms Scheme area, Western Australia. *J. Dep. Agric. W. Aust.* **16**, 202-30.

Teakle, L. J. H., Southern, B. L. and Stokes, S. J. (1940). A soil survey of the Lakes District, Western Australia. *J. Dep. Agric. W. Aust.* **17**, 251-94.

United States Department of Agriculture, Soil Conservation Service (1973). 'Guide for Interpreting Engineering Uses of Soils.' (U.S. Dep. Agric. Soil Cons. Serv.: Washington, D.C.)

van Dijk, D. C. (1959). Soil features in relation to erosional history in the vicinity of Canberra. CSIRO Aust. Soil Publ. No. 13.
van Dijk, D. C., Riddler, A. H. M. and Rowe, R. K. (1968). Criteria and problems in ground surface correlations with reference to a regional correlation in south eastern Australia. Ninth. Internat. Congr. Soil Sci. Trans. Vol. 4, pp. 131-8.
Walker, P. H. (1963a). Soil layers on hill slopes. A study at Nowra, New South Wales. *J. Soil Sci.* **13**, 167-77.
Walker, P. H. (1963b). Terrace chronology and soil formation on the South Coast, New South Wales. *J. Soil Sci.* **13**, 178-86.
Webster, R. (1971). Wilkes' criterion: A measure for comparing the value of general purpose soil classifications. *J. Soil Sci.* **22**, 254-60.
Webster, R. and Butler, B. E. (1976). Soil classification and survey studies at Ginninderra. *Aust. J. Soil. Res.* **14**, 1-24.

Physical and chemical characteristics of Australian soils
(Parts 1-3)

Part 1

Mineralogy

Chapters 22-24

Chapter 22
Clay minerals

K. Norrish and J. G. Pickering

When the various particle–size separates of soils are analyzed mineralogically the minerals of the coarser fractions are usually those that are residual from the parent rock. This is not true of the fine fraction; very often the minerals of this particle size are not present in the rock, but have formed during weathering. The fine-grained minerals found in different soils vary but are related in that they are usually layer lattice silicates. Natural clay deposits are similar and contain the same minerals that occur in soils, i.e. kaolin, illite, smectite, etc. This situation has led to these fine-grained layer lattice silicates being called *the clay minerals*. To examine the clay minerals in a soil it is usual to sample the $< 2\,\mu m$ fraction, because it is in this fraction that they are abundant, while in coarser fractions they are absent or in low concentration. Although the clay minerals usually dominate the clay fractions of soils other minerals are often present and may contribute significantly to soil properties. The most abundant non-clay minerals found in the clay fractions of soils are the oxides and oxyhydroxides of iron; hematite, maghemite, goethite and lepidocrocite: the aluminium hydroxides and oxyhydroxides; gibbsite and boehmite: the silicon oxides; quartz, opaline silica and some of the disordered forms of silica: and the carbonates; calcite, dolomite and aragonite. Clay fractions may also contain various rock-forming minerals e.g. feldspars, amphiboles, etc.

Clays and other minerals in the fine fractions of soils not only have a large surface area, but the surfaces are chemically active and can hold a variety of cations and anions by various mechanisms. It is the presence of cations on clay mineral surfaces that gives them their characteristic properties, plasticity and the ability to absorb water and swell. The fine mineral fraction of the soil, together with organic matter, is important in holding nutrient elements so that they are retained against leaching and yet are available to plants.

In this chapter the various clay minerals and their properties are discussed. The treatment accorded the various minerals is not even, but is meant to reflect the importance of the minerals in Australian soils; e.g. illite is given a fuller discussion than other minerals because it is the dominant mineral of the agricultural soils of southern Australia.

The fine-grained oxides and hydroxides are considered in Chapter 23, and carbonates in Chapter 10.

Kaolin

Structure and chemistry

The kaolin minerals are 1:1 layer structured aluminosilicates, one octahedrally co-ordinated aluminium layer combined with one tetrahedrally co-ordinated silicon layer (Brindley 1961; Grim 1968; Dixon, 1977). Only two thirds of the octahedral cation sites are occupied by aluminium. The serpentine minerals chrysotile and antigorite have a kaolin structure but all three of the octahedral sites are filled with magnesium. Other minerals of this group may have iron (Fe^{3+} and Fe^{2+}) replacing aluminium, and Fe^{3+} can replace some silicon. Garnierites are often kaolin type minerals high in nickel (Brindley and Hang 1973).

The kaolin minerals commonly have a somewhat disordered structure arising from disorder in stacking of the unit 7Å sheets, or disorder in the occupancy of the octahedral sites. Kaolinite is an ordered structure while halloysite (7Å) is disordered: minerals intermediate between the two are common in soils. Kaolinite usually has a hexagonal platy habit (Fig. 22.1*a*); halloysite (7Å) particles can have a variety of shapes, tubular, rosette-like, spheroidal, tabular or even platy (Dixon 1977). Commonly in soils kaolin minerals have ill-defined shapes (Fig. 22.1*b*), and the identification of specific kaolin minerals from the morphology of particles is difficult, as although there is a tendency for the more ordered structures to give hexagonal plates, there are exceptions. Halloysite (7Å) can have a platy habit (Souza Santos *et al.* 1966) while not all well ordered kaolinites exhibit hexagonal plates.

Mineralogical analyses of soils high in kaolin show that the crystallinity of kaolin varies, but in Australia there has been no systematic study of this. In other parts of the world attempts have been made to assess crystallinity and to relate it to soil forming factors. Muir *et al.* (1957) and Webster (1965) suggest profile drainage as the cause of variability; Jungerius and Levelt (1964) and West and Dumbleton (1970) associate the variability with parent material; Alwis and Pluth (1976) attribute it

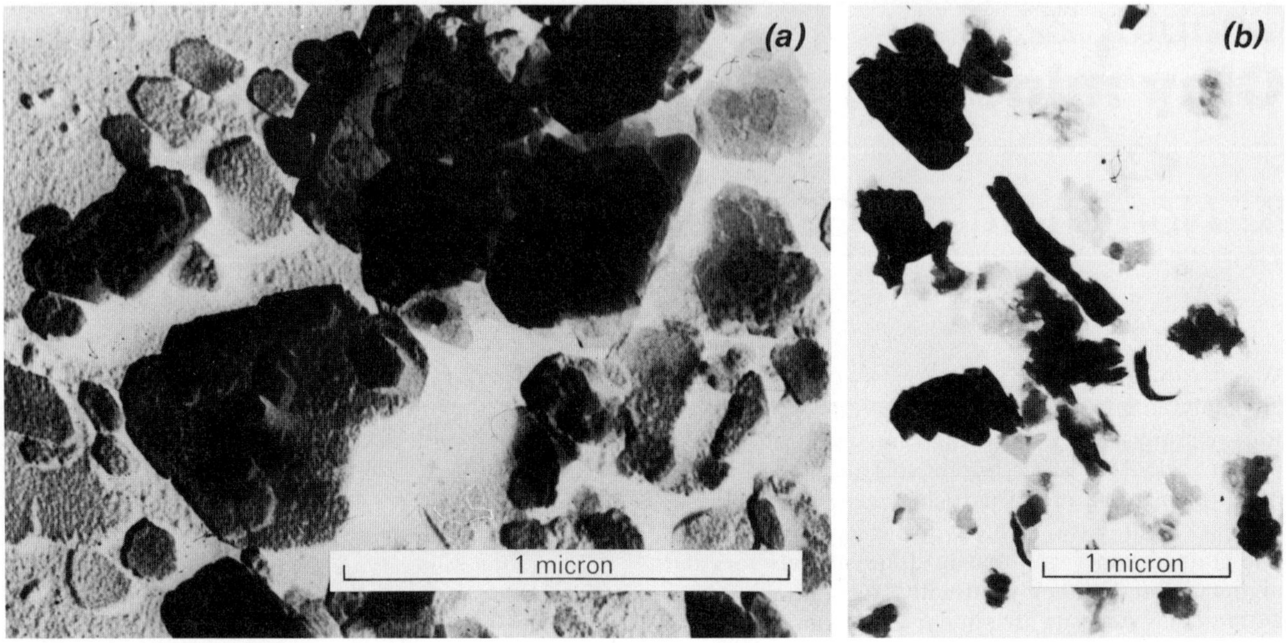

Fig. 22.1. (*a*) Kaolinite. Clay fraction of a *lateritic podzolic soil* from Rocky Gully, W. A. (pallid zone material). Free iron oxides removed. (*b*) Halloysite. Clay fraction of hand-picked specimen in a *krasnozem* from Wollongbar, N.S.W. (From Norrish 1952.)

to depth in the profile; and Hughes and Brown (1979) suggest crystallinity may be influenced by the age of the soil or by periods of drying.

Halloysite (10 Å) has a single layer of water molecules between the unit layers so that the interlayer spacing increases from 7 to 10 Å. The morphology of halloysite (10 Å) is as variable as that of halloysite (7 Å). Simple air drying of halloysite (10 Å) will usually remove the water layer to collapse the structure irreversibly to that of halloysite (7 Å). Because it is so easily dehydrated, halloysite (10 Å) is not found in soils, especially surface soils that are subject to drying. However, it is found in the soils of humid regions. Often the methods of handling soil will preclude its identification, e.g. in many laboratories soils are air dried prior to dispersion and separation of clay, and this may dehydrate the halloysite (10 Å) irreversibly. Therefore halloysite (10 Å) is probably a more common soil constituent than the literature would indicate.

The kaolin minerals are relatively pure aluminium silicates, having a composition near to that indicated by their formula $Al_2Si_2O_5(OH)_4$. However there is recent good evidence that small amounts (~2%) of iron may replace aluminium in the structure (Maldon and Meads 1967; Angel and Hall 1973; Rengasamy *et al.* 1975; Jefferson *et al.* 1975; Jepson and Rowse 1975; Herbillon *et al.* 1976). The soil clay shown in Fig. 22.1*a* is from a laterite in Western Australia, and contains Fe_2O_3. After dithionite treatment to remove free iron oxides electron microprobe analyses give 1.4% Fe_2O_3 in particles that appear to be pure single crystals. In principle there seems to be no reason why small amounts of other ions might not be present in kaolin. Weiss and Range (1966) report titanium in the structure, but the study of Herbillon *et al.* (1976) casts doubt on their finding. The interpretation to be placed on the presence of such elements in kaolin can be confused by the presence of small amounts of other minerals that may escape detection by most analytical techniques. Analysing single kaolin particles does not give unambiguous results since mica can be present in kaolin particles (Altschuler *et al.* 1963; Lee *et al.* 1975; Jepson and Rowse 1975).

Compared to other clay minerals, kaolins have low cation exchange capacities, 1–10 mequiv.%; at least part of the exchange capacity is pH dependent, indicating that some of the charge on the clay arises from dissociation of H^+ and OH^- at the crystal edges, rather than to isomorphous replacement that gives a permanent charge that is independent of pH. Whether kaolins have any charge due to isomorphous substitution is still in doubt. Some experimental observations indicate the presence of a permanent charge (Schofield and Samson 1954), but it could be due to impurities or to mixed layering of other minerals with the kaolin.

Occurrence and genesis

Some kaolin is present in most soils of the world and this is true of Australia (see Table 22.8). The highly weathered soils, *krasnozems*, *euchrozems*, *yellow* and *red earths*, mostly have kaolin as their dominant clay mineral. The same is true of most podzolic soils and the *earthy* and *siliceous sands*. The clays of pallid zones of laterites are often almost pure kaolin and in some areas of Australia such horizons are sources of ceramic clay. Some soils such as *red* and *yellow podzolic soils*, may be formed from lateritic materials; they retain kaolin as their major clay constituent. Gibbsite, hematite and goethite are usually present in appreciable amounts in such soils. Either dioctahedral chloritized smectite or vermiculite may be present with kaolin in certain highly weathered soils; these will be discussed later.

The amount of kaolin in the clay of the less weathered soils is variable and undoubtedly reflects the parent material and the history of the soil. The soils of southern Australia commonly have 20–30% kaolin in their clay fractions. The only soil clays in which kaolin is absent or very low are those associated with a particular parent material and they are usually restricted in area.

Kaolins undoubtedly form by weathering in the soil. Although some sedimentary rocks may contain kaolin, and contribute this to the soil on weathering, many rocks that do not contain kaolin weather to soils high in the mineral. Basalt is a good example as it normally contains no layer lattice silicates, yet basalts commonly weather to *krasnozems*, soils in which kaolin is the dominant mineral (see Table 22.8*e*). Aluminium and silicon are the essential constituents of kaolin and are abundant in most rocks. If elements such as Mg, Fe and K that are necessary for the formation of other clay minerals are absent from the rock, or if they are removed rapidly by intense weathering, then kaolin formation is favoured. Severe weathering can remove the silicon from kaolin and then gibbsite forms.

Illite

Structure and chemistry

Illites are clay micas. They are 2:1 structures, the sheet of octahedrally co-ordinated cations being sandwiched between tetrahedrally co-ordinated silica sheets. The unit layers have potassium ions holding them together, these positive ions balancing a negative charge brought about by isomorphous substitution of cations in the layer, e.g. Al replacing Si, or Mg replacing Al (Bradley and Grim 1961; Grim 1968; Fanning and Keramidas 1977). As in the coarser grained micas, the interlayer spacing is about 10 Å. In coarse-grained micas usually about one in four silicon atoms is replaced by aluminium; even when there is substitution in the octahedral sheet the overall charge on the layer remains the equivalent of that due to replacement of one in four silicon atoms by aluminium. This charge requires complete occupancy of the interlayer positions with potassium to balance the charge. Muscovite contains 11% K_2O, equivalent to 234 mequiv.%, but most of this potassium is not readily exchangeable, so the actual cation exchange capacity of a mica is much lower, and depends on particle size.

As with micas, Na^+ can occupy the interlayer positions in illites, and brammallite is such a mineral. The published analyses of illites show Na_2O in the range 0–1%. Almost invariably illites have some of the octahedral Al replaced by Mg and Fe, and in this respect the dioctahedral members resemble the mica phengite rather than muscovite. Illites usually contain a few percent of MgO, with a range of 0–4%; most illites have an Fe_2O_3 content in the range 0–8% with a small fraction of this present as FeO; some illites, however have a higher Fe_2O_3 content, ~ 15%, and can be regarded as intermediates between aluminous illites and glauconites. Because soil illites usually occur mixed with other minerals definitive data on their chemistry is lacking. Indications are that they usually contain significant concentrations of MgO and Fe_2O_3. Illites high in Fe_2O_3 are common in Australian soils, and are possibly more common than varieties low in Fe_2O_3.

Trioctahedral clay micas, in which the octahedral positions are mostly filled with Mg^{2+} and/or Fe^{2+}, do occur, and they are found in soils. In Australian soils they are comparatively rare and are only found in young soils forming from rocks containing the trioctahedral micas biotite or phlogopite.

Relationship to micas

Some clay micas resemble coarse grained micas in every respect except particle size, i.e. some clays are fine muscovites. Normally, however, there are significant differences, the most consistent being a decreased K_2O content and an increased H_2O content compared to micas (Fig. 22.2). Simply increasing the surface area of a mica will decrease its K_2O content as the exposed K^+ ions become exchangeable. The exchangeable cations will hold water of hydration so that the decreased K_2O

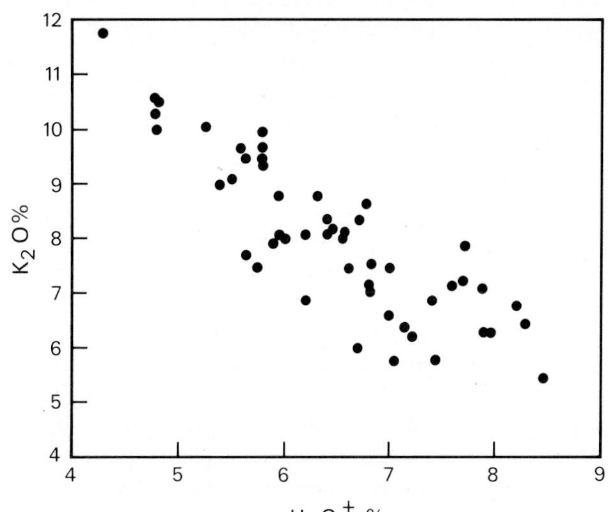

Fig. 22.2. Plot of K_2O versus H_2O^+ for hydrous micas.

Table 22.1. Charges on various micas before and after removal of potassium
The table is comprised of unpublished data from T. Kato and K. Norrish

Mica	Before treatment K_2O (%)	Before treatment Charge (mequiv. %)	After exchange[A] K_2O (%)	After exchange[A] Charge (mequiv. %)
Muscovite 0.2–2 μm	9.27	222	5.53	216
Illite, Tumut 0.5–2 μm	10.24	218	7.08	218
Sericite, Yoji Pass 0.2–2μm	8.82	194	4.48	193
Illite, Marble Head <0.2 μm	9.00	220	4.82	213
Illite, Missouri <0.2 μm	8.41	204	4.84	200
Illite, Ouachita 2–10 μm	7.28	163	4.52	164
Illite, Fithian <2 μm	6.15	156	2.67	145
Illite, Morris 0.2–2 μm	6.21	157	2.90	163

[A]15 days treatment with sodium tetraphenylboron.

and increased water are due, to some extent, to the decreased particle size, but this effect accounts for only a portion of the observed differences from mica. A variety of explanations have been put forward to account for these chemical differences from mica.

If the potassium ions are removed from some of the interlayers of mica and replaced with hydrated exchangeable cations, the overall effect is a decreased K_2O content with an increase in H_2O and cation exchange capacity. The resultant mineral is one in which mica and smectite (or vermiculite) layers are mixed and because of different inter-layer spacings it gives a diffraction pattern different from that of mica. Hower and Mowatt (1966) examined illites and mixed layer minerals and concluded that for many illites the low K_2O was partly due to the presence of expanding (smectite) layers, but that even when no expanding layers were present the illite had a K_2O content only 0.8 that of a true mica. Weaver (1965) obtained an average of 8.75% K_2O for 18 Palaeozoic illites, which is about 0.8 that of muscovite. Gaudette *et al.* (1966) also found that some illites are low in K_2O when free from impurities or interstratification.

Raman and Jackson (1966) and Alexiades and Jackson (1966) claim the low K_2O and high H_2O levels are due to mineral impurities and they estimate the mica contents of clays on the basis of 10% K_2O for the pure mineral. Bailey (1966) has a similar view, regarding illites as mixtures of micas and mixed layer minerals. Calculation of clay mica on the basis of 10% K_2O can give ambiguous results, as some specimens have a low K_2O content yet give no diffraction evidence of impurities or mixed layering.

Brown and Norrish (1952) proposed that in illites H_30^+ replaced K^+ in the interlayer positions. Structural formulae calculated on this basis give illites the same layer charge as true micas, and account quantitatively for the decrease in K_2O and the accompanying increase in H_2O (see Fig. 22.3). Hower and Mowatt (1966) considered that their data did not fit the above model, and Raman and Jackson (1966) could find no evidence for H_3O^+ when Na^+ was used to displace the interlayer cations of illite. In fact the pH rose during the reaction whereas it should have dropped if H_3O^+ was being released from interlayer positions. If sodium tetraphenylboron is used to remove the interlayer ions from illite (Scott and Reed 1962) the validity of the Brown and Norrish model can be tested directly, since the layer charge can then be measured directly as the cation exchange capacity. If illite has the full charge of mica, balanced by K^+, H_3O^+ and exchangeable cations, then as exchange proceeds and H_3O^+ is displaced, the sum of the exchangeable cations and K^+ should rise and reach that of mica (~250 mequiv.%) when all interlayer positions are accessible. Table 22.1 shows unpublished data of Norrish and Kato in which such measurements were made. With fine-grained illite it is often difficult to replace all the K^+, but the data indicate that as K^+ was displaced the sum of exchangeable cations and K^+ was

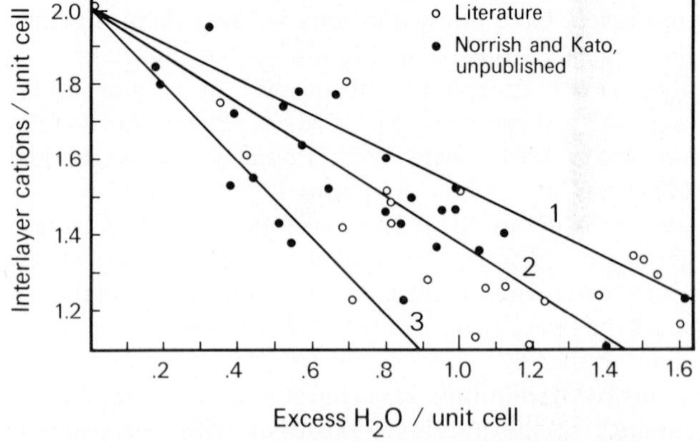

Fig. 22.3. A comparison of interlayer cation and excess H_2O per unit cell, compared to proposed models. Unit cell contents on the basis of 22 oxygens for ignited clay. (1) Mica plus kaolin. (2) H_3O^+ replacing K^+ (Brown and Norrish 1952). (3) H_2O occupying interlayer vacancies (Hower and Mowatt 1966).

constant within experimental error. The sum does not show the increase expected if H_3O^+ was being displaced. From such data it must be concluded that H_3O^+ is not normally in interlayer positions, and that illites have a lower layer charge than true micas. Yet for very many illites, when structural formulae are calculated as proposed by Brown and Norrish the sum of K^+ and H_3O^+ is equal to the number of interlayer positions in mica, and this observation requires explanation even if the basis of the calculation is wrong. Rosenquist (1963) used tritium to study proton exchange in illites and concluded that essentially H_3O^+ replaces K^+, but that the H^+ attaches itself to an oxygen of the silicate sheet leaving H_2O in the K^+ position. Such a model would predict the same relationship between K_2O and H_2O as that of Brown and Norrish (1952). Hower and Mowatt (1966) proposed that illites had a low layer charge and that H_2O molecules were trapped in the holes vacated by K^+ (Fig. 22.3). This model predicts only slightly less structural H_2O than that of Brown and Norrish, but the difficulty of measuring structural H_2O in illites precludes distinguishing the models on the basis of H_2O contents.

Another possibility that we have considered is that some illite might be a randomly interstratified mica–kaolin. Although we have no direct evidence for the presence of kaolin layers in illite, interstratifications of kaolin–smectite are known to occur (Shimoyama *et al.* 1969; Schultz *et al.* 1971); mica flakes commonly weather to kaolin aggregates (Gilkes and Suddhiprakarn 1979) and it seems likely that during this process there might be an intimate association between the layers of the two minerals. Lee *et al.* (1975) have observed mica layers in kaolin by high resolution microscopy. Kaolin contains no K_2O and considerably more structural H_2O than mica, and its incorporation into mica would affect the chemistry in the desired way (Fig. 22.3). Diffraction effects from mixed layer smectite–kaolin clays are weak and broad, and if mica–kaolin gives similar diffraction then such minerals could easily go undetected, particularly in a mixture containing mica and kaolin.

In terms of K_2O and H_2O contents, the structural relationships between illite and mica must be regarded as uncertain. It is very difficult to determine the structural water content of fine-grained minerals such as illite as there is a continuous loss of water with temperature, with no region that could be regarded as dividing the structural from the hygroscopic water. Distinguishing the various possibilities by diffraction is difficult. Although some minerals low in K_2O give effects to be expected from mixed layer mica–smectite, others do not. However, there are many possibilities in terms of mixed layering that have not been treated theoretically; kaolin could be involved, the mixing could be of more than two interlayer spacings, considerable potassium can be lost from small mica particles by edge replacement (Jackson *et al.* 1952), and in this zone where spacings are increasing to greater than 10 Å to accommodate the hydrated cations, spacings are variable and unit mica layers

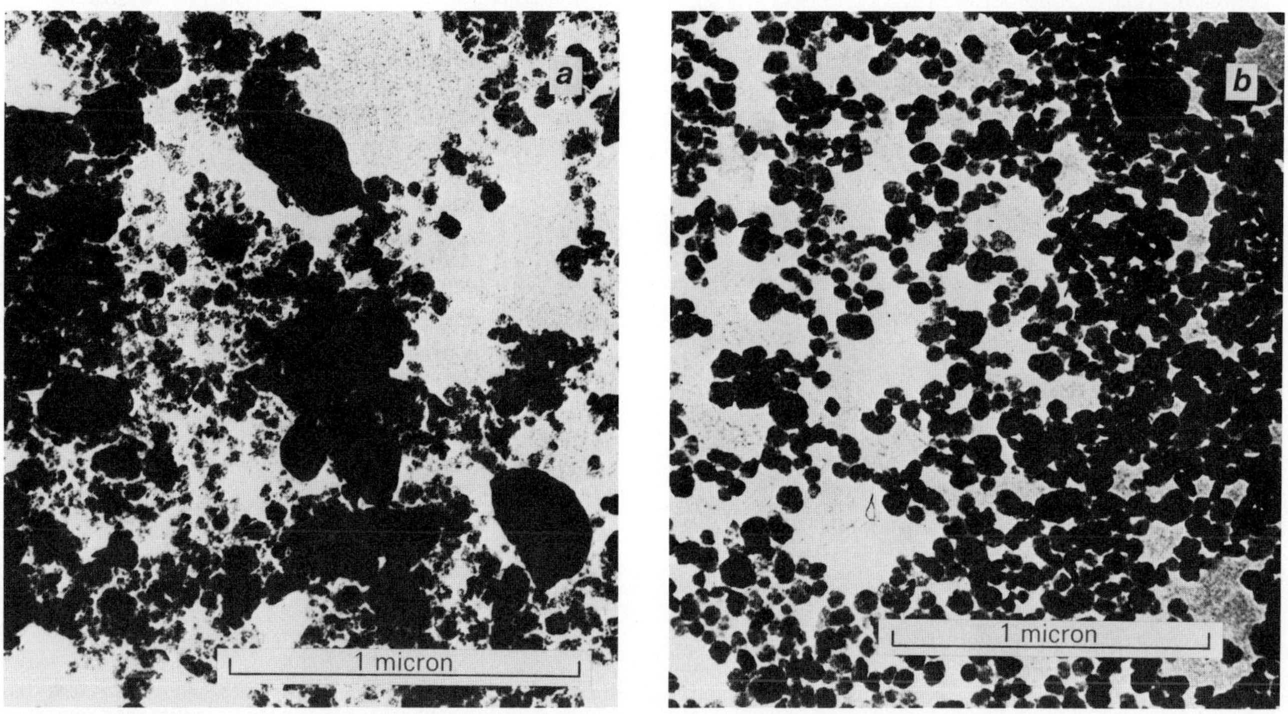

Fig. 22.4. (*a*) The <2 μm fraction of a *solodized solonetz* soil from Willalooka, S. A. This clay is mainly illite with ~ 20% kaolin. Its chemical analysis is given in Table 22.4. (*b*) The <2 μm fraction of clay from Muloorina Station, S. A. Note the small (~ 0.07 μm) but uniform particle size. Its chemical analysis is given in Table 22.4.

are not parallel. Such edge zones could form an appreciable fraction of fine grained illites, and the diffraction effects from them would be weak and lack specificity, particularly if lack of parallelism caused neighbouring sheets to scatter independently.

Particle size and shape

In the electron micrograph, illites generally appear to have a platy habit and sometimes show hexagonal outlines. Several samples have been described as having lath-shaped particles (Weaver 1953; Rex 1966). When present in soils they are usually thin flakes but with indefinite shapes.

Fanning and Keramidas (1977) consider that illite is usually most abundant in the coarse clay, 0.2–2 μm. This is not generally true of Australian soils where the illite is often very fine, < 0.1 μm, and can be concentrated relative to kaolin by taking a very fine fraction. Fig. 22.4*a* shows an electron micrograph of a typical illitic soil clay. It seems likely that this small particle size is due to its origin (see p. 290).

Exchange properties

Illites commonly have a cation exchange capacity (CEC) of 10–30 mequiv.%. As mentioned above many Australian soils contain very fine-grained illite and these have a lower K_2O content and a higher CEC than would be expected of illite. In studying the clays of the *solonized brown soils* (Mallee soils) Norrish (1949) found a complementary relationship between K_2O and CEC. Fig 22.5 shows that for these soils the CEC increases as the K_2O content decreases. Exchange capacities and K_2O contents measured on clay separates showed a similar trend and Norrish concluded that for the illite component the sum of the CEC and K_2O (expressed as mequiv.%) was constant at 150 mequiv.%. Radoslovich (1958) found a similar relationship for the clays from *red brown earths.*

To some extent the complementary relationship between CEC and K_2O could be due to particle size variations, since as illite particles become thinner, potassium ions exposed at surfaces are replaced by exchangeable cations. However, to account for the high exchange capacities in this way the crystallites would have to be only two or three layers thick, 20–30 Å, but the width of (001) diffraction lines suggests that for most clays the illite particles are at least 100 Å thick. If the crystals are 10 layers thick, only about 20 mequiv.% of the exchange capacity would be due to exposed surfaces. For those clays with the higher CEC the additional exchange capacity is approximately equal to the non-exchangeable potassium (expressed as mequiv.%). Norrish (1949) suggested that the additional CEC was due to the replacement of K^+ from the outer annulus of the disc-like crystals, and that the areas of the unaltered core and the weathered edge would then be equal. If D is the diameter of a crystal and l the depth of replacement of K^+, i.e. the width of the annulus, then the two areas are equal if $l = 0.14D$. If illite particles are 700 Å in diameter, then replacement of edge K^+ to a depth of 100 Å would halve its potassium content and give the required exchange capacity. Such a model is consistent with the X-ray diffraction data in that the annulus would not be expected to give specific lines of another phase, but would account for the observed low angle scatter.

Further work will be required to check the validity of the above model; detailed X-ray diffraction studies should be coupled with internal and external surface area measurements as made by Greenland and Quirk (1964).

Although the illites of many clays are degraded in the sense of being low in K_2O with a high exchange capacity, there is no X-ray diffraction evidence that these properties are due to interstratification with a smectite component. The small particle sizes and high exchange capacities of these illites give the soil containing them properties that are usually associated with smectite.

Although when no smectite is present the clays can show considerable swelling and have undesirable engineering properties (Aitchison and Holmes 1953; Aitchison *et al.* 1954), not all illitic soils behave in this way. There are areas in Australia where illitic clay soils exhibit a very strong aggregation; physically they behave as sands or gravels in not swelling and in resisting dispersion (Norrish and Tiller 1976).

Potassium release

The interlayer potassium of micas is not normally exchangeable but intensive work on this subject over the last decade or so shows that this fixed potassium can be

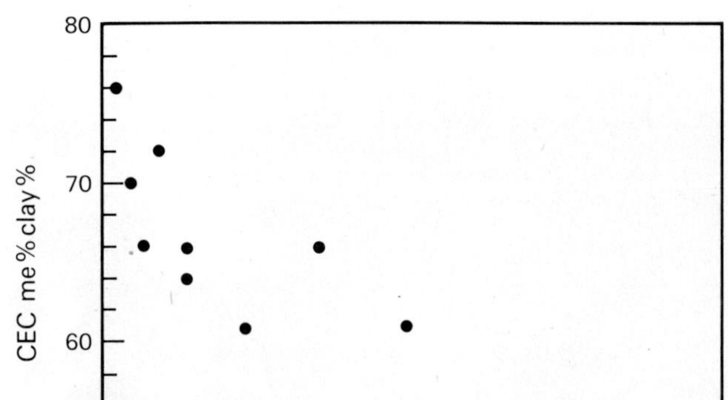

Fig. 22.5. Data from Norrish (1949) for Mallee soils showing the complementary relationship between CEC and K_2O of the clay fractions. CEC and K_2O were measured on the soil. These clays do not appear to have a smectite component but are dominantly illite (~ 70% illite and 30% kaolin).

replaced in short times under appropriate conditions. Norrish (1973) and Fanning and Keramidas (1977) have reviewed the transformation of micas to vermiculite, a reaction involving the replacement of potassium in mica.

If the potassium concentration in solution is sufficiently low, replacement of the interlayer potassium proceeds from the edge of the mica crystal (Fig. 22.6). The exchange is controlled by diffusion, the rate depending on the particular mica, the replacing cation and its concentration, temperature, and the concentration of potassium in solution (Rausell-Colom *et al.* 1965; Norrish 1973). The most important feature within the mica structure that affects the replacement seems to be the presence of the proton attached to an oxygen of the octahedral layer. This hydroxyl ion faces onto the hole in the silicate sheet that contains the potassium ion, and when its proton is vertically orientated away from the octahedral layer, as in trioctahedral micas, the proton is nearer the potassium and exchange is easiest. The mechanism seems to be one of the interlayer ion being repelled by the like-charged proton (Bassett 1960). In those mica structures where the proton is inclined away from the vertical, as in dioctahedral micas and oxidized biotite, or where the hydroxyl is replaced by fluorine, the replacement of interlayer potassium is more difficult.

For exchange to proceed, the potassium in solution must be kept at a low concentration. Potassium in solution will slow the exchange and above a certain critical level the reaction will cease. The critical concentration depends on all the factors discussed above. Barium is more efficient than the other common exchangeable cations in replacing potassium and at elevated temperatures with 1 M salt solutions trioctahedral micas require critical potassium concentrations of several hundred ppm in solution (Rausell-Colom *et al.* 1965) to prevent exchange. With Ca^{2+} as the exchange cation this value

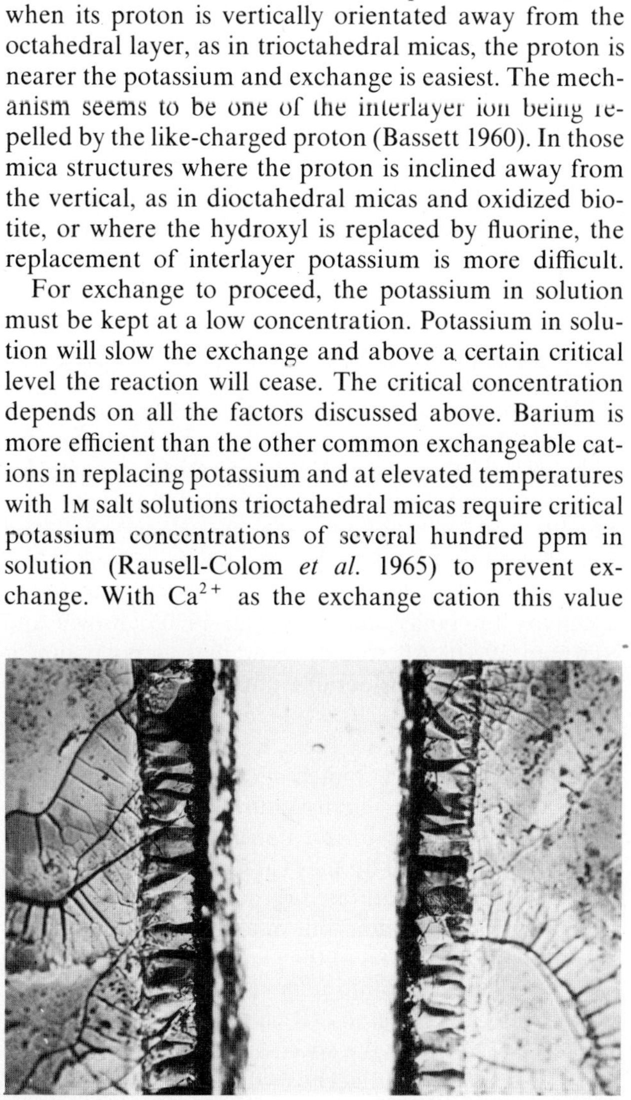

Fig. 22.6. Biotite flake after immersion in $SrCl_2$. Prior to immersion a cut was made in the mica surface and this appears in the photograph as the clear central area. At the two edges of the cut the interlayer K^+ has been replaced by Sr^{2+}.

drops, and if weaker salt solutions are used the critical concentration will be only a few ppm or less (Rausell-Colom *et al.* 1965; Scott and Smith 1966). All data indicate that trioctahedral micas (biotites and phlogopites) in soils are to be regarded as sources of potassium for plant growth. Mortland *et al.* (1956) used biotite as the sole source of potassium in growing wheat and Gilkes and Young (1974) used various oxidized biotites to supply potassium to clover.

The dioctahedral micas, muscovite and illite, require very low levels of potassium in solution for exchange to take place. The potassium content of even high quality laboratory reagents may be too high, and this has led to the use of sodium tetraphenyl boron as an exchange reagent, since by precipitation of potassium tetraphenyl boron, it maintains very low potassium concentrations in solution, i.e. about 0.005 ppm (Scott and Smith 1966). Even then the exchange rate is slow, being about 0.005 times that for trioctahedral micas (Scott and Smith 1966; Leonard and Weed 1970).

Because of the slow reaction rates, and the low concentrations involved, it is difficult to obtain reliable values for the critical potassium concentration for muscovites and illites. Scott and Smith (1966) indicate a value of 0.1 ppm for these minerals in the presence of 1 M NaCl. Quirk and Chute (1968) were able to exchange potassium from illite using 0.3 M NaCl containing 0.3 ppm potassium, but they observed very low diffusion rates that were 8–10 orders of magnitude lower than those reported by Rausell-Colom *et al.* (1965) for trioctahedral micas, and these low rates might have been partly due to the potassium content of their NaCl solution. For illitic soil clay, Tucker (1967*a*, *b*) found that, once the easily exchangeable potassium was removed, solution equilibrium concentrations were <0.2 ppm K.

As would be expected, initial exchange is more rapid as the particle size of the mica decreases (Scott 1968; Scott and Reed 1962), but there are several unexpected effects associated with small particles. Reichenbach and Rich (1969) and Reichenbach (1973) have shown for biotite, phlogopite and muscovite that the critical concentrations of potassium required to stop exchange *decrease* with *decreasing* particle size. Coupled with this is the observation that *less* of the total potassium is exchanged in the *fine* fractions. Ground muscovite, phlogopite and biotite all show this effect (Scott 1968; Reichenbach and Rich 1969; Reichenbach 1973). It does not appear to be an artefact of grinding as particle size separates of illite behave in the same way (Scott 1968). The illites of soils show the same effect (Smith *et al.* 1968) in that the finest fractions release potassium more slowly than the coarser fractions, and the proportion of potassium that can be replaced is less for the fine fractions. Doll *et al.* (1965) found that wheat plants removed potassium preferentially from the coarse clay.

As yet there is no satisfactory explanation for this

unexpected behaviour of small particles of mica. Ross and Rich (1973*a*,*b*) have proposed that differences between the behaviour of coarse and fine particles are due to differences in bonding as expansion takes place. Sometimes incomplete exchange of potassium is associated with interstratification; some interlayer regions lose all their potassium while others are unaffected. Scott (1968) suggested that potassium retention by small particles was due to interstratification, as Bassett (1959), in order to account for the occurrence of regularly interstratified minerals, had proposed that removal of potassium from one interlayer region caused potassium to be held more strongly in adjacent interlayer regions. Norrish (1973) suggested a model based on changes in the orientation of the OH group that could explain such behaviour. However, potassium can be easily replaced from coarse-grained regularly interstratified hydrobiotite (Sawhney 1967; Norrish 1973) so the above explanations do not account for the behaviour of fine biotite or phlogopite particles, though they may account for retention by interstratified dioctahedral micas.

There is agreement amongst most researchers that OH orientation is the main factor governing the ease of potassium replacement (Bassett 1960; Rausell-Colom *et al.* 1965; Quirk and Chute 1968; Newman 1969; Mackintosh *et al.* 1971), and the behaviour of fine particles could be understood if size influenced OH orientation. Calculations by Giese (1973) show that OH orientation is influenced by polytypism; in the 2M and 3T muscovite structures the OH dipoles make an angle of 14.5° with the mica surface; for the 1M structure the angle is 0°. Giese points out that for the 1M muscovite there is almost no interaction between the proton of the OH and potassium. On this basis the removal of potassium should be more difficult from the 1M than from the common 2M structure. The 1M polytype should behave like high fluorine micas, in which the potassium–proton interaction is absent; such micas are the most stable in regard to potassium exchange (Rausell-Colom *et al.* 1965; Newman 1970*a*). Illites commonly, but not always, have 1M or 1Md structures (Levinson 1955). In a mixture of 1Md and 2M micas the 2M component is likely to be over estimated because it gives more defined and stronger diffraction effects (Bradley and Grim 1961).

Whatever the explanation for the lack of reactivity of interlayer potassium in the fine mica particles, illites certainly owe their stability to this effect. If their potassium was as easily removed as that from coarse-grained micas, then illites would weather and lose their potassium very quickly indeed.

In soils there are factors that will assist potassium exchange from micas. Large cations have a higher competing power than the simple inorganic cations, and organic cations may play an important role in soils. Mackintosh and Lewis (1968) have shown that alkylammonium ions are very effective in replacing potassium. Compared to inorganic cations, exchange is quicker, more complete, and will continue with comparatively high concentrations of potassium in solution. Using 0.025 M dodecylammonium chloride they found critical concentrations of 80 and 20 ppm of potassium for two illites.

Muscovite in highly leached soils is often converted to dioctahedral chlorite; hydroxy–aluminium polymers replace the interlayer potassium. This suggests that large cationic polymers are also more effective in replacing potassium than simple inorganic cations, and in this respect they behave like large organic cations. Norrish and Wells (1970) observed such a reaction in the laboratory. They found potassium replacement occurred when mica was immersed in 'aged' distilled water. No cations were detectable in the water, yet exchange proceeded even when the potassium in solution had reached a level that would prevent exchange in strong salt solutions! They were unable analytically to distinguish the 'aged' reactive water from fresh unreactive water but the difference could be due to very small amounts of organic material produced by microbial action. Microprobe analyses of the mica showed high aluminium in the exchanged edge of the mica flakes, consistent with hydroxy–aluminium ions. The source of aluminium appeared to be the mica itself. Serstevens *et al.* (1978) have recently observed a similar reaction with distilled water, and one in which aluminium from the mica structure became the replacing interlayer cation.

Illites are considered to be the minerals responsible for potassium fixation in soils. When other hydrated cations, such as Ca^{2+} replace K^+, the interlayer spacing increases, converting the mica to vermiculite. If the vermiculite is K^+-saturated the spacing collapses to about that of the original mica, but the whole process is not reversible in that the structure does not regain its full selectivity for potassium (Newman 1970*b*; Brown and Newman 1970). All studies show that as potassium is replaced there is considerable physical distortion of the edge areas (Fig. 22.6) (Rausell-Colom *et al.* 1965; Wells and Norrish 1968; Scott 1968; Reichenbach 1973), and these prevent a perfect match of opposing silicate sheets when the mineral is again saturated with potassium. Most micas are well ordered structures, but once the structures are opened up they cannot be perfectly reconstituted. Potassium-saturated vermiculite has a moderate selectivity for potassium, but one that is very much reduced compared to that of the parent mica. If a mica has only part of its potassium removed there will be wedge-shaped interlayer regions (Rich 1964; Sawhney 1972; Tucker 1967*b*) where the spacing changes from 10 to 14 Å. In these regions adjacent to the unaltered mica, the layers will retain their register, so over a small distance the mica can be reconstituted to fix potassium with a selectivity comparable to the original mica (Newman 1970*b*; Reichenbach 1973). These wedge-shaped sites

Table 22.2. Analysis of saturated soil extracts from illitic soils

Values are calculated from the data of Jackson (1962) and expressed as mequiv. l^{-1}

	K	Na	Ca	Mg
Mean of 35 samples	0.39	22.1	8.6	9.0
Standard deviation	0.24	39.7	13.3	17.8
Minimum:	0. 1	0.3	0.3	0.2
Maximum:	1. 2	190.0	56.0	80.0

that are capable of fixing potassium will be only a small fraction of the exchange sites, and Bolt *et al.* (1963) found they were about 4% of the exchange capacity of an illitic soil.

Illite is generally a major component of the clays of the agricultural soils of Southern Australia, and the absence of potassium deficiencies in these soils is probably due to their illite content. Analyses of saturated extracts from soils show remarkably little variation in potassium concentration. Table 22.2 shows that whereas other cations have concentration variations of many hundreds, that for potassium is less than 10, and confirms the view that micas are to be regarded as soil potassium buffers (Newman 1969).

As discussed above, very fine-grained micas do not readily release their potassium. Many Australian soils contain illite that is very fine grained and the potassium-supplying ability of these soils over a long period might be less than expected when compared to some overseas soils. There have been no comparable experiments on Australian illitic soils to those experiments made on U.S. soils by Smith *et al.* (1968) who found that potassium exchange was greatest in the < 2 μm fraction; using NaTPB, least potassium was replaced in the < 0.08 μm and the < 20 μm fractions. Of the total soil potassium 20–47% was replaceable in 20 weeks.

Asher and Ozanne (1967) found that plant growth in solution culture was not limited by potassium supply when the solution concentration was > 0.4 ppm; in the range 0.4–0.04 ppm growth was affected, with the lowest levels giving plants with severe deficiency symptoms. Illites are capable of maintaining potassium concentrations of a fraction of a part per million in solution, and it seems that plants have adapted their requirements to what illites can supply. The potassium in illite is to be regarded as a reservoir that is available over a long period, but illite cannot be regarded as a good source when potassium is required at a fast rate.

Stevenson (1959) and Stevenson and Dhariwal (1959) have shown that micas in soils and rocks may contain nitrogen, and in some subsoils much of the fixed ammonium in the soil is due to ammonium ions in illite. The ammonium ion is the same size as the K^+ ion, and in mica the two ions occupy the same structural positions. Not surprisingly then the reactions of the ammonium ion with illite are much the same as those of K^+; ammonium added as fertilizer can be fixed in the interlayer positions of illite, and native ammonium can be slowly released from the structure.

Martin *et al.* (1970) and Gilkes and Little (1972) found very high correlation coefficients (*r* values as high as 0.99) between the fixed ammonium and total potassium ions for soils formed from phyllites in Queensland. Their soils were very high in fixed ammonium ions in the subsoils, ~ 1000 ppm nitrogen, and they showed that, in the different profiles, from 1 in 8 to 1 in 18 of the potassium ions in illite were replaced by ammonium.

Occurrence and genesis

Illite occurs very widely as the dominant mineral in the clays of many Australian soils. *Red brown earths* (southern), *solonized brown soils, calcareous red earths, rendzinas, terra rossas,* and *desert loams* often have illite as their dominant clay mineral (see Tables 22.8 and 22.9). With certain exceptions, illite is a major clay mineral of the soils of southern Australia and those of the arid regions. The exceptions are the highly weathered soils, *podzols* and laterites and soils that have as their parent material highly weathered rocks. In tropical soils illite is less common, and is only occasionally the dominant mineral of the clay fraction. Potassium is an essential constituent of illite, and rocks with a low potassium content, e.g. basic igneous rocks, weather to clays in which illite is absent or present in only small amounts. Published clay mineral analyses of North American and European soils, and those from desert areas, commonly show illite as a major constituent.

The source of much of the illite in soils is mica from the parent rocks. Mica is a constituent of many igneous, metamorphic and sedimentary rocks and it can be the dominant mineral in schists, shales and slates; it is often the dominant mineral in the insoluble residue from limestones. Where weathering is not severe, i.e. in arid and semi-arid soils, illite appears to be relatively stable, as it is present in many very old soils. Chlorites, vermiculites and trioctahedral micas that may be present in the parent material with the dioctahedral micas are much less stable.

The dominance of illite in soils over such large areas raises questions regarding the mechanism and extent of illite formation in soils. They certainly form from micas, but such a transformation may involve little more than a particle size reduction. There has been speculation as to whether illites form in soils from structurally unrelated minerals. All other clay minerals can form in this way so it would be surprising if illite did not.

In many soils in Australia it is common for the illite content of the clay to increase towards the surface and this is most easily observed by the increased potassium

Table 22.3 Some soils that have increased potassium in the clay fraction of the surface horizons

Location	Great Soil Group	Parent material	% K_2O in clay[A] 0-10 cm	~100 cm
(A1133) South of Hahndorf township, S.A.	*yellow podzolic*	Laterite and dark pyritic shales and slates	4.16	1.02
(A1126) Mt Osmond, top of ridge, S.A.	*red podzolic*	Calcareous siltstones	2.65	0.95
(A1140) Mt Osmond, bottom of valley, S.A.	*non-calcic brown*	Calcareous siltstones	2.74	1.08
(A1116) Overlooking Onkaparinga valley West of Hahndorf, S.A.	*yellow podzolic*	Feldspathic sandstone	2.13	1.14

[A]Pickering and Falk (1975).

Table 22.4. Chemical composition of soil illites compared to some sedimentary illites

Values expressed as % composition (ignited-basis). **1**. <2 μm fraction of Willalooka sand, a *solodized-solonetz* soil, from Hundred of Laffer, S.A. Analysis corrected for the presence of 20% kaolin. **2**. <2 μm fraction of *red brown earth* from Tharbogang, N.S.W. Analysis corrected for the presence of 33% kaolin. 2.1% free Fe_2O_3 removed prior to analysis, using sodium dithionite. **3**. <0.2 μm fraction of green clay from Muloorina Station, S.A., a lacustrine deposit of Oligocene age. **4**. Glauconitic mica from Uravan, U.S. (Keller 1958). **5**. <0.5 μm fraction of Aardebrug clay, a non-marine deposit of Oligocene age (Porrenga 1968)

	1	2	3	4	5
SiO_2	54.8	60.4	55.2	52.2	58.0
Al_2O_3	20.1	16.4	17.9	19.1	15.2
Fe_2O_3	13.5	11.8	11.9	14.0	13.0
FeO	–	–	0.85	1.39	2.17
TiO_2	1.1	1.44	0.32	1.13	0.24
MgO	3.1	3.67	4.33	2.97	4.29
K_2O	5.4	3.98	7.88	8.34	6.50
Na_2O	–	–	0.0	0.11	0.04
CaO	1.86[A]	2.25[A]	0.84[A]	0.41	0.73

[A]Ca saturated.

content of the clay in the surface horizons (Table 22.3). Similar observations have been made by Majhoory (1975) on the soils of Iran and by Nettleton *et al.* (1973) for some semi-arid soils of North America. The high illite in the surface is considered by these workers to be due to illite formation in the soil, the accumulation of potassium in the surface, by biocycling, favouring illite formation at the surface. However, there are other possible explanations of the observations and more detailed work is required to determine if illite is forming pedogenically in these cases.

Studies of the chemical composition of Australian illitic clays by the authors indicate that not all illites are derived ultimately from the micas of hard rocks. Table 22.4 shows analyses of illites from some Australian soils. Their notable feature is a relatively high ferric iron content, and in this respect the illites do not resemble common hardrock micas, but they do resemble the mica of certain green clays described as glauconitic illite or ferric illite (Porrenga 1968). A similar green clay occurs at Muloorina Station, near Lake Eyre in Australia, and X-ray diffraction patterns show that the < 2 μm fraction of this sample is very pure illite. Its chemical composition is given in Table 22.4 and apart from its higher K_2O content it resembles the soil illites. Porrenga (1968) points out that such green clays have been found in the United States and Europe, and are Oligocene in age; they are non-marine in origin and have formed in a lagoonal or lacustrine environment; they are often associated with gypsum or carbonates. The green clays are considered to have formed by the alteration of detrital smectite or illite, although Jung *et al.* (1954) considered that the mica formed by chemical precipitation.

The Muloorina clay sample was from the Billa Kalina Clay Member of the Etadunna Formation. It occurs as a soft green clay layer up to 6m thick; it is lacustrine in origin and Stirton *et al.* (1961) consider it to be Oligocene in age. The clay contains gypsum, and at Billa Kalina Station the < 2 μm fraction contains palygorskite as well as illite. Dr W. Compston of the Australian National University has measured the age of the Muloorina illite as 25 million years using its rubidium/strontium ratio. This agrees well with its stratigraphic age and argues against the illite being formed from mica detritus derived from surrounding rocks which are much older. Particle size studies indicate that almost all of the clay is < 1 μm in size and electron micrographs show an almost uniform particle size of ~0.07 μm (Fig. 22.4*b*). High magnification micrographs show that despite their small particle size, the particles tend to be hexagonal in shape. All the above data indicate that this clay formed by chemical precipitation rather than as a result of alteration of detrital clays.

As shown in Table 22.4 the soil illites resemble the Muloorina clay chemically, and Fig. 22.4 shows a general similarity in particle size in that much of the clay from the Willalooka soil is < 0.1 μm. As mentioned earlier, hard rocks do not generally contain micas high in ferric iron that could weather to the iron-rich illites. The above considerations indicate that soil illite is probably authigenic, forming in the soil. In arid soils the soil solution can resemble that of saline lakes, and if illite can precipitate in such lakes it is likely also to form in the soil. An alternative explanation is that all the high-iron illite was of sedimentary origin, forming under lacustrine conditions. Subsequent dispersal of the sediments led to its presence in soils. Though this could be the origin of the illite, it probably has a very wide distribution and we consider its formation in the soil is more likely.

There are insufficient data to delineate those soils in which the mica is pedogenic from those in which the mica is residual. If illite is forming pedogenically the phenomenon is probably widespread in Australian arid and semi-arid soils. Many soils will contain a mixture of the two types of illite.

Smectites

Structure and chemistry

The silicate layer of smectites is a 2:1 structure, and resembles mica in this respect. The layer has a negative charge due to isomorphous substitution, but it is only about half that of micas. As in micas, this charge is balanced by interlayer cations, but for smectite the interlayer cations are readily exchangeable and the charge on the silicate layer is equivalent to the cation exchange capacity of the clay (MacEwan 1961; Grim 1968; Borchardt 1977).

The smectite minerals can have a variety of chemical compositions, giving rise to the different minerals within the group.

Montmorillonite has aluminium in the octahedral sheet. A small amount of magnesium substituting for this aluminium gives rise to the layer charge. In some literature the term montmorillonite is also used as a group name for the smectite minerals.

Beidellite also has aluminium in the octahedral sheet but in this mineral the charge originates by substitution of aluminium for silicon in the tetrahedral sheet.

Nontronite contains mostly ferric iron in the octahedral sheet, the charge arising by aluminium substituting for silicon in the tetrahedral sheet.

Saponite has magnesium in octahedral co-ordination, the charge originating in the tetrahedral sheet by aluminium replacing silicon.

Montmorillonite, beidellite and nontronite all have trivalent ions in octahedral co-ordinations, so only two thirds of these structural positions are occupied, i.e. the minerals are dioctahedral.

Saponite is trioctahedral with octahedral positions filled. There are a variety of other smectite species (MacEwan 1961) but they are unlikely to occur in soils.

Soil smectites are often intermediate in composition between the above species, and Table 22.5 shows analyses of several soil smectites. All of these minerals have a considerable aluminium-for-silicon substitution in the tetrahedral position, and must be regarded as members of a beidellite-nontronite series.

In electron micrographs, montmorillonite, beidellite and saponite particles usually appear as irregularly shaped sheets; in lateral dimensions they can be ~ 1 μm, but are distinguished by being very thin. Nontronite appears as fine fibres or laths.

Table 22.5. Chemical composition of some soil smectites

Values expressed as % composition ignited basis, Ca saturated. **1**. Drayton, Qld. Basalt parent material. <2 μm fraction, pure smectite (beidellite-nontronite). **2**. Pelican clay, Lismore, N.S.W. Basalt parent material. <2 μm fraction, smectite plus very little kaolin. **3**. Black Hill, S.A. Norite parent material. < 0.5 μm fraction, pure smectite (beidellite-nontronite) (Hutton *et al.* 1977). **4**. Plaine des Gaiacs, 253 km from Noumea on Route Coloniale No. 1, New Caledonia. Ultrabasic parent material. High magnesium black clay soil. <0.5 μm fraction, pure nontronite

	1	2	3	4
SiO_2	56.5	56.1	60.5	59.4
Al_2O_3	19.9	18.2	14.1	10.4
Fe_2O_3	10.6	19.5	17.1	21.8
TiO_2	0.30	1.59	0.31	0.54
MgO	9.0	2.36	4.52	4.91
K_2O	0.02	0.36	0.28	0.05
CaO	4.03	2.42	3.15	3.46
Na_2O	–	0.20	–	–
CEC[A]	144		113	124

[A]Cation exchange capacity (mequiv. per 100 g)

Swelling behaviour

Smectites have cation exchange capacities varying from about 80 to 150 mequiv.%. The exchangeable cations are located between silicate layers and are normally hydrated. The presence of the relatively large hydrated cations increases the interlayer spacing of smectites above the 10Å of mica. With sodium as the exchangeable cation smectites may swell greatly on hydration and the interlayer spacing may increase to at least several hundred Angstroms (Norrish 1954*a*). In soil such swelling is unlikely because sodium-saturated clays are rare, and even a small soil overburden pressure would limit their swelling. In soils where calcium or magnesium is the dominant cation the interlayer spacing is likely to vary between 14 and 20 Å depending on water tension, and in this range the swelling pressure is great. So it is not surprising that many soils with swelling problems contain smectite.

In the Melbourne area, the soils that swell and shrink, causing failure of building foundations, are those derived from basalt and the clay is predominantly smectite. Other soils in Melbourne overlie Silurian shale and Pliocene sandstone and have illite and kaolin respectively as their clay minerals (Norrish 1954*b*). Certain Adelaide soils have undesirable engineering properties, but here the influence of clay mineralogy is not so clear-cut; all the clays are predominantly illite; in the problem areas there is little or no smectite present as a distinct mineral phase, but the illite can be interstratified, i.e. the potassium has been removed from some interlayers to form interstratified mica–smectite.

Landslides and soil creep are often associated with soils that contain smectite.

Smectites usually have a cation exchange capacity of about 100 mequiv.%, but sometimes it can be as high as 150 mequiv.% (Table 22.5); such clays may exhibit a physical behaviour very different from that expected of smectite (Norrish 1973*a*; Norrish and Tiller 1976). Swelling is limited, and the clays can be difficult to disperse.

The swelling of smectites can be restricted also by hydroxy-aluminium polymers that replace the exchangeable cations in the interlayer region (Tamura 1957).

Distribution and genesis

No Australian Great Soil Group has smectite consistently as the dominant clay mineral. Some soil groups such as *desert loams, grey, brown and red clays,* and *black earths* that have illite as their dominant clay mineral in southern Australia have smectite as their dominant clay in northern Australia (see Tables 22.8b and c) ('northern' implies north of approximately Lat. 31°S). Other soils having large amounts of smectite are of restricted occurrence, usually a result of local parent material or drainage conditions. In many parts of the world basic igneous rocks weather to red soils high in kaolin, or to black soils rich in smectite. Often, but not always, the black soils form in poorly drained areas. In Australia the *black earths* of Queensland are high in smectite and these soils are commonly derived from basalt which usually contains no layer lattice silicates as primary minerals. The smectite forms authigenically during weathering. Veen (1973) studied some *grey, brown,* and *red clays* in south-eastern Queensland. The clay of these soils was dominantly smectite in the surface horizons, while kaolin dominated in the subsoil. Veen considered that the present climate favoured the formation of smectite, and that smectite was forming from the remnants of old lateritic soils that had kaolin as their clay mineral.

Table 22.5 shows that soil smectites commonly contain appreciable concentrations of magnesium and iron. Harder (1977) found he could not synthesize pure aluminium-smectites at 20°C, but that minerals having a mixed population in the octahedral layer were easily synthesized. Table 22.4 would indicate that the same is probably true for illites.

In some cases smectite can form by the removal of potassium from the interlayer region of micas, and this is accompanied by small changes in the octahedral and tetrahedral sheets.

Vermiculites

Structure and chemistry

Vermiculites resemble smectites in structure, and differ from them primarily in having a higher charge (intermediate between smectite and mica) on the silicate layer. The charge in vermiculite is usually due to the substitution of aluminium for silicon in the tetrahedral positions. As with smectite, the charge is balanced by interlayer cations (Walker 1961; Douglas 1977).

Usually vermiculites are trioctahedral, with magnesium as the dominant ion in the octahedral position of the silicate sheet.

Because of their higher charge, the swelling of vermiculites is more restricted than that of smectites and usually the interlayer spacing does not go beyond 14 Å.

When viewed in the electron microscope, vermiculite particles are seen as large irregularly shaped sheets. Where vermiculite occurs in the clay it is often present also in the sand fraction of the soil.

The cation exchange capacity of vermiculites ranges from about 140 to 245 mequiv.%, on an ignited basis (Norrish 1973*b*). The exchange properties of vermiculites have been the subject of intensive studies as for many years it was thought that vermiculites were responsible for potassium fixation in soils (Barshad 1948, 1950; Krishnamoorthy and Overstreet 1950; Wear and White 1951). Although in laboratory experiments vermiculite will fix K^+ against exchange by ammonium ions, it is now realised (Norrish 1954*a*) that this fixation is due to the use of ammonium as exchanging cation. Potassium and ammonium ions have a low hydration energy; they cannot bring water into the interlayer region, so the silicate sheets collapse to give a 10-Å spacing. Exchange between these ions is then exceedingly slow since it relies on solid state diffusion. If an ion of higher hydration energy (Ca^{2+}, Mg^{2+}, Na^+) is used as exchanging cation, the ions carry a water shell that opens up the silicate sheets; the potassium is then easily exchanged.

Sawhney (1972) found that the ability of some vermiculites to fix potassium was due to microscopic zones of mica within the vermiculite.

Distribution and genesis

Vermiculite is not a major component of any soil group. It occurs in variable amounts in different soils, and usually its presence is related to mica or chlorite in the parent material.

Phlogopites and biotites easily weather to vermiculite. In many soils this alteration can be observed on sand-sized grains that are unaltered mica in the deeper soil horizons, and pass to golden coloured vermiculite flakes in the higher horizons of the soil. This conversion is easily duplicated in the laboratory (Norrish 1973*b*).

Vermiculites are high in magnesium and relatively unstable compared to clay minerals that are high in aluminium. It is probable that vermiculites rarely form authigenically from unrelated minerals during weathering, as even where leaching is mild the removal of magnesium will favour the destruction of vermiculite rather than its formation. Where vermiculite is present in soils it is usually the result of weathering of related minerals, i.e. micas or chlorites. Kittrick (1973), in laboratory studies, could not find a stability area for vermiculite and considered that it was present in soils as an unstable intermediate. However, Ildephonse *et al.* (1979) report vermiculite forming in a soil from a meta-gabbro that lacked a layer lattice silicate precursor, and this indicates that given the appropriate conditions, which may occur only rarely, vermiculite can crystallize from solution.

Dioctahedral vermiculites are common in certain highly weathered soils, and usually have polymerized aluminium hydroxide in the interlayer region. These minerals will be discussed under chlorites.

Chlorites

Structure and chemistry

The silicate sheet of chlorite resembles that of mica, vermiculite and smectite; chlorite differs from vermiculite and smectite in that the interlayer region is occupied with magnesium, iron, or aluminium hydroxides instead of hydrated cations (Brindley, 1961). The interlayer spacing is about 14 Å and in this respect is similar to natural vermiculites, but for chlorite the interlayer spacing is fixed and usually the interlayer region is not involved in exchange reactions.

Distribution and genesis

Two distinct chlorites are now recognized as occurring in soils.

The normal chlorite present in rocks such as slates and shales contains much magnesium and sometimes ferrous iron. The interlayer region consists of a brucite-like (magnesium hydroxide) layer. These chlorites are sometimes observed in soil clays, and in such cases their presence can usually be attributed to the mineral in the parent rock. Because they are high in magnesium and ferrous iron they are relatively unstable and weather rapidly; they may form a trioctahedral vermiculite or smectite, but these minerals also weather easily. Very often these chlorites may be present in the deeper part of the soil profile but diminish in abundance towards the surface.

By contrast, the other chlorite tends to occur in highly weathered soils. This mineral is an almost pure aluminium silicate; the mica sheet is dioctahedral and aluminium hydroxide occupies the interlayer region. The pure end-member has a chemical composition almost identical to that of kaolin. Rich (1968) has reviewed the occurrence and properties of these minerals. In soils literature this chlorite is referred to by a variety of names: dioctahedral chlorite; chloritized vermiculite; inhibited vermiculite. This mineral is common in certain highly weathered soils and may be the dominant clay mineral in them. Often its presence in a soil can be related to the occurrence of mica in the deeper horizons of the soils; the interlayer potassium has been removed from the mica, and replaced by charged aluminium hydroxide polymers. Usually the interlayer region consists of hydrated cations together with the hydroxide polymers. In these cases the mineral is intermediate in properties between vermiculite and chlorite; the polymers prevent the crystal swelling that is characteristic of vermiculite and smectite, but the interlayer cations may still be exchangeable. In these cases treatment with sodium citrate, ammonium fluoride, etc. (Brown 1953; Rich and Obenshain 1955; Klages and White 1957) can remove the interlayer aluminium and then the mineral behaves as vermiculite or smectite. Where chloritization has proceeded further it is often not possible to remove the interlayer aluminium.

These dioctahedral chlorites do not occur in rocks or as coarse-grained crystals, but minerals closely resembling those occurring in soils can be synthesized by precipitating aluminium in the interlayer region of vermiculite or smectite (Rich 1968; Barnhisel, 1977).

In soil clays where smectite or vermiculite is chloritised, the swelling of the clay will be reduced and the cation exchange capacity will be reduced depending on the amount of hydroxy polymer (Hsu and Bates 1964;Turner 1965; Brydon and Kodama 1966; Hsu 1968; Barnhisel 1977). The aluminium of the interlayer region is in a more reactive form than the aluminium of clay mineral sheets, and could be available to react with phosphate in soils.

Where dioctahedral chlorite is observed in Australian soils its presence can generally be attributed to the weathering of mica, and this is in agreement with observations in the U.S. (Douglas 1977). It is worth noting that *krasnozems* rarely contain significant amounts of this mineral. They are highly weathered soils but usually mica is absent from the parent basalt.

Interstratified clay minerals

Structure and chemistry

As mentioned earlier in this chapter, the interlayer potassium of mica can be exchanged in the laboratory. If the exchange is carried out in very dilute solutions (Rausell-Colom *et al.* 1965) the potassium is removed from some layers and not from others. Where the potassium is retained the mica has its normal interlayer spacing of 10 Å; where the potassium is replaced the hydrated cation replacing potassium will cause a lattice expansion, e.g. to 14 Å. The mineral now has a mixture of 10 and 14 Å spacings, and is regarded as a mixed layer mica–vermiculite. Note that in this alteration the mica silicate sheet undergoes little or no change; the interstratification is brought about by changes in the interlayer region. It is easy to understand then that interstratification of mica, vermiculite, smectite and chlorite occur, as the main difference between these minerals is in the composition of the interlayer region. In recent years there have been reports of kaolin forming interstratified mixtures with some of the 2:1 minerals. A great many two-component interstratifications have been observed and studied in clays. The mixing of layers can be random, regular, or partially ordered. In the mica–vermiculite example cited above, a regular interstratification is when mica and vermiculite layers (interlayers) alternate regularly; if there is a tendency for the layers to alternate, but there are mistakes, then the mixing is partially ordered; where there is no tendency for order, the mineral is randomly interstratified.

In a two-component interstratification, the proportion of the two components can vary.

There is thus a bewildering array of possibilities. Different components can be mixed in different proportions and with various degrees of order. In principle there is no reason to suppose that such minerals are restricted to two components, and then the number of possibilities increases considerably. Weaver (1956) and Foscolos and Kodama (1974) report the occurrence of three-component mixed layering.

Where interstratified minerals occur in clays their chemical composition and properties will be some average of those of the component minerals.

It is not intended here to discuss the various minerals in this group that have been reviewed by MacEwan *et al.* (1961) and Sawhney (1977).

Occurrence and genesis

In soil clays interstratification is very common; rarely is the interlayer region of micas completely intact; commonly, smectite gives diffuse spacings indicative of the presence of an interlayer material other than simple inorganic cations, probably hydroxy polymers. The components and ordering of interstratification are often difficult to identify, as soil clays are complex mixtures, often with the interstratified mineral as a minor component.

Regularly interstratified minerals are usually found in slightly weathered rocks or in clays of hydrothermal origin. Such minerals do not survive in highly weathered soils. Even the randomly interstratified minerals in soil clays seem less well defined than those of clay deposits. The diffraction effects from the soil clays are usually more diffuse; small particle size may contribute to the diffuseness of the diffraction pattern, but it is probably also due to an inherent variety in the clay. In a soil clay, even if there is only two-component interstratification, the proportions and the degree of order may vary from particle to particle.

Soils formed from rocks high in biotite or phlogopite commonly contain mixed layer mica-vermiculite minerals in the clay fraction. The same minerals are often observed in the weathered mica grains of sand size.

Where dioctahedral chlorite has formed from mica in highly weathered soils, the intermediate soil horizons may contain mica interstratified with smectite, vermiculite or chlorite

Interstratified kaolin–smectite

This mineral has only recently been recognized; Shimoyama *et al.* (1969), Schultz *et al.* (1971), Wiewiora (1971) and Sakharov and Drits (1973) have described the material in clay deposits. The only report of its presence in soils is from some Scottish brown forest soils (Wilson and Cradwick 1972), where it was a minor component, usually in the C horizons.

This mineral is discussed here because it occurs in Australian soil clays, and is possibly a very common major component.

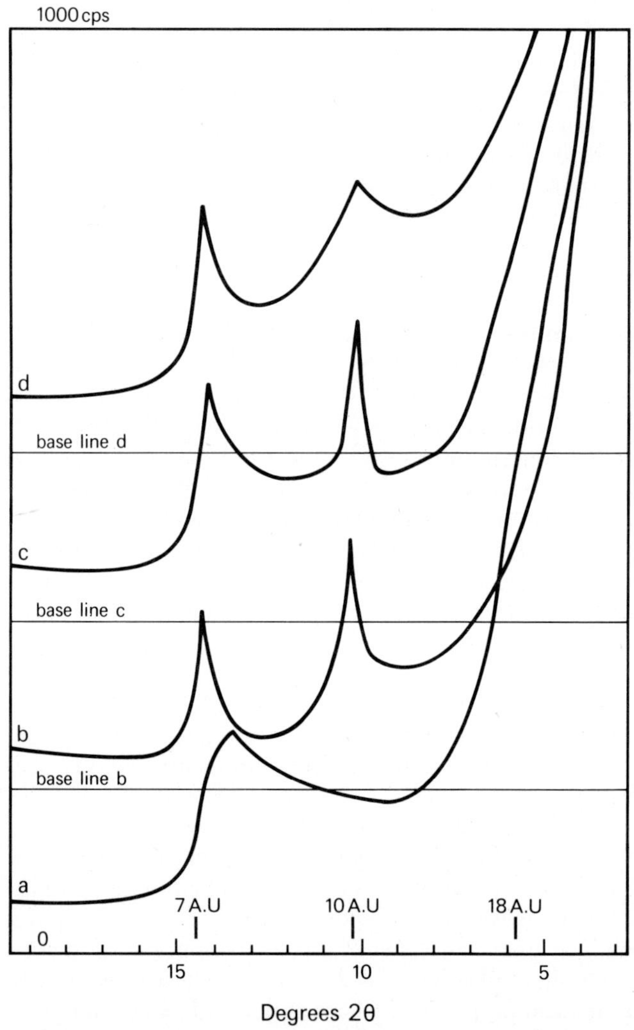

Fig. 22.7. Diffraction patterns, using CoKα radiation, of interstratified kaolin-smectite. (*a*) Typical kaolin-smectite from the $<2\mu m$ fraction of a *lithosol* from Darling Downs, Queensl. (CSIRO Div. Soils No. B171). (*b*) Kaolin and illite from the $<2\ \mu m$ fraction of a *solodized solonetz* from Wanilla, S. Aust. (CSIRO Div. of Soils sample SA/68/C4/30–40 cm). (Reproduced with permission from Norrish 1979). (*c*) A mixture of equal parts of samples a and b. (*d*) The $<2\ \mu m$ fraction of a *solodic soil* from near Woorndoo, Vic. (CSIRO Div. Soils sample VIC/72/P11 35–40 cm). This is a diffraction pattern typical of many soil clays. The clay is composed of illite and kaolin and appears to contain an appreciable amount of interstratified kaolin-smectite.

Some soil clays give no basal diffraction lines, but the clays are not amorphous as the non-basal lines are normal. In the region of the basal reflections (40–7 Å) they give intense central scatter that decreases with increasing angle (decreasing spacing) until at about 7 Å the scatter drops suddenly to a normal value (Fig. 22.7). These diffraction effects resemble those described for the mixed-layer kaolin–smectite minerals. Heating and other tests (Schultz *et al.* 1971), and a direct comparison with the Mexican clays described by Schultz *et al.*, confirm that these clays are mixed-layer kaolin–smectite minerals. Table 22.6 shows analyses of several of these minerals from soils that are geographically widely separated. Although the clays of Table 22.6 are all derived from basic igneous rocks, similar clays form from different parent materials. These clays appear to be *pure* and do not contain significant amounts of any of the normal clay minerals; we have identified such clay minerals from some 40 soil profiles. Although this is not many occurrences, these are specimens in which the kaolin–smectite appears to be *pure*, and it has only been identified because no other clay minerals are present. As described above, the diffraction effects from this mineral are diffuse, and even if the mineral is present in large amounts it will probably go undetected in the presence of clay minerals that give defined basal diffraction peaks (Fig. 22.7). Thus it is likely that interstratified kaolin–smectite occurs quite widely.

Over very large areas of southern Australia, illite and kaolin are the two dominant clay minerals, and usually their diffraction peaks are superimposed on a high background, indicative of disorder in the interlayer spacings. As discussed for illites, these clays usually have a high cation exchange capacity and a lower than expected potassium content. The diffraction effects are not consistent with the presence of mixed-layer illite–smectite, but are more like those of interstratified kaolin–smectite or kaolin–smectite–illite. The analyses of Table 22.6 show appreciable potassium, and so the clays should be described as mixed-layer kaolin–smectite–illite, even though the illite component is small. Further work will be required to establish whether such minerals explain the chemical properties and diffraction effects of the illitic soil clays described above.

If interstratifications involving kaolin are confirmed, it is probable that these minerals are present in considerable amounts over large areas of Australia. The effect of these minerals on the chemical and physical properties of soils warrants further study.

Palygorskite

Structure and chemistry

Palygorskite, sometimes called attapulgite, does not have a layer lattice structure like the above minerals. However, its structure resembles that of smectite in that the 2:1 structure exists within it, but is limited to one direction. These 2:1 ribbons are the width of five octahedral positions in one direction and continue indefinitely in the other direction. At the edge of a ribbon the silica tetrahedra reverse and form another ribbon of the same width, after which the silica tetrahedra reverse again to give a ribbon identical to the first. The overall structure is one that is extended in one direction; at right angles to this the ribbons have a chain-like structure with definite channels parallel to the ribbons. For more details of the structure Caillere and Henin (1961), Grim (1968) or Zelazny and Calhoun (1977) should be consulted.

Because of its structural similarity to smectite, the composition of palygorskite is rather similar to that of smectite. There appears to be little or no tetrahedral substitution in palygorskites; the octahedral positions are occupied by aluminium and magnesium with small amounts of iron. The magnesium oxide content of palygorskite (6–14%) is higher than that of smectite and the charge on palygorskite arises from the substitution of Mg^{2+} for Al^{3+} in the octahedral positions. Table 22.7 gives analyses of two soil palygorskites. Exchangeable

Table 22.6. Chemical composition of mixed-layer kaolin–smectite minerals in soil clays

Values expressed as % composition (ignited basis, Ca saturated). Analyses were carried out on <2 μm fractions after the removal of free iron oxides. **1**. *Brown earth* on Basalt, Launceston Basin, Tas. CSIRO Division of Soils, Sample H27.2. **2**. *Prairie soil* on basalt, 30–45 cm, Glen Innes, N.S.W. **3**. *Non-calcic brown soil* west of Dysart, Tas. CSIRO Division of Soils, sample TG 336.

	1	2	3
SiO_2	58.6	55.9	59.1
Al_2O_3	26.8	27.2	28.0
Fe_2O_3	7.1	10.3	8.2
TiO_2	2.47	2.49	0.40
MgO	0.65	1.21	1.07
K_2O	0.89	0.20	0.35
CaO	0.95	1.15	1.48
CEC[A]	34	41	53

[A]Cation exchange capacity (mequiv. per 100 g)

Table 22.7. Chemical composition of palygorskite skins and that of the adjacent clay

The red earth was from near Alice Springs, N.T. The paleosol came from a *red earth/red brown earth*, eastern Riverina, N.S.W. The data are from Singer and Norrish 1974 and the table is reproduced with the kind permission of The Royal Australian Chemical Institute (Norrish 1979)

	Calcareous red earth		Paleosol	
	Palygorskite skin	Clay matrix	Palygorskite skin	Clay matrix
SiO_2	60.29	46.48	62.77	50.75
Al_2O_3	11.86	18.2	12.33	23.95
Fe_2O_3	2.22	16.14	0.41	7.30
TiO_2	0.10	0.69	0.09	0.88
MgO	10.26	2.29	9.62	1.61
K_2O	0.39	1.75	0.62	2.32
CaO	0.14	0.57	0.24	0.39
Na_2O	0.15	0.18	0.33	0.64

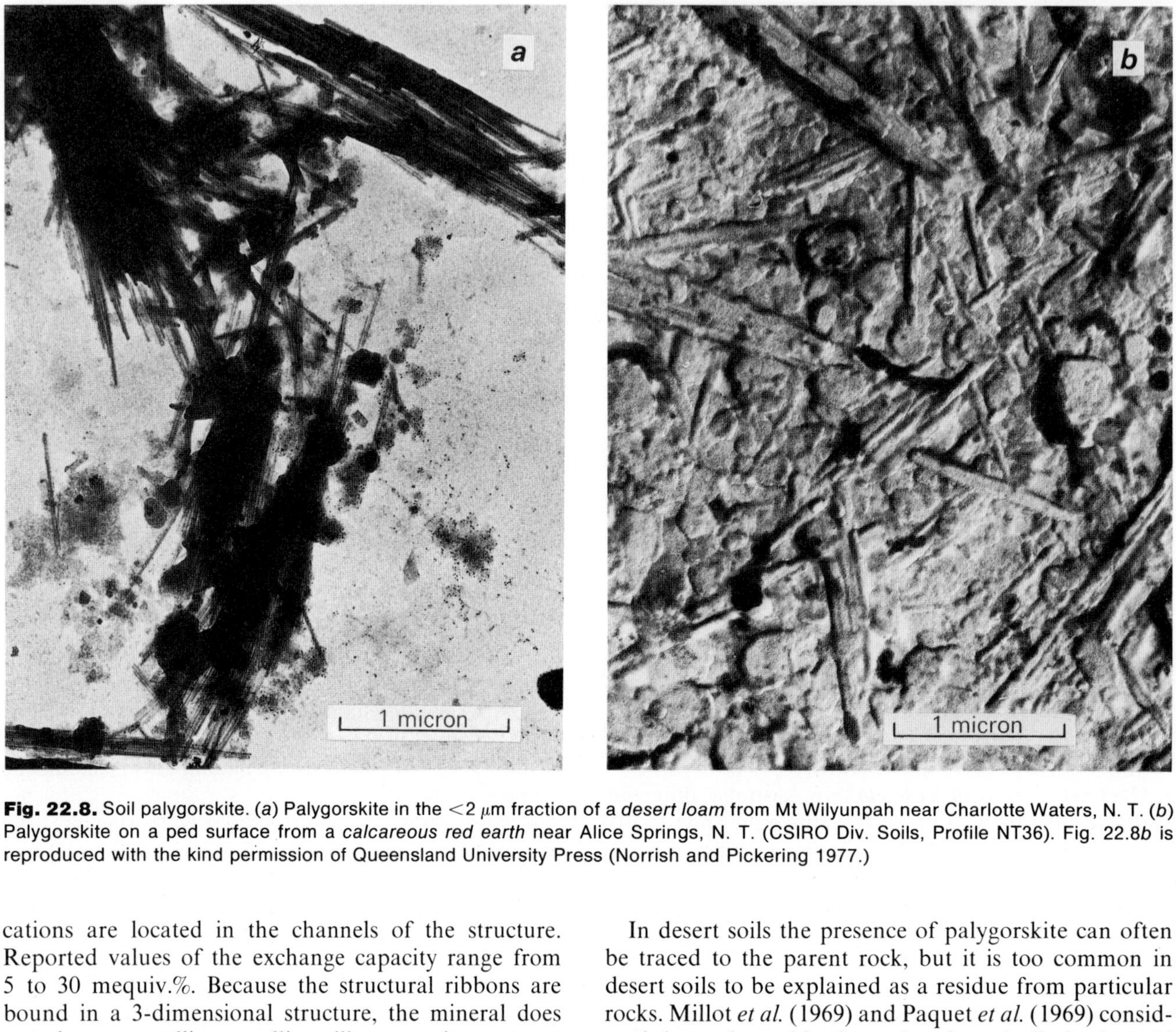

Fig. 22.8. Soil palygorskite. (*a*) Palygorskite in the $<2\ \mu m$ fraction of a *desert loam* from Mt Wilyunpah near Charlotte Waters, N. T. (*b*) Palygorskite on a ped surface from a *calcareous red earth* near Alice Springs, N. T. (CSIRO Div. Soils, Profile NT36). Fig. 22.8*b* is reproduced with the kind permission of Queensland University Press (Norrish and Pickering 1977.)

cations are located in the channels of the structure. Reported values of the exchange capacity range from 5 to 30 mequiv.%. Because the structural ribbons are bound in a 3-dimensional structure, the mineral does not show crystalline swelling like smectite or vermiculite.

The structure of palygorskite is reflected in its morphology in that the mineral invariably occurs as fibres (Fig. 22.8*a, b*).

Occurrence and genesis

Palygorskite is a common component of the arid and semi-arid soils of Australia (Stace *et al.* 1968) (Table 22.8a, b, e). It occurs in the desert soils of North Africa and the Middle East (Muir 1951; Yaalon 1955; Barshad *et al.* 1956; Ravikovitch *et al.* 1960; Elgabaly 1962; Singer 1971; Aba-Husayn and Sayegh 1977) and the U.S.S.R. (Chukhrov 1968).

Palygorskite is often a major component of sedimentary rocks of lacustrine origin. The weathering of these rocks gives soils that are high in palygorskite, but in regions of higher rainfall the palygorskite is present only in the deeper horizons of the soil.

In desert soils the presence of palygorskite can often be traced to the parent rock, but it is too common in desert soils to be explained as a residue from particular rocks. Millot *et al.* (1969) and Paquet *et al.* (1969) considered that palygorskite formed pedogenically in the arid soils of Morocco since it appeared to be absent from the parent material, and Vanden Heuvel (1966) postulated a pedogenic origin for palygorskite in some New Mexico soils.

The palygorskite in some Australian soils is present as thin, white, translucent clay skins on soil aggregates (Beattie and Haldane 1958; Singer and Norrish 1974), but may be absent from the clay within the aggregates. Table 22.7 gives the composition of the palygorskite and compares it with the adjacent clay matrix. Singer and Norrish (1974) considered that this palygorskite formed in the soil and showed that the composition of the soil solutions is compatible with its formation. Their study shows that palygorskite formation is favoured by high pH and high Mg^{2+} and $Si(OH)_4$ concentrations. Such conditions are likely in drier areas of poor drainage where evaporation exceeds precipitation.

Table 22.8. The clay mineralogy of Australian soils

Abbreviations used in this table: K, kaolin; I, illite; M, smectite; V, vermiculite; Vi, chloritized vermiculite; C, chlorite; P, palygorskite; RIM, randomly interstratified material; T, talc; Py, pyrophyllite; F, feldspar; H, hematite; Go, goethite; L, lepidocrocite; Ma, maghemite or magnetite; G, gibbsite; B, boehmite; An, anatase; Il, ilmenite; Pl, a phosphate mineral of the plumbogummite group; Ho, amphibole Ti, rutile. An asterisk against a mineral name indicates an isolated occurrence of the particular mineral

Great Soil Group	Location (with no. of profiles)	Major components	Minor components
(a) Soils showing no profile development			
Solonchaks	Canterbury, Qld (1)	kaolin, smectite, illite	
	Lakes District, Co. Russell, S.A. (1)	RIM, kaolin	
Alluvial soils	Pemberton, W.A. (8)	kaolin	H/Go, G, Vi, M, I
	Launceston Basin, Tas. (6)	kaolin, smectite, RIM, goethite*	I, Vi, L, Go
	Barossa Valley, S.A. (1)	illite (trioctahedral), kaolin	V, RIM, T
	Innisfail, Qld (1)	gibbsite, kaolin	
Lithosols	Darling Downs, Qld (1)	interstratified kaolin-smectite	M, H/Go, F
	Barossa Valley, S.A. (1)	illite (trioctahedral)	K
Calcareous sands	Port Augusta, S.A. (1)	illite, kaolin, RIM	P, H/Go
Siliceous sands	S.W. of W.A. (3)	kaolin, boehmite*, gibbsite*	H/Go, G, Vi
	Northern Qld (3)	illite, kaolin	H/Go
	Kununurra, W.A. (1)	kaolin	I, RIM, H/Go
	Alice Springs, N.T. (1)	kaolin, illite	RIM, H/Go
Earthy sands	S.W. of W.A. (2)	kaolin, gibbsite*	G, B, H/Go
	Merredin, W.A. (1)	kaolin	H/Go
(b) Soils showing minimal profile development			
Grey-brown and red calcareous soils	Alice Springs, N.T. (1)	illite, RIM, kaolin	H/Go
	Jamestown, S.A. (1)	illite, RIM, kaolin	H/Go
	Whyalla, S.A. (1)	illite, kaolin, RIM	P, H/Go
Desert loams	S.W. of Qld (7)	kaolin, smectite, illite*, RIM*	I
	Whyalla, Port Augusta, Oodnadatta, S.A. (6)	kaolin, illite, RIM	P (quite often), H/Go
	Others in N.T. and S.A. (2)	smectite, illite, kaolin	H/Go
Red and brown hardpan soils	N.W. of S.A. (1)	kaolin, illite	RIM, H/Go
Grey-brown and red clays	Barossa Valley; lower north of S.A. (5)	variable amounts of illite, kaolin and RIM	
	Riverina, N.S.W. (5)	illite, kaolin	RIM
	Namoi floodplain, N.S.W. (3)	smectite, illite	K, RIM, V, F
	N.W. of Vic. (43)	illite, RIM, kaolin, variable smectite*	
	S.E. of Qld (14)	smectite, kaolin, illite*	I
	Western Qld (12)	smectite, kaolin, illite*	I, F, H/Go
	Northern Qld (4)	smectite	K, V, Vi, I, F
	Oodnadatta, S.A. (2)	kaolin, smectite, illite	H/Go
	Alice Springs, N.T. (1)	smectite, kaolin	I, H/Go
	Kununurra, W.A. (1)	smectite, kaolin	I, H/Go
(c) The mildly leached dark soils			
Black earths	near Adelaide, S.A. (5); Western Vic. (2); Namoi, N.S.W. (1)	illite, kaolin, RIM, smectite*	F in Vic. profiles
	Southern Qld (13); Northern Qld (1)	smectite, kaolin, illite*	F, H/Go
	East coast and S.E. of Tas. (4)	smectite	F, T or Py
Rendzinas	near Adelaide, S.A. (8)	illite, kaolin, RIM	C*
	S.E. of S.A. (13)	illite, kaolin, RIM, smectite*, palygorskite*	
Chernozems	S.E. of S.A. (1)	kaolin, RIM	I
	Hunter Valley, N.S.W. (1)	smectite, illite	K
	Inverell, N.S.W. (9)	smectite, variable kaolin, illite*	
	Liverpool Plains, N.S.W. (2)	smectite	K
	Monaro area, N.S.W. (2)	smectite	I

Table 22.8. *continued.* The clay mineralogy of Australian soils

Great Soil Group	Location (with no. of profiles)	Major components	Minor components
Prairie soils	South coast of N.S.W. (5)	kaolin, with the other major components, comprising several of the following: smectite, vermiculite, feldspar, RIM. Illite is a major in only one of the profiles.	H/Go, Vi
	Southern Qld (5)	kaolin, smectite*, vermiculite*, feldspar*	H/Go, RIM, Vi, V, C, F
	Western Vic. (1)	kaolin, RIM	I, H/Go
	Northern Tablelands, N.S.W. (1)	kaolin, illite (trioctahedral)	RIM
	North-west slopes, N.S.W. (1)	smectite, kaolin	
	Launceston Basin, Tas. (1)	interstratified kaolin–smectite	K
Weisenböden	near Adelaide, S.A.	illite, RIM	K
(d) The mildly leached soils			
Solonetz	Upper S.E. of S.A. (1)	illite, kaolin, RIM	
	Yudnapinna, S.A. (1)	kaolin, RIM	Go
	Far north of S.A. (1)	illite, kaolin	RIM
	Narrogin, W.A. (1)	kaolin	F, I, RIM
	Lower Burdekin Valley, Qld (1)	smectite, kaolin	I, F
Solodized solonetz and solodic soils	Eyre Peninsula, S.A. (9)	kaolin, variable illite	RIM, H/Go
	Barossa Valley, S.A. (7)	kaolin, illite (trioctahedral)	RIM, H/Go, T
	S.E. of S.A. (18); Darwin, Alice Springs, N.T. (2)	variable kaolin, illite and RIM	H/Go, RIM, F, M, V, Vi
	Western Vic. (46)	kaolin, variable illite, smectite*, interstratified kaolin–smectite*	RIM, H/Go
	Southern Qld (15)	kaolin, smectite	RIM, H/Go, F, Vi
	Northern Qld (9)	kaolin, smectite, variable illite	RIM
	Upper Hunter Valley, N.S.W. (1)	smectite, RIM	K, F
	Riverina district, N.S.W. (2); Benalla, Bairnsdale, Vic. (2)	illite, kaolin	RIM
Soloths	Southern Qld (15)	kaolin, smectite, RIM*	I, H/Go, RIM, F
	Western Vic. (3)	kaolin	RIM, H/Go, I, F
	S.E. of S.A. (1)	illite, RIM	K
	Eyre Peninsula, S.A. (2)	kaolin, RIM	I, Go
Solonized brown soils	Merredin, W.A. (7)	kaolin, illite*	I, RIM, F, An
	Eyre Peninsula, Yorke Peninsula, lower north, S.E. and Mallee areas of S.A. (27)	illite, kaolin with variable RIM	
	Moonta, S.A. (3)	kaolin, RIM	H/Go
	Riverina district, N.S.W. (3)	illite, kaolin, RIM	
Red-brown earths	Merredin, W.A. (2); lower north and Adelaide Plains, S.A. (34); Wimmera and northern Vic. (39); Riverina and central N.S.W. (13)	illite, kaolin, with variable RIM	H/Go, M in a few Vic. profiles
	Southern Qld (7)	kaolin, variable illite	M, RIM
	Barkly Tableland, Qld (3)	kaolin (usually halloysite)	M, RIM, I
	East Central Coast, Qld (9)	kaolin, variable illite	M, RIM, H/Go, F
Non-calcic brown soils	Avon Valley, W.A. (1)	kaolin, illite	H, F, T, Ho
	Northern Qld (6)	kaolin, variable illite	M, RIM, H/Go, F
Chocolate soils	Glenelg region, Vic. (2)	kaolin, feldspar, magnetite	H, Vi
	Southern Tablelands, N.S.W. (4)	smectite, variable illite	K
	Central Tablelands, N.S.W. (1)	illite, kaolin	M
	North-west slopes, N.S.W. (1)	smectite	K, H
	Northern Tablelands, N.S.W. (8)	*either*; smectite, variable kaolin, illite; *or* kaolin, gibbsite	H/Go, Vi, Ma, Ti
Brown earths	Launceston Basin and other areas of Tas. (6)	smectite, kaolin, interstratified kaolin–smectite*	H/Go, RIM
	Barossa Valley, S.A. (6)	illite (trioctahedral), variable kaolin, RIM	H, V, T
	Mt Lofty Ranges, S.A. (1)	illite, chlorite	RIM, H/Go
	Coastal Lowlands, northern N.S.W. (1)	kaolin, vermiculite, (illite at depth)	RIM, H/Go
	South coast, N.S.W. (1)	amorphous material, kaolin	F, M, V

Table 22.8. *continued.* The clay mineralogy of Australian soils

Great Soil Group	Location (with no. of profiles)	Major components	Minor components
(e) Soils dominated by sesquioxides			
Calcareous red earths	Whyalla, S.A. (4)	illite, kaolin, palygorskite*	RIM, H/Go
	Alice Springs, N.T. (2)	kaolin, illite	RIM, H
Red earths	S.W. of W. A. (15) } Darwin, N.T. (1) }	kaolin, with variable hematite, goethite and gibbsite	Vi, F, Ma
	Katherine, N.T. (1); Alice Springs N.T. (1)	kaolin, illite	RIM, H/Go, Ma
	Southern Qld (6)	kaolin	I, RIM, H/Go
	Leichardt–Gilbert area, Qld (5)	kaolin, with variable illite (trioctahedral)	H, M, RIM
	Northern Qld (7)	kaolin, with variable hematite	H/Go, RIM, I
	South Coast (1), Northern Tablelands, N.S.W. (1)	kaolin, with variable gibbsite	V or C, Vi, I, H/Go, RIM
	South–west slopes, N.S.W. (1)	illite, kaolin	H
	Ballarat, Vic. (1)	kaolin	I, RIM, H/Go
	Daly Basin, N.T. (3)	variable amounts of kaolin, illite and hematite	V*
Yellow earths	S.W. of Qld (2)	kaolin with variable smectite	H/Go, I
	S.E. of Qld (6)	variable kaolin, illite and gibbsite	H/Go, RIM
	Northern Qld (13)	kaolin	H/Go, I, Vi, G
	S.W. of W.A. (3)	kaolin	Vi, H/Go
	Kununurra, W.A. (1)	illite, kaolin	H/Go
	Tennant Creek, N.T. (1)	kaolin, illite	H/Go, RIM
	South coast of N.S.W. (1)	kaolin, gibbsite	Vi, H/Go
Grey earths	Northern Qld (4)	kaolin	I, RIM
Terra rossa soils	S.E. of S.A. (7)	illite, kaolin, with variable RIM	C, H (M in one profile)
	near Adelaide, S.A. (7)	illite, kaolin, with variable RIM	H/Go, C, V
	Mt Larcom, Qld (1)	kaolin, illite, RIM	
Euchrozems	Kununurra, W.A. (1)	smectite, kaolin, hematite	I
	Toowoomba, Qld (1)	smectite, kaolin	H
	Jondaryan, Qld (1)	kaolin	H/Go
	Northern Tablelands, N.S.W. (4)	kaolin	H, RIM
	Northern Qld (5)	kaolin (10Å halloysite*), hematite*	H/Go, Ma, M
Brown basaltic soils	Northern Qld (6)	kaolin, with variable RIM	H/Go, Ma
Xanthozems	Northern Qld (2)	kaolin	Vi, H/Go
	Gympie, Qld (1)	illite, kaolin	Vi, H/Go
Krasnozems	North coast, Darling Downs, south coast and others Qld (30) }	kaolin, with variable amounts of gibbsite, hematite and goethite; smectite*	Il, Ma, F, Vi*, Pl*
	Northern rivers, northern tablelands and others N.S.W. (31) }	kaolin, with variable amounts of gibbsite, hematite and goethite	Ma, Vi*, Pl* Cristobalite*, Tridymite*
	Launceston Basin, (5) } Burnie area Tas. (8) }	kaolin, with variable gibbsite, hematite and goethite	H/Go, F, V
(f) The mildly to strongly leached soils			
Grey-brown podzolic soils	Launceston Basin and others, Tas. (6)	kaolin, with variable smectite and RIM	H/Go, Vi, F, L*
	Eyre Peninsula, S.A. (1)	kaolin, illite	H/Go
	Mt Lofty Ranges, S.A. (9)	kaolin, with variable illite (trioctahedral), RIM	H/Go, Pl*
Red podzolic soils	Swan coastal plain and S.W. of W.A. (7)	kaolin, gibbsite, with variable hematite	H/Go, Vi, I, F
	Barossa Valley, S.A. (1)	kaolin	I, V or C, H
	Eyre Peninsula, S.A. (1)	kaolin	I, Go
	Canberra, A.C.T. (1)	kaolin, illite	R, Vi, H/Go
	Ebor–Dorrigo, N.S.W. (5)	kaolin, variable illite, chloritized vermiculite and gibbsite	
	South coast, N.S.W. (1)	illite, kaolin	H/Go, Vi, RIM
	Northern Qld (2)	kaolin	I, R, H/Go, G
	Beenleigh, Qld (1)	interstratified illite–smectite and kaolin	H/Go
	Gympie, Qld (1)	illite and interstratified illite/chloritized vermiculite	K

Table 22.8. *continued.* The clay mineralogy of Australian soils

Great Soil Group	Location (with no. of profiles)	Major components	Minor components
Yellow podzolic soils	near Perth and S.W. coast of W.A. (20)	kaolin, (illite in some profiles from York)	Vi, H/Go, G, F
	Barossa Valley, S.E., near Adelaide, Fleurieu Pen., S.A. (12)	kaolin, variable illite	Vi, H/Go, G, RIM, T, Pl*
	Northern Tas. (11)	kaolin, illite, chloritized vermiculite	RIM
	Preston, Vic. (2)	kaolin, illite*	Go, G
	A.C.T. and N.S.W. (6)	kaolin, illite, variable chloritized vermiculite*	H/Go, RIM, F, G, Pl*
	South coast, Qld (3)	kaolin	RIM, I, H/Go, Vi
	Leichardt–Gilbert area, Qld (1)	kaolin	I, H/Go
Brown podzolic soils	York area, W.A. (3)	kaolin, illite (trioctahedral)	M at depth, H/Go, F
	Glenelg region, Vic. (1)	kaolin	Vi, G, H/Go
Lateritic podzolic soils	East Darling Range, Merredin, near Perth and South Coast, W.A. (22)	kaolin, variable illite	H/Go, G, Vi, M, B
	Eyre Peninsula, Fleurieu Pen., Mt Lofty Ranges, S.A. (9)	kaolin, illite	H/Go, Vi, G, RIM
	Launceston Basin, Tas. (10)	kaolin	Vi, RIM, I, Go
	Wimmera, Ballarat, Preston, Vic. (3)	kaolin, illite, variable smectite	RIM, H/Go, V, Ma
	Northern Tablelands, N.S.W. (1)	smectite, kaolin, RIM	I, V, Go
	Darwin, N.T. (1)	kaolin	I, Vi, H/Go
Gleyed podzolic soils	South coast, Qld (2)	kaolin	H/Go, Vi, RIM
	Eastern Darling Ranges, W.A. (1)	kaolin	V, G, B, H/Go
Podzols	King Is., Tas.	kaolin, illite	F, H/Go
	Mt Burr, S.A. (4)	kaolin, RIM	Vi, H/Go, An
	Barossa Valley, S.A. (1)	kaolin, RIM	H
	Beerwah, Qld (2)	(only quartz)	
Humus podzols	Barossa Valley, S.E. of S.A. (2)	kaolin, variable feldspar and vermiculite	I, T, An
	Launceston Basin, Tas. (4)	kaolin, variable illite and RIM	H/Go, Vi
(g) The organic soils			
Alpine humus soils	Mt Kosciusko, N.S.W. (6)	kaolin, hydrobiotite, gibbsite	I
	Monaro, N.S.W. (1)	illite, kaolin, RIM	G, V or C
	Corryong, Vic. (2)	kaolin, illite, vermiculite, smectite	H/Go
	Northern Tablelands, N.S.W. (2)	kaolin, gibbsite	H, Ma, F
Humic gleys	South coast, Qld (5)	kaolin, illite, smectite	Vi, H/Go
	Northern Tablelands, N.S.W. (2)	interstratified kaolin–smectite	Ti

Eswaran and Barzanji (1974) concluded that palygorskite was forming in some alluvial and co-alluvial soils of Iraq; Yaalon and Wieder (1976) found palygorskite concentrated in the B_3 horizons of some Arid Brown soils in Israel and they considered that it formed there.

The available data indicate that palygorskite is a common component of desert soils and can form during soil formation; if palygorskite is present in a sedimentary parent material it can persist in the soil in arid climates; where there is more leaching, palygorskite is rapidly removed from the soil (Yaalon *et al.* 1966).

Sepiolite

Sepiolite is a fibrous mineral resembling palygorskite. Structurally the two minerals are made up of ribbons of 2:1 structures, but the ribbon widths are different, eight octahedral positions for sepiolite. Sepiolite differs from palygorskite in containing more magnesium and less aluminium.

Sepiolite has been reported in one South Australian soil (Rogers *et al.* 1956), where in that it appeared to be a residue from the parent material. Mann and Deutscher (1978) found sepiolite in soils formed on calcrete near Lake Way in Western Australia. Sepiolite with palygorskite is found in some New Mexico soils (Vanden Heuval 1966; McLean *et al.* 1972) but there is disagreement as to whether the minerals are forming in the soil or are entirely residual from the parent material. Sepiolite is less stable than palygorskite, but given the right conditions, which may be rare, there would seem to be no reason why it could not persist in soils or even form in them; it is easily synthesized at room temperature (Isphording 1973).

Allophane and related minerals

Structure and chemistry

Hydrated aluminium silicates that are amorphous can occur as major constituents of soils and clays derived from volcanic rocks. Because they have a disordered structure, such materials give ill-defined data when studied by X-ray diffraction, infra-red spectrometry (IR), differential thermal analysis (DTA) or the thermogravimetric analysis (TGA). This makes it exceedingly difficult to say whether the amorphous constituents of different clays are the same.

Allophane has a SiO_2/Al_2O_3 mole ratio between 1.3 and 2 (for kaolin this ratio is 2). High resolution electron microscopy (Wada 1977) suggests that allophane consists of hollow spherules, 35–40 Å in diameter.

Allophane has a very high surface area, but its value depends on how it is measured; Aomine and Otsuka (1968) obtained values ranging from $\sim 50 m^2 g^{-1}$ using N_2, to ~ 900 $m^2 g^{-1}$ using ethylene glycol. Soils containing much allophane can have very high moisture contents, > 100% on an oven dry basis, but these soils do not swell and shrink, and on drying the loss of moisture is largely irreversible.

The cation exchange capacity of allophanes depends on pH, concentration, and the ionic species used (Wada 1977), but it is usually in the range 20–50 mequiv.%.

Allophane has a high anion sorption capacity and soils containing it retain large amounts of phosphate.

Occurrence and genesis

Soils high in allophane are generally formed from recent volcanic ash. The allophanic soils of Japan, New Zealand and New Guinea are such cases, where often allophane is the only clay constituent. As would be expected, allophane occurs in the soils of Australia that are associated with recent volcanic activity. Such soils in western Victoria and some near Mount Gambier in South Australia have appreciable allophane in the clay, but other clay minerals are also present (Briner and Jackson 1969).

Allophane is generally absent in older volcanic soils and Fieldes (1955) suggested that it transforms with time to halloysite and kaolinite; it can also transform to gibbsite (Wada and Aomine 1966; Wada and Greenland 1970).

In recent volcanic ash soils, where allophane is a major clay component, it can be identified by a variety of techniques. However, as mentioned earlier, the common techniques of clay identification are not very satisfactory for identifying small amounts of allophane, so there is doubt as to whether allophane occurs, commonly and in significant amounts in non-ash soils. Because the physical methods of identification are so unsatisfactory, chemical dissolution techniques have been developed to estimate its abundance in soils. These usually involve equating allophane to the silica and alumina dissolved in hot 2% sodium carbonate or 0.5 M NaOH (Jackson 1956; Hashimoto and Jackson 1960). Although these reagents may be effective in dissolving amorphous material in clay, they also attack crystalline clays, particularly if they are fine-grained (Langston and Jenne 1971; Wada and Tokashiki 1972). Any clay mineral, no matter how well crystallized, will yield some 'amorphous material' when attacked by hot NaOH. For such estimates to be meaningful they must be coupled with very careful X-ray diffraction and IR measurements; even then, when the amounts are small, the results may be ambiguous.

In reviewing the amorphous consitituents of soils, Wada and Harward (1974) concluded that, although such material has been postulated as a clay constituent of soils derived from parent materials other than volcanic ash and pumice, the data were inconclusive. In Australia, there is no evidence that many soils contain significant amounts of allophane. The great age of most of the Australian landscape would favour this viewpoint.

Aluminium-organic matter complexes in soil can be a source of amorphous material in soils. The organic matter of certain soils can be high in aluminium; electron microprobe analyses of organic matter, made in this laboratory, give Al_2O_3 contents of up to 20%, which is consistent with the chemical estimates of Martin and Reeve (1957*a*, *b*) and Schnitzer and Khan (1972). Many of the extractants used to remove amorphous material from soil will extract this aluminium. In clay mineral studies it is usual to destroy organic matter with H_2O_2 prior to separation of the clay, and this will result in precipitation of aluminium hydroxide. In certain soils where the organic matter is high compared to the clay content, e.g. *podzols*, this precipitated aluminium may be a major component of the separated clay. The podzolic B horizons studied by Martin and Reeve are in this category. They showed that a large fraction of the total soil aluminium was complexed with organic matter. Destruction of the organic matter in such soils would lead to clay very high in amorphous aluminium hydroxide.

Although aluminium, iron or silicon combined with organic matter is, strictly speaking, amorphous, it is misleading to describe such materials as amorphous clays; it is certainly more fruitful to consider them as metal-organic complexes.

Imogolite

Imogolite is a clay mineral often associated with allophane in volcanic ash soils. Imogolite consists of very fine threads having diameters of 100–300 Å and lengths of several micrometres (Yoshinaga and Aomine 1962). High resolution electron microscopy has shown that each thread is made up of finer (20 Å diameter) hollow tubes (Wada *et al.* 1970). Because of the very small diameter of the threads imogolite can be regarded as a one dimensional structure. Despite this it gives more definitive X-ray diffraction and IR data than does allophane (Yoshinaga and Aomine 1962; Cradwick *et al.* 1972; Wada 1977).

Imogolite has not been identified in Australian soils.

Table 22.9. The distribution of clay minerals in Australian soils

% dominant, percentage of samples in which the mineral is the dominant clay mineral; % major, percentage of samples in which the mineral is >20%; average %, average % composition,

Great Soil Group	Total No. of samples	Kaolin % dominant	Kaolin % major	Kaolin Average %	Illite % dominant	Illite % major	Illite Average %	Smectite % dominant	Smectite % major	Smectite Average %	RIM % dominant	RIM % major	RIM Average %	H+ Go %
(a) Soils showing no profile development														
Solonchaks	2	50	100	30	–	50	13	–	50	13	50	50	23	–
Alluvial soils	13	46	100	40	23	31	15	7.7	31	13	–	7.7	1.9	5.3
Lithosols	3	–	–	–	33	33	30	33	33	12	2	2	27	5.3
Calcareous sands	2	–	100	25	100	100	40	–	–	–	–	100	25	1.5
Siliceous sands	12	92	100	58	–	67	23	–	–	–	–	–	–	2.1
Earthy sands	2	100	100	80	–	–	–	–	–	–	–	–	–	3.0
Average	*(34)*	*59*	*91*	*44*	*18*	*47*	*19*	*5.9*	*1.8*	*6.6*	*8.8*	*18*	*5.9*	*3.5*
(b) Soils showing minimal profile development														
Grey-brown and red calcareous soils	8	13	100	28	75	100	36	–	–	–	50	100	31	2.8
Desert loams	31	55	100	37	26	61	20	23	35	16	16	55	15	1.4
Red and brown hardpan soils	2	100	100	53	–	50	30	–	–	–	–	–	–	15
Grey, brown and red clays	190	21	78	25	54	65	29	27	36	17	5.3	36	10	0.5
Average	*(231)*	*26*	*82*	*27*	*50*	*66*	*28*	*25*	*34*	*16*	*8*	*41*	*12*	*0.8*
(c) The mildly leached dark soils														
Black earths	47	28	60	24	19	21	9	49	72	39	17	21	10	0.3
Rendzinas	38	2.6	47	12	87	100	54	5.3	7.9	3	2.6	42	12	0.2
Chernozems	5	20	60	21	–	100	23	40	40	21	40	60	31	–
Prairie soils	24	42	67	35	21	21	14	13	21	6	83	17	6	4.5
Wiesenböden	2	–	50	13	100	100	53	–	–	–	–	100	30	–
Average	*(116)*	*22*	*57*	*22*	*42*	*52*	*26*	*26*	*38*	*19*	*11*	*30*	*11*	*1.1*
(d) The mildly leached brown soils														
Solonetz	14	57	93	38	43	43	20	21	21	11	21	29	10	0.2
Solodized-solonetz and solodic soils	262	78	95	53	17	41	15	5.3	12	4	6.5	23	8	3.6
Soloths	43	74	93	44	2.3	9.3	3.3	14	40	15	12	28	10	1.7
Solonized brown soils	132	32	82	31	53	77	33	0.8	1.5	0.5	4.5	44	12	0.8
Red-brown earths	168	32	94	32	80	94	44	–	2.4	0.7	1.8	41	11	1.6
Non-calcic brown soils	2	50	50	23	–	100	25	–	–	–	–	–	–	–
Chocolate soils	8	50	75	26	–	–	–	–	–	–	50	50	28	8.7
Brown earths	21	9.5	62	22	62	62	29	29	29	15	–	4.8	4.8	5.9
Average	*(650)*	*54*	*90*	*41*	*41*	*59*	*26*	*4.6*	*9.4*	*3.6*	*6*	*32*	*9.8*	*2.5*
(e) Soils with profiles dominated by Sesquioxides														
Calcareous red earths	13	46	100	30	54	100	42	–	–	–	–	7.7	1.9	5.3
Red earths	71	94	100	67	4.2	8.5	2.4	–	–	–	–	2.8	0.70	12
Yellow earths	82	94	98	74	3.7	7.3	2.9	–	1.2	0.4	–	–	–	1.0
Terra rossa soils	37	46	95	33	86	100	48	–	–	–	5.4	8.1	3.6	2.8
Euchrozems	32	94	100	75	–	–	–	6.3	19	6	3	6.3	2.5	11
Brown soils on basalt in North Queensland	16	100	100	69	–	–	–	–	–	–	–	56	15	5.7
Xanthozems	7	100	100	75	29	29	14	–	–	–	–	–	–	2.3
Krasnozems	90	78	93	52	–	–	–	1.0	2	0.9	–	–	–	21
Average	*(348)*	*83*	*97*	*61*	*14*	*19*	*8.1*	*0.9*	*2.6*	*0.9*	*0.6*	*4.9*	*1.5*	*10*

Table 22.9. (Continued) The distribution of clay minerals in Australian soils

% dominant, percentage of samples in which the mineral is the dominant clay mineral; % major, percentage of samples in which the mineral is >20%; average %, average % composition,

Great Soil Group	Total No. of samples	Kaolin % dominant	Kaolin % major	Kaolin Average %	Illite % dominant	Illite % major	Illite Average %	Smectite % dominant	Smectite % major	Smectite Average %	RIM % dominant	RIM % major	RIM Average %	H+Go %
(f) The mildly to strongly leached highly differentiated soils														
Grey-brown podzolic soils	24	58	96	49	33	46	21	4.2	8.4	3.3	8.4	21	9.3	3.3
Red podzolic soils	35	83	97	53	20	29	10	—	—	—	2.9	5.8	2.4	7.4
Yellow podzolic soils	91	86	99	56	18	37	14	2.2	2.2	1.4	—	11	2.8	3.9
Brown podzolic soils	19	100	100	44	53	68	24	—	21	5.8	—	—	—	9.7
Lateritic podzolic soils	89	94	100	76	4.5	10	3.5	—	—	—	1.1	3.4	1.2	0.5
Podzols	4	100	100	56	—	—	—	—	—	—	—	—	—	4.2
Gleyed podzolic soils	5	80	100	58	—	—	—	—	—	—	—	—	—	3.2
Humus podzols	9	100	100	38	11	11	3.9	—	—	—	—	22	5.5	2.1
Average	*(276)*	*87*	*99*	*60*	*20*	*28*	*11*	*1.1*	*2.9*	*1.2*	*1.5*	*8.0*	*2.6*	*5.0*
(g) The organic soils														
Humic gleys	11	55	64	29	—	27	6.8	36	45	24	18	73	23	—
Alpine humus soils	16	44	100	35	38	44	16	—	—	—	6	63	19	4
Average	*(27)*	*48*	*85*	*33*	*22*	*37*	*12*	*15*	*19*	*9.8*	*11*	*67*	*21*	*2.4*
Average of all samples	(1682)	59	89	45	32	45	20	7.7	13	5.6	4.8	24	7.3	4.1

RIM, randomly interstratified material; H + Go, hematite plus goethite

Clay mineralogy of Australian soils

The data presented in Tables 22.8 and 22.9 are summaries of analyses made in the Adelaide laboratories of the CSIRO Division of Soils since 1948, and more recently, in the CSIRO's Brisbane laboratory. Although some of these data are published as clay mineral studies, much is in internal reports, or as incidental information in publications not concerned primarily with clay mineralogy. In addition, the results of other workers (Shearer and Cole 1939*a*, *b*; Cole 1943; Hosking 1940, 1957; Hosking *et al.* 1957; Turton *et al.* 1962; Hallsworth *et al.* 1952, 1953; Costin *et al.* 1952; Simonett and Bauleke 1963; and Day 1977) have been included where sufficient information on Great Soil Group and locality is available.

X-ray techniques

Results are presented for approximately 900 profiles, of which 10% have been extracted from the publications of other workers, listed previously. For the approximately 800 profiles analysed in the CSIRO laboratories the following procedure was adopted. The $< 2\mu m$ fraction of the soil was obtained after the removal of carbonate. Organic matter was not always removed prior to dispersion. X-ray diffraction powder photographs of oven-dry, salt-free clay, and diffractometer traces of glycerol-wet, magnesium-saturated, oriented clay were obtained for each sample using CoK α radiation. Additional diagnostic tests were made where necessary (see Brown 1961). The diffraction data were supplemented by measurements of the cation exchange capacity and total potassium made on barium-saturated clay using X-ray fluorescence techniques.

Interpretation

Semi-quantitative estimates of all components, except the layer-lattice silicates, were made by visual comparison with powder photographs of standard mixtures. The clay minerals were estimated by careful consideration of both the diffraction data and the exchange capacity and potash content. The ratio of illite to kaolin was determined by comparison (from diffractometer traces) of the peak heights of their basal reflections weighted in favour of illite by a factor of two. Using this ratio and the estimates of non-clay minerals the amounts of illite and kaolin can be estimated. At this stage reference is made to the CEC determination and to average values of the CEC for the various clay minerals (after Grim 1968). Generally the measured CEC is in excess of the calculated CEC. This excess is attributed to randomly interstratified material, smectite or vermiculite. The presence of any of these was of course confirmed from the diffraction data, and the previous percentages were then modified. Because of the problems associated with quantitative clay mineral analysis, these estimates were originally presented as percentage ranges rather than as specific percentages.

Distribution of clay minerals

In summarizing results for a Great Soil Group it is often found that the mineralogy varies with locality (climate, parent material, etc.). Useful information is lost in averaging such a group, so in Table 22.8, under each Great Soil Group, results are presented for a number of localities, each of which have reasonably consistent mineralogies. This arrangement highlights differences that may occur within one Great Soil Group. For example, Norrish and Pickering (1977) have pointed out that various Great Soil Groups are dominated by illite in southern Australia, while their northern counterparts tend to be high in smectite (Table 22.8*b*, *c*, *d*).

In Table 22.8, the minerals are divided into major and minor components, where major represents > 20%. Percentage figures are not given, as there may be considerable variation between the individual clays. However, for the major components, the minerals are listed in decreasing order of average abundance. For the minor components the minerals are listed in decreasing order of the frequency of occurrence. An asterisk in either column indicates an isolated occurrence of the particular mineral.

In Table 22.8, quartz is not recorded for any soil as it is present in almost all soil clays. Carbonate minerals are also absent because of the previously mentioned pretreatment of the soil. The distinction between hematite and goethite in clay samples is made difficult because of mutually coincident diffraction lines and the masking effect by some of the stronger clay lines. In Table 22.8 'H/Go' is used to indicate the presence of hematite and/or goethite. In many soils both of these minerals are probably present. The data is based on average profile mineralogy.

By contrast, Table 22.9 is based on the results of individual soil clay samples. Results are only given for the most common and abundant clay minerals (kaolin, illite, smectite and randomly interstratified material) and free iron oxides. Under each clay mineral heading the first column lists the percentage of samples in which that mineral is the dominant mineral; the second column lists the percentage of samples in which that mineral is a major component; the third column gives the average for that mineral.

The two tables follow the arrangement of Great Soil Groups adopted in 'A Handbook of Australian Soils' (Stace *et al.* 1968), with two additions. Included in Table 22.8*e*, following *yellow earths*, are results of several profiles from northern Queensland which Isbell and Smith (1976) described as 'grey earths'. In the same table, following the data on *euchrozems* appears a group from the same general area described by Isbell *et al.* (1977) as 'brown basaltic soils'.

From Table 22.9 it is obvious that the grouping of Great Soil Groups is not related to their average clay mineralogy. *Yellow earths* and *terra rossa soils* are in the same group yet in the former 94% of samples are dominantly kaolin with a low average illite content, while 86% of the latter have illite as the dominant mineral.

Table 22.9 shows the extent to which kaolin is present in clays; it is the dominant mineral of 60% of the samples examined and is a major component of 90%. Kaolin is consistently the dominant mineral of the podzolic soils.

Illite is the dominant clay mineral of about one third of the samples and a major component of 45% of samples.

Smectite is the dominant mineral in only about 8% of samples and a major component in only 12%.

Although not shown in Table 22.9, palygorskite was dominant in only two samples, i.e. ~ 0.1% of those examined, and it was a major in ~ 0.2%. The data of Table 22.9 do not represent a true average for the continent as most samples examined have been taken from the higher rainfall and agricultural areas. On an area basis, the arid and semi-arid soils are under-represented, and these soils tend to be higher in illite and montmorillonite than the overall averages of 19 and 6% given for these minerals in Table 22.9.

References

Aba-Husayn, M. M. and Sayegh, A. H. (1977). Mineralogy of Al-Hasa desert soils (Saudi Arabia). *Clays Clay Miner.* **25**, 138-47.

Aitchison, G. D. and Holmes, J. W. (1953). Aspects of swelling in the soil profile. *Aust. J. Appl. Sci.* **4**, 244-59.

Aitchison, G. D., Sprigg, R. C. and Cochrane, G. W. (1954). The soils and Geology of Adelaide and Suburbs. Geol. Surv. S. Aust. Bull. No. 32.

Alexiades, C. A. and Jackson, M. L. (1966). Quantitative clay mineralogical analysis of soils and sediments. *Clays Clay Miner.* **14**, 35-52.

Altschuler, Z. S., Dwornick, E. J. and Kramer, H. (1963). Transformation of montmorillonite to kaolinite during weathering. *Science* **141**, 148-52.

Alwis, K. A. and Pluth, D. J. (1976). The red latosols of Sri Lanka: I. Macro-morphological, physical and chemical properties, genesis and classification. II. Mineralogy and weathering. *Soil Sci. Soc. Am. Proc.* **40**, 912-28.

Angel, B. R. and Hall, P. L. (1973). Electron spin resonance studies of kaolins. Proc. 4th Internat. Clay Conf. (Madrid), Vol. 1, pp. 47-60.

Aomine, S. and Otsuka, H. (1968). Surface of soil allophanic clays. Trans. 9th Internat. Congr. Soil Sci. (Adelaide), Vol. 1, pp. 731-7.

Asher, C. J. and Ozanne, P. G. (1967). Growth and potassium content of plants in solution cultures maintained at constant potassium concentrations. *Soil Sci.* **103**, 155-61.

Bailey, S. W. (1966). The status of clay mineral structures. *Clays Clay Miner.* **14**, 1-23.

Barnhisel, R. L. (1977). Chlorites and hydroxy interlayered vermiculite and smectite. *In* 'Minerals in Soil Environments' (Eds J. B. Dixon, *et al.*) pp. 331-356. (Soil Science Society of America: Madison.)

Barshad, I. (1948). Vermiculite and its relation to biotite as revealed by base exchange reactions, X-ray analysis, differential thermal curves and water content. *Am. Mineral.* **33**, 655-78.

Barshad, I. (1950). The effect of the interlayer cation on the expansion of the mica type of crystal lattice. *Am. Mineral.* **35**, 225-38.

Barshad, I., Halevy, E., Gold, H. A. and Hagin, J. (1956). Clay minerals in some limestone soils of Israel. *Soil Sci.* **81**, 423-37.

Bassett, W. A. (1959). The origin of the vermiculite deposit at Libby, Montana. *Am. Mineral.* **44**, 282-99.

Bassett, W. A. (1960). Role of hydroxyl orientation in mica alteration. *Bull. Geol. Soc. Am.* **71**, 449-56.

Beattie, J. A. and Haldane, A. D. (1958). The occurrence of palygorskite and barytes in certain parna soils of the Murrumbidgee region, New South Wales. *Aust. J. Sci.* **20**, 274-5.

Bolt, G. H., Sumner, M. E. and Kamphorst, A. (1963). A study of the equilibria between three categories of potassium in an illitic soil. *Soil Sci. Soc. Am. Proc.* **27**, 294-9.

Borchardt, G. A. (1977). Montmorillonite and other smectite minerals. *In* 'Minerals in Soil Environments' (Eds J. B. Dixon, *et al.*) pp. 293-330. (Soil Science Society of America: Madison.)

Bradley, W. F. and Grim, R. E. (1961). Mica clay minerals. *In* 'The X-ray Identification and Crystal Structures of Clay Minerals' (Ed. G. Brown) pp. 208-41. (Mineralogical Society: London.)

Brindley, G. W. (1961). Chlorite minerals, *In* 'The X-ray Identification and Crystal Structures of Clay Minerals' (Ed. G. Brown) pp. 242-96. (Mineralogical Society: London.)

Brindley, G. W. (1961). Kaolin, serpentine and kindred minerals. *In* 'The X-ray Identification and Crystal Structures of Clay Minerals' (Ed. G. Brown) pp. 51-131. (Mineralogical Society: London.)

Brindley, G. W. and Hang, D. T. (1973). The nature of garnierites: 1. Structures, chemical compositions and colour characteristics. *Clays Clay Miner.* **21**, 27-40.

Briner, G. P. and Jackson, M. L. (1969). Allophanic material in Australian soils derived from Pleistocene basalt. *Aust. J. Soil Res.* **7**, 163-9.

Brown, G. (1953). A dioctahedral analogue of vermiculite. *Clay Miner. Bull.* **2**, 64-70.

Brown, G. (ed.) (1961). 'The X-ray Identification and Crystal Structures of Clay Minerals'. (Mineralogical society: London.)

Brown, G. and Newman, A. C. D. (1970). Cation exchange properties of micas. III. Release of potassium sorbed by potassium depleted micas. *Clay Miner.* **8**, 273-8.

Brown, G. and Norrish, K. (1952). Hydrous micas. *Miner. Mag.* **29**, 929-32.

Brydon, J. E. and Kodama, H. (1966). The nature of aluminium hydroxide–montmorillonite complexes. *Am. Mineral.* **51**, 875-89.

Caillere, S. and Henin, S. (1961). Palygorskite. In 'The X-ray Identification and Crystal Structures of Clay Minerals' (Ed. G. Brown) pp. 343-53. (Mineralogical Society: London.)

Chukhrov, F. V. (1968). Some results of the study of clay minerals in the USSR. *Clays Clay Miner.* **16**, 3-14.

Cole, W. F. (1943). X-ray analysis of some soil colloids from Gingin, Western Australia. *Soil Sci.* **56**, 153-71.

Costin, A. B., Hallsworth, E. G. and Woof, M. (1952). Studies in pedogenesis in N.S.W. III. The Alpine Humus Soils. *J. Soil Sci.* **3**, 190-218.

Cradwick, P. D. G., Farmer, V. C., Russell, J. D., Masson, C. R., Wada, K. and Yoshinaga, N. (1972). Imogolite, a hydrated aluminium silicate of tubular structure. *Nature Phys. Sci.* **240**, 187-9.

Day, K. J. (1977). Fertility Studies on Three Red Earth Soils of the Daly Basin, Northern Territory. Dep. N. Territory (Aust.) Industry Agric. Branch, Tech. Bull. No. 22.

Dixon, J. B. (1977). Kaolinite and serpentine group minerals. *In* 'Minerals in Soil Environments' (Eds J. B. Dixon *et al.*) pp. 357-403. (Soil Science Society of America: Madison.)

Doll, C. E., Mortland, M. M., Lawton, K. and Ellis, B. G. (1965). Release of potassium from soil fractions during cropping. *Soil Sci. Soc. Am. Proc.* **29**, 689-702.

Douglas, L. A. (1977). Vermiculites. *In* 'Minerals in Soil Environments' (Eds J. B. Dixon *et al.*) pp. 259-92. (Soil Science Society of America: Madison.)

Elgabaly, M. M. (1962). The presence of attapulgite in some soils of the western desert of Egypt. *Soil Sci.* **93**, 387-90.

Eswaran, H. and Barzanji, A. F. (1974). Evidence for the neoformation of attapulgite in some soils of Iraq. Trans. 10th Internat. Congr. Soil Sci. (Moscow), Vol. 7, pp. 154-61.

Fanning, D. S. and Keramidas, V. Z. (1977). Micas. *In* 'Minerals in Soil Environment's. (Eds J. B. Dixon *et al.)* pp. 195-258. (Soil Science Society of America: Madison.)

Fieldes, M. (1955). Clay mineralogy of New Zealand soils. Part 2. *N. Z. J. Sci. Tech.* **37B**, 336-50.

Foscolos, A. E. and Kodama, H. (1974). Diagenesis of clay minerals from lower cretaceous shales of North Eastern British Columbia. *Clays Clay Miner.* **22**, 319-35.

Gaudette, H. E., Eades, J. L. and Grim, R. E. (1966). The nature of illite. *Clays Clay Miner.* **13**, 33-48.

Giese, R. F. (1973). *In* General discussion on 'K-exchange in micas'. Internat. Clay Conf. Proc. (Madrid), Vol. 1, pp 494-5.

Gilkes, R. J. and Little, I. P. (1972). Weathering of chlorite and some associations of trace elements in Permian Phyllites in southeast Queensland. *Geoderma* **7**, 233-46.

Gilkes, R. J. and Suddhiprakarn, P. (1979). Biotite alteration in deeply weathered granite. I. Morphological, mineralogical and chemical properties. *Clays Clay Miner.* **27**, 349-60.

Gilkes, R. J. and Young, R. C. (1974). Artificial weathering of oxidized biotite: III Potassium uptake by subterranean clover. *Soil Sci. Soc. Am. Proc.* **38**, 41-3.

Greenland, D. J. and Quirk, J. P. (1964). Determination of the total specific surface areas of soils by adsorption of cetyl pyridinium bromide. *J. Soil Sci.* **15**, 178-91.

Grim, R. E. (1968). 'Clay Mineralogy', 2nd edn. (McGraw-Hill Book Co.: New York.)

Hallsworth, E. G., Costin, A. B., Gibbons, F. R. and Robertson, G. (1952). Studies in pedogenesis in N.S.W. II. The Chocolate Soils. *J. Soil Sci.* **1**, 103-24.

Hallsworth, E. G., Colwell, J. D. and Gibbons, F. R. (1953). Studies in pedogenesis in N.S.W. V. The euchrozems. *Aust. J. Agric. Res.* **3**, 305-25.

Harder, H. (1977). Clay mineral formation under lateritic weathering conditions. *Clay Miner.* **12**, 281-88.

Hashimoto, F. and Jackson, M. L. (1960). Rapid dissolution of allophane and kaolinite-halloysite after dehydration. *Clays Clay Miner.* **7**, 102-13.

Herbillon, A. J., Mestdagh, M. M. Vielvoye, L. and Derouane, E. G. (1976). Iron in kaolinite with special reference to kaolinite from tropical soils. *Clay Miner.* **11**, 201-20.

Hosking, J. S. (1940). The soil clay mineralogy of some Australian soils developed on granitic and basaltic parent material. *J. CSIR Aust.* **13**, 206-16.

Hosking, J. S. (1957). Clay Mineralogy and Parent Material. 2nd Aust. Conf. Soil Sci. (Melbourne), Pap. No. 7.

Hosking, J. S., Neilson, M. E. and Carthew, A. R. (1957). A study of clay mineralogy and particle size. *Aust. J. Agric. Res.* **8**, 45-74.

Hower, J. and Mowatt, T. C. (1966). The mineralogy of illites and mixed layer illite/montmorillonites. *Am. Mineral.* **51**, 825-54.

Hsu, Pa Ho (1968). Heterogeneity of montmorillonite surface and its effect on the nature of hydroxy-aluminium interlayers. *Clay Miner.* **16**, 303-11.

Hsu, Pa Ho and Bates , T. F. (1964). Fixation of hydroxy-aluminium polymers by vermiculite. *Soil Sci. Soc. Am. Proc.* **28**, 763-9.

Hughes, J. C. and Brown, G. (1979). A crystallinity index for soil kaolins and its relation to parent rock, climate and soil maturity. *J. Soil Sci.* **30**, 557-63.

Hutton, J. T., Lindsay, D. S. and Twidale, C. R. (1977). The weathering of norite at Black Hill, South Australia. *J. Geol. Soc. Aust.* **24**, 37-50.

Ildefonse, P., Copin, E. and Velde, B. (1979). A soil vermiculite formed from a meta-gabbro, Loire-Atlantique, France. *Clay Miner.* **14**, 201-10.

Isbell, R. F., Gillman, G. P., Murtha, G. G. and Jones, P. N. (1977). Brown basaltic soils in North Queensland. CSIRO Aust. Div. Soils Tech. Pap. No 34.

Isbell, R. F. and Smith, G. McL. (1976). Some properties of red, yellow and grey massive earths in North Queensland. CSIRO Aust. Div. Soils Tech. Pap. No. 30.

Isphording, W. C. (1973). Discussion of the occurrence and origin of sedimentary palygorskite–sepiolite deposits. *Clays Clay Miner.* **21**, 391-401.

Jackson, E. A. (1962). Soil studies in central Australia; Alice Springs – Hermannsburg – Rodinga Areas. CSIRO Aust. Soil Pub. No. 19.

Jackson, M. L. (1956). 'Soil Chemical Analysis — Advanced Course'. (Published by the author: Madison, Wisconsin).

Jackson, M. L., Hseung, Y., Corey, R. B., Evans, E. J. and Vanden Heuvel, R. C. (1952). Weathering of clay size minerals in soils and sediments. II Chemical weathering of layer silicates. *Soil Sci. Soc. Am. Proc.* **16**, 3-6.

Jefferson, D. A., Tricker, M. J. and Winterbottom, A. P. (1975). Electron-microscopic and Mössbauer spectroscopic studies of iron-stained kaolinite minerals. *Clays Clay Miner.* **23**, 355-60.

Jepson, W. B. and Rowse, J. B. (1975). The composition of kaolinite—an electron microscope microprobe study. *Clays Clay Miner.* **23**, 310-17.

Jung, J., Alexanian, C., Blot, P., Oberlin, A. and Rey, M. (1954). Les illites du bassin oligocene de Salins (Cantal). *Bull. Soc. Fr. Mineral. Cristallogr.* **77**, 1231-49.

Jungerius, P. D. and Levelt, T. W. M. (1964). Clay mineralogy of soils over sedimentary rocks in Eastern Nigeria. *Soil Sci.* **97**, 89-95.

Keller, W. D. (1958). Glauconitic mica in the Morrison formation in Colorado. *Clays Clay Miner.* **5**, 120-8.

Kittrick, J. A. (1973). Mica-derived vermiculites as unstable intermediates. *Clays Clay Miner.* **21**, 479-88.

Klages, M. G. and White, J. L. (1957). A chlorite-like mineral in Indiana soils. *Soil Sci. Soc. Am. Proc.* **21**, 16-20.

Krishnamoorthy, C. and Overstreet, R. (1950). An experimental evaluation of ion exchange relationships. *Soil Sci.* **69**, 41-53.

Langston, R. B. and Jenne, E. A. (1971). NaOH dissolution of some oxide impurities from kaolins. *Clays Clay Miner.* **19**, 633-47.

Lee, S. Y., Jackson, M. L. and Brown, J. L. (1975). Micaceous occlusions in kaolinite observed by ultramicrotomy and high resolution electron microscopy. *Clays Clay Miner.* **23**, 125-9.

Leonard, R. A. and Weed, S. B. (1970). Mica weathering rates as related to mica type and composition. *Clays Clay Miner.* **18**, 187-95.

Levinson, A. A. (1955). Polymorphism among illites and hydrous micas. *Am. Mineral.* **40**, 41-9.

MacEwan, D. M. C. (1961). Montmorillonite minerals. *In* 'The X-ray Identification and Crystal Structures of Clay Minerals'. (Ed. G. Brown). pp. 143-207. (Mineralogical Society: London.)

MacEwan, D. M. C., Ruiz-Amil, A. and Brown, G. (1961). Interstratified clay minerals. *In* 'The X-ray Identification and Crystal Structures of Clay Minerals'. (Ed. G. Brown). pp. 393-445. (Mineralogical Society: London.)

Mackintosh, E. E. and Lewis, D. G. (1968). Displacement of potassium from micas by dodecylammonium chloride. Trans 9th Internat. Congr. Soil Sci. (Adelaide), Vol II, pp. 695-703.

Mackintosh, E. E., Lewis, D. G. and Greenland, D. J. (1971). Dodecylammonium-mica complexes. 1. Factors affecting the exchange reaction. *Clays Clay Miner.* **19**, 202-18.

McLean, S. A., Allen, B. L. and Craig, J. R. (1972). The occurrence of sepiolite and attapulgite on the southern High Plains. *Clays Clay Miner.* **20**, 143-49.

Majhoory, R. A. (1975). Clay mineralogy, physical and chemical properties of some soils in arid regions of Iran. *Soil Sci. Soc. Am. Proc.* **39**, 1157-64.

Maldon, P. J. and Meads, R. E. (1967). The solid state substitution by iron in kaolinite. *Nature* (London) **215**, 844-6.

Mann, A. W. and Deutscher, R. L. (1978). Hydrochemistry of a calcrete-containing aquifer near Lake Way, Western Australia. *Hydrology* **38**, 357-77.

Martin, A. E., Gilkes, R. J. and Skjemstad, J. O. (1970). Fixed ammonium in soils developed on some Queensland phyllites and its relation to weathering. *Aust. J. Soil Res.* **8**, 71-80.

Martin, A. E. and Reeve, R. (1957*a*). Chemical studies on podzolic illuvial horizons. I. The extraction of organic matter by organic chelating agents. *J. Soil Sci.* **8**, 268-78.

Martin, A. E. and Reeve, R. (1957*b*). Chemical studies of podzolic illuvial horizons. II The use of acetylacetone as extractant of translocated organic matter. *J. Soil Sci.* **8**, 279-86.

Millot, G., Paquet, H. and Ruellan, A. (1969). Neoformation de l'attapulgite dans les sols à carpaces calcaires de la Basse Moulouya (Maroc oriental). *C. R. Acad, Sci. Paris* **268-D**, 2771-4.

Mortland, M. M., Lawton, K. and Uehara, G. (1956). Alteration of biotite to vermiculite by plant growth. *Soil Sci.* **82**, 477-81.

Muir, A. (1951). Notes on Syrian soils. *J. Soil Sci.* **2**, 163-81.

Muir, A., Anderson, B. and Stephens, E. (1957). Characteristics of some Tanganyika soils. *J. Soil Sci.* **8**, 1-18.

Nettleton, W. D., Nelson, R. E. and Flach, K. W. (1973). Formation of mica in surface horizons of dryland soils. *Soil Sci. Soc. Am. Proc.* **37**, 473-8.

Newman, A. C. D. (1969). Cation exchange properties of micas. I. The relation between mica composition and potassium exchange in solutions of different pH. *J. Soil Sci.* **20**, 357-73.

Newman, A. C. D. (1970*a*). The synergetic effect of hydrogen ions on the cation exchange of potassium in micas. *Clay Miner.* **8**, 361-73.

Newman, A. C. D. (1970*b*). Cation exchange properties of micas. II. Hysteresis and irreversibility during potassium exchange. *Clay Miner*, **8**, 267-72.

Norrish, K. (1949). Soil colloids of the solonized brown soils. Paper presented to Aust. N.Z. Assoc. Adv. Sci., Hobart. 1949.

Norrish, K. (1952). A study of the mineralogy of some Australian soils formed from basic igneous rocks. Ph.D. Thesis, University of London.

Norrish, K. (1954*a*). The swelling of montmorillonite. *Disc. Faraday Soc.* **18**, 120-34.

Norrish, K. (1954*b*). Relationship of foundation characteristics and the clay mineralogy of some Melbourne soils. CSIRO Aust. Div. of Soils Tech. Memo. No. 5/54.

Norrish, K. (1973*a*). Forces between clay particles. Proc. Internat. Clay Conf. (Madrid), Vol. 1, pp. 375-83.

Norrish, K. (1973*b*). Factors in the weathering of mica to vermiculite. Proc. Internat. Clay Conf. (Madrid), Vol. 1, 417-32.

Norrish, K. (1979). Mineralogy and properties of soil colloids. *In* 'Colloids in Soils — Principles and Practice' (Ed. D. E. Yates). (The Royal Australian Chemical Institute: Melbourne.)

Norrish, K. and Pickering, J. G. (1977). Clay mineralogic properties. *In* 'Soil Factors in Crop Production in a Semi-arid Environment' (Eds. J. S. Russell and E. L. Greacen). (University of Queensland Press: Brisbane.)

Norrish, K. and Tiller, K. G. (1976). Subplasticity in Australian soils. V. Factors involved and techniques of dispersion. *Aust. J. Soil Res.* **14**, 273-89.

Norrish, K. and Wells, C. B. (1970). Chemical changes associated with the artificial weathering of mica. Proc. Aust. Clay Min. Ceramics Conf. Vol. C1, pp. 1-4.

Paquet, H., Ruellan, A., Tardy, Y. and Millott, G. (1969). Geochimie d'un bassin versant au Maroc oriental. Evolution des argiles dans les sols des montagnes et des plaines de la Basse Moulouya. *C. R. Acad. Sci. Paris* **269-D**, 1839-42.

Pickering, J. G. and Falk, H. (1975). Soils and landforms of the Mt. Lofty Ranges, South Australia. X-ray diffraction analyses of clay fractions. CSIRO Aust. Div. Soils Tech. Memo. No. 22/1975.

Porrenga, D. A. (1968). Non-marine glauconitic illite in the Lower Oligocene of Aardedrug, Belgium. *Clay Miner.* 7, 421-30.

Quirk, J. P. and Chute, J. H. (1968). Potassium release from mica-like clay minerals. 9th Trans. Internat. Cong. Soil. Sci. (Adelaide), Vol II, pp. 671-81.

Radoslovich, E. W. (1958). Clay mineralogy of some red brown earths. *J. Soil Sci.* **9**, 242-51.

Raman, K. V. and Jackson, M. L. (1966). Layer charge relations in clay minerals of micaceous soils and sediments. *Clays Clay Miner.* **14**, 53-68.
Rausell-Colom, J. H., Sweatman, T. R., Wells, C. B. and Norrish, K. (1965). Studies in the artificial weathering of mica. *In* 'Experimental Pedology' (Eds. E. G. Hallsworth and D. V. Crawford). pp. 40-72. (Butterworths: London.)
Ravikovitch, S., Pines, F. and Ben-Yair, M. (1960). Composition of the colloids in the soils of Israel. *J. Soil Sci.* **11**, 82-91.
Reichenbach, H. G. V. (1973). Exchange equilibria of interlayer cations in different particle size fractions of biotite and phlogopite. Proc. Internat. Clay Conf. (Madrid), Vol. 1, 457-66.
Reichenbach, H. G. V. and Rich, C. I. (1969). Potassium release from muscovite as influenced by particle size. *Clays Clay Miner.* **17**, 23-9.
Rengasamy, P., Krishna Murti, S. R. and Sarma, V. A. K. (1975). Isomorphous substitution of iron for aluminium in some soil kaolinites. *Clays Clay Miner.* **23**, 211-4.
Rex, R. W. (1966). Authigenic kaolinite and mica as evidence for phase equilibria at low temperature. *Clays Clay Miner.* **13**, 95-104.
Rich, C. I. (1964). Effect of cation size and pH on potassium exhange in Nason soil. *Soil Sci.* **98**, 100-6.
Rich, C. I. (1968). Hydroxy interlayers in expansible layer silicates. *Clays Clay Miner.* **16**, 15-30.
Rich, C. I. and Obenshain, S. S. (1955). Chemical and clay mineral properties of a red-yellow podzolic soil derived from muscovite schist. *Soil Sci. Soc. Am. Proc.* **19**, 334-9.
Rogers, L. E. R., Quirk, J. P. and Norrish, K. (1956). Occurrence of an aluminium-sepiolite in a soil having unusual water relationships. *J. Soil Sci.* **7**, 177-84.
Ross, G. J. and Rich, C. I. (1973*a*). Effect of particle thickness on potassium exchange from phlogopite. *Clays Clay Miner.* **21**, 77-81.
Ross, G. J. and Rich, C. I (1973*b*). Effect of particle size on potassium sorption by potassium-depleted phlogopite. *Clays Clay Miner.* **21**, 83-7.
Rosenquist, I. Th. (1963). Studies in position and mobility of the H atoms in hydrous micas. *Clays Clay Miner.* **11**, 117-35.
Sakharov, B. A. and Drits, V. A. (1973). Mixed-layer kaolinite-montmorillonite: a comparison of observed and calculated diffraction patterns. *Clays Clay Miner.* **21**, 15-7.
Sawhney, B. L. (1967). Interstratification in vermiculite. *Clays Clay Miner.* **15**, 75-84.
Sawhney, B. L. (1972). Selective sorption and fixation of cations by clay minerals: A review. *Clays Clay Miner.* **20**, 93-100.
Sawhney, B. L. (1977). Interstratification in layer silicates. *In* 'Minerals in Soil Environments' (Eds J. B. Dixon *et al.*) pp. 405-34. (Soil Science Society of America: Madison.)
Schnitzer, M. and Khan, S. (1972). 'Humic Substances in the Environment'. (Marcel Dekker Inc.: New York.)
Schofield, R. K. and Samson, H. R. (1954). Flocculation of kaolinite due to the attraction of oppositely charged crystal faces. *Disc. Faraday Soc.* **18**, 135-45.
Schultz, L. G., Shepard, A. O., Blackman, P. D. and Starkey, H. C. (1971). Mixed layer kaolinite-montmorillonite from the Yucatan Peninsula, Mexico. *Clays Clay Miner.* **19**, 137-50.
Scott, A. D. (1968). Effect of particle size on interlayer potassium exchange in mica. Trans. 9th Internat. Congr. Soil Sci. (Adelaide), Vol. II, 649-60.
Scott, A. D. and Reed, M. G. (1962). Chemical extraction of potassium for soils and micaceous minerals with solutions containing sodium tetraphenyl boron: III. Illite. *Soil Sci. Soc. Am. Proc.* **26**, 45-58.
Scott, A. D. and Smith, S. J. (1966). Susceptibility of interlayer potassium in micas to exchange with sodium. *Clays Clay Miner.* **14**, 69-81.
Serstevens, A. T., Rouxhet, P. G. and Herbillon, A. J. (1978). Alteration of mica surfaces by water and solutions. *Clay Miner.* **13**, 401-10.
Shearer, J. and Cole, W. F. (1939*a*). Analysis of soil colloids by X-ray diffraction methods. *J. R. Soc. West Aust.* **26**, 121-31.
Shearer, J. and Cole, W. F. (1939*b*). X-ray analysis of some Tasmanian soil colloids. *J. R. Soc. West Aust.* **26**, 133-37.
Shimoyama, A., Johns, W. D. and Sudo, T. (1969). Montmorillonite-kaolin clay in acid clay deposits from Japan. Internat. Clay Conf. Proc. (Tokyo), Vol. 1, pp. 225-31.
Simonett, D. S. and Bauleke, M. P. (1963). Mineralogy of soils on basalt in North Queensland. *Soil Sci. Soc. Am. Proc.* **27**, 205-12.
Singer, A. (1971). Clay minerals in the soils of the Southern Ramat Hagolan. *Israel J. Earth Sci.* **20**, 105-12.
Singer, A. and Norrish, K. (1974). Pedogenic palygorskite occurrences in Australia. *Am. Mineral.* **59**, 508-17.
Smith, S. J., Clark, L. J. and Scott, A. D. (1968). Exchangeability of potassium in soils. Internat. Trans. 9th Congr. Soil Sci. (Adelaide), Vol. II, pp. 661-69.
Souza Santos, P., Souza Santos, H. and Brindley, G. W. (1966). Mineralogical studies of kaolinite-halloysite clays. Part IV. A platy mineral with structural swelling and shrinking characteristics. *Am. Mineral.* **51**, 1640-8.
Stace, H. C. T., Hubble, G. D., Brewer, R., Northcote, K. H., Sleeman, J. R., Mulcahy, M. J. and Hallsworth, E. G. (1968). 'Handbook of Australian Soils'. (Rellim Technical Publications: South Australia.)
Stevenson, F. J. (1959). On the presence of fixed ammonium in rocks. *Science* **130**, 221-2.
Stevenson, F. J. and Dhariwal, A. P. S. (1959). Distribution of fixed ammonium in soils. *Soil Sci. Soc. Am. Proc.* **23**, 121-5.
Stirton, R. A., Tedford, R. H. and Miller, A. H. (1961). Canozoic stratigraphy and vertebrae palaeontology of the Tirari Desert, South Australia. *Rec. S. Aust. Mus.* **14**, 19-61.
Tamura, T. (1957). Identification of the 14Å clay mineral component. *Am. Mineral.* **42**, 107-10.
Tucker, B. M. (1967*a*). The solubility of potassium from soil illites. III Reactivity towards other ions. *Aust. J. Soil Res.* **5**, 173-90.
Tucker, B. M. (1967 *b*). The solubility of potassium from soil illites. IV Rates of reaction and exchange constants. *Aust. J. Soil Res.* **5**, 191-201.
Turner, R. C. (1965). Some properties of aluminium hydroxide precipitated in the presence of clays. *Can. J. Soil Sci.* **45**, 331-6.
Turton, A. G., Marsh, N. L., McKenzie, R. M. and Mulcahy, M. J. (1962). The chemistry and mineralogy of laterite soils in the southwest of Western Australia. CSIRO. Soil Publ. No. 20.
Vanden Heuvel, R. C. (1966). The occurrence of sepiolite and attapulgite in the calcareous zone of a soil near Las Cruces, New Mexico. *Clays Clay Miner.* **13**, 193-207.
Veen, A. W. L. (1973). Evaluation of clay mineral equilibria in some clay soils (Usterts) of the Brigalow lands. *Aust. J. Soil Res.* **11**, 167-84.
Wada, K. (1977). Allophane and imogolite. *In* 'Minerals in Soil Environments' (Eds J. B. Dixon *et al.*). pp. 603-38. (Soil Science Society of America: Madison.)
Wada, K. and Aomine, S. (1966). Occurrence of gibbsite in weathering of volcanic materials at Kuroishibaru, Jumamato. *Soil Sci. Plant Nutr. (Tokyo).* **12**, 151-7.
Wada, K. and Greenland, D. J. (1970). Selective dissolution and differential infrared spectroscopy for characterization of 'amorphous' constituents in soil clays. *Clay Miner.* **8**, 241-54.
Wada, K. and Harward, M. E. (1974). Amorphous clay constituents of soils. *Adv. Agron.* **26**, 211-60.
Wada, K. and Tokashiki, Y. (1972). Selective dissolution and difference infrared spectroscopy in quantitative mineralogical analysis of volcanic-ash soil clays. *Geoderma* **7**, 199-213.
Wada, K., Yoshinaga, N., Yotsumoto, H., Ibe, K. and Aida, S. (1970). High resolution electron micrographs of imogolite. *Clay Miner.* **8**, 487-9.
Walker, G. F. (1961). Vermiculite minerals. *In* 'The X-ray Identification and Crystal Structures of Clay Minerals'. (Ed. G. Brown). pp. 297-324. (Mineralogical Society: London.)
Wear, J. I. and White, J. L. (1951). Potassium fixation in clay minerals as related to crystal structure. *Soil Sci.* **71**, 1-14.

Weaver, C. E. (1953). A lath shaped non-expanded dioctahedral 2:1 mineral. *Am. Mineral.* **38**, 279-87.
Weaver, C. E. (1956). The distribution and identification of mixed-layer clays in sedimentary rocks. *Am. Mineral.* **41**, 202-21.
Weaver, C. E. (1965). Potassium content of illite. *Science.* **147**, 603-5.
Webster, R. (1965). A catena of soils on the Northern Rhodesia Plateau. *J. Soil Sci.* **16**, 31-43.
Weiss, A. and Range, K. J. (1966). Übber titan im gitter von kaolin. Internat. Clay Conf. Proc. (Israel), Vol. 1, 53-66.
West, G. and Dumbleton, M. J. (1970). The mineralogy of tropical weathering illustrated by some west Malaysian soils. *J. Eng. Geol.* **3**, 25-40.
Wells, C. B. and Norrish, K. (1968). Accelerated rates of release of interlayer potassium from micas. 9th Trans Internat. Congr. Soil. Sci. (Adelaide), Vol II, pp. 683-94.
Wiewiora, A. (1971). A mixed layer kaolinite-smectite from Lower Silesia, Poland. *Clays Clay Miner.* 19, 415-6.
Wilson, M. J. and Cradwick, P. D. (1972). Occurrence of interstratified kaolinite-montmorillonite in some Scottish soils. *Clay Miner.* **9**, 435-7.
Yaalon, D. H. (1955). Note on the clay mineralogy of the major soil types in Israel. *Bull. Res. Counc. Israel.* **58**. 168-73.
Yaalon, D. H., Nathan, Y., Koyumdjisky, H. and Dan, J. (1966). Weathering and catenary differentiation of clay minerals in soils of various parent materials in Israel. Proc. Internat. Clay Conf. (Jerusalem), Vol. 1, pp. 187-98.
Yaalon, D. H. and Wieder, M. (1976). Pedogenic palygorskite in some arid brown (calciorthid) soils of Israel. *Clay Miner.* **11**, 73-80.
Yoshinaga, N. and Aomine, S. (1962). Allophane in some Ando soils. *Soil Sci. Plant Nutr., Tokyo,* **8**, 6-13.
Zelazny, L. W. and Calhoun, F. G. (1977). Palygorskite, (attapulgite), sepiolite, talc, pyrophyllite, and zeolites, *In* 'Minerals in Soil Environments' (Eds J. B. Dixon *et al.*). pp. 435-66. (Soil Science Society of America: Madison.)

Chapter 23
Oxide minerals

R. M. Taylor, R. M. McKenzie, A. W. Fordham and G. P. Gillman

THE GROUP OF MINERALS REFERRED TO AS SOIL OXIDES IS generally taken to include hydroxides and hydrous oxides. Several metal elements often occur in these relatively stable forms, in particle size ranges extending from clay size (< 2 μm) to cemented aggregates or concretions of a few centimetres diameter.

In relation to the uptake or release of nutrients, these minerals differ from the layer lattice silicates, in which the surface charge generally arises from isomorphous substitutions, in that adsorption is often specific or controlled by pH even though there may often be many foreign ions incorporated in the crystal lattice.

These minerals are widespread in soils of all humid regions although the metallic oxides and hydroxides of Fe, Mn, Ti and Al occur at highest concentrations in the soils of warm, humid climates (Segalen 1971). These oxidic soils are generally located within the tropics (see world map of Ganssen and Hädrich 1965) or in tropical, subtropical equatorial, and mediterranean climatic zones, and are therefore subject to wide variations in other soil-forming factors, such as topography, vegetation, parent material and rainfall. The oxides do not form in arid zones within these geographical regions, though they may be present there; if they do occur, they are remnants from older climatic conditions. In wetter areas, aluminium oxides occur, if there is also good drainage.

The oxides of Fe, Mn, Al, Si and Ti are the most common in soils, and will be discussed separately in this chapter.

Iron oxides

Iron oxides can be distributed evenly throughout a profile, or preferentially concentrated in particular horizons, and may constitute over 50% of the soil minerals present. They can exist as discrete fine particles, as coatings on sand grains and crack faces, or as concentrations in cemented aggregates. In most forms of occurrence, they were formed, distributed, and concentrated by soil-forming processes.

The soil-forming factors responsible for these distributions, as well as for the mineral species present, are in many cases the factors also responsible for many of the morphological features upon which the whole profile is classified; for example, due to the high pigmenting power of the two most common oxides, goethite and hematite, quite low concentrations can impart so much colour as to be reflected in the naming and classification of soils in many countries e.g. red-yellow podzolic, Braunerde, sols rouges tropicaux, *terra rossa* and *krasnozem*.

Iron is mobile within a profile because of its ease of reduction under anaerobic conditions and its susceptibility to be complexed. It can influence the structure and fabric, by the cementation of aggregates or by the development of its own zones of concentration, e.g. concretionary layers, ironstone cappings etc.

Although some of the aggregates or concretions grow to several centimetres in diameter, the crystallites of their constituent iron oxides are generally quite small and of large specific surface area, which facilitates adsorption of anions and cations, some of which are incorporated into the structure. Many of these elements so bound are essential for plant growth, so that these oxides may act as a source or sink, depending on the prevailing environmental factors.

The large range of climatic conditions and associated soil-forming factors encountered in the Australian continent, from the wet and alpine conditions to the arid regions, from the humid temperate to the tropical, enable all the common pedogenic iron oxides to form. From each, some information about earlier genetic factors can generally be gained.

Forms of iron oxides in soils

The most common soil iron oxides are goethite (α-FeOOH), lepidocrocite (γ-FeOOH), hematite (α-Fe_2O_3), maghemite (γ-Fe_2O_3), ferrihydrite ($HFe_5O_8.4H_2O$) and magnetite (Fe_3O_4). All these minerals occur in different Australian soils, many of which contain more than one form. With the exception of magnetite, all are products of pedogenesis and derived either directly from primary sources of iron, or from the transformation of earlier pedogenic forms under new environmental conditions. Magnetite generally exists as a residual primary mineral, which can weather via solution to other forms, or be oxidized in a solid state transformation to maghemite; such a transformation is favoured by smaller particle sizes and a moist environment (Feitknecht 1965). The formula γ-Fe_2O_3 for soil maghemites is not strictly correct because they generally contain a few percent of ferrous iron (Taylor and Schwertmann 1974*a*), but not sufficient to classify them as a magnetite.

Goethite is the most common mineral, and can be found in most soil types and in most climatic regions. It is the 'yellow ochre' pigment which imparts a yellow-brown colour in many soils e.g. to *yellow earths*, and produces the yellow mottles of laterites (Davey *et al.* 1975). It is also present, generally with hematite, in the consolidated ironstone capping of most laterites. In massive mineral deposits, goethite crystals are often acicular, but this morphology is less pronounced, or even absent, in soils, possibly due to the effects of foreign ions in the environment during nucleation and crystal formation of soil goethite.

Lepidocrocite has the same composition, but a different crystalline structure, to goethite, and is less common. However, it is a common mineral in hydromorphic soils of the humid temperate regions, where it is often associated with goethite. This association is quite common throughout the world (Schwertmann and Fitzpatrick 1977), yet lepidocrocite is rarely, if ever, formed contemporaneously with hematite. Lepidocrocite has been found in the periodically flooded terraces of the South Esk river near Longford in Tasmania, and at the emergence of ground water and its exposure to oxidising conditions on the side of Mt Wellington near Hobart, where it is associated with goethite. It is often recognised as a bright orange or rusty-colored mottle, band, or fleck, in gley or pseudogley soils, especially those high in clay.

Electron micrographs of natural and synthetic lepidocrocites generally exhibit the same elongated serrated lath morphology, and sometimes a twinned star shape (Schwertmann, 1973).

Hematite is also a very common mineral in the Australian landscape, particularly in the soils of the arid and tropical regions where the decomposition of organic matter is so rapid that it has little effect in either promoting goethite or inhibiting hematite formation (Schwertmann 1971). It occurs often in association with goethite, in other climatic zones. It is the 'red ochre' responsible for the red coloration of many soils and soil features e.g. *terra rossas, red earths*, red mottles in laterites etc. In many of these occurrences the hematite may be less abundant than the goethite with which it is associated, but, because of its high pigmenting power, minor amounts may give a distinct red colour to a soil matrix.

Hematite is often a major component of soil concretions, especially of those that are magnetic due to their maghemite content, e.g. in *krasnozems* and podzolic soils. Whereas synthetic hematites often exhibit an hexagonal morphology, natural soil hematites are generally isodimensional particles with a poorly expressed morphology. From the sharpness of their respective X-ray diffraction patterns soil hematites generally appear to be more crystalline than soil goethites.

Maghemite is common in very weathered soils of tropical and subtropical climates, especially in soils formed on basic igneous rocks in these regions. Normal methods of mineral analysis are not sensitive to small amounts of maghemite; it may be more prevalent in many soil types than published analyses reveal. Besides being concentrated in the magnetic concretions of many soils, it can also occur in a finely dispersed form in soils (Oades and Townsend 1963). Although fully oxidized synthetic maghemite is red-brown, natural occurrences, or the products of syntheses at pH 7, 20°C and low rates of oxidation (Taylor and Schwertmann 1974*b*), may be darker (brown-black), possibly due to additional amounts of ferrous iron in the spinel structure. As has been stated, this so-called γ-ferric oxide generally contains a few percent of ferrous iron, but less than its isomorph, magnetite (Fe_3O_4) which generally contains close to the theoretical 33.3% of its total iron in the ferrous state (Fasiska 1967).

Ferrihydrite was previously regarded as an amorphous phase, but has recently been accepted as a crystalline mineral (Chuckhrov *et al.* 1973) with a definite short-range crystal order. In synthetic preparations its colour can vary from yellow-brown to red-brown. As a pure phase it can be identified, but in association with other soil minerals positive detection is difficult. Ferrihydrite is found often in drainage ditches, and has been identified in pipes and bore holes from the Hume reservoir and Mt Isa areas. It is often considered as having formed by microbial decomposition of the soluble Fe-organic compounds, and subsequent hydrolysis and precipitation. Its persistence may be due to the inhibition, by adsorption of foreign ions from the environment, of further crystallization to other phases (Schwertmann and Fischer 1973; Fischer and Ottow 1972). For example, it transforms to hematite (Chukhrov *et al.* 1973) or, via solution, to goethite, depending on environmental conditions.

Origin and formation of iron oxides in soils

The iron in soils arises originally from the weathering of primary iron-bearing minerals, particularly the silicates, in which the iron is generally in the divalent state. In this form it is more easily dissolved and more mobile than the trivalent form. The dissolved ferrous iron is subsequently precipitated and oxidized, or oxidized and precipitated. These processes may take place either *in situ*, or, after transportation in subsurface waters, in other parts of the profile or landscape. The sequence of the oxidation and precipitation processes is important. In general, where oxidation precedes hydrolysis the α-iron oxides, goethite and hematite, form; when, however, hydrolysis and precipitation occur before oxidation, the γ-oxides, lepidocrocite and maghemite, often occur.

Divalent iron can also be utilized directly as an energy source by some bacteria, leading to the ultimate accumulation of ferric hydroxides. These biogenic deposits are transformed, in time, into ortsteins and eventually, after complete breakdown of organic matter, into goethite and/or hematite.

Iron already oxidized and precipitated within a soil profile may be mobilized subsequently without undergoing reduction, by the formation of complexes with the soil organic matter. These processes may result from direct reaction of microbial slimes upon the iron in mineral lattices (Claus *et al.* 1958), or from reaction with microbial enzymes or with the various organic and mineral acids present as microbial metabolites in a soil (Aristovskaya, 1974). Adsorption of $Si(OH)_4$ can also stabilize colloidal ferric hydroxide and permit further mobilization.

The type of organic matter may determine in what form the complexed iron is bound, e.g. Pauli (1968) observed that organic matter rich in nitrogen complexed Fe(II) in preference to Fe(III). In addition, the type of organic ligand, and the conditions under which the complex is broken down, could determine the mineral form in which the new precipitation occurs, (McHardy *et al.* 1974). Readers are referred to articles on organometallic interactions in soils by Schnitzer and Skinner (1964).

Maghemite

The oxidation and hydrolysis of a ferrous system can produce a ferrous–ferric–hydroxy anion complex called 'green rust' (Feitknecht and Keller 1950; Bernal *et al.* 1959) from a reaction between the ferric hydroxide and the soluble ferrous hydroxy species still in solution. This compound undergoes simultaneous oxidation and dehydration, and, depending on the relative rates of these two processes, maghemite or lepidocrocite may dominate. The lepidocrocite (hydrated) is favoured over maghemite (dehydrated) by higher rates of oxidation (Feitknecht 1959; Misawa *et al.* 1973), whereas maghemite is favoured by both higher temperatures and ferrous concentrations in the soil solution (Taylor and Schwertmann 1974*b*).

Maghemite, which requires ferrous iron for its formation, is often found in association with hematite, which is derived from hydrolysis of trivalent iron. If such an association could be explained, it might offer some insight into the genesis of maghemite.

A further factor that appears important in the occurrence of maghemite is a low level of soluble aluminium. Taylor and Schwertmann (1974*a*), observed in an examination of magnetic and non-magnetic concretions from a variety of Australian soils, that the magnetic samples had much less layer lattice silicate in their matrix than did the morphologically similar non-magnetic concretions with which they were associated. These non-magnetic samples generally contained goethite, with or without hematite, rather than maghemite and hematite which are the major components of the magnetic concretions. This has also been observed in morphologically similar concretions of different sizes from different horizons of grey and *red earth* profiles from Queensland (Taylor and Coventry unpublished data). In both cases the magnetic samples always contained much lower (or undetectable) amounts of kaolin than the morphologically similar non-magnetic concretions, which generally contained increased amounts of goethite. Laboratory syntheses have confirmed, that a small amount of Al present during the slow oxidation of a ferrous system at pH 7 (Taylor and Schwertmann 1978) favours the formation of α-phases and depresses the formation of the γ-phases which, under the same conditions in the absence of aluminium, would normally form. This will be discussed again later.

Maghemite can also form in nature from the oxidation of primary magnetite, indicated by pseudomorphs of maghemite after magnetite (Bonifas, 1959), from the dehydration of lepidocrocite and from the thermal transformation of other oxides, e.g. goethite, in the presence of organic matter (van der Marel 1951; Schwertmann and Heinemann 1959). However, its intimate association with other soil iron oxides, of pedogenic origin and presumably formed from solutions of mobile Fe, suggests that in most soils the latter methods of formation are not so important (Oades and Townsend 1963; Schwertmann and Taylor 1977).

Lepidocrocite

Although lepidocrocites have been synthesised from inorganic Fe (III) compounds (Fordham 1970; Murphy *et al.* 1975), and from the decomposition of ferric chelates (McHardy *et al.* 1974), Schwertmann and Thalman (1976) maintain that, generally, lepidocrocite will not form in soils unless there is Fe(II) present. This claim is supported by the morphological similarity between naturally occurring lepidocrocite and lepidocrocite synthesized from the oxidation of a ferrous system, (Schwertmann 1973), and also by its almost exclusive occurrence in hydromorphic soils, where microbially induced anaerobic conditions lead to the formation of Fe(II). Subsequent oxidation and hydrolysis of Fe(II) may lead to the formation of the 'green rust' compound previously mentioned.

Where the rate of oxidation in syntheses is sufficiently rapid this 'green rust' phase may not be detected because it becomes fully oxidized as soon as it is formed. Although lepidocrocite can be synthesized in many ways, its common association with goethite in soils requires that a mode of formation be postulated that can lead, by such slight changes in environmental conditions as could be expected in soils, to either or both phases. Although lepidocrocite can form from ferrous iron derived from the weathering of primary minerals, it is often the result of microbial reduction, under hydromorphic conditions, of other soil iron oxides formed during an earlier weathering cycle under different soil-forming conditions. For example, in the Launceston Tertiary Basin in Tasmania, the flood terrace soils of the South Esk river show lepidocrocite to be present in profiles whose parent material was derived from a widespread area (Taylor and Pickering 1963).

The carbon dioxide regime in hydromorphic soils is generally higher than in more aerated soils, and the mobility of Fe(II) in these hydromorphic soils may be sustained by the solubility of ferrous carbonate, which is much more soluble than the ferric oxide. This has been supported by laboratory synthesis. It was previously shown that the carbonate anion favours goethite formation over lepidocrocite (Schwertmann 1959*a*; Schwertmann and Fitzpatrick 1977). Recent syntheses (Taylor and Schwertmann 1978), which show that either or both phases can form from a ferrous carbonate solution, give a ready explanation for their common association in soils: lepidocrocite was found to be favoured by higher aeration rates and lower Fe(II) concentrations.

Normally, the oxidation of a ferrous salt solution at a fixed pH requires alkali to be added to compensate for the pH drop during hydrolysis. However, in the carbonate system the reaction is similar to the buffered reaction expected in soils; alkali addition becomes unnecessary because the pH drop is balanced to some extent by a pH increase caused by the loss of carbon dioxide from the system, depending on the rate of aeration. Experimentally, the presence of small amounts of aluminium in a chloride, sulphate or carbonate system suppresses formation of the γ-phase, causing the formation of goethite instead. The counterpart of these restrictions to lepidocrocite formation can readily be found in nature, and provides additional reasons for its limited occurrence in soils.

Schwertmann and Thalmann (1976) found also that silicate in solution can inhibit the formation of lepidocrocite, causing ferrihydrite to form. They explain this by assuming that the ferrihydrite formed from the oxidation of ferrous hydroxide $Fe(OH)_2$ adsorbs silicate from solution, thus precluding adsorption of further Fe(II), and preventing the formation of a 'green rust' phase.

Ferrihydrite

Fast hydrolysis of Fe(III) in aqueous solution yields ferrihydrite (formerly called amorphous ferric hydroxide). Many naturally occuring ferrihydrites are rich in organic compounds which, through adsorption, can often inhibit further crystallization (Schwertmann 1966). Ferrihydrite can be synthesized by microbial oxidation of ferric citrate solutions (Fischer and Ottow 1972), and such a mechanism has been suggested to explain the occurrence of carbon-rich samples. Organic compounds carried in water percolating the acidic surface horizons of some soils can complex and mobilize Fe(III). On exposure to an oxidising environment, such as in drainage ditches and springs, the organic chelate is attacked by microorganisms, and the Fe(III) is hydrolyzed and precipitated. Rapid oxidation of the ferrous carbonate which forms in carbon dioxide-enriched water can also cause ferrihydrite to form (Taylor and Schwertmann 1978), which transforms to goethite or hematite on ageing. Krause *et al.* (1970) also produced ferrihydrite from a ferrous carbonate system, and showed that, under the conditions used, its formation is influenced by the presence of foreign cations such as Ca, Mg, Co or Pb.

The presence of inorganic (as well as organic) anions, such as phosphate (Scheffer *et al.* 1957) and silicate (Schellmann 1959), especially those with hydroxyl groups, can not only inhibit further crystallization of ferrihydrite allowed to age in their presence, but also can cause it to form, probably by preventing nucleation of other species. For example, 150 ml of 0.016 M $FeSO_4$ solution, oxidized at pH 7 with air at 5 $cm^3\ min^{-1}$, gives lepidocrocite with a trace amount of goethite; the addition of silicate to the system (5.6×10^{-4}M) is sufficient to cause only ferrihydrite to form (R. M. Taylor unpublished data). The products from ageing ferrihydrite depend only on the environmental conditions and the concentration of ferrihydrite. Suggested further references on this subject are Feitknecht and Michaelis (1962), Sylva (1972), Feitknecht *et al.* (1973), Knight and Sylva (1974), and Murphy *et al.* (1976) from which many earlier references can be obtained.

Goethite

This most common soil iron oxide forms from either ferrous or ferric systems. During the slow hydrolysis of ferric solution, the comparatively low solubility product of goethite, compared with that of ferrihydrite, may be exceeded; goethite nucleates and grows from soluble hydroxy species such as $Fe(OH)_2^+$ (Knight and Sylva 1974). It can also form from precipitated ferrihydrite, in an environment where these soluble species are still present. Its formation from ferrihydrite is enhanced by the presence of Fe^{2+} in the soil solution (Fischer 1973), or of organic matter. This transformation to goethite via solution is competitive with the coalescence and aggregation of the ferrihydrite particles leading to hematite nucleation. Therefore any reaction or foreign ion that retards this aggregation will enhance the formation of goethite.

Although goethite can be synthesized easily, its formation path in soils will be determined by the soluble forms of iron most commonly found in those soils. Formation through oxidation and hydrolysis of soluble ferrous salts, such as chloride, sulphate or nitrate, would be relatively uncommon. A more likely reaction would be precipitation following oxidation and hydrolysis of soluble ferrous carbonate in the soil solution. Goethite synthesis from these solutions is favoured by higher concentrations of Fe(II) and lower rates of aeration (Taylor and Schwertmann 1978). The common inclusion of aluminium, with an atomic radius comparable to Fe(III), may not merely reflect an incidental presence of aluminium in the soil solution, but may be an indication of its positive influence on the formation of goethite. As has been previously stated, the presence of aluminium in small amounts (i.e. the ratio Al/(Al + Fe) is 0.1 – 0.3) favours goethite over the γ-phases, under conditions which, in the absence of aluminium would have led to maghemite or lepidocrocite formation: excessive levels of aluminium cause ferrihydrite formation. This influence was observed in the carbonate, chloride and sulphate media (Taylor and Schwertmann 1978). The underlying effect of this added aluminium, and possibly of other cations which promote goethite formation, appears to be the slowing down of the oxidation of Fe(II), so that nucleation of goethite from soluble ferric hydroxy species can occur (Stauffer and Lovell 1968). The ability of many anions and cations to influence reaction kinetics may be the reason for the prevalance of goethite as the most common oxide formed in the complex soil system.

The recent observation of Taylor and McKenzie (1980) that aluminium hydroxy species can induce the precipitation and hydrolysis of Fe(II) in carbonate solutions with low amounts of dissolved oxygen may be important in the formation of aluminium substituted goethites. A green Fe(II) Al(III) hydroxy-carbonate, isostructural with the pyroaurite group of minerals is formed. This is very unstable, and if allowed to oxidize in water gives substituted goethite. This pathway requires no pH control and occurs around neutral pH and ambient temperatures, and utilizes components that could be expected in the soil system.

Hematite

Ferrihydrite is a necessary pre-stage for hematite, which forms by the aggregation and subsequent dehydration of the fine particles. Acceleration of the aggregation process, e.g. by higher ferrihydrite concentrations and higher temperature (Schwertmann and Fischer 1966) or by the presence of certain anions, e.g oxalate (Fischer and Schwertmann 1975), increases the rate of hematite formation. Although hematite is formed when goethite dehydrates at temperatures up to 300°C (depending on goethite particle size), this method of formation is not considered to be responsible for soil hematites or intimate goethite–hematite associations.

Certain anions, e.g. calcium and magnesium, also influence hematite formation during the ageing of ferrihydrite (Schellmann 1959). Gastuche *et al.* (1964) found that small amounts of Al co-precipitated with Fe (III) hydroxide inhibit goethite but favour hematite formation. The Al can substitute for Fe in the hematite structure, as in goethite, modifying its lattice parameters. Schwertmann *et al.* (1979) note, that Al-hematite is formed from a ferrihydrite in which Al is substituted for Fe, rather than from a simple co-precipitate of the two hydroxides. Small amounts of Al appear to have a stabilising effect on the hematite; at higher levels of substitution, there is retardation of crystal growth in the *c*-axis direction, and/or structural disorder occurs. At still higher levels of Al, inhibition or retardation of hematite crystallization results (Schwertmann *et al.* 1977).

Effect of environment on formation and properties of iron oxides

Subtle changes in parameters during laboratory syntheses can often produce marked variations in the final products. In syntheses from Fe solutions, a chosen set of conditions may produce a particular morphology or an association of mineral phases that does not occur, or at least is not common, in nature. The explanation possibly lies in the fact that in soils we are rarely free from the variable influence of the chemical environment, which may exert a control that can override the general controlling parameters viz. pH, Eh, temperature, concentration and valence of Fe in solution. This influence can extend to:

(i) Inhibition of crystallization e.g. the effects of added silicate, phosphate, citrate etc. (Schellmann 1959; Chukhrov *et al.* 1973; Schwertmann 1970).

(ii) Change in mineral species, e.g. the effects of Al (III) and Cr(III) in promoting goethite rather than lepidocrocite and maghemite (Taylor and Schwertmann 1978), or the catalytic effect of Cu in the formation of maghemite rather than lepidocrocite (R. M. Taylor unpublished data).

(iii) The modification of crystal morphology, e.g. the effects of added citrate or silicate, (Schwertmann *et al.* 1968; Schwertmann and Taylor 1972*a*).

The effect of foreign ions on the crystallite size and morphology is no doubt a major factor controlling the generally small particle size of soil goethites which form from solution, and offers an explanation as to why the marked acicular morphology, so easily produced in syntheses from pure systems, is uncommon in actual soil samples. Electron micrographs show that synthetic goethites prepared in the presence of Al from an Fe(II) system resemble, both in morphology and particle size, an aluminiferous goethite extracted from a soil (Taylor and Schwertmann 1978).

In many cases the foreign ions present in the environment can be incorporated into the crystal structure of the oxide during its formation. Aluminium substitution for Fe is now commonly recognized in soil goethites (Norrish and Taylor 1961; Thiel 1963), hematites (Perinet and Lafont 1972; Schwertmann *et al.* 1977), and possibly in lepidocrocites (de Villiers and van Rooyen 1967). Often, due to the incorporation of the smaller Al cation, a change in the X-ray diffraction spacings can be observed, and the influences of this substitution on the infrared spectra have been reported: Jonas and Solymar (1970) report, that in synthetic and natural Al-goethites, increasing Al substitution causes the 900 cm^{-1} goethite band to shift linearly to higher wavenumbers, and the extinction coefficient to decrease linearly. The effects of Al substitution on Mössbauer spectra are also being examined in many laboratories. However, many other elements in the soil environment can also be incorporated, and some produce an increase in unit cell size, due to their ionic radii being greater than Fe(III), e.g. the presence of small amounts of Ti(IV) in hematites. Norrish (1975) has found Si, Ti, V, Zn, Cu, Mo and P included within the goethite structure, but it is not clear how these elements are fixed or what positions they occupy within the lattice.

It is likely that the properties of synthetic iron oxides, for instance point of zero charge (pzc), anion and cation adsorption capacity, specific surface, and ease of dissolution, may reflect the method of formation, and may not give a true picture as to the contribution that natural oxides make to the soil properties.

Stability of mineral forms of iron oxides in soils

Topotactic transformations, other than the oxidation of magnetite to maghemite, between the common soil iron oxides do not generally take place under pedogenic conditions; for example, under normal conditions, dehydration of lepidocrocite to maghemite, or the dehydration of goethite to hematite, and its converse, the hydration of hematite to goethite, do not appear to occur.

Transformations however do occur between the different mineral forms, but are generally the result of solution and precipitation under new environmental conditions which determine the new mineral form.

Thermodynamic considerations are often of little help in predicting the mineral form which may occur in a soil, or whether a metastable form will persist. For example, published data for the standard free energies of formation of hematite and goethite are similar, and for the reaction

$$\text{hematite} + H_2O = \text{goethite}$$

small values, both positive, 1.6 kJ mol^{-1} (Berner 1969), and negative, −0.4 to −0.8 kJ mol^{-1} (Mohr *et al.* 1972) have been obtained. However, these values apply only to the pure systems investigated, and not to systems where inhibition of crystallization, isomorphous substitutions, limitations to crystal growth, *etc.* may occur as a result of the effects of the foreign ions in the environment.

For similar reasons, mineral forms stable under the environmental conditions of their formation may be regarded, under a new environment, as metastable, yet they do not revert to the thermodynamically more favoured form. Their persistence in a metastable state may possibly be due to the high activation energy needed to initiate the transformation. One such example is the stability of lepidocrocite under conditions where goethite is considered more stable. The transformation should be accelerated by the presence of goethite (with which the lepidocrocite is commonly associated), but in fact is generally hindered by the low solubility of the Fe(III) from the lepidocrocite, coupled with the inhibition of further goethite nucleation by Si in the soil solution (Schwertmann and Taylor 1972*a,b*).

Removal of iron oxides from soils

The Fe content of a soil can vary markedly down a profile, as can the relative amounts of the mineral phases in which the Fe is distributed. In soil analyses it is sometimes necessary to know the amounts of Fe distributed between these pedogenically formed oxides and the other Fe-bearing minerals present, and so dissolution treatments have been developed to remove the oxides.

These chemical treatments, commonly referred to as 'deferration', cannot be used to estimate the proportions of the different minerals present, as no single technique is specific for one mineral. Moreover, a particular dissolution treatment might remove goethite and hematite effectively from one soil, but prove ineffective in removal of the same two minerals from a different soil. Segalen (1971) has critically examined the parameters that influence the efficiency of extraction by the sodium dithionite method of Deb (1950), from which many modified treatments currently in use have evolved. Many of these modifications appear to have their greatest efficacy on the soils for which they were developed. This is understandable in view of the variations in particle size, degree of crystallinity, and composition, that can be expected in these common soil iron minerals. Segalen (1971) concludes that: the efficiency of Fe removal is dependent (for a given number of treatments) on the total free iron content of the soil; the efficiency is not associated with the pedological classification except where this affects the total amount of Fe; and whereas gibbsite, kaolinite, and illite are only slightly attacked by the treatment, as is magnetite, Fe can be removed from octahedral positions of montmorillonite.

Segalen also emphasized that many workers investigating efficiencies of iron removal techniques embrace different groups of minerals in their use of the term 'free iron oxide'. Most techniques rely on the reduction of the Fe(III) in the oxides to a soluble Fe(II) or a subsequently complexed form. Maghemite appears less sensitive to dithionite reduction treatment than hematite, with which it is commonly associated (Taylor and Schwertmann 1974*a*), but it is preferentially dissolved in hot 0.2 M oxalic acid (Schwertmann and Taylor 1977).

Although it is not generally possible to differentiate between crystalline forms on the basis of their rate of dissolution, a partition between the Fe in crystalline oxides, ferrihydrite, and amorphous Fe–organic complexes, can be made by extractions with selected reagents (McKeague *et al.* 1971; Weber *et al.* 1974). The organically bound Fe is first removed by treatment with 0.1 M sodium pyrophosphate (Aleksandrova 1960). A second dissolution removes the poorly crystalline ferrihydrite, the so-called amorphous inorganic fraction, and is accomplished by an acid ammonium oxalate (pH = 3) treatment in the absence of light (Schwertmann 1964). In the presence of natural light, the ultraviolet component causes photochemical reduction of the more crystalline Fe(III) oxides, whereas in the dark, the dissolution of the ferrihydrite depends on its greater surface area and high solubility. However, in the presence of a source of Fe(II), e.g. siderite or maghemite, crystalline iron oxides may also be dissolved, even in the absence of ultraviolet wavelengths (Schwertmann 1959*b*; Pawluk 1972). The crystalline Fe(III) oxides remaining are then removed by the dithionite–citrate–bicarbonate method of Mehra and Jackson (1960).

Care must be exercised in attributing trace metals in a dithionite extract to the iron oxides originally present in the soil, because Mn oxides, which contain large quantities of these metals as impurities, are also dissolved during reduction with dithionite. Moreover, the presence of citrate can cause the complexing of many other trace metals which may be associated with other mineral phases.

Iron oxide mineral identification and estimation

Identification of soil iron oxides can be made by X-ray diffraction, infra-red spectroscopy, differential thermal analysis (DTA), and Mössbauer spectroscopy. In most cases the efficiency and limitation of the method will depend on the amount of Fe, the number of its different mineral phases present, and the nature of the remaining soil minerals.

Not all minerals give reactions with DTA, and generally this technique can be used only as corroborative evidence for minerals already found or suspected from other techniques. The Fe oxides possess distinctive infrared spectra but positive mineral identification can generally be made only when sufficient amounts of the mineral are present in a soil (Kodama *et al.* 1977).

Mössbauer spectroscopy is currently being developed for the identification of Fe minerals in soils, and, according to Kodama *et al.* (1977), the limits of detection are lower than for X-ray diffraction; 1% hematite can be detected, whereas by diffraction the limit is around 10%, provided there is no marked interference from other minerals. These authors also state that this method offers a potential to quantify analyses of oxide minerals without chemical and/or heat treatment. Analysis by Mössbauer is made more difficult by the supermagnetism caused by the very fine particle size of soil goethites and hematites. However, in spite of this limitation, Gangas *et al.* (1973) used Mössbauer to identify hematite crystals of only 40 Å from a red soil in Greece. Readers are also referred to the work of Karpachevskii and Babanin (1974), who suggest that Mössbauer may be useful in studying changes in the Fe mineral crystal lattices, and the rate of oxidation and reduction of Fe in soils, as well as in the identification of their various Fe compounds.

Crystal morphology as determined by electron microscopy may give an indication as to the mineralogy of the oxide, but only in special cases, because the crystal shape and size are so susceptible to environmental influences.

At present, however, X-ray diffraction is still the most widely used and most useful technique for Fe-oxide mineral identification. In this technique it is important that a radiation be used that does not cause fluorescence of the Fe in the sample. Cu K-α, for example, increases the background in a powder diffraction photograph, and small amounts of Fe minerals cannot be detected. In recorder diffractometry, this difficulty is overcome, and peak-to-background ratios are restored by the use of monochromators before the detector stage. If however, Co K-α radiation is used, the other normal soil components, *viz* quartz and clays, have mass absorption coefficients similar to the common iron oxides, and relative proportions of minerals can often roughly be estimated by comparisons of line densities in photographs, or of peak intentisites in diffractometer traces.

In X-ray diffraction, the presence of goethite or hematite, which are commonly associated in Australian soils, can be recognized by a common line at 2.69–2.70 Å. With low amounts of total free iron (up to 10% Fe_2O_3), this line may be the only one clearly visible outside the normal suite of lines arising from other common soil components. The major goethite line (110) at 4.18 Å is often unresolved from the quartz 4.26 Å line, and also occurs on the tail of the (Okl) bands arising from the clay minerals present. If however, sufficient goethite is present, the 2.69 Å (130) line is visible above the weaker lines from other minerals in this area, and a distinctive doublet with the 2.45 Å (111) reflection is observed. If the sample is free from hematite, a semiquantitative estimation of the goethite can be made from the integrated peak intensity and measured mass absorption coefficient, by reference to the values they have for a pure goethite (Norrish and Taylor 1962). However, inaccuracies are introduced by the peak broadening caused by the small particle sizes of these oxides, and also by the change in diffraction intensity caused by isomorphous substitution.

If hematite is also present, the increase in intensity of the 1.69 Å line after heating the sample at 230°C for 1 hour is used to give a rough partition between goethite and hematite (Taylor and Graley 1967). Although the presence of hematite may be suspected from the reddish colour it imparts to the soil, it has no really distinctive line which is free from the influences of other soil minerals. At higher concentrations its 2.69 and 2.52 Å reflections make a distinctive doublet: the former line being the more intense when hematite is the only oxide present. The 3.68 Å line cannot be resolved satisfactorily from the (002) kaolinite line.

Lepidocrocite is generally better crystallized than soil goethites, and gives a distinctive diffraction line at 6.25 Å at concentrations of only a few percent. However this (020) line often appears poorly expressed in relation to the weaker (120) line at 3.29 Å, perhaps because of influence from foreign ions (Taylor and Schwertmann 1980), or formation paths (Schwertmann and Taylor 1979).

Maghemite and residual primary *magnetite* cannot be differentiated except by their Fe(II) content: magnetite contains close to the theoretical 33%, whereas maghemite is generally much lower. Although this change in Fe(II) content produces, due to a shrinkage of the unit cell, a change in X-ray diffraction spacings, this displacement is often too small to measure accurately, and furthermore may be offset by the incorporation of other cations of different sizes. The major line coincides with the 2.52 Å hematite line, but at higher proportions a distinctive line at 2.95 Å is visible. Although certain soil phosphate minerals, e.g. the gorceixite family, have a line in this position, concentration of the maghemite and/or magnetite by a hand magnet can confirm the identification. Maghemite and hematite are commonly associated in Australian soils, especially in concretions. In such an association, the intensity of the 2.52 Å line increases relative to the 2.68 Å as the maghemite content is increased.

Ferrihydrite can only be detected by X-ray diffraction in high concentrations. However, it has been observed (R.M. Taylor, unpublished data), that where ferrihydrite exists predominantly with some other forms of oxides, then the tangent of the slope of the background of the X-ray diffraction trace between 28 and 33° 2θ (using Co K-α radiation) has a linear relationship with the amount of noncrystalline material present. The common method of identifying and quantifying this component is by measuring the amount of Fe dissolved by acid ammonium oxalate in the dark at pH 3 (Schwertmann 1964, 1959b). Another method (R.M. Taylor unpublished data) of chemical detection is by solution in a 10:1 mixture of 5 M NaOH and ethylene glycol; the amorphous or poorly crystalline iron oxides, because of their large surface areas and greater solubility, dissolve more readily at the higher pH than do the more crystalline oxides: and subsequent complex formation with the glycol facilitates the reaction.

Surface charges of Fe oxides

The oxide minerals are capable, sometimes in a significant way, of contributing to the surface charge characteristics of a soil. The hydroxylated surface participates in proton dissociation–association reactions, so that the surface potential is governed by the adsorption of the potential-determining ions H^+ and OH^-. The system is considered to be of the 'constant potential' type, because once the solution concentration of H^+ and OH^- (i.e. the pH) is fixed, so is the surface potential. The surface charge (δ_o) is related to the surface potential (ψ_o) through the Gouy-Chapman equation

$$\sigma_o = \frac{\kappa\epsilon}{4\pi}\,\psi_o \qquad \text{(for } \psi_o << 25\text{mV)},$$

where κ is the reciprocal of the double layer thickness, and ϵ is the dielectric constant of water. Because κ is directly related to the electrolyte concentration, the surface charge is not dependent only upon the sign and magnitude of the surface potential, but also upon the ionic strength.

The surface properties of such minerals are in contrast to the silicate clay minerals, the latter deriving most of their surface charge from isomorphous substitution, within the clay lattice, of ions of a particular valency by others of another (usually lower) valency, thus creating a charge imbalance; the surface charge in this case is fixed by the extent of substitution, and is not dependent upon the pH and ionic strength of the electrolyte.

Most soils undoubtedly contain mixtures of these variable-charge and constant-charge colloids, and it is their relative proportions which will control the net surface charge characteristics of a soil. Much of the pioneering work in soil surface chemistry was carried out on soils dominated by the constant-charge minerals; there being greater incidence of constant-charge minerals in temperate regions, where such work was largely conducted. As interest in soils from humid tropical regions developed, it was found that many soils contain appreciable quantities of oxides, and as a result, the net surface charge characteristics relate to the variable charge system. There is no intention to draw a sharp contrast between soils from tropical and temperate regions, as will be explained below. Van Raij and Peech (1972) and Morais *et al.* (1976) have shown, that in Brazil, some oxisols, ultisols, and alfisols have surface charge properties akin to metal oxides; similar experience has been reported by Keng and Uehara (1974) and El Swaify and Sayegh (1975) for some Hawaiian oxisols and ultisols, and by Gallez *et al.* (1976) for some Nigerian alfisols and ultisols.

In Australia, Gillman and Bell (1976, 1978) examined samples from highly weathered soils in tropical north Queensland (*krasnozem, xanthozem, red podzolic soils, red earths*, and *yellow earths*), and showed, that these soils, although dominated by kaolin and sesquioxides in the clay fraction, possess constant-potential surface charge characteristics. Though such work has been conducted only on soils from tropical Australia, oxidic soils with similar properties will certainly exist in the temperate regions of the country.

At some pH value, an oxide will have adsorbed equal quantities of H^+ and OH^-, so that its net surface charge will be zero, in the absence of any specifically adsorbed ions. This isoelectric point for the common soil oxides is usually between pH 7 and 9, so that in the acid range of pH, an oxide is positively charged, and can be present in sufficient quantities to give the soil an appreciable anion exchange capacity. A *krasnozem* subsurface soil, studied by Gillman and Bell (1976), containing about 50% goethite and hematite, has an anion exchange capacity of 4.0 mequiv. per 100 g and a cation exchange capacity of only 0.5 mequiv. per 100 g.

The ability of an oxide to adsorb OH^- imparts a high buffering capacity to soils containing oxides. Gillman and Bell (1976) found buffering capacity to be as high as 5.3 mequiv. per 100 g for a north Queensland *red podzolic soil*, indicating that attempts to raise the soil pH of such a soil would require an enormous amount of liming material.

Oxidic soils normally possess a net negative surface charge in the top-soil because the organic matter present is negatively charged in the soil pH range. If however, organic matter levels are considerably reduced by cultivation, the positively charged oxide surfaces could become dominant. In this situation, the soil has little capacity to retain nutrient cations against leaching losses. This could be an important problem in tropical Australia where such soils occur under a warm climate with high annual rainfall. It then becomes difficult to conserve organic matter under some cropping systems, and other ways may have to be found for increasing the cation exchange capacity: for instance, silicate amendments might be useful, since it is well known that the silicate anion can be specifically adsorbed onto oxides, with a resultant increase in negative charge.

Specific adsorption by iron oxides in soils

Adsorption of heavy metals by iron oxides is an important process in soils. Although the iron oxides are generally positively charged below about pH 8, adsorption of some anions (such as molybdate, phosphate and selenite) and weakly hydrolyzed cations (such as Cu, Co, Ni, Zn and Pb) may occur in excess of the exchange capacity, as determined for example by Cl^- or Na^+. This process is referred to as specific adsorption. The ions are thought to be chemisorbed, i.e. they exchange with surface protons or hydroxyl groups.

Hingston *et al.* (1970, 1972) studied the adsorption of various anions, including molybdate and selenite, and found that the maximum adsorption, or a break in the curve of adsorption versus pH, occurs near the pK of the conjugate acid of the anion concerned. On the other hand, specific adsorption of cationic species increases with pH. Grimme (1968) studied the adsorption of Cu, Zn, Co and Mn on goethite, and found a linear relation between the pH at which 50% of the added metals are adsorbed and the solubility product of the hydroxide; Forbes *et al.* (1976) found a similar order for the adsorption of Cd, Co, Cu, Pb and Zn on goethite.

Such reactions lead to the accumulation of heavy metals in iron oxides in soils. Taylor and Giles (1970) found that vanadium and molybdenum accumulate in the iron oxides in the fine fractions in soils; Turton *et al.* (1962) found that Mo, Ga, V and Pb accumulate in the ferruginous gravels in lateritic soils in Western Australia, and McKenzie (1975) found that Cu, Zn and Ni accumulate in the iron oxides in the fine fractions of soils.

Jenne (1968) proposed that iron and manganese oxides act as sinks for the heavy metals, and that the presence of only small amounts of these oxides may be adequate to control the distribution of the heavy metals between earth materials and the associated aqueous phase. Jones and Smith (1972) showed that the uptake of molybdenum by plants is reduced by the addition of freshly precipitated ferric oxide to the soil, and such reactions may well control the uptake of many heavy metals by plants.

Sorption of phosphate and other anions by iron oxides

Because iron oxides in most soils are positively charged, they attract anions and often react strongly with them, e.g. phosphate, arsenate, silicate and organic anions. This has both beneficial and detrimental effects. For instance, under the extensive leaching conditions of the Tertiary era in Australia (Wild 1958), phosphate was removed from soils low in iron oxides, aluminium oxides and clay minerals, and from those subject to the seasonal reducing phase of lateritization in which iron oxides and phosphate came into solution. As a result, Australian soils generally contain less phosphate than those in the United States of America and in Europe. On the other hand, some exceptional soils like *krasnozems* retained phosphate against strong leaching because iron oxides developed and remained *in situ*. Again, eutrophication of lakes and streams would be far greater and more widespread if phosphate were not relatively immobile in the soil. Sewage and other pollution problems would be worse than they now are if many of the polluting elements drained freely through the soil into surface waters instead of being retained by iron oxides and other soil components.

The most important detrimental aspect arises when attempts are made to improve the agricultural productivity of soils by addition of phosphate fertilizers. Plant responses to added fertilizer are usually low because much of the phosphate becomes unavailable to plants; it is fixed in acidic soils, at least in part by iron oxides.

Because of the agricultural importance of phosphate fixation, a considerable amount of work (reviewed by Wild 1950 and Smith 1965) has been done to determine which soil components are primarily responsible for it. The outstanding contenders in acidic soils are iron oxides and aluminium oxides. Much evidence has accumulated in favour of one or the other, but it is confused mainly because the solutions designed to extract the oxides from soils do not do so sufficiently specifically to reveal the true relative contributions. Synthetic oxides of both Fe and Al take up phosphate from model systems, but these studies are of limited value when applied to soils, where much depends on the types and relative amounts of the oxides present; their reaction in competition with each other and with other components over short and long periods; the size and crystallinity of the particles; and the nature of other ions in the environment. A comparative study of several Australian soils by autoradiography and electron probe microanalysis (Fordham and Norrish 1974, and unpublished data) indicates that the reaction of phosphate with iron oxides dominates over that with aluminium oxides, even when gibbsite is present in much greater amounts; that iron oxides occur as compact crystals or micro-aggregates and as fine-grained material on other mineral and clay surfaces (both morphological forms being active (Fig. 23.1)), and that, in some circumstances, titanium oxides compete effectively with iron oxides for phosphate.

Phosphate is initially taken up quickly by iron oxides, but then is removed from solutions at a much slower rate for long periods. The initial reaction is decidedly dependent on pH (Hingston *et al.* 1972; Breeuwsma and Lyklema 1973; Yates and Healy 1975) and not easily elucidated, because pH affects the nature of both the iron oxide surface and the ionic species of phosphate in solution. Several competing reactions are therefore possible, each contributing differently to net uptake at different pH values and different degrees of surface coverage. However, it is clear that phosphate ions (and others, like silicate and arsenate) can be held very firmly at the iron oxide surface by strong bonds with covalent characteristics, in contrast to ClO_4^-, I^-, Cl^-- and NO_3^-, which are bound by electrostatic forces and thus easily displaced. Phosphate ions react with $-OH_2^+$ and $-OH$ groups in the iron oxide surface, causing expulsion of H_2O molecules which, under some conditions, include a proton derived from the phosphate ion itself. The result of these reactions is a strong Fe–O–P bond at the surface. Parfitt *et al.* (1975) believe that an even stronger bidentate bond is formed when each phosphate ion reacts with adjacent surface groups, but McLaughlin *et al.* (1977*a*) dispute that such a bond would be formed under normal soil conditions. To explain the almost irreversible uptake which is most prominent at the end of the reaction, these authors favour a mechanism involving diffusion of phosphate into the interior of iron oxide crystals or microaggregates. McLaughlin *et al.* (1977*b*) also emphasise that the phosphate-sorbing characteristics of iron oxides are very similar to those of many soils. This implies that reactions of phosphate with iron oxides may often predominate in soils, particularly in the initial stages; they may even be dominant in some alkaline soils in which $CaCO_3$ is present (Holford and Mattingly 1975).

The strong affinity which iron oxides have for organic anions (Evans and Russell 1959; Levashkevich 1966) has important consequences in several aspects of soil science.

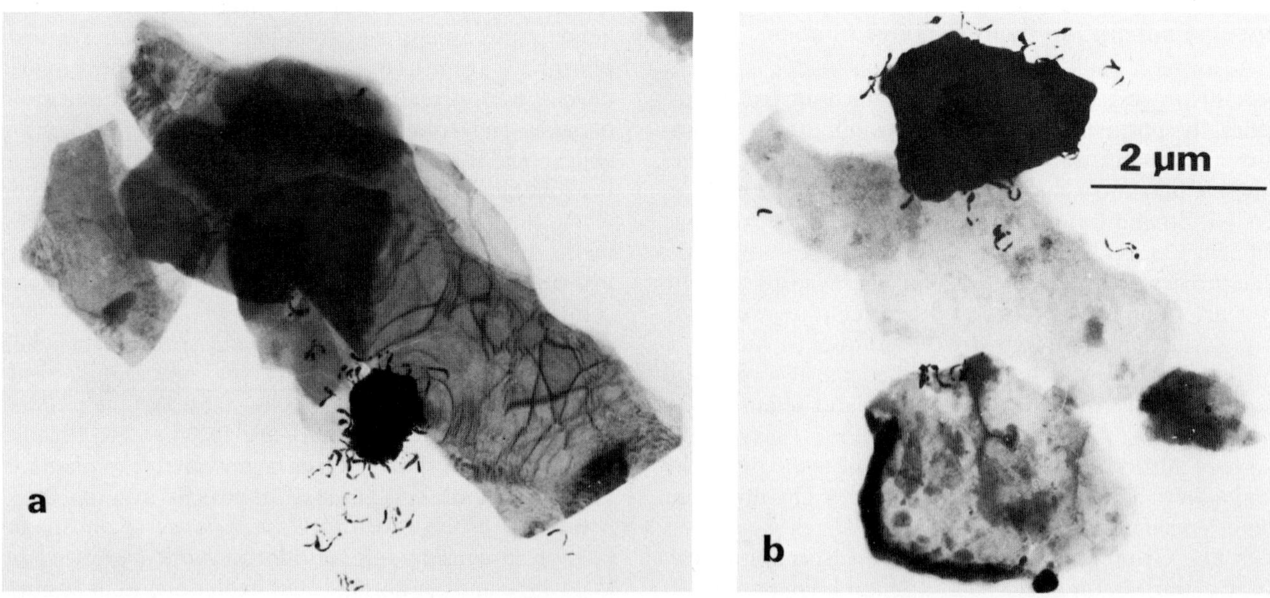

Fig. 23.1. (*a*) Hematite pellet, and (*b*) kaolinite flake carrying heavy deposits (not evident at this magnification) of finely divided goethite. Both particles show accumulation of silver grains (whorls) after auto-radiographic exposure, indicating strong uptake of phosphate. Clean kaolinite flakes included in the figure are free of silver.

It is involved in the maintenance of good soil structure, and in the mobilization of iron during soil genesis. It affects availability of nutrients, as in the case when phosphate fixation is reduced by the competitive influence of organic matter. The cation exchange capacity of organic matter, and therefore the ability of surface soils to retain nutrient cations, can be considerably decreased by interaction of the organic matter with iron oxides. Microbial decomposition by sorbed organic matter is retarded, and soil enzymes may be rendered inactive. Herbicides and pesticides lose their efficiency by reaction with iron oxides and persist undecomposed for long periods in soils.

Because much of the organic matter in soils is in anionic form, reaction mechanisms similar to those previously discussed for phosphate with iron oxides can be expected. Indeed, the beneficial effect which organic matter has on release of phosphate to plants is due in part to direct competition for sorbing sites on iron oxides and other components. But additional reactions also occur, because of the large number of different types of functional groups on organic molecules; hydrogen bonding and various types of electrostatic and physical interactions, as well as steric effects, have been observed (Greenland 1971). The ability of organic molecules to make multiple contacts with the surface results in strong association, as in the case of fulvic acid on goethite (Parfitt *et al.* 1977*a*). Carboxylate groups of each fulvic acid ion form covalent bonds by displacing $-OH_2^+$ and $-OH$ groups from the iron oxide surface, but, because of steric hindrance, fewer surface sites are available to fulvic acid than to smaller phosphate ions; other less specific interactions occur at higher surface coverage. Watson *et al.* (1973) showed that the hydrophilic nature of goethite can be altered by adsorption of the herbicide 2,4-D.

Negatively charged inorganic polymers probably react with iron oxides in a similar manner to organic anions. Thus Krishna Murti *et al.* (1976) suggest that mixed Si–Al polymers adsorb on iron oxide, and extend three-dimensionally to produce a network of gel which contains iron oxide domains. Follett *et al.* (1965) also found a close association between iron oxide and poorly ordered aluminosilicates. Structures of an iron oxide core within a shell of organic macromolecules may be relatively common in nature; they have been shown to exist in ferritin (see Spiro and Saltman 1969) and in iron-dextran compounds (Marshall and Rutherford 1971).

Association of iron oxides with other inorganic soil components

Finely divided iron oxides (often X-ray amorphous) occur in close association with clay minerals and other soil components. In so doing, they exert a strong influence on soil properties, not only because of their high surface areas and physico-chemical reactivity, but also because of the modifying effect they have on the host particles.

Although iron oxide-clay associations were previously proposed by several authors from indirect evidence, Fripiat and Gastuche (1952) were the first to demonstrate them by electron microscope studies. They showed that the surfaces of soil kaolinites are much cleaner after treatment with reagents which extract iron oxides; the disappearance of localized accumulations being particularly noticeable. The removal of iron oxides is accompanied by changes of surface area – a decrease as iron oxides on the clay surface dissolve, followed by an increase as previously inaccessible internal surfaces of microaggregates of clay are exposed. Changes in reverse order are obtained when iron oxides are precipitated onto deferrated clays. Working with similar synthetic preparations, Greenland and Oades (1968) distinguished differences between kaolinites in their ability to retain surface deposits of iron oxide, as well as differences due to the type of iron oxide and pH of the system. Thus, ferric hydroxide gel precipitated (with positive charge) at low pH values is held firmly at the clay surface, whereas goethite precipitated (with negative charge) at high pH values produces a separate phase. These results were later confirmed by measurements of positive and negative charges in the mixtures as the type and content of iron oxide were altered (Greenland 1975).

Iron oxides on clay surfaces were originally described in terms such as 'amorphous skins and coatings', but the oxide is probably present in most cases in discrete particulate form. Several electron micrographs to show this in synthetic iron oxide-clay mixtures have been published (Follet 1965; Greenland *et al.* 1968; Greenland and Oades 1968; Greenland and Wilkinson 1969; Greenland 1975). Using very dilute solutions containing ^{55}Fe, Fordham (1973) showed that iron oxide is dispersed evenly over the surface of dickite but, because insufficient Fe is available to form a continuous polymeric film, it must be present as discrete units. Mössbauer spectroscopy has proved very useful in defining the nature of particles too small to be characterized by X-ray diffraction. Thus, Nalovic *et al.* (1975) found that dried ferric hydroxide gel is composed partly of ultracrystalline particles ($<$25 Å) and partly of larger but poorly crystalline particles. The same material after leaching with water at 60°C contains mostly $>$250 Å particles of hematite, together with a small amount of finely crystalline iron oxide ($<$100 Å). In their view, true amorphous iron oxides do not exist in soils. Similarly, van der Giessen (1966) and van der Giessen *et al.* (1968) concluded from Mössbauer and other studies that undried ferric hydroxide gel consists of a network of small particles (20–30 Å) interlaced with pores containing water. Even at this size, the iron oxide particles have some crystalline characteristics and a unit cell dimension could be measured. Yassoglou and Peterson (1969) found evidence for more than incidental association between iron oxide and kaolinite, since the Mössbauer spectra are altered by their combination, probably as a result of increased bond rigidity. Babanin *et al.* (1975) studied synthetic iron oxide-clay preparations and concluded that iron oxide particles <70 Å are held at clay surfaces partly by chemical and partly by van der Waal's forces.

Mössbauer studies of soil samples also reveal the small but discrete nature of iron oxide particles. Gangas *et al.* (1973) examined the clay fraction of a soil from Greece in which no crystalline iron oxides are detected by X-ray diffraction. They found iron oxide is present as hematite particles with a mean diameter of about 130 Å, calculated on the assumption that the particles are spherical. Karpachevskii *et al.* (1972) identified most of the iron oxide in several Russian soils as goethite, occurring as small particles <150 Å either collected together in aggregates (1 μm) or combined with larger particles of other soil components. In deep-sea iron–manganese nodules which show no X-ray diffraction characteristics of iron oxides, Mössbauer spectra (Gager 1968) indicate the presence of lepidocrocite or geothite in a finely divided state (<70 Å).

Several Australian soils examined by electron microscopy and electron probe microanalysis (Fordham and Norrish unpublished data) also contain iron oxides in two size ranges. There are single crystals or micro-aggregates of about 1 μm in size as well as much smaller particles <200 Å. The latter form localized deposits on larger particles of other soil components, particularly kaolinite and illite, although not all these clay particles in any one soil carry them. Deposits are also found on rutile, anatase and some vermiculites, but ilmenite and quartz consistently have clean surfaces and contain very little iron other than that within the ilmenite itself. In addition, small iron oxide particles occur aggregated with kaolinite and/or gibbsite, all the particles being of comparable size.

The persistence of such small particles in soil is most probably due to inhibition of crystal growth by organic compounds (Schwertmann 1966; Schwertmann *et al.* 1968), by trace elements (Nalovic *et al.* 1975), by silicic acid (Herbillon and Tran Vinh An 1969; Schwertmann and Thalmann 1976), by aluminium (Taylor and Schwertmann 1978), or by phosphate (Krause and Borkowska 1963). All of these substances are abundant in soil solutions.

Association of iron oxides with clay surfaces alters the charge characteristics of the particles (Greenland 1975), and so reduces the electrostatic repulsion between these particles and other unaffected ones. Consequently, it promotes clay aggregation. This has been observed in many soils and is particularly prominent in subsoils (Harris *et al.* 1966; Krishna Murti *et al.* 1977). Other mechanisms such as polycation bridging and cementation have been proposed (Rengasamy and Oades 1977), although the latter process is usually applied to the rather exceptional case of formation of laterites (Sivarajasingham *et al.* 1962). It seems likely that iron compounds by themselves might lead only to micro-aggregates, and that organic molecules have to be involved to produce larger aggregates. Greenland (1971) reviews these aspects. Further discussion on the topic of aggregation is found in Chapter 31.

Management of soils with appreciable oxide content

The correct management of soils containing significant quantities of oxidic material has yet to be elucidated. This is an important area of study because the poorest nations in the world are situated in or near the tropics where such soils are prevalent. As Uehara (1977) has pointed out, there are large tracts of level land in Latin America, Africa, and south-east Asia with deep, adequately watered and well-drained soils that are nevertheless impoverished and under-utilized. The high oxide content of many of these soils reflects earlier soil-forming conditions which would have removed nutrient elements, as well as creating an environment which can render residual or further applied nutrients unavailable. Both these factors could be responsible for the infertility of some of these soils.

The aggregation mechanisms discussed above are important in bestowing beneficial physical properties, in that they result in well aerated soils with good internal drainage. Unfortunately however, the surface charge characteristics of the oxides are such that these soils often have a low capability for retaining nutrient cations, and at the same time have a tendency to strongly sorb nutrient anions and render them unavailable to plants. This is particularly true in the deeper layers of these soils. Consequently, root penetration into the subsoil could be inhibited, and plants could therefore be denied access to an important water source.

Future studies could profitably be devoted to ways of overcoming subsoil infertility, and could include deep tillage accompanied by the introduction of amendments at depth. As mentioned previously, the addition of silicate, and even rock phosphate, at depth should increase the soil cation exchange capacity, at the same time reducing subsoil acidity, which would also alleviate any toxicity due to aluminium at low pH values.

Manganese oxides

Manganese occurs in a number of oxidation states, but only the cationic oxidation states 2, 3, and 4 occur in nature (Day 1963). A large number of oxides and hydroxides exists, in which there is extensive substitution: of Mn(II) and Mn(III) for Mn(IV), and of OH^- for O^{2-}. The manganese ions may be oxidized, or reduced, without changing position, and when the valency of a sufficient number has been changed, the structure becomes mechanically unstable, and rearranges to that of a new phase (Feitknecht *et al.* 1960; Wadsley and Walkley 1951). There is, therefore, a continuous series of compositions from MnO to MnO_2, within which there are several stable and metastable arrangements of atoms to form the well-known minerals, many of which may cover a wide range of compositions. In addition to this series, there are oxides which contain cations other than manganese as an essential component of the structure.

The manganese oxide minerals

Some of the more important oxides of manganese are listed in Table 23.1. The composition of each mineral listed is an ideal formula, and subject to some degree of variation. The oxides may be classified according to their structures (Burns and Burns 1975*b*).

Chain structures

The structures of pyrolusite and ramsdellite are based on chains of MnO_6 octahedra; single chains in pyrolusite and double chains in ramsdellite. These oxides are regarded as the only true varieties of MnO_2; all the others are non-stoichiometric, and exhibit defect structures to varying degrees. Nsutite includes a group of oxides consisting of structural intergrowths of pyrolusite and ramsdellite (Giovanoli 1969), and has the general composition

$$Mn^{4+}_{(1-x)} Mn^{2+}_{x} O_{(2-2x)} (OH)_{2x}$$

(Zwicker *et al.* 1962). Groutite is isostructural with ramsdellite (Dent Glasser and Ingram 1968), and the structure of manganite resembles that of pyrolusite (Dachs 1963).

Framework structures

Minerals of the hollandite group contain large foreign cations as an essential part of the structure, and are characterized by tunnels, running in the *c*-axis direction, in which the foreign cations are situated (Bÿstrom and Bÿstrom 1950). The foreign ion is barium in hollandite, potassium in cryptomelane, and lead in coronadite. Psilomelane has a similar structure, with larger tunnels which accommodate barium atoms and water molecules (Wadsley 1953).

Layer structures

Minerals of the birnessite group have a layer structure, and were previously known as manganous manganite or 7 Å-manganite. The composition of the type mineral was given by Jones and Milne (1956) as $(Na_{0.7}\ Ca_{0.3})\ Mn_7 O_{14} . 2.8H_20$, but it is now known that the calcium and sodium are not essential constituents. δ-MnO_2 is a very fine-grained disordered birnessite in which only the X-ray diffraction spacings at 2.40 Å and 1.42 Å are present. Lithiophorite also has a layer structure in which layers of MnO_6 octahedra alternate with layers of $(Al, Li)(OH)_6$ octahedra (Wadsley 1952), but it does not belong to the birnessite group (Giovanoli *et al.* 1973).

Todorokite is a common mineral of unknown structure, occurring as fibrous aggregates of small needle-shaped crystals (Burns and Burns 1975*b*).

The manganese oxides in soils are very fine grained, often poorly crystalline, and non-stoichiometric; they are often reported to be amorphous (Childs 1975; McKeague *et al.* 1968), but crystalline varieties have been identified on a number of occasions. Up to the present time birnessite has been identified 18 times, lithiophorite 14, hollandite 4, todorokite twice, and pyrolusite, psilomelane and manganite once each (Eswaran and Raghu Mohan 1973; Gallaher *et al.* 1973; Koljonen *et al.* 1976; Ross *et al.* 1976; Taylor 1968; Taylor *et al.* 1964; Tiller 1963). Thus birnessite and lithiophorite are by far the most common in soils.

Mode of occurrence of manganese oxides in soils

The manganese oxides occur in soils as deposits in cracks and veins, as coatings and stains on soil aggregates, or as a cementing material; they may be dispersed throughout the soil as microscopic accumulations, or concentrated as nodules up to about 2 cm in diameter. The nodules are roughly spherical in shape, and many have a concentric layered structure suggestive of seasonal growth; they contain a considerable concentration of iron oxides, as well as other soil minerals. Nodules occur in greatest numbers and in largest sizes in horizons which are subject to seasonal waterlogging, where restricted drainage results in slow removal of the products of weathering. Manganese is reduced and dissolved during waterlogging, and, when the soil dries out under aerobic conditions, the reduced manganese compounds are oxidized and precipitated around localized nuclei (Drosdoff and Nikiforoff 1940; Winters 1938).

Whereas birnessite occurs in both acid and alkaline horizons, it is more common in alkaline surface horizons. Lithiophorite is most commonly found in acidic subsurface horizons (Taylor *et al.* 1964).

Table 23.1. Some manganese oxide minerals

Mineral name	Other names	Composition
Pyrolusite	β-MnO_2, polianite	MnO_2
Ramsdellite	—	MnO_2
Nsutite	γ-MnO_2, ρ-MnO_2	$Mn(O, OH)_2$
Birnessite	δ-MnO_2, manganous manganite, 7 Å manganite	variable
Todorokite	—	variable
Cryptomelane	α-MnO_2	$K_2Mn_8O_{16}$
Hollandite	α-MnO_2	$Ba_2Mn_8O_{16}$
Coronadite	α-MnO_2	$Pb_2Mn_8O_{16}$
Psilomelane	romanèchite	$BaMn_9O_{16}(OH)_4$
Lithiophorite	—	$(Al, Li)MnO_2(OH)_2$
Groutite	α-MnOOH	MnOOH
Manganite	γ-MnOOH	MnOOH
Hausmannite	—	$Mn^{2+}Mn^{3+}_2 + O_4$

Oxidation of manganese in soils

Manganese is oxidized rapidly by air under strongly alkaline conditions. In slightly alkaline conditions, Bricker (1965) observed that $Mn(OH)_2$ is oxidized to hausmannite, and, after prolonged oxidation, to manganite, but further oxidation is not achieved under these conditions. From the equation,

$$Mn^{2+} + H_2O + \tfrac{1}{2}O_2 = MnO_2 + 2H^+ \qquad \Delta G = -0.88 \text{kJ},$$

it can be calculated that the concentration of Mn^{2+} in water saturated with air should be <0.1 ppm at pH >3.7. The failure of this oxidation to proceed at a measureable rate is shown by the stability of Mn^{2+} solutions in the laboratory, and may be due to a high activation energy for the oxidation of the manganous ion, or for the reduction of oxygen.

Oxidation of manganese is catalyzed by fine particles in soils (Morgan and Stumm 1964*b*), and Hem (1964) reported that the oxidation is catalyzed by MnO_2 at pH 7.8 or higher, but not at pH 7.2. Hem (1964) also discussed the oxidation of manganese by precipitated $Fe(OH)_3$, and concluded that this is possible in slightly acid solution, provided the activity of Fe^{2+} is low enough. The higher oxides are also formed by the disproportionation of hausmannite and manganite under acid conditions.

The biological oxidation of manganese has been demonstrated many times (Bromfield 1958; Harriya and Kikuchi 1964; Schweisfurth and Gattow 1966; Zavarzin 1968), and a number of the bacteria and fungi responsible for this oxidation have been identified.

Origin and formation of manganese minerals in soils

A number of studies on the transformation of synthetic birnessite to nsutite and finally to pyrolusite, and on the formation of nsutite or pyrolusite by hydrolysis of $MnCl_4$ (Buser *et al.* 1954; Cole *et al.* 1947; Gattow and Glemser 1961; Glemser *et al.* 1961; McKenzie 1971; Zwicker *et al.* 1962) show that the presence of large amounts of foreign ions prevents the formation of nsutite, while even small amounts of foreign ions prevent the formation of pyrolusite. In soils, foreign ions derived from the weathering of other minerals will always be present, and would probably limit the formation of nsutite and pyrolusite.

Hausmannite and manganite would not be expected to persist in soils because, under acid conditions, they would disproportionate to give higher oxides, and under alkaline conditions they would be oxidized. Hausmannite has not been found in soils, and the only occurrence of manganite (Eswaran and Raghu Mohan 1973) was in the interior of an ironstone nodule, where it would have been protected from the soil environment.

Birnessite has been shown to be formed by bacterial oxidation (Schweisfurth and Gattow 1966), and also by disproportionation of hausmannite under mildly acid conditions (Bricker 1965). These processes, as well as direct or catalytic oxidation, probably contribute to its formation in soils.

Birnessites containing suitable concentrations of a large foreign cation are readily converted to cryptomelane-type minerals by boiling (Gattow and Glemser 1961; Glemser *et al.* 1961; Maxwell and Thirsk, 1954; McKenzie 1971). Lithiophorite has been synthesized by substitution of aluminium and lithium into a birnessite-type mineral followed by hydrothermal treatment of the product (Giovanoli *et al.* 1973; Wadsley 1950*a,b*): and todorokite, by treatment of a birnessite-type mineral with transition metal cations (McKenzie 1971). Burns *et al.* (1974) have also demonstrated the recrystallization of δ-MnO_2 to todorokite under water at high pressure at room temperature. It seems likely for soils therefore, that either δ-MnO_2 or birnessite are formed first, and that these minerals may recrystallize subsequently, under suitable conditions, to lithiophorite or todorokite or the cryptomelane group.

The growth of manganese nodules

Burns and Burns (1975*a*) suggest that structural similarities between ferrihydrite ($HFe_5O_8.4H_2O$) and δ-MnO_2 make these minerals amenable to epitaxial intergrowth. They consider that the nucleation and growth of many manganese nodules is initiated by a deposit of ferrihydrite on a nucleating agent, followed by epitaxial growth of δ-MnO_2 and ferrihydrite on one another.

It is known that ferrous iron is oxidized much more rapidly in the presence of MnO_2 than in pure solution or in the presence of goethite (McKenzie unpublished data). If, as was suggested by Hem (1964), the precipitated ferrihydrite can in turn catalyse the oxidation of manganese, then, under conditions of fluctuating Eh and pH, the two processes would occur alternately, leading to the growth of nodules.

Observations in soils, using the electron probe microanalyser (McKenzie 1975), suggest that while iron may deposit on the manganese oxides as well as on the purer iron oxides, manganese tends to deposit at sites where both manganese and iron already exist. This is also evident from the electron probe microanalyzer pictures in Fig. 23.2, which show that iron has accumulated with manganese, but mostly interlayered with it in a manner suggestive of incipient nodule formation.

Methods of identification of manganese oxides

The manganese oxides in soils are identified by X-ray diffraction, using manganese-filtered iron radiation. Due to the minerals being present in such small amounts this method can be applied only where natural segregations of the minerals occur, such as in nodules and veins. Even then, the diffuse nature of the X-ray diffraction patterns of some of the minerals, and the coincidence of the

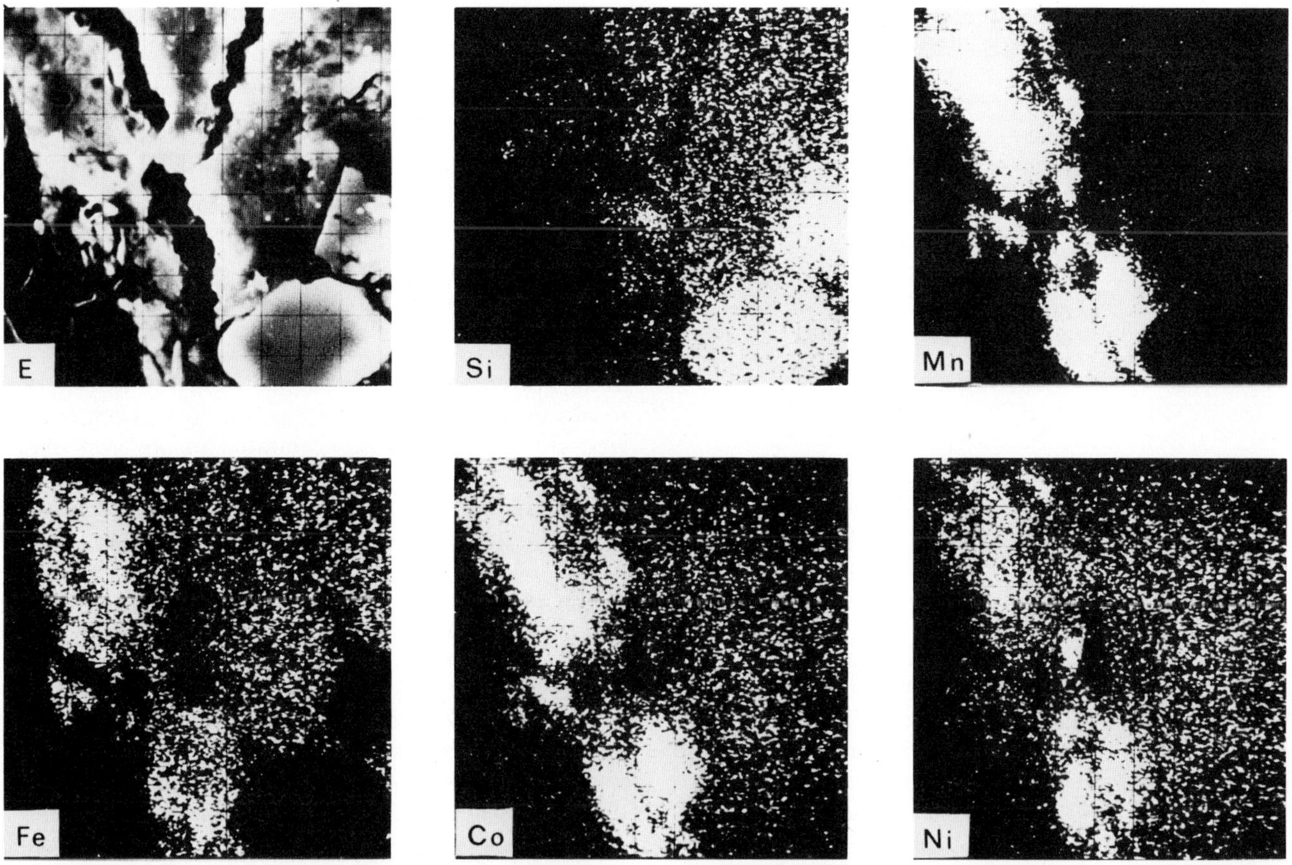

Fig. 23.2. Back-scattered electron image (E) and distribution of elements in an area of manganese accumulation in a black earth. (Reproduced from McKenzie (1975).

diagnostic lines with those of other minerals, necessitates some pretreatment to concentrate the minerals to permit positive identifications to be made.

Taylor *et al.* (1964) removed kaolin by boiling the sample in sodium hydroxide; when quartz was the main contaminant, the manganese mineral could sometimes be concentrated by magnetic separation. They also used a flotation method in which the ground sample was treated with hydrogen peroxide; particles of the manganese dioxide were carried to the surface by bubbles of oxygen, and were collected by elutriation. This method was also used by Ross *et al.* (1976). In many cases samples were treated with H_2O_2 at pH 3 to remove the manganese oxides: the comparison of the diffraction patterns before and after treatment led to positive identification in most cases.

The diffraction patterns of the birnessite and nsutite groups show considerable variation. Positive identification is difficult in some cases because of changes in unit cell dimensions and line broadening brought about by substitution, non-stoichiometry, disorder and small crystallite size.

The birnessite group is characterized by a strong basal reflection at about 7.0 Å, and its higher orders. The patterns of the more crystalline varieties contain many lines, but in the more disordered forms usually found in soils, only four lines are present. When the disordered varieties occur in a very finely divided form, the basal reflections disappear altogether, and only two lines at 2.4 Å and 1.4 Å are left (Bricker 1965).

Chemical composition of manganese oxides

The chemical composition of manganese oxides in soils cannot be determined directly, as they occur intimately mixed with iron oxides and other soil minerals. Taylor *et al.* (1964) extracted the manganese oxides from finely ground soil nodules by treatment with hydrogen peroxide at pH 3. This treatment reduces and dissolves manganese dioxide, with little attack on other soil constituents; other elements extracted with manganese are assumed to be associated with the manganese oxide phase in the nodules. Table 23.2 shows mean analyses of a number of samples. There is more aluminium and lithium in lithiophorite, and barium in hollandite, because these elements are essential constituents of the respective minerals. Individual samples show a considerable range of concentrations, thus chemical composition is not a useful means of differentiating between the various minerals.

Table 23.2. Mean analyses of manganese minerals in soils Extracted with acidified H_2O_2; parts percent

	Lithiophorite	Birnessite	Hollandite
Mn	52.7	59.9	47.4
Al	6.1	0.61	3.0
Si	1.8	0.65	3.1
Fe	3.3	0.65	4.6
Co	1.2	0.45	1.1
Ni	0.10	0.34	0.17
Ca	0.14	2.3	4.1
Mg	0.15	0.66	0.60
Ba	3.2	4.0	5.4
Li	0.07	0.03	0.05
Na	0.15	0.07	0.10
K	0.27	0.18	0.31

Minor elements feature commonly in the chemical composition. Manganese oxides in soils are finely divided, have large surface areas, and their points of zero charge range from pH 1.5 for birnessite to pH 4.6 for the cryptomelane group (Healy *et al.* 1966). Because they are negatively charged in all but the most acid soils, they have high sorptive capacities, particularly for heavy metals. Loganathan and Burau (1973) give, for adsorption onto MnO_2 at pH 4.0, Langmuir capacities of 1.4 mmol g^{-1} for cobalt and 1.0 for zinc. Similarly, McKenzie (1970) gives 1.8 mmol g^{-1} for copper at pH 5 Morgan and Stumm (1964*a*) give 0.5 mol per mol of MnO_2 (5.7 mmol g^{-1}) for manganese at pH 7.5.

This adsorption leads to the accumulation of relatively high concentrations of heavy metals in the manganese oxides in soils. Taylor and McKenzie (1966) have compared analyses of whole soils and manganese nodules separated from them; some of their results, in Table 23.3, show concentration factors ranging from 16 to 19 for cobalt between nodules and whole soil, whereas concentration factors for nickel are 2.8 to 5. The greater concentration factor for cobalt is quite typical for manganese nodules in soils. Data presented by Childs (1975) show concentration ratios between nodules and surrounding soil which range from 9 to 140 for cobalt and 1 to 5.5 for nickel.

The accumulatoin of cobalt in manganese oxides in soils has been studied extensively (McKenzie 1970, 1975; Taylor 1968; Taylor and McKenzie 1966). Taylor and McKenzie (1966) found that an average of about 80% of the total soil cobalt is associated with the manganese oxides, whereas only a small fraction of the total nickel is in this form. In another study based on spot analyses with the electron probe microanalyser, McKenzie (1975) found that 30–53% of cobalt, but only 0.3–19% of the total nickel, in the soil was associated with the manganese oxides. The electron probe microanalyser pictures in Fig. 23.2 show a very close relation between cobalt and manganese, and a less close relation between nickel and manganese. Although the highest concentrations of nickel are found in the manganese oxides, nickel also occurs in the iron oxides, and the nickel in this form can account for 18–80% of the total. Copper and zinc are also strongly concentrated in the iron oxide phase in the soils.

Table 23.3. Concentrations of elements in whole soils and in manganese nodules separated from them

Values as ppm. (Reproduced with permission from McKenzie (1980)

	Mn	Co	Ni	V	Pb	Ba	Cr	Zn	Cu
Whole soil	92	5	22	47	14	180	86	12	12
Nodules	3600	82	67	97	100	140	120	30	24
Whole soil	2500	20	8	49	10	220	26	12	8
Nodules	72000	380	41	110	50	2300	30	30	23
Whole soil	1300	6	14	49	14	180	37	20	5
Nodules	40000	110	39	88	34	1600	31	33	20

The fixation of cobalt by manganese oxides is of considerable practical importance; Adams *et al.* (1969) found, that the manganese content of the soil is one of the factors controlling the availability of cobalt applied as a fertilizer, and McKenzie (1978) showed that the addition of manganese dioxide to soils reduces the uptake of cobalt by plants.

Lead has also been observed to accumulate in the manganese oxides in soils. Taylor and McKenzie (1966) reported values up to 2500 ppm in the manganese oxide phase of soil nodules, and Norrish (1975) reported values of up to 2% of lead in manganese oxide in a surface soil. McKenzie (1978) showed that the addition of manganese dioxide to soils contaminated with lead reduces the uptake of lead by plants.

Experiences with cobalt and lead uptake by plants show that adsorption is an important factor in the incorporation of heavy metals into the manganese oxide phase in soils, and that at least some of these metals are thereby rendered unavailable to plants.

Aluminium oxides

Many different Al oxides and hydroxides can be synthesized and identified by their distinctive X-ray diffraction patterns; a brief summary is given by Rooksby (1961), in which many references to earlier work are made. There are, however, only about five naturally occurring mineral forms, and of these, only two are considered to be commonly formed during pedogenesis. Many of the Al oxide minerals have isostructural Fe counterparts, which are often more likely to form in soils. Readers are referred to a recent review article by Hsu (1977) covering the crystallography, infrared, X-ray diffraction and DTA data of the Al oxides, and the solution chemistry that leads to their formation in soils.

Whereas Fe(II) is generally liberated as the more mobile divalent ion during the weathering of primary minerals, Al exists in these minerals in its trivalent state, and, if the pH is greater than about 5, hydrolysis and precipitation occur when it weathers from its parent material.

According to Segalen (1971), hydrolysis is the main driving force in chemical weathering. The removal, by solution, of alkalis and alkali earth metals from silicate structures, and their replacement by the smaller H^+ ion are sufficient to weaken the structure of the alumino-silicate mineral; the Si–O–Al bond is affected, and a separation of Si and Al results. The continuous supply of H^+ ions needed comes from the conversion of dissolved CO_2 into HCO_3^-.

Depending on the environmental pH and drainage, the breakdown products may be removed together, may react together to form clay minerals, or the Al may be precipitated while the silica is removed. The accumulation of aluminium hydroxide as a first stage in the further weathering of clay minerals is largely dependent on the soluble silica content (silicic acid) of the weathering soil solution. According to Kittrick (1969) the level of silica in solution is not controlled so much by quartz as by the rate of dissolution of unstable silicates, removal of the soluble silica in the soil solution by leaching, rate of precipitation of stable silicates and uptake of silica in solution by plants.

It has been shown (Gardner 1970) that if the silica in solution is <1 ppm, desilication occurs and aluminium hydroxide can be accumulated under suitable pH conditions (4–9).

Forms of aluminium oxide minerals in soils

The common aluminium oxide minerals include gibbsite (α-$Al(OH)_3$), nordstrandite ($Al(OH)_3$), boehmite (γ-AlOOH), diaspore (α-AlOOH), and corundum (α-Al_2O_3). Bayerite, designated as γ-$Al(OH)_3$, is regarded as a metastable form, easily produced during synthesis involving precipitation of the hydroxide by alkali addition to an Al salt solution. Reports of natural occurrences of bayerite are, however, rare. Gibbsite and boehmite are the most common Al oxides formed during pedogenesis. Diaspore is not at all as common as the other hydroxide forms or as its isomorphous Fe analogue, goethite; there is uncertainty as to whether it can form under the normal pressures and temperatures associated with soil formation (Deer *et al.* 1962). Likewise, corundum is not as common as its Fe counterpart hematite (α-Fe_2O_3), and is generally regarded as being residual from the weathering of thermally metamorphosed or of igneous rocks; it has been found in Australian soils inside ferruginous concretions of a laterite profile (K. Norrish personal communication), but no evidence is available as to whether it formed during pedogenesis, or was a residual from an earlier period of weathering. Readers are also referred to an article by Schoen and Robertson (1970) on the structures of Al hydroxide, and geochemical implications.

Gibbsite has been found in the arid-region soils of Australia, where it is considered to be a relic of a past and more severe weathering cycle (Norrish and Pickering 1977). These authors also point out that the particle size of soil gibbsite is often in the micron (μm) range, which gives it a much lower relative surface area than other soil oxides, e.g. those of Fe, with which it must often compete for adsorption of anions and cations present in the soil environment.

From the mineralogy given for many soils described in 'A Handbook of Australian Soils' (Stace *et al.* 1968), gibbsite is detectable in the following soil types: *alluvial soils, chocolate soils, yellow earths,* the *red, yellow,* and *lateritic podzolic soils, krasnozems* and *alpine humus soils.* These occurrences are in agreement with the general climatic requirements for gibbsite formation reported by Segalen (1971), and by other workers who also noted it, that well-drained conditions in the wetter areas of warm humid zones favour its formation. Kittrick (1969) suggests, from thermodynamic data, that gibbsite should be found only in association with kaolinite, and should not occur with amorphous silica or montmorillonite. In general, the mineral associations presented in the Australian handbook, agree with this view.

Boehmite is less common in nature than gibbsite, with which it is often associated in bauxites. In Australia it has been detected in *earthy sands* and *siliceous sands, lateritic* and *gleyed podzolic soils* (Stace *et al.* 1968). Kittrick (1969) considers, from thermodynamic studies, that it is more stable than gibbsite, and observations on bauxite deposits indicate that the proportions of boehmite to gibbsite increase with the age of the formation. Although gibbsite does not dehydrate to boehmite in the laboratory at temperatures below 130° C, Schoen and Robertson (1970) consider that the transformation could take place at lower temperatures in soil profiles; they suggest that the presence of dissolved salts in the soil solution may reduce the activity of water sufficiently to facilitate the transformation. Boehmite forms under the same conditions as gibbsite, but in less well drained areas (Segalen 1971); it is not a transformation product of pseudoboehmite, which generally crystallizes to gibbsite.

Bauxites at Weipa in Queensland contain boehmite as well as gibbsite, and Schellmann (personal communication) has noted that the association of trace elements with the two mineral varieties is different, which suggests other than thermal transformation; for example, the Ni and Ti concentrations in bauxite are correlated with its boehmite content, but the Ga and Th are correlated with gibbsite.

Nordstrandite, a further polymorph of $Al(OH)_3$ has been found in nature, and differs in its stacking sequence from gibbsite. Hsu (1977) states, that the nature and concentration of environmental anions may influence nordstrandite formation. Since its X-ray diffraction pattern can be quite variable (Hsu 1977), and its major reflections are close to those of gibbsite, its identification may be uncertain. It has not been positively identified in Australian soils.

Properties of aluminium oxides in soils

As has been remarked, many of the reactions of the Fe oxides in soils, particularly their adsorption and fixation of anions, apply also to the Al oxides. The affinity of both species of oxides for P has been extensively investigated, and is well established (Wild 1950). Muljadi *et al.* (1966) suggest that its adsorption takes place at specific sites at edge faces of the gibbsite, by ligand exchange of two singly co-ordinated –OH groups for oxygen attached to the phosphate anion.

Various laboratory experiments have been carried out (e.g. Haseman *et al.* 1950) demonstrating that the Al oxides fix phosphate to a greater degree than the iron oxides, but electron probe microanalysis of thin sections of soils (Norrish, personal communication) and soil concretions (Taylor, unpublished data) in which both gibbsite and goethite are present, shows that the P is generally associated with the Fe oxide. The discrepancy between these observations may arise from differences in morphology and particle size of the materials used in laboratory experiments and those reacting in soils.

Besides phosphate, many other inorganic and organic anions can be specifically adsorbed by the aluminium oxides. Most studies have been carried out on gibbsite, which is justifed in view of its more frequent occurrence than other Al oxides in soils. Sulfate is specifically adsorbed (Chao *et al.* 1964), but can be replaced by the more tightly held phosphate. Parfitt *et al.* have studied the uptake of fulvic and humic acids (1977a) and oxalate and benzoate (1977*b*) on gibbsite. These studies revealed that only singly coordinated –OH groups at the edge surfaces of gibbsite are involved, as suggested for phosphate adsorption by Muljadi *et al.* (1966). Un-ionized fulvic acid was also found to be slightly adsorbed by the doubly coordinated –OH groups on the 001 face of gibbsite, and Jones and Handreck (1967) observed the adsorption of silicic acid from solution. Because of the similarity of the hydroxylated surfaces of the Fe and Al oxides, most other anions and cations specifically adsorbed on Fe oxide surfaces could be expected to display some degree of affinity to the surfaces of the Al oxides.

Al oxides may, like Fe oxides, coat other soil constituents, and can influence physical properties of the soil, e.g. the stabilization of aggregates (Kroth and Page 1947), or the dispersability and swelling characteristics (El Swaify and Emerson 1975).

Removal of aluminium oxides from soils

The sodium dithionite extraction designed to remove the iron oxides from soils does not generally attack gibbsite or boehmite. However, many modifications of this extraction, for instance the addition of other reagents such as citrate, could indeed cause an attack on these minerals (Saunders 1959). It must be remembered however, that in some cases much of the Si and Al found in soil extracts following deferration treatment, may come from the iron oxides themselves, as they are known to incorporate Al (Norrish and Taylor 1961) and Si (Norrish 1975). Extraction with acid ammonium oxalate will also cause solution of aluminium from Al oxides and possibly clay minerals.

Often, free Al oxides can be removed by boiling in 5 M NaOH, but this treatment also dissolves the kaolin present (Norrish and Taylor 1961). Depending on the degree of crystallinity, this may be avoided by using lower strength NaOH and lower temperatures. However, even if effective, this treatment is again not specific, because at pH $>$ 10 in NaOH solutions the solubility of Fe(III) increases, due to the formation of $Fe(OH)_4^-$.

Techniques of identification of aluminium oxides in soils

Gibbsite and boehmite both lose water on heating, and give endothermic DTA peaks at around 300 and 450–580°C respectively, but, as with the Fe oxides, these cannot be used alone for positive identification. A similar limitation would apply to infrared spectra, because Al–OH bonds in the oxides may not be distinguished from those of the associated clay minerals.

At present, X-ray diffraction offers the only reliable method of identification for both gibbsite and boehmite.

Gibbsite is reasonably easily identified, even at low concentrations, by the distinctive doublet formed by the two most intense and highest-spacing lines at 4.83 Å and 4.34 Å. Boehmite is difficult to recognize separately from its Fe analogue, lepidocrocite, as in most soils it can only be recognized by the innermost (020) line around 6.23 Å. The soil colour and type should give some indication as to whether lepidocrocite could be expected. Repeated diffraction after successive deferration treatments may confirm identifications. Gibbsite has no isomorphous Fe analogue, and is not subject to the same uncertainty.

Silicon oxides

Oxides of silicon in soils

Silica is the most abundant oxide in the lithosphere, occurring in combined and free forms in most igneous, metamorphic and sedimentary rocks, and this abundance is reflected in the soils derived from these rocks. Most soils indeed contain free silica in one or more of its polymorphic forms, or as an amorphous or poorly crystalline hydrated form, $SiO_2.nH_2O$.

The forms of silica commonly found in soils are quartz, cristobalite, tridymite, and the poorly crystalline opals: and each variety can to some extent reflect the type of parent material and/or formation conditions. According to Wilding *et al.* (1977), cristobalite and tridymite reflect, to some extent, magmatic volcanic or marine origins, whereas quartz is useful as a reference constituent, and as an index of parent material uniformity and origin. The amorphous or poorly crystalline hydrated silica, besides existing on the outer surfaces of the more crystalline quartz, is often biogenic, originally having been deposited in the form of phytoliths in plant cells, and now remaining as residues after decomposition of the organic matter.

Liebau *et al.* (1969) suspect that organo-silica complexes may account for the transport of dissolved silica in the body fluids of plants and animals, but to date no such compounds have been isolated. Griffith and Schnitzer (1975) reported Si in association with Fe and Al in the fulvic acid fraction of soil organic matter, but this could have come from the dissolution, in the NaOH extract, of biogenic silica. Jones and Handreck (1965), however, showed that the levels of biogenic silica in plants could be adequately explained by the concentrations of monomeric silicic acid in the soil solution.

Thermodynamically, quartz is the only stable variety at ordinary temperatures. Although various laboratory workers have found that the crystallisation of amorphous silica → quartz is extremely slow, and in many cases not observed, Jackson and Sayin (1975) report evidence for a silica gel → opal → cristobalite → quartz transformation in some young Wisconsin soils. This is supported by laboratory experiments of other workers: Liebau *et al.* (1969) state that impurities which are found in natural waters can hasten gel formation from soluble silica species; Harder and Flehmig (1970) identified quartz formed one month after coprecipitation of silica gel with hydroxides of Al, Fe and other metals; Brümmer (personal communication) has stated that if precipitated silica gel is mixed with precipitated ferrihydrite and organic matter (to produce reducing conditions) quartz is formed, whereas the silica gel precipitate remains amorphous if Fe is not added. The interaction between Si and Fe in soils, therefore, is two-fold; in one case, soluble silica can be adsorbed on to ferrihydrite (Hingston *et al.* 1967; Herbillon and Tran Vinh An 1969) inhibiting hematite formation (Chukhrov *et al* 1973) whereas the adsorption of soluble Fe onto precipitated silica gel promotes recrystallization to quartz (Hazel *et al.* 1949)).

The surface of quartz is hydroxylated, and resembles that of amorphous silica. The surface contains ions that are not fully coordinated, and are hence charged. On exposure to water, hydroxide groups, which function as protonic acids, are formed (Cotton and Wilkinson 1962; Fripiat 1972). Such hydroxylated silica surfaces can readily adsorb hydrolyzed Fe species, which may be oxidized subsequently to ferrihydrite, and then crystallize to goethite and/or hematite under suitable environments. Thus, hematite often occurs in cracks and on faces of quartz agglomerates and quartz grains in many soils and situations, e.g. desert varnish. An alternative explanation for this association may be the adsorption of positively charged Fe(III) hydroxide colloids onto the negatively charged hydroxylated silica surface, the negative charge arising from migration of H^+ from the hydroxylated surface. The absence of these ferruginous coatings in some soils is indicative of certain soil-forming processes, e.g. podzolization.

The observations of Brümmer that quartz forms from the amorphous silica precipitate in the presence of Fe, and the similar ideas of Liebau *et al.* (1969) and Harder and Flehmig (1970), reported above, suggest that the adsorbed Fe species may have a greater affinity for the shared surface hydroxyl groups than has the Si to which they were originally co-ordinated, hence resulting in the observed alteration of the amorphous silica to quartz.

Quartz often comprises the bulk of the sand and silt fractions of soils, although the clay fraction may also contain several percent. It is dissolved by HF, but other soil components are also attacked: it is not readily removed by boiling 5M NaOH. Douglas and Platt (1977) report from scanning electron microscopy studies that quartz grains show both solution and precipitation phenomena, and that the degree of alteration can be related to soil texture and age of the parent material, and is more pronounced in the B rather than the A horizons.

This process of solution and deposition no doubt contributes to the marked siliceous cementation observed in many soils, e.g. *red and brown hardpan soils* of Western Australia (Stace *et al.* 1968, p. 76), and the *humic podzols* of the coastal lowlands of Queensland (Stace *et al.* 1968, p. 377).

Although not generally expected, free quartz has been observed in many Australian soils derived from basic rocks (CSIRO Div. Soils, unpublished diffraction data) e.g. *krasnozems* on basalt. However, there is no evidence as to whether the quartz originated from the soil parent material or from a vug of a more acid character occurring within the basic rock. There is no definite association between the presence of quartz and other common soil minerals.

A form of opal or, more correctly, random layering of cristobalite and tridymite, often identified as cristobalite, is often found in association with montmorillonite in soils (Norrish, personal communication). Quartz can contain trace amounts of other elements either interstitially or as isomorphous substitutions. In their treatise on silica in soils, Wilding *et al.* (1977) quote Frondel (1962) and Dennen (1966) who report that OH and trace amounts of elements such as Al, Ti, Fe, Na, Li, K, Mg, Ca and H can occur as occlusions in quartz, either interstitially or as isomorphous replacements.

Because of its generally larger particle sizes and its relatively unreactive surfaces, quartz plays a negligible part in the nutritional function of soils, other than its contribution to the soil structure.

Identification of oxides of silicon in soils

Amorphous or opaline silica is not readily detected in association with other soil minerals because, although it is readily removed by treatment with hot alkaline solutions, there is always the uncertainty that the dissolved silica may have arisen from dissolution of other soil components.

Quartz displays many well defined X-ray diffraction peaks and, although many of these occur in positions that can be masked by reflections of other common soil components, the major reflection (101) occurring at 3.34 Å is clearly visible, even when only a few per cent quartz is present. In purer samples, the $\alpha \rightarrow \beta$ inversion, which gives a sharp DTA endothermic peak at 573°C and a similar exothermic peak on cooling, is very specific for quartz.

The X-ray diffraction pattern of cristobalite can vary according to the conditions of its formation, often showing disorder when formed in the presence of foreign ions (Brown 1961). In soils its presence may be confused with orthoclase feldspar because of the near coincidence of their respective X-ray peaks, and positive identification can only be made by comparing relative intensities of lines, and observing what lines of either orthoclase or cristobalite are absent (Wilding *et al.* 1977). In the absence of feldspars the presence of a reflection at around 4.04 Å, where interference from other common soil minerals is not expected, would suggest that cristobalite could be present. Because of the variations that can be obtained in the X-ray diffraction patterns of tridymite, Wilding *et al.* (1977) suggest that this is not a satisfactory technique for the identification of this mineral.

Opal has been regarded previously as amorphous, but some X-ray diffractograms show diffuse broad bands, suggesting a disordered cristobalite, and other variations in the pattern, resulting from different modes of formation (see references cited by Wilding *et al.* 1977). Since heat treatment can cause alteration to more crystalline varieties, there are no inversion peaks on DTA; the same consequences apply after ashing plant matter in an effort to isolate phytoliths.

Some opals isolated from soils may show diffraction patterns indicating more crystalline varieties of silica, the presence of which in some cases can be attributed to diagenesis on ageing (Jones *et al.* 1964).

Titanium oxides

There are six common oxides of Ti found in soils: three are nominally polymorphic forms of TiO_2, viz. rutile, anatase and brookite, and three are combined oxides or titanates, viz. ilmenite ($FeTiO_3$), sphene ($CaTiSiO_5$) and perovskite ($CaTiO_3$). Other mineralogical species e.g. pseudorutile ($FeTi_3O_9$) which is an intermediate product in the weathering of ilmenite (Teufer and Temple 1966), and leucoxene, embracing all fine-grained titania and sphene minerals, will not be discussed.

Titanium minerals have little effect on the nutritional status of a soil; as Hutton (1977) emphasises, their main value is in studies of soil development.

The first six minerals can be present in primary igneous and metamorphic rocks and, because of their resistance to weathering, they persist as residual minerals in the derived soils. Although authigenic alteration of these minerals to other phases does occur, the main source of secondary titania in soils is possibly via the Ti-rich minerals, such as biotites, garnets, amphiboles and augites, in which Ti is generally substituting for Al^{3+} or Fe^{3+} in octahedral positions (Tillmanns and Correns 1969).

These secondary formations and alterations indicate dissolution and reprecipitation, but the form of the soluble species is not identified. Loughnan (1969) believes that $Ti(OH)_4$, which is not precipitated till about pH 5, still persists after Al in solution has been precipitated, and gives rise to one of the polymorphs of TiO_2 on precipitation and subsequent dehydration. Sherman (1952), in a study on Hawaiian soils, stated that Ti moves as a colloidal hydrated oxide and, on dehydration, precipitates as TiO_2. According to Cotton and Wilkinson (1962), no definite hydroxide $Ti(OH)_4$ exists, and initial precipitates are best regarded as hydrous TiO_2. However, they suggest that hydrolyzed species such as $Ti(OH)_3^+$ and $Ti(OH)_2^{2+}$ do exist under certain conditions, and that these mobile soluble forms can be further hydrolyzed.

According to Jackson (1964), anatase and/or rutile are the main Ti minerals in the fine fraction, and concentrations in clays are between 1 and 3%. Similar values for clays have been noted by Loughnan (1969) in the soils around Sydney. In certain soil formations, e.g. lateritic crusts, and under particular combinations of weathering and parent material, e.g. tropical weathering of igneous rocks, the TiO_2 content can be as high as 20–30% (Loughnan 1969; Sherman 1952). Stace *et al.* (1968) gave the average Ti content of 70 Australian soils (surface and subsurface) as 0.6%, whereas in some Queensland soils under tropical conditions the average content of Ti is around 3.4%, again showing the influence of weathering on the Ti content.

Fitzpatrick *et al.* (1978) produced goethite, hematite, pseudorutile, and anatase in ageing precipitates of varying Ti/(Ti + Fe) ratios at pH 5.5 and 70°C. In the molar range $0.7 < Ti/(Ti + Fe) < 1$, anatase and ferriferous anatase form, whereas in the lower concentration range 0.2–0.7, microcrystalline pseudorutile occurs, and in the range 0–0.2, goethite and hematite are formed instead. Their results suggest that 5 mole per cent Fe actually causes increased anatase crystallization, as seen by an increase in the integrated intensity of the anatase X-ray diffraction lines, again demonstrating the influence of foreign ions on the mineral pedogenesis. These workers also observed that amorphous Ti oxides are completely soluble in acid ammonium oxalate extraction carried out in the absence of light, a dissolution treatment designed for the extraction of amorphous iron oxides (Schwertmann 1959*b*). A comparison of the amount of Ti extracted from various soils by this technique, their total Ti content and their oxalate- and dithionite-soluble Fe, suggested that Ti and Fe follow similar crystallization processes in soils, and that crystallization may be somewhat inhibited in cooler climatic conditions.

Rutile is the most common form of TiO_2. It occurs in many igneous and metamorphic rocks, and persists as a residual mineral in the soil during weathering. It is a common alteration product of other Ti minerals, such as ilmenite, titaniferous magnetite, sphene and perovskite, and often contains Fe^{3+} and Fe^{2+} (Deer *et al.* 1962). Ziv (1956) reports that ilmenite can convert to rutile under supergene conditions, so that in some cases its occurrence in soils may be of pedogenic origin.

Anatase, similarly, occurs in igneous and metamorphic rocks, or can be a product of alteration of other Ti minerals. It can form under pedogenic conditions, and then persist as a residual mineral during further soil weathering (Deer *et al.* 1962). However, Loughnan (1969) states that anatase is generally formed from the titania released during the weathering of primary minerals as, for example, from the partial loss of Ca^{2+}, Mg^{2+} and Fe^{2+} during the weathering of pyroxene. Iron and Sn are often minor impurities in anatase (Deer *et al.* 1962).

Brookite is less common than anatase or rutile in soils, and is considered to be of secondary origin, occurring in metamorphic rocks (Hutton 1977).

The factors responsible for the preferential formation of either anatase or rutile from the Ti liberated into solution during pedogenesis are not known. It is suggested that the ionic environment could influence the final crystalline form, as in the case of the iron oxides, but this is a matter for future research.

Perovskite occurs as an accessory mineral in basic igneous rocks and in some thermally metamorphosed calcareous rocks in contact with basic igneous rocks. Rare earths and alkalis can substitute for Ca (Deer *et al.* 1962).

Ilmenite is isostructural with hematite, and occurs in soils as a detrital mineral from the weathering of igneous and metamorphic rocks, but the experiments of Fitzpatrick *et al.* (1978) suggest that it could form by co-precipitation of Fe and Ti hydroxides. In a Ti-poor environment titaniferous hematite is the more likely product.

Sphene occurs generally as a residual from the weathering of igneous and metamorphic rocks, although Stephen (1952) observed that in the weathering of biotite-rich primary rocks, the early formation of chlorite is accompanied by the formation of sphene from segregations of calcium carbonate, titania and silica.

Identification of titanium oxides in soils

At the generally low concentrations of Ti minerals found in clay fractions (1–3%), and the still lower amounts in whole soils, they are not readily detected by X-ray diffraction techniques, without prior concentration by dissolution of the silicate minerals by cold HF and HCl (Jackson 1964*a*).

J. G. Pickering (personal communication) has successfully concentrated these minerals by a dissolution in 24% HF for 10 minutes; the reaction was terminated by dilution with water. When present in sufficiently high amounts after the concentration procedure, there is little ambiguity between the diffraction patterns of the various minerals, although some interference arises from residual coarse-grained iron oxides and quartz (Hutton 1977).

Titanium causes a high fluorescence background in X-ray diffraction with radiations normally used in soil studies. When large amounts of Ti are present, a monochromator should be used to increase resolution.

There is no published work on possible amorphous phases in soils; if present, they would possibly be at a still lower level of concentration.

References

Adams, S. N., Honeysett, J. L., Tiller, K. G. and Norrish, K. (196). Factors controlling the increase of cobalt in plants following the addition of cobalt fertilizer. *Aust. J. Soil Res.* **7**, 29-42.

Aleksandrova, L. N. (1960). The use of sodium pyrophosphate for isolating free humic substances and their organic mineral compounds from soil. *Pochvovedenie* **2**, 190-97.

Aristovskaya, T. V. (1974). Role of microorganisms in iron mobilisation and stabilisation of soils. *Geoderma* **12**, 145-50.

Babanin, V. F., Veronin, A. D., Karpachevskii, L. O., Manucharov, A. S., Opalenko, A. A. and Pochatkova, T. N. (1975). Some modes of transformation of Fe compounds in soils. *Pochvovedenie* **2**, 35-40.

Bernal, J. O., Dasgupta, D. R. and Mackay, A. L. (1959). The oxides and hydroxides of iron and their structural inter-relationships. *Clay Min. Bull.* **4**, 15-30.

Berner, R. A. (1969). Goethite stability and the origin of red beds. *Geochim. Cosmochim. Acta* **33**, 267-73.

Bonifas, M. (1959). Contributions a l'étude géochimique de l'altération latérique. *Mem. Serv. Carte Geol. Alsace Lorraine* **17**, 159.

Breeuwsma, A. and Lyklema, J. (1973). Physical and chemical adsorption of ions in the electrical double layer on hematite (α Fe_2O_3). *J. Colloid Interface Sci.* **43**, 437-48.

Bricker, O. P. (1965). Some stability relations in the system $Mn-O_2$ $2-H_2O$ and one atmosphere total pressure. *Am. Mineral.* **50**, 1296-354.

Bromfield, S. M. (1958). The properties of a biologically formed manganoxide, its availability to oats and its solution by root washings. *Plant Soil* **9**, 325-37.

Brown, G. (1961). Other minerals. *In* 'The X-ray Identification and Crystal Structures of Clay Minerals'. (Ed. G. Brown.) 2nd ed. (Mineralogical Society: London.)

Burns, R. G., Burns, V. M., Sung, W. and Brown, B. A. (1974). Ferromanganese nodule mineralogy: suggested terminology of the principal manganese oxide phases. Geol. Soc. Am. Ann. Meet. Miami, Abstr. Vol. 6, pp. 1029-31.

Burns, R. G. and Burns,V. M. (1975*a*). Mechanism for nucleation and growth of manganese nodules. *Nature* (London) **255**, 130-31.

Burns, R. G. and Burns, V. M. (1975*b*). Structural relationships between the manganese (IV) oxides. *In* 'Proc. Intern. Symp. Manganese Dioxide, Cleveland 1975'. (Eds. A. Kozawa and R. J. Brodd.) pp. 306-27. (Electrochem. Soc. Publication.)

Buser, W., Graf, P. and Feitknecht, W. (1954). Beitrag zur Kenntnis der Mangan (II) - manganite und des δ-MnO_2. *Helv. Chim. Acta* 37, 2322-33.

Býstrom, A. and Býstrom, A. M. (1950). The crystal structure of hollandite, the related manganese oxide minerals, and α-MnO_2. *Acta Crystallogr.* **3**, 146-54.

Chao, T. T., Harward, M. E. and Fang, S. C. (1964). Iron or aluminium coatings in relation to sulphate adsorption characteristics of a soil. *Soil Sci. Soc. Am. Proc.* **28**, 632-35.

Childs, C. W. (1975). Composition of iron-manganese concretions from some New Zealand soils. *Geoderma* **13**, 141-52.

Chuckhrov, F. V., Zvyagin, B. B., Ermilova, L. P. and Gorshkov, A. I. (1973). New data on iron oxides in the weathering zone. Proc. Internat. Clay Conf. 1972 (Madrid), Vol. 1, pp. 397-404.

Claus, D., Wittmann, A. and Rippel-Baldes, A. (1958). Untersuchungen über Zusammensetzung von Bakterienschleimen und deren Lösungsvermögen gegenüber schwerlöslichen anorganischen Verbindungen. *Arch. Mikrobiol.* **29**, 169-78.

Cole, W. F., Wadsley, A. D. and Walkley, A. (1947). An X-ray diffraction study of manganese dioxide. *Trans. Electrochem. Soc.* **92**, 133-58.

Cotton, F. A. and Wilkinson, G. (1962). *In* 'Advanced Inorganic Chemistry' (Interscience Publishers: London.)

Dachs, H. (1963). Neutron-und Röntgenuntersuchungen am Manganit. *Z. Kristallogr.* 118, **303**-26.

Davey, B. G., Russell, J. D. and Wilson, M. J. (1975). Iron oxide and clay minerals and their relation to colours of red and yellow podzolic soils near Sydney, Australia. *Geoderma* **14**, 125-38.

Day, F. H. (1963). 'The Chemical Elements in Nature'. (Harrap: London.)

Deb, B. C. (1950). Estimation of free iron oxides in soils and clays and their removal. *J. Soil Sci.* **1**, 212-30.

Deer, W. A., Howie, R. A. and Zussman, J (1962). Rock Forming Minerals. Vol. 5. Non Silicates. (Longmans: London.)

Dennen, W. A. (1966). Stoichiometric substitution in natural quartz. *Geochim. Cosmochim. Acta.* **30**, 1235-41.

Dent Glasser, L. S. and Ingram, L. (1968). Refinement of the crystal structure of groutite α-MnOOH. *Acta Crystalallogr. Sect. B.* **24**, 1233-36.

Douglas, L. A. and Platt, D. W. (1977). Surface morphology of quartz and age of soils. *Soil Sci. Soc. Am. J.* **41**, 641-45.

Drosdoff, M. and Nikiforoff, C. C. (1940). Iron-manganese concretions in Dayton soils. *Soil Sci.* **49**, 333-45.

El Swaify, S. A. and Sayegh, A. H. (1975). Charge characteristics of an oxisol and an inceptisol from Hawaii. *Soil Sci.* **120**, 49-56.

El Swaify, S. A. and Emerson, W. W. (1975). Changes in the physical properties of soil clays due to precipitated aluminium and iron hydroxides: I. Swelling and aggregate stability after drying. *Soil Sci. Soc. Am. Proc.* **39**, 1056-63.

Eswaran, H. and Raghu Mohan, N. G. (1973). The microfabric of petroplinthite. *Soil Sci. Soc. Am. Proc.* **37**, 79-82.

Evans, L. T. and Russell, E. W. (1959). The adsorption of humic and fulvic acids by clays. *J. Soil Sci.* **10**, 119-32.

Fasiska, E. J. (1967). Structural aspects of the oxides and oxyhydrates of iron. *Corros. Sci.* **7**, 833-39.

Feitknecht, W. (1959). Uber die Oxydation von festen Hydroxyverbindungen des Eisens in wassrigen Lösungen. *Z. Elektrochem.* **63**, 34-43.

Feitknecht, W. (1965). Oxidation of Fe_3O_4 and Fe. *Colloq. Int. CNRS.* **122**, 121-26.

Feitknecht, W., Giovanoli, R., Michaelis, W. and Müller, M. (1973). Uber die Hydrolyse von Eisen (III) Salzlösungen. I. Die Hydrolyse der Lösungen von Eisen (III) Chlorid. *Helv. Chim. Acta* **56**, 2847-56.

Feitknecht, W. and Keller, G. (1950). Uber die dunkelgrünen Hydroxyverbindungen des Eisens. *Z. Anorg. Allg. Chem.* **262**, 61-8.

F itknecht, W. and Michaelis, W. (1962). Uber die Hydrolyse von Eisen (III) — perchlorat Lösungen. *Helv. Chim. Acta* **45**, 212-24.

Feitknecht, W., Oswald, H. R. and Feitknecht-Steimann, V. (1960). Uber die topochemische einphasige Reduktion von γ-MnO_2. *Helv. Chim. Acta* **48**, 1947-50.

Fischer, W. R. (1973). Die Wirkung von zweiwertigem Eisen auf Lösung und Umwandlung von Eisen (III) — hydroxiden. 'Pseudogley and Gley'. Int. Soc. Soil Sci. Comm. V and VI Trans. Stuttgart, 1971. pp. 37-44. (Verlag Chemie: Weinheim.)

Fischer, W. R. and Ottow, J. C. G. (1972). Abbau von Eisen (III) citrat in durchlüfteter wässriger Lösung durch Bodenbakterien. *Z. Pfanzenernaehr. Bodenkd.* **131**, 243-53.

Fischer, W. R. and Schwertmann, U. (1975). The formation of hematite from amorphous iron (III) — hydroxide. *Clays Clay Miner.* **23**, 33-7.

Fitzpatrick, R. W., Le Roux, J. and Schwertmann, U. (1978). Amorphous and crystalline titanium and iron-titanium oxides in synthetic preparations, at near ambient conditions and in soil clays. *Clays Clay Miner.* **3**, 189-201.

Follett, E. A. C. (1965). The retention of amorphous colloidal ferric hydroxide by kaolinites. *J. Soil Sci.* **16**, 334-41.

Follett, E. A. C., McHardy, W. J., Mitchell, B. D. and Smith, B. F. L. (1965). Chemical dissolution techniques in the study of soil clays: Part II. *Clay Miner.* **6**, 35-43.

Forbes, E. A., Posner, A. M. and Quirk, J. P. (1976). The specific adsorption of divalent Cd, Co, Cu, Pb and Zn on goethite. *J. Soil Sci.* **17**, 154-66.

Fordham, A. W. (1970). Sorption and precipitation of iron on kaolinite. III. The solubility of iron (III) hydroxides precipitated in the presence of kaolinite. *Aust. J. Soil Res.* **8**, 107-22.

Fordham, A. W. (1973). The location of iron-55, strontium-85 and iodide-125 sorbed by kaolinite and dickite particles. *Clays Clay Miner.* **21**, 175-84.

Fordham, A. W. and Norrish, K. (1974). Direct measurement of the composition of soil components which retain added arsenate. *Aust. J. Soil Res.* **12**, 165-72.

Fripiat, J. J. (1972). Time concepts in surface chemistry. Proc. Int. Clay Conf. Madrid, pp. 527-47.

Fripiat, J. J. and Gastuche, M. C. (1952). Physico-chemical study of clay surfaces. Nat. Inst. Belg. Congo Agronomical Studies. Sci. Ser. No. 54.

Frondel, C. (1962). Dana's System of Mineralogy. *Vol. III, Silica Minerals.* (John Wiley: New York.)

Gager, H. M. (1968). Mössbauer spectra of deep-sea iron-manganese nodules. *Nature* (London) **220**, 1021-23.

Gallaher, R. N., Perkins, H. F., Tan, K. H. and Radcliffe, D. (1973). Soil concretions. II. Mineralogical analysis. *Soil. Sci. Soc. Am. Proc.* **37**, 469-72.

Gallez, A., Juo, A. S. R. and Herbillon, A. J. (1976). Surface and charge characteristics of selected soils in the tropics. *Soil Sci. Soc. Am. J.* **40**, 601-8.

Gangas, N. H., Simopoulos, A., Kostikas, A., Yassoglou, N. J. and Filippakis, S. (1973). Mössbauer studies of small particles of iron oxides in soil. *Clays Clay Miner.* **21**, 151-60.

Ganssen, R. and Hädrich, F. (1965). 'Atlas zur Bodenkunde'. (Hochschulatlanten Bibliog. Inst., 85: Mannheim.)

Gardner, L. R. (1970). A chemical model for the origin of gibbsite from kaolinite. *Am. Miner.* **55**, 1380-89.

Gastuche, M. C., Bruggenwert, T. and Mortland, M. M. (1964). Crystallization of mixed iron and aluminium gels. *Soil Sci.* **98**, 281-89.

Gattow, G., and Glemser, O. (1961). Darstellung und Eigenschaften von Braunsteinen. II. Die γ-und η-Gruppe der Braunsteine. *Z. Anorg. Allg. Chem.* **309**, 20-36.

Giessen, A. A. van der (1966). The structure of iron (III) oxide-hydrate gels. *J. Inorg. Nucl. Chem.* **28**, 2155-59.

Giessen, A. A. van der, Rensen, J. G. and Wieringen, J. S. van (1968). A study of the constitution and freezing behaviour of iron oxide-hydrate gels by means of the Mössbauer effect. *J. Inorg. Nucl. Chem.* **30**, 1739-44.

Gillman, G. P. and Bell, L. C. (1976). Surface charge characteristics of six weathered soils from tropical north Queensland. *Aust. J. Soil Res.* **14**, 351-60.

Gillman, G. P. and Bell, L. C. (1978). Soil solution studies on weathered soils from tropical north Queensland. *Aust. J. Soil Res.* **16**, 67-77.

Giovanoli, R. (1969). A simplified scheme for polymorphism in the manganese oxides. *Chimia* **23**, 470-72.

Giovanoli, R., Buhler, H. and Sokolowaka, K. (1973). Synthetic lithiophorite: electron micrography and X-ray diffraction. *J. Mikrosc.* (Paris) **18**, 271-84.

Glemser, O., Gattow, G. and Meisiek, H. (1961). Darstellung and Eigenschaften von Braunsteinen. I. Die δ-Gruppe der Braunsteine. *Z. Anorg. Allg. Chem.* 309, 1-19.

Greenland, D. J. (1971). Interactions between humic and fulvic acids and clays. *Soil Sci.* **111**, 34-40.

Greenland, D. J. (1975). Charge characteristics of some kaolinite-iron hydroxide complexes. *Clay Miner.* **10**, 407-16.

Greenland, D. J. and Oades, J. M. (1968). Iron hydroxides and clay surfaces. Trans. 9th Int. Congr. Soil Sci. (Adelaide), Vol. 1, pp. 657-68.

Greenland, D. J., Oades, J. M. and Sherwin, T. W. (1968). Electron-microscope observations of iron oxides in some red soils. *J. Soil Sci.* **19**, 123-26.

Greenland, D. J. and Wilkinson, G. K. (1969). Use of electron microscopy of carbon replicas and selective dissolution analysis in the study of the surface morphology of clay particles from soils. Proc. Int. Clay Conf. (Tokyo), Vol. 1, pp. 861-70.

Griffith, S. M. and Schnitzer, M. (1975). The isolation and characterisation of stable metal-organic complexes from tropical volcanic soils. *Soil Sci.* **120**, 126-31.

Grimme, H. (1968). Die Adsorption von Mn, Co, Cu and Zn durch Goethit aus verdünnten Lösungen. *Z. Pflanzenernaehr. Bodenkd.* **121**, 58-65.

Harder, H. and Flehmig, W. (1970). Quarzsynthese bei tiefen Temperaturen. *Geochim. Cosmochim. Acta.* **34**, 295-305.

Harris, R. F., Chesters, G. and Allen, O. A. (1966). Dynamics of soil aggregation. *Adv. Agron.* **18**, 107-69.

Harriya, Y. and Kikuchi, T. (1964). Precipitation of manganese by bacteria in mineral springs. *Nature* (London) **202**, 416-17.

Haseman, J. F., Brown, E. H. and Whitt, C. D. (1950). Some reactions of phosphate with clays and hydrous oxides of iron and aluminium. *Soil Sci.* 70, 257-71.

Hazel, F., Schock, R. V. and Gordon, M. (1949). Interactions of ferric ions with silicic acid. *J. Am. Chem. Soc.* **71**, 2256-57.

Healy, T. W., Herring, A. P. and Fuerstenau, D. W. (1966). The effect of crystal structure on the surface properties of a series of manganese dioxides. *J. Colloid Interface Sci.* **21**, 435-44.

Hem, J. D. (1964). Deposition and solution of manganese oxides. U.S. Geol. Survey Water Supply Paper No. 1667B.

Herbillon, A., and Tran Vinh An (1969). Heterogeneity in silicon-iron mixed hydroxides. *J. Soil Sci.* **20**, 223-35.

Hingston, F. J., Posner, A. M. and Quirk, J. P. (1970). Anion binding at oxide surfaces – the adsorption envelope. *Search* **1**, 324-27.

Hingston, F. J., Atkinson, R. J., Posner, A. M. and Quirk, J. P. (1967). Specific adsorption of anions. *Nature* (London) **215**, 1459-61.

Hingston, F. J., Posner, A. M. and Quirk, J. P. (1972). Anion adsorption by goethite and gibbsite. I. The role of the proton in determining adsorption envelopes. *J. Soil Sci.* **23**, 177-92.

Holford, I. C. R. and Mattingly, G. E. G. (1975). The high- and low-energy phosphate adsorbing surfaces in calcareous soils. *J. Soil Sci.* **26**, 407-17.

Hsu, Pa Ho. (1977). Aluminium hydroxides and oxyhydroxides. *In* 'Minerals in Soil Environments'. pp. 99-143. (*Soil Sci. Soc. Am.)*

Hutton, J. T. (1977). Titanium and zirconium minerals. *In* 'Minerals in Soil Environments'. (Soil Sci. Soc. Am.)

Jackson, M. L. (1964*a*). Soil clay mineralogical analysis. *In* 'Soil Clay Mineralogy'. (Univ. North Carolina Press: Chapel Hill.)

Jackson, M. L. (1964*b*). Free oxides, hydroxides, and amorphous aluminosilicates. *In* 'Methods of Soil Analysis'. Agron. Monograph 9, p. 257, (Am. Soc. Agron.)

Jackson, M. L. and Sayin, M. (1975). The occurrence of quartz in the fine clay fraction of soils. Agronomy Abstracts. p. 178. 1975 Annu. Meet. Am. Soc. Agron., Crop Sci. Soc. Am., Soil Sci. Soc. Am.

Jenne, E. A. (1968). Controls on Mn, Fe, Co, Ni, Cu and Zn concentrations in soils and water: the significant role of hydrous Mn and Fe oxides. *In* 'Trace Inorganics in Water'. Adv. Chem. Series. Vol. 73, pp. 337-87.

Jónás, K. and Solymar, K. (1970). Preparation, X-ray, derivographic and infrared study of aluminium-substituted goethites. *Acta Chim. Acad. Sci. Hungar.* **66**, 383-94.

Jones, J. B., Sanders, J. V. and Segnit, E. R. (1964). Structure of opal. *Nature* (London) **198**, 119.

Jones, L. H. P., and Milne, A. A. (1956). Birnessite, a new manganese oxide mineral from Aberdeenshire, Scotland. *Miner. Mag.* **31**, 283-88.

Jones, L. H. P. and Handreck, K. A. (1965). Studies of silica in the oat plant. III. Uptake of silica from soils by the plant. *Plant Soil* **23**, 79-96.

Jones, L. H. P. and Handreck, K. A. (1967). Silica in soils, plants and animals. *Adv. Agron.* **19**, 107-49.

Jones, L. H. P. and Smith, B. H. (1972). The effect of sesquioxides on the availability of molybdenum to *Trifolium subterraneum*. *Plant Soil* **37**, 649-55.

Karpachevskii, L. O., Babanin, V. F., Gendler, T. S., Opalenko, A. A. and Kuz'min, R. N. (1972). Identification of ferruginous soil minerals by Mössbauer spectroscopy. *Sov. Soil Sci.* **4**, 639.

Karpachevskii, L. O. and Babanin, V. F. (1974). Forms of iron compounds in soils and methods of studying them. *Pochvovedenie* **29**, 54-63.

Keng, J. C. W. and Uehara, G. (1974). Chemistry, mineralogy, and taxonomy of oxisols and ultisols. *Proc. Soil Crop Sci. Soc. Fla.* **33**, 119-26.

Kittrick, J. A. (1969). Soil minerals in the Al_2O_3–SiO_2–H_2O system and a theory of their formation. *Clays Clay Miner.* **17**, 157-67.

Knight, R. J. and Sylva, R. N. (1974). Precipitation in hydrolysed iron (III) solutions. *J. Inorg. Nucl. Chem.* **36**, 591-97.

Kodama, H., McKeague, J. A., Tremblay, R. J., Gosselin, J. R. and Townsend, M. G. (1977). Characterisation of iron oxide compounds in soils by Mössbauer and other methods. *Can. J. Earth Sci.* **14**, 1-15.

Koljonen, T., Lahermo, P. and Carlson, L. (1976). Origin, mineralogy and chemistry of manganiferous and ferruginous precipitates found in sand and gravel deposits in Finland. *Bull. Geol. Soc. Finl.* **48**, 111-35.

Krause, A. and Borkowska, A. (1963). Der Einfluss von PO_4^{3-} Ionen auf die Oxydation des $Fe(OH)_2$ mit Luftsauerstoff. *Mh. Chem.* **94**, 460-65.

Krause, A., Ignasiak, J. and Kostrzewa, E. (1970). Uber die Bildung von röntgenamorphem Eisen (III) – hydroxid durch Luftoxydation von Eisen (II) – hydroxide und Eisen (II) – carbonat. *Z. Anorg. Allg. Chem.* **378**, 210-12.

Krishna Murti, G. S. R., Sarma, V. A. K. and Rengasamy, P. (1976). Amorphous ferri-aluminosilicates in some tropical ferruginous soils. *Clay Miner.* **11**, 137-45.

Krishna Murti, G. S. R., Singh, G. and Rengasamy, P. (1977). The nature of soil clays and the stability of microaggregates. *Aust. J. Soil Res.* **15**, 115-19.

Kroth, E. M. and Page, J. B. (1947). Aggregation formation in soils with special reference to cementing substances. *Soil Sci. Soc. Am. Proc.* **11**, 27-34.

Levashkevich, G. A. (1966). Interaction of humic acids with iron and aluminium hydroxides. *Sov. Soil Sci.* 422-27.

Liebau, F., Condie, K. C. and Siever, R. (1969). Silicon. *In* 'Handbook of Geochemistry'. Vol. II-1. (Ed. K. H. Wederpohl). (Springer-Verlag: Berlin.)

Loganathan, P. and Burau, R. G. (1973). Sorption of heavy metals by a hydrous manganese oxide. *Geochim. Cosochim. Acta.* **37**, 1277-93.

Loughnan, F. C. (1969). 'Chemical Weathering of the Silicate Minerals'. (Elsevier: New York.)

McHardy, W. J., Thompson, A. P. and Goodman, B. A. (1974). Formation of iron oxides by decomposition of iron-phenolic chelates. *J. Soil Sci.* **25**, 471-82.

McKeague, J. A., Brydon, J. E. and Miles, N. M. (1971). Differentiation of forms of extractable iron and aluminium in soils. *Soil Sci. Soc. Am. Proc.* **35**, 33-8.

McKeague, J. A., Damman, A. W. H. and Heringa, P. K. (1968). Iron-manganese and other pans in some soils of Newfoundland. *Can. J. Soil Sci.* **48**, 243-53.

McKenzie, R. M. (1970). The reaction of cobalt with manganese dioxide minerals. *Aust. J. Soil Res.* **8**, 97-106.

McKenzie, R. M. (1971). The synthesis of cryptomelane and some other oxides and hydroxides of manganese. *Miner. Mag.* **38**, 493-502.

McKenzie, R. M. (1975). An electron microprobe study of the relationships between heavy metals and manganese and iron in soils and ocean floor nodules. *Aust. J. Soil Res.* **13**, 177-88.

McKenzie, R. M. (1978). The effect of two manganese dioxides on the uptake of lead, cobalt, nickel, copper and zinc by subterranean clover. *Aust. J. Soil Res.* **16**, 209-14.

McKenzie, R. M. (1980). The manganese oxides in soils. In 'Geology and Geo-chemistry of Manganese'. Vol I, pp 259-269. (Eds I. M. Varentsov and Gy. Grasselly). (Hungarian Academy of Sciences: Budapest.)

McLaughlin, J. R., Ryden, J. C. and Syers, J. K. (1977*a*). Development and evaluation of a kinetic model to describe phosphate sorption by hydrous ferric oxide gel. *Geoderma* **18**, 295-307.

McLaughlin, J. R., Ryden, J. C. and Syers, J. K. (1977*b*). Mechanisms of phosphate sorption by soils and hydrous ferric oxide gel. *J. Soil Sci.* **28**, 72-92.

van der Marel, H. W. (1951). Gamma ferric oxide in sediments. *J. Sedimentol. Petrol.* **21**, 12-21.

Marshall, P. R. and Rutherford, D. (1971). Physical investigations on colloidal iron-dextran complexes. *J. Colloid Interface Sci.* **37**, 390-402.

Maxwell, K. H. and Thirsk, H. R. (1954). 'Electromotive Force and Structural Relations in the MnO_2 system'. Int. Comm. Electrochem. Thermodyn. Kinetics. Proc. 6th Meet., Poitiers, 1954. (Butterworth: New York.)

Mehra, O. P. and Jackson, M. L. (1960). Iron oxide removal from soils and clays by dithionite-citrate system buffered with sodium bicarbonate. *Clays Clay Miner.* **7**, 317-27.

Misawa, T., Hashimoto, K. and Shimodaira, S. (1973). Formation of Fe (II), Fe (III) intermediate green complex on oxidation of ferrous iron in natural and slightly alkaline sulphate solutions. *J. Inorg. Nucl. Chem.* **35**, 4167-74.

Mohr, E. C. J., van Baren, F. A. and van Schuylenborgh, J. (1972). *'Tropical Soils'.* 3rd Edn. (Monton-Ichtior Baru-Van Hoeve: The Hague.)

Morais, F. I., Page, A. L. and Lund, L. J. (1976). The effect of pH, salt concentration, and nature of electrolytes on the charge characteristics of Brazilian tropical soils. *Soil Sci. Soc. Am. J.* **40**, 521-7.

Morgan, J. J. and Stumm, W. (1964*a*). Colloid chemical properties of manganese dioxide. *J. Colloid Sci.* **19**, 347-59.

Morgan, J. J. and Stumm, W. (1964*b*). The role of multivalent metal oxides in limnological transformations as exemplified by iron and manganese. 2nd Int. Conf. Water Pollution Res., Tokyo. pp. 103-18.

Muljadi, D., Posner, A. M. and Quirk, J. P. (1966). The mechanism of phosphate adsorption by kaolinite, gibbsite and pseudoboehmite. II. The location of the adsorption sites. *J. Soil Sci.* **17**, 230-37.

Murphy, P. J., Posner, A. M. and Quirk, J. P. (1975). Chemistry of iron in soils. Ferric hydrolysis products. *Aust. J. Soil Res.* **13**, 189-201.

Murphy, P. J., Posner, A. M. and Quirk, J. P. (1976). Characterization of hydrolysed ferric ion solutions. A comparison of the effects of various anions on the solutions. *J. Colloid Interface Sci.* **56**, 312-19.

Nalovic, Lj., Pedro, G and Janot, C. (1975). Demonstration by Mössbauer spectroscopy of the role played by transitional trace elements in the crystallogenesis of iron hydroxides (III). Proc. Int. Clay Conf., Mexico City. pp. 601-10.

Norrish, K. (1975). Geochemistry and mineralogy of trace elements. *In* 'Trace Elements in Soil-Plant-Animal systems'. (Eds D. J. D. Nicholas and A. R. Egan). pp. 55-81. (Academic Press: New York.)

Norrish, K. and Pickering, J. G. (1977). Clay mineralogic properties. *In* 'Soil Factors in Crop Production in a Semi-Arid Environment'. (Eds J. S. Russell and E. L. Greacen). (University of Queensland Press: Brisbane.)

Norrish, K. and Taylor, R. M. (1961). The isomorphous replacement of iron by aluminium in soil goethites. *J. Soil Sci.* **12**, 294-306.

Norrish, K. and Taylor, R. M. (1962). Quantitative analysis by X-ray diffraction. *Clay Miner. Bull.* **5**, 98-109.

Oades, J. M. and Townsend, W. N. (1963). The detection of ferromagnetic minerals in soils and clays. *J. Soil Sci.* **14**, 179-87.

Parfitt, R. L., Atkinson, R. J. and Smart, R. St. C. (1975). The mechanism of phosphate fixation by iron oxides. *Soil Sci. Soc. Am. Proc.* **39**, 837-41.

Parfitt, R. L., Frazer, A. R. and Farmer, V. C. (1977*a*). Adsorption on hydrous oxides. III. Fulvic acid and humic acid on goethite, gibbsite, and imogolite. *J. Soil Sci.* **28**, 289-96.

Parfitt, R. L., Frazer, A. R., Russel, J. D. and Farmer, V. C. (1977*b*). Adsorption on hydrous oxides. II. Oxalate, benzoate and phosphate on gibbsite. *J. Soil Sci.* **28**, 40-47.

Pauli, F. W. (1968). Some recent developments in biogeochemical research. *Geol. Assoc. Canada Proc.* **19**, 45-49.

Pawluk, S. (1972). Measurement of crystalline and amorphous iron oxides. *Can. J. Soil Sci.* **52**, 119-23.

Perinet, G. and Lafont, R. (1972). Sur la présence d'hématite alumineuse désordonée dans les bauxites du Var. *C. R. Acad. Sci.* **274**, 272-74.

Van Raij, B. and Peech, M. (1972). Electrochemical properties of some oxisols and ultisols of the tropics. *Soil Sci. Soc. Am. Proc.* **36**, 587-93.

Rengasamy, P. and Oades, J. M. (1977). Interaction of monomeric and polymeric species of metal ions with clay surfaces. I. Adsorption of iron (III) species. *Aust. J. Soil Res.* **15**, 221-33.

Rooksby, H. P. (1961). Oxides and hydroxides of aluminium and iron. *In* 'The X-ray Identification and Crystal Structures of Clay Minerals'. (Ed. G. Brown). 2nd edn. (Mineralogical Society: London.)

Ross, S. J., Franzmeier, D. P. and Roth, C. B. (1976). Mineralogy and chemistry of manganese oxides in some Indianna soils. *Soil Sci. Soc. Am. J.* **40**, 137-43.

Saunders, W. M. H. (1959). Aluminium extracted by neutral citrate-dithionite reagent. *Nature* (London) **184**, 2037.

Scheffer, F. Welte, E. and Ludwieg, F. (1957). Zur Frage der Eisenoxidhydrate im Boden. *Chem. Erde.* **19**, 51-64.

Schellmann, W. (1959). Experimentelle Untersuchungen über die sedimentäre Bildung von Goethit und Hämatit. *Chem. Erde* **20**, 104-35.

Schnitzer, M. and Skinner, S. I. M. (1964). Organo-metallic interactions in soils. 3. Properties of iron – and aluminium-organic – matter complexes, prepared in the laboratory and extracted from a soil. *Soil Sci.* **98**, 197-203.

Schoen, R. and Robertson, C. E. (1970). Structures of aluminium hydroxide and geochemical implications. *Am. Miner.* **55**, 43-77.

Schweisfurth, R. and Gattow, G. (1966). Untersuchungen über Röntgen-Struktur und Zusammensetzung mikrobiell gebildeter Braunsteine. *Z. Allg. Mikrobiol.* **6**, 303-8.

Schwertmann, U. (1959*a*). Über die Synthese definierter Eisenoxyde unter verschiedenen Bedingungen. *Z. Anorg. Allg. Chem.* **298**, 337-48.

Schwertmann, U. (1959*b*). Die fraktionerte Extraktion der freien Eisenoxyde unter verschiedenen Bedingungen. *Z. Pflanzenernaehr. Dueng. Bodenkd.* **84**, 194-204.

Schwertmann, U. (1964). Differenzierung der Eisenoxide des Bodens durch photo-chemische Extraction mit saurer Ammoniumoxalat-Lösung. *Z. Pflanzenernaehr. Dueng. Bodenkd.* 105, 194-202.

Schwertmann, U. (1966). Inhibitory effect of soil organic matter on the crystallization of amorphous ferric hydroxide. *Nature* (London) **212**, 645-46.

Schwertmann, U. (1970). Der Einfluss organischer Anionen auf die Bildung von Goethit und Hämatit aus amorphem Fe (III)-hydroxid. *Geoderma* **3**, 207-14.

Schwertmann, U. (1971). Transformation of hematite to goethite in soils. *Nature* (London) **232**, 624-25.

Schwertmann, U. (1973). Electron micrographs of soil lepidocrocites. *Clay Minera.* **10**, 59-60.

Schwertmann, U. and Fischer, W. R. (1966). Zur Bildung von α-Fe_2O_3 aus amorphem Eisen (III)-hydroxid. *Z. Anorg. Allg. Chem.* **346**, 137-42.

Schwertmann, U. and Fischer, W. R. (1973). Natural 'amorphous' ferric hydroxide. *Geoderma* **10**, 237-47.

Schwertmann, U., Fischer, W. R. and Papendorf, H. (1968). The influence of organic compounds in the formation of iron oxides. Trans. 9th Int. Congr. Soil Sci. (Adelaide), Vol. 1, 645-55.

Schwertmann, U. and Fitzpatrick, R. W. (1977). Occurrence of lepidocrocite and its association with goethite in Natal soils. *Soil Sci. Soc. Am. J.* **41**, 1013-18.

Schwertmann, U., Fitzpatrick. W. and Le Roux J. (1977). Al substitution and differential disorder in soil hematites. *Clays Clay Miner.* **25**, 377-403.

Schwertmann, U. and Heinemann, B. (1959). Uber das Vorkommen und die Enstehung von Maghemit in nordwestdeutschen Böden. *Neues Jahrb. Miner. Monatsh.* 174-81.

Schwertmann, U., Fitzpatrick, R. W., Taylor, R. M. and Lewis, D. G. (1979). The influence of aluminium on iron oxides. Part II. Preparation and properties of Al substituted hematites. *Clays Clay Miner.* **27**, 105-12.

Schwertmann, U. and Taylor, R. M. (1972*a*). The influence of silicate on the transformation of lepidocrocite to goethite. *Clays Clay Miner.* **20**, 159-64.

Schwertmann, U. and Taylor, R. M. (1972*b*). The transformation of lepidocrocite to goethite. *Clays Clay Miner.* **20**, 151-58.

Schwertmann, U. and Taylor, R. M. (1977). Iron oxides, *In* 'Minerals in Soil Environments'. pp. 145-80. (Soil Sci. Soc. Am.: Madison.)

Schwertmann, U. and Taylor, R. M. (1979). Natural and synthetic poorly crystalline lepidocrocites. *Clay Miner.* **14**, 285-293.

Schwertmann, U. and Thalmann, H. (1976). The influence of [Fe(II)], [Si] and pH on the formation of lepidocrocite and ferrihydrite during oxidation of aqueous $FeCl_2$ solutions. *Clay Miner.* **11**, 189-200.

Segalen, P. (1971). Metallic oxides and hydroxides in soils of the warm and humid areas of the world: formation, identification evolution. *In* Natural Resources Research XI. Soils and Tropical Weathering. 'Proc. Bandung Symposium'. pp. 15-24. (UNESCO: Paris.)

Sherman, G. D. (1952). The titanium content of Hawaiian soils and its significance. *Soil Sci. Soc. Am. Proc.* **16**, 15-18.

Sivarajasingham, S., Alexander, L. T., Cady, J. G., and Cline, M. G. (1962). Laterite. *Adv. Agron.* **14**, 1-60.

Smith, A. N. (1965). Aluminium and iron phosphates in soils. *J. Aust. Inst. Agric. Sci.* **31**, 110-26.

Spiro, T. G. and Saltman, P. (1969). Polynuclear complexes of iron and their biological implication. *Struct. Bonding (Berlin)* **6**, 116-56.

Stace, H. C. T., Hubble, G. D., Brewer, R., Northcote, K. H., Sleeman, J. R., Mulcahy, M. J. and Hallsworth, E. G. (1968). 'A Handbook of Australian Soils'. (Rellim Tech. Publications: Adelaide.)

Stauffer, T. E. and Lovell, H. L. (1968). The oxygenation of Fe (II) – relationship to coal mine drainage treatment. Coal Res. Board, Comm. of Pennsylvania, Special Research Dept. SR-69.

Stephen, I. (1952). A study of rock weathering with reference to the soils of the Malvern Hills. *J. Soil Sci.* **3**, 20-33.

Sylva, R. N. (1972). The hydrolysis of iron (III). *Rev. Pure Appl. Chem.* **22**, 115-31.

Taylor, R. M. (1968). The association of manganese and cobalt in soils – further observations. *J. Soil Sci.* **19**, 77-80.

Taylor, R. M. and Giles, J. B. (1970). The association of vanadium and molybdenum with iron oxides in soils. *J. Soil Sci.* **21**, 203-15.

Taylor, R. M., Graley, A. M. (1967). The influence of ionic environment on the nature of iron oxides in soils. *J. Soil Sci.* **18**, 341-48.

Taylor, R. M. and McKenzie, R. M. (1966). The association of trace elements with manganese minerals in Australian soils. *Aust. J. Soil Res.* **4**, 29-39.

Taylor, R. M., McKenzie, R. M. and Norrish, K. (1964). The mineralogy and chemistry of manganese in some Australian soils. *Aust. J. Soil Res.* **2**, 235-48.

Taylor, R. M. and McKenzie, R. M. (1980). The influence of aluminium on iron oxides. VI. The formation of Fe(II) Al(III) hydroxy-chlorides, -sulphates and -carbonates as new members of the pyroaurite group and their possible significance in soils. *Clays Clay Miner.* **28**, 179-87.

Taylor, R. M. and Pickering, J. G. (1963). Clay mineralogy of soils of the Launceston Tertiary basin, Tasmania. CSIRO Aust. Div. Soils Div. Rep. 9/63.

Taylor, R. M. and Schwertmann, U. (1974*a*). Maghemite in soils and its origin. I. Properties and observations in soil maghemites. *Clay Miner.* 10, 289-98.

Taylor, R. M. and Schwertmann, U. (1974*b*). Maghemite in soils and its origin. II. Maghemite syntheses at ambient temperature and pH 7. *Clay Miner.* **10**, 299-310.

Taylor, R. M. and Schwertmann, U. (1978). The influence of aluminium on iron oxides. I. The influence of Al on Fe oxide formation from the Fe(II) system. *Clays Clay Miner.* **26**, 373-83.

Taylor, R. M. and Schwertmann, U. (1980). The influence of aluminium on iron oxides. VII. Substitution of Al(III) for Fe(III) in synthetic lepidocrocites. *Clays Clay Miner.* **14**, 285-293.

Teufer, G. and Temple, A. K. (1966). Pseudorutile – a new mineral intermediate between ilmenite and rutile in the natural alteration of ilmenite. *Nature* (London) **211**, 179-81.

Thiel, R. (1963). Zum System – αAlOOH. *Z. Anorg. Allg. Chem.* **326**, 70-8.

Tiller, K. G. (1963). Weathering and soil formation on dolerite in Tasmania, with particular reference to several trace elements. *Aust. J. Soil Res.* **1**, 79-90.

Tillmanns, E. and Correns, C. W. (1969). Titanium. *In 'Handbook of Geochemistry'.* Vol. II-2. (Ed. K. H. Wedepohl). (Springer-Verlag: Berlin.)

Turton, A. G., Marsh, N. L., McKenzie, R. M. and Mulcahy, M. J. (1962). The chemistry and mineralogy of lateritic soils in the southwest of Western Australia. CSIRO Aust. Div. Soils, Soils Publ. No. 20.

Uehara, G. (1977). An overview of the soils of the arable tropics. College of Tropical Agriculture (Univ. Hawaii) Misc. Publ. No. 145.

de Villiers, J. M. and van Rooyen, T. H. (1967). Solid solution formation of lepidocrocite – boehmite and its occurrence in soils. *Clay Miner.* **7**, 229-35.

Wadsley, A. D. (1950*a*). A hydrated manganese oxide with exchange properties. *J. Am. Chem. Soc.* **72**, 1881-84.

Wadsley, A. D. (1950*b*). Synthesis of some hydrated manganese minerals. *Am. Miner.* **35**, 485-99.

Wadsley, A. D. (1952). The structure of lithiophorite (Al, Li)MnO_2 $(OH)_2$. *Acta Crystallogr.* **5**, 676-80.

Wadsley, A. D. (1953). The crystal structure of psilomelane. $(Ba,H_2O)_2Mn_5O_{16}$. *Acta Crystallogr.* **6**, 433-38.

Wadsley, A. D. and Walkley, A. (1951). The structure and reactivity of the oxides of manganese. *Rev. Pure Appl. Chem.* **1**. 203-13

Watson, J. R., Posner, A. M. and Quirk, J. P. (1973). Adsorption of the herbicide 2,4-D on goethite. *J. Soil Sci.* **24**, 503-11.

Weber, M. D., McKeague, J. A., Raad, A. T., De Kimpe, C. R., Chang Wang, Haluschak, P., Stonehouse, H. B., Pettapiece, W. W., Osborne, V. E. and Green, A. J. (1974). A comparison among nine Canadian laboratories of dithionite-, oxalate-, and pyrophosphate-extractable Fe and Al in soils. *Can. J. Soil Sci.* **54**, 293-98.

Wild, A. (1950). The retention of phosphate by soil. A review. *J. Soil Sci.* **1**, 221-38.

Wild, A. (1958). The phosphate content of Australian soils. *Aust. J. Agric. Res.* **9**, 193-204.

Wilding, L. P., Smeck, N. E. and Drees, L. R. (1977). Silica in soils: quartz, cristobalite, tridymite and opal. *In* 'Minerals in Soil Environments'. pp. 471-552. (Soil Sci. Soc. Am.: Wisc.)

Winters, E. (1938). Ferromanganiferous concretions from some podzolic soils. *Soil Sci.* **46**, 33-40.

Yassoglou, N. J. and Peterson, J. B. (1969). Mössbauer effect in clay-iron oxide complexes. *Soil Sci. Soc. Am. Proc.* **33**, 967-70.

Yates, D. E. and Healy, T. W. (1975). Mechanism of anion adsorption at the ferric and chromic oxide/water interfaces. *J. Colloid Interface Sci.* **52**, 222-28.

Zavarzin, G. A. (1968). Bacteria in relation to manganese metabolism. In 'International symposium on Ecology of soil Bacteria'. (Ed. T. R. T. Gray and D. Parkinson) (1966). pp. 612-31. (Liverpool University Press: Liverpool.)

Ziv, E. F. (1956). Rutilization of ilmenite under supergene conditions. *Izv. Akad. Nauk Uzb. SSR Ser. Geol.* **12**, 57.

Zwicker, W. K., Meijer, W. O. G. J. and Jaffe, H. W. (1962). Nsutite, a wide-spread manganese oxide mineral. *Am. Miner.* 47, 246-66.

Chapter 24
Mineral phosphate

K. Norrish and H. Rosser

THE LEVEL OF PHOSPHORUS IN SOILS, MORE THAN THAT OF any other single nutrient, governs soil fertility.

Although some soil profiles have an increased phosphorus content in the surface soil, phosphorus is a relatively immobile element in the landscape and usually the phosphorus content, and hence the fertility of a soil, is inherited from its parent material. The other major soil nutrients, N, S, Mg, Ca and K are more mobile and parent material is not such a controlling factor with them; such elements move in waters and can be introduced in appreciable quantities from rain.

Lindsay and Vlek (1977) state that phosphorus in soils usually ranges from 0.02 to 0.5% with an average content of about 0.05%. A wheat crop requires the equivalent of only several ppm of phosphorus in the top 15 cm of the soil, an amount that is only a small fraction of the phosphorus in even a low-phosphorus soil. Despite this, many soils supply insufficient phosphorus for plant growth, indicating that not all the phosphorus in soils is available to plants.

When phosphatic fertilizers are added to soils only some of the P is utilized by plants; some of the P is retained in the soil but is unavailable for plant growth and this phosphorus is regarded as fixed.

Phosphate fertilizers are used in large amounts throughout the world (about 84 million tonnes of phosphate rock annually (Emigh 1973), of which Australia uses about 2 million tonnes). Much of the world's agricultural production would suffer if phosphate fertilizers were not used, and this is particularly true of Australia.

Considerable efforts have been made to render native phosphorus more available and to improve the efficiency of use of that added as fertilizer. But on the whole such experiments have not been very successful. Such efforts have been made difficult by our lack of knowledge of the forms and reactions of phosphorus in soils. Soil scientists have recognized the need for this knowledge, but despite intensive work on the subject for some 50 years the nature and chemistry of soil phosphate is still inadequately understood.

Despite the complexities and problems associated with soil phosphate studies, progress is being made, and in this chapter we consider the phosphate minerals in soils and the reactions of phosphate with soil minerals. In Chapter 29 the sorption of phosphorus by soil is discussed.

Distribution

In relation to parent material

As indicated above, parent material is the dominant factor governing soil phosphorus contents. Table 24.1 gives data from McKelvery (1973) on the phosphorus

Table 24.1. Average phosphorus contents of rocks
Data from McKelvery (1973)

Source	%P_2O_5	Source	%P_2O_5
Lithosphere	0.22		
Igneous rocks			
Plutonic		Volcanic	
Alkali granite	0.14	Alkali rhyolite	0.07
Peralkaline granite	0.07	Peralkaline rhyolite	0.03
Calc-alkali granite	0.18	Calc-alkali rhyolite	0.07
Adamellite	0.20	Dellenite	0.12
Granodiorite	0.21	Rhyodacite	0.17
Tonalite	0.21	Dacite	0.17
Alkali syenite	0.19	Alkali trachyte	0.18
Peralkaline syenite	0.17	Peralkaline trachyte	0.12
Calc-alkali syenite	0.55	Calc-alkali trachyte	0.20
Monzonite	0.44	Latite	0.49
Calc-alkali mangerite	0.43	Calc-alkali doreite	0.33
Alkali mangerite	0.63	Alkali doreite	0.66
Diorite	0.35	Andesite	0.35
Gabbro	0.24	Alkali andesite	0.52
Alkali gabbro	0.44	Tholectic basalt	0.23
Pyroxenite	0.09	Alkali basalt	0.39
Alkali pyroxenite	0.59	Phonolite	0.17
Peridotite	0.05	Nepheline latite	0.37
Alkali peridotite	0.05	Nepheline ordanchite	0.60
Nepheline syenite	0.19	Nepheline tephrite	0.35
Malignite	0.71	Analcite tephrite	0.67
Nepheline monzonite	0.45	Nephelinite, etc.	0.35-1.02
Essexite	0.48		
Glenmuirite	0.66		
Theralite	0.31		
Teschenite	0.50		
Ultra alkaline rocks	0.02-1.68		
Metamorphic rocks		*Sedimentary rocks*	
Leptites and hellaflintas	0.1	Sandstone	0.10
Quartz feldspathic gneiss	0.2	Shale	0.16
Mica schists	0.2	Carbonate rocks	0.06
Slates	0.2		
Phyllites	0.2		
Amphibolites	0.3		

Table 24.2. Average phosphorus contents of Australian rocks
Calculated from the data of Joplin (1963, 1965, 1975)

Source	Mean % P_2O_5	S.D.[A]	Number of analyses
Igneous rocks			
Granites	0.12	0.11	666
Diorites	0.28	0.19	208
Norites and gabbros	0.15	0.20	74
Rhyolites	0.12	0.11	226
Andesites	0.31	0.30	174
Trachytes	0.25	0.22	86
Basalts	0.46	0.34	908
Serpentinites and associated ultramafics	0.08	0.10	47
Metamorphic rocks			
Metamorphic mafics	0.24	0.20	196
Metamorphics of granitic composition	0.15	0.08	111
Metamorphics of dioritic composition	0.28	0.18	45
Metamorphosed pelitics (non-calcic)	0.16	0.12	192
Metamorphosed psammitics	0.11	0.08	59
Metamorphosed calcic rocks	0.12	0.10	37
Metasomatized and deuterically altered mafic rocks	0.49	0.44	119
Sedimentary rocks			
Quartz sandstones	0.10	0.09	22
Clays	0.12	0.16	88
Shales	0.19	0.27	54

[A] Standard deviation.

content of various rocks. Joplin (1963, 1965, 1975) has tabulated analyses of Australian rocks, and Table 24.2 shows the average phosphorus contents calculated from her data. Although the various rocks have different averages, and there is a big range of contents for any particular rock type, there is not a consistent relationship between parent material and phosphorus content of soils. The variability is evident from Table 24.2 where the standard deviation for most rocks is comparable to the mean, and it may exceed the mean. (The standard deviations were calculated in the usual way although the phosphorus concentrations do not have a gaussian distribution.) Soils derived from siliceous sands are usually very low in phosphorus, often $< 0.005\%$; calcareous sands and limestones give soils with contents of about 0.03%; soils derived from other sedimentary rocks, shales, slates, and mudstones have variable contents but are usually in the range 0.01–0.05%; soils from metamorphic rocks and granites have a similar range. Usually, but not always, basaltic soils are higher and average $> 0.1\%$. The basaltic soils of Queensland are consistently high and the average of about 30 such soils is 0.24% P. The basalts of northern New South Wales and southern Queensland have a comparable value; the average of the data given by Johnson (1976) is 0.25% P for these rocks. By contrast, the Antrim Plateau Volcanics of the Kimberley region are low, averaging $\sim$ 0.05% P (Johnson 1976), and the various soils near Kununurra, that are formed on these basalts, are correspondingly low in P (Stace *et al.* 1968; Rutherford 1962; R. F. Isbell personal communication).

An important feature of basic igneous rocks is their high iron content; soils formed from them are usually high in free iron oxides and the soils have the mechanism for retaining phosphorus (see p. 343), even where these soils develop under strongly leaching conditions. By contrast, soils formed from rocks that are lower in iron can lose much phosphorus during soil formation. Table 24.3 shows results for two soils that are high in free iron oxides; the P_2O_5 content of these soils changes little with depth. The two soils with lower free iron oxide contents show a considerable reduction in phosphorus in passing from the subsoil to the surface.

In Great Soil Groups

The phosphorus content of different Great Soil Groups is largely dependent on parent material, but phosphorus may be lost from strongly leached soils.

Table 24.4 compares the results given by Wild (1958) with the averages calculated from the data of Stace *et al.* (1968). If surface soil contents are compared, the latter are usually higher than the former and it may be significant that the data of Stace *et al.* was obtained by X-ray fluorescence methods that give total phosphorus contents. Wild used HCl to extract the phosphorus, a technique that extracts only about 70% of the total. However, overall the agreement is good,and where there is disagreement, it can often be traced to one or two very high values raising the average of a group.

Siliceous sands and the various *podzolic soils, podzols,* and *yellow earths* are those lowest in phosphorus, usually $< 0.02\%$ and often $< 0.005\%$. At the other end of the scale are soils that often have basic igneous rocks as parent materials, i.e. *chocolate soils, euchrozems, krasnozems, xanthozems, chernozems* and *prairie soils*; these soils usually contain $> 0.1\%$ P, but may be much lower if the parent rock is low.

Of the other soils, *solonetz, solodized-solonetz, soloths, solodized brown soils, calcareous red earths, red earths* and *terra rossas* are low, ranging beween 0.02 and 0.04% P.

There are insufficient data to give meaningful means for individual great soil groups, but the remaining soils that constitute much of the agricultural land of Australia usually contain from 0.02 to 0.08% P; these soils include, *black earths, red-brown earths, red earths, calcareous red earths, grey, brown and red clays, desert loams* and *calcareous sands.* Wild (1957) gave the phosphorus contents of many Australian soils, and after correcting for the area occupied by the various soils, Wild (1958) estimated the Australian average as 0.030% P, and pointed out that this is significantly lower than the averages for the major soils of the United States.

Table 24.3. The phosphorus contents of soils with high and low levels of free iron oxides

	Krasnozem		Krasnozem		Red earth		Non-calcic brown soil	
Location:	Malanda, Qld[A]		Lismore, N.S.W.[B]		Kulgera, N.T.[C]		Charters Towers, Qld[D]	
Parent material:	Basalt		Basalt		Granite		Granodiorite	
	% P_2O_5	% Fe_2O_3[E]	% P_2O_5	% Fe_2O_3[E]	% P_2O_5	% Fe_2O_3[E]	% P_2O_5	% Fe_2O_3[E]
Depth (cm)								
0–10	0.44	24.1	0.89	23.9	0.23	5.2	0.048	3.6
~ 50	0.37	26.2	0.73	25.3	0.23	5.9	0.037	7.9
~ 100	0.32	24.2	0.62	30.0	0.32	4.9	0.31	6.6
Rock	0.44	11.0	–	–	0.62	6.4	–	–

[A] Site 9 of Isbell et al. (1976). [B] Data from Nicholls and Tucker (1956), Profile 1. [C] Data from J. T. Hutton, pers. comm.
[D] Profile T10 of Crack and Isbell (1971). [E] Total Fe_2O_3.

Table 24.4. Phosphorus contents of Australian soils

Values are % P. The grouping of soils is the same as that of Stace *et al.* (1968). The grouping is split where there appears to be a difference in P content between Great Soil Groups within the major grouping. Handbook data calculated from the data of Stace *et al.* (1968).

Great Soil Groups	Number	Handbook data Surface	Handbook data Subsoil	Wild (1958) Surface
1. *Alluvial soils; lithosols; calcareous sands*	4	0.045	0.030	
Siliceous and *earthy sands*	5	0.014	0.010	
2. *Grey-brown* and *red calcareous soils; desert loams; red and brown hardpan soils; grey, brown and red clays*	23	0.036	0.027	0.027
3. *Black earths; rendzinas; chernozems; prairie soils; wiesenboden*	13	0.078	0.079	0.087
4. *Solonetz; solodized-solonetz; solodic soils; soloths; solonized brown soils; red-brown earths; non-calcic brown soils; brown earths*	27	0.033	0.023	0.023
Chocolate soils	3	0.163	0.143	
5. *Calcareous red earths; red earths; terra rossa soils*	12	0.028	0.023	0.021
Yellow earths	3	0.013	0.010	
Euchrozems; xanthozems; krasnozems	8	0.169	0.106	0.134
6. *Grey-brown, red, yellow* and *brown podzolic soils*	12	0.020	0.011	
Lateritic podzolic soils; gleyed podzolic soils; podzols; humus podzols; peaty podzols	17	0.011	0.006	0.013
7. *Alpine humus soils; humic gleys; neutral* to *alkaline peats; acid peats.*	11	0.125	0.096	0.042

Within the profile

Almost all soils show increased phosphorus contents in their surface horizons. Where the phosphorus content of the soil is low and there is organic matter accumulation at the surface, the increase may be to several times that of the sub-soil. Available phosphorus, which is usually only a small fraction of the total, may increase 10 or 20 times in the surface soil. Averages for all the data of Stace *et al.* (1968) give means of 0.055% P for surface soils and 0.038% for sub-soils (Table 24.4).

The increase in the surface is due to a biological pumping action, i.e. plants obtain phosphorus from the whole soil profile and deposit much of it on the surface as litter. Not surprisingly then, the increased phosphorus in surface horizons is often present in organic matter, but this is not always the case. Though the increase may be via organic matter, phosphorus can be mineralized and retained inorganically in the surface.

Fertilizer added to the soil will increase the phosphorus content of the surface soil, but under normal agricultural practice the increase is slight. About 10 kg ha^{-1} of P is applied in wheat production, and if this is distributed over a depth of about 15 cm then it represents an increase of 5 ppm. A fraction of this increase is removed with the crop, and the phosphate additions may be made less frequently than annually. Under these conditions the increase amounts to several ppm per year, an amount that is only a very small fraction of the total phosphorus content of most soils.

Where larger amounts of phosphate are added the phosphorus content of the surface soil may increase considerably. Some of the heavily fertilized New Zealand ash soils show a considerable increase of P, as much as 0.2%.

Nature of the inorganic phosphorus

The amount of phosphorus available for plant growth is usually only a small fraction of the total in the soil, and there is no simple relationship between the amount available and the total. Soils that are high in total phosphorus may require phosphate fertilizers for adequate plant growth. This is true of many of the *krasnozems* that have a total of 0.2 to 0.5% phosphorus. Grange and Fox (1953) found that lateritic soils of the Lower Cook Group were deficient in phosphorus, even those whose totals ranged from 0.5 to 2%! This rather unpredictable behaviour of soil phosphorus is obviously due to some variability in the nature of the phosphorus.

For several decades attempts have been made to identify the inorganic fractions of soil phosphorus in the hope of understanding their availability in soils. This problem is by no means solved though modern analytical instruments are aiding by providing data that were unobtainable a few years ago.

There are particular problems in identifying the forms of phosphorus in soils. Usually it is in the fine particle size fractions so that the conventional petrographic methods of identification cannot be used. X-ray diffraction methods are the most reliable for identifying fine-grained minerals, but powder diffraction techniques are not very sensitive and rarely can a phase be identified in a complex mixture if it is less than one per cent of the mixture. Phosphorus is present in most soils at a concentration one or two orders below this limit of detection.

An additional complication, which many researchers suspected but lacked proof of, was that the phosphorus was not necessarily present as a crystalline phase. This would invalidate many physical methods of identification.

Because of these difficulties, indirect methods of identification have been attempted based on solubility or exchange data (Clark and Peech 1955; Lindsay *et al.* 1959*b*). Differential dissolution has also been used, and Chang and Jackson (1957) have presented a series of chemical treatments that classifies the inorganic phosphorus into six mineral categories. These indirect methods give useful information on the behaviour of soil phosphate but the interpretation of such data in terms of specific minerals is questionable. Phosphate chemistry, particularly in soils, is too complex to be interpreted with certainty. A large number of phosphate minerals have been 'identified' by such methods, but mostly such identifications have remained unconfirmed; apatite and various iron and aluminium phosphates are those most commonly postulated to explain solubility data.

Despite the above problems a variety of phosphorus containing minerals have been identified, with certainty, in soils.

Some of the results given below are unpublished data obtained by the authors in studying soils with an electron probe micro-analyser. This instrument is capable of analysing very small volumes, of the order of a cubic micron, and so can give information that was previously unobtainable. The probe gives chemical information only, but generally the mineral phase present can be identified without ambiguity from a complete chemical analysis.

Residual minerals

In this section those phosphate minerals native to soils are discussed; phases formed from reaction with applied fertilizer are considered later.

A variety of phosphate minerals that are found commonly are residues from parent rock. Their continued presence in soils suggests that their phosphorus is almost completely unavailable for plant growth. Usually, but there are exceptions, such minerals account for only a small fraction of the total soil phosphorus.

Apatite

Most of the phosphorus in rocks occurs as apatite. In igneous and metamorphic rocks it is usually fluorapatite, $Ca_5(PO_4)_3F$. In sedimentary phosphorites, rocks that are often sources of rock phosphate, the apatite is one in which some of the phosphate is replaced by carbonate. Fluorine can be replaced by chlorine (chlorapatite), or by –OH (hydroxyapatite). Carbonate–apatite is invariably fine-grained and is more soluble than the other apatites.

In calcareous soils and soils that are not highly leached, apatite can persist and it is often observed microscopically in the sand fractions of such soils (Fry 1913; Leahey 1935; Plummer 1915; Skipp and Matelski 1960; Quershi *et al.* 1969). Apatite often occurs in rocks as micro-inclusions in other minerals, and in such cases the apatite is protected and can persist even where weathering is more severe (Syers *et al.* 1967; Cescas and Tyner 1970). Obviously such phosphorus will be completely unavailable to plants.

In our microprobe studies of Australian soils, apatite grains are often observed in the less weathered soils, but rarely in amounts that would constitute an appreciable fraction of the total phosphorus.

In the two granitic soils of Table 24.3, apatite accounts for virtually all the phosphorus in the rock and for most of it in the deeper soil horizons. There is little apatite in the surface horizons.

Monazite and xenotime

Both monazite, (Ce, La, Th)PO_4, and xenotime, YPO_4, are rare earth phosphates and can be present in soils from most parent materials.

These minerals are very resistant to weathering and are often identifiable in the heavy mineral fraction of soils. The presence of monazite and xenotime in heavy mineral beach sands is evidence of their stability. In our microprobe studies, grains of these minerals are observed in a great variety of soils, including those that are highly weathered. Usually their contribution to total soil phosporus is trivial.

Plumbogummite minerals

This group of minerals will be discussed fully under pedogenic minerals, since plumbogummite minerals often form in the soil. But these minerals may be present in sedimentary rocks and then they can persist into the soil, as at least certain members of the group are very resistant to weathering.

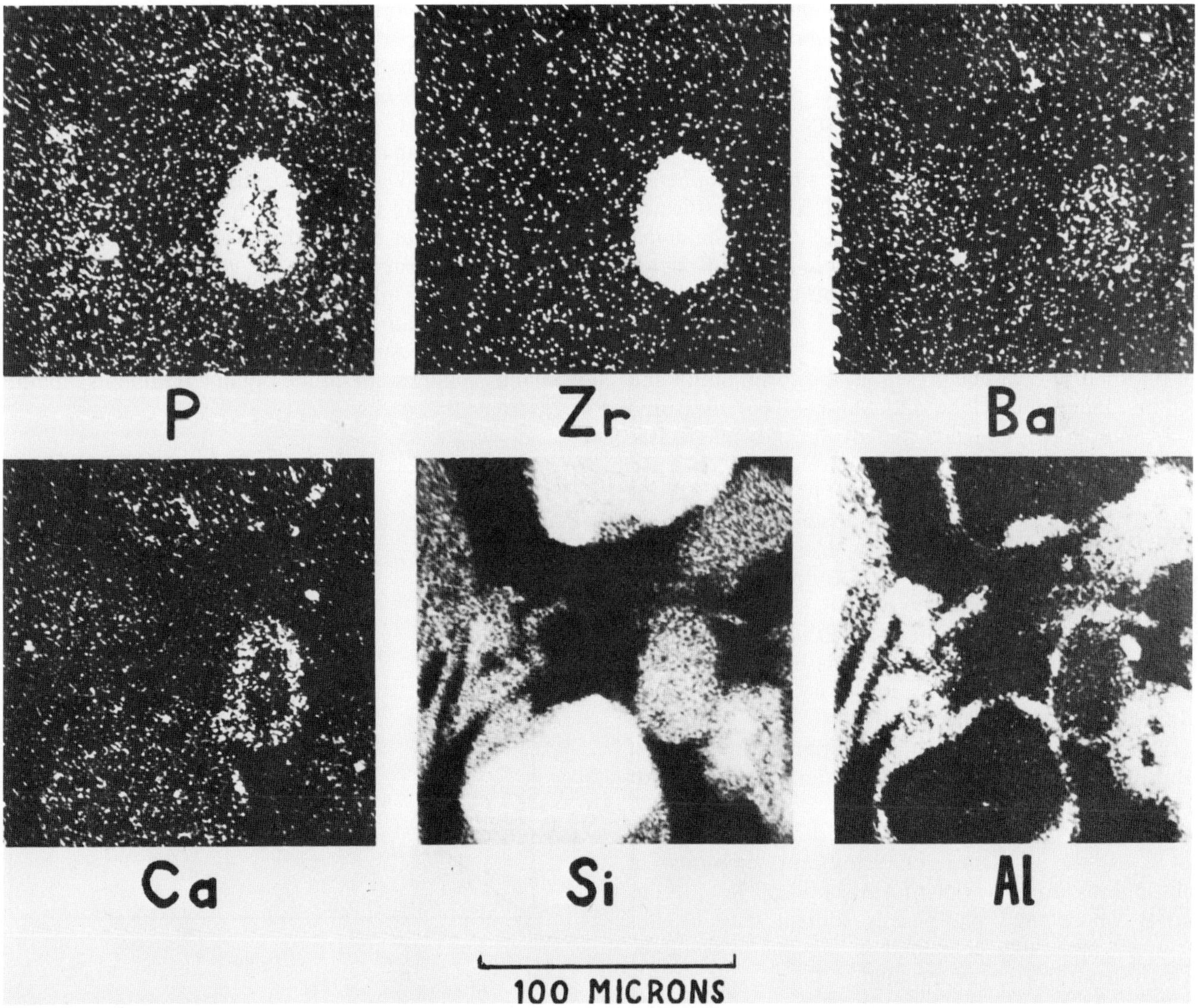

Fig. 24.1. Phosphorus-bearing minerals in a *yellow podzolic soil* from the Barossa Valley, S.A. Large grain is zircon. The smaller grain high in P and Ba is gorceixite.

Table 24.5. Phosphorus contents of common silicates
Data from Koritnig (1965)

Mineral	Phosphorus concentration Average (ppm)	Range (ppm)
Feldspars	131	5–1000
Amphiboles	77	24–180
Pyroxenes	81	18–630
Olivines	223	48–517
Micas	86	48–332
Garnets	185	41–560
Quartz	9	0.005–42
Chalcedony	35	23–47
Opal	62	31–173

Silicates

It is well known that in certain silicates phosphorus can replace silicon in tetrahedral co-ordination. Zircon (zirconium silicate) is a resistant mineral and in it phosphorus can replace silicon to a considerable extent. Fig. 24.1 shows microprobe photographs of a zircon grain in which the outer rim is high in phosphorus. Zircons often contain appreciable phosphorus, but in most soils it would constitute a very small fraction of the total, and it would be unavailable to plants.

Koritnig (1965) has shown that small amounts of phosphorus replace silicon in most common silicates. He examined 140 specimens representing 45 minerals and the data of Table 24.5 are from his paper. Koritnig calculated that from 2 to 26% of the total phosphorus in rocks is present in this form, and it is obvious that in soils that are not highly weathered significant amounts of phosphorus could be present in this form. Smith and Ribbe (1966) and Ribbe and Smith (1966) measured the phosphorus contents of alkali and plagioclase feldspars, with the electron microprobe, and obtained values in each group as high as 1000 ppm. Cescas and Tyner (1970) observed that in some New Zealand soils phosphorus was uniformly distributed throughout the plagioclase feldspars at a concentration of 160 ppm.

In our microprobe studies of soils we often observe phosphorus in feldspars, e.g. the potash feldspars of the Urrbrae loam, a *red brown earth*, from Adelaide, South Australia contain ~ 500 ppm P.

Volcanic glass can contain variable amounts of phosphorus, e.g. glass fragments in soils near Mt Gambier contain 0.23%.

The most surprising result in Koritnig's data (Table 24.5) is the presence of significant amounts of phosphorus in the various forms of silica. In siliceous sands whose total phosphorus content may be less than 50 ppm, that contained in the quartz could be a significant fraction of the total.

Most Australian soils are low in primary minerals, apart from quartz, and so such minerals will not contribute much to total soil phosphorus, but in some soils an appreciable fraction of the total could reside in silicates.

Pedogenic minerals

Under this heading are discussed those minerals that form during weathering and soil formation. Those that may be formed from applied fertilizer will be discussed later, but it could well be argued that the ultimate fate of applied phosphorus should be in the pedogenic phosphates.

Apatite

As apatite is the major phosphate constituent of most rocks it is usually difficult to tell whether apatite is residual or pedogenic. Certainly apatites are easily synthesized at low temperatures and solubility data indicate that they should form in alkaline soils (Lindsay and Vlek 1977).

A puzzling feature of our microprobe studies is that analyses corresponding to apatite are not common, even in calcareous soils. Some apatite grains are observed but these are usually comparatively large crystals and are probably residual.

If pedogenic apatite is present as very small crystals, $< 1\ \mu m$, it would escape recognition in the microprobe, particularly where there is a high background Ca level from carbonate or even exchangeable cations.

That apatite can form in soils is evidenced by results on a *chernozem*; microprobe analyses showed high Ca and P within organic matter fragments. The chemistry of these points corresponded to apatite (Fig. 24.2), and after hand picking some of the organic fragments, apatite

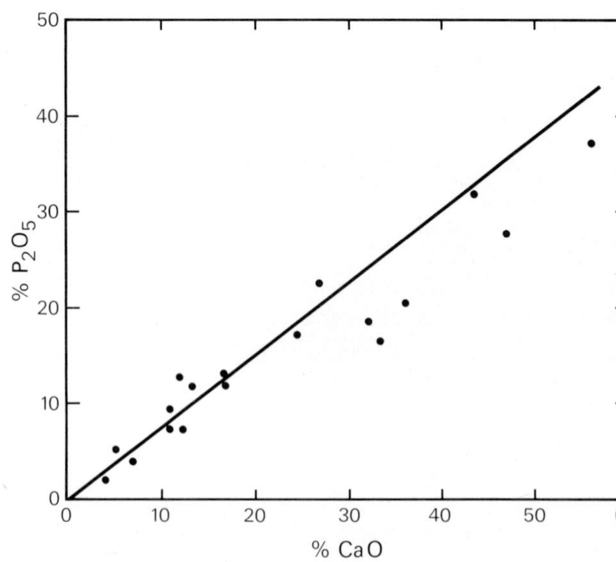

Fig. 24.2. Showing the relationship between P_2O_5 and CaO within organic matter fragments in a *chernozem* from Mt Gambier, S.A. The line gives the ratio for apatite. (Microprobe analyses.)

Table 24.6. Plumbogummite type, and structurally related, minerals
Minerals with asterisks are those that have been found in soils

Mineral	Formula
Crandallite*	$CaAl_3(PO_4)_2(OH)_5H_2O$
Goyazite*	$SrAl_3(PO_4)_2(OH)_5H_2O$
Gorceixite*	$BaAl_3(PO_4)_2(OH)_5H_2O$
Florencite*	$CeAl_3(PO_4)_2(OH)_6$
Plumbogummite*	$PbAl_3(PO_4)_2(OH)_5H_2O$
Hinsdalite*	$(Pb,Sr)Al_3(PO_4)(SO_4)(OH)_6$
Svanbergite	$SrAl_3(PO_4)(SO_4)(OH)_6$
Woohouseite	$CaAl_3(PO_4)(SO_4)(OH)_6$
Alunite*	$KAl_3(SO_4)_2(OH)_6$
Jarosite*	$KFe_3(SO_4)_2(OH)_6$
Carphosiderite	$(H_3O)Fe_3(SO_4)_2(OH)_6$

was confirmed by X-ray diffraction. Whether apatite crystallized in the organic debris, or whether it was present in the live plant is not known, but McManus *et al.* (1979) have shown that apatite is a constituent of plant ashes and they considered the mineral was present in the plant prior to ashing. Even if apatite is formed in plants and deposited in the soil it is pedogenic in the broad sense of the term.

The extent to which phosphorus is present in Australian alkaline soils as apatite is uncertain, but evidence will be given later that at least in some calcareous soils part of the phosphorus is present in other forms.

Plumbogummite minerals

Norrish (1957, 1968) first identified minerals of this group in 40 soils from Australia and elsewhere, and showed that in certain soils the phosphates had formed

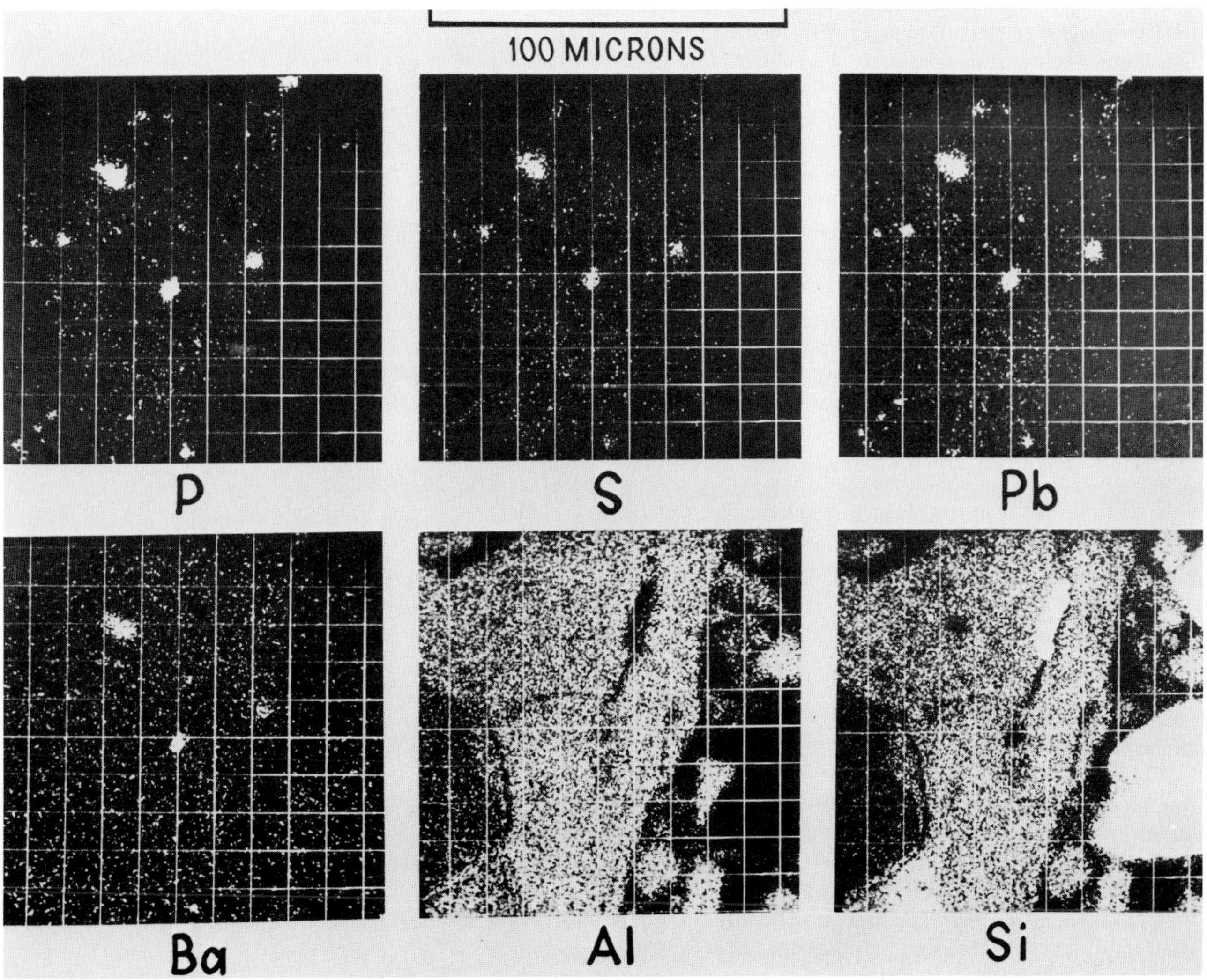

Fig. 24.3. Microprobe results showing phosphorus associated with S, Pb, Ba and Al in a *lateritic podzolic soil,* Mount Torrens area, S.A.

Table 24.7. Percentage composition of some soil plumbogummite minerals

1. Average of 6 grains. Laterite, W.A. Profile 31 of Norrish (1968). Data from Norrish (1974). 2. Average of 5 grains. *Lateritic podzolid soil* near Mt Torrens, S.A. Data from Norrish (1974). 3. Average of 9 grains. Andosol, Efate Is., New Hebrides. Site 374 of Lee (1974).

	1	2	3
P_2O_5	23.8	14.6	25.1
SO_3	2.9	7.8	1.11
Al_2O_3	29.5	21.5	28.2
Fe_2O_3	3.2	10.1	0.67
PbO	1.3	35.2	-
BaO	11.9	2.9	19.0
La_2O_3	4.0	-	-
CeO_2	3.8	-	1.63
CaO	1.2	-	0.78
SrO	2.7	0.3	1.27
K_2O	-	-	0.18
SiO_2	2.6	0.8	1.84

during weathering. These minerals are also present in Pacific island soils (Fieldes *et al.* 1960; Schroo 1963), in Christmas Island soils (Trueman 1965) and Bermuda (Ruhe *et al.* 1961). Adams *et al.* (1973) reported a plumbogummite mineral in a strongly weathered New Zealand soil formed on granite and T. Gudmundsson and K. Stahr (personal communication) found gorceixite in a podzol from the southern Black Forest in Germany. Altschuler (1973), in discussing the weathering of phosphate deposits, points out that apatite, and other phosphate minerals, react with clay minerals to form plumbogummite type minerals. These minerals are associated with phosphate rock deposits in the U.S., Senegal, Brazil and Africa, and are the products of quarternary weathering.

Table 24.6 gives the idealized formulae of plumbogummite and structurally related minerals. Within this structure the tetrahedral position may be occupied by P, S or As; Al or Fe can be in octahedral coordination. The third structural position can accommodate a great range of cations, more than those shown in Table 24.6. For a full description of these minerals see Palache *et al.* (1951). Rarely are these minerals found as pure end members and usually the cation position is occupied by a variety of elements; this is true also of the minerals in soils (Norrish 1968). Table 24.6 indicates those minerals that have been recognized in soils using electron microprobe analysis. In most cases this identification was confirmed by X-ray diffraction after concentrating the mineral (Norrish 1968).

A gorceixite particle is evident in Fig. 24.1 (high P and Ba), and Fig. 24.3 shows microprobe photographs of a *lateritic podzolic soil* where the phosphate mineral is high in Pb, and about one third of the P is replaced with S. Table 24.7 gives analyses of some of these minerals from soils, and it can be seen that they are chemically complex. In all cases there is some replacement of P by S. It seems most likely that other members of this structural type will be found in soils. One aspect of their chemistry

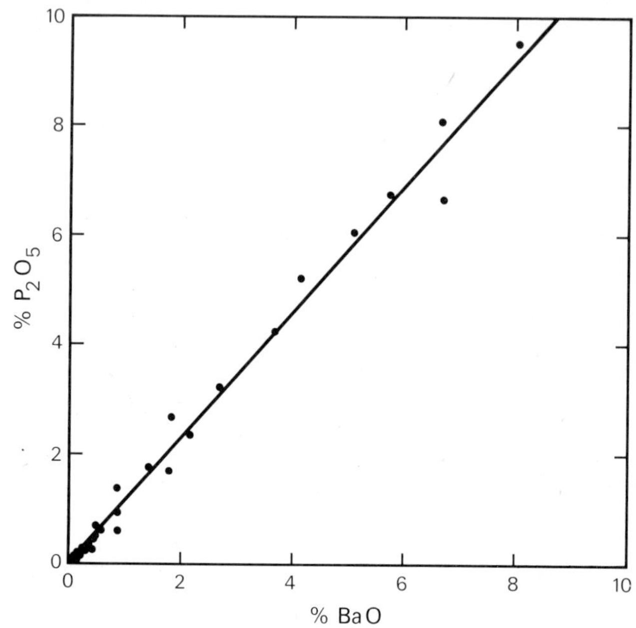

Fig. 24.4. The relationship between P_2O_5 and BaO in an andosol from Efate Is., New Hebrides. See Table 24.7. Phosphorus is present as gorceixite.

warrants comment; amongst these phosphates we have not analysed particles in which iron exceeds aluminium.

Norrish (1968) showed that these minerals are present in very different soils overlying different parent materials. The soils occurred under a wide range of climates and ranged from calcareous and mildly leached soils (*terra rossa* and *rendzina*) to strongly weathered soils (*krasnozems, lateritic podzolic soils*).

Subsequent microprobe studies have confirmed that plumbogummite minerals are present in many soils, but usually only in small amounts. However, in some soils much of the phosphorus can be accounted for by these minerals; Norrish (1968) found at least 12 to 60% of the total P was present in plumbogummite minerals in certain soils and Fig. 24.4 shows microprobe data obtained on a New Hebrides soil that contains gorceixite. When random points were analysed, the correlation between phosphorus and barium was high (R^2 = 0.996), and this is good evidence that virtually all the phosphorus is present as gorceixite.

Minerals of this group are usually fine grained in soils, 0.5 to 5 μm, but despite this particle size they resist breakdown and persist in highly weathered soils. Norrish (1957, 1968) was able to concentrate these minerals by dissolving most of the clay minerals in hydrofluoric acid!

It is not surprising then that the evidence indicates that these minerals are poor sources of phosphorus for plant growth. Their presence explains the deficiencies of some soils that are high in total phosphorus. A microbiological assay of a soil gorceixite yielded less P than any of the 62 phosphate minerals tested by Swaby (1956).

Although there are no data on the relative solubilities of the different members of the plumbogummite group, it is the barium and rare earth members that are present in highly weathered soils, suggesting that these are less soluble than the Sr and Ca minerals.

When heated to $\sim$ 400°C, these minerals lose structural water and become amorphous and in this form their phosphorus is more soluble (Norrish 1957). Their presence in Cook Island soils explains the observation of Arbuckle (Grange and Fox 1953, Appendix) that preheating the soils caused a considerable increase in phosphorus availability. Christmas Island C-grade rock phosphate is high in crandallite, the Ca member of the plumbogummite group, and the above behaviour is the basis of rendering its phosphate more available by heating.

Preheating soil to 550°C forms the basis of one method of estimating organic phosphorus in soils (Saunders and Williams 1955). By this method much of the phosphorus in plumbogummite minerals ($\sim$ 50%) would be estimated as organic phosphorus (Norrish 1968; Adams *et al.* 1973).

Other minerals

Koch (1956) identified vivianite, $Fe_3^{2+}(PO_4)_2 \cdot 8H_2O$, in alluvial beds in New South Wales; the vivianite had formed by the reduction of iron in the presence of organic matter and so it might be expected to occur in swampy soils (Palache *et al.* 1951).

Perhamite has only recently been described as a new mineral (Dunn and Appleman 1977). For some years we have observed an unknown phosphate-silicate phase in a variety of rocks. This mineral was always intimately intergrown with plumbogummite minerals, usually crandallite, and gave a characteristic X-ray diffraction spacing at 20Å. The chemical and diffraction data confirmed the presence of perhamite, $3CaO \cdot 3.5Al_2O_3 \cdot 3SiO_2 \cdot 2P_2O_5 \cdot 18H_2O$. Although we have not identified this mineral in soils we have observed it in tonsteins and often as a superficial weathering product associated with goethite and crandallite. Such occurrences suggest that it might form in soils, but our data indicate that it is less resistant to weathering than the plumbogummite minerals.

In our microprobe studies we have examined about 150 soils, with representatives of all the great soil groups of Stace *et al.* (1968). An important negative finding has been that the two minerals commonly postulated as occurring in soils, strengite and variscite, have not been found. Though these minerals might yet be identified in soils, it seems that they are unlikely to be common or very significant, and this is in accord with the solubility studies of Bache (1963).

Dispersed forms

In most soils the above identifiable mineral grains account for only a small portion of the total soil phosphorus. Microprobe studies show that most of the phosphorus is present dispersed in the clay matrix of the soil. If this phosphorus is associated with particular minerals then their particle size is very small, $< 1\ \mu m$, much too small to be analysed as a distinct phase with the microprobe.

To try and obtain meaningful information on this dispersed form of phosphorus, we have used the following technique. Many points (100–200) selected at random were analysed for those elements considered to be significant. In addition to phosphorus, Ca, Al, Fe, Si, Ti and S were usually measured, and depending on the particular sample, other elements such as K, Ba, or rare earths were added. Statistical methods using simple and multiple regressions were then used to determine with which elements phosphorus is associated. To obtain the most meaningful results, considerable manipulation of the data is necessary; e.g. such mineral phases as quartz are low in phosphorus and all elements but Si, so their presence tends to give correlations, between P and other elements, which have little chemical significance. Quartz points are therefore deleted from the statistical analysis, and the same is done with certain other identifiable minerals such as feldspars. Analytical points that happen to hit only the impregnating plastic have the same effect as quartz and are eliminated. In addition to excluding such points, it may be desireable to reject also those points that correspond to mixtures of plastic and the clay matrix, but in which the clay occupies only a small part of the analysed volume. Such points give low totals for the sum of the major components ($SiO_2 + Al_2O_3 + Fe_2O_3$ etc.), and the inclusion of such data has the effect of forcing a meaningless correlation between all the elements, including phosphorus, in the clay matrix. For example, Figs 24.5, 6, 8 and 9 do not include points for which the totals of the major components are less than 50%. If rare phosphate mineral grains are encountered such points are rejected, as several grains of apatite, for instance, will give analyses very high in P and Ca, and these points will produce a very high correlation between P and Ca, but will suppress relationships that might exist within the clay matrix between P and other elements.

Iron oxides and oxyhydroxides

The iron oxide hematite, Fe_2O_3, and the oxyhydroxide goethite, FeOOH, are the most abundant and the most frequently occurring free iron oxides in soils. Here, and subsequently in this chapter, the term 'free iron oxides' is used to include both minerals, although strictly speaking goethite is an oxyhydroxide. Very often both minerals are present in the one soil and often, particularly when they are in low concentrations, it may be very difficult to identify the mineral phases specifically.

Fig. 24.5 shows the relationship between P_2O_5 and Fe_2O_3 for the deep subsoil of a *krasnozem* from

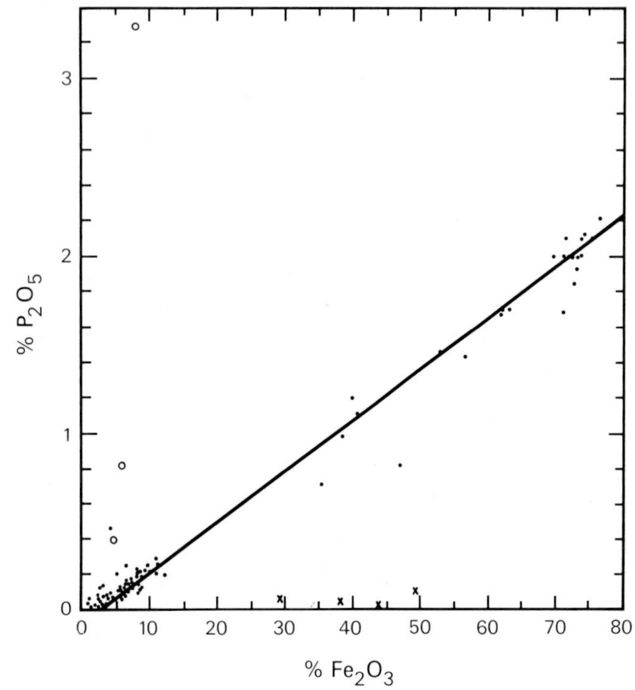

Fig. 24.5. The relationship between P_2O_5 and Fe_2O_3 in the subsoil of a *krasnozem* from Malanda, Queensland. x , these points are high in TiO_2 and correspond to ilmenite; o , these points are high in Ba and rare earths and indicate gorceixite.

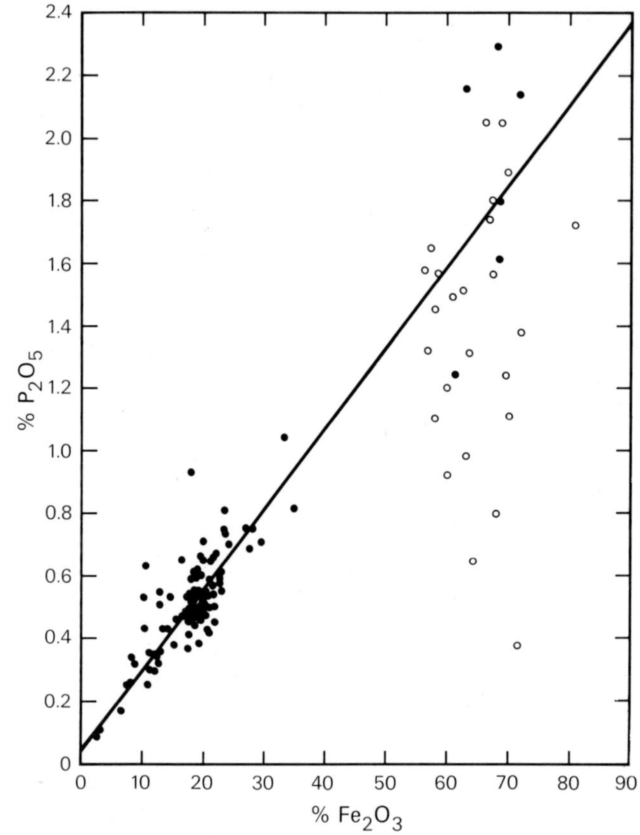

Fig. 24.6. The relationship between P_2O_5 and Fe_2O_3 in the surface soil of a *krasnozem* from near Innisfail, Queensland. ● ; points selected at random: o : points selected for high Fe_2O_3 (goethite grains).

Malanda, north Queensland. The analyses high in Fe and low in P are high in Ti and are of ilmenite grains; those low in Fe and high in P contained barium and rare earths, indicative of gorceixite. Several of the latter points have P_2O_5 contents too high to appear on the graph. After deleting the points associated with ilmenite and gorceixite, P_2O_5 is highly correlated with Fe_2O_3 (R^2 = 0.97).

The line of best fit is given by

$$P_2O_5 = 0.0285\ Fe_2O_3 - 0.0316$$

In the multiple regression none of the other elements, S, Ti, Si and Al were significant.

The average of the 116 points analysed was 0.57% P_2O_5 while a determination on the soil gave 0.59% P_2O_5, and a similar good agreement is found in the Fe_2O_3 contents. Such agreement suggests that despite the total very small volume analysed by the microprobe it is reasonably representative of the soil.

The Fe_2O_3 contents of the points in Fig. 24.5 range from zero to 75% and the variation is indicative of the volume of iron oxide (number of micro-crystals) in the analysed volume. Those points near the origin are representative of the clay matrix; the cluster of points at ∼ 70% Fe_2O_3 represents very small iron oxide aggregates. It is interesting that both sets of points lie on the same straight line, indicating that the association of P_2O_5 and Fe_2O_3 is constant, and the same for the iron oxide aggregates as for the clay matrix.

The fact that the line of best fit passes very close to the origin indicates that phosphorus is associated with the iron oxides only, not with the clay. Kaolin was the only clay mineral identified. Both goethite and hematite were present in the soil and the results indicate that the relationship to phosphorus is independent of mineral form.

The P/Fe ratio is indicative of an adsorption complex rather than an iron phosphate compound. If the phosphate existed as iron phosphate there would not necessarily be a correlation between P and *total* Fe. To have such a correlation it would be necessary to postulate an iron phosphate precipitating on the surface of iron oxide; in fact such a mechanism has been postulated (Hemwall 1957), but if the phosphate phase does not crystallize then the distinction from an adsorption complex is a matter of terminology.

Rarely is the association of P and Fe as close as shown in Fig. 24.5, and Fig. 24.6 shows results for another *krasnozem*, this time for a virgin surface soil. For the sake of clarity, the ilmenite (high Ti) points are not shown. The P_2O_5 content of the soil was 0.62% and the average of the

120 microprobe analyses was 0.59%. For the random points there is a highly significant correlation between P_2O_5 and Fe_2O_3 (R^2 = 0.85), but the points are more spread out than those of Fig. 24.5. In the multiple regressions, TiO_2 is significant but with a negative coefficient that is approximately equal to the positive coefficient for Fe_2O_3. Despite the fact that the high Ti points were rejected, the results imply that low concentrations (very small particles) of ilmenite are still present. Ilmenite, FeO.TiO_2, can vary in composition but in it $Fe_2O_3 \simeq TiO_2$, so the negative effect of TiO_2 can be removed if instead of Fe_2O_3 the free iron oxide is expressed as $Fe_2O_3 - TiO_2$. Then phosphorus was not significantly correlated with other elements; if only those points with $Fe_2O_3 < 50\%$, i.e. those of the clay matrix, are considered then Fe, Al and Ca are all significant and,

$$P_2O_5 = 0.020\ \text{Go} + 1.03\ \text{CaO} + 0.004\ \text{Gi}$$

Mean P_2O_5 = 0.54%. Error in estimate = 0.079% P_2O_5.

$$\text{Go} = Fe_2O_3 - TiO_2 \quad \text{(see above)}.$$

$$\text{Gi} = \text{gibbsite} = Al_2O_3 - SiO_2 - 0.18\ \text{Go}.$$

The above equation aims to estimate gibbsite, expressed as free Al_2O_3, from the Al_2O_3 content; where SiO_2 is present it is assumed to be present as kaolin in which $Al_2O_3 \simeq SiO_2$; where Fe_2O_3 is present it is assumed to contain 18% Al_2O_3 substituting for Fe_2O_3 (this determined on pure goethite points).

In the above equation for P_2O_5, CaO has the highest coefficient with a one to one relationship between CaO and P_2O_5. But the soil is low in CaO, average 0.14%, and the phosphorus associated with calcium accounts for only about 24% of the total. It is probable that all or some of this phosphorus is that associated with organic matter.

The coefficient for gibbsite is much less than that of iron oxide. It agrees well with the coefficient derived from analyses of pure, or almost pure, gibbsite points which gave a coefficient of 0.0048 for Al_2O_3. Gibbsite is only two thirds Al_2O_3, so the coefficient for the mineral would be about 0.003, i.e. about one seventh of that for iron oxide, and this agrees well with the results of Fordham and Norrish (1979*b*), who found that gibbsite retained about one tenth the amount of arsenate as iron oxide.

The high Fe_2O_3 points in Fig. 24.6 correspond to pure goethite aggregates and those observed in the random sampling showed a considerable variation in phosphorus contents. To try and understand the reason for this variation, additional high Fe_2O_3 points were found and analysed, and these are shown also in Fig. 24.6: they confirm the wide variation in phosphorus content of the goethite particles.

Fig. 24.7 shows that for the goethite particles the phosphorus content is related to their Al_2O_3 content. This Al_2O_3 is not free, it is that substituting for Fe_2O_3 in the goethite structure (Norrish and Taylor 1961). Note that

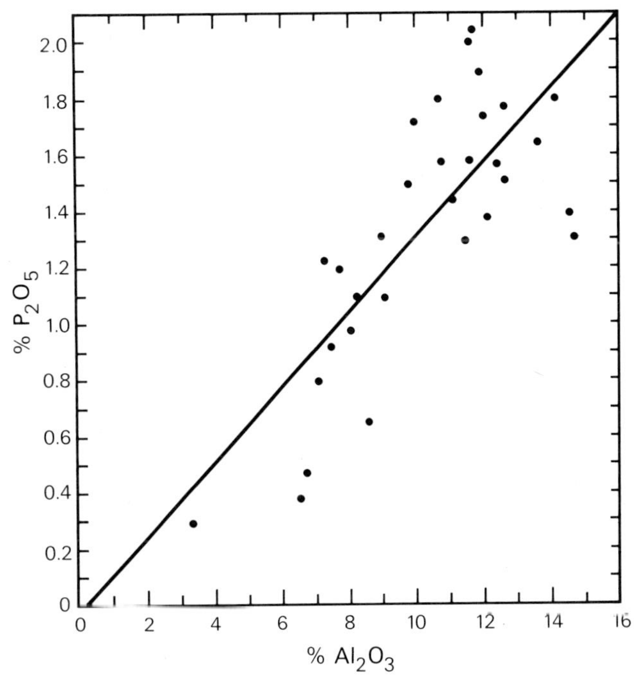

Fig. 24.7. The relationship between P_2O_5 and Al_2O_3 in goethite grains. These are the high Fe_2O_3 points of Fig. 24.6.

the goethites showing this variation in Al_2O_3 content all occur within one soil. The larger goethite aggregates can be handpicked from the soil and X-ray diffraction patterns of individual grains confirm that they are mineralogically pure goethite (no gibbsite) and their lower interlayer spacings are in accord with Al_2O_3 replacing Fe_2O_3 (Norrish and Taylor 1961).

The result shown in Fig. 24.7 is not a 'freak' result, as a similar dependence of P_2O_5 on the Al_2O_3 content of goethite has been observed in about ten other soils. All are soils formed on basalt and are *krasnozems* or *euchrozems*. In about half of the ten soils, several horizons have been studied, and in each case the above effect is more marked in the surface horizons. Both the P_2O_5 and the Al_2O_3 contents of the goethite grains are relatively uniform in the deeper horizons of the soils. It is not known whether this variation of the goethite with depth is a pedogenic effect, but it would seem to be. Although the absolute phosphorus contents of the goethite in the different soils vary, the dependence on Al_2O_3 content resembles that shown in Fig. 24.7. As Al_2O_3 increases, the P_2O_5 content rises steeply and reaches a maximum at about 12% Al_2O_3. At higher Al_2O_3 contents, and some go to 21%, the P_2O_5 content does not increase, and it may even decrease slightly. Usually the goethites containing about 6% Al_2O_3 contain a half to one third of the P_2O_5 content of the 12% Al_2O_3 grains.

There are two obvious ways in which the presence of Al_2O_3 in the goethite structure might increase phosphorus uptake. The presence of Al could increase the specific activity of the goethite surface towards the

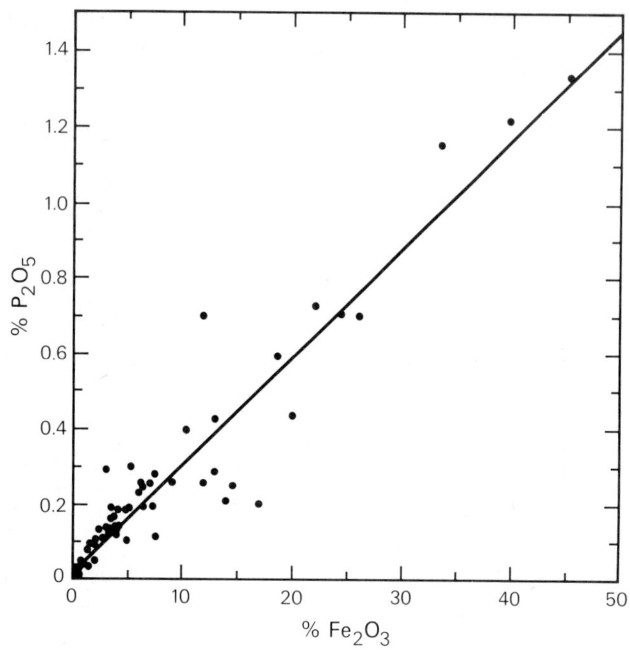

Fig. 24.8. The relationship between P_2O_5 and Fe_2O_3 in a young *podzol* from Queensland.

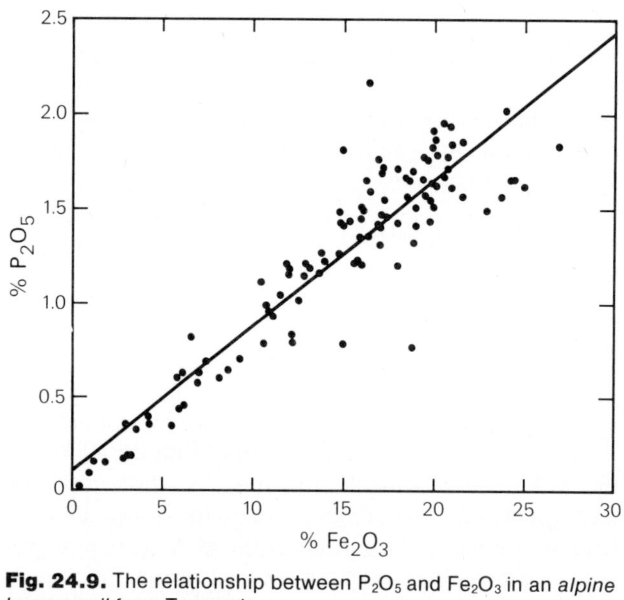

Fig. 24.9. The relationship between P_2O_5 and Fe_2O_3 in an *alpine humus soil* from Tasmania.

phosphorus; alternatively the Al might restrict the particle size of the goethite so that the specific area of the goethite increases with Al content. As yet we have not been able to determine which explanation is correct.

The relationship between P and Al in the goethite, and particularly as it varies with depth, is the more puzzling since almost all the micro-aggregates of goethite are pseudomorphs after ferro-magnesium minerals, generally olivine. The particles are pseudosingle crystals, as optically they behave as single crystals but consist

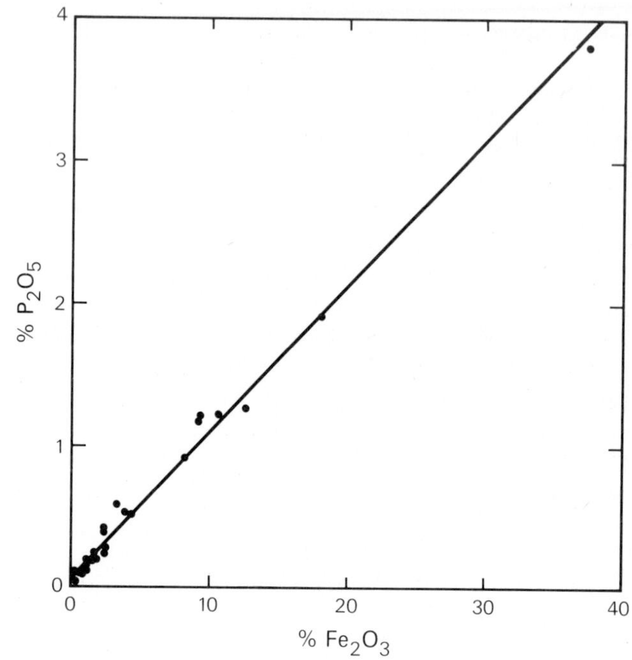

Fig. 24.10. The relationship between P_2O_5 and Fe_2O_3 in organic fragments in a *chernozem* from South Australia.

of three-dimensionally orientated micro-crystals. The grains are similar to those described by Brown and Stephen (1959), but since goethite was the only phase detectable in powder and single crystal X-ray diffraction patterns, the grains appear to be more weathered than those studied by Brown and Stephen.

It is obvious by comparing Figs. 24.5 and 24.6 that the relationship of P_2O_5 to Fe_2O_3 is poorer for Fig. 24.6 than Fig. 24.5, but the above discussion shows that this is not because the P_2O_5 is associated with other phases; it is mainly due to variations within the goethite.

The above examples are for tropical soils that are high in P and Fe_2O_3. Figs. 24.8 and 24.9 show similar relationships for very different soils. The *podzol* of Fig. 24.8 contains only 0.2% Fe_2O_3 yet this accounts for much of the phosphorus in the soil. Sulphur, presumably in organic matter, also gave a significant correlation with phosphorus so that,

$$P_2O_5 = 0.029\ Fe_2O_3 + 0.42\ SO_3 - 0.03$$

The data of Fig. 24.9 were obtained from an *alpine humus soil* from Tasmania, and Fe_2O_3 accounts for practically all the phosphorus. The soils contained 17% organic matter, and the clay contained both hematite and goethite. In this case

$$P_2O_5 = 0.076\ Fe_2O_3 + 0.41\ SO_3 + 0.06$$

Fig. 24.10 shows the very close relationship between P and Fe in the organic fragments present in a *chernozem*; the P_2O_5 content is precisely one tenth that of Fe_2O_3. The very high Fe_2O_3 content of some fragments indicates that the iron is probably not present as an organic complex, but for some reason has precipited as iron oxide in the organic matter.

Many other great soil groups give data where there are highly significant correlations between P_2O_5 and Fe_2O_3, including *prairie soils, black earths, chernozems, red* and *brown podzolic soils, red earths, red-brown earths, calcareous red earths, desert loams, solonized brown soils, non-calcic brown soils, chocolate soils* and *humic gleys.*

In these soils a very significant fraction of the total phosphorus is associated with Fe_2O_3; often the fraction decreases towards the surface as more is associated with sulphur (organic matter).

Thus, it is safe to say that in many Australian soils most of the inorganic phosphorus is retained by the iron oxides and hydroxides. There have been many laboratory measurements made studying the fixation of phosphorus by iron oxides (Hingston *et al.* 1968; Ryden *et al.* 1977*a,b*; Parfitt *et al.* 1975), and it is interesting to compare the results given here with those. Hingston *et al.* (1968) found that at pH 5 goethite adsorbed about 2 μmole P per m^2; Parfitt *et al.* (1975) found at pH 3.5 the corresponding figures for lepidocrocite, hematite and iron hydroxide gel were, 8, 7.7 and 3.7; at pH 6 Ryden *et al.* (1977*a*) found iron hydroxide gel retained about 5 μmole P per m^2.

A triangle enclosing three basal oxygen atoms of a PO_4 tetrahedron would have an area of 23 $Å^2$, so that a perfectly ordered monolayer coverage of the surface would give 7 μmole P per m^2. This estimate is in reasonable agreement with the above laboratory measurements of phosphorus retention.

The amount of phosphorus retained by soil iron oxide will depend on its surface area, but usually such measurements are not available for the iron oxide of soil. However, estimates of particle size of the goethite of some soils have been made by measuring broadening of the X-ray diffraction lines, and these measurements give values of $\sim$ 200 Å diameter. For spheres, surface area and particle size are related by $A = 6/\rho.D$, where A is the surface area in $m^2\ g^{-1}$; ρ is the mineral density, (4.3 g cm^{-3} for goethite); D is the particle diameter in μm. Goethite with 200 Å particles would have a surface area of 70 $m^2\ g^{-1}$, and on the basis of 7 μmole P per m^2 the goethite would be expected to retain $\sim$ 3.5% P_2O_5, or 3.9% P_2O_5 relative to pure Fe_2O_3. The above linear regression equations have coefficients for Fe_2O_3 that range from 0.02 to 0.076, i.e. the iron oxide component retains 2 to 7.6% P_2O_5. For the soils studied, where there is a significant correlation of P_2O_5 with Fe_2O_3, the coefficient for Fe_2O_3 generally lies between 0.005 and 0.04; the 0.076 for the *alpine humus soil*, and the 0.1 of the organic fragments of the *chernozem*, are exceptions in this respect. Both are high in organic matter and this could result in the iron oxides having a higher than usual surface area. Schwertmann (1966) has shown that organic molecules prevent the growth of iron oxide crystals.

Bennema (1980) found that for a variety of latosols in Brazil the P_2O_5 content of the soils could be expressed as,

$$P_2O_5 = 0.007\ Fe_2O_3 + 0.009C + 0.0016,$$

where C = carbon content.

Taylor and Schwertmann (1974) studied ferruginous concretions and the coefficient relating P_2O_5 to Fe_2O_3 averaged 0.003 for concretions from Australian soils; for European concretions from pseudo-gley and loess soils it was 0.01; for concretions from European bog iron ores it was 0.024.

The phosphorus retained by iron oxides will depend on the amount available, and this may be the overriding factor in most cases, but it would also be influenced by surface area of the oxides, pH, competition for the surface by organic matter and various anions, as well as other things, e.g. Al content of goethite. It is surprising then that for a very wide range of soils the retention of phosphorus by iron oxides varies only by a factor of about eight.

From the above considerations it can be seen that for many soils the surfaces of the iron oxides are fractionally saturated with phosphate, and sometimes this fraction might be near unity. Calculation of the coverage in a particular soil depends on determining the surface area of the iron oxide in that particular soil and this is not easy.

Deshpande *et al.* (1968) and Gallez *et al.* (1976) have measured the surface areas of clays prior to, and after, dithionite treatment and have obtained estimates of iron oxide surface areas in this way. Such estimates are liable to positive error as the removal of any other material (organic matter, fine clay, etc) by the dithionite treatment will give too high a surface area for the iron oxide. Gallez *et al.* (1976) obtained results as high as 324 $m^2\ g^{-1}$ Fe_2O_3 on a tropical basaltic soil not dissimilar to those discussed above. An area of 324 $m^2\ g^{-1}$ implies goethite particles of 40 Å size, and such particles are unlikely to be identified by X-ray diffraction methods, yet goethite was identified. By the same methods, Bigham *et al.* (1978*a*) obtained values up to 200 $m^2\ g^{-1}$ for a soil goethite (Norfolk B21t) and this would indicate a particle size of $\sim$ 70 Å; the diffraction pattern of this goethite (Bigham *et al.* 1978*b*) shows line broadening that indicates the particles are $\geqslant$ 200 Å.

The surface area of iron oxides will vary from soil to soil, but since in most soils these minerals can be identified by X-ray diffraction, their surface areas will generally be less than 200 $m^2\ g^{-1}$ (i.e. $>$ 70 Å diameter). This value could be exceeded in swampy soils or in those very high in organic matter. It has been established that the high phosphorus retention by reoxidised flooded soils is due to amorphous iron oxides being formed

(Khalid *et al.* 1977; Willett and Higgins 1978; Kuo and Mikkelsen 1979). As would be expected, such iron oxides show higher than usual retentive capacities, and Khalid *et al.* found that for Louisiana rice soils P_2O_5 uptake was linearly related to oxalate extractable (amorphous) iron oxide with a coefficient of 0.10 for Fe_2O_3. For two such soils Willett and Higgins (1978) obtained coefficients of 0.18 and 0.27 for Fe_2O_3, where the Fe_2O_3 was that extracted by acid ammonium oxalate.

It must be pointed out that in comparing the uptake of phosphorus in laboratory experiments with that present on iron oxide in soils, very different solution concentrations of P are involved. Soil solution concentrations in virgin soils are likely to be less than 0.1 ppm and possibly less than 0.01 ppm, whereas sorption measurements are often made using concentrations of 100 ppm or more. This large discrepancy would seem to invalidate a comparison, but the sorption experiments are usually short term and it is known that phosphate binding increases with time (Munns and Fox 1976). In addition, the isotherms appear to be irreversible (Hingston *et al.* 1974) so that in the desorption cycle the iron oxides can retain phosphorus almost without loss at very low solution concentrations.

Aluminium hydroxides

Gibbsite is the common aluminium hydroxide found in soils and it does not have the very wide distribution that iron oxides have in soils, though gibbsite can be a major component of highly weathered soils.

When analysing the data from random point analyses, it is only rarely that Al gives a significant correlation with P. The contribution from gibbsite is usually only seen by finding and analysing high Al points, and always the coefficient for Al_2O_3 is smaller than that for Fe_2O_3, so that in most soils the latter accounts for most of the phosphorus. When iron oxides and gibbsite are present in the same soil, the coefficient for Fe_2O_3 ranges from 4 to 20 times that for Al_2O_3. As mentioned earlier, Fordham and Norrish (1979*b*) estimated that gibbsite was about one tenth as effective as iron oxides in adsorbing arsenate. Bennema (1980) did not find that free Al_2O_3 contributed to the phosphorus content of Brazilian latosols. There should be some soils, high in gibbsite and low in iron, in which most of the phosphorus is held by gibbsite, but as yet we have not studied such a soil.

Laboratory studies (Cabrera *et al.* 1977) indicate that aluminium oxides and hydroxides are considerably more reactive than the iron oxides in adsorbing phosphorus, yet this is not observed in soils. Gibbsite particles are plate-shaped and the edges are considered to be the only active points for phosphorus retention (Muljadi *et al.* 1966*b*). The comparison made by Cabrera *et al.* (1977) was on the basis of total area and if the surface area of gibbsite was corrected for the non-reactive faces the difference between it and iron oxides would be accentuated. In some soils the reduced activity of gibbsite can be explained by its increased particle size, as, in contrast to goethite, whose particles rarely exceed ~ 200 Å in soils, gibbsite crystals grow to coarse clay, silt, and even fine sand sizes. But in most of the soils examined by us the gibbsite was fine-grained, and Fordham and Norrish (1979*a,b*) used a soil with very small gibbsite particles. Organic matter is known to compete with phosphate for surfaces (Moshi *et al.* 1974), and Nagarajah *et al.* (1970) have shown that various organic molecules are much more effective in competing with phosphate on gibbsite than on goethite.

The high uptake of phosphorus by gibbsite, in laboratory experiments, could be explained, if at high solution concentrations of phosphorus, reactions start that do not take place at the low concentrations found in virgin soils.

Boehmite occurs only rarely in soils, and corundum is present in some West Australian laterites; laboratory experiments indicate that these minerals (see p. 000) would behave the same as gibbsite in retaining phosphorus (Cabrera *et al.* 1977).

In acid soils, in addition to the crystalline aluminium minerals, 'amorphous' aluminium hydroxide occurs; this is generally determined as the Al extracted with Tamm's acid oxalate solution. Most of this aluminium is probably present as exchangeable Al or as charged hydroxy polymers (Rich 1968) that are in exchange positions but are not easily replaced (Hsu and Rich 1960). These forms of Al are associated with the clay minerals, and the lack of association between P and clay minerals (see p. 354) indicates that in virgin soils these forms of Al are not associated with phosphorus. Tamm's acid oxalate solution also extracts the Al bonded to organic matter, but again our evidence is that in most virgin soils this Al is not associated with phosphorus.

From the above it seems that the various forms of Al play only a minor role in retaining native phosphorus, but this does not mean that these forms are unreactive to added phosphate (see p. 351).

Titanium oxides

Anatase and rutile are the titanium oxides found in soils, and laboratory studies show that they can retain phosphorus (Cabrera *et al.* 1977); Fordham and Norrish (1979*b*) also observed that added arsenate was retained by these minerals.

In our microprobe studies of soils, P is observed associated with TiO_2 in some soils, but usually this represents only a small part of the total phosphorus. A *lateritic podzolic soil* from Western Australia is an exception to this in that TiO_2 gave a significant correlation with P and this association accounted for ~ 40% of the total phosphorus. This soil had the pallid zone of a laterite as its parent material and contained almost no free iron oxide.

We have insufficient data to compare the activity of titanium oxides with those of Al or Fe, but the generally

Table 24.8. Microprobe analyses of high phosphorus areas in allophane soils from New Zealand

The sample of silt loam was supplied by the Soil Bureau, D.S.I.R. Wellington, New Zealand. Parent material, andesitic ash. The sample of sandy loam was supplied by Dr W. H. M. Saunders, Ruakura Agricultural Research Centre, Hamilton. This sample had not received phosphate fertilizer. Parent material rhyolitic alluvium with some andesitic ash.

	Egmont black silt loam. B_3 horizon. 46–60 cm. pH = 7					Waihou sandy loam. Surface sample. pH = 5.6			
	Soil Composition (%)								
P_2O_5	0.40	3.38	17.8	25.2	30.4	3.12	9.54	12.9	18.5
Al_2O_3	19.9	21.4	18.0	21.0	19.1	23.3	25.8	23.9	23.5
SiO_2	28.2	26.6	12.8	8.3	4.6	32.2	22.9	19.3	14.3
SO_3	0.27	0.70	0.61	0.36	0.28	1.56	1.45	1.38	0.90
K_2O	0.93	1.20	0.62	0.77	0.56	1.94	0.45	0.48	0.94
Fe_2O_3	6.1	6.54	3.62	7.22	1.23	3.5	4.2	2.4	11.7
CaO	0.46	1.26	3.57	3.69	3.67	1.8	2.6	1.35	2.66
TiO_2	0.82	0.64	0.37	0.52	0.14	0.42	0.33	0.18	0.34
	Number of atoms relative to Al = 1								
P	0.01	0.11	0.71	0.86	1.14	0.096	0.27	0.39	0.57
Si	1.20	1.06	0.60	0.34	0.20	1.18	0.76	0.69	0.52
P+Si	1.21	1.17	1.31	1.20	1.34	1.27	1.03	1.08	1.09

low concentration of Ti in soils would mean that the TiO_2 minerals would have a low capacity for retaining phosphorus.

Ilmenite ($FeTiO_3$) is a common soil component, but microprobe analyses of such grains show low phosphorus contents. It is doubtful if ilmenite forms in the soil, and it is probably always a residue from parent material. Although ilmenite can occur in the clay fractions of soil it occurs in the coarse clay, so even if it had reactivity per unit area comparable to anatase or rutile it would adsorb less phosphorus. Fordham and Norrish (1979*b*) found that ilmenite grains were almost inert as far as arsenate uptake was concerned.

Clay minerals

Although laboratory experiments show that the clay minerals can retain phosphorus (Haseman *et al.* 1950; Low and Black 1950; Muljadi *et al.* 1966*a*), microprobe results indicate that they do not hold it in virgin soils. The reasons for this discrepancy are the same as those discussed under aluminium oxides.

Allophane

Amorphous material is relatively rare in Australian soils, and our studies indicate that such material in not of importance in holding phosphorus in Australian soils. But the same is not true of the high allophane soils of New Zealand, and Table 24.8 shows analyses of some high phosphorus points in such soils. Although the Egmont soil had received fertilizer, it is unlikely that this phosphate would have penetrated to 50 cm. Rajan (1975*a*) and Rajan and Watkinson (1976) have observed for allophanic clays that Si is released as P is fixed. Their experiments involved only 3-hour reaction times and the replacement of Si by P occurred only at relatively high solution concentrations, > 100 ppm P. The data of Table 24.8 indicate that the same reaction may occur, over long periods, at much lower P levels, since the soil solution (0.01 M $CaCl_2$) of the Waihou soil was 0.001 ppm!

When the substitution is small, comparable to the first analysis of Table 24.8, P may substitute directly for surface Si, but at higher levels of P it is likely that a new aluminium phosphate phase forms; if both the allophane and the aluminium phosphate are amorphous then there is no clear-cut criterion for saying which process is operative.

Not all the phosphorus of these New Zealand soils is present in this form, as the results of the random analyses show significant correlations of P with S (organic matter), and Fe_2O_3. Apatite is the phosphate mineral of the rocks and occasional apatite grains were also observed. Presumably they are weathering and releasing P to react with allophane.

Reactions of applied phosphate

When phosphate fertilizers are applied to soils some of the phosphorus is fixed and unavailable for plant growth. This reaction with the soil to render phosphorus unavailable was studied over a century ago and those early workers concluded that phosphorus reacted to form insoluble components with iron and aluminium oxides, and to form calcium phosphate in calcareous soils (Wild 1949). Despite intensive research for the last 50 years, our understanding of phosphorus–soil reactions is inadequate, and Hemwall (1957) in reviewing the subject says 'The present literature on the fixation of phosphorus by soils gives a somewhat confusing picture regarding the reactions involved and the compounds formed'. The picture is almost as confused at the present time. In this section the reaction between phosphate and minerals is discussed. The more chemical aspects of the sorption of phosphate by soils are considered in Chapter 29.

There have been two main methods used in attempting to identify the soil components that retain added phosphorus. The first method involves measuring the capacity of the soil to retain phosphorus and seeing if this is correlated with other soil properties. The second method involves reacting pure minerals with phosphate solutions to see which minerals can retain phosphorus.

In both methods a wide range of phosphate concentrations have been used. Very few studies are made at normal soil solution concentrations, as low as < 0.01 ppm P in solution. Very often the adsorption capacity of a soil is measured by the P it retains when the soil solution is 0.2 ppm (Beckwith 1965); anion exchange capacity is commonly measured using 0.3 M $NH_4H_2PO_4$ at pH 4, i.e. 10^4 ppm P, and other experiments have been made using solutions comparable to those in a wet fertilizer granule, i.e. 4.5 molar in P and pH 1, equals $\sim 10^5$ ppm P. Various intermediate concentrations have been used, and very often adsorption isotherms are determined, i.e. phosphorus uptake as a function of P in solution (Chapter 29).

The studies have indicated that the following soil constituents are important in phosphorus retention.

Calcium

In neutral and alkaline soils, apatite is considered to form from added phosphate, the source of calcium being exchangeable cations, soluble salts or calcium carbonate. However, the initial reaction at low P concentrations appears to be one of adsorption.

Holford and Mattingly (1975*a*) determined adsorption isotherms for 24 calcareous soils, and fitted a two-surface Langmuir equation to their data; they were able to relate the capacity of the soil for low energy adsorption to the surface area of the $CaCO_3$. The low energy adsorption capacity per unit area corresponded to that observed for pure limestones (Holford and Mattingly 1975*b*). Initial solution concentrations ranged from 0.78 to 31 ppm P. The high energy adsorption surfaces in these soils were iron oxides. Fitter and Sutton (1975) fitted Freundlich adsorption isotherms to their data and found that for soils with pH > 5, one of the constants of the isotherm was related to exchangeable Ca. One P atom was adsorbed for each 6 Ca atoms. The isotherms were made using solutions whose initial concentrations ranged from 0.6 to 9 ppm P.

Cole *et al.* (1953) found that phosphate was adsorbed onto the surface of pure $CaCO_3$ for solution concentrations up to 10 ppm P, but that above this concentration di-calcium phosphate, $CaHPO_4.2H_2O$, precipitated. They considered that the latter would slowly, in 1–2 years, alter to hydroxyapatite releasing phosphorus at the same time. Griffin and Jurinak (1973) also studied the adsorption of phosphate on pure $CaCO_3$, and they fitted their data to a two-surface Langmuir equation. High energy adsorption took place at soil solution concentrations of < 0.6 ppm P, and they considered that other adsorption studies using higher concentrations would not have detected this region. They considered, on the basis of solubilities, that their first region represented the formation of hydroxyapatite nuclei on the surface of the carbonate, while the second region commenced when octa-calcium phosphate was nucleating. Amer and Ramy (1971) also considered that octa-calcium phosphate formed when phosphate solutions, 1–2 ppm P, reacted with $CaCO_3$. Holford and Mattingly (1975*b*) also fitted their data, for phosphate adsorption onto $CaCO_3$, to a two-surface Langmuir equation, and they found that the high adsorption region was for soil solutions < 0.5 ppm, a result that agrees well with that of Griffin and Jurinak (1973). However Holford and Mattingly (1975*b*) interpreted the high and low energy adsorption regions differently; they considered the first region corresponded to chemisorption and that the second region was due to physical adsorption, possibly of hydroxyapatite or octa-calcium phosphate on the surface. On the basis of solubility products they considered that above ~ 3 ppm P in solution, octa-calcium phosphate was precipitating; their solutions were undersaturated with respect to di-calcium phosphate.

Larsen (1967) considered that octa-calcium phosphate is unlikely to occur in soils and pointed out that it is not known as a natural mineral and is difficult to synthesize. It is generally postulated because the soil solutions of fertilized soils are supersaturated with respect to hydroxyapatite. He considered that the apparent supersaturation is due to using pure hydroxyapatite in simple solution as a model. The solubilities of fluor-, hydroxy- and carbonate-apatites vary considerably, and each will be influenced by the impurities it is likely to contain. Larsen further points out that the presence of small amounts of calcium carbonate have the effect of making the equilibrium solutions of apatite appear to be supersaturated. Discrepancies in the interpretation of how the P is held in the various adsorption regions are due to the fact that data can be fitted to the equation for an adsorption isotherm even when precipitation is taking place (Hsu and Rennie 1962*b*; Veith and Sposito 1977). Whatever the form of phosphate, octa-calcium, di-calcium or adsorbed, it should slowly convert to apatite, and this is consistent with the findings of Nagelschmidt and Nixon (1944), who found apatite in the carbonate nodules of soil with a long fertilizer history, but that apatite was absent from similar nodules from unfertilized soil.

When strong phosphate solutions at low pH react with

soils, a variety of Ca compounds are formed, some containing Al or Fe (Huffman 1962). Such compounds may form in the contact zone between a fertilizer granule and the soil. The amount of phosphate incorporated into such compounds will depend on the soil and on its moisture content subsequent to fertilization, but most of these compounds are relatively soluble and will be relatively short-lived. However, in soils, little is known of the actual time required for these compounds to revert to less soluble forms.

Exchangeable and active aluminium

Data presented by many workers indicate that the phosphorus sorbing capacities of acid soils are related to some form of active Al in the soils. In very acid soils some Al^{3+} will be present on the exchange complex, but above pH 4.5 hydroxy species will be dominant. The monomers $Al(OH)^{2+}$ and $Al(OH)_2^+$ tend to polymerize to species such as $Al_6(OH)_{15}^{3+}$ (Hsu and Rich 1960), but on the clay surface there can be further polymerization until gibbsite islands are formed (Rich 1968). As the hydroxy polymers grow in size they become more difficult to replace in simple cation exchange reactions. Also, as the polymers grow the effective charge on the aluminium decreases, and Barnhisel (1977) presents data showing that the Al in polymers on clays had an effective charge of only 0.3. This drop in charge is important since it means that the equivalent weight of Al is increased to 90, so that soils of low cation exchange capacity can retain appreciable amounts of Al in this way, $\sim$ 1% Al for each 10 mequiv.%.

Organic matter has a high cation exchange capacity, often > 100 mequiv.%, and in oxidic soils it may dominate the exchange complex. In acid soils Al hydroxy polymers will occupy these exchange sites also, the hydroxy species being preferred by organic matter (Bloom *et al.* 1979). As with clays, this Al is not readily displayed.

The phosphorus fixing powers of soils are often attributed to the presence of amorphous aluminium hydroxide in soils, but there is little evidence that such materials exist in soils as a separate entity, and that which is found probably exists as described above, on the exchange complex of minerals and organic matter. Much of this Al would not readily exchange with simple salt solutions, but complexing agents, such as the Tamm acid oxalate solution, will remove the aluminium.

Hydroxy-aluminium species on clays and organic matter are in the active form and would be able to react with phosphate. This would be particularly true of the aluminium bound to organic matter. Much of the Al on clays can be in interlayer positions which would be less active, although Kodama and Webber (1975) have prepared smectite with interlayers of $AlPO_4 . nH_2O$.

Working with the subsoil of an acid dark clay from Queensland, Beckwith (1965) found that leaching the exchangeable Al from the soil resulted in a large decrease in its phosphate sorption. Coleman *et al.* (1960) attributed P retention in their soils to exchangeable Al and found $\sim$ 1 P atom was fixed for each 1.3 Al. Jai Prakash and Bhasker (1974) measured the P sorption at zero time and this corresponded to $\sim$ 1 P atom per 2 Al atoms on the exchange complex. With time the P sorption increased but by then the relationship with exchangeable Al was poorer. Fitter and Sutton (1975) related one of the constants of the Langmuir isotherm to exchangeable Al, 1 P atom being fixed for 6 Al.

Williams *et al.* (1958) and Williams (1960) found that the P sorption capacities of 40 acid Scottish soils were closely related to the Al extracted by Tamm acid oxalate solution, and after removal of this Al the soils lost about 90% of their sorption capacity. There were significant correlations of sorption capacities with organic carbon and loss on ignition, both of which were correlated with Tamm Al, so they concluded that most of the aluminium responsible for fixing phosphate was complexed with organic matter, 1 P atom was fixed for 1 Al.

Bromfield (1965), working with acid soils from eastern Australia found that Tamm Al gave the best correlation with P sorption, but the correlation was improved when Tamm Al + Fe was used. Also he found that after the Tamm extraction the reduction in sorption capacity was much less than for Scottish soils. The explanation of this difference would seem to be in the higher iron oxide contents of the Australian soils ($\sim$ 8%, whereas the Scottish soil averaged 1.4% Fe_2O_3). The soils showing considerable reduction after Tamm extraction were the two *alpine humus soils* and the soil that was the lowest in free Fe_2O_3 (0.5%). Bromfield's results are consistent with Tamm Al and free Fe_2O_3 both contributing to P sorption. Bromfield found that when he progressively removed free iron, using dithionite with an acetate buffer, the adsorption capacity of most soils increased! He attributed this to 'activation' of aluminium by the treatment. It is now well known that the soil iron oxides can be high in Al, and if free iron removal treatments are carried out in the absence of an ion that will complex Al, the Al is likely to precipitate, and this will enhance the adsorption capacity of the soil.

For a range of New Zealand soils Saunders (1965) found a good relation between Tamm Al and phosphorus sorption capacity, and as with Scottish soils, Tamm Al was correlated with organic matter, indicating that most of the active Al was complexed with the organic matter. Saini and MacLean (1965) found for New Brunswick soils that P sorption capacity was highly correlated with Al removed by dithionite extraction; organic matter and Al were correlated and a multiple regression analysis indicated that both contributed to P retention; 1.9 P atoms were adsorbed for each Al. Vijayachandran and Harter (1975) obtained similar results with soils from Puerto Rico in that extractable Al and organic matter were significantly correlated with P

Table 24.9. Phosphate adsorption by a cation exchange resin partially saturated with hydroxy-aluminium polymers. Data from Robarge and Corey (1979)

P concn in solution (ppm)	Al/P (molar ratio)	Al^{3+} (%) on exchange complex
0.004	17.4	–
0.009	10.2	26
0.015	6.8	22
0.023	4.5	25
0.033	3.6	25
0.052	2.7	22
0.21	1.9	18
20.0	1.3	16
96.0	1.2	10

adsorption, but they considered that the Al was associated with clay rather than organic matter.

All the above results indicate that in many acid soils, hydroxy-aluminium polymers attached to clay or organic matter are responsible for most of the P sorption of these soils. Robarge and Corey (1979) have investigated the P sorption by a cation exchange resin prepared with hydroxy-aluminium polymers on the exchange complex.

Table 24.10. Microprobe analyses of soil organic matter fragments

Values are percentages. **1.** *Brown earth*, N.S.W. Average of 4 fragments whose compositions were variable. **2.** *Lateritic podzolic soil*, W.A. Average of 3 fragments whose compositions were variable. **3.** *Podzol*, Fraser Is., Qld. Analysis of spores shown in Fig. 24.11. The composition of spores and other organic matter of this soil was reasonably uniform. **4.** *Red earth*, N.T. Average of 15 fragments of reasonably uniform composition. **5.** From the same soil as 4, but sampled near a fertilizer pocket. Average of 20 fragments.

	1	2	3	4	5
P_2O_5	0.12	0.48	0.02	0.006	9.30
SO_3	0.59	0.29	0.10	0.087	0.097
K_2O	0.33	0.51	0.31	0.29	0.38
CaO	2.50	4.77	0.09	0.57	3.33
SiO_2	0.09	0.10	0.74	0.29	0.26
Al_2O_3	1.93	1.46	15.2	8.30	7.67
Fe_2O_3	0.18	0.25	0.99	0.57	0.53

Overall the resin showed many of the characteristics of soils in terms of the effects of pH and salt concentration. Table 24.9 shows some data taken from the paper by Robarge and Corey. For mole ratios of Al/P of > 2, the

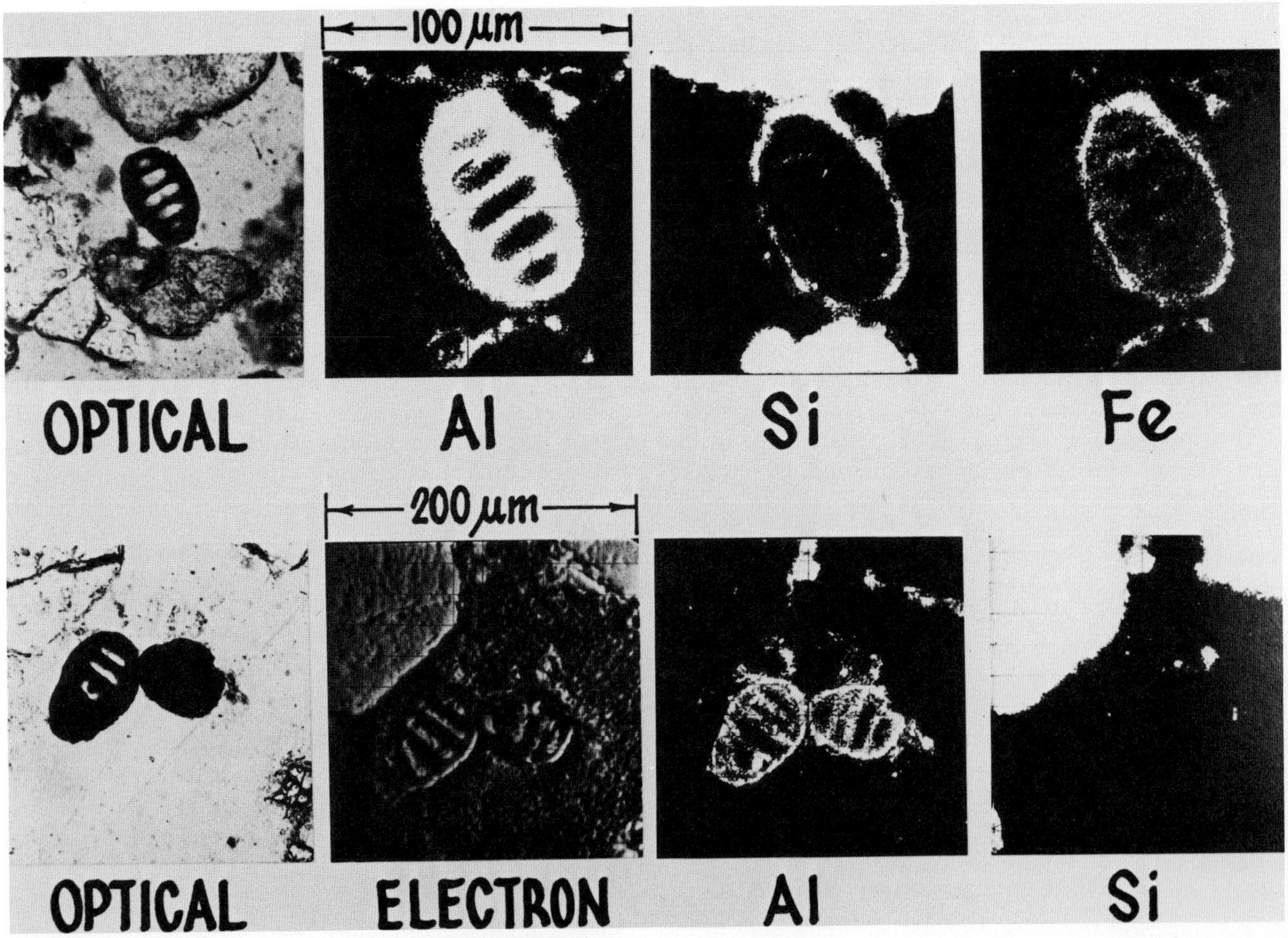

Fig. 24.11. Microprobe results showing the high Al content of spores in a *podzol* from Fraser Is., Queensland. See Table 24.10. Other organic matter is high in Al.

adsorption complex maintains a concentration in solution of < 0.2 ppm P. It is of interest to recall that 0.2 ppm P in solution is that suggested by Beckwith (1965) and others as the concentration above which phosphorus will not limit growth. As can be seen from Table 24.9, once the Al/P ratio drops below 2 the phosphorus in solution increases rapidly and the authors suggest that precipitation is then taking place and Al/P tends to unity. The constancy of exchangeable Al^{3+} on the resin during phosphate adsorption (Table 24.9), indicates that P reacts preferentially with the hydroxy-aluminium polymers, and only reacts with Al^{3+} at the higher P concentrations when precipitation is taking place.

Hsu and Rennie (1962*b*) reacted phosphate with a cation exchange resin saturated with Al^{3+}, and they present evidence that precipitation was taking place even though phosphorus retention conformed to Langmuir and Freundlich adsorption isotherms. The lowest concentrations used were ~ 1 ppm P and even at this concentration Robarge and Corey (1979) recorded precipitation. The Al/P ratio of the precipitate was 1 and this ratio agrees with that obtained on soils by Williams *et al.*

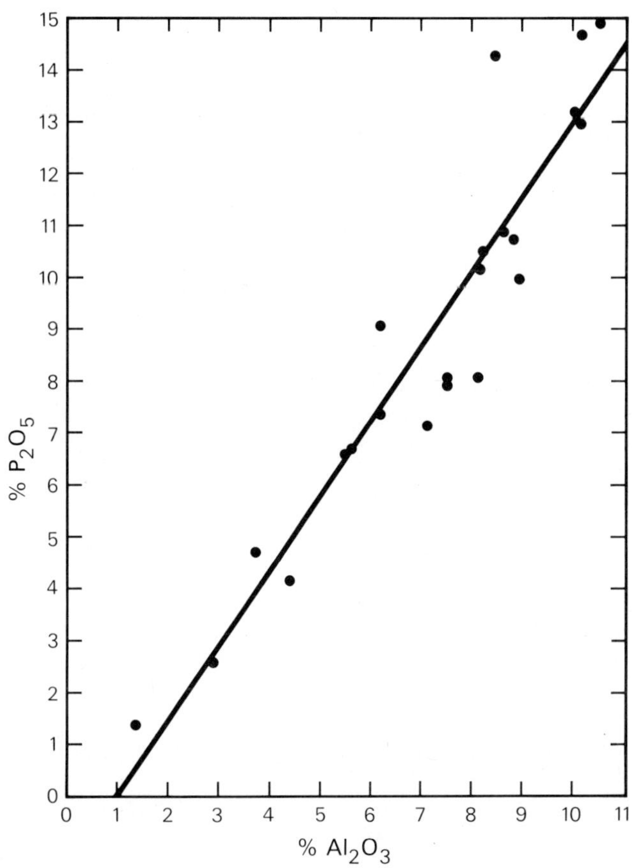

Fig. 24.12. Showing the relationship between P_2O_5 and Al_2O_3 in organic fragments near a fertilizer pocket. See Table 24.10. *Red earth* from the Northern Territory.

(1958) and Coleman *et al.* (1960), who used high concentrations of P in measuring sorption. Saini and MacLean (1965) obtained an Al/P ratio of ~ 0.5, but this could be due to using a gentle extractant in estimating reactive Al. Jai Prakash and Bhasker (1974) obtained a ratio of 2 when phosphorus adsorption was measured at zero time.

If all the reactive aluminium is available to react with phosphorus, then the Al/P ratio of 2 observed by Robarge and Corey (1979), for low solution concentrations, would correspond to the phosphate being held as a bridging complex (Kyle *et al.* 1975; Rajan 1975*b*; Parfitt *et al.* 1977), each P atom being bonded via oxygen to two Al atoms.

Much of the above data indicate that Al complexed with organic matter is important in reacting with added phosphorus. Martin and Reeve (1957) found large amounts of Al complexed with organic matter in podzolic soils from Australia, Scotland and New Zealand. The organic matter of the four soils they analysed averaged ~ 16% Al_2O_3. Microprobe studies confirm the high aluminium content of organic matter in many soils. It is not possible to analyse the fine humus that is intimately associated with clay, but larger organic fragments can be analysed, and some analyses are shown in Table 24.10. Even when organic fragments with recognizable structures are analyzed there can be some doubt as to whether there is contamination from very fine soil minerals. Fig. 24.11 shows Al in organic matter where there is no ambiguity in the results. The spores in Fig. 24.11 were intact until cut during specimen preparation, yet despite their large size they are high in Al, up to 15% Al_2O_3, throughout their volume (sample 3, Table 24.10). Other organic matter in this soil, root fragments and amorphous organic matter, had a similar composition.

In Table 24.10, analyses 4 and 5 are of particular interest as they are from the same soil, but the latter was near a fertilizer pocket into which superphosphate and other fertilizers had been placed. The samples were taken a year later, when 1 m of rain had fallen.

Fig. 24.12 shows that the amount of phosphorus retained by the organic fragments, near the fertilized pocket, was related to the aluminium content of the organic matter. Analysis 5 gives 1.1 Al atoms for each P, and this is consistent with the results of other workers discussed above. If the Al of organic matter can fix P on a 1:1 basis, then the organic matter represented by analyses 1–3 (Table 24.10) can retain appreciable phosphorus, and would make a significant contribution to the sorption capacity of a soil even where the organic matter is not high; it is likely to be the major contributor in many soils.

When Robarge and Corey (1979) studied phosphate adsorption using an exchange resin they stated, 'Of course, the resin model cannot be directly applied to soils', but where P sorption is by Al attached to organic matter, the resin model would seem to be a very apt one.

If phosphorus concentrations in the soil are high

enough (> 2 ppm) to produce precipitation, it is likely that the aluminium phosphate will be sufficiently soluble to release phosphorus for plant growth. The availability of the P adsorbed on an Al–organic matter complex will depend on the degree of saturation by phosphorus, but the results in Table 24.9 indicate that as the degree of saturation drops the complex would not maintain soil solution at a level that is adequate for good plant growth.

Aluminium hydroxides

Gibbsite, $Al(OH)_3$, is the common aluminium hydroxide that occurs in soils, but boehmite, AlOOH, is also found. As indicated in the previous section we do not consider that amorphous aluminium hydroxide occurs in soils as an independent phase.

Hsu and Rennie (1962*a*) studied the adsorption of phosphate by amorphous aluminium hydroxide and found that there was an initial very rapid surface adsorption, and after a time this was followed by precipitation. Bache (1964), working with gibbsite, found a surface adsorption at low solution concentrations of phosphorus; at higher solution concentrations he concluded, on solubility grounds, that precipitation had commenced. Muljadi *et al.* (1966*a,b,c*) also studied gibbsite, and they divided their isotherms into three regions and considered that phosphate reacted only at the edge faces of the platy crystals. High energy adsorption, Region 1, was considered to be due to the phosphate being attached to an Al atom at the crystal edge, but subsequently it was proposed that this phosphate was held as a bridging complex attached to two Al atoms (Kyle *et al.* 1975; Parfitt *et al.* 1977). Muljadi *et al.* (1966*a*) considered that this high energy part of the isotherm extended to ~ 3 ppm P in solution. Region III, that of low energy adsorption, took place at solution concentrations in the range 30–3000 ppm P, and although this region showed some of the characteristics of precipitation, Muljadi *et al.* (1966*b*) rejected this possibility in favour of one involving penetration of phosphate into less crystalline regions of the gibbsite surface. Chen *et al.* (1973) using α-Al_2O_3, and van Riemsdijk *et al.* (1975, 1977) using α-Al_2O_3 and amorphous $Al(OH)_3$ observed a new solid phase when the compounds were reacted with solutions initially containing 10 ppm P. Both detected the new phase with the electron microscope, and van Riemskijk *et al.* (1975, 1977) also obtained electron and X-ray diffraction patterns from it. In both cases the precipitation continued, at pH 5, when solution concentrations had dropped to a few ppm P. Although gibbsite was not used, it seems likely that it would react in the same way to form a precipitate as found by Bache (1964). The experiments of Chen *et al.* and of van Riemsdijk *et al.* were conducted for several months, whereas Muljadi *et al.* (1966*a*) used a reaction time of 48 hours.

It seems that in a general way, and perhaps even in detail, the reactions of gibbsite with phosphate resemble those of the hydroxy polymers discussed in the previous section. At low solution concentrations, phosphate will be adsorbed on the edges of gibbsite crystals, while at higher concentrations precipitation will occur. In the sorption region the capacity of gibbsite will be unfavourable when compared to the polymers. Muljadi *et al.* (1966*a*) used a gibbsite with a surface area of 48 $m^2\ g^{-1}$, and although the particle size of gibbsite can vary greatly from soil to soil, the specimen would approximate in surface area those occurring in many soils.

In region 1 this specimen retained 15 mmole of P per 100 g; the estimated edge area was 19 $m^2\ g^{-1}$, so that each phosphate ion occupied ~ 21 Å^2 indicating a monolayer coverage of the edge surface. However, for polymers in which all Al ions are accessible, and on the basis of each PO_4 bonding to two Al atoms, the polymer would have an adsorption capacity of 600 mmole of P per 100 g, i.e. a capacity about 40 times greater than the gibbsite. The soil conditions that produce gibbsite are those that favour the Al hydroxy-polymers, so in many soils, and particularly those with appreciable organic matter, the capacity of the hydroxy-polymers will exceed that of gibbsite.

In solutions of higher phosphorus concentration, Al from gibbsite will form a precipitate with the phosphate and then the capacity, per unit mass, of the gibbsite will be comparable to the Al-hydroxy-polymers. So in a soil containing gibbsite, the capacity of gibbsite is likely to exceed that of the polymers. However, with gibbsite the reactions leading to precipitation are much slower, and so its capacity to react with phosphate is likely to go undetected in most (rapid) methods of measuring sorption capacity. Because of its crystallinity, the transition from surface adsorption to reactions giving precipitates, is likely to occur at somewhat higher solution concentrations for gibbsite than for the hydroxy-polymers which seem to be at ~2 ppm. The species formed as a precipitate will influence the concentration at which precipitation commences, and so the presence of other ions will be important in this respect.

To what extent gibbsite will react with phosphate to form precipitates will depend on the time the soil solution concentrations are maintained at levels of a few ppm or more, but it seems likely that in a fertilized soil such concentrations could persist for weeks or even months, and then gibbsite would exhibit a high sorption capacity; as mentioned above this capacity is unlikely to be measured by rapid laboratory determinations of sorption capacity.

From the above it is not surprising that there are almost no data implicating gibbsite in phosphorus retention, but Sree Ramulu *et al.* (1967) and Sree Ramulu and Pratt (1970) found that the A_1 horizon of the Cecil soil, which contained gibbsite, had an anomolously high sorption capacity.

Little is known yet of the crystalline phase that forms when phosphate reacts with aluminium hydroxide, but it is likely to be sufficiently soluble to supply phosphorus for plant growth. In the contact zone between the fertilizer granule and the soil, gibbsite will react quickly with phosphate to form new phases that will depend on the soil and fertilizer chemistries (Haseman *et al.* 1950). The evidence is that such phases are comparable to monocalcium phosphate as fertilizers (Lindsay and Taylor 1960; Haseman *et al.* 1950) and are likely to be short lived.

Clay minerals

If clay minerals have Al^{3+} or hydroxy-aluminium polymers as part of their exchangeable cations, then as discussed above, phosphorus can be retained by this aluminium (Muljadi *et al.* 1966*b*).

Low and Black (1950) studied the reactions of phosphate with kaolin. At high phosphate concentrations the kaolin precipitated aluminium phosphate and released SiO_2. At low solution concentrations they concluded that phosphorus was retained by surface adsorption, the phosphate ions replacing surface hydroxyl ions.

Muljadi *et al.* (1966*a, b, c*) studied the adsorption of phosphate onto kaolinite and, as with gibbsite, they divided the adsorption isotherm into three regions. However, Chen *et al.* (1973), using solutions < 10 ppm P and pH <7, considered that kaolinite underwent a similar slow reaction with phosphate as does α-Al_2O_3, to form an aluminium phosphate precipitate. Kafkafi (1968) and Kafkafi and Bar-Yosef (1969) reported that silica was released during phosphate adsorption by kaolinite, and concluded that dissolution of the kaolinite was taking place even in dilute solutions. The experiments of Chen *et al.* and Kafkafi involved much longer reaction times (several months) than the adsorption experiments of Muljadi *et al.* (1966*a*), which were for 24 hours. Rajan and Fox (1975), using a 6-day reaction period, found that structural silicon was released from a kaolinitic soil at concentrations of ~ 50 ppm P.

It seems that kaolinite behaves like gibbsite in that at very low solution concentrations, ~ <3 ppm P, there is surface adsorption of phosphate bound to the aluminum of the edge faces of the crystals (Muljadi *et al.* 1966*a*). At higher solution concentrations there is slow dissolution of the kaolinite with the formation of aluminium phosphates that may incorporate other cations.

Illite and smectite adsorb phosphorus at low, < 10 ppm P, solution concentrations; Pissarides *et al.* (1968) found that in the pH range 6–7, kaolinite, illite and smectite retained comparable amounts of phosphate; clays saturated with divalent ions sorbed about twice as much as those with monovalent cations, and they attributed this to diffuse double layer effects. Edzwald *et al.* (1976) also studied the adsorption of phosphorus by kaolinite, illite, and smectite in the pH range 7–8 and found that the capacity for sorption was greatest for illite, being ~ 28 times that of kaolinite, and 3.4 times that of smectite. Although the above workers did not find evidence of precipitation, Edzwald *et al.* (1976) related the sorption capacity of the clays to the acid extractable iron. There is ample evidence (see below) that in high phosphorus concentrations dissolution of all the clay minerals occurs, but there are no data to indicate whether illite and smectite form aluminium phosphate at low concentrations when the reaction time is long.

Rajan and his co-workers (Rajan 1975*a*; Rajan and Perrott 1975; Rajan and Fox 1975; Rajan and Watkinson 1976) have shown that the initial adsorption of phosphate onto allophane is by displacement of hydroxyl ions and adsorbed silicate. At higher solution concentrations of phosphorus, the retention of phosphate is accompanied by the release of structural silicon from the allophane. This release of structural silicon appeared to commence when solution concentrations were ~ 100 ppm P, and it would seem to be indicative of the formation of an aluminium phosphate phase. Rajan used equilibrium times of 3 hours, and it seems very likely that with much longer times aluminium phosphate would form at lower concentrations. Veith and Sposito (1977) reacted allophanes with phosphate for up to 20 days and considered equilibrium had not been reached in that time. Amorphous 'variscite' formed with the release of $Si0_2$.

In strong phosphate solutions there is ample evidence that all the clay minerals are dissolved to form aluminium phosphates. Low and Black (1947, 1950) using solutions ranging from 0.25 to 1.5 M in phosphate, showed that kaolinite decomposed with release of $Si0_2$ to form aluminium phosphate. Haseman *et al.* (1950) reacted kaolinite, illite and smectite with 0.01, 0.1 and 1 molar solutions at various temperatures with pH from 3 to 7. They concluded that there was an initial rapid reaction (adsorption) and that this was followed by a slower reaction in which the clay was decomposed. The speed of the latter reaction was increased by increasing the phosphate concentration in solution, by increasing the temperature, or by lowering the pH. Haseman *et al.* obtained X-ray diffraction data and chemical analyses of the crystalline phosphates formed. To simulate the environment at the contact zone of the soil and a fertilizer granule, Lindsay *et al.* (1959*a*) reacted soils with solutions 4.5 molar in P and pH 1.0. Most of the soil constituents were dissolved to some extent, and there was subsequent precipitation of a variety of calcium, aluminium and iron phosphates, some containing Na, K or NH_4. In later experiments Lindsay *et al.* (1962) identified 30 crystalline compounds from the reaction of soil with strong phosphate solutions. When phosphate reacts with clays to precipitate iron or aluminium phosphates, the species that precipitate will depend on the presence of other cations (Ca, Mg, K, etc.) derived from the fertilizer itself or from the clay.

In the absence of other cations, clays and gibbsite react with phosphate to give an amorphous aluminium phosphate (Taylor and Gurney 1962; Gebhardt and Coleman 1974). Veith (1978) found the compound formed in the pH range 2.5 to 8.6 and had a solubility product pK_{so} = 28.06, and although he considered it to be stable over a wide pH range, in the higher pH region, 5–7, it would be stable only in the presence of relatively high concentrations of phosphorus in solution; approximately 50 ppm at pH 5 and much greater than this at pH 6.

All the phosphate compounds formed by precipitation are sufficiently soluble to act as good fertilizers (Lindsay and Taylor 1960).

When the phosphate sorption capacity of soils is measured quickly in the laboratory, the conditions are usually such that there will be very little retention due to clay decomposition, and the result will depend on surface adsorption. In many soils, the hydroxy aluminium polymers on the organic matter, or attached to the clay itself, will contribute more to adsorption than the edges of the clay crystals *per se*.

Iron oxides and oxyhydroxides

Unlike aluminium, at normal soil pH values, iron is not stable as Fe^{3+} or as hydrolyzed species on the exchange complex of clays, and laboratory attempts to prepare iron-saturated clays usually result in the precipitation of iron hydroxide. Although iron may complex with organic matter, microprobe analyses (Table 24.10) usually show much less iron than aluminium, and this is consistent with the results of Martin and Reeve (1957) whose data give an average of 16% Al_2O_3, but only 1.1% Fe_2O_3 for the organic matter of four soils.

Goethite (α-FeOOH) and hematite (α-Fe_2O_3) are the common and most abundant iron oxides in soils, but lepidocrocite (γ-FeOOH), maghemite (γ-Fe_2O_3) and amorphous, or almost amorphous iron oxides (ferrihydrite) also occur in soils (see Chapter 23). Magnetite (Fe_3O_4) and ilmenite ($FeO.TiO_2$) in soils are often residual from the parent rock; these minerals are usually coarser-grained than other iron oxides, and there is little evidence that they retain phosphorus in soils.

There is ample evidence that iron oxides control or contribute to the sorption capacity of some soils; Holford and Mattingly (1975*a*) showed that the high energy adsorption capacity of calcareous soils was correlated with dithionite extractable iron; Bigham *et al.* (1978*b*) found that the surface area of the iron oxide controlled phosphate retention, and because the goethites in their soils were very fine-grained, they adsorbed more than hematite; Saunders (1965) and Lopez-Hernandez and Burnham (1974), studying a range of soils, observed that free iron oxides were important in only some soils; Coleman *et al.* (1960) found for most of their soils that exchangeable Al was the main factor controlling sorption, but free Fe_2O_3 was important in soils that were high in iron oxides and these soils retained their sorption capacity when exchangeable Al was removed; Sree Ramulu and Pratt (1970) related the speed of dissolution of dicalcium phosphate dihydrate to the free iron oxide contents of their soils, and the phosphate needed to saturate the soils was also related to free iron oxide contents.

Soils that are subjected to prolonged periodical flooding show increased sorption capacities on oxidation; the reducing conditions during flooding dissolve goethite and hematite, and amorphous iron oxides precipitate with high surface areas when the soils are drained to produce oxidizing conditions. Amorphous iron oxides are extracted with acid ammonium oxalate, so the iron extracted by this reagent correlates well with the sorption capacity of the soils (Khalid *et al.* 1977; Willet and Higgins 1978; Kuo and Mikkelsen 1979).

Fordham and Norrish (1974, 1979*b*), using radioactive arsenate as a proxy for phosphate, combined autoradiography, electron microscopy and microprobe analysis to locate and analyze the sorbing particles dispersed on an electron microscope grid. In this way they showed quite unambiguously in a variety of acid soils that iron oxides were responsible for most of the sorption, with gibbsite and TiO_2 making small but significant contributions.

Posner and Quirk and their co-workers have studied the adsorption of phosphate by goethite (Hingston *et al.* 1968, 1972, 1974; Atkinson *et al.* 1972); phosphate was strongly and irreversibly adsorbed and to explain this they postulated that it formed a binuclear bridging complex the PO_4 being attached to two iron atoms, the same as with gibbsite; desorption was more difficult from goethite than from gibbsite. Russell *et al.* (1974) and Parfitt and Russell (1977) confirmed the bridging complex using infrared spectrometry. Ryden *et al.* (1977*a, b*) studied phosphate sorption by ferric oxide gel and concluded that three mechanisms operated; at very low solution concentrations, $<$ 0.02 ppm P, chemisorption was important (this could be the binuclear bonding referred to above); bonding was weaker in the concentration range from 0.02 to 1 ppm P, with phosphates replacing hydroxyls; from 1 to 20 ppm P in solution the phosphate was adsorbed as a potential-determining ion. Ryden and Syers (1977) found that with the ferric oxide gel, irreversibility increased with time, and they atttributed this to phosphate moving from region III (physical adsorption) to regions I or II (chemisorption). Parfitt *et al.* (1975) and Cabrera *et al.* (1977) studied adsorption by goethite, hematite and lepidocrocite, the former workers studying $Fe(OH)_3$ gel as well. Hematite has hydroxylated surfaces, and in a general way all the iron oxides behave the same, although Parfitt *et al.* (1975) found the adsorption per unit area was greatest for hematite, and Cabrera *et al.* (1977) showed that while the binding energies for various goethites and lepidocrocite were similar, that of hematite was greater.

In strong phosphate solutions iron oxides react to form various iron phosphates (Kittrick and Jackson 1955; Lindsay and Stephenson 1959; Lindsay *et al.* 1962), many of which are sufficiently soluble to be comparable to monocalcium phosphate as a fertilizer.

There are no data showing that goethite or hematite react to form iron phosphates in solutions of low P concentration, and the indications are that the iron oxides are more stable than the aluminium oxides in this respect.

As discussed earlier in this chapter, the iron oxides are largely responsible for the retention of native phosphorus in most Australian soils, and although the sorption capacities of soils are sometimes related to their free iron oxide contents, it is surprising that this relationship is not a lot more common.

To some extent it is the fact that iron oxides retain native phosphorus that restricts their contribution to laboratory sorption capacity measurements. In native soils, as shown earlier, the iron oxides can retain several per cent P_2O_5 while gibbsite and the hydroxy aluminium polymers are very low in P_2O_5. When the soil solution is increased in phosphorus, the iron oxides' capacity is limited due to the fact that they are already partly, or even largely, saturated. Because the various forms of aluminium are unsaturated in the soil, they exhibit their full capacity for sorption when the phosphorus level is raised.

As discussed earlier, each aluminium atom of hydroxy aluminium polymers on organic matter or on clay is available, whereas with iron oxides only those iron atoms exposed at particle surfaces can react. As pointed out for gibbsite, the capacity for adsorption on small crystals is likely to be about one fortieth that of an equal weight of active aluminium hydroxide.

Viewed in this way it is possible to reconcile many of the differing results. Williams *et al.* (1958) worked with soils low in iron oxides and high in organic matter; the soils were acid and so the organic matter would be high in aluminium and this would dominate phosphorus sorption. Holford and Mattingly (1975*a*) found free iron oxides were important, but they were working with calcareous soils whose high pH would not favour the presence of hydroxy aluminium polymers. Fordham and Norrish (1974, 1979*b*) observed little uptake (of arsenate) by any minerals other than iron oxides, but their sorption measurements were made with very low concentrations, comparable to those in the soil solution (equivalent ~ 0.03 ppm P), and under these conditions the iron oxides will remain unsaturated. Then uptake will be governed by the energy of adsorption, rather than by capacity.

General discussion of phosphorus retention

Although it is perhaps a depressing thought to the researcher looking for a simple explanation, the retention of phosphate by soils is very complex. Solution concentrations, moving from a fertilizer granule into the soil, can change by seven or eight orders of magnitude with respect to phosphorus, and under these circumstances it is not surprising that a variety of different reactions take place. Also, with the exception of quartz, it seems that all the common clay components of soils can react with phosphorus. Most soils retain phosphorus no matter how, or in what concentration, it is applied, and it is the variety of reactions, in terms of mechanisms and the different minerals involved, that makes this possible. The only soils that show appreciable losses of phosphorus from fertilizer are siliceous sands with almost no clay minerals (Ozanne *et al.* 1961; Gillman 1973).

A great many detailed mechanisms have been proposed to explain phosphate retention by the various minerals of soils, however, a certain pattern of behaviour seems to be common for all the minerals studied, as follows.

A. At very low solution concentrations there is rapid and strong chemisorption of phosphate. As solution concentrations increase, larger amounts are adsorbed, but the energy of adsorption decreases.

B. Even at moderately low phosphate concentrations there can be slow reactions leading to some dissolution of minerals, and the precipitation of a phosphate. At high solution concentrations there is rapid dissolution of minerals, and they simply become sources of Al and Fe for producing phosphates.

The solution concentrations at which stage B commences will depend on the mineral involved as well as pH and other soil factors, but, as discussed above, precipitation may commence at low concentrations where adsorption is still active; i.e. regions A and B can overlap.

Iron oxides have a very high energy of adsorption for phosphate, and this results in the native mineral phosphate of many soils, acid and alaline, being retained by iron oxides. Despite the strong retention of phosphate by iron oxides, it seems probable that in several respects iron oxide with adsorbed phosphate is an ideal native phosphate fertilizer. The first requirement of such a fertilizer is that phosphate is retained in the soil, even when leached for geological periods. Any mineral grains that persist under such conditions will be so insoluble that they will not act as sources of P for plant growth, e.g. monazite, xenotime and gorceixite.

The phosphate is on the surface of iron oxides of very high surface area and yet is retained under even severe leaching conditions. Such retention would imply almost complete unavailability of the P for plant growth, yet plants do seem to have access to this P even if the amounts available are short of optimum. It seems likely that plants have special mechanisms for releasing the phosphate bound to iron oxides. Changes of pH, and particularly of Eh, around a root could act to make phosphorus more available. Roots excrete organic matter, and some fractions of this might displace phosphate from iron oxide surfaces; even if the organic matter does not compete for the iron oxide surfaces, its presence in the soil could lead to increased bacterial activity with consequent complexing or reduction of the iron oxides, and release of P, in the vicinity of a root. If plants use one or more of the above mechanisms the high surface area of the iron oxides would aid in the release of phosphate.

In Australian soils aluminium hydroxides and aluminium hydroxy polymers retain very little native phosphate compared to iron oxides. Phosphate is observed associated with Al in some of the allophanic soils of New Zealand, but these soils are very low in free iron oxides. When the concentration of phosphate in solution is raised, more phosphate is bound to Al, so Al can be dominant in retaining the P of added fertilizer. But this P is held less strongly than that adsorbed by iron oxide. Barrow (1980) observed that despite the very high capacity of the allophanic soils to retain P, they showed a higher residual value than that of Australian soils.

In laboratory sorption experiments the use of high phosphorus concentrations and short times simulates the reactions near a fertilizer grain. The products of such reactions are likely to be comparatively soluble and short lived.

Phosphorus fixation in soils will usually involve low phosphorus concentrations and long reaction times. Most phosphorus adsorption studies have been made at comparatively high phosphorus concentrations; there is need to make these at much lower concentrations that are comparable to soil solutions. The recent studies of Chen *et al.* (1973) and van Riemedijk *et al.* (1977) have shown that a crystalline aluminium phosphate forms when aluminium oxide and aluminium hydroxide react with surprisingly low, $<$ 10 ppm, concentrations of phosphorus. There is a need to identify the product of this reaction and to establish the minimum phosphorus concentration at which precipitation occurs, and to determine how these are affected by the presence of other cations. Similar information is needed for the other soil minerals that react with phosphate, i.e. clay minerals, iron oxides and carbonates.

References

Adams, J.A., Hawarth, D. T. and Campbell, A. S. (1973). Plumbogummite minerals in a strongly weathered New Zealand Soil. *J. Soil. Sci.* **24**, 224-31.

Altschuler, Z. S. (1973). The Weathering of phosphate deposits — geochemical and environmental aspects. *In* 'Environmental Phosphorus Handbook' (Eds E. J. Griffith, A. Beeton, J. M. Spencer and D. T. Mitchell). pp.33-96. (John Wiley and Sons: New York.)

Amer, F. and Ramy, A. (1971). On the possibility of characterizing calcium phosphates in calcareous soils by isotopic exchange. *J. Soil Sci.* **22**, 267-74.

Atkinson, R. J., Posner, A. M. and Quirk, J. P. (1972). Kinetics of isotopic exchange of phosphate at the α-FeOOH–aqueous solution interface. *J. Inorg. Nucl. Chem.* **34**, 2201–11.

Bache, B. W. (1963). Aluminium and iron phosphate studies relating to soils. I. Solution and hydrolysis of variscite and strengite. *J. Soil Sci.* **14**, 1-123.

Bache, B. W. (1964). Aluminium and iron phosphate studies relating to soils. II. Reactions between phosphate and hydrous oxides. *J. Soil Sci.* **15**, 110-6.

Barnhisel, I. R. (1977). Chlorites and hydroxy interlayered vermiculite and smectite. *In* 'Minerals in Soil Environments' (Ed. R. C. Dinauer). (Soil Sci. Soc. Am.: Madison.)

Barrow, J. (1980). Differences amongst a wide ranging collection of soils in the rate of reactions with phosphate. *Aust. J. Soil Res.* **18**, 215-24.

Beckwith, R. S. (1965). Sorbed phosphate at standard supernatant concentration as an estimate of the phosphate needs of soils. *Aust. J. Exp. Agric. Anim. Husb.* **5**, 52-8.

Bennema, J. (1980). Total phosphorus in relation to total carbon in Brazilian oxisols. *Int. Soc. Soil. Sci.*, CLAMATROPS Conf., Kuala Lumpur. In press.

Bigham, J. M., Golden, D. C., Bowen, L. H., Buol, S. W. and Weed, S. B. (1978*a*). Iron oxide mineralogy of well-drained ultisols and oxisols: I. Characterisation of iron oxides in soil clays by Mössbauer spectroscopy, X-ray diffractometry, and selected chemical techniques. *Soil Sci. Soc. Am. J.* **42**, 816-24.

Bigham, J. M., Golden, D. C., Buol, D. W., Weed, S. B. and Bowen, L. H. (1978*b*). Iron oxide mineralogy of well-drained ultisols and oxisols: II. Influence on color, surface area and phosphate retention. *Soil Sci. Soc. Am. J.* **42**, 825-30.

Bloom, P. R., McBride, M. B. and Weaver, R. M. (1979). Aluminium organic matter in acid soils: Buffering and solution aluminium activity. *Soil Sci. Soc. Am. J.* **43**, 488-93.

Bromfield, S. M. (1965). Studies of the relative importance of iron and aluminium in the sorption of phosphate by some Australian soils. *Aust. J. Soil Res.* **3**, 31-44.

Brown, G. and Stephen I. (1959). A structural study of iddingsite from New South Wales, Australia. *Am. Mineral.* **44**, 251-60.

Cabrera, F., Madrid, L. and Arambarri, P.De. (1977). Adsorption of phosphate by various oxides: Theoretical treatment of the adsorption envelope. *J. Soil Sci.* **28**, 306-13.

Cescas, M. P. and Tyner, E. H. (1970). Distribution of apatite and other mineral inclusions in rhyolitic pumice ash and beach sands from New Zealand: An electron-microprobe study. *J. Soil Sci.* **21**, 79-84.

Chang, S. C. and Jackson, M. L. (1957). Fractionation of soil phosphorus. *Soil Sci.* **84**, 133-44.

Chen, Y. S. R., Buttler, J. N. and Stumm, W. (1973). Kinetic study of phosphate reaction with aluminium oxide and kaolinite. *Environ. Sci. Technol.* **7**, 327-32.

Clark, J. S. and Peech, M. (1955). Solubility criteria for the existence of calcium and aluminium phosphates in soils. *Soil Sci. Soc. Am. Proc.* **19**, 171-74.

Cole, C. V., Olsen, S. R. and Scott, C. O. (1953). The nature of phosphate sorption by calcium carbonate. *Soil Sci. Soc. Am. Proc.* **17**, 352-6.

Coleman, N. T., Thorup, J. T. and Jackson, W. A. (1960). Phosphate and sorption reactions that involve exchangeable Al. *Soil Sci.* **90**, 1-7.

Crack, B. J. and Isbell, R. F. (1971). Studies on some neutral red duplex soils (Dr 2.12) in north-eastern Queensland. 1. Morphological and chemical characteristics. *Aust. J. Exp. Agric. Anim. Husb.* **11**, 328-35.

Deshpande, T. L., Greenland, D. J. and Quirk, J. P. (1968). Changes in soil properties associated with the removal of iron and aluminium oxides. *J. Soil Sci.* **19**, 108-22.

Dunn, P. J. and Appleman, D. E. (1977). Perhamite, a new calcium aluminium silico-phosphate mineral, and a re-examination of viseite. *Mineral. Mag.* **41**, 437-42.

Edzwald, J. K., Toensing, D. C. and Leung, M. C. (1976). Phosphate adsorption reactions with clay minerals. *Environ. Sci. Technol.* **10**, 485-90.

Emigh, G. D. (1973). Economic phosphate deposits. *In* 'Environmental Phosphorus Handbook' (Eds E. J. Griffith, A. Beeton, J. M. Spencer and D. T. Mitchell). pp.97-116. (John Wiley and Sons: New York.)

Fields, M., Bealing, G., Claridge, G. G., Wells, N. and Taylor, N. H. (1960). Mineralogy and radioactivity of Niue island soils. *N.Z. J. Sci.* **3**, 658-75.

Fitter, A. H. and Sutton, C. D. (1975). The use of the Freundlich isotherm for soil phosphate adsorption data. *J. Soil Sci.* **26**, 241-46.

Fordham, A. W. and Norrish, K. (1974). Direct measurement of the composition of soil components which retain added arsenate. *Aust. J. Soil Res.* **12**, 165-72.

Fordham, A. W. and Norrish, K. (1979*a*). Electron microprobe and electron microscope studies of soil clay particles. *Aust. J. Soil Res.* **17**, 283-306.

Fordham, A. W. and Norrish, K. (1979*b*). Arsenate-73 uptake by components of several acidic soils and its implications for phosphate retention. *Aust. J. Soil Res.* **17**, 307-16.

Fry, W. H. (1913). The condition of the soil phosphoric acid insoluble in hydrochloric acid. *J. Ind. Eng. Chem.* **5**, 665-6.

Gallez, A., Juo, A. S. R. and Herbillon, A. J. (1976). Surface and charge characteristics of selected soils in the tropics. *Soil Sci. Soc. Am. J.* **40**, 601-8.

Gebhardt, H. and Coleman, N. T. (1974). Anion adsorption by allophanic tropical soils: III. Phosphate adsorption. *Soil Sci. Soc. Am. Proc.* **38**, 263-6.

Gillman, G. P. (1973). Studies on some deep sandy soils in Cape York Peninsula, North Queensland. 3. Losses of applied phosphorus and sulphur. *Aust. J. Exp. Agric. Anim. Husb.* **13**, 418-22.

Grange, L. I. and Fox, J. P. (1953). Soils of the Lower Cook Group. *New Zealand D.S.I.R. Soil Bur. Bull. No. 8.*

Griffin, R. A. and Jurinak, J. J. (1973). The interaction of phosphate with calcite. *Soil Sci. Soc. Am. Proc.* **37**, 847-50.

Haseman, J. F., Brown, E. H. and Whitt, C. D. (1950). Some reactions of phosphate with clays and hydrous oxides of iron and aluminium. *Soil Sci.* **70**, 257-71.

Hemwall, J. B. (1957). The fixation of phosphorus by soils. *Adv. Agron.* **9**, 95-112.

Hingston, F. J., Atkinson, R. J., Posner, A. M. and Quirk, J. P. (1968). Specific adsorption of anions on goethite. Trans. 9th Int. Congr. Soil Sci. (Adelaide). Vol. 1, pp.669-78.

Hingston, F. J., Posner, A. M. and Quirk, J. P. (1972). Anion adsorption by goethite and gibbsite. I. The role of the proton in determining adsorption envelopes. *J. Soil Sci.* **23**, 177-92.

Hingston, F. J., Posner, A. M. and Quirk, J. P. (1974). Anion adsorption by goethite and gibbsite. II. Desorption of anions from hydrous oxide surfaces. *J. Soil Sci.* **25**, 16-26.

Holford, I. C. R. and Mattingly, G. E. G. (1975*a*). The high- and low-energy phosphate adsorbing surfaces in calcareous soils. *J. Soil Sci.* **26**, 407-17.

Holford, I. C. R. and Mattingly, G. E. G. (1975*b*). Phosphate sorption by Jurassic oolitic limestones. *Geoderma* **13**, 257-64.

Hsu, P. H. and Rennie, D. A. (1962*a*). Reactions of phosphate in aluminium systems. I. Adsorption of phosphate by X-ray amorphous 'aluminium hydroxide'. *Can. J. Soil Sci.* **42**, 197-209.

Hsu, P. H. and Rennie, D. A. (1962*b*). Reactions of phosphate in aluminium systems. II. Precipitation of phosphate by exchangeable aluminium on a cation exchange resin. *Can. J. Soil Sci.* **42**, 210-21.

Hsu, P. H. and Rich, C. I. (1960). Aluminium fixation in a synthetic cation exchanger. *Soil Sci. Soc. Am. Proc.* **24**, 21-5.

Huffman, E. O. (1962). Reactions of phosphate in soils: Recent research by TVA. *Fert. Soc. Proc.* **71**, 5-35.

Isbell, R. F., Stephenson, P. J., Murtha, G. G. and Gillman, G. P. (1976). Red basaltic soils in North Queensland. *CSIRO Aust. Div. Soils Tech. Pap. No. 28.*

Jai Prakash, B. S. and Bhasker, T. D. (1974). Exchangeable aluminium and phosphorus sorption of some acid soils of Mysore state. *Soil Sci.* **118**, 243-6.

Johnson, R. W. (Ed). (1976). 'Volcanism in Australia'. (Elsevier Scientific Publishing Co: New York.)

Joplin, G. A. (1963). Chemical Analysis of Australian Rocks. Part I: Igneous and Metamorphic. Aust. Bur. Miner. Resour. Geol. Geophys. Bull. No. 65.

Joplin, G. A. (1965). Chemical Analyses of Australian Rocks. Part II: Sedimentary Rocks. Aust. Bur. Miner. Resour. Geol. Geophys. Bull. No. 78.

Joplin, G. A. (1975). Chemical Analyses of Australian Rocks. Part III: Igneous and Metamorphic, Supplement 1961-1969. Aust. Bur. Min. Resour. Geol. Geophys. Bull. No. 146.

Kafkafi, U. (1968). Hydrogen consumption and silica release during initial stages of phosphate adsorption on kaolinite at constant pH. *Israel J. Chem.* **6**, 367-73.

Kafkafi, U. and Bar-Yosef, B. (1969). The effect of pH on the adsorption and desorption of silica and phosphate on and from kaolinite. Proc. 3rd Int Clay Conf. (1969). Vol. 1, pp.691-6.

Khalid, R. A., Patrick, W. H. and De Laune, R. D. (1977). Phosphorus sorption characteristics of flooded soils. *Soils Sci. Soc. Am. J.* **41**, 305-10.

Kittrick, J. A. and Jackson, M. L. (1955). Rate of phosphate reaction with soil minerals and electron microscope observations on the reaction mechanism. *Soil Sci. Soc. Am. Proc.* **19**, 292-5.

Koch, L. E. (1956). Phosphate mineralization in alluvials near Cobargo. *Aust. J. Sci.* **18**, 200-2.

Kodama, H. and Webber, M. D. (1975). Clay-inorganic studies. II. Hydroxy aluminium phosphate-montmorillonite complex. *Can. J. Soil Sci.* **55**, 225-33.

Koritnig, S. (1965). Geochemistry of phosphorus. I. The replacement of Si^{4+} by P^{5+} in rock-forming silicate minerals. *Geochim. Cosmochim. Acta* **29**, 361-71.

Kuo, S. and Mikkelsen, J. S. (1979). Distribution of iron and phosphorus in flooded and unflooded soil profiles and their relation to phosphorus adsorption. *Soil Sci.* **127**, 18-25.

Kyle, J. H., Posner, A. M. and Quirk, J. P. (1975). Kinetics of isotopic exchange of phosphate adsorbed in gibbsite. *J. Soil Sci.* **26**, 32-43.

Larsen, S. (1967). Soil phosphorus. *Adv. Agron.* **19**, 151-210.

Leahey, A. (1935). Mineralogical and chemical studies on some of the inorganic phosphorus compounds in the soil. *Sci. Agric.* **15**, 704-12.

Lee, K. E. (1974). Royal Society and Percy Sladen Expedition to the New Hebrides, 1971. Collection data. 77 pages. (Royal Society: London.)

Lindsay, W. L. and Stephenson, H. F. (1959). Nature of the reactions of monocalcium phosphate monohydrate in soils: II. Dissolution and precipitation reactions involving iron, aluminium, manganese and calcium. *Soil Sci. Soc. Am. Proc.* **23**, 18-22.

Lindsay, W. L. and Taylor, A. W. (1960). Phosphate reaction products in soil and their availability to plants. Proc. 7th Int. Congr. Soil Sci. (Madison). Vol. 3, pp.580-5.

Lindsay, W. L. and Vlek, P. L. G. (1977). Phosphate minerals. *In* 'Minerals in Soil Environments' (Ed. R. C. Dinauer). (Soil Sci. Soc. Am.: Madison.)

Lindsay, W. L., Frazier, A. W. and Stephenson, H. F. (1962). Identification of reaction products from phosphate fertilizers in soils. *Soil Sci. Soc. Am. Proc.* **26**, 446-52.

Lindsay, W. L., Lehr, J. R. and Stephenson, H. F. (1959*a*). Nature of the reactions of monocalcium phosphate monohydrate in soils: III. Studies with metastable triplepoint solution. *Soil Sci. Soc. Am. Proc.* **23**, 342-5.

Lindsay, W. L., Peech, M. and Clark, J. S. (1959*b*). Solubility criteria for the existance of variscite in soils. *Soil Sci. Soc. Am. Proc.* **23**, 357-60.

Lopez-Hernandez, D. and Burnham, C. P. (1974). The covariance of phosphate sorption with other soil properties in some British and tropical soils. *J. Soil Sci.* **25**, 196-206.

Low, P. F. and Black, C. A. (1947). Phosphate-induced decomposition of kaolinite. *Soil Sci. Soc. Am. Proc.* **12**, 180-4.

Low, P. F. and Black, C. A. (1950). Reactions of phosphate with kaolinite. *Soil Sci.* **70**, 273-90.

Martin, A. E. and Reeve, R. (1957). Chemical studies of podzolic illuvial horizons. II. The use of acetylacetone as extractant of translocated organic matter. *J. Soil Sci.* **8**, 279-86.

McKelvery, V. E. (1973). Abundance and distribution of phorphorus in the lithosphere. *In* 'Environmental Phosphorus Handbook' (Eds. E. J. Griffith, A. Beeton, J. M. Spencer and D. T. Mitchell). pp.13-31. (John Wiley and Sons: New York.)

McManus, W. R., Anthony, R. G., Grout, L. L., Malin, A. S. and Robinson, U. N. E. (1979). Biocrystallization of mineral material on forage plant cell walls. *Aust. J. Agric. Res.* **30**, 635-49.

Moshi, A. O., Wild, A. and Greenland D. J. (1974). Effect of organic matter on the charge and phosphate adsorption characteristics of kikuyu red clay from Kenya. *Geoderma* **11**, 275-85.

Muljadi, E., Posner, A. M. and Quirk, J. P. (1966*a*). The mechanism of phosphate adsorption by kaolinite, gibbsite and pseudoboehmite. Part I. The isotherms and the effect of pH on adsorption. *J. Soil Sci.* **17**, 212-29.

Muljadi, D., Posner, A. M. and Quirk, J. P. (1966*b*). The mechanism of phosphate adsorption by kaolinite, gibbsite and pseudoboehmite. Part II. The location of the adsorption sites. *J. Soil Sci.* **17**, 230-7.

Muljadi, D., Posner, A. M. and Quirk, J. P. (1966*c*). The mechanism of phosphate adsorption by kaolinite and gibbsite and pseudoboehmite. Part III. The effect of temperature on adsorption. *J. Soil Sci.* **17**, 238-47.

Munns, D. N. and Fox, R. L. (1976). The slow reaction which continues after phosphate adsorption: Kinetics and equilibrium in some tropical soils. *Soil Sci. Soc. Am. J.* **40**, 46-51.

Nagarajah, S., Posner, A. M. and Quirk, J. P. (1970). Competitive adsorption of phosphate with polygalacturonate and other organic anions on kaolinite and oxide surfaces. *Nature* (London) **228**, 83-5.

Nagelschmidt, G. and Nixon, H. L. (1944). Formation of apatite from superphosphate in the soil. *Nature* (London) **154**, 428-9.

Nicolls, K. D. and Tucker, B. M. (1956). Pedology and chemistry of the basaltic soils of the Lismore district N.S.W. CSIRO Aust. Soil Publ. No. 7.

Norrish, K. (1957). Some phosphate minerals of soils. Proc. 2nd Aust. Conf. Soil Sci. Vol 1(1), pp.17/1-15.

Norrish, K. (1968). Some phosphate minerals of soils. Trans. 9th Int. Congr. Soil Sci., (Adelaide). Vol. 2, pp.713-23.

Norrish, K. (1974). Geochemistry and mineralogy of trace elements. *In* 'Trace Elements in Soil and Plant-Animal Systems' (Eds D. J. D. Nicholas and A. R. Egan). (Academic Press: London.)

Norrish, K. and Taylor, R. M. (1961). The isomorphous replacement of iron by aluminium in soil goethites. *J. Soil Sci.* **12**, 294-306.

Ozanne, P. G., Kirton, D. J. and Shaw, T. C. (1961). The loss of phosphorus from sandy soils. *Aust. J. Agric. Res.* **12**, 409-423.

Palache, C., Berman, H. and Frondel, C. (1951). 'The System of Mineralogy II'. (John Wiley: New York.)

Parfitt, R. L. and Russell, J. D. (1977). Adsorption on hydrous oxides. IV. Mechanisms of adsorption of various ions on goethite. *J. Soil Sci.* **28**, 297-305.

Parfitt, R. L., Atkinson, R. J. and Smart, R. St. C. (1975). The mechanism of phosphate fixation by iron oxides. *Soil Sci. Soc. Am. Proc.* **39**, 837-41.

Parfitt, R. L., Fraser, A. R., Russel, J. D. and Farmer, V. C. (1977). Adsorption on hydrous oxides. II. Oxalate, benzoate and phosphate on gibbsite. *J. Soil Sci.* **28**, 40-7.

Pissarides, A., Stewart, J. W. B. and Rennie, D. A. (1968). Influence of cation saturation on phosphorus adsorption by selected clay minerals. *Can. J. Soil Sci.* **48**, 151-7.

Plummer, J. K. (1915). Petrography of some North Carolina soils and its relation to fertilizer requirements. *J. Agric. Res.* **5**, 569.

Quershi, R. H., Jenkins, D. A., Davies, R. I. and Rees, J. A. (1969). Applications of microprobe analyses to the study of phosphorus in soils. *Nature* (London) **221**, 1142-3.

Rajan, S. S. S. (1975*a*). Phosphate adsorption and the displacement of structural silicone in an allophane clay. *J. Soil Sci.* **26**, 250-66.

Rajan, S. S. S. (1975*b*). Adsorption of divalent phosphate on hydrous aluminium oxide. *Nature* (London) **253**, 434-6.

Rajan, S. S. S. and Fox, R. L. (1975). Phosphate adsorption by soils: II. Reactions in tropical acid soils. *Soil Sci. Soc. Am. Proc.* **39**, 846-51.

Rajan, S. S. S. and Perrott, K. W. (1975). Phosphate adsorption by synthetic amorphous aluminosilicate. *J. Soil Sci.* **26**, 257-66.

Rajan, S. S. S. and Watkinson, J. H. (1976). Adsorption of selenite and phosphate on an allophane clay. *Soil Sci. Soc. Am. J.* **40**, 51-4.

Ribbe, P. H. and Smith, J. V. (1966). X-ray-emission microanalyses of rock forming minerals. IV. Plagioclase feldspars. *J. Geol.* **74**, 217-31.

Rich, C. I. (1968). Hydroxy interlayers in expansible layer silicates. *Clays Clay Miner.* **16**, 15-30.

Robarge, W. P. and Corey, R. B. (1979). Adsorption of phosphate by hydroxy-aluminium species on a cation exchange resin. *Soil Sci. Soc. Am. J.* **43**, 481-7.

Ruhe, R. V., Cady, J. G. and Gomez, R. S. (1961). Paleosols of Bermuda. *Bull. Geol. Soc. Am.* **72**, 1121-42.

Russell, J. D., Parfitt, R. L., Fraser, A. R. and Farmer, V. C. (1974). Surface structures of gibbsite goethite and phosphated goethite. *Nature* (London) **248**, 220-1.

Rutherford, G. K. (1962). Chemical and morphological data and some observations on the pedogenesis of the soils of the West Kimberley area, Western Australia. CSIRO Aust. Div. Land Res. Reg. Sur. Div. Rep. 62/3.

Ryden, J. C. and Syers, J. K. (1977). Desorption and isotopic exchange relationships of phosphate sorbed by soils and hydrous ferric oxide gel. *J. Soil Sci.* **28**, 596-609.

Ryden, J. C., McLaughlin, J. R. and Syers, J. K. (1977*a*). Mechanisms of phosphate sorption by soils and hydrous ferric oxide gel. *J. Soil Sci.* **28**, 72-92.

Ryden, J. C., McLaughlin, J. R. and Syers, J. K. (1977*b*). Time-dependent sorption of phosphate by soils and hydrous ferric oxides. *J. Soil Sci.* **28**, 585-95.

Saini, G. R. and MacLean, A. A. (1965). Phosphorus retention capacities of some New Brunswick soils and their relationship with soil properties. *Can. J. Soil Sci.* **45**, 15-18.

Saunders, W. M. H. (1965). Phosphate retention by New Zealand soils and its relationship to free sesquioxides, organic matter, and other soil properties. *N.Z. J. Agric. Res.* **8**, 30-57.

Saunders, W. M. H. and Williams, E. G. (1955). Observations on the determination of total organic phosphorus in soils. *J. Soil Sci.* **6**, 254-67.

Schroo, H. (1963). A study of highly phosphatic soils in a karst region of the humid tropics. *Neth. J. Agric. Sci.* **11**, 209-31.

Schwertmann, U. (1966). Inhibitory effect of soil organic matter on the crystallization of amorphous ferric hydroxide. *Nature* (London) **212**, 645-6.

Skipp, R. F. and Matelski, R. P. (1960). A microscopic determination of apatite and a study of phosphorus in some Nebraska soil profiles. *Soil Sci. Soc. Am. Proc.* **24**, 450-2.

Smith, J. V. and Ribbe, P. H. (1966). X-ray emission microanalysis of rock-forming minerals. III. Alkali feldspars. *J. Geol.* **74**, 197-216.

Sree Ramulu, U. S. and Pratt, P. F. (1970). Dissolution of dicalcium phosphate in relation to iron oxide content of acid soils. *Soil Sci.* **109**, 35-9.

Sree Ramulu, U. S., Pratt, P. F. and Page, A. L. (1967). Phosphorus fixation by soils in relation to extractable iron oxides and mineralogical composition. *Soil Sci. Soc. Am. Proc.* **31**, 193-6.

Stace, H. C. T., Hubble, C. D., Brewer, R., Northcote, K. H., Sleeman, J. R., Mulcahy, M. J. and Hallsworth, E. G. (1968). 'A Handbook of Australian Soils'. (Rellim Technical Publications: South Australia.)

Swaby, R. J. (1956). Microbiological assay of available phosphates. Aust. Plant Nut. Conf. Proc. Vol. 1, pp.193-9.

Syers, J. K., Williams, J. D. H., Campbell, A. S. and Walker, T. W. (1967). The significance of apatite inclusions in soil phosphorus studies. *Soil Sci. Soc. Am. Proc.* **31**, 752-6.
Taylor, A. W. and Gurney, E. L. (1962). Solubility of amorphous aluminium phosphate. *Soil Sci.* **93**, 241-5.
Taylor, R. M. and Schwertmann, U. (1974). The association of phosphorus with iron in ferruginous soil concretions. *Aust. J. Soil Res.* **12**, 133-45.
Trueman, N. A. (1965). The phosphate, volcanic and carbonate rocks of Christmas Island, Indian Ocean. *J. Geol. Soc. Aust.* **12**, 261-83.
Vijayachandran, P. K. and Harter, R. D. (1975). Evaluation of phosphorus adsorption by a cross section of soil types. *Soil Sci.* **119**, 119-26.
van Riemsdijk, W. H., Westrate, F. A. and Bolt, G. H. (1975). Evidence for a new aluminium phosphate phase from reaction rate of phosphate with aluminium hydroxide. *Nature* (London) **257**, 473-4.
van Reimsdijk, W. H., Westrate, F. A. and Beek, J. (1977). Phosphates in soils treated with sewerage water. III. Kinetic studies on the reaction of phosphate with aluminium compounds. *J. Environ. Qual.* **6**, 26-9.
Veith, J. A. (1978). Formation of X-ray amorphous aluminium o-phosphates from precipitation and secondary precipitation. *Z. Pflanzenernaehr. Bodenkd.* **141**, 29-42.
Veith, J. A. and Sposito, G. (1977). On the use of the Langmuir equation in the interpretation of 'adsorption' phenomena. *Soil Sci. Soc. Am. J.* **41**, 697-702.
Wild, A. (1949). Retention of phosphate by soil. A review. *J. Soil Sci.* **4**, 231-8.
Wild, A. (1957). The phosphate content of Australian soils: Tabulated data. CSIRO Aust. Div. Soils Div. Rep. No. 10/57.
Wild, A. (1958). The phosphate content of Australian Soils. *Aust. J. Agric. Res.* **9**, 193-204.
Willet, I. R. and Higgins, M. L. (1978). Phosphate sorption by reduced and reoxidized rice soils. *Aust. J. Soil Res.* **16**, 319-26.
Williams, E. G. (1960). Some aspects of phosphate retention and availability in soils. Proc. 7th Int. Congr. Soil Sci. (Madison). Vol. 3, pp.604-10.
Williams, E. G., Scott, N. M. and McDonald, M. J. (1958). Soil properties and phosphate sorption. *J. Sci. Food Agric.* **9**, 551-9.

Part 2
Chemistry

Chapters 25-29

Chapter 25
Micronutrients

K. G. Tiller

Soils contain small amounts of many elements. Sometimes the actual concentrations are difficult to measure accurately. For all elements there is a limit below which reliable analyses cannot be obtained, but these limits are being continually lowered by new methods and modern instrumentation. In past decades, many elements which were detected with difficulty were often quoted in analysis as 'trace amounts'. The resultant term 'trace elements' continues in use, even though the trace amounts can now often be measured quickly and accurately. Many of these trace elements in soils are essential for the growth and reproduction of living things. In this chapter, in accordance with common practice, those trace elements known to be essential for either animal or plant nutrition will be referred to as *micronutrients*; the term trace elements will be retained as a general term for elements present in soils and used by animals and plants in small amounts, without consideration of their physiological function, if any. The micronutrients of importance for plant growth are Cu, Zn, Fe, Mn, Co, B, Mo, Cl, Na and possibly V (Nicholas 1975) For animal production, the elements I, Se, Cr, Sn, Si and Ni should also be added, and B be omitted (Underwood 1975).

The discovery of the importance of these micronutrients is a continuing process. Iron was shown in about 1860 to be needed by plants, but the essentiality of Co for symbiotic bacteria was demonstrated one hundred years later. The list of essential elements is probably not yet complete. However, future additions may be of less practical importance to animal or plant production; they are likely to be elements, such as was the case with Cl, which are ubiquitous and needed in small amounts. Many micronutrients can affect living processes, both at low concentrations, as a deficiency, and at high concentrations, when toxicity or nutrient imbalance may occur. Readers are referred to reviews, such as by Nicholas (1975) and Underwood (1975), for discussions of the physiological functions of the micronutrients.

Studies on micronutrients in Australian soils by Australian researchers, are emphasized in the following sections.

Sources of micronutrients

The micronutrient content of soils is derived mainly from parent materials, but significant contributions, especially to surface soils, may arise from industrial and urban pollution, agricultural sprays, and incidentally as contaminants in fertilizers and irrigation waters.

Soil parent materials as a source of micronutrients

The micronutrient content of soils that is inherited from the parent rock depends upon the kind and intensity, and duration of weathering processes operating since exposure of the present land surface. The type and intensity of weathering processes are related to the climatic zone, whereas the duration of the processes depends on land stability, because glaciation, erosion and sedimentation and other processes, may generate new soil surfaces. In regions of comparatively young and weakly weathered land surfaces, a close relation is evident between composition of parent rock and associated soils. This association was well documented in the excellent review by Mitchell (1955), which stressed such studies on Scottish soils, and the studies by Wells (1960) on soils of the south-east Pacific region. The successful use in some countries of soil sampling as a basis of geochemical prospecting for minerals in underlying rocks, depends on the close relationship between rocks and soil composition. In young landscapes, the main soil process which affects the distribution of trace elements in soil profiles is surface accumulation by plants. However, in older landscapes, especially under conditions of intense weathering, pedogenic factors assume a much greater role than parent material in determining the trace metal status of soils. In fact, Oertel (1961) has shown statistically for many Australian soils that the trace element content of soil profiles is not closely related to the composition of parent materials, but rather to undetermined pedogenic factors. Of course, extremes in rock geochemistry, e.g. basic igneous rocks and siliceous sand, will always be reflected in the composition of their derived soils. The range in the composition of some micronutrients in rocks is indicated in Table 25.1 (data from Wedepohl 1969-78).

Table 25.1. Micronutrient concentrations in rocks

Values are expressed in μg per gram, from Wedepohl (1969-78)

Rock	Co	Cu	Mn	Zn	Mo	B
Granite	4	13	325	48	1.1	12
Diorite, andesite	21	53	1390	70	0.8	—
Gabbro, basalt	45	90	1390	100	1.2	8
Ultrabasic	110	47	1050	56	0.3	—
Limestone	∿1	6	550	20	0.4	20
Bituminous shales	10	95	150	200	70	—
Shales and clays	19	35	∿600	110	∿1	130
Upper continental crust	—	24	480	60	1.2	∿10

Agricultural chemicals and fertilizers as sources of micronutrients

Although micronutrients are often mixed with fertilizers as required, the major nutrient fertilizers often contain appreciable amounts as incidental impurities. The extent of this contamination in the various fertilizer sources of the world has been reviewed by Swaine (1962). Table 25.2 summarizes some information for superphosphate, an important fertilizer source. These impurities arise from the raw materials, and are also introduced during their processing: e.g. for superphosphate, from the phosphate rock and the sulphuric acid used in the manufacture. Some fertilizer sources contain sufficient of some micronutrients to benefit crop growth (Ozanne *et al.* 1965); unfortunately, undesirable impurities such as Cd may also be present (Williams and David 1973). Superphosphate application of 100 kg ha^{-1} would only contribute, assuming the mean concentrations shown in Table 25.2, the Cu and Zn required by 1 tonne of dry matter of normal composition, and part of this contribution would be lost by leaching, or fixation by soil colloids.

Metal-containing sprays are often used to control pests and plant diseases of horticultural crops, and, when continued over several decades, can increase the concentrations of metals in surface soils appreciably (Frank *et al.* 1976); for example, sprays of copper oxychloride may increase the Cu concentrations in soil by several hundred micrograms per gram of soil, to the extent that establishment of new crops after orchard clearance can be difficult (Delas 1963).

Industrial pollution as a source of micronutrients

The implications of the accession of toxic by-products of industrial and urban activity to local soils are now well recognized; appreciable amounts of the micronutrients may also arise from the same sources. Urban and industrial pollution contributes toxic and trace elements by many means: e.g. from burning refuse, oil and coal, by volatilization from heating natural metal-containing materials such as clay, by the abrasion of car tyres, and by emissions from metal-processing plants and smelters. The effects of industrial contamination on crops and soils, air and water, has been reported adequately during recent years, and will not be reviewed here. However, it is worth stressing that many plant nutrients can arise from the same source. This contamination can be a mixed blessing. Sulphur dioxide, for example, may be a health hazard close to its source or cause deleterious environmental effects over wide areas due to acidification of rain water; yet sulphur dioxide from industry may contribute significantly to the sulphur requirements of crops in some regions. In parts of England and Wales, addition of sulphur from the atmosphere and rainfall is of such importance to grassland production that further decreases in sulphur dioxide pollution of the atmosphere may require adjustment of fertilizer practice (Jones *et al.* 1972). Purves and McKenzie (1969) and Purves (1972) have reported that urban soils and their produce can have elevated levels of Cu, Zn, Mo and B. The atmospheric accession of Cu and Zn in one year onto soils several kilometres from the metal works at Wollongong, N.S.W. (Beavington 1977), if available to plants, would be more than adequate for the requirements of plant growth. The accession of such metals from industries can be quite widespread, but the areas affected can often not be defined because of overlapping sources. However, Cartwright *et al.* (1977) showed that emissions from the lead-zinc smelter at Port Pirie in S. Australia had contaminated over 3000 km^2 of land; much of this area would have received an accumulated application of Zn which greatly exceeded that used in commercial fertilizer practice in Zn deficient areas.

Concentrations of micronutrients in soils

The multitude of weathering processes arising from different combinations of factors of weathering and acting on an immense variety of parent materials derived from rocks of different composition (see Table 25.1), must lead inevitably to great variety in composition and concentration of micronutrients in the soils of the world;

Table 25.2. The range in concentration (μg per gram) of some micronutrients in superphosphate

Co	Cu	Mo	Zn	B	Mn	Reference
0.2-13	4-270	0.5-35	4-3000	0-1000	20-3000	Swaine (1962), overseas superphosphates
ND[A]	10-43	1-4	270-750	ND	27-180	Williams (1974), Australian superphosphates
ND	14-115 (49 ± 32)	ND	380-630 (470 ± 80)	ND	23-370 (120 ± 90)	Tiller (unpublished) superphosphate purchased in South Australia, 1962–77, mean ± S.D. in brackets

[A]ND, no data.

Table 25.3. Total concentrations of trace elements in normal soils of the world

Ranges (and means, where available, in brackets) are expressed as μg per gram. Swaine estimated ranges for normal soils, mainly surface horizons, based on numerous sources (often thousands) of data from the world literature up to 1955. The values of Vinogradov include all horizons of profiles studied, based on about 350–500 samples for Cu and Mn, and about 50 for Co, Zn and Mo. Results from Shacklette *et al.* come from 863 subsurface horizons collected at about 20 cm depth. Tiller assessed surface horizons only (commonly about 0–10 cm) basing his values on data produced since 1955. The numbers of values used by Tiller are indicated in the Table. Values for B in soils are reported to range from 7 to 80 μg per gram (Krauskopf 1972) or 2 to 100 μg per gram (Swaine 1955)

Country	Reference	Co	Cu	Mn	Zn	Mo
World-wide	Swaine (1955)	1–40	2–100	200–3000	10–300	0.2–5.0
U.S.S.R.	Vinogradov (1959)	0.1–15 (6)	1.6–200 (17)	100–12000 (1000)	20–120 (48)	1–12 (2.6)
U.S.A.	Shacklette *et al.* (1971)	<3–70 (10)	<1–300 (25)	<1–7000 (560)	<25–2000 (54)	<3–7
Australia	Tiller (this study)	<2–170 (11) n=232	<1–190 (22) n=479	4–5100 (780) n=779	<2–180 (34) n=459	<1–20 (3.2) n=286

Table 25.4. Ranges of total concentrations of micronutrients in some Australian surface soils

ND, no data. ( ), no. of samples, if fewer than indicated for other elements. References: 1. Giles (1959). 2. Giles *et al.* (1962). 3. McKenzie (1957). 4. McKenzie (1958). 5. Giles (1960). 6. Tiller (1958). 7. McKenzie (1960). 8. Turton *et al.* (1961). 9. McKenzie (1966). 10. Tiller (1957). 11. McKenzie (pers. comm.). 12. Isbell and Smith (1976). 13. Isbell *et al.* (1976). 14. Isbell (pers. comm.). 15. Isbell and Gillman (1973). 16. Thompson and Moore (pers. comm.).

Soil groups	Region	No. samples	Range in concentrations (μg per gram of soil) Co	Cu	Mn	Zn	Mo	Reference
Alluvial soils	Qld/Vic.	9	14–44	45–92	940–2400	32–150	0.8–20	1,2
Lithosols	Qld	12	ND	16–32	500–4700	66–151	0.8–1.5 (3)	16
Calcareous sands	Vic.	3	ND*	4–9*	ND*	ND*	3–6	9
Arid zone soils	N.T.	21	<2–23	4–27	75–900	12–66	1.5–3.6	7
Grey, brown and red clays	Qld	24	3–26	17–64*	360–2800	30–140*	1.7–7.9	1,2
Black earths	Qld	17	24–170	28–110*	900–3000*	43–150*	2.3–5.5	1,2
Black/brown earths on dolerite	Tas.	20	7–60	31–121	670–3100	31–65	1.3–6.0	2,6
Prairie soils	Qld	4	ND	35–136	2000–4700	79–177	0.5–1.4	16
Rendzina	Qld/S.A.	26	<2–7*	1–46*	36–1100*	5–91*	0.3–5.0	1,2,4,5
Terra rossa	Qld/S.A.	21	<2–30*	1–65*	78–1400*	1–36*	<1-5.0	1,2,4,5
Solodized solonetz and *solodic soils*	Qld	15	<2–59*	1–74*	59–2000*	6–70*	1.9–7.0*	1,2
	Vic.	53	ND	4–58*	ND	3–104*	0.3–11	9
Soloths	Qld	5	ND	4–50	300–1900	20–73		-16
Solonized brown soils	S.A.	6	ND*	5–20*	5–110*	ND*	<2–3	11
Red-brown earths	Qld/S.A.	22	<2–39	7–63	240–2800*	ND*	1.3–5.1*	1,2,3
Non-calcic brown soils	Qld	13	ND	13–82	–	22–89	ND	14
Red earths	Qld/W.A.	24	1–51 (12)	17–93	110–2100	9–110	3.9–6.8 (12)	1,2,12,13
Yellow earths	Qld/W.A.	12	ND	5–80	15–230	4–42	ND	12,13,14
Euchrozems	Qld/W.A.	32	ND	43–97	420–3530	50–151	ND	14,15
Xanthozems	Qld	17	ND	4–35	200–2200 (11)	17–101	0.1–2.4 (3)	14,16
Krasnozems	Qld/Vic./Tas.	28	5–110*	22–190*	280–5100	25–180	2–12*	2,6,10
Grey-brown and *brown podzolic soils*	W.A./Tas.	15	<2–23*	11–52*	320–3050	13–86*	1.6–5.0*	2,6,8
Red podzolic soils	Qld	8	3–8 (4)	4–19	80–1700	24–56	3.1–5.7 (4)	1,2,14
Lateritic podzolic soils	W.A./Qld	16	all <2*	0.5–24*	6–1300*	5–38*	<0.5–9*	1,2,8
Podzols and *deep sands*	Vic.	20	ND	2–8	ND	2–8	1.8–7.2	9
	Qld	29	<2–5 (4)	2–19	40–530 (4)	1.5–164	3.3–5.3 (4)	1,2,15

* Nutrient deficiencies reported in this soil group (Stephens and Donald 1958).

representative data from several sources are assembled in Table 25.3. The range and mean values from the well-separated areas of Australia, U.S., and U.S.S.R. agree very closely, and illustrate the variation in composition which can be expected from the interaction of the various factors of soil formation and parent materials. However, appreciable variation in micronutrient composition is found between soils and within many soil groups, as exemplified by the more detailed information, given in Table 25.4, for Australian soils. The soil groups are presented in an approximate order of decreasing degree of weathering, following the somewhat subjective arrangement used in 'A Handbook of Australian Soils' (Stace *et al.* 1968). Although the ranges in concentrations of some micronutrients do vary appreciably between some soil groups, especially those representing the

Table 25.5. Ranges of concentrations of trace elements extracted from Australian surface soils

Unpublished data, Division of Soils, CSIRO, Regions *a* and *c*, Tiller *et al.*; *b*, Cartwright *et al.*; *d* and *f*, Isbell *et al.*; *e*, Thompson and Moore. Regions as follows: *a* = Field trials in southern Australia; *b* = Eyre Peninsula, S.A.; *c* = Mt Lofty Ranges, S.A.; *d* = North Queensland; *e* = South Queensland; *f* = Kimberleys, W.A. Co extracted by DTPA (Lindsay and Norvell 1978) followed by concentration using APDC/MIBK. ( ), no. of samples, if fewer than indicated for other elements. Cu, Mn and Zn extracted with 0.1 M EDTA, pH 6 (Clayton and Tiller 1979).

Soil group	Regions	No. of samples	Range in concentrations (μg per gram of soil)			
			Co	Cu	Mn	Zn
Lithosols	*e*	12	0.29–2.6	2.0–11	170–3420	3.9–47
Grey, brown and red clays	*af*	36	0.39–2.6 (4)	2.5–11	190–1010 (4)	0.15–16
Prairie soils	*e*	12	0.58–6.3	2.7–23	133–3530	2.0–38
Solodized solonetz and *solodic soils*	*abe*	89	0.33–1.8 (18)	0.2–5.7	4–668 (25)	<.05–9.1
Soloths	*ace*	128	0.40–5.2 (12)	0.2–16	4.8–1080 (12)	0.7–37
Solonized brown soils	*ab*	307	0.04–1.15	0.2–9.3	6–288	0.2–4.7
Red-brown earths	*ab*	76	0.52–1.12 (8)	0.5–9.3	18–52 (8)	0.7–10
Non-calcic brown soils	*d*	12	0.82–3.0	1.3–12	108–2280	0.8–21
Calcic red earths	*a*	10	–	2.3–4.7	–	1.2–8.1
Red earths	*ade*	25	<0.01–4.3 (15)	<0.1–9.3	0.5–1060	<0.05–23
Yellow earths	*df*	17	<0.01–1.6	<0.1–9.1	0.3–940	<0.05–0.8
Euchrozems	*def*	27	1.4–6.5	2.4–45	360–3170	1.0–29
Xanthozems	*de*	19	<0.01–14	2–11	13–2870	1.5–15
Krasnozems	*d*	15	0.21–2.4	4.3–13	59–3330	1.7–18
Red podzolic soils	*ade*	53	0.02–6.7 (27)	0.2–9.0	7–2560 (27)	0.5–23
Yellow podzolic soils	*acde*	40	–	0.3–10	–	<0.05–19
Lateritic podzolic soils	*ac*	56	–	0.2–13	–	1.1–16
Podzols/siliceous sands	*def*	26	≤0.01–1.6	<0.1–0.8	0.3–510	<0.05–4.4
Organic soils		7	–	2–13	10–2100	5–20

extremes of weathering and parent material, e.g. *black earths* and *brown earths* on dolerite *vs podzols* and *deep sands*, most soil groups have similar ranges in total micronutrient concentration. This means that broad-scale pedological surveys are of very limited value in predicting micronutrient concentration of most soils. Surveys within restricted regions may be more useful since regional differences may outweigh differences between soil groups (McKenzie 1959). A further complication is the limited extent to which total analysis can indicate the soil's ability to supply micronutrients to plants; sometimes a negative correlation is found. Some soil groups are strongly associated with a particular micronutrient deficiency, e.g. Co deficiency in animals on *krasnozem* landscapes, and Zn deficiency in cereals on *black earths* and *grey, brown and red clays*, yet the total concentration of the deficient element in these soils is often at the upper end of the range; the generally high Mo contents of *lateritic podzolic soils* is another example. This situation arises because the very chemical processes which promote low availability to plants also promote accumulation in the soil during soil formation or following fertilization. Intense weathering processes could well lead to either high or low total micronutrient concentrations, depending on the element concerned and the soil conditions. Thus these values generally give little insight into the possible limitation to crop or animal production, or the likely response to micronutrient fertilizers. The only exceptions are soils such as *podzols* and *deep sands*, which have very low concentrations in relation to the amounts being required for crop production.

For these reasons, as analytical techniques have improved, chemical assessment of the micronutrient status of soils has been progressively based on empirical extraction procedures. Some of the philosophy and practice of soil testing for micronutrient status will be presented later. Some Australian state authorities and private fertilizer companies carry out testing programs on farmers' soil samples, but these surveys are generally unpublished. Although the DTPA procedure (Lindsay and Norvell 1978), following American experience, is now being widely accepted in Australia as the most useful method for the micronutrient metals Cu, Zn, Fe and Mn, published data for a variety of soils are still not available. No widespread testing of other micronutrients in farmers' soils is carried out in Australia. Table 25.2 shows the ranges in concentrations of several micronutrient metals extracted by EDTA from representative soil groups (Division of Soils, CSIRO, unpublished data). This EDTA procedure extracted amounts of Zn similar to the standard DTPA procedure, but up to several-fold more Cu and Mn; the difference depended on soil type and decreased when small amounts of Cu and Mn were present (Clayton and Tiller 1979). The values of Tables 25.4 and 25.5 are not derived from an identical set of samples, but clearly only a small proportion of the total content of most soils has been extracted by the complexing agents. In the highly leached sandy soils, in which much of the micronutrients is associated with organic and oxide surfaces rather than with constituents of mineral lattices, the amounts removed by complexing agents may approximate that found by total analysis.

Almost half of the soil groups listed in Table 25.5 have members with insufficient Cu, Zn, and sometimes Mn, for healthy crop growth. These soils were probably taken from virgin sites in which nutrition is maintained by cycling within a closed system. The soil groups in which individual micronutrient deficiencies have been recorded by Australian research workers have been discussed by Stephens and Donald (1958), and are indicated in Table 25.4. Since then, studies of the *red* and *yellow earth* soils of northern Queensland have shown them to be low in micronutrients and to respond to applications of Cu and Zn (Isbell *et al.* 1976).

Chemical and mineral forms

The forms and associations of micronutrients in the soil solution

In the soil solution, like all aqueous environments, micronutrient ionic species are often strongly associated with other ions and molecules, including water. The stability of these associations depends on the size, charge, and electronic structures of the species involved. The transition metal ions have especially strong affinities for some ligands.

Soil solution contains both organic and inorganic ligands. The common inorganic non-cationic constituents of soil solution are silicic acid, nitrate, phosphate, chloride, sulphate, and bicarbonate. There is some evidence of ion-pair formation involving Cu, Zn and Mn with phosphate, and Cu and Mn with bicarbonate (Sillén and Martell 1964, 1971). Sulphate and chloride, however, undoubtedly form stable complex ions with these metals under the normal conditions of many soils. Complexes do not form appreciably between trace metals and nitrates or monomeric silicic acid of soil solution. Hydrolyzed ionic species of the form MOH^+ and $M(OH)_2^0$ also form in solution, especially at higher pH values. The few published constants available for solutions with ionic strengths similar to those found in soil solution ($< \sim 0.1$) suggest that $M(OH)_2^0 = M^{2+}$ at about pH 7.5 for Cu and Zn, and at about pH 9 for Co. However, if the solubility of soil Zn and Cu, as shown experimentally by Norvell and Lindsay (1969), has general applicability in the form they propose (Lindsay 1972), e.g. $[Zn^{2+}] = 10^6 \times [H^+]^2$, then the actual concentrations of $Zn(OH)_2^0$ and $Cu(OH)_2^0$ species in soil solution should be respectively about 10^{-9} and 10^{-12}M, and relatively independent of pH within the range of pH 6–8 covered by Norvell and Lindsay's data. The formation constants of some complexes with some inorganic and organic ligands are summarized in Table 25.6.

The naturally occurring organic ligands are of much greater significance yet much harder to study because of the difficulty in isolating and defining the chemical composition and properties of the active components. The organic components of soil solution which are responsible for metal complexing will be mainly aliphatic acids, polycarboxylic acids, amino acids, and the less well defined polymeric compounds of higher molecular weight which are similar, possibly, to the fulvic acid fractions separated chemically from soil organic matter. For a full discussion of naturally occurring organic substances which can complex metals, the reader is referred to Stevenson and Ardakani (1972).

To interpret the relative significance of the various ligands, organic and inorganic, values of ligand concentrations are needed. These values are often unavailable; if measured, they may only represent a transient state, especially for the biochemical products. However, we do know that the maximum concentration of SO_4^{2-} in soil solution, which occurs in soils containing free gypsum, will not exceed 10^{-2}M (Novogamsky *et al.* 1976). In normal soils from arid areas it will be $\leqslant 10^{-3}$M. As far as the organic ligands are concerned, the soil solution displaced from a New York soil was found to contain about 10^{-7}M of total amino acids and $\sim 10^{-2}$ M acetic acid (Geering and Hodgson 1969). Stevenson and Ardakani (1972) have proposed concentrations of acetic acid and polybasic acids of the order of 10^{-3}M in soil solutions of incubated soils. Similarly, free amino acid concentrations of about 10^{-4}M in the bulk soil, and possibly up to about 10^{-3}M in the rhizosphere, were suggested. These values were based on extraction data, and recalculated assuming that amounts extracted were in water-soluble forms. This assumption will probably often lead to overestimates since many organic acids will be adsorbed to some extent by soil colloids. Most of these organic acids are strongly dissociated in the pH range of normal soils.

Table 25.6. Formation constants (log K_1) of some micronutrient complexes

Ligand (L)[A]	Metal			
	CoII	CuII	ZnII	MnII
Chloride (L^-)	−0.5	0.5	0.5	0
Hydroxyl (L^-)	4.0	6.5	5.0	3.5
Sulphate ($L^=$)	2.5	2.5	2.5	2.5
Acetate (L^-)	1.5	2.0	1.5	1.0
Oxalate (L^-)	6.5	9.0	8.0	6.0
Citrate ($H_2L_2^=$)	3.0	7.5	3.0	2.0
Alanate (L^-)	4.5	8.5	5.0	3.0
Valate (L^-)	NA[D]	8.5	5.0	3.0
Fulvate (pH 5)[B]	3.5	8.5	2.5	4.0
Non-dialysable organic fraction[C]	NA	5.5	4.5	NA

[A] A consensus of values determined at low ionic strength where $K_1 = [(ML)^{2-n}]/[(M^{2+})(L^{n-})]$. Values were taken from Sillén and Martell (1964, 1971) and rounded to nearest 0.5.
[B] Organic matter from a podzol B horizon, soluble in both sodium hydroxide and acid (Schnitzer and Skinner 1966, 1967).
[C] Separated, without chemical treatment, from the soil solution of the A horizon of a silt loam soil (Geering and Hodgson 1969).
[D] Data not available.

A consideration of the formation constants, and of some of the likely concentrations of ligands shows that the stable complexes of metals with natural organic compounds must dominate the solution chemistry of trace metals, especially Cu, in the soil solution of most soils. Hodgson *et al.* (1965, 1966) have shown that Cu and Zn in soil solutions are present mainly as metal-organic complexes; the organic complexes, of Cu at least, are anionic (Hodgson *et al.* 1966, Beckwith 1973). However, the inorganic ligands are more important in soils from more arid regions because of increased concentrations of Cl^- and especially of SO_4^{2-} combined with probable lower concentrations of organic ligands associated with reduced biological activity. The formation of uncharged complexes, especially $ZnSO_4^0$, may dominate solubility in such soils.

Thus the total metal concentration (M_T) in solution, as would be measured by direct flaming in atomic absorption spectrophotometry, should be viewed as a summation of different species of the form:

$$M_T = M^{2+} + \Sigma ML_i^{2-n_i}$$

where L_i represents different ligands with a negative charge n_i. Since all these complexes are interrelated through equilibria with the various ligands present, a change in any one of them, through change in ligand concentration, or through the involvement of a particular species in either adsorption or precipitation of a separate phase, will result in a corresponding change in all other species present. Hence the specific reaction of one, possibly minor, component of soil solution, say M^{2+} or MOH^+, may moderate the concentration of them all. The possible equilibria or quasi-equilibria describing micronutrient reactions between solid and solution phases in soils will be discussed in later sections.

Much less is known of the chemistry of the micronutrient anions in soil solution, but, from available formation constants (Sillén and Martell 1964), the following species are expected:

B, generally as undissociated $H_3BO_3^0$ with borate ($H_2BO_3^-$) in the most alkaline soils ($pK = 9.2$);

Mo, mainly as monomeric MoO_4^{2-} above pH 4.2, and $H_2MoO_4^0$ below pH 4.0, with the possibility of mixed and polymeric species; and

Se, as SeO_4^{2-} in oxidized soils.

Although Mo concentrations in saturation extracts of some Indiana soils have been reported in the range 2–8 x 10^{-8}M (Lavy and Barber 1964), even less seems to be known of the levels of the micronutrient anions in soil solution than of the micronutrient cations.

The micronutrient metals Cu, Zn and Co only occur in soil solution in the divalent form. The metals Mn and Fe are subject to the same kinds of equilibria as discussed earlier, but their reactions are complicated further by redox processes leading to the occurrence in soil solution of these metals in two valence states, Fe (II) $\rightleftarrows$ Fe (III), and Mn (II) $\rightleftarrows$ Mn (IV). Reducing conditions have the dominant role in determining the concentrations of Fe and Mn in soil solution. Conditions which change the stability of Fe and Mn oxide surfaces in soils can indirectly have a major influence on the mobility of elements associated with these surfaces, even though the elements themselves are unaffected by oxidation – reduction reactions under these conditions. For example, the distribution of Co between solid and solution would be markedly affected by appreciable loss of Mn oxide surface; similarly, trace metals and molybdate would be sensitive to variations in the stability of iron oxides. The mobilization of Zn, Cu and P by gleying of basaltic soils in south-east South Australia was ascribed by Tiller (1958) to instability of Fe oxides. Oxidation–reduction reactions are discussed more fully in Chapter 28.

Adsorption and micronutrients

Adsorption refers to the accumulation of the reacting species at the interface of the solid and liquid or gaseous phases. Sorption is a more general term, which includes the possibility that some of the reacting species may be absorbed within the solid phase, either in pores or in lattice positions. Micronutrients may be sorbed by specific and/or non-specific adsorption reactions. Non-specific adsorption refers to the bonding, by weak electrostatic interactions, between ions and surface groups of opposite charge. These ions are associated with the outer, or diffuse, part of the double layer, and comprise the bulk of what are usually referred to as the exchangeable cations of soils. The term 'specific adsorption' was first introduced by Stern (1924) and has been widely used in various contexts since. Specific adsorption may be viewed as the adsorption which cannot be accounted for by diffuse double layer theory (Parsons 1968). Hence specific adsorption implies stronger bonding, and is thought to include the formation of chemical bonds with the surface (chemisorption) and other interactions arising from the short-range forces associated with the inner part of the double layer surrounding soil colloids. Experimentally, specific and non-specific sorption reactions can be separated by use of a large excess of appropriate electrolyte, to effectively eliminate non-specific adsorption, and by carrying out reactions under conditions where the surface and adsorbate do not have opposite charge. Because of the heterogeneity between and within soil surfaces, sorption at very low surface coverage will tend to be of higher energy than at higher surface coverage. This situation will apply to all adsorbates, whether Ca, K, Zn or Mo ionic species for example, but because micronutrients occur in very low concentrations in soils, the higher energy specific sorption reactions would be favoured, quite apart from any tendency arising from their intrinsic chemical properties.

Most views of the sorption of micronutrients by soils

have not been derived from studies of soils but from simpler model systems of silicate clays, carbonates, iron or manganese oxides, silica and organic matter separated from soils. Such studies have limitations, apart from a lack of realism arising, for example, from the substitution of a simple type of surface for the heterogeneous soil surfaces. There is firstly the difficulty in measuring the very low concentrations applicable to soil solutions, and secondly the severe restriction of time scale, which limits interpretation to the initial reaction without providing any guidelines for the possible slow transformation which can occur in Nature's own time scales.

Layer silicate clays as a model system of micronutrient adsorption

Studies of Tiller and Hodgson (1962), and later unpublished studies by the author, have considered the reactions of micronutrient metals, especially Zn, with layer silicate clays in the presence of excess salt. The reactions were characterized as being endothermic and slow (equilibrium taking several weeks at 20°C), the reaction rates being dependent on temperature and concentration of zinc in solution. The reaction was not reversible, except to pH. Protons were released during the specific sorption of Co, Zn, Cu and Ni by several minerals (Tiller 1961). Sorption of Zn, Co and Ni by hectorite resulted in an approximately equimolar release of H^+ (Tiller 1968*c*). Similar results were found for the specific sorption of Zn by montmorillonite (Tiller unpublished data). Other cations which also hydrolyze, such as Cu and Co, competed strongly with the reaction, and could also displace readily any Zn adsorbed in the presence of Ca salts (Tiller *et al.* 1979). Magnesium salts displaced sorbed Co and Zn much more readily than did Ca, Li or Na salts, because of the greater hydrolysis of Mg relative to the other ions (Tiller 1961; Hodgson *et al.* 1964). The above properties are more characteristic of chemisorption than simple coulombic attraction. The preferential sorption of different trace metals, in the presence of Ca salts, by montmorillonite hectorite and kaolinite was in the order Cu > Zn > Co > Mn > Ni (Tiller 1968*c* and unpublished data).

Tiller (1961) pointed out that although this order approximated that of the values of the hydrolysis constants of these metals, the ability to polarize the coordinated water molecules in hydrolysis would just as well facilitate reactions with the mineral surface by enhancing the polarization of surface hydroxyl groups. Since the more complex clays are not readily amenable to physical chemical interpretation, others have used pure silica as a model of clays and oxide surfaces generally. Dugger *et al.* (1964) and Schindler (1975) have made comparative studies of the reactions of many trace metals with silica, and have shown that the equilibrium constant of each reaction was closely related to the hydrolysis constant of the corresponding metal. Fordham (1969) showed that the chemisorption of Fe(III) by kaolinite was correlated with particular hydrolysed species of Fe(III), the species being dependent on the pH of reaction.

The anions Cl^- and SO_4^{2-} decreased the specific sorption of Zn by montmorillonite, relative to that found in the pure NO_3^- system. However, when the amount of Zn sorbed was measured as a function of $ZnOH^+$ concentration in equilibrium solution, the Cl^- and SO_4^{2-} and NO_3^- systems have identical results. Apart from demonstrating the control of Zn sorption by $ZnOH^+$ or Zn^{2+}/H^+ in solution, this showed that $ZnCl^+$ and $ZnSO_4^0$ species are not involved in the specific sorption of Zn by this model system (Tiller unpublished data).

However, monomeric silicic acid ($Si(OH)_4$), another common inorganic component of soil solution, increased the specific sorption of Co, Cu and Zn by kaolinite and montmorillonite but not by alumina (Tiller 1967, 1968*a* and unpublished data). As a complement to the latter reactions, Tiller (1968*c*) found that the sorption of heavy metals by hectorite led to an increase in subsequent sorption of silicic acid. From these two studies it was proposed that heavy metals may be sorbed specifically at the edges of clay minerals. The sorbed silicic acid increases the number of sites involving silanol groups that exist naturally around edges and other defects of layer silicate clays. Earlier studies had also suggested that Co was only specifically sorbed at the edges of vermiculite clay (Hodgson and Tiller 1962). Fordham (1973) showed that Fe(III) was chemisorbed at the edges of kaolinite. In another approach, the etching of clay edge-surfaces by pretreatments involving Li salts at low temperatures also resulted in increased specific sorption by heavy metals (Tiller unpublished data). The proposed sorption of heavy metals at the edges of layer silicate clays has been considered as a partial extension of the crystal lattice (Tiller 1968*a*); this view is consistent with the irreversibility of these reactions as neutrality is approached. In the studies described above, the adsorption sites were not identified but were assumed to be the silanol groups.

The exact mechanism of the specific sorption reactions by silicate clays is not clear. Bower and Truog (1940) obtained anomalously high values for exchange capacities of soil clays when using easily hydrolyzable cations. Although it now seems likely that precipitation occurred under the conditions of their measurements, their work is important because it alerted researchers to the possible role of hydrolyzed metals in reactions with soils (Wiklander 1955; Menzel and Jackson 1951; Spencer and Gieseking 1954; Hodgson 1960; and others since). Matejevic (1967) concluded that charge reversal of oxides arises from the adsorption of soluble hydroxo-metal ions. The high affinity of MOH^+ is not simply explained. Reduced charge and its effect on hydration energy of cations generally, and thus on their adsorption affinity, has long been recognized. Jurinak and Bauer's (1954) view of specifically adsorbed ions was of them being held adjacent to the mineral surface in a partially dehydrated state. The role of solvation energy has been incorporated into the model of adsorption by James and Healy (1972) for reactions of heavy metals with silica and titania surfaces.

Their model is based on the proposition that the energy of adsorption,

$$\Delta G^{o}_{ads} = \Delta G^{o}_{coul} + \Delta G^{o}_{solv} + \Delta G^{o}_{chem}$$

where the components of energy favouring adsorption (ΔG^{o}_{coul} arising from coulombic attraction, and ΔG_{chem} arising from undefined short-range chemical interactions) are opposed to ΔG_{solv}, which arises from the secondary solvation sheath of water molecules surrounding the adsorbing cations. This outer sheath, or sphere, of water needs to be partially displaced to allow a sufficiently close approach to the surface. By their model, the dependence of adsorption on pH is not directly connected with the specific reactions of hydrolyzed ionic species; rather it is the decreased solvation energy, related quadratically to the average charge of adsorbing cations, which favours the adsorption of the hydrolyzed species (MOH^+) relative to M^{2+}. Charge of the adsorbing species, and thus ΔG_{solv}, may also be reduced by complexation with ligands such as OH^-. This model and its associated discussion provide a very useful approach to the understanding of these reactions, but it remains to be tested as to how well it can predict the reactions of trace metals with soil colloids.

MacNaughton and James (1974) have discussed some limitations of the James and Healy model in terms of the specific noncoulombic interactions of hydroxo species which favour their adsorption by oxide surfaces. Reactions in soil solution, at higher ionic strengths than used for the James and Healy model, and with a variety of heterogeneous surfaces available, will tend to maximize the role of the ΔG^{o}_{chem} term (see equation above). The latter term may well dominate the reactions in the more complex soil colloid system. Even in relatively simple systems of 'pure' clays, there are (Tiller 1961) great differences in affinity by different surfaces for a particular trace metal cation, under given conditions of pH and ionic strength; also, for different minerals, the pH at which sorption increases sharply may vary as much as two units under identical experimental conditions and a given cation, e.g. Co or Zn. Similarly, James and Healy (1972) also found that the pH-dependent sorption of Co depended on the particular oxide surface involved. On the other hand, cations like Zn(II) or Ni(II), with very similar stability constants and hydration energy, have appreciably different affinities for particular mineral surfaces in the presence of excess salts (Tiller 1968*c* and unpublished data). The reduction in solvation energy (ΔG_{solv}) of Zn^{2+} arising from complex formation with Cl^-, which is predicted by the James and Healy model to favour the reaction with oxide surfaces, resulted in less specific sorption of Zn by montmorillonite (Tiller unpublished data). However, sorption of Zn by goethite was slightly favoured in a Cl^-, compared with an NO_3^- system (Forbes *et al.* 1974), as would be predicted by James and Healy (1972).

The above observations imply a strong influence of a specific interaction term, related to the particular surface involved. This creates problems in developing adequate models for heterogeneous systems like soils, unless one type of surface dominates all sorption reactions.

Oxide minerals as model systems for micronutrient adsorption

Iron and Mn oxides are ubiquitous in Australian soils and probably provide a significant adsorption surface for micronutrients, even in soils in which these minerals cannot be detected by X-ray diffraction.

The major contribution to our understanding of adsorption by iron oxides has been the work of Posner and Quirk and associates, at the University of Western Australia. Their early emphasis was on anion adsorption (reviewed by Hingston *et al.* 1970). Experimentally, adsorption isotherms were measured on synthetic goethite using dilute sodium salts as supporting electrolyte. Inflections in the curve describing maximum adsorption as a function of pH, referred to as adsorption envelopes, occurred at pH values corresponding to the pK value of anion dissociation. This led to the proposition that specific adsorption requires the involvement of both the dissociated anion and its corresponding protonated product. At pH values more alkaline than the point of zero charge (PZC), the surface has a negative charge and little adsorption of anions of strong acids is expected. They suggested that the ease of dissociation of a conjugate acid increases with pH in solution, and thus the possibility of surface-promoted dissociation of the acid will increase in the same way. The H^+ ion thus released combines with an adjacent OH^- in the surface coordination layer of goethite, providing a positive site for anion adsorption. There does not appear to be any chemical basis, however, for their implication that the bonding energy of acid dissociation is a function of pH, and thus for their explanation for maximum adsorption at pH = pKa. Implicit in their hypothesis also, is the assumption that the mechanism controlling the reaction at high solution concentrations and surface saturation will apply to reactions in soils under quite different conditions. These provisos do not detract, however, from their major contribution in stressing the specific nature of anion adsorption, the importance of ionic species and surface charge in relation to pH, and the change in surface charge due to specific sorption. These features have provided the basis for the theoretical model developed by Bowden *et al.* (1973) for adsorption on oxide surfaces.

Hingston *et al.* (1970) have also suggested that the specific sorption of some metal cations on positively charged oxide surfaces may be due to the dissociation of the hydrolysed species (MOH^+) at the surface, the released hydroxyl ion thus removing a proton from the mineral surface as water and permitting the metal ion to react with the resulting negative site. Later, Quirk and Posner (1975) have preferred to view heavy metal sorption as surface promoted hydrolysis, resulting in mono- and bi-dentate bonding of metals through hydroxyl bridges. However, in the end result, the bonding of the metal as proposed would be indistinguishable from the highly specific sorption of the hydrolyzed species, MOH^+ or $M(OH)_2$. The proposed mechanism of Posner and Quirk is analogous to the 'increased hydrolysis' of Al(III) reported by Raglan and Coleman (1960), in which the sorption of each mole of Al(III) by clay, under conditions which precluded precipitation, was accompanied by the release of 3 moles of H^+.

Forbes *et al.* (1976) continued the study of the sorption of several metals on synthetic goethite (PZC at pH 7.6). The relative affinities for goethite of the micronutrient metals studied were in the order Cu > Zn > Co. The reaction was pH dependent, with 1.5 to 2.0 moles H^+ being released per mole of metal sorbed, and increasing over the pH range 4.5–5.0 (Forbes 1973, as quoted by Quirk and Posner 1975). A similar variation in molar ratio (i.e. H^+ released per M(II) sorbed) from 1.5 to 2.0 was found by Kalbasi *et al.* (1978) in the pH range 6–7 for the sorption of Zn(II) by reagent grade hematite and corundum; they attributed the variation to a pH dependent contribution of nonspecifically sorbed $ZnCl^+$. Grimme (1968) found that the adsorption of each mole of Cu, Zn, Co and Mn, with affinity in that order, by synthetic goethite (PZC at pH > 7) was accompanied by the release of one mole of protons in solution, over the pH range 4–7. The variations in stoichiometry found by different authors may arise from different chemical conditions used, but could also reflect the surface properties of the various synthetic preparations. Of special note in these studies is the high affinity of trace metal cations for positively charged iron oxide surfaces.

The reactions of trace metals with Mn oxides differ in the important respect that Mn oxides are negatively charged in the normal pH range of soils (McKenzie 1977). Sorption reactions with Mn oxides are also more complex under normal conditions because of possible occupation, in addition, of lattice sites. When Cu, Co and Ni were reacted with ground Mn nodules separated from soils, and with synthetic Mn minerals, the sorption of Cu was initially favoured over Mn and Co and Ni, in that order, but the slower subsequent reactions, involving lattice sites, favoured Co (McKenzie 1967, 1970). As assessed by acid extraction, Co was more strongly retained than other trace metals. Later experiments demonstrated variations in the behaviour of different Mn oxide phases. This was ascribed by McKenzie (1972) to the presence of Mn(III) in the crystal lattice, and its replacement by Co(III), due to crystal field stabilization. Murray (1975) also studied the interaction of heavy metals with a synthetic Mn oxide. Metals had the relative affinity Cu > Co > Mn > Zn > Ni. He found that approximately one mole of H^+ was released per mole of metal sorbed, over the pH range of 3.5–7. Cobalt and Zn have also been reported to release 2 moles H^+ per mole of metal sorbed at pH 4 (Loganathan and Burau 1973). These variations in experimental data parallel partially the situation found with studies with Fe oxide. Hence some caution must be exercised in the detailed application of model systems to soils, especially since the oxide phase cannot often be identified, even if the bulk mineralogical composition is known.

Comments on the choice of model systems for micronutrient adsorption

Jenne (1968) has proposed that the hydrated oxides of Fe and Mn provide the main matrix for the control of heavy metal concentrations in natural solutions, through coprecipitation during formation, or adsorption later. Many researchers would now support the generality of this point of view, but it should not be concluded that these oxides are the exclusive means by which heavy metal chemistry is controlled. Strong associations of a micronutrient with segregations of a particular solid phase may only indicate conditions operative during an early stage of pedogenesis, or a system at saturation which no longer controls the partition of metals added more recently. Tiller *et al.* (1963) and Hodgson *et al.* (1969) showed that at low concentrations ($\sim 10^{-6}$M initially), Co sorption by all but one of a range of soil clays was not decreased by the removal of Fe and Mn oxides or organic or amorphous materials.

Iron and manganese segregations often contain higher concentrations of some micronutrients than their surrounding matrices do, yet sometimes only account for a small part of the total micronutrient content. There are clearly exceptions to this, *viz.* Co in Mn oxides. Even if a large proportion of the micronutrient content of soils is associated with the oxide minerals, it is still open to question as to whether these oxides or the surrounding organic–clay complex or carbonate minerals control the soil-solution equilibria of micronutrients in the solution of many soil environments. Individual studies using model systems, e.g. carbonate minerals or silicate clays or oxides of Fe, Si and Mn, as discussed above, and various investigations on soil organic matter and peats, all demonstrated high affinity of trace metals for the surface being studied. These different approaches cannot be compared because of variation in experimental conditions and in surface area and crystallinity of the mineral being used. These studies often demonstrate common factors in the reactions being studied, *viz.* similar order of affinity of trace metal for each surface, a strong pH dependence, sorption accompanied by release of protons, and a tendency for desorption hysteresis especially as the pH of reaction approaches pK_1^* (first hydrolysis constant). The OH group is the common denominator of most soil surfaces, whether silicate clays or mineral oxides or organic matter. The strong tendency of trace metals to combine with the hydroxyl ligand in solution, or on the mineral surface, provides the basis for similar reactions with different surfaces. Differences in affinity of metals for different sites and surfaces will be due to factors affecting bond strength of OH groups, such as crystal structure and defects, e.g. local ionic substitutions.

A problem common to all model systems is how to distinguish confidently between reaction mechanisms for reactions of known stoichiometry. This question was raised by Hodgson *et al.* (1964), but developed fully by MacNaughton and James (1974). One set of experimental results could variously be interpreted in terms of simple ion exchange, or adsorption with hydrolysis of the metal at the surface, or adsorption with hydrolysis in solution, or surface-encouraged precipitation. There is a risk that the preferred interpretation could be largely intuitive and supported by an arbitrary choice of model parameters chosen to maximize agreement between predictive model and experimental results. Schindler (1975), in discussing a quantitative model for the removal of metals from sea water, has rather delightfully admitted that the reader 'was manipulated by the author' because of the difficulty in choosing parameters which are meaningful in the field sense, and because different combinations of parameters may appear equally suitable. The main strength of a predictive model is not the degree with which a particular set of data can be predicted, but rather a frank appraisal of its limitations. This would assist research into refinement of the model and meanwhile act as a caution against unjustifiable predictions.

Model systems provide a developing basis for interpretation and understanding, but all suffer from the weakness of relevance to real soil conditions by reason of unrealistic surface areas and solution concentrations and time scale, and frequently by absence of competition of other adsorbates or surfaces. Improvement in the understanding and prediction of micronutrient reactions in soils will require research which includes studies on actual soils, or on their active fractions, provided the fractions can be separated without introducing unacceptable artefacts.

Mineral surface disorder and micronutrient adsorption

If defects in the surfaces of minerals provide the main reaction sites for specific sorption of micronutrient cations, then the macro mineralogical properties become much less important.

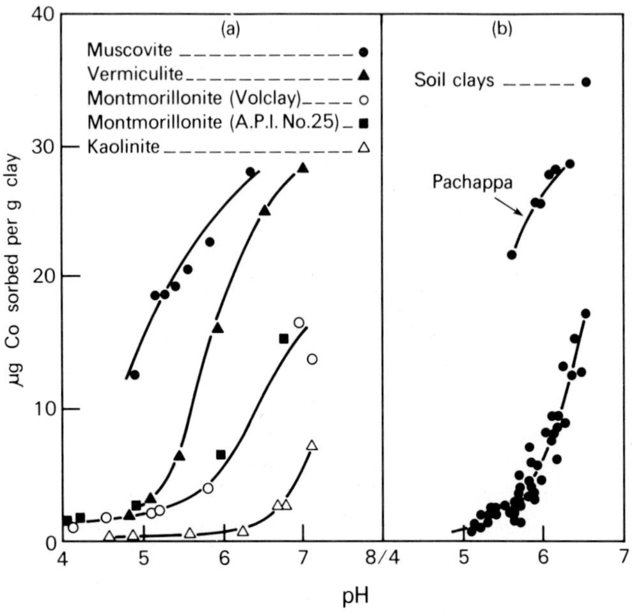

Fig. 25.1. Sorption of Co by untreated soil clays (<2 μm) $CaCl_2$ as a function of pH (Tiller *et al.* 1963).

Tiller *et al.* (1963) found that the specific sorption of Co by soil clays of widely ranging composition could be described, with one exception, by a single curve, whereas sorption by pure minerals varied considerably under identical conditions (Fig. 25.1). These soil clays were selected for having a dominance respectively of iron oxides, kaolinite, illite, montmorillonite or, the exception referred to above, vermiculite; the latter soil also contained an appreciable Mn content (Jenne 1968). Associated studies showed that the reactivity of different mineral surfaces to retention of Co and Zn is related to their ease of weathering, or their attack by dilute acids (Tiller and Hodgson 1962). From these two studies was developed the hypothesis that surface defects of natural soil minerals provide the reaction sites of specific sorption. This special role of surface defects is consistent with Jurinak and Bauer's (1954) proposal that the sorption of Zn by magnesium carbonate surfaces involves the 'healing' of the surface defects associated with missing Mg atoms. In clays, the increased specific sorption of heavy metals (Tiller 1968*a* and unpublished data) which results from adsorption of silicic acid by the clay could be viewed as being due to increased surface disorder.

Surface defects of silicate clays will behave like an oxide surface. The dominance of one mineral or another in trace metal reactions will then depend on the relative surface areas, the crystallinity as expressed at the surface, and the levels of micronutrients involved. The emphasis on defect structures means that the mineral species *per se* may be of much less influence in micronutrient reactions than is their weathering environment; by this view, mineral species would be important only in relation to their susceptibility to weathering.

A full understanding of these reactions remains a challenge requiring studies on soils, and their fractions, using integrated approaches, including solution chemistry and microprobe analysis; the recent studies by Fordham and Norrish (1974) on the soil components responsible for sorption of As are a good example.

Micronutrients in the solid phase

Micronutrient minerals

The various mineral forms of the micronutrients found in nature have been reviewed in the text books of geochemistry, e.g. by Rankama and Sahama (1950), Goldschmidt (1958) and Wedepohl (1969–78), and also in reviews in the soils context by Krauskopf (1972) and Norrish (1975). Many of the main mineral forms occurring in the earth's crust and in mineral deposits have not been detected in soils, and are unlikely to be found.

In general geochemical terms, Cu, Co, Mo and Zn are associated with sulphide minerals (chalcophile), and Zn, Mn and B with the silicate and oxide minerals (lithophile). Of great significance to the distribution of trace constituents is their ability to proxy for major constituents in minerals by isomorphous substitution; thus Co, Zn and Mn are often closely associated with Mg and Fe(II) in minerals. However, some soil minerals containing major concentrations of micronutrients have been identified and isolated, e.g. gorceixite (Norrish 1968), and undoubtedly more will be in the future as techniques improve. The other main source of micronutrients in the solid phase will be that associated with surfaces, either adsorbed or coprecipitated, on the outer layers of some of the more unstable oxide and carbonate phases whose solubility is determined by pCO_2 or pH and/or reduction potential. The geochemistry of the micronutrient elements in soils will be discussed briefly in relation to the above considerations.

Cobalt in soils is mainly associated with the oxide minerals, especially the Mn oxide minerals. Tiller (1959, 1963) showed that Co was closely associated with Mn in some *krasnozem* soils of northern Tasmania, and that a segregation of lithiophorite contained about 1% Co. Oertel and Giles (1964) found that Co and Mn are correlated in some Queensland soils. Taylor and McKenzie (1966) separated Mn from many soils and found Co to be universally concentrated in the Mn oxide fraction. In microprobe studies of some Australian soils McKenzie (1975) showed that 30–50% of the total soil Co was associated with manganese oxides in soils with soil Mn contents ranging from 110 to 360 μg per gram of soil. However, in many studies reviewed by Jenne (1968), Co has also been found associated with the Fe oxide component of soils.

Manganese in rocks is commonly associated with silicate minerals. In soils, it occurs mainly as complex oxide minerals which can act as host to many minor components. These minerals are discussed in detail elsewhere. The remaining Mn is mainly associated with silicate clays and organic matter. Manganese carbonate is thought to precipitate in flooded rice soils under conditions of high pCO_2 (I.R.R.I. 1970).

Zinc occurs in rocks as a minor replacement of Mg and Fe(II) in silicate and in sulphide minerals. Separate zinc silicate minerals, hemimorphite and willemite, also exist in rocks but they have not been found in soils. Zinc is to be expected in soil clay minerals by isomorphous replacement of Mg and Fe(II). Zinc sulphide possibly forms under very reduced conditions in soils, e.g. in some flooded rice soils (I.R.R.I. 1972). Manganese minerals in soils often contain higher concentrations of Zn than their surrounding soil matrix (Taylor and McKenzie 1966) but investigations by Tiller *et al.* (1972) on a range of Australian soils showed the Zn to be associated with Mn oxides only through adsorption. Resistant soil minerals, such as ilmenite and magnetite, can have a high Zn content (Norrish 1975). Studies based on extraction procedures suggested that Zn is strongly associated with iron oxides (White 1957). Recent microprobe studies of McKenzie (1975) showed that the Zn of some Australian soils is especially accumulated in the iron oxide segregations but not in the manganese ones.

Copper in soils has been associated most commonly with soil organic matter. Geochemically, it is associated with sulphides rather than silicates, yet soil profile studies (see later) often show a concentration of Cu in the clay-rich B horizon, and probably largely in sorbed forms. Copper sulphide can form under the reduced conditions of some flooded soils (I.R.R.I. 1972). Some of the soil Cu will be accounted for by isomorphous replacement of Ca or Mg and Fe(II) in silicate minerals; the conservation of Cu in the A horizon of some soils on dolerite results from the Cu in the felspar fraction (Tiller 1963). Copper is also enriched in manganese oxide concretions in soils (see Table 23.3), but direct investigation of soils using the microprobe analyser, shows that from 33 to 75% of the total soil Cu content may be accounted for in the iron oxides and only 1–6% in the manganese oxides (McKenzie 1975). Copper will be strongly associated with all soil surfaces by adsorption. This factor alone could account for the low concentrations in soil solutions (Krauskopf 1972).

Molybdenum is strongly associated with the ironstone gravels in soils (Oertel and Prescott 1944; and others since). In igneous rocks, it occurs as the sulphide; the forms of its relatively high concentrations in shales are generally less well understood, but Norrish (1975) showed, by microprobe analysis, a strong association with Fe, as in soils. Molybdenum in soils is probably mainly associated with iron oxide surfaces.

Boron in soils is expected to occur mainly in the adsorbed state and on oxide minerals and clays, but it may also occur as tourmaline inherited from the parent rocks. Tourmaline is very resistant to weathering, and it occurs as an accessory mineral in granites and pegmatites. Higher contents of B are found in shales, especially those of marine origin.

The use of solubility criteria in micronutrient studies

The solubility relationships of compounds of the micronutrients have been described in detail by Lindsay (1972). His review is a full and clear account of phase equilibria of known compounds; the reader is referred to it for background information and discussion. For solubility relationships in flooded (reduced) soils, the reader is referred to Ponnamperuma (1972).

It soon becomes clear that few of the micronutrient compounds for which thermodynamic data are available are of any relevance to soil chemistry. The main contribution, and an important one, has been in a negative sense. Thus we now know that under normal circumstances, $Zn(OH)_2$ and $Zn(CO_3)$ and $Zn_3(PO_4)_2 \cdot 4H_2O$ are all too soluble, as are many simple compounds of other micronutrients, to control the known concentrations in soil solutions. The solubility products of known compounds may have most relevance in heavily fertilized soils and very polluted soils and under some extreme conditions, e.g. in flooded soils.

Although the phase equilibria of individual simple compounds will continue to provide a framework for thinking about soils, and a valuable working tool under some conditions, the generally ill-defined nature of micronutrients at soil surfaces, and the lack of appropriate stability constants, will restrict the usefulness of the approach for many years. A recognition of the high solubilities of zinc carbonate, phosphate and hydroxide, led Tiller (1967) to investigate the possible role of monomeric silicic acid in controlling metal ion concentration, through either enhanced adsorption or the formation of insoluble silicate phases. Subsequently Tiller and Pickering (1974) spent some years trying to synthesize layer lattice zinc silicates, under conditions which might encourage the same phase in soil, of sufficient purity to permit the direct measurement of stability constants. Meanwhile Norvell and Lindsay (1970), using available thermodynamic data, showed that zinc silicate was too soluble to account for Zn^{2+} in soil solution. The 2:1 layer lattice zinc silicate synthesized at 20°C by Tiller and Pickering (1974), and aged for several years, had a solubility, near pH 7, which could be described by:

$$K = \frac{(Zn^{2+})\,(Si(OH)_4)}{(H^+)^2} = 10^{-5 \cdot 5}$$

for the reaction:

$$ZnSiO_3 + H_2O + 2H^+ = Zn^{2+} + Si(OH)_4$$

Tiller (unpublished data). The solubility in terms of $(Zn^{2+})\,(OH^-)^2$ was about 10^{-18}, that is, at least an order of magnitude less than the pure $Zn(OH)_2$ phase, but still far too soluble to explain the Zn^{2+} concentrations measured in several Colorado soils (Norvell and Lindsay 1969). However, whether this means that zinc silicates are irrelevant to Zn solution chemistry, or whether we haven't the appropriate constants for the as yet unidentified silicate phase, remains for the future.

The application of solubility criteria to the soil chemistry of trace constituents has several major limitations.

(i) The soil conditions being specified tend at best to a quasi-equilibrium condition; this is so because of continual fluctuations in temperature, variable removal and dilution of reaction products by leaching or capilliary rise, and variations in kind and concentrations of reactants due to biological activity (organic ligands and pCO_2). Superimposed on the short-term fluctuations mentioned above is the long-term reversion of metastable minerals to more stable forms. Some equilibria are effectively permanently poised in a non-equilibrium condition; two examples are: amorphous iron oxides formed in the presence of various sorbed foreign ions, such as silicate, humic substances etc. (see Chapter 23): and quartz and amorphous silica phases which are never likely to exert their theoretical solubility in soils because of either disordered outer layers (quartz) or adsorbed cations (amorphous silica).

Many of the limitations discussed above are not restricted of course to micronutrients, nor for that matter to the application of solubility products. Models of micronutrient reactions in soils, in terms of adsorption isotherms or adsorption equations, have their own limitations, e.g. the tendency for hysteresis on desorption, and variations in bonding energy with surface coverage.

(ii) Because of the small amounts involved, formation of separate phases of micronutrients are unlikely to be pure compounds but rather members of a solid solution series. The control of chemical reactions will be very hard to assess if we must use the indeterminate stability constants arising from ionic substitutions.

(iii) Finally, the levels of trace constituents in solution may be determined by the stability of an incipient separate solid phase, which occurs in the interface of the soil solution and main solid phases of the soil system, *viz* clays and oxides. Here the problem is less chemical and more philosophical. Two examples are presented: (a) Integrated sorption of hydrolyzed metal ions and silicic acid at low concentrations are proposed (Tiller 1968*a*, 1968*c*) to constitute a partial extension of the clay edge. Depending on the metal concerned, and thus the bonding strength, this reaction has affected to varying degrees the solubility of hectorite, as expressed by Mg in solution as a function of pH (Fig. 25.2). These data could be viewed in terms of either sorption stabilization, since solubility (Mg or Si as function of pH) is directly related to metal affinity by sorption, or a rudimentary layer of heavy metal-magnesium silicate exercising its own

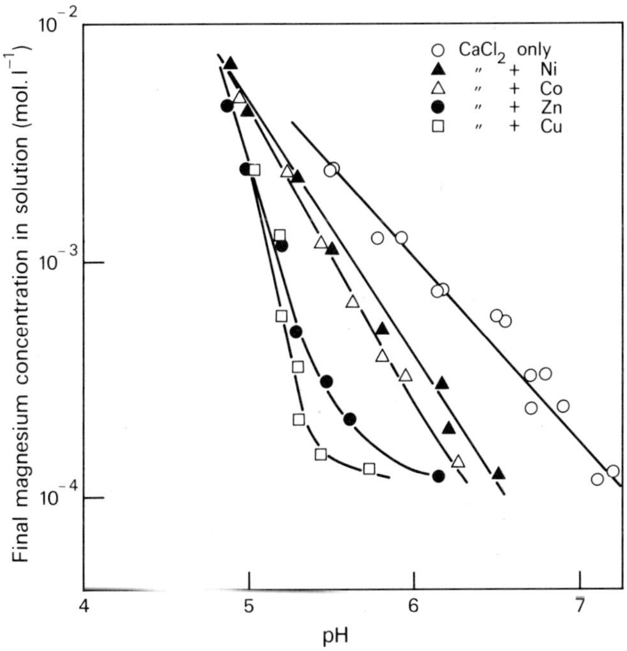

Fig. 25.2. The effect of heavy metal cations on hectorite solubility, expressed as magnesium in solution as a function of pH (Tiller 1968*c*).

intrinsic solubility. (b) As the second example, the solubility of zinc hydroxide precipitated in the presence of montmorillonite (~ 1mmol. Zn per g clay) agreed with published values for $Zn(OH)_2$ as a separate phase (Bingham *et al.* 1964). The solubility of similar amounts of Fe(III) precipitated in the presence of kaolinite was also unaffected by the presence of the clay (Fordham 1970), but lower and irregular solubility products, in terms of $Fe(OH)_3$, were obtained in clay samples containing less Fe. Similarly, when Zn and Co were precipitated in low concentrations in the presence of montmorillonite (~ 10 μmol. metal per gram of clay), the solubility products, measured as $(Zn)(OH)^2$ and $(Co)(OH)^2$, were respectively 3 to 4 orders of magnitude less soluble than these hydroxides precipitated in the absence of clay but otherwise under identical conditions (Tiller 1968*a*); precipitation in the presence of clay was indicated by marked upward inflections of adsorption isotherms. The actual solubility of Zn in the montmorillonite suspension, containing 10 μmol. precipitated Zn per gram of clay, was $[Zn][OH]^2 = 10^{-18.5}$, which agrees closely with the values of 10^{-18} to 10^{-19} quoted by Udo *et al.* (1970) for some Arizona soils of pH < 7.5. The solubility of Zn, as $Zn(OH)_2$, on the surface of montmorillonite (Tiller 1968*a*) is also the same as the Zn of the synthetic Zn silicate (Tiller and Pickering 1974) in which the Zn occurs effectively as a brucite-like $Zn(OH)_2$ layer. Since the apparent solubility has been markedly reduced, there must be a strong interaction between the hydroxide molecules and the clay surface silanol groups with which they are associated, probably at edges. The strong interaction raises the question as to whether the control of solution concentrations in these circumstances should be treated in terms of solubility of a separate phase, or a sorption reaction. If this kind of reaction is treated as solubility of a particular phase, then clearly, the stability constants which are normally available, and which refer to the 'free' phases, are not applicable. This may apply to many soil reactions.

Progress in soil chemistry in the near future will depend for its guidance on empirical relationships generated from studies on real soils. The experiments of Norvell and Lindsay (1969) and Lindsay and Norvell (1969) have shown a lead in this subject. Measurements of metal activity in soil systems containing complexing agents of known stability have led to the following relationships for a small group of American soils (Lindsay 1972):

$$(Cu^{2+}) = 10^{3.2} \times (H^+)^2 \text{ and } (Zn^{2+}) = 10^6 \times (H^+)^2$$

This approach deserves extension to other groups of soils and soil conditions, and will provide a sounder basis for research into the reactions which control micronutrient concentrations in soil solutions.

Reactions of micronutrients with soils

Many researchers have reported on the reactions of Zn and other trace metals with soils. Most report that sorption follows the Langmuir adsorption equation and that sorption capacity is correlated with clay and/or organic matter contents, together with interactions with calcium carbonate, when present (Udo *et al.* 1970; Shuman 1975; Sidhu *et al.* 1977; Singh and Silhon 1977). Although most published studies of the sorption of micronutrients have used concentrations greatly in excess of what is expected in soil solution, many have used concentrations and conditions more appropriate to studies of the macronutrients. The high zinc concentrations used in many investigations may well obscure the more relevant reactions taking place at much lower concentrations.

There have been few studies of the adsorption of micronutrients by Australian soils. Kanwar (1954) studied the adsorption of Cu by a soil from Kangaroo Island in South Australia which was associated with many nutrient deficiencies; the strong retention of Cu was ascribed mainly to the organic matter content. The research by Beckwith (1955) into the complexing of metals by soil organic matter, and his discussion of the implications for uptake of micronutrients by plants, provided the basis for many developments in this area of research. Tiller *et al.* (1969), using isotope techniques, studied the adsorption of Co by 25 Australian soils suspended in 0.05 M $CaCl_2$. For twenty of these soils, representing four soil groups, the sorption capacity was highly correlated with the manganese mineral content and with surface area. This correlation, however, did not apply to the fifth soil group, containing appreciable amounts of montmorillonite clay. The affinity of Co for soils was also relatively constant for all soils in which sorption capacity was dominated by Mn; the actual values were found to vary with the degree of site saturation for low surface coverage (<5%). These experiments with soils confirmed the special role of Mn minerals in the chemistry of Co, as discussed earlier (Taylor *et al.* 1964; McKenzie 1967), but also remind us that other soil surfaces are of importance in some soils.

The measurement of adsorption of Zn by a range of Australian soils suspended in $CaCl_2$ (Tiller *et al.* 1972) confirmed some of the conclusions made earlier, using model systems. The logarithm of the initial isotherm slope, a measure of affinity of adsorbate for the adsorption surface, had a linear relation with pH. These measurements were restricted to total zinc concentrations in solution of less than about 10^{-7} M for calcareous soils and less than about 2 x 10^{-6} M for weakly acidic soils. The generalization was developed from this relation that the initial reactions of low concentrations of Zn with all soils are determined by adsorption reactions of constant bonding energy in terms of the $ZnOH^+$, the ionic species controlling the reaction.

An attempt was also made to describe quantitatively the solubility relationships of Zn in these soil suspensions after shaking for 2 weeks. Soil solutions were not separated but an equilibrium concentration of Zn in solution was measured under conditions in which there was no net transfer between soil and solution. From these total concentrations of Zn, referred to as Zn_Q by Tiller *et al.* (1972), can be computed the concentration of Zn^{2+} in the equilibrium solution if we know the proportion of the Zn in solution that is complexed. These Zn_Q values were highly correlated with pH measured in the same suspensions of 22 soils (r^2 = 0.92). Three sandy soils dominated by organic matter were excluded. The regression equation (Tiller, unpublished data) was

$$\log Zn_Q = -2.20\text{–}0.75 \text{ pH}$$

for the pH range 5–7.4 measured in 0.05 M $CaCl_2$ or 6–8.5 in water. Similar regression coefficients were reported by Saeed and Fox (1977) who found a linear correlation between logarithm of total Zn concentration and pH in solutions that had been separated, after serial titration and treatment with constant amounts of $ZnSO_4$, from a series of soil suspensions from individual soils. Although constant conditions of ionic strength etc. were not maintained, slopes of the relation:

$$\log Zn_T = a - b \text{ pH}$$

averaged about 0.6 for the acidic soils and about 1 for calcareous soils, (by inspection of diagrams published by Saeed and Fox) in the pH range 5–7. Hodgson *et al.* (1966) showed, for a range of acidic and calcareous soils that about 40% of Zn in soil solution was complexed. If this applies approximately to these Australian soils then the following relationship controls the Zn ion activity in the soil suspension:

$$(Zn^{2+}) \simeq 10^{-2.8} \text{ x } (H^+)^{0.75}$$

It has a similar form to the one describing the solubility of hectorite in the absence of added Mg^{2+} and silicic acid (Tiller 1968*b*, Fig. 1), that is,

$$\text{pMg} = 1.00 \text{ pH} - 2.55$$

which on transformation becomes:

$$(Mg^{2+}) = (\text{constant})(H^+)$$

It is consistent with the results of Sinha *et al.* (1978), who used an equilibrium method based on EDTA complexes to study the solubility of Zn and other metals in some Indian soils; they found that:

$$(Zn^{2+}) = 10^{-2.2} \text{ x } (H^+)$$

over a narrow range of pH (7.14–7.76). Hence, the proposed relation describing Zn^{2+} and H^+ in soil solution may well result from the dissolution of talc- or hectorite-like rudimentary layers of Zn silicates at the edges of aluminosilicate clays, as discussed earlier. The relation $[Zn^{2+}] \simeq 10^{-3} \text{ x } [H^+]^{0.75}$ is essentially different from that published by Norvell and Lindsay (1969), as discussed earlier, but predicts very similar values of $[Zn^{2+}]$ for near neutral and alkaline soils. Although these relationships between Zn^{2+} and H^+ concentrations in solution are generally consistent with the different studies based on model systems, further detailed investigation is needed of the chemical equilibria of micronutrient ionic species in soil solution.

Early studies (Stout *et al.* 1951; Barshad 1951; Jones 1957) showed that Mo reacts strongly with soils, clays and oxides. Williams and Moore (1951) suggested that the availability of soil Mo to plants is related to pH and iron oxide content, probably through an adsorption process involving molybdate ions and free iron oxides. Jones' work showed that the affinity of Mo for Fe oxides is much greater than for other oxides and silicate clays. The high affinity for Fe oxides was also demonstrated by comparing the sorptions of Mo, before and after mild deferration treatment, by a northern N.S.W. soil rich in iron oxide. These experimental expressions of the exceptional affinity of Mo for iron oxides are consistent with the high concentrations of Mo found in ironstone nodules separated from soils (Oertel and Prescott 1944; Wells 1956). Reisenauer *et al.* (1962) measured adsorption isotherms of Mo on various oxides and several Californian soils; all exhibited similar adsorption curves as a function of pH, which suggested a common reaction, involving chemical bonding and the release of hydroxyl ions and water.

Smith and Leeper (1969) suggested that the initial adsorption of Mo by oxide minerals is followed by slow conversion to less soluble forms. The difficulty in

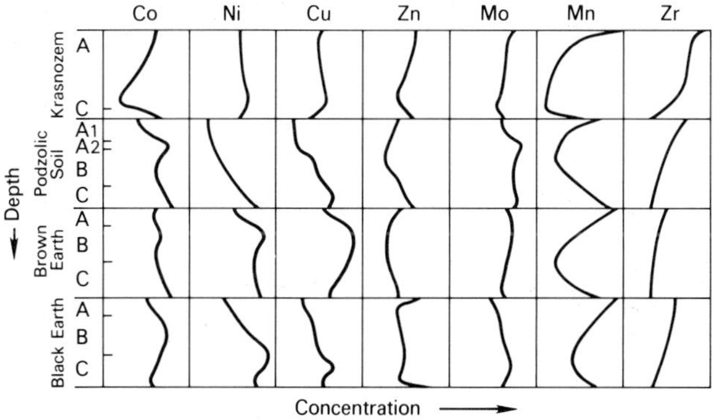

Fig. 25.3. Trends in the amounts of trace elements in four soil groups developed on dolerite (Tiller 1963).

distinguishing between formation of new phases and chemisorbed monolayers of micronutrients has been discussed earlier. Reisenauer *et al.*, Jones, and Smith and Leeper all recognized the possibility of formation of insoluble molybdates in soils, but their experimental evidence was equivocal in this regard. Later studies by Hingston *et al.* (1972), using Fe and Al oxides, supported the view that specific sorption of Mo is the preferred fixation mechanism in soils. According to their proposals, the sorption of molybdate takes place by ligand exchange of MoO_4^{2-} for surface-coordinated water, with a corresponding decrease in positive charge. At pH values around the pK_2 of molybdic acid (~ 4.5), sorption is enhanced by dissociation of $HMoO_4^-$ at the surface. The released proton removes a hydroxyl from the surface, thus generating a positive site for the deprotonated anion (Hingston *et al.* 1970). The validity of the application of the theories of the latter authors to soils has been supported by a comparative study of the sorption of sulphate and molybdate and phosphate by soil suspensions (Barrow 1970). Little and Kerridge (1978) have recently shown that the Mo sorption by some Queensland soils follows the form of the Freundlich isotherm, and have provided an improved basis for prediction of molybdenum fertilizer needs.

Vertical distribution of micronutrients in soil profiles

A major reorganization of micronutrients often takes place during pedogenesis. Soils may contain more or less trace elements than the original parent materials and may vary in composition from horizon to horizon, and also between different components of one horizon. The processes leading to this redistribution of elements during soil formation are complex, because of the interactions of parent material composition, geomorphic processes, and chemical weathering, as a function of time. Fig. 25.3 gives examples of the variations in vertical distribution of several trace elements in four soil groups formed on one fairly uniform parent material (Tiller 1963). These curves are based on somewhat subjective assessments of the general trend with depth, for typical residual profiles of each soil group. The trends in relative abundances of a group of elements within a soil profile may be viewed as a fingerprint of the soil's genetic history, and in particular, of the mobilities of the particular elements concerned. These profile trends are a very important consideration in the planning and interpretation of geochemical surveys for mineral deposits. These topics will be developed further below. The mobility of an element in the weathering zone is inversely related to the chemical stability of its solid phase forms whether sorbed or as a compound. The movement of inorganic and organic colloids through soils, as in illuviation, is an exception; generally, it is the tendency to partition into the solution-phase that controls the movement of ions and molecules through soils and sediments by diffusion or mass transport in solution.

Chemical weathering and micronutrient profiles

A consideration of elemental mobility includes much of the subject of chemical weathering of rocks and minerals, which is beyond the scope of this chapter. However, it should be recognized that under any given different environmental conditions, minerals differ greatly in their chemical and physical stability. Hence, the release of micronutrients from host minerals during weathering is related to the chemical properties of the major ions which constitute the crystal structure, rather than to the properties of the elements occurring in trace amounts by isomorphous substitution. Another general concept often discussed in weathering studies, and of importance in determining trace element distribution in soil profiles, is that of *relative* and *absolute* accumulation of weathering products. Some losses of all elements released by weathering processes are to be expected. Quantitative evaluation of these losses is possible if inert minerals or very immobile elements are present because they can be used as indices of weathering. If large proportions of the more soluble weathering products, *viz.* Na, Ca, Mg and Si are removed by leaching, then the less mobile elements will appear to have accumulated (relative accumulation), even though no inputs of these elements from other weathering zones or soil horizons have taken place. The zone of greatest apparent concentration of an element, as measured by chemical analyses, may well be the horizon of its greatest loss by leaching. Since Zr is considered to be a relatively immobile element, by reason of its mineralogy and its strong sorption affinity, the decreasing concentrations of Zr with depth for each soil group (Fig. 25.3) give an indication of the decreasing weathering intensity, and losses of soluble products with depth. Absolute accumulation results from the transfer of weathering products and their partial retention in another horizon; examples are the accumulation of some nutrient elements in surface soils by biotic cycling, the scavenging effect of Fe and/or Mn oxides in concretions near the surface of the water table, and the retention of some elements by calcium carbonate, clays and oxide minerals in soil B and C horizons. The mobility of micronutrients in soil profiles, and their retention in zones of accumulation, are controlled by the degree and kind of complexing in solution, and by sorption and precipitation reactions as discussed in earlier sections. The parent material also affects these secondary reactions because of its influence on the nature of sorption surfaces being formed, and on inorganic ligands being released, e.g. Cl^-, SO_4^{2-}, HCO_3^-.

Biological activity and micronutrient profiles

Biological activity greatly influences the mobility and distribution of micronutrients in soil profiles. Organic ligands capable of forming stable complexes with micronutrients are most important by-products of this activity. Soil and sediment pH may be modified by microbial activity, especially in the vicinity of plant roots and in situations where sulphide minerals are exposed to oxidizing conditions.

In waterlogged soils, and possibly some microenvironments, oxygen stress resulting from microbial activity may lower *Eh* sufficiently to affect the stability of solid phases of Fe(III) and Mn(IV), which are normally stable in aerated environments. On the other hand, new phases, e.g. trace metal sulphides, may form if *Eh* becomes sufficiently low. Carbon dioxide produced by microbial respiration can affect micronutrient mobility through its effect on pH, and on carbonate equilibria and degree of complexing in solution by HCO_3^-. Finally, the absorption of metals in deeper soil horizons by plant roots results in the translocation of micronutrients to the above-ground parts of the plant (including trees). Regular leaf fall, and the loss of micronutrients by exudation from living leaves, leaching from senescent leaves, and death of the whole plant, can all contribute to the redistribution of elements capable of absorption by plants; the elements need not be plant nutrients. The organic debris is broken down and incorporated in the surface horizon by soil organisms. However, any accumulation of micronutrients found at the soil surface may not remain associated chemically with organic matter but could be strongly sorbed on finely divided Fe and Mn oxides which themselves resulted from biological degradation of plant remains. Of the micronutrients, Mn has the highest concentration in plants and is commonly accumulated in soil surface horizons (see Figs 25.3 and 25.4).

Geomorphic processes and micronutrient profiles

Landscape history has a major influence on micronutrient status and concentration trends down soil profiles. These trends may be accentuated or attenuated by geomorphic processes. New land surfaces may develop on colluvium comprised of materials from surface horizons of soils, or on the deeper horizons of soils exposed by erosion. The levels of micronutrients in the new soils will vary greatly, depending on whether the truncation of the antecedent soil occurred in a zone of depletion, or in a zone of accumulation of a particular element. Since the zone of accumulation of one element may well be a zone of maximum depletion of another, there is a potential for great variation in micronutrient status, even in one climatic region on a uniform parent material. The cobalt status of some *krasnozem* soils on dolerite is an example in this variation; whereas in the case of four widely separated residual *krasnozems* a range of about 10-fold could be found for the Co content of the horizons of one profile and only a 2-fold range (25–51 μg per gram of soil) in their surface horizons, the Co concentrations of surface soils of *krasnozems* in areas of landscape instability were much more variable, with values from 5 to 90 μg per gram.

Downslope movement and aerial accession of soil material results in subadjacent horizons that are unrelated genetically; this situation can lead to complex variations in depth–concentration functions of trace elements, and problems of interpretation unless such discontinuities are recognized. These factors are of practical significance in geochemical prospecting, and need to be sorted out in orientation surveys. In studies of soil genesis, the well defined trends of some trace elements with depth in soil profiles can provide independent evidence for fossil or buried soils. If the overlying and underlying soil materials are of similar origin, the degree of expression of the profile trend may indicate relative exposure to the two land surfaces. Examples of these trends in relation to buried soils are given by Tiller (1958) in studies of the geochemistry of soils on old basaltic ash showers in south eastern South Australia.

Water and micronutrient profiles

Water is the major agent involved in the redistribution of elements during pedogenesis. Besides its essential role as a reactant in chemical weathering reactions such as hydration and hydrolysis, and in biological activity, it has other important functions; it provides the medium for mass transfer of reaction products, soluble and colloidal, by vertical and lateral leaching; water also provides the lubricant which facilitates the effect of gravity on many geomorphic processes, e.g. solifluction and other forms of erosion which effect transfers of micronutrients in the landscape.

Examples of trace metal trends down soil profiles

The distributions of total Co, Cu, Mn, Mo, and % clay (<2 μm) are shown in Fig. 25.4 for four profiles. The data were prepared by A. C. Oertel (Division of Soils, unpublished data) using reduced variables of depth and concentration. These curves illustrate some of the points discussed in the sections above; they emphasize the complexity of the processes involved, and they caution against simple generalizations.

(i) Trends with depth for many elements are strongly dependent on the kind of soil, even if on the same parent materials; for example, the two soils on dolerite had quite different trends for Cu and Co.

(ii) The variation in clay content in a profile is a poor basis for predicting trace metal concentrations. In soil H128, the trends of Mo and Cu mirror that of clay content; in soil B233 the peaks in Co and Mo content coincide with the top of the horizon of clay accumulation, but the trends of Cu and other elements are unrelated to the clay trend; furthermore, the upper part of soil H165 has a high, and relatively constant, clay content, yet Cu, Co and Mn exhibit marked changes with depth.

(iii) Strong geochemical associations exhibited in one soil, e.g. Co and Mn in soil H165, are no guarantee that they will occur in another soil, even on the same parent materials (e.g. soil H128).

(iv) Some soils have minimal changes in trace element composition with depth.

Micronutrient availability and assessment

Concepts of plant nutrient availability

The term 'availability' lacks an agreed, exact definition but is nevertheless widely used; it serves to describe the ability of plants to utilize the nutrient resources of the soil. This topic has been reviewed by Viets (1962), Beckwith (1963), Loneragan (1975) and many others.

Earlier researchers treated available nutrients in soils in terms of their solubility, specified by chemical extraction or equilibrium procedures. Availability has also been viewed in terms of chemical pools (Viets 1962) which describe the amounts of nutrient in defined chemical states; examples of such pools for micronutrients are: water soluble, nonspecifically adsorbed, specifically adsorbed, and held within crystal lattices. Each of these chemical components is viewed as having its own rates of solution and desorption, and equilibria with other pools; all contribute eventually to the uptake of micronutrients by plants. According to the particular chemical state involved, each pool can be defined by specific chemical extraction procedures. Most soil tests used in diagnostic work are based, implicitly or otherwise, on Viet's

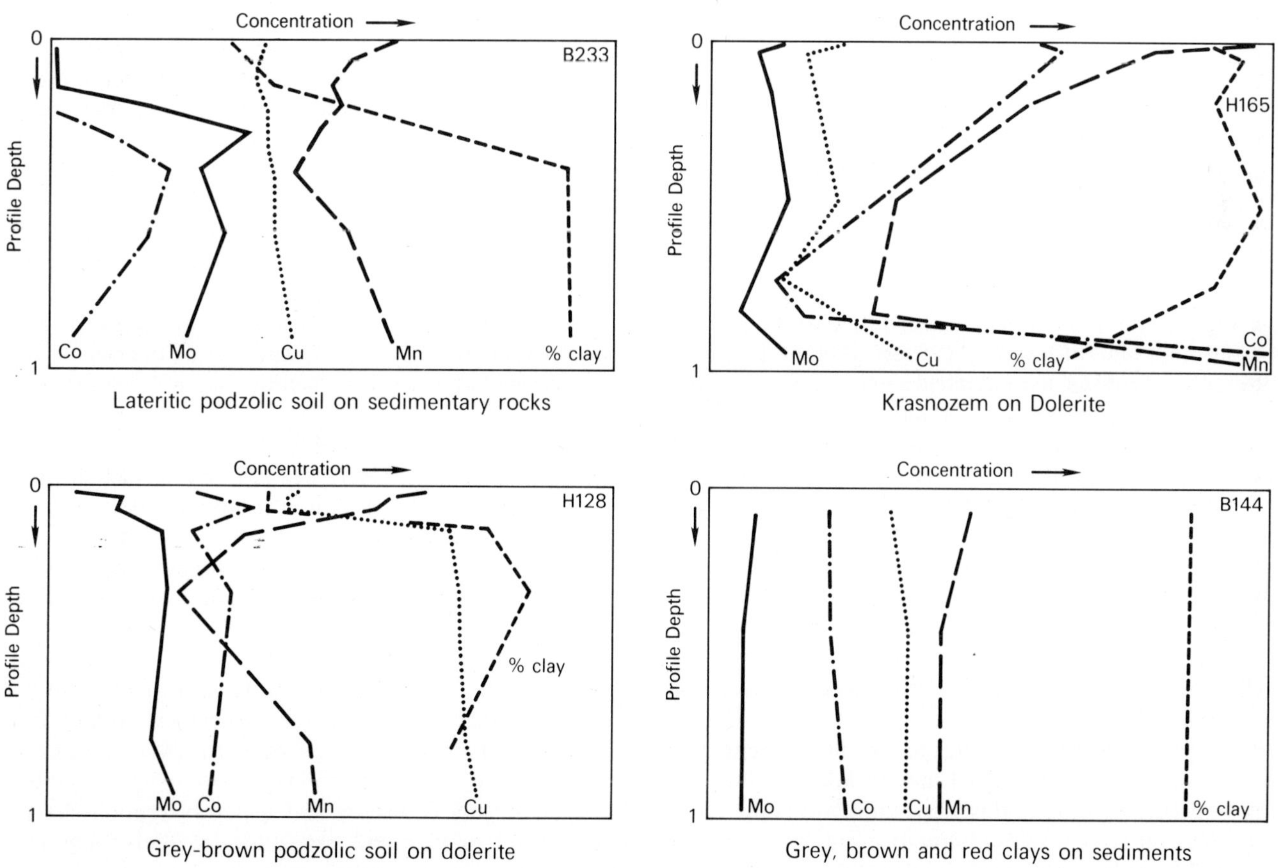

Fig. 25.4. Examples of total trace element distribution in four soil profiles.

approach. These chemical extraction procedures may provide a useful guide to the ability of the soil to sustain the needs of a specified crop for a particular nutrient, even though the processes involved in the absorption of micronutrients by plants are incompletely understood. For this latter reason, the values obtained by chemical and extraction procedures should not be termed 'available' nutrient, but defined only in terms of the extraction procedure used.

A complementary approach to the question of the nutrient status of a soil, and one more amenable to theoretical treatment, emphasizes the chemical interactions between solid and solution phases in soils and the factors which affect this equilibrium. In the short term, the concentrations of micronutrients in soil solution in the root zone will determine the extent of nutrient absorption by plants. However, soil solution contains such low concentrations of micronutrients that it can only supply a very small part of a crop's seasonal requirements. Under these conditions, the rate of desorption and transport of each micronutrient will determine its availability to the plant. Since micronutrients are usually strongly bound, the ability of a soil to supply them to plant roots will be limited by the ease of desorption and by the need to rely on diffusion in response to concentration gradients near plant roots as the main means of nutrient transport. Hysteresis of micronutrient-colloid reactions may complicate predictions of nutrient supply. Rate of supply of micronutrients to plants in many deficient soils, and hence vigour of plant growth, may be thus much more dependent on the partition between the solid and solution phases of soils than on the total amount of labile micronutrient present as measured by selective extraction or isotopic dilution procedures.

The degree of complexing and the pH status are the two key variables which can affect this equilibrium and thus the levels of micronutrients in solution and the rates of desorption and diffusion in response to plant demand. Differences between the chemical environments of the bulk soil and of the root zone should also be recognized. Plant exudates and the by-products of microbial activity in the vicinity of plant roots (Rovira 1969) can decrease pH and increase the concentrations of complexing substances. Thus, in the root zone the reversibility of reactions of micronutrient cations with soil surfaces may be more readily achieved (Quirk and Posner 1975) and diffusion to the root surface facilitated (Elgawhary *et al.* 1970), but the extent to which these processes are encouraged may be difficult to predict using criteria based on plant-free soil.

Many other researchers prefer to use the nutrient status of plants as the indicator of the availability of nutrients in soil. According to this view, the plant is the only acceptable judge of a soil's ability to supply nutrients to a plant. This is hard to dispute: the real problem is then one of communication with the plant, to determine which factors were considered in the judgement, and if any are suited for the task of routine assessments. In practice, the micronutrient concentration in the plant, often with the plant part and maturity specified, are commonly and successfully used in the diagnosis of deficiencies. Other people (e.g. Beckwith 1963) consider that the total uptake of nutrients, rather than the concentration in the plant, provides a better basis for comparing nutrient availability in soils. Such comparisons require plants to be grown under standardized conditions which permit full exploitation of the soil's nutrients. This approach is recommended for calibration or other screening procedures but is not itself suitable for routine testing of soils from farms. The main limitation of plant analysis as a means of assessing nutrient availability in soils and the need for fertilizer amendments results from the effects of differences in climatic conditions, maturity, kind of plant, and interactions with other nutrients.

Assessment of micronutrient deficiency

Leaf symptoms and micronutrient deficiency

The visual recognition of leaf symptoms is an important means of identifying micronutrient disorders; it has been reviewed for various plants by Wallace (1964) and Chapman (1966).

Confident diagnosis by this approach requires much experience in the nutrition of particular crop varieties in relation to local soils, since the symptoms of some nutrient deficiencies are difficult to differentiate. Another limitation of this technique is that yield decreases may occur before nutrient levels are low enough to induce visual symptoms; this has been referred to as the region of 'hidden hunger'. There is no doubt however, that leaf symptoms and growth characteristics of many crops will readily define instances of extreme deficiency. A variant of the general application of leaf symptoms is the use of indicator plants which are sensitive to the status of certain nutrients in soils. One advantage of the above approaches is that analytical facilities are not required.

Plant analysis for assessing micronutrient deficiency

Foliar analysis is widely used as a means of detecting micronutrient deficiency and as a basis for determining fertilizer amendments; the subject has been well reviewed by Chapman (1966) and Jones (1972). This approach was stimulated by the modern developments in spectrometry, both X-ray fluorescence and atomic absorption. Its most successful application has been in areas of more intensive agriculture and in monitoring trends in nutrient levels in the sufficiency ranges. For example, under-fertilization leading to reduced nutrient reserves, or over-fertilization leading to toxicity or nutrient antagonisms, can soon be detected by regular checking of leaf composition, especially in tree crops, and remedial measures put into practice.

Although foliar analysis has been an invaluable aid in

Table 25.7. Approximate concentrations in mature leaf tissue

Actual values will vary for different crops, plant parts and times of sampling. After Jones (1972)

Micro-nutrient	Concentration (μg per gram of tissue)		
	Deficient	Sufficient	Excess or toxic
B	<15	20–100	>300
Cu	<4	5–20	>20
Fe	<50	50–250	–
Mn	<20	20–500	>500
Mo	<0.1	0.5–	–
Zn	<20	25–150	>400

improving crop nutrition, the critical levels or ranges are often poorly defined, since variations occur between species and varieties. The critical value is the concentration below which yield responses should result from fertilization, if no other factors limit growth. Chapman (1966) has published a comprehensive list of critical values for many crops. Approximate concentrations of micronutrients, corresponding to deficient, sufficient, and toxic levels in plants are given in Table 25.7 (after Jones 1972). Another limitation of plant analysis is the relative insensitivity of plant composition to the extent of deficiency, when in the deficiency and incipiently deficient ranges. An extreme manifestation of this is the tendency for extremely deficient plants to concentrate nutrients (Steenbjerg 1951). An important bonus of plant analyses is that some data may also be useful in studies of human and animal nutrition.

Soil analysis for assessing micronutrient deficiency

Many chemical extraction procedures have been used to estimate the micronutrient requirement of plants; most commonly, dilute acids and complexing agents have been used. These methods have been reviewed for many micronutrients by Cox and Kamprath (1972). Soil tests are usually calibrated in glasshouse studies; critical soil extraction values are assessed, either in relation to the nutrient concentrations in reference crop plants, or in terms of the yield response to additions of the micronutrient under investigation. Glasshouse calibration of soil tests often gives good correlations because, in contrast to field experiments, the plants are grown under conditions of maximum soil–plant root interaction and of minimized variation in environmental conditions. The relevance of parameters derived in the glasshouse to field-grown plants has often been questioned, however.

Soil extraction tests for micronutrients are rarely used in practice for estimating an application rate of micronutrient fertilizer, but many organizations have found routine soil testing a valuable guide as to whether micronutrients need be applied. The prediction is often improved when other soil characteristics, such as organic matter or pH etc. according to the particular soil type and micronutrient, are also taken into consideration. Advantages of soil extraction for diagnostic testing are the greater simplicity of the analysis and its ready adaptation to routine large-scale surveys. Care is required in the interpretation of data obtained, because critical values for well defined extraction procedures may vary between different soil groups and different crops. Analyses of either the soil or the plant for diagnosis of availability suffer from the same limitations, although the factors involved differ. Acceptance of critical values from soil extraction procedures derived in other regions or countries should be done with caution because there may be important differences in climate or crop varieties etc., but especially because minor variations in extraction procedure may result in significantly different amounts of nutrient extracted (Soltanpour *et al.* 1976). The same applies to some extent to critical values based on plant analysis.

General comment on assessment of micronutrient availability

Micronutrient deficiencies result from the interaction of soil conditions, genetic variation between plants, climatic effects, and nutrient imbalance. Deficiencies are thus often difficult to predict with confidence by simple procedures based on either soil or plant analysis. Because of the many causes, micronutrient deficiencies may be spotty within a field, and spasmodic in time, occurring in one season and not another; deficiencies may be evident early in the growing season, with a possible set-back in growth, but not evident in the mature plant. Interactions with fertilizers, e.g. phosphate with zinc, are not always consistently expressed, and may vary from year to year.

In circumstances of extreme deficiency most methods would suffice, whether based on visual symptoms or on analysis of the plant or soil. Plant analysis has definite advantages in monitoring nutrient trends due to fertilizer practice in areas of intensive high-production agriculture where growth stresses due to other nutrients or water etc. are minimized; soil-testing procedures have advantages in broad-scale surveying of soil nutrient resources. Plant and soil analysis are much more effective diagnostic tools when used in conjunction with each other.

Finally, it is most surprising that Australian soil scientists and agriculturists, famous for their early research into the field recognition of micronutrient deficiencies, have done so little in the development and practice of soil testing for micronutrients. The reason lies mainly in the effectiveness of the *ad hoc* simple solutions devised some decades ago to overcome the problems. But the lack of general acceptance and enthusiasm for micronutrient testing is also related to the multiplicity of organizations and departments administering agricultural research and extension, as well as to the inadequacy of analytical techniques in earlier years, and possibly to unreal expectations both of the precision of prediction, and of the benefits. There is today, still, plenty of scope for the development of good micronutrient testing programmes; land is still being developed for agriculture, especially in northern Australia; existing agricultural land is being used for new purposes; new fertilizer practices are being introduced, and new crop varieties, with possibly different demands for micronutrients, are being bred and introduced. Some fertilizer companies and government organizations have micronutrient testing programmes, but the approach should be more widespread and based as far as possible on standardized procedures. The application and usefulness of data should not be limited by farm or state boundaries. Donald and Prescott (1975) have estimated that about 15 million hectares of agricultural production in Australia have benefited from micronutrient fertilization. The main soil types and situations associated with micronutrient deficiencies were reviewed by Stephens and Donald (1958). The various means of correction of micronutrient deficiencies, and the associated problems of fertilizer technology, have been reviewed by Silverberg *et al.* (1972), Lehr (1972), Giordano and Mortvedt (1972), and Reuter (1975).

Farmers and their advisers deserve a reliable basis on which to decide whether deficits in micronutrients are limiting production, whether current fertilizer practice may lead to nutrient imbalance, thus affecting future plant and/or animal production, and whether nutrients are being added unnecessarily. Donald and Prescott (1975), in summarizing the progress in trace element studies in Australia, have stated that, "the most pressing need in the use of trace elements in Australian agriculture is for effective tests of deficiency".

References

Barrow, N. J. (1970). Comparison of the adsorption of molybdate, sulfate and phosphate by soils. *Soil Sci.* **109**, 282-288.

Barshad, I. (1951). Factors affecting the molybdenum content of pasture plants. 1. Nature of soil molybdenum, growth of plants, and soil pH. *Soil Sci.* **71**, 297-313.

Beavington, F. (1977). Trace elements in rainwater and dry deposition around a smelting complex. *Environ. Poll.* **13**, 127-131.

Beckwith, R. S. (1955). Metal complexes in soils. *Aust. J. Agric. Res.* **6**, 685-698.

Beckwith, R. S. (1963). Chemical extraction of nutrients in soils and uptake by plants. *Agrochimica* **7**, 296-313.

Beckwith, R. S. (1973). The availability of micronutrient metals to plants. Ph.D. Thesis, University of Western Australia.

Bingham, F. T., Page, A. L. and Sims, J. R. (1964). Retention of Cu and Zn by H-montmorillonite. *Soil Sci. Soc. Am. Proc.* **28**, 351-354.

Bowden, J. W., Bolland, M. D. A., Posner, A. M. and Quirk, J. P. (1973). Generalized model for anion and cation adsorption at oxide surfaces. *Nature* (London), *Phys. Sci.* **245**, 81-83.

Bower, C. A. and Truog, E. (1940). Base exchange capacity determination as influenced by nature of cation employed and formation of basic exchange salts. *Soil Sci. Soc. Am. Proc.* **5**, 86-89.

Cartwright, B., Merry, R. H. and Tiller, K. G. (1977). Heavy metal contamination around a lead smelter at Port Pirie, South Australia. *Aust. J. Soil Res.* **15**, 69-81.

Chapman, H. D. (1966). 'Diagnostic Criteria for Plants and Soils'. (University of California: Berkeley.)

Clayton, P. M. and Tiller, K. G. (1979). A chemical method for the determination of the heavy metal content of soils in environmental studies. CSIRO Aust. Div. Soils Tech. Pap. No. 41.

Cox, F. R. and Kamprath, E. J. (1972). Micronutrient soil tests. *In* 'Micronutrients in agriculture.' pp. 289-318. (Soil Sci. Soc. Am.: Madison.)

Delas, J. (1963). La toxicité du cuivre accumulé dans les sols. *Agrochimica* **7**, 258-288.

Donald, C. M. and Prescott, J. A. (1975). Trace elements in Australian crop and pasture production. *In* 'Trace Elements in Soil-Plant-Animal Systems.' pp.7-38. (Academic Press: New York.)

Dugger, D. L., Stanton, J. H., Irby, B. N., McConnell, B. L., Cummings, W. W. and Maatman, R. W. (1964). The exchange of twenty metal ions with the weakly acidic silanol group of silica gel. *J. Phys. Chem.* **68**, 757-760.

Elgawhary, S. M., Lindsay, W. L. and Kemper, W. D. (1970). Effect of EDTA on the self-diffusion of zinc in aqueous solution and in the soil. *Soil Sci. Soc. Am. Proc.* **34**, 66-93.

Forbes, E. A. (1973). Ph.D. Thesis, University of Western Australia. (quoted by Quirk and Posner 1975).

Forbes, E. A., Posner, A. M. and Quirk, J. P. (1974). The specific sorption of inorganic Hg(II) species and Co(III) complex ions on goethite. *J. Coll. Interface Sci.* **49**, 403-409.

Forbes, E. A., Posner, A. M. and Quirk, J. P. (1976). The specific sorption of divalent Cd, Co, Cu, Pb and Zn on goethite. *J. Soil Sci.* **27**, 154-166.

Fordham, A. W. (1969). Sorption and precipitation of iron on kaolinite. II. Sorption isotherms and their interpretation in terms of iron (III) ionic equilibria. *Aust. J. Soil Res.* **7**, 185-197.

Fordham, A. W. (1970). Sorption and precipitation of iron on kaolinite. III. The solubility of iron (III) hydroxides precipitated in the presence of kaolinite. *Aust. J. Soil Res.* **8**, 107-122.

Fordham, A. W. (1973). The location of iron-55, strontium-85 and iodide-125 sorbed by kaolinite and dickite particles. *Clays Clay Miner.* **21**, 175 184.

Fordham, A. W. and Norrish, K. (1974). Direct measurement of the composition of soil components which retain added arsenate. *Aust. J. Soil Res.* **12**, 165-172.

Frank, R., Braun, H. E., Ishida, K. and Suda, P. (1976). Persistent organic and inorganic pesticide residues in orchard soils and vineyards of southern Ontario. *Can. J. Soil Sci.* **56**, 463-484.

Geering, H. R. and Hodgson, J. F. (1969). Micronutrient complexes in soil solution: III. Characterization of soil solution ligands and their complexes with Zn^{2+} and Cu^{2+}. *Soil Sci. Soc. Am. Proc.* **33**, 54-59.

Giles, J. B. (1959). Trace element content of some Queensland surface soils. CSIRO Aust. Div. Soils Div. Rep. No. 1/59.

Giles, J. B. (1960). Trace element profiles of some ground water rendzina and terra rossa soils from County Grey, S.E. South Australia. CSIRO Aust. Div. Soils Div. Rep. No. 9/60.

Giles, J. B., McKenzie, R. M. and Tiller, K. G. (1962). Revised data for molybdenum in some Australian soils. CSIRO Aust. Div. Soils Div. Rep. No. 12/61.

Giordano, P. and Mortvedt, J. J. (1972). Agronomic effectiveness of micronutrients in macronutrient fertilizers. *In* 'Micronutrients in Agriculture.' pp.505-524. (Soil Sci. Soc. Am.: Madison.)

Goldschmidt, V. M. (1958). 'Geochemistry.' (Ed. A. Muir). (Clarendon Press: Oxford.)

Grimme, H. (1968). Die Adsorption von Mn, Co, Cu and Zn durch Goethit aus verdünnten Lösungen. *Z. Pflanzenernaehr. Bodenkd.* **21**, 58-65.

Hingston, F. J., Posner, A. M. and Quirk, J. P. (1970). Anion binding at oxide surfaces – the adsorption envelope. *Science* **1**, 324-327.

Hingston , F. J., Posner, A. M. and Quirk, J. P. (1972). Anion adsorption by goethite and gibbsite. I. The role of the proton in determining adsorption envelopes. *J. Soil Sci.* **23**, 177-192.

Hodgson, J. F. (1960). Cobalt reactions with montmorillonite. *Soil Sci. Soc. Am. Proc.* **24**, 165-168.

Hodgson, J. F. and Tiller, K. G. (1962). The location of bound cobalt on 2:1 layer silicates. *Clays Clay Miner.* **9**, 404-411.
Hodgson, J. F., Tiller, K. G. and Fellows, M. (1964). The role of hydrolysis in the reaction of heavy metals with soil-forming materials. *Soil Sci. Soc. Am. Proc.* **28**, 42-46.
Hodgson, J. F., Geering, H. R. and Norvell, W. A. (1965). Micronutrient cation complexes in soil solution. I. Partition between complexed and uncomplexed forms by solvent extraction. *Soil Sci. Soc. Am. Proc.* **29**, 665-669.
Hodgson, J. F., Lindsay, W. L. and Trierweiler, J. F. (1966). Micronutrient cation complexing in soil solution: II. Complexing of zinc and copper in displaced solution from calcareous soils. *Soil Sci. Soc. Am. Proc.* **30**, 723-726.
Hodgson, J. F., Tiller, K. G. and Fellows, M. (1969). Effect of iron removal on cobalt sorption by clays. *Soil Sci.* **108**, 391-396.
I.R.R.I. (1970). Annual Report Int. Rice Res. Inst., Los Banos, Philippines.
I.R.R.I. (1972). Annual Report Int. Rice Res. Inst., Los Banos, Philippines.
Isbell, R. F. and Gillman, G. P. (1973). Studies on some deep sandy soils in Cape Yorke Peninsula, North Queensland. I. Morphological and chemical characteristics. *Aust. J. Exp. Agric. Anim. Husb.* **13**, 81-87.
Isbell, R. F., Jones, R. K. and Gillman, G. P. (1976). Plant nutrition studies on some yellow and red earth soils in northern Cape Yorke Peninsula. *Aust. J. Exp. Agric. Anim. Husb.* **116**, 532-541.
Isbell, R. F. and Smith, G. M. (1976). Some properties of red, yellow, and grey massive earths in Nth Queensland. CSIRO Aust. Div. Soils Tech. Pap. No. 30.
James, R. O. and Healy, T. W. (1972). Adsorption of hydrolysable metal ions at the oxide-water interface. III. A thermodynamic model of adsorption. *J. Coll. Interface Sci.* **10**, 65-81.
Jenne, E. A. (1968). Controls on Mn, Fe, Co, Ni, Cu and Zn concentrations in soils and water; the significant role of hydrous Mn and Fe oxides. *In* 'Trace Inorganics in Water.' Adv. Chem. Series No. 73 (Am. Chem. Soc.)
Jones, L. B. (1972). Plant tissue analysis for micronutrients. *In* 'Micronutrients in Agriculture.' pp.319-346. (Soil Sci. Soc. Am.: Madison.)
Jones, L. H. P. (1957). The solubility of molybdenum in simplified systems and aqueous soil suspensions. *J. Soil Sci.* **8**, 313-326.
Jones, L. H. P., Cowling, D. W. and Lockyer, D. R. (1972). Plant-available and extractable sulfur in some soils of England and Wales. *Soil Sci.* **114**, 104-114.
Jurinak, J. J. and Bauer, N. (1954).Thermodynamics of zinc adsorption on calcite, dolomite and magnesite-type minerals. *Soil Sci. Soc. Am. Proc.* **20**, 466-471.
Kalbasi, M., Racz, G. J. and Loewen-Rudgers, L. A. (1978). Mechanism of zinc adsorption by iron and aluminium oxides. *Soil Sci.* **125**, 146-150.
Kanwar, J. S. (1954). Influence of organic matter on copper fixation in soil. *J. Ind. Soc. Soil Sci.* **2**, 73-80.
Krauskopf, K. B. (1972). Geochemistry of micronutrients. *In* 'Micronutrients in Agriculture'. pp.7-40. (Soil Sci. Soc. Am.: Madison.)
Lavy, T. L. and Barber, S. A. (1964). Movement of molybdenum in the soil and its effect on availability to the plant. *Soil Sci. Soc. Am. Proc.* **28**, 93-97.
Lehr, J. R. (1972). Chemical reactions of micronutrients in fertilizers. *In* 'Micronutrients in Agriculture.' pp.459-504. (Soil Sci. Soc. Am.: Madison.)
Lindsay, W. L. and Norvell, W. A. (1969). Equilibrium relationships of Zn^{2+}, Fe^{3+}, Ca^{2+} and H^+ with EDTA and DTPA in soils. *Soil Sci. Soc. Am. Proc.* **33**, 62-68.
Lindsay, W. L. (1972). Inorganic phase equilibria of micronutrients in soils. *In* 'Micronutrients in Agriculture.' pp.41-58. (Soil Sci. Soc. Am.: Madison.)
Lindsay, W. L. and Norvell, W. A. (1978). Development of a DTPA soil test for zinc, iron, manganese and copper. *Soil Sci. Soc. Am. J.* **42**, 421-428.
Little, I. P. and Kerridge, P. C. (1978). A laboratory assessment of the molybdenum status of nine Queensland soils. *Soil Sci.* **125**, 102-106.
Loganathan, P. and Burau, R. G. (1973). Sorption of heavy metal ions by a hydrous manganese oxide. *Geochim. Cosmochim. Acta* **37**, 1277-1293.
Loneragan, J. F. (1975). The availability and absorption of trace elements in soil-plant-animal systems and their relation to movement and concentration of trace elements in plants. *In* 'Trace Elements in Soil-Plant-Animal Systems.' pp.109-134. (Academic Press: New York.)
Matejevic, E. (1967). Charge reversal of lyophobic colloids. *In* 'Principles and Applications of Water Chemistry.' pp.328-369. (Wiley: New York.)
McKenzie, R. M. (1957). Trace element distribution in some South Australian red-brown earths. CSIRO Aust. Div. Soils Div. Rep. No. 9/56.
McKenzie, R. M. (1958). The trace element status of some South Australian terra rossa and rendzina soils. CSIRO Aust. Div. Soils Div. Rep. No. 12/57.
McKenzie, R. M. (1959). Trace elements in some South Australian terra rossa and rendzina soils. *Aust. J. Agric. Res.* **10**, 52-57.
McKenzie, R. M. (1960). Trace elements in some soils from the Todd River area, Central Australia. CSIRO Aust. Div. Soils Div. Rep. No. 6/60.
McKenzie, R. M. (1966). The relation of laboratory analyses for copper, zinc, and molybdenum in some Victorian soils to the results of field trials. *Aust. J. Exp. Agric. Anim. Husb.* **13**, 89-97.
McKenzie, R. M. (1967). The sorption of cobalt by manganese minerals in soils. *Aust. J. Soil Res.* **5**, 235-246.
McKenzie, R. M. (1970). The reaction of cobalt with manganese oxide minerals. *Aust. J. Soil Res.* **8**, 97-106.
McKenzie, R. M. (1972). The sorption of some heavy metals by the lower oxides of manganese. *Geoderma* **8**, 29-35.
McKenzie, R. M. (1975). An electron microprobe study of the relationships between heavy metals and manganese and iron in soils and ocean floor nodules. *Aust. J. Soil Res.* **13**, 177-188.
McKenzie, R. M. (1977). Manganese oxides and hydroxides. *In* 'Minerals in Soil Environments.' pp.181-193. (Soil Sci. Soc. Am.: Madison.)
MacNaughton, M. G. and James, R. O. (1974). Adsorption of aqueous mercury (II) complexes at the oxide/water interface. *J. Coll. Interface Sci.* **47**, 431-440.
Menzel, R. G. and Jackson, M. L. (1951). Sorption of copper from acid systems by kaolinite and montmorillonite. Trans. 5th Congr. Int. Soil Sci. Soc. (Amsterdam) Vol. 1, pp.1-4.
Mitchell, R. L. (1955). Trace elements. *In* 'Chemistry of the Soil.' pp. 253-285. (Van Nostrand Reinhold: New York.)
Murray, J. W. (1975). The interaction of metal ions at the manganese dioxide-solution interface. *Geochim. Cosmochim. Acta.* **39**, 635-647.
Nicholas, D. J. D. (1975). The functions of trace elements in plants. *In* 'Trace Elements in Soil-Plant-Animal Systems.' pp.181-198. (Academic Press: New York.)
Norrish, K. (1968). Some phosphate minerals of soils. Trans. 9th Congr. Int. Soc. Soil Sci. (Adelaide). Vol. 2, pp.713-723.
Norrish, K. (1975). The geochemistry and mineralogy of trace elements. *In* 'Trace Elements in Soil-Plant-Animal Systems.' pp.55-82. (Academic Press: New York.)
Norvell, W. A. and Lindsay, W. L. (1969). Reactions of EDTA complexes of Fe, Zn, Mn and Cu with soils. *Soil Sci. Soc. Am. Proc.* **3**, 86-91.
Norvell, W. A. and Lindsay, W. L. (1970). Lack of evidence for $ZnSiO_3$ in soils. *Soil Sci. Soc. Am. Proc.* **34**, 360-361.
Novogamsky, I., Beek, J. and Bolt, G. H. (1976). Chemical equilibria. *In* 'Soil Chemistry, A. Basic Elements.' p.27. (Elsevier: Amsterdam.)
Oertel, A. C. and Prescott, J. A. (1944). A spectrochemical examination of some ironstone gravels from Australian soils. *Trans. R. Soc. S. Aust.*, **68**, 173-176.

Oertel, A. C. (1961). Relation between trace-element concentrations in soil and parent material. *J. Soil Sci.* **12**, 119-128.
Oertel, A. C. and Giles, J. B. (1964). A study of some brigalow soils based on trace-element profiles. *Aust. J. Soil Res.* **2**, 162-172.
Ozanne, P. G., Shaw, T. C. and Kirton, D. J. (1965). Pasture responses to traces of zinc in phosphate fertilizers. *Aust. J. Exp. Agric. Anim. Husb.* **5**, 29-33.
Parsons, R. (1968). Some problems of the electrical double layer. *Rev. Pure Appl. Chem.* **18**, 91-104.
Ponnamperuma, F. N. (1972). Chemistry of submerged soils. *Adv. Agron.* **24**, 29-88.
Purves, D. and McKenzie, E. J. (1969). Trace element contamination of parklands in urban areas. *J. Soil Sci.* **20**, 288-290.
Purves, D. (1972). Consequences of trace-element contamination of soils. *Environ. Poll.* **3**, 17-24.
Quirk, J. P. and Posner, A. M. (1975). Trace element adsorption on mineral surfaces. *In* 'Trace Elements in Soil-Plant-Animal Systems.' pp.95-108. (Academic Press: New York.)
Raglan, J. L. and Coleman, N. T. (1960). The hydrolysis of aluminum salts in clay and soil systems. *Soil Sci. Soc. Am. Proc.* **24**, 457-460.
Rankama, K. and Sahama, T. G. (1950). 'Geochemistry.' (Univ. Chicago: Chicago.)
Reisenauer, H. M., Tabikh, A. A. and Stout, P. R. (1962). Molybdenum reactions with soils and hydrous oxides of iron, aluminum, and titanium. *Soil Sci. Soc. Am. Proc.* **26**, 23-27.
Reuter, D. J. (1975). The recognition and correction of trace element deficiencies. *In* 'Trace Elements in Soil-Plant-Animal Systems.' pp.291-324. (Academic Press: New York.)
Rovira, A. D. (1969). Plant root exudates. *Bot. Rev.* **35**, 35-57.
Saeed, M. and Fox, R. L. (1977). Relations between suspension pH and zinc solubility in acid and calcareous soils. *Soil Sci.* **124**, 199-204.
Schindler, P. W. (1975). Removal of trace elements from the oceans: a zero order model. *Thalassia Jugosl.* **11**, 101-111.
Schnitzer, M. and Skinner, S. I. M. (1966). Organo-metallic interactions in soils: 5. Stability constants of Cu^{++}-, Fe^{++}-, and Zn^{++}-fulvic acid complexes. *Soil Sci.* **102**, 361-365.
Schnitzer, M. and Skinner, S. I. M. (1967). Organo-metallic interactions in soils: 7. Stability constants of Pb^{++}-, Ni^{++}-, Mn^{++}-, Co^{++}-, Ca^{++}-, and Mg^{++}- fulvic acid complexes. *Soil. Sci.* **103**, 247-252.
Shacklette, H. T., Hamilton, J. C., Boerngen, J. G. and Bowles, J. M. (1971). Elemental composition of surficial materials in the conterminous United States., U.S. Geol. Surv. Prof. Paper 574-D.
Shuman, L. M. (1975). The effect of soil properties on zinc adsorption by soils. *Soil Sci. Soc. Am. Proc.* **39**, 454-458.
Sidhu, A. S., Randhawa, N. S. and Sinha, M. K. (1977). Adsorption and desorption of zinc in different soils. *Soil Sci.* **124**, 211-218.
Sillén, L. G. and Martell, A. E. (1964). 'Stability Constants of Metal-Ion Complexes.' Special Publ. No. 17. (Chem. Soc.: London.)
Sillén, L. G. and Martell, A. E. (1971). 'Stability Constants of Metal-Ion Complexes.' Supplement No.1. Special Publ. No. 25. (Chem. Soc.: London.)
Silverberg, J., Young, R. D. and Hoffmeister, G. (1972). Preparation of fertilizers containing micronutrients. In 'Micronutrients in Agriculture.' pp.431-458. (Soil Sci. Soc. Am.: Madison.)
Singh, B. and Silhon, G. S. (1977). The effect of soil properties on adsorption and desorption of zinc by alkaline soils. *Soil Sci.* **124**, 366-369.
Sinha, M. K., Dhillon, S. K., Dhillon, K. S. and Dyanand, S. (1978). Solubility relationships of iron, manganese, copper and zinc in alkaline and calcareous soils. *Aust. J. Soil Res.* **16**, 19-26.
Smith, B. H. and Leeper, G. W. (1969). The fate of applied molybdate in acidic soils. *J. Soil Sci.* **20**, 247-254.
Soltanpour, P. N., Khan, A. and Lindsay, W. L. (1976). Factors affecting DTPA-extractable Zn, Fe, Mn and Cu from soils. *Commun. Soil Sci. Plant. Anal.* **7**, 797-821.
Spencer, W. F. and Gieseking, J. E. (1954). Cobalt adsorption and release in cation-exchange systems. *Soil Sci.* **78**, 267-276.
Stace, H. C. T., Hubble, G. D., Brewer, R., Northcote, K. H., Sleeman, J. R., Mulcahy, M. J. and Hallsworth, E. G. (1968). 'A Handbook of Australian Soils.' (Rellim: Glenside.)
Steenbjerg, F. (1951). Yield curves and chemical plant analyses. *Plant Soil* **3**, 97-109.
Stephens, C. G. and Donald, C. M. (1958). Australian soils and their responses to fertilizers. *Adv. Agron.* **10**, 167-256.
Stern, O. (1924). Zur Theorie der electrolytischem Doppelschicht. *Z. Electrochem.* **30**, 508-516.
Stevenson, F. J. and Ardakani, M. S. (1972). Organic matter reactions involving micronutrients in soils. *In* 'Micronutrients In Agriculture.' pp.79-114. (Soil Sci. Soc. Am.: Madison.)
Stout, P. R., Meagher, W. R., Pearson, G. A. and Johnson, C. M. (1951). Molybdenum nutrition of crop plants. I. *Plant Soil* **3**, 51-87.
Swaine, D. J. (1955). 'The Trace-element Content of Soils.' Commonw. Bur. Soil Sci., Tech. Commun. No. 48.
Swaine, D. J. (1962). 'The Trace-element Content of Fertilizers.' Commonw. Bur. Soils Tech. Commun. No. 5'.
Taylor, R. M., McKenzie, R. M. and Norrish, K. (1964). The mineralogy and chemistry of manganese in some Australian soils. *Aust. J. Soil Res.* **2**, 235-248.
Taylor, R. M. and McKenzie, R. M. (1966). The association of trace elements with manganese minerals in Australian soils. *Aust. J. Soil Res.* **4**, 29-39.
Tiller, K. G. (1957). The geochemistry of the basaltic materials and associated soils of the S.E. of South Australia. MSc. Thesis, University of Adelaide.
Tiller, K. G. (1958). The geochemistry of basaltic materials and associated soils of south-eastern South Australia. *J. Soil Sci.* **9**, 225-241.
Tiller, K. G. (1959). Distribution of some trace elements in soils developed on dolerite in Tasmania. CSIRO Aust. Div. Soils Div. Rep. No. 6/58.
Tiller, K. G. (1961). Specific sorption of some heavy metal cations by pure minerals and soil clays. Ph.D. Thesis, Cornell University.
Tiller, K. G. and Hodgson, J. F. (1962). The specific sorption of cobalt and zinc by layer silicates. *Clays Clay Miner.* **9**, 393-403.
Tiller, K. G. (1963). Weathering and soil formation in Tasmania with particular reference to several trace elements. *Aust. J. Soil Res.* **1**, 74-90.
Tiller, K. G., Hodgson, J. F. and Peech, M. (1963). Specific sorption of cobalt by soil clays. *Soil Sci.* **95**, 392-399.
Tiller, K. G. (1967). Silicic acid and the reaction of zinc with clays. *Nature* (London) **214**, 852.
Tiller, K. G. (1968*a*). The interaction of some heavy metal cations and silicic acid at low concentrations in the presence of clays. Trans. 9th Congr. Int. Soil Sci. Soc. Adelaidel. Vol. 2, pp.567-575.
Tiller, K. G. (1968*b*). Stability of hectorite in weakly acidic solutions. II Studies of the chemical equilibrium and the calculation of free energy. *Clay Miner.* **7**, 262-270.
Tiller, K. G. (1968*c*). Stability of hectorite in weakly acidic solutions. III. Adsorption of heavy metal cations and hectorite solubility. *Clay Miner.* **7**, 409-419.
Tiller, K. G., Honeysett, J. L. and Hallsworth, E. G. (1969). The isotopically exchangeable form of native and applied cobalt in soils. *Aust. J. Soil Res.* **7**, 43-56.
Tiller, K. G., Honeysett, J. F. and de Vries, M. P. C. (1972). Soil zinc and its uptake by plants. II. Soil chemistry in relation to prediction of availability. *Aust. J. Soil Res.* **10**, 151-164.
Tiller, K. G. and Pickering, J. G. (1974). The synthesis of zinc silicates at 20°C and atmospheric pressure. *Clays Clay Miner.* **22**, 409-416.
Tiller, K. G., Nayyar, V. K. and Clayton, P. M. (1979). Specific and nonspecific sorption of cadmium by soil clays as influenced by zinc and calcium. *Aust. J. Soil Res.* **17**, 1-12.
Turton, A. G., Marsh, N. L. and McKenzie, R. M. (1961). Morphological, chemical, physical, clay mineralogy and minor element data on lateritic profiles from south western Australia. CSIRO Aust. Div. Soils Div. Rep. No. 8/6.

Udo, E. J., Bohn , H. L. and Tucker, T. C. (1970). Zinc adsorption by calcareous soils. *Soil Sci. Soc. Am. Proc.* **34**, 405-407.

Underwood, E. J. (1975). Trace elements and their physiological roles in the animal. *In* 'Trace Elements in Soil-Plant-Animal Systems.' pp.227-242. (Academic Press: New York.)

Viets, F. G. (1962). Chemistry and availability of micronutrients in soils. *Agric. Food Chem.* **10**, 174-178.

Vinogradov, A. P. (1959). 'The Geochemistry of Rare and Dispersed Chemical Elements in Soils.' (Consultants Bureau: New York.)

Wallace, T. (1964). 'The Diagnosis of Mineral Deficiencies in Plants by Visual Symptoms.' (HMSO: London.)

Wedepohl, K. H. (Ed.) (1969-78). 'Handbook of Geochemistry.' (Springer-Verlag: Berlin.)

Wells, N. (1956). Soil studies using sweet vernal for assessing element availability. II. *N.Z. J. Sci. Technol.* **337**, 482-502.

Wells, N. (1960). Total elements in topsoils from igneous rocks: an extension of geochemistry. *J. Soil Sci.* **11**, 409-424.

White, M. L. (1957). The occurrence of zinc in soil. *Econ. Geol.* **52**, 645-651.

Wiklander, L. (1955). Cation and anion exchange phenomena. *In* 'Chemistry of the Soil.' pp.107-148. (Van Nostrand Reinhold: New York.)

Williams, C. H. and Moore, C. W. E. (1951). The effect of stage of growth on the copper, zinc, manganese, and molybdenum contents of Algerian oats grown on thirteen soils. *Aust. J. Agric. Res.* **3**, 343-361.

Williams, C. H. and David, D. J. (1973). The effect of superphosphate on the cadmium content of soils and plants. *Aust. J. Soil Res.* **11**, 43-56.

Williams, C. H. (1974). Heavy metals and other elements in fertilizer–environmental considerations. *In* 'Fertilizers and the Environment.' pp.123-130. Proc. Symp. Univ. Sydney (A.I.A.S.: Sydney.)

Chapter 26
Soil solution and silicate clay mineral reactions

M. Raupach

MINERAL REACTIONS IN THE SOIL SOLUTION ARE STUDIED for two reasons, both of which relate to the transport of dissolved material. The first reason is that the soil solution has a major role in soil profile development. Over geological time large quantities of materials such as alumina and silica have been transported by the soil solution during periods when conditions were favourable for leaching. Such large scale movements of soil constituents have had profound effects on the development of Australian soils. The second reason for studying the soil solution is its importance in the transport of nutrients to plant roots. If conditions in the soil solution do not favour a well balanced supply of nutrients, plant growth may become restricted.

The soil solution contains dissolved species of molecules and ions. Where does this dissolved material come from? Sometimes the solutes that enter the soil are already dissolved, for example as salts in rain. More frequently the soluble species result from the flow of matter across gas–liquid or solid–liquid interfaces in the soil. The dissolved species are simple or complexed, charged or uncharged, and monomeric or polymeric. The concentrations of the dissolved species in any unit volume of the soil solution depend on inflows and outflows of solutes from the surrounding volume elements of soil solution, and on the sources and sinks for the various dissolved species in the volume element.

Among the sources of dissolved species are inorganic soil components including salts; materials from organic soil components such as litters and decomposing debris and excreta, and added materials from fertilization, irrigation or pollution. Among the sinks for dissolved species are solid phases which separate or crystallize out (e.g. hydroxides, salts, phosphates and carbonates); exchange complexes whose surfaces are charged either positively (mainly oxides and hydroxides) or negatively (mainly silicate minerals and organic matter); other sites capable of fixing nutrients, and also assimilative sinks in the form of plants, microbes or animals.

The composition of the soil solution is variable in both space and time. Although many detailed accounts of sorption, ion exchange and solubility phenomena are available for systems containing relatively simple solid phases, it is difficult to generalize about the soil solution because of this variability and because the contributions of the various sources and sinks are both complicated and difficult to quantify.

The aim of this chapter is to give a treatment of silicate mineral equilibria in the soil solution as applied to soil systems. To do this, emphasis is given to chemical reactions in the soil solution and to the inclusion of experimental data on systems containing soils. As a result, new mechanisms are proposed for the reaction of silicate clays and the soil solution. This allows a number of previous studies on the soil solution to be consolidated.

Solubility and adsorption concepts

Reactions in the soil solution have been considered from various points of view. The soil may be thought of as a system at stable or metastable equilibrium; or it may be regarded as a steady state system in which there is sorption of ions on a range of sites (at best similar to each other, and at worst different) with a spectrum of sorption energies which varies from soil to soil. These views imply control of ionic concentrations in the soil solution, either by solubility or by adsorption phenomena.

The soil solution shares much of its chemistry with ground waters, stream waters, and other aqueous systems containing estuarine and marine deposits or sediments. Sillen's (1967) view, that in the ocean the things that we observe are caused not by equilibrium but by lack of it, holds for soils just as for the sea. His argument that it is wise to work on the equilibrium model as a useful guide to actual processes also applies to soils with equal force.

The evidence examined in this chapter favours both the equilibrium concept and also the control of the behaviour of at least some of the ions by solubility phenomena. The result is that a more uniform view of the soil solution is now possible.

Silicon

Relation between pH and the silica in the soil solution

Raupach (1957) and Raupach and Piper (1959) were the first to point out that, over a wide pH range, the concentration of soluble silica in the soil solution is

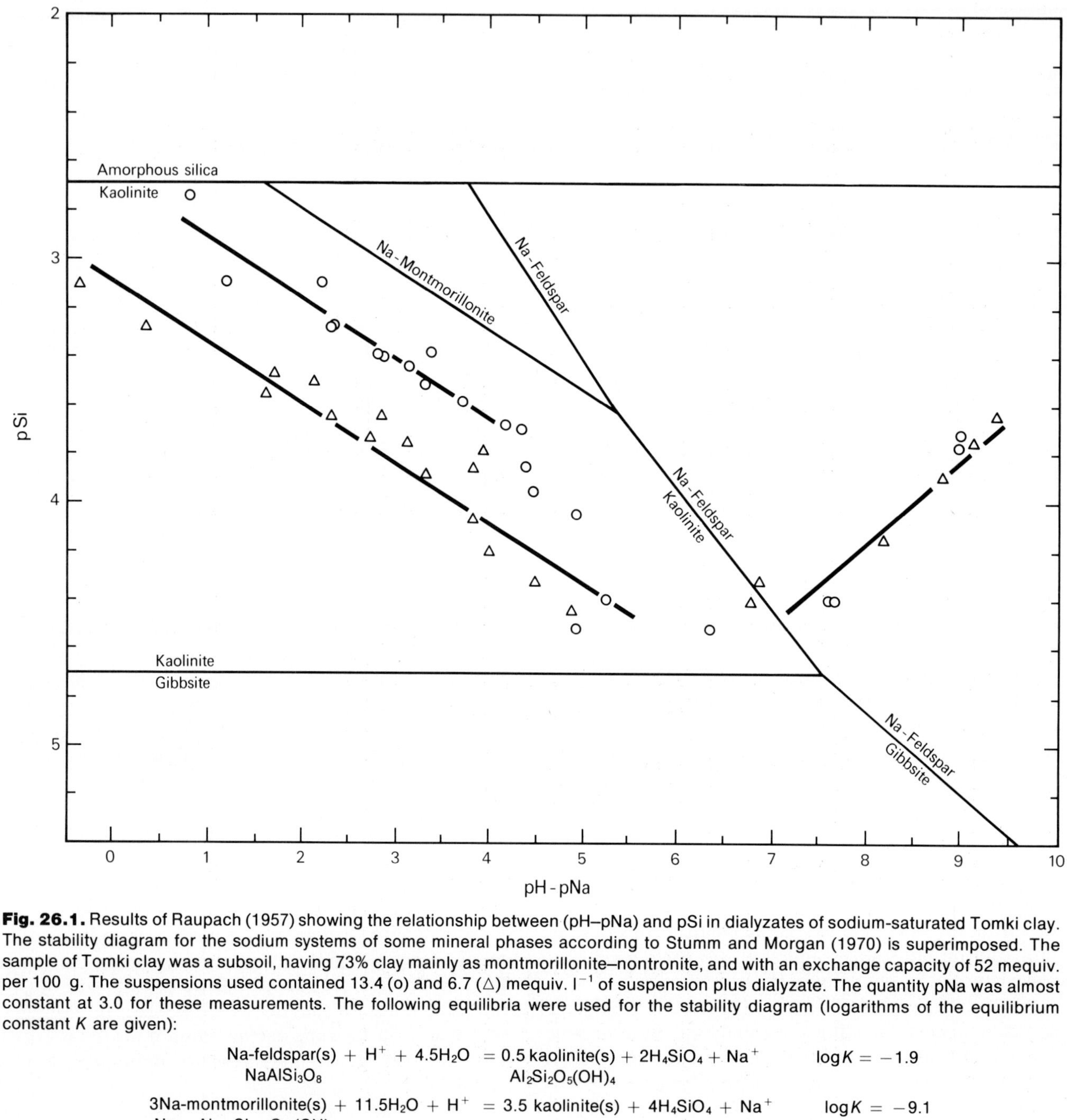

Fig. 26.1. Results of Raupach (1957) showing the relationship between (pH–pNa) and pSi in dialyzates of sodium-saturated Tomki clay. The stability diagram for the sodium systems of some mineral phases according to Stumm and Morgan (1970) is superimposed. The sample of Tomki clay was a subsoil, having 73% clay mainly as montmorillonite–nontronite, and with an exchange capacity of 52 mequiv. per 100 g. The suspensions used contained 13.4 (o) and 6.7 (△) mequiv. l^{-1} of suspension plus dialyzate. The quantity pNa was almost constant at 3.0 for these measurements. The following equilibria were used for the stability diagram (logarithms of the equilibrium constant K are given):

$$\text{Na-feldspar(s)} + H^+ + 4.5H_2O = 0.5\,\text{kaolinite(s)} + 2H_4SiO_4 + Na^+ \qquad \log K = -1.9$$
$$NaAlSi_3O_8 \qquad Al_2Si_2O_5(OH)_4$$

$$\text{3Na-montmorillonite(s)} + 11.5H_2O + H^+ = 3.5\,\text{kaolinite(s)} + 4H_4SiO_4 + Na^+ \qquad \log K = -9.1$$
$$Na_{0.33}Al_{2.33}Si_{3.67}O_{10}(OH)_2$$

$$\text{kaolinite(s)} + 5H_2O = \text{gibbsite(s)} + 2H_4SiO_4 \qquad \log K = -9.4$$
$$Al_2O_3.3H_2O$$

$$SiO_2\ \text{(amorphous)} + 2H_2O = H_4SiO_4 \qquad \log K = -2.7.$$

related to its pH. Aluminium ions were believed to be responsible for the relationships between the negative logarithm of the silica concentration (expressed as pSi) and the pH of the solution phase of soil suspensions. This contrasts with the higher silica solubility in silica-water systems which showed no such relationship below pH 9.

Experimentally, below pH 7, the relationship between the soluble silica and the pH took the form pSi $= k$.pH $+$ constant, where $k = 0.25$; Fig. 26.1 illustrates this relationship. Note that silicate was not added. Above pH 7, the concentration of soluble silica decreased to a minimum at about pH 9, and then increased again at higher pH values. The increase in soluble silica at high pH was presumed to occur because of the dissolution of

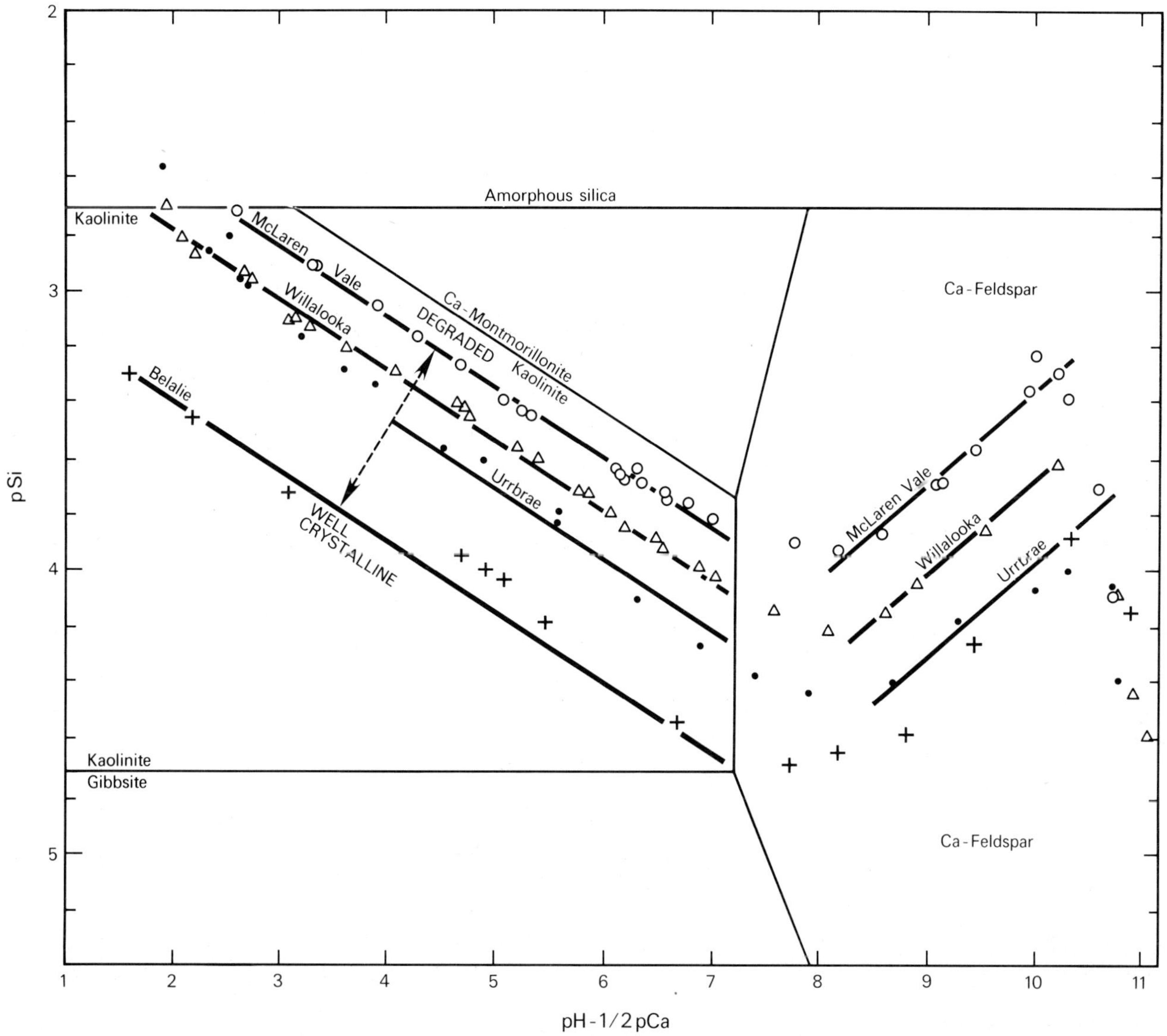

Fig. 26.2. Results of Tucker (1964) showing the relationship between (pH–0.5pCa) and pSi. The following four illitic clay soils were used in order of increasing degradation: Belalie (+) slight (well crystalline); Willalooka (△) slight, Urrbrae (●) moderate; McLaren Vale (○) high (poorly crystalline). The suspensions used were in 0.0033 M $CaCl_2$ and contained 10 millimoles of total potassium per litre. The stability diagram from Stumm and Morgan (1970) is given. Equilibrium constants and equations additional to those in Fig. 26.1 for this diagram are:

$$\underset{CaAl_2Si_2O_8}{\text{Ca-feldspar(s)}} + 2H^+ + H_2O = \text{kaolinite(s)} + Ca^{2+} \qquad \log K = 14.4$$

$$\underset{Ca_{0.33}Al_{4.67}Si_{7.33}O_{20}(OH)_4}{\text{3Ca-montmorillonite(s)}} + 2H^+ + 23H_2O = \text{7kaolinite(s)} + 8H_4SiO_4 + Ca^{2+} \qquad \log K = -15.4$$

monovalent and divalent silicate ions from the clay.

It was concluded that under acid conditions the pH arises from a silicic acid compound with some aluminium attached, but the exact mechanism could not be ascertained at the time when the work was first reported.

On adding chlorides of various cations to clay–water systems, the above authors found relationships between the number of equivalents (M) of exchangeable cation (mainly as chloride) in solution and the pH. These took the form: pH $= C.pM +$ constant, where C was a constant which varied with the pH value in the absence of salt (pH_o). The constant C decreased linearly with pH_o, from $C = 1$ at pH_o6.4 to $C = 0.2$ at pH_o3. Reactions involving aluminium were thought to casue the decrease.

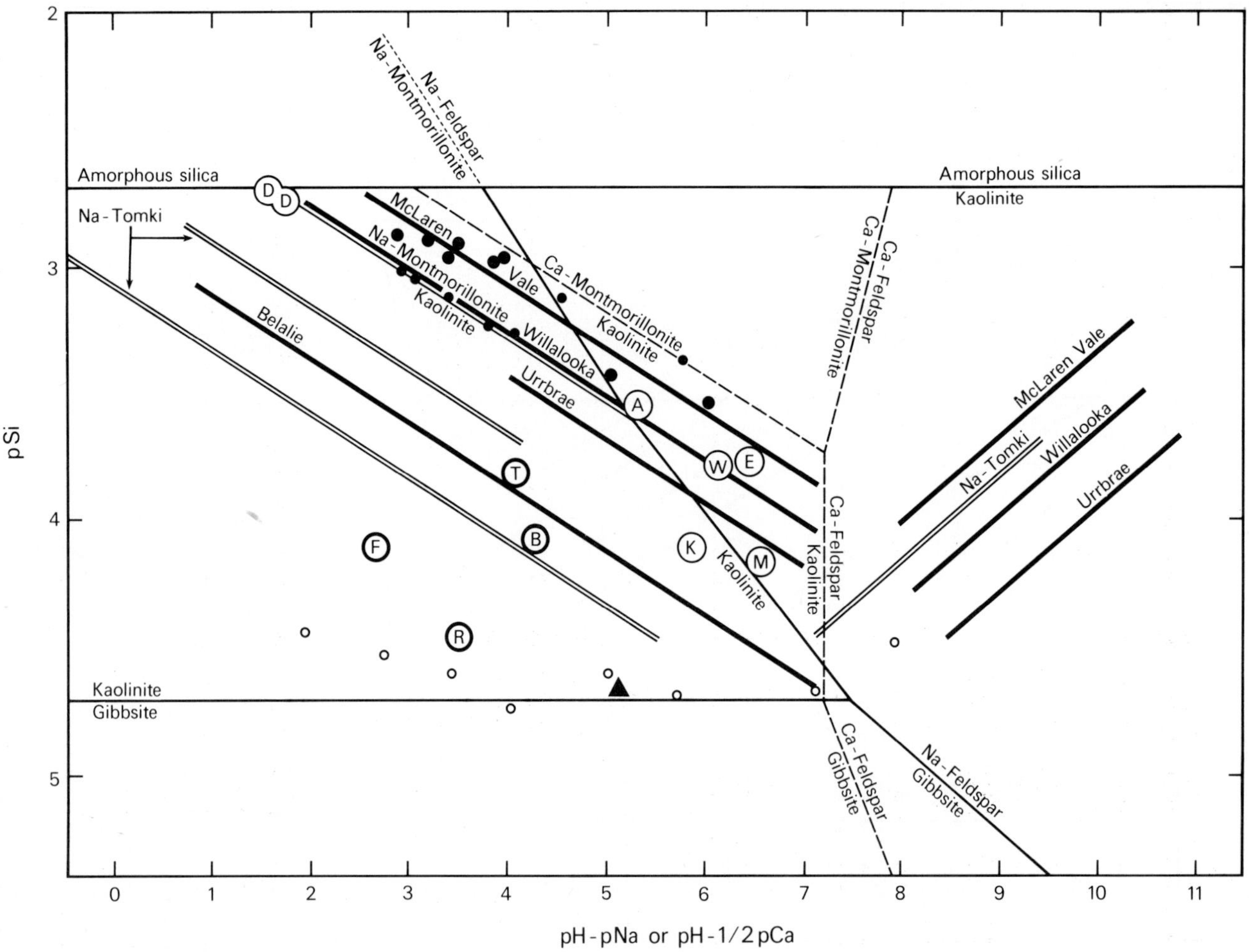

Fig. 26.3. Results from the literature used to calculate the free energy of formation of various solid phases from mineral equilibria. The results are represented by encircled symbols (heavy for the muscovite type minerals and light for the montmorillonites) which give a single point lying on the appropriate boundary of the stability diagram according to the phases present. Symbols are: A, Arizona montmorillonite, W, Wyoming montmorillonite, B, Beavers Bend illite and F, Fithian illite from Huang and Keller (1973); M, Mg-beidellite and K, K-beidellite from Misra and Upchurch (1976); T, muscovite from Huang and Keller (1972); E, Belle Fourche and D, Aberdeen montmorillonites, from both supersaturation and undersaturation, from Kittrick (1971*a*, *b*); R, Marblehead illite from Reesman (1974) and ▲, chlorite from Weaver *et al.* (1971). Data are also given from Raupach and Piper (1959) for a *lateritic podzolic* soil (○) which contained kaolinite and gibbsite and from Veen (1973) for saturation extracts of samples taken at increasing depths from profiles of a grey clay (●) and a brown clay (●), both from brigalow soils and having montmorillonite, kaolinite, and quartz in the clay fraction.

Adsorption of silica

McKeague (1962) and McKeague and Cline (1963*a*, *b*, *c*) confirmed the relationship in Fig. 26.1 between pH and pSi in aqueous soil extracts. However, they suggested that the reaction was governed by pH-dependent adsorption that followed the Freundlich equation; thus they argued for sites with a range of energies under conditions which favoured the adsorption of dissolved silica and the precipitation of aluminium hydroxide. Beckwith and Reeve (1963, 1964) reasoned similarly from experiments which followed the fate of silicic acid added to soil systems, including its release into solutions of common extracting reagents.

Results on soil illites

Tucker (1964). in the second of a series of papers on the reaction of potassium in illites, gave data on the silicic acid, potassium, and aluminium contents of solutions reacted with four soil illitic clays possessing varying degrees of degradation. His experiments were over a very wide pH range and lasted for 21 and 125 days. Tucker obtained relationships between pH and pSi which confirmed the results of Raupach (1957), for as shown in Fig. 26.2 the slopes of the line pSi = k.pH + constant were almost constant at k = 0.25 for the different illites. The alkaline side of the pH range also indicated the possibility of linear relationships.

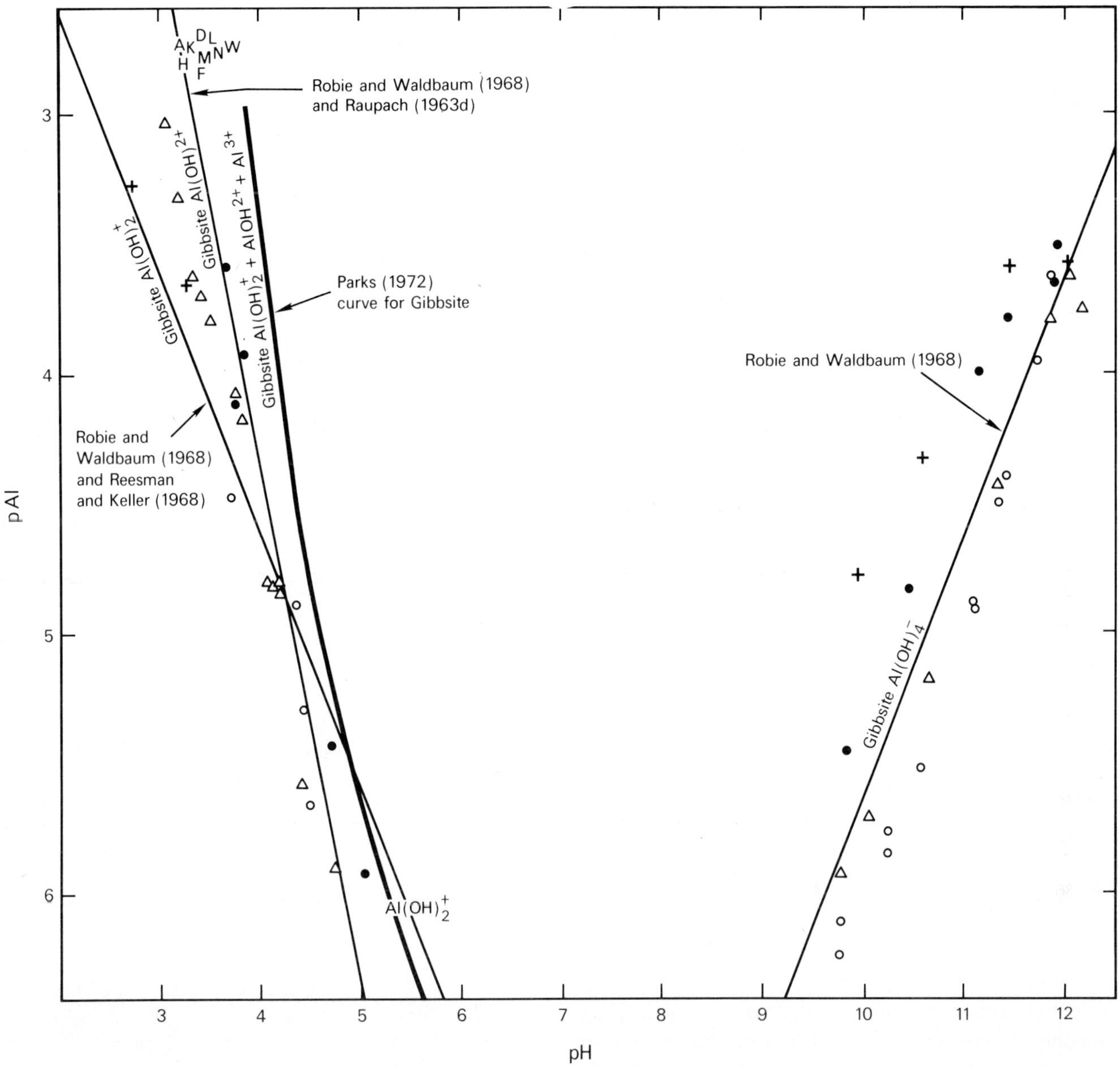

Fig. 26.4. Relationship between the negative logarithm of the number of moles of aluminium per litre and the pH of solutions from the four soil clays used by Tucker (1964), with symbols as in Fig. 26.2, and from the following nine soil clays used by Marion *et al.* (1976): Ava (A) Davidson (D), Fanno (F), Nicholson (N), Kalkaska (K), Molokai (L) Mohave (M), Anthony (N) and Wagram (W). The species of aluminium ions are indicated for the various theoretical curves calculated from thermodynamic data taken from the references given.

Further Results

Solutions reacted with a *lateritic podzolic* soil from Kangaroo Island by Raupach and Piper (1959) and saturation extracts of Veen (1973) from profiles of grey, brown and red clays, gave results as shown in Fig. 26.3.

Aluminium in the soil solution

Solubility of aluminium hydroxide

Raupach (1957, 1963*a, b, c, d*) examined the solubility of simple aluminium hydroxides and oxyhydroxides in an attempt to define further the species of aluminium likely to be found in soil solutions, including solutions where phosphate was present. The conclusion was reached that the soil solutions reacted as though simple aluminium ions and solid phases were present, and that this was consistent with soil data. Graphical representation enabled the course of the reactions to be followed with increasing time. The reactions either approached equilibrium, or they proceeded from one set of metastable equilibrium conditions to another. Some reconstitution of the aluminium hydroxide solids occurred with increasing reaction times. As shown in Fig. 26.4. the species $AlOH^{2+}$ was favoured in the range $3 < pH < 4.5$, and $Al(OH)_2^+$ in the range of pH from 4.5 up towards 7 where higher polymers were present.

Solubility results were used to argue that, above pH 4, phosphate reacted with aluminium hydroxide to form a product which was not as simple as variscite ($Al(OH)_2H_2PO_4$) thus favouring either adsorption, e.g. as proposed by Hingston *et al.* (1972, 1974), or the formation of some other insoluble phosphate. The $AlOH^{2+}$ was suggested to be potential determining (i.e. to give rise to surface charges) in the aluminium hydroxides. Parks (1965) took up and generalized the suggestion that crystalline modification of solid surfaces accompanied the dissolution of aluminium hydroxides and that particular hydrolyzed species may be potential determining. Parks (1972) used the experimental data of Raupach (1963*a*) as a basis for extensive additional calculations and discussion; the results of some of Parks's calculations are given in Fig. 26.4.

Hem and Robertson (1967) were the first to use similar approaches to study the chemistry of aluminium in natural water systems.

Marion *et al.* (1976), working on aluminium and silica solubility in soils, concluded in agreement with Raupach (1963*a, b, d*) that the monomeric model for soluble aluminium can describe both the electrophoretic behaviour and the solubility of aluminium hydroxide (for some of their data see Fig. 26.4). In studies on rice-growing soils, Nhung and Ponnamperuma (1966), see also Ponnamperuma (1972), confirmed a relationship $pAl = 2pH - 4.41$ of the form found by Raupach (1963*c*). Bache (1974) advocated that deductions about aluminium ions in soils or pure mineral systems should preferably be based on the solution chemistry of aluminium, and sought to explain the aluminium concentration in 0.01 M calcium chloride extracts of limed and acidified soils in terms of the hydrolysis of Al^{3+}. However, his results, extremes of which are presented in Fig. 26.5, can also be explained by solubility phenomena. This is because the data lie in a region of the pH–pAl diagram where hydrolysis and solubility phenomena virtually overlap. The overlap is seen from the way curves drawn through results on the hydrolysis of salt solutions (curves C, D, E of Fig. 26.5; Raupach 1963*b*) abut those from the dissolution of gibbsite (e.g. curve B of Fig.26.5; Raupach 1963*d*). Bache's soils gave points lying between curve C (slope +2) and a further curve A (slope +1). Curves C and A, respectively, express the solubility equilibria $[AlOH^{2+}][OH^-]^2$ = constant, and $[Al(OH_2^+)][OH^-]$ = constant. Small changes in the free energy of the solid phase, such as those occurring with variations in particle size, cause the results to lie parallel to these curves, but to one side or the other of them.

The aluminium ion concentrations measured by Tucker (1964) on illitic soil clays are also in agreement with the solubility of divalent and monovalent aluminium species or their polymers from aluminium hydroxide. These data are presented in Fig. 26.4. and do not appear to support Parks's (1972) calculated solubility for gibbsite, for which he used Al^{3+} as well as the lower-charged species. Tucker's results tend rather to follow curves of lower slope, such as may be calculated from$[AlOH^{2+}][OH^-]^2$ = constant, using the free energy values from Robie and Waldbaum (1968) for gibbsite and from Raupach (1963*a*) for $Al(OH)_2^+$. It should be noted that Raupach (1963*d*), using conductance measurements to study aluminium hydroxide solubility equilibria, confirmed the presence of monovalent and divalent aluminium species rather than the trivalent species. He proposed a reaction akin to a catalysed hydrolysis of the Al^{3+} at the surface.

In the illitic soil clays he examined, Tucker (1964) concluded that the release of potassium was independent of the dissolution of silicic acid and of aluminium. This leads us to consider the use of stability diagrams for clays.

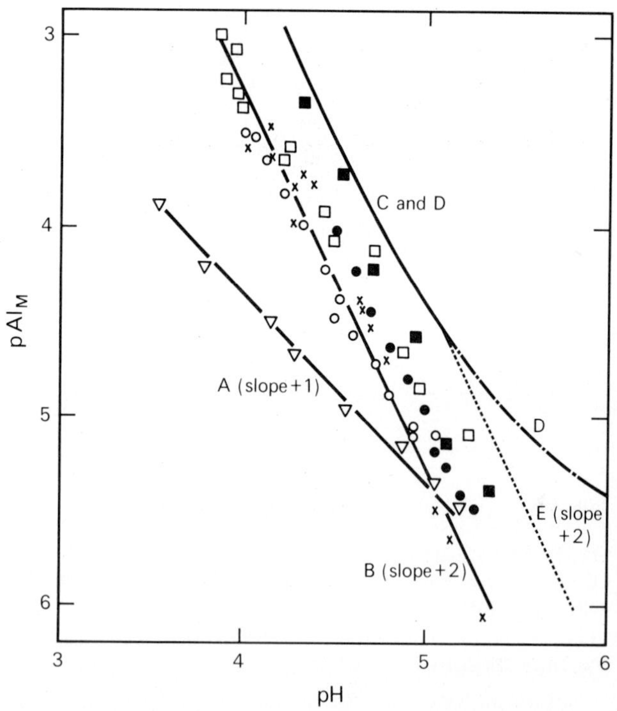

Fig. 26.5 Relationship between pH and pAl from data on soils, the solubility of gibbsite and the hydrolysis of dilute aluminium salt solutions: acid surface soils of Bache (1974), Insch (○) Countesswells (□) and Cawdor (▽), the respective infilled symbols representing subsoils and curve A being drawn through the data for Cawdor; gibbsite reacted with 0.01M K_2SO_4 or 0.03M KCl for 1230 to 8480 hr from data of Raupach (1963*d*) (x) and curve B; dilute aluminium salt solutions (Raupach 1963*b*): theoretical curve (C) and experimental curves with (D) and without (E) admitting atmospheric CO_2.

Interpretation of stability diagrams

Much of the behaviour of certain ions, like aluminium, in soil solutions can be explained by solubility principles. This simple view does not hold for other ions, like phosphate, capable of strong attachment to soil surfaces and frequently present in small amounts. Stability diagrams which use solubility principles have been used for silicate and other mineral systems. What kinds of reactions do these diagrams imply?

Stumm and Morgan (1970) summarized the stability relations for the mineral phases of aqueous systems in weathering situations such as those found in the presence of ground water. Some of the stability zones are indicated for monovalent and divalent cation systems in Figs 26.1 and 26.2.

Among the relationships illustrated, the following applies to a solution in the presence of montmorillonite and kaolinite:

$$3Ca_{0.33}Al_{4.67}Si_{7.33}O_{20}(OH)_4(s) + 23H_2O + 2H^+$$

(montmorillonite)

$$= 7Al_2Si_2O_5(OH)_4(s) + 8H_4SiO_4 + Ca^{2+}.$$

(kaolinite)

Here the equilibrium constant K is given by $\log K = -15.4$. The equilibrium constant may be expressed in terms of solution activities: $4pSi - (pH - 0.5pCa) = 7.7$.

Tucker's range of illites, with their varying degrees of degradation as judged from potassium contents, X-ray scatter and surface area measurements, extended across the kaolinite zone from the kaolinite–montmorillonite boundary to the kaolinite–gibbsite boundary. The points (Fig. 26.2) for the Belalie soil, which contained the illite having the highest degree of crystallinity, were nearest to the kaolinite–gibbsite boundary where the lateritic soil was situated. This gives the important conclusion that for illitic clays the degree of degradation influences the free energy and the equilibrium constant for dissolution.

Deductions from the analyses of soil and clay suspensions

The literature of the last decade contains analyses of solutions 'in equilibrium' with various clays. Calculations of free energies from these analyses give the boundaries of the stability diagrams. The boundaries in turn reflect, at least semi-quantitatively, the composition of the soil solution.

For equilibrium to be attained over increasing times of reaction, the measurements should fall along boundaries for the appropriate solid phases in the stability diagrams. Sometimes reactions were monitored from both undersaturated and supersaturated conditions (e.g. Kittrick 1971*a,b*), and results which agreed under both these conditions were presumed to represent the equilibrium state. Variations in free energy cause the measurements to fall along lines parallel to the boundaries. Against this, Reesman (1974) has expressed a rather extreme view about illites, that 'to assume true equilibrium in these mineral systems is folly. No such condition has been proven, nor can it be'.

Nevertheless, 'equilibrium' solution concentrations (expressed as activities) have been used to calculate the free energies of formation for various minerals; even Reesman carried his calculations through to this stage.

Once the free energies of formation have been calculated, equilibrium constants can be used to delineate the boundaries within which various solid phases are stable. This allows diagrams like Fig. 26.1 to be drawn. Such diagrams are useful to examine experimental data on soil solutions (see Figs 26.1–3).

Equilibrium is not the only requirement for the use of analyses from clay–water systems to calculate free energy values. A valid equation for the reaction is required which identifies the solid phases and the ionic species in solution. It is sometimes possible to set up an equation using ideal clay formulae, but real clays with complicated compositions give difficulties.

As an example, Huang and Keller (1973) gave the following reaction for the dissolution of Wyoming montmorillonite at pH 8.2:

$$(Na_{0.27}Ca_{0.10}K_{0.02})(Al_{1.52}Fe^{3+}_{0.19}Mg_{0.22})(Si_{3.94}Al_{0.06})O_{10}(OH)_2$$

$$+ 10.46\ H_2O = 0.27Na^+ + 0.10Ca^{2+} + 0.02K^+$$

$$+ 1.58Al(OH)_4^- + 0.19Fe(OH)^{2+}$$

$$+ 0.22Mg^{2+} + 3.94H_4SiO_4 + 0.46H^+.$$

Their calculations using this equation and inserting the activities of the ions present in the equilibrium solution gave a free energy of formation of -5222.5 kJ mol^{-1} (-1248.2 kcal mol^{-1}) for Wyoming montmorillonite. With the further assumption that $Fe(OH)_3$ was consumed and amorphous $Al(OH)_3$ was produced, a value of -5196.9 kJ mol^{-1} (-1242.1 kcal mol^{-1}) was returned. If the $Al(OH)_3$ was gibbsite, a range of free energies up to -5221.6 kJ mol^{-1} (-1248.0 kcal mol^{-1}) resulted, depending on the free energy of formation attributed to the gibbsite. Further literature values for the free energy of Wyoming montmorillonite are -5314 and -5164.3 kJ mol^{-1} (-1270 and -1234.3 kcal mol^{-1}) from Reesman and Keller (1968) and Kittrick (1971*c*) respectively. Such variations are sufficiently large to place predicted values of pSi above and below the bounds plotted in Fig. 26.1.

Apart from these variations, the experimental results below pH 7 have been found to be somewhat sensitive to the quality of clay in suspension (Raupach 1957; Tucker 1964; Kittrick 1971*b*). There should be no such dependence if the reaction only involves the solubility products of simple phases such as $Al(OH)_3$, and if the equilibrium condition holds.

Formulae of clays and mechanism of reaction

In recent careful work on muscovite, Mattigod and Kittrick (1979) observed that their activity ratios involving soluble silica were not compatible with a dissolution reaction based on the original chemical composition of the muscovite used. They established that their system was at equilibrium but could not detect any new phases by X-ray diffraction or by infrared spectroscopy, so that, in their opinion, the presence of a new muscovite phase was unlikely. Mattigod and Kittrick's (1979) results all fell on the line in Fig. 26.2 expressing the relationship between pH $-$ ½pCa and pSi for Willalooka illite if pK is used instead of ½pCa.

The reasons for the observed pH–pSi relationship are not obvious if the formula for the clay is complicated. Wyoming bentonite would, for example, have the reaction:

$$(Na_{0.27}Ca_{0.10}K_{0.02})(Al_{1.52}Fe^{3+}_{0.19}Mg_{0.22})(Si_{3.94}Al_{0.06})O_{10}(OH)_2$$
$$+\ 3.52H_2O\ +\ 0.22H^+\ =$$
$$(Na_{0.27}Ca_{0.10})(Al_{1.52}Fe^{3+}_{0.19})(Si_{3.06}Al_{0.06})O_{6.26}(OH)_{5.30}$$
$$+0.88H_4SiO_4\ +\ 0.20Na^+\ +\ 0.02K^+\ +\ 0.22Mg(OH)_2.$$

Since only formulae with non-fractional subscripts should be used in solubility equilibria (Lippmann 1977) such a reaction is quite unmanageable. Lippmann suggested that the dissolving solid could be thought of as a mixture of end-member states, some of which may be hypothetical. The solubility would then be expressed by a number of partial solubility products, each product having integral exponents. These views, which are based on the assumption that average mineral formulae should be used to represent the reaction, also lead to an extremely complicated treatment of solubility.

The experimental results in Figs 26.1–3 support the use of integral exponents for dissolution reactions. Consequently, there ought to be a simple way of representing the dissolution of clays having complicated formulae.

The basis now proposed for the simplification is to represent the reactions using formulae for unit cells of the mineral structure. Some cells have isomorphous replacements. If a particular cell has a substituted ion, chance favours the neighbouring cells being unsubstituted. For example, ≡Al-0-Si≡ bonds can be somewhat common in the tetrahedral ≡Si-O-Si≡ sheet of clay minerals, but ≡Al–O–Al≡ bonds are rare (Barrar and Klinowski 1977). It is now proposed that the decomposition of illite or montmorillonite occurs as a result of the preferred reaction of one hydrogen ion with one isomorphously substituted positive ion in the lattice. The valency of the ion attacked is one less than the normal occupancy. This attack is followed by the release of aluminium and silicon from vulnerable points in the clay lattice, including regions near the edges.

For pH $<$ 9.2, the attack by a single proton on octahedral and tetrahedral substitutions is proposed to give the following reactions in which the enclosed portions of the equations are not in solution but combined as part of the clay:

(i) octahedral substitution of magnesium for aluminium:

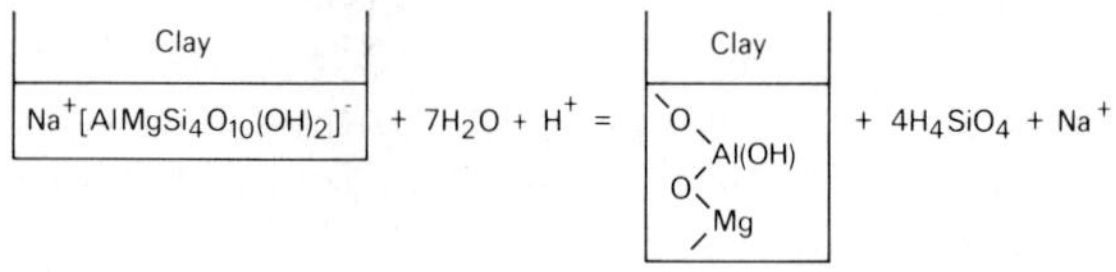

(ii) tetrahedral substitution of aluminium for silicon:

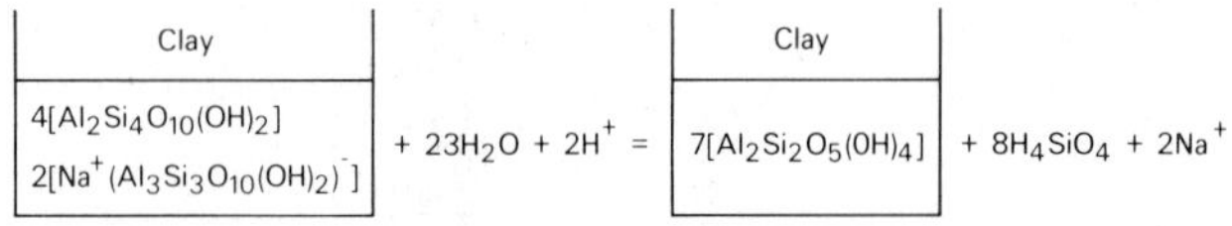

These reactions give the required relation in the solution phase: 4pSi $-$ (pH $-$ pNa) $=$ constant.

According to reaction (i), four silicons are released for each magnesium ion attacked, because there are four silicons attached via oxygen bridges to each octahedral ion, as shown in Fig. 26.6*a*. The space thus created in the octahedral layer houses a magnesium ion (see Fig. 26.7).

According to reaction (ii), and as shown in Fig. 26.8, four silicons are released on the removal of a tetrahedrally substituted aluminium ion by stripping the silicons from the lattice to leave the kaolinite structure as a residue in that locality. Thus, areas of kaolinite are created by the action of hydrogen ions on montmorillonite by either of the reactions of Fig. 26.8 in agreement with many studies on weathering. This brings such a clay to the montmorillonite–kaolinite boundary of the stability diagram without any addition of kaolinite to the system being required.

Since the number of exchange positions enters into the left hand side of the equations, these two reactions will be somewhat sensitive to the concentration of clay in the system.

It is of interest that at pH$<$7, Tiller (1968) found that hydrogen ions decomposed hectorite by attack on the octahedral layer which was partly lithium substituted:

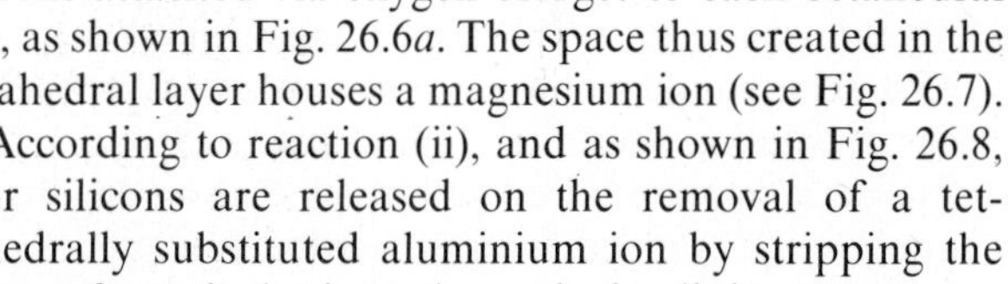

This corresponds to the apparent solubility of a hydrated $MgSiO_3$.

The reactions when 9.2 $<$ pH $<$ 11.2 are less certain and possible explanations are therefore less detailed. Figs 26.1 and 26.2 show that a slope of $-1/3$ is found over

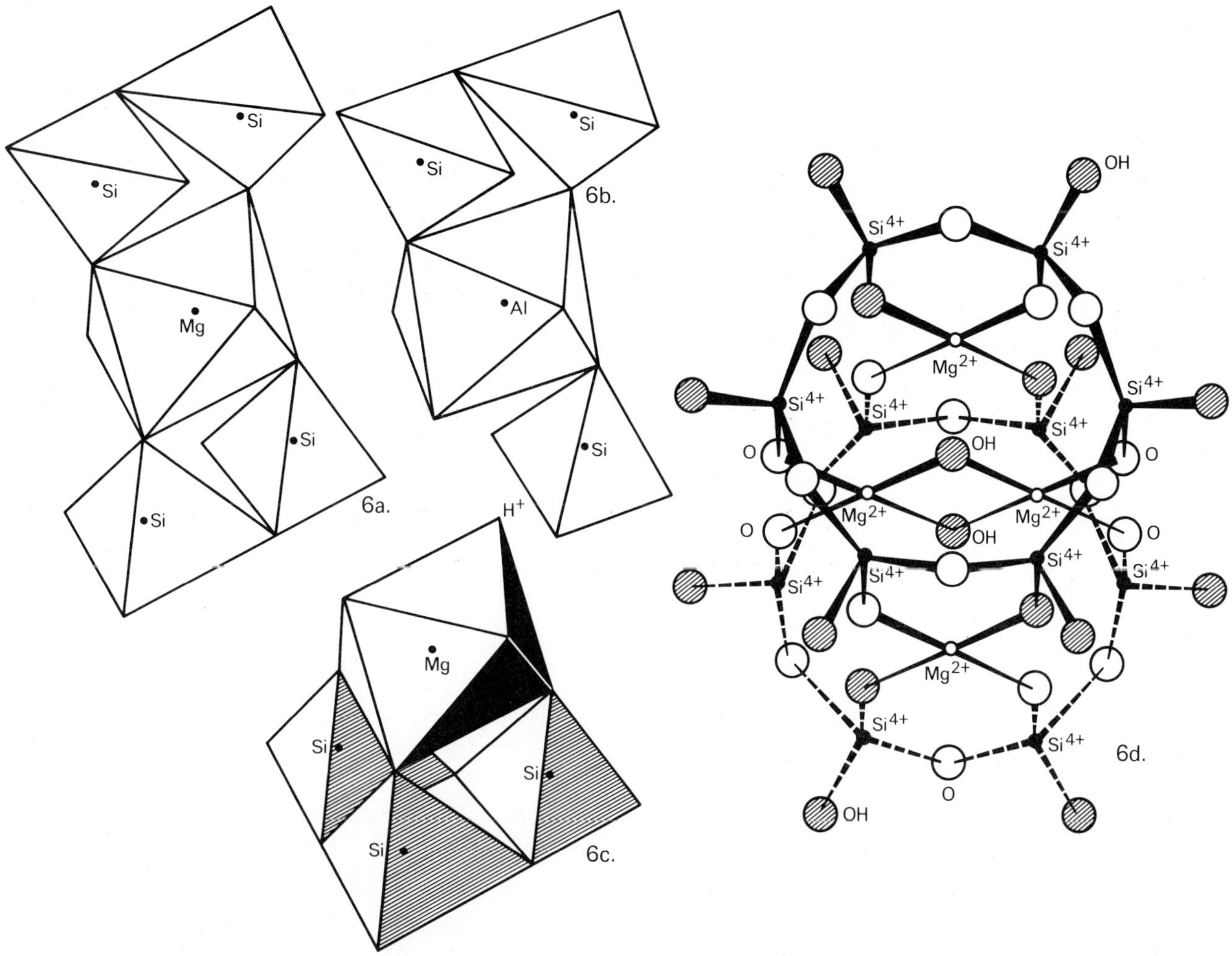

Fig. 26.6. Diagrams of units thought to be excised from clays or present in solution upon reaction with hydrogen or hydroxyl ions: *a.* four silicon atoms released for each octahedral magnesium ion attacked by a hydrogen ion; *b.* three silicon atoms released for each tetrahedral aluminium ion attacked by hydroxyl; *c.* and *d.* two possible forms of magnesium silicate present in alkaline solution. In Fig. 26.5.*d*, the bonds of the upper and lower rings of the six silicon tetrahedra are shown by full and broken lines respectively.

this pH range; thus for pH + 3pSi = constant, a net release of three silicon atoms by one hydroxyl ion is required. This happens as an octahedral aluminium ion (contained in a unit shown in Fig. 26.6*b*) reverts to an aluminate ion, with the release of the three attached silicons in a 2:1 layer silicate lattice.

Furthermore, for pH > 9.2, an equivalent unit, with three silicon tetrahedra attached to both a magnesium ion and a hydrogen ion (to neutralize the excess negative charge, see Fig. 26.6*c*), is suggested from the relationship pMg - pH = constant, found by Tucker (see Chapter 27). Gac *et al.* (1977) examined the evaporation of waters in the Lake Chad system, and the results from their more dilute concentrates (the first 11) support 3pSi + pH = constant for systems containing magnesium silicate under alkaline conditions, suggesting the same unit. The reaction basically involves the solubility relationship:

$$Mg(H_3O)\,(SiO_2\,(OH))_3 + 6H_2O = 3Si(OH^-)_5 + Mg^{2+} + H_3O^+.$$

This reaction could involve dissolution of part of a saponite structure, present as portion of an existing clay phase or as a separate structural unit. Four of these units together make up a compact segment of magnesium montmorillonite, having two rings each of six silicon tetrahedra above and below the four octahedral magnesium ions, and stabilized by four hydrogen ions (see Fig. 26.6*d*).

It is believed that data used to illustrate this chapter, in which a series of reactions are considered over a wide range of pH, will be helpful for future work on the soil solution and for the discovery of soil processes.

Many other sources of ions contribute to the soil solution. Organic moieties, soluble salts, and exchangeable cations provide pools from which nutrients and salts are

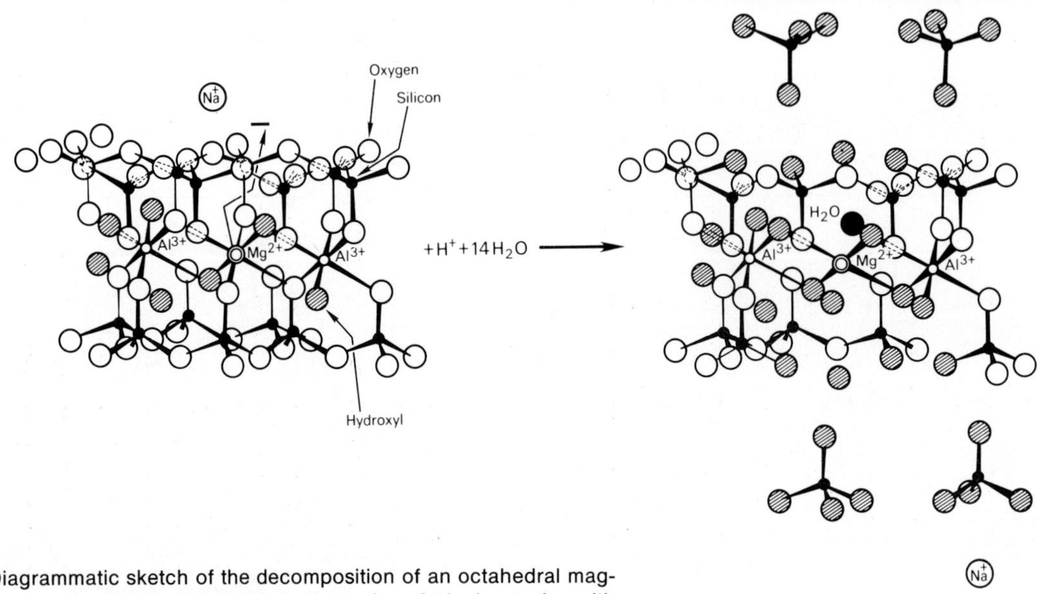

Fig. 26.7. Diagrammatic sketch of the decomposition of an octahedral magnesium ion in a montmorillonite type lattice by the action of a hydrogen ion with the release of four silicic acid molecules.

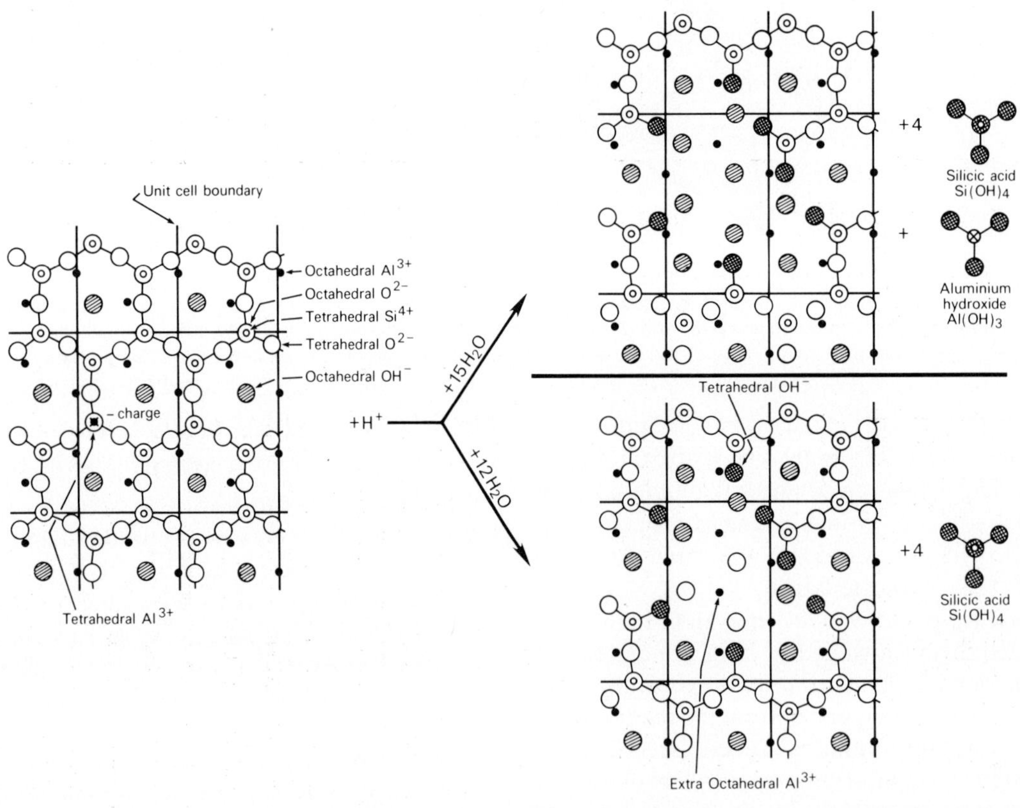

Fig. 26.8 A plan view of montmorillonite type lattice, showing the attack on a tetrahedral aluminium ion by a hydrogen ion with the release of four silicic acid molecules. Two alternative courses are given for the reaction, one releasing aluminium hydroxide into the system, and the other incorporating the aluminium ion into the octahedral layer.

supplied to plants. These ions also influence soil structure and permeability, as described in other chapters. Little information is available on many of the solubilities and adsorption equilibria involved, although the order in which the various salts are expected to separate on concentration of the soil solution is available from computer programs (Helgeson 1968; Helgeson *et al.* 1969; Truesdell and Jones 1974) which apply for aqueous systems. The solubility of some mineral phases, e.g. calcium sulphate and calcium and magnesium carbonates (e.g. Doner and Lynn 1977) and the hydroxides and phosphates, appear to be straightforward in many instances. In other cases however, redox phenomena and complex formation with organic compounds make the situation so complicated that it has not been examined in detail. The capacity of the soil to act as a reservoir for dissolved gases, whose release may favour or hinder plant growth, awaits a thorough investigation.

Most proposals made about the above systems depend to some extent on the equilibria of the ubiquitous silicate structures which are the main consideration of this chapter. Until these equilibria are clarified, the behaviour of many of the less abundant ions in the soil solution will continue to remain rather obscure.

Conclusion

For free energy data and stability diagrams to be used to study ionic concentrations in the soil solution, two important conditions are required. The first is that there should be equilibrium. The second is that the reactions taking place should be formulated correctly. The equilibrium condition appears to be met without undue difficulty, but the formulation of the reactions is more difficult to accomplish.

In this chapter, a formulation is advanced which differs from the ones used previously; it is based on formula units, each unit being equivalent to an unsubstituted unit cell, or to a cell having an isomorphously substituted lattice ion which gives rise to a cation exchange position. This formulation appears to be consistent with existing data on soil–water systems and saturation extracts, as well as with data in the literature used for the calculation of free energy values.

This gives a rather different interpretation to many reactions, e.g. compare Veen (1973). One large difference is that on dissolution of the montmorillonite lattice by hydrogen ions portion of the montmorillonite structure decomposes to give kaolinite. For this reason, montmorillonite clays go into solution as though both montmorillonite and kaolinite are present together.

References

Bache, B. W. (1974). Soluble aluminium and calcium–aluminium exchange in relation to the pH of dilute calcium chloride suspensions of acid soils. *J. Soil Sci.* **25**, 320-32.

Barrar, R. M. and Klinowski, J. (1977). Theory of isomorphous replacement in aluminosilicates. *R. Soc. Lond. Phil. Trans.* **285A**, 637-76.

Beckwith, R. S. and Reeve, R. (1963). Studies on soluble silica in soils. I. The sorption of silicic acid by soils and minerals. *Aust. J. Soil Res.* **1**, 157-68.

Beckwith, R. S. and Reeve, R. (1964). Studies on soluble silica in soils. II. The release of monosilic acid from soils. *Aust. J. Soil Res.* **2**, 33-45.

Doner, H. E. and Lynn, W. C. (1977). 'Minerals in Soil Environments.' (Soil Sci. Soc. Am.: Madison.)

Gac, J. I., Droubi, A., Fritz, B. and Tardy, Y. (1977). Geochemical behaviour of silica and magnesium during the evaporation of waters in Chad. *Chem. Geol.* **19**, 215-28.

Helgeson, H. C. (1968). Evaluation of irreversible reactions in geochemical processes involving minerals and aqueous solutions. I. Thermodynamic relations. *Geochim. Cosmochim. Acta* **32**, 853-77.

Helgeson, H. C., Garrels, R. M. and Mackenzie, F. T. (1969). Evaluation of irreversible reactions in geochemical processes involving minerals and aqueous solutions. II. Applications. *Geochim. Cosmochim. Acta* **33**, 455-81.

Hem, J. D. and Robertson, C. E. (1967). Form and stability of aluminium hydroxide complexes in dilute solution. U.S. Geol. Surv. Water-supply Pap. No. 1827-A.

Hingston, F. J., Posner, A. M. and Quirk, J. P. (1972). Anion adsorption by goethite and gibbsite. I. The role of the proton in determining adsorption envelopes. *J. Soil Sci.* **23**, 175-92.

Hingston, R. J., Posner, A. M. and Quirk, J. P. (1974). Anion adsorption by goethite and gibbsite. II. Desorption of anions from hydrous oxide surfaces. *J. Soil Sci.* **25**, 16-26.

Huang, W. H. and Keller, W. D. (1972). Standard free energies of formation calculated from dissolution data using specific mineral analyses. *Am. Mineral.* **57**, 1152-62.

Huang, W. H. and Keller, W. D. (1973). Gibbs free energies of formation calculated from dissolution data using specific mineral analyses. III. Clay minerals. *Am. Mineral.* **58**, 1023-8.

Kittrick, J. A. (1971*a*). Montmorillonite equilibria and the weathering environment. *Soil Sci. Soc. Am. Proc.* **35**, 815-20.

Kittrick, J. A. (1971*b*). Stability of montmorillonites. II. Aberdeen montmorillonite. *Soil Sci. Soc. Am. Proc.* **35**, 820-3.

Kittrick, J. A. (1971*c*). Stability of montmorillonites. I. Belle Fourche and Clay Spur montmorillonites. *Soil Sci. Soc. Amer. Proc.* **35**, 140-45.

Lippmann, F. (1977). The solubility products of complex minerals, mixed crystals, and three-layer clay minerals. *Neves Jahrb. Mineral. Abh.* **130**, 243-63.

McKeague, J. A. (1962). Silica in solution in the presence of soil and related materials: forms, concentrations and reactions. PhD thesis, Cornell University.

McKeague, J. A. and Cline, M. G. (1963*a*). Silica in soil solutions. I. The form and concentration of dissolved silica in aqueous extracts of some soils. *Can. J. Soil Sci.* **43**, 70-82.

McKeague, J. A. and Cline, M. G. (1963*b*). Silica in soil solutions. II. The adsorption of monosilicic acid by soil and by other substances. *Can. J. Soil Sci.* **43**, 83-96.

McKeague, J. A. and Cline, M. G. (1963*c*). Silica in soils. *Adv. Agron.* **15**, 339-96.

Marion, G. M., Hendricks, D. M., Dutt, G. R. and Fuller, W. H. (1976). Aluminium and silica solubility in soils. *Soil Sci.* **121**, 76-85.

Mattigod, S. V. and Kittrick, J. A. (1979) Aqueous solubility studies of muscovite: apparent nonstoichiometric solute activities at equilibrium. *Soil Sci. Soc. Am. Proc.* **43**, 180-7.

Misra, U. K. and Upchurch, W. J. (1976). Free energy of formation of beidellite from apparent solubility measurements. *Clays Clay Mineral.* **24**, 327-31.

Nhung, M. M. and Ponnamperuma, F. N. (1966). Effects of calcium carbonate, manganese dioxide, ferric hydroxide and prolonged

flooding on chemical and electrochemical changes and growth of rice in a flooded acid sulphate soil. *Soil Sci.* **102**, 29-41.
Parks, G. A. (1965). The isoelectric points of solid oxides, solid hydroxides and aqueous hydroxo complex systems. *Chem. Rev.* **65**, 177-98.
Parks, G. A. (1972). Free energies of formation and aqueous solubilities of aluminium hydroxides and oxide hydroxides at 25°C. *Am. Mineral.* **57**, 1163-89.
Ponnamperuma, F. N. (1972). The chemistry of submerged soils. *Adv. Agron.* **24**, 29-96.
Raupach, M. (1957). Investigations into the nature of soil pH. CSIRO Aust. Soil Publ. No. 9.
Raupach, M. (1963*a*). Solubility of simple aluminium compounds expected in soils. I. Hydroxides and oxyhydroxides. *Aust. J. Soil Res.* **1**, 28-35.
Raupach, M. (1963*b*). Solubility of simple aluminium compounds expected in soils. II. Hydrolysis and conductance of Al^{3+}. *Aust. J. Soil Res.* **1**, 36-45.
Raupach, M. (1963*c*). Solubility of simple aluminium compounds expected in soils. III. Aluminium ions in soil solutions and aluminium phosphates in soils. *Aust. J. Soil Res.* **1**, 46-54.
Raupach, M. (1963*d*). Solubility of simple aluminium compounds expected in soils. IV. Reactions of aluminium hydroxide under acid conditions. *Aust. J. Soil Res.* **1**, 55-62.
Raupach, M. and Piper, (1959). Interactions of silicate and phosphate in a lateritic soil. *Aust. J. Agric. Res.* **10**, 818-31.
Reesman, A. L. (1974). Aqueous dissolution studies of illite under ambient conditions. *Clays Clay Miner.* **22**, 443-54.
Reesman, A. L. and Keller, W. D. (1968). Aqueous solubility studies of high-alumina and clay minerals. *Am. Mineral.* **53**, 929-42.
Robie, R. A. and Waldbaum, D. R. (1968). Thermodynamic properties of minerals and related substances at 298.15°K (25.0°C) and one atmosphere (1.013 bars) pressure and at higher temperatures. U.S. Geol. Surv. Bull. No. 1259.
Sillen, L. G. (1967). *In* 'Equilibrium Concepts in Natural Water Systems'. Advances in Chemistry Series No. 67. (Am. Chem. Soc.: Washington.)
Stumm, W. and Morgan, J. J. (1970). 'Aquatic Chemistry.' (Wiley-Interscience: New York.)
Tiller, K. G. (1968). Stability of hectorite in weakly acidic solutions. II. Studies of the chemical equilibrium and the calculation of free energy. *Clay Mineral.* **7**, 261-70.
Truesdell, A. H. and Jones, B. F. (1974). WATEQ: A computer program for calculating chemical equilibria of natural waters. *J. Res. U.S. Geol. Surv.* **2**, 233-48.
Tucker, B. M. (1964). The solubility of potassium from soil illites. II. Mechanisms of potassium release. *Aust. J. Soil Res.* **2**, 67-75.
Veen, A. W. L. (1973). Evaluation of clay mineral equilibria in some clay soils (Usterts) of the brigalow lands. *Aust. J. Soil Res.* **11**, 167-84.
Weaver, R. M., Jackson, M. L. and Syers, J. K. (1971). Magnesium and silicon activities in matrix solutions of montmorillonite-containing soils in relation to clay mineral stability. *Soil Sci. Soc. Am. Proc.* **35**, 823-30.

Chapter 27
Basic exchangeable cations

B. M. Tucker

CATION EXCHANGE IS ONE ASPECT OF SOIL CHEMISTRY that has received a great deal of attention, ranging from theoretical studies on idealized model clay systems to empirical work on practical problems in soil management. Insufficient attention has been given to the actual phenomena in cation exchange so that the various studies have been too simplistic and often not well related to the properties of field soils. In theoretical work considerable reliance has been placed on comparisons with synthetic ion exchangers or on electrical analogues, and the derived mathematical models have been very useful in providing a guide to the empirical work which is necessary for practical application of exchange theory. The danger is that modification of such models may be thought sufficient to fit them to the actual behaviour of soils instead of pursuing the more difficult task of devising the better models which need to be developed now on the basis of the new concepts of soil properties.

Rather less is known of the behaviour of soluble salts in clay soils. The conventional approach has been to treat them as a mechanical mixture with other soil constituents, but increasingly in recent years this has been recognized as too simple a view. The surface charges, the structure and the swelling of soil components are able to modify the mobility of salts in the soil solution under some circumstances and it is then possible for salts to simulate exchangeable ions (see below). The importance of these factors cannot be judged from the small amount of research done so far.

Cation exchange in soils takes the form of ionic reactions with mixtures of various clay minerals, some metal oxides, and variable amounts of organic compounds, and generally involves more than two cations simultaneously. In addition it is clouded by the presence of sparingly soluble minerals and soluble salts. In field soils, a cation exchange in one part of the solum is necessarily followed by some further exchange in another part when the displaced cations are transferred by movement of the soil solutions – a simple fact of great importance when improvement of the exchangeable cation suite is attempted in the management of the soil system.

Less attention therefore will be given to the electrochemical aspects of cation exchange; instead some directions, in which new principles for understanding the role of cation exchange in the whole of soil chemistry may be found, will be suggested. Discussion of the theory is already available in reviews and recent texts (e.g. Bolt and Bruggenwert 1976). Double-layer theory will be dealt with in Chapter 31, Interparticle bonding. Basic exchangeable cations include the cations of the alkali and alkaline earth metals and magnesium and ammonium. Cations of the trace metals like zinc, copper or cobalt, may take part in ion exchange reactions, but since other reactions, such as chelation, are so much more important for them in soils they will not be treated in this chapter.

Ion exchange capacity

Ion exchange in soils is understood in terms of the charges on surfaces accessible to ions in the soil solution. The exchangeable ions are pictured as lying in a layer of solvent at the surface of the soil exchanger particles, the greatest concentration of ions being near the surface and decreasing exponentially away from this surface. Associated with this layer is a volume in which ions of the same charge as the surface are at a lower concentration than in the bulk of the soil solution – an effect called 'negative absorption' – and it is conveniently known as a double layer volume.

In clay minerals, part of the surface charge is due to imperfections in the layer structures. These imperfections result in negative charges on the clay particles which are balanced by cations situated either between the structural layers within the particles, or at external planar surfaces. They are called permanent charges because they are not discharged under normal conditions in soils. In addition further charges may develop on the edges of clay crystals, on organic materials and on the surfaces of other minerals (e.g. metal oxides) by the ionization of surface groups. This ionization is controlled mainly by the pH and the concentration and nature of salts in the soil solution, and to a slight degree by other conditions such as temperature. These additional charges are called pH-dependent.

Metal oxide minerals are important among the sources of pH-dependent charge because they may develop positive surface charges, with exchangeable anions as counter-ions, in neutral or acid soil solutions, in contrast to negative charges under alkaline conditions. There are well-established techniques for finding the solution pH at which such surfaces have no charge (point of zero charge: PZC) and are unable to hold either anions or cations at their surface with the soil solution. Some of these techniques may be applied to soils but because the soil contains a mixture of charged surfaces the PZC as measured has a rather different meaning; the soil at this pH retains equivalent quantities of anions and cations. These quantities nevertheless are small and are easily lost into percolating soil solutions when the salt concentrations are low, so that the PZC is a useful measurement on soils with large contents of metal oxides.

The concept of ion exchange capacity is a natural consequence of viewing the process as the result of surface charges and it is understood in two ways. First, what may be called a *standard* ion exchange capacity is a measure of the ability of a soil to hold ions under defined conditions of pH and concentration of the soil–solution mixture containing only ions of specified kinds. It is necessary to define these conditions because all natural soils have at least some proportion of pH-dependent charges. The pH chosen is usually either 7 (corresponding to a neutral soil acidity) or 8.5 (a non-saline soil containing carbonates of calcium or magnesium); the final salt concentration is perhaps 0.1 M, a low salinity under natural conditions; the cation is usually calcium, sodium, or ammonium, chosen according to some reasoned argument, and the anion is nearly always chloride or acetate. The conditions are artificial but they do allow comparisons of exchange capacities between different soils, and the relating of capacities to the content of clay minerals, organic materials, and other soil components.

Secondly, we may use an *actual* ion exchange capacity, which is a measureof the ability of a soil to retain ions at its own pH and with a soil solution close to its natural composition. This is rather simply measured as the sums of the exchangeable ions present in the soil – it is then necessary to distinguish carefully between the ions held by the surface charges and others which might be derived from soluble salts or dissolving minerals. (This difficulty is usually avoided in the methods designed to measure standard capacities.) An alternative is to measure the actual capacities in an unbuffered salt solution, at a concentration comparable to that in the natural soil, allowing the soil itself to dictate the eventual pH. This is one step away from field conditions.

When a soil is able to retain both cations and anions at charged sites it is not possible to measure the two capacities in an absolute sense, and it is necessary to make some assumptions about the content of soluble ions after equilibration of the soil under standard conditions, or in the soil as it was actually sampled, as follows.

(1) In the standard methods the wet soil can be weighed to calculate the volume of solution present and then it is assumed that the concentration of the salt is equal to that in the equilibrating solution.

(2) Some soil solution can be expressed from the soil and analysed, assuming the concentration of the whole soil solution in the natural sample to be the same as that of the portion analysed.

These two assumptions are inexact because the negative adsorption of anions and cations at negatively and positively charged soil particle surfaces makes the salt concentration in the soil solution non-uniform, and on average less than that calculated on the basis of these assumptions.

(3) The soil solution may be removed by washing the soil sample with some solvent in which ion exchange is minimized (i.e. no salt impurities and a low dielectric constant). The cations and anions remaining in the soil can be displaced and analysed to measure the cation and anion exchange capacities. Here it is assumed that the exchangeable ions are not disturbed by this reduction of solution concentrations below those in the standard conditions or in the natural soil sample. This introduces a small, variable and unknown error into the measurements.

The alternative to making these assumptions is to measure the difference between the sums of cations and those of anions, both soluble plus exchangeable in the natural soil or after equilibration under standard conditions. This difference is called the *net charge*. For soils with negligible anion exchange capacity this net (negative) charge is equivalent to the cation exchange capacity. In soils with some anion exchange capacity, notably those with a high content of iron oxide minerals accompanied by silicate clay minerals of low negative charge and organic materials of strongly pH-dependent charge, the distinction between net charge (negative or positive) and ion exchange capacities is an important one. In these circumstances the net charge will underestimate the ability of the soil to retain cations and anions against loss by leaching. Both exchange capacities may have small non-zero values when the net charge as measured is zero, i.e. in a soil at its point of zero charge.

Ion exchange equilibria

Under laboratory conditions, for example in a stirred suspension of clay in water, exchange reactions are rapid. Since they involve only an interchange of ions (low heat of reaction) their rates are determined by the movement of the ions to the exchange site rather than by the reaction itself. It is therefore reasonable to describe ion exchange in terms of equilibria existing instantaneously and responding immediately to changes in composition of the soil solution at the sites of exchange. (Exchanges of acidic exchangeable cations, hydrogen and aluminium, seem to be more complex and equilibrium is approached very slowly.) In field soils actual ion exchanges are limited in rate by relatively slow diffusions of ions in the soil solution, slow solution processes of soil minerals and slow movement of the soil solution through the soil body.

A commonly considered cation exchange reaction is that between sodium and calcium ions, which may be written in simple terms as

$$2Na^{+}_{o} + Ca^{2+}_{s} \leftrightarrow 2Na^{+}_{s} + Ca^{2+}_{o}$$

where the subscripts *o* and *s* refer the ions to the solution and solid exchanger phases respectively. Such a reaction never occurs in isolation. Other ions, which enter into parallel reactions of their own, are present; and in addition changes in the soil solution which accompany such an exchange may make small changes in the pH-dependent component of the actual cation exchange capacity. These capacity changes are necessarily accompanied by minor adjustments in the anion distribution near the charged surfaces and by an exchange of hydrogen ions with the soil solution.

The equation for this Na–Ca exchange may be written in terms of cation activities as

$$\frac{(Na^{+}_{s})^2(Ca^{2+}_{o})}{(Na^{+}_{o})^2(Ca^{2+}_{s})} = K$$

where the exchange constant K is a function of the (four) different standard activity states of the two cations in the solution and exchanger phases. For such an equation to be useful the ion activities have to be calculated from measurable quantities. Normally these quantities are cation molalities expressed as mol. kg^{-1} of the soil solution and of the solid exchanger in some specified standard condition e.g. dried to remove all free water.

The standard state for salts in solution is accepted as that expected for an ideal solution (no interactions between the ions, nor with the solvent) at molal concentration. The activity of a salt, whether found by experiment or by calculation, is the product of the ionic activities. For a salt $M^{n+}_{m}A^{m-}_{n}$, its activity

$$(M_mA_n) = \{(M)^m (A)^n\},$$

and is relative to that of the standard state for that salt. Individual ionic activities are not experimentally accessible and the problem of their calculation is resolved by the convention of assuming that for potassium chloride solutions the ionic activities are equal $(KCl)^{1/2} = (K^+) = (Cl^-)$, and then assuming that the activity coefficient γ = (activity/concentration) of an ion species is the same in all solutions of the same ionic strength I in the same solvent. The ionic strength is defined as

$$I = \tfrac{1}{2}\Sigma_i c_i z_i^2$$

where c_i is the concentration and z_i the charge of the ion species i, the summation being made for all cations and anions in solution. The concentration is for actual ion concentrations so that the total (analytical or prepared) concentrations have to be corrected for ion-pair formation as in the important cases

$$Ca^{2+} + SO_4^{2-} \leftrightarrow CaSO_4^{o}$$

$$\text{and } Ca^{2+} + HCO_3^{-} \leftrightarrow CaHCO_3^{+}$$

The ion-pair formation constants available in literature tables are based on ion activities so that iterative calculations of ionic strength from concentrations corrected for ion-pairs based on activities based on ionic strength, and so on, are required. Conventionally, for uncharged ion-pairs, the activity coefficient is taken as unity since uncharged species should have zero interaction with the solvent or ions in solution.

For dilute soil solutions the Debye-Hückel equation, or some modification of it, can be used to calculate ion activity coefficients; for example in the formulation

$$-\log \gamma_i = Az_i^2 \sqrt{I}/(1 + å_i B \sqrt{I})$$

where $å_i$ is the mean diameter of the solvated ion, i, in angström units. The constants A, B and $å_i$ are available in tables. This calculation may be used for soil solutions up to total cation concentrations of about 0.1 molal. For higher concentrations, and this is usually the case for soils in the field, tabulated data for activities of specific salts have to be used, obtaining ion activities by the KCl convention mentioned above. Allowance still has to be made for ion-pair formation where significant, i.e. between the cations Ca^{2+}, Mg^{2+} (and Al) and the anions SO_4^{2-}, HPO_4^{2-} or HCO_3^- (Adams 1971).

These considerations are based on the chemistry of aqueous electrolytes. They are assumed to apply to that part of the soil solution beyond the range of the charges on the soil particle surfaces. At low soil moisture and low salt concentrations this part of the soil solution may be small or even absent. As a guide, for a soil with a particle surface area of 50 m^2 g^{-1} and an effective double-layer thickness of say 10 nm as might be found in a soil solution of 0.01 M NaCl, the volume affected by the surface charges would be 0.5 cm^3 g^{-1}. Under such circumstances the use of exchange equations based upon laboratory experiments with rather dilute soil suspensions is not appropriate. A direct experimental approach, simulating field conditions as closely as practicable, is advisable.

The standard states for cations in the solid soil exchanger phases are less easily specified. Gaines and Thomas (1953) used for each cation the exchange capacity completely filled with that cation in equilibrium with an infinitely dilute solution of the cation. The problem still remains of relating the activities of mixtures of exchangeable cations to the composition of the exchangeable cation suite. There is no principle, comparable to the concept of ionic strength for ion activities in solution, which can be used to allow for the effect of exchangeable cations on the activity coefficients of their companion cations. For exchangers containing only one kind of clay mineral an analogy has been made between the cation mixture and a solid solution where the activities of the components are taken as proportional to their mole fraction, x, in the solid. For this model the exchange equation becomes

$$\frac{(\mathrm{Ca}_o^{2+}).\, x_{\mathrm{Na}}^2}{(\mathrm{Na}_o^+)^2\, x_{\mathrm{Ca}}} = K_m$$

where x is the mole fraction of each exchangeable cation and K_m incorporates the proportionality factor between mole fraction and solid phase activity. The total moles of all exchangeable cations changes as the proportions of cations of different charges in the suite change. It is not equal to the cation exchange capacity.

More usually the mole fractions are replaced by the equivalent fractions q. This has the advantages that $\Sigma_+ q$ is a constant under fixed conditions in the soil solution, and that equivalents are the customary units for exchangeable ions. In the exchange equation q replaces x and another exchange constant K_e is needed. However, K_m and K_e are both found to vary with composition of the cation suite, so such equations are only useful for predicting comparatively small changes in cation status. Various attempts have been made to adapt the simple exchange equations so that they can be used over a wider range of conditions, especially for exchange of cations of different charges. The proposal of Krishnamoorthy and Overstreet (1950), based on the ideas of Davis (1950), has attracted much attention.

The exchange sites on a clay particle are envisaged as a regular array on a plane surface. The probability of a double-charged ion, Ca^{2+}, replacing a single-charged ion, Na^+, depends on the number of adjoining sites also occupied by Na^+; and this is determined not only by the proportion of sodium in the exchange suite but also by the geometrical arrangement of the sites. For Na–Ca exchange the equation then becomes

$$\frac{(\mathrm{Ca}_o^{2+})}{(\mathrm{Na}_o^+)^2} \cdot \frac{x_{\mathrm{Na}}^2}{x_{\mathrm{Ca}}\{x_{\mathrm{Na}} + \frac{2(n-1)}{n}\, x_{\mathrm{Ca}}\}} = K'_m$$

where n is the maximum number of nearest-neighbour charged sites. This virtually has the effect of modifying the exchange constant so that it is adjusted to favour entry of calcium as the proportion of exchangeable sodium increases (and *vice versa*) especially if n is more than 2, that is, for two-dimensional arrays of sites. This agrees with the generally observed fact that, on clays, cation species in the smaller proportion tend to be held more tightly at exchange sites, other factors being equal. Consequently this reformed equation can be used for larger exchanges of cations. The model proposed by Davis is a simplified one and takes no account of the important possibility that double-charged ions may be attracted by pairs of sites on separate charged surfaces. Instead of deciding on a value of n by structural reasoning, one might adopt the suggestion of Bolt (1967) of empirically determining n to give a best fit to experimental data. Such a device undermines the basis of the Davis approach to the exchanger phase activity problem.

These descriptions of cation exchange have a theoretical rationale, but they can only be applied as they stand to simple equilibrium situations involving only two cation species and a single type of exchanger. For mixtures of several exchangers with different individual exchange constants the application of these equations necessarily finds the apparent general constant to change rather rapidly with composition of the cation suite. Another real problem in soils is that usually more than two cation species interact. The three-cation problem has been studied and graphical solutions presented by André (1975), Elprince and Babcock (1975), Helfferich (1962), Jangg (1958), and Galkina *et al.* (1962); Dranoff and Lapidus (1957) have also discussed this problem. No straight-forward mathematical solution to this or to more complicated equilibria seems likely at the present time; but it is probable that computer models based on empirical equations (Carlson and Buchanan 1973) will be useful.

A practical expedient that is often used to cope with multiple cation equilibria is to ignore some cations and to lump others together to give an apparent two-cation equilibrium which can be treated satisfactorily with simple equations. In the common Ca, Mg, Na ternary exchange, K is ignored, which is not objectionable if exchangeable potassium is low or the soil (clay) has a high selectivity for potassium so that it is little disturbed by other exchange reactions, hydrogen is neglected if pH changes are expected to be small, and magnesium is put together with calcium, the sum being treated as one double-charged cation. This is acceptable in many situations since the exchange constants for calcium and magnesium are nearly equal on various exchange materials and soils very often contain much more calcium than magnesium (Beckett 1965). This expedient may prove to be less satisfactory for Australian soils, which often contain more exchangeable Mg than Ca in important horizons. In addition Mg may have an adverse effect on clay dispersion in opposition to the effect of Ca, for example in the presence of appreciable exchangeable Na (Bakker and Emerson 1973), so that combining Mg with Ca may give misleading predictions of clay behaviour after cation exchange.

An exchange equation due to Gapon (1933) is widely used, especially because of its application to the reclamation of sodium-affected soils (Richards 1954). Gapon's main assumption is that the rate of exchange of a cation from a soil is proportional to the fraction of the cation exchange capacity occupied by that cation, expressed in equivalents, not moles. The composition of the exchange suite then enters the exchange equation in the form q_{Na}/q_{Ca}, where q is the equivalent fraction of the cation, instead of the form x^2_{Na}/x_{Ca} of other equations. Gapon writes the exchange reaction in a barbaric form which can be represented as

$$\tfrac{1}{2}Ca_s^{2+} + Na_o^+ \leftrightarrow Na_s^+ + \tfrac{1}{2}Ca_o^{2+}$$

For this reaction we have the variables $[Na_o^+]$ for Na_o^+ and $[Ca_o^{2+}]$ for Ca_o^{2+}, solution concentrations replacing the activities used in other equations, q_{Na} proportional to Na_s, and q_{Ca} proportional to $\tfrac{1}{2}Ca_s$.

Then

$$\frac{q_{Na}[Ca_o^{2+}]^{1/2}}{q_{Ca}[Na_o^+]} = K_G$$

where K_G is the Gapon exchange constant for the exchanger. In spite of its theoretical deficiencies, this equation has been found to fit experimental data rather well. Sposito and Mattigod (1977) give a justification for Gapon's main assumption on the relative reactivity of cations at exchange sites. The exponents of the activity of each species in the equilibrium equation must equal the number of that species involved in the reaction if the equation is to have a theoretical as opposed to an empirical foundation. This requirement is not met by their justifications of the Gapon equation – a step in the mathematical argument is missing.

The Gapon reaction may be viewed as the first step in the reaction sequence formally written as

$$Ca_o^{2+} + Na_s \leftrightarrow Ca_s^+ + Na_o^+,$$

followed by

$$Ca_s^+ + Na_s \leftrightarrow Ca_s + Na_o^+.$$

To make the nature of the reaction sequence more clearly seen, the charges on the cations which bind them to the exchange sites are omitted. The disagreement between the molecularity of the Gapon reaction and the exponents in the Gapon equation suggest that it may be possible to justify the equation by considering some such complex reaction mechanism, in which the equilibrium is controlled by only one of the steps.

Kittrick (1976) has introduced a formulation, similar to Gapon's equation, but based on the idea of a separation factor (see Helfferich 1962). Equivalent fraction ratios are used to express the composition of the cation suite as Gapon did, but activities are used by Kittrick for solution ions and there are no ion-charge exponents in the equation. Using the symbols of this chapter his formula reads

$$\frac{q_{Na}(Ca_o^{2+})}{q_{Ca}(Na_o^+)} = \alpha^{Na}_{Ca}$$

where α is the separation factor for this pair of cations. He is able to obtain linear plots of data by this formula in its logarithmic form over a wide range of cation ratios. The separation factor may depend on solution concentration and equivalent fraction ratio so that experimental tests of Kittrick's formula would normally be needed before applying it to particular soils.

It is disappointing that the traditional approach to exchange equilibrium theory has not been more successful, but this was inevitable for the mixture of ion exchange carriers in soils. The failure to produce a suitable expression for ion activity ratios in the exchanger phase has meant that the equations are not really very useful even for single exchange carriers. Meanwhile simple formulas like that of Gapon, and perhaps that of Kittrick, provide substitutes which are easy to apply.

Partition of exchangeable cations

The study of cation exchange in soils in terms of exchange equations and exchange constants has a long history; but its main application has been almost entirely in the management of soils under irrigation and in the reclamation of soils damaged by excess soluble and exchangeable sodium. In other areas, particularly that of plant nutrition, the concepts of exchange capacity and relative composition of the cation suite have been interpreted in terms of nutrient reserves, cation balance and availability to plants.

Soils are mixtures of exchange carriers which bind ions in differing degrees, and it is probable that exchangeable ions will have different effects on soil properties according to the kind of site to which they are bound. Recent years have seen attempts to separate cations into different categories and this movement gained some impetus from the discovery that exchangeable sodium in some soils could be divided into two classes, namely sodium which did or did not affect clay dispersability (Schulz *et al.* 1964).

Classes of exchange sites

The total exchange capacity of soils has been subdivided in several ways; into organic and inorganic capacities, or into permanent and pH-dependent charges, or into specific and nonspecific charge sites or according to different exchange constants.

Organic and inorganic capacities. This may be done by determining standard exchange capacities before and after destruction of the organic material by an oxidising agent like hydrogen peroxide. This method is objectionable because clay minerals can be altered by such chemical treatment and organic matter is not entirely removed. A better way is to make statistical analyses of exchange capacities against functions of organic and inorganic components within groups of closely related soils (Curtin and Smillie 1976). This kind of partition is useful information because changing the organic content of soils is the most effective way of altering their exchange properties either for better or for worse. It may be noted here that cation adsorption methods of estimating surface area, used as a measure of inorganic components, are essentially disguised measurements of exchangeable cations or surface charge, so that correlations based on them may be unrealistic.

Permanent and pH-dependent charges. This is done by measuring exchange capacities at low pH (pH 5: permanent charge only) and high pH (pH 8: permanent plus pH-dependent charges). This assumes that pH-dependent charges do not develop at pH 5 or below, perhaps not a justifiable assumption; but at lower pH values some decomposition of mineral exchange carriers is likely. Sawhney *et al.* (1970) have devised a method to split the pH-dependent charge between its organic and its metal-oxide components. These measurements are useful when the pH of the soil needs to be changed nearer to an optimum for plant growth, or where pH changes are used to alter the soil's ability to retain or release nutrient ions. The pH-dependent capacity is a guide to the amounts of alkaline or acidic amendments required to change the soil pH. An older measurement known simply as 'lime requirement' is just the part of the pH-dependent charge found between the natural pH of the soil and some desirable pH value (usually pH7).

Specific-nonspecific charge sites. A specific site is one at which a particular cation is bound much more strongly than any other cations that may be present. A separation of sites into these classes may be made by measuring exchange capacities with a non-specifically adsorbed cation before and after saturation of the specific sites with the specifically adsorbed cation, which by its nature is not replaceable by the other cation.

These measurements are important where cations on the two kinds of sites contribute unequally to some soil property. An example is the work of Bower and Rhoades (1972) on the different exchanges of sodium with non-specific and potassium-specific sites in some soils under irrigation. Such differentiation is necessarily related to some special feature of the soil, in this case the presence of vermiculite in the soil clay. The designation of sites as specific or non-specific can only be made in terms of particular cations.

Different exchange constants. If the ratios of the amounts of two cations on exchange sites in a soil are plotted graphically against their ratio in solution according to some exchange equation (e.g. for the Gapon equation, equivalent ratios against reduced solution concentration, or activity, ratios) the points then fall on a curve which in some cases appears to be a composite of two or more linear segments. These may be interpreted in terms of several kinds of sites with different exchange constants and it is possible, if the constants are sufficiently different, to produce a compound exchange equation to fit the experimental curve. For Ca–K exchange on an illitic soil clay, van Schouwenburg and Schuffelen (1963) were able to define three kinds of sites on the faces, edges and interlayer surfaces of the mineral, each with its own exchange capacity and Ca–K Gapon exchange constant. These are rather special circumstances. More usually a continuous curve, not obviously composed of linear segments, would be obtained. The method requires considerable experimental work for each clay sample, and its value lies in understanding the mechanisms of cation exchange, as for example in the work of Ehlers *et al.* (1967*a,b*), rather than in pedological studies. There is no guarantee that the separate kinds of sites apply to cations other than the actual pair investigated.

Classes of exchangeable cations

It is possible to fractionate the cations actually present in soils according to two distinct criteria, ease of displacement, or the conditions of exchange.

Ease of displacement. If a soil is brought to equilibrium with increasing volumes, or increasing concentrations, of a displacing salt solution the amount of each cation displaced can be plotted against the ratio of its activity to that of the displacing cation in the equilibrium solution. By including graded amounts of the displaced cation in the displacing reagent, it is possible to extend the curve above the point where no cation has been displaced from the soil (instead, the cation is taken up by the soil from the reagent). This is the basis for the quantity–intensity curves of Beckett (1964). These curves are comparable with the exchange constant curves of van Schouwenburg and Schuffelen (1963), and may appear to be composed of linear segments, but they are not interpreted in terms of exchange constants. It is possible to relate these Q/I curves, at least for potassium, with some confidence to the types of exchange carriers in soils (Beckett and Nafady 1967). The construction of a complete Q/I curve is laborious, and Moss (1969) used shortened procedures to measure either the intensity with which the cation is supplied by the solid phase of the soil to the soil solution – that is the ratio of its solution activity to that of the displacing cation at zero displacement from the soil – or the amount of cation held by the soil above or between arbitrary activity ratios in the soil solution.

Munns (1976) has been able to divide exchangeable calcium and magnesium into loosely and tightly bound fractions by incremental leaching or equilibration with potassium chloride solutions. This is also a method which could be widely used if suitable shortened procedures can be devised.

Conditions of exchange. The displacement of a cation from a soil may depend on certain conditions being met by the displacing reagent, so that it is possible to fractionate the cation suite by using different reagents which do or do not meet these conditions. A first reagent is used to completely displace all the cations for which the exchange conditions have been met by the reagent. Other classes of cations are left in the soil until another reagent which can fulfil their exchange requirements is used, and so on in a logical sequence of reagents. This is an alternative to separating the cations according to the different degrees of tightness of binding in a single medium that meets the exchange conditions for all the exchangeable cations present in the soil. Some of the conditions which affect exchanges are as follows.

(i) Presence of water. The activity of water affects the selectivity coefficients for competing cations (Laudelout *et al.* 1972). Water also affects cation activity coefficients, and the swelling of clays and organic materials.

(ii) Size of displacing cation. Small cations are able to enter restricted sites e.g. interlayer sites in illites (Tucker 1967).

(iii) Acidity. Some cations form hydroxide cations above certain pH values; these would be more tightly bound than the normal, neutral cations.

(iv) Mixtures of displacing cations. There is some evidence (Tucker 1974*a*) that such mixtures reduce the effects of specific adsorption of cations.

(v) Use of double-charged displacing cations.

Reagents can be devised which fulfil or do not fulfil one or more of these conditions in order to displace different classes of cation; the possible combinations are numerous. Taking the first two conditions, Tucker (1975) was able to divide the exchangeable cations in soils into discrete classes, which he called subsets, depending on the combination of conditions required for their displacement. As in conventional methods rather concentrated reagents (0.5 M) were used, so no discrimination on the basis of exchange constants is made within the subsets.

Even with the simplest partition into two subsets by using a sequence of two reagents – a primary subset (the reagent contains no water and has a large displacing cation) and a conditional subset (the reagent contains water and a small cation) – the exchange scene at once becomes more complex, quite apart from the possibility of choosing other cations and solvents which also meet these conditions. These subsets behave as if belonging to separate exchange carriers; they have exchange capacities, measured as the sum of all cations in the subset, which are stable, expanding or contracting slightly depending on whether they are filled with more or less tightly bound cations just as the whole exchange capacity of a soil does; and pairs of cations have separate exchange constants for each subset. When cations enter (or leave) subsets they show a preference, for one subset over another, which decreases as the amount in the preferred subset increases. These changes are illustrated in Fig. 27.1, which summarizes four experiments on the entry of single cations into one soil. Points to be noted are the small entry of sodium but its high preference for the primary subset; a strong but decreasing preference of potassium for the conditional subset because the soil has an illitic clay, and the rather erratic values for

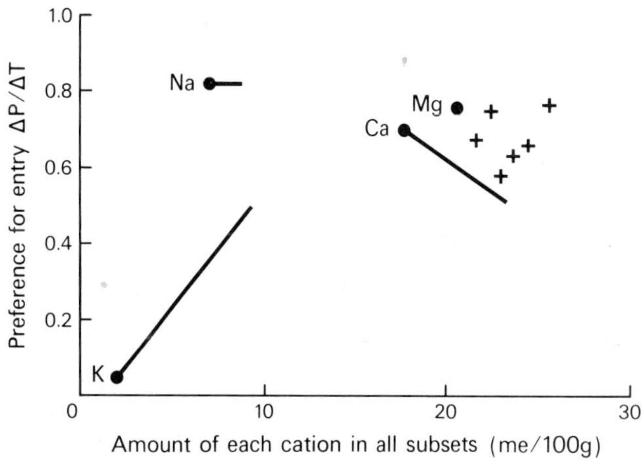

Fig. 27.1. Preference of cations for entry into the primary subset. The lines represent the results of four separate experiments in which single cation species were applied in solution to samples of the natural soil in amounts ranging from 1 to 10 me/100g soil (left to right along the line). Except for magnesium the experimental points all fall close to the lines shown. ●, cation status in natural soil; +, magnesium additions; ΔP, entry of added cation into primary subset P; ΔT, entry of added cation into all subsets T. This Claremont soil (pH 8.5) contains degraded illite in the clay fraction (47%). Actual exchange capacities are primary 20, conditional 25, total 45 me/100g at pH 8.5. The subset capacity ratio P/T = 0.44.

magnesium – the soil is at pH 8.5, the lower end of the range where $MgOH^+$ cations play a significant role; compare Fig. 27.2*a* which is discussed in a later section.

Generally in soils the primary subset is preferred for entry and release. Exceptions are for potassium in clay soils with illite, and in organic soils where the conditional subset is preferred for entry of cations.

One advantage of this second way of fractionating exchangeable cations, on the basis of changing exchange conditions, is that values for a wide range of soils are more easily obtained than when differences in exchange constants are sought, and it should be less difficult to relate these subsets to agricultural and pedological properties of soils.

Measuring exchange properties

There have been many proposals for measuring exchangeable ions and exchange capacities, and they continue to multiply, so that it is only possible to discuss the methodology of these measurements and indicate a few examples of valid techniques. Sources of error in these measurements are noted in several places. Generally it is not possible to estimate the size of these errors as there are no independent, non-chemical, methods which can give reference values for comparison. The best advice is to use methods which minimize the likely errors in the hope of obtaining accurate values and which give values of acceptable precision. The precision required for applying exchange measurements to the management of soils is largely unknown. For a ratio such as exchangeable sodium percentage, a precision of 10% would seem to be adequate. Single laboratory measurements can be made to a precision (relative standard deviation) of 2% (Tucker 1974*b*) but the difficulty of securing representative soil samples is a much greater source of error.

Exchange capacity

The purpose of this measurement is to estimate the surface charge, particularly for clay minerals, or to evaluate the ability of a soil to retain ions especially those which are plant nutrients.

The actual exchange capacities, measured under the conditions obtaining in the soils, are probably best estimated as the sum of all exchangeable cations, or anions, in the soil, provided that only exchangeable forms of these ions are included in the total. An alternative to this practical approach is to use ions of a strong electrolyte to measure actual capacities starting with concentrated solutions to remove the natural exchangeable ions and then reducing the concentration to a level comparable to the salt concentration in the original sample, for example to about the same electrolytic conductivity. Using an unbuffered solution in principle allows the soil sample itself to dictate the pH so that this source of variation in pH-dependent charge is restrained. The measuring ions are then displaced by another salt solution, and analysed. The difference between the equivalents of cations and anions displaced is the net (negative) charge of the soil. It is equal to the cation exchange capacity only if the anion exchange capacity is zero. Otherwise it is necessary to estimate the amounts of cations and anions derived from the residue of soluble measuring salt so that these can be subtracted from the total displaced cations and anions to calculate the separate cation and anion exchange capacities (see next section).

Standard exchange capacities are determined under defined conditions of pH and of measuring salt concentration at the final stage of displacement of the natural cations. It is necessary to choose the measuring salt carefully according to the purpose of the measurements. Calcium may be chosen because it is usually the major cation in soils; sodium is used where reactions of sodium salts with soils are being studied; chloride is a frequent choice of anion because it is often the commonest soil anion and is least liable to specific adsorption. On the other hand where capacity measurements are to be combined with analysis of natural ions, it is usual to use a cation, like barium or ammonium, which is only a minor member of the natural ions and is not intended to be measured.

Generally, it must be said, there are no ideal measuring salts; some compromise has to be made between desirable and undesirable properties of possible reagents when the soil has pH-dependent charges to be included in the exchange capacity. A single-charged cation is best for measurement of surface charge but some cations may be undesirable for certain minerals e.g. K and NH_4 for micas and vermiculites. On the other hand it may be very difficult to saturate soils with weakly bound cations like sodium and lithium; five or more treatments with 2 M solutions are generally needed for soils of medium or high capacity. Some double-charged cations may be unsuitable under alkaline conditions where they form hydroxide cations (e.g. $MgOH^+$ above pH 8.5, cf. Fig. 27.2*a*), or if they undergo equimolar rather than equivalent exchange (Papanicolaou and Overstreet 1969). Calcium is unsuitable for soils containing calcite or gypsum since these minerals dissolve in most displacing solutions; barium exchanges with calcium in calcite so that it may overestimate capacities in calcareous soils.

Concentration control is usually obtained by finishing the saturation step with relatively dilute reagents, say 0.1 M. These are rather more concentrated than would be found naturally in the soil solution of non-saline soils: 0.1 M corresponds roughly to an electrolytic conductivity of 1 mS cm^{-1} in a saturation extract. However it is hardly possible to go much lower than this without losing control of pH and also needing several dilution steps. Concentration is not a strong factor in determining exchange capacities so the compromise at about 0.1 M is a reasonable one. Sometimes washing with a solvent such as aqueous ethanol is used to remove excess measuring salt before the measuring ions are displaced. This has an advantage that direct evaluation of cation and anion exchange capacities can be made by analysis of the displaced ions, but this useful device is usually avoided because measuring ions may be lost in the washing step, and because, after washing, both concentration and pH are uncontrolled.

The pH at which capacities are measured is controlled by buffering the measuring-salt solution at the required pH.

pH 5: typically a buffer of acetic acid with sodium, calcium or ammonium acetate. This value is used to measure permanent negative charges on clay minerals. In soils some small component of pH-dependent charge is likely to be present also, notably on organic materials.

pH 7: no simple buffers exist at this pH; most common weak acids or weak bases are too strong to buffer at pH 7. Sometimes neutralized ammonium chloride or acetate solutions are used instead, relying on repeated treatments to bring the soil sample near to pH 7. This is not normally successful unless the soil has a low capacity or is already close to pH 7. Capacities at pH 7 are useful because this is a suitable pH for the growth of many economic plants.

pH 8.5: buffers containing a weak base, for example ammonia-ammonium chloride, are used here. Capacities at pH 8.5 are relevant to soils containing calcium or magnesium carbonates, since it is a common pH for such soils when exchangeable sodium is low. It represents an upper practical limit to the cation exchange capacity attainable under field conditions since serious agricultural problems occur in soils more alkaline than this.

Sometimes a small proportion of a weak base is added to the measuring salt solution in the first steps of the cation saturation to raise the pH to 8.5. This is followed by a more dilute measuring salt solution (without the base to avoid the complication of having to analyse for two cations for exchange capacity) before displacing the measuring ions in another salt solution. This is the basis of Mehlich's method (1948) using triethanolamine and barium chloride. One possible problem with using organic-N bases is that many otherwise suitable ones are complexing agents likely to affect some exchange sites; the complexing of aluminium by triethanolamine is an example.

Natural exchangeable cations

In order to measure the amounts of natural exchangeable ions which are true to name – that is, are held in the soil as ions at charged sites – it is necessary to separate them from ions present as soluble salts, and the extraction of ions through non-exchange reactions has to be prevented. If these steps are not possible, then these unwanted ions have to be allowed for.

Removal of soluble salts. To do this the soil sample is washed with a suitable solvent; but this procedure is far from straightforward. Dilution of the soil solution leads to a change in the suite of cations. If sodium is as usual the major soluble cation, some exchangeable sodium will be lost in the washing procedure and this loss is aggravated by sodium being the most loosely bound of the main cations in soils. This error is minimized by using a solvent such as a mixture of ethanol and water, which has just sufficient power to dissolve the soluble salts present; this works because sodium and other cations are more tightly bound to the exchanger in such solvents than in water. Some loss of clay by dispersion may occur on washing out the salts from saline soils, but this may be reduced by including some conditioning agent such as polyvinyl alcohol in the solvent.

It is now known that, in some clay soils, pores with charged surfaces may occlude soluble ions so that they are only slowly released from soil aggregates when the soil solution concentration is low, but are released readily into less dilute solutions especially if they contain double-charged cations (Blackmore 1976). It is then possible that such occluded salts may simulate exchangeable ions, and be wrongly measured as them, by resisting removal by aqueous solvents, yet entering readily the displacing solution. This possibility remains to be tested, but the difficulty should be lessened by using wash solvents of lower dielectric constants, thereby decreasing the thickness of the double layer at the charged sites which occlude the salts.

Another problem arises with soluble chlorides in clay soils when non-aqueous solvents are used to remove these salts. Solvents such as glycol–ethanol mixtures are capable of dissolving sufficient amounts of chlorides, as free salts, to remove them completely from all but quite saline soils, but they may fail to do so in some cases. The amounts of chloride retained are positively correlated to the cation exchange capacities of these soils. Because of this phenomenon, all wash solvents should contain some water to ensure complete removal of all soluble salts.

An alternative to the removal of the salts is to estimate their concentration in the soil solution; for example in a saturation extract of the soil. This is a reasonable procedure where an analysis of the salt composition is desired. A difficulty with this method is in knowing how to divide the water in the wet soil between free solution and the double layer volume. This difficulty is also present if more dilute soil–water mixtures are used but it is less serious because of the larger volume of free solution; instead, the disturbance of the natural cation suite replaces it as the main source of error.

Exclusion of nonexchangeable cations. Apart from soluble salts, the chief errors from these unwanted ions arise from the solution of calcite, dolomite, and gypsum in the displacing solution, with minor contributions from less soluble minerals e.g. felspars and micas. The calcium and magnesium solubility from carbonate minerals can be reduced to a low level by using aqueous ethanol as the solvent and making the reagent alkaline. For example, the error in exchangeable calcium or magnesium of calcareous soils is reduced to +0.5me/100g less in molal ammonium chloride in 60% ethanol: 40% water solvent made to pH 8.5 with ammonia (Tucker 1954). This reagent can also be used for non-calcareous soils and has been suggested as a general reagent for both capacity and exchangeable cation measurements (Tucker 1974*b*). Alkaline reagents containing other displacing cations have also been used: Mehlich (1948) proposed barium chloride with triethanolamine in water and Yaalon *et al.* (1962) use lithium chloride–acetate in water. No direct method for gypseous soils has yet been developed. Calcium sulfate is less soluble if the reagent contains ethanol, so it is possible to obtain good values by washing the soil first with aqueous ethanol to remove other sulfates, then extracting the exchangeable cations with a displacing reagent in aqueous ethanol and finally analysing the extract for sulfate as well as the cations. The preliminary washing ensures that all sulfate in the extract is due to dissolved calcium sulfate and a reliable correction to the exchangeable calcium can be made on that basis.

Choice of displacing salt. For the general purpose extraction of basic exchangeable cations, there is no entirely satisfactory reagent, but some comments will be made on those in common use.

(i) Cation. It is convenient to use as a displacing cation

one which is not intended to be measured, or is not already present in the sample. Ammonium salts are often chosen; they are obtainable as quite pure reagents and do not generally cause analytical problems. They have been avoided by some workers either because ammonium releases specifically adsorbed potassium or because it is itself specifically adsorbed and hinders the release of potassium from some soil minerals. The writer believes the hindrance effect may be found in some Australian soils which contain illite minerals so that specifically adsorbed potassium is underestimated; such potassium can be displaced by hydrogen ions and is available to plants. Apart from hydrogen no other cation will displace this potassium as well as ammonium does, so there is no alternative if this potassium is to be included in the measured cation suite e.g. for available potassium studies.

Lithium is sometimes used but it is a weak displacer of double-charged cations and it is not easy to obtain a pure reagent. Barium (and occasionally strontium) is a popular choice; it is a strong displacer so that relatively dilute reagents can be used, but it may displace calcium and magnesium from their carbonate minerals when they occur in the silt and clay fractions of the soil.

Organic-N cations offer some possibilities, but their large ionic sizes and, in the case of most quaternary bases, their low solubility restrict their usefulness. Methylammonium is one likely cation but it has not been accepted for general use. For special purposes, the particular properties of some cations may be utilized; for example large organic cations may be used to separate cations held at open sites from those at less accessible sites (Tucker, 1975).

(ii) Anion. The anions of strong acids are necessary. Weak acid anions, such as acetate, favour the dissolution of cations from non-exchange sources in minerals. Chloride is the most usual choice; it is easily analysed when anion exchange capacities are to be measured after extraction of the exchangeable anions but natural exchangeable chloride cannot then be estimated. Nitrate is a useful alternative to chloride.

(iii) Solvent. Water is the usual solvent, but there is an advantage in using alcohol/water mixtures because the dissolution of minerals is greatly reduced in such solvents. Ion exchange occurs quite readily in solvents such as 60% ethanol/40% water. Some non-aqueous solvents may be used for special purposes, e.g. in the partition procedures, but the number of such solvents with a sufficient ability to hold the displaced cations in solution and to permit ionic reactions is rather limited. Methanol, ethanol, ethanediol (glycol) and their mixtures are among the few to be used successfully.

(iv) Acidity. Hydrogen ions are very effective displacers of exchangeable cations since their entry to all kinds of sites seems to be unrestricted. They are never used where the intention is to relate the extracted cations to surface charges because of the inevitable large amounts of cations they dissolve from minerals. Cations displaced from soils by hydrogen ions (dilute acids) are, however, measured in plant nutrition research. Reagents for extracting basic exchangeable cations are always prepared in the range pH 7 to 8.5 mainly to avoid solution of mineral cations. Above pH 8.5, magnesium is more difficult to displace.

Complications in cation exchange

From what has been said so far, it will be seen that theories of ion exchange are based on ideal models, and that measurements of exchange properties are usually made with a straightforward view of the reactions involved but taking proper precautions against obvious sources of error. There are however some complications to this simple scene, which have not been studied extensively but which may be useful in the management of the cation status of field soils.

Hysteresis of exchange

It has been observed in laboratory experiments that when a soil sample, already saturated with cation A, is treated with cation B then this second cation, B, is more easily replaced by other cations than it would be if a B-saturated soil had been treated with cation A, for similar proportions of the two adsorbed cations. Put in another way the equilibrium solution concentration ratio B/A will be greater when B has entered the exchange sites after A than when the entry order is reversed. Marshall (1964) says that this imperfect reversibility is to be expected when each cation can be held with a range of bonding energies, because then more than one composition of the cation suite can be in true equilibrium with a given soil solution. However this is a complex phenomenon; it has bee n observed for pairs of cations with different charges, but it apparently does not occur between pairs of basic exchangeable cations of the same charge.

This phenomenon does not appear to have been studied recently; but if it occurs in field situations it would be important there. The availability and retention of a fertilizer cation in a soil would then depend on whether it was applied separately or with other cations. The effect of a given level of exchangeable sodium on the physical properties of soils may depend on whether it was naturally present, had been derived from irrigation waters, or was a residue after treatment with calcium amendments.

Some aspects of cation exchange revealed by the partition of cations into subsets have a similarity to hysteresis. Sodium has been found to have a larger proportion in the conditional subset if the soil is saturated with sodium before partial replacement by calcium than if sodium is added to a calcium-saturated soil. This agrees with the observations that these two cations prefer to enter the primary subset and to displace previously adsorbed cations from that subset rather than from the conditional subset (see above). As in hysteresis, the exchange properties depend on the sequence of cation entry; but the property here is the conditions under which a cation may be exchanged instead of relative equilibrium concentration in the soil solution.

Demixing of cations

In a review of cation exchange, Wiegner (1935) noted that for ions of equal charge, their ease of displacement from inorganic exchangers was related to their degree of hydration in solution. This is a useful general principle; highly hydrated ions tend to be weakly bound by clays. Wiegner's explanation of site strength in terms of particle surface morphology (edges, planes, flaws, etc.) is not favoured now, although Kelley (1948) did dismiss his concept too easily.

Jarusov's rule that ions with the highest charge prefer to occupy the strongest sites, and that weaker sites tend to be occupied by cations with lesser charge is based on ideas expressed in his paper (Jarusov 1937). Actually he views positions close to the surface as being the strong sites with weaker sites further away, which is reminiscent of the double-layer picture of ion distribution. This is an important rule which is consistent with the requirement that at equilibrium the cation distribution should have a minimum of free energy. Its consequence is a segregation of cations with each species tending to occupy different zones of the clay surfaces.

Recently Fink *et al.* (1971) have been able to demonstrate that such demixing of cations does occur over some ranges of cation composition on sodium-calcium bentonite. Their work is based on the differential swelling of sodium and calcium-bearing interlayers. This is not necessarily due to intrinsic differences in site strength. It may be that interlayer regions with Ca^{2+} already in place will favour the entry of more Ca^{2+} and reject Na^+; an effect analogous to, but weaker than, the interstratification observed in mica-type minerals (Sawhney 1972). There the adsorption of a strongly bound ion (K^+) in one interlayer is thought to weaken the bonding in adjacent interlayer spaces which then tend to take up other weaker-bound, more hydrated cations (Ca^{2+}, Mg^{2+}).

Complex reactions

There is a little evidence that the release of specifically bound cations may be through a sequence of reactions rather than through a simple exchange of one cation for another. This was suggested for the release of potassium from soil illites (Tucker 1964) the sequence being an essential first reaction:

$$H_3O^+ + K_s \rightarrow K^+ + H_3O_s$$

followed by n reactions of the form

$$Ca^{2+} + 2K_s \rightarrow 2K^+ + Ca_s$$

when n appears to be 3 in accord with an hexagonal arrangement of K in the interlayer region.

A comparable reaction appears to be involved in the release of magnesium from a mineral vermiculite and from a soil clay containing chlorite and illite. Here one hydrogen ion is associated with the release of four Mg^{2+} ions (Figs 27.2*a* and *b*). Above pH 8.5 or 9 the release and uptake of magnesium is dominated by the reaction with $MgOH^+$ cations.

In considering complex ionic reactions like these which are determined by structural arrangement of the cations on crystalline mineral surfaces, we come close to thinking of cation exchange as the solubility of a surface mineral phase, a topic dealt with in Chapter 26. In ion exchange, the carrier of the exchange sites is considered

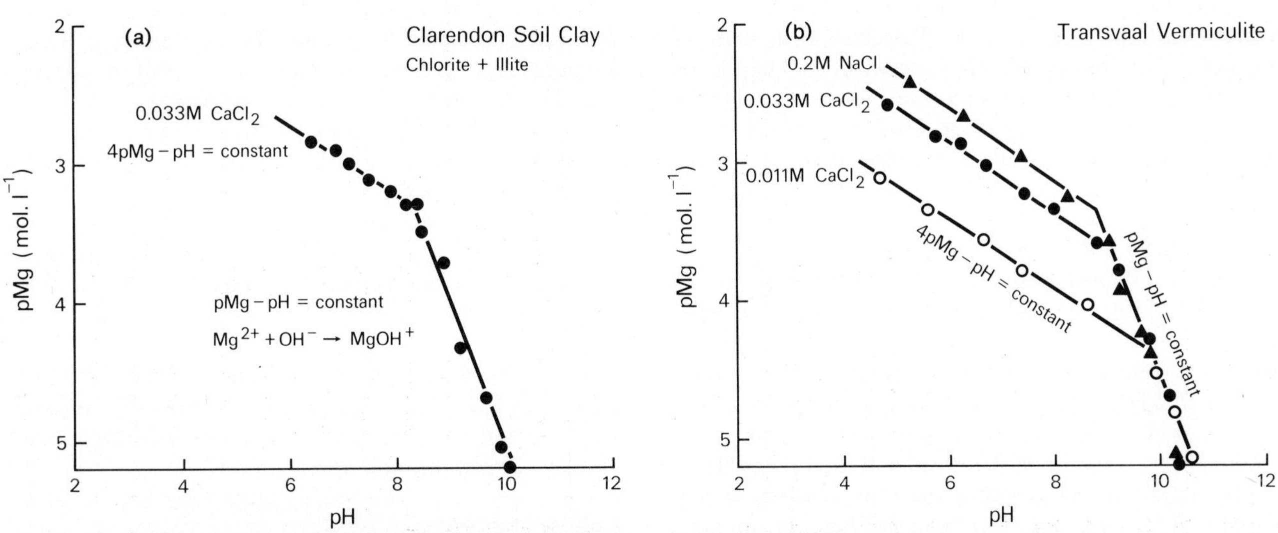

Fig. 27.2. Magnesium equilibria with hydrogen and calcium or sodium ions. *a.* (left) clay from Clarendon soil containing chlorite and illite. *b.* (right) mineral vermiculite from the Transvaal.

to be unaffected by the reaction; in solubility, the solid phase as a whole enters the solution. A third kind of reaction, called by the indefinite term, cation release, and due to a subtle change in the solid phase such as a loss of charge or a structural rearrangement, may also cause a change in the concentration of cations in solution without the solid actually being dissolved at the point of release. In practice it may be difficult to distinguish between these three kinds of reaction in soil minerals.

Exposition of exchange measurements

Much effort continues to go into devising new ways of measuring exchange properties of soils and perhaps the time has come to accept current methods. Attention must now be given to the more difficult problem of handling the mass of information so easily accumulated. Using conventional methods for each soil analysed, consisting of say five horizon samples, one obtains a table containing 25 values (4 cations, 1 capacity per sample), even two or three times that if partition of the cations is carried out. If Beckett's *Q/I* approach is used then one has for each cation, a set of diagrams containing one curve for each horizon. There is as yet no way of fully utilizing so much information.

One fundamental difficulty lies in the lack of a unifying concept which can express the interdependence of soil horizons. This means that we cannot expect to have a well-founded procedure for handling chemical information on soil profiles.

One possible expedient would be to use a code similar to that worked out for the morphology of Australian soils by Northcote (1971), which is based on the useful idea of profile trends (texture and pH) as well as other important features. However such a code ought to be able to rank soil profiles for selected cation exchange properties so that correlations with other soil properties such as nutrient availability or permeability are feasible. The cluster analysis procedures also offer a possible way of handling exchange information. But not too much should be expected of either approach in the absence of a rational basis for their use.

Cation ratios

Instead of considering values for the whole profile, it is usual to select one important horizon e.g. for plant nutrition, the surface horizons; for physical and engineering properties, a major clay horizon. In such an horizon a cation ratio is often used as an indicator of soil properties. The use of the ratio of a prime cation to the other cations is based on the ideas that the other cations oppose or dilute the action of the prime cation, or that they provide a reference level in a soil equilibrium state from which its action may be judged. These ideas can be seen as an application of the proposition, referred to in the section on ion exchange equilibria, that activities of cations in the solid exchanger are proportional to their equivalent or mole fractions. If demixing of cations is normal in soils it might be argued that actual values of the prime cations are likely to be more closely related to other soil properties than the ratios are.

A ratio of exchangeable sodium to cation exchange capacity (by Na at pH 8.2) of 0.15, or above, was used to define sodic soils (Richards 1954) without specifying horizons. Recently the same index has been used to define natric horizons (Soil Survey Staff 1975). The Na/CEC (cation exchange capacity) ratio, usually expressed as the exchangeable sodium percentage (ESP), has been used quite widely for assessing the likely physical behaviour of soil horizons under irrigation since there is a close relation, based on Gapon's equation, between this ratio and the concentration ratio $[Na^+]/[Ca^{2+} + Mg^{2+}]^{1/2}$ in moles per litre of the soil solution. Within a group of related or similar soils there is usually a close relation between soil permeability and clay dispersion, and the ESP. The values of ESP at which the adverse effects of exchangeable sodium become serious differ between groups of soil depending on the clay minerals and the amounts of organic matter which are present. In some soils Schulz *et al.* (1964) found that not all exchangeable sodium contributes to the sodic effects so that it is necessary to divide the sodium into effective and ineffective fractions before calculating an effective Na/CEC ratio. On the other hand Bower and Rhoades (1972) found that in vermiculite soils those exchange sites able to fix potassium are less involved in soil swelling. In such cases it was better to use a reduced exchange capacity that excluded such sites when relating ESP to hydraulic conductivities. It is also known that the relation between ESP and physical properties of the soils is considerably modified if the other cations have more magnesium than calcium (Bakker and Emerson 1973), but more work is needed to define this Ca–Mg–Na interaction. An analogous ratio for potassium, K/CEC, would be less useful in assessing the effect of potassium on soil properties because in many Australian soils this cation is bound at sites with greatly differing exchange constants. Instead, much work has been done on solutions in equilibrium with the soil using especially the potassium activity ratio $(K^+)/(Ca^{2+} + Mg^{2+})^{1/2}$, which will be recognized as the solution side of a Gapon-type equation. In addition graphical plots of $(K^+)/(Ca^{2+})^{1/2}$ in solution against changes in exchangeable potassium as K is added to or taken from the soil – the *Q/I* plots of Beckett (1964) – have been used to evaluate the availability of potassium to plants with more success than the use of simple activity ratios in the natural soil solution.

Cation cascades

Exchange relationships in soils may also be presented as cascade diagrams (Chorley and Kennedy 1971) which

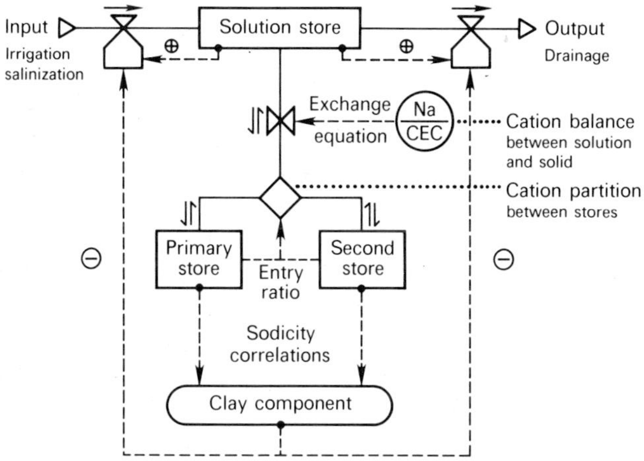

Fig. 27.3. Sodium cascade. ———, sodium ion flows; - - - - - -, information flows. Exchange equations express balance between sodium and its competitor cations as activities in solution and in the solid phases. Entry ratios express distribution of sodium between the available cation exchange stores (subsets) as a function of the relative content of sodium already in the stores. Sodicity correlations express the relation between sodium in the various stores and clay dispersion effects. ⊖, negative feedback, increasing sodicity of clay decreases flow of cations in solution. ⊕ positive feedback, increasing solution concentration increases solution flow.

show the connexions between ion exchange and other soil processes, and indicate the most needed areas of research, or the most effective points for soil management. The unit of observation of a cascade system would be initially a soil structural unit or an horizon, broadening eventually to solums and soil landscapes. A simple example is shown in Fig. 27.3. In this cascade the exchange capacities of the subsets are treated as stores of cations; the controls on the movement of cations between these stores and the soil solution are described as information transfers customarily modelled as exchange equations and subset distribution functions. The presence of cations in the stores affects the physical properties of the solid components of the system, especially of some clay minerals by a sodicity correlation in the case of exchangeable sodium and in turn these properties affect the movement of cations into and out of the soil system unit as an information feedback. For sodium this feedback is negative, more sodium in the stores leading to less sodium entering the soil. For calcium the feedback is positive.

The sodium cascade in Fig. 27.3 is only a starting point. It is brought nearer to the field situation by recognizing delays and restrictions in the flow of cations – slow diffusion of cations, slow movement of soil solution. In other systems, delays lead to overshooting and hunting for steady states; in soil units they could lead to locking of the system in a non-uniform state e.g. external parts of structural units having different composition to their interiors as may be observed in some features of the soil fabric (Chapter 30) and in the trapping of salts within soil aggregates (Blackmore 1976). These system diagrams are open to modification as knowledge and concepts of ion exchange are improved; but at present they offer a possible way of relating soil chemistry to other areas of soil science.

Australian soils

It is part of the folklore of Australian soil science that our soils contain, in general, relatively large proportions of magnesium and sodium compared with soils in other parts of the world. The comparison, naturally enough for people endeavouring to establish a European style of land use, is with the soils of Europe and North America, lands with a very different geological character and climatic history. The first observations of a distinctive cation status appear in Prescott's essay (1931), and have been repeated since then in general discussions of Australian soils e.g. Stephens and Donald (1958). Impressions of this kind were based on limited and selective data; it is only a decade since a mapping of Australian soils on a continental scale has been made (Northcote *et al.* 1960–68). This was based on profile morphology and land form and has not yet been matched by a comparable study of physical and chemical soil properties.

It is worthwhile comparing the cation status of Australian and United States soils. For Australia we can use the collection of analyses in the Handbook of Australian Soils (Stace *et al.* 1968); three quarters of the profiles presented lie to the south-east of a line from Rockhampton to Adelaide, under warm temperate climates, while the soils of tropical, arid, and western parts of Australia are less well represented. For the U.S. we use the data given in 'Soil Taxonomy' (Soil Survey Staff, 1975) but exclude pedons from Alaska, Hawaii, Puerto Rico, South America and Europe. This gives a comparable group of soils to that in the Handbook with a similar climatic range except for a few pedons from cool climates.

For comparison of calcium and magnesium, the profiles (pedons) or horizons containing carbonates have been left out, mainly because of some uncertainty on the reliability of the data in some of these soils. Fig. 27.4*a* and *b* shows little difference in the totals of basic exchangeable cations between the soils of the two countries either in the surface samples or in the profiles as a whole. Fig. 27.4*c* shows that Australia does have more soils with a comparatively high exchangeable magnesium content in the surface horizons, i.e. Mg is more nearly equal to Ca, compared with United States soils. Fig. 27.4*d* illustrates the universal trend for the proportion of magnesium to increase with depth in the soil. This trend is greater in Australian soils than in the U.S.

The effect of exchangeable sodium in soils is usually described in terms of sodium as a percentage of the cation exchange capacity. Here we compare the ESP in the A or B horizons having the maximum clay content

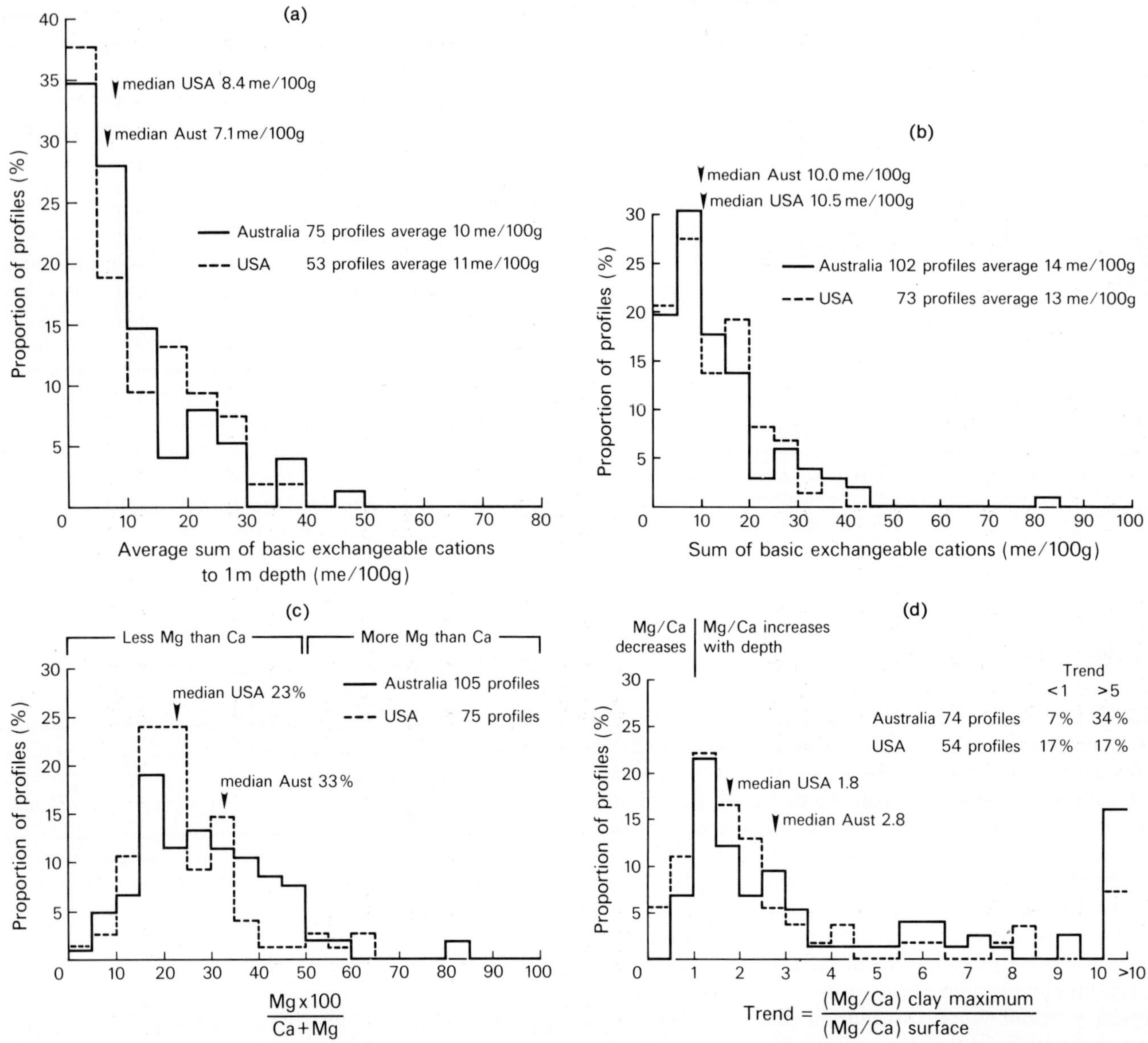

Fig. 27.4. Comparisons of cation exchange properties of Australian and United States non-carbonate soils. *a*. Average content of basic exchangeable cations to 1 metre depth. *b*. Sum of basic exchangeable cations in surface horizons. *c*. Exchangeable magnesium as a fraction of calcium plus magnesium in surface horizons. *d*. Trend of the magnesium/calcium ratio from the surface to the horizon of maximum clay content.

since the major effect of excessive sodium is on the physical behaviour of clay horizons. The exchange capacity values have been used as quoted – the pH range may be 7 to 8.5; the measuring cations are usually ammonium in Australia, but may also be sodium or barium in the U.S.A. For better comparability, ammonium values have been used where they were available, although the standard definition of ESP in America is based on sodium exchange capacities at pH 8.2. Fig. 27.5*a* shows that high-sodium soils are much more common in Australia than in the U.S.

It is the experience of Australian pedologists that the effects of sodium are apparent in profile morphology and in physical properties of soils when the ESP of an horizon is 7% or more (Northcote and Skene 1972). This agrees with the observations of Emerson (1967) that soil aggregates begin to disperse in water at similar ESP values. On this basis, half of the Australian soils listed in the Handbook should show some evidence of sodicity in at least one horizon. American pedologists set a higher value of ESP = 15 as one of the criteria for the appearance of significant sodicity (definitions of natric horizons, Soil Survey Staff 1975). One reason for the lower limit in Australian soils may be the generally higher level of

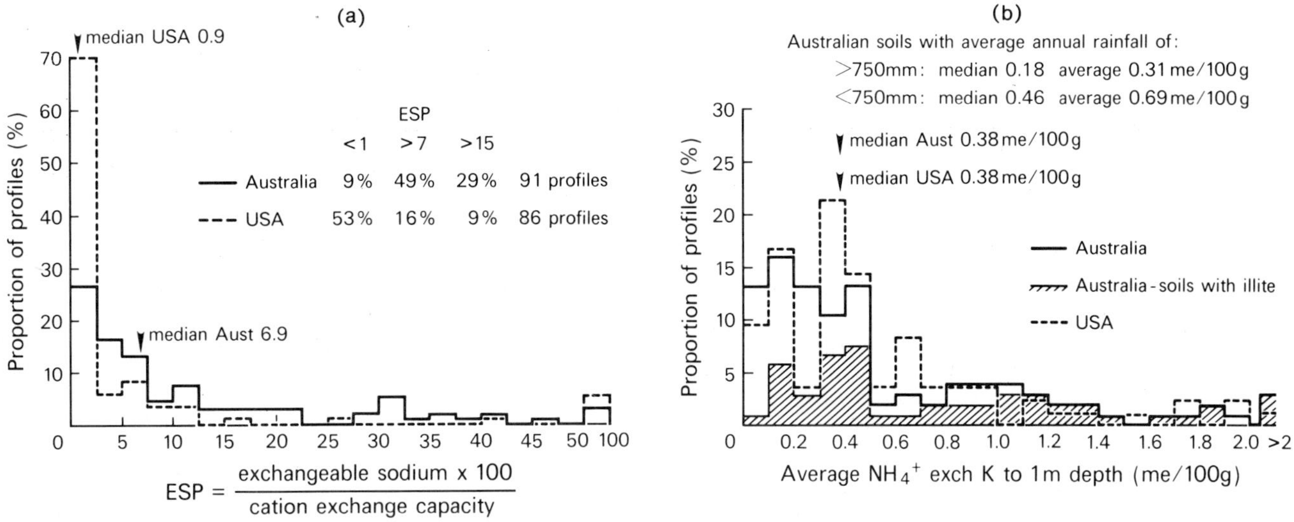

Fig. 27.5. Comparisons of exchangeable sodium and potassium in Australian and U.S. soils. *a.* Exchangeable sodium percentage at maximum clay content in A + B horizons. *b.* Average content of potassium exchanged by ammonium to 1 metre depth.

exchangeable magnesium, which is known to affect soils adversely in the presence of sodium (Bakker and Emerson 1973). Emerson believes the effect of magnesium on clay dispersion is due to its greater affinity for water so that it is less strongly bound to exchange sites than calcium is, and not to the presence of $MgOH^+$, which in any case is not likely to be significant below pH 8 in soils (see Fig. 27.2). The high levels of sodium and magnesium in many soils are a real handicap to Australian agriculture and the problems they raise in soil management will be discussed in a later chapter.

The appearance of abnormal sodium and magnesium in the duplex soils of south-eastern Australia, notably those with bleached A_2 horizons characteristic of a history of sodium effects, has been explained by Downes (1954) as being due to the accumulation of wind-borne salts from the oceans during an arid part of the Quaternary period, followed by a period of wet climates when these salts were removed to leave soils with moderate exchangeable sodium and magnesium, and strongly developed texture-contrast profiles. Elsewhere in southern Australia, incursions of the sea in Miocene times (the Murravian and Eucla Gulfs) left salts in the soils which now have highly sodic gradational profiles instead of the duplex soils of the higher rainfall areas. In addition the contemporary lack of drainage to the sea in much of Australia, and the redistribution of terrestrial salts through the landscape by the wind, tend to maintain a high level of sodicity in our modern soils.

The Australian and U.S. soils have very similar contents of ammonium-exchangeable potassium in the upper metre of the solum. The median and average values are virtually the same (Fig. 27.5*b*). For the third of the Australian profiles in the Handbook with average annual rainfalls of 750 mm or more, the potassium levels are much lower than in soils of the drier areas. The medians are 0.18 and 0.46 me/100g, and the averages 0.31 and 0.69 me/100g.

In the Australian soils high levels of exchangeable potassium, averaging above 1 me/100g in the one-metre depth profile, are only found where illite clay minerals are present, except in one soil which contains partly decomposed basalt at shallow depth. However the presence of illite does not guarantee that much exchangeable potassium will be found in the soil. Some illites may contain fair amounts of potassium which is not easily displaced even by ammonium ions. Optimum exchangeability occurs in illites which have moderate amounts of potassium and moderately degraded structures. Potassium is less exchangeable from illites with lower contents and more degraded structures, and from well-crystallized illites with higher potassium contents. The tables given by Tucker (1967) show some of the complex interactions of displacing cation, potassium content, and illite crystallinity for some Australian soil clays.

References

André, J. P. (1975). Graphical representation of tricationic exchanges where one of the cations is the proton. *J. Soil Sci.* **26**, 81-92.

Adams, F. (1971). Ionic concentrations and activities in soil solutions. *Soil Sci. Soc. Am. Proc.* **35**, 420-6.

Bakker, A. C. and Emerson, W. W. (1973). The comparative effects of exchangeable calcium, magnesium and sodium on some physical properties of red-brown earth subsoils. III. The permeability of Shepparton soil and comparison of methods. *Aust. J. Soil Res.* **11**, 159-65.

Beckett, P. H. T. (1964). Studies on soil potassium 2. The 'immediate' Q/I relations of labile potassium in the soil. *J. Soil Sci.* **15**, 9-23.

Beckett, P. H. T. (1965). The cation exchange equilibria of calcium and magnesium. *Soil Sci.* **100**, 118-23.

Beckett, P. H. T. and Nafady, M. H. M. (1967). Potassium-calcium exchange equilibria in soils: the location of non-specific (Gapon) and specific exchange sites. *J. Soil Sci.* **18**, 263-81.

Blackmore, A. V. (1976). Salt sieving within clay aggregates. *Aust. J. Soil Res.* **14**, 149-58.

Bolt, G. H. (1967). Cation-exchange equations used in soil science – A review. *Neth. J. Agric. Sci.* **15**, 81-103.
Bolt, G. H. and Bruggenwert, M. G. M. (eds) (1976). 'Soil Chemistry. A. Basic Elements'. Developments in Soil Science 5A. (Elsevier: Amsterdam.)
Bower, C. A. and Rhoades, J. D. (1972). Fractionation of cation exchange capacity for assessing soil and water sodicity. *Soil Sci. Soc. Am. Proc.* **36**, 174-5.
Carlson, R. M. and Buchanan, J. R. (1973). Calcium-magnesium-potassium equilibria in some California soils. *Soil Sci. Soc. Am. Proc.* **37**, 851-5.
Chorley, R. J. and Kennedy, B. A. (1971). 'Physical Geography. A Systems Approach'. (Prentice-Hall International: London.)
Curtin, D. and Smillie, G. W. (1976). Estimation of components of soil cation exchange capacity from measurements of specific surface and organic matter. *Soil. Sci. Soc. Am. J.* **40**, 461-2.
Davis, L. E. (1950). Ionic exchange and statistical thermodynamics. I. Equilibria in simple exchange systems. *J. Colloid. Sci.* **5**, 71-9.
Downes, R. G. (1954). Cyclic salt as a dominant factor in the genesis of soils in south-eastern Australia. *Aust. J. Agric. Res.* **5**, 448-64.
Dranoff, J. and Lapidus, L. (1957). Equilibrium in ternary ion exchange systems. *Ind. Eng. Chem.* **49**, 1297-302.
Ehlers, W., Gebhardt, H. and Meyer, B. (1967a). K-Austausch und kristallo-graphisches Verhalten der Dreischicht-Tonminerale. *Z. Pflanzenernähr. Bodenkd.* **117**, 29-53.
Ehlers , W., Meyer, B. and Scheffer, F. (1967b). K-Selectivität und Fraktionierung des Austauschkaliums. *Z. Pflanzenernähr. Bodenkd.* **117**, 1-29.
Elprince, A. M. and Babcock, K. L. (1975). Prediction of ion-exchange equilibria in aqueous systems with more than two counter-ions. *Soil Sci.* **120**, 332-8.
Emerson, W. W. (1967). A classification of soil aggregates based on their coherence in water. *Aust. J. Soil Res.* **5**, 47-57.
Fink , D. H., Nakayama, F. S. and McNeal, B. L. (1971). Demixing of exchangeable cations in free-swelling bentonite clay. *Soil Sci. Soc. Am. Proc.* **35**, 552-5.
Gaines, Jr, G. L. and Thomas, H. C. (1953). Adsorption studies on clay minerals. II. A formulation of the thermodynamics of exchange adsorption. *J. Phys. Chem.* **21**, 714-8.
Galkina, N. K., Rubinshtein, R. N. and Senyavin, M. M. (1962). Exchange of ions in mixtures under static conditions. *Russ. J. Phys. Chem.* **36**, 1006-11.
Gapon, E. N. (1933). Theory of exchange adsorption in soils. *J. Gen. Chem. USSR* **3** (2) 144-58.
Helfferich, F. (1962). 'Ion Exchange'. (McGraw-Hill: New York.)
Jangg, G. (1958). Berechnung von ternären lonenaustausch-Gleichgewichten. *Österr. Chem.* **59**, 331-4.
Jarusov, S. S. (1937). On the mobility of exchangeable cations in the soil. *Soil Sci.* **43**, 285-303.
Kelly, W. P. (1948). 'Cation Exchange in Soils'. p. 60. (Reinhold: New York.)
Kittrick, J. A. (1976). The separation factor applied to some soil ion exchange equilibria. *Soil Sci. Soc. Am. J.* **40**, 147-48.
Krishnamoorthy, C. and Overstreet, R. (1950). An experimental evaluation of ion exchange relationships. *Soil Sci.* **69**, 41-53.
Laudelout, H., van Bladel, R. and Robeyns, J. (1972). Hydration of cations adsorbed on a clay surface from the effect of water activity on ion exchange selectivity. *Soil Sci. Soc. Am. Proc.* **36**, 30-4.
Marshall, C. E. (1964). 'The Physical Chemistry and Mineralogy of Soils'. Vol. 1. Soil Materials. (Wiley: New York.)
Mehlich, A. (1948). Determination of cation and anion exchange properties of soils. *Soil Sci.* **66**, 429-45.
Moss, P. (1969). A comparison of potassium-activity ratios derived from equilibration procedures and from measurements on displaced soil solution. *J. Soil Sci.* **20**, 297-306.
Munns, D. N. (1976). Heterovalent cation exchange equilibria in soils with variable and heterogeneous charge. *Soil Sci. Soc. Am. J.* **40**, 841-5.
Northcote, K. H. (1971). 'A Factual Key for the Recognition of Australian Soils'. 3rd edn. (Rellim: Adelaide.)
Northcote, K. H. and Skene, J. K. M. (1972). Australian soils with saline and sodic properties. CSIRO Aust. Soil Publ. No. 27.
Northcote, K. H. *et al.* (1960–1968). 'Atlas of Australian Soils'. (CSIRO and Melbourne Univ. Press, Melbourne.)
Papanicolaou, E. P. and Overstreet, R. (1969). The determination of cation exchange capacity over a wide pH range using various index cations. *Z. Pflanzenernähr. Bodenkd.* **123**, 205-11.
Prescott, J. A. (1931). The soils of Australia in relation to vegetation and climate. CSIR Aust. Bull. No. 52.
Richards, L. A. (ed.) (1954). Diagnosis and improvement of saline and alkali soils. U.S. Dept. Agric. Hbk 60.
Sawhney, B. L. (1972). Selective sorption and fixation of cations by clay minerals: a review. *Clays Clay Miner.* **20**, 93-100.
Sawhney, B. L., Frink, C. R. and Hill, D. E. (1970). Components of pH dependent cation exchange capacity. *Soil Sci.* **109**, 272-8.
Schulz, R. K., Overstreet, R. and Barshad, I. (1964). Some unusual ionic exchange properties of sodium in certain salt-affected soils. *Soil Sci.* **99**, 161-5.
Soil Survey Staff (1975). 'Soil Taxonomy'. U.S. Dept. Agric. Hbk 436.
Sposito, G. and Mattigod, S. V. (1977). On the chemical foundation of the sodium adsorption ratio. *Soil Sci. Soc. Am. J.* **41**, 323-9.
Stace, H. C. T. *et al.* (1968). 'A Handbook of Australian Soils'. (Rellim: Adelaide.)
Stephens, C. G. and Donald, C. M. (1958). Australian soils and their response to fertilizers. *Adv. Agron.* **10**, 167-256.
Tucker, B. M. (1954). The determination of exchangeable calcium and magnesium in carbonate soils. *Aust. J. Agric. Sci.* **5**, 706-15.
Tucker, B. M. (1964). The solubility of potassium from soil illites. II. Mechanisms of potassium release. *Aust. J. Soil Res.* **2**, 67-75.
Tucker, B. M. (1967). The solubility of potassium from soil illites. III. Reactivity towards other ions. *Aust. J. Soil Res.* **5**, 173-90.
Tucker, B. M. (1974*a*). Displacement of ammonium ions for cation exchange capacity measurements. *J. Soil Sci.* **25**, 333-7.
Tucker, B. M. (1974*b*). Laboratory procedures for cation exchange measurements on soils. CSIRO Aust. Div. Soils Tech. Pap. No. 23.
Tucker, B. M. (1975). The partition of basic exchangeable cations. *Aust. J. Soil Res.* **13**, 77-89.
van Schouwenburg, J. Ch. and Schuffelen, A. C. (1963). Potassium-exchange behaviour of an illite. *Neth. J. Agric. Sci.* **11**, 13-22.
Wiegner, G. (1935). Ionenumtausch und Struktur. Trans. 3rd Int. Cong. Soil Sci. (Oxford). Vol. 3, pp. 5-28.
Yaalon, D. H., van Schuylenborgh, J. and Slager, S. (1962). The determination of cation-exchange characteristics of saline and calcareous soils. *Neth. J. Agric. Sci.* **10**, 217-22.

Chapter 28
Oxidation–reduction reactions

I. R. Willett

SOILS BECOME WATERLOGGED WHEN WATER CANNOT drain quickly enough from the soil profile because of its impervious nature, or because of the presence of a shallow watertable. The watertable may be permanent, as in riverine floodplains, or temporarily perched over an impermeable horizon (e.g. fragipans, iron pans, or dense clay B horizons) in periods of heavy rainfall. Soils are deliberately waterlogged for paddy rice cultivation, and excessive irrigation of other crops may also cause waterlogging.

Under these hydrological conditions, chemical reactions involving the transfer of electrons from organic matter to other soil components (oxidation – reduction, or redox, reactions) are of particular importance in soil formation and in the supply of nutrients to plants.

The diffusion of O_2 through soils is drastically reduced when they are waterlogged (Lemon and Kristensen 1960). Aerobic microorganisms reduce O_2 at a faster rate than it can diffuse into the soil, and an anaerobic environment, relatively rich in CO_2, develops. Facultative anaerobic bacteria then turn to oxidized soil components to replace O_2 as an electron acceptor, thus initiating soil reduction. Recent research (Smith 1976*a,b*) indicates that redox reactions may be of greater significance than previously accepted in soils normally considered well aerated, because of anaerobic micro-sites within the soil fabric.

Soil redox potentials

A redox reaction involves the transfer of electrons from a donor to an acceptor. The redox potential (*Eh*) is a quantitative measure of the tendency of a particular system to accept or donate electrons (reduce or oxidize) and at equilibrium can be expressed as:

$$Eh = E_o + \frac{RT}{nF} \ln \frac{(Ox)}{(Red)}$$

where (Ox) and (Red) are the activities of the oxidized and reduced species of a redox couple; E_o the standard potential (voltage when $(Ox) = (Red)$); n the number of electrons tranferred; F the Faraday constant; R the gas constant, and T the absolute temperature.

Large positive values indicate strongly oxidizing systems, whereas negative values represent reducing systems. The term 'reduced' has been applied to oxygen-free soils with small positive *Eh* values, as well as to very highly reduced soils with large negative *Eh* values. Patrick and Mahapatra (1968) presented the following scheme to define soil redox conditions in terms of *Eh* values: oxidized soils, $Eh > 0.4$ V; moderately reduced soils, $Eh = 0.4$ to 0.2 V; reduced soils, $Eh = 0.1$ to –0.1 V; and highly reduced soils, $Eh = -0.1$ to –0.3 V.

There has been a recent trend to use the minus $\log_{10}$ of the electron activity (pe) instead of *Eh*, where $pe = Eh \times 0.059$ (V) at 25°C. The term *pe* is more convenient than *Eh* when used in chemical calculations; however, the more familiar *Eh* values have been adopted for this discussion.

Since *Eh* is governed by the ratio of (Ox) and (Red), it represents an intensity measurement, and is analogous to pH in reactions involving transfer of protons. The resistance to change in *Eh* is termed 'poise', and is analogous to pH buffering; poise increases with (Ox) and (Red). In aerated soils, (Ox), and (Red) are low, and the poise is low, and this makes reliable *Eh* measurements difficult. On the other hand in highly reduced soils (Red) is relatively high, and *Eh* measurements are easier to make and are more reproducible.

Soil *Eh* is measured as the potential (E) between an inert platinum electrode and a standard reference electrode (usually the saturated calomel electrode) inserted into the soil. The potential of the inert electrode should be governed only by its role as an electron acceptor or donor, and is usually measured by a pH meter. The *Eh* is obtained by adding the value of the potential of the standard calomel electrode relative to the potential of the standard hydrogen electrode (at the temperature of the measurement), to E.

In principle, *Eh* measurements should quantitatively define the redox status of soils. However, several severe limitations are found when applying *Eh* concepts to soils and other natural systems. Bohn (1971) concluded that soil *Eh* values are nonequilibrium potentials and that this invalidates any strict thermodynamic interpretation of them. He also pointed out that for a redox couple to exert itself at an inert electrode surface, (Ox) and (Red) must be present at relatively high concentrations. This is frequently not the case in soils, where the bulk of the compounds participating in redox reactions are solids (e.g. $Fe(OH)_3$ and $Fe(OH)_2$) and have no influence on inert electrodes. He also considered soil *Eh* values to be mixed potentials, i.e. the (*Ox*) and (*Red*) species influencing *Eh* do not belong to the same redox couple; this also makes quantitative treatment of soil *Eh* values extremely complex.

Soil *Eh* measurements are subject to liquid junction potential errors, which Ponnamperuma *et al.* (1967) suggested are minimized if measurements are made in filtered soil solutions. This method also helps to overcome *Eh* variations in soils, which are caused by the uneven distribution of organic matter and bacteria. However, it is extremely difficult to prevent O_2 contamination of reduced soil solutions, and this may account for the observation that *in situ Eh* values of soils are frequently 0.10–0.30 V lower than those measured in extracted soil solutions (van Breemen 1969).

In accordance with these limitations, many workers have found that *Eh* measurements could not be quantitatively interpreted in soil systems (Bohn 1969; Brümmer 1974). However, in reduced soils in which Fe(III)–Fe(II) systems were dominant, Jeffery (1960) and Ponnamperuma *et al.* (1967) found *Eh* measurements to be quantitatively related to pH and Fe^{2+} as predicted by equilibria equations. In these studies, the activities of Fe(III) and Fe(II) species must have been high enough to dominate the reactions involving both electrons and protons, without interference from other redox couples nor buffering of pH by other soil components.

Further information on the application of *Eh* to soils may be found in Quispel (1947), Aomine (1962), Garrels and Christ (1965), Stumm (1965) and Bohn (1971). Examples of *Eh* measurements in some Australian soils are given by Williams (1972*a*, *b*) and Willett (1978).

The sequential reduction of soils

When a soil is flooded, there is a rapid decline in *Eh* as facultative and obligate anaerobic microorganisms use soil components as final electron acceptors in place of O_2. The stability of individual soil redox couples under the prevailing *Eh* is determined by their standard potential (E_o). Table 28.1. presents values of E_o for some redox couples expected in soils. After a soil is flooded, the couples should reduce in the sequence given in the table. In practice, O_2 and NO_3^- should rapidly disappear, followed by the appearance of soluble Mn^{2+} and Fe^{2+}, since the divalent oxides of these metals are more soluble than their respective higher oxides; sulfate should be reduced to sulfide, and CO_2 to CH_4, at much lower potentials. This sequence has been experimentally demonstrated although some overlapping does occur (Turner and Patrick 1968). The results presented in Fig. 28.1

Note: 0–15 cm plugs of soil were removed from a flooded intact core of Willbriggie clay loam (Dr 2.13). NO_3^- was anaerobically extracted with KCl, and Fe^{2+} and Mn^{2+} with ammonium acetate, (pH 4.8).

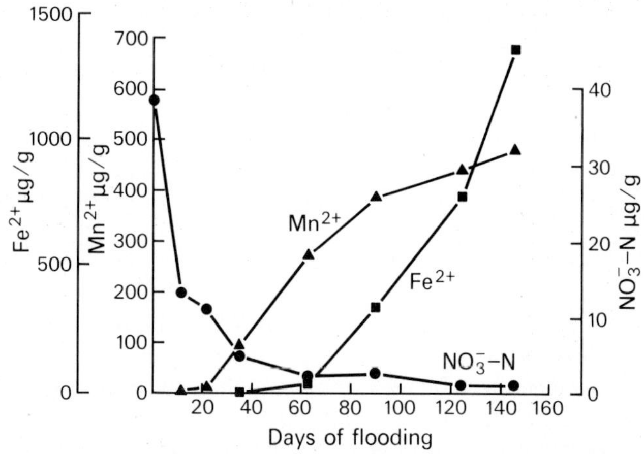

Fig. 28.1. Sequence of reduction of nitrate, Fe (III) and Mn (IV) in a rice soil of N.S.W.

Table 28.1. The standard potentials (E_0) of some soil redox systems (Ponnamperuma 1977*a*)

System	E_0(V) (pH = 7)
$O_2 + 4H^+ + 4e \rightleftharpoons 2H_2O$	0.814
$2NO_3^- + 12H^+ + 2e \rightleftharpoons N_2 + 8H_2O$	0.741
$MnO_2 + 4H^+ + 2e \rightleftharpoons Mn^{2+} + 2H_2O$	0.401
$Fe(OH)_3 + 3H^+ + e \rightleftharpoons Fe^{2+} + 3H_2O$	−0.185
$SO_4^{2-} + 10H^+ + 8e \rightleftharpoons H_2S + 4H_2O$	−0.214
$CO_2 + 8H^+ + 8e \rightleftharpoons CH_4 + 2H_2O$	−0.244
$2H^+ + 2e \rightleftharpoons H_2$	−0.413

show the sequence of the reduction of NO_3^-, Mn(IV) and Fe(III) in a typical rice-growing soil of the irrigation areas of New South Wales.

As reduction proceeds, each redox couple, if present in sufficient quantities in soluble form, tends to poise *Eh* at a value characteristic for that couple. Nitrate tends to stabilize *Eh* above 0.25 V and prevents further reduction (Gotoh 1973). Mn(IV) and Fe(III) compounds have similar effects, but they are less marked because of their low solubility. However, soils rich in Fe(III), and relatively low in oxidizable organic matter frequently show *Eh* poise between 0.10 and 0.20 V, corresponding to an Fe(III)–Fe(II) redox couple (Jeffery 1960; Willett 1978). Fe(III) and Mn(IV) compounds added to soils retard the normal decline in *Eh*, and prevent sulfate reduction (Nhung and Ponnamperuma 1966).

Redox reactions of soil components

Soil components subjected to redox reactions are limited by the redox reactions of water: i.e., the oxidation of water to O_2 and the reduction of H^+ to H_2. Between

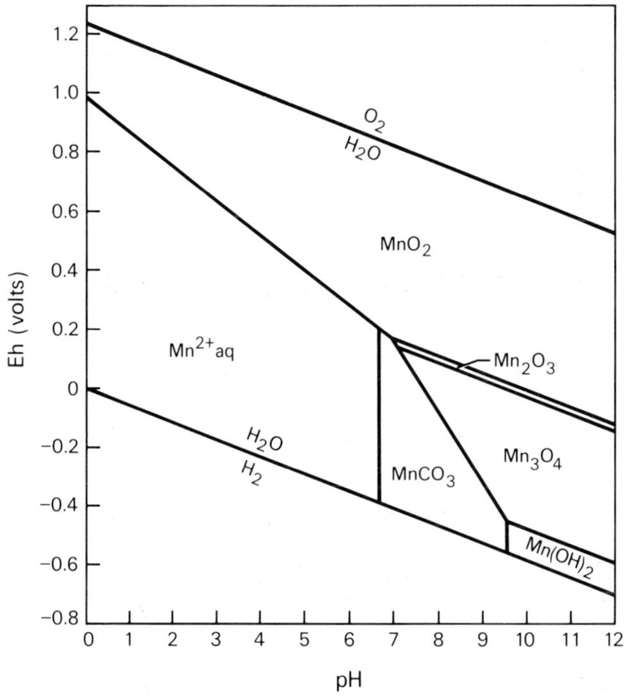

Fig. 28.2. Stability diagram of hypothetical soil manganese oxides. (Reproduced with permission: © 1969 Williams and Wilkins Co., Baltimore.)

Table 28.2. Redox reactions of ideal manganese oxides

Pyrolusite–Mn^{2+}	
$MnO_2 + 4H^+ + 2e \rightleftharpoons Mn^{2+} + 2H_2O$	(1)
$Eh = 0.123 - 0.030 \log (Mn^{2+}) - 0.118$ pH	
Partridgeite–Mn^{2+}	
$Mn_2O_3 + 6H^+ + 2e \rightleftharpoons 2Mn^{2+} + 3H_2O$	(2)
$Eh = 0.145 - 0.059 \log (Mn^{2+}) - 0.177$ pH	
Hausmannite–Mn^{2+}	
$Mn_3O_4 + 8H^+ + 2e \rightleftharpoons 3Mn^{2+} + 4H_2O$	(3)
$Eh = 0.182 - 0.089 \log (Mn^{2+}) - 0.236$ pH	

these extremes lie the couples listed in Table 28.1. In this section, redox reactions of couples important in soils will be briefly discussed.

Redox reactions of nitrate in soils

The transformations of nitrogen in soils are discussed in Chapter 37. Of these, only the reactions in which NO_3^- serves as an electron acceptor in place of O_2 in anaerobic soils (denitrification) will be considered here. Table 28.1 indicates that after microbial consumption of O_2 present in soils, NO_3^- should be the next compound to act as an electron acceptor for anaerobic bacteria; this is found to be the case (e.g. Takai and Kamura 1966; Turner and Patrick 1968). Since NO_3^- is converted to gaseous N_2 and N_2O, it represents the loss from the soil of a major plant nutrient, hence the process has received considerable attention (reviewed by Alexander 1961; Broadbent and Clark 1965; Patrick and Mahapatra 1968; and Ponnamperuma 1972).

Soils contain large numbers of microorganisms capable of denitrification, and the rate at which this occurs is mainly dependent on the supply of oxidizable organic matter. Consequently, denitrification is slow in soils low in organic matter (MacRae *et al.* 1968). Increasing the temperature accelerates denitrification, at least on the initial submergence of a soil, as discussed by Ponnamperuma (1972).

There appears to be no quantitative relationship between *Eh* and the behaviour of NO_3^- in soils, possibly because of its biological lability. However, Pearsall (1938) could not detect NO_3^- in soils with *Eh* values less than 0.32 V at pH 5.0, and Patrick (1960) found NO_3^- to be unstable at 0.34 V at pH 5.1. The presence of NO_3^- in waterlogged rice-growing soils with lower bulk redox potentials can be attributed to local areas of oxygenation at the surface of rice roots, or at the water – soil interface. The latter probably accounts for the low concentration of NO_3^- coexisting with Fe^{2+} in the example shown in Fig. 28.1, and was discussed in relation to N losses in rice paddies by Patrick and Mahapatra (1968).

Redox reactions of manganese in soils

Manganese in oxidized soils occurs as complex oxides of Mn(IV) and Mn(III) which were described in Chapter 23 and by Leeper (1947) and McKenzie (1972*b*, 1977). Table 28.2 shows the redox reactions of Mn oxides with ideal compositions, a simplification of those found in soils. A stability diagram (Fig. 28.2) for soil Mn systems was presented by Ponnamperuma *et al.* (1969) based on experimental values of the standard free energies of Mn oxides and carbonate. Discrepancies shown between the standard free energies of pure Mn oxides and those derived from equilibria studies of soil solutions indicate that soil Mn oxides are complex non-stoichiometric oxides of variable composition. Nevertheless, Olumu *et al.* (1973) found some evidence for the operation of eqn (2) of Table 28.2 in soils.

Fig. 28.2 indicates that in oxidized soils (*Eh* 0.6–0.8 V), solid MnO_2 should be the stable phase at the pH values of most soils. However, only very mildly reducing conditions (e.g. *Eh* 0.40 V at pH 5.0) are required for the stability of Mn^{2+}. Studies with soil suspensions at controlled levels of *Eh* and pH tend to support these pre- dictions, at least qualitatively (Gotoh and Patrick 1972).

When a soil is waterlogged, and the *Eh* declines, there is a release of Mn^{2+} into solution. However, on prolonged waterlogging, CO_2 accumulates causing the precipitation of $MnCO_3$ and a decrease in Mn^{2+} in solution (Ponnamperuma *et al.* 1969). Additional organic matter accelerates the release of soluble Mn^{2+} in soils, either indirectly by accelerating the decline in *Eh*, or possibly by the direct reduction of MnO_2 by organic reducing agents (Cotter and Mishra 1968).

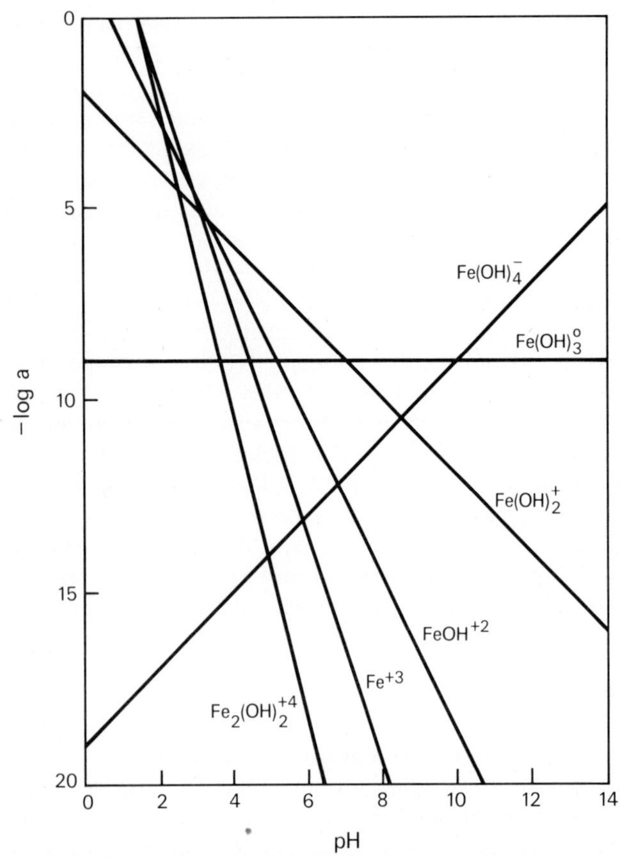

Fig. 28.3. The activity of Fe (III) species in relation to pH. (Reproduced with permission from Norozamsky *et al.* 1978.)

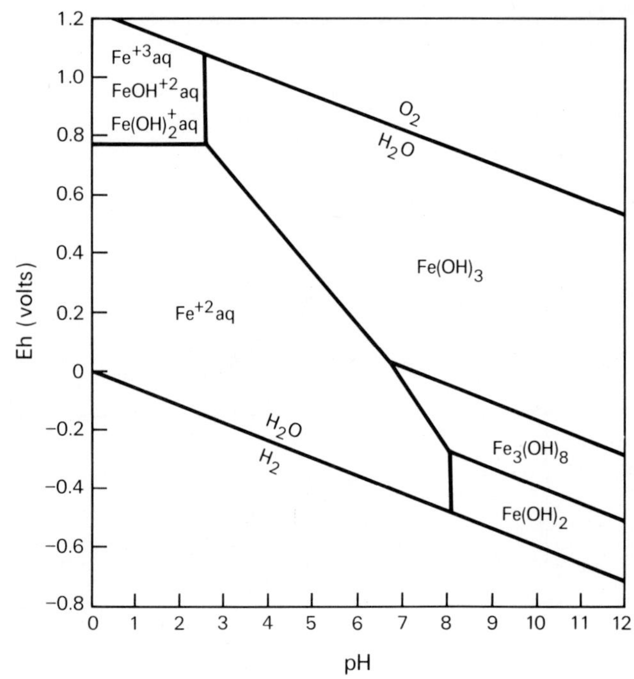

Fig. 28.4. Stability diagram of iron species. (Reproduced with permission: © 1967 Williams & Wilkins Co., Baltimore.)

Redox reactions of iron in soils

In aerated soils, iron occurs as ferric oxides and hydroxyoxides, which have been described in Chapter 23. The activity of Fe^{3+} in soil solution should be governed by the solubility of the solid oxides, and by hydrolysis reactions which can be represented as follows (ignoring hydration):

$$Fe(OH)_3 \rightleftharpoons Fe^{3+} + 3\ OH^- \ (\log K^{o}_{SO}\ -37.5)$$

$$Fe^{3+} + OH^- \rightleftharpoons FeOH^{2+}$$

$$Fe^{3+} + 2OH^- \rightleftharpoons Fe(OH)_2^+$$

$$2Fe^{3+} + 2OH^- \rightleftharpoons Fe_2(OH)_2^{4+}$$

$$Fe^{3+} + 3OH^- \rightleftharpoons Fe(OH)_3^o$$

$$Fe^{3+} + 4OH^- \rightleftharpoons Fe(OH)_4^-$$

Fig. 28.3 shows the relationship between the activity of the various Fe(III) species and pH; in oxidizing solutions with pH values less than 3, Fe^{3+} is the most abundant species; between pH 3.5 and 7, $Fe(OH)_2^+$ is the dominant species. The hydrolysis reactions were reviewed by Sylva (1972) and the precipitation reactions were investigated by Hsu (1972) and Murphy *et al.* (1975).

Bohn (1967) equilibrated an acidic soil in aerated $CaCl_2$ solution, and determined the activities of Fe^{3+} and H^+ in the supernatant solution. He found that the Fe^{3+} activity is determined by the solubility of $Fe(OH)_3$, and that between pH 4 and 8 the predominant iron species is $Fe(OH)_2^+$, in accordance with the above equilibria systems. In clay suspensions to which Fe(III) was added, Fordham (1969) found that the activity of Fe^{3+} in solution is controlled by the solubility of $Fe(OH)_3$. However, in some soils, formation of complexes of Fe^{3+} with organic compounds may increase the concentration of soluble Fe^{3+} to values greater than expected from the equilibria equations.

In addition to the reactions of Fe(III) with OH^- in aerated soils, Fe(III) compounds participate in redox reactions when the soil is subjected to high electron activity (low *Eh*), as for example, when soils are waterlogged in the presence of organic matter. The stability diagram (Fig. 28.4) indicates that, at pH 7 (the value for most highly reduced soils), $Fe(OH)_3$, $Fe_3(OH)_8$ or Fe^{2+} are the stable Fe phases as *Eh* decreases from +0.85 to -0.40 V.

Possible redox reactions of Fe(III) in soils are listed in Table 28.3, and include those used to construct Fig. 28.4. In each reaction, protons as well as electrons are transferred, so solution pH, as well as *Eh*, determines the stable phase.

Ponnamperuma *et al.* (1967) attempted to determine which of the reactions in Table 28.3 operate in flooded

Table 28.3. Possible redox reactions of iron in soils

Ferric hydroxide–ferrous ion	
$Fe(OH)_3 + 3H^+ + e \rightleftharpoons Fe^{2+} + 3H_2O$	(1)
$Eh = 1.058 - 0.177\ pH - 0.059 \log (Fe^{2+})$	
Ferric hyrdoxide–ferrosoferric hydroxide	
$3Fe(OH)_3 + H^+ + e \rightleftharpoons Fe_3(OH)_8 + H_2O$	(2)
$Eh = 0.429 - 0.059\ pH$	
Ferrosoferric hydroxide–ferrous ion	
$Fe_3(OH)_8 + 8H^+ + 2e \rightleftharpoons 3Fe^{2+} + 8H_2O$	(3)
$Eh = 1.373 - 0.0885 \log (Fe^{2+}) - 0.236\ pH$	
Ferrosoferric hydroxide–ferrous hydroxide	
$Fe_3(OH)_8 + 2H^+ + 2e \rightleftharpoons 3\ Fe(OH)_2 + 2H_2O$	(4)
$Eh = 0.195 - 0.059\ pH$	
Goethite–ferrous ion	
$\alpha FeOOH + 3H^+ + e \rightleftharpoons Fe^{2+} + 2H_2O$	(5)
$Eh = 0.736 - 0.177\ pH - 0.059 \log (Fe^{2+})$	
Hematite–ferrous ion	
$\alpha Fe_2O_3 + 6H^+ + 2e \rightleftharpoons 2\ Fe^{2+} + 3H_2O$	(6)
$Eh = 0.728 - 0.177\ pH - 0.059 \log (Fe^{2+})$	

soils. They measured *Eh* and pH and determined the elemental composition of the solutions of flooded soils, and attempted to fit the data to the equilibria equations. Their *Eh* and pH and (Fe^{2+}) data for the initial 17 weeks of flooding fitted eqn (1) of Table 28.3; after a decrease in the values of (Fe^{2+}) on prolonged flooding, eqns (2) and (3) also fitted the data. It was concluded, that in the rice-growing soils they studied, $Fe(OH)_3$ and $Fe_3(OH)_8$ are the active iron oxides participating in the equilibria.

In a similar experiment with non-rice-growing soils, van Breemen (1969) found that the stability of the Fe(III) oxides present was intermediate between that of $Fe(OH)_3$ and $\alpha FeOOH$. The Fe(III) oxides controlling the equilibria could not be defined.

It is possible that previous cycles of reduction and oxidation, which are characteristic of soils used for rice production, brought about an increase in activity of $Fe(OH)_3$ (at the expense of the more crystalline Fe(III) oxides) in the soils studied by Ponnamperuma *et al.* (1967). Willett and Higgins (1978*b*) have demonstrated that a flooding-drying cycle in previously non-flooded soils causes an increase in the amount of amorphous ferric oxide as determined by oxalate extraction. Therefore, amorphous $Fe(OH)_3$ may have a greater influence on redox equilibria in rice-growing soils than in soils which have not previously undergone frequent cycles of oxidation and reduction. It is possible that in non-rice-growing soils, reactions (5) and (6) of Table 28.3 may participate in redox equilibria.

On flooding an aerobic soil, there is a large increase in soluble and exchangeable Fe^{2+}, corresponding to the Fe^{2+} stability region that lies between *Eh* 0.77 and 0.03 V at pH values less than 6.8 (Fig. 28.4). Accordingly, the amount of Fe^{2+} in solution increases markedly with decreasing *Eh* at pH values 5 and 6 (Gotoh and Patrick 1974). The rate of Fe(III) reduction also depends on the type and amount of organic matter present (as the

Table 28.4. Redox reactions of sulfate

$SO_4^{2-} + 10H_2O + 8e \rightleftharpoons H_2S + 4H_2O$	(1)
$Eh = 0.303 - 0.074\ pH + 0.007 \log \frac{(SO_4^{2-})}{(H_2S)}$	
$SO_4^{2-} + 9H^+ + 8e \rightleftharpoons HS^- + 4\ H_2O$	(2)
$Eh = 0.252 - 0.066\ pH + 0.007 \log \frac{(SO_4^{2-})}{(HS^-)}$	
$SO_4^{2-} + 8H^+ + 8e \rightleftharpoons S^{2-} + 4H_2O$	(3)
$Eh = 0.148 - 0.059\ pH + 0.007 \log \frac{(SO_4^{2-})}{(S^{2-})}$	

electron donor) and on the temperature (IRRI 1964). On further reduction to *Eh* values between 0.1 and -0.1 V, and a rise in pH to 7 (because of the proton-consuming reactions listed in Tables 28.1–4), there is a decrease in the concentration of Fe^{2+} corresponding to the stability region for the precipitation of $Fe_3(OH)_8$ in Fig. 28.4 (Ponnamperuma *et al.* 1967).

Further decreases in Fe^{2+} in flooded soils may be caused by the precipitation of $FeCO_3$ and FeS. The slow diffusion of CO_2 from flooded soils favours its accumulation (IRRI 1964) and the precipitation of $FeCO_3$ (and $MnCO_3$), as discussed by Ponnamperuma (1972). Since S^{2-} may be expected in only very highly reduced soils, precipitation of Fe(II) by S^{2-} is significant only in organic-rich soils containing SO_4^{2-}.

Redox reactions of sulfate in soils

Inorganic sulfur is present in aerated soils at its highest oxidation state of SO_4^{2-}, but in very reducing soils it is subjected to the redox reactions listed in Table 28.4, some of which were used to construct the stability diagram shown in Fig. 28.5. The stability field for SO_4^{2-} is large; in solutions of pH greater than 8, extremely low values of *Eh* are required for sulfide stability; at lower pH values (e.g. 5 to 7), small negative values of *Eh*, similar to those found in organic-rich waterlogged soils, result in H_2S being the stable phase; as the prevailing pH rises at low *Eh* values, dissociation of H_2S to HS^- and S^{2-} increases.

In common with NO_3^-, Fe(III) and Mn(IV) reduction, SO_4^{2-} reduction is largely a microbiological process. When the numbers of sulfate-reducing bacteria are not limiting, the rate of SO_4^{2-} reduction depends on the availability of readily oxidizable organic matter (Ogato and Bower 1965). However, the activity of the main sulfate-reducing bacteria (*Desulfovibrio desulfuricans*) is sensitive to pH (Cappenberg 1975) and temperature (Kaplan and Rittenberg 1964), so that the kinetics of SO_4^{2-} reduction depends on the soil as a microbiological environment, as well as on physico-chemical factors.

Implications of redox reactions in soils

Redox reactions and soil formation

The principal redox reactions involved in the genesis of soils are those of the Fe(III)–Fe(II) systems and, to a

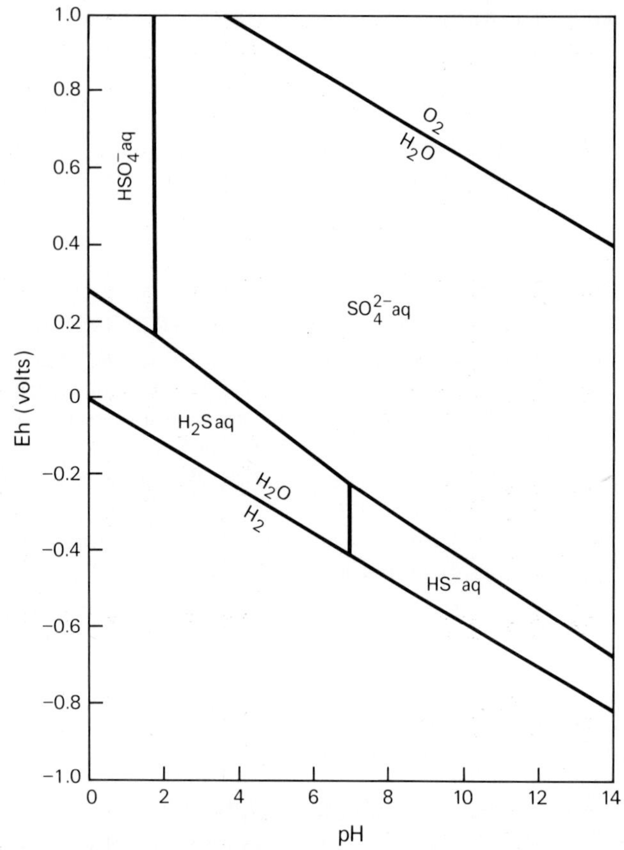

Fig. 28.5. Stability diagram of sulphur species. (Reproduced with permission from Garrels and Christ 1965.)

lesser degree, the Mn(IV)–Mn(II) systems. These are reflected in soil colour, which is strongly influenced by the state of iron in particular. In the various classification systems for soils, colour is usually regarded as a significant differentiating attribute.

In terms of soil formation, it is important to distinguish between soil horizons which are periodically reduced, and those which are permanently reduced. Permanently waterlogged horizons are characterized by blue-green colours, which have been ascribed to possible mixed Fe(II)–Fe(III) hydroxy compounds (R. M. Taylor, personal communication.). Sulfide-rich soils are frequently intensely black, due to the presence of FeS. If continuously reduced horizons are also subjected to leaching, Fe, as Fe(II), may be leached out to unmask the colour of the Fe-free matrix (Bloomfield 1973). In Australia, permanently waterlogged horizons are found in marshy depressions in areas of high rainfall (e.g. in *humic gleys* derived from clay alluvium) and below peats in areas of low temperature and high rainfall (Stace *et al.* 1968). Organic matter accumulates in permanently or frequently waterlogged surface horizons to form a layer of peat, because anaerobic oxidation is slower than aerobic oxidation.

Soils subjected to alternating oxidation and reduction are characterized by localized concentrations of Fe(III) and Mn(IV) compounds. Fe(III) stains, tubules, and concretions, are commonly found in the surface horizons of duplex soils of eastern Australia. The surface horizons may be well drained, but temporarily waterlogged during periods of heavy rainfall because of perched watertables overlying impermeable clay B horizons. Soils may also be subjected to alternating oxidizing – reducing conditions when overlying a permanent, but fluctuating, watertable. Both hydrological regimes produce horizons showing marked mottlings of coloured Fe(III) oxides in a greyish Fe-depleted matrix. This is due in both cases to the reduction of very sparingly soluble Fe(III) oxides to more mobile Fe(II) species, and their subsequent transport and reoxidation. Mn(II) reoxidizes at a higher *Eh* than Fe(II) (Table 28.1), so that if a gradient in *Eh* exists in the soil fabric, as for example from the interior to the surface of a ped, separation of the elements may occur.

On soil reduction, Fe^{2+} displaces other cations from the exchange complex and increases the soil solution concentration of Ca^{2+}, Mg^{2+}, K^+, and Na^+. These cations may be slowly leached from the profile, so that on reoxidation an acid Al-clay is produced. This process has been termed 'ferrolysis' (Brinkman 1970), and leads to the acidification of periodically flooded soils. Ferrolysis has not been specifically reported from Australia; small decreases in soil pH associated with periodic flooding for rice cultivation appear to be caused by the leaching of salts (Hallsworth *et al.* 1952) and, possibly, by the accumulation of organic matter (Blackmore *et al.* 1956). However, ferrolysis could be involved in the genesis of the acid peats of upland areas of eastern Australia.

The drainage and oxidation of soils and sediments rich in sulfide also leads to acidification. Sulfide-rich sediments occur in marine and fossil estuarine environments in which sulfate from seawater has been reduced. A plentiful source of organic matter is required for this SO_4^{2-} reduction (Moormann 1963). Whilst reduced, sulfide and Fe(II) react to form FeS and ultimately FeS_2 (pyrites). On drainage, the sulfides are oxidized to sulfate, oxides, and jarosite; in the absence of carbonate, the pH may decrease to 2.5 (e.g. Walker 1972), or even lower (Moormann 1963). The chemistry and genesis of acid sulfate soils have been thoroughly discussed (Dost 1973).

In New South Wales, those estuarine sediments rich in sulfide, and exposed at the surface in back swamps, have very little agricultural potential, especially if drained (Walker 1972). However, the soils nearer the river are derived from alluvial sediments (overlying the sulfide-rich sediments), and are not acidified.

In northern Australia, large areas of mangroves provide an organic matter and sulfate-rich environment with potential for the formation of acid sulfate soils. However, no attempt appears to have been made to drain them for agricultural production.

Redox reactions and the mineral nutrition of plants

As stated earlier, the behaviour of nitrate, manganese, iron and sulfur in soils is strongly influenced by the prevailing *Eh* and pH. The reactions discussed affect directly the plant availability of these elements. However, redox reactions indirectly affect the availability of nutrient elements which do not normally undergo valence changes in soil environments, particularly P, Mo, Co, Cu, Zn and Si.

The 'availability' of NO_3^- is decreased by its reduction to gaseous N_2O and N_2 and their diffusion from the soil. Because nitrate is the next substance to be reduced after the consumption of O_2 (Table 28.1), it is highly vulnerable to reduction in soils. Nitrate is unstable in very mildly reducing environments, and, in soils rich in organic matter, may be reduced a few hours after waterlogging. In addition, significant denitrification occurs in soils normally considered aerated, because of micro-sites depleted in O_2 by intense microbial activity, e.g. in peds containing fresh plant residues (Smith 1976*a*; reviewed by Stolzy and Flühler 1978). The clay soils of the Murrumbidgee Irrigation Area of New South Wales develop *Eh* values low enough to cause denitrification during normal flood irrigations (Willett 1978). However, denitrification does not occur in clay soils from the Darling Downs at moisture contents less than 90% w/w; the soil moisture content required to cause denitrification depends on the availability of easily oxidizable organic matter (Craswell and Martin 1974).

Nitrate is denitrified when soils are flooded for paddy rice cultivation. Nitrogen fertilizers containing NO_3^- are therefore unsuitable for this crop. Additional N is also lost from rice paddies because of nitrification in oxygenated parts of the soil (at the soil–water interface, and possibly in the rhizosphere of rice) and the subsequent diffusion of NO_3^- to reduced parts of the soil, where it is denitrified (Patrick and Mahapatra 1968).

Manganese uptake by plants is in the form of Mn^{2+}, whereas most manganese in aerated soils is in the form of insoluble Mn(IV) oxides. The effects of soil *Eh* and pH on the Mn(IV)–Mn(II) equilibria, described above, therefore directly influence the availability of Mn to plants.

Our understanding of the availability of soil manganese is based on the extensive studies of G. W. Leeper and his co-workers at the University of Melbourne. In aerated soils, it was found that Mn availability depends not only on the amount of Mn(II) in solution, or on the exchange complex, but also on the nature of the Mn(IV) oxides present (Leeper 1947). Structurally disordered Mn(IV) oxides of small particle size are very easily reduced to soluble Mn(II), and also contribute to available Mn. On the other hand, well structured oxides with relatively larger particle size are unavailable in aerated soils (Jones and Leeper 1951*a, b*).

Intense reduction of soils (by waterlogging), followed by oxidation, causes an increase in easily reducible Mn(IV) oxides, and an increase in Mn availability. Conversely, ageing of the oxides increases structural order, and decreases their availability (Jones and Leeper 1951*b*).

Mn deficiency is associated with aerated alkaline soils, because the formation of insoluble Mn(IV) oxides is favoured, e.g. the calcareous soils of the Yorke Peninsula, South Australia (Reuter *et al.* 1973). However, Mn availability in alkaline soils may be greater than expected from the *Eh* – pH considerations, because of the stabilization of Mn^{2+} by organic complexes (Beckwith 1955), and the reduction and complex formation of Mn(IV) with organic ligands from root exudates (Passioura and Leeper 1963). In alkaline soils, Mn availability is increased by waterlogging, but not to levels of toxicity. However, waterlogging can cause Mn toxicity in near neutral soils (Grasmanis and Leeper 1966).

Fig. 28.2 shows that, in acid environments, Mn^{2+} is the stable form of Mn, even at relatively high values of *Eh*. Mn^{2+} is therefore found in solution and on the exchange complex of acid soils, and is readily available to plants. In aerated acid soils derived from Mn-rich materials, Mn toxicity to plants has frequently been reported (e.g. Fergus 1954; Lovett and Johnson 1968). The waterlogging of acidic soils further increases the availability of Mn, and results in severe Mn toxicity (Adams and Honeysett 1964).

The concentration of Fe^{3+} in the soil solution is largely controlled, as indicated previously, by the solubility of the most soluble soil Fe(III) hydroxide, $Fe(OH)_3$. Since its solubility product is very small (log K°_{SO} –37.5), the concentration of Fe^{3+} (or $FeOH^{2+}$) in the solutions of aerated soils is extremely low; and because the total concentration of Fe(III) species in solution decreases with increasing pH, iron deficiency in plants is associated with alkaline soils. However, the solubility of Fe(III), and hence its availability to plants, is increased by the formation of complexes, especially with organic ligands from soil organic matter and root and microbial exudates.

The reduction of Fe(III) hydroxides to Fe^{2+} markedly increases Fe availability. Low *Eh* values are required for Fe(III) reduction, and it is retarded by the presence of NO_3^- and Mn(IV). Iron toxicity is therefore associated with acidic soils rich in organic matter, relatively low in Mn(IV), and subject to waterlogging. In Australia, reports of iron toxicity appear to be confined to acidic soils of the Northern Territory that have been flooded for rice cultivation (Strickland 1968). On the other hand, alkaline rice-growing soils of northern Australia release very little Fe^{2+} (Williams 1972*a*), and

Fe toxicity is not found in the rice-growing soils of New South Wales.

Plants take up sulfur mainly in the form of sulfate, but in highly reduced soils sulfate is reduced to H_2S, which is highly toxic even at low concentrations. Because sulfate is reduced at lower *Eh* values (Table 28.1), than are NO_3^-, Mn(IV) or Fe(III), these redox systems may poise *Eh* at values too high for sulfate reduction. However, sulfate reduction occurs in soils rich in organic matter after prolonged waterlogging. As sulfate can only be reduced at *Eh* values lower than required for the formation of Fe^{2+} and Mn^{2+} free H_2S only occurs at concentrations of the order 10^{-8}M, due to the precipitation of MnS and FeS. However, in some Asian paddy soils low in Fe, H_2S toxicity is a serious problem. The precipitation of insoluble sulfides of Cu and Zn by H_2S may be involved in deficiencies of these metals in some Asian paddy soils (Ponnamperuma 1977*b*), but precipitation by H_2S is not considered to contribute towards the relatively low levels of available Zn and Cu in rice-growing soils of New South Wales (Beckwith *et al.* 1975).

The availability of phosphorus is influenced indirectly by the redox status of the soil. Even though P can exist in valence states from +5 to -3, in field soils the reduction of orthophosphate to phosphine is of no significance (Patrick *et al.* 1975). The waterlogging of soils tends to increase the solution concentration of P, by the reduction of Fe(III) phosphate to the more soluble Fe(II) phosphate (Islam and Elahi 1954). The reduction of Fe(III) oxides may release occluded P, and increase its availability (Patrick 1964). However, P released during reduction may be re-sorbed on to the soil surfaces. Recent studies have shown that the reduction of soils (in which the pH was allowed to rise to near neutrality) increases their P sorption capacity and bonding energy, presumably because Fe(II) oxides have greater surface area and affinity for P than Fe(III) oxides (Willett and Higgins 1978*a*, *b*; Holford and Patrick 1979). In reduced soils of pH 5, Fe(II) oxides do not precipitate, and Fe(II) and P accumulate in the soil solution (Holford and Patrick 1979). However, most reduced soils have near neutral pH values, so that Fe(II) oxides precipitate, and P should be strongly re-sorbed and its availability decreased. Consequently, some soils show no increase, or only a transitory increase, in P availability on reduction (see examples presented by Ponnamperuma 1972). Williams *et al.* (1958) found that anaerobic incubation increases the availability of Fe(III) phosphate in soil, but decreases the availability of Al or Ca phosphates. The overall effects of soil reduction on P availability appear to depend on the relative influences of the reduction on the release of P from Fe(III) compounds and on the re-sorption of P by Fe(II) oxides.

The reduction of Fe(III) oxides may also liberate Cu, B, Mo and Co into solution, but changes in pH and complexation by organic compounds probably have a greater influence on their availability (see Chapter 25). Co, Ni, Cu and Zn tend to accumulate in soil Mn oxides (McKenzie 1972*a*), so these elements may be more available for plant uptake in reduced soils.

Fe(III) oxides formed by the oxidation of Fe(II) species are less crystalline, and sorb P with greater energy, than oxides which have not undergone a cycle of reduction and oxidation (Willett and Higgins 1978*b*; Kuo and Mikkelsen 1979; Willett 1979). This can lead to P deficiency in crops sown into soils which have been flooded and drained (Simpson and Williams 1970; Willett 1979).

Molybdenum availability is also related to the crystallinity of soil Fe(III) oxides (Jarrell and Dawson 1978) so that a cycle of reduction and oxidation may also depress its availability to subsequent crops; boron could be similarly affected.

Conclusion

Studies on soil redox reactions have contributed greatly to defining the soil conditions necessary for paddy rice production. Further research is required on the effects of waterlogging on P availability to rice, in particular the relative effects of reduction on P mobilization from Fe(III) compounds and its immobilization by Fe(II) compounds. The forms of Fe(II) in reduced soils have not been positively determined.

The effects of cycles of reduction and oxidation on nutrient mobility have been more recently studied. These studies are of practical significance, not only to soil fertility but also in relation to the soil as a source or sink of pollutants. It is now becoming accepted that redox reactions are of significance in well-drained soils, at least in micro-sites protected from oxygenation. Such soils may consist of a mosaic of zones of differing redox status, and may be considered to be undergoing continual cycles of reduction and oxidation. Such a process would be expected to affect the behaviour of the two major plant nutrients – nitrogen and phosphorus. Further advances must await a technique for determining the redox status of very small volumes of intact soil.

References

Adams, S. N. and Honeysett, J. L. (1964). Some effects of soil waterlogging on the cobalt and copper status of pasture plants grown in pots. *Aust. J. Agric. Res.* **15**, 357-67.

Alexander, M. (1961). 'Introduction to Soil Microbiology'. (Wiley: New York.)

Aomine, S. (1962). A review of research on redox potentials of paddy soils in Japan. *Soil Sci.* **94**, 6-13.

Beckwith, R. S. (1955). Metal complexes in soils. *Aust. J. Agric. Res.* **6**, 685-98.

Beckwith, R. S., Tiller, K. G. and Suwadji, E. (1975). The effects of flooding on the availability of trace metals to rice in soils of differing organic matter status. *In* 'Trace Elements in Soil-Plant-Animal Systems.' (Eds D. J. D. Nicholas and A. R. Egan). (Academic Press: New York.)

Blackmore, A. V., Talsma, T., and Hutton, J. T. (1956). The cumulative effects of rice growing on two soil types in the Murrumbidgee Irrigation Area. CSIRO Aust. Div. Soils Div. Rep. No. 1/56.

Bloomfield, C. (1973). Some chemical properties of hydromorphic soils. *In* 'Pseudogley and Gley.' (Eds E. Schlichting and U. Schwertmann). Trans. Comm. V and VI. I.S.S.S.

Bohn, H. L. (1967). The (Fe)(OH)^3 ion product in suspension of acid soils. *Soil Sci. Soc. Am. Proc.* **31**, 641-44.

Bohn, H. L. (1969). EMF of platinum electrodes in dilute solutions and its relation to soil pH. *Soil Sci. Soc. Am. Proc.* **33**, 639-40.

Bohn, H. L. (1971). Redox potentials. *Soil Sci.* **112**, 39-45.

Brinkman, R. (1970). Ferrolysis, a hydromorphic soil forming process. *Geoderma* **3**, 199-206.

Broadbent, F. E. and Clark, F. E. (1965). *In* 'Soil Nitrogen'. (Eds W. V. Bartholomew and F. E. Clark). (Am. Soc. Agron: Madison.)

Brümmer, G. (1974). Redoxpotentiale und Redoxprozesse von Mangan-, Eisen- und Schwefelverbindungen in hydromorphen Böden und Sedimenten. *Geoderma* **12**, 297-22.

Cappenberg, T. E. (1975). Relationships between sulphate-reducing and methane producing bacteria. *Plant Soil* **43**, 125-39.

Cotter, D. J. and Mishra, U. N. (1968). The role of organic matter in soil manganese equilibrium. *Plant Soil* **29**, 439-48.

Craswell, E. T. and Martin, A. E. (1974). Effect of moisture content on denitrification in a clay soil. *Soil Biol. Biochem.* **6**, 127-29.

Dost, H. (1973). 'Acid Sulphate Soils'. International Institute for Land Reclamation and Improvement. Wageningen. Public. 18, pt. 1-2.

Fergus, I. F. (1954). Manganese toxicity in an acid soil. *Queensl. J. Agric. Sci.* **11**, 15-27.

Fordham, A. W. (1969). Sorption and precipitation of iron on kaolinite, II. *Aust. J. Soil Res.* **7**, 199-212.

Garrels, R. M. and Christ, C. L. (1965). 'Solutions, Minerals and Equilibria.' (Harper and Row: New York.)

Gotoh, S. (1973). Reduction processes in waterlogged soils with special reference to transformation of nitrate, manganese and iron. *Bull. Kyushu Agric. Exp. Sta.* **16**, 669-714.

Gotoh, S. and Patrick, W. H. (1972). Transformation of manganese in a waterlogged soil as affected by redox potential and pH. *Soil Sci. Soc. Am. Proc.* **36**, 738-42.

Gotoh, S. and Patrick, W. H. (1974). Transformation of iron in a waterlogged soil as influenced by redox potential and pH. *Soil Sci. Soc. Am. Proc.* **38**, 68-71.

Gramanis, V. O. and Leeper, G. W. (1966). Toxic manganese in near neutral soils. *Plant Soil* **25**, 41-48.

Hallsworth, E. G., Lemerle, T. M. and Rayner, H. V. (1952). Studies in pedogenesis in New South Wales. I. *J. Soil Sci.* **3**, 89-102.

Holford, I. C. R. and Patrick, W. H. (1979). Effects of reduction and pH changes on phosphate sorption and mobility in an acid soil. *Soil Sci. Soc. Am. J.* **43**, 292-97.

Hsu, P. H. (1972). Nucleation, polymerisation and precipitation of FeOOH. *J. Soil Sci.* **23**, 409-19.

I.R.R.I. (1964). Annual Report of the International Rice Research Institute. (Los Banos: Philippines.)

Islam, M. A. and Elahi, M. A. (1954). Reversion of ferric iron to ferrous iron under waterlogged conditions and its relation to available phosphorus. *J. Agric. Sci.* **45**, 1-2.

Jarrell, W. M. and Dawson, M. D. (1978). Sorption and availability of molybdenum in soils of Western Oregon. *Soil Sci. Soc. Am. J.* **42**, 412-15.

Jeffery, J. W. O. (1960). Iron and the Eh of waterlogged soils with particular reference to paddy. *J. Soil Sci.* **11**, 140-48.

Jones, L. H. P. and Leeper, G. W. (1951*a*). The availability of various manganese oxides to plants. *Plant Soil* **3**, 141-53.

Jones, L. H. P. and Leeper, G. W. (1951*b*). Available manganese oxides in neutral and alkaline soils. *Plant Soil* **3**, 154-59.

Kaplan, I. R. and Rittenberg, S. C. (1964). Microbiological fractionation of sulphur isotopes. *J. Gen. Microbiol.* **34**, 195-212.

Kuo, S. and Mikkelsen, D. S. (1979). Distribution of iron and phosphorus in flooded and unflooded soil profiles and their relation to phosphorus adsorption. *Soil Sci.* **127**, 18-25.

Leeper, G. W. (1947). The forms and reactions of manganese in the soil. *Soil Sci.* **63**, 79-94.

Lemon, E. and Kristensen, J. (1960). An edaphic expression of soil structure. Trans. 7th Int. Congr. Soil Sci. Vol. 1, pp. 232-40.

Lovett, W. K. and Johnson, A. D. (1968). Manganese uptake by tobacco and bean plants grown on the soils of the Mareeba-Dimbulah Irrigation Area, North Queensland. *Aust. J. Exp. Agric. Anim. Husb.* **8**, 466-69.

MacRae, I. C., Ancajar, R. R. and Salandanan, S. (1968). The fate of nitrate nitrogen in some tropical soils following submergence. *Soil Sci.* **105**, 327-34.

McKenzie, R. M. (1972*a*). The sorption of some heavy metals by the lower oxides of manganese. *Geoderma* **8**, 29-35.

McKenzie, R. M. (1972*b*). The manganese oxides in soils – a review. *Z. Pflanzernernaehr. Bodenkd.* **131**, 221-42.

McKenzie, R. M. (1977). Manganese oxides. In 'Minerals in Soil Environments.' (Soil Sci. Soc. Am.: Madison.)

Moormann, F. R. (1963). Acid sulphate soils (cat-clays) of the tropics. *Soil Sci.* **95**, 271-75.

Murphy, P. J., Posner, A. M. and Quirk, J. P. (1975). Chemistry of iron in soils. Ferric hydrolysis products. *Aust. J. Soil Res.* **13**, 189-201.

Nhung, M. M-T., and Ponnamperuma, F. N. (1966). Effects of calcium carbonate, manganese dioxide, ferric hydroxide, and prolonged flooding on chemical and electrochemical changes and growth of rice in a flooded acid sulfate soil. *Soil Sci.* **102**, 29-41.

Norozamsky , I., Beek, J. and Bolt, G. H. (1978). Chemical equilibria. In *'Soil Chemistry. A. Basic Elements.'* (Eds G. H. Bolt and M. G. M. Bruggenwert.) (Elsevier: Amsterdam.)

Ogata, G. and Bower, C. S. (1965). Significance of sulphate reduction in soil salinity. *Soil Sci. Soc. Am. Proc.* **29**, 23-25.

Olumu, M. O., Racz, G. J. and Cho, C. M. (1973). Effects of flooding on the Eh, pH and concentrations of Fe and Mn in several Manitoba soils. *Soil Sci. Soc. Am. Proc.* **37**, 220-24.

Passioura, J. B. and Leeper, G. W. (1963). Soil compaction and manganese deficiency. *Nature* (London) **200**, 29-30.

Patrick, W. H. (1960). Nitrate reduction rates in a submerged soil as affected by redox potential. Trans. 7th Int. Congr. Soil Sci. Vol. II. pp. 494-500.

Patrick, W. H. (1964). Extractable iron and phosphorus in a submerged soil at controlled redox potentials. Trans. 8th Int. Congr. Soil Sci. Vol. IV. pp. 605-09.

Patrick, W. H., de Laune, R. D. and Antie, D. A. (1975). Transformation of added phosphate in flooded soil. Trans. 10th Int. Congr. Soil Sci. Vol. IV. pp. 294-304.

Patrick, W. H. and Mahapatra, I. C. (1968). Transformation and availability to rice of nitrogen and phosphorus in waterlogged soils. *Adv. Agron.* **20**, 323-59.

Pearsall, W. H. (1938). The soil complex in relation to plant communities. I. *J. Ecol.* **26**, 180-93.

Ponnamperuma, F. N. (1972). The chemistry of flooded soils. *Adv. Agron.* **24**, 29-96.

Ponnamperuma, F. N. (1977*a*). Physicochemical properties of submerged soils in relation to fertility. International Rice Research Institute, Research Paper No.5.

Ponnamperuma, F. N. (1977*b*). Behaviour of minor elements in paddy soils. International Rice Research Institute, Research Paper No.8.

Ponnamperuma, F. N., Tianco, E. M. and Loy, T. A. (1967). Redox equilibria in flooded soils. I. *Soil Sci.* **103**, 374-82.

Ponnamperuma, F. N., Loy, T. A. and Tianco, E. M. (1969). Redox equilibria in flooded soils. II. *Soil Sci.* **108**, 48-57.

Quispel, A. (1947). Measurement of the oxidation-reduction potentials of normal and inundated soils. *Soil Sci.* **63**, 265-75.

Reuter, D. J., Heard, T. G. and Alston, A. M. (1973). Correction of manganese deficiency in barley crops on calcareous soils. I. *Aust. J. Exp. Agric. Anim. Husb.* **13**, 434-39.

Simpson, J. R. and Williams, C. H. (1970). The effects of fluctuations in soil moisture content on the availability of recently applied phosphate. *Aust. J. Soil Res.* **8**, 209-19.

Smith, A. M. (1976*a*). Ethylene production by bacteria in reduced microsites in soil and some implications to agriculture. *Soil Biol. Biochem.* **8**, 293-98.

Smith, A. M. (1976*b*). Anaerobic microsites in the rhizosphere of plants as a mechanism for increasing phosphate availability. Reviews in Rural Science, (University of New England) Vol. 3, pp. 119-21.

Stace, H. C. T., Hubble, G. D., Brewer, R., Northcote, K. H., Sleeman, J. R., Mulcahy, M. J. and Hallsworth, E. G. (1968). 'A Handbook of Australian Soils.' (Rellim Technical Publications: Adelaide.)

Stolzy, L. H. and Flühler, H. (1978). Measurement and prediction of anaerobiosis in soils. *In* 'Nitrogen in the Environment.' Vol. 1. Nitrogen Behaviour in Field Soil. (Eds D. R. Nielson and J. G. MacDonald). (Academic Press: New York.)

Strickland, R. W. (1968). The effect of drainage on physiological disorders of rice grown on the sub-coastal plains of the Adelaide River, N.T. *Aust. J. Exp. Agric. Anim. Husb.* **8**, 212-22.

Stumm, W. (1965). Redox potentials as an environmental parameter, conceptual significance and operational limitations. Advances in Water Pollution. Proc. 3. International Conf. on Water Pollution Research. Vol. 1. (Pergamon: New York.)

Sylva, R. N. (1972). The hydrolysis of iron (III). *Rev. Pure Appl. Chem.* **22**, 115-32.

Takai, Y. and Kamura, T. (1966). The mechanisms of reduction in waterlogged paddy soil. *Folia Microbiol.* **11**, 304-13.

Turner, F. T. and Patrick, W. H. (1968). Chemical changes in waterlogged soils as a result of oxygen depletion. Trans. 9th Int. Congr. Soil Sci. Vol. IV, pp. 53-65.

van Breemen, N. (1969). The effect of ill-defined ferric oxides on the redox characteristics of flooded soils. *Neth. J. Agric. Sci.* **17**, 256-60.

Walker, P. H. (1972). Seasonal and stratigraphic control in coastal floodplain soils. *Aust. J. Soil Res.* **10**, 127-42.

Willett, I. R. (1978). Redox potentials and pH of two Murrumbidgee Irrigation Area rice soils during flooding for rice culture. CSIRO Aust. Div. Soils Div. Rep. No. 31.

Willett, I. R. (1979). The effects of flooding for rice culture on soil chemical properties and subsequent maize growth. *Plant Soil* **52**, 373-83.

Willett, I. R. and Higgins, M. L. (1978*a*). An evaluation of methods for characterizing phosphate sorption in rice soils. *Comm. Soil Sci. Plant Anal.* **9**, 559-69.

Willett, I. R. and Higgins, M. L. (1978*b*). Phosphate sorption by reduced and reoxidized rice soils. *Aust. J. Soil Res.* **16**, 319-26.

Williams, B. G. (1972*a*). Flooded rice soils of Northern Australia I. *Aust. J. Soil Res.* **10**, 43-51.

Williams, B. G. (1972*b*). Flooded rice soils of Northern Australia, II. *Aust. J. Soil Res.* **10**, 53-60.

Williams, R. F., Bromfield, S. M. and Williams, C. H. (1958). Studies in soil fertility with special reference to organic manures, IV. *Aust. J. Agric. Res.* **9**, 640-63.

Chapter 29
The sorption of phosphate by soils
M. E. Probert

WHEN PHOSPHATE SOLUTIONS COME INTO CONTACT WITH soil the concentration of phosphate in solution decreases. This phenomenon, which here will be referred to as sorption, has been much studied in recent years. It is an important characteristic of soils because it affects the availability of phosphate to plants, the effectiveness of phosphate fertilizers, the concentration of phosphate in lake waters (and thus eutrification problems) and the usefulness of soils for the renovation of waste waters containing phosphate.

Some aspects of the reaction of phosphate with soil components are discussed in Chapter 25. Here the emphasis will be on low phosphate concentrations, of the order of that necessary for optimum plant nutrition.

Studies have been carried out using both soils and soil components such as clays, calcite, and sesquioxides of iron and aluminium. However, despite the attention the subject has attracted, there is little consensus of opinion about it in the literature, indeed one might say there is some confusion, and for this reason, it is appropriate to devote some space to it in this book.

Factors affecting phosphate sorption

When phospate sorption is studied under laboratory conditions, the experimental variables can have considerable influence upon the results obtained. It is worthwhile stating in general terms some of these observed effects at the outset.

The reactions between phosphate solutions and soil continue for long periods of time. Although the rate of change of the concentration of phosphate in solution declines, it is doubtful whether studies ever reflect a true equilibrium situation.

It is usual to observe marked differences between adsorption and desorption; this has been taken as indicative of the non-reversibility of the phenomenon and referred to as hysteresis, but it may merely reflect the non-equilibrium that exists.

Sorption of phosphate by soil increases the negative charge on the surface of the solid phase.

The nature and ionic strength of the support electrolyte used have major influences on the extent of sorption observed.

Other experimental variables such as temperature, soil:solution ratio, vigour of shaking, even the size of container used for shaking the suspensions, all influence the results obtained.

No doubt it is this complexity of the phenomenon that has led to the diversity of approaches and conclusions one finds in the literature.

Principles of reaction of phosphate between the liquid and solid phases

The processes that can be thought of as controlling the phosphate ion concentration in solution (Larsen 1967) may be summarized as:

$$\text{P in minerals} \rightleftharpoons \text{P in solution} \rightleftharpoons \text{P adsorbed on surfaces.}$$

On the left hand side it is postulated that sparingly soluble phosphates control the concentration in solution in accordance with solubility product principles, whilst on the right hand side the processes considered are those between the phosphate ions in solution and the surface of solid soil particles. If discrete phosphate compounds were present in soil and if a true equilibrium could be attained, it is obvious that the chemical potential in all three phases would be equal, and that both the reactions would be satisfied. Disturbances to the system, either by withdrawal of phosphate or through additions, would then be expected to cause changes in all three phases until the system readjusted to some new equilibrium level. It should be obvious that if the discrete phosphate phase remained unchanged, it would have a constant chemical potential and ultimately the original solution concentration would be re-attained provided that other ion activities remained unchanged; thus the basic reactions would be dissolution and precipitation. This simple situation is not usually encountered in a heterogenous system such as soil.

The sorption process that is observed in the laboratory has aspects that can be associated with both reactions and it is not usually possible to positively identify which is the controlling process. In the 1950s and after, there was much emphasis on the solubility products of possible soil phosphate constituents. This work stemmed largely from workers at Tennessee Valley Authority, but except for showing the probable occurrence of dicalcium phosphate in some soils following enrichment with phosphate, conformity with other discrete calcium, iron and aluminium phosphates was more tenuous. Some aspects of studies to examine whether solution phosphate concentration is controlled by discrete phosphate compounds are to be found, for example, in the papers of Murrmann and Peech (1968, 1969*a*). In a further paper (1969*b*) the same authors concluded '... regardless of whether or not any highly insoluble crystalline phosphate ... is in equilibrium with the labile phosphate of the soil, it is the labile phosphate that determines the concentration of phosphate in solution and the rate at which this equilibrium is replenished upon leaching or removal of phosphate by plant roots.' In this context the labile phosphate is essentially surface phosphate that is capable of isotopic exchange. Other observations which show increases in negative charge on the soil surfaces with phosphate sorption also strongly implicate the greater importance of the surface reaction, since a precipitation type of reaction cannot account for such behaviour whereas a sorption process involving ligand exchange does.

More recently there has been a trend towards treating the sorption process in terms of adsorption equations. The rationale for seeking to describe the observations in the form of mathematical equations lies in the need to summarise this characteristic of a soil in terms of a few parameters, which may in turn be related to other soil properties or used in computations concerning phosphate uptake or movement through soil. For the latter purpose mathematical equations are easier to manipulate than hand fitted curves, though in some instances this simpler approach is adequate. For example, where sorption curves are used as an aid to fertilizer recommendations (for example Ozanne and Shaw 1969; Fox and Kamprath 1970) all that is required is the phosphate sorbed, or in some instances the slope of the curve, at a particular solution concentration and the overall shape of the rest of the curve is ignored.

Phosphate sorption isotherms

Barrow (1978) has recently published an excellent critique of the fitting of equations to describe the phosphate sorption characteristics of soils and readers are referred to this for further details.

The forms of equations that have been used are predominantly the following.

(a) The Freundlich equation, which can be written

$$x = ac^b$$

where x is the amount of sorbed phosphate per unit weight of soil and c is the phosphate concentration in solution; a and b are coefficients.

This equation was probably the first to be fitted to sorption data and it is still widely used; some examples of its use are to be found in the work of Barrow (1973), Fitter and Sutton (1975) and White and Taylor (1977).

(b) The Langmuir isotherm, which has the form

$$x = x_m kc/(1 + kc).$$

It has been usual to ascribe physical meaning to the coefficients of this equation: x_m is the adsorption maximum and k is an affinity term which reflects the relative rates of adsorption and desorption.

The first workers to apply the Langmuir isotherm to phosphate sorption were Olsen and Watanabe (1957) and it has been widely used since. However, it is usually found that this equation only describes sorption over a narrow range of concentration so that the advantage claimed for it, that coefficients having physical meaning can be determined, does not stand up since the 'constants' evaluated depend upon the concentration range considered.

(c) Multi-surface Langmuir equations. A basic assumption of the Langmuir equation is that discrete sorption sites are present that have uniform affinity for phosphate. This can be extended by postulating that more than one discrete surface is present, each having its own affinity term and sorption maximum so that one can write

$$x = \frac{x_{m1}k_1c}{1 + k_1c} + \frac{x_{m2}k_2c}{1 + k_2c} + \cdots$$

This approach has gained considerable popularity of late and examples are to be found in the literature of workers using two- and three-surface equations (e.g. Holford *et al.* 1974; Ryden *et al.* 1977).

(d) The Temkin equation can be reduced to the form

$$x = a \ln (bc)$$

where a and b are coefficients.

Plotting sorption against a logarithmic function of concentration is a convenient means of presenting sorption curves since a large range of concentration can be shown. However, such plots are usually smooth curves, and this equation has been little used in describing the sorption process.

(e) Mechanistic equations. When phosphate, or other anions, is specifically adsorbed onto soil, the negative charge on the soil surface increases. This effect means that it becomes increasingly difficult for each increment of phosphate to be adsorbed, but this change in the nature of the sorbing surface is not allowed for in the models above; the Langmuir equation assumes a constant affinity for each site, whilst the Temkin and Freundlich equations imply that the affinity term decreases linearly and exponentially, respectively, with surface coverage.

Bowden *et al.* (1973, 1977) have formulated equations which describe the effects on adsorption of not only changes in concentration but also those of pH and the concentration of the support electrolyte, both of which also affect the net charge on the surface. The model is of necessity complex and to date has been little applied to sorption data, but it does offer real possibilities of advancing our understanding of the process in ways that have not been possible with the other equations, which fail to adequately describe the nature of all the effects which occur.

In brief, it is necessary to solve a series of equations describing the charging behaviour of the adsorbing material to derive the charge density and electrostatic potential at the surface and in the various planes of adsorption. This involves all the ions in the system. The adsorption of a single ion, *i*, can be written:

$$x_i = \frac{x_m k_i c_i \exp(-z_i F\psi_a/RT)}{1 + \sum_i k_i c_i \exp(-z_i F\psi_a/RT)}.$$

x, *c* and *k* are defined above, but now have different values denoted by the subscript *i* for each ion present; *z* indicates valency (including sign); x_m is the adsorption maximum but is unique for the system and independent of such variables as pH; *F* is the Faraday, ψ_a the electrostatic potential in the plane of adsorption, *R* the gas constant and *T* the temperature in degrees Kelvin.

An ion that adsorbs in the plane of potential ψ_a thus both affects and depends upon the magnitude of that potential.

The sorption equation itself is based on Langmuir adsorption statistics but the exponential terms take account of the electrostatic interactions which must be present when ionic species adsorb at charged surfaces. If these are ignored, that is $\psi_a = 0$, the equation becomes equivalent to the Langmuir equation given above.

Fitting and interpreting sorption equations

When experimental data are found to be fitted by some mathematical expression, it is logical to assume that the premises on which it is based are true. By the same token, if a good fit is not obtained the premises will be rejected. However, there are inherent dangers in this approach. As mentioned previously, the operational details can have considerable effects on the observed sorption characteristics. Such effects may contribute to the lack of fit of some particular model. In this section some of these effects are considered further.

Phosphate initially present in the soil

Irrespective of the model chosen, there is a need to take account of phosphate already present in the soil before it is brought into contact with a phosphate solution: if allowance is not made this can be the cause of apparent non-linearity. Fitter and Sutton (1975) and Barrow (1978) have shown how this can affect the fit of the Freundlich isotherm. In this case the equation that should be fitted is

$$x + Q = ac^b$$

where Q is the phosphate initially present, for which an independent estimate is needed or else it can be fitted as an extra unknown.

Soil : solution ratio

The sorption isotherm, if it represents a genuine equilibrium situation, ought to be independent of the soil:solution ratio. However, there have been numerous reports that the soil:solution ratio does have an effect on the amount of phosphate sorbed. Barrow and Shaw (1979*a*) have shown that the effect of soil:solution ratio can interact with the vigour of shaking, and have attributed this effect to the breakdown of soil particles during the shaking period. If the shaking was sufficiently gentle, the effect of soil : solution ratio was shown to become non-existent.

Support electrolyte and pH

Both the nature and the ionic strength of the medium influence the charge characteristics of the soil surfaces, and whilst this can be accommodated in a model such as that due to Bowden *et al.* (1977) this is not so for the other models. The pH of the system is also a variable and Hingston *et al.* (1972) have used the term 'adsorption envelopes' to define the relationship between adsorption and pH. Thus an experimentor has to decide whether to adjust the pH of the suspension to some fixed value or carry out the sorption measurements in unbuffered solutions. If this latter approach is adopted a further complication arises because the pH of the system changes as phosphate is sorbed, presumably due to the release of OH^- ions from the surface as this ligand is replaced by phosphate. Often the change in pH from this source is not very large, but it clearly is a source of variation which could influence the goodness of fit of a chosen equation. Rajan *et al.* (1974) have devised means whereby the pH of the system is controlled during sorption studies.

Many support electrolytes have been used for sorption studies. Most recent workers tend to use dilute solutions of $CaCl_2$, NaCl or KCl. The effects of varying the support electrolyte have been reported by, amongst others, Rajan and Fox (1972), Ryden and Syers (1975) and Helyar *et al.* (1976*a*, *b*).

The usual observation is that sorption increases with increasing concentration of support electrolyte; also at a given ionic strength there are specific effects of the cation present. Helyar *et al.* (1976a) explained the latter in terms of a complex being formed between calcium and adjacent adsorbed phosphate ions, thus reducing the repulsion force between them. In a soil, which would contain exchangeable calcium, increasing the concentration of say sodium or potassium in the support electrolyte would displace calcium into solution and thus could cause a similar effect. Barrow and Shaw (1979*b*) have questioned this explanation; they discount the idea of a complex between calcium and the adsorbed phosphate

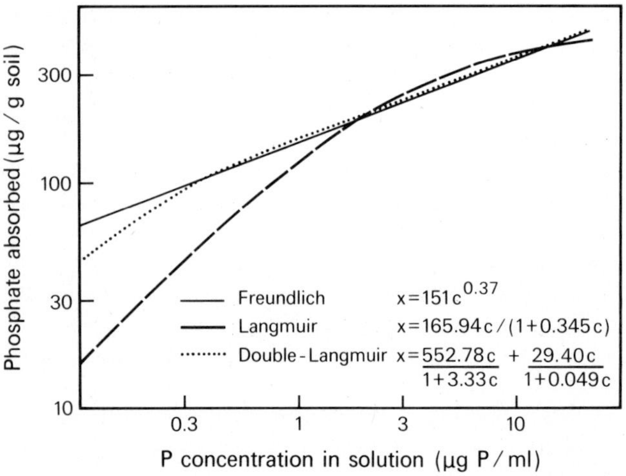

Fig. 29.1. Comparison of Freundlich, Langmuir and Double-Langmuir equations fitted to phosphate sorption data for a Western Australian soil. The log sorption vs log concentration presentation is used to illustrate the poorer agreement between the three equations at low concentrations of phosphate. The residual variance of the original data points about the fitted curves was least for the Freundlich equation. (Reproduced with permission from Barrow 1978.)

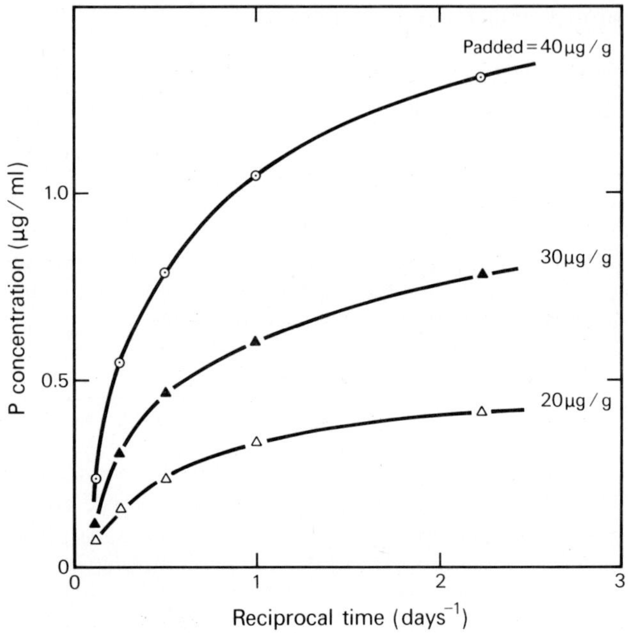

Fig. 29.2. Phosphate concentration remaining in solution after various periods of shaking for a north Queensland *non-calcic brown* duplex surface soil. The plots at three levels of added phosphate are made against reciprocal time as proposed by Ryden and Syers (1975) and show no indication of the linearity they obtained. Conditions for shaking: 2 g soil to 20 ml 10^{-2} M $CaCl_2$ at 25°C. (M.E. Probert, unpublished data.)

ions and explain the effects of ionic strength and nature of the cation in terms of distribution of charge at the adsorbing surface according to the model of Bowden *et al.* (1977).

Other workers (Ryden and Syers 1975) have suggested that the observed effects of support electrolyte are not so much associated with the total phosphate sorbed as with the rate of approach to equilibrium. Thus, using their method of extrapolation, they have suggested that the 'equilibrium isotherm' is independent of electrolyte and its concentration at least for solutions of $CaCl_2$ and NaCl.

Limitations of the Langmuir isotherm

The Langmuir isotherm, as it is usually applied to sorption studies, should not be taken as sufficient evidence that a sorption mechanism is necessarily operating. Veith and Sposito (1977) have discussed this aspect and have shown how good fits to the Langmuir isotherm can be obtained for reactions that are clearly not sorption but precipitation. Furthermore, an equation of similar form to the Langmuir equation is appropriate for ion exchange reactions (Harter and Baker 1977; Sposito 1979). In this application of the equation it is necessary to take account of the concentration of the desorbed species in solution. Griffin and Au (1977) have shown that using such an approach the apparent effect of soil: solution ratio on the adsorption of lead by montmorillonite disappeared. It has been suggested that this form of equation ought to be applicable to phosphate sorption but to date it has not been so used. Its use could possibly resolve some of the observed effects on phosphate sorption of soil:solution ratio and ionic strength of the support electrolyte.

Similarly, the use of multi-surface Langmuir equations could be little more than a curve fitting exercise, and it is obvious that the extra coefficients should be expected to lead to improved fits to data sets. Fig. 29.1, after Barrow (1978), illustrates the similarity between the shape of the Freundlich and Double-Langmuir equations.

If such multi-surface equations are to be used it is essential that they be justified by some criteria other than the goodness of fit of the equations to the experimental points. Some evidence of this nature is available (Ryden *et al.* 1977; Rajan and Fox 1975) and the indications are that the reactions occurring may be different for each of the surfaces. For the high energy sites the surface becomes more negative with no change in pH, for the lower energy sites the charge is unchanged but pH increases, whilst at yet higher concentrations of phosphate the surface may be destroyed and Si released (Rajan and Watkinson 1976) or a weaker, more physical sorption of P (as distinct from chemisorption) is observed (Ryden *et al.* 1977). Ryden and Syers (1977) have further suggested that it is this more physically sorbed P that is amenable to desorption and isotopic exchange. Results of this nature at least give some credence to the use of the multi-surface equations.

The variable surface charge–variable surface potential model

Whilst the model of Bowden *et al.* has not been applied to phosphate sorption by soil, workers in Western Australia have had some spectacular successes in using it to describe the sorption of phosphate on goethite (Bowden *et al.* 1980). In particular, many of the aspects of phosphate sorption mentioned above can be at least qualitatively predicted by the model. These include the effects of pH, ionic strength and nature of the support electrolyte. Furthermore Posner and Bowden (1980) have shown that the model, which is based on a single type of adsorption site, generates theoretical isotherms which behave very similarly to experimental data for which other workers have invoked the concept of two or more types of sorption sites. Posner and Bowden state '(their) continuous model shows the futility of splitting isotherms into two or more regions and hence we think this procedure should not be undertaken . . .'

From their work to date it appears that although the model requires that all the individual ionic species be considered, for the sorption of phosphate onto goethite it is only the divalent ion that is sorbed.

Kinetics of phosphate sorption

Although discussion of the effect of time on phosphate sorption has been left to last, it is probably the most significant aspect in that it is crucial to our understanding of the phenomenon and assumes dominance when the movement into soil of solutions containing phosphate is considered (Novak and Adriano 1975; Enfield *et al.* 1976). It is probably the one aspect of sorption studies on which there is widespread agreement in that equilibrium is approached very slowly in the systems under study and it is doubtful whether a true equilibrium has ever been obtained.

The fact that one is not dealing with an equilibrium situation should effectively disqualify any of the adsorption isotherms as being genuine descriptions of the phosphate sorption process, since these all assume the system is at equilibrium. For example, the affinity term of the Langmuir equation is in fact the ratio of the rates of the forward and reverse reactions and thus only has physical meaning at equilibrium: this is also true of the affinity terms for the individual species of the Bowden model.

The manner in which various workers handle the effects of time indicate how poor our understanding of this aspect really is; there are basically four approaches found in the literature.

Firstly, it is assumed that after some suitable period equilibrium is effectively attained, and then soils are compared at this uniform period of contact. Times chosen for this purpose vary, but often one day (e.g. Helyar *et al.* 1976a) or six days (e.g. Fox and Kamprath 1970) are used. Where the studies are aimed at understanding fertilizer responses, such comparisons between soils are probably adequate, though it would seem dubious whether this approach is good enough where considerable effort is to be devoted to fitting adsorption isotherms and arguing their significance.

A second approach is advocated by Syers and his coworkers (Ryden and Syers 1975). They propose that by plotting the concentration in solution against reciprocal time, the 'sorption' process can be extrapolated to an equilibrium situation ($1/t \rightarrow 0$, i.e. $t = \infty$). These extrapolated values of concentration are then used to fit an equilibrium isotherm. Such an approach assumes that the continuing slow reactions are part of the adsorption process. Experience in attempting to obtain plots of this type indicate that linearity is rarely a feature (see Fig. 29.2). Furthermore, it might be remarked that it would be rather peculiar chemical kinetics that would give rise to such a relationship.

Barrow and Shaw (1975*a*) adopt a third approach that is conceptually very different. Their model of the behaviour of phosphate in soil has three compartments: the phosphate in solution is controlled by the amount present in the adsorbed form at all times and the third compartment contains phosphate which has been converted into a form which is not in direct equilibrium with the soil solution.

Thus the model assumes that the sorption process itself is a rapid one, and that subsequent changes in solution concentration arise from slower reactions which do not interfere with the sorption equilibrium. This amounts to an extrapolation to $t = 0$ rather than to $t = \infty$ as above, although the actual equations used become indeterminate at $t = 0$. By seeking equations to describe the slow reactions whereby phosphate is lost from the adsorbed form they have had considerable success in modelling changes with time in the phosphate concentration in solution, isotopically exchangeable phosphate, extractable phosphate, and the decrease in effectiveness of applied phosphate for plant growth. In using this approach, Barrow and Shaw have found that the Freundlich isotherm gives an adequate description of that part of the overall reaction which they refer to as adsorption.

In a fourth category we can lump the various other treatments of the kinetic aspects of phosphate sorption that have been attempted. A wide variety of models has been tried with varying degrees of success; some examples are found in the work of Griffin and Jurinak (1973, 1974), Chen *et al.* (1973), Kuo and Lotse (1972, 1974*a*, *b*), Novak and Adriano (1975), Enfield *et al.* (1976) and Overman and Chu (1977). Some of the approaches tried are based upon the solution of the Langmuir equation at finite times, a fuller discussion of which is given by Loebenstein (1962, 1963). Others (e.g. Kuo and Lotse 1974*a*, *b*) have similarities with the approach of Barrow and his co-workers in that the Freundlich isotherm is used to describe the basic sorption behaviour. However, none of them would appear to be as successful as the model of Barrow and Shaw (1975*a*) in describing the observed behaviour of phosphate when contacted with soil.

The failure to adequately understand the effect of time on phosphate sorption makes it extremely difficult to ascertain the true extent of its reversibility. The frequent references to the marked hysteresis involved with phosphate adsorption-desorption could well be due to a lack of equilibrium during the adsorption process, so that both forward and reverse reactions continue to operate during the desorption period (Munns and Fox 1976; White and Taylor 1977). Barrow and Shaw (1975b) have shown how their model, which assumes rapid adsorption followed by slow reactions of the adsorbed phosphate, can explain such behaviour.

Outlook for furthering our understanding

It is to be regretted that no conclusion can be reached here as to which model or models should be used to describe the phenomenon of phosphate sorption. Despite the attention the subject has received there are still major gaps in our understanding. The speculation that has been advanced on the choice of the appropriate isotherm may count for little when the kinetics of the process have largely been ignored. Trying to decide the merits of the various isotherms without regard to whether one is in fact dealing with an equilibrium system would seem rather futile.

There does seem to be something of a picture that emerges in terms of treating the overall process in terms of a rapid initial adsorption reaction followed by slower reactions. This is explicit in the model of Barrow and Shaw (1975a) and implied in the work of Chen *et al.* (1973), Helyar *et al.* (1976*a*) and Munns and Fox (1976). Where they differ is in how the slow reactions should be treated: Barrow and Shaw describe the rate of the slow reaction in terms of the proportion of the added phosphate remaining in equilibrium with the solution phase, i.e. the adsorbed phosphate, whereas the others have remarked upon the apparent linearity of semi-logarithmic plots of phosphate concentration against time of contact suggesting that the rate of the slow reactions depend upon the solution phosphate concentration, not on sorbed phosphate. Munns and Fox (1976) remark upon the possibility of extrapolation of such plots to zero time in order to derive the equilibrium state of the adsorption reaction. An example of this method of treatment is shown in Fig. 29.3, from which it could be concluded that for this soil the sorption data after shaking for one day is a close approximation to the extrapolated values assumed to be free from the effects of the slow reactions.

It is axiomatic that any complete model for describing the phosphate sorption phenomenon should be capable of predicting all of the observations. This includes the changes in charge upon the surfaces which occur. Only the mechanistic approach of Bowden *et al.* (1973, 1977) promises such progress, and it would appear that for the simpler, non-soil systems that have been studied to date this model can be tuned to accurately describe the sorption and charge data.

In this chapter, no attempt has been made to deal with the literature concerned with understanding the nature of the sorption sites. This aspect has been reviewed by Parfitt (1978) and he describes the evidence available from infrared spectroscopy regarding the nature of the bonding between the adsorbed anion and the surface. It would be expected that the mathematical expressions of sorption would accord with such facts.

However, for some purposes a less than complete model may be adequate to advance our understanding. A good example of this is to be found in the use of sorption data as an aid to fertilizer recommendations. Here the detailed mathematical equation for the whole isotherm is of little consequence: users of sorption data in this manner would be well advised to stop worrying whether their data fit single or multi-surface Langmuir or Freundlich isotherms, since it is not to be expected that they will, and concern themselves rather with the operating variables under which their data are obtained, as these are likely to have an influence and thus affect the usefulness of the predictions that can be made.

Barrow (1978), in discussing some of the factors to be considered in choosing between different equations for describing sorption data, advocates the 'simplest model' unless some *a priori* reason exists for the more complex. The 'simplest' is likely to be that adequate for the immediate needs.

Phosphate sorption by Australian soils

No systematic study has been made of the phosphate sorption characteristics of a wide range of Australian soils, and differences in methodology make it a dubious practice to try to compare results obtained by various workers.

Beckwith (1965), when he suggested that sorption data could be interpreted to indicate the phosphate needs of soils, reported data for a range of south-eastern Queensland soils and sub-soils. Whilst the use of 0.1 M sodium acetate at pH 5.1–5.2 as the support electrolyte would not now be the normal practice, the results do highlight the differences that are to be found. The soils studied included *krasnozems*, which sorbed large amounts of phosphate, grey and brown clays of the brigalow lands, which sorbed much smaller amounts of phosphate, and those developed from phyllite in the Gympie district

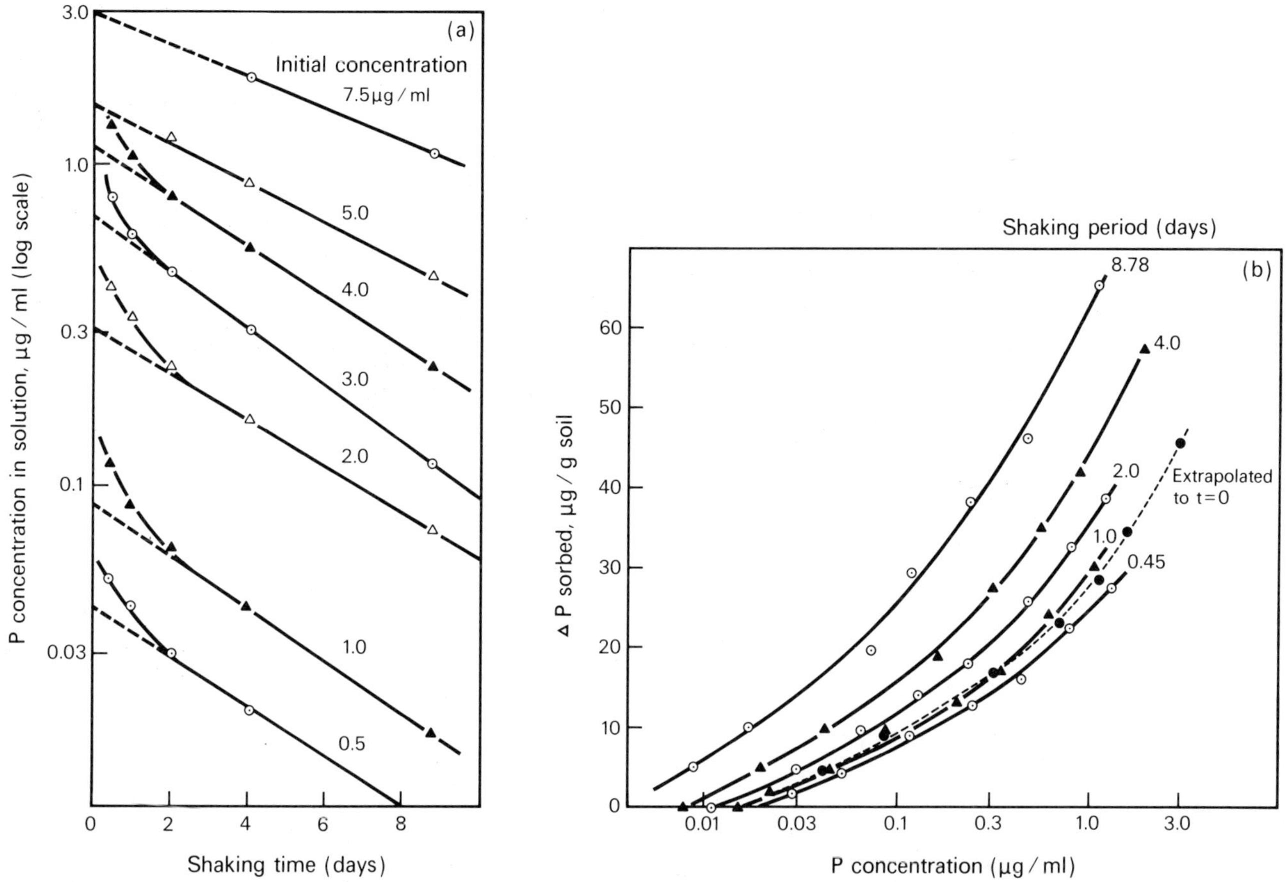

Fig. 29.3. The effect of shaking time on phosphate sorption by a north Queensland *non-calcic brown* duplex surface soil. (M.E. Probert, unpublished data.) *a*, The close approximation to linearity of the log concentration vs shaking time plots for various initial concentrations. These plots could be extrapolated to $t = 0$, thus giving an indication of the adsorbed phosphate free of effects from the slower reactions. *b*, Same data plotted as sorption isotherms for the different shaking periods. The values derived by extrapolation to $t = 0$ are shown by the dashed line. All curves are hand drawn.

(dominantly *red podzolic soils* and *xanthozems*) which were intermediate between the extremes of the *krasnozems* and soils of the brigalow lands.

A compilation of data for a wider range of soils is reported by Barrow (1980). For a collection of 16 Australian surface soils from five localities, the sorption data were fitted to the model:

$$x = ac^{b_1}\, t^{b_2}$$

where x is the phosphate sorbed (μg P g^{-1} soil), c is the solution concentration (μg P ml^{-1}) and t is time in days; a, b_1 and b_2 are coefficients.

For the soils studied, a, which represents the phosphate sorbed under the standard conditions of 1 μg P ml^{-1} in solution after one day, ranged from 12 to 930 μg P g^{-1} soil, whilst the coefficient b_1, indicating the effect of concentration at constant time, was in the range 0.35 to 0.58, and b_2, summarizing the effect of time at constant concentration, was in the range 0.14 to 0.27. The wide variation in both b_1 and b_2 was something of a surprise in view of earlier experience with Western Australian soils (Barrow and Shaw 1975*a*), but the two coefficients were found to be correlated.

Barrow (1980) suggests that the range in the values of the coefficients is less when soils from a restricted locality are considered, and it is to be presumed that a still narrower range would be encountered if a single soil group was considered. Some data for a range of virgin north Queensland soils show this effect (Table 29.1).

The alfisols sorb little phosphate in their surface layers whilst the basaltic oxisols (*krasnozems*) sorb large amounts. The non-basaltic oxisols, ultisols and inceptisols were somewhat variable in the amounts of P sorbed but were intermediated between the extremes of the alfisols and oxisols derived from basalt. Invariably the subsoils sorbed more phosphate than the surface soils.

These phosphate sorption data again indicate the diversity that exists. However, they do accord reasonably well with experience in establishing legumes into the native pastures of north Queensland. For example, Kerridge *et al.* (1972) have reported on the high phosphate requirements for *krasnozems*; Winter and Gillman

Table 29.1. Phosphate sorption characteristics for some soils from tropical Queensland

The characteristics were determined by shaking soil with 10^{-2} M $CaCl_2$ solutions (soil:solution ratio = 1:10) containing graded amounts of KH_2PO_4 on an end-over-end shaker for 24 h at 25°C

Soil taxonomy	Great soil group	Site	A_1 horizon		B_2 horizon	
			P sorbed[A] (μ g^{-1} soil)	PBC[B] (μ g^{-1} 0.1 ppm^{-1})	P sorbed[A] (μ g^{-1} soil)	PBC[B] (μ g^{-1} 0.1 ppm^{-1})
Oxisols		Malanda	225	64	667	88
	Krasnozem	Gregory Falls	239	53	742	119
		Clump Point	239	49	964	129
		Tolga	133	43	445	52
	Red podzolic (?)	Innisfail	110	42	219	36
	Xanthozem	Lake Barrine	54	24	122	24
		Tully	100	27	287	45
		Mt Spec.	134	21	nd	nd
Ultisols	*Red podzolic*	Lamb Range	155	55	314	43
		Cardwell Range	26	11	113	20
	Red earth	Mt Tyson	40	14	484	46
		McDonnell	101	14	128	15
	Yellow earth	McDonnell	66	12	163	19
Inceptisols	*Euchrozem*	Meadowbank	26	34	358	66
		Talavera	53	22	522	59
		Hillgrove	80	15	nd	nd
		Boomerang	63	28	nd	nd
Alfisols	*Yellow earth*	Laura	9	2	93	12
		Redlands	14	3	91	11
	Red earth	Redlands	14	4	58	10
		Pentland	10	4	58	10
	Non-calcic brown	Virginia Park	13	4	nd	nd
	Solodic	Manton	18	4	nd	nd

[A] P sorbed at final concentration of 0.2 ppm.
[B] P sorbed in increasing final concentration from 0.2 to 0.3 ppm.

(1976) found that at least 110 kg ha^{-1} of P was required to obtain maximum growth of *Stylosanthes guianensis/Brachiaria decumbens* pastures on the *yellow earth* at McDonnell, Cape York Peninsula, in contrast to recent studies on the *red* and *yellow earths* at 'Redlands' west of Charters Towers (M. E. Probert and J. Williams, unpublished data) which have shown maximum yields of *Stylosanthes hamata* with less than 20 kg ha^{-1} of P. The absence of a response to applied phosphate on the *euchrozems* (Miller and Jones 1977) may appear to be anomolous in view of the values reported for them in Table 29.1. However, for these soils, the isotherms indicate that the final concentration at which P sorbed is zero is in the range 0.05 to 0.1 ppm; this is much higher than in the alfisols and is probably above the external P requirement for *Stylosanthes* species. The isotherms suggest, then, that these soils are well buffered against changes in solution concentration when phosphate is withdrawn through uptake by plants.

Similar comments regarding the phosphate requirements of his soils were made by Beckwith (1965). The use of sorption data for predicting the phosphate requirements of pastures has been studied in greater detail in Western Australia (Ozanne and Shaw 1967, 1969), where it has been shown that the method offers considerable improvement over the use of traditional extraction techniques.

It would seem desirable to have better information on the phosphate sorption characteristics of our soils in as much as this is likely to be a unifying parameter when comparing results of phosphate fertilizer response experiments. In general terms, phosphate sorption would be expected to be high in acid soils, in those with large surface areas, and in those with an abundance of 'reactive' iron and aluminium. These properties are likely to be found in the more highly weathered soils such as *krasnozems*, *xanthozems*, *red* and *yellow earths* and *earthy sands*; the general distribution of these soils is discussed in Chapter 3.

References

Barrow, N. J. (1973). Relationship between a soil's ability to adsorb phosphate and the residual effectiveness of superphosphate. *Aust. J. Soil Res.* **11**, 57-63.

Barrow, N. J. (1978). The description of phosphate adsorption curves. *J. Soil Sci.* **29**, 447-62.

Barrow, N. J. (1980). Differences amongst a wide-ranging collection of soils in the rate of reaction to phosphate. *Aust. J. Soil Res.* **18**, 215-24.

Barrow, N. J. and Shaw, T. C. (1975*a*). The slow reactions between soil and anions: 2. Effect of time and temperature on the decrease in phosphate concentration in the soil solution. *Soil Sci.* **119**, 167-77.

Barrow, N. J. and Shaw, T. C. (1975*b*). The slow reactions between soil and anions: 5. Effects of period of prior contact on the desorption of phosphate from soils. *Soil Sci.* **119**, 311-20.

Barrow, N. J. and Shaw, T. C. (1979*a*). Effects of solution:soil ratio and vigour of shaking on the rate of phosphate adsorption by soil. *J. Soil Sci.* **30**, 67-76.

Barrow, N. J. and Shaw, T. C. (1979*b*). Effects of ionic strength and nature of the cation on desorption of phosphate from soil. *J. Soil Sci.* **30**, 53-65.

Beckwith, R. S. (1965). Sorbed phosphate at standard supernatant concentration as an estimate of the phosphate needs of soils. *Aust. J. Exp. Agric. Anim. Husb.* **5**, 52-8.

Bowden, J. W., Bolland, M. D. A., Posner, A. M. and Quirk, J. P. (1973). Generalized model for anion and cation adsorption at oxide surfaces. *Nature* (London) **245**, 81-3.

Bowden, J. W., Nagarajah, S., Barrow, N. J., Posner, A. M. and Quirk, J. P. (1980). Describing the adsorption of phosphate, citrate and selenite on a variable-charge mineral surface. *Aust. J. Soil Res.* **18**, 49-60.

Bowden, J. W., Posner, A. M. and Quirk, J. P. (1977). Ionic adsorption on variable charge mineral surfaces. Theoretical charge development and titration curves. *Aust. J. Soil Res.* **15**, 121-36.

Chen, Y. S. R., Butler, J. N. and Stumm, W. (1973). Kinetic study of phosphate reaction with aluminium oxide and kaolinite. *Environ. Sci. Technol.* **7**, 327-32.

Enfield, C. G., Harlin, C. C. and Bledsoe, B. E. (1976). Comparison of five kinetic models for orthophosphate reactions in mineral soils. *Soil Sci. Soc. Am. J.* **40**, 243-9.

Fitter, A. H. and Sutton, C. D. (1975). The use of the Freundlich isotherm for soil phosphate sorption data. *J. Soil Sci.* **26**, 241-6.

Fox, R. L. and Kamprath, E. J. (1970). Phosphate sorption isotherms for evaluating the phosphate requirements of soils. *Soil Sci. Soc. Am. Proc.* **34**, 902-7.

Griffin, R. A. and Au, A. K. (1977). Lead adsorption by montmorillonite using a competitive Langmuir equation. *Soil Sci. Soc. Am. J.* **41**, 880-2.

Griffin, R. A. and Jurinak, J. J. (1973). Test of a new model for the kinetics of adsorption-desorption processes. *Soil Sci. Soc. Am. Proc.* **37**, 869-72.

Griffin, R. A. and Jurinak, J. J. (1974). Kinetics of the phosphate interaction with calcite. *Soil Sci. Soc. Am. Proc.* **38**, 75-9.

Harter, R. D. and Baker, D. E. (1977). Applications and misapplications of the Langmuir equation to soil adsorption phenomenon. *Soil Sci. Soc. Am. J.* **41**, 1077-80.

Helyar, K. R., Munns, D. N. and Burau, R. G. (1976*a*). Adsorption of phosphate by gibbsite. I. Effects of neutral chloride salts of calcium, magnesium, sodium, and potassium. *J. Soil Sci.* **27**, 307-14.

Helyar, K. R., Munns, D. N. and Burau, R. G. (1976*b*). Adsorption of phosphate by gibbsite. II. Formation of a surface complex involving divalent cations. *J. Soil Sci.* **27**, 315-23.

Hingston, F. J., Posner, A. M. and Quirk, J. P. (1972). Anion adsorption by goethite and gibbsite. I. The role of the proton in determining adsorption envelopes. *J. Soil Sci.* **23**, 177-92.

Holford, I. C. R., Wedderburn, R. W. M. and Mattingly, G. E. G. (1974). A Langmuir two-surface equation as a model for phosphate adsorption by soils. *J. Soil Sci* **25**, 242-55.

Kerridge, P. C., Andrew, C. S. and Murtha, G. G. (1972). Plant nutrition status of soils of the Atherton Tableland, North Queensland. *Aust. J. Exp. Agric. Anim. Husb.* **12**, 618-27.

Kuo, S. and Lotse, E. G. (1972). Kinetics of phosphate adsorption by calcium carbonate and Ca-kaolinite. *Soil Sci. Soc. Am. Proc.* **36**, 725-9.

Kuo, S. and Lotse, E. G. (1974*a*). Kinetics of phosphate adsorption and desorption by hematite and gibbsite. *Soil Sci.* **116**, 400-6.

Kuo, S. and Lotse, E. G. (1974*b*). Kinetics of phosphate adsorption and desorption by lake sediments. *Soil Sci. Soc. Am. Proc.* **38**, 50-4.

Larsen, S. (1967). Soil phosphorus. *Adv. Agron.* **19**, 151-210.

Loebenstein, W. V. (1962). Batch adsorption from solution. *J. Res. Nat. Bur. Stand. Sect. A.* **66A**, 503-14.

Loebenstein, W. V. (1963). Adsorption, desorption and resorption. *J. Res. Nat. Bur. Stand. Sect. A.* **67A**, 615-24.

Miller, C. P. and Jones, R. K. (1977). Nutrient requirements of *Stylosanthes guianensis* pastures on a euchrozem in North Queensland. *Aust. J. Exp. Agric. Anim. Husb.* **17**, 607-13.

Munns, D. N. and Fox, R. L. (1976). The slow reaction which continues after phosphate adsorption: kinetics and equilibrium in some tropical soils. *Soil Sci. Soc. Am. J.* **40**, 46-51.

Murrmann, R. P. and Peech, M. (1968). Reaction products of applied phosphate in limed soils. *Soil Sci. Soc. Am. Proc.* **32**, 493-6.

Murrmann, R. P. and Peech, M. (1969*a*). Effect of pH on labile and soluble phosphate in soils. *Soil Sci. Soc. Am. Proc.* **33**, 205-10.

Murrmann, R. P. and Peech, M. (1969*b*). Relative significance of labile and crystalline phosphates in soil. *Soil Sci.* **107**, 249-55.

Novak, L. T. and Adriano, D. C. (1975). Phosphate movement in soils: soil-orthophosphate reaction kinetics. *J. Environ. Qual.* **4**, 261-6.

Olsen, S. R. and Watanabe, F. S. (1957). A method to determine a phosphate adsorption maximum of soil as measured by the Langmuir isotherm. *Soil Sci. Soc. Am. Proc.* **21**, 144-9.

Overman, A. R. and Chu, R. L. (1977). A kinetic model of steady state phosphorus fixation in a batch reaction — I. Effect of soil/solution ratio. *Water Res.* **11**, 771-5.

Ozanne, P. G. and Shaw, T. C. (1967). Phosphate sorption by soils as a measure of the phosphate requirements for pasture growth. *Aust. J. Agric. Res.* **18**, 601-12.

Ozanne, P. G. and Shaw, T. C. (1969). Advantages of the recently developed phosphate sorption test over the older extraction methods for soil phosphate. Trans. 9th Int. Cong. Soil Sci. (Adelaide) Vol. II, pp. 273-80.

Parfitt, R. L. (1978). Anion adsorption by soils and soil materials. *Adv. Agron.* **30**, 1-50.

Posner, A. M. and Bowden, J. W. (1980). Adsorption isotherms: should they be split? *J. Soil Sci.* **31**, 1-10.

Rajan, S. S. S. and Fox, R. L. (1972). Phosphate adsorption by soils. I. Influence of time and ionic environment on phosphate adsorption. *Comm. Soil Sci. Plant Anal.* **3**, 493-504.

Rajan, S. S. S. and Fox, R. L. (1975). Phosphate adsorption by soils. II. Reactions in tropical acid soils. *Soil Sci. Soc. Am. Proc.* **39**, 846-51.

Rajan, S. S. S., Perrott, K. W. and Saunders, W. H. M. (1974). Identification of phosphate-reactive sites of hydrous alumina from proton consumption during phosphate adsorption at constant pH values. *J. Soil Sci.* **25**, 438-47.

Rajan, S. S. S. and Watkinson, J. H. (1976). Adsorption of selenite and phosphate on allophane clay. *Soil Sci. Soc. Am. J.* **40**, 52-4.

Ryden, J. C., McLaughlin, J. R. and Syers, J. K. (1977). Mechanisms of phosphate sorption by soils and hydrous ferric oxide gel. *J. Soil Sci.* **28**, 72-92.

Ryden, J. C. and Syers, J. K. (1975). Rationalization of ionic strength and cation effects on phosphate sorption by soils. *J. Soil Sci.* **26**, 395-406.

Ryden, J. C. and Syers, J. K. (1977). Desorption and isotopic exchange relationships of phosphate sorbed by soils and hydrous ferric oxide gel. *J. Soil Sci.* **28**, 596-609.

Sposito, G. (1979). Derivation of the Langmuir equation for ion exchange reactions in soils. *Soil Sci. Soc. Am. J.* **43**, 197-8.

Veith, J. A. and Sposito, G. (1977). On the use of the Langmuir equation in the interpretation of 'Adsorption' phenomena. *Soil Sci. Soc. Am. J.* **41**, 697-702.

White, R. E. and Taylor, A. W. (1977). Reactions of soluble phosphate with acid soils: the interpretation of adsorption-desorption isotherms. *J. Soil Sci.* **28**, 314-28.

Winter, W. H. and Gillman, G. P. (1976). Plant nutrient studies on some yellow and red earth soils in northern Cape Yorke Peninsula. 2. Phosphorus — plant response and soil retention. *Aust. J. Exp. Agric. Anim. Husb.* **16**, 542-8.

Part 3

Physics and Hydrology

Chapters 30-33

Chapter 30

The fabric of Australian soils

R. Brewer, J. R. Sleeman and R. C. Foster

IT IS NOT THE PURPOSE OF THIS CONTRIBUTION TO SET out in detail the principles and practice of fabric analysis, but it seems advisable to summarize briefly the concepts and methods. A comprehensive treatment of mineral soils and biological materials can be found in Brewer (1964), Bal (1973) and Babel (1975).

Fabric deals with the size, shape and spatial arrangements of ultimate particles, and of compound units and amorphous constituents. Descriptions of the arrangement of the solid components automatically describe the characteristics of the voids, but patterns of voids are often described positively by treating them as individuals (Brewer 1964; Beckmann and Geyger 1967). Most soil materials consist of mineral grains and/or rock nodules, plant fragments in varying stages of decomposition, and amorphous material that lacks describable shape and dimensions. They may contain biological units such as spores, hyphae, pollen grains, living plant roots, microfossils, and so on, and they are inhabited by a variety of fauna. In its broadest sense, soil fabric deals with the spatial distribution of all these units except, perhaps, for living fauna and, to a lesser extent, living plant tissue, whose mobility is high and/or persistence low. Even so, the present relationship of even short-lived units can be described.

Methods of analysis

Ideally, studies of fabric require three-dimensional observations at all levels from macroscopic to ultramicroscopic as appropriate for the size of the units in the specific sample. Most of the data on soil fabric have been obtained however from thin sections of undisturbed samples using a polarizing light microscope; more recently, some data are accumulating from use of the electron microscope. So far, these data have not been correlated, to any significant degree, with macroscopic and mesoscopic descriptions of soil structure in the field.

The major limitations of using thin sections with a light microscope are that they are two-dimensional and the limit of resolution is about 2 μm. Therefore, three-dimensional spatial relationships must be inferred from two-dimensional data, and units smaller than 2 μm cannot be observed directly. However most of the important characteristics of fabric are repeated many times in the one thin section, so a reasonably reliable three-dimensional model can be built up from a large number of observations. Also, the spatial relationships of grains smaller than 2μm can sometimes be inferred from optical properties, as for the anisotropic layer-lattice clay minerals for which orientation patterns can be inferred from extinction patterns using crossed polarizers.

Less important difficulties arise from the thickness (25 μm) of the sections. Units are often stacked on top of each other, and the more so with decreasing size; voids smaller than 25 μm are often embedded in the section, and voids of any size may only partially penetrate through the section. As a secondary effect, finely dispersed organic matter and iron oxides may mask the optical properties of other constituents. These difficulties can be minimized with thinner sections.

Transmission electron microscopy enables soil components with a high intrinsic electron density to be detected almost to the limits of resolution of the electron microscope (0.3 nm). However, it is necessary to enhance the electron density of organic components by deposition of heavy metals onto or into them, and this limits the resolution to about 2–5 nm. Because of the low penetrating power of electrons, specimens must be less than 100 nm thick.

Tovey (1974) and Smart (1974) reviewed the methods and limitations of the techniques available to produce specimens of soil materials for electron microscopy. More recently, by stabilizing soil materials with agar, and fixing the organics with aldehydes and divalent metals, and introducing heavy metal stains, Foster (1978) has been able to demonstrate the detailed interrelationships between both organic and inorganic components *in situ* in ultrathin sections of undisturbed soil materials. Although the larger sand and silt-size particles are usually torn out of the embedding medium during sectioning, leaving electron transparent holes, any surrounding cutans may remain in place (Fig. 30.1). Higher resolutions are attainable in this way than are possible by the replica or scanning electron microscope methods and, in general, larger volumes can be examined.

Such ultramicroscopic observations provide information concerning the fate of soil constituents down to the

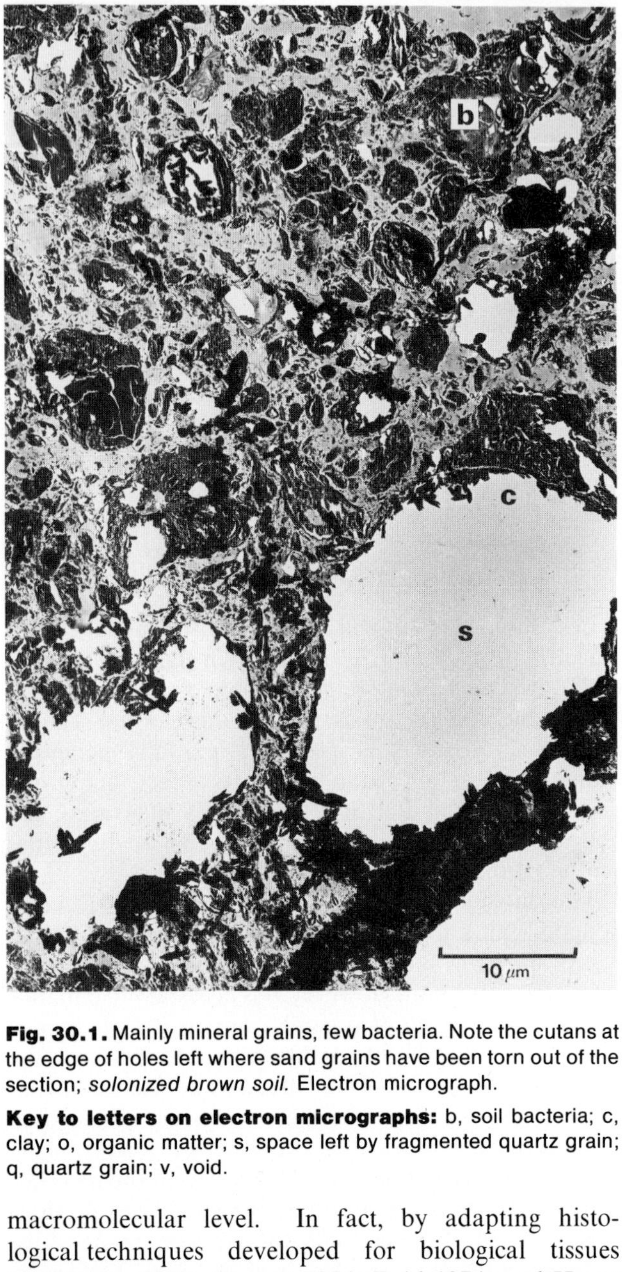

Fig. 30.1. Mainly mineral grains, few bacteria. Note the cutans at the edge of holes left where sand grains have been torn out of the section; *solonized brown soil.* Electron micrograph.

Key to letters on electron micrographs: b, soil bacteria; c, clay; o, organic matter; s, space left by fragmented quartz grain; q, quartz grain; v, void.

macromolecular level. In fact, by adapting histological techniques developed for biological tissues (see Pearse 1972; Glauert 1974; Reid 1974; and Hayat 1975), sites of specific kinds of carbohydrates and proteins (e.g. enzymes) have been recognized in soils (Foster 1978). As more advanced techniques, such as STEM, become more common, it should be possible to look at thicker sections and so avoid edge effects. By using the electron probe micro analyser to check the identity of electron-dense deposits seen after histological tests, the location of specific biochemicals in fabrics can be made more certain. At present, histological techniques are limited by the fact that heavy metals must be used in the stains, and, in general, these are adsorbed non-specifically by many soil organics; since it is impossible to treat successive sections differently, the preparation of adequate controls is limited.

Description and interpretation of soil fabric

Soil materials are commonly highly heterogeneous, both in composition and fabric. Therefore, complete descriptions of fabric are complex, unless divided into logical subdivisions each of which can be described independently. A number of devices have been used for this purpose (Brewer 1964); they are generally referred to as levels of fabric or organization. The most useful of these levels are specific related distribution patterns, plasmic fabrics, and pedological features or, broadly, any features of particular interest. These three are the levels discussed here because they have the most relevance to field description, genesis, and distribution of constituents analysed in the laboratory.

Brewer (1972) discussed the basis of interpretation of fabrics and pedological features; it depends ultimately on consideration of the properties and distribution of the constituents in relation to the processes that are most likely to have operated in the formation of the profile, having regard to its present and possible past environments. Two important processes must always be considered: inheritance of the fabric or feature from the parent material, and pedoturbation which may change the distribution patterns of features subsequent to their formation.

Clearly, the correctness of such interpretations is not generally demonstrable; their reliability depends on patient accumulation of data on fabric in relation to kinds of constituents, and to soil profiles and their environments.

Related distribution patterns

Terminology and principles of occurrence

The concepts used here are those presented by Brewer and Pawluk (1975), as modified from proposals by Stoops and Jongerius (1975). The terminology is that used by Brewer (1977); *f-matrix* and *f-members* are general terms for materials that exhibit recognizable, characteristic, related distribution patterns with regard to each other; *f-clasts* are occasional, exceptionally large units in the soil material that are not an essential part of particular arrangements exhibited by the f-matrix and f-members. The f-members are usually the larger units, whether simple or compound, and the f-matrix the finer-grained material.

Brewer (1979) examined the relationships between related distribution patterns of f-matrix to f-members and particle size distribution of the soil material. He proposed three fabric sequences: the chlamydic (originally chitonic), iunctic and matrigranic sequences. Differences

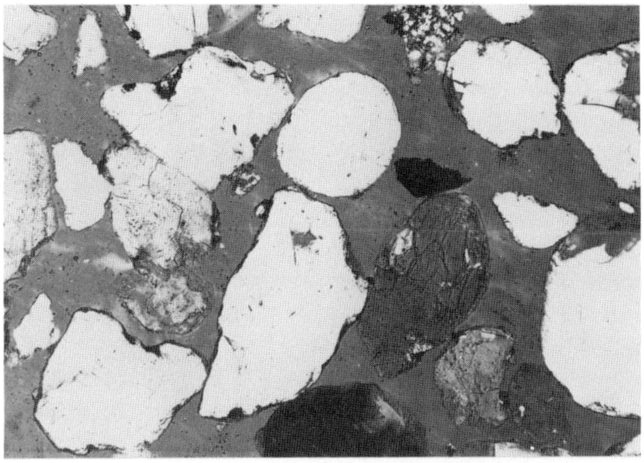

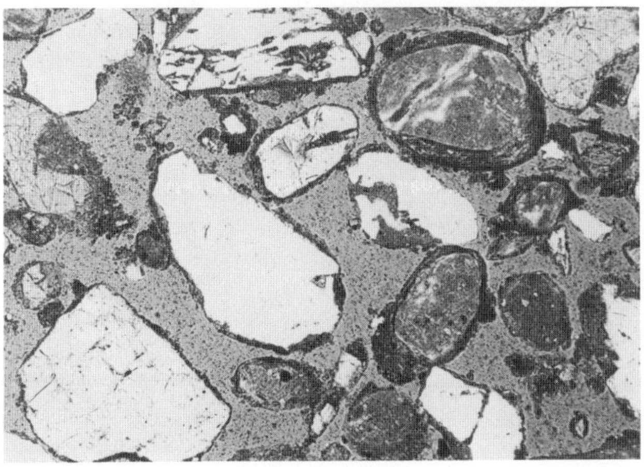

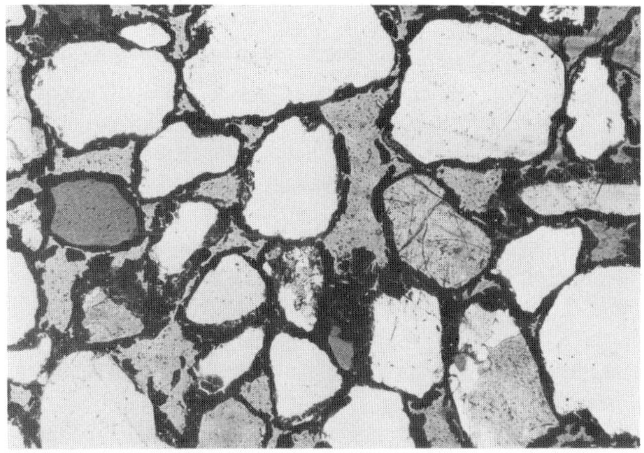

Fig. 30.2. Orthogranic fabric, orthogranic stage of the chlamydic sequence; B_1 horizon of *podzol*; depth 64 cm. Frame length 1.3 mm. Partially crossed polarizers.

Fig. 30.3. Chlamydic fabric; surface horizon of *earthy sand*; depth 13 cm. Frame length 1.3 mm. Partially crossed polarizers.

Fig. 30.4. Plectic fabric; buried B_2 horizon below *podzol*; depth 909 cm. Frame length 1.3 mm. Partially crossed polarizers.

Fig. 30.5. Plectic porphyric fabric; top of B horizon of *podzol*; depth 45 cm. Frame length 1.3 mm. Partially crossed polarizers.

in fabric between the first two sequences, and between stages within them, are related to particle size distribution, although this relationship can be affected by composition and specific soil-forming processes. The matrigranic sequence is only remotely related to particle size, but more strongly to composition and process. Brewer's study was confined to materials with less than 33% silt plus clay, so detailed examination of the large group of porphyric materials and the full range of granic types was not undertaken. Nor was such an examination made of the fragmic group (Brewer and Pawluk 1975) that occurs mostly in finer-grained materials.

Chlamydic sequence

As a general rule, the members of the chlamydic sequence have a ratio of silt to clay of less than one, and successive stages correlate with increasing proportions of silt plus clay. The modal stages are as follows.

Orthogranic (Fig. 30.2)

Loosely packed, mainly bare, skeleton grains and/or rock nodules (ortho-f-members); rare, partially decomposed, intergranular organic fragments; distinguished from the same stage in other sequences by the almost universal occurrence of a minor chlamydic component.

Chlamydic (Fig. 30.3)

Coatings of f-matrix on f-members; the latter are usually, but not exclusively, ortho-f-members; intergranular material as for orthogranic fabrics; usually in soil materials with less than 12% silt plus clay.

Plectic (Fig. 30.4)

Coatings of f-matrix on f-members, broadening and coalescing where adjacent f-members approach each other; the coatings may include skeleton grains and

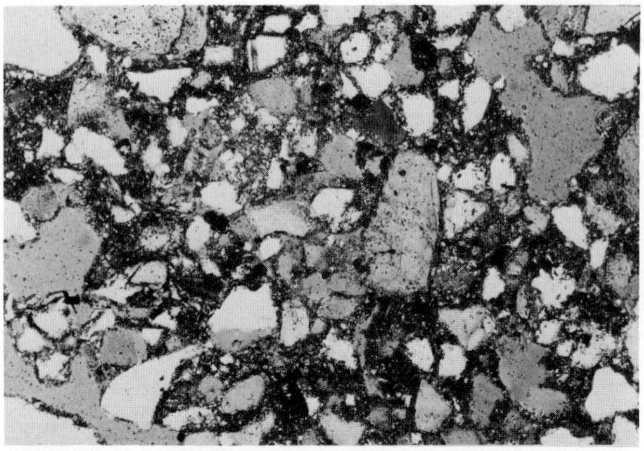

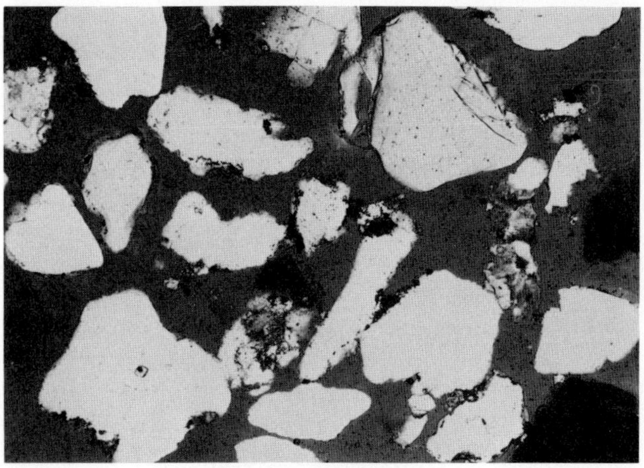

Fig. 30.6. Porphyric fabric, porphyric stage of the chlamydic sequence; surface horizon of *desert loam*; depth 2 cm. Frame length 1.3 mm. Partially crossed polarizers.

Fig. 30.7. Orthogranic fabric, orthogranic stage of the iunctic sequence; A_{21} horizon of *podzol*; depth 50 cm. Frame length 1.3 mm. Partially crossed polarizers.

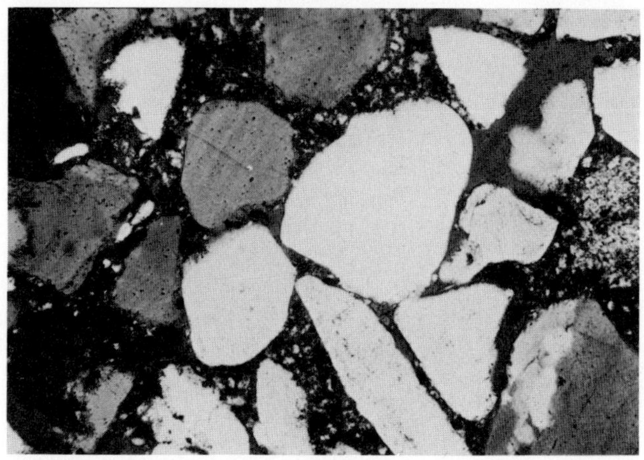

Fig. 30.8. Iunctic fabric; A_2 horizon of *yellow podzolic soil*; depth 17 cm. Frame length 1.3 mm. Partially crossed polarizers.

Fig. 30.9. Iunctic porphyric fabric; B_2 horizon of *podzol*; depth 1112 cm. Frame length 1.3 mm. Partially crossed polarizers.

glaebules; there may be some intergranular aggregates; usually in soil materials with 12-20% silt plus clay.

Plectic porphyric (Fig. 30.5)

Intergrades between plectic and porphyric, not as dense as porphyric; f-members subjacent to voids usually coated; usually in soil materials with 20–23% silt plus clay.

Porphyric (Fig. 30.6)

F-members embedded in a relatively continuous f-matrix; f-members subjacent to visible voids characteristically coated with f-matrix which has a relatively smooth dense appearance due to the dominance of clay over silt; usually in soil materials with more than 23% silt plus clay, but less characteristic as the percentage increases.

Iunctic sequence

As a general rule, the members of the iunctic sequence have a ratio of silt to clay that is greater than one, and successive stages correlate with increasing proportions of silt plus clay. The modal stages are as follows.

Orthogranic (Fig. 30.7)

As for the chlamydic sequence; distinguished by the occurrrence of a minor iunctic component.

Iunctic (Fig. 30.8)

F-matrix mostly packed around points of contact and in the spaces between adjacent f-members; otherwise, surfaces of f-members are essentially bare; usually in soil materials with less than 15% silt plus clay.

Iunctic porphyric (Fig. 30.9)

Intergrades between iunctic and porphyric; interstices between adjacent f-members tend to be filled, with

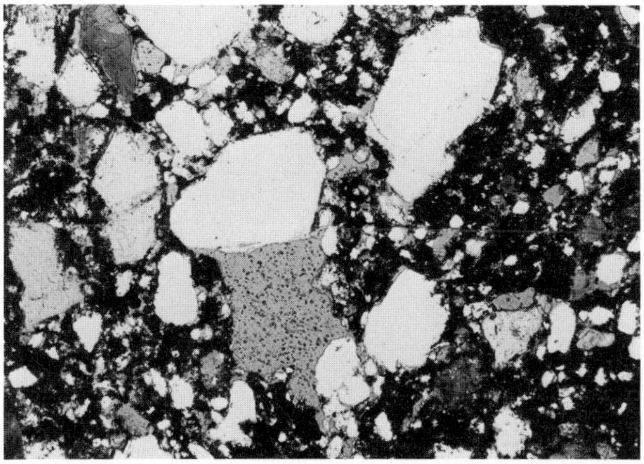

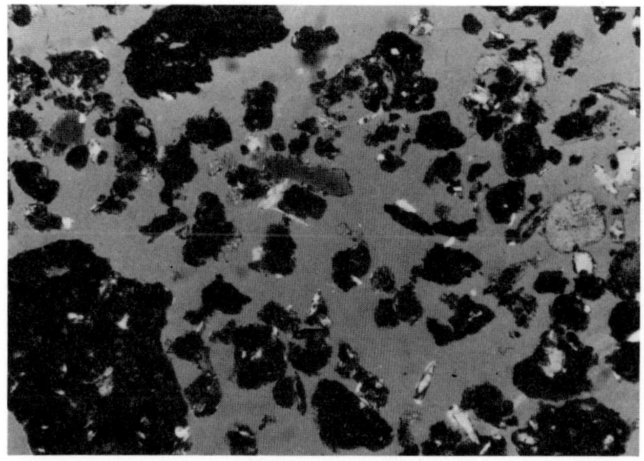

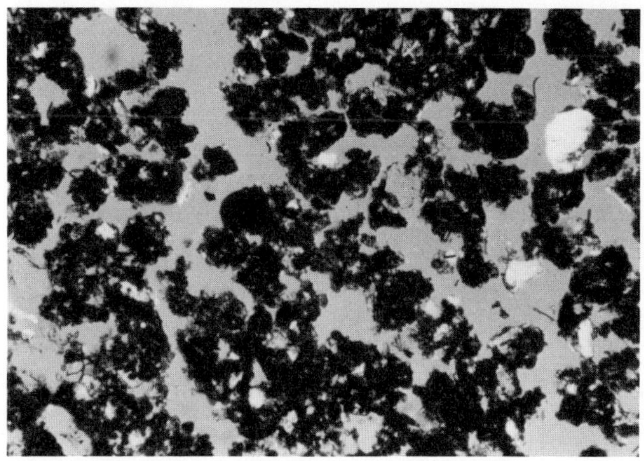

Fig. 30.10. Porphyric fabric, porphyric stage of the iunctic sequence; surface horizon of *humic gley*; depth 8 cm. Frame length 1.3 mm. Partially crossed polarizers.

Fig. 30.11. Orthogranic fabric, orthogranic stage of the granic sequence; sedimentary stratum below minimal *prairie soil*; depth 473 cm. Frame length 3.3 mm. Plain light.

Fig. 30.12. Mullgranic fabric; surface horizon of *alpine humus soil*; depth 4 cm. Frame length 1.3 mm. Partially crossed polarizers.

Fig. 30.13. Mullgranoidic fabric; subsurface horizon of *alpine humus soil*; depth 11 cm. Frame length 1.3 mm. Partially crossed polarizers.

minimal formation of coatings, and the f-members are commonly more widely spaced; usually in soil materials with 15-23% silt plus clay.

Porphyric (Fig. 30.10)

As for the chlamydic sequence, but surfaces of f-members adjacent to voids are mostly bare; the characteristic dominance of silt over clay in the f-matrix gives it a rough appearance; usually in soil materials with more than 23% silt plus clay, but less characteristic as the percentage increases. In materials with a low proportion of clay, the f-matrix itself may have an iunctic fabric in which the silt grains behave as f-members.

Granic sequences

The matrigranic sequence does not correlate with the ratio of silt to clay or with the percentage of silt plus clay except that a lower limit is required to give a dominantly matrigranic fabric. The successive stages correlate with degree of coalescence of the matri-f-members. The modal stages are as follows.

Orthogranic (Fig. 30.11)

As for the chlamydic sequence; distinguished by the occurrence of a minor matrigranic or matrigranoidic component and, sometimes, a minor chlamydic component.

Matrigranic (Fig. 30.12)

Loosely packed, apparently discrete, compound units of plasma plus skeleton grains, sometimes with included pedological features (matri-f-members); there is no f-matrix at this level of description.

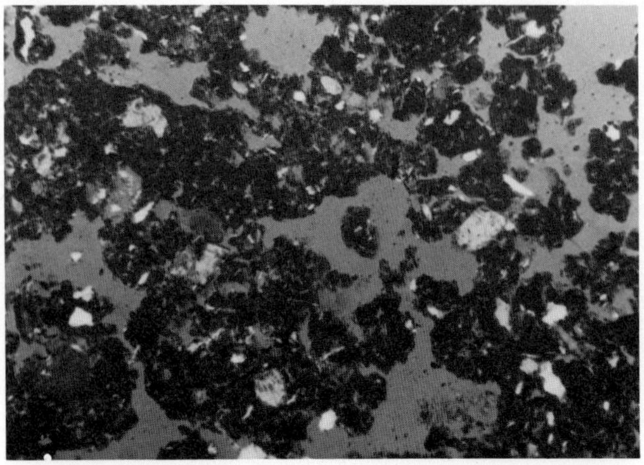

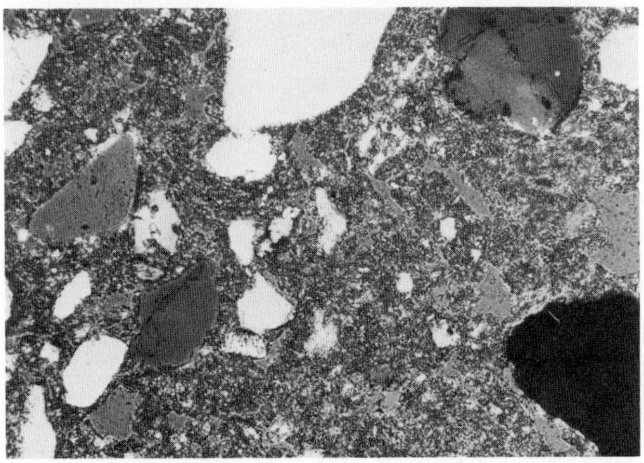

Fig. 30.14. Mullgranoidic porphyric fabric; subsurface horizon of *alpine humus soil*; depth 26 cm. Frame length 1.3 mm. Partially crossed polarizers.

Fig. 30.15. Porphyric fabric, porphyric stage of the granic sequence; lower B_{Ca} horizon of a *solonized brown soil*; depth 84 cm. Frame length 1.3 mm. Partially crossed polarizers.

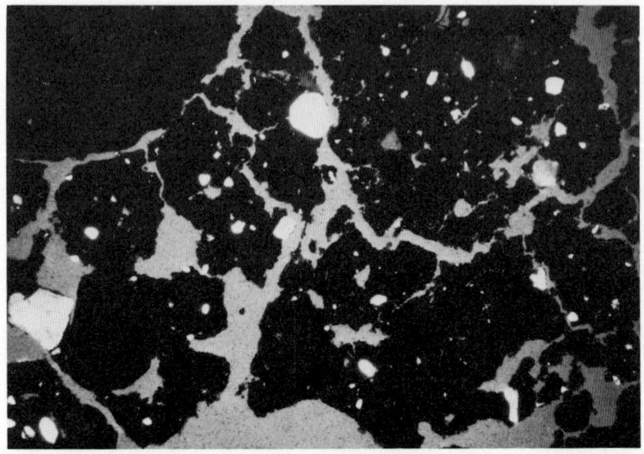

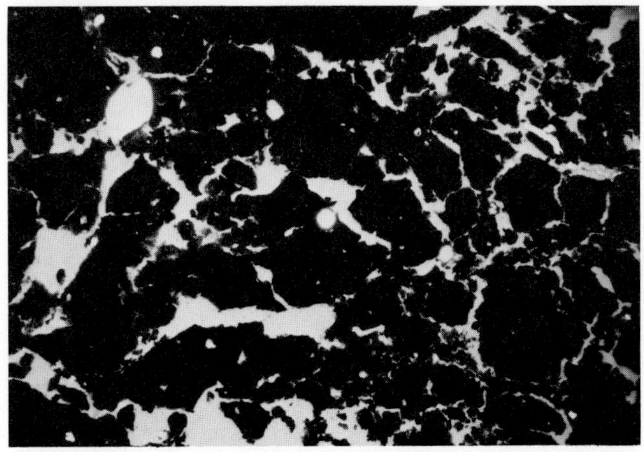

Fig. 30.16. Fragmic fabric; subsurface horizon of *euchrozem*; depth 11 cm. Frame length 3.3 mm. Partially crossed polarizers.

Fig. 30.17. Fragmoidic fabric; subsoil of *euchrozem*; depth 42 cm. Frame length 3.3 mm. Plain light.

Matrigranoidic (Fig. 30.13)

Partially coalesced matri-f-members; there is no f-matrix at this level of description.

Matrigranoidic porphyric (Fig. 30.14)

Intergrades between matrigranoidic and porphyric; the material apparently consists of strongly coalesced matri-f-members.

Porphyric (Fig. 30.15)

As for the chlamydic sequence; characteristically, distinct compound units cannot be recognized consistently, but the form of coalesced units is suggested by the conformation of the walls of voids and by indistinct forms within the f-matrix.

Other kinds of granic materials, not discussed by Brewer (1979), have been observed. The *mullgranic* and *humigranic sequences* are analogous to the matrigranic sequence; the former has mull-like units of plasma plus skeleton grains in which the birefringence of the plasma is masked by finely divided organic matter, and the latter has dark, usually isotropic, moder-like (highly humified organic matter) units. *Phytogranic fabric* consists of loosely packed, partially decomposed plant fragments; it does not form a sequence, and is usually associated with an orthogranic component of the iunctic sequence. In the following discussion, the matrigranic and mullgranic sequences are grouped together as matrigranic, since they both consist basically of plasma plus skeleton grains, unless it is necessary to specify the mullgranic sequence; other granic materials are specified as ortho-, humi-, and phytogranic.

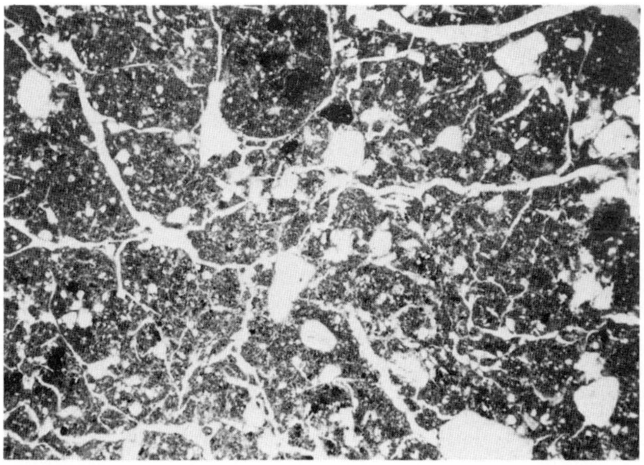

Fig. 30.18. Fragmoidic porphyric fabric; subsoil of *grey brown and red clay*; depth 35 cm. Frame length 3.3 mm. Plain light.

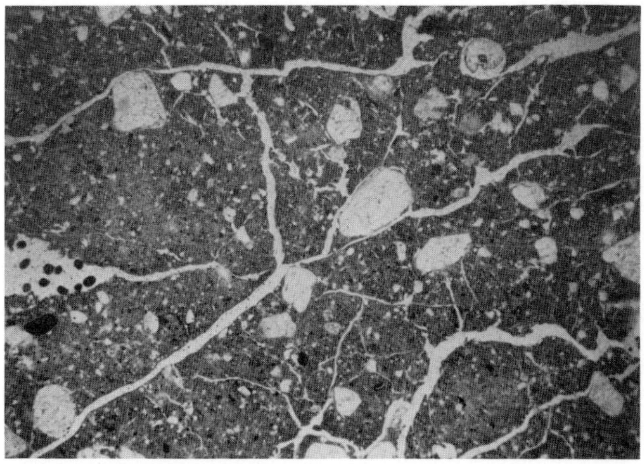

Fig. 30.19. Porphyric fabric, porphyric stage of the fragmic sequence; subsoil of *grey brown and red clay*; depth 35 cm. Frame length 3.3 mm. Plain light.

Fragmic sequence

The fragmic sequence is analogous to the matrigranic sequence except there is no orthogranic stage because of the requirement that the compound units must, at least partially, accommodate each other (Brewer and Pawluk 1975). The successive stages correlate with the degree of coalescence of the matri-f-members rather than particle size distribution. The modal stages are as follows.

Fragmic (Fig. 30.16)

Relatively densely packed, accommodated, or partially accommodated, discrete units without coatings on, or bridges between, units; there is no f-matrix at this level of description.

Fragmoidic (Fig. 30.17)

Forms like those in fragmic fabric but the units are not discrete – appearing in thin section to be fused (united) at their contacts; that is, where the units are fully accommodated there are frequent skewplanes.

Fragmoidic porphyric (Fig. 30.18)

Intergrades between fragmoidic and porphyric.

Porphyric (Fig. 30.19)

As for the chlamydic sequence; characteristically, distinct compound units cannot be recognized consistently.

Porphyric fabrics

Porphyric fabrics vary widely in composition, fabric of the f-matrix, and characteristics of the voids. The characteristic porphyric stages of the various sequences have been discussed. However, above about 30% silt plus clay these fabrics are less characteristic, and other attributes must be used to subdivide the large group of porphyric fabrics. Data of this kind are not yet available for Australian soils. In the following discussion, reference is made to a few specific types that seem to correlate with classification groupings.

Heterogeneity of soil fabrics

Most soil materials, or soil horizons, are heterogeneous at the level of related distribution patterns of f-matrix and f-members; that is, they have mixed or separated complex fabrics (Brewer and Pawluk 1975). *Mixed complex fabrics* consist of inextricably intermixed fabric components: for example, coated skeleton grains randomly distributed among uncoated compound units. *Separated complex fabrics* consist of repeating areas of two or more fabric components. Mixed complex fabrics are recorded by using a hyphen (-), and separated complex fabrics by using a slash (/), between the names of the fabric components.

Effects of composition and specific processes

The relationships between related distribution patterns of f-matrix and f-members, and particle size distribution may be modified by composition and specific processes (Brewer 1979). Most important are organic matter, fine carbonate crystallites, faunal activity and illuviation. Fabrics will be denser than expected from silt plus clay analysis where the f-matrix includes organic matter and/or carbonate crystallites. Likewise fabrics will be less dense than expected where the particles of the f-matrix are closely packed, as in illuviation argillans, or where f-members yield silt and/or clay on analysis, for example clay mineral f-members. Relatively high proportions of organic matter in a soil climate suitable for faunal activity are inducive to formation of members of the mullgranic and humigranic sequences, neither of which correlate with percentages of silt plus clay; in fact, the latter may have virtually no silt or clay.

Genesis of the fabric sequences

It is, of course, dangerous to impute the same genesis to all occurrences of the fabrics of the various sequences. The stages within and between the sequences that have been defined are broad, and not specific in terms of some important characteristics such as the composition of the f-matrix, conformation of the coatings in the chlamydic sequence, and internal fabric of the f-matrix and f-members. However, the data suggest that some broad generalizations concerning genesis can be made, with the proviso that each example should be studied in detail before firm interpretations are made.

Clearly there are two groups of fabric sequences: the chlamydic and iunctic sequences which have f-members and f-matrix, and the granic and fragmic sequences which consist of compound units of soil material packed in different ways. In the first group, interpretation of genesis is concerned with the reasons why the f-matrix forms coatings on f-members in the chlamydic sequence but is packed in the interstices between the f-members in the iunctic sequence. In the second group, it is concerned with the formation and packing of the compound units.

Chlamydic and iunctic fabrics

Some members of the chlamydic sequence, especially the chlamydic and plectic stages, can be interpreted with some confidence as resulting from illuviation: for example, in some B_t horizons where the f-members are coated with strongly oriented clay minerals (free grain argillans), and in the B horizons of some *podzols* where the f-members have strong dark brown organic coatings. Neither of these examples conforms to the general relationships between particle size distribution and fabric: the former has too high a proportion of clay (up to 30% in a plectic fabric) due to the close packing of oriented clay particles, and the latter has too low a proportion of silt plus clay (down to 2% in a plectic fabric) due to the f-matrix being dominantly organic matter with very low bulk density.

In the majority of the members of the chlamydic sequence however, the f-matrix consists of plasma plus fine skeleton grains (matri-types), sometimes masked by finely dispersed organic matter (mull-types). In these, the coatings are commonly irregular in thickness and in conformation of the outer boundary. There is no evidence suggesting illuviation: such fabrics occur to the surface in many profiles, particle sizes are not well differentiated, and sometimes the coatings contain randomly distributed carbonate crystallites. It seems therefore, that stages of the chlamydic sequence are usually the result of the universal process of wetting and drying causing f-matrices with relatively high clay to silt ratios to coat the f-members (coarser particles). This hypothesis is commonly supported by the orientation within the f-matrix, viz. a degree of preferred orientation parallel to the surfaces of the f-members it coats. The data also suggest that this process is more effective the higher the clay to silt ratio.

The f-matrix of the stages of the iunctic sequence, with its relatively low clay to silt ratio, behaves differently to that of the chlamydic sequence. The data suggest that the principal factor of formation is rather slight gravitational movement under the influence of wetting and drying, causing packing of the f-matrix between the f-members and around their points of contact. In some instances, such fabrics may be inherited in sedimentary deposits, but they occur in eluviated surface horizons where loss of clay causes the formation of a silt-enriched material, and in silt- and clay-deficient materials where the f-matrix consists of particulate melanized organic material derived from plant fragments and aged organic faunal excrement. In some soil materials, partially aged, inorganic mineral, faecal pellets may form bridges between f-members that are not in contact.

The orthogranic stage of both the chlamydic and iunctic sequences, even those with low percentages of silt plus clay, is restricted in occurrence. This is because even very low proportions of colloidal material almost invariably tend to coat ortho-f-members to form chlamydic fabrics (low proportions of organic matter can form coatings of remarkably low bulk density and so of unexpected thickness (Brewer and Thompson, 1980) and because fine, particulate organic matter, especially when melanized, behaves in the same manner as silt and tends to form members of the iunctic sequence.

Granic and fragmic fabrics

The most common process of formation of the stages of the matrigranic and mullgranic sequences in Australian soils is the activity of soil fauna. In some soil materials, a restricted variety of fauna produce matrigranic and mullgranic fabrics having a limited range of sizes and shapes of units that are, in fact, faecal pellets; in others, a larger variety produce a wide range of sizes and shapes. Gradual aging of the pellets produces the granoidic stage, and finally the porphyric stage: in which case the coalescence of the units is rather extreme, and the original presence of faecal pellets is deduced mainly from the rounded shapes of the walls of voids. At this stage, the interpretation of faunal activity is often open to question.

An interesting observation is the common association of a dominantly separated granoidic/granic fabric with the presence of moderate proportions of fine carbonate crystallites, with little evidence for faunal activity; the almost spherical shape of the matrigranic units is unusual for faecal pellets. The reason for the association is unexplained. Larger proportions of crystallites form a porphyric fabric by effectively increasing the proportion of f-matrix.

Granic fabrics can also be formed by fracturing and flaking of organic coatings in the upper B horizons of *podzols* (Flach 1960). Such fabrics have not been observed as yet in Australian *podzols*, although fractured coatings are common. Finally, granic fabrics can be inherited, as in the deeper horizons of soils formed on dune parnas (Butler 1974).

The stages of the fragmic sequence can be formed by a variety of processes. In many soil materials, they can be attributed to faunal activity, especially where they are associated with matrigranic and mullgranic fabrics in a separated complex fabric: as in the rather specific granic/fragmic fabrics referred to earlier, which have areas of various proportions of granic, granoidic, fragmic, and fragmoidic fabrics. The larger soil fauna commonly excrete material in the form of pedotubules, especially isotubules, rather than as pellets. These isotubules are often dense, but tend to fracture to a fragmic or fragmoidic fabric; the tubular form may be obvious or rather indistinct, with rounded surfaces common, and separate pedotubules may adjoin each other, enhancing the fragmic fabric. Part of the rationale of this interpretation is the logical deduction, that if a high proportion of a thin section can be interpreted with confidence as being the result of faunal activity, then the whole soil material probably has been worked, since aging tends to obliterate the evidence. In fact, the degree of faunal activity is often underestimated because of this effect of aging, which, in fine-grained soil materials tends to produce areas with porphyric fabric.

The fragmoidic porphyric, especially, and the fragmoidic stages can be produced by simple cracking due to wetting and drying. In some Canadian soils, fragmic and fragmoidic fabrics have been attributed to freezing and thawing (Brewer and Pawluk 1975); such processes have not yet been observed in Australian soils.

Porphyric fabrics

Most porphyric fabrics result from the normal packing of grains in materials with a high proportion of f-matrix; often correlated with the proportion of silt plus clay, although porphyric fabrics with an f-matrix of dominantly fine sand have been recorded. In the porphyric stage of the chlamydic sequence, the f-members adjoining large voids are coated, as in the less dense stages; in the iunctic sequence they are bare, as in the less dense stages. This suggests that the specific processes are still operative to some degree. The porphyric stage of the matrigranic and mullgranic sequences are due to aging of faecal pellets, but they are often difficult to recognize as such. The porphyric stage of the fragmic sequence is probably most often due to cracking caused by shrinkage, but some may be aged faunal features. In some soil materials, high proportions of finely disseminated organic matter or carbonate crystallites enhance the proportion of f-matrix, and tend to form porphyric fabrics. The void patterns and other details of specific kinds of porphyric fabrics have not been examined in sufficient detail as yet to suggest processes of formation.

Distribution of the fabric sequences

Since fabric, at the level of related distribution of f-matrix and f-members, is not used consistently as a parameter in soil classification, it cannot be expected that particular fabrics will occur consistently in the soils of particular classes. As shown in the preceding discussion, only soil classes that happen to be defined on characteristics of particle size distribution, composition and genesis can be expected to have consistent fabrics.

In the following discussion, an attempt is made to relate fabrics to the major soil groups set out in 'A Handbook of Australian Soils' (Stace *et al.* 1968) and to kinds of soil horizons, and to discuss variability. The data available are inadequate for a full discussion; there are no data for some soil groups and very few for most.

Soils showing no profile development

No data are available for the *lithosols* and *solonchaks*.

The *alluvial soils* conform to the general relationships of particle size distribution with fabric. They often have areas of members of the matrigranic or mullgranic sequences where fauna have been active.

The three groups of sands are defined primarily on particle size characteristics and, as expected, have a dominant, or important, chlamydic component. The *calcareous sands* have a mixed complex matrigranic-chlamydic fabric throughout their depth, with fine carbonate crystallites abundant in the matrigranic units and the f-matrix of the chlamydic component; f-members include frequent shell fragments. In profiles with lower proportions of f-matrix, the skeleton grains and rock nodules are coated in preference to the shell fragments, but, with higher proportions, the shell fragments are also coated, and the chlamydic component grades into plectic fabric. In the *siliceous sands*, all the samples examined are dominantly chlamydic. In materials with low silt plus clay (down to 2%), the coatings are thin and incomplete, except those in surface horizons which are enhanced by additional organic matter. In these profiles, materials with more clay (up to 11%) usually have little or no silt; the fabric is chlamydic with a significant plectic component. Comminuted melanized plant fragments sometimes occur in the upper horizons in the intergranular spaces; the finer-grained fragments tend to produce a weak iunctic component similar to that in the A horizons of many *podzols*. In spite of a higher than expected proportion of silt plus clay, the *earthy sands* are also dominantly chlamydic, with plectic and matrigranoidic components more important with increasing depth. The higher proportions of silt plus clay are due partly to the matrigranoidic component, and, in the profile examined, to the argillaceous f-members that partially disperse during particle size analysis.

Soils showing minimal profile development

The fabrics of the soils grouped under this heading conform to the general relationships of fabric and particle size distribution within the chlamydic sequence, and to the common association of mixed matrigranic-granoidic fabrics with fine carbonate crystallites.

In the *grey-brown and red calcareous soils*, coarser-grained surface horizons are dominantly plectic with significant porphyric and minor matrigranic components. Finer-grained surfaces and subsurface horizons are porphyric with variable plectic and matrigranic components. The matrigranic component usually is more important with increasing depth, and is sometimes associated with a chlamydic component in a mixed complex fabric. Deeper, even finer-grained horizons are more dominantly porphyric. Carbonate crystallites are abundant in the matrigranic units and the f-matrix of the other fabric components.

The *desert loams* are dominantly porphyric throughout, except that thin coarser-grained surface horizons are dominantly plectic with a significant porphyric component, or porphyric with a plectic porphyric component, according to particle size distribution. A minor plectic, or even chlamydic, component due to local clustering of the ortho-f-members, may be present in any part of the profile.

For *red and brown hardpan soils*, the only data are for those with a silica-cemented hardpan. The coarser-grained types have a dominantly chlamydic to plectic surface, sometimes with a minor porphyric component. Otherwise they are dominantly porphyric, with variable proportions of a chlamydic and/or plectic component due to varying localized deficiencies in f-matrix. All have a minor matrigranic component.

The *grey brown and red clays* are dominantly porphyric throughout, correlating with high proportions of silt plus clay; there may be restricted chlamydic or plectic areas due to localized clustering of ortho-f-members, and mixed matrigranic-granoidic areas due to faunal activity. The red types commonly have mixed fragmic-fragmoidic fabrics in the upper horizons, and the browner types have a tendency towards those fabrics.

The mildly leached dark soils

Most profiles available in this group have much more than 23% silt plus clay, and surface horizons relatively high (6–11%) in organic matter (mostly as part of the plasma). Therefore, porphyric fabrics would be expected to dominate, but in most soil groups the upper horizons have been modified by fauna.

The *black earths* are dominantly porphyric throughout, with visible voids decreasing with increasing depth; there may be some fragmoidic areas and smaller matrigranoidic areas, probably due to faunal activity.

The *rendzinas* are fragmoidic porphyric in the upper horizons, as is common in soil materials with high silt and clay and relatively high organic matter; some fragmoidic units have rounded surfaces suggestive of aged faunal excrement. This fabric is closely related to separated granic/fragmic, and is more so with increasing depth.

The *chernozems* are dominantly metafragmoidic (fragmoidic with smoothed surfaces on the units) in the surface, due to faunal activity. With increasing depth, they have a mixed chlamydic-matrigranoidic component, and are dominantly so where carbonate crystallites are abundant in the f-matrix.

The *prairie soils* with high silt plus clay are similar to the *rendzinas* in the upper horizons; some have a strong separated granic/fragmic fabric, but are less so with increasing depth; others are essentially porphyric, tending towards fragmoidic. Those with less than about 20% silt plus clay in the surface are iunctic porphyric in the upper horizons, but porphyric (>20% silt plus clay) in the subsoil.

There are no data for the *wiesenboden*.

The mildly leached brown soils

The soils of this group differ in degree of textural differentiation. Their fabrics conform to the relationships with particle size distribution and composition rather than to soil group, except in so far as some soil groups tend to have similar particle size distributions and compositions in equivalent horizons.

Solonetz, solodized solonetz and solodic soils, and *soloths* are texturally strongly differentiated. The B_2 horizons have high silt plus clay, and more clay than silt: they are dominantly porphyric, except in one unique B_2 horizon that has 33% silt plus clay (30% clay) as free grain argillans in a plectic fabric. Most A horizons have relatively high silt to clay ratios, so the fabrics tend to belong to the iunctic sequence. In the *solonetz* and *soloths*, surface horizons with low silt plus clay are iunctic, sometimes with small porphyric areas of the iunctic type. Those with high silt plus clay are porphyric. Melanons (Bal 1973) are commonly frequent, and behave as silt and sand size particles. In the *solodized solonetz and solodic soils*, A horizons with low silt plus clay have a mixed chlamydic–iunctic fabric; one has a matrigranoidic component (organic-rich aggregates). Those with high silt plus clay are dominantly porphyric with minor areas of mixed chlamydic-iunctic fabric. One A horizon with a low silt to clay ratio has a separated chlamydic/plectic fabric.

The *solonized brown soils*, generally are less strongly differentiated texturally, but there is considerable variation between profiles. Most have low silt to clay ratios throughout the profile, but the proportion of silt plus clay varies between profiles. Carbonates are commonly present, particularly in the lower horizons, as fine randomly distributed crystallites. In horizons without significant proportions of carbonates, the fabrics are members of the chlamydic sequence: dominantly chlamydic, plectic, plectic porphyric, or porphyric, according to the proportions of silt plus clay. Horizons with carbonates are of two kinds: those with high carbonates and high silt plus clay are dominantly porphyric, often with a matrigranoidic component; those with moderate proportions of carbonates are dominantly matrigranoidic with a chlamydic component. The carbonates occur as fine crystallites throughout the matrigranoidic units and the f-matrix of the associated chlamydic component.

The *red-brown earths* are texturally strongly differentiated. All profiles examined have moderate to high proportions of silt plus clay, and most have low silt to clay ratios. They are dominantly porphyric throughout the profiles, but the A horizons with the lowest proportions of silt plus clay have a plectic component; two A horizons with high silt to clay ratios are porphyric because of their high silt plus clay. The A horizon of one profile is unusual in that it is dominantly porous matrigranoidic porphyric, perhaps due to faunal activity. Minor mixed matrigranic-granoidic components in any horizon are generally identifiable as due to faunal activity; the carbonates present in many deeper horizons are not generally associated with mixed matrigranic-granoidic fabrics.

The *non-calcic brown soils* are similar to the majority of *red-brown earths*, except there are no carbonates in the profile, and the surface has a separated granic/fragmic fabric due to faunal activity.

The *chocolate soils* and *brown earths* are essentially texturally undifferentiated profiles with considerable organic matter in their surfaces. They have a wide range of silt to clay ratios and high silt plus clay; the lower horizons are porphyric; the upper horizons are commonly separated granic/fragmic due to faunal activity, otherwise they are porphyric.

Soils with profiles dominated by sesquioxides

Most *calcareous red earths* and *red earths* examined have more clay than silt throughout the profiles; in the one exception, difficulty was experienced in dispersing the clay for particle size analysis. The fabrics of the two groups are similar above the carbonate-rich horizons of the *calcareous red earths*: the A horizons are dominantly chlamydic, plectic, or porphyric, according to the proportion of silt plus clay; all profiles become porphyric with increasing depth and proportion of silt plus clay, except one particularly coarse *red earth* that does not become denser than plectic porphyric with minor areas of plectic and porphyric material. The porphyric fabric in the carbonate-free lower horizons in both groups is rather characteristic: a porous type that appears to result from an increase in volume and coalescence of irregular coatings. The carbonate horizons of the *calcareous red earths* commonly have a matrigranoidic or matrigranoidic porphyric component, with occasionally significant areas with mixed chlamydic-matrigranoidic fabric.

The surface of the one *yellow earth* examined has slightly more silt than clay, and a plectic iunctic fabric; the sub-surface has more clay than silt, and is dominantly plectic, but becoming porphyric with increasing depth and proportion of silt plus clay.

Fine-grained *terra rossa soils* are dominantly porphyric throughout, with a tendency towards separated granic/fragmic fabric (cf. *rendzinas*), with matrigranoidic areas. A sandy (22% silt plus clay) profile examined is dominantly plectic porphyric throughout, with minor plectic areas and some intergranular aggregates.

Euchrozems, xanthozems and *krasnozems* have much in common. They are texturally weakly differentiated, with gradual boundaries, and are high in silt plus clay (>50%). The upper horizons have a separated complex fabric of the granic/fragmic type to various depths in different profiles, probably due to faunal activity; compaction of this fabric in some profiles results in a thin surface horizon with a highly vughy porphyric fabric. One profile that is denser in the upper horizons has a surface cover of basalt stones, and may be an eroded phase. With increasing depth, the compound units gradually lose their form, distinct units are restricted to pedotubules, and a rather characteristic porous porphyric fabric develops. In deeper horizons, the porphyric fabric is denser, but short planar voids and irregular vughs are often abundant. The fabric of deep horizons is often a dense porphyric type with skew planes.

The mildly to strongly leached highly differentiated soils

There are no data for the *grey-brown podzolic soils* nor for the *peaty podzols*.

The A horizons of all the *red podzolic, yellow podzolic, brown podzolic* and *gleyed podzolic soils* examined have more silt than clay; their fabrics are dominantly members of the iunctic sequence, the particular fabric, or complex of fabrics, depending primarily on the proportion of silt plus clay. In the immediate surfaces however, finely comminuted organic matter may cause a denser stage than expected from the proportion of silt plus clay; iunctic porphyric or porphyric of the iunctic type is most common, the underlying material often being less dense. Apart from this feature, most A horizons become denser with increasing depth, usually porphyric of the iunctic type immediately above the prominent B horizons. The A horizon of one *brown podzolic soil* is essentially fragmoidic porphyric. The B_2 horizons of all these profiles are porphyric types in which there is more clay than silt.

The *lateritic podzolic soils* with more silt than clay are similar to the other podzolic soils. Those with more clay than silt in the A horizons still have a iunctic or iunctic porphyric fabric in the immediate surface. In the subsurface, one profile with low silt plus clay has a poorly developed chlamydic fabric with a minor iunctic component; with increasing depth, the chlamydic fabric is weaker and abruptly overlies a porphyric B_2 horizon. Another profile has a poorly developed plectic fabric in the subsurface, which becomes gradually denser with increasing depth to plectic porphyric; all boundaries between horizons are diffuse, except that the plectic porphyric material abruptly overlies a vesicular laterite.

All the *podzols* and *humus podzols* examined have been formed on well sorted windblown sands. The A horizons have extremely low silt plus clay; their fabric is dominantly a mixed complex orthogranic-iunctic type in which the iunctic component is due to finely comminuted, melanized, organic material. In restricted areas, concentrations of these opaque fragments, behaving as f-matrix, cause an iunctic porphyric, a plectic, or even a porphyric fabric. There may also be a minor chlamydic component, similar to that in the B horizons, which is more important in weakly developed profiles with relatively thin A horizons. Most B horizons have a strongly developed chlamydic and/or plectic fabric, in spite of the usually very low silt plus clay. The coatings are dominantly organic with some sesquioxides, and are generally thickest at the top of the B horizons below a thin A_2/B_1 transitional horizon, becoming thinner and less complete with increasing depth. In the younger, thinner less strongly developed profiles, the fabric is chlamydic, and the coatings are not fractured. In the older, thicker profiles, the upper B horizon is plectic with darker fractured coatings. In the most strongly developed profiles, the upper B horizon is porphyric.

Some *podzols* have more complex B horizons, for example, one on a sand sheet overlying sandy sediments at Jervis Bay in the A.C.T. The B horizons of this profile are dominantly plectic or chlamydic, but the upper B horizon is a complex of small areas with mixed humigranic-chlamydic, plectic and porphyric fabrics. Also, the plectic component is compound, that is, two layers of differently coloured materials (darker and paler browns); the darker outer coatings can be considered as void cutans (organans) overlying a simple plectic material.

Some *podzols*, particularly the older ones, have a transitional A–B horizon that consists of an f-matrix of varying proportions of very fine sand, silt, clay and very fine particulate organic matter. The proportion of f-matrix varies considerably between profiles and within a single horizon. The fabric varies from dense porphyric with a high proportion of f-matrix, to dominantly iunctic with a low proportion of f-matrix.

The organic soils

All the *alpine humus soils* examined have more than 23% silt plus clay, less silt than clay, and high organic matter in the upper horizons. The fabric of the upper horizons however, can be confidently attributed to faunal activity, which has an overriding effect on the relationship between fabric and particle size distribution. The profiles are essentially texturally undifferentiated, with the proportion of organic matter gradually decreasing with increasing depth. The immediate surface has an ortho-mull-phytogranic mixed complex fabric. Below this, the upper horizons have various expressions of a strongly developed separated granic/fragmic fabric on a much finer scale than in soils with similar fabrics in other groups; that is, the compound units are smaller. Even the larger units often have an internal separated fine granic/fragmic fabric, and the porphyric component is very minor or absent. In deeper horizons with less organic matter, the fabric changes gradually to dominantly porphyric with minor separated granic/fragmic areas, and then into the basically porphyric fabric of the colluvial parent material. Even in the transition zone from high to low organic matter, there is a significant separated granic/fragmic component.

The *humic gleys* have various degrees of textural differentiation. Most profiles examined have very high silt plus clay throughout, and more silt than clay. However, the surface horizons vary from dense porphyric with large isolated vughs, through porphyric with areas of separated rather coarse granic/fragmic fabric, to separated granic/fragmic, according to the degree of faunal activity. Most profiles have dark melanized organic fragments in the surface horizons. With increasing depth, the profiles with a separated granic/fragmic component in the surface become dense porphyric with small areas, confined to pedotubules, of separated granic/fragmic fabric, and eventually a uniform dense porphyric fabric. One profile with relatively low silt plus clay in the surface horizons (15%) has more silt than clay, and frequent dark melanized organic fragments that behave as silt and fine sand grains; the surface horizons are iunctic porphyric; with depth, where silt plus clay is higher and silt less than clay, the fabric is dense porphyric.

There are no data for the *peats*.

Fabric of the f-members and f-matrix

In many soil materials, the f-members are skeleton grains and/or rock nodules (ortho-f-members), whose internal fabrics are beyond the scope of this discussion. In others, the f-members are compound units with specific kinds of internal fabrics: such as the wide variety of glaebules, and the compound soil units that participate in the various stages of the granic sequences – these last are regarded as f-members, there being no f-matrix at this level of description.

In most instances, both the compound f-members and

the f-matrix have a porphyric fabric, or more specifically, a porphyroskelic fabric in which skeleton grains, rock nodules, and pedological features occur within a relatively continuous matrix of plasma. Sometimes the f-matrix and the f-members may consist entirely of plasma. The most important characteristic of the internal fabric of the f-members and f-matrix, therefore, is the plasmic fabric, which is usually expressed as orientation patterns of the plasma, as interpreted from extinction patterns observed with the polarizing microscope.

Where the f-members and f-matrix do not have a porphyroskelic fabric, their internal fabric can be described in terms of the fabric sequences already discussed: for example, the f-matrix itself may have a fine-grained iunctic fabric, and large granic/fragmic units may have a matrigranoidic internal fabric. Such two-level fabrics can be described at some level however, in terms of plasmic fabric, unless they entirely lack plasma.

The following discussion deals with the various fabric sequences and the porphyric group, whose distribution has already been discussed, in terms of the internal fabric of f-members and f-matrix, with emphasis on the plasmic fabric.

Chlamydic sequence

The coarser stages of the chlamydic sequence have less than 20% silt plus clay, more clay (or plasma) than silt, and a usually fine-grained f-matrix coating the f-members. It is not surprising, therefore, that the dominant fabric of the f-matrix is a striated orientation pattern parallel to the surfaces of the f-members, except where the f-matrix is unusually coarse. Wetting and drying, with slight movement of the material, would tend to produce such a fabric. The degree of orientation varies from weak to moderately strong in non-illuviated materials, to strong continuous orientation in the B_2 horizon of the *solodic soil* in which the coatings are illuviated clay. The thicker coatings of the denser stages (plectic) are usually less strongly oriented, unless formed by illuviation, but are commonly better oriented adjacent to the surfaces of the f-members.

Composition also affects the plasmic fabric. Some chlamydic materials have an undulic f-matrix: for example, in a *red earth* on impure sandstone, and a *terra rossa soil* on calcarenite; undulic fabrics are usually associated with basic parent materials (see Undulic and inundulic fabrics p.454). Fine carbonate crystallites tend to produce weaker orientation patterns: as in the *calcareous sands*, the *grey-brown and red calcareous soils*, and the lower horizons of the *chernozems* and some *solonized brown soils*. Finely dispersed organic matter tends to mask the birefringence of the plasma, and so the plasmic fabric of the f-matrix; this occurs most commonly in surface horizons, particularly where fauna have been active and where the chlamydic component is associated with stages of the mullgranic sequence. Chlamydic components associated with stages of the humigranic sequence usually have an isotic f-matrix. In the B horizons of the *podzols*, the dominantly organic f-matrix of the chlamydic fabric of weakly developed profiles has a weak orientation pattern parallel to the surfaces of the ortho-f-members; the organic f-matrix of plectic and porphyric fabrics of the more strongly developed profiles is essentially isotic.

Iunctic sequence

The coarser stages of the iunctic sequence have less than 23% silt plus clay, more silt than clay, and the f-matrix is packed between adjacent f-members and around their points of contact. These characteristics militate against orientation of the plasma. The plasmic fabric, therefore, is usually silasepic or argillasepic, according to the ratio of silt to clay. Birefringence of the plasma may be masked by finely dispersed organic matter. In some instances, where there is essentially no plasma, the f-matrix has a fine-grained orthogranic fabric, as in the A horizons of some *red podzolic* and *yellow podzolic soils* formed on old beach ridges, or a mixed granic-iunctic fabric in which the f-matrix consists of finely comminuted melanized organic fragments, as in the A horizons of the *podzols*.

Granic sequences

Apart from phytogranic, humigranic, and orthogranic fabrics, the granic units usually have a porphyric or porphyroskelic internal fabric. In the matrigranic and mullgranic sequences, the units are argillasepic, masked by finely dispersed organic matter in the mullgranic sequence. Some mullgranic materials, especially, have coarse units with a mullgranoidic internal fabric (see *Alpine humus soils* p.450); the mullgranoidic units have a masked argillasepic internal fabric. Many matrigranic materials have a high proportion of randomly distributed, fine carbonate crystallites that give a calciasepic fabric (see Calciasepic fabric p.453). Some matrigranic components are undulic or inundulic, particularly when associated with basic parent materials. In the humigranic sequence, the units are isotic, and lack skeleton grains.

Some granic materials have inherited f-members that may have a wide variety of internal fabrics. The *earthy sands* have kaolinitic f-members, as well as ortho-f-members, with a characteristic neostrian with striated orientation around their margins; the *calcareous sands* have shell fragments as f-members.

Fragmic sequence

As in the granic sequences, the various stages of the fragmic sequence consist of units with a porphyric, or porphyroskelic fabric; most are argillasepic (masked to varying degrees by organic matter) undulic, or inundulic. Argillasepic units occur in surface horizons of *prairie soils*, *rendzinas*, *chernozems*, *chocolate soils*, *brown earths*,

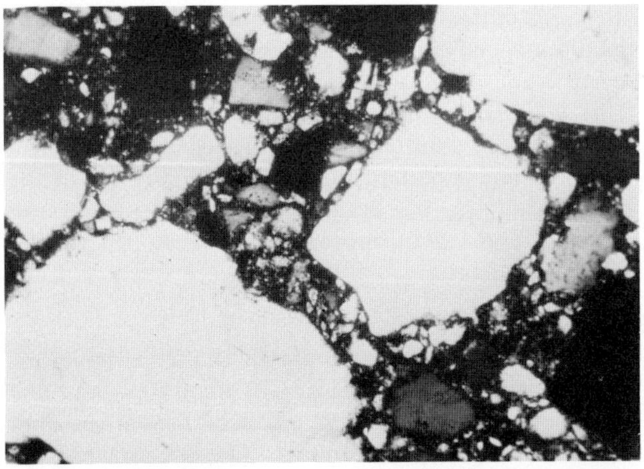

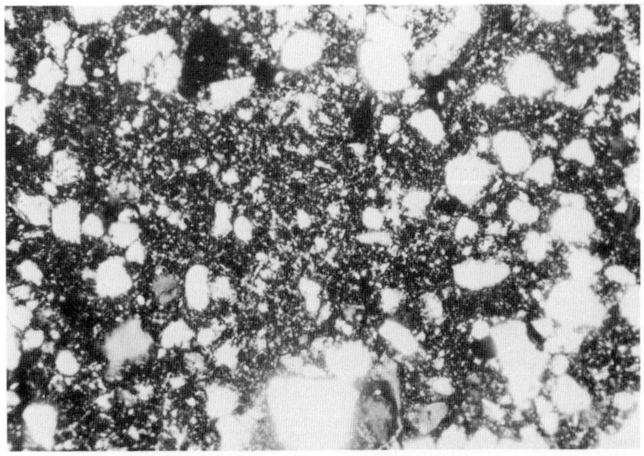

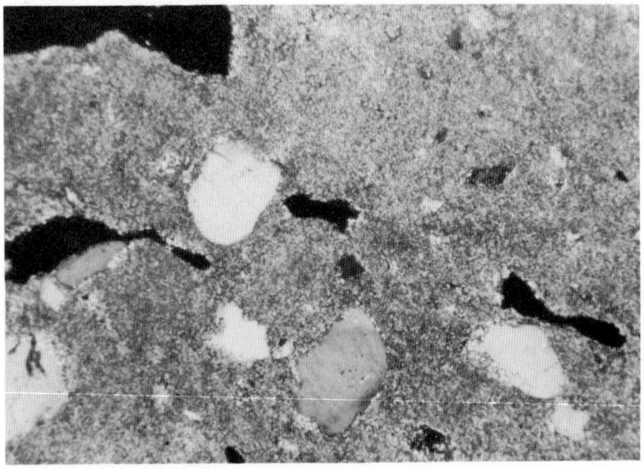

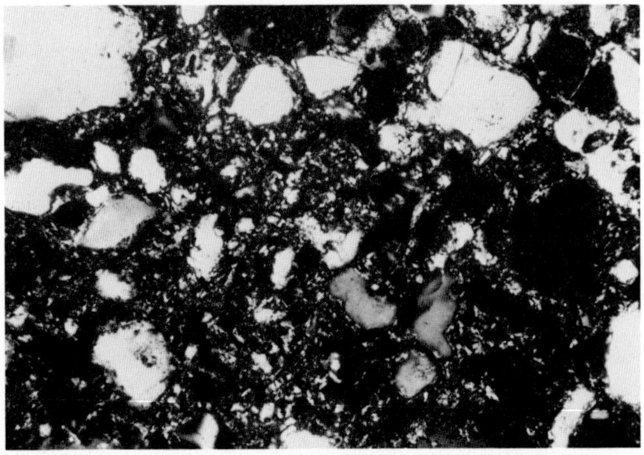

Fig. 30.20. Silasepic fabric; A_2 horizon of *solonetz*; depth 17 cm. Frame length 0.9 mm. Crossed polarizers.

Fig. 30.21. Calciasepic fabric; *grey brown and red clay*; depth 132 cm. Frame length 0.9 mm. Crossed polarizers.

Fig. 30.22. Argillasepic fabric; B horizon of *red earth*; depth 75 cm. Frame length 0.9 mm. Crossed polarizers.

Fig. 30.23. Insepic fabric component; B horizon of *yellow earth*; depth 64 cm. Frame length 0.9 mm. Crossed polarizers.

Key to letters on electron micrographs: b, soil bacteria; c, clay; o, organic matter; s, space left by fragmented quartz grain; q, quartz grain; v, void.)

terra rossa soils, and *xanthozems*; and undulic or in-undulic types particularly in *euchrozems* and *krasnozems*, but also in *alpine humus soils*, *chernozems*, *prairie soils*, *non-calcic brown soils* and *chocolate soils* formed on basic parent materials, especially basaltic materials. In all these soils, the internal fabric of the units grades into inundulic or argillasepic with increasing depth, except that the *chernozems* become calciasepic.

Porphyric fabrics

Most porphyric materials with more than 23% silt plus clay are also porphyroskelic. They are most common in textural B horizons, fine-grained A horizons of texture-contrast profiles, and the lower horizons of profiles with matrigranic or separated granic/fragmic surface horizons. They are generally absent from the three groups of sands, coarser-grained A horizons, and all except the most strongly developed *podzol* B horizons.

The plasmic fabrics of such materials have various degrees and patterns of preferred orientation of the clay size constituents (plasma separations) as interpreted from extinction patterns. For purposes of discussion, they can be regarded as having *non-linear fabric components* and *linear plasma separations* in various combinations. The former either lack plasma separations or the plasma separations are essentially equi-dimensional. The latter are plasma separations that appear elongated, often curved, in thin section.

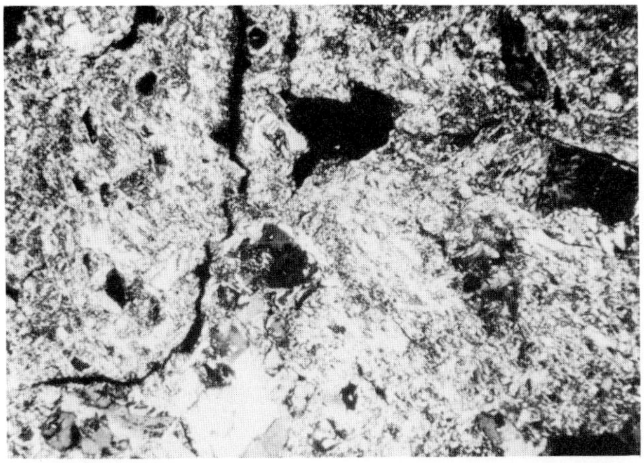

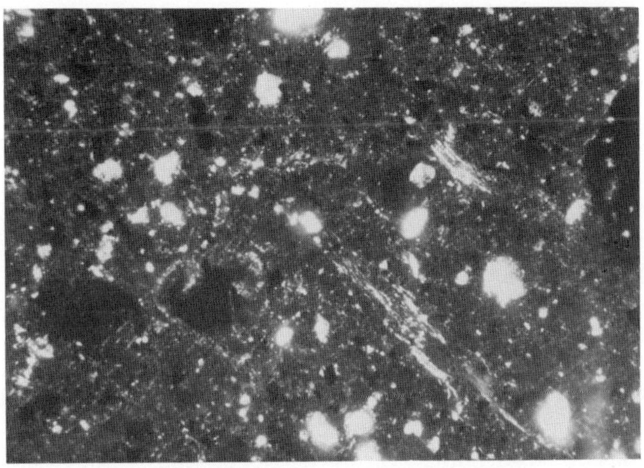

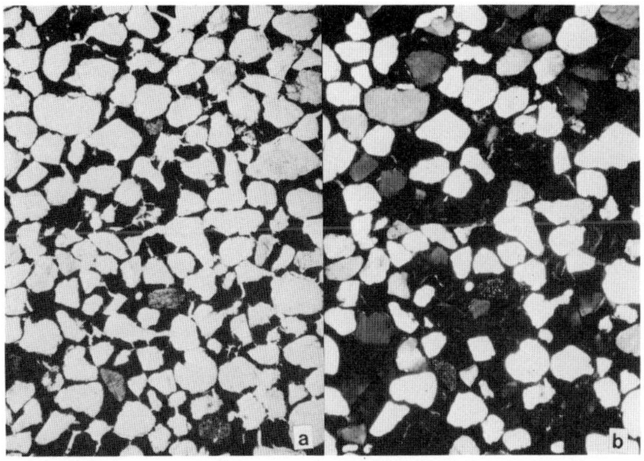

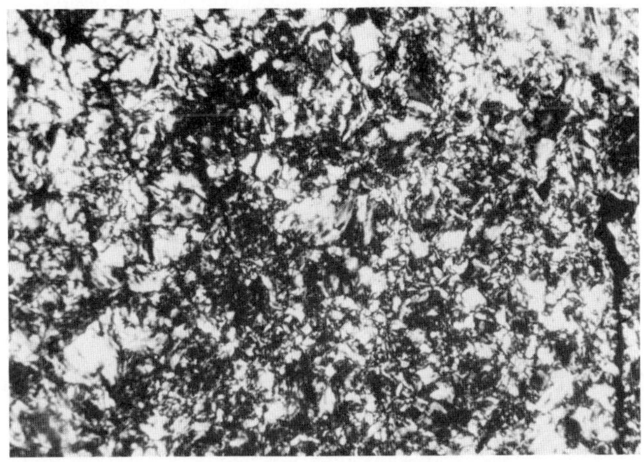

Fig. 30.24. Mosepic fabric; II B_2 horizon below *red podzolic soil*; depth 82 cm. Frame length 3.3 mm. Crossed polarizers.

Fig. 30.25. Inundulic fabric; A horizon of *krasnozem*; depth 4 cm. Frame length 0.9 mm. Crossed polarizers.

Fig. 30.26. Isotic fabric; B_2 horizon of *podzol*; depth 161 cm. Frame length 3.3 mm. (*a*) Plain light. (*b*) Crossed polarizers.

Fig. 30.27. Papulic fabric; C horizon below *red podzolic soil*; depth 82 cm. Frame length 3.3 mm. Crossed polarizers.

Non-Linear fabric components

Silasepic fabric (Fig. 30.20)

This is associated with moderate to high proportions of fine sand (30–60%), low clay content, and more silt than clay; linear plasma separations are absent or rare. It is most common in the porphyric members of the iunctic sequence, as in the A horizons of *solonetz, solodized solonetz and solodic soils, soloths, red podzolic soils, lateritic podzolic soils* and *gleyed podzolic soils*; it is also associated with some sedimentary deposits.

Calciasepic fabric (Fig. 30.21)

A variety of silasepic fabric in which abundant carbonate crystallites behave essentially as silt. It occurs in the finer-grained *grey-brown and red calcareous soils, calcareous sands, solonized brown soils, calcareous red earths, black earths,* and *chernozems*, and less commonly in some horizons of the *desert loams, grey brown and red clays, solodized solonetz and solodic soils,* and *red-brown earths.*

Argillasepic fabric (Fig. 30.22)

This occurs with and without linear plasma separations. It is common in the porphyric surface horizons of many soil groups, except those formed on basaltic parent material, other than *black earths*. Clay is usually less than 30%, but may be as high as 70%, and birefringence of the plasma is usually masked by a dark stain, probably finely dispersed organic matter. It is usually an important component in finer-textured B horizons, in association with the equant plasma separations of insepic fabric, or with linear plasma separations.

Insepic fabric (Fig. 30.23)

In association with linear plasma separations, insepic fabric is very common in finer-grained porphyric B horizons of a wide range of soil groups. Without linear plasma separations, it is confined to porphyric surface horizons of some *red-brown earths, yellow earths, solonized brown soils* and *calcareous red earths.*

Fig. 30.28. Crystic fabric; within a gypsum crystal chamber; subsoil of a *grey brown and red clay*; depth 126 cm. Frame length 3.3 mm. Crossed polarizers.

Mosepic fabric (Fig. 30.24)

It has been recorded, with associated linear plasma separations, only in the B_2 horizons of some *red-brown earths, brown earths, red earths, yellow earths, red podzolic soils* and *gleyed podzolic soils*. These materials usually have high clay contents (40–60%) and some evidence of intermittent waterlogging, in the form of segregations of iron and manganese.

Undulic and inundulic fabrics (Fig. 30.25)

These are closely related. They occur most commonly as the internal fabric of the separated granic/fragmic types (see Fragmic sequence p.445). With increasing depth, most of these materials grade, either directly or through inundulic fabric, into asepic or sepic porphyroskelic fabrics. Surface horizons usually lack linear plasma separations, but the deeper materials may have associated plasma separations. These fabrics seem to be associated with basic parent materials, especially basalt, and with moderate to strong weathering and leaching.

Isotic fabric (Fig. 30.26)

This has been observed only in the uppermost horizon of a *gleyed podzolic soil*, the B horizon of *podzols* and throughout a *humic gley*. Elsewhere, it is common in soils formed on volcanic ash that weathers to produce allophane.

Papulic fabric (Fig. 30.27)

Consists of abundant, randomly arranged papules that could be considered as pedological features; but because they comprise most of the plasma of the soil material, they are dealt with here as a specific plasmic fabric. This has been observed only in weathering pyroclastic rock at the base of a *red podzolic soil*, and in the B horizon of a *desert loam* on alluvium. In the former, some of the papules are clearly weathered micas, and others are kaolinite books; such features are common in weathered granite-like rocks. In the latter, they may be due to illuviation, with subsequent pedoturbation causing disruption of the argillans, or they may be sedimentary features formed during depositon of the alluvial parent material.

Crystic fabric (Fig. 30.28)

Rare, except in pedological features. However, random crystic fabric observed in the base of *desert loams* is composed of gypsum crystals of variable size that are regarded as sedimentary deposits. Such fabrics can be regarded as specific kinds of granic fabric in which the f-members are gypsum crystals.

Linear plasma separations

Linear plasma separations are generally *in situ* features formed by differential swelling and shrinking, but they may be inherited from the parent rock. They are characteristic of the sepic fabrics: skelsepic, vosepic, masepic and omnisepic. The first three are frequently associated with argillasepic or insepic components, and much less frequently with undulic, inundulic, and mosepic components. Below the shallowest occurrence of linear plasma separations in a profile, their degree of development with increasing depth usually either remains uniform or it increases, sometimes with a subsequent decrease; however, in some *desert loams, solonetz, red-brown earths* and *euchrozems* it decreases with increasing depth.

The depth to the shallowest occurrence of linear plasma separations varies between, and within, soil groups. To generalize rather broadly: they occur in the surface or subsurface horizons of clay soils such as *grey brown and red clays, black earths, humic gleys, prairie soils, xanthozems, chocolate soils,* and *brown earths*, and of gradational soils such as *red earths* and *yellow earths*; they occur in the horizon immediately above the clay maximum in texture-contrast soils such as *solonetz, solodized solonetz and solodic soils, red podzolic soils, yellow podzolic soils, lateritic podzolic soils,* and *soloths*, and of gradational soils such as *solonized brown soils*; they occur in the horizon of maximum clay content in some *red-brown earths* and *calcareous red earths*; and in the base of *non-calcic brown soils*, some *red earths*, and *krasnozems*. They are developed best in or immediately above the clay maximum at depths up to 60 cm in some *grey brown and red clays, solonetz, solodized solonetz and solodic soils, red-brown earths, red podzolic soils* and *lateritic podzolic soils*; and in the base (70–90 cm) in some *chocolate soils, xanthozems,* and *red earths*. The former are most likely to be orthic features, and the latter truncated buried soils or orthic weathering features.

Omnisepic fabrics (Fig. 30.29) are extreme expressions of linear plasma separations, usually associated with weathered rocks: for example, the C horizon of a *yellow podzolic soil* formed on shale has an extreme form

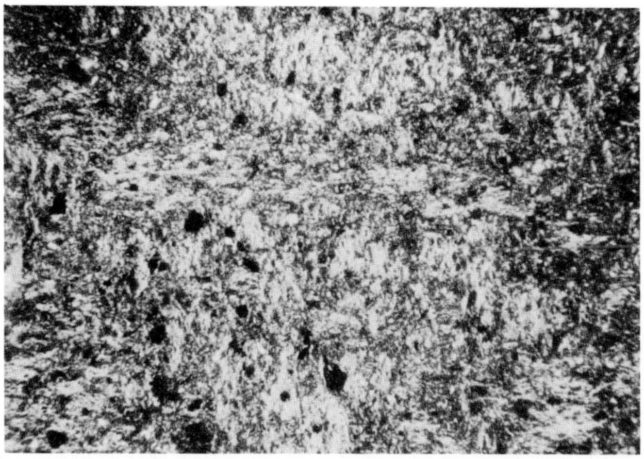

Fig. 30.29. Omnisepic fabric; C horizon below *yellow podzolic soil*; depth 123 cm. Frame length 3.3 mm. Crossed polarizers.

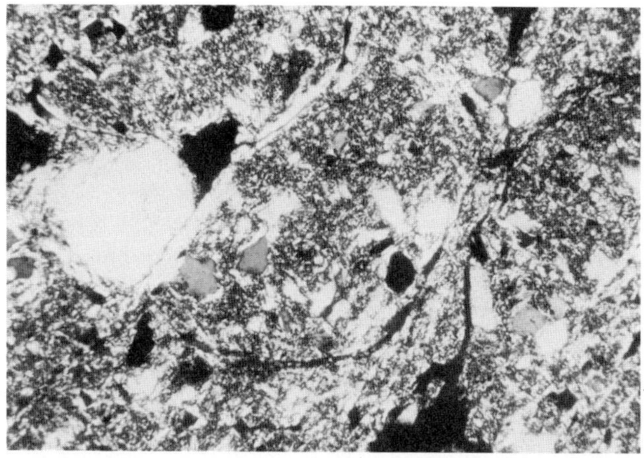

Fig. 30.30. Vo-ma-skel-insepic fabric; *grey brown and red clay*; depth 14 cm. Frame length 0.9 mm. Crossed polarizers.

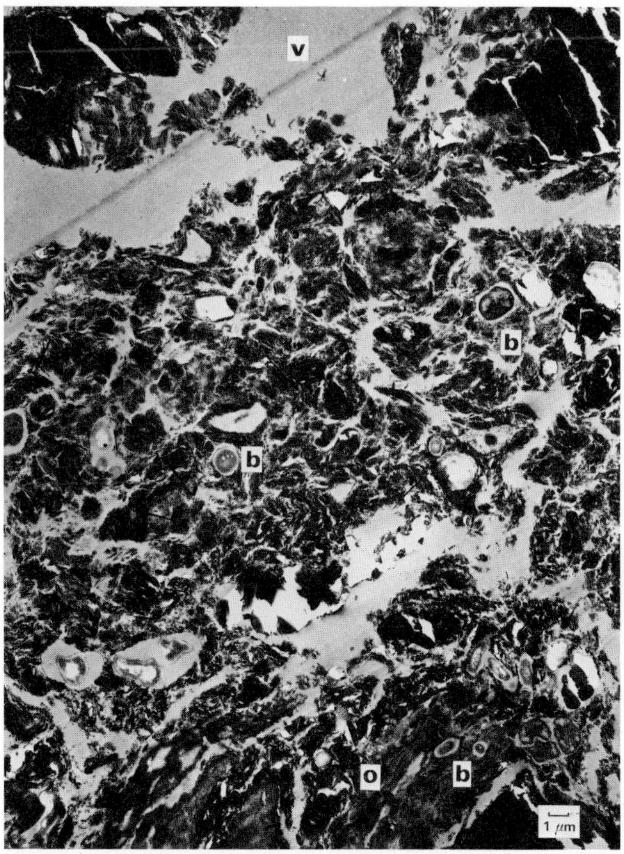

Fig. 30.31. Cutan-lined cryptovoids, quartz grains, clay minerals, organic matter and a few bacteria; *black earth*. Electron micrograph.

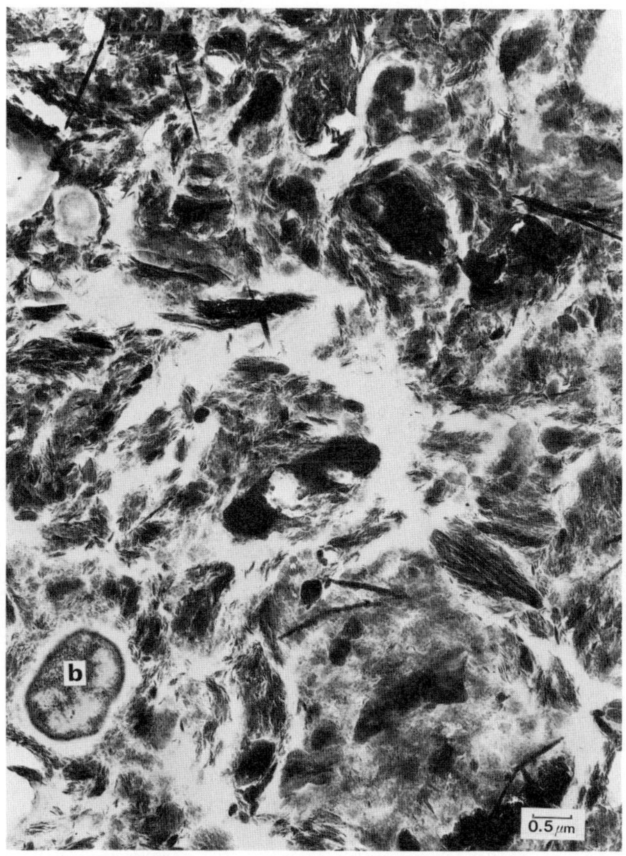

Fig. 30.32. Packets and stacks of clay platelets, acicular crystals, bacteria with capsule; *black earth*. Electron micrograph.

of masepic fabric, and one formed on granodiorite an extreme form of skelsepic fabric.

Whilst there is no specific relationship between the form of linear plasma separations and soil groups, the following generalizations are based on observed dominance within groups. Variants of skel-vomasepic fabric (Fig. 30.30) are by far the most common; they occur in *grey brown and red clays, red-brown earths, prairie soils, solonetz, solodized solonetz and solodic soils, solonized brown soils, chocolate soils, brown earths,*

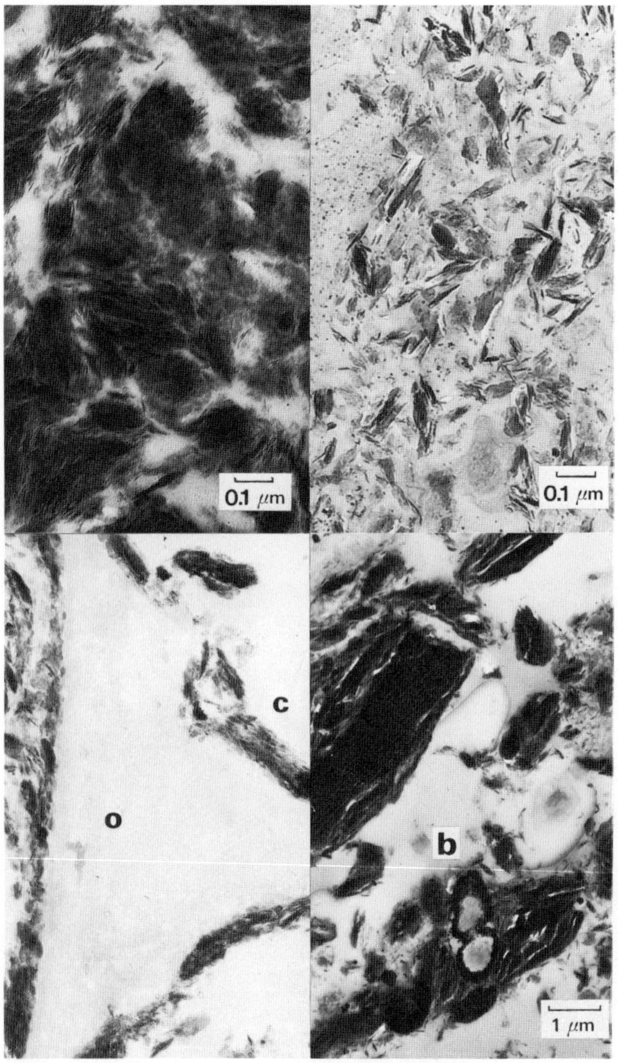

Fig. 30.33. (Top, left) Details of stacks, domains, and bundles of clay lamellae; *black earth.* Electron micrograph.

Fig. 30.34. (Top, right) Plasmic fabric; honeycomb arrangement of platelets. Electron micrograph.

Fig. 30.35. (Bottom, left) Cutan on middle lamellar fragment; *calcareous sand.* Electron micrograph.

Fig. 30.36. (Bottom, right) Cutan on lamellate grain; *solonized brown soil.* Electron micrograph.

euchrozems, and *humic gleys*. Ma-skelsepic fabric occurs in podzolic soils (red, yellow, brown, lateritic, and gleyed), *yellow earths*, and *xanthozems*. Skelsepic fabric occurs in *calcareous red earths, grey-brown and red calcareous soils, desert loams, non-calcic brown soils*, and as a striated orientation pattern within the f-matrix in many chlamydic and plectic materials; it tends to be better developed in materials with more than 30% fine sand, and commonly occurs to the surface. Masepic fabric occurs in *soloths, red earths,* and *krasnozems.* A number of fine-grained profiles formed on basalt (*krasnozems* and *euchrozems*) have a characteristic masepic component in their lower horizons; this has the form of numerous striotubules with indistinct, sometimes interfering, boundaries.

Ultramicroscopic features

Ultrathin sections of mineral soils (Figs 30.1, 30.31–35) show a wide range of material in the size range from 10 µm to 10 nm. Particles may consist of single minerals or compound units. Most particles larger than 100 nm are randomly arranged within the s-matrix, but they may display a preferred orientation near roots (see Ultrastructure of root-soil interface, Chapter 41). Smaller particles may be aligned parallel to the surfaces of adjacent mineral and organic fragments. Most of the voids in Figs 30.1, 30.31 and 30.32 fall into the cryptovoid category. Voids smaller than 500 nm in diameter may be filled with plasma composed of small acicular platelets about 40 nm long (Fig. 30.1). Plasma that may fill much larger voids forms a loose but stable network of particles (Fig. 30.34), with a honeycomb rather than a card house structure (O'Brien 1971), of which only a small part is seen because of the thinness of the section (Smart 1975).

At higher magnifications (Figs 30.32–36), the fine structure of such micromorphological features as domains, packets, and bundles (Smart 1975) can be seen in detail, especially where the section passes at right angles to the lamellae of which they are composed (Figs 30.32 and 30.33). The clay particles may be separated by a network of cryptovoids ranging from 300 to 30 nm or less in diameter. Acicular crystals as well as clay minerals occur, each about 1 µm long and 30 nm thick (Fig. 30.32). Fig. 30.33 shows a complex domain about 500 nm across consisting of packets of platelets at different angles to each other. Some of the smaller packets are composed of only 4 or 5 lamellae, and are only 30 nm across. The individual lamellae are about 70 nm long and 5–10 nm wide; they are typically spindle-shaped in cross section. They are common in most of the soils examined, and may occur in small clusters of 3 or 4, or singly (Fig. 30.34). They often become adsorbed on to other soil constituents to form ultramicroscopic argillans (Brewer 1964, p. 212) (Figs 30.35 and 30.36). Because of the low electron density of much of the organic matter after conventional fixation, such cutans are the only way of recognizing the organics contained in them. It seems that small mineral particles have a propensity for adsorption on to surfaces.

These data can be related to microscopic plasmic fabrics. The larger units are logically those that cause the flecked extinction pattern of argillasepic fabric components, while the smaller units are unlikely to appear birefringent in microscopic thin sections; these latter probably constitute the indeterminate material that is common in many soil materials (Brewer 1964, p. 170). Similar studies are required to relate other microscopic plasmic fabrics to ultramicroscopic features.

Table 30.1 Distribution of segregations of constituents

Soil group	No. of profiles	Clay minerals					Iron								Manganese				Al	Carbonates						Sulfates			Silica							Silt			Sand				Soil material						
		Argillans	Laminae	Papules	Nodules	Concretions	Nodules	Concretions	Haloes	Mottling	Ferrans	Neoferrans	Quasiferrans	Iwatoka	Nodules	Haloes	Mangans	Neomangans	Gibbsans	Crystallites	Nodules	Concretions	Calcans	Neocalcans	Crystallaria	Crystallaria	Nodules	Gypsans	Silans	Neosilans	Nodules	Plasma	Phytoliths	Diatoms	Sponge Spicules	Nodules	Siltans	Laminae	Granotubules	Skeletans	Laminae	Nodules	Nodules	Aggrotubules	Isotubules	Striotubules	Matrans	Neomatrans	Laminae
Alluvial soils	1																																1																
Calcareous sands	1	1		1	1										1					1	2		1	1		1								1		1							1						
Siliceous sands	2	2		1			1																										1						1				1		1				
Earthy sands	1	1		1	1	1	1								1																		1	1									1						
Grey-brown and red calc. soils	2	2		1			2								1			1		2	2		2		1								2	1									1	1					
Desert loams	5	5	3	5			5					1			5		1	4		4	5		3		1	5		3					4	1		2							4	3	1				
Red and brown hardpan soils	3	3				3	3								2		2												2	2	3	1																	
Grey brown and red clays	25	20	10	24	7		17	2		6	1	1			19	1	2	12		5	17		7		1	5	1	1					22	10	1	13		7			3	3	14	13	2	3			3
Black earths	1	1		1	1		1								1		1	1		1	1					1							1																
Rendzinas	4																				3												4		1									1					
Chernozems	1																			1													1																
Prairie soils	3	3	1	3	1		3			1		1			2			1															3	2		1		1					2	1	1				
Solonetz	1	1	1	1	1										1						1												1	1		1	1						1						
Solodized solonetz and solodic soils	6	6	3	6			6	1	1			2			5		1	1		1	4					1						1	6	4	2	2	1	1					5	2	1	1			1
Soloths	7	7	2	5			5			1		1			2			1															7	4	1	1					2		5	2	1	3			1
Solonized brown soils	6	6	2	6	2	1	5								6	1	1	5		6	6	1	6	2	5								4	1		2		1	1				5	3	1	1			1
Red-brown earths	9	9	1	9	1		9	1			1	2	1		8		3	3		5	9		6	2									9	4		5		2					9	3	4	1			
Non-calcic brown soils	2	2		2			1								2																		2										2						
Chocolate soils	3	1					2																										2										1	1					
Brown earths	1			1			1		1																								1													1			
Calcareous red earths	6	6	2	6			4								6	1		5		3	5		3	1	1	1							4	1		1		1					2	1	1			1	1
Red earths	16	11	1	9	3	3	10	1						2	6			4															9	1			1		1				6	5	5	3			
Yellow earths	1	1	1	1			1																										1																
Terra rossa soils	1	1					1					1																					1		1														
Euchrozems	3	1		1			3					1			1																		3										2	3	1	2			
Xanthozems	1	1	1	1			1								1																		1											1		1			
Krasnozems	2				1		2	1							1																		2										2	2		1			
Red podzolic soils	1	1		1			1								1																		1										1						
Yellow podzolic soils	5	3	1	4			5	2	2	1	2	3			3			2															3	2	1	1		1		1			1		1				1
Brown podzolic soils	1	1		1	1		1							1																			1										1						
Lateritic podzolic soils	4	3	3	4	1		4			2		2	1					1															4	2						1			4	1	1	1	1	1	
Gleyed podzolic soils	1	1	1	1			1					1		1																			1																
Podzols	5																																1	1	1														
Humus podzols	3																																3																
Alpine humus soils	6						1																										6	1										4	2	2			
Humic gleys	5	3	2	4			4			1	1	4	1																				5	4	2	1		1					1						1
Indurated lateritic horizons	4	4				2	4				2						1		1							1			2			1																	
Pseudogley subsola	5	5	5	5	2		4	2		4	1	3		1	3			3									1					2	4	1		2	2	5	2			2	3	4		1			2

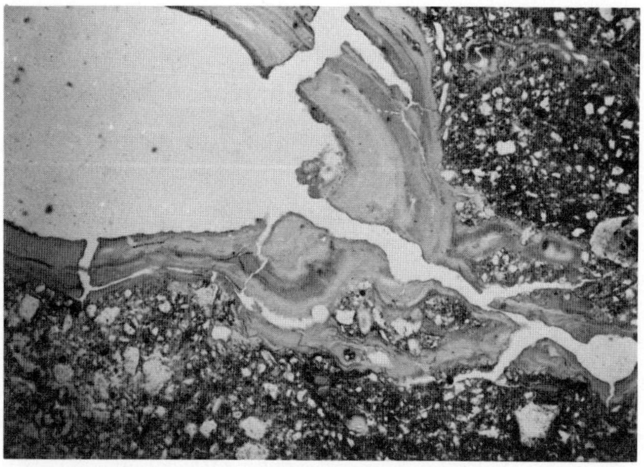

Fig. 30.37. Thick smooth yellow argillan with strong continuous extinction pattern; II B_2 horizon (pseudogley) below a *red earth*; depth 132.5 cm. Frame length 3.3 mm. Plain light.

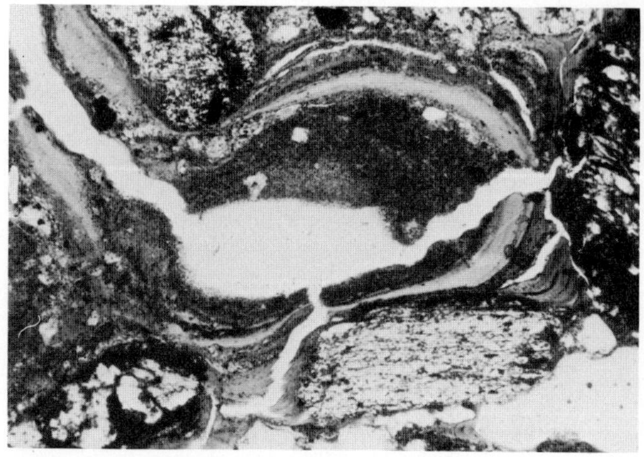

Fig. 30.38. Compound argillan, smooth yellow with strong continuous extinction pattern and dusty argillans with striated/flecked extinction pattern; II B_2 horizon (pseudogley) below a *red earth*; depth 261 cm. Frame length 0.9 mm. Plain light.

Key to letters on electron micrographs: b, soil bacteria; c, clay; o, organic matter; s, space left by fragmented quartz grain; q, quartz grain; v, void.

Segregations of constituents

The following discussion is concerned with segregations of various constituents, their distribution in relation to great soil groups, and their genesis. Their broad distribution in terms of presence or absence within profiles of each great soil group is shown in Table 30.1.

Segregations of clay minerals

Argillans

Most argillans are composed of the same general mixture of clay minerals as in the plasma of the s-matrix of the soil material in which they occur, often with somewhat higher proportions of iron oxides (ferri-argillans). Argillans of pure clay minerals are rare, and have only been recorded as palygorskans or compound palygorskan/kaolinans in old, buried, truncated profiles developed in parna (Beattie and Haldane 1958), and in the base of a single *solonized brown soil.* Argillans usually have a strong continuous extinction pattern, indicating strong orientation of the clay mineral particles parallel to the surfaces on which they occur, but the degree of orientation is variable. Most are smooth red or yellow in plain transmitted light (Figs 30.37 and 30.38), but some have a dusty appearance or dusty zones parallel to the surfaces they coat (Fig. 30.38). The dusty zones are usually less strongly oriented, and electron microprobe analysis shows that they are generally higher in silica than the clear zones, suggesting accumulation of secondary silica or very fine quartz.

The distribution of argillans, if considered in terms of frequency, suggests they are generally less abundant in well structured soils, especially those formed on basalt. The dusty types tend to occur in silasepic A horizons or the top of B horizons in texture-contrast profiles, and in pseudogley subsolum horizons (van Dijk 1969) in association with thick (up to 1 cm) compound ferri-argillan-siltans, clay plugs, and void fillings; they also occur in some *red-brown earths, yellow podzolic soils,* and *gleyed podzolic soils.*

The maximum of void argillans may occur in any section of the soil profile, including the underlying weathered rock, even within the one soil group. The only tentative generalization that can be made is that where they occur in profiles with carbonate horizons, they tend to reach a maximum above the carbonate horizon.

Most argillans are orthic features formed by illuviation; they are sometimes weathering features formed, especially in coarse micaceous rocks, by re-alignment of weathered mica grains. Wustenquarz (mineral grains coated with red ferri-argillans, Fig. 30.39) is an inherited feature indicative of provenance; it tends to be concentrated in surface horizons, especially in *red-brown earths* and *grey brown and red clays* developed in parna in the Riverine Plain of south-western N.S.W. Similar features occur sporadically in soils from other parts of Australia.

In texture-contrast soils, the amount of illuviated clay in the B horizons, as assessed from the content of strongly oriented clay (usually argillans plus papules), is usually less than 5% of the soil material, occasionally between 5 and 10%, rarely up to 15%, and extremely rarely up to 30%. The last figure was obtained from the B horizon of a *solodic soil,* which consisted entirely of sand grains with free grain argillans. The amount of illuviated clay estimated in this way is rarely sufficient to account for the difference in clay content between the A and B horizons. Part of this difference is probably due to differential weathering between the A and B horizons, and/or sedimentary layering.

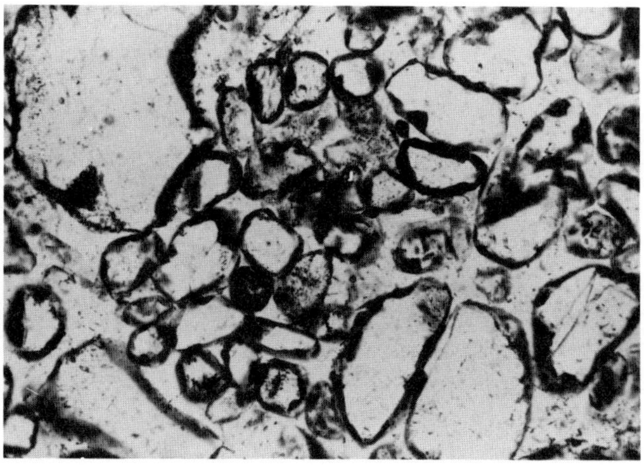

Fig. 30.39. Wustenquarz; *red earth*. Frame length 0.9 mm. Plain light.

Laminae

Laminated or elongated zones of strongly oriented clay (Fig. 30.40) usually occur in the lower section of profiles formed on unconsolidated and consolidated sediments, except for two profiles formed on basalt and one on pallid zone material of a laterite. This general association with sedimentary bodies suggests they are usually sedimentary features, but in swelling soils, such as *desert loams* and *grey brown and red clays*, they may be engulfed void and ped argillans, especially where they have been recorded as neo-argillans and quasi-argillans because of their relationships with voids. Neo-argillans have been recorded in *grey brown and red clays, red-brown earths, calcareous red earths, xanthozems*, and *yellow podzolic soils*, and quasi-argillans in a *solodized solonetz and solodic soil*, and pseudogley subsolum horizons. Compound clay-silt laminae also occur, particularly in *grey brown and red clays* and in buried soils and/or pseudogley subsolum horizons.

Papules (Fig. 30.41)

These have a wide distribution, similar to argillans, but are not generally found in well-structured soils formed on basalt or limestone. This suggests that most papules are derived from argillans disrupted by shrink-swell processes. However, many materials containing papules have little capacity to swell; in these, the papules are either weathering features as in the base of some *red podzolic soils* and pallid zones on granite-like rocks, or are transported inherited features as in lake and lunette parnas, or argillans disrupted by faunal activity or rock weathering. In others, such as the subplastic clays of the Riverine Plain of N.S.W., the stability of the clays postdates the formation of both argillans and papules (Brewer and Blackmore 1976).

Papules with lamellar fabric (Fig. 30.42) are confined to profiles formed on sedimentary parent materials, either unconsolidated alluvium or aeolian clay, or on consolidated fine-grained sediments. This suggests that they are sedimentary relicts, and certainly transported inherited features in the aeolian clays.

Nodules and concretions

Clay nodules (Fig. 30.43) are not common. Some of the profiles in which they occur are formed *in situ* on basalt, laterite, lake parna, and transported material derived from laterite. The evidence suggests that apart from in the lake parnas, the nodules are weathering features in lateritic profiles and in soils developed on basalt, and that they have the capacity to be transported and deposited without destruction.

Clay concretions (Fig. 30.44) are rare. Some are orthic features characteristic of some lateritic profiles, but the most abundant are detrital f-members in sandy soils formed in colluvium or windblown deposits derived from lateritic materials. They are dominantly kaolinitic, well rounded, with internal fabrics similar to fabric features in the pallid zones of lateritic profiles.

Segregations of iron

Segregations of iron oxides and hydroxides are generally assumed to result from some degree of hydromorphism, possibly with micro-organisms playing a role, but they are also commonly transported inherited features.

Nodules

Ferruginous nodules (Fig. 30.45) are widespread. Because they are common transported inherited features, they may occur in any profile formed on sediments. In polished or thin section, they are mainly red in reflected light (rarely black or bronze) and irregular (rarely normal), with sharp or diffuse boundaries. Diffuse boundaries indicate formation *in situ* due to periodic waterlogging, while sharp boundaries suggest inheritance as an alternative to formation *in situ*. Their internal fabric is highly variable, ranging from rather weak concentrations impregnating the s-matrix of the soil material so that the fabric of the s-matrix is still recognizable, to opaque units. Internally, there may be complex patterns of different degrees of concentration; they usually contain skeleton grains, but may not; they may have cores of mineral grains or rock fragments, or decomposed organic fragments, or no apparent core; compound nodules suggesting more than one cycle of formation are common.

The distribution of ferruginous nodules within profiles even of the one soil group is somewhat erratic. Most commonly, they occur throughout the *grey brown and red clays, chocolate soils, red earths, yellow earths, terra rossa soils*, and *krasnozems*, also in the lower A and upper B

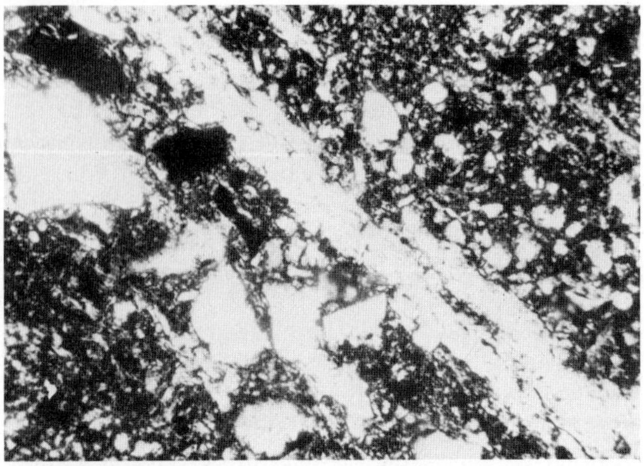

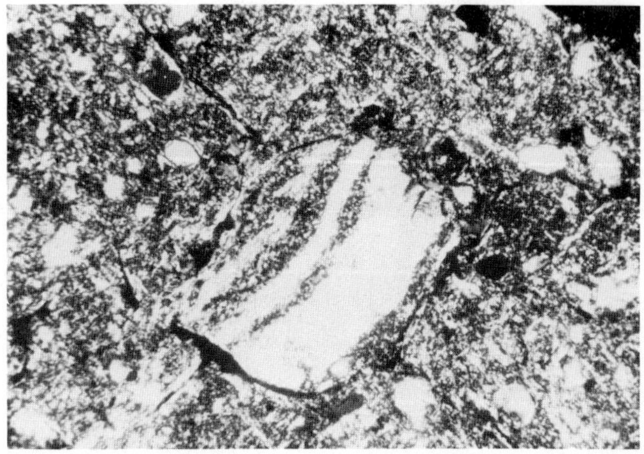

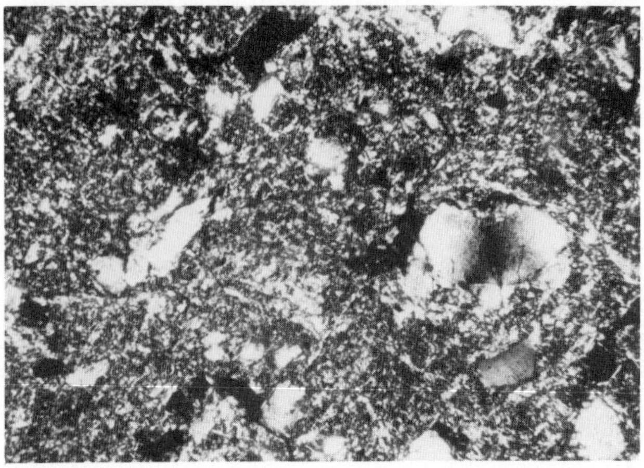

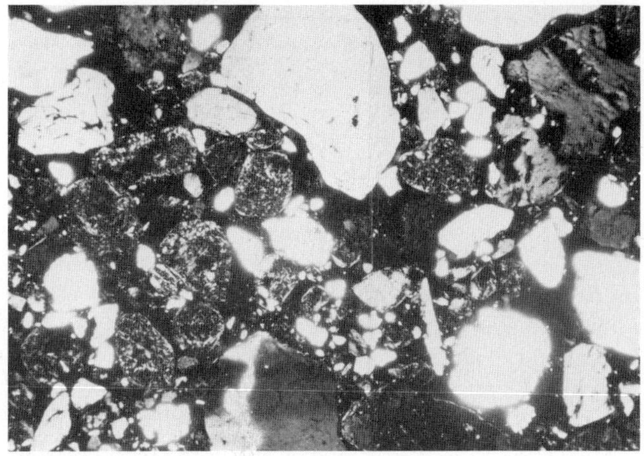

Fig. 30.40. Clay lamina, strongly oriented; A_2 horizon of *solodized solonetz;* depth 28 cm. Frame length 0.9 mm. Crossed polarizers.

Fig. 30.41. Papules; *grey brown and red clay*; depth 85 cm. Frame length 0.9 mm. Crossed polarizers.

Fig. 30.42. Papules with lamellar fabric; *grey brown and red clay;* depth 85 cm. Frame length 0.9 mm. Crossed polarizers.

Fig. 30.43. Clay nodules; A horizon of *earthy sand*; depth 2.5 cm. Frame length 3.3 mm. Crossed polarizers.

horizons of texture-contrast soils such as *solodized solonetz and solodic soils, soloths, red-brown earths, brown podzolic soils*, and *lateritic podzolic soils*, and in the surface horizon of soils of arid regions; they are most abundant in some *chocolate soils, red earths*, and *krasnozems*. In the field descriptions, they are commonly reported as abundant in soils formed on basalt, but such nodules are often identifiable microscopically as weathered basaltic lithorelicts. Nodules that are bronze in reflected light have been observed in indurated lateritic horizons, *red and brown hardpan soils*, and *yellow podzolic soils*. In the first group, they are interpreted as *in situ* features, and in the last two as transported inherited features. Nodules with diffuse boundaries tend to be more common in deeper horizons: for example, in some *red earths, red-brown earths*, and *lateritic podzolic soils*.

There are insufficient data to ascribe particular kinds of ferruginous nodules to horizons within specific soil groups. The erratic distribution within and between profiles and soil groups is probably related to local variation in micro-environment, such as redox potential and organic matter and micro-organisms, and to the fact that many occurrences are inherited features.

Concretions

Ferruginous concretions (Fig. 30.46) are not common. There is some evidence that they are associated with horizons that are subjected to periods of marked desiccation, hence their concentric fabric.

Glaebular haloes (Fig. 30.47)

These are confined to lower horizons. They are orthic features and reflect the effects of intermittent waterlogging in the base of these profiles.

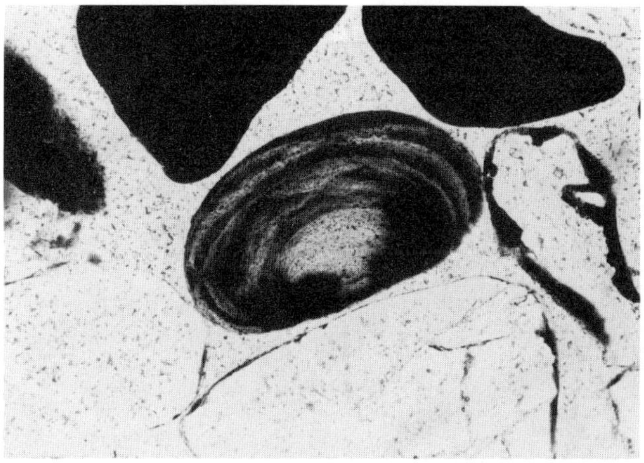

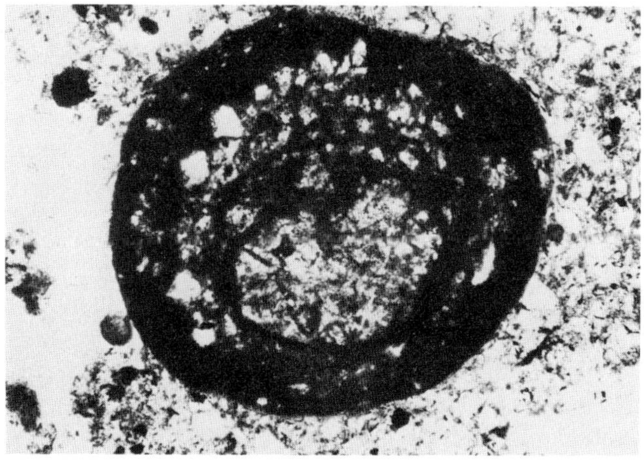

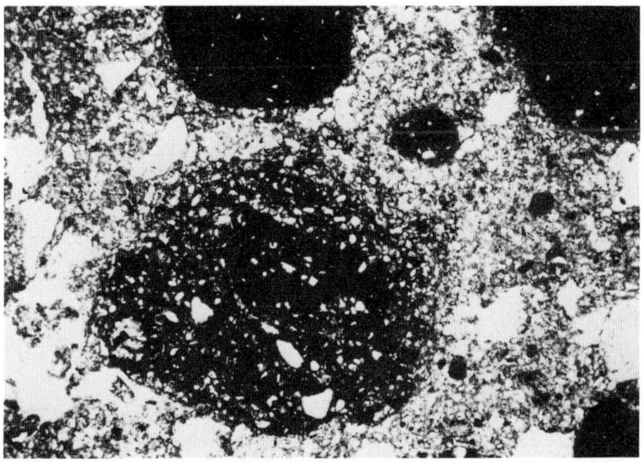

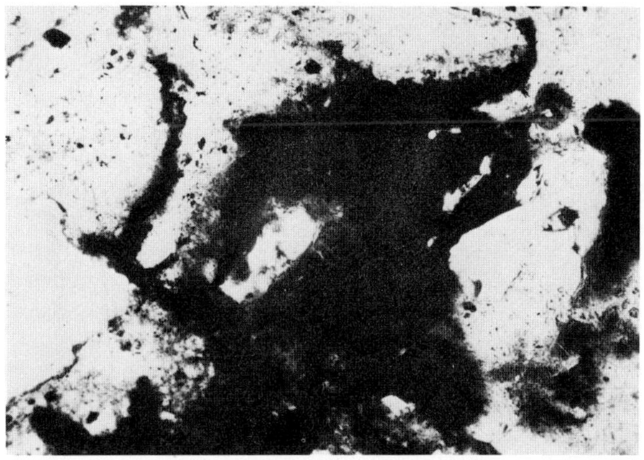

Fig. 30.44. Abraded kaolinitic concretion; *siliceous sand.* Frame length 0.9 mm. Plain light.

Fig. 30.45. Ferruginous nodules; strongly separated with diffuse and sharp boundaries; A horizon of *red earth*; depth 12 cm. Frame length 3.3 mm. Plain light.

Fig. 30.46. Normal ferruginous concretion; A horizon of *red earth*; depth 12 cm. Frame length 3.3 mm. Plain light.

Fig. 30.47. Ferruginous glaebular halo; lower B horizon of *yellow podzolic soil*; depth 162 cm. Frame length 0.9 mm. Plain light.

Mottling (or stain formation) (Fig. 30.48)

This is mainly red, rarely black or yellow. It is associated with poorly drained horizons or strata.

Ferrans (Fig. 30.49)

These have been observed rarely in lower horizons. They are orthic features indicative of periodic waterlogging.

Neoferrans and quasiferrans

Neoferrans (Fig. 30.49) are not common, and quasiferrans (Fig. 30.50) are even less common. Both of these features usually occur in subsoils associated with intermittent waterlogging, but this does not appear to be so in the apparently well drained *euchrozems* and *terra rossa soils*; these two soil groups, however, have high proportions of iron oxides.

Iwatoka (Fig. 30.51)

These are concentrations of iron oxides that occur as clusters of extremely small units, sometimes crystallites, frequently grouped together in the form of a neoferran. They are usually associated with horizontally layered materials that are distinct from the overlying profile, presumably buried soils or intermittently waterlogged subsola.

Segregations of manganese

Manganiferous segregations are most common in alkaline materials, especially in those that include carbonate segregations, or in horizons with some evidence for intermittent waterlogging. Chemical oxidation of manganese occurs in the pH range 8.0–9.0, whereas microbial oxidation is effective in the pH range 5.0–8.5 (Mulder and van Veen 1968).

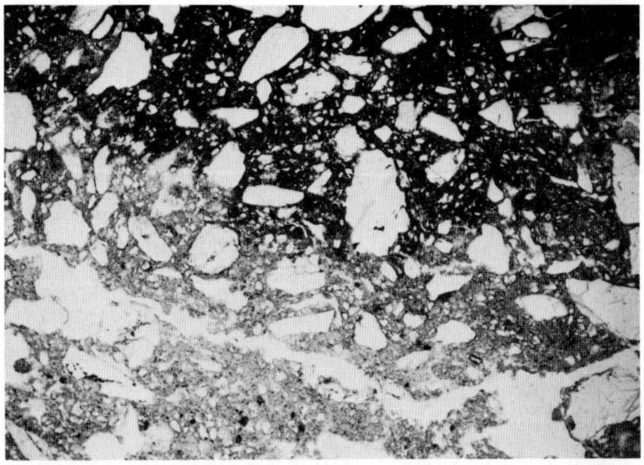

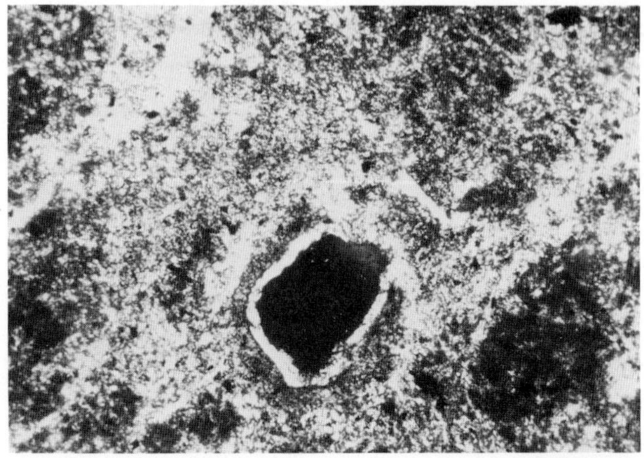

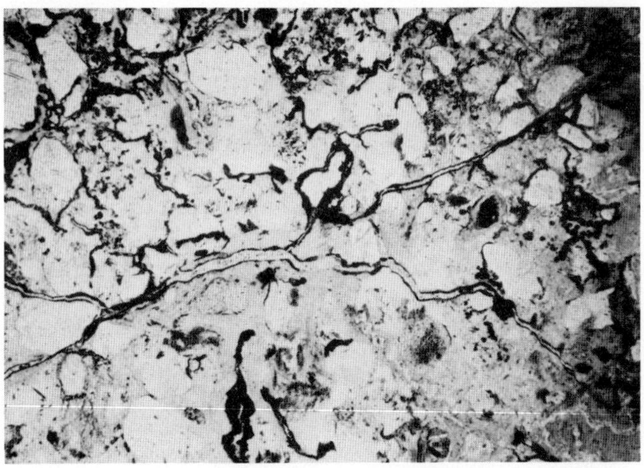

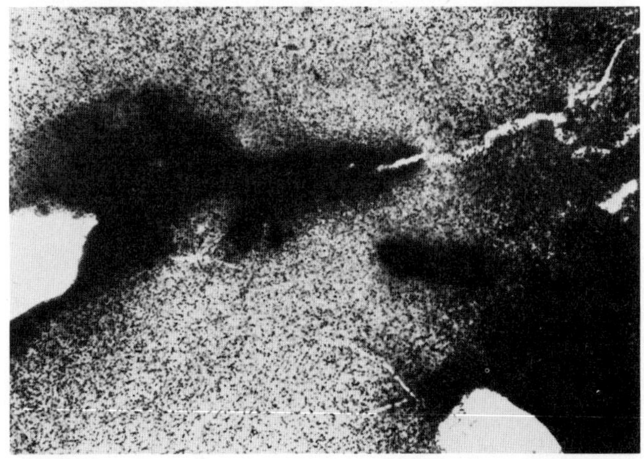

Fig. 30.48. Ferruginous mottling; B horizon of a *yellow podzolic soil*; depth 61 cm. Frame length 3.3 mm. Plain light.

Fig. 30.49. Void ferrans and neoferrans; lower B horizon of *yellow podzolic soil*; depth 174 cm. Frame length 3.3 mm. Plain light.

Fig. 30.50. Void quasiferrans; B/C horizon of *humic gley*; depth 188 cm. Frame length 0.9 mm. Crossed polarizers.

Fig. 30.51. Iwatoka — in part clustered to form neoferrans; ferruginous zone below *red earth*; depth 72 cm. Frame length 0.4 mm. Plain light.

Nodules

Manganiferous nodules (Fig. 30.52) are common. They are usually irregular with sharp boundaries, rarely diffuse. They may be inherited features in some profiles.

Glaebular haloes

Manganiferous glaebular haloes (Fig. 30.53) are very rare. They have been observed only in subsoils of materials with high exchangeable sodium (>10%) and total soluble salts (>0.3%).

Mangans (Fig. 30.54)

These have been observed only in subsoils, sometimes, as in the base of a *black earth*, as compound mangan-argillans.

Neomangans

These occur mainly as void neomangans, with a characteristic dendritic form associated with fine pore spaces, in subsoils and lower subsoils; but occasionally they occur throughout some *grey brown and red clays*. Embedded grain neomangans are rare (Fig. 30.55). All these occurrences are in calcareous materials, or in horizons with features characteristic of intermittent waterlogging.

Segregations of aluminium oxides

Segregations of aluminium oxide most commonly occur in mottled and pallid zones of lateritic profiles as pseudomorphs after felspars and as void gibbsans.

Segregations of carbonates

Most, but not all, segregations of carbonates consist of calcium carbonate. In the following record, no distinction is made between the various carbonate minerals. Lebedeva and Ovechkin (1975) discussed the factors governing the microstructure and form of carbonate segregations. A wide range of forms occurs in the *calcareous*

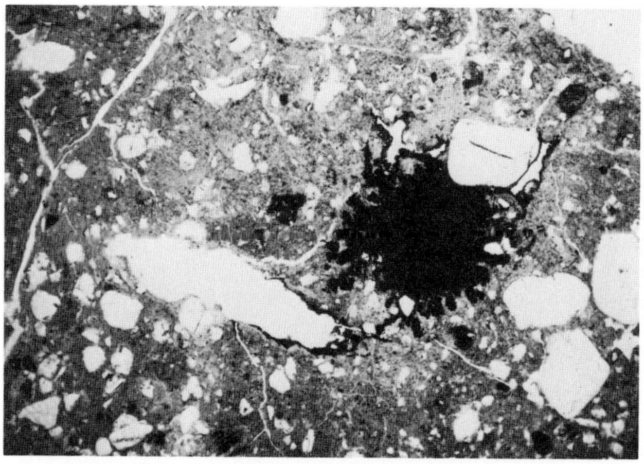

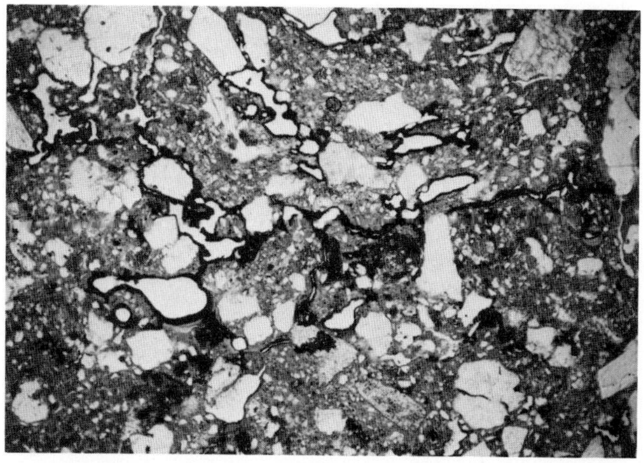

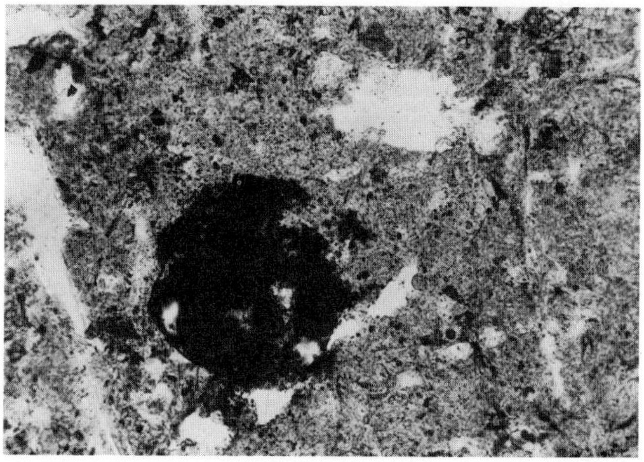

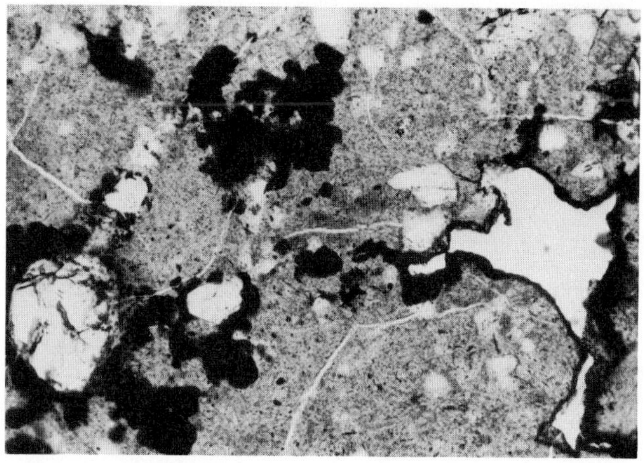

Fig. 30.52. Irregular manganiferous nodule and void neomangans; deep subsoil of *grey brown and red clay*; depth 104 cm. Frame length 3.3 mm. Plain light.

Fig. 30.53. Irregular manganiferous nodule with glaebular halo; subsoil of *grey brown and red clay*; depth 65 cm. Frame length 0.4 mm. Plain light.

Fig. 30.54. Void mangans; B horizon of *yellow podzolic soil*; depth 71 cm. Frame length 3.3 mm. Plain light.

Fig. 30.55. Neomangans (grain and void) and clustered irregular manganiferous nodules; deep subsoil of *grey brown and red clay*; depth 31 cm. Frame length 0.9 mm. Plain light.

sands, grey-brown and red calcareous soils, desert loams, grey brown and red clays, solonized brown soils, red-brown earths, and *calcareous red earths*; a much more restricted range, usually nodules only, occurs in the *black earths, rendzinas, solonetz* and *solodized solonetz and solodic soils*; in the *chernozems*, carbonates usually occur only as crystallites in the s-matrix.

Crystallites in the s-matrix

Fine crystallites, randomly distributed throughout the s-matrix (Fig. 30.56) are common in carbonate horizons; when present in high proportions, as in the base of a *solonized brown soil*, they may form a pan. The proportion of crystallites usually increases with increasing depth. In calcareous soils with mixed matrigranic-granoidic fabrics, the crystallites occur within the compound units and in the f-matrix of associated minor fabric components.

Nodules

Carbonate nodules (Fig. 30.57) are quite common in calcareous horizons; the proportion usually increases with increasing depth. The nodules are almost invariably irregular, with diffuse or sharp boundaries. They commonly include skeleton grains and manganiferous segregations, but may also include argillans and papules, and may have neostrians or neocalcans associated with their surfaces. They are usually orthic features, but may be inherited: for example, the fine nodules in some parnas. Bal (1977) suggested that they may, in some instances, be formed by earthworms.

Concretions

Carbonate concretions (Fig. 30.58) have been recorded only throughout a single profile in the *solonized brown soil* group, in which they are inherited features derived

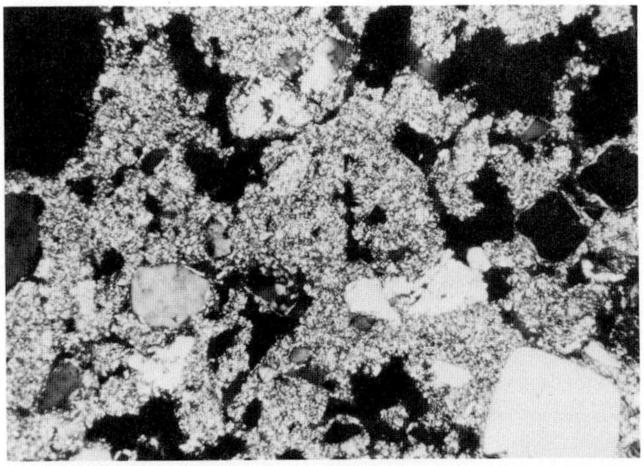

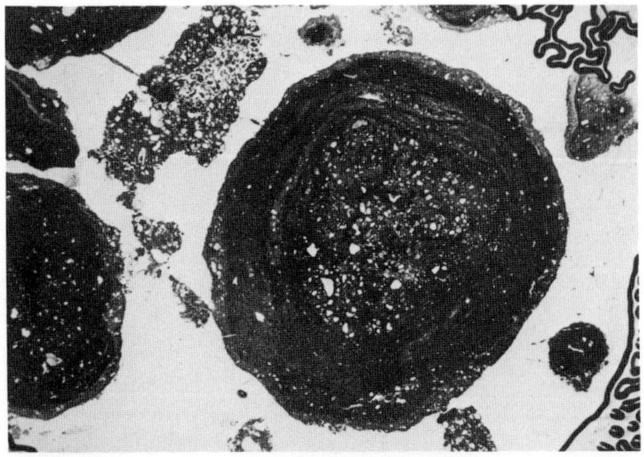

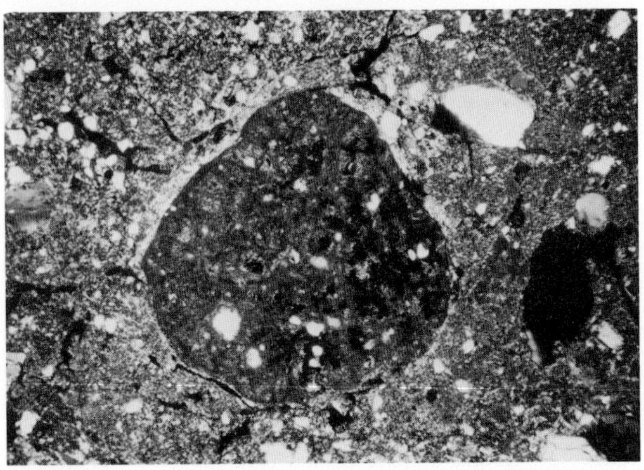

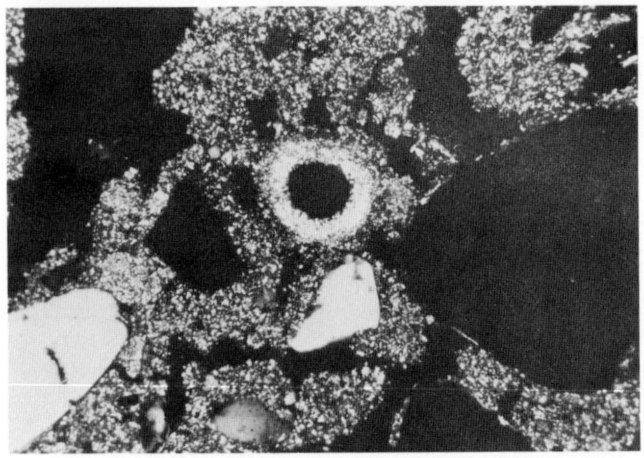

Fig. 30.56. Carbonate crystallites through the s-matrix (calciasepic fabric); lower subsoil of *grey-brown and red calcareous soil*; depth 46 cm. Frame length 0.9 mm. Crossed polarizers.

Fig. 30.57. Normal carbonate nodule with included skeleton grains, mangans, and clay nodules; subsoil of *grey brown and red clay*; depth 81 cm. Frame length 3.3 mm. Crossed polarizers.

Fig. 30.58. Carbonate concretion; subsoil of *solonized brown soil*; depth 24 cm. Frame length 28 mm. Plain light.

Fig. 30.59. Calcan/carbonate crystal tube and calciasepic plasmic fabric; lower subsoil of *grey-brown and red calcareous soil*; depth 46 cm. Frame length 0.4 mm. Crossed polarizers.

from kunkar in an old soil from which the parent material of the present soil was derived.

Calcans and neocalcans

Calcans (Fig. 30.59) and neocalcans (Fig. 30.60) are not common; compound void argillan-calcans occur in some *calcareous red earths* and *red-brown earths.*

Crystallaria

Crystal chambers (Fig. 30.61) have been recorded in the *solonized brown soils, desert loams, grey brown and red clays*, and *calcareous red earths*; crystal tubes (Fig. 30.62), in *solonized brown soils* and *grey-brown and red calcareous soils*; and intercalary crystals in *calcareous red earths*. The crystal tubes referred to above are of very fine diameter (10–20 μm), and are developed mainly on void surfaces or as tangled masses embedded in the s-matrix. Much coarser crystal tubes have been recorded in sands. Bal (1977) suggested that certain crystallaria may be formed by earthworms.

Segregations of sulfates

Crystallaria

Gypseous crystallaria occur as crystal chambers and intercalary crystals (Fig. 30.63) throughout, or in the subsoil of, some *grey brown and red clays*, and in the subsoil or basal portion of *desert loams, calcareous red earths* and *solodized solonetz and solodic soils*; they occur as crystal chambers in the subsoil of *calcareous sands*; and as intercalary crystals in the subsoil of *black earths*. Barite crystal chambers occur in a ferruginous zone of laterite.

Gypsans (Fig. 30.64)

These are confined to the subsoil of some *desert loam* and *grey brown and red clay* profiles.

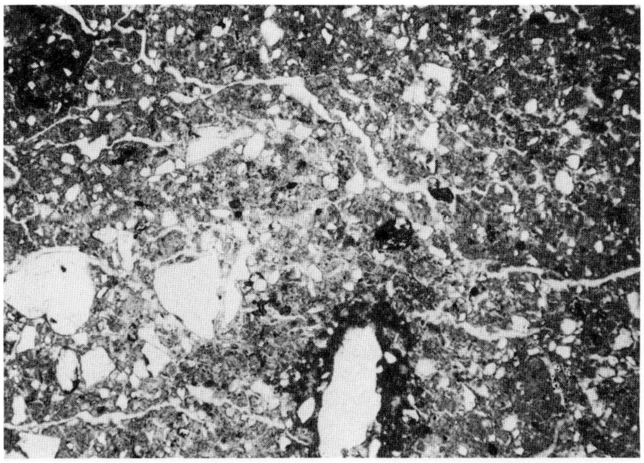

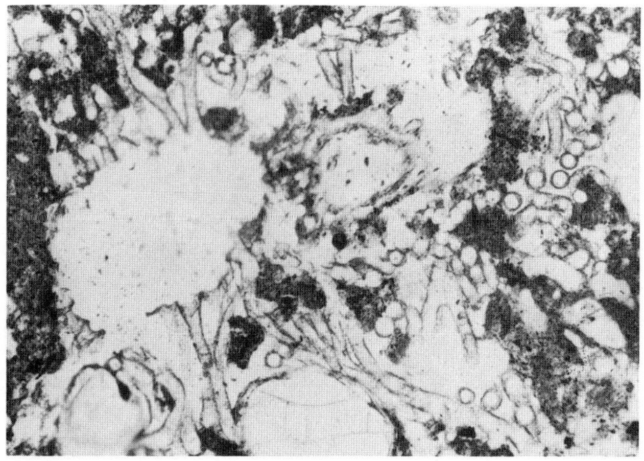

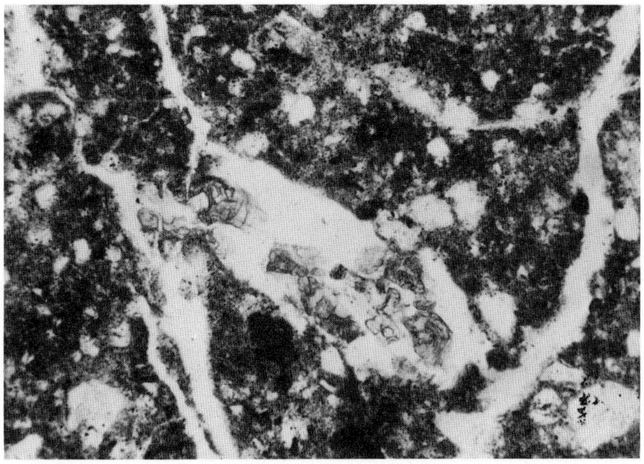

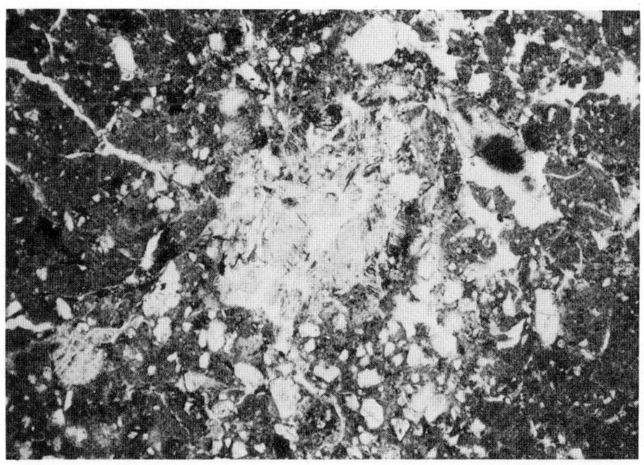

Fig. 30.60. Channel neocalcan; buried B_2 horizon below *red earth*; depth 50 cm. Frame length 3.3 mm. Plain light.

Fig. 30.61. Carbonate crystal chamber; surface of *grey brown and red clay*; depth 3 cm. Frame length 0.9 mm. Plain light.

Fig. 30.62. Carbonate crystal tubes; surface horizon of *solonized brown soil*; depth 14 cm. Frame length 0.9 mm. Plain light.

Fig. 30.63. Gypsum crystal chamber and intercalary crystals; subsoil of *grey brown and red clay*; depth 95 cm. Frame length 3.3 mm. Plain light.

Nodules

Barite/celestite nodules (Fig. 30.65) occur in the subsoils of *grey brown and red clays*, in pseudogley subsola and a buried soil (K_3) at Wagga.

Segregations of silica

Chemical segregations

These segregations of silica occur as silans (Fig. 30.66) (frequently compound argillan-silans), and as inherited nodules (Fig. 30.67) of clear or cloudy or somewhat laminated secondary silica. Neosilans occur in hardpan horizons of *red and brown hardpan soils*. Sometimes secondary silica appears to impregnate the s-matrix of some soil materials as part of the plasma, as in some indurated materials, in some pseudogley subsola, and in the surface horizons of some profiles classified as solodized soils. In the last example, the silica apparently accumulated in a shallow depression or lake, because the material is very rich in diatom skeletons.

Biogenic segregations

Opal phytoliths (Fig. 30.68) are almost ubiquitous in surface soils. Usually, they are most frequent in the upper 0.5 m of the soil profile, but they may occur throughout some profiles.

Diatom skeletons (Fig. 30.69) have been recorded in some surface horizons. They may be inherited from the parent materials, but may also result from temporary waterlogging.

Sponge spicules (Fig. 30.70) are rare. They are almost certainly inherited features.

Segregations of silt

Nodules

Silt nodules (Fig. 30.71) are rare. All the profiles in which they occur are formed on transported parent material : mainly alluvium, but also colluvium, parna and aeolian clay; the single exception is a *prairie soil*

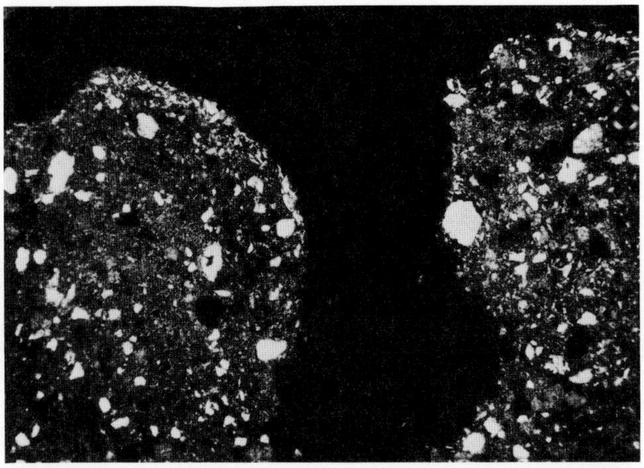

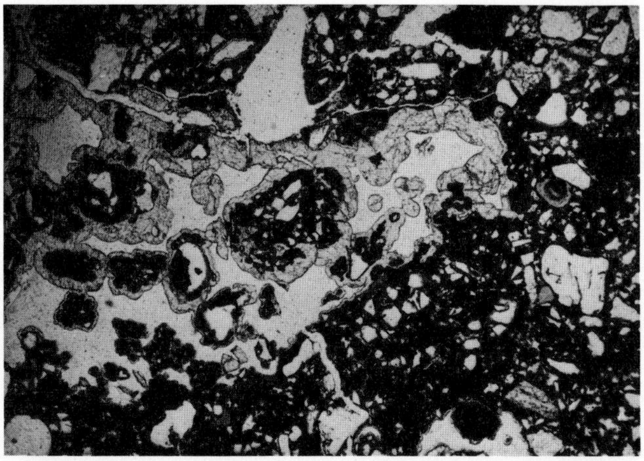

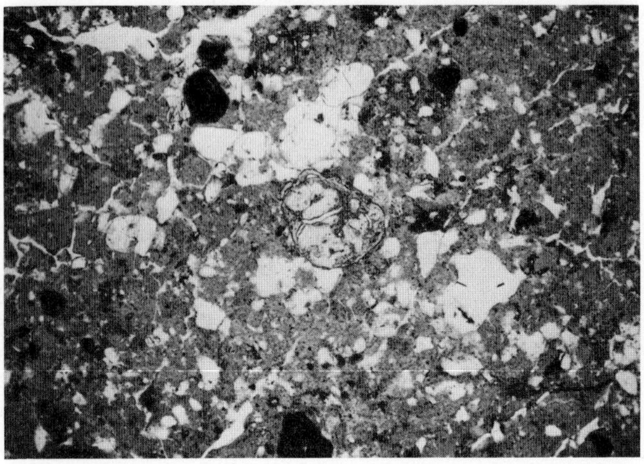

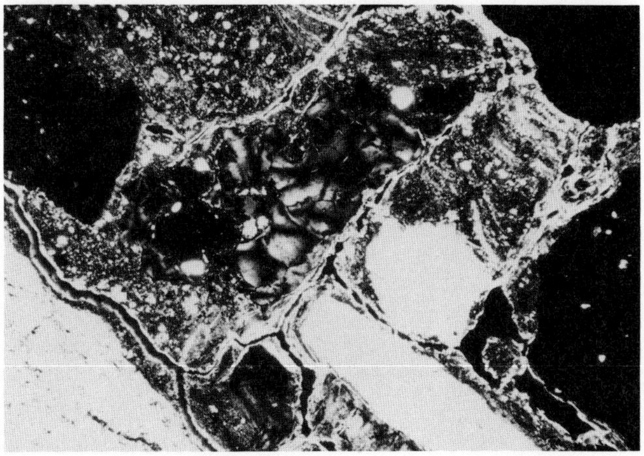

Fig. 30.64. Gypsans in transitional zone between buried *red earth* and underlying kopi; depth 72 cm. Frame length 3.3 mm. Crossed polarizers.

Fig. 30.65. Barite/celestite and halloysite nodule with included intercalary carbonate crystallites; central; at 47 cm in a *grey brown and red clay*. Frame length 3.3 mm. Plain light.

Fig. 30.66. Void and grain silans; in hardpan of *red and brown hardpan soil*. Frame length 3.3 mm. Plain light.

Fig. 30.67. Silica nodule; central; in hardpan of *red and brown hardpan soil*. Frame length 3.3 mm. Crossed polarizers.

reputed to be formed on basalt. The nodules are inherited features, commonly associated with silty laminae; as such, they may occur in almost any of the soil groups.

Siltans (Fig. 30.72)

These have been observed only in the A_2 horizon of a *solonetz*, and of a *solodized solonetz and solodic* profile; in pseudogley subsola, and weakly developed in a transitional *alpine humus soil*. They are usually associated with rounded or mammillated vughs, but also occur as plane siltans in the sediments below one pseudogley subsolum; the s-matrix usually has a silasepic plasmic fabric.

Laminae

Silt laminae (Fig. 30.73) are usually associated with silt nodules. They commonly have a silasepic unistrial fabric, and may occur as compound clay–silt laminae; they are sedimentary features.

Segregations of sand

Granotubules (Fig. 30.74)

These have been observed in some subsoils. They commonly have an associated neo-organan in the adjacent s-matrix, which is usually sandy except in pseudogley subsola. The neo-organan indicates that faunal activity or root development is responsible for the tubular-shaped void; the sandy void filling is probably due to faunal activity rather than gravity, since most occurrences are in subsoils.

Skeletans (Fig. 30.75)

These have been observed only in upper B horizons of some profiles with sandy A_2 horizons. They may be depositional or degradational features.

Laminae

Sand laminae (Fig. 30.73) are sedimentary features that occur rarely in profiles formed on sandy alluvium and colluvium.

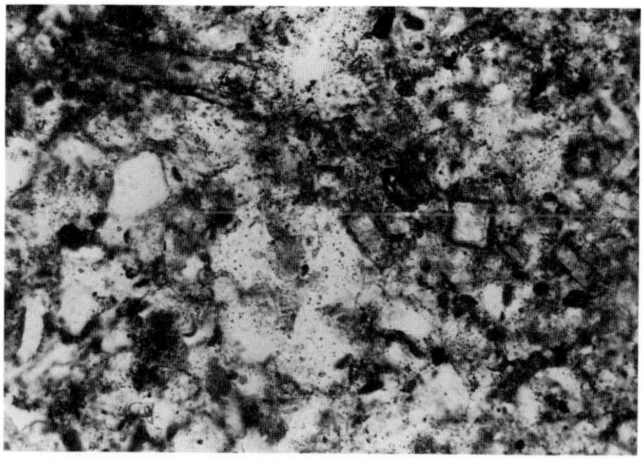

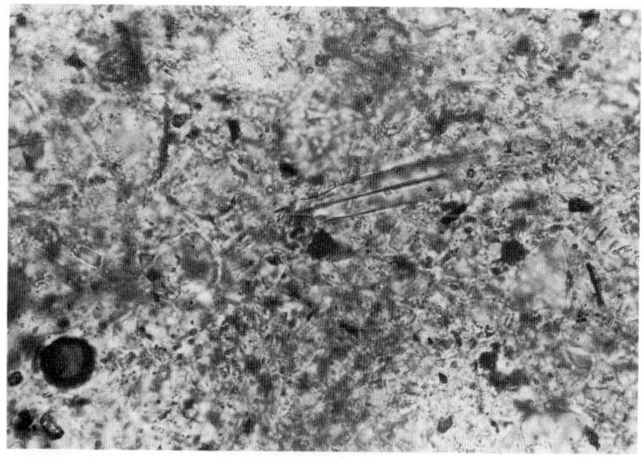

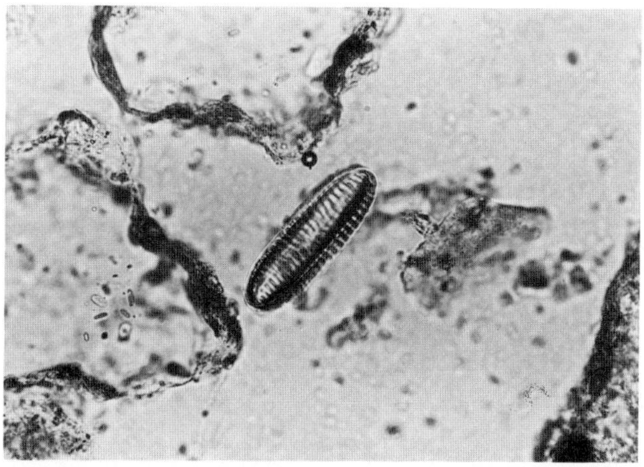

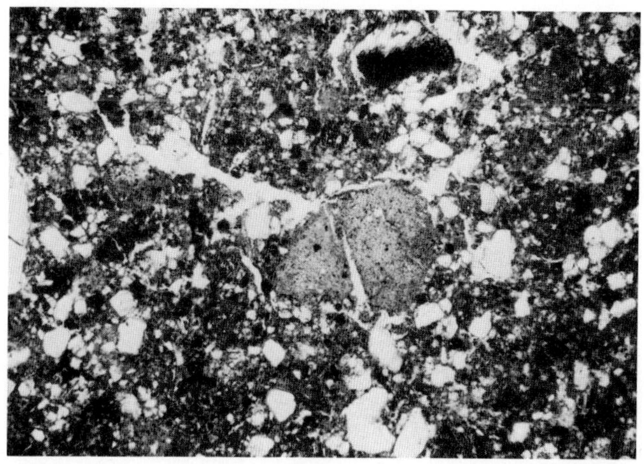

Fig. 30.68. Opal phytoliths; A horizon of *red earth*; depth 4 cm. Frame length 0.4 mm. Plain light.

Fig. 30.69. Diatom skeleton; A horizon of *grey brown and red clay*; depth 1 cm. Frame length 0.2 mm. Plain light.

Fig. 30.70. Portion of sponge spicule; right central; A horizon of *solodized solonetz*; depth 2 cm. Frame length 0.2 mm. Plain light.

Fig. 30.71. Silt nodule; central; A horizon of *grey brown and red clay*; depth 3 cm. Frame length 3.3 mm. Plain light.

Nodules

Sand nodules have been observed only in pseudogley subsola formed in colluvium, and *grey brown and red clays* formed in alluvium.

Segregations of soil material

In this context, segregations of soil material are recognizable features consisting of skeleton grains and plasma, perhaps with included pedological features.

Nodules

The occurrence of compound units of soil material has been considered earlier in the treatment of members of the matrigranic (Fig. 30.12) and fragmic (Fig. 30.16) sequences; these are fabrics composed almost entirely of variously packed soil nodules of different kinds, quite commonly faecal pellets, but sometimes without evidence of faunal activity. In other soils, soil nodules occur as more or less isolated features, in a similar manner to skeleton grains and rock nodules (Fig. 30.76). Such features have a widespread distribution. They may be inherited transported features, inherited *in situ* features (pedorelicts), or inherited sedimentary features (sedimentary relicts); as such, they may occur in virtually any soil formed on transported material. Apparent discrete soil nodules (pseudonodules) noted in field studies of some *red earth* profiles from northern Australia have, in thin section, the same fabric as the s-matrix and are not micromorphologically recognizable as such unless surrounded by voids, commonly with an associated argillan.

Aggrotubules (Fig. 30.77)

These have an internal fabric that belongs to the matrigranic or mullgranic sequence. They may occur within material of very different or very similar fabric; in the latter instance they are recognized because of their distinct boundaries. They can be treated as component

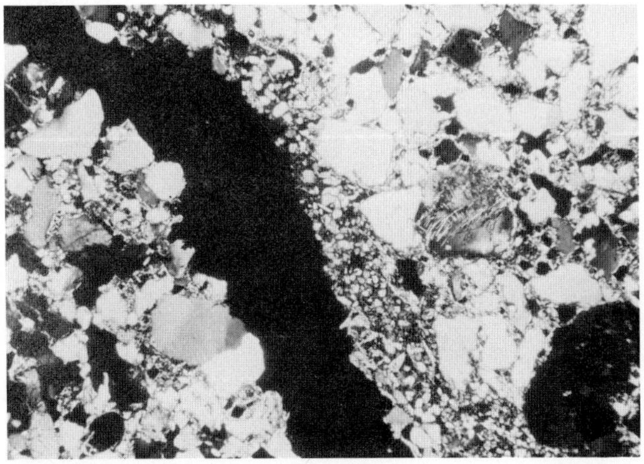

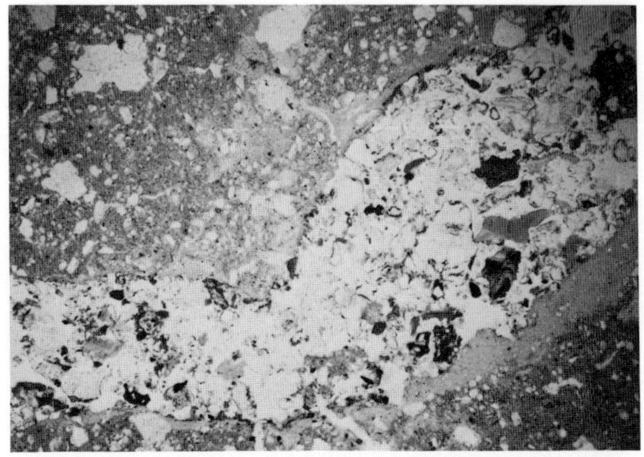

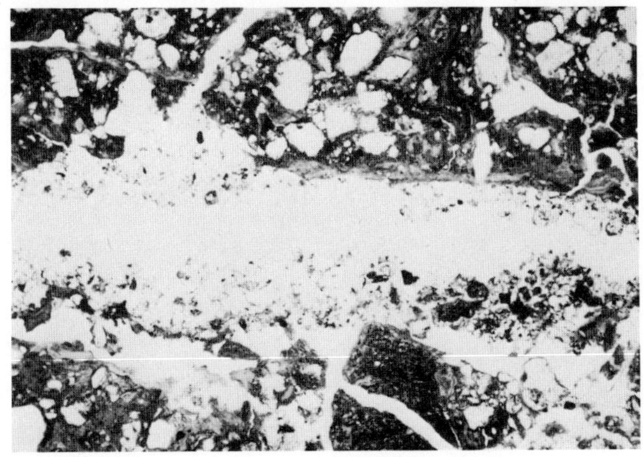

Fig. 30.72. Plane siltan; base of pseudogley subsolum; depth 294 cm. Frame length 3.3 mm. Crossed polarizers.

Fig. 30.73. Sand/silt/clay laminae; A horizon of *yellow podzolic soil*; depth 4 cm. Frame length 3.3 mm. Plain light.

Fig. 30.74. Granotubule; base of pseudogley subsolum; depth 246 cm. Frame length 3.3 mm. Plain light.

Fig. 30.75. Skeletan-siltan; B_1 horizon of *yellow podzolic soil*; depth 58 cm. Frame length 3.3 mm. Plain light.

areas of the overall fabric of the soil material in which they occur, but it is useful to treat them more positively as pedological features, especially as their composition may be different from the surrounding material.

Occasionally, they may be enclosed by a neo-organan, as in a *solonized brown soil*. Most of these features, which occur widely, are formed by faunal activity, but an interesting occurrence is in the base of some of the red soils formed on basalt, the *krasnozems* and *euchrozems*, where faunal activity would not be expected.

Isotubules (Fig. 30.78)

These have a porphyric, or porphyroskelic, internal fabric. Remarks concerning recognition of aggrotubules apply also to isotubules. Some isotubules have an enclosing neomatran or argillan. Most are due to faunal activity or the infilling of biotic voids. They are relatively rare.

Striotubules

These are recognized because of a saucer-shaped internal fabric that is expressed by the arrangement of voids, by materials of different composition (Fig. 30.79), or by preferred orientation of the constituents; they may have sharp or indistinct boundaries. They occur in clay-rich subsoils, commonly associated with red colour mottling. They also occur in the A_2 horizon of *soloths*. In some basaltic soils, especially the *krasnozems* and *euchrozems*, they are abundant in the deeper horizons immediately above the weathering basalt, where they have rather indistinct boundaries and form a special kind of sepic fabric that has been discussed earlier (see information on masepic plasmic fabric under Linear plasma separations, p. 445).

Matrans and neomatrans

Matrans are usually the free grain type. They are common in soil materials whose fabrics belong to the

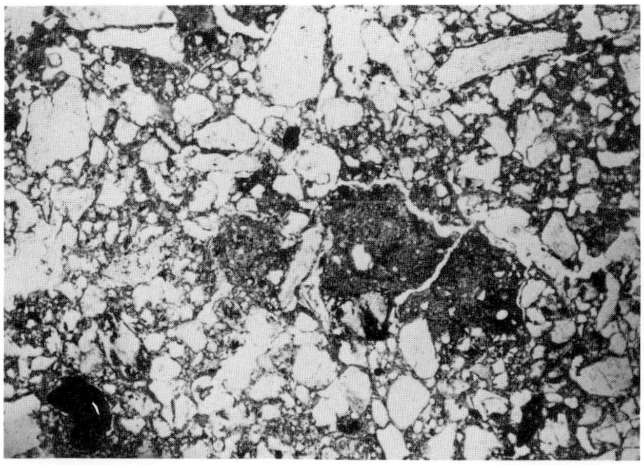

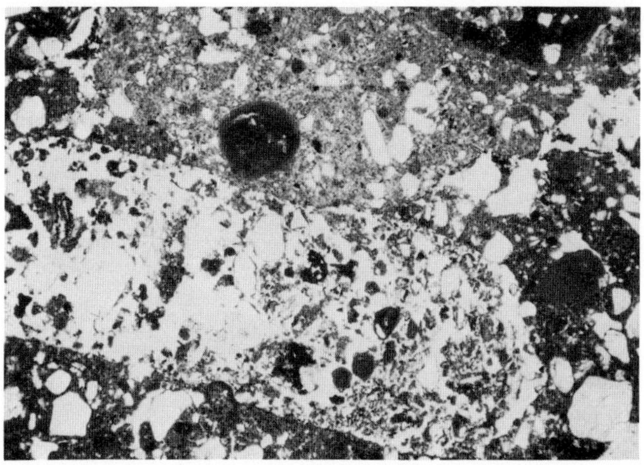

Fig. 30.76. Soil nodule; central; A horizon of *yellow podzolic soil*; depth 4 cm. Frame length 3.3 mm. Plain light.

Fig. 30.77. Grano-aggrotubule; subsurface horizon of a *grey-brown and red calcareous soil*; depth 18 cm. Frame length 3.3 mm. Plain light.

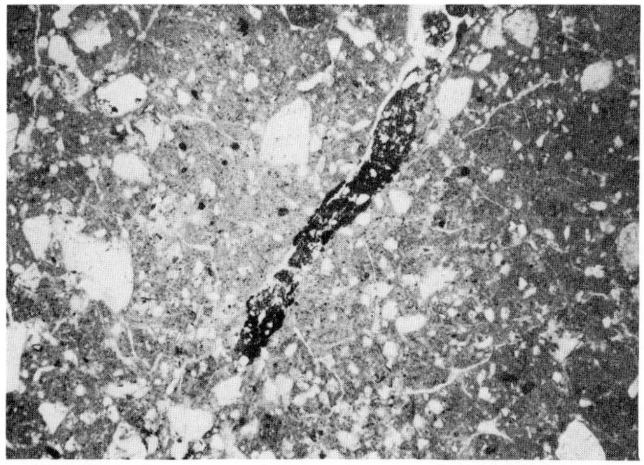

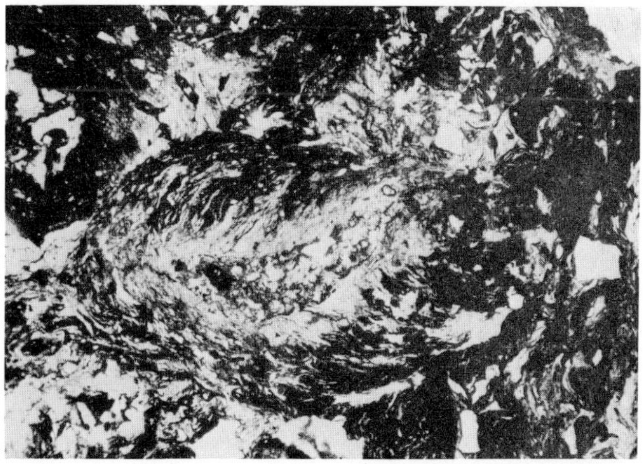

Fig. 30.78. Para-isotubule; in a buried soil underlying a *grey brown and red clay*; depth 232 cm. Frame length 3.3 mm. Plain light.

Fig. 30.79. Striotubule; saucer-shaped iron rich and iron deficient striated clay zones; base of *brown earth*; depth 127 cm. Frame length 3.3 mm. Plain light.

chlamydic sequence discussed earlier, particularly the chlamydic and plectic stages (Figs 30.3 and 30.4). Neomatrans are usually confined to the walls of faunal voids, or enclosing aggrotubules and isotubules. In soils of arid regions they may be due to termite activity.

Laminae

In a number of soil profiles, soil material with different proportions of plasma occurs in bands. All these soils are formed on alluvium, and the banding may reflect original sedimentary stratification. It is not comparable with the banded fabric reported particularly from periglacial soils in Canada (Acton and St Arnaud 1963).

Organic matter

Microscopic features

Organic matter occurs as finely dispersed, highly decomposed material intimately mixed with mineral soil and forming part of the plasma, as plant parts or fragments in various stages of decomposition by micro-organisms (Figs 30.80–82), as animal parts (Fig. 30.83), as fungal tissue (Figs 30.84 and 30.85), and as faecal pellets composed entirely of highly decomposed organic matter (Fig. 30.80) or of comminuted plant fragments (Fig. 30.86).

There are no exhaustive microscopic studies of organic matter in Australian soils; for a general introduction to the subject, see Bal (1973) and Babel (1975). General observation suggests, that apart from the peats, most Australian soils in virgin undisturbed sites have only a thin layer of litter consisting of partially decomposed and comminuted plant fragments (phytogranic material), and pass suddenly into dominantly mineral soil in which the organic matter occurs as finely dispersed plasma and/or dark melanized fragments. In

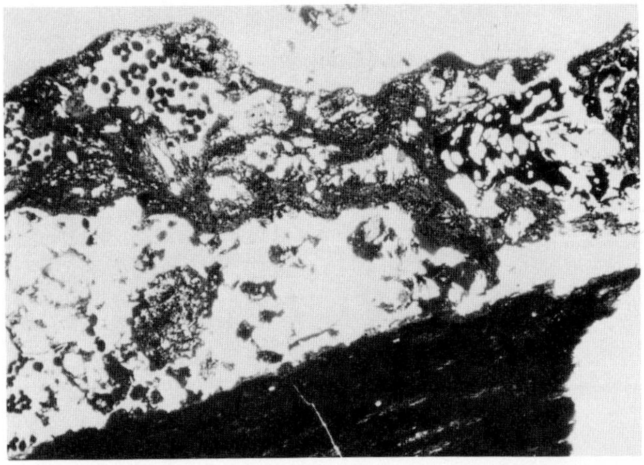

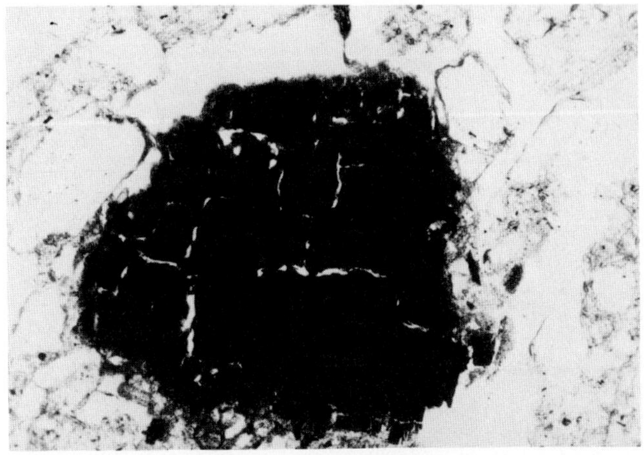

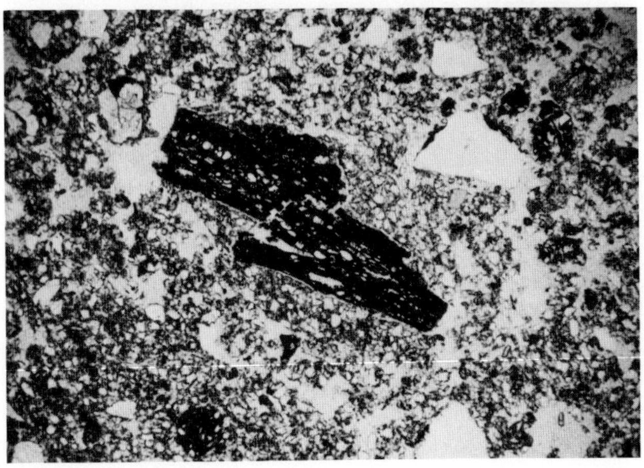

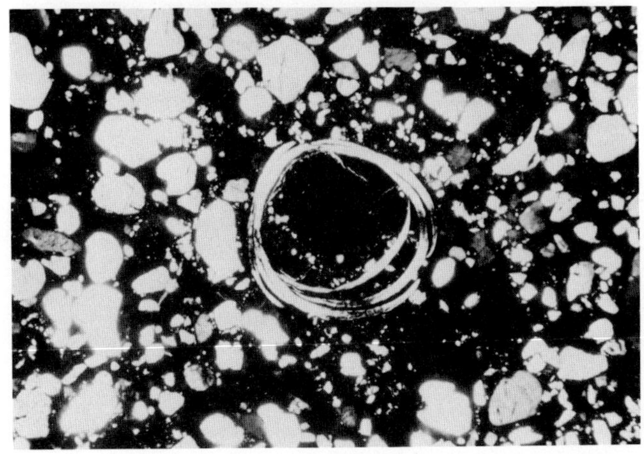

Fig. 30.80. Reddened, isotropic plant fragment, melanized tissue, large modexi (faecal pellets) and fine, ovoid modexi (faecal pellets) composed of decomposed organic matter in cavity; B_1 horizon of a *solonized brown soil*; depth 20 cm. Frame length 3.3 mm. Plain light.

Fig. 30.81. Charcoal or melanized tissue; B horizon of *red earth*; depth 26 cm. Frame length 3.3 mm. Plain light.

Fig. 30.82. Black opaque organic fragment; BC horizon of *yellow podzolic soil*; depth 98 cm. Frame length 0.9 mm. Plain light.

Fig. 30.83. Ground pearl pupal case; A horizon of *lateritic podzolic soil*; depth 10 cm. Frame length 3.3 mm. Crossed polarizers.

Key to letters on electron micrographs: b, soil bacteria; c, clay; o, organic matter; s, space left by fragmented quartz grain; q, quartz; v, void.

most of the soil groups, the proportion of organic matter decreases rapidly from the surface downward, except in some profiles of the *alluvial soils, prairie soils, chocolate soils, brown earths, xanthozems, peaty podzols*, and *alpine humus soils*.

The depth sequence of organic types in coarse-textured soils such as *podzols*, starts with a thin surface phytogranic layer, often with faecal pellets of comminuted plant fragments, over A horizons of mineral soil with fine particulate melanized organic material as the f-matrix, with reddened organic fragments associated with fungi in the deeper horizons. Faecal pellets composed entirely of highly decomposed organic matter may occur at any level, associated with plant fragments. In finer-textured soils, the horizons underlying the phytogranic layer have various proportions of melanized particles embedded in the s-matrix, but much of the organic matter occurs as finely dispersed plasma within the s-matrix. Some also occurs as channel organans (Fig. 30.87), free grain organans in the chlamydic and plectic fabrics of the B horizons of *podzols* discussed earlier (see p.450), and as neo-organans on voids, or enclosing pedotubules. Organic-rich striotubules and isotubules have also been recorded, and many aggrotubules have finely dispersed organic matter as part of the plasma.

Ultramicroscopic features

Recently, using ultrathin sections with a transmission electron microscope, one of the authors has studied the ultramicroscopic characteristics of organic soils that consist largely of variously decomposed plant material.

In one such study, samples taken a few centimetres below the surface show little cellular structure, and the

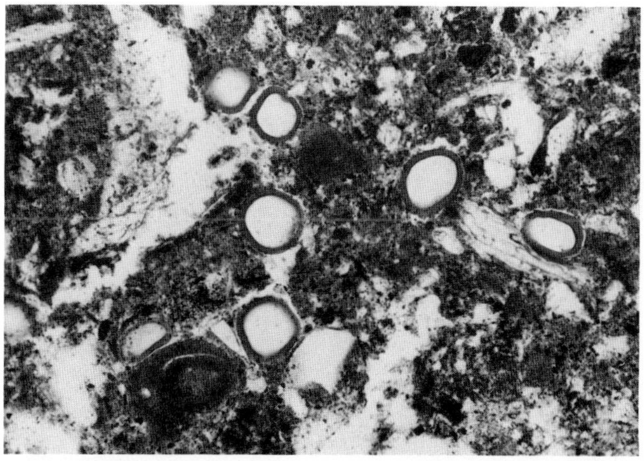

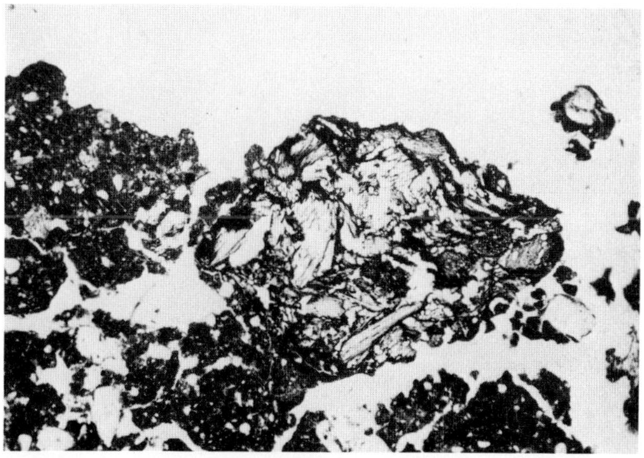

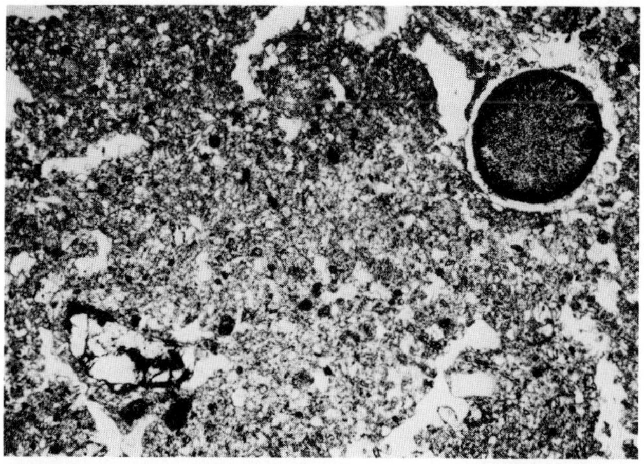

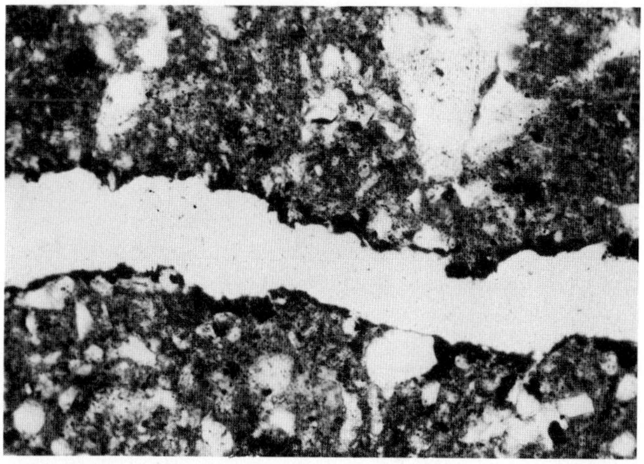

Fig. 30.84. Fungal spores; subsurface horizon of a *grey-brown and red calcareous soil*; depth 18 cm. Frame length 0.9 mm. Plain light.

Fig. 30.85. Sclerotium; A_2 horizon of *yellow podzolic soil*; depth 27 cm. Frame length 3.3 mm. Plain light.

Fig. 30.86. Scarab beetle modex (faecal pellet) composed of plant fragments; at the surface of a *grey brown and red clay*. Frame length 3.3 mm. Plain light.

Fig. 30.87. Channel organan; deep subsoil of a *grey brown and red clay*; depth 85 cm. Frame length 0.9 mm. Plain light.

fabric is composed of an intricate, convoluted, lamellate mass of cell wall remnants (Fig. 30.88). Polyphenol-rich layers are persistent and very electron dense, whilst those with some carbohydrate remaining show various shades of grey (Bland *et al.* 1971). The mass of cell wall remnants is embedded in a fine, amorphous, but electron dense, organic matrix, possibly composed of humic materials (Visser 1963). Nearer the surface of the soil, undecomposed tissue can be recognized as histons (Fig. 30.89). The lumena and the cell walls of the tissue can be recognized, but in some cases, the lumen may be filled with osmiophyllic polyphenolic or resinous materials, especially in soils from radiata pine forests. It is possible however, that in some residues, electron dense humic materials are redeposited in histons. Bacteria occur even in the most heavily impregnated tissues, and there are lysis holes of bacterial size even in the polyphenolic masses. Some bacteria secrete extensive capsules which further occlude the lumena of histons. At more advanced stages of decomposition (Fig. 30.90), the partially digested cell walls collapse, so that the bacteria are trapped between the cell-wall remnants. Heavily lignified middle-lamellar cell-wall corners may persist unchanged, but the carbohydrate-rich cell-wall lamellae are gradually reduced to a fibrous network. As the carbohydrate is removed, osmiophyllic polyphenolic groups are unmasked, so that as the cell-wall remnants become progressively thinner, they become progressively more electron dense, and much more crushed and folded, giving rise to the sort of multilamellate structures seen in Fig. 30.90. Living cells contain proteins, some of which, such as enzymes, may survive even to advanced stages of cellular decomposition, especially where they are protected by surrounding cell-wall materials. Fig. 30.91 shows, for instance, that the enzyme peroxidase is present in the middle lamella of a cell-wall fragment,

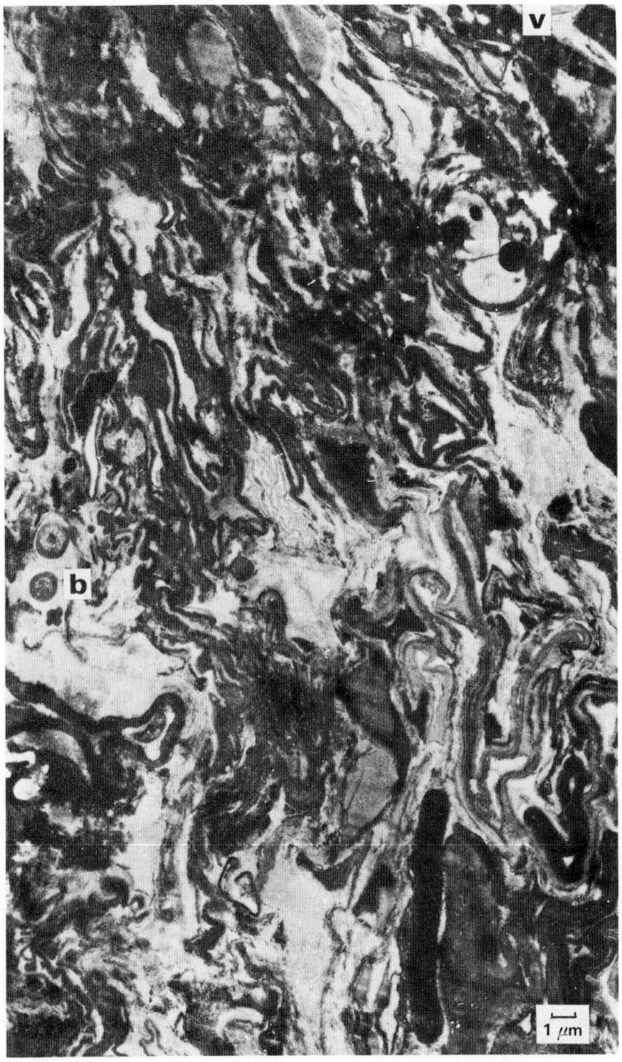

Fig. 30.88. Highly convoluted cell wall remnants containing dark polyphenol-rich lamellae. Electron micrograph.

Key to letters on electron micrographs: b, soil bacteria; c, clay; o, organic matter; s, space left by fragmented quartz grain; q, quartz grain; v, void.

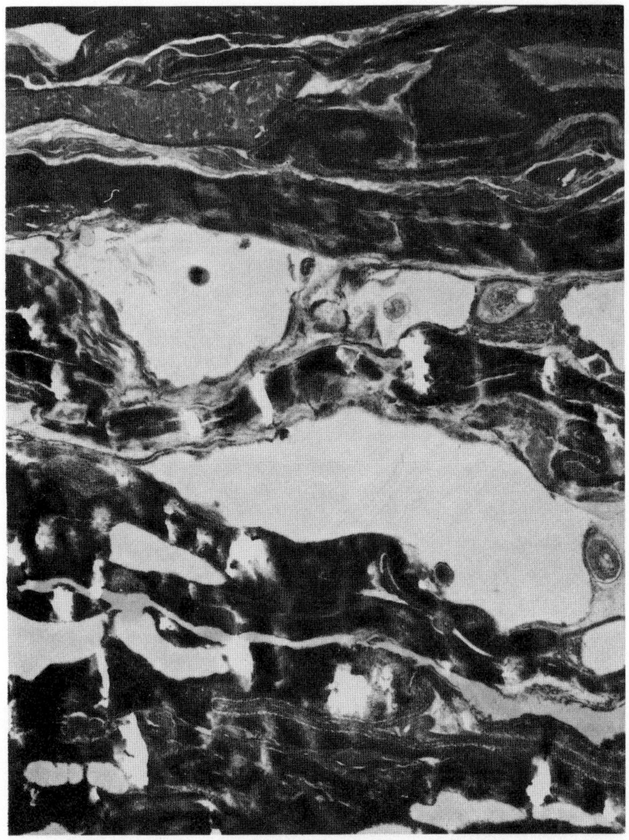

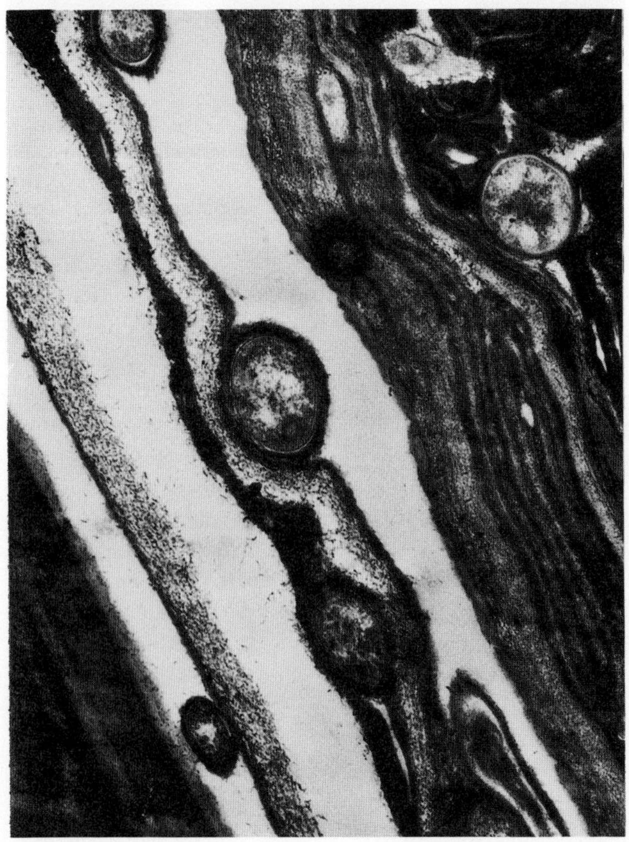

Fig. 30.89. Organic fragments from radiata-pine forest; *podzol*. Electron micrograph.

Fig. 30.90. Highly decomposed cell wall material; *solonized brown soil*. Electron micrograph.

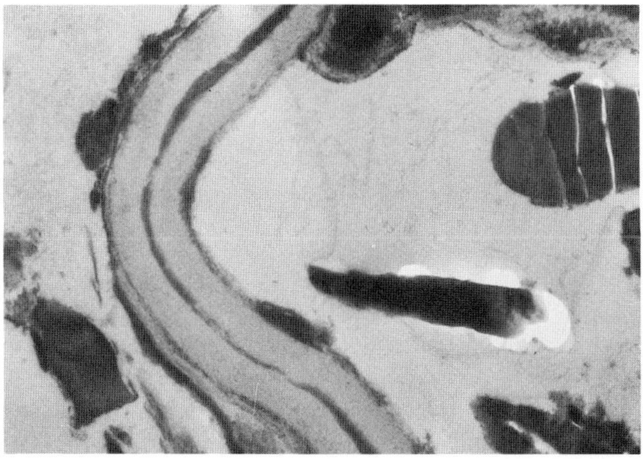

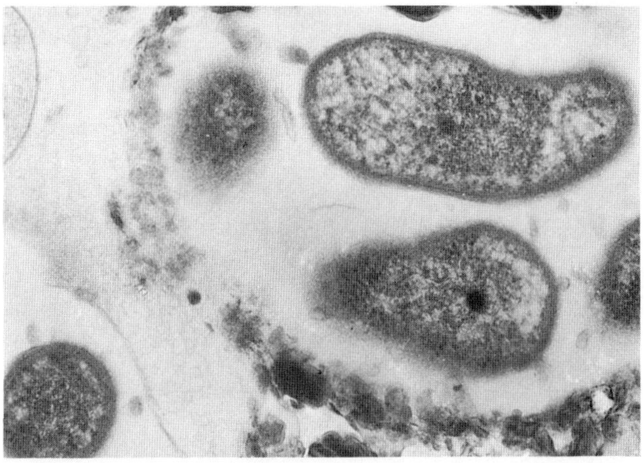

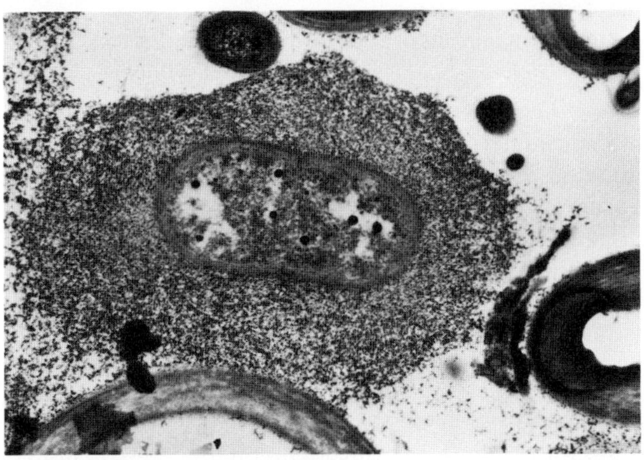

Fig. 30.91. Peroxidase in the middle lamella of a plant cell wall; *podzol*. Electron micrograph.

Fig. 30.92. Bacteria with electron transparent capsule; *solonized brown soil*. Electron micrograph.

Fig. 30.93. Bacterial capsule stained with ruthenium red; *podzol*. Electron micrograph.

and in Fig. 30.36 the enzyme acid phosphatase is demonstrated in the capsule of bacteria. These surviving

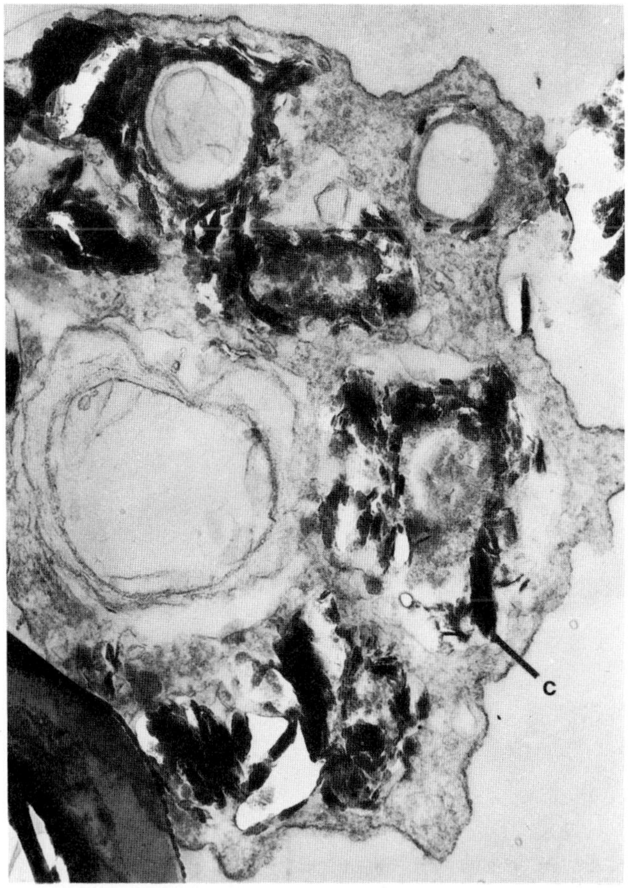

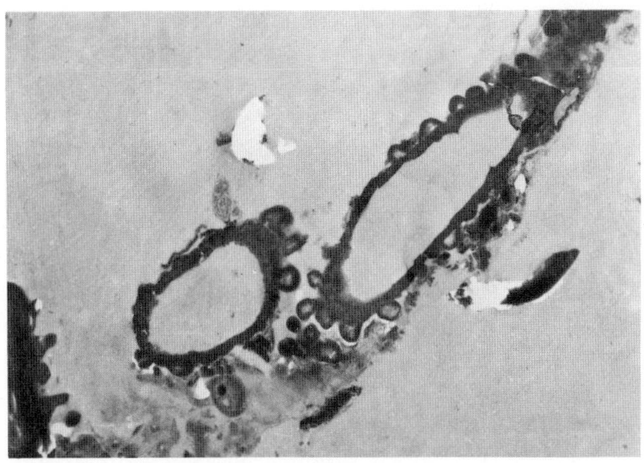

Fig. 30.94. Animal histon; TS small invertebrate with ingested clay in gut; *grey brown and red clay*. Electron micrograph.

Fig. 30.95. Invertebrate cuticular remains; *podzol*. Electron micrograph.

enzymes may play an important role in soil biochemical processes.

Although the metabolic activities of soil microorganisms are the main cause of the breakdown of organic matter in soils, bacteria and fungi do give rise to

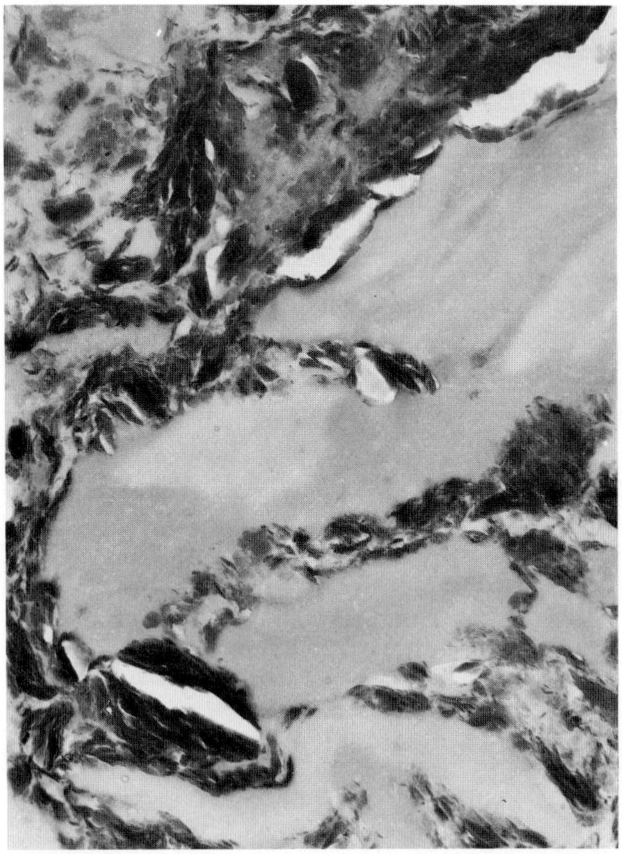

Fig. 30.96. Mucilage with clay embedded; possibly earthworm cast; *calcareous sand*. Electron micrograph.

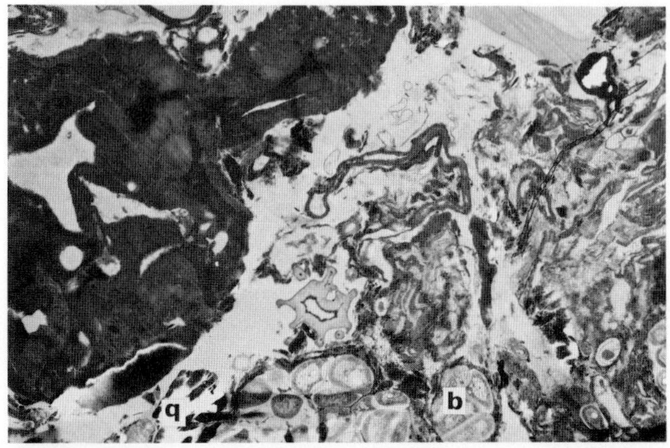

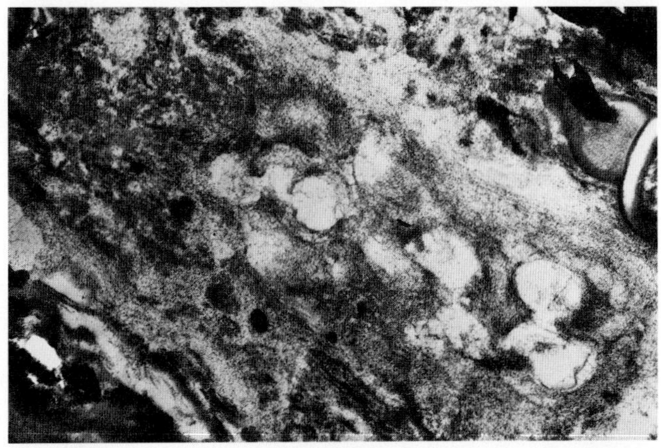

Fig. 30.97. Faecal pellet with bacterial lysis holes, remnants of cellular membranes supporting a bacterial colony in the adjacent soil. Electron micrograph.

Fig. 30.98. Remnants of a bacterial colony showing fibrils persisting and binding soil constituents. Electron micrograph.

extracellular biochemicals unique to these organisms. Fungi secrete slimes, and bacteria form extensive fibrillar capsules. Both of these are composed of carbohydrates which do not react with conventional fixatives, so that they are recognized as an electron transparent space between the organisms and surrounding soil materials (Figs 30.32 and 30.92); the carbohydrates can be demonstrated by using ruthenium red or lanthanum salts (Fig. 30.93).

Animals also contribute to soil organic matter. Occasionally, small invertebrates are trapped in the s-matrix and appear as histons in ultrathin sections (Fig. 30.94). They also leave cast fragments of exoskeleton, such as cuticular plates (Fig. 30.88) or hollow appendages (Fig. 30.95). Earthworms secrete slimes, and these appear as electron transparent masses outlined by adsorbed clay fragments (Fig. 30.96). They also leave faecal pellets, which have characteristic morphologies, are usually electron dense (Fig. 30.97), and in which recognizable remnants of partially digested tissues can be seen. Microorganisms invade the pellets and leave lysis holes; older faecal deposits may support large bacterial colonies in the adjacent soil.

As decay progresses, all histological features of the organic materials are obliterated and the highly comminuted remnants can then only be recognized by their appearance after heavy metal staining. Three kinds of organic deposits occur at the ultramicroscopic level in Australian soils. The first is plant cell walls and bacterial capsules which both contain fibrous materials; the individual fibres are about 7–10 nm in diameter, and where large bacterial colonies occur, the fibrillar material may be recognized even after the bacteria themselves have perished (Fig. 30.98). In some agricultural soils, large quantities of these deposits occur, binding the soil minerals together in crumbs (Martin 1971). Secondly, roots secrete mucigels into soils (see The plant root environment, chapter 41), and these may persist where they are protected by cutans (Fig. 30.99); they are recognized by their granularity. Thirdly, most soils contain amorphous, slightly electron dense materials (Fig. 30.100); these organic materials act as the cement holding the various soil constituents together and conferring on the soil its characteristic properties.

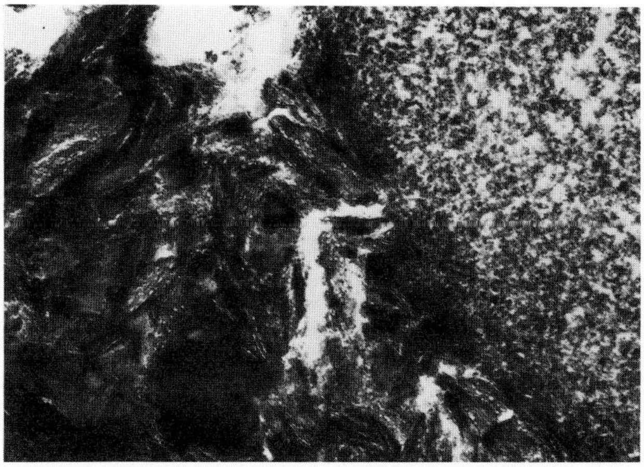

Fig. 30.99. Granular organic remnants, possibly mucigel, protected by mineral fabric; *podzol*. Electron micrograph.

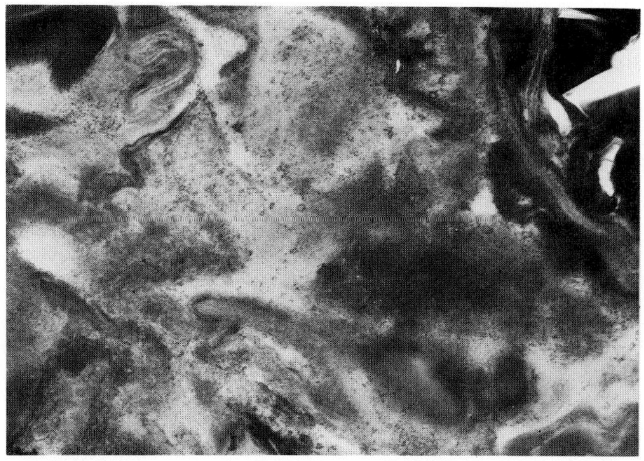

Fig. 30.100. Amorphous organic matrix; *podzol*. Electron micrograph.

Micro-organisms are uncommon in mineral soils (Fig. 30.1), except in the vicinity of roots (see Ultrastructure of root-soil interface, chapter 41), and where they do occur, they are often associated with organic fragments (Fig. 30.31). They may be surrounded by an electron-transparent organic capsule (Figs 30.32 and 30.92) to which clay particles attach, either face-or edge-on (Foster 1978) depending on surface charges on the bacterial cell wall. Organic matter is similarly adsorbed on to minerals, but this is only distinguished readily from amorphous mineral matter where the organics are fibrous. Humic materials are said to be either particulate (Visser 1963) or fibrous, depending on the local pH (Chen and Schnitzer 1976).

Such studies of the ultrastructural features of soils are being used to investigate the fine details of the processes involved in the breakdown of litter and root systems in forest soils, and of the role of microbial products, such as carbohydrate microfibrils and amorphous gels, in building and stabilizing soils under pastures and crops by binding soil components to form crumbs.

References

Acton, D. F. and St Arnaud, R. J. (1963). Micropedology of major profile types of the Weyburn catena. *Can. J. Soil Sci.* **43**, 377-86.

Babel, U. (1975). Micromorphology of soil organic matter. *In* 'Soil Components, Vol. 1, Organic Components.' (Ed. J. E. Gieseking). (Springer-Verlag: Berlin.)

Bal, L. (1973). 'Micromorphological Analysis of Soils.' (Soil Survey Inst.: Wageningen, The Netherlands.)

Bal, L. (1977). The formation of carbonate nodules and intercalary crystals in the soil by the earthworm *Lumbricus rubellus*. *Pedobiologia* **17**, 102-6.

Beattie, J. A. and Haldane, A. D. (1958). The occurrence of palygorskite and barytes in certain parna soils of the Murrumbidgee region, New South Wales. *Aust. J. Sci.* **20**, 274-5.

Beckmann, W. and Geyger, E. (1967). Entwurf einer Ordnung der natürlichen Hohlraum-, Aggregat- und Struktur-formen im Boden. *In* 'Die Mikromorphometrische Bodenanalyse.' (Ed. W. L.
Bal, L. (1977). The formation of carbonate nodules and intercalary crystals in the soil by the earthworm *Lumbricus rubellus*. *Pedobiologia* **17**, 102-6.

Beattie, J. A. and Haldane, A. D. (1958). The occurrence of palygorskite and barytes in certain parna soils of the Murrumbidgee region, New South Wales. *Aust. J. Sci.* **20**, 274-5.

Beckmann, W. and Geyger, E. (1967). Entwurf einer Ordnung der natürlichen Hohlraum-, Aggregat- und Struktur-formen im Boden. *In* 'Die Mikromorphometrische Bodenanalyse.' (Ed. W. L.
Kubiena). (Ferdinand Enke Verlag: Stuttgart.)

Bland, D. E., Foster, R. C. and Logan, A. F. (1971). The mechanism of permanganate and osmium tetroxide fixation and the distribution of lignin in the cell wall of *Pinus radiata*. *Holzforchung* **25**, 137-43.

Brewer, R. (1964). 'Fabric and Mineral Analysis of Soils.' (John Wiley and Sons: New York.)

Brewer, R. (1972). The basis of interpretation of soil micromorphological data. *Geoderma* **8**, 81-94.

Brewer , R. (1977). Micromorphology of a euchrozem from Talavera, Queensland. CSIRO Aust. Div. Soils Div. Rep. No. 18.

Brewer, R. (1979). Relationships between particle size, fabric and other factors in some Australian soils. *Aust. J. Soil Res.* **17**, 29-41.

Brewer, R. and Blackmore, A. V. (1976). Relationships between subplasticity rating, optically oriented clay, cementation and aggregate stability. *Aust. J. Soil Res.* **14**, 237-48.

Brewer, R. and Pawluk, S. (1975). Investigations of some soils developed in hummocks of the Canadian sub-arctic and southern-arctic regions. I. Morphology and micromorphology. *Can. J. Soil Sci.* **55**, 301-19.
Bland, D. E., Foster, R. C. and Logan, A. F. (1971). The mechanism of permanganate and osmium tetroxide fixation and the distribution of lignin in the cell wall of *Pinus radiata*. *Holzforchung* **25**, 137-43.

Brewer, R. (1964). 'Fabric and Mineral Analysis of Soils.' (John Wiley and Sons: New York.)

Brewer, R. (1972). The basis of interpretation of soil micromorphological data. *Geoderma* **8**, 81-94.

Brewer , R. (1977). Micromorphology of a euchrozem from Talavera, Queensland. CSIRO Aust. Div. Soils Div. Rep. No. 18.

Brewer, R. (1979). Relationships between particle size, fabric and other factors in some Australian soils. *Aust. J. Soil Res.* **17**, 29-41.

Brewer, R. and Blackmore, A. V. (1976). Relationships between subplasticity rating, optically oriented clay, cementation and aggregate stability. *Aust. J. Soil Res.* **14**, 237-48.

Brewer, R. and Pawluk, S. (1975). Investigations of some soils developed in hummocks of the Canadian sub-arctic and southern-arctic regions. I. Morphology and micromorphology. *Can. J. Soil Sci.* **55**, 301-19.

Brewer, R. and Thompson, C. H. (1980). Morphology of two subtropical podzols formed on a siliceous dune in coastal Queensland. Proceedings Conference on Classification and Management of Tropical Soils 1977. (Ed. K. T. Joseph). (Malaysian Soc. Soil. Sci.; Kuala Lumpur.)

Butler, B. E. (1974). Contribution towards the better specification of parna and some other aeolian clays in Australia. *Z. Geomorphol. Suppl.* **20**, 106-16.

Chen, Y. and Schnitzer, M. (1976). Scanning electron microscopy of a humic acid and of a fulvic acid and its metal and clay complexes. *Soil Sci. Soc. Am. J.* **40**, 682-6.

Flach, K. W. (1960). Sols bruns acides in the northeastern United States. Genesis, morphology, and relationships to associated soils. Unpubl. Ph.D. Thesis, Cornell, New York.

Foster, R. C. (1978). Ultramicromorphology of some South Australian Soils. *In* 'Modification of Soil Structure.' (Eds W. W. Emerson, R. D. Bond and A. R. Dexter). (John Wiley and Sons: New York.)

Glauert, A. M. (1974). Fixation, dehydration and embedding of biological specimens. *In* 'Practical Methods in Electron Microscopy.' (Ed. A. M. Glauert). (North-Holland: Amsterdam.)

Hayat, M. A. (1975). 'Positive Staining for Electronmicroscopy.' (Van Nostrand Reinhold: New York.)

Lebedeva, I. I. and Ovechkin, S. V. (1975). New carbonate formations in the chernozems of the Ukraine east of the Dnieper. *Soviet Soil Sci.* **7**, 637-53.

Martin, J. P. (1971). Decomposition and binding action of polysaccharides in soil. *Soil Biol. Biochem.* **3,** 33-41.

Mulder, E. G. and van Veen, W. L. (1968). Effect of microorganisms on the transformations of mineral fractions in soil. Trans. 9th Int. Congr. Soil Sci. (Adelaide) Vol. 4, pp. 651-661.

O'Brien, N. R. (1971). Fabric of kaolinite and illite floccules. *Clays Clay Miner.* **19**, 353-59.

Pearse, A. G. E. (1972). 'Histochemistry: Theoretical and Applied.' Vol. 2, 3rd Edn. (Churchill: London.)

Reid, N. (1974). Ultramicrotomy. *In* 'Practical Methods in Electron Microscopy.' (Ed. A. M. Glauert). (North-Holland: Amsterdam.)

Smart, P. (1974). Electron microscope methods in soil micromorphology. *In* 'Soil Microscopy.' (Ed. G. K. Rutherford). (The Limestone Press: Kingston, Ontario.)

Stace, H. C. T., Hubble, G. D., Brewer, R., Northcote, K.H., Sleeman, J. R., Mulcahy, M. J. nad Hallsworth, E. G. (1968). 'A Handbook of Australian Soils.' (Rellim Technical Publications: Adelaide)

Stoops, G. and Jongerius, A. (1975). Proposal for a micromorphological classification of soil materials. I. A classification of the related distribution of fine and coarse particles. *Geoderma* **13**, 189-99.

Tovey, N. K. (1974). Some applications of electron microscopy to soil engineering. *In* 'Soil Microscopy.' (Ed. G. K. Rutherford). (The Limestone Press: Kingston, Ontario.)

van Dijk, D. C. (1969). Psuedogley in Gundaroo subsola, Southern Tablelands, N.S.W. *Aust. J. Soil Res.* **7**, 143-61.

Visser, S. A. (1963). Electron-microscopic and electron-diffraction patterns of humic acids. *Soil Sci.* **96**, 353-6.

Chapter 31
Inter-particle bonding

W. W. Emerson

THE PHYSICAL PROPERTIES OF THE SOIL SUCH AS ITS hydraulic conductivity are largely determined by the arrangement of the clay particles in the soil and by the stability of that arrangement to osmotic and mechanical stresses. Both of these in turn, depend mainly on the strength of the bonds between the particles.

The first part of this chapter summarizes the current knowledge of the forces acting between particles of pure clay minerals. For condensed systems, which soils are, it is the short-range forces which are the most important. Unfortunately, these forces are also the most difficult to measure. In fact, their nature has often to be inferred from the properties of suspensions of the particles, or from measurements of macroscopic swelling. The difficulty is accentuated with coarse-grained clay particles, such as kaolinite, where the average distance apart cannot be measured by x-rays.

In experiments on inter-particle forces, it is important to establish the precise composition of the ions balancing the electric charges of the clay particles. This point is highlighted in an initial section on kaolinite.

In the discussion of the forces, emphasis is given to the conditions under which the net force between the particles in a given solution becomes repulsive, i.e. the clay disperses. Such conditions can induce large changes in the physical properties of a soil, such as a marked reduction in the saturated hydraulic conductivity. Correspondingly, at small applied suctions the excessive swelling can lead to reduction in unsaturated hydraulic conductivity, with its implication for impeded drainage.

It will become evident that at present our quantitative knowledge of the bonding between clay particles is limited. Further, in a soil, the clay fraction not only usually contains more than one clay mineral, but other particles such as carbonates and sesquioxides may also be present. In addition, the proportion of clay in a soil relative to the content of coarser particles has to be considered.

Simple tests have been developed therefore, based on the results obtained with the pure clay minerals, to qualitatively rank the strengths of the bonds between the clay particles in soil aggregates. Simple tests have also been devised to determine whether there are additional bonds, such as organic matter bonds, holding domains of clay particles together. These tests are described in the second part of the chapter. Lastly, the use of the tests for deciding how the physical properties of a soil can be optimized for a particular purpose is discussed.

Bonding between clay mineral particles

Bonding in kaolinite suspensions

The negative charge of the 2:1 clay minerals can be shown by chemical analysis to be due to isomorphous substitution within the crystal structure. It is much more difficult to show this in the case of kaolinite, because the cation exchange capacity (CEC) of a pure ceramic kaolinite is only 2–3 mequiv. per 100 g, corresponding to the substitution of about one Si atom in 400 by Al. Approximate agreement between CEC and chemical analysis has been found for a fine-grained kaolin (Robertson *et al.* 1954). Nevertheless, the presence of a permanent negative charge usually has to be demonstrated indirectly, such as by its invariance with pH.

Schofield and Samson (1953, 1954) prepared Na-kaolinite by washing with NaCl acidified to pH 3; HCl was added so that any aluminium impurities would be removed. The clay was then washed with water, and finally dried. This whole process will be referred to as the NaCl–HCl method. Some of their results are plotted in Fig. 31.1. The Na-clay flocculated in water (pH$\sim$5), as shown by its high viscosity; it also sorbed a small amount of Cl from a dilute NaCl solution. The sorption was interpreted to mean that the edge surfaces of the particles were positively charged. Then flocculation was ascribed to electrostatic attraction between positively charged edges and negatively charged basal surfaces, the latter charge being due to isomorphous replacement. On raising the pH, Cl sorption became negative (Fig. 31.1), and at the same time, the clay dispersed, as shown by the low viscosity in water. In a separate experiment, Na-kaolinite was suspended in water and sufficient polyphosphate added to cause the clay to disperse. The clay was further washed with water and then dried. Suspensions of the clay in water then dispersed (low viscosity, Fig. 31.1), and Cl sorption from dilute NaCl was negative. However, if acid was added to the suspension, then Cl sorption again became positive. This

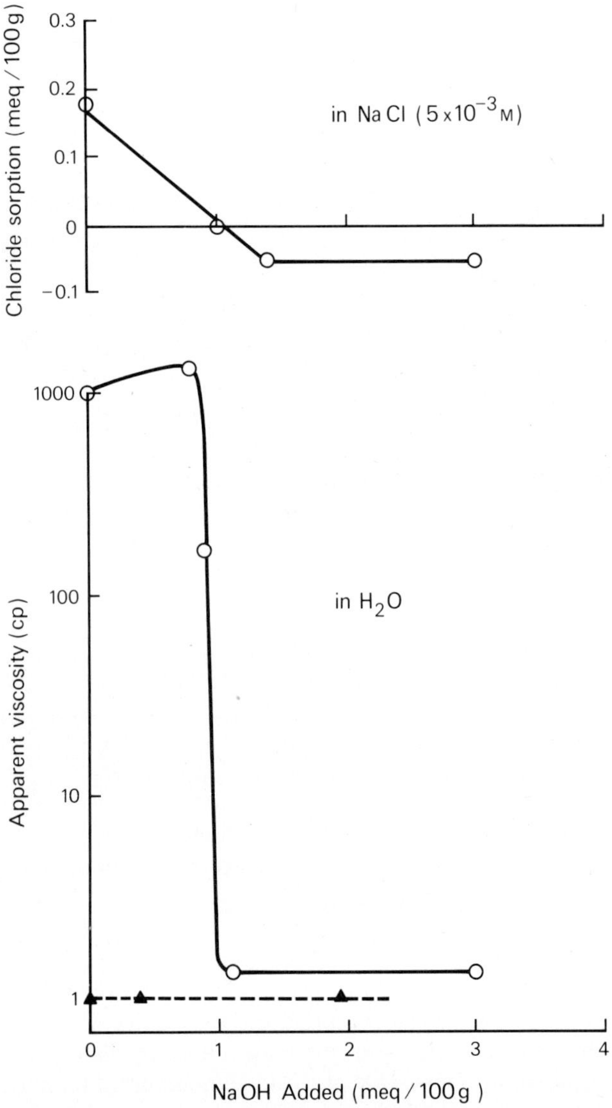

Fig. 31.1. The effect of alkali on the sorption of chloride by Na-kaolinite from an NaCl solution and on the viscosity in water of kaolinite, as prepared (o) or dispersed with polyphosphate (▲) (after Schofield and Samson 1954).

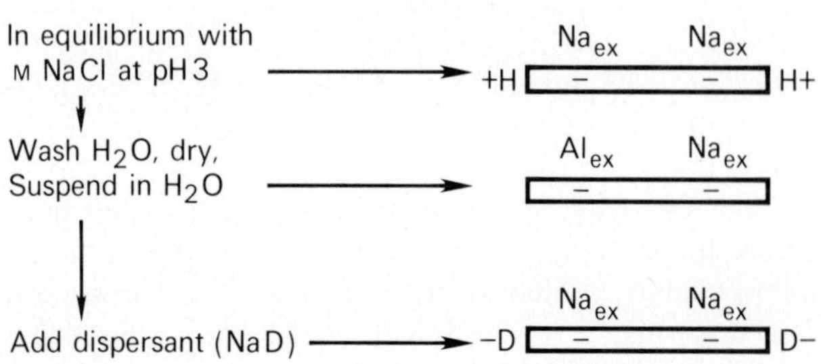

Fig. 31.2. The effect of different treatments on the electric charges of kaolinite surfaces and associated ions.

was interpreted to mean that the polyphosphate neutralizes the positive edge charges of the clay as prepared, but lowering the pH allows additional positive charges to be developed.

Subsequently Cashen (1959, 1961) examined the electric charge of a similar kaolinite, given the same NaCl–HCl pre-treatment, using a qualitative electro-osmotic technique. On the basis of his results, he suggested that during the final stages of preparation of the NaCl–HCl clay, the edges which initially were positively charged become discharged; a corresponding amount of aluminium ions in an unspecified cationic form is transferred to the basal surfaces. The subsequent addition of dispersant removes this exchangeable Al_{ex}, and replaces it with Na_{ex}. The subscript ex indicates cations occupying exchange sites. Smith and Emerson (1976) extended Cashen's work. They looked at the effect of oxalate in more detail, and confirmed that its initial reaction with Al_{ex} on the basal surfaces could be distinguished from a secondary reaction, presumably with edge Al atoms. The reactions are shown diagrammatically in Fig. 31.2.

The important conclusions to be deduced from the work on kaolinite are as follows.

It is difficult to prepare homoionic kaolinite without some of the negative charge of the clay being balanced by Al_{ex} (Bolland *et al.* 1976). In making deductions from viscosity measurements as to the nature of the forces between kaolinite particles, Rand and Melton (1977) stressed that their kaolinite was 'aluminium free'; in fact it was prepared by the NaCl–HCl method. Similar considerations should apply to other clay minerals prepared in the same way, e.g. Na–montmorillonite, as shown by Cashen (1961).

Flocculation of kaolinite particles in salt-free suspensions may occur due to bonding of the particles by Al_{ex}; there is no necessity to invoke positive edge to negative face bonds. The small positive sorption of Cl by NaCl–HCl washed clay (Fig. 31.1) could be due to the displacement of OH groups from the Al_{ex}. This is not to deny that positive charges are developed when kaolinite is acidified. A striking illustration of this is given by electron micrographs of kaolinite from an acidified suspension to which colloidal gold has been added; the negatively charged gold particles are sorbed at the crystal edges (Follett 1965).

Soils in which the dominant clay mineral is kaolinite are usually acid and strongly leached. Therefore the edge surfaces of the kaolinite particles are likely to be uncharged. Note that in Western Australia, the salt present in lateritic kaolinitic soils has been brought in since the formation of the laterite (Dimmock *et al.* 1974).

Bonding in lithium–vermiculite gels

It is convenient when discussing the forces between clay particles to distinguish three states of a clay. The

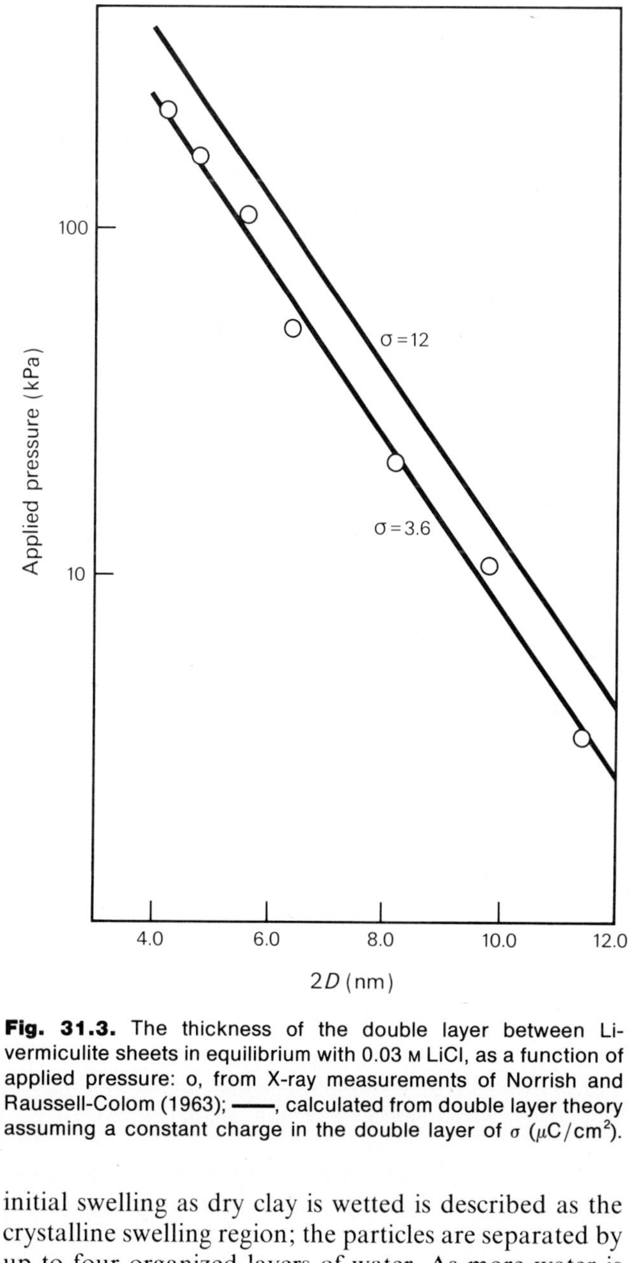

Fig. 31.3. The thickness of the double layer between Li-vermiculite sheets in equilibrium with 0.03 M LiCl, as a function of applied pressure: o, from X-ray measurements of Norrish and Raussell-Colom (1963); ——, calculated from double layer theory assuming a constant charge in the double layer of σ (μC/cm^2).

initial swelling as dry clay is wetted is described as the crystalline swelling region; the particles are separated by up to four organized layers of water. As more water is added, the system enters the more swollen gel state, and lastly the suspension stage.

The bonding in Li-vermiculite gels is discussed first, since most is known about it, starting with the work of Walker (1960). Norrish and Rausall-Colom (1963) applied a dead load to a Li-vermiculite crystal immersed in a LiCl solution, and then measured with X-rays the modal spacing (d) of the individual sheets. This spacing was found to be almost reversible with respect to applied load. Some of the results are shown in Fig. 31.3. The effective thickness of the double layer $2D$ has been derived from d by subtracting 2.1 nm. The latter includes the 1 nm thickness of a sheet and allows 0.5–0.6 nm for

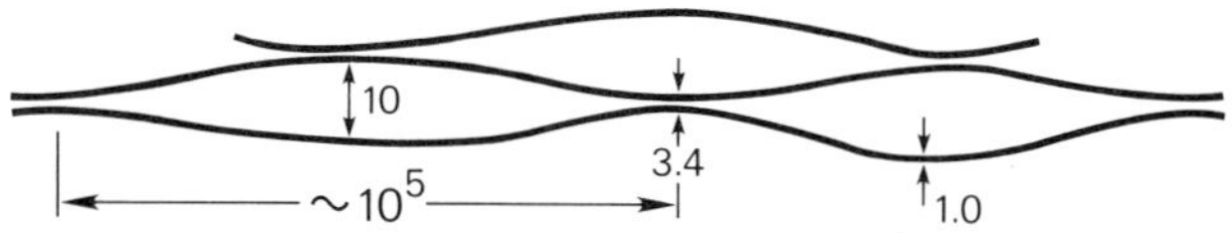

Fig. 31.4. Model of Li-vermiculite swollen in a dilute LiCl solution (numbers indicate distances in nm).

the thickness of the organized water network on each side. The two lines are the repulsive pressures calculated from Gouy double layer theory (Verwey and Overbeek 1948) assuming that the charge in the double layer is constant and equal either to that of the vermiculite or 30% of it; note that the line using the lower charge density gives the better fit to the data. However, the fit does not prove that the surface charge remains constant; lines obtained assuming constant surface potentials (152 mV and 92 mV respectively) would be indistinguishable from them.

Adams and Israelachvili (1978) measured the repulsive force and the spacing between two mica sheets immersed in dilute KNO_3 solutions. Mica has an even greater surface density of charge (30 μC cm^{-2}) than vermiculite. Nevertheless there was good agreement between the force calculated from Gouy theory and the measured force, if a constant surface potential of about 75 mV was assumed. The results of both experiments are therefore explained apparently by the presence of repulsive forces only.

However, Li-vermiculite does not disperse in very dilute LiCl solutions, so there must be some small attractive force. Norrish and Rausell-Colom (1963) found by extrapolation that this force was equivalent to an applied pressure of 2 kPa; moreover, this force was independent of both LiCl concentration and spacing. Measurement of the macroscopic swelling of Li-vermiculite crystals in free solution by Emerson (1963) showed the swelling at a given LiCl concentration to be largely reversible and independent of pH. A suggested model for swollen Li-vermiculite is shown in Fig. 31.4. The sheets are considered to be anchored together locally, probably where the charge density approaches that of mica (c.f. blisters on the surface of Li-vermiculite, Sridhar *et al.* 1972). Barshad (1950) has shown that expanded spacings are not formed by Li-mica. Swelling would be reversible, even in the absence of an applied load, for, as the LiCl concentration is increased, and therefore the Gouy repulsive force between the sheets reduced, the sheets unbend.

Bonding in sodium–montmorillonite gels

Similar experiments have been carried out using Na-montmorillonite. In order to arrange montmorillonite sheets as parallel as possible, an aliquot of a suspension of dispersed salt-free Na-montmorillonite is usually dried on a plane surface to form a flake. When such a flake is immersed in NaCl solution, it swells progressively as the salt concentration is reduced, and eventually the flake disperses; at neutral pH, this occurs in the range 1–2 x 10^{-2}M. It should be noted that this is the same concentration range required to flocculate a suspension of Na-montmorillonite (Emerson 1963); in other words the equilibrium is reversible.

The X-ray diffraction pattern of the swollen flakes in NaCl concentrations < 0.5M varies with the method of preparation of the clay. Treatments able to remove Al_{ex}, such as alkali or polyphosphate, result in more diffuse patterns and larger spacings for a given NaCl concentration (Norrish and Rausell-Colom 1963). Sharp spacings were obtained by Foster *et al.* (1955), who washed their clay with HCl and water, and then titrated to pH 7 with NaOH. The Al_{ex} present after the acid wash (Cashen 1966) would have been converted to hydroxy cations only.

Measurements of the macroscopic swelling of flakes in NaCl solutions (Emerson 1963), show that swelling increases with pH. Flakes also dispersed in neutral sodium pyrophosphate solutions at Na^+ concentrations of < 0.5 mol. l^{-1}. Unlike Li-vermiculite in LiCl solutions, swelling is almost irreversible with respect to NaCl concentration.

It is suggested that in a montmorillonite gel the links holding the particles together normal to their average orientation arise from Al_{ex} and from areas of higher than average charge density. The latter are proposed, as for Li-vermiculite, because stable sodium gels are formed in dilute NaOH solutions. The equilibrium in 1–2 x 10^{-2} M NaCl is attributed to a balance, at the edge surfaces of the individual 1-nm-thick sheets, between the Gouy repulsive force and the van der Waals attractive force. In the plane of the particles the van der Waals force only decays as $1/d^3$, as compared with $1/d^5$ normal to the particles (Verwey and Overbeek 1948). As pointed out originally by Norrish (1954), the van der Waals force normal to the particles is inadequate, at the measured spacings, to balance the Gouy repulsive force.

On this tactoid model, swelling would be expected to be irreversible. For if the NaCl concentration is increased, and the Gouy force reduced, there is no attractive force normal to the sheets to bring them back together. In Li-vermiculite, as pointed out above, this is supplied by the bending of the continuous sheets. Pyrophosphate anions could disperse the flakes, both by removing Al_{ex} and also by increasing the repulsive force between the edge surfaces of the particles (see Fig. 31.2).

Bonding in sodium–montmorillonite suspensions

Two main types of experiments have been carried out to see how far particle interaction in suspension is described by Gouy theory. In the first, pressure is applied to a suspension of the clay in contact, through a membrane, with a reservoir containing a given NaCl concentration; the average spacing between each silicate sheet is calculated from the equilibrium water content of the clay; and the applied pressure is then compared with the repulsive pressure calculated from Gouy theory for an infinite parallel plate model. Reasonable agreement has been found for Na-montmorillonite dispersed in 10^{-4} M NaCl e.g. Warkentin *et al.* (1957) and Callaghan and Ottewill (1974). However, in such a dilute solution, the calculated pressure is insensitive to the actual proportion of the cations dissociated from the clay surface.

The second type of experiment is to measure the negative adsorption of anions, usually Cl^-, for an initally dry and salt-free Na-clay that has been shaken in a dilute NaCl solution. The results are compared with values predicted by Gouy theory, again on the basis of uniformly spaced infinite sheets. Edwards *et al.* (1965), using NaCl–HCl washed illite (Posner and Quirk 1964), found less adsorption than would be predicted. To reconcile this with Gouy theory, they suggested that 90–99% of the exchangeable cations are tightly bound to the clay surface (ion pair formation), the proportion increasing with decreasing salt concentration. A similar deduction was made by Mattson (1929) by applying simple Donnan theory to his results for negative adsorption by Na-bentonite.

A value for the surface potential may be deduced from measurements of the electrophoretic mobility of Na-montmorillonite particles. Using the Smoluchowski equation, it works out at about 50 mV (e.g. Mattson 1929; Callaghan and Ottewill 1974) – a value very much less than would be calculated from Gouy theory if the charge in the double layer were equal to the permanent negative charge of the clay. This is in agreement with the deduction from the measurements of negative adsorption. The presence of Al_{ex} would cause some reduction; it may also affect the negative adsorption of anions (see earlier discussion of kaolinite suspensions).

Bonding in calcium–montmorillonite suspensions

Similar experiments to those of the first type have been carried out by Blackmore and Miller (1961) using Ca-montmorillonite suspended in 10^{-4}M $CaCl_2$. They found reasonable agreement between the applied pressure and the calculated Gouy repulsive pressure between the particles in suspension, provided they made allowance for aggregation of the individual sheets. Aggregation, as estimated from the width of the (001) spacing measured, increased with the maximum pressure applied; the spacing corresponded to a sheet separation of 0.92 nm. Previously, Norrish and Quirk (1954) had measured a maximum separation of 0.95 nm for Ca-montmorillonite flakes expanded in water.

A similar model was used by Shainberg (1968) to reconcile his measurement of the negative adsorption of Cl^- by Ca-montmorillonite suspensions with that

Fig. 31.5. Ultra-thin section of a Ca-montmorillonite gel (x28 000). (Courtesy Dr M. Arnold, and Dr R. C. Foster).

calculated using Gouy theory. The particles were assumed to consist of 4–6 sheets held 0.95 nm apart.

The concentration of $CaCl_2$ required to flocculate a suspension of the Ca-clay is 3-4 x 10^{-4}mol. l^{-1} Flocculation is again considered to be due to a balance between the Gouy repulsive force and the van der Waals attractive force between the edges of individual sheets of different particles. Then it can be calculated (Verwey and Overbeek 1948, p. 118) that the flocculating concentration (mol. l^{-1}) should be 1/64th of the value for the Na-clay, as is found. The links between clay particles, normal to the silicate sheets, are also thought to be due to variations in the surface density of charge of the clay.

Swelling of calcium–montmorillonite, dry to gel

When montmorillonite is dried, cations such as Ca^{2+} are dehydrated and accommodated within the spaces between the oxygen atoms of the silicate sheets. If the relative vapour pressure (p/p_o) is increased, the cations hydrate, causing the sheets to move apart. This crystalline swelling is prevented, if Ca^{2+} is replaced by cations which do not hydrate, or if the isomorphous replacement charge of the clay is neutralized by heating.

At p/p_o = 0.985, the (001) spacing of an orientated flake has attained its maximum value, corresponding to the sheet separation of 0.95 nm mentioned earlier. The volume of a flake is also increased correspondingly, by about 70%. However, there is at least a further equal increase in volume when the flake is immersed in 0.1 M $CaCl_2$ (Emerson 1962*a*). Further swelling may also occur as the equilibrium $CaCl_2$ concentration is reduced, but this depends on the source of the montmorillonite

(Emerson 1962*b*). The additional swelling beyond the value at $p/p_o = 0.985$ must be due to the opening up of gaps between the packets of sheets. This would be caused, as $p/p_o \rightarrow 1.0$, both by the reduction in the restraining surface-tension force, and then by the development of the Gouy repulsive force.

An ultra-thin section of a Ca–montmorillonite (API-25) gel, water content 46% by weight, in equilibrium with a pressure of about 100 kPa is shown in Fig. 31.5. Ca–clay, initially in the form of a slurry, was repeatedly cycled under loads of from about 10^2 to 10^4 kPa, before impregnation with plastic, and then sectioning. The individual 1-nm-thick sheets are aggregated into larger continuous sheets (or packets) about 50 nm thick. These larger sheets are linked at intervals. The regularity of the linkage, and the degree of orientation of the sheets, are much less than in the simple tactoid model suggested earlier (Emerson 1962*b*).

The most important difference in the bonding between Ca- and Na-montmorillonite sheets is that in dilute salt solutions the spacing between most of the former is restricted to 0.95 nm, whereas the spacing of the latter is unlimited. MacEwan (1954), using the Madelung equation, has calculated the electrostatic energy of attraction between the exchangeable cations and the negative charges that arise from isomorphous replacement. He showed that, for a sheet separation of 0.95 nm, the attractive energy was about 2.5 times as large for Ca^{2+} as for Na^+. This supported an earlier suggestion, by Russell (1934), based on the idea that divalent cations can 'bridge' charges on two adjacent surfaces. The energy available to move the sheets apart at this spacing is the loss of free energy of the hydrated cations, due to sorption on the sheets. A value for the energy has not yet been worked out.

Bonding between illite particles

The illite present in the clay fraction of Willalooka subsoil (Quirk 1955; Chi *et al.* 1977) has been taken to be typical of the fine-grained illite that occurs commonly in Australian soils (Norrish and Pickering 1977). The particles have an average diameter of 45 nm and a thickness of 6 nm (surface area to N_2 of 180 m^2 g^{-1}). The dispersion characteristics of this clay are similar to those of montmorillonite. The concentration of NaCl at which orientated flakes of the Na–clay disperse (Emerson 1967) is equal to that required to flocculate a suspension of the Na–clay (Quirk and Schofield 1955). Equally, dried, orientated flakes of the Ca–clay show limited swelling in water, but the clay disperses if shaken in suspension. The flocculating concentrations for the Na- and Ca-clay, at neutral pH, are similar to those given respectively, for Na- and Ca-montmorillonite.

The volumetric swelling of flakes of Ca–Willalooka illite, from dry to $p/p_o = 0.985$, is 26%. This is comparable to that of Ca–montmorillonite, allowing for the difference in the total surface areas of the two clays. There is a further approximately equal amount of swelling when the illite flakes are equilibrated with 0.1 M $CaCl_2$, but no more when the $CaCl_2$ concentration is reduced (Emerson 1964; see also, Aylmore and Quirk 1962). The structure of a swollen Ca–flake is thought to be similar to that of Ca–montmorillonite, with the particles linked edge to edge in the form of a tactoid (Emerson 1977).

Emerson and Chi (1977) have found recently, that coarser-grained illites disperse more easily than the Willalooka illite. Orientated flakes of the Ca-form of the coarsest illite used (surface area to N_2 of 26 m^2 g^{-1}), from the Tumut shale in New South Wales, even dispersed slightly in water. This is probably due to the poorer packing of the larger, irregularly shaped particles, which allows fewer opportunities for particle bridging.

Bonding between kaolinite particles

Soil kaolinites are much coarser than soil illites; a typical example, from Rocky Gully in Western Australia (B horizon, Kent sand), has a surface area to N_2 of only 36 m^2 g^{-1} (Aylmore and Sills 1978). The volumetric swelling of kaolinite, from dry to $p/p_o = 0.985$, is small: Aylmore and Quirk (1966), using compressed cores of Rocky Gully kaolinite, found 3.5%; Emerson (1964) using flakes of an even coarser ceramic kaolinite, found 2%. By comparison with Ca–montmorillonite, the swelling is about the same as the ratio of total surface areas of the two clays. Rocky Gully kaolinite and the ceramic kaolinite swelled a further 26% and 11% when equilibrated with 0.1 M $CaCl_2$ or lower, and, in the case of Rocky Gully, equally with 0.1 M NaCl or below.

As kaolinite particles are thick, compared with montmorillonite, their van der Waals force normal to the particles will no longer be negligible compared to that between edge surfaces. The decisive factor in determining particle linkages will be the presence of Al_{ex} on the periphery of the basal surfaces. For a Na–kaolinite, the most stable arrangement is likely to be an overlapping one, with Al_{ex} bridging the particles (Fig. 31.6*a*).

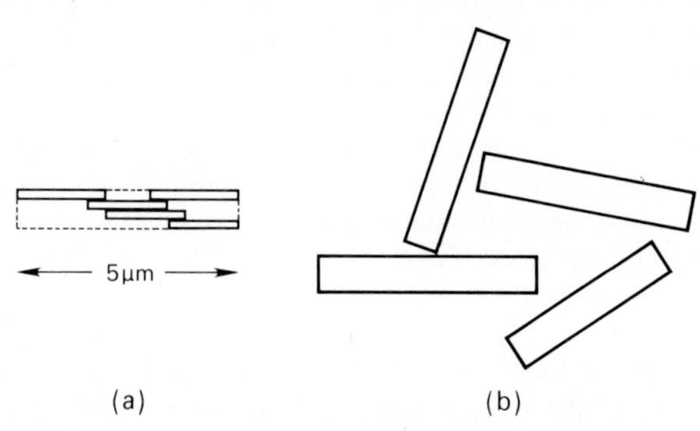

Fig. 31.6. Model of the structure of a kaolinite flake. (*a*) arrangement of individual particles into aggregates (*b*) arrangement of aggregates.

Clusters or domains settle out of suspension, and form a poorly orientated flake on drying, as shown in Fig. 31.6*b*. In a previous model (Emerson 1977), the importance of Al_{ex} in determining the stacking pattern was overlooked. Stacking within the aggregates now resembles the stepped face to face arrangement of kaolinite particles seen by Smalley and Cabrera (1969) in the electron microscope.

Conclusions

In the case of 2:1 layer silicate particles with monovalent cations on the exchange sites, and having distances of separation of the particles > 3 nm, the repulsive force between the particles may be explained on the basis of Gouy double layer theory. However, allowance has to be made for the fact that only a small fraction of the exchangeable cations is actually present in the double layers. It is suggested that the attractive forces between the basal surfaces of the particles are due to areas of higher than average charge density and to Al_{ex}. Flocculation–dispersion is a reversible process with respect to salt concentration, and is apparently associated with the balance of forces between the edge surfaces of the particles.

In the case of Ca–clays, starting from dry clay, most of the individual particles remain at a small distance of separation, even in water. Once the particles are forced into suspension by external work, then the interaction between the packets of particles (or domains) seems to be similar to that for the single particles of the Na–clays.

The swelling of a Ca–clay, from dry to $p/p_o = 0.985$, can be followed in the case of Ca–montmorillonite as a progressive increase in the spacing between the individual sheets, up to the maximum separation of 0.95 nm. The volumetric swelling of other Ca–clays over this vapour pressure range, as compared with that of Ca–montmorillonite, is approximately in the ratio of their total surface areas. The amount of further swelling of Ca–clays when the clays are immersed in water cannot be predicted yet.

These conclusions will now be used to help explain the swelling of soils and the conditions under which clay particles disperse from soils immersed in aqueous solutions. Dispersion will depend not only on the clay minerals present and their associated cations, but also on other soil components, such as organic matter and carbonate.

Swelling of soils

Greene-Kelly (1974) brought remoulded samples of 63 clay subsoils into equilibrium with $p/p_o = 0.99$, and then measured their shrinkage on drying; the samples had previously been through one drying and wetting cycle. Only three of the subsoils contained more than 0.7 mequiv. per 100 g of Na_{ex}. He found a good correlation between shrinkage and the total surface area of the clay present, the latter being measured using ethylene glycol.

He also showed that shrinkage of samples from the same 63 soils, equilibrated at 10 kPa suction and then dried at $p/p_o = 0.99$, was not correlated with total surface area; these results would be anticipated from the results presented for Ca–clays. Earlier, Gallavan and Greene-Kelly (1973) had confirmed that the individual sheets in two montmorillonitic clay soils aggregate in the same way as reported here for flakes of Ca–montmorillonite.

The crystalline swelling of the clay present is one of the main reasons for the macroscopic break-up (slaking) of dry soil aggregates on wetting with water. Such slaking can only be prevented if the rate of wetting is very slow, or if the domains of clay particles are bonded together by organic matter (Emerson 1959*a*).

It is unfortunate that the magnitude of the swelling of Ca–clays from $p/p_o = 0.985$ to free solution cannot yet be predicted. This region of swelling (wilting point, $p/p_o \simeq 0.98$) strongly influences the amount of water available for plant growth in clay soils; it also fixes the extent of the soil movement caused by the use of this water. In soils that are clays to the surface, the cracks caused by the removal of water by plants can be beneficial in allowing the entry of subsequent irrigation water (Emerson 1977, p. 98). However, the vertical movement of clay soils, resulting from the withdrawal of water by plant roots, can cause damage to adjacent structures such as roads and buildings. This damage may be accentuated by additional swelling near saturation when sufficient Na_{ex} is present so that the potential swelling of a soil is not finite (see Fig. 31.14) and additional shrinkage in the crystalline swelling region, if airdrying occurs along cracks.

The stability in water of initially dry soil aggregates

Exchangeable cation composition and aggregate dispersion

Dry soil aggregates which disperse when immersed in water have always been found to contain some Na_{ex} on the clay. Exceptions to this may however occur; as mentioned earlier, two pure illites have recently been shown to disperse from dry without Na_{ex} being present (Emerson and Chi 1977). It is convenient to express the amount of Na_{ex} present as a percentage of the net negative charge (ESP). Here the charge at pH 7 is used, unless otherwise stated.

To illustrate how dispersion increases with increasing ESP results for a group of soils are plotted in Fig. 31.7*a* (Emerson 1972); these are the 0–10 cm samples of 51 soils used by Loveday (1974). Here the ESP is expressed as a percentage of the sum of the exchangeable basic cations, plus the buffering to pH 8.4, determined using *m*-nitrophenol (Piper 1950). Dispersion was estimated by eye on the scale 0–4, where 4 means apparently complete

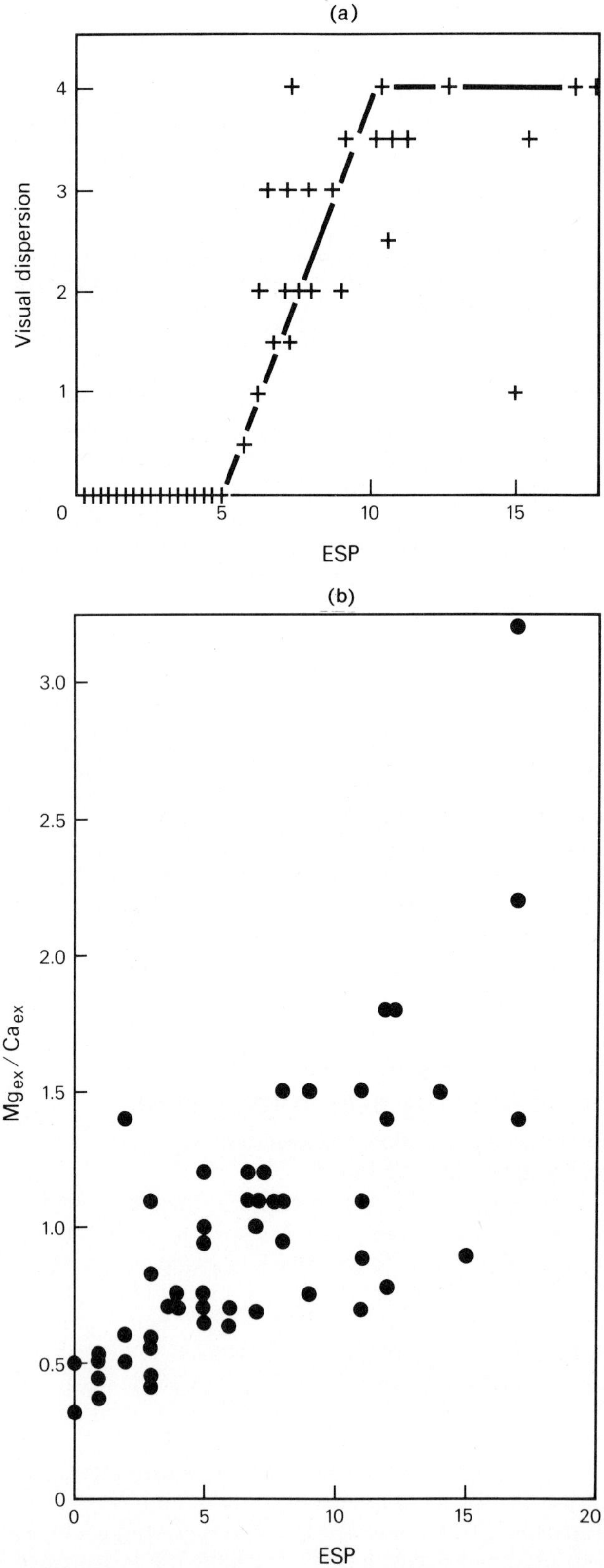

Fig. 31.7. (*a*) The degree of dispersion in water of dry aggregates from a group of surface soils from the Riverina, N.S.W. (*b*) The relation between exchangeable Na, Ca, and Mg for a group of surface soils from the Riverina.

dispersion. Dispersion started at an ESP of 5, and was complete in most samples at an ESP of 10; the exceptional point at ESP 15 is from a soil which contained 0.3% $CaCO_3$.

In general, the ESP required before a dry soil disperses in water will be increased either as the strength of the osmotic repulsive force between the clay particles is reduced, or if the clay particles are bonded together by other soil components, such as organic matter and oxides. The presence of Al_{ex} will reduce the repulsive force because the number of ions dissociated from the clay surface will be correspondingly reduced; also there is the possibility of interparticle bridging by polymeric hydroxylated Al_{ex} cations. Cashen (1959) showed that one kaolinite, with an ESP of 83 (net negative charge equal to the isomorphous replacement charge), did not disperse in water when the remaining cations consisted of Al_{ex} in the nominal form $Al(OH)_{2.5}$. The difficulty of determining the role of Al_{ex} in the dispersion of soil clays is that of evaluating precisely the amount and form of the Al_{ex} present.

An example of Al_{ex} acting to reduce dispersion is given by a comparison of samples from two depths of the acid clay subsoil of a Dy3.41 soil (Northcote 1971), Caniambo loam (Downes 1949). The composition of the clay in both was the same, but the pH of the upper sample was 6.4 and of the lower, 5.0. The percentage of the negative charge at pH 7 balanced by exchangeable basic cations was 95, with an ESP of 20, for the upper sample, and was 85, with an ESP of 24, for the lower one. The main difference was in the lower sample having a three-fold increase in the proportion of the charge presumed to be balanced by Al_{ex}. This was reflected in the dispersion of the dry aggregates; those from the upper sample dispersed completely in water, whereas the others did not disperse at all.

The influence of Mg_{ex} will now be considered. It has been found for illitic *red-brown earth* (Dr2.33) subsoils around Tatura in Victoria (Emerson and Bakker 1973) that the ESP required for dispersion is about halved when Mg^{2+} is the complementary exchangeable cation rather than Ca^{2+}. This result has been confirmed for pure illites (Emerson and Chi 1977). It is likely that the 20% greater hydration energy of Mg^{2+} (Bernal and Fowler 1933), leads to an increase in the percentage of ions that are dissociated from exchange sites on the clay surface. Jenny and Reitemeier (1935) found an increase of 2.5% in the ζ-potential of Mg- compared with Ca-Putnam clay. The importance of discriminating between Mg_{ex} and Ca_{ex} lies in the fact that in many Australian subsoils Mg_{ex} exceeds Ca_{ex}; further, there is often a corresponding increase in Na_{ex}. To illustrate this, the ratio Mg_{ex}/Ca_{ex} has been plotted against ESP in Fig. 31.7*b* for the 51 samples used already for Fig. 31.7*a*. Much larger values of Mg_{ex}/Ca_{ex} do occur: e.g. in the Caniambo loam mentioned previously, the ratio varies from 10-30.

A comparison of the effect of Mg_{ex} and Na_{ex} on the dispersion in water of an otherwise Ca–illite (Muloorina) showed that Mg_{ex} was 10% as effective on an equivalence basis as Na_{ex} in causing dispersion (Emerson and Chi 1977).

The last major cation on soil clays, K^+, usually only occurs in the interlayer positions and is not readily exchangeable (Tucker 1975); it therefore does not usually contribute towards the repulsive force. In principle, K_{ex} might be expected to cause dispersion in a similar manner to Na_{ex}, both ions being monovalent, except that, the hydration energy of a K ion being 18% smaller (Bernal and Fowler 1933), the proportion of exchangeable cations contributing to the osmotic repulsive force will be reduced, and this in turn will lead to a reduction in the Cl^- concentration at which dispersion occurs. In fact the KCl concentrations at which initially dry flakes of a K–montmorillonite and a K–illite disperse are about one fifth of the NaCl concentrations for the corresponding Na–clays (Emerson 1967). Jenny and Reitemeier (1935) reported a 2% reduction in the ζ-potential of K-Putnam clay as compared with Na–Putnam clay.

Organic bonds, aggregate dispersion and stability

Emerson (1954) showed that the organic bonds in surface soil aggregates (crumbs) from permanent grassland are able to prevent dispersion of the aggregates in water, even after the aggregates have been first washed with a NaCl solution. In contrast, aggregates of similar inorganic composition, from a permanent arable field, washed in the same way, dispersed when the NaCl concentration was reduced to 3.5×10^{-2} M NaCl. In such experiments, mechanical disturbance of the wet aggregates must be kept to a minimum. (cf. a later section on organic matter).

Methods of estimating the extent of interparticle bonding have been developed on the basis of these results (Greenland *et al.* 1962; Clapp and Emerson 1965). Most of the exchangeable cations of the aggregates are replaced by Na^+, using NaCl solution. Next, the aggregates are equilibrated with a lower concentration of NaCl, one in which the clay separated from the aggregates should *just* disperse completely, i.e. clay free of organic matter, and in the Na-form at the same pH. Then, either the amount of clay actually released, or the decrease in permeability of a bed of aggregates, can be measured as an index of the degree of bonding. If the aggregates do not disperse, then various extractants can be tried. The aim is to attain, under the same conditions, complete dispersion both of the aggregates and of the separated clay, given the same extraction treatments.

Using these methods, the relative contributions that periodate-oxidizable polymers (such as polysaccharides) and polymers complexed with metal ions make to the organic bonds holding particles together can be determined (Clapp and Emerson 1965; Stefanson 1971; Hamblin and Greenland 1977). The Na-method can also be used for the rapid qualitative screening of potential soil conditioners. The conditioner is added to a soil; then the dried soil aggregates are washed with a sodium chloride solution, followed by immersion in water. Aggregates are considered stabilized if there is no dispersion. Observation of the swelling of stabilized aggregates also helps to understand bonding mechanisms; aggregates treated with polyvinyl alcohol or other non-ionic polymer do not swell, aggregates stabilized by carboxylated polymers swell progressively as the salt concentration is reduced (Fig. 31.8). This is presumably because polyvinyl alcohol collapses down onto the surfaces of the particles, whereas the unbonded segments of the carboxylated polymers progressively uncoil (Emerson 1956; Greenland 1972).

The Na-method is particularly suited to investigating the bonding between the clay and silt sized particles in aggregates. At the macroscopic level (2–0.2 mm), many measurements of the slaking of dry aggregates in water have been made in order to show that the water stability of surface soil aggregates is increased after the soil has been under grass for a period (Clarke and Marshall 1947; Low 1972). In laboratory experiments, Tisdall *et al.* (1978) added readily decomposable organic matter to the surface soil of a *red-brown earth* (30% silt); the resultant biological activity led to a marked reduction in slaking. Further, this was largely maintained, despite repeated 'cultivation'. Giovannini and Sequi (1976) have shown, that the organic matter which links the soil particles together to form aggregates > 0.2 mm diam. does so through metal ions extractable by acetylacetone in benzene. This organic matter could be fungal hyphae bonded to the particles by bacterial gums (Emerson 1977). Equally, once a crust has formed on the surface of a soil, the growth of filamentous algae could increase the strength of the crust (Bond and Harris 1964).

Iron oxide, aggregate dispersion and stability

Surface soils with a high content of uniformly distributed iron oxide usually have a stable structure (Russell 1971). However it is difficult to prove that iron oxide as such contributes to this stability. The soils, e.g. *krasnozems*, are acid, and their clay fraction consists of: kaolin, a little vermiculite, about 20% haematite plus goethite, and a variable amount of gibbsite (Stace *et al.* 1968). Aggregates from these soils, including subsoil aggregates with low organic matter content, do not disperse in water after washing with 1 M NaCl; however, this can be due to Al_{ex} on the kaolin, rather than any bonding by Fe_2O_3 particles. Subsoil aggregates disperse completely in water after extraction with a neutral sodium pyrophosphate solution. Unfortunately, this treatment not only removes Al_{ex} from the kaolin, but also increases the negative charge of the oxides (Hingston *et al.* 1972), so that its effect is not specific. Similarly, attempts to determine the

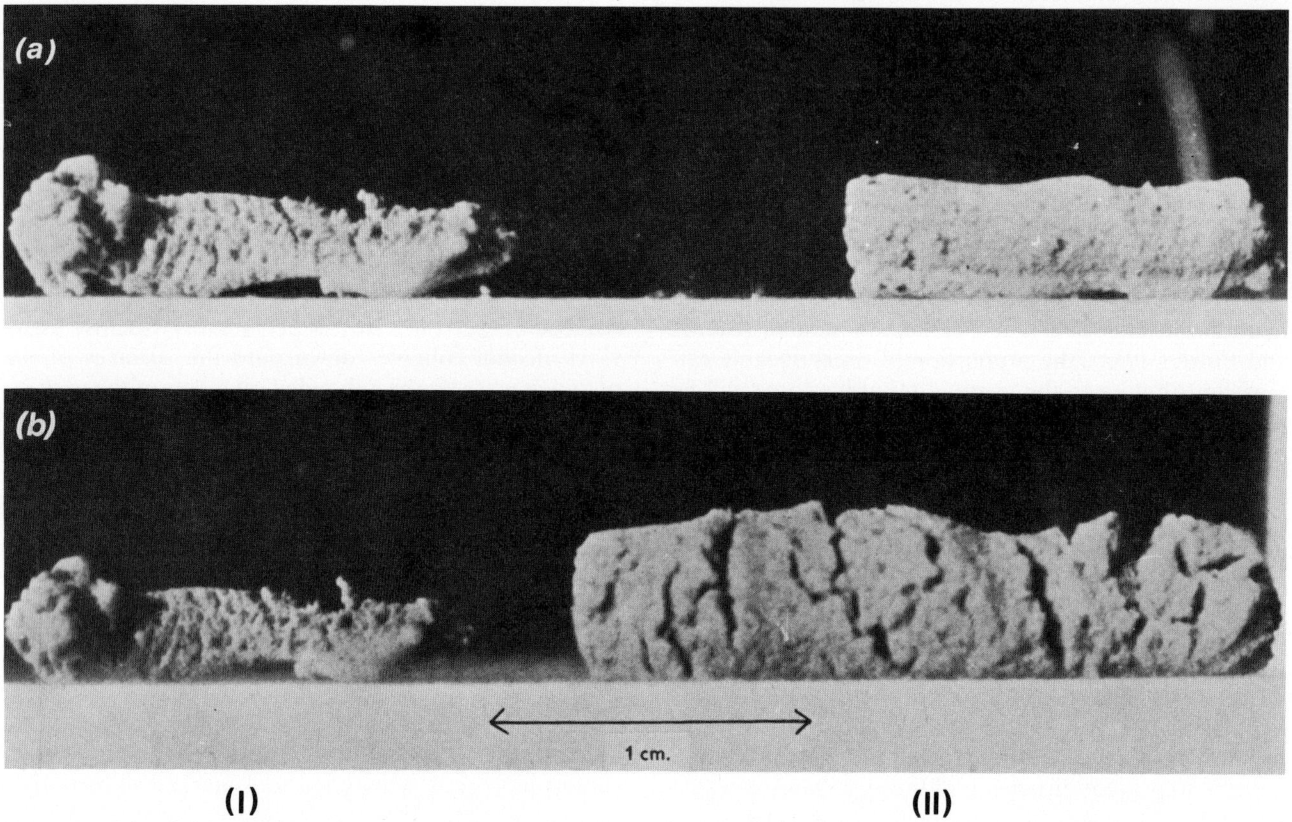

Fig. 31.8. Synthetic aggregates with 0.25% PVA (I) or 0.25% sodium alginate (II) added (*a*) in 0.5 M NaCl; (*b*) in H_2O.

effect of Fe_2O_3 by a process of subtraction are usually complicated by secondary reactions: for example, the citrate ion in the buffer used in the Mehra and Jackson (1960) extraction method reacts in the same way as the pyrophosphate ion.

In nearly neutral subsoils, with a lower content of iron oxide, and illite a major component of the clay fraction, e.g. the *red-brown earths* (Stace *et al.* 1968), no evidence has been found for additional bonding due to the oxide particles. For example, the dispersion in water of *red-brown earth* (Lemnos loam) subsoil aggregates, with 6% free iron oxide in the clay fraction, has been studied (Emerson and Bakker 1973); an ESP of only 5 was required to initiate dispersion of otherwise Ca-aggregates, as compared with 12 for a fine-grained illite, Willalooka (Emerson 1967). The latter value also applied to aggregates from the subsoil of another *red-brown earth*, which happened to contain appreciable Al_{ex} (Emerson and Bakker 1973). Electronmicrographs (Fig. 31.9) of the clay fraction of Lemnos subsoil dispersed in water show the iron occurring as clusters of particles, the individual particles being < 5 nm diam. Similar clusters were observed by Greenland *et al.* (1968) in a range of soils. In the natural soil, the clusters are probably associated with the basal surfaces of some of the clay particles (Follett *et al.* 1965).

In studies using synthetic iron oxides, Schahabi and Schwertmann (1970) showed that only the addition of particles sufficiently small as to be amorphous to X-rays is able to increase the water stable aggregation of a silty soil. El-Swaify and Emerson (1975) added alkali to suspensions of a Na–illite (Willalooka) and a Na–kaolinite (Rocky Gully) and then added an equivalent amount of $FeCl_3$; the Fe precipitated as particles about 4 nm diam. onto the basal surfaces of the kaolinite. It was inferred from surface area measurements that similar particles were also present on the basal surfaces of the illite. The illite no longer slaked in water or dispersed in water after washing with NaCl; however, both clays still dispersed completely in dilute neutral sodium pyrophosphate solutions. It is considered that the freshly precipitated oxide particles may act as large bridging cations.

In the field it is likely that most of the positive charges of the oxide particles will be rapidly neutralised by the uptake of organic anions. In the very young Hilo soil (hydrandept), which consists mainly of precipitated iron and aluminium, the sesquioxide particles are strongly bonded together by organic matter (Tama and El-Swaify 1978). Equally, in *krasnozems*, the surfact horizons show negative sorption of anions, while the corresponding subsoils show positive sorption (Black and Waring 1979).

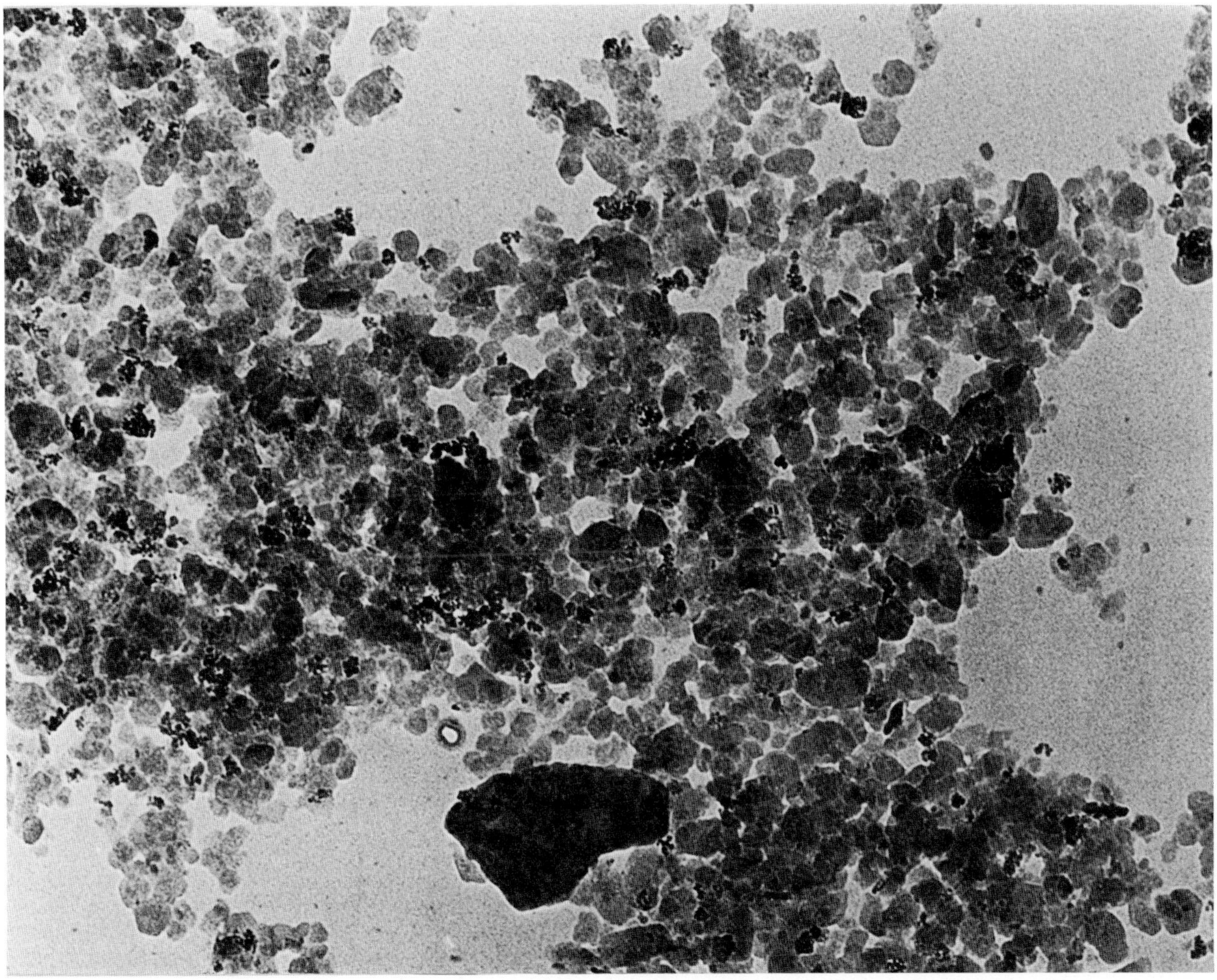

Fig. 31.9. Electron micrograph of the fine fraction of the subsoil of Lemnos loam, Vic. (Dr2.33, Skene and Poutsma 1962) (x 54 000).

Stace (1956) measured some properties of a group of *rendzinas* and a group of *terra rossa soils*. For these neutral to alkaline soils, as will be seen from Fig. 9 of Stace (*ibid*), there is a good correlation in each group between the nitrogen content of the soil and the content of free iron oxide in the clay fraction. McIntyre (1956) selected samples from the group with more than 26% clay (and therefore more than 0.14% N). He showed that the macroporosity of these soils increases sympathetically with the content of free iron oxide in the clay fraction. He suggested that this is due to the iron interacting with the organic matter. It is likely that other positive correlations between indexes of structural stability, such as water stable aggregation (Kemper and Koch 1966, cf. their Figs 4 and 6), and iron oxide content, are mainly due to interaction between iron and organic matter.

Aluminium oxide and aggregate dispersion

When aluminium oxide occurs in soils as discrete particles, it is in the form of gibbsite (Norrish and Pickering 1977). Soils containing gibbsite and $< 5\%$ Fe_2O_3 are acid, and kaolin is the only clay mineral present, e.g. pallid zones of laterites. Therefore, as discussed already for the case of iron oxides, the contribution of gibbsite particles as such to bonding between clay particles cannot be determined using a sodium method. Direct evidence of such bonding was obtained by Fordham and Norrish (1979). These authors washed aggregates from krasnozems with NaCl and then applied sufficient mechanical action to break the organic matter bonds linking the clay-sized particles together; the particles were then shown to consist of aggregates of kaolin and gibbsite.

Planar sheets of Al-hydroxy cations can be formed in solution, with a structure similar to that of gibbsite (Rengasamy and Oades 1978). This was probably the form of the Al precipitated onto illite and kaolinite by El-Swaify and Emerson (1975) in experiments parallel to those already described for Fe. It was deduced that the precipitate was present as discontinuous films about 0.7 nm thick on the basal surfaces of the particles. The clays plus precipitated Al dispersed in pyrophosphate solutions but not in NaCl solutions, in a similar way to the clays plus Fe. The precipitated Al was found to be more effective than the Fe particles in bonding the clay together.

Carbonate, aggregate dispersion and stability

One important function of carbonate in soil is to act as a source of Ca^{2+}; this helps to reduce the dispersion of sodic soils (cf. Fig. 31.7*a*), and also, as will be discussed later, increases the resistance to dispersion of sheared wet, effectively calcium soils. Abedi and Talibudeen (1974) measured the specific surface area of carbonate present in soils from Iran and Britain; for carbonate contents of < 1% by weight, this was about 300 $m^2 g^{-1}$, as compared with 15 $m^2 g^{-1}$, when 10–20% carbonate was present. It follows, that small amounts of carbonate in a soil can be effective sources of Ca^{2+}. This is important for example in self-mulching soils (Butler 1955; Emerson 1977) which may contain only about 1% by weight of carbonate, e.g. Yooroobla clay (Stace *et al.* 1968, p. 106).

Rimmer and Greenland (1976) studied how the removal of carbonate affected the bonding between the clay particles in soil aggregates taken down the profile of a surface water gley. They measured the water uptake by the NaCl-washed clay from water at a suction of 1 kPa; this is a variation of the Na-method where swelling (dispersion) is observed at zero suction. The swelling of the untreated Na–clay was about half of that of the clay from which carbonate had been removed. The authors considered that this was largely due to the Ca^{2+} concentration maintained by the carbonate. A small increase was also measured in the surface area of the clay to nitrogen after treatment; however this could be the result of breaking up the clay domains by the repeated shaking and centrifuging of the clay during the treatment, rather than to the removal of carbonate coatings. In arid environments, actual engulfing of the soil particles occurs by deposition of carbonate (Gile *et al.* 1966; Wetherby and Oades 1975).

Crystal growth and clay dispersion

Hardpans can be formed in soils by the growth of carbonate, silica, or iron oxide crystals through a soil (Flach *et al.* 1969). Portions of the pans are usually of rock-like consistence. Intermediate in consistence between this and normal clay subsoils, are the highly subplastic or SPIII+ heavy clay subsoils, (Butler 1955; van Dijk 1958). These soils texture initially in the field as grits; their texture even after 15 min manipulation changes only to a medium clay with grit (McIntyre 1976).

Such SPIII+ subsoils occur in the Riverina, and contain carbonate (McIntyre, *ibid*). The soil aggregates do not disperse in water after immersion in 1M NaCl, nor even after extraction with neutral sodium pyrophosphate. The latter distinguishes them from the acid subsoil aggregates bonded together by Al_{ex} as described previously, and, the synthetic clay aggregates bonded together by amorphous iron oxide. Complete visual dispersion of the aggregates can be obtained by immersing them first in 1 M HCl, followed by decanting with a dilute solution of NaOH or neutral sodium pyrophosphate, followed by water (W. W. Emerson, unpublished). Using a combination of acid plus alkali extraction, followed by shaking in Calgon, Norrish and Tiller (1976) were able to disperse the clay fraction of the aggregates completely.

A characteristic feature of the Riverina subsoils is that cation exchange reactions are not affected by the degree of subplasticity of the soil (Beattie 1970). Similar observations were made by Flach *et al.* (1969) in studying a non-calcareous kaolinitic soil profile with a presumed silica hardpan. They found that, as compared with the upper part of the profile, very little clay was released from hardpan aggregates when these were shaken in a polyphosphate solution, and the CEC of the aggregates was unchanged. Sills (1975) reported the same two properties for glaebules in the sand fraction of a kaolinitic subsoil.

Norrish and Tiller (1976) suggested that the bonding between the clay particles in the Riverina subsoils is due to the authigenic growth of the clay minerals themselves, rather than to the precipitation of silica, for example. Another possibility is that the domains of clay particles could be bonded together by a precipitated aluminosilicate. This is similar to a suggestion made earlier to explain the bonding of soil aggregates by Portland cement, the exchangeable cations in the clay domains being free to hydrate normally (Emerson 1959*a*).

In a clay shale, such as Tumut, strong mechanical forces are needed to release the clay present (Chi *et al.* 1977). Negligible clay is released in water after extraction by any combination of acid, alkali, and polyphosphate. In this case, a bonding mechanism such as crystal growth is required.

The dispersion of clay from wet sheared aggregates

Water content and aggregate dispersion

Soil aggregates that do not disperse in water when immersed dry may do so after shearing sufficiently wet. In order to characterize the ease of dispersion of wet sheared soil in water, a water content for dispersion has been defined (Emerson and Smith 1970); it is the maximum content at which a soil can be uniformly remoulded with a spatula, and then *immediately* immersed in water without there being subsequent diffusion of any clay particles into the water. It will be referred to here as

θ_D (% by weight oven-dry soil); and θ_D^{Ca}, for example, refers to the value for the Ca-form of the soil.

The aim of remoulding the soil is to achieve a uniform separation of the clay particles; and the criterion used is that of an apparently uniform paste. Usually, just sufficient water is added to a single sample of dry soil so that it can be uniformly remoulded within the half minute prescribed (Emerson 1968); the force applied to the spatula in the process is not controlled, so that the drier the soil relative to its clay content, the greater the pressure applied. Aliquots are then taken for checking dispersion, and for measurement of water content. Next, further water is added to the main sample, which is again remoulded, followed by subsampling. The process is repeated as required. An alternative method is to add increasing amounts of water to separate samples of dry soil, so that each sample is only remoulded once (Emerson 1967). The two methods are compared in a later section.

In relating θ_D to field conditions, θ_{10}, the water content of unsheared <2 mm soil wetted at a suction of 10 kPa, has been used (Emerson 1967). This is a convenient approximation to field capacity, where this concept is applicable.

Clay composition, soil texture and aggregate dispersion

The composition and form of the clay minerals present in a soil considerably affect the value of θ_D (Emerson 1967). In the cases of a fine-grained soil illite (Willalooka), θ_D^{Ca} is not much greater than θ_{10}, respectively, 52 and 45. On the other hand, a suspension of Ca-kaolinite in water flocculates, so that the clay cannot be dispersed by adding water and remoulding, however wet it is made. Emerson (1968) studied synthetic mixtures of the Willalooka illite and two kaolinites. The value of θ_D^{Ca} remains constant and equal to the value for the pure illite, as kaolinite is added initially, but then increases rapidly. The increase occurs approximately when the amount of illite present is less than sufficient to cover the surfaces of the kaolinite particles, as calculated from the nitrogen-determined surface areas of the clays (Emerson 1968); for Rocky Gully kaolinite, this corresponds to two thirds kaolinite in the mixture. Ca-montmorillonite (Redhill, UK) disperses similarly to the illite. In mixtures of fine sand and silt with the illite or montmorillonite, θ_D is reduced in value, approximately in proportion to the amount of clay present (Emerson 1967, 1977). In other words, after remoulding, all the water is effectively retained by the clay only.

Ca_{ex} and Mg_{ex} and aggregate dispersion

Dispersion–water content curves obtained by the two methods outlined above are shown in Fig. 31.10; they are for a surface soil which had been converted to the Ca or Mg form by washing with corresponding solutions of the chlorides (0.1 M down to 1 mM). This soil is one of nine non-carbonate samples from the group of soils used previously to check dispersion dry (Fig. 31.7*a*). Using a single sample, the value of θ_D increases slightly but the main difference is the slower increase in the rate of dispersion as the water content exceeds θ_D. This means that the water is distributed more efficiently between the individual clay particles in using the single sample method. Otherwise, some clay particles remain aggregate in packets, particularly after remoulding very wet samples. However, as complete dispersion is apparently attained, these packets must be < 2 μm diam. The single sample method is preferred for simplicity, and also for pure clays, when attempting to relate average spacing (i.e. water content) to dispersion.

The values of θ_D for the Ca and Mg forms of the nine soils (pH range, 5.1–7.3; %C, 0.6–1.7) determined by the single sample method have been plotted in Fig. 31.11 against clay content. The points lie approximately on straight lines through the origin, indicating that the composition of the clay minerals in this group of soils is similar. The average ratio $\theta_D^{Mg}/\theta_D^{Ca}$ from the slopes of the lines is 0.62. Previous values are: 0.72 for the surface of a kaolinitic soil (Emerson and Smith 1970), 0.50 for a *red-brown earth* subsoil (Bakker *et al.* 1973), and 0.67 for Willalooka illite (Emerson and Chi 1977). Presumably

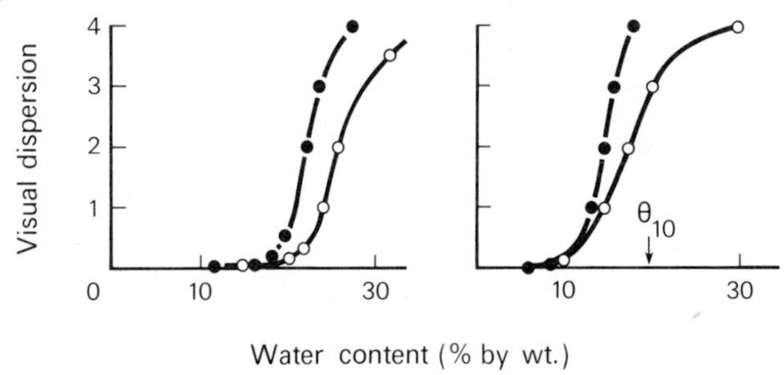

Fig. 31.10. The visual dispersion in water of a wet remoulded soil (0–10 cm of a Ug3.2, 50% <2 μm), washed with $CaCl_2$ or $MgCl_2$. The arrowed water content was determined on unsheared soil. ●, separate samples; ○, single samples.

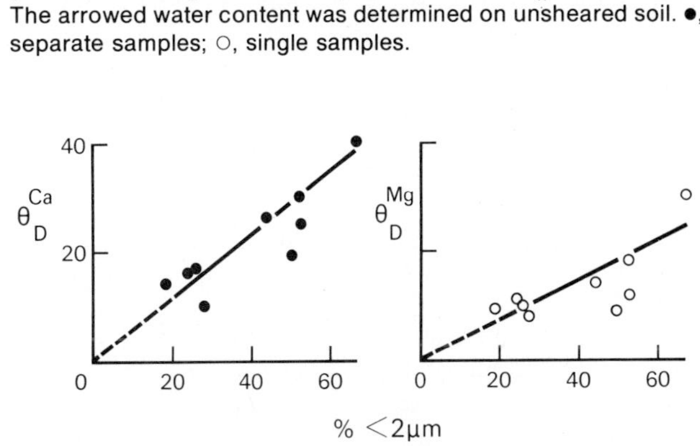

Fig. 31.11. The water contents for dispersion of nine surface soils from the Riverina, washed with $CaCl_2$ or $MgCl_2$.

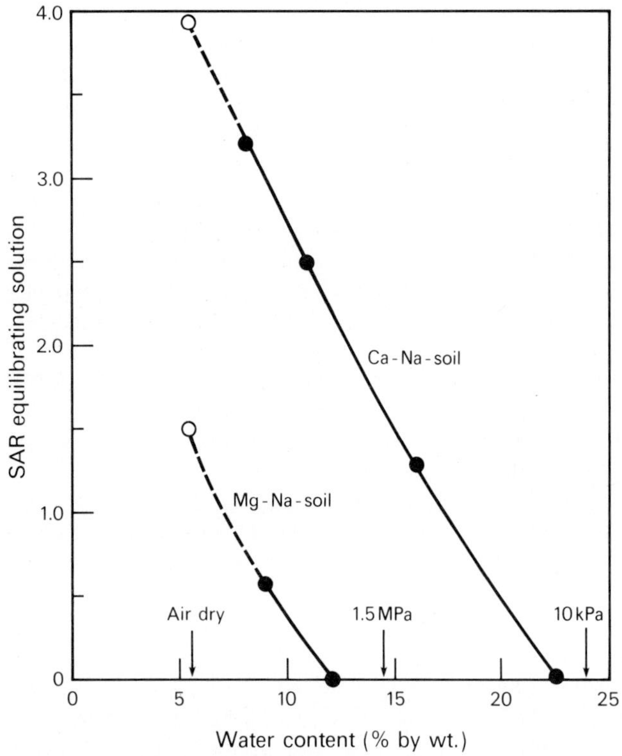

Fig. 31.12. Subsoil aggregates of a *red-brown earth* equilibrated with $CaCl_2$–NaCl or $MgCl_2$–NaCl solutions of known SAR: ●, water content for dispersion for a given SAR; ○, values of SAR for dispersion of air-dry aggregates. Arrows indicate water contents of unsheared, Ca- and Mg-aggregates, air-dried and then equilibrated with water at the stated suctions.

soils with even smaller ratios may exist, since, as noted earlier, the Mg form of one of the illites studied by Emerson and Chi (1977) dispersed dry in water. The ratio for this clay would be zero.

Na_{ex} and aggregate dispersion

As some of the Ca_{ex} and Mg_{ex} is replaced by Na_{ex} there is a rapid reduction in θ_D, until the soil becomes too dry to remould uniformly. Finally, sufficient Na_{ex} is present for dispersion to occur dry, that is, without external work being required; this is shown in Fig. 31.12, following Emerson (1977), for subsoil aggregates of a *red-brown earth*, Shepparton fine sandy loam, Dr2.23, (Skene and Poutsma 1962). The aggregates were prepared by equilibration with solutions of a given sodium adsorption ratio (SAR), with a final divalent ion concentration of 1 mM. An SAR of 2 corresponds approximately here to an ESP of 3. It will be noted that there is no threshold value of Na_{ex}; θ_D decreases immediately any Na_{ex} is present. The increased ESP required to cause dispersion in water of dry, otherwise Ca soils, as compared with Mg soils, discussed previously, is seen to be reflected in a corresponding increase in the value of θ_D.

Carbonate and aggregate dispersion

For water contents in excess of θ_D, dispersion of a wet remoulded clay or soil will be inhibited if it is immersed in a sufficiently strong salt solution. The greater θ_D, the nearer this salt concentration approaches to the flocculating concentration for a suspension of the clay or soil. For example, a sample of Ca-illite (Willalooka), remoulded at a water content of 80% by weight compared with its θ_D^{Ca} value of 52, required a $CaCl_2$ concentration of 0.3 mmol. l^{-1} to inhibit dispersion (Emerson 1967); the concentration required to flocculate a suspension of the Ca-illite is 0.5 mmol l^{-1}. Similarly, there will be a reduction in the degree of dispersion of wet remoulded soil if it is immersed in a solution of $CaCl_2$ rather than in water. For example, the value of θ_D^{Ca} for the surface soil aggregates of a *red-brown earth* was known to be 8 (Emerson 1968); when the aggregates were remoulded wetter, at $\theta_{10} = 18$, the percentages of the clay content that dispersed were 95 in water, 40 in 0.3 mM $CaCl_2$, and 0 in 1 mM $CaCl_2$ (Emerson 1977); the concentration required to flocculate a suspension of the soil is 0.5 mmol l^{-1}.

These results do not mean that the presence of Ca^{2+} will increase the net attractive forces between the clay particles sufficiently to prevent a soil being brought to the puddled or dispersed condition when worked wet; Olson and Mitronovas (1960) showed, that the strength of a remoulded illite is not affected by $CaCl_2$ concentration. However, the presence of carbonate, by maintaining the Ca^{2+} concentration in the range 0.3–1 mM will have two effects: it will minimize the risk of soil actually dispersing in subsequent rain or irrigation water, after cultivating wet; and further, the puddled wet soil will tend to crack, that is self-mulch, on drying out (Emerson 1977).

Organic matter and aggregate dispersion

When considering the dispersion of dry soil aggregates in water, the contribution of organic bonds in preventing dispersion has been emphasized. However, once the bonds are broken by forcing the clay particles sufficiently far apart, the organic matter helps to stabilize the separated particles in suspension. The most striking examples occur in acid kaolinitic soils: the surface soil, if shaken into suspension in water, forms a dispersed suspension, whereas a suspension of the subsoil aggregates in water, flocculates.

The input of energy required to break the organic bonds in soil aggregates depends on the numbers of bonds, as well as on the water content of the aggregates. For example, weakly bonded aggregates from the surface soil of a *red earth* (C content, 1.3%) dispersed severely in water after remoulding with a spatula at a water content equal to θ_{10} (Emerson and Smith 1970). On the other hand, aggregates from the surface soil of a *krasnozem* (C content, 4.6%) showed no dispersion in water after remoulding with a spatula at a water content equal to θ_{10}. Further, prolonged sonification in water only released about half the clay particles present (Emerson 1971). Subsoils of both these acid kaolinitic soils formed flocculated suspensions in water.

Field applications of inter-particle bonding

Soil aggregate classification

Soil aggregates have been classified according to their dispersion characteristics in water (Emerson 1967; Australian Standard 1980). Class 1 aggregates disperse apparently completely when immersed dry in water. By contrast, class 6 aggregates yield suspensions in water that flocculate rapidly; i.e. there is a net attractive force between the clay particles, irrespective of their distance apart. Class 4 aggregates are distinguished by the presence of a source of divalent ions sufficient to prevent dispersion of the sheared wet soil.

Within the framework of the classification, given schematically in Fig. 31.13, additional sub-divisions can be made: for example, more degrees of dispersion of aggregates may be distinguished. In tests on the dispersion of dry aggregates in water, Loveday and Pyle (1973) use four degrees of dispersion, rather than the three as in Fig. 31.13. Engineers, who refer to such tests as 'crumb' tests, also use four grades (e.g. Marshall and Workman 1977). Similarly Loveday and Pyle (1973) distinguish four degrees of dispersion of aggregates remoulded at θ_{10}, rather than two. Another possibility for dividing class 3 aggregates would be according to the relation of θ_D to θ_{10}. Such subdivision is justified if it reflects significant differences in the physical property being investigated in the field.

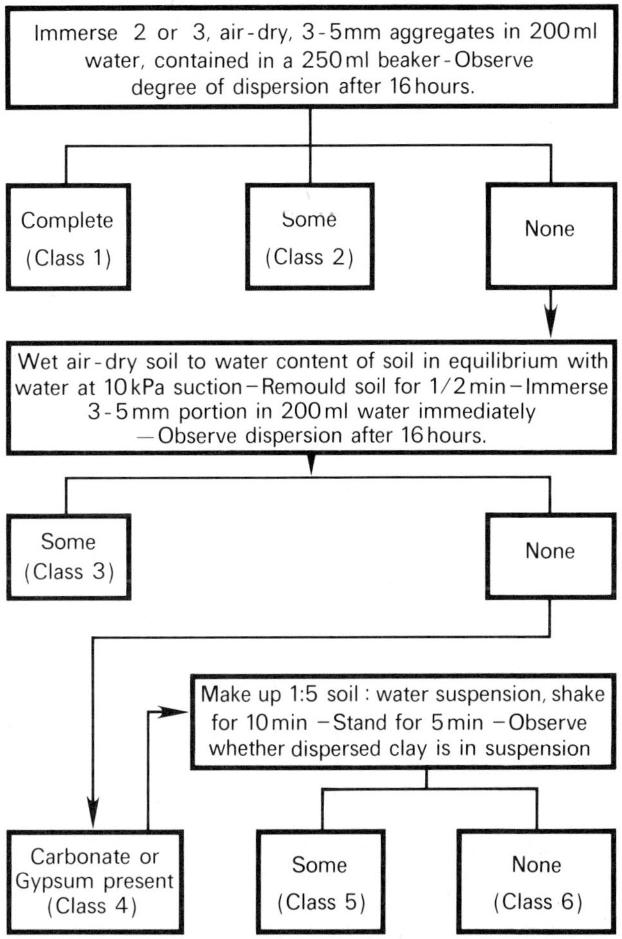

Fig. 31.13. A classification of soil aggregates based on their dispersion in water.

Classifying aggregates

The time taken for dry aggregates to disperse, after immersion in water, is dependent on the rate of diffusion of soluble salts out of the aggregates. This in turn depends on the sizes of the fragments the aggregates slake into, and the amount the fragments swell before clay particles are released. Therefore, the time will be longest for aggregates from sodic montmorillonitic soils, which have a coarse slaking pattern, and also swell greatly (Emerson 1964). Such aggregates occur in the acid subsoils of the Brigalow Belt in Queensland. These Class 1 aggregates have an ESP of 25–35, and contain more than 0.5% by weight of soluble salts (Stace *et al.* 1968, pp. 88–90); it can take up to 4 h before the *first* clay particles diffuse out of the swollen slaked fragments. Rallings (1966), when checking the dispersion of subsoils from this area, used an observation period of 1 h only. He suggested the use of 0.001 M NaOH as a reagent instead of water, because it speeds up the dispersion of the aggregates. In such a solution, the bonds between the clay particles are weakened in a way that will not ordinarily occur in the field. Some soils would therefore be classed erroneously as dispersive, as pointed out by Sherard *et al.* (1976*b*).

Ingles and Metcalf (1972) suggested that dispersion of dry aggregates should be assessed after 5–10 min. The assessment is based on the appearance of halos round the periphery of the slaked particles. This will work satisfactorily only for aggregates which slake, and which also disperse appreciably. Aggregates should not be oven dried; this may give an additional resistance to dispersion (Holmgren and Flanagan 1977), which is not appropriate for field conditions.

Sherard *et al.* (1976*b*) also adopted an observation period of 5–10 min, but used aggregates at their field water content. An objection to using wet aggregates is the possibility of mechanically shearing the outsides of the aggregates while handling the aggregates prior to immersion in water. This could lead to confusion between class 3 and class 2 aggregates, particularly with the short period of observation. The use of wet aggregates

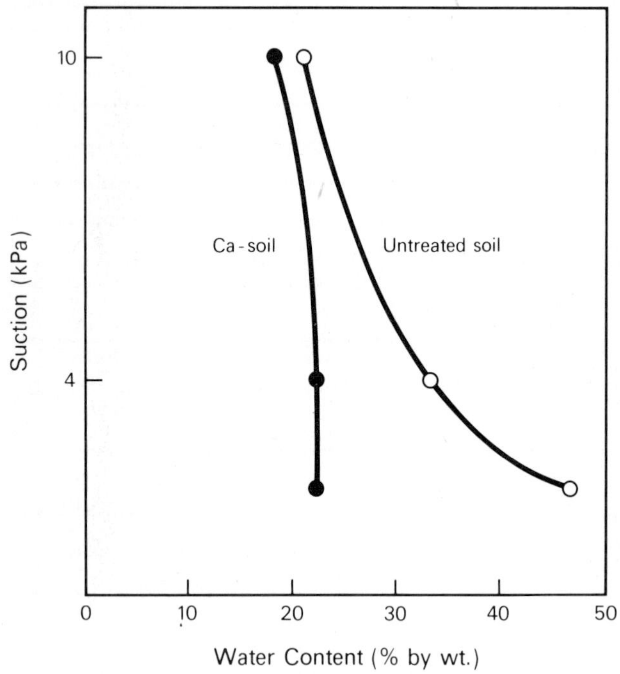

Fig. 31.14. The equilibrium water uptake of Ca-aggregates from Kondinin, W.A., and the water uptake, after 144 h, of untreated 4.0–5.6 mm aggregates.

would be appropriate for normally consolidated sediments, e.g. for estuarine muds that are not dried before incorporation in an earth structure.

In distinguishing class 3 aggregates, the present method of remoulding, although it prescribes the time of remoulding, does not control the force applied to the soil. However, the method has the virtue of simplicity. It may be more convenient to bring the soil to a water content corresponding to its plastic limit, rather than to the 10 kPa suction value (Fig. 31.13); the former would usually be a slightly drier condition (Emerson 1978).

Properties of class 1 aggregates

In situ, class 1 aggregates contain a flocculating concentration of soluble salts. The result is that before the soluble salts are removed, some properties of the aggregates may differ little from those of aggregates converted to the Ca form; the water uptake (swelling) of the natural aggregates from a solution equivalent to the saturated extract is the same as that of Ca aggregates from water (e.g. both 22%, at 2 kPa suction, for the soil used in Fig. 31.14).

The liquid limit used by engineers to classify soils is a poor index of a soil's potential dispersibility. The water content at the liquid limit may not change appreciably even when the soluble salts are removed, depending on the clay minerals present (White 1958). As an example, results obtained using class 1 aggregates taken at Kondinin in W.A. from a dam wall (Cole and Lewis 1960) will be given. The soil contains 27% clay and 0.35% soluble salts; it has an ESP of 22, and a plastic limit of 15.5. The value of the liquid limit of the soil washed with $CaCl_2$ is slightly greater than that of the natural soil aggregates: 44 and 39; the latter value was also obtained for soil from which the salts had been removed by washing with EtOH. In contrast, the water uptake of natural aggregates at a small suction (Emerson and Bakker 1973) is much increased compared with that of aggregates washed with $CaCl_2$ (Fig. 31.14). The actual magnitude of the increase at a given suction depends on the rate of loss of the soluble salts by diffusion, and hence both on the size of the aggregates and the elapsed time.

The particular soil from Western Australia was also chosen for illustration because the clay fraction consists predominantly (> 80%) of kaolin. Aggregates can be in class 1 without montmorillonite being the dominant clay mineral present, as suggested by Ingles and Metcalf (1972). It is only necessary that there should be sufficient dispersible particles, on a surface area basis (cf. earlier discussion of the dispersion of kaolinite-illite mixtures). In this soil, there is a small amount of randomly interstratified fine clay.

The dispersion of class 1 aggregates will be prevented if the water in which they are immersed contains sufficient salts. Here, the concentration of divalent cations in the water is the critical factor. Quirk and Schofield (1955) and Rowell *et al.* (1969) washed subsoil aggregates with NaCl–$CaCl_2$ solutions of decreasing concentration, while maintaining the ESP of the aggregates at a constant value within the range 0–54. From their results it may be deduced, that the concentration of $CaCl_2$ required to prevent dispersion over this ESP range is 0.25–0.4 mM.

Piping failures in dams

In 1937, Volk reported that the failure of the walls of some small dams and dikes in the semi-arid areas of New Mexico and Arizona was associated with soils which, after gentle shaking in water, dispersed severely. He considered that failure occurred through arches formed above the wetted soil due to slumping of the wetted soil on first filling. The general significance of this work was not realized until after the detailed study in Australia on the causes of piping failures in dams. Piping commonly occurs along a temporary phreatic surface established through a dam wall during first filling. Such dams have usually been built too dry to avoid appreciable settlement on wetting (Emerson 1959*b*; Aitchison *et al.* 1963; Ritchie 1963; Rallings 1966). Other factors emphasized were the ESP of the soil in contrast to the low salt content of the water, and rapid filling of the dams. The difficulty, in the field, of adding water uniformly to such soils to bring them to the optimum water content for compaction was also pointed out (Cole and Lewis 1960).

Soil samples from over 40 breached dams filled by run-off in Australia have been examined. All contained some aggregates in class 1. Such aggregates came either from the B-horizons of sodic duplex soils (Northcote and Skene 1972), or from sedimentary layers underlying the soil profile. For example, the Blanchetown Clay (Firman 1965) that covers most of the Murray Valley has an ESP of about 55 and contains 0.8% by weight of NaCl. If a sample from the borrow area of a proposed dam disperses appreciably, (level 3 in the scale 0–4 used in Fig. 13.7*a*), then the presence of class 1 aggregates in the area should be suspected.

Technically, dispersion should be checked in the actual water used to fill the dam. In practice with run-off water, the concentration of Ca^{2+} plus Mg^{2+} will be < 0.2 mM, insufficient to prevent dispersion of class 1 aggregates. Grant *et al.* (1977), to prevent possible piping of a dam, placed gypsum in the inlet, so that the Ca^{2+} concentration in the water was increased to 1 mM. Methods of dam construction using class 1 aggregates are discussed in Edwards (1964).

Dispersion and erosion

Sherard *et al.* (1976*a*) introduced a new test for detecting dispersive soils in relation to piping failures and erosion. Soil at a water content near the plastic limit is compacted into a core, 3.8 cm diam., 3.8 cm long. A 1-mm diam. hole is punched through the centre of the core. Distilled water is then allowed to flow through, at a pressure of 0.5 kPa initially, then increased to 10 kPa in stages. The degree of dispersion of the soil is deduced both from the increase in flow rate and from the turbidity of the effluent.

Using this test, class 1 and some class 2 aggregates, corresponding to grades 4 and 3 in the crumb test, will be definitely classed as dispersive (Sherard *et al.* 1976*b*). Some class 3 aggregates will also be included; it depends on how much the strength of the aggregates increases in the curing period, i.e. during a time interval between the shearing, due to compaction and insertion of the pinhole, and subsequent testing. The curing period can be up to one month (Sherard *et al.* 1976*b*; Riley 1977); the thixotropic increase can be sufficient to inhibit dispersion (Emerson 1978).

The classical index of a soil's susceptibility to erosion is the 'dispersion ratio' of Middleton (1930). The ratio, modified by Volk (1937) and termed the 'dispersion percentage', has since been standardized (Sherard *et al.* 1972). Soil at its field water content is mechanically shaken end-over-end for 1 min., the amount of $< 5\mu m$ particles released is expressed as a percentage of the total amount of $< 5\mu m$ particles present. Ritchie (1963) proposed a similar method, but using dry soil and shaking end-over-end for 10 minutes; he then expressed his results as a 'dispersal index', the inverse of the 'dispersion percentage'.

It has been found independently that soils with a dispersion percentage of > 30, (Sherard *et al.* 1972) or a dispersal ratio of < 3 (Rosewell 1977), are potentially erosive in the field. The disadvantage of this general method is that it does not distinguish dispersion caused simply by a reduction in electrolyte concentration and that for which an energy input is required.

The trigger for erosion is often severe crusting. The resultant surface flow of water can erode the surface soil, then cut into the subsoil and initiate gullies. Gulleying will be most severe if the subsoil contains class 1 aggregates, since these will disperse in the low-salt run-off water, e.g. the so-called 'sugary' soils (Dy3.4) of Downes (1946). Soils with surface aggregates in classes 1 and 2 will crust under rainfall, or during irrigation by water of low salt content, due to dispersion of the clay present. Surface soils consisting of class 3 aggregates will disperse, and hence develop a severe crust, if there is an energy input when the soil is wet; this input will usually come through cultivation or trampling by stock. However, if θ_D approaches the air-dry water content of a soil, then it is likely that the mechanical action of raindrops alone will cause the soil to disperse. This may be the case with the silty loess soils of Mississippi (plasticity index $\sim$ 5); here shallow vertical tunnels are formed along the tops of the banks of dams, apparently due to dispersion of the soil in the rainwater (Perry 1977).

Aggregate dispersion and plant growth

A large area of sodic clay soil irrigated from the Murray and Murrumbidgee rivers contains aggregates in classes 1 and 2. It has been shown, starting with Barley and Hutton (1956) and Davidson and Quirk (1961), that where flood irrigation is used, gypsum has to be added either to the soil or to the water to prevent severe surface crusting, and to ensure good germination of seeds such as clover. With furrow irrigation, most of the surface soil is wetted from below, under suction; surface crusting is not so important, water entry is determined by the crack patterns; deep tillage can be used as an adjunct to gypsum to increase further water entry and hence the yield of a crop such as cotton (Muirhead *et al.* 1970).

Where wheat is grown under dryland conditions on *red-brown earths* that have hard-setting A and bleached A_2 horizons (Dr2.33, Northcote 1971), gypsum has been used to reduce surface crusting, and this leads to improved germination and presumably increased rainfall acceptance. Results showing the increased yields of wheat in the Wimmera district of Victoria have been summarized by Sims and Rooney (1965). The ESP of a typical surface soil from this area increases from 3 in the surface soil to 40 at depth (Stace *et al.* 1968, p. 218). The aggregates of the surface soil are in class 3, but a concentration of 1 mM $CaCl_2$ will prevent dispersion (Emerson 1977). Rudd (1974) reported a 14% increase in the yield of wheat on a similar sodic *red-brown earth* at Mallala in S.A., following the application of gypsum.

If class 3 soil aggregates are brought to a dispersed or puddled state by cultivating wet, then very dense lumps may be formed on drying, because there are no continuous air-filled voids to initiate cracking. The surface soil of a neutral *red earth* from Katherine, in the Northern Territory, consists of class 3 aggregates; clods resulting from remoulding the soil aggregates at θ_{10}, and then drying, had a bulk density of 2.0 g cm^{-3} (Emerson 1977). Such clods are water stable because of their slow rate of wetting. Arndt (1963) formed a gravel-like mulch by cultivating this soil when wet, and was thus able to prevent the severe crusting that leads to poor germination of cotton (Arndt 1965).

If subsoils consist of class 1 aggregates, waterlogging can be expected in wet winters. Around Shepparton, in Victoria, this has led to the death of peach trees. The injection of gypsum into these subsoils has been one of the factors responsible for a four-fold increase in the yield of peaches (Cockroft and Tisdall 1978).

In contrast, soils consisting of class 4 aggregates containing clay-sized carbonate can be manipulated at water contents up to field capacity without subsequent dispersion. Such soils occur around Narrabri, in N.S.W., where extensive land-forming is carried out successfully for the production of irrigated cotton. Similarly, class 4 subsoils can be used to form unlined mole drainage channels, as on the Gault clay in England (Emerson 1967).

Aggregate dispersion and playing turf

The surface soil aggregates of turf should preferably not disperse under wet conditions. If the aggregates are in class 3, the addition of gypsum will help prevent disaggregation of the clay. Thus water entry and turf growth will be maintained (Harris 1978).

However, the soils preferred for cricket pitches are black clay subsoils with appreciable Na_{ex} (Piper 1932); aggregates of such soils are in class 2, and also disperse completely after remoulding at the plastic limit (W. W. Emerson, unpublished). The first rolling of the cricket pitch is given when the clay is wetter than the plastic limit (Harris 1961). This will eliminate the airfilled pore space, and also tend to align the clay particles horizontally. As the clay dries out towards the plastic limit, rolling is repeated with progressively heavier rollers. The final drying-out gives a hard pavement with a few widely spaced vertical cracks (Harris 1961).

Aggregate dispersion and farm dams

The aggregate dispersion classification was developed originally to predict the suitability of soils for dams (Emerson 1959*b*). The information given in Table 31.1 is a summary of the results of a combination of laboratory measurements on samples of the subsoils used for building dams, and of field experience on the performance of the dams (Emerson 1978). The hydraulic conductivity given in the second column is an average of values determined in the laboratory for a number of samples by using a bed of aggregates and a standard compaction treatment; 1 mM $CaCl_2$ is used as the measuring fluid in order to avoid any reduction in conductivity, due to dispersion of the clay. For run-off dams, the smaller the class numbers of the aggregates, the easier seepage is to control. If a dam is filled from a bore or an ephemeral creek, the concentration of salts present will usually be sufficient to prevent dispersion of aggregates in classes 1–3. This means that compaction as wet as possible is required during construction; subsequently it is important to try and maintain some water in the dam to avoid irreversible shrinkage cracks.

The seepage loss through compacted subsoils consisting of class 5 or class 6 aggregates will normally be too great for satisfactory dams. These soils include subsoils of *red* and *yellow podzolic soils* (class 5 aggregates), and *krasnozems* and pallid zones under laterite (class 6). Aggregates in these classes can be transformed into class 2 aggregates by the addition of a dispersant, such as sodium tripolyphosphate, and satisfactory dams constructed in this way (Wickham 1978). Pepper (1978) has also cured the seepage problem in a dam constructed partly of class 6 aggregates by adding sodium tripolyphosphate to the water.

A problem with run-off dams in low rainfall areas is to ensure that the dams are filled every year. In the wheat-belt of Western Australia, e.g. around Tambellup and Corrigen, roaded catchments are used successfully to enhance run-off. The subsoils in this area are similar to that discussed at Kondinin (Fig. 31.14), in that, although the clay fraction is dominantly kaolin ($> 80\%$), many of the aggregates are in class 1. Not only will such aggregates form a surface seal, but with the low surface area of the clay (about 45 m^2 g^{-1}), cracking due to shrinkage will be minimal.

Conclusions

Dry soil aggregates that contain Na_{ex}, when immersed in water, may disperse as the soluble salts diffuse out. If the exchange sites are occupied by divalent ions only, then wet soil aggregates may disperse in water, after being worked mechanically. The second process has been characterized by a water content for dispersion; it is defined as the maximum water content at which soil can be remoulded uniformly, and then immediately immersed in water, without there being any clay particles subsequently diffusing into the water.

Simple tests based on these two types of dispersion have been used to investigate bonding in soil aggregates. The bonds between clay particles are weaker when Mg^{2+} is the dominant exchangeable cation rather than Ca^{2+}: for example, the water content for dispersion

Table 31.1. The hydraulic conductivity of beds of compacted aggregates, and seepage losses from dams, in relation to the dispersion class number of the soil aggregates

Soil wetted at θ_{10}, remoulded, loaded uniaxially at 340 kPa for a short time; measuring fluid, 1 mM $CaCl_2$. Excessive seepage regarded as a loss of 3 mm per day under an average head of 5 m, i.e. a compacted layer, 150 mm thick with a conductivity of 1.10^{-9} m s^{-1}.

Class No. subsoil aggregates	Laboratory hydraulic conductivity (m s^{-1})	Field requirements to avoid excessive seepage: dam filled by; Run-off	From creek etc.
1	$< 1.10^{-10}$	Water content at compaction not critical. Note danger of piping if conductivity of wall $> 1.10^{-8}$	Compaction near θ_{10} required.
2	$< 1.10^{-9}$	Water content at compaction not critical	Recompaction necessary if soil dries out
3	$< 1.10^{-9}$	Water content must be near θ_{10}, unless stock dam	
4	1.10^{-9}	Water content must be near θ_{10}. Good compaction essential, sheepsfoot roller or prolonged puddling with stock. Any *seams* of gypsum have to be covered.	
5	1.10^{-7}	Conductivity can be reduced to 1.10^{-8} using very heavy compaction at θ_{10}.	
6	1.10^{-8}	Conductivity can be reduced only temporarily to 1.10^{-8} even if soil compacted wetter than θ_{10}.	

of a Mg–soil is smaller than that of the corresponding Ca-soil. Starting with Mg- or Ca-aggregates, their respective water contents for dispersion decrease continuously as Na_{ex} is introduced. The curves of water content for dispersion against Na_{ex} extrapolate to the ESP values required to cause dispersion of the dry aggregates in water.

Organic matter has been found to have a dual role. It can bond the clay surface particles together in soil aggregates so strongly that unworked aggregates do not disperse in water even if most of the exchangeable cations are first replaced by Na^+. On the other hand, some surface soil aggregates may disperse completely in water after being sheared wet, whereas the corresponding subsoil aggregates do not. This is particularly important for kaolinitic soils with a relatively small organic matter content.

Al_{ex} increases the resistance of aggregates to dispersion. On the other hand, there is no clear evidence that iron oxides strengthen the bonds between clay particles in soil aggregates. The divalent ion concentration derived from any carbonate present can be important in preventing the dispersion of wet sheared aggregates.

Aggregates have been formally divided into six classes based on dispersion of their clay fraction in water either due to the induced osmotic stress only, or in combination with prior mechanical stress. Class 1 aggregates, for example, disperse completely, starting from dry, whereas class 6 aggregates yield flocculated suspensions in water. Using the classification, it can be predicted whether a subsoil will form a satisfactory dam, and in particular, whether piping is a potential hazard. It can also be used to predict whether a soil can safely be manipulated wet, e.g. for land forming or drainage, without the risk of producing a dispersed structure. Testing aggregates directly for dispersion is much easier than trying to predict possible dispersion from other soil properties.

In the two other current tests of aggregate dispersion in water, the classical end-over-end shaking method and the recent pin-hole test, dispersion which only occurs after an energy input is not specifically distinguished from that which occurs simply due to the reduction in the external salt concentration.

For most agricultural crops, an extremely strong soil structure is not required. It is necessary only to prevent aggregates becoming dispersed or swelling excessively at small suctions so that severe crusting and poor drainage can be avoided. This can usually be achieved by the addition of gypsum or $CaCO_3$, and by not cultivating or grazing a soil when it is too wet. Conversely, when selecting soils for water retention structures, those soils whose aggregates are very difficult to disperse in water should be avoided.

References

Abedi, M. J. and Talibudeen, O. (1974). The calcareous soils of Azerbaijan. I. Catena development related to the distribution and surface properties of soil carbonate. *J. Soil Sci.* **25**, 357-72.

Adams, G. E. and Israelachvili, J. N. (1978). Measurement of forces between two mica surfaces in aqueous potassium nitrate solutions. *In* 'Modification of Soil Structure.' (Eds W. W. Emerson, R. D. Bond, and A. R. Dexter.) pp. 27-33. (John Wiley: London.)

Aitchison, G. D., Ingles, O. G. and Wood, C. C. (1963). Post-construction deflocculation as a contributory factor in failure of earth dams. Proc. 4th Aust.-N.Z. Conf. Soil Mech. Found. Eng. pp. 275-81.

Arndt, W. (1963). Investigations of some physical problems of Katherine soils. CSIRO Aust. Div. Land Res. Reg. Surv. Tech. Memo No. 64/3.

Arndt, W. (1965). The impedance of soil seals and the forces of emerging seedlings. *Aust. J. Soil Res.* **3**, 55-68.

Australian Standard (1980). Determination of Emerson class number of a soil. Standards Association of Australia. No. 1289. C8.1.

Aylmore, L. A. G. and Quirk, J. P. (1962). The structural status of clay systems. 9th. Nat. Conf. Clays and Clay Mineral. pp. 104-30.

Aylmore, L. A. G. and Quirk, J. P. (1966). Adsorption of water and electrolyte solutions by kaolin clay systems. *Soil Sci.* **102**, 339-45.

Aylmore, L. A. G. and Sills, I. D. (1978). Pore structure and mechanical strength of soils in relation to their constitution. *In* 'Modification of Soil Structure.' (Eds W. W. Emerson, R. D. Bond and A. R. Dexter.) pp. 53-8. (John Wiley: London.)

Bakker, A. C., Emerson, W. W. and Oades, J. M. (1973). The comparative effects of exchangeable calcium, magnesium, and sodium on some physical properties of red-brown subsoils. I. Exchange reactions and water contents for dispersion of Shepparton soil. *Aust. J. Soil Res.* **11**, 143-50.

Barley, K. P. and Hutton, J. T. (1956). Effects of a lowland rice crop and of gypsum on a saline alkali clay. *Aust. J. Agri. Res.* **7**, 110-26.

Barshad, I. (1950). The effect of interlayer cations on the expansion of the mica type of crystal lattice. *Am. Mineral.* **35**, 225-38.

Beattie, J. A. (1970). Peculiar features of soil development in parna deposits in the eastern Riverina, N.S.W. *Aust. J. Soil Res.* **8**, 145-56.

Bernal, J. D. and Fowler, R. H. (1933). A theory of water and ionic solution, with particular reference to hydrogen and hydroxyl ions. *J. Chem. Phys.* **1**, 515-48.

Black, A. S. and Waring, S. A. (1979). Adsorption of nitrate, chloride and sulfate by some highly weathered soils from south-east Queensland. *Aust. J. Soil Res.* **17**, 27-82.

Blackmore, A. V. and Miller, R. D. (1961). Tactoid size and osmotic swelling in calcium montmorillonite. *Soil Sci. Soc. Am. Proc.* **25**, 169-73.

Bolland, M. D. A., Posner, A. M. and Quirk, J. P. (1976). Surface charge on kaolinites in aqueous suspension. *Aust. J. Soil Res.* **14**, 197-216.

Bond, R. D. and Harris, J. R. (1964). The influence of the microflora on the physical properties of soils. I. Effects associated with filamentous algae and fungi. *Aust. J. Soil Res.* **2**, 111-22.

Butler, B. E. (1955). A system for the description of soil structure and consistence in the field. *J. Aust. Inst. Agric. Sci.* **21**, 239-49.

Callaghan, I. C. and Ottewill, R. H. (1974). Interparticle forces in montmorillonite gels. *Faraday Disc. Chem. Soc.* **57**, 110-8.

Cashen, G. H. (1959). Electric charges of kaolin. *Trans. Faraday Soc.* **55**, 477-86.

Cashen, G. H. (1961). Electric charges of clays. *Chem. Ind.* **43**, 1732-7.

Cashen, G. H. (1966). Electric charges of clays. *J. Soil Sci.* **17**, 303-16.

Chi, C. L., Emerson, W. W. and Lewis, D. G. (1977). Exchangeable calcium, magnesium and sodium and the dispersion of illites in water. I. Characterization of illites and exchange reactions. *Aust. J. Soil Res.* **28**, 243-53.

Clapp, C. E. and Emerson, W. W. (1965). The effect of periodate oxidation on the strength of soil crumbs. *Soil Sci.Soc. Am. Proc.* **29**, 127-34.

Clarke, G.B. and Marshall, T. J. (1947). The influence of cultivation on soil structure and its assessment in soils of variable mechanical composition. *J. Counc. Sci. Ind. Res. Aust.* **20**, 165-75.

Cockroft, B. and Tisdall, J. M. (1978). Soil management, soil structure and root activity. *In* 'Modification of Soil Structure.' (Eds W. W. Emerson, R. D. Bond and A. R. Dexter.) pp. 387-92. (John Wiley: London.)

Cole, D. C. H. and Lewis, J. G. (1960). Piping failures of earthen dams built of plastic materials in arid climates. Proc. 3rd. Aust.-N.Z. Conf. on Soil Mech. and Found. Eng. pp. 93-9.

Davidson, J. L. and Quirk, J. P. (1961). The influence of dissolved gypsum on pasture establishment on irrigated sodic clays. *Aust. J. Agric. Res.* **12**, 100-10.

Dimmock, G. M., Bettenay, E. and Mulcahy, M. J. (1974). Salt content of lateritic profiles in the Darling Range, Western Australia. *Aust. J. Soil Res.* **12**, 63-70.

Downes, R. G. (1946). Tunnelling erosion in North-Eastern Victoria. *J. Coun. Sci. Ind. Res. Aust.* **19**, 283-310.

Downes, R. G. (1949). Soil, land-use, and erosion survey around Dookie, Victoria. CSIR Aust. Bull. No. 243.

Edwards, D. G., Posner, A. M. and Quirk, J. P. (1965). Repulsion of chloride ions by negatively charged clay surfaces. Part 2 — Monovalent cation montmorillonites. *Trans. Faraday Soc.* **61**, 2816-24.

Edwards, J. A. (1964). Storing water in farm dams. Dept. Agric. S. Aust. Bull. No. 467.

El-Swaify, S. A. and Emerson, W. W. (1975). Changes in the physical properties of soil clays due to precipitated aluminium and iron hydroxides I. Swelling and aggregate stability after drying. *Soil Sci. Soc. Am. Proc.* **39**, 1056-63.

Emerson, W. W. (1954). The determination of the stability of soil crumbs. *J. Soil Sci.* **5**, 235-50.

Emerson, W. W. (1956). A comparison between the mode of action of organic matter and synthetic polymers in stabilizing soil crumbs. *J. Agric. Sci.* **47**, 117-21.

Emerson, W. W. (1959*a*). The structure of soil crumbs. *J. Soil Sci.* **10**, 235-44.

Emerson, W. W. (1959*b*). The sealing of earth dams. CSIRO Aust. Div. Soils Tech. Memo No. 4/59.

Emerson, W. W. (1962*a*). The swelling of Ca-montmorillonite due to water absorption. I. Water uptake in the vapour phase. *J. Soil Sci.* **13**, 31-9.

Emerson, W. W. (1962*b*). 2. Water uptake in the liquid phase. *J. Soil Sci.* **13**, 40-5.

Emerson, W. W. (1963). The swelling of Na-montmorillonite due to water absorption. *Aust. J. Soil Res.* **1**, 128-43.

Emerson, W. W. (1964). The slaking of soil crumbs as influenced by clay mineral composition. *Aust. J. Soil Res.* **2**, 211-17.

Emerson, W. W. (1967). A classification of soil aggregates based on their coherence in water. *Aust. J. Soil Res.* **5**, 47-57.

Emerson, W. W. (1968). The dispersion of clay from soil aggregates. Trans. 9th Int. Congr. Soil Sci. Vol. 1, pp. 617-26.

Emerson, W. W. (1971). Determination of the contents of clay-sized particles in soils. *J. Soil Sci.* **22**, 50-9.

Emerson, W. W. (1972). Soil tilth and aggregation. Symp. on Swelling Clay Soils, Univ. of New England, pp. 39-44.

Emerson, W. W. (1977). Physical properties and structure. *In* 'Soil Factors in Crop Production in a Semi-arid Environment.' (Eds J. S. Russell and E. L. Greacen). pp. 78-104. (Qld Univ. Press: Brisbane.)

Emerson, W. W. (1978). Aggregate classification and the hydraulic conductivity of compacted subsoils. *In* 'Modification of Soil Structure.' (Eds W. W. Emerson, R. D. Bond and A. R. Dexter.) pp. 239-248. (John Wiley: London.)

Emerson, W. W. and Bakker, A. C. (1973). The comparative effects of exchangeable calcium, magnesium and sodium on some physical properties of aggregates from subsoils of red-brown earths. II. The spontaneous dispersion of aggregates. *Aust. J. Soil Res.* **11**, 151-57.

Emerson, W. W. and Chi, C. L. (1977). Exchangeable calcium, magnesium and sodium and the dispersion of illites in water. II. Dispersion of illites in water. *Aust. J. Soil Res.* **15**, 255-62.

Emerson, W. W. and Smith, B. H. (1970). Magnesium, organic matter and structure. *Nature* (London) **228**, 453-54.

Firman, J. B. (1965). Late Cainozoic lacustrine deposits in the Murray Basin, South Australia. Adelaide Geological Survey of South Australia, Quarterly Geol. Notes, 16.

Flach, K. W., Nettleton, W. D., Gile, L. H. and Cady, J. G. (1969). Pedocementation: Induration by silica, carbonates and sesquioxides in the Quarternary. *Soil Sci.* **107**, 442-53.

Follett, E. A. C. (1965). The retention of amorphous, colloidal 'ferric

hydroxide' by kaolinites. *J. Soil Sci.* **16**, 334-41.
Follett, E. A. C., McHardy, W. J., Mitchell, B. D. and Smith, B. F. L. (1965). Chemical dissolution techniques in the study of soil clays. Part I. *Clay Mineral.* **6**, 35-43.
Fordham, A. W. and Norrish, K. (1979). Electron microprobe and electron microscope studies of soil clay particles. *Aust. J. Soil Res.* **17**, 283-306.
Foster, W. R., Savins, J. G. and Waite, J. M. (1955). Lattice expansion and rheological behaviour relationships in water–montmorillonite systems. 3rd. Nat. Conf. Clays Clay Mineral. pp. 296-316.
Gallavan, R. C. and Greene-Kelly, R. (1973). Structural studies on clay soils. Rep. Rothamsted Exp. Station, 1972, Pt 1, pp. 72-3.
Gile, L. H., Peterson, F. F. and Grossman, R. B. (1966). Morphological and genetic sequences of carbonate accumulation in desert soils. *Soil Sci.* **101**, 347-60.
Giovannini, G. and Sequi, P. (1976). Iron and aluminium as cementing substances of soil aggregates. 1. Acetylacetone in benzene as an extractant of fractions of soil iron and aluminium. *J. Soil Sci.* **27**, 148-53.
Grant, K., Ingles, O. G., Lawrence, I. A. and Sommerville, P. J. (1977). Protecting a dispersive soil by water treatment. *In* 'Dispersive Clays, Related Piping and Erosion in Geotechnical Projects.' (Eds J. L. Sherard and R. S. Decker.) ASTM Spec. Tech. Publ. No. 623, pp. 191-201.
Greene-Kelly, R. (1974). Shrinkage of clay soils: A statistical correlation with other soil properties. *Geoderma* **11**, 243-57.
Greenland, D. J. (1972). Interactions between organic polymers and inorganic soil particles. *Meded. Fac. Landbouwwet. Rijksuniv. Gent* **37**, 897-914.
Greenland, D. J., Lindstrom, G. R. and Quirk, J. P. (1962). Organic materials which stabilize natural soil aggregates. *Soil Sci. Soc. Am. Proc.* **26**, 366-71.
Greenland, D. J., Oades, J. M. and Sherwin, T. W. (1968). Electron-microscope observations of iron oxides in some red soils. *J. Soil Sci.* **19**, 123-26.
Hamblin, A. P. and Greenland, D. J. (1977). Effect of organic constituents and complexed metal ions on aggregate stability of some East Anglian soils. *J. Soil Sci.* **28**, 410-17.
Harris, J. R. (1961). The crumbling of cricket pitches. *Aust. Sci.* **1**, 173-78.
Harris, J. R. (1978). Maintenance of soil structure under playing turf. *In* 'Modification of Soil Structure.' (Eds W. W. Emerson, R. D. Bond and A. R. Dexter.) pp. 289-303. (John Wiley: London.)
Hingston, F. J., Posner, A. M. and Quirk, J. P. (1972). Anion adsorption by goethite and gibbsite. 1. The role of the proton in determining adsorption envelopes. *J. Soil Sci.* **23**, 177-92.
Holmgren, G. G. S. and Flanagan, C. P. (1977). Factors affecting spontaneous dispersion of soil materials as evidenced by the crumb test. *In* 'Dispersive Clays, Related Piping and Erosion in Geotechnical Projects.' (Eds J. L. Sherard and R. S. Decker.) ASTM Spec. Tech. Publ. No. 623, pp. 218-239.
Ingles, O. G. and Metcalf, J. B. (1972). Soil Stabilization. (Butterworths: Australia.)
Jenny, H. and Reitemeier, R. F. (1935). Ionic exchange in relation to the stability of colloidal systems. *J. Phys. Chem.* **39**, 593-604.
Kemper, W. D. and Koch, E. J. (1966). Aggregate stability of soils from Western United States and Canada. U.S. Dept. Agric. Tech. Bull. No. 1355.
Loveday, J. (1974). Recognition of gypsum-responsive soils. *Aust. J. Soil Res.* **12**, 87-96.
Loveday, J. and Pyle, J. (1973). The Emerson dispersion test and its relationship to hydraulic conductivity. CSIRO Aust. Div. Soils Tech. Pap. No. 15.
Low, A. J. (1972). The effect of cultivation on the structure and other physical characteristics of grassland and arable soils (1945-1970). *J. Soil Sci.* **23**, 363-80.
MacEwan, D. M. C. (1954). Short-range electrical forces between charged colloidal particles. *Nature* (London) **174**, 39-40.
McIntyre, D. S. (1956). The effect of free ferric oxide on the structure of some Terra Rossa and Rendzina soils. *J. Soil Sci.* **7**, 302-6.
McIntyre, D. S. (1976). Subplasticity in Australian soils. I. Description, occurrence and some properties. *Aust. J. Soil Res.* **14**, 227-36.
Marshall, A. F. and Workman, J. P. (1977). Identification of dispersive clays in the Texas Gulf Coast Area. *In* 'Dispersive Clays, Related Piping and Erosion in Geotechnical Projects.' (Eds J. L. Sherard and R. S. Decker.) ASTM Spec. Publ. No. 623, pp. 274-86.
Mattson, S. (1929). The laws of soil colloidal behaviour. *Soil Sci.* **28**, 179-220.
Mehra, O. P. and Jackson, M. L. (1960). Iron oxide removal from soils and clays by a dithionite – citrate system buffered with sodium bicarbonate. 7th Nat. Conf. Clays Clay Mineral. pp. 317-27.
Middleton, H. E. (1930). Properties of soil which influence erosion. U.S. Dept. Agric. Tech. Bull. No. 178.
Muirhead, W. A., Loveday, J. and Saunt, J. E. (1970). Soil and cotton responses to tillage and ameliorant treatments in a brown clay soil. 2. Growth, yield and quality of cotton. *Aust. J. Exp. Agric. Anim. Husb.* **10**, 325-33.
Norrish, K. (1954). The swelling of montmorillonite. *Disc. Faraday Soc.*, **18**, 120-34.
Norrish, K. and Pickering, J. G. (1977). Clay mineralogic properties. *In* 'Soil Factors in Crop Production in a Semi-arid Environment.' (Eds J. S. Russell and E. L. Greacen.) pp. 33-53. (Qld Univ. Press: Brisbane.)
Norrish, K. and Quirk, J. P. (1954). Crystalline swelling of montmorillonite. *Nature* (London). **173**, 225-26.
Norrish, K. and Rausell-Colom, J. A. (1963). Low angle diffraction studies of the swelling of montmorillonite and vermiculite. 10th Nat. Conf. Clays Clay Mineral. pp. 123-49.
Norrish, K. and Tiller, K. G. (1976). Subplasticity in Australian soils. V. Factors involved and techniques of dispersion. *Aust. J. Soil Res.* **14**, 273-90.
Northcote, K. H. (1971). 'A Factual Key for the Recognition of Australian Soils.' (Rellim Tech. Publ.: Adelaide)
Northcote, K. H. and Skene, J. K. M. (1972). Australian soils with saline and sodic properties. CSIRO Aust. Soils Publ. No. 27.
Olson, R. E. and Mitronovas, F. (1960). Shear strength and consolidation characteristics of calcium and magnesium illite. *Proc. Am. Soc. Civil Eng.* **85**, SM4.
Pepper, R. G. (1978). Surface seals in earth dams. *In* 'Modification of Soil Structure.' (Eds W. W. Emerson, R. D. Bond and A. R. Dexter.) pp. 257-65. (John Wiley: London.)
Perry, J. P. (1977). Lime treatment of dams constructed with dispersive soils. Am. Soc. Agric. Eng. Tech. Paper 76-2093.
Piper, C. S. (1932). Some characteristics of soils used for turf wickets in Australia. *Trans. R. Soc. S. Aust.* **56**, 15-18.
Piper, C. S. (1950). 'Soil and Plant Analysis.' (Univ. of Adelaide: Adelaide.)
Posner, A. M. and Quirk, J. P. (1964). The adsorption of water from concentrated electrolyte solutions by montmorillonite and illite. *Proc. R. Soc.* **278**, 35-56.
Quirk , J. P. (1955). Significance of surface areas calculated from water vapor sorption isotherms by use of the B.E.T. equation. *Soil Sci.* **80**, 423-30.
Quirk, J. P. and Schofield, R. K. (1955). The effect of electrolyte concentrations on soil permeability. *J. Soil Sci.* **6**, 163-78.
Rallings, R. A. (1966). An investigation of the causes of failure of farm dams in the Brigalow Belt of Central Queensland. Water Res. Found. Aust. Bull. No. 10.
Rand, B. and Melton, I. R. (1977). Particle interactions in aqueous kaolinite suspensions. I. Effect of pH and electrolyte upon the mode of particle interaction in homionic sodium kaolinite suspensions. *J. Coll. Interface Sci.* **60**, 308-20.
Rengasamy, P. and Oades, J. M. (1978). Interaction of monomeric and polymeric species of metal ions with clay surfaces. III. Aluminium (iii) and chromium (iii). *Aust. J. Soil Res.* **16**, 53-66.

Rimmer, D. L. and Greenland, D. J. (1976). Effects of calcium carbonate on the swelling behaviour of a soil clay. *J. Soil Sci.* **27**, 129-39.

Riley, P. B. (1977). Dispersion in some New Zealand clays. *In* 'Dispersive Clays, Related Piping and Erosion in Geotechnical Projects.' (Eds J. L. Sherard, and R. S. Decker.) ASTM Spec. Publ. No. 623, pp. 355-59.

Ritchie, J. A. (1963). Earthwork tunnelling and the application of soil testing procedures. *J. Soil Conserv. Serv. N.S.W.* **19**, 111-29.

Robertson, R. H. S., Brindley, G. W. and Mackenzie, R. C. (1954). Mineralogy of kaolin clays from Pugu, Tanganyika. *Am. Mineral.* **39**, 118-38.

Rosewell, C. J. (1977). Identification of susceptible soils and control of tunnelling. *In* 'Dispersive Clays, Related Piping and Erosion in Geotechnical Projects.' (Eds J. L. Sherard and R. S. Decker.) ASTM Spec. Publ. No. 623, pp. 362-69.

Rowell, D. L., Payne, D. and Ahmad, N. (1969). The effect of the concentration and movement of solutions on the swelling, dispersion and movement of clay in saline and alkali soils. *J. Soil Sci.* **20**, 176-88.

Rudd, C. L. (1974). More about gypsum. *Agric. S. Aust.* **77**, 70-4.

Russell, E. W. (1934). The interaction of clay with water and organic liquids as measured by specific volume changes and its relation to the phenomena of crumb formation in soils. *Phil. Trans. R. Soc.* **A233**, 361-89.

Russell, E. W. (1971). Soil structure: Its maintenance and improvement. *J. Soil Sci.* **22**, 137-51.

Schahabi, S. and Schwertmann, U. (1970). Der Einfluss von synthetischen Eisenoxiden auf die Aggregation zweier Lussbodenhorizonte. *Z. Pflanzenernaehr. Bodenkd.* **125**, 193-204.

Schofield, R. K. and Samson, H. R. (1953). The deflocculation of kaolinite suspensions and the accompanying change-over from position to negative chloride adsorption. *Clay Mineral. Bull.* **2**, 45-51.

Schofield, R. K. and Samson, H. R. (1954). Flocculation of kaolinite due to the attraction of oppositely charged crystal faces. *Disc. Faraday Soc.* **18**, 135-45.

Shainberg, I. (1968). Electrochemical properties of Na^+ and Ca^{++} montmorillonite suspensions. Trans. 9th Int. Congr. Soil Sci. Vol. 1, pp. 578-86.

Sherard, J. L., Decker, R. S. and Ryker, N. L. (1972). Hydraulic fracturing in low dams of dispersive clay. Proc. Spec. Conf. on Performance of Earth and Earth-supported Structures, ASCE, pp. 589-626.

Sherard, J. L., Dunnigan, L. P., Decker, R. S. and Steele, E. F. (1976*a*). Pinhole test for identifying dispersive soils. *J. Geotech. Eng. Div. ASCE*, **102**, 69-85.

Sherard, J. L., Dunnigan, L. P. and Decker, R. S. (1976*b*). Identification and nature of dispersive soils. *J. Geotech. Eng. Div. ASCE* **102**, 287-301.

Sills, I. D. (1975). Internal surface area and porosity of kaolinitic glaebules. *Aust. J. Soil Res.* **13**, 235-38.

Sims, H. J. and Rooney, D. R. (1965). Gypsum for difficult clay wheat-growing soils. *J. Dept. Agric. Vic.* **63**, 401-9.

Skene, J. K. M. and Poutsma, T. J. (1962). Soils and land use in the Goulburn Valley, Victoria. Dept. Agric. Vic. Tech. Bull. No. 14.

Smalley, I. J. and Cabrera, J. G. (1969). Particle association in compacted kaolinite. *Nature* (London) **222**, 80-1.

Smith, B. H. and Emerson, W. W. (1976). Exchangeable aluminium on kaolinite. *Aust. J. Soil Res.* **14**, 43-53.

Sridhar, K., Jackson, M. L. and Syers, J. K. (1972). Cation and layer charge effects on blister-like osmotic swelling of micaceous vermiculite. *Am. Min.* **57**, 1832-48.

Stace, H. C. T. (1956). Chemical characteristics of Terra Rossas and Rendzinas of South Australia. *J. Soil Sci.* **7**, 280-93.

Stace, H. C. T., Hubble, G. D., Brewer, R., Northcote, K. H., Sleeman, J. R., Mulcahy, M. J. and Hallsworth, E. G. (1968). 'A Handbook of Australian Soils.' (Rellim Tech. Pub.: Adelaide)

Stefanson, R. C. (1971). Effect of periodate and pyrophosphate on the seasonal changes in aggregate stabilization. *Aust. J. Soil Res.* **9**, 33-41.

Tama, K. and El-Swaify, S. A. (1978). Charge, colloidal and structural stability interrelationships for oxidic soils. *In* 'Modification of Soil Structure.' (Eds W. W. Emerson, R. D. Bond and A. R. Dexter.) pp. 41-9. (John Wiley: London.)

Tisdall, J. M., Cockroft, B. and Uren, N. C. (1978). The stability of soil aggregates as affected by organic materials, microbial activity and physical disruption. *Aust. J. Soil Res.* **16**, 9-18.

Tucker, B. M. (1975). Primary and conditional basic exchangeable cations in Australian soil profiles. *Aust. J. Soil Res.* **13**, 91-100.

van Dijk, D. C. (1958). Soil distribution in the Griffith-Yenda District, N.S.W. CSIRO Aust. Soils Land Use Ser. No. 40.

Verwey, E. J. W. and Overbeek, J.Th.G. (1948). 'Theory of the Stability of Lyophobic Colloids.' (Elsevier: Amsterdam.)

Volk, G. M. (1937). Method of determination of the degree of dispersion of the clay fraction of soils. *Soil Sci. Soc. Am. Proc.* **2**, 561-65.

Walker, G. F. (1960). Macroscopic swelling of vermiculite crystals in water. *Nature (*London). **187,** 312.

Warkentin, B. P., Bolt, G. H. and Miller, R. D. (1957). Swelling pressure of montmorillonite. *Soil Sci. Soc. Am. Proc.* **21**, 495-97.

Wetherby, K. G. and Oades, J. M. (1975). Classification of carbonate layers in highland soils of the Northern Murray Mallee, S. A., and their use in stratigraphic and land-use studies. *Aust. J. Soil Res.* **13**, 119-32.

White, W. A. (1958). Water sorption properties of homoionic clay minerals. Illinois Geol. Survey Report 208.

Wickham, H. G. (1978). Sodium tripolyphosphate and seepage from dams. *In* 'Modification of Soil Structure.' (Eds W. W. Emerson, R. D. Bond and A. R. Dexter.) pp. 263-65. (John Wiley: London.)

Chapter 32
Physical properties and water relations

Soil mechanical properties and water movement E. L. Greacen

Soil hydrology John Williams

AUSTRALIA HAS THE DOUBTFUL DISTINCTION NOT ONLY of being the driest continent but also of having extensive areas of very old landscape and, as a consequence, many soils that are hydrologically complex. This situation has resulted in a population that is sensitive to water supply and management, and soil physicists who are inordinately concerned with soil-water relations. It is not surprising, then, that Australian scientists have always been numbered amongst the world leaders in soil-water relations: Green and Ampt (1911) on infiltration, Prescott (1934) on climate, Philip (1954) on theoretical analyses of water movement in soil, Marshall (1958) on unsaturated conductivity, Holmes (1956) on water measurement, and Aitchison (1960) on soil mechanical properties and water relations. Today water balance models developed in Australia for studies of plant growth and crop production rank among the best available.

The combination of age and semi-aridity in soils over the greater part of the continent has led to strongly developed texture contrasts and hydrologically complex profiles. Low rainfall has meant that large areas contain massive amounts of salt in the subsoil and in underground water systems that are shallow enough to be critical factors in land use.

Any land use practice, engineering, water resources, recreation or whatever, depends largely on water relationships and the soil water regime. The first section of this chapter deals with relationships between water content and the soil physical and mechanical properties of practical concern; some important processes and mechanisms of water movement and uptake by vegetation are then discussed. The next section deals with hydrological properties and soil morphology; information and ideas on the comparative hydrology of some important Australian soils are also reviewed.

Soil mechanical properties and water relations

The morphological properties of soils are discussed in relation to two broad aspects of land use: water supply in agriculture and soil strength in engineering. The first case is concerned with how soil properties alter the water regime and the second with how water content affects the physical and mechanical properties of soil in such applications as structure stability, erosion, tillage and root penetration. These processes all depend basically on the mechanical strength of the soil, and in a field assessment based on morphological features some simple method of measuring strength, e.g. consistence, almost always as it is affected by water content, is usually employed. In this section we develop an analytical basis for relationships between morphological criteria, particularly as they are affected by water content, and engineering concepts of soil strength.

Mechanical strength of soil

The theory of soil mechanics has been developed by engineers who are interested mainly in the maximum force that a soil can withstand without undergoing large deformation; this is important in building and dam construction, roadways and the bearing capacity and trafficability of soils, so that in effect the engineer is concerned with deformation before failure. On the other hand, the agriculturalist is interested to know the force required to create a loose tilth or seed bed, the resistance of the soil to erosion by wind or water, and its resistance to penetration by roots or implements. These largely involve stressing and deforming the soil to failure and frequently beyond failure. This deformation can be by tensile failure, for example, when a clod bursts under pressure, by shear failure without compression in front of a tyne, or by shear failure with compression under traffic or during root penetration.

Determination of strength parameters

In Fig. 32.1*a* the shear strength τ is given by the classical strength parameters c and ϕ according to Coulomb's equation

$$\tau = c + \sigma_n \tan \phi$$

where σ_n is the pressure acting on the plane of shear, c is the cohesion and ϕ is the angle of internal friction. The relationships between these parameters and measures of strength arising from the other types of deformation discussed above are demonstrated in a Mohr diagram in Fig. 32.1*b*.

A satisfactory characterization of soil strength can be obtained by determining the failure envelope on a

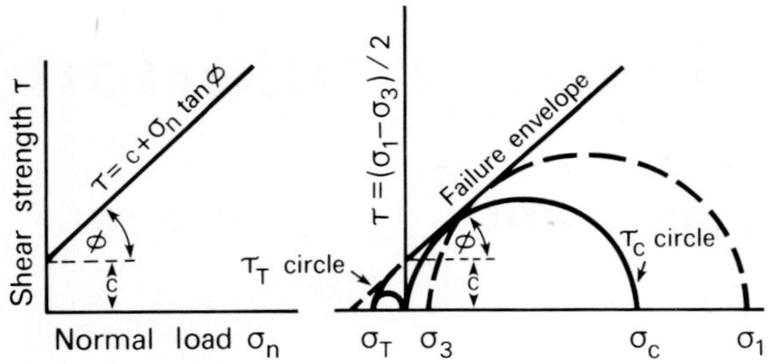

Fig. 32.1. (*a*) Shear strength of soil as measured in a direct shear box and described by the Coulomb equation. (*b*) Mohr diagram with the failure envelope described by c and ϕ; σ_1 and σ_3 are the principal stresses, which in a triaxial test are the axial and allround radial stress respectively. Mohr circles for the compressive strength σ_c and the tensile strength σ_T are also shown.

Mohr diagram with one of the recognized techniques. The torsion box (Payne and Fountaine 1952) or the direct shearbox (Terzaghi and Peck 1948) are often employed, the former being useful for small samples or peds. The most versatile method for soil cores is the triaxial test as discussed by Bishop and Henkel (1962). The diagram in Fig. 32.1*b* can be considered as a plot of triaxial test data where σ_1 is the axial pressure required to fail in shear a cylinder of soil subjected to radial stress σ_3; shear strength is given on the vertical axis.

Where the deformation involves local shear failure with compression (e.g. where failure occurs under foundations, tractor wheels, or around plant roots) analytical estimates of mechanical resistance require values for the parameters c and ϕ together with a measured compressibility curve. The compressibility characteristics can be expressed as Young's modulus and as the slopes of the failure and rebound curves for compression with shear. The parameters c and ϕ and the compressibility characteristics are important in deformation under tractor wheels and by penetrating roots. As Farrell and Greacen (1966) have shown, they can be measured with sufficient accuracy by means of the triaxial cell.

Mohr circles for the unconfined compressive strength σ_c and for tensile strength σ_T are also shown in Fig. 32.1*b*; it can be seen that both σ_c and σ_T depend on c and ϕ. Farrell *et al.* (1967) have shown that at soil water potentials as high as −30 kPa, i.e. almost as wet as field capacity, compact loams behave as brittle materials, for which $\sigma_c = 8\,\sigma_T$ (Griffith 1924). Where the sample is in the form of a core, either natural or remoulded, σ_T can be measured indirectly by means of the so-called Brazilian test (Kirkham *et al.* 1959) and σ_c can be measured by an unconfined loading test. Both tests are made in a compression test machine: in the first test the lateral load required to burst the core in tension is measured, and in the second, the axial load required to rupture the core in shear is measured.

The above methods measure the bulk strength of the soil, which is often limited by inter-aggregate strength. In contrast to this the strength of the individual aggregates might be more important in controlling processes of deformation where actual crushing or disruption of the soil peds is involved. Wet sieving is an attempt to measure aggregate stability, but a more rigorous, soundly based method is required.

Rogowski (1964) has developed a rupture test for natural aggregates. The crushing strength σ_z is given as

$$\sigma_z = \frac{6F}{\pi d^2} - \left(\frac{14+5\mu}{7+5\mu}\right) F \tag{2}$$

where F is the load, μ is the ratio of transverse to longitudinal strain (Poisson ratio) and d is the aggregate diameter. Tensile strength is given as $\sigma_T = 0.1727\sigma_z$. Rogowski also concluded for soils in an air-dry or moderately dry condition (−1500 kPa water potential) that brittle strength theory applied. He found that tensile strength varied inversely as the square root of the diameter of the aggregates, indicating Griffith cracks of the same dimension as the aggregate. However, Braunack and Dexter (1978) found that for *red brown earth* aggregates over the range of available water, σ_T was independent of aggregate diameter; they concluded that maximum crack size was independent of aggregate diameter.

The effect of water content on soil strength

The effect of water content on soil strength is most easily demonstrated in the case of a saturated soil during consolidation. For saturated soil, the void ratio e is equal to about two times the volumetric water content θ as $e = \theta/(1\text{-}\theta)$, and so can be used as a good index of water content. During consolidation for the first time, 'virgin' or 'normal' consolidation, (Fig. 32.2) the relation between e and the consolidation pressure σ is given by

$$e = e_o - I_c \ln(\sigma/\sigma_o) \tag{3}$$

where I_c is a constant for a particular soil called the compression index and e_o is the void ratio at an arbitrary pressure σ_o.

In soil that has been consolidated (curve ab) and then unloaded (curve bc), the void ratio depends on the maximum consolidation pressure σ_b, and the extent of swelling that occurs on unloading. The effect of a previous cycle of compression on the compressibility of such 'over-consolidated' soil is evident from the reloading curve (cd). The soil at c is in an 'over-consolidated' state, similar to the state of soil peds in the field. Thus, for any water content in saturated soil, ignoring hysteresis, there is an equivalent pressure on the virgin consolidation curve which is a useful index of equivalent strength, i.e. it is the consolidation pressure that the soil can withstand without further deformation (Greacen 1959, 1960*a*).

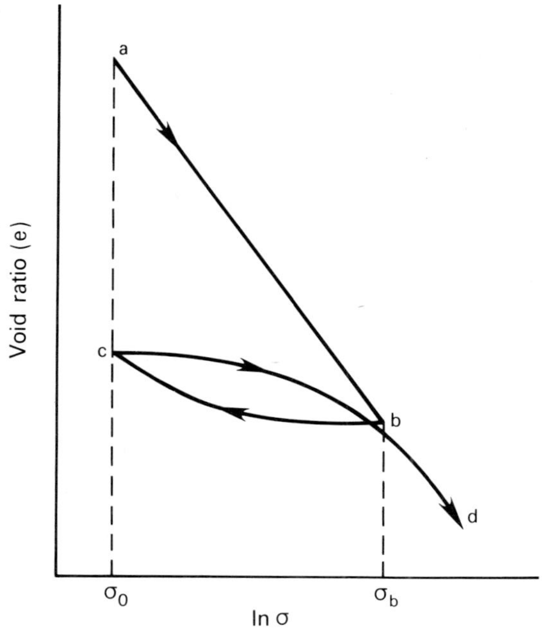

Fig. 32.2. Void ratio e as a function of the logarithm of consolidation pressure (ln σ) for a saturated soil.

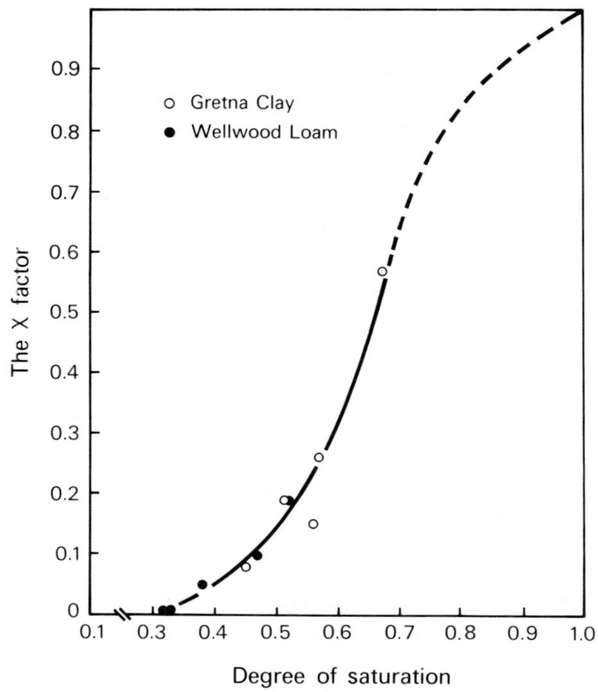

Fig. 32.3. The relationships between the χ factor and the degree of saturation for two unsaturated soils. After Williams and Shaykewich (1970).

The equivalent strength of an over-consolidated soil is expressed as the cohesion of the clay peds in a fully wet condition. Hvorslev (1937) attempted to express it when he modified Coulomb's equation given above as

$$\tau_f = c_o \exp(-e_f/I_c) + \sigma_f' \tan \phi_o, \qquad (4)$$

where τ_f is the shear strength at failure, e_f the void ratio at failure and σ_f' the effective normal pressure acting on the plane of failure, and c_o and ϕ_o are soil constants. The first term on the right of Eqn (4) is the expression for cohesion modified by the void ratio and the compressibility. The effect of water content on shear strength for this saturated condition is expressed through e_f.

Soil water suction and effective stress

In a saturated soil, an increase in soil water suction has the same effect on strength as an increase in an externally applied load (Childs 1955). Terzaghi (1923) showed experimentally for a saturated soil that the amount of one-directional consolidation depended on the effective pressure. The effective pressure is defined as $\sigma' = \sigma - u_w$, where σ is the applied pressure and u_w is the soil water pressure (suction being accepted as a negative pressure, i.e. below atmospheric datum). For this reason the second term on the right of Eqn (4) allows for effective pressure σ' rather than applied pressure σ and c_o and ϕ_o are defined as intrinsic soil properties.

In unsaturated soil, where the pores contain both air and water, the suction is regarded as acting over an effective area χ per unit area of soil. The effective pressure is then $\sigma' = \sigma - \chi u_w$.

At saturation, $\chi = 1$ and σ' is identified with Terzaghi's expression above. Bishop (1960) showed experimentally that χ is a non-linear function of the degree of saturation, and work by Williams and Shaykewich (1970) suggests an S-shaped relationship not differing greatly for a loam and a clay soil (Fig. 32.3). The effect of soil water suction on effective stress and consequently on soil strength depends on the nature of the soil moisture characteristic. Soils which develop high strength with increasing soil water suction are those which maintain a high degree of saturation as suction increases. Towner and Childs (1972) successfully use the moisture characteristic and intrinsic soil properties to predict soil strength as a function of soil water suction.

Empirical measurement of soil strength

Soil friability

The successful application of Griffith's theory of brittle fracture even to soils in the very wet condition (Farrell *et al.* 1967) is relevent to the concept of friability in soils. Friability has been described as being concerned with the ease of crushing, crumbling or rubbing apart. Marshall and Quirk (1950) at Adelaide, attempted to obtain an index of friability or aggregate strength using a drop-shatter test. They found that break-up was related to cumulative drop height, the

Table 32.1. Effect of water content on soil strength parameters related to friability of a *red-brown earth*

Data from Farrell *et al.* 1967

Water content (%)	Matric potential (−kPa)	Tensile strength (kPa)	Tensile strain energy (kg m^{-2})	Cumulative drop height (m)	$\frac{CDH}{TSE}$
2	–	158	2.15	5.0	2.3
4	2500	101	2.00	3.6	1.8
6	650	68	1.88	3.2	1.7
8	330	48	1.83	3.0	1.6
10	180	31	1.78	2.8	1.6
12	50	16	1.66	2.5	1.5

sum of the heights of repeated drops, and within limits, was independent of drop height. This suggests a storage of drop energy and, in an air-dry material, this storage must be in the form of microscopic cracks. For loam from a *red-brown earth*, Farrell *et al.* (1967) found a direct relationship over most of the range of available water between the tensile strain energy (TSE) required to rupture a remolded core and the cumulative drop height (CDH) required to shatter a clod to a given degree of break-up. Tensile strain energy is defined as the area under the stress-strain curve up to failure under tension (Table 32.1).

Subplasticity

A morphological feature of certain subsoils in Australia is that of subplasticity (Butler 1955, 1976). This relates to the field assessment of soil consistence and is an indicator of the amount of work that is expended in kneading the soil to a final steady consistence. Soils with a high degree of subplasticity require a considerably greater expenditure of energy than other soils to achieve breakdown of aggregates. Greacen (1960*b*) has shown that subplastic soils do not swell greatly on wetting up. Using the plastic limit and liquid limit as points on the consolidation-with-shear curve, Greacen quantified the strength at saturation in terms of equivalent compressive strength as described above. Values of the tensile strength of individual aggregates of subplastic soils measured as crushing strength are about an order of magnitude lower than the equivalent compressive strength. This suggests that the strength is related directly to the degree of consolidation and that there is only an indirect or second-order effect of intercrystalline growth (Norrish and Tiller 1976).

The resistance of aggregates to a crushing force (Rogowski 1964) offers what appears to be a useful method for characterizing the strength of individual aggregates. Braunack and Dexter (1978) suggest that mathematical expressions describing the compaction of f eds of brittle aggregates might be made more generally applicable by normalizing the compaction load with the tensile strength of the aggregates. Their universal compaction equation takes the form

$$H/H_i = 0.4 + 0.6\ \exp[0.017(P/\sigma_T) - 0.38\ (P/\sigma_T)^{1/2}] \quad (5)$$

where P is the compaction load on an aggregate bed of initial height H_i. From Table 32.1 there is a 10-fold effect of water content on σ_T in remoulded soil, suggesting large differences in compressibility. Braunack and Dexter (1978), however, found only a small effect of water content on σ_T for natural aggregates from a *red-brown earth* similar to that used by Farrell *et al.* (1967).

Resistance to penetration

The resistance offered to growing plant roots by the mechanical strength of the soil is an important feature of the physical environment of the root (Greacen 1976). Resistance to penetration by a metal probe is often used to assess this aspect of mechanical strength, with the probe being used as an analogue of the root. From an analysis of the types of deformation caused by different shapes of probes, Farrell and Greacen (1966) and Greacen *et al.* (1968) were able to predict the forces acting on the probe in terms of the classical strength parameters described above. Although critical penetrometer resistance at times approaches an order of magnitude greater than the maximum pressure that roots are able to exert, Greacen *et al.* (1968) have been able to reconcile this difference on the basis of their analysis and offer an explanation of the predictive performance of the penetrometer.

Greacen and Oh (1972) showed that, in soils of moderate strength, the turgor pressure developed by the root growing through the soil could be reconciled with the forces assessed from penetrometer data. While their results were for experimental conditions that were rigorously controlled, the penetrometer test is a useful method for correlative measurements in the field. Greacen and Hignett (1976) found that the seminal axes of wheat grew vertically downwards at a rate determined by the temperature and mechanical resistance of the soil. Fig. 32.4 gives penetrometer resistance measurements for a range of wheatland soils in South Australia measured at a soil water suction close to field capacity several days after the cessation of rain.

There is evidence also of a high correlation between root growth of *Pinus radiata* and penetrometer resistance of sandy soils in the south-east of South Australia: Sands *et al.* (1979) found that in the surface 1 m of soil, rooting density was negatively correlated with penetrometer resistance, with few roots occurring in areas of penetrometer resistance greater than 3000 kPa. But high penetrometer resistance is often associated with low organic matter content and less favourable water relations and it is not easy to establish causal relationships.

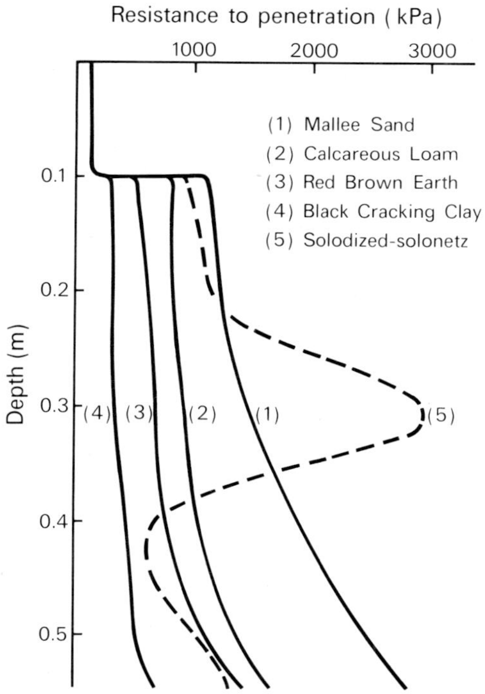

Fig. 32.4. Profiles of mechanical impedance to a metal penetrometer for five wheatland soils near field capacity.

Water movement in soil

Infiltration

Solutions to hydrological problems involving water transfer in the soil, such as prediction of infiltration or estimation of recharge to an aquifer, depend on an understanding of the physics of soil water movement. The basic equation for water movement in the vertical direction in a porous salt-free medium was given by Klute (1952) as

$$\partial\theta/\partial t = \partial(D\partial\theta/\partial z)/\partial z - \partial K/\partial z \qquad (6)$$

where z is the vertical coordinate positive downwards; K, the hydraulic conductivity, is defined as the flux per unit gradient ($d\psi/dz$) in the soil water potential ψ, and D, the diffusivity, is the flux per unit gradient ($d\theta/dz$) in water content θ. It remained for Philip (1957) to obtain a useful solution to this equation. This was the beginning of an extensive series of papers by Philip giving an exhaustive mathematical treatment of infiltration.

Prior to this, empirical expressions such as that of Horton (1933) were used:

$$i = i_\infty + (i_o - i_\infty)\exp(-bt) \qquad (7)$$

where i is the infiltration rate at any time t, i_o is the initial rate and i_∞ is the infiltration rate at large t. The constant b depends in an unknown way on the physical properties of the soil. More recently Boughton (1968), for greater ease in computing, has modified the Horton type of equation to give daily infiltration rate f as a function of soil water content rather than time, as

$$f = f_c + (f_o - f_c)\exp(-aW_B) \qquad (8)$$

f_c and f_o are the daily infiltration rates when the B horizon is at field capacity and permanent wilting point respectively, W_B is the water stored in subsoil and a is a constant. Berndt and White (1976) further modified Boughton's equation in a water balance study in southern Queensland. They assumed that surface storage capacity of the soil and rainfall amount, rather than intensity, largely determine the infiltration pattern. They determined weekly infiltration capacity F as a function of soil water content as $F = F_O \exp(-aW)$, where F_o is the weekly infiltration capacity (mm) when the storage is empty. W is the amount of stored soil water and a is a constant determined by measuring ΔW after a runoff-producing rainfall. Weekly surface drainage or runoff D_s is given as $D_s = P_w - F \tanh(P_w/F)$; P_w is precipitation for the week. Such formulae as these are sometimes useful, although they have the serious disadvantage that the parameters require measurement for each set of conditions.

Philip's solution of the flow equation provided a more intellectually satisfying basis for infiltration. It results in an infinite series of powers of $t^{1/2}$ as

$$I = St^{1/2} + At + Bt^{3/2} + \ldots \qquad (9)$$

where I is the cumulative infiltration at time t, and S, and A and B are parameters which can be determined from the hydraulic characteristics of the soil (Philip 1957). Philip termed S the sorptivity of the soil. In the initial minutes of infiltration in non-swelling soils the process is dominated by the first term, which describes the sorption of water by the matric forces of the soil; at this stage the other terms may be neglected. As infiltration proceeds, the contribution of the other terms, which express the effect of gravity and hydraulic conductivity, become more important. At intermediate times, one to two hours, the approximation $I = St^{1/2} + At$ is used, A being commonly approximated as $K_o/3$ (Talsma 1969). For long times, infiltration depends solely on the saturated hydraulic conductivity of the soil (K_o), and the analogous empirical expression $I = St^{1/2} + K_o t$ is often used in practice. An extension of this empirical model has been proposed which describes infiltration at all times (Collis-George 1977). While much of the appeal of Philip's equation derives from its sound physical basis, its application owes much to the contribution of Talsma (1969) in developing simple and rapid field methods for measuring the two soil properties sorptivity and hydraulic conductivity.

The above treatment is restricted to describing water movement in rigid soils. However, for water movement in swelling soil, Philip (1969) has given an expression similar to Eqn (6) but with a mass coordinate (m) in place of z, thus

$$\partial v/\partial t = \partial(D\partial v/\partial m)\partial m - \partial((1-\rho_w)K)/\partial m \quad (10)$$

where ρ_w is the wet specific gravity and v is the water content ratio. Eqns (6) and (10) suggest that the mechanisms of infiltration will have different emphases in rigid and swelling soils. The importance of the gravity term will be reduced in swelling soil by the $(1\text{-}\rho_w)$ factor. This appears to be so for Yandera loam, a typical non-swelling soil from south-eastern New South Wales; sorptivity dominates infiltration only during the first few minutes and by 300 minutes, the final steady infiltration rate of 31.4 mm h^{-1} has been reached; K_o for this soil is 31 mm h^{-1} (Talsma 1976). For swelling soils from the Murrumbidgee Irrigation Area, Talsma and Van der Lelij (1976) suggest sorptivity or linear $t^{1/2}$ infiltration for about a week with an infiltration rate of 1–2 mm day^{-1} after 40–50 days. In cracking clays, the filling of the shrinkage cracks complicates the initial infiltration rate (Fig. 32.5). The second part of the process, as water is absorbed from the cracks, again follows the $t^{1/2}$ relationship. Gardner (1978) demonstrates the application of Philip's solution to cracking clays in the Emerald Irrigation Area in central Queensland.

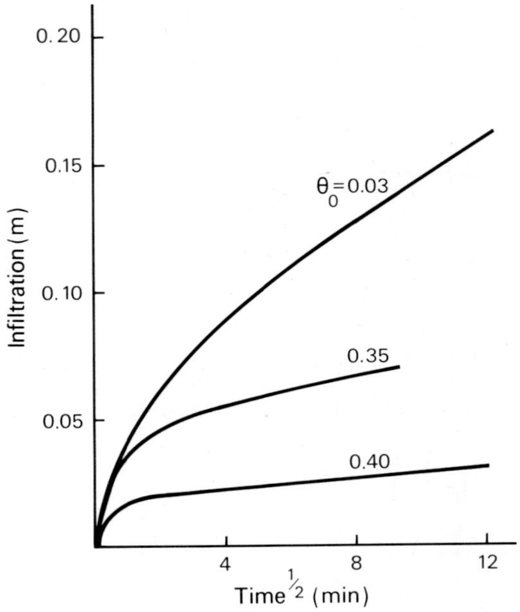

Fig. 32.5. Infiltration (I) in a swelling clay at different initial water contents θ_o.

An old infiltration equation for new models

The problem of predicting water movement into field soils can be solved, even for multiple layered soils and difficult boundary conditions, by application of the basic equation for water movement using numerical methods with computers. The cost in computer time and data can be enormous and is generally not warranted. Neverthelesss solutions are still required. Answers to this problem have recently been found in the application of a very early work by two Australians, Green and Ampt (1911). They described infiltration as water being moved by an effective suction h_f at the wetting front together with gravity, through a depth of wetted soil L with constant conductivity K, as

$$i = dI/\mathrm{d}t = (h_o + h_f \cos\alpha + L - p)K/L \quad (11)$$

L, of course, increases with I as $I = \Delta\theta L$, where $\Delta\theta = \theta_i - \theta_f$. Gardner (1977) has modified the original equation by reducing the effective hydraulic potential by the pressure p of air trapped in the soil ahead of the wetting front (Peck 1965) and by including the wetting angle α which has been shown to be critical in non-wetting sands in southern Australia (Bond 1964). In this case cos α may be negative and high values of h_o are required to force water into the soil.

It is deserving of comment that Philip arrived at the Green and Ampt infiltration equation independently in 1953, 40 years after the original work, and that 20 years later still, people are discovering the advantages of its simplicity in computer simulation. The Green and Ampt equation has been used for investigating infiltration into layered soils (Bouwer 1969), time to runoff (Mein and Larsen 1973) and surface crusting effects (Hillel and Gardner 1970).

The Green and Ampt equation is particularly useful when infiltration is determined by the hydraulic conductivity of the least permeable layer. Commonly, the least permeable layer is the cultivated surface layer that has slumped or been sealed over by crusting. Depending on the wet strength of aggregates, an unprotected cultivated layer may collapse under the impact of raindrops and seriously reduce infiltration. MacIntyre (1958) has shown that in some hard-setting duplex soils (*red-brown earths*) of southern Australia, crusting can reduce the hydraulic conductivity to less than 0.02 m day^{-1}. In northern Australia where summer rainstorms are common, soils may be classified on the basis of their resistance to disruption and compaction (W. Arndt pers. comm.). Class A soils, which include the basaltic *krasnozems*, will accept 0.5 m day^{-1} even with puddling by traffic when wet; Class B soils accept the same high rate at first, but this decreases to 0.1 m day^{-1} following heavy mechanical puddling; Class C soils gradually deteriorate from a moderate 0.1 m day^{-1} to a low value under repeated rainfalls with no effect of puddling, while Class D soils degrade under the first rainfall to accept only 0.025 m day^{-1}. The physico-chemical properties of such soils are discussed more fully in Chapter 31.

Redistribution and drainage

Downward movement of water in the profile following infiltration is important agronomically in that it determines the distribution of water, soluble plant nutrients and salinity relative to that of the roots of plants. This movement of water within the root zone is generally termed redistribution, while movement of water beyond the influence of roots and evaporation is treated as loss by deep drainage.

A rigorous description of redistribution is essentially an exercise in applying Eqn (6) above, but with the important requisite that, since both wetting and drainage are simultaneously involved (Peck 1971), full account must be taken of the hysteresis in the hydraulic properties of the soil, the moisture characteristic $\psi(\theta)$ and the conductivity-potential function $K(\psi)$. Talsma (1974, following Peck 1971), gives the draining rate q at the transition from the draining to the wetting zone Z_t at any time t as

$$q = -\left(D\frac{d\theta_m}{dz} - K\right) \qquad (12)$$

where D and K are the wetting values of diffusivity and conductivity at θ_m, the maximum value of θ reached at Z_t.

An interesting outcome from the analysis is that while it might be expected that the rate of redistribution varies considerably between soils, the effect of the amount of water and the initial water content on the draining rate, apart from being different, might even be reversed from one soil to another. For example, at one week after infiltration, the flow rate at the transition point for a loam varied from 0.02 m day^{-1} for an initial water content of 0.02 to 0.01 m day^{-1} for a wetter initial water content of 0.07. In contrast to this, however, the flow rate for a sand increased from 1×10^{-3} to 2×10^{-3} m day^{-1} for about the same two values of initial water content.

Another important aspect of the above results is the magnitude of the draining rates at long time, seven days, after the cessation of infiltration. While these results do not deny the usefulness of the concept of field capacity in budgeting soil water under field conditions, they indicate areas where caution must be observed. In simple water balance models, based on the field capacity concept, a maximum water storage capacity is assumed and rainfall in excess of this is considered as loss by drainage. For coarse-grained soils, where infiltration is non-limiting, these models provide an acceptable prediction of total drainage, but their reliability decreases in fine textured soils. In any case, they do not provide a continuous simulation of the drainage process.

To overcome the limitations of these storage models, particularly where a simulation of drainage is required, e.g. in plant nutrient or salt leaching situations, drainage models have been developed on the basis of Darcy's law applied to the whole profile as a unit,

$$v = -K\left(\frac{d\psi}{dz} + 1\right) \qquad (13)$$

where v is the drainage flux (m day^{-1}). In a uniform soil during drainage, $d\psi/dz$ is close to zero and the profile drains uniformly under unit hydraulic gradient of gravity. Combinations of the field capacity model and the drainage model based on Darcy's law are providing a useful compromise. Greacen and Hignett (1976), in a two-layer model of the water balance for wheat in South Australia, use an exponential $K(\theta)$ function based on Rogowski's (1971) model as an alternative to the field capacity model. These represent the simplest methods of accounting for drainage in water balance models. Other solutions to the drainage problem are discussed later.

Changes in hydraulic conductivity also occur in the soil immediately below the cultivated layer and even deeper; it may decrease due to the compacting effect of tractor wheels and plough shares or it may increase due to the meliorative effect of swelling and shrinking, and, probably more importantly in our pasture-wheat rotation agriculture, to the creation of pores by plant roots. In Fig. 32.6, evidence of the cycling of hydraulic conductivity is shown in the subsurface layer (100–175 mm) and the subsoil layer (300–375 mm) of a *red-brown earth* in the long-term rotation experiment at the Waite Agricultural Research Institute in Adelaide (E. L. Greacen, unpublished). The low value of hydraulic conductivity of the soil in the continuous wheat plots after 50 years, corresponds to the hydraulic conductivity that would be expected from the radius and density of channels left by wheat roots at this depth (Greacen *et al.* 1976). Hydraulic

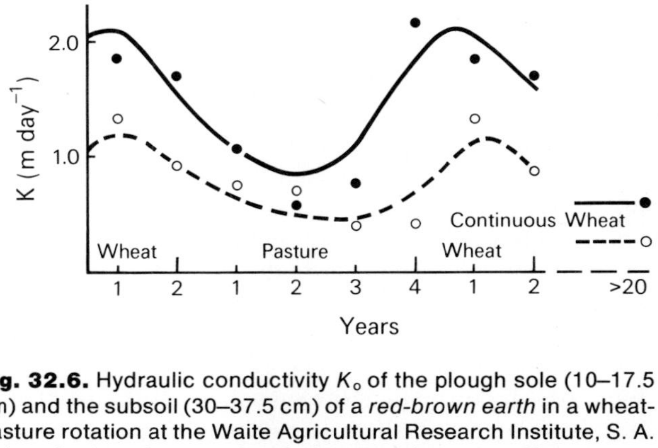

Fig. 32.6. Hydraulic conductivity K_o of the plough sole (10–17.5 cm) and the subsoil (30–37.5 cm) of a *red-brown earth* in a wheat-pasture rotation at the Waite Agricultural Research Institute, S. A. Values for 'continuous wheat' plots are given at > 20 years.

conductivity values obtained on other duplex soils under pasture on a catchment in the Adelaide Hills changed from 1.5 m day^{-1} in late summer, the same time as the rotation results, to 0.1 m day^{-1} in spring (S. West pers. comm.). A possible explanation of this decrease is that new roots are blocking pores and temporarily causing a decrease in the hydraulic conductivity.

Water uptake by roots

The rate at which water is required by vegetation depends on the evaporative demand of the atmosphere. While water may be available in the soil store, the rate at which it can be taken up depends on the hydraulic properties of the soil and the roots and on the geometry of the root system. Thus, if the conductance of the soil-plant system is high enough and water is available, evaporation will be determined by the environment at the potential rate. If the conductance of any part of the system falls below the requirements for potential evaporation and a minimum leaf water potential, evaporation will be determined by the hydraulic properties of the system.

Resistance to flow in the soil-root system

Resistance to flow in the soil-root system may be defined as the difference in water potential that must develop between the base of the stem and the bulk of the soil to cause a flow of water per unit ground area of 1 cm day^{-1}. This resistance is conveniently considered as three separate components: (i) resistance per unit length of root to radial flow in the cylinder of soil immediately surrounding the root, sometimes called rhizosphere resistance R_s (kPa cm^{-2} day); (ii) resistance per unit length of root to radial flow in the root R_r (kPa cm^{-2} day) i.e. flow from the root surface to the xylem vessels in the centre of the root; and (iii) resistance per unit length of root to axial flow R_x (kPa cm^{-4} day) up through the xylem to the base of the stem.

It is possible to measure these root resistances directly on excised segments of root or to calculate them from more basic measurements (Ponsana 1975). In the seminal roots of cereals, for example, the main vertical axes or tap roots possess a single central vessel having a diameter of 40–100 μm. Using this value in the Poiseuille equation, and knowing the number of axial roots per unit ground area, it is possible to calculate a value for axial resistance over any part of the root system. Buscher and Fritton (1978) claim that xylem resistance for the whole root system of oats is negligible, less than 1 kPa cm^{-3} day. But this depends on the number of main roots and they were dealing with plants with as many as 12 main axes growing in solution. Under field conditions in Australia the number of axial roots in wheat at 0.3 m depth passing through to the subsoil is less than 400 m^{-2} of ground area and there is evidence that this value is largely independent of plant density (Ponsana 1975). Under these conditions the potential drop required to overcome axial resistance and supply subsoil water to meet a transpiration rate of 10 mm per day can be as high as 300 kPa (Greacen *et al.* 1976).

Resistance to radial flow in the root is considered by some to be the critical resistance in the root–soil pathway (Newman 1969), but unfortunately there is no completely acceptable way of measuring it; direct measurements have been made on excised axial roots but their reliability is uncertain. Also, in cereals, most of the uptake is probably by first order laterals, which may make up 80% of the total root length; these roots are very small, 0.2 mm in diameter, and direct measurement is difficult. Indirect measurements on the whole root system (Newman 1973) are difficult to interpret because radial resistance depends strongly on the age of the root. For cereal roots, Meyer *et al.* (1978) give an estimated value of radial resistance of 10 mPa cm^{-2} day, that is, a 10 mPa drop in pressure is required to overcome radial resistance to a flow rate of 1 cm^3 per cm length of root per day.

Resistance to flow of water through the soil to a root can be calculated from the hydraulic conductivity–water content function of the soil. As will be discussed below, the conductivity function is an important basic hydrologic property of the soil and depends in a complex way on the pore space relations and the soil moisture characteristic, i.e. the water retention curve. Over the past decade there has been the contention (Newman 1969; Williams 1974) that for many surface soils, soil resistance does not play a major part in restricting the supply of water to the root and that often it only approaches root resistance in magnitude when high suctions, approaching wilting point, have been reached. This conclusion has been based largely on low rates of flow in a soil with a hydraulic conductivity-water content function representative of highly conducting soils.

This dependence of soil resistance on the conductivity function is readily demonstrated. An analysis of flow towards a root has been given by Whisler *et al.* (1970) for soils with a conductivity function $K = a|\psi|^{-n}$, where a is a constant and n is always greater than 1. The flux towards unit length of root (Q) in units of cm^2 day^{-1} is given by

$$Q = \frac{2\Pi a}{\ln(r_2/r_1)\,(n\text{-}1)\,(\psi_2)^{n-1}}\left(1 - \left|\frac{\psi_2}{\psi_1}\right|^{n-1}\right) \qquad (14)$$

in which r_1, r_2, ψ_1 and ψ_2 are the radii and water potentials at the root surface and at the mid point between roots respectively.

The soil resistance as a function of soil matric potential is illustrated in Fig. 32.7. for three Australian soils in which the conductivity function is given by $K = 4800\,|\psi|^{-3.2}$ for a *siliceous sand*; $K = 54\,|\psi|^{-2.0}$ for a loam from a *red-brown earth* and $K = 11.2\,|\psi|^{-1.6}$ for a *black earth*. The flow rate was set high at about 0.01 cm^2

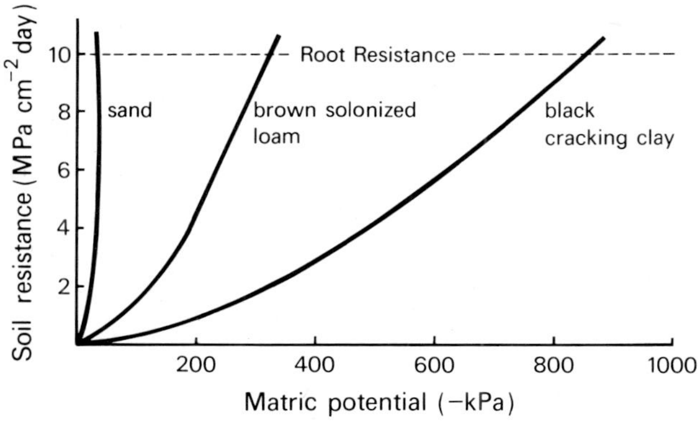

Fig. 32.7. Soil resistance to radial flow of water towards a root as a function of the matric potential of the soil water for three field soils. Root radius 0.01 cm and rooting density L_v of 4 cm^{-2}.

day^{-1}, at and below which R_s changes little with flow rate. The soil water potential at which R_s becomes equal to R_{root} varies from -850 kPa for the clay, through -350 kPa for the medium textured soil to -35 kPa for the sand. For very coarse Bassenden Sand (Carbon 1973) soil resistance dominates water uptake for ψ as high as -4 kPa. For values of Q greater than 0.1 cm^2 day^{-1}, soil resistance becomes dominant at much higher values of ψ (Williams 1974).

These estimates of the resistance of different soils to the flow of water to plant roots depend on the basic hydrologic properties, the conductivity – and the potential – water content functions which will be described later in the chapter. An important factor is soil texture but, as discussed below, probably more as it affects soil structure than through the effect of texture *per se.*

Root Geometry

It is sometimes convenient to consider root systems as being composed of two layers, the upper or surface layer, often corresponding to the A horizon of the soil profile (Greacen and Hignett 1976) with high rooting densities concerned with the immediate use of current light rainfalls, and the lower layer extending to the bottom of the root zone, frequently concerned with subsistence supply of water from the subsoil. A good example of the two-layer system is that of cereals where in dry seasons the nodal roots are confined to the upper layer and the lower root system is made up almost entirely of seminal roots (Ponsana 1975). In the root systems of many species, for example *Eucalyptus regnans, Pinus radiata* and vines, members of the two layers are distinguished as spreader and sinker roots. Ashton (1975) gives a good description of the change of the root system of *Eucalyptus regnans* with age.

Detailed information on root geometry is required to quantify resistance to flow and hence the ability of the soil–root system to supply water to the crop without inducing stress. For example, radial root resistance and soil resistance, and hence uptake of water from a layer of soil, depend on the total length of root in that layer (Greacen *et al.* 1976). This may be obtained from measured values of rooting density L_v (cm cm^{-3}) as a function of depth, or better, the hair-bearing roots L_H or white roots, which are more realistic indices of absorbing root length. In wheat, L_H is commonly 0.2–0.4 L_v but may be as high as 0.7 L_v in zones of younger roots (Greacen 1977). High values of rooting density are required in surface soils to take advantage of full water supply in good seasons.

Efficiency of water uptake from the subsoil layer depends on axial resistance, which in turn depends on the number of axes between that layer and the surface, and the equivalent radii of the xylem vessels. In regions, such as southern Australia, which enjoy a mediterranean climate, perennial crops such as *Pinus radiata* forests require an axial root system with resistance to flow low enough to transport water up through the dry surface layer in summer to carry the crop through periods of drought.

Hydrological properties and soil morphology

Soil morphology has been traditionally described by those properties detected by sight and handling: orientation, dimension, texture, structure, colour, and the consistence of the horizons, which make up the soil profile. Morphology aims to provide an integrated view of the solum which, as a hydrological unit, sheds, absorbs, stores and conducts water. The hydrology of the profile is dependent not only on the hydraulic properties of each relatively homogenous horizon but also on the depth, position and sequence in which these horizons occur within the profile. Profile morphology is therefore of consequence to soil hydrology because it provides, by way of texture and structure descriptions, some insight into the hydraulic properties of each horizon as well as a description of the dimensions and spatial arrangement of these morphological units.

The principles governing water entry, storage and movement in a homogeneous soil system are well established (Childs 1969; Hillel 1971; Baver *et al.* 1972), and can generally be derived from a knowledge of two important soil hydraulic properties, namely the $\psi(\theta)$ relationship or pF curve, originally termed by Childs (1940) the soil *moisture characteristic*, and the $K(\theta)$ relationship, where ψ is the water potential, K the hydraulic conductivity and θ the volumetric soil water content. The challenge before soil physics presently is the application (Klute 1973) of these principles to real world soil profiles using perhaps the computer simulation techniques explored by Hillel (1977) or the classical analytical procedures as represented by Childs and Bybordi (1969), Hillel and Gardner (1969, 1970) and others.

It is the integration of such knowledge to obtain a comprehensive understanding of the way the whole profile operates which is the essential step if we are to generalize experience (Hillel 1977) for a particular profile morphology under specific conditions and predict its behaviour under different rainfall and geomorphic settings. In our Australian experience this objective is yet to be achieved. At the moment our understanding must therefore be limited to a qualitative discussion of the relationships that exist between soil horizon morphology, whole profile morphology and water dynamics.

The moisture characteristic

The moisture characteristic is related to the morphological properties of texture and structure, but the nature of the relationship is not quantitative and is at best empirical. To date no satisfactory theory exists for the prediction of the $\psi(\theta)$ relationship from basic morphological and other soil properties. The effects due to surface adsorption of water and the capillary uptake by pores appear to be too complex to be described from first principles using simple models (Marshall 1958).

Effect of texture and structure on the moisture characteristic

The obvious effect of texture and structure on the nature of the moisture characteristic is of considerable hydrological importance and has often been expressed quantitatively in empirical regression-type relationships (Salter and Williams 1965). It is unlikely that unique analytical relationships exist. The position and shape of the moisture characteristic is determined by the surface area of solid to which water is adsorbed and the pore size distribution. Colloidal solids, either clay or organic matter, exert a dominant influence on the solid surface area with the result that increasing clay and organic matter displace the curve towards higher water contents at all matric potentials. The arrangement of the solid material, and the resultant texture and structural pore space (Childs 1969), determines the shape of the function from matric potential 0 to −100 kPa, while the desorption of water from solid surfaces is reflected in the shape of the curve from approximately −100 to −2000 kPa.

Several empirical equations have been proposed for the $\psi(\theta)$ relationship and are useful, provided application is restricted to situations for which they were developed. Perhaps the best known is the simple logarithmic function in which $\ln|\psi| = a + b\,\theta$, where a and b are derived constants (Cowan 1965). Rogowski (1971), Jakobsen (1973) and McQueen and Miller (1974) provide good evidence that this form can yield a useful description of the moisture characteristic, although Jakobsen (1973) suggests that for ψ greater than −50 kPa, the power function $\ln|\psi| = a + b \ln\theta$ describes the characteristic better.

The draining moisture characteristic of large cores determined by Prebble (1970*a*) for 17 soil groups in Australia can be adequately described by either of the above empirical models. Despite wide variation in texture, structure and clay mineralogy of the 78 horizons investigated, the proportion of the variation accounted for by $\ln|\psi| = a + b \ln\theta$ was always in excess of 0.910 with a mean of 0.941 for all 78 soil horizons.

In Fig. 32.8 an empirical function of this form is used to describe the position and general shape of the moisture characteristic for the six texture groups (Northcote 1971), using the data of Prebble (1970*a*) and that presented by Stace *et al.* (1968) for the same profiles. Clearly, the heavier the texture, the more water is held by the soil over the whole range of matric potential from −4 to −1500 kPa.

The functions for clay loam, light clays and heavy clays are based on soils with structure of different degrees and types. In these textures the influence of structure is minimal compared with its influence in the lighter texture groups. In light to medium textured soils, presence or absence of structure can have a marked effect on the position and shape of the moisture characteristic as is evident in Fig. 32.9, in which an apedal fine sandy loam from the surface of a *solodized–solonetz* (B558, Stace *et al.* 1968 p. 158) is contrasted with the fine sandy loam surface of a *humic gley* (B565, Stace *et al.*, 1968 p. 158) exhibiting a moderate crumb structure and increased organic matter. The organic matter not only absorbs more water but it also contributes to the development of structural pore water which in turn determines the shape of the moisture characteristic. Further influence of structure in light to medium textured material (Fig. 32.9) is illustrated by the difference in shape and position between the moisture characteristic for the loam of massive structure from a *soloth* (B551, Stace *et al.* 1968 p. 177) and a loam with a well developed crumb structure from a *krasnozem* (B551, Stace *et al.* 1968 p. 302). Sharma and Uehara (1968*a*) illustrate that much of the influence of structure on the form of the moisture characteristic in the range 0 to −30 kPa can be attributed to the inter-ped fabric whilst at matric potential below −30 kPa the intra-ped fabric dominates the form of the characteristic. Both of the structured soils of Fig. 32.9 exhibit a form consistent with this thesis.

Effect of texture and structure on the available water storage capacity

Despite the theoretical and practical difficulties in defining and estimating the available water storage capacity (AWSC), it is a useful hydrologic concept which is directly derived from the draining moisture characteristic. The water retained by intact soil cores between matric potentials of −10 kPa and −1500 kPa is generally

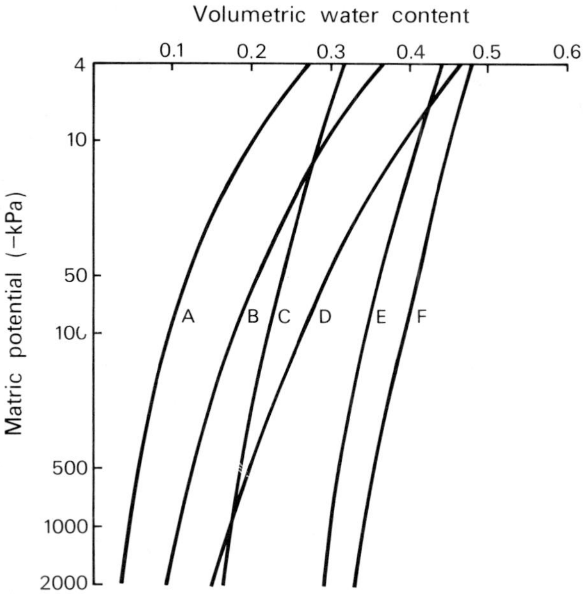

Fig. 32.8. Empirical curves for the draining moisture characteristic for six textural and structural groupings. These curves are described by $\ln \psi = a + b \ln \theta$, where ψ has units of kPa, θ is volumetric water content expressed as a fraction. (*A*) Sands, including fine sands, loamy sands and clayey sands. a = −2.28, b = −3.19, n = 9, r^2 = 0.876. (*B*) Sandy loams and loams exhibiting minimal structural development. a = −3.24, b = −4.56, n = 5, r^2 = 0.805. (*C*) Massive sandy clay loams, and sandy clays. a = −9.19, b = −9.21, n = 7, r^2 = 0.822. (*D*) Light to medium clays exhibiting strong fine self-mulching properties. a = −2.74, b = −5.48, n = 5, r^2 = 0.689. (*E*) Light to medium clays exhibiting coarse structure and silty clay loams. a = −9.75, b = −13.64, n = 23, r^2 = 0.858. (*F*) Medium to heavy clays exhibiting coarse structure and silty clays. a = −10.93, b = −16.78, n = 18, r^2 = 0.776.

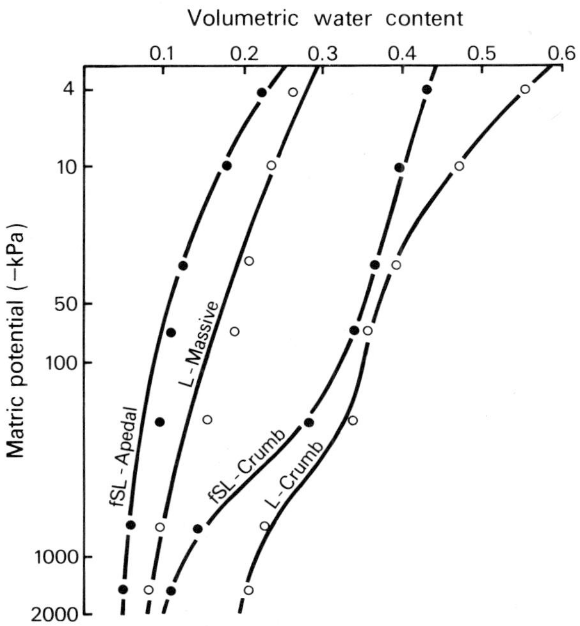

Fig. 32.9. Influence of structure on the shape and position of the draining soil moisture characteristic for light and medium textured soils. (Data from Prebble 1970*a*).

considered to give some approximation to the soil water available for extraction by the plant root. Empirical relationships between the textural composition, the upper and lower limits of available water, and the AWSC have been established (Wilcox and Spilsbury 1941; Stirk 1957; Jamison and Kroth 1958; Salter *et al.* 1966; Shaykewich and Zwarich 1968; Coughlan 1969).

Initially, consideration will be restricted to the textural and structural influences on the available water storage capacity for a homogenous horizon, or sample, out of context of the profile. This will allow some qualitative principles to emerge. In later sections these influences will be considered in the context of the whole profile.

Strong positive correlations exist between the lower limit of available water (i.e. −1500 kPa) and clay content. The slope of the relationship is dependent on clay mineralogy (Stirk 1957; Coughlan 1969). Although it is general experience that the upper limit of available water can be positively correlated with the content of fine sand and organic matter, the correlation is often weak, reflecting the confounding influence of structural pore space which dominates water retention near the upper limit. The compensating nature of the relationships between texture and the limits of available water result in the AWSC being positively correlated with fine sand and negatively correlated with clay (Jamison and Kroth 1958; Salter *et al.* 1966).

The predictive value of these empirical equations is perhaps adequate for many purposes. Salter and Williams (1967) were able to predict to within ±10% the AWSC for some 39 British soils using 21 field texture classes in conjunction with their regression equations.

The AWSC as estimated from water retained between matric potentials of −10 and −1500 kPa for a wide range of texture classes for structured and structureless Australian soils is reported in Table 32.2. Although the AWSC varies from 9.2 mm dm^{-1} for a structureless silty clay to 26.5 mm dm^{-1} for a structured fine sandy loam, the AWSC for many texture classes are similar and fall in the range 12 to 16 mm dm^{-1}. The generally held view (Salter and Williams 1965) that the medium textured soils have the largest AWSC is supported; however the influence of structure, particularly in the loam and sandy loam group, is of greater importance. The differences between cracking clays, which exhibit self mulching properties, and those that do not, again illustrate the importance of structural pore space in determining the available water storage capacity.

As seen from Fig. 32.8, increasing clay content shifts the moisture characteristic towards high water contents at all potentials, usually without a change in the slope of the curve, thus having minor influence overall on the AWSC. It is the morphological property which influences the slope of the characteristic which must exert the dominant influence on the AWSC. Fine sand, coarse silt and structural pore space appear to be important in

Table 32.2. Influence of field texture and structure on water retained between −10 and −1500 kPa for a range of Australian soils

Values are means of the number of horizons enclosed in brackets. Data derived from 244 soil horizons for which moisture characteristic information and morphology is reported by Stace *et al.* (1968) and Prebble (1970*a*).

Field texture class	Water retained ($mm\ dm^{-1}$) Structured soils	Structureless soils	All soils in texture group
	Sands		13.6(35)
Sand		14.1(6)	
Fine sands		15.3(6)	
Loamy sands	15.5(4)	12.3(12)	
Clayey sands	14.0(2)		
	Sandy loams		15.5(28)
Sandy loams		11.9(18)	
Fine sandy loams	26.5(7)	14.6(3)	
	Loams		15.8(37)
Loam	24.3(9)	13.6(11)	
Sandy clay loam		12.7(17)	
	Clay loams		16.9(26)
Clay loams	17.7(11)	18.5(8)	
Silty clay loams	12.7(4)	12.7(3)	
	Light clays		13.8(41)
Sandy clays		16.7(22)	
Silty clays		9.2(7)	
Light clays	11.5(12)	14.0(5)	
	Medium to heavy clays		11.5(67)
Medium clays	12.2(33)		
Heavy clays	11.5(34)		
	Self-mulching clays		
	21.4(10)		21.4(10)

increasing the slope of the moisture characteristic, particularly in the range of matric potential between −10 and −100 kPa, where the variation in AWSC between soils occurs.

At this point it is essential to note that these comments on AWSC relate to a homogenous sample or perhaps a uniform soil profile, both of which are free draining. The nature of AWSC in field profiles and the relationship to morphology will be dealt with later.

The hydraulic conductivity of soil

The majority of experiments indicate that the hydraulic conductivity K is effectively a single-valued function of the volumetric water content, in contrast to the $K(\psi)$ relationship, which exhibits hysteresis as does the moisture characteristic. The $K(\theta)$ relation has meaning only for a homogenous horizon in the profile. For a rigid soil the nature of $K(\theta)$ will depend on the pore space dimensions and distribution as reflected in the moisture characteristic. Consequently, Childs and Collis-George (1950) were able to construct a model for $K(\theta)$ based on the form of the moisture characteristic. Other workers (Marshall 1958; Millington and Quirk 1961; Lalibarte *et al.* 1966) have subsequently developed models which, by using a matching factor for K at some value of θ, generates a $K(\theta)$ function from the moisture characteristic.

The direct measurement of $K(\theta)$ over the whole range of θ using laboratory methods (Klute 1972) is expensive and of necessity involves slow field procedures to collect the 'undisturbed' sample. Rose *et al.* (1965), van Bavel *et al.* (1968) and Hillel *et al.* (1972) describe field methods which have received much attention.

Effect of texture and structure on the conductivity function

Many of the $K(\theta)$ relationships available relate to artificial porous media or are restricted to a narrow range of θ for soil samples for which the texture and structure descriptions are inadequate. Apart from the data of Rose *et al.* (1965) and Olsson and Rose (1978) our knowledge of the $K(\theta)$ function for Australian soils is restricted to that which can be derived from the moisture characteristic (Greacen *et al.* 1974). Fig. 32.10 illustrates the general influence of field texture on the

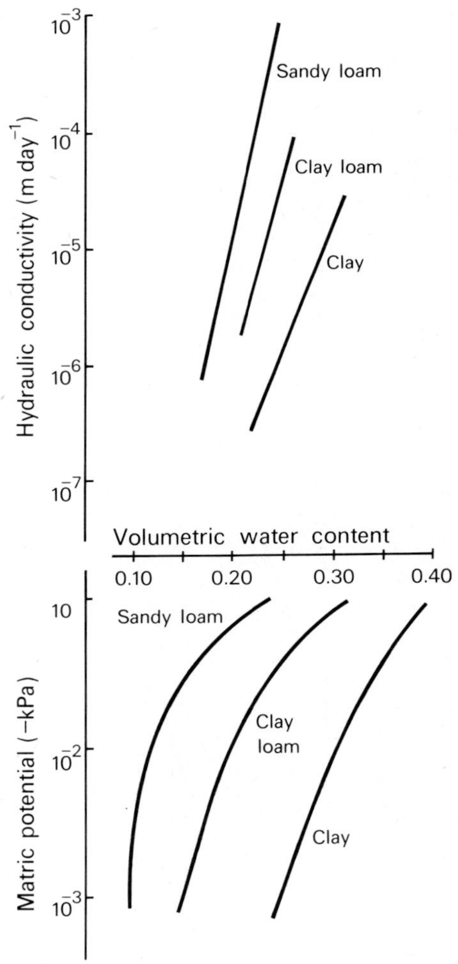

Fig. 32.10. Unsaturated hydraulic conductivity and matric potential as function of volumetric water content for clay, clay loam and sandy loam soil textures. (Data redrawn from Rose *et al.* 1965.)

form and nature of $K(\theta)$. At a given θ, K can be reduced by about three orders of magnitude as the field texture increases from a sandy loam to a clay. Alternatively, at $\psi = -60$ kPa, the clay, clay loam and sandy loam would have K values of 2×10^{-5}, 2×10^{-6} and 5×10^{-7} m day^{-1}. In Fig. 32.11, where K is drawn as a function of matric potential, it is clear that for a light textured soil, K decreases very much more rapidly than it does for heavier textured soil as matric potential decreases with drying of the soil. The slope of the $K(\psi)$ function has very important hydrological consequences. As will be seen from subsequent discussion, the effect of texture cannot be readily separated from that of structure. The magnitude of K between saturation and -10 kPa appears to be largely determined by structure. The slope of the function in drier conditions appears to be related to texture. Structure as reflected in ped size, shape, arrangement and ped orientation determines the nature of the macropore space and consequently saturated hydraulic conductivity K_o (Bouma and Anderson 1973). It is by determining both K_o and the retention of water with decreasing θ and matric potential that structure exerts its influence on the $K(\psi)$ function.

Bouma and Anderson (1973) demonstrate the influence of seven basic soil structures on the $K(\psi)$ function down to -10 kPa. Although confounded by the influence of texture, their data clearly show the substantial influence of structure.

Fig. 32.12 illustrates the effect of structural development on the $K(\psi)$ function for the A and B horizons of two important texture-contrast profiles in Australia, namely the *red podzolic* and the *red-brown earth.* The texture of the A and B horizons of both are sandy loam and medium clay respectively. The Samford *red podzolic* (Stace *et al.* 1968, p. 323; Prebble 1970*b*) has well developed fine granular structure in both horizons whilst the Conargo *red-brown earth* (Churchward 1957; Bridge 1968) is practically structureless in the surface horizon with a very coarse prismatic structure in the B horizon.

Although both surface horizons are sandy loams, the

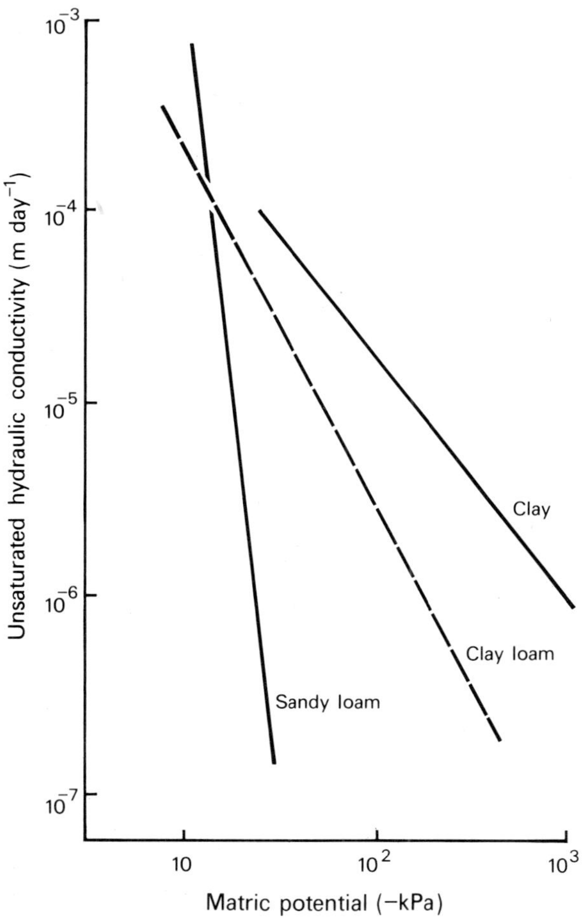

Fig. 32.11. Unsaturated hydraulic conductivity as function of matric potential for clay, clay loam and sandy loam. (Data redrawn from Rose *et al.* 1965.)

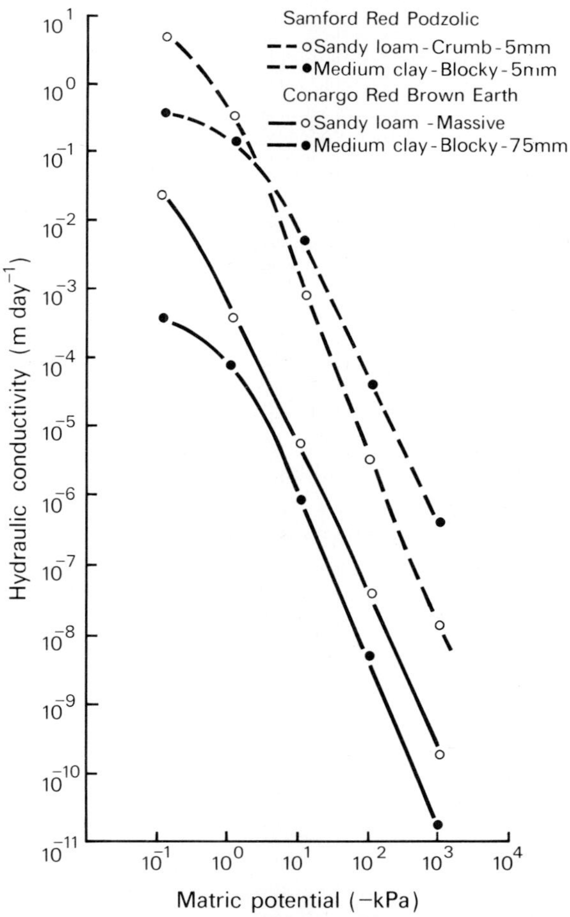

Fig. 32.12. Hydraulic conductivity–matric potential relationship illustrating the influence of structure at similar textures for two important Australian soils. (Based on data from Prebble 1970*b*, Bridge 1968.)

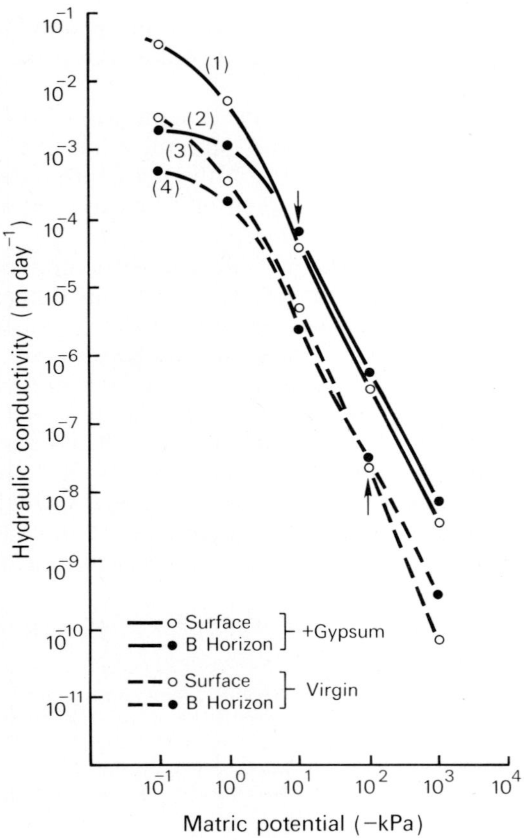

Fig. 32.13. Effect of gypsum and structural stability on the hydraulic conductivity–matric potential relationship for Billabong grey cracking clay. (Johnston 1952; Bridge 1968.)

red podzolic, with a well developed 5 mm crumb ped, has a K_o of 9.0 m day^{-1} whilst the *red-brown earth*, with a weak, platy structure, has a K_o of 0.04 m day^{-1}. The moisture characteristic of the *red podzolic* is similar to the *red-brown earth* for matric potentials less than −100 kPa but dramatically different in the region from −20 kPa to saturation. The thesis for the B horizon is much the same where the 5 mm granular peds in the *red podzolic* behave very differently to the 50–75 mm peds of the *red-brown earth*, although the greater swelling properties of the *red-brown earth* may confound the influence. Morphologically, the main difference between these two B horizons is the size, shape and proportion of peds which make up the fabric. Clearly these structural differences have important hydrological implications with regard to water movement in the profile.

Bridge (1968) presents data for the effect of gypsum on K_o and the moisture characteristic for the surface and B horizons of grey cracking clay (Billabong clay, Johnston 1952, p. 34). The effect of gypsum is understood to be through the effect on structural stability (Loveday and Scotter 1966). Using the procedure of Jackson (1972) and matching at K_o, a $K(\psi)$ function for both horizons and the influence of structural change as brought about by gypsum application is set out in Fig. 32.13. Although the original assumptions (Childs and Collis-George 1950; Marshall 1958) are invalid for swelling soils, the use of a matching factor in effect reduces the technique to a curve fitting procedure. Comparisons of these $K(\psi)$ functions with *in situ* measurement (Olsson and Rose 1978) indicate that the calculated functions are of the correct order and form.

The addition of gypsum and the associated improvement in structural stability not only increased K_o by an order of magnitude in both horizons but brought important changes in the nature of the $K(\psi)$ relationship.

The shift in $K(\psi)$ which can be attributed to structure in both Figs 32.12 and 32.13 is of the same order as the shift in $K(\psi)$ attributable to texture differences. Structural differences between soils of similar texture have equal, if not greater, hydrological importance than do differences in texture. In addition, structure has a major influence on the hydraulic properties at matric potentials where most water movement takes place because at these potentials K is sufficiently large (10^{-3} to 10^{-4} m day^{-1}). This emphasises the value of *in situ* measurements. Disturbance exerts greatest influence at high matric potentials and produces lower K values compared to those for undisturbed soils. At low matric potentials the reverse appears to be true (Shaykewich 1970).

The effect of texture and structure on saturated hydraulic conductivity

The saturated hydraulic conductivity K_o of soil depends on its porosity characteristics. Pore size distribution and total porosity are determined by the size, shape and arrangement of the individual particles. Many investigations have been carried out in the attempt to formulate a relationship between porosity and saturated hydraulic conductivity (Engelund 1953). For sands and rock material they have been reasonably successful but in natural soils this has not been the case.

Commencing with Aronovici (1940) empirical relationships between texture and hydraulic conductivity have been found and utilized. In a large data set examined by Mason *et al.* (1957) saturated hydraulic conductivity decreased with increasing silt and clay content and increased as one might expect with percent large pores. Talsma and Flint (1958) found that K_o decreased with increasing clay for subsoils in the Murrumbidgee Irrigation Area, New South Wales, provided that compaction as a consequence of overburden depth was considered. Increasing overburden with depth shifted the relationship to lower K_o for all clay contents. In these subsoils the level and variation in structural development was small and consequently texture was the dominant variable. They demonstrated further that in the cemented layers, which occur in some of the subsoils, K_o increased linearly with increasing aggregate stability as reflected in the 'apparent clay content'. Some cemented aggregates are completely stable, wet little internally, and consequently K_o in these subsoils is about 1 m day^{-1}. In layers where cementation of the aggregates was weak, resulting in an increased apparent clay content, K_o was reduced by at least an order of magnitude from that where the cemented aggregate was stable.

The pioneering work of O'Neal (1949) demonstrated that structure is the most significant factor in evaluating saturated hydraulic conductivity. In fact O'Neal (1949), using both texture and structure, was able to group correctly 95% of horizons examined into seven hydraulic conductivity classes. However, K_o cannot be accurately evaluated on the basis of type or degree of pedality alone; the nature of the ped and the orientation between peds must be considered. For example, the relationship between the vertical and horizontal axes, and the direction and amount of overlap between peds is important.

Although many difficulties arise in the development of quantitative relationships between K_o and morphology, there is sufficient evidence to encourage further collaboration between soil physicists and pedologists. The possibility of using morphology to group soils in terms of their hydraulic properties was explored by Bouma and Anderson (1973) and more of this type of work is needed. Baker (1978) showed that it is feasible to group soil series in terms of their $K(\psi)$ and K_o in such a way that general functions for a group or soil series can be defined with a level of definition which is perhaps adequate for many hydrologic and agronomic purposes.

The heterogeneity of soil morphology and consequently the variability of K_o is notoriously large, and currently there is much interest in the nature and analysis of this variability. Clearly much of this variability is associated with variation in texture and, particularly, structure. There is some hope that better definition of structural properties of consequence to K_o will assist in setting up a stratified sampling procedure to cope with the large variability (Dylia and Grutjens 1970). Talsma and Flint (1958) and others since, showed that, within a soil unit, variation in K_o over several metres is of the same order of magnitude as that between sites several kilometres apart.

Data presented by Bridge (1968) and redrawn in Fig. 32.14 for a *grey clay* and a *red-brown earth*, show that K_o is log normally distributed over five orders of magnitude. These are data from large undisturbed soil cores and are many times more variable than those reported by Talsma and Flint (1958) and Baker and Bouma (1976) for *in situ* measurements. Baker and Bouma (1976) also demonstrate that K_o data are log normally distributed.

Marshall (1959) presents a table which groups K_o into various classes, not only to place values in perspective but also to illustrate that K_o varies over many orders of magnitude from perhaps 10–50 m day^{-1} for well structured *krasnozem* surface horizons or a coarse sand, to

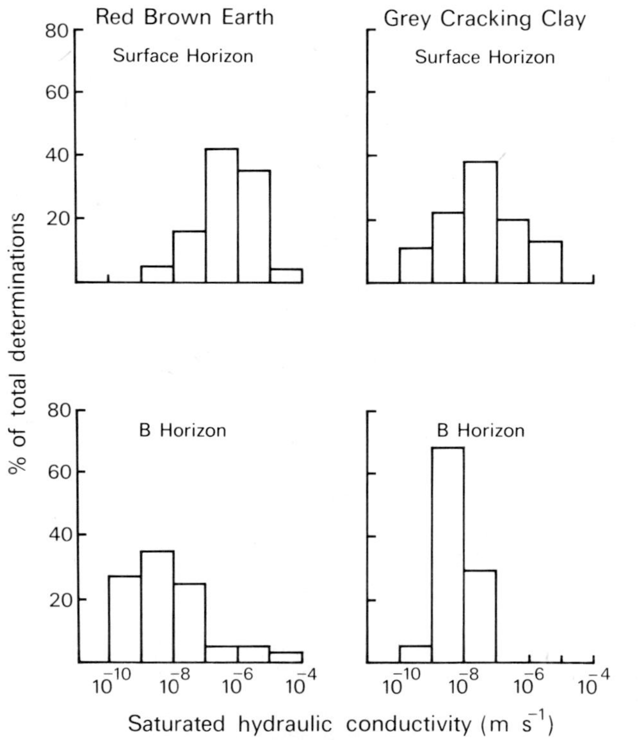

Fig. 32.14. Variability of saturated hydraulic conductivity for two important Australian soils. (Redrawn from Bridge 1968.)

10^{-4} m day^{-1} for the B horizon of a *red-brown earth* or a cracking *grey clay*. It is important to note that clay content and field texture are poorly related to K_o for surface soils. For example, the massive, dense, sandy loam surface of a *red earth* may have K_o values considerably lower than that of a well structured medium clay from the surface of a *krasnozem*.

The form of the relationship between hydraulic conductivity and matric potential

Wesseling (1957) was one of the first to describe the hydraulic conductivity by

$$K = a|\psi|^{-n}$$

where a and n are empirical constants for the soil. For matric potential less than -10 kPa there is good evidence (Rijtema 1965; Wesseling and Wit 1966) that the model describes the relationship quite well. In the region between saturation and -10 kPa, Baker and Bouma (1976) demonstrate that

$$\ln K = \ln(a|\psi|^{-n})$$

fits the data for extensive field measurements in silty clay loam subsoils.

However, for some soils in which the hydraulic conductivity remains relatively constant or changes slowly at matric potentials near saturation the model can be written

$$K = K_o e^{-a(|\psi| - |\psi_e|)}$$

where ψ_e is the matric potential at which air entry takes place. Brooks and Corey(1964) find good agreement using the models:

$$K = K_o|\psi|^{-n} \quad \text{for } |\psi| > |\psi_e|$$

and

$$K = K_o \quad \text{for } |\psi| < |\psi_e|$$

Although many samples prepared in the laboratory demonstrate unambiguous air entry values, these are less common in naturally occurring soils, particularly those exhibiting substantial ped development.

For $|\psi| > |\psi_e|$ the form $K = a|\psi|^{-n}$ has had considerable use and it appears that n has values of 1 to 2 for clays, 2 to 3 for loams and greater than 3 for sands. The value of n reflects the slope of the function in Figs. 32.11 and 32.12. The effect of structure does not appear to change the value of n greatly in this range of ψ, but structural differences are strongly reflected in the value of a. This might be expected, as structure has a dominant influence on K_o and a smaller influence on water retention in this range.

There is insufficient information available on the conductivity functions at different textures and structures to attempt to find a useful grouping of a and n values for classes of texture and structure.

Soil texture and the hydrology of a uniform profile

The consequences of the effect of texture on the hydraulic properties in the profile are documented in the literature. The lower loss by evaporation and more rapid drainage of light textured soils compared to equivalent heavier textured profiles has been observed by Alizai and Hulbert (1970). Although these observations are consistent with the influence of texture on the $\psi(\theta)$ and $K(\theta)$ functions, I am not aware of any quantitative comparative data in the literature which brings together the integrated influence of texture on the profile water balance, despite its importance to agriculture, hydrology and ecology. Hillel and van Bavel (1976) demonstrated the integrated influence of texture on the water balance of the soil profile. They used the $\psi(\theta)$, $K(\theta)$ and $K(\psi)$ relations for a sand, a loam, and a clay to simulate the influence of texture on infiltration, drainage, evaporation and water balance for a uniform non-vegetated profile under a series of rainfall inputs. Their results are redrawn in Figs. 32.15 and 32.16.

The simulation generates more rapid infiltration, lower evaporation, and higher drainage from light textured profiles than from equivalent heavier textured profiles. This is consistent with the empirical

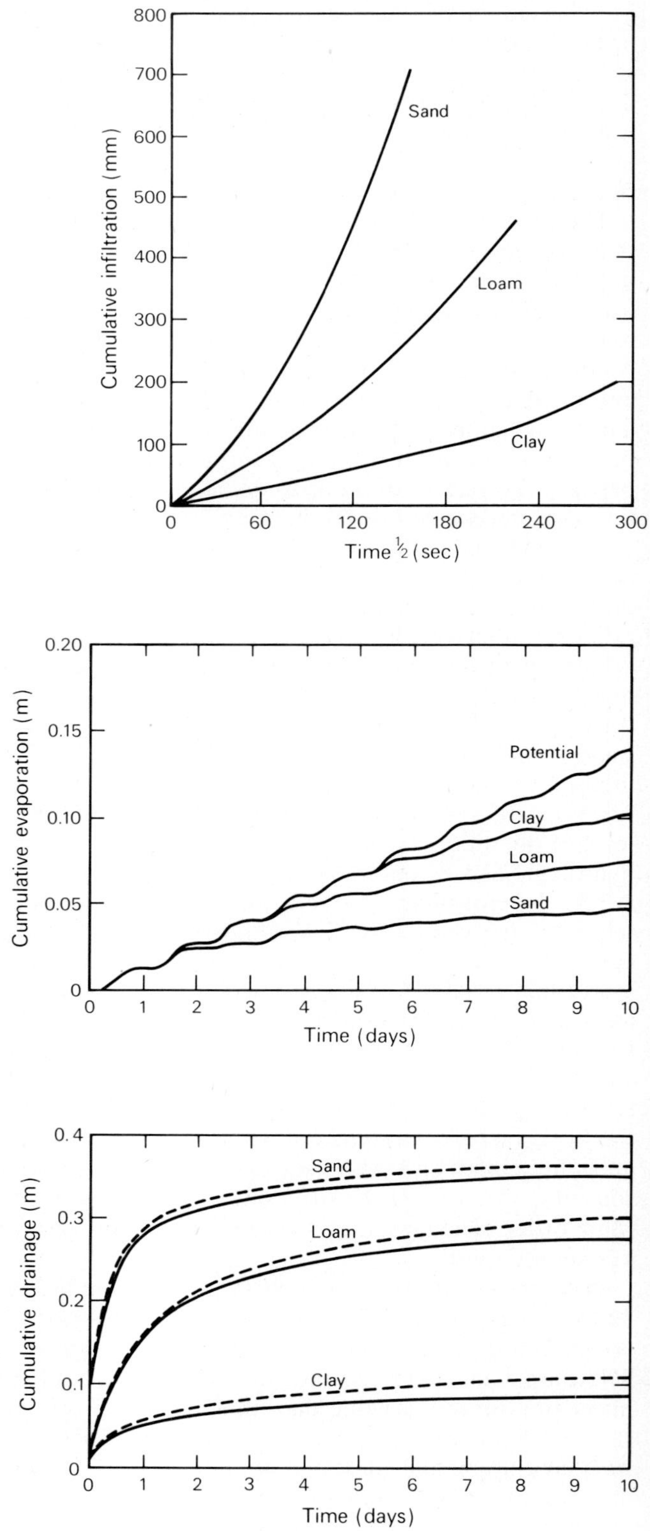

Fig. 32.15. The influence of texture on infiltration, evaporation and drainage in a uniform profile. (From Hillel and Van Bavel 1976.)

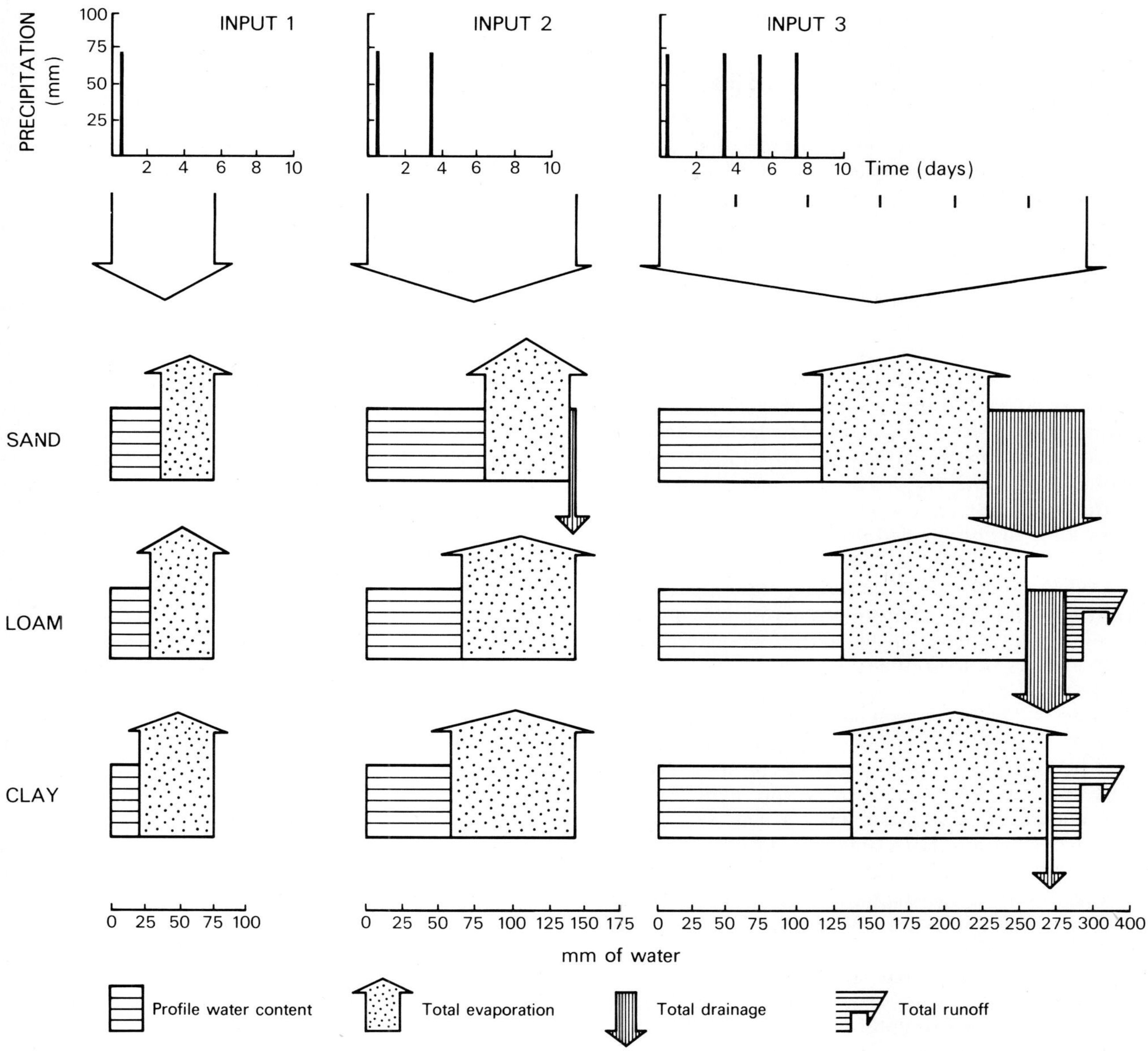

Fig. 32.16. The influence of texture on the water balance of a uniform profile. (From data of Hillel and Van Bavel 1976.)

observations of Alizai and Hulbert (1970). In Fig. 32.16 the hydrology of 1.20 m of uniform sand, loam and clay profiles is depicted at the tenth day following three rainfall regimes.

For rainfall amounts of the order of 75 mm distributed two or three days apart, as is common during the wet season in semi-arid Australia, the storage of water in the profile is optimal in the light textured soils because the evaporation term is minimized. Under larger rainfall inputs the drainage loss from the sand exceeds the runoff which begins to occur in the heavier loams and clays. Evaporation is least for light textured soils but if the inputs are large enough, increasing amounts of water drain through the profile. The relative amounts of runoff and evaporation in the heavier textured profile compared to drainage in the light textured profile are the crucial factors determining which profile stores the most water.

In general terms, the simulation demonstrates and confirms the empirical observation of ecologists that in semi-arid environments light textured profiles store more water (and more of that store is available to plants) than do heavy textured soils. In humid environments, heavy textured soils store more water although less of the stored water will be available to plants. Further, the proportion of the available water held at matric potentials greater than −100 kPa is much greater in light textured than in heavy textured soil. As discussed above, this benefit could well be compensated for by the fact that soil resistance to water flow (a consequence of the $K(\psi)$ function of the soil) becomes a constraint to water uptake by roots at potentials usually below -100 kPa in sands whereas in clay soils, soil resistance does not appear to be important until below -1000 kPa.

It must be emphasized that the simulation does not consider transpiration, and the effect of different profile textures on transpiration could well produce very different water balances to those depicted in Fig. 32.16. The lower evaporation in sands may be compensated by higher rates of transpiration because more of the water stored is held at higher matric potentials.

The higher evaporation rate of the clay profile simulated above does not consider the consequence of the self mulching behaviour often associated with cracking clays. Any such mulch on the surface will substantially reduce evaporation (Talsma 1976). The simulation applies to a uniform profile but, once a mulch develops with drying and cracking, the profile no longer has uniform hydraulic properties.

Fig. 32.16 illustrates the overriding importance of the interaction between the nature of the rainfall input and the hydrologic response of the soil profile with a given texture and set of hydraulic properties. It follows that there is considerable room for compensating influences to operate over a wet season such that the influence of texture on the annual profile water balance may be less than, equal to, or greater than the differences displayed in Fig. 32.16. In one rainfall event, the light textured profile may store more water than a heavier textured profile, whilst in another event, the position could be reversed. The simulation demonstrates that, although profile texture has a substantial influence on the fate of rainfall arriving at the soil surface, the hydrologic response of the profile is to a great extent determined by the time series nature of the rainfall.

The field capacity concept

It is interesting to note in Figs 32.10 and 32.11 that the $K(\psi)$ functions for the different soil textures tend to converge and to cross over as saturation is approached (see also Olsson and Rose 1978). In the saturated condition, sandy loams have a much higher K value than the clay and clay loams. This is in contrast to the range of ψ depicted in Fig. 32.10.

The fact that $K(\psi)$ tends to converge in the region of -5 to -20 kPa at K values of 10^{-3} to 10^{-4} m day^{-1} (Baker and Bouma 1976) points to the reasons why, for many freely draining soils, θ at $\psi = 10$ kPa tends to coincide with the observed field capacity. When these matric potentials are reached in a major section of the profile during redistribution, K has values of the order of 10^{-3} to 10^{-4}m day^{-1} and this is reflected in a reduced rate of drainage. This reduced rate of drainage is associated with a slowly changing water content in the profile. It is this slowly changing water content in a major section of the profile that has been termed 'field capacity'.

Hillel and van Bavel (1976) simulate θ as a function of time for three different textures and illustrate that θ changes very slowly when ψ approaches -10 kPa with K values of the order of 10^{-3} m day^{-1}. In heavy textured soils, however, K at $\psi = -10$ kPa is often in excess of 10^{-3} m day^{-1} with the consequence that the field capacity does not coincide with $\psi = -10$ kPa. For clays, the $K(\psi)$ relation is such that more negative matric potentials must be reached before a very low rate of drainage and small changes in θ are observed. The application of the concept of field capacity in heavy textured soils is difficult, if not inappropriate, because of this slow drainage over long time intervals during which plant water use can be substantial. As will be discussed later, it is often water accession rather than retention that determines the value of the upper limit to the water store in these soils.

Profile depth and hydrological behaviour

Profile and horizon depths are obvious and important morphological characteristics which have a significant influence on hydrological behaviour. Profile depth usually refers to the distance from the surface to the interface with a weathered rock substrate for soils formed *in situ* or the depth to a barrier to water movement and/or plant root penetration.

The thickness of each horizon and the moisture characteristic determine the profile water store and the potential plant-available water store. As seen from Table 32.1, the amount of water held between -10 kPa and -1500 kPa may vary by a factor of two over the extremes of field texture and structure. For many Australian profiles the variation in plant-available water per unit depth is small (Prebble 1870*a*; Loveday *et al.* 1978), while the variation in profile depth or rooting depth can be quite large (Greacen and Hignett 1976). Tennant (1976) showed that the available water in the root zone of wheat in five different soil textures depended on rooting depth more than it did on the moisture characteristic of the horizons within the profile. McCown (1971, 1973, 1976) indicated that depth of the A horizon in the *solodic soils* near Townsville, Queensland was the major variable determining the plant-available water store; these differences in water store were responsible for significant differences in pasture and animal production in these environments.

The depth to impermeable material at the base of *yellow earths* (Stace *et al.* 1968) constrains the profile water store to between 70 and 150 mm (Williams and Coventry 1979) with the consequence that free water surfaces develop during the wet season in tropical, semi-arid environments. This is in contrast to the adjacent deep *red earths* which have similar moisture characteristics yet behave as freely draining profiles with profile water stores in excess of 480 mm.

The thickness of the surface horizon is important to water entry; this is particularly true for texture contrast soils but there is also evidence that for many gradational or uniform texture soils the thickness of the surface horizon is important to the generation of overland flow (Holtan and Minshall 1968). The surface horizons normally have K_o values greater than those of the horizons below, often because of biological activity and increased structural development. In such situations the depth of these more permeable surface horizons is important to the fate of rainfall.

Comparative hydrology of some important Australian soils

The morphological properties of texture and structure exert their influence on the hydrologic behaviour of the profile via their influence on the basic soil hydraulic properties, namely the moisture characteristic and the hydraulic conductivity–water content function. In order that the principles of soil physics be used to predict profile hydrology it is essential to be able to group soil hydraulic properties in terms of profile and horizon morphology. The use of morphology to stratify sampling of soil physical properties is an important tool in dealing with soil heterogeneity, which presently limits quantitative application of soil physics to problems in hydrology (Philip 1975) and agronomy. Work by Baker (1978), coupled with that of Warrick *et al.* (1977) and Peck *et al.* (1977) provides evidence that the relationships between hydraulic properties and morphology are sufficiently substantial and important to be further pursued.

The hydrology of a soil profile is dependent not only on its hydraulic properties and physical dimension, but also on the time series and amount of rainfall, root distribution and water extraction by the plant in response to the atmospheric environment. Although an attempt will be made to describe the hydrology of the major groups in Australian soils, it must be emphasized that the water balance is the result of a dynamic interaction between soil, plant and the atmospheric environment.

For comparative purposes, four generalized groups (Butler and Hubble 1977) will be discussed. It is recognized that within each group there is considerable heterogeneity, yet the groups represent the major differences in hydrological behaviour that exist between Australian soils.

The deep sands

These soils have sand to sandy loam textures with little profile development or variation in texture with depth (the Uc profiles of Northcote (1971)). These soils include the *earthy sands, calcareous sands* and the *siliceous sands*; they are all briefly described in Chapter 3 and are reported to occupy up to 20% of the Australian continent. The hydrology of these soils appears to be relatively simple and amenable to analysis by established soil physical principles and methods. Carbon and Galbraith (1975) demonstrate that a water balance model based on soil physical parameters of sands gives good agreement with observed soil hydrology and plant water use. Carbon (1975) has shown that redistribution in coarse sands can be largely predicted from the moisture characteristic.

The deep sands are characterized by rapid infiltration, low water holding capacity, and under most rainfall conditions generate little runoff and considerable deep drainage. Minimum infiltration rates into sands are often in excess of 50 mm h^{-1} although there are many fine sands and loamy sands in which the rate may be of the order of 15 mm h^{-1}. Low infiltration rates on fine sands and loamy sands are often associated with water repellence (Bond 1964) which is characterized by high spatial variability in infiltration and redistribution.

Although it is generally true that sands have low water holding capacity (less than 10 mm dm^{-1}), there are many earthy sands and siliceous sands with fine sand surface horizons that hold in excess of 15 mm dm^{-1} (see Table 32.2). Most of this available water is retained between -10 and -30 kPa (Prebble 1970*a*). This may well account for the very low available water reported by Tennant (1976) who based his estimates of field capacity on -30 kPa. There is good evidence in Day (1977) that for these textures 'field capacity' is associated more closely with -10 kPa than -30 kPa. Therefore it is important to distinguish between the coarse sands as described by Carbon (1973) and fine sands (Stace *et al.* 1968, pp. 42–46) in terms of hydraulic properties.

The importance of rooting depth as well as water holding capacity on the profile water store can be seen in Fig. 32.17. The deep sand and the *red earth* are similar in this respect and contrast with the *grey cracking clay* and the *black earth*. In the study of Holmes (1960) the profile water store of these deep sands for six sites varied from 273 mm to 473 mm. From the data presented it did not appear that complete recharge of the profile took place each year under rainfed conditions in the 500 mm per year environment. It is important to note that in this environment and under deep rooting heath and legume pasture, deep drainage did not occur during the study interval. In other rainfall environments, particularly those with a marked wet season, deep drainage is no doubt an important term in the water balance of deep sands under rainfed

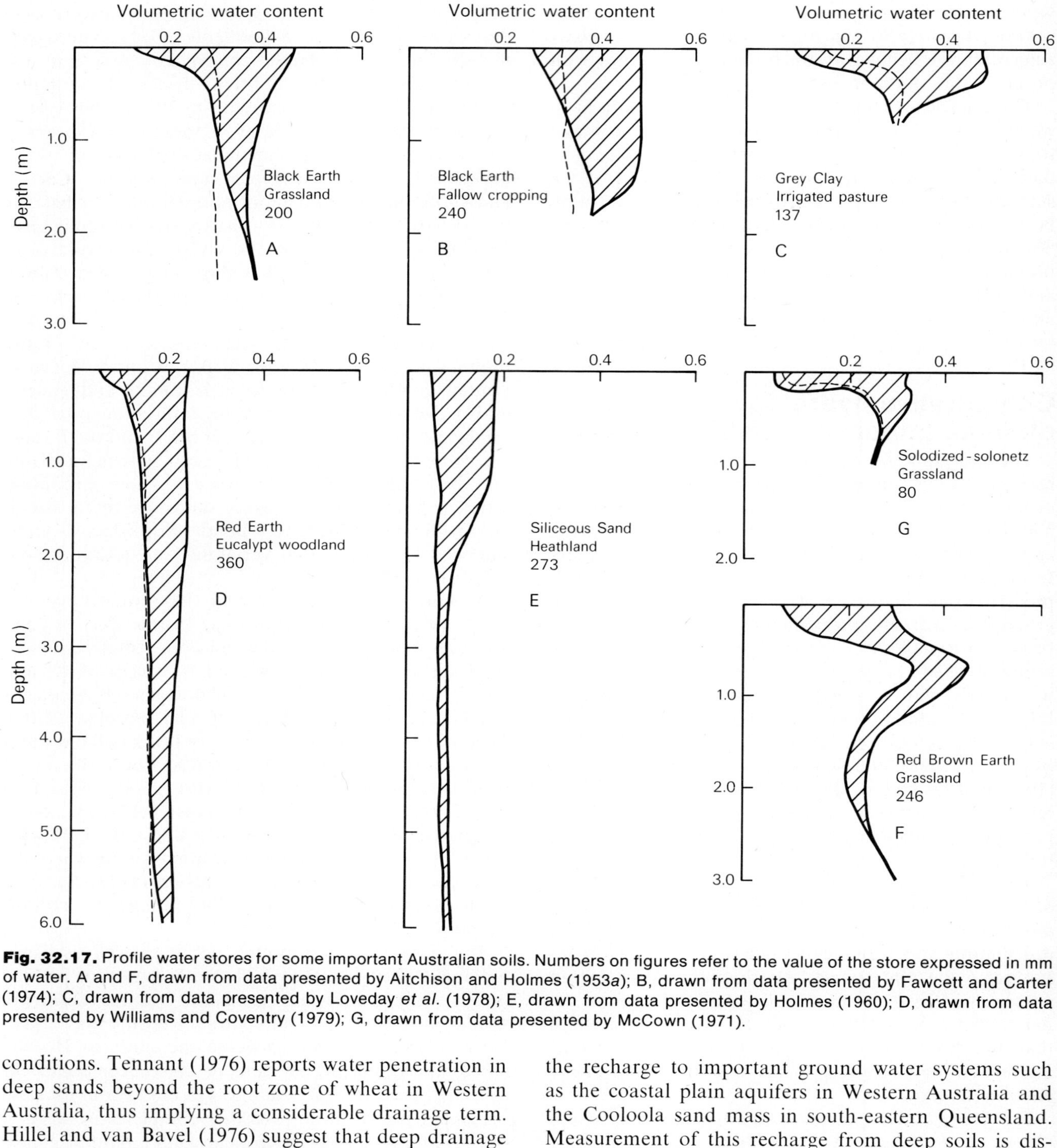

Fig. 32.17. Profile water stores for some important Australian soils. Numbers on figures refer to the value of the store expressed in mm of water. A and F, drawn from data presented by Aitchison and Holmes (1953*a*); B, drawn from data presented by Fawcett and Carter (1974); C, drawn from data presented by Loveday *et al.* (1978); E, drawn from data presented by Holmes (1960); D, drawn from data presented by Williams and Coventry (1979); G, drawn from data presented by McCown (1971).

conditions. Tennant (1976) reports water penetration in deep sands beyond the root zone of wheat in Western Australia, thus implying a considerable drainage term. Hillel and van Bavel (1976) suggest that deep drainage will be an increasing part of the hydrology of sand as rainfall increases or is concentrated over short 'wet' seasons. The water use patterns and rooting morphology of vegetation are equally important as is the pattern and amount of rainfall in determining deep drainage from deep sands.

Deep drainage from profiles of this form provides the recharge to important ground water systems such as the coastal plain aquifers in Western Australia and the Cooloola sand mass in south-eastern Queensland. Measurement of this recharge from deep soils is discussed in Chapter 33.

The coherent earths

Coherent earths (Butler and Hubble 1977) bring together profiles which exhibit gradual or no horizon differentiation and behave as essentially homogenous, rigid profiles in which kaolin and iron oxides dominate the clay mineralogy. In chapter 3 they are described as the massive and structured sesquioxidic soils and include *red, yellow* and *grey earths, krasnozems, xanthozems* and *euchrozems.* Although they encompass a wide range of texture, from loamy sands to clays, these soils generally behave as homogenous rigid profiles; it is anticipated that their water dynamics can be analyzed by the classical principles of soil physics. Hydrological studies of these soils have been restricted to the *red earths, yellow earths* and *krasnozems.* The *red earths* and *krasnozems* both behave as freely draining profiles and provide an interesting contrast of the effect of structure on hydrological behaviour. Compare the moisture characteristic of a structureless loam horizon of a *red earth* with the well structured loam horizon from a *krasnozem* in Fig. 32.9.

Water entry and surface runoff

Infiltration into a *krasnozem* is characteristically very rapid with minimum infiltrating rates of 10–50 mm h^{-1} (Stirk 1948*b*) for soils under pasture near Lismore, New South Wales. There are some estimates of saturated K_o for *krasnozems* under rainforest (Bonell and Gilmour 1978) which range as high as 1200 mm h^{-1}; however, these values are restricted to the top 0.2 m where biological activity is high. Further down the profile K_o has values of the order of 13–64 mm h^{-1} and it tends to decline with depth.

The *red earths* have only moderate minimum infiltration rates of approximately 5–20 mm h^{-1} (Arndt and Norman 1959; Prebble and Stirk 1961; Winkworth 1961; Slatyer 1962; Day 1977). Day (1977) reports very much higher infiltration rates for the light textured *red earths*, with values of 110 mm h^{-1} after 4 h of ponding. The $i(t)$ function appears to be very stable for all the textures of *red earths* examined (Slatyer 1962; Day 1977).

Some *red earth* landscapes are characterized by patterns of microrelief (0.1–0.3 m) which recur at intervals of 3–6 m (Arndt and Norman 1959) and to which water entry is related. This microrelief is usually associated with vegetation patterns, as demonstrated by Slatyer (1962), who measured a minimum infiltration rate of 25 mm h^{-1} in the depression near the shrub *Acacia aneura* compared with 8 mm h^{-1} on the ridges between the shrub groves. Similarly, Torssell (1973) found that sorptivity was highest (9 mm $h^{-1/2}$) in the depression and lower (3 mm $h^{-1/2}$) on the slopes and ridges. This microrelief plays a very important role in the hydrology of *red earths.* During monsoonal rain, ponding in the depressions to depths of 20–30 mm, without generation of surface runoff, is common when the microrelief is not disturbed in the native woodland (J. Williams unpublished data). This is consistent with the lack of an integrated stream network in the low relief *red earth* landscapes. In contrast to this behaviour for native woodlands, Arndt and Norman (1959) and Ive *et al.* (1976) recorded considerable runoff from *red earths* which had been cropped continuously and subsequently sown to pasture. The heavier textured *red earths* are subject to surface sealing under cropping (Arndt *et al.* 1963) whilst the lighter textured profiles tend to erode under cropping at Katherine, Northern Territory, despite their very high infiltration properties. Recently Mott *et al.* (1979) have demonstrated that the surface hydraulic properties of *red earths* are fragile. Heavy grazing and burning can cause an order of magnitude decrease in both S and K_o of the surface horizon, resulting in the development of degraded scald areas within the grassland, and considerable surface redistribution of incident rainfall.

Krasnozems and *xanthozems* by contrast, are strongly structured with very stable peds (W. Arndt, unpublished data). Consequently water entry, although apparently reduced, does not degenerate under sustained cultivation, and minimum infiltration rates of 20 mm h^{-1} are reported under puddled conditions.

Deep drainage

Movement of water beyond the zone of water extraction is to be expected in many of the coherent earths under appropriate rainfall and topographic conditions (Hillel and Van Bavel 1976). Chapman (1963) estimated an average 400 mm per year drained from a *red earth* profile in the Cape York Peninsula, North Queensland, to provide direct recharge to an unconfined aquifer. Despite the high rainfall and its intensity there was little runoff and the large drainage term dominated the water balance. In more arid environments, deep drainage is observed in the *red earths* at Katherine, Northern Territory, under introduced legume pastures (R. L. McCown unpublished data). Wetselaar (1962) reports considerable movement of nitrate beyond the root zone of sorghum and millet under rain fed conditions in the same locality. Similar observations confirming deep drainage have been made on *red earths* under native woodlands and introduced pastures near Charters towers, North Queensland (Williams and Coventry 1979, 1981).

A substantial drainage term is expected in *krasnozems* under tropical rainfall conditions although detailed studies are lacking. Drainage from *krasnozems* in the Coranderrk Catchment near Melbourne contributes perhaps 230 mm annually to the streamflow (Langford *et al.* 1977).

There are many coherent earths, however, which do not exhibit freely draining profiles. These include the *yellow* and *grey earths* and some *xanthozems.* During the wet season in tropical environments, free water accumulates in the *yellow* and *grey earths* in most years, to yield fluctuating perched water tables (Williams and Coventry 1979, 1981). Lateral drainage from these profiles is observed and may well be common behaviour where soil and rock allow.

The profile water store

The freely draining coherent earths represent profiles in which it is appropriate to use the well established upper and lower limits of the soil store, namely 'field capacity' and the water content at −1500 kPa. The principal difficulty in estimating the profile water store is the determination of the depth of water extraction for the given land use. Without direct observation of soil water regimes, estimates using rooting depth are at best estimates of the potentially available water in the profile (Tennant 1976). Fig. 32.17(*d*) illustrates the profile water store for a *red earth* under eucalypt woodland in North Queensland. The profile is deep and, although the water holding capacity is relatively low (approximately 11 mm dm^{-1}), the profile water store is approximately 335 mm, although the potentially available water to 5 m is nearly 500 mm. Dagg (1968) reports similar sized profile water stores for deep rooting grasslands and woodlands on similar soils in East Africa. The full recharge of these profiles is not necessarily an annual event (Williams and Coventry 1979, 1981). Tennant (1976) reported a potentially available profile store for a deep *yellow earth* to be 200 mm under wheat cropping in Western Australia. Based on the data of Arndt *et al.* (1963) and Day (1977) it appears that *red earths* under cropping of sorghum and millet at Katherine, Northern Territory, have potentially available profile stores which range from 146 mm to 269 mm for the maximum rooting depths observed. *Red earths* and *yellow earths* have a relatively low (approximately 11 mm dm^{-1}) water holding capacity, although it does vary considerably (Prebble and Stirk 1961; Arndt *et al.* 1963; Day 1977) to yield a range from 7–19 mm dm^{-1}. Much of the variation is associated with texture as depicted in Table 32.2. However, the higher water holding capacities are often associated with high bulk densities (1.7–1.8 g cm^{-3}), and the value of this water to the root system may be constrained by high soil strength and/or poor aeration. Prebble and Stirk (1961) noted that in *red earths* a large proportion of the available water is retained at matric potentials between −10 and −30 kPa (see moisture characteristics (A) and (B) in Fig. 32.8).

The water holding capacity of some *krasnozems* is exceptionally high (B551, Stace *et al.* 1968 p. 305) and may be 30 mm dm^{-1} in the surface horizons; however, the majority of soils examined (Stirk 1948*a*; Stace *et al.* 1968) hold approximately 15 mm dm^{-1} with a range of 9–20 mm dm^{-1}. K. J. Langford (pers. comm.) reports profile water stores of the order of 300 mm for *krasnozems* in the Coranderrk Catchments north of Melbourne, and Langford *et al.* (1977) emphasize the important role that these large soil stores play in the water supply to Melbourne.

The texture-contrast soils

The texture-contrast soils described in Chapter 3 refer to a wide grouping of soils which are characterized by a sharp increase in texture between the A and B horizons. The depth of the A horizon varies from 0.03 to 0.6 m whilst the texture ranges from loamy sand to clay loam. The light to heavy clay B horizons exhibit a wide variation in grade of structure and size of ped. The presence or absence of sodic conditions and the nature of the clay mineral in the B horizon, along with the size, grade and stability of the ped, are the principal factors which appear to determine or correlate with the different soil hydrologies observed within the group. The important soils in this group include *red-brown earths, non-calcic brown soils, solodics, solodized-solonetz* and many of the *red* and *yellow podzolics*.

The hydrological significance of a sharp increase in texture is dependent on the absolute and relative difference in hydraulic properties between the two horizons. In some texture-contrast soils, such as the *red podzolics* and the *non-calic brown soils*, the texture contrast has a relatively small hydrologic consequence and many behave essentially as freely draining profiles (see Soil No. 18, McCown *et al.* 1976). The increase in clay is often offset by the presence of a large proportion of relatively small (10–20 mm) stable peds. In the *solodized-solonetz, solodics* and many sodic *red-brown earths*, the hydraulic properties of the two horizons are very different. The increase in clay in these soils is associated with a low grade of structure, low stability

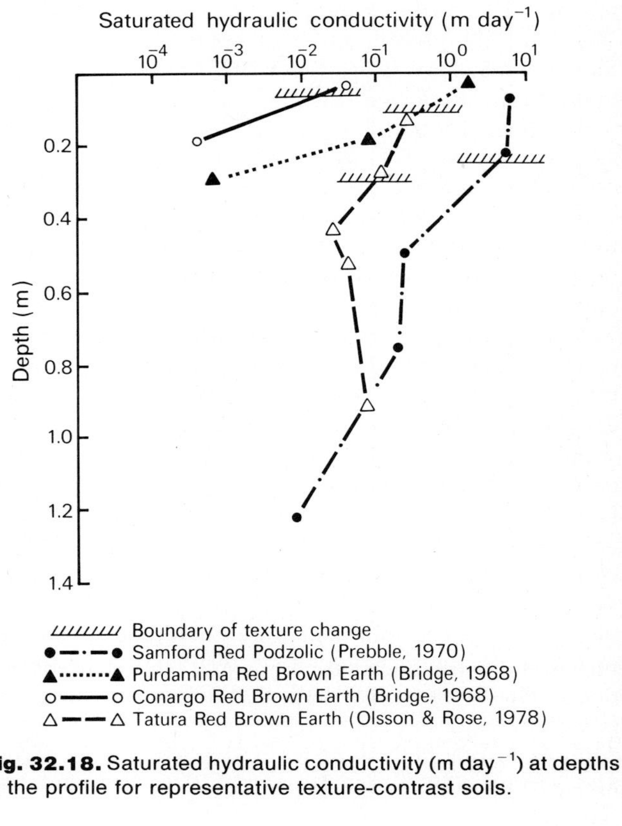

Fig. 32.18. Saturated hydraulic conductivity (m day^{-1}) at depths in the profile for representative texture-contrast soils.

and usually a very large ped size (100 mm or more), often with high levels of salts and exchangeable Na and Mg; all of these factors contribute to a low saturated hydraulic conductivity in the B horizon.

Data from Bridge (1968), Prebble (1970*b*) and Olsson and Rose (1978) brought together in Fig. 32.18, illustrate that at least an order of magnitude decrease in K_o usually occurs in Australian texture-contrast soils between the A and B horizons. In the sodic *red-brown earths* the decline in K_o from the A to the B horizon is often many orders of magnitude (Bridge 1968). Fig. 32.18 also illustrates the reason for the general observation of different hydrologies between sodic *red-brown earths* and *red podzolics*, both with a marked texture contrast. The K_o values in the A and B horizons are at least an order of magnitude larger in the *red podzolics*. The consequence of this is that under most conditions the *red podzolics* behave as freely draining profiles. But the differences in K_o values between these soils is closely correlated with soil structure and its stability, not texture. The clay mineralogy of the B horizons of the *red-brown earths* is usually dominated by illite with montmorillonite present as a minor constituent (Norrish and Pickering 1977); these B horizons usually exhibit shrink/swell behaviour. The B horizons of the *red* and *yellow podzolics, solodized-solonetz* and the *solodics* by contrast are usually dominated by kaolinite (Norrish and Pickering 1977) and generally behave as rigid soils, although the presence of small amounts of montmorillonite cause some to swell and shrink.

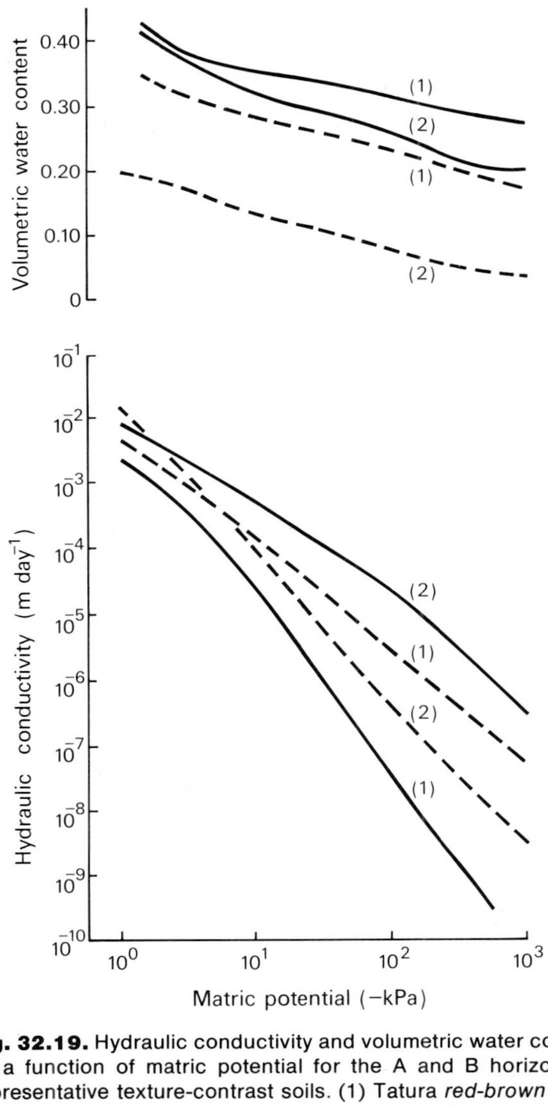

Fig. 32.19. Hydraulic conductivity and volumetric water content as a function of matric potential for the A and B horizons of representative texture-contrast soils. (1) Tatura *red-brown earth* (Based on data from Olssen and Rose 1978). (2) Narayen *gleyed podzolic* (G. B. Stirk and R. E. Prebble, unpublished data). Dashed line, A horizon; solid line, B horizon.

Not only is the K_o much lower in the sodic form of the *red-brown earths* but it appears to be a characteristic of these soils for the $K(\psi)$ function of the B horizon to locate below the $K(\psi)$ function of the A horizon for the whole range of ψ. At all potentials, K for the B horizon is less than K for the A horizon. A feature of many texture contrast soils is that the $K(\psi)$ function of the A horizons cross over at some potential to reverse the order of hydraulic conductivity. This is apparent in Figs. 32.12 and 32.19 where the hydraulic properties of the *red-brown earths* are contrasted with those of *red* and *gleyed podzolics.* It is well known that there is considerable variation in the hydraulic properties of the B horizon of the *red-brown earths* although little is documented in the literature. The differences appear to be associated with aggregate stability as determined by the cation balance of the exchange complex. Once dispersion of the colloid commences, there is a dramatic decline in the saturated hydraulic conductivity (Emerson and Bakker 1973). Many of the South Australian *red-brown earths* (Aitchison and Holmes 1953*a*) have B horizons which, although slowly permeable, do not generally develop perched water tables and saturation in the A and upper B horizons as is common with sodic forms distributed in the Riverina of New South Wales and the Goulburn Valley of Victoria.

Analysis of water entry and movement in texture-contrast soils

Childs and Bybordi (1969) and Bybordi (1973) have applied the Green and Ampt (1911) analysis to water entry into layered or texture-contrast soils. Hanks and Bowers (1962) used numerical methods to simulate the water dynamics. Field confirmation of these analyses is unfortunately restricted to that of Bruce and Whisler (1973) and although encouraging, the simulation of infiltration was by no means completely satisfactory. The observations of Miller and Gardner (1962) are of the form anticipated from both analytical and numerical solutions of the infiltration process. After the wetting front enters the B horizon, a zone of positive pressure head develops as the A horizon becomes saturated, and the infiltration rate rapidly declines. If the conductivity of the B_2 is substantially less than the B_1, the B_1 will tend to saturate following entry of the wetting front into the B_2 and a second period of rapid decay in the infiltration rate will occur. It is the depth and magnitude of the change in the $K(\psi)$ function between the A and B horizons which determines the shape of the infiltration rate/time curve ($i(t)$) for a profile. The analysis of Bybordi (1973) shows that for a layered soil with various combinations of K_o and horizon thickness, it is possible for the $i(t)$ relation to take almost any form. In practice, therefore, the form of the $i(t)$ curve for a texture contrast profile may be indistinguishable from that of a given uniform profile, particularly if the peds disperse during infiltration.

The analyses of Swartzendruber (1960) and Bybordi (1973) clearly demonstrate that it is not only the hydraulic properties of the texture-contrast horizons but also the depth of each horizon relative to total profile depth which is critical to water entry, storage and movement in the profile. The effective saturated hydraulic conductivity of a layered profile is therefore not equal to the K_o for the least permeable horizon but has a value larger than that of the least permeable horizon by an amount dependent on the thickness of the horizons.

The simulation by Hillel and Talpaz (1977) of the water dynamics for layered profiles can give some insight to the hydrology of some texture-contrast soils in Australia. The clay horizon in their simulation is at least an order of magnitude more permeable than that encountered in many of the B horizons in sodic *red-brown earths, solodized-solonetz* and *solodics*. However, it is of a similar order to many of the B horizons of the *podzolics* and perhaps the *non-calcic brown soils.*

The simulation again indicates the strong interaction between the nature of the rainfall intercepted and the profile water balance. As has been widely observed, the simulation predicts impeded infiltration; yet for subsoils with reasonable hydraulic properties (K_o 200 mm day^{-1}), the subsoil may in many instances promote drainage of the A horizon. Evaporative loss from a texture-contrast profile is reduced below that which would occur if either texture occupied a uniform profile of the same dimension. This is only true if the hydraulic properties of the B allow substantial water to enter the B horizon. The coarser the texture of the A horizon the larger is the reduction in the evaporation. However, the serious restriction of the simulation to non-vegetated soil must be kept in mind.

Deep drainage under rainfall and irrigation

Drainage from a texture-contrast profile is simulated (Hillel and Talpaz 1977) to be very small and this is consistent with the observations (Dunin 1970; McCown 1971 and McCown *et al.* 1976) for *solodic soils* and *solodized-solonetz* under rain-fed conditions and long intervals of ponding. A similar behaviour appears to be true for the *red-brown earths* with sodic B horizons examined by Loveday *et al.* (1978). There was zero drainage beyond 2 m after 120 days ponding in a transitional *red-brown earth*; however, substantial deep drainage was observed when the hydraulic properties were stabilized by heavy applications of gypsum.

There are many texture-contrast soils, however, which do exhibit a significant deep drainage term under rain-fed conditions. Texture-contrast soils with B horizons which possess hydraulic properties similar to the Narayan *gleyed podzolic* (see Fig. 32.11) appear to yield significant drainage terms under rain-fed conditions. Holmes and Colville (1970) measured drainage in the range 25–134 mm year^{-1} for a *gleyed podzolic* (Stace *et al.* 1968, Profile B p. 364) in the Gambier Plain of South Australia. Aston and Dunin (1977) measured drainage from a *red podzolic* to be as high as 2.5 mm day^{-1} during winter in the Southern Tablelands of New South Wales. There is evidence that the *non-calcic brown soils* (Crack and Isbell 1971) exhibit a significant drainage term under a semi-arid tropical rainfall environment, particularly when the eucalypt woodland is removed. The salt and sulphate distribution (Probert and Jones 1977) in these soils would suggest that drainage from these texture-contrast profiles occurs sufficiently frequently to remove sulphate from the solum.

Lateral drainage along the top of the B horizon is a common observation in many texture-contrast soils (G. B. Stirk unpublished data) often resulting in salt seeps at breaks in slope or at lower regions of the landscape (Cope 1958; Conacher 1975). This is particularly true in texture-contrast soils with sodic B horizons in which change in land use has resulted in an increase in infiltration and a reduction in evapotranspiration, with the consequence that there is a significant increase in the vertical and horizontal drainage term of the water balance (Peck 1978).

Profile water store

The simulation of Hillel and Talpaz (1977) suggests that a texture-contrast profile consisting of a sand or loam A horizon over a clay B horizon will store considerably more of the incident rainfall than the equivalent uniform profile of either of the three textures. This is consistent with the findings of Tennant (1976), although the agreement may be fortuitous. The differences in the simulation were principally due to reduced evaporation in the case of the clay, and reduced drainage in the case of sand or loam. It is important to recall that their simulation was for a non-vegetated profile in which the hydraulic properties of the B horizon allow substantial water entry.

Many Australian texture-contrast soils have very impermeable B horizons (McCown *et al.* 1976; Loveday *et al.* 1978). In such soils the above conclusions are inappropriate. Texture-contrast soils with sodic B horizons are characterized by a small profile water store (25–100 mm, see also Fig. 32.17*g*) which is largely determined by the depth and texture of the A horizon and the depth of water penetration into the B horizon (McCown *et al.* 1976).

It is common, however, for non-sodic texture-contrast soils such as many *red-brown earths, podzolics* and *non-calcic brown soils* to have profile water stores of 120–230 mm (Aitchison and Holmes 1953*a,b*; Johns and Smith 1975; McCown *et al.* 1976). Water recharge and extraction occurs to perhaps 2.5 m under native pasture and to shallow depths under cropping and irrigation. Fig. 32.17*g* and *f* contrast the form and magnitude of the profile water store for a *solodized-solonetz* and a *red-brown earth*. They illustrate that in these soils the water store is determined by the depth of the A horizon and water entry properties of the B horizon to a much greater extent than the water holding capacity of either horizon as set out in Table 32.2. For texture-contrast soils, it is the water entry properties of horizons rather than their water holding properties that govern the magnitude of the profile water store.

Water entry and surface runoff

Except for Talsma (1976) and Eastgate *et al.* (1979), there appears to be little published work which attempts to relate rainfall-runoff behaviour to either profile morphology or hydraulic properties of Australian soils, although the approach has much to recommend it (Turner 1963).

There is however, considerable data on the minimum infiltration rate for many soils, usually obtained during assessment for irrigation (Stirk 1948*a*, 1952). Talsma (1969, 1976) reports $i(t)$ functions for a number of texture contrast soils in the Southern Tableland of New South Wales and suggests that except for a *yellow podzolic* most of these soils are unlikely to generate surface runoff from the A horizon for rainfalls up to 60 mm under the rainfall intensity experienced in this environment. Dunin and Reyenga (1978) conclude that surface runoff is a negligible component in the water balance of a *Themeda* grassland on a *red podzolic* at Krawarree near Canberra. Texture-contrast soils with structured A horizons are unlikely to generate runoff before the wetting front arrives at the B horizon, under commonly occurring rainfall intensities (less than 50 mm h^{-1}). However, once the A horizon approaches saturation, surface runoff can usually be expected from most texture-contrast soils. In the provisional grouping of minimum infiltration rates for Australian soils devised by G. B. Stirk and G. D. Hubble and reported by Eastgate *et al.* (1979), the texture-contrast soils have low rates ranging from 8.5 to less than 1.0 mm h^{-1}. Fig. 32.20 provides a comparison of minimum infiltration rates for the four major groupings of Australian soils. The cracking clays and the texture-contrast soils with sodic B horizons have low minimum infiltration rates, usually less than 5 mm h^{-1}, but the coherent earths and deep sands usually have rates in excess of 10–15 mm h^{-1}.

It is general experience that many texture-contrast soils, particularly those with shallow, massive, hard setting A horizons which exhibit dispersion and surface sealing (Matheson 1969), generate substantial runoff under rain-fed conditions (Dunin and Downes 1962; McCown 1973). The huge peak discharge of the Burdekin River, North Queensland appears to result from relatively high intensity rain falling on large areas of shallow solodics in the upper reaches of the catchment.

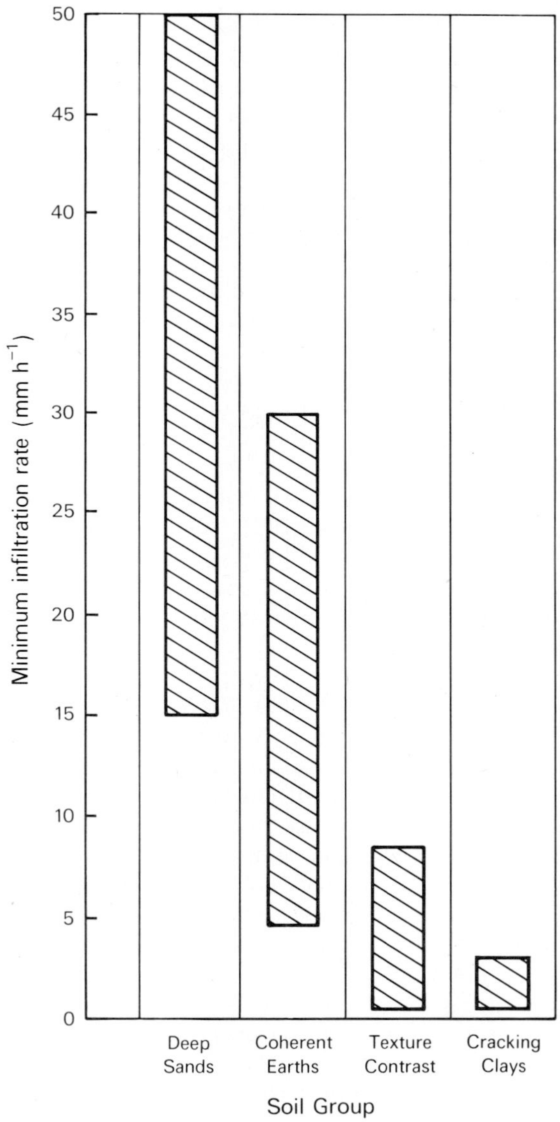

Fig. 32.20. Generalized range of minimum infiltration rates for four major groups of Australian soils.

The cracking clays

The term cracking clay brings together all uniform, fine textured soils which crack seasonally (the Ug profiles of Northcote 1971) and encompass *grey, brown* and *red clays* and *black earths* (Stace *et al.* 1968). These soils occupy large areas of eastern Australia (see Chapter 3, Fig. 3.7) where they not only support productive grasslands, but also are used for both dryland and irrigated cropping. They are widespread in most of the irrigation areas in New South Wales, Queensland and Western Australia. The climate in which they are distributed is invariably semi-arid, although the nature of the rainfall distribution can vary from mediterranean to monsoonal. The range of morphology encompassed by this extensive group is described by Stace *et al.* (1968) and in Chaper 3 of this book. In general these soils are relatively uniform, heavy textured profiles which exhibit three dimensional swelling and shrinking to produce cracking and various degrees and depths of surface self-mulching.

As discussed above, water movement in swelling soil is quite different from that in a rigid soil. The importance of overburden potential as well as matric and osmotic potential in determining the total water potential in a swelling clay is the principal difference. Thus for a swelling profile in equilibrium with a water table, the matric potential decreases with depth down the profile. In the case of a rigid soil, it increases with depth towards the water table. In contrast to the case of a rigid profile in equilibrium with a water table, the volumetric water content of a swelling soil decreases with increasing depth, and water content decreases beneath the water table. Although field experimental data is scant, Talsma (1973) concluded that the hydric and xeric water profile forms proposed by Philip (1969) are found in field soils. There is, however, a need for a great deal more experimental work to confirm the theory.

Infiltration under ponding

The initial infiltration rate following ponding on a dry cracked clay is usually very high, with up to 70% of the water increment entering the profile within the first 10 minutes (Quirk and Blackmore 1955). Under crack entry conditions, initial water uptake is limited by the availability of crack volume (Swartz 1966; Gardner 1978). Once the crack volume is full, infiltration proceeds as a linear function of $t^{1/2}$ until the time (days after ponding for some soils and 2–3 h on others, depending on the relative values of S and K) when deviation from linearity occurs as gravity induced flow starts to influence the process.

There is some controversy as to the nature and importance of gravity induced flow during infiltration in field cracking clays. The data of Talsma and van der Lelij (1976) show that infiltration after one week proceeds as for a rigid soil with the net effect of gravity making a positive contribution to flow. For gravity to make a positive contribution in a swelling soil it implies that α values (Philip 1969) must be small, and/or permanent preferential flow paths must exist (Ritchie *et al.* 1972) between the large structural units.

Gardner (1978) presents data which show that two-thirds of the water increment instantly enters the soil by filling of crack volume. This is consistent with normal three dimension, equidimensional volume change. Subsequent water entry, equal to about one third of the water increment, is determined by the vertical surface area of the cracks and in this period it is a sorption dominated process, in which cumulative infiltration is linearly related to the square root of time. Gardner's finding and analysis using current infiltration theory is consistent with the earlier findings of Quirk and Blackmore (1955).

The volume and particularly the depth of cracking is a consequence of water extraction by plant roots and direct evaporation. As water entry under flooding depends on the cracking depth and volume, the entry of water to a cracking clay depends on the volume and extent of the previous water extraction process (Swartz 1966; Loveday *et al.* 1978).

Following very rapid water entry as crack volume is filled, the declining infiltration rate in a cracking clay proceeds at rates determined by S and K in equations presented by Collis-George (1977). Based on data provided by Talsma and van der Lelij (1976) and Collis-George (1977) for field measurements on cracking clays, S, and K range from 50–352 mm day$^{-1/2}$ and 2–30 mm day^{-1} respectively. After long periods (days or weeks) following ponding, very low infiltration rates of the order of 0.1 mm day^{-1} are reported for soils in which throttles develop within or at the surface of the profile because of structural instability. For cracking clays with a more stable structure which maintain cracks that provide preferential pathways (Talsma and van der Lelij 1976; Loveday *et al.* 1978), long term infiltration rates of about 1–10 mm day^{-1} appear to be common, although rates of 35 mm day^{-1} have been reported by Ritchie *et al.* (1972).

Except for those of Collis-George (1977), the empirical infiltration equations are unable to describe the initially very rapid water entry into cracked clays. At present I am not aware of an infiltration equation which describes infiltration into a cracking clay from fundamental physical properties. Quantitative description of water entry into cracking clays remains a challenging area in soil physics research.

Infiltration under rainfall

Under rainfall on clay with open cracks, the amount of water which enters the cracks directly depends on rainfall intensity and the stability of the surface. Rainfall in excess of the infiltration rate of the clay between cracks runs off to enter the open crack system. This water travels under gravity to the base of the cracking system where it enters the soil matrix in accordance with swelling media infiltration theory (Philip 1969). The stability and nature of the cracking system in conjunction with rainfall intensity determines water entry. If rainfall intensity is low, the surface cracks may close by swelling of the soil and subsequent entry will be much reduced. In fact any process which tends to reduce crack openings at the surface will obviously reduce water entry into a cracking clay. Stirk (1954) demonstrated that cultivation of the surface soil to form a mulch tends to eliminate any contribution to conduction by small cracks which are present at high water contents.

The soil water store

The cracking clays are one group of profiles in which the soil water store is determined by the 'water entry capacity' rather than by the form and nature of the moisture characteristic expressed as a 'water holding capacity' (Farbrother 1972; Loveday *et al.* 1978). Bridge (1968) and Loveday *et al.* (1970) have demonstrated that gypsum increases water entry and subsequently the available water store in swelling soils. Consequently, the definition of the upper limit of the profile water store presents real difficulty with cracking clays. The concept of field capacity as the upper limit of the profile water store is inappropriate in a cracking clay. A number of approaches have been adopted. Loveday *et al.* (1978) use the water content profile following extended ponding as the upper limit, whilst Gardner (1978) calculates maximum water content from theory (Philip 1969) for a swelling profile. The duration of ponding used is usually an attempt to simulate the conditions experienced in the land use under examination. Van der Lelij and Talsma (1978) found that water entry, although small (0.1 to 1.0 mm day^{-1}), continued for many weeks after ponding, with the consequence that the upper boundary of the profile store was dependent on ponding duration. This is in contrast to a rigid, freely draining profile.

There also appears to be some difficulty, although this is not restricted to cracking clays, in using the water content at −1500 kPa as the lower limit of the soil store. Under native and introduced pastures following long drying regimes (Winkworth 1961; Scotter 1964; McCown 1971), and as seen in Fig. 32.17*a*, *c* and *g*, much of the profile is dried well beyond −1500 kPa. In cropping situations, Fawcett and Carter (1974), McIntyre *et al.* (1976) and Gardner (1978) report that, except for the surface 0.2 m or so, crops such as sorghum or wheat rarely dry the profile beyond −1500 kPa. However, W. Berndt (pers. comm.) has data for a *black earth* in which wheat cropping dried the profile beyond −1500 kPa to a depth of 1.5 m. Clearly, application of arbitrary upper and lower limits based on water contents derived from the moisture characteristic are of limited value in describing the water store for a cracking clay. This is of course equally true for many texture-contrast (duplex) soils: there is ample evidence (Loveday and McIntyre 1966; Loveday and Scotter 1966; Farbrother 1972; McCown *et al.* 1976) to demonstrate that it is water entry rather than the water holding capacity which determines the available soil water store. Therefore it is essential to monitor *in situ* the profile water regime over a large number of infiltration and water extraction events to obtain a reliable estimate of the profile water store.

An important measure relating to the available soil water store for a cracking clay is the water increment following irrigation. The water added, usually under ponded conditions, is strongly dependent on the previous drying history as reflected in the depth and volume of cracking. Much of the variation in water increment for cracking clays reported in the literature is a consequence of this. Water increment in cracking clay profiles which have been dried beyond −1500 kPa for much of the profile, ranges from 80 to 150 mm (Loveday *et al.* 1978; Gardner 1978; Farbrother 1972). Values above 120 mm are usually, although not always, those of *black earths* rather than *grey clays*.

Other work (Loveday and Scotter 1966; Tisdall 1951; Bridge and Kleinig 1968; Quirk and Blackmore 1955) with irrigated cracking clays shows much smaller water increments, ranging from 34 to 58 mm. In these instances it appears that cracking was much reduced or water entry was limited by very low saturated hydraulic conductivity in an upper zone of the profile as a consequence of structural instability or other mechanisms. For many cracking clays it appears that the depth of recharge following irrigation reflects rooting behaviour and water extraction rather than a basic soil hydraulic property (Farbrother 1972). Soil strength and aeration, however, may well be factors which restrict rooting depth and extraction and subsequently the water increment. The water store in a *grey clay* under irrigated pasture is depicted in Fig. 32.17*c*. Note that the soil store of 137 mm is consistent with the water increment (Loveday *et al.* 1978) but very different from that estimated using data in Table 32.2. Using the traditional values for the upper and lower limits and a depth of 70 cm, medium clay yields a store of approximately 80 mm. In this situation this is an underestimate but in other situations it could equally well be an overestimate.

Water store under rainfall

Recharge of a swelling clay under rainfall is a very different process from that under ponding. Fig. 32.17*a* illustrates the nature of the soil store for a *black earth* under two different environments. Under fallow and wheat cropping near Narrabri, New South Wales the profile has recharged to a greater extent than that under pasture near Adelaide where the profile has undergone considerably more drying, particularly at depths less than 0.5 m. From the data of Aitchison and Holmes (1953*a*), Waring *et al.* (1958), Swartz (1966), and Fawcett and Carter (1974), *black earths* under rainfall have a profile store of the order of 180–250 mm of water. Some recent data (R. D. Berndt, pers. comm.) suggest that *black earth* profiles under fallow cropping of wheat on the Darling Downs of Southern Queensland can recharge to a depth of 2 m and dry beyond –1500 kPa over most of this depth to yield in excess of 330 mm of water. This behaviour, however, appears to be restricted to the Waco clay (Thompson and Beckman 1959), which exhibits extensive self mulching and a stable fine structure to considerable depths. The type of behaviour depicted in Fig. 32.17*a* and *b* represents the more common behaviour of the *black earths*.

The *grey, brown* and *red clays* are extremely variable in the amount of water they store under rainfall. Although clay content, mineralogy and exchangeable Na are highly variable in these soils, their soil water store results not from the water holding capacity but rather from the water entry properties and the rooting depth and subsequent cracking pattern, which are of course interrelated. The structural instability of the surface and the low hydraulic conductivity of the throttle associated with sodic and low electrolyte conditions are the principal variables controlling water entry (Loveday 1976). R. E. Winkworth (pers. comm.) reports profile water stores for *grey clays* on the grassland of the Barkly Tablelands of the Northern Territory of 200–260 mm. In contrast, Tennant (1976) reports less than 60 mm for a *grey clay* in the central wheat belt of Western Australia. *Grey clays* with a small water store can often be recognized by the absence of, or presence of, only a very thin self-mulching surface horizon, underlain by generally large peds in excess of 100 mm when dry. The *grey clays* with water stores in excess of 150 mm under dryland conditions tend to be correlated with fine peds of 10–20 mm which readily break to peds of 3 mm or less (Butler and Hubble 1977).

From this discussion, it is clear that within a soil group such as the cracking clays, there is considerable variation in the profile water store. Much of this variation is the result of an interaction between rainfall and vegetation which determines water entry and extraction and may not reflect extreme heterogeneity in hydraulic properties within the group. Studies to establish the factors contributing to this variability in soil water entry and storage seem warranted if we are to use soil physical approaches to practical problems in agriculture, hydrology and engineering involving the cracking clays.

Deep drainage under irrigation and rainfall

Studies by Stern (1965) on irrigated safflower in the Ord River Valley of north-western Australia show that drainage in a cracking clay with hydraulic properties similar to that for the clay in Fig. 32.11 ranged from 0.3 to 0.8 mm day^{-1}. Holmes and Watson (1967) report drainage from irrigated pasture on a cracking clay at Murray Bridge in South Australia to be considerably higher, ranging from 1.6 mm day^{-1} in winter to 3.3 mm day^{-1} during summer. This is similar to the 2 mm day^{-1} recorded by van der Lelij and Talsma (1978) under flooded rice in the Murrumbigee Irrigation Area of New South Wales. Observations on cracking clays with a sodic horizon show that deep drainage is very much smaller, ranging from 0.05 to 0.16 mm day^{-1}. Farbrother (1972) and Loveday *et al.* (1978) indicate that deep drainage may penetrate to 14 m after 20 to 30 years cropping under irrigation.

The presence of sodic conditions and the influence of electrolytes on colloidal behaviour and subsequent hydraulic properties (Loveday 1976) are important factors which account for the very different behaviour reported in the literature. The presence of deep, stable cracks as described by Ritchie *et al.* (1972) and Bonn and Wösten (1979) may well be a factor associated with the relatively high rates of deep drainage reported by Holmes and Watson (1967) and van der Lelij and Talsma (1978). It cannot be assumed that deep drainage in cracking clays under flood irrigation will be insignificant and of little hydrological importance in the long term water balance of the regional landscape.

Under rain-fed conditions, studies of Aitchison and Holmes (1954) and Winkworth (1969) indicate that little, if any, water movement takes place beyond the root zone in cracking clay under semi-arid and arid grasslands. These studies are by no means comprehensive and under higher monsoon rainfall conditions and under vegetation such as fallow cropping, water may well move beyond the zone of extraction by plants.

References

Aitchison, G. D. (1960). Relationships of moisture stress and effective stress functions in unsaturated Soil. *In* 'Pore Water Pressure and Suction in Soils.' Conf. Int. Soc. Soil Mech. Found. Eng. (Butterworths: London.)

Aitchison, G. D. and Holmes, J. W. (1953*a*). Seasonal changes of soil moisture in a red brown earth and black earth in South Australia. *Aust. J. Appl. Sci.* **4**, 260-73.

Aitchison, G. D. and Holmes, J. W. (1953*b*). On water stores in soils. CSIRO Aust. Div. Soils Tech. Mem. No. 11/53.

Alizai, H. U. and Hubert, L. C. (1970). Effects of soil texture on evaporative loss and available water in semi-arid climates. *Soil Sci.* **110**, 328-32.

Arndt, W. and Norman, M. J. T. (1959). Native pastures in Tipperary clay loam at Katherine, N. T. CSIRO Aust. Div. Land Res. Regional Survey Tech. Pap. No.3.

Arndt, W., Phillips, L. J. and Norman, M. J. T. (1963). Comparative performance of crops on three soils of the Tipperary Region, N.T. CSIRO Aust. Div. Land Res. Regional Survey Tech. Pap. No. 23.

Aronovici, V. S. (1940). The mechanical analysis as an index of subsoil permeability. *Soil Sci. Soc. Am. Proc.* **11**, 137-41.

Ashton, D. H. (1975). The root and shoot development of *Eucalyptus regnans* F. Muell. *Aust. J. Bot.* **23**, 867-87.

Aston, A. R. and Dunin, F. X. (1977). Empirical model for drainage from soil under rain-fed conditions. *Aust. J. Soil Res.* **15**, 205-10.

Baker, F. G. (1978). Variability of hydraulic conductivity within and between nine Wisconsin soil series. *Water Resour. Res.* **14**, 103-8.

Baker, F. G. and Bouma, J. (1976). Variability of hydraulic conductivity in two subsurface horizons of two silt loam soils. *Soil Sci. Soc. Am. J.* **40**, 219-22.

Bakker, A. C. and Emerson, W. W. (1973). Comparative effects of exchangeable calcium, magnesium and sodium on some physical properties of red brown earth subsoils. III. Permeability of Shepparton soil and comparison of methods. *Aust. J. Soil Res.* **11**, 159-65.

Baver, L. D., Gardner, W. H. and Gardner, W. R. (1972). 'Soil Physics.' (John Wiley and Son: New York.)

Berndt, R. D. and White, B. J. (1976). A simulation-based evaluation of three cropping systems on cracking-clay soils in a summer rainfall environment. *Agric. Meteorol.* **16**, 211-31.

Bishop, A. W. (1960). The measurement of pressure in the triax test. *In* 'Pore Water Pressure and Suction in Soils.' Conf. Int. Soc. Soil Mech. Found. Eng. (Butterworths: London.)

Bishop, A. W. and Henkel, D. J. (1962). 'The Measurement of Soil Properties in the Triaxial Test.' (Arnold: London.)

Bond, R. D. (1964). The influence of the microflora on the physical properties of soils. *Aust. J. Soil Res.* **2**, 123-31.

Bonell, M. and Gilmour, D. A. (1978). Development of overland flow in a tropical rainforest catchment. *J. Hydrology* **39**, 365-82.

Boughton, W. C. (1968). A mathematical catchment model for estimating runoff. *J. Hydrol. (N.Z.)* **7**, 173-8.

Bouma, J. and Anderson, J. L. (1973). Relationships between soil structure characteristics and hydraulic conductivity. *In* 'Field Water Regime.' (Eds R. R. Bruce *et al.*), pp. 77-105. SSSA special publication No. 5. Wisconsin, USA.

Bouma, J. and Wösten, J.H.M. (1979). Flow patterns during extended saturated flow in two undisturbed swelling clay soils with different macrostructure. *Soil Sci. Soc. Amer. J.* **43**, 16-22.

Bouwer, H. (1969). Infiltration of water into non-uniform soil. *J. Irrig. Drain. Div. Am. Soc. Civil Eng.* **95**, 451-62.

Braunack, M. V. and Dexter, A. R. (1978). Compaction of aggregate beds. *In* 'Modification of Soil Structure.' (Eds W. W. Emerson, R. D. Bond and A. R. Dexter). (Wiley: London.)

Bridge, B. J. (1968). Soil physical properties as influenced by gypsum. M.Sc. Agric. Thesis, University of Sydney, Sydney, Australia.

Bridge, B. J. and Kleinig, C. R. (1968). The effect of gypsum on the ground water storage in a sandy loam under an irrigated perennial pasture. Proc. 9th Int. Cong. Soil Sci. Adelaide Vol. 1, pp. 313-24.

Brooks, R. H. and Corey, A. T. (1964). Hydraulic properties of porous media. Hydrology paper No. 3. Colorado State University, Fort Collins, Colorado.

Bruce, R. R. and Whisler, F. D. (1973). Infiltration of water into layered field soils. *In* 'Physical Aspects of Soil Water and Salts in Ecosystems.' (Springer-Verlag: Berlin.)

Buscher, W. J. and Fritton, D. D. (1978). Simulated flow through the root zylem. *Soil Sci.* **125**, 1-6.

Butler, B. E. (1955). A system for the description of soil structure and consistence in the field. *J. Aust. Inst. Agric. Sci.* **21**, 239-49.

Butler, B. E. (1976). Subplasticity in Australian soils. *Aust. J. Soil Res.* **14**, 225-7.

Butler, B. E. and Hubble, G. D. (1977). Morphologic properties. Chapter 2. *In* 'Soil Factors in Crop Production in a Semi-arid Environment.' (Eds J. S. Russell and E. L. Greacen). (University of Queensland Press: St Lucia, Qld.)

Bybordi, M. (1973). Infiltration of water into non-uniform soils. *In* 'Physical Aspects of Soil Water and Salts in Ecosystems.' (Springer-Verlag: Berlin.)

Carbon, B. A. (1973). Diurnal water stress in plants grown on a coarse soil. *Aust. J. Soil Res.* **11**, 33-42.

Carbon, B. A. (1975). Redistribution of water following precipitation on previously dry sandy soils. *Aust. J. Soil Res.* **13**, 13-19.

Carbon, B. A. and Galbraith, K. A. (1975). Simulation of the water balance for plants growing on coarse-textured soils. *Aust. J. Soil Res.* **13**, 23-31.

Chapman, T. G. (1963). Groundwater hydrology of the Weipa Peninsula. *J. Inst. Eng. Aust.* July-Aug. 1963.

Childs, E. C. (1940). The use of soil moisture characteristics in soil studies. *Soil Sci.* **55**, 317-27.

Childs, E. C. (1955). The physical aspect of some concepts in soil mechanics. Proc. Natl. Acad. Sci. (India) **24A**, 86-92.

Childs, E. C. (1969). 'The Physical Basis of Soil Water Phenomena.' (Wiley: London.)

Childs, E. C. and Bybordi, M. (1969). The vertical movement of water into stratified porous material. I. Infiltration. *Water Resour. Res.* **5**, 446-59.

Childs, E. C. and Collis-George, N. (1950). The permeability of porous materials. *Proc. R. Soc. London* **A201**, 391-405.

Churchwood, H. M. (1957). The soils and land use of the Denimein Irrigation District of New South Wales. CSIRO Aust. Div. Soils, Soils Land Use Ser. No. 27.

Collis-George, N. (1977). Infiltration equations for simple soil systems. *Water Resour. Res.* **13**, 395-403.

Conacher, A. J. (1975). Throughflow as a mechanism responsible for excessive soil salinization in non irrigated, previously arable lands in the West Australian wheat belt. A field study. *Catena* **2**, 31-67.

Cope, F. R. (1958). Catchment salting in Victoria. M. Agric. Sci. Thesis, University of Melbourne, 1957. *J. Aust. Inst. Agric. Sci.* **24**, 364-5.

Coughlan, K. J. (1969). Prediction of the moisture holding characteristics of Queensland soils — Preliminary study. *Queensl. J. Agric. Anim. Sci.* **26**, 465-73.

Cowan, I. R. (1965). Transport of water in the soil-plant-atmosphere systems. *J. Appl. Ecol.* **2**, 221-39.

Crack, B. J. and Isbell, R. F. (1971). Studies on some neutral red duplex soils (Dr 2.12) in north-eastern Queensland. I. Morphological and chemical characteristics. *Aust. J. Exp. Agric. Anim. Husb.* **11**, 328-35.

Dagg, M. (1968). Hydrological implications of grass roots studies at a site in East Africa. *J. Hydrol.* **9**, 438-44.

Day, K. J. (1977). Fertility studies on three red earth soils of the Daly Basin, Northern Territory. Aust. Dept. N. Territory Tech. Bull. No. 22.

Dunin, F. X. (1970). Changes in water balance components with pasture management in south-eastern Australia. *J. Hydrol.* **10**, 90-102.

Dunin, F. X. and Downes, R. G. (1962). The effect of subterranean clover and Wimmera ryegrass in controlling runoff from four acre catchments near Bacchus Marsh, Victoria. *Aust. J. Exp. Agric. Anim. Husb.* **2**, 148-52.

Dunin, F. X. and Reyenga, W. (1978). Evaporation from a Themeda grassland. I. Controls imposed on the process in a sub-humid environment. *J. Appl. Ecol.* **15**, 317-25.

Dylla, A. S. and Grutjens, J. C. (1970). Hydraulic conductivity sampling for confidence. *Trans. Am. Soc. Agric. Eng.* **13**, 485-88.

Eastgate, W. I., Swartz, G. L. and Briggs, H. S. (1979). Estimation of runoff rates from small Queensland catchments. Department of Primary Industries, Queensland, Division of Land Utilisation, Tech. Bull. No. 15.

Emerson, W. W. and Bakker, A. C. (1973). Comparative effect of exchangeable calcium, magnesium and sodium on some physical properties of red brown earth subsoils. II. The spontaneous dispersion of aggregates in water. *Aust. J. Soil Res.* **11**, 151-7.

Engelund, F. (1953). On the laminar and turbulent flows of groundwater through homogeneous sand. *Trans. Danish Acad. Tech. Sci.* **3**, 53.

Farbrother, H. G. (1972). Field behaviour on Gezira clay under irrigation. *Cotton Grow. Rev.* **49**, 1-27.

Farrell, D. A. and Greacen, E. L. (1966). Resistance to penetration of fine probes in compressible soil. *Aust. J. Soil Res.* **4**, 1-17.

Farrell, D. A., Greacen, E. L. and Larson, W. E. (1967). The effect of water content on axial strain in a loam soil under tension and compression. *Soil Sci. Soc. Am. Proc.* **31**, 445-50.

Fawcett, R. G. and Carter, O. G. (1974). Utility of a simple soil water budget in agronomic research. 2. Estimates of available soil water in relation to profile change of soil water content and potential. *Aust. J. Exp. Agric. Anim. Husb.* **14**, 80-4.

Gardner, E. A. (1978). Techniques for evaluating suitability for irrigation of cracking clay soils in the Emerald Irrigation Area. M.Sc. Agric. Sci. Thesis, University of Queensland, Australia.

Gardner, W. R. (1977). Water Entry and Movement in Relation to Erosion. *In* 'Soil conservation and management in the humid tropics.' (Eds D. J. Greenland and R. Lal). (John Wiley and Sons: Chichester.)

Greacen, E. L. (1959). The strength of cultivated soil. *J. Agric. Eng. Res.* **4**, 60-61.

Greacen, E. L. (1960*a*). Water content and soil strength. *J. Soil Sci.* **11**, 313-33.

Greacen, E. L. (1960*b*). Aggregate strength and soil consistence. Proc. 7th Int. Conf. Soil Sci. Madison. Vol 1, pp. 256-64.

Greacen, E. L. (1976). Soil conditions affecting crop water supply. CSIRO Aust. Div. Soils Biennial Rep. 1974-75, pp. 10-14.

Greacen, E. L. (1977). Mechanisms and models of water transfer. *In* 'Soil Factors in Crop Production in a Semi-arid Environment.' (Eds J. S. Russell and E. L. Greacen). (Univ. Queensland Press: Brisbane.)

Greacen, E. L., Barley, K. P. and Farrell, D. A. (1968). The mechanics of root growth in soils with particular reference to implications for root distribution. *In* 'Root Growth.' (Ed. W. J. Whittington). Proc. 15th Eastern School Agric. Sci. Univ. Nottingham.

Greacen, E. L. and Hignett, C. T. (1976). A water balance model and supply index for wheat in South Australia. CSIRO Aust. Div. Soils Tech. Pap. No. 27.

Greacen, E.L. and Oh, J. S. (1972). Physics of root growth. *Nat. New Biol.* **235**, 24-5.

Greacen, E. L., Ponsana, P. and Barley, K. P. (1976). Resistance to water flow in the roots of cereals. *In* 'Water and Plant Life.' (Eds O. L. Lange, L. Kappen and E. D. Schulze). (Springer-Verlag: Berlin.)

Greacen, E. L., Ponsana, P. and Forrest, J. A. (1974). The use of Marshall's model for describing the field hydraulic conductivity of soil. Trans. 10th Int. Congr. Soil Sci. Moscow. Vol. 1, pp. 319-26.

Green, W. H. and Ampt, G. A. (1911). Studies in soil physics. I. The flow of air and water through soils. *J. Agric. Sci.* **4**, 1-24.

Griffith, A. A. (1924). The theory of rupture. 1st Int. Congr. Appl. Math. Proc. pp. 55-63.

Hanks, R. J. and Bowers, S. A. (1962). Numerical solution of the moisture flow equation for infiltration into layered soils. *Soil Sci. Soc. Am. Proc.* **26**, 530-4.

Hillel, D. (1971). 'Soil and Water – Physical Principles and Processes.' (Academic Press: New York).

Hillel, D. (1977). 'Computer Simulation of Soil-water Dynamics: A Compendium of Recent Work.' (IDRC Ottawa: Canada.)

Hillel, D. and Gardner, W. R. (1969). Steady infiltration into crust-topped profiles. *Soil Sci.* **108**, 137-42.

Hillel, D. and Gardner, W. R. (1970). Transient infiltration into crust-topped profiles. *Soil Sci.* **109**, 69-76.

Hillel, D., Krentos, V. D. and Stylianou, Y. (1972). Procedure and test of an internal drainage method for measuring soil hydraulic characteristics *in situ*. *Soil Sci.* **114**, 395-400.

Hillel, D. and Talpaz, H. (1977). Simulation of soil water dynamics in layered soils. *Soil Sci.* **123**, 54-62.

Hillel, D. and van Bavel, C. H. M. (1976). Simulation of profile water storage as related to soil hydraulic properties. *Soil Sci. Soc. Am. Proc.* **40**, 807-15.

Holmes, J. W. (1956). Calibration and use of the neutron scattering method for measuring soil water content. *Aust. J. Appl. Sci.* **7**, 45-58.

Holmes, J. W. (1960). Water balance and the water table in deep sandy soils of the upper south-east, South Australia. *Aust. J. Agric. Res.* **11**, 970-88.

Holmes, J. W. and Colville, J. W. (1970). Grassland hydrology in a karstic region of southern Australia. *J. Hydrol.* **10**, 38-58.

Holmes, J. W. and Watson, C. L. (1967). The water budget of irrigated pasture land near Murray Bridge, S.A. *Agric. Meteorol.* **4**, 177-87.

Holtan, H. N. and Minshall, R. J. (1968). Plot samples of watershed hydrology. USDA, ARS 41-133, January 1968.

Horton, R. E. (1933). The role of infiltration in the hydrologic cycle. *Trans. Amer. Geophys. Union* **14**, 446-60.

Hvorslev, M. J. (1937). Uber die Festigkeitseigenschaften gestorter bindiger Boden. *Ingenioervidensk. Skr.* **A45**, 1-159.

Ive, J. R., Rose, C. W., Wall, B. H. and Torssell, B. W. R. (1976). Estimation and simulation of sheet runoff. *Aust. J. Soil Res.* **14**, 129-38.

Jackson, R. A. (1972). On the calculation of hydraulic conductivity. *Soil Sci. Soc. Am. Proc.* **36**, 380-3.

Jakobsen, B. F. (1973). Inter-relations of soil physical characteristics. *Acta Agric. Scand.* **23**, 165-72.

Jamison, V. C. and Kroth, E. M. (1958). Available moisture storage capacity in relation to textural composition and organic matter content of several Missouri soils. *Soc. Soil Sci. Am. Proc.* **22**, 189-92.

Johns, G. G. and Smith, R. G. C. (1975). Accuracy of soil water budgets based on a range of relationships for the influence of soil water availability on actual water use. *Aust. J. Agric. Res.* **26**, 871-83.

Johnston, E. J. (1952). Soils of the Deniboota Irrigation District and their classification for Irrigation. CSIRO Aust. Div. Soils, Soils Land Use Ser. No. 5.

Kirkham, D., De Boodt, M. F and de Leenheer, L. (1959). Modulus of rupture determination on undisturbed soil core samples. *Soil Sci.* **87**, 141-4.

Klute, A (1952). A numerical method for solving the flow equation for water in unsaturated materials. *Soil Sci.* **73**, 105-16.

Klute, A. (1972). The determination of the hydraulic conductivity and diffusivity of unsaturated soils. *Soil Sci.* **113**, 264-77.

Klute, A. (1973). Soil water flow theory and its application in field situations. *In* 'Field Water Regime.' (Eds R. R. Bruce *et al.*) Soil Sci. Soc. Am. Special Publication No. 5, Wisconsin, USA, pp. 9-36.

Lalibarte, G. E., Corey, A. T. and Brooks, R. H. (1966). Properties of unsaturated porous media. Colorado State University. Hydrology Paper No. 17.

Langford, K. J., Duncan, H. P. and Heeps, D. P. (1977). Evaluation and use of a water balance model. Institution of Engineers, Hydrology Symposium, Proceedings, Brisbane, June 1977, pp. 151-155.

Loveday, J. (1976). Relative significance of electrolyte and cation exchange effects when gypsum is applied to a sodic clay soil. *Aust. J. Soil Res.* **14**, 361-71.

Loveday, J. and McIntyre, D. S. (1966). Soil properties influencing the growth of subterranean clover in the Coleambally Irrigation Area of N.S.W. *Aust. J. Exp. Agric. Anim. Husb.* **6**, 283-95.

Loveday, J., McIntyre, D. S. and Watson, C. L. (1978). Aspects of the hydrology of riverine plain soils. *In* 'Hydrogeology of the Riverine Plain of South East Australia.' Proc. Symp. Aust. Soil. Sci. Soc. Riverina Branch, Griffith, N.S.W. 99-115.

Loveday, J., Saont, J. E., Fleming, P. M. and Muirhead, W. A. (1970). Soil and cotton responses to tillage and ameliorant treatments in a brown clay soil. I. Soil responses and water use. *Aust. J. Exp. Agric. Anim. Husb.* **10**, 313-24.

Loveday, J. and Scotter, D. R. (1966). Water entry, water use and seasonal-moisture regimes in flood-irrigated Riverine soils. *Aust. J. Exp. Agric. Anim. Husb.* **6**, 396-402.

McCown, R. L. (1971). Available water storage in a range of soils in north-eastern Queensland. *Aust. J. Exp. Agric. Anim. Husb.* **11**, 343-8.

McCown, R. L. (1973). An evaluation of the influence of available soil water storage capacity on growing season length and yield of tropical pastures using simple water balance models. *Agric. Meteorol.* **11**, 53-63.

McCown, R. L., Murtha, G. G. and Smith, G. D. (1976). Assessment of available water storage capacity of soils with restricted subsoil permeability. *Water Resour. Res.* **12**, 1255-9.

McIntyre, D. S. (1958). Permeability measurements of soil crusts formed by rainfall impact. *Soil Sci.* **85**, 185-9.

McIntyre, D. S., Watson, C. L. and Loveday, J. (1976). Field measurement of water and salt movement in a swelling clay soil. Proc. Symp. 'Water in Heavy Soils.' Bratislava, Czechoslavakia, Sept. 1976.

McQueen, I. S. and Miller, R. F. (1974). Approximating soil moisture characteristics from limited data: Empirical evidence and tentative model. *Water Resour. Res.* **10**, 521-7.

Marshall, T. J. (1958). A relation between permeability and size distribution of pores. *J. Soil Sci.* **9**, 1-8.

Marshall, T. J. (1959). Relations between water and soil. Common. Agric. Bur. Tech. Comm. No. 50.

Marshall, T. J. and Quirk, J. P. (1950). Stability of structural aggregates of dry soil. *Aust. J. Agric. Res.* **1**, 266-75.

Mason, D. D., Lutz, J. F. and Petersen, R. G. (1957). Hydraulic conductivity as related to certain soil properties in great soil groups. *Soil Sci. Soc. Am. Proc.* **21**, 554-60.

Matheson, W. E. (1969). Gypsum for hard setting soils. *J. Dep. Agric. (S. Aust.)* **72**, 402-6.

Mein, R. G. and Larsen, C. L. (1973). Modelling infiltration during a steady rain. *Water Resour. Res.* **9**, 384-94.

Meyer, W. S., Greacen, E. L. and Alston, A. M. (1978). Resistance to water flow in the seminal roots of wheat. *J. Exp. Bot.* **29**, 1451-61.

Miller, D. E. and Gardner, W. A. (1962). Water infiltration into stratified soil. *Soil Sci. Soc. Am. Proc.* **26**, 115-9.

Millington, R. J. and Quirk, J. P. (1961). Permeability of porous solids. *Trans. Faraday Soc.* **57**, 1200-6.

Mott, J., Bridge, B. J. and Arndt, W. (1979). Soil seals in tropical tall grass pastures of Northern Australia. *Aust. J. Soil Res.* **17**, 483-94.

Newman, E. I. (1969). Resistance to water flow in soil and plant. I. Soil resistance in relation to amounts of root: theoretical estimates. *J. Appl. Ecol.* **6**, 1-12.

Newman, E. I. (1973). Permeability to water of the roots of five herbaceous species. *New Phytol.* **72**, 547-55.

Northcote, K. H. (1971). 'A Factual Key for the Recognition of Australian Soils.' (Rellim Technical Publications: Adelaide, S. Aust.)

Norrish, K. and Pickering, J. G. (1977). Clay mineralogic properties. *In* 'Soil Factors in Crop Production in a Semi-arid Environment.' (Eds J. S. Russell and E. L. Greacen). (Univ. Queensland Press: Brisbane.)

Norrish, K. and Tiller, K. G. (1976). Subplasticity in Australian soils. V. Factors involved and techniques of dispersion. *Aust. J. Soil Res.* **14**, 273-89.

Olsson, K. A. and Rose, C. W. (1978). Hydraulic properties of a red-brown earth determined from *in situ* measurements. *Aust. J. Soil Res.* **16**, 169-80.

O'Neal, A. M. (1949). Soil characteristics significant in evaluating permeability. *Soil Sci.* **67**, 403-9.

Payne, P. C. J. and Fountaine, E. F. (1952). A field method of measuring the shear strength of soils. *J. Soil Sci.* **3**, 136-44.

Peck, A. J. (1965). Moisture profile development and air compression during water uptake by porous bodies. *Soil Sci.* **100**, 44-51.

Peck, A. J. (1971). Redistribution of water after infiltration. *Aust. J. Soil Res.* **9**, 59-71.

Peck, A. J. (1978). Salinization of non-irrigated soils and associated streams. A review. *Aust. J. Soil Res.* **16**, 157-68.

Peck, A. J., Luxmore, R. J. and Stolzy, J. L. (1977). Effects of spatial variability of soil hydraulic properties in water budget modelling. *Water Resour. Res.* **13**, 348-54.

Philip, J. R. (1954). An infiltration equation with physical significance. *Soil Sci.* **77**, 153-7.

Philip, J. R. (1957). The theory of infiltration. I. The infiltration equation and its solution. *Soil Sci.* **83**, 345-57.

Philip, J. R. (1969). Hydrostatics and hydrodynamics in swelling soils. *Water Resour. Res.* **5**, 1070-7.

Philip, J. R. (1975). Some remarks on science and catchment prediction. *In* 'Prediction in Catchment Hydrology.' (Eds T. G. Chapman and F. X. Dunin). (Aust. Acad. Sci.: Canberra.)

Ponsana, P. (1975). Drainage and water uptake terms in the water balance. Ph.D. Thesis. University of Adelaide.

Prebble, R. E. (1970*a*). Physical properties of 17 soil groups from Queensland. CSIRO Aust. Div. Soils Tech. Memo. No. 10/1970.

Prebble, R. E. (1970*b*). Soil physical measurements and grass root distributions on a red podzolic at Samford, South East Queensland. CSIRO Aust. Div. Soils Tech. Memo No. 13/1970.

Prebble, R. E. and Stirk, G. B. (1961). Some physical measurement on soils from the Tobacco Research Institute, Mareeba, North Queenskand. CSIRO Aust. Div. Rep. No. 3/61.

Prescott, J. A. (1934). Single value climatic factors. *Trans. R. Soc. S. Aust.* **58**, 48-61.

Probert, M. E. and Jones, R. K. (1977). The use of soil analysis for predicting the response to sulphur of pasture legumes in the Australian tropics. *Aust. J. Soil Res.* **15**, 137-46.

Quirk, J. P. and Blackmore, A. V. (1955). Effect of electrolyte concentration of the irrigation water on the infiltration of several Riverina soils. CSIRO Aust. Div. Soils Div. Rep. No. 12/55.

Ritchie, J. T., Kissel, D. E. and Burnett, E. (1972). Water movement in undisturbed swelling clay soil. *Soil Sci. Soc. Am. Proc.* **36**, 874-9.

Rijtema, P. E. (1965). An analysis of actual evapotranspiration. Thesis Wageningen, V.L.O. 659 Pudoc. Wageningen.

Rogowski, A. S. (1964). Strength of soil aggregates. Ph.D Thesis. Iowa State University of Science and Technology. (Univ. Microfilms, Inc.: Arbor, Michigan.)

Rogowski, A. S. (1971). Watershed physics: model of the soil moisture characteristic. *Water Resour. Res.* **7**, 1575-82.

Rose, C. W., Stern, W. R. and Drummond, J. E. (1965). Determination of hydraulic conductivity as a function of depth and water content for soil *in situ*. *Aust. J. Soil Res.* **3**, 1-9.

Salter, P. J. Berry, G. and Williams, J. B. (1966). The influence of texture on the moisture characteristic of soils. III. Quantitative relationships between particle size,scomposition and available water capacity. *J. Soil Sci.* **17**, 93-8.

Salter, P. J. and Williams, J. B. (1963). The effect of farm yard manure on the moisture characteristic of a sandy loam soil. *J. Soil Sci.* **14**, 73-81.

Salter, P. J. and Williams, J. B. (1965). Influence of texture on the moisture characteristic of soils. II. Available water capacity and moisture release characteristics. *J. Soil Sci.* **16**, 310-17.

Salter, P. J. and Williams, J. R. (1967). The influence of texture on the moisture characteristic of soils. IV. A method of estimating the available water capacities of profiles in the field. *J. Soil Sci.* **18**, 174-81.

Salter, P. J. and Williams, J. B. (1969). The influence of texture on the moisture characteristic of soil. V. Relationships between particle size composition and moisture content at the upper and lower limits of available water. *J. Soil Sci.* **20**, 126-31.

Sands, R., Greacen, E. L. and Gerard, C. J. (1979). Compaction of sandy soils in *Radiata pine* forests. I. Penetrometer study. *Aust. J. Soil Res.* **17**, 101-13.

Scotter, P. R. (1964). A study of soil water content and potential, and their edaphic significance on a range of irrigated Riverina soils. M.Sc. (Agric.) University of Sydney, Australia.

Sharma, M. L. and Uehara, G. (1968*a*). Influence of soil structure on water relations in low humic latosola. I. Water retention. *Soil Sci. Soc. Am. Proc.* **32**, 765-70.

Sharma, M. L. and Uehara, G. (1968*b*). Influence of soil structure on water relations in low humic latosola. II. Water movement. *Soil Sci. Soc. Am. Proc.* **32**, 770-4.

Shaykewich, C. F. (1970). Hydraulic properties of disturbed and undisturbed soils. *Can. J. Soil Sci.* **50**, 431-7.

Shaykewich, C. F. and Zwarich, M. A. (1968). Relationships between soil physical constants and soil physical components of some Manitoba soils. *Can. J. Soil Sci.* **48**, 199-204.

Slatyer, R. O. (1962). Methodology of a water balance study conducted on a desert woodland (*Acacia aneura* F. Muell) community in central Australia. *UNESCO Arid Zone Res.* **16**, 15-26.

Stace, H. C. T., Hubble, G. D., Brewer, R., Northcote, K. H., Sleeman, J. R., Mulcahy, M. J. and Hallsworth, E. G. (1968). 'A Handbook of Australian Soils.' (Rellim Technical Publications: Adelaide, S. Aust.)

Stern, W. R. (1965). Evapotranspiration of safflower at three densities of sowing. *Aust. J. Agric. Res.* **16**, 961-71.

Stirk, G. B. (1948*a*). The entry of water into several soils at Deniliquin, N.S.W. CSIRO Aust. Div. Soils Div. Rep. No. 5/48.

Stirk ,G. B. (1948*b*). Studies on the basaltic soils of the Northern Rivers area, N.S.W. – Part IV. Infiltration characteristics. CSIRO Aust. Div. Soils Div. Rep. No. 6/48.

Stirk, G. B. (1952). Water entry characteristics of certain soils of the lower Burdekin Valley, North Queensland. CSIRO Aust. Div. Soils Div. Rep. No. 11/52.

Stirk, G. B. (1954). Some aspects of soil shrinkage and the effect of cracking upon water entry into the soil. *Aust. J. Agric. Res.* **5**, 270-90.

Stirk, G. B. (1957). Physical properties of soils of the lower Burdekin Valley, North Queensland. CSIRO Aust. Div. Soils Div. Rep. No. 1/57.

Swartz, G. L. (1966). Water entry into a black earth under flooding. *Queensl. J. Agric. Anim. Sci.* **23**, 407-22.

Swartzendruber (1960). Water flow through a soil profile as affected by the least permeable layer. *J. Geophys. Res.* **65**, 4037-42.

Talsma, T. (1969). *In situ* measurement of sorptivity. *Aust. J. Soil Res.* **7**, 269-76.

Talsma, T. (1973). Hydrology of swelling soils. *In* 'Physical Aspects of Swelling Clay Soil.' Symposium, University of New England, Armidale, Australia.

Talsma, T. (1974). The effect of initial water content on infiltration quantity or redistribution of soil water. *Aust. J. Soil Res.* **12**, 15-26.

Talsma, T. (1976). Soil water movement, infiltration, redistribution and ground water movement. *In* 'Watershed Management of Range and Forest Lands.' (Eds H. F. Heady *et al.*). (Utah State University: Logan.)

Talsma, T. and Flint, S. E. (1958). Some factors determining the hydraulic conductivity of subsoils with special reference to tile drainage problems. *Soil Sci.* **58**, 198-207.

Talsma, T. and van der Lelij, A. (1976). Infiltration and water movement in an *in situ* swelling soil during prolonged ponding. *Aust. J. Soil Res.* **14**, 337-49.

Tennant, D. (1976). Wheat root penetration and total available water on a range of soil types. *Aust. J. Exp. Agric. Anim. Husb.* **16**, 570-7.

Terzaghi, K. (1923). Die Berechnung der Durchlassigkeitsziffer des Tones aus dem Verlauf der hydrodynamischen Spannungserscheinungen. *Sitzungber. Akad. Wiss. Wien, Math.-Naturwiss. Kl. Abt.* 2A, **12**, 105-124.

Terzaghi, K. and Peck, R. B. (1948). 'Soil Mechanics in Engineering Practice.' (Wiley: New York.)

Thompson, C. H. and Beckmann, G. G. (1959). Soils and land use in the Toowoomba, Darling Downs Queensland. CSIRO Aust. Div. Soils, Soil Land Use Ser. No. 28.

Tisdall, A. L. (1951). Antecedent soil moisture and its relation to infiltration. *Aust. J. Agric. Res.* **2**, 342-8.

Torssell , B. W. R. (1973). Microrelief and edaphic factors on a lateritic red earth in a dry monsoonal climate. CSIRO Aust. Div. Land Use Res. Tech. Memo No. 73/5.

Towner, G. D. and Childs, E. C. (1972). The mechanical strength of unsaturated porous granular material. *J. Soil Sci.* **23**, 481-98.

Turner, A. K. (1963). Infiltration runoff and soil classification. *J. Hydrol.* **1**, 129-43.

van Bavel, C. H. M., Stirk, G. B. and Brust, K. J. (1968). Hydraulic properties of a clay loam soil and the field measurement of water uptake by roots. I. Interpretation of water content and pressure profiles. *Soil Sci. Soc. Am. Proc.* **32**, 301-7.

van der Lelij, A. and Talsma, T. (1978). Infiltration and water movement in Riverina Plain soils used for rice growing. Proc. Symp. 'Hydrogeology of the Riverine Plain of south-east Australia.' (Ed. R. R. Storrier.) (ASSS Inc.: Riverina Branch, Australia.)

Waring, S. A., Fox, W. E. and Teakle, L. J. H. (1958). Fertility investigations on the black earth wheatlands of the Darling Downs, Queensland. *Aust.J. Agric. Res.* **9**, 717-29.

Warrick, A. W., Mullen, G. J. and Nielsen, D. R. (1977). Scaling field-measured soil hydraulic properties using a similar media concept. *Water Resour. Res.* **13**, 355-462.

Wesseling, J. (1957). Enige asperten van de water-beheersing in hand bouwgronden. *Versl. Landb. Ondeiz.* 63-5.

Wesseling, J. and Wit, K. E. (1966). An infiltration method for the determination of the capillary conductivity of undisturbed soil cores. Inst. of Land and Water Management Research Misc. Reprint No. 81.

Wetselaar, R. (1962). Nitrate distribution in tropical soils. III. Downward movement and accumulation in the subsoil. *Plant Soil* **16**, 19-31.

Whisler, F. D., Klute, A. and Millington, R. J. (1970). Analysis of radial, steady-state solution, and solute flow. *Soil Sci. Soc. Am. Proc.* **34**, 382-7.

Wilcox, J. C. and Spilsbury, R. H. (1941). Soil moisture studies. II. Some relationships between moisture measurements and mechanical analysis. *Sci. Agric.* **21**, 459-72.

Winkworth, R. E. (1961). Soil water relationships in an *Eragrostis eripoda* community in Central Australia. CSIRO Aust. Div. Land Res. Regional Survey-Tech. Memo No. 61/26.

Winkworth, R. E. (1969). The soil water regime of an arid grassland (*Eragrostis eripoda*) in Central Australia. *Agric. Meteorol.* **7**, 387-94.

Williams, J. and Shaykewich, C. F. (1970). The influence of soil water matric potential on the strength properties of unsaturated soil. *Soil Sci. Soc. Am. Proc.* **34**, 835-40.

Williams, J. (1974). Root density and water potential gradients near the plant root. *J. Exp. Bot.* **25**, 669-74.

Williams, J. and Coventry, R. J. (1979). The contrasting soil hydrology of red and yellow earths in landscapes of low relief. *In* 'The Hydrology of Areas of Low Precipitation.' Proc. Symp. Canberra, 1979. IAHS-AISH Pub. No. 128 pp. 385-96.

Williams, J. and Coventry, R. J. (1981). The potential for groundwater recharge through red, yellow and grey earths in central north Queensland. Aust. Water Resources Council, Conference Series, No. 2. (Aust. Govt. Printing Service: Canberra.)

Chapter 33
Water balance and groundwater studies

G. B. Allison, J. S. Colville and E. L. Greacen

In recent years, investigations of the behaviour of water in the soil–plant–atmosphere continuum have been directed towards the development of predictive simulation models of subsystems or processes within the hydrologic cycle. The use of these models may be qualitative, where an attempt is made only to determine which are the important parameters determining some particular process such as soil erosion, or they may be predictive and attempt to estimate actual quantities, for example, the yield of water from a catchment. In this chapter we look at the philosophy behind the development of many of the models currently in use in Australia, and then describe some systems to which they have been applied.

Following this, some of the more important factors which determine the water balance are discussed. Rainfall and evapotranspiration are usually the major components of the water balance. Considerable effort has been made to understand and predict evapotranspiration, because water is usually limiting for plant growth. Of considerable interest in Australia, being the driest continent, are rainfall distribution and drought.

Replenishment of groundwater systems occurs after the rainfall input to the earth's surface has been modified by the processes of runoff, evapotranspiration and soil storage. While the amount of water involved is usually small compared with rainfall, quantitative assessment is important, as this allows estimation of the amount of water which can be safely withdrawn from the system. We discuss some of the techniques which can be used to measure this parameter.

In many regions of economic importance, groundwater is used to support industry and agriculture. In the final section, some of the more important aspects of groundwater hydrology are discussed.

Water balance models and applications in Australia

Philosophy and development

Simple models of the soil water balance have been used in Australia for preparing irrigation schedules for almost 50 years. Briggs and Shantz (1917) showed that evapotranspiration from well watered plants closely follows the rate of water loss from an evaporimeter. Following this, and on the basis of experience at the CSIRO Irrigation Research Station at Griffith in New South Wales, orchard trees were watered at intervals of 100 mm evaporation from an evaporimeter.

Modification of the rate of evapotranspiration by the drying out of the soil was described by West and Perkman (1953) and Slatyer (1956). Slatyer (1960*a*) developed what might be considered one of the first modern models of the soil water balance. He used this to categorize and describe the water supply factor in landscapes in the Northern Territory and New South Wales on the basis of their effective water supply.

Water is generally accepted as the dominant factor in pasture and crop production, but it is surprising to learn, even from broad general data, just how much the water supply influences productivity in a semi-arid environment. Cornish (1950) has demonstrated this for the wheatlands of South Australia. Using a simple regression model he related wheat yields over the whole of South Australia to three rainfall variates: rainfall for the three periods April through May, June through August, and September through October, for the years 1896–1941. His conclusions are worth quoting: 'In 267 "hundreds"*, out of a total or 296, 50 percent or more of the observed variation in yield is expressible in terms of the rainfall variates and time, a fact sufficient to excite remark when it is recalled that all other causes of variation without exception are included in the remaining 50 percent or less'.

This result is even more exciting when we consider that no explicit account is taken of any effect of the soil itself in modifying the supply of water to the crop. Incorporating the soil as a simple storage unit, and modifying evapotranspiration for crop stage and soil water content, almost invariably improves the prediction to account for a further 24% of the observed variation in yield (Fitzpatrick and Nix 1969; Greacen and Hignett 1976).

This dominant role of the water supply factor in determining agricultural and pastoral productivity and its importance in major engineering projects, such as irrigation, land development schemes, watershed hydrology

*A hundred, is the terminology used in South Australia for a subdivision of a county, being 100 square miles or 259 km^2.

and soil conservation, has called for considerable effort in modelling the soil–water regime. As described in the preceding chaper, substantial progress has been made over the last 50 years towards understanding soil water processes and describing them as analytic mathematical models or equations. Because of increasing pressure for a more efficient use of our soil and water resources and a growing awareness of their vulnerability to abuse, there is an increasing need to incorporate this analytical knowledge in predictive models. Because of the dominance of the water supply factor in determining plant growth, simulation of the soil–water regime should be a major part of any agronomic project.

Besides their possible use in decision making and prediction in agriculture, simulation models provide a disciplined theoretical basis for research. In a dynamic system, such as the soil–water system, mathematical models can organize the interactions and interrelations of variables into manageable subsystems, can give leads to possible lines of research and identify the parameters to be measured.

Mathematical models of the soil–water regime

Early attempts to model the water supply factor were based on observed empirical relationships between the variables. These relationships were usually expressed as a regression equation, which is a form of *statistical model.* Although the act of seeking such a regression implies that a functional or causal effect is anticipated, no attempt is made in a statistical model to describe the mechanisms concerned. Where functional relationships exist and the mechanisms relating the variables are known or assumed and are incorporated mathematically, the model is called a *mechanistic model.* Where the functional relationships depend on chance or random variables, the model is called a *stochastic model,* but when random variables are not recognized and the relationship is exact we have a *deterministic model.*

Many of the mechanisms determining the water balance are known, and numerous simulation models of the soil–water regime have been developed in recent years. Depending on the objectives of the program, whether it is the prediction of agronomic yield, catchment yield of water, surface runoff, or deep drainage, relevant aspects of the model are treated in greater detail while other aspects may be grossly simplified. Such simplifications however must be allowed by the peculiarities of the particular environment. For example, Slatyer (1960*a*) found that in southern New South Wales, with a predominantly winter rainfall of moderate intensity, a model that accepted all rainfall until the soil storage was filled and after that discarded any excess rainfall, gave an acceptable simulation of infiltration. Fitzpatrick *et al.* (1967) modified the simple overflow model to accept only a fraction of the rainfall, (depending on the quantity of rainfall received during the interval) and this enabled them to simulate infiltration in the Northern Territory where a summer monsoonal climate prevails. Thus, while detail in a simulation model is determined largely by the overall objective, the degree of simplification must not ignore other aspects of the system. Instances of this fine structuring can be seen in the development of models for different hydrological applications.

Agronomic applications of hydrological models

The water balance model WATBAL (Keig and McAlpine 1974) has been used for simulating the water regime, and thus defining periods of growth, over a wide range of pastoral country in Australia. The popularity of this model derives from its simplicity, the input requirements being weekly rainfall and evaporation and an estimate of the water storage capacity of the soil (Isbell and McCown 1978). Estimates of weekly evaporation are usually made from temperature records (Fitzpatrick 1963), and the water storage capacity of the soil is either measured or assumed. The coefficient which determines the actual evapotranspiration from the crop as a fraction of the potential evaporation varies with the water content of the soil and has to be adjusted for different regions. Thus for Canberra the coefficient is 1.0 if storage is more than 50% full, and 0.5 if less than 50%; for Alice Springs the corresponding values are 0.5 and 0.25. McAlpine (1970) showed the effectiveness of this simple model by comparing 100 observations of predicted and measured soil water at Canberra ($r = 0.94$) and at Alice Springs ($r = 0.84$).

The simplicity of the soil term in WATBAL, namely a single bucket system, often with an assumed capacity, gives some cause for concern. McCown (1973) used a more sophisticated model with a two-layer storage in a water balance study on Townsville stylo (*Stylosanthes humilis* L.) in northern Queensland. Over four seasons, WATBAL prediction was only marginally less accurate than the two-layer model. Using WATBAL, McCown compared the effect of soil water storage capacity on productivity over a 60 year period. Three soils were used with storage capacities of 75, 150 and 180 mm. He found that for years of moderately low rainfall, one year in eight, there was no effect of soil, but for 76% of years, increasing storage capacity from 75 mm to 180 mm gave a 34% increase in plant yield. It is generally considered that the one-layer model is not as widely applicable as the two-layer one but comparisons that have been made do not show a pronounced difference. On the other hand, we are not aware of any extensive comparison and, as McCown's work has shown, this might be required.

The problem of the number of soil layers to be used in a model of the water balance relates to processes that depend on the soil water content, e.g. evapotranspiration or water movement in the profile. If a limited rainfall is added to the whole profile, as in a single storage model, the relative water content will be much lower than if this rainfall is added only to a shallow surface layer in a multi-layered model. Fleming (1964), Flinn (1969) and Greacen and Hignett (1976) try to overcome this problem with a rain balance term, which is a sub-model of the shallow surface layer, run independent of the main program, but Johns and Smith (1975) concluded that the rain balance term confers no benefit in work on pastures. Johns and Smith (1975) also discuss the problem of estimating the storage capacity of the soil when the effective root zone is changing. Rose *et al.* (1972) and Byrne and Tognetti (1969) overcome this problem for pastures by using a multi-layered profile and an assumed rate of root growth. Greacen and Hignett (1976) use a two-layer model for wheat in South Australia, but keep track of the depth of the root zone and adjust the maximum available storage accordingly. In their model the root zone deepens at a rate in accord with soil temperature and mechanical impedance. Rooting density is increased as a function of simulated crop growth.

Estimation of available storage is complicated in dry situations where the soil dries to a water content below the wilting point or − 1500 kPa water potential, which is often taken as the lower limit of available water. A common solution has been to take the driest recorded profile as the lower limit for deeper layers, and, for the soil surface, to discard rainfall events below a certain threshold. Fitzpatrick *et al.* (1967) assume two soil stores, as opposed to two layers; water held above − 1500 kPa is referred to as Store A, and that below − 1500 kPa, but above the driest limit, as Store B. They assume maximum values of storage of 120 mm for A and 40 mm for B, and rain input to the two stores is in the ratio of these values.

In early water balance models developed for agronomy, infiltration and deep drainage have generally been treated by the overflow or bucket technique. This is often satisfactory, but in some environments and for some soils, as discussed in the previous chapter, models have had to be adjusted to take more account of these processes. Thus for infiltration, Fitzpatrick *et al.* (1967) had to allow for a limited rainfall absorption for periods of heavy rain in Northern Australia. For restricted drainage, Greacen and Hignett (1976) drain the profile at a rate depending on the hydraulic conductivity of the subsoil. Rosenthal *et al.* (1976) have drawn on many early crop models (Fitzpatrick and Nix 1969; Keig and McAlpine 1974; Boughton 1968) to produce an extremely versatile model for agronomic application in the sub-tropics. Two options are available in their model for calculating infiltration and runoff; firstly, the simple bucket or overflow technique, where runoff occurs only when maximum storage is reached, and secondly, a more realistic simulation of infiltration using a surface retention storage along with Boughton's infiltration model described in Chapter 32 of this book.

Watershed and surface runoff applications

In contrast to the simulation models of the soil water balance for agronomic application, models developed for catchment hydrology, soil conservation, or deep drainage are concerned with a more accurate simulation of runoff and redistribution (movement) of water in and through the profile. An ambitious general-purpose model of catchment hydrology is being developed for the Representative Basins Programme (Australian Water Resources Council 1969) for which data are being collected from over 80 catchments representing the major hydrological units within Australia. In Fig. 33.1 the general arrangement of the Representative Basins model (Fleming 1973) is shown. Runoff is generated in the wetting cycle of the model. Foliage is represented as an interception store which overflows into a depression store, representing depressions due to surface roughness and litter. Water is lost from this store by infiltration to the upper soil layer and, as overflow, to generate surface runoff. Special attention has been given to the infiltration function. A method based on the Philip infiltration constants (see Chapter 32) and the time compression approximation (Reeves and Miller 1975) is being investigated.

Functions for redistribution of water between the two soil layers and for drainage from the lower layer are based on Darcy's law, using simplified boundary conditions. The model is mechanistic and does not allow for any stochastic effects. It is a 'lumped parameter' or 'point water balance' model, i.e. the calculations are carried out for a single hypothetical point using averaged parameters and characteristics.

Aston and Dunin (1979) also concentrate on the infiltration process in a watershed model (WATSIM) for a 5 ha catchment under natural pasture near Canberra. The Philip infiltration constants are again used, but with a modified Green and Ampt approach (Mein and Larson 1973). WATSIM is a single store model and weekly drainage is modelled empirically as a function of the maximum volumetric water content for the week (Aston and Dunin 1977). The program was run treating the catchment both as a lumped parameter and as three stratified zones; the upper, middle, and lower zones were considered hydrologically uniform with time. Present indications (F.X. Dunin, pers. comm.) are that it is probably worth including spatial variability in the model for the prediction of small to intermediate flows, but for extreme events the improvement of the stratified over the lumped parameter approach is only marginal.

Spatial variability in the hydraulic properties of soils can be high. For example, saturated hydralic conductivity K_o varies by a factor of about 10 for uniform soils

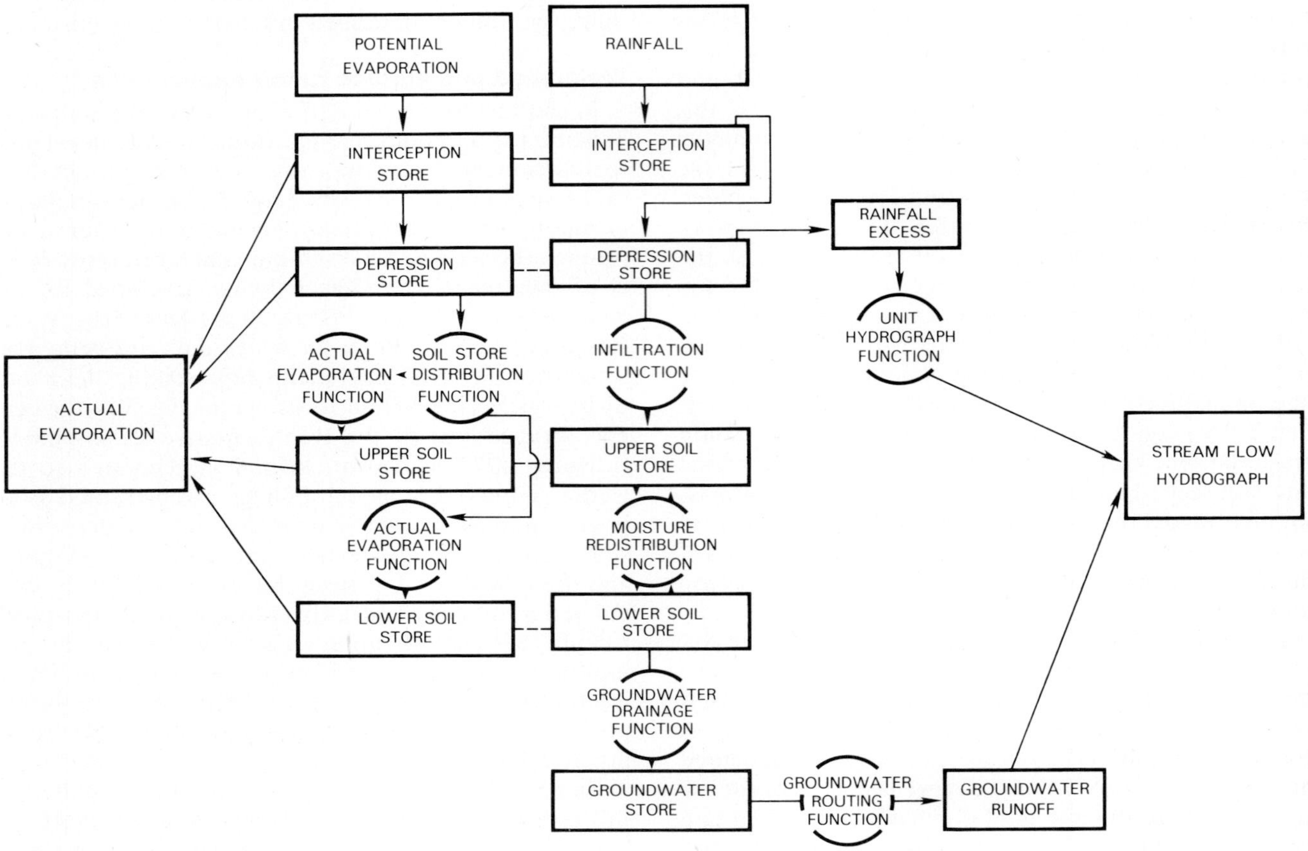

Fig. 33.1. Schematic flow diagram for the initial mathematical model for the Australian Representative Basins Program (Fleming 1973).

(Rogowski 1972), to three orders of magnitude for many forest soils (T. Talsma, unpublished data). Currently, attention is being focussed on modelling the effect of this variabiltiy on catchment behaviour. Peck *et al.* (1977) described the spatial variability of infiltration, conductivity and the moisture characteristic of a catchment using the standard deviation (s.d.$_\alpha$) of a scaling factor α. Partial areas of the catchment are then determined for class intervals of α and the model is run as a point water balance for these partial areas. For s.d.$_\alpha$ = 0.25, the model predicts essentially the same as a lumped parameter model for a limited climatic input. Sharma and Luxmore (1979) consider that 0.6 is a more realistic value for s.d.$_\alpha$ for the catchment they studied and find that this magnitude of variabiltiy has quite strong effects on runoff prediction.

The problem of synthesizing analytic information into simulation models of the soil–water regime is largely a problem of simplifying these functions without losing their basic integrity, and of finding techniques for dealing with spatial and temporal variability of the soil and the climate. Not the least important part of this latter aspect is the measurement of the parameters used in the model. As will be discussed later, other techniques are available for studying individual components of the soil–water regime.

Rainfall in Australia

Distribution of rainfall

The Australian continent lies between latitudes 11°S and 44°S and is therefore subject to a range of wind systems which control the regional rainfall characteristics.

The characteristic feature of the whole wind system is a series of anticyclones which form over the Indian Ocean and move eastwards over the whole continent. An individual anticyclonic system can cover the whole continent but at other times two distinct systems can be seen.

The high pressure centres of the anticyclones are formed by streams of air which have descended from a height of about 10 km and are therefore very dry. However, between successive anticyclones are low-pressure troughs often containing moist air from which rain can be deposited if conditions are suitable.

On average, anticyclones pass every nine days throughout the year, but their path varies with the season. In summer they travel along latitudes 37–38° S, i.e. the latitude of Bass Strait, but in winter they move northward to latitudes 29 – 32° S, i.e. the latitudes of northern New South Wales. The seasonal shift of the anticyclone path allows rain-bearing air systems to arrive in the northern or southern parts of the continent.

During the summer the high pressure regions in the south keep the frontal rains to the fringe of the continent but allow tropical monsoons and associated systems to reach northern Australia between October and March. A land area of about 4 x 10^6 km^2 (about half of the continent) receives predominantly summer rain, but only about 1.6 x 10^6 km^2 has a median annual rainfall of more than 500 mm. The rain is only partly monsoonal; trade winds bring heavy falls to a narrow coastal strip near Tully in Queensland; low pressure troughs between anticyclones provide rain in parts of central Queensland.

Spectacular mid-summer events are the passage of tropical cyclones which begin over warm seas and may cross the coast anywhere north of 20° S. Their wind speed, which increases as they move across the sea, is reduced over land, and they eventually degenerate into large 'tropical depressions' when their average wind speed drops to 'gale force' (63 km h^{-1}). Cyclone Tracy produced wind speeds of 140 km h^{-1} at Darwin in December 1974; the rain gauge was destroyed, but, after allowing for reduced interception at high wind speeds and attempting to reconstruct the record, the estimated rainfall was over 300 mm in 24 h (Australian Bureau of Meteorology 1977). The heaviest Australian recorded rainfall over 24 h was 907 mm at Crohamhurst near Brisbane in 1893; this also was associated with a tropical cyclone.

In its dying phase an individual cyclone can bring heavy rain to large areas; Brunt (1966) presents data for some eastern cyclones, and quotes areas as large as 10^6 km^2 and runoff in excess of 3 x 10^{10} m^3.

About six tropical cyclones a year have an effect on rainfall in Australia and therefore are responsible for extreme variability in rainfall. One frequently quoted example is Onslow on the northwest coast, which has recorded annual rainfalls from 14 to 997 mm, and for the years 1921 to 1924 received 565, 64, 681, and 53 mm.

During winter the anticyclones move north, and the frontal rains move over the southern part of the continent. By May and June about 3 x 10^6 km^2 have received winter rain but only about 20% of this area receives a median rainfall greater than 300 mm.

There is a large region in the centre and west of the continent which receives only spasmodic rain in all seasons. Much more important is a region of about 4 x 10^5 km^2 which receives a median rainfall greater than 300 mm without a large seasonal variation. This is in the south-east of the continent and extends across the Dividing Range into central New South Wales. Seasonal rainfall zones are given in Map Set 5 of the Climatic Atlas of Australia, prepared by the Australian Bureau of Meteorology (1975). Another valuable map, showing the first, fifth, and ninth deciles for each month at some forty stations, is Map No. 6 in the Atlas of Australian Resources published by the Department of National Development (1970). Finally, the climatological survey of Gentilli (1971) is very useful.

Comparison with other continents

Australia is generally known as the driest continent, and comparison with other regions is instructive.

Some statistical index is needed to allow for variation between years. Conventionally the thirty year arithmetical mean of annual rainfall is quoted as the 'rain normal'. This is confusing because the rainfall distribution is not statistically normal but is skewed, especially if the rainfall is small. Accordingly, a median value evaluated over a longer period is preferable. This can be considerably less than the mean if there is little rain; a standard example is Hall's Creek, which in July has a mean of 6 mm and a median of zero.

According to the 'Official Year Book of Australia' (1976, p.33), 29.6% of Australia receives a median annual rainfall less than 200 mm, 52.5% less than 300 mm, 63.7% less than 400 mm, and 71.3% less than 500 mm. Comparable statistics for the rest of the world are not available. However, some collections of mean rainfall are available. Plate 4 in the 'Times Atlas' (1975) shows the mean annual rainfall throughout the world classified into thirteen divisions. The lowest division, i.e. less than 25 mm, is not represented in Australia, but it does occur in North Chile in the southern hemisphere, and in the Sahara and Gobi deserts of the northern hemisphere. The second driest division, 25 to 100 mm, includes a region near Lake Eyre and also occurs in Saudi Arabia and Somalia. The third division, 100 to 200 mm, covers a large inland region of Australia, and also occurs in Iraq, Iran, and the deserts of the western United States.

At the other extreme, wetter regions can be considered. About 85% of New Zealand receives more than 750 mm, and the lowest individual station mean is 335 mm. Most of Europe receives more than 500 mm, although one very small region in Spain receives less than 200 mm. Although about one tenth of Australia receives more than 800 mm of rain, much of this is along the north coast where the rains fall in the summer when evaporation is high. Accordingly, most of the agricultural production comes from drier southern regions which have winter rainfall.

Drought

It is surprisingly difficult to define the occurrence of drought. The effect of a water deficit will depend on land use. Also, the effect of a moderate water deficit will worsen with increasing duration of drought. Accordingly, Foley (1957), in his survey of Australian droughts, limited himself to a collection of historical records of crop failures.

A more precise but arbitrary definition can be based on rainfall records; Gibbs and Maher (1967) showed that occurrence of rain within the first decile was generally acceptable. However, on this criterion, over a long period, every site has one drought year in ten and nine drought free years. For the 30% of the Australian continent with a median of 200 mm or less, this would seem to be unduly optimistic. Gibbs and Maher published maps showing drought affected areas from 1885 to 1965. They concluded that on their criterion, a drought over the whole of Australia would be impossible, but a drought over half the continent would occur once in 50 years, and the whole continent would be drought free for about one year in five years. There is a quarterly publication 'Drought Review' (Bureau of Meteorology) which continues this survey.

A drought could be defined as some abnormally long period of plant water stress. Usually no evaporation measurements are available, and therefore various indices based on temperature as well as rainfall have been proposed. Gentilli (1971) discusses several such indices, but finds that they are seldom reliable outside the region in which they were first derived. As described earlier, computer models which predict plant-water stress are currently popular.

The common impression is that droughts show persistence, i.e. they extend longer than would be expected from a random series. This impression could be the result of several mediocre years in succession producing a very poor final crop yield. Anderson (1957) examined Melbourne rainfall records from 1856 to 1956 and showed that long runs of rainfall below the median value occurred as often as could be expected in a random series. However, Maher (1973) quotes unpublished calculations by D. N. Body for 31 Australian stations, which show that long runs of monthly or annual rainfall within the first or first three deciles do occur significantly often.

At the other extreme, daily rainfall records for Melbourne between 1859 and 1965 were examined by Watterson and Legg (1967). They found that the chance that a day would be dry, i.e. receiving less than 0.25 mm of rain, depended upon the time of year and the rainfall of the preceding day but there was no influence of earlier days.

Long-term variations of rainfall

The reliable seasonal estimation of rain, say a year in advance, would obviously be very valuable. The search for long-term trends and recurrent cycles has therefore fascinated generations of observers but no accepted useful method of prediction has yet been found. Long-term predictions which are commercially available have been criticized for their vague and ambiguous wording.

Basic knowledge of physical meteorology is still inadequate to predict the occurrence of rain, still less to estimate it. It is not even certain which subsidiary processes are important in the control of rain some months hence. Therefore proposed methods of prediction have always been highly empirical and generally are based on statistical correlations between rainfall and some other meteorological variables. A relation between rainfall and air pressure at some earlier period has always appeared possible and many apparently useful correlations have been derived in the past and recommended for prediction. The common experience is that the correlations once thought useful have become less significant for later periods. Treloar and Grant (1953) list various regressions of southern rainfall on Darwin pressure which have become worthless: one correlation coefficient changed from −0.87 in one 20-year period to −0.14 in the next period and this last is quite useless for prediction; some regression coefficients have even changed sign. Similar changes have occurred in other parts of the world.

The search for trends and cycles in rainfall records themselves is essentially confined to discussions of rainfall statistics. To establish any trend in rainfall it is necessary to use long records of rainfall, which are rare. Furthermore, there is always the possibility that some gradual changes in exposure of the gauge occurred, and were never recorded. Some mathematical techniques are needed to reveal a concealed pattern among the random fluctuations. Unfortunately, some of the techniques which have been widely used are unreliable; Gani (1975) lists various examples of dubious procedures. A valuable discussion of meteorological statistics and of some precautions to be taken in their use has been given by the World Meteorological Organization (1966).

Cornish (1954) found no significant trend in annual rainfall at Adelaide between 1839 and 1950, but detected a cyclic change in the date at which one quarter of the year's rain had fallen; the cycle was considered to be the same as the double sun spot cycle. However, Leeper (1957) could not find this cycle in the records of other stations. He also rejected some other claims that summer rainfall was increasing and winter rainfall was decreasing.

There is indirect evidence of long-term changes in rainfall, but it cannot be verified from existing raingauge records because the random changes conceal any trend.

Many attempts have been made to find approximately periodic changes in rainfall. The random fluctuations are again so large that some statistical manipulation is needed. Time series analysis (Kendall 1973) is statistically safer than crude moving averages. Unfortunately, most of the periods detected for one station do not occur for other stations or for other lengths of records. Also, the effects, even if significant, are always small, and their value is not in prediction but to suggest possible meteorological research.

O'Mahony (1961) found some evidence for 2–3-year cycles and also for a seven-year cycle in some but not all of the seasonal records of eight stations examined. The short 2–3-year cycle is thought to be related to some natural oscillation of the earth's atmosphere; interestingly he found no evidence for the 11-year solar cycle. In his regional analysis, Coughlan (1975) found marginally significant evidence for 2–3-year periods but not for O'Mahony's 6–7-year cycle nor for any solar cycles. He mapped regions where seasonal and annual variations are significant but they seldom coincide.

Rain controls beyond the atmosphere

Many attempts have been made to relate rainfall characteristics to causes outside the earth's atmosphere but none are at present generally accepted.

Sunspot cycles were long ago related to changes in the ionosphere, but claims of correlations with rainfall have repeatedly been disputed. A sunspot cycle is not periodic; individual cycles have varied between 8 and 17 years, with an average of 11 to 12 years. The greatest difficulty in accepting any influence of solar cycles is that no physical mechanism is known; in particular, the solar radiation received by the earth is not significantly altered by sunspots. Some tentative mechanisms are outlined by King (1975).

Bowen (1953) readmitted meteors into meteorology when he suggested that meteoric dust acts as rain nuclei in clouds and is responsible for heavy rains. However, Martyn (1954) and Swinbank (1954) have criticized this hypothesis; also Canadian data examined by Maybank and Qureshi (1966) do not support the hypothesis. Subsequently, this suggestion became entwined with possible lunar effects (Bowen 1963). The incidence of large falls of rain at Sydney, Rockhampton, Perth, and Adelaide from about 1861 to 1960 was examined for any lunar influence by O'Mahony (1965); the only evidence he found was a small, statistically insignificant rainfall minimum at full moon for each of the four stations.

Evaporation and Evapotranspiration

Evaporation involves the presence of an energy supply to vaporize water, a process to remove the vapour into the atmosphere and usually a process to move water to the evaporating surface. If plants are present each of these processes can be modified and it is then usual to use the term evapotranspiration. Linacre (1963) distinguishes four rates of water loss by evaporation: the potential evaporation rate E_0, which is the evaporation rate from a large flat free water surface, the evaporation rate measured from a pan or tank evaporimeter, E_{pan} or E_{tank}, the potential evapotranspiration E_t, which applies to a crop (plants and soil surface) for which water is not limiting, and the actual rate of evaporation E_a, which is the rate obtaining in reality and determined by soil, plant and climate.

Measurement of Evaporation

It is convenient to illustrate details of the various processes by discussing methods of measurement. All measurements of evaporation are troublesome and only modest accuracy is possible.

Evaporimeters

Evaporation from free water surfaces can be measured by recording the water loss from an evaporimeter, either a sunken tank or a surface pan. Evaporimeters of various types have been used at some Australian stations for over a century. Hounam (1961) edited the records and later (1964) published their statistical parameters. Class A pans of the U.S. Weather Bureau are now standard and evaporation contours from them have been published by the Australian Bureau of Meteorology (1975 Map 3). Very few of these should be regarded as relevant for actual evaporation, e.g. the annual pan evaporation in most of central Australia is about ten times the annual rainfall, and over most of the continent the rainfall is less than the pan evaporation in every month.

Some soils are so impermeable that crops grown on them can be treated as though they were in evaporimeter tanks. Thus Talsma and van der Lelij (1976) found that for rice growing in fine-textured soils near Griffith, drainage at 1 mm day^{-1} was a relatively constant and small part of the water balance. Usually, however, drainage is a large and essential part of the water balance for irrigated crops and some artificial barrier like a deep lysimeter is needed to measure drainage.

Lysimeters

The simplest lysimeters serve only to collect drainage. Unless the change in soil water content is monitored, only annual evaporation can be estimated. A neutron moisture meter installed in the lysimeter can yield monthly evaporation loss; Holmes and Colville (1970) used this technique to measure evapotranspiration from pasture near Mt Gambier in South Australia. Where more frequent and more accurate measurements of evapotranspiration were required on the River Murray flats in South Australia, Holmes and Watson (1967) used household buckets as weighable lysimeters. To simulate natural drainage patterns and root growth, lysimeters must be quite large, and elaborate engineering is usually required to obtain accurate sensitive weighing; McIlroy (1975) has developed such a large weighing lysimeter with good sensitivity but of low cost and simple construction. Puckridge (1978) has developed this instrument further.

Oasis effect

If possible, a representative crop should be planted to protect pans or lysimeters from the oasis effect, i,e, the abnormal sensible heat flux present at the junction of dryland and growing crop. Pans placed outside irrigated crops can lose a quarter more than pans inside the crop. The minimum width of the protective belt or fetch is usually accepted as 100 m, but the evaporation decreases for a longer fetch.

Lang *et al.* (1973) measured evaporation at four sites within a flooded rice crop near Griffith. The fetch varied with different wind directions for each site, and values from 14 to 497 m were used. Evaporation near the edge was about a fifth greater than for the largest fetch, and was considered to vary as a negative power of the fetch; their relation predicts a reduction of only one third in the variable component for a tenfold increase of fetch.

Estimation of evaporation

Lysimeters are difficult to install and operate; therefore much effort has been given to using meteorological data to estimate evaporation. A recent review of various approaches based on turbulence theory is given by Webb (1975). The energy balance approach apportions the difference in net radiation R and thermal ground flow G between sensible heat flux and evaporative flux E from the evaporating surface. An alternative aerodynamic method requires knowledge of wind speeds and air moisture. Penman (1948) combined these two approaches to avoid some difficult measurements. Various modifications and reformulations of Penman's equation have been suggested. Thus Slatyer and McIlroy (1961) write a generalized form as

$$LE = \frac{s}{s+\gamma}(R-G) + h\,(D_z - D_o), \tag{1}$$

where L is the latent heat, s is the slope of the saturation vapour pressure temperature curve at the temperature measured at height z above the evaporating surface, γ is the psychometric 'constant', which in fact varies with pressure, h is an atmospheric conductance term depending upon wind speed at height z, and D_z and D_o are wet bulb depressions at height z and at the evaporating surface. The term $s/(s+\gamma)$ is well tabulated, but the expression for h varies between different formulations.

Potential evapotranspiration

Potential evapotranspiration is defined as the evaporation from well watered short grass. The term D_o in Eqn(1) then vanishes and the equation becomes a direct modification of the original Penman equation. Although originally discussed for English conditions, Penman's potential evaporation equation has been used successfully for several Australian stations: Katherine in the Northern Territory, Yass (Slatyer 1960*a*, *b*), and Griffith in New South Wales (Fleming 1964), and Aspendale in Victoria (McIlroy and Angus 1964). Potential evaporation depends only on meteorological conditions. Thus van Bavel (1966) found good agreement with lysimeter results even on an hourly basis for bare wet soil, free water and short well watered lucerne.

Because of differences in reflection between vegetation and a water surface (0.25 for crops versus 0.05 for water), net radiation can be 25% higher for water than for grass. Under conditions of low advection, in European environments, E_{grass} is about 0.8 E_{tank}. Sometimes though, under Australian conditions, while the Penman equation describes E_{tank} surprisingly well, E_{grass} is often 5% higher than E_o. This is generally explained on the basis of surface roughness and advection.

The need for considerable micrometeorological data has always been a limit on the use of Penman's equation, e.g. early formulations used hours of bright sunshine instead of net radiation. Linacre (1977) has given two approximate formulae, one for free water and one for a green crop, which both require only mean temperatures and mean dewpoint in addition to station latitude and elevation. Comparisons of monthly estimates with lysimeter values show discrepancies of only 0.5 mm day^{-1} or 10%, which are acceptable.

An alternative simplification has been to calibrate evaporimeters against Penman estimates. Thus Fleming (1964) at Griffith found that Penman estimates agreed with evaporation from sunken tanks, but were only 80% of pan evaporation. Pruitt *et al.* (1972) quote a series of conversions for different crops and wind conditions in California.

Approximate methods for estimating evaporation

It is often necessary to estimate evaporation when only routine meteorological data are available. Various formulae have been proposed, but none is suitable for providing the daily estimates desirable in managing large irrigation projects.

McGuinness and Bordne (1972) listed fourteen formulae for estimating potential evaporation and gave all numerical procedures. All formulae were compared with the smoothed results of 15 years observations on the weighable lysimeter at Coshocton, Ohio, which has a humid climate. Four of these formulae involve only temperature data, and one of these, that of Blainey-Criddle, gave an annual total 10% less than the lysimeter result; this was achieved by systematic underestimation in the winter and overestimation in the summer. For comparison, the van Bavel form of the combination formula gave the closest annual total of 5% in excess, but showed the opposite pattern through the year.

Daily results for 70 summer days at the same site were considered by Parmele and McGuinness (1974). The simple Blainey–Criddle formula accounted for only 10% of the variance, which should be compared with 66% for Penman or van Bavel's formulae and 64% for the simple net radiation.

Monthly total evaporation from the same site was examined by Hargreaves (1974), who obtained correlations (r^2) as large as 90% for various elements, even for daylength.

Potential evapotranspiration with crop development

The original idea of potential evapotranspiration of a pasture, which applied to a uniform and well-covered turf, has been extended to potential evapotranspiration for the crop at different stages of development. While Penman allowed for a winter and summer season for pasture, Slatyer (1960*a*) considered potential evaporation E_t for four stages of crop development: emergence, establishment, period of maximum growth, and finally, maturation, with a crop factor f of 0.4, 0.6, 0.8 and 0.2 respectively, where $f = E_t/E_o$. This coefficient, in one value, takes account of leaf area of the crop, evaporation from the soil, and changes in structure and morphology of the crop. More detailed models (Paltridge 1970; Ritchie 1972) separate E_t into evaporation from the soil and transpiration from the crop on the basis of leaf area or row spacing index.

The stage of development of the crop can sometimes be predicted from simple meteorological parameters. Fitzpatrick and Nix (1969) for day-neutral varieties of wheat found that accumulated potential evaporation, which correlates well with total radiation and temperature over the growing season, predicts anthesis better than day-degrees summation or photothermal units. The same approach was used by Berndt and White (1976) for comparing the water balance aspect of cropping strategies involving wheat and sorghum on the Darling Downs in southern Queensland.

Actual evaporation

While the practice of first estimating potential evapotranspiration for the crop has been retained, in more recent water balance models there have been significant departures from the original Penman model in estimating actual evaporation loss. Where Penman originally assumed that E_t was maintained until the water supply in the root zone was exhausted, i.e. water content less than wilting point, present models modify evaporation loss according to the soil water content. McIlroy (1965, 1968) first calculates E_t using a modified Penman equation and then modifies it according to soil water content as

$$E_a = E_t/(1 + \frac{\gamma}{s+\gamma}\frac{h}{h_i}) \qquad (2)$$

The internal transfer coefficient h_i of the soil–root system depends on soil water content as $h_i = c(\theta - \theta_c)$, where c and θ_c are constants for the particular system, θ_c being the critical water content below which there is no significant uptake. These parameters can be determined experimentally for a particular soil–crop system; Dunin and Costin (1970) give an example of their evaluation in a water study on pastures.

The concept of a maximum evapotranspiration rate based on the conductance of the soil–root system was implicitly demonstrated by Denmead and Shaw (1962) and was defined by Linacre (1963) as the maximum rate E_m at which the soil conductivity and the concentration and hydraulic properties of the root allow water to be supplied to the evaporating surface. Hydraulic conductivity of the soil decreases rapidly with decreasing soil water content and E_m decreases accordingly.

Thus there is an upper limit of crop evaporation E_m, the maximum evapotranspiration rate. While E_t for a particular growth stage is less than E_m, E_a will equal E_t. As the soil store is depleted, E_m falls, or if E_t exceeds E_m due to a weather change, then E_a will fall below E_t and become equal to E_m. Johns and Smith (1975) conclude that of six functions for predicting E_a for four temperate pastures of clovers and grasses at Armidale on the New England tableland, a linear E_m function (Linacre 1963) and a quadratic E_m function (Linacre 1973) gave close to the best prediction. The quadratic function $E_m = E_{mm} W^2$ (mm day^{-1}), E_{mm} is the maximum value of E_m at $W = 1$ and W is the proportion of the available water currently present, was preferred because of the markedly low sensitivity to assumed values of both E_{mm} and surface soil capacity. For modelling water extraction by wheat from the subsoil, Greacen and Hignett (1976) used the linear E_m function, $E_m = E_{mm}W$. They estimated E_{mm}, for predicted length of root in the subsoil, from a measured mean maximum rate of uptake of 0.002 cm^3 day^{-1} for each cm length of root. Their data, however, (their Fig. 7) probably favour the quadratic function.

The use of a measured mean rate of uptake per unit length of root is a simplification in that it assumes that axial resistance to flow in the root is negligible, and that radial resistance is uniform and can be integrated in parallel over the entire length of the root system. Landsberg and Fawkes (1978) assume uniform radial resistance of a main root, on the basis of possible uniformity along the lateral roots. Experimental evidence consistent with uniform total resistance to uptake throughout the root zone, reported by Taylor and Klepper (1975) and Burch (1979), support this assumption, but a strong case can be made against their conclusions. Fig. 33.2 (composed from Greacen 1977) gives predicted density of young root (which may be read as 'low resistance root') as a function of depth. If the assumption is made that a significant resistance to axial flow exists, it is not difficult to ascribe, through a function of root age, a reasonable radial resistance that will produce apparent uniform total resistance throughout the profile.

Specification of regions of lower resistance in the roots of plants means that effective rooting densities are reduced, flow rates in the rhizosphere are increased and, as a consequence, soil resistance increases in importance. This raises the whole problem of the significance of the various components of resistance to flow in the soil–root system. It will only be resolved by direct or indirect estimates of the important parameters as they occur under field conditions.

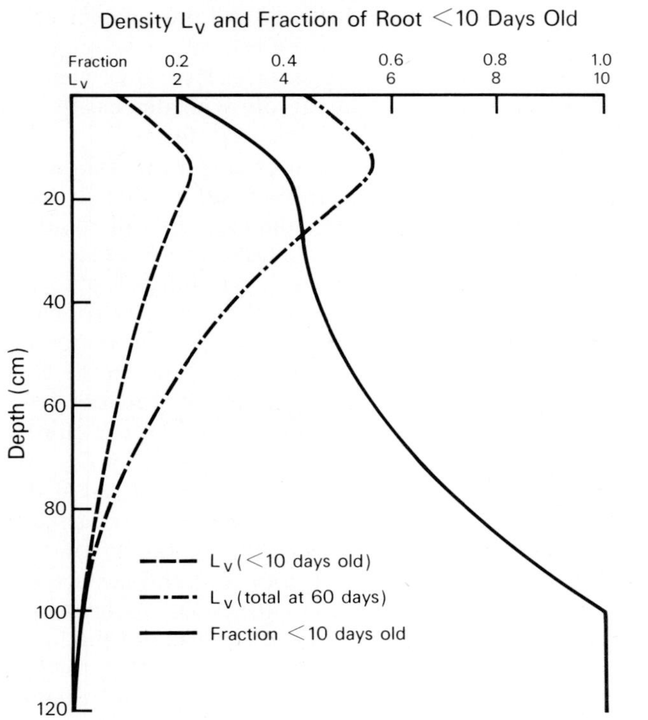

Fig. 33.2. Distribution of the rooting density for different ages of root (Greacen 1977).

Recharge to groundwater systems

Water moving to a groundwater system is often what remains after plant and soil water storage requirements are satisfied. In many situations, recharge can be identified with the drainage component of the water balance, as discussed in Chapter 32. Assessment of the mean annual recharge to a groundwater system is important as this is the parameter most often required for water resource evaluation.

Techniques for estimating recharge

This section is primarily devoted to a description of the techniques available for the estimation of mean annual recharge.

Recharge can occur either through a large portion of the material overlying an aquifer (local recharge) or at well defined geomorphic locations as, for example, through river beds or surface outcrops of aquifer material. In Australia, examples of the former are the coastal-plain sand aquifer in Western Australia, and the Gambier Limestone aquifer in southern Australia. The Great Artesian Basin aquifer provides an example of the latter. Confined aquifers usually have restricted sources of recharge, whereas unconfined aquifers can be recharged by either of the mechanisms given above.

Recharge through river beds is generally best estimated by determining changes in river flow at two or more locations, or by observation of water table fluctuations. Recharge through such localised areas will not be discussed further here; discussion will be centred on local recharge, that is, drainage of water through the unsaturated soil profile to a water table.

Several techniques have been developed for estimating recharge; the earlier ones by engineers, but recently, soil physicists have made an important contribution to its evaluation.

Engineering methods of estimating recharge

One of the simplest techniques for estimating recharge is to use the equation $R = \Sigma O_i + \Delta S$, where R is recharge, ΣO_i the total outflow from the aquifer, and ΔS the change in storage. However, the difficulty with this method is that a component of the discharge of many groundwater systems occurs through lake floors or coastal margins, making ΣO_i almost impossible to estimate. Unless there are strong reasons for assuming that all outflow occurs to a spring or river system, this approach is not usually profitable.

Empirical expressions of the type $R = k_1(P - k_2)$, where P is the precipitation and k_1 and k_2 are constants for a particular area, have been used with varying success in estimating approximate water budgets for an area.

Another approach used to determine local recharge is to solve the water balance equation. However, evapotranspiration is usually the largest loss component and, since the techniques avai lable for estimating this often give rise to quite large errors, values of recharge can be in error by a factor of 2 or more.

Lysimetry, as discussed in the previous section, should also be included in this group. If the land use and soil of the lysimeter is identical with that of the surrounding area, and if the lysimeter does not disturb the water regime, the local recharge occurring each year is equal to the amount of drainage collected at the base of the lysimeter.

Soil physics methods of estimating recharge

Solution of the drainage equation (Chapter 32) is possible if both the $\psi(\theta)$ and $K(\psi)$ relationships are known. However, in real soils, solution of this equation is impracticable because K and ψ are difficult to measure, so it is not often used to estimate recharge.

Various analytical and numerical techniques are available for determining recharge using water table behaviour, as demonstrated by Childs (1969). Analytical methods have been used successfully to determine recharge to aquifers underlying irrigated areas, but they have been less successful in modelling large aquifer systems. Numerical models are readily applicable to the estimation of recharge, but such models often require detailed data input, such as reasonably accurate estimates of hydraulic conductivity and specific yield.

Tracer methods for estimating recharge

Natural and injected tracers have both been used to evaluate mean annual recharge, but, as natural tracers are injected with precipitation over the whole surface of the aquifer, they seem to be a more attractive proposition, and only these will be discussed here.

During infiltration, only that portion of the water in the large pores of the soil is mobile. However, where the distance between large and small pores is small, diffusion will cause the concentration of a perfect tracer (one whose properties are identical to those of water) in the water moving in the large pores to rapidly become equal to that in the small pores (Zimmermann *et al.* 1967). As a consequence of this, all molecules of a perfect tracer move downward with the same apparent velocity. It also follows that water which reaches the surface at different times will be layered in the soil profile, with the most recent near the top and the oldest at the bottom.

Since this book is primarily concerned with soils, little of the work using tracers in the saturated zone will be discussed. The natural tracers, tritium, ^{14}C-labelled carbon, and chloride, have all been used in modelling aquifer behaviour and recharge with varying degrees of success. In an early paper, Nir (1964) proposed several useful models. Examples of Australian work are studies of the Mereenie Sandstone aquifer near Alice Springs using ^{14}C (Calf 1978), of a shallow aquifer near Mt Gambier in southern Australia using tritium (Allison and Hughes 1975) and of catchments in the Darling Ranges near Perth in Western Australia using chloride balance (Peck and Hurle 1973).

The remainder of this section will be devoted primarily to a discussion of the usefulness of natural tritium and chloride in the unsaturated zone for evaluating mean annual recharge (MAR).

(1) Natural tritium

In most soils, tritiated water, $^3H^1HO$, moves in almost the same way as ^1H_2O and is therefore a near-perfect tracer. For the reasons given above, a series of pulses of $^3H^1HO$ applied at the surface of a soil will move down layer upon layer as piston flow. Depending on the rate of flow, diffusion and dispersion will cause some mixing between layers.

Thus, in the unsaturated zone of soils where recharge to groundwater occurs, each year's recharge will lie above that of the preceeding year. Since tritium concentrations of rainfall have varied from year to year it is possible to identify a particular year's recharge within the soil profile, provided that the unsaturated zone is capable of holding an amount of water large enough to ensure that many years' recharge are represented.

In the northern hemisphere a large peak in tritium fallout in precipitation occurred in 1963. Smith *et al.* (1970) showed that this peak was preserved in the English chalk. Their tritium profile is shown in Fig. 33.3 together with a curve showing the way in which the tritium concentration of rainfall has varied with time in the northern hemisphere.

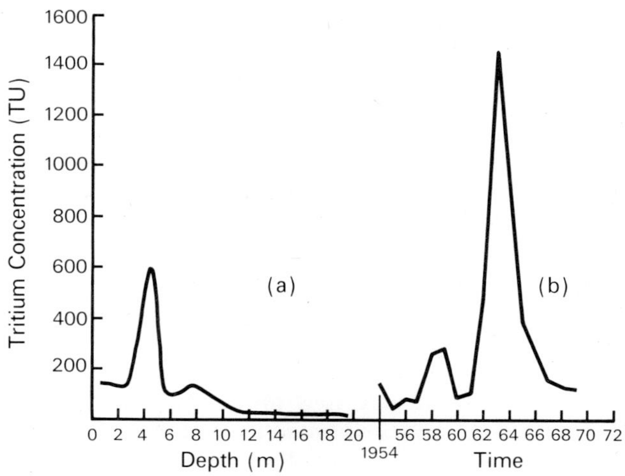

Fig. 33.3. Variation of the tritium concentration of (a) soil water with depth for a chalk profile and (*b*) rainfall with time for the U.K. (from Smith *et al.* 1970). 1TU is one atom of tritium in 10^{18} atoms of hydrogen.

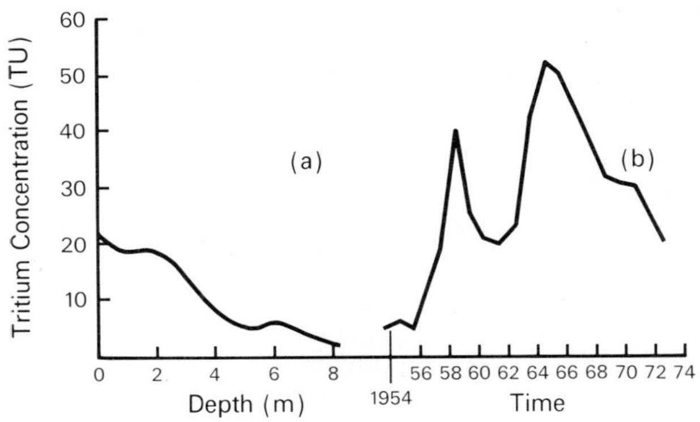

Fig. 33.4. Variation of the tritium concentration of (*a*) soil water with depth for a loam overlying clay profile (from Allison and Hughes 1974) and (*b*) rainfall with time for southern Australia (from Allison and Hughes 1977).

Although such a marked peak in tritium fallout has not occurred in the southern hemisphere (Taylor 1971; Allison and Hughes 1977), it is possible to find the tritium fallout record preserved in the soil profile. For example, Bredenkamp *et al.* (1974) have reported tritium profiles for an area in southern Africa, and Allison and Hughes (1974) for sites in southern Australia. A tritium profile obtained for a clay soil in southern Australia is shown in Fig. 33.4.

Three techniques are available for estimating MAR from such a profile: from the position of the tritium peak, from the shape of the tritium peak, and from the total amount of tritium stored in the soil profile.

From the position of the tritium peak. If water moves by

piston flow, all the water above the peak of tritium concentration in the profile must have entered the soil after the peak in tritium concentration of rainfall. Thus, in principle, recharge may be evaluated simply by dividing the total amount of water in the profile above the peak by the time elapsed since the fallout peak. Problems with this technique arise because roots extract differing amounts of soil water at different depths; these problems and possible ways of overcoming them are discussed by Allison and Hughes (1974).

The technique is most suited to areas having shallow-rooted vegetation; it is unsuitable where a significant amount of water moves through fissures, so partially invalidating the piston flow model. In climates where rainfall is high and distributed fairly uniformly over the year, the technique is probably applicable fairly generally because only small soil moisture deficits are generated and rooting depths are shallow.

From the shape of the tritium peak. A more reliable technique for estimating MAR than that described in the previous section is to develop a model for the infiltration and redistribution of water in the root zone, and hence calculate tritium profiles for different recharge rates, and compare them with the observed profile.

Because there are differences in the concentration of tritium in rainfall from year to year, vertical diffusion will occur between the annual increments. The effect of diffusion can be calculated as shown in Schmalz and Polzer (1969) and Smith *et al.* (1970). Discrepancies between observed and calculated profiles can give an indication of the importance of fissure flow or other non-ideal flow mechanisms.

Schmalz and Polzer attempted to describe tritium profiles in a low rainfall region in the United States by assuming that a fixed percentage of rainfall moved into the soil each year and that loss of water occurred from the soil surface only. Smith *et al.* also assumed loss of water (and hence $^3H^1HO$) from the surface layers of the soil only; in the rather humid English climate this seems to be a reasonable assumption. They used the difference between rainfall and a modified Penman evaporation as their recharge input, rather than using a fixed percentage of rainfall, and obtained an excellent match between observed and calculated profiles when they assumed that 15% of the water moved rapidly through fissure systems. This shows the importance of matching profiles; the method based on the position of the peak would have overestimated recharge by 15%.

The type of model used by Smith *et al.* requires estimates of evapotranspiration from the soil. In climatic regions where water becomes limiting over extended periods, as in most of Australia, continuous evaluation of evapotranspiration is difficult. The following technique may be used to overcome this problem and to enable estimates of mean annual recharge to be made in areas where limited meteorological data are available. Annual recharge R is written as $R = \omega_i \bar{R}$, where $\bar{R}$ is mean annual recharge and ω_i is a weighting function for year i, which accounts for year to year variation in recharge (Allison and Hughes 1974). Tritium profiles can then be calculated using a range of values of $\bar{R}$. In practice, a range of weighting functions should be tried in order to test the sensitivity of the calculated profile to differences in ω_i. Weighting functions which have been used are: $\omega_i = (P_{\omega i} - c)/(\bar{P}_{\omega} - c)$, where $P_{\omega i}$ is winter plus spring rainfall in year i, $\bar{P}_{\omega}$ the mean value, and c is a constant, $\omega_i = \Delta h_i / \overline{\Delta h}$ where Δh_i is the annual watertable rise and $\overline{\Delta h}$ is the mean value. A comparison of several weighting functions is given in Allison and Hughes (1978).

From the total amount of tritium stored in the soil profile. The total amount of tritium stored in the soil profile, T_s, is given as

$$T_s = \int_0^\infty T(z)\theta(z)\mathrm{d}z, \tag{3}$$

where $T(z)$ is the tritium concentration of the soil water at distance z beneath the soil surface, and θz the volumetric water content. Provided that sampling has been carried out to such a depth that the soil water being intersected fell as rain before 1950 ($T \simeq 0$), then the integral may be approximated by

$$T_s \simeq \int_0^d T(z)\theta(z)\mathrm{d}(z), \tag{4}$$

where d is the depth of sampling. MAR can then be estimated from T_s using

$$T_s = \bar{R} \sum_{i=1}^{\infty} \omega_i \, T_{pi} \exp(-i\lambda) \tag{5}$$

where T_{pi} is the tritium concentration of recharge water i years before the present and λ is the decay constant for tritium.

An advantage this technique has over the earlier methods is that the estimate of MAR is independent of the distribution of tritium in the profile; thus, if some water is moving in a non-ideal fashion, making it difficult to model the profile shape, it is still possible to obtain an estimate of MAR from such a profile.

This technique seems to be the most useful for routine resource evaluation. However, in applying it, the assumption must be made that none of the tritium that is in the profile at the driest time of the year will be lost subsequently. When sampling does not take place at the driest time of the year a correction term based on the net accession of tritium since that time should be subtracted from T_s.

The three techniques described so far for estimating MAR from tritium profiles have been concerned only with an unsaturated zone that can store at least 20 years of recharge water, and therefore lateral movement of water can be ignored. However, tritium profiles can be extended beneath the water table. For this case Atakan *et al.* (1974) have demonstrated that lateral flow, which now becomes important, can be modelled readily, and estimates of local recharge obtained.

(2) Natural chloride

Although chloride does not have all the properties of a perfect tracer, it moves as water does through most soils. It has the advantages over tritium that it is concentrated by evapotranspiration and is not lost from the soil as is tritium.

Chloride can provide information about recharge rates, provided there are no sources of chloride in the soil matrix. However, surprisingly little use has been made of chloride for this purpose; Anderson (1945) recognized the potential of the method and used it to study several groundwater systems in southern Australia, but the work was hampered by lack of data for input of chloride. Since that time, the work of Hutton and Leslie (1958), Hingston and Gailitis (1976), and Hutton (1976) has greatly improved our knowledge of chloride fallout in Australia.

One approach has been to measure the ratio of the mean annual concentration of chloride in rainfall to that in groundwater; provided runoff is negligible, this ratio should be equal to the ratio of annual recharge to total precipitation. The technique has been used to evaluate the water resources of portions of the coastal aquifer in Western Australia (Balleau 1973; Forth 1973). Eriksson and Khunakasem (1969) have used essentially the same technique, but with more detailed data, in a model which allowed them to compute recharge distribution from variations in groundwater chloride. A major difficulty with this approach is that it is usually assumed that no long-term changes in recharge have occurred. Unfortunately, land use has often changed considerably in the past 100 years. Such changes usually give rise to changes in recharge and most sizeable groundwater systems will not have reached a new equilibrium in this time. For example, Peck and Hurle (1973), in studying the chloride balance of several small catchments (0.4–380 x 10^3 ha) undergoing changes in land use, estimated that the time for chloride equilibrium to be re-established under new regimes was 20–400 years.

Since many groundwater systems of interest have a small ratio of recharge to total volume, the time to reach chloride equilibrium will be large. Thus it seems profitable to consider alternative methods of estimating recharge from chloride. One possibility is to sample the top of the groundwater table, as this would represent recent recharge; tritium analysis could be used to confirm this. Provided a depth of aquifer is intersected such that several years' recharge is sampled, an estimate of recharge $\bar{R}$ may be made using

$$\overline{PC_p} = \bar{R}C_G,$$

where $\bar{P}$ is the mean annual precipitation, C_P is the mean chloride concentration including dryfall, and C_G is the chloride concentration of the groundwater. In very active groundwater systems lateral flow may have to be taken into account.

Another possibility is to determine the chloride concentration of soil water in the unsaturated material beneath the root zone. Provided information on chloride fallout is available at the site, recharge can be estimated on the following basis (Allison and Hughes 1978).

On average, the flux of chloride through the soil surface is equal to the flux of chloride beneath the root zone, thus

$$\overline{PC_p} = \overline{RC_s},$$

where C_s is the chloride concentration of soil water, and C_p takes fertilizer additions of chloride, if any, into account. However, because correlation is expected between R and C_s,

$$\overline{RC_s} = \bar{R}\bar{C}_s + \overline{\Delta R \Delta C_s},$$

where Δ refers to deviation from the mean.

Combining these we obtain

$$\bar{R} = (\overline{PC_p} - \overline{\Delta R \Delta C_s})/\bar{C}_s.$$

To obtain estimates of $\overline{\Delta R \Delta C_s}$, the various weighting schemes discussed earlier may be used to obtain estimates of ΔR; an estimate of ΔC_s can be obtained from the variability in C_s beneath the root zone. Allison and Hughes found that $\overline{\Delta R \Delta C_s}$ was in the range $(0.10\text{–}0.15)\overline{RC_s}$, but in more arid areas this term may be considerably larger. In a study in Israel, Eriksson and Khunakasem (1969) found that neglecting the correlation term led to errors of less than 10%.

Considerable interest is currently centred on methods for evaluating the water balance of forested catchments. The technique outlined here may have considerable application in estimating recharge beneath deep-rooting species. However, in areas of low recharge it may be important to choose sites where the water table is well beneath the root zone since, if higher or lower recharge occurs upgradient, chloride may be gained or lost to the unsaturated zone by diffusion, which becomes important at low flow rates. The possible importance of diffusion can be seen in salinity profiles presented by Dimmock *et al.* (1974) for some Western Australian soils.

Comparison of estimates of mean annual recharge made using different techniques

The use of the tracers tritium and chloride has been emphasized in this section primarily because of their potential for estimating MAR as the average of a number of years' recharge from a single profile. Little work has been published on the comparison of estimates of recharge made using tracers and other techniques. Holmes and Colville (1970) obtained values for MAR of 47 mm and 79 mm from two lysimeters for the period 1961–65; at a nearby site, Allison and Hughes (1974) found from tracer work that MAR was 49 ± 9 mm. As rainfall for the period 1961–65 was higher than the 30-year mean, the agreement is reasonable.

To assess the usefulness of the tritium profile technique in evaluating recharge over a relatively large area, studies have been carried out by Bredenkamp *et al.* (1974) in southern Africa and by Allison and Hughes (1978) in southern Australia. In both of these studies quite close agreement was obtained between estimates of total recharge made using tritium profiles and those made using standard methods. However, it should be pointed out that in both cases considerable uncertainties exist in the estimates.

A comparison of estimates of recharge made using chloride and tritium has been made in southern Australia by Allison and Hughes. At the 12 sites they studied they found excellent agreement over a range of mean annual recharge from 50 to 250 mm (Fig.33.5).

Since the tritium method relies on the measurement of the total amount of tritium in the profile, the precision of estimates of recharge increases as the amount of recharge increases. In contrast, the chloride technique relies on the estimation of concentrations; and since recharge is inversely related to chloride concentration in the unsaturated zone, sampling and analytical errors become less important at low recharge rates. This indicates that a combination of both techniques should be used, where possible, to obtain reliable estimates over a wide range of values of mean annual recharge.

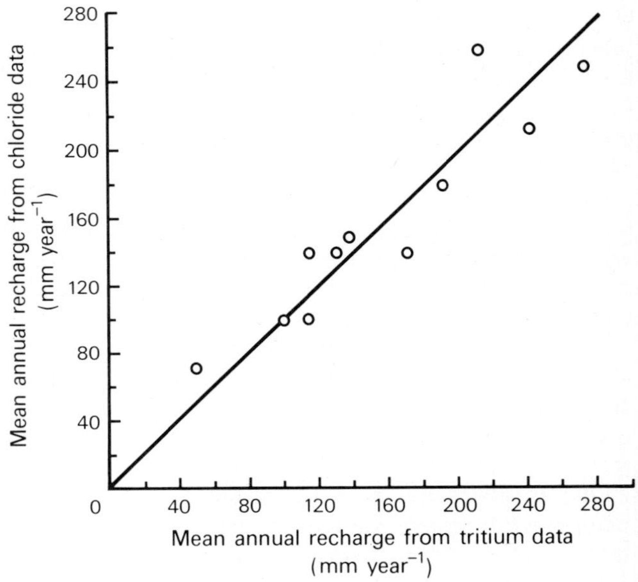

Fig. 33.5. Comparison of recharge estimates obtained from chloride and from tritium data (Allison and Hughes 1978).

Saturated groundwater flow

Water infiltrates a soil profile until eventually it reaches some impermeable material and causes a zone of saturation to develop. There is a vast literature on flow within the saturated groundwater phase, and much of this has been condensed into book form. Accordingly this brief guide can do little more than list some topics and indicate some helpful references.

Saturated groundwater flow is usually regarded as water flowing within the pores of an inert soil fabric. For water-saturated material this can vary from about 60% by volume for some clays to less than 0.1% for some rocks. Davis (1969) has reviewed this subject.

If no fluid motion occurs, the ordinary laws of hydrostatics should apply, i.e. the pressure decreases with height measured above some reference level. Alternatively, a potential defined as $\phi = p + z$, where p is the pressure measured in metres of water and z is the height in metres above some reference level, will be constant throughout the system.

Calculation of water flow using Darcy's law

If fluid flows in a stable saturated soil, experiments indicate that the velocity $\bar{v} = -K \text{ grad } \phi$, where K is a constant for a given soil. This is the original form of Darcy's law, and it is much simpler than the modification discussed in the previous chapter, where K must be a function of water content. Numerical values of K vary from about 0.1 m s^{-1} for a clean gravel to 10^{-9} m s^{-1} for a fine silt or clay and even less for unweathered clay.

The combination of Darcy's law with the equation of continuity for an incompressible aquifer shows that the potential function must satisfy Laplace's equation.

The most tractable problems involve steady state water flow within a confined surface, such as occurs in an aquifer bounded above and below by impermeable layers, or in flow below dam works. The interest in the latter case is to predict pressure distributions along the base of the dam and to predict any excessive flow velocity below the dam works.

In these problems, any aquifer surface in contact with a large reservoir will be an equipotential surface, any seepage surface exposed to the air will be at zero pressure and hence the potential will be a known function of position, and impermeable surfaces will have a vanishing normal potential gradient. Some boundary surfaces can leak water into another system and these can be described by a linear relation between potential and its normal gradient. Thus quite varied boundary conditions occur even in the simplest problems.

If the pressure varies significantly with time, then the very small compressibility of the aquifer material must be considered, and the potential must satisfy the diffusion equation.

It is rather more common to find that the water system is not confined and two conditions hold along the free surface of the watertable. By definition it is a surface of zero pressure. Also, some boundary condition must be defined by the problem, e.g. the boundary may be a streamline if no recharge occurs, or alternatively the recharge must be specified and a non-linear boundary condition will result along the watertable. A part of the calculation will be to locate this watertable.

Traditionally, the hydraulic conductivity of the unsaturated zone above the watertable has been ignored, e.g. any recharge has been assumed to descend vertically from the surface. More recent numerical treatments, e.g. Vauclin *et al.* (1976), have considered the influence of this zone for some transient problems.

Detailed solution of these problems involves mathematical difficulties, and approximations are often essential. Three methods of attack are available, viz. analytical, numerical, and analogue methods.

Analytic methods are usually restricted to two-dimensional problems. Special techniques exist for the Laplace equation, and these are described in the books of Bear (1972), Haar (1962), and Polubarinova–Kochina (1962). For problems involving the diffusion equation, the book of Carslaw and Jaeger (1959) is a standard reference.

Numerical methods have been widely used since high speed computers became readily available. Most of the published solutions have used finite difference methods applied to a regular grid of points. Enormous machine storage can be required, especially for multi-dimensional problems. Special care is also needed if the original boundaries do not coincide with the grid. The newer method of finite elements treats the potential separately in each of many small regions, or 'finite elements', which can be chosen at will. The principal advantages claimed for the finite element method are that it is well adapted to irregular boundaries because the elements can be shaped accordingly, that it is easier to select small elements in critical regions than it is to reduce the grid size for finite differences, and that usually less computer capacity is required than is needed for finite difference calculations. An introductory discussion of both methods applied to groundwater is given by Remson *et al.* (1971). A bibliography of applications of the finite element method to flow in porous media is given by Desai (1975); it includes references to some modifications of Darcy's law.

Another numerical method, which is restricted to solution of Laplace's equation, is the method of boundary integral equations. Liggett (1977) has described its application to some unconfined flow problems.

The analogue solution of groundwater flow problems has been overshadowed recently by numerical methods. Direct scale reduction in a sand tray has often been used, but the influence of capillary action can be excessive for unconfined systems. Other analogue methods are based on different physical processes which can be described by the same differential equations as the original water flow and which are easy to measure. For a detailed discussion of various analogue methods the reader should refer to the text book of Bear (1972) and the review article of Prickett (1975), which includes a very large bibliography.

Hydraulic theory of groundwater flow

Most calculations of flow of groundwater with a free surface avoid the Laplace equation altogether by replacing Darcy's law with a simpler 'hydraulic' assumption introduced by Dupuit. This assumes that the horizontal pressure gradient is independent of height and can be replaced by the watertable slope which is always small in field problems. The most general formulation is that the total flux in the direction of a coordinate x will be

$$q = -Kh\frac{\partial h}{\partial x}$$

where h is the height above the horizontal bed.

There are some configurations (e.g. Young (1974)) for which it is known that the Dupuit approximation correctly estimates the flux but predicts the wrong watertable shape. The configurations involve groundwater flow between two free water surfaces. It is known experimentally that the watertable is ultimately connected to the lower free water surface by a seepage zone of zero pressure, but this cannot be included in the Dupuit approximation. Therefore the predicted watertable is always too low near the outflow.

Although the Dupuit assumption was formulated in 1867 for a horizontal bed, the extension to a sloping bed is still being debated a century later. Childs (1971) emphasized that earlier calculations implied horizontal flow and accordingly he revised the assumption to include water flow parallel to a sloping bed. Marie and Towner (1975) tested the two formulas with a Hele–Shaw apparatus and found that the revised assumption was significantly better at large slopes. However, Jaiswal and Chauhan (1975) found very little difference in the calculation of watertable position, and actually preferred the original formula for calculations of flux.

A further simplification is usually made to the Dupuit assumption, especially in time-dependent problems, by neglecting explicit changes in the flux expression, i.e. by treating Kh as a constant transmissivity. This reduces time-dependent problems once again to a diffusion equation instead of a troublesome nonlinear partial differential equation.

Groundwater modelling

In many regions, groundwater is used for irrigation and also as a domestic and industrial water supply. Accordingly, it is often important to predict the effect of any proposed removal of groundwater or the effect of any proposed change in recharge from recharge basins or a different crop cover. It is often important to predict changes in water quality, such as pollutant or sea water intrusion in coastal aquifers.

A simplified description or model of the real aquifer can be either a physical analogue model or a mathematical model. If the description is adequate, the model can be used for prediction on a greatly reduced time scale. However, no real aquifer is uniform, and the irregularities are seldom obvious. The boundary conditions are at best only approximately known along the aquifer's upper surface and are almost unknown along the lower surface. Therefore it is always necessary to compare the response of a proposed model with historical records, and to successively adjust details of the model until good agreement is obtained.

For mathematical models the simplified hydraulic theory is usually thought to be adequate, in view of the other uncertainties. Analytical solutions are only available for highly idealized problems and can therefore only supply approximate estimates for real problems.

An Australian example of an analytic solution is a discussion (Colville and Holmes 1972) of possible groundwater recharge under pine trees at Mt Gambier. The approximately circular symmetry of groundwater flow and forest was idealized to allow the oscillation of the water table to be calculated. This was adequate to confirm that some recharge occurred under the forest but only about a third of that under pasture. An even smaller estimate of recharge under part of this forest was obtained by tritium analyses of the groundwater (Allison and Hughes 1972).

A discussion (Volker 1977) of the groundwater in the delta of the Burdekin river shows the greater power and complexity of numerical models. The model is formulated as a finite difference approximation of the hydraulic theory, but the values used for transmissivity are allowed to vary according to the height of groundwater above the sloping bedrock and other geological information which also supplied storage coefficients. The greatest uncertainties were the estimation of recharge and of pumping loss from the aquifer.The solution adopted was to use a subsidiary model which estimates net recharge by maintaining a daily water budget for the soil at each node. A calibration process was then performed which adjusted all the parameters until a satisfactory agreement was obtained with historical records. The final model accounted for 93% of the total equivalent water-level change. One unexpected result of this process was to reveal the very large recharge of groundwater from the Burdekin river itself.

References

Allison, G. B. and Hughes, M. W. (1972). Comparison of recharge to groundwater under pasture and forest using environmental tritium. *J. Hydrol.* **17**, 81-95.

Allison, G. B. and Hughes, M. W. (1974). Environmental tritium in the unsaturated zone: estimation of recharge to an unconfined aquifer. Proc. Symp. Isotope Tech. in Groundwater Hydrol. Vol. 1, 57-72. (IAEA: Vienna.)

Allison, G. B. and Hughes, M. W. (1975). The use of environmental tritium to estimate recharge to a South Australian aquifer. *Aust. J. Soil Res.* **13**, 245-54.

Allison, G. B. and Hughes, M. W. (1977). History of tritium fallout in Southern Australia as inferred from rainfall and wine samples. *Earth Planet Sci. Lett.* **36**, 334-40.

Allison, G. B. and Hughes, M. W. (1978). The use of environmental chloride and tritium to estimate total local recharge to an unconfined aquifer. *Aust. J. Soil Res.* **16**, 181-95.

Anderson, G. N. (1957). Are there climatic cycles? – Melbourne's rainfall examined. *J. Aust. Inst. Agric. Sci.* **23**, 235-37.

Anderson, V. G. (1945). Some effects of atmospheric evaporation and transpiration on the composition of natural waters in Australia, 4. Underground waters in riverless areas. *J. Aust. Chem. Inst.* **12**, 83-98.

Aston, A. R. and Dunin, F. X. (1977). An empirical model for drainage from soil under rain-fed conditions. *Aust. J. Soil Res.* **15**, 205-10.

Aston, A. R. and Dunin, F. X. (1979). Coupled infiltration and surface runoff on a 5 ha experimental catchment. *Aust. J. Soil Res.* **17**, 53-64.

Atakan, Y., Roether, W., Münnich, K. and Matthess, G. (1974). The Sandhaussen shallow groundwater experiment. Proc. Symp. Isotope Tech. in Groundwater Hydrol. 1974. Vol. 1, 21-43 (IAEA: Vienna.)

Australian Bureau of Meteorology (1975). 'Climatic Atlas of Australia Map Set 5.' (Aust. Govt. Publ. Serv.: Canberra.)

Australian Bureau of Meteorology (1977). Report on Cyclone Tracey, December 1974.

AustralianWater Resources Council (1969). The Representative Basins concept in Australia. AWRC Hydrol. Series No.2. (Dept. Natl. Dev.: Canberra.)

Balleau, W. P. (1973). Flow-through of water and chloride ion in a watertable aquifer of Bassendean Sands, Perth Basin. Hydrol. Symp. 1973, Perth, Publ. No. 73/3, 55-60 (Inst. Eng.: Aust.)

Bear, J. (1972). 'Dynamics of Fluids in Porous Media.' (American Elsevier: New York.)

Berndt, R. D. and White, B. J. (1976). A simulation-based evaluation of three cropping systems on cracking-clay soils in a summer-rainfall environment. *Agric. Meteorol.* **16**, 211-29.

Boughton, W. C. (1968). A mathematical catchment model for estimating runoff. *N.Z. J. Hydrol.* **7**, 75-100.

Bowen, E. G. (1953). The influence of meteoric dust on rainfall. *Aust. J. Phys.* **6**, 490-97.

Bowen, E. G. (1963). A lunar effect on the incoming meteor rate. *J. Geophys. Res.* **68**, 1401-03.

Bredenkamp, D. B., Schutte, J. M. and Du Toit, G. J. (1974). Recharge of a dolomitic aquifer as determined from tritium profiles. Proc. Symp. Isotope. Tech. in Groundwater Hydrol. 1974. Vol. 1, 73-94 (IAEA: Vienna.)

Briggs, L. J. and Shantz, H. L. (1917). Comparison of hourly evaporation rate of atmometers and free water surfaces with the transpiration rate of Medicago sativa. *J. Agric. Res.* **9**, 277-82.

Brunt, A. T. (1966). Rainfall associated with tropical cyclones in the northeast Australian region. *Aust. Meteorol. Mag.* **14**, 85-109.

Burch, G. (1979). Soil and plant resistances to water absorption by plant root systems. *Aust. J. Agric. Res.* **30**, 279-93.

Bureau of Meteorology (Australia). (Quarterly). 'Drought Review' (Aust. Publ. Ser.: Canberra.)

Byrne, G. F. and Tognetti, K. (1969). Simulation of a pasture-environment interaction. *Agric. Meteorol.* **6**, 151-63.

Calf, G. E. (1978). The isotope hydrology of the Mereenie sandstone aquifer, Alice Springs, N.T. Australia. *J. Hydrol.* **38**, 343-55.

Carslaw, H. S. and Jaeger, J. C. (1959). 'Conduction of Heat in Solids.' (Oxford Univ Press: London.)

Childs, E. C. (1969). 'An Introduction to the Physical Basis of Soil Water Phenomena.' (Wiley Interscience: London.)

Childs, E. C. (1971). Drainage of groundwater resting on a sloping bed. *Water Resour. Res.* **7**, 1256-63.

Colville, J. S. and Holmes, J. W. (1972). Water table fluctuations under forest and pasture in a karstic region of southern Australia. *J. Hydrol.* **17**, 61-80.

Cornish, E. A. (1950). The influence of rainfall on the yield of wheat in South Australia. *Aust. J. Sci. Res. Series B – Biol Sci.* **3**, 178-218.

Cornish, E. A. (1954). On the secular variation of rainfall at Adelaide. *Aust. J. Phys.* **7**, 334-46.

Coughlan, M. J. (1975). The evidence for climatic change in some Australian meteorological data. Australian Conference on climate and climatic change, Monash University Dec. 1975. To be published by Royal Meteorological Society.

Davis, S. N. (1969). Porosity and permeability of natural materials. *In* 'Flow through Porous Media'. (Ed. R. J. M. de Wiest). (Academic Press: New York.)

Denmead, O. T. and Shaw, R. H. (1962). Availability of soil water to plants as affected by soil moisture content and meteorological conditions. *Agron. J.* **54**, 385-90.

Department of National Development. (1970). 'Atlas of Australian Resources', Second Series Map 6. (Commonw. Gov. Printer: Canberra.)

Desai, C. S. (1975). Finite element methods for flow in porous media. *In* 'Finite Elements in Fluids.' (Eds R. H. Gallagher, J. T. Oden, C. Taylor and O. C. Zienkiewica). Vol. 1. (Wiley: London.)

Dimmock, G. M., Bettenay, E. and Mulcahy, M. J. (1974). Salt content of lateritic profiles in the Darling Range, Western Australia. *Aust. J. Soil Res.* **12**, 63-9.

Dunin, F. X. and Costin, A. B. (1970). Analytical procedures for evaluating the infiltration and evapotranspiration components of the water balance. IASH-UNESCO Symp. on Representative and Experimental Basins. Wellington, N.Z. Vol. 1, pp. 39-54.

Eriksson, E. and Khunakasem, V. (1969). Chloride concentrations in ground water, recharge rate and rate of deposition of chloride in the Israel Coastal plain. *J. Hydrol.* **7**, 178-97.

Fitzpatrick, E. A. (1963). Estimates of pan evaporation from mean maximum temperature and vapour pressure. *Appl. Meteorol.* **2**, 780-92.

Fitzpatrick, E. A. and Nix, H. (1969). A model for simulating soil water regime in alternating fallow-crop systems. *Agric. Meteorol.* **6**, 303-19.

Fitzpatrick, E. A., Slatyer, R. O. and Krishnan, A. I. (1967). Incidence and duration of periods of plant growth in Central Australia as estimated from climatic data. *Agric. Meteorol.* **4**, 389-404.

Fleming, P. M. (1964). A water budgeting method to predict plant response and irrigation requirements for widely varying evaporative conditions. Trans. 6th. Int. Cong. Agric. Engs. Lausanne, Switzerland, pp.66-77.

Fleming, P. M. (1973). Modelling of total hydrologic systems. Aust. Water Resour. Counc. Tech. Committee Underground Water Seminar, Manly Vale.

Flinn, J. C. (1969). Allocation of water resources; a derived demand function for irrigation water and efficiency allocation. Farm Manag. Bull., Univ. New England, Armidale No.1.

Foley, J. C. (1957). Droughts in Australia. Aust. Bur. Meteorol. Bull. No.43.

Forth, J. R. (1973). Albany Coastal aquifer – long term safe yield. Hydrol. Symp. 1973, Perth, Pub. No. 73/3, pp.119-26. (Inst. Eng. Aust.).

Gani, J. (1975). The use of statistics in climatological research. *Search* **6**, 504-08.

Gentilli, J. (1971). Climate of Australia and New Zealand. *In* 'World Survey of Climatology.' (Ed H. E. Landsberg). (Elsevier: Amsterdam.)

Gibbs, W. J. and Maher, J. V. (1967). Rainfall deciles as drought indicators. Aust. Bur. Meteorol. Bull. 48.

Greacen, E. L. (1977). Mechanisms and models of water transfer. *In* 'Soil Factors in Crop Production in a Semi-arid Environment.' (Eds J. S. Russell and E. L. Greacen) pp. 164-96. (Univ. Queensland Press: Brisbane.)

Greacen, E. L. and Hignett, C. T. (1976). A water balance model and supply index for wheat in South Australia. CSIRO Aust. Div. Soils Tech. Pap. No.27.

Haar, M. E. (1962). 'Groundwater and Seepage.' (McGraw-Hill: New York.)

Hargreaves, G. H. (1974). Estimation of potential and crop evapotranspiration. *Trans. A.S.A.E.* **17**, 701-04.

Hingston, F. J. and Gailitis, V. (1976). The geographic variation of salt precipitated over Western Australia. *Aust. J. Soil Res.* **14**, 319-35.

Holmes, J. W. and Colville, J. S. (1970). Grassland hydrology in a karstic region of southern Australia. *J. Hydrol.* **10**, 38-58.

Holmes, J. W. and Watson, C. L. (1967). The water budget of irrigated pasture land near Murray Bridge, S.A. *Agric. Meteorol.* **4**, 177-88.

Hounam, C. E. (1961). Evaporation in Australia. Bur. Meteorol. Bull. No. 44.

Hounam, C. E. (1964). The variability of Australian tank evaporation. *Aust. Meteorol. Mag.* **47**, 26-39.

Hutton, J. T. (1976). Chloride in rainwater in relation to distance from the ocean. *Search* **7**, 207-08.

Hutton, J. T. and Leslie, T. I. (1958). Accession of non-nitrogenous ions dissolved in rainwater to soils in Victoria. *Aust. J. Agric. Res.* **9**, 492-507.

Isbell, R. F. and McCown, R. L. (1978). Land. *In* 'Tropical Pasture Research – Principles and Methods.' (Eds N. H. Shaw and W. W. Bryan). CAB Bull. No. 51. Hurley, Berkshire, U.K.

Jaiswal, C. S. and Chauhan, H. S. (1975). A Hele-Shaw model study of steady state flow in an unconfined aquifer resting on a sloping bed. *Water Resour. Res.* **11**, 595-600.

Johns, G. G. and Smith, R. C. G. (1975). Accuracy of soil water budgets based on a range of relationships for the influence of soil water availability on actual water use. *Aust. J. Soil Res.* **26**, 871-85.

Keig, G. and McAlpine, J. R. (1974). WATBAL – A computer system for the estimation and analysis of soil moisture regimes from simple climatic data (2nd edn). CSIRO Aust. Div. Land Use Res. Tech. Memo. No. 74/4.

Kendall, M. G. (1973). 'Time Series.' (Griffith: London.)

King, J. W. (1975). Sun-weather relationships. *Astronaut. Aeronaut.* **13**(4), 10-19.

Landsberg, J. J. and Fawkes, N. D. (1978). Water movement through plant roots. *Ann. Bot.* **42**, 493-508.

Lang, A. R. G., Evans, G. N. and Ho, P. Y. (1973). Evaporation from an irrigated rice crop in a semi-arid region. 1st. Aust. Conf. Heat and Mass Trans. Monash, Melbourne. Sect. 3, 9-14.

Leeper, G. W. (1957). Changing climates and agriculture. *J. Aust. Inst. Agric. Sci.* **23**, 232-35.

Liggett, J. A. (1977). Location of free surface in porous media. *J. Hydraul. Proc. Am. Soc. Civ. Eng.* **103**, 353-65.

Linacre, E. T. (1963). Determining evapotranspiration rates. *J. Aust. Inst. Agric. Sci.* **29**, 165-77.

Linacre, E. T. (1973). A simpler expression for actual evapotranspiration rates – a discussion. *Agric. Meteorol.* **11**, 451-52.

Linacre, E. T. (1977). A simple formula for estimating evaporation rates in various climates, using temperature data alone. *Agric. Meteorol.* **18**, 409-24.

McAlpine, J. R. (1970). Estimating pasture growth periods and droughts from simple water balance models. Proc. 11th Int. Grassld. Cong. Surfers Paradise, Queensland. Univ. Press, Brisbane.

McCown, R. L. (1973). An evaluation of the influence of available soil water storage capacity on growing season length and yield of tropical pastures using simple water balance models. *Agric. Meteorol.* **11**, 53-63.

McGuinness, J. L. and Bordne, E. F. (1972). A comparison of lysimeter-derived potential evapotranspiration with computed values. U.S.D.A. Technical Bulletin No. 1452.

McIlroy, I. C. (1965). Routine measurements of crop evaporation. *Aust. Met. Mag.* **50**, 45-8.

McIlroy, I. C. (1968). Evaporation and its measurement, Part B. Energy balance and combination methods. *In* 'Agricultural Meteorology,' Proc. W.M.O. Seminar, Melbourne 1966, Bur. Met. Aust., 409-31.

McIlroy, I. C. (1975). A sensitive low-cost weighing system. *In* 'Recent Advances in Measurement Technology.' ISCA Natl. Symp. Melb. 21-1 to 21-6.

McIlroy, I. C. and Angus, D. E. (1964). Grass, water and soil evaporation at Aspendale. *Agric. Meteorol.* **1**, 201-24.

Maher, J. V. (1973). Meteorological aspects of drought. *In* 'The Environmental, Economic and Social Significance of Drought.' (Ed J. V. Lovett). (Angus and Robertson: Sydney.)

Marie, S. M. and Towner, G. D. (1975). A Hele-Shaw analog study of the seepage of groundwater resting on a sloping bed. *Water Resour. Res.* **11**, 589-94.

Martyn, D. F. (1954). Comments on paper by E. G. Bowen entitled 'The influence of meteoric dust on rainfall'. *Aust. J. Phys.* **7**, 358-64.

Maybank, J. and Qureshi, M. M. (1966). An alternative explanation of rainfall anomalies. *J. Atmos. Sci.* **23**, 13-24.

Mein, R. G. and Larson, C. L. (1973). Modelling infiltration during a steady rain. *Water Resour. Res.* **9**, 384-94.

Nir, A. (1964). On the interpretation of tritium 'age' measurements of groundwater. *J. Geophys. Res.* **69**, 2589-95.

Official Yearbook of Australia (1976). No.61 Australian Bureau of Statistics, Canberra.

O'Mahony, G. (1961). Time series analysis of some Australian rainfall data. Aust. Bureau of Meteorology. Meteorological Study No. 14.

O'Mahony, G. (1965). Rainfall and moon phase. *Quart. J. Roy. Meteorol. Soc.* **91**, 196-208.

Paltridge, G. W. (1970). A model of a growing pasture. *Agric. Meteorol.* **7**, 93-130.

Parmele, L. H. and McGuinness, J. L. (1974). Comparisons of measured and estimated daily potential evapotranspiration in a humid region. *J. Hydrol.* **22**, 239-51.

Peck, A. J. and Hurle, D. H. (1973). Chloride balance of some farmed and forested catchments in southwestern Australia. *Water Resour. Res.* **9**, 648-57.

Peck, R. J., Luxmore, R. J. and Stolzy, J. L. (1977). Effects of spatial variability of soil hydraulic properties in water budget modelling. *Water Resour. Res.* **13**, 348-55.

Penman, H. L. (1948). Natural evaporation from open water, bare soil and grass. *Proc. R. Soc.* **A193**, 120-46.

Polubarinova-Kochina, P. Y. (1962). 'Theory of Groundwater Movement.' Trans. R. J. M. De Wiest (Princeton University Press).

Prickett, T. A. (1975). Modelling techniques for groundwater evaluation. *Adv. Hydrosci.* **10**, 1-143.

Pruitt, W. O., Lourence, F. J. and von Oettingen, S. (1972). Water use by crops as affected by climate and plant factors. *Calif. Agric.* (October) 10-14.

Puckridge, D. W. (1978). A comparison of evapotranspiration measurements of crop communities using lysimeters and assimulation chambers. *Aust. J. Soil Res.* **16**, 229-37.

Remson, I., Hornberger, G. M. and Molz, F. J. (1971). 'Numerical Methods in Subsurface Hydrology.' (Wiley-Interscience: New York.)

Reeves, M. and Miller, E. E. (1975). Estimating infiltration for erratic rainfall. *Water Resour. Res.* **11**, 102-10.

Ritchie, J. T. (1972). Model for predicting evaporation from a row crop with incomplete cover. *Water Resour. Res.* **8**, 1204-13.

Rogowski, A. S. (1972). Watershed physics: Soil variability criteria. *Water Resour. Res.* **8**, 1015-23.

Rose, C. W., Begg, J. E., Byrne, G. F., Torssell, B. W. F. and Goncz, J. H. (1972). A simulation model of growth-field environment relationships for Townsville stylo *(Stylosanthes humilis* 4 B.K.) pasture. *Agric. Meteorol.* **10**, 161-83.

Rosenthal, K. M., White, B. J. and Berndt, R. D. (1976). A versatile computer model for simulating the soil water balance of cropping systems. Qld. D.P.I. Div. Land Utilis. Tech. Bull. No. 17.

Schmalz, B. L. and Polzer, W. L. (1969). Tritiated water distribution in unsaturated soil. *Soil Sci.* **108**, 43-7.

Sharma, M. L. and Luxmore, R. J. (1979). Soil spatial variability and its consequences on simulated water balance. *Water Resour. Res.* **15**, 1567-74.

Slatyer, R. O. (1956). Evapotranspiration in relation to soil moisture. *Neth. J. Agric. Sci.* **4**, 73-6.

Slatyer, R. O. (1960*a*). Agricultural climatology of the Yass Valley. CSIRO Aust. Div. Land Res. Reg. Surv. Tech. Pap. No. 6.

Slatyer, R. O. (1960*b*). Agricultural climatology of the Katherine area, N.T. CSIRO Aust., Div. Land Res. Reg. Surv. Tech. Pap. 13.

Slatyer, R. O. and McIlroy, I. C. (1961). 'Practical Climatology,' (CSIRO: Melbourne.)

Smith, D. B., Wearn, P. L., Richards, H. J. and Rowe, D. C. (1970). Water movement in the unsaturated zone of high and low permeability strata by measuring natural tritium. Proc. Symp. Isotope Hydrol. 1970. pp. 73-87 (IAEA: Vienna).

Swinbank, W. C. (1954). Comments on a paper by E. G. Bowen entitled 'The influence of meteoric dust on rainfall'. *Aust. J. Phys.* **7**, 354-58.

Talsma, T. and van der Lelij, A. (1976). Interpretation and water movement in an *in situ* swelling soil during prolonged ponding. *Aust. J. Soil Res.* **14**, 337-49.

Taylor, C. B. (1971). Influence of 1968 French thermonuclear tests on tritium fallout in the southern hemisphere. *Earth Planet. Sci. Lett.* **10**, 196-198.

Taylor, H. M. and Klepper, B. (1975). Water uptake by cotton root systems: an examination of assumptions in the single root model. *Soil Sci.* **120**, 57-67.

Times Atlas of the World. (1975). Comprehensive edition. (Times Book in collaboration with John Bartholomew and Son: Edinburgh.)

Treloar, H. M. and Grant, Alison M. (1953). Correlation studies of Australian rainfall. *Aust. J. Agric. Res.* **4**, 423-29.

van Bavel, C. H. M. (1966). Potential evaporation: the combination concept and its experimental verification. *Water Resour. Res.* **2**, 455-67.

Vauclin, M., Khanji, D. and Vachaud, G. (1976). Étude expérimentale et numérique du drainage et de la recharge des nappes à surface libre, avec prise en compte de la zone non-saturée. *J. Méc.* **15**, 307-48.

Volker, R. E. (1977). Numerical modelling of an aquifer system with intermittent recharge. A.W.R.C. Tech. Paper No.26 (Aust. Gov. Publ. Serv: Canberra.)

Watterson, G. A. and Legg, M. P. C. (1967). Daily rainfall patterns at Melbourne. *Aust. Meteorol. Mag.* **15**, 1-12.

Webb, E. K. (1975). Evaporation from catchments. *In* 'Prediction in Catchment Hydrology.' (Eds T. G. Chapman and F. X. Dunin). (Aust. Acad. Sci. Griffin Press: Netley, S. Aust.)

West, E. S. and Perkman, O. (1953). Effect of soil moisture on transpiration. *Aust. J. Agric. Res.* **4**, 326-33.

World Meteorological Organization (1966). Climatic change. Tech. Note No. 79.

Youngs, E. G. (1974). Seepage rates and the horizontal flow approximation. *Water Resour. Res.* **10**, 874-76.

Zimmermann, U., Münnich, K. O. and Roether, W. (1967). Downward movement of soil moisture traced by means of hydrogen isotopes. Geophys. Monograph No. 11 (Ed G. E. Stout) Amer. Geophys. Union, pp. 28-35.

Organic matter and its decomposition

Chapter 34
Soil organic matter

A. V. Spain, R. F. Isbell and M. E. Probert

SOIL ORGANIC MATTER IS A KEY COMPONENT OF ANY ecosystem, and variation in its abundance and nature has profound effects on many of the processes that occur in the system. Discussion of specific aspects of the importance of soil organic matter in chemical, physical and biological soil processes is to be found in several chapters of this book. However, it is appropriate to consider soil organic matter in its own right and to attempt to describe the levels of, and aspects of variation in, the organic matter status of Australian soils.

The amount present in soils is the result of input of organic materials and their breakdown in soil. The basic source of this organic matter and its contained energy is the autotrophic section of the plant community. In, or on, the soil it is subjected to progressive comminution, decomposition and redistribution. The final, organic end-product of the decomposer cycle is microbially derived 'humified' material, chemically complex and with no recognizable characteristics of its origins.

In functional terms, it is neither simple nor desirable to distinguish between the organic fraction of the litter layer (see Chapter 38) and that of the soil. However, such a distinction is traditionally made on the basis of profile morphology, and soil organic matter is normally considered to exclude the O horizon. We shall follow this convention. Furthermore, it is also normal practice to remove fresh root material and the larger soil fauna during sample preparation so that these too are excluded from what is here considered to be soil organic matter.

Given the importance of soil organic matter (Allison 1973), it is pertinent to consider the question 'How much organic matter is necessary in soil?'. The answer of course depends on the form and intensity of land use. The most crucial consideration is likely to be the maintenance of soil structure and land stability; a reduction in soil organic matter can lead to loss of structure and erosion in many soils.

In terms of agricultural productivity, if organic matter is the only source of nutrients, especially nitrogen, available to the crop the answer will be 'Enough to supply the needs of the crop.' However, such a constraint does not usually apply in more intensive farming systems where fertilizer application is a regular part of agricultural practice. An inquiry in the United Kingdom (Anon. 1970) found that there was little evidence that low organic matter is associated with reduced yields except on unstable soils, although such experiments have not usually investigated differences in flexibility of cultivation, in draught and in wear and tear of equipment, or on earliness and uniformity of ripening. However, the inquiry recognized that there may be some conflict between experimental results and commercial farm practice in that small field plots can be worked when the soil is at its most suitable state, whereas on a large scale of operation this may not be possible. The report considered that the effects of organic matter in soils and the practical value it has in improving structural stability, porosity, infiltration of water, consistency and workability require further study.

In Australia, where the greater part of the landscape can never be subjected to intensive agricultural exploitation, the situation is somewhat different. Soil organic matter levels must be maintained at appropriate levels to ensure ecosystem stability. Maintenance of an effective ground cover will help to minimize soil erosion and thus loss of the important A_1 horizon with its higher organic matter and nutrient status. Management must be of the total environment, rather than of the soil system alone as is practised in the more intensive forms of agriculture.

Factors controlling organic matter levels in soils

Soils in the context of an ecosystem are the product of the joint activities of five soil-forming factors, *viz.* climate, parent material, time, organisms and topography (Jenny 1941, 1961*a*).

Climate is possibly the most important factor determining soil organic matter levels in that it most strongly influences the productivity of the vegetation and the level and perhaps the type of decomposer activity. In relation to soil organic matter levels, Birch and Friend (1956), Jenny *et al.* (1948) and Jenny *et al.* (1968) have shown that precipitation and temperature are the most important variables although they have different effects on carbon and nitrogen levels, and the nature of the relationships appears to be different in tropical and temperate climates (Laudelout *et al.* 1960).

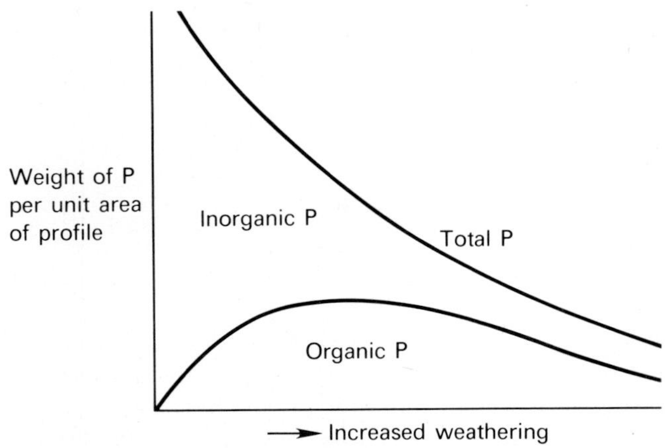

Fig. 34.1. Schematic representation of changes in forms and amounts of soil phosphorus during a weathering process (after Walker and Syers 1976)

The influence of parent material on soil organic matter status has been the subject of some debate. It is clear from the work of Jenny *et al.* (1968) and many others that the capacity of a soil to grow plants is related to the nature and mineral constitution of its parent material. The effect of the variations in parent material on the level of organic matter in the soil (other factors being held equal) is less clear. Walker (1965), and more recently Walker and Syers (1976), have discussed the special role of phosphorus in pedogenesis. Phosphorus is the one major constituent of soil organic matter which must be supplied almost entirely from the parent material, and therefore is a key element in pedogenesis. It has been shown that during weathering processes, organic phosphorus increases and acid-extractable forms of the element decrease. This may be associated with a loss of total phosphorus from the system and/or a conversion of phosphorus to forms unavailable to plants. Although maximum levels of soil organic matter attained during pedogenesis may not always be determined solely by initial levels of phosphorus in the parent material, organic matter would begin to decline once the reserves were severely depleted, either by conversion to organic phosphorus or to very insoluble inorganic phosphorus compounds, or through loss from the system (Fig. 34.1). Walker (1965) considers that no more nitrogen could accumulate in an ecosystem once this stage had been reached. However, continuing soil development would probably result in organic phosphorus levels declining faster than organic nitrogen or carbon, and thus there would be a widening of N:P and C:P ratios (Walker 1965).

This hypothesis thus suggests that the organic matter content of a soil depends to an important degree upon the phosphorus content of the parent material and the degree of weathering, which includes the effects of both climate and time. Furthermore, it suggests that 'young' soils (towards the left hand side of Fig. 34.1) are less likely to be responsive to phosphorus. At the other extreme, strongly weathered soils would be expected to be grossly deficient in phosphorus, and to have a high proportion of their phosphorus present in organic form or as very insoluble inorganic phosphorus. Such soils may be rejuvenated to create conditions under which organic matter may again increase by inputs of additional phosphate and other limiting elements, or through geological processes.

There has been little study in Australia of sequences of soils such as those reported by Walker (1965) in New Zealand. However, the widespread occurrence of acute phosphorus deficiency, the loss of phosphorus during soil formation (Wild 1958; Beadle 1962), and the response in organic matter levels when legumes fertilized with superphosphate are introduced (Chapter 37) all suggest that many of our soils are towards the right hand side of Fig. 34.1 and have passed beyond their peak organic matter contents.

Time is thus particularly relevant to soil nutrient status, and therefore to organic matter levels, in the predominantly old landscapes of the Australian continent. Stark (1978) has described the inevitable loss of productivity suffered by soils with the passage of time. Continuing loss of nutrient elements from the soil ecosystem, and the consequent reduction in vegetation biomass necessarily leads to a reduction in soil organic matter input. Jenny (1961*b*) claims that the levels of organic components in a soil reflect those of the 'current' ecosystem state factors. However, the great age of much soil carbon as indicated by radiocarbon dating suggests that this must be true at a gross level only.

The effects of living organisms on soil organic matter are substantial. Of these, vegetation is the primary source of soil organic matter and is thus the major influencing factor on the amount present. It not only sets the amount of organic matter supplied to the soil but Jenny *et al.* (1968) have suggested that the nature of this organic matter is influenced by the C:N ratio of the vegetation. The presence of a wide variety of secondary plant compounds in living plant material, litter and leachates (Chapter 38) suggests that the effect is a complex one.

The nature of the populations of consumer animals may circumscribe the paths of carbon flow in an ecosystem. Earthworms have long been known to condition soil organic matter status (Darwin 1881) but, in Australia, termites probably play at least as important a role in this respect (Chapter 39).

Topography influences soil organic matter through the effects of slope and drainage. Slope influences soil stability through creep effects and interacts with drainage in partitioning the incident precipitation. The effect of impeded drainage in reducing microbial activity is well known and in extreme conditions leads to the formation of peat.

The age of soil organic matter

The resistance of some fractions of soil organic matter to decomposition has long been realised but was considerably underestimated prior to the advent of radiocarbon dating studies (Allison 1973). It is now known that the mean residence time of the carbon present in the surface horizon of soils, where new organic matter is continually being added, is less than that deeper in the profile. The greatest ages are recorded from organic matter deep in the profile or adsorbed onto clay particles.

Few age determinations of soil profile organic matter have been carried out in Australia. Blackburn *et al.* (1979) have examined the ages of carbon in studies of gilgai phenomena and the ^{14}C ages (as mean residence times) of the humus fractions of eight *krasnozems* from south-eastern Queensland have been recorded by Beckmann and Hubble (1974), who found that the age of the surface humus fraction was of the order of a few to several hundred years; maximum ages of 4000 years were found at depths greater than 60 cm. These values are at the low end of the range recorded from many other areas of the world (Scharpenseel and Schiffmann 1977). O'Brien and Stout (1978) working on New Zealand soils have developed a means of partitioning soil organic carbon on the basis of its age. In their model, 'old' carbon (5700 to > 15000 years old) is uniformly distributed throughout the profile while 'modern' (< 100 years) carbon decreases exponentially with depth. For the soils they studied, old carbon was estimated to be 16% of the total soil organic carbon of the profile.

Many studies and most models of soil organic carbon distribution and flow treat this element as though all carbon compounds in soil organic matter were equally susceptible to breakdown. This is patently untrue for the various forms of organic matter derived directly from plant tissue but may well be true for at least some fractions of the materials that have undergone microbial transformation. The models of Jenkinson and Rayner (1977) and O'Brien and Stout (1978), which incorporate information derived from radiocarbon dating studies, go some way towards the evolution of more realistic formulations of soil organic matter age, its vertical distribution in soil profiles and its dynamics.

Soil organic matter concentrations

Few attempts have been made to describe the organic matter contents of Australian soils on a continental basis. Prescott (1931) presented a frequency distribution for the surface nitrogen concentration of 817 eastern Australian soils. Hosking (1935) examined 319 surface, 191 subsurface and 108 subsoil samples irregularly distributed throughout the whole of Australia and presented frequency distributions for C:N ratios for a number of major soils, together with mean carbon and nitrogen concentrations. Since these early studies, there has been no attempt to look systematically at organic matter levels in Australian soils as a whole although Hubble and Martin (unpublished 1960) examined the nitrogen content of major Queensland soils. Stace *et al.* (1968) presented data showing carbon and nitrogen concentrations for several profiles of each of the recognized Australian Great Soil Groups.

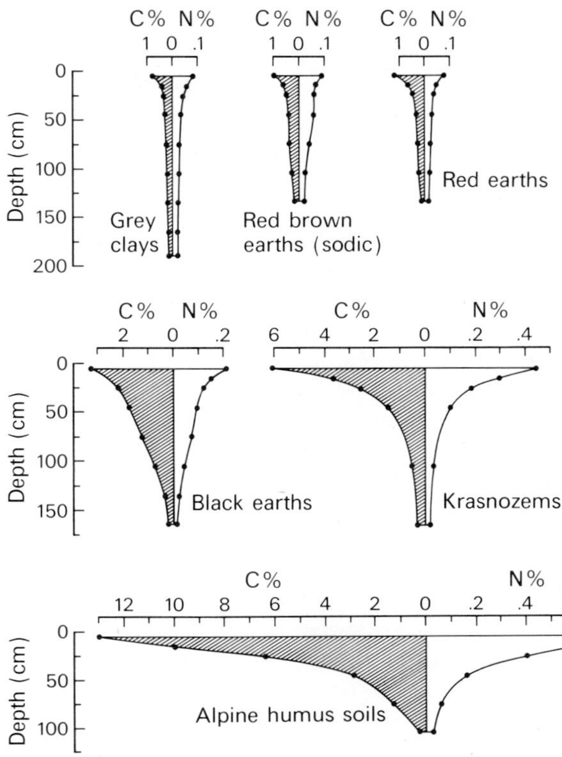

Fig. 34.2. Distribution of organic carbon and nitrogen with depth in selected Australian soils based on a mean of ten profiles (five for the *alpine humus soils*). The sodic *red-brown earths* from South Australia are all cleared and/or cultivated (mean annual rainfall 350—450 mm) but all others are virgin soils. The *grey clays* are from arid to semi-arid grasslands (250—450 mm), the *red earths* from arid to sub-humid shrublands and woodlands (250—1000 mm), the *black earths* from subtropical grasslands (650 mm), the *krasnozems* from tropical rainforests (1500—3000 mm) and the *alpine humus soils* from high-elevation woodlands with winter snow (1500—2300 mm).

Organic matter concentration in soil usually declines rapidly with depth. Typical depth functions of organic carbon and nitrogen are shown for a selection of great soil groups in Fig. 34.2; profiles for nitrogen in a wider range of great soil groups are shown in Fig. 37.7 of Chapter 37. Of interest is the relatively slow decline of organic carbon in the grassland *black earths*. Fig. 34.2 may be compared with similar diagrams for extra-Australian soils shown in Parsons and Tinsley (1975).

In some Australian soils, as elsewhere, there is an increase in the organic matter concentration of the subsoil. This is most pronounced in *podzols*, but in some *solodized-solonetz* the marked organic staining on the B horizon columns also gives a very slight subsoil increase to the organic matter profile. In such soils and in others with a pronounced bleached A_2 horizon, the decrease with depth is more rapid in the A_2 than in the lower horizons.

In discussing the organic matter content of soils, it would be most appropriate to consider it in terms of the total present in the solum on a unit area basis, as was attempted by Hubble and Martin (unpublished 1960). However, this requires not only a knowledge of the distribution of carbon or nitrogen with depth, but also of the bulk density of the soils; such information is not readily available. Whilst it may be less informative, there are no such difficulties in discussing the concentrations of organic carbon and nitrogen in surface soils.

Organic Matter in Surface Soils

Over the years many data have been accumulated. It seemed appropriate that this information be collated and examined in order to define the levels of organic matter in Australian soils, and its variation with geographic location and a range of environmental factors.

Sources in the literature and many unpublished records have yielded information on 3652 soils distributed somewhat unevenly throughout Australia. Because of inconsistencies in the depths of sampling and analytical methods, direct comparison of all samples was not possible. For this reason, the data have been edited to remove obvious errors and all Walkley-Black carbon values have been scaled by an arbitrary (but widely used) factor of 1.3 to convert them to a level comparable with that obtained by dry combustion (Little *et al.* 1962).

The data set finally assembled refers to A_1 horizons (ranging in depth from 5 to 15 cm) in terms of carbon, nitrogen, total and 'available' phosphorus, pH and clay content, but data for one or more attributes were frequently missing in individual samples. This has resulted in differing numbers of records being available for the various analyses that follow. In addition, a simplified structural vegetation class (rainforest, open forest and tall woodland, open woodland, shrubland and heathland, native grassland, cleared land with native pasture, cleared land with exotic pastures or crops, conifer plantations) (Specht 1970), mean annual precipitation; and its seasonality, and mean annual temperature class (0–6, 6–12, 12–18, 18–24, 24–30°C) were recorded for each site. The soils were classified according to Stace *et al.* (1968), but some great soil groups were combined and others subdivided for analytical purposes (see Fig. 34.3, 4 and 5).

The representation of samples on a geographical basis is: Northern Territory 106; Northern Queensland (north of 23°S) 674; Southern Queensland 1270; New South Wales 408; Victoria 428; Tasmania 163; South Australia 369; Western Australia 234.

As would be expected, the sample distribution is far from uniform. The best coverage on areal, soil group and environmental bases is from Queensland and Victoria. The largest gaps exist in the arid areas of South Australia, Western Australia and New South Wales, although these are partly compensated for by the extensive coverage available for arid western Queensland. In the tropics, representative data are available for Queensland but the coverage is much poorer in the Northern Territory and the north of Western Australia.

The need to attempt an assessment of sample site vegetation is emphasized by the fact that whereas all Queensland samples are from virgin sites, a great many from South Australia and Victoria are from areas with long histories of improved pastures and/or cultivation. The data set is available as CSIRO Aust. Div. Soils Tech Mem. No. 13, 1981.

Organic matter in the Great Soil Groups

Frequency distributions were generated for each of the Great Soil Groups for carbon, nitrogen and the C:N ratio.

The frequency distributions for carbon and nitrogen were rather similar: all showed some degree of non-normality. The most common form of distribution showed marked positive skewness, others were platykurtic or leptokurtic and others yet were polymodal. The great soil groups showing positive skewness were generally those with low organic carbon and nitrogen levels. Where levels were higher, skewness was less evident but platykurtic and polymodal distributions were more in evidence. For the C:N ratio, samples for each great soil group appeared more normally distributed although platykurtic and positively skewed distributions were common here also.

Because of the nature of the frequency distributions of the variables, it is considered preferable to present summaries of the data as medians and interquartile ranges. The data for the great soil groups for organic carbon, nitrogen and the C:N ratios are presented in Figs 34.3–34.5; in each case Great Soil Groups are arranged in descending order of median and the number of records for each is given. In addition, the frequency distribution for organic carbon for the entire data is included in Table 34.1. Because of the relatively high correlation between organic carbon and nitrogen ($r = 0.87$ for the total data set) the order of decreasing median is much the same both for organic carbon and for nitrogen.

Based on the data of Northcote *et al.* (1975), it may be estimated that about 75% of Australia has surface soils with a median organic carbon content of less than 1%. In arriving at this figure it has been assumed that the widespread shallow stony soils – *lithosols* – (see Fig. 3.4 in Chapter 3) have median carbon contents of less than 1%. Data for these soils are extremely limited, and hence the group was not included in the data set. However, available evidence would suggest that this is a reasonable

Table 34.1. Distribution of organic carbon records with mean annual precipitation

Blanks indicate no records

Precipitation (mm)	Organic carbon (%) 0–1	1–2	2–3	3–4	4–5	5–6	6–7	7–8	8–9	9–10	10–11	11–12	> 12
> 4000		1	1	3	1								
3500–4000			2	1	1	2	1						
3000–3500		2	2	4	3	1	3	1					
2500–3000				2		1	2	1					
2000–2500		1	1	4	5	7	3	5		4	1	1	2
1500–2000	12	33	12	21	16	11	3	6	10	3	4	1	9
1000–1500	61	72	50	49	43	22	28	22	16	6	5	3	3
500–1000	448	630	328	144	102	50	25	9	12	7	3	1	5
< 500	601	252	44	5	3					1			
No. of records	1122	991	440	233	174	94	65	44	38	21	13	6	19
% of total records	34.4	30.4	13.5	7.1	5.3	2.9	2.0	1.3	1.2	0.6	0.4	0.2	0.6

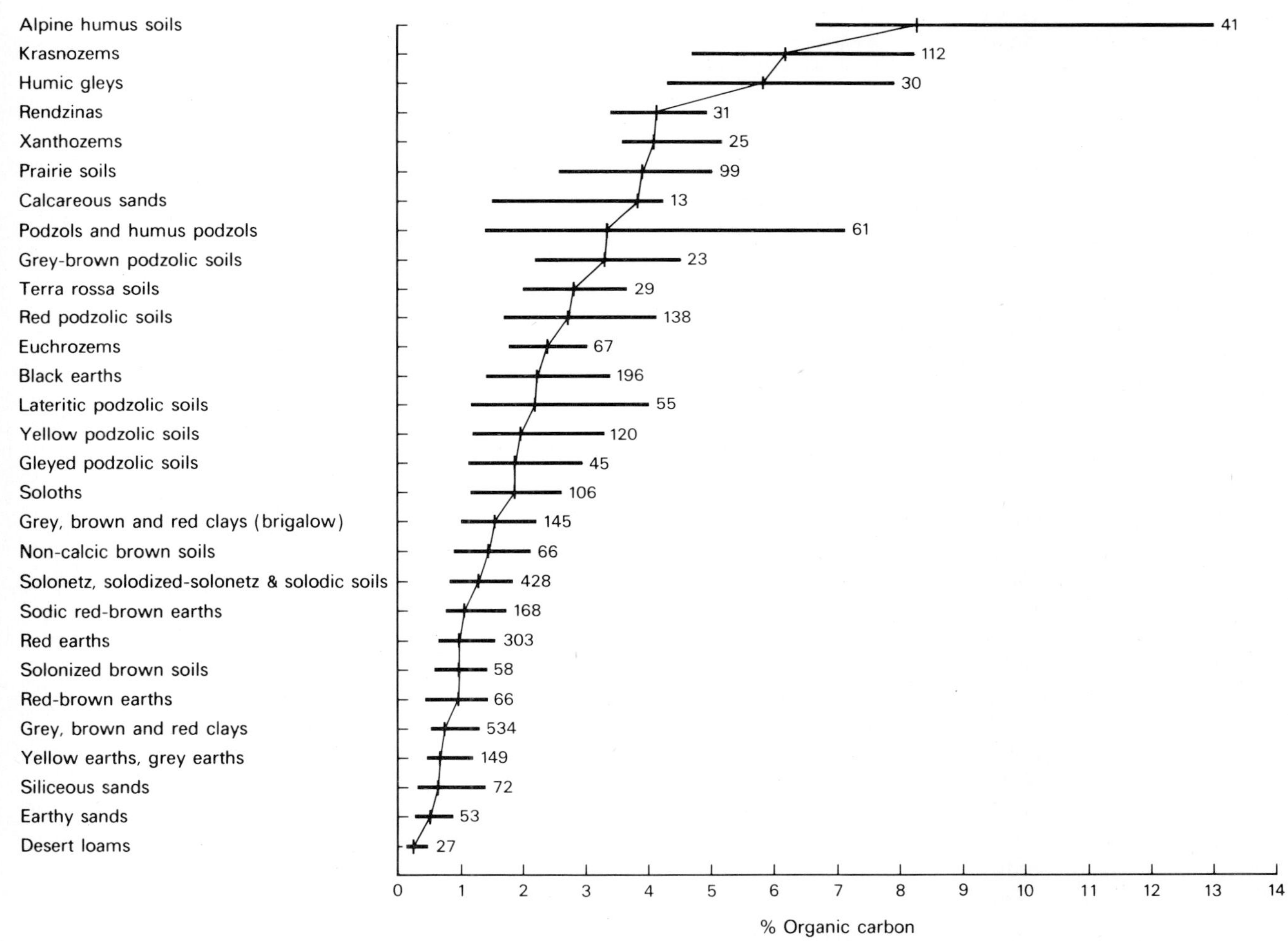

Fig. 34.3 Organic carbon levels in Australian soils. Great Soil Group medians, interquartile ranges and the number of records available.

assumption, particularly where the soils occur in drier areas.

The low-carbon soils include some that are of considerable importance in dryland agriculture, e.g. the *red-brown earths, solonized brown soils* and the *grey, brown* and *red clays*. It may be noted here that the *grey, brown* and *red clays* supporting brigalow (*Acacia harpophylla*) forests in eastern Queensland and New South Wales were treated separately. This leguminous species is known to nodulate, and it has been suggested that the higher carbon and nitrogen values for the clay soils where this species is dominant are probably due to

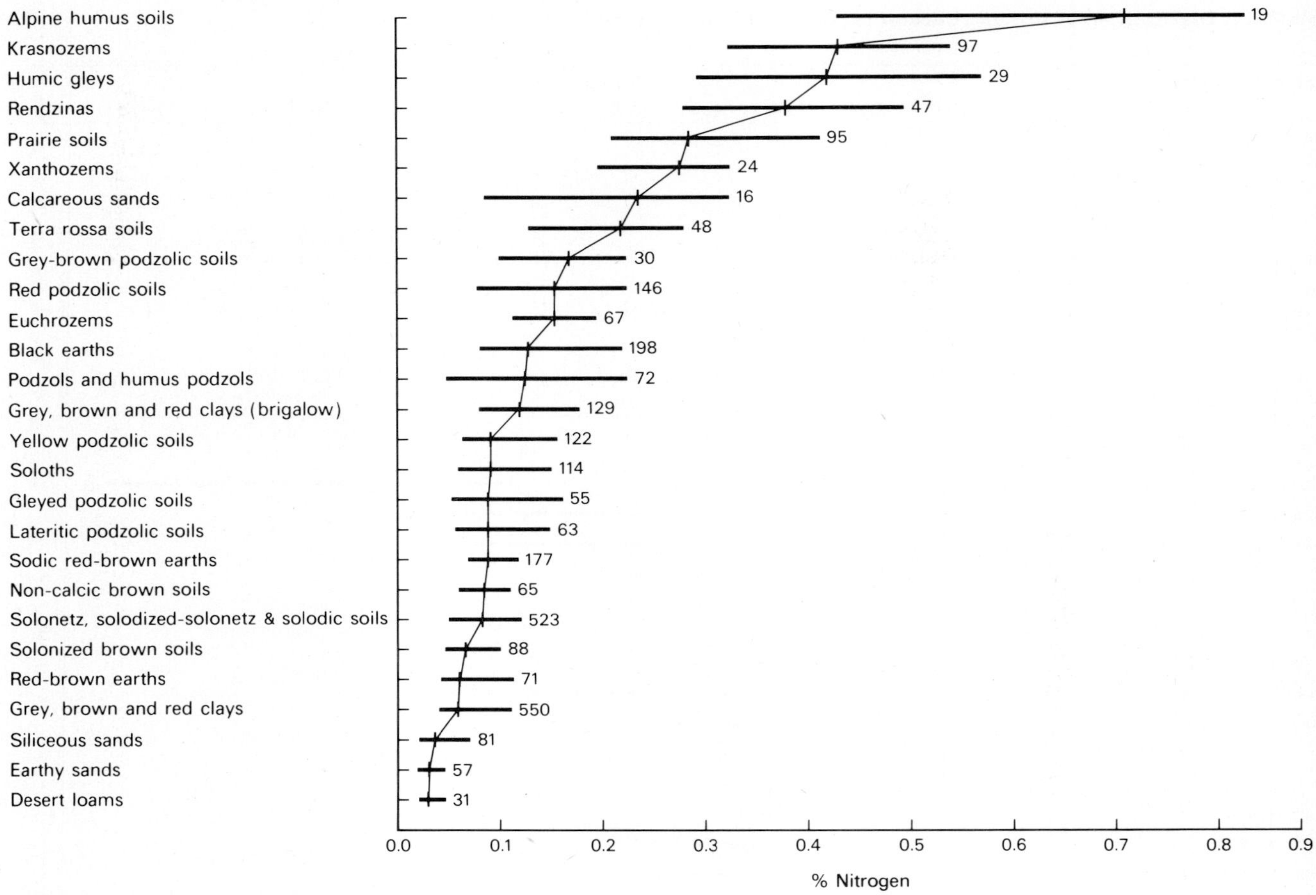

Fig. 34.4 Nitrogen levels in Australian soils. Great Soil Group medians, interquartile ranges and the number of records available

symbiotic nitrogen fixation (Hubble and Martin unpublished 1960; Isbell 1966). Some other agriculturally-important soils, e.g. the *black earths*, have moderate organic levels (Fig. 34.3) but these only occupy about 0.7% of the continent.

Not surprisingly, at the higher end of the scale, the *alpine humus soils* show the greatest carbon and nitrogen values, while *krasnozems* and *humic gleys* also have median levels greater than 5% organic carbon and 0.4% nitrogen respectively. However, the areal extent of these soils is small (less than 2% of the Australian landscape). Similarly, soils with between 3 and 5% organic carbon occupy only relatively small areas of Australia.

The considerable range in values of the C:N ratios for the great soil groups is demonstrated in Fig. 34.5. A problem with ratios is that they are subject to extreme variability resulting from errors in the determination of either variable. Thus, considerable editing of the data set was required prior to analysis. Another factor that could distort C:N ratios is the presence of fixed ammonium. Although there are few studies reported for Australian soils, Martin *et al.* (1970) found that up to 12% of the total nitrogen in some southern Queensland surface soils was present in this form, with very much higher proportions (up to 90%) in some subsoils.

Nearly half of the great soil groups have a median C:N ratio greater than 15, and on an area basis this would probably account for at least 50% of Australia. Three soil groups (*podzols* and *humus podzols, lateritic podzolic soils* and *yellow podzolic soils*) have median ratios greater than 20. However, the modal classes are lower than the medians due to positively skewed distributions in the *podzols* and *humus podzols* and *yellow podzolic soils*. In the former the frequency distribution shows one modal class ranging from 15 to 27 and a lesser mode at 33 to 36.

Organic matter by regions

Fig. 34.6 presents the medians and interquartile ranges for organic carbon, nitrogen and the C:N ratio for various regions of Australia. The grouping of records is arbitrary but useful comparisons may be made on a latitudinal basis and contrasts drawn between tropical and temperate areas where appropriate.

For both carbon and nitrogen, the frequency distributions for the data from each region were again strongly

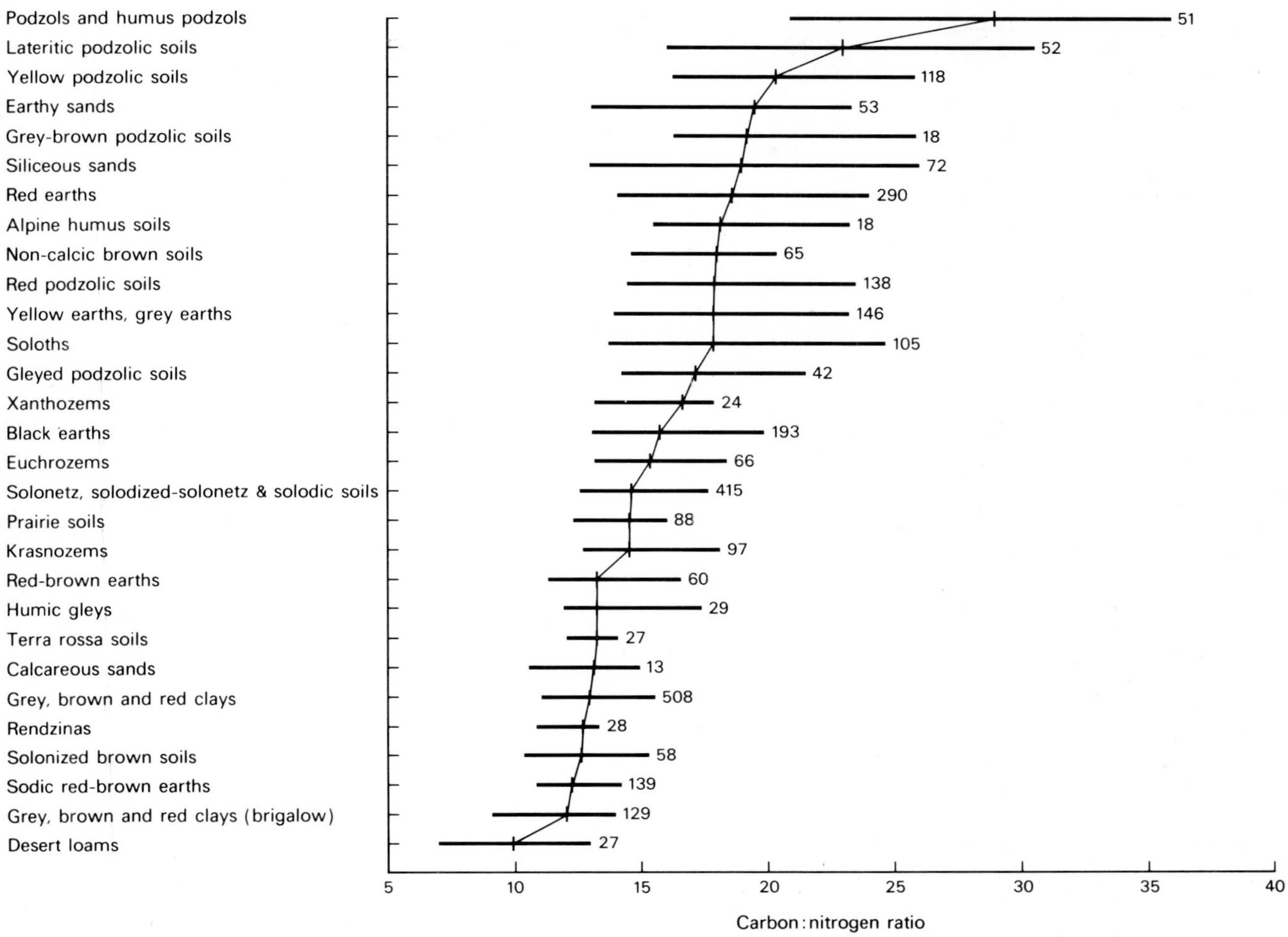

Fig. 34.5. Carbon:nitrogen ratios of Australian soils. Great Soil Group medians, interquartile ranges and sample sizes

positively skewed and those for Tasmania were bimodal with a major mode at 3–5% organic carbon and a minor mode at 7–9% organic carbon. The frequency distributions for the C:N ratios for all localities were also positively skewed, although less markedly so than those for carbon and nitrogen.

In terms of the geographical distribution of median values, carbon and nitrogen had very similar patterns. For both elements, median concentrations were lowest in the tropical areas and showed a general increase with latitude. The median concentrations of carbon and nitrogen for New South Wales were both somewhat higher than for locations to the north or south of them. Concentrations of both elements are substantially higher in Tasmanian soils.

Median C:N ratios show a decrease with increasing latitude with the exception of records from temperate Western Australia and Tasmania. The median nitrogen recorded from soils in the former area was less than 0.1% which was similar to those from Queensland and the Northern Territory. In contrast, carbon appears commensurate with latitude, leading to an elevated C:N ratio. The median C:N ratio of approximately 16 for Tasmania appears somewhat higher than would be expected from its location.

Variability of carbon and nitrogen in surface soils

No estimate of site variability can be obtained from our data set, and in view of the known high variability of many soil chemical properties (e.g. Beckett and Webster 1971) it is of interest to look at this in the Australian context using other sources. Some data are available for virgin north Queensland soils. These include *non-calcic brown soils* (Crack and Isbell 1971), *siliceous sands* (Isbell and Gillman 1973), *krasnozems* and *euchrozems* (Gillman 1976), *yellow earths* (Isbell *et al.* 1976) and *euchrozems* (Isbell *et al.* 1977).

These soils span a rainfall environment from about 600 mm to over 2000 mm. Vegetation ranges from open woodland to open forest, rainforest in the case of the *krasnozems*, and heath in the case of the *yellow earths*. Both within- and between-site coefficients of variation

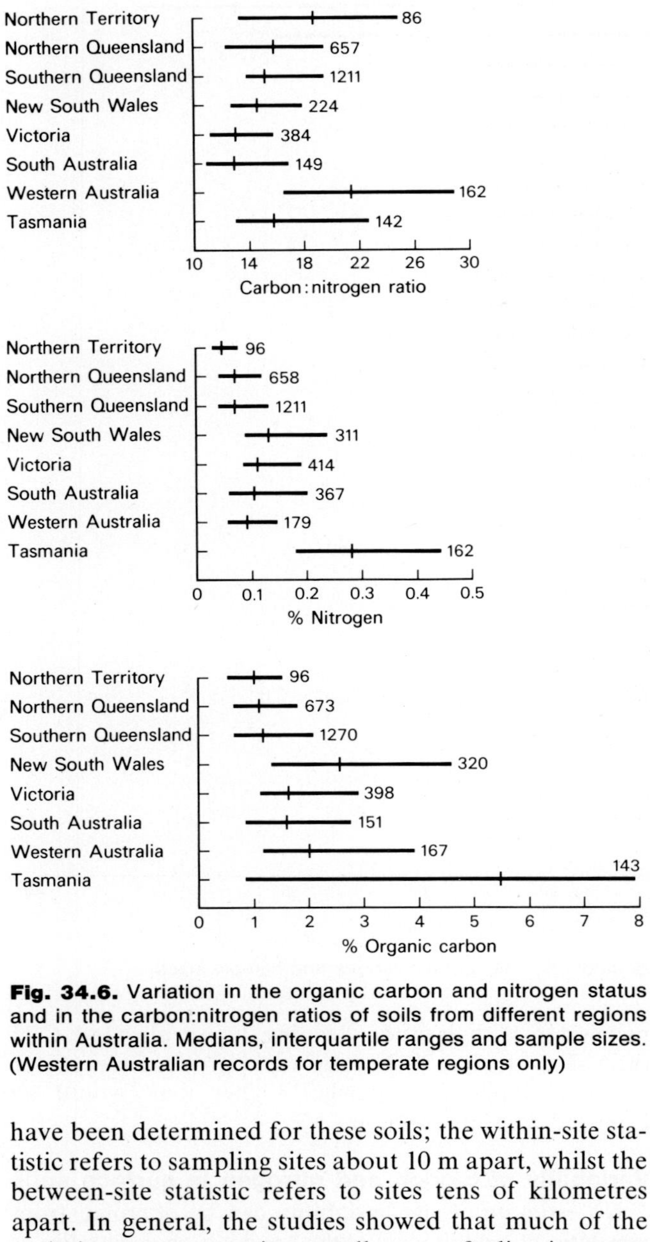

Fig. 34.6. Variation in the organic carbon and nitrogen status and in the carbon:nitrogen ratios of soils from different regions within Australia. Medians, interquartile ranges and sample sizes. (Western Australian records for temperate regions only)

have been determined for these soils; the within-site statistic refers to sampling sites about 10 m apart, whilst the between-site statistic refers to sites tens of kilometres apart. In general, the studies showed that much of the variation was present in a small area, a finding in agreement with that of Beckett and Webster (1971). Within-site coefficients of variation for both carbon and nitrogen were generally in the range of 10 to 20%, with between-site coefficients of variation generally between 25 and 40%.

Relationships with selected environmental factors

The approaches found in the literature for investigating the relationships between soil organic matter and environmental factors have often been based on simple linear regression models (e.g. Birch and Friend 1956; Jones 1973; Kadeba 1978) although the work of Jenny and his co-workers has shown that such models may be inappropriate in that the relationships are often far from linear.

For the data set we have compiled, the diversity of edaphic, vegetation and environmental variables make it most unlikely that simple universal relationships will be present for these Australian soils. Thus we encompass a range almost as diverse as the Californian and Colombian soils studied by Harradine and Jenny (1958) and Jenny *et al.* (1948) for which it has been shown that the coefficients of the models used were very different for these two situations.

From other studies, it might be expected that rainfall would be the single most important variable. For example, Prescott (1931) found nitrogen levels to be significantly related to rainfall, although Hubble and Martin (unpublished data 1960) found that rainfall only accounted for some 8% of the variation in total nitrogen in their Queensland soils.

In Table 34.1 the total data set for organic carbon is depicted as a function of rainfall. It is very obvious that no simple single relationship exists between organic carbon and rainfall in our data. Indeed, there is some suggestion that there are two contrasting patterns present: there are a number of soils represented that have very high organic carbon contents at intermediate to low rainfall whilst, on the other hand, many of the high rainfall soils have only moderate organic carbon contents. The pattern shown in Table 34.1 confounds the effects of soil group, vegetation and temperature. Thus, the high organic matter soils are predominantly *alpine humus soils* whilst the soils from areas of high rainfall (> 2400 mm) are largely *krasnozems, xanthozems* and *red podzolic soils* under rainforests.

A possible aid to the interpretation of the information contained in Table 34.1 is to restrict the data set on the basis of soil taxonomy or geography. Fig. 34.7*a* and *b* shows scatter diagrams of the relationships between rainfall and organic carbon for *solodic soils* and *krasnozems* respectively. These two soil groups were chosen because they occur over a wide latitudinal range. In both cases, the general pattern is very similar to that shown by the total data set. In Fig. 34.7 the soils from the tropical environments and from the southern-most locations (Victoria and Tasmania) have been distinguished. It is clear that there is a very strong latitudinal effect present. A plausible explanation of this effect can be offered in terms of the influence of the individual contributions of the input and decomposition rates on the organic carbon present. In the tropical soils, high rainfall and, probably more importantly, high temperatures will encourage decomposition, thus resulting in soils that have lower levels of organic carbon than might be otherwise expected. On the other hand, at higher latitudes, and especially at higher altitudes, temperature may be a constraint to decomposition, thus giving soils with high organic carbon contents. In the extreme, the latter situation will result in

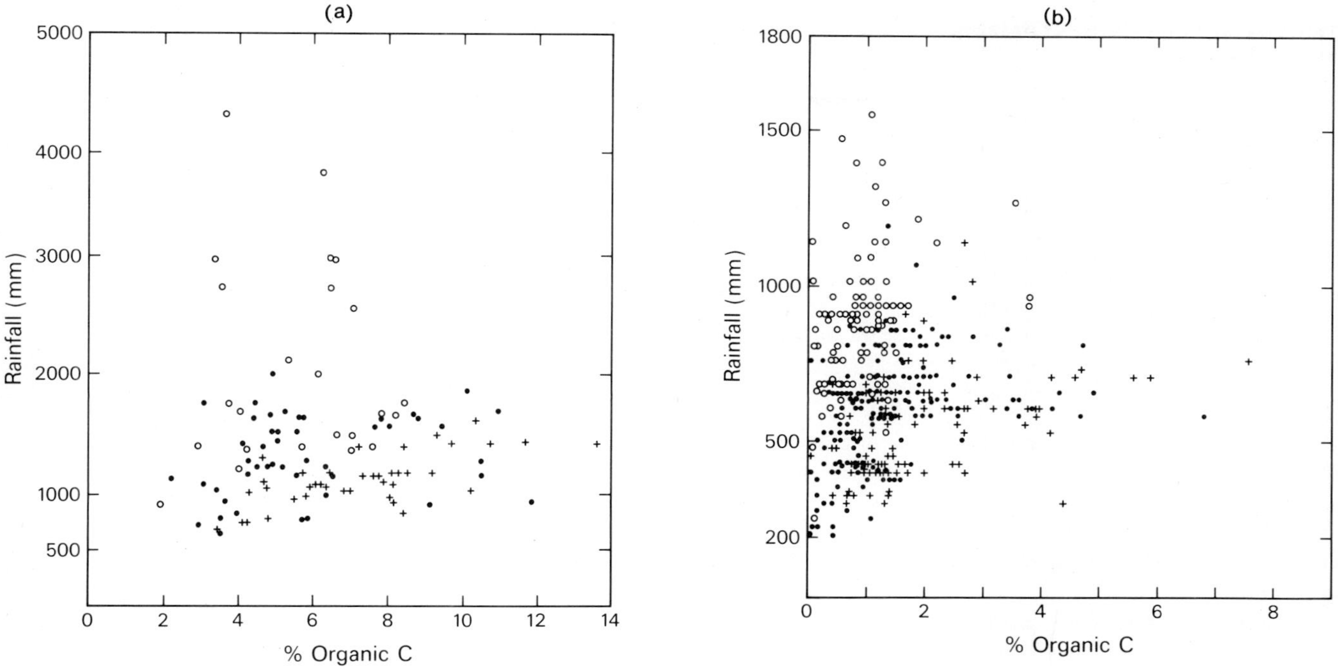

Fig. 34.7. Relationships between organic carbon levels and precipitation for (a) *solodic soils;* (b) *krasnozems.* ○, tropical locations; +, Victoria and Tasmania; •, other.

peats, the more so in conditions of impeded drainage.

There are alternative ways of stratifying the data: indeed, because climate is itself a soil forming factor, it is not particularly appropriate to use soil groups as the stratifying factor when investigating the effects of climatic and environmental variables.

By using regional subsets of the data, closer correlation between rainfall and organic carbon concentrations ($r = 0.72$ for New South Wales soils, $r = 0.81$ for Victorian soils) were obtained.

In order to further examine the relationships between soil organic matter and the environmental factors measured, the records were divided somewhat arbitrarily into those from the tropical areas (northern Queensland, Northern Territory and the tropical areas of Western Australia), the warm temperate area of Queensland south of the Tropic of Capricorn, and the cool temperate areas of New South Wales, Victoria, Tasmania, South Australia and Western Australia. For these areas, multiple correlation analyses were carried out relating both soil organic carbon and nitrogen to precipitation, annual temperature class, pH and clay content. Table 34.2 presents the analyses for each of the environmental factors separately. In an initial attempt to account for nonlinearity in the relationships, the squares of the environmental factors were included where appropriate in these analyses.

Using a stepwise procedure, the multiple correlations of soil carbon and nitrogen with the environmental factors and their squares were computed. The results of these analyses are presented in Table 34.3.

Precipitation is undoubtedly the most important of the environmental factors and is positively correlated with soil organic matter. Both organic carbon and nitrogen correlate better with this factor than with any other variable (Table 34.2) and, with the exception of nitrogen for the entire data set, it was the first of the environmental factors to be entered in the stepwise fitting procedure. In all of the analyses presented, the relationships of both organic carbon and nitrogen with precipitation were modelled as quadratics. As shown in Table 34.1, this must only be regarded as a first approximation.

Temperature is negatively correlated with organic carbon and nitrogen in the analyses for all regions. However, as shown in Table 34.2, the coefficients of

Table 34.2. Coefficients of determination for the relationships between organic carbon, nitrogen and selected environmental factors

Values are single factor coefficients,

	Precipitation, (Precipitation)^2	Temp.	Temp. (Temp.)^2	pH, (pH)^2	% Clay	% Clay (% Clay)^2
			Tropical areas			
C	36.5	4.9	—	17.0	—	14.0
N	30.2	7.1	—	15.1	—	18.0
			Southern Queensland			
C	33.7	—	13.6	3.3	—	3.7
N	25.1	10.5	—	1.6	—	6.9
			Cool temperate areas			
C	51.8	—	21.9	29.2	3.7	—
N	28.5	—	14.7	7.3	—	13.8

Table 34.3. Coefficients of determination for the relationships between organic carbon, nitrogen and selected environmental factors

Maximum subsets of variables accounting for 'useful' proportions of total sums of squares.

	Variables	R^2 (%)
	Tropical areas	
C	Precipitation, $(\text{precipitation})^2$, clay, $(\text{clay})^2$, pH, mean annual temperature	50.0
N	Precipitation, $(\text{precipitation})^2$, clay, $(\text{clay})^2$, pH, $(\text{pH})^2$, mean annual temperature	50.3
	Southern Queensland	
C	Precipitation, $(\text{precipitation})^2$, mean annual temperature, $(\text{mean annual temperature})^2$, $(\text{clay})^2$, pH, $(\text{pH})^2$	52.3
N	Precipitation, $(\text{precipitation})^2$, mean annual temperature, clay, $(\text{clay})^2$, pH, $(\text{pH})^2$	45.6
	Cool temperate areas	
C	Precipitation, $(\text{precipitation})^2$, mean annual temperature, $(\text{mean annual temperature})^2$, clay, $(\text{clay})^2$, pH, $(\text{pH})^2$	59.8
N	Precipitation, mean annual temperature, clay, $(\text{clay})^2$	44.6

determination are not high. This may be partly due to the restricted range of temperature classes present in some of the regional groupings of records. Except for the tropical records, organic carbon concentration is more highly correlated with mean annual temperature than is nitrogen.

Soil pH was negatively correlated with both organic carbon and nitrogen in all of the analyses and showed quadratic trends. For organic carbon in the cool temperate areas and also for nitrogen in the tropical areas, this factor accounted for a substantial proportion of the total sums of squares. Again, the coefficients of determination were much greater for organic carbon than for nitrogen except for the tropical records.

Nitrogen and organic carbon were both positively correlated with clay content, although correlations with nitrogen were invariably higher. In the multiple correlation analyses clay content was selected earlier in the stepwise fitting process for the tropical area than for the other areas studied, and may be more important as a factor controlling organic matter in tropical soils than elsewhere. Such a contrast may reflect the differing clay mineralogies which are known to exist between tropical and temperate zones (Millot 1979). Relationships between clay and organic matter in tropical soils have been found by Birch and Friend (1956), Jones (1973), and Kadeba (1978).

Thus, for organic carbon content, precipitation is the most important factor. Except for the tropics, pH and mean annual temperature class are next with only a small contribution to the total sums of squares being made by clay content. In contrast, both pH and clay content attain a relatively greater importance in the tropics. For nitrogen, the order of importance of the environmental variables is, for the temperate areas, precipitation, mean annual temperature class, clay content and soil pH. In the tropics, mean annual temperature class is of much less importance than clay content or pH.

In spite of the general correlation between organic carbon and nitrogen, it is obvious that different relationships exist between the nature and levels of soil organic matter between the different regions and probably also within the arbitrary groupings of records considered here.

Many other factors will of course be operating. For example, Harradine and Jenny (1958) showed the effects of vegetation and parent material on soil organic matter in California. Others have suggested relationships with soil phosphorus content (Prescott 1931). Hubble and Martin (unpublished data 1960) found for Queensland soils that, using partial regressions with rainfall as the other variable, nitrogen and phosphorus were highly correlated, with phosphorus accounting for 47% of the variation.

There are problems in attempting to quantitatively define individual effects. The system is essentially multivariate and interactive. Above, we have demonstrated that precipitation, temperature regime, clay content and pH all influence organic matter status. However, we do not feel that these are sufficiently well modelled that other minor factors can be identified.

The relationship with phosphorus is likely to be particularly hard to define. Because organic phosphorus in soil is strongly correlated with the other constituents of soil organic matter (carbon, nitrogen and sulphur) it follows that total soil phosphorus is likely to be high in soils rich in organic matter. This tells little about whether it is causative. It does seem likely that part of any parent material effect may be attributable to its phosphorus component but the analytical data normally available for soils are unlikely to be appropriate to reveal such relationships. This is particularly true for our data set as there is no uniform measure of 'available' phosphorus, while 'total' phosphorus has been determined by chemical methods using either boiling HCl extracts or, more recently, by X-ray fluorescence analysis of solid samples. Available evidence indicates that the former method recovers between 65 and 95% of the amount of phosphorus recovered by the latter. In their work on New Zealand soil sequences, Walker and Syers (1976) found it necessary to fractionate the total soil phosphorus into several more meaningful components.

The influence of vegetation and land use on soil organic matter

Due to the confounding effects of other environmental factors, it is difficult to completely isolate the role of vegetation in the data set we have examined. Previous studies have examined the effects of vegetation on soil organic matter in particular localities where there is a wide variety of vegetation types or where the vegetation has been drastically modified by man.

In the former case, it is still difficult to separate cause and effect even when the paired site technique is used. Isbell (1966) found that brigalow (*Acacia harpophylla*) forests had substantially higher organic carbon and nitrogen contents than adjacent grasslands in north central Queensland, both communities being on very similar *grey* and *brown clays*. Similarly, Gillman (1976) found significantly higher carbon and nitrogen levels in *krasnozems* under rainforest in north Queensland than under adjacent eucalypt forest. C:N ratios were much wider under the latter. In contrast, Stocker (1968) found no significant differences between similar soils under a dry rainforest variant and eucalypt forests in the Northern Territory.

Although exotic conifer plantations are widespread in parts of Australia, few studies have been published on soil changes resulting from their establishment. Hamilton (1965) compared the top 2.5 cm of soil under eucalypt forest and 20–30 year old *Pinus radiata* plantations near Canberra, A.C.T. Although statistical comparisons were not made, five of the six paired sites showed a decline in organic carbon and nitrogen under the exotic species. C:N ratios increased at all sites. Waring (1968) has reported that under *Pinus elliottii* growing on coastal sands in New South Wales there was a 25% reduction in the concentration of total nitrogen in the surface soil compared to that in the soil under the original stand of eucalypts.

A more recent study by Turner and Kelly (1977) compared soils under a 50-year-old Douglas-fir (*Pseudotsuga menziesii*) plantation and an adjacent regenerated eucalypt forest near Batlow in southern New South Wales. The mean values from 15 paired sites of the A_1 horizon (0–7 cm) for organic carbon and nitrogen were significantly lower under the exotic species.

Although information is available for organic matter changes following cultivation and the use of grass–legume leys (see Chapter 37), there is surprisingly little information available for the effects of clearing and the establishment of pure grass pastures. Further, the evidence is conflicting.

Teakle (1950) suggested that after removing the rainforest from north Queensland *krasnozems*, organic carbon content could be halved after about 20 or 30 years under *Paspalum dilatatum* sward. Nicholls and Tucker (1956) examined *krasnozems* in the Lismore district of New South Wales but found that the evidence for organic matter decline after clearing the rainforest was inconclusive, a result similar to that obtained by Colwell (1958) for the same general area. However, McGarity (1959), again for Lismore *krasnozems*, concluded that surface horizons had suffered a decline of 20% in total nitrogen since clearing of the rainforest some 60 to 80 years previously. It would appear that these few studies are not sufficiently comprehensive for definitive statements to be made.

Similarly, the effects on organic matter when brigalow (*Acacia harpophylla*) forests are cleared has been documented in central Queensland. Webb *et al.* (1977) found higher organic carbon contents in virgin sites on *grey clays*, while in a more comprehensive examination of *solodized-solonetz* and *solodic soils*, Graham (1978) found that in 8 out of 17 paired sites a significant decrease in organic carbon occurred on clearing. The period since clearing in these instances ranged from 7 to 27 years, with percentage change ranging from 17 to 41% of the virgin level.

In the data set assembled, it is possible to examine the effects of imposing the vegetation classification on soil organic carbon levels by treating the categories of vegetation as factors in a regression analysis. This was done by first fitting temperature and precipitation and then imposing the vegetation categories. The effects of fitting the vegetation classification were examined by analysis of deviance.

Regressions were computed separately for data records from the tropical, warm temperate and cool temperate regions with soil organic carbon as the dependent variable. For the cool temperate region, imposing the vegetation classification on the regression gave a highly significant reduction in the deviance ($P < 0.001$). This was also found in the other two regions although in both of these a further contrast could be defined. Comparing the rainforest soils, which are higher in organic matter, with those of the other vegetation classes, large reductions in the deviances were observed; smaller, but still significant, reductions in the deviances could be obtained by separating the other vegetation classes.

For many years, it has been recognized that vegetation exerts a considerable influence on both the nature and amount of organic matter in soils. This is borne out for Australia by the analyses discussed here and by the few examples presented. Characterization of the nature and status of soil organic matter levels both under indigenous vegetation types and under conditions of altered land use has considerable implications for future productivity and needs further study in Australia.

Concluding remarks

There are very real problems involved both in the analysis of the data described in this chapter and in drawing inferences from these analyses. Firstly, the data were not randomly drawn from either the soil groups or regions under study. Secondly, the data gathered for carbon, nitrogen and the C:N ratio characteristically have non-normal frequency distributions. In view of this, the conclusions we have reached must necessarily be regarded as somewhat tentative.

Much yet remains to be learned about the nature and environmental relationships of organic matter in Australian soils. More representative data on a regional, climatic and soil taxonomic basis are required if statements of more general application are to be made. In particular, little is understood of how the joint action of the complex set of interrelated environmental factors effects a dynamic balance of the various components of organic matter. Thus, except in the coarsest sense, we are not yet in the desirable position of being able to predict the long term effects of manipulating the soil system. If sustained future productivity is to be assured in our agricultural and other terrestrial ecosystems, it appears that a deeper understanding of the nature, distribution, role and behaviour of soil organic matter is necessary.

References

Allison, F. E. (1973). 'Soil Organic Matter and its Role in Crop Production.' (Elsevier: Amsterdam.)

Anon. (1970). 'Modern Farming and the Soil. Report of the Agricultural Advisory Council on Soil Structure and Fertility.' (H.M.S.O.: London.)

Beadle, N. C. W. (1962). An alternative hypothesis to account for the generally low phosphate content of Australian soils. *Aust. J. Agric. Res.* **13**, 434-42.

Beckett, P. H. T. and Webster, R. (1971). Soil variability: a review. *Soils Fert.* **34**, 1-15.

Beckmann, G. G. and Hubble, G. D. (1974). The significance of radiocarbon measurements of humus from *krasnozems* (Ferralsols) in subtropical Australia. Trans. 10th Int. Congr. Soil Sci. Vol. 6, pp. 362-71.

Birch, H. F. and Friend, M. T. (1956). The organic matter and nitrogen status of East African soils. *J. Soil Sci.* **7**, 156-67.

Blackburn, G., Sleeman, J. R. and Scharpenseel, H. W. (1979). Radiocarbon measurements and soil micromorphology as guides to the formation of gilgai at Kaniva, Victoria. *Aust. J. Soil Res.* **17**, 1-15.

Colwell, J. D. (1958). Observations on the pedology and fertility of some *krasnozems* in northern New South Wales. *J. Soil Sci.* **9**, 46-57.

Crack, B. J. and Isbell, R. F. (1971). Studies on some neutral red duplex soils (Dr2.12) in north-eastern Queensland. I. Morphological and chemical characteristics. *Aust. J. Exp. Agric. Anim. Husb.* **11**, 328-35.

Darwin, C. R. (1881). 'The Formation of Vegetable Mould Through the Action of Worms with Observations on their Habits.' (John Murray: London.)

Gillman, G. P. (1976). Red basaltic soils in north Queensland. II. Chemistry. CSIRO Aust. Div. Soils Tech. Paper No. 28.

Graham, T. W. G. (1978). Soil nitrogen status in relation to land development and pasture productivity in the brigalow region of central Queensland. Unpublished M.Agric. Sci. Thesis, University of Queensland.

Hamilton, C. D. (1965). Changes in the soil under *Pinus radiata*. *Aust. For.* **29**, 275-89.

Harradine, F. and Jenny, H. (1958). Influence of parent material and climate on texture and nitrogen and carbon contents of virgin California soils. I. Texture and nitrogen contents of soils. *Soil Sci.* **85**, 235-43.

Hosking, J. S. (1935). The carbon-nitrogen ratios of Australian soils. *Soil Res.* **4**, 253-68.

Isbell, R. F. (1966). Soils of the east Bald Hill area, Collinsville district, north Queensland. CSIRO Aust. Div. Soils, Soils Land Use Ser. No. 48.

Isbell, R. F. and Gillman, G. P. (1973). Studies on some deep sandy soils in Cape York Peninsula, north Queensland. I. Morphological and chemical characteristics. *Aust. J. Exp. Agric. Anim. Husb.* **13**, 81-8.

Isbell, R. F., Gillman, G. P., Murtha, G. G. and Jones, P. N. (1977). Brown basaltic soils in north Queensland. CSIRO Aust. Div. Soils Tech. Pap. No. 34.

Isbell, R. F., Jones, R. K. and Gillman, G. P. (1976). Plant nutrition studies on some yellow and red earth soils in northern Cape York Peninsula. I. Soils and their nutrient status. *Aust. J. Exp. Agric. Anim. Husb.* **16**, 532-41.

Jenkinson, D. S. and Rayner, J. H. (1977). The turnover of soil organic matter in some of the Rothamstead classical experiments. *Soil Sci.* **123**, 298-305.

Jenny, H. (1941). 'Factors of Soil Formation.' (McGraw-Hill: New York.)

Jenny, H. (1961*a*). Derivation of state factor equations of soils and ecosystems. *Proc. Soil Sci. Soc. Am.* **25**, 385-8.

Jenny, H. (1961*b*). Comparison of soil nitrogen and carbon in tropical and temperate regions. University of Missouri, College of Agriculture. Agricultural Experiment Station Research Bulletin No. 765, 5-31.

Jenny, H., Bingham, F. and Padilla-Sarava, B. (1948). Nitrogen and organic matter contents of equatorial soils of Colombia, S.A. *Soil Sci.* **66**, 173-86.

Jenny, H., Salem, A. E. and Wallis, J. R. (1968). Interplay of soil organic matter and soil fertility with state factors and soil properties. *In* 'Study Week on Organic Matter and Soil Fertility.' pp. 5-36, Pontificae Academiae Scientiarum, Scripta Varia 31. (North-Holland Publishing Co.: Amsterdam.)

Jones, M. J. (1973). The organic matter content of the savanna soils of West AFrica. *J. Soil Sci.* **24**, 42-53.

Kadeba, O. (1978). Organic matter status of some savanna soils of northern Nigeria. *Soil Sci.* **125**, 122-7.

Laudelout, H., Meyer, J. and Peeters, A. (1960). Les relations quantatives entre la teneur en matière organique du sol et le climat. *Agricultura, Louvain* **8**, 103-40.

Little, I. P. ,Haydock, K. P. and Reeve, R. (1962). The correlation of Walkely-Black organic carbon with the value obtained by total combustion in Queensland soils. CSIRO Aust. Div. Rep. No. 3/62.

Martin, A. E., Gilkes, R. J. and Skjemstad, J. O. (1970). Fixed ammonium in soils developed on some Queensland phyllites and its relation to weathering. *Aust. J. Soil Res.* **8**, 71-80.

McGarity, J. W. (1959). The influence of sod-seeded legumes on the nitrogen economy of grassland soils at Lismore, N.S.W. *J. Aust. Inst. Agric. Sci.* **25**, 287-93.

Millot, G. (1979). Clay. *Sci. Am.* **240**, (4), 76-84.

Nicholls, K. D. and Tucker, B. M. (1956). Pedology and chemistry of the basaltic soils of the Lismore district, N.S.W. CSIRO Aust. Soil Publ. No. 7.

Northcote, K. H., Hubble, G. D., Isbell, R. F., Thompson, C. H. and Bettenay, E. (1975). 'A Description of Australian Soils.' (CSIRO: Melbourne.)

O'Brien, B. J. and Stout, J. D. (1978). Movement and turnover of soil organic matter as indicated by carbon isotope measurements. *Soil Biol. Biochem.* **10**, 309-17.

Parsons, J. W. and Tinsley, J. (1975). Nitrogenous substrates. *In* 'Soil Components, 1. Organic Components.' (Ed. J. E. Gieseking). pp. 263-304. (Springer-Verlag: New York.)

Prescott, J. A. (1931). The soils of Australia in relation to vegetation and climate. Bull. Coun. Sci. Ind. Res. No. 52.

Scharpenseel, H. W. and Schiffmann, H. (1977). Radiocarbon dating of soils, a review. *Z. Pfl-Ernahr. Düng. Bodenk.* **140**, 159-74.

Specht, R. L. (1970). Vegetation. *In* 'The Australian Environment.' (Ed. G. W. Leeper) 4th Edn, pp.43-67. (Melbourne University Press: Melbourne.)

Stace, H. C. T., Hubble, G. D., Brewer, R., Northcote, K. H., Sleeman, J. R., Mulcahy, M. J. and Hallsworth, E. G. (1968). 'A Handbook of Australian Soils.' (Rellim Technical Publications: Glenside, S.A.)

Stark, N. (1978). Man, tropical forests, and the biological life of a soil. *Biotropica* **10**, 1-10.

Stocker, G. C. (1968). Fertility differences between the surface soils of monsoon and eucalypt forests in the Northern Territory. *Aust. For. Res.* **4**, 31-8.

Teakle, L. J. H. (1950). Notes on the soils of coastal Queensland and portions of the hinterland with special reference to the tropical latitudes. *Pap. Fac. Agric. Queensl.* **1** (1).

Turner, J. and Kelly, J. (1977). Soil chemical properties under naturally regenerated *Eucalyptus* spp. and planted Douglas fir. *Aust. For. Res.* **7**, 163-72.

Walker, T. W. (1965). The significance of phosphorus in pedogenesis. *In* 'Experimental Pedology.' (Eds E. G. Hallsworth and D. V. Crawford). pp.285-316. (Butterworths: London.)

Walker, T. W. and Syers, J. K. (1976). The fate of phosphorus during pedogenesis. *Geoderma* **15**, 1-19.

Waring, J. D. (1968). The effect of cultural techniques on the growth of *Pinus eliottii*, Engelm. near Jervis Bay, New South Wales. Trans. 9th Int. Congr. Soil Sci. Vol. 3, pp.447-54.

Webb, A. A., Crack, B. J. and Gill, J. Y. (1977). Studies on the gilgaied clay soils (Ug5.2) of the Highworth land system in east-central Queensland. I. Chemical characteristics. *Queensl. J. Agric. Anim. Sci.* **34**, 53-65.

Wild, A. (1958). The phosphate content of Australian soils. *Aust. J. Agric. Res.* **9**, 193-204.

Chapter 35

Aspects of the chemistry of soil organic matter

R. S. Beckwith and J. H. A. Butler

ORGANIC MATTER IN SOILS INCLUDES PART OF THE LIVING biomass, animal, vegetable and microbial, as well as decomposing residues from the biomass. It includes resistant debris as well as compounds which may occur only transitorily unless stabilized by association with one or more other soil constituents. Other organic materials, minute in quantitative terms, nevertheless have important biological activities. The root exudates (Hale *et al.* 1971) and phytotoxins (Chapter 47) are examples of such materials. Much of this complexity and diversity will be necessarily disregarded in this short chapter.

The term 'soil organic matter' is less inclusive but rather non-specific in at least two respects. Thus physically, in undisturbed natural soils, it denotes a somewhat arbitrary part of an ecósystem (Jenny 1961) and, chemically, the bulk of the carbon compounds in soils have not been identified. Several defined organic constituents make up only a small proportion of the whole. Those containing nitrogen, sulfur or phosphorus are considered in other chapters.

Most organic carbon occurs as dark-coloured, seemingly stable organic material designated by such terms as 'humified material', 'humic substances' or simply 'humus'. Residues with recognizable morphology are usually excluded, together with particulate mineral matter, no matter how intimately mixed. Even this degraded material contains components of all ages including some from recently dead litter and organisms (Jenkinson 1971) which, particularly if mobile, can react with solutes, with other organic moieties or with minerals in a great variety of micro-environments. These include an animal gut (Greenland 1965), root channel or rhizosphere, or at a sorption site on an activating mineral or organic surface.

Beneficial effects of organic matter in soils have been appreciated since the times when manures and plant residues were the only fertilizers (Allison 1973). The understanding of its roles, and of changes to added organic materials, may become even more important in future in relation to organic waste disposal (Anon. 1976; Loehr 1977), as well as to recycling various plant nutrients in wastes. The protection of the quality of natural waters should be one important aspect.

Despite their growing practical importance, the above topics are outside the scope of this chapter. Here we attempt to summarize present understanding of humified materials, the methods used to study them, their importance to some soil properties and plant growth and selected aspects of Australian work from which further progress may come.

Environmental factors and soil organic matter

External factors

Soil is part of an ecosystem and its organic content is a function of time and of all the external environmental factors which influence that ecosystem (Jenny 1941, 1961). These factors (see Chapter 34) include the soil parent material, which also has a notable role as an internal factor.

Internal factors

Besides influencing the nutrient status and biomass production of the ecosystem, the parent material is the source, direct or after weathering, of the persistent minerals in soils. These can, to a greater or lesser degree, adsorb and/or stabilize organic soil constituents. Still less directly, parent material influences aeration and other aspects of the internal soil environment that can affect rates of metabolism of, and/or rates of removal of, organic substrates in soils. Although humus is an inherently stable organic material, as discussed later, internal soil properties appear to superimpose secondary differences in stability. Hence soil environment can influence the amount of organic matter which occurs in soils and the changes in content which will result from given soil management.

Severely impeded drainage is especially significant in decreasing the rate of decomposition of organic residues and appears to be a characteristic of all soils in which the content of organic matter reaches very high levels (Stephens 1949).

Acidity is known to slow the growth of soil organisms, especially actinomycetes, and generally slows transformation of residues to humus and turnover of organic matter in soils. However, production of plant residues may be much less from acid soils (Allison 1973) and the net effect of acidity can vary.

Possibly related, but less often discussed, is the large and rapid buildup of organic content in allophanic soils. For example, Jackman (1960) noted an increase in soil carbon in Taupo pumice soil under pasture from 4.6 to 12.0% in 22 years. This represents accumulation at a rate of approximately 1100 kg carbon per hectare per year and may be contrasted with Barrow's (1969) values from Western Australia of 440 kg carbon per hectare per year during 30-40 years. Jackman discounted low pH or abnormally large yields of pastures as contributing causes to this rapid buildup.

There are probably a number of interactions between soil minerals and organic constituents which confer increased stability on the latter (Allison 1973). On many occasions, aluminium has appeared to be implicated through chelation and coprecipitation reactions (Martin and Reeve 1958; Russell 1973) or reactions resembling tanning of leather (Jackman 1960) and perhaps also in contributing to some effects of soil acidity. Carbon in *krasnozem* surface soils, but not in some others, correlated strongly with their contents of aluminium soluble in 0.005 M sulphuric acid (Beckwith *et al.* 1962).

Aluminium is the predominant polyvalent cationic element in soils; it is immune to valence change, and so it would not be surprising if, as the preceding evidence collectively suggests, this metal was found to be uniquely active in stabilizing humus in acidic soils. Apart from its aluminium content, allophane might be expected to be especially active by virtue of its exceptionally small particle size and large surface which would enable it more efficiently to protect physically the soil organic moieties. Random physical mixing and coprecipitation of comparably-sized mineral and organic polymers may be more relevant to humus stability than the often quoted protection of organic molecules by trapping within clay lattices.

Humic substances

Extraction from soils

Sodium hydroxide, the classical extractant of soil humic materials, remains the most frequently used reagent (Russell 1973) because of its effectiveness (Table 35.1; Schnitzer *et al.* 1958; Hayes *et al.* 1975). Pretreatment of soils is now recommended to remove raw undecomposed plant fragments (Ford *et al.* 1969) and (using hydrochloric acid) to enhance the extraction of organic matter and lessen mineral contamination of the extracted organic materials. As an alternative to the latter, inclusion of pyrophosphate in the sodium hydroxide extractant has been advocated (Kononova and Bel'chikova 1961).

It is not possible to diagnose the state of combination in soils of organic matter extracted with sodium hydroxide. The minerals contained in such extracts may be adventitious, while mineral matter which in the soil was combined with organic moieties, may be left in the insoluble residue due to hydrolysis and/or precipitation. Again, organic matter probably undergoes change during extraction (Dubach *et al.* 1961; Swift and Posner 1972). The question of the extent of depolymerization and oxidation of organic components during extraction with alkali still appears to be unresolved (Hayes *et al.* 1975), but it seems certain that the organic polymers undergo change of molecular shape during this treatment.

Table 35.1. Amounts of carbon extracted by various reagents from the A horizon of a *podzol*

Data from Schnitzer *et al.* (1958)

Extractant	Conc.	Carbon extracted (% of total)
$Na_4P_2O_7.10H_2O$	0.1 M	6.1
$Na_3PO_4.12H_2O$	0.1 M	5.5
$Na_2B_4O_7.10H_2O$	0.1 M	7.7
NaF	0.5 M	6.6
NaCl	0.5 M	2.5
Na_2CO_3	0.5 M	8.3
NaOH	0.5 M	24.3
Na_2EDTA	0.2 M	Trace

A major difficulty in devising milder but efficient extractants for separating soil organic matter arises from the lack of understanding of the mechanisms by which organic matter is retained in soils (e.g. Dubach and Mehta 1963; Greenland 1971). Bremner *et al.* (1946) observed that solutions of pyrophosphates, or metal-complexing organic anions such as citrate, extracted organic matter quite effectively, and considered that iron had an important role in precipitating humus in soils. However, such reagents compete strongly for calcium ions and so organic matter bound in neutral and alkaline soils can also be extracted.

The eluviated organic matter of podzol B horizons is believed to have a smaller molecular size than much of the humus of A horizons (Butler and Ladd 1971). Martin and Reeve (1957) found that 80–90% of the NaOH-extractable organic matter of some groundwater podzol B horizons could be extracted under less extreme chemical conditions. They showed that the removal of aluminium from this organic matter, through competitive chemical reaction with acetylacetone, was sufficient to promote its dispersal in water. It can thence be deduced that one important action of sodium hydroxide in dissolving organic matter from podzol B horizons, and probably from other soils also, is the freeing of Al-linked carboxylate groups through aluminate formation, to produce ionized sodium carboxylate groups. Simultaneously, hydrolysis reactions may also free some carboxylate groups bound to Fe, Ca and other metals.

Sodium silicate is another general contaminant of alkali-extracted organic matter. As yet, there is no

Table 35.2. Extraction of soil carbon

Carbon was extracted by two solutions using ultrasonic dispersion for 5 min followed by shaking for 120 h at a 1:20 soil:extractant ratio. R. S. Beckwith and V. K. Nayyar, unpublished data.

Extractant	Percentage of total soil carbon extracted Clay loam pH 5.7	Loam *(red-brown earth)* pH 6.3
0.05 M $Na_4P_2O_7$ in 8 M urea	66	57
0.05 M $Na_4P_2O_7$ in 0.5 M NaOH	60	58

conclusive evidence as to whether all the silicon present is adventitious or whether silicic acid bonds to humus in soils.

Emerson (1956) suggested that the binding of humus in soils probably involved hydrogen bonding. Beckwith (1973) applied this idea and noted that an extracting solution containing both 8 M urea and 0.3 M Na_2HPO_4 removed much more organic matter from two calcareous soils than either reagent alone. Subsequently Beckwith and Nayyar (Table 35.2) found that 0.1 M $Na_4P_2O_7$ in 8 M urea extracted at least as much carbon from a range of soils as did 0.5 M NaOH with or without added $Na_4P_2O_7$.

A solution of 0.3 M $NaHCO_3$ containing 8 M urea appeared as good an extractant of organic matter as the orthophosphate reagent. Thus the need for separation of polyvalent metals may have been exaggerated as a requirement for successful extraction of organic matter. When using an extractant that does not break down the minerals binding the organic matter, what is needed may be a strongly adsorbed anion, to replace anionic groups (carboxylate) from adsorption sites on minerals, together with a reagent which disrupts hydrogen bonds. The superiority of pyrophosphate over other anions could arise from its more avid attraction to anion sorption sites. The special effectiveness of sodium hydroxide may be due to its ability to disrupt hydrogen bonds, by replacing the H of phenolic groups with Na, as well as the other reactions discussed above.

Properties of extracted humic substances

The recorded properties of some 'organic matter' preparations have probably been influenced by the contaminants which they contained. Humus preparations which have been separated by extraction with alkali contain varying amounts of silicon and polyvalent metals which may not be readily removed. The classical fractionation of organic matter into 'fulvic' and 'humic' acids can be materially affected by the extent of purification of the alkaline soil extract (Sequi *et al.* 1975) and by conditions of acidification e.g. when HF is included much less 'humic' acid is usually precipitated. The size of the 'humin' fraction appears to decrease progressively as more raw plant fragments are removed (Oades and Ladd 1977) and the separation of fulvic acid from the humic fraction depends on the amount of contamination with polyvalent metals. Also, Acton *et al.* (1963) reported that 12% of the total soil carbohydrate was coprecipitated with the humic acid when an alkaline soil extract was acidified quickly.

Nevertheless, the distinction betwen humic and fulvic acids appears to have some reality and, allowing the importance of experimental conditions (Goh and Reid 1975), fulvic acid is generally agreed to contain a larger proportion of low molecular material with a larger percentage of carboxyl carbon (Stevenson and Butler 1969; Butler and Ladd 1971). The absorbance in the ultraviolet region at wavelengths below about 300 nm rises more steeply for fulvic than for humic solutions (Butler and Ladd 1969).

The structures and properties of humic and fulvic acid preparations have been reviewed by Dubach and Mehta (1963), by Schnitzer and Khan (1972) by Flaig *et al.* (1975) and by Schnitzer (1978).

Fractionation of humic substances

As most preparations of soil organic matter contain materials with a wide range of characteristics, preliminary fractionation should give more sensible values for measurements of structure and properties. Most extracted soil organic matter is polymeric and of a molecular weight up to a million. Thus fractionation procedures can be based, to at least some degree, on molecular size.

The traditional method using acid and alcohol as precipitating agents gives fractions with different characters, but the quantities of each fraction depend very much on the exact procedure used.

Theng *et al.* (1968) salted out organic matter with ammonium sulphate solutions. As the salt concentration was increased the precipitated humic fractions became gradually more oxidized, of smaller molecular weight and less aromatic. However, there were no abrupt changes in properties.

Other workers (Posner 1963; Ladd 1969; Butler and Ladd 1971; Swift and Posner 1971), have used gel exclusion chromatography and ultra-filtration as methods of fractionation without achieving sharp separations. Indeed there is evidence that Sephadex gel can give anomalous results, particularly with fulvic acids (Lindquist 1967; Butler and Ladd 1969). Ladd (1969) studied soil humic acids fractionated by gel exclusion chromatography and found a contrast with humic acids fractionated by solubilities; the larger molecular size fractions had smaller optical extinction coefficients in the ultraviolet and visible spectra.

Butler and Ladd (1969) studied humic acids extracted with different extractants from several soils, and a series of humic acids obtained by fractionating material from one soil, on Sephadex gels. As the molecular size of these fractions increased, both aliphatic C–H and acid-hydrolyzable amino acid contents increased, whereas the extinction values at 260 nm and carboxyl contents decreased. Swift *et al.* (1970) fractionated humic acid by agar gel chromatography and also found that the extinction coefficients increased with decreasing molecular size and that the higher molecular size fractions more closely resembled lignin. In contrast to other workers (Kumada and Kawamura 1968), Swift *et al.* (1970) proposed that as humification of organic matter proceeds the degree of aromaticity increases and the molecular weight decreases. This latter relationship was confirmed by Butler and Ladd (1971).

Functional groups of humic substances

One of the characteristics of humic materials, and probably the major reason for their beneficial effects in soil, is the presence of a considerable number of functional groups on the polymers.

Those groups demonstrated with considerable certainty are carboxyl, carbonyl and phenolic hydroxyl (Stevenson and Butler 1969).

The acidic groups (carboxyl and phenolic hydroxyl) can be titrated (Martin and Reeve 1958; Beckwith 1959; Posner 1964, 1966). However, assessing what the titration curves mean in terms of numbers of these groups is virtually impossible as the p*K* of a particular group depends strongly on its electronic environment. Thus the p*K* values of the carboxyl and phenolic hydroxyl groups overlap and cannot be distinguished by titration.

Butler and Ladd (1969) used a calcium acetate exchange procedure (measures all acidic hydrogens replaceable by Ca^{2+}) to measure carboxyl groups. They also methylated the humic compounds and measured the methanol released by alkaline hydrolysis as another method of determining carboxyl groups. Both procedures gave similar results. This may be fortuitous as again acidity rather than structure (carboxyl versus phenolic hydroxyl) determines whether or not a particular group will react.

Phenolic hydroxyl groups have been estimated usually as the difference between total acidity and carboxyl acidity. This method is likely to be unreliable because of limitations with both of these measurements.

Carbonyl groups are usually determined via derivative formation using reagents such as hydroxylamine, phenylhydrazine, 2,4-dinitrophenylhydrazine and semicarbazide. Theng and Posner (1967) reacted three soil humic acids with a number of carbonyl reagents and studied their infrared spectra. The spectral changes led them to conclude that these humic acids contained conjugated ketonic and H-bonded conjugated ketonic carbonyl groups. Quinone groups were not revealed by any tests.

The oxygen accounted for in functional groups is usually much less than the total oxygen of humic acids (Table 35.3). It seems either that estimates of numbers of various functional groups are low or that there are some oxygen-containing groups, perhaps internal ether bonds, which are not being measured (Stevenson and Butler 1969).

If some means of quantifying the infrared spectra of humic materials were available, better estimates of the numbers of some functional groups might be obtained. Some infrared absorbance measurements have been made by Butler and Ladd (1969), Theng *et al.* (1967) and Posner *et al.* (1968) but results are only semiquantitative and need calibrating. Great care must be taken when determining infrared spectra of humic materials to avoid anomalous results from impurities present in the preparation (Theng *et al.* 1966; Russell and Anderson 1977*a*, *b*).

Molecular size of humic materials

Cameron and Posner (1974) used the sedimentation velocity technique with a preparative ultracentrifuge to determine the molecular weight distribution of some humic acids. Four humic acids extracted from various plant parts, (rotted for 68 weeks), had weight-average molecular weights ranging from 4.9 x 10^4 to 1.1 x 10^5 while that of a humic acid extracted from a highly humified muck soil was measured as 2.2 x 10^5.

One significant advance in humus chemistry over recent years has been the use of gel exclusion chromatography to fractionate humic materials and to give some estimate of the molecular weight of the resultant fractions. Swift *et al.* (1970) determined the molecular sizes of a number of humic extracts using agar gel exclusion chromatography. The gel column was previously calibrated using humic acid fractions the molecular weights of which had been determined by osmometry (Wake and Posner 1967). The humic acid fractions had number-average molecular weights ranging from 1.36 x 10^3 to 6.2 x 10^4. In a later study Cameron *et al.* (1972), also using humic and fulvic acids fractionated by agar gel exclusion chromatography, found the molecular weight to range from 2.6 x 10^3 to 1.4 x 10^6. Comparison of the actual frictional coefficients of the molecules with that of a spherical molecule suggested that the higher molecular weight fractions were randomly coiled polymers with some branching.

Structure and origin of humic materials

The various organic matter fractions, particularly the humic fractions, are probably all more or less heterogeneous. Undoubtedly the humic fractions are acidic, polyfunctional polymers, with considerable ranges of molecular weights and most likely are made up from a variety of structural units. Further possibilities are as follows.

(1) Humic substances may be merely demethylated and artly oxidized lignin, but the supporting evidence for

Table 35.3. Distribution of oxygen in some humic and fulvic acids from terrestrial soil

Reproduced with permission from Stevenson and Butler (1969)

Oxygen (%)	COOH (% of O)	Acidic OH (% of O)	Weakly acidic plus alcoholic OH (% of O)	C=O (% of O)	Recovered (%)
			Humic acids		
35.6	13.6	19.0	12.2	4.1	48.9
36.7	26.2	24.8	14.2	7.8	74.0
			Fulvic acids		
44.7	65.1	11.8	12.9	11.1	100.9
47.3	57.5	19.3	11.5	5.8	94.1
47.0	41.5	22.8	11.2	10.2	85.7

this is poor; lignin is a relatively stable material in soil and the general composition and infra-red spectrum of humic acid resembles that of lignin to some degree (Flaig *et al.* 1975; Swift *et. al.* 1970). Gottlieb and Hendricks (1945) applied the alkaline nitrobenzene oxidation procedure to soil organic matter (unmethylated and methylated) and found only traces of vanillin and no syringaldehyde. The same procedure applied to wood samples yielded 21–43% of these lignin aldehydes. Swaby and Ladd (1962) tried the same technique on Australian soils and also found only traces of lignin aldehydes. However, oxidized lignin in soil might behave differently, and so the contribution of lignin to soil organic matter still remains uncertain.

(2) As proposed by some other workers (Swaby and Ladd 1962; Flaig 1967), humic materials are formed by polymerization of individual phenols, derived from lignin breakdown, with amino acids (to produce the nitrogen content of humic substances). However, if lignin were degraded to its constituent phenols, these would be generally expected to metabolize rapidly, leaving few phenolics able to repolymerize. Also, Ladd and Brisbane (1967) and Brisbane *et al.* (1972) demonstrated convincingly that humic acids contain peptide bonds, as they showed that protein-hydrolyzing enzymes can liberate 28–39% of the amino acids released from humic acids by hot acid hydrolysis.

(3) Humic substances may be a conglomeration of different polymers of various sizes and origins. This is all that would be expected to result from extraction and purification procedures which are so non-specific. Any polymeric material in soil with a proportion of acidic groups should be included.

(4) Haider and Martin (1967) amongst others, have suggested that as fungal cultures produce black humic-like substances, fungi contribute to the synthesis of humic acid. Results obtained by Piper and Posner (1972*a*, *b*) support this idea. After sodium amalgam reduction of a number of humic acids they obtained phenolic compounds that they considered characteristic of fungal polymers and of lignin.

Kimber and Searle (1970) pyrolyzed a series of humic acids and measured the released benzene and toluene by gas chromatography. Yields of benzene and toluene were best correlated with amino acid content but also with molecular weight and inversely with extinction at 260 nm and 460 nm (Butler and Ladd 1969). These results, together with those of Swift *et al.* (1970), suggest a change in the nature of the humic polymer as the molecular weight increases. The higher molecular weight material appears less cross-linked, and more lignin-like whereas other constituents are either more degraded or originate from other sources.

Jackson *et al.* (1972) used an interesting procedure to degrade humic acid to identifiable fragments viz. heating with phenol in the presence of *p*-toluene sulphonic acid or BF_3 as catalyst. The following reaction was postulated.

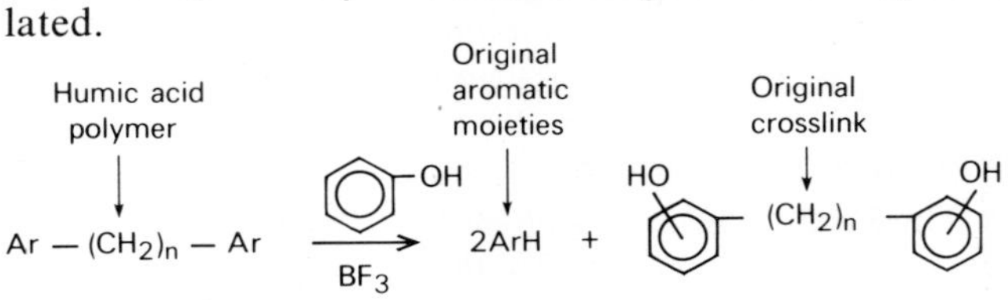

After methylation, the compounds of the mixture were separated and identified by the usual organic chemical procedures. Both the crosslink and the aromatic units were supposedly identified but the above mechanism still needs confirmation with extensive studies on model compounds. Accepting the above reaction, three- and four-point linkages appeared to predominate, suggesting that the humic polymer has a highly crosslinked, three-dimensional structure.

Biological effects of humic materials

Limited work has been done on the biological effects of humic substances. Some workers (e.g. Schnitzer and Poapst 1967; Lineman 1977) found that humic materials in solution stimulated root initiation. However, this effect could be nutritional e.g. by mobilizing metal ions, in particular Fe^{3+}.

Knowledge of enzymes in soils has recently been extensively reviewed (Ladd 1978; Burns 1978; Kiss *et al.* 1975). Ladd *et al.* (1968), Ladd and Butler (1969*a*, *b*) and Butler and Ladd (1971) showed that soil humic and fulvic acids in solution at concentrations of 2–20 $\mu g\ ml^{-1}$

increased the activity of some proteolytic enzymes and decreased the activity of others. Monovalent and divalent cations, at concentrations of 10^{-1} and 10^{-2} mol. l^{-1} respectively (Ladd and Butler 1970), and methylation with CH_2N_2 (Butler and Ladd 1969) both prevented the effect of the humic acids on the enzymes. Butler and Ladd (1971) also noted that high molecular weight fractions of either humic or fulvic acids from a podzol B horizon brought about the largest changes in enzyme activity. A humic acid had more effect than a fulvic acid fraction of the same size range obtained by Diaflo Ultrafiltration. Whether humic acid or organic matter has a significant effect on enzymes in the soil is not known but organic matter levels far exceeding 10 μg ml^{-1} occur in some soil solutions. Extracted humic polymers can readily bond to free amino acids or protein (Ladd and Butler 1975). The latter could be protected from microbial breakdown if this reaction occurs in soils and Lloyd (1975) suggested that urease in alpine humus soil is protected from breakdown by protease enzymes in such a way.

Polysaccharides

Soil polysaccharides have been recently reviewed by Swincer *et al.* (1969), Greenland and Oades (1975) and Lowe (1978). They originate from plant, animal and microbial saccharides present in debris added to the soil and from microbes decomposing this debris. Thus a very wide range of these compounds are no doubt present in soil, but some only in trace amounts.

Nature of saccharides

The saccharides in soils pose a problem different from that of humic substances; generally we know the structures involved, but procedures for extracting, identifying and measuring the various saccharides present are very imperfect. Probably most of the methods for estimating the various classes of saccharides in soil are not specific and give variable recoveries. Thus figures quoted can only be approximate. Some problems associated with the different methods are discussed by Greenland and Oades (1975). Table 35.4 shows the structures of the saccharides common in soil and the amounts found as a percentage of the total organic matter.

Extraction of saccharides

Soil saccharides have been extracted with a variety of reagents, the most common probably being water, followed by 0.5 M NaOH. A comprehensive table published by Greenland and Oades (1975) showed yields against methods of extraction used. Swincer *et al.* (1968*a*) used a number of different procedures on a *red brown earth* (Table 35.5). Acetic anhydride with H_2SO_4 was clearly

Table 35.4. Nature and amounts of saccharides found in some soils

Values from Greenland and Oades (1975). *n* (in uronic acid formula) = 1 or 2

	Structure	Amount (% of organic matter)
Free saccharides	$CHO-(CHOH)_n-CH_2OH$	< 1
Total reducing sugars	Carbohydrates with no free aldehyde or ketone group	5–45
Hexoses	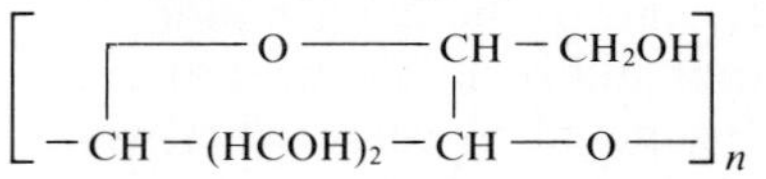	5–20
Pentoses	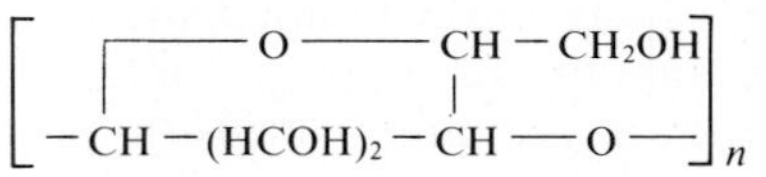	< 5
Uronic acids	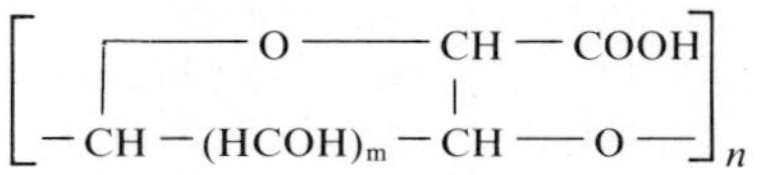	1–5
Hexo-samines	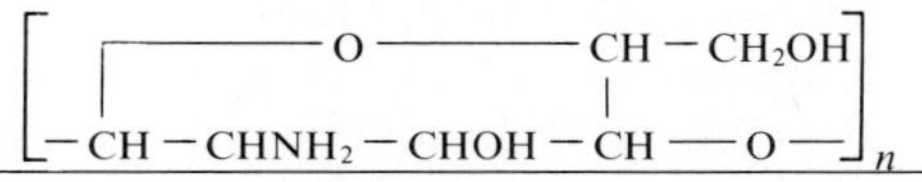	2–5

Table 35.5. Yields of carbohydrate extracted from Urrbrae soil by various procedures
Adapted from Swincer *et al.* (1968*a*)

Extractant	Time and temperature	Yield: % of soil	% of organic matter	% of total soil carbohydrate
0.5 M H_2SO_4	16 h, 20°C	0.026	0.45	6
1 M HCl	16 h, 20°C	0.040	0.70	9
1 M HF	16 h, 20°C	0.076	1.3	18
Water	16 h, 70°C	0.059	1.0	14
Phosphate buffer, pH7.0 (+ 5% KCl)	16 h, 20°C	0.022	0.39	5
0.5 M sodium pyrophosphate	16 h, 20°C	0.046	0.80	11
0.5 M NaOH	16 h, 20°C	0.13	2.2	30
	1 h, 100°C	0.16	2.7	37
Dimethyl formamide (DMF) containing 0.4 M boric acid, 0.4 M oxalic acid, and 0.2 M LiCl	1 h, reflux	0.12	2.1	29
Acetic anhydride/pyridine 1:1 (v/v)	16 h, reflux	0.05	0.88	12
Acetic anhydride/conc. H_2SO_4 20:1 (v/v)	10 min, 60°C	0.32	5.5	76

the most efficient, being twice as effective as hot 0.5 M NaOH. In their final procedure they used a stepwise extraction with first 1.0 M HCl then 0.5 M NaOH and finally an acetic anhydride–H_2SO_4 mix, and claimed removal of 63–94% of the total soil saccharide content from a series of eleven soils.

Fractionation of extracted saccharides

Fractionation techniques have generally given poor results with soil saccharides. Only gel filtration (Roulet *et al.* 1963; Barker *et al.* 1965; Swincer *et al.* 1968*b*) and chromatography on DEAE-Sephadex (Barker *et al.* 1965) or on DEAE cellulose (Thomas 1963; Roulet *et al.* 1963) have produced fractions containing varying proportions of different soil saccharides.

Swincer *et al.* (1968*b*) fractionated polysaccharides from a *red brown earth* with Sephadex G25 and Biogel P100 and found they ranged widely in molecular weight, with appreciable proportions less than 4000 and greater than 100 000. They showed that as the molecular weight of the fraction increased, the proportion of amino acids present decreased, contrasting with the amino acid content of humic acid fractions of different molecular weights (Butler and Ladd 1971).

The high molecular weight polysaccharide fractions contained more deoxyhexoses, suggesting that they are of microbial origin. These fractions are considered likely to have the greatest effect on aggregate stability (Swincer *et al.* 1968*b*).

Lipids

Soil lipids have been comprehensively reviewed by Morrison (1969) and Braids and Miller (1975). They include a wide range of different chemical compounds of biological origin, soluble in fat solvents. They normally make up only a small fraction (1–5%) of soil organic matter (Braids and Miller 1975).

Lipids are generally extracted from soils with benzene and ethanol mixtures using a soxhlet extractor.

Butler *et al.* (1964) investigated the lipids and dye in an Australian green-pigmented soil. They found that the soil containerd a polynuclear chloroquinone (Cameron and Sidell 1978) and about 2% of wax, five times the soil concentration of the surrounding non-pigmented soil. Saponification, separation and gas chromatography revealed long-chain acids and alcohols with even numbers of carbon atoms predominating, like other biolipids. This fits the suggested origin of the pigment and wax, i.e. microbial breakdown of tree roots. However, the hydrocarbons also present showed no such biological pattern.

Forms and reactions of organic matter in soils

The solid phase

Organic matter is only one component of the soil solids but it shows great versatility in interacting with other constituents. Organic components have a greater variety of reactive groups than mineral surfaces and include moieties which are more mobile than most particulate minerals. The possibility of cleavage of organic polymers by enzymes, and redeposition and reaction of some fragments at other sites on the soil solid surface, explains a further part of the greater reactivity of soil organic solids. Also, organic polymers with a long-chain configuration will be likely to increase aggregate stability in a way impossible for mineral particles (Emerson 1977).

The concept of adsorbed organic matter (Greenland 1965, 1971), supported by preceding observations on its extractability from soils, allows a further possibility. In soils containing adequate levels of organic matter, mineral surfaces may become coated and their properties changed. For example, Leaver and Russell (1957) diminished the phosphate-sorbing capacities of soils by pretreating them with extracted humic acid. Coated mineral particles could assume the character of giant organic moieties having some organic reactivity combined with size commensurable with soil mineral particles.

Structural effects

The structural properties of mineral soils are predominantly determined by the nature and orientation of *mineral* particles, especially clay minerals. However, organic matter appears to be disproportionately important in stabilizing such configurations of particles and the deteriorating structure of cultivated soils has long been attributed to loss of organic matter. Even so, it is somewhat difficult to try to isolate the role of organic matter for discussion because nearly all ecological and environmental factors (e.g. parent material, climate, biology) influence soil structure, sometimes indirectly as well as directly. It may be because of this complexity that direct demonstrations of the effects of soil organic matter have often proved difficult.

The only chemically identified organic constituents shown to stabilize soil aggregates are polysaccharides, especially microbially produced polysaccharides (Greenland 1965). Solvent extractions of soils to remove waxes and lipids did not affect aggregate stability (Stefanson 1968). Swaby (1949) considered that the effects of microorganisms, through polysaccharide production, were too temporary to account for the crumb structures of many soils. However, many natural soil aggregates lose stability after destruction of any carbonates and treatment with dilute periodate (Greenland *et al.* 1962*b*; Clapp and Emerson 1965) and this indicates that polysaccharides are of predominant importance in some soils. Possibly under field conditions polysaccharides are added repeatedly to the same soil crumbs and so their effectiveness increased. Aggregates from surface horizons of *red-brown earth*, a *solonized brown soil* and a *rendzina* were increasingly stable, in that order, towards periodate treatment. This reagent completely destroyed the aggregates of a *red brown earth* subsoil (Greenland *et al.*

1962*b*). Aggregates produced under long-term pasture or woodland were little affected by periodate and appear to be stabilized by other mechanisms.

For a *red brown earth*, Stefanson (1971) found that the contribution from polysaccharides to aggregate stability was near-constant throughout the year and was not diminished by cultivations, even though total aggregate stability varied. Soil structure was thoroughly reviewed very recently by Emerson (1977) and here we note only that, as a moist soil dries out, non-adsorbed soluble constituents will tend to deposit at the zones of contact between adjacent soil particles which the soil solution last occupied. If the material of the resulting interparticle 'bridge' is for any reason made less soluble, e.g. by virtue of polymerization or coprecipitation processes, the interparticle cement can grow in size and strength during successive wettings. Such repeated deposition of organic matter from solution could quite conceivably contribute to aggregate stability even without any tendency to bond specifically to mineral surfaces. It would then at least augment deposits of other constituents such as monosilicic acid (Beckwith and Reeve 1962), calcium carbonate and re-oxidized iron.

Although there has been no unequivocal demonstration of their formation, sesquioxide, silica and aluminosilicate particles appear to be capable, together with precipitated silicic acids and humic materials, of building into an enormous variety of multilayered microconcretions, interparticle cements and compound aggregates of these. A multiplicity of compositions of aggregates with these ingredients may account for the difficulty in achieving clarification of some macrophysical soil properties.

Chemical effects

The precise nature of the mineral-organic complexes in soils has not yet been fully resolved (Mortland 1970) but some carboxylic groups of the organic moieties appear to be involved (e.g. Schnitzer 1978). Also, semi-specific bonding probably occurs at the sites of attachment to minerals because anions which are not specifically adsorbed are very poor at displacing humic materials from soils. Anions such as fluoride, orthophosphate and citrate are very much better (Table 35.6). Similar results were obtained when ultraviolet absorbances at 256–270 nm were used to measure the organic matter contents of soil extracts with ammonium salts of nitrate, sulphate and phosphate (Beckwith 1973).

In addition to linkages between carboxylate groups and minerals, there appear to be reinforcing attachments through hydrogen bonds because of the large increase, noted previously, in extracted organic matter in the presence of 8 M urea. These hydrogen bonds might link an organic moiety directly to a mineral surface or to another adjacent, aligned organic moiety, or both. These linkages between organic moieties may sometimes persist during

Table 35.6. The extraction of nitrogen from soils by various sodium salt solutions

Adapted from Bremner and Lees (1949)

Extractant	Percent of total N extracted from: Burwell peat	Broadbalk soil
0.5 M sodium bromide	1.2	2.4
0.5 M sodium chloride	1.2	2.5
0.2 M sodium acetate	0.8	2.2
0.2 M sodium succinate	0.8	2.2
0.1 M sodium orthophosphate	3.8	5.6
0.5 M sodium fluoride	21.9	7.6
0.2 M sodium citrate	19.5	8.6
0.1 M sodium pyrophosphate	22.0	11.0

extraction of humic materials. The involvement of silicic acids (and perhaps of minute mineral precipitates or fragments) in these associations also seems possible as considerable silicon is commonly extracted with organic matter, and as polysilicic acids have a propensity to link to hydroxy compounds by hydrogen bonding (Iler 1955).

The carboxylate groups of the solid phase organic matter are able to retain sodium and the cation macronutrients potassium, calcium and magnesium. In acid soils the dissociation of hydrogen from carboxylic groups is decreased and thus exchange capacity is diminished (e.g. Coleman and Thomas 1967). Also, many carboxylate groups bind iron and aluminium preferentially (Schnitzer and Skinner 1963), which further limits ion exchange reactions. Thus to retain a useful cation exchange capacity of humus it is advisable to maintain soil pH above about 5, at least. Except in soils which experience quite prolonged waterlogging, this acidity should allow only limited dissolution of aluminium and iron from minerals in soils. Sufficient carboxylate groups should then be present to retain nutrient cations against leaching.

The importance of the cation exchange capacity of organic origin differs greatly even between soils of like pH; it contributes a major part of the capacities of some *krasnozems* (Nicholls and Tucker 1956), of peaty soils and of sands and sandy loams but probably accounts for less than one fourth of the capacities of some fine-textured surface soils (e.g. Brewer and Butler 1953). In most Australian subsoils, its contribution is very small; at Kybybolite the texturally similar 5–10-cm soil horizon had only about half the exchange capacity of the 0–5-cm horizon, presumably due to its smaller organic content (Russell 1961). The exchange capacities of soils are commonly less than the sum of the capacities of separated organic and clay mineral components, (Schnitzer 1965). There is considerable uncertainty in estimating these separate values (Brewer and Butler 1953) but nevertheless it seems that, particularly under acid conditions, there is loss of some exchange capacity of humus by involvement of its acidic groups in linkages to polyvalent metals, especially aluminium (Martin and Reeve 1958), to sesquioxides and to clay mineral edges.

Many trace metals, including the micronutrients Cu, Zn, Co, Fe and Mn and the toxic metals Pb, Hg, Cd and Ni, react avidly with humic material isolated from soils (Schnitzer 1978) and so also, presumably, with organic matter in soils. In so far as checks have yet been made, small amounts of these metals often cannot be exchanged easily or completely from soils. Larger recoveries may be achieved by extracting soils with solutions of metal chelating agents (pyrophosphate, acetylacetone, citrate, EDTA etc.) or with extractants which disperse organic matter (Bremner *et al.* 1946). As a further example, 1 M ammonium chloride solution recovered only 0.1 mg of 2 mg Cu added per gram to a sample of peat but subsequent extraction with 2 M NH_3 solution recovered an additional 1.45 mg Cu together with much organic matter (R. S. Beckwith unpublished). It seems that many of the carboxylic groups in organic matter have neighbouring carboxyls or hydroxyls which together can bind iron, aluminium and transition group metals in chelate compounds or bond strongly to hydrated oxide surfaces (e.g. Aleksandrova 1954; Aleksandrova *et al.* 1968; Schnitzer and Skinner 1963). In untreated soil, many such chelating sites on organic matter may be occupied by iron and aluminium atoms or utilized in bonding organic moieties to mineral particles (Emerson 1977).

The soil solution

Part of the organic matter in moist soils occurs in the soil solution and, frequently, carbon is one of the more abundant soluble elements. The molecular species which account for most of this carbon have not been identified but this material may be closely related to that in freshwater streams and lakes. The distribution of organic matter between soil and solution could be important in the control of pollution and eutrophication of our water supplies, and therefore deserves more study.

Organic solutes affect both the amounts and forms of other elements in solution. The structurally bound nutrients, N, P and S are among these elements (Gieseking 1975) but are discussed elsewhere in this volume. The importance of water solubility to mineralization of such organic-bound nutrients does not seem to have been fully evaluated. The simple, poorly soluble compound oxamide is mineralized only slowly (Smith 1977) and so very complex insoluble humus will be even more stable to enzymic breakdown.

Hodgson *et al.* (1965) showed that soluble organic soil components held transition metals such as Cu and Zn in solution as complexes. Lindsay (1972) reviewed additional research on soil Zn. Beckwith (1973) reported that, for two calcareous soils, the soluble copper was mostly anionic and was largely recoverable from 1/2.5 water extracts with an anion exchange resin.

The occurrence of soluble organic complexes can increase the amounts of transitional metals in soil solutions, especially for non-acid soils. Thus, accepting the value for peat of 1×10^{-6} (Coleman *et al.* 1956) as the general dissociation constant of Cu–(organic matter) in soils, the increase in concentration of Cu^{2+} in the soil solution of Robe calcareous sand can be estimated when a corrective fertilizer application of 2 kg Cu per hectare is made. If the field capacity of the soil is 20% and $CuSO_4$ is mixed uniformly to 5 cm, this added fertilizer would establish a concentration of Cu^{2+} in the soil solution of only about 13 parts in 10^{12}. This is more than 200-fold smaller than the minimal copper concentration required by oats in culture solutions (Piper 1942) and would be an *over* estimation of Cu^{2+} concentration if small additions are retained more avidly than the larger ones employed to calculate stability constants (Davies *et al.* 1969). The total native copper content of Robe soil is also only about 2 mg per kg and so would not add enough soluble Cu^{2+} to support healthy plant growth by passive uptake (Hodgson *et al.* 1966; Oliver and Barber 1966). The calculated Cu^{2+} concentration is also much smaller than the measured 0.02 ppm *total* Cu content of 1/2.5 water extracts of Cu-fertilized, Robe soil (Beckwith 1973). This concentration is within the range found by Hodgson *et al.* (1965) for American, Cu-sufficient soils and so is not abnormal. Apparently soluble organic matter can increase the level of total Cu in solution by two or more orders of magnitude and, in some non-acidic soils, probably has qualitatively similar effects on ferric iron, zinc (Lindsay 1972) and other nutrient metals.

Extracts containing larger concentrations of soluble organic matter also generally contain more metals (Bremner *et al.* 1946). An extract of Robe soil (1/5) with 0.2 M diammonium phosphate, rather than water, contained 0.18 ppm Cu (cf. 0.02) while its organic content was about 7-fold higher judged by absorbances at 256–313 nm (Beckwith 1973). Concentrations of total Fe and Mn were also higher in the phosphate-containing extract, which was strongly coloured by humus. Greater recovery of Cu with NH_4OH than with NH_4Cl provides a further example as NH_4OH extracts much more humic material.

The predominantly anionic Cu extracted from calcareous soils by non-acidic extractants, in the absence of added metal chelating agents, probably occurs in the soil solid phase as pre-existing Cu complexes (Beckwith 1973). The displacement of this Cu into solution increases when sodium salts of avidly sorbed anions are added to the soil suspension. Many such anions are also metal chelating agents (e.g. pyrophosphate, citrate) and, used as extractants, these can free some metals from natural complexes, and so extract more from soils than expected from their displacement of organic matter (e.g. EDTA). On the other hand, sodium hydroxide probably precipitates some of the iron which was combined with the organic matter before it was extracted with alkali.

Organic matter in the soil solution has so far been accorded little significance. It constitutes only a small part of the total in soils and has apparently been regarded as a discrete material largely unrelated to solid phase organic matter (e.g. Birch 1958; Hodgson *et al.* 1966). With the Robe soil and another calcareous soil from Dongara, Western Australia, Beckwith (1973) studied the release of ultraviolet-absorbing materials in soil suspensions. The absorbances at 256 nm to 313 nm (370 nm for nitrate-containing extracts) reflected relative concentrations of organic carbon determined by dichromate oxidation of residues from evaporated soil extracts. The ratio of absorbances *A*256/*A*313 was used to monitor suspensions for any sharp change in the nature of the organic matter extracted under different conditions (Theng *et al.* 1967) and the aqueous suspensions were kept saturated with chloroform to inhibit microbial changes to extracted organic materials (see below).

Successive extracts of the soils with a given extractant showed decreasing absorbances. For example, the third (1/5) water extract of Robe soil (72 h extraction periods) gave only about half the absorbance of the first. Extraction did not remove a finite amount of 'soluble organic matter' and leave an 'insoluble' fraction, nor was there evidence of solid organic matter with limited solubility. Rather, the amount of organic matter in solution depended on the soil/solution ratio so that extracts at a wider ratio contained more *total* organic matter, but at a smaller concentration, than extracts at a narrower ratio (Table 35.7). The amount of 'soluble' organic matter was greatly influenced by added anions. With Robe and Dongara soils the extraction efficiencies of 0.5 M $(NH_4)_2HPO_4$, 0.5 M $(NH_4)_2SO_4$ and 1.0 M NH_4NO_3 decreased in that order with absorbances in the ratios 6.43 : 3.57 : 1.0 and 6.90 : 3.45 : 1.0 respectively. Finally, as evidenced later in Table 35.8, extracted organic matter inhibited release of more organic matter from a second portion of the same soil.

In total this behaviour is characteristic of that of an adsorbed solute. Hence it can be proposed that the amount of humified material in solution in moist calcareous soils, and so of contained chelated metals, is governed by a desorption equilibrium under the prevailing conditions of moisture, temperature, anion concentrations and microbial activity. While this hypothesis might have been inferred from past discussions of organic matter (e.g. Greenland 1965), specific experimental evidence for it does not seem to have been obtained previously.

Under different experimental conditions, Jenkinson (1966) found that pretreatment of calcareous soils with chloroform vapour stimulated mineralization of carbon in inoculated soils and related this mineralized carbon to that of the biomass. Jenkinson recognized that 'more than one fraction of soil organic matter serves as substrate after partial sterilization'. The present results, obtained in the *continuous* presence of chloroform, are not readily interpretable in terms of the biomass as the predominating source of soluble, ultraviolet-absorbing compounds. In particular, the effects of soil/solution ratio, of repeated extraction and of exposing fresh soil to soil extract (Table 35.8) are not readily explicable with such a finite, major source of soluble carbon. Even if such a source accounted for most of the carbon in extracts with 1.0 M NH_4NO_3 solutions it would seem to account for only 10–15% of the larger amounts in 0.5 M $(NH_4)_2HPO_4$ extracts.

Table 35.7. Effect of soil:water ratio on the relative concentration and total organic matter in Robe soil extracts as measured by ultraviolet absorbance at 263 nm

Soil:water ratio	Relative concentration of organic matter	Total organic matter, relative to 1:1.5 extract
1:5	0.563	1.89
1:3	0.750	1.51
1:1.5	0.993	1.00

Table 35.8. Ultraviolet absorbances of 1:5 water extracts of soils before and after re-equilibration with fresh soil at ratio 1:10

Wave length (nm)	Water + soil at 1:10 ratio	Water + soil at 1:5 ratio	1:5 extract + soil at 1:10 ratio
	Robe Soil		
263	0.375	0.575	0.670
313	0.150	0.250	0.250
*A*256/*A*313	2.57	2.44	2.80
	Dongara soil		
263	0.192	0.310	0.378
313	0.084	0.150	0.176
*A*256/*A*313	2.65	2.20	2.27

The postulated release of soluble organic matter by a desorption reaction helps to explain a number of phenomena. As soluble material is metabolized under nonsterile conditions, more organic matter will come into solution from adsorbed reserves. The immediately mineralizable organic carbon would then not be just that which comes into solution during leaching (Birch 1958) or sterile extraction with water. Further, if this mode of release is general for soils, desorption would be an important step in the mineralization of organic-bound nutrients and in bringing substrates into contact with enzymes able to degrade them (Oades and Ladd 1977). Soluble organic matter would then have a significance analogous to that of soluble phosphate relative to phosphate availability. However, study of calcareous soils was facilitated by their inbuilt pH stability. Different techniques may be required to allow recognition of sorption and desorption phenomena with humic materials in more poorly-buffered, acid soils. Nevertheless, the extractability of organic matter in acid and calcareous soils seems no different with stronger reagents (Bremner and Lees 1949).

Control of Cu in solution in calcareous soils by desorption of pre-existing chelate complexes accounts for some previously unexplained aspects of Cu availability. Hodgson *et al.* (1965) supposed that soluble chelating ligands competed for metal atoms held at avid sites on solid organic and/or mineral surfaces. Such a model does not readily explain why large top-dressed applications of Cu salts should be poorly available (Piper 1942) nor why mixing Cu fertilizer throughout calcareous soils should greatly increase its availability to cereals (Beckwith 1973). From the desorption model, Cu fertilizer can come into solution as a complex only as solid phase organic matter desorbs in the Cu-enriched soil zone. *Localized* Cu fertilizer will only influence the Cu content of the small proportion of the organic matter desorbed in a restricted soil zone. By contrast, *mixed* Cu affects the Cu content of *all* organic matter desorbed within the plant root zone and this much larger amount of organic matter will solubilize more Cu. As transport of Cu to roots in soils is mostly passive (Oliver and Barber 1966) a larger total amount of Cu in the soil solution will strongly increase the uptake by plants.

Dynamics

In the text above we have stressed the relationship of soil organic matter to the ecosystem of which soil forms a part. Until the ecosystem is disturbed, approximate balance is maintained between, on one hand, accessions of organic materials from dead and living roots (Martin 1977), from litter and from residues of fauna and microorganisms and, on the other hand, losses by erosion, leaching and decomposition (Jenny 1961). The balance is a complex one because the added organic materials undergo continuous modification, and there may be loss of constituents, and interactions between different constituents, at all stages (Jenkinson 1971). The added materials have different resistances to decomposition depending on their solubility, chemical structure and complexity as well as on physical, chemical and microbiological soil environment factors. They can be stabilized by association with other soil components (Jenkinson 1977), or their breakdown can be stimulated transiently by accessions of other energy sources, causing microbial proliferation (Oades and Ladd 1977).

Changes to the ecosystem, caused by land clearing and cropping, initiate a trend towards a new organic matter balance which will depend on the nature and magnitude of those changes. Cropping and fertilization introduce particularly important new factors to the balance.

Cultivation (Rovira and Greacen 1957) and artificial diminution of soil aggregates (Crasswell and Waring 1972) stimulate decomposition of organic matter, presumably by exposing additional soil surfaces and so associated organic moieties, to microbial attack. Repeated wetting and drying of soils, the latter intensified by 'burning off' and by leaving soils bare during summer, also promote decomposition of organic matter (Birch 1958; Sauerbeck 1966; Jenkinson 1966). The effect of drying may depend in part on fracturing soil aggregates by shrinkage, and on release of fresh substrate from the killed biomass, but could also be effected through increasing the subsequent desorption of organic matter from newly exposed soil surfaces. It may be relevant that adsorbing surfaces in soils are capable of acting in this way; the amounts of monosilicic acid desorbed from soils during a few days equilibration with water can be considerably increased by preheating the soils (R. Reeve and R. S. Beckwith, unpublished data).

Fertilization of cleared land and subsequent build up of soil organic matter under legume-based pasture (Donald and Williams 1954) can be viewed, in general terms, as a reversal of the prolonged weathering process which depleted the soil nutrient supplies and thus reduced the plant productivity of the ecosystem. Compared to the return of litter from dry, sclerophyll forest (Lee and Correll 1978) annual pasture producing 8000 kg dry matter per hectare adds three times as much organic matter per year to the system. The ratio would almost certainly be larger if returns from roots were included.

The critical importance of plant inputs was illustrated by Wetselaar's (1967) results at Katherine. A soil under Townsville lucerne pasture lost N when the pasture was harvested but gained nitrogen if the pasture was grazed. Carbon correlated closely with soil nitrogen in another experiment at this site (Wetselaar and Norman 1960).

In all producing ecosystems very recent biological products coexist with aged, humified materials. As contents leak from recently dead cells, reactions can occur between previously isolated constituents and between these and soil components. Extracellular enzymes appear to be responsible for breakdown of humus and of polymeric plant constituents, as these molecules are too large to enter microbial cells (Ladd and Butler 1975). Such breakdown requires an appropriate enzyme to contact a poorly mobile organic substrate. But the former may only survive in soil if stabilized by bonding to a clay particle or organic soil component and this association will largely immobilize the enzyme (Burns *et al.* 1972). Chance contacts leading to enzymic breakdown of soil organic ingredients might occur in solution or at a soil surface. In either case the opportunities for enzymic breakdown of organic matter will be increased when conditions favour proliferation of the microflora. Effects of temperature, moisture, manuring and fertilization generally support this supposition. Enzymic breakdown may also be favoured by regular wetting and drying of soils. During drying, reactants could be deposited on surfaces in close proximity, thus favouring their interaction with each other or with a third, mobile ingredient.

Leaching of organic matter has not been much considered although its reality is established by the occurrence of illuvial organic B horizons in profiles and by the visible presence of organic matter in stream and drainage waters. A removal rate of 50 kg carbon per hectare would result from 25 cm drainage water containing 20 ppm carbon. This may seem trivial but the lost organic material, if predominantly stable humified matter, would require a much larger input of fresh residues to balance its removal (Jenkinson 1965).

The selective removal down profiles of the lower molecular constituents of humus was suggested by some results of Beckwith (1973). During studies of two calcareous soils, a fresh aliquot of soil was equilibrated at 1:10 ratio with a water extract of the same soil previously obtained at 1:5 ratio (Table 35.8). Compared with a fresh 1:10 water extract of the soil (Column 2) and with a fresh 1:5 extract (Column 3), the 1:5 extract re-equilibrated with fresh soil at 1:10 (Column 4) contained less total organic matter as estimated by respective ultraviolet absorbances. This can readily be seen by comparing the sum of absorbances in Columns 2 and 3 with that in Column 4, Table 35.8. Re-equilibration of the 1/5 extract with fresh soil increased the *A*256/*A*313 ratio of the supernatant. This ratio increases with the steepness of slope of the ultraviolet absorbance curve, which is larger for more readily extractable organic matter (Theng *et al.* 1967). Among polymers with similar chemical characteristics, smaller molecular species would be expected to be more readily extractable (Greenland *et al.* 1962).

If 'polyanion exchange', such as suggested by the above results, occurs in acid free-draining soils it would result in the selective leaching of smaller organic moieties from A horizons. Subsoil organic matter, apart from that derived from organisms and roots in that horizon, should then contain a larger proportion of smaller moieties. Drainage and runoff waters should also contain a higher proportion of low molecular humic materials than the surface soils of the catchment whence they originated.

Organic matter and soil fertility

Soil fertility integrates the effects on plant growth of all relevant physical, chemical and biological soil properties. As organic compounds strongly influence all three groups of soil properties, their importance to soil fertility is profound. Only some major aspects can be mentioned here but it should be observed, in advance, that additional organic matter does not necessarily benefit a soil, physically or chemically.

Physical properties

Organic constituents appear able to improve soil structure by acting as building blocks in soil aggregates and by stabilizing coalesced mineral particles (Giovannini and Sequi 1976). Polysacchardides have an identified role, as described above.

Soil aeration and water economy are both affected by soil structure (Butler and Hubble 1977). Soils of good structure not only absorb rain readily with minimal runoff, but also mulch themselves when dry, thus slowing moisture loss. Hence through conservation the proportion of a given application of water which is utilized by plants can be significantly increased in soils with larger organic matter contents. Since organic matter holds water very tightly, it must be increased considerably before available water capacity increases appreciably. Thus Russell and Shearer (1964) found that the available water capacity of a solonetzic soil under various pasture treatments was strongly related to the organic matter content, but an increase of 157% in soil nitrogen increased available water capacity by only 10–18%.

Interactions between soil physical conditions, water supply and nutrient uptake seem to have received less than their due attention considering the shortage of water in most Australian crop lands (Beckwith 1963). Many such interactions seem likely, although unproven. For example, improved aeration, brought about by increased organic matter, might aid uptake of nutrients in several ways. In some soils root proliferation would be encouraged and a larger volume of soil would be explored. In some soils mineralization of N, S and P might proceed faster if aeration was improved.

A problem caused by organic matter is water repellence in sandy-surfaced soils in southern Australia (Bond 1969). This organic matter coating the sand grains appears to be 'stable humic-like material' produced by fungi or plants (Roberts and Carbon 1972). Physical benefits conferred by additional organic matter may have a price as, for example, loss of availability of nutrients retained in resistant organic residues or build up of phytotoxins (Kimber 1973).

Nutrient supplies

The major soil reserves of nitrogen, sulphur and often phosphorus are present in organic matter. In very general terms, deficiencies of these nutrients may occur either because the total organic content is insufficient or because its decomposition is too slow to meet plant demands. The former aspect is discussed below and the latter in other chapters.

Besides reserves of N, S and P, organic matter provides cation exchange sites, as mentioned previously. This has been documented in several instances for the organic matter built up under temperate, subclover-based pastures (e.g. Williams and Donald 1957; Russell 1960). Data from Barrow (1969) are shown in Table 35.9 and similar organic matter increases have now been found with tropical and sub-tropical leguminous pastures (Henzell *et al.* 1966). Despite all the soil improvements accompanying such humus buildup – N, S and P contents and cation retention capacity all increased – sandy soils remained susceptible to K, Mg and Ca deficiencies if cropped or used for hay production

Table 35.9. Some properties of two samples of Coolup sand selected for detailed study
Reproduced with permission from Barrow (1969)

	Soil 5		Soil 8	
	Virgin	33-year pasture	Virgin	40-year pasture
Total carbon (%)	0.77	2.03	0.84	2.20
Total nitrogen (%)	0.038	0.122	0.038	0.158
Total sulphur (%)	0.005	0.018	0.006	0.022
Water at 0.1 bar (%)	10.8	15.0	10.7	15.5
Water at 15 bar (%)	1.7	3.2	1.7	3.0
Organic phosphorus (ppm)	14	44	14	80
Phosphorus soluble in 0.1 M sulphuric acid (ppm)	6	106	5	54
Exchangeable cations (mequiv. per 100 g)				
Calcium	0.72	1.68	0.62	1.38
Magnesium	0.38	0.25	0.31	0.31
Sodium	0.16	0.33	0.42	0.42
Potassium	0.11	0.15	0.10	0.14
Exchange acidity (mequiv per 100 g)	2.72	5.87	3.11	6.59
Cation exchange capacity (mequiv per 100 g)	4.09	8.28	4.56	8.84

without addition of fertilizers. Also, the increased acidity, due to organic matter accumulation in sandy soils under pasture, increased the availabilities of some trace metals (Williams and David 1976) and so toxicities could occur over a long term.

The mobility, and hence the availability of some metals to plants, can be increased in neutral and alkaline soils by the organic matter in solution (Lindsay 1972). By contrast the solid phase organic matter without soluble organics adsorbs metals avidly under these pH conditions (Schnitzer and Skinner 1963; Schnitzer 1978). Some organic moieties may carry chelated metals from the biomass during their incorporation into humus polymers but there is no evidence for this.

The occurrence of some micronutrient metals in calcareous soils as adsorbed organic, anionic complexes was discussed previously. Ferric iron and Cu in particular are believed to occur in such forms in many soils. If so, a proportion of these metal-containing moieties seem always to be accessible to water and their desorption and migration in response to changes in the soil solution appears to allow continuing transfer of metals to plant roots even in neutral and alkaline soils. This mechanism apparently suffices to supply plants, even though complex metal ions, such as $(Cu\ EDTA)^{2-}$, may be less readily absorbed from solution by roots than metal cations (Loneragan 1975).

Uptake by plants of Zn, Mn and Co may involve similar mechanisms, at least in part in some soils. These metals, like Fe, more frequently persist in inert minerals (Norrish 1975), so that their total amounts in soils are often a quite misleading guide to plant supplies. The proportions of these metals bound organically in different soils have not been determined.

Soil organic constituents affect availability of phosphate, and probably other anionic nutrients, in yet another way. These nutrients are all minor constituents of soils but, even when added in excess, are not very effective in displacing humic materials from soils. Phosphate has been most studied and less is adsorbed by soils in the presence of soluble humic materials (Leaver and Russell 1957). Native soil organic matter presumably occupies the sorption sites which would otherwise compete with roots for this anion, and probably others. Beckwith (1965) noted the much larger phosphate sorption capacity, at constant supernatant concentration, of a krasnozemic subsoil whose mineralogical composition differed little from that of the overlying surface soil. Organic matter in the latter apparently blocked many phosphate sorption sites. It seems that because of their larger contents of organic matter, surface soils will usually contain the largest reactive concentrations of most nutrients and also provide the zone in the profile in which mineral surfaces compete least avidly for anionic nutrients.

In waterlogged soils, organic substrates may be metabolized more slowly and to different products depending on the duration of submergence (Allison 1973). Thus in paddy soils the availabilities of many nutrients can be affected (Ponnamperuma 1972). In particular, organic matter usually promotes the reduction of iron and manganese oxides, increasing the availabilities of these metals. Under appropriate prolonged anaerobic conditions, one can speculate that the sulphides of Fe and Mn may assume the role played by their oxides in aerobic soils i.e. sulphide surfaces may reversibly sorb a range of ions including plant nutrients in both inorganic and organically complexed forms. If so, the effects of organic matter in paddy soils are likely to be as many and as complex as in well aerated soils.

From the viewpoints of maintaining soil nutrient reserves, especially of N, S and P, and of providing good conditions for water entry, root proliferation and nutrient uptake, prevention of loss by erosion of the organic-rich surface soil horizon is generally a first priority for maintaining soil fertility.

Management of soil organic matter

Apart from erosion prevention there is need to balance the level of organic matter built up and maintained in soils. Use is made of the effect of cultivation and cropping on organic matter to balance that of the pasture phase (Chapter 37). The aim should be to avoid both excessive waste of fertilizer residues in stable humus and serious deterioration of soil physical conditions from which can result erosion losses and poor conditions for root growth. The extension of such management practices to subtropical and tropical environments in Australia has been a major endeavour during the last 30 years. Its success waited on the introduction and breeding of appropriate pasture legumes with effective root nodulation.

Ladd and Russell (Chapter 37) have summarized estimates of nitrogen and carbon accumulated in various soils under legume-based pastures and Williams (1970) calculated that the monetary value of equivalent fertilizer nitrogen exceeded $A250 million at that time. On permanent pasture lands much of this accumulated nitrogen remains unused by plants.

However, there have been some instances where soil nitrogen and carbon were not measurably increased under pastures (e.g. Wetselaar 1967; Myers 1976; Kohn *et al.* 1977) so there is need for some caution. At the other extreme, Barrow (1969) found a near 3-fold increase of organic matter, during 40 years, in two sandy soils in Western Australia (Table 35.9). The difference made at Katherine by harvesting and removing pasture, rather than grazing it (Wetselaar 1967) was noted previously and can complicate measurement of the accumulation of nitrogen and humus in soils. Clearly a number of factors will influence the ease and extent of organic build-up under pastures including temperature and climatic environment (Jenny 1961), the soil mineral characteristics (Jackman 1960), the productivity of the biomass under the previous eco-system and the extent to which it is increased under pasture.

Under continuous cropping the run down of organic matter may continue for many years before a new soil equilibrium is approached (Clarke and Russell 1977). The new, depleted, equilibrium levels of nitrogen and carbon attained in cropped soils in the absence of fertilizer usually have only academic interest. However, at any time, the amount of soil nitrogen above this depleted equilibrium level represents the maximum which can be recovered by crops.

Because of the relatively short period of use and study of pasture improvement, optimal levels of organic build-up remain to be established for many Australian soils and climates. It is the aim that, after pastureland is cultivated, sufficient nitrogen will be recoverable by one or two crops to provide them with near optimal nitrogen supply without nitrogen fertilizers. There are large land areas which were developed after 1946 and have been since used continuously for improved pasture. Much biologically fixed nitrogen has now accumulated in many of these and its recovery for crops, together with phosphorus and sulphur, will become increasingly attractive.

Cropping by broadacre techniques would not be possible on some such lands (e.g. hilly country with and without rock outcrops) nor perhaps worthwhile in some climates. However, more attention could be given to utilizing the nutrient reserves of others, especially those which can be provided with an assured water supply and which have favourable aspect close to population centres. Where cultivation alone fails to achieve rapid mineralization of nutrients, the use of lime on acidic soils, and other fertilizer amendments, deserves closer consideration (e.g. Barrow 1965). Even where yields are not extremely high, cropping such lands is likely to become increasingly worthwhile in future because fertilizer and transport costs will tend to make produce from distant centres steadily less competitive.

A changed farming system will eventually lead to a different soil organic content. Such longer-term changes, and their possible implications for fertilizer requirements and other cropping costs, need to be carefully considered. A topical example is the suggested adoption of 'chemical ploughing' and other reduced tillage techniques. Increased build up of organic nitrogen, phosphorus and sulphur under such systems could conceivably result in the need for larger fertilizer inputs for crops. If so, the proper economic assessment of such a basic change to farming will need considerable time.

References

Acton, C. J., Paul, E. A. and Rennie, D. A. (1963). Measurement of the poly-saccharide content of soils. *Can. J. Soil Sci.* **43**, 141-50.

Aleksandrova, L. N. (1954). The nature and properties of the interaction of humic acids and humates with sesquioxides. Pochvovedenie (1954). No. 1. 14-29. Abstract. *Soils Fert.* **17**, 243.

Aleksandrova, L. N., Arshavskay, T. Th., Dorfman, E. M., Lyuzin, M. F. and Yurlova, O. V. (1968). Humus acids and their organo-mineral derivatives in soil. Trans. 9th Congr. Int. Soc. Soil Sci. Vol. 3, pp. 143-52.

Allison, F. E. (1973). 'Soil Organic Matter and its Role in Crop Production.' (Elsevier: Amsterdam.)

Anon. (1976). Land application of waste materials. Soil Conservation Society of America. Proceedings of National Conference, Aukeny, Iowa, 1976.

Barker, S. A., Finch, P., Hayes, M. H. B., Simmonds, R. G. and Stacey, M. (965). Isolation and preliminary characterization of soil poly-saccharides. *Nature (London)* **205**, 68-9.

Barrow, N. J. (1965). Further investigations of the use of lime on established pastures. *Aust. J. Exp. Agric. Anim. Husb.* **5**, 442-9.

Barrow, N. J. (1969). The accumulation of soil organic matter under pasture and its effect on soil properties. *Aust. J. Exp. Agric. Anim. Husb.* **9**, 437-44.

Beckwith, R. S. (1959). Titration curves of soil organic matter. *Nature (London)* **184**, 745.

Beckwith, R. S., (1963). Chemical extraction of nutrients in soils and uptake by plants. *Agrochimica* **7**, 296-313.

Beckwith, R. S. (1965). Sorbed phosphate at standard supernatant concentration as an estimate of the phosphate needs of soils. *Aust. J. Exp. Agric. Anim. Husb.* **5**, 52-8.

Beckwith, R. S. (1973). The availability of micronutrient metals to plants. Ph.D. Thesis, University of Western Aust. 1973.

Beckwith, R. S. and Reeve, R. (1962). Effects of added silicic acids on the physical properties of soils. 3rd Aust. Conf. Soil Science, Canberra 1962. Paper No. 110.
Beckwith, R. S., Haydock, K. P., Little, I. P. and Reeve, R. (1962). CSIRO Aust. Div. Soils Div. Rep. No. 2/62.
Birch, H. F. (1958). The effect of soil drying on humus decomposition and nitrogen availability. *Plant Soil* **12**, 81-96.
Bond, R. D. (1969). Factors responsible for water repellence of soils. *In* 'Water-Repellent Soils.' (Eds L. F. Debano and J. Letey). (University of California: Riverside, Calif.)
Braids, O. C. and Miller, R. H. (1975). Fats, waxes, and resins in soil. *In* 'Soil Components.' (Ed. J. E. Gieseking). Vol. 1. (Springer-Verlag: Berlin.)
Bremner, J. M. and Lees, H. (1949). Studies on soil organic matter. Part 2. The extraction of organic matter from soil by neutral reagents. *J. Agric. Sci.* **39**, 274-9.
Bremner, J. M., Heintze, S. G., Mann, P. J. G. and Lees, H. (1946). Metallo-organic complexes in soils. *Nature (London)* **158**, 790.
Brewer, R. and Butler, B. E. (1953). Pedology of the Upper Hunter Valley, N.S.W. CSIRO Aust. Soil Publ. No. 2.
Brisbane, P. G., Amato, M. and Ladd, J. N. (1972). Gas chromatographic analysis of amino acids from the action of proteolytic enzymes on soil humic acids. *Soil Biol. Biochem.* **4**, 51-61.
Burns, R. G. (1978). Enzyme activity in soil: Some theoretical and practical considerations. *In* 'Soil Enzymes.' (Ed. R. G. Burns). (Academic Press: New York.)
Burns, R. G., El-sayed, M. H. and McLaren, A. D. (1972). Extraction of an urease-active organo-complex from soil. *Soil Biol. Biochem.* **4**, 107-8.
Butler, B. E. and Hubble, G. D. (1977). Morphologic properties. *In* 'Soil Factors in Crop Production in a Semi-arid Environment.' (Eds J. S. Russell and E. L. Greacen). (Univ Queensland Press: St Lucia.)
Butler, J. H. A. and Ladd, J. N. (1969). Effect of extractant and molecular size on the optical and chemical properties of soil humic acids. *Aust. J. Soil Res.* **7**, 229-39.
Butler, J. H. A. and Ladd, J. N. (1971). Importance of the molecular weight of humic and fulvic acids in determining their effects on protease activity. *Soil Biol. Biochem.* **3**, 249-57.
Butler, J. H. A., Downing, D. T. and Swaby, R. J. (1964). Isolation of a chlorinated pigment from green soil. *Aust. J. Chem.* **17**, 817-9.
Cameron, D. W. and Sidell, M. D. (1978). 1,3,6,8,11,13-Hexachloro-4, 10-dihydroxydinaphtho(2,1-6:1'2'-*d*)furan-5,9-dione. A polychloroquinone from green soils. *Aust. J. Chem.* **31**, 1323-33.
Cameron, R. S., Thornton, B. K., Swift, R. S. and Posner, A. M. (1972). Molecular weight and shape of humic acid from sedimentation and diffusion measurements on fractionated extracts. *J. Soil Sci.* **23**, 394-408.
Cameron, R. S. and Posner, A. M. (1974). Molecular weight distribution of humic acid from density gradient ultra-centrifugation profiles corrected for diffusion. Trans. 10th Congr. Int. Soil Sci. Soc. Moscow. Vol. 2, pp. 325-31.
Clapp, C. E. and Emerson, W. W. (1965). The effect of periodate oxidation on the strength of soil crumbs. *Proc. Soil Sci. Soc. Am.* **29**, 127-30.
Clarke, A. L. and Russell, J. S. (1977). Crop sequential practices. *In* 'Soil Factors in Crop Production in a Semi-arid Environment'. (Eds J. S. Russell and E. L. Greacen). (Univ. Queensland Press: St Lucia.)
Coleman, N. T., McLung, A. C. and Moore, D. P. (1956). Formation constants for Cu^{2+}— peat complexes. *Science* **123**, 330-1.
Coleman, N. T. and Thomas, G. W. (1967). The basic chemistry of soil acidity. *In* 'Soil Acidity and Liming.' (Eds R. W. Pearson and F. Adams). Amer. Soc. Monograph No. 12. (Madison: Wisconsin.)
Craswell, E. T. and Waring, S. A. (1972). Effect of grinding on the decomposition of soil organic matter. I. The mineralization of organic nitrogen in relation to soil type. *Soil Biol. Biochem.* **4**, 427-33.
Davies, R. I., Cheshire, M. V. and Graham-Bryce, I. J. (1969). Retention of levels of copper by humic acid. *J. Soil Sci.* **20**, 65-71.
Donald, C. M. and Williams, C. H. (1954). Fertility and productivity of a podzolic soil as influenced by subterranean clover and superphosphate. *Aust. J. Agric. Res.* **5**, 664-87.
Dubach, P. and Mehta, N. C. (1963). The chemistry of soil humic substances. *Soils Fert.* **26**, 293-300.
Dubach, P., Mehta, N. and Deuel, H. (1961). Extracktion von Huminstroffen aus dem B-Horizont eines Podsols mit Athylendiamintetraessigsaure. *Z. PflErnähr. Dung* **95**, 119.
Emerson, W. W. (1956). A comparison between the mode of action of organic matter and synthetic polymers in stabilising soil crumbs. *J. Agric. Sci.* **47**, 350-3.
Emerson, W. W. (1977). Physical properties and structure. *In* 'Soil Factors in Crop Production in a Semi-Arid Environment.' (Eds J. S. Russell and E. L. Greacen). (Univ. Queensland Press: St Lucia.)
Flaig, W. (1967). Veranderungen amlignin und Anlagerung von stickstoffhaltigen Verburdungen zu Beginn der Bildung und bei der Nutzung von Torf. *Landbauforsch Volkenrode* **17**, 1-16.
Flaig, W., Beutelspatcher, H. and Rietz, E. (1975). Chemical composition and physical properties of humic substances. In 'Soil Components.' (Ed. J. E. Gieseking). Vol. 1. (Springer Verlag: Berlin.)
Ford, G. W., Greenland, D. J. and Oades, J. M. (1969). Separation of the light fraction from soils by ultrasonic dispersion in halogenated hydrocarbons containing a surfactant. *J. Soil Sci.* **20**, 291-6.
Gieseking, J. E. (Ed.) (1975). 'Soil Components'. Vol. 1. (Springer-Verlag: Berlin.)
Giovannini, G. and Sequi, P. (1976). Iron and aluminium as cementing substances of soil aggregates. II. Changes in stability of soil aggregates following extraction of iron and aluminium by acetylacetone in a non polar solvent. *J. Soil Sci.* **27**, 148-53.
Goh , K. M. and Reid, M. R. (1975). Molecular weight distribution of soil organic matter as affected by acid pretreatment and fractionation into humic and fulvic acids. *J. Soil Sci.* **26**, 207-22.
Gottlieb, S. and Hendricks, S. B. (1945). Soil organic matter as related to newer concepts of lignin chemistry. *Proc. Soil Sci. Soc. Am.* **10**, 117-25.
Greenland, D. J. (1965). Interaction between clays and organic compounds in soils. II. Absorption of soil organic compounds and its effect on soil properties. *Soils Fert.* **28**, 521-32.
Greenland, D. J. (1971). Interactions between humic and fulvic acids and clays. *Soil Sci.* **111**, 34-41.
Greenland, D. J., Laby, R. H. and Quirk, J. P. (1962*a*). Adsorption of glycerine and its di-, tri- and tetra-peptides by montmorillonite. *Trans. Faraday Soc.* **58**, 829-41.
Greenland, D. J., Lindstrom, G. R. and Quirk, J. P. (1962*b*). Organic materials which stabilize natural soil aggregates. *Proc. Soil Sci. Soc. Am.* **26**, 366-71.
Greenland, D. J. and Oades, J. M. (1975). Saccharides. *In* 'Soil Components.' Vol. 1. (Ed. J. E. Gieseking). (Springer-Verlag: Berlin.)
Haider, K. and Martin, J. P. (1967). Synthesis and transformation of phenolic compounds by *Epicoccum nigrum* in relation to humic acid formation. *Proc. Soil Sci. Soc. Am.* **31**, 766-72.
Hale, M. G., Foy, C. L. and Shay, F. J. (1971). Factors affecting root exudation. Adv. Agron. **23**, 89-109.
Hayes, M. H. B., Swift, R. S., Wardle, R. E. and Brown, J. K. (1975). Humic materials from an organic soil. A comparison of extractants and of properties of extracts. *Geoderma* **13**, 231-45.
Henzell, E. F., Fergus, I. F. and Martin, A. E. (1966). Accumulation of soil N and C under a *Desmodium uncinatum* pasture. *Aust. J. Exp. Agric. Anim. Husb.* **6**, 157-60.
Hodgson, J. F., Geering, H. R. and Norvell, W. A. (1965). Micronutrient cation complexes in soil solution. *Proc. Soil Sci. Soc. Am.* **29**, 665-9.
Hodgson, J. F., Lindsay, W. L. and Trierweiler, J. F. (1966). Micronutrient cation complexing in soil solution. 2. Complexing of zinc and copper in displaced solution from calcareous soils. *Proc. Soil Sci. Soc. Am.* **30**, 468-71.

Iler, R. K. (1955). 'The Colloid Chemistry of Silica and Silicates.' (Cornell University Press: Ithaca, New York.)

Jackman, R. H. (1960). Organic matter stability and nutrient availability in Taupo pumice. *N.Z. J. Agric. Res.* **3**, 6-23.

Jackson, M. P., Swift, R. S., Posner, A. M. and Knox, J. R. (1972). Phenolic degradation of humic acid. *Soil Sci.* **114**, 75-8.

Jenkinson, D. S. (1965). Studies on the decomposition of plant material in soil. I. Losses of carbon from ^{14}C labelled ryegrass incubated with soil in the field. *J. Soil Sci.* **16**, 104-15.

Jenkinson, D. S. (1966). Studies on the decomposition of plant material in soil. II. Partial sterilization of soil and the soil biomass. *J. Soil Sci.* **17**, 280-302.

Jenkinson, D. S. (1971). Studies of the decomposition of ^{14}C-labelled organic matter in soil. *Soil Sci.* **111**, 64-70.

Jenkinson, D. S. (1977). Studies on the decomposition of plant material in soil. V. The effects of plant cover and soil type on the loss of carbon from ^{14}C-labelled ryegrass decomposing under field conditions. *J. Soil Sci.* **28**, 424-34.

Jenny, H. (1941). 'Factors of Soil Formation.' (McGraw-Hill Book Co. Inc.: New York.)

Jenny, H. (1961). Comparison of soil nitrogen and carbon in tropical and temperate regions. Missouri Agric. Exp. Stn. Res. Bull. No. 765, pp. 5-31.

Kimber, R. W. L. (1973). Phytotoxicity from plant residues. III. The relative effect of toxins and nitrogen immobilization on the germination and growth of wheat. *Plant Soil* **38**, 543-5.

Kimber, R. W. L. and Searle, P. L. (1970). Pyrolysis gas chromatograhy of soil organic matter. 2. The effect of extractant and soil history on the yields of products from pyrolysis of humic acids. *Geoderma* **4**, 57-71.

Kiss, S., Dragan-Bulada, M. and Radulescu, D. (1975). Biological significance of enzymes accumulated in soils. *Adv. Agron.* **27**, 25-87.

Kohn, G. D., Osborne, G. J., Batten, G. D., Smith, A. N. and Lill, W. J. (1977). The effect of top-dressed superphosphate on changes in nitrogen: carbon: sulphur: phosphorus and pH on a red earth soil during a long term grazing experiment. *Aust. J. Soil Res.* **15**, 147-58.

Kononova, M. M. and Bel'chikova, N. P. (1961). Quick methods of determining the humus composition of mineral soils. Soviet Soil Science Eng. Trans. p. 1112.

Kumada, K. and Kawamura, Y. (1968). On the fractionation of humic acids by a fractional precipitation technique. *Soil Sci. Plant Nutr.* **14**, 198-500.

Ladd, J. N. (1969). The extinction coefficients of soil humic acids fractionated by "Sephadex" gel filtration. *Soil Sci.* **107**, 303-6.

Ladd, J. N. (1978). Origin and range of enzymes in soil. *In* 'Soil Enzymes.' (Ed. R. G. Burns). (Academic Press: New York.)

Ladd, J. N. and Brisbane, P. G. (1967). Release of amino acids from soil humic acids by proteolytic enzymes. *Aust. J. Soil Res.* **5**, 161-71.

Ladd, J. N., Brisbane, P. G. and Butler, J. H. A. (1968). The susceptibility of nitrogenous components of humic acids to enzyme attack – inhibition of pronase activity. Trans. 9th Congr. Int. Soc. Soil Sci. Adelaide. Vol. 3, pp. 319-27.

Ladd, J. N. and Butler, J. H. A. (1969*a*). Inhibitory effect of soil humic compounds on the proteolytic enzyme pronase. *Aust. J. Soil Res.* **7**, 241-51.

Ladd, J. N. and Butler, J. H. A. (1969*b*). Inhibition and stimulation of proteolytic enzyme activities by soil humic acids. *Aust. J. Soil Res.* **7**, 253-61.

Ladd, J. N. and Butler, J. H. A. (1970). The effect of inorganic cations on the inhibition and stimulation of protease activity by soil humic acids. *Soil Biol. Biochem.* **2**, 33-40.

Ladd, J. N. and Butler, J. H. A. (1975). Humus-enzyme systems and synthetic, organic polymer-enzyme analogs. *In* 'Soil Biochemistry 4.' (Eds E. A. Paul and A. D. McLaren). (Dekker: New York.)

Leaver, J. P. and Russell, E. W. (1957). The reaction between phosphate and phosphate-fixing soils. *J. Soil Sci.* **8**, 113-26.

Lee, K. E. and Correll, R. L. (1978). Litterfall and its relationship to nutrient cycling in a South Australian dry sclerophyll forest. *Aust. J. Ecol.* **3**, 243-52.

Lindsay, W. L. (1972). Zinc in soils and plant nutrition. *Adv. Agron.* **24**, 147-86.

Lineman, D. J. (1977). Growth promoting properties of soil fulvic acid and a synthetic polycarboxylic acid. *Soil Biol. Biochem.* **9**, 427-8.

Lindquist, I. (1967). Adsorption effects in gel filtration of humic acid. *Acta Chem. Scand.* **21**, 2564-6.

Lloyd, A. B. (1975). Extraction of urease from soil. *Soil Biol. Biochem.* **7**, 357-8.

Loehr, R. C. (Ed.) (1977). 'Land as a Waste Management Alternative.' (Ann Arbor Science Publishers Inc.: Ann Arbor.)

Loneragan, J. F. (1975). The availability and adsorption of trace elements in soil-plant systems and their relation to movement and concentration of trace elements in plants. *In* 'Trace Elements in Soil-Plant-Animal Systems.' (Eds D. J. D. Nicholas and A. R. Egan). (Academic Press Inc.: New York.)

Lowe, L. E. (1978). Carbohydrates in soil. *In* 'Soil Organic Matter.' (Eds M. Schnitzer and S. U. Khan). (Elsevier: Amsterdam.)

Martin, A. E. and Reeve, R. (1957). Chemical studies of podzolic illuvial horizons. 2. The use of acetylacetone as extractant of translocated organic matter. *J. Soil Sci.* **8**, 279-86.

Martin, A. E. and Reeve, R. (1958). Chemical studies of podzolic illuvial horizons. 3. Titration curves of organic matter suspensions. *J. Soil Sci.* **9**, 1-7.

Martin, J. K. (1977). The chemical nature of the carbon-14-labelled organic matter released into soil from growing wheat roots. *In* 'Soil Organic Matter Studies.' Vol. 1. (International Atomic Energy Agency: Vienna.)

Morrison, R. I. (1969). Soil lipids. *In* 'Organic Geochemistry.' (Eds G. Eglington and M. T. J. Murphy). (Springer-Verlag: New York.)

Mortland, M. M. (1970). Clay-organic complexes and interactions. *Adv. Agron.* **22**, 75-117.

Myers, R. J. K. (1976). Nitrogen accretion and other soil changes in Tindall clay loam under Townsville stylo/grass pastures. *Aust. J. Exp. Agric. Anim. Husb.* **16**, 94-8.

Nicholls, K. D. and Tucker, B. M. (1956). Pedology and chemistry of the basaltic soil of the Lismore district, N.S.W. CSIRO Aust. Soil Publ. No. 7.

Norrish, K. (1975). Geochemistry and mineralogy of trace elements. *In* 'Trace Elements in Soil-Plant-Animal Systems.' (Eds D. J. D. Nicholas and A. R. Egan). (Academic Press Inc.: New York.)

Oades, J. M. and Ladd, J. N. (1977). Biochemical properties: Carbon and nitrogen metabolism. *In* 'Soil Factors in Crop Production in a Semi-Arid Environment.' (Eds J. S. Russell and E. L. Greacen). (Univ. Queensland Press: St Lucia.)

Oliver, S. and Barber, S. A. (1966). Mechanisms for the movement of Mn, Fe, B, Cu, Zn, Al and Sr from one soil to the surface of soybean roots. *Proc. Soil Sci. Soc. Am.* **30**, 468-71.

Piper, C. S. (1942). Investigations on copper deficiency in plants. *J. Agric. Sci.* **32**, 143-78.

Piper, T. J. and Posner, A. M. (1972*a*). Sodium amalgam reduction of humic acid. I. Evaluation of the method. *Soil Biol. Biochem.* **4**, 513-23.

Piper, T. J. and Posner, A. M. (1972*b*). Sodium amalgam reduction of humic acid. II. Application of the method. *Soil Biol. Biochem.* **4**, 525-31.

Ponnamperuma, F. N. (1972). The chemistry of submerged soil. *Adv. Agron.* **24**, 29-96.

Posner, A. M. (1963). Importance of electrolyte in the determination of molecular weights by 'Sephadex' gel filtration with especial reference to humic acid. *Nature (London)* **198**, 1161-3.

Posner, A. M. (1964). Titration curves of humic acid. Trans. 8th Congr. Int. Soil Sci. Bucharest. Vol. 3, pp. 161-73.

Posner, A. M. (1966). The humic acids extracted by various reagents from a soil. Part I. Yield, inorganic components, and titration curves. *J. Soil Sci.* **17**, 65-78.

Posner, A. M., Theng, B. K. G. and Wake, J. R. H. (1968). The extraction of soil organic matter in relation to humification. Trans. 9th Congr. Int. Soc. Soil Sci., Adelaide, Vol. 3, pp. 153-62.

Roberts, R. J. and Carbon, B. A. (1972). Water repellence in sandy soils of South-western Australia. II. Some chemical characteristics of hydrophobic skins. *Aust. J. Soil Res.* **10**, 35-42.

Roulet, N., Mehta, N. C., Dubach, P. and Deuel, H. (1963). Abtrennung von Kohlehydraten und Stickstoffverbindungen aus Huminstoffen durch Gelfiltration und Ionenaustausch-Chromatographie. *Z. Pflanzenernaehrung Bodenkd.* **103,** 1-9.

Rovira, A. D. and Greacen, E. L. (1957). The effect of aggregate disruption on the activity of microorganisms in the soil. *Aust. J. Agric. Res.* **8**, 659-73.

Russell, E. W. (1973). 'Soil Conditions and Plant Growth.' 10th Ed. (Longmans Group: London.)

Russell, J. D. and Anderson, H. A. (1977*a*). Comment on 'Contamination of humic acid by silica gel and sodium bicarbonate' by K. H. Tan. *Plant Soil* **47**, 263-4.

Russell, J. D. and Anderson, H. A. (1977*b*). Comment on 'Spectroscopie infra-rouge de quelques fraction d'acides humique obtennes sur Sephadex'. *Plant Soil* **48**, 547-8.

Russell, J. S. (1960). Soil fertility changes in long term experimental plots at Kybybolite, South Australia. I. Changes in pH, total N, organic carbon and bulk density. *Aust. J. Agric. Res.* **12**, 902-26.

Russell, J. S. (1961). *Ibid.* 3. Changes in cation exchange capacity and exchangeable cations. *Aust. J. Agric. Res.* **12**, 273-85.

Russell, J. S. and Shearer, R. C. (1964). *Ibid.* 4. Changes in certain moisture characteristics. *Aust. J. Agric. Res.* **15**, 91-9.

Sauerbeck, D. (1966). A critical evaluation of incubation experiments on the priming effect on green manure. *In* 'The Use of Isotopes in Soil Organic Matter Studies.' p. 209. (FAO/IAEA Tech. Meeting). (Pergamon: New York.)

Schnitzer, M. (1965). Contribution of organic matter to the cation exchange capacity of soils. *Nature (London)* **207**, 667-8.

Schnitzer, M. (1978). Humic substances: Chemistry and reactions. *In* 'Soil Organic Matter.' (Eds M. Schnitzer and S. U. Khan). (Elsevier: New York.)

Schnitzer, M. and Khan, S. V. (1972). 'Humic Substances in the Environment.' (M. Dekker: New York.)

Schnitzer, M. and Poapst, P. A. (1967). Effects of a soil humic compound on root initiation. *Nature (London)* **213**, 598-9.

Schnitzer, M. and Skinner, S. I. M. (1963). Organo-metallic interactions in soils. 2. Reactions between different forms of iron and aluminium and the organic matter of a podzol Bh horizon. *Soil Sci.* **96**, 181-6.

Schnitzer, M., Wright, J. R. and Desjardins, J. G. (1958). A comparison of the effectiveness of various extractants for organic matter from two horizons of a podzol profile. *Can. J. Soil Sci.* **38**, 49-53.

Sequi, P., Guidi, G. and Petruzelli, G. (1975). Influence of metals on solubility of soil organic matter. *Geoderma* **13**, 153-61.

Smith, A. N. (1977). Fertilization. *In* 'Soil Factors in Crop Production in a Semi-Arid Environment.' (Eds J. S. Russell and E. L. Greacen). (Univ. Queensland Press: St Lucia.)

Stefanson, R. C. (1968). Factors determining seasonal changes in the stabilities of soil aggregates. Trans. 9th Congr. Int. Soil Sci. Soc. Adelaide. Vol. 2, pp. 395-402.

Stefanson, R. C. (1971). Effect of periodate and pyrophosphate on the seasonal changes in aggregate stabilization. *Aust. J. Soil Res.* **9**, 33-41.

Stephens, C. G. (1949). Comparative morphology and genetic relationships of certain Australian, North American and European soils. *J. Soil Sci.* **1**, 123-49.

Stevenson, F. J. and Butler, J. H. A. (1969). Chemistry of humic acids and related pigments. *In* 'Organic Geochemistry. Methods and Results.' (Eds G. Eglington and M. T. J. Murphy). (Springer-Verlag: Berlin.)

Swaby, R. J. (1949). The relationship between micro-organisms and soil aggregation. *J. Gen. Microbiol.* **3**, 236-54.

Swaby, R. J. and Ladd, J. N. (1962). Chemical nature, resistance and origin of soil humus. International Soil Conference, New Zealand. Vol. A5, pp. 3-8.

Swift, R. S. and Posner, A. M. (1971). Gel chromatography of humic acid. *J. Soil Sci.* **22**, 237-49.

Swift, R. S. and Posner, A. M. (1972). Autoxidation of humic acid under alkaline conditions. *J. Soil Sci.* **23**, 381-93.

Swift, R. S., Thornton, B. K. and Posner, A. M. (1970). Spectral characteristics of a humic acid fractionated with respect to molecular weight using an agar gel. *Soil Sci.* **110**, 93-9.

Swincer, G. D., Oades, J. M. and Greenland, D. J. (1968*a*). Studies on soil polysaccharides. I. The isolation of polysaccharides from soil. *Aust. J. Soil Res.* **6**, 211-24.

Swincer, G. D., Oades, J. M. and Greenland, D. J. (1968*b*). Studies on soil polysaccharides. II. The composition and properties of polysaccharides in soils under pasture and under fallow–wheat rotation. *Aust. J. Soil Res.* **6**, 225-35.

Swincer, G. D., Oades, J. M. and Greenland, D. J. (1969). The extraction, characterization and significance of soil polysaccharides. *Adv. Agron.* **21**, 195-235.

Theng, B. K. G. and Posner, A. M. (1967). Nature of the carbonyl groups in soil humic acid. *Soil Sci.* **104**, 191-201.

Theng, B. K. G., Wake, J. R. H. and Posner, A. M. (1966). The infrared spectrum of humic acid. *Soil Sci.* **102**, 70-2

.Theng, B. K. G., Wake, J. R. H. and Posner, A. M. (1967). The humic acids extracted by various reagents from a soil. II. Infra-red, visible, and ultra-violet absorption spectra. *J. Soil Sci.* **18**, 349-63.

Theng, B. K. G., Wake, J. R. H. and Posner, A. M. (1968). The fractional precipitation of soil humic acid by ammonium sulphate. *Plant Soil* **29**, 305-16.

Thomas, R. L. (1963). Fractionation and characterization of the polysaccharides extracted from the Brookston soil. Ph.D. Thesis, Ohio State University.

Wake, J. R. H. and Posner, A. M. (1967). Membranes for measuring low molecular weights by osmotic pressure. *Nature (London)* **213**, 692-3.

Wetselaar, R. (1967). Estimation of nitrogen fixation by four legumes in a dry monsoonal area of north-western Australia. *Aust. J. Exp. Agric. Anim. Husb.* **7**, 518-22.

Wetselaar, R. and Norman, M. J. T. (1960). Recovery of available soil nitrogen by annual fodder crops at Katherine, Northern Territory. *Aust. J. Agric. Res.* **11**, 693-704.

Williams, C. H. (1970). Pasture nitrogen in Australia. *J. Aust. Inst. Agric. Sci.* **36**, 199-205.

Williams, C. H. and David, D. J. (1976). Effects of pasture improvement with subclover and superphosphate on the availability of trace elements to plants. *Aust. J. Soil Res.* **14**, 85-93.

Williams, C. H. and Donald, C. M. (1957). Changes in organic matter and pH in a podzolic soil as influenced by subterranean clover and superphosphate. *Aust. J. Agric. Res.* **8**, 179-89.

Chapter 36
Organic phosphorus and sulfur

M. E. Probert

SOIL ORGANIC MATTER CONTAINS BOTH PHOSPHORUS and sulfur and there is a considerable body of evidence from diverse parts of the world that indicates relationships between the amounts of these two elements and those of nitrogen and carbon in organic forms. The proportions of carbon, nitrogen and sulfur found in most cases are rather similar (C:N:S about 140:10:1.3) (Williams 1967), but appreciable variations in these ratios do occur between individual soil samples and between soil groups. Some soils, for reasons that are unknown, contain rather high concentrations of carbon: however, the relation between nitrogen and sulfur does not appear to be subject to wide variations and most surface soils have N:S ratios in the range 6.6–10:1. It thus appears that the association between nitrogen and sulfur in soil organic matter is closer than their association with carbon.

The content of organic phosphorus, relative to 10 parts of nitrogen, is not as uniform as that of sulfur; it is usually in the range 0.33 to 3.0. No doubt part of the reason for this is that some of the organic phosphorus in soils is present in compounds that contain no nitrogen or sulfur. A further indication that phosphorus should be treated as a variable, rather than as a characteristic, fraction of soil organic matter is afforded by the observations that a very high proportion of the organic phosphorus in soils can be removed by relatively mild extraction procedures that certainly do not remove all of the organic C, N or S present, and that organic P is largely found in the fulvic fraction of soil organic matter.

Studies of organic phosphorus and sulfur in soil are hampered by the fact that it is not fully known in what chemical compounds they are present, and also by analytical difficulties. Although separation techniques such as gel chromatography (Steward and Tate 1971) and anion-exchange chromatography (Martin 1970) by means of which soil extracts can be separated into their inorganic P and organic P components have been developed, such methods are at best tedious. For routine determinations it has been more usual to estimate the total organic fraction as the difference between the total element concentration and the inorganic fraction. For phosphorus the methodology may be based on either the difference in acid-extractable phosphorus of soil samples before and after ignition or the difference found in an extract. Steward and Oades (1972) discuss some of the limitations of these methods and the factors affecting the amount of organic P measured. For sulfur, the organic fraction is again taken as the difference between total S present and the inorganic forms, though often the latter can be determined fairly simply by extraction with a phosphate solution, which removes sorbed and soluble sulfate. Where carbonates are present some sulfate is likely to be present as a coprecipitate and an acid extraction is appropriate, whilst occasionally it will be necessary to allow for reduced forms of inorganic sulfur. Methodologies also exist whereby discrete fractions of organic S can be determined chemically and these are discussed below where the nature of organic sulfur is considered.

Using the crude fractionation into organic and inorganic fractions, some 25–80% of the total phosphorus in soils is found to be present in organic form. In most surface soils the proportion of the sulfur present in organic form is very high, but due to the accumulation of sorbed sulfate in some subsoils, the proportion may decrease with depth (for example soils 1 and 10 of Fig. 36.1). Concentrations of both organic P and organic S usually decrease with depth in the soil profile, reflecting organic matter levels. However, changes in C:N:S:P ratios are often observed with depth, the phosphorus content of organic matter tending to increase with depth (Barrow 1961), as is the tendency with organic sulfur relative to nitrogen (Williams 1974; Probert 1977).

The nature of soil organic phosphorus

Most of the reliable information on the quantities and composition of soil organic phosphorus has accrued during the last 20 years; it has been the subject of several recent reviews, for example Cosgrove (1977), Dalal (1977) and Halstead and McKercher (1975). The same constituents have been found in soils from diverse parts of the world: from the information presented by Halstead and McKercher (1975) it appears that the phospholipid fraction may contain approximately 1% of the organic P, nucleic acids some 5% of the organic P, with up to 60% being identified as an inositol polyphosphate

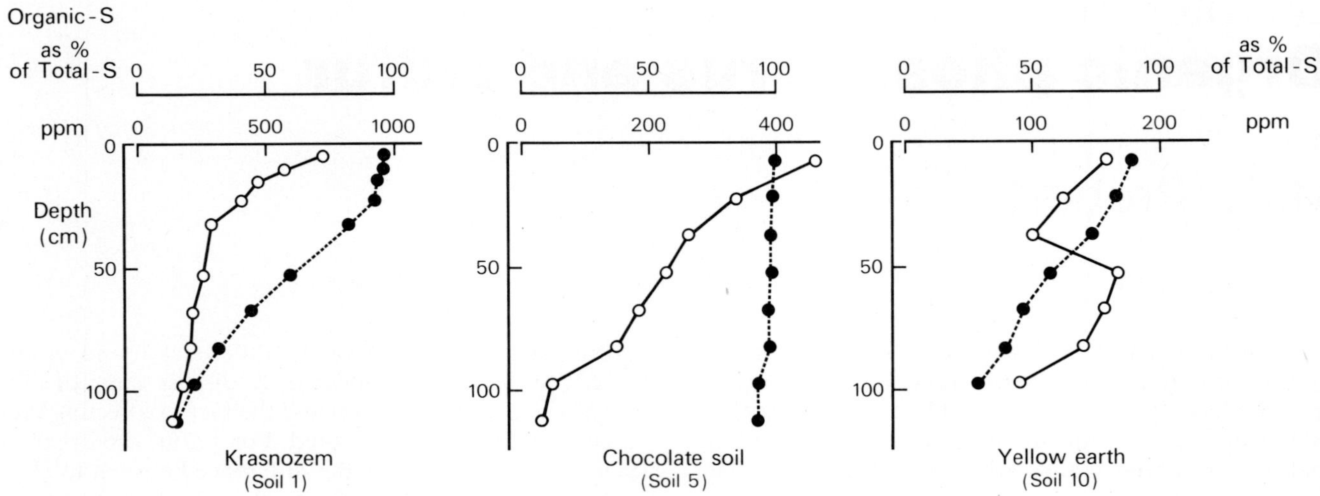

Fig. 36.1. Variation of organic sulfur with depth in three contrasting soil profiles, when expressed in terms of concentration (○——○) and as a proportion of the total sulfur present (●----●). There is a marked increase in sulfate sulfur with depth in soils 1 and 10, but not in soil 5. (Data from Williams 1974.)

fraction. In addition to these three main groups there is some evidence that phosphoproteins, sugar phosphates, and glycerol phosphates are present in soils.

Not all of the organic phosphorus in soil has been identified. However, virtually all of the organic P in a number of Australian and New Zealand soils can be accounted for in two major fractions; (i) inositol polyphosphates, and (ii) a high molecular weight component not as yet characterized (Martin 1970; Steward and Tate 1971). For eight Australian soils, Steward and Tate (1971) found the inositol polyphosphates to represent between 2 and 31% of the organic P extracted.

The inositol polyphosphate is present chiefly as the penta and hexa-phosphates of the various inositol isomers. These are often referred to as 'soil phytate' or 'soil phytin' although strictly speaking phytic acid is *myo*-inositol hexaphosphate and phytin is its calcium/magnesium salt. Smaller amounts of the lower esters (mono-, di-, tri-, and tetraphosphates) have also been detected. Although it is predominantly the esters of *myo*-inositol that are present in soil, the *scyllo*-, D-*chiro*- and *neo*-inositol esters have also been detected.

Besides the uncertainty regarding the nature of the organic phosphorus present, doubt also surrounds its origin. Thus of the inositol polyphosphate esters, only the *myo*-isomer is found in plants and it is generally assumed that the other isomers found in soil are synthesized by soil microorganisms, though the organisms responsible have not been identified.

Nucleic acids, or their derivatives, and phospholipids enter the soil in significant quantities from plant and animal remains and by *in situ* microbial synthesis. Since they comprise only a small part of the soil organic P it could be inferred that their degradation in soil is fairly rapid. However, as pointed out by Cosgrove (1977), the possibility shouldn't be overlooked that our current estimates for nucleic acids and phospholipids in soil may need revising upwards if improved methods of extraction are found.

The exact nature of the recycling of organic P components from plant and animal tissue to soil is not fully understood. Thus Martin and Cunningham (1973) studied the release of organic phosphorus from wheat roots and concluded that it can be released directly into the soil solution as inorganic phosphate as a result of autolysis. Air drying of the root material was particularly effective in causing the degradation of organic P in this manner.

Phosphatases produced by living organisms and enzymes present in soil are usually considered to be the means by which organic P in soil is mineralized. Cosgrove (1967) concluded that soils contain a wide range of microorganisms capable of dephosphorylating all known organic phosphorus compounds of plant origin. It has been observed that enzyme activity is greater in the rhizosphere than elsewhere in soils, and moreover that P-deficiency in plants increases phosphatase activity. Also there is evidence for phosphatase activity associated with mycorrhizas, but there are only negative results for the contention that organic phosphorus is the source of the extra phosphorus taken up by mycorrhizal roots.

The reason for the relative stability of organic phosphorus in soil is generally assumed to be due to its low solubility in the soil solution. For the most part it exists as insoluble complexes with other organic materials, as relatively insoluble salts, or sorbed onto soil surfaces. As an example, Anderson *et al.* (1974) have reported the strong sorption of inositol hexaphosphate; in the same

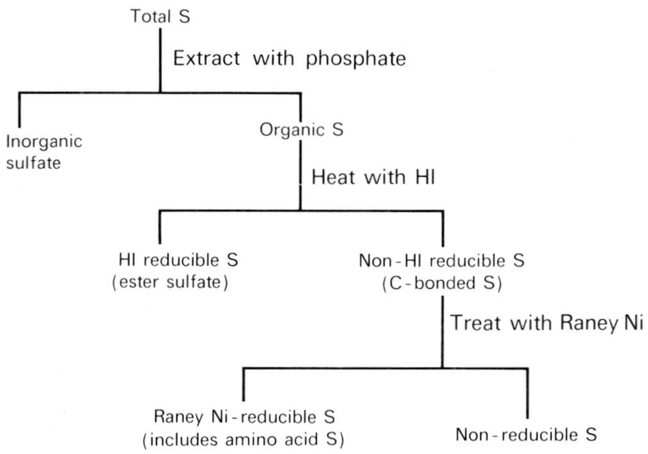

Fig. 36.2. The fractionation of organic sulfur (after Freney *et al.* 1975).

context Martin and Cartwright (1971) observed high availability of inositol hexaphosphate to plants when added at high concentration to a soil of low P retention, but not when it was added at low concentration, nor in soils with high P retention.

However, some organic phosphorus is found in the soil solution, and its concentration can be greater than that of the inorganic P present (Wild and Oke 1966). These authors detected the presence of the lower esters of inositol, including the monophosphate, and furthermore demonstrated the uptake of at least some of this organic phosphorus by plants. That this organic phosphorus was present in solution does not negate the suggestion above that the stability of most soil organic phosphorus is derived from its low solubility. There is evidence for the existence of phosphatases that can dephosphorylate lower esters of inositol but that are unable to attack the penta- and hexa-phosphates.

The nature of soil organic sulfur

Many organic compounds containing sulfur are produced by organisms growing in or on the soil, and some examples of these are given by Freney (1967). Those occurring in greatest amounts in plants and microorganisms are the proteins.

In soils, despite the fact that most of the sulfur is present in organic form, few compounds have been isolated in the free state. This is at least partly attributable to the propensity of sulfur compounds to decompose, and it has proved impossible to date to find extraction techniques that remove organic sulfur compounds from soil in an unaltered form. Of the compounds that have been detected, trithiobenzaldehyde has been found in the greatest quantity but it has not been found in all soils (Anderson 1975). The amino acids cystine and methionine, and derivatives such as methionine sulfoxide, methionine sulfone and cysteic acid have been obtained from soil hydrolyzates (Freney 1967, Anderson 1975); they presumably are present in soil as proteins or polypeptides.

Whilst it is true therefore that most of the organic sulfur in soil can not be positively identified, progress has been made in subdividing it into broad categories that give some insight as to its nature. Such a scheme is shown in Fig. 36.2.

The two main fractions identified are the hydriodic acid-reducible sulfur, often referred to as ester sulfate, and the remainder, which is not reduced by HI and is thought to be present in compounds with the sulfur directly bonded to carbon (carbon-bonded S). A subfraction of the carbon-bonded S is reduced by Raney nickel: some organic compounds containing a C–S bond, including the amino acids methionine and cystine, are reduced to inorganic sulfide by this reagent, but others such as alkyl sulfones are not. Tin and hydrochloric acid has also been studied as a reactant for fractionating sulfur in soil, but was not found to be very satisfactory in that whilst it can reduce some organic sulfur compounds, those with a C=S bond being particularly susceptible, it is not specific for this grouping (Melville *et al.* 1971).

Data from such a fractionation scheme for Australian soils have been reported by Williams (1974) and discussed in greater detail in relation to observations for other soils by Williams (1975). In general, the HI-reducible sulfur accounts for approximately 50% of the total organic sulfur, whilst the Raney nickel-reducible fraction averages some 50% of the carbon-bonded S, although fairly wide variations are found. The relationship with nitrogen has been found to be closer for the carbon-bonded S than for the HI-reducible fraction (Williams 1975; Scott and Anderson 1976).

The transformations of sulfur in soil are mainly microbiological. During the decomposition of organic compounds some of the sulfur is used for the synthesis of cell material as the microbial population grows and only that not needed is converted to inorganic forms. The end products of the decomposition will depend upon the substrate, the organisms, and the conditions, though where net mineralization occurs it is the increase in sulfate that is usually determined. The decomposition of several organic S compounds has been studied by Frederick *et al.* (1957), who showed that in some instances decomposition was accompanied by formation of volatile sulfur products. The possible formation and loss of hydrogen sulfide from soils especially under anaerobic conditions has been the subject of much speculation. Interesting studies on the nature of volatile sulfur compounds formed in soil have been reported by Banwart and Bremner (1975,1976); a range of volatile products were detected, including methyl mercaptan, dimethyl sulfide, dimethyl disulfide and carbon disulfide, but not hydrogen sulfide. Up to 50% of S added as amino acid could be lost by volatilization, but losses from sludges, manures and plant materials were relatively small even under waterlogged conditions.

Since the conversion of organic sulfur to inorganic sulfur is mainly the result of microbiological activity, the factors which affect mineralization are largely those that influence the growth of microorganisms. Temperature, moisture, pH and the availability of food supply have all been shown to affect the rate of mineralization, whilst mineralization can be suppressed by additions of formaldehyde or toluene; these effects have been discussed in greater detail by Freney and Swaby (1975).

The assimilation of sulfate into organic forms has been studied by Freney *et al.* (1971) by adding radioactive ^{35}S-labelled sulfate to soils. Following sterilization by autoclaving, very little labelled sulfate was incorporated into the organic fractions, showing that isotopic exchange does not occur between the inorganic and organic fractions. However, without sterilization, sulfate was incorporated into both the HI-reducible and carbon-bonded sulfur fractions; the addition of glucose increased the amount of sulfur incorporated. Differences in the specific activity of the organic fractions were found, the HI-reducible fraction being more highly labelled than the sulfur bonded to carbon.

The same authors (Freney *et al.* 1975) have also investigated changes in soil organic sulfur fractions during cropping. The greatest change was found in the carbon-bonded fraction, with decreases in both the Raney nickel-reducible and non-reducible fractions being involved. However, the earlier paper (Freney *et al.* 1971) had shown that there must have been conversion of carbon-bonded sulfur to HI-reducible sulfur during incubation. Thus all fractions undergo change during uptake of sulfur by plants and the authors were forced to conclude that none of the fractions studied are likely to be of value for predicting the sulfur requirements for plants.

The results of these relatively short-term studies are not in full accord with longer term effects seen in the field. Thus in contrast to the initial incorporation of freshly added sulfate into predominantly the HI-reducible fraction, organic sulfur accumulating under subterranean clover (*Trifolium subterraneum*) pastures is largely accounted for in the carbon-bonded fraction (Williams 1975).

Organic matter as a source of phosphorus and sulfur for plant nutrition

It is generally accepted that plants obtain their nutrients as ions from the soil solution. Thus the constituents of soil organic matter are in forms that are unavailable, and the organic matter must be mineralized to release its nutrients prior to their being taken up by plants. Mineralization of organic P and S occurs concurrently with the reverse process, immobilization, through which inorganic forms are converted to organic compounds. In incubation studies, it is the difference between these two processes, 'net mineralization', that is usually measured and this is often found to be rather small. However, this is not to say that the two processes are in themselves necessarily slow, but rather that they often tend to occur at comparable rates. Using a radiotracer technique it has been observed that 40% of the P applied as superphosphate to soil under pasture appeared as organic P within 28 days (unpublished data of G. J. Blair, cited by Dalal (1977)).

Thus both phosphorus and sulfur are involved in dynamic cycles in soil and are continually turning over between inorganic and organic forms. Notwithstanding the fact that the relative rates of mineralization and immobilization may vary at different times of the year, the organic fractions of P and S can only be viewed as sources of these materials in the sense that they are capable of giving increased production that can be harvested, if organic matter levels are declining. This of course is practised under crop-pasture rotations where the organic matter built up under the pasture phase is utilized during the cropping sequence. During the build-up phase however, the organic fractions are essentially a sink for P and S, and sources of P, S and N are required in order for organic matter to accumulate in soils. It is presupposed here that the close relationship between C, N, S and P in organic matter is maintained when organic matter is accumulating and whilst it is being depleted. Williams (1967) suggests that this is so, especially in regard to the N:S ratio, though changes in the C:N ratio are found following clearing and establishment of improved pastures; he cites an instance where the C:N ratio declined from 33 to 11 but with no change in the N:S ratio.

Conclusions

Whilst our knowledge of the organic forms of phosphorus and sulfur in soils has expanded, their study and identification remain a challenge to chemists. A large proportion of these nutrients in soils, both native and from fertilizers, is present in organic forms.

The possible shortage of phosphate on a world scale, and certainly its escalating cost to the agricultural community, will attract attention to the feasibility of utilizing the existing forms of phosphate in soil for plant nutrition, whilst other man-made problems such as the disposal of wastes and the eutrophication of waters also involve the cycling of phosphorus in the environment. Progress in tackling such problems is likely to be dependent in no small part upon a better understanding of the cycling of phosphorus in biological cycles.

Similarly, a better understanding of the role of sulfur as a constituent of soil organic matter and of its contribution to plant nutrition is only likely if our knowledge of the sulfur cycle becomes more comprehensive.

Current fractionation procedures would seem too crude to achieve these purposes, and further advances are likely to depend upon refinements in methodology for separating better defined fractions of organic matter. To expect complete characterization of the compounds involved is probably unrealistic.

References

Anderson, G. (1975). Sulfur in soil organic substances. *In* 'Soil Components.' Vol. 1 (Ed J. E. Gieseking) pp. 333-41. (SpringerVerlag: Berlin).

Anderson, G., Williams, E. G. and Moir, J. O. (1974). A comparison of the sorption of inorganic orthophosphate and inositol hexaphosphate by six acid soils. *J. Soil Science* **25**, 51-62.

Banwart, W L. and Bremner, J. M. (1975). Formation of volatile sulphur compounds by microbial decomposition of sulphur-containing amino acids in soils. *Soil Biol. Biochem.* **7**, 359-64.

Banwart, W. L. and Bremner, J. M. (1976). Evolution of volatile sulphur compounds from soils treated with sulphur-containing organic materials. *Soil Biol. Biochem.* **8**, 439-43.

Barrow, N. J. (1961). Phosphorus in soil organic matter. *Soil Fert.* **24**, 169-73.

Cosgrove, D. J. (1967). Metabolism of organic phosphates in soil. *In* 'Soil Biochemistry.' Vol. 1. (Eds A. D. McLaren and G. H. Peterson). pp. 216-28.(Marcel Dekker: New York).

Cosgrove, D. J. (1977). Microbial transformations in the phosphorus cycle. *In* 'Advances in Microbial Ecology.' Vol. 1. (Ed M. Alexander). pp. 95-134.(Plenum Press: New York).

Dalal, R. C. (1977). Soil organic phosphorus. *Adv. Agron.* **29**, 83-117.

Frederick, L. R., Starkey, R. L. and Segal, W. (1957). Decomposability of some organic sulfur compounds in soil. *Soil Sci. Soc. Am. Proc.* **21**, 287-92.

Freney, J. R. (1967). Sulfur-containing organics. *In* 'Soil Biochemistry.' Vol. 1. (Eds A. D. McLaren and G. H. Peterson). pp. 229-59. (Marcel Dekker: New York).

Freney, J. R., Melville, G. E. and Williams, C. H. (1971). Organic sulphur fractions labelled by addition of ^{35}S-sulphate to soil. *Soil Biol. Biochem.* **3**, 133-41.

Freney, J. R., Melville, G. E. and Williams, C. H. (1975). Soil organic matter fractions as sources of plant-available sulphur. *Soil Biol. Biochem.* **7**, 217-21.

Freney, J. R. and Swaby, R. J. (1975). Sulphur transformations in soils. *In* 'Sulphur in Australasian Agriculture.' (Ed. K. D. McLachlan). pp. 31-9. (Sydney Univ. Press: Sydney).

Halstead, R. L. and McKercher, R. B. (1975). Biochemistry and cycling of phosphorus. *In* 'Soil Biochemistry'. Vol. 4. (Eds E. A. Paul and A. D. McLaren). pp. 31-63. (Marcel Dekker: New York).

Martin, J. K. (1970). The measurement of inositol polyphosphates in soil extracts by isotope dilution analysis. *N.Z. J. Agric. Res.* **13**, 930-9.

Martin, J. K. and Cartwright, B. (1971). The comparative plant availability of ^{32}P myo-inositol hexaphosphate and $KH_2{}^{32}PO_4$ added to soils. *Commun. Soil Sci. Plant Anal.* **2**, 375-81.

Martin, J. K. and Cunningham, B. R. (1973). Factors controlling the release of phosphorus from decomposing wheat roots. *Aust. J. Biol. Sci.* **26**, 715-27.

Melville, G. E., Freney, J. R. and Williams, C. H. (1971). Reaction of organic sulfur compounds in soil with tin and hydrochloric acid. *Soil Sci.* **112**, 245-8.

Probert, M. E. (1977). The distribution of sulphur and carbon-nitrogen-sulphur relationships in some North Queensland soils. CSIRO Aust. Div. Soils Tech. Pap. No. 31.

Scott, N. M. and Anderson, G. (1976). Organic sulphur fractions in Scottish soils. *J. Sci. Fd. Agric.* **27**, 358-66.

Steward, J. H. and Oades, J. M. (1972). The determination of organic phosphorus in soils. *J. Soil Sci.* **23**, 38-49.

Steward, J. H. and Tate, M. E. (1971). Gel chromatography of soil organic phosphorus. *J. Chromatogr.* **60**, 75-82.

Williams, C. H. (1967). Nitrogen, sulphur and phosphorus, their interactions and availability. Trans. Int. Soc. Soil Sci., Comm. II and IV, pp. 93-111.

Williams, C. H. (1974). The chemical nature of sulphur in some New South Wales soils. *In* 'Handbook on Sulphur in Australian Agriculture.' (Ed K. D. McLachlan). pp. 16-23. (CSIRO Aust.: Melbourne).

Williams, C. H. (1975). The chemical nature of sulphur compounds in soils. *In* 'Sulphur in Australasian Agriculture.' (Ed. K. D. McLachlan). pp. 21-30. (Sydney Univ. Press: Sydney).

Wild, A. and Oke, O. L. (1966). Organic phosphate compounds in calcium chloride extracts of soils: identification and availability to plants. *J. Soil Sci.* **17**, 356-71.

Chapter 37

Soil nitrogen

J. N. Ladd and J. S. Russell

Forms and distribution

Concentrations and profile distribution

In Australia there is a great diversity of soil micro-environments ranging from tropical to alpine and from arid to humid, resulting in a wide range of organic nitrogen profiles. Soil nitrogen concentrations are generally low, since most of the soils are old, and the widespread climatic conditions of low rainfall and high temperature do not favour organic matter accumulation. However in the humid areas small areas of soils such as *krasnozems* have high N concentrations, and in the sub-humid areas, localized soils with high calcium status such as *rendzinas* have moderate N levels.

One distinguishing feature of Australian soils in the semi-arid to humid areas is the extensive occurrence of *Eucalyptus* spp.; more than 500 species of this tree have been identified, and the ecological range of the genus is very wide. In an early survey of Australian soils, Prescott (1931) referred to the possible effects of *Eucalyptus* spp. on soil N status. Although climatic and intrinsic soil factors were seen as having important roles in determining the amount and composition of soil organic matter, Prescott believed that the non-deciduous nature of the vegetation contributes to the low soil organic matter concentrations. However, data on litter production of *Eucalyptus* forests (Bray and Gorham 1964) suggest that annual falls are similar to those of other warm temperate areas, but that the proportion of non-leaf litter is higher, and that litter-fall occurs throughout the year.

The early work of Prescott also suggested that low soil P concentrations may have affected those of soil organic N; subsequent studies have emphasized the low P content of Australian soils generally (Wild 1958). Low S status has also been implicated by Williams and Steinbergs (1958) as a factor affecting the low N status of some Australian soils; these authors have found a close relationship between organic N and non-sulfate S, and similar C:N:S:P ratios in the organic matter of a range of soils. Beneficial effects of P and S applications on soil organic matter concentrations (Donald and Williams 1954; Russell 1960) support the suggestion that the low organic N status of many Australian soils is related to their low P and S status.

Substantial changes in the organic N concentrations of many Australian soils have occurred as a result of man's activities and are still occurring. On the one hand, many of the low fertility soils in the higher rainfall areas now have considerably higher soil N contents than in their natural state, due to the addition of deficient soil nutrients, to the selection of adapted pasture legumes which can contribute N to the soil–plant system, and to the use of grazed grass–legume pastures. On the other hand, N concentrations of some of the more fertile soils decreased initially with cultivation for crop production. Satisfactory equilibrium levels have subsequently been achieved on some soils due to the use of conservative rotations of crops and ley pastures; on others, particularly in the sub-tropical areas, soil N concentrations are still decreasing (Clarke and Russell 1977).

A considerable amount of data has accumulated on the amounts and profile distribution of organic N of soils in Australia. Stace *et al.* (1968) have presented some of these data for the main Great Soil Groups; the surface-soil organic N data have been further summarized by Williams and Colwell (1977) and in Chapter 34 of this book.

The mean profile distribution of organic N of some of the more important Great Soil Groups in Australia are illustrated in Fig. 37.1, calculated from data from Stace *et al.* (1968). Mean depth trends for several Great Soil Groups are also given in Fig. 34.4 of Chapter 34. More than half of the groups considered in Fig. 37.1 had N contents of topsoils (0–10 cm) of less than 0.12%, and of subsoils, at a depth of 50 cm, of less than 0.05%. Four of the groups, *grey brown and red clays*, *red-brown earths*, *prairie soils*, and *black earths*, ranged from 0.12 to 0.22% N at the surface. Only *krasnozems* and *humic gleys* had surface-soil N contents greater than 0.3%, and even with these soils there was a marked decline in organic N content with depth.

Fractionation

Most of the nitrogen of topsoils is organic. Most plant and animal nitrogenous residues are decomposed rapidly and extensively in soils through the activities, predominantly, of successive populations of soil microflora. Hence soil organic N is considered to be mainly of microbial origin (Bartholomew 1965), and will

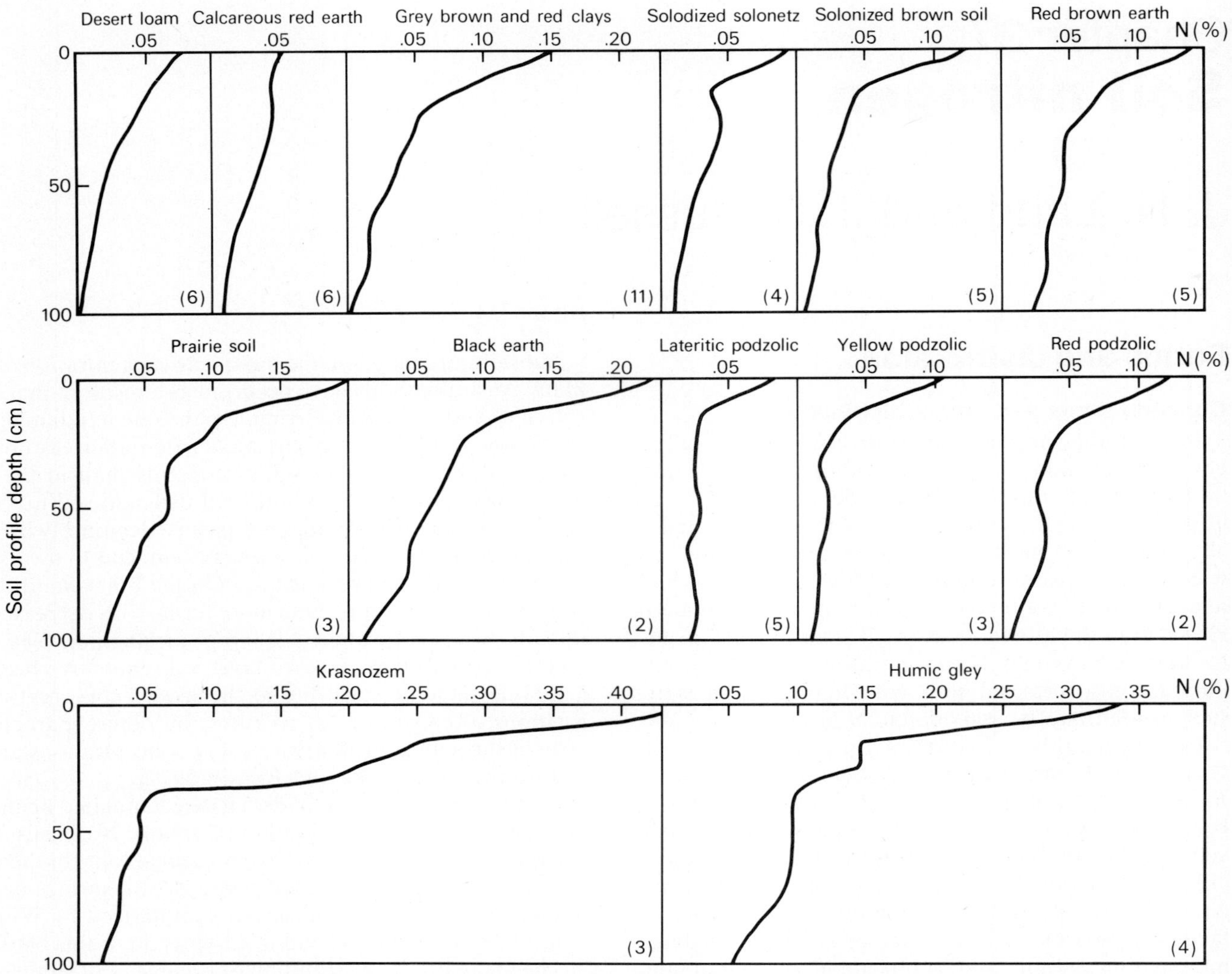

Fig. 37.1. Distribution of organic N in some Australian Great Soil Group profiles. Profile data are from Stace *et al.* (1968). Numbers in brackets refer to the numbers of profiles for each group.

be distributed within microbial cells and cell debris and the more resistant products of microbial metabolism.

Ideally, fractionation of soil N would reveal both its chemical nature and its physical distribution in intact plant and animal and microbial cells, particulate cell debris, soil solution, and in complexes with organic and inorganic soil colloids (Oades and Ladd 1977). In practice, techniques for fractionating soil N into components of biological significance are limited. Chemical procedures are based mainly on hydrolyzing soil N with hot 6 M hydrochloric or 3 M sulfuric acid, with the certainty that similar products will be formed from a variety of sources. Physical techniques are generally applied to dispersed soils in aqueous or organic solvents. Destruction of cells during dispersion may give rise to artefacts through adsorption of soluble and particulate organic N to soil colloids.

Biomass N fraction is estimated indirectly, using biomass carbon. Jenkinson and Powlson (1976) and Ayanaba *et al.* (1976) have determined biomass C in soils. Their method is based on calculations of the net formation of carbon dioxide C or mineral N after nearly complete sterilization of soils by chloroform fumigation, then inoculation and incubation under standard conditions. The method requires that the flush of mineral C and mineral N is derived exclusively from decomposing microbial cells and animal tissue, and is a constant percentage of biomass C. Soil biomass N may be approximately calculated by assuming that the average C:N ratio of the soil biomass is 5:1 (Ladd *et al.* 1977*c*).

Applying the technique to several South Australian soils, biomass N accounted for 3.5–5.5% of the total N of a range of calcareous soils, and for 1.6–3.1% of the total N of a *red-brown earth* with a long-term history of

Table 37.1. Fractionation of nitrogen for some South Australian soils

Biomass N, total N, and chemical fraction N, determined on five calcareous soils (0–10 cm) sampled under pasture, and on Urrbrae *red-brown earth* (0–5 cm) sampled 1976 before wheat cropping, and sampled 1977 (permanent pasture). Light fraction N determined by Ford (1968) on Urrbrae soil (0–6.3 cm), sampled 1965. W, wheat; P, pasture; F, fallow; N.D., not determined.

Great Soil Group	Northcote key[A]	Soil location	Rotation	Total N (% by wt)	Fractionation (%) of total soil Nitrogen			
					Biomass N	Acid-soluble N	Acid-soluble amino acid N	Light fraction N
Red-brown earth	Dr2.23	Urrbrae	P	0.216	3.0	76.0	36.8	26
			WWPPPP(P4)	0.182	3.1	70.3	39.8	20
			WPPF(P2)	0.138	1.7	65.2	34.1	21
			WPF(P)	0.080	1.6	60.1	34.6	N.D.
			WF(F)	0.071	1.7	61.9	36.7	19
Solonized brown soil	Gc1.12	Avon		0.188	4.9	64.0	34.2	N.D.
Black earth	Ug5.16	Northfield		0.136	5.0	72.2	41.8	N.D.
Solonized brown soil	Gc1.2	Two Wells		0.125	3.5	70.1	37.2	N.D.
Solonized brown soil	Gc1.22	Caliph (a)		0.089	5.5	63.8	37.6	N.D.
Calcareous sand	Uc5.11	Caliph (b)		0.042	5.0	55.2	25.9	N.D.

[A] Soils categorized according to Northcote (1965).

different sequences of cereal cropping and pasture phases (Table 37.1). Rotations with increasing frequency of pasture increased total N and biomass N of the soil. The proportion of biomass N to soil N was greatest in sequences where cultivation was absent or infrequent.

Values for biomass N in Table 37.1 are approximate; they will be influenced not only by C:N ratios of organisms killed by fumigation, and by those of newly synthesized microorganisms and their metabolic products formed during incubation, but also by the C:N ratios of extraneous organic materials available in the soil at the time of sampling. The effect of season and phase of rotation on biomass N has not been determined. Nevertheless an effect of soil management on biomass N is apparent, and is consistent with the results of Powlson and Jenkinson (1976), who found that biomass C of a soil declines in the order: grassland > cropped arable > bare fallow.

Chemical fractionation is most frequently done by hot acid hydrolysis, which dissolves 60–80% of the organic N of surface soils. Amino acids account, as a minimum, for 20–40% of topsoil N, the amino sugars glucosamine and galactosamine for 5–10%, and purines and pyrimidines for less than 1% (Bremner 1965). Most of the amino acids are typical of those formed from proteins under similar hydrolytic conditions, and are probably released in soils by cleavage of peptide bonds. Amino sugars are present in soils as high molecular weight compounds of unknown structure. Purine and pyrimidine bases are probably derived from nucleic acids.

Chemical fractionation of Australian soils has been less detailed. Moore and Russell (1968, 1970), Osborne (1977) and Amato, *et al.* (Table 37.1) have examined a range of soils, differing in chemical and physical properties and vegetative cover, and sampled from widely different climatic environments. Percentages of soil N hydrolyzed, and accounted for as acid-soluble, alkali-distillable N, or as acid-soluble amino acid N, were similar to those recorded elsewhere. For some calcareous soils and a *red-brown earth*, the percentage of soil N hydrolyzed increased with the N content of the soil; for the *red-brown earth* this proportionate increase corresponded with the increasing frequency of the pasture phase (Table 37.1). Acid-soluble amino acid N accounted for a relatively constant percentage of soil N, irrespective of the soil or of its previous agronomic history.

The percentages of soil N dissolved with hot acid are not fixed; they may be increased (to 90%) by pretreating soils with hydrofluoric acid, then hydrolyzing the soils, and alkaline extracts of the soil residues, with 3 M sulfuric acid (Freney 1968). These studies, and those of Freney and Miller (1970), support the contention that soil N is partly attributable to low molecular weight organic compounds entrapped and protected within the lattices of clay minerals (Bremner 1965).

The classical soil fractions, humic acid, fulvic acid and humin, contain organic N, partly accounted for, after hot acid hydrolysis, as amino acids and amino sugars (Bremner 1965; Oades and Ladd 1977). Humic acids extracted with neutral sodium pyrophosphate solutions are of lower average molecular weight, and they contain less N with lower proportions of acid hydrolysable N and amino acid N than do humic acids extracted with dilute alkali (Bremner 1965; Butler and Ladd 1969). Humic acid fractions of increasing molecular size contain increasing amounts of total N, attributable to increasing

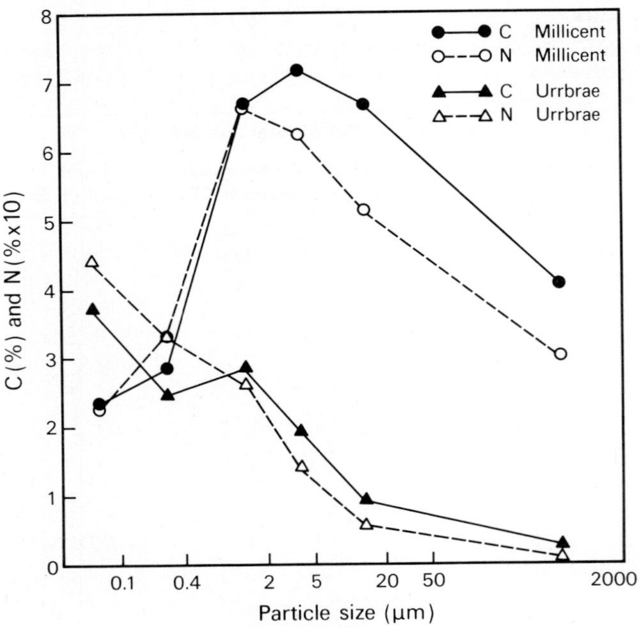

Fig. 37.2. Organic C and N contents of particle size fractions of two South Australian soils. (Adapted from Turchenek and Oades 1979.)

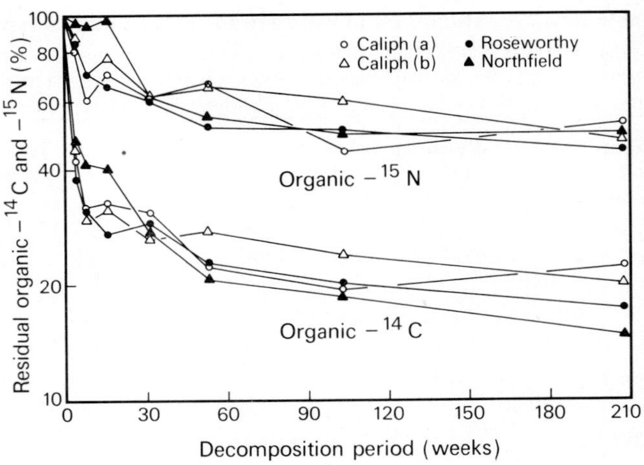

Fig. 37.3. Decomposition of ^{14}C, ^{15}N-labelled medic residues at four South Australian field sites. Percentage clay and total N contents, respectively, of topsoils were Caliph (*a*), 5, 0.05; Caliph (*b*), 20, 0.09; Roseworthy, 12, 0.09; Northfield, 42, 0.14. (From Ladd *et al.* 1981*a*).

contents of acid-soluble amino acid N, and derived possibly from peptide-bound components of greater chain-length (Ladd and Butler 1966; Piper and Posner 1968; Butler and Ladd 1969; Swift and Posner 1972*a*).

The range of amino acids released by hot acid hydrolysis of humic acids from Australian soils, and elsewhere, is typical of that from protein hydrolysis (Piper and Posner 1968; Brisbane *et al.* 1972). The view that acid hydrolysis releases amino acids from soil humic acids by cleavage of peptide bonds is supported by studies with model compounds (Ladd and Butler 1963, 1966) and with proteinases (Scharpenseel and Krausse 1962; Ladd and Brisbane 1967; Brisbane *et al.* 1972), by employing mild acid hydrolyses (Piper and Posner 1968), and by isolation of proteinaceous components (Simonart *et al.* 1967; Biederbeck and Paul 1973).

Peptides or proteins may be hydrogen bonded in the humic acid complexes, or may be covalently linked via terminal α- and ε-amino groups to aromatic components. N- Phenylamino acids have not been identified in humic acid hydrolyzates, despite their resistance to hot acids (Haider *et al.* 1965; Ladd and Butler 1966). If present, such components may account for some of the non-hydrolyzable N of humic acids. Amino acids are readily released from quinoneimines, formed from N-phenylamino acids under oxidative conditions (Ladd and Butler 1966). Such reactions may occur when amino acids are released by alkaline hydrolysis of humic acids previously hydrolyzed with 6 M hydrochloric acid (Piper and Posner 1972).

Physical fractionation involves, generally, soils being dispersed with minimal losses of organic N, and then fractionated according to nominal particle size and/or particle density. Particle size fractions are obtained by sedimentation techniques similar to those described for soils with organic matter removed. The organic N (and C) concentrations of size fractions usually conform to one of two patterns (Fig. 37.2). Turchenek (1975) and Turchenek and Oades (1979) have shown that with a *red-brown earth* the C and N contents of fractions increase with decreased particle size. Similar patterns are obtained with Ando soils and with leached acid or slightly acid soils of low clay content. Alternatively, with a (ground-water) *rendzina*, the highest C and N contents occur in the coarse-clay size and silt size fractions, a pattern typically obtained with *chernozems* and chestnut soils of high base status. Ladd *et al.* (1977*a*) observed a similar pattern with particle size fractions from calcareous clay and sandy soils. C:N ratios generally decrease with particle size (Swift and Posner 1976; Turchenek and Oades 1979).

Organic N of clay size particles is more extensively hydrolyzed with hot acid than that of larger particles (Kyuma *et al.* 1969; McGill 1971), but the percentage distribution of 'defined' nitrogenous hydrolytic products shows no consistent trend with the size of particle hydrolyzed.

Densimetric techniques have been extensively used in Australia to fractionate soils (Greenland and Ford 1964; Ford 1968; Ford *et al.* 1969), or to further fractionate particle size fractions (Turchenek 1975; Turchenek and Oades 1979; Ladd *et al.* 1977*a*). Soils or soil fractions have been dispersed in organic solvents ranging in specific gravity from 1.6 to 2.6.

Nitrogen contents of light fractions may be more than ten times those of the soil or particle size fraction from which they were obtained. The proportion of soil N accounted for in the light fraction depends upon soil type and agronomic history, and has ranged from 40 to 79% (Ford 1968; Greenland and Ford 1964). Light fraction N from a *red-brown earth* is extensively dissolved by heating with 6 M hydrochloric acid; most of the hydrolyzate N is accounted for as amino acids (Ford 1968).

Turchenek and Oades (1979) have shown that organic C and N are concentrated in low-density fractions, irrespective of the size of particle fractionated. Light fractions from sand and silt size particles have the highest N contents. (Ladd *et al.* 1977*a*; Turchenek and Oades 1979).

Mineralization and immobilization

Decomposition of plant and microbial residues

Unground ^{14}C-,^{15}N-labelled medic (*Medicago littoralis*) residues incorporated in topsoils at four wheat growing field sites in South Australia followed decomposition patterns typical of those established elsewhere with other ^{14}C-labelled plant materials (Jenkinson 1971; Oberlander 1973; Sauerbeck and Gonzalez 1977 Ladd *et al.* 1981*a*). The initially rapid carbon decomposition phase lasted 4–8 weeks at all sites (Fig. 37.3). After 8 weeks, residual organic ^{14}C averaged 34% of the amounts incorporated; after 2 years, 21.5%; and after 4 years, 19%. However, rates of decomposition differed between soils, especially within 16 weeks of incorporation, and seasonal effects were also evident. During the slower phase of decomposition, the concentrations of residual organic ^{14}C decreased at rates similar to those established for temperate (UK) locations (Jenkinson and Ayanaba 1977).

Rates of C mineralization from ground leaves or stems or roots of *M. littoralis*, *M. truncatula* and *Lupinus angustifolius* have been determined during 4 weeks laboratory incubation of amended soils (Jackson, Ladd and Amato, unpublished). In these short-term experiments, percentage decomposition rates differed slightly, but significantly, with plant type and part. In longer term field experiments, however, percentage rates of decomposition of plant C in soils have been found generally to be similar, irrespective of plant type and plant part (Jenkinson 1971), although some differences have been reported (Nyhan 1975).

In the field experiments illustrated in Fig. 37.3, the average C:N ratio of the added medic residues was low (15.3:1), and mineral ^{15}N formation was evident soon after incorporation. Residual organic ^{15}N concentrations declined less rapidly during early summer, accompanying drying of topsoils.

Rates of formation of mineral N have been shown, during 4-week laboratory incubations of ^{15}N-labelled medic and lupin residues in soils, to vary with plant type and part; net N immobilization occurred in those soils amended with stem material and lupin roots (Jackson, Ladd and Amato, unpublished data); and all the results are explicable in terms of the varying C:N ratios of the plant tissue used. However, Vallis and Jones (1973) have attributed the lower release of mineral N from *Desmodium intortum* residues, when compared with that from *Macroptilium atropurpureum*, to a protection of proteins in the former tissue, possibly due to the formation of resistant polyphenol-protein complexes (Handley 1961).

Field studies on the rates of decomposition of nitrogen-poor, ^{14}C-labelled plant residues, e.g. wheat stubble, are in progress at Avon in South Australia (Amato *et al.*, unpublished data). The decomposition of incorporated wheat residue C after one year is similar to that of medic residue C, *viz.* approximately 75% of input. Thus formation of decomposition products of C:N ratio say 10:1 would create a demand for N of 25 kg ha^{-1}, per tonne of plant residue C available. Such an amount of N immobilized would be equivalent to that exported in grain from crops yielding approximately 1.2 t ha^{-1}, a common yield under South Australian conditions (Schultz and French 1976).

The total return of residue C to the soil will depend on agricultural practice and climatic conditions. Estimates of wheat root C at flowering approximate 0.4 t ha^{-1} in South Australia (Schultz 1974). Such C sources will be augmented by return of some wheat stubble, and by C released to the soil from the growing crop prior to flowering (see Chapter 42). We have shown in unpublished work that in a field experiment in which wheat grain yields were < 0.9 t ha^{-1}, of the 50 kg ha^{-1} of ^{15}N-labelled fertilizer applied at seeding, approximately 15 kg ha^{-1} were immobilized by flowering. Total mineral nitrogen immobilized during crop growth would be derived partly from the decomposing crop residues themselves, partly from the mineralization of soil nitrogen, and partly from fertilizer nitrogen.

Nitrate and ammonium formation from tropical soils has been described as a function of temperature (Myers 1975) and soil water potential (Wetselaar 1968). Ammonification had a higher temperature optimum, and continued at a higher water potential, than nitrification. In two tropical soils, a clay loam and a sand, percentage mineralization *in situ* of topsoil N during the first year after clearing was, respectively, 4.8% and 12.5%, the latter decreasing to 5.0–5.9% in following years. The extent of organic N mineralization was based on the amounts of nitrate in the soil profile at the end of the wet season. No nitrate was formed during the dry season, and no losses of mineral N occurred over extended periods (Wetselaar 1967*a*).

Mineralization of N from natural soil aggregates from clay soils of the Darling Downs in Queensland increases with decreasing aggregate diameter (Craswell *et al.* 1970); smaller aggregates are considered to contain higher proportions of readily mineralizable organic N. Mineralization rates increase when aggregates are ground, especially with aggregates of high montmorillonitic clay content, and when disruptive treatments are sufficiently severe to reduce their size to < 0.18 mm (Craswell and Waring 1972*a, b*). The effects of grinding are attributed to increased accessibility of organic N to the soil microflora. Organic N protected within aggregates is only slowly rendered decomposable by field tillage practices (Craswell and Waring 1972*b*).

Incubation of two calcareous soils, that had undergone drying and leaching intermittently, increased mineralization of both native unlabelled soil N and of organic ^{15}N formed from immobilized $^{15}NO_3^-$ (Ladd *et al.* 1977*b*). The more recently formed organic ^{15}N was the more readily mineralized, irrespective of the incubation conditions, but the increased mineralization due to drying and leaching of the soils was proportionately greater from unlabelled residues. The effects of wetting and drying are not fully understood, but they may be due to enhanced accessibility of older stabilized residues to microbial attack. Studies (by Ladd *et al.* 1981*b*) suggest that only small percentages of the N of wheat crops in South Australia are derived from the decomposition of the previous year's medic pasture; most of the wheat N is derived from the mineralization of soil organic N, the decomposition of which occurs partly during summer and autumn under conditions of intermittent drying and wetting and leaching of topsoils.

Nevertheless, most of the organic N of soils is very stable to biological attack; $< 3\%$ of the total organic N is mineralized annually (Bremner 1965). Pure, biochemically resistant nitrogenous compounds have not been isolated from soils, and about 50% of soil N may be characterized as compounds which typically are readily decomposed when added to soils. Thus biological stability is thought to be conferred mainly by adsorption of extracellular, high molecular weight, nitrogenous constituents onto clay minerals, or by reactions of the nitrogenous compounds with resistant organic compounds (lignin, tannin, polyphenolic heterocondensates), thus limiting accessibility of substrates to the appropriate extracellular enzymes (Bremner 1965; Oades and Ladd 1977; Swaby and Ladd 1962).

Rates of hydrolysis of peptide or amide derivatives by soils are less affected by seasonal changes (Ladd *et al.* 1976), or by heating or drying of soils, or by conditions causing marked fluctuations in microbial populations, than are rates of hydrolysis of a protein substrate (Ladd and Paul 1973). The comparative stability of activities towards the low molecular weight substrates may be due possibly to reactions of extracellular proteinases with soil colloids. Formation of clay-proteinase or organo-proteinase complexes may both protect the enzyme, and limit its accessibility to substrates of low molecular weight only (Ladd and Butler 1975; Rowell *et al.* 1973). In an analogous fashion, soil urease is thought to be stabilized as humic–enzyme complexes (Burns *et al.* 1972*a, b*; McLaren *et al.* 1975). By contrast, soil proteinases active against high molecular weight substrates may be vulnerable not only to adsorption reactions but to autolysis and attack by other proteinases. Their activities may be more dependent upon the availability of energy sources for continued enzyme synthesis.

A small percentage (up to 5%, Table 37.1) of soil N resides in cellular tissue. Mineralization of cellular N will be influenced by cell type and by cell viability under the changing conditions induced within the soil environment by climate and management.

Changes in the organic nitrogen of soil fractions

The concept of 'active' and 'passive' nitrogenous pools within soils was introduced by Jansson (1958). Subsequently, soils have been fractionated both chemically and physically in attempts to define more clearly the nature of readily mineralizable nitrogen. Comparisons have been made of the distribution of organic N in cultivated and uncultivated soils, and of the relative changes of organic ^{14}N and organic ^{15}N of soil fractions during long periods of net N mineralization. In addition, measurements have been made of changes in concentration or availability of non-exchangeable (fixed) ammonium in soils (Nommik 1965). The native fixed ammonium of a montmorillonitic clay soil in Queensland has been found to be almost unavailable to wheat (Black and Waring 1972), which is in agreement with previous Australian work (Martin *et al.* 1970) and with results generally obtained elsewhere (Bremner 1965; Nommik 1965). Recently added fixed ammonium, however, is more available: but even so, under field conditions, less than half of it is likely to be available to the first crop (Black and Waring 1972).

Chemical fractionation based on hot acid hydrolysis of soils shows that the non-distillable, acid-soluble N (amino acid N) appears to be a major source of mineral N, and in some soils the percentage decrease of this fraction with time and cultivation slightly exceeds that of other organic N fractions (Freney and Simpson 1969; Moore and Russell 1968,1970). On the other hand, Osborne (1977) found no consistent differences in the percentage N decline of various chemical fractions from a range of New South Wales soils. The results of the Australian workers are typical of those reported elsewhere (Bremner 1965): viz. all chemical fractions obtained by acid hydrolysis contain some readily mineralizable N, and although acid-soluble amino acid N often shows marginally the greatest percentage decline, particularly during short-term mineralization experiments under intermittent drying and wetting conditions, no organic N fraction is consistently more susceptible to decomposition than any other. Similar inconsistencies have been demonstrated with soils fractionated classically into fulvic acid, humic acid, and humin components (reviewed by Oades and Ladd 1977).

Changes in physical-fraction N have been less extensively investigated, and any generalizations would be premature. Mineralization of ^{15}N-labelled plant tissue (Amato and Ladd 1980) and of immobilized $^{15}NO_3^-$ (Ladd *et al.* 1977*b*) in calcareous soils was accompanied by disproportionate changes in the organic ^{15}N contents of soil components of different particle size or density; most consistent was the decline in the organic ^{15}N of the fine clay (< 0.2 μm) and very fine clay (< 0.04 μm) fractions. This decline was partly associated with decreases in the amounts of newly formed microbial biomass of these fractions. The organic ^{15}N of coarse clay (0.2–2 μm) fractions may increase during short-term periods of net N mineralization, but all evidence suggests that in the very long term the silt size (2–50 μm) fraction will accumulate an increasing proportion of the more stable nitrogenous residues. Intermittent drying and wetting of soils may stimulate N mineralization, and thus may magnify changes in the organic ^{15}N of soil fractions, but qualitatively the pattern of change is similar to that of soils incubated under constantly moist conditions (Ladd *et al.* 1977*b*).

Sand size particles contained very little immobilized ^{15}N, except that associated with relatively coarse, decomposing straw residues. These plant residues were readily separated as a light fraction from the sand size particles after dispersion in an organic solvent, specific gravity 1.6. The organic ^{15}N of this light fraction appeared to be readily decomposed, especially during periods of intermittent drying and wetting of the soils. However, the decline in the organic ^{15}N of this macro-organic matter fraction could be due not only to conversion to mineral ^{15}N, but also to transfer of organic ^{15}N to finer or heavier soil fractions (Parsons *et al.* 1975). Any assumptions that mineralization only has occurred in the light fractions, e.g. Ford and Greenland (1968), may overestimate light fraction N availability.

Most evidence suggests that the percentage mineralization of the organic N of light fractions, obtained by dispersing whole soils in solvents of specific gravity 2.0, is not greater than that of the unfractionated soil (Greenland and Ford 1964; Ford 1968). Much of the light fraction N separated from soils after dispersion in solvents of specific gravity as high as 2.0 was previously associated with silt size particles. Ladd and Amato (1980) have shown that the percentage changes in the organic ^{15}N of light and heavy subfractions of silt size components are similar, whether during periods of net N immobilization in soils, or of net N mineralization in soils, or following destruction of soil biomass ^{15}N after chloroform fumigation.

Changes in the amounts of organic ^{15}N of soil physical fractions (e.g. fine clay fraction, light fraction of specific gravity < 1.6) will reflect changes in the size and nature of the newly formed microbial biomass and of its metabolic products. Soil biomass ^{15}N (as estimated from the flush of mineral ^{15}N following soil fumigation) declines disproportionately more rapidly than total organic ^{15}N during a 4–5-month period of net mineralization of recently immobilized $^{15}NO_3^-$ (Ladd *et al.* 1977*c*). Although biomass N appears to have been a relatively labile nitrogenous pool during this period, its magnitude at the commencement of the net mineralization phase, and its subsequent decline, did not reflect the capacities of the soils to release $^{15}NO_3^-$. Similarly, Ayanaba *et al.* (1976) showed that mineral N released by fumigation of a range of Nigerian soils is poorly correlated with mineral N release in unfumigated soils in standard incubation tests. Mineral N concentrations in soils will be influenced by many, often opposing, reactions, including those determined by the amounts of available C and N substrates, and by the activities and viabilities of the soil microflora. Measurements of the fluxes of N between the biomass, the non-biomass and the inorganic N pools, rather than net changes in pool sizes, are basic to an understanding of nitrate accumulation in soils.

Nitrogen balance and recovery

A change in the total N content of a soil is the net result of the actions of various mechanisms for N gain and loss. Mechanisms for N gain include symbiotic and non-symbiotic N_2 fixation, accretion from the atmosphere by adsorption and in dust and rain, and application of fertilizers, manures, and seeds. Mechanisms for N loss include volatilization, erosion, leaching below the root zone, and removal of plant and animal products. Measurements of N gains and losses by each process permit the drawing up of a N balance sheet or equation; the net N balance should equal in magnitude and sign the measured changes in the total N of the system.

In practice, especially under field conditions, changes in total N are relatively small, and measurements with adequate precision are not easily achieved; also, individual components (e.g. N_2 fixation, losses by volatilization, erosion) of the balance equation are rarely measured directly. Vallis (1969) has described the relative importance of various factors in the N balance equation in assessing total N changes in grazed pastures at several Queensland sites.

Quantitative assessments of seasonal gains or losses of soil N by specific established mechanisms have not been made with certainty for Australian soils under natural field conditions. Gains and losses may be expected to vary markedly in response to large variations in plant and soil type, climatic conditions, and management practices. Some field studies have shown that losses of N from added ^{15}N-labelled fertilizers or plant residues are negligible. In other studies where deficits have been noted, an assumed loss of N has sometimes been attributed to a particular mechanism(s), mainly by eliminating unlikely alternatives. However, the relative extents to which favoured mechanisms affect soil N levels in a season or a year are difficult to measure.

Gains

The use of legumes to maintain or increase the N status of Australian soils is widespread, and has been studied extensively (see Chapter 44). Nevertheless, accurate measurements, under practical management conditions, of the amounts of nitrogen fixed symbiotically in each season by legume-based pastures have been seldom achieved. Net changes in plant N plus soil N must be adjusted for any N gains by non-symbiotic processes, and for any N losses (e.g. Wetselaar 1967*b*). Estimates based on yields of legume N require a further correction for the proportions of legume N derived from the uptake of mineralized soil organic N. Estimates of the latter have been made (Vallis *et al.* 1977) from the relative ^{15}N atom % enrichments of legumes and grasses grown for 2 years in soils which were supplied every 4 weeks with small amounts of N (0.3 kg ha^{-1}), as $^{15}NO_3^-$ or $^{15}NH_4^+$ fertilizer. In mixed legume-grass pastures after 2 years, legume ^{15}N accounted for approximately 25% of the total fertilizer ^{15}N uptake. The proportions of legume N derived from N_2 fixation averaged 92% and 94% at two sites. To be valid, the method requires either that the added mineral ^{15}N has equilibrated with the unlabelled mineral N formed from soil between fertilizer additions, or that the time-curves for N uptake by the legumes and grasses are similar. Mineral ^{15}N, more uniformly labelled in both space and time, may possibly be released using soils containing old, slowly decomposing organic ^{15}N residues (Ladd 1981) incorporated to greater depth in the soil profile.

Comparatively few studies have been made of non-symbiotic nitrogen fixation in Australian soils. Early work by Swaby (1939), Jensen (1940*a*, *b*) and Jensen and Swaby (1940), with laboratory incubated soils, showed

Table 37.2. Recovery of applied ^{15}N from plant and soil in glasshouse or lysimeter experiments

Soil	Plant	Plant part	Applied ^{15}N recovered in plant (%)	Applied ^{15}N in plant+soil+atmosphere (where applicable) (%)	References
Red podzolic sandy loam (pH 5.6; 0.11% N)	Rhodes grass (*Chloris gayana* Kunth)	Tops Roots	60–63 13–14	96–97	Martin *et al.* (1963) Henzell *et al.* (1964)
	Rhodes grass Townsville lucerne (*Stylosanthes humilis* H.B.K.)	Whole	32–35 63–65	not published	Vallis *et al.* (1967)
	Rhodes grass Siratro (*Macroptilium atropurpureum*)	Whole	4	not published	Henzell *et al.* (1968)
3 soils of grano-diorite origin (pH 4 9–5.9; 0.14–0.36% N)	Rye grass (*Lolium perenne* L.)	Whole	80–85	>90	Simpson and Freney (1967)
Sandy loam (pH6.5; 0.13% N)	Rhodes grass	Tops Roots	14–34 6–8	>90	Moore (1974)
Black earth (pH 8.5; 74% clay; 0.10% N)	Wheat (*Triticum aestivum* var. Timgalen)	Grain Straw	45–50 13.5–15.5	>97	Craswell and Martin (1975*a*)
'*Prairie*-like *soil*' (0–10 cm; pH 5.9; 0.33% N; 42% clay)	Rhodes grass	Tops + Litter	34–60	97–102	Catchpoole (1975)
Black earth	Wheat	Grain Straw	34–53 15	99	Saffigna (1977)
Red-brown earth	Wheat	Grain Straw	60 15	99	
Acid-washed silica sand	Rhodes grass	Whole	63	not published	Moore and Craswell (1976)
Washed calcareous dune sand	Medic (*Medicago littoralis* var. Harbinger)	Tops Roots	56 17	82	Parsons, Ladd and Amato (unpublished)

little or no gains in soil N; numbers of the classical N_2-fixing microorganisms were low. Nevertheless, appreciable net gains in soil N have occurred under grass pastures at several sites in temperate southern Australia, indicative of non-symbiotic fixation (reviewed by Clarke and Russell 1977). With the availability of new and sensitive methods for estimating N_2 fixation rates, interest in non-symbiotic fixation has been rekindled in Australia, especially in soils under tropical grasses in Queensland, where no suitable perennial or self-regenerating pasture legumes have yet been found for clay soils in the sub-humid and semi-arid areas.

In addition to measuring organic N changes in soils under leguminous and grass pastures, studies have been carried out of N_2 fixation in coniferous forests (Richards 1964; Richards and Voigt 1965*a*; Richards and Bevege 1967; Bevege and Richards 1970), and in the rhizospheres of *Araucaria* (Bevege and Richards 1971; Richards 1973) and *Pinus* seedlings (Richards 1973; Richards and Voigt 1964).

Biological N_2 fixation appears to be the major mechanism for N accretion in Australian soils. However, estimates of annual amounts fixed range widely with season and locality, and virtually no measurements have been made of the amounts of N added from the atmosphere to soils, in soluble or particulate forms or by ammonia absorption. Eriksson (1952) has concluded that N entering the soil in precipitation each year is < 10 kg ha^{-1} in temperate and tropical regions. Wetselaar and Hutton (1963) have shown that at Katherine in a monsoonal area of northern Australia (average rainfall 925 mm $year^{-1}$), nitrate N added to the soil in rainwater each year is < 1 kg ha^{-1}.

Losses

The magnitude of soil N losses by established mechanisms might be anticipated to vary markedly according to season, soil type, and management of soil and plant cover. Complete, or nearly complete ($> 90\%$), recoveries of ^{15}N have been achieved in Australian soils under pasture or cereal in experiments conducted both in the glasshouse (Simpson and Freney 1967; Moore 1974; Catchpoole 1975; Craswell and Martin 1975*a*; Saffigna 1976) and in the field (Simpson 1968; Henzell 1971; Vallis *et al.* 1973; Craswell and Martin 1975*b*; Ladd *et al.* 1981*b*), (Tables 37.2 and 37.3); soils have ranged widely

Table 37.3. Recovery of applied ^{15}N from plant and soil in field experiments

Applied ^{15}N	Soil	Location	Plant	Plant part	Applied ^{15}N recovered in plant (%)	Applied ^{15}N recovered in plant + soil (%)	References
NH_4^+ NO_3^-	*Red podzolic* sandy loam	Samford, Qld	Rhodes grass	Whole	37–42	57–86	Henzell (1971)
NH_4^+	'*Prairie*-like *soil*' (0–15 cm; pH 5.2; 0.12% N)	Samford, Qld	Rhodes grass	Tops Roots	38 27	83	Vallis *et al.* (1973)
NO_3^-	*Solodic soil* (0–15 cm; pH 6.4; 0.07% N)	Rodd's Bay, Qld	Townsville stylo-spear grass (*Heteropogon contortus*)	Tops Roots	17 30	77	Vallis *et al.* (1973)
Urea	*Yellow* and *gleyed podzolic soils* (0–15 cm; 0.12% N)	Samford, Qld	Setaria (*Setaria anceps*)	Tops Roots	35 16	98	Vallis *et al.* (1973)
NO_3^-	*Black earth* (0–10 cm; pH 8.5; 74% clay; 0.10%N)	Waco, Qld	Wheat (*Triticum aestivum*, var. Timgalen)	Grain Straw	26–38 17–25	75–94	Craswell and Martin (1975*b*)
NH_4^+ NO_3^-	'*Prairie*-like *soil*' (0–10 cm; pH 5.9; 0.33% N; 42% clay)	Samford, Qld	Rhodes grass	Tops Litter + stubble	0.2–20 4–7	29–48	Catchpoole (1975)
Medic residues	*Solonized brown soil* (0–10 cm; pH 8.5; 12% clay; 0.09% N)	Roseworthy, S.A.	Wheat (*Triticum aestivum*, var. Condor)	Whole Grain	17.3 11.8	92	Ladd *et al.* (1981*b*)
Medic residues	*Black earth* (0–10 cm; pH 8.4; 42% clay; 0.14% N)	Northfield, S.A.	Wheat (*Triticum aestivum*, var. Condor)	Whole Grain	10.9 7.6	87	Ladd *et al.* (1981*b*)
Medic residues	*Solonized brown soil* (0–10 cm; pH 8.6; 20% clay, 0.11% N)	Caliph, S.A.	Wheat (*Triticum aestivum*, var. Condor)	Whole Grain	13.8 7.1	93	Ladd *et al.* (1981*b*)
NO_3^-	*Solonized brown soil* (0–10 cm; pH 8.6; 0.18% N)	Avon, S.A.	Wheat (*Triticum aestivum*, var. Condor)	Whole Grain	42.6 18.7	92	Ladd *et al.* (unpublished)

in properties, and have been located in different climatic zones. Wetselaar (1962*b*) has demonstrated that nitrate added to a bare-fallow tropical soil at Katherine is leached, but can be accounted for completely 5 years after application.

However, deficits in the amounts of recovered ^{15}N have also been obtained, indicative of losses of N from the soil–plant systems (Simpson and Freney 1967; Simpson 1968; Henzell *et al.* 1970; Henzell 1971, 1972; Vallis *et al.* 1973; Wetselaar *et al.* 1973; Catchpoole 1975; Craswell and Martin 1975*b*). Mechanisms of N loss were not specifically determined in most of these experiments.

By denitrification in soils

This has been demonstrated under a variety of experimental conditions (McGarity 1961; McGarity and Myers 1968; Myers and McGarity 1971, 1972; Myers 1972; Burford and Stefanson 1973; Craswell and Martin 1974; Freney *et al.* 1978), and the conditions favouring it have been defined (Myers and McGarity 1972; Craswell and Martin 1974; 1975*a, b*). Denitrification is an anaerobic process; if soils contain NO_3^- and readily oxidizable carbon sources and denitrifying microorganisms, then, under conditions where O_2 diffusion to the potential sites of reaction is limiting, e.g. in soils of high moisture content, denitrification rates will be increased. Nevertheless, even in soils of low water content which are apparently aerobic overall, aerobic catabolism of readily decomposable substrates may induce anaerobic zones, thus creating conditions suitable for denitrification, given that NO_3^- and denitrifiers are also present. Reported rates of denitrification in Australian soils vary widely, but conditions of assay or incubation also vary, thus limiting comparisons of potential activities.

Burford and Millington (1968) and Burford and Greenland (1970) obtained indications of seasonal trends in the N_2O present in the atmosphere of a well-aerated *red-brown earth*. However, the extent to which N_2O losses are due to denitrification, and the extent to which potential activity is realized under natural field conditions, are difficult to assess; some N_2O may be formed during NH_4^+ oxidation. Quantitative measurements of denitrification can be attempted using sealed gas lysimeters, permitting plant growth on undisturbed soil cores, and allowing direct analyses of plant and soil, and atmosphere (Ross *et al.* 1964, 1968; Stefanson 1970). Stefanson (1972*a,b,c,d,* 1973, 1974) has determined the relative and absolute amounts of N_2 and N_2O evolved in these gas-tight chambers, under a range of conditions. Variables included soil type and agronomic history, soil organic matter and water contents, forms and location of applied N, planted and unplanted soils, and disturbed and undisturbed soils. However, no assessment was made of the magnitude of denitrification losses under field conditions, or of the likely frequency and duration of field conditions that favour denitrification. Craswell and Martin (1975*a*) found no evidence for denitrification in gas lysimeter experiments, despite using a *black earth* of high clay and adequate nitrate content, which contained actively growing plant roots, thus creating conditions which were expected to favour denitrification. The soil had a high potential denitrification activity, but this was not expressed. Waterlogging of the soil appears to have been necessary for significant denitrification to occur (Craswell and Martin 1974). Similarly, Catchpoole (1975) found no evidence for denitrification in cores taken from a *'prairie-like soil'* under pasture, installed in a gas lysimeter and run at or below field capacity. Deliberate waterlogging of the cores caused 27% of applied fertilizer N to be lost to the atmosphere.

By ammonia volatilization

This is potentially an important mechanism for N loss from plants and soils, especially from the drying surfaces of alkaline soils of low cation exchange capacity (e.g. Fenn and Kissel 1976; Fenn and Escarzaga 1976). Calcareous soils under crop or pasture are widespread in Australia, and climatic conditions are such that significant losses of NH_3 from fertilizer, urine and dung may be anticipated. Watson and Lapins (1969) estimated a 54% loss of N from urine applied to the surface of a sandy soil. Simpson (1968) showed that 60% of urea ^{15}N applied to undisturbed pasture soils was unaccounted for, presumably due to NH_3 volatilization after decomposition of urea by soil urease; N losses were related to urease activities of soil layers to 1 cm below the respective surfaces of topsoils and subsoils to which urea was applied. By contrast, Denmead *et al.* (1977) have shown, by direct analysis of the atmosphere, that NH_3 volatilization in the field may account for less than 1% of anhydrous NH_3 injected at 12.4 cm depth into a moist clay soil, pH 8.2. Baker *et al.* (1959) have shown that fertilizer NH_3 losses are determined by the depth of application.

Denmead *et al.* (1974, 1976) have calculated the hourly rates of N lost as NH_3 to the atmosphere from a grazed and an ungrazed pasture to be respectively 13 and 2 g ha^{-1}. The greater loss from the grazed pasture is attributed partly to NH_3 volatilization from dung and urine, and partly to less efficient absorption of NH_3 by the more open vegetation of the grazed area; Denmead *et al.* (1977) have demonstrated that in a mature ungrazed pasture NH_3 is almost completely absorbed by the plant cover. Losses of NH_3 to the atmosphere from wheat crops also have been detected by gas lysimetry (Craswell and Martin 1975*a*); the NH_3 is believed to have been lost from the crop during leaf senescence, rather than directly from the soil.

By leaching

Rate of leaching in a soil profile will depend on the concentrations of soluble N compounds (of nitrate N principally), and on the downward flux of infiltrating water (Simpson and Freney 1974). Thus, losses of nitrate N by leaching may be negligible or large, and will be influenced by rainfall, by rates of evaporation at or near the soil surface, by the porosity of and channelling within the soil profile, by adsorption reactions, and by the capacity of plant roots to explore the soil profile and utilize NO_3^- at depth; for example, Wetselaar and Norman (1960) have shown that, at a tropical site near Katherine, bullrush millet growing during the wet season effectively utilizes NO_3^- leached to 152 cm in the profile, whereas sorghum and Sudan grass deplete NO_3^- mainly from the 0–60 cm zone. In the same region, Wetselaar (1962*a*) has shown that rates of NO_3^- leaching are dependent upon soil type, and that the mean downward movement of NO_3^- in a clay loam during one wet season was highly correlated with rainfall. NO_3^- and Cl^- are equally mobile, and are leached 1.075 mm per mm of rainfall. Movement of anions is enhanced by channels left by partly decomposed roots. During the dry winter season, NO_3^- (and Cl^-) accumulate near the soil surface, due to upward capillary movement (Wetselaar 1961*a*, *b*).

Movement of mineral N from banded ammonium nitrate has been followed vertically and laterally in a *krasnozem*, kept bare or cropped using a horticultural rotation, at Redland Bay in Queensland (Black and Waring 1976*a*). Ammonium N remains within 40 cm of the soil surface, and is considered available to most horticultural crops, whereas considerable amounts of NO_3^- are leached beyond the root zone; prolonged cover cropping is recommended to ensure recovery of NO_3^- in the deep subsoil. The relative movement of NO_3^- with water in the soil profile decreases with depth, and is concentration dependent. Further experimentation with soil columns (Black and Waring 1976*b*,*c*) indicates that NO_3^- leaching rates are markedly influenced by NO_3^- adsorption reactions. These would increase with increasing positive charges on soil surfaces, and would be associated with decreasing organic matter contents, increasing kaolinite concentrations, and decreasing pH of soils, with profile depth. Fractionation of nitrate by differential adsorption rates during leaching has been suggested to explain the observed increases in the ^{15}N abundance of NO_3^- with increasing depth of sample (Black and Waring 1977).

In southern Australia, seasonal changes in the concentrations of mineral N have been followed in profiles of a range of N-fertilized or unfertilized soils, kept fallow and/or cropped to wheat (e.g. Parry 1971; Hingston 1974; Storrier 1965; McGarity and Myers 1973).

By erosion

The extent of loss by this means varies with rainfall, soil type, topography, vegetative cover and management practices. In the United States, annual losses of N due to wind and water erosion from cropped soils averaged 27.1 kg ha^{-1}, which may in some areas be similar to the amounts of N removed in the crop itself (Allison 1965). The concentrations of N in eroded soil may be five times that in the topsoil remaining. No overall quantitative assessment of N losses by wind and water erosion has been made in Australia. Catchpoole (1975), using ^{15}N fertilizer in field microplots with facilities for collecting surface run-off, recovered < 50% of the applied N in the soil–plant system; generally < 5% was accounted for in surface run-off water.

By transfer into plants

The percentage of uptake of applied N will be determined partly by the extent to which mineral N is immobilized, or is lost from the soil–plant system by various competing mechanisms, and partly by other factors which may limit or stimulate plant growth and the capacity to utilize the mineral N available. Thus N uptake by crops and pastures will be influenced by plant type, climate, soil type, and C:N ratios of extraneous available organic substrates, and also by the nature and amounts of added fertilizer or plant residues, and by the time and location of their application.

Wide variation in the recovery of applied N by plants has been reported by Hauck (1971); a similar variation has been found in field experiments in Australia. In dryland environments N recovery is usually low: Russell (1964) found that, in wheat grain, the recovery of N applied at rates of 13, 26 and 52 kg ha^{-1} in 52 experiments over a period of 6 years, was 19.9, 16.7 and 12.6%; maximum values recorded at these rates were respectively, 55, 45 and 35% recovery. With summer-growing pastures under favourable conditions, moderate N recoveries have been found; Henzell (1963) reported that with pasture grasses in south-east Queensland N recovery increased to a value of 47% when N had been applied at 448 kg ha^{-1}. Nitrogen recovery values of 52% were found with irrigated forage sorghum at the Ord River when fertilized with N at 180 kg ha^{-1} (Blunt and Fisher 1973). Under irrigation, N recoveries of 45, 28 and 29% were recorded for irrigated wheat, safflower and linseed (Wetselaar and Beech 1968) and 41% for oilseed *Brassica* varieties (Beech and Norman, 1964).

Several ^{15}N tracer experiments have been conducted in the glasshouse and field in Australia (Tables 37.2 and 37.3). Uptake of ^{15}N by Rhodes grass was directly related to the amount of $^{15}NH_4NO_3$ applied (to provide N up to 448 kg ha^{-1}), and was distributed between tops and roots in a constant ratio, 4.5:1 (Henzell *et al.* 1964). Rhodes grass and Townsville stylo, grown both separately and together, took up about one third of ammonium ^{15}N and two thirds of nitrate ^{15}N, applied regularly over a 13-week period (Vallis *et al.* 1967); however, for both fertilizers, only 6.5% of the ^{15}N taken up by the two plant species, when grown in association, was present in the legume (averaged over three harvests). The Townsville stylo in mixed culture fixed N_2 sooner than in pure stands, because of the competition from the Rhodes grass for applied and soil N. In a similar experiment, in which ^{15}N-labelled Rhodes grass residues were substituted for the ^{15}N-labelled fertilizers (Henzell *et al.* 1968), the C:N ratio of the ^{15}N-labelled plant residues was 44:1, and their incorporation resulted in some net immobilization of unlabelled soil mineral N. Nevertheless, the total ^{15}N uptake of Rhodes grass (*Chloris gayana*) and Siratro (*Macroptilium atropurpureum*), grown separately or together in pots containing ^{15}N-labelled plant residues, was similar in all combinations; although when the grass and legume were grown together, about one third only of the ^{15}N uptake was associated with the legume. After 3 weeks' growth, about 50% of the N of the Siratro plants was derived from the soil, but after 15 weeks, only 2–4% of the legume N was derived from this source, reflecting large increases in the amounts of N_2 fixed symbiotically. Utilization by Rhodes grass of ^{15}N released from decomposing Rhodes grass residues was further examined by Moore (1974); the effects of mineral N addition, of plant part, and of the age and size and placement of the residues, on ^{15}N uptake by two crops over a 16 month period were determined. For most treatments the percentage uptakes by Rhodes grass tops ranged from 23 to 34%, least utilized (14%) was ^{15}N from surface litter, and Rhodes grass roots contained 6–8% of the plant residue ^{15}N. Periodical additions of mineral N to the soil during growth of Rhodes grass increased its uptake of ^{15}N, whereas grinding and age (C:N ratios) of added plant residues had little effect.

Wheat has been grown on a *black earth* amended with ^{15}N-labelled NH_4^+, NO_3^- (Craswell and Martin 1975*a*), NH_3, or urea (Saffigna 1976). Wheat straw accounted for about 15% of the applied N, irrespective of fertilizer type or its N concentration (22–84 kg ha^{-1}) or the moisture content of the soil during crop growth; wheat grain accounted for approximately 50% of the applied ^{15}N for most treatments, but the percentage uptake declined at the higher fertilizer application rates. Higher percentage uptakes were achieved with wheat grown on a *red-brown earth*; wheat grain contained 60% of the ^{15}N of applied NH_3 or urea, irrespective of application rate. Following wheat crops took up 3.5–8% and 3–12% of the residual fertilizer ^{15}N in the *red-brown earth* and *black earth*.

Uptake of applied ^{15}N by wheat or pasture has been determined in the field using open or enclosed microplots (Table 37.3). Fertilizer concentration did not significantly affect the percentage uptake of ^{15}N by Rhodes grass (Henzell 1971). The percentage uptake by pasture or wheat from applied fertilizers exceeded the uptake by wheat from decomposing ^{15}N-labelled medic residues, but not when expressed as percentages of the mineral ^{15}N released from the medic residues during the eight months prior to seeding (Ladd, *et al.* 1981*b*). Portions of the mineral ^{15}N in the soils at seeding were re-immobilized during wheat growth. Current field experiments in South Australia are establishing those periods during cereal cropping when the net N immobilization takes place.

Mathematical models of nitrogen cycling

Attempts to express the nitrogen cycle in mathematical terms have arisen from a recognition of the important effects that changes in soil nitrogen with time have on land use, and of the complexity of the nitrogen cycle. The models proposed can be considered in two categories: simple analytical models, and detailed process models.

Simple analytical models have a long history in evaluating the changes in soil organic matter with time (Jenny 1933; Leeper 1938; Greenland and Nye 1959; Bartholomew and Kirkham 1960; Russell 1960; Greenland 1971). The differential equation used in these studies is that originally proposed as

$$\frac{dN}{dt} = -K_1 N + K_2$$

by Jenny (1933), where N is the soil organic nitrogen content, K_1 is a decomposition coefficient, K_2 is an addition term, and t is time. The solution to this equation can be obtained by fitting appropriate field data to it, and estimating the coefficients. Aspects of this approach include the concept of equilibrium soil N levels and half-life periods of increase or decrease in soil N content; half life of equilibrium for instance is $t_{1/2} = 0.693/K_1$.

Calculation of this value for a number of environments suggests that organic matter change in field situations takes place over a period of decades. Thus analysis of the soil N increase of the surface soil under temperate climate pastures at Kybybolite in South Australia has given K_1 values of 2.9 percent, indicating a half life of 24 years (Russell 1962).

There have been a number of variations and additions to the above equation (Henin and Dupuis 1945; Woodruff 1949; Dawson 1950; Henin *et al.* 1959). The two common variations have been to include plant yield in the addition term, and to consider different forms of organic N. With increased computing power available, it has also been possible to consider different crop sequences and different crop yields over time, and to calculate coefficients from crop and pasture rotation experiments, where data are available (Russell 1975).

There is a number of limitations to the analytical approach. Firstly, soil organic matter information, where available, has usually been restricted to organic N or C levels, whereas additional data on other components of soil organic matter would be useful. Secondly, changes in soil organic N generally occur over a period of decades; continuity of interest, consistency of sampling procedures, and the maintenance of the integrity of previous samples, can usually only be achieved at well established research stations. In almost all cases, insufficient past soil sampling represents a major limitation to the analytical approach.

Detailed process models to simulate the N cycle are theoretically attractive. The concept of a model linking the various processes affecting N transformations and transfers, and simulating changes as influenced by factors such as soil properties, temperature, rainfall, and plant growth, would appear to be of considerable value in understanding soil N changes in ecosystems.

Although in the long term the approach is attractive, its practical application in the short term is limited by the diversity of N forms, the large number of processes affecting N in soil, and the lack of knowledge of these processes. Allison (1955) referred to the 'enigma' of soil N balance sheets; this enigma concerned our inability to balance even the crude N inputs and outputs of many systems. In a later paper, Allison (1966) indicated that some progress had been made in obtaining more accurate balance data, and also in a better definition of transformation mechanisms. Nevertheless, whilst considerable progress has been made in recent years, vital information, particularly of loss pathways and quantitative information concerning fluxes, is still not available for many systems. Also, the recent discovery that under moderately high soil temperature conditions biological N_2 fixation can occur with the association of *Azospirillum* and certain non-leguminous plants (Dobereiner 1974; Dobereiner and Day 1975), has uncovered an additional pathway that has to be evaluated in some soil systems.

Some soil N models have been published (Frere *et al.* 1970; Dutt *et al.* 1972; Myers 1974). Even though quantitative information is not available to allow these models to operate over other than a very limited range, they do indicate areas where information and research is required. There have been attempts to model parts of the soil N system. Ross *et al.* (1972) developed a model to examine the effect of soil N on the vegetative growth of a grass–legume pasture; the authors believed that this submodel could be used as part of a larger model if required.

Thus although the concept of a soil N process model is sound, the main limitation concerns knowledge of the processes themselves, and of quantitative information of the factors affecting them. Much more intensive study of sub-models, that is, of specific parts of the N cycle is needed. With adequate mathematical representation and validation of these sub-models, it should be possible to combine sub-models, and eventually develop a comprehensive model of the soil N cycle.

Manipulation of nitrogen in field systems

Changes in soil organic nitrogen amounts and composition are consequences of man's manipulation of soil ecosystems. In the last 180 years, land use changes in Australia have been substantial; large areas have been cultivated, and even larger areas have been cleared of forest and savannah vegetation to be replaced by pastures grazed by sheep and cattle.

Two main effects of these changes can be recognized. Firstly, soil organic nitrogen levels in some of the higher fertility soils have declined where cropping and cultivation have been carried out. Secondly, increases in soil organic nitrogen have occurred in many soils associated with pastures, in most cases with the removal of limiting factors to plant growth by the application of phosphorus and sulphur and, in some soils, of trace elements such as copper, zinc and molybdenum. Studies of changes in soil organic nitrogen content with time suggest that equilibrium has not yet been attained in many of these systems because rates of change are low. Grass-legume ley farming, using particular sequences of crops, has been successfully developed in many of the more humid winter rainfall areas to attain equilibrium at adequate soil organic nitrogen levels. However, cultivation of summer rainfall areas is more recent, and, in the subtropical semi-arid regions in particular, large decreases in soil organic nitrogen levels have been reported (Martin and Cox 1956). Satisfactory land use systems have not yet evolved in these areas.

Effects of cultivation and cropping

Cultivation invariably results in decreased soil organic N levels. Clarke and Marshall (1947) showed that 16–20 years of cultivation in a fallow–cereal rotation on *red-brown earths* near Adelaide in South Australia may reduce the soil organic N to less than 65% of its original level. There is evidence that even greater losses occur over larger time periods. In the permanent rotation experiment begun in 1926 at the Waite Institute, the original soil N content of the 0–22.5 cm layer of seven treatments was 0.175% N; in 1961 (Greenland 1971) mean soil N levels in these plots were recorded as 0.102% N, and by 1973 (Russell, J. S. and Oades, J. M., unpublished data) the mean N content was found to be 0.084%. Thus losses of more than 50% had occurred in less than half a century. Although soil N losses in this environment are higher than in many other areas (Clarke and Russell 1977, Table 7), there is little question that cropping and cultivation lead to declines in soil N status of initially fertile soils. Fallowing, in particular, causes rapid losses of soil organic matter. This practice has been widely used in Australia for more than 100 years (Sims 1977), and involves soil cultivation many months prior to crop planting; its purpose is to allow soil water accumulation, and the build- up of available soil N for the subsequent crop. However, the practice of fallowing increases the rate of soil organic matter decomposition, particularly when it is accompanied by frequent soil cultivation for weed control.

Numerous studies of the effect of different cultivation and cropping treatments on soil organic N levels have been made in Australia (reviewed by Clarke and Russell 1977), particularly in southern winter-rainfall areas. In all cases, while fallow–wheat rotations have maintained relatively high crop yields in comparison with other treatments, they have resulted in substantial losses of soil organic N. In contrast, although yields of continous wheat have been lower, the decline in soil organic N levels has been less severe. Thus, taking the mean of six different experiments (Clarke and Russell 1977, Table 6), the mean suface soil N content of the fallow–wheat treatments was 0.053%, compared with 0.066% for the continuous wheat treatments. The recent development of minimum tillage technology offers considerable promise in both the conservation of soil organic nitrogen and its more efficient utilization.

The use of grass–legume leys has been found to offset declines in soil N due to cultivation and cropping (Clarke and Russell 1977). This has been due to both the decrease in cultivation intensity associated with fewer crops, and an increase in N input associated with the growth of legumes. For many environments the maintenance of a stable soil organic N level, with adequate available N for the crop phase, is possible through the use of crop rotations that involve both periods of crop and periods of pasture ley.

The practicability of grass–legume leys in different environments depends firstly on the availability of suitable pasture legumes, and secondly on the development of crop–livestock enterprises that allow a lower cropping intensity. Satisfactory pastures, based on *Trifolium* and *Medicago* spp., are available in winter rainfall areas, and leys are usually used. However, in subtropical summer rainfall areas there is a lack of perennial or self-regenerating annual legumes which can persist in these environments and make contributions to soil organic nitrogen. In areas such as the *black earths* of the Darling Downs, therefore, leys are rarely used.

There is, however, increasing interest in the use of grain legumes in crop sequences in these areas, and in their possible role as a source of N input to cultivated areas. The possiblity of using *Stylosanthes* species, such as *S. humilis* (Townsville stylo), as pasture leys in rotation with crops such as sorghum is also being examined in northern Australia. The possible combination of *Azospirillum* in association with tropical grasses, such as *Panicum maximum*, remains to be determined. The bacterium *Azospirillum* has been shown to be present in a *grey clay* from south east Queensland (Date, R.A. pers. comm.).

The willingness to accept a low cropping intensity also depends on the economic viability of crop–livestock enterprises. In recent years there has been a tendency for cropping intensity to increase, resulting in increased pressure on soil N levels. If cropping intensity increases further in the future, soil organic N is likely to decrease to lower equilibrium levels. This would result in inadequate available N for crop growth, and the application of N fertilizers would be necessary to maintain crop yield;

Table 37.4 Increases in soil nitrogen content under pastures of different soils in Australia

Great Soil Group	Soil nitrogen (%) original	final	Period (Years)	Depth (cm)	Principal pasture legume	Reference
Lacteritic podzolic	0.052	0.158	6	0–5	*Desmodium uncinatium*	Henzell *et al.* (1966)
Lateritic sand	0.057	0.083	5	0–10	*Trifolium subterraneum*	Watson (1963)
Podzolic	0.085	0.139	26	0–10	*Trifolium subterraneum*	Donald and Williams (1954)
Transitional *podzolic*	0.054	0.086	8	0–15	*Trifolium subterraneum*	Mullaly *et al.* (1967)
Red-brown earth	0.084	0.130	10	0–15	*Trifolium subterraneum*	Mullaly *et al.* (1967)
Red-brown earth	0.068	0.078	3	0–12.5	*Trifolium subterraneum*	Clarke (1970)
Solonized solonetz	0.060	0.238	40	0–5	*Trifolium* spp.	Russell (1960)
Solonized brown soil	0.056	0.059	2	0–10	*Medicago truncatula*	Parker (1957)
Solonized brown soil	0.032	0.041	6	0–15	*Medicago truncatula*	Mullaly *et al.* (1967)
Grey brown soil of *heavy texture*	0.090	0.128	8	0–15	*Medicago truncatula*	Mullaly *et al.* (1967)
Grey brown soil of *heavy texture*	0.092	0.109	3	0–15	*Medicago truncatula*	Meagher and Rooney (1966)

there could also be a deterioration in soil physical conditions, with effects on soil structure, infiltration characteristics and erodibility. The magnitude of these effects will depend very much on intrinsic soil characteristics, and these require better definition.

Effects of pasture

Although short-term pastures are important as components of crop–pasture systems on soils of moderate fertility, long-term pastures have played a very significant role in the development and use of low-fertility soils. A key factor has been the evolution of practices that increase organic N levels in the soil to well above their original levels.

The concept of using the biological fixation of N_2 by legumes to increase the amount of N in soil–plant systems is not new, and has been common since early agriculture. However, unique features of the Australian approach have been: its application to soils of very low fertility; the systematic search for pasture legumes and associated effective *Rhizobium* spp. which are adapted to these soils, and the use of scientific methods to identify and overcome limiting major and minor soil nutrient factors as they affect legume growth. Although the soils have often been acid, the use of lime has been rare; rather, legumes such as *Trifolium subterraneum* that tolerate these conditions have been used.

A consequence of the increased amount of N in the soil–plant system has been substantial increases in soil organic N levels. The original and final N contents obtained in a series of studies are shown in Table 37.4. In most cases where substantial increases have been found, the original soil levels were low; also, increases have invariably been associated with the application of P, and in some cases by the introduction of pasture legumes; time of accumulation has also been important.

When expressed in terms of rates of change, the mean annual increment of N has been as high as 182 kg ha^{-1}, for short term studies (Clarke and Russell 1977); however, for studies exceeding 20 years the annual values become 47-67 kg ha^{-1}.

Accompanying these changes in organic N have been decreases in soil pH, with increases in cation exchange capacity and exchangeable cations (Donald and Williams 1954; Williams and Donald 1957; Russell 1960, 1961), and changes in soil moisture characteristics (Russell and Shearer 1964).

While considerable success has been achieved in the past in increasing soil organic N levels, there has been less success in making use of the accumulated fertility. More thought now needs to be given to the development of short cropping periods interspersed between long ley periods. Although large areas of the improved soils have arable disabilities such as steepness, or have excessive wetness, or are susceptible to erosion, there does appear to be scope for intermittent cultivation of some of these soils and for the development of agronomic practices that can make productive use of high soil N levels.

References

Allison, F. E. (1955). The enigma of soil nitrogen balance sheets. *Adv. Agron.* **7**, 213-51.

Allison, F. E. (1965). Evaluation of incoming and outgoing processes that affect soil nitrogen. *In* 'Soil Nitrogen.' (Eds W.V. Bartholomew and F.E. Clark). pp.573-606. (Am. Soc. Agron.: Madison.)

Allison, F. E. (1966). The fate of nitrogen applied to soils. *Adv. Agron.* **18**, 219-59.

Amato, M. and Ladd, J. N. (1980). Studies of nitrogen immobilization and mineralization in calcareous soils. V. Formation and distribution of isotope-labelled biomass during decomposition of ^{14}C, ^{15}N-labelled plant material. *Soil Biol. Biochem.* **12**, 405-11.

Ayanaba, A., Tuckwell, S. B. and Jenkinson, D. S. (1976). The effects of clearing and cropping on the organic reserves and biomass of tropical forest soils. *Soil Biol. Biochem.* **8**, 519-25.

Baker, J. H., Peech, M. and Musgrave, R. B. (1959). Determination of application losses of anhydrous ammonia. *Agron. J.* **51**, 361-2.

Bartholomew, W. V. (1965). Mineralization and immobilization of nitrogen in the decomposition of plant and animal residues. *In* 'Soil Nitrogen.' (Eds W.V. Bartholomew and F.E. Clark). pp.285-306 (Am. Soc. Agron.: Madison.)

Bartholomew, W. V. and Kirkham, D. (1960). Mathematical descriptions and interpretation of culture induced soil nitrogen changes. Trans. 7th Int. Congr. Soil Sci. Vol. 2, pp.471-6.

Beech, D. F. and Norman, M. J. T. (1964). The effect of time of planting and nitrogen fertilizer on yield attributes of brassica oilseed crops in the Ord River Valley. *Aust. J. Exp. Agric. Anim. Husb.* **4**, 206-14.

Bevege, D. I. and Richards, B. N. (1970). Nitrogen in the growth of *Araucaria cunninghamii* Ait. underplanted in *Pinus* stands. *Ecology* **51**, 134-42.

Bevege, D. I. and Richards, B. N. (1971). Nitrogen fixation in the rhizosphere of mycorrhizal *Auracaria* seedlings. XV. IUFRO Congress, Gainsville, Florida.

Biederbeck, V. O. and Paul, E. A. (1973). Fractionation of soil humate with phenolic solvents and purification of the nitrogen-rich portion with polyvinylpyrrolidone. *Soil Sci.* **115**, 357-66.

Black, A. S. and Waring, S. A. (1972). Ammonium fixation and availability in some cereal producing soils in Queensland. *Aust. J. Soil Res.* **10**, 197-207.

Black, A. S. and Waring, S. A. (1976*a*). Nitrate leaching and adsorption in a krasnozem from Redland Bay, Qld. I. Leaching of banded ammonium nitrate in a horticultural rotation. *Aust. J. Soil Res.* **14**, 171-80.

Black, A. S. and Waring, S. A. (1976*b*). Nitrate leaching and adsorption in a krasnozem from Redland Bay, Qld. II. Soil factors influencing adsorption. *Aust. J. Soil Res.* **14**, 181-88.

Black, A. S. and Waring, S. A. (1976*c*). Nitrate leaching and adsorption in a krasnozem from Redland Bay, Qld. III. Effect of nitrate concentration on adsorption and movement in soil columns. *Aust. J. Soil Res.* **14**, 189-95.

Black, A. S. and Waring, S. A. (1977). The natural abundance of ^{15}N in the soil–water system of a small catchment area. *Aust. J. Soil Res.* **15**, 51-7.

Blunt, C. G. and Fisher, M. J. (1973). Production and utilization of fodder and grain sorghum as forage for cattle in the Ord River Valley, Western Australia. *Aust. J. Expt. Agric. Anim. Husb.* **13**, 234-37.

Bray, J. R. and Gorham, E. (1964). Litter production in forests of the world. *Adv. Ecol. Res.* **2**, 100-57.

Bremner, J. M. (1965). Organic nitrogen in soils. *In* 'Soil Nitrogen.' (Eds W.V. Bartholomew and F.E. Clark). pp.93-149. (Am. Soc. Agron.: Madison.)

Brisbane, P. G., Amato, M. and Ladd, J. N. (1972). Gas chromatographic analysis of amino acids from the action of proteolytic enzymes on soil humic acids. *Soil Biol. Biochem.* **4**, 51-61.

Burford, J. R. and Greenland, D. J. (1970). Denitrification under an annual pasture. 11th Int. Grassland Congr. pp.458-61. (Univ. of Queensland Press, St. Lucia.)

Burford, J. R. and Millington, R. J. (1968). Nitrous oxide in the atmosphere of a red-brown earth. Trans. 9th Int. Congr. Soil Sci. Adelaide. Vol. 2, pp.505-11.

Burford, J. R. and Stefanson, R. (1973). Measurement of gaseous losses of nitrogen from soils. *Soil Biol. Biochem.* **5**, 133-41.

Burns, R. G., El-Sayed, M. H. and McLaren, A. D. (1972*a*). Extraction of a urease-active organo-complex from soil. *Soil Biol. Biochem.* **4**, 107-8.

Burns, R. G., Pukite, A. H. and McLaren, A. D. (1972*b*). Concerning the location and persistence of soil urease. *Soil Sci. Soc. Am. Proc.* **36**, 308-11.

Butler, J. H. A. and Ladd, J. N. (1969). Effect of extractant and molecular size on the optical and chemical properties of soil humic acids. *Aust. J. Soil Res.* **7**, 229-39.

Catchpoole, V. R. (1975). Pathways for losses of fertilizer nitrogen from a Rhodes grass pasture in South-eastern Queensland. *Aust. J. Agric. Res.* **26**, 259-68.

Clarke, A. L. (1970). Nitrogen accretion by an impoverished red-brown soil under short-term leys. Proc. 11th Int. Grassl. Congr. pp.461-65.

Clarke, A. L. and Russell, J. S. (1977). Crop sequential practices. *In* 'Soil Factors in Crop Production in a Semi-arid Environment.' (Eds J.S. Russell and E.L. Greacen.) pp.279-300. (Univ. of Queensland Press: St. Lucia.)

Clarke, G. B. and Marshall, T. J. (1947). Influence of cultivation on soil structure and its assessment in soils of variable mechanical composition. *J. Counc. Sci. Ind. Res. Aust.* **20**, 162-75.

Craswell, E. T. and Martin, A. E. (1974). Effect of moisture content on denitrification in a clay soil. *Soil Biol. Biochem.* **6**, 127-29.

Craswell, E. T. and Martin, A. E. (1975*a*). Isotopic studies of the nitrogen balance in a cracking clay. I. Recovery of added nitrogen from soil and wheat in the glasshouse and gas lysimeter. *Aust. J. Soil Res.* **13**, 43-52.

Craswell, E. T. and Martin, A. E. (1975*b*). Isotopic studies of the nitrogen balance in a cracking clay. II. Recovery of nitrate ^{15}N added to columns of packed soil and microplots growing wheat in the field. *Aust. J. Soil Res.* **13**, 53-61.

Craswell, E. T., Saffigna, P. G. and Waring, S. A. (1970). The mineralization of organic nitrogen in dry soil aggregates of different sizes. *Plant Soil* **33**, 383-92.

Craswell, E. T. and Waring, S. A. (1972*a*). Effect of grinding on the decomposition of soil organic matter. I. The mineralization of organic nitrogen in relation to soil type. *Soil Biol. Biochem.* **4**, 427-33.

Craswell, E. T. and Waring, S. A. (1972*b*). Effect of grinding on the decomposition of soil organic matter. II. Oxygen uptake and nitrogen mineralization in virgin and cultivated cracking clay soils. *Soil Biol. Biochem.* **4**, 435-42.

Dawson, J. E. (1950). Soil organic matter. *In* 'Agricultural Chemistry. Vol. 1 Principles of Agricultural Chemistry.' (Ed. D.E. Frear.) pp.778-97. (Van Nostrand: New York.)

Denmead, O. T., Simpson, J. R. and Freney, J. R. (1974). Ammonia flux into the atmosphere from a grazed pasture. *Science* **185**, 609-10.

Denmead, O. T., Freney, J. R. and Simpson, J. R. (1976). A closed ammonia cycle within a plant canopy. *Soil Biol. Biochem.* **8**, 161-64.

Denmead, O. T., Simpson, J. R. and Freney, J. R. (1977). A direct field measurement of ammonia emission after injection of anhydrous ammonia. *Soil Sci. Soc. Am. Proc.* **41**, 1001-04.

Dobereiner, J. (1974). Nitrogen fixing bacteria in the rhizosphere. *In* 'The Biology of Nitrogen Fixation.' (Ed. A. Quispel.) pp.86-120. (North-Holland: Amsterdam.)

Dobereiner, J. and Day, J. M. (1975). Nitrogen fixation in the rhizosphere of tropical grasses. *In* 'Nitrogen Fixation by Free-living Microorganisms'. (Ed. W. D. P. Stewart.) (Camb. Univ. Press: Cambridge.)

Donald, C. M. and Williams, C. H. (1954). Fertility and productivity of a podsolic soil as influenced by subterranean clover (*Trifolium ubterraneum L.*) and superphosphate. *Aust. J. Agric. Res.* **5**, 664-87.

Dutt, G. R., Shaffer, M. J. and Moore, W. J. (1972). Computer simulation model of dynamic bio-physic chemical processes in soils. Ariz. Agr. Expt. Sta. Tech. Bull. No. 196.

Eriksson, E. (1952). Composition of atmospheric precipitation. I. Nitrogen compounds. *Tellus* **4**, 145.

Fenn, L. B. and Escarzago, R. (1976). Ammonia volatilization from surface applications of ammonium compounds on calcareous soils. V. Soil water content and method of nitrogen application. *Soil Sci. Soc. Am. J.* **40**, 537-41.

Fenn, L. B. and Kissel, D. E. (1976). The influence of cation exchange capacity and depth of incorporation on ammonia volatilization from ammonium compounds applied to calcareous soils. *Soil Sci. Soc. Am. J.* **40**, 394-98.

Ford, G. W. (1968). Partly humified organic matter in soils. Its contribution to mineralizable nitrogen. Ph.D. Thesis, Univ. of Adelaide.

Ford, G. W. and Greenland, D. J. (1968). The dynamics of partly humified organic matter in some arable soils. Trans. 9th Int. Congr. Soil Sci. Adelaide. Vol. 2, pp.403-10.

Ford, G. W., Greenland, D. J. and Oades, J. M. (1969). Separation of the light fraction from soils by ultrasonic dispersion in halogenated hydrocarbons containing a surfactant. *J. Soil Sci.* **20**, 291-96.

Freney, J. R. (1968). The extraction and partial characterization of non hydrolysable nitrogen in soil. Trans. 9th Int. Congr. Soil Sci. Adelaide. Vol. 3, pp.531-39.

Freney, J. R. and Miller, R. J. (1970). Investigation of the clay mineral protection theory for non-hydrolysable nitrogen in soil. *J. Sci. Fd. Agric.* **21**, 56-61.

Freney, J. R. and Simpson, J. R. (1969). The mineralization of nitrogen from some organic fractions in soil. *Soil Biol. Biochem.* **1**, 241-51.

Freney, J. R., Denmead, O. T. and Simpson, J. R. (1978). Soil as a source or sink for atmospheric nitrous oxide. *Nature (London)* **273**, 530-32.

Frere, M. H., Jensen, M. E. and Carter, J. N. (1970). Modelling water and nitrogen behaviour in the soil-plant system. Proc. 1970 Summer Computer Simulation Conf. Denver. Vol. II, pp.746-50.

Greenland, D. J. (1971). Changes in the nitrogen status and physical conditions of soils under pastures with special references to the maintenance of the fertility of Australian soils growing wheat. *Soils Fert.* **34**, 237-51.

Greenland, D. J. and Ford, G. W. (1964). Separation of partially humified organic materials from soils by ultrasonic dispersion. Trans. 8th Int. Cong. Soil Sci. Bucharest. Vol. 3, pp.137-48.

Greenland, D. J. and Nye, P. H. (1959). Increases in the carbon and nitrogen contents of tropical soils under natural fallows. *Soil Sci.* **10**, 284-99.

Haider, K., Frederick, L. R. and Flaig, W. (1965). Reactions between amino acid compounds and phenols during oxidation. *Plant Soil* **22**, 49-64.

Handley, W. R. C. (1961). Further evidence for the importance of residual leaf protein complexes in litter decomposition and the supply of nitrogen for plant growth. *Plant Soil* **15**, 37-73.

Hauck, R. D. (1971). Quantitative estimates of nitrogen-cycle processes. *In* 'N^{15} in Soil and Plant Studies.' pp. 65-80. (IAEA: Vienna.)

Henin, S. and Dupuis, M. (1945). A tentative balance sheet for soil organic matter. *Ann. Agron.* **15**, 17-29.

Henin, S., Monnier, G. and Turc, L. (1959). Un aspect de la dynamique des matieres organiques du sol. *C. R. Acad. Sci.* **248**, 138-41.

Henzell, E. F. (1963). Nitrogen fertilizer responses of pasture grasses in south-eastern Queensland. *Aust. J. Exp. Agric. Anim. Husb.* **3**, 290-99.

Henzell, E. F. (1971). Recovery of nitrogen from four fertilizers applied to Rhodes grass in small plots. *Aust. J. Exp. Agric. Anim. Husb.* **11**, 420-30.

Henzell, E. F. (1972). Loss of nitrogen from a nitrogen-fertilized pasture. *J. Aust. Inst. Agric. Sci.* **38**, 309-10.

Henzell, E. F., Fergus, I. and Martin, A. E. (1966). Accumulation of soil nitrogen and carbon under a *Desmodium uncinatum* pasture. *Aust. J. Exp. Agric. Anim. Husb.* **6**, 157-60.

Henzell, E. F., Martin, A. E. and Ross, P. J. (1970). Recovery of fertilizer nitrogen by Rhodes grass. Proc. 11th Int. Grassl. Congr. pp. 411-14.

Henzell, E. F., Martin, A. E., Ross, P. J. and Haydock, K. P. (1964). Isotopic studies on the uptake of nitrogen by pasture grasses. II. Uptake of fertilizer nitrogen by Rhodes grass in pots. *Aust. J. Agric. Res.* **15**, 876-84.

Henzell, E. F., Martin, A. E., Ross, P. J. and Haydock, K. P. (1968). Isotopic studies on the uptake of nitrogen by pasture plants. IV. Uptake of nitrogen from labelled plant material by Rhodes grass and Siratro. *Aust. J. Agric. Res.* **19**, 65-77.

Hingston, F. J. (1974). Seasonal distribution of mineral nitrogen with particular reference to leaching. *Aust. J. Exp. Agric. Anim. Husb.* **14**, 815-21.

Jansson, S. L. (1958). Tracer studies on nitrogen transformation in soil with special attention to mineralization–immobilization relationships. *Lantbrukshoegsk. Ann.* **24**, 101-361.

Jenkinson, D. S. (1971). Studies on the decomposition of C^{14}-labelled organic matter in soil. *Soil Sci.* **111**, 64-70.

Jenkinson, D. S. and Ayanaba, A. (1977). Decomposition of carbon-14 labelled plant material under tropical conditions. *Soil Sci. Soc. Am. J.* **41**, 912-15.

Jenkinson, D. S. and Powlson, D. S. (1976). The effects of biocidal treatments on metabolism in soil. V. A method for measuring soil biomass. *Soil Biol. Biochem.* **8**, 209-13.

Jenny, H. (1933). Soil fertility losses under Missouri conditions. Mo. Agr. Exp. Sta. Bull. No. 324.

Jensen, H. L. (1940*a*). Contribution to the nitrogen economy of Australian wheat soils with particular reference to New South Wales. *Proc. Linn. Soc. N. S. W.* **65**, 1-122.

Jensen, H. L. (1940*b*). Nitrogen fixation and cellulose decomposition by soil microorganisms. I. Aerobic cellulose-decomposers in association with *Azotobacter*. *Proc. Linn. Soc. N. S. W.* **65**, 543-56.

Jensen, H. L. and Swaby, R. J. (1940). Further investigations on nitrogen-fixing bacteria in soil. *Proc. Linn. Soc. N. S. W.* **65**, 557-64.

Kyuma, K., Hussain, A. and Kawaguchi, K. (1969). The nature of organic matter in soil organo-mineral complexes. *Soil Sci. Plant Nutr.* **15**, 149-55.

Ladd, J. N. (1981). The use of ^{15}N in following organic matter turnover, with specific reference to rotation systems. *Plant Soil* **58**, 401-11.

Ladd, J. N. and Amato, M. (1980). Studies of nitrogen immobilization and mineralization in calcareous soils. IV. Changes in the organic nitrogen of light and heavy subfractions of silt- and fine clay-size particles during nitrogen turnover. *Soil Biol. Biochem,* **12**, 185-189.

Ladd, J. N., Parsons, J. W. and Amato, M. (1977*a*). Studies of nitrogen immobilization and mineralization in calcareous soils. I. Distribution of immobilized nitrogen amongst soil fractions of different particle size and density. *Soil Biol. Biochem.* **9**, 309-18.

Ladd, J. N., Parsons, J. W. and Amato, M. (1977*b*). Studies of nitrogen immobilization and mineralization in calcareous soils. II. Mineralization of immobilized nitrogen from soil fractions of different particle size and density. *Soil Biol. Biochem.* **9**, 319-25.

Ladd, J. N., Amato, M. and Parsons, J. W. (1977*c*). Studies of nitrogen immobilization and mineralization in calcareous soils. III. Concentration and distribution of nitrogen derived from the soil biomass. 'International Symposium on Soil Organic Matter Studies', FAO, IAEA, Agrochimica Braunschweig, Vol. 1, pp. 301-10.

Ladd, J. N., and Brisbane, P. G. (1967). Release of amino acids from soil humic acids by proteolytic enzymes. *Aust. J. Soil Res.* **5**, 161-71.

Ladd, J. N., Brisbane, P. G., Butler, J. H. A. and Amato, M. (1976). Studies on soil fumigation. III. Effects on enzyme activities, bacterial numbers and extractable ninhydrin reactive compounds. *Soil Biol. Biochem.* **8**, 255-60.

Ladd, J. N. and Butler, J. H. A. (1963). Comparison of properties of synthetic and natural humic acids. *In* 'The Use of Isotopes in Soil Organic Matter Studies.' pp. 143-51. (FAO/IAEA Technical Meeting Braunschweig, Pergamon Press, London.)

Ladd, J. N. and Butler, J. H. A. (1966). Comparison of some properties of soil humic acids and synthetic phenolic polymers incorporating amino derivatives. *Aust. J. Soil Res.* **4**, 41-54.

Ladd, J. N. and Butler, J. H. A. (1975). Humus-enzyme systems and synthetic, organic polymer-enzyme analogs. *In* 'Soil Biochemistry'. (Eds E. A. Paul and A. D. McLaren.) Vol. 4, pp. 143-93. (Marcel Dekker: New York.)

Ladd, J. N. and Paul, E. A. (1973). Changes in enzymic activity and distribution of acid-soluble amino acid-nitrogen in soil during nitrogen immobilization and mineralization. *Soil Biol. Biochem.* **5**, 825-40.

Ladd, J. N., Oades J. M. and Amato, M. (1981*a*). Microbial biomass formed from ^{14}C, ^{15}N-labelled plant material decomposing in soils in the field. *Soil Biol. Biochem,* **13**, 119-26.

Ladd, J. N., Oades, J. M. and Amato, M. (1981*b*). Distribution and recovery of nitrogen from legume residues decomposing in soils sown to wheat in the field. *Soil Biol. Biochem.* **13**, 251-6.

Leeper, G. W. (1938). Organic matter of soil as determined by climate. *J. Aust. Inst. Agric. Sci.* **4**, 145-47.

Martin, A. E. and Cox. J. E. (1956). Nitrogen studies on black soils from the Darling Downs Queensland. Seasonal variations in moisture and mineral nitrogen fractions. *Aust. J. Agric. Res.* **1**, 169-83.

Martin, A. E., Gilkes, R. J. and Skjemstad, J. O. (1970). Fixed ammonium in soils developed on some Queensland phyllites and its relation to weathering. *Aust. J. Soil Res.* **8**, 71-80.

Martin, A. E., Henzell, E. F., Ross, P. J. and Haydock, K. P. (1963). Isotopic studies on the uptake of nitrogen by pasture grasses. I. Recovery of fertilizer nitrogen from the soil : plant system using Rhodes grass in pots. *Aust. J. Agric. Res.* **1**, 169-84.

McGarity, J. W. (1961). Denitrification studies on some South Australian soils. *Plant Soil* **14**, 1-21.

McGarity, J. W. and Myers, R. J. K. (1968). Denitrifying activity in solodized solonetz soils of Eastern Australia. *Soil Sci. Soc. Am. Proc.* **32**, 812-17.

McGarity, J. W. and Myers, R. J. K. (1973). Seasonal trends in the content of mineral nitrogen in solodized solonetz soil. *Aust. J. Exp. Agric. Anim. Husb.* **13**, 423-29.

McGill, W. B. (1971). Turnover of microbial metabolites during nitrogen mineralization and immobilization in soil. Ph. D. Thesis, Univ. of Saskatchewan.

McLaren, A. D., Pukite, A. H. and Barshad, I. (1975). Isolation of humus with enzymatic activity from soil. *Soil Sci.* **119**, 178-80.

Meagher, J. W. and Rooney, D. R. (1966). The effect of crop rotations in the Victorian Wimmera on the cereal rust nematode (*Heterodera avenae*), nitrogen fertility and wheat yield. *Aust. J. Exp. Agric. Anim. Husb.* **6**, 425-36.

Moore, A. W. (1974). Availability to Rhodes grass (*Chloris gayana*) of nitrogen in tops and roots added to soil. *Soil Biol. Biochem.* **6**, 249-55.

Moore, A. W. and Craswell, E. T. (1976). Non-uniformity of ^{15}N-labelling in plant material. *Commun. Soil Sci. Plant Anal.* **7**, 335-44.

Moore, A. W. and Russell, J. S. (1968). Relative constancy of soil nitrogen fractions with varying total soil nitrogen. Trans. 9th Int. Congr. Soil Sci. Adelaide. Vol. 2, pp. 557-66.

Moore, A. W. and Russell, J. S. (1970). Changes in chemical fractions of nitrogen during incubation of soils with histories of large organic matter increase under pasture. *Aust. J. Soil Res.* **8**, 21-30.

Mullaly, J. V., McPherson, J. B., Mann, A. P. and Rooney, D. R. (1967). The effect of length of legume and non-legume leys on quantitative soil nitrogen at some locations in the Victorian wheat areas. *Aust. J. Exp. Agric. Anim. Husb.* **7**, 568-71.

Myers, R. J. K. (1972). The effect of sulphide on nitrate reduction in soil. *Plant Soil* **37**, 431-33.

Myers, R. J. K. (1974). Soil processes affecting nitrogenous fertilizer. Proc. Symp. Ecol. Aspects Fert. Tech. Aust. Inst. Agric. Sci., Sydney, pp. 3-25.

Myers, R. J. K. (1975). Temperature effects on ammonification and nitrification in a tropical soil. *Soil Biol. Biochem.* **7**, 83-6.

Myers, R. J. K. and McGarity, J. W. (1971). Factors influencing high denitrifying activity in the subsoil of solodized solonetz. *Plant Soil* **35**, 145-60.

Myers, R. J. K. and McGarity, J. W. (1972). Denitrification in undisturbed cores from a solodized solonetz B horizon. *Plant Soil* **37**, 81-9.

Nommik, H. (1965). Ammonium fixation and other reactions involving a non-enzymatic immobilization of mineral nitrogen in soil. *In* 'Soil Nitrogen.' (Eds W. V. Bartholomew and F. E. Clark). pp. 198-258. (Am. Soc. Agron. : Madison.)

Northcote, K. H. (1965). A Factual Key for the Recognition of Australian Soils, CSIRO Aust. Div. Soils Div. Rep No. 2/65.

Nyhan, J. W. (1975). Decomposition of carbon-14 labelled plant materials in a grassland soil under field conditions. *Soil Sci. Soc. Am. Proc.* **39**, 643-48.

Oades, J. M. and Ladd, J. N. (1977). Biochemical properties: Carbon and nitrogen metabolism. *In* 'Soil Factors in Crop Production in a Semi-arid Environment.' (Eds J. S. Russell and E. L. Greacen). pp. 127-60. (Univ. of Queensland Press: St. Lucia.)

Oberlander, H. E. (1973). The fate of organic manures in soil as traced by means of radiocarbon. *Pontif. Accad. Sci. Scripta Varia* **38**, 1001-71.

Osborne, G. J. (1977). Chemical fractionation of soil nitrogen in six soils from southern New South Wales. *Aust. J. Soil Res.* **15**, 159-65.

Parker, C. A. (1957). Non symbiotic N-fixing bacteria in soil. III. Total nitrogen changes in a field soil. *J. Soil Sci.* **8**, 48-59.

Parry, D. J. (1971). The distribution, availability and measurement of subsoil nitrogen. M. Agric. Sci. Thesis, Univ. of Adelaide.

Parsons, J. W., Ladd, J. N. and Amato, M. (1975). Mineralization and transformation of ^{15}N labelled plant residues in soil *J. Sci. Fd. Agric.* **27**, 1426-27.

Piper, T. J. and Posner, A. M. (1968). On the amino acids found in humic acid. *Soil Sci.* **106**, 188-92.

Piper, T. J. and Posner, A. M. (1972). Humic acid nitrogen. *Plant Soil* **36**, 595-98.

Powlson, D. S. and Jenkinson, D. S. (1976). The effects of biocidal treatments on metabolism in soil. II. Gamma irradiation, autoclaving, air-drying and fumigation. *Soil Biol. Biochem.* **8**, 179-88.

Prescott, J. A. (1931). The soils of Australia in relation to vegetation and climate. C.S.I.R. Aust. Bull. No. 52.

Richards, B. N. (1964). Fixation of atmospheric nitrogen in coniferous forests. *Aust. For.* **28**, 68-74.

Richards, B. N. (1973). Nitrogen fixation in the rhizosphere of conifers. *Soil Biol. Biochem.* **5**, 149-52.

Richards, B. N. and Bevege, D. I. (1967). The productivity and nitrogen economy of artificial ecosystems comprising various contaminations of perennial legumes and coniferous tree species. *Aust. J. Bot.* **15**, 467-80.

Richards, B. N. and Voigt, G. K. (1964). Role of Mycorrhiza in nitrogen fixation. *Nature (London)* **201**, 310-11.

Richards, B. N. and Voigt, G. K. (1965). Nitrogen accretion in coniferous forest ecosystems. Proc. 2nd North American Forest Soils Conf. Corvallis, Oregon, pp. 105-15.

Ross, P. J. Henzell, E. F. and Ross, D. R. (1972). Effects of nitrogen and light in, grass–legume pastures — a systems analysis approach. *J. Appl. Ecol.* **9**, 535-56.

Ross, P. J., Martin, A. E. and Henzell, E. F. (1964). A gas-tight growth chamber for investigating gaseous nitrogen changes in the soil:plant: atmosphere system. *Nature (London)* **204**, 444-47.

Ross, P. J., Martin, A. E. and Henzell, E. F. (1968). Gas lysimetry as a technique in nitrogen studies on the soil:plant:atmosphere system. Trans. 9th Int. Congr. Soil Sci. Adelaide. Vol. 2, pp. 487-94.

Rowell, M. J., Ladd, J. N. and Paul, E. A. (1973). Enzymically active complexes of proteases and humic acid analogues. *Soil Biol. Biochem.* **5**, 699-703.

Russell, J. S. (1960). Soil fertility changes in the long-term experimental plots at Kybybolite, South Australia. I. Changes in pH, total nitrogen, organic carbon and bulk density. *Aust. J. Agric. Res.* **11**, 902-26.

Russell, J. S. (1961). Soil fertility changes in the long-term experimental plots at Kybybolite, South Australia. III. Changes in cation exchange capacity and exchangeable cations. *Aust. J. Agric. Res.* **12**, 273-85.

Russell, J. S. (1962). Estimation of the time factor in soil organic matter equilibrium under pastures. Trans. Int. Soil Conf. N. Z. Comm. Vol. IV, pp. 3-8.

Russell, J. S. (1964). Some observations on the influence of nitrogen fertilizer applied at seeding on the nitrogen content of the wheat grain. *Aust. J. Exp. Agric. Anim. Husb.* **4**, 345-51.

Russell, J. S. (1975). A mathematical treatment of the effect of cropping system on soil organic nitrogen in two long-term sequential experiments. *Soil Sci.* **120**, 37-44.

Russell, J. S. and Shearer, R. C. (1964). Soil fertility in the long-term experimental plots at Kybybolite, South Australia. IV. Changes in certain moisture characteristics. *Aust. J. Agric. Res.* **15**, 91-9.

Saffigna, P. G. (1976). Fertilizer nitrogen balance and transformations in Queensland wheat soils using ^{15}N. M. Agric. Sci. Thesis, Univ. of Queensland.

Sauerbeck, D. R. and Gonzalez, M. A. (1977). Field decomposition of ^{14}C-labelled plant residues in different soils of Germany and Costa Rica. *In* 'International Symposium on Soil Organic Matter Studies.' (FAO, IAEA, Agrochimica, Braunschweig.)

Scharpenseel, H. W. and Krausse, R. (1962). Aminosäureuntersuchungen an verschiedenen organischen Sedimenten, besonders Grau-und Braunhuminsäure-fraktionen verschiedener Bodentypen. *Z. Pflanzernaehr. Dueng. Bodenkd.* **96**, 11-34.

Schultz, J. E. (1974). Root development of wheat at the flowering stage under different cultural practices. *Agric. Record.* **1**, 12-17.

Schultz, J. E. and French, R. J. (1976). Mineral content of herbage and grain of Halberd wheat in South Australia. *Aust. J. Exp. Agric. Anim. Husb.* **16**, 887-92.

Simonart, P., Batistic, L. and Mayaudon, J. (1967). Isolation of protein from humic acid extracted from soil. *Plant Soil* **27**, 153-61.

Simpson, J. R. (1968). Losses of urea nitrogen from the surface of pasture soils. Trans. 9th Int. Congr. Soil Sci. Adelaide. Vol. 2, pp. 459-66.

Simpson, J. R. and Freney, J. R. (1967). The fate of labelled mineral nitrogen after addition to three pasture soils of different organic matter contents. *Aust. J. Agric. Res.* **18**, 613-23.

Simpson, J. R. and Freney, J. R. (1974). The fate of fertilizer nitrogen under different cropping systems. *In* 'Fertilizers and the Environment.' Symp. Proc., Sydney. (Ed. D. R. Leece). pp. 27-33.

Sims, H. J. (1977). Cultivation and fallowing practices. *In* 'Soil Factors in Crop Production in a Semi-arid Environment.' (Eds J. S. Russell and E. L. Greacen). pp. 243-61. (Univ. of Queensland. Press: St Lucia.)

Stace, H. C. T., Hubble, G. D., Brewer, R., Northcote, K. H., Sleeman, J. R., Mulcahy, M. J. and Hallsworth, E. G. (1968). 'A Handbook of Australian Soils'. (Rellim: Adelaide.)

Stefanson, R. C. (1970). Sealed growth chambers for studies of the effects of plants on the soil atmosphere. *J. Agric. Eng. Res.* **15**, 295-301.

Stefanson, R. C. (1972*a*). Soil denitrification in sealed soil–plant systems. I. Effect of plants, soil water content and soil organic matter content. *Plant Soil* **37**, 113-27.

Stefanson, R. C. (1972*b*). Soil denitrification in sealed soil–plant systems. II. Effect of soil water content and form of applied nitrogen. *Plant Soil* **37**, 129-40.

Stefanson, R. C. (1972*c*). Soil denitrification in sealed soil–plant systems. III. Effect of disturbed and undisturbed soil samples. *Plant Soil* **37**, 141-49.

Stefanson, R. C. (1972*d*). Effect of plant growth and form of nitrogen fertilizer on denitrification from four South Australian soils. *Aust. J. Soil Res.* **10**, 183-95.

Stefanson, R. C. (1973). Evolution patterns of nitrous oxide and nitrogen in sealed soil–plant systems. *Soil Biol. Biochem.* **5**, 167-69.

Stefanson, R. C. (1974). Denitrification from fertilizer-N placed at various depths in the soil–plant system. Trans. 10th Int. Congr. Soil Sci. Moscow. pp. 141-50.

Storrier, R. R. (1965). The leaching of nitrogen and its uptake by wheat in a soil from southern New South Wales. *Aust. J. Exp. Agric. Anim. Husb.* **5**, 323-28.

Swaby, R. J. (1939). The occurrence of *Azotobacters* and *Clostridium butyricum* in Victorian soils. *Aust. J. Exp. Biol. Med. Sci.* **17**, 401-23.

Swaby, R. J. and Ladd, J. N. (1962). Chemical nature, microbial resistance, and origin of soil humus. Trans. Int. Soc. Soil Sci. Soil Conference, New Zealand Commissions IV and V, pp. 3-8.

Swift, R. S. and Posner, A. M. (1972*a*). Nitrogen, phosphorus and sulphur contents of humic acids fractionated with respect to molecular weight. *J. Soil Sci.* **23**, 50-7.

Swift, R. S. and Posner, A. M. (1972*b*). The distribution and extraction of soil nitrogen as a function of soil particle size. *Soil Biol. Biochem.* **4**, 181-86.

Turchenek L. W. (1975). Organo-mineral associations in soils. Ph. D. Thesis, Univ. of Adelaide.

Turchenek, L. W. and Oades, J. M. (1979). Fractionation of organo-mineral complexes by sedimentation and density techniques. *Geoderma* **21**, 311-43.

Vallis, I. (1969). The measurement of gains and losses of nitrogen in grazed pastures. Ph. D. Thesis, Univ. of Queensland.

Vallis, I., Haydock, K. P., Ross, P. J. and Henzell, E. F. (1967). Isotopic studies on the uptake of nitrogen by pasture plants. III. The uptake of small additions of ^{15}N-labelled fertilizer by Rhodes grass and Townsville lucerne. *Aust. J. Agric. Res.* **18**, 865-77.

Vallis, I., Henzell, E. F. and Evans, T. R. (1977). Uptake of soil nitrogen by legumes in mixed swards. *Aust. J. Agric. Res.* **28**, 413-25.

Vallis, I., Henzell, E. F., Martin, A. E. and Ross, P. J. (1973). Isotopic studies on the uptake of nitrogen by pasture plants. V. ^{15}N balance experiments in field microplots. *Aust. J. Agric. Res.* **24**, 693-702.

Vallis, I. and Jones. R. J. (1973). Net mineralization of nitrogen in leaves and leaf litter of *Desmodium intortum* and *Phaseolus atropurpureum* mixed with soils. *Soil Biol. Biochem.* **5**, 391-98.

Watson, E. R. (1963). The influence of subterranean clover pasture on soil fertility. I. Short-term effects. *Aust. J. Agric. Res.* **14**, 796-807.

Watson, E. R. and Lapins, P. (1969). Losses of nitrogen from urine on soils from south-western Australia. *Aust. J. Exp. Agric. Anim. Husb.* **9**, 85-91.

Wetselaar, R. (1961*a*). Nitrate distribution in tropical soils. I. Possible causes of nitrate accumulation near the surface after a long dry period. *Plant Soil* **15**, 110-20.

Wetselaar, R. (1961*b*). Nitrate distribution in tropical soils. II. Extent of capillary accumulation of nitrate during a long dry period. *Plant Soil* **15**, 121-33.

Wetselaar, R. (1962*a*). Nitrate distribution in tropical soils. III. Downward movement and accumulation of nitrate in the subsoil. *Plant Soil* **16**, 19-31.

Wetselaar, R. (1962*b*). The fate of nitrogenous fertilizers in a monsoonal climate. Trans. Joint Meeting of Comm. 4 and 5, I. S. S. S., New Zealand. pp. 588-95.

Wetselaar, R. (1967*a*). Determination of the mineralization coefficient of soil organic nitrogen on two soils at Katherine, N. T. *Aust. J. Exp. Agric. Anim. Husb.* **7**, 266-74.

Wetselaar, R. (1967*b*). Estimation of nitrogen fixation by four legumes in a dry monsoonal area of north-western Australia. *Aust. J. Exp. Agric. Anim. Husb.* **7**, 518-22.

Wetselaar, R. (1968). Soil organic nitrogen mineralization as affected by low soil water potentials. *Plant Soil* **18**, 9-17.

Wetselaar, R. and Beech, D. F. (1968). The effect of wet season land treatment and nitrogen fertilizer on safflower linseed and wheat in the Ord River valley. I. Soil and crop nitrogen. *Aust. J. Exp. Agric. Anim. Husb.* **8**, 59-65.

Wetselaar, R. and Hutton, J. T. (1963). The ionic composition of rainwater at Katherine, N. T. and its part in the cycling of plant nutrients. *Aust. J. Agric. Res.* **14**, 319-29.

Wetselaar, R. and Norman, M. J. T. (1960). Recovery of available soil nitrogen by annual fodder crops at Katherine, Northern Territory. *Aust. J. Agric. Res.* **11**, 693-704.

Wetselaar, R., Jakobsen, P. and Chaplin, G. R. (1973). Nitrogen balance in crop systems in tropical Australia. *Soil Biol. Biochem.* **5**, 35-40.

Wild, A. (1958). The phosphate content of Australian soils. *Aust. J. Agric. Res.* **9**, 193-204.

Williams, C. H. and Donald, C. M. (1957). Changes in organic matter and pH in a podzolic soil as influenced by subterranean clover and superphosphate. *Aust. J. Agric. Res.* **8**, 179-89.

Williams, C. H. and Colwell, J. D. (1977). Inorganic chemical properties. *In* 'Soil Factors in Crop Production in a Semi-arid Environment.' (Eds J. S. Russell and E. L. Greacen). pp. 105-26. (Univ. of Queensland. Press: St Lucia.)

Williams, C. H. and Steinbergs, A. (1958). Sulphur and phosphorus in some eastern Australian soils *Aust. J. Agric. Res.* **9**, 483-91.

Woodruff, C. M. (1949). Estimating the nitrogen delivery of soil from the nitrogen determination as reflected by Sanborn field. *Soil Sci. Soc. Am. Proc.* **14**, 208-12.

Soil fauna

Chapter 38
Dynamics and fauna of the litter layers

A. V. Spain and B. R. Hutson

THE STUDY OF ENERGY FLOW AND THE CYCLING OF nutrients and other substances through natural and man-modified ecosystems has achieved considerable prominence in recent years. Initially, this resulted from appreciation of the need to better understand the function of agricultural ecosystems. While this aspect remains dominant, the principles of energy flow and nutrient cycling have a general application. They provide a basis for understanding the functioning of ecosystems and, from this, principles for the long term conservation of ecosystems as functional units may be derived. This will hopefully allow the development of a conceptual framework against which perturbations resulting from agricultural or industrial activity may be viewed.

Over much of Australia, plant biomass and productivity, and thus soil organic matter, are restricted by unfavourable environmental characteristics (Chapter 34). The landscapes in many areas are old and eroded and soil fertility is frequently low. Given these constraints, the roles of biological nutrient cycling and conservation mechanisms in productivity assume greater importance than in regions where the soils are inherently more fertile. Climatic factors do not favour high levels of productivity over much of the continent because of low effective rainfall (Chapter 1). The net effect of both these sets of restrictions is reflected in the distribution of the natural vegetation (Carnahan 1976), the most productive of which is confined to the eastern coastline and a few other (mainly coastal) areas of higher rainfall.

To date, several aspects of nutrient cycling have been studied in a number of Australian ecosystems (Bevege 1978). However, no large scale, fully integrated studies have yet been conducted and in terms of understanding the resources of our ecosystems, this work is in its infancy. In spite of this, selected aspects of the unique features of nutrient cycling and energy flow in a number of Australian ecosystems have been at least partly characterized. In particular, the litter layers of indigenous forests have been the subject of research aimed at the assessment of fire risk. There are, however, other reasons for studying the litter layers and recent publications edited by Anderson and Macfadyen (1976) and Dickinson and Pugh (1974) summarize much of the general information available from other countries.

Dead organic matter falling from the aerial parts of plants plays an important role in conditioning the behaviour of other components of the ecosystem to which it belongs. It is the major source of particulate organic matter accessed by the mineral soil and, where it forms a surface layer, modifies soil temperature and moisture regimes and provides some degree of physical protection against erosion.

As a source of energy, nutrient elements and physical protection, this dead organic matter at once provides a nutritional substrate and the habitat for a diverse biota allied to that of the mineral soil and which has as its base the lower members of the decomposer cycle.

Dead organic matter also modifies and conditions the chemical status of the underlying mineral soil, and, particularly in coarse-textured soils and in older landscapes, efficient degradation may be critical in maintaining fertility levels. Variation in land use usually implies changes in the nature, location and quantity of the organic matter accessed by the soil and the long-term consequences of this are generally unknown.

The purpose of this review is to summarize our current knowledge of the characteristics and dynamics of the litter layers of Australian ecosystems and, where known, their interactions with the mineral soil horizons. It is virtually confined to woodlands because of the almost complete lack of information on the nature and magnitudes of litters under grassland and cropping situations.

The system of vegetation classification employed is that of Specht (1970).

Nature of the litter layers

Nomenclature

Although the presence of some dead organic matter is characteristic of all ecosystems, it is on the forest floor of woodlands that this becomes particularly noticeable. Because of its conspicuous development in certain forest types, the phenomenon has been the subject of considerable study in the forest environment and concepts and nomenclature have tended to reflect this. However, the development of a layer of dead organic matter on the surface of the predominantly mineral soil is of wide occurrence in grasslands, cropping systems and other vegetation types. The term 'litter layers' has been recommended because of its broader connotations (Medwecka-Kornas 1971) and will be used here.

Progress in the study of litter layers in general has been impeded by lack of a standardized system of nomenclature. A relatively simple approach to naming the litter layers has been proposed (Soil Science Society of America 1975) which considers the layers of dead organic matter to comprise the O horizon. This places the litter layers in what is conceptually a better ecological perspective as an integral part of the soil profile and thus emphasizes their important role in influencing soil processes.

Under this system, the O horizon is defined as the horizon forming the upper part of the soil above the mineral part and dominated by fresh or partly decomposed organic matter. The relationship to the underlying mineral soil is recognized by requiring that the organic matter levels be greater than 30% if the clay fraction of the mineral soil is greater than 50% and greater than 20% if the mineral fraction has no clay. The horizon is subdivided into the O_1 in which the nature of the original vegetation is apparent to the naked eye and the O_2 in which discrete parts are no longer recognizable.

Stratification is a marked feature of the litter layers of conifer forests and this is reflected in their nomenclature. In contrast, most broadleaf forests have litter layers in which progressive fragmentation and loss of leaf area occur with depth in the O horizon. In these situations stratification is rarely obvious.

Other more specialized systems have been devised to classify the various layers of the forest floor. One of the earlier and better known of these is reviewed by Millar (1974) and is particularly appropriate to certain conifer forest types. In this, the A horizon is considered to comprise up to three layers above the predominantly mineral soil. The topmost is called the A_{000} or L layer and comprises freshly fallen, undecomposed needles. The second layer is the A_{00} and has also been called the fermentation or F layer. It is characterized by the presence of recognizable needles, or parts thereof, extensively colonized by fungal hyphae, and the presence of animal parts and faeces. It may often be compressed and the mesophyll is collapsed. The A_0, or layer adjacent to the mineral soil, is also known as the H or humus layer and is characterized by the presence of an amorphous mass of organic matter, together with small fragments of arthropod exoskeleton and microbial fragments. The O_1 corresponds to the combined A_{000} and A_{00}, the O_2 to the A_0 layers.

Broadleaf forests contrast sharply with those of conifers in the nature of their forest floors. Many broadleaf forests have a relatively small litter mass and often no clear dividing line is present between the litter layers and the A horizon (Ashton 1975). This is the typical mull situation found in many eucalypt and rainforests. It may be contrasted with the mor type, typically found in conifer forests, where a deep mat of decomposing vegetation and fungal mycelia may form. Characteristic of mor litters is a sharp boundary between the litter layers and the mineral soil (Spain 1972; Turner and Kelly 1977). Furthermore, these litters retain their cellular integrity for relatively long periods. Moder litters are somewhat intermediate between the two and many eucalypt litters appear to belong to this category.

Mass

The masses of the litter layers of Australian forests are not well known. Current studies in a number of major forest types are going some way towards rectifying the deficiencies in our knowledge of indigenous forests (Bevege 1978).

Variation in the mass and other characteristics of the litter layers is substantial. The causes of this are undoubtedly multiple and complex. This variation may be partly attributed to the non-random nature of both tree spacing and litterfall accession and partly to variation in soil characteristics. Frequency distributions of random samples taken to estimate litter layer mass tend to exhibit positive skews (Spain 1972). This is partly because of the presence of certain structures, such as fruits or cones which tend to remain as large discrete objects substantially longer than most of the other components. Other factors tending to increase sample variability include the nature of the distribution of litterfall components in relationship to the size of and distance from each tree (Spain 1973*b*). The presence of bark 'haloes' around trees in many forests is visual evidence for one aspect of this source of variability. Various forms of 'clumped' distributions have also been recorded for a range of litter faunal categories (Usher 1976) and mineral elements (Usher 1970).

In eucalypt and other forests, the mass of the litter layers is related to that of the above ground components of the stand and is thus dependent on the stage of development of a forest. It has been shown with *Pinus radiata* in South Australia that litter mass varies with soil characteristics and it appears that this largely results from the capacity of certain soils to maintain an elevated breakdown rate (Florence and Lamb 1974), possibly because they possess more favourable moisture retention capacities. O'Connell *et al.* (1978) have also shown that substantial differences in litter layer mass and nature are present in jarrah (*Eucalyptus marginata*) forests growing on different soils.

Litter layer mass is also directly related to the length of time a stand has been protected from fire (Hatch 1955; Peet 1971). Thus, in many areas the litter layers are unlikely to ever reach an approximate equilibrium state. Fox *et al.* (1979), however, did find a steady state reached in the litter of a New South Wales eucalypt forest 10 years after fire. Eucalypt forests rarely form single species stands and a shrub stratum and other low vegetation is usually present. The nature of the plant material which forms the litter layer is therefore heterogeneous. In many cases, the shrub layer may have a legume component whose foliage will have quite different breakdown characteristics compared with those of the dominant eucalypts. In addition, the various woody, reproductive and other litter fractions all have different breakdown characteristics and further contribute to the diversity of the organic matter in the litter layers.

As stated above, the mass of the litter layers may range from virtually nothing after an intense fire to quite substantial levels. The maximum levels so far recorded for eucalypt forests are those of Meakins (1966) for communities dominated by *Eucalyptus fastigata* in the Australian Capital Territory and *E. signata* on North Stradbroke Island (Rogers and Westman 1977). Both had litter layers with total masses of 27 t ha^{-1}. In the latter community, the litter layers consisted of 32% wood, 33% leaves, 19% bark and 16% 'residue'. The proportions vary for different ecosystems and leaves may comprise up to 60% of the total mass (Anon. 1969). In general, the proportion of leaves appears to be higher on harsher sites (Anon. 1969) and to decrease proportionally in favour of the woody components with increasing years of protection from fire (Hatch 1955; Fox *et al.* 1979) and development towards a presumptive equilibrium condition.

In closed-canopy rainforests, the masses of litter layers are usually somewhat lower although few studies have been reported. In Victoria, Howard (1973) recorded a litter layer mass of 15.5 t ha^{-1} from a *Nothofagus cunninghamii* association. Bale (1974) records masses of 7 and 9 t ha^{-1} for two forest types in the New England National Park.

In tropical areas, rainforests may also build up litter layers of some magnitude. Recent studies of the forest floor in the species-rich rainforests of tropical north-eastern Queensland have shown maximum masses ranging from 7.4 t ha^{-1} in a high rainfall situation to 11.1 t ha^{-1} at a drier site (CSIRO Aust. Div. Soils Tech. Mem. No. 29/1976). The proportions of leaf ranged from 40% to 60%, that of wood from 24% to 53% and reproductive structures comprised 1% to less than 3% of the total mass. Stocker (1966) has reported a total mass of 12.5 t ha^{-1} from a monsoon forest in the Northern Territory.

Other indigenous forest systems, including the low open forests, usually t ave lower biomass levels in their above-ground parts and also tend to have a lower litter layer mass. One notable exception is a brigalow (*Acacia harpophylla*) low open forest in southern Queensland studied by Moore *et al.* (1967). The total mass of the litter layers was estimated at 75.5 t ha^{-1}, with a further 58 t ha^{-1} of standing dead material. These components (i.e. the dead material) comprised an estimated 46% of the total mass of dry matter in the system. The litter layers in this woodland were composed of 19% leaves and twigs, 6% bark, 22% branches and 53% trunks. The substantial mass of dead woody material present in this situation may be ascribed to the lack of an efficient xylophagous termite fauna on these soils (A.W. Moore, pers. comm.).

Among the exotic conifers, *Pinus radiata* is without doubt the most widely grown plantation species in Australasia although others assume dominance where, for climatic or other reasons, *P. radiata* is not grown. The nature of the litter layers in the conifer forests of Europe and North America has been the subject of considerable study over the last century and similar litter layers form under the exotic conifers grown in Australia. These comprise, in most situations, typically deep layers of decomposing needles and other plant structures with a high fungal biomass and have a sharp boundary with the mineral soil horizons below. The masses of these layers may be considerable (Table 38.1). Forrest and Ovington (1970) estimated that the litter layers formed about 12% of the above-ground organic mass of one nearly mature *P. radiata* stand.

The indigenous conifer *Araucaria cunninghamii* is grown in Queensland and elsewhere as a plantation species (Hall *et al.* 1970). This species also forms substantial litter layers and an annual range in mass of approximately 11.0 to 22.5 t ha^{-1} has been recorded in a mature stand (CSIRO Aust. Div. Soils Tech. Mem. 36/1978). The litter forms a diffuse layer mostly composed of the third and fourth order branchlets which, together with their attached leaves, are the normal unit of litterfall.

Table 38.1. Mass of the litter layers in selected Australian conifer forests

Species	Mass (t ha^{-1})	Author
Araucaria cunninghamii	11.0-22.5	A.V. Spain, unpublished
Larix decidua	26.3	A.V. Spain and M.T. Tanton, unpublished
Pinus banksiana	16.0	A.V. Spain and M.T. Tanton, unpublished
P. jeffreyi	30.0	A.V. Spain and M.T. Tanton, unpublished
P. lambertiana	27.0	A.V. Spain and M.T. Tanton, unpublished
P. nigra	28.0	A.V. Spain and M.T. Tanton, unpublished
P. ponderosa	30.0	A.V. Spain and M.T. Tanton, unpublished
P. radiata	16.4-32.9	Anon (1969)
P. radiata a	17.9-27.7	Florence and Lamb (1974)
P. radiata b	19.5-22.6	Florence and Lamb (1974)
P. radiata c	15.2-19.1	Florence and Lamb (1974)
P. radiata d	11.8	Florence and Lamb (1974)
Pseudotsuga menziesii	22.8	A.V. Spain and M.T. Tanton, unpublished
Ps. menziesii	35.7	Turner and Kelly (1977)

Little information is available for Australian conditions on the magnitudes of the litter layers in grassland or cropping situations. One feature of grazed pastures is the large area of ground covered by domestic animal faeces present in the litter layers. Where the stocking rate is high, these faeces may occupy a considerable area and have an inimical effect on the palatability of the sward surrounding them. Marsh and Campling (1970) have reviewed in general terms the problem of pasture fouling.

Australia's indigenous coprophagous biota is generally not efficient in breaking down ungulate faeces (Waterhouse 1974; Bornemissza 1976) although some groups are sporadically active (Ferrar and Watson 1970; Hughes 1975). In addition to their important effects on pasture production, ungulate faeces provide a breeding medium for muscid flies of considerable public health and veterinary importance. The rapid burial of dung is therefore of importance and a number of coprophagous beetles from Africa have been introduced with this aim in mind (Bornemissza 1976).

Vertebrate carrion is a unique substrate with its own sequence of specialized decomposer organisms. Bornemissza (1957) has outlined the effects of carrion on the soil fauna and the sequence of communities of organisms involved in the various stages of reduction of guinea pig carcasses. It appears, though, that the nature of carrion breakdown and the organisms responsible for it vary between environments and that more work is necessary to delineate their characteristics. Less is known of the breakdown of invertebrate carrion. However, chitin from the exoskeletons of arthropods may form a considerable proportion of the organic matter in the lower part of certain litters.

Faunal communities

In a large proportion of Australian soils, organic matter production is low and of limited variety because of high temperatures, low rainfall, plant nutrient limitations and the nature of the vegetation. Soil organic matter determines the pattern of distribution of soil animals and the relative abundance of the dominant groups. It has been shown (Castri 1973) that in the drier regions of Italy and Chile, changes in faunal density are more directly correlated with soil organic matter levels than with moisture content.

Castri (1973) summarizes the results of faunal investigations in several mediterranean areas (including eucalypt woodlands) in Chile. He concludes that the dominant fauna in the more arid areas includes prostigmatid mites, Olpiidae (Chelonethi or pseudoscorpions) and Formicidae whereas included in the dominant fauna of the true mediterranean areas are cryptostigmatid mites, Psocoptera, Diplopoda, Chthoniidae (Chelonethi), Collembola (Isotomidae and Entomobryidae in the litter, Poduridae and Onychiuridae in the deeper soil) as well as Protura and Symphyla. In humid areas, cryptostigmatid and mesostigmatid mites, onychiurid and podurid Collembola, larval Diptera, Copepoda and Enchytraeidae are included in the dominant fauna. Much information concerning the distribution and feeding habits of these soil animals outside Australia is reviewed in Dickinson and Pugh (1974) and Wallwork (1970, 1976). Their role in litter breakdown is discussed on p.622 and the general ecology of the Australian soil fauna is discussed in Chapter 40.

Indigenous vegetation sites

In Australia, a similar trend of faunal dominance and distribution to that described above is now becoming apparent from the results of limited investigations of the litter and soil communities. Unfortunately, the majority of samples taken by field investigators have been of the litter layer (if present) combined with some of the mineral soil immediately below. Thus, it is impossible to discuss only the litter fauna from these results. Table 38.2 gives the mean annual densities of mites and Collembola calculated from results of several surveys (abbreviated in future to 'S'), progressing from the driest to the wettest sites.

In the arid sites, the densities of soil microarthropods only increase slightly from arid grassland (S1) to low shrubland (S4). A significant increase has been shown to occur in the open scrub sub-formations between semi-arid mallee (S5) and sclerophyll mallee (S6) areas as the mean annual rainfall increases from less than to greater than 500 mm and the plant litter layer changes from discontinuous to more or less continuous. In the more arid areas, Collembola and mites are found in about equal numbers, with Isotomidae dominating the collembolan, and Prostigmata the mite, populations. As the sites become less arid and the amount of soil organic matter increases, the proportion of cryptostigmatid and mesostigmatid mites increases and the other groups of Collembola become more frequent. The only other microarthropods found in relatively large numbers in the drier areas have been harpacticoid Copepoda (Wood 1971*b*), a form of Crustacea which are more often found in tall open forest formations and rainforests.

In open scrubland and low open forested areas, mean annual population densities of mites and Collembola have been shown to range between 30 and 70 x 10^3 m^{-2} (S6–12) with the proportion of mites generally rising with increasing rainfall.

A study in three low open forest sites in South Australia (S8, 9 and 12) has shown that prostigmatid mites still dominate the drier sites throughout the year whereas at the colder, wetter site (S12), except for the summer months, Cryptostigmata have become dominant, particularly in the litter (which generally consists of an O_1 layer only). A tendency in the litter of these sites has been for isotomid Collembola to be dominant for most

Table 38.2. Mean annual densities (× 10^3 m^{-2}) of mites and Collembola observed in soil and litter by various authors at selected indigenous sites

Survey	S1	S2	S3	S4	S5	S6	S7	S8	S9	S10	S11	S12	S13	S14	S15	S16	S17
Mites (M)	0.7	1.2	3.7	4.3	8.3	33.6	28.4	35.9	59.1	—	60.1	46.4	—	56.1	67.0	30.3	—
Collembola (C)	1.2	1.0	1.5	2.5	4.1	32.2	12.7	4.9	10.0	—	11.6	6.5	—	30.2	14.1	2.3	—
M + C	1.9	2.2	5.2	6.8	12.4	65.8	41.1	40.8	69.1	69.4	71.7	52.9	129.1	86.3	81.1	32.6	117.5
C:M	0.5	1.1	2.5	1.7	2.1	1.0	2.2	7.3	5.9	—	5.2	7.1	—	1.8	4.7	13.1	—

Survey	Author	Site Vegetation	Rainfall (mm)	Comments	
S1	Wood (1971*b*)	Desert grassland	76–204	No litter	(S + W)[A]
S2	Wood (1971*b*)	Desert steppe	76–204	Short shrubs (46 cm), very little litter	(S + W)
S3	Wood (1971*b*)	Low layered woodland	127–254	Litter only under trees	(S + W)
S4	Wood (1971*b*)	Low shrubland	152–406	Litter only under shrubs	(S + W)
S5	Wood (1971*b*)	Open scrub (mallee, semi-arid)	204–432	Litter discontinuous	(S + W)
S6	Wood (1971*b*)	Open scrub (mallee, sclerophyll)	320–660	Litter continuous	(S + W)
S7	Wood (1971*b*)	Open woodland	457–762	Litter continuous	(S + W)
S8	B.R. Hutson (unpubl.)	Low open forest (S.A.)	610	*Eucalyptus obliqua–E. baxteri*	(M)
S9	B.R. Hutson (unpubl.)	Low open forest (S.A.)	762	*E. obliqua–E. baxteri*	(M)
S10	Wood (1971*a*)	Low open forest (N.S.W.)	889	*E. pauciflora–E. dalrympleana*	(Q)
S11	Springett (1976*b*)	Low open forest (W.A.)	886	*Eucalyptus, Banksia, Hakea*	(M)
S12	B.R. Hutson (unpubl.)	Low open forest (S.A.)	1016	*E. obliqua–E. baxteri*	(M)
S13	Wood (1971*a*)	Tall open forest (N.S.W.)	1144	*E. delegatensis* assoc.	(Q)
S14	Plowman (1979)	Closed forest (Qld.)	1788	Complex notophyll vine forest	(Q)
S15	Plowman (1979)	Tall open forest (Qld.)	1788	*E. saligna–Tristania conferta*	(Q)
S16	Ahern and Yen (1977)	Tall open forest (Vic.)	1910	*E. regnans*	(X)
S17	Wood (1971*a*)	Alpine herbfield (N.S.W.)	2160	*Poa caespitosa–Celmisia longifolia*	(Q)

[A]S + W = mean of summer and winter samples; M = mean of monthly samples; Q = mean of quarterly samples; X = one day's sample taken in May.

of the year, supported by a large entomobryid population. In the summer months, however, when their density was low, Hypogastruridae became dominant (not in very large numbers) for a short period. As might be expected, only a small population of the deeper-dwelling Onychiuridae were found in the litter layer at any time, but a relatively large population occurred in the soil in the autumn months.

Large mite and collembolan populations were found in similar sites near Mt. Kosciusko (S10) and in Western Australia (S11). At the site in Western Australia, nearly 50% of the mites were Cryptostigmata. Similarly, in the soil and litter of a closed forest (S14) and nearby tall open forest (S15) in Queensland, large populations of both mites and Collembola occurred, with Cryptostigmata forming 41% and 68% respectively of the mite populations. Samples of soil and litter taken in Victorian tall open forest sites (S16) one day in May contained a very high ratio of mites to Collembola whereas samples taken on four occasions throughout the year at the Queensland sites (S14 and 15) contained a very much lower ratio overall. The maximum mean annual densities of these two groups so far recorded in Australia are from tall open forest and alpine herbfield sites at Mt. Kosciusko (S13 and 17).

The densities of the mites and Collembola occurring in the litter layer and upper and lower soil layers of three low open forest sites in South Australia (S8, 9 and 12) are summarized in Table 38.3. Throughout the year, an average of only 7–12% of the population occurred in the litter layers, between 17 and 23% occurred in the lower soil layers and the majority of the population (66–76%) inhabited the upper soil layers. A similar proportion (9%) occurred in the thin and discontinuous litter layer of site S14 in Queensland, but a far greater proportion (26%) was found to inhabit the thicker, continuous litter layer at site S15 (Table 38.3). In the two drier South Australian sites (S8 and 9), the proportion of mites to

Table 38.3. Mean annual densities of mites and Collembola in various soil horizons

Values in the table are mean annual densities (× 10^3 m^{-2}) of mites (M) and Collembola (C) in the litter (L), upper soil (0–4 cm, SU) and lower soil (4–8 cm, SL) layers of three South Australian low open forest sites, and the litter (L) and upper soil (0–5 cm, SU) layers of a closed forest and tall open forest in Queensland.

Survey[A]:	S8			S9			S12			S14		S15	
	L	SU	SL	L	SU	SL	L	SU	SL	L	SU	L	SU
Mites (M)	3.6	25.7	6.6	5.9	38.7	14.5	3.4	34.9	8.1	5.2	50.9	17.4	49.6
Collembola (C)	0.7	3.7	0.5	1.9	6.9	1.2	0.5	4.9	1.1	2.5	27.7	3.5	10.6
M + C	4.3	29.4	7.1	7.8	45.6	15.7	3.9	39.8	9.2	7.7	78.6	20.9	60.2
% in each layer	10.5	72.1	17.4	11.3	66.0	22.7	7.4	75.2	17.4	8.9	91.1	25.8	74.2
C:M	5.1	6.9	13.2	3.1	5.6	12.1	6.8	7.1	7.4	2.1	1.8	5.0	4.7

[A] According to Table 38.2.

Table 38.4. Mean annual densities of 'other' fauna in litter and soil at various indigenous sites

Values in the table are mean annual densities (actual number m^{-2}) of 'other' fauna in the litter and upper 0–8 cm of soil in three South Australian low open forest sites, and the litter and upper 0–5 cm layer of soil in a closed forest and tall open forest in Queensland, with percentages occurring in the litter layer.

	Mean annual density m^{-2} in litter and soil					% occurring in litter layer				
Survey[A]:	S8	S9	S12	S14	S15	S8	S9	S12	S14	S15
Chelonethi[B]	78.4	189.2	149.4	640.6	63.2	5.3	24.1	5.7	9.8	13.7
Araneida[C]	27.4	28.7	60.4	246.2	474.1	32.8	84.0	35.3	47.4	38.7
Copepoda	0	3.0	9.1	0	44.8	–	23.3	74.7	–	79.7
Amphipoda	0	0	13.6	98.5	59.3	–	–	66.2	21.2	38.8
Isopoda	11.6	27.3	78.2	248.2	664.6	40.5	41.0	14.6	58.2	16.6
Pauropoda	944.6	1319.3	1769.7	1209.6	459.1	0.3	0.5	0.9	2.3	1.0
Diplopoda	31.3	13.9	61.7	1486.4	1835.7	19.2	20.1	19.4	7.1	5.4
Chilopoda[B]	97.7	204.9	154.1	863.8	987.7	3.8	1.6	1.4	5.1	8.9
Symphyla	203.0	327.0	235.3	535.7	484.8	0.6	1.1	0.2	0.2	0.7
Protura	58.1	102.2	344.3	370.9	213.5	0.2	0	1.3	0	74.5
Collembola (Symphypleona)	361.8	276.5	503.3	–[D]	–[D]	18.2	37.7	29.7	–[D]	–[D]
Thysanoptera	62.0	104.6	84.8	104.6	93.9	44.3	53.8	37.6	25.8	70.9
Lepidoptera larvae	25.7	111.4	49.7	384.7	847.8	55.2	35.9	30.6	43.9	51.8
Diptera larvae[C]	378.3	221.0	279.4	2811.7	4107.4	6.5	21.2	12.0	14.4	14.8
Formicidae[B]	995.2	275.1	507.0	997.0	2678.0	5.8	4.6	4.2	7.4	13.1
Coleoptera larvae[C]	101.9	88.8	97.7	1505.0	825.4	11.8	24.8	10.4	22.0	24.0
Coleoptera adults[C]	60.4	50.8	79.8	1152.7	1660.2	13.4	23.8	13.1	25.2	22.3

[A] According to Table 38.2 [B] All predators. [C] Predators included in the group. [D] No record.

Collembola increased from the litter layer to the lower soil layer to a much greater extent than in the wetter South Australian site (S12), and in the even wetter Queensland sites (S14 and 15) this proportion tended to decrease from the litter to soil layers.

Very few estimates of the mean annual densities of populations of the 'other' soil fauna have been made in Australia, so there are few comparisons to make. Earthworms and termites are discussed in the next chapter.

On one occasion, Wood (1971*b*) sampled the nematodes in sites S1–S3 and their densities ranged from 17.6 to 313.5 x 10^3 m^{-2}, showing a positive correlation with moisture content. Nematode populations are much higher in the sub-humid temperate regions of Europe (1–20 x 10^6 m^{-2}) and sites S10, 13 and 17 in Australia (2.5–7.5 x 10^6 m^{-2}) where their density decreased with increasing moisture content. In the latter areas, the enchytraeid population increased with increasing moisture content from 250 m^{-2} (S10) to 3250 m^{-2} (S17). The mean annual densities of several other members of the litter and soil fauna from the South Australian sites S8, 9 and 12 are given in Table 38.4. A population increase from the drier to the wetter sites is shown by the Araneida, Copepoda, Amphipoda, Isopoda, Pauropoda, and Protura. Amphipoda were in fact entirely absent from the two drier sites. Only larval Diptera, larval Coleoptera and Formicidae were fewer at the wettest than at the driest site. The litter fauna in these low open forest sites is generally not as abundant throughout the year as might be expected, due basically to the very low moisture content of the litter for several months. Only larval Lepidoptera (S8), Thysanoptera and Araneida (S9), and Amphipoda and Copepoda (S12) show an overall dominance in the litter rather than in the soil. In contrast, less than 2% of the Pauropoda, Symphyla and Protura inhabit the litter layer.

The density of most 'other' faunal groups was far greater in the litter and soil of the Queensland sites, S14 and 15, than in that of the South Australian sites (Table 38.4). Although the Queensland sites were established in a much higher rainfall area, the proportion of the fauna inhabiting the litter layers was not, however, greater than at the South Australian sites. Only Isopoda at site S14 and Copepoda, Protura, Thysanoptera and Lepidoptera larvae at site S15 showed an overall dominance in the litter, rather than in the soil. This suggests that some factor other than moisture content was operating to either reduce the density of the litter population or increase the density of the soil population. As was found at the South Australian sites, the proportions of Pauropoda, Symphyla and Protura in the litter layers were extremely low.

In Victoria, Ashton (1975) estimated the crustacean population in the litter of *E. regnans* tall open forests (rainfall 1300 mm) and recorded mean densities of individuals per square metre in mature and pole forests respectively as follows: Amphipoda 258, 790; Isopoda 666, 169, and Diplopoda 59, 73. These densities are much greater than those recorded in the litter layers only of the warmer and drier site S12 (Amphipoda 9, Isopoda 11, and Diplopoda 12) and, except for Diplopoda, in the warmer and wetter sites S14 and 15 (Amphipoda 21, 23; Isopoda 145, 110, and Diplopoda 106, 100 m^{-2} respectively).

Coniferous forests

Densities of mites and Collembola in some Australian *Pinus* sites are given in Table 38.5. Except for P8 (results

Table 38.5. Densities of mites and Collembola observed in soil and litter by various authors at *Pinus* sites
Values are expressed as $\times$ 10^3 m^{-2}.

Survey:	P1	P2	P3	P4	P5	P6	P7	P8
Mites (M)	51.0	125.0	131.0	78.6	79.2	79.3	25.9	37.8
Collembola (C)	11.8	12.0	13.0	7.2	11.0	16.3	35.6	7.4
C + M	62.8	137.0	144.0	85.8	90.2	95.6	61.5	45.2
C:M	4.3	10.4	10.1	10.9	7.2	4.9	0.7	4.5

Survey:	Author	Site Vegetation		Rainfall (mm)	Comments	
P1	Howard (1967)	*Pinus radiata*	(S.A.)	785	Orig. *E. baxteri–E. obliqua*	(N)[A]
P2	Springett (1971)	*P. pinaster*	(W.A.)	886	Unburnt	(W)
P3	Springett (1971)	*P. pinaster*	(W.A.)	886	Burnt 3 years before sampling	(W)
P4	Springett (1976*b*)	*P. pinaster*	(W.A.)	912	Orig. as Table 38.2, S11. 18 years old	(B)
P5	Springett (1976*b*)	*P. pinaster*	(W.A.)	912	Orig. as Table 38.2, S11. 23 years old	(B)
P6	Springett (1976*b*)	*P. pinaster*	(W.A.)	912	Orig. as Table 38.2, S11. 31 years old	(M)
P7	Spain (1974)	*P. nigra* var. *maritima*	(N.S.W.)	1450	35 years old	(Q)
P8	Ahern and Yen (1977)	*P. radiata*	(Vic.)	1910	Orig. as Table 38.2. S14. Mature	(X)

[A] N = mean of nine months sampling; W = mean of three winter biased samples (May, July, Sept.); B = mean of bimonthly samples; M = mean of monthly samples; Q = mean of quarterly samples; X = one day's sample taken in May.

of a single sampling in May), the figures given are basically an estimate of the annual mean.

Howard (1967) sampled sites of three differing age classes of *P. radiata*: mature (25–35 years), canopy just closed (9–12 years) and young (2–5 years). As he found no significant differences in mite and collembolan densities between the three classes he combined his results (P1). The mites were the most abundant group and were dominated by Cryptostigmata (85–91%) throughout the year. Isotomidae generally dominated the collembolan population, but up to 96% Hypogastruridae were recorded in the summer. Their density was low, however, and with the Cryptostigmata being more resistant to the dry period, the ratio Collembola:mites was as low as 1:62 in the summer.

Springett (1971) surveyed two 15-year-old unburnt *P. pinaster* sites and two adjacent plots three years after a 'cold' burn (P2, 3). She recorded no significant difference between the total numbers of arthropods but concluded that the fauna of the burnt sites had not recovered fully since the proportion of fungal feeders present was significantly lower than on the unburnt sites.

Springett (1976*b*) also examined the populations of microarthropods in three *P. pinaster* stands 18, 23 and 31 years old (P4–6) and in the indigenous open forest site, S11. As on Howard's (1967) sites, densities on the three *Pinus* sites were relatively similar (varying between 90 and 101 x 10^3 m^{-2}). These densities were also similar to those of the indigenous site. Species diversity (mainly of mites and Collembola) varied, however, with least occurring under the young pine stand and an increasing amount occurring with increasing crop maturity. Greatest diversity was found under the eucalypt forest stand. The mites were dominated (between 60–70%) by Cryptostigmata on all sites.

A rather different situation was found by Spain (1974) in the cooler and wetter *P. nigra* site (P7). The mean annual density of Collembola was greater than that of mites, of which 48% were Mesostigmata and only 43% were Cryptostigmata. Thus, the density of Cryptostigmata of about 10 x 10^3 m^{-2} was low when compared with the densities of this group on Howard's (P1) and Springett's (P4–6) sites (43–46 x 10^3 m^{-2} and 47–55 x 10^3 m^{-2} respectively).

On the day that Ahern and Yen (1977) sampled the soil and litter in *P. radiata* plots in a tall open forest (P8) they found a greater invertebrate density than in the eucalypt sites in the same area (S16); species diversity was less, however, as was found by Springett (1976*b*).

Very little information is available concerning the 'other' litter and soil fauna in *Pinus* sites. The mean annual densities of individuals per square metre of some 'other' fauna obtained from Spain's P7 site were: Coccoidea 405; Coleoptera adults 353, larvae 105; Diptera 399; Araneida 270; Chilopoda 146, and Symphyla 474. Thus, compared with Hutson's wettest eucalypt site (S12) a much greater population of Coleoptera adults and Araneida was found, but the numbers of larval Coleoptera and Chilopoda were similar.

Springett (1976*b*) recorded Protura between 0.9 and 4.1 x 10^3 m^{-2} compared with 4.6 x 10^3 m^{-2} in her eucalypt forest site. In a *P. radiata* site (rainfall 900 mm) in the Australian Capital Territory, Elliott (1970) recorded the density of immature Diplopoda in the late summer and autumn as up to 35 m^{-2}, and that of the adults of one large species as up to 4 m^{-2}.

The litter layers as a pool of energy and nutrients

The often substantial amounts of dry matter present in the litter layers of many ecosystems represent a considerable store of energy, of elements and of a range of organic compounds. The total energy contained in a litter layer of 25 t ha^{-1} containing 8% ash and organic matter with an energy level of 18 kJ g^{-1} is approximately 4.1 x 10^8 kJ ha^{-1}. Ovington and Heitkamp (1960) have shown that, in a range of English conifer forests, the energy tied up in the litter layers may be as much as 14% of the total for the ecosystem. Gosz *et al.* (1978) record approximately 18% for the same value in a North American temperate broadleaf forest. This is equivalent to about half the energy in the living plant biomass.

No work has been published in Australia that directly provides this sort of estimate. However, using conservative values of 18.8 kJ g^{-1} for the organic matter and estimating average values of 7 and 10% ash for the above-ground biomass and litter layers respectively, it may be calculated that the eucalypt forest ecosystem described by Rogers and Westman (1977) and Westman and Rogers (1977*a*, *b*) also had approximately 14% of its total above-ground energy tied up in the litter layers.

Litterfall input may also be quantified in the same way. In a tropical rainforest with an annual litterfall of 10 t ha^{-1} of dry matter, the energy input will be of the order of 1.58×10^8 kJ ha^{-1} $year^{-1}$ given an energy level of 16.8 kJ g^{-1} (Golley 1969) and 6% ash. The figures for energy content of the forest floor and annual litterfall may be combined to give an approximate estimate of the turnover rate of energy by the forest floor. In low open forest in South Australia (Table 38.2: S8, 9 and 12) the energy input ranged from 0.3 to 0.5×10^8 kJ ha^{-1} $year^{-1}$ and the mean energy content of the litter was 1.5 to 2.2×10^8 kJ ha^{-1}. Thus, the approximate turnover time is estimated at 4–5 years.

In a similar way, the litter layers may be considered as a pool of elements, both nutrient and non-essential. Ovington and Madgwick (1959) demonstrate that an appreciable proportion of the total ecosystem pool for an element may be tied up in the litter layers. In an English stand of *Pinus sylvestris*, the litter layers and dead trees together account for 82% of the nitrogen, 66% of the phosphorus, 42% of the potassium and 59% of the calcium in the phytomass of the ecosystem.

Few published data on Australian ecosystems are available to allow comparison with these figures although the proportions of total ecosystem nutrient stocks tied up in Australian litter layers are also substantial. Westman and Rogers (1977*b*) have estimated that in an ecosystem dominated by *Eucalyptus signata* on a sandy soil on North Stradbroke Island, the masses of nitrogen, phosphorus and calcium in the litter layers were 195, 6.8 and 142 kg ha^{-1} respectively. These values represent 30, 28 and 29% of the respective totals for these elements in the phytomass of the ecosystem and approximately 0.6, 7.7 and 14.3% of the estimated total ecosystem nutrient stocks. Nitrogen was not considered to be limiting in this ecosystem although calcium is in short supply. There is, however, considerable variation between ecosystems and in a Brigalow (*Acacia harpophylla*) low open forest ecosystem in southeastern Queensland, Moore *et al.* (1967) record large pools of nutrient elements in the litter and standing dead components. For example, the masses of nitrogen and phosphorus in the litter layers are 368 and 10 kg ha^{-1} and these represent 23% and 26% of the ecosystem phytomass totals respectively. Hingston *et al.* (1979) recorded the nitrogen and phosphorus contents of the litter layers of both a karri (*Eucalyptus diversicolor*) and a mixed karri-marri (*E. calophylla*) stand in Western Australia as 224 and 7 kg ha^{-1} (47 and 26% of the total phytomass stock), and 105 and 5 kg ha^{-1} (23 and 19% of the total stock) respectively. Arid zone communities have a much lower biomass and in the *Eremophila gilesii* shrubland community studied by Burrows (1972) in south western Queensland only 15.8 and 0.87 kg ha^{-1} of nitrogen and phosphorus were present in the litter. These levels represent 26.3% and 27.8% of that present in the phytomass of the ecosystem. In the *Atriplex* grassland community studied by Charley and Cowling (1968) the litter layer levels are proportionally much smaller and account for only 1.9% and 3.8% of the nitrogen and phosphorus in the ecosystem phytomass.

As with energy, the masses of the elements accessed annually in litterfall may be combined with those of the litter layer levels to give approximate estimates of turnover rate. Bevege (1978) summarizes what is known of the masses of elements present in litterfall and in the litter layers for the indigenous forests of Australia.

Dynamics of the litter layers

Inputs

While the qualitative nature of the litter layers is largely dependent on that of the plant community present, the mass at any given time is dependent on the balance between past and present rates of input and breakdown. Accession of organic and inorganic material by the litter layers occurs principally through litterfall but inputs from bark and foliar washing, from leaching and aeolian deposition, and from faeces derived from arboreal herbivores may contribute appreciable quantities.

Litterfall

The rate of input of litterfall to the litter layers is dependent in a general way on the biomass of the aerial parts of the plant community. This, in turn, varies with climate, latitude, exposure, soil fertility, moisture (Bray and Gorham 1964), stage of development of the plant community and cultural treatments. Thinning has an obvious effect by temporarily reducing the standing crop biomass of leaves. In closed canopy forests, though, no consistent relationships have been demonstrated between stem density and litterfall although there is a relationship between the latter variable and stand basal area (Bray and Gorham 1964). Fertilization may be expected to have an indirect effect on litterfall mass by increasing the standing crop of leaves, particularly in areas where nutrition is limiting (Miller *et al.* 1976).

(i) *Litterfall in eucalypt forests.* In Australia, eucalypt forests and woodlands dominate the landscape in terms of area, and it is in this vegetation type that most studies of litterfall have been conducted.

The reported total litterfall mass varies widely from a low figure of 0.64 t ha^{-1} $year^{-1}$ for a *Eucalyptus piperita* association on sandstone near Sydney (Hannon 1958) to values of more than 8 t ha^{-1} $year^{-1}$ for an *E. regnans* stand in Victoria (Ashton 1975).

In low open forests near Adelaide, two studies have been conducted over sufficient time to allow some assessment of seasonal and annual variation in mass and composition. Lee and Correll (1978) studied an *E. obliqua–E. baxteri* stand whose total litterfall accessions ranged from 2.10 to 2.65 t ha^{-1} $year^{-1}$ over a 5-year period of study. These values correspond well with those obtained from a similar study conducted on three sites in the same area (CSIRO Aust. Div. Soils Tech. Mem. 5/1977) in which total mass inputs ranged from 1.41 to 2.59 t ha^{-1} $year^{-1}$.

The physical composition of the litterfall accessed was quite constant in these studies. The proportion of leaf ranged from 77% to 81% while woody material varied from 15% to 18% of total litterfall mass. The seasonality of leaf fall is particularly marked in this environment and most occurs from December to March. The amount falling in this latter period is eight to twelve times that in the later winter. Sticks fall at a relatively low and constant rate (Lee and Correll 1978) with the suggestion of a slight increase at the time of maximum leaf fall (CSIRO Aust. Div. Soils Tech. Mem 5/1977).

Tall open forests have a greater standing crop biomass in their above ground parts and thus a heavier litterfall. The maximum litterfall mass recorded from these ecosystems is a 5-year average of 7.76 t ha^{-1} $year^{-1}$ (Ashton 1975) although Bevege (1978), in his compilation, lists one stand having an annual litterfall in excess of 10 t ha^{-1} $year^{-1}$. A number of other authors have published litterfall data for this forest type and mean annual values for total litterfall mass are mostly of the order of 6–7 t ha^{-1} $year^{-1}$ for well developed forests (see, for example, Stoate 1958; Florence 1961; McColl 1966; Webb *et al.* 1969; Rogers and Westman 1977). The total litterfall mass in a mature *E. regnans* forest ranged from 7.05 to 9.95 t ha^{-1} $year^{-1}$ over a 5-year period. In a pole stand the range was from 5.40 to 7.45 t ha^{-1} $year^{-1}$ over the same period, and for a 4-year period, litterfall mass in a spar stand ranged from 6.99 to 9.74 t ha^{-1} $year^{-1}$ (Ashton 1975). Most other studies have not been conducted over a sufficient length of time to permit realistic evaluation of year to year variation.

The distribution of physical components in tall open forests differs from those of the low open forests in having proportionally less leaf and more woody fall. Rogers and Westman (1977), for example, record 44% leaf and 47% bark and wood from a *E. signata–E. umbra* forest on North Stradbroke Island. Similar levels of 43% leaf and 48% wood are recorded by McColl (1966) for a *E. maculata* forest in New South Wales. Ashton (1975) records of 51% for leaf and 44% for bark and wood combined. Stoate (1958) records 49% leaf and 51% wood.

The understorey vegetation of an open forest in Queensland has been shown to contribute 8 to 15% of the total litterfall (Birk 1979*a*). In Ashton's (1975) study the understory contributed 25% of the total litterfall in mature and 11% in spar and pole stage tall open forests, reflecting the greater development of the understory in mature stands. Reproductive structures represented 8% of the total mass in certain tall open forests in the Australian Capital Territory (Anon. 1975) while both McColl (1966) and Rogers and Westman (1977) record 7% for the same fraction. Proportionately lower figures are found in younger stands.

Overall, eucalypt forests show a variation in litterfall mass consistent with their structure. The low open forests have a lower leaf biomass and a low litterfall (Bray and Gorham 1964) although they have a higher proportion of leaves than tall open forests. Litterfall in the tall open forests is high by world standards and the proportion of woody material is considerably greater than average (Bray and Gorham 1964). In particular, gum barked species of *Eucalyptus* have a high annual production of bark (Rogers and Westman 1977).

(ii) Litterfall in closed forests. Australian closed forest communities return quite variable amounts of organic matter to the forest floor. This is related to the considerable degree of structural and floristic variability they display associated with the latitudinal change from the cool temperate climate of Tasmania to the tropical northern parts of the country (Specht 1970). In fact, few detailed studies of litterfall accessions have been conducted in Australian closed forest ecosystems.

In the cool temperate *Nothofagus cunninghamii* rainforest of Victoria, Howard (1973) has recorded an annual litterfall mass of 6.5 t ha^{-1} $year^{-1}$. Watson (1977) recorded a 4 year mean accession of 4.99 t ha^{-1} $year^{-1}$ from a *Nothofagus moorei* ecosystem of a similar type. Webb *et al.* (1969) estimated litterfall from complex and simple notophyll vine forests in northern New South Wales over a two year period and recorded values ranging from 5.94 to 8.77 t ha^{-1} $year^{-1}$ and from 3.44 to 5.51 t ha^{-1} $year^{-1}$ respectively, in the two ecosystems. Litterfall mass in three tropical mesophyll vine forest sites in north-eastern Queensland ranged from 7.79 t ha^{-1} $year^{-1}$ at a relatively xeric site to 10.97 t ha^{-1} $year^{-1}$ at the most mesic site (CSIRO Aust. Div. Soils Tech. Mem. No. 28/1976). Brasell *et al.* (1980) have estimated annual litterfall accessions of 9.02 and 9.87 t ha^{-1} $year^{-1}$ at two sites in the same area over 4 and 3 years respectively.

Tropical closed forests return somewhat greater amounts of organic matter to the litter layers than the tall open forests. However, values given for warm temperate

Table 38.6. Estimates of annual litterfall (t ha^{-1} year^{-1}) in Australian conifer forests

Species	Total mass	Leaf	Wood	Female cones	Male cones	Author
P. lambertiana	5.50	4.87	0.53	0.01	0.07	Spain (1973*a*)
P. nigra	5.19	4.33	0.77	0.05	0.01	Spain (1973*a*)
P. ponderosa	5.64	4.96	0.55	0.09	0.03	Spain (1973*a*)
P. radiata	5.3, 8.2					Anon. (1969)
P. radiata	2.9, 4.6					Anon. (1969)
P. radiata	3.6, 4.8					Anon. (1969)
P. radiata	3.5					Forrest and Ovington (1970)
P. radiata	4.27					Pawsey (1959)
P. radiata	3.79, 3.79					Florence and Lamb (1974)
P. radiata	3.26, 4.12					Florence and Lamb (1974)
P. radiata	3.95, 3.36					Florence and Lamb (1974)
P. radiata	4.08, 4.19					Florence and Lamb (1974)
Ps. menziesii	3.95	3.39	0.47	0.05	0.005	Spain (1973*a*)
P. radiata	4.50					Tanton (1970)
A. cunninghamii	6.3–10.2[A] 5.1–12.7[B]					Brasell *et al.* (1980)

[A] Range of 4-year collections. [B] Range of 3-year collections.

forests do not appear substantially larger than those for many tall open forests. The maximum levels recorded for Australian tropical closed forest ecosystems are similar to those of Nye (1961) from Ghana and those of some secondary Malaysian forests (Bray and Gorham 1964) but are lower than the maximum value of 15.3 t ha^{-1} year^{-1} recorded by Laudelout and Meyer (1954) from Zaire, even allowing for the fact that the last of these results was presented on an air-dry weight basis. The lower levels of litterfall mass recorded from other tropical sites accord with values from other forests of the tropics and subtropics (Klinge and Rodriques 1968).

Little information is available on annual variation in litterfall mass for Australian closed forests although it is apparently low. Watson (1977) records a range of variation of 4.8 to 5.2 t ha^{-1} year^{-1} over a 4-year period Similarly low levels of annual variation of less than 10% are also recorded from rainforest communities in north-eastern Queensland (CSIRO Aust. Div. Soils Tech. Mem. No. 14/1978; Brasell *et al.* 1980). Bray and Gorham (1964) have reported similar levels from a four-year study of a mixed secondary forest in Malaysia and normal levels of variation appear to be low in this forest type. However, storm damage and other climatic factors have a considerable influence on litterfall, especially in areas subject to tropical revolving storms.

Proportions of the different physical components present in closed forest leaf litter have not often been reported. Howard (1973) records that leaf comprised 41% of total litterfall in a cool temperate forest. Leaf mass ranged from 54% to 64% of the total in samples from three closed forest ecosystems in north-eastern Queensland. In the same study, woody material comprised 12% to 17% and reproductive structures from 6% to 11% of the total mass (CSIRO Aust. Div. Soils Tech. Mem. No. 28/1976).

Peak litterfall rates occur during the period from late spring to early autumn in the Australian closed forests studied to date.

(iii) Litterfall in conifer plantations. In conifer forests, the litterfall derived from the planted species rises progressively to a peak at canopy closure and appears not to change markedly thereafter, as described by Forrest and Ovington (1970) for an age series of *Pinus radiata* stands in inland New South Wales.

P. radiata is widely grown in the temperate areas of Australia and several litterfall studies have been conducted (Table 38.6). The mass of litterfall produced varies markedly with site quality. Estimates extending from a low of 2.9 t ha^{-1} year^{-1} to a maximum of 8.2 t ha^{-1} year^{-1} have been made from contrasting sites within the Australian Capital Territory (Anon. 1969). Most other studies carried out in *P. radiata* plantations have recorded litterfall mass levels that range from 3.5 to 5.0 t ha^{-1} year^{-1} (Pawsey 1959; Forrest and Ovington 1970; Florence and Lamb 1974) although higher values have been recorded overseas. For example, Will (1959) estimated a maximum value of 10.31 t ha^{-1} year^{-1} for a highly productive stand in New Zealand. Other coniferous species have been little studied in Australia. Spain (1973*a*) estimated litterfall mass to range from 5.19 to 5.64 t ha^{-1} year^{-1} for three species of *Pinus* and to be 3.95 t ha^{-1} year^{-1} for a stand of *Pseudotsuga menziesii* in the Snowy Mountains region of New South Wales. Litterfall in plantations of the indigenous species *Araucaria cunninghamii* has been studied by Brasell *et al.* (1980) in tropical north-eastern Queensland. The mass of annual litterfall accessed is substantially higher than that of the other coniferous species so far studied in Australia (Table 38.6).

Seasonal variation in the mass of conifer litterfall is usually large. The major variation is in the fall of needles and the late spring fall of male cones. In *Pinus radiata*, maximum needle fall occurs in autumn (Pawsey 1959). Maximum needle fall also occurs in autumn for the other *Pinus* species studied by Spain (1973*a*) and for *P. radiata* and *P. nigra* in New Zealand (Will 1959). *Pseudotsuga menziesii* stands showed no marked seasonal pattern in needle fall in both the two latter studies. Substantial annual variation is evidenced in conifer stands. Total litterfall mass of one year in one stand may be as much as twice that of an adjacent year in the same stand (Will 1959; Brasell *et al.* 1980). The proportion of needles in litterfall for the species studied by Spain (1973*a*) varied from 83% to 88% which is higher than the 60% to 76% suggested by Bray and Gorham (1964).

(iv) Litterfall in other woodlands. Bevege (1978) lists a number of studies of nutrient cycling in arid woodland and shrubland. In most of these ecosystems, the aboveground living biomass of the community is relatively small and, consequently so is the mass of litterfall accessed annually.

Other inputs

A wide range of organic and inorganic substances have been found in washings and leachates from the aerial parts of plants (Tukey 1970). For elements such as calcium, manganese, phosphorus, potassium and sodium the amounts involved may account for considerable proportions of the total annual loss from the aerial parts of the plant (Ovington and Madgwick 1959; Will 1959; Attiwill 1966). Except for the studies of Attiwill (1966), Attiwill *et al.* (1978), Guthrie *et al.* (1978) and Westman (1978), few data have yet been published for Australian ecosystems.

Feller (1978) has measured inputs of cations into the soil from combined litter and canopy leachates and compared indigenous forests with plantations of exotic conifers.

High rates of defoliation by insects may occur in eucalypt forests (Morrow and LaMarche 1978; Springett 1978) and have been ascribed to the low levels of nitrogen in the leaves (Fox and Macauley 1977). This defoliation leads to a heavy fall of frass and Jacobs (1955) has recorded hourly rates of fall as high as 64 kg ha^{-1} in a *Eucalyptus moluccana* stand.

Considerable amounts of dust and dissolved elements may be transported by wind and rain and deposited at some distance from their origin. This phenomenon has been described by Walker and Costin (1971) in the south-eastern Australian ranges where substantial deposition of soil material has been recorded.

The amount of plant nutrient and other elements in rainfall may be appreciable although Australian records of this phenomenon are sparse. Concentrations of sodium and chloride in rainwater are related to the distance of the collecting site from the coast (Hutton 1976) although other elements are present which are not derived from marine sources. Australian studies of this phenomenon are, *inter alia*, those of Drover and Barrett-Lennard (1956), Hingston and Gailitis (1976), Probert (1976) and Guthrie *et al.* (1978).

Breakdown of organic matter

Efficient breakdown of the organic matter accessed by the litter layers is necessary to avoid the lengthy immobilization of large amounts of nutrients and other substances resulting from the formation of unduly heavy litter layers. Breakdown of litterfall is mediated through the activity of the decomposer cycle and by fire. Both breakdown routes involve a loss of structure and conversion of the organic matter to simpler compounds with a loss of energy (Odum 1971). Fire is particularly important in the Australian environment and is considered further below.

The products derived from the breakdown of plant material through the decomposer cycle usually become available to the plant community after incorporation into the mineral soil horizons. However, a direct cycling mechanism has been shown to occur in South American rainforests where fine roots penetrate the litter layers to feed directly on products of the decaying vegetation (Went and Stark 1968). This phenomenon has also been noted in north-eastern Australian rainforests and probably occurs in certain other vegetation types.

Breakdown through activity of the decomposer cycle

(i) Processes and agents of breakdown. In the context of reduction of plant and animal residues in the litter layers, breakdown must be considered as distinct from decomposition, which is the simplification of chemical compounds. Breakdown, as defined here, refers to the loss of material from the litter layers. It involves decomposition but also includes the processes of particulation and comminution of plant and animal remains and their incorporation in the upper surface horizons of the mineral soil. Incorporation in the upper mineral horizons may result from the direct feeding activity of animals such as earthworms on the surface and excretion of the partly broken down organic matter as faeces in the mineral soil horizons. It may result from translocation of particulate organic matter or faeces into the mineral horizons and, in some areas, the burial of organic remains through feeding and other activities of higher animals such as lyre birds (Ashton 1975) which 'cultivate' the upper soil layers.

Decomposition is mediated principally through the activity of the microflora, since few groups of soil animals possess the required enzymes to decompose the more resistant structural plant polysaccharides (Nielsen 1963). Termites may be a partial exception to this in that they can at least partly decompose lignin (Lee and Butler 1977), although the ecological significance of this remains to be demonstrated. The processes of leaf decomposition start during senescence and prior to abcission with their colonization by weak fungal saprovores (Dickinson 1976). In the litter layers, a succession of substrate-specific fungi is associated with the changing chemical composition that occurs during the course of decomposition. Macauley and Thrower (1966) have followed the fungal succession in the leaves of *Eucalyptus regnans* and the topic is reviewed by Hudson (1968). Leaching of the litter layers by throughfall and stemflow at once adds a range of compounds and allows passage into the mineral soil of the substances solubilized by decomposition processes.

Although the role of the soil and litter fauna is not completly defined, all the faunal groups discussed above play some part in litter breakdown and decomposition, although predators (Table 38.4) have an indirect influence through their regulatory effect on the numbers of other organisms.

Earthworms have long been recognized as important in organic matter breakdown in the temperate climates at least and have been the subject of considerable study in grassland and other situations (Chapter 39). Little appears to be known of the ecology of the Australian earthworm fauna although their numbers and activity are probably both restricted in drier areas.

Termites are widespread in the Australian environment and are important soil formers (Lee and Wood 1971) and decomposers with some unusual patterns of nutrient cycling and energy flow. These are discussed in relation to termite activity in Chapter 39 and by Wood (1976).

Within the Collembola, the wide range of feeding habits found indicates that they consume virtually the complete range of organic matter available, including the faeces of larger arthropods. The majority of mesostigmatid and some prostigmatid mites are predatory, feeding on soft-bodied arthropods and their eggs. These two groups also include some fungivorous forms but little is known of the feeding habits of small Prostigmata. Many Cryptostigmata are unspecialized feeders; fungi and bacteria as well as organic matter are common foods and, where dominant, this group may play an important role in community food webs (Harding and Stuttard 1974).

Free-living nematodes do not contribute directly to decomposition but feed on microorganisms and, where abundant, they may provide an important food source for other groups (Nicholas 1975).

It is considered that the main contribution of the saprovores (including Isopoda, Symphyla, Diplopoda, Diptera, larval Lepidoptera and other groups as well as many of the mites and Collembola) is to act in a loose form of symbiosis with the microflora to produce a synergistic effect on the rate and extent of breakdown and decomposition. This is achieved through particulation, comminution and conditioning of dead plant remains during their passage through the animal gut. These activities increase the surface area of the substrate for the microflora to work on and facilitate further decomposition. Direct effects of the soil fauna on the activity of the soil and litter microflora include an increase in spore dissemination throughout the substrate and the stimulation of senescent colonies through browsing. Reciprocal effects have also been noted in the sense that certain substrates, particularly wood, need to be softened externally by microfloral activity before they will become attractive to the litter and soil animals.

As with any biological community, the activity of the saprobiota is subject to environmental constraints. Temperature, moisture, and interactions between the two, are perhaps the most important variables controlling the breakdown rate of organic substrates (Wood 1974; Bunnell *et al.* 1977*a*). Moisture, particularly, limits decomposer activity for considerable periods in many Australian ecosystems. Even in the tropical rainforest of north-eastern Queensland, breakdown of litter and soil organic matter slows markedly in the dry season leading to a substantial build up of the litter layers (CSIRO Aust. Div. Soils Tech. Mem. No. 29/1976). With the onset of the summer rains, rapid breakdown rates supervene to reduce the mass of the litter layers to a low level.

The physical and chemical nature of the substrate has an important bearing on both the pathway and the rate of breakdown. Hard substrates usually break down more slowly than soft and the concentrations of a range of both organic and inorganic compounds exert a major influence. In Australia, replacement of the indigenous vegetation with exotic conifers leads to a redistribution of organic matter within the profile and to the build up of substantial litter layers. The modified environment of these conifer litters possesses a depauperate litter fauna with a reduced ability to break down organic materials (Spain 1974; Springett 1976*b*). It is not clear, however, whether these effects are wholly nutritionally based.

Minderman (1968) has demonstrated that the breakdown of litter may be modelled as the sum of a series of negative exponentials corresponding to the various organic fractions present. Also, the differential mobility of organic elements in decomposing plant tissue has been recognized for some years with sodium and potassium being lost rapidly and other elements, such as calcium, being lost relatively much more slowly (Attiwill 1968). It thus appears that it should be possible to characterize decomposing plant material at its various stages of breakdown by a mathematical combination of the levels of the above substances. Van Cleve (1974) and Bunnell *et al.* (1977*b*) have related breakdown rates to the concentration of particular substances or groups of substances although this work requires extension. In particular, a better characterization of the major organic components of plant material in terms of their differential susceptibilities to breakdown is required.

In field studies, it has been shown that the concentrations of resistant fractions, such as lignins, increase with time in litters (Alexander 1965). However, as litter does not appear to build up indefinitely, a stage of dynamic equilibrium must be attained where resistant fractions are broken down at a rate approximating their accession. It is likely that, with time, a saprobiota develops that has the capacity to degrade these resistant fractions.

Of particular relevance in influencing the susceptibility of organic matter to breakdown is the important role of the secondary compounds present in many plant tissues. These compounds have been evolved by plants as part of their defence against herbivores, and perhaps pathogens, and are known to actively inhibit faunal and microbial activity in living and dead tissues. The ecological consequences of the presence of these secondary compounds are considerable and they are known to exert a protective effect on various substrates otherwise susceptible to enzymic attack. This protective effect extends long past the death of the plants and affects the breakdown of plant litter and soil organic matter.

Janzen (1974) discusses some of the ecological relationships of these compounds and the consequences of their redistribution in the environment. The chemical nature of the compounds involved and aspects of their activity have been recently reviewed by Levin (1976) and Swain (1977).

The concentrations of secondary compounds in Australian vegetation are not well known. Hingston (1963) presented data on the levels of selected groups of secondary compounds from the plant tissues of several eucalypts and other genera and concluded that polyphenols, in particular, play an important role in soil genesis. Fox and Macauley (1977) and Macauley and Fox (1980) have reported high and variable levels of polyphenols in the foliage of several species of *Eucalyptus*. In addition, eucalypt foliage is known to have high concentrations of a wide range of essential oils (Penfold and Willis 1961). The detailed ecological relationships of these secondary compounds are not yet known for any Australian ecosystem.

(ii) Measurement of breakdown rate. The various methods used to assess rates of breakdown of organic substrates may be considered as direct or indirect. In general, though, most provide indices rather than absolute levels of breakdown.

The direct methods attempt to assess breakdown by confining litter, leaves or other substrates in inert containers such as polyester mesh bags or by tethering or marking the substrate in such a way that it may be readily recovered. Such factors as the rate of loss of leaf area, dry weight and energy, changes in the concentrations of a range of plant nutrient elements and other substances are measured after varying periods of exposure to provide an index of the rate of breakdown.

There has been some variation in the way litter bag studies have been carried out, and this has led to problems in making comparisons. Wood (1974) used bags with meshes of different sizes to estimate the effects of selected environmental factors and various decomposer groups on the rate of breakdown of *Eucalyptus delegatensis* foliage. He found yearly rates of loss of dry matter from coarse mesh bags that ranged from 40 to 84% over a range of sites. The variation in these results was considered to be due to the different rainfall and temperature regimes pertaining at the exposure locations. In contrast, it was found that in a similar experiment effectively all of a sample of the leaves of *Castanospermum australe* had broken down over a period of ninety days during the wet season in the warm humid environment of a tropical rainforest near Innisfail (latitude 17°30′S, longitude 146°E) (CSIRO Aust. Div. Soils Tech. Mem. No. 34/1978).

By using fine or medium mesh litter bags, Wood (1974) excluded, selectively, nearly all soil animals or only the larger soil animals from his litter samples and showed that depressed rates of breakdown of substrate resulted. Macauley (1975) exposed leaves of *Eucalyptus pauciflora* treated with two fungicides and two insecticides and found depressed rates of loss of dry weight, phosphorus and nitrogen content of the leaves commensurate with reduction of the target organism populations.

Litter bag studies are often carried out with the aim of assessing the relative susceptibilities of different substrates to breakdown or differences between environments. Wood (1971*a*) used litter bag techniques to show that the leaves of *Eucalyptus* species break down more slowly than those of other broadleafed species and that marked differences occur between species within the genus. In similar experiments Springett (1976*b*) demonstrated reduced breakdown rates following replacement of the indigenous eucalypt forest with plantations of *Pinus pinaster*. A comparison of the first year rates of disappearance from litter bags of overstorey eucalypt and understorey shrub leaves has indicated that the litter fauna may be more responsible for the breakdown of overstorey than understorey leaves, whereas leaching and microbial activity may be more important for understorey litter (Birk 1979*b*).

Confinement methods, including the use of litter bags, suffer from a degree of artificiality since the regime within the bag is unlikely to be the same as that without. However, with standardization of substrate, bags and exposure techniques, they provide a useful index of breakdown.

There are several indirect methods of obtaining indices of decomposition. Springett (1976*b*) exposed cotton threads for selected periods and used the reduction in tensile strength as a measure of decomposer activity. A widely used method is to measure the rate of input of organic matter to the litter layer and relate this to the mass present, often assuming that a steady state exists in the forest floor (Olson 1963). This method and its variants must be regarded as only providing first approximations since they imply that both litter fall and the forest floor are homogeneous in nature and that an improbable steady state exists.

Measurement of substrate respiration rates has been used to assess microbial activity for many years. Although it is a useful tool and provides much valuable information, it must be used in conjunction with other techniques in litter breakdown studies because of the inter-related effects of the soil fauna and microflora. Microfloral populations can degrade plant material to only a limited extent in the absence of faunal elements and then only at a much reduced rate. In decomposition studies Howard and Howard (1974) found that, in the absence of soil animals, the amount of litter remaining tended to a limiting value and that an asymptotic model was appropriate in this situation.

Breakdown through the action of fire

Fires vary considerably in intensity, depending on the amount and quality of the fuels available and on environmental conditions. The enormous areas affected and substantial losses of life and property resulting from severe bushfires have been discussed by Cheney (1976). To reduce the damage, prophylactic or prescribed burning has been advocated (Vines 1975) in an effort to reduce the mass of the litter layers so that when bushfires occur they will be low-intensity, slow-moving fires confined to the forest floor. A problem with this approach is that it may provide a rationale for an unduly high frequency of burning.

Fire and its consequences have shaped and conditioned the nature and distribution of the vegetation of Australia since long before the arrival of Aboriginal man. Gill (1975) and papers in Gill *et al.* (1981) review our knowledge of the effects of fire on the Australian biota and Luke and McArthur (1978) provide a useful summary of bushfire distribution, behaviour and control in Australia.The effects of high-intensity fire on the rate of litterfall and nutrient concentration in the litter for two years after burning a jarrah forest have been examined by O'Connell *et al.* (1979). Raison (1980) reviews the effect of fire on nutrient cycling processes in Australia.

When the action of fire – rather than the normal decomposer cycle – causes the partial or complete decomposition of the litter layers, the available energy is dissipated rapidly with the litter mass to leave an ash bed on the surface of the mineral soil horizon.

Although the ecological effects of regular burning are not well known, some effects of burning have been characterized. Low temperature prescribed burns are usually patchy in their distribution and do not markedly affect the crowns of the dominant tree species. Tanton (1974) states that 70% or less of the area covered by a prescribed burn is actually burned although Campbell and Tanton (1979) estimated 73 to 97% and Row and Hagel (1974) estimated that 80 to 90 % of their experimental plots were covered by fire. Intense wildfires, though, may bare the mineral soil almost completely over substantial areas and kill the standing vegetation by burning the crowns.

When heavy rains follow firing before vegetation regrowth can stabilize the soil surface, severe soil erosion often results (Brown 1972). Mineral elements, are removed through leaching, in runoff and in ash lost in the updraught of any fire (Rowe and Hagel 1974). Loss of substantial amounts of plant nutrient and other elements in smoke has been reported (Harwood and Jackson 1975; King *et al.* 1977). Of these, nitrogen loss must be considered the most important since it frequently limits plant growth. Ash bed effects may, however, enhance subsequent regrowth (Humphreys and Lambert 1965).

Populations of ground-dwelling fauna are substantially reduced for a period following extensive burning although some species are affected less than others (Newsome *et al.* 1975). Perhaps of more importance is the reduction in the biological activity of the decomposer cycle following firing. However, the extent and duration of this suppression are not well known. Watson (1977) recorded a temporary fall in soil respiration following firing and showed that breakdown (as measured by loss in weight of bagged litter) was also reduced. Similar findings are reported by Springett (1976*a*), who noted that the saprobiota had a reduced ability to degrade the needles of *Pinus pinaster* four years after a prescribed burn in a Western Australian plantation of this species. Springett (1976*a*, 1979) also noted a reduction in the species diversity and population levels of microfauna in eucalypt forest one and five years after prescribed burning. In contrast, Campbell and Tanton (1981) found that, following a light burn, the reduction in litter invertebrate populations was temporary and that communities of these animals were essentially similar to those of unburned control areas within eighteen months of firing.

Comparative studies of the soil microflora in regularly burned and unburned areas of low open forest (Jones and Richards 1977) have shown that total soil bacterial numbers are reduced although fungal populations appear unaffected.

In the long term, the effects of periodically removing a major proportion of the litter layers through firing must be considered in relation to the long term fertility of forest soils. This is, manifestly, one area that requires additional study, considering the large areas burned each year, and is particularly cogent where, as in parts of northern Australia, annual or near annual burning occurs (Stocker 1966).

Conclusions

An understanding of the general nature of the biological productivity of a particular ecosystem requires a knowledge of the stocks of organic matter and nutrients present in the major compartments and the rates of transfer between them. It also involves a knowledge of energy flows into the system and of the pathways of its eventual dissipation. In addition, the functioning of an ecosystem must be seen in the light of given climatic and edaphic regimes and of its biogeographical history.

As stated earlier in this review, integrated approaches to nutrient cycling studies are still in their infancy in Australia. A few studies, such as that reported by Rogers and Westman (1977), Westman and Rogers (1977*a*, *b*) and Westman (1978) have made a substantial contribution to our understanding of particular types of ecosystems. However, these studies must be extended both in depth and nature to a wider range of environments before general conclusions may be drawn as to the special nature of Australian nutrient cycling and energy flow regimes. Work of this nature could lead to a functional classification of Australian ecosystems which would have considerable scientific value and be of benefit in land use management and planning.

Of the studies reported in the Australian literature, a distinct geographical bias to the south-eastern part of the continent is noticeable. The remaining parts of Australia, particularly the large areas falling within the tropics, are considerably less well known and possess a range of qualitatively different ecosystems such as those of the arid lands and the wetter tropical rain and mangrove forests. These and most other tropical ecosystems are poorly known in terms of their structure, biomass and nutrient cycling characteristics. It is to be hoped that, in future, some greater effort will be expended in attempting to characterize these areas, given the large areas involved and their national and scientific importance.

Within ecological systems, flow of energy and the accompanying movement of mineral elements may become constrained at some point in the cycle such that accession by one compartment exceeds its rate of loss for a suitable period. The resulting build up has a far reaching influence on the activities of other ecosystem components. This normally occurs during development of a crop or after harvest while the biomass of the major autotrophs builds up. It may also occur when the nature of the plant community is altered or when biologically active chemicals are added.

The litter layers are potentially large compartments in many forested ecosystems and in all ecosystems are seen as a dynamic reservoir of nutrients and non-essential elements within a complex body of energy-rich organic materials. Their magnitudes are generally assumed to be low in grassland and cropping situations although little Australian information is available.

Within the litter layers, a characteristic biota develops that is related to and overlaps that of the underlying mineral soil. To date, there have been very few detailed surveys of the soil and litter fauna of different Australian ecosystems and our knowledge of the nature and activity of this biota is minimal.

In general, the low sampling intensity and variety of methods used have rendered the results of many studies sufficient only for tentative conclusions to be reached on the nature of the differences between populations. In addition, many faunal studies have methodological deficiencies in sample handling and extraction and little credence can be placed in the results for certain groups. The results of adequately intensified and standardized sampling programmes would enable more positive conclusions to be reached on soil faunal population structure and variation, their relationships to the physical and chemical state of the environment and their effects on the fertility of the soil.

A feature of litters is their degree of variation in mass and nature over both small and large time scales. Variation in mass results from an imbalance between the rates of accession and breakdown of litterfall and other inputs from the dominant community autotrophs. Climatic factors considerably influence this as litterfall is often seasonal. Sporadic inputs may result from storm effects and can substantially modify the normal patterns of physical and chemical inputs from leaf abcission. In addition, both seasonal and extended drought are a feature of many Australian environments and markedly restrict the activity of the decomposer cycle. Adequate assessment of annual variation thus requires several years sampling; a time span not achieved in most studies.

As an intermittent feature of many Australian landscapes, fire exerts a modifying influence on many ecosystems and maintains the litter layers in a state of disequilibrium. It is, however, a valuable management tool and the use of prescribed burning to protect human life and resources may be expected to increase. Little is yet known of the long-term ecological effects of regular burning and more information is needed particularly on the effects on subsequent biological activity and on the surface of the mineral soil.

Litterfall is a distinctive feature of terrestrial communities and its timing, magnitude, and physical and chemical nature are an important combination of properties useful in a functional classification of ecosystems. In particular, the chemical characteristics of both litterfall and the litter layers of Australian ecosystems are poorly known. Several studies have assessed levels of a number of major and minor nutrient elements (Bevege 1978). This work should be extended to include a larger range of elements, both essential and non-essential, together with selected physical characteristics. Characterisation of the organic components of litterfall from the often dominant eucalypts has not proceeded far and may be of benefit in helping to explain the nature of the reported differences in breakdown characteristics between these and other broadleafed plants (Wood 1971*a*).

There is a particular need to define the development stage of the community in ecological studies. While this is a particularly obvious feature of plantation crops it is less marked in many indigenous communities. Nevertheless, it is important to assess the structural characteristics associated with development as Ashton (1975) has shown in his studies of communities dominated by *Eucalyptus regnans*. In many circumstances, it may also be desirable to partition litterfall from the various strata and species of the plant community where an understorey or shrub layer has developed (Birk 1979*a*). This is particularly relevant in studies of the extensive open forests and woodlands, where the relationships between the grass stratum and the canopy trees have not been delineated satisfactorily.

It is clear that little is yet known of breakdown and incorporation processes in Australian ecosystems. Breakdown studies have been restricted in nature and, in general, are not directly comparable due to a lack of standardization of both aims and methodology. It is perhaps time for standardized comparative studies to be instigated with the aim of characterizing breakdown processes and rates in relation to climatic factors within the soil and litter layers of the major Australian ecosystems. Further, there is also a need for studies that will integrate the somewhat disparate faunal, microbial and biochemical approaches to this work.

The capacity of the saprobiota of an ecosystem to degrade its own or other organic matter may be assessed quite readily using litter bags and related techniques. To understand the faunal aspects of the processes whereby this organic matter degrades is more difficult since we have little knowledge of the nature and roles of the saprobiota of many Australian environments. This is particularly true of the soil and litter fauna.

References

Ahern, L. D. and Yen, A. L. (1977). A comparison of the invertebrate fauna under *Eucalyptus* and *Pinus* forests in the Otway Ranges, Victoria. *Proc. R. Soc. Vic.* **89**, 127-36.

Alexander, M. (1965). Biodegradation: problems of molecular recalcitrance and microbial fallibility. *Adv. Appl. Microbiol.* **7**, 35-80.

Anderson, J. M. and Macfadyen, A. (Eds) (1976). 'The Role of Terrestrial and Aquatic Organisms in Decomposition Processes.' (Blackwell Scientific Publications: Oxford.)

Anon (1969). Forestry and Timber Bureau, Annual Report for 1968. (Aust. Govt. Printer: Canberra.)

Anon (1975). Forestry and Timber Bureau, Annual Report for 1973/74. (Aust. Govt. Printer: Canberra.)

Ashton, D. H. (1975). Studies of litter in *Eucalyptus regnans* forests. *Aust. J. Bot.* **23**, 413-33.

Attiwill, P. M. (1966). The chemical composition of rainwater in relationship to the cycling of nutrients in mature *Eucalyptus* forests. *Plant Soil* **24**, 390-406.

Attiwill, P. M. (1968). The loss of elements from decomposing litter. *Ecology* **49**, 142-5.

Attiwill, P. M., Guthrie, H. B. and Leuning, R. (1978). Nutrient cycling in a *Eucalyptus obliqua* (L. Hérit.) forest. I. Litter production and nutrient return. *Aust. J. Bot.* **26**, 79-91.

Bale, C. L. (1974). Topographic studies in the New England National Park, B. Sc. (Hons.) Thesis, University of New England, Armidale.

Bevege, D. I. (1978). Biomass and nutrient distribution in indigenous forest ecosystems. Queensl. For. Dep. Tech. Pap. No. 6.

Birk, E. M. (1979*a*). Overstory and understory litterfall in a eucalypt forest: spatial and temporal variability. *Aust. J. Bot.* **27**, 145-56.

Birk, E. M. (1979*b*). Disappearance of overstory and understory litter in an open eucalypt forest. *Aust. J. Ecol.* **4**, 207-22.

Bornemissza, G. E. (1957). An analysis of arthropod succession in carrion and the effect of its decomposition on the soil fauna. *Aust. J. Zool.* **5**, 1-12.

Bornemissza, G. E. (1976). The Australian dung beetle project 1965-1975. Australian Meat Research Committee Review No. 20.

Brasell, H. M., Unwin, G. L. and Stocker, G. C. (1980). The quantity, temporal distribution and chemical composition of litterfall in two forest types at two sites in tropical Australia. *J. Ecol.* **68**, 123-39.

Bray, J. R. and Gorham, E. (1964). Litter production in forests of the world. *Adv. Ecol. Res.* **3**, 102-57.

Brown, J. A. H. (1972). Hydrologic effects of a bushfire in a catchment in south-eastern New South Wales, *J. Hydrol.* **15**, 77-96.

Bunnell, F. L., Tait, D. E. N., Flanagan, P. W. and Van Cleve, K. (1977*a*). Microbial respiration and substrate weight loss. I. A general model of the influence of abiotic variables. *Soil Biol. Biochem.* **9**, 33-40.

Bunnell, F. L., Tait, D. E. N. and Flanagan, P. W. (1977*b*). Microbial respiration and substrate weight loss. II. A model of the influences of chemical composition. *Soil Biol. Biochem.* **9**, 41-7.

Burrows, W. H. (1972). Productivity of an arid zone shrub (*Eremophila gilesii*) community in south-western Queensland. *Aust. J. Bot.* **20**, 317-29.

Campbell, G. and Tanton, M. T. (1981). The effects of fire on the invertebrate fauna of soil and litters of a eucalypt forest. *In* 'Fire and the Australian Biota.' (Eds A. M. Gill, R. H. Groves and I. R. Noble) pp. 215-41. (Australian Academy of Science: Canberra.)

Carnahan, J. A. (1976). Natural vegetation. Atlas of Australian Resources, 2nd series. (Dep. Nat. Res. : Canberra.)

Castri, F. di (1973). Soil animals in latitudinal and topographical gradients of mediterranean ecosystems. *In* 'Mediterranean Type Ecosystems.' (Eds F. di Castri and H. A. Mooney) pp. 171-90. (Chapman and Hall: London.)

Charley, J. L. and Cowling, S. W. (1968). Changes in soil nutrient status resulting from overgrazing and their consequences in plant communities of semi-arid areas. *Proc. Ecol. Soc. Aust.* **3**, 28-38.

Cheney, N. P. (1976). Bushfire disasters in Australia, 1945-1975. *Aust. For.* **39**, 245-68.

Dickinson, C. H. (1976). Fungi on the aerial surfaces of higher plants. *In* 'Microbiology of Aerial Plant Surfaces.' (Eds C. H. Dickinson and T. F. Preece) pp. 293-324. (Academic Press: London.)

Dickinson, C. H. and Pugh, G. J. F. (Eds) (1974). 'Biology of Plant Litter Decomposition.' (Academic Press: London.)

Drover, D. P. and Barrett-Lennard, I. P. (1956). Accessions of nitrogen (ammonia, nitrate and nitrite) in Western Australia wheat belt rains. *J. Aust. Inst. Agric. Sci.* **32**, 193-7.

Elliot, H. J. (1970). The role of millipedes in the decomposition of *Pinus radiata* litter in the Australian Capital Territory. *Aust. For. Res.* **4**, 3-10.

Feller, M. C. (1978). Nutrient movement into soils beneath eucalypt and exotic conifer forests in southern central Victoria. *Aust. J. Ecol.* **3**, 357-72.

Ferrar, P. and Watson, J. A. L. (1970). Termites (*Isoptera*) associated with dung in Australia. *J. Aust. Entomol. Soc.* **9**, 100-2.

Florence, R. G. (1961). Studies in the ecology of blackbutt (*Eucalyptus pilularis* Sm.) Unpublished Ph. D. Thesis, University of Sydney.

Florence, R. G. and Lamb, D. (1974). Influence of stand and site on Radiata pine litter in South Australia. *N. Z. J. For. Sci.* **4**, 502-10.

Forrest, W. G. and Ovington, J. D. (1970). Organic matter changes in an age series of *Pinus radiata* plantations. *J. Appl. Ecol.* **7**, 177-86.

Fox, B. J., Fox, M. D. and McKay, G. M. (1979). Litter accumulation after fire in a eucalypt forest. *Aust. J. Bot.* **27**, 157-65.

Fox, L. R. and Macauley, B. J. (1977). Insect grazing on *Eucalyptus* in response to variation in leaf tannins and nitrogen. *Oecologia* **29**, 145-62.

Gill, A. M. (1975). Fire and the Australian flora: a review. *Aust. For.* **38**, 4-25.

Gill, A. M., Groves, R. H. and Noble, I. R. (Eds) (1981). 'Fire and the Australian Biota.' (Australian Academy of Science: Canberra.)

Golley, F. B. (1969). Calorific values of wet tropical vegetation. *Ecology* **50**, 517-9.

Gosz, J. R., Holmes, R. T., Likens, G. E. and Bormann, F. H. (1978). The flow of energy in a forest ecosystem. *Sci. Am.* **238**, 92-103.

Guthrie, H. B., Attiwill, P. M. and Leuning, R. (1978). Nutrient cycling in a *Eucalyptus obliqua* (L'Hérit.) forest. II. A study in a small catchment. *Aust. J. Bot.* **26**, 189-201.

Hall, N., Johnston, R. D. and Chippendale, G. M. (1970). 'Forest Trees of Australia.' 3rd edition. (A. G. P. S. : Canberra.)

Hannon, N. J. (1958). The status of nitrogen in the Hawkesbury sandstone soils and their plant communities in the Sydney district. II. The distribution and circulation of nitrogen. *Proc. Linn. Soc. N. S. W.* **83**, 65-85.

Harding, D. J. L. and Stuttard, R. A. (1974). Microarthropods. *In* 'Biology of Plant Litter Decomposition.' (Eds C. H. Dickinson and G. J. F. Pugh) Vol. 2, pp. 489-532. (Academic Press: London.)

Harwood, C. E. and Jackson, W. D. (1975). Atmospheric losses of four plant nutrients during a forest fire. *Aust. For.* **38**, 92-9.

Hatch, A. B. (1955). The influence of plant litter on the Jarrah forest soils of the Dwellingup region – Western Australia. For. Timb. Bur. Leaflet No. 70.

Hingston, F. J. (1963). Activity of polyphenolic constituents of leaves of *Eucalyptus* and other species in complexing and dissolving iron oxide. *Aust. J. Soil Res.* **1**, 63-73.

Hingston, F. J. and Gailitis, V. (1976). The geographic variation of salt precipitated over Western Australia. *Aust. J. Soil Res.* **14**, 319-35.

Hingston, F. J., Turton, A. G. and Dimmock, G. M. (1979). Nutrient distribution in karri (*Eucalyptus diversicolor* F. Muell.) ecosystems in southwest Western Australia. *For. Ecol. Manag.* **2**, 133-58.

Howard, G. W. (1967). Arthropods inhabiting pine litter in the south-east of South Australia. Unpublished M. Sc. thesis, University of Adelaide (Waite Institute).

Howard, P. J. A. and Howard, D. M. (1974). Microbial decomposition of tree and shrub leaf litter. I. Weight loss and chemical composition of decomposing litter. *Oikos* **25**, 341-52.

Howard, T. M. (1973). Studies in the ecology of *Nothofagus cunninghamii* Oerst. II. Phenology. *Aust. J. Bot.* **21**, 79-92.

Hudson, H. J. (1968). The ecology of fungi on plant remains above the soil. *New Phytol.* **67**, 837-74.

Hughes, R. D. (1975). Assessment of the burial of cattle dung by Australian dung beetles. *J. Aust. Entomol. Soc.* **14**, 129-34.

Humphreys, F. R. and Lambert, M. J. (1965). An examination of a forest site which has exhibited the ash-bed effect. *Aust. J. Soil Res.* **3**, 81-94.

Hutton, J. T. (1976). Chloride in rainwater in relation to distance from the ocean. *Search* **7**, 207-8.

Jacobs, M. R. (1955). 'Growth Habits of the Eucalypts.' (Commonwealth Gov. Printer: Canberra.)

Janzen, D. H. (1974). Tropical blackwater rivers, animals, and mast fruiting by the Dipterocarpaceae. *Biotropica* **6**, 69-103.

Jones, J. M. and Richards, B. N. (1977). Changes in the microbiology of eucalypt forest soils following reafforestation with exotic pines. *Aust. For. Res.* **7**, 229-40.

King, N. K., Packham, D. R. and Vines, R. G. (1977). On the loss of selenium and other elements from burning forest litter. *Aust. For. Res.* **7**, 265-8.

Klinge, H. and Rodrigues, W. A. (1968). Litter production in an area of Amazonian Terra Firma forest. Part I. Litterfall, organic carbon and total nitrogen contents of litter. *Amazoniana* **1**, 287-302.

Laudelout, H. and Meyer, J. (1954). Les cycles d'éléments minéraux et de matière organique en forêt équatoriale congolaise. *Int. Congr. Soil Sci. Trans.* **2**, 267-72.

Lee, K. E. and Butler, J. H. A. (1977). Termites, soil organic matter decomposition, and nutrient cycling. *In* 'Soil Organisms as Components of Ecosystems.' (Eds U. Lohm and T. Persson). 6th Int. Colloq. Soil Zool., Uppsala, 1976. *Ecol. Bull.* (Stockholm) **25**, 544-8.

Lee, K. E. and Correll, R. L. (1978). Litterfall and its relationship to nutrient cycling in a South Australian dry sclerophyll forest. *Aust. J. Ecol.* **3**, 243-52.

Lee, K. E. and Wood, T. G. (1971). 'Termites and Soils.' (Academic Press: London.)

Levin, D. A. (1976). The chemical defenses of plants to pathogens and herbivores. *Ann. Rev. Ecol. Syst.* **7**, 121-59.

Luke, R. H. and McArthur, A. G. (1978). 'Bushfires in Australia.' (A. G. P. S. : Canberra.)

Macauley, B. J. (1975). Biodegradation of litter in *Eucalyptus pauciflora* communities. I. Techniques for comparing the effects of fungi and insects. *Soil Biol. Biochem.* **7**, 341-4.

Macauley, B. J. and Fox, L. R. (1980). Variation in total phenols and condensed tannins in *Eucalyptus*: leaf phenology and insect grazing. *Aust. J. Ecol.* **5**, 31-5.

Macauley, B. J. and Thrower, L. B. (1966). Succession of fungi in leaf litter of *Eucalyptus regnans*. *Trans. Br. Mycol. Soc.* **49**, 509-20.

McColl, J. G. (1966). Accession and decomposition of litter in spotted gum forests. *Aust. For.* **30**, 191-8.

Marsh, R. and Campling, R. C. (1970). Fouling of pastures by dung. *Herbage Abstr.* **40**, 123-30.

Meakins, J. (1966). The effect of understory species on soil properties under mountain eucalypts. B. Sc. For. (Hons.) Thesis, Aust. Nat. Univ., Canberra.

Medwecka-Kornas, A. (1971). Plant litter. *In* 'Methods of Study in Quantitative Soil Ecology: Population, Production and Energy Flow.' (Ed. J. Phillipson). pp. 24-33. I. B. P. Handbook No. 18. (Blackwell Scientific Publications: Oxford.)

Millar, C. S. (1974). Decomposition of coniferous leaf litter. *In* 'Biology of Plant Litter Decomposition.' (Eds C. H. Dickinson and G. J. F. Pugh). Vol. 1, pp. 105-28. (Academic Press: London.)

Miller, H. G., Cooper, J. M. and Miller, J. D. (1976). Effects of nitrogen supply on nutrients in litterfall and crown leaching in a stand of corisican pine. *J. Appl. Ecol.* **13**, 233-48.

Minderman, G. (1968). Addition, decomposition and accumulation of organic matter in forests. *J. Ecol.* **56**, 355-62.

Moore, A. W., Russell, J. S. and Coaldrake, J. E. (1967). Dry matter and nutrient content of a subtropical semiarid forest of *Acacia harpophylla* F. Muell. (Brigalow). *Aust. J. Bot.* **15**, 11-24.

Morrow, P. A. and LaMarche, V. C. (1978). Tree ring evidence for chronic suppression of productivity in subalpine *Eucalyptus*. *Science* **201**, 1244-6.

Newsome, A. E., McIllroy, J. and Catling, P. (1975). The effects of an extensive wildfire on populations of twenty ground vertebrates in south-east Australia. *Proc. Ecol. Soc. Aust.* **9**, 107-23.

Nicholas, W. L. (1975). 'The Biology of Free-Living Nematodes.' (Oxford University Press: London.)

Nielsen, C. O. (1963). Carbohydrates in soil and litter invertebrates. *Oikos* **13**, 200-15.

Nye, P. H. (1961). Organic matter and nutrient cycles under moist tropical forest. *Plant Soil* **13**, 333-46.

O'Connell, A. M., Grove, T. S. and Dimmock, G. M. (1978). Nutrients in the litter on jarrah forest soils. *Aust. J. Ecol.* **3**, 253-60.

O'Connell, A. M., Grove, T. S. and Dimmock, G. M. (1979). The effects of a high intensity fire on nutrient cycling in jarrah forest. *Aust. J. Ecol.* **4**, 331-7.

Odum, E. P. (1971). 'Fundamentals of Ecology'. 3rd edn. (W. B. Saunders: Philadelphia.)

Olson, J. S. (1963). Energy storage and the balance of producers and decomposers in ecological systems. *Ecology* **44**, 322-31.

Ovington, J. D. and Heitkamp, D. (1960). The accumulation of energy in forest plantations in Britain. *J. Ecol.* **48**, 639-46.
Ovington, J. D. and Madgwick, H. A. I. (1959). Distribution of organic matter in a plantation of Scots Pine. *For. Sci.* **5**, 344-55.
Pawsey, C. K. (1959). The seasonal fall of litter in Monterey pine in the south-east of South Australia. *Inst. For. Aust. Newsl.* **2**, 16-17.
Peet, G. B. (1971). Litter accumulation in jarrah and karri forests. *Aust. For.* **35**, 258-62.
Penfold, A. R. and Willis, J. L. (1961). 'The Eucalypts.' (Leonard Hill: London.)
Plowman, K. P. (1979). Litter and soil fauna of two Australian subtropical forests. *Aust. J. Ecol.* **4**, 87-104.
Probert, M. E. (1976). The composition of rainwater at two sites near Townsville Qld. *Aust. J. Soil Res.* **14**, 397-402.
Raison, R. J. (1980). A review of the role of fire in nutrient cycling in Australian native forests, and of methodology for studying the fire-nutrient interaction. *Aust. J. Ecol.* **5**, 15-21.
Rogers, R. W. and Westman, W. E. (1977). Seasonal nutrient dynamics of litter in a subtropical eucalypt forest, North Stradbroke Island. *Aust. J. Bot.* **25**, 47-58.
Rowe, R. K. and Hagel, U. (1974). Leaching of plant nutrient ions from burned forest litter. *Aust. For.* **36**, 154-63.
Soil Science Society of America (1975). Glossary of soil science terms. (Soil Science Society of America.)
Spain, A. V. (1972). A study of aspects of the nutrient cycle in a New South Wales conifer forest. Unpublished Ph. D. Thesis, Aust. Nat. Univ. Canberra.
Spain, A. V. (1973*a*). Litterfall in a New South Wales conifer forest: A multivariate comparison of plant nutrient element status and return in four species. *J. Appl. Ecol.* **10**, 527-56.
Spain, A. V. (1973*b*). A preliminary study of spatial patterns in the accession of conifer litterfall. *J. Appl. Ecol.* **10**, 557-67.
Spain, A. V. (1974). The effects of carbaryl and DDT on the litter fauna of a Corsican pine (*Pinus nigra* var. *maritima*) forest: A multivariate comparison. *J. Appl. Ecol.* **11**, 467-81.
Specht, R. L. (1970). Vegetation. *In* 'The Australian Environment.' (Ed G. W. Leeper) 4th edn. pp. 43-67. (Melbourne University Press: Melbourne.)
Springett, B. P. (1978). On the ecological role of insects in Australian eucalypt forests. *Aust. J. Ecol.* **3**, 129-39.
Springett, J. A. (1971). The effects of fire on litter decomposition and on soil fauna in a *Pinus pinaster* plantation. IV. Colloquium pedobiologiae. *Ann. Zool.* **17**, 257-63.
Springett, J. A. (1976*a*). The effect of prescribed burning on the soil fauna and on litter decomposition in Western Australian forests. *Aust. J. Ecol.* **1**, 77-82.
Springett, J. A. (1976*b*). The effect of planting *Pinus pinaster* Ait. on populations of soil microarthropods and on litter decomposition at Gnangara, Western Australia. *Aust. J. Ecol.* **1**, 83-7.
Springett, J. A. (1979). The effects of a single hot summer fire on soil fauna and on litter decomposition in jarrah (*Eucalyptus marginata*) forest in Western Australia. *Aust. J. Ecol.* **4**, 279-91.
Stoate, T. N. (1958). Karri silviculture – Litterfall. Western Australia Forests Dep., Annual Report, pp. 25-6.
Stocker, G. C. (1966). Effects of fires on vegetation in the Northern Territory. *Aust. For.* **30**, 223-30.
Swain, T. (1977). Secondary compounds as protective agents. *Annu. Rev. Plant Physiol.* **28**, 479-501.
Tanton, M. T. (1970). The biology of radiata pine stands. The biological community. Paper presented to '*Radiata Pine Symposium*' held by the Radiata Pine Association of Australia, 25-28 August, 1970 at the Australian National University, Canberra.
Tanton, M. T. (1974). Bushfires and the forest. pp. 58-9. *In* 'Proceedings of the 1971 Tropical and Arid Fire Symposium.' (Bush Fires Council of the Northern Territory: Darwin.)
Tukey, H. B. (1970). The leaching of substances from plants. *Annu. Rev. Plant Physiol.* **21**, 305-24.
Turner, J. and Kelly, J. (1977). Soil chemical properties under naturally regenerated *Eucalyptus* spp. and planted Douglas-fir. *Aust. For. Res.* **7**, 163-72.
Usher, M. B. (1970). Pattern and seasonal variability in the environment of a Scots pine forest soil. *J. Ecol.* **58**, 669-79.
Usher, M. B. (1976). Aggregation responses of soil arthropods in relation to the soil environment. *In* 'The Role of Terrestrial and Aquatic Organisms in Decomposition Processes.' (Eds J. M. Anderson and A. Macfadyen). pp. 61-94. (Blackwell Scientific Publications: Oxford.)
Van Cleve, K. (1974). Organic matter quality in relation to decomposition. *In* 'Soil Organisms and Decomposition in Tundra.' (Eds A. J. Holding, O. W. Heal, S. F. Maclean and P. W. Flanagan). pp. 311-24. (Tundra Biome Steering Committee: Stockholm.)
Vine, R. G. (1975). Bushfire research in CSIRO. *Search* **6**, 73-8.
Walker, P. H. and Costin, A. B. (1971). Atmospheric dust accession in south-eastern Australia. *Aust. J. Soil Res.* **9**, 1-5.
Wallwork, J. A. (1970). 'Ecology of Soil Animals.' (McGraw-Hill: London.)
Wallwork, J. A. (1976). 'The Distribution and Diversity of Soil Fauna.' (Academic Press: London.)
Waterhouse, D. F. (1974). The biological control of dung. *Sci. Am.* **230**, 100-9.
Watson, G. W. (1977). Metabolism of forest floors. Unpublished M. Sc. Thesis. University of New England, Armidale.
Webb, L. J., Tracey, J. G., Williams, W. T. and Lance, G. W. (1969). The pattern of mineral return in leaf litter of three subtropical Australian forests. *Aust. For.* **33**, 99-110.
Went, F. W. and Stark, N. (1968). Mycorrhizae. *BioScience* **18**, 1035-9.
Westman, W. E. (1978). Inputs and cycling of mineral nutrients in a coastal subtropical eucalypt forest. *J. Ecol.* **66**, 513-31.
Westman, W. E. and Rogers, R. W. (1977*a*). Biomass and structure of a subtropical eucalypt forest, North Stradbroke Island. *Aust. J. Bot.* **25**, 171-91.
Westman, W. E. and Rogers, R. W. (1977*b*). Nutrient stocks in a subtropical eucalypt forest, North Stradbroke Island. *Aust. J. Ecol.* **2**, 447-60.
Will, G. M. (1959). Nutrient return in litter and rainfall under some exotic-conifer stands in New Zealand. *N. Z. J. Agric. Res.* **2**, 719-34.
Wood, T. G. (1971*a*). The effects of soil fauna on the decomposition of *Eucalyptus* leaf litter in the Snowy Mountains, Australia. IV. Colloquium Pedobiologiae. *Ann. Zool.* **17**, 349-58.
Wood, T. G. (1971*b*). The distribution of *Folsomides deserticola* (Collembola: Isotomidae) and other microarthropods in arid and semi-arid soils in southern Australia, with a note on nematode populations. *Pedobiologia* **11**, 446-68.
Wood, T. G. (1974). Field investigations on the decomposition of leaves of *Eucalyptus delegatensis* in relation to environmental factors. *Pedobiologia* **14**, 343-71.
Wood, T. G. (1976). The role of termites (*Isoptera*) in decomposition processes. *In* 'The Role of Terrestrial and Aquatic Organisms in Decomposition Processes.' (Eds J. M. Anderson and A. Macfadyen). pp. 145-68. (Blackwell Scientific Publications: Oxford.)

Chapter 39
Soil animals and pedological processes

K. E. Lee

PEDOGENESIS AND SOIL FERTILITY RESULT FROM A combination of *abiotic factors*, including the composition of the parent rock, physical and chemical rock weathering, temperature regimes, hydrology and landscape stability, and biotic factors, especially uptake by plants of water and nutrients, photosynthesis, litter fall, and decomposition of litter. The *biotic factors* are cyclic processes involving the elaboration of complex organic molecules by plants, death of plant tissue, utilization of its contained energy by decomposer organisms, fixation of atmospheric nitrogen, mainly by soil microorganisms, eventually resulting in recirculation of nutrients back to plants, and further photosynthesis. Man's use of the soil for food production consists of siphoning off a proportion of the energy captured in the photosynthetic phase of the cycle. The obvious importance of photosynthesis as the basic source of energy for all animal life has led to much scientific investigation and considerable understanding of the processes involved. The less obvious but equally important reciprocal process of decomposition of dead plant tissue and the release of the plant nutrients it contains for further plant growth has received much less attention; this is particularly so in Australia, where soil science has always been dominated by physicists, mineralogists, and inorganic chemists, whose main efforts have been devoted very largely to abiotic aspects of soil processes. Decomposition processes are the causal agents of mineral cycling; they involve the whole range of microorganisms, fungi and soil animals that live in soils.

Maldague (1970), in a comprehensive study of the fertility of tropical and temperate forest soils, concluded that fertility results from the maintenance of soil conditions such that decomposition processes operate at a level adequate to release plant nutrients from the litter at a rate that will sustain optimum growth. Maldague (1970) proposed that the greater the flow of energy the more fertile the soil, because energy flow rate is proportional to the rate of circulation of nutrients in the ecosystem and so to the rate of plant growth that can be maintained. Reichle (1977), taking saprovores (feeders on dead plant material), herbivores and carnivores separately, has developed this proposition further; he shows that saprovores have a low assimilation efficiency, i.e. the amount of energy assimilated is low compared with the energy available in the food ingested, but they have a high production efficiency, i.e. of the energy assimilated a high proportion (11–17%) is converted into new tissue, compared with the proportion converted into new tissue by herbivores (2–7%) and carnivores (*c*. 2%). Although it is apparent that metabolic activity, as measured by energy consumption, is a key to the interpretation of rates of nutrient cycling, we are not yet able to open the lock and relate metabolism to nutrient cycling in quantitative terms for all of the major heterotrophic levels of soil organisms. It is, however, reasonable to distinguish biological fertility, which depends on the rate of circulation of the pool of available nutrients and of energy flow, from inherent mineralogical fertility, due to the store of nutrients available in the weathering soil horizons and the intensity of the physical and chemical processes that make them available for plant growth. There is always a balance between the two, not a steady state, but a dynamic equilibrium that varies continuously within limits set largely by the physical environment. An extreme case of the dominance of biological fertility is found in tropical rainforests, where very rapid turnover of organic matter and plant nutrients in a warm, constantly humid environment maintains vigorous plant growth on very strongly leached and weathered soils whose plant nutrient content is frequently very low. Australian examples have been discussed in the previous chapter; similar conditions in the nearby Solomon Islands are described by Lee (1969) and many other examples have been described for other tropical regions.

Soil animals affect pedological processes in two main ways. Firstly, through the modification of soil profiles. Burrowing and construction involve movement of material within and between soil horizons, with accompanying mixing of organic and inorganic soil constituents, disintegration and reformation of aggregates, and changes in porosity, aeration and water infiltration.

The second modification comes about through decomposition processes and nutrient cycling. The basic food source of soil animals is plant litter, which is comminuted, more or less decomposed, and incorporated into the soil, resulting in release of energy and plant nutrients to be recycled in the ecosystem.

Table 39.1. Types of food taken by some Australasian termites
Data from Gay and Calaby (1970); Ferrar and Watson (1970)

No. of species considered	No. of species known to take various types of food							
	Wood			Grass	Various plant debris	Living herbaceous and woody plants	Dung of herbivores	Carton of other termites
	Living	Sound dead	Rotten dead					
116	18	76	30	25	23	9	47[A]	4

[A] Of those that feed on dung, only four are recorded from dung of native Australian herbivores, whereas 46 are recorded from the dung of introduced domesticated or feral herbivores.

Soil animals include representatives of a very wide range of animal groups, many of which have significant effects on soils; the most important in terms of pedogenesis are the larger burrowing animals, especially earthworms and termites, which dominate the macrofauna of most of the world's soils. Both are very successful animal groups; about 3000 species of earthworms and 2000 species of termites are known. Only a brief summary of their ecology and relationship to soils can be presented here. For more detailed treatments the reader is referred to Lee (1959), Satchell (1967), Lee and Wood (1971*a*) and Bouché (1972).

Termites are social insects, making up the Order Isoptera, that live in nests of their own construction. They are known to have existed since the early Tertiary, but may be much older. Some species do not live in the soil, but nearly all are soil animals that forage for food on or under the soil surface and transport it to a central nest.

Earthworms are segmented annelid worms, belonging to several families of the Oligochaeta. They are probably among the most ancient groups of terrestrial animals. Nearly all species pass most of their lives in the soil, but some have secondarily adopted aquatic or arboreal habits.

Termites are confined to tropical and warm temperate regions, approximately between 45°N and 45°S; between these latitudes lie about two-thirds of the earth's land surface. Earthworms are more widely distributed and are entirely lacking only in regions of extreme cold or drought. Both groups are well represented in Australia, where there are about 200 species of termites and several hundred species of earthworms.

A few cosmopolitan termite species have been introduced by man since European settlement of Australia. They are mainly species that are particularly suited to life in wooden structures and furniture, and are mainly confined to towns and cities. The remainder of Australian termites are endemic species. Land development has greatly affected the distribution of some species, especially those that build above-ground mounds and nests and some that have special physiological requirements that are not compatible with man-made environmental changes; for example, the alarm pheromone of *Nasutitermes exitiosus*, a common southern Australian species that feeds mainly on timber of *Eucalyptus* spp., is largely composed of α and β pinenes, which are present in the essential oils of many plants, especially *Pinus* spp.; therefore when native vegetation is cleared and replaced by pine forests, *N. exitiosus* rapidly dies out. Mound-building species have been eliminated or much reduced by cultivation in cropland areas, but the distribution and abundance of many Australian termites appears not to have been much affected by the changes made by European man; common grass-, litter- and wood-feeding species continue to thrive in pastures, croplands, planted forests and gardens. Some, that apparently fed at least partly on the dung of marsupials, have readily adopted the dung of ruminants as a food supply, and may very likely have increased in abundance since the introduction of these animals (Table 39.1).

Wood *et al.* (1977*a*) found that grazing, cultivation and the growing of maize in former savanna woodlands in Nigeria resulted in a reduction from 23 species in undisturbed sites to 20 in grazed pasture, 8 in first year maize and 4 in maize grown on land cultivated for up to 24 years. Mound-inhabiting species and some subterranean species disappeared, but *Amitermes evuncifer* and a group of five species of *Microtermes* increased in number under continuous cultivation and became major pests of the maize crop; the population of *Microtermes* spp. was 820 m^{-2} in grazed pasture, 925 m^{-2} in first year maize, and 4425 m^{-2} after 24 years of maize cultivation.

Co-evolution of plants and insects has led to the development in plants of chemical and physical defenses against common insect pests (see Levin 1976). Mielke (1978) has pointed out that in tropical Africa the indigenous crop sorghum grows especially well on termite mounds and is not apparently damaged by the termites, while non-indigenous crops such as maize, rice, potatoes and groundnuts, commonly suffer much damage from termites, leading to an escalation in use of pesticides. Non-indigenous pasture legumes have been successfully introduced into northern Australia, where termites are common, and new introductions are in progress. Screening of new candidate plants for susceptibility to termite attack and possibly breeding of termite-resistant strains may be desirable to enhance the possibility of successful introductions.

European man has profoundly affected the distribution and species composition of the Australian earthworm fauna. It is generally true that endemic earthworms, which belong to a group of megascolecoid families that are distributed widely throughout the southern land masses of the world, are now confined to areas in Australia where there has been little or no interference with the vegetation by European man. The soils of planted pastures and cultivated lands in temperate regions are inhabited by a small group of species of the family Lumbricidae. They are mainly of the genera *Allolobophora*, *Eisenia*, *Lumbricus* and *Octolasion*, of European origin but now established throughout the temperate regions of the world. In tropical regions, man-modified soils are inhabited by introduced species of the *Pheretima* group of megascolecid genera, originally probably of Oriental and Indo-Malaysian origin but now pantropical, and *Pontoscolex corethrurus*, a pantropical species of the family Glossoscolecidae that probably originated in Africa. Primitive man may have been responsible for the introduction of some or all of the tropical species, as appears to be the case for this group of species in Pacific islands (Lee 1975). The introduction and very rapid spread of lumbricids were partially documented in New Zealand by Smith (1893, 1894,); the worms were apparently introduced with soil used as ballast in ships and in soil that came with imported English trees that were propagated and then spread throughout the country from nurseries. Their initial spread in Australia was probably achieved in the same way. Much of the work done in Europe on lumbricid earthworms is relevant to the same species in Australia, and I have drawn on European work in preparing this chapter.

Earthworm populations, and their activity, are limited in Australia mainly by the need for moisture. Earthworms have no special respiratory organs; gas exchange takes place through a dense epidermal network of capillary blood vessels beneath the thin outer cuticle, which is highly permeable to water and must be kept constantly moist to maintain respiratory exchange. The limits of earthworm distribution in Australia are not accurately known. Pickford (1937) found that in South Africa earthworms are rare or absent in regions with mean annual rainfall below 250 mm, while Ljungström *et al.* (1973) in Argentina found that earthworms are rare or absent where soil moisture is consistently below 15%. Though some earthworms have been recorded around isolated springs and watercourses, even in the driest parts of Australia, the general situation is probably similar to that in South Africa and South America.

The lethal upper temperature limit for earthworms is comparatively low, between about 25° and 34°C for most species (Bodenheimer 1935; Satchell 1967; Reynolds 1973; Byzova 1975 and many other authors). Death probably results from inability of the respiratory system to supply oxygen and remove waste gases at the higher metabolic rate that results from increasing body temperature. Some degree of tolerance of high ambient temperatures can be achieved by evaporative cooling, but only when moisture is sufficient for the earthworm to avoid death by desiccation. Hogben and Kirk (1944) found that in the laboratory the body temperature of *Lumbricus rubellus* did not exceed that of a wet-bulb thermometer by more than 2°C at relative humidities higher than 37%. El-Duweini and Ghabbour (1965) found that, in Egypt, *Allolobophora caliginosa*, an introduced European lumbricid, could tolerate temperatures of 39–40°C, whereas Grant (1955) measured an upper lethal limit of 26.3°C for the same species in America. Various authors have shown that acclimatization at steadily increasing temperatures raises the lethal limit; Mangum (1978) discussed the possibility of control by neuroendocrine secretions that are believed to trigger the synthesis of enzymes affecting glucose metabolism and possibly the activity of other enzymes, thus affecting oxygen utilization. He also suggested the possibility of temperature-induced changes in the rate of blood flow that might account for an increase in efficiency of oxygen uptake. It is known that the haemoglobin content of the blood varies within some lumbricid species, increasing in environments with low concentrations of oxygen (Byzova 1975); a similar adaptation might be used by earthworms to increase their rate of oxygen uptake and consequently their range of temperature tolerance. Any or all of these mechanisms could account for acclimatization, but there is no conclusive evidence for them.

Termites individually have little resistance to desiccation, but because of their social organization and nest construction they are able to isolate themselves for most of the time in a small space, which enables them to control temperature and humidity. Consequently, they can live in all but the driest and hottest of desert areas, and those that build above-ground or arboreal nests can avoid seasonal flooding of the soil. The main limitation on their distribution is low ambient temperature; in Australia they are not found on the mainland mountains at altitudes above about 1200 m. In southern Australia they remain within the nest during the winter months, in a torpid state, crowded together to make the best use of metabolic heat output. A more detailed treatment of their relationships with, and control over, their environment will be found in Lee and Wood (1971*a*).

Drummond (1886) and many authors subsequently have described termites as the analogues in tropical soils of earthworms in the soils of temperate regions. It is implied that the one group replaces the other, but this is not so. Termites are common in the dry tropics while earthworms are few or absent; earthworms are common in cool temperate and even in sub-arctic regions where termites are virtually unknown. There is, however, a broad region of the more humid tropics and warm temperate regions where both groups are present in large numbers; they fill rather different niches, and have very different effects on soil morphology and on the processes of organic matter decomposition and nutrient cycling.

Table 39.2. Abundance and biomass of earthworms in selected habitats

Habitat	Location	Earthworm group	Abundance (no. m^{-2})	Biomass (g m^{-2} wet wt)	Reference
Sown pasture	South Australia	Lumbricidae	460–625	62–78	Barley 1959
Sown pasture	New Zealand	Lumbricidae	740–1240	140–323	Waters 1955
Old pasture	England	Lumbricidae	390–470	52–110	Svendsen 1955
Fallow	South Australia	Lumbricidae	210–460	16–76	Barley 1959
Wheat	South Australia	Lumbricidae	20–25	2–2.5	Barley 1959
Natural grassland	South Africa	Glossoscolecidae	74	96	Ljungström and Reinecke 1969
Alpine herb field	New South Wales	Megascolecidae	8–135	1.2–82	Wood 1974
Orchard	Holland	Lumbricidae	300–500	75–122	van Rhee and Nathans 1961
Savanna woodland	New South Wales	Megascolecidae	13.8	10.2	Wood 1974
Sclerophyll forest	New South Wales	Megascolecidae	7.2–76	1.3–48	Wood 1974
Subalpine woodland	New South Wales	Megascolecidae	15–106	5.7–36	Wood 1974
Maple forest	Canada	Lumbricidae	240–780	38–109	Maldague 1970
Beech forest	Germany	Lumbricidae	164–192	101–126	Nordström and Rundgren 1976
Pine woodland	England	Lumbricidae	40	17	Edwards and Lofty 1977

Table 39.3. Abundance and biomass of termites in selected habitats

Habitat	Location	Abundance (no. m^{-2})	Biomass (g m^{-2} wet wt)	Reference
Semi-arid tropical savanna	Senegal	229	1.0	Lepage 1974
Tropical savanna	Nigeria	4400	11.1	Wood and Sands 1977
Grazed pasture	Nigeria	2010	2.8	Wood and Sands 1977
Maize (8–24 years)	Nigeria	6825	18.9	Wood and Sands 1977
Dry sclerophyll forest	South Australia	600	3.0	Lee and Wood 1968
Riverine forest	Zaire	1000	11.0	Maldague 1964
Tropical rainforest	Trinidad	4450	–	Strickland 1944
Tropical rainforest	Malaysia	1340	3.5	Matsumoto 1976
Various habitats	Nigeria and Ghana	1000–10000	5–50	Lee and Wood 1971*a* (from data of Sands 1965, Bodot 1967)

Populations

Abundance and biomass of soil animals

Soil animal populations are nearly always highly aggregated; they live in a three-dimensional space that is itself non-uniform and most of its variations are not apparent on the surface. A basic requirement for quantitative studies of the activities in, and influence on, soils of soil animals is to obtain a reasonable assessment of their abundance and biomass. This requires many samples to be taken at a time, with the size of individual samples, the sampling methods, and the frequency of sampling matched to the size, population density and behaviour of the animal group sampled, and to seasonal or shorter-term variations in numbers, vertical distribution and activity.

Figures for abundance and biomass of particular groups of soil animals can only be regarded as reliable if sampling methods suitable for those animals have been used, and even then they should be treated as probably underestimating the true numbers. A variety of sampling methods has been used for earthworms, but the most reliable methods are those based on hand-sorting or sieving (usually with the assistance of washing) of soil samples, or those that depend on the careful use of chemical repellants to drive earthworms out of the soil. Termites that build above-ground nests can be sampled fairly easily, but populations of subterranean nesting species are very difficult to sample reliably. Josens (1974), Wood *et al.* (1977*b*), Wood and Sands (1977) and Lepage (1974, quoted in Wood and Sands 1977) have recently provided reasonably reliable estimates of populations of entirely subterranean species in western Africa, based on laborious handsorting of many large soil samples.

A representative selection of estimates of abundance and biomass of earthworms is presented in Table 39.2 and of termites in Table 39.3. These population figures illustrate that earthworms and termites are very widespread and numerous. Detailed summaries of population biology of earthworms may be found in Satchell (1967) and of termites in Lee and Wood (1971*a*), and will not be discussed here.

Vertical distributions in soils

Earthworm populations are typically stratified into species associations that are adapted for, and live most of the time within, specific depth zones in the soil. Three groups are recognizable in forest soils: litter species (Lee 1959), usually small, heavily pigmented, very active, with well developed body-wall musculature and relatively high respiratory rates, that have no burrows, but move freely in the litter layer ('formes épigées' of Bouché 1972); topsoil species (Lee 1959), coloured red, red-brown to brownish-black, of medium to large size, fairly active and with well-developed musculature (but relatively less than litter species), that have burrows (often vertical) in the soil, but feed in the litter layer and frequently cast at the surface ('formes anéciques' of Bouché 1972); subsoil species (Lee 1959), varying in size from very small to the largest known earthworms, usually unpigmented, rather sluggish and usually with weakly-developed body wall musculature, that make extensive burrows, mainly horizontal, and do not usually cast at the surface, feeding on dead roots or organic matter contained in soil ingested during burrowing ('formes endogées' of Bouché 1972). Various combinations of all or some of these associations have been recorded from continental Europe (Graff 1953, Bouché 1972), New Zealand (Lee 1959), Japan (Nakamura 1972), Australia (Wood 1974) and England (Phillipson *et al.* 1976).

Bouché (1977) has developed the concept of the three associations further, based on the Lumbricidae of France; he represented the three groups, épigés, anéciques and endogés, as the apices of a triangle, with various species distributed over the area of the triangle on the basis of morphological, physiological and behavioural parameters. Lavelle (1979) found that Bouché's tripolar concept was useful for qualitative description of the earthworm communities (Megascolecidae and Eudrilidae) of a west African savanna, but he defined a 'demographic profile' for each species, which takes account of the time required to grow to maximum weight, mean life expectancy and annual cocoon production. The demographic profile correlates with a number of ecological parameters, including efficiency of energy utilization, and Lavelle considered it to be a more useful concept than the tripolar concept of Bouché.

Perel (1977) distinguished two associations among European Lumbricidae: the humus-formers that feed on plant debris that is only slighty decomposed, and the humus-feeders that feed on plant debris that is already much decomposed and dispersed in the soil. She related some morphological features of the species, especially features of the digestive system, and some behavioural characteristics to the feeding groups.

As with most work on earthworms, the majority of these studies relate to Lumbricidae in Europe, and it must be appreciated that these are only about 10% of known earthworm species.

The two, three, or perhaps more groups combine to bury, ingest, comminute and partially digest plant litter from the surface and dead roots within the soil. Lee (1959) and Maldague (1970) have related the stage of differentiation of the three groups to the morphology of litter layers and concluded that, in forest soils, the rates of organic matter incorporation and decomposition, and hence the fertility of the soil, are greatly enhanced by the activities of a well developed stratified earthworm fauna.

Termites are not stratified in the same way. The humus feeders (*Cubitermes* spp.) of Africa apparently burrow continuously in the soil, ingesting soil and digesting its included organic matter, in the manner of many earthworms. Most termites though, including nearly all Australian species, use the soil, above-ground mounds or arboreal nests as refuges, and forage on the surface for plant material, living, dead, and in various stages of decay, as food (Table 39.1). The nest and associated galleries that give access to food, building materials in the soil, and water, together form a closed system largely isolated from the external environment. Feeding galleries are usually more or less horizontal and not very deep, while those that give access to soil for building and to water are usually more or less vertical, and in the latter case may go to very great depths; Hill (1921), in northern Australia, found them at depths of more than 3 m, Ghilarov (1962), in central Asia, at 10–15 m, and Yakushev (1968), in Africa, at 70 m.

Modification of soil profiles

Earthworms

Effects of earthworms on soil profiles result from the construction of burrows and also ingestion and casting of soil mixed with plant debris in various stages of decomposition.

Burrows

The burrows of earthworms are more or less circular in cross-section, and vary in diameter with the diameter of the earthworms, from about 1 mm to several centimetres. They are made in two ways, either by elongation and extension of the anterior segments of the body between aggregates and into cracks, followed by contraction and lateral expansion of the segments, pushing the soil sideways, or by ingestion of soil which is passed through the gut and voided, often on the surface, as casts. The walls of burrows are lined with mucus secreted by gland cells in the worm's epidermis which acts as a lubricant and facilitates movement through the soil. Jeanson (1961, 1964, 1971) found that burrow walls are lined with a mixture of mineral grains, oriented parallel to the wall, and partially digested organic matter. She noted visible concentrations of iron and manganese, probably deposited by evaporation from percolating water, surrounding the burrows, diminishing gradually to disappear at about 0.6 mm from the burrow walls; in moist soils burrows had reddish linings up to 2–3 mm thick, high in ferric iron, apparently due to oxidation of ferrous iron in the presence of air in the burrows. Kretzschmar (1978) made a very detailed study of the burrows of lumbricid earthworms in a French pasture soil. He studied a 2-mm-thick layer around the burrows, which he called the 'drilosphere'; this zone included a volume of 10 l of soil for each m^2 of surface area, or about 0.9% of the total soil volume, and in this volume of soil he found 43% of the total nitrogen-fixing micro-organisms in the soil and 15% of the total soil microflora. The biomass of earthworms was 150 g m^{-2}, which is in the normal range of biomass in the soils of temperate pasture lands (Table 39.2).

Earthworm burrows frequently make up a high proportion of the large pore space of soils. Raw (1959) counted a mean of 50 burrows m^{-2} of *Lumbricus terrestris*, with a total cross-sectional area of 14 cm^2 m^{-2}, at the soil surface in an English apple orchard; Hoeksema and Jongerius (1959) counted 200 earthworm burrows m^{-2} at 25 cm depth in a Dutch orchard; Bouché (1971) counted 800 m^{-2}, with a total cross-sectional area of *c.* 57 cm^2 m^{-2}, in a French pasture. Burrows have frequent intersections, and Bouché poured 100 l of water into one burrow opening without it overflowing. Various authors, e.g. Slater and Hopp (1947), Hopp and Slater (1949), Teotia *et al.* (1950), Guild (1952), Scharpenseel and Gewehr (1960), and Stockdill (1966), have shown that soils with earthworms drain two to ten times faster than soils without earthworms. Ehlers (1975) investigated the effect, on some German soils derived from loess, of zero-tillage compared with normal tillage practices. One of the results of zero-tillage was a great increase in earthworm population; after four years zero-tilled areas had about twice as many earthworm burrows per unit surface area as did the tilled areas. Because of the large diameter of earthworm burrows (2–11 mm in the experimental areas) water will run down them only when it is in a tension-free state, so the burrows can function as water channels only during heavy rain or under irrigation. Simulation of such conditions showed that they do function in this way. In the zero-tilled soil, water infiltration due to earthworm burrows alone was calculated to be 6150 ml m^{-2} per 5 min while for tilled soil the total infiltration rate was 1110 ml m^{-2} per 5 min. Ehlers concluded that earthworm burrows were a very important additional drainage system within the soil matrix during heavy rain or irrigation. In contrast Barley (1959) concluded that the burrows of *Eisenia rosea* and *Allolobophora caliginosa* in the A horizon of a *red-brown earth* under permanent pasture in a Mediterranean environment near Adelaide were too wide to conduc water at the tensions usual during infiltration, and that the small population of earthworms present in the soil probably had little influence on water penetration or aeration. This may be so in much of Australia, where there is little rain and the amount of water infiltrating into the soil is low, and where the long hot summers limit the numbers and the period of activity of earthworms, but it is unlikely to be true of the soils of the wetter eastern coastal regions of Australia, where no work has been done on earthworms in relation to soils. The evidence overwhelmingly implies that where earthworm activity is not specially restricted by seasonal aridity, their burrows are very important in promoting water penetration into the soil.

Earthworm burrows are probably not very significant in promoting aeration of the soil. Bouché (1971) calculated that, with the very large number of 800 burrows m^{-2}, assuming that the burrows averaged 20 cm deep, the additional surface area represented by burrow walls was only about 1.5 m^2 m^{-2} of soil surface, and this is a very small additional surface area compared with the total surface area of soil aggregates.

Casts

These consist of soil mixed with residues of comminuted and digested plant tissue, excreted by earthworms. Many species, especially those that make predominantly horizontal burrows, excrete their casts in their own burrows or in other sub-surface spaces, but many of those that feed at the surface and live in predominantly vertical burrows excrete their casts on the ground surface adjacent to or over the surface opening of the burrow, and it is these species that most affect soil profile morphology. There are two basic forms of casts: sub-spherical to sub-cylindrical granules, ranging from less than 1 mm to about 1 cm in diameter, varying with the size of the species that excretes them; and paste-like slurries that form generally rounded but less regular shapes. Composite casts, consisting of aggregated masses of the two basic forms, are common. Some species whose casts are semi-liquid when excreted build up massive composite casts over their burrow entrances: e.g. *Archipheretima tumulifaciens*, in the Sepik Plains, New Guinea, whose casts are up to 50 mm high and 50 mm wide, with an oven-dry weight of 25–45 g (Lee 1967); much larger casts, about 100 mm high and 100 mm wide, weighing more than 400 g, produced by an undescribed megascolecid species in north Queensland rainforests, and similar casts, weighing up to 1.6 kg, reported from Burma (Gates 1961).

Annual surface cast production in cool temperate regions, under pastures, natural grasslands and forests, has been estimated by various authors to be 1–50 kg m^{-2}, dry weight; most estimates are about 4–5 kg m^{-2}, representing a depth of 3–4 mm of soil. Barley (1959) found that a population of *Allolobophora caliginosa*, near Adelaide, deposited only 0.25 kg m^{-2} $year^{-1}$ of surface casts, representing a depth of 0.18 mm of soil, and concluded that it would take 60 years for complete turnover of the top 15 cm of the soil. For tropical regions there are few estimates of rates of surface cast production; in Indian grasslands Roy (1957) measured cast production of 0.02–0.25 kg m^{-2} $year^{-1}$, while Dash and Patra (1979) measured production of 7.79 kg m^{-2} $year^{-1}$; in west African savanna Lavelle (1974) found a total surface cast production of 20–30 kg m^{-2} $year^{-1}$; Edwards and Lofty

(1977) estimated cast production as 210–260 kg m^{-2} $year^{-1}$ in the Cameroons and the Nile valley.

Casting activity is seasonal, and in both temperate and tropical regions is at a maximum during warm wet periods of the year and at a minimum during cold or dry periods.

Most surface-casting species are active only in the top 10–20 cm of the soil. Their effect is to stir the upper soil horizons and mix organic matter from the surface into the soil. They are selective in the size range of the soil particles they ingest; casts contain little or no gravel and usually have more clay and less sand than the soil from which they are derived. Thus worm-worked soils generally have a fine-textured surface horizon overlying sandy or stony horizons and stone-lines, resulting from burial of former stony surface material, and have been described from many parts of the world.

Various authors have demonstrated that worm casts contain more water-stable aggregates than non-cast soil. It has been suggested that their stability is due to cementing materials (calcium humate) produced in the gut (Meyer 1943), gums produced by bacteria in the gut (Swaby 1950), the reinforcing effect of included plant fibres (Ponomareva 1953) and the growth of fungal hyphae in the casts after deposition (Parle 1963*a,b*). Jeanson (1960*a,b*) found that casts with low organic matter content are less stable and those with high organic matter content are more stable than unworked soil; Parle (1963*a,b*) found that casts are semi-liquid when produced, increase in stability for about 15 days, then gradually lose their stability.

Termites

Lee and Wood (1971*a,b*) discussed in detail the effects of termites on soil profiles; they were obliged to confine their discussion mainly to the effects of soil-inhabiting species that build above-ground mounds, but pointed out that the effects of subterranean species may also be significant. Wood and Sands (1977) have summarized more recent work that estimates the effects of some subterranean species.

Mounds are constructed from mixtures of soil, plant debris, termite excreta and salivary secretions, in varying proportions. They range in height from a few centimetres to about 10 m and in diameter from a few centimetres to about 60 m.

Soil used for construction is carried in the mandibles in small pieces, moulded, and stuck together with excreta or salivary secretions. Detailed description of the fabric of mounds and nests of some Australian species may be found in Lee and Wood (1971*a*) and in Sleeman and Brewer (1972). Gillman *et al.* (1972) investigated the organic constituents of the very hard cemented mound wall of *Coptotermes acinaciformis* from northern Queensland. They identified the two major components as a polysaccharide fraction, which they termed hemicellulose, predominantly composed of xylose and galactose, which they considered was probably included in the faeces and derived from the termites' food, and a glycoprotein fraction, probably a glandular secretion of the termite. They considered that the hemicellulose probably functions as a space-filling mortar, while the glycoprotein probably functions as an adhesive in cementing soil particles together.

Lee and Wood (1971*b*) have shown that the soil used for mound construction by a variety of Australian termites is usually B-horizon material, excavated and transported to the surface by the termites, and that it nearly always contains a higher proportion of fine particles (clay and silt) and a lower proportion of sand than the horizon from which it is derived. Various authors (Hesse 1955; Kemp 1955; Nye 1955; Harris 1956; Maldague 1959; Stoops 1964; Watson 1974, 1976; Miedema and van Vuure 1977) in various parts of Africa, Joachim and Kandiah (1940), in Ceylon, and Pathak and Lehri (1959) in India, have come to the same conclusions. It has been shown in several investigations that much material in mounds comes from weathering rock, often several metres below the surface, e.g. Watson (1962) has found *Odontotermes* galleries at 23 m, and Boyer (1971) found that *Macrotermes subhyalinus* brought building material from depths of 12–15m.

Boyer (1973, 1975*a,b*) made a very detailed study of the effects on pedogenesis and hydrological properties of the soil of mounds of *Macrotermes bellicosus* and *M. natalensis* in the seasonally wet/dry savanna lands of the Central African Republic. Mounds of *M. bellicosus* were up to 5–6 m in height and 30–60 m in diameter, while those of *M. natalensis* were 3–4 m in height and 3–5 m in diameter. He found that they were constructed of sandy-argillaceous materials derived from deep in the soil profile, 10–12 m and about 2 m deep for *M. bellicosus* and for *M. natalensis* respectively. The very large mounds of *M. bellicosus* were found at densities up to 10–12 ha^{-1} between about 6°N and 6°S latitude, over a total land area that Boyer described as two to three times that of France. The soils of the region are highly ferruginous and between the mounds there are compact clay-rich subsurface horizons that interfere with water infiltration. Beneath the mounds of *M. bellicosus* the soil horizons are much modified, with a rejuvenation of pedogenesis at depths down into the underlying decomposing rock. The increase in permeability to water of the soil beneath the mound relative to surrounding soils has led to rejuvenation of mineral weathering and formation of new clay minerals, with additional illite formed from mica by prolonged chewing and moistening with saliva by the termites. The hydrological regime of the region has been much modified; the soil beneath the mounds allows water to percolate readily into the deep aquifers that underlie the savanna lands, and also provides access for water held in temporary water tables on the compact horizons at higher levels in the soil to drain directly into the deep aquifers. In the rainy season, when the regional water table level rises into the soil horizons, the mounds have a water table higher than the surrounding soils, and this is considered to be a major factor in the accumulation of the high concentrations, and sometimes layers of concretions, of calcareous minerals that are often found in *Macrotermes* mounds (see discussion in Lee and Wood 1971*a*). Watson (1962, 1969, 1974, 1976) reported on soils in and under mounds in Rhodesia. He described (1962) concentric zones of whitened soil extending to a depth of about 6 m beneath a mound on leached ferrallitic (non-calcareous) soil; the white horizons contained 1.7% (a total of about 20 000 kg) calcium carbonate. He also showed (1969) that radioactive tracers were retained through the wet season in termite mounds but were leached from adjacent soils. De Leenheer *et al.* (1952) in Zaire, noted great differences in soils under high concentrations of similar mounds that occupied 30% of the soil surface.

Termite mounds are constantly eroded and the soil from them is spread on the surface; the termites repair them, bringing more material from deep horizons. Williams (1968) described soil profiles at Brock's Creek, in Northern Territory, with an upper fine-textured (M) horizon about 15 cm deep, overlying a stony (S) horizon about 30 cm deep, on deeply weathered hornblende granite. He considered that mound-building termites (*Tumulitermes hastilis*, *T. pastinator* and *Nasutitermes triodiae*) were responsible for the formation of the M horizon, which was derived from fine material in quartz veins in the weathered granite, and that the S horizon was a former surface horizon, buried by the termites. He calculated that 6000 $m^3\ ha^{-1}$ of fine material had been moved to the surface, that about 75% of this had been lost by erosion, and that the present M horizon had accumulated in about 12 000 years, at a rate of 0.0125 mm $year^{-1}$. Lee and Wood (1971*b*) calculated rates of accumulation of 0.02 mm $year^{-1}$ and 0.10 mm $year^{-1}$ at two sites in the Northern Territory. In western Africa, Nye (1955) calculated an accumulation rate of 0.025 mm $year^{-1}$, and de Ploey (1964) estimated that a 1–2 m thick soil mantle had been formed in 10 000 years, at a rate of 0.1–0.2 mm $year^{-1}$.

Many subterranean-nesting termite species that forage for food on the soil surface build covered runways or sheetings of soil providing access to, or covering, sources of food; others that feed on logs, standing trees or other plants, dung, or other food sources above ground, pack old feeding galleries with soil. These structures are well known in Australia, but no estimates have been made of the quantities of soil moved to the surface for these purposes, nor of their effects on pedogenesis. It is well known that the runways and soil sheetings are destroyed by rain and are rapidly replaced. Wood and Sands (1977) comment that Lepage (1974) found that of the 2000 kg ha^{-1} of soil brought to the surface each year by *Macrotermes subhyalinus* in Senegal, 675–950 kg ha^{-1} was used to construct foraging runways and sheetings, and that in Nigerian savanna lands *Microtermes* spp. attacking maize bring 250 kg ha^{-1} of subsoil to the surface each year and pack it within the eaten-out stems of the maize plants.

Much of the area of northern Australia, and to a lesser extent of southern Australia, probably owes its present soil mantle to termite-transported material; many duplex soil profiles, with a sharp texture boundary, resemble African profiles for which there is good evidence that termites are responsible. Apart from the work of Williams (1968) and Lee and Wood (1971*b*) reported above, which was of very limited extent and was confined to the effects of mound-building termites, very little is known of the contribution that these animals make to pedogenesis and to soil fertility in Australia.

Watson (1970, 1974) has reported on the use of termite mounds in geochemical prospecting in Rhodesia, where Kalahari sand overlies mineral-bearing rocks, and termites sometimes penetrate into the underlying material to find construction material. This technique might be useful in northern Australia.

The rates of accumulation of soil brought to the surface by termites are low compared with rates attributed to earthworm casting, but the building behaviour of termites involves physical inversion of selected components of soil profiles, with accumulation on the surface of material that was formerly deep in the profile or was derived from partially weathered material below the soil profile. In general this is not so of earthworms.

Microrelief features

The mounds of termites are prominent features of landscapes throughout the warm temperate and tropical regions of the earth. Most termite colonies are comparatively short lived, and their mounds are eroded away fairly quickly after the colony dies, but some mounds, notably those built by some African and Asian Macrotermitinae, remain as permanent landscape features, although their original inhabitants and builders occupy them for only a short time.

Large macrotermitid mounds often support vegetation different from surrounding soils. Fuller (1915) recognized that the 'park-formation' of inland Natal resulted from the growth of trees being more or less confined to old termite mounds, while Troll (1936) described 'termiten-savannen', a similar formation that is widespread in western Africa. Kang (1978) recorded *Macrotermes* mounds occupying 0.10–0.84% of the ground surface in the forest zone of south-western Nigeria, and found highest mound densities on well drained and lowest densities on poorly drained sites, and Sys (1955) recorded similar mounds occupying 4.3 to 7.8% of the ground surface in Zaire. The mounds of Australian termites usually cover not more than about 2% of the ground surface and do not usually support the growth of any higher plants (Lee and Wood 1971*a*). They are ephemeral structures, most of them lasting for only a few years, although the very large mounds of *Nasutitermes triodiae* may last 100 years or more (Watson 1972), and the strongly cemented slab-type nests of *Drepanotermes rubriceps* and *D. perniger* in south-western Queensland continue to resist water penetration and seed lodgement for up to 30 years (Watson and Gay 1970).

Haantjens (1965) described two types of microrelief on the grassy plains of the lower Sepik valley in New Guinea that he attributed to earthworms. He distinguished 'pitted soils', with steep-sided irregular rounded holes, varying to elongate trenches, up to about 60–70 cm deep, and 'sorted stripes', an irregular pattern of low gravel strips separated by earthen rises. Large earthworms (*Archipheretima tumulifaciens*) that produce very large surface casts (see p.634) were concentrated around the rims of the pits and trenches and on the earthen rises, and Haantjens proposed that the earthworms move fine soil materials out of the pits or stone stripes and build higher areas in which they can live above the water table during the wet season, when lower ground is saturated or flooded. Lee (1967) considered this explanation unlikely, because earthworm burrows are concentrated in the raised areas with very few under the lower ground, and the casts, which consist of fine material, are eroded by rain and must wash onto the lower ground, tending to eliminate rather than produce the microrelief. He considered that the microrelief might be initiated by uneven burning of grass tussocks in frequent dry season fires, followed by uneven wind- and water-erosion of severely burnt patches, but Haantjens (1969) concluded that this was an unlikely explanation. Rather similar hollows and trenches are known from wet coastal areas of north Queensland, and it has been suggested that they may be due to earthworms, but no work has been done on them. Very similar cup-shaped depressions (kommetjes), 30–100 cm deep and about 100 cm in diameter, on seasonally flooded grassed flats in South Africa have been attributed by Pickford (1926) and Ljungström and Reinecke (1969) to earthworms that cast around their edges; mound- and trench-type depressions in grassed swamps in Uganda were similarly attributed to earthworms by Wasawo and Visser (1959). More investigation is needed to resolve the question of whether earthworms are indeed responsible for these microrelief features.

Organic matter decomposition

Organic matter inputs to the soil consist of the plant litter that falls on the ground surface and dead roots and root exudates below the ground surface. The nature and dynamics of litter layers, and quantities involved are discussed in Chapter 38.

Litter decomposition, i.e. the release of the energy it contains and its included plant nutrients, is due to the 'decomposer industry' (Macfadyen 1963), a complex of soil animals, fungi and microorganisms. The traditional view of the place of soil animals in the complex is that they physically incorporate surface litter into the soil, ingest it, comminute it, use a small proportion of the energy available in the organic compounds elaborated by plants, and excrete most of it relatively unchanged chemically, but with a much greater surface area available as a substrate for microbial decomposition. This is more or less true of earthworms, but is not so for termites, which are the dominant group of soil animals in the soils of perhaps half the land surface of the earth, including most of the productive and potentially productive soils of Australia.

The principal structural components of plant tissue are cellulose and lignin. Woody tissues contain something of the order of 50–60% cellulose and 15–25% lignin, while leaves of trees and grasses have $>60\%$ cellulose and 20% or less lignin. The digestive capabilities of earthworms and termites, and therefore the amount of energy that they can derive from a given weight of plant tissue, differ very greatly.

Earthworms feed on litter when it is already partially decomposed by bacteria and fungi. Study of the enzymes in the gut indicates that they probably digest a small amount of the cellulose but none of the lignin in plant tissue, and are apparently more or less restricted to digesting the simpler plant carbohydrates and proteins, and probably some of the fungi and bacteria living on the partially decomposed litter; current knowledge of the digestive enzymes of earthworms is summarized by Michel and de Villez (1978). It has been shown by various authors (summary in Satchell 1967) that annual food consumption by earthworms is more or less equal to the total input from the vegetation. No work has been done in Australia on the dynamics of organic matter decomposition by earthworms. Results of an investigation by Lavelle (1974), on earthworms in a sparsely wooded savanna at Lamto, Ivory Coast, are summarized in Table 39.4. The energy consumed by the earthworms, expressed as a percentage of the total energy available in the organic matter of the soil ingested, is the assimilation efficiency (= 9%). Most of the energy derived from food is used in respiration, but a small proportion is used in production, i.e. increase in earthworm biomass. The few other measurements available indicate that the assimilation efficiency of earthworms is generally less than 10%, and that they do not, directly, greatly influence decomposition. The most important contribution of earthworms to decomposition processes appears to be in the comminution and mixing of plant litter into the soil, and the effects that this process has on further organic matter decomposition by the soil microflora.

Table 39.4. Soil consumption and energy budget of earthworms in a sparsely wooded savanna ecosystem at Lamto, Ivory Coast
Based on data from Lavelle (1974)

Earthworm population data	
Numbers	20 m^{-2}
Biomass	20.5 g m^{-2} (fresh wt)
Production	43 g m^{-2} $year^{-1}$ (fresh wt)
Total soil ingested	50 kg m^{-2} $year^{-1}$
Energy budget	
Total available in ingested soil	11 750 kJ m^{-2} $year^{-1}$
Earthworms	
Biomass	35 kJ m^{-2}
Production	73 kJ m^{-2} $year^{-1}$
Ratio production:biomass	2.1
Earthworm food consumption	1050 kJ m^{-2} $year^{-1}$
Earthworm faeces	10 700 kJ m^{-2} $year^{-1}$
Loss via earthworm respiration	975 kJ m^{-2} $year^{-1}$ = 8.3% of energy available in food
Assimilation efficiency[A]	9%

[A] $\frac{\text{Total energy utilized (or respiration + production)}}{\text{Total energy in food}}$

The accumulation of thick layers of slowly decomposing plant litter on the soil urface is characteristic of soils that have no earthworms or where earthworms are very few. Usually their absence can be attributed to unfavourable soil moisture, temperature, pH or some other physical environmental condition; sometimes the environment can be modified, e.g. by irrigation, drainage, addition of lime, and earthworms may invade from surrounding areas, or earthworms may be deliberately introduced. Noble *et al.* (1970) studied irrigated pasture lands at Deniliquin (New South Wales) where lumbricid earthworms were apparently lacking until the introduction of *Allolobophora* spp. in 1955 (Barley and Kleinig 1964). Noble *et al.* (1970), in 1968, found that in control plots, where *Allolobophora* was not introduced, there was a surface mat of pasture roots and other plant debris with a dry weight of about 72 000 kg ha^{-1}, while in the earthworm-inoculated plots, which had about twelve times the earthworm biomass of controls, the weight of the surface mat was about 7600 kg ha^{-1}; they were not, however, able to show a significant difference in pasture production between control and inoculated plots. Stockdill (1966) introduced *Allolobophora caliginosa* and *Lumbricus rubellus* to pasture soils that lacked earthworms at Hindon, in New Zealand, and showed that surface root mats were incorporated into the soil and that 5 years after the earthworms were introduced pasture production was about 72% higher in earthworm-inoculated than in control plots without earthworms. Waters (1951), also in New Zealand pasture land, has shown that the lumbricid earthworms common in pasture soils feed primarily on dead pasture root material.

Parle (1959, 1963*a*,*b*) showed that the microflora of the gut in some lumbricid earthworms did not differ qualitatively from that of surrounding soils, but that numbers of bacteria and actinomycetes increased up to 1000-fold; yeasts and fungi, which were found mainly as spores in the intestine, germinated and grew in the asts, reaching a maximum when the casts were about 15 days old; oxygen uptake by microorganisms fell as the casts aged, but was still higher than that of the surrounding soil in casts 50 days old; there was a steady decline in the ammonia-nitrogen and an increase in nitrate-nitrogen content of casts, but with an overall increase in nitrogen content relative to surrounding soil. Barley and Jennings (1959), Kozlovskaya (1969) and Syers *et al.* (1979) have also noted increases in nitrogen retention in earthworm casts and Nowak (1975) considered this to be the most important contribution of earthworms to the fertility of grassland soils. The physical grinding of soil as it passes through the gizzard of an earthworm is probably an important factor in exposing organic matter to further decomposition by microorganisms. Foster (Chapter 41) illustrates particles of soil organic matter that are coated with clay particles which appear to shield the organic matter from microbial attack; rupture of such clay coatings may explain the significance of the comminution of aggregates of organic matter and soil in accelerating organic matter breakdown.

The digestive processes of termites are very different from, and much more efficient than those of earthworms. With the help of a symbiotic microfauna or microflora in the gut, very like that of ruminant mammals, they can digest cellulose, and are also able to digest some of the lignin in plant tissue. A comprehensive account of their digestive processes and their effects on soil organic matter may be found in Lee and Wood (1971*a*).

Field and laboratory experiments with *Nasutitermes exitiosus*, a common wood-feeding termite in southern Australia, show an assimilation efficiency of 54%, with > 90% assimilation of the cellulose and 14–23% assimilation of the lignin content of the wood (Lee and Wood 1971*a*; Lee and Butler 1977). Other wood-feeding termites have been shown to have assimilation efficiencies of 50–85% and to digest up to 97% of the cellulose in their food.

Many Australian termites are grass-feeders, but no work has been done on their assimilation rates and significance in ecosystem processes. Wood (1976) has provided information on assimilation of grass by the termite *Hodotermes mossambicus* in Nigeria. He has determined the amount of various classes of compounds contained in the faeces of *H. mossambicus*, and expressed the apparent loss, as a ratio of the amount contained in the faeces to that in the food (*Themeda triandra*), in the form of a coefficient of digestibility (Table 39.5). *H. mossambicus* has an assimilation efficiency of 61% and digests 80–90% of the polysaccharide content of its food.

It apparently digests only 0.3% of the lignin and virtually none of the soluble carbohydrates.

Wood *et al.* (1977*b*) have calculated the total energy used by termites in a Nigerian savanna ecosystem (Table 39.6). They consume 24–42% of the total dry matter production of the ecosystem, and from that return about 7.3% as faeces, 7% as dead termite tissue, and dissipate about 86% in respiration. The faeces consist largely of material that is not readily decomposed further, whereas the dead termite tissue would be rapidly decomposed.

The faeces of termites are used very largely in mound or nest construction, and contain, as is shown by analyses of the organic matter in the nests of Australian termites (Table 39.7), much lignin, little carbohydrate, and much polyphenolic material with properties similar to the humic fraction of soils. There is little difference in the lignin and carbohydrate contents, or in the ratio lignin: carbohydrate, between the excreta of termites feeding on wood, which has a high lignin content, and those feeding on grass, which has a low lignin content, illustrating the generally high level of cellulose digestion in all termites. The production of humic substances by termites, directly from undecomposed plant material, is of interest because it is generally accepted that humic materials are the residual products of the whole process of organic matter decomposition by the full range of soil animals and microorganisms.

French and Bland (1975) fed ^{14}C-labelled eucalypt wood to *Coptotermes lacteus* and *Nasutitermes exitiosus* and were able to show that the ^{14}C was incorporated into body tissue of the termites. Lee and Butler (1977), fed ^{14}C-labelled synthetic lignin to *N. exitiosus* and found that 14–23% of the labelled lignin was decomposed. Butler and Buckerfield (1979) fed ^{14}C-labelled natural and synthetic lignin and a number of other ^{14}C-labelled phenolic compounds related to lignin to laboratory cultures of *N. exitiosus*. The experiments were done under carefully controlled conditions to allow for the possibility that decomposition of the lignin and other

Table 39.5. Utilization of redgrass (*Themeda triandra*) by *Hodotermes mossambicus*
Published with permission from Wood (1976)

	Food Dry wt (g)	Food Dry wt (%)	Faeces Dry wt (g)	Faeces Dry wt (%)	Coefficient of apparent digestibility
Ash	3.96	12.09	4.04	31.60	−0.022
Lignin	5.69	17.36	5.67	44.30	0.003
Soluble carbohydrate	0.55	1.68	0.76	5.91	−0.374
Hexosans	12.34	37.66	1.15	8.99	0.907
Pentosans	7.10	21.66	1.14	8.87	0.840
Other solubles	0.36	1.11	0.10	0.80	0.819
Nitrogen	0.25	0.75	0.13	0.99	0.486
Unaccounted for	2.52	7.70	–	–	–
Total		100.00		101.46	0.609

Table 39.6. Organic matter turnover and energy budget of termites in a Nigerian savanna ecosystem
Data from Wood *et al.* (1977*b*)

Primary production (dry wt)	400–700 g m^{-2} $year^{-1}$
Termite population data[2]	
Numbers	3500 m^{-2}
Biomass	9.4 g m^{-2} (fresh wt)
	= 2.4 g m^{-2} (dry wt)
Food consumption by termites	167.5 g m^{-2} $year^{-1}$
	= 24–42% of primary production
Returns to soil	
Faeces	12.3 g m^{-2} $year^{-1}$
Alates + Predation	11.7 g m^{-2} $year^{-1}$
Energy budget	
Primary production	8000–14 000 kJ m^{-2} $year^{-1}$
Termites	
Biomass	50 kJ m^{-2}
Production	240 kJ m^{-2} $year^{-1}$
Ratio production:biomass	4.8
Termite food consumption	3300 kJ m^{-2} $year^{-1}$
Termite faeces	250 kJ m^{-2} $year^{-1}$
Loss via termite respiration	c. 2800 kJ m^{-2} $year^{-1}$ = 20–35% of primary production
Assimilation efficiency	92%

Table 39.7. Composition of nest materials of some Australian termites
Data from Lee and Wood (1971*b*) calculated on the basis of loss on ignition

	N (%)	Ratio C:N	Total carbohydrate (%)	Lignin (%)	Ratio lignin: carbohydrate	Polyphenolic (humic) material (%)
Wood-feeding species						
Mastotermes darwiniensis	0.53	92.0	11.9	64.1	5.4	5.6
Coptotermes acinaciformis	0.60	89.3	17.0	40.0	2.4	12.2
Coptotermes lacteus	1.07	50.0	19.2	46.1	2.4	8.1
Amitermes sp.	0.98	50.0	8.2	41.8	5.1	11.1
Microcerotermes nervosus	1.11	32.4	10.8	44.2	4.1	3.9
Nasutitermes exitiosus	0.83	57.8	14.5	53.1	3.7	6.6
Grass-feeding species						
Nasutitermes magnus	2.16	21.3	15.1	47.9	3.2	10.7
Nasutitermes triodiae	2.10	22.9	11.2	40.3	3.6	15.9
Tumulitermes pastinator	2.17	22.1	12.1	41.3	3.4	11.7

compounds was due to microbial or fungal contaminants of the termite cultures and not to the termites and their symbiotic gut microflora. They showed that the termites respired $^{14}CO_2$, derived from lignin decomposition by the termites, and that a small proportion of the ^{14}C from the lignin was incorporated into the termites' tissue. The ^{14}C label was attached to specific sites in the lignin monomers; the experiments showed that for synthetic lignin 23% of the total ^{14}C incorporated into lignin as a methoxyl side chain was released, implying demethoxylation of the lignin molecules, 15% was released when the ^{14}C was incorporated as the C-2 of the aliphatic alcohol side chain, implying depolymerization of the lignin molecules, and 16.5% was released when the ^{14}C was incorporated in the six carbon atoms of the phenolic ring, implying depolymerization and decomposition of the monomers; the corresponding figures for ^{14}C-labelled natural lignin (from maize) were 63%, 32% and 15% respectively. They concluded that *N. exitiosus* is a significant contributor to lignin decomposition in field populations. The mechanism of lignin decomposition is not known, but Butler and Buckerfield (1979) suggested that it may be partly due to the intestinal symbionts and partly to the termites' own digestive processes.

The rate at which lignin decompostion by *N. exitiosus* occurs is very rapid compared with known lignin decomposition rates due to fungi. If termites are generally capable of decomposing lignin, and since it is well established that they are extremely efficient decomposers of cellulose, their very widespread occurrence and high populations in tropical regions may be an important contributing factor to the generally low organic matter content of tropical soils. Further investigation is necessary to quantify their role in organic matter turnover and its impact on Australian and other similar soils.

Nutrient cycling

Nutrients contained in plant tissue ingested by earthworms and termites are excreted or incorporated into body tissues, and are eventually returned to the soil to become available for further plant growth.

No work has been done in Australia on the effects of earthworms on nutrient cycling, but various workers in Europe have investigated the effects of lumbricid species that are now common in Australia. Graff (1971) compared the weight of some plant nutrients in the annual cast production of seven earthworm species with the weight of these nutrients in the casts plus the A_h (humified) horizon of a pasture soil in Germany; a summary of his findings is in Table 39.8. The earthworms ingest and cast about 25% of the total A_h horizon each year but, because they select organic in preference to inorganic matter, their casts contain a disproportionately large percentage of plant nutrients.

Sharpley and Syers (1976, 1977) have shown that in New Zealand pastures the release of inorganic phosphorus from earthworm casts is about four times greater than that from surface soil. Data for phosphatase activity in casts indicated that 0.1 M NaCl-extractable phosphorus was converted to inorganic phosphorus, and that this was released to solution. The released phosphorus is readily available for plant growth, but is also readily lost in solution in surface runoff.

It is not easy to assess the role of earthworms in nitrogen turnover because much of the nitrogenous output of earthworms is excreted in urine or is contained in the mucus secreted on the body surface and is discharged directly into the soil, while the remainder is excreted in casts. On the basis of nitrogen in casts only, Nowak (1975) estimated that 42% of the total nitrogen available annually in plant litter and dung in a sheep fold passed through earthworms; Raw (1962), Perel and Sokolov (1964) and Satchell (1967) concluded that lumbricid earthworms in several European habitats were capable of ingesting all the plant litter available each year, so would have the capacity to contribute to the cycling of the total pool of available nutrients, including all the nitrogen. Syers *et al.* (1979) found that casts of *Lumbricus rubellus* in a New Zealand pasture contained 73% of the total ingested nitrogen from organic sources and commented that earthworm activity largely results in a 'pass through' of organic nitrogen with a low efficiency of utilization; this is in accordance with the generally low energy assimilation efficiency of earthworms (see above). Nitrogen excreted in urine is mainly in the form of ammonia, with some urea and other compounds, including allantoin. The proportions of ammonia and urea vary with the source of food, and also with variations in environmental temperature, water availability and soil texture (see El-Duweini and Ghabbour 1971).

The casts of earthworms are deposited fairly uniformly over or under the ground surface, but termites forage widely from the nest for food and use their faeces for nest construction, so the nutrients they contain are concentrated and withdrawn from circulation in the ecosystem for as long as nests persist. Lee and Wood (1971*b*)

Table 39.8. Contribution of earthworms to nutrient cycling in a pasture soil
Data from Graff (1971).

Nutrient element	Nutrient content A	Nutrient content B	Amount cycled by earthworms annually (A/B%)
C	2137	5081	42.1
N	134	396	33.8
P (total)	26	77	33.8
P (lactic acid sol.)	3.42	5.07	67.5
K (total)	274	873	31.4
K (lactic acid sol.)	11.15	16.40	68.0

Earthworm biomass (7 species), 96.5 g m^{-2}; weight of A_h horizon of soil, *c.* 100 kg m^{-2}; annual cast production 25.75 kg m^{-2}. A, annual cast production (g m^{-2}); B, casts plus Ah horizon (g m^{-2}).

Table 39.9. Organic carbon, nitrogen and exchangeable cations in termite mounds as a percentage of total quantities in the mound/A_1 horizon systems in two northern Australian ecosystems

From Lee 1974. Since the mounds will eventually be destroyed by erosion and become part of the A_1 horizon of the soil, the weights of nutrients etc. in the mounds are calculated as a percentage of the total in the mounds and the A_1 horizon

Mound-building termites sampled	Depth of A_1 horizon (cm)	Proportion of soil in mounds as % of mound/A_1 system (%)	Quantities in mounds as % of total in mound/A_1 horizon system						
			Org. C	N	P	Exch. Ca	Exch. Mg	Exch. K	Exch. Na
Nasutitermes triodiae	8	2.0	9.6	5.3	5.0	9.1	22.0	13.1	2.7
Amitermes vitiosus } *Tumulitermes hastilis* }	10	3.1	6.3	4.6	3.0	5.2	5.2	6.6	3.5

Table 39.10. Contribution of the wood-feeding termite *Nasutitermes exitiosus* to nutrient cycling in a *Eucalyptus obliqua/E. baxteri* forest at Engelbrook, South Australia

Data from Lee and Butler 1977

Mean annual litter fall		233 g m^{-2}
Consisting of: Wood		43 g m^{-2}
Other litter		190 g m^{-2}
Nasutitermes exitiosus population		
Numbers		600 m^{-2}
Biomass		3 g m^{-2} (fresh wt)
	=	0.65 g m^{-2} (dry wt)
Food consumed by *N. exitiosus*		11.6 g m^{-2} $year^{-1}$
Total food assimilated		6.27 g m^{-2} $year^{-1}$
Faeces		5.33 g m^{-2} $year^{-1}$
Assimilation efficiency		54%
Total food consumed	=	26% of available wood
	=	5% of total litter fall

Nutrient content of annual food consumption

Nutrient element	Amount in wood eaten (mg m^{-2})	% of total available in wood	% of total available in total litter fall
N	29.6	26.9	3.0
P	0.8	21.6	2.6
K	4.9	25.8	1.8
Ca	53.7	25.6	3.1
Mg	9.1	23.9	2.1
Zn	0.2	33.3	3.6
Mn	1.0	28.6	2.5
Fe	2.5	25.0	5.8
Cu	0.05	25.0	5.5

have calculated the quantities of nutrients withheld in mounds in some northern Australian wooded savanna ecosystems (Table 39.9). Lee and Butler (1977) discuss litter fall rates and the contribution of the wood-feeding termite *Nasutitermes exitiosus* to nutrient cycling in a South Australian dry sclerophyll eucalypt forest (Table 39.10). This species builds mounds largely composed of its faeces; colonies and their mounds survive for many (perhaps 20–40) years, and the plant nutrients incorporated in their excreta are withdrawn from circulation for that time. Further information on the same forest site and its termites may be found in Lee and Wood (1968, 1971*a*), Wood and Lee (1971), and Lee and Correll (1978). The very large mounds of *Macrotermes* spp., found especially in Africa, sometimes contain high concentrations of plant nutrients. Accumulation is partly due to concentration from excreta, as described above for *N. exitiosus*, but unlike the mounds of Australian termites those of some species of *Macrotermes* become permanent landscape features and pedological and hydrological processes that ensue may result in accumulations of plant nut ients (see Boyer 1973, 1975*a*, *b*). In some areas in Africa soil from the mounds is spread over the surrounding landscape as a valuable fertilizer for crop production (see Watson 1977, Kang 1978, Mielke 1978); in other areas it has been found that *Macrotermes* mounds are particularly low in plant nutrient content, and that spreading the soil from them over surrounding land can cause crop failure. This apparent contradiction is probably to be understood in terms of differences in local hydrological and pedological regimes, and of differences in fertility between surface soil horizons that are used for growing crops and underlying horizons from which material is transported by the termites to build their mounds.

Conclusions

Soils, as dynamic systems that support the growth of plants, are driven by cyclic biological processes that draw on solar energy, and soil animals contribute much to these processes. Soil animals also contribute directly to the physical organizatio and reorganization of soil profiles. The very widespread distribution and large populations of termites in Australian soils and the great impact that these insects have on soil profile development, hydrology, landscape dynamics, organic matter decomposition, and nutrient cycling make them a particularly significant factor in pedogenesis in Australia. Some work has been done, but much more is necessary, and though it is possible to infer a good deal from work in Africa, where termites are also widespread and numerous and where soils and environmental conditions are often rather similar to those found in Australia, the termites of Australia differ considerably in their building and feeding behaviour from those of Africa, and there is still much to learn of African as well as Australian termites.

Earthworms, because of their needs for water, are not

common over much of the area of Australia, but they are widespread and numerous in the higher rainfall areas. Very little work has been done on their ecology and relationships to soils in Australia. It is known that the common species in pastures and cultivated lands of southern Australia are introduced lumbricid earthworms of European origin, and it is possible to draw some inferences on their probable effects on soils from the results of European work and experience; it is also known that the common earthworms of similar lands in northern Australia include megascolecid and glossoscolecid species, probably all of them introduced from tropical regions of the world, but nothing is known of their ecology and effects on soils.

The last 50 years have seen great progress in Australia in the understanding of the physical and chemical characteristics of our soils, and their geographical distribution, as is illustrated in other chapters of this book. Perhaps it is not too much to hope that in the next 50 years more support might be found for research on the biological processes that make soils what they are and maintain their fertility.

References

Barley, K. P. (1959). Earthworms and soil fertility. IV. The influence of earthworms on the physical properties of red-brown earths. *Aust. J. Agric. Res.* **10**, 371-76.

Barley, K. P. and Jennings, A. C. (1959). Earthworms and soil fertility. III. The influence of earthworms on the availability of nitrogen. *Aust. J. Agric. Res.* **10**, 364-70.

Barley, K. P. and Kleinig, C. R. (1964). The occupation of newly irrigated lands by earth worms. *Aust. J. Sci.* **26**, 290-91.

Bodenheimer, F. S. (1935). Soil conditions, which limit earthworm distribution. *Zoogeographica* **2**, 572-78.

Bodot, P. (1967). Etude écologique des termites des savanes de Basse Côte d'Ivoire. *Insectes Soc.* **14**, 229-58.

Bouché, M. B. (1971). Relations entre les structures spatiales, et fonctionelles des écosystèmes, illustrées par le rôle pédobiologique des vers de terre. *In* 'La Vie dans les Sols.' (Ed. C. Delamare Deboutteville). pp. 189-209. (Gauthiers-Villars: Paris.)

Bouché, M. B. (1972). Lombriciens de France. Ecologie et Systématique. I.N.R.A. Publ. 72-2. (INRA: Paris).

Bouché, M. B. (1977). Stratégies lombriciennes. *In* 'Soil Organisms as Components of Ecosystems.' (Eds U. Lohm and T. Persson). Ecol. Bull. (Stockholm) **25**, pp. 122-32.

Boyer, P. (1971). Les différents aspects de l'action des termites sur les sols tropicaux. *In* 'La Vie dans les Sols – Aspects Nouveaux. Etudes Expérimentales.' (Ed. P. Pesson). pp. 279-334. (Gauthiers-Villars: Paris.)

Boyer, P. (1973). Action de certains termites constructeurs sur l'évolution de sols tropicaux. I. Les termites et le sol. *Annal. Sci. Nat. Zool. Paris Ser. 12* **15**, 329-498.

Boyer, P. (1975*a*). Action de certains termites constructeurs sur l'évolution des sols tropicaux. II. Etude particulière de trois termitières de *Bellicositermes* et de leur action sur les sols tropicaux. *Annal. Sci. Nat. Zool. Paris Ser.12* **17**, 273-446.

Boyer, P. (1975*b*). Action de certains termites constructeurs sur l'évolution des sols tropicaux. III. Les différents aspects de l'action des *Bellicositermes* sur les sols tropicaux. *Annal. Sci. Nat. Zool. Paris Ser.12*, **17**, 447-504.

Butler, J. H. A. and Buckerfield, J. C. (1979). Digestion of lignin by termites. *Soil Biol. Biochem.* **11**, 507-13.

Byzova, J. B. (1975). Specific haemoglobin content as an ecological characteristic in earthworms (Lumbricidae). *In* 'Progress in Soil Zoology.' (Ed. J. Vanek). Proc. 5th Int. Colloq. Soil Zool. Prague, 1973. pp.559-62.

Dash, M. C. and Patra, V. C. (1979). Wormcast production and nitrogen contribution to soil by a tropical earthworm population from a grassland site in Orissa, India. *Rev. Ecol. Biol. Sol.* **16**, 79-83.

de Leenheer, L., d'Hoore, J. and Sys, K. (1952). Cartographie et caractérisation pédologique de la catena de Yangambi. Les termitières. *Publ. Inst. Natl. Etude Agron. Congo Ser. Sci.* **55**, 53-4.

de Ploey, J. (1964). Nappes de gravats et couvertures argilo-sableuses au Bas-Congo; leur génèse et l'action des termites. *In* 'Etudes sur les Termites Africains'. (Ed. A. Bouillon). pp.399-414. (Leopoldville Univ. Press: Leopoldville.)

Drummond, H. (1886). On the termite as the tropical analogue of the earthworm. *Proc. R. Soc. Edinburgh* **13**, 137-46.

Edwards, C. A. and Lofty, J. R. (1977). 'Biology of Earthworms'. (2nd edn). (Chapman and Hall: London.)

Ehlers, W. (1975). Observations on earthworm channels and infiltration on tilled and untilled loess. *Soil Sci.* **119**, 242-49.

El-Duweini, A. K. and Ghabbour, S. I. (1965). Population density and biomass of earthworms in different types of Egyptian soils. *J. Appl. Ecol.* **2**, 271-87.

El-Duweini, A. K. and Ghabbour, S. I. (1971). Nitrogen contribution by live earthworms to the soil. *In* 'IV Co loquium Pedobiologiae'. (Ed. J. d'Aguilar). I.N.R.A. Publ. No. 71-7, pp.495-501. (INRA:Paris.)

Ferrar, P. and Watson, J. A. L. (1970). Termites associated with dung in Australia. *J. Aust. Entomol. Soc.* **9**, 100-2.

French, J. R. J. and Bland, J. E. (1975). Lignin degradation in the termites *Coptotermes lacteus* and *Nasutitermes exitiosus*. *Mater. Org.* **10**, 281-88.

Fuller, C. (1915). Observations on some South African termites. *Ann. Natal Mus.* **3**, 329-505.

Gates, G. E. (1961). Ecology of some earthworms with special reference to seasonal activity. *Am. Midl. Nat.* **66**, 61-86.

Gay, F. J. and Calaby, J. H. (1970). Termites from the Australian region. *In* 'Biology of Termites.' (Eds K. Krishna and F. M. Weesner), Vol.2, pp.393-448. (Academic Press: New York.)

Ghilarov, M. S. (1962). Termites of the USSR, their distribution and importance. *In* 'Termites in the Humid Tropics.' New Delhi Symposium, pp.131-35. (UNESCO: Paris.)

Gillman, L. R., Jefferies, M. K. and Richards, G. N. (1972). Non-soil constituents of termite (*Coptotermes acinaciformis*) mounds. *Aust. J. Biol. Sci.* **25**, 1005-13.

Graff, O. (1953). Die Regenwürmer Deutschlands. *Schrift. Forsch. Land. Braunschweig – Volk.* **7**, 81pp.

Graff, O. (1971). Stickstoff, Phosphor und Kalium in der Regenwürmlosung auf der Wiesenversuchs fläche des Sollingprojektes. *In* 'IV Colloquium Pedobiologiae.' (Ed. J. d'Aguilar), pp.503-11. I.N.R.A. Publ. No.71-7. (INRA: Paris.)

Grant, W. C. (1955). Temperature relationships in the megascolecid earthworm *Pheretima hupeiensis*. *Ecology* **36**, 412-17.

Guild, W. J. McL. (1952). Variation in earthworm numbers within field populations. *J. Anim. Ecol.* **21**, 169-81.

Haantjens, H. A. (1965). Morphology and origin of patterned ground in a humid tropical lowland area, New Guinea. *Aust. J. Soil Res.* **3**, 111-29.

Haantjens, H. A. (1969). Fire and wind erosion, or earthworms as the cause of microrelief in the Lower Sepik Plains, New Guinea. *Aust. J. Sci.* **32**, 52-54.

Harris, W. V. (1956). Termite mound building. *Insectes Soc.* **3**, 261-65.

Hesse, P. R. (1955). A chemical and physical study of the soils of termite mounds in East Africa. *J. Ecol.* **43**, 449-61.

Hill, G. F. (1921). The white ant pest in Northern Australia. *Bull. Advis. Coun. Sci. Ind. Melbourne* **21**, 1-26.

Hoeksema, K. J. and Jongerius, A. (1959). On the influence of earthworms on the soil structure in mulched orchards. Proc. Int. Symp. Soil Struct. Ghent, 1958, pp.188-94.

Hogben, L. and Kirk, R L. (1944). Studies on temperature regulation. I. The Pulmonata and Oligochaeta. *Proc. R. Soc. Ser. B* **132**, 239-52.

Hopp, H. and Slater, C. S. (1949). The effect of earthworms on the productivity of agricultural soil. *J. Agric. Res.* **78**, 325-39.

Jeanson, C. (1960a). Étude expérimentale de l'action de *Lumbricus herculeus* (Savigny) (Oligochète Lumbricide) sur la stabilité structurale des sols. *C.R. Acad. Sci. Ser. D.* **250**, 3041-43.

Jeanson, C. (1960*b*). Évolution de la matière organique du sol sous l'action de *Lumbricus herculeus* (Savigny) (Oligochète Lumbricidae). *C.R. Acad. Sci. Ser. D.* **250**, 3500-02.

Jeanson, C. (1961). Sur une méthode d'étude du comportement de la faune du sol et de sa contribution à la pédogenèse. *C.R. Acad. Sci. Ser. D.* **253**, 2571-73.

Jeanson, C. (1964). Micromorphology and experimental soil zoology; contribution to the study, by means of giant-sized thin sections, of earthworm-produced artificial soil structure. *In* 'Soil Micromorphology.' Proc. 2nd. Int. Wk. Mtg. Soil Micromorph. Arnhem. pp.47-55.

Jeanson, C. (1971). Structure d'une galerie de lombric à la microsonde électronique. *In* 'IV Colloquium Pedobiologiae' (Ed. J. d'Aguilar). I.N.R.A. Publ. No.71-7, pp.513-25. (INRA: Paris.)

Joachim, A. W. R. and Kandiah, S. (1940). Studies on Ceylon soils. XIV. A comparison of soils from termite mounds and adjacent land. *Trop. Agric. Mag. Ceylon Agric. Soc.* **95**, 333-38.

Josens, G. (1974). Étude fonctionelle de quelques groupes animaux: les termites de la savane de Lamto. Bull. de Liaison des Chercheurs de Lamto, No. spéc. 1974, Fasc. 5, pp.91-131.

Kang, B. T. (1978). Effect of some biological factors on soil variability in the tropics. III. Effect of *Macrotermes* mounds. *Plant Soil* **50**, 241-51.

Kemp, P. B. (1955). The termites of north-eastern Tanganyika: their distribution and biology. *Bull. Ent. Res.* **46**, 113-35.

Kozlovskaya, L. S. (1969). Der Einfluss der Exkremente von Regenwurmen auf die Aktivierung der mikrobiellen Prozesse in Torfböden. *Pedobiologia* **9**, 158-64.

Kretzschmar, A. (1978). Quantification écologique des galeries de lombriciens. Techniques et premières estimations. *Pedobiologia* **18**, 31-8.

Lavelle, P. (1974). Les vers de terre de la savane de Lamto. *In* 'Analyse d'un Écosystème Tropical Humide: la Savane de Lamto'. Bull. de Liaison des Chercheurs de Lamto, No. spéc. 1974, Fasc. 5, pp. 133-66.

Lavelle, P. (1979). Relations entre types écologiques et profils démographiques chez les vers de terre de la savane de Lamto (Côte d'Ivoire). *Rev. Écol. Biol. Sol.* **16**, 85-101.

Lee, K. E. (1959). The earthworm fauna of New Zealand. N.Z. Dep. Sci. Ind. Res. Bull. No.130.

Lee, K. E. (1967). Microrelief features in a humid tropical lowland area, New Guinea, and their relation to earthworm activity. *Aust. J. Soil Res.* **5**, 263-74.

Lee, K. E. (1969). Some soils of the British Solomon Islands Protectorate. *Phil. Trans. R. Soc. London* Ser. B. **255**, 211-57.

Lee, K. E. (1974). The significance of soil animals in organic matter decomposition and nutrient cycling in tropical forest and savanna ecosystems. Trans. 10th Int. Congr. Soil Sci. Moscow. Vol.3, pp.43-51.

Lee, K. E. (1975). Results of the 1971 Royal Society-Percy Sladen Expedition to the New Hebrides. 13. Conclusions. *Phil. Trans. R. Soc. London Ser. B.* **272**, 477-86.

Lee, K. E. and Butler, J. H. (1977). Termites, soil organic matter decomposition and nutrient cycling. *In* 'Soil Organisms as Components of Ecosystems.' (Eds U. Lohm and T. Persson). 6th Int. Colloq. Soil Zool., Uppsala, 1976. *Ecol. Bull. (Stockholm)* **25**, 544-48.

Lee, K. E. and Correll, R. (1978). Litter fall and its relationship to nutrient cycling in a South Australian dry sclerophyll forest. *Aust. J. Ecol.* **3**, 243-52.

Lee, K. E. and Wood, T. G. (1968). Preliminary studies of the role of *Nasutitermes exitiosus* (Hill) in the cycling of organic matter in a yellow podzolic soil under dry sclerophyll forest in South Australia. Trans. 9th. Int. Congr. Soil Sci. Adelaide, 1968. Vol. 2, pp.11-18.

Lee, K. E. and Wood, T. G. (1971*a*). 'Termites and Soils.' (Academic Press: London.)

Lee, K. E. and Wood, T. G (1971*b*). Physical and chemical effects on soils of some Australian termites and their pedological significance. *Pedobiologia*, **11**, 376-409.

Lepage, M. (1974). Les termites d'une savane sahélienne (Ferlo Septentrional, Sénégal): peuplements, consommation, rôle dans l'écosystème. Doctoral Thesis, University of Dijon, Dijon.

Levin, D. A. (1976). The chemical defenses of plants to pathogens and herbivores. *Annu. Rev. Ecol. Syst.* **7**, 121-59.

Ljungström, P. O., Orellana, J. A. de and Priano, L. J. J. (1973). Influence of some edaphic factors on earthworm distribution in Santa Fe Province, Argentina. *Pedobiologia* **13**, 236-47.

Ljungström, P. O. and Reinecke, A. J. (1969). Ecology and natural history of the microchaetid earthworms of South Africa. 4. Studies on influence of earthworms upon the soil and the parasitological question. *Pedobiologia* **9**, 152-57.

Macfadyen, A. (1963). The contribution of the fauna to total soil metabolism. *In* 'Soil Organisms.' (Eds J. Doeksen and J. van der Drift). pp.3-17. (North-Holland: Amsterdam.)

Maldague, M. E. (1959). Analyses de sols et matériaux de termitières du Congo. *Insectes Soc.* **6**, 343-59.

Maldague, M. E. (1964). Importance des populations de termites dans les sols équatoriaux. Proc. 8th Int. Congr. Soil Sci., Bucharest. Vol. 3, pp.743-52.

Maldague, M. E. (1970). Rôle des animaux édaphiques dans la fertilité des sols forestiers. Publ. Inst. Natl. Etude Congo, Ser. Sci. No.112.

Mangum, C. P. (1978). Temperature adaptation. *In* 'Physiology of Annelids'. (Ed. P. J. Mill). pp.447-78. (Academic Press: London.)

Matsumoto, T. (1976). The role of termites in an equatorial rain forest ecosystem of West Malaysia. *Oecologia* **22**, 153-78.

Meyer, L. (1943). Experimental study of microbiological effects on humus and soil formation. *Bodenkd. Pflanzenernaehr.* **29**, 119-40.

Michel, C. and DeVillez, E. J. (1978). Digestion. *In:* 'Physiology of Annelids.' (Ed. P. J. Mill). pp.509-53. (Academic Press: London.)

Miedema, R. and van Vuure, W. (1977). The morphological, physical and chemical properties of two mounds of *Macrotermes bellicosus* (Smeathman) compared with surrounding soils in Sierra Leone. *J. Soil Sci.* **28**, 112-24.

Mielke, H. W. (1978). Termitaria and shifting cultivation: the dynamic role of the termite in soils of wet-dry Africa. *Trop. Ecol.* **19**, 117-22.

Nakamura, Y. (1972). Ecological studies of the family Lumbricidae from Hokkaido. I. Ecological distribution. *Japan J. Appl. Ent. Zool.* **16**, 18-23.

Noble, J. C., Gordon, W. T. and Kleinig, C. R. (1970). The influence of earthworms on the development of mats of organic matter under irrigated pastures in southern Australia. Proc. 11th Int. Grassl. Conf., Brisbane, 1970. pp.465-68.

Nordström, S. and Rundgren, S. (1976). *Lumbricus polyphemus* in the Kaiserstuhl area, southwestern Germany. *Pedobiologia* **16**, 373-78.

Nowak, E. (1975). Population density of earthworms and some elements of their production in several grassland environments. *Ekol. Polska* **23**, 459-91.

Nye, P. H. (1955). Some soil-forming processes in the humid tropics. IV. The action of the soil fauna. *J. Soil Sci.* **6**, 73-83.

Parle, J. N. (1959). Activities of micro-organisms in soil and influence of these on soil fauna. Ph. D. Thesis, London Univ.

Parle, J. N. (1963*a*). Micro-organisms in the intestines of earthworms. *J. Gen. Microbiol.* **31**, 1-13.

Parle, J. N. (1963*b*). A microbiological study of earthworm casts. *J. Gen. Microbio*l. **31**, 13-22.

Pathak, A. N. and Lehri, L. K. (1959). Studies on termite nests. I. Chemical, physical and biological characteristics of a termitarium in relation to its surroundings. *J. Indian Soc. Soil Sci.* **7**, 87-90.

Perel, T. S. (1977). Differences in lumbricid organization connected with ecological properties. *In* 'Soil Organisms as Components of Ecosystems.' (Eds U. Lohm and T. Persson). *Ecol. Bull. (Stockholm)* **25**, 56-63.

Perel, T. S. and Sokolov, D. F. (1964). Quantitative evaluation of the participation of the earthworm *Lumbricus terrestris* Linné (Lumbricidae-Oligochaeta) in the transformation of forest litter. *Zool. Zh.* **53**, 1618-25.

Phillipson, J., Abel, R., Steel, J. and Woodell, S. R. J. (1976). Earthworms and the factors governing their distribution in an English beechwood. *Pedobiologia* **16**, 258-85.

Pickford, G. E. (1926). The Kommetjie Flats. Blythswood Review No.3, pp.57-58. Cape Province, South Africa.

Pickford, G. E. (1937). 'A Monograph of the Acanthodriline Earthworms of South Africa.' (Heffer: Cambridge.)

Ponomareva, S. I. (1953). The influence of the activity of earthworms on the creation of a stable structure in a sod-podzolized soil. *Trud. Pochv. Inst. Dokuchaeva* **41**, 304-78.

Raw, F. (1959). Estimating earthworm populations by using formalin. *Nature (London)* **184**, 1661-62.

Raw, F. (1962). Studies of earthworm populations in orchards. I. Leaf burial in apple orchards. *Ann. Appl. Biol.* **50**, 389-404.

Reichle, D. E. (1977). The role of soil invertebrates in nutrient cycling. *In* 'Soil Organisms as Components of Ecosystems.' (Eds U. Lohm, and T. Persson). Proc. 6th Int. Coll. Soil Zool. *Ecol. Bull. (Stockholm)* **25**, 145-56.

Reynolds, J. W. (1973). Earthworm (Annelida: Oligochaeta) ecology and systematics. Proc. 1st Soil Microcommunities Conf., Syracuse, N.Y. 1971 (Ed. D. Dindal), pp.95-120 (USAEC, Office of Information Services.)

Roy, S. K. (1957). Studies on the activities of earthworms. *Proc. Zool. Soc, Calcutta* **10**, 81-98.

Sands, W. A. (1965). Mound population movements and fluctuations in *Trinervitermes ebenerianus* Sjöstedt (Isoptera, Termitidae, Nasutitermitinae). *Insectes Soc.* **12**, 49-58.

Satchell, J. E. (1967). Lumbricidae. *In* 'Soil Biology.' (Eds A. Burges and F. Raw). pp.259-322. (Academic Press: London.)

Scharpenseel, H. W. and Gewehr, H. (1960). Studien zur Wasserbewegung im Boden mit Tritium-Wasser. *Z. Pflanzenernaehr. Dueng.* **88**, 35-49.

Sharpley, A. N. and Syers, J. K. (1976). Potential role of earthworm casts for the phosphorus enrichment of run-off waters. *Soil Biol. Biochem.* **8**, 341-46.

Sharpley, A. N. and Syers, J. K. (1977). Seasonal variation in casting activity and in the amounts and release to solution of phosphorus forms in earthworm casts. *Soil Biol. Biochem.* **9**, 227-31.

Slater, C. S. and Hopp, H. (1947). Relation of fall protection to earthworm populations and soil physical conditions. *Soil Sci. Soc. Am. Proc.* **12**, 508-11.

Sleeman, J. R. and Brewer, R. (1972). Micromorphology of some Australian termite nests. *Pedobiologia* **12**, 347-73.

Smith, W. W. (1893). Further notes on New Zealand earthworms with observations on the known aquatic species. *Trans. N.Z. Inst.* **25**, 111-46.

Smith, W. W. (1894). Further notes on New Zealand earthworms. *Trans. N.Z. Inst.* **26**, 155-75.

Stockdill, S. M. J. (1966). The effects of earthworms on pastures. *Proc. N.Z. Ecol. Soc.* **13**, 68-75.

Stoops, G. (1964). Application of some pedological methods to the analysis of termite mounds. *In* 'Etudes sur les Termites Africains.' (Ed. A. Bouillon). pp. 379-98. (Leopoldville Univ: Leopoldville.)

Strickland, A. H. (1944). The arthropod fauna of some tropical soils, *Trop. Agric. Trin.* **21**, 107-14.

Svendsen, J. A. (1955). Earthworm population studies: a comparison of sampling methods. *Nature (London)* **175**, 864.

Swaby, R. J. (1950). The influence of earthworms on soil aggregation. *J. Soil Sci.* **1**, 195-97.

Syers, J. K., Sharpley, A. N. and Keeney, D. R. (1979). Cycling of nitrogen by surface-casting earthworms in a pasture ecosystem. *Soil Biol. Biochem.***11**, 181-85.

Sys, C. (1955). The importance of termites in the formation of latosols. *Afr. Soils* **3**, 393-95.

Teotia, S. P., Duley, F. L. and McCalla, T. M. (1950). Effect of stubble mulching on number and activity of earthworms. *Bull. Nebraska Agric. Exp. Stn.* **165**, 1-20.

Troll, C. (1936). Termitensavannen. *In* 'Landekunliche Forschrift Festschrift für Norbert Krebs'. pp.275-312. (Engelhorn: Stuttgart.)

van Rhee, J. A. and Nathans, S. (1961). Observations on earthworm populations in orchard soils. *Neth. J. Agric. Sci.* **9**, 94-100.

Wasawo, D. P. S. and Visser, S.A. (1959). Swampworms and tussock mounds in the swamps of Teso, Uganda. *East Afr. Agric. J.* **25**, 86-90.

Waters, R. A. S. (1951). Earthworms and the fertility of pasture. Proc. 13th Conf. N.Z. Grassl. Assoc. New Plymouth. pp. 168-75.

Waters, R. A. S. (1955). Numbers and weights of earthworms under a highly productive pasture. *N.Z. J. Sci. Tech.* **A36**, 516-25.

Watson, J. A. L. (1972). An old mound of the spinifex termite, *Nasutitermes triodiae* (Froggatt). (Isoptera: Termitidae). *J. Aust. Ent. Soc.* **11**, 78-80.

Watson, J. A. L. and Gay, F. J. (1970). The role of grass-eating termites in the degradation of a mulga ecosystem. *Search* **1**, 43.

Watson, J. P. (1962). The soil below a termite mound. *J. Soil Sci.* **13**, 46-51

Watson, J. P. (1969). Water movement in two termite mounds in Rhodesia. *J. Ecol.* **55**, 663-69.

Watson, J. P. (1970). Contribution of termites to development of zinc anomaly in Kalahari sand. Trans. Am. Inst. Min. Metall. Pet. Eng. (Sect. B: Appl. Earth Sci.). pp.B53-B59.

Watson, J. P. (1974). Termites in relation to soil formation, groundwater, and geochemical prospecting. *Soils and Fert.* **37**, 111-14.

Watson, J. P. (1976). The composition of mounds of the termite *Macrotermes falciger* (Gerstäcker) on soil derived from granite in three rainfall zones of Rhodesia. *J. Soil Sci.* **27**, 495-503.

Watson, J. P. (1977). The use of mounds of the termite *Macrotermes falciger* (Gerstäcker) as a soil amendment. *J. Soil Sci.* **28**, 664-72.

Williams, M. A. J. (1968). Termites and soil development near Brock's Creek, Northern Territory. *Aust. J. Sci.* **31**, 153-54.

Wood, T. G. (1974). The distribution of earthworms (Megascolecidae) in relation to soils, vegetation and altitude on the slopes of Mt. Kosciusko, Australia. *J. Anim. Ecol.* **43**, 87-106.

Wood, T. G. (1976). The role of termites (Isoptera) in decomposition processes. *In* 'The Role of Terrestrial and Aquatic Organisms in Decomposition Processes.' (Eds J. M. Anderson and A. Macfadyen). pp.145-68. (Blackwell: Oxford.)

Wood, T. G. and Lee, K. E. (1971). Abundance of mounds and competition among colonies of some Australian termite species. *Pedobiologia* **11**, 341-66.

Wood, T. G. and Sands, W. A. (1977). The role of termites in ecosystems. *In* 'Production Ecology of Ants and Termites.' (Ed. M. V. Brian). pp. 245-92. (Cambridge Univ. Press: Cambridge.)

Wood, T. G., Johnson, R. A. and Ohiagu, C. E. (1977*a*). Populations of termites (Isoptera) in natural and agricultural ecosystems in southern Guinea savanna near Mokwa, Nigeria. *Geo-Eco-Trop.* **1**, 139-48.

Wood, T. G., Johnson, R. A., Ohiagu,C. E., Collins, N. M. and Longhurst, C. (1977*b*). Ecology and importance of termites in crops and pastures. Min. Overseas Devel./Inst. agr. Res., Ahmadu Bello Univ., Nigeria (ODM Res. Scheme R2709). (Centre for Overseas Pest Research: London.)

Yakushev, V. M. (1968). Influence of termite activity on the development of laterite soil. *Soviet Soil Sci.* **1968**, 109-11.

Chapter 40
Ecology of soil invertebrates

P. J. M. Greenslade and Penelope Greenslade

In the previous chapter Lee has described the pedological influences of termites and earthworms. On any given site the density of individuals and/or population biomass of only one or a few species in either of these groups may be such that their activities have a direct measureable effect on soil processes. Spain and Hutson in Chapter 38 outlined the composition of the invertebrate fauna of the litter layer and discuss its contribution to the breakdown and decay of organic matter. In this chapter we survey some aspects of the ecology of soil invertebrates, emphasizing their diversity. We use the term not in a technical sense but to describe the condition of being varied or diversified. The variety of higher taxonomic groups, life forms and species of invertebrates that compose the soil fauna is one of its most striking features and is the reason for much of its inherent biological interest. At the same time this diversity is responsible for some formidable problems in studying and understanding the role of the fauna in the soil system. One might differentiate between the soil fauna in the strictest sense living in a protected equable environment where movement is physically constricted and the mobile surface fauna subject to a widely fluctuating and often harsh climate. However, movement and interaction between species across the air–soil interface is extensive and we include in 'soil fauna', surface, organic horizon and mineral soil components. We are mainly concerned with arthropods and, within the arthropods, take many of our examples from the two groups with which we are most familiar, the ants (Hymenoptera, Formicidae) and Collembola. In a recent paper (P.J.M. Greenslade and P. Greenslade, in press) we have dealt with a number of aspects of ant and collembolan biology in the arid zone and these topics are not covered again in detail here.

Diversity of soil invertebrates

Higher taxa and life forms

Figs 40.1–6 include representatives of the main taxonomic groups that we refer to or are mentioned in Chapters 38 and 39 and they give some idea of the range of size and morphology covered by soil arthropods. They also illustrate simple but important relationships between gross morphology and mode of locomotion, migratory ability and capacity for penetrating narrow passages in soil or actively burrowing. These attributes are, in turn, primary factors that determine limits to distribution. Kevan (1962) gives a detailed classification of different types of locomotion by soil animals and a series of studies of the functional morphology of terrestrial arthropods by Manton (1978) allows their locomotory mechanisms to be related to habits and habitat.

Soil porosity generally decreases with depth, resulting in vertical patterns of zonation in the size and morphology of arthropods that are unable to burrow (Haarlov 1955). In Australian pastures reduction of pore space at high stocking rates contributes to a reduction in the abundance of the soil fauna (King and Hutchinson 1976). Poor structure is also associated with low organic content of most soils in the arid zone. Here the survival of many soil animals depends on the ability to escape down the soil profile away from high temperatures and saturation deficits at the surface. Deep coarse-textured soils in red earth, sand plain and sand hill landscapes provide a favourable medium for excavation and support a rich fauna of burrowing arthropods (Matthews 1976), although many species are excluded from the mobile crests of dunes if they cannot maintain permanent, subterranean nests or burrows there. For arthropods that cannot burrow to escape from the surface climate, arid zone survival depends on secondary behavioural, morphological or physiological adaptations (P. J. M. Greenslade and P. Greenslade, in press). Commonly this involves 'anhydrobiosis'. In a number of isotomid collembolan species, for example *Proisotoma brisbanensis* (Fig. 40.4*b*), inactive anhydric morphs are resistant to desiccation and are able to survive dry periods, rapidly becoming rehydrated and active again after rain. Inability either to burrow or to withstand desiccation at the surface is one of the factors that seems to exclude certain groups from the arid zone and contributes to the reduction of diversity of higher taxa there that was noted by Spain and Hutson in Chapter 38.

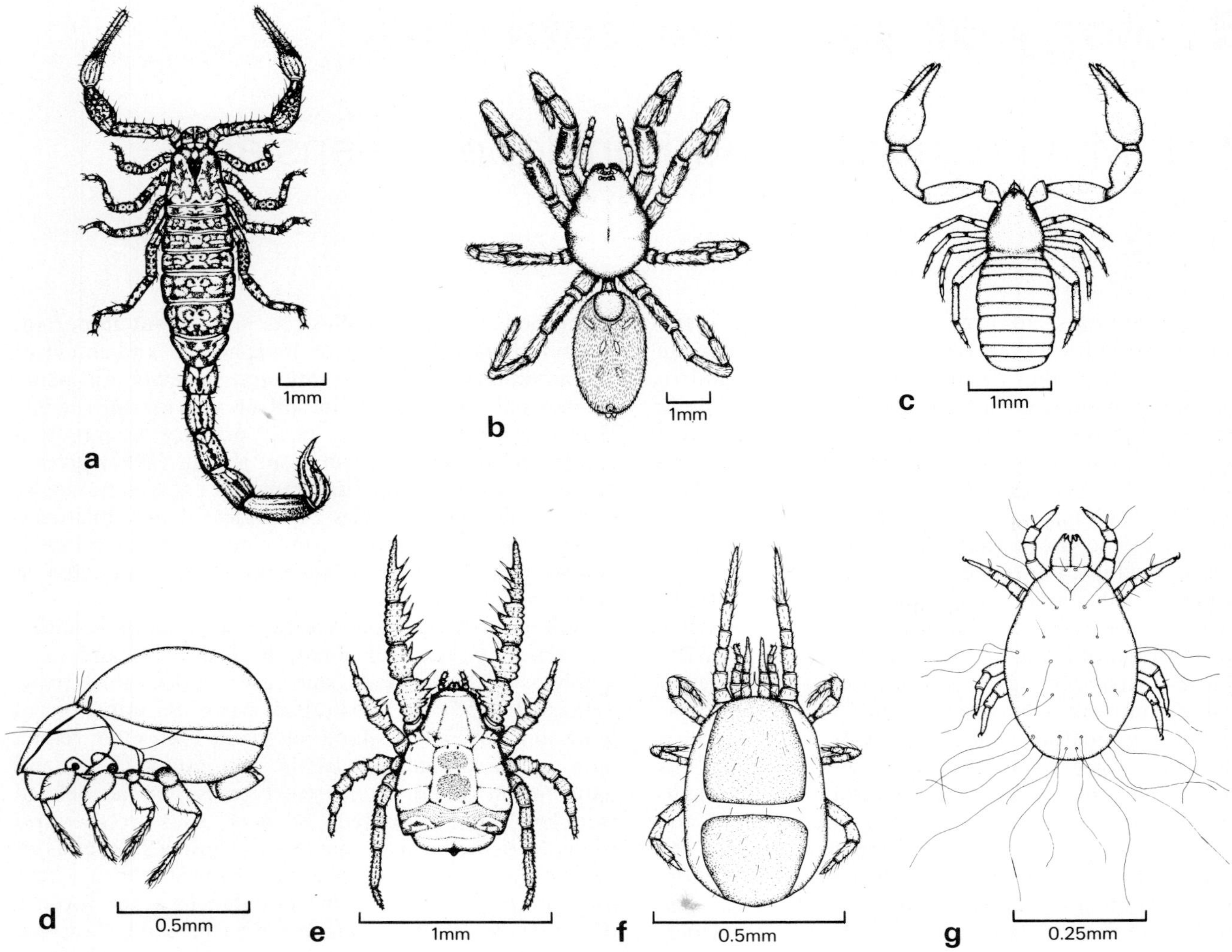

Fig. 40.1. Examples of soil invertebrates, Arachnida: (*a*) immature scorpion, *Lychas marmoreus* (Boothidae, Scorpionida); (*b*) immature female spider, *Lampona* sp., (Gnaphosidae, Araneida); (*c*) pseudoscorpion, *Xenolpium* sp., (Olpiidae, Pseudoscorpiones); (*d*) cryptostigmatid mite, *Scheloribates* sp., (Oribatulidae, Acarina); (*e*) prostigmatid mite (Caeculidae, Acarina); (*f*) mesostigmatid mite, *Acugamasus semipunctatus*, (Rhodacaridae, Acarina); (*g*) astigmatid mite (Acarina).
Apart from the pseudoscorpion, collected in the Great Victoria Desert, all these arachnids come from one site at Bridgewater in the Mt Lofty Ranges, South Australia. The spider and scorpion (*a*, *b*) are general predators on the soil surface. Many species in both these groups excavate burrows but their relatively large size and convex build are not suited to activities within mineral soil. Pseudoscorpions are also predators but they are smaller, with a more flexible morphology. Olpiids (*c*) are moderately compressed and occur in litter and the surface layers of the mineral soil. Chthoniid pseudoscorpions, often abundant in forest litter (Chapter 38), are more convex and spider-like with longer legs. Members of the other pseudoscorpion families are smaller and elongate and are found mainly in soil, or they are strongly dorso-ventrally compressed and live under bark. Representatives of the four main suborders of mites that inhabit soil are illustrated (*d*–*g*). Cryptostigmata (*d*) are particularly abundant in soils rich in organic matter where they feed on microorganisms and dead plant material. The large Prostigmata are surface predators and may gain protection from ants by spines, scales and a heavily armoured integument (*e*), or a dense covering of long hairs. Very small Prostigmata are active after rain in the arid zone but little is known of their biology. One family, the Nanorchestidae, can be found in a variety of very harsh environments from deserts to the Antarctic where they feed on algae (D. E. Rounsevell, personal communication). The Mesostigmata (*f*) include predators on other microarthropods, nematodes and insect eggs and larvae, and species that feed on fungi. The predators are frequent in temporary habitats such as dung and compost heaps where the immature stages of many species disperse to new habitats by phoresy, attaching themselves to flying insects such as adult Diptera and Coleoptera. Many Astigmata are very small and weakly pigmented and feed in the mineral soil on fungi and other microorganisms and on roots (Krantz 1970).

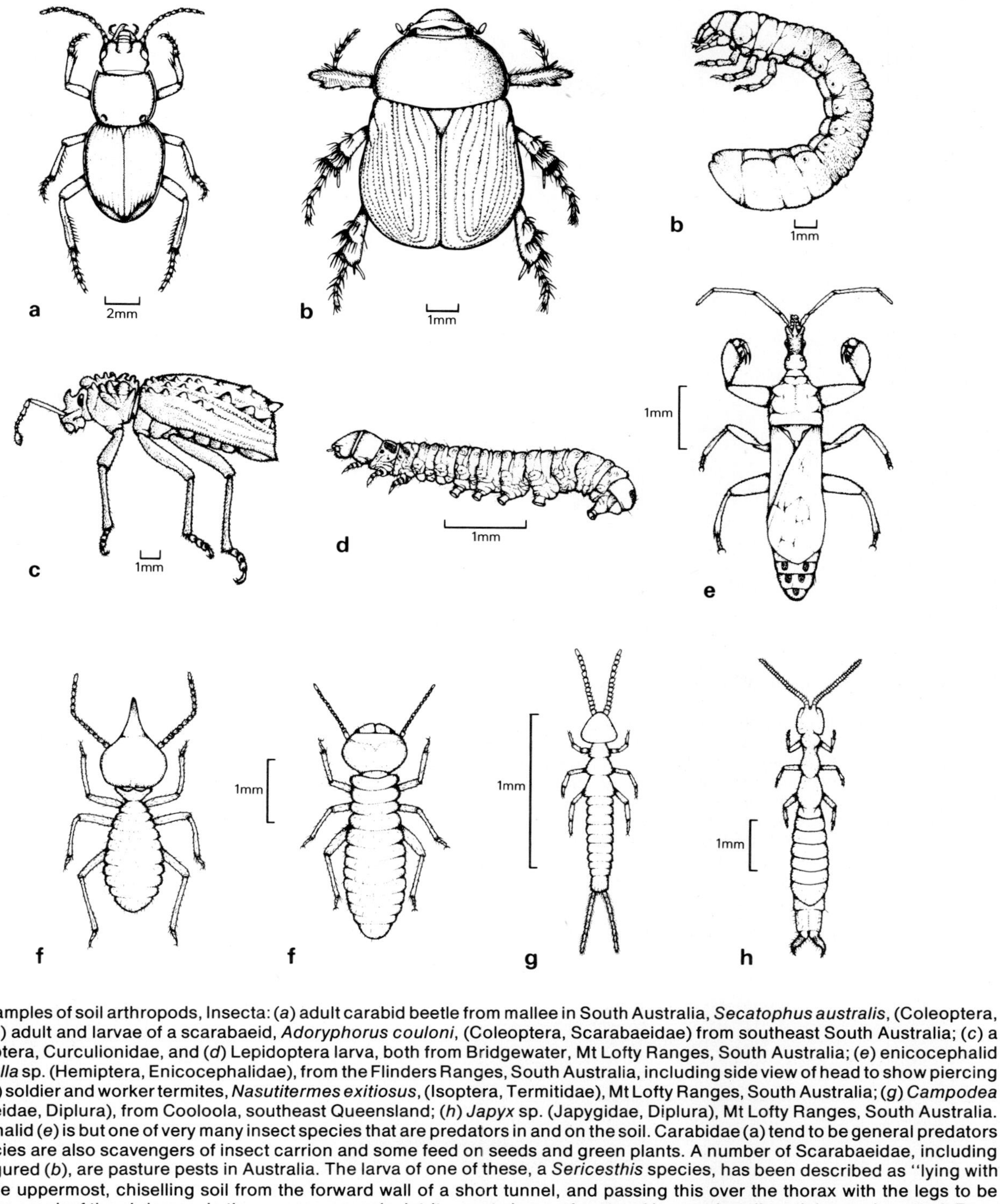

Fig. 40.2. Examples of soil arthropods, Insecta: (*a*) adult carabid beetle from mallee in South Australia, *Secatophus australis*, (Coleoptera, Carabidae); (*b*) adult and larvae of a scarabaeid, *Adoryphorus couloni*, (Coleoptera, Scarabaeidae) from southeast South Australia; (*c*) a weevil, Coleoptera, Curculionidae, and (*d*) Lepidoptera larva, both from Bridgewater, Mt Lofty Ranges, South Australia; (*e*) enicocephalid bug, *Henschiella* sp. (Hemiptera, Enicocephalidae), from the Flinders Ranges, South Australia, including side view of head to show piercing mouthparts; (*f*) soldier and worker termites, *Nasutitermes exitiosus*, (Isoptera, Termitidae), Mt Lofty Ranges, South Australia; (*g*) *Campodea* sp., (Campodeidae, Diplura), from Cooloola, southeast Queensland; (*h*) *Japyx* sp. (Japygidae, Diplura), Mt Lofty Ranges, South Australia. The enicocephalid (*e*) is but one of very many insect species that are predators in and on the soil. Carabidae (a) tend to be general predators but many species are also scavengers of insect carrion and some feed on seeds and green plants. A number of Scarabaeidae, including the species figured (*b*), are pasture pests in Australia. The larva of one of these, a *Sericesthis* species, has been described as "lying with the ventral side uppermost, chiselling soil from the forward wall of a short tunnel, and passing this over the thorax with the legs to be deposited in the crook of the abdomen. In the process, organic detritus, cut pieces of root and inorganic soil particles are ingested. From time to time the larvae somersaults and drops the soil at the rear of the tunnel then rams it with the abdomen" (Davidson *et al.* in press). The family Scarabaeidae also includes the native dung beetles of the genus *Onthophagus*, adapted to the dry dung pellets of marsupials. Exotic dung beetles are being introduced to Australia in order to dispose of larger dung pads of cattle (Bornemissza 1976). Larvae of many curculionid beetles (*c*) and of some Lepidoptera (*d*) feed on roots. Lepidoptera larva are also abundant in litter in Australia where representatives of at least six families feed on fallen leaves (Common 1970). *Nasutitermes exitiosus* is a frequent mound-building termite that feeds on wood in south-eastern Australia (see Table 39.10). Campodeidae and Japygidae are predators and Manton (1972) has related their locomotory achievements to their skeletal morphology and musculature. The campodeids with long, fragile appendages are fast runners confined to litter and very porous soils. The more robust Japygidae reserve their agility for penetrating poorly structured soils where manoeuvring in narrow crevices is facilitated by powerful leg movements and great flexibility, especially of the thorax.

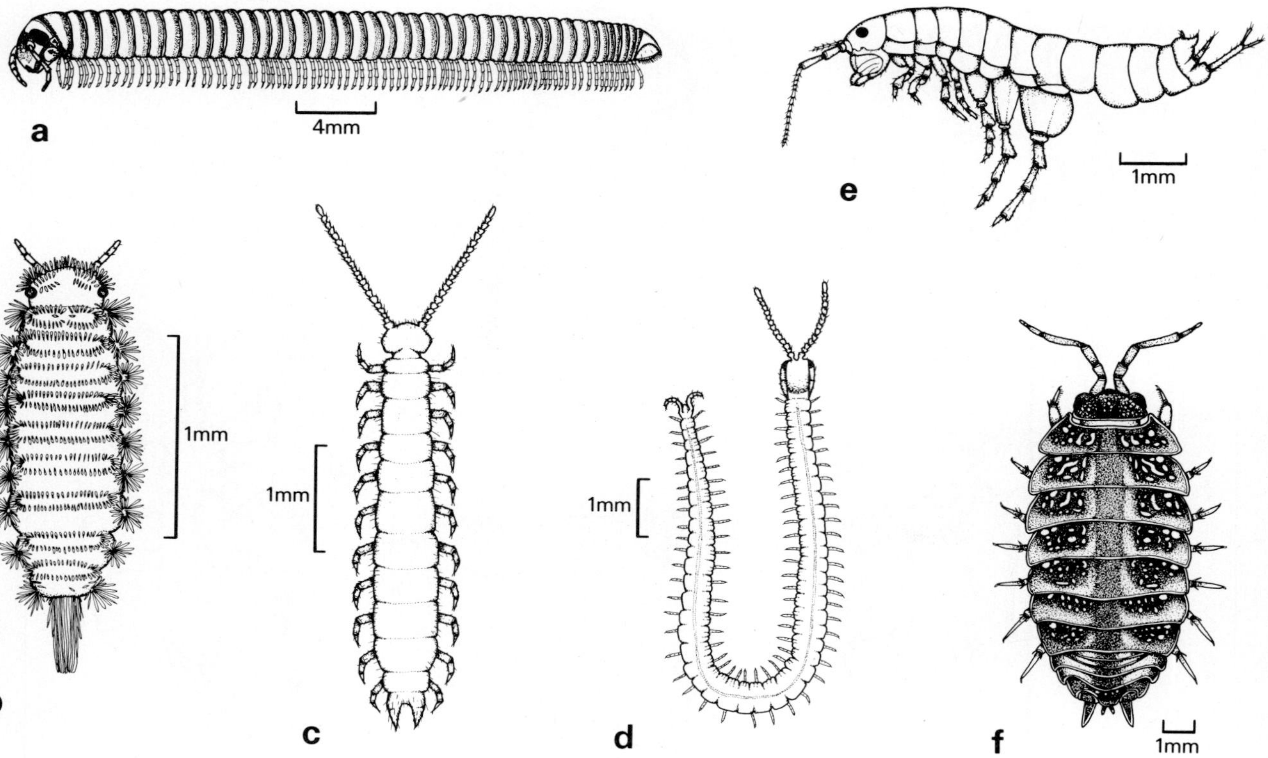

Fig. 40.3. Examples of soil invertebrates, Diplopoda, Symphyla, Chilopoda, Crustacea: (*a*) a millipede, *Ommatoiulus moreletii* (Iulidae, Diplopoda); (*b*) a polyxenid, *Polyxenus* sp. (Polyxenidae, Diplopoda); (*c*) a symphylan (Symphyla); (*d*) geophilomorph centipede (Chilopoda); (*d*) the land-hopper, *Talitrus sylvaticus* (Talitridae, Amphipoda); (*f*) woodlouse, *Porcellio leavis* (Oniscidae, Isopoda).
The members of these groups are all more or less elongate and many legged. Apart from the chilopod (*d*), a predator of Oligochaeta (small earthworms, Enchytraeidae) and probably many other invertebrates, they feed primarily on decomposing plant material. The introduced millipede, *Ommatoiulus moreletii* may become so abundant as to be a household pest in parts of southeast Australia (Baker 1976). It has a smooth, hard integument and is able to force its way into refuges in grass tussocks and under stones and logs. In contrast, the more fragile Polyxenidae (*b*), are restricted to litter and the upper horizons of well structured soils. Symphyla and geophilomorph centipedes are dorso-ventrally compressed and very flexible and they move extensively up and down the soil profile through narrow passages (e.g. Edwards 1955). Most Crustacea (*e, f*) have little control over water loss and are much less successful on land than insects or arachnids. The bilaterally compressed morphology of terrestrial Amphipoda (*e*) is ill adapted to penetrating lower soil horizons or humid refuges on the surface. They are limited to areas of high rainfall and in dry weather are further restricted to accumulations of moist organic matter. The dorso-ventrally compressed Isopoda (*f*) are also confined to the soil surface but they are able to retreat to refuges under stones and, where such refuges are available, extend into the arid zone. The species illustrated (*f*) is exotic and is common in disturbed habitats. (Figs 40.1–3 by Lesley Howard.)

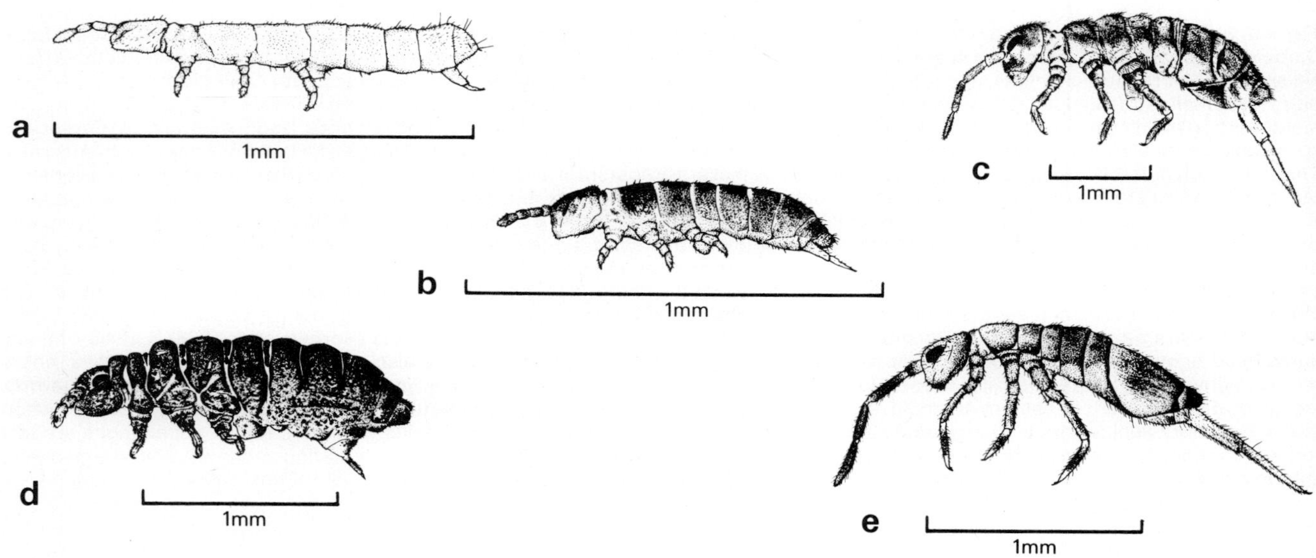

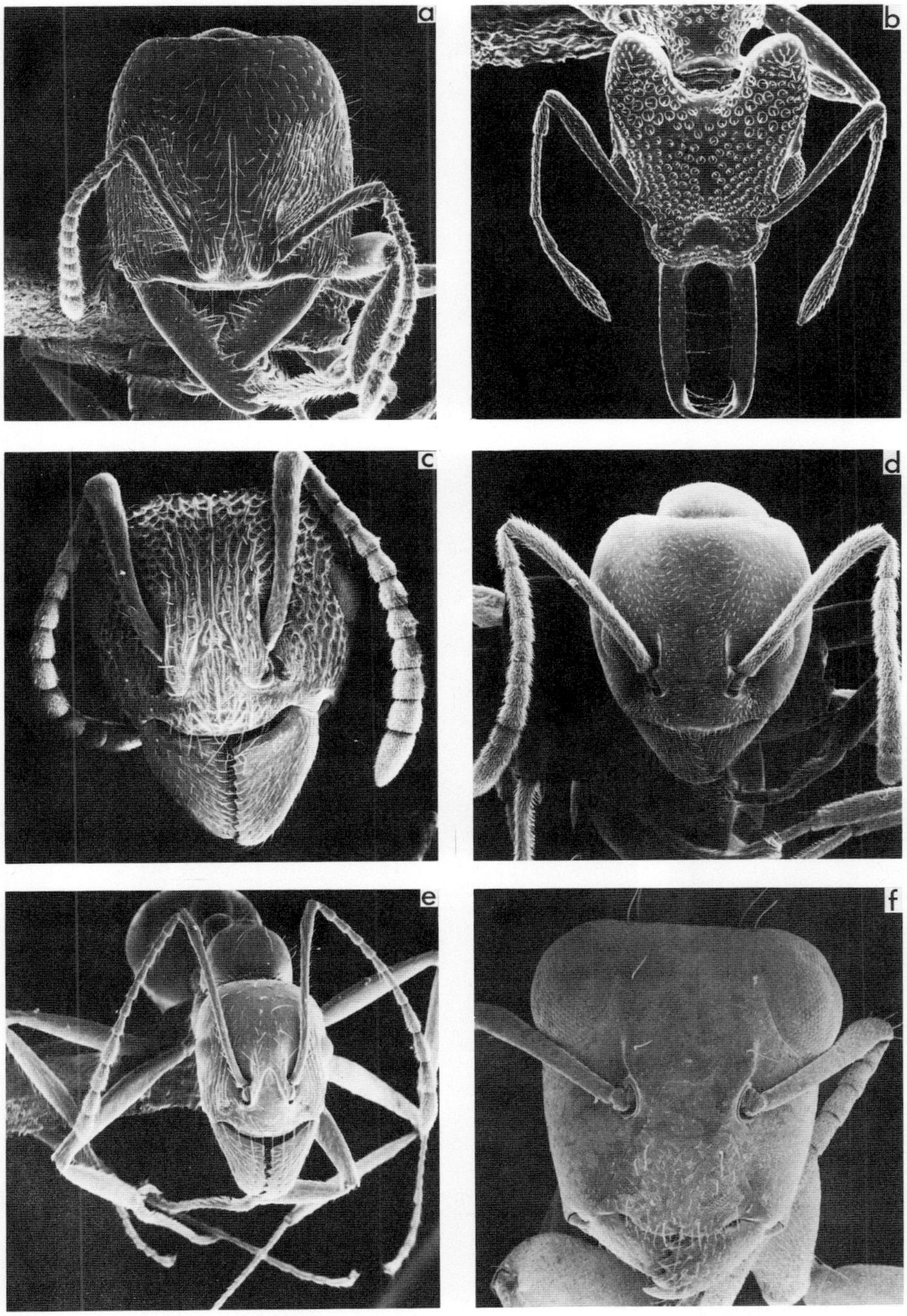

Fig. 40.5. (Left) Scanning electron micrographs of the heads of some commonly encountered Australian ants that are active in soil, litter and on the ground surface. (*a*) *Amblyopone australis*, a predaceous soil inhabitant in humid climates; (*b*) *Orectognathus robustus* from the litter layer of moist forests where it preys on microarthropods; (*c*) *Rhytidoponera metallica*, this, and very closely related species, are opportunist predators and scavengers and can be found throughout Australia; (*d*) *Iridomyrmex nitidus*, a species belonging to the dominant genus on the soil surface; (*e*) *Aphaenogaster nitidiceps*, members of this genus often nest in loose sands and sandy soils where they construct characteristic crater-like nest entrances; (*f*) *Opisthopsis haddoni*, a tropical species that nests under stones and in termite mounds. Photographs by R. W. Taylor and C. D. Beaton, CSIRO, Division of Entomology.

Fig. 40.4. (opposite) Collembola. With Fig. 40.6 this figure illustrates some of the main collembolan families and life forms. (*a*) *Isotomodes productus* (Isotomidae); (*b*) *Proisotoma brisbanensis* (Isotomidae); (*c*) *Acanthomurus* sp., (Isotomidae); (*d*) *Australella* sp., (Neanuridae); (*e*) *Drepanura* sp., (Entomobryidae). Scale lines, 1 mm. (*a*) – (*c*) illustrate a morphosequence within the Isotomidae. *Isotomodes productus* (Fig. 4*a*) is 'euedaphic' and is found in mineral soil; it is blind, white, with a cylindrical body and very short appendages. The *Proisotoma* (Fig. 4*b*) 'hemiedaphic', living higher in the soil profile, is pigmented, has ocelli and a stouter body and longer appendages. *Acanthomurus* sp., (Fig. 4*c*) which is 'epigaeic', living on the soil surface, is much larger with long appendages. (Fig. 40.4 by Jenni Thurmer)

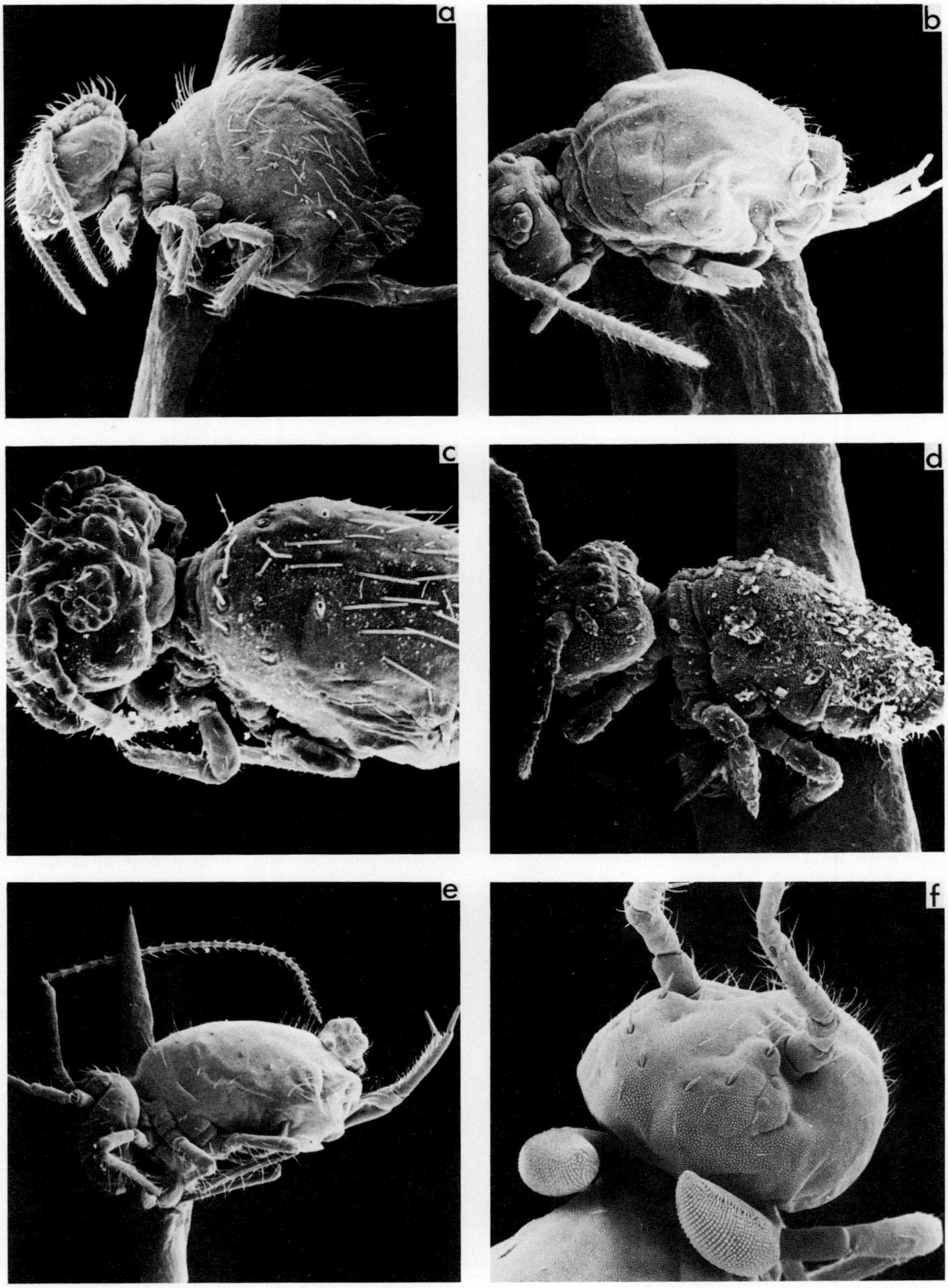

Fig. 40.6. Scanning electron micrographs of sminthurid Collembola from leaf litter. The actual specimens are 1–2 mm in length. (*a*) *Katianna australis*, found in the upper litter layer and on low-growing vegetation in southern humid climates; (*b*) Katiannini, an undescribed species belonging to a new genus found in leaf litter and on the soil surface in subhumid woodlands; (*c*) *Spinotheca santiagoi*, from lowland tropical vine forests; (*d*) *Neosminthurus* sp., *cf. mirabilis* from high altitude tropical vine forests, with Lepidoptera scales, microflora and mineral particles embedded in mucilaginous patches on the body surface and perhaps acting as camouflage; (*e*) *Temeritas regalis* the long antennae appear to be permitted by life on the surface of litter and logs in moist forests; (*f*) *Adelphoderia regina*, from the litter of very wet *Nothofagus* forests with neck organs that may contribute to respiration in waterlogged anaerobic environments.

Selective reduction in the diversity of the soil fauna also follows disturbances such as cultivation (Wallwork 1976) and again it is strongly influenced by gross morphology. Taxa that can survive the disturbance by retreating down the profile, for example Symphyla (Fig. 40.3) and some Collembola (Fig. 40.4*c*), exploit opportunities presented by improvements in soil structure and organic content caused by cultivation and may be abundant in arable land. Populations that are tied by their morphology to litter or the soil surface are liable to be destroyed by cultivation and their subsequent re-establishment depends on mode of locomotion and migratory ability (P. Greenslade and P. J. M. Greenslade, 1980).

Numbers of species

It is a truism that a few square metres of soil and litter under moderately productive vegetation will support several hundred arthropod species. This applies in Australia as elsewhere but here the resulting problems of species identification are accentuated by a lack of taxonomic work. The Australian invertebrate fauna is a continental one and most higher taxa contain a substantial proportion of total world species. An estimated 1000–2000 Australian collembolan species represent 10–15% of the world total. In the Coleoptera, the largest arthropod order, 70 of the families recognized in Australia by Britton (1970) contain species that are associated either with soil or with decomposition habitats that are an integral part of the soil system: dung, carrion, fallen fruits, decaying wood and fungal fruiting bodies. More than 19 000 species of Coleoptera have been described from Australia (Britton 1970), and allowing for undescribed ones, the total has been estimated at 33 500 species (Taylor 1976), 5–10% of the world total. Some families of Coleoptera are particularly well developed here; for example there are more than 2000 Australian species of Scarabaeidae (12% of the world total), the larvae of nearly all of which live in dung or the soil (Fig. 40.2*b*). Other families are not so well represented. For example, the Carabidae (Fig. 40.2*a*) which are important ground layer predators and scavengers in cool temperate climates seem to be limited by the great abundance of ants in Australia. Brown and Taylor's (1970) estimate of 1000–1500 Australian ant species or a tenth of the world total was a conservative one and on the basis of recent collections it could be increased three-fold. In almost all parts of Australia, excluding only the coldest climates and most disturbed habitats, more than 100 ant species are readily found in an area of a few square kilometres. Extrapolation of cumulative ant species plotted against cumulative samples suggests that local ant faunas, including all the rarer species, usually consist of 200–300 species (P. J. M. Greenslade and P. Greenslade, in press; Greenslade and Mott, 1979; Greenslade and Thompson, 1981). This compares with some 40 ant species in the whole of the U.K. In semi-arid mallee in South Australia 30–40 ant species can be recorded from an area of only a few square metres (although falling to a handful of species in adjacent arable land). Similar densities of species occur in other major groups of soil invertebrates. Plowman (1977) recognized totals of 93 apparent species or 'morphotypes' of cryptostigmatid mites (Fig. 40.1*d*) and 76 Mesostigmata (Fig. 40.1*f*) in two forest plots, each *c.* 150 m², near Brisbane.

The lack of previous taxonomic work on most groups of Australian soil invertebrates can be illustrated by two examples. The Meat ant complex, treated by Greaves (1971) as three colour forms of one species, *Iridomyrmex purpureus* (F. Smith), proves to consist of ten sibling species (Halliday 1978; P. J. M. Greenslade and R. B. Halliday, unpublished results). Among the Collembola, Salmon (1964) listed five Australian species in the sminthurid genus *Corynephoria*; Wallace (1974) mentioned at least 10 undescribed species in northern Australia and the total is now approaching 50 species (P. Greenslade 1977 and unpublished results). According to Taylor's (1976) assessment, only 45% of Australian insect species have been named and about the same proportion still remains to be collected (Fig. 40.7).

It is only possible to suggest partial solutions to the identification problem. To a certain extent the taxonomic process can be accelerated by omitting the alpha-stage of species description (Matthews 1975). Familiarity with one's species is essential and the ecologist may have to become his own taxonomist, although at the cost of narrowing the range of taxa that are studied. However, when the ecologist or soil zoologist is thoroughly familiar with a group this knowledge and a supporting reference collection form an invaluable instrument with which to carry out further research. When no names are available species can be referred to by code numbers. These should relate to a single national reference collection (Imai *et al.* 1977). All published ecological observations involving soil invertebrates need to be supported by adequate voucher material deposited in an institutional collection; this material also adds to that which is available for taxonomic study.

High levels of species diversity are found only in the native or near native vegetation that predominates over most of Australia despite grazing by rabbits, sheep and cattle. In areas more strongly modified by man, the 'culture steppe' (Matthews 1976), including urban areas, improved pastures, plantations etc., a reduction in diversity that we have already commented on in arable land is accompanied by the entry of exotic species. In various disturbed habitats near Adelaide exotic species formed 18–54% of the insect fauna (Gross 1954). A few native soil invertebrates increase in abundance in disturbed areas and some attain pest status, especially in improved pasture (Wallace 1970).

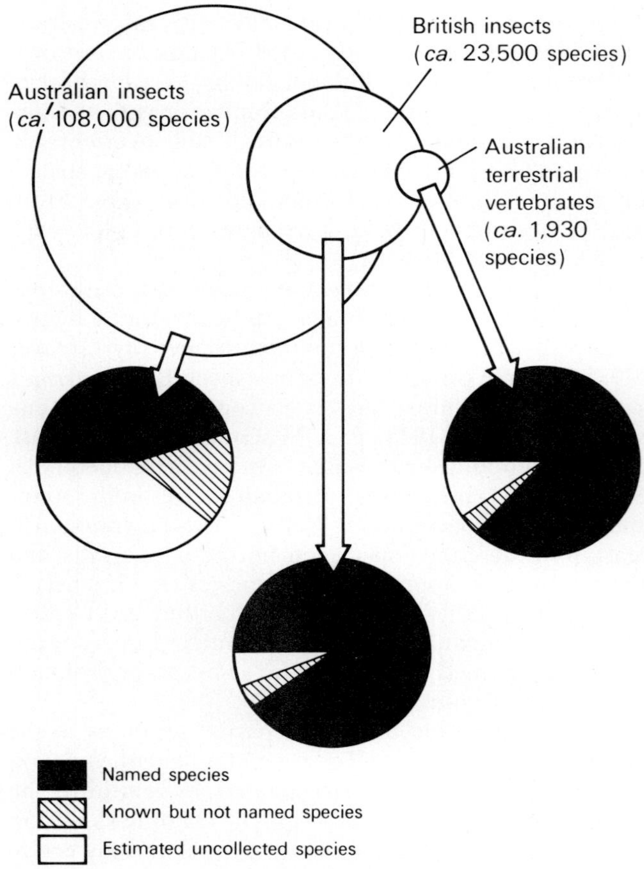

Fig. 40.7. Comparison of estimated overall sizes of Australian insect, British insect and Australian terrestrial vertebrate faunas (above), and levels of taxonomic knowledge concerning them (below). Knowledge of the taxonomy of most soil-inhabiting groups of insects in Australia is below the average for Australian insects as a whole. Other, non-insect arthropod groups living in the soil stand with the poorest known insects. Redrawn with permission from Taylor (1976).

Much of the ecological work that has been carried out on soil invertebrates has been centred on western Europe. Most of this area has to be regarded as culture steppe. Concepts and techniques developed from species-poor faunas there may be applicable in the Australian culture steppe but they are not necessarily relevant to less disturbed, species-rich soils under native vegetation.

Estimating populations

Sampling methods developed by insect ecologists (Southwood 1966) are generally suitable for ecological studies of soil invertebrates involving the macrofauna (e.g. larger soil insects, Diplopoda, Crustacea etc.), for intensive studies on single species and for most economic work in the culture steppe. But for extensive studies in native vegetation the diversity of soil invertebrates causes difficulties in estimating populations precisely. Despite research over many years (e.g. see Edwards and Fletcher (1971) and authors in P. W. Murphy (1962)) there have been few real improvements in the techniques that are available and they all possess serious deficiencies. The most widely used methods are extraction of the fauna from mineral soil or litter samples by flotation, a mechanical process used mainly for soil, or in various forms of Tullgren funnel (Edwards and Fletcher 1971), where invertebrates actively move out of the sample under the desiccating influence of heat (Vannier 1970). The labour of sorting, counting and identifying the extracted fauna always limits the scope of extensive investigations. As a rule, funnel methods are preferred since they are, on average, at least as efficient as flotation, the fauna is obtained in better condition for identification and the extraction process is less laborious. These funnels yield large numbers of individuals belonging to many different invertebrate groups but relative efficiency varies with such factors as soil type, season, taxon and life history stage, and the age structure of individual populations (Edwards and Fletcher 1971; Lasebikan *et al.* 1978; Takeda 1979; Tamura 1976). Efficient extraction is claimed for some groups by some methods (e.g. Petersen 1978) but efficiency can be extremely low (Fig. 40.8). It is also difficult, and usually impracticable, to determine any method's absolute efficiency when a soil fauna contains many species. Known numbers of a species can be added to and then extracted from sterile samples (e.g. Niijima 1971) but the effects of handling on behaviour, activity and mortality may lead to over or under-estimation of actual efficiency. Treated samples can be searched for individuals that failed to emerge (Takeda 1979) but this requires an additional check on the accuracy of search (Lussenhop 1971). This second method will tend to overestimate efficiency, while the first method usually underestimates it (Kempson *et al.* 1963; Marshall 1972); therefore the two used together should give a reasonable indication of absolute efficiency and a correction factor can then be applied to the results. Nevertheless, in a rich fauna these procedures would need to be repeated for many species in turn.

In soil and litter samples from arid areas, facultatively anhydrobiotic invertebrates present a special problem. During the normal process of extraction in Tullgren funnels some active individuals are extracted, while others become anhydric and immobile and are not extracted (P. J. M. Greenslade and P. Greenslade, in press). The best population estimate that can be obtained, which also distinguishes between active and inactive individuals, requires flotation of some samples and funnel extraction of others, both dry and moistened.

Alternative methods of assessing soil animal populations also have their limitations. Microscopical examination of soil sections impregnated with agar or gelatine

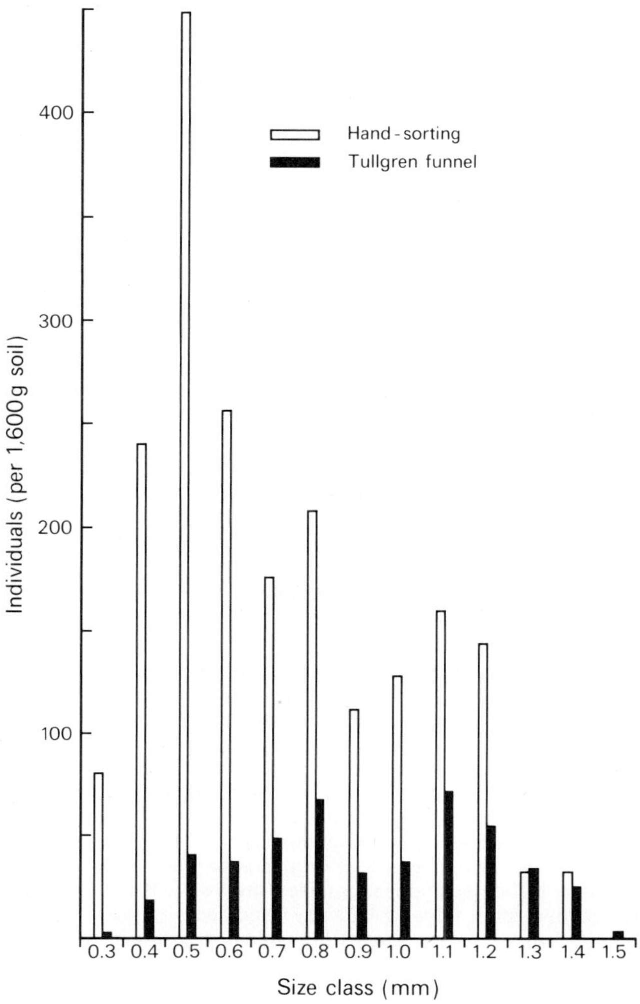

Fig. 40.8. Effect of method on efficiency of extraction. Frequency distribution of each size class at 0.1 mm intervals of a *Folsomina* species (Collembola, Isotomidae), per soil sample extracted in simple Tullgren funnels or sorted by hand. This shows how Tullgren funnels may greatly underestimate the actual population present in soil. And since funnels are biased towards the larger individuals, extraction errors will vary seasonally according to the age and size structure of the population being sampled. More elaborate Tullgren funnels may be more efficient but, strictly, the onus is on the user of any apparatus to demonstrate that this is the case. Taken with permission from Tamura (1976).

(Haarlov and Weis-Fogh 1953; Anderson and Healey 1970; Pande and Berthet 1973*b*) allows direct observations of the distribution of invertebrates in relation to the fine structure of the habitat. The method has been used in an ecological study of Amphipoda in Tasmania (Friend and Richardson 1977). Soil sections have shown that despite high densities of invertebrates per unit area (Chapter 38), individuals can be widely dispersed in relation to the total surface area of cavities in the upper part of the soil profile (Haarlov 1955). Using this method it may be possible to examine species' distribution in relation to microbial populations since almost the same techniques are used in microbial ecology (e.g. Chapter 41). However, the method is time consuming and limited by the range of soils for which it can be used (Pande 1975).

Many invertebrate groups that are active on the soil surface are effectively sampled with a suction apparatus (Wallace 1972). This method should give estimates of absolute active populations but results are subject to any hour to hour vertical migrations to and from the surface and a sorting or extraction process is still needed to separate animals from plant debris.

As a method for sampling surface fauna, pitfall trapping (P. Greenslade and P. J. M. Greenslade 1971; Majer 1978*a*) has numerous defects, but these have to be set against the disadvantages of, and time consumed by, other methods. Pitfall traps are still a commonly used technique for obtaining relative population estimates of surface fauna. Catches vary with the nature of the ground cover and the extent to which it impedes movement but this is not a serious problem in many Australian landscapes where a patchy vegetation exposes much bare soil. The catch of any one species also depends on the size of the active population, level of locomotor activity, size of animal and its behaviour. But it is claimed that by integrating these factors, catches of pitfall traps are a measure of a species' relative importance on the surface of the soil (P. J. M. Greenslade 1973; Thiele 1977).

Distributions in space

Patterns related to climate and soil landscapes

The gradient of declining rainfall from the coast to arid central Australia is accompanied by simplification of the soil fauna. Some higher taxa are lost and in most of the groups that do penetrate the arid zone the number of species is much reduced compared with areas of higher rainfall. This applies to Collembola (P. Greenslade 1978; P. J. M. Greenslade and P. Greenslade, in press), and to cryptostigmatid mites where the number of species on an arid site in South Australia was only a quarter of the number found in humid, coastal localities (D.C. Lee, unpublished results). These trends run parallel with a fall in primary production and, if we accept Spain and Hutson's figures (Chapter 38) as a broad indication of actual changes, a fall in the density of soil invertebrates.

Two important groups, the ants and termites, do not show these patterns. The density of ant species is rather constant over most of Australia (P. J. M. Greenslade and P. Greenslade, in press; Greenslade and Mott, 1979), although in the arid zone generic diversity is low and the number of species per genus is correspondingly high. Termites are abundant in central Australia (Watson *et al.* 1978) but are most prominent in open forest and woodland in the semi-arid and subhumid tropics (Lee and

Wood 1971). Apart from some information on ants and termites, very little has been published concerning the ecology of soil invertebrates in tropical northern Australia (see below). The diversity of ants and termites is reduced at the cool, wet end of the climatic range, especially on mountains in south-east Australia. And both groups are rather poorly represented in Australian vine forests, in contrast to their abundance in similar forests in the equatorial tropics to the north of Australia. In the ants this can be related to altitudinal and latitudinal temperature lapse coupled with low temperature maxima and a small amplitude of temperature fluctuation on the forest floor (Greenslade and Thompson, 1981; Taylor 1972).

A few invertebrate species, which appear to be closely adapted to certain soil types, have extraordinarily wide distributions on the appropriate soils throughout Australia. A notable example is the ant *Brachyponera lutea* found in coherent sandy soils. Its distribution extends from the Torres Strait Islands through eastern, central and southern Australia to south-west Western Australia.

Soils with extreme physical properties (e.g. cracking clays, loose sands, lithosols, gleyed soils and others subject to seasonal waterlogging) exclude many species but may support their own special faunas. Elsewhere, distributions are determined ultimately by soil acting through vegetation production and structure and its effects on the climate of the soil surface (Fig. 40.9). Within any locality the diversity of ants and Collembola generally increases with primary production although, as we have noted, ant diversity declines in the densest forests. Geographical and local patterns are interrelated. In humid climates for example, species of Collembola with arid affinities are restricted to the most open, best insolated habitats (P. Greenslade 1974).

Small-scale horizontal pattern

Within any one combination of soil and vegetation type the distributions of soil invertebrates are almost invariably aggregated in the horizontal plane. Aggregation reaches an extreme form among ants and those termites whose colonies are centred on mounds and subterranean nests, and among the many arthropod species with populations divided between discrete habitat units such as decaying logs. In open vegetation (hummock and tussock grassland, open woodland and shrubland), the alternation of litter and bare ground determines the distribution of surface-active and litter-inhabiting invertebrates (P. J. M. Greenslade and P. Greenslade, in press). On the floor of forest and woodland with a complete cover of litter, its heterogeneity (Chapter 38) is reflected in complex patterns of dispersion in the litter fauna and, to a lesser extent, in the fauna of the mineral soil. The nature of these patterns, their causes and the behavioural mechanisms by which they are maintained have been studied in the northern hemisphere (Joose and Verhoef 1974; Usher 1976). According to Usher, most aggregated distributions are vectorial in Hutchinson's (1953) system, determined by external influences such as moisture (Joose and Groen 1970) or the distribution of food or refuges.

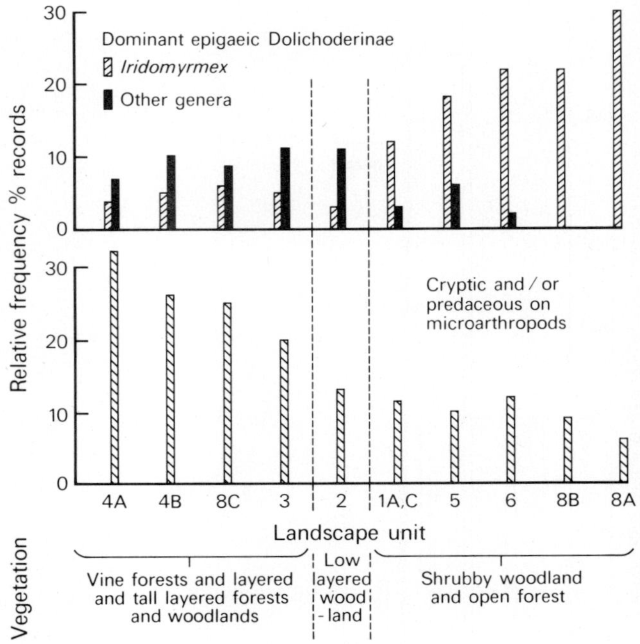

Fig. 40.9. Distributions of some major elements of ant communities in relation to vegetation structure on well-drained sites in the Cooloola area, southeast Queensland. Vegetation structure depends on soil and position in landscape. In turn it determines input of litter to the soil and the climate of the soil surface (Greenslade and Thompson, 1981). These landscape units fall into two main groups. In vine forests and the taller eucalypt forests and woodlands there is a continuous litter layer and soil surface temperatures are relatively low. Cryptic ants (below), involved as predators in the decomposer complex, are frequent and *Iridomyrmex* (above) tends to be replaced by allied genera *(Leptomyrmex, Dolichoderus)*. This represents the typical structure of ant communities in dense vegetation in humid, eastern Australia. In the second group of landscape units with lower, more open shrubby woodland and open forest, both tree canopies and litter layers give incomplete cover, soil surface temperatures are higher and the litter is liable to dry out. Cryptic ants are much less frequent and habitats are more suited to epigaeic *Iridomyrmex* species, active on the soil surface and vegetation. This type of ant community prevails in open vegetation in humid climates and throughout the rest of the continent.

Pattern may also be social, determined by signalling which can lead to spacing or clumping. It is probable that communication, and hence control over dispersion, often depends on pheromones, volatile secretions that modify the activity of other individuals of the same species, but the subject has hardly been investigated among soil animals other than social insects. In one pyrrochorid heteropteran, scent glands are known to produce pheromones that affect aggregation (Youdeowei and Calam 1969). In this and allied heteropteran families the degree of development of a metathoracic scent gland is greater among ground-living species than those living on plants (Schaeffer 1972). Grüm (1971*a,b*) has shown that population dispersion of a number of European Carabidae results from the interaction of locomotor activity, hunger and satiation, the availability of prey and the mosaic of physically more or less favourable habitat patches. The dispersion of Carabidae also depends on levels of activity as affected by the presence or absence of other individuals of the same species (P. J. M. Greenslade 1963) and in at least one species a pheromone is responsible for aggregation (Wautier 1970). Parisi (1972) and Verhoef *et al.* (1977) have reported aggregation pheromones in Collembola.

Other patterns are coactive, determined by interaction between competing species. They are well illustrated by ants. The workers are large in comparison with many soil invertebrates and much activity takes place on the soil surface. Interaction between different species is overt and frequently consists of hostile or evasive behaviour and patterns of exclusion and coexistence can be observed. Most of Australia is covered by a patchwork of colonies of mutually antagonistic, dominant species in the genus *Iridomyrmex*. Different dominants are associated with different subordinate ant species and these varying combinations affect the composition and diversity of the surface fauna as a whole (P. J. M. Greenslade and P. Greenslade, in press). Similar coactive patterns are described for Collembola in a subsequent section.

Vertical distributions

In humid climates many soil invertebrates are concentrated in the lower part of the organic horizon and in organic rich, well-structured mineral soil. The dependence of the fauna on the structure of the upper part of the profile has been mentioned previously. In many poorly structured soils of the arid zone a large part of the soil fauna's activity takes place at the surface, on and in a crust of fungal hyphae and lower plants. These soils could prove to be a valuable research tool since it is possible to observe and manipulate processes *in situ* that are normally distributed throughout a volume of soil and litter in areas of higher rainfall (P. J. M. Greenslade and P. Greenslade, in press). Soil samples can be stored in a dry state in the laboratory and biological activity can be induced by moistening the samples when required.

Much ecological work on soil invertebrates has focused on their role in the breakdown of leaf litter and other plant residues that are deposited on the soil surface. Apart from studies of pests, especially nematodes and the larvae of Coleoptera and Diptera, the invertebrates associated with roots have received relatively little attention, even though production by roots may exceed that of the above-ground parts of plants (Coleman 1976). Davidson (1978) describes root systems as 'the forgotten component of pastures'. Roots are now receiving more attention (see Coleman (1976), Estey (1977) and other authors in Lohm and Persson (1977)).

In Australia, the woody tissues of the older roots of trees and shrubs are attacked by larvae of Coleoptera allied to the wood-boring fauna of the aerial parts of plants, and by externally feeding larvae of hepialid and cossid moths (Common 1970; Matthews 1976). Australia has a large and poorly known curculionid fauna and adults of many species are found in litter (Fig. 40.2*c*). The larvae are probably associated with roots but nothing is known of their biology. The root-feeding larvae of two introduced curculionids (*Cosmopolites*, *Sitona* spp.) are pests of crops. Finer, actively growing roots, especially under grassland, pasture and annual crops, provide a variety of resources for soil invertebrates. The younger roots are a food source for ectoparasitic nematodes and chewing insect larvae. Soluble and mucilaginous exudates from root tips (see Chapters 41 and 42) form the base of a trophic system in which bacteria are prominent and they seem to be preyed upon extensively by Protozoa (Wallwork 1976). Higher levels of predators include nematodes, Collembola and mites. Many of the individuals in these groups are small in size and are active in a water film. The size distribution of the fauna of roots that penetrate poorly structured mineral soil may differ from that of the upper part of the profile. Here size of individual and abundance are inversely related and a double logarithmic plot of number of individuals against body length results in a straight line (Rapoport 1970). The size distribution of the fauna of root apices tends to be bimodal with very small forms on one hand and large, burrowing insect larvae on the other.

One question to emerge is the extent to which invertebrates that are dependent on roots and litter, and the processes to which they contribute, are functionally separate in ecosystems. Persson and Lohm (1977) suggest that soil-inhabiting predators form a link. This probably applies in grasslands where the soil surface consists of a mat of live and dead roots, stem bases and litter. Scarabaeid larvae that feed on soil organic matter may form an earlier link if their food is derived both from above-ground litter and from roots. In forests, roots often ramify in the litter layer and in vine forests they extend above the soil surface and penetrate logs. Unspecialized invertebrates may feed equally on leaf litter and its associated microorganisms or on roots and rhizosphere organisms. This factor should be taken into account when populations of invertebrates thought to be responsible for the breakdown of litter are related to litter input. There is likely to be a much clearer distinction between these two elements of the soil fauna (i.e. depending either on roots or on litter) in arid ecosystems, especially open shrublands where leaf litter is concentrated on the soil surface immediately below the plant that produces it and there is little organic matter in the soil apart from roots. Vertical migration by the soil fauna is negligible except for movement to and from refuges.

The only root-litter link will be any predators, probably ants, that feed on the litter and surface fauna and on adults of root-feeding insects that leave the soil to migrate (Fig. 40.12*b*).

Distributions in time

Seasonal rhythms and the influence of rainfall

The seasonal life histories of a number of soil invertebrates that are pests of crops and pasture have been described in Australia, but little is known about seasonal change in the density and composition of the soil fauna as a whole. The only detailed studies are those of King and Hutchinson (1976) in pastures and of B. R. Hutson (unpublished) in woodlands near Adelaide. The climate of the latter area is Mediterranean with winter rainfall and hot summers. On one of Hutson's sites Baker (1976) regularly sampled a population of the millipede *Ommatoiulus moreletii* (Fig. 40.3), and some information is available on ants (Fig. 40.10). The predictable climate of this area is reflected in predictable seasonal fluctuations in the abundance and activity of the soil fauna. However, there is month to month variation about the overall seasonal trend, for example in density of microarthropods, which is still seen in Fig. 40.10*c*, *d*, when 2 year's results are averaged. In studies such as this, sampling must be carried out with intervals of no more than a month between samples. Less frequent sampling, say quarterly, would reveal statistically significant differences between sampling occasions but could lead to quite misleading conclusions as to the nature of seasonal fluctuation. Fig. 40.10 also shows how different sampling methods

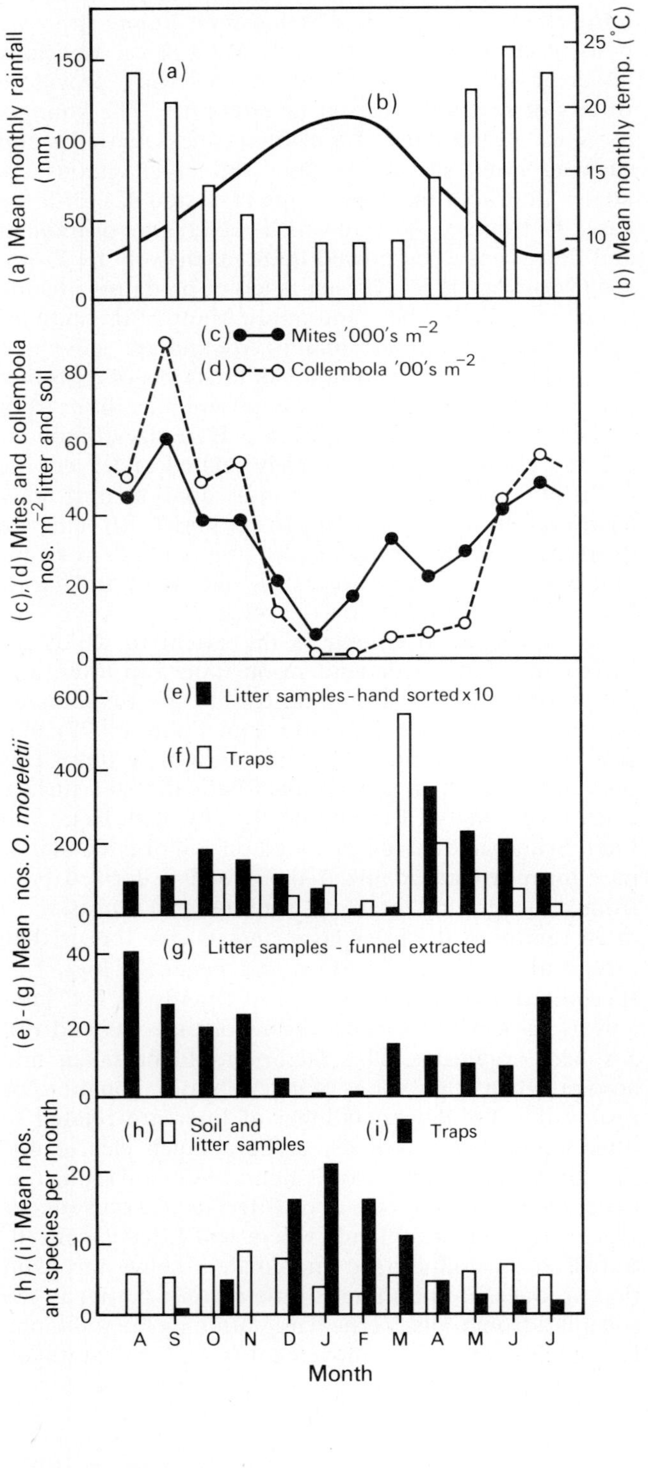

Fig. 40.10. Seasonal trends in the numbers and activity of some invertebrates in open woodland at Bridgewater, Mt Lofty Ranges, South Australia, in relation to mean monthly rainfall (*a*) and mean temperature (*b*). (*c*), (*d*) Mean numbers (per m^2) of mites and Collembola in litter and soil samples (0–8 cm), March 1975 – January 1977. (*e*) Mean numbers of the millipede *Ommatiulus moreletii* per m^2 (from hand-sorted litter samples, 10 x 1 m^2 samples per month), September 1974 – November 1975. (*f*) Numbers of *O. moreletii* in pitfall traps (224 trap days per month), October 1972 – January 1974. (*g*) Mean number of *O. moreletii* per m^2 in Tullgren funnel-extracted litter samples (12 x 0.1 m^2 samples per month), March 1975 – January 1977. (*h*) Mean number of ant species in soil and litter samples per sampling occasion, March 1975 – January 1977. (*i*) Number of ant species per 224 trap-days, October 1972 – January 1974.

Trends in the density of microarthropods (mites (c), and Collembola (*d*)) are closely related to regular seasonal changes in rainfall (*a*), temperature (*b*) and soil moisture; maxima and minima in soil moisture follow rainfall with a lag of about a month. The highest microarthropod densities were obtained in the winter months and lowest during the summer. In contrast, the numbers of ant species active on the surface and taken in pitfall traps (*i*), varies with temperature although there was only slight variation in the number of species extracted from soil and litter (*h*) by Tullgren funnels. Changes in the density of the millipede *O. moreletii*, in funnel-extracted litter samples (*g*), were similar to those shown by the microarthropods with a winter maximum and summer minimum. But in hand-sorted litter (*e*), there was an autumn maximum and a winter decline; very similar trends were indicated by pitfall traps (*f*) with a major burst of locomotor activity in the early autumn. It is unlikely that this species' seasonal pattern changed between 1972–75 and 1975–77. More probably, funnel extraction and hand sorting differ in the efficiency with which they sample different age or size classes or populations differing in levels of activity. Whatever the explanation, this Figure illustrates the contrasting results that can be obtained by using different sampling methods. We are grateful to Dr G. F. Baker and Dr B. R. Hutson for access to unsorted ant samples and unpublished results.

produce different results, in this case both for millipedes (Fig. 40.10(*e*)(*g*) and ants (Fig. 40.10(*h*)(*i*). Fig. 40.10 distinguishes between ants whose population activity on the surface varies with temperature and the microarthropods in soil and litter whose numbers vary with rainfall. This distinction is important in the drier parts of Australia, particularly the central and southern arid zone. Mean annual rainfall is rather evenly distributed over the year but the actual amount received in a locality in any one year and its distribution are unpredictable. Part of the soil fauna (e.g. ants, spiders, some Collembola and predatory mites), is independent of rainfall in that activity on the soil surface occurs throughout the year. The rest of the surface fauna, associated with litter, includes anhydrobiotic nematodes and Collembola and is only activated by rain which also stimulates microbial activity. The scarcity of specialized, cryptic predators in arid areas can be seen as the product of three factors: the filtering effect of a severe climate on some groups; the narrow resource base of the decomposer food web caused by low levels of primary production; unpredictability in the availability of prey. Thus decomposer invertebrates escape the attention of specialized predators.

Daily rhythms

The timing of above-ground activity is an important aspect of community organization among ants and differences in time of foraging allow potential competitors to coexist in the same area (P. J. M. Greenslade and P. Greenslade, in press). In warm to hot weather three sets of species can be recognized: diurnal ants active in the morning and at dusk including dominant *Iridomyrmex*; exclusively diurnal species whose activity requires high temperatures; nocturnal species. In cool weather, the second group and many species in the third group are inactive, limited by low maximum and night temperatures respectively. The first group and part of the third group consist of species that are more active in the middle of the day. There has been very little work on species' water and temperature relations in the context of foraging activity and it is not known whether most rhythms are innate or dependent on day to day weather changes.

Succession

Usher and Parr (1977) distinguished two types of successional change among soil invertebrates. The first accompanie seral change in plant communities. Included here are the effects of change in the plant community that follow profile development on newly exposed surfaces such as sand blows or during rehabilitation of landscapes. This topic is mentioned again later. The second type of succession is change in the invertebrate community in the course of decomposition of organic material that is induced by the decomposers themselves. A third type can be added, the succession of organisms on aging roots.

The main successional habitats of the second type are logs and sticks, dung, carrion, leaf litter and dead roots. Others include fallen fruits and decaying fruiting bodies of the higher fungi. In Australia the activities of termites often dominate decomposition of wood. Their role has been studied by Lee and Wood (1971) but not in a successional context, and it is not known how the termites interact with other organisms. Invertebrate succession in logs has been described briefly for vine forests in New Guinea and the Pacific (Wilson 1959; P. J. M. Greenslade 1972*a*). In the latter area, change in the log habitat was rapid at first and then proceeded more slowly; an initially specialized and rapidly changing log fauna was eventually replaced by species that also occurred in soil. The replacement of log specialists by more generalized species as decomposition proceeds has also been commented on by Wallwork (1976). A substantial amount of information has accumulated on the biology of invertebrates that inhabit dung in Australia (e.g. see Matthews 1976), although succession has not been emphasized. Succession in vertebrate carrion has been studied by Bornemissza (1957). Invertebrate carrion is usually preempted by ants and all but heavily sclerotized tissues are immediately consumed. The scavenging efficiency of ants was demonstrated by Doube (1967) who showed that workers of the Meat ant, *Iridomyrmex purpureus sens. strict.*, visit all parts of the colony's territory several times a day. Consequently, most invertebrate carrion is discovered by ants within hours or even minutes.

In the northern hemisphere, faunal succession on fallen leaves consists of initial colonization by a few fungal-feeding species that do not directly attack the litter itself. Subsequently, the diversity of invertebrates and their activities increase (Harding and Stuttard 1974; Wallwork 1976) and there are similar changes among microorganisms (Swift 1976). However, there are indications that early colonizers among isotomid Collembola may be able to feed on dead plant material directly or at a very early stage of microbial succession, and this is discussed below. In a deep extensive litter layer the sequence of decomposition and succession forms a vertical zonation. Where the litter layer is sparse because of rapid decomposition (as in tropical vine forests at low latitudes and elevations), or low rates of litter fall (as in some arid zone ecosystems), the fauna is distributed in a mosaic in a horizontal plane.

Wallwork (1976) and Martin (Chapter 42) suggest that as growing roots become older there is a succession of rhizosphere microorganisms and invertebrates. As the root ages, specialists at the mucilage-exudate stage are replaced by forms that are more widely distributed through the soil. Wallwork notes the analogy to replacement of specialists with time in decaying logs.

The soil fauna can be seen as a horizontally layered matrix of widely distributed decomposers and their

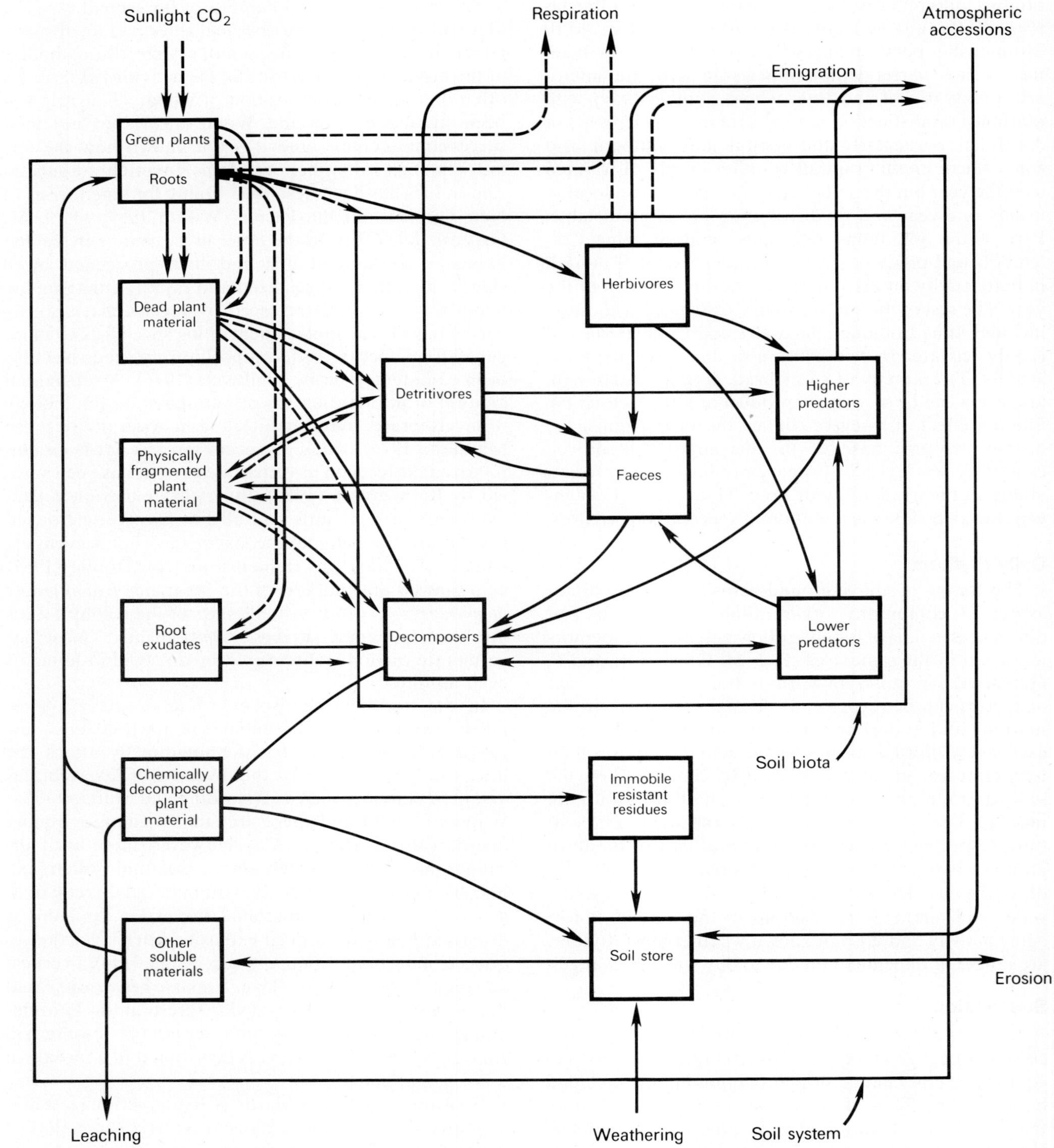

Fig. 40.11. Conventional diagrammatic representation of soil processes involving biota and organic matter: boxes, stores; arrows, flows of energy (– – –) and material and plant nutrients (——). The chief decomposition pathway is thought to be ingestion of dead plant material by soil fauna, egestion in a physically fragmented condition and eventual chemical breakdown by microorganisms (Chapter 38). Such diagrams are unsatisfactory simplifications in that they fail to emphasise several important points: the necessity for prior attack by microorganisms before most dead plant material is acceptable to the majority of 'detritivores'; the operation by many species simultaneously at several different trophic levels (here represented by biota 'boxes'), or their ability to move between levels; the series of steps which has been condensed to 'higher' and 'lower' predators; the distinction between live and dead food of animal and microbial origin (Weigert *et al.* 1970). However, such diagrams form a basis for comparing the functional roles of different species or higher taxa (see Fig. 40.12.)

predators adapted to later stages of decomposition at increasing depths. It is interrupted by discrete units such as wood fragments and roots that change with time at various rates and are continually disappearing and being replaced. Initially they alter rapidly and support specialized faunas. These are gradually supplanted by widely distributed species from the surrounding matrix adapted to middle and later stages of succession.

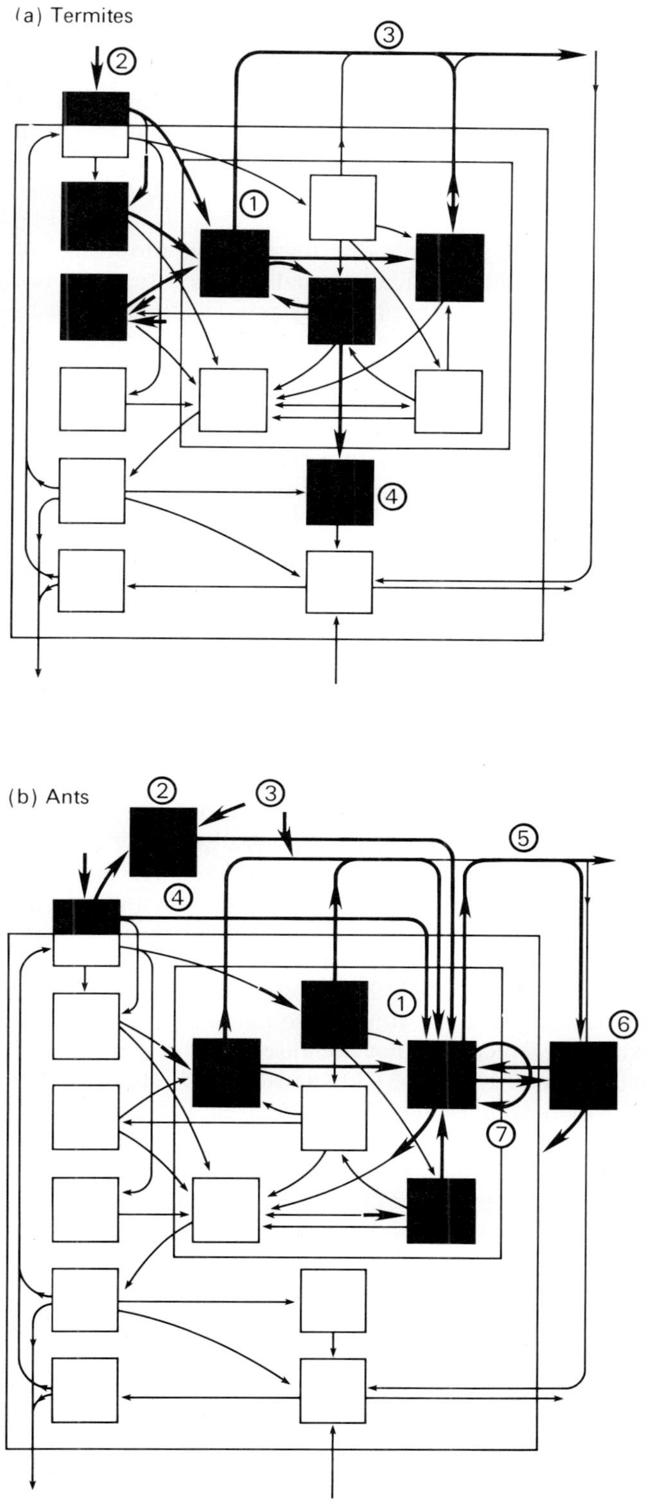

Fig. 40.12. Transfers of materials and plant nutrients involving (*a*) termites, (*b*) ants, (bold arrows and solid boxes): (*a*) 1, Termites; 2, seeds and dry grasses; 3, release of winged sexuals; 4, organic matter bound up in the structure of nests and galleries. (*b*) 1, Ants; 2, prey captured on vegetation; 3, migrants from other habitats; 4, seeds, honeydew, nectar and other exudates; 5, release of winged sexuals; 6, other higher predators; 7, predation by ants on other ants.

Structure of communities

Trophic relationships

Fig. 40.11 is a much simplified summary of the process of breakdown and decomposition of plant material as described by Spain and Hutson (Chapter 38). It stresses the central importance of invertebrate faeces for it is thought that organic matter becomes progressively modified as it undergoes a series of ingestions and egestions by different species (Crossley 1977; Webb 1977). Very often we are uncertain as to what food items are actually digested by soil animals and this is dealt with in the next section. Only total soil respiration is shown, in part to simplify the diagram but also to stress the difficulties of separately measuring or estimating metabolic activity due to roots, fauna and microorganisms.

Fig. 40.12*a* illustrates Lee's account of the activities of termites in the soil system. It shows the chief outputs from a typical termite population as faecal material incorporated in nest structures and transfer of biomass to higher predators, both by way of worker termites and through the very large production of winged sexual forms. Wood (1976) cites figures for sexual production equivalent to *c*. 50% of the total biomass and energy content of parent colonies.

Whereas many termites delay the process of decomposition by locking up organic matter in long-lived mounds and nest structures, ants have an accelerating effect. As dominant predators they can influence virtually all other arthropod groups in and above the soil (Fig. 40.12*b*), although herbivores such as Scarabaeidae are preyed upon mainly as emigrating adults. With predation on other emigrants such as termite sexuals and on immigrants, this represents retention and capture of invertebrate biomass by the ecosystem. Petal (1978) considers that through their scavenging efficiency, ants accelerate the return to the soil of plant nutrients contained by invertebrate biomass. Ants also accelerate the transfer of material from the aerial parts of plants to the ground layer by preying on above-ground herbivores and by feeding extensively on honeydew and plant exudates. D. H. Murphy (pers. comm.) emphasized the importance of exudates as an alternative means of transfer of material to the decomposer complex both above and below the soil surface and in aquatic as well as terrestrial systems. A major source of loss from ant populations is the release of winged sexuals, as it is in termite populations, and the biomass and energy output from individual colonies may be of the same order of magnitude (Nielsen and Josens 1978). In a stable ant population of long-lived colonies that saturate their habitat, none of these sexuals will be involved in the successful founding of new colonies. Most are captured by predators (Greaves and Hughes 1974). Although ants must be a significant element in the diet of many insectivorous vertebrates in Australia, predation by ants on other ants is also significant.

From Fig. 40.12 it is clear that ants and termites have very different and indeed complementary roles in the soil system and they depend on quite different resources. The only shared resource is seeds but we do not know to what extent the supply of seeds is limiting to either group. Ants certainly prey on termites but we have no evidence that they exert any quantitative influence on Australian termite populations. There may be some interaction when ants nest in termite mounds but the indications are that neither group is in any way dependent on or much affected by the other (Holt and Greenslade 1979). Together, ants and termites are capable of dominating most of the biological processes shown in Fig. 40.11 but with the exception of those involving roots (Fig. 40.12.) Roots do not appear to be an important food source for termites in Australia (Lee and Wood 1971) and although a few ants, notably species of *Acropyga*, culture Homoptera on roots, they are infrequent. Thus over much of the northern half of Australia, where both ants and termites are abundant, the soil fauna seems to consist of three main elements which hardly interact: ants, termites and, presumably, a third element associated with roots about which we have no information.

Specialization of feeding habits

There is doubt as to the exact nature of the food of many soil invertebrates and the relationship between what is ingested and what is digested. This can be illustrated by the Scarabaeidae. Larval food ranges from mainly soil organic matter in *Rhopaea* and *Anoplognathus* species (Davidson and Roberts 1968*a, b*), which are therefore detritivores in Fig. 40.11, to fresh leaves of pasture species in the later larval instars of *Aphodius tasmaniae*; larvae of this species excavate burrows in the soil from which they emerge to forage on the surface at night (Carne 1956). *Sericesthis* larvae occupy an intermediate position and are soil herbivores, feeding on the roots of pasture grasses. However, they can survive and develop in the absence of living plants, selectively ingesting organic matter (Ridsdill-Smith 1975). There is also evidence which indicates that microorganisms in the rhizosphere and decomposing organic matter contribute to larval nutrition. The materials that are ingested rank in digestibility: microorganisms > roots > other organic matter (R. L. Davidson pers. comm.). If microorganisms do prove to be a significant element in the diets of many of these scarabaeid larvae they could be placed in the 'lower predator' box in Fig. 40.11 rather than the 'herbivore' and 'detritivore' boxes. The differences in feeding habits that, according to Matthews (1976), may account for the coexistence of several Scarabaeidae at any given time and place, are then resolved into preferences for particular microorganisms.

The food of the mites, especially Crypostigmata (Fig. 40.1*d*), and Collembola that form a large proportion of total invertebrate individuals and species in soil has been studied in the northern hemisphere. The main objects have been to define these invertebrates' functional role and to determine whether coexisting species are separated by specialized feeding habits. Much of this work has been based on analysis of gut contents and on laboratory trials of species' preferences, survival and reproductive performance on different foods. However, feeding trials may have little relevance to the more complex field situation (Mitchell and Parkinson 1976), and gut contents commonly reveal what is ingested but cannot be digested (McMillan and Healey 1971). Consequently results have been contradictory.

Many Cryptostigmata only seem able to utilize microbial growth on the plant material that they ingest (Wallwork 1967), but some are equipped with enzymes to decompose plant structural polysaccharides and lack the enzymes to exploit microorganisms (Luxton 1972). Feeding by Cryptostigmata has been described and reviewed within the last 10 years by Wallwork (1970), Luxton (1972), Pande and Berthet (1973*a*), Harding and Stuttard (1974), Anderson (1975), Mitchell and Parkinson (1976) and Behan and Hill (1978). Luxton's account remains the most comprehensive.

Hale (1967) stated that Collembola ingest organic matter only for the fungi it contains and it has been widely accepted that they are unselective in their feeding behaviour (Paclt 1956; Christiansen 1964; Healey 1970; McMillan and Healey 1971; Petersen 1971; McMillan 1975). On the other hand, there are indications that many soil-inhabiting Collembola are specific in their food requirements and that some Isotomids feed directly on dead plant material. There is considerable uncertainty as to the actual food of many Collembola and here we examine their feeding habits in some detail.

There are several reasons for the idea that soil Collembola are generalized feeders, primarily on microorganisms.

(1) With the exception of an aberrant, predacious isotomid from North America, most species (i.e. the members of all but one of the seven collembolan families, the Neanuridae) have very similar mouthparts.

(2) In the laboratory, many Collembola will feed on a great variety of materials and are often observed to ingest the culture substrate. Field populations commonly contain mineral particles in the gut, implying that feeding is quite indiscriminate.

(3) Gut contents can be particularly misleading if digestion takes place rapidly in the foregut. This is suggested by lack of change in the structure of the gut contents from one end to the other. Gilmore (1970) killed and examined individuals of a *Sinella* species immediately after feeding them with nematodes, but in only a few individuals was he able to detect the nematodes in the gut. It would appear that rapid action of digestive enzymes had altered the optical properties of the nematode cuticle in the foregut.

(4) Most work on feeding preferences has used common species of the European and northern American culture steppe. They tend to be widely distributed and are likely to be widely adapted in other respects such as food preferences.

(5) There have been misconceptions as to the action of the mouthparts and these are slowly being resolved by detailed studies of their functional morphology and musculature (Manton 1964; Goto 1972; Adams and Salmon 1972). In most Collembola, i.e. all except Neanuridae, an expansion of the base of the mandibles is commonly referred to as the 'grinding mandibular plate'. However, gut contents generally consist of discrete particles such as whole pollen grains, fungal spores, unicellular algae, yeasts and lengths of fungal hyphae, and Goto (1972) proposed that the mandibular plates have a raking rather than a grinding action. Only in a few Isotomidae (*Folsomides*, *Proisotoma* spp.), are the gut contents exceptionally finely divided.

From a survey of the literature and our own observations, making allowance for the five points above, the following broad generalizations can be made on the food of the main families of Collembola other than the Neanuridae.

Isotomidae: probably plant litter (see notes below), some live higher and lower plant material, to a lesser extent fungal hyphae and spores.

Entomobryidae: mainly fungi, pollen and algae; nematodes in a few species.

Onychiuridae: variable, including nematodes, fungi, plant roots.

Hypogastruridae: very variable between species; algae, fungal hyphae and spores, bacteria associated with vertebrate and invertebrate carrion, autolysing fungi and dung.

Sminthuridae: live plants and the microflora on them, pollen. Species of *Sphaeridia* and other litter and soil-inhabiting genera may feed on fungi.

There is no reliable information on the food of Neelidae or Paronellidae.

Reasons for suspecting that some Isotomidae feed on litter include the physically fragmented condition of material in the guts of *Proisotoma* and *Folsomides* species and observation of direct attack on leaf litter in culture. Species of these two genera seem to be the only organisms in at least some arid landscapes that are both sufficiently abundant and apparently capable of disposing of the annual production of leaf litter. The gut microflora of the isotomid *Proisotoma minuta* produces a cellulase (von Torne 1976) and, using litter bags on land reclamation sites in England, Hutson (1980) correlated rates of litter disappearance with the population biomass of *Isotoma* species.

There are a number of records of Collembola feeding on plant roots (e.g. Spencer and Stracener 1929) but most accounts fail to demonstrate that healthy roots are attacked and that plant material rather than micro-organisms is actually digested.

Fungi are clearly an important food for many Collembola and under experimental conditions different fungal species vary in attractiveness and different fungal-feeding collembolan species vary in their abilities to reproduce on different species of fungi (Mills and Sinha 1971). Booth and Anderson (1979) have shown, in culture, how the quality (nitrogen content) of fungal food can affect the growth and reproduction of an isotomid.

These Collembola, whose mouthparts all share the same basic morphology, represent many life forms and they occur in a variety of different habitats which must differ greatly in the food that is available. In contrast, all neanurid species are active in essentially the same habitat (amongst moist, finely divided, decomposing organic material under the bark of logs and sticks and in the soil) but they exhibit a much greater range of variation in the structure of the mouthparts. Commonly, the mandibles are reduced or absent, while the maxillae may be elaborated and bear complex fringed lamellae or toothed plates. The mouthparts of some species with styliform maxillae could be described as piercing. Many if not all species feed through a water film and, following Wolter (1963), their mouthparts have often been described as having a sucking action. However, Adams and Salmon (1972) describe a scooping-lapping action in *Brachystomella*. There have been few studies of the feeding biology of the Neanuridae, probably because of the difficulties that exist in maintaining cultures on standard laboratory diets such as yeast. This implies specialized food requirements (Adams and Salmon 1972). Brachystomellini appear to feed on fungal spores (Womersley 1939) and possibly bacteria, nematodes and Protozoa. Frieseini are predators of nematodes, rotifers and arthropod eggs and may be cannibals in culture (Cassagnau 1952; Macnamara 1924; Petersen 1971). Maritime species of *Anurida* and *Pseudanurida* feed in or on decomposing animals such as molluscs in the intertidal zone (Joosse 1976; D.H. Murphy 1971). In culture a *Neanura* species would not feed on sterilized organic matter and preferred fungal spores to hyphae and decayed wood to yeast, sugar or starch (Singh 1969). When *Ceratrimeria* and *Neanura* species are kept amongst the material from which they are collected their feeding behaviour appears to involve selective searching.

Soil inhabiting Collembola as a whole exploit a wide variety of foods: living higher and lower plant material, litter, microorganisms and invertebrate prey. At least some species and higher taxa are specialized feeders. One factor that contributes to the diversity of insects is coevolution with food plants (Matthews 1976). This will also account for some of the diversity of soil invertebrates, such as Collembola, providing they feed on specific microorganisms (Pimental 1968). Were microarthropods indiscriminate feeders, this route to diversification would be closed.

Indirect relationships

Vlijm *et al.* (1974) distinguished two types of relationship between species of soil animals: direct and indirect. Direct, trophic relationships have been discussed in the two preceding sections. Vlijm *et al.* regard direct relationships between 'key species', the most abundant members of the fauna, as the community's 'skeleton' (Elton 1966). Indirect relationships, described by Elton as the community's 'girder system', 'damping down fluctuation and slowing down deviation from the norm' are interfering activities, notably competition, by which different species affect one another so that food chains are inter-connected. Typically, communities of organisms have a structure of a few abundant key species and a larger number of scarcer ones (Williams 1964). As a rule, the key species contribute 15–40% of total individuals (Elton 1966). It is through indirect relationships that the less abundant species can play a part in the functioning of the community. Abundance and functional significance are not necessarily correlated and many successful agents of biological control show how a rare species can determine the structure of a community. A familiar example outside the soil system is provided by scarce moths of the genus *Cactoblastis* which maintain control over Prickly Pear.

One would expect interspecific competition to be important in population processes of soil invertebrates living in species-rich communities in an equable environment. Lee and Wood (1971) inferred that there is competition between some Australian termite species from the dispersion of their mounds. Interspecific competition is certainly important amongst Australian ants. An emerging picture of this competition (Greenslade 1979) reveals two elements of the community's girder system: diffuse competition and the activities of unspecialized, opportunist species.

In diffuse competition (Terborgh 1971) a constellation of species can more easily outcompete a ‡d displace a competitor than one species alone (MacArthur 1972). The success in competition of dominant *Iridomyrmex* species depends on their foraging efficiency and high population densities within colonies. Access to a large supply of resources is needed to maintain high densities and this entails generalized feeding habits. Ultimately this will bring a dominant species into competition with other less abundant but more specialized species. Individually they may be inferior competitors but within their specializations they will be more efficient than the unspecialized dominant, and together will outcompete it. The Meat ants, *Iridomyrmex purpureus* and allies, the leading dominants, show specializations of size and daily foraging rhythm that can be seen as devices to avoid diffuse competition with other ants without reducing the range of resources that are used. The workers are large, 8–9 mm in length, whereas workers of the majority of Australian ants are smaller, only 2–5 mm long. A bimodal foraging rhythm avoids interaction with species that are exclusively nocturnal or active in the hottest part of the day. Diffuse competition seems to be one of Elton's (1966) 'strong ecological and evolutionary forces against monopoly'. Without this constraint, excessive dominance could result in whole communities being affected by population fluctuation by one or a few species.

The other element of the girder system is the presence of opportunist species. An ideal opportunist ant is unspecialized, with a flexible foraging rhythm, variable worker size and wide food range, often including seeds. In established ant communities opportunists rarely become so abundant as dominant *Iridomyrmex* despite the variety of resources to which they have access, presumably because of diffuse competition. They often nest between territories of dominants, are found in temporary and disturbed habitats, especially in the culture steppe, and expand their populations in response to the ephemeral increase in the availability of food that follows effective rain in the arid zone. In these ways opportunists exploit under-utilized resources. They also bear the brunt of intensified competition if the supply contracts. Their presence allows the community as a whole to accommodate environmental change with little disturbance of its integrated core of dominant and associated species.

Similar behaviour by opportunists in the rest of the soil invertebrate community would allow it to keep pace with changing patterns in the distribution of resources. Anderson and Healey (1972) postulated that this instability of the habitat might allow 'fugitive' species (Hutchinson 1951) of Collembola to coexist with superior competitors by moving between temporarily unexploited patches of their habitat. Most studies of competition between soil invertebrates, other than ants, have been carried out in Europe and north America on simple, two-species interactions among Collembola (Christiansen 1967; Culver 1974; Longstaff 1976) and cryptostigmatid mites (Anderson 1978) under controlled experimental conditions.

It is difficult to apply the results of this work to a diverse field community where any one species is liable to interact with a patchwork of others at varying densities and in varying combinations. Intuitively the idea of diffuse competition seems more useful. Two sets of field observations on Collembola include both opportunist activity and diffuse interactions.

In Polish pine forests, Kaczmarek (1975*a,b*) found that the distributions and densities of 'oligovalent' collembolan species, i.e. specialized in respect to habitat, depended to a large extent on moisture. The pattern of density of 'polyvalent' species, with less specialized habitat requirements, varied inversely with the density of the more specialized ones. Kaczmarek concluded that the oligovalent species had a limiting effect on the unspecialized species. Their distributions could not be accounted for in terms of any one oligovalent species and so diffuse competition is indicated.

We have made similar observations on isotomid Collembola in the Solomon Islands (P. Greenslade and P. J. M. Greenslade 1980). On each of three sites there were four isotomid species forming a morpho-sequence of increasing adaptation to life in mineral soil. They all appear to be opportunists. They have wide geographical distributions and are usually found in species-poor, disturbed habitats where they sometimes reach very high densities. Over the three sites, the relative abundances of each of the four species could be explained by two processes, neither of which provided a sufficient explanation by themselves: the same influences that affected the rest of the soil fauna (e.g. disturbance, the nature of the vegetation cover); diffuse interaction with other soil invertebrates. Whenever the rest of the community was poorly developed, these isotomids' different, but overlapping, habitat requirements allowed one or more of them to exploit an opportunity at any level in the upper part of the soil profile. The term 'interaction' is more appropriate than competition here since we cannot distinguish between various activities of other species that may be injurious to isotomid populations, the continuum ranging from predation through interference to exploitation competition for resources that are in limited supply.

Habitat and selection processes

Both the fugitive and opportunist species referred to in the preceding section are synonymous with *r*-strategists in MacArthur and Wilson's (1967) *r–K* spectrum (Southwood 1977). MacArthur and Wilson recognized two selection processes operating in different types of habitats. In temporary habitats that are not crowded with competitors, *r*-selection favours capacity for rapid population increase and migratory ability; interspecific competition is occasional, usually between a pair of species and it may be intense. Under *K*-selection in more predictable and crowded habitats, interactions between species, including diffuse competition, are more frequent; selection is for specialization and mechanisms to ensure individual and progeny survival; fecundity and migratory ability are reduced. A third selection process operates in habitats that are predictable but very unfavourable. It was originally described as 'beyond *K*-selection' (P. J. M. Greenslade 1972*b*) since it was recognized in an altitudinal series in a montane zone above the region in which *K*-selection predominated. A similar process was referred to by Whittaker (1975) as 'adversity-selection' and the term is adopted here, abbreviated to **A**-selection. In continuously harsh environments, diversity is low and there is little interaction between species. Selection is first for adaptation to the physical environment and then for conservation of adaptation. This commonly involves parthenogenesis and other simplifications of life histories and further reduction of migratory ability.

For Australian soil invertebrates, *r*-selection operates in temporary, unpredictable environments, including transient decomposition habitats such as fallen fruits and the early stages of decomposition of logs, ephemeral vegetation in the arid zone, the early stages of plant succession and the culture steppe. *K*-selection should be frequent in most undisturbed native vegetation although the previous section shows how *r*-selected species can occur in diverse communities in *K* environments. **A**-selection will apply on the tops of high mountains in south-eastern Australia (Matthews 1976). For some groups, cold, wet forests are an **A** environment, for others the arid zone. However, arid areas are not unfavourable for ants and their activities seem to maintain a sufficiently high level of interaction between species for *K*-selection to occur as well. This is suggested by the prevalance of *K*-strategies such as ant mimicry in zodariid and salticid spiders. An idealized soil profile can be seen as forming an *r–K–***A** sequence with increasing depth (Fig. 40.13).

Southwood (1977) condensed a variety of quantitative characters by which habitats can be defined into two axes equivalent to favourableness and predictability. The three types of selection processes are related to these axes in Fig. 40.14. Southwood emphasized external properties of the habitat, measuring one aspect of favourableness for example in terms of the length of time a habitat remains suitable for breeding in relation to an organism's generation time. This, incidentally, explains how a habitat can select for *r*-strategies in one group and *K*- in another. Adaptive strategies are a product of evolution and interaction between species is perhaps the most important single influence on the evolutionary process (Bock 1972), especially competitive interaction (Franklin 1978). In Fig. 40.14 we include a third

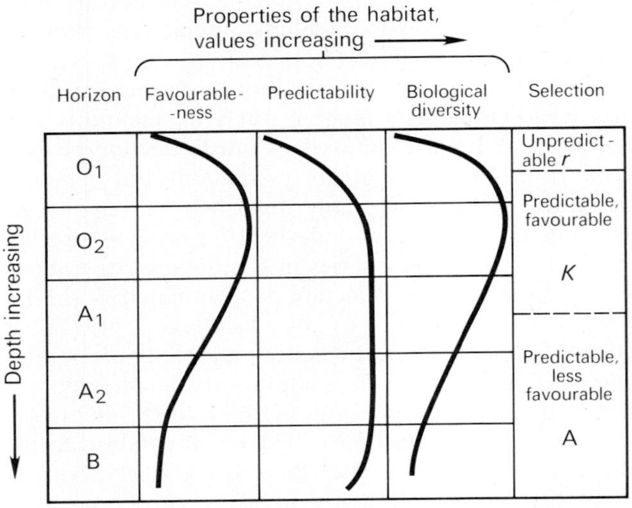

Fig. 40.13. Selection processes in an idealized soil profile. Habitat favourableness is low in the upper litter layer which is subject to high rates of evaporation while microbial succession on dead plant material is at an early stage. Favourableness increases as decomposition proceeds and microbial variety and biomass increase (see Fig. 8.7 in Swift 1976), reaching a maximum in the sheltered, well structured environment at the base of the litter layer and in the organic-rich, upper A horizon. At greater depths, favourableness declines as soil structure and drainage deteriorate. Predictability is low near the soil surface where the microclimate is variable and the initial rate and course of decomposition depends both on the physical environment and the composition of the surface biota. At deeper levels the soil environment is stable and highly predictable although finally it may become less predictable if movements of ground water affect the fauna.

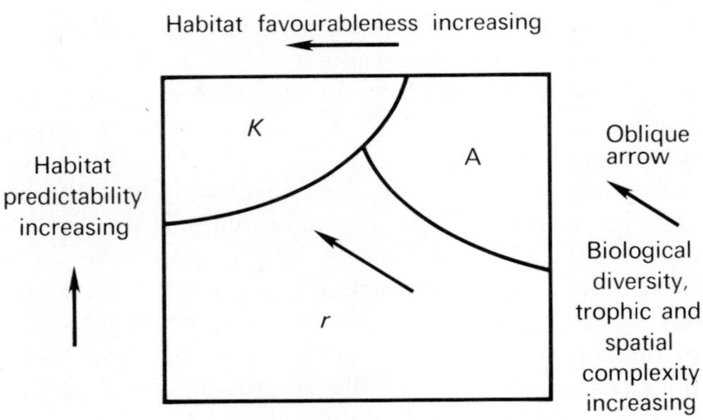

Fig. 40.14. The habitat templet: relationships between habitat predictability, habitat favourableness, biological unpredictability and selection processes. This is schematic and the precise form of the diagram depends on the scaling of the axes. The zones of *K*, *r* and **A** selection are areas in which each type of selection tends to predominate; they intergrade and there are no clear boundaries between them.

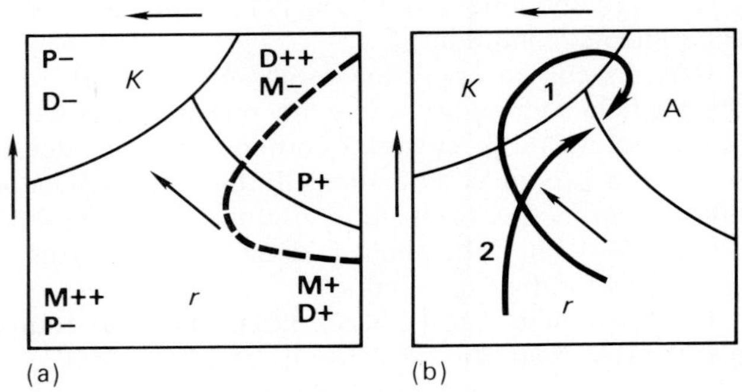

Fig. 40.15. Applications of relationships between habitat and selection: (*a*) Areas of maximum (++,+), and minimum (−) selection for ecological strategies: P, parthenogenesis; D, dormancy; M, migratory ability. The area in which parthenogenesis is favoured is enclosed within the broken line; it is selected against in areas of high biotic or abiotic predictability to the left of the figure. Under **A**-selection it reduces genetic variability and departures from the adapted type; under *r*-selection it effectively doubles the potential reproductive rate. Dormancy will be selected for in unfavourable habitats, most strongly in predictably unfavourable ones. It is likely to be selected against in the most favourable, predictable and competitive environments where resources are continuously available and must be continuously occupied. Migratory ability is correlated with unpredictability of the habitat. It will be selected against very strongly under **A**-selection since it leads to outcrossing and genetic variability. For the same reason it should be selected for to some extent under extreme *K*-selection to allow variation in response to a biologically varying environment. (*b*) Selection processes during decomposition; 1, leaf litter or a moist log; 2, a fleshy fruit (see text).

dimension for diversity and biological complexity which, as Southwood points out, increase with predictability and favourableness. All three dimensions are measurable. On an Australia-wide scale predictability will often be a function of the predictability of rainfall; a measure of favourableness would need to take into account rainfall and temperature. Density of species or rate of increase in number of species with increase in area can be used to describe biological complexity. Fig. 40.14, essentially Southwood's (1977) 'habitat templet', is a basis for correlating and predicting several different aspects of the biology of soil invertebrates (Fig. 40.15*a*), for instance migratory ability and capacity for dormancy. Thus species able to recolonize by migration are selected for in temporary unpredictable habitats such as arable land, and harsh environments in the arid zone select for facultatively dormant, anhydrobiotic Collembola. Glesener and Tilman (1978) correlated parthenogenesis in terrestrial animals with environmental predictability consisting of abiotic and biotic components. The first is simply the vertical axis in Fig. 40.14. Biotic unpredictability can be equated with the third dimension, diversity and complexity. Transposing Glesener and Tilman's Fig. 1 to Fig. 40.14 here shows parthenogenesis to be most frequent in unfavourable **A**- and *r*-selection environments (Fig 40.15*a*). This accounts for its occurrence amongst Australian Collembola. Parthenogenesis is frequent amongst the Isotomidae of temporary habitats and the arid zone. It is found also in species of *Arrhopalites*, *isotomodes* and other isotomids, *Onychiurus* and *Tullbergia* adapted to life in mineral soil (see Fig. 40.13). A facultatively parthenogenetic species of *Cyphoderus* inhabits ant and termite nests. These provide equable predictable habitats that are clearly very unfavourable for most arthropods.

If 'time' is substituted for 'depth' in Fig. 40.13 it describes changes that take place in the course of decomposition of leaf litter or a moist log. These can be expressed as a trajectory on Fig. 40.14, where the radius of the curve at a point reflects rate of decomposition. On any one site all decomposing organic matter will tend to converge towards the same area in Fig. 40.15*b* as widespread soil inhabitants eventually replace specialists. This figure also shows the course of decomposition of a fleshy fruit, favourable from the start because it can be exploited by many invertebrates without microbial intervention. Predictability is low and decomposition rapidly goes through to completion within the area of *r*-selection. In contrast, a log that dries out decomposes slowly. As a habitat it is predictable but unfavourable for most log specialists. Wood-feeding termites can be thought of as originally **A**-selected but the strategy has been so successful as to alter the dimensions of the habitat templet. Termites themselves have become abundant enough to form a resource for other organisms and through interaction with them are secondarily *K*-selected. If arid zone Isotomidae do feed on fresh litter they have also acquired an adaptation to an unfavourable resource in a severe environment. Because of its climatic severity and unpredictability these Collembola appear to have avoided coevolution of predators.

Ecology of soil invertebrates and land use

Many agricultural pest problems involve soil invertebrates. Although they are the province of the specialist economic entomologist and nematologist, current approaches to control require an understanding of the ecology of the pest and the community of which it is a part. It may then be possible to manipulate the pest population by modifying cultural practices. To take but one example, work on the ecology of soil invertebrates in New England pastures has led to trials of a system of managing larval scarabaeid pests by alternating high and low stocking rates (Roberts 1979). This depends on earlier observations on the relationships between grazing pressure and the relative densities of pest species and those that ingest only dead plant material.

Soil invertebrates show considerable promise in the assessment of land condition. Ants, for example, are sensitive to any disturbance to their nests in soil and to changes in vegetation structure (Greenslade and Thompson, 1981). Groups that are unable to burrow are influenced by changes in soil structure. The fauna of grassland litter may be affected by trampling at levels lower than those required to produce changes in the vegetation (Duffey 1975). Consequently, soil invertebrates may act as indicators of land deterioration before changes in vegetation or soil properties are evident. Here the diversity of the soil fauna is an asset rather than an obstacle. Without it being necessary to determine absolute densities, quantitative data can be obtained from the frequency of species classed by their geographical affinities, ecological roles or adaptive strategies. However tentative, the initial assessment of species as *r*, *K* or **A** strategists, the process of classifying them leads to further hypotheses as to their biology and properties of the habitat.

As yet the Australian soil fauna has hardly been exploited in this way. Ants are being used to monitor the rehabilitation of land after mining (Majer 1978*b*; Lentfer 1978), and there is scope for using soil invertebrates to follow and possibly anticipate the effects of recreational use of land. At present the greatest potential lies in the arid zone in relation to grazing on soils that are liable to surface erosion and scalding. Since much of the soil fauna's activity takes place on the surface of the soil, it is sensitive to change in surface conditions.

If changes in soil invertebrate populations are to be studied in landscapes disturbed by man, it has to be remembered that these communities are simplified departures from more complex natural systems. They are best described and understood by comparison with the original state.

Conclusions

Lee, and Spain and Hutson in Chapters 39 and 38 comment on how little we know about the Australian soil fauna. We can add further areas in which information is scarce or absent: geographical distributions, composition of soil faunas in the northern half of the continent, effects of the seasons, nature and control of activity rhythms, communication and dispersion in non-social groups, the ecological physiology of water and temperature relations, and nutrition. Some of the reasons for this lack of basic knowledge are immediately evident from Figs 40.7 and 40.8: a large, taxonomically little-known fauna and very unsatisfactory sampling methods. As a result it is not practicable for work on soil invertebrates in most Australian ecosystems to follow two major trends in contemporary ecology: towards formal theoretical descriptions of processes and rigorous quantitative investigations of actual systems. For example, the emphasis on energy flow as a unifying process in ecosystem analysis is in conflict with the realities of accurately measuring populations of soil animals (Macfadyen 1970). The possible solution to this difficulty is to accumulate energy budgets for each of the more abundant species (Richards 1974; Weigert *et al.* 1970), i.e. the key species of Vlijm *et al.* (1974), when it should be feasible to correct for the errors involved in estimating absolute populations. However, if indirect relationships are as important as we have implied, the rarer members of the soil fauna cannot be ignored. There is a real dilemma here and, although it is not peculiarly Australian, the problem becomes acute when one has to deal with a diverse, poorly known continental fauna.

There is little merit in estimating abundance for its own sake and with the exception of some termite and earthworm populations, the soil fauna's contribution to total ecosystem biomass and respiration is small (Reichle *et al.* 1975). If each step from one trophic level to the next in the sequence: green plants → microbial decomposers → lower predators, detritivores, herbivores → higher predators, is accompanied by a reduction in biomass of an order of magnitude the soil fauna's direct contribution is negligible.

Alternative or complementary approaches to a purely quantitative one need be considered. Southwood (1977) suggests that experimental, theoretical and observational approaches should be combined to bring qualitative observations into a quantitative framework. But the framework itself is weak if the activities, functions and relationships of soil invertebrates are not understood. Work in a number of currently unfashionable fields is required. They include taxonomy at least to the point where diagnoses and keys for Australian genera and species groups are available, functional morphological studies in relation to habits and habitat, and taxonomically oriented natural history.

References

Adams, E. C. G. and Salmon, J. T. (1972). The mouth-parts and feeding methods of *Brachystomella parvula* (Schaeffer) Collembola: (Brachystomellidae). *Trans. R. Entomol. Soc. London* **124**, 269-86.

Anderson, J. M. (1975). Succession, diversity and trophic relationships of some animals in decomposing leaf litter. *J. Anim. Ecol.* **44**, 475-95.

Anderson, J. M. (1978). Competition between two unrelated species of soil Cryptostigmata (Acari) in experimental microcosms. *J. Anim. Ecol.* **47**, 787-803.

Anderson, J. M. and Healey, I. N. (1970). Improvements in the gelatine embedding technique for woodland soil and litter samples. *Pedobiologia* **10**, 108-20.

Anderson, J. M. and Healey, I. N. (1972). Seasonal and inter-specific variations in major components of the gut contents of some woodland Collembola. *J. Anim. Ecol.* **41**, 359-68.

Baker, G. H. (1976). The ecology and life history of the introduced millipede *Ommatoiulus moreletii* (Lucas 1860), in South Australia. Ph. D. thesis, University of Adelaide.

Behan, V. M. and Hill, S. B. (1978). Feeding habits and spore dispersal of Oribatid mites in the north American arctic. *Rev. Ecol. Biol. Sol.* **15**, 497-516.

Bock, W. J. (1972). Species interactions and macroevolution. *Evol. Biol.* **5**, 1-24.

Booth, R. G. and Anderson, J. M. (1979). The influence of fungal food quality on the growth and fecundity of *Folsomia candida* (Collembola:Isotomidae). *Oecologia* **64**, 317-23.

Bornemissza, G. E. (1957). An analysis of arthropod succession in carrion and the effect of its decomposition on the soil fauna. *Aust. J. Zool.* **5**, 1-12.

Bornemissza, G. E. (1976). The Australian dung beetle project 1965-1975. Australian Meat Research Committee Review No. 30, 1-30.

Britton, E. B. (1970). Coleoptera (Beetles). *In* 'Insects of Australia'. pp. 495-621. (Melbourne University Press: Melbourne.)

Brown, W. L., and Taylor, R. W. (1970). Hymenoptera Superfamily Formicoidea. *In* 'Insects of Australia'. pp. 951-59. (Melbourne University Press: Melbourne.)

Carne, P. B. (1956). An ecological study of the pasture scarab *Aphodius howitti* Hope. *Aust. J. Zool.* **4**, 259-314.

Cassagnau, P. (1952). Sur le régime alimentaire des *Friesea* dalla Torre (Collemboles Poduromorphes). *Vie Milieu* **3**, 304-05.

Christiansen, K. (1964). Bionomics of Collembola. *Annu. Rev. Entomol.* **9**, 142-78.

Christiansen, K. (1967). Competition between Collembola species in culture jars. *Rev. Ecol. Biol. Soc.* **4**, 439-62.

Coleman, D. C. (1976). A Review of Root Production Processes and Their Influence on Soil Biota in Terrestrial Ecosystems. *In* 'The Role of Terrestrial and Aquatic Organisms in Decomposition Processes'. (Eds J. M. Anderson and A. Macfadyen). pp. 417-34. (Blackwell Scientific Publications: Oxford.)

Common, I. F. B. (1970). Lepidoptera (moths and butterflies). *In* 'Insects of Australia'. pp. 765-866. (Melbourne University Press: Melbourne.)

Crossley, D. A. (1977). The Roles of Terrestrial Saprophagous Arthropods in Forest Soils: Current Status of Concepts. *In* 'The Role of Arthropods in Forest Ecosystems'. (Ed. W. J. Mattson). pp. 49-56. (Springer-Verlag: New York.)

Culver, D. (1974). Competition between collembolan species in a patchy environment. *Rev. Ecol. Biol. Sol.* **11**, 533-40.

Davidson, R. L. (1978). Root Systems —the Forgotten Component of Pastures. *In* 'Plant Relations in Pastures'. (Ed. J. R. Wilson). pp. 86-94. (CSIRO: Melbourne.)

Davidson, R. L. and Roberts, R. J. (1968*a*). Species differences in scarab-pasture relationships. *Bull. Entomol. Res.* **58**, 315-24.

Davidson, R. L. and Roberts, R. J. (1968*b*). Influence of plants, manure and soil moisture on survival and live-weight gain of two scarabaeid larvae. *Entomol. Exp. Appl.* **11**, 305-14.

Doube, B. M. (1967). The foraging behaviour of the Meat ant, *Iridomyrmex detectus* (Smith). Honours Thesis, University of Adelaide.

Duffey, E. (1975). The effects of human trampling on the fauna of grassland litter. *Biol. Conser.* **7**, 255-74.

Edwards, C. A. (1955). Soil Sampling for Symphylids and a Note on Populations. *In* 'Soil Zoology'. (Ed. D. K. Mc. E. Kevan). pp. 152-56. (Butterworths Scientific Publications: London.)

Edwards, C. A. and Fletcher, K. E. (1971). A comparison of Extraction Methods for Terrestrial Arthropods. *In* 'Methods of Study in Quantitative Soil Ecology'. IBP Handbook No. 18. (Ed. J. Phillipson). pp. 150-85. (Blackwell Scientific Publications: Oxford.)

Elton, C. S. (1966). 'The Pattern of Animal Communities.' (Methuen: London.)

Estey, R. H. (1977). Roots, Rhizomes and Rhizophaga. *In* 'Soil Organisms as Components of Ecosystems'. (Eds U. Lohm and T. Persson). Proc. 6th Int. Coll. Soil Zool. *Ecol. Bull (Stockholm)* **25**, 293-98.

Franklin, I. R. (1978). Population Genetics and Rates of Evolution. *In* 'Biology and Quaternary Environments'. (Eds D. Walker and J. C. Guppy). Australian Academy of Sciences: Canberra.)

Friend, J. A. and Richardson, A.M.M. (1977). A preliminary Study of Niche Partition in Two Tasmanian Terrestrial Amphipod Species. *In* 'Soil Organisms as components of Ecosystems'. (Eds U. Lohm and T. Persson). Proc. 6th Int. Coll. Soil Zool. *Ecol. Bull. (Stockholm)* **25**, 24-35.

Gilmore, S. K. (1970). Collembola predation on nematodes. *Search Agric.* **1**, 1-12.

Glesener, R. R. and Tilman, D. (1978). Sexuality and the components of environmental uncertainty: clues from geographic parthenogenesis in terrestrial animals. *Am. Nat.* **112**, 659-73.

Goto, H. E. (1972). On the structure and function of the mouthparts of the soil-inhabiting collembolan *Folsomia candida. Biol. J. Linn. Soc.* **4**, 147-68.

Greaves, T. (1971). The distribution of three forms of the meat ant *Iridomyrmex purpureus* (Hymenoptera: Formicidae) in Australia. *J. Aust. Entomol. Soc.* **10**, 15-21.

Greaves, T. and Hughes, R. D. (1974). The population biology of the meat ant. *J. Aust. Entomol. Soc.* **13**, 329-51.

Greenslade, P. (1974). Ecological and zoogeographical notes on Collembola of Kangaroo Island, South Australia. *Pedobiologia* **14**, 256-65.

Greenslade, P. (1977). A re-examination of the genus *Corynephoria*, Absolon (Collembola, Sminthuridae). *Rev. Ecol. Biol. Sol.* **14**, 241-56.

Greenslade, P. (1978). The physical and biological features of Kunoth Paddock in central Australia. VIID. Collembola. CSIRO (Aust.) Div. Land Res. Manag. Tech. Pap. No. 4, pp. 114-23.

Greenslade, P. and Greenslade, P. J. M. (1971). The use of baits and preservatives in pitfall traps. *J. Aust. Entomol. Soc.* **10**, 253-60.

Greenslade, P. and Greenslade, P. J. M. (1980). Relationships of some Isotomidae (Collembola) with habitat and other soil fauna. Proc. VII Int. Coll. Soil Zool., pp. 491-506.

Greenslade, P. J. M. (1963). Further notes on aggregation in Carabidae (Coleoptera), with special reference to *Nebria brevicollis* (F.). *Entomol. Mon. Mag.* **99**, 109-14.

Greenslade, P. J. M. (1972*a*). Distribution patterns of *Priochirus* species (Coleoptera: Staphylinidae) in the Solomon Islands. *Evolution* **26**, 130-42.

Greenslade, P. J. M. (1972*b*). Evolution in the staphylinid genus *Priochirus* (Coleoptera). *Evolution* **26**, 203-20.

Greenslade, P. J. M. (1973). Sampling ants with pitfall traps: digging-in effects. *Insects Soc.* **20**, 343-53.

Greenslade, P. J. M. (1979). 'A Guide to Ants of South Australia'. (South Australian Museum: Adelaide.)

Greenslade, P. J. M. and Greenslade, P. (in press). 'Some Insects of the Soil Surface in the Australian Arid Zone.' (Australian Museum: Sydney.)

Greenslade, P. J. M. and Mott, J. J. (1979). Ants (Hymenoptera, Formicidae) of Native and Sown Pastures in the Katherine Area, N. T., Australia. Proc. 2nd Australasian Conference on the Ecology of Grassland Invertebrates, pp. 153-6.

Greenslade, P. J. M. and Thompson, C. H. (1981). Soils, Vegetation and Ant Distributions in the Cooloola-Noosa river Area, Queensland. *In* 'Vegetation Classification in Australia.' (Eds A. N. Gillson and D. J. Anderson). (CSIRO and ANU Press: Canberra.)

Gross, G. F. (1954). The insect communities of the Brompton Pugholes in relation to those of some nearby climax and disclimax areas. M. Sc. thesis, University of Adelaide. (Cited by Matthews (1976)).

Grüm, L. (1971*a*). Spatial differentiation of the *Carabus* L. (Carabidae, Coleoptera) mobility. *Ekol. Pol.* **14**, 1-34.

Grüm, L. (1971*b*). Remarks on the differentiation in Carabidae mobility. *Ekol. Pol.* **14**, 47-56.

Haarlov, N. (1955). Vertical Distribution of Mites and Collembola in Relation to Soil Structure. *In* 'Soil Zoology'. (Ed. D. K. McE. Kevan). pp. 167-78. (Butterworths Scientific Publications: London.)

Haarlov, N. and Weis-Fogh, T. (1953). A microscopical technique for studying the undisturbed texture of soils. *Oikos* **4**, 44-57.

Hale, W. G. (1967). Collembola. *In* 'Soil Biology'. (Eds A. Burges and F. Raw). pp. 397-411. (Academic Press: London.)

Halliday, R. B. (1978). Genetic studies of Meat ants (*Iridomyrmex purpureus*). Ph. D. thesis, University of Adelaide.

Harding, D. J. L. and Stuttard, R. A. (1974). Microarthropods. *In* 'Biology of Plant Litter Decomposition'. (Eds C. H. Dickinson and G. J. F. Pugh). pp. 489-532. (Academic Press: London.)

Healey, I. N. (1970). The Study of Production and Energy Flow in Populations of Soft-bodied Microarthropods. *In* 'Methods of Study in Soil Ecology'. (Ed. J. Phillipson). pp. 175-81. (UNESCO: Paris.)

Holt, J. A. and Greenslade, P. J. M. (1979). Ants (Hymenoptera: Formicidae) in mounds of *Amitermes laurensis* (Isoptera: Termitidae). *J. Aust. Entomol. Soc.* **18**, 349-61.

Hutchinson, G. E. (1951). Concluding remarks. *Cold Spring Harbour Symp. Quant. Biol.* **22**, 415-27.

Hutchinson, G. E. (1953). The concept of pattern in ecology. *Proc. Acad. Nat. Sci. Philadelphia* **105**, 1-12.

Hutson, B. R. (1980). The influence on soil development of the invertebrate fauna colonising industrial reclamation sites. *J. Appl. Ecol.* **17**, 277-86.

Imai, H. T., Crozier, R. H. and Taylor, R. W. (1977). Karyotype evolution in Australian ants. *Chromosoma* **59**, 341-44.

Joose, E. N. G. (1976). Littoral Apterygotes. *In* 'Marine Insects'. (Ed. L. Cheng) pp. 151-86. (North-Holland: Amsterdam.)

Joose, E. N. G. and Groen, J. B. (1970). Relationship between saturation deficit and the survival and locomotory activity of surface dwelling Collembola. *Neth. J. Zool.* **20**, 315-28.

Joose, E. N. G. and Verhoef, H. A. (1974). On the aggregational habits of surface dwelling Collembola. *Pedobiologia* **14**, 245-49.

Kaczmarek, M. (1975*a*). An analysis of Collembola communities in different pine forest environments. *Ekol. Pol.* **23**, 265-93.

Kaczmarek, M. (1975*b*). Influence of Humidity and Specific Interactions on Collembolan Populations in Pine Forests. *In* 'Progress in Soil Zoology'. pp. 333-39. (Ed. J. Vanek). (Dr. W. Junk: The Hague; and Academia: Prague.)

Kempson, D., Lloyd, M. and Ghelardi, R. (1963). A new extractor for woodland litter. *Pedobiologia* **3**, 1-21.

Kevan, D. K. McE. (1962). 'Soil Animals'. (H. F. and G. Witherby: London.)

King, K. L. and Hutchinson, K. J. (1976). The effects of sheep stocking intensity on the abundance and distribution of mesofauna in pastures. *J. Appl. Ecol.* **13**, 41-56.

Krantz, G. W. (1970). 'A Manual of Acarology'. (O. S. V. Book Stores Inc. : Corvallis, Oregon.)

Lasebikan, B. A. Belfield, W. and Gibson, N. H. E. (1978). Comparison of relative efficiency of methods for the extraction of soil arthropods. *Rev. Ecol. Biol. Sol.* **15**, 39-65.

Lee, K. E. and Wood, T. G. (1971). 'Termites and Soils'. (Academic Press: London.)

Lentfer, C. (1978). The effects of sand mining on the ant fauna of the Holocene dunes in the Myall Lakes National Park. Honours thesis, University of New England.

Lohm, U. and Persson, T. (1977). Eds. 'Soil Organisms as Components of Ecosystems'. Proc. 6th Int. Coll. Soil Zool. *Ecol. Bull. (Stockholm)* **25**.

Longstaff, B. C. (1976). The dynamics of Collembolan populations: competitive relationships in an experimental system. *Can. J. Zool.* **54**, 948-62.

Lussenhop, J. (1971). A simplified canister-type soil arthropod extractor. *Pedobiologia* **11**, 40-5.

Luxton, M. (1972). Studies on the oribatid mites of a Danish beech wood soil. I. Nutritional biology. *Pedobiologia* **12**, 434-63.

MacArthur, R. H. (1972). 'Geographical Ecology, Patterns in the Distribution of Species'. (Harper and Row: New York.)

MacArthur, R. H. and Wilson, E. O. (1967). 'The Theory of Island Biogeography'. (Princeton University Press: Princeton, New Jersey.)

Macfadyen, A. (1970). Soil Metabolism in Relation to Ecosystem Energy Flow and to Primary and Secondary Production. *In* 'Methods of Study in Soil Ecology'. (Ed. J. Phillipson) pp. 167-70. (UNESCO: Paris.)

McMillan, J. H. (1975). Interspecific and seasonal analyses of the gut contents of three Collembola (Family Onychuridae). *Rev. Ecol. Biol. Sol.* **12**, 449-57.

McMillan, J. H. and Healey, I. N. (1971). A quantitative technique for the analysis of the gut contents of Collembola. *Rev. Ecol. Biol. Sol.* **8**, 295-300.

Macnamara, C. (1924). The food of Collembola. *Can. Entomol.* **56**, 100-05.

Majer, J. D. (1978*a*). An improved pitfall trap for sampling ants and other epigaeic invertebrates. *J. Aust. Entomol. Soc.* **17**, 261-62.

Majer, J. D. (1978*b*). The Importance of Invertebrates in Successful Land Reclamation with Particular Reference to Bauxite Mine Rehabilitation. *In* 'Rehabilitation of Mined Lands in Western Australia'. pp. 47-61. (Western Australian Institute of Technology: Bentley, W. A.)

Manton, S. H. (1964). Mandibular mechanisms and the evolution of arthropods. *Phil. Trans. R. Soc.* **247B**, 1-183.

Manton, S. M. (1972). The evolution of arthropodan locomotory mechanisms. Part 10. Locomotory habits, morphology and evolution of the hexapod classes. *Zool. J. Linn. Soc.* **51**, 203-400.

Manton, S. M. (1978). 'The Arthropoda: Habits, Functional Morphology and Evolution'. (Clarendon Press: Oxford.)

Marshall, V. G. (1972). Comparison of two methods of estimating efficiency of funnel extractors for soil microarthropods. *Soil Biol. Biochem.* **4**, 417-26.

Matthews, E. G. (1975). Museum collections as an alternative to publication in insect taxonomy. *Syst. Zool.* **24**, 223-25.

Matthews, E. G. (1976). 'Insect Ecology'. (University of Queensland Press: Brisbane.)

Mills, J. T. and Sinha, R. N. (1971). Interactions between the springtail, *Hypogastrura tullbergi*, and soil-borne fungi. *J. Econ. Entomol.* **64**, 398-401.

Mitchell, M. J. and Parkinson, D. (1976). Fungal feeding of Oribatid mites (Acari: Cryptostigmata) in an Aspen woodland soil. *Ecology* **57**, 302-12.

Murphy, D. H. (1971). Revision of the tropical marine littoral genus *Pseudanurida* Schott (Collembola: Pseudachorutidae) *Pacific Insect* **13**, 49-63.

Murphy, P. W. (1962). Ed. 'Progress in Soil Zoology'. (Butterworths: London.)

Nielsen, M. G. and Josens, G. (1978). Production by Ants and Termites. *In* 'Production Ecology of Ants and Termites'. (Ed. M. V. Brian). pp. 45-53. (Cambridge University Press: Cambridge).

Niijima, K. (1971). Seasonal changes in Collembolan populations in a warm temperate forestof Japan. *Pedobiologia* **11**, 11-26.

Paclt, J. (1956). 'Biologie der Primar Flugellosen Insekten'. (Gustav Fischer Verlag: Jena.)

Pande, Y. D. (1975). Scope of the Gelatin-embedding Technique for Studying Soil Arthropod Fauna with Particular Reference to Oribatei. *In* 'Progress in Soil Zoology'. (Ed. J. Vanek). pp. 547-58. (Dr. W. Junk: The Hague, and Academia: Prague.)

Pande, Y. D. and Berthet, P. (1973*a*). Studies on the food and feeding habits of soil Oribatei in a black pine plantation. *Oecologia* **12**, 413-26.

Pande, Y. D. and Berthet, P. (1973*b*). Comparison of the Tullgren funnel and soil section methods for surveying oribatid populations. *Oikos* **24**, 273-77.

Parisi, V. (1972). La distanza individuale nei Collemboli e la sua difesa attiva. *Atti. Conv. Entomol.* **9**, 245-53. (Cited by Tosi *et al.* (1977)).

Persson, T. and Lohm, U. (1977). The Energetical Significance of the Annelids and Arthropods in a Swedish Grassland Soil. *Ecol. Bull. (Stockholm)* **23**, 1-211.

Petal, J. (1978). The Role of Ants in Ecosystems. *In* 'Production Ecology of Ants and Termites'. (Ed. M. V. Brian). pp. 293-325. (Cambridge University Press: Cambridge.)

Petersen, H. (1971). Collembolernes ernaeningsbiologi og dennes ikologiskes betydning. *Ent. Meddr.* **39**, 97-118.

Petersen, H. (1978). Some properties of two high-gradient extractors for soil micro-arthropods, and an attempt to estimate their extraction efficiency. *Jutlandica* **20**, 95-122.

Pimental, D. (1968). Population regulation and genetic feedback. *Science N. Y.* **159**, 1432-37.

Plowman, K. P. (1977). A study of the litter and soil faunas of two Australian sub-tropical forests with particular reference to the Cryptostigmata and the Mesostigmata (Acarina). Ph. D. thesis, University of Queensland.

Rapoport, E. H. (1970). Pyramid of Numbers in Soil Micro- and Mesofauna. *In* 'Methods of Study in Soil Ecology'. (Ed. J. Phillipson). pp. 183-86. (UNESCO: Paris.)

Reichle, E., McBrayer, J. F. and Ausmus, S. (1975). Ecological Energetics of Decomposer Invertebrates in a Deciduous Forest and Total Respiration Budget. *In* 'Progress in Soil Zoology. (Ed. J. Vanek). pp. 283-92. (Dr. W. Junk, The Hague and Academia: Prague.)

Richards, B. N. (1974). 'Introduction to the Soil System'. (Longman: Essex.)

Ridsdill-Smith, T. J. (1975). Selection of living grass roots in the soil by larvae of *Sericesthis nigrolineata* (Coleoptera: Scarabaeidae). *Entomol. Exp. Appl.* **18**, 75-86.

Roberts, R. J. (1979). The theory and use of alternating stocking rates to control pasture pests. Proc. 2nd Australasian Conference on the Ecology of Grassland Invertebrates. pp. 179-80.

Salmon, J. T. (1964). An index to the Collembola. R. Soc. N. Z. Bull. No. 7.

Schaeffer, C. W. (1972). Degree of metathoracic scent-gland development in the trichophorous Heteroptera (Hemiptera). *Ann. Entomol. Soc. Am.* **65**, 810-21.

Singh, S. B. (1969). Preliminary observations on the food preference of certain Collembola (Insecta). *Rev. Ecol. Biol. Sol.* **6**, 461-67.

Southwood, T. R. E. (1966). 'Ecological Methods with Particular Reference to the Study of Insect Populations'. (Methuen: London.)

Southwood, T. R. E. (1977). Habitat, the templet for ecological strategies? *J. Anim. Ecol.* **46**, 337-65.

Spencer, H. and Stracener, C. L. (1929). Soil animals injurious to sugar-cane roots. *Ann. Entomol. Soc. Am.* **22**, 641-49.

Swift, M. J. (1976). Species Diversity and the Structure of Microbial Communities. *In* 'The Role of Terrestrial and Aquatic Organisms in Decomposition Processes'. (Eds J. M. Anderson and A. Macfadyen). pp. 185-222. (Blackwell Scientific Publications: Oxford)

Takeda, H. (1979). On the extraction process and efficiency of Macfadyen's high gradient extractor. *Pedobiologia* **19**, 89-98.

Tamura, H. (1976). Biases in extracting Collembola through Tullgren Funnel. *Rev. Ecol. Biol. Sol.* **13**, 21-34.

Taylor, R. W. (1972). Biogeography of Insects of New Guinea and Cape York Peninsula. *In* 'Bridge and Barrier: the Natural and Cultural History of Torres Strait'. (Ed. D. Walker). pp. 213-30. (Research School of Pacific Studies, Australian National University: Canberra.)

Taylor, R. W. (1976). A submission to the enquiry into the impact on the Australian Environment of the current inquiry programme. Aust. Senate Official Hansard Rep. (Ref. Current Woodchip Enquiry), Melbourne, 12th August, 1976. pp. 3724-31.

Terborgh, J. (1971). Distribution on environmental gradients: theory and a preliminary interpretation of distributional patterns in the avifauna of the Cordillera Vilcabamba, Peru. *Ecology* **52**, 23-40.

Thiele, H. V. (1977). 'Carabid Beetles in their Environments'. (S pringer-Verlag: Berlin.)

Tosi, L., Parisi, V. and Nieder, L. (1977). Analysis of the feeding behaviour of *Sinella loeia* (Schött) (Collembola). *Rev. Ecol. Biol. Sol.* **14**, 483-92.

Usher, M. B. (1976). Aggregation Responses of Soil Arthropods in Relation to the Soil Environment. *In* 'The Role of Terrestrial and Aquatic Organisms in Decomposition Processes'. (Eds J. M. Anderson and A. Macfadyen). (Blackwell Scientific Publications: Oxford.)

Usher, M. B. and Parr, T. W. (1977). Are there successional changes in arthropod decomposer communities? *J. Environ. Manag.* **5**, 151-60.

Vannier, G. (1970). Réactions des microarthropodes aux variations de l'état hygrique du sol. Techniques relatives a l'extraction des Arthropodes du sol. *Rev. Ecol. Biol. Sol.* **7**, 289-309.

Verhoef, H. A., Nagelkerke, C. J. and Joosse, E. N. G. (1977). Aggregation responses in Collembola (Apterygota); a biotic cause of aggregation. *Rev. Ecol. Biol. Sol.* **14**, 21-25.

Vlijm, L., Van der Kraan, C. and Van Wingerden, W. K. R. E. (1974). Significance of functional and structural relationships in the soil and litter layer. *Rev. Ecol. Biol. Sol.* **11**, 511-18.

von Torne, E. (1967). Beispele fur indirekte Enflusse von Bodentieren auf die Rotte von Zellulose. *Pedobiologia* **7**, 220-27.

Wallace, M. M. H. (1970). Insects of Grasslands. *In* 'Australian Grasslands'. (Ed. R. M. Moore). pp. 361-70. (Australian National University Press: Canberra.)

Wallace, M. M. H. (1972). A portable power operated apparatus for collecting epigaeic Collembola and Acari. *J. Aust. Entomol. Soc.* **11**, 261-63.

Wallace, M. M. H. (1974). The Entognathous Hexapods. *In* 'The Insects of Australia, supplement 1974'. pp. 31-32. (Melbourne University Press:Melbourne.)

Wallwork, J. A. (1967). Acari. *In* 'Soil Biology'. (Eds N. A. Burges and F. Raw). pp. 363-95. (Academic Press: London.)

Wallwork, J. A. (1970). 'Ecology of Soil Animals'. (McGraw-Hill: London.)

Wallwork, J. A. (1976). 'The Distribution and Diversity of Soil Fauna'. (Academic Press: London.)

Watson, J. A. L., Barrett, R. A. and Lendon, C. (1978). The physical and biological features of Kunoth Paddock. VIII B. Termites. CSIRO (Aust.) Div. Land Res. Managmt Tech. Pap. No. 4, pp. 101-08.

Wautier, V. (1970). Existence d'une interattraction olfactive chez les Brachinides (Col., Carabidae). *Ann. Soc. Entomol. Fr. (N. S.)* **6**, 433-36.

Webb, D. P. (1977). Regulation of Deciduous Forest Litter Decomposition by Soil Arthropod Faeces. *In* 'The Role of Arthropods in Forest Ecosystems'. (Ed. W. J. Mattson). pp. 57-69. (Springer-Verlag: New York.)

Weigert, R. G., Coleman, D. C. and Odum, E. P. (1970). Energetics of the Litter-soil Subsystem. *In* 'Methods of Study in Soil Ecology). (Ed. J. Phillipson). pp. 93-98. (UNESCO: Paris.)

Whittaker, R. H. (1975). The Design and Stability of Some Plant Communities. *In* 'Unifying Concepts in Ecology'. (Eds W. H. van Dobben and R. H. Lowe-McConnell). pp. 169-81. (Dr W. Junk: The Hague.)

Williams, C. B. (1964). 'Patterns in the Balance of Nature'. (Academic Press: London.)

Wilson, E. O. (1959). Some ecological characteristics of ants in New Guinea forests. *Ecology* **40**, 437-47.

Wolter, H. (1963). Vergleichende Untersuchungen zur Anatomie und Functionsmorphologie der stechendsangendenden Mundwerksenge der Collembolen. *Zool. Jahrb. Abt. Anat. Ontog. Tiere* **81**, 27-100.

Womersley, H. (1939). 'Primitive Insects of South Australia'. (Government Printer: Adelaide.)

Wood, T. G. (1976). The Role of Termites (Isoptera) in Decomposition Processes. *In* 'The Role of Terrestrial and Aquatic Organisms in Decomposition Processes'. (Eds J. M. Anderson and A. Macfadyen). pp. 145-68. (Blackwell Scientific Publications: Oxford.)

Youdeowei, A. and Calam, D. H. (1969). The morphology of the scent glands of *Dysdercus intermedius* Distant (Hemiptera: Pyrrochoridae) and a preliminary analysis of the scent gland secretions of the fifth instar larvae. *Proc. R. Entomol. Soc. London Ser. (A)* **44**, 38-44.

Soil / microbial / plant relationships

Chapter 41
The plant root environment

R. C. Foster

MANY STUDIES OF SOILS ARE MADE WITH ONE PURPOSE IN mind, to improve the growth of crops. In practice, this means investigating methods of improving the nutrient status of the plants, or of combating root diseases. In both these approaches, the crucial site of interaction is the root–soil interface. It is here that nutrients enter the plant, and it is here that root pathogens first encounter their host.

As they grow, roots take in nutrients *from*, and secrete materials *into*, their surroundings. This means that the soil at the root surface is very different from the bulk soil, physically, chemically and biologically, and these differences may extend 1 mm or more from the root surface. This specialized zone near roots is called the *rhizosphere* (Hiltner 1904), while the actual surface between the root and soil is the *rhizoplane* (Clark 1949).

The precise nature of the rhizosphere depends on three factors: the species of plant, the age of the root surface, and the nature of the soil. Because the particular physiological nature of many species limits them to particular types of environment and particular types of soil, we will begin with a consideration of the physiological morphological and ultrastructural characters at the root surface, which change with time, and with the effects these changes impose on the nature of the rhizosphere. We will then discuss how the nature of the soil modifies these changes.

Physical changes in the rhizosphere soil induced by root growth

As a root pushes its way into a soil, it must of necessity displace its own volume, and this causes mechanical compaction of the soil fabric near the root surface; it will be slight in the case of a fine lateral root 1 mm in diameter, but near to a major root which has undergone secondary thickening, considerable compression may occur. In either case this means that near the surface of the root there will be a cylinder of soil with increased bulk density, resulting in a zone of minimal voids surrounding the root. This local reduction in void diameter may be important in enhancing the flow of capillary water to the root, and the flow of exudates away from it. It may also restrict movement of both beneficial and deleterious microorganisms to the rhizosphere (Baker and Cook 1974). The mechanical compaction induced even by primary roots of annual plants such as wheat may be sufficient to cause the reorientation of soil minerals (Blevins *et al.* 1970), so that, instead of being randomly dispersed as in the bulk soil, particles of micron and sub-micron size in the rhizosphere may come to lie with their long axis parallel to the root, and with their narrowest dimension tangential to it (see Fig. 41.4) (Greaves and Darbyshire 1972; Foster and Rovira 1976). Where adjacent grains are thereby caused to overlap, the root surface may be protected from lysis by microorganisms, and mucilages may accumulate locally, and persist after the death of the root (see Chapter 30).

Chemical changes in rhizosphere soil induced by root metabolism

Although only a few of the chemical changes can be detected directly by conventional transmission electron microscopy, they are responsible for much of the subsequent change in rhizosphere biology, and are therefore of supreme importance in rhizosphere ecology. Firstly, because the root is primarily an absorbing organ, nutrients, especially insoluble ones like phosphate, are soon removed from the rhizosphere to create a zone of depletion around the root. Removal of particular ions, and the secretion of balancing ions by the root may change the pH of the rhizosphere by as much as two pH units compared with the bulk soil (Baker and Cook 1974), and this may affect the solubility of particular nutrients near the root. Similarly, in the neighbourhood of a transpiring plant, the water potential near the root may be quite different from that in the bulk soil where measurements of water potentials are made. Water potentials of -20 bar may be generated by plant roots on hot days with high transpirational loss, so that bacteria at the rhizoplane experience much lower potentials than those in the outer rhizosphere. This effect may be especially severe when the soil has a low hydraulic conductivity, and where, as in the case of forest trees, the roots are widely spaced.

In normal soil, respiration by the root may deplete the rhizosphere of oxygen, and enrich it with carbon dioxide; this may result in the precipitation of carbonates in

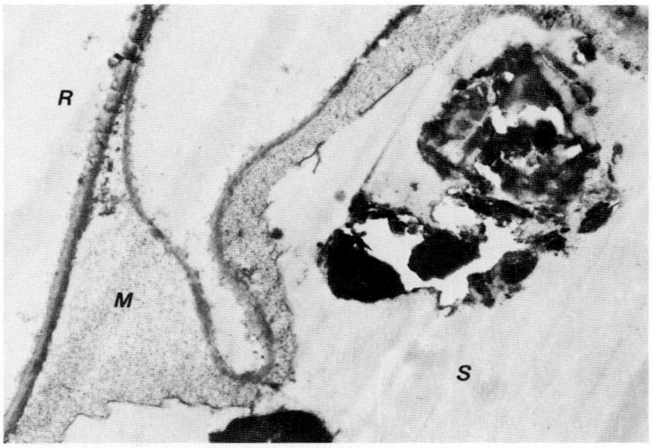

Fig. 41.1.* Continuous layer of mucigel on *Pinus radiata* root. Sterile soil.

the rhizosphere in some soils. Conversely, plants such as rice, adapted to anaerobic waterlogged soils may supply oxygen to the nearby soil, resulting in the precipitation of iron oxides in the rhizosphere.

However, the main chemical and morphological changes in the soil fabric at the rhizosphere are due directly to the energy-rich organic materials which escape from roots into the surrounding soil (see below).

The type and volume of root-derived organic materials in the rhizosphere depend on the age of the adjacent root i.e. on the distance from the root apex, since roots extend in length only at their tips. There are four major plant-derived sources of organic materials at the root surface viz. mucilages, exudates, lysates, and discarded cell wall materials. Firstly, the embryonic cells at the root tip are protected by a cap of tissue which is continuously replaced from within. Both the root cap and epidermal cells within it secrete a copious *mucilaginous gel* which lubricates the surface of the root, easing its passage through the soil. At the same time, the intercellular layer between the outer root cap cells may be enzymatically broken down, and so the cells can slide past each other. This means, that as the root presses forward into the soil, the outer cap cells are rubbed off, so that the root apex glides through a lubricated tube of its own making. In this process, adjacent soil particles are smeared with mucilage (Rovira 1978), and discarded cap cells are left in the soil fabric. More mucilage is secreted by the epidermal cells behind the root tip, so that in certain situations a gelatinous sheath up to 10–50μm thick may be secreted around the root (Figs 41.1, 41.2 and 41.12). Oades (1978) reviews the data on the chemical composition of root mucilage, and the cytological processes involved in its secretion.

Because of their internal turgor pressure, epidermal cells typically arch outwards into the soil (Old and Nicolson 1975), so that deep fissures are formed between the cells (Fig. 41.7). At the root apex the fissures become completely filled with mucilage, so that in scanning electron micrographs (Campbell and Rovira 1973), the root cap and adjacent regions appear externally as a smooth cone. In transmission electron micrographs of conventionally fixed material, the mucilage has a fine granular texture (Fig. 41.1), and apparently consists of two populations of particles differing markedly in size; it also may contain cellulose microfibrils (Leppard 1974). It is bounded on its outer surface by an electron-dense membrane (Fig. 41.1) which has a tri-lamellate substructure (Foster, 1981). The outer membrane is usually observed only in plants grown in axenic culture, or in radiation-sterilized soils, but in natural soils it is occasionally observed where the mucilage stretches over an air gap between mineral grains, from which bacteria are excluded. Usually, d e to root pressure, the mucilage is forced deep into the soil fabric, and is normally in such intimate contact with soil particles that its outer boundary cannot be easily resolved (Fig. 41.12).

In order to provide a mechanical support against which the apex may thrust its way into the soil fabric, and to increase the surface area available for the adsorption of nutrients, the epidermal cells a few millimetres behind the apex produce hairs at right angle to the root axis; the hairs penetrate between and around the soil particles and, like the root apex, also produce a protective mucilage, especially at their growing tips. This means, that in the root hair zone the surrounding soil receives organic materials to a distance of up to 2–3 mm from the root surface, so that in dry soils, soil may be removed from the root hair zone only with some difficulty (Greaves and Darbyshire 1972). Because they function as absorbing organs, the root hairs and epidermal cells are necessarily porous, and this makes them permeable not only to nutrients passing *into* the root from the soil, but also to organic compounds synthesized by the root in the normal processes of metabolism. This means that, in addition to those substances such as mucilage and 'balancing ions' (Rovira 1978) which are specifically secreted by the root into the rhizosphere, the root also passively loses metabolites, which leak out of the surface cells into the soil. These *exudates* contain most of the common metabolites found in living cells (Rovira 1965, 1969), ranging from simple sugars to complex vitamins. Being of low molecular weight, most of these substances are invisible in the electron microscope, but some, e.g. certain amino acids and phenolics, may contribute to the electron density of the mucilage on fixation with heavy metals (Bland *et al.* 1971; Bowen and Rovira 1976).

Although soluble nutrients continuously reach the rhizosphere in the mass flow of water pulled to the root

**Symbols used* A, actinomycete; B, bacterium; C, clay mineral; Ca, capsule material; M, mucigel; O, organic matter; P, polyphenolics; R, root; S, soil; V, bacteriophage.

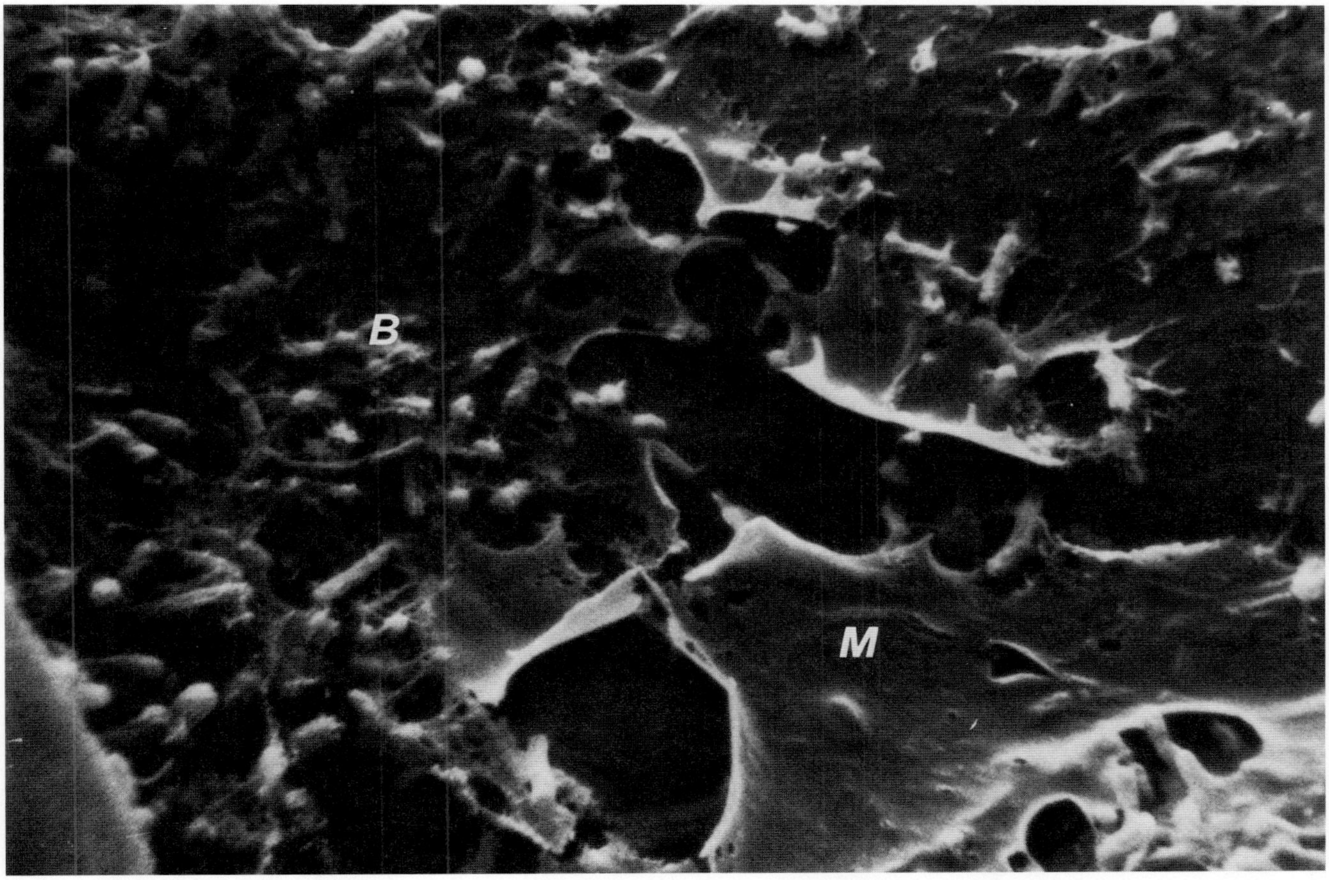

Fig. 41.2. Mucigel layer attacked by bacteria, wheat root. Scanning electron micrograph by A.D. Rovira.

surface by the plant's transpiration stream (Barber and Ozanne 1970), the concentration of some ions (e.g. potassium) in this water is not sufficient to support plant growth, and most of the ions reach the rhizoplane by diffusion. Insoluble nutrients (e.g. phosphates) cannot move through the soil by either process, and are obtained by the root by direct contact. This means, that insoluble nutrients soon become exhausted at any one point in the soil, so that, whenever the plant is metabolically active, it is necessary for the root to grow, and to explore new soil. The root hairs and epidermal cells at any one point of the root therefore normally function only for a short period, probably only a few days, and are continually replaced at the root tip. Behind the tip, root hairs and epidermal cells which are no longer functional undergo lysis, and the resulting organic breakdown products are also released into the rhizosphere. These may differ chemically from the mucilages and exudates, and have been called *lysates* by Martin (1977). Martin showed, that wheat roots may lose into the soil as much as 30% of the metabolites reaching them from the leaves. In older roots, most of the cortex may become invaded by microorganisms (Old and Nicholson 1975; Foster and Rovira 1973), so that by flowering time, the functional region of the root may be limited to the central stele, and the endodermis becomes the functional rhizoplane. The cell wall materials of the root cortex, and of root hairs and sloughed-off root cap cells, therefore constitute a fourth source of organic matter deposited in the rhizosphere (Figs 41.4 and 41.5).

Structure of the rhizosphere outside the root

This combination of energy-rich organic materials (the carbohydrates of the mucilages and cell wall materials), supplemented with metabolites in the exudates and lysates, provides a rich medium for the growth of soil microorganisms. Microorganisms are abundant in soils (populations of 10^8 cm^{-3} are not uncommon), but they are not uniformly distributed in the soil fabric. Because they require energy, most non-rhizosphere bacteria are associated with organic matter, such as the remains of old root systems, or litter which has become incorporated into the soil. They therefore show a clumped distribution in the soil fabric (Foster 1978). The rapid growth of root tips (1–5 cm day^{-1}) means that, despite the irregular distribution of bacteria

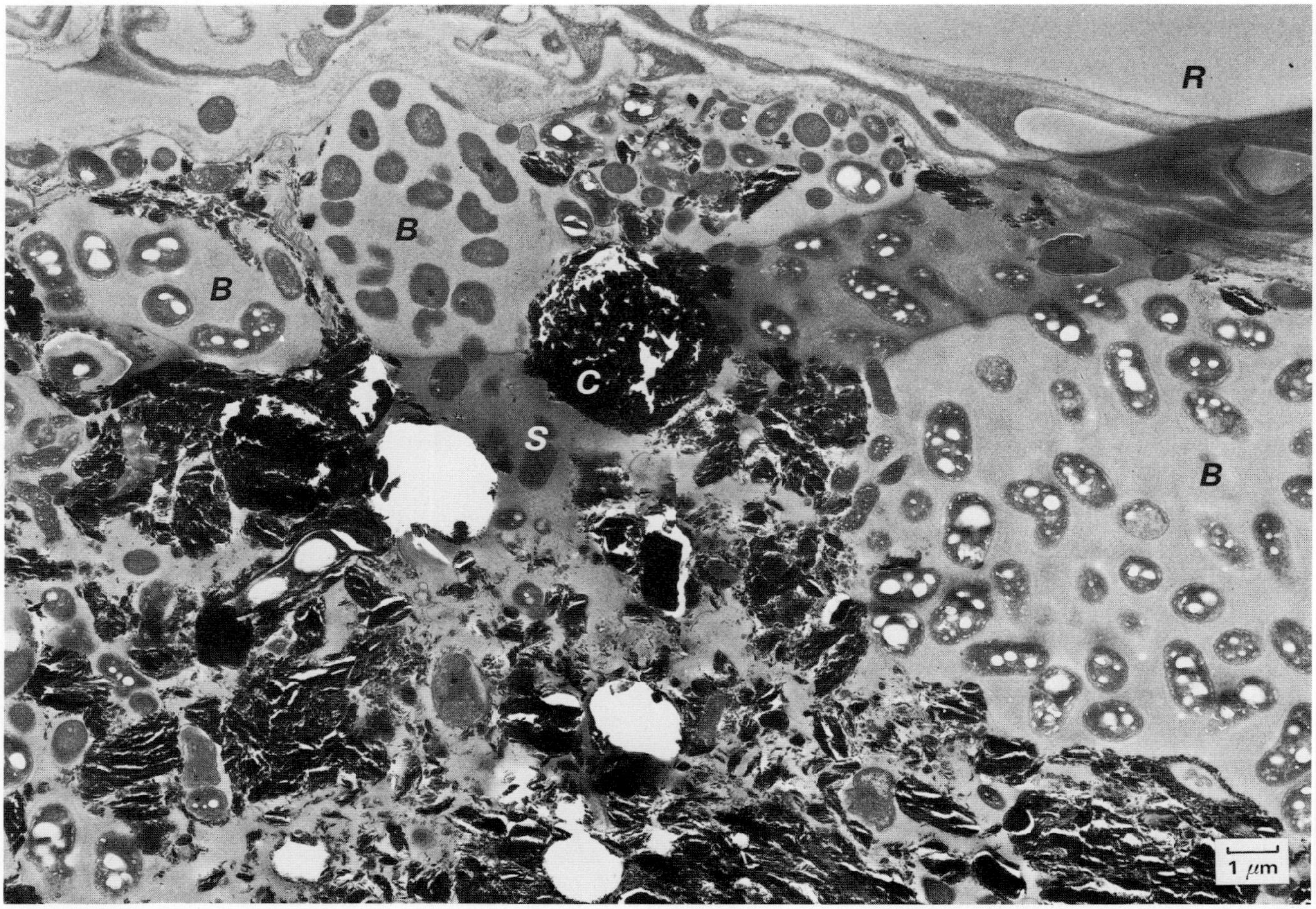

Fig. 41.3.* Bacterial colonies at the rhizoplane of *Trifolium subterraneum* L.

in soils, it is inevitable that they encounter microbially-infected organic matter from time to time, and that these microorganisms become incorporated into the rhizosphere, where they act as localized sites of infection. Bacteria therefore also show non-random distribution patterns on the root surface (Malajczuk and Bowen 1974; Newman and Bowen 1974), resulting in the formation of distinct colonies round the founder individuals (Figs 41.3 and 41.6).

The mucilage on the root surface is quickly colonized by bacteria, many of which produce degradative enzymes, so that, except where it is protected by clay cutans, the mucilage is quickly broken down by microorganisms (Figs 41.2, 41.3 and 41.12). The bacteria are often observed to attach themselves end-on (Bowen and Foster 1978), and to produce lysis zones in the mucilage (Fig. 41.12). These zones gradually coalesce, so that quite soon the epidermal cells behind the root apex are exposed to the soil, and in scanning electron micrographs the surfaces of their arched outer walls are then freely visible (Rovira and Campbell 1974, 1975; Old and Nicolson 1975). Bacterial lysis of mucilage is particularly severe in roots infected with pathogens (Rovira and Campbell 1975). As the mucilage is removed from the outer surfaces of the cells, bacterial colonization is progressively restricted to the grooves between the epidermal cells (Bowen and Foster 1978), where secretion of mucilage continues (Fig. 41.7). This means that, whereas at the apex the rhizoplane is located at the outer surface of the mucilage, a centimetre or two behind the apex the outer tangential walls of the epidermal cells function as the rhizoplane, except where mucilage persists in the interstices between the cells.

Because of the abundance of nutrients in the rhizosphere, the generation time of the bacteria at the rhizoplane may fall to less than one-tenth of its value in in the bulk soil (e.g. 5.2 h compared with 77 h for *Pseudomonas*, Bowen and Foster 1978), so bacterial numbers at the root surfaces rapidly increase. Even so, on normal healthy roots within a few mm of the apex, less than 10%

**Symbols used:* A, actinomycete; B, bacterium; C, clay mineral; Ca, capsule material; M, mucigel; O, organic matter; P, polyphenolics; R, root; S, soil; V, bacteriophage.

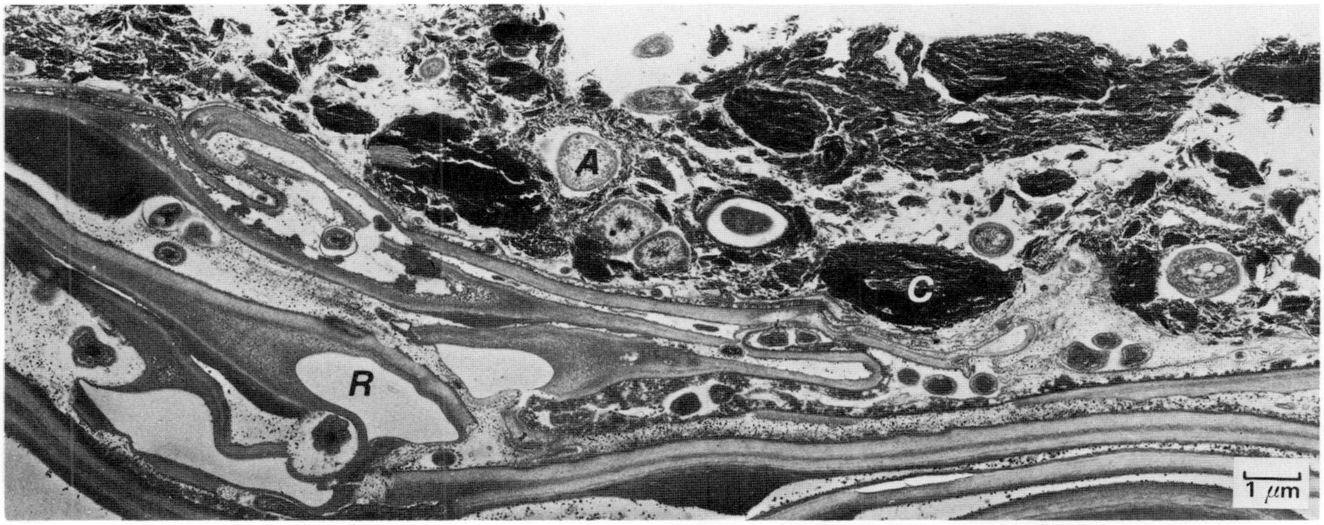

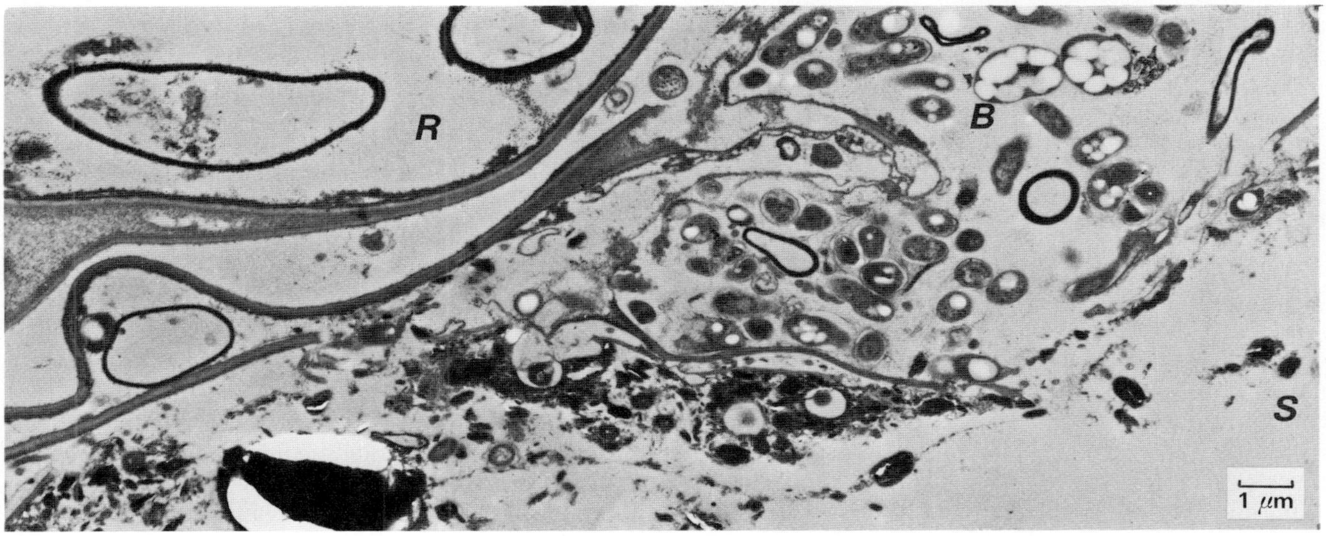

Fig. 41.4. (Top) Early stage in colonization of mature wheat root.

Fig. 41.5. (Bottom) Late stage of wheat root colonization. Bacteria with polyhydroxybutyrate.

of the root surface may be occupied by bacteria (Bowen and Theodorou 1973; Rovira *et al.* 1974). This may rise in older parts of the root (Fig. 41.3), but after 90 days exposure to the soil microflora, the bacterial cover may still be less than 40%, although it may be several cells deep in some places. As in other energy-rich environments, such as tropical forests and coral reefs, rhizospheres are remarkable for the abundance of organisms and the richness of forms present (Figs 41.3–6). Bacteria are typically arranged in colonies 5–15μm in diameter (Fig. 41.3), and species can be distinguished on the basis of their morphology, cytology, wall structure, and storage compounds (Figs 41.3–6) (Foster and Rovira 1978). Although it is not easy to obtain high resolution ultrathin sections of soils containing mineral grains, it is sometimes possible to distinguish between Gram positive and Gram negative forms on the basis of wall structure. By comparing bacteria in micrographs of natural rhizospheres with those of known species reinoculated into otherwise sterile roots, individual species of microorganisms, including actinomycetes (Fig. 41.4 (A)), can sometimes be identified (Foster and Rovira 1978). Thus in Fig. 41.3, forms with polyhydroxybutyrate resemble *Rhizobium*, and others with dense nuclear regions resemble fluorescent pseudomonads. Some forms of microorganisms seen in natural rhizospheres of field crops are quite bizarre in appearance (Fig. 41.6; Foster and Rovira 1978; Old *et al.* 1975), and have not been brought into culture, so that they are known only from electron micrographs. Some have been found only in the grooves between the root cells or very close to the rhizoplane, so that they may be obligate

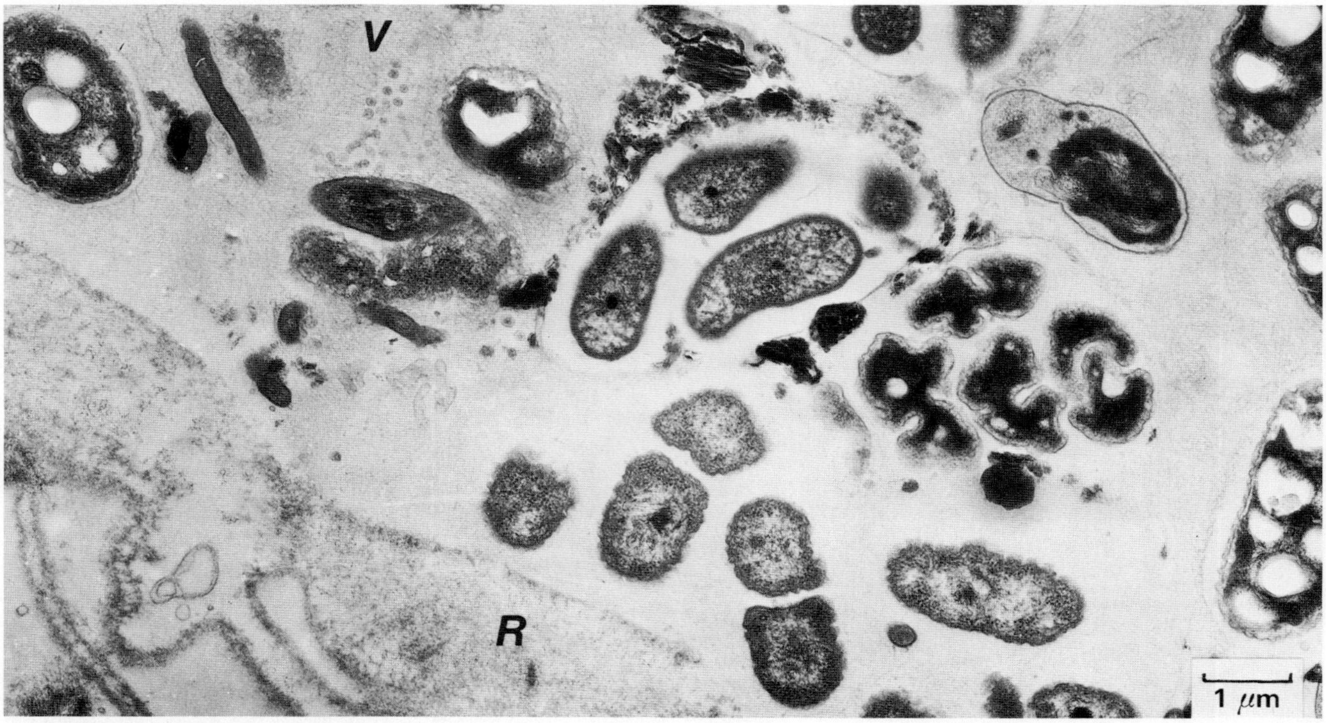

Fig. 41.6.* Bacterial colonies on *Trifolium* root. Note variety of forms.

rhizosphere inhabitants with very specific nutrient requirements.

Bacteria are most numerous in a zone 5–10 μm from the rhizoplane, where numbers may reach 200 x 10^9 cm^{-3}, but numbers often fall to one-tenth of this value only 10–15 μm from the root surface. There is also a much greater diversity of forms at the root surface than in the outer rhizosphere (Foster and Rovira 1978), and it is possible that the rhizoplane constitutes a very specific ecological niche. In the outer rhizosphere, Gram negative forms such as pseudomonads are present in great abundance (Foster and Rovira 1978), so that rhizosphere soils have a different microbiological composition to that of the bulk soil, and this is confirmed by the normal plating techniques. This probably results from the fact that the energy-poor conditions of the bulk soil favour the survival of spore-forming species such as *Bacillus*, whereas the rhizosphere favours species such as *Pseudomonas*, which have wide biochemical capabilities and rapid rates of cell division.

Routine plating techniques commonly detect less than one-tenth of the numbers of bacteria seen in electron micrographs of similar rhizospheres. This is partially due to the fact that bacteria on roots are bound together in groups by their common capsule materials, so that several give rise to one colony on the agar plate, but it is possible that routine media only detect the less fastidious organisms, either in terms of nutrients or of physical or chemical requirements. It has been suggested also that the usual laboratory media favour fast-dividing, capsuleless forms of bacteria found in natural environments, so that these techniques give a biased view of the nature and rates of division of microorganisms *in situ*.

The richness of the rhizoplane microflora appears to be influenced by both soil and host type. Thus legumes such as *Trifolium* (Fig. 41.3) appear in electron micrographs, to have a richer flora, both in numbers and types, than wheat growing adjacent to it in the same soil (Figs 41.4 and 41.5), and this is supported by plate counts (Rovira 1978). Similarly, in an irrigated pasture *Paspalum* roots (Fig. 41.12) secrete more mucilage into the rhizosphere than *Lolium* roots. Again, the rhizosphere of a dry forest soil (Fig. 41.11) is much less rich than those of dry wheatland soils. Grasses both of wetland and dryland species often secrete an extensive mucilage (Fig. 41.12) which may become infected with elongate spirillar bacteria. This is of particular interest since *Spirillum lipoferum* has been shown by Döbereiner and Day (1974) to be associated with grass roots, and to be capable of fixing atmospheric nitrogen. Wetland soils contain abundant cell wall debris because microbial decomposition is slowed down by the generally anaerobic conditions. Despite this abundance of energy-rich material,

**Symbols used* A, actinomycete; B, bacterium; C, clay mineral; Ca, capsule material; M, mucigel; O, organic matter; P, polyphenolics; R, root; S, soil; V, bacteriophage.

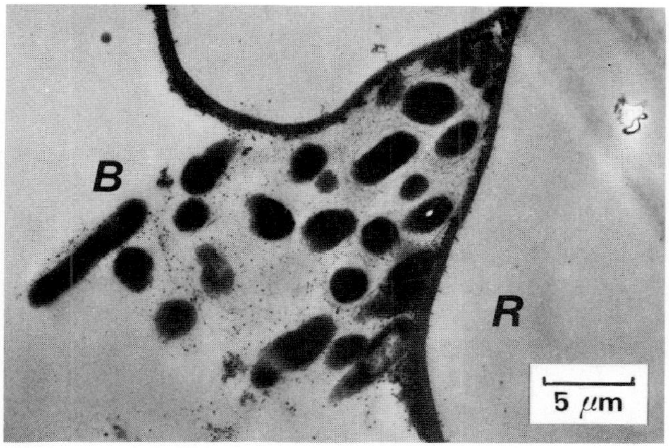

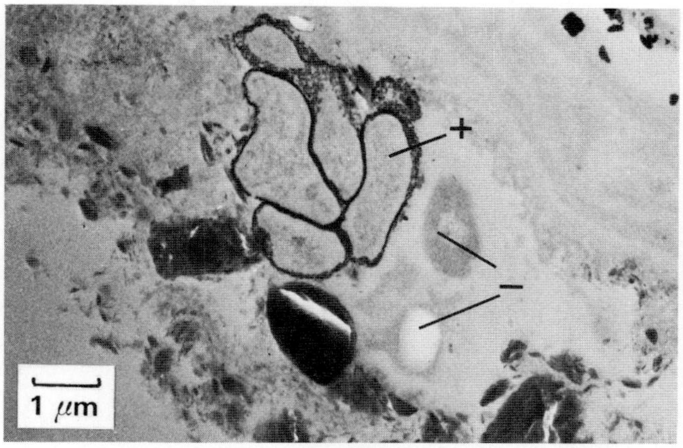

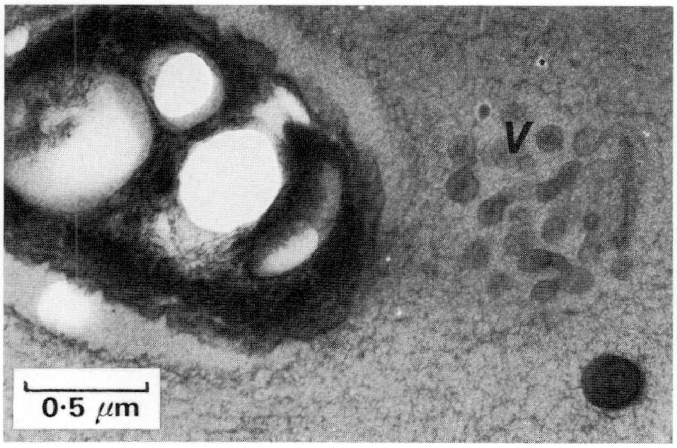

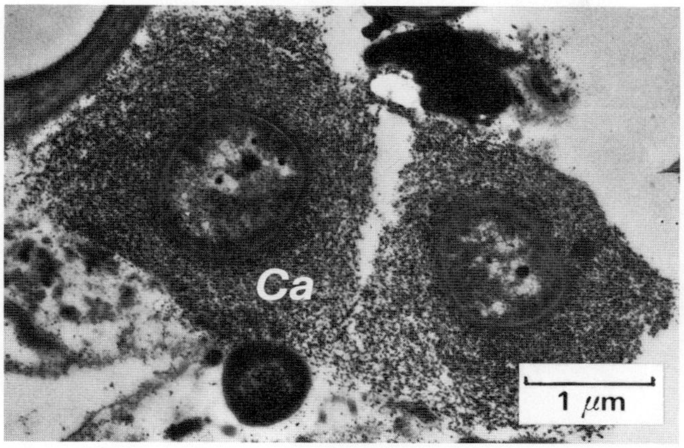

Fig. 41.7. Bacteria in intercellular groove of *Pinus radiata* root.

Fig. 41.8. Rhizosphere of wheat tested for acid phosphatase. Some bacteria reactive, adjacent ones negative.

Fig. 41.9. Bacteriophage near microorganism in *Trifolium* rhizosphere.

Fig. 41.10. Bacteria from *Pinus radiata* mycorrhizosphere, specifically stained for capsule material.

rhizospheres of plants growing in wetlands often contain fewer bacteria than those of similar species which grow in a calcareous grassland, possibly due to soil compaction, slight anaerobiosis, the toxic effects of humus leachates, or to low pH values.

In forest soils, the structure of the rhizosphere is complicated by the fact that most trees support ectotrophic mycorrhizal fungi; these are usually basidiomycetes, although ascomycete associations have been described (Scannerini 1968). In *Pinus radiata* and *Pseudotsuga menziesii*, the hyphae of the outer mantle are often devoid of cytoplasm, and may be partially collapsed, and, as with the epidermal cells of non-mycorrhizal roots, groups of bacteria may be associated with the lysed cells (Foster and Marks 1967). Rhizosphere bacteria are known to influence the development of mycorrhizas, and the phosphate metabolism of roots (Bowen and Rovira 1966), and they might be involved in nitrogen fixation in pine forests (Richards and Voigt 1964). Since mycorrhizas grow in the surface layers of soils, and pine leaves are rich in polyphenolics and resins, tree rhizospheres contain much osmiophilic organic matter.

Evidence is accumulating that mycorrhizas inhibit the breakdown of litter under pines, perhaps because the mycorrhizal fungi prevent the escape of energy-rich exudates into the rhizosphere, so depriving the normal litter-destroying bacteria of nutrients essential to the degradative processes.

Although rhizosphere organisms break down the mucilage produced by the root, many themselves secrete extensive coats of carbohydrate (Figs 41.6 and 41.10), so that in the older rhizosphere it is impossible to distinguish between root mucilage and bacterial slime (Fig. 41.3). Moreover, as well as these colloidal materials secreted by the root and by the associated microorganisms, the rhizosphere contains authigenic mineral fragments of colloidal size and colloidal organic matter, derived from the soil. This complex mixture of colloids was

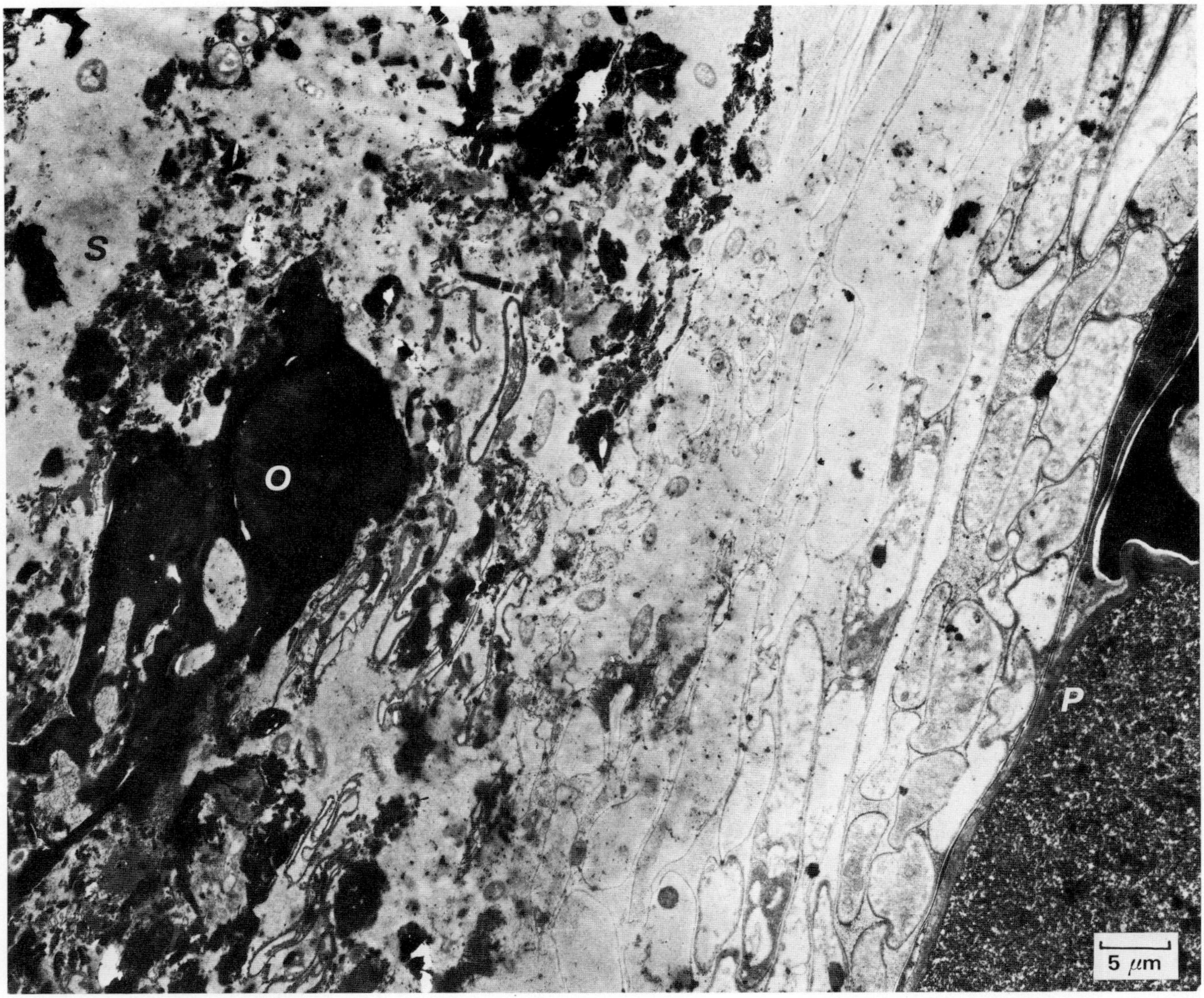

Fig. 41.11. Mycorrhizosphere of *Pinus radiata* showing polyphenol secretion by root; mantle hyphae and soil bacteria.

called *mucigel* by Jenny and Grossenbacher (1963), and, except where roots or bacteria are grown separately in axenic culture, or where the root mucilage can be demonstrated to be uninfected by microorganisms, their term *mucigel* would appear to be the most appropriate one for surface slimes on roots. Because carbohydrates do not react with the heavy metals used in conventional fixation (Bland *et al.* 1971), the mucigel normally appears to be only slightly electron-dense (Figs 41.3-6). Treatment with lanthanum hydroxide increases the electron density of root-derived mucilage (Fig. 41.12), whilst ruthenium red may be used to demonstrate the capsule material secreted by rhizosphere bacteria (Fig. 41.10). The capsules of some rhizosphere bacteria may be very extensive (Figs. 41.3, 41.6, 41.10), and it is possible that, as well as protecting the organisms against desiccation induced by low water potentials at the root surface, capsule carbohydrates may be used as a food source in times of dearth. As well as clay-protected pockets of mucilage, extensive remnants of bacterial capsule material may survive in soils, even after the death of the organisms that secreted them. The microfibrils of the capsule materials often bind soil components together to form crumbs, especially under pastures, thus playing an important role in maintaining soil stability (Martin 1971). Bacterial carbohydrate may also be important in the ion-exchange properties of soils (Martin 1971). Some rhizosphere microorganisms secrete extra-cellular enzymes such as acid phosphatase into the rhizosphere (Fig. 41.8), and these may increase the mineralization of

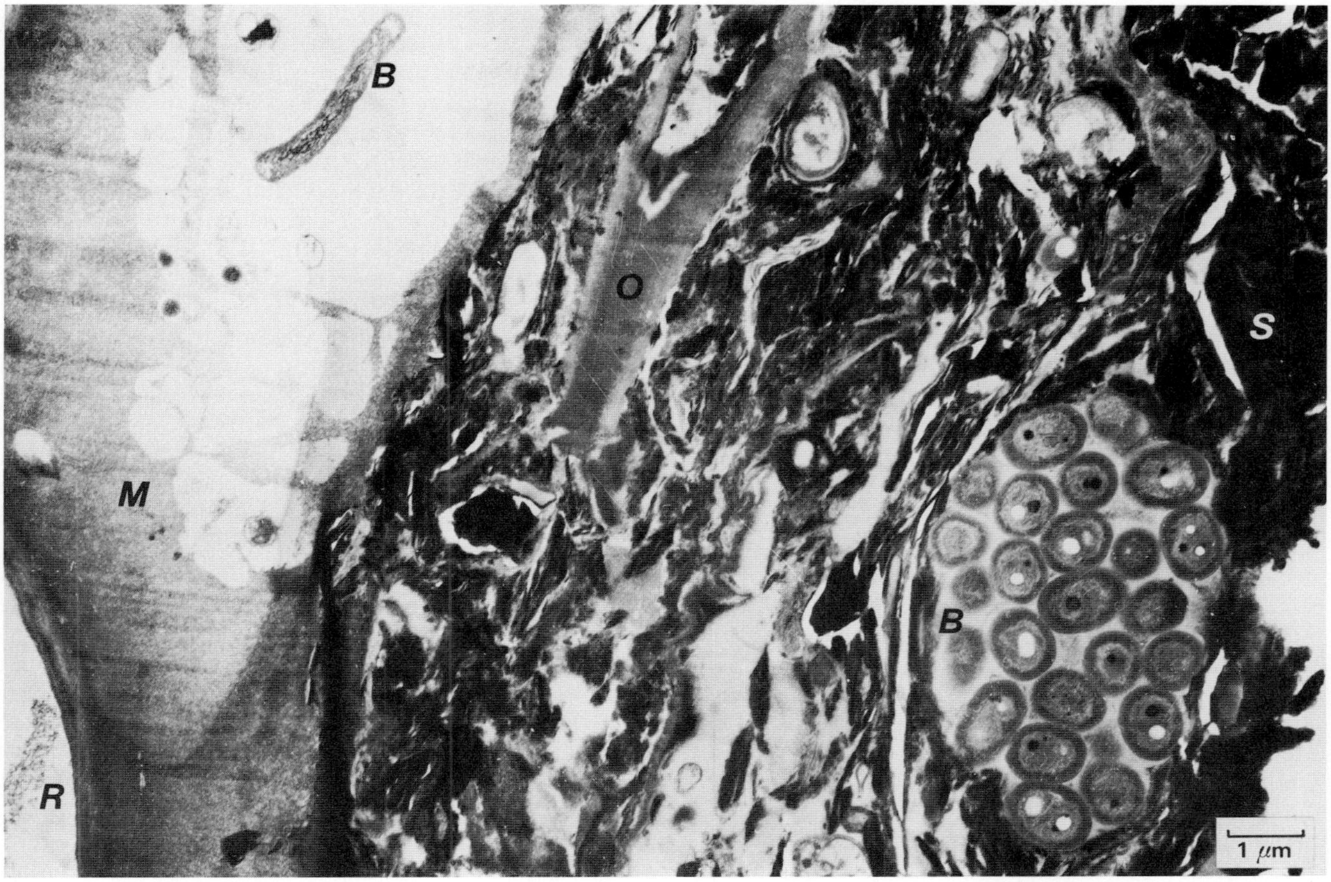

Fig. 41.12. Rhizosphere of *Paspalum* from irrigated pasture. The thick mucigel contains elongated bacteria. A bacterial colony is rich in polyhydroxybutyrate and polyphosphate reserves.

nutrients contained in organic residues and perhaps even in minerals.

The rhizosphere within the root

In the warm dry conditions towards the end of the growing season in the calcareous soils of the South Australian wheat belt, the outer cortical cells soon collapse and autolyse, and behind the root hair zone the cells themselves are progressively colonized by bacteria and fungi (Figs 41.4, 41.5). By anthesis, most of the cortex may be occupied by microorganisms, and at this stage there is probably no identifiable rhizoplane, though the outer wall of the cells of the endodermis may function as such (Foster and Rovira 1976; Martin 1977). Within the endodermis the cytoplasmic structure is still quite normal, so it is probable that even in heavily infected roots the stele is still functional. The occupation of the cortex of apparently healthy roots by a wide variety of microorganisms, combined with the indications that there is abundant available energy (capsule materials, polyhydroxybutyrate in cells), is of particular interest, since Patriquin and Dobereiner (1978) have suggested that it may be here, in what has been called the *endorhizosphere*, rather than in the mucigel or in the soil external to the root, that 'non symbiotic' nitrogen fixation takes place in tropical grasses.

Rhizosphere ecology

Two techniques have traditionally been used to study the micro-ecology of the rhizosphere. The roots may be washed in sterile water and the washings plated out onto a nutrient agar; after a few days the different kinds of colonies can be identified and counted. Where counts of endorhizosphere organisms are required, the roots may be shaken with glass beads to release organisms from the root surface; the roots are then ground and suspended in water, and aliquots of the suspension are transfered to agar media. The bacteria may then be classified; many systems of classification are used, based, for example, on the Gram stain, on morphology, and on biochemical tests.

The second traditional method consists of gently washing roots, to remove the soil, and then staining them with dyes which specifically react with the microorganisms but not with the cells of the root. Using these techniques, the distribution patterns of microorganisms in the rhizosphere can be investigated (Bowen and Malajczuk 1974; Rovira *et al.* 1974).

Because bacteria are so small (0.3–3μm long), only gross differences in morphology can be detected by light microscopy, so that studies ofthe distributions and interrelationships of only the most distinctive organisms can be made. As we have seen, using electron microscopy it is possible to recognize many more types on the basis of their wall structure, capsule structure, or details of their cytology (Bae *et al.* 1972; Foster and Rovira 1976, 1978). Unfortunately, the ultra-thin sections necessary for conventional electron microscopy sample a very small volume of soil: the sections are about 0.25 mm^2 in area, and only 10^{-4} mm thick. Scanning electron microscopy can be used to examine larger areas of root (Fig. 41.2), but the method is limited to surfaces, and, as in conventional light microscopy, different kinds of bacteria cannot be distinguished unless they differ markedly in shape and size.

Many published studies of rhizosphere ultrastructure, unfortunately, have not been made in plants growing in natural soils. Some have been made on plants grown in sterile sands (Greaves and Darbyshire 1972; Dart and Mercer 1964), radiation-sterilized soils (Greaves and Darbyshire 1972), or even quite artificial supporting media such as vermiculite (Dart and Mercer 1964) or Ca-permutitite and Ca-bentonite (Jenny and Grossenbacher 1963). Some observations are on roots growing in continuous liquid culture (Dart and Mercer 1964). Where soils have been used, they have been reconstituted from sieved air-dry materials (Greaves and Darbyshire 1972), and plants grown in the soils in plant pots (Leppard 1974; Leppard and Ramamoorthy 1975), and even boiling-tubes (Greaves and Darbyshire 1972); these have been incubated in controlled environments, with ample nutrient supplies, and in some cases specified microorganisms have been deliberately introduced into the system (Greaves and Darbyshire 1972; Darbyshire and Greaves 1971; Dart and Mercer, 1964). In most cases therefore, only very young plants, and even seedlings, were used, so that it is very unlikely that the rhizospheres observed were at all similar to those likely to be found on mature plants from natural soils in field environments. On the other hand, Foster and Rovira (1973, 1976, 1978) and Old and Nicolson (1975) examined plants from natural soils: from wheat fields and irrigated pastures in the case of Foster and Rovira, and from dune sands in the case of Old and Nicolson.

The conventional plating techniques suffer from the disadvantages that, although the biochemical capabilities of the various microorganisms can be determined, one cannot know whether they actually do perform these reactions in the very different conditions existing in the rhizosphere, and that one cannot know precisely where in the rhizosphere particular forms occur in nature. A combination of histochemistry and transmission electron microscopy is needed to provide the link between the biochemical characteristics used in the identification of rhizosphere microorganisms, and the morphological and cytological characteristics observed in electron micrographs of roots fixed *in situ*. For example, when rhizospheres such as that in Fig. 41.3 are treated for the enzyme acid phosphatase (Fig. 41.8), only 25% of the organisms gave a positive reaction; particular colonies were very reactive, whilst neighbouring cells a micron or two away gave no reaction. Attempts to extend this kind of biochemical electron microscopy in natural soils and rhizospheres, however, have been somewhat inhibited by the fact that the reagents of enzyme-specific media, especially the heavy metals necessary to confer electron density on the reaction products, tend to be adsorbed on the colloidal components of the soil fabric (Foster and Martin 1981).

Despite these difficulties of technique, enzyme tests have suggested that even in the rhizosphere not all the bacteria with cytologically intact protoplasts may be metabolically active, and this may partially account for the observation that plate counts are much lower than direct observations would suggest. But there must be competition between rhizosphere organisms for nutrients and space; in electron micrographs, colonies appear to be closely pressed against themselves and the adjacent soil, so that the space between the root and the soil is filled with bacteria and their capsule materials (Figs 41.3, 41.6). Electron micrographs also reveal the presence of pathogenic microbes such as *Bdellovibrio* and bacteriophage (Figs 41.6, 41.9) in natural rhizospheres. These are associated with several types of bacteria, and they may be involved in the biological control of species in the rhizosphere (Foster and Rovira 1978).

One of the most striking differences between rhizosphere and non-rhizosphere microorganisms is that in the former there are large deposits of the bacterial energy reserve polyhydroxybutyrate, which occurs as electron transparent (white) droplets in the cells (Figs 41.3, 41.5, 41.6); this is true whether the organisms are inside the root (Fig. 41.5) or external to it (Fig. 41.3). Some organisms also contain polyphosphate, detectable as small electron dense crystals inside the cells (Fig. 41.3); this may act as a phosphate reserve for the whole rhizosphere complex, including eventually the root. Bacteria in the bulk soil have very little polyhydroxybutyrate or polyphosphate; this supports a view that many bacteria remote from living roots may be in a largely dormant state (Clark and Paul 1970), and only those few fortunates which happen to lie near organic matter, either debris or capsular material, survive to give rise to the transient population explosions observed when a root happens to pass by.

Although in pastures the bulk of the surface soil is probably influenced by roots (Rovira 1978), it is likely that in the arid regions occupying most of Australia, where the organic matter content of the soil is very low, conditions for soil microorganisms are very severe indeed.

Consequences of changes in the rhizosphere

Ultrastructural studies such as these, show that the precise nature of the root/soil boundary depends on the species and soil type, and that on any one root it changes with the age of the adjacent root cells. Within a few weeks, the *functional* surface of a given point on the root of an annual crop like wheat will migrate from the mucigel surface (at the root apex) to the epidermal cell surface (in the older root hair zone) and then progressively deeper and deeper into the root, so that by anthesis the endodermis is probably best regarded as the root-soil interface of roots near the crown. For much of this time the *morphological* surface may remain unchanged. It is interesting to note in this regard the similarities between the cell wall structures of the outer tangential faces of the epidermis and of the endodermis. As the cortex is invaded by microorganisms, the cell walls are progressively reduced to the thin persistent polyphenol-enriched middle and terminal lamellae, and there appears to be some division of labour between the microorganisms in this process (Foster and Rovira 1976). Under the stress of diurnal shrinkage and swelling of the root and surrounding soil, the epidermis may break down, so that not only the soil microflora but even soil minerals may enter space formerly occupied by root tissue (Figs 41.4, 41.5). In root nutrition experiments, the consequences of these changes in the root/soil interfaces may be very significant.

The importance of ultrastructural studies of the rhizosphere lies in the information they provide on the extent of the mucigel, and on the distribution of the various types of microorganism at the root surface. Mucigel is not chemically inert, nor are the microoorganisms merely passive epiphytes occupying space at the rhizosphere. Microorganisms may compete with the root for inorganic ions, since water entering the root must first traverse the rhizosphere. Moreover, some organisms secrete biochemicals which profoundly influence plant growth and the growth of pathogens and symbionts (Rovira 1978); they affect the growth and longevity of root hairs, and thus influence nutrient uptake by the root; in some instances they actually control root morphology, as is the case in members of the *Proteaceae*. The presence of certain microorganisms has been shown to influence the uptake by plant roots not only of trace elements such as molybdenum and manganese but even of major nutrients such as phosphorus and potassium (see Rovira 1978 for review); they may enhance or inhibit the establishment of mycorrhizal associations (Bowen and Theodorou 1979), and may detoxify soils, enabling plants to grow where they otherwise could not, as in the case of the removal of hydrogen sulfide by sulfur bacteria in rice rhizospheres (Joshi and Hollis 1977). Conversely, rhizosphere organisms may compete with pathogens for root exudates, and inhibit them by the production of antibiotics. Again, the function of mucigel, whether composed mainly of mucilage produced by the roots or of slimes secreted by bacteria, may not be restricted to its lubricating properties; many of the ion exchange properties of roots may in fact lie in the mucigel, but its major function may be to maintain continuity between the minimal voids zone of the outer rhizosphere and the symplast of the root. Carbohydrate gels are able to swell as they are wetted and to contract as they dry out, *without* cracking. In drying soils, this property may be vital to the function of the root. By preventing the formation of an air gap between the root and the soil during the diurnal shrinkage and swelling of roots caused by transpiration-induced water stress cycles, mucigel may ensure the continuity of water and nutrient uptake in drying soils.

References

Bae, H. C., Cota-Robles, E. H., and Casida, L. E. (1972). Microflora of soil as viewed by transmission electron microscopy. *Appl. Microbiol.* **23**, 637-48.

Baker, K. F. and Cook, R. J. (1974). Biological Control of Plant Pathogens. pp. 218-220. (W. H. Freeman and Co. : San Francisco.)

Barber, S. A. and Ozanne, P. G. (1970). Auto-radiographic evidence for the differential effect of four plant species in altering the calcium content of the rhizosphere soil. *Soil Sci. Soc. Am. Proc.* **34**, 635-8.

Bland, D. E., Foster, R. C. and Logan, A. F. (1971). The mechanism of permanganate and osmium tetroxide fixation and the distribution of lignin in the cell wall of *Pinus radiata*. *Holzforschung* **25**, 137-43.

Blevins, R. L., Holowaychuk, N. and Wilding, L. P. (1970). Micromorphology of soil fabric at the tree-root-soil interface. *Soil Sci. Soc. Am. Proc.* **34**, 460-5.

Bowen, G. D. and Foster, R. C. (1978). Dynamics of Microbial Colonization of Plant Roots. *In* 'Symposium on Soil Microbiology and Plant Nutrition'. (Eds W. J. Broughton and C. K. John). (University of Malaya.)

Bowen, G. D., and Malajczuk, N. (1974). Microbial colonisation of root surfaces. *Aust. Plant. Pathol. Soc. Newsl.* **3**, 49-50.

Bowen, G. D. and Rovira, A. D. (1966). Microbial factors in short term phosphate uptake studies with plant roots. *Nature (London)* **211**, 665-6.

Bowen, G. D. and Rovira, A. D. (1976). Microbial colonization of plant roots. *Annu. Rev. Phytopathol.* **14**, 121-44.

Bowen, G. D. and Theodorou, C. (1973). Growth of Mycorrhizal Fungi around Seeds and Roots. *In:* 'Ectomycorrhizae – Their Physiology and Ecology'. (Eds T. T. Kozlowski and G. Marks). pp. 107-150. (Academic Press: New York.)

Bowen, G. D. and Theodorou, C. (1979). Interaction between bacteria and ectomycorrhizal fungi. *Soil Biol. Biochem* **11**, 119-126.

Campbell, R. and Rovira, A. D. (1973). The study of the rhizosphere by scanning electron microscopy. *Soil Biol. Biochem.* **5**, 747-52.

Clark, F. E. (1949). Soil microorganisms and plant roots. *Adv. Agron.* **1**, 241-88.

Clark, F. E. and Paul, E. A. (1970). The microflora of grassland. *Adv. Agron.* **22**, 375-435.

Darbyshire, J. F. and Greaves, M. P. (1971). The invasion of pea roots, *Pisum sativum* L., by soil microorganisms, *Acanthamoeba palestinensis* (Reich) and *Pseudomonas* sp. *Soil Biol. Biochem.* **3**, 151-5.

Dart, P. J. and Mercer, F. V. (1964). The legume rhizosphere. *Arch. Mikrobiol.* **47**, 344-78.

Döbereiner, J. and Day, J. M. (1974). Associative symbiosis in tropical grasses: characterization of microorganisms and dinitrogen fixation sites. Proc. Int. Symp. on N_2-fixation. Interdisciplinary discussions. pp. 518-538. (Washington State Univ. : Pullman.)

Foster, R. C. (1978). Ultramicromorphology of some South Australian Soils. *In* 'Modification of Soil Structure'. (Eds W. W. Emerson, R. D. Bond and A. R. Dexter). pp. 103-109. (Wiley: London.)

Foster, R. C. (1981). The ultrastructure and histochemistry of the rhizosphere. *New Phytol.* **89**, 263-73.

Foster, R. C. and Marks, G. C. (1967). Observations on the mycorrhizas of forest trees. The Rhizosphere of *Pinus radiata. Aust. J. Biol. Sci.* **20**, 915-26.

Foster, R. C. and Martin, J. K. (in press). *In situ* analysis of soil components of biological origin. Soil Biochemistry. Vol. 5 (Eds E. H. Paul and J. N. Ladd).

Foster, R. C. and Rovira, A. D. (1973). The rhizosphere of wheat roots studied by electron microscopy. *Bull. Ecol. Res. Comm (Stockholm).* **17**, 93-102.

Foster, R. C. and Rovira, A. D. (1976). Ultrastructure of wheat rhizosphere. *New Phytol.* **76**, 343-52.

Foster, R. C. and Rovira, A. D. (1978). The Ultrastructure of the Rhizosphere of *Trifolium subterraneum* L. *In* 'Microbial Ecology'. (Eds M. W. Loutit and J. A. R. Miles) pp. 278-90. (Springer Verlag: Berlin.)

Greaves, M. P. and Darbyshire, J. F. (1972). The ultrastructure of the mucilaginous layer on plant roots. *Soil Biol Biochem.* **4**, 443-9.

Hiltner, L. (1904). Uber neuere Erfahrungen und Probleme auf dem Gebiet der Bodenbakteriologie und unter besonderer Berücksichtigung der Grundungung und Brache. *Arb. dt. Landw-Ges.* **98**, 59-78.

Jenny, H. and Grossenbacher, K. (1963). Root-soil boundary zones as seen in the electron microscope. *Soil Sci. Soc. Am. Proc.* **27**, 273-7.

Joshi, M. M. and Hollis, J. P. (1977). Interaction of *Beggiatoa* and rice plant: Detoxification of hydrogen sulfide in the rice rhizosphere. *Science* **195**, 179-80.

Leppard, G. G. (1974). Rhizoplane fibrils in wheat: Demonstration and Derivation. *Science* **185**, 1066-7.

Leppard, G. G. and Ramamoorthy, S. (1975). The aggregation of wheat rhizoplane fibrils and the accumulation of soil bound cations. *Can. J. Bot.* **53**, 1729-35.

Malajczuk, N. and Bowen, G. D. (1973). Proteoid roots are microbially induced. *Nature (London).* **251**, 316-7.

Martin, J. K. (1977). Factors influencing the loss of organic carbon from wheat roots. *Soil Biol. Biochem.* **9**, 1-9.

Martin, J. P. (1971). Decomposition and binding action of polysaccharides in soil. *Soil Biol. Biochem.* **3**, 33-41.

Newman, E. I. and Bowen, H. J. (1974). Patterns of distribution of bacteria on root surfaces. *Soil Biol. Biochem.* **6**, 205-11.

Oades, J. M. (1978). Mucilages at the root surface. *J. Soil Sci.* **29**, 1-17.

Old, K. M. and Nicolson, T. H. (1975). Electron microscopical studies of the microflora of roots of sand dune grasses. *New Phytol.* **74**, 51-8.

Old, K. M., Hallam, S. and Nicolson, T. H. (1975). Helically lobed bacteria in plant roots. *Soil Biol. Biochem.* **1(1)** 73-5.

Patriquin, D. G. and Döbereiner, J. (1978). Light microscopy observations of tetrazolium-reducing bacteria in the endorhizosphere of maize and other grasses in Brazil. *Can. J. Microbiol.* **24**, 734-42.

Richards, B. N. and Voigt, G. K. (1964). Role of mycorrhizas in nitrogen fixation. *Nature (London)* **201**, 310-11.

Rovira, A. D. (1965). Plant Root Exudates and Their Influence upon Soil Microorganisms. *In* 'Ecology of Soil-borne Plant Pathogens – Prelude to Biological Control'. (Eds K. F. Baker and W. C. Snyder). pp. 170-186. (Univ. Calif. Press: Berkeley.)

Rovira, A. D. (1969). Plant root exudates. *Bot. Rev.* **35**, 35-57.

Rovira, A. D. (1978). Microbiology of Pasture Soils and some Effects of Microorganisms on Pasture Plants. *In* 'Plant Relations in Pastures'. (Ed. J. R. Wilson). pp. 95-110. (CSIRO: Melbourne.)

Rovira, A. D. and Campbell, R. (1974). Scanning electron microscopy of microorganisms on the roots of wheat. *Microbial. Ecol.* **1**, 14-22.

Rovira, A. D. and Campbell, R. (1975). A scanning electron microscope study of interactions between microorganisms and *Gaeumannomyces graminis* (Syn. *Ophiobolus graminis*) on wheat roots. *Microbial Ecol.* **2**, 177-85.

Rovira, A. D., Newman, E. I. and Bowen, H. J. (1974). Quantitative assessment of rhizoplane microflora by direct microscopy. *Soil Biol. Biochem.* **6**, 211-6.

Scannerini, S. (1968). Sull ultrastruttura delle ectomicorrize. II. Ultrastruttura di una micorriza di Ascomycete *'Tuber albidum' X Pinus strobus. Allionia* **14**, 77-95.

Chapter 42
Biology of the rhizosphere

J. K. Martin

HILTNER OBSERVED IN 1904 THAT THERE WERE MORE bacteria in soil surrounding plant roots than there were in soil free from plant roots. He concluded that the accumulation of bacteria was due to the activity of the roots, and proposed the term *rhizosphere* to describe the zone of soil where the microbial population was subject to the effects of plant roots.

Thus the concept of the rhizosphere arose from the recognition of the influence of plant roots on the microbial population in the surrounding soil. Consequently, much of the early work involved measurement of the numbers and types of microorganisms in the soil surrounding the roots of different plant species at different stages of growth, together with studies of the organic matter released from the roots. However, the economic and agronomic significance of the converse effect, the influence of soil microflora on plant growth, was also recognized. Apart from the obvious adverse effects of root pathogens and beneficial effects of mycorrhiza and nitrogen fixing organisms (see Chapters 43–45), there are four broad groupings of microbial activity directly or indirectly influencing plant growth. These arise from effects on the availability of the major nutrients nitrogen and phosphorus and sulfur, the availability of trace elements, plant physiology, and soil structure.

Initially, information on the rhizosphere was obtained with standard microbiological and biochemical techniques: spreading dilutions of soil suspensions on selective or non-selective media, and counting and identifying microbial colonies; determining the nutritional requirements and metabolic activities of microorganisms isolated from soil sampled at different distances from root surfaces; paper and ion-exchange chromatography of organic compounds present in nutrient solutions supporting growth of plants with or without microorganisms present; measuring the decomposition of organic material added to soil, frequently in the absence of plants, and with constant temperature and moisture conditions. However, since about 1970, techniques have been used which allow *in situ* measurements of soil microflora and estimates of the amount and nature of the organic material released into soil from growing plant roots.

This chapter will be divided into two sections: a review of the major facts established by traditional methods, followed by a review of recent work, with a discussion of the implications of the new information for future research on the rhizosphere in Australia.

Rhizosphere studies 1904–1970

Most of the work from this period has been extensively reviewed, so only a summary and some representative data are presented here.

Microbiology of the rhizosphere

The validity of Hiltner's proposal that there is a gradient of microbial numbers in soil around plant roots, with numbers decreasing with increasing distance from the root, was soon confirmed. However, it was evident that the extent to which plant roots influence microbial populations varies widely with plant species and soil type and soil moisture, and that there is frequently a selective action on different groups of soil organisms. The selective stimulation of microorganisms by roots was referred to as the *rhizosphere effect*, and the concept of the $R:S$ ratio was introduced (Katznelson 1946) in an attempt to quantify the rhizosphere effect. The $R:S$ ratio was defined as, the ratio of the microbial numbers per unit weight of rhizosphere soil (R) to the numbers per unit weight of adjacent non-rhizosphere soil (S). Since the greatest numbers of microorganisms in a rhizosphere soil are present on the root surface, wide variations are obtained for the $R:S$ ratio for a particular soil-plant association, dependent on the relative amounts of soil and roots contained in the unit weight of rhizosphere soil. Clark (1949) stated that the two components should be separated, and proposed the term *rhizoplane*, which he defined as being the external surface of plant roots together with any closely adhering particles of soil or debris. There would thus be two overlapping microbial populations influenced by roots; on the rhizoplane and in the adjacent rhizosphere soil.

A number of attempts have been made to determine the dimensions of the rhizosphere; it has been found that, although the outer limits show considerable variation, there is a consistent stimulation of microflora within distances of 3–5 mm from the root surface. Since

Barley (1970) considered that the half distance between axial parts of neighbouring roots in topsoils under dense crops or pastures is usually less than 2 mm, it follows that most of such top-soil will be rhizosphere.

Tables 42.1–3 present selected data which illustrate three major components of the rhizosphere effect: the differential stimulus of different classes of organisms in the rhizosphere, the difference in the rhizosphere populations of different plants grown in the same soil, and the effect of plant age on rhizosphere populations.

The work of Rouatt and co-workers showed that there are also differences in the metabolic activity, the nutritional requirements, and the morphology of bacteria from a wheat rhizosphere, compared with bacteria from root-free soil. The numbers of ammonifying and denitrifying bacteria were significantly ($P < 0.01$) higher in the rhizosphere than in the control soil (Rouatt *et al.* 1960). The rhizoplane flora from wheat plants was dominated by Gram-negative bacteria, predominantly *Pseudomonas* spp., and there was also a dominance in the rhizoplane of bacteria requiring amino acids for optimal growth. In the rhizoplane and the rhizosphere and in root-free samples respectively, the percentage incidence of Gram-negative bacteria was 79, 62, and 40, and of bacteria requiring amino acids, it was 58, 45, and 9.

More detailed coverage of rhizosphere microbiology will be found in reviews by Clark (1949), Krasil'nikov (1958), Rovira (1965*a*) and Rovira and Davey (1974).

Table 42.1. A comparison of the numbers of various groups of organisms in the rhizosphere of spring wheat and in control soil

From Rouatt *et al.* (1960)

Organisms	Organisms in oven-dry soil	
	Rhizosphere soil ($\times 10^6$ g^{-1})	Control soil ($\times 10^6$ g^{-1})
Bacteria	1200	53
Actinomycetes	46	7
Fungi	1.2	0.1
Protozoa	0.002	0.001
Algae	0.005	0.027

Table 42.2. A comparison of bacterial counts in the rhizoplane and rhizosphere of different plants and in root-free soil

From Rouatt and Katznelson (1961)

Plant	Cells ($\times 10^6$) per gram of:		
	Rhizosphere root (dry)	Rhizosphere soil (oven-dry)	Root-free soil (oven-dry)
Red clover	3844	3255	134
Flax	2450	1015	184
Oats	3588	1090	184
Maize	4500	614	184
Barley	3216	505	140
Wheat	4119	710	120

Table 42.3. The ratio of numbers of protozoa and bacteria in the rhizosphere (*R*) and unplanted soil (*S*) in relation to the age of *Sinapis alba* L.

Calculated from data of Derbyshire and Greaves (1967)

Plant age (days)	Amoebae (*R*:*S*)	Flagellates (*R*:*S*)	Total bacteria (*R*:*S*)
8	0.61	1.01	1.14
15	2.79	3.17	2.37
22	5.38	9.18	9.22
29	5.29	12.91	40.91
36	12.46	87.67	5.80
57	0.48	1.6	3.95

Organic material entering the rhizosphere

Starkey (1929) identified a number of factors that contributed to the stimulation of microorganisms by plant roots: 'The plants withdraw considerable amounts of substances from the rhizosphere, modify the soil structure by the penetration of roots, exert solvent effects by the excretion of carbon dioxide and eventually introduce large amounts of organic matter to the soil in the form of their dead tissues which undergo disintegration through the agency of the microorganisms. These changes all affect the development of the soil organisms'.

However, there was agreement that the major factor is the plant-derived organic matter introduced into the soil as excretion products or as sloughed and moribund epidermal and cortical tissue. It was proposed (Rovira 1965*b*), that root exudates are more important than cell debris in the young seedling, but that, as the plant develops, exudation declines, and decomposition of moribund tissue contributes a substantial portion of the nutrients to the rhizosphere population. There was a heavy emphasis on the nature and amount of organic material in root exudates, with most of the studies employing axenic plants grown in nutrient or sand culture. Rovira and McDougall (1967) summarized the results of more than 120 publications, and presented a composite table of the compounds exuded from wheat roots as reported in 10 different studies. The compounds identified included nine sugars, eighteen amino acids, ten organic acids, nucleotides, and enzymes. The amounts and ranges of compounds exuded were shown to vary with plant species, plant age, the conditions of growth (nutrition, temperature, light), and the presence or absence of microorganisms. Although there are wide variations, the total amount of carbon released from roots of plants grown in nutrient solution has been estimated to be $< 1\%$ of plant dry weight (Rovira and Davey 1974).

Electron microscopy provided evidence for an additional source of organic carbon for soil microorganisms, with the demonstration by Jenny and Grossenbacher (1963) that the root is surrounded by a mucilage filling the space between the root cell wall and mineral soil particles. Dart and Mercer (1964) showed that the mucilaginous layer surrounding legume roots could be heavily colonized by rhizobia, and suggested that the microorganisms utilize the mucilage as a carbon source.

During this period to 1970 there were no direct estimates of the amount of carbon released from roots as mucilage or from decomposing root tissue. Shamoot *et al.* (1968) measured the deposition of organic matter in soil by several grasses and legumes grown for 212 days in an atmosphere containing $^{14}CO_2$; they reported values for plant-derived carbon in root-free soil equivalent to 5–20% of the total dry weight of the plant, but their study gave no estimate of the relative amounts of soil carbon arising from the growing roots and from root residues after the plants reached maturity.

Additional material will be found in reviews by Rovira (1969), Hale *et al.* (1971), Rovira and Davey (1974), and in a colloqium 'La Rhizosphere' (Mangenot 1975).

Effects of rhizosphere microorganisms on plant growth

Availability of nitrogen

The activities of rhizosphere microorganisms determine the processes of nitrification, denitrification, mineralization of organically combined nitrogen, the immobilization of soil ammonium and nitrate nitrogen, and nitrogen fixation. These topics have been comprehensively reviewed (Bartholomew 1965) and discussions of recent work, particularly with Australian soils, can be found in Chapters 37 and 44.

Availability of phosphorus

A high proportion of the microorganisms isolated from rhizosphere soil release inorganic from organic phosphates, and many also have the ability to dissolve insoluble inorganic phosphates incorporated into agar media (reviewed by Hayman 1975). Hence there have been many proposals that the rhizosphere flora increases the plant availability of insoluble inorganic and organic soil phosphate. Mycorrhizal associations undoubtedly influence the phosphorus status of the host (Chapter 43), but otherwise, the probable situation is as stated by Hayman (1975): 'except when microbes find utopia as in litter in springtime, the liberation of soluble phosphates surplus to their own requirements probably contributes only slightly to the phosphorus nutrition of plants'.

Evidence for the converse effect, competition by microorganisms with roots for available phosphorus, has been presented by Barber (1969). There was apparent conflict with the results of Bowen and Rovira (1966), who reported higher uptake and translocation of phosphate by roots of non-sterile than by roots of sterile tomato and clover plants; however, it was subsequently shown (Barber *et al.* 1976) that microorganisms enhance absorption and translocation in young seedlings during short periods of uptake, but cause decreases over longer periods and in older plants.

Availability of sulphur

Most of the sulphur in the surface horizon of well-drained, non-calcareous soils is in organic combination. As it does with nitrogen and phosphorus, the soil microflora mediates the transformation of inorganic to organic forms, and the mineralization of organic sulphur. However, there is even more uncertainty for sulphur than for nitrogen and phosphorus as to the nature of the organic compounds in soil, and the extent to which they contribute to plant nutrition. Reviews on sulphur transformations in soil have been published by Freney and Swaby (1975), Till (1975) and Fitzgerald (1976), the former two being related to Australian soils.

Availability of trace elements

There is evidence for an involvement of soil microorganisms in a zinc deficiency of fruit trees and a manganese deficiency of oats (Barber 1968). The influence of microorganisms on the plant availability of manganese has also been discussed by Bromfield and David (1976). Loutit and Brooks (1970) have reported an effect of rhizosphere microflora on the molybdenum content of vegetables; they showed that plants of the same variety, grown under the same environmental conditions in two different soils, supported qualitatively different rhizosphere populations, and that these populations influence the molybdenum concentration in the plants.

Effects on plant physiology

Inhibitory effects on root development by non-invasive soil microflora have been reported by Bowen and Rovira (1961), and Darbyshire and Greaves (1970). Stimulatory effects on plant growth by the rhizosphere microflora have also been claimed, particularly by Russian workers, as the result of seed inoculation with *Azotobacter chroococcum* or *Bacillus megaterium*. It now seems likely that the production of growth-regulating substances by the bacteria is responsible for the stimulation, rather than it being caused by an increase in plant available nitrogen or phosphorus as originally claimed (Brown 1975).

Stabilization of soil structure

The mechanisms responsible for the aggregation of soil particles and for the maintenance of such aggregates are only partially understood, but there is little doubt that the soil microflora is heavily involved. Plants, particularly perennial grasses, are very active in promoting a desirable soil structure, and there is substantial evidence that the rhizosphere flora contributes to the structure, partly by the growth of fungal and actinomycete hyphae over soil particles and partly by the production of polysaccharide gums. Root mucilage and autolytic products from the decomposition of plant and microbial tissue also seem to be involved (Russell 1973; Hepper 1975).

Soil enzymes

Nearly all the transformation of organic molecules occurring in soil are enzyme-mediated, and most of the enzyme reactions occur within cells. However, a number of processes are thought to occur outside cells, e.g. polysaccharides and proteins are degraded by cellulases and proteinases released by microorganisms which then assimilate the decomposition products. Enzymes will also be released into soil by lysis of senescent plant and microbial cells, and there is substantial evidence that many of these enzymes are stabilized, and that their activity may persist for many years, e.g. Skujins and McLaren (1968) reported phosphatase and urease and dehydrogenase activities in permafrost soils of geological age > 8000 years.

There has been considerable progress in the qualitative recognition of *free* enzyme activity in soils, and Skujins (1976) stated that over 50 different enzymes, representative of all enzyme classification groups, have been detected in soils. However, quantitative estimation of extracellular enzyme activity has not made the same advances, primarily because of problems in methodology, e.g. during an assay some intracellular enzyme activity may be released, or substrate or product may be lost by sorption onto soil particles or assimilation by the microflora. Comparison between soils is uncertain because of the influence of soil variables such as pH value and humus and clay content, or enzyme activity and the accessibility of substrates.

Descriptions of the enzyme activities found in soil, and discussion of the methods used and their shortcomings, can be found in reviews by Skujins (1967, 1976), Kuprevich and Shcherbakova (1971), Ladd and Butler (1975), and Burns (1977).

Rhizosphere studies since 1970

Rhizosphere studies received a major impetus in the middle to late 1960s from two distinct sources: the United States space exploration program involving the development of methods to detect extra-terrestrial life, and the International Biological Programme (1964–74) (IBP), particularly the section on terrestrial productivity. There was a critical need to have values for the energy and nutrient fluxes below ground for all of the ecosystems under study. This resulted in a major effort to quantify microbiological processes in the field, with special emphasis on microbial biomass and the associated topics of energy flux and nutrient cycling.

Measurements of microbial biomass in the soil

Microbial biomass was the most critical area requiring new techniques, since the deficiencies of the standard plate count for soil microflora had been widely recognized, e.g. Skinner *et al.* (1952) showed that estimates of microbial numbers obtained by microscopic counts on stained films prepared from suspensions of soil in agar can be 40–50 times higher than the numbers obtained from plate counts. There was also evidence for a qualitative difference, and it was suggested that the plate count estimates an element of the soil population that reacts to external factors differently from the total population. Three different approaches were used in attempts to obtain better quantitative estimates of the soil microflora.

Improvements to light microscopy

These were sought because the original method for direct counts of soil organisms (Jones and Mollison 1948) had the disadvantages of not distinguishing live from dead cells, and it was extremely difficult to differentiate small microbial cells from soil particles. The fluorescent dyes, acridine orange and fluorescein isothiocyanate and europium chelate (Anderson and Slinger 1975), have improved the recognition of microbial cells in soil smears, but there is still doubt as to their ability to distinguish live from dead cells.

Scanning and transmission electron microscopy

These have been used to obtain information on the morphology of the soil microflora and its spatial distribution on the surfaces of soil particles and roots. Quantitative estimates of the numbers of organisms can be obtained by serial sectioning of resin-impregnated blocks of soil. This technique has been employed by Foster to obtain numbers of microorganisms at the soil-and-root interface (see Chapter 41).

Indirect methods of estimating soil biomass

These involved the measurement of organic molecules characteristic of viable microbial tissue, such as muramic acid or adenosine 5-triphosphate, ATP. It seems probable that extracellular muramic acid persists in soil, so that microbial biomass would be overestimated. ATP should give a better measure, but there are methodological problems associated with extracting ATP quantitatively from viable cells, and with preventing sorption of the extracted ATP by soil particles. A system based on the CO_2 flush from partially sterilized soils (Jenkinson 1966) has given values in agreement with biomass estimated from direct counts (Jenkinson *et al.* 1976); because the small fraction of soil organic matter rendered decomposable by treatment with chloroform vapour is postulated to arise from the biomass, the amount of soil biomass can be estimated from the flush of CO_2 which follows reinoculation of the treated soil with soil microflora.

Further details of methods and some of the results obtained will be found in Parkinson *et al.* (1971) and Rosswall (1973).

Jenkinson and Powlson (1976) have estimated that the biomass (microflora and microfauna) in the topsoil

(0–23 cm) from long-term experiments at Rothamsted represents 2–3% of the organic carbon in these soils. Values for biomass carbon, measured by the fumigation technique, ranged from 440 kg ha^{-1} under fallow soil to 2020 kg ha^{-1} under permanent grass, with an average of 1107 kg ha^{-1} for the nine soils studied. Clark and Paul (1970) have reported similar values for the IBP native grassland site at Matador in Saskatchewan; the microbial biomass in the topsoil (0–30 cm), measured by light microscopy, was estimated to be, as dry matter, 200 g m^{-2}, which is equivalent to a carbon content of 1000 kg ha^{-1}, assuming that microbial dry matter contains 50% carbon.

Energy flux through the rhizosphere

Measurements of the respiratory activity of soils and litters have been used by ecologists to estimate the energy flow through terrestrial ecosystems (Macfadyen 1971). Long-term data obtained for native grasslands as part of the IBP have emphasized the very high proportion of dry matter production decomposed by the soil microflora and microfauna, at least in these ecosystems. Coleman (1973) estimated that CO_2 respired from soil in a successional grassland represented 65–76% of the 4.7 t ha^{-1} annual carbon input into the soil, after allowing for root respiration. Similarly, de Jong *et al.* (1974) estimated that CO_2 evolution from untreated native grassland plots at the Matador site represented a carbon production of 750 g m^{-2} for a 4-year period, compared with a total productivity of carbon for the same period of 500 g m^{-2} above ground, and an equal amount below ground, i.e. respired CO_2 was equivalent to all the below-ground and half of the above-ground production.

Measured as carbon, the value for respired CO_2 from the Matador site of 750 g m^{-2} during 4 years is equivalent to an annual value of some 1900 kg ha^{-1}; this is less than twice the carbon content of the microbial biomass in the topsoil from the same site, estimated by Clark and Paul (1970) to be equivalent to 1000 kg ha^{-1}. Since active microbial cultures produce several times their weight of CO_2, it was concluded by Clark and Paul, and independently by Gray and Williams (1971), that a lack of available nutrients causes the bulk of the microbial biomass in soil to be dormant for long periods of time.

A new dimension to the apparent disparity between microbial biomass and available carbon has come from recent studies of microbial numbers at the root-and-soil interface,and of the amounts of organic carbon released into soil from actively growing cereal roots.

Foster has estimated (Chapter 41) that microbial numbers per cubic centimeter adjacent (< 10 μm) to root surfaces may be as high as 200×10^9, compared with earlier estimates of 2×10^9 in a productive agricultural soil (Clark and Paul 1970). Foster has also observed that the microorganisms close to the root surface contain large amounts of inclusion material, probably poly-hydroxybutyrate and polyphosphate, whereas there is little evidence of reserve material in bacterial cells from the bulk soil.

Barber and Martin (1976) showed that organic carbon equivalent to 12–18% of the net dry matter production may be released into soil from the roots of 3-week-old wheat and barley plants. Additional experiments (Martin 1977*a*) with wheat grown continuously in an atmosphere containing $^{14}CO_2$ at temperatures of 10°C or 18°C for periods from 3 to 8 weeks showed the organic carbon released into the soil to be equivalent to 29–44% of the ^{14}C translocated from shoots to the roots; it was proposed, that most of the organic carbon resulted from lysis of epidermal and cortical tissue and from sloughed root cap material. Tissue degeneration, which apparently starts before tillering and continues past flowering, appears to be part of normal root development for wheat, but may be accelerated by the soil microflora, without cell invasion (Martin 1977*a*). There is also evidence (Martin 1977*b*) that localized water stress increases lysis of root tissue and increases the secretion of root mucilage.

Thus the direct evidence, from transmission electron microscopy, for large and active microbial populations adjacent to root surfaces is supported by the ^{14}C data, which shows that release of organic carbon from active roots is substantial and in amounts representing up to 20% of net photosynthate. A partial characterization of the ^{14}C-labelled material remaining in root-free soils shows (Martin 1977*c*) that, as well as high molecular weight humic materials, there are materials present behaving as simple sugars and amino acids. It has been proposed (Martin 1978) that there is a succession of different classes of plant-derived material which could support different groups of the rhizosphere microflora: the first are the root exudates and root mucilage, which could support a population having very specific growth requirements and capable only, perhaps, of growth at root surfaces; the next source of organic nutrient arises from autolysis of epidermal and cortical tissue, and is probably responsible for the stimulation of Gram-negative organisms observed in the rhizosphere of wheat plants from early tillering to flowering; finally, subsequent to cell lysis there will be extensive colonization of the cell wall by fungi and actinomycetes and bacteria.

There are two explanations which could reconcile the transmission micrographs of a very large active microbial population at root surfaces to the apparent low growth rate averaged over the whole microbial biomass in soils. Many of the microorganisms which have grown at the root surfaces may remain dormant until new root growth stimulates further multiplication, or many of the microbial cells in the bulk soil, though non-viable, may nevertheless have a real, though low, respiration, and retain intact osmotic barriers and an ATP pool (Postgate 1973).

Further evidence for a substantial loss of carbon from growing roots came from experiments by Sauerbeck and Johnen (1977), who grew wheat to maturity in an atmosphere containing $^{14}CO_2$. These experiments showed the amount of root carbon mineralized to CO_2 during growth to be nearly three times the amount of root carbon recovered at harvest. More direct evidence has come from Jenkinson and Rayner (1977) who used the increase, resulting from thermonuclear tests, in the ^{14}C content of soil organic matter to estimate the amount of fresh organic carbon entering the soil each year in the permanent wheat rotation at Rothamsted. For the unmanured plot, they calculated values of organic carbon of 1200 kg ha $^{-1}$ entering the top 23 cm of soil; there were no estimates of root production, but other experiments at Rothamsted (Welbank *et al.* 1974) gave values per hectare for root dry weight 6 weeks before harvest equivalent to 300 kg of (organic) carbon. There would be some carbon addition to the soil from the crown and the residual stubble, but it is evident that there was a substantial input of organic carbon into the soil from the wheat roots prior to maturity.

Nutrient cycling within the rhizosphere

Recognition of the very considerable amount of carbon released into the soil from cereal roots in the early stages of growth has added emphasis to the point made by Barber (1968) that rhizosphere organisms will be in competition with plant roots for nutrients. The possible magnitude of the effect is shown by Cole *et al.* (1977), who estimated a phosphorus uptake by microbial and other soil decomposers of plant material as 2.5 and 3.2 g m^{-2} for two IBP grassland sites compared with plant uptakes of 0.5 and 1.2 g m^{-2} for the corresponding sites. Similarly, Hiltbold *et al.* (1950) estimated that, for every two parts of nitrogen ammonified from soil organic matter, one part is re-utilized by the soil microflora.

Calculations based on the data of Jenkinson and Rayner (1977) show that there will be an annual incorporation of carbon of some 400 kg ha^{-1} into the soil organic matter in the unmanured continuous wheat plot at Rothamsted. Accepting a value for the ratio of organic carbon: nitrogen: phosphorus in topsoils of 120:10:1 (Barrow 1961), there will be an annual nitrogen incorporation of 33 kg ha^{-1}, and of phosphorus 3 kg ha^{-1}, into the soil organic matter. These values are comparable to the amounts of nitrogen and phosphorus in the grain harvested from this plot, 20.6 kg ha^{-1} for N and 3.8 kg ha^{-1} for P (Johnston 1969). An immobilization of nitrogen and phosphorus into soil organic matter equal to the nitrogen and phosphorus content of the harvested grain could also be true for those areas of southern Australia where the wheat yield is 1000 kg ha^{-1} or less.

However, climatic differences between England and Australia could alter the net incorporation of nitrogen and phosphorus into soil organic matter in the two areas. Studies in the African savanna (Jones and Wild 1975) have shown that the intense drying of the surface soil during the dry season leads to a very rapid flush of decomposition on re-wetting, and there is evidence that the longer and hotter the dry season the greater the breakdown of organic material at the start of the next wet season; the same is presumably true for southern Australia. The higher air temperatures in Australia than England would also result in different rates of breakdown of organic matter. Jenkinson and Ayanaba (1977) showed, that the decomposition of ^{14}C-labelled plant material in Nigerian soils followed the same pattern as that observed at Rothamsted, but was four times faster. The difference in rates is reasonably in accord with the difference between the mean annual air temperatures at the Nigerian (26.1°C) and Rothamsted sites (8.9°C). The rate of breakdown of soil carbon in southern Australia, where the mean annual air temperature ranges between 15 and 18°C, would therefore be expected to be approximately double the rate at Rothamsted. The organic carbon content of soil cropped continuously with wheat in the Waite Institute trials was comparatively constant from 1927–1967 (Steward and Oades 1972) at a value comparable to the unfertilized plot in the Broadbalk continuous wheat experiment, suggesting an input of organic carbon into the Waite soil approximately double the annual carbon input of 1200 kg ha^{-1} estimated for the Broadbalk soil by Jenkinson and Rayner (1977).

Soil management will also strongly influence the extent to which fertilizer nitrogen or phosphorus or sulphur accumulates in the soil in organic combination. Steward and Oades (1972) estimated that, of the phosphorus added as superphosphate to a permanent legume-grass pasture in the period 1929–1967, 30% had accumulated as soil organic phosphorus compared with values of 12% and 6% for the conversion to organic phosphorus of fertilizer phosphorus added to the continuous wheat and the wheat/fallow rotations during the same period. Table 42.4 shows a comparison between the ratio of the values for organic phosphorus accumulated in the continuous wheat, the wheat/fallow, and the wheat/pasture/fallow rotations at the Waite Institute between

Table 42.4. A comparison of organic phosphorus accumulated in soil under different crop rotations, and the phosphorus content of wheat harvested from the rotations, in the period 1929–67

Rotation	P content of harvested grain[A] (kg ha^{-1})	Increase in soil P(org)[B] (kg ha^{-1})	$\frac{\Delta \text{ soil P(org)}}{\text{grain P}}$
Continuous wheat	79.0	72.4	0.92
Wheat/fallow	95.7	17.3	0.18
Wheat/pasture/fallow	79.6	24.1	0.30

[A] Calculated from data of Norton (1976).
[B] Calculated from data of Steward and Oades (1972).

1929–1967 (Steward and Oades 1972), and the phosphorus content of the grain harvested from these plots in the same period (Norton 1976), assuming a phosphorus content of 0.26% of the dry weight of grain. The values for the continuous wheat rotation are in agreement with the proposal that the annual formation of soil organic phosphorus in the Broadbalk unfertilized plot is equivalent to phosphorus in the harvested grain. However, the introduction of a fallow into the rotation causes a marked reduction in the accumulation of soil organic phosphorus, presumably due to mineralization during the fallow period.

Conclusions

The studies on microbial biomass referred to above, together with studies on the microbial colonization of roots (Bowen and Rovira 1976), have emphasized the discontinuities of microbial populations in soil, both with respect to numbers of cells and to their metabolic activity. There has also been the realization that the rhizosphere continues past the root surface, into the epidermal and cortical tissue.

The amount of carbon flowing through the rhizosphere, probably some ten times greater than earlier estimates, must have a significant effect on the formation and mineralization of organic nitrogen and phosphorus and sulfur, as well as possibly representing a diversion of photosynthate sufficient to influence grain yield. Conversely, a substantial input of organic carbon into the rhizosphere increases the potential for nitrogen fixation by non-symbiotic microorganisms. Although this is probably relatively unimportant in temperate cereal and grassland agriculture, there is the possibility of significant nitrogen fixation in the soils of northern Australia, as in other tropical soils (Dart and Day 1975).

Particular emphasis has been given in this chapter to studies of total biomass and energy flow, with little discussion of detailed studies of individual microorganisms or of specific microbial–plant associations. Such studies have not been discounted by the recent work; rather, they have been placed in the context of the overall activity in natural systems over whole seasons.

There is the necessity for continuing studies in Australia at two levels: broad population changes and their relation to energy flux and nutrient turnover and soil structure; and narrow population balances and their influence on the colonization of plant roots by pathogens or symbionts, and the survival of these organisms in soil. Both are highly relevant to the major goals in Australian cereal and pasture production, i.e. maintaining or increasing yields with low fertilizer input, and reducing losses from soil-borne root diseases.

References

Anderson, J. R. and Slinger, J. M. (1975). Europium chelate and fluorescent brightener staining of soil propagules and their photomicrographic counting. II. Efficiency. *Soil Biol. Biochem.* **7**, 211-15.

Barber, D. A. (1968). Microorganisms and the inorganic nutrition of higher plants. *Annu. Rev. Plant Physiol.* **19**, 71-88.

Barber, D. A. (1969). *In* 'Ecological Aspects of the Mineral Nutrition of Plants'. (Ed. I. Rorison). (Blackwell: Oxford.)

Barber, D. A., Bowen, G. D. and Rovira, A. D. (1976). Effects of microorganisms on absorption and distribution of phosphate in barley. *Aust. J. Plant Physiol.* **3**, 801-08.

Barber, D. A. and Martin, J. K. (1976). The release of organic substances by cereal roots into soil. *New Phytol.* **76**, 69-80.

Barley, K. P. (1970). The configuration of the root system in relation to nutrient uptake. *Adv. Agron.* **22**, 159-201.

Barrow, N. J. (1961). Phosphorus in soil organic matter. *Soils Fert.* **24**, 169-73.

Bartholomew, W. V. (1965). *In* 'Soil Nitrogen'. (Eds W. V. Bartholomew and F. E. Clark). (Am. Soc. Agron. : Madison.)

Bowen, G. D. and Rovira, A. D. (1961). The effects of microorganisms on plant growth. I. Development of roots and root hairs in sand and agar. *Plant Soil* **15**, 166-88.

Bowen, G. D. and Rovira, A. D. (1966). Microbial factor in short-term phosphate uptake studies with plant roots. *Nature (London)* **211**, 665-6.

Bowen, G. D. and Rovira, A. D. (1976). Microbial colonization of plant roots. *Annu. Rev. Phytopathol.* **14**, 121-44.

Bromfield, S. M. and David, D. J. (1976). Sorption and oxidation of manganous ions and reduction of manganese oxide by cell suspensions of a manganese oxidizing bacterium. *Soil Biol. Biochem.* **8**, 37-43.

Brown, M. E. (1975). *In* 'Soil Microbiology'. (Ed. N. Walker). (Butterworths: London.)

Burns, R. G. (1977). Soil enzymology. *Sci. Prog. (Oxford)* **64**, 275-85.

Clark, F. E. (1949). Soil microorganisms and plant roots. *Adv. Agron.* **1**, 241-88.

Clark, F. E. and Paul, E. A. (1970). The microflora of grassland. *Adv. Agron.* **22**, 375-435.

Cole, C. V., Innis, G. S. and Stewart, J. W. B. (1977). Simulation of phosphorus cycling in semi-arid grasslands. *Ecology* **58**, 1-15.

Coleman, D. C. (1973). Soil carbon balance in a successional grassland. *Oikos* **24**, 195-9.

Darbyshire, J. F. and Greaves, M. P. (1967). Protozoa and bacteria in the rhizosphere of *Sinapis alba* L., *Trifolium repens* L., and *Lolium perenne* L. *Can. J. Microbiol.* **13**, 1057-68.

Darbyshire, J. F. and Greaves, M. P. (1970). An improved method for the study of the interrelationships of soil microorganisms and plant roots. *Soil Biol. Biochem.* **2**, 63-71.

Dart, P. J. and Day, J. M. (1975). *In* 'Soil Microbiology'. (Ed. N. Walker). (Butterworths: London.)

Dart, P. J. and Mercer, F. V. (1964). The legume rhizosphere. *Arch. Microbiol.* **47**, 344-78.

de Jong, E., Schappert, H. J. V. and MacDonald, K. B. (1974). Carbon dioxide evolution from virgin and cultivated soil as affected by management practices and climate. *Can. J. Soil Sci.* **54**, 299-307.

Fitzgerald, J. W. (1976). Sulphate ester formation and hydrolysis: a potentially important yet often ignored aspect of the sulphur cycle of aerobic soils. *Bacteriol. Rev.* **40**, 698-721.

Freney, J. R. and Swaby, R. J. (1975). *In* 'Sulphur in Australasian Agriculture'. (Ed. K. D. McLachlan). (Sydney Univ. Press: Sydney.)

Gray, T. R. G. and Williams, S. T. (1971). *In* 'Microbes and Biological Productivity'. 21st Symp. Soc. Gen. Microbiol. (Eds D. E. Hughes and A. H. Rose). (Cambridge Univ. Press: Cambridge.)

Hale, M. G., Foy, C. L. and Shay, F. J. (1971). Factors affecting root exudation. *Adv. Agron.* **23**, 89-109.

Hayman, D. S. (1975). *In* 'Soil Microbiology'. (Ed. N. Walker). (Butterworths: London.)

Hepper, C. M. (1975). *In* 'Soil Microbiology'. (Ed. N. Walker). (Butterworths: London.)

Hiltbold, A. E., Bartholomew, W. V. and Werkman, C. H. (1950). The use of tracer techniques in the simultaneous measurement of mineralization and immobilization of nitrogen in soil. *Soil Sci. Soc. Am. Proc.* **15**, 166-73.

Hiltner, L. (1904). Uber neuere Erfahrungen und probleme auf dem Gebiete der Bodenbakteriologie und unter besonderer Berück – sichtigung der Gründüngung und Brache. *Arb. Deut. Landw. Ges.* **98**, 59-78.

Jenkinson, D. S. (1966). Studies on the decomposition of plant material in soil. II. Partial sterilization and the soil biomass. *J. Soil Sci.* **17**, 280-302.

Jenkinson, D. S. and Ayanaba, A. (1977). Decomposition of carbon-14 labelled plant material under tropical conditions. *Soil Sci. Soc. Am. J.* **41**, 912-5.

Jenkinson, D. S. and Powlson, D. S. (1976). The effects of biocidal treatments on metabolism in soil. V. A method for measuring soil biomass. *Soil Biol. Biochem.* **8**, 209-13.

Jenkinson, D. S., Powlson, D. S. and Wedderburn, R. W. M. (1976). The effects of biocidal treatments on metabolism in soil. III. The relationship between soil biovolume, measured by optical microscopy and the flush of decomposition caused by fumigation. *Soil Biol. Biochem.* **8**, 189-202.

Jenkinson, D. S. and Rayner, J. H. (1977). The turnover of soil organic matter in some of the Rothamsted classical experiments. *Soil Sci.* **123**, 298-305.

Jenny, H. and Grossenbacher, K. (1963). Root–soil boundary zones as seen in the electron microscope. *Soil Sci. Soc. Am. Proc.* **27**, 273-7.

Johnston, A. E. (1969). The plant nutrients in crops grown on Broadbalk. Rothamsted Exp. St. Rep. Pt. 2, pp. 50-62.

Jones, M. J. and Wild, A. (1975). Soils of the west African Savanna. Commonw. Bur. Soils Harpenden Tech. Commun. No. 55.

Jones, P. C. T. and Mollison, J. E. (1948). A technique for the quantitative estimation of soil microorganisms. *J. Gen. Microbiol.* **2**, 54-69.

Katznelson, H. (1946). The rhizosphere effect of mangels on certain groups of soil microorganisms. *Soil Sci.* **62**, 343-54.

Krasil'nikov, N. A. (1958). 'Soil Microorganisms and Higher Plants'. (Acad. of Sci. U. S. S. R. English translation I. P. S. T. Press: Jerusalem.)

Kuprevich, V. F. and Shcherbakova, T. A. (1971). *In* 'Soil Biochemistry'. Vol. 2. (Eds A. D. McLaren and J. Skujins). (Marcel Dekker: New York.)

Ladd, J. N. and Butler, J. H. A. (1975). *In* 'Soil Biochemistry'. Vol. 4 (Eds E. A. Paul and A. D. McLaren) (Marcel Dekker: New York.)

Loutit, M. W., and Brooks, R. R. (1970). Rhizosphere organisms and molybdenum concentrations in plants. *Soil Biol. Biochem.* **2**, 131-5.

MacFadyen, A. (1971). *In* 'Methods of Study in Quantitative Soil Ecology'. IBP Handbook No. 18. (Ed. J. Phillipson). (Blackwell: Oxford.)

Mangenot, F. (1975). La rhizosphère. *Bull. Soc. Bot. Fr.* **122**, 1-215.

Martin, J. K. (1977*a*). Factors influencing the loss of organic carbon from wheat roots. *Soil Biol. Biochem.* **9**, 1-7.

Martin, J. K. (1977*b*). Effect of soil moisture on the release of organic carbon from wheat roots. *Soil Biol. Biochem.* **9**, 303-04.

Martin, J. K. (1977*c*). *In* 'Soil Organic Matter Studies'. Vol. 1. (IAEA: Vienna.)

Martin, J. K. (1978). *In* 'Microbial Ecology'. (Eds M. W. Loutit and J. A. R. Miles). (Springer-Verlag: Berlin.)

Norton, R. (1976). Permanent rotation experiment. Waite Agric. Res. Inst. Bienn. Rep. 1974-1975.

Parkinson, D., Gray, T. R, G. and Williams, S. T. (1971). 'Methods for Studying the Ecology of Soil Microorganisms'. IBP Handbook No. 19. (Blackwell: Oxford.)

Postgate, J. R. (1973). *In* 'Modern Methods in the Study of Microbial Ecology'. (Ed. T. Rosswall). (Ecological Research Committee/NFR: Stockholm.)

Rosswall, T. (1973). *In* 'Modern Methods in the Study of Microbial Ecology'. (Ed. T. Rosswall). (Ecological Research Committee/NFR: Stockholm.)

Rouatt, J. W., Katznelson, H. and Payne, T. M. B. (1960). Statistical evaluation of the rhizosphere effect. *Soil Sci. Soc. Am. Proc.* **24**, 271-3.

Rouatt, J. W. and Katznelson, H. (1961). A study of the bacteria on the root surface and in the rhizosphere soil of crop plants. *J. Appl. Bacteriol.* **24**, 164-71.

Rovira, A. D. (1965*a*). Interactions between plant roots and soil microorganisms. *Annu. Rev. Microbiol.* **19**, 241-66.

Rovira, A. D. (1965*b*). *In* 'Ecology of Soil-borne Plant Pathogens – Prelude to Biological Control'. (Eds K. F. Baker and W. C. Snyder). (Univ. Calif. Press: Berkeley.)

Rovira, A. D. (1969). Plant root exudates. *Bot. Rev.* **35**, 35-57.

Rovira, A. D. and Davey, C. B. (1974). *In* 'The Plant Root and its Environment'. (Ed. E. W. Carson). (Univ. Press Virginia: Charlottsville.)

Rovira, A. D. and McDougall, B. M. (1967). *In* 'Soil Biochemistry'. Vol. 1. (Eds A. D. McLaren and G. H. Peterson). (Marcel Dekker: New York.)

Russell, E. W. (1973). 'Soil Conditions and Plant Growth'. 10th edn. p. 493. (Longmans: London.)

Sauerbeck, D. R. and Johnen, B. G. (1977). *In* 'Soil Organic Matter Studies'. Vol. 1. (IAEA: Vienna.)

Shamoot, S., McDonald, I. and Bartholomew, W. V. (1968). Rhizodeposition of organic debris in soil. *Soil Sci. Soc. Am. Proc.* **32**, 817-20.

Skinner, F. A., Jones, P. C. T. and Mollison, J. E. (1952). A comparison of a direct-and a plate-counting technique for the quantitative estimation of soil microorganisms. *J. Gen. Microbiol.* **6**, 261-71.

Skujins, J. J. (1967). *In* 'Soil Biochemistry'. Vol. 1. (Eds A. D. McLaren and G. H. Peterson). (Marcel Dekker: New York.)

Skujins, J. J. (1976). Extracellular enzymes in soil. *Crit. Rev. Microbiol.* **4**, 383-421.

Skujins, J. J. and McLaren, A. D. (1968). Persistence of enzymatic activities in stored and geologically preserved soils. *Enzymologia* **34**, 213-25.

Starkey, R. L. (1929). Some influences of the development of higher plants upon the microorganisms in the soil. 1. Historical and introductory. *Soil Sci.* **27**, 319-34.

Steward, J. H. and Oades, J. M. (1972). The determination of organic phosphorus in soils. *J. Soil Sci.* **23**, 38-49.

Till, A. R. (1975). *In* 'Sulphur in Australasian Agriculture'. (Ed. K. D. McLachlan). (Sydney Univ. Press: Sydney.)

Welbank, P. J., Gibb, M. J., Taylor, P. J. and Williams, E. D. (1974). Root growth of cereal crops. Rothamsted Exp. Stn Rep. Pt. 2, pp. 26-66.

Chapter 43

Microbial determinants of plant nutrient uptake

G. D. Bowen

IT IS BECOMING INCREASINGLY CLEAR THAT SOIL micro-organisms associated with roots can have very large effects on nutrient uptake by almost all species of plants, and can markedly change the productivity of soils, especially those of low nutrient status which are so common in Australia. Microbial associations with roots thus play an important role in the nutrition of all Australian indigenous ecosystems such as eucalypt forests, semi-arid grasslands, and impoverished heath lands, and in optimizing productivity of managed agricultural and forestry systems. This opens new avenues for optimizing plant production with much lower addition of fertilizers. Correct management of the soil microflora can also suppress soil borne disease organisms (Chapter 45) but here I discuss mainly the microbial systems affecting ion uptake, and especially mycorrhizal stimulation of plant growth, as nitrogen fixation is treated in Chapter 44. Most research on mycorrhizal associations has focussed on their greater nutrient uptake in the traditional monoculture systems of agriculture and forestry, but they have almost certainly been extremely important also in the evolution and composition of mixed communities. Furthermore, if we regard mycorrhizas as an alternate root strategy we can perceive roles they may have in several situations deleterious to 'normal' root growth and function. Some of these situations are discussed.

The microbial systems

Soil organisms affecting ion uptake can be non-infective (restricted to the rhizosphere) or infective. The main examples of the latter are nitrogen-fixing bacteria and actinomycetes on legumes and non legumes such as *Casuarina*, and symbiotic fungal infections of the root, called *mycorrhizas*, causing large increases in uptake of ions from soils.

Mycorrhizas concerned with inorganic nutrition of plants are of three main types.

Ectomycorrhizas

Tree species in some eight families, including the genera *Pinus* and *Eucalyptus*; all pines and eucalypts are usually infected, at least to some extent. A fungus sheath (mantle) grows around short lateral roots, and hyphae usually grow between the cortical cells but rarely enter them (Fig. 43.1*a*). Mycorrhiza formation is usually accompanied by increased branching of the infected lateral roots, and extensive mycelial strand growth into soil can occur (Fig. 43.1*b*). However, penetration of the root is not necessary for the mycorrhizal response, as highly beneficial associations have been recorded in which growth of the fungi was restricted to the rhizosphere e.g. *Pisolithus tinctorius* and *Endogone eucalypti* on some eucalypts (Neumann 1959; Warcup 1975) and unidentified fungi on pine (Lamb and Richards 1971).

Ericaceous mycorrhizas

These occur on members of Ericaceae, Epacridaceae and Empetraceae, in which a fine weft of hyphae on the outside of the root is connected with hyphae entering the cortical cells (hence called 'endotrophic') to form compact intracellular masses.

Vesicular-arbuscular mycorrhizas

Referred to here as VA mycorrhizas, these are also endotrophic. In this case the fungus grows extensively between cortical cells, where it produces storage vesicles and penetrates the cells to produce vesicles and multi-branched arbuscles (Fig. 43.2). The arbuscle increases the surface area of plasmalemma available for transfers of nutrients between fungus and plant by 2–3 fold and a large stimulation of cytoplasm formation in the infected root cells also occurs (Cox and Tinker 1976). These mycorrhizas occur on most angiosperms (including most agricultural, pasture, forest, and horticultural species) but only rarely in Chenopodiaceae, Cruciferae, Cyperaceae and Juncaceae.

The inescapable conclusion is that almost all Australian native higher plants and almost all introduced species of plants used in farming and forestry are mycorrhizal, or have the potential to be mycorrhizal. Indeed some of our native plants, e.g. *Leptospermum*, some *Acacia* and some other native legumes can be both endomycorrhizal and ectomycorrhizal (Baylis 1971; J. H. Warcup, pers. comm.). This then leads to questions such as: How many different organisms cause these intimate associations? How do they function? Is one organism better than another? Can plant benefit be increased by

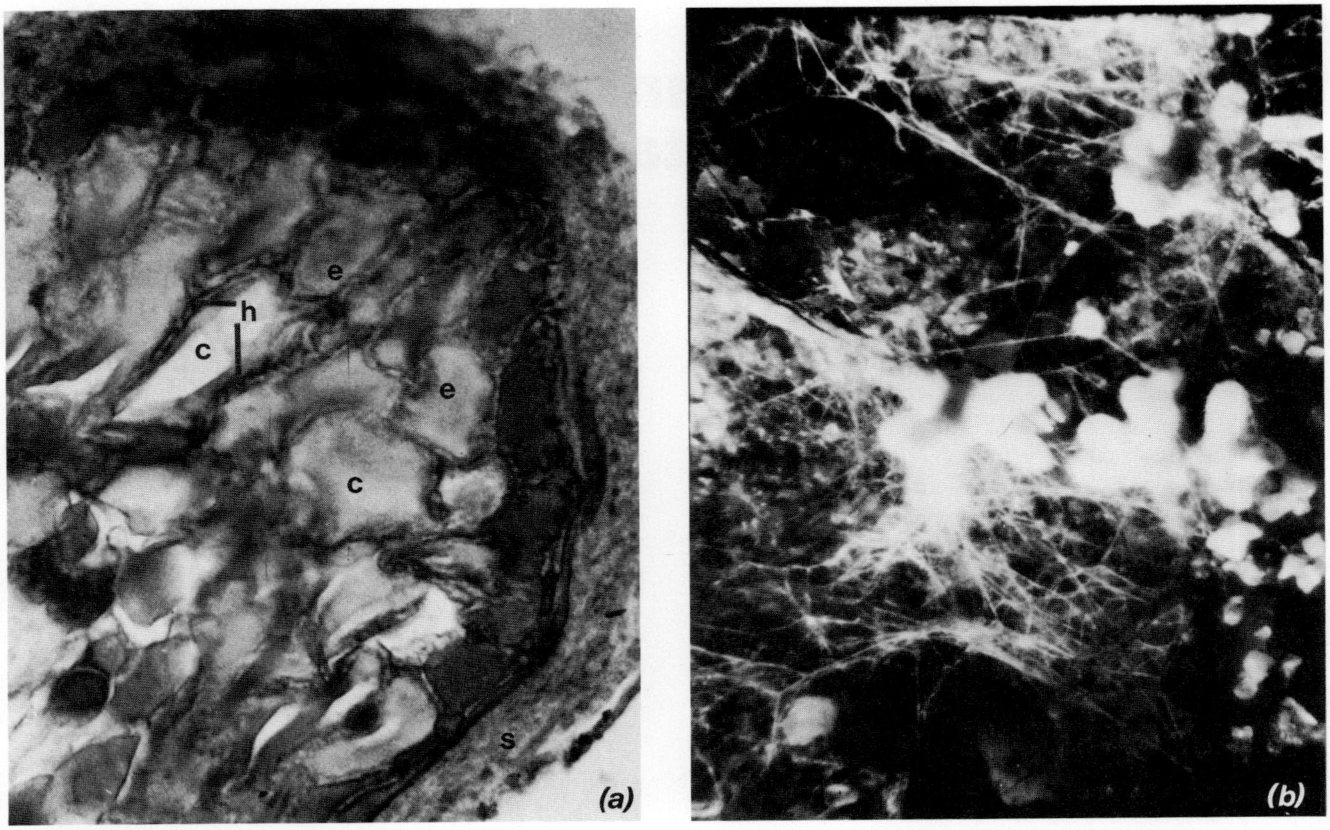

Fig. 43.1. (*a*) Transverse section of an ectomycorrhiza of *Pinus radiata* showing a well developed fungus sheath (s) penetrating between root epidermal (e) and outer cortical (c) cells to form a Hartig net (h); **(*b*)** extensive mycelial strand growth from ectomycorrhizas of *P. radiata* into soil.

changing the infecting organisms at a site or by increasing the amounts of roots being infected?

Ectomycorrhizas are usually thought to be formed mainly by several genera of Basidiomycetes, such as *Amanita*, *Suillus*, *Cortinarius*, *Russula*, *Rhizopogon*, *Pisolithus* and *Scleroderma* (Bowen 1969; Chilvers 1973; Ashton 1956), all very common in Australian forests. However, isolation of fungi from roots reveals that emphasis on fungi which form fruit bodies readily has been somewhat misplaced (Bowen 1969; Lamb and Richards 1970). Many other Basidiomycetes form fruit bodies readily, and a number of Ascomycetes (Trappe 1969; J. H. Warcup, pers. comm.) and at least two Phycomycetes, *Endogone flammicorona* and *E. eucalypti*, form ectomycorrhizae on *Pinus* or *Eucalyptus* (Fassi *et al.* 1969; Warcup 1975). The dominant mycorrhizal fungi on a plantation may change with time; Lamb (1974) isolated 48 probable mycorrhizal fungi from plantations of *P. elliottii* in New South Wales, most of which did not occur on seedlings in the nursery. There is thus a much greater reservoir of fungi to select from than had previously been thought.

Some ectomycorrhizal fungi infect a wide range of tree species but others infect only a very narrow range (Bowen 1969). There are conflicting reports on fungus-tree compatibility e.g. in studies by Lamb and Richards (1974), *Pisolithus tinctorius* formed mycorrhizas on *P. radiata* but in studies by Chilvers (1973), an isolate forming mycorrhizas on *E. st-johnii* did not form them on *P. radiata*. Under conditions used by Pryor (1956), *Scleroderma flavidum* formed mycorrhizas on three species of the *Eucalyptus* sub-genus Renantherae but not on *E. bicostata* (Macrantherae). In some cases specificity may be determined by the particular fungus strain used, for variation exists within mycorrhizal fungus species in many properties (Trappe 1977). In other cases, differences in infection may be due to the particular environmental conditions preventing infection, e.g. the absence of mycorrhizal infection in highly fertilized soils, has been correlated with low levels of sucrose in the root (Marx *et al.* 1977). [Such correlations however may not indicate cause and effect, for the fertility level may also affect root production of polyphenols or terpenes, which could play a major role in compatibility or incompatibility of the fungus and the root (Hillis *et al.* 1968; Krupa and Fries 1971). Another possible reason for failure of a fungus to produce mycorrhizas is that some environmental factors operate by preventing or

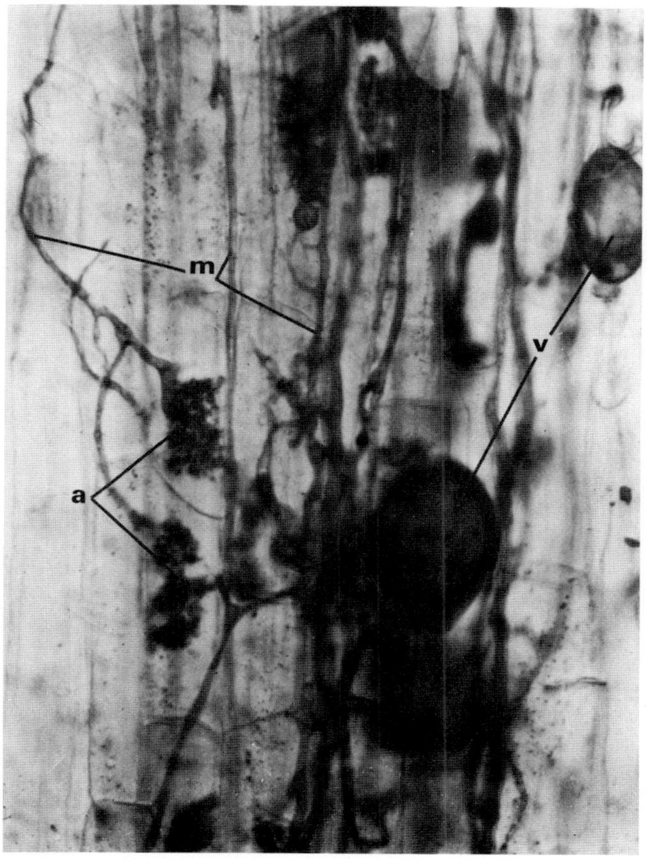

Fig. 43.2. Mycelium, (m), arbuscules (a) and vesicles (v) in a vesicular- arbuscular mycorrhiza of clover.

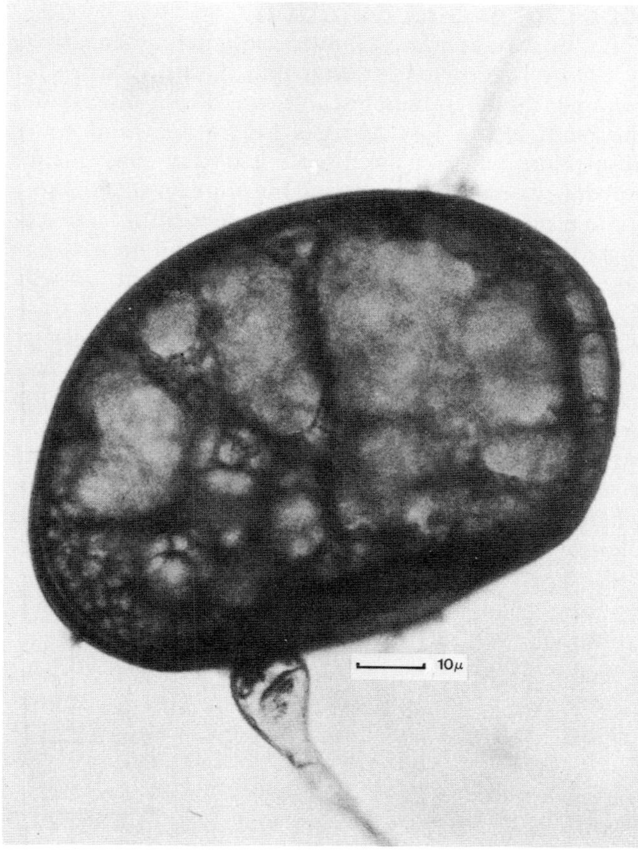

Fig. 43.3. An external spore of a vesicular–arbuscular mycorrhizal fungus (*Gigaspora* sp.) obtained from soil.

retarding fungus growth in the rhizosphere (Bowen and Theodorou 1973)].

Vesicular- arbuscular mycorrhizas are formed by several species of *Gigaspora*, *Acaulospora*, *Glomus* and *Sclerocystis*, all members of the Endogonaceae in the Phycomycetes. The fungi of these mycorrhizas are often collectively referred to as 'Endogone', although the genus *Endogone* itself has so far been implicated only in ectomycorrhizas. All of these genera occur in Australia under a wide range of conditions in indigenous ecosystems and in cultivated soils. They occur in soil in sporocarps or as external vesicles (Fig. 43.3), and can be recovered by wet sieving processes (Gerdemann and Nicolson 1963) or flotation methods. Using various stable spore characteristics, Mosse and Bowen (1968*a*) recognized eight different VA mycorrizal fungi in 117 of 141 Australian sites over a very wide range of soils and plant communities, but many more types certainly exist (e.g. Bevege 1971; Koske 1975; Sward *et al.* 1978). Soil type and agricultural practice play an important role in determining which species occur in particular soil samples (Mosse and Bowen 1968*b*; Abbott and Robson 1977*a*) and sometimes three or more of these fungi occur in the one soil. Some indications of specificity in the association occur but this is usually low and involves plant response rather than infection (Bevege 1971; Hall 1978*a*).

The same species of VA mycorrhizal fungi occur throughout the world (Mosse and Bowen 1968*a*; Gerdemann and Trappe 1974); rhizomes of the earliest land plants were infected with them (Nicolson 1975). Their wide distribution, their ability to infect a wide range of plants, their early evolution, and their known ability to increase nutrient uptake from soils low in nutrition point to their being critical factors in the evolution and success of land plants. Further study may show a similar argument can be sustained for other mycorrhizal types, for there is increasing evidence of a continuum of mycorrhizal types rather than sharp distinctions between them. For example the genus *Endogone,* closely related to VA fungi, cause ectomycorrhizas on eucalypts and pines (Fassi *et al.* 1969; Warcup 1975) and Zak (1976) has suggested arbutoid mycorrhizas of some ericaceous plants are formed by fungi causing ectomycorrhizas of forest trees.

Detailed discussions of the biology of ectomycorrhizas are given in Harley (1969) and Marks and Kozlowski (1973), and of endomycorrhizas in Harley (1969) and Sanders *et al.* (1975).

Root form and function

It is important that the influence of microorganisms on plants be interpreted in terms that the plant 'sees': for example the microbiologist may see a root disease but the plant sees a loss of roots for nutrient and water absorption; whether the disease is of any consequence will depend on the nutrient and moisture level of the soil, for in a rich soil the plant may be able to afford a loss of some roots, but the same loss in a poor soil may lead to large reductions in ion uptake and plant growth. Fig. 43.4 (discussed in more detail by Bowen and Cartwright (1977) and Bowen 1979), indicates processes involved in nutrition and growth of plants. This shows root growth and function are closely integrated with top growth, for the root absorbs nutrients and water and produces cytokinins (and possibly other growth compounds) which are transported to the tops of the plant where they markedly affect leaf size, leaf longevity, and photosynthetic rates. Photosynthates are then translocated to the root and provide energy for further root growth. Some of this energy is diverted to the use of micro-organisms associated with the root.

Ion uptake from soil is dominated by two factors: transport of ions to the root and the absorbing ability of the root. With conditions of high soil moisture some ions, e.g. nitrate and sometimes potassium, move rapidly to the root by convection in water and/or diffusion, and their absorption is determined mainly by the uptake ability of the root. At the other extreme some ions are immobile in soil, e.g. zinc, copper, molybdenum, or they diffuse to the root only very slowly, e.g. phosphate, and for such ions movement to the root is the limiting factor in uptake. Even a fairly sparse root system can quickly deplete soil of available mobile ions such as nitrate (Andrews and Newman 1970) but with ions such as phosphate, zinc and to a lesser extent potassium and ammonium, uptake is closely related to root growth and root hair production.

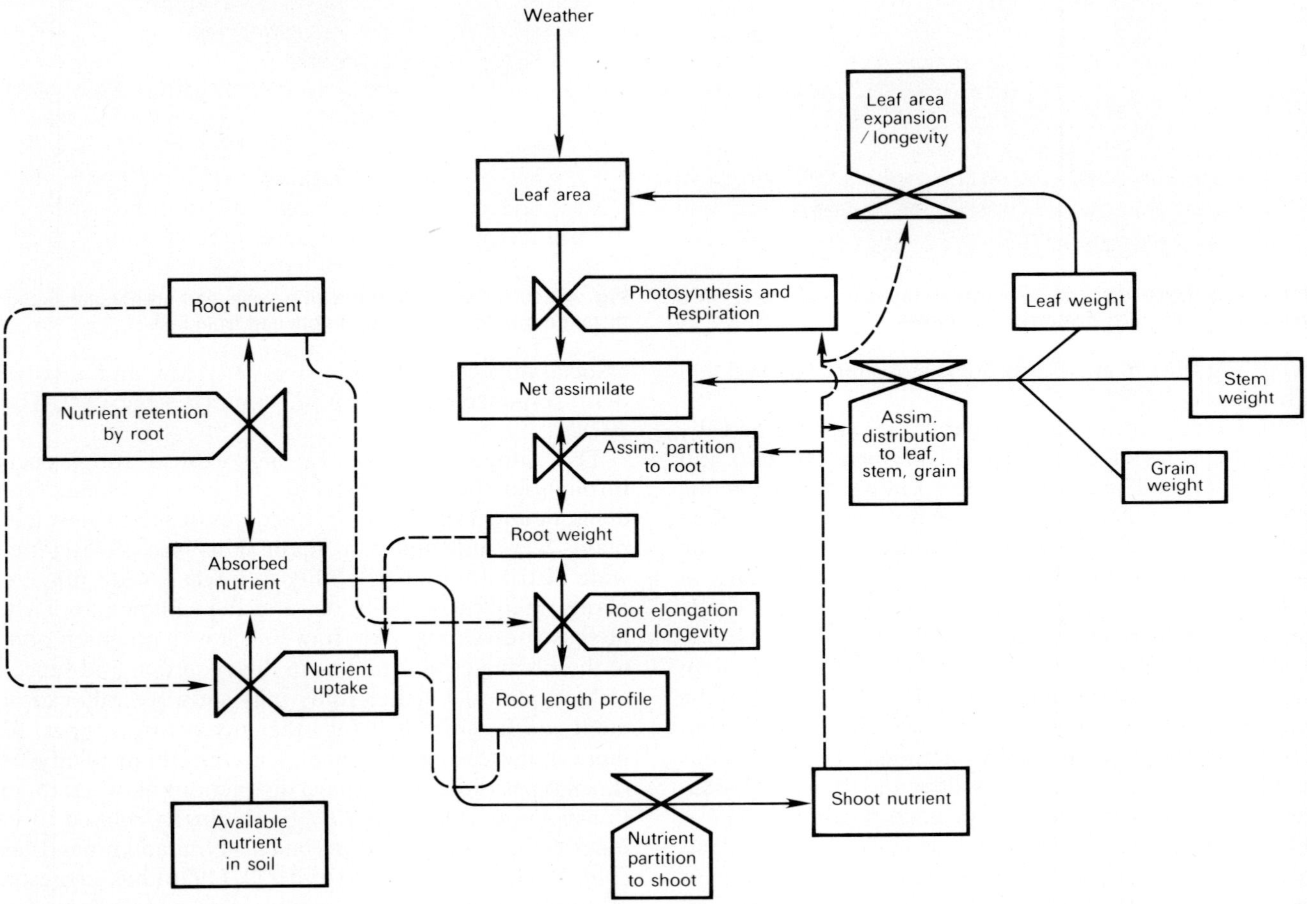

Fig. 43.4. Assimilation processes in plants. A scheme to represent the main compartments for assimilate and nutrient distribution in a simple model (after Bowen and Cartwright 1977). Rectangular boxes are used for measurable variables whose values at any time describe the state of the system at that instant. These boxes are connected by continuous lines to represent material (or energy) flows. The 'valve' symbols represent processes which affect the use and distribution of dry matter and nutrients. Process rates may be output values of the state variables on which they operate or they may be affected by variables elsewhere in the system. Such controls are shown by broken lines.

As soil dries, the tortuosity of diffusion paths increases, and apparent diffusion coefficients of ions decrease markedly (Marriott and Nye 1968); then transfer of ions to the root can become limiting even for relatively mobile ions such as potassium. The plant sometimes compensates partly by producing longer root hairs; Reid and Bowen (1979*a*) found in a sandy soil that root hairs on *Medicago truncatula* averaged 80 μm long at 22% soil moisture but 150 μm long at 15% and 220 μm long at 11% soil moisture.

For monoculture systems, uptake of poorly mobile ions particularly is governed largely by the geometry of the root system. However, in mixed systems such as mixed pastures, inter-row cropping, and in almost all natural ecosystems such as Australian eucalypt forests, the sharing of soil resources of even highly mobile ions between different species is governed by root length and morphology of the different species (Trenbath 1977).

The key components of root geometry are root length and root hair development. The former determines the speed and extent of colonization of new soil zones, while the latter affects the efficiency of uptake of poorly mobile ions and the extent of utilization of these ions from soils. It is seen from Fig. 43.4 that assimilation by the plant top (affected by plant species, climate, nutrition and disease), and the partitioning of assimilate between tops and roots, are important controls on root growth. However, given equal availability of assimilate for root growth, conversion of assimilate to root length varies considerably between species. Baylis (1975) recognized 'magnolioid' root systems at one end of this spectrum and 'graminoid' roots at the other. Magnolioid roots are usually relatively thick, with diameters usually more than 0.5 mm, coarsely branched, and produce few or no root hairs. Typical magnolioid root systems occur with many trees e.g. *Magnolia*, *Persea* (avocado), *Laurus*, *Liriodendron*, *Araucaria* (hoop pine), and citrus. Graminoid roots are formed by plants such as rushes, sedges and grasses; these roots are fine, often less than 0.1 mm diameter, much branched and often have profuse development of root hairs (1–2 mm long). The specific root length (*Lv*, the length of root per cm^3 of soil) of magnolioid roots is low compared with that of grasses and Barley (1970) reported *Lv* for the surface 10 cm of soil to be 30–50 cm cm^{-3} for grasses, whereas for woody plants, e.g. *P. radiata* and tea, it is only 2–4 cm cm^{-3}; herbs and legumes commonly fall between these extremes. Thus, graminoid roots and their root hairs occupy the upper layers of the soil very effectively whereas magnolioid root systems have very large volumes of unused soil between the roots. The *Lv* declines markedly with depth; cereals may have an *Lv* of only 2–4 at 25–100 cm depth (Barley 1970). Furthermore, nutrient deficiencies lead to much smaller root systems and thus a much lower *Lv* of most plants: e.g. Bowen *et al.* (1974) found total length of first order laterals on the third seminal root of wheat in phosphate deficient soil was only one fifth of that in soil fertilized with phosphate.

Root hairs can grow to 1.5 mm within one day, and a root of 400 μm diameter can increase the volume of soil it occupies by twenty times by producing root hairs 700 μm long. Such root hairs, some 100 μm apart, can deplete 60–100% of the available phosphate in soil contained in the root hair cylinder in 3–5 days (Lewis and Quirk 1967; Bhat and Nye 1973); Barley and Rovira (1970) found that peas with root hairs absorbed 70% more phosphate from a clay soil than peas without root hairs. Depletion of poorly mobile ions is quite restricted beyond the root hair zone and depends on the ion transmission characteristics of the soil and the time for which the root is active at that point (Bhat *et al.* 1976).

Large differences occur between plant species in root hair production; some legumes have well developed root hairs but others, e.g. *Stylosanthes* species, produce very few root hairs. The traditional picture of many well developed root hairs on most plants is far from correct and even in species capable of good root hair production environmental conditons such as calcium deficiency, soil compaction, high soil moisture and diurnal environmental fluctuations can drastically reduce their numbers and length (Cormack 1962; Champion and Barley 1969; Reid and Bowen 1979*a*). Unfortunately there is almost no good quantitative data on root geometry and morphology of Australian flora, and we must extrapolate from the generalizations from other plants (above): viz. that tree species tend to have low (but persistent) root densities, consisting largely of a stout structural framework forming smaller and more ephemeral roots, and too often have poor development of root hairs and thus a low intensity of soil occupation. Considering that most Australian soils are nutritionally deficient, one might have expected more Australian research on roots and their function, for this is obviously the key to efficient use of meagre soil nutrients by plants.

It follows that the major effects of microorganisms on ion uptake will be exerted partly through effects on ion absorption ability at the root surface but particularly through effects on root geometry. They affect root geometry both by their effects on root morphology itself and by adding their own characteristics to those of the roots. Microorganisms also affect nutrient transport to the tops of the plants. These matters are discussed in the following sections. Foliage diseases and insect pests affect root growth indirectly by reducing the assimilation by the plant, and virulent root diseases will directly reduce root length considerably with subsequent effects on nutrient uptake.

Effects of rhizosphere microorganisms on nutrition

Organisms in the general rhizosphere microflora can affect most of the parameters important in ion uptake.

They may, for example, affect the availability of ions in soil either by solubilizing poorly soluble forms of nutrients such as phosphate or by changing the oxidation state of ions such as manganese (Alexander 1961). The soil microflora can sometimes reduce root length and root hair production to such an extent that the volume of soil contained in the 'rhizocylinder' (and contributing poorly mobile ions such as phosphate) can be reduced by some 60% (Bowen and Rovira 1961) and it is possible that some of the large plant responses occurring with soil fumigation (Chapter 45) are due to reduction in the levels of sub-clinical pathogens which retard root growth. In laboratory studies some organisms have also stimulated root growth (Otto 1965; Rovira and Bowen 1971) but these have not been studied in field situations and their successful manipulation poses some problems.

The rhizosphere microflora can affect ion uptake at the root surface and ion translocation to the shoot in two ways. Micro-organisms may compete with the root for nutrients; this can be demonstrated in solution culture where microbial populations in the rhizosphere are very high (Barber 1966), but it is not likely to be a major factor in soil as only some 7–10% of the root is covered with micro-organisms and these take some 3–4 days to develop to that level on any part of the root (Bowen and Rovira 1976). The second way is that micro-organisms can greatly stimulate ion uptake and translocation of phosphate, potassium, manganese, zinc and iron, especially with very young plants (Bowen and Rovira 1966; Barber and Lee 1974; Barber *et al.* 1976; Williamson and Wyn Jones 1973); up to a quadrupling of potassium and phosphate transport to the tops has been recorded. These studies have been performed only in solution culture and need to be extended to soil. Growth regulators may be implicated, as in the hastening of flowering caused by inoculating soil with certain soil micro-organisms (Rovira 1963; Brown *et al.* 1964).

Proteoid roots

An extreme and important example of the stimulation of nutrient uptake by rhizosphere micro-organisms occurs with the production of proteoid roots on many Australian native plants. These roots are caused by non-infecting rhizosphere micro-organisms (Lamont and McComb 1974; Malajczuk and Bowen 1974), and occur on many plants of the Proteaceae such as *Banksia* and waratah, and on the native legume *Viminaria juncea* (Lamont 1972) and lupins (Trinick 1977). Intense lateral root formation is stimulated (Fig. 43.5*a*), and dense layers of proteoid roots are formed just below the surface of the soil. These roots absorb nutrients very efficiently from the soil solution (Jeffrey 1967) and possibly those in leachings from foliage and surface litter. Thus they help to form a tightly coupled nutrient cycling system which is very important in conserving nutrients in the extremely poor soils in which these plants grow in Australia. They also increase nutrient uptake from soil itself: inoculation of young *Banksia grandis* plants to form proteoid roots doubled plant growth in low phosphate soils in studies by Malajczuk and Bowen 1974). Roots of *Banksia serrulata* and *Hakea ulicina* with well formed proteoid roots stored (as polyphosphates) some 20–30% of a 24 h pulse of orthophosphate (Jeffrey 1968); this is reminiscent of storage of phosphate as polyphosphate in mycorrhizas, and is probably a further phosphate conservation mechanism.

Physiological effects on ion uptake also occur, for proteoid roots have an absorbing capacity per square centimeter of root surface several times that of non-proteoid roots (Malajczuk and Bowen 1974). They are considerably more active metabolically than the non-proteoid roots; ^{14}C-labelled assimilates are rapidly and preferentially translocated to the proteoid roots (Fig. 43.5*b*). However, these roots are very susceptible to the soil-borne root disease fungus *Phytophthora cinnamomi*, and their critical role in nutrient (and possibly water) uptake is indicated by the rapid decline of the plant after proteoid roots are attacked by this fungus. Indeed, in the jarrah forests of West Australia the proteaceous understorey is the first component of the forest to succumb to *P. cinnamomi*.

It is quite possible that further study of the native flora, especially those growing in low nutrient soils, will produce other instances of microbial modification of roots leading to increased nutrient uptake, e.g. roots of sedges often have swollen lateral roots thickly covered with long root hairs, 'dauciform' roots (Lamont 1974), and it is tempting to speculate that these might also be microbially induced. So far, studies on proteoid roots have focussed on increased phosphate uptake, but these roots also play a key role in uptake and recirculation of other nutrients such as zinc and other trace elements which are frequently deficient in Australian soils. There is certainly now little doubt that this microbial association has been critical to the evolution and establishment of proteaceous and possibly several other types of plants.

Mycorrhizal functions

Nutritional responses to mycorrhizas

There are many reports of responses of several plant species to inoculation with mycorrhizal fungi. They have often been glasshouse studies using sterilized soils but data exist also for field studies, where responses are still striking but are sometimes smaller because of naturally occurrring mycorrhizal fungi and/or difficulty in establishing inoculum.

Table 43.1 (from Lamb and Richards 1971) shows 7–8-fold increases in growth of *P. elliottii* in pots, and 10-fold increases in uptake of N, P and K, following inoculation with ectomycorrhizal fungi. There are also

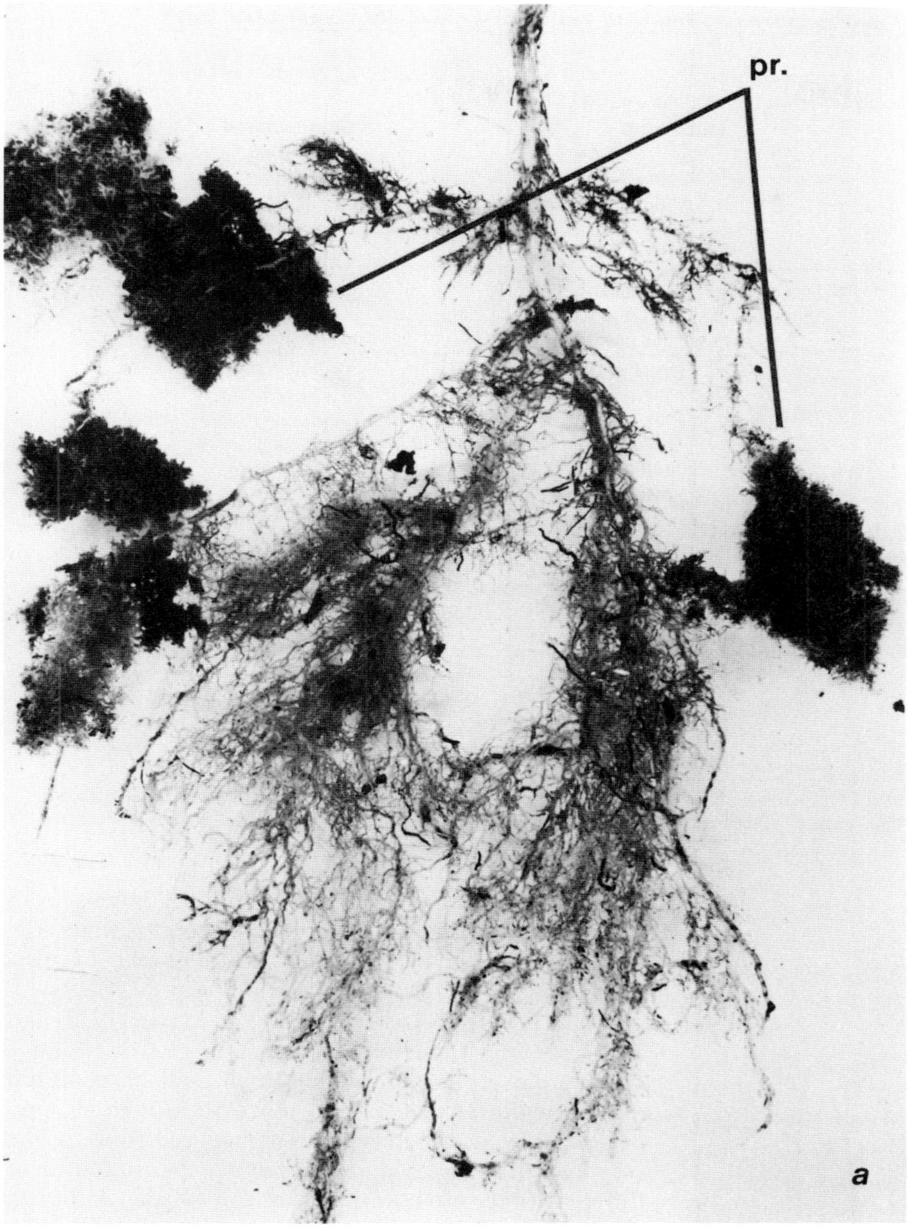

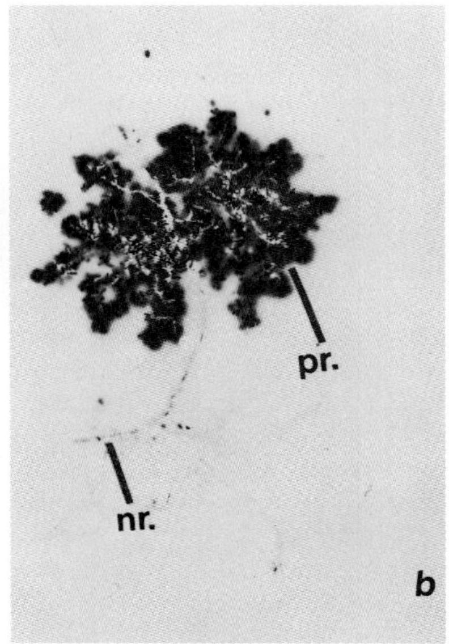

Fig. 43.5. (At actual size) **(*a*)** Proteoid root development (pr) in *Banksia grandis*. The thick clumps of roots, caused by microorganisms form a mat in the surface layers of soil. **(*b*)** Preferential translocation of ^{14}C-labelled assimilates to proteoid roots (pr) compared with non-proteoid roots (nr); 24 h after feeding tops with $^{14}CO_2$ (G. D. Bowen and N. Malajczuk, unpublished).

large differences between mycorrhizal fungi in the responses they give; Table 43.2 (from Bowen *et al.* 1972) shows that growth of *P. radiata* after five years in the field in South Australia was 50% greater with certain

Table 43.1. Differential response of *P. elliotii* var. *elliotii* to mycorrhizal fungi (from Lamb and Richards 1971)

Inoculum	Dry Wt (g)	Nutrient content (mg)		
		N	P	K
Nil	0.26	2.44	0.22	2.27
Isolate R. 8.35[A]	1.11	11.14	1.12	10.72
Rhizopogon sp.	1.96	25.15	2.20	23.77
Isolate R. 8.22[A]	2.81	38.92	3.86	35.60

[A] Unidentified isolates from mycorrhizas.

Table 43.2. Field response of *Pinus radiata* to different mycorrhizal fungi 60 months after planting (from Bowen *et al.* 1972)

Inoculum	Mean height[A] (cm)
Suillus granulatus #5	482
Rhizopogon luteolus #17D	454
Rhizopogon luteolus #10C	448
Suillus luteus #8	414
Uninoculated	334
LSD	40 $P=0.05$
	58 $P=0.01$

[A] Mean of 27 trees

Table 43.3. Responses of plants to inoculation with vesicular-arbuscular mycorrhizal fungi

Araucaria cunninghamii (Bevege, 1971)
At 26 months; the soil was a fertile red loam (total P 1600 ppm) and steam sterilized to eliminate naturally occurring VA mycorrhizal fungi. 20 t ha^{-1} of manure was applied before sowing.

Plant variable	Inoculated	Uninoculated
Dry wt (g)	76.1	7.2
Shoot N%	1.25	1.88
Root N%	0.96	0.94
Shoot P%	0.079	0.042
Root P%	0.109	0.053

Onion and *Paspalum notatum* (Mosse 1972)
Onion

Endophyte[A]	Dry wt per plant (mg)[B]	
	Soil 11 (pH 4.6)	Soil 7 (neutral)
Laminate	737 (m)	370 (m)
E3[C]	708 (a)	441 (a)
Honey coloured	689 (m)	475 (m)
Small brown clusters	649 (m)	631 (a)
Yellow vacuolate[D]	246 (s)	410 (m)
E. microcarpa[E]	151 (s)	134 (s)
Bulbous reticulate	139 (s)	153 (v.s.)

Paspalum notatum

Inoculum	Dry wt of shoots (mg)	
	+ lime, pH 5.8	− lime, pH 4.8
E3[c]	115	58
Honey coloured	41	65
Indigenous strain	79	14
Laminate	33	16
Nil	10	11

[A] After Mosse and Bowen (1968 *a*). [B] Plants were grown in irradiation sterilized soil and harvested at 11 weeks. Infection Level: a, abundant; m, moderate; s, slight; v.s., very slight. [C] = *Glomus fasciculatus*. [D] = *G. mosseae*. [E] = *G. microcarpa*.

Fig. 43.6. Mycorrhizal stimulation of phosphorus uptake and growth of barley in a sterilized lateritic podzolic soil (1 ppm available P). At 6 weeks the plants inoculated with *Glomus fasciculatus* (right) were 3 times the weight and had 20 times the phosphate content of the uninoculated (left). (D. A. Barber and G. D. Bowen, unpublished).

fungal strains than that of uninoculated plants even though mycorrhizal fungi occurred naturally. Similarly, growth of eucalypts is aided considerably by mycorrhizas (Malajczuk *et al.* 1975).

Table 43.3, from Bevege (1971) and Mosse (1972) and Fig. 43.6 show: large increases occur in growth of native species, e.g. *Araucaria cunninghamii*, and horticultural and agricultural plants from inoculation with VA mycorrhizal fungi; differences occur between fungi in the response they elicit; and the relative performance of different fungi often varies from soil to soil. Although naturally occurring mycorrhizal fungi are adapted to the particular soil habitat, they may not be able to stimulate growth of plants as much as other fungi which could be introduced.

Further insight into the nature of responses is given by Fig. 43.7, from Abbott and Robson's work (1977*b*) on subterranean clover: plant growth responses and enhanced phosphate uptake occurred with the intermediate applications of phosphate fertilizers but were largely abolished with high fertilizer applications. Indeed there is some evidence (Mosse 1973*b*) that at supra-optimal levels of added phosphate the increased phosphate uptake by mycorrhizas may induce phosphate toxicity. It is thus possible that high additions of fertilizer in the field could sometimes be counterproductive.

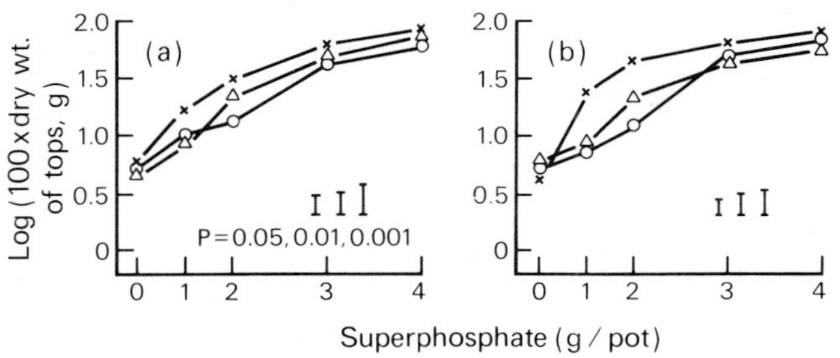

Fig. 43.7. Effect of vesicular–arbuscular endophytes and phosphorus application on dry weight of tops of subterranean clover in (*a*) untreated and (*b*) steamed soils. From Abbott and Robson (1977*b*).

The response shown in Fig. 43.7 was mainly a phosphate response but increased absorption of potassium, sulphate, zinc and ^{90}Sr (a calcium tracer) by mycorrhizal plants has also been shown experimentally (Powell 1975; Gray and Gerdemann 1973; Gilmore 1971; Jackson *et al.* 1973). The increase in uptake of potassium (a moderately mobile ion in soil) shown by Powell (1975), some 23%, was not as dramatic as many recorded increases in absorption of phosphate (a poorly mobile ion). Mycorrhizas are probably also involved in absorption of ammonium ions from soil and litter (Bowen and Smith 1980). Differences have been shown sometimes between mycorrhizal and non-mycorrhizal plants in their concentration of N, Ca, Na, Mg, Fe, Mn, Cu, B, and Al (Mosse 1973*a*). On general grounds (see below) one might expect increases in uptake of most poorly mobile ions such as Mo, Fe, Cu, Co and sometimes Mn. Research emphasis has been on ions essential for plant growth, but it has been suggested (Bowen 1979) that increased uptake of potentially harmful elements, such as cadmium and some organic residues, may also occur and that this should be studied further, even though danger to animal and human health from this may arise only in very special soil and crop situations.

The important conclusions that can be drawn regarding the mycorrhizal stimulation of plant growth as indicated by Tables 43.1–3 and Fig. 43.6 are summarized as follows:

Infection with an appropriate mycorrhizal fungus can radically change the estimate of production potential of a soil and fertilizer required to produce a good crop;

Nutritional studies on plant species, e.g. ranking their relative fertilizer requirements, can be badly in error by neglecting the mycorrhizal factor. Almost no nutritional studies have ensured adequate mycorrhizal infection with highly efficient fungal strains;

The search for plant genotypes capable of tolerating deficiency conditions, now being carried out by many scientists, must recognize the importance of using appropriate mycorrhizal fungi in test programmes. These fungi may short-circuit and facilitate the selection-breeding programmes. Selection for productivity under deficiency conditions involves selection for efficient uptake of nutrients in low supply, and for efficient use of nutrients once they have been absorbed. It may be that the former of these can be achieved by selecting appropriate mycorrhizal fungi;

The endomycorrhizal increase in ion uptake can have large consequences for nitrogen fixation by legumes and by nonlegumes such as *Casuarina* spp. VA mycorrhizas have been demonstrated to increase the uptake of phosphate, sulphate and zinc, and probably would increase uptake of cobalt, molybdenum, copper and iron, all of which are involved in nitrogen fixation. Mosse *et al.* (1976) found clover, *Centrosema* and *Stylosanthes* did not nodulate in grossly phosphate-deficient soil if they were not also mycorrhizal. Mycorrhizal infection is thus important for efficient fixation of nitrogen and therefore for sustained production in our managed pasture systems, and in our natural ecosystems.

Mechanisms for increased nutrient uptake

It has been deduced theoretically and experimentally demonstrated for ectomycorrhizas and endomycorrhizas, that the main mode of increased uptake of poorly mobile ions is by fungal growth into soil, absorption of ions and their transport back to the higher plant (Bowen 1968, 1973; Skinner and Bowen 1974*a*; Sanders and Tinker 1971; Hattingh *et al.* 1973). Sanders and Tinker (1973) found with endomycorrhizal onions that there could be 80 cm of hyphae per cm of infected root, thus increasing the 'rooting length' enormously. Greatly increased longevity of the mycorrhiza (up to several months for ectomycorrhizas) in the one position in soil can also increase radii of diffusion of ions to the root (Bowen 1968), but this is likely to be of less importance than the penetration of soil (and litter) by mycelia under most soil conditions. Longevity in function would be important particularly in sandy soils with low buffering power and moderate levels of phosphate. Sanders *et al.* (1977) found that the extra inflow of phosphate into mycorrhizal onion roots was closely related to the percentage of root infected by the fungi, which in turn was correlated with production of external hyphae of the VA mycorrhizas.

Because of their poor rooting intensity and poor production of root hairs, magnolioid roots respond most markedly to mycorrhizal infection, and some such plants, e.g. tree crops such as citrus, (Kleinschmit and Gerdemann 1972) coffee and *Araucaria* (Table 43.3), may be almost dependent on the mycorrhizal system for normal growth because the mycorrhizal response can sometimes not be eliminated by fertilizer additi- n. The response of various species tends to decline as roots become more graminoid (Baylis 1975).

Table 43.4, from Crush (1974) illustrates well the much greater responses to mycorrhizal infection when root hairs are poorly developed. However, many Australian soils are sufficiently low in available phosphate for even graminaceous species to respond markedly to mycorrhizal inoculation, e.g. Fig. 43.6. Even where surface soil concentrations of grass roots are high, they are frequently low enough in sub-surface horizons for mycorrhizas to have a possible supplementary nutritional function by acting there. Mycorrhizal infection can occur to 50 cm depth, although sometimes the level of infection is much reduced (Sutton 1973; Smith 1978; G. D. Bowen unpublished).

We can now conclude that the pattern of uptake of poorly mobile ions by plants is one of intensive depletion of a narrow zone around the root, delineated by the root hairs if they occur and limited depletion around them depending on the ion and the soil (Bhat *et al.* 1976) then followed by quite extensive exploration by mycorrhizal fungi further out into the soil.

A body of evidence so far indicates that mycorrhizal plants use the same sources of inorganic phosphate as do plant roots (Hayman and Mosse 1972; Barrow *et al.* 1977), but have greater access to them via the fungus hyphae than do plant roots. This also appears to be the way in which mycorrhizal responses to phosphates such as rock phosphate occur. Access to poorly soluble forms of phosphate is of major interest in some tropical soils especially where fixation of phosphate by iron and aluminium occurs but use of these poorly soluble phosphates by plants seems to be small e.g. Sanni (1976) did not find mycorrhizal rice able to use strengite, a major form of insoluble phosphate in Nigerian soils.

The use of organic phosphate by mycorrhizas has received rather less attention than their use of inorganic phosphates, but it is pertinent to recycling of nutrients in litter in perennial pasture and in indigenous ecosystems. The demonstration of phytases and various phosphatases associated with ectomycorrhizas and VA mycorrhizas (Theodorou 1968; Bowen 1973; Williamson and Alexander 1975; R. C. Foster, unpublished) raises a good case for mycorrhizas aiding the plant in uptake of organic phosphates. Phosphatase activity occurs also with uninfected plant roots growing in sterile conditions and it is probable that the uptake of phosphates from litter occurs not by a unique property of the fungi but by their advantageous spatial positioning (compared with the root), i.e. in penetrating litter and organic matter (see below).

Table 43.4. Response to VA mycorrhizal infection with respect to root characters (from Crush 1974).

All plants were inoculated with *Rhizobium* and harvested at 7 weeks. For the mycorrhizal experiment the plants were inoculated with endophyte E3 (= *Glomus fasciculatus*). Phosphate was added as 0.4 g monocalcium phosphate per kg soil.

	Centrosema	Stylosanthes	Trifolium	Lotus
Weight of tops (g fresh wt per plant)				
Control	1.67	0.47	1.56	2.54
Mycorrhizal	3.88	1.63	2.57	2.01
Added phosphate[A]	4.95	0.91	3.97	3.86
Root characters				
Root diameter (μm)	288	285	221	229
Root hair length (μm)	106	108	213	809
Roots with root hairs (%)	17	6	79	99

[A]Not inoculated

Different responses to various ectomycorrhizas and VA mycorrhizal fungi may be related to the relative ease and spread of infection, to differences in rate of spread through the root system and to differences in growth of the hyphae into soil. Ectomycorrhizal fungi can differ considerably in growth into soil and they behave differently in different soils; Bowen (1973) indicated that the more efficient ectomycorrhizal fungi readily produced mycelial strands. Despite the importance of growth of mycorrhizal fungi into soils, surprisingly little experimental study has been made of factors affecting this growth. Rhodes and Gerdemann (1975) observed hyphal growth from VA mycorrhizas of onions over and just under the surface of soil for 8 cm from the root. In a plantation of *P. radiata*, Skinner and Bowen (1974*a*) demonstrated translocation of phosphate to ectomycorrhizas, via mycelial strands, from soil 12 cm away from the root. Using an experimental approach to mycelial strand growth into soil, Skinner and Bowen (1974*b*) showed up to 100-fold differences in penetration of different soils by mycelial strands. Compacting a South Australian forest soil from 1.2 g cm^{-3} to 1.6 g cm^{-3} reduced mycelial growth from mycorrhizas of *P. radiata* by up to 90%, and this would be expected to reduce nutrient uptake considerably. The effects of agricultural and silvicultural practices on fungus growth into soil require detailed study if response to mycorrhizal fungi is to be optimized.

In examining the mechanism of response to mycorrhizas, the growth of mycorrhizal and non-mycorrhizal plants experimentally adjusted to have the same nutrient content should be compared. If growth is similar (as is often the case) it can be concluded that the *main* mechanism of response to mycorrhizas is increased efficiency of nutrient uptake. However in some cases, e.g. Bowen (1973) and Abbott and Robson (1977*b*), they are different and this suggests factors other than uptake are also operating: Schönbeck (1979) reported an increased susceptibility of endomycorrhizal plants to foliage diseases but decreased susceptibility to root diseases; some unusual nematode – endomycorrhiza interactions occur (Schenck *et al.*, 1975). These are areas needing further study and emphasize that in many ways the microbe–plant system may be a very different system from that of the plant alone.

Other mycorrhizal functions

There is no reason why we should restrict our thinking on mycorrhizas to their being important only in soils of low nutrient status. Once we eliminate this 'tunnel' vision and consider the mycorrhizal root as an alternate strategy for producing 'root tissue' other modes of function (sometimes related to nutrition) are suggested. Some of these are described here.

Absorption of water

The main interactions between mycorrhizas and water are likely to be as follows. First, fungal growth into soils may increase the rate of water movement to the root, especially with low soil moisture in sandy soils and especially with low rooting intensities. In a sandy soil, Sands *et al.* (1979) have calculated that the hydraulic conductivity of soil falls by 5–10 orders of magnitude between field capacity (–10 kPa matric potential) and wilting point (–1500 kPa); in such soils this transport factor often becomes the rate limiting step and fungus growth into soil from mycorrhizas may thus be important in water uptake. Secondly, fungal growth into soil may give a greater opportunity for root–soil contact, compensating for any shrinkage of roots from soil with drying. Thirdly, and probably most important, mycorrhizas should increase absorption of ions which are relatively highly mobile in water-saturated soils but poorly mobile in unsaturated soils, e.g. K^+. Reid and Bowen (1979*b*) found uptake processes for phosphate at the root surface of mycorrhizal and non-mycorrhizal roots were little affected by water potentials commonly found in soil from field capacity to permanent wilting point.

Counteracting toxicities

Many soil and environmental conditions may decrease root growth, and hence ion uptake. However, provided the mycorrhizal fungus present is not as susceptible to the deleterious factor as the roo t is, mycorrhizas may then take over the main nutrient and water uptake roles. This may be the reason for the much greater mycorrhizal stimulation of plants at high soil temperatures recorded by Moawad (1979).

A main effect of high acidity on plants can be *via* aluminium toxicity, one effect of which is to injure root tips and prevent root growth. Theoretically, fungal growth into soil from mycorrhizas could compensate for this loss of root growth, provided the fungi are not also affected. Aluminium-tolerant varieties of plants often are tolerant of low soil phosphate situations (Sanchez 1976) – could there be a mycorrhizal component in this? Some aluminium-tolerant species have a lower translocation of aluminium to the tops – could mycorrhizal fungi act similarly by being a sink for aluminium? R. C. Foster (unpublished) has found lead and chloride to be accumulated in some ectomycorrhizal and VA mycorrhizal fungi, and they may buffer the plant to some extent from toxins.

Similarly, the mycorrhizal factor in highly saline avd alkaline soils, so common in Australia, needs further study. G. D. Bowen (unpublished) has found high populations of VA mycorrhiza spores in soils with greater than 500 ppm Cl^- and at least some infection under these extreme conditions. In other soils typical mycorrhizas were found at pH 9.2. Whether mycorrhizal infection substantially assists plant tolerance of high alkalinity or high salinity is a question well worth experimental study.

Effects on soil structure

Nicolson (1960) found high VA mycorrhizal infection of many grasses on developing sand dunes. It probably assisted ion uptake by the grasses, but could the fungal growth itself have assisted in stabilizing the dunes? Recently Sutton and Sheppard (1976) showed, by inoculating pots of sterilized dune sand with *Glomus*, that endomycorrhizal *Phaseolus vulgaris* had some five times (by weight) more sand aggregates per kilogram than did non-mycorrhizal plants of similar size, the sand grains being held together by the fungal hyphae. Tisdall and Oades (1979) ascribed the greater efficiency of ryegrass roots than of clover in stabilizing soil aggregates in a loam, to the ryegrass supporting a larger population of the hyphae of VA mycorrhizas in soil. The hyphae were covered with a layer of amorphous material, probably polysaccharide, to which clay particles were firmly attached. Possible soil structure effects of VA mycorrhizas warrant further examination in poorly structured soils, especially in some high rainfall areas of the tropics where intensive agriculture could lead to loss of structure.

Nutrient conservation

In natural communities even in the high rainfall areas of the tropics, losses of nutrient from soils are remarkably small, despite high temperatures and other conditions for high decomposition rates of litter. In forests and in grasslands, it is probable that mycorrhizal associations play an important part in the conservation of nutrients and in nutrient cycling. Fungal hyphae readily penetrate litter and decomposing organic matter, and can spatially compete with other soil micro-organisms for organic and inorganic nutrients far more efficiently than can plant roots (Bowen 1973). This is probably an important part of the tight nutrient cycle of rain forests and other systems. This factor is likely to be of far more general importance in nutrition from organic matter than only the production of enzymes such as phosphatases by the fungi. A possible expression of this competition phenomenon is seen in the work of Gadgil and Gadgil (1975) who found litter decomposition of *P. radiata* was decreased by the presence of mycorrhizas; they hypothesized that absorption of nutrients by the mycorrhizal fungi suppressed decomposition by removing nutrients needed by decomposer micro-organisms.

Roles in plant competition

A further dimension of plant and mycorrhizal ecophysiology is in maximizing productivity from, and in stabilizing the composition of, mixed natural plant communities such as tropical rain forests and eucalypt forests and savannah lands, or planted communities such as mixed pastures. The nutrient demands of two or more species growing concurrently are usually considerably in excess of that of a single species, and a competitive situation often occurs.

In some communities competition is minimized by stratification of root zones of different species: e.g. ground-nut roots occupy the superficial layers of soil, while maize plants root at greater depth. In other cases different species have their major nutrient demand at different times and avoid intensive competition. However, competition intensifies when demands are concurrent and roots are occupying essentially the same location in soil. In an excellent essay on plant interactions in mixed communities, Trenbath (1977) pointed out that competition begins when there is an overlap of depletion zones for nutrients by roots of different components of a mixed community. Therefore competition begins first for the more mobile ions, e.g. nitrate, sulphate, potassium; for such ions a good correlation is observed between root abundance and nutrient absorbed. In contrast to monocultures, in this situation mycorrhizal fungus growth into soil is critical for uptake of mobile ions as well as for poorly mobile ions because sparsely rooting species such as trees in grassland, or legumes in a grass pasture would be greatly disadvantaged in the absence of mycorrhizas. For this reason it is probable that mycorrhizas have been key factors in the evolution of mixed communities in nutrient-deficient soils. However, the situation is not simple for the densely rooting species are probably also mycorrhizal, and it is possible that these could absorb luxury amounts of the elements in limited supply and disadvantage the sparsely rooting species even further. Any theoretical treatment or experimental study of competition for limited soil nutrients must necessarily include mycorrhizal systems.

Few experimental studies have been made on the roles of mycorrhizas in competition between plants; Crush (1974) found that when *Trifolium repens* and *Lolium perenne* were grown together, mycorrhizas preferentially stimulated the growth of the legume in phosphorus deficient soil; Hall (1978*b*) recorded that in a soil deficient in phosphorus the growth of mycorrhizal white clover in ryegrass was forty times that of non-mycorrhizal clover and that the ryegrass gave a poor response both to phosphorus and to VA mycorrhizas; Fitter (1977) found mycorrhizal inoculation of the combination of two grasses, *Holcus lanatus* and *Lolium perenne*, gave considerable competitive advantage to *H. lanatus* in uptake of both potassium and phosphate.

Ameliorating and avoiding root diseases

Plant growth is often reduced by root pathogens, especially in nutritionally poor soil, e.g. a spectacular example of root disease damage is that due to *Phytophthora cinnamomi* in eucalypt forests of Australia. Mycorrhizas can ameliorate root disease in several ways:

(i) Antibiosis. Marx (1973) has indicated that some ectomycorrhizas can produce antibiotics towards pathogens such as *Phytophthora cinnamomi*, e.g. mycorrhizas produced by *Leucopaxillus cerealis* var. *piceina*. He also found that a well developed fungus sheath of ectomycorrhizas is less able to be penetrated by fungi such as *P. cinnamomi*. Similary, Malajczuk (1975) found that some mycorrhizas of eucalypts were resistant to *P. cinnamomi*.

(ii) Improved nutrition caused by mycorrhizal development can indirectly affect disease, e.g. increased arginine content of mycorrhizal tobacco roots may be the reason for suppression of chlamydospore formation by *Thielaviopsis basicola* (Schönbeck and Schlösser, 1976).

(iii) Compensation. Diseased, ineffective roots will lead to reduced yield in a nutritionally poor soil but abundant mycorrhiza formation and hyphal growth into soil can compensate for such losses. Furthermore, this compensating tissue, ie. the fungal hyphae, is not susceptible to the same diseases as the root; and is a nice example of the many insurance systems plants have developed during evolution. V. Bumbieris (pers. com.) found that mycorrhizal infection of *P. radiata* by *Rhizopogon luteolus* compensated for damage which would have occurred from *P. cryptogea* had the plant been nonmycorrhizal. (However, mycorrhizal infection in the absence of *P. cryptogea* gave an 80% growth increase.)

(iv) Avoidance. Suberization or cutinization of newly formed roots confers a degree of disease protection, but it also causes a very large decline in ion uptake by that part of the root. However, if the root has become mycorrhizal before suberization it can maintain an efficient uptake function *via* the mycorrhiza while enjoying the disease protection of suberization: e.g. in studies by Bowen *et al.* (1975) suberized uninfected lateral roots of the sub-tropical tree *Araucaria cunninghamii*, exposed to ortho[^{32}P]phosphate had an activity of 3.8 counts per second, most of which was physical adsorption to the surface (i.e. there was not true phosphate uptake into the root) but similar mycorrhizal roots (VA mycorrhizas) had 12.7 counts per second, some three quarters of which was real uptake.

It is also possible that mycorrhizal infection may elicit production of plant defence compounds against other micro-organisms for this occurs with orchid mycorrhizas (see Smith 1974) but this needs further study. Foster and Marks (1967) found enhanced polyphenol production in the root cortex of mycorrhizal *Pinus radiata*, and Krupa and Fries (1971) found up to eight-fold increases in volatile terpenes and sesquiterpenes (often fungistatic) with ectomycorrhizal infection of *P. sylvestris* by *Boletus variegatus*.

Claims for nitrogen fixation by mycorrhizas

Claims for nitrogen fixation by mycorrhizas must be viewed with the greatest caution; in no cases have the mycorrhizas been free from other organisms, and significant nitrogen fixation has so far only been recorded with procaryotic systems such as bacteria and blue green algae. It has also been suggested that the observed nitrogen fixation in laboratory experiments with mycorrhizal plants has been due to fixation by free-living bacteria encouraged by the mycorrhizas, but few quantitative or qualitative studies of nitrogen fixing bacteria associated with mycorrhizas and non-mycorrhizal roots have been carried out. Studies on the hyphosphere of mycorrhizas, while often showing increased general microbial activity, are inadequate in the present context.

A very disturbing facet of claims for nitrogen fixation by mycorrhizas or their associated microflora is the quite unusual conditions under which the hypothesis is tested; plants are sometimes placed in a sealed tube of limited volume under high humidity conditions for several days (e.g. Bevege *et al.* 1978). This would give good conditions for abnormal development of bacteria, including nitrogen fixers, on roots and on foliage. Great difficulty can be experienced in washing bacteria from the rhizoplane so that 'root incorporation' of ^{15}N in such experiments may be really in micro-organisms. The tendency not to use similar non-mycorrhizal plants as controls is a further very disturbing facet of many such studies.

Growth depression by mycorrhizas

The main energy source for mycorrhizas is assimilates from the higher plant and although the mycorrhizal demand can be considerable, the benefits from the association usually far outweigh the energy cost by the fungus (see Bowen 1978). Detailed analyses of the energy relations of these symbioses have yet to be made; perhaps the plant may be able to compensate for an energy drain to the fungus by increasing photosynthesis. However, there are cases where growth depressions have been recorded with mycorrhizas. Some examples follow.

Temporary and relatively small depressions of plant growth (due to fungal growth requirements) before the mycorrhiza response occurs have been recorded by a number of workers, e.g. on maize (Khan 1972).

In situations where fertility is high enough to make mycorrhizas superfluous, but not sufficiently high to inhibit mycorrhizal development, depressions may occur due to the fungal demand for assimilate: data of Hepper (1977) show that some 17% of the root weight can be fungal material in a heavily VA mycorrhizal plant; Crush (1976) found decreases in growth of mycorrhizal *Medicago sativa* and *Trifolium hybridum* of 3.5–16.2% in fertile soils. These situations are quite unusual in agricultural practice.

Mosse (1973*b*) found growth depressions of up to 40% in mycorrhizal onion with supra-optimal additions of phosphate, possibly caused by phosphate toxicity resulting from increased phosphate uptake by the mycorrhiza.

The scope for microbial manipulation

I have indicated above that the microbial associations of roots can have very great effects on plant growth, that they can change our concept of productivity of a soil, that they give plants (and plant scientists) an alternate strategy for optimizing productivity, and they have probably been critical in the evolution of species and ecological systems in our infertile soils. This leads to questions such as: to what extent are organisms presently in soil contributing to plant growth, and what is the scope for increasing productivity by introduction of more efficient organisms and judicious management of these microbial systems?

Considerations in the foregoing sections indicate that responses to inoculation with beneficial micro-organisms would be expected when the soil is nutritionally sub-optimal, and when naturally occurring beneficial organisms for that species are absent or low in number, or where poorly efficient naturally occurring organisms can be replaced by more efficient ones.

Most Australian soils are nutritionally impoverished, especially in phosphate, nitrogen, and several trace elements. This prerequisite for inoculation response is therefore met; most plant species are likely to benefit from a highly efficient microbial associate, but certain plant species will benefit much more than others. Increases in nutrient uptake by mycorrhizas will greatly reduce the fertilizer needed for good growth, but often do not eliminate the need for some fertilizer application. It is sometimes claimed that when sorption sites for phosphate in soil are all satisfied, (and if there is no loss of nutrients by leaching), only maintenance levels of phosphate need be added to soil for maximum productivity and mycorrhizas will not be needed. While this may be so for species with well developed root systems, species with coarse root systems (magnolioid) respond to inoculation in even highly fertile soils and it is highly likely that several species of plant will always benefit from mycorrhizas.

Some 83–95% of Australian soils studied by Mosse and Bowen (1968*b*) and Abbott and Robson (1977*a*) contained spores of the VA mycorrhizal fungi, but often only few: e.g. they were abundant in only 38% of the soils Mosse and Bowen studied. Spore numbers are frequently low in virgin soils (Mosse and Bowen 1968*b*; Abbot and Robson 1977*a*). Relationships between infectivity of soil and plant responses are complex bu tMosse (1977) found responses to inoculation in nine soils from different parts of the world to be greater in soils containing few indigenous endophytes. Thus it would be expected that with so many Australian soils low in mycorrhizal fungi, crops in many would respond greatly to inoculation. The study of management techniques to increase and maintain numbers of spores or infectivity of soil would be worthwhile. Similarly, although ectomycorrhizal fungi beneficial to eucalypts, and possibly pine trees, occur in most eucalypt stands, as one goes into areas with grasses dominating, the population of mycorrhizal fungi declines markedly (Bowen 1963) and spectacular responses to inoculation would be more frequent.

The naturally occurring mycorrhizal fungi have evolved to be most compatible with the existing soil microflora and natural soil conditions (they have been selected primarily for survival) and are not necessarily the most stimulative. They may be able to be replaced to some extent by more efficient fungi (see below). This was the reason for the inoculation responses of Table 43.2.

We can thus predict that many Australian soils fulfil two or more of the three conditions under which inoculation responses to mycorrhizal fungi would be expected. Where inoculation of a highly efficient strain does not result in plant growth increases, it may be that the naturally occurring fungi are already highly efficient or that the inoculated organisms could not establish under those particular conditions. In inoculation studies on *P. radiata* in South Australia, West Australia and Victoria growth increases of 50% have been obtained in some places following inoculation with mycorrhizal fungi but only 10% growth increase occurred in others, due to the naturally occurring fungus in those soils (which often had been previously planted with pine) being of high efficiency in plant stimulation.

How then can one successfully introduce highly beneficial micro-organisms in the face of competition from other soil organisms (and other mycorrhizal fungi of lesser efficiency)? Unfortunately there has been little study of the dynamics of microbial growth in the rhizosphere but useful experimental approaches are now developing in this field (Bowen 1979; Bowen and Theodorou 1973; Bowen and Rovira 1976). It is possible that some of the different responses with different mycorrhizal fungi, e.g. in Table 43.2, are due to ease of establishment of some fungi and difficulty of establishment of others. Table 43.5 shows that even with different strains of the one ectomycorrhizal fungus, considerable differences occur in the colonization of the rhizosphere at soil temperatures commonly encountered in the field in southern Australian forest soils during the growing season (e.g. 16°C) and Table 43.6 shows that large differences exist between mycorrhizal fungi in their colonization of the rhizosphere in the presence and absence of other micro-organisms; *Suillus luteus* was the most sensitive to the other organisms and *Thelephthora terrestris* was one of the least sensitive. Such factors have to be considered in the selection of inoculants.

Table 43.5. Effects of soil temperature on rhizoplane colonization by ectomycorrhizal fungi (from Theodorou and Bowen 1971)

Growth in the rhizoplane for 4 weeks in sterilized soil. In laboratory media all fungi grew approximately half as well at 16°C as at 25°C.

Fungus	Length colonized (mm) P. radiata seedling root		Inert glass fibre	
	16°C	20°C	16°C	20°C
Suillus luteus #1	6.2	24.9	4.5	6.9
S. luteus #3	18.6	31.8	5.1	9.1
Rhizopogon luteolus	4.1	27.8	4.4	3.2
L.S.D.	*Suillus*	$P < 0.05$	5.1	
		$P < 0.01$	6.7	
	Rhizopogon	$P < 0.01$	7.4	

Table 43.6. Length of root colonized (mm) by ectomycorrhizal fungi with bacteria present (from Bowen and Theodorou 1979)

Fungus	Expt I			Expt II	
	Nil	Pseudomonas sp.	Bacillus sp. 1	Nil	Bacillus sp. 2
Rhizopogon luteolus	27.6	8.3	23.7	30.7	34.1
Suillus luteus	16.2	3.2	8.7	19.1	26.2
Thelephthora terrestris	22.4	12.9	18.1	12.1	11.2
Corticium bicolor	17.2	3.2	15.3	15.8	27.1
Pisolithus tinctorius	6.9	5.7	4.2	6.9	1.9
L.S.D.	$P < 0.05$, 5.4		$P < 0.05$, 6.2		
	$P < 0.01$, 7.1.		$P < 0.01$, 8.2		

Indigenous soil organisms are so well adapted to the existing environment that it is often difficult to superimpose another organism unless the natural system is changed in some way. In the most extreme case this would involve soil sterilization; successful establishment of inoculated mycorrhizal fungi is considerably easier in partially sterilized soil than in non-sterile soil, especially when the natural soil already contains mycorrhizal fungi (Theodorou 1967). In other instances, changing the cropping system or applying fertilizer causes sufficient change of the soil environment to move the microbial equilibrium to a new position and introduction of selected micro-organisms suitable to the new environment may then be more easily accomplished: e.g. Mosse (1977) recorded that on adding phosphate to some phosphate-deficient soils, the indigenous VA mycorrhizal fungi appeared to compete poorly with introduced fungi which were apparently suited to higher phosphate. Some of these fungi are almost restricted to arable or cultivated soils (Mosse and Bowen 1968*b*; Abbott and Robson 1977*a*). Introduction and proliferation of particular fungi can be aided considerably by plant management.

Rhizosphere manipulation

Successful inoculation with organisms inducing proteoid roots occurs with the practice of some nurserymen of adding to seedlings, some soil from beneath vigorously growing proteaceous plants. Unfortunately, such methods could also transmit plant diseases and there is a need to develop inoculation methods using pure cultures of the bacteria causing proteoid roots.

Some eucalypt forest soils are suppressive toward the root disease organism *Phytophthora cinnamomi* and Malajczuk (1975) showed that this is due to the presence of soil micro-organisms antagonistic to the fungus. The best way to obtain biological control of disease organisms in such extensive systems as forests whould be by using forestry practices encouraging antagonists to the disease organisms. Shea and Malajczuk (1977) have shown in pot studies that *Acacia pulchella* and *A. extensa* reduce the population of *P. cinnamomi*. They suggested that the use of an understorey burning regime to encourage such legumes (rather than disease susceptible proteaceous plants) may help control this disease: This imaginative approach is currently under field evaluatio n(S.R. Shea, unpublished).

Ectomycorrhizal inoculation

Suitable inoculation techniques have now been devised to inoculate forest tree nurseries at seeding with ectomycorrhizal fungi selected for high efficiency in tree response (Theodorou and Bowen 1973; Lamb and Richards 1974; Marx 1976). The simplest methods involve application of spores of selected fungi to the seed. The spores can be collected from fruit bodies in the forest, stored under freeze dried or refrigerated conditions (Theodorou 1971, Theodorou and Bowen 1973; Marx 1976), and applied as a slurry in water to the seeds. This method restricts one to organisms which form fruit bodies, but methods also exist for inoculation of limited areas using mycelial cultures produced in the laboratory (Theodorou 1967; Marx and Bryan 1975). The desirability and practicality of altering the mycorrhizal component on *existing* extensive natural forests or plantations has not been studied. However, it may be necessary to examine this if we are to optimize productivity and use of our native forests. It is almost certain in our nutrient deficient soils that root symbioses play an extremely important role in the productivity of our native forests.

Vesicular-arbuscular mycorrhizal inoculation

The major problem at present in large scale inoculation with VA mycorrhizal fungi is that it has not been possible to grow the organisms in laboratory culture. However, this and management procedures encouraging high populations in soil are now receiving considerable attention by researchers. At present, inoculation experiments employ natural inocula, such as soils with a high population of efficient strains of the VA mycorrhizal fungi, spores which have been separated from soil or roots infected under controlled conditions. It is practicable to mass-produce inoculated glasshouse or nursery plants, and soil and infected roots produced in this way could readily be used to inoculate labour-intensive crops such as horticultural plants, tea, rubber, and forest nurseries. However, the ultimate aim should be inoculation on an extensive scale. Crush and Pattison (1975) and Hall (1979) have demonstrated some success in inoculation by pelleting of seeds with inoculum of lyophilized soil sievings containing VA mycorrhizal fungal spores, or with dried soil pellets containing infected roots and spores of a VA mycorrhizal fungus. If further studies confirm the hopes that large field responses can be obtained routinely, it is likely that large scale methods such as those above will be developed vigorously. If responses are to be obtained over wide areas, they will be most significant in places such as Australia with its widespread nutrient deficiencies and low fertilizer input agriculture.

Conclusion

Soil conditions are such in Australia that it is most likely that microbial root associations have played a very large role in agricultural production and in growth and survival of our indigenous forests. We now know enough about variation between fungi and factors affecting host response to be confident that there is scope in many Australian soils for improving plant production very considerably with proper micro-organism selection, plant selection and management techniques.

A considerable amount is known about how the organisms act and in some cases, e.g. ectomycorrhizal fungi, efficient inoculation methods have been devised for pines and eucalypts growing in plantations. However, much less is known about the efficient management and inoculation of these associations in native forests. Similarly, the immediate tasks in VA mycorrhizal research are of the ecological kind. In particular, a task of prime importance is to increase inoculation studies in the field considerably, using a range of mycorrhizal fungi, to ascertain the soils and conditions under which the biggest responses will be obtained. We can see the great potential of manipulating microbial–plant systems to increase productivity; to convert such potential into actuality is the challenge ahead.

References

Abbott, L. K. and Robson, A. D. (1977*a*). The distribution and abundance of vesicular arbuscular endophytes in some Western Australian soils. *Aust. J. Bot.* **25**, 515-22.

Abbott, L. K. and Robson, A. D. (1977*b*). Growth stimulation of subterranean clover with vesicular arbuscular mycorrhizas. *Aust. J. Agric. Res.* **28**, 639-49.

Alexander, M. (1961). 'An Introduction to Soil Microbiology' (John Wiley: New York.)

Andrews, R. E. and Newman, E. I. (1970). Root density and competition for nutrients. *Oecologia Plantarum* **5**, 319-34.

Ashton, D. H. (1956). Autecology of *Eucalyptus regnans*, Ph. D. Thesis, Univ. Melbourne.

Barber, D. A. (1966). Effect of micro-organisms on nutrient absorption by plants. *Nature (London)* **212**, 638-40.

Barber, D. A., Bowen, G. D. and Rovira, A. D. (1976). Effects of micro-organisms on absorption and distribution of phosphate in barley roots. *Aust. J. Plant Physiol.* **3**, 801-8.

Barber, D. A. and Lee, R. B. (1974). The effect of micro-organisms on the absorption of manganese by plants. *New Phytol.* **73**, 97-106.

Barley, K. P. (1970). The configuration of the root system in relation to nutrient uptake. *Adv. Agron.* **22**, 159-201.

Barley, K. P. and Rovira, A. D. (1970). The influence of root hairs on the uptake of phosphate. *Soil Sci. Plant Anal.* **1**, 287-92.

Barrow, N. J., Malajczuk, N. and Shaw, T. C. (1977). A direct test of the ability of vesicular arbuscular mycorrhiza to help plants take up fixed soil phosphate. *New Phytol.* **78**, 269-76.

Baylis, G. T. S. (1971). Endogonaceous mycorrhizas synthesized in *Leptospermum* (Myrtaceae). *N. Z. J. Bot.* **9**, 293-6.

Baylis, G. T. S. (1975). The Magnolioid Root and Mycotrophy in Root Systems Derived From it. *In* 'Endomycorrhizas'. (Eds F. E. Sanders, B. Mosse and P. B. Tinker). pp. 373-89. (Academic Press: London.)

Bevege, D. I. (1971). Vesicular-arbuscular mycorrhizas of *Araucaria*: Aspects of their ecology and physiology and role in nitrogen fixation. Ph. D. Thesis, Univ. New England, Armidale, Australia.

Bevege, D. I., Richards, B. N. and Moore, A. W. (1978). Nitrogen fixation associated with conifers. CSIRO Aust. Div. Soils Div. Rep. No. 26.

Bhat, K. K. S. and Nye, P. H. (1973). Diffusion of phosphate to plant roots in soil. I. Quantitative autoradiography of the depletion zone. *Plant Soil.* **38**, 161-75.

Bhat, K. K. S., Nye, P. H. and Baldwin, J. P. (1976). Diffusion of phosphate to plant roots in soil. IV. The concentration distance profile in the rhizosphere of roots with root hairs in a low-P soil. *Plant Soil* **44**, 63-72.

Bowen, G. D. (1963). The natural occurrence of mycorrhizal fungi for *Pinus radiata* in South Australian soils. CSIRO Aust. Div. Soils Div. Rep. No. 6/63.

Bowen, G. D. (1968). Phosphate uptake by mycorrhizas and uninfected roots of *Pinus radiata* in relation to root distribution. Trans. 9th Int. Cong. Soil Sci, Adelaide, Vol. 2, pp. 219-28.

Bowen, G. D. (1969). The Roles of Mycorrhizas and Root Nodules in Tree Nutrition. *In* 'Collected Papers of the Post-graduate School for Foresters' (University of New England).

Bowen, G. D. (1973). Mineral Nutrient Relations of Ectomycorrhizae. *In* 'Ectomycorrhizae – Their Ecology and Physiology'. (Eds G. C. Marks and T. T. Kozlowski). pp. 151-205. (Academic Press: New York.)

Bowen, G. D. (1978). Dysfunction and Shortfalls in Symbiotic Response. *In* 'Plant Disease – An Advanced Treatise'. (Eds J. G. Horsfall and E. B. Cowling). Vol. III. pp. 231-26. (Academic Press: New York.)

Bowen, G. D. (1979). Integrated and Experimental Approaches to the Study of Growth of Organisms Around Roots. *In* 'Soil-Borne Plant Pathogens'. (Eds B. Schippers and W. Gams). pp. 209-27. (Academic Press: London.)

Bowen, G. D., Bevege, D. I. and Mosse, B. (1975). Phosphate Physiology of Vesicular-arbuscular Mycorrhizas. *In* 'Endomycorrhizas'. (Eds. F. E. Sanders, B. Mosse and P. B. Tinker). pp. 241-60. (Academic Press: London.)

Bowen, G. D., Cartwright, B. and Mooney, J. R. (1974). Wheat Root Configuration under Phosphate Stress. *In* 'Mechanisms of Regulation of Plant Growth'. (Eds R. L. Bieleski, A. R. Ferguson and M. M. Creswell). pp. 121-25. Bull. No. 12. Roy. Soc. New Zealand.

Bowen, G. D. and Cartwright, B. (1977). Mechanisms and Models of Plant Nutrition. *In* 'Crop Growth in Semi-arid Environments'. (Eds J. E. Russell and E. L. Greacen). pp. 197-223. (Queensland University Press: Brisbane.)

Bowen, G. D. and Rovira, A. D. (1961). The effects of micro-organisms on plant growth. I. Development of roots and root hairs in sand and agar. *Plant Soil 15*, 166-88.

Bowen, G. D. and Rovira, A. D. (1966). Microbial factors in short-term phosphate uptake studies with plant roots. *Nature (London)* **211**, 665-66.

Bowen, G. D. and Rovira, A. D. (1976). Microbial colonization of roots. *Annu. Rev. Phytopathol.* **14**, 121-44.

Bowen, G. D. and Smith, S. E. (1980). The Effects of Mycorrhizas on Nitrogen Uptake by Plants. *In* 'Nitrogen Cycling in Terrestrial Ecosystems'. (Eds F. E. Clark and T. Rosswall). *Ecol. Bull. (Stockholm).* **33**, 237-47.

Bowen, G. D., Theodorou, C. and Skinner, M. F. (1972). Towards a Mycorrhizal Inoculation Programme. *In* 'Australian Conference in Tree Nutrition'. (Ed R. Boardman). pp. 68-81. (Australian Forestry Council: Canberra.)

Bowen, G. D. and Theodorou, C. (1973). Growth of Mycorrhizal Fungi around Seeds and Roots. *In* 'Ectomycorrhizae – Their Physiology and Ecology' (Eds G. C. Marks and T. T. Kozlowski). pp. 147-50. (Academic Press: New York.)

Bowen, G. D. and Theodorou, C. (1979). Interactions between bacteria and ectomycorrhizal fungi. *Soil Biol. Biochem.* **11**, 119-26.

Brown, M. E., Burlingham, S. K. and Jackson, R. M. (1964). Studies of *Azotobacter* species in soil. III. Effects of artificial inoculation on crop yields. *Plant Soil* **20**, 194-214.

Champion, R. A. and Barley, K. P. (1969). Penetration of clay by root hairs. *Soil Sci.* **108**, 402-7.

Chilvers, G. A. (1973). Host range of some eucalypt mycorrhizal fungi. *Aust. J. Bot.* **21**, 103-11.

Cormack, R. G. H. (1962). Development of root hairs in angiosperms. II. *Bot. Rev.* **28**, 446-64.

Cox, G. and Tinker P. B. (1976). Translocation and transfer of nutrients in vesicular–arbuscular mycorrhizas. I. The arbuscule and phosphorus transfer: a quantitative ultrastructural study. *New Phytol.* **77**, 371-8.

Crush, J. R. (1974). Plant growth responses to vesicular–arbuscular mycorrhizas. VII. Growth and nodulation of some herbage legumes. *New Phytol.* **73**, 743-9.

Crush, J. R. (1976). Endomycorrhizas and legume growth in some soils of the Mackenzie Basin, Canterbury, New Zealand. *N. Z. J. Agric. Res.* **19**, 473-76.

Crush, J. R. and Pattison, A. C. (1975). Preliminary Results on the Production of Vesicular Arbuscular Mycorrhizal Inoculum by Freeze Drying. *In* 'Endomycorrhizas'. (Eds F. E. Sanders, B. Mosse and P. B. Tinker). pp. 485-93. (Academic Press: London.)

Fassi, B., Fontana, A. and Trappe, J. M. (1969). Ectomycorrhizae caused by *Endogone lactiflua* with species of *Pinus* and *Pseudotsuga. Mycologia* **61**, 412-4.

Fitter, A H. (1977). Influence of mycorrhizal infection on competition for phosphorus and potassium by two grasses. *New Phytol.* **79**, 119-25.

Foster, R. C. and Marks, G. C. (1967). Observations on the mycorrhizas of forest trees. II. The rhizosphere of *Pinus radiata* D. Don. *Aust. J. Biol. Sci.* **20**, 915-26.

Gadgil, R. L. and Gadgil, P. D. (1975). Suppression of litter decomposition by mycorrhizal roots of *Pinus radiata. N. Z. J. For. Sci.* **5**, 33-41.

Gerdemann, J. W. and Nicolson, T. H. (1963). Spores of mycorrhizal *Endogone* species extracted from soil by wet sieving and decanting. *Trans. Br. Mycol. Soc.* **46**, 235-44.

Gerdemann, J. W. and Trappe, J. M. (1974). The Endogonaceae in the Pacific North-west. *Mycol. Mem.* **5**, 1-76.

Gilmore, A. E. (1971). The influence of endotrophic mycorrhizae on the growth of peach seedlings. *J. Am. Soc. Hort. Sci.* **96**, 35-8.

Gray, L. E. and Gerdemann, J. W. (1973). Uptake of sulphur-35 by vesicular arbuscular mycorrhizae. *Plant Soil* **39**, 687-9.

Hall, I. R. (1978*a*). Effect of vesicular-arbuscular mycorrhizas on two varieties of maize and one of sweetcorn. *N. Z. J. Agric. Res.* **21**, 517-9.

Hall, I. R. (1978*b*). The effects of endomycorrhizas on the competitive ability of white clover. *N. Z. J. Agric. Res.* **21**, 509-15.

Hall, I. R. (1979). Soil pellets as a method of introducing vesicular-arbuscular mycorrhizal fungi into soil. *Soil Biol. Biochem.* **11**, 85-6.

Harley, J. L. (1969). 'The Biology of Mycorrhiza'. (Leonard Hill: London.)

Hattingh, M. J., Gray, L. E. and Gerdemann, J. W. (1973). Uptake and translocation of ^{32}P-labelled phosphate to onion roots by mycorrhizal fungi. *Soil Sci.* **116**, 383-7.

Hayman, D. S. and Mosse, B. (1972). Plant growth responses to vesicular-arbuscular mycorrhizas. III. Increased uptake of labile P from soil. *New Phytol.* **71**, 41-7.

Hepper, C. M. (1977). A colorimetric method for estimating vesicular-arbuscular mycorrhizal infection in roots. *Soil Biol. Biochem.* **9**, 15-8.

Hillis, W. E., Ishikura, W., Foster, R. C. and Marks, G. C. (1968). The role of extractives in the formation of ectotrophic mycorrhizae. *Phytochemistry* **7**, 409-10.

Jackson, N. E., Miller, R. H. and Franklin, R. E. (1973). The influence of vesicular-arbuscular mycorrhizas on uptake of ^{90}Sr from soil by soybeans. *Soil Biol. Biochem.* **5**, 205-12.

Jeffrey, D. W. (1967). Phosphate nutrition of Australian heath plants. I. The importance of proteoid roots in *Banksia* (Proteaceae). *Aust. J. Bot.* **15**, 403-11.

Jeffrey, D. W. (1968). Phosphate nutrition of Australian heath plants. II. The formation of polyphosphate by five heath species. *Aust. J. Bot.* **16**, 603-13.

Khan, A. G. (1972). The effect of vesicular-arbuscular mycorrhizal associations on the growth of cereals. I. Effects on maize growth. *New Phytol.* **71**, 613-9.

Kleinschmidt, G. D. and Gerdemann, J. W. (1972). Stunting of citrus seedlings in fumigated nursery soils related to the absence of endomycorrhizae. *Phytopathol.* **62**, 1447-53.

Koske, R. E. (1975). *Endogone* spores in Australian sand dunes. *Can. J. Bot.* **53**, 668-72.

Krupa, S. and Fries, N. (1971). Studies on the ectomycorrhizae of pine. I. Production of volatile organic compounds. *Can. J. Bot.* **49**, 1425-31.

Lamb, R. J. (1974). Autecology of ectomycorrhizal fungi of *Pinus*. Ph. D. Thesis, University of New England, Armidale, Australia.

Lamb, R. J. and Richards, B. N. (1970). Some mycorrhizal fungi of *Pinus radiata* and *P. elliottii* var. *elliottii* in Australia. *Trans. Br. Mycol. Soc.* **54**, 371-8.

Lamb, R. J. and Richards, B. N. (1971). Effect of mycorrhizal fungi on the growth and nutrient status of slash and radiata pine seedlings. *Aust. For.* **35**, 1-7.

Lamb, R. J. and Richards, B. N. (1974). Inoculation of pines with mycorrhizal fungi in natural soils. II. Effects of density and time of application of inoculum and phosphorus amendment on seedling yield. *Soil Biol. Biochem.* **6**, 173-7.

Lamont, B. B. (1972). 'Proteoid' roots in the legume *Viminaria juncea*. *Search* **3**, 9-12.

Lamont, B. B. (1974). The biology of dauciform roots in the sedge *Cyathochaete avenacea*. *New Phytol.* **73**, 985-96.

Lamont, B. B. and McComb, A. J. (1974). Soil micro-organisms and the formation of proetoid roots. *Aust. J. Bot.* **22**, 681-8.

Lewis, D. G. and Quirk, J. P. (1967). Phosphate diffusion in soil and uptake by plants, III. ^{31}P movement and uptake by plants as indicated by ^{32}P autoradiography. *Plant Soil* **26** 445-53.

Malajczuk, N. (1975). Ph. D. Thesis, Univ. of Western Australia.

Malajczuk, N. and Bowen, G. D. (1974). Proteoid roots are microbially induced. *Nature (London)* **251**, 316-17.

Malajczuk, N., McComb, A. J. and Loneragan, J. F. (1975). Phosphorus uptake and growth of mycorrhizal and uninfected seedlings of *Eucalyptus calophylla. Aust. J. Bot.* **23**, 231-8.

Marks, G. C. and Kozlowski, T. T. (1973). 'Ectomycorrhizae – Their Ecology and Physiology'. 444 pp. (Academic Press: New York.)

Marriott, F. H. C. and Nye, P. H. (1968). The importance of mass flow in the uptake of ions by roots from soil. Trans. 9th Int. Cong. Soil Sci. Adelaide. Vol. 1, pp. 127-34.

Marx, D. H. (1973). Mycorrhiza and Feeder Root Diseases. *In* 'Ectomycorrhizae – their Ecology and Physiology'. (Eds G. C. Marks and T. T. Kozlowski). pp. 351-82. (Academic Press: New York.)

Marx, D. H. (1976). Synthesis of ectomycorrhizae on loblolly pine seedlings with basidiospores of *Pisolithus tinctorius. For. Sci.* **22**, 13-20.

Marx, D. H. and Bryan, W. C. (1975). Growth and ectomycorrhizal development of loblolly pine seedlings in fumigated soil infested with the fungal symbiont *Pisolithus tinctorius. For. Sci.* **21**, 245-54.

Marx, D. H., Hatch, A. B. and Mendicino, J. F. (1977). High soil fertility decreases sucrose content and susceptibility of loblolly pine roots to ectomycorrhizal infection by *Pisolithus tinctorius. Can. J. Bot.* **55**, 1569-74.

Moawad, M. (1979). Ecophysiology of VA Mycorrhiza in the Tropics. *In* 'The Soil-Root Interface'. (Eds J. L. Harley and R. S. Russell). pp. 197-209 (Academic Press: London.)

Mosse, B. (1972). The influence of soil type and *Endogone* strain on the growth of mycorrhizal plants in phosphate deficient soils. *Rev. Ecol. Biol. Sol.* **9**, 529-37.

Mosse, B. (1973*a*). Advances in the study of vesicular-arbuscular mycorrhiza. *Annu. Rev. Phytopathol.* **11**, 171-96.

Mosse, B. (1973*b*). Plant growth response to vesicular-arbuscular mycorrhiza. IV. In soil given additional phosphate. *New Phytol.* **72**, 127-36.

Mosse, B. (1977). Plant growth responses to vesicular-arbuscular mycorrhiza. X. Responses of *Stylosanthes* and maize to inoculation in unsterile soils. *New Phytol.* **78**, 277-88.

Mosse, B. and Bowen, G. D. (1968*a*). A key to the recognition of some *Endogone* spore types. *Trans. Br. Mycol. Soc.* **51**, 469-83.

Mosse, B. and Bowen, G. D. (1968*b*). The distribution of *Endogone* spores in some Australian and New Zealand soils and in an experimental field soil at Rothamsted. *Trans. Br. Mycol. Soc.* **51**, 485-92.

Mosse, B., Powell, C. L. and Hayman, D. S. (1976). Plant growth responses to vesicular-arbuscular mycorrhiza. IX. Interactions between VA mycorrhiza, rock phosphate and symbiotic nitrogen fixation. *New Phytol.* **76**, 331-42.

Neumann, R. (1959). Relationships between *Pisolithus tinctorius* and *Eucalyptus camaldulensis (E. rostrata). Bull. Res. Council Israel* **7D**, 116.

Nicolson, T. H. (1960). Mycorrhiza in the Gramineae. II. Development in different habitats, particularly sand dunes. *Trans. Br. Mycol. Soc.* **43**, 132-45.

Nicolson, T. H. (1975). Evolution of Vesicular-arbuscular Mycorrhizas. *In* 'Endomycorrhizas'. (Eds F. E. Sanders, B. Mosse and P. B. Tinker). pp. 25-34. (Academic Press: London.)

Otto, G. (1965). The Influence Exerted by Fungi in the Rhizosphere on the Density of Ramification of Cucumber Roots. *In* 'Plant Microbes Relationships'. (Eds J. Macura and V. Vancura). pp. 209-19. (Czech, Acad. Sci: Prague.)

Powell, C. L. (1975). Potassium Uptake by Endotrophic Mycorrhizas. *In* 'Endomycorrhizas'. (Eds F. E. Sanders, B. Mosse and P. B. Tinker). pp. 461-68. (Academic Press: London.)

Pryor, L. D. (1956). Ectotrophic mycorrhiza in renatherous species of *Eucalyptus, Nature (London)* **177**, 587-88

Reid, C. P. P. and Bowen, G. D. (1979*a*). Effects of Soil Moisture on VA Mycorrhiza Formation and Root Development in *Medicago. In* 'The Root-Soil Interface'. (Eds J. L. Harley and R. S. Russell). pp. 211-9. (Academic Press: London.)

Reid, C. P. P. and Bowen, G. D. (1979*b*). Effect of water stress on phosphorus uptake by mycorrhizas of *Pinus radiata. New Phytol.* **23**, 103-07.

Rhodes, L. H. and Gerdemann, J. W. (1975). Phosphate uptake zones of mycorrhizal and non-mycorrhizal onions. *New Phytol.* **75**, 555-61.

Rovira, A. D. (1963). Microbial inoculation of plants. I. Establishment of free-livng nitrogen-fixing bacteria in the rhizosphere and their effects on maize, tomato and wheat. *Plant Soil* **19**, 304-14.

Rovira, A. D. and Bowen, G. D. (1971). Microbial Effects on Nutrient Uptake by Plants. *In* 'Recent Advances in Plant Nutrition'. (Ed. R. M. Samish) Vol. 1, pp. 307-16. (Gordon and Breach: New York.)
Sanchez, P. A. (1976). Properties and Management of Soils in the Tropics. (John Wiley and Sons: New York.)
Sanders, F. E., Mosse, B. and Tinker, P. B. (1975). 'Endomycorrhizas'. (Academic Press: London.)
Sanders, F. E. and Tinker, P. B. (1971). Mechanism of absorbtion of phosphate from soil by *Endogone* mycorrhizas. *Nature (London)* **233**, 278-79.
Sanders, F. E. and Tinker, P. B. (1973). Phosphate flow into mycorrhizal roots. *Pestic. Sci.* **4**, 385-95.
Sanders, F. E., Tinker, P. B., Black, R. L. B. and Palmerley, S. M. (1977). The development of endomycorrhizal root systems. I. Spread of infection and growth-promoting effects with four species of vesicular-arbuscular endophyte. *New Phytol.* **78**, 257-68.
Sands, R., Greacen, E. L. and Gerard, C. J. (1979). Compaction of sandy soils in radiata pine forests. I. A penetrometer study. *Aust. J. Soil Res.* **17**, 101-13.
Sanni, S. O. (1976). Vesicular-arbuscular mycorrhiza in some Nigerian soils: The effect of *Gigaspora gigantea* on the growth of rice. *New Phytol.* **77**, 673-74.
Schenck, N. C., Kinlock, R. A. and Dickson, D. W. (1975). Interaction of Endomycorrhizal Fungi and Root-knot Nematodes on Soybean. *In* 'Endomycorrhizas' (Eds. F. E. Sanders, B. Mosse and P. B. Tinker). pp. 607-17. (Academic Press: London.)
Schönbeck, F. (1979). Endomycorrhiza in relation to disease. *In* 'Soil Borne Plant Pathogens'. (Eds B. Schippers and W. Grams). pp. 271-80. (Academic Press: London.)
Schönbeck, F. and Schlösser, E. (1976). Preformed Subtances as Potential Protectants. *In* 'Physiological Plant Pathology'. (Eds R. Heitefuss and P. H. Williams). pp. 653-78. (Springer-Verlag: Berlin.)
Shea, S. R. and Malajczuk, N. (1977). Potential for Control of Eucalypt Dieback in Western Australia. *In* 'Australian Nursery Assoc. Conference, Hobart, 1977'. (Ed. K. F. Baker.) pp. 13-9.
Skinner, M. F. and Bowen, G. D. (1974*a*). The uptake and translocation of phosphate by mycelial strands of pine mycorrhizas. *Soil Biol. Biochem.* **6**, 53-6.
Skinner, M. F. and Bowen, G. D. (1974*b*). The penetration of soil by mycelial strands of pine mycorrhizas. *Soil Biol. Biochem.* **6**, 57-61.
Smith, T. F. (1978). A note on the effect of soil tillage on the frequency and vertial distribution of spores of vesicular-arbuscular endophytes. *Aust. J. Soil Res.* **16**, 359-61.
Smith, S. E. (1974). Mycorrhizal fungi. *Crit. Rev. Microbiol.* **2**, 275-313.
Sutton, J. C. (1973). Development of vescular-arbuscular mycorrhizae in crop plants. *Can. J. Bot.* **51**, 2487-93.
Sutton, J. C. and Sheppard, B. R. (1976). Aggregation of sand dune soil by endomycorrhizal fungi. *Can. J. Bot.* **54**, 326-33.
Sward, R. J. Hallam, N. D. and Holland, A. A. (1978). *Endogone* spores in a heathland area of south-eastern Australia. *Aust. J. Bot.* **26**, 29-43.
Theodorou, C. (1967). Inoculation with pure cultures of mycorrhizal fungi of radiata pine growing in partially sterilized soil. *Aust. For.* **31**, 303-09.
Theodorou, C. (1968). Inositol phosphates in needles of *Pinus radiata* and the phytase activity of mycorrhizal fungi. Trans. 9th Int. Cong. Soil Sci., Adelaide. Vol. 3, pp. 438-93.
Theodorou, C. (1971). Introduction of mycorrhizal fungi into soil by spore inoculation of seed. *Aust. For.* **35**, 23-6.
Theodorou, C. and Bowen, G. D. (1971). Influence of temperature on the mycorrhizal associations of *Pinus radiata*. *Aust. J. Bot.* **19**, 13-20.
Theodorou, C. and Bowen, G. D. (1973). Inoculation of seeds and soil with basidiospores of mycorrhizal fungi. *Soil Biol. Biochem.* **5**, 763-71.
Tisdall, J. M. and Oades, J. M. (1979). Stabilization of soil aggregates by the root systems of rye grass. *Aust. J. Soil Res.* **17**, 429-41.
Trappe, J. M. (1969). Mycorrhiza Forming Ascomycetes. *In* 'Mycorrhizae'. (Ed. E. Hacskaylo). pp. 19-37. (U. S. Government Printing Office: Washington, D. C.)
Trappe, J. M. (1977). Selection of fungi for ectomycorrhizal inoculation in nurseries. *Annu. Rev. Phytopathol.* **15**, 203-22.
Trenbath, B. R. (1977). Plant Interactions in Mixed Crop Communities. *In* 'Multiple Cropping'. (Ed. M. Stelley). pp. 129-69. American Soc. Agron. Special Publ. No. 27. (Am. Soc. Agron. : Madison.)
Trinick, M. J. (1977). Vesicular-arbuscular infection and soil phosphorus utilization in *Lupinus* spp. *New Phytol.* **78**, 297-304.
Warcup, J. H. (1975). A Culturable *Endogone* Associated with Eucalypts. *In* 'Endomycorrhizas'. (Eds F. E. Sanders, B. Mosse and P. B. Tinker). pp. 53-63. (Academic Press: London.)
Williamson, B. and Alexander, I. J. (1975). Acid phosphatase localized in the sheath of beech mycorrhiza. *Soil Biol. Biochem.* **7**, 195-8.
Williamson, F. A. and Wyn Jones, R. G. (1973). The influence of soil micro-organisms on growth of cereal seedlings and on potassium uptake. *Soil Biol. Biochem.* **5**, 569-75.
Zak, B. (1976). Pure culture synthesis of Pacific madrone ectendomycorrhizae. *Mycologia* **68** 362-9.

Chapter 44

Rhizobium as a factor in soil fertility

R. A. Date

LEGUMES IN ASSOCIATION WITH THEIR APPROPRIATE root-nodule bacteria, *Rhizobium*, make a major contribution to soil fertility by converting atmospheric nitrogen into readily assimilable organic forms. This association is especially important in Australia, where many of our agricultural soils are either naturally low in, or have been depleted of organic nitrogen reserves. Because our indigenous legume flora are unable to make good this nitrogen deficiency, more suitable legume species have been introduced deliberately from other parts of the world. The earliest introductions were the traditional crop and pasture legumes of Europe and included the accidental entry of such important species as subterranean clover and Townsville stylo.

Early investigations into the need for inoculation of the introduced grain legumes, cowpea, soybean and pea revealed mixed responses. Work with pasture species followed later. One of the first critical research reports concerning legumes and *Rhizobium* was that of Strong (1937), of the CSIR Division of Soils, Adelaide. Strong's report on the specialization between rhizobia from clovers and their hosts for effectiveness in nitrogen fixing ability pioneered the way in understanding the specificity relationships within the clover group.

Australian contributions to research in legume bacteriology have been reviewed by Vincent (1962) and, more recently, in general state-of-the-art reviews, by Brockwell (1977), Date and Roughley (1977), Dudman (1977), Gibson (1977), Norris and Date (1976), and Parker *et al.* (1977). The outstanding work of Emeritus Professor J. M. Vincent and his contributions to the understanding of the *Rhizobium*/legume symbiosis have been of major importance in the use of legumes in Australian agriculture.

The need for seed inoculation

The subject of inoculation of legumes with nitrogen-fixing bacteria was introduced to the farmers of New South Wales by Guthrie (1896) in an article in the Agricultural Gazette of New South Wales, and to the scientific community by Grieg-Smith (1899, 1901, 1911) in a series of papers concerning the *Rhizobium*/legume association and its contribution to soil fertility. The first field trial comparing inoculated with non-inoculated seed was set up in 1905 at Hawkesbury Agricultural College (Marks 1905), using a broth culture prepared in the United States. Similar experiments were conducted by McAlpine (1906) in New South Wales and by Easterby (1910) in Queensland, using locally prepared Professor Bottomly's nitrobacterine culture. Other attempts to inoculate seed were made by Donnan (1912) and Hamblin (1920) in New South Wales and Pound (1910) in Queensland, but growth and nitrogen fixation responses in most were variable and inconclusive.

The form of inoculum used in these early experiments varied considerably. Liquid cultures were frequently mixed with soil and spread over the area to be sown or the liquid was added directly to the soil without prior mixing. In some cases, seed inoculation was tried by 'wetting' the seed with skimmed milk or sugar solution, then coating it with soil impregnated with a liquid suspension of *Rhizobium*. Results were still variable, although some workers began to realize that failure to obtain a response to inoculation was due to natural nodulation, i.e., uninoculated controls were nodulated effectively. The relatively wide distribution of naturally occurring strains of *Rhizobium* continued to confuse results during the 1930s and 1940s, but no doubt some situations were complicated by such factors as poor quality inoculant cultures, and by limitations imposed by mineral deficiencies affecting the plant directly, or infection of its roots by rhizobia.

A striking example of the sparseness of rhizobia in acid soils, and of the ineffectiveness of the few rhizobia that did occur, was observed by Vincent (1954*a*) in red basaltic soils of the Lismore district of New South Wales; subterranean and crimson clovers and barrel medic and vetches failed to establish without inoculation with effective strains of *Rhizobium*. In the same area several years later, Ireland and Vincent (1968) observed, for subterranean clover, that high levels of inoculum are required to overcome competition from a population of naturalized white clover rhizobia ineffective on the subterranean clover. Even at high inoculum levels, effective nodulation was achieved only in the first year, unless the test host was the cultivar Woogenellup (Gibson *et al.* 1976). This cultivar forms nodules only very sparsely with white clover rhizobia and thus provides greater opportunity for the inoculum strain to do so.

Variable responses, from ineffective to fully effective, were obtained for medics (Purchase *et al.* 1951*a,b*; Hely and Brockwell 1962; Brockwell and Hely 1961) in Queensland and New South Wales. Pittman (1935), Cass-Smith (1938) and Cass-Smith and Pittman (1939) demonstrated responses to inoculation of legumes in certain eastern Australian soils, and Strong (1938) in South Australia found it advantageous to inoculate polished (cleaned) subterranean clover seed.

The importance of inoculation for first-year establishment, even though inoculation was not essential in some near-neutral (pH6–7), soils, was observed by Harris (1961). In the newly cleared sandy soils of Western Australia it was essential to inoculate clovers and medics to ensure good first year establishment. Specific strains were required to obtain good persistence of the reseeding annual subterranean clover (see review by Parker *et al.* 1977).

In tropical and subtropical areas, Steindl (1941) and McKnight (1949) obtained variable responses to inoculation of cowpea, soybeans and lucerne associated with the sugar cane areas of Queensland. In special environments, e.g. the coastal lowlands of south-eastern Queensland, inoculation was found to be necessary for successful early establishment of a range of legumes (Bryan and Andrew 1955). More recently, responses have been obtained with the legume *Lotononis bainesii* (Norris 1970), *Trifolium semipilosum* (Jones and Date 1975) and soybeans (Diatloff and Brockwell 1976).

Development of legume inoculants and their quality control in Australia

One of the most significant achievements of legume bacteriologists in Australia in recent years has been an exercise in applied bacteriology, in which fundamental principles have been used successfully for the development of quality legume-seed inoculants at a commercial level.

Roughley (1962) records that the New South Wales Department of Agriculture commenced distribution of cultures to farmers in 1914, and that demand increased to such an extent by 1932 that a charge was made to defray costs. Similar developments took place in the Western Australian Department of Agriculture in the 1920s and 1930s (Pittman 1933, 1935; Cass-Smith and Pittman 1939), in South Australia in 1938 (by the soil microbiology group of the then Council for Scientific and Industrial Research), in the Queensland Department of Primary Industry in 1937–38 (A. Diatloff, personal communication), and in the Victorian and Tasmanian Departments of Agriculture at about that time also. In New South Wales, progress in pasture improvement, especially the sowing of subterranean clover in newly cleared lands, caused an increase, to a total of 2029, in the demand for cultures supplied by the Department of Agriculture. Meanwhile, experimental programs had developed a peat-based culture, similar to today's inoculant, which was despatched by the New South Wales Department of Agriculture in 1953, when 20 188 cultures were distributed (Roughley 1962). Demand by this stage had reached such proportions, that commercial interests were invited to take over the preparation of cultures. Commercial production began in 1954 (Date 1969), but many nodulation failures were reported and attributed to poor-quality cultures (Brockwell 1954; Waters 1954). Difficulties were highlighted by the use of inoculants in newly cleared areas, on soils of low fertility, and in dry situations. Also there was a generally increased awareness of the need for good early and effective nodulation, along with early recognition of the signs of failure. The need for better production procedures, and for the establishment of some control over quality of cultures (Vincent 1954*b*), led to the establishment of the University of Sydney–New South Wales Department of Agriculture Laboratory Service, or U–DALS as it became popularly known (Vincent 1958*a*). This resulted in rapid improvement in quality, through the combined effort of manufacturers and U–DALS, so that by the 1964/65 season the quality of cultures had reached 500 x 10^6 rhizobia per gram, guaranteed over the 6-month shelf life of the cultures. The first 10 years of quality control have been summarized by Date (1969). The second decade of activity has seen a further improvement in quality, and, although standard test procedures have been reduced and the final product changed from non-sterile to sterilized peat (i.e. pure cultures), the consumer is still guaranteed a product of high quality that will ensure nodulation under most field situations (Date and Roughley 1977). The fact that few failures attributable to inoculant quality have been reported in recent years (J. A. Thompson, personal communication) stands as tribute to the efforts of those concerned, both on the manufacturing side and in the quality-control and research aspects.

Since 1971 the quality control group has been supported by government resources and quality control has become obligatory. The group's functions remained the same but its name changed to the Australian Inoculant Research and Control Service (AIRCS).

Associated with the development of quality control procedures was the equally important launching of Rhizobium Newsletter in 1956. This news sheet has provided, and still provides, scope for rapid and informal exchange of information between *Rhizobium* researchers. Initially, Rhizobium Newsletter was distributed to research workers in Australia and New Zealand only, but it now includes contributions from, and is received by, *Rhizobium* workers around the world; it ranks as a major contribution to legume bacteriology research in Australia.

Improved inoculation technology and the era of pelleting

All early cultures used to inoculate seed were based on agar or broth suspensions, and it was not until the 1950s and 1960s that improved survival of rhizobia on the seed, and thus a greater chance of early infection and nodule formation, could be achieved by using the peat form of the inoculum (Vincent 1958*b*; Date 1959). Further improvement in survival on seed was obtained by use of a sticker, such as gum arabic or methyl cellulose, and a coating of lime or rock phosphate (known as pelleting) (Cass-Smith and Goss 1958; Bergersen *et al.* 1958; Brockwell 1962; Date 1968).

The idea of lime-pelleting seed for promoting legume nodulation comes from Snieszko (1941) in the U.S., who aimed for improved nodulation in field peas in acid soils. It was first used in Australia by Loneragan *et al.* (1955), who were primarily interested in overcoming soil acidity for pasture species in acid (pH 5.5) soils of the southern tablelands of New South Wales; it was only later that the technique was used as a general inoculating procedure. Initially, Loneragan *et al.* (1955) incorporated the inoculum in a sugar suspension under the coating-material with subsequent rapid loss of viability of the rhizobia. Dusting the inoculum on the outside proved satisfactory, even when seed was in contact with acid fertilizer (D. Meyer, J. F. Loneragan and K. C. Marshall, personal communication). However, with the use of methyl cellulose and gum arabic as stickers, the inoculum was again incorporated in the pellet (under the coating), with highly successful results (Cass-Smith and Goss 1958; Brockwell 1962); in fact, it is the adhesive or sticker in pelleted seed that promotes survival of the rhizobia (Date 1968). The coating of lime or rock phosphate provides protection from acid soils and fertilizers, during sowing and germination. Thompson (1960) suggested that improved survival is due to physical separation of the inoculum from water-soluble toxic substances in the seed coat.

The potential of the technique as a means of pre-inoculation was first reported in New Zealand by Hastings and Drake (1960), but was soon reported here in Australia by Brockwell (1962), Brockwell and Whalley (1962), and Goss and Shipton (1965). Lime-pelleting is in fact now an established practice in temperate Australia; it has the advantage of protecting acid-sensitive strains against excessively acid soils and acid fertilizers, allowing nodulation without lime amendment; it also protects against toxic chemicals such as fungicides, allows for delays in sowing due to weather conditions or mechanical breakdown, facilitates aerial sowing, ensures that a high proportion of established seedlings are nodulated effectively (Murguia and Date 1965), and provides protection from seed-harvesting ants (Champ and Sillar 1961).

Pelleting of tropical legume seed began during the 1960s, amid considerable controversy as to whether lime or some other material should be used as a coating. Norris (1967) discussed arguments for and against lime-pelleting in relation to legume species, and suggested that those having fast-growing acid-producing *Rhizobium* should be lime pelleted, and species having slow-growing alkali-producing *Rhizobium* should be pelleted with relatively inert substances, such as rock phosphate. The rationale for this concept is that the legume/*Rhizobium* symbiosis originated in the wet tropics, under low pH and leached soil conditions, where legumes adapted to soils of low calcium status, and their associated rhizobia developed the alkali-producing characteristic as a survival mechanism. As legumes migrated from the tropics and into non-acid soils, the selection pressures changed, so that plants adapted to alkaline soils now had an associated *Rhizobium* producing acid as a survival mechanism. Thus, most tropical legumes of today represent the original species, and their rhizobia the archetype nodulating organism. Exceptions to this generalization, such as *Leucaena* or *Lotus*, represent adaptation to special environmental conditions; i.e. *Leucaena* is naturally adapted to alkaline soils, and has a fast-growing acid-producing *Rhizobium* partner, yet is generally considered a tropical plant; *Lotus*, on the other hand, is a temperate species, growing frequently in acid soils, and has a slow-growing alkali-producing *Rhizobium*.

The use of lime or rock phosphate as a coating material has little effect on the survival of inoculated and pelleted seeds whilst stored (Date 1968 and unpublished); however lime can depress subsequent nodulation in some species, e.g. *Desmodium uncinatum, Lotoninis bainesii* (Norris 1971*a*, 1972) and *Lupinus luteus* and *Ornithopus sativus* (Parker and Oakey 1965), but is without effect on others, e.g. *Macroptilium atropurpureum* and *Centrosema pubescens* (Norris 1971*a*, 1973*b*). The rule-of-thumb guide, suggested by Norris, is to use lime for species requiring fast growing acid-producing *Rhizobium,* and rock phosphate (or similar inert material) on species with slow growing alkali-producing *Rhizobium.* Experience in recent years suggests, that unless pelleting is necessary to counteract adverse conditions, such as acid soils or fertilizers, or protection against fungicides, or because the seed is being oversown, then simple dilute (quarter to half strength) sticker, without coating, ensures satisfactory adhesion of inoculum to the seed, and survival of the *Rhizobium* (Date 1968), and nodulation in the field (Norris 1971*a*, *b*, 1972, 1973*a*, *b*).

Serological techniques

One of the outstanding technical contributions to legume bacteriology in the past 50 years has been the development and use of serological techniques for identification of *Rhizobium.* Identification of strains forming nodules is a basic necessity in any ecological assessment of *Rhizobium* and its contribution to fertility in soils systems. For many years ecological studies were not attempted because of the enormity of the task of isolating rhizobia from nodules for identification. The first serological typing of *Rhizobium* in Australia was done by Vincent (1941) for rhizobia from species of *Medicago*. He recognized the difference between flagellar (H) and somatic (O) antigen reaction and, by destroying the less specific flagellar material, was able to increase specificity or discriminatory ability between strains. Using this technique and agglutinin absorption to further increase specificity of an antiserum, Vincent and coworkers (Vincent 1941, 1942; Hughes and Vincent 1942; Purchase and Vincent 1949; Purchase *et al.* 1951*a*, *b*) described the antigenic relationships between naturalized strains of *Rhizobium* for clovers and medics, and the performance of inoculants for these species in the field.

The agglutination test, however, required isolation of the *Rhizobium* into pure culture, until direct typing of nodule juice or bacterioid antigen (Means *et al.* 1964; Parker and Grove 1970) widened the scope and capacity of the technique.

Further improvements in techniques were made by Dudman (1964, 1971, 1977), who developed the Ouchterlony (1958) double diffusion, or gel immunodiffusion, precipitin band technique. The increased precision of this technique, combined with the use of bacteroid antigen, has provided a means to expand ecological studies, and to assess more accurately the success of applied inocula to form nodules when in competition with existing soil populations of rhizobia; in this laboratory one operator can routinely serotype 500–600 nodules per day. Because asepsis is not required, and non-viable material is used as test antigen, nodules sampled from the field can be stored dry or frozen for years, if necessary, before assay. The technique does have one disadvantage, however; typing is restricted to one or two antisera, because of the limited amount of antigen available from a single nodule. Obviously the technique does not work for small nodules (< 2 mm), but fortunately, modification of the fluorescently labelled antibody (F.A.) technique (Trinick 1969; R. A. Date, unpublished) has overcome this problem. Nodules of < 0.5 mm provide sufficient antigen for several tests by this method.

Serological checking remains important in quality control of commercial inoculants (Date 1969), since it is the only rapid specific identification test available for immediate trueness-to-type assessment.

Ecological studies with *Rhizobium* for subterranean clover

Although an accidental introduction to the Australian pastoral scene, subterranean clover has played an important role in the development of pastures in temperate regions. There have been several major studies related to its nodulation and nitrogen fixation with *Rhizobium*. The first detailed research undertaking appears to be that of Strong (1937, 1938, 1940) who observed: that the rhizobia from subterranean clover are not mutually interchangeable with those of red and white clover for effectiveness in nitrogen fixation; that significant responses to inoculation could be obtained from cleaned seed; that uncleaned seed, still in the burr with soil attached, nodulated effectively but more slowly; and that there is a likelihood of competition between effective and ineffective strains for nodulation.

Pittman (1935) also made the observation, that established soil rhizobia may offer sufficient competition to keep out applied strains. More detailed observations on the existence of specificity relationships between naturally occurring strains and the sparseness of these rhizobia were reported for the acid soils of the Lismore District of New South Wales by Vincent (1954*a*), in New England by Baird (1955), and in north-western New South Wales by Vincent and Crofts (1958). Harris (1961) reported, that in relatively non-acid soils, inoculation is still worthwhile in the first year, but not vital to successful establishment of the clover.

In south-eastern Australia, large areas of subterranean clover have established, either by natural spread following widespread application of superphosphate, or as a component in specific pasture improvement programs. In some areas, seed was inoculated, but the inoculum strain tended to be replaced by a naturalized population (Dudman and Brockwell 1968). The effectiveness of the naturalized population was studied by Gibson *et al.* (1975), who observed that effectiveness of the populations sampled varied from 62–93% of that of the standard strain, TA1. There was wide variability within sites and between regions, and no apparent relation to any soil character, or experimental treatment.

More recent studies (Gibson *et al.* 1976; Roughley *et al.* 1976) have confirmed that inoculation is advantageous in the first year, but thereafter the inoculum strains tend to disappear, forming smaller proportions of the nodule population. There is no evidence to indicate whether these strains persist in the soil but fail to form nodules because they are not strong competitors, or whether they simply decline in number. There are, however, strain-of-*Rhizobium* and host-cultivar differences; for example, in both these studies, *Rhizobium* strain WU95 persisted longer (i.e. formed more nodules in subsequent years) than other strains, and was found more frequently in the nodules of cultivar Woogenellup than Mt Barker. The degree of persistence (or recovery from nodules) was influenced also by the presence or absence of rhizobia in the soil before sowing, e.g., in an initially *Rhizobium*-free site, inoculum strains formed larger proportions of the nodule population than at sites with initially high levels of native strains (Roughley *et al.* 1976).

Although the legume exercises a distinct preference for strains in nodule formation, the success of an applied inoculum in competition with soil strains is influenced by the relative numbers of inoculum- and soil rhizobia. Ireland and Vincent (1968) found, that strains which occur naturally in the soil, and form ineffective nodules on white clover, prevented the formation of nodules by an effective strain introduced on the seed of subterranean clover, unless the number introduced per seed was greater than the number of ineffectives per gram of soil. Increasing the level of inoculum on the seed by a factor of 10 increased the proportion of effectively nodulated plants by a factor that was greater than the inhibitive effect of an equal increase in the number or ineffective soil strains. Quantitative and qualitative differences between strains have been studied in the laboratory by Marques Pinto *et al.* (1974), Labanderra and Vincent (1975) and Franco and Vincent (1976). Using a mathematical function to obtain a competitive index, these workers were able to demonstrate that inoculum strain ratios were not always reflected in initial root surface ratios; strains differed in their ability to colonize the root surface; dependence of nodulation success on root surface representation existed in a limited number of situations; there was no relation between symbiotic effectiveness and competitive ability; and environment exerted a marked influence on interstrain competition (see review Date and Brockwell 1978).

The problem of second-year clover mortality in the coastal sand plain soils of Western Australia is now almost a classical story in practical *Rhizobium* ecology. A microbiological problem was reported by Parker (1962) and described by Marshall *et al.* (1963), who observed that, although successful nodulation was readily obtained in the year of sowing by inoculation and lime-pelleting of seed (Cass-Smith and Goss 1958), there were few inoculum rhizobia in the soil by the second year; lupins and serradella did not however suffer from this problem. They referred to this as 'second-year clover mortality'. In a series of laboratory and field experiments, this was observed to be primarily due to the very low (< 5 per gram of soil) number of clover rhizobia in the soil at the end of the growing season (Chatel and Parker 1973*a*). Studies on colonization of soil and rhizophere showed that lupin rhizobia; established more readily and in higher numbers than clover rhizobia, i.e. *Rhizobium* species differences existed (Chatel and Parker 1973*b*), and that there were significant differences in colonization for strains of *Rhizobium* for subterranean clover (Chatel and Greenwood 1973*a*). These tended to survive from season to season in 'pockets' associated primarily with old nodules, whereas lupin rhizobia were distributed throughout the soil (Chatel and Greenwood 1973*b*). Finally, strains of rhizobia isolated from nodules of the occasional healthy plant from originally 'failed' fields, when used as inocula, demonstrated superior persistence. The inclusion of these strains brought about a solution of the second-year mortality problem (Chatel *et al.* 1973). The characteristic(s) so included in such strains has been described as 'saprophytic competence' (Chatel *et al.* 1968), and might be somewhat analogous to Harris's (1954) 'incursiveness', described some years earlier; these terms refer to what other authors encompass in the terms 'persistence', and 'nodulating and competitive abilities'. It has been suggested that the failure of the original strain, TA1, to survive into the second year was an intolerance of high-dry soil temperatures during summer in the absence of the growing plant. Survival was improved by inclusion of colloidal materials (montmorillonite, flyash), with improvement in the proportion of plants effectively nodulated in the second year (Marshall 1964; Marshall and Roberts 1963).

Experience in northern Australia

Except for isolated studies (Steindl 1941; McKnight 1949), there had been little work prior to the 1950s on the nodulation of legumes, especially of pasture legumes, for the tropics and subtropics of Australia. When the now CSIRO Division of Tropical Crops and Pastures began in 1952, Dr D. O. Norris started a systematic study of the nodulation and nitrogen fixing requirements of legumes suited to the tropics and sub-tropics. His primary task was to assemble and test the effectiveness of a collection of rhizobia suited to the legumes of interest to Divisional workers. Very little of this routine information has been reported in the literature. However, the Division does maintain a bank of tested strains of *Rhizobium*, for a very wide range of indigenous and imported legume species. In a similar study, Bowen (1956*a*) recorded the nodulation of 83 indigenous species from 36 genera, and obtained inoculation responses for *Glycine max, Neonotonia wightii, Pueraria phaseoloides, Centrosema pubescens, Clitorea ternatea,* and *Indigofera hirsuta,* but not for *Vigna unguiculata, Macroptilium lathyroides* and *Stylosanthes guianensis* on the coastal sand plain soil at Coolum, near Brisbane (Bowen 1956*b*).

More recently, inoculation responses have been demonstrated in *Trifolium semipilosum* (Jones and Date 1975), *Lotononis bainesii* (Diatloff 1977), some *Stylosanthes* species and, at some sites, *Centrosema pubescens* and *Desmodium intortum* (R. A. Date, unpublished). Species such as *Macroptilium atropurpureum* cv. Siratro, *Stylosanthes humilis* (Townsville stylo), *Stylosanthes hamata* cv. Verano (but not other *hamatas*), *S. fruticosa* and *S. scabra* have not responded to inoculation, and appear to be nodulated effectively by native strains of *Rhizobium.*

Field experiments to evaluate the success of selected strains to form nodules on freely nodulating species, in competition with existing soil populations, revealed that, like the situation already described for subterranean clover, response is dependent on the nature and number of rhizobia at a particular site, and on host. For example, at Samford (near Brisbane, Qld) *Desmodium intortum* did not form nodules with the inoculum strain at any stage during five years of sampling, whereas at Tumbulgum (northern NSW) and Beerwah (near Brisbane, Qld), 85–90% of nodules were formed by the same inoculum strain in the year of sowing, but had declined to less than 5% of the nodule population by the end of 5 years. *Macroptilium atropurpureum* at these same three sites did not form nodules with the inoculum strain at any time, but was effectively nodulated by native strains (R. A. Date, unpublished).

From the many studies now completed, a general picture has emerged, that in most cases inoculum strains will not persist. Reasons for the failure of such strains to persist, or to compete with native populations, are as yet undefined. However, some clue may be obtained from detailed studies of associations between *Lotononis bainesii* (Diatloff 1977) and *Trifolium semipilosum* (Jones and Date 1975), where the inoculum strain is the one found in 80% or more of nodules, even after 6–12 years.

Likely future research

The impetus for development of new lands, for either cropping or pasture improvement, has been severely curtailed in the latter part of the post-World War II period because of high costs of production and lack of export markets. Therefore the trend in any program for improvement, or even maintenance, has been toward low-cost systems. One outcome of this has been the demand for legumes adapted to, or at least tolerant of, low phosphorus, low pH, high soluble aluminium soils; another, has been the attempt to find legumes suited to specific environment situations, such as the dry tropics of nothern Australia. Many of the legume species of interest need specific or elite strains of *Rhizobium,* that will form nodules in acid soils or in heavy clay alkaline soils, or under conditions of high soil temperature, frequently at low soil moisture levels, or of high soluble aluminium or manganese or iron, and in the presence of competing naturally occurring strains of *Rhizobium.*

In an era of energy conservation and of pollution control, biological nitrogen in general, and legume/*Rhizobium*-fixed nitrogen in particular, will assume greater and greater importance. Such attributes as retention of the organic nitrogen fraction in the soil (i.e. resistance to leaching), the potential for high levels of fixation, and cycling of nitrogen in the biosphere, will no doubt receive much closer scrutiny as researchers make closer examination of the gains and loss of nitrogen available for plant growth.

Seed inoculation will receive special emphasis. As a commercial practice, preinoculation of seed has not been successful in Australian conditions, and has not been accepted by Departments of Agriculture or other research establishments, because of the poor survival of inoculum in samples tested (Brockwell *et al.* 1975). The principle of preinoculation is an attractive one. Its main advantages are that it would relieve the farmer of the responsibility of inoculating legume seed, and it would eliminate problems associated with the use of peat cultures by individual farmers; it would be amenable to quality control procedures, so that a guaranteed product would be available, as are peat inoculants today. The problem to be resolved is one of improved survival of rhizobia on inoculated and stored seed. A minimum of twelve months storage life would be required, so that unsold seed prepared in one sowing season could be used without risk in the next. Some investigation of the factors affecting survival will need close attention. Indications so far suggest that relative humidity, water retention by the organism, and oxygen level are key factors. Strain differences have been observed, and interactions between oxygen and humidity levels during storage noted (H. V. A. Bushby, personal communication).

Associated with this, it might be reasonable to expect inoculant cultures to improve in storage life, so that these also would have almost indefinite shelf-life, as do lyophilized cultures. This would be a significant advance in practical application of knowledge to industry.

Finally, the ecology of *Rhizobium* in soil will demand a large amount of study, particularly aspects of factors or characteristics involved in enabling an organism (i.e. select strain) to establish and persist in the presence of existing soil rhizobia. Some headway in this direction may be achieved by specific selection and /or breeding of new legume lines that nodulate sparsely, if at all, with naturally occurring strains but which are effectively nodulated by a previously selected inoculum strain. This kind of situation was observed by Gibson *et al.* (1976) in the Woogenellup cultivar of subterranean clover and *Rhizobium* strain WU95. This strain persisted over several years in red basaltic soils in the presence of other clover rhizobia and accounted for the majority of nodules on each occasion of sampling.

Some consideration of direct *Rhizobium* to *Rhizobium* competition may reveal reasons for strains not persisting in certain soils or not competing strongly with other strains. There is a suggestion that bacteriocin production by a strain is advantageous in soil and rhizophere colonization, and in competition for nodule-forming sites (Schwinghamer 1971, 1975).

References

Baird, K. J. (1955). Clover root-nodule bacteria in the New England Region of New South Wales. *Aust. J. Agric. Res.* **6**, 15-26.

Bergersen, F. J., Brockwell, J. and Thompson, J. A. (1958). Clover seed pelleted with bentonite and organic material as an aid to inoculation with nodule bacteria. *J. Aust. Inst. Agric. Sci.* **24**, 159-60.

Bowen, G. D. (1956*a*). Nodulation of legumes indigenous to Queensland. *Queensl. J. Agric. Sci.* **13**, 47-60.

Bowen, G. D. (1956*b*). Responses of tropical legumes to inoculation. *Queensl. J. Agric. Sci.* **13**, 61-2.

Brockwell, J. (1954). Non-infective nodule bacteria in peat cultures. *J. Aust. Inst. Agric. Sci.* **20**, 242-6.

Brockwell, J. (1962). Studies on seed pelleting as an aid to legume seed inoculation. *Aust. J. Agric. Res.* **13**, 638-49.

Brockwell, J. (1977). Application of Legume Seed Inoculants. *In* 'A Treatise on Dinitrogen Fixation. Section IV. Agronomy and Ecology'. (Eds R. W. F. Hardy and A. H. Gibson). Chapt. 8, pp.277-309. (Wiley Interscience.)

Brockwell, J. and Hely, F. W. (1961). Symbiotic characteristics of *Rhizobium meliloti* from brown acid soils of the Macquarie Region of N.S.W. *Aust. J. Agric. Res.* **12**, 630-43.

Brockwell, J., Herridge, D. F., Roughley, R. J., Thompson, J. A. and Gault, R. R. (1975). Studies on seed pelleting as an aid to legume seed inoculation. 4. Examination of preinoculated seed. *Aust. J. Exp. Agric. Anim. Husb.* **15**, 780-7.

Brockwell, J. and Whalley, R. D. B. (1962). Incorporation of peat inoculant in seed pellets for inoculation of *Medicago tribuloides* sown into dry soils. *Aust. J. Sci.* **24**, 458.

Bryan, W. W. and Andrew, C. W. (1955). Pasture studies on the coastal lowlands of subtropical Queensland. II The Interrelation of legumes, *Rhizobium* and calcium. *Aust. J. Agric. Res.* **6**, 291-8.

Cass-Smith, W. P. (1938). The inoculation of tangier peas. *J. Agric. West. Aust.* **15**, (2nd Ser.) 36-9.

Cass-Smith, W. P. and Goss, O. M. (1958). A method of inoculating and lime pelleting legume seed. *J. Agric. West. Aust.* **7** (3rd Ser.) 119-21.

Cass-Smith, W. P. and Pittman, H. A. (1939). The influence of methods of planting in the effective inoculation and establishment of subterranean clover. *J. Agric. West. Aust.* **16** (2nd Ser.) 61-73.

Champ, B. R. and Sillar, D. I. (1961). Pellet your Buffel seed and thwart ants. *Queensl. Agric. J.* **87**, 583.

Chatel, D. L. and Greenwood, R. M. (1973*a*). The colonization of host root and soil by rhizobia. II. Strain differences in the species *Rhizobium trifolii. Soil Biol. Biochem.* **5**, 433-40.

Chatel, D. L. and Greenwood, R. M. (1973*b*). The location and distribution in soil of rhizobia under senesced annual legume pastures. *Soil Biol. Biochem.* **5**, 799-808.

Chatel, D. L., Greenwood, R. M. and Parker, C. A. (1968). Saprophytic competence as an important character in the selection of *Rhizobium* for inoculation. XI Int. Congr. Soil Sci. Trans. Vo. II, pp.65-73.

Chatel, D. L. and Parker, C. A. (1973*a*). Survival of field grown rhizobia over the dry summer period in Western Australia. *Soil Biol. Biochem.* **5**, 415-23.

Chatel, D. L. and Parker, C. A. (1973*b*). The colonization of host-root and soil by rhizobia. I. Species and strain differences in the field. *Soil Biol. Biochem.* **5**, 425-32.

Chatel, D. L., Shipton, W. A. and Parker, C. A. (1973). Establishment and persistence of *Rhizobium trifolii* in Western Australian soils. *Soil Biol. Biochem.* **5**, 815-24.

Date, R. A. (1959). Growth and survival of the root-nodule bacteria. M.Sc. Agric. Thesis Univ. of Sydney, Australia.

Date, R. A. (1968). Rhizobial survival on the inoculated legume seed. IX Int. Cong. Soil Sci. Trans. Vol. II, pp.75-83.

Date, R. A. (1969). A decade of legume inoculant quality control in Australia. *J. Aust. Inst. Agric. Sci.* **35**, 27-37.

Date, R. A. and Brockwell. J. (1978). *Rhizobium* Strain Competition and Host Interaction for Nodulation. *In* 'Plant Relations in Pastures'. (Ed. J. R. Wilson). Chapt. 14. pp 202-16. (CSIRO: Melbourne.)

Date, R. A. and Roughley, R. J. (1977). Preparation of Legume Seed Inoculants. *In* 'A Treatise on Dinitrogen Fixation. Section IV. Agronomy and Ecology'. (Eds R. W. F. Hardy and A. H. Gibson). Chapt. 7, pp.243-75. (Wiley Interscience: New York.)

Diatloff, A. (1977). Ecological studies of root-nodule bacteria introduced into field environments. 6. Antigenic and symbiotic stability in *Lotononis* rhizobia over a 12 year period. *Soil Biol. Biochem.* **9**, 85-8.

Diatloff, A. and Brockwell, J. (1976). Ecological studies of root-nodule bacteria introduced into field environments. 4. Symbiotic properties of *Rhizobium japonicum* and competitive success in nodulation of two *Glycine max* cultivars by effective and ineffective strains. *Aust. J. Exp. Agric. Anim. Husb.* **16**, 514-21.

Donnan, A. V. (1912). Experiment in inoculation of lucerne. *Agric. Gaz. N.S.W.* **23**, 953-8.

Dudman, W. F. (1964). Immune diffusion analysis of the extra-soluble antigens of two strains of *Rhizobium meliloti. J. Bacteriol.* **88**, 782-94.

Dudman, W. F. (1971). Antigenic analysis of *Rhizobium japonicum* by immuno-diffusion. *Appl. Microbiol.* **21**, 973-85.

Dudman, W. F. (1977). Serological Methods and their Application to Dinitrogen Fixing Organisms. *In* 'A Treatise on Dinitrogen Fixation. Section IV. Agronomy and Ecology'. (Eds R. W. F. Hardy and A. H. Gibson). Chapt. 13, pp.487-508. (Wiley Interscience: New York.)

Dudman, W. F. and Brockwell, J. (1968). Ecological studies of root-nodule bacteria introduced into field environments. I. A survey of field performance of clover inoculants by gel immune diffusion serology. *Aust. J. Agric. Res.* **19**, 739-47.

Easterby, H. T. (1910). Experiments with green manure seed by inoculation with Professor Bottomley's nitro-bacterine culture. Bur. Sug. Exp. Stations. Annl. Rep. 1910, pp.10-11.

Franco, A. A. and Vincent, J. M. (1976). Competition amongst rhizobial strains for the colonization and nodulation of two tropical legumes. *Plant Soil* **45**, 27-48.

Gibson, A. H. (1977). The Influence of the Environment and Managerial Practices on the Legume-*Rhizobium* Symbiosis. *In* 'A Treatise on Dinitrogen Fixation. Section IV. Agronomy and Ecology'. (Eds R. W. F. Hardy and A. H. Gibson). Chap. 11, pp.393-450. (Wiley Interscience: New York.)

Gibson, A. H., Curnow, B. C., Bergersen, F. J., Brockwell. J. and Robinson, A. C. (1975). Studies of field populations of *Rhizobium:* Effectiveness of *Rhizobium trifolii* associated with *Trifolium subterranean* pastures in south eastern Australia. *Soil Biol. Biochem.* **7**, 95-102.

Gibson, A. H., Date, R. A., Ireland, J. A. and Brockwell, J. (1976). A comparison of competitiveness and persistence amongst five strains of *Rhizobium trifolii. Soil Biol. Biochem.* **8**, 395-401.

Goss, O. M. and Shipton, W. A. (1965). Nodulation of legumes on new light lands. I. Survival of rhizobia on inoculated pelleted seed held for varying periods before sowing into dry soil. *J. Agric. West. Aust.* **6** (4th Ser.) 616-21.

Greig-Smith, R. (1899). The nodule organism of the leguminosae. *Proc. Linn. Soc. NSW* **24**, 653-73.

Greig-Smith, R. (1901). The nature of the bacteriods of the leguminous nodule and the culture of *Rhizobium leguminosarum. Proc. Linn. Soc. NSW* **26**, 152-5.

Greig-Smith, R. (1911). Contributions to our knowledge of soil fertility. II The determination of rhizobia in soil. *Proc. Linn. Soc. NSW* **36**, 492-503.

Guthrie, F. B. (1896). Inoculation of the soil for leguminous crops. *Agric. Gaz. NSW* **7**, 690-4.

Hamblin, C. O. (1920). To infect lucerne seed with nodule organisms. *Agric. Gaz. NSW* **31**, 466.

Harris, J. R. (1954). Rhizophere relationships of subterranean clover. I. Interactions between strains of *Rhizobium trifolii. Aust. J. Agric. Res.* **5**, 247-70.

Harris, J. R. (1961). Rhizobial inoculation and fertilizer rates as factors in the establishment of subterranean clover in South Australia. *Aust. J. Agric. Res.* **12**, 84-99.

Hastings, A. and Drake, A. D. (1960). Inoculation and pelleting of clover seed. *N.Z. J. Agric.* **101**, 619-20.

Hely, F. W. and Brockwell, J. (1962). An exploratory survey of the ecology of *Rhizobium meliloti* in inland N.S.W. and Queensland. *Aust. J. Agric. Res.* **13**, 864-79.

Hughes, D. Q. and Vincent, J. M. (1942). Serological studies of the root nodule bacteria. III Tests of neighbouring strains of the same species. *Proc. Linn. Soc. NSW.* **67**, 142-52.

Ireland, J. A. and Vincent, J. M. (1968). A qualititative study of competition for nodule formation. IXth Int. Cong. Soc. Sci. Trans. Vol. II, pp.85-93.

Jones, R. M. and Date, R. A. (1975). Studies in the nodulation of Kenya White Clover (*Trifolium semipilosum*) under field conditions in south-east Queensland. *Aust. J. Exp. Agric. Anim. Husb.* **15**, 519-26.

Labanderra, C. A. and Vincent, J. M. (1975). Competition between an introduced strain and native Uruguayan strains of *Rhizobium trifolii. Plant Soil* **42**, 327-47.

Loneragan, J. F., Meyer, D., Fawcett, R. G. and Anderson, A. I. (1955). Lime pelleted clover seeds for nodulation on acid soils. *J. Aust. Inst. Agric. Sci.* **21**, 264-5.

McAlpine, D. (1906). Experiments with nitrogen fixing bacteria in connection with lucerne. *Agric. Gaz NSW* **18**, 349-53.

McKnight, T. (1949). Efficiency of isolates of *Rhizobium* in the cowpea group, with proposed additions to this group. *Queensl. J. Agric. Sci.* **6**, 61-76.

Marks, G. (1905). Experiments with nodule culture. *Agric. Gaz NSW* **17**, 136-41.

Marques Pinto, C., Yao, P. Y. and Vincent, J. M. (1974). Nodulating competitiveness amongst strains of *Rhizobium meliloti* and *R. trifolii. Aust. J. Agric. Res.* **25**, 317-39.

Marshall, K. C. (1964). Survival of root-nodule bacteria in dry soils exposed to high temperatures. *Aust. J. Agric. Res.* **15**, 273-81.

Marshall, K. C., Mulcahy, M. J. and Chowdhury, M. S. (1963). Second-year clover mortality in Western Australia — a microbiological problem. *J. Aust. Inst. Agric. Sci.* **29**, 160-4.

Marshall, K. C. and Roberts, F. J. (1963). Influence of fine particle materials on survival of *Rhizobium trifolii* in sandy soil. *Nature (London)* **198**, 410-11.

Means, U. M., Johnson, H. W. and Date, R. A. (1964). Quick serological method of classifying strains of *Rhizobium japonicum* in nodules. *J. Bacteriol.* **87**, 547-53.

Murguia, J. L. and Date, R. A. (1965). The use of pellets in the oversowing of pasture legumes. Proc. IX Int. Grasslds. Cong. Sao Paulo, Brazil. Vol. 1. pp.279-82.

Norris, D. O. (1967). The intelligent use of inoculants and lime pelleting for tropical legumes. *Trop. Grasslands* **1**, 107-21.

Norris, D. O. (1970). The contribution of research in legume bacteriology to the development of Australian pastures. Proc. XI Int. Grasslds. Cong. Australia, 1970. pp.A22-A30.

Norris, D. O. (1971*a*). Seed pelleting to improve nodulation of tropical and subtropical legumes. 2. The variable response to lime and rock phosphate pelleting of eight legumes in the field. *Aust. J. Exp. Agric. Anim. Husb.* **11**, 282-9.

Norris, D. O. (1971*b*). Seed pelleting to improve nodulation of tropical and subtropical legumes. 3. A field evaluation of inoculant survival under lime and rock phosphate. *Aust. J. Exp. Agric. Anim. Husb.* **11**, 677-83.

Norris, D. O. (1972). Seed pelleting to improve nodulation of tropical and subtropical legumes. 4. The effects of various mineral dusts on nodulation of *Desmodium uncinatum. Aust. J. Exp. Agric. Anim. Husb.* **12**, 152-8.

Norris, D. O. (1973*a*). Seed pelleting to improve nodulation of tropical and subtropical legumes. 5. The contrasting response to lime pelleting of two *Rhizobium* strains on *Leucaena leucocephala. Aust. J. Exp. Agric. Anim. Husb.* **13**, 98-101.

Norris, D. O. (1973*b*). Seed pelleting to improve nodulation of tropical and subtropical legumes. 6. The effects of dilute sticker and of bauxite pelleting on the nodulation of six legumes. *Aust. J. Exp. Agric. Anim. Husb.* **13**, 700-4.

Norris, D. O. and Date, R. A. (1976). Legume Bacteriology. *In* 'Tropical Pasture Research. Principles and Methods'. (Eds N. H. Shaw and W. W. Bryan), CAB Bulletin 51, Chapt. 7, pp.134-74. (Alden Press: Oxford.)

Ouchterlony, O. (1958). Diffusion in gel methods for immunological analysis. *Prog. Allergy* **5**, 1-78.

Parker, C. A. (1962) Light lands in Western Australia. 3. Microbiological problems in the establishment of legumes on light lands. *J. Agric. West. Aust.* **3** (4th Ser.), 713-6.

Parker, C. A. and Grove, P. L. (1970). The rapid serological identification of rhizobia in small nodules. *J. Appl. Bacteriol.* **33**, 248-52.

Parker, C. A. and Oakey, A. E. (1965). Reduced nodulation of lupins and serradella due to lime pelleting. *Aust. J. Exp. Agric. Anim. Husb.* **5**, 144-6.

Parker, C. A., Trinick, M. J. and Chatel, D. L. (1977). Rhizobia as Soil and Rhizosphere Inhabitants. *In* 'A Treatise on Dinitrogen Fixation. Section IV. Agronomy and Ecology'. (Eds R. W. F. Hardy and A. H. Gibson), Chap. 9, pp.311-52. (Wiley Interscience: New York.)

Pittman, H. A. (1933). Leguminous crops and their place in the building up of soil fertility. *J. Agric. West. Aust.* **10** (2nd Ser.), 470-8.

Pittman, H. A. (1935). Leguminous crops and their place in the building up of soil fertility. *J. Agric. West. Aust.* **12** (2nd Ser.), 105-16.

Pound, C. J. (1910). Nitrogen fixing bacteria. Queensl. Dep. Agric. and Stock Annu. Rep. 1909-10, p.24.

Purchase, H. F. and Vincent, J. M. (1949). A detailed study of the field distribution of strains of clover nodule bacteria. *Proc. Linn. Soc. NSW* **74**, 227-36.

Purchase, H. F., Vincent, J. M. and Ward, L. M. (1951*a*). The field distribution of strains of nodule bacteria from species of *Medicago. Aust. J. Agric. Res.* **2**, 261-72.

Purchase, H. F., Vincent, J. M. and Ward, L. M. (1951*b*). Serological studies of the root-nodule bacteria. IV Further analysis of isolates from *Trifolium* and *Medicago. Proc. Linn. Soc. NSW* **76**, 1-6.

Roughley, R. J. (1962). *Rhizobium* research and service in the N.S.W. Department of Agriculture. *Agric. Gaz NSW* **73**, 260-2.

Roughley, R. J., Blowes, W. M. and Herridge, D. F. (1976). Nodulation of *Trifolium subterranean* by introduced rhizobia in competition with naturalized strains. *Soil Biol. Biochem.* **8**, 403-7.

Schwinghamer, E. A. (1971). Antagonism between strains of *Rhizobium trifolii* in culture. *Soil Biol. Biochem* **3**, 355-63.

Schwinghamer, E. A. (1975). Properties of some bacteriocins produced by *Rhizobium. J. Gen. Microbiol.* **91**, 403-13.

Snieszko, S. F. (1941). The influence of lime on nodulation and growth of canning peas. *Canning Age* **22**, 49-50.

Steindl, D. R. L. (1941). Legume inoculation in Queensland canefields: Efficiency of strains of nitrogen fixing bacteria. Bur. Sug. Exp. Stations Tech. Comm. No. 2, pp.29-55.

Strong, T. H. (1937). The influence of host plant species in relation to the effectiveness of the *Rhizobium* of clovers. *J. Counc. Sci. Indust. Res.* **10**, 12-6.

Strong, T. H. (1938). Legume establishment and its function in relation to the development of some of the poor soils of Kangaroo Island. *J. Agric. South Aust.* **41**, 542-50.

Strong, T. H. (1940). Non-effective associations of nodule bacteria and legumes. *J. Aust. Inst. Agric. Sci.* **6**, 14-20.

Thompson, J. A. (1960). Inhibition of nodule bacteria by an antibiotic from legume seed coats. *Nature (London)* **187**, 619-20.

Trinick, M. J. (1969). Identification of legume nodule bacteria by the fluorescent antibody reaction. *J. Appl. Bacteriol.* **32**, 181-6.

Vincent, J. M. (1941). Serological studies of the root-nodule bacteria. I. Strains of *Rhizobium meliloti Proc. Linn. Soc. N.S.W.* **66**, 145-54.

Vincent, J. M. (1942). Serological studies of the root-nodule bacteria. II. Strains of *Rhizobium trifolii. Proc. Linn. Soc. N.S.W.* **67**, 82-6.

Vincent, J. M. (1954a). The root-nodule bacteria as factors in clover establishment in the red basaltic soils of the Lismore district, N.S.W. I. A survey of 'native' strains. *Aust. J. Agric. Res.* **5**, 55-60.

Vincent, J. M. (1954*b*). The control of the quality of legume seed inoculants. *J. Aust. Inst. Agric. Sci.* **20**, 247-9.

Vincent, J. M. (1958*a*). Towards better legume inoculants. *J. Aust. Inst. Agric. Sci.* **24**, 226-8.
Vincent, J. M. (1958*b*). Survival of the Root-nodule Bacteria. *In* 'Nutrition of Legumes'. (Ed. E. G. Hallsworth). Chapt. 8, pp. 108-23. (Butterworths Scientific Publications: Sydney.)
Vincent, J. M. (1962). Australian studies of the root-nodule bacteria. II. A review. *Proc. Linn. Soc. N.S.W.* **87**, 8-38.
Vincent, J. M. and Crofts, F. C. (1958). Pasture improvement in the Pilliga. University of Sydney, School of Agriculture. Report No. 3.
Waters, L. M. (1954). Tests of commercial peat inoculants. *J. Aust. Inst. Agric. Sci.* **20**, 250-1.

Chapter 45

Soil-borne root diseases and wheat

A. D. Rovira and E. H. Ridge

SOIL BIOLOGICAL FACTORS WHICH RESTRICT ROOT growth and root function influence the uptake of nutrients and water by plants. This is particularly so in many Australian wheat soils which are low in nutrients and which are in areas of low rainfall compared with many other wheat-growing areas of the world.

Although rainfall is a major determinant of wheat yield in Australia (Greacen and Hignett 1976), the supply of nutrients and the incidence of root pathogens are two important factors determining whether or not the rainfall is used to maximum efficiency. The responses in yield to the introduction of superphosphate and later to improved rotations with pasture legumes are demonstrated in Fig.45.1. Despite the introduction of improved varieties and improved nutrition, there has been no significant increase in yield per hectare over the past 10 years and we believe this is very much due to limitations imposed by biological factors in the soil.

Studies with soil fumigation at several sites in southern Australia demonstrated that in most years the potential yield for a particular soil under a given rainfall has not been reached (Table 45.1).

The major factor responsible for these yield increases following fumigation is the elimination of soil-borne root

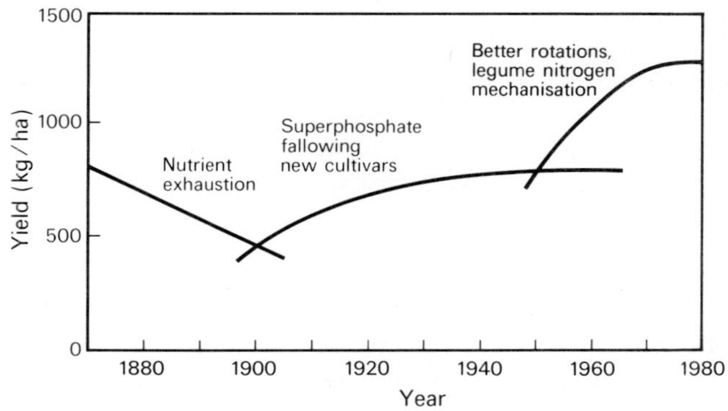

Fig. 45.1. Trends in Australian wheat yields (after Donald 1964).

Table 1. Effect of soil fumigation on wheat yields

All plots received superphosphate at 200 kg ha^{-1} with trace elements where required. Fumigation was with either methyl bromide or chloropicrin at 200 to 400 kg ha^{-1}. Abbreviations, soil-borne root pathogens: Ggtr, *Gaeumannomyces graminis* var. *tritici*; Rhiz, *Rhizoctonia solani*; CCN, cereal cyst nematode (*Heterodera avenae*); L, light; M, moderate; H, High. *, **, ***, significance at $P = 0.05$, $P = 0.01$ and $P = 0.001$.

Site	Soil description	Year	Grain yield (t ha^{-1})		Levels of root pathogens		
			Untreated	Fumigated	Ggtr	Rhiz	CCN
Reeves Plains	Solonized brown soil	1971	2.4	3.3**	L	M	M
Roseworthy		1971	2.2	3.9**	M	0	L
Agricultural College	Solonized brown soil	1972	1.6	2.4**	L	0	M
Sea Lake[A]	Solonized brown soil	1971	2.1	3.2*	L	M	H
		1973	0.9	3.6***	L	M	H
Streaky Bay	Highly calcareous sandy loam	1974	0.5	2.7***	M	M	H
Nunjikompita	Highly calcareous sandy loam	1974	0.8	1.4*	M	M	H
Coonalpyn[B]	Deep siliceous sand	1975	0.2	0.5**	Detailed information not available, but at all sites each of the above pathogens was in the moderate to severe range		
	Sand over clay	1975	0.5	1.5***			
	Shallow sandy loam over clay	1975	1.4	2.8***			
	Deep siliceous sand	1975	0.3	0.9***			
Coonalpyn[B]	Sand over clay	1976	2.1	2.4*			
	Deep siliceous sand	1976	1.0	1.7**			
	Shallow sand over clay	1976	0.9	2.7***			

[A] Joint CSIRO–Victorian Department of Agriculture.
[B] Trials by P. King, S.A. Dep. Agric. Fish.

pathogens, although the release of mineral nitrogen as NH_4^+-N is another factor contributing to the higher yields.

Since the earliest days of cereal growing in Australia, farmers have been troubled by diseases which attacked the roots. The problem was so serious, indeed, in the early days of colonization that the Government of South Australia set up a Commission to investigate 'Diseases in Cereals' (Anon. 1868). The report stated:

"This disease, Take-all, so destructive in many districts of South Australia, is not so well understood as rust. Farmers and chemists are alike at sea – ploughmen and microscopical observers differ *in toto* as to its nature and causes."

Even at that early stage of agriculture in Australia, the low nutrient status of the soils was recognized and experiments were conducted with ashes and crushed bones in an attempt to reduce losses to wheat crops from Take-all (Mücke 1870).

The aim of this chapter is to discuss the major soil-borne root diseases of wheat in southern Australia and to relate the extent and severity of these diseases to the soils and climates of Australia in comparison with some of the overseas experiences in this area of research.

The major soil-borne root pathogens of wheat in Australia are as follows.

Fungi

Gaeumannomyces graminis–commonly called take-all,
Rhizoctonia solani (purple or bare patch),
Pythium spp. (root rots),
Gibberella zeae or *Fusarium roseum* (crown rot), and
Cochliobolus sativus or *Helminthosporium sativum* (common root rot).

G. graminis and *Rhizoctonia* appear to be the major soil-borne fungal pathogens of cereals in southern Australia and will be discussed further. *Pythium* is widespread, while *Gibberella* and *Cochliobolus* are more important in northern New South Wales and Queensland (Burgess *et al.* 1980), but these three genera are mentioned here only for completeness.

Nematodes

Heterodera avenae (cereal cyst nematode or cereal eelworm) which is prevalent in Victoria and South Australia,
Pratylenchus thornei which is important in certain crop rotations in the Darling Downs, Queensland,
Pratylenchus minyus which is widespread, but considered of minor importance, and
Tylenchorhynchus brevidens which is also widespread, but considered of minor importance.

In this chapter we will discuss *H. avenae* because its effects on growth are better understood than those caused by other nematodes.

Take-all or hay-die (*Gaeumannomyces graminis* var. *Tritici*)

G. graminis var. *tritici* is one of three varieties distinguished by Walker (1972) on the basis of morphology and host range. Earlier, the species had been known as *Ophiobolus graminis*, later renamed as *G. graminis*. Turner (1940) had distinguished the take-all fungus causing disease of oats as *Ophiobolus graminis* var. *avenae*, and Walker (1975) defined varieties of *Gaeumannomyces graminis* as:

var. *graminis*, which is common on grasses and is the cause of crown sheath rot on rice,
var. *avenae*, the cause of take-all of oats and other cereals, and
var. *tritici*, the cause of take-all of wheat and other cereals, but not of oats.

There is evidence (Chambers and Flentje 1967*a*; Yeates 1980) that there is a form which is pathogenic on wheat, that also attacks oats, but in many respects it is intermediate between var. *tritici* and var. *avenae*. Such a strain has been found in 11 to 56% of wheat paddocks sampled in Western Australia, the higher frequency being in soils in which oats had been grown. The presence of this oat-attacking strain would negate the use of oats as a 'cleansing' crop.

In a recent historical review on this pathogen, Garrett (1981) wrote: 'in its historical origins, Take-all is very much an Australian disease, and it is to this country that we must look for records of its early history'.

The name 'take-all' was derived from the symptom of seedling death which was the manifestation of the disease in the early days of wheat growing in Australia. In 1870 Mücke wrote, 'The first signs of it (take-all) were noticed about fourteen years ago . . . Small patches were observed here and there, where the crops died off not long after their appearance above the ground. These patches increased in number and size, and were taken possession of by weeds hitherto unknown in the district. . . . The people, very appropriately, called the mysterious disease "Take-all", . . .'

Garrett (1942) ranked take-all as second in importance to stem rust as a cause of yield losses in wheat on a world-wide basis; Butler (1961) stated that 'throughout the cereal-growing areas of the world the root- and foot-rot complex has constituted a serious problem' and 'the most important of the root rot diseases is Take-all'.

With the introduction of superphosphate and legume pastures and consequent improvement in soil fertility, the fungus more often causes a premature ripening of wheat, producing symptoms of 'whiteheads', 'dead heads' or 'hay-die'. Examination of roots at this stage shows black lesions on the seminal roots, spreading to the subcrown internode and, in very severe cases, blackening of the nodal roots and base of the stem, so that such plants are easily pulled from the ground. These are also the symptoms of the disease in Europe, the United States, Brazil and Chile, where it causes serious losses in yield.

It is difficult to assess yield losses in wheat caused by take-all, especially when subclinical attacks may produce premature ripening, but estimates of overall losses in New South Wales are 1 to 10%, with individual losses of up to 70%; in Victoria, South Australia, and Western Australia the estimates of overall losses are 5 to 10% of the crop. Parker (1977) stated that take-all is widespread, almost universal in the clover belt, in the eastern wheat belt, as well as in the south and south-east of Western Australia.

A relationship between grain yield and percentage of infected crowns per site or percentage of infected roots per core was found by MacNish and Dodman (1973*a*), but more detailed studies over several seasons are required to obtain data to calculate the disease incidence-yield regression more precisely in relation to seasonal conditions.

Influence of season

Banyer (1966*b*) and H. J. Sims (pers. comm.) reported that take-all is favoured by wet conditions in August–October, especially after a dry summer. This has been confirmed experimentally by a technique in which a standard number of disease propagules (*G. graminis* grown on dead oats) is introduced into soils with the wheat seed. By this technique, Rovira (1978*b*) found that in a wet season loss of grain due to take-all was 50%, compared with 10% in drier seasons for the same number of propagules.

Influence of soil conditions

Soil conditions affect both the survival of the pathogen and its effectiveness in initiating infection and promoting disease in the host. The effects may be directly upon the pathogen or upon the host, but indirect effects, mediated through the microflora of soil and rhizosphere, may also be important.

Take-all is more often a problem in sandy soil than in heavy clay soil. Loose, light-textured, well-aerated soils of alkaline reaction are the most favourable and Garrett (1936) considered that such soils provided a lower concentration of carbon dioxide which favoured the growth of runner hyphae along the roots. However, take-all is by no means confined to sandy soils and can be a serious problem in the red brown earths and self-mulching black clay soils of Australia, the phosphate-fixing allophanic soils of Chile and the acidic lateritic soils of Brazil.

Effects of fertilizers

A link between take-all and the nutrient status of the soil was established even before the cause of the disease was known. Mücke (1870) reported that the disease could be checked by application of 'a dilute solution of alkalies and phosphates, which are to be obtained best from night-soil'. A further link with plant nutrition was proposed by the analyst of the Victorian Department of Agriculture (Hopwood 1877) who recommended 'a thorough course of treatment for "take-all" patches—(*a*) dressing with slacked lime and ploughing well in; (*b*) fallowing; (*c*) manuring with bone-dust or human excrement and ashes before sowing'.

Reports conflict on the effects of nitrogen on the disease in the plant, but differences in the form of nitrogen, in the timing of application or in seasonal and soil conditions influence the result. Addition of nitrogen or NPK to barley in the field generally reduces disease (Garrett 1948) because, with added fertilizer, new crown roots are produced faster than the fungus can infect them. Earlier, Garrett (1941) had reported that nitrogen-deficient plants in pots had least root infection and he postulated that the addition of nitrogen renders existing (seminal) roots more susceptible, but that it also encourages the production of new (crown) roots. Shipton (1972), in work on a succession of wheat and barley crops in English field soil carrying *G. graminis* inoculum, found that in the first 4 years applications of up to 125 kg ha^{-1} nitrogen did not produce a consistent effect on yield. In later years, barley and then wheat (after the seventh year) suffered less severe attacks when nitrogen was applied.

Rovira (1978*b*) reported that the application of ammonium sulfate reduced the effect of the disease; MacNish (1977) found that the application of nitrogen in the ammonium form reduced loss due to take-all, whereas nitrate did not reduce take-all. Similar results had been found in the United States by Huber and Watson (1972) and Smiley and Cook (1973), especially when the ammonium nitrogen was initially protected from nitrification by the inhibitor N-Serve [2-chloro-6-(trichloromethyl) pyridine]. Smiley and Cook reported that control of disease severity was mediated by the fall in rhizosphere pH, caused by root uptake of ammonium nitrogen and hence that this lower pH probably increased the proportion of bacteria antagonistic to *G. graminis* in the rhizosphere (Smiley 1978*a*, *b*).

There is conflicting evidence from England, where it was reported that a higher ammonium nitrogen:nitrate nitrogen ratio or ammonium nitrogen alone favoured the disease (Hornby and Goring 1972; Brown *et al.* 1973; Hornby and Brown 1977).

Early studies in Australia showed that phosphorus deficiency favours infection and that there is a cumulative effect of superphosphate in reducing disease (Samuel 1924; Butler 1948). The decline in severe take-all in light soils of southern Australia could be due to the accumulation of available phosphate after years of applying superphosphate to these originally phosphate-deficient soils. However, later studies by Syme (1966) and Chambers (1971*a*) have not substantiated these early findings. Clearly, this is an area requiring further research.

Trace elements appear to influence take-all. Millikan (1938) reported a reduction of root rot following the application of zinc and Toms (1958) found that there was a reduction in incidence and severity of root diseases when a copper deficiency was satisfied.

In Brazil, severe take-all occurs when lime is added to acidic lateritic soils to raise the pH from 4.4 to 5.2, and crops are destroyed after higher rates of liming which raise the pH to 6.6 (Anon. 1976).

Survival of take-all inoculum in soil

The infective propagules of *G. graminis* survive in crop residues, principally root fragments and stem bases. Survival is influenced by soil temperature, moisture, aeration, nitrogen level, and the presence of a susceptible host.

Increased soil nitrogen favours the survival of *G. graminis* in wheat straw (Garrett 1938, 1940, 1944; Chambers 1971*a*) and this is related to the effect of added nitrogen on the cellulolytic ability of the fungus (Garrett 1967). Garrett (1938) had earlier found that fungal survival on straw is decreased when grass meal is added to the sand or soil used for incubation, and he ascribed this to non-availability to the fungus of the nitrogen locked up in the decomposition of the grass. Lucas (1955) also found that nitrogen, added in the pretreatment of straws before burial in inoculated soil, hastened the saprophytic colonization of the straws by *G. graminis* from the soil.

The effect of soil microorganisms on survival was shown by Garrett (1934*a*, *b*, *c*,) when he found less survival in moist soil at 25°C than in moist soil at 12°C. Similar results were obtained by MacNish and Dodman (1973*b*) from undisturbed cores of calcareous sandy loam containing the roots and crowns of field-grown wheat infected with take-all. Survival of take-all in naturally infected residues for up to 45 weeks in hot dry soil is greater than in laboratory-infected straw (MacNish and Dodman 1973*a*; MacNish 1976). These results indicate how well the pathogen is adapted to survive the hot dry summers of southern Australia.

Chambers and Flentje (1967*b*) found that the survival of weakly pathogenic *G. graminis* from ascospores developed in culture is much less than for those from living host tissue, indicating that the more virulent forms will have a better chance of survival in the field.

Cultivation

With the increasing trend to reduced cultivation, generally after herbicide treatment, there are reports of less (Brooks and Dawson 1968), more (Moore 1978; Moore and Cook 1978) and equal (Yarham 1979) incidence of take-all compared with conventional cultivation. The beneficial effects of direct drilling were claimed to be due to limiting the spread of the fungus in undisturbed soil (Brooks and Dawson 1968). However, Moore (1978) claimed that lack of cultivation led to more disease because of the presence of a greater amount of inoculum in the less fragmented debris from the previous season. Ellington and Reeves (1981) reported less lodging from take-all in wheat sown by direct drilling compared with conventional cultivation, whether in continuous wheat or in lupin–wheat rotations. *G. graminis* was isolated from the stem bases of the diseased plants but the field symptom of lodging is not typical of severe take-all; no other root pathogen was isolated from affected plants (Kollmorgen, pers. comm.).

Our own experience over 2 years on a sandy loam soil with a 350 mm annual rainfall is that direct drilling has led to greater incidence of take-all and lower wheat yields compared with conventional cultivation (Rovira 1979). This may be due to the persistence, despite the herbicide treatments, of a proportion of barley grass (*Hordeum glaucum*) which is a common carrier of *G. graminis*. In a different type of soil, or with grasses completely controlled, or with direct drilling practised over a longer period, the results could well be different.

The varying results on the influence of tillage emphasize the point that, although there are some general principles applicable to the effects of soil-borne diseases on crops, e.g. the role of alternate hosts, there are other factors which we cannot expect to produce similar results over different soils, climates and cropping systems. We must accept apparently contradictory results and interpret them within the framework of site specificity.

The infection process and effects on wheat

Infection of the host root surface is preceded by growth from the propagule through soil towards the root, followed by clumping of mycelia among the root hairs, and then ectotrophic growth along the root (Brown and Hornby 1971; Wildermuth 1977).

The fungus does not greatly damage the cortex, which it penetrates *en route* to the stele, where the hyphae first disrupt the phloem and thereafter occupy the xylem vessels (Clarkson *et al.* 1975). This causes a reduction in movement of assimilate to root tips, with a resultant reduction in elongation, and also impedes the entry and transport of ions via the xylem. The blocking of the xylem vessels of wheat by the fungus and the production of mucilage within the vessels are shown in Fig.45.2 (Foster, R. C., pers. comm.).

The difficulty of obtaining the 'white head' symptom (i.e. premature ripening) in glasshouse experiments led Price and Stubbs (1963) to doubt the evidence that associated *G. graminis* with premature ripening. Subsequently, Chambers (1971*b*) produced the white heads in wheat grown in pots with *G. graminis*, with reduced soil water and at temperatures of 20°C rather than 28°C. The higher soil temperature (which does not occur in the

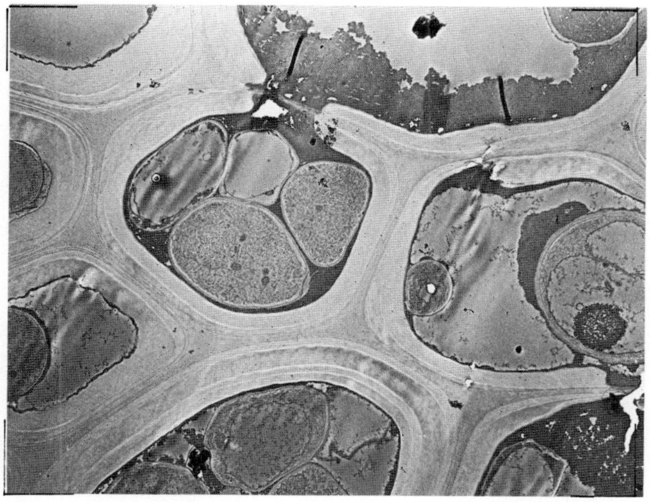

Fig. 45.2. Transmission electron micrograph of wheat root xylem vessels occupied by hyphae of *G. graminis* and mucilage showing as dark material. (R. C. Foster, pers. comm.)

field during the growing season) probably encouraged biological control of the disease reported to occur above 20°C (Henry 1932).

Field trials with introduced inoculum of *G. graminis* have demonstrated that the white head symptom occurred only where the disease has established on the roots; the proportion of white heads in the crop at a given level of inoculum was greatest in the year when the presence of adequate soil moisture up to flowering time favoured the pathogen. In drier seasons, *G. graminis* reduced yields by 10% without the appearance of white heads (Rovira 1978*b*).

Role of Rotation

Allied to the problem of survival of *G. graminis* in soil is the colonization of roots of plant species other than wheat. Nilsson (1969) lists 400 species of graminaceous plants that are alternate hosts to the pathogen. Butler (1959) found 50% of the grass component of clover leys to be infected with *G. graminis* and this factor, together with higher nitrogen which promoted the saprophytic survival of the pathogen, was responsible for the high level of take-all in the succeeding wheat crop. Adam (1951) and Price (1970) showed that a grass-free fallow of 1 year or an oat crop helps to control take-all.

In Victoria, Sims (1958) and Sims *et al.* (1961) in papers on the links between rotations and 'deadheads' in wheat, reported on the complexity of the problem, with apparent inconsistencies of experimental results pointing to the need for far more precise information on the factors *in the field* responsible for the development and incidence of deadheads.

Chambers and Flentje (1968) found less take-all in wheat following some grasses, e.g. *Ehrharta calycina* (perennial veldt grass), *Vulpia myuros* (silver grass) and *Lolium rigidum* (Wimmera rye grass) than following *Hordeum leporinum* (barley grass); and the differences in yield were large, viz. 1.2 t ha^{-1} after *E. calycina* and 0.24 t ha^{-1} after *H. leporinum.*

Chambers (1971*c*) found that in artificially infected sods of grass species *G. graminis* survived better on *Bromus gussonii* and *Hordeum hystrix* than on *Phalaris minor* and *Avena sterilis*.

The value of a grass-free legume crop before barley or wheat has been shown by Holden (1977), who found less disease and higher yields after a grass-free lupin crop than after sown clover with grass or the normal grass-dominant pastures.

Varietal resistance

To date, commercial wheat varieties with resistance to *G. graminis* have not been found (Simmonds 1953; Butler 1961; Nilsson 1969). However, it is likely that much of the natural variation from more primitive wheats has been lost in breeding for desirable commercial attributes and so Rovira with Simon and Rathjen (unpubl.) have screened 3000 wheats from the Australian Wheat Collection in the field using added take-all inoculum. Fourteen of these 3000 lines have shown resistance over several seasons and sites and are now being used as breeding material.

Chemical control

After using six systemic fungicides, including benomyl, as seed dressings, Jenkyn and Prew (1973) found no control of take-all in the field but Whitehead (1976) found in pots that PCNB (pentachloronitro-benzene) or benomyl, mixed through the soil, controlled the disease—a finding which could be useful in field studies. Bateman (1980) also found that benomyl was most effective in controlling take-all in soil in pots, but that the effectiveness of benomyl was less in soils with high organic matter.

Biological control

Most soils contain microorganisms which reduce the level of disease caused by *G. graminis,* an effect termed 'general suppression' and increased by the addition of organic matter (Baker and Cook 1974). The suppression reported by Henry (1932) at soil temperatures above 20°C during the growth of wheat in pots was due to microorganisms which survive 70°C but are killed by autoclaving (121° for 20 min). Suppression under warm, moist conditions may explain why take-all is not the serious problem in northern New South Wales and Queensland that it is in southern Australia.

The importance of soil microorganisms in suppressing take-all has been shown in some of our own work in which fumigation of the soil with methyl bromide before seeding greatly increased the yield loss due to take-all when the fungus re-infested the fumigated soil.

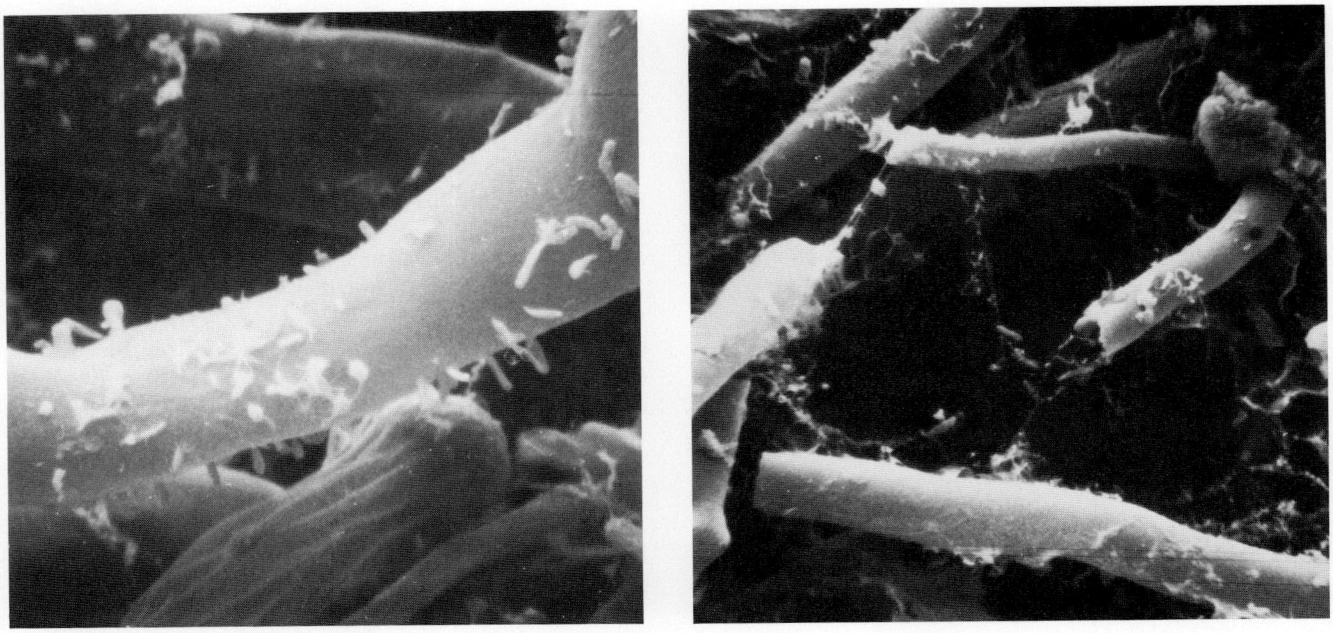

Fig. 45.3. Scanning electron micrograph of *G. graminis* on wheat roots in suppressive soil showing attachment of bacteria to hypha, lysis of hypha showing bacteria, and remnants of hyphae with bacteria. (Rovira and Campbell 1975).

'Take-all Decline' is a phenomenon reported to occur in Europe and north-western United States after several successive wheat or barley crops (Glynne *et al.* 1956; Slope and Cox 1964; Lester 1967; Gerlagh 1968; Vojinovic 1972; Pope and Jackson 1973; Baker and Cook 1974; Shipton 1975). The suppression found in take-all decline has been termed 'specific suppression'; it is transferable (Pope and Hornby 1975) but is destroyed by 60°C for 30 min (Shipton *et al.* 1973), and is thought to be due to specific fluorescent pseudomonads (Cook and Rovira 1976). Vojinovic (1972) found small non-sporing rods on roots in suppressive soil; Rovira and Campbell (1975) showed that non-sporing bacteria adhere to hyphae of *G. graminis* and are responsible for their lysis (Fig.45.3).

Pope and Jackson (1973) reported that the presence of suppressive soil reduced the trophic response of *G. graminis* towards roots and Wildermuth and Rovira (1977) found a lower density of hyphae of *G. graminis* on roots as early as 7 days after seeding in a suppressive soil (Fig.45.4).

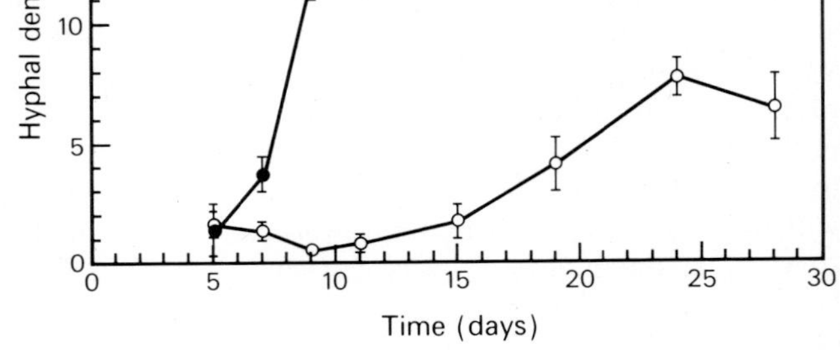

Fig. 45.4. Density of runner hyphae of *G. graminis* var. *tritici* on wheat roots, estimated by the line intercept method. Plants were grown for 28 days in fumigated soil infested with 0.1% oat inoculum without a suppressive soil (●——●) or with 1% suppressive soil (○——○).

Subsequent studies by Wildermuth *et al.* (1978) reported that the site of suppressive action is around the pathogen propagule in the soil rather than on the root surface, that in suppressive soil fewer propagules give rise to hyphae and that there is greater lysis of hyphae in suppressive than in non-suppressive soils.

The relative importance of 'specific' and 'general' suppression in wheat soils in Australia has yet to be assessed, but two factors which may reduce the practical importance of 'specific' suppression are (*a*) the inadvisability of growing wheat each year without an alternate crop/pasture to maintain soil nitrogen and structure and to reduce infestation by cereal cyst nematode, and (*b*) high summer soil temperatures which will be destructive to the non-sporing bacteria where they are the major active component of the suppressive microflora.

To what extent can the phenomenon of suppressiveness be used in the field? Shipton *et al.* (1973) accelerated the onset of take-all decline in Washington State, U.S.A., by thoroughly mixing (rotovating) small quantities of soil, taken from a field showing this effect, into soil fumigated with methyl bromide. Pot experiments by Rovira and Cook (unpubl.) demonstrated that the property of disease suppression could be transmitted to fumigated soil but not to natural (unfumigated) soil. Natural soil apparently resists the introduction of a new microflora and, even without suppressive effects, is itself more antagonistic to *G. graminis* than is fumigated soil.

Parker and Fang (1971) failed in attempts to control take-all in the field by inoculating wheat seed with 20 bacterial isolates selected from the rhizosphere for the production of antibiotics against *G. graminis.* They stressed the need for greater knowledge of the qualities required of the microorganisms concerned before seed inoculation could be effective in the control of take-all.

Sivasithamparam (1977) found that several *Phialophora* spp. protected wheat from the pathogen, and one, *P. hoffmanni* 481, which produced no antibiotic against *G. graminis* in agar culture, significantly reduced the number of dead plants when the fungus was introduced on infected oat kernels sown with the wheat seed, but pelleting seed with mixtures of selected bacteria and actinomycetes gave no significant protection in the same field trial.

Colonization of roots by the non-pathogenic *G. graminis* var. *graminis* reduces take-all infection of glasshouse-grown plants (Wong 1975). The rate of growth of the protective fungus along the root will be important if competition for sites is the mode of its activity against the pathogen. Root elongation exceeds fungal growth. Hence if a fungus with potential biological control is introduced on or near the seed, it has little chance of occupying sites along the root before the take-all fungus, which is distributed throughout the soil, colonizes the root. Soil texture is also a factor in this competitive spread of fungi along the root, for Balis (1970) reported that addition of sand to soil increases two-fold the growth of *G. graminis* along roots but scarcely affects that of *Phialophora radicicola,* a fungus also proposed as a means of controlling the pathogen (Deacon 1973).

Both Wong (1975) and Sivasithamparam (1977) postulated that the protection given by introduced organisms may not be due entirely to competition for sites on the roots, but could also be host-induced, since the extension of the pathogen in the vascular tissue is also reduced.

When inoculation of seeds with microorganisms is being considered as a possible aid to better growth, either directly by their influence upon the plant, or indirectly to control plant pathogens, attention must be given to attaining maximum inoculum size, to its survival on the seed, and to the subsequent distribution of the organisms along the roots. If the microorganisms are to benefit plants, the inoculum must survive, firstly, between inoculation and sowing time and, secondly, in the soil. Studies in sterile soil are of little value in predicting the colonizing ability of desirable microorganisms, nor the antagonistic activity of these organisms in natural soil (Ridge, unpubl.).

Failures in attempting to control take-all by seed inoculation emphasize the difficulty of employing this technique in the field.

The most effective biological control will occur when the suppressive organisms are distributed through the top 20 cm of soil. Thus root residues occupied by the take-all fungus and forming the infection propagules can be colonized by suppressive organisms and emergence of the fungus prevented (Wildermuth *et al.* 1980). Likewise, if suppression occurs in the rhizosphere as on the root surfaces, the strategic distribution of suppressive organisms through the soil will ensure early colonization of roots by these protecting organisms. Such distribution of suppressive organisms has been achieved by growing continuous wheat in the presence of the take-all fungus; the consequent build-up of suppressive organisms has been responsible for a decline in the number of take-all organisms.

Rhizoctonia solani

Samuel (1928) and Samuel and Garrett (1932) first recorded *R. solani* as the cause of 'bare patch' in cereals in sandy alkaline soils of South Australia. The patches of stunted seedlings are 1–20 m across, covering 3–30% of the crop (Kerr 1955; de Beer 1965). The root-attacking strain of this fungus produces brown lesions in the cortex with exposure of the stele; the root breaks to leave a brown, pointed stub. Another strain of *R. solani* attacks the stem, causing 'eye spot' lesions, distortion of the coleoptile and pre-emergence damping-off.

Although Kerr and de Beer found greater amounts of hyphae of *Rhizoctonia* within patches of poor plant growth than outside in more normal areas, it is difficult to reproduce the symptoms of the disease in the glasshouse unless undisturbed cores are used (R. L. Dodman, unpubl.). The problem may arise from the normal process of preparing field soil for pot experiments in which the plant debris, which carries the *Rhizoctonia,* is sieved out.

R. solani, with a wide host range and having a saprophytic ability to colonize plant residues, is difficult to control. Banyer (1966*c*) advocated control by (*a*) improving soil fertility; (*b*) avoiding deep seeding; and (*c*) preparing a firm seed bed, because the disease is encouraged by loose soil.

Cereal cyst nematode (*Heterodera avenae*)

The first Australian reports of field symptoms and losses from cereal cyst nematode were by Davidson (1930) and Hickinbotham (1930) in South Australia, and by Millikan (1938) in Victoria, although the Plant Research Institute Herbarium at Burnley (Victoria) contains a sample of wheat sent from South Australia in 1904 with typical symptoms and cysts produced by this nematode on the roots (Meagher 1972). The pathogen was probably introduced from Europe in the mid 1800s, and subsequently spread across South Australia and Victoria as farmers moved stock and equipment to new lands. Meagher (1968*a*) reported finding hundreds of cysts in soil adhering to tractor wheels, and also has evidence of the cysts being carried by wind during dust storms.

Banyer (1966*a*), Stynes (1974), A. Dubé (pers. comm.) and Rovira *et al.* (1981) showed that the cereal cyst nematode occurs throughout the cereal belt in South Australia in an area of over 1.5 million ha. Meagher (1968*b*) found it in Victoria in 71% of Mallee solonized brown soils (0.75 million ha sown to cereals in 1962) and 46% of Wimmera soils (0.5 million ha sown to cereals in 1962). This nematode has also been reported from New South Wales (McLeod 1968) and Western Australia (Parkin and Goss 1968), although in neither State does it cause widespread disease.

Life cycle

The nematode is less than 0.5 mm long, and attacks the roots of wheat, barley and oats by invading the zone of elongation immediately behind the root tip. This invasion is followed by a proliferation of lateral roots which, in turn, may become infested. The coralloid appearance of an infested root system is illustrated in Fig.45.5, in which the effect on top growth is also shown. Invasion of the roots can occur for some 6 weeks following germination, and in severe attacks there can be up to 200 'strikes' or galls per root system on a 6-week-old plant. Secondary or nodal roots are seldom as heavily infested as the seminal roots. After 12–16 weeks the fertilized females swell to 0.5 to 1 mm in diameter and adhere to the root; they are referred to as 'white cysts'. As the plants mature, the cysts, each containing from 200 to 400 eggs, become brown and survive the summer in this state (see Fig.45.6 for the life cycle).

The eggs in the cysts remain dormant through the hot dry summer until the autumn rains when, under the influence of moist soil and temperatures below 20°C, hatching commences. Eggs of other *Heterodera* species respond to host root exudates by hatching, but there are conflicting reports with respect to the effects of roots on hatching of eggs of *H. avenae*. Winslow (1955) reported that the eggs do not respond to exudates, whereas Williams and Beane (1972), working at lower temperatures, viz. 10 and 15°C, compared with 20 and 25°C used by Winslow, found that, in solution, the presence of roots of oats and barley increases hatching. Recently, Williams and Beane (pers. comm.) at Rothamsted found that root exudates double the rate of hatching when cysts are stored at 5°C for 2 months and then exposed to 15°C—conditions which exist in the field in England during the growth of winter wheat or prior to sowing of

Fig. 45.5. Effect of pre-seeding conditions on the influence of *H. avenae* on root and shoot growth of wheat. Dry, soil kept dry until 24 h before seeding so there were no free nematodes in the soil; wet, soil at field capacity for 8 weeks at 15°C before seeding to promote hatching of eggs; C, superphosphate only; T, aldicarb 'Temik' (nematocide), P, PCMB (fungicide); B, Benlate (nematocide and fungicide).

spring wheat. Such conditions never occur in southern Australia, where spring wheats are sown in the autumn and early winter and nematode hatching and attack occur when soil temperatures are falling. Banyer (1970) reported no such effect in southern Australia with either root exudate or roots in soil on hatching of *H. avenae*.

Although Banyer and Fisher (1971) found maximum hatch (46%) occurs after exposure to 7°C for 4 weeks followed by 15°C—a temperature sequence unlikely in southern Australia at seeding—they also noted a 10% hatch after soil temperatures of 20°C for 4 weeks followed by 15°C—the latter conditions approximating those in Meagher's field assessments in 1968.

Soil in the field is subject to diurnal temperature fluctuations, with night temperatures falling to 10°C in March in the 0–10-cm zone where the cysts produced by the nematode occur, but no studies on the effects of such variations on hatching have yet been conducted in Australia.

Field studies by Meagher and Rooney (1966) showed that in a single season 56–75% of eggs hatched. These figures, which are higher than those of Banyer and Fisher (1971), would mean a faster decline of *H. avenae* in the absence of host plants—a factor which would influence the strategy of using crop rotations to control this pathogen.

Influence on plant yield

Meagher and Brown (1974) reported that as few as two eggs g^{-1} of soil reduced yields by 10% and 10 eggs g^{-1} caused a 50% loss of yield. These yield losses for given population densities will depend greatly upon soil conditions such as temperature and fertility. Simon (1980) found that, in soils with 2–5 eggs g^{-1}, rooting depth was reduced by an average of 21%. O'Brien (1976) showed that extension of the main seminal root axes was reduced by 14 and 50% when two and eight larvae respectively penetrated the root tip.

It is reported from the British Isles that 10 eggs g^{-1} cause a loss of only 188 kg ha^{-1} in wheat crops of 4–5 t ha^{-1}(Duggan 1961; Dixon 1969). Such small losses compared with those in Australia reflect the much more favourable conditions for plant growth. Dixon commented that losses were most common when a wet spring precedes a dry summer.

Barker and Olthof (1976) in their general review of nematode population densities and crop yields stated that 'most field and microplot research with natural or artificial infestations has provided data which show that yields are lower with increasing densities of nematodes'.

Wallace (1965) stated that it was impossible to be precise about annual losses in cereals caused by the cereal cyst nematode in Australia, but the monetary toll must then have been of the order of several million pounds.

From field surveys and the use of nematicides in field trials by Brown *et al.* (1970) and Brown (1972, 1973) in Victoria, Rovira (1978*a*), Rovira *et al.* (1981) and A. Dubé (unpubl.) in South Australia, it has been calculated that cereal cyst nematode causes annual losses of up to 250 000 t wheat in each State (Brown, pers. comm.).

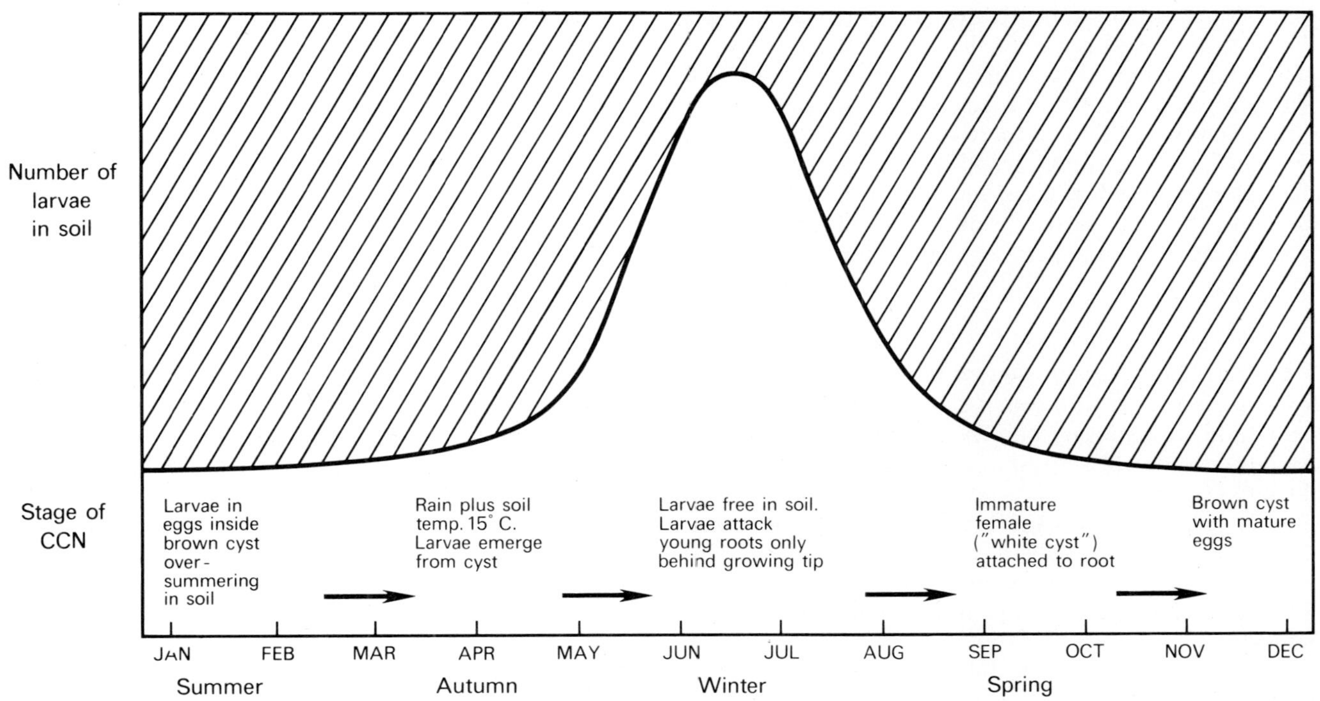

Fig. 45.6. Life cycle of cereal cyst nematode.

Influence of soil type

Wallace (1963) showed that soils with pore diameters from 30–100 μm provide the best conditions for *Heterodera* spp., and thus sandy soils support larger populations than clay soils. Meagher (1968*b*) found the cereal cyst nematode in sandy soils and grey self-mulching clays but not in hard-setting red-brown earths, and demonstrated in pot experiments how its numbers increased at a faster rate in sandy soils. In South Australia the nematode is most widespread in calcareous sandy loams, but, in contrast to the Victorian results, can cause yield losses in soils classified as red-brown earths (Rovira *et al.* 1981).

Effect of fertilizers

In their review, Barker and Olthof (1976) stress that plant nutrition plays a major role in the damage caused by nematodes, e.g. high levels of potassium tend to protect crops from nematode damage, while ammonium nitrate at high rates suppresses the development of *Heterodera glycines* on soybeans and reduces damage.

Banyer (1966*a*) reported that in South Australia nitrogenous fertilizer often increased yields in sandy soils infested with cereal cyst nematode, while Barry *et al.* (1974) in Victoria concluded that, although nitrogen gave higher yields, these increases were not economic. However, the outcome depends upon season.Thus in a wet year, Meagher *et al.* (1978) found that at a site with a high population of *H. avenae* nitrogenous fertilizer at seeding increased grain yield from 0.95 to 1.71 t ha^{-1}, while aldicarb increased yield to 1.48 t ha^{-1} and combined nitrogen and aldicarb gave a yield of 2.57 t ha^{-1}. Barry *et al.* (1974) reported that ammonium sulfate at 100 kg ha^{-1} increased cyst populations by 50% compared with soil receiving no nitrogen, whereas urea had no effect. Simon and Rovira (pers. comm.) have found that the application of superphosphate in the field increases the numbers of 'white cysts' on roots of wheat 2–3-fold and attribute this to the proliferation of lateral roots which occurs when superphosphate is banded with the seed.

Control

Garrett (1934*d*) demonstrated that if peas or bare fallow are alternated with cereals, the effects of cereal cyst nematode are reduced; later Banyer (1966*a*) wrote: 'For a long-term control, steps should be taken to increase soil fertility and prevent a build-up of eelworm numbers. On heavily infested paddocks, cereals should not be grown for at least two years. Every encouragement should be given to establishing or maintaining a predominantly legume pasture during this time, with a long clean fallow . . .'

Studies in Victoria by Millikan (1938), Meagher and Rooney (1966), and Meagher and Brown (1974) have demonstrated the beneficial effects of fallow, Avon oats, peas, or medic pasture in reducing the population of this nematode.

The use of chemicals that control *H. avenae* has demonstrated the magnitude of the losses in yield owing to this pathogen (Brown *et al.* 1970; Brown 1972, 1973; Rovira 1978*a*; Rovira *et al.* 1981). Research conducted by the Division of Soils, CSIRO, has shown that economic control of *H. avenae* can be achieved by the use of the insecticides Counter or Terbufos (*0*, 0-diethyl-S-t-betylthiomethyl phosphoroditioate) applied as a granule at 400–600 g ha^{-1} and with Vydate or Oxamyl (methyl N'N'-D, methyl-N[(methylcarbamoyl)oxy]-1-thiooxamimidate) applied to the seed at 5 g kg^{-1} seed (Roget and Simon 1980). Gurner (pers. comm.) has achieved economic control of *H. avenae* in Australia by applying Nemadi (ethylene dibromide) at 3.7 litres ha^{-1}in the drill row with the seed. With the development by Simon (1980) of a plant assay of soil to assess the potential damage in fields from *H. avenae*, farmers can now have their soils tested to help devise appropriate strategies to control this pathogen.

The finding of resistant varieties by O'Brien (1976) has opened the way for wheat breeders to incorporate resistance into commercial wheats. Sparrow (pers. comm.) has bred a barley (var. Galleon) which is resistant to *H. avenae* and as such does not increase the levels of the pathogen.

Field experiments in 1978 showed that a current commercial wheat variety, Festiguay, is more resistant to the cereal cyst nematode than other commercial varieties, and in severely infested fields will have only 5–10 cysts per plant compared with 80–100 cysts per plant on other wheat varieties (Rovira, Dillon and Simon, unpubl.). This low cyst population on roots of Festiguay explains why other varieties of wheat grown in parts of fields which had grown Festiguay 2 or 3 years previously yielded 0.4–1 t ha^{-1} more than other parts of the same fields which had grown susceptible wheat varieties.

Biological control

It often happens that when pests, weeds or diseases are introduced into a country the predators of these problem organisms are not introduced and so the problems caused are greater than in the country of origin. *H. avenae* is no exception to this general phenomenon and the natural decline in populations reported in Britain as being due to biological control (Kerry 1975) has not yet been shown to operate in Australia. Tribe (1977) reviewed the pathology of cyst nematodes and reported a range of fungi capable of attacking them, generally at the 'white cyst' (actually an immature female) or mature cyst stages. It was shown in Britain that decline in *H. avenae* occurs with continuous cropping of spring oats, wheat or barley and that treatment of the soil with formalin or calcium cyanide causes a considerable increase in the nematode population with further cereal cropping possibly because of the destruction of fungi parasitic on *H. avenae*. Kerry (1975) found an *Entomophthora*-like fungus, now named *Nematophthora gynophila* (Kerry and Crump 1980), and another fungus, *Verticillium chlamydosporium*, to parasitize eggs and females of *H. avenae* in 'decline' fields. Studies in South Australia by Stirling (1980) have shown a number of fungi and mites which parasitize *H. avenae* but these do not include *Nematophthora gynophila*. The great capability of *H. avenae* to increase in numbers (because of the large number of eggs per cyst) means that successful biological control requires a very high rate of 'kill' of *H. avenae* by the controlling organism (Tribe 1977).

There appear to be possibly two factors which could limit the effectiveness of biological control of *H. avenae* in Australia: firstly, some of these parasites move as motile zoospores through water films in soil to attack the nematode—such wet conditions seldom prevail in Australian wheat fields where CCN occurs; secondly, the fungi that control CCN appear to operate at a threshold population of *H. avenae* higher than that found under Australian conditions. However, there could be fields in Australia in which parasites of *H. avenae* more suited to the local conditions are operating, and this could be a fruitful area of research.

Fungus–nematode associations

Benedict and Mountain (1956) reported a consistent association between root rot of winter wheat and root-lesion nematode. In view of the damage nematodes cause to roots and the accompanying leakage of nutrients (Meagher 1974), it would be surprising if associations of fungi and nematodes causing root disease did not occur. Meagher and Chambers (1971) reported that in glasshouse trials the combined effect on root stunting was greater than when each organism was applied separately. However, at the time of writing this review there was no published work to confirm this effect in the field, although both pathogens have been found to occur in the same fields.

Conclusions

The efficient utilization of soil and water in production of a great many crops can be reduced by soil-borne root diseases. We have described how two such pathogens of wheat interact with soil conditions.

Modification of microbial populations is difficult to achieve, especially on a broad-acre basis in low input agricultural systems such as cereal farming in southern Australia. However, research on cereal cyst nematode, in Australia has developed economic methods, e.g. rotations, resistant varieties and chemicals, by which the level of this pathogen can be managed to minimize crop losses.

Despite over 100 years of research on take-all, this disease has proved far more recalcitrant than cereal cyst nematode. One of the major difficulties with the former is the definition of a propagule. In contrast to the cereal cyst nematode, whose eggs may be accurately defined as propagules, a propagule of take-all can be of sizes varying from minute fragments of infected root residue to whole crowns of plants from previous wheat crops or weed grasses. Superimposed upon this problem are the requirements of the fungus for high water content of soil, early in the life of the host plants, the interaction with the nutritional status of the host plant and with the water stress placed upon the plants after infection.

However, success with cereal cyst nematode in the field offers hope of development of control measures for root diseases such as take-all caused by soil fungi.

Economic pressures, including especially the soaring cost of fuel, are leading to increasing use of minimal cultivation and to the inclusion of more cereals in rotation sequences. Hence continuing investigations are required to ensure that, over the longer term, any economic gains from the newer practices will not be eroded by an increase in losses caused by soil-borne diseases. The development of continuous cropping systems with grain legumes and wheat in alternate years, together with minimal cultivation, in the wetter parts of the cereal belt has been one innovation that has reduced losses from root diseases, while maintaining soil nitrogen levels.

Factors such as soil type and climate are of such importance in the incidence of these diseases and in their effects on yield that experience gained in southern Australia might be applicable to other areas of the world having a similar Mediterranean climate.

References

Adam, D. B. (1951). The control of take-all in wheat after a long period in pasture. *Aust. J. Agric. Res.* **2,** 273-82.

Anon. (1868). Report of Commission on Diseases in Cereals. (Gov. Printer: Adelaide.)

Anon. (1976). EMBRAPA annu. tech. rep. National Wheat Research Centre, Passo Fundo, Brazil, 1975-76.

Baker, K. F., and Cook, R. J. (1974). 'Biological Control of Plant Pathogens'. (W. H. Freeman: San Francisco.)

Balis, C. (1970). A comparative study of *Phialophora radicicola,* an avirulent fungal root parasite of grasses and cereals. *Ann. Appl. Biol.* **66,** 59-73.

Banyer, R. J. (1966*a*). Cereal root diseases and their control. Part I. Cereal root eelworm. *J. Dep. Agric. S. A.* **69,** 310-5.

Banyer, R. J. (1966*b*). Cereal root diseases and their control. Part II. Take-all. *J. Dep. Agric. S. A.* **69,** 372-6.

Banyer, R. J. (1966*c*). Cereal root diseases and their control. Part III. *Rhizoctonia. J. Dep. Agric. S. A.* **69,** 415-7.

Banyer, R. J. (1970). Hatching and survival of eggs of *Heterodera avenae* Wollenweber, 1924. Ph.D. Thesis, Univ. Adelaide.

Banyer, R. J., and Fisher, J. M. (1971). Seasonal variation in hatching of eggs of *Heterodera avenae. Nematologica* **17,** 225-36.

Barker, K. R., and Olthof, T. H. A. (1976). Relationship between nematode population densities and crop responses. *Ann. Rev. Phytopathol.* **14,** 329-52.

Barry, E. R., Brown, R. H., and Elliott, B. R. (1974). Cereal cyst nematode *(Heterodera avenae)* in Victoria: influence of cultural practices on grain yields and nematode populations. *Aust. J. Exp. Agric. Anim. Husb.* **14,** 566-71.

Bateman, G. L. (1980). Prospects for fungicidal control of take-all of wheat. *Ann. Appl. Biol.* **96,** 275-82.

de Beer, J. F. (1965). Studies on the ecology of *Rhizoctonia solani* Kühn. Ph.D. Thesis, Univ. Adelaide.

Benedict, W. E., and Mountain, W. R. (1956). Studies on the etiology of a root rot of winter wheat in south-western Ontario. *Can. J. Bot.* **34,**159-285.

Brooks, D. H., and Dawson, M. G. (1968). Influence of direct drilling of winter wheat on incidence of take-all and eyespot. *Ann. Appl. Biol.* **61,** 57-64.

Brown, M. E., and Hornby, D. (1971). Behaviour of *Ophiobolus graminis* on slides buried in soil in the presence or absence of wheat seedlings. *Trans. Br. Mycol. Soc.* **56**, 95-103.

Brown, M. E., Hornby, D., and Pearson, V. (1973). Microbial populations and nitrogen in soil growing consecutive cereal crops infected with take-all. *J. Soil Sci.* **24,** 296-310.

Brown, R. H. (1972). Chemical control of cereal cyst nematode *(Heterodera avenae)* in Victoria. A comparison of systemic and contact nematicides. *Aust. J. Exp. Agric. Anim. Husb.* **12,** 662-7.

Brown, R. H. (1973). Chemical control of cereal cyst nematode (*Heterodera avenae*) – a comparison of methods and rates of application of two systemic nematicides. *Aust. J. Exp. Agric. Anim. Husb.* **13**, 587-92.

Brown, R. H., Meagher, J. W., and McSwain, N. K. (1970). Chemical control of the cereal cyst nematode *(Heterodera avenae)* in the Victorian mallee. *Aust. J. Exp. Agric. Anim. Husb.* **10,** 172-3.

Burgess, L. W., Dodman, R. L., Pont, W., and Mayers, P. (1980). *Fusarium* diseases of wheat, maize and grain sorghum in eastern Australia. *In* 'Fusarium: Biology and Taxonomy', ed. Paul E. Nelson. (Pennsylvania State Univ. Press). (In press.)

Butler, F. C. (1948). Take-all of wheat. *Agric. Gaz. N.S.W.* **59,** 248-50.

Butler, F. C. (1959). Saprophytic behaviour of some cereal root-rot fungi. IV. Saprophytic survival in soils of high and low fertility. *Ann. Appl. Biol.* **47,** 28-36.

Butler, F. C. (1961). Root and foot rot diseases of wheat. N.S.W. Dep. Agric. Sci. Bull. No. 77.

Chambers, S. C. (1971*a*). Some factors affecting the survival of the cereal root pathogen *Ophiobolus graminis* in wheat straw. *Aust. J. Exp. Agric. Anim. Husb.* **11,** 90-3.

Chambers, S. C. (1971*b*). Glasshouse studies on *Ophiobolus graminis* as a cause of whiteheads in wheat. *Phytopathol. Z.* **71,** 169-82.

Chambers, S. C. (1971*c*). Some factors affecting the relative importance of hosts in the survival of *Ophiobolus graminis. Aust. J. Agric. Res.* **22,** 111-21.

Chambers, S. C., and Flentje, N. T. (1967*a*). Studies of oat-attacking and wheat-attacking isolates of *Ophiobolus graminis* in Australia. *Austr. J. Biol. Sci.* **20,** 927-40.

Chambers, S. C., and Flentje, N. T. (1967*b)* Studies on variation with *Ophiobolus graminis. Aust. J. Biol. Sci. 20,* 941-51.

Chambers, S. C., and Flentje, N.T. (1968). Saprophytic survival of *Ophiobolus graminis* on various hosts. *Aust. J. Biol. Sci.* **21,** 1153-61.

Clarkson, D. T., Drew, M. C., Ferguson, I. B., and Sanderson, J. (1975). The effect of the take-all fungus, *Gaeumannomyces graminis,* on the transport of ions by wheat plants. *Physiol. Plant Pathol.* **6,** 75-84.

Cook, R. J., and Rovira, A.D. (1976). The role of bacteria in the biological control of *Geaumannomyces graminis* by suppressive soils. *Soil Biol. Biochem.* **8,** 269-73.

Davidson, J. (1930). Eelworms *(Heterodera schachtii* Schm.) affecting cereals in South Australia. *J. Dep. Agric. S. A.* **34,** 378-85.

Deacon, J. W. (1973). Control of take-all fungus by grass leys in intensive cereal cropping. *Plant Pathol.* **22,** 88-94.

Dixon, G. M. (1969). The effect of cereal cyst eelworm on spring-sown cereals. *Plant Pathol.* **18,** 109-112.

Donald, C. M. (1964). The progress of Australian agriculture and the role of pastures in environmental change. *Aust. J. Sci.* **27,** 187-8.

Duggan, J. J. (1961). The effect of cereal root eelworm on its hosts. *Ir. J. Agri. Res.* **1,** 7-16.

Ellington, A., and Reeves, T. G. (1981). Direct drilling – long- and-short-term effects on the soil. Proc. Nat Spray-Seed Conf., Albury, N.S.W. 1981. (I.C.I. Australia Ltd).

Garrett, S. D. (1934*a*). Factors affecting the severity of take-all. 1. The importance of soil microorganisms. *J. Dep. Agric. S. A.* **37,** 664-74.

Garrett, S. D. (1934*b*). Factors affecting the severity of take-all. 2. Soil temperature. *J. Dep. Agric. S. A.* **37,** 799-805.

Garrett, S. D. (1934*c*). Factors affecting the severity of take-all. 3. The climatic factor. *J. Dep. Agric. S. A.* **37,** 976-83.

Garrett, S. D. (1934*d*). The effect of crop rotation on the eelworm *(Heterodera schachtii)* disease of cereals. *J. Dep. Agric. S. A.* **37,** 984-7.

Garrett, S. D. (1936). Soil conditions and the take-all disease of wheat. *Ann. Appl. Biol.* **23,** 667-99.

Garrett, S. D. (1938). Soil conditions and the take-all disease of wheat. III. Decomposition of the resting mycelium of *Ophiobolus graminis* in infected wheat stubble buried in the soil. *Ann. Appl. Biol.* **25,** 742-66.

Garrett, S. D. (1940). Soil conditions and the take-all disease of wheat. V. Further experiments on the survival of *Ophiobolus graminis* in infected wheat stubble buried in the soil. *Ann. Appl. Biol.* **27,** 199-204.

Garrett, S. D. (1941). Soil conditions and the take-all disease of wheat. VI. The effect of plant nutrition upon disease resistance. *Ann. Appl. Biol.* **28,** 14-18.

Garrett, S. D. (1942). The take-all disease of cereals. Tech. Comm. Bur. Soil Sci. Harpenden, No. 41.

Garrett, S. D. (1944). Soil conditions and the take-all disease of wheat. VIII. Further experiments on the survival of *Ophiobolus graminis* in infected wheat stubble. *Ann. Appl. Biol.* **31,** 186-91.

Garrett, S. D. (1948). Soil conditions and the take-all disease of wheat. IX. Interaction between host plant nutrition, disease escape and disease resistance. *Ann. Appl. Biol.* **35,** 14-7.

Garrett, S. D. (1967). Effect of nitrogen level on survival of *Ophiobolus graminis* in pure culture on cellulose. *Trans. Br. Mycol. Soc.* **50,** 519-24.

Garrett, S. D. (1981) Introduction. *In* 'Biology and Control of Take-all', eds, P. J. Shipton and M. J. C. Asher. (Academic Press: London.) (In press.)

Gerlagh, M. (1968). Introduction of *Ophiobolus graminis* into new polders and its decline. *Neth. J. Plant Pathol.* **74,** Suppl. No. 2.

Glynne, M. D., Salt, G. A., and Slope, D. B. (1956). Rep. Rothamsted Exp. Stn for 1955, **32,** 103-4.

Greacen, E. L., and Hignett, C. T. (1976). A water balance model and supply index for wheat in South Australia. CSIRO (Aust.) Div. Soils Tech. Pap. No. 27.

Henry, A. W. (1932). Influence of soil temperature and soil sterilization on the reaction of wheat seedlings to *Ophiobolus graminis* Sacc. *Can. J. Res.* **7,** 198-203.

Hickinbotham, A. R. (1930). Eelworm and 'no growth' patches. *J. Dep. Agric. S. A.* **34,** 386-92.

Holden, K. J. (1977). The role of grain legumes in the mixed farming areas. S. A. Dep. Agric. Fish., Agronomy Branch Rep. No. 95.

Hopwood, G. M. (1877). The take-all – letter to Bendigo Agric. Soc. *Bendigo Advertiser* (Victoria) 30 Sept. 1877.

Hornby, D., and Brown, M. E. (1977). Nitrate and ammonium in the rhizosphere of wheat crops and concurrent observations of take-all. *Plant Soil* **48,** 455-71.

Hornby, D., and Goring, C. A. I. (1972). Effects of ammonium and nitrate nutrition on take-all disease of wheat in pots. *Ann. Appl. Biol.* **70,** 225-31.

Huber, D. M., and Watson, R. D. (1972). Nitrogen form and plant disease. *Down to Earth* **27,** 14-5.

Jenkyn, J. F., and Prew, R. D. (1973). Activity of six fungicides against cereal foliage and root diseases. *Ann. Appl. Biol.* **75,** 241-52.

Kerr, A. (1955). Studies on the parasitic and saprophytic activities of *Pellicularia filamentosa* (Pab.) Rogers and *Sclerotinia homeocarpa* Bennett. Ph.D. Thesis, Univ. Adelaide.

Kerry, B. R. (1975). Fungi and the decrease of cereal cyst nematode populations in cereal monoculture. *E. P. P. O. Bull.* **5,** 353-61.

Kerry, B. R., and Crump, D. H. (1980) Two fungi parasitic on females of cyst-nematode *(Heterodera* spp.). *Trans. Br. Mycol. Soc.* **74,** 119-25.

Lester, E. (1967). Disease problems in intensive cereal growing. Proc. 4th Br. Insectic. Fungic. Conf. Vol. **2,** 510-6.

Lucas, R. L. (1955). A comparative study of *Ophiobolus graminis* and *Fusarium culmorum* in saprophytic colonization of wheat straw. *Ann. Appl. Biol.* **43,** 134-43.

McLeod, R. W. (1968). Cereal cyst nematode, a nematode new to New South Wales. *Agric. Gaz. N.S.W.* **79,** 293-5.

MacNish, G. C. (1976). Survival of *Gaeumannomyces graminis* var. *tritici* in artificially colonized straw buried in naturally infested soil. *Aust. J. Biol. Sci.* **29,** 163-74.

MacNish, G. C. (1977). Effect of nitrogen form on take-all severity. Proc. Aust. Cer. Agron. Conf. Perth, 1977.

MacNish, G. C., and Dodman, R. L. (1973*a*), Relation between incidence of *Gaeumannomyces graminis* var. *tritici* and grain yield. *Aust. J. Biol. Sci.* **26,** 1289-99.

MacNish, G. C., and Dodman, R. L. (1973*b*). Survival of *Gaeumannomyces graminis* var. *tritici* in the field. *Aust. J. Biol. Sci.* **26,** 1309-17.

Meagher, J. W. (1968*a*). The distribution of cereal cyst nematode *(Heterodera avenae)* in Victoria and its relation to soil type. *Aust. J. Exp. Agric. Anim. Husb.* **8,** 637-40.

Meagher, J. W. (1968*b*). The cereal eelworm — an increasing pest in the Mallee and the Wimmera. *J. Agric. Vic.* **66,** 230-3.

Meagher, J. W. (1972). Cereal cyst nematode *(Heterodera avenae* Woll.). Studies on ecology and control in Victoria. Dep. Agric. Vic. Tech. Bull. No. **24.**

Meagher, J. W. (1974). The biology of the cereal cyst nematode *(Heterodera avenae* Woll.) and its interaction with a fungus in a soil-borne disease of wheat. Ph.D. Thesis, Univ. Melbourne.

Meagher, J. W., and Brown, R. H. (1974). Microplot experiments on the effect of plant hosts on populations of the cereal cyst nematode *(Heterodera avenae)* and on the subsequent yield of wheat. *Nematologica* **20,** 337-46.

Meagher, J. W., Brown, R. H., and Rovira, A. D. (1978). The effects of cereal cyst nematode *(Heterodera avenae)* and *Rhizoctonia solani* on the growth and yield of wheat. *Aust. J. Agric. Res.* **29,** 1127-37.

Meagher, J. W., and Chambers, S. C. (1971). Pathogenic effects of *Heterodera avenae* and *Rhizoctonia solani* and their interaction on wheat. *Aust. J. Agric. Res.* **22,** 189-94.

Meagher, J. W., and Rooney, D. R. (1966). The effect of crop rotations in the Victorian Wimmera on the cereal cyst nematode *(Heterodera avenae),* nitrogen fertility and wheat yield. *Aust. J. Exp. Agric. Anim. Husb.* **6,** 425-31.

Millikan, C. R. (1938). Eelworm *(Heterodera schachtii* Schmidt) disease of cereals. *J. Agric. Vic.* **36,** 452-68.

Moore, K. J. (1978). The influence of no-tillage on take-all of wheat. Ph.D. Thesis, Washington State Univ.

Moore, K. J., and Cook, R. J. (1978). Influence of no-tillage on take-all of wheat. Abstr. 3rd Int. Congr. Plant Pathol. Munich.

Mücke, C. (1870). The take-all *(Zenodochius cerealium)* — the corn disease of Australia. Prize entry pub. Board of Agric. Vic.

Nilsson, H. E. (1969). Studies of root and foot rot diseases of cereals and grasses. 1. On resistance to *Ophiobolus graminis* Sacc. *Ann. Agric. College Sweden* **35,** 275-807.

O'Brien, P. C. (1976). Studies on the relation of *Heterodera avenae* to susceptible and resistant wheat. Ph.D. Thesis, Univ. Adelaide.

Parker, C. A. (1977). Spray-seeding, soil microbes and root disease. Proc. Minimum Cultivation Conv. , Perth, W.A., 1977. (I.C.I. Australia Ltd.).

Parker, C. A., and Fang, C. S. (1971). Problems in the biological control of root disease in cereals. Proc. Aust. Plant Pathol. Conf. Hobart, Nov. 1971, 3(e), 43-5.

Parkin, R. J., and Goss, O. M. (1968). Cereal eelworm: a new disease of cereal crops in the Geraldton area. *J. Agric. W.A.* **8,** 259-60.

Pope, A. M. S., and Hornby, D. (1975). Decrease of take-all by a transmissible factor in take-all decline soils. *Ann. Appl. Biol.* **81,** 145-60.

Pope, A. M. S., and Jackson, R. M. (1973). Effects of wheat field soil on inocula of *Gaeumannomyces graminis* (Sacc.) Arx and Olivier var. *tritici* J. Walker in relation to take-all decline. *Soil Biol. Biochem.* **5,** 881-90.

Price, R. D. (1970). Stunted patches and deadheads in Victorian cereal crops. Dep. Agric. Vic. Tech. Publ. No. 23.

Price, R. D., and Stubbs, L. L. (1963). An investigation of barley yellow dwarf virus as a primary or associated cause of premature ripening of 'dead heads' in wheat. *Aust. J. Agric. Res.* **14,** 154-64.

Roget, D. K., and Simon, A. (1980). Cereal cyst nematode in wheat — 1979 field trial results. CSIRO (Aust.) Div. Soils. Tech. Memo. No. 12/80.

Rovira, A. D. (1978*a*). Assessment of yield losses in wheat due to cereal cyst nematode *(Heterodera avenae)* by the use of selective chemicals.. *In* 'Epidemiology and Crop Loss Assessment', eds. R. C. Close *et al.* Proc. A.P.P.S. Workshop, Lincoln College, August 1977. Publ. Lincoln College, 27-1 to 27-5.

Rovira, A. D. (1978*b*). Manipulation of the level of take-all disease in the field by inoculation. *In* 'Epidemiology and Crop Loss Assessment', eds. R. C. Close *et al.* Proc. A.P.P.S. Workshop, Lincoln College, August 1977. Publ. Lincoln College, 11-1 to 11-4.

Rovira, A. D., (1979). Direct drilling — friend or foe to soil-borne root diseases of cereals? Proc. Natl Direct Drilling Conf., Adelaide, 1979. (I.C.I. Australia Ltd.)

Rovira, A. D., and Campbell, R. (1975). A scanning electron microscope study of interactions between microorganisms and *Gaeumannomyces graminis* (Syn. *Ophiobolus graminis)* on wheat roots. *Microbiol. Ecol.* **2,** 177-85.

Rovira, A. D., Brisbane, P. G., Simon, A., Whitehead, D. G., and Correll, R. L. (1981). Influence of cereal cyst nematode *(Heterodera avenae)* on wheat yields in South Australia. *Aust. J. Exp. Agric. Anim. Husb.* **21,** 516-23.

Samuel, G. (1924). Take-all investigation. II. *J. Dep. Agric. S.A.* **27,** 1134-47.

Samuel, G. (1928). Two stunting diseases of wheat and oats. *J. Dep. Agric. S.A.* **32,** 40-3.

Samuel, G., and Garrett, S. D. (1932). *Rhizoctonia solani* on cereals in South Australia. *Phytopathol.* **22,** 827-36.

Shipton, P. J. (1972). Take-all in spring-sown cereals under continuous cultivation: disease progress and decline in relation to crop succession and nitrogen. *Ann. Appl. Biol.* **71,** 33-46.

Shipton, P. J. (1975). Take-all decline during cereal monoculture. *In* 'Biology and Control of Soil-borne Plant Pathogens', ed. G. W. Bruehl, pp. 137-44. (Am. Phytopathol. Soc.: St. Paul, Minn.)

Shipton, P. J., Cook, R. J., and Sitton, J. W. (1973). Occurrence and transfer of a biological factor in soil that suppresses take-all of wheat in eastern Washington. *Phytopathol.* **63,** 511-17.

Simmonds, P. M. (1953). Root rots of cereals. II. *Bot. Rev.* **19,** 131-46.

Simon, A. (1980). A plant assay of soil to predict potential damage to wheat by *Heterodera avenae. Plant Dis.,* **64,** 917-19.

Sims, H. J. (1958). Deadheads in wheat — their relationship to crop rotation in Mallee Research Station Trials. *J. Agric. Vic.* **56,** 282-4.

Sims, H. J., Meagher, J. W., and Millikan, C. R. (1961). Deadheads in wheat — field studies in Victoria. *Aust. J. Exp. Agric. Anim. Husb.* **1,** 99-108.

Sivasithamparam, K. (1977). The rhizosphere microflora of the seminal and secondary roots of wheat and the effect of selected isolates on the take-all fungus and its potential for disease. Ph.D. Thesis, Univ. Western Australia.

Slope, D. B., and Cox, J. (1964). Rep. Rothamsted Exp. Stn for 1963.

Smiley, R. W. (1978*a*). Antagonists of *Gaeumannomyces graminis* from the rhizoplane of a wheat in soils fertilized with ammonium- or nitrate-nitrogen. *Soil Biol. Biochem.* **10,** 169-74.

Smiley, R. W. (1978*b*). Colonization of wheat roots by *Gauemannomyces graminis* inhibited by specific soils, microorganisms and ammonium-nitrogen. *Soil Biol. Biochem.* **10,** 175-80.

Smiley, R. W., and Cook, R. J. (1973). Relationship between take-all of wheat and rhizosphere pH in soils fertilized with ammonium *vs* nitrate-nitrogen. *Phytopathol.* **63,** 882-90.

Stirling, G. R. (1980). Parasites and predators of cereal cyst nematode *(Heterodera avenae)* in South Australia. *In* Proc. 4th Natl Plant Pathol. Conf. Perth, W.A.

Stynes, B. A. (1964). A synoptic study of wheat. Ph.D. Thesis, Univ. of Adelaide.

Syme, J. R. (1966). Fertilizer and varietal effects on take-all in irrigated wheat. *Aust. J. Exp. Agric. Anim. Husb.* **6,** 246-9.

Toms, J. (1958). The use of copper and zinc in the cereal-growing districts of Western Australia. *J. Agric. W.A.* (3rd Ser.) **7,** 197-203.

Tribe, H. T. (1977). Pathology of cyst nematodes. *Biol. Rev.* **52,** 477-507.

Turner, E. M. (1940). *Ophiobolus graminis* Sacc. var. *avenae* var. n., as the cause of take-all or whiteheads of oats in Wales. *Trans. Br. Mycol. Soc.* **24,** 269-81.

Vojinović, Z. (1972). Antagonists from soil and rhizosphere to phytopathogens. Tech. Rep., Inst. Soil Sci., Beograd, Yugoslavia.

Walker, J. (1972). Type studies on *Gaeumannomyces graminis* and related fungi. *Trans. Br. Mycol. Soc.* **58,** 427-57.

Walker, J. (1975). Take-all diseases of Gramineae: a review of recent work. *Rev. Plant Pathol.* **54,** 13-44.

Wallace, H. R. (1963). 'The Biology of Plant Parasitic Nematodes'. (Edward Arnold Ltd.: London.)

Wallace, H. R., (1965). The ecology and control of cereal root nematode. *J. Aust. Inst. Agric. Sci.* **31,** 178-86.

Whitehead, D. G. (1976). Studies in the control of *Gaeumannomyces graminis* var. *tritici* ('Hay-die' or 'Take-all') and of *Heterodera avenae* (cereal cyst nematode), by the application of fungicides and a nematicide to the soil, in pot experiments. CSIRO (Aust.), Div. Soils Tech. Memo. No. 12/1976. (Available on request.)

Wildermuth, G. B. (1977). Studies on suppressive soils in relation to the growth of *Gaeumannomyces graminis* var. *tritici* and other root pathogens of wheat. Ph.D. Thesis, Univ. Adelaide.

Wildermuth, G. B., and Rovira, A. D. (1977). Hyphal density as a measure of suppression of *Gaeumannomyces graminis* var. *tritici* on wheat roots. *Soil Biol. Biochem.* **9,** 203-5.

Wildermuth, G. B., Rovira, A. D., and Warcup, J. H. (1980). Mechanism and site of suppression of *Gaeumannomyces graminis* var. *tritici* in soil. *In* 'Soil-borne Plant Pathogens', eds. B. Schippers and W. Gams, pp. 157-64. (Academic Press: London.)

Williams, T. D., and Beane, J. (1972). Cereal cyst nematode. Effect of temperature and root exudates on *H. avenae* hatch. *Ann. Rep. Rothamsted Exp. Stn* 1971.

Winslow, R. D. (1955). The hatching responses of some root eelworms of the genus *Heterodera. Ann. Appl. Biol.* **43,** 19-36.

Wong, P. T. W. (1975). Cross-protection against wheat and oat take-all fungi by *Gaeumannomyces graminis* var. *graminis. Soil Biol. Biochem.* **7,** 189-94.

Yarham, D. J. (1979). The effect on soil-borne diseases of changes in crop and soil management. *In 'Soil-borne Plant Pathogens',* eds. B. Schippers and W. Gams. (Academic Press: London.)

Yeates, J. S. (1980). The distribution and importance of an oat-attacking form of the take-all fungus *(Gaeumannomyces graminis)* in the Western Australian wheat belt. (In preparation.)

Chapter 46
Effect of fire and sun-baking on the soil microflora and seedling growth in forest soils

J. H. Warcup

AUSTRALIAN SOILS, IN BOTH THE TROPICAL AND temperate regions, are subject to prolonged dry and often hot conditions. Furthermore, much of the vegetation is prone to fire, which is considered a natural environmental variable over most of Australia (Gill 1975). This paper considers the effect of heat on soil organisms and on seedling growth especially in natural forest soils.

Fire

The common occurrence of fire in Australia is generally regarded as preceding the arrival of European man and even that of Aboriginal man. Lightning, which is still responsible for the burning of significant areas, doubtless burnt larger areas of the drier vegetation types before the arrival of man. Fire frequency probably increased in Aboriginal hunting grounds (Gill 1975), where patches of country were burnt to provide succulent growth for game and as a barrier to assist in its capture (Wallace 1966).

Most of Australia is subject to the influence of fire (McArthur 1972). The relatively restricted areas of rain forest rarely support fire (Francis 1970; Gilbert 1959; Jackson 1968), and in semi-arid areas fires tend to occur only after exceptional rainfall that allows sufficient growth to later carry a fire (McArthur 1972). Many of the areas subject to fire are dominated by a largely endemic flora (Burbridge 1960) especially by the genera *Eucalyptus* and *Acacia*. Specht (1970) comments that all open (dry sclerophyll) forests and all tall open (wet sclerophyll) forests are subject to periodic bush fires. Fires in the tall open forest tend to be severe and, as most of the *Eucalyptus* dominants are sensitive to fire, widespread destruction of the original forest results. Dense regeneration from seed rapidly follows the fire, the seedlings gradually thinning to form an even-aged stand. Unlike the eucalypts of the tall open forests, however, most of the *Eucalyptus* species of the open forests are fire-insensitive; after a fire a mass of shoots rapidly arises from numerous epicormic buds buried in the stem. While fire-sensitive shrubs do occur in the understorey, many others have underground root stocks which survive, even though the tops are burnt, and which enable regeneration to occur.

Gilbert (1959) showed that *Eucalyptus regnans* F. Muell., a tall open forest species, when in association with rain forest species could not regenerate without fire. Ashton (1956) and Cunningham (1960) working with tall open forests in eastern Australia found that eucalypts do not regenerate naturally without fire. Mount (1964) considered that all forest eucalypts probably require fire for effective natural regeneration.

In forests, depending on the fuel available at different sites, the effects of a fire may vary from transient heating of surface soil to the production of 'ashbeds' under fallen trees or logs, where surface soil temperatures may be several hundred degrees (Beadle 1940; Roberts 1965; Cromer and Vines 1966; Floyd 1966; Tunstall *et al.* 1976). Floyd (1966), working in tall open forest in northern New South Wales, found that a light fuel burn gave a surface maximum of 510°C with the temperature staying above 100°C for 15 minutes, and at 2.5 cm depth the temperature reached 44°C; a medium fuel burn gave 67°C at 5 cm, and a heavy fuel burn similar temperatures to 10 cm depth. Tunstall *et al.* (1976) measured temperatures under and beside a windrow fire, an extreme case of the modification of the soil environment by fire; the soil surface in the centre of the windrow reached 780°C and temperatures above 80°C occurred at 40 cm depth. More interestingly, soil at 2.5 cm deep 2 metres from the heap was above 60°C, and this temperature was recorded 4 metres away on the windward side.

The ashbeds formed after heaps of slash or logs are burnt have a pronounced stimulatory effect on seedling growth of many forest trees. In Western Australia, *Eucalyptus diversicolor* F.v.M. (karri) seedlings are stimulated considerably in ashbeds (Hatch 1960), and growth of young karri in non-ashbeds is generally so poor that deliberate creation of ashbeds has become part of karri forest regeneration (Loneragan and Loneragan 1964). Pryor (1976) mentions that fire-sensitive species of *Eucalyptus* that regenerate well from seed after fire include *E. pilularis* Sm., *E. grandis* Hill ex Maiden, *E. sieberi* L. Johnston, *E. regnans*, *E. delegatensis* R. T. Baker, and *E. diversicolor*. All these species are important timber trees.

Hatch (1960) studied the chemical changes in soil in ashbeds, and found that they are most pronounced in the first 2.5 cm and decrease rapidly with depth; the most important are increases in pH and total soluble salts, and formation of $CaCO_3$. By contrast the soil organic matter is markedly reduced by high temperature burns. A period of at least 20 years was required before the ashbed soil returned to its previous condition. The dominant cations of the ash were Ca, Mg, Na and K, and the dominant anions were carbonate and phosphate. Hatch commented, that although total N of the ashbed soil was decreased there was apparently an increase in availability.

Loneragan and Loneragan (1964) found that when karri seedlings were grown in soil from karri forest they showed remarkable responses, up to twenty-fold, to N and P fertilizers added simultaneously. Responses to ash and to heat treatment, 12 h at about 150°C, were also obtained. The major part of the ash response was shown to be due to the supply of phosphate in the ash; the heat response was also believed to be largely due to an increased supply of nutrients including N, but they made no attempt to resolve the effects of heat treatment.

While changes in soil nutrients after forest fires can be attributed to ash, the effect of heat must also be considered. Griffith (1946) showed, in India, that there was little or no difference in growth response of seedlings of several trees between ash plus heat and heat alone. Pryor (1960) reported that the addition of ash produced a 25% increase in the growth of eucalypts, but that baking the soil at 150°C for several hours produced a 100% increase over plants grown in unheated soil; the addition of ash to the baked soil produced no further growth increase. Florence and Crocker (1962) reported a roughly fourfold increase in *E. pilularis* seedlings grown in soil heated at 160°C for 48 h, but poor growth in autoclaved soil.

Attiwell (1962) grew seedlings of *Eucalyptus obliqua* L'Herit. in surface soil from a mature *E. obliqua* forest and in the same soil after heating to 150, 300 and 500°C for 10 h. He found an initial temporary toxicity in soils preheated to 150 and 300°C, but subsequently, significantly better growth in soils treated at 300 and 500°C than in untreated soil. Since addition to phosphate also increased growth, he suggested that the effect of soil heating is to release previously unavailable phosphate in a form available to plants and that this is a large part of the ashbed effect. Attiwell (1962, 1972) concluded that the level of available soil P under mature *Eucalyptus obliqua* forest is limiting to the growth of seedlings of that species.

Renbuss *et al.* (1973) investigated the microbial population of an ashbed soil produced by burning a log pile. The temperatures recorded under the ashbed varied from about 650°C at the surface to about 100°C at 20 cm and they stayed high for several hours. Microbiological assays showed the soil to be apparently sterilized to 25 cm, but bacteria rapidly recolonized the soil, and shortly their numbers exceeded those in unheated soil, and stayed high for several weeks. Actinomycetes and fungi recolonized the soil more slowly. The early bacterial and fungal recolonizers included many types not detected in unheated soil. Species typical of unheated soil soon reappeared and replaced the ashbed species, so that 48 weeks after the burn none of the latter could be detected.

Renbuss *et al.* (1973) considered that the ashbed effect owes much to the sterilizing event. They studied the growth of eucalypt seedlings planted in ashbed and untreated soil collected from different depths immediately after the burn. Increased growth occurred in all samples to 25 cm, except the top 0–5 cm of soil which had an increased pH and was toxic to seedlings. Further, soil which was heated dry in an oven at 150°C to simulate the original burn gave enhanced growth similar to that of ashbed soil.

Most work on seedling response of eucalypts to fire has considered ashbed sites where temperatures have been above 100°C, and often 150 to 500°C. But after a forest fire there is likely to be a mosaic of sites, ranging from those where soil temperatures are little changed from normal to others, such as the ashbeds, where temperatures have been high. In the latter sites all organisms in the heated zone will have been killed; in the former most will have survived. However, except in very severe fires, ashbed sites are likely to be in a minority; nevertheless much of the surface soil will have had some heat-treatment.

The effect of a fire on soil is analogous to heat-treatment or fumigation of glasshouse or agricultural soils. While such soil treatments are usually undertaken to control soil-borne pests or pathogens, it was early realized that even in the absence of such organisms this treatment often markedly improved plant growth (Wilhelm 1966; Jenkinson 1966). Commercial heat-treatments have dropped over the years from 121°C for several hours in an autoclave, to 100°C for 30 min in flowing steam, to 60° for 30 min with aerated steam (Baker 1970). Such relatively low temperatures have been little considered in relation to the effect of fire on soil, soil organisms, or subsequent plant growth in forest soils. Another factor which has been neglected is that plants in forest communities, where available nutrients are generally low, are almost invariably mycorrhizal. The effect of fire on these organisms has generally not been taken into consideration.

Aerated steam treatment of forest soil

A pot culture study has been made (J. H. Warcup, unpublished data) of the effects heat-treatment with aerated steam has on a eucalypt forest soil and on the subsequent growth of eucalypt seedlings. Meadows fine sand (Rix and Hutton 1953) from Kuitpo, South Australia was used; the vegetation, following clearing of the original open forest, was mature planted *Eucalyptus maculata* Hook. with a sparse shrub layer invaded by introduced grasses and herbs. In most experiments *Eucalyptus regnans*, a species showing a marked seedling response to fire in the field (Ashton 1956) and growing relatively faster on good than on poor soil (Ellis 1971), was used but other eucalypts were occasionally grown.

Table 46.1 compares the growth of *E. regnans* seedlings in untreated, fumigated (chloropicrin 220 kg ha^{-1}) or in soil treated with aerated steam at 60°C and 71°C for 30 min, or in autoclaved soil (121°C for 30 min). Growth of seedlings in autoclaved soil was significantly worse, and in all other treatments significantly better than in untreated soil. While treatment at 71°C was significantly better than that at 60°C, subsequent experiments have shown that treatment at these temperatures usually gives a similar response. Seedlings in autoclaved soil scarcely grew and were non-mycorrhizal; in contrast all other seedlings were mycorrhizal.

Analyses of Kuitpo soil immediately after treatment with aerated steam have shown (J. H. Warcup, unpublished data) that the fungal flora is markedly reduced by heat treatment at 60 and 71°C, but that many of the surviving organisms are Ascomycetes, which are not recorded on soil plates of unheated soil. Ascomycetes are often resistant to heat treatment of soil (Warcup 1951, 1957*a*; Bollen 1973), and Warcup and Baker (1963) have shown that some ascospores from soil show heat-stimulated germination. Analysis of soil dilution plates showed that 96 and 98% of the fungal propagules in Kuitpo soil were killed by heat treatment at 60 and 71°C.

Analyses of the forms of available N in Kuitpo soil after aerated steam treatment or after autoclaving showed (Table 46.2) that heating gives an immediate increase in the ammonium N level of this soil and that this increase becomes higher 14 days after treatment. It has often been demonstrated that heat or soil fumigation results in an immediate and a continuing release of ammonium N, and a temporary disruption of nitrification (Wilhelm 1966; Jenkinson and Powlson 1976; Rovira 1976).

It was noted that the mycorrhizas present on roots of plants grown in heated Kuitpo soil appeared different from those on plants from untreated soil. Table 46.3 records the types of mycorrhizas on, and the dry weights of the aerial parts of, six seedlings of *E. regnans* grown in untreated soil, soil treated with aerated steam at 60°C for 30 min or sun-baked soil. The data show fewer kinds of mycorrhizas present on plants grown in treated soils. In particular, basidiomycete mycorrhizas were relatively abundant on seedlings in untreated soil, and black mycorrhizas caused by *Cenococcum graniforme* (Sow.) Ferd & Winge were sparse; but both were absent from plants grown in treated soil. All plants in all treatments were heavily mycorrhizal.

In the aerated-steamed soil the common mycorrhizal fungi in addition to an unidentified Ascomycete were a species of *Endogone* that forms ectomycorrhizas on eucalypts (Warcup 1975); *Endogone* also occurred, though more sparsely, in untreated soil. Examination of soil and seedlings from other experiments with steamed Kuitpo and other soils have shown that a range of larger Ascomycetes grow in soil treated with aerated steam. Some, like species of *Anthracobia* which fruit occasionally in pots of steamed soil, were saprophytes; others were mycorrhizal. Their identities are as yet unknown. While Ascomycetes are not generally considered to be mycorrhizal fungi (Harley 1972), they have been found commonly as mycorrhizas on seedling eucalypts; they were recognized by their simple septal pores and other hyphal characters. It was also noticeable that in many cases both the *Endogone* and the mycorrhizal Ascomycetes ramified extensively through the heated soil. Several of these mycorrhizal Ascomycetes have been

Table 46.1. Mean dry weight (mg) of *Eucalyptus regnans* seedlings grown in untreated, heat-treated or chloropicrin-fumigated Kuitpo soil

Plant part	None	Heat 60/30[A]	71/30	121/30	Chloropicrin
Tops	43	293	539	15	253
Roots	18	85	129	7	57
Plant	61	378	668	22	310

[A] Temperature (°C) and time (min) of treatment.

Table 46.2. Amino-nitrogen, ammonium and nitrate concentrations in untreated and heat-treated Kuitpo soil immediately and 14 days after treatment

Treatment	Amino 0	Amino 14[A]	NH_4 0	NH_4 14	NO_3 0	NO_3 14
Untreated	5.0	6.0	3.6	3.9	1.0	1.4
60°/30[B]	8.0	8.0	18.0	29.6	2.5	1.1
71°/30	8.0	6.2	20.0	30.2	1.0	1.2
121°/30	6.0	7.0	39.0	32.0	0.0	0.7

[A] Days after treatment.
[B] Temperature (°C) and time (min) of heat treatment.

Table 46.3. Mean top dry weight (mg), and types of mycorrhizas on *Eucalyptus regnans* seedlings grown in untreated, sun-baked or aerated-steamed Kuitpo soil.

	Untreated	Sun-baked	Steamed 60°/30[A]
Top weight	165	697	1255
Mycorrhizal type			
Black	6[B]	0	0
curly bas.[C]	5	0	0
bas. 1	2	0	0
bas. 2	1	0	0
asco. 1	4	0	0
Endogone N191	3	6	4
asco. 4	0	5	6

[A] temperature (°C) and time (min) of treatment.
[B] No. of plants.
[C] bas. = basidiomycete; asco. = ascomycete.

isolated from eucalypt seedlings growing after regeneration burns or wild fires in Western Australia, Victoria, South Australia and Tasmania.

Eucalypt seedlings grown in untreated Kuitpo soil generally looked healthy, but their growth was very slow; however, some experiments that were continued for 4–8 months instead of 2 months showed that the rate of growth of seedlings may gradually improve with time. This poor growth of seedlings was not necessarily associated with fungal root pathogens nor with attack by ectoparasitic nematodes, though both types of damage were noted in untreated but not heated soil. Attacked seedlings often showed signs of damage such as withering of tips of leaves; such seedlings were often smaller than the usual size for untreated soil. Also, seedlings from untreated Kuitpo soil did not have stunted blackened roots, as did those reported for seedling *E. pilularis* (blackbutt) grown in untreated soil from under blackbutts (Florence and Crocker 1962) and ascribed to the toxin nectrolide produced by strains of *Cylindrocarpon radicicola* Wollenw. (Evans *et al.* 1967.) In freshly collected soil from under blackbutt Florence and Crocker (1962) found that none of the levels of N or P, or both, they used overcame the inhibiting factors operating to restrict normal seedling growth; however, heat treatment at 35, 70 or 160°C, or irradiation of soil gave healthier seedlings.

Table 46.1 shows that the growth of *E. regnans* was extremely poor in autoclaved soil; it is probable that this was due to lack of available P, for seedlings grew well in autoclaved soil if inoculated with a mycorrhizal fungus (but not a saprophytic one) or if soluble P was added to the soil (Warcup 1975). Barrow *et al.* (1977) comment that when phosphate is added to soil it slowly becomes more firmly bound and less available to plants, and that this change is accelerated by high temperatures. Florence and Crocker (1962) also considered that the poor growth of blackbutt seedlings observed in autoclaved soil was probably related to some form of heat-induced immobilization of available P. Whether this is the mechanism involved with autoclaved soil here has not been investigated, but the poor growth encountered would seem different from that due to the toxin reported by Attiwell (1962) that killed transplanted seedlings and that disappeared with time, or that removed by leaching from soil baked at 150°C (Pryor 1963).

The data in Tables 46.1 and 46.3, and those obtained from other experiments, show that the growth of eucalypt seedlings in soils treated with aerated steam in the range 60–71°C for 30 min is markedly stimulated compared with growth in untreated forest soils. This would appear to be a typical partial sterilization effect, such as occurs with many nursery and agricultural soils (Wilhelm 1966, Jenkinson 1966), but the reasons for such increases in growth are still not fully understood. In these experiments, although control of parasitic nematodes and pathogenic fungi occasionally occurred, the roots of most plants in unheated soil were healthy in appearance and were usually mycorrhizal.

In part, the increased growth may be due to the initial and continuing release of ammonium N from the heated soil, for the plants usually looked well fed with N. However, in autoclaved soil, although the level of ammonium N was relatively high, practically no growth of seedlings occurred unless a mycorrhizal fungus was present or soluble P added. Plant analyses for P, K, Mg, S, Ca, Cu and Zn (Warcup 1975) showed that P alone was markedly increased by the addition of superphosphate to the soil, but in mycorrhizal plants both P and K were markedly higher. Analysis of tops of seedlings grown in aerated-steamed soil also showed major differences in P and K, but not Ca, S, Mg and Na, compared with plants from untreated soil. The percentage occurrence of P in plants from untreated soil and from soil steamed at 71°C for 30 min was 0.07 and 0.264, yet both series of plants were mycorrhizal. Thus a major effect of the heat-treatment is on P uptake, but it is not known how much is due to change in chemical availability or how much to changed mycorrhizal uptake. Chemically available P (bicarbonate-extraction) in untreated soil was about 3 $\mu g\ g^{-1}$, and in heated soil was slightly but not significantly higher. However, plants of a *Juncus* species which grew from seed present in the Kuitpo soil and were non-mycorrhizal grew better in soil steamed at 60°C than in untreated soil, so that the availability of P in a heated soil must have changed. The difference in P uptake of mycorrhizal plants of *Eucalyptus* in untreated and heated soil has not been investigated. In heated soil there was a change in mycorrhizae, at least in part due to the treatment killing the species common in untreated soil. It is possible that the fungi from heated soil were more efficient in P uptake than those in untreated soil, but since the latter group contain the usual mycorrhizal fungi, this seems unlikely. The ability of a mycorrhizal fungus to obtain phosphate may differ markedly under different soil conditions. Mosse (1972), for instance, showed that an endomycorrhizal fungus might give markedly different growth responses with the same plant species growing in different soils or in the presence or absence of lime. Certainly, in heated soil the mycorrhizal fungi quickly ramified extensively throughout the soil mass; it is doubtful if such extensive growth occurred in untreated soil. Irrespective of other factors, this extensive growth would bring a mycorrhizal fungus into greater contact with soil P (Skinner and Bowen 1974).

Sun-baking

Besides heat from a fire, forest soil is often exposed to direct sunlight afterwards, in the absence of canopy and litter. Surface soil temperatures (Marshall 1964; Quinlivan 1971) in southern Australia in areas originally covered with open forest may reach 60°C or higher and be above 50°C for several hours a day. These temperatures are in the range where one might expect a partial sterilization effect from heat as well as from desiccation. Dry heat is less lethal than moist heat at the same temperature (Meynell and Meynell 1970) but loss of viability is a temperature/time interaction, and, while the temperature in sun-baked soil may be relatively low, the time may be long.

The flush of microbial growth on wetting an air-dry soil is well known (Birch 1958; 1959; Jagers and Bruins 1975; Powlson and Jenkinson 1976), though Birch (1959) considered that the effect of temperature was more important than the effect of drying. Both heat and air-drying render some non-biomass C decomposable (Powlson and Jenkinson 1976), as well as killing an appreciable fraction of a biomass (Jenkinson and Powlson 1976).

It has long been known that drying a soil improves its fertility. Prescott (1920) showed that the Egyptian practice of sheraqi, a fallow period during extreme dryness and surface temperatures of up to 55°C, is associated with an increased availability of nutrients. Greenland (1958) considered that the rapid increase in nitrate N at the start of the rains in Ghana is related to the intense drying to which the soil is subjected during the dry season. Nitrate N accumulation, as well as moisture conservation, is one of the results of bare-fallowing wheat (Sims 1977).

To investigate the effect of exposing dry forest soil to direct sunlight, surface soil from Kuitpo was collected in mid-summer, coarse-sieved, and a subsample exposed in the sun for 6 days as a layer 2.5 cm deep on a wooden tray. Air temperature maxima at the time varied from 27 to 35°C; the temperature of the exposed soil reached 62°C and was above 50°C for several hours most days. At the start the soil was just below wilting point, but after sunbaking it was much closer to oven-dry than air-dry; most of the water was lost in the first 2 days of drying. A further sub-sample of the soil was treated with aerated steam at 71°C for 30 min. The growth of seedling *E. regnans* in untreated, sun-baked, and steamed soil is compared in Table 46.3. Early growth of seedlings was best in sun-baked replicates but after about 4 weeks those in steamed soil surpassed them and at 8 weeks were significantly larger; both treatments gave significantly larger seedlings than untreated soil. Table 46.3 also shows that the mycorrhizal fungi on plants in sun-baked soil were similar to those in steamed rather than to those in untreated soil.

Untreated and treated soils were also examined for fungi. Soil dilution plates gave counts of 470,000 units g^{-1} in untreated soil, 46,000 in sun-baked and 11,000 in steamed soil. Apart from lower numbers, there seemed little difference in the fungi obtained from baked or untreated soil. The main difference was a marked increase in the number of colonies of *Trichoderma* from baked soil; the significance of this is not known, though *Trichoderma* is moderately resistant to heat and fumigants (Warcup 1957*a*). Two fungi known to have heat-activated ascospores, *Aspergillus fischeri* var. *glaber* Fennell and Raper and *Anthracobia*, were isolated only from aerated-steamed soil. However, there were fewer colonies of Ascomycetes than usual in this sample of Kuitpo soil, and if heat activation in sun-baked soil is a mild effect the number of activated spores may have been below sampling level.

Three plants that normally germinate more abundantly after heat-treatment of soil, *Trifolium dubium* Sibth., *Geranium solanderi* Carolin and *Dichondra repens* Forst. et f. (J. H. Warcup, unpublished data), germinated in both sun-baked and aerated-steamed soil but not in untreated soil. Quinlivan (1971) considered that in environments with hot dry summer periods, such as most of southern Australia, the daily temperature cycle in summer is the principal if not the only mechanism causing the softening of impermeable (hard) seeds of many legumes, provided the treatment was continued for several weeks or months.

Warcup (1957*b*) studied the occurrence of soil fungi in a wheat-field for 18 months including two summer dry periods. He noted that fungal numbers were low, about 25 percent of winter figures, during the dry periods, and that the fall in number was general and not confined to particular organisms. Although a number of fungi were recorded that are now known to have heat-activated spores (Warcup and Baker 1963), re-examination of the data has failed to show that these fungi are isolated more frequently during, or just after the summer dry periods. However, further work on the microbial and nutritional aspects of sun-baking soil seems desirable.

While evidence is lacking for heat stimulation of spores of fungi by summer dry periods, exposure of forest soil to full sunlight markedly lowered fungal number in the soil, and stimulated growth of eucalypt seedlings. Whether the effect was due to the extreme dryness, the heat, or both, remains unresolved. The growth of eucalypt seedlings on bare soil along forest roads (Gilbert 1959) or in larger gaps in forests (Jackson 1968) but not on the forest floor may be due to soil factors as well as to the seedlings being light-demanding.

References

Ashton, D. H. (1956). 'Studies on the autecology of *Eucalyptus regnans* F.v.M.' Ph.D. Thesis, Univ. Melbourne.

Attiwell, P. M. (1962). The effect of heat pre-treatment of soil on the growth of *Eucalyptus obliqua* seedlings. 3rd Gen. Conf. Inst. For. Australia.

Attiwell, P. M. (1972). On the cycling of elements in mature *Eucalyptus obliqua* forest. The Australian Forest-tree Nutrition Conf. 1971. (Forestry and Timber Bureau: Canberra.)

Baker, K. F. (1970). Selective Killing of Soil Microorganisms by Aerated Steam. *In* 'Root Diseases and Soil-borne Pathogens'. (Eds T. A. Toussoun, R. V. Bega and P. E. Nelson). pp. 234-9. (Univ. Calif. Press: Berkeley.)

Barrow, N. J., Malajczuk, N. and Shaw, T. C. (1977). A direct test of the ability of vesicular-arbuscular mycorrhiza to help plants take up fixed soil phosphate. *New Phytol.* **78**, 269-76.

Beadle, N. C. W. (1940). Soil temperatures during forest fires and their effect on the survival of vegetation. *J. Ecol.* **28**, 180-92.

Birch, H. F. (1958). The effect of soil drying on humus decomposition and nitrogen availability. *Plant Soil* **10**, 9-31.

Birch, H. F. (1959). Further observations on humus decomposition and nitrification. *Plant Soil* **11**, 262-86.

Bollen, G. J. (1973). Fungal recolonization of heat-treated glasshouse soils. *Agro-ecosystems* **1**, 139-55.

Burbridge, N. T. (1960). Phytogeography of the Australian region. *Aust. J. Bot.* **8**, 72-211.

Cromer, R. N. and Vines, R. G. (1966). Soil temperatures under a burning windrow. *Aust. For. Res.* **2**, 29-34.

Cunningham, T. M. (1960). The natural regeneration of *Eucalyptus regnans*. Univ. Melb. Sch. For. Bull. No. 1.

Ellis, R. G. (1971). Growth of *Eucalyptus* seedlings on four different soils. *Aust. For.* **35**, 107-18.

Evans, G., Cartwright, J. B. and White, N. H. (1967). The production of a phytotoxin, nectrolide, by some root surface isolates of *Cylindrocarpon radicicola* Wr. *Plant Soil* **26**, 253-60.

Florence, R. G. and Crocker, R.L. (1962). Analysis of blackbutt (*Eucalyptus pilularis* Sm.) seedling growth in a blackbutt forest soil. *Ecology* **43**, 670-9.

Floyd, A. G. (1966). Effect of fire on weed seeds in wet sclerophyll forest of northern N.S.W. *Aust. J. Bot.* **14**, 243-56.

Francis, W. D. (1970). 'Australian Rain-forest Trees'. Forestry and Timber Bureau, Canberra.

Gilbert, J. M. (1959). Forest succession in the Florentine Valley, Tasmania. *Pap. Proc. R. Soc. Tas.* **93**, 129-51.

Gill, A. M. (1975). Fire and the Australian flora – a review. *Aust. For.* **38**, 4-25.

Greenland, D. J. (1958). Nitrate fluctuations in tropical soils. *J. Agric. Sci.* **50**, 82-92.

Griffith, A. L. (1946). The effects of burning on the soil as a preliminary to artificial regeneration. Indian For. Bull. No. 130. Dehra Dun.

Harley, J. L. (1972). 'The Biology of Mycorrhiza'. (2nd edn). (Leonard Hill: London.)

Hatch, A. B. (1960). Ashbed effects in Western Australian forest soils. Bull. For. Dep. West. Aust. No. 64.

Jackson, W. D. (1968). Fire, air, water and earth – an elemental ecology of Tasmania. *Proc. Ecol. Soc. Aust.* **3**, 9-16.

Jagers, G. and Bruins, E. H. (1975). Effect of repeated drying at different temperatures on soil organic matter decomposition and characteristics and on the soil microflora. *Soil Biol. Biochem.* **7**, 153-9.

Jenkinson, D. S. (1966). Studies on the decomposition of plant material in soil – II. Partial sterilization of soil and soil biomass. *J. Soil Sci.* **17**, 280-302.

Jenkinson, D. S. and Powlson, D. S. (1976). The effects of biocidal treatments on metabolism in soil – I Fumigation with chloroform. *Soil Biol. Biochem.* **8**, 167-77.

Loneragan, O. W. and Loneragan, J. F. (1964). Ashbed and nutrients in the growth of seedlings of karri (*Eucalyptus diversicolor* F.v.M). *J. R. Soc. W.A.* **47**, 75-80.

Marshall, K. C. (1964). Survival of root-nodule bacteria in dry soils exposed to high temperatures. *Aust. J. Agric. Res.* **15**, 273-81.

McArthur, A. G. (1972). Fire Control in the Arid and Semi-arid Lands of Australia. *In* 'The Use of Trees and Shrubs in the Dry Country of Australia'. pp. 488-575. (Aust. Govt Publ. Service: Canberra.)

Meynell, G. G. and Meynell, E. (1970). 'Theory and Practice in Experimental Bacteriology'. (Cambridge Univ. Press: Cambridge.)

Mosse, B. (1972). The influence of soil type and *Endogone* strain on the growth of mycorrhizal plants in phosphate deficient soils. *Rev. Ecol. Biol. Soc.* **IX**, **3**, 529-37.

Mount, A. B. (1964). The interdependence of eucalypts and forest fires in southern Australia. *Aust. For.* **28**, 166-72.

Powlson, D. S. and Jenkinson, D. S. (1976). The effects of biocidal treatments on metabolism in soil – II. Gamma irradiation, autoclaving, air-drying and fumigation. *Soil Biol. Biochem.* **8**, 179-88.

.Prescott, J. A. (1920). A note on the sheraqi soils of Egypt. A study in partial sterilization. *J. Agric. Sci.* **10**, 177-81.

Pryor, L. D. (1960). The ashbed effect in eucalypt ecology. *Inst. For. Aust. Newsletter* **2** (9), 23-6.

Pryor, L. D. (1963). Ashbed growth response as a key to plantation establishment on poor sites. *Aust. For.* **27**, 48-51.

Pryor, L. D. (1976). 'Biology of Eucalypts'. (Edward Arnold: Melbourne.)

Quinlivan, B. J. (1971). Seed coat impermeability in legumes. *J. Aust. Inst. Agric. Sci.* **37**, 283-95.

Renbuss, M. A., Chilvers, G. A. and Pryor, L. D. (1973). Microbiology of an ashbed. *Proc. Linn. Soc. N.S.W.* **97**, 302-10.

Rix, C. E., and Hutton, J. T. (1953). A soil survey of the Hundred of Kuitpo in the Mt Lofty ranges of South Australia. S. Aust. Land Tax Dep. Bull. No. 1.

Roberts, W. B. (1965). Soil temperatures under a pile of burning logs. *Aust. For. Res.* **1**(3), 21-3.

Rovira, A. D. (1976). Studies on soil fumigation – I. Effects on ammonium, nitrate and phosphate in soil and on growth, nutrition and yield of wheat. *Soil Biol. Biovhem.* **8**, 241-7.

Sims, H. J. (1977). Cultivation and Fallowing Practices. *In* 'Soil Factors in Crop Production in a Semi-arid Environment'. (Eds J. S. Russell and E. L. Greacen). pp. 243-61 (Univ. Queensland Press: Brisbane.)

Skinner, M. F. and Bowen, G. D. (1974). The penetration of soil by mycelial strands of ectomycorrhizal fungi. *Soil Biol. Biochem.* **6**, 57-61.

Specht, R. L. (1970). Vegetation. *In* 'The Australian Environment'. (Ed G. W. Leeper). pp. 44-67. (CSIRO: Melbourne.)

Tunstall, B. R., Martin, T., Walker, J., Gill, A. M. and Aston, A. (1976). Soil temperatures induced by an experimental logpile fire: preliminary data analysis. CSIRO Aust. Div. Land Use Res. Tech. Mem. No. 76/20.

Wallace, W. R. (1966). Fire in the jarrah forest environment. *J. R. Soc. W. Aust.* **49**, 33-44.

Warcup, J. H. (1951). Soil-steaming: a selective method for the isolation of Ascomycetes from soil. *Trans. Br. Mycol. Soc.* **34**, 515-18.

Warcup, J. H. (1957*a*). Chemical and biological aspects of soil sterilization. *Soils Fert.* **20**, 1-5.

Warcup, J. H. (1957*b*). Studies on the occurrence and activity of fungi in a wheat-field soil. *Trans. Br. Mycol. Soc.* **40**, 237-62.

Warcup, J. H. (1975). A Culturable *Endogone* Associated with Eucalypts. *In* 'Endomycorrhizas'. (Eds F. E. Sanders, B. Mosse and P. B. Tinker). pp. 53-63. (Academic Press: London.)

Warcup, J. H. and Baker, K. F. (1963). Occurrence of dormant ascospores in soil. *Nature (London)* **197**, 1317-8.

Wilhelm, S. (1966). Chemical treatments and inoculum potential of soil. *Ann. Rev. Phytopathol.* **4**, 53-78.

Chapter 47
Phytotoxins in soil

J. R. Harris and R. W. L. Kimber

Infertility, allelopathy and toxicity

Plants do not grow equally well in all soils for a variety of reasons. Fertile soils have an abundant resource of chemical nutrients and provide a good physical environment for root growth, for moisture supply and for aeration. Infertile soils are generally thought of as being deficient in one or other of these respects.

Nevertheless, not all soils are infertile because they are deficient. Plant growth may be retarded by excessive amounts of substances that are toxic. In this chapter we discuss how biological products accumulate in soil, and that some of these substances may be toxic to plants. We do not deal with the infertility problems which follow from heavy metal excesses from mineralization or from spoilage, with pollution from industrial wastes or agricultural chemicals, or with salinization.

Early agricultural writings contain abundant observations that some plants do not grow well in the vicinity of others, whereas some are not affected by, or are even stimulated by, the proximity of their neighbours. Where there is either retardation or stimulus, the condition is described as allelopathy. It has been a powerful force in determining the stability and the composition of ecosystems in some plant communities.

More correctly, allelopathy describes the situation where byproducts from plant growth accumulate and affect the growth of neighbours. The magnitude of the effect will depend upon whether the environment will allow a potentially active metabolite to persist and to accumulate into a dose-effective quantity.

Soil toxicity describes a situation where the accumulation of allelochemicals in soil measurably reduces the growth of a plant. The rather older terms 'soil sickness' and 'soil fatigue' reflect the reductions in growth rates that are observed either in existing plants or in succeeding crops.

Irrespective of what we call it, this problem has received attention because it has long been part of intensive agricultural systems of the Old World. But it is no stranger to the low intensity farming systems of Australia. In many cases it is expressed most severely in the first stages of developing new land for agriculture when the native vegetation has been cleared and a sown pasture or a crop is introduced for the first time. We will cite some examples of this in due course.

Modern agricultural science has tended to disregard soil toxicity as a major constraint to plant growth. This has been a reaction against the single factor organic chemical hypotheses produced early this century where the notable contributors were Schreiner and Reed (1907*a*, *b*, 1908) and Schreiner and Shorey (1909, 1910) and their co-workers of the Bureau of Soils of the United States Department of Agriculture. In a series of studies of soils in which the fertility has been depleted by overcropping, they consistently isolated toxic substances in appreciable quantities. The most common four were identified as dihydroxystearic acid, picolinic acid, salicylaldehyde and vanillin. Quite a number of other substances were isolated in lesser amounts or occurred irregularly.

When successive crops of wheat were grown in the same container, Schreiner and Shorey (1908*a*, *b*) showed that impaired root growth accompanied yield decline. Finally, germination became uncertain, and thermolabile toxic compounds were found to have accumulated. These compounds were isolated and tested for their effects upon plant growth in water-culture experiments. Root function was impaired early (Schreiner and Sullivan 1909). Nutrient uptake became inefficient and relatively greater amounts were needed to maintain plant growth at the level of the controls (Schreiner and Skinner 1912). Dihydroxystearic acid and vanillin specifically impeded the uptake of nitrogen; quinone decreased the uptake of potassium; coumarin caused decreased uptake of phosphorus.

Although these toxins clearly depressed plant growth, the most effective way to overcome soil fatigue under field conditions was to increase nutrient availability by increasing the rate of application of fertilizers (Schreiner and Skinner 1912; Skinner and Noll 1916; Skinner 1918). It was arguable whether the fertilizer salts reacted with the organic compounds to detoxicate them, or whether the effect was nothing more than improved plant nutrition in an impoverished soil. In the latter case, less significance was attached to the fact that certain organic products had had a tendency to accumulate. The general concensus of opinion was that soil toxins played a lesser role than nutrient depletion in infertility problems.

This was reinforced by Loehwing's (1937) critical review which attacked most of the earlier studies that had attributed soil sickness to substances excreted by plants. He maintained that only rarely had the substances been identified chemically, that their origin was mostly uncertain, and that there were always other cultural factors, notably nutritional impoverishment, which could not be excluded as major influences upon plant growth.

In the same year, Molisch's (1937) book was published. It brought together much of the European studies on plant interactions, and it placed a new emphasis upon the importance of volatile substances such as ethylene. The term allelopathy emerged with a new concept; the old catch-all terms of soil fatigue and soil sickness were abandoned because they were imprecise.

Allelopathy became recognized as a powerful ecological factor capable of altering the structure, function and composition of many plant communities. Went's (1942, 1970) studies of desert shrubs demonstrated that they possessed a chemistry which precluded the establishment of annuals close to them. A water-soluble leaf toxin from the shrub *Encelia farinosa* was identified by Gray and Bonner (1948*a,b*) as 3-acetyl-6- methoxybenzaldehyde. In Southern California the soft shrubs of the chapparal were shown by Muller (1966, 1969) and coworkers (Muller and Muller 1956; McPherson and Muller 1969) to exert strongly allelopathic effects upon the adjacent grassland vegetation by both water-soluble leaf toxins and by volatile terpenes. Australian eucalypts introduced into California were found by del Moral and Muller (1969, 1970) to possess similar allelopathic influences. In the Australian coastal heath of southern Victoria, del Moral *et al.* (1978) demonstrated a zone of suppression beneath the canopies of *E. baxteri* where leaf litter accumulated. The tree species of the humid regions of the Pacific North-West coast of the United States were shown (del Moral and Gates 1971) to possess widespread allelopathic potential; but under a humid environment allelopathy did not develop the ecological pressure that characterized the more highly stressed arid zone vegetations.

Terminology in the field of allelopathy has developed largely from European influences. The term 'phytotoxicity' loosely embraces adverse reactions to substances of botanical origin. Thus a phytotoxin may refer either to a substance inhibitory to plant growth or to a substance of plant origin with an effect upon animal physiology. Where greater precision is needed, Grümmer (1955) and Rademacher (1957, 1959) introduced terms which account for the origin of the inhibitor and the subject of the reaction. They used the term '*choline*' to describe those substances produced by higher plants which have effects upon higher plants; '*marasmine*' for substances produced by microorganisms which have effects upon higher plants; '*antibiotics*' (in the strict sense) for substances produced by microorganisms which affect other microorganisms; and '*phytoncides*' for substances produced by higher plants which affect microorganisms. We will follow this terminology here.

A wider classification of interactive biological chemicals has been proposed by Whittaker and Feeny (1971) to separate allelochemic effects (i.e. those affecting individuals or populations of a different species from that of the source) from intraspecific chemical effects (i.e. those operating within individuals of the same species). This classification attempts to combine both plant and animal behaviour patterns with responses to chemical stimulus.

The current position is that a large number of substances of plant or microbial origin can accumulate in the environment in sufficient quantities to affect plant growth adversely. If they accumulate in soil and are able to persist there for long enough, unthrifty growth or poor establishment of plants will follow. This is the condition described as soil toxicity. It is likely to be manifested more severely in infertile soils than it would be in otherwise fertile soils.

The toxic allelochemicals

A large number of substances have been described as allelochemicals. The diversity of their origins, their chemistry and their physiological activities may be appreciated from consulting the wide-ranging reviews of Evanari (1949, 1961), Bonner (1950), Krasilnikov (1958), Goodman (1959), Nickell (1959), Börner (1960), Boullard and Moreau (1962), McCalla and Haskins (1964), Savory (1966), Moje (1966), Tukey (1966, 1969), Muller (1969, 1970, 1974), Whittaker and Feeny (1971), Rice (1974), Putnam and Duke (1978), and Wood (1960).

Although a wide range of allelochemicals of plant or microbial origin affect the growth of higher plants, there are a number of chemical groups which not only are regularly well-represented, but are the most common causes of soil toxicity. Six such groups are as follows.

Phenols and phenolcarboxylic acids have been isolated from a large number of soils, plant tissues and residues. The cholines most frequently reported to be associated with toxic soils are *p*-hydroxybenzoic, gentisic, benzoic, salicylic, *p*-coumaric, syringic, ferulic, cinnamic and the chlorogenic acids. This is the group most frequently implicated in agricultural and forestry ecosystems where an inhibition of root growth is a prime feature (Börner 1960; Moje 1966; DeBell 1970; Rice 1974).

Aldehydes were isolated by Skinner and Noll (1916) and Skinner (1918) in the early Bureau of Soils studies of arable soils and were found to be potent soil toxins. Salicaldehyde, benzaldehyde, 3-acetyl-6-methoxy benzaldehyde (Gray and Bonner 1948*a, b*) and vanillin have been regarded as the most toxic cholines within this group.

Coumarins and unsaturated lactones received early attention as germination inhibitors (Evanari 1949) and as fungistatic agents. Most frequently reported as active are coumarin, the furanocoumarins, esculentin and scopoletin.

Glucosides from woody plants received early recognition as providing a wide range of potent cholines (Savory 1966; Moje 1966). Juglone from walnut and amygdalin and phlorizin from apple, peach and almond roots are examples which have been implicated in causing well-known replant problems (Koch 1955; Patrick 1955; Ward and Durkee 1956; Mountain and Boyce 1958; Mountain and Patrick 1959; Hendrix and Powell 1970; Israel *et al.* 1973). Whilst glucosides themselves are moderately toxic, their breakdown in soils often releases products of even greater toxicity (Börner 1960), e.g. amygdalin hydrolyses to hydrocyanic acid and benzaldehyde, each of which is more toxic than its precursor.

Terpenes have been recognized as the class of volatile compounds which are most frequently associated with allelopathic relationships of woody shrubs in dry Mediterranean and in arid zone vegetations, e.g. in the chaparral of California (Went 1942, Gray and Bonner 1948 *a*, *b*; Muller 1953, 1966; Muller and Muller 1956; McPherson and Muller 1969) and in the spinifex tussock grass communities of arid Australia (Went 1956). Dry soils appear to act as reservoirs for terpene accumulation (Muller 1970, 1974). Camphene, cineole, camphor, and α- and β-pinene have been recovered from toxic soils and from plant parts, mostly leaves.

Antibiotics (sensu stricto) have been indicated in cases where fungi dominate in a major colonization of plant residues, and particularly of surface litter. Many of the antibiotics show some degree of phytotoxicity (Goodman 1959), but in soil they are generally sorbed too strongly to retain marasmine activity. The best documented example of crop damage due to an antibiotic is that of patulin produced by the fungus *Penicillium urticae* on stubble mulched corn in Nebraska (McCalla and Haskins 1964; Guenzi and McCalla 1962; McCalla *et al.* 1964). In Western Australia, a fungal colonization of litter gives rise to a problem of antagonism to rhizobia associated with forage legumes in pastures sown on new land after clearing scrub for agricultural development (Holland 1962, 1966; Holland and Parker 1966).

Factors influencing allelopathy

Before considering the likelihood of a soil to develop toxicity, we must bear in mind a number of general points about the phenomenon of allelopathy and the factors which affect its manifestation. Some of these are as follows.

Widespread occurrence

Allelopathy occurs in a wide range of agricultural, forest and natural ecosystems. These range from desert shrub communities (Went 1942, 1970; Muller 1953, 1966, 1969, 1974; Muller and Muller 1956) to rainforest (Webb *et al.* 1961, 1967; Canon *et al.* 1961; Bevege 1968) and in the monoculture of many crop plants (Moje 1966); and in geographical regions which extend from the tundras to the tropics (Krasilnikov 1958; Lodhi 1976). It follows that soil toxicity can occur in many climates, under many vegetations, and in many soil types.

Variety of routes of release

Cholines may originate from almost any plant part, leaf surfaces, glands, stones, floral parts, fruits, roots, wood and bark. They may be released via a number of routes, e.g. by volatilization (Molisch 1937; Tukey 1966, 1969) or by rainwash and fog-drip (del Moral and Muller 1969, 1970) from aerial parts; by litter fall (del Moral *et al.* 1978); and by excretion from living roots (Woods 1960; Börner 1960; Tukey 1966; 1969). A metabolite may be released from living tissue and possess choline activity, or it may be shed as detritus and become the precursor of a marasmine produced during litter decomposition. To quote Whittaker and Feeny (1971), in general 'the poison will out: accumulations of potentially toxic materials in plants will find their way into the soil sooner or later, by one route or another'. And, of course, it is well known that many plants are sources of highly poisonous substances.

Plant community

The plant community tends to determine the rate of accession of the potential choline. A monoculture is more likely to produce an excess of a single substance than is a mixed plant community. In plant succession, the more dominant a species or the closer the approach to a climax community, the greater is the likelihood that a single plant species will determine the organic matter composition of the soil (Bevege 1968; Lodhi 1976). In this way plant succession is determined, and in extreme cases even extends to autotoxicity (Curtis and Cottam 1950; McNaughton 1968). In some cultivated plants, e.g. cucumber, some cultivars are more allelopathic than are others in their ability to control weed invasion (Lockerman and Putnam 1979).

Soil microflora

Most cholines which find their way into the soil become subject to microbial decomposition. Their accumulation to a dosage level which affects plant growth will occur only when:

–the soil microflora is relatively inactive at the time; the substance is relatively refractory to further decomposition; or

–the rate of accession is unusually great; or

–the choline (or marasmine) is biologically active in very low concentrations.

There are many observations that allelopathy tends to

flare up at seasonal or irregular intervals, or to follow major disturbance of a plant community. This may reflect a fluctuation in activity of the soil microflora as much as a fluctuation in plant growth.

Effects upon nutrient cycles

If a soil toxin has phytoncidal activity it may affect the growth of micro-organisms which control mineralization cycles and thereby cause a secondary effect upon the availability of plant nutrients. In infertile soils that have low nutrient resource levels this phytoncidal effect may be just as important as the direct effect of a choline upon plant growth.

This situation has been highlighted in the studies by Rice and his colleagues (Rice 1964, 1965, 1974; Floyd and Rice 1967; Abdul-Wahab and Rice 1967; Neill and Rice 1971; Lehman and Rice 1972) of old field succession in Oklahoma. There the abandoned farmlands are colonized only very slowly by a succession of broad-leaved weeds which rarely reach the climax association of prairie grasses. The soil resources of both nitrogen and phosphorus in abandoned fields are too low to support the growth of the climax grasses and forbs which have higher nutrient requirements than the early weedy invaders. If the bacteria associated with the nitrogen fixation and nitrification cycles are inhibited, there is no accession of nitrogen into the system and the soil nitrogen status remains too low to permit the next stage of succession. Chlorogenic acids and gallotannins from the weeds *Ambrosia*, *Euphorbia* and *Helianthus* have been shown (Rice 1964, 1965) to inhibit the growth of *Azotobacter*, *Rhizobium*, *Nitrosomonas* and *Nitrobacter*.

In the Willochra Plain of South Australia there are farmlands which were abandoned at the turn of the century. Even now they carry depauperate barley grass and weeds and are unable to succeed to the chenopodiaceous plants of the climax shrub steppe vegetation. Allelopathic effects upon nutrient cycles appear to have the greatest impact and the longest standing repercussions in fragile plant communities such as those of the arid zone where the soils have low nutrient and limited moisture resources.

In the Hunter Valley of the tablelands of New South Wales, circles of sparse grasses occur under the canopies of the trees which comprise the open woodland of the partially cleared vegetation. These circles often extend out for considerable distances beyond the canopies of the *Acacia*, *Casuarina* and *Eucalyptus* species of the region. On aerial photographs, plant growth appears both paler and sparser than the surrounding ground cover. Story (1967) has demonstrated that this growth retardation cannot be explained in terms of effects upon soil water regime. It is tempting to postulate that this is another example of a phytoncidal interference with mineralization cycles which affect the availability of nutrients to the ground cover plants.

Table 47.1. Effect of the proximity of a desert shrub canopy on nitrogen fixation.
The data are reported as mean ethylene production (in nmoles) per 24 h from acetylene reduction (Rychert and Skujins 1974) in soil cores with intact algal–lichen crusts collected from three plant communities.

Plant community	Between canopies	Under canopy
Atriplex confertifolia	78.7	20.7
Eurotia lanata	4.5	2.3
Artemesia tridentata	79.5	10.8

The data of Table 47.1 show a likely phytoncidal inhibition of nitrogen fixation in the algal–lichen crusts which form biocoenoses over desert floors. Rychert and Skujins (1974) demonstrated how this delicate balance is disrupted under shrub canopies and even well beyond them in the case of the more toxigenic plant species of the Great Basin Desert of Utah.

Enhanced pathogenesis

Upon exposure to a choline, the root system is generally the part of the plant which reacts most strongly. It is not unusual for an attack by root pathogens to follow inhibition of root growth, and a disease syndrome ensues. Two examples of diseases of vegetable crops may be cited where this complication follows the incorporation of readily decomposable plant residues, viz. *Fusarium* root-rot of bean demonstrated by Toussoun and Patrick (1963) and Toussoun *et al.* (1968) and the *Pythium* corky root-rot of lettuce demonstrated by Amin and Sequeira (1966*a*, *b*). The replant problems of pome and stone fruits, citrus and vine are examples where the root residues of the same preceding horticultural crop produces a comparable effect (Savory 1966). In this case the cholines are usually glycosides or their breakdown products. Although glycosides are markedly phytoncidal to some soil micro-organisms (Hendrix and Powell 1970; Israel *et al.* 1973), the attack by tylenchoid nematodes exacerbates the replant syndrome into a disease syndrome in both Canada (Koch 1955; Patrick 1955; Ward and Durkee 1956; Mountain and Boyce 1958; Mountain and Patrick 1959) and Australia (Colbran 1953; Anon. 1960). Attempts to ascribe all the causal symptoms to any one of these factors alone tend to be inconclusive.

The soil as a reservoir of toxins

For a soil to develop toxicity it is necessary that the active metabolites should be retained in the root zone in a dose-effective concentration. This has to be maintained without appreciable decline from leaching, microbial decomposition or sorbtion.

There is a root-rot of tobacco in which seedling damage is associated with the presence of freshly incorporated plant residues. The toxins are water-soluble, thermostable and relatively non-specific in action.

Experiments which attempted to assay the toxicity of soil extracts gave inconsistent results until it was realised that the toxicity was associated with particulate organic matter and that this was irregularly distributed throughout the soil mass. The highest levels of activity occurred around the residue fragments, and toxicity diminished with distance beyond (Patrick and Koch 1958; Patrick *et al.* 1963; Patrick *et al.* 1964).

When the soils were sampled, toxicity was reduced if the procedures increased aeration, allowed the soil to dry out, or held it for periods of time (in excess of 25 days) under conditions in which decomposition could continue. If held in the cold (0 to 2°C) or under an atmosphere of nitrogen, the phytotoxic activity of the soil remained undiminished.

The tobacco root-rot syndrome provides an example which parallels many other root-rots associated with soil toxicity (Cochrane 1948). It is most severe when an active decomposition of plant residues occurs under conditions of poor aeration. And it emphasizes that the soil does not behave as a uniform mass, but rather that within it are anaerobic microsites which are ecologically and allelopathically different from the background. Damage may occur when an exploring root encounters an active microsite. The overall damage to the root system reflects the number of contacts with active microsites.

Plant damage will depend upon how much of the root system is exposed to the choline. The primary root of seedling tobacco was especially susceptible; the growth of a radical was either partially or completely inhibited whereas the hypocotyl was little affected. The effect on the apical meristem of the root was rapid; within hours it became discoloured by browning and was easily killed. The growth of root hairs was inhibited (Patrick and Koch 1958); Patrick *et al.* 1964).

Plant behaviour in the field may contrast greatly with what happens in an experiment in which plant response is assayed in water culture or in which seed germination is assayed in a Petri dish, or in a nursery flat. It has been relatively easy to extract a whole range of cholines from all sorts of soils, litters and plant parts for the demonstration of their activity in a sensitive laboratory experiment. There have been many cautions issued by European workers (Grümmer 1961; Winter 1961) that a successful laboratory experiment may stop short of establishing analogous behaviour in a field soil. The experiences of Cannon *et al.* (1961) and of Webb *et al.* (1961) during the investigation of allelopathy in Australian plant communities highlight the problems of choosing a representative experimental approach. Their dilemma arose from an enormous discrepancy between the amounts of dehydroangustione from *Backhousia angustifolia* leaves which seemed to inhibit the germination of hoop-pine in a forest site compared to laboratory germination tests on filter paper.

Studies in which pure chemical compounds and plant extracts have been added to soil have demonstrated frequently that activity may be short-lived. Several reasons account for this, viz. that readily soluble substances may be leached out of the root-zone; many active metabolites will be attacked by those members of the soil microflora that are not inhibited; polar compounds may become sorbed onto the colloidal fraction of soil and thereby be inactivated (Schönbeck 1968).

The soil microflora might be expected to have some difficulty in decomposing antibiotics. Investigations by Brian (1957) have shown that they are not likely to be produced in soil other than at discrete microsites. Even then, it has been consistently difficult to demonstrate antibiotic activity and persistence in soil because they become adsorbed onto and are often held tenaciously by, clay minerals (Simonoff and Gottlieb 1951; Martin and Gottlieb 1952). In fact, special extraction procedures are required to assay soils for the presence of the basic antibiotics such as streptomycin and carbomycin, and the amphoteric antibiotics such as terramycin, aureomycin and bacitracin (Pinck *et al.* 1961; Soulides *et al.* 1961, 1962; Soulides 1964). In the bound condition these antibiotics are biologically inactive against the soil microflora and against sensitive plants.

In addition to the sorption of antibiotics on clay minerals, Simonoff and Gottlieb (1951) and Martin and Gottlieb (1952) have produced evidence which showed that some were decomposed by non-susceptible members of the microflora. One example of this is the antibiotic griseofulvin produced by *Penicillium nigricans*. Wright and Grove (1957) demonstrated its persistence in sterile soil; but in natural soil it was inactivated readily by bacteria of the genus *Pseudomonas*.

A parallel situation exists for the phenolic acids, which are probably the commonest agents associated with soil toxicity. These compounds are widespread in nature, being found in both living plants and plant residues (Collison and Conn 1925; Börner 1955*b*, Bates-Smith 1956; Guenzi and McCalla 1962, 1966*a*; McCalla *et al.* 1964) as well as in agricultural soils which exhibit decline syndromes or which possess residue toxicity problems (Schreiner and Shorey 1908*a*, *b*; Whitehead 1964; Rice 1964, 1965; Guenzi and McCalla 1966*b*; Lehman and Rice 1972). They are sparingly soluble in water, but may be extracted from soil readily by mild alkalis. Even the relatively low concentrations in which they can be found in many normal soils exceed the amounts that would cause significant retardation of the growth of plants in water-culture experiments (Guenzi and McCalla 1966*b*; Yang *et al.* 1966; Wang *et al.* 1967; Kuwatsuka and Shindo 1973; Rao *et al.* 1975; Shindo *et al.* 1978). Why then do they not cause considerable damage to plants more often?

In the soil, phenolic acids are at least partially adsorbed. The amounts depend upon both the chemically bonded state of the compound itself and upon the

Table 47.2 Phenolic acids and organic matter contents of soils

Mean concentrations (ppm) of phenolic acids (PA) extracted by 2 M NaOH from an arable Sharpsburg soil under two cultivation treatments, together with the proportion (%) of the organic matter content (OM) of each soil. After Guenzi and McCalla (1966*b*).

	Subtilled soil		Ploughed soil	
Phenolic acid	**PA**	**OM**	**PA**	**OM**
Ferulic	7.6 ± 4.0	0.021	3.7 ± 1.5	0.010
p-Coumaric	14.4 ± 2.3	0.040	9.4 ± 2.9	0.026
Syringic	< 1.0	–	< 1.0	–
Vanillic	1.5 ± 0.7	0.004	1.5 ± 0.5	0.004
p-Hydroxybenzoic	1.2 ± 0.3	0.003	1.3 ± 0.1	0.003

composition of the clay mineral fraction. In arable soils in Japan (Shindo and Kuwatsuka 1976), allophanic soils adsorbed more phenolic acids than did kaolinitic or montmorillonitic soils, and polyphenolic acids complexed to soil minerals more readily than monophenolic compounds. Where phenolics are not adsorbed, but remain in the soil solution, they appear to be readily leached and disappear from the ecosystem.

For plant damage to occur, the concentration of phenolic acids in the soil solution must rise at least transiently to an abnormally high level, or the high concentration may occur localized in microsites, or suboptimal conditions for growth must heighten the sensitivity of the plant (Guenzi and McCalla 1966*b*; Kuwatsuka and Shindo 1973; Shindo and Kuwatsuka 1976; Shindo *et al.* 1978). The data of Table 47.2 show the concentrations of phenolic acids which occurred in an arable soil under stubble mulching in Nebraska. The samples were taken at a time when concentrations were low and when the crops showed no symptoms of phytotoxicity. The data indicate the low normal background concentrations of phenolic acids and also a response to past cultivation treatments.

Another similarity between phenolic acids and antibiotics is that they act not only as cholines, but that their phytoncidal effects are strong enough to disturb the balance of the microbiological population of the soil (Knoesel 1959; Hatsuda *et al.* 1963, 1965). The old field succession studies of Rice and his associates (Rice 1964, 1965; Neill and Rice 1971; Lehman and Rice 1972) indicate that this may be of greater importance in impoverished soils where the phytoncidal effect is a major factor in maintaining homeostasis in a pioneer plant community.

Although a great many cholines and phytoncides accrue in or are produced from discrete decomposing residues and react with clay minerals, there is very little evidence available which suggests an interaction with the humic fractions of soil organic matter. These probably occur, but we know very little about them. Certainly the highest concentrations of phenolic acids in Japanese soils occur in forest soils where the organic matter fractions are highest (Shindo *et al.* 1978). We may assume that sorption phenomena occur, and that this will influence both the level of activity and the persistence of cholines and phytoncides (Shindo and Kuwatsuka 1976).

One piece of experimental evidence that emphasizes the importance of sorption of toxic substances in the rootzone follows from the observations of Eliasson (1959) that roots of wheat plants growing in solution cultures could be stimulated strongly by the addition to the medium of strongly absorbing activated charcoal.

We may summarize the factors which determine whether the soil will become a reservoir of toxins under these items: accession rate and kind of residues; microbial decomposition rate and diversity; bonding to clay minerals and sorption phenomena; leaching from the root-zone; and extreme soil reaction and environmental conditions.

Residual toxicity in changing ecosystems

In some natural ecosystems the expression of soil toxicity provides homeostasis to the plant community as well as imposing a selective pressure on soil organisms. The Soviet microbiologists Krasilnikov and Garkina (1946) Krasilnikov *et al.* (1955) Krasilnikov (1958) and Yuzhina (1958*a*, *b*) have demonstrated this homeostasis in the strongly acid turf-podzol soils of the Kola Peninsula. Not only is plant growth affected, but the microflora of the soil is so severely inhibited that it has provided the basis for a method for assaying soil toxicity by measuring the survival rate of introduced populations of micro-organisms (Zhukova 1959, 1960; Yevdokimova 1962).

Acid podzols with humic or ferruginous hard-pans tend to possess a microflora dominated by fungi whose decomposition products from grassland residues are sources of toxicity (Mirchink 1957; Mirchink and Greshnyk 1961; Mirchink *et al.* 1966). Such soils do not provide particularly favourable physico-chemical environments for plant growth. During part of the year they become water-logged and anaerobic conditions prevail. Their low nutrient resource status, proclivity to strong leaching and the low base exchange capacity predispose plants growing on them to assimilate persistent marasmines. This often leads to unbalanced plant nutrition, particularly in relation to nitrogen metabolism (Mirchink *et al.* 1962; Zhukova 1964).

Whether *podzols* carry grassland, open woodland or forest vegetation in the natural state, they seem to possess sufficient background soil toxicity to single them out as providing a real difficulty in development for arable agriculture in all parts of the world. In Australia we have a number of examples where residual toxicity from a natural ecosystem is a problem when podzolic soils are sown to pastures as the first stage of introducing an agroecosystem.

Pasture establishment on the *lateritic podzols* of

Western Australia became difficult when introduced subterranean clover (*Trifolium subterraneum* L) failed to nodulate. The failure was not associated with nutrient deficiency. Holland (1962, 1966) ascribed the inhibition of nodulation to marasmine production by fungi which colonized the copious accession of particulate organic residues which resulted from clearing the scrub. The residues could amount to as much as 8% by weight of the soil. The primary colonists were not the fungi which previously inhabited the rhizosphere of the native vegetation, but a new succession of species of *Mortierella*, *Penicillium* and asporogenous Phycomycetes. Later they gave way to a succeeding microflora of Ascomycetes and Basidiomycetes which appeared to be far less toxigenic (Holland and Parker 1966).

Only in a few cases were truly antibiotic species able to be demonstrated by Holland and Parker (1966). But where they were present, the active toxin was assimilated by the plant and became concentrated in the clover leaves. This induced an increase in anthocyanin and a decrease in chlorophyll production. The plants remained stunted and distinctively anthocyanosed to a purplish-red colour in both leaves and petioles. Although the primary root system was not seriously stunted, secondary lateral branching and root hair development was markedly inhibited. Nodulation was inhibited. Some of the primary fungal colonists of the debris were antagonistic to *Rhizobium trifolii*. Thus both partners of the nitrogen-fixing symbiosis could become affected adversely.

A similar establishment problem which occurs in South Australia on both *lateritic* and sandy *podzols* has been described by Harris (1961). Following emergence, a large number of subterranean clover plants remain unthrifty, anthocyanosed and poorly nodulated. The proportions of poor plants can be reduced and the nodulation can be improved by increasing the rate of application of phosphatic fertilizer, by liming, and by artificial inoculation of the clover seed with rhizobial culture. The ratio of poor plants to good plants was not reduced by applying nitrogenous fertilizer, so that the condition of the poor plants could not be ascribed to nitrogen starvation following nodulation failure.

There is little doubt that this unthriftiness in subterranean clover was associated with a residual soil toxicity which heightened following the clearing of the native vegetation. The abnormal clover nutrition could be relieved by increased rates of application of fertilizers which supplied the nutrients deficient in these soils. But the plants were left with an inefficient root system and suffered from a high incidence of damage by root pathogens. Ludbrook *et al.* (1953) have described the association of root-attacking fungi and nematodes which incited bare patches and unthriftiness in the new pastures of the Coonalpyn Downs of South Australia. The condition tended to abate after several years and to have been a problem associated mainly with the establishment phase of the pasture.

The peaks of soil toxicity provide something of a chicken and egg dilemma. This may be illustrated from a study by Geller and Dobrotvorskaya (1967) of sugar-beet production on grey-podzolic and chernozem soils in a Russian agricultural system. Soil toxicity depressed the germination and the growth of sugar-beet as well as the mineralization cycles of the soil microflora on soils which were virgin or rarely cultivated. Soils which were regularly fertilized and which were cropped in rotation showed little of this effect or it was much reduced.

This is the opposite situation to the soil fatigue associated with over-cropping in agroecosystems. But it does serve to emphasize that circumstances which set limits to available plant nutrients or which set limits to microbiological activity will exacerbate the apparent effects of soil toxicity in both types of situation.

Sometimes the growth of a plant is dependent upon the associated growth of micro-organisms. The symbiotic association of rhizobia in the leguminous rhizosphere or of mycorrhizal fungi with many plant species are cases where limits to plant growth will be set if the microbial component is adversely affected by soil toxicity. It is not always easy to determine which partner of the symbiosis is the most sensitive. The subterranean clover establishment problems in Western Australia and South Australia exemplify situations in which the major impact of soil toxicity has been upon the host plant. But in a different soil the effect could be expressed most severely upon the microbial partner with equally disastrous results.

In the New England district of New South Wales, poor nodulation of subterranean clover on *yellow podzolic soils* is due to the rhizobia inoculated onto the seed-coat failing to establish. This was ascribed by Hely *et al.* (1957) to microbial antagonism in the rhizosphere. An interesting aspect of this problem is that it did not occur on sites at which logs had been heaped and burned. Experimentally, it could be overcome by incorporating charcoal into the soil. The suggestion was made that the charcoal adsorbed unidentified factors toxic to the rhizobia. This provides an interesting parallel to Eliasson's (1959) experience with toxin adsorption by charcoal when added to culture solutions containing wheat roots. But whether the source of the toxicity was from rhizosphere micro-organisms, from the soil solution or from plant parts remains an open question in view of the findings of Thompson (1961) that the seed-coat of subterranean clover itself produces a phytoncidal exudate during germination. An important attribute of the success of lime pelleting of clover seed has been that it physically separates the inoculated rhizobia from the testa during this critical phase of germination, and thereby a better establishment of bacteria is achieved.

The problems of establishing agriculture have counterparts in the regeneration of forests. The rainforest ecosystems of Australia are natural plant communities from which certain tree species are selectively harvested. It is important that the timber resource be sustained by regeneration of the wanted species. There are examples in which this regeneration does not proceed readily and in which allelopathy is a significant component.

Hoop pine (*Araucaria cunninghamii*) is the dominant gregarious species of a number of rainforest formations in eastern Australia and in New Guinea. The regeneration of hoop pine in rainforest is extremely slow. The leachates from soil under a number of tree species were found by Bevege (1968) to reduce dry matter production and even cause mortality in hoop pine seedlings. The yield data of one of his experiments are given in Table 47.3. The cholines responsible were of unknown origin and were rather non-specific. This contrasts with the specific autotoxicity described by Webb *et al.* (1967) which occurred under another rainforest tree, the silky oak (*Grevillea robusta*). In both cases more than one choline may be involved. Not only was plant growth rate severely reduced (Cannon *et al.* 1961; Bevege 1968), but there were nutritional effects in which nitrogen, phosphorus and potassium metabolism appeared to be abnormal.

In one example of a forest regeneration problem from New South Wales, the source of soil toxicity has been identified by Evans *et al.* (1967) as a marasmine produced by a saprophytic fungus commonly found as a root-surface inhabitant of many plant species. The marasmine nectrolide is produced by *Cylindrocarpon radicicola* (with perfect stage *Nectria radicicola*). It may be identical with the antibiotic brefaldin A produced by a *Penicillium*. It arrested the growth of blackbutt (*Eucalyptus pilularis*) seedlings, including a severe stunting and blackening of the roots. The symptoms observed conformed to what has been described by Florence and Crocker (1962) to happen in a mature blackbutt forest.

The soil microflora and microfauna are integral components of natural edaphic ecosystems. Specific examples of constraints to microbiological activity incited by phytoncidal metabolites produced by higher plants are widespread and may be referred to in the papers of the Russian workers Krasilnikov and Garkina (1946), Krasilnikov *et al.* (1955), Krasilnikov (1958), Mirchink (1957), Mirchink and Greshnykh (1961), Mirchink *et al.* (1966). Yuzhina (1958*a*, *b*), Zhukova (1959, 1960), Yevdokimova (1962) and Geller and Dobrotvorskaya (1967); of the German workers Knoesel (1959) and Schönbeck (1968); of the British workers Brian *et al.* (1945) and Robinson (1972); of the United States workers Nickell (1959), Rice (1964, 1965), Floyd and Rice (1967), Abdul-Wahab and Rice (1967), Menzies and Gilbert (1967), Gilbert *et al.* (1969), Hendrix and Powell (1970), Israel *et al.* (1973) and Rychert and Skujins (1974); of the Canadian workers Basaraba (1964) and Neal (1969); of the Japanese workers Hatsuda *et al.* (1965); and of the Australian workers Hely *et al.* (1957) and Thompson (1961). This list is by no means exhaustive; but it is sufficient to pose the question of how often the limits to microbial growth in ordinary soils are set by phytoncidal allelopathy. This is an important question because it highlights the long-standing observation that fertile soils are microbiologically active compared to the much lower order of activity of infertile soils (Harris 1968).

Table 47.3. Mean dry weights (g per pot) of seedling growth of hoop pine (*Araucaria cunninghamii*)
Data from a pot experiment in which leachates from three rainforest tree species were applied to three soils. After Bevege (1968).

Leachate source	Basaltic red loam	Lateritic podzol	Sterilised quartz sand	Mean
Hoop-pine	1.90	1.26	2.27	1.81
Slash-pine	1.88	1.46	1.43	1.59
Crows ash	1.36	1.15	0.72	1.08
Control	2.74	3.32	4.25	3.44
Mean	1.97	1.80	2.17	

The first studies of soil toxicity to be carried out in Australia pursued the hypothesis that bacteriotoxins set the limits to growth. The soil microbiologist who conducted this research was Greig-Smith. He was the first appointee as Macleay Bacteriologist to the Linnean Society of New South Wales. His long series of papers (e.g. Greig-Smith 1910, 1911, 1915, 1918) recount his investigations of the soils of the Hawksbury Agricultural College and make fascinating reading. He demonstrated clearly that there was a build-up and a decline of toxins in the soil. This was associated with microbiological activity and he conducted experiments to determine which organisms might be most active in toxin production. He was interested in the formation of waxy layers which spread over the surface of soil particles and organic matter. This waterproofed them and rendered decomposition difficult. He called this layer the 'agricere'. Both toxin and agricere production fluctuated seasonally with accession of organic residues, with cultivation, with aeration, and with moisture regime. The removal of the agricere by organic solvents and the elimination of toxigenic bacteria by fumigants restored the soil microbiological activity to normal.

Greig-Smith introduced a population of *Bacillus prodigiosus* (*Serratia marcescans*) into soils to assess the toxicity. In toxic soils the introduced organisms were short-lived. The data of Table 47.4 are taken from one of his early experiments which demonstrated that thermolabile soil toxins could be inactivated by boiling a soil extract. After boiling they supported a higher population of the test organism. In the untreated extracts, many arable soils were relatively toxic to the test organism.

Fifty years later, microbiological techniques were still in use (Harris 1968) because they provide the most sensitive test organisms for equating the biological activity of

Table 47.4. An early experiment to demonstrate soil toxicity to bacteria

Serratia marcescens was introduced into aqueous extracts of Hawkesbury soils before and after heating at 94°C for 2h. Data are colony counts per 10 introduced cells. After Greig-Smith (1910).

	After 1 day		After 4 days	
	Unheated	Heated	Unheated	Heated
Soil 1 (rich alluvial soil)	16	741	5	2
Soil 2	19	56	10	2
Soil 3	15	712	87	594
Soil 4 (poor sandy soil)	33	19	20	117
Soil 5	36	119	30	66
Garden Soil	41	4015	125	650
Sand	75	—	52	—

a soil with its fertility. They remain the best basis for the assay of soil toxicity.

Induced toxicity in an agroecosystem

Monocultural agroecosystems provide many examples of toxicities which are induced by heavy accessions of the one sort of plant material, by invasion of aggressive weeds, by replant problems in vineyards and orchards, and by autotoxicities. The references of the chapter list a number of these. But in this section we will confine attention to a single important example, viz. straw toxicity.

The problem of coping with the straw residues from a cereal crop has prompted innumerable investigations into a many-sided question. Since the advent of the combine harvester, cereal crop residues have normally been returned to the soil. On erosion-prone soils and with short grain-fallow rotations, it is advantageous to retain the straw as a surface mulch. In continuous cropping or in grain-pasture rotations, the retention of stubbles and surface mulch makes it difficult to prepare a seed-bed, so that incorporation or heavy grazing or burning becomes necessary. In some parts of the world clean air regulations restrict or preclude burning as an option.

Whilst incorporation has the obvious advantage that it returns the nutrients of the straw into a mineralization cycle, this is difficult to achieve mechanically with the current trends towards minimum tillage farming operations or with heavy stubbles. If weed control requires the use of a pre-emergent herbicide which needs to be incorporated efficiently, trash in the surface layers becomes a serious disadvantage. There are also hazards that straw return will tie-up essential nutrients, or that it will release toxins.

The deleterious effects of the latter two hazards occur sporadically in most farming systems. There is considerable variation in the severity of the effects depending upon soil type, climate, the nature of the straw or stover, and the degree of decomposition it has undergone prior to incorporation. The greatest hazards are associated with the early stages of straw rotting.

The classical experiments of Winter ahd Schönbeck (1953*a*, *b*, 1954) had shown that cold water extracts of cereal straw were toxic for seed germination and for the development of cereal plants. These substances were liberated into the soil under stubbles and from incorpoated residues. Börner (1955*a*, *b*, 1956) demonstrated that the active substances were phenolic acids.

Phenolic constituents are widely distributed throughout the plant kingdom (Bates-Smith 1956), but the heavy residues of cereal crops may provide high accession rates of phenolics to arable soils and could become a hazard if toxic constituents persist. Their occurrence has now been reported from many arable soils, e.g. Whitehead (1964), Guenzi and McCalla (1962, 1966*a*, *b*), Wang *et al.* (1967), Rao *et al.* (1975) and Shindo *et al.* (1978). The hazard to gramineous crops is well authenticated and covers the range of the temperate cereals (McCalla *et al.* 1964; Börner 1956) through rice (Kuwatsuka and Shindo 1973; Rao *et al.* 1975); Shindo *et al.* (1978) to the tropical gramineous crops such as sugar-cane (Yang *et al.* 1966; Wang *et al.* 1967). Some data of the phenolic acid content of tropical soils under a range of crops is given in Table 47.5. These are all young crops so that the differences between phenolic acids within rows and between rows is ascribed by Wang *et al.* (1967) to phenolic acid production by the current crop as well as to residual phenolic acid from a previous crop.

The individual phenolic acid which occurs in the greatest amount is generally *p*-coumaric acid (Bates-Smith 1956; Guenzi and McCalla 1966*b*; Kuwatsuka and Shindo 1973), and this is probably because it occurs as the dominant phenolic constituent of cereal straw (Guenzi and McCalla 1966*a*).

The total amount of phenolic acids in a standing crop can be quite high. One estimate which has been given by Guenzi and McCalla (1966*a*) for the *p*-coumaric acid equivalents in the residue returns from four crops is shown in Table 47.6. Whilst the majority of phenolic acids do not exist free in the plant residues, it does seem

Table 47.5. Phenolic acid content of Taiwan surface soils

Data are the phenolic acid contents (10^{-2} μmole per 100 g soil) of surface soils in Taiwan plantations carrying four crops and sampled within and between the crop rows. After Wang *et al.* (1967).

	p-Hydroxy-benzoic	Vanillic	*p*-Coumaric	Total
Tobacco				
within rows	2.86	3.00	22.64	28.50
between rows	3.66	3.00	10.99	17.65
Sweet potato				
within rows	0.40	0.27	0	0.67
between rows	3.66	2.03	1.97	7.66
Pineapple				
within rows	2.86	1.37	28.31	32.54
between rows	3.66	1.60	44.62	49.88
Banana				
within rows	0.83	0.80	0	1.63
between rows	1.00	0.33	0	1.73

that the risks attendant upon the return of these amounts to the soil could be substantial under particular circumstances (Kuwatsuka and Shindo 1973). It has been pointed out by Cochran *et al.* (1977) that there are few valid estimates of the reduction in yield of field crops from phytotoxicity despite the amount of experimental work that has been carried out in glasshouse and laboratory experiments to substantiate the role of phenolic acids and to isolate them from 'fatigue' soils.

The data of Tables 47.5 and 47.6 suggest also that the toxicity hazard from some crops will be greater than from others, and that in some soils it will be greater than in others. This is borne out both by field observations (Cochran *et al.* 1977; Shindo *et al.* 1978) and by water culture experiments (Yang *et al.*, 1966; Wang *et al.*, 1967; Kimber, 1967, 1973*a*, *b*). The order in which preceding crops may be ranked for hazard will depend upon the amount of residues they produce, the conditions under which they are grown, whether or not they contain toxins other than phenolic acids, and the amount of rotting or weathering that takes place in the standing stubble prior to incorporation.

A number of observations show that weathered wheat straw is less toxic than fresh straw (Horricks 1964; Wang *et al.* 1967; Kimber 1967, 1973*a*, Ridky 1977; Cochran *et al.* 1977); but the peaks of toxicity that have been reported vary with the conditions of decomposition (Kuwatsaka and Shindo 1973; Shindo *et al.* 1978) and may be cyclic (Kimber 1973*a*, *b*; Cochran *et al.* 1977). In general, the peak occurs in the first day or two of decomposition, and will persist for some weeks. Not all observations reconcile easily because it is not always certain whether the demonstrated toxicity is due to the release of phenolic acids, to other toxins of unknown composition, or to byproducts of microbial growth.

Table 47.6. Total amounts of extractable phenolic acids in *p*-coumaric acid equivalent in the residues of four gramineous crops, assuming the indicated yields

Crop	Yield and residue	Phenolic acid
Sorghum	3.2 t ha^{-1} grain with 7.5 t ha^{-1} stover	79 kg ha^{-1}
Maize	5.4 t ha^{-1} grain with 7.5 t ha^{-1} stover	64 kg ha^{-1}
Oats	2.8 t ha^{-1} grain with 5.0 t ha^{-1} stover	21 kg ha^{-1}
Wheat	2.2 t ha^{-1} grain with 3.7 t ha^{-1} stover	7 kg ha^{-1}

The susceptibility of the subsequent crop has to be considered also. Ridky (1976) showed that barley residues had lost all of their toxicity to a subsequent planting of winter wheat after a period of three months. Nevertheless, such a soil still remained sufficiently toxic to reduce the germination of a more sensitive crop of spring barley.

Buried layers of wheat straw may affect different plants in different ways. The data of Table 47.7 from Iswaran and Harris (1968) showed that both the roots and the tops of subterranean clover were equally inhibited in terms of dry matter production over the range of rates of incorporation of decomposing straw. In lucerne, root growth tends to be inhibited more than top growth, whilst the reverse is true for oats.

When wheat straw is buried in a layer, toxicity has been demonstrated by Iswaran and Harris (1968) to be associated with an early flush of bacterial growth in which members of the families Enterobacteriaceae, Achromobacteriaceae and Bacillaceae dominate. An aerogenic fermentation supported by a suite of facultatively anaerobic bacteria takes place within the layer of straw. The environment for toxin production is within the straw layer; the soil type is immaterial. If the straw is rich in simple carbonaceous substrate, and Australian cereal straws generally are, the fermentation reaction is vigorous, and the straw layer becomes anaerobic within 24 h.

This is the condition which the Russian workers Mishustin and Erofeev (1965, 1966) have pointed out corresponds to maximum toxicity, both by failure to decompose phenolic acids and by the production of volatile aliphatic acids. This conforms also to the North American observations of Cochran *et al.* (1977) that toxicity is greatest in the low-lying and wetter portions of fields where high residue accessions are found.

Another type of microbially induced toxicity occurs in stubble-mulch farming systems where the toxicity follows the colonization of surface residues by a predominantly fungal population. This has been described by McCalla *et al.* (1963) and Norstadt and McCalla (1968, 1969) to be associated with production of the antibiotic patulin by the fungus *Penicillium urticae*. Maximum toxicity is reached more slowly in the stubble mulch than

Table 47.7. Reduction in dry matter production of lucerne, subterranean clover and oats.
Reduction caused by planting above layers of wheat straw at different rates. Mean weights in mg per pot (and percentage reduction) for plants growing in sand watered with nutrient solution.

Equivalent straw incorporation rate	Lucerne		Sub clover		Oats	
	Tops	Roots	Tops	Roots	Tops	Roots
Nil	3700	4566	6169	2295	4599	9859
490 kg ha^{-1}	2287 (38%)	2279 (50%)	2161 (65%)	785 (66%)	1852 (63%)	6015 (39%)
980 kg ha^{-1}	1245 (66%)	1401 (60%)	1746 (72%)	644 (72%)	1505 (70%)	7560 (23%)
1470 kg ha^{-1}	719 (81%)	739 (84%)	1252 (80%)	524 (77%)	1282 (74%)	3411 (65%)
1960 kg ha^{-1}	418 (89%)	631 (86%)	779 (87%)	342 (85%)	1006 (80%)	4119 (58%)

would occur in buried layers which support bacterial fermentations. The peak toxicity may not occur for 3–4 weeks, although initial cycling may occur. Again, toxicity tends to be highest at high soil moisture levels, and to persist as long as the residues sustain microbial growth.

There is another sort of toxicity which can occur when residues are high in nitrogen; that is ammonia toxicity. High concentrations of gaseous ammonia have been reported to be released in the vicinity of decomposing layers of forage legume stover, and lesser concentrations from corn stover and barley straw. The risk has been shown by Smith (1964, 1967) and Smith and Burns (1965) to arise when the nitrogen content of the residue is high enough to support vigorous ammonification and when decomposition is rapid.

Nutrient immobilization, particularly of nitrogen and of sulphur, has been regarded as a risk attendant upon incorporation of cereal straw residues. The role of nitrogen and the C:N ratio in straw decomposition have been reviewed in depth by Harmsen and van Schreven (1955) and by Allison (1966). Often there have been fears expressed that low-nitrogen residues in low-nitrogen status soils could cause nitrogen starvation in a succeeding crop. Much of the evidence for this comes from laboratory experiments. There have not been many field trials; but what evidence is available from them (Smith and Douglas 1971; Smith *et al.* 1973) supports the view that environmental conditions in the field are rarely ideal for rapid decomposition. Thus when decomposition proceeds slowly wheat straw low in nitrogen is able to decompose without requiring further nitrogen, and moreover is more likely to release nitrogen than to immobilize it. Under these conditions, toxicity could still be a greater hazard than immobilization (Kimber 1973*b*; Smith *et al.* 1973).

The evidence which has been brought together here suggests that it is sounder to consider straw toxicity as a suite of toxicities rather than as one specific sort of toxicity. The features that the toxicities have in common are that the hazards are greatest with: fresh (unweathered) residues and high-nitrogen residues; heavy stubbles or high rates of incorporation of residues; rapid microbial decomposition; anaerobic conditions which follow burial or arise from excessive moisture; and a sensitive succeeding crop following too close after incorporation, or planted in the incorporated layer.

These features characterize the hazards that are associated with almost any monocultural agroecosystem. It would seem that man's attempt to produce food, fibre and forage crops will always be attended by a discount attributable to soil toxicity. No longer do we consider it to be the major cause of infertility. Nevertheless, it constitutes a 'negative feed-back' in decomposition and mineralization cycles which is too great to be ignored. Occasionally it becomes a major determinant in soil processes. But only rarely does it meet the requirements of a single factor explanation in observed plant behaviour.

References

Abdul-Wahab, A.S. and Rice, E.L. (1967). Plant inhibition by Johnson grass and its possible significance in old field succession. *Bull. Torrey Bot. Club* **94**, 486-97.

Allison, F.E. (1966). The fate of nitrogen applied to soils. *Adv. Agron.* **18**, 219-58.

Amin, K.S. and Sequeira, L. (1966*a*). Role of certain soil factors in the etiology of corky root rot in lettuce. *Phytopathology* **56**, 1047-53.

Amin, K.S. and Sequeira, L. (1966*b*). Phytotoxic substances from decomposing lettuce residues in relation to the etiology of corky root rot of lettuce. *Phytopathology* **56**, 1054-61.

Anon. (1960). Peach and prune replanting problem. Ann. Rep. N.S.W. Dep. Agric. 1959-60, 82.

Basaraba, J. (1964). Influence of vegetable tannins on nitrification in soil. *Plant Soil* **21**, 8-16.

Bates-Smith, E.C. (1956). The commoner phenolic constituents of plants and their systematic distribution. *R. Dublin Sci. Soc. Proc.* **27**, 165-76.

Bevege, D.I. (1968). Inhibition of seedling hoop pine (*Araucaria cunninghamii* Ait.) on forest soils by phytotoxic substances from the root zones of *Pinus*, *Araucaria* and *Flindersia*. *Plant Soil* **29**, 263-73.

Bonner, J. (1950). The role of toxic substances in the interactions of higher plants. *Bot. Rev.* **16**, 51-65.

Börner, H. (1955*a*). Die Ausscheidung organischer Verbindungen aus den Samen von Roggen (*Secale cereale* L.), Weizen (*Triticum aestivum* L.) und Gerste (*Hordeum vulgare* L.) Während der Quellung. *Naturwiss.* **42**, 48.

Börner, H. (1955*b*). Untersuchungen über phenolische Verbindungen aus Getreidestroh und Getreiderückständen. *Naturwiss.* **42**, 583-4.

Börner, H. (1956). Die Abgabe organischer Verbindungen aus den Karyopsen, Wurzeln und Ernterückständen von Roggen (*Secale cereale.* L.), Weizen (Triticum aestivum L.) und Gerste (*Hordeum vulgare* L.) und ihre Bedeutung bei der gegenseitigen Beeinflussung der höhren Pflanzen. *Beitr. Biol. Pflanz.* **33**, 33-83.

Börner, H. (1960). Liberation of organic substances from higher plants and their role in the soil sickness problem. *Bot. Rev.* **26**, 393-424.

Boullard, B. and Moreau, R. (1962). 'Sol, Microflore et Vegetation'. (Masson et Cie: Paris.)

Brian, P. W. (1957). The Ecological Significance of Antibiotic Production. *In* 'Microbial Ecology'. 7th Symposium, Society of Microbiology. (University Press: Cambridge.)

Brian, P. W., Hemming, H. G. and McGowan, J. C. (1945). Origin of a toxicity to mycorrhiza in Wareham Heath soil. *Nature (London)* **155**, 637-8.

Cannon, J. N. R., Corbett, N. H., Haydock, K. P., Tracey, J. G. and Webb, L. J. (1961). Antibiotic effect in plant communities. *Nature (London)* **190**, 189-90.

Cochrane, V. W. (1948). The role of plant residues in the etiology of root rot. *Phytopathology* **38**, 185-96.

Cochran, V. L., Elliott, L. F. and Papendick, R. I. (1977). The production of phytotoxins from surface crop residues. *Soil Sci. Soc. Am. Proc.* **41**, 903-8.

Colbran, R. C. (1953). Problems in tree replacement. I. The root-lesion nematode *Pratylenchus coffeae* Zummerman as a factor in the growth of replant trees in apple orchards. *Aust. J. Agric. Res.* **4**, 384-9.

Collison, R. C. and Conn, H. J. (1925). The effect of straw on plant growth. New York (Geneva) Agric. Exp. Sta. Tech. Bull. No. 114. 35pp.

Curtis, J. T. and Cottam, G. (1950). Antibiotic and autotoxic effects in prairie sunflower, *Bull. Torrey Bot. Club* **77**, 187-91.

DeBell, D. S. (1970). Phytotoxins. New problems in forestry? *J. For.* **68**, 335-7.

del Moral, R. and Muller, C. H. (1969). Fog drip: a mechanism of toxin transport from *Eucalyptus globulus*. *Bull. Torrey Bot. Club* **96**, 467-75.

del Moral, R. and Muller, C. H. (1970). The allelopathic effects of *Eucalyptus camaldulensis*. *Am. Mid. Nat.* **83**, 254-82.

del Moral, R. and Gates, R. G. (1971). Allelopathic potential of the dominant vegetation of Western Washington. *Ecology* **52**, 1030-7.

del Moral, R., Willis, R. J. and Ashton, D.H. (1978). Suppression of coastal heath vegetation by *Eucalyptus baxteri*. *Aust. J. Bot.* **26**, 203-19.

Eliasson, L. (1959). Inhibition of the growth of wheat roots in nutrient solutions by subtances exuded from roots. *K. Lantbrukshögsk Ann.* **25**, 285-93.

Evanari, M. (1949). Germination inhibitors. *Bot. Rev.* **15**, 153-94.

Evanari, M. (1961). Chemical Influences of Other Plants (Allelopathy). Handbuch Pflanzenphysiologie Vol. 16, pp.691-736. (Springer Verlag: Berlin.)

Evans, G., Cartwright, J. B. and White, N.H. (1967). The production of a phytotoxin, nectrolide, by some root-surface isolates of *Cylindrocarpon radicicola*, Wr. *Plant Soil* **26**, 253-60.

Florence, R. G. and Crocker, R. L. (1962). Analysis of Blackbutt (*Eucalyptus pilularis* Sm.) seedling growth in blackbutt forest soil. *Ecology* **43**, 670-9.

Floyd, G. L. and Rice, E. L. (1967). Inhibitors of higher plants by three bacterial growth inhibitors. *Bull. Torrey Bot. Club* **94**, 125-9.

Geller, I. A. and Dobrotvorskaya, K. M. (1967). (Effect of soil toxicity on the vital activity of soil micro-organisms and productivity of sugar beets. – Eng. summary) *Mikrobiol. Zh. (Kiev)* **29**, 13-7.

Gilbert, R. G., Menzies, J. D. and Griebel, G. E. (1969). The influence of volatiles from alfalfa upon growth and survival of soil micro-organisms. *Phytopathology* **59**, 992-5.

Goodman, R. N. (1959). The Influence of Antibiotics on Plants and Plant Disease Control. *In* 'Antibiotics – Their Chemistry and Non-medical Use.' (Ed. H. S. Goldberg). pp.322-448. (Van Nostrand Company Inc.: New Jersey.)

Gray, R. and Bonner, J. (1948*a*). An inhibitor of plant growth from the leaves of *Encelia farinosa*. *Am. J. Bot.* **35**, 52-7.

Gray, R. and Bonner, J. (1948*b*). Structure determination and synthesis of a plant growth inhibitor, 3-acetyl-6-methoxybenzaldehyde, found in the leaves of *Encelia farinosa*. *J. Am. Chem. Soc.* **70**, 1249-53.

Greig-Smith, R. (1910). Contributions to our knowledge of soil fertility. I. The action of wax-solvents and the presence of thermolabile bacteriotoxins in soil. *Linn. Soc. N.S.W. Proc.* **35**, 808-22B.

Greig-Smith, R. (1911). Contributions to our knowledge of soil fertility. IV. The agricere and bacteriotoxins of soil. *Linn. Soc. N.S.W. Proc.* **36**, 679-99.

Greig-Smith, R. (1915). Contributions to our knowledge of soil fertility. XIII. The toxicity of soils. *Linn. Soc. N.S.W. Proc.* **40**, 631-45.

Greig-Smith, R. (1918). Contributions to our knowledge of soil fertility. XVI. The search for toxin-producers. *Linn. Soc. N.S.W. Proc* **43**, 142-90.

Grümmer, G. (1955). 'Die gegenseitige Beeinflussung höhrer Pflanzen, Allelopathie'. (Gustav Fischer Verlag: Jena.)

Grümmer, G. (1961). The Role of Toxic substances in the Interrelationships Between Higher Plants. *In* 'Mechanisms of Biological Competition'. (Ed. F.L. Milthorpe). S.E.B. Symposium No. 15, pp.219-28. (Cambridge University Press: Cambridge.)

Guenzi, W.D. and McCalla, T.M. (1962). Inhibition of germination and seedling development by crop residues. *Soil Sci. Soc. Am. Proc.* **26**, 456-8.

Guenzi, W.D. and McCalla, T.M. (1966*a*). Phenolic acids in oats, wheat, sorghum, and corn residues and their phytotoxicity. *Agron. J.* **58**, 303-4.

Guenzi, W.D. and McCalla, T.M. (1966*b*). Phytotoxic substances extracted from soil. *Soil Sci. Soc. Am. Proc.* **30**, 214-6.

Harmsen, G.W. and van Schreven, D.A. (1955). Mineralisation of organic nitrogen in soil. *Adv. Agron.* **7**, 299-398.

Harris, J.R. (1961). Rhizobial inoculation and fertilizer rates as factors in the establishment of subterranean clover in South Australia. *Aust. J. Agric. Res.* **12**, 84-99.

Harris, J.R. (1968). Techniques for assessing the net effect of microbial activities in soil. 9th Int. Cong. Soil Sci., Trans. Vol. 3, pp. 665-74.

Hatsuda, Y., Hamasaki, T., Nishimura, S., and Renbutsu, M. (1963). On the phytotoxic substances in pea roots. *J. Agric. Chem. Soc. (Japan)* **37**, 262-4.

Hatsuda, Y., Nishimura, S., Hamasaki, T., and Renbutsu, M. (1965). Inhibitory effects of some phytotoxic substances produced by plant roots upon the growth of higher plant seedlings and micro-organisms. *J. Fac. Agric. Tottori Univ.* **4**, 183-8.

Hely, F.W., Bergersen, F.J., and Brockwell, J. (1957). Microbial antagonism in the rhizosphere as a factor in the failure of inoculation of subterranean clover. *Aust. J. Agric. Res.* **8**, 24-44.

Hendrix, F.F. and Powell, W.W. (1970). Control of root pathogens in peach tree decline sites. Univ. Georgia Agric. Exp. Sta. Bull. No. 77, pp.28-9.

Holland, A.A. (1962). The Effect of Indigenous Saprophytic Fungi upon Nodulation and Establishment of Subterranean Clover. *In* 'Antibiotics in Agriculture'. (Ed. M. Woodbine). pp.147-64. (Butterworths: London.)

Holland, A.A. (1966). Studies on microbial antagonism in the establishment of clover pasture. I. The number of species and percentage frequency of occurrence of the major fungi. *Plant Soil* **25**, 238-48.

Holland, A.A. and Parker, C.A. (1966). Studies on microbial antagonism in the establishment of clover pasture. II. The effect of saprophytic soil fungi upon *Rhizobium trifolii* and the growth of subterranean clover. *Plant Soil* **25**, 329-40.

Horricks, J.S. (1964). Effect of wheat residue in soil on plant growth. Canad. Phytopathol. Soc. Proc. 30th Session, No. 31, p.13.

Israel, D.W., Giddens, J.E. and Powell, W.W. (1973). The toxicity of peach tree roots. *Plant Soil* **39**, 103-12.

Iswaran, V. and Harris, J.R. (1968). The bacteria of decomposing cereal straw, and their effects on plant growth. Ninth Int. Congr. Soil Sci. Trans. Vol. 3, pp.501-11.

Kimber, R.W.L. (1967). Phytotoxicity from plant residues. I. The influence of rotted wheat straw on seedling growth. *Aust. J. Agric. Res.* **18**, 361-74.

Kimber, R.W.L. (1973*a*). Phytotoxicity from plant residues. II. The effect of time of rotting of straw from grasses and legumes on the growth of wheat seedlings. *Plant Soil* **38**, 347-61.

Kimber, R.W.L. (1973*b*). Phytotoxicity from plant residues. III. The relative effect of toxins and nitrogen immobilization on the germination and growth of wheat. *Plant Soil* **38**, 543-55.

Knoesel, D. (1959). Über die Wirkung aus Pflanzenresten, freiwerden, phenolischer Substanzen auf Micro-organismen des Boden. *Z. Pflanzenernaehr. Düng Bodenkd.* **85**, 225.

Koch, L.W. (1955). The peach replant problem in Ontario. I. Symptomatology and distribution. *Can. J. Bot.* **33**, 450-60.

Krasilnikov. N.A. (1958). Soil micro-organisms and higher plants. Translated and published. National Science Foundation, Washington, (1961).

Krasilnikov, N.A. and Garkina, N.R. (1946). (Microbiological factors of soil exhaustion). *Mikrobiologiya* **15**, 109-14.

Krasilnikov, N.A., Korenyako, A.E., and Mirchink, T.G. (1955). (On the toxicosis of podsol soils). *Izv. AN. SSR, Ser. Biol.* **3**, 34-48.

Kuwatsuka, S. and Shindo, H. (1973). Behaviour of phenolic substances in the decaying process of plants. I. Identification and quantitative determination of phenolic acids in rice straw and its decayed produce by gas chromatography. *Soil Sci. Plant Nutr. (Japan)* **19**, 219-27.

Lehman, R.H. and Rice, E.L. (1972). Effect of deficiencies of nitrogen, potassium and sulfur on chlorogenic acids and scopoletin in sunflower. *Am. Midl. Natur.* **87**, 71-80.

Lockerman, R.H. and Putnam, A.R. (1979). Evaluation of allelopathic cucumbers (*Cucumis sativus*) as an aid to weed control. *Weed Sci.* **27**, 54-7.

Lodhi, M.A.K. (1976). Role of allelopathy as expressed by dominating trees in a lowland forest in controlling the productivity and pattern of herbaceous growth. *Am. J. Bot.* **63**, 1-8.

Loehwing, W.F. (1937). Root interactions of plants. *Bot. Rev.* **3**, 195-239.

Ludbrook, W.V., Brockwell, J. and Riceman, D.S. (1953). Bare-patch disease and associated problems in subterranean clover pastures in South Australia. *Aust. J. Agric. Res.* **4**, 403-14.

McCalla, T.M., Guenzi, W.D. and Norstadt, F.A. (1963). Microbial studies of phytotoxic substances in the stubble-mulch system. *Z. Allg. Mikrobiol.* **3**, 202-10.

McCalla, I.M. Guenzi, W.D. and Norstadt, F.A. (1964). Phytotoxic substances in stubble mulching. Eighth Int. Congr. Soil Sci., Bucharest, Rom. Vol. 3, pp.933-43.

McCalla, T.M. and Haskings, F.A. (1964). Phytotoxic substances from soil micro-organisms and crop residues. *Bacteriol. Rev.* **28**, 181-207.

McNaughton, S.J. (1968). Autotoxic feedback in relation to germination and seedling growth in *Typha latifolia. Ecology* **49**, 367-9.

McPherson, J.K. and Muller, C.H. (1969). Allelopathic effects of *Adenostoma fasciculatum*, Chamise, in the California chaparral. *Ecol. Monogr.* **39**, 177-98.

Martin, N. and Gottleib, D. (1952). The production and role of antibiotics in soils. III. *Phytopathology* **43**, 294-6.

Menzies, J.D. and Gilbert, R.G. (1967). Responses of the soil microflora to volatile components in plant residues. *Soil Sci. Soc. Am. Proc.* **31**, 495-6.

Mirchink, T.G. (1957). On the fungi causing toxicity of turf-podzol soil in various stages of cultivation. *Microbiology* (Eng. trans. *Mikrobiologiya*) **26**, 83-90.

Mirchink, T.G. and Greshnykh, K.P. (1961). Toxin formation in the soil by some fungal species of the genus *Penicillium. Microbiology* (Eng. trans. *Mikrobiolgiya*) **30**, 851-4.

Mirchink, T.G., Blagoveshchenskii, V.S. and Fedorov, V.A. (1966). Production and characteristics of the toxin of *Penicillium* sp. no. 19. *Microbiology* (Eng. trans *Mikrobiologiya*) **35**, 222-6.

Mirchink, T.G., Kopysskaya, F.G. and Greshnykh, K.P. (1962). Effect of soil fungi toxins on nitrogen and amino-acid content of plants. *Microbiology* (Eng. Trans. *Mikrobiologiya*) **31**, 546-51.

Mishustin, E.N. and Erofeev, N.S. (1965). Elimination of the nitrogen deficiency in soil by using straw as an organic fertilizer, *Microbiology* (Eng. trans. *Mikrobiologiya*) **34**, 927-32.

Mishustin, E.N. and Erofeev, N.S. (1966). Nature of the toxic compounds accumulating during the decomposition of straw in soil. *Microbiology* (Eng. trans. *Mikrobiologiya*) **35**, 126-8.

Moje, W. (1966). *In* 'Diagnostic Criteria for Plants and Soils'. (Ed. H.D. Chapman). Chapt. 37, pp.533-69 (Univ. Calif.: Berkeley.)

Molisch, H. (1937). 'Der Einfluss einer Pflanze auf die andere – Allelopathie' 106pp. (Gustav Fischer Verlag: Jena.)

Mountain, W.B. and Boyce, H.R. (1958). The peach replant problem in Ontario. VI. The relation of *Pratylenchus penetrans* to the growth of young peach trees. *Can. J. Bot.* **36**, 135-51.

Mountain, W.B. and Patrick, Z.A. (1959). The peach replant problem in Ontario. VII. The pathogenicity of *Pratylenchus penetrans* (Cobb, 1971) Filip and Stek 1941. *Can. J. Bot.* **37**, 459-70.

Muller, C.H. (1953). The association of desert annuals with shrubs. *Am. J. Bot.* **40**, 53-60.

Muller, C.H. (1966). The role of chemical inhibition (allelopathy) in vegetational composition. *Bull. Torrey Bot. Club* **93**, 332-51.

Muller, C.H. (1969). Allelopathy as a factor in ecological process. *Vegetatio* **18**, 348-57.

Muller, C.H. (1970). Phytotoxins as plant habit variables. *Recent Adv. Phytochem.* **3**, 105-21.

Muller, C.H. (1974). Allelopathy in the Environmental Complex. *In* 'Handbook of Vegetation Science, Part 6. Vegetation and Environment'. (Eds B.R. Strain and W.D. Billings). pp.71-85. (Junk BV.: The Hague.)

Muller, W.H. and Muller, C.H. (1956). Association patterns involving desert plants that contain toxic products. *Am. J. Bot.* **43**, 354-61.

Neal, J.L., Jr. (1969). Inhibition of nitrifying bacteria by grass and forb root extracts. *Can. J. Microbiol.* **15**, 633-5.

Neill, R.L. and Rice, E.L. (1971). Possible role of *Ambrosia psilostachya* on pattern succession in old fields. *Am. Midl. Natur.* **86**, 344-57.

Nickell, L.G. (1959). Antimicrobial activity in vascular plants. *Econ. Bot.* **13**, 281-318.

Norstadt, F.A. and McCalla, T.M. (1968). Microbially induced phytotoxicity in stubble-mulched soil. *Soil Sci. Soc. Am. Proc.* **32**, 241-5.

Norstadt, F.A. and McCalla, T.M. (1969). Microbial populations in stubble-mulched soil. *Soil Sci.* **107**, 188-93.

Patrick, Z.A. (1955). The peach replant problem in Ontario. II. Toxic substances from microbial decomposition products of peach root residues. *Can. J. Bot.* **33**, 461-86.

Patrick, Z.A. and Koch, L.W. (1958). Inhibition of respiration, germination and growth by substances arising during the decomposition of certain plant residues in the soil. *Can. J. Bot.* **36**, 621-47.

Patrick, Z.A., Toussoun, T.A. and Snyder, W.C. (1963). Phytotoxic substances in arable soils associated with decomposition of plant residues. *Phytopathol.* **53**, 152-61.

Patrick, Z.A., Toussoun, T.A. and Koch, L.W. (1964). Effect of crop residue decomposition products on plant roots. *Annu. Rev. Phytolpathol.* **2**, 267-92.

Pinck, L.A., Soulides, D.A. and Allison, F.E. (1961). Antibiotics in soils. II. Extent and mechanism of release. *Soil Sci.* **91**, 94-9.

Putnam, A.R. and Duke, W.B. (1978). Allelopathy in agroecosystems. *Annu. Rev. Phytopathol.* **16**, 431-51.

Rademacher, B. (1957). Die Bedeutung allelopathischer Erscheinungen in der Pflanzenpathologie. *Z. Pflanzenkr. (Pflanzenpathol.) Pflanzenschutz.* **64**, 427-39.

Rademacher, B. (1959). Gegenseitige Beeinflussung höhrer Pflanzen. *In* 'Handbuch Pflanzenphysiologie'. Vol. 11, pp. 655-706. (Springer Verlag: Berlin.)

Rao, A.V., Pajaram, K.P., Siddaramappa, R., and Sethunathan, N. (1975). Phenols in rice soils. *Soil Biol. Biochem.* **7**, 227-9.

Rice, E.L. (1964). Inhibition of nitrogen-fixing and nitrifying bacteria by seed plants (I). *Ecology* **45**, 824-37.

Rice, E.L. (1965). Inhibition of nitrogen-fixing and nitrifying bacteria by seed plants (II). Characterization and identification of inhibitors. *Physiol. Plant.* **18**, 255-68.

Rice, E.L. (1974). 'Allelopathy'. (Academic Press: New York.)

Ridky, K. (1976). Fytotoxicita zbytku jecmene pri ruznem zpracovani pudy. *Rostlinna Vyroba, Praha*, **22**, 525-35.

Robinson, R.K. (1972). The production by roots of *Calluna vulgaris* of a factor inhibitory to growth of some mycorrhizal fungi. *J. Ecol.* **60**, 219-24.

Rychert, R.C. and Skujins, J. (1974). Nitrogen fixation by blue-green algae-lichen crusts in the Great Basin Desert. *Soil Sci. Soc. Am. Proc.* **38**, 768-71.

Savory, B.M. (1966). Specific replant diseases causing root necrosis and growth depression in perennial fruit and plantation crops. Res. Rev. No. 1. (Cwlth. Bur. Hort. Plant. Crops; Cwlth. Agr. Bur.: Farnham Royal.)

Schönbeck, F. (1968). Untersuchungen zur Bedeutung von Hemmstoffen im Boden. I. Originäre Pflanzenstoffe. *Z. Pflanzenkrank (Pflanzenpathol.) Pflanzenschutz* **75**, 193-213.

Schreiner, O. and Reed, H.S. (1907*a*). The production of deleterious excretions by roots. *Bull. Torrey Bot. Club* **34**, 279-303.

Schreiner, O. and Reed, H.S. (1907*b*). Some factors influencing soil fertility. U.S. Dep. Agric. Bur. Soils Bull. No. 40.

Schreiner, O. and Reed, H.S. (1908). The toxic action of certain organic plant constituents. *Bot. Gaz.* **45**, 73-102.

Schreiner, O. and Shorey, E.C. (1908*a*). The isolation of picoline carboxylic acid from soils and its relation to soil fertility. *J. Am. Chem. Soc.* **30**, 1295-307.

Schreiner, O. and Shorey, E.C. (1908*b*). The isolation of dihydroxystearic acid from soils. *J. Am. Chem. Soc.* **30**, 1599-607.

Schreiner, O. and Shorey, E.C. (1909). The isolation of harmful organic substances from soils. U.S. Dep Agric. Bur. Soils Bull. No. 53.

Schreiner, O. and Shorey, E.C. (1910). Chemical nature of soil organic matter. U.S. Dep. Agric. Bur. Soils Bull. No. 74.

Schreiner, O. and Skinner, J. J. (1912). The toxic action of organic compounds as modified by fertilizer salts. *Bot. Gaz.* **54**, 31-48.

Schreiner, O. and Sullivan, M.X. (1909). Soil fatigue caused by organic compounds. *J. Biol. Chem.* **6**, 39-50.

Shindo, H. and Kuwatsuka, S. (1976). Behavoiur of phenolic substances in the decaying process of plants. IV. Adsorption and movement of phenolic acids in soils. *Soil Sci. Plant Nutr. (Japan)* **22**, 23-33.

Shindo, H., Ohta, S., and Kuwatsuka, S. (1978). Behaviour of phenolic substances in the decaying process of plants. IX. Distribution of phenolic acids in soils of paddy fields and forests. *Soil Sci. Plant Nutr. (Japan)* **24**, 233-43.

Simonoff, P. and Gottlieb, D. (1951). The production and role of antibiotics in soils. I. The fate of streptomycin. *Phytopathology* **41**, 420-30.

Skinner, J.J. (1918). Soil aldehydes – a scientific study of a new class of soil constituents unfavourable to crops, their occurrence, properties and elimination in practical agriculture. *J. Franklin Inst.* **186**, 165-86, 289-316, 547, 584, 723-41, 1111-6.

Skinner, J.J. and Noll, C.F. (1916). Field tests of fertilizer action on soil aldehydes. *J. Am. Soc. Agron.* **8**, 273-98.

Smith, J.H. (1964). Relationships between cation-exchange capacity and the toxicity of ammonia to the nitrification process. *Soil Sci. Soc. Am. Proc.* **28**, 640-4.

Smith, J.H. (1967). Nitrogen gradients and nitrification associated with decomposing corn plants and barley straw in soil. *Soil Sci. Soc. Am. Proc.* **31**, 377-9.

Smith, J.H. and Burns, G.R. (1965). Ion gradients and nitrification associated with decomposition of a plant material layer in soil. *Soil Sci. Soc. Am. Proc.* **29**, 179-81.

Smith, J.H. and Douglas, C.L. (1971). Wheat straw decomposition in the field. *Soil Sci. Soc. Am. Proc.* **35**, 269-72.

Smith, J.H. Douglas, C.L. and LeBaron, J.J. (1973). Influence of straw application rates, plowing dates, and nitrogen applications on yield and chemical composition of sugarbeets. *Agron. J.* **65**, 797-800.

Soulides, D.A. (1964). Antibiotics in soils. VI. Determination of microquantities of antibiotics in soils. *Soil Sci.* **97**, 286-9.

Soulides, D.A., Pinck, L.A. and Allison, F.E. (1961). Antibiotics in soils. III. Further studies on release of antibiotics from clays. *Soil Sci.* **92**, 90-3.

Soulides, D.A., Pinck, L.A. and Allison, F.E. (1962). Antibiotics in soils. V. Stability and release of soil-adsorbed antibiotics. *Soil Sci.* **94**, 239-44.

Story, R. (1967). Pasture patterns and associated soil water in partially cleared woodland. *Aust. J. Bot.* **15**, 175-87.

Thompson, J.A. (1961). Studies on nodulation responses to pelleting of subterranean clover seed. *Aust. J. Agric. Res.* **12**, 578-92.

Toussoun, T.A. and Patrick, Z.A. (1963). Effect of phytotoxic substances from decomposing plant residues on root rot of bean. *Phytopathology* **53**, 265-70.

Toussoun, T.A., Weinhold, R.A., Linderman, R.G. and Patrick, Z.A. (1968). Nature of phytotoxic substances produced during plant residue decomposition in soil. *Phytopathology* **58**, 41-5.

Tukey, H.B. Jr. (1966). Leaching of metabolites from above-ground plant parts and its implications. *Bull. Torrey Bot. Club* **93**, 395-401.

Tukey, H.B. Jr. (1969). Implications of allelopathy in agricultural plant science. *Bot. Rev.* **35**, 1-15.

Wang, T.S.C., Yang, T.K., and Chuang, T.T. (1967). Soil phenolic acids as plant growth inhibitors. *Soil Sci.* **103**, 239-46.

Ward, G.M. and Durkee, A.B. (1956). The peach replant problem in Ontario. III. Amygdalin content of peach tree tissues. *Can. J. Bot.* **34**, 419-22.

Webb, L.J., Tracey, J.G. and Haydock, K.P. (1961). The toxicity of *Eremophila mitchelii* Benth. leaves in relation to the establishment of adjacent herbs. *Aust. J. Sci.* **24**, 244-5.

Webb, L.J., Tracey, J.G. and Haydock, K.P. (1967). A factor toxic to seedlings of the same species associated with living roots of the non-gregarious subtropical rainforest tree *Grevillea robusta* A. Cunn. *J. Appl. Ecol.* **4**, 13-25.

Went, F.W. (1942). The dependence of certain annual plants on shrubs in Southern California deserts. *Bull. Torrey Bot. Club* **69**, 100-14.

Went, F.W. (1956). Some Aspects of Plant Research in Australia. A Report on a Visit to Australia. (CSIRO: Melbourne.)

Went, F.W. (1970). Plants and the Chemical Environment. *In* 'Chemical Ecology'. (Eds E. Sondheimer and J.B. Simone). pp.71-82. (Academic Press: New York.)

Whitehead, D.C. (1964). Identification of p-hydroxybenzoic, vanillic, p-coumaric and ferulic acids in soils. *Nature (London)* **202**, 417-8.

Whittaker, R. H. and Feeny, P. P. (1971). Allelochemics: chemical interactions between species. *Science* **171**, 757-70.

Winter, A.G. and Schönbeck, F. (1953*a*). Untersuchungen über die Beeinflussung der Keimung und Entwicklung von Getreidesamen durch Kaltwasserauszüge aus Getreidestroh. *Naturwiss.* **40**, 168-9.

Winter, A.G. and Schönbeck, F. (1953*b*). Untersuchungen über den Einfluss von Kaltwasserextrakten aus Getreidestroh und anderer Blattstreu auf Wurzelbildung und -wachstum. *Naturwiss.* **40**, 513-4.

Winter, A.G. and Schönbeck, F. (1954). Untersuchungen über wasserlosliche Hemmstoffe aus Getreideboden. *Naturwiss.* **41**, 145-6.

Winter, A.G. late, (1961). New Physiological and Biological Aspects in the Inter-relationships Between Higher Plants. *In* 'Mechanisms of Biological Competition'. (Ed. F.L. Milthorpe). S.E.B. Symposium No. 15, pp.229-44. (Cambridge University Press: Cambridge.)

Woods, F.W. (1960). Biological antagonisms due to phytotoxic root exudates. *Bot. Rev.* **26**, 546-69.

Wright, J.M. and Grove, J.F. (1957). The production of antibiotics in soil. V. Breakdown of griseofulvin in soil. *Ann. Appl. Biol.* **45**, 36-43.

Yang, T.K., Wang, T.S.C. and Tsai, M.B. (1966). Effect of soil phenolic acids upon young plant growth. *Rep. Taiwan Sugar Exp. Sta.* No. 41, pp.9-18.

Yevdokimova, T.I. (1962). Toxicity of some mosses and lichens to *Azotobacter* in Karelian soils. *Soviet Soil Sci.* (Eng. trans. *Pochvovedeniye*) **8**, 852-5.

Yuzhina, Z.I. (1958*a*). Survival of *Azotobacter* in cultivated and virgin soils of Kola Peninsula. *Microbiology* (Eng. trans. *Mikrobiologiya*) **27**, 199-203.

Yuzhina, Z.I. (1958*b*). Relationship between toxic properties of soil of Kola Peninsula and numbers of bacterial antagonists of *Azotobacter*. *Microbiology* (Eng. trans. *Mikrobiologiya* (**27**, 452-6.)

Zhukova, R.A. (1959). Toxicity of Kola Peninsula soils for aerobic cellulose bacteria. *Microbiology* (Eng. trans. *Mikrobiologiya*) **28**, 844-50.

Zhukova, R.A. (1960). Role of the biological factor in toxicity of Kola Peninsula soil for aerobic cellulose bacteria. *Microbiology* (Eng. trans. *Mikrobiologiya*) **29**, 220-8.

Zhukova, R.A. (1964). Soil toxicity for plants on Kola Peninsula. *Soviet Soil Science* (Eng. trans. *Pochvovedeniye*) **9**, 957-63.

Soils and land management

Chapter 48
Soils and land use

C. H. Thompson, A. W. Moore and K. H. Northcote

SOIL IS THE MAJOR COMPONENT OF LAND. CONSEQUENTLY it influences land use and any changes from one land use to another. Conversely land use, and changes in it, affect soil, and different soils respond differently to land use change.

In this chapter, some interrelations between soils and land use are considered briefly. In a schematic, conceptual (and very simplistic) way the relations between classes of land use are indicated in Fig. 48.1. 'Use' has an anthropomorphic connotation as far as land is concerned; however, it is obvious that land in a natural, unused state is being used by elements of the biota other than man.

The boundaries between the land use classes of Fig. 48.1 tend to be tension zones, where drastic changes in use have a correspondingly large effect on soils. Although changes may occur in either direction across any boundary, the progression from natural to rural to urban tends to occur as a chronological trend that is largely irreversible because of the increasing input of capital (= energy). Within each broad land use, other changes also occur at what is usually considered to be management level.

In this chapter, we are concerned largely with aspects of this natural to rural to urban progression; in the past, soils have played a lesser role in recreational and extractive uses, so these are only mentioned in passing. In the brief history of European settlement in Australia, the natural to rural transition was the dominant feature until 30–40 years ago when the rural to urban transition began to assume considerable importance. In the section of this chapter where we are concerned with the influence of soils on land use, it is largely the first transition that is involved. In the following section on the effect of changing land use on soils, both transitions are considered.

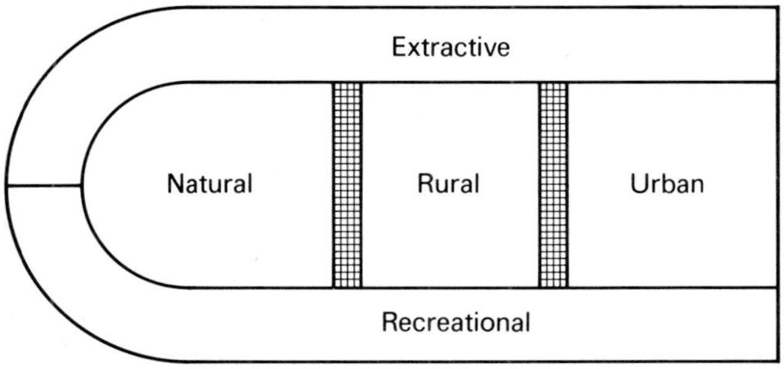

Fig. 48.1. General relationships between classes of land use.

Man's capacity to change soils has increased enormously since European settlement began in Australia. In the beginning, the areas affected were small and the equipment used was limited to hand implements. Horse-drawn implements were ushered in with the plough in 1795, and broadacre cultivation began with the development of new machinery for wheat-growing after 1840. The tractor era began in the 1920s, culminating in the bulldozer and large earth-moving machinery of the present day. We now have the capacity to drastically change or destroy thousands of years of natural soil development in a few minutes, and over large areas in a few months. We can also make, and have made, large changes to our soils by application of chemicals and fertilizers and through the introduction of new plants and animals. Some of these changes are briefly discussed in this chapter.

Because the area of climatically favoured soils in Australia is limited, and because our capacity to destroy soils is so great, we must take greater care in land use decisions in the future than we have in the past. Protection of our soil and water resources, and the establishment of soil priorities for some forms of use, are necessary if future generations are to have the opportunity to enjoy reasonable living standards. Making better use of these resources in both rural and urban areas is discussed.

Soil effects on land use

Environmental background

Australia extends from 10^0 to 43^0S latitude, reaches elevations of 1–2000m along the ranges of the eastern coast, and has an arid or semi-arid environment over 70% of its area. Combined thermal and moisture environments favourable to improved pasture and rainfed agriculture are restricted to a narrow coastal belt, less than 150km wide, extending from Cairns to Adelaide and from Perth to Albany (Fitzpatrick and Nix 1970).

Soils over a large part of the continent have been formed and severely leached under past warm wet climates, and large areas of deeply weathered materials have persisted with little modification by subsequent soil formation; it is only where the streams have cut through to fresh rock that reasonably fertile soils have formed. This is particularly evident through the hilly country of eastern Australia, where a great diversity of soils occurs (Northcote *et al.* 1975), having their character and inherent fertility closely related to the underlying rocks. The same rocks are also indirectly responsible for the fertile soils of the floodplains and younger terraces of the coastal streams, and of the large floodplains, levees and fans of some inland rivers, e.g. the Darling and Murray River systems.

Widespread low phosphorus levels have severely restricted biological activity in many of the soils; as a consequence the surface horizons over large areas are low in organic matter, and are massive and hard-setting when dry. In addition, many of the subsoils have high clay contents, coarse structure and low permeability, and may contain appreciable amounts of sodium salts.

Large tracts of poor soils and an absence of outstanding indigenous food plants are not conducive to settled agriculture, and the Australian aborigines remained nomadic. Their widespread use of fire was probably the major effect on the land during 40 000 years of occupation; it accelerated erosion by destroying surface cover, and so further reduced the low inherent fertility.

Although many of the ecosystems appear to have reached near-equilibrium with the Australian environment before settlement, most of the hillslopes were graded to higher base levels of earlier periods and their stability was largely dependent on their vegetative cover. Thus much of the landscape was poised for rapid change (latent instability, Peltier 1950) when the protecting vegetation was removed or the plant cover altered.

This then was the environmental situation into which the first white settlers stepped in 1788, bringing with them a very limited knowledge of agriculture, for few had rural backgrounds, and the new advances in English agriculture initiated by the industrial revolution were only just beginning (King 1948).

Soil constraints on early settlement

Most early settlements along the Australian coast were made on soils of low fertility, the main exception being Adelaide where a large area of moderately fertile *red-brown earths* extends to the coast. Very low levels of available phosphorus and nitrogen were the main plant nutrient constraints, although deficiencies of trace elements and an inadequate supply of potassium restricted farming in some areas. There were no fertilizers, and animal manures were not available, so the early settlers were heavily dependent on what little fertility the virgin soils possessed.

Agriculture was also constrained by the poor physical properties of the soils, the hard-setting surfaces and tough clay subsoils of many *podzolic soils* and *soloths* impeding water entry and cultivation. The lateral rooting habit of most Australian hardwoods, and the persistence of tree stumps, seriously restricted the use of ploughs. Although the plough was first used in 1795, hand hoeing was the normal form of cultivation until about 1820; in fact the axe, hoe and mattock remained important agricultural implements on pioneering coastal farms well into the twentieth century.

Farming began in 1788 with the sowing of about four hectares of barley and some wheat on podzolic *lithosols* and *podzols* at Farm Cove, Sydney. The crop failed and a second Public Farm was set up at Rosehill near Parramatta on *red podzolic soils*. It initially produced good crops of wheat, barley and vegetables, but grain yields soon declined, and within three years it was producing little more than the seed sown. James Ruse, the first individual farmer, was also settled at Parramatta in 1789, and successfully grew wheat, maize and vegetables. His opinion of the *red podzolic soil* was that "it was middling . . . it will make do with manure but without cattle it will fail" (King 1948). By 1795, the small area of reddish *chocolate* and *prairie soils* on dolerite at Prospect, the *alluvial* and *prairie soils* along the Hawkesbury-Nepean River, and the *krasnozems* of Norfolk Island, had been found to be the most productive soils in the colony. Five zones of land were soon recognized between the coast and the Nepean River (Wentworth 1819), approximating the *lithosol/podzol*, *yellow podzolic*, *lateritic podzolic*, *red podzolic* and *alluvial* soil associations of Walker (1960).

By 1840, trial and error farming in the coastal areas had established marked variations in productivity between soils on different rock types and under different vegetation. Some of the soils were incapable of producing crops without added plant nutrients and well-distributed rain, e.g. *podzols* and podzol intergrades around Sydney, Perth, and Port Phillip. On other soils, yields had markedly declined within two to three years of cropping, e.g. *red and yellow podzolic soils* and *soloths* around Sydney, Melbourne and Brisbane. The only fertile soils capable of producing a succession of good crops had been found to be the *alluvial* and *prairie soils* of the river floodplains, e.g. Hawkesbury, Hunter and Yarra Rivers, and the *krasnozems*, *chocolate* and *prairie soils*, and *chernozems* on basic volcanic rocks, e.g. Prospect Hill, Norfolk Island, Derwent Valley and Portland. Even on these soils productivity was declining with continued cropping.

The widespread low fertility of the coastal soils had several effects on early agriculture. It was largely, but not solely, responsible for the settlement at Sydney failing to become self-sufficient in grain. Farmers on soils with low inherent fertility could take two or three crops only before yields declined and they were forced to clear new land or change to grazing; many moved to new districts, thus practising a form of shifting cultivation. Trial-and-error farming built up experience with different soils so, as settlement advanced, soils of proven productivity were in great demand, while those of low fertility were neglected or left to grazing and forestry. The result was a patchwork of settlement, with agriculture along most of the river valleys being interspersed with large areas of pastoral and unused land.

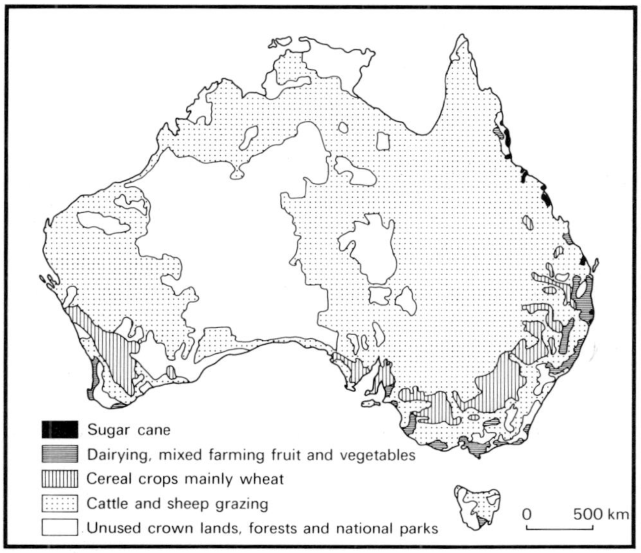

Fig. 48.2. Distribution of main farmland and grazing areas. (Based on "Land Use" in Atlas of Australian Resources, 2nd Ser., Dep. Minerals and Energy, Canberra, 1973.)

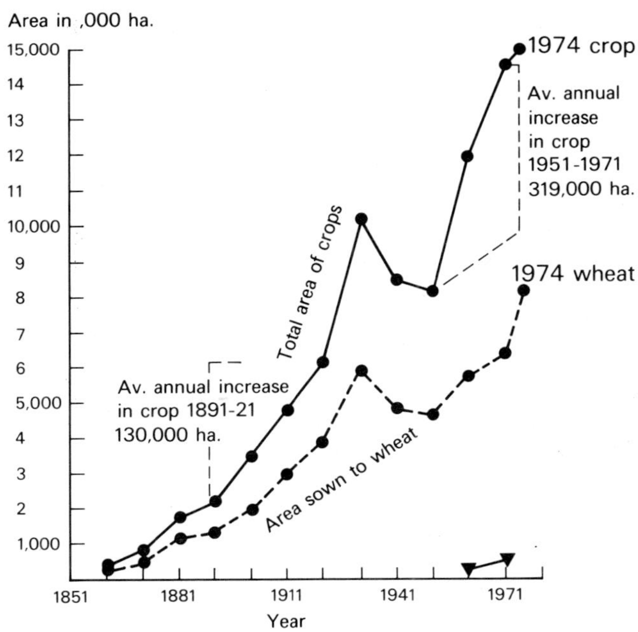

Fig. 48.3. Areas sown to wheat (●— —●) and to all crops (●——●) (including pastures for both hay and seed) in Australia 1861-1971, and in tropical Australia 1961-71. (▼——▼). (From Official Year Books, Commonwealth of Australia 1953, 1963, 1975-6; Dunsdorfs 1956; and Courtenay 1978.)

Because of the widespread low fertility and lack of fertilizers, gross production during the first hundred years could be increased only by expanding the area of crop, so early agriculture was wholly exploitative. Small amounts of wood ashes, seaweed and shellgrit were applied to some areas in attempts to maintain productivity. As domestic stock numbers increased, some farmyard manure was used, and sheep were folded on some fields used for crops. Small amounts of local guano and basic slags were tried from 1850, but it was not until superphosphate came into general use after 1910 that the farmers were able to increase production by means other than expanding the area of cultivation.

The greater part of Australia was first settled for sheep and cattle grazing on native grasses, and most of the semi-arid grassy woodlands and grasslands have remained in pastoral use (Fig. 48.2). However, the clearing and cultivation of the farmlands of eastern and southern Australia, beginning with about four hectares in 1788 and reaching some 15 million hectares in 1978, have been mainly associated with cereal crops, dairying and mixed farming, and sugar cane. Slow expansion in these three industries occurred in the 1840–90 period with the establishment of export markets and the development of new technology in cultivation, harvesting, processing and transport. Although several other industries are locally important, e.g. horticulture, these three have dominated the development of our farmlands in the areas suited to rainfed agriculture.

The total area sown to crops (Fig. 48.3) shows an average annual increase of 130 000 ha for the 1891–1921 period, due mainly to the expanding wheat industry and its improved machinery. It also shows the marked increase in cultivation associated with post-war expansion (1921–1931) and the pronounced decline during the subsequent depression and Second World War periods. A very marked average increase of 320 000 ha annually is evident since 1951, due to the widespread use of tractors and bulldozers, as large areas of the wheat belt in Western Australia, Eyre Peninsula, north-western New South Wales, and the brigalow lands in Queensland came under the plough. Even so, 96% of the cultivated lands still lie south of the Tropic of Capricorn. Soil, climate and transport problems have made farming in tropical Australia impracticable, except for favoured areas along the Queensland coast and adjoining hinterland where some half million hectares were cultivated in 1971 (Courtenay 1978).

Wheat

Since the early years of settlement wheat has been Australia's most important agricultural crop (Fig. 48.3).

In 1974, 8.3 million hectares were sown in the six states, representing 55% of the total area used for all crops (Anon. 1977). It is grown extensively in the semi-arid to sub-humid areas of southern and eastern Australia that lie between the 250 and 625 mm rainfall isohyets south of the Tropic and receive 150–400 mm of rain during the May to October growing season (Nix 1975). Wheat is grown on a great variety of soils, ranging from sands to cracking clays, encompassing a very wide range of physical characteristics and inherent fertility. It is grown on soils of most of the Australian Great Soil Groups, but eight – *red-brown earths, solodized solonetz and solodic soils, solonized brown soils, yellow earths, red earths, black earths, grey, brown* and *red clays*, and *earthy sands* – are of major importance (McGarity 1975). Economic production from such a variety of soils is dependent on varying cultivation practices, fertilizer inputs, rotational cropping systems, farm machinery used and size of the farming unit, according to the soil properties controlling storage and availability of soil water.

Wheat growing began in the coastal districts around each of the new settlements, generally on poor sandy soils; the crops failed or did poorly. As farming experience grew, there was a move to the more fertile areas along the river flats, to the friable soils on volcanic rocks, and to *red podzolic soils, non-calcic brown soils* and *red-brown earths* on sedimentary and metamorphic rocks in the coastal districts. The levels of organic matter and available phosphorus in these soils were low, and soon declined with continuous cropping, leading to severely reduced yields.

The settlement on *red-brown earths* in South Australia provided entry near the coast to the semi-arid to sub-humid lands of the present Australian wheat belt. With suitable soils and climate, and the development of new machinery for harvesting and cultivation, the industry expanded rapidly, bringing prosperity to the farming community, and for the first time demonstrating the viability of Australian agriculture. With the development of railways, wheat growing spread across the *red-brown earths* of south-eastern Australia and, as machinery improved, onto the *grey, brown and red clays* and *black earths* of Victoria, New South Wales and Queensland. It also moved into drier, climatically marginal areas, e.g. on *solonized brown soils* of the mallee lands and onto *desert loams* in northern South Australia, the *earthy sands* in Western Australia, the *red earths* in western New South Wales, and onto *red-brown earths* of the Western Downs in Queensland. In many of these areas, the rainfall was too low or too unreliable for farming to be successful using management techniques then available, and many farms were abandoned and later amalgamated for grazing.

Under dry farming, the highest yields and best quality grain came from the *black earths* and *grey, brown and red clays*. With continuous cropping, both quality and yields declined on most other soils, and farming techniques were changed to counter falling yields. Thus from 1890, bare fallowing, to store moisture and increase available nitrogen through mineralization, and after 1910, the use of superphosphate, led to a marked improvement. The introduction of fertilized grass-legume leys into the farm rotation, using subterranean clover on acid soils and lucerne or medics on calcareous soils and in drier areas, has made dramatic changes in soil fertility and wheat yields since 1930 (Donald 1965). Yield and quality of grain are still improving through most of the southern wheatlands as the crop rotation system becomes better understood, more widely used, and modified to suit a wider range of soil properties.

Using fertilized ley-crop rotation systems and modern machinery, wheat growing has returned and become consolidated in many of the areas previously regarded as marginal (Dahlke 1975). In recent years, it has also spread further into drier areas, especially on the very light sandy soils of Wester n Australia, Eyre Peninsula, and the western margin of the riverine plain in New South Wales. Since the middle 1950s, large areas of cracking clays, *black earths* and *grey, brown and red clays*, have been cultivated for wheat in north-western New South Wales, and through the brigalow lands of Queensland reaching into the tropics in the Central Highlands. Here, the winter rainfall is both low and unreliable, and wheat growing is dependent on the ability of the clay soils to store water from the summer rains, whereas in southern areas with more reliable winter rainfall expansion onto light soils is possible.

Dairying

Commercial dairying began in Australia about 1808 with the Blaxland brothers supplying Sydney with dairy products from their Parramatta farm (King 1948). However, because of transport and storage problems, the industry developed slowly, and for much of the nineteenthcentury remained localized around the centres of population. From about 1890, advances in technology, access to British markets, and encouragement from colonial governments provided tremendous stimulation to the industry. It then expanded rapidly along the coast, forming the basis for the mixed farming settlement of most of the near-coastal lands of eastern and southern Australia.

Productive dairying is dependent on good pastures and thus on at least moderately fertile soils in areas with preferably more than 750 mm of rain annually. In New South Wales, soils and slopes strongly delineate the areas developed for dairying (Holmes 1962) and largely confine it to the coastal river valleys. Similar valley development is evident in Queensland, but with a greater spread into the rainforest lands on basaltic plateaux, and a marked westward extension into drier areas with highly fertile soils, e.g. the Darling Downs. Dairying is widely distributed throughout Victoria (Wilson 1961), except for the mountainous areas of the north-east and the dry mallee lands of the north-west. In Tasmania, the main dairy lands are along the north coast, concentrated on the basaltic hills and in the river valleys. The principal dairying areas of South Australia are the southern slopes of the Mt Lofty Ranges, the Lower Murray Swamps and the Mt Gambier district, while in Western Australia the industry stretches along the coastal plain from Perth to Albany.

The fertile *alluvial* and *prairie soils* of the river flats and the *krasnozems* and *xanthozems* of the rainforests were soon recognized as prime dairy lands, and vigorously sought by prospective farmers despite almost inaccessible locations. Soils with low inherent fertility were unable to support good pastures and were by-passed, e.g. the infertile *lateritic podzolic soils* and *gleyed podzolic soils* of the coastal lowlands in Queensland, and the *podzols* and *acid peats* of coastal eastern Australia. In Western Australia, infertile soils severely impeded dairying until after 1920 when subterranean clover pastures and superphosphate came into general use.

Because of the great diversity and local variablility of the soils of eastern Australia, many soils of different quality were used for dairying in any one district, and frequently on any one farm. At the peak of development in the 1930s, the industry had extended over a very wide range of soils, representing most Australian Great Soil Groups except those occurring only in arid areas. It then occupied most of the better soils in the area most suited to rain-fed agriculture.

Indigenous pasture species of the river valleys and coastal areas are not highly productive, so from the beginning, dairying was largely dependent on introduced grasses. These make heavier demands on nutrient supply, particularly of nitrogen and phosphorus, than the native species, and within a few years growth declined on most soils as the organic matter and available phosphorus levels fell. The decline was most serious on soils of borderline inherent fertility, such as *red* and *yellow podzolic soils* on sedimentary and metamorphic rocks where the poorer producing farms went out of dairying, e.g. those in the Dorrigo area of New South Wales (McArthur 1964) and the Wide Bay District in Queensland (Thompson 1965). Even the more fertile *krasnozems* and *prairie soils* of the basalt areas showed marked fertility decline under grass-dominant dairy pastures, e.g. in the Lismore District (Nicolls *et al.* 1953). In southern Australia the fertility decline was countered by the use of superphosphate and pasture legumes, but no suitable legumes were available for the warm drier subtropical areas before the breeding of siratro (*Macroptilium atropurpureum*) (Hutton 1962).

Although the interfaces between dairying and other rural industries have fluctuated over the years with changing economic circumstances, the overall trend since 1950 has been the withdrawal of dairying from low-productivity areas, e.g. in the 1960–70 period the number of dairy farmers declined by 34% (Anon. 1975*a*). The biggest changes have occurred in the hotter and drier areas and on soils with acute fertility problems, because the costs of overcoming the soil and moisture constraints are uneconomic.

The growth and decline of the dairying industry illustrates the way rural industries in Australia extend into climatically marginal areas and onto marginal soils during boom periods, and retreat as economic conditions become less favourable, often leaving problems for subsequent land use. Changes of this kind clearly indicate the relatively small area of fertile soils that can support stable mixed farming in our coastal areas using existing techniques.

Sugar cane

Sugar cane was first grown for the production of sugar at Port Macquarie in 1826, and is now cultivated over some 290 000 ha in the more fertile valleys along 2000 km of the eastern coast from the Clarence River in New South Wales to Mossman in North Queensland. The first commercial crop in Queensland was grown on the fertile *krasnozems* at Redland Bay in 1863. Within five years, about 800 ha were planted to cane in the nearby districts, mainly on *alluvial soils*, *humic gleys* and *prairie soils* of the river valleys.

The industry spread northwards from one river valley to the next, suitable areas being determined by the navigability of the coastal streams and the extent of their fertile alluvial flats. By 1880 the industry had spread to Bundaberg and Mackay and into the tropical wet lowlands around Ingham and Innisfail. Areas of vine scrub were in great demand, since their soils were found to have higher fertility than forest areas, and after clearing and burning the stumps decayed within a few years, permitting the use of ploughs.

Early development of the sugar industry was constrained by the low phosphorus and nitrogen levels of many of the soils, and by the droughtiness of others, e.g. sandy *alluvial* and *solodic soils*. Early attempts to use the infertile *gleyed podzolic* and *lateritic podzolic soils* of the relatively moist coastal lowlands were soon abandoned. The droughty and infertile *solodized solonetz and solodic soils* of the coastal areas between Bundaberg and Mackay and around Bowen and Townsville were also avoided. Even in the more fertile valleys, soil differences caused piece-meal development and wide variations in yields. Production from old lands soon declined as levels of organic matter fell, and nitrogenous and phosphatic fertilizers were recommended as early as 1874 (Voelckers 1874).

Research by the Bureau of Sugar Experiment Stations has solved many of the early problems related to nutrition, disease and poor varieties. This has led to the economic expansion of cane onto large areas of soils previously regarded as unsuitable, e.g. the *solodized solonetz and solodic soils* and *grey clays* of the lower Burdekin Valley, and the *lateritic podzolic* and *gleyed podzolic soils* of the coastal lowlands near Bundaberg.

An average crop of sugar cane in Queensland produces about 100 t ha^{-1} of leaves, stalks and roots during the growing season (Molnar 1974), uses about 4600 m^3 ha^{-1} of water in transpiration (Kerr and Bell 1939), and makes heavy demands on the soil for nitrogen, phosphorus and potassium.

The nutritional needs are largely met by the use of various fertilizer mixtures providing N at 50–100, P at 10–40 and K at 60–130 kg ha^{-1} annually (calculated from Molnar 1974), according to the nutrient status of different soils. Irrigation, deep ploughing, the use of leguminous green manure crops and the provision of shallow drains have overcome some of the physical and soil water problems. Although sugar cane is now grown on a wide range of soils, spanning more than fifteen Australian Great Soil Groups and thus implying that it is not soil-sensitive, continued high production is maintained only by varying the cultural practices and the inputs of water and nutrients, according to the different soils. The two soil constraints restricting further expansion in the rain-fed areas are droughtiness, due to poor water entry and limited soil water storage and, in the wetter situations, inadequate drainage of *humic gleys* and *acid peats*.

Summary

Although the regional pattern of different types of agricultural use has been largely determined by climate, and to some extent by demand and economics, differences in soil nutrient status and water regime have mostly determined the local areas used for particular crops within the broad climatic regions. The successful establishment of early agriculture was strongly dependent on inherent soil fertility, because the very limited reserves of nitrogen, phosphorus and sulphur in so many Australian soils have either proved inadequate for agricultural crops or have been exhausted rapidly by continuous cropping. Differences in water entry and storage, due to variations in porosity, organic matter, clay content and depth of soil, have also been important in determining crop performance under Australia's unreliable rainfall, controlling the magnitude of the crop or causing its failure in droughty soils. Soils of moderate or better fertility and with good water entry and storage, such as *alluvial soils*, *prairie soils*, *krasnozems*, *xanthozems*, *non-calcic brown soils*, *red-brown earths* and *black earths*, were therefore in demand, and their distribution largely controlled the pattern of agricultural development in the subhumid and humid lands.

Although agriculture in the coastal areas is largely confined to the valley floors, topography had a lesser effect on early agriculture than soil quality because of the small areas involved, due to use of manual equipment rather than large mechanical implements. Quite steep slopes were cleared and cropped where the soils were suitable, e.g. *krasnozems* in many rainforest areas. Soils and topography are inter-related so that steep slopes with very shallow soils were largely avoided, but so were quite gentle slopes with moderately deep infertile soils, e.g. *podzols* and some *podzolic* and *solodic soils*. With the modern trend to broadacre farming, topography has become more important, mainly because the complex pattern of small areas of different soils associated with hilly terrain cannot be worked economically with large machinery.

Modern agricultural technology, using a range of fertilizers, large and improved farm machinery and irrigation has erased much of the patchwork pattern of early agriculture in many districts by bringing previously unsuitable soils under crop. However, the high productivity of these soils is mostly dependent on large inputs of fertilizer and/or water and on managerial skills; consequently their continued use will come under greatest pressure during down-turns in market prices. Much of the soil variation will remain, and its effect on land use patterns will reappear in the future, at different times in different industries, as economic changes force the reduction of various inputs.

Effects on soils of changing land use

Introduction

European settlement brought profound changes to Australian landscapes as both urban and rural areas were developed. The land was subdivided on a rectilineal system and assigned to individual management; increasingly larger areas were cleared, burnt and cultivated; exotic plants and animals were introduced; roads and towns built; and larger and more powerful machines used. Thus the new population set about restructuring the landscape, initially to resemble rural England and then, in the present century, mechanized North America.

These developments have markedly changed soil environments, by destroying native vegetation and replacing it with introduced species having different canopies, root systems and life cycles, or with urban and engineering structures. Soil horizons have been loosened and mixed, and exposed to new patterns of solar radiation, rain-drop energy and surface water flow. The most obvious consequences have been large increases in runoff and soil erosion from sloping lands, and in frequency of flooding and deposition along valley bottoms.

Many facets of European settlement have contributed to changes in Australian soils, but the frequent interdependence between them makes the isolation of individual effects virtually impossible: for example, the build-up of soil nitrogen, through ley rotations, and its utilization in the wheatlands result from interactions between fertilizers, introduced plants, microorganisms and tillage.

In this section, changes in Australian soils following the change to rural, urban and extractive uses are treated separately, and effects are discussed under the practice considered to have a major influence.

Changing to rural land use

Cadastral influences

In Australia, as elsewhere, land has been subdivided on a rectilineal system because of advantages in survey and in distances travelled. Subdivision within farms has largely followed the same system because of ease and cost of fencing. However, straight lines are rarely the result of natural processes and where they have been imposed on the landscape they influence man's activities, particularly those that affect infiltration and erosion, e.g. cultivation parallel to fence lines, and concentration of traffic adjacent to fences irrespective of slope.

The main indirect effects on Australian soils have been reduced water entry and accelerated soil loss from sloping sites. These are most marked in soils with low infiltration rates, such as cracking clays and texture-contrast soils with hard-setting loamy surfaces and dense clay subsoils, particularly where the clays are sodium-affected and disperse readily, e.g. *solodized solonetz and solodic soils*. Rectilineal farm and paddock design, coupled with individual ownership of different parts of hillslopes and first-order stream catchments, has also seriously hindered management of land as natural drainage units in soil erosion control programmes.

Clearing and burning

Clearing affects soils by decreasing the protection of the surface, by changing the hydrological status, by accelerating loss of nutrients, particularly those stored in the vegetation, and by altering soil fauna populations.

Increased exposure of its surface to the elements usually results in degradation of the soil. With cover gone or reduced, increased desiccation and higher temperatures result and these are usually deleterious to plant growth in Australia. Even more serious is the resulting acceleration of erosion by wind and water; this is likely to be high in the period immediately following clearing, before a new cover has had a chance to become established. Subsequent to this, the amount of cover is largely a function of other factors.

Removal of native vegetation usually changes the evapotranspiration rate and thus the hydrology of the soil. Lowered evapotranspiration may cause watertables to rise adversely affecting the soil when they reach the upper metre. In other situations, clearing may result in increased movement of water through the profile and lead to salinization of soils in lower parts of the landscape, e.g. the increased salinity in some valleys of the south-west of Western Australia following clearing of native forest (Wood 1924). Drainage-way salinity also has often increased in cleared areas, e.g. on Kangaroo Island in South Australia, and in the Yass Valley of New South Wales (van Dijk and Woodyer 1961) (Fig. 48.4).

Fig. 48.4. Drainage-way salinity, Yass Valley, N.S.W. (Photograph D. C. van Dijk.)

Clearing changes soil fauna populations by destroying their primary food source and altering their habitat. Many indigenous species are unable to survive in the altered environment, even when an alternative food source of exotic plants is available. Ants and earthworms are known to be severely affected, and similar changes in other populations have almost certainly occurred.

Clearing is usually accompanied by burning which causes massive redistribution of nutrients in the soil-plant system; much of the nitrogen and sulphur is volatilized and other nutrients are probably lost to some extent in air-borne ash.

The Australian aborigines have long used burning to drive game, and fires from lightning strikes have been relatively common. Since European settlement, fire has also been used widely as a management tool in the pastoral industry. Thus frequent extensive burning has been a feature of the Australian environment for a long time, removing the combustible materials from the soil surface and preventing the accumulation of litter and the development of duff layers on the soils. While hot burns are generally deleterious, there is evidence that cool burns may be beneficial in some cases by increasing available nutrient levels in the soil.

Tillage and traffic.

The main effects of tillage are to loosen, fracture, mix and eventually homogenize the surface soil to the depth of the cultivation. It has destroyed the separate pedological identity of the upper horizons of many Australian soils by incorporating the thin O horizons into the underlying mineral soil, by mixing A_1 and A_2 horizons and, in deep cultivation or where the A horizons are thin, has often incorporated the top of the clay subsoil in the

Fig. 48.5. Water and sediment movement following heavy rain on contoured and untreated cultivated *black earths*, Darling Downs, Queensland. (Photograph Dep. Primary Industries, Qld.)

Fig. 48.6. Wind-blown sand due to crop failure on *solonized brown soils* during the 1977 drought, Mannum, South Australia. (Photograph W. E. Matheson, Dep. Agriculture and Fisheries, South Australia.)

newly formed anthropic A horizon. Tillage has increased water intake and aeration, and in most soils has extended the depth of major biological activity; it has also increased the availability of some nutrients through enhanced mineralization and by the exposure of new soil faces to root exploitation.

While the homogenization of surface horizons by cultivation may be beneficial to plant growth, the greatest effect of tillage on Australian soils has been deleterious, through increased erodibility. Tillage destroys plant cover and the binding effect of roots so that the resulting fragmented soil is readily transported by surface wash, leading to rill erosion and gullying. Enormous losses of soil have occurred where bare cultivated fields have received high intensity rains, e.g. the *black earths* of the Darling Downs (Cummins and Bass 1978) (Fig 48.5), or have been exposed to strong winds, e.g. the *solonized brown soils* of the mallee areas (Fig. 48.6).

The development of fine tilth under cultivation also favours the formation of crusts and seals in some soils. Rate of structural deterioration varies with speed of cultivation and type of machine used, and there is a pronounced interaction with soil surface condition varying with moisture and organic matter contents and type of soil, e.g. pedality of *black earths*, *prairie soils* and *krasnozems* deteriorates more slowly than that of the hardsetting *red-brown earths*, *podzolic* and *solodic soils*.

Both animal and mechanical traffic compact soil surfaces. The introduction of hard-hooved livestock into Australia has undoubtedly reduced rainfall acceptance in many soils. On early pastoral holdings, livestock were concentrated on selected areas to increase runoff to the creeks. It was found that 'on the spongy surface of virgin country, untrodden by any hoof, there was little "run" off the surface after rain but when hardened by the tread of stock the creeks received a fairer share of the downpour' (Anon. 1909). Similar effects can be demonstrated experimentally; simulated trampling of surface soils from south-eastern Queensland had little effect on *xanthozems* but markedly reduced water acceptance rates in *red podzolic soils* (G. G. Beckmann, C. H. Thompson and W. Arndt, unpublished data). Livestock may also puddle the surface layer during wet periods, effectively reducing water entry; this technique is used to advantage in the paddy fields of Asia where buffalo are used to puddle the surface to prevent soakage losses.

The wheels of tractors and implements also compact the surface; and ploughs exert pressure and a shearing action on the underlying soil, leading to plough pans which restrict penetration of water and roots. Significant differences in soil properties have been observed in the bands between and beneath tractor wheel tracks on *red earths* at Katherine, Northern Territory (Arndt 1966).

Introduced plants.

Introduced plants have played an important role in changing Australian soils, mainly through their influence on the organic matter cycle, but also through their effect on erosion and through their greater demand for nutrients and soil water.

Introduced legumes, particularly in fertilized pastures, have markedly increased organic matter contents, soil nitrogen levels and mineralization rates over large areas of Australian soils. The outstanding example is subterranean clover, estimated to cover more than 11 million hectares of southern Australia (Donald 1964), but other temperate legumes, e.g. white clover, lucerne, medics and lupins, are grown extensively in the south, and several tropical legumes, e.g. siratro, stylo and glycine, have been introduced to large areas in the moist subtropics and tropics. All have made important changes to a wide range of soils, through increased levels of soil nitrogen, faster nutrient cycling, changed soil biota and improved surface condition, the latter leading in many cases to improved water entry.

On the other hand, agricultural crops and introduced grasses have higher growing season demands for nutrients and soil water than is general for the native grass species. In the absence of legumes and mineral fertilizers they have been highly exploitative, often rapidly depleting the inherent reserves of nitrogen and phosphorus, resulting in seriously reduced yields, e.g. the marked decline in Australian wheat yields prior to 1910. Even on the fertile *black earths*, cereal cropping over a 25-year period reduced total soil nitrogen by about 1% per year (Martin and Cox 1956). Because of their more vigorous growth, introduced pasture species may dry out the soil profile earlier, and to greater depths, than equivalent native gramineous species (Donald 1970).

Differences in form, structure and canopy of introduced plants have altered erosion rates on many soils. In the south, temperate grass/legume pastures have reduced soil loss where they have replaced native tussock grasses, by increasing soil cover and increasing water entry; similar effects have been achieved in the north through the stoloniferous habit of several of the tropical legumes. In other cases, the very rapid growth of introduced plants, e.g. *Dolichos lablab*, has provided urgently needed short-term soil protection. Against this, many introduced plants have been associated with soil losses from sloping sites, e.g. orchards, tropical fruit plantations, and row crops such as maize, sunflower and sugar cane, particularly where inter-row cultivation is practised.

The effects of many species have never been measured and are difficult to assess. Deep-rooting plants such as lucerne and siratro possibly recycle plant nutrients from greater depths than many indigenous species. Folklore has it that some species, e.g. lantana, improve fertility and structure, and that the destruction of enormous quantities of prickly pear by the cactoblastis insect is recorded in the thin, dark, highly organic surface layer of many clays soils of the brigalow lands.

Introduction of new fauna

A unique fauna evolved in Australia following its separation from other land masses, and presumably reached equilibrium with other components of the ecosystems. The introduction of the dingo by the aborigines probably did little to upset this equilibrium and had virtually no effect on the soils. European settlers, however, brought a large number of exotic animals with them and these have considerably changed Australian soils.

The main effect of domestic animals has been the large increase in soil erosion following extensive over-grazing of pastoral lands. Extensive compaction and localized puddling of soil surfaces by hard-hooved animals, particularly cattle and sheep, have further aggravated the problem by reducing water entry and so increasing run-off. Changes in soil fauna, populations of microorganisms, nutrient cycling rates and in the distribution of plant nutrients have also occurred, due to the large and changing numbers of domestic stock.

Wild rabbits and feral pigs have mixed soil surface horizons and spread pathogens. Because of its particular requirements, physiologically and for burrowing, the wild rabbit has had its greatest effect on the coarse-textured soils of southern Australia, e.g. *solonized brown soils*, *red earths*, *desert loams* and sandy *lateritic podzolic soils*. From semi-arid western New South Wales into the wet tropics, the feral pig has rooted and wallowed in a large range of soils, and has recently been found to spread the die-back fungus (*Phytophthora cinnamomi*) in some northern rainforests. Other feral animals, particularly goats and donkeys, have heavily over-grazed parts of the semi-arid grassy woodland belt, thereby accelerating erosion.

Many of the native earthworms have apparently been unable to adapt to agricultural and urban environments; species of European origin are now more common in many southern farming areas, and species from South America and Indonesia are becoming common in warm coastal urban areas (G. Dyne pers. comm.). It is likely that these species, in association with introduced plants, are more active and promote faster nutrient cycling than the native species. Other additions to the soil fauna may also have had significant effects but these have not been documented. An interesting example of trying to alter an effect of one exotic animal by the introduction of another is the use of dung beetles from Africa to break down cattle pats and incorporate them in the soil, so speeding up the release of nutrients immobilized in the manure.

Introduction of microorganisms

Most soils are biologically well buffered; in spite of continual input of inoculum, from the air and by surface soil movement and by transit of animals over and through the soil, the nature and size of a particular population of soil microflora remain fairly constant. It is only when appreciable interference with the soil system occurs that this buffering may be overcome and a change in microflora initiated. While it is certain that drastic changes in land use, such as clearing of native vegetation, cultivation, introduction of new plants, fertilizer application or fumigation of greenhouse soils, have a pronounced effect on the soil microflora, we often cannot be sure whether the overall effect is beneficial or otherwise. Such perturbations may change the numbers or relative proportions of native microorganisms, and may allow establishment of exotic microorganisms.

The deliberate introduction of specific microorganisms from outside faces the problem of buffering also. It is almost mandatory that a favourable microenvironment be created for the introduced microorganism to persist and carry out its function.

Observable changes to Australian soils following introduced microorganisms are difficult to detect but indirect effects through organism/plant associations are well known. The best-known example of a deliberately introduced microorganism is *Rhizobium*. In this case, a favourable environment is provided by the presence of the host legume roots, and much research on *Rhizobium* has been concerned with host specificity of various strains of the bacteria. Of equal importance, but receiving less attention, has been the need to provide a suitable soil microenvironment for multiplication of the *Rhizobium*; pelletting of legume seed with mixtures of peat, lime and concentrated *Rhizobium* inoculum is one technique used to achieve this.

Another example of deliberate use of a microorganism is that of soil *Thiobacillus* strains used to oxidize the sulphur of biosuper, a mixture of rock phosphate and elemental sulphur, to sulphuric acid, so producing biological superphosphate in the soil (Swaby 1975).

The widespread occurrence of ectomycorrhizae on the roots of forest trees has led to attempts to introduce particular strains by inoculation of seed with spores of mycorrhizal fungi (e.g. *Pinus radiata* seed with *Rhizopogon luteolus*). More recently, recognition that most plants carry endomycorrhizae on their roots, and that these are important in increasing nutrient availability, has resulted in research on inoculation of seed with strains of endogone selected for their ability to increase plant growth.

Fertilizers and other agricultural chemicals

Applications of mineral fertilizers change the composition and concentration of the soil solution and thus the availability of plant nutrients. Applications which correct soil deficiencies usually result in marked increases in biological activity at most levels from the microbial to higher plants, although very heavy dressings may initially suppress some populations, e.g. earthworms.

Mineral fertilizers have been used in Australia for about 90 years, firstly superphosphate, and subsequently nitrogen, potash and trace elements. The main observable effects on the soils are those which have been accomplished indirectly through enhanced growth of introduced plants; the most spectacular have been those of the well-known superphosphate/legume combinations, viz. large increases in soil organic matter, pedality, water entry and storage, and changes in soil animal populations. This combination has restored fertility to worn-out farmlands and allowed agriculture to extend into areas previously regarded as unsuitable. It obviously has wide application in improving the surface condition of vast areas of hard-setting surface soils characteristic of much of the Australian farmlands.

Not all of the effects of fertilizer application are beneficial; deleterious effects on soils are often blanketed by short-term increases in productivity. Under continuous cropping where legumes and green manure crops are not used, mineral fertilizer application has accelerated the depletion of organic matter and consequent deterioration of surface condition. The adverse effects of ammonium salts on the cation exchange capacity of soils is also well-known, inducing the loss of metallic cations, particularly calcium, from many soils; with time, soil acidity is markedly increased, and manganese and aluminium toxicities may result, e.g. induced manganese toxicity following continued use of ammonium sulphate for pineapples on *krasnozems* in Queensland (Fergus 1954).

Residues of pesticides and weedicides used in agriculture are known to accumulate and persist in soils but little is known of their effects on the micro-fauna and -flora. In some soils, chemical compounds have reached toxic levels affecting subsequent productivity; examples are copper build-up in some horticultural soils, from fungicides such as Bordeaux mixture, and arsenic toxicity, following its use to control harrisia cactus (*Eriocereus martinii*), on *grey clays* in Queensland. The magnitude and persistence of these effects depend on the compounds used and the nature of the soil colloids.

Organic matter and clay, particularly that with 2 : 1 lattice structure, have a high adsorptive capacity for many organic pesticides, so retarding their loss from the soil. The availability of adsorbed pesticide to plants and other members of the food chain determines the extent of its effect on the soil. Very large applications of pesticides, e.g. DDT to cotton soils and Dieldrin to many sugar soils, probably markedly change soil fauna and microflora populations, and so may have long-term effects on the organic cycle.

A recent review of the effects of soil fumigants, fungicides and one soil insecticide on endomycorrhizae (Smith 1978) shows a marked reduction in root infection or spore numbers or both. In such cases, the advantages of chemical treatments in reducing pathogen attack may be partly offset by reduced yields, unless additional nutrients are provided to compensate for the reduction in mycorrhizal effectiveness.

Irrigation and drainage

Apart from maintaining soil moisture at favourable levels for plant growth, the main effects of irrigation are waterlogging and salting.

Irrigation almost invariably supplies more water than is needed for crop production. Initially this leads to deep percolation and leaching but, where lateral water transmission rates are slow, ground waters begin to rise and carry dissolved salts towards the root zone. Significant parts of Australian irrigation areas have become waterlogged and salted by this process, e.g. Kerang and Wakool Districts in the Murray Valley, and much more is threatened (Pels and Stannard 1977). Depending on topographic situation, soil and substrata permeability, and intensity of land use, some form of artifical drainage, e.g. surface, tile or tube-well drains, may be used to alleviate waterlogging and salting of the root zone. Heavy flood irrigation also reduces aeration, leading to the mobilization of iron and manganese and to the development of pseudo-gley features in some soils, e.g. *grey, brown and red clays* and *solodic soils* under rice.

Irrigation has probably also had a considerable effect on soil fauna since, in conjunction with other agricultural activities, it drastically changes native habitats; for example, the build-up of organic mats under irrigated pastures may be due, not only to the inability of the native biota to handle the large increase in plant material, but also to their poor survival in the new environment.

Artificial drainage has been used in Australia to bring wetlands into cultivation and to control watertables in irrigation areas. The main effects are those associated with improved aeration and reduced saturation in the root zone.

The draining of *acid peats* has led to shrinkage and irreversible drying of the peat granules so that they now behave like pieces of coarse sand or gravel, e.g. in parts of Mowbray Swamp, Tasmania. Where sulphides are present, acidification to $pH < 3.0$ has occurred as the sulphides have oxidized, e.g. in the swamp lands of Western Australia and New South Wales (Teakle and Southern 1937; Walker 1963). The fire hazard increases as peats are drained, and reflooding has sometimes been necessary to extinguish bad peat fires. Drainage has also led to the immobilization of sesquioxides as hardened aggregates and nodules in *humic gleys* and *gleyed podzolic soils* of coastal Queensland.

Changing to urban land use

Urbanization

Urban developments commit land, and therefore soils, to uses which are virtually irreversible because of the capital costs involved and because the drastic changes made to the soil mantle greatly reduce the options for other forms of use subsequently. The most marked changes are those associated with truncation of soils and the mixing and inversion of soil horizons and underlying rock; with the additions of paving and foundation materials, industrial and household wastes, and garden top-dressings and fertilizers; and with changing moisture regimes due to increased runoff from impermeable installations and to additional water from gardens and household drainage. The overall effect of urban use is to produce extreme local variability in the soil mantle so that any subsequent use by current agricultural systems is impracticable.

Loss of agricultural soils to urban development varies with a city's location and planning; those located in areas of good agricultural soils can often only expand at the expense of agriculture; for example, several of the Queensland sugar-producing centres are land-locked in coastal valleys, and urban growth in these valleys results in loss of sugar-growing soils which can no longer be replaced by bringing new land into production (Anon. 1978*a*). Since its inception, Adelaide has continued to expand over good farmlands; the vegetable, orchard and vine areas of the Adelaide plains have been reduced from 3070 ha to 1050 ha in the 1957/8–1973/4 period (Anon. 1976) and it is anticipated that by 1990 only areas designated as historical, such as some of the early vineyards, will remain in production there.

However, many of the cities of the eastern coast are situated on, and surrounded by, soils of varied agricultual quality; here land use planning might have prevented the loss of the better agricultural soils. For example, more than 80% of the Brisbane city area consists of soils of poor agricultural quality, and *red earths* and *krasnozems*, prized for horticultural crops, occupy less than 8% of the area; yet most of these high quality soils have been committed to housing.

In yet other cities where even the best soils are of only poor to moderate agricultural quality and occupy relatively small areas they may be used to advantage for urban development; for instance, in Townsville development is now recommended on and around the areas of better soils because they have little agricultural potential, but are important for parklands, sports grounds and gardens (Murtha and Reid 1976).

Not only have good agricultural soils been lost as a result of urban development but many of the soils of local government areas adjacent to the larger cities have been markedly changed by dormitory suburb and hobby farm development, extractive industries, waste disposal, water storage, feed lots and transport and communication facilities. Redland Shire adjoining Brisbane to the south-east has long provided the city market with fruit and vegetables grown on its *krasnozems*, which occupy less than 12% of its area; half of this red soil area has already been lost to dormitory suburb housing (Fig. 48.7) and it is almost

Fig. 48.7. The urban development/horticulture interface on a *krasnozem*, Redland Shire, Queensland.

certainly only a matter of time before the remainder is also engulfed. Conversely, Redland Shire is benefiting from the sub-division for hobby farms of the much larger area of poor *podzolic soils*, which are now receiving fertilizer and other inputs previously considered to be uneconomic.

High population densities associated with individual home ownership generally result in high economic inputs per unit area and a general fertility improvement in soils around dwellings. A common pattern in the past exploitative phase of Australian agriculture has been a fertility shift from rural to urban areas but the increasing rural use of mineral fertilizers now tends to offset this trend.

Waste disposal

Wastes are those materials returned to the environment as a result of the activities of living organisms, including man. Agriculture provides large volumes of waste in the form of straw, bagass, manure, etc., but where this is returned to the soil it is usually in a sufficiently diluted form (i.e. spread over large areas) to be beneficial. Where the human population is dense, natural wastes become concentrated, and wastes that formerly may have been considered harmless, or even beneficial, become detrimental to the soil. The nature of these wastes changes also, most notable being the increasing proportion of synthetic polymers that are resistant to breakdown by the natural soil micro-flora.

Solid wastes (garbage) when disposed of by landfill result in the loss of the soil that is covered but this is often of little consequence since many landfill sites are chosen because they are far from ideal for other purposes. However, leakage from these sites has sometimes been a problem in adjacent soils. Liquid wastes are usually discharged into water courses, estuaries and the sea, and consequently have little effect on soil. However, the disposal of effluent from sewage treatment is often accomplished by passing it into soil, by sprinkler, flood irrigation or evaporation trenches. In most cases this effluent contains high levels of organic matter, inorganic nutrients and some heavy metals. Many soils have the capacity to decompose or inactivate this material with little detriment to themselves. The best known example in Australia is the Melbourne sewage farm at Werribee, which has been operating for about 80 years and handles sewage at $> 500\,000$ m^3 day^{-1}; flood irrigation is used to apply about 10 cm of sewage to legume/grass pastures on *red-brown earths* at about three-weekly intervals. The main changes to the soils have been: the development of a thick organic loam surface; large increases in soil nitrogen, organic carbon and phosphorus; and significant increases in many trace element levels, particularly nickel, zinc, cadmium and copper (Johnson *et al.* 1974). However, the trace elements have not reached toxic levels for the pastures or the beef cattle grazing them.

Aerial wastes also fall on soils, and gaseous discharges from mining and manufacturing industry are usually distributed over large areas of land unless prevailing winds blow them out to sea. In some instances they may be beneficial, e.g. the accretion of sulphur dioxide originating from industrial areas may improve plant growth in soils with low contents of indigenous sulphur (Whitehead 1964). On the other hand, if the concentration of sulphur dioxide is high, pH in nearby soils and streams may be lowered, with detrimental effects (Overrein 1976).

Heavy metals may also accumulate in soils around smelters. Thus, soils surrounding Port Pirie in South Australia show high levels of cadmium, lead and zinc in the plough layer, although concentrations of these metals in the crops grown on these soils do not exceed levels considered to constitute a potential health hazard (Tiller and de Vries 1977).

Transport and water storage

Transport facilities serve both rural and urban areas but most new constructions connect and service towns and cities. The main effects are increasing loss of soil for other uses, increased erosion hazard and pollution of adjacent soils.

Roads and railways cut across landscapes, largely ignoring differences in topography and soils unless these present serious engineering problems, e.g. steep terrain or wetlands with low bearing strength. The engineer's predilection for straight lines, usually imposed on a curved landscape, leads to interference with natural drainage and to erosion and siltation problems on many soils, particularly those with low permeability. This is most obvious where serious gully erosion has developed at points of water dispersal from roads on *solodized solonetz and solodic soils* in the summer rainfall areas, but the problem is also widespread on hillslopes throughout Australia. Severe gully erosion also occurs where runoff waters are concentrated by built-up roads into large drains and culverts in the valley bottoms, e.g. on *black earths* of the Darling Downs. Eroding gullies also create further problems by depositing sediment on lower land.

Australia has some 845 000 km of roads (Anon. 1977), representing about 0.4% of its total area, assuming an average road reserve width of 40 m. However, roads use larger proportions of land as development intensifies and population per unit area increases; for instance, road reserves occupy about 0.5%, 1.0% and 2.9% respectively of Queensland, New South Wales and Victoria.

Modern divided highways, with multiple traffic lanes, entrances, exits and intersections designed to maintain fast traffic flow, are large users of land (Fig. 48.8). New areas of land are being used at increasing rates, while abandoned routes are of little value except for urban development. Pollution of adjacent lands is widespread;

apart from rubbish dumped by highway users, trace elements, such as lead from modern fuels, accumulate in the nearby soils. Another interesting phenomenon has been the occurrence of grey speck disease of oats bordering roads in south-eastern South Australia, due to drift of limestone dust from the roads raising the soil pH (Samuel and Piper 1928).

Surface water storage has alienated good agricultural land by inundation of flood plain and terrace soils of the valley floors (Fig. 48.9). Storage dams alter sediment transport, resulting in siltation and increasing flooding above the dam, scouring of the stream bed immediately downstream of the structure, and reduced sedimentation on the lower floodplains thus largely depriving them of natural nutrient additions. Water distribution channels and pipelines for rural and domestic use have led to further loss of agricultural land and, in some cases, to waterlogging and salting of soils adjacent to the distribution channels, e.g. in the Murray Valley.

Changing to extractive land use

Extractive industries, viz. mines, quarries, borrow pits and surface scrapings, result in destruction of soil through removal and burial and through increased erosion due to removal of vegetation and changes in natural drainage. The mixing of soil horizons and underlying materials creates new soils in regeneration areas.

Fig. 48.8. Modern divided highway north of Brisbane, Queensland, on mainly *lateritic podzolic soils, red earths* and *gleyed podzolic soils*. (Photograph Courier-Mail, Brisbane.)

Borrow pits and surface scrapings (Fig. 48.10)

These are widely used in both urban and rural areas to provide road base, construction materials and top-dressing. Borrow pits are very common in gravelly soils, such as *lateritic podzolic soils*, lateritic *red earths* and *alluvial soils*, where they occur adjacent to roads and

Fig. 48.9. North Pine River Dam, Queensland, inundating *alluvial, prairie* and *podzolic soils* of the river flats and lower hill slopes previously used for dairying and beef production. (Photograph Courier-Mail, Brisbane.)

Fig. 48.10. Roadside scraping of the gravelly sandy loam surface horizons of *lateritic podzolic soils*, Hervey Bay, Queensland.

railways. In addition, thick gravelly, sandy or loamy surface horizons of many *alluvial, prairie* and *podzolic soils* have been scraped to provide top-dressing and construction materials for nearby cities. In some rural districts, the thick sandy surface horizons of even some *solodic soils* have been scraped to provide sub-grade material for roads over adjacent clay soils, leaving bare dispersive subsoils exposed to erosion. Altogether, borrow pits and surface scrapings probably affect larger areas than other extractive industries, but few have received regenerative treatment, except along new highways where the trend is to roadside development by contouring, top-dressing and re-vegetating disturbed areas to produce a pleasing landscape.

Quarries and clay pits

Construction materials, such as clay, gravel, limestone and hard and weathered rock, are usually mined as close as possible to user facilities; soils are destroyed in the process but the areas involved are relatively small.

Underground and open-cut mines

Underground mining, in itself, affects insignificant areas of soils but the disposal of mullock, slag and ore residues often creates problems through burial of soil and, on occasion, through toxic components. Open-cut mining is being increasingly used, and locally may destroy large areas of valuable soils; although the national total is not yet great, it will increase as more of the extensive shallow deposits of coal are worked and the overburden of high quality agricultural land is removed, e.g. *black earths* and *grey, brown and red clays* of the Darling Downs and Central Highlands of Queensland (Fig.48.11). Regeneration after open-cut mining is now common practice, and new soils of highly variable quality are formed by returning topsoil, which has been stockpiled or removed ahead of current mining, over a highly variable substrate made up of mixed soils and underlying rock. Regeneration problems associated with the acidity of some materials and with differentials in water penetration and availability have occurred, and may well affect subsequent crop performance. Spatial variability in these new soils is very high and may make subsequent agricultural use, other than growing pastures, uneconomic.

Fig. 48.11. Open-cut coal mining on *grey clays* of the Central Highlands, Queensland. (Photograph Courier-Mail, Brisbane.)

Replacing topsoil, applying fertilizer and revegetating disturbed areas are also established practices in the heavy-mineral industry on *siliceous sands* and *podzols*, and following bauxite mining of *lateritic podzolic soils* and lateritic *red earths*. In these cases, some of the new soils so formed, e.g. in the coastal sands and on the coastal plain at Weipa, may prove more productive after mining than before, because of inputs of fertilizers and other agronomic practices.

Making better use of land and soil resources

Much of the present Australian approach to land use stems from the pioneering period and is characterized by three attitudes: the overwhelming desire to modify land and environment towards that of the northern hemisphere, the demand by commercial interests for exploitation of natural resources for short-term monetary gain, and the desire to expand production at ever-increasing rates. These attitudes have led to the degradation of our soil landscapes and have jeopardized their future productivity. Together with low population density, large area, and lack of appreciation of environmental processes, these attitudes have led to the general acceptance by the community that, in Australia, land is expendable, a resource to be used for immediate needs without consideration of the longterm consequences and justified by a belief that there is plenty more over the hill for the next generation.

The climatically-favoured coastal zone of Australia contains most of its forest and water resources, its rain-fed agriculture and horticulture, its secondary industries, and about 90% of its population. Consequently, competition for land in this coastal strip is high, with urban, industrial, transport and extractive industries continually encroaching on rural lands and water catchments. Almost every industry in Australia is looking for either increased areas of land or relocation on new land. The modern trend to larger unit areas in horticulture, agriculture and forestry has brought changing demands within the rural lands, while the expanding population has increased the pressure on them for additional areas for industrial, urban, water storage, hobby farming and recreational use.

Although the local land needs of a community change with population numbers and their prosperity and technology, the three basic needs of man, viz. water, food and shelter, must always be given first consideration in land use decisions. Changes in technology, economics and population mean that few land use decisions can be truly permanent, but need to be reviewed as the demand changes. Because of this, and because we are temporary occupiers of the land with a moral responsibility to future generations, all land use decisions must be made with care and responsibility. A rational assessment of the various use options for each piece of land is needed; this requires an adequate inventory of land resources at both national and district levels, and a clear understanding of the natural processes controlling sediment and water movement through soils and landscapes.

As far as a resource itself is concerned, the best use decision is one which minimizes its degradation or destruction. However, such a decision may either be uneconomic or fail to meet a required objective, so that in some cases conservation of the resource may have to be forgone to meet the immediate needs of the community. We must however be careful to see that suitable pieces of land are both available and used for our diverse requirements; nothing is gained by having to increase management inputs simply to overcome constraints arising from poor allocation of land for various purposes. The modern community needs land for its many activities; all of these must, of course, be considered but priorities are necessary if we are to make the best continuing use of our lands.

Rural land use

If rural productivity in Australia is to be maintained or increased in the future, a number of current attitudes towards land use and soil management must be changed, and several soil constraints need to be reduced.

Firstly, all land use decisions will need to take account of the very limited total area of prime agricultural soils available to the nation, and give priority to food production over other potential uses, particularly in areas adjacent to urban centres. There is also the need to evaluate the likely effects of particular uses on adjacent areas, before decisions are made.

Secondly, the emphasis must shift to less exploitative farming, and quite marked changes will be required in both farm systems and management to achieve high production without cumulative degradation of soil resources. Again, the likely consequences of farming systems on different soils need to be evaluated ahead of general implementation, and this demands a better understanding of soil and landscape processes.

Thirdly, soil technology should give increasing attention to enhancing plant growth by modifying the rooting zone of our agricultural soils to reduce present constraints; these are principally poor water entry and storage, inefficient use of plant nutrients and soil water due to inadequate fertilizer use or poor placement, and water-table and salinity effects associated with irrigation and clearing.

Limited area of prime agricultural land

Because of erratic rainfall and unfavourable terrain, only about 77 million hectares of Australia are suitable for rainfed agriculture. About one quarter of this is already cultivated, one third carries sown pasture, and one tenth is in non-agricultural use. This leaves about 25 million hectares of unimproved land that might still be developed (Hallsworth *et al.* 1976), but which generally has greater constraints than existing farmlands. In the south, rainfed agriculture has moved inland to the rainfall limit (around the 250 mm isohyet); most of the water available for irrigation is already committed to existing irrigation districts; and areas of undeveloped soils are generally small, discontinuous, and subject to severe constraints for current agriculture. Land for future expansion is available in quantity only in the summer rainfall zone of the north, where rainfall is much less reliable and erosion hazards are generally greater. Increased production is therefore dependent either on expansion into this area, with its additional farming hazards and transport problems, or on increasing yields in the present farmlands. In both cases, greater technological inputs will be needed and production costs will probably be higher.

Within the present farmlands, virtually all of the arable soils with few constraints are in agricultural use. Many areas are badly eroded and may have to be retired from cultivation; other areas are being needlessly lost to urban and industrial development. Any reduction in the present cultivation area will put greater pressure on that remaining, and continued reductions will force farming onto poorer soils and into areas with greater hazards, where returns will be less reliable and where the risks of landscape damage are considerably greater. Loss of high quality soils is particularly important in the horticultural areas of the rural/urban interface where housing developments are still given precedence over other forms of use. Urban subdivision of these districts is often excused on the grounds that the small individual farm areas are no longer economic units for modern mechanization, but restructuring of farms to increase unit size would overcome this difficulty in many areas if we were determined to keep them in horticulture. Much more detailed information about soils surrounding urban areas is urgently needed so that priorities can be established for particular soils, e.g. those that are very suitable for horticulture, such as *prairie soils, red earths* and *krasnozems.*

Conserving soil resources

Large areas of Australian farmlands are severely eroded and, across the country as a whole, soil losses are still increasing faster than conservation measures can be implemented (Anon. 1978*b*). The low permeabilities of most of our soils make them highly susceptible to erosion, particularly the cracking clays and those soils with hard-setting surface horizons and dense clay subsoils. Many soil conservation practices imported from the northern hemisphere are unsatisfactory for erodible Australian soils, especially under the higher rainfall intensities of the summer rainfall zone. In this environment, considerable modification of existing techniques and new approaches based on an understanding of landscape processes are needed. Sizable areas of badly eroded farmland may have to be retired from cultivation and used for pastures or forestry, or be cultivated less frequently under ley-farming systems.

Our early settlers used farming techniques familiar to them from the northern hemisphere; many of these were not well-suited to Australian environments, and modifications and change have come slowly; instead, we have tended to reinforce old techniques by using improved agricultural machinery. There is a real need to fit farming systems into the landscape to achieve harmony between various farming activities and geomorphic processes. This can only be attained through a better understanding of soil and geomorphic processes and of the pressures imposed on the soils by different agricultural crops and farming systems.

While the general principles are probably well enough known, there are few actual data on the geomorphic response of different Australian soil landscapes to varying agricultural techniques. Simple techniques for monitoring various landscape processes are needed to provide base data for different natural ecosystems, against which the effects of various types of future use may be judged. Better knowledge will permit manipulation of both farming and soil landscape systems in ways that will lead to reduced soil losses in the future.

Modifying the root zone

While traditional cultivation techniques have been gradually modified to better suit Australian conditions and the main nutrient deficiencies have been defined, much still remains to be done to enhance plant growth by understanding and manipulating the chemical, physical and biological processes occurring in the root zone.

One of the most pressing problems of many Australian soils is the hard-setting character of their surfaces (Fig. 48.12) associated with low organic matter and weak pedality. Under native vegetation these soils have poor rainfall acceptance and in many cases probably less than two-thirds of the annual rain is effective, e.g. hard-setting surface soils from the Gympie district of Queensland. Agricultural use has led to further deterioration of surface structure and to increased runoff and erosion.

The hard-setting characteristic probably stems from the low phosphorus status and low biological activity under which many Australian soils have developed; both will need to be overcome to effect lasting improvement. In southern Australia, grass/legume leys and superphosphate have proved to be one way of improving surface structure and water entry, and the technique is being extended northwards as suitable legumes become available. Chemical ameliorants have generally proved uneconomic, with the exception of gypsum used to reduce surface crust formation in some soils. Most changes in the past have been associated with the use of new plants and fertilizers and little attention has been given to introducing new soil animal populations. Perhaps future improvement also lies in this direction, e.g. the introduction of dung beetles may have significant long-term effects on soil surfaces in northern Australia.

Fig. 48.12. Rainwater lying on the hard-setting surface of a *solodic soil* near Tenterfield, New South Wales.

Many currently farmed Australian soils have clay subsoils which restrict root penetration at shallow depths. Although research on deep tillage and ripping has given conflicting yield results, there seems little doubt that it leads to enhanced root development and better plant growth in many soils. Gradually increasing the plough depth so as to extend the main root zone, as has been done in many of the thin-surfaced soils used for growing sugar cane, would probably increase yields in many crops. However, since tillage affects water relationships, different soils will respond differently, and response will vary with changing moisture regime, e.g. deep tillage of light sandy soils may reduce yields in the drier areas.

Australian soils generally are of low fertility, and the plant nutrients they contain are mostly concentrated in the surface horizons while the subsoils contain low to very low amounts. Cultivation depth for many crops, and particularly for cereals, seldom exceeds 10 cm so, even where fertilizers are used, the less mobile nutrients such as phosphorus are concentrated in this layer. Nutrition of rainfed crops may be affected in dry years if the cultivation layer dries out within a few weeks of germination. In these instances, deep placement of fertilizer, especially phosphorus, at depths of about 50 cm may prove beneficial in some soils, although in others the fertilizer may be rapidly adsorbed onto the subsoil clays; here slow-release fertilizers may prove useful.

Urban land use

Urban growth in Australia has increased dramatically in the last thirty years due to the combined effects of high migrant intake, the pronounced movements of rural population to the cities and larger country towns, and natural increase. We are now a highly urbanized nation and, more than ever before, we need better management of urban land development in order to meet the future demands of an expanding population. Most of the problems of past urban development have arisen from the inappropriate use of natural resources, from excessive real estate speculation, from the nature of land tenure and from poor planning. We are concerned here only with the natural resources; it is fairly obvious that the best use has not been made of these, both within and adjacent to most cities.

Most Australian cities have grown almost haphazardly from initial points of settlement and government. Few sites were deliberately chosen and planned as cities from the outset. Most have expanded as population increased, engulfing the surrounding countryside in repetitive suburban development irrespective of intrinsic differences in soils and landscapes. Although development has been controlled and, in the last half century at least, generally guided by town planning, its direction appears to have arisen mainly from the need for housing, transport, water reticulation and sanitary disposal, rather than by any long-term principles of rational land use. No-one can deny the importance of these needs or the urgency with which they should be met; however, where cities have engulfed land with high values for non-urban use, e.g. first-class agricultural land or land containing deposits of constructional materials or sizable underground aquifers, one may question development for economic reasons as well as commonly stated aesthetic ones.

Modern city developmental costs are high and, basically, cities grow on their deposits of constructional materials: hard rock; limestone; gravels; sands; clays and soils. The modern city, with its increasing use of bricks and concrete and its provision of motorways, generates an enormous demand for these resources, while the costs of providing them increase greatly with each additional kilometre of transport. Better use must be made of these resources in the future by locating and characterizing them ahead of urban development, so that planning can allow for their continued extraction while the resource is viable. They must also be protected from urban encroachment, by allocating adjoining lands for some non-conflicting use such as recreation, and access roads of sufficient quality must be provided to allow rapid transport to consumer facilities. All too frequently in the past, such deposits have been built over, or their extraction prevented by encroaching suburban development. The same situation applies to energy resources; for example, there are reputedly 600 million tonnes of high-quality coal beneath the stored waters of the Sydney Metropolitan Water, Sewerage and Drainage Board (Anon. 1979). We cannot afford such misuse in the future.

Australian cities have been built on a wide range of soils, from sands to cracking clays, which includes most of the Australian Great Soil Groups. Sixteen of these, *red, yellow, lateritic* and *gleyed podzolic soils, red-brown earths, soloths, solodized solonetz and solodic soils, humus podzols, podzols, prairie soils, black earths, grey, brown and red clays, krasnozems, red earths* and *lithosols*, have been commonly used, and vary widely in their suitability for urban development. Differences in soil engineering, corrosive and drainage properties affect the choice of constructional materials, type of foundation needed, road and building design, drainage and sewerage layout, and the durability of cables and pipes. High incidence of failure or short service life has been experienced on particular soils, e.g. the failure of footings, walls and earthenware pipes in swelling clays, and the high corrosion rates of unprotected metal pipes in *solodized solonetz and solodic soils*.

Rather less attention has been given to lateral subsurface drainage problems in texture-contrast soils with slowly permeable clay subsoils, yet these are the most common soils of the cities of the east coast. During periods of protracted rain large amounts of water seep downslope in the A_2 horizons, above the clay subsoils, and build up behind retaining walls and house foundations causing failure or basement dampness, unless adequate cut-off drains have been installed. In these soils, soakage trenches for household sullage and septic tanks also become flooded during wet periods, and are consequently ineffective.

Soils also vary in their suitability for sports grounds, parks and gardens, yet the location of these has rarely been determined by soil type. Sizable areas of first-class agricultural and horticultural soils have been lost to urban development because their long-term importance to the community has not been sufficiently stressed by agriculturists; as a consequence second-class soils are being cropped at greater expense in adjacent areas. In the past, urban planning has taken little cognizance of intrinsic soil differences, yet these are obviously important in determining type of use and in the design and costs of development. Detailed soil surveys have rarely been prerequisites for urban expansion but it is unlikely that we can afford to continue ignoring soil information in the future.

Little attention has been given to groundwater resources in past urban development and cities have frequently extended over sizable aquifers. These waters are rarely used in city reticulation systems because of contamination from the urban population; in some cases they provide water for industry and recreational areas and could be used more extensively if their extent and recharge rates were better known. Future planning should consider better ways of using these resources and the possibility of recharge from the large amounts of rainwater lost from city roofs and streets; even if only 10–20% of this could be diverted into aquifers, it would still represent significant recharge. A country as dry as Australia cannot continue to ignore these resources while drowning good agricultural land by surface storage for urban water supply.

Australian cities will continue to grow as the population increases, making further demands on the surrounding rural lands. Rational land use dictates that our future urban needs for water, food, shelter, employment, transport and recreation be interfaced with and directed by the natural resources available to each city. Our planners face an exciting challenge to achieve the best possible use of our resources while still improving the quality of urban life. Future development should be planned around the natural resources available, so that the various urban uses are located on the most suitable pieces of land. Priorities should be given to extractive industries, water supply and horticulture, allowing urban development to proceed around these and some industries to be located above or close to underground aquifers. This can only happen if full characterization of the natural resources, their use options and their likely response to various forms of use are available to the planner.

Conclusions

Soils are dynamic systems, with continually changing water regimes, chemistry and biological populations, delicately poised in landscapes controlled by the geomorphic processes governing the movements of solids and solutions. As a soil develops over a long period of time it comes to dynamic equilibrium with its particular environment and, in the absence of man, short-term changes are mainly repetitive and in response to seasonal and biological variations. Every action of man, however, alters some of the environmental conditions and evokes responses from landscapes and soils as both endeavour to adjust to the changed circumstances. Consequently, all land use makes tremendous demands on both soils and landscapes, with use involving major changes initiating large, and often far-reaching responses.

Since European settlement, our land use has made, and will continue to make, major changes to Australian soils. In the past, rural use has been largely exploitative, resulting initially in widespread decline in soil fertility and subsequently in immense soil losses by erosion. Future rural use and management must aim at least at protecting our soil resources against further degradation, and preferably at improving their agricultural characteristics. This can only be achieved if all agricultural scientists and landowners have a fuller knowledge of our soils than in the past and a better appreciation of the consequences of various land use practices. As far as soil is concerned, we can no longer afford to farm on a trial-and-error basis; quite apart from the human heartbreak it usually entails, we cannot afford the loss of soil associated with failure of farming in marginal conditions. In the past, such situations have invariably led to the abandonment of farms, and have usually been accompanied by marked decline in soil fertility and severe erosion losses.

In the future, we need to learn much more about our soils, and in more detail, than has been possible in the past. While our knowledge of soil geography at the continental level is reasonably well covered by the Atlas of Australian Soils (Northcote *et al.* 1960-68), and at regional level by reconnaissance surveys by CSIRO and various state organizations, the detailed characterization of soils necessary for sound land use planning is generally not available outside of irrigation areas. This must be rectified in the near future to provide the resource base for more rational land use.

A better understanding of the natural landscape processes is also needed, and must be integrated with this improved soil knowledge and with better appreciation of the pressures exerted on the landscapes by different forms of use. Data at all levels are necessary for predictive models; some are available for nutrient and water balance modelling but there is a general lack of hard data relating to landscape processes. The recent recognition of cycloidal slope profiles (Bridge and Beckmann 1977) may provide the first step towards quantitative prediction of landscape response to man-made changes in base level.

Urban developments must be planned to make more effective use of natural resources than has been the case in the past. Again, this can only occur if the natural resources are carefully characterized and the natural processes controlling them elucidated. Such data are vital to the town planner if good use decisions are to be made.

Increasing areas of land will be needed for water catchments, recreation, and educational and scientific studies. Satisfactory management of these areas is also largely dependent on much better information being assembled on their landscapes, soils, ecosystems and land use potential, to permit assessment, e.g. the North Pine Dam Catchment Study, Queensland (Anon. 1975*b*).

In sum, rational land use is dependent on better information at all levels, beginning on a base of much better knowledge of the natural resources, of their distribution and of the processes responsible for their formation and persistence, and ending with a balanced appraisal of both the present and likely future needs of the community.

References

Anon. (1909). Our First Half Century. (Govt Printer: Brisbane, Qld.)
Anon. (1953). Yearbook of Commonwealth of Australia. (Commonw. Bureau Census Statistics: Canberra.)

Anon. (1963). Yearbook of Commonwealth of Australia. (Commonw. Bureau Census Statistics: Canberra.)

Anon. (1973). Land use. *In* 'Atlas of Australian Resources'. 2nd Ser. (Dep. Minerals and Energy: Canberra.)

Anon. (1975*a*). 'The Australian Dairy Farming Industry.' (Bureau Agric. Economics: Canberra.)

Anon. (1975*b*). 'North Pine Catchment Area Land Use Study.' (Dep. Local Govt: Queensland.)

Anon. (1976). Report of the committee on the preservation of land for horticultural and viticultural use (J. Harris, Chairman). (Govt Printer: Adelaide.)

Anon. (1977). 'Official Year Book of Australia No. 61, 1975 and 1976.' (Aust. Bureau Statistics: Canberra.)

Anon. (1978*a*). Annual Report, Development Planning Branch. (D.P.I. Queensland: Brisbane.)

Anon. (1978*b*). A basis for soil conservation in Australia. Commonwealth and State Government Collaborative Soil Conservation Study 1975-1977. Rep. No. 1. (Aust. Govt. Publ. Service: Canberra.)

Anon. (1979). Top coal reserves frozen by NSW urban planning. The Australian, 6th March, 1979, p. 12.

Arndt, W. (1966). Traffic compaction and tillage requirements. IV. The effect of traffic compaction on a number of soil properties. *J. Agric. Eng. Res.* **11**, 182-7.

Bridge, B. J. and Beckmann, G. G. (1977). Slope profiles of cycloidal form in south-east Queensland. *Science* **198**, 6102.

Courtenay, P. P. (1978). Tropical Australia. *In* 'Australia–A Geography'. (Ed. D. N. Jeans). (Sydney Univ. Press: Sydney.)

Cummins, V. G. and Bass, G. A. (1978). The Pittsworth plains and associated upland catchments, Darling Downs, Queensland. (Div. Land Utilisation, D.P.I. Qld.)

Dahlke, J. (1975). Evolution of the wheat belt in Western Australia. *Aust. Geogr.* **13**, 3-14.

Donald, C. M. (1964). The legacy of a legume – the Howard Memorial Appeal Opening. *J. Aust. Inst. Agric. Sci.* **30**, 4-5.

Donald, C, M, (1965). The progress of Australian agriculture and the role of pastures in environmental change. *Aust. J. Sci.* **27**, 187-97.

Donald, C. M. (1970). Temperate pastures species. *In* 'Australian Grasslands'. (Ed. R. M. Moore). pp. 303-20. (ANU Press: Canberra.)

Dunsdorfs, E. (1956). 'The Australian Wheat-growing Industry 1788-1948.' (Melbourne Univ. Press: Melbourne.)

Fergus, I. F. (1954). Manganese toxicity in an acid soil. *Q. J. Agric. Sci.* **11**, 15-27.

Fitzpatrick, E. H. and Nix, H. A. (1970). The climatic factor in Australian grassland ecology. *In* 'Australian Grasslands'. (Ed. R. M. Moore). pp. 3-26. (ANU Press: Canberra.)

Hallsworth, E. G., Martin, A. E., Millington, R. J. and Perry, R. A. (1976). 'Principles of a Balanced Land Use Policy for Australia'. (CSIRO Aust., Land Res. Labs: Adelaide.)

Holmes, J. (1962). The distribution of dairying in coastal New South Wales. *Aust. Geogr.* **8**, 207-20.

Hutton, E. M. (1962). Siratro as a tropical pasture legume bred from *Phaseolus atropurpeus* D.C. *Aust. J. Exp. Agric. Anim. Husb.* **2**, 117-25.

Johnson, R. D., Jones, R. L., Hinesley, T. O. and David, D. J. (1974). Selected chemical characteristics of soils, forages, and drainage water from the sewage farm serving Melbourne, Australia. (U.S. Dep. Army, Corps of Engineers: Washington, D.C.)

Kerr, H. W. and Bell, A. F. (1939). 'The Queensland Cane Growers' Handbook.' (Bur. Sug. Exp. Stats & Dept. Agric.: Brisbane.)

King, C. J. (1948). The first fifty years of agriculture in New South Wales. *Rev. Market. Agric. Econ. Sydney* **16**, 387–417.

McArthur, W. M. (1964). Soils and land use in the Dorrigo-Ebor-Tyringham Area, New South Wales. CSIRO Aust. Div. Soils, Soils and Land Use Ser. No. 46.

McGarity, J. W. (1975). Soils of the Australian wheat growing areas. *In* 'Australian Field Crops'. (Eds A. Lazenby and E. M. Matheson). Vol. 1. (Angus and Robertson: Sydney.)

Martin, A. E. and Cox. J. E. (1956). Nitrogen studies on black soils from the Darling Downs, Queensland. 1. Seasonal variations in moisture and mineral nitrogen fractions. *Aust. J. Agric. Res.* **7**, 169-83.

Molnar, I. (1974). 'A Manual of Australian Agriculture.' (3rd edn). (Heinemann: Melbourne.)

Murtha, G. G. and Reid, R. (1976). Soils of the Townsville area in relation to urban development. CSIRO Aust. Div. Soils, Rep. No. 11.

Nicolls, K. D., Colwell, J. D. and Tucker, B. M. (1953). A survey of soils, and some aspects of soil fertility, in the Lismore District, New South Wales. CSIRO Aust. Div. Soils, Soils and Land Use Ser. No. 11.

Nix, H. A. (1975). The Australian climate and its effects on grain yield and quality. *In* 'Australian Field Crops'. (Eds A. Lazenby and E. M. Matheson). Vol. 1. (Angus and Robertson: Sydney.)

Northcote, K. H., Beckmann, G. G., Bettenay, E., Churchward, H. M., van Dijk, D. C., Dimmock, G. M., Hubble, G. D., Isbell, R. F., McArthur, W. M., Murtha, G. G., Nicolls, K. D., Paton, T. R., Thompson, C. H., Webb, A. A. and Wright, M. J. (1960-68). 'Atlas of Australian Soils.' Sheets 1-10, with explanatory booklets. (CSIRO and Melbourne University Press: Melbourne.)

Northcote, K. H., Hubble, G. D., Isbell, R. F., Thompson, C. H. and Bettenay, E. (1975). 'A Description of Australian Soils.' (CSIRO: Melbourne.)

Overrein, L. N. (1976). General introduction and synthesis. *In* 'Impact of Acid Precipitation on Forest and Freshwater Ecosystems in Norway'. (Ed. F. H. Braekke). Fagrapport FR6/76 SNSF-project, NISK, Norway.

Pels, S. and Stannard, M. E. (1977). Environmental changes due to irrigation development in semi-arid parts of New South Wales. *In* 'Arid Land Irrigation in Developing Countries, Environmental Problems and Effects'. (Ed. E. Worthington). (Pergamon Press: London.)

Peltier, L. C. (1950). The geographic cycle in periglacial regions as it is related to climatic geomorphology. *Ann. Ass. Am. Geogr.* **40**, 214-36.

Samuel, G. and Piper, C. S. (1928). Grey speck (manganese deficiency) disease of oats. *J. Agric. South Australia* **31**, 696-705, 789-99.

Smith, T. F. (1978). Some effects of crop protection chemicals on the distribution and abundance of vesicular–arbuscular endomycorrhizas. *J. Aust. Inst. Agric. Sci.* **44**, 82-8.

Swaby, R. J. (1975). Biosuper – biological superphosphate. *In* 'Sulphur in Australasian Agriculture'. (Ed. K. D. McLachlan). pp. 213-20. (Sydney Univ. Press: Sydney.)

Teakle, L. J. H. and Southern, B. L. (1937). The peat soils and related soils of Western Australia. II. A soil survey of Herdsman Lake. *J. Dep. Agric. W. Aust.* **14**, 404-24.

Thompson, C. H. (1965). Soils and soil problems of the Wide Bay District. *Proc. Trop. Grasslds Soc. Aust.* **6**, 7-12.

Tiller, K. G. and de Vries, M. P. C. (1977). Contamination of soils and vegetables near the lead-zinc smelter, Port Pirie, by cadmium, lead and zinc. *Search* **8**, 78-9.

van Dijk, D. C. and Woodyer, K. D. (1961). The soils of the Yass River Valley. Regional Research and Extension Study, Southern Tablelands, N.S.W., Rep. No. 6.

Voelckers, A. (1874). Report *in* Daintree, Richard 'Analyses of Queensland Sugar and Wheat Soils with Dr Voelckers Report on their Fertility.' (Agent-General for Queensland: London 1874.)

Walker, P. H. (1960). Soil survey of County of Cumberland. Soil Surv. Unit Dep. Agric. N.S.W. Bull. No. 2.

Walker, P. H. (1963). A reconnaissance of soils in the Kempsey District, N.S.W. CSIRO Aust. Div Soils, Soils and Land Use Ser. No. 44.

Wentworth, W. C. (1819). 'Statistical, Historical and Political Description of the Colony of New South Wales and its Dependent Settlements in Van Diemans Land'. (Whittaker: London 1819. Facsimile Edition, Griffin Press: Adelaide 1978.)

Whitehead, D. C. (1964). Soil and plant-nutrition aspects of the sulphur cycle. *Soils Fert.* **27**, 1-8.

Wilson, R. K. (1961). Distribution of dairying in Victoria. *Aust. Geogr.* **8**, 51-64.

Wood, W. E. (1924). Increase of salt in soils and streams following the destruction of the native vegetation. *J. Roy. Soc. W. Aust.* **10**, 35-47.

Chapter 49
Plant nutrients in Australian soils

C. H. Williams and M. Raupach

AUSTRALIAN SOILS VARY GREATLY IN THEIR CAPACITIES to provide the chemical nutrients essential for plant growth. Although many are of at least moderate fertility, the areas of highly fertile soils are small, and there is a remarkably high proportion of extremely poor soils.

Soil weathering processes, discussed in earlier chapters, have favoured the depletion of several nutrient elements in many regions of the continent and have resulted in large areas of infertile soils. Acute deficiencies of nitrogen and phosphorus are common, and frequently there are associated deficiencies of other nutrients, especially sulfur and potassium and the trace elements molybdenum, copper, zinc and manganese. In addition deficiencies of cobalt, copper and selenium in stock are common, and there are records of more limited occurrences of deficiencies of calcium, magnesium, boron, iron and cobalt in plants. Toxicity problems also occur but are of minor importance compared with deficiency problems; most are aluminium and manganese toxicities associated with soils of low pH (Parberry 1943; Anderson and Spencer 1949; Munns 1965*a*, *b*; Russell 1966; Fergus 1954; Siman *et al.* 1971); small areas of nickel, cobalt and chromium toxicities, associated with soils formed on ultra basic parent materials, have also been reported (Anderson *et al.* 1973).

The original native vegetation was, in general, well adapted to these low fertility conditions. An ability to grow and persist at low levels of available phosphorus seems to have been particularly important and Beadle (1954, 1962*a*, 1966) has postulated that soil phosphate level was primarily responsible for the delineation of many important native plant communities. Several investigations (Specht and Rayson 1957; Specht 1963; Groves 1965; Jones 1968; Specht *et al.* 1977) have shown phosphorus to be the main nutrient limiting growth of natural plant communities, and some of these investigations have also shown that application of phosphate to the soil not only increases plant production but may also greatly change the floristics of the community. Specht and Groves (1966) have shown that Australian heath species possess several characteristics related to efficient phosphorus utilization which give them decided advantages over introduced economic species for survival under low phosphorus conditions.

The areas of Australia suitable for intensive agricultural development have been determined primarily by climatic factors affecting moisture supply. Less than 20% of the continent has natural rainfall that is adequate and reliable enough to support growth of agricultural crops and improved pastures; the areas are largely confined to the coastal fringe of the continent. Within them the correction of nutrient deficiencies has been essential to agricultural production and the widespread occurrence of phosphorus and nitrogen deficiencies has played a major part in the development of farming practice in many regions of the country.

Phosphatic fertilizers, mostly superphosphates, have been used extensively to correct the almost universal phosphorus deficiency; the more costly nitrogen fertilizers have not been greatly used except for high-production crops such as sugar, cotton, fruit and vegetables.

Farming practices have been developed which minimize the need for nitrogen fertilizer. For cereal cropping, nitrogen deficiency was initially met by bare-fallowing, which made use of and depleted the limited soil nitrogen reserves, but subsequently, soil nitrogen has been maintained and crop requirements met by the use of leguminous pasture leys. For pastures, the deficiency has been met almost entirely by the use of legumes. The most important legume is subterranean clover (*Trifolium subterraneum*) which has been sown in over 80 per cent of the improved pastures in southern Australia, mainly on acid to neutral soils. Other important legumes include barrel medic (*Medicago truncatula*), mainly on soils of high pH, white clover (*Trifolium repens*), lucerne (*Medicago sativa*) and, in northern Australia, Townsville stylo (*Stylosanthes humilis*) and siratro (*Macroptilium atropurpureum*).

Of the land now developed for intensive agriculture, approximately 27 x 10^6 ha have been sown to improved leguminous pasture, and 17 x 10^6 ha are used for cropping. Over 80% of the cropped land is used for cereal growing, about two-thirds of which is wheat.

The importance of phosphatic fertilizers, together with the successful use of the legume-superphosphate strategy for meeting the nitrogen requirements of both pastures and crops, is reflected in the relative proportions of nitrogen, phosphorus and potassium fertilizers used. For Australia the N:P:K ratio of fertilizers used in 1972–73 was 0.43:1.0:0.18, compared with 3.9:1.0:1.0 for the U.K. and 3.7:1.0:1.7 for the United States. Beyond the regions suitable for intensive agriculture large areas of arid and semi-arid natural rangelands are used for grazing by sheep and cattle. Plant production is limited by low moisture supply so that stocking rates are of necessity low and there is no economic potential for the use of fertilizers. These areas, however, account for only about 30% of the total Australian sheep and cattle population.

Phosphorus

The phosphorus content of Australian soils varies greatly; P values ranging from 1 to more than 5000 ppm have been reported for unfertilized surface soils, but the majority fall within the range 10–400 ppm. Wild (1958) estimated an overall average of 300 ppm P for Australian surface soils, thus indicating clearly a generally low content by world standards. The majority of American soils, for example, have phosphorus contents appreciably in excess of 500 ppm (Parker 1953), and English soils an average of 650 ppm (Cooke 1958).

Exceptional for their higher values are the *black earths* of eastern Australia, with phosphorus contents usually ranging from 300 to 5000 ppm, and limited areas of *krasnozems*, *euchrozems*, and *chocolate soils*, with phosphorus contents of 200–4500 ppm. These groups of soils are commonly, although not exclusively, associated with basalt or its alluvial derivatives which provide phosphorus-rich parent material.

The occurrence of such a high proportion of low-phosphorus soils has been attributed by Wild (1958, 1962) largely to the leaching of phosphate under poor drainage conditions during the mid- to late-Tertiary periods of lateritization. Beadle (1962*b*) however has rejected leaching as a major factor. He attributed the low phosphorus contents largely to the occurrence of extensive areas of phosphate-low parent materials (mainly sedimentary rocks or unconsolidated material of similar origin) which had been built up of coarser particles left behind after finer phosphate-rich material has been removed by wind and water. He also suggested that substantial losses of phosphorus could have occurred through the agencies of vegetation, either in plant material blown or washed away, or when surface soil, enriched in phosphorus by plant growth, was eroded. Both these theories emphasize different aspects of intensive weathering, which undoubtedly was a major factor leading to the low phosphorus content of so many of our soils.

As might be expected, the level of plant-available phosphorus varies widely from soil to soil, although, as already indicated, most soils are phosphorus deficient. In general, the *black earths* are the least responsive to phosphorus application. Important areas of them in south eastern Queensland and northern New South Wales have been extensively cropped, mainly for wheat and sorghum, for over 80 years without the use of phosphate fertilizer. Consistent with this, of the 460 000 ha sown to wheat in the Northern Plains districts of New South Wales in 1970–71 only 12.9% received fertilizer (Anon. 1973). In contrast, phosphorus deficiency is so acute in many of the sandy and gravelly soils of Western Australia that no yield at all is obtained if cereals are sown without superphosphate (Teakle and Carriss 1943).

In the majority of mineral soils, organic forms of phosphorus make up 20–60% of the total phosphorus of surface soils. In southern Australian soils this organic phosphorus generally has only low availability to plants, but under intensive fallowing and cropping systems its mineralization has contributed small amounts of phosphate to the available supply (Williams and Lipsett 1961).

The generally low available phosphorus status of the soils is reflected in the phosphorus content of many Australian crops. For example, an average phosphorus content of 0.26% has been found for wheat grown in southern Australia, even through phosphorus fertilizer is applied in all cases; this compares with average P values ranging form 0.36 to 0.42% for American and English grain (Woodroffe and Williams 1953). Only Queensland wheat grown on the *black earths* of the Darling Downs has a phosphorus content (0.38%) comparable with that of overseas wheats.

Residual fertilizer phosphate

Although the quantity of phosphorus taken up by plants and permanently removed from the soil in farm produce varies greatly depending on many different factors, the amounts so removed under most Australian conditions are small compared with fertilizer accessions. Teakle and Carriss (1943), for example, estimated that the average recovery of phosphorus by wheat crops in Western Australia is less than one third of that added in the fertilizer. Data of Woodroffe and Williams (1953) indicate that the average recovery of phosphorus by wheat crops in South Australia is little more than 20% of that applied, and Donald and Williams (1954) estimated that the removal of phosphorus in farm produce from pastures grazed by sheep in the Crookwell district of New South Wales is less than 5% of that applied. Ample opportunity therefore exists for the accumulation of residual phosphorus from fertilizers added to most agricultural soils.

The rate of phosphate accumulation is usually greater in pasture soils than in cropped soils because of more frequent and somewhat heavier applications of fertilizers and because of the removal of appreciably smaller amounts of phosphorus in farm produce. The availability to plants of this accumulated phosphorus is determined, however, by its ultimate chemical form.

Pasture conditions frequently favour the retention of inorganic phosphate in forms more easily available to plants. Under permanent pasture the phosphate fertilizer is broadcast onto the soil surface, where it accumulates because of its relative immobility; as a result, a zone at the soil surface becomes enriched with phosphate which is readily available to plants. If cultivation is practised, however, the phosphate is mixed throughout the entire cultivated depth of soil; this exposes the residual phosphate to chemical reaction with a large volume of soil and results in a reduced availability to plants (Williams and Simpson 1965). Thus in pasture soils, the residual inorganic phosphate retained by the soil usually has a higher relative availability to plants than in cultivated soils.

At the same time, particularly in soils of low initial organic matter content, accumulation of organic matter often occurs under pasture. This accumulation influences the level of availability of the residual phosphorus, since, especially under pasture, organic phosphorus has only low availability to plants. Thus the incorporation of applied phosphorus into the organic matter results in considerable reduction in phosphorus available to plants. Losses of available phosphorus due to this immobilization can sometimes be as serious as those resulting from grain removal under cereal cropping. The build-up in phosphorus status in pasture soils as a result of fertilizer application is thus also determined by the proportion accumulating in organic forms.

The surface accumulation also has certain disadvantages in that the residual phosphorus is exposed to a risk of ready loss should erosion take place, and surface drying may, at times, deny the plant roots full access to the readily available phosphate.

Availability of residual phosphate

The availability of the residual fertilizer phosphate varies considerably from soil to soil, depending on the capacity of the soil to sorb phosphate and the nature of the chemical reactions involved. In spite of the high incidence of phosphorus deficiency in Australian soils, most have only moderate phosphate sorption capacities and in the majority of soils the residual fertilizer phosphate retains some degree of plant availability. The most notable exceptions include some of the *krasnozems* and some *red* and *yellow earths*. These soils have high capacities to adsorb phosphate and show poor residual effects from applied phosphate (e.g. Hughes and Searle 1964). In some very light-textured soils the phosphate sorption capacity is too low to retain soluble phosphate against leaching, and serious losses of fertilizer phosphate have been reported (Hingston 1959; Paton and Loneragan 1960; Russell 1960*a*; Ozanne *et al.* 1961; Gillman 1973). However, the areas of soils with very high or very low sorption capacity are small. In the majority of soils the accumulation of residual fertilizer phosphate has led to substantial improvements in levels of plant-available phosphorus, both in cultivated soils (Teakle and Carriss 1943; Woodroffe and Williams 1953; Sims 1953) and in pasture soils (Anderson and McLachlan 1951; Donald and Williams 1954; McLachlan and Norman 1962; Cameron and McGowan 1969).

In many soils, accumulation of residual fertilizer phosphate seems likely to raise the available phosphorus level eventually to a point where the current rate of fertilizer application could be reduced to a maintenance level without adverse effects on plant yield. The point at which such a stage is reached, and the size of the maintenance requirement, vary considerably according to soil type, farming enterprise, and rainfall (e.g. Cameron *et al.* 1974). For many soils it seems possible that this level may be reached after the application of 150–200 kg of phosphorus per ha to pastures, and 200–250 kg per ha for cereal crops. Estimates of the total amounts of phosphorus added to Australian soils in fertilizers are shown in Fig. 49.1.

Nitrogen

The total nitrogen content of Australian soils varies widely, ranging from 0.01 to 0.8% in the surface horizon

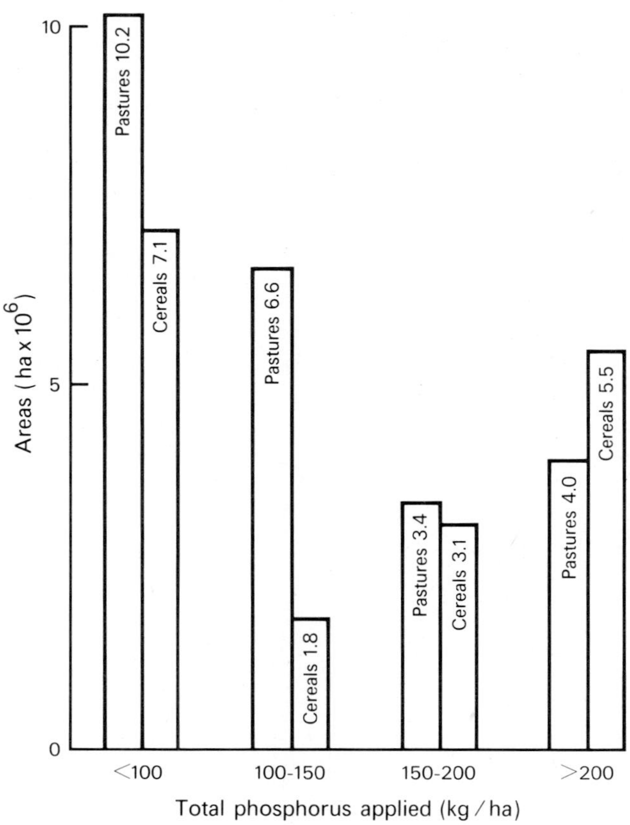

Fig. 49.1 Estimated cumulative amounts of phosphorus added in superphosphate to areas under pasture and cereals in Australia up to 1971 (ha 10^6).

of normal mineral soils and up to 2% in swamp and fen soils. Soil nitrogen is discussed in detail in Chapter 37. The majority of mineral soils have nitrogen contents of less than 0.2% in the surface horizon. The nitrogen content of surface soils is broadly correlated with rainfall (Prescott 1931) and is generally between 0.01 and 0.05% in the drier regions. Parent material is also important; where *black earths*, *euchrozems* or *krasnozems* are formed on basalt, their nitrogen content generally ranges from 0.1 to 0.8%.

Prescott (1931) also suggested that soil nitrogen may be correlated with phosphorus content. Similarly, Hubble and Martin (1960) found in Queensland soils a significant correlation between nitrogen and rainfall but they also found that nitrogen and phosphorus were more highly correlated. Partial correlation revealed that phosphorus separately accounted for 47% of the variation in total nitrogen, whereas rainfall accounted for only 8%. They considered, however, that the relationship between nitrogen and phosphorus could be indirect rather than causal.

The early practice of fallowing and cropping resulted in substantial losses of soil organic matter from all cultivated soils. The rate of organic matter breakdown varies considerably, depending on soil and environmental conditions, resulting in average annual losses of total nitrogen ranging from about 0.5% from wheat-growing soils of southern New South Wales (Williams and Lipsett 1961) to 2% from the wheat soils of the north-west of the State (Hallsworth *et al.* 1954). Commonly, the rate of organic matter loss is greatest during the first few years of cultivation and decreases with time (Clarke and Marshall 1947).

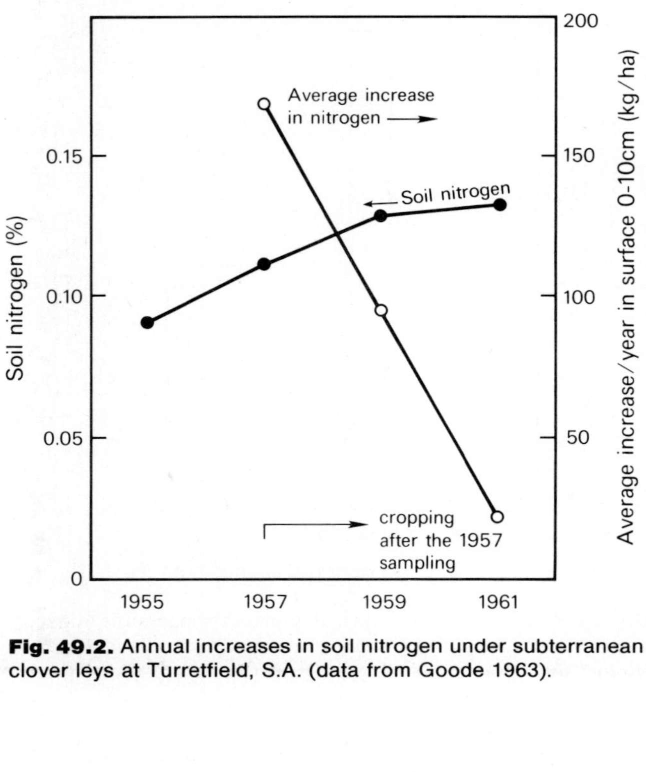

Fig. 49.2. Annual increases in soil nitrogen under subterranean clover leys at Turretfield, S.A. (data from Goode 1963).

Introduction of clover leys into cropping rotations reverses these trends, resulting in accumulation of nitrogen in the soil and restoration of fertility levels. An example of such an effect is shown in Fig. 49.2, which indicates the accumulation of nitrogen in a *red-brown earth* at Turretfield, South Australia, following the introduction of subterranean clover leys into a cropping rotation (Goode 1963). It should be noted that the 1955 sampling represents the soil nitrogen level before subterranean clover was sown and that the increases after 1957 occurred under all phases of the system, including cropping and grazing. Wheat yields for the period 1958–62 under the ley system increased to 2990 kg ha^{-1}, compared with 1690 kg ha^{-1} for the period 1953–57 under fallow-wheat.

Mineralization of nitrogen under fallow

Although fallowing has other beneficial effects such as suppression of weeds, control of pests, and seed-bed preparation, its two main functions on most Australian cereal-growing soils are to conserve moisture, and to assist the accumulation of mineral nitrogen. Under the initial conditions of low soil nitrogen status, the rainfall during the growing season in the winter rainfall areas of southern Australia was generally sufficient to satisfy crop needs (Prescott 1933), and the main function of the fallow was shown to involve the release and accumulation of mineral nitrogen (Prescott and Piper 1930; Richardson and Fricke 1931; Penman and Rountree 1932; Teakle and Samuel 1936). Fallowing enabled nitrification over and above that normally taking place during the growing season and ensured the accumulation of a store of available nitrogen to meet the early demands of the crop. In addition, the appreciable amounts of organic nitrogen mineralized during the growth of the crop contribute to the crop nitrogen supply (Storrier 1965*a*; Kohn and Storrier 1970).

It is a characteristic of arable soils generally that most of the mineral nitrogen both at the end of the fallow and during the growing season, is present in nitrate form, and that the amounts of ammonium nitrogen are usually quite small (Martin and Cox 1956; Waring and Teakle 1960; Storrier 1962). In southern Australia this nitrate may be leached into deeper subsoil horizons by autumn and winter rains (Prescott and Piper 1930; Richardson and Fricke 1931; Storrier 1962, 1965*c*). The extent of this leaching, and the possible loss of nitrate from the root zone of the crop, are determined by seasonal conditions and the soil type. On *red-brown earths* it has been found both in South Australia (Prescott and Piper 1930) and in southern New South Wales (Storrier 1962, 1965*c*) that there is usually little leaching beyond the root zone under the climatic conditions which normally prevail. Except in years of heavy rainfall, most of the leached nitrogen is retained in the surface 75–90 cm of the soil and much of it is subsequently taken up by the crop.

However, on the more sandy soils of Western Australia considerable leaching of nitrate often occurs because the water-holding capacity of the soils is low and the area has an intense winter rainfall (Halse *et al.* 1969). In northern Australia, intense summer-autumn rains may leach nitrate to 150 cm or more, i.e. well beyond the root zone of some species (Wetselaar and Norman 1960).

The early investigations into the mineralization of nitrogen were carried out under farming systems involving fallow–wheat, or similar rotations, which deplete the soil of organic matter. Subsequent investigations have shown that the patternof nitrogen release in the soil is essentially unchanged by a clover ley (Storrier 1962, 1965*a*, *c*). At the higher available nitrogen status prevailing after clover ley however, the accumulation of nitrate by the fallow may have little, or no, value in increasing crop yields (Fisher 1962; French 1963; Kohn *et al.* 1966). Indeed, under these conditions the high level of available nitrogen may sometimes result in an over supply of nitrogen, leading to depression in grain yield and haying-off of the crop (Storrier 1962; 1965*a*; Colwell 1963). This loss in potential yield at high levels of available nitrogen has been attributed by Storrier (1965*b*) to moisture stress and competition for light which, in turn, causes increased tiller mortality. In some cases yield depression at high nitrate levels could be brought about by inadequate molybdenum availability (Lipsett and Simpson 1971).

Although the losses of nitrogen from the soil by leaching are often small, leaching of mineral nitrogen from the surface soil can sometimes affect crop growth adversely. For example, Halse *et al.* (1969) showed that wheat on sandy loams in Western Australia suffered considerable nitrogen stress during the first 8–9 weeks of growth, because leaching of nitrate decreased the available soil nitrogen; the plants subsequently recovered rapidly due, probably, to increased root growth that enabled the recovery of nitrogen previously leached below the root zone. Storrier (1965*c*) considers that a similar leaching of nitrate from the surface horizon may adversely affect the growth of wheat on soils of low nitrogen status in southern New South Wales. On *red-brown earths* in South Australia, Piper and de Vries (1964) found that when wheat is grown under conditions of approximately uniform phosphorus status, yield is negatively correlated with the amount of rain falling before seeding; they attributed this to leaching of nitrate.

In addition to possible losses of nitrogen by leaching, there is evidence of gaseous loss, as N_2O and N_2, and NH_3, by denitrification processes and volatilization. Burford and Millington (1968), Burford and Greenland (1970), McGarity (1961) and McGarity and Myers (1968) have demonstrated denitrification losses from *red-brown earths*, *black earths* and *solodized solonetz soils* under field conditions in South Australia, New South Wales and Queensland. Substantial losses of nitrogen by the volatilization of ammonia from animal excreta, especially from decomposition of urea, are likely under the intermittently dry conditions commonly encountered in Australian pastures (Gillard 1967; Simpson 1968; Watson and Lapins 1969). Under sheep grazing, daily ammonia losses averaging 260 g ha^{-1} have been reported from lucerne pastures on the southern tablelands of New South Wales during autumn (Denmead *et al.* 1974), and similar losses have been found from subterranean clover pastures in the same region during spring (Denmead *et al.* 1976). Insufficient information is yet available to assess gaseous losses accurately.

Accumulation of nitrogen under pasture ley

The accumulation of nitrogen under a ley is influenced by many soil and environmental factors, and the amount of nitrogen accumulated in any one season depends largely upon whether the seasonal conditions favour the growth of the pasture legume. Clarke (1970), for example, found that under a subterranean clover sward approximately 220 kg nitrogen accumulated per ha in the surface 12.5 cm of a *red-brown earth* during a 3-year period, but that most of this increase occurred in one favourable season. Simpson *et al.* (1974) also found that nitrogen accumulation under subterranean clover pasture varies according to season, and, in fact, they recorded decreases in soil nitrogen in a season unfavourable for clover growth; Vallis (1972) has found similar seasonal effects on soil nitrogen changes under legume–grass pastures in subtropical coastal Queensland.

Average annual accessions of soil nitrogen under leguminous pasture have been reported for a wide range of soils in southern Australia to range from 28 to 170 kg ha^{-1} (Williams 1970). These figures, however, indicate only the average net gain in soil nitrogen under continuous pasture, and do not represent the total nitrogen being contributed by the legume since they do not include the nitrogen removed in farm produce, or the soil nitrogen lost by volatilization, leaching and denitrification. Further, they do not include litter nitrogen, which may provide a substantial source of readily available nitrogen for the initial crop. The total amount of nitrogen contributed by the legume under normal field conditions is probably about 50% greater than the average annual increment in soil organic nitrogen (Williams 1970).

Sulfur

The total sulfur content of Australian soils varies widely, and ranges from less than 20 to 1200 ppm in the surface horizons of mineral soils and up to 2000 ppm in peat and swamp soils. A high proportion of soils have low natural sulfur contents, with total sulfur in the range 40–200 ppm. As with nitrogen, many of the soils of higher sulfur content are associated with basaltic parent materials, e.g. the *black earths* and *krasnozems* (see Table 49.1). The higher sulfur content of these soils is due largely to their higher organic matter content and, in some cases, to the retention of sulfate by adsorption. Highly calcareous, arid, or poorly drained soils may also have high total sulfur contents due to the presence of inorganic sulfates.

Although less widespread than phosphorus or nitrogen deficiencies, sulfur deficiency is common, and the application of sulfur fertilizer is essential in many regions of the country, particularly for pasture soils. The occurrence of sulfur deficiency has been almost entirely suppressed by the widespread use of single superphosphate (containing over 11% S) for the correction of phosphorus deficiency. Use of this fertilizer has always corrected sulfur deficiency along with the phosphorus deficiency for which it was initially applied.

Until recently, single superphosphate was the main fertilizer used in Australia, but over the past 10 years there has been increasing use of higher analysis fertilizers, especially double and triple superphosphates. This has accentuated the possible occurrence of sulfur deficiency in crops and pastures and has greatly increased the need for precise recognition of sulfur-deficient soils.

The chemical nature of available sulfur in Australian soils

Sulfur occurs in soils as both organic and inorganic forms, but the proportion of organic to inorganic sulfur varies widely according to soil type and depth in the profile. In the surface horizons of most soils the sulfur is present mainly in organic forms.

The inorganic sulfur occurs almost entirely as sulfate, which may be adsorbed on the soil colloids, or be present as either water-soluble salts or as insoluble sulfate associated with calcium carbonate. Sulfate has also been reported to occur as barytes in nodular material in some soils (Beattie and Haldane 1958). Except in water-logged soils, the amounts of inorganic sulfur at lower oxidation states (e.g. sulfide) are negligible.

Adsorbed sulfate is readily available to plants. Significant amounts of adsorbed sulfate may be present in acid soils of moderate and high clay contents, such as *krasnozems*. In such soils the amounts of adsorbed sulfate adsorbed are usually adequate to ensure satisfactory levels of available sulfur for plant growth.

Sulfate, water-soluble or adsorbed or associated with calcium carbonate, is often present in the subsoil, with the amounts usually increasing with depth in the profile (Probert 1974, 1977; Williams 1974). The occurrence of gypsum is common in the subsoil horizons of many soils, including *black earths*, *red-brown earths*, *desert loams*, and *grey*, *brown and red clays*. Subsoil sulfate is often available to deep-rooted species, and can play an important role in decreasing the incidence of sulfur deficiency, especially in cereal crops and in lucerne or phalaris pastures (Lipsett and Williams 1971).

Light-textured soils, and heavier-textured soils with pH greater than 6.0, have little capacity to adsorb sulfate; the sulfate in them is largely water-soluble and subject to leaching. Consequently, except where drainage is poor, the surface horizons of these soils generally have low concentrations of soluble sulfate-sulfur, and contain most of their sulfur in organic forms. The availability of the sulfur is thus largely determined in a great many soils by the mineralization of organic sulfur, which is in turn influenced by environmental factors such as moisture, temperature and aeration, as for nitrogen.

Calcareous soils often contain appreciable amounts (up to 1000 ppm) of insoluble sulfate associated, as an impurity, with the calcium carbonate. This form of sulfate is totally unavailable to plants (Williams and Steinbergs 1964) and the availability of sulfur in these soils is also largely determined by the amounts and mineralization of organic forms of sulfur.

Sulfur is closely associated with carbon and nitrogen in the soil organic matter, therefore sulfur deficiency is not very common in crops grown without the addition of fertilizer nitrogen (Table 49.2), because adequate

Table 49.1. Sulfur content of Australian unfertilized surface soils

Figures in brackets indicate number of soils. Data taken from: Crack and Isbell (1970); Crack and Isbell (1971); Gillman (1976); Isbell and Gillman (1973); Isbell and Smith (1976); Murtha (1975); Spencer and Freney (1960); Williams, C. H. (published and unpublished data).

Parent material or soil type	Sulfur content (ppm) Range	Mean
Basalt (110)	106–1100	370
Acid igneous rocks (141)	24– 417	89
Sedimentary and metamorphic rocks (33)	32– 340	135
Transported materials (alluvium, parna, etc.) (128)	40– 430	133
Organic soils (9)	390–1860	848
Calcareous sands (9)	560–1030	818[A]

[A] 520 ppm soluble in acid.

Table 49.2. Field experiments with sulfur on cereal crops

State	Number of sites	Number of sites responding to applied sulfur	Reference
Vic.	5	0	Sims (1959)
Vic.	26	1	Briner *et al.* (1974)
Vic.	38	3	Briner *et al.* (1974)
N.S.W.	4	2	Blair & Crofts (1970)
N.S.W.	14	0	Lipsett & Williams (1971)
S.A	4	0	Rudd (1970)
S.A.	33	0	Clarke & Lewis (1974)
W.A.	11	2	Cox (1968)
W.A.	86	10	Colwell & Grove (1976)
Total	221	18	

sulfur for plant growth is usually mineralized from the soil organic matter concurrently with the mineralization of nitrogen. However, the application of nitrogen fertilizer increases the crop requirement for sulfur, and is likely to increase the incidence of sulfur deficiency; many of the responses of crops to sulfur reported in Australia have been recorded in the presence of applied nitrogen. Sulfur deficiency is common in pasture soils however, because accumulation of organic matter, rather than its mineralization, is favoured under pasture conditions. Whereas the legume component of the pasture is independent of the soil for its nitrogen requirement, it is almost entirely dependent on it for sulfur. Further, many pasture species are shallow-rooted and thus unable to exploit available subsoil sulfate.

The accession of sulfate in rainfall, common in most industrialized regions of the world, is low in most agricultural areas of Australia (e.g. Hutton and Leslie 1958) because of lower rainfall and generally smaller emissions by industry and because much of the industry is concentrated at a relatively few localities on the coast or remote from agriculture. This accentuates the likelihood of sulfur deficiency in pastures and forests established on soils of low total sulfur content at sites remote from industrial activity.

Residual fertilizer sulfur

The residual effects of sulfur applied in fertilizers are determined by many factors: they include soil type, rainfall, chemical and physical nature of the fertilizer, and the amount of fertilizer applied. Because of the mobility of soluble sulfate in most soils, immediate residual effects are usually poor because sulfate not taken up by plants during the growing season may be leached from the root zone. Hilder (1954), for example, found little residual effect in the following season from sulfate in superphosphate applied to pasture on a *black earth* on the New England tableland (predominantly summer rainfall, 750 mm); he attributed this largely to leaching. McLachlan and De Marco (1971) however, also working with a *black earth* on the southern tablelands of New South Wales (predominantly winter rainfall, 580 mm), concluded that leaching has no significant influence on the residual value of calcium sulfate; they considered that residual effects are largely determined by plant uptake during the year of application. Barrow (1966) found, on a *siliceous sand* from Western Australia (winter rainfall, 950 mm), that leaching of autumn-applied calcium sulfate could lead to sulfur deficiency in the pasture in the following spring. Gillman (1973) showed heavy losses of sulfur from the 100 cm profile of deep *siliceous sands* in north Queensland due to leaching, but Townsville stylo, because of its deep rooting ability, was able to make adequate use of the leached sulfate.

As already mentioned, the majority of improved leguminous pastures have been established on soils of low intitial fertility. Substantial increases in soil organic matter, with consequent increases in organic sulfur, have been recorded on many soils (Donald and Williams 1954; Williams and Donald 1957; Barrow 1969). Much of this organically bound sulfur originated from fertilizer top-dressings (Donald and Williams 1954) and its accumulation has resulted in higher levels of available sulfur, with a consequent reduction in the incidence of sulfur deficiency. This reflects a long-term residual effect of fertilizer sulfur. The importance of fertilizer sulfur in reducing the incidence of sulfur deficiency in this way has been clearly indicated by Briner *et al.* (1974) in a summary of sulfur responses obtained in 337 field experiments conducted by the Victorian Department of Agriculture. These results (Table 49.3) indicate a lower incidence of sulfur deficiency in soils previously treated with superphosphate. In addition, the incidence of the deficiency decreases with increase in the total quantity of superphosphate applied.

Potassium

Most Australian soils contain adequate levels of available potassium. Although areas of potassium deficiency occur throughout the continent, these are small in comparison with those deficient in phosphorus, nitrogen or sulphur (see e.g. Stephens and Donald 1958). The main cereal-growing soils for example, the *red-brown earths, solonized brown soils, grey, brown and red clays* and the *black earths*, all have adequate available potassium levels, and have been cropped now for 80–90 years without any need to apply potassium. Responses to potassium have been reported, however, in clover leys (Carroll 1961) and wheat grain (Colwell and Grove 1976) on some of the light-textured soils in the cereal-growing districts of Western Australia. This undoubtedly reflects a generally lower potassium status of these soils.

The potassium content of Australian soils, as indicated by potassium soluble in boiling 6 M hydrochloric acid, varies greatly and ranges from less than 0.01 to over 3.0% in surface soils. Low potassium levels are common in highly leached soils such as light-textured podzolized soils, in which acid-soluble potassium levels of less than 0.04% are frequent, and *krasnozems*, which often have potassium contents within the range 0.04–0.20%.

Most soils in south-eastern Australia have been shown to contain appreciable amounts of illite (Norrish and

Table 49.3. Response to sulfur by Victorian pastures — 337 experiments (Briner *et al.* 1974).

	Total superphosphate applied (kg ha^{-1})			
	Nil	<625	625-1875	>1875
Sites with no response	24	39	80	96
Sites with response to sulfur	46	19	22	10
% of sites responding to sulfur	66	33	22	9

Rogers 1956; Radoslovich 1958; Norrish 1967); this has probably ensured satisfactory levels of potassium for plant growth, and, at the same time, prevented serious loss of potassium by leaching and weathering. Deficiencies are more likely on soils formed from basalt, on lateritic or laterite-derived soils which contain little or no illite (Norrish 1967) and in a few areas such as parts of the LaTrobe Valley in Victoria where soils have developed on siliceous sands or other parent materials of low potassium content.

Potassium deficiency was first reported in pastures in 1933 in Victoria (Anon. 1933) and since then the amounts of potassium applied to pastures have increased annually. By 1966 application of potassium fertilizers to pastures accounted for almost 40% of the potassium fertilizers used. As indicated above, the deficiency is common in pastures on very light-textured soils such as siliceous and podzolized sands, and application of potassium to them is often essential for pasture establishment. On many other soils potassium deficiency reflects a deficiency induced by the depletion of available soil reserves by intensive hay making, cereal cropping or dairying.

Almost without exception, soils deficient in potassium are also deficient in phosphorus. In many soils naturally low in available potassium, the potassium deficiency has initially been masked by a more acute phosphorus deficiency. As the phosphorus status of soils is raised by the continued use of phosphorus fertilizers there is likely to be an increasing recognition of limitations to pasture growth due to these marginal potassium deficiencies.

Other nutrients

Trace elements

Trace elements have played a vital role in the development of Australian agriculture and clearly no other country has derived so much benefit from their use. The first report of a trace element deficiency in Australia was of manganese, which Samuel and Piper (1928) identifed in oats growing on *rendzina* soils in the Mt Gambier district of South Australia; by 1942 deficiencies of manganese, copper, cobalt, zinc, boron and molybdenum had been identified in crops, grazing animals or pastures. Although no accurate records are available, it seems likely that at least 16 x 10^6 ha, about 33% of the land that has been developed for agriculture, have now received at least one application of trace elements.

The deficiencies occur over a wide range of soils (Stephens and Donald 1958) and multiple deficiencies involving two or more elements are common. Most important are molybdenum, copper and zinc, which together account for over 80% of all trace element deficiencies. Their importance is indicated in Table 49.4 which presents an estimation by Donald and Prescott (1975) of the average areas of land treated annually with trace elements over the 5-year period 1967/68–1971/72. In addition to the areas listed in Table 49.4, at least 100 000 ha would have been treated with cobalt to satisfy animal requirements. Since the occurrence of trace element deficiencies is usually associated with phosphorus deficiency, it has proved relatively simple, in most cases, to correct them using additives to superphosphates. Approximately 9% of all superphosphate used in Australia contains trace elements (Williams and Andrew 1970) and about three-quarters of this is applied to pastures.

The total trace element contents of Australian soils vary widely. Table 49.5 summarizes the copper, zinc, manganese and molybdenum contents of surface soils from three regional studies (McKenzie 1966; Blackburn and Giles 1963; Giles 1959); most of the soils included in the Victorian and South Australian studies were known to be deficient in at least one of the nutrient elements.

The widespread occurrence of trace element deficiency reflects once more the prolonged or intense weathering and leaching to which so many Australian soils, or their parent materials, have been subjected. This has resulted in the removal of most of the plant-available forms from the soils or their immobilization into chemical forms of extremely low availability. Oertel and Prescott (1944) for example, showed that molybdenum has been concentrated in the ironstone gravels of many Australian soils and that substantial amounts of zinc, copper and manganese have also been immobilized in the same way. Taylor *et al.* (1964) and Taylor and McKenzie (1966) have shown that cobalt, and possibly zinc and molybdenum, are immobilized in the minerals birnessite and lithiophorite associated with manganiferous nodules and deposits in many soils. Adsorption of the limited reserves of molybdate by soil colloids also seems to be an important factor that contributes to the occurrence of molybdenum deficiency in many soils. Application of lime usually releases available molybdenum, and in

Table 49.4. Mean annual area treated with trace elements 1967/68–1971/72
(Donald and Prescott 1975)

Element	Area (ha)	Element	Area (ha)
Molybdenum	874 000	Manganese	46 000
Copper	556 000	Iron	34 000
Zinc	374 000	Boron	1 000

Table 49.5. Copper, zinc, manganese and molybdenum (ppm) in surface soils from Victoria, South Australia and Queensland

Element	Victoria[A] Range	Victoria[A] Mean	South Australia[B] Range	South Australia[B] Mean	Queensland[C] Range	Queensland[C] Mean
Copper	2.0–5.0	9.9	0.9–130	11.8	1.0–190	45.6
Zinc	1.5–100	15.7	4.0–110	23.0	6.0–180	64.7
Manganese			4–2000	293	6–3800	1222
Molybdenum	0.3–9.3	3.0	0.3–4.0	1.7	1.2–32	8.7

[A]McKenzie (1966). [B] Blackburn and Giles (1963). [C]Giles (1959)

some soils this may be adequate to fully overcome the deficiency.

In many calcareous and alkaline soils, such as aeolian *calcareous sands, solonized brown soils, grey, brown and red clays* and *black earths* of South Australia, Victoria and Queensland, manganese, copper and zinc are often present in chemical forms of very low plant-availability and deficiencies of these three nutrients are common. In many of these soils conversion of soluble manganese, copper or zinc to insoluble unavailable forms is so rapid that an application of the deficient nutrient is usually necessary at the sowing of each crop and must be supplemented by foliar spraying in some cases in order to ensure a satisfactory yield (e.g. Reuter *et al.* 1973). In acid soils good residual effects from the application of molybdenum, copper, zinc and manganese are usually found and generally it is only necessary to reapply them at intervals of 5 to 10 years. Cobalt shows poorer residual effects, and reapplication at intervals of at least every 3 years is frequently necessary.

Magnesium and calcium

Magnesium deficiency is uncommon in Australian soils. It has been recorded on only a few very acid soils and in some cases has been induced by the application of heavy dressings of potassium or ammonium fertilizers.

The most important occurrences have been in fruit and vegetable crops in New South Wales and Western Australia (Parberry 1937; Halse 1959), although Von Stieglitz (1951) recorded the deficiency in pastures on *krasnozems* formed on rhyolite in southern Queensland. The relative unimportance of magnesium deficiency in Australian soils is illustrated by results reported by Hosking (1970); in experiments conducted over a wide range of soils, mainly in areas with rainfall in excess of 500 mm, only one transitory response was recorded from the application of magnesium (as $MgSO_4$) to pastures in 190 experiments in Victoria.

Many Australian soils contain high proportions of magnesium in the exchangeable cations, especially in the subsoil and this undoubtedly is a factor contributing to the low incidence of the deficiency. These unusually high levels of exchangeable magnesium probably result, at least in part, from the influence of cyclic salts during soil development (Prescott 1931).

The occurrence of calcium deficiency in soils is often accompanied by effects on plant growth of associated factors such as low pH and metal toxicities. Liming materials are thus generally preferred to correct the deficiency, although their use in Australia has been surprisingly small.

Simple calcium deficiency has been reported as affecting the growth of flax on acid silty clays and on acid clay loams in Victoria (Millikan 1944), the seed-set in subterranean clover on a *meadow podsol* in Western Australia (Ozanne and Howes 1973), and as being the cause of bitter-pit disorder in apples in Victoria, Western Australia and Tasmania (Baxter 1960; Melville *et al.* 1964; Martin *et al.* 1965). However, the most important occurrences of the deficiency are in pasture soils where low calcium levels and associated soil properties may affect both the nodulation and growth of legumes.

Thus most of the lime used in Australia is applied to pasture in order to secure satisfactory nodulation for the establishment of legumes, to overcome deficiency of calcium, to increase soil pH, or to moderate toxicities of manganese and aluminium (Anderson and Spencer 1949; Loneragan and Dowling 1958; Munns 1965*a*, *b*; Russell 1966). The main use is on acid soils (pH < 5.5) at pasture establishment in order to provide a suitable environment for legume nodulation (Anderson and Moye 1952). This is usually achieved using quite low levels of application (200–300 kg ha^{-1}) by drilling the lime with the seed. The amount of lime used is often reduced still further by pelleting inoculated seed in a coating of lime (Loneragan *et al.* 1955).

Lime is also sometimes necessary for the correction of manganese toxicity in vegetable crops (Parberry 1943; Fergus 1954; Siman *et al.* 1966; Siman *et al.* 1971); in many cases the toxicity has been induced by heavy use of acidifying nitrogen fertilizers.

As mentioned above, liming may often release available molybdenum in deficient soils and many early reports of pasture responses to lime were later found to have been due to this effect. In such areas liming has been replaced by the simpler direct practice of molybdenum application.

Soil and fertilizer requirements of Australian crops

Wheat, barley and oats

The winter cereals wheat, oats and barley together account for approximately 85% of the total area of crops in Australia. Wheat, the most important, accounts for 55% of the cropped land (see Fig. 49.3). It is grown on a wide range of soils, the most important of which are *red-brown earths*, *solodized solonetz* and *solodic soils*, *solonized brown soils*, *yellow earths*, *red earths*, *black earths*, and *grey, brown* and *red clays* (McGarity 1975). As described above, all of these, except the *black earths*, are phosphorus deficient and require the application of phosphatic fertilizer.

In addition, the application of copper, zinc or manganese is sometimes necessary on some of the more alkaline of the *solonized brown soils* and the *grey*, *brown* and *red clays*, and deficiencies of copper, zinc and molybdenum are common in the *yellow* and *red earths*, especially in the *earthy sands*, of the sandplain soils of Western Australia. Some of the *black earths* respond to zinc.

The nutritional requirements and cultural factors governing the growth of oats and barley are similar to those of wheat. These crops are thus generally grown on the same soils as wheat, and often in rotation with it. Barley tends to be grown more frequently on lighter, more alkaline soils in low-rainfall districts, whereas oats is often grown on the more acid, moister and less well drained soils. Fertilizer practice for these two crops is closely similar to that described for wheat.

Sorghum

The growing of sorghum has undergone rapid development in recent years, the area sown having expanded from 209 000 ha in 1968/69 to almost 700 000 ha in 1972/73. The greater part of this crop is now grown on the *black earths* and *grey, brown* and *red clays* of northern New South Wales and the Darling Downs of Queensland, most commonly without the use of fertilizers. Important areas of sorghum are also grown on similar soils in the Burnett and Dawson–Callide regions of Queensland.

Rice

Rice-growing in Australia is confined to the Murray and Murrumbidgee Irrigation Areas in New South Wales, the Burdekin area of Queensland, and the Ord River area of Western Australia. In all cases it is grown as a fully irrigated crop.

In the Murray and Murrumbidgee Irrigation Areas, which in 1975/76 accounted for over 95% of the total Australian rice production, the rice is grown mainly on *grey, brown* and *red clays* and on *red-brown earths* or transitional *red-brown earths*. Yields are among the highest in the world, averaging over 5 t ha^{-1}. Much of the rice is grown in 3- to 7-year rotations with pastures and cereals and the crop is usually sown following several years of leguminous pasture (generally Wimmera ryegrass and subterranean clover). Although the application of superphosphate is necessary for satisfactory pasture growth, nitrogen is the only fertilizer required for the rice crop and its application depends upon the previous history of the soils; in many cases the nitrogen fixed by the legume is adequate and no fertilizer is necessary.

Recently there has been a trend toward growing successive crops of rice and other cereals, eliminating the pasture phase of the rotation. Consequent problems have been experienced in the growth of summer crops after rice on some soils, and may be due to low phosphorus availability following increases in phosphate sorption and buffer capacity of the soil during waterlogging (Willett 1979).

Sugarcane

Sugarcane is cultivated on a range of soils along the coastal regions of Queensland and New South Wales. Climatic factors have determined the suitability of these areas for sugar growing; the more important soils include *gleyed podzolic soils, humic gleys, podzols, krasnozems* and *red earths*, but many other soil groups are also involved. The high nutrient requirements of the crop are met by extensive use of nitrogen, potassium and phosphorus fertilizers, and, although this crop only accounts for 2% of the total area of cropped land (Fig 49.3), it

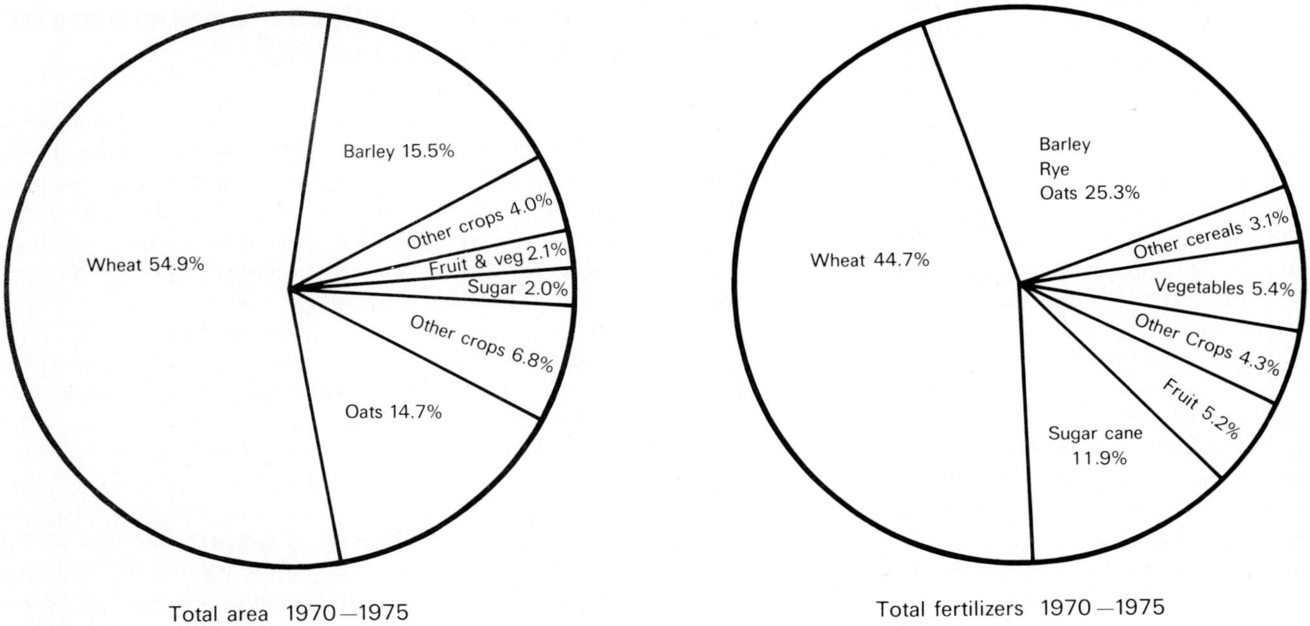

Fig. 49.3. Percentages of the total area under various crops and total fertilizer applied to them in Australia over the years 1970–75 (data from Year Book Australia 1975–76, No. 61).

accounts for 40, 25 and 1.5% of the total nitrogen, potassium and phosphorus fertilizer usage.

Fruit and vegetables

Fruit is grown on a wide range of soils because the choice of site may have been determined by suitability of soil or climate to the particular crop, or by the availability of irrigation water; the soils include *red-brown earths*, *solonized brown soils*, *brown podzolic soils*, *gleyed podzolic soils*, *krasnozems*, *terra rossa soils*, *rendzinas* and *alluvial soils*. The initial fertility of the soils thus ranged from poor to moderate and high, so that fertilizer practice has been determined by the soil selected and the nutrient requirement of the particular crop. Generally, nitrogen rather than phosphorus is the most important nutrient, potassium deficiency is common on some soils, and deficiencies of calcium, magnesium, copper, zinc, manganese, iron and boron have been reported in specific cases.

Fruit crops account for about 1.3% of the total area under crops in Australia, but they use about 15% of the nitrogen, 10% of the potassium and only 1.1% of the phosphorus fertilizers.

Vegetables are also grown on a wide range of soils; the choice of which has been determined, as with fruit crops, by soil or climatic factors essential to the particular crop, availability of irrigation water, or nearness to market. Potatoes, for example, require deep friable soils, but are grown on many different soil types; the crop itself has a high requirement for phosphorus and, since many of the preferred soils have high capacities to fix phosphate into unavailable forms, high annual rates of phosphate application are normal. The application of the 1000–2000 kg of superphosphate per ha per crop is common, nitrogen application is also usual and potassium is required on some soils.

Cotton

Approximately three-quarters of the Australian cotton crop is now grown under irrigation on *black earths* in the Namoi and Macquarie Valleys of northern New South Wales. The phosphorus status of these soils is quite adequate and only nitrogenous fertilizer (mainly anhydrous ammonia) is used. The remainder of the crop is grown on similar soils at several centres in Queensland.

Nitrogen, phosphorus, potassium and sulfur in forest soils

Australian native forests have grown useful stands of hardwood on a number of relatively impoverished soils in areas, frequently mountainous, where the rainfall is assured. Also, eucalypts and other native species have been re-established successfully, with the addition of only token amounts of fertilizer, in areas which have been clear felled or used for catchments or for other strategic purposes. Because it has now become apparent that there are profitable returns in growth when more liberal amounts of fertilizer are used, a number of field trials based mainly on nitrogen and phosphorus have been established (Cromer 1971). In at least one case the responses of eucalypts in such a trial have been described as dramatic (Cromer *et al.* 1975).

By contrast, Australia's softwood resources are man-made forests, mostly conifers, and predominantly *Pinus radiata* in the southern half of the continent. A review of the soil and fertilizer requirements of this species has been given by Raupach (1967*a*).

As a result of a somewhat similar philosophy to that adopted for hardwoods, that fertilizers were not necessary even on very poor soils, a number of softwood plantations either failed or performed very poorly during the early decades of this century.

Practice has shown that on poor soils the addition of fertilizers at establishment frequently ensures survival, presumably because the young trees are enabled to explore a sufficient volume and depth of soil to obtain supplies of nutrients and moisture in situations which would otherwise prove restrictive. By this means, growth is sustained until nutrient cycling accelerates due to canopy closure. This also means that, if fertilizers are added early, subsequent annual increments of wood are laid down on a stem of appreciable dimensions. An eventual return on the cost of establishment is thus made more likely. For these reasons the regeneration of stagnated trees by fertilizer additions is not favoured, except when growth has slowed down in spite of the previous stimulation effect of earlier fertilizer additions.

The soils on which *P. radiata* is grown comprise *humus podzols*, *podzols*, *yellow podzolic soils*, *red podzolic soils*, *lateritic podzolic soils*, *gleyed podzolic soils*, *krasnozems*, and *terra rossa soils*. The first six have a wide range of parent materials, the *krasnozems* are restricted to basalt and the *terra rossa soils* to sedimentary and dune limestones. Plantations on these soils fail where they are shallow or excessively stony or have low moisture storage capacity or where their surface pH is less than 4.5. Responses to phosphorus or zinc or to both of these nutrients are particularly common.

In addition, deficiencies have been reported for nitrogen, sulfur, potassium, boron, copper, and manganese, as well as toxicities due to excess sodium or chloride.

Many of the soils requiring fertilizer present difficulties either because added fertilizers are fixed and so rendered unavailable to trees (as for phosphorus in the laterite-rich horizons of *lateritic podzolic soils*), or because leaching of soluble nutrients occurs very easily (as for the bleached siliceous sand horizons in some deep *podzols* and *yellow podzolic soils*). Various designs of phosphate fertilizers (Raupach 1968) and sequences of cultural practices (Woods 1976) have been proposed to combat such problems.

To assist management decisions on the timing of fertilizer additions and on the nutrients to be added to the forests, monitoring schemes have been proposed (see for example Richards and Bevege 1972). While these may be based on either soil or plant analyses, foliar analysis seems in practice to be superior, except that both types of analysis are useful in isolated cases of extreme nutrient deficiency. The conditions for monitoring by using foliar analyses have been defined by Raupach (1967*a*).

For *P. radiata*, sampling criteria and foliar critical levels suitable for diagnostic use in the field have been established for potassium (Fig. 49.4); Raupach and Hall 1974), for phosphorus and nitrogen (Raupach 1967*a*; Raupach *et al.* 1969), and for some of the trace elements (Raupach 1975; Raupach *et al.* 1978; Raupach and Clarke 1978). The various errors entering into foliar sampling have been estimated by Raupach and Hall (1974), using foliar potassium as an example.

Raupach (1967*a*, *b*) and Raupach *et al.* (1969) found that the productivity of plantations of *P. radiata* was related to the phosphorus (p) and nitrogen (n) foliar levels. The interaction term $n \times p$ was not significant, and the relationships found took the form: volume increment $= k + k_1 n + k_2 p$ where k, k_1 and k_2 are coefficients estimated from the data. This relationship means that the critical foliar levels of these two nutrients depend on one another. In field experiments with this species at Belanglo State Forest in New South Wales, Waring (1969) found that the addition of nitrogen and phosphorus together causes the trees to take up about six times the nitrogen taken up by trees receiving nitrogen alone or by untreated trees; tree growth is proportionately greater. After 3.5 years the nitrogen content of the surface soil (0–8 cm depth) of the nitrogen-plus-phosphorus plots had increased by an amount equivalent to 30% of the nitrogen (0.5 t ha^{-1}) applied as urea, and 40% of the fertilizer nitrogen had been recovered by the aerial parts of the trees. Similar responses to the combined application of nitrogen and phosphorus have been found for the following species on *lateritic podzolic soils: P. pinaster* (Keay *et al.* 1969; Turton and Keay 1970), *P. taeda* (Richards and Bevege 1967) and *Agathis robusta* (Kauri pine) (Richards and Bevege 1968).

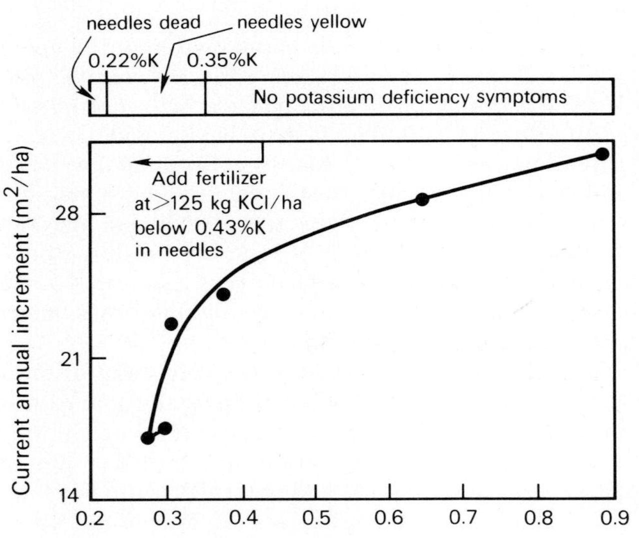

Fig. 49.4 Relationship between the current annual increment in wood volume of *P. radiata* and the potassium content of the foliage (Raupach and Hall 1974). Samples of foliage were from mature needles of the current season's growth on the first laterals down from the top of the trees. Foliar needle symptoms are also indicated.

Sulfur requirements under forestry conditions are low (about 1 kg ha^{-1} $year^{-1}$) and growth responses to sulfur by trees are uncommon. There are, however, several places in eastern Australia, particularly on the tablelands, where symptoms identical with those of sulfur deficiency have been observed; they occur most commonly in fast-growing plantations on soils derived from granite or basalt at sites remote from coastal or industrial influence (Humphreys *et al.* 1975).

Effect of cultural practices on tree growth

Some site properties of the soil and some cultural practices of forestry also influence tree growth and nutrition.

The responses of tree growth to phosphate fertilizer are greater on wet soils in the southeast of South Australia than on dry sites, but the foliar concentrations of phosphorus tend to be less than on the dry sites (Raupach *et al.* 1969).

Cultivation increased the rate of height growth of three pine species, including *P. radiata* on three soils (two *podzols*, one shallow and the other deep, and a 'shallow ironstone' soil) over the first 3–5 years of tree growth (Raupach *et al.* 1975).

Ashbeds are a common feature of Australian softwood plantations. When the land is prepared for planting, the cleared vegetation is heaped into windrows which are then burnt. Quantities of surface soil are inevitably scraped up with the vegetation and so incorporated in the ashbeds. Compared with the normal soil, ashbed soils contain higher amounts of soluble salts, nitrogen, phosphorus, potassium and manganese, and they support a superior rate of tree growth (in terms of both height and basal area) for the first 12 years after planting, but not subsequently (Cromer *et al.* 1970; Waring 1973).

The second and possibly subsequent crops of *P. radiata* produce less timber than the first crop (Keeves 1966; Bednall 1968). The reasons for this decline are not well understood and probably differ among plantation soils.

Effects of pasture improvement on soil fertility

When leguminous pastures were first established their sole purpose was to provide better grazing for increased animal production. They have also proved to be a valuable means for improving the fertility of poor soils, and have become an essential factor in Australian agriculture. Their development, however, has in all cases depended upon the correction of phosphorus deficiency by the use of phosphate fertilizers.

When such a pasture is established, it initiates a series of soil changes that lead to a general build-up in the available supply of several plant nutrients. Residual fertilizer phosphate is accumulated, and soil nitrogen and sulfur increase as a consequence of gains in soil organic matter. The importance of these elements has already been discussed, but other factors are also involved, some of which contribute to fertility improvement and some of which may prove disadvantageous.

The increases in organic matter can lead to improvements in soil structure. In some soils under cereal cropping rotations, the introduction of clover leys has arrested deterioration in soil structure and associated erosion losses, and has restored the soil structure to satisfactory levels (Greacen 1958; Smith 1969).

The increases in soil organic matter have also resulted in appreciable increases in the water-holding capacity of many pasture soils (Williams and Donald 1957; Russell and Shearer 1964; Barrow 1969), although Russell and Shearer (1964) have shown that the resultant increases in available water capacity are small in relation to the increases in organic matter.

Compared with the mineral constituents of the soil, soil organic matter has a high cation exchange capacity, so that the increases in organic matter result in substantial increases in the cation exchange capacity of many soils (Williams and Donald 1957; Russell 1961; Barrow 1969). Associated with these increases there have been changes in the amounts of the exchangeable cations; in most cases increases in exchangeable calcium and hydrogen have been recorded, the changes in exchangeable calcium reflecting the retention, at least in part, of calcium added in superphosphate. In *yellow podzolic soils* from the Crookwell district of New South Wales, Williams and Donald (1957) also recorded increases in exchangeable potassium, which were subsequently shown (Williams and Lipsett 1960) to result in significant improvements in the potassium status of these soils. Russell (1961) found that exchangeable potassium had remained unchanged in a *meadow podzol* at Kybybolite in South Australia, and Barrow (1969), although he recorded increases in exchangeable potassium in *meadow podzolic soils* in Western Australia, concluded that the buffering capacity of the soil for potassium is only slightly affected by increases in organic matter. Simpson *et al.* (1971) recorded substantial increases in exchangeable potassium on a *podzolic soil* near Canberra in the Australian Capital Territory after 6 years under subterranean clover–phalaris pasture.

A further consequence of the increase in cation exchange capacity, and the accompanying increases in exchangeable hydrogen, is that in many soils the pH decreases as organic matter builds up under the pasture (Donald and Williams 1954; Russell 1960*b*). The decrease in pH has been shown (Williams and David 1976) to increase the availability and plant uptake of the metals manganese, copper, zinc, iron and aluminium. Although the increase in availability of copper, zinc, and manganese may be advantageous in marginally deficient soils and may prolong the residual effects of trace metal fertilizers applied to deficient soils, the amounts of available manganese and aluminium may be raised to toxic levels in some soils (Flemons and Siman 1970; Williams and David 1976). It is clear that as a result of the decline in pH some soils have now reached a stage where corrective measures, such as liming, are becoming necessary and it seems likely that increasing areas of soils will be similarly affected in the future.

Future trends

Economic projections suggest that major changes in the patterns of agricultural production in Australia are unlikely in the immediate future. At the same time, they indicate that substantial increases in crop and pasture productivity will be essential if overseas markets are to be maintained.

It is also unlikely that there will be any major change in the general pattern of fertilizer use. Phosphorus, nitrogen and sulfur will remain the main nutrient deficiencies in Australian soils. In spite of the long history of phosphate fertilizer usage, phosphorus is still the most important nutrient deficiency, and, although the phosphorus status of many soils has been raised considerably by past fertilizer practice, most soils are still responsive. Over 40% of all developed land has received, as yet, a total of less than 100 kg of phosphorus per ha, and only about 20% has had more than 200 kg. In addition, there will be an increasing need for phosphorus fertilizer on the limited areas of naturally phosphorus-adequate soils as their phosphorus reserves are depleted by continued cropping. Moreover, it is reasonable to assume that all new land brought into cultivation will require phosphorus and nitrogen and sulfur in amounts at least equal to those that have been required on land already developed.

There will probably be decreasing needs for sulfur, especially on pastures, as the sulfur status of soils is built up by fertilizers, but a need will remain for sulfur on many soils, especially if there are increases in the use of nitrogen fertilizers and high-analysis phosphate fertilizers. The need for potassium fertilizers seems likely to increase.

Legumes are likely to remain the most satisfactory means for meeting nitrogen deficiency in pastures in southern Australia, both because of fertilizer cost and because of the difficulities associated with maintaining satisfactory stock feed with fertilizer nitrogen under the intermittently dry conditions which prevail over large areas. Continued development of leguminous pastures in northern Australia will require both phosphorus and sulfur fertilizer as at present. Although there is considerable potential for the use of nitrogen fertilizers on grass pastures in some regions of northern Australia (Donald 1960), their full role is yet to be defined.

Henzell (1970) considers that in Queensland most nitrogen fertilizer will be used for selective or strategic applications to overcome temporary feed shortages and that its extensive use on grass is unlikely in the immediate future. In his view the main use in the future will be principally for dairy production in the coastal areas south of the Tropic of Capricorn receiving more than 900–1000 mm annual rainfall.

In areas of southern Australia where rainfall is more reliable, it seems possible that the use of fertilizer nitrogen on cereal crops will increase, especially on second or third crops in the rotation.

Increased productivity must thus be attained by more efficient use of fertilizers, by more precise recognition of nutrient deficiencies, and by more precise estimation of the amounts of fertilizer nutrients required for optimum production. Changes in fertilizer form, as for example more extensive use of ground rock phosphate or rock phosphate-containing fertilizers on pastures, may be required to reduce costs.

References

Anderson, A. J. and McLachlan, K. D. (1951). The residual effect of phosphorus on soil fertility and pasture development on acid soils. *Aust. J. Agric. Res.* **2**, 377-400.

Anderson, A. J., Meyer, D. R. and Mayer, F. K. (1973). Heavy metal toxicities. Levels of nickel, cobalt and chromium in the soil and plants associated with visual symptoms and variation in growth of an oat crop. *Aust. J. Agric. Res.* **24**, 557-71.

Anderson, A. J. and Moye, D. V. (1952). Lime and molybdenum in clover development on acid soils. *Aust. J. Agric. Res.* **3**, 95-110.

Anderson, A. J. and Spencer, D. (1949). Molybdenum and sulphur in symbiotic nitrogen fixation. *Nature (London)* **164**, 273-4.

Anon. (1933). Pasture improvement. A summary of a progress report on the investigations initiated in 1932 by the Department of Agriculture in cooperation with the Victorian Pasture Improvement League. *J. Agric. Victoria* **31**, 235-7.

Anon. (1973). Official Year Book N.S.W. No. 62.

Barrow, N. J. (1966). The residual value of the phosphorus and sulphur components of superphosphate in some Western Australian soils. *Aust. J. Exp. Agric. Anim. Husb.* **6**, 9-16.

Barrow, N. J. (1969). The accumulation of soil organic matter under pasture and its effect on soil properties. *Aust. J. Exp. Agric. Anim. Husb.* **9**, 437-44.

Baxter, P. (1960). Bitter pit of apples. Effect of calcium sprays. *J. Agric. Victoria* **58**, 801-11.

Beadle, N. C. W. (1954). Soil phosphate and the delimitation of plant communities in eastern Australia. *Ecology* **35**, 370-5.

Beadle, N. C. W. (1962*a*). Soil phosphate and the delimitation of plant communities in eastern Australia. II *Ecology* **43**, 281-8.

Beadle, N. C. W. (1962*b*). An alternate hypothesis to account for the generally low phosphate content of Australian soils. *Aust. J. Agric. Res.* **13**, 434-42.

Beadle, N. C. W. (1966). Soil phosphate and its role in molding segments of the Australian flora and vegetation with special reference to xeromorphy and sclerophylly. *Ecology* **47**, 992-1007.

Beattie, J. A. and Haldane, A. D. (1958). The occurrence of palygorskite and barytes in certain parna soils of the Murrumbidgee region, New South Wales. *Aust. J. Sci.* **20**, 274-5.

Bednall, B. H. (1968). The problem of lower volumes associated with second rotations in *Pinus radiata* plantations in South Australia. Woods and Forests Dep. South Australia Bull. No. 17.

Blackburn, G. and Giles, J. B. (1963). Trace element concentrations in certain soils of the lower south-east of South Australia. CSIRO (Aust.) Div. Soils Div. Rep. No. 2/63.

Blair, G. J. and Crofts, F. C. (1970). The effect of sulphur fertilization on the yield and chemical composition of forage oats on the central tablelands of New South Wales. *Aust. J. Exp. Agric. Anim. Husb.* **10**, 768-73.

Briner, G. P., Hosking, W. J. and Rooney, D. R. (1974). Importance of Sulphur as a Nutrient for Victorian Agricultural Industries. *In* 'Handbook on Sulphur in Australian Agriculture'. (Ed. K. D. McLachlan). p.24. (CSIRO: Melbourne.)

Burford, J. R. and Greenland, D. J. (1970). Denitrification under an annual pasture. Proc. XI Int. Grassl. Cong. pp.458-61.

Burford, J. R. and Millington, R. J. (1968). Nitrous oxide in the atmosphere of a red-brown earth. Trans. 9th Int. Cong. Soil Sci. Adelaide. Vol. 2, pp.505-11.

Cameron, I. H., Douglas, M. H. and Hosking, W. J. (1974). How much super for pastures now? *J. Agric. Victoria* **72**, 220-4.

Cameron, I. H. and McGowan, A. A. (1969). Responses by pasture to current and previous applications of superphosphate. *Aust. J. Exp. Agric. Anim. Husb.* **9**, 617-21.

Carroll, M. D. (1961). Potassium deficiency in the wheat belt. *J. Agric. West. Aust.* **2**, 457-8.

Clarke, A. L. (1970). Nitrogen accretion by an impoverished red-brown earth soil under short-term leys. Proc. XI Int. Grassl. Congr. p. 461.

Clarke, A. L. and Lewis, D. C. (1974). South Australia. *In* 'Handbook on Sulphur in Australian Agriculture'. (Ed. K. D. McLachlan). p.42. (CSIRO: Melbourne.)

Clarke, G. B. and Marshall, T. J. (1947). The influence of cultivation on soil structure and its assessment in soils of variable mechanical composition. *J. Counc. Sci. Ind. Res. Aust.* **20**, 162-75.

Colwell, J. D. (1963). The effects of fertilizers and season on the yield and composition of wheat in southern New Southern Wales. *Aust. J. Exp. Agric. Anim. Husb.* **3**, 51-61.

Colwell, J. D. and Grove, T. S. (1976). Assessments of potassium and sulphur fertilizer requirements of wheat in Western Australia. *Aust. J. Exp. Agric. Anim. Husb.* **16**, 748-54.

Cooke, G. W. (1958). The nations food larder. *J. Sci. Food Agric.* **3**, 761-72.

Cox, W. J. (1968). The phosphorus fertilizers – how they compare. *J. Agric. West. Aust.* **9**, 434-6.

Crack, B. M. and Isbell, R. F. (1970). Studies on some solodic soils in north-western Queensland. 1. Morphological and chemical characteristics. *Aust. J. Exp. Agric. Anim. Husb.* **10**, 334-41.

Crack, B. M. and Isbell, R. F. (1971). Studies on some neutral red duplex soils (Dr 2.12) in north-eastern Queensland. 1. Morphological and chemical characteristics. *Aust. J. Exp. Agric. Anim. Husb.* **11**, 328-35.

Cromer, R. N. (1971). Fertilizer trials in young plantations of eucalypts. *Aust. For. Res.* **5**, 1-10.

Cromer, R. N., Raupach, M. and Clarke, A. R. P. (1970). An examination of ashbed soils growing radiata pine in Gippsland, Victoria. *Aust. For.* **34**, 1-10.

Cromer, R. N., Raupach, M., Clarke, A. R. P. and Cameron, J. N. (1975). Eucalypt plantations in Australia – the potential for intensive production and utilization. *Appita* **29**, 165-73.

Denmead, O. T., Simpson, J. R. and Freney, J. R. (1974). Ammonia flux into the atmosphere from a grazed pasture. *Science* **185,** 609-10.

Denmead, O. T., Freney, J. R. and Simpson, J. R. (1974). A closed ammonia cycle within a plant canopy. *Soil Biol. Biochem.* **8,** 161-4.

Donald, C. M. (1960). The impact of cheap nitrogen. *J. Aust. Inst. Agric. Sci.* **26,** 319-38.

Donald, C. M. and Prescott, J. A. (1975). Trace Elements in Australian Crop and Pasture Production, 1924-1974. *In* 'Trace Elements in Soil-plant-animal Systems'. (Eds D. J. D. Nicholas and A. R. Egan). p.7. (Academic Press: New York.)

Donald, C. M. and Williams, C. H. (1954). Fertility and productivity of a podzolic soil as influenced by subterranean clover *(Trifolium subterranean L.)* and superphosphate. *Aust. J. Agric. Res.* **5,** 664-87.

Fergus, I. F. (1954). Manganese toxicity in an acid soil. *Queensl. J. Agric. Sci.* **11,** 15-27.

Fisher, H. M. (1962). Farming without fallow. *J. Dep. Agric. West. Aust.* **3,** 172-84.

Flemons, K. and Siman, A. (1970). Goulburn lucerne failures linked to induced manganese toxicity. *Agric. Gaz. N.S.W.* **81,** 662-3.

French, R. J. (1963). New facts about fallowing. *J. Dep. Agric. South Aust.* **67,** 42-8.

Giles, J. B. (1959). Trace element content of some Queensland surface soils. CSIRO (Aust.) Div. Soils Div. Rep. No. 1/59.

Gillard, P. (1967). Coprophagous beetles in pasture ecosystems. *J. Aust. Inst. Agric. Sci.* **33**, 30-4.

Gillman, G. P. (1973). Studies on some deep sandy soils in Cape Yorke Peninsula, North Queensland. 3. Losses of applied phosphorus and sulfur. *Aust. J. Exp. Agric. Anim. Husb.* **13**, 418-22.

Gillman, G. P. (1976). Red basaltic soils in north Queensland. 2. Chemistry. CSIRO (Aust.) Div. Soils Tech. Pap. No. 28, p.23.

Goode, J. R. (1963). Clover leys lift wheat and wool output. *J. Dep. Agric. South Aust.* **66**, 358-65.

Greacen, E. L. (1958). The soil structure profile under pastures. *Aust. J. Agric. Res.* **9**, 129-37.

Groves, R. H. (1965). Growth of heath vegetation. II. The seasonal growth of a heath on a ground-water podsol at Wilson's Promontory, Victoria. *Aust. J. Bot.* **13**, 281-9.

Hallsworth, E. G., Gibbons, F. R. and Lemerle, T. H. (1954). The nutrient status and cultivation practices of the soils of the north-west wheat belt of New South Wales. *Aust. J. Agric. Res.* **5**, 422-47.

Halse, N. J. (1959). Magnesium deficiency of apples and oranges. *J. Agric. West. Aust.* **8**, 417-19.

Halse, N. J., Greenwood, E. A. N., Lapins, P. and Boundy, C. A. P. (1969). An analysis of the effects of nitrogen deficiency on the growth and yield of a Western Australian wheat crop. *Aust. J. Agric. Res.* **20**, 987-98.

Henzell, E. F. (1970). Use of nitrogenous fertilizers on subtropical pastures in Queensland. *J. Aust. Inst. Agric. Sci.* **36**, 206-13.

Hilder, E. J. (1954). Some aspects of sulphur as a nutrient for pastures in New England soils. *Aust. J. Agric. Res.* **5**, 39-54.

Hingston, F. J. (1959). The loss of applied phosphorus and sulphur from soils under pasture in W.A. *J. Aust. Inst. Agric. Sci.* **25**, 209-13.

Hosking, J. (1970). Dolomite. The facts on magnesium as a fertilizer. *J. Agric. Victoria* **68**, 281.

Hubble, G. D. and Martin, A. E. (1960). The nitrogen status of Queensland soils. Symposium on nitrogen Part 1. Cunningham Laboratory, CSIRO, Brisbane, p.5-1.

Hughes, J. D. and Searle, P. G. E. (1964). Observations on the residual value of accumulated phosphorus in a red loam. *Aust. J. Agric. Res.* **15**, 377-83.

Humphreys, F. R., Lambert, M. J. and Kelly, J. (1975). The Occurrence of Sulphur Deficiency in Forests. *In* 'Sulphur in Australasian Agriculture'. (Ed. K. D. McLachlan). p.154. (Sydney Univ. Press: Sydney.)

Hutton, J. T. and Leslie, T. I. (1958). Accession of non-nitrogenous ions dissolved in rainwater to soils in Victoria. *Aust. J. Agric. Res.* **9**, 492-507.

Isbell, R. F. and Gillman, G. P. (1973). Studies on some deep sandy soils in Cape York Peninsula, North Queensland. 1. Morphological and chemical characteristics. *Aust. J. Exp. Agric. Anim. Husb.* **13**, 81-8.

Isbell, R. F. and Smith, R. M. (1976). Some properties of red, yellow and grey massive earths in north Queensland. CSIRO (Aust.) Div. Soils Tech. Pap. No. 30.

Jones, R. (1968). Productivity studies on heath vegetation in southern Australia. The use of fertilizer in studies of production processes. *Folia Geobot. Phytotaxon.* **3**, 355-62.

Keay, J., Turton, A. G. and Campbell, N. A. (1969). Fertilizer response to maritime pine on a lateritic soil. *Aust. For.* **33**, 248-58.

Keeves, A. (1966). Some evidence of loss of productivity with successive rotations of *Pinus radiata* in south-east South Australia. *Aust. For.* **30**, 51-63.

Kohn, G. D. and Storrier, R. R. (1970). Time of sowing and wheat production in southern New South Wales. *Aust. J. Exp. Agric. Anim. Husb.* **10**, 604-9.

Kohn, G. D., Storrier, R. R. and Cuthbertson, E. G. (1966). Fallowing and wheat production in southern New South Wales. *Aust. J. Exp. Agric. Anim. Husb.* **6**, 233-41.

Lipsett, J. and Simpson, J. R. (1971). Wheat responses to molybdenum in southern N.S.W. *J. Aust. Inst. Agric. Sci.* **37**, 348-51.

Lipsett, J. and Williams, C. H. (1971). The sulphur status of wheat in red-brown earths in southern New South Wales. *Aust. J. Exp. Agric. Anim. Husb.* **11**, 59-63.

Loneragan, J. F. and Dowling, E. J. (1958). The interaction of calcium and hydrogen ions in the nodulation of subterranean clover. *Aust. J. Agric. Res.* **9**, 464-72.

Loneragan, J. F., Meyer, D., Fawcett, R. G. and Anderson, A. J. (1955). Lime-pelleted clover seeds for nodulation on acid soils. *J. Aust. Inst. Agric. Sci.* **21**, 264-5.

McGarity, J. W. (1961). Denitrification studies on some South Australian soils. *Plant Soil* **14**, 1-21.

McGarity, J. W. (1975). Soils of the Australian Wheat-growing Areas. *In* 'Australian Field Crops 1. Wheat and Other Temperate Cereals'. (Eds A. Lazenby and E. M. Matherson). p.227. (Angus and Robertson: Sydney.)

McGarity, J. W. and Myers, R. J. K. (1968). Denitrifying activity in solodized solonetz soils of eastern Australia. *Soil Sci. Soc. Am. Proc.* **32**, 812-17.

McKenzie, R. M. (1966). The relation of laboratory analyses for copper, zinc and molybdenum in some Victorian soils to the results of field trials. *Aust. J. Exp. Agric. Anim. Husb.* **6**, 170-4.

McLachlan, K. D. and De Marco, D. G. (1971). Current and residual effects of calcium sulphate applications on a sulphur deficient basaltic soil. *Aust. J. Exp. Agric. Anim. Husb.* **11**, 64-70.

McLachlan, K. D. and Norman, B. W. (1962). Effects of previous superphosphate applications to the pasture environment and the response by pasture to a current dressing. *Aust. J. Agric. Res.* **13**, 836-52.

Martin, A. E. and Cox, J. E. (1956). Nitrogen studies on black soils from the Darling Downs, Queensland. 1. Seasonal variations in moisture and mineral nitrogen fractions. *Aust. J. Agric. Res.* **7**, 169-83.

Martin, D., Lewis, T. L. and Cerny, J. (1965). Experiments with orchard spray treatments for the control of bitter pit in apples in Tasmania. CSIRO (Aust.) Div. Plant Ind. Tech. Pap. No. 22.

Melville, F., Hardisty, S. E. and Shorter, N. S. (1964). The control of bitter pit in apples. *J. Agric. West. Aust.* **5**, 938-44.

Millikan, C. R. (1944). 'Wither top' (calcium deficiency) disease in flax. *J. Agric. Victoria* **42**, 79-91.

Munns, D. N. (1965*a*). Soil acidity and growth of a legume. 1. Interactions of lime with nitrogen and phosphate on growth of *Medicago sativa* L. and *Trifolium subterranean* L. *Aust. J. Agric. Res.* **16**, 733-41.

Munns, D. N. (1965*b*). Soil acidity and growth of a lucerne. III. Interactions of lime and phosphate on growth of *Medicago sativa* L. in relation to aluminium toxicity and phosphate fixation. *Aust. J. Agric. Res.* **16**, 757-66.

Murtha, G. G. (1975). Soils and land use on the northern section of the Townsville coastal plain, North Queensland. CSIRO (Aust.) Soils Land Use Series No. 55.

Norrish, K. (1967). Clay Mineralogy and Sources of Natural Potassium. *In* 'Potassium in South-eastern Australia'. p.6. (Aust. Inst. Agric. Sci., Vic Branch: Parkville, Victoria.)

Norrish, K. and Rogers, L. E. R. (1956). The mineralogy of some terra rossas and rendzinas of South Australia. *J. Soil Sci.* **7**, 294-301.

Oertel, A. C. and Prescott, J. A. (1944). A spectrochemical examination of some ironstone gravels from Australian soils. *Trans. R. Soc. South Aust.* **68**, 173-6.

Ozanne, P. G. and Howes, K. M. W. (1973). The effect of calcium supply on burr set and seed formation in subterranean clover (*Trifolium subterranean* L). *Aust. J. Agric. Res.* **24**, 839-50.

Ozanne, P. G., Kirton, D. J. and Shaw, T. C. (1961). The loss of phosphorus from sandy soils. *Aust. J. Agric. Res.* **12**, 409-23.

Parberry, N. (1937). Magnesium deficiency in cabbages and cauliflowers. *Agric. Gaz. N.S.W.* **48**, 55-77.

Parberry, N. H. (1943). The excessive uptake of manganese by beans showing scald and magnesium deficiency. *Agric. Gaz. N.S.W.* **54**, 14-17.

Parker, F. W. (1953). Phosphorus Status and Requirements of Soils in the United States. *In* 'Soil and Fertilizer Phosphorus in Crop Nutrition'. (Eds W. H. Pierre and A. G. Norman). p.401. (Academic Press: New York.)

Paton, D. F. and Loneragan, J. F. (1960). An effect of lime on residual phosphorus in soil. *Aust. J. Agric. Res.* **11**, 524-9.

Penman, F. and Rountree, P. M. (1932). Influence of a wheat crop on accumulation of soil nitrate. *J. Agric. Victoria* **30**, 496-504.

Piper, C. S. and de Vries, M. P. C. (1964). The residual value of superphosphate on a red-brown earth in South Australia. *Aust. J. Agric. Res.* **15**, 234-72.

Prescott, J. A. (1931). The soils of Australia in relation to vegetation and climate. CSIR (Aust.) Bull. No. 52.

Prescott, J. A. (1933). The nitrogen problem in the Australian wheat belt. Proc. 5th Pacif. Sci. Cong. Canada Vol. 4, pp.2657-67.

Prescott, J. A. and Piper, G. R. (1930). Nitrate fluctuations in a South Australian soil. *J. Agric. Sci.* **20**, 517-31.

Probert, M. E. (1974). The Sulphur Status of Some North Queensland Soils. *In* 'Handbook on Sulphur in Australian Agriculture'. (Ed. K. D. McLachlan). p. 64. (CSIRO: Melbourne.)

Probert, M. E. (1977). The distribution of sulphur and carbon-nitrogen-sulphur relationships in some north Queensland soils. CSIRO (Aust.) Div. Soils Tech. Pap. No. 31.

Radoslovich, E. W. (1958). Clay mineralogy of some Australian red-brown earths. *J. Soil Sci.* **9**, 242-51.

Raupach, M. (1967*a*). Soil and fertilizer requirements for forests of *Pinus radiata*. *Adv. Agron.* **19**, 307-53.

Raupach, M. (1967*b*). The growth of Radiata pine on a lateritic soil at Second Valley, South Australia. *Aust. For.* **31**, 246-62.

Raupach, M. (1968). Nutrient release from coated fertilizer pellets. Trans. 9th Int. Cong. Soil Sci. Adelaide. Vol. 2, pp.755-63.

Raupach, M. (1975). Trace Element Disorders in *Pinus* and Their Correction. *In* 'Trace Elements in Soil-plant-animal Systems'. (Eds D. J. D. Nicholas and A. R. Egan). p.353. (Academic Press: New York.)

Raupach, M., Boardman, R. and Clarke, A. R. P. (1969). Growth rates of *Pinus radiata* D. Don. in relation to foliar levels of nitrogen and phosphorus for plantations in the south-east of South Australia. CSIRO (Aust.) Soil Publ. No. 26.

Raupach, M. and Clarke, A. R. P. (1978). Soil-tree relationships in a forest of *Pinus radiata* with micronutrient deficiencies. *Aust. J. Soil Res.* **16**, 121-35.

Raupach, M., Clarke, A. R. P. and Cellier, K, M. (1978). Disorder symptoms of a forest of *Pinus radiata* in relation to foliar nutrient levels. *Aust. For. Res.* **8**, 1-11.

Raupach M., Clarke, A. R. P., Gibson, B. F. and Cellier, K. M. (1975). Cultivation and fertilizer effects on the growth and foliage nutrient concentrations of *P. radiata, P. pinaster* and *P. caribaea* on three soil types at Anglesea (Victoria). CSIRO (Aust.) Div. Soils Tech. Pap. No. 25.

Raupach, M. and Hall, M. J. (1974). Foliar levels of potassium in relation to potassium deficiency symptoms in radiata pine. *Aust. For.* **36**, 204-13.

Reuter, D. J., Heard, T. G. and Alston, A. M. (1973). Correction of manganese deficiency in barley crops on calcareous soils. 1. Manganous sulphate applied at sowing and as foliar sprays. *Aust. J. Exp. Agric. Anim. Husb.* **13**, 434-9.

Richards, B. N. and Bevege, D. I. (1967). Effects of cultivation and felling on potential yield of pulpwood from loblolly pine. *Aust. For.* **31**, 202-10.

Richards, B. N. and Bevege, D. I. (1968). Nutrient requirements of kauri pine on a lateritic podzolic soil in southern Queensland. *Aust. For.* **32**, 55-62.

Richards, B. N. and Bevege, D. I. (1972). Principles and practice of foliar analysis as a basis for crop-logging in pine plantations. *Plant Soil* **36**, 109-19.

Richardson, A. E. V. and Fricke, E. F. (1931). Effect of nitrogenous fertilizers on the growth and yield of wheat and barley. *J. Agric. South Aust.* **35**, 57-86.

Rudd, C. L. (1970). Wheat responses to individual nutrients in superphosphate. Dep. Agric. South Aust. Exp. Record No. 5, p.28.

Russell, J. S. (1960*a*). Soil fertility changes in the long-term experimental plots at Kybybolite, South Australia. 1. Changes in pH, total nitrogen, organic carbon and bulk density. *Aust. J. Agric. Res.* **11**, 902-26.

Russell, J. S. (1960*b*). Soil fertility changes in the long-term experimental plots at Kybybolite, South Australia. II. Changes in phosphorus. *Aust. J. Agric. Res.* **11**, 927-47.

Russell, J. S. (1961). Soil fertility changes in the long-term experimental plots at Kybybolite, South Australia. III. Changes in cation exchange capacity and exchangeable cations. *Aust. J. Agric. Res.* **12**, 273-85.

Russell, J. S. (1966). Plant growth on a low calcium status sodic soil in a sub-tropical environment. I. Legume species, calcium carbonate, zinc and other minor element interactions. *Aust. J. Agric. Res.* **17**, 673-86.

Russell, J. S. and Shearer, R. C. (1964). Soil fertility changes in the long-term experimental plots at Kybybolite, South Australia. IV. Changes in certain moisture characteristics. *Aust. J. Agric. Res.* **15**, 91-9.

Samuel, G. and Piper, C. S. (1928). Grey speck (manganese deficiency) disease of oats. *J. Dep. Agric. South Aust.* **31**, 696-705.

Siman, A., Cradock, F. and Autry-Hall, F. (1966). Manganese toxicity in beans at Terranora. *Agric. Gaz. N.S.W.* **77**, 169-72.

Siman, A., Craddock, F. W., Nicholls, P. J. and Kirton, H. C. (1971). Effects of calcium carbonate and ammonium sulphate on manganese toxicity in an acid soil. *Aust. J. Agric. Res.* **22**, 201-14.

Simpson, J. R. (1968). Losses of urea nitrogen from the surface of pasture soils. Trans. 9th Int. Cong. Soil Sci. Adelaide. Vol. 2, pp.459-66.

Simpson, J. R., Bromfield, S. M. and Jones, O. L. (1974). Effects of management on soil fertility under pasture. 3. Changes in total nitrogen, carbon, phosphorus and exchangeable cations. *Aust. J. Exp. Agric. Anim. Husb.* **14**, 487-94.

Sims, H. J. (1953). Some aspects of soil fertility in the cereal areas of Victoria. *J. Aust. Inst. Agric. Sci.* **19**, 89-98.

Sims, H. J. (1959). Cereal trials with rock phosphate in Victoria. *J. Agric. Victoria* **57**, 759-65.

Smith, S. T. (1969). The long-term effects of frequent cropping. *J. Dep. Agric. West. Aust.* **10**, 530-3.

Specht, R. L. (1963). Dark Island Heath (Ninety-mile Plain, South Australia). VII. The effect of fertilizers on composition and growth 1950-60. *Aust. J. Bot.* **11**, 67-94.

Specht, R. L., Connor, D. J. and Clifford, H. T. (1977). The heath-savannah problem. The effect of fertilizers on sand-heath vegetation of North Stradbroke Island, Queensland. *Aust. J. Ecol.* **2**, 179-86.

Specht, R. L. and Groves, R. H. (1966). A comparison of the phosphorus nutrition of Australian heath plants and introduced economic plants. *Aust. J. Bot.* **14**, 201-21.

Specht, R. L. and Rayson, P. (1957). Dark Island Heath (Ninety-mile Plain, South Australia). I. Definition of the ecosystem. *Aust. J. Bot.* **5**, 52-85.

Spencer, K. and Freney, J. R. (1960). A comparison of several procedures for estimating the sulphur status of soils. *Aust. J. Agric. Res.* **11**, 948-59.

Stephens, C. G. and Donald, C. M. (1958). Australian soils and their responses to fertilizers. *Adv. Agron.* **10**, 167-256.

Storrier, R. R. (1962). The availability of mineral nitrogen in a wheat soil from southern New South Wales. *Aust. J. Exp. Agric. Anim. Husb.* **2**, 185-92.

Storrier, R. R. (1965*a*). The influence of water on wheat yield, plant nitrogen uptake and soil mineral nitrogen concentration. *Aust. J. Exp. Agric. Anim. Husb.* **5**, 310-16.

Storrier, R. R. (1965*b*). Excess soil nitrogen and the yield and uptake of nitrogen by wheat in southern New South Wales. *Aust. J. Exp. Agric. Anim. Husb.* **5**, 317-22.

Storrier, R. R. (1965*c*). The leaching of nitrogen and its uptake by wheat in a soil from southern New South Wales. *Aust. J. Exp. Agric. Anim. Husb.* **5**, 323-8.

Taylor, R. M. and McKenzie, R. M. (1966). The association of trace elements with manganese minerals in Australian soils. *Aust. J. Soil Res.* **4**, 29-39.

Taylor, R. M., McKenzie, R. M. and Norrish, K. (1964). The mineralogy and chemistry of manganese in some Australian soils. *Aust. J. Soil Res.* **2**, 235-48.

Teakle, L. J. H. and Samuel, L. W. (1936). The value of nitrogenous fertilizers for wheat in Western Australia. *J. Dep. Agric. West. Aust.* **13**, 74-89.

Teakle, L. J. H. and Carriss, H. G. (1943). Superphosphate requirements for growing wheat in Western Australia. *J. Dep. Agric. West. Aust. 20,* 1-28.

Turton, A. G. and Keay, J. (1970). Changes in dry-weight and nutrient distribution in maritime pine after fertilization. *Aust. For.* **34**, 84-96.

Vallis, I. (1972). Soil nitrogen changes under continuously grazed legume–grass pastures in subtropical coastal Queensland. *Aust. J. Exp. Agric. Anim. Husb.* **12**, 495-501.

Von Stieglitz, C. R. (1951). Investigation of plant and animal nutrition problems on Springbrook Plateau. *Queensl. Agric. J.* **73**, 41-9.

Waring, H. D. (1969). The role of nitrogen in the maintenance of productivity in conifer plantations. *Commw. For Rev.***48**, 226-37.

Waring, H. D. (1973). Early fertilization for maximum production. Proc. Int. Symp. on For. Fertilization. p. 215. (FAO-IUFRO: Paris.)

Waring, S. A. and Teakle, L. J. H. (1960). Fertility investigations on the black earth wheatlands of the Darling Downs, Queensland. III. Mineral nitrogen in the soil and its relation to the wheat crop. *Aust. J. Agric. Res.* **11**, 27-41.

Watson, E. R. and Lapins, P. (1969). Losses of nitrogen from urine on soils of south-western Australia. *Aust. J. Exp. Agric. Anim. Husb.* **9**, 85-91.

Wetselaar, R. and Norman, M. J. T. (1960). Recovery of available soil nitrogen by annual fodder crops at Katherine, Northern Territory. *Aust. J. Agric. Res.* **11**, 693-704.

Wild, A. (1958). The phosphate content of Australian soils. *Aust. J. Agric. Res.* **9**, 193-204.

Wild, A. (1962). A pedological study of phosphorus in 12 soils derived from granite. *Aust. J. Agric. Res.* **12**, 286-99.

Willett, I. R. (1979). The effect of flooding for rice culture on soil chemical properties and subsequent maize growth. *Plant Soil* **52**, 373-83.

Williams, C. H. (1970). Pasture nitrogen in Australia. *J. Aust. Inst. Agric. Sci.* **36**, 199-205.

Williams, C. H. (1974). The Chemical Nature of Sulphur in Some New South Wales Soils. *In* 'Handbook on Sulphur in Australian Agriculture'. (Ed. K. D. McLachlan). p.16. (CSIRO: Melbourne.)

Williams, C. H. and Andrew, C. S. (1970). Mineral Nutrition of Pastures. *In* 'Australian Grasslands'. (Ed. R. M. Moore).p.321. (Aust. Nat. Univ. Press: Canberra.)

Williams, C. H. and David, D. J. (1976). Effects of pasture improvement with subterranean clover and superphosphate on the availability of trace metals to plants. *Aust. J. Soil Res.* **14**, 85-93.

Williams, C. H.and Donald, C. M. (1957). Changes in organic matter and pH in a podzolic soil as influenced by subterranean clover and superphosphate. *Aust. J. Agric. Res.* **8**, 179-89.

Williams, C. H. and Lipsett, J. (1960). The build-up of available potassium under subterranean clover pastures on a podzolic soil. *Aust. J. Agric. Res.* **11**, 473-84.

Williams, C. H. and Lipsett, J. (1961). Fertility changes in soils cultivated for wheat in southern New South Wales. *Aust. J. Agric. Res.* **12**, 612-29.

Williams, C. H. and Simpson, J. R. (1965). Some effects of cultivation and waterlogging on the availability of phosphorus in pasture soils. *Aust. J. Agric. Res.* **16**, 413-27.

Williams, C. H. and Steinbergs, A. (1964). The evaluation of plant-available sulphur in soils. II. The availability of adsorbed and insoluble sulphates. *Plant Soil* **21**, 50-62.

Woodroffe, K. and Williams, C. H. (1953). The residual effect of superphosphate in soils cultivated for wheat in South Australia. *Aust. J. Agric. Res.* **4**, 127-50.

Woods, R. V. (1976). Early silviculture for upgrading productivity on marginal *Pinus radiata* sites in the south-eastern region of South Australia. Woods and For. Dep. South Aust. Bull. No. 24.

Year Book Australia (1975-1976). No. 61. Aust. Bureau of Statistics, Canberra, Australia.

Chapter 50
Fertilizer requirements

J. D. Colwell

FERTILIZERS PROVIDE THE SIMPLEST AND MOST OBVIOUS means for increasing crop production when there are soil nutrient deficiencies and accordingly a very considerable amount of research is being carried out in most parts of the world relating to their use. Nutrient deficiencies are being diagnosed for various crops, and fertilizing programs are being formulated for agricultural regions. The work has been in progress for many years, and has become largely routine in many instances, merely applying standard and accepted procedures to new regions and crops. Nevertheless, there are still problems for original research which are both challenging because solutions continue to be elusive, and perplexing because they involve overlapping considerations from economics, agronomy and soil science. Some of these, important to Australia, are described in this chapter.

The Australian work on the use of fertilizers commenced in the 1880s with the introduction of superphosphate and has continued since with numerous fertilizer experiments and related studies on soil as described in many comprehensive reviews (Russell 1957; Stephens and Donald 1958; Smith 1965, 1971, 1977; Jackson 1966; Colwell 1967*a,b*; McLachlan 1975; Blair 1976; Williams and Colwell 1977; Chapter 49). This chapter is not intended to repeat or extend these reviews but rather to describe our present state of knowledge in relation to current problems. Accordingly, we commence by considering the basic information that is required, and the necessary computations, for the calculation of economic-optimal fertilizer rates. The representation of this information in a form suitable for the computations requires the estimation of mathematical functions, expressing yield as a function of nutrient rates and the choice of a mathematical model for this purpose can present quite a problem, as described subsequently. Then fertilizer requirements have to be predicted on the basis of regional information and site variables, including the so-called soil tests, and this requires special computational procedures which are also detailed here. The effects of fertilizers often extend however, beyond the crop to which they have been applied and allowances then have to be made for residual effects on future crops, and this leads to the set of problems considered in the penultimate section. Finally, the problems reviewed are summarized, together with some suggestions on directions for research.

Calculation of optimal fertilizer rates

The basic procedures for calculating optimal rates of application of fertilizer is derived from simple application of the differential calculus and has been known for a long time (e.g. Mitscherlich 1909; Baum *et al.* 1956; Anderson 1967). In the past, however, application of the procedure has been in general very limited, presumably because of the practical problems involved in carrying out the necessary statistical and mathematical computations with appropriate allowances for non-ideal response forms, extrapolation, data variability, etc. Thus most of the descriptions of fertilizer experiments published in the past (Agricultural Gazettes, Technical Bulletins etc. of State Departments of Agriculture) rely heavily on 'common sense' interpretations of data with at most some very simple supporting calculations. The computational difficulties have, however, now been largely removed by the development of modern computing facilities and of 'software packages' for computing optimal fertilizer rates for a variety of circumstances (Colwell 1978). It is unfortunate therefore that some of the pre-computer methods (% yields and such) are still often used to obtain estimates that may be unnecessarily misleading. For this reason we consider the basic principles of computing optimal rates in some detail, indicating the difficulties and hence current problems for research.

The definition of optimal rate

Given a suitable yield function expressing yield as a function of rate of application of a nutrient in fertilizer, the rates for maximum yield and maximum profit are simply calculated by standard applications of the differential calculus. If, for example, the relationship between the yield of wheat and the rate of application of phosphorus in fertilizer at a particular site* is represented by the equation

$$Y = 447.8 + 445.4P^{0.5} - 42.60P \qquad (1)$$

where Y = kg wheat ha^{-1} and P = kg phosphorus ha^{-1}, then the rate $P = 27.3$ required for maximum yield or yield response is calculated in the usual way by solving the equation $dY/dP = 0$. Similarly, if the wheat has the value $V = \$0.08$* kg^{-1} and the cost of applying the phos-

*Functions derived by least squares regression with data for the National Soil Fertility Project, site N68/W26 (Colwell 1977).

phorus in fertilizer is $C_P = \$0.85$ ha^{-1} then the rate $P = 17.5$ for maximum profit is calculated from the corresponding profit equation

$$\pi = V(447.8 + 445.4P^{0.5} - 42.60P) - C_P P - Q$$

by solving $d\pi/dI_P = 0$ where π = profit, $I_P = C_P P$ = the investment in phosphatic fertilizer and Q = fixed costs not affected by fertilizer use.

Now at first sight it might seem obvious that fertilizer should be applied for maximum profit and that the rate 17.5 kg phosphate ha^{-1} for the example function should be described as optimal, but this is not so, as will be seen from a consideration of the implications of the 'diminishing return' form of the yield function and the fact that there is a delay, usually of at least 1 year while the crop is sown, grown, harvested and sold, before the farmer receives a return on his investment in fertilizer. Yield and profit both increase with increase in fertilizer rate, but in such a way that yield increments from successive fertilizer increments become progressively smaller. Consequently at some point below that for maximum profit, returns or increases in profit from further investments in fertilizer cease to be attractive relative to alternative investment opportunities. In mathematical terms, $d\pi/dI_P$ decreases with increase in I_P and if there is an alternative investment opportunity yielding a rate of return R then rates of return from further investments in fertilizer are not attractive when $d\pi/dI_P < R$. The optimal rate is defined accordingly as 'that for which $d\pi/dI_P = R$ where R is a marginal (or minimal) rate of return nominated from a consideration of alternative investment opportunities' (Colwell 1978).

In more general terms the definition of optimal rate follows from the need to allow for a delay of n interest or conversion periods between the time of the investment and the ultimate return. If a return of profit π is received after n interest periods then this value must be discounted back to the time of the investment by

$$\pi_D = \pi(1+R)^{-n} \quad (2)$$

to determine the optimum (maximum) economic investment, by $d\pi_D/dI = 0$ where I = investment. Since $\pi = VY - CX - Q$, as in the above example, and $I = CX$, where V = value of a unit of Y = yield, C = cost of a unit of X = nutrient, and Q = fixed costs, it follows that the optimal rate is the value of X satisfying the equation

$$\frac{dY}{dX} = \frac{C}{V}(1+R)^n \quad (3)$$

For annual crops where return is about 1 year after the investment, number of interest periods will be $n = 1$.

*Australian dollars.

For the non-mathematician the rationale for equation (2) is easily demonstrated by calculating yields profit increases and fertilizer costs (investments) for a range of application rates, and then comparing as in Table 50.1 for equation (1). Comparing for example the lines for $R = 0.2$ and $R = 0.0$ in Table 50.1, the cost of increasing profit from \$74.45 for a 20% marginal rate of return to \$74.56 for maximum profit is \$14.88 − 13.76 = \$1.12 ha^{-1}. In other words the cost of chasing the last \$0.11 ha^{-1} of profit is an additional investment of \$1.12 ha^{-1}. An investment of \$1.12 ha^{-1} for a return of \$0.11 ha^{-1} about 1 year later, when the crop has been produced and sold, would be hardly attractive to most farmers.

An important point to note about the definition is that optimum simply means *best* and that economically best does not always mean maximum yield or maximum profit. Because of the time delay in obtaining the return from the fertilizer investment, optimum means maximum only with respect to the time-discounted profit π_D. Failure to distinguish between optimum and maximum in this context can lead to confusion as shown by the controversy caused by Jardine (1975) in attempting to clarify the point by proposing the term 'sub-optimal'.

Note also that the optimal rate depends on the value nominated for R, as well as the magnitude of the yield response, the value of the crop and the cost of the fertilizer application. The nomination of a value for R may not be simple in practice because of lack of information on rates of return from the many alternative investment opportunities of a complex farm situation. The range of possible values can, however, usually be reduced by some general considerations. If, for instance, there are safe investment opportunities (Government Debentures, Building Societies, etc.) yielding about 10% then clearly $R = 0.1$ would be an absolute minimum value and a value of about $R = 0.2$ would seem more appropriate for farm operations. Simple considerations such as these have led the author to use $R = 0.2$ for fertilizer requirement studies, on the basis that this value cannot be too far wrong for the average reasonably prosperous Australian farmer, and that calculations are easily repeated anyway for any other values that may seem to be more appropriate.

Table 50.1 Yields, profit increases and fertilizer costs for various nominated rates of return (*R*)

R	Phosphorus (kg ha^{-1})	Yield response (kg ha^{-1})	Investment (\$A ha^{-1})	Profit (\$A ha^{-1})
2.0	8.9	951	7.60	68.49
1.0	12.2	1035	10.34	72.49
0.5	14.5	1078	12.30	73.94
0.2	16.2	1103	13.76	74.45
0.0	17.5	1118	14.88	74.56
−0.46	21.2	1148	18.05	73.79
−1.0	27.3	1164	23.23	69.92

Simultaneous rates and the ideal mix

Soils are commonly deficient in more than one nutrient and this introduces a difficulty to the estimation of optimal fertilizer rates. Optimal rates for each nutrient cannot simply be determined separately, and then combined to obtain an ideal mix, because crop yield responses to each nutrient are usually affected by the rate of application of the others, or in statistical terms the responses to the applied nutrients are not independent of each other and there are interaction effects. This interaction effect is represented for example by the $(NP)^{0.5}$ term in the two nutrient yield function, for nitrogen and phosphorus,

$$Y = 447.8 + 44.78N^{0.5} + 445.4P^{0.5} + 4.050\,(NP)^{0.5} - 5.639N - 42.60P \qquad (4)$$

Given a multi-nutrient function such as this, optimal rates are calculated by application of the same mathematical principles that are used for a single nutrient, solving now a simultaneous set of equations having the general form

$$\frac{\partial \pi}{\partial I_i} = R, \quad \text{or} \quad \frac{\partial \pi_D}{\partial I_i} = 0, \quad i = 1, 2, \ldots, k$$

or equivalently

$$\frac{\partial Y}{\partial X_i} = \frac{Ci}{V}(1+R)^{\pi}, \quad i = 1, 2, \ldots, k$$

for k different nutrients X_i with costs C_i, and where π = profit, I_i = investment in X_i, V = crop value and R = marginal rate of return as for equation (3), with n = number of interest periods. Solution of the simultaneous equations

$$\frac{\partial Y}{\partial N} = \frac{C_n}{V}(1+R) \quad \text{and} \quad \frac{\partial Y}{\partial P} = \frac{C_P}{V}(1+R)$$

for equation (4), with $C_n = \$0.40$ kg^{-1}N, $C_p = \$0.85$ kg^{-1}P, $V = \$0.08$ kg^{-1} wheat and $n = 1$, for the range of R values used in Table 50.1, gives the corresponding set of simultaneous rates in Table 50.2. These may be compared directly with those in Table 50.1 derived for equation (1), since the functions have been derived from the same data, the equation (1) being that for phosphorus with a nil application of nitrogen, obtained by putting $N = 0$ in equation (4). The P rates, yield, investment and profit are all reduced by not using N as well as P. Corresponding reductions can be demonstrated due to P by calculating optimal rates with the function derived from equation (4) by substituting $P = 0$. The reductions in profit due to non-use of N or P, obtained in this way and given in Table 50.2, serve as measures of the relative importance of these nutrients in an ideal mix. This method of calculation is necessary because the interacting effect of the two nutrients precludes the simple apportioning of yield response or profit for each.

A further example of these computations is given in Table 50.3 for the three-nutrient function

$$Y = 447.8 + 44.78N^{0.5} + 445.4P^{0.5} + 23.24S^{0.5} + 4.050\,(NP)^{0.5} + 0.7385(NS)^{0.5} + 1.793(PS)^{0.5} - 5.639N - 42.60P - 3.377S \qquad (5)$$

Table 50.2 Yields, profit increases and fertilizer costs for simultaneous applications of nitrogen and phosphorus at various rates of return (*R*)

R	N (kg ha^{-1})	P (kg ha^{-1})	Yield Response kg ha^{-1}	Investment ($A ha^{-1})	Profit Increase ($A ha^{-1})	Reductions[A] N ($A ha^{-1})	Reductions[A] P ($A ha^{-1})
2.0	1.9	9.2	1026	8.56	73.54	5.06	70.66
1.0	3.6	12.6	1136	12.13	78.73	6.24	75.35
0.5	5.3	15.1	1197	14.95	80.80	6.86	77.17
0.2	7.0	17.0	1235	17.22	81.59	7.14	77.85
0.0	8.5	18.5	1261	19.10	81.77	7.21	78.00
−0.46	14.8	22.7	1318	25.23	80.24	6.44	76.75
−1.0	35.4	30.4	1361	39.97	68.87	−1.04	68.07

[A]Reductions in profit due to non-use of N or P.

Table 50.3 Profit increases and nutrient application rates at various nominated rates of return (*R*) for simultaneous applications of nitrogen, phosphorus and sulphur

R	N (kg ha^{-1})	P (kg ha^{-1})	S (kg ha^{-1})	Profit increase ($A ha^{-1})	Reductions[A] N ($A ha^{-1})	Reductions[A] P ($A ha^{-1})	Reductions[A] S ($A ha^{-1})
2.0	2.0	9.2	0.8	75.37	5.23	71.37	1.83
1.0	3.7	12.7	1.7	81.11	6.52	77.58	2.38
0.5	5.5	15.3	2.7	83.49	7.21	78.33	2.69
0.2	7.3	17.3	3.7	84.43	7.53	79.09	2.84
0.0	9.0	18.8	4.7	84.65	7.61	79.27	2.88
−0.46	16.0	23.3	9.4	82.56	6.66	77.70	5.81
−1.0	40.6	31.9	31.8	63.15	−4.17	64.28	-5.72

[A]Reductions in profit due to non-use of N, P or S.

representing the effect of sulphur (S) as well as nitrogen (N) and phosphorus (P), with C_s = \$0.34. The reductions for the three nutrients, calculated from profits from optimal applications of the nutrients less each in turn, show that S is relatively unimportant for the soil, crop and site represented by this function.

Compound fertilizers

The application of simultaneous optimal rates of fertilizers requires in practice a blending of the appropriate single nutrient fertilizers into an ideal mix by a fertilizer blending plant, as provided in some regions by fertilizer manufacturers and distributors. The alternative is to use a compound fertilizer containing the required nutrients in some fixed ratio, approximating in composition and cost to the ideal mix. However the choice of the most appropriate compound from a fertilizer price and composition list, such as that in Table 50.4, is not obvious. Given, for example, the expected responses to N, P and S represented by equation (5) and the optimal simultaneous rates of Table 50.3, what should a farmer do? If he can purchase an ideal mix from a blending plant, is it going to cost more or less than using one of the compounds A, B or C at an optimal rate? Which compound is best and what is the optimal rate for its application? Would it be worth applying the urea separately, say with the last cultivation, and then applying the P and S with superphosphate B or C, but again at what rates? Such questions make a fertilizer list rather bewildering to a farmer or advisory officer. The calculations required to answer these questions are rather tiresome but can be carried out by a computer, as by the subroutine COMP (Colwell 1978). Optimal rates are calculated for each of the fertilizers in the list and for mixes of the fertilizers, with appropriate allowances for composition and price, as given in Table 50.5 for equation (5) and Table 50.4 fertilizers. The choice of the most appropriate fertilizer or mixture is then made by comparing profits. Thus for the example, compound A is the best of the range of fertilizers, there is only a slight advantage in using the mix (urea + superphosphate A + gypsum), and highest returns would result from the use of urea + superphosphate B given, of course, a convenient means for applying the urea.

An example of optimal rates of compound fertilizers for a region in southern New South Wales is given by Colwell (1973*a*), but such lists are soon out of date because of changes in prices and the fertilizers that are locally available for purchase. It is, however, simple to compute annual listings for individual regions, to guide farmers.

Other calculations

The above calculation procedures are all for a simple crop such as wheat and a very simple farming system in which fertilizers are purchased before sowing and the return on the investment obtained about 1 year later when the crop is sold. Many cropping systems are, however, much more complicated than this, with the benefits from fertilizer use spread over much longer periods and perhaps many crops. Many variations and developments of the basic calculation procedures are therefore possible as described in many references (e.g. Heady *et al.* 1961, 1963, 1966; Anderson 1967; Dillon 1968; Colwell 1973*a*, *b*, 1975, 1978). All represent applications of standard mathematics and the difficulties arise not from the mathematics but rather from the fact that few of the systems can be modelled with much accuracy. Problems exist for example in modelling the relationships between fertilizer rates and yields from long-term, continuing crops such as forests, or for dairy farming where fertilizer is applied to pasture to increase milk production. These represent current and recurring research challenges relating to the optimal use of fertilizers.

Table 50.4. Cost and composition of several fertilizers

Fertilizer	Composition (%)			Cost
	N	P	S	($SA t^{-1}$)
Urea	46	–	–	184
Superphosphate A	–	20	–	170
Gypsum	–	–	16	54.4
Superphosphate B	–	8.6	11.5	83
Superphosphate C	–	17.5	5	165
Compound A	12.4	22.2	3	248
Compound B	15.0	13.0	10.3	204
Compound C	11.0	14.5	5.1	226

Table 50.5. Optimal application rates and associated profit increases for various fertilizers and mixes of fertilizers.

Fertilizer	Optimal rate (kg ha^{-1})	Profit increase (\$A ha^{-1})
Urea	8.1	3.74
Superphosphate A	81.0	74.45
Gypsum	11.9	1.40
Superphosphate B	183.1	78.00
Superphosphate C	93.0	76.95
Compound A	76.7	84.24
Compound B	107.8	81.41
Compound C	96.8	79.03
Ideal mixes of:		
Urea	15.9	
Superphosphate A	86.5	84.43
Gypsum	23.1	
Urea	16.8	
Superphosphate B	193.8	85.95
Urea	16.0	
Superphosphate C	98.0	80.77

The yield function

Estimates of the rate at which fertilizer should be applied to a crop are based essentially on estimates of the effects that different rates will have on yield and hence on profit. Accordingly our first task in studies of fertilizer requirements is to specify a functional form or model which can be used to represent the relationship between yield and rate of fertilizer application for particular situations. But a problem is immediately apparent. There is no fundamental theory to indicate a suitable model, nor does such even seem possible given the many variable factors that contribute to the yield of crops between the time when fertilizer is applied and when they are harvested. Many models are possible as for example:

	Model
$Y = b_0 + b_1 X + b_2 X^2$	1
$Y = b_0 + b_1 X^{0.75} + b_2 X^{1.5}$	2
$\log Y = b_0 + b_1 X^{0.5} + b_2 X$	3
$Y = b_0 + b_1 X^{0.5} + b_2 X$	4
$Y = b_0 + b_1 e^{b_2 X}$	5
$\log Y = b_0 + b_1 e^{b_2 X}$	6
$Y = b_0 + b_1 \{(X-b_2) - \sqrt{(X-b_2)^2 + g}\}$	7
$Y = b_0 + b_1 \{X-b_2 - \lvert X-b_2 \rvert\}$	8
$Y = b_0 + b_1 e^{b_2 X^2}$	9

where Y is yield and X is nutrient rate. The source of the problem lies in the essential variability of the biological world, so that deterministic or mechanistic models simply cannot be derived from differential equations with as much confidence as in the physical sciences. Rather, empirical or statistical models must be evolved by observation, experience and interaction with real data, and it must be accepted that these models are appropriate only in a probabilistic statistical sense i.e. they are satisfactory only most of the time and exceptions must be expected. The unfortunate consequence is that, in the absence of rigid theory, many seemingly suitable models can be (and are) formulated (reviews by Russell 1937; Johnson 1953; National Soil Test Group 1956; Heady *et al.* 1961, 1963, 1966; Fried and Broeshart 1967; Willey and Heath 1969; Mead and Pike 1975). Moreover, there are endless opportunities for debate on the merits of alternative models since supporting data can be found for almost any of them. Models with a theoretical basis in other contexts may be adapted for the purpose, so that biological or physical meanings can be attached to their parameters (e.g. models by Mitscherlich 1907, 1909; Balmukand 1928; Hagin 1960; Nelder 1966; Campbell and Keay 1970; Greenwood and Cleaver 1971; Russell 1972), but this must be done with caution lest the model become some sort of esoteric device conveying particular notions or prejudices. As pointed out by Mead and Pike (1975) 'all models are essentially empirical; the point at issue is to what extent, for a particular objective, is it necessary to use biologically reasonable models'.

One way to avoid the problem is to restrict the range of data so that the relationships can be treated as being linear. It has been argued for example that the main source of trouble with yield functions is the inclusion of the nil rate in the data range and accordingly that it should be omitted (Anderson 1957). A lowest treatment rate for a fertilizer experiment might thus be a maintenance rate, chosen to maintain the level of fertility by adding about the amount of nutrient lost in crop products, by leaching, etc. The difficulty is, however, not really avoided in this way, since the functions are required to calculate optimal rates and this depends on an estimation of curvature, and hence on at least an appreciable range of data. Moreover, nil-treatment rates may be needed for series of experiments within a region to cover the situation of high levels of fertility, where it is appropriate to exploit (excessive) soil nutrient reserves.

Criteria for the choice of a model

Given a series of possible models, such as the models 1 to 9, a choice must be made on the basis of practicality, potential usefulness and standard statistical considerations (e.g. Daniel and Wood 1971). The more important criteria in these respects are:

Computational convenience

The ease with which actual functions can be estimated from data, their suitability for standard analyses of variance, their suitability for generalization for two or more nutrients, and the ease of calculation of optimal rates, all vary enormously between models. The polynomial models 1 and 4 are undoubtedly the most convenient in this respect. Equation-fitting and analysis of variance is simple, particularly with factorial design experiments which allow the use of orthogonal polynomials and also the calculation of optimal rates by solving equations of the type $dY/dX = (C/V)(1+R)^n$ (Table 50.6). Most of the others become near impossibly complex, with highly unstable parameters (see discussion by Ross 1970) when extended to two or more nutrients. Such difficulties preclude the general use of many models despite the appeal of some of their mathematical features.

Number of constants

Economy in the use of constants (parameters) in the model facilitates their statistical estimation from data. Thus the number of treatment levels in fertilizer experiments must be greater than the number of parameters to be estimated, and elaborate models require correspondingly elaborate experiments which become impracticable, especially for experiments with more than two nutrients. Moreover, a restriction in the number of parameters imposes a degree of inflexibility on the models that can be useful for detecting and smoothing out aberrations. The model 7 with four parameters is thus less suitable than the others, unless some general value for g,

Table 50.6. Models representing the relationship between yield and fertilizer applications

Model	$\frac{dY}{dX} = R^*$	Solution method for $\frac{dY}{dX} = R^*$
1. $Y = b_0 + b_1X + b_2X^2$	$X = \frac{R^* - b_1}{2b_2}$	Direct
2. $Y = b_0 + b_1X^{0.75} + b_2X^{1.5}$	$1.5b_2X^{0.75} - R^*X^{0.25} + 0.75b_1 = 0$	Cardan's method for cubic
3. $Y = \exp(b_0 + b_1X^{0.5} + b_2X)$	$(\frac{b_1}{2X^{0.5}} + b_2)\exp(b_0 + b_1X^{0.5} + b_2X) = R^*$	Newton's successive approximation method
4. $Y = b_0 + b_1X^{0.5} + b_2X$	$X = \left[\frac{b_1}{2(R^* - b_2)}\right]^2$	Direct
5. $Y = b_0 + b_1\exp(b_2X)$	$X = \frac{1}{b_2}\log(\frac{R^*}{b_1b_2})$	Direct
6. $Y = \exp\{b_0 + b_1\exp(b_2X)\}$	$b_1b_2\exp\{b_0 + b_2X + b_1\exp(b_2X)\} = R^*$	Newton's method
7. $Y = b_0 + b_1\{(X-b_2) - \sqrt{(X-b_2)^2+g}\}$	$b_1\left\{1 - \frac{(X-b_2)}{\sqrt{(X-b_2)^2+g}}\right\} = R^*$	Newton's method
8. $Y = b_0 + b_1\{X - b_2 - \lvert X - b_2\rvert\}$	$X = b_2$, if $2b_1 \geqslant R^*$ $X = 0$, if $2b_1 < R^*$	Direct
9. $Y = b_0 + b_1\exp(b_2X^2)$	$b_2X^2 + \log X - \log(\frac{R^*}{2b_1b_2}) = 0$	Newton's method

$(R^* = \frac{C}{V}[1 + R]$, or similar)

determining rate of curvature, is specified on the basis of previous experience.

Biological meaning of parameters

It is obviously convenient to be able to attribute direct meaning to the magnitude of the coefficients (parameters) in fitted functions, and many models are proposed mainly on this basis. Thus the classical exponential growth model 5 originally proposed as a fertilizer model by Mitscherlich (1907), may be written as $Y = A(1 - e^{B-CX})$ so that 'biological meanings' can be attached to the parameters: A is the asymptotic yield maximum, e^B is the overall response as a proportion of the maximum i.e. $e^B = (A - Y_0)/A$ where Y_0 is the yield for nil fertilizer, and C is a curvature parameter determing rate of response (the ratio of successive yield increments for fertilizer increments ΔX is $e^{-C\Delta X}$. Similarly in the Griffiths and Miller (1973) models 7 and 8, b_0 is the asymptotic maximum yield, b_1 is half the slope or rate of response to the maximum in model 8, and half the slope of the response asymptote in model 7, b_2 is the turning point between the response and maximum lines or asymptotes, and g is a parameter determing the curvature in the vicinity of the intersection of the maximum and slope parameters in model 7. There is a danger however in reading too much algebraic meaning into statistical coefficients. Statistical estimates of parameters are not independent, except for special orthogonal models (e.g. orthogonal polynomials), so that small variations in data can sometimes cause large compensatory variations in the parameter values, as of b_2 and b_1 in the exponential model 5 i.e. there is instability. Such problems become very much more serious with generalizations of such 'non-orthogonal models' so that any algebraic or biological significance of the parameters becomes largely lost in statistical estimates from real world data containing the inevitable error aberrations. Although, for example, parameters for generalizations of the exponential function used in model 5

$$Y = b_0(1 - a_1e^{a_2X_1})(1 - b_1e^{b_2X_2})(1 - c_1e^{c_2X_3}) \ldots \text{etc.}$$

can be estimated by an iterative procedure (Carmer 1964), the biological significance of the estimated, highly unstable parameters is likely to be very doubtful.

Residual error variance

The relative suitability of models can be judged simply by comparisons of 'goodness of fit' of the model to real data, based on comparisons in some form or other of the residual mean square or 'error' variance in an analysis of variance. Residual mean squares are provided by standard analyses of variance, or may be computed indirectly from the sum of square of deviations of data points about the fitted model. This latter procedure is necessary for example with models 3 and 6, if they are fitted to data by transforming to $\log Y = b_0 + b_1X^{0.5} + b_2X$ and $\log Y = b_0 + b_1\exp(b_2X)$, since interest lies in yield and not logarithm yield. Lower residual mean squares are obtained in these instances if the models are fitted as

$$Y = \exp(b_0 + b_1X^{0.5} + b_2X)$$

and $$Y = \exp[b_0 + b_1\exp(b_2X)].$$

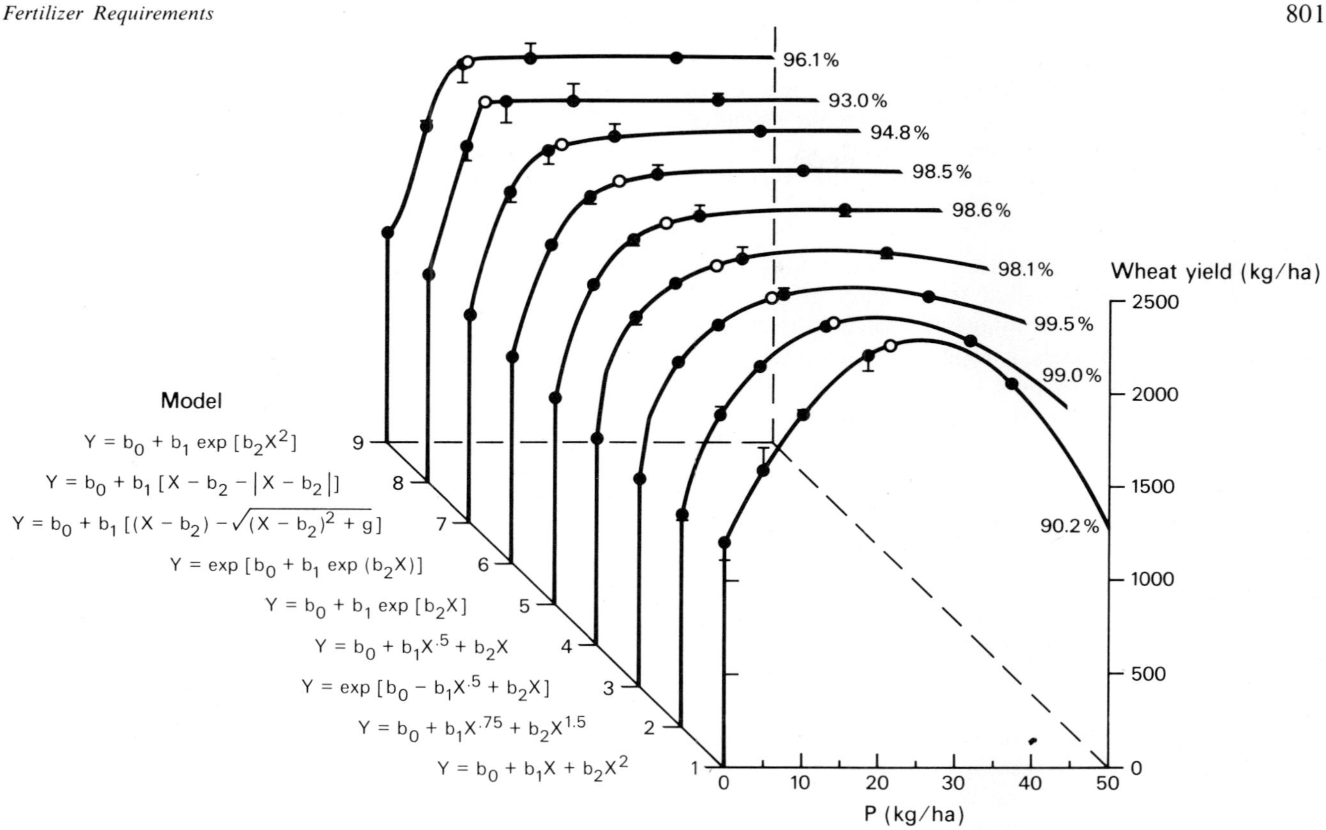

Fig. 50.1. Yield functions for the yield function models 1 to 9 for the Wagga Wagga data (Table 50.7) with deviations of data indicated by the vertical crossed bars. R_a^2 values for goodness of fit are indicated as percentages and the point for $(dY/dX) = (C_p/V)(1 + R) = 12.75$ corresponding to optimal fertilizer rates are marked by **O**.

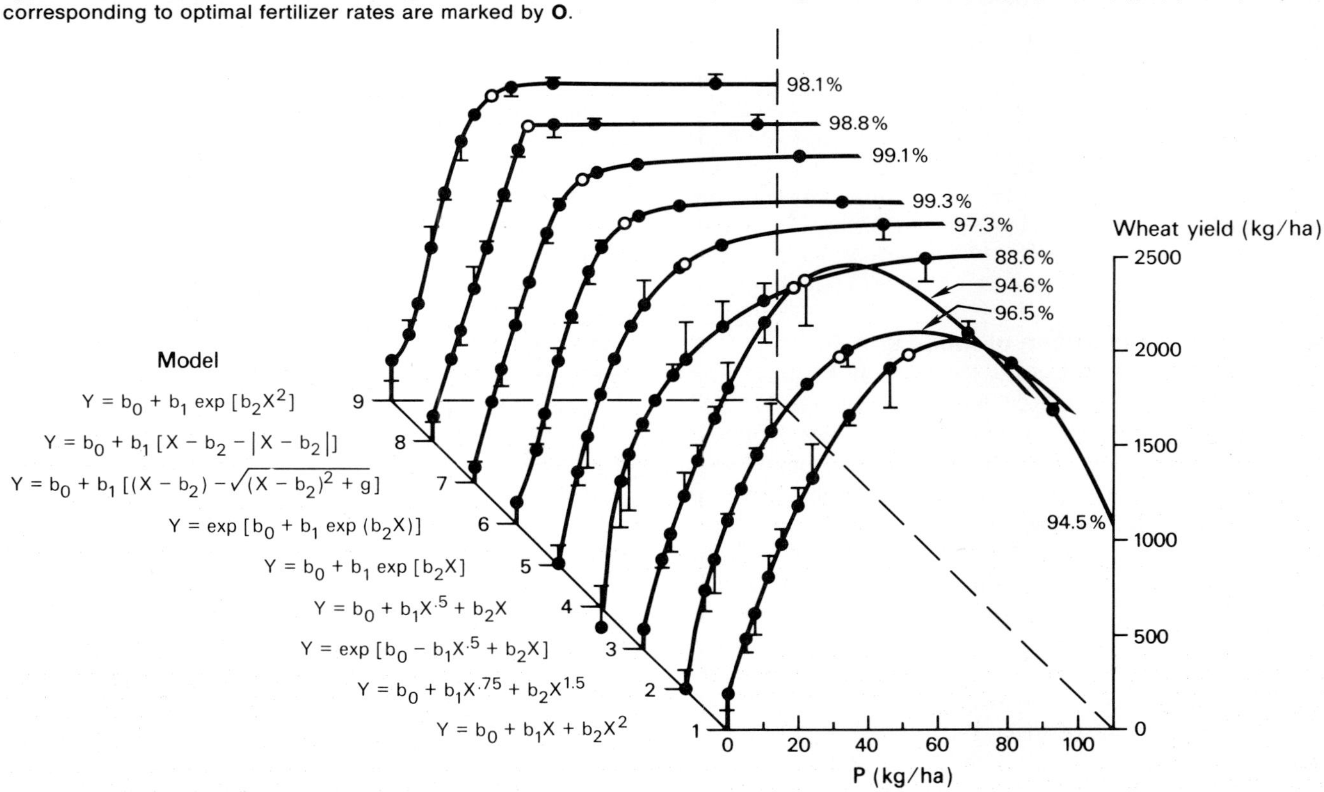

Fig. 50.2. Yield functions for Lake Grace data (Table 50.7) as in Fig. 50.1.

A convenient scaleless measure of goodness of fit based on the residual mean square (RMS) is the so-called 'adjusted R^2 value', being the proportion of variance (Var) of the Y variable accounted for by the regression,

$$R_a^2 = 1 - \frac{\text{RMS}}{\text{Var}} \quad (6)$$

This value relates to the square of the correlation coefficient (R^2), sometimes called the coefficient of determination, but has the advantage of allowing direct comparisons between models, irrespective of the number of model parameters. For poor models R_a^2 values become small and even negative when the residual mean square is greater than the original variance, and for good models and little error, values approach 1. The values do depend however on a ratio, RMS/Var, so that other things being equal high R_a^2 values are favoured by wide ranges of data (high Var values). For this reason R_a^2 values should be used to compare different models against the same set of data rather than to compare functions for the same model with different sets of data.

Deviations

The form of deviations of data about fitted functions should be studied, as recommended by Draper and Smith (1968) and Daniel and Wood (1971), and as illustrated in the yield-fertilizer model simulation study of Cady and Laird (1969). Such studies become particularly simple with precise computer plots of fitted functions and deviations as illustrated in Figs 50.1 and 50.2. A non-random distribution of data deviations may indicate the unsuitability of a model, at least for certain ranges of the data values, or may indicate a need for the use of the more general weighted least squares method, weighting each point inversely by its variance. A weighting which may prove excessive is applied incidentally by the transformation of Y to $\log Y$ for the fitting of a model, e.g. the direct fitting of $\log Y = b_0 + b_1X^{0.5} + b_2X$ and $\log Y = b_0 + b_1 \exp(b_2X)$ rather than models 3 and 6.

Table 50.7. Selection of a suitable model using experimental data

Data used are from fertilizer experiments with wheat.

Wagga Wagga data:

X =	0.0	5.0	10.0	18.6	37.3
Y =	1109	1714	1915	2124	2070

Model	Fitted function	Optimal rate
1	$Y = 1201 + 85.18X - 1.670X^2$	21.7
2	$Y = 1129 + 189.1X^{0.75} - 8.432X^{1.5}$	19.0
3	$\log Y = 7.008 + 0.2538X^{0.5} - 0.02461X$	17.0
4	$Y = 1095 + 371.3X^{0.5} - 34.07X$	15.7
5	$Y = 2103 - 993.8 \exp(-0.1837X)$	14.5
6	$\log Y = 7.647 - 0.6348 \exp(-0.2215X)$	14.0
7	$Y = 2111 + 86.25 \{(X-5.1) - \sqrt{(X-5.1)^2+16}\}$	11.6
8	$Y = 2037 + 68.15 \{X - 6.8 - \lvert X - 6.8\rvert\}$	6.8
9	$Y = 2050 - 932.9 \exp(-0.03784X^2)$	10.3

Lake Grace data:

X =	0.0	5.2	7.8	11.6	15.3	20.1	23.9	34.4	46.1	92.2
Y =	100	420	510	920	1060	1270	1500	1610	1700	1720

Model	Fitted function	Optimal rate
1	$Y = 188.5 + 58.41X - 0.4580X^2$	49.9
2	$Y = -5.171 + 164.1X^{0.75} - 3.574X^{1.5}$	43.2
3	$\log Y = 4.585 + 0.7917X^{0.5} - 0.0518X$	46.0
4	$Y = -111.3 + 376.2X^{0.5} - 18.07X$	37.3
5	$Y = 1804 - 1805 \exp(-0.05951X)$	35.8
6	$\log Y = 7.437 - 2.815 \exp(-0.1218X)$	30.8
7	$Y = 1736 + 34.31 \{(X-23.4) - \sqrt{(X-23.4)^2+70}\}$	30.2
8	$Y = 1677 + 29.41 \{X-26.4 - \lvert X-26.4\rvert\}$	26.4
9	$Y = 1668 - 1465 \exp(-0.003909X^2)$	28.9

An example comparison of models

The procedure for choosing a model can be illustrated with the two differing types of data given in Table 50.7 and the models 1 to 9. Both sets of data are from phosphatic fertilizer experiments with wheat, the data from Wagga Wagga, New South Wales (Colwell 1963*a*) being chosen to represent the diminishing return form of response commonly assumed for fertilizer experiments and those from Lake Grace, Western Australia (unpublished data provided by Bowden 1977) to represent the so-called 'broken stick' response form (Mead and Pike 1975) suggesting two intersecting straight lines. This latter form was perhaps first suggested for fertilizer experiments by Blackman (1905), and again more recently by Boyd (1970), Anderson and Nelson (1975) and Nelson and Anderson (1977). The fitted functions listed in Table 50.7 are illustrated by the computer-drawn curves in Figs 50.1 and 50.2.

Comparison of the curves in these figures shows the subtle range of functional forms that can be estimated from a set of experimental data, simply by choice of model. At one extreme the function gives a well defined maximum curvature for the region of maximum yield (model 1) and at the other a sigmoidal response form, with maximum curvature at relatively low rates (model 9).

The R_a^2 values given in Figs 50.1 and 50.2 provide a simple but crude basis of choice. Thus models 1, 7 and 8 are clearly inferior in Fig. 50.1 and also models 1, 2, 3 and 4 in Fig. 50.2, but apart from this broad separation of good from bad, the values are not very helpful because for the better models the values are all much the same. Moreover the values can be affected by the number and distribution of data points as can be seen from the effect that the number and distribution of data points would have on the sum of squares of deviations and hence on the RMS term and the R_a^2 values for the examples. Thus a predominance of data between the 20 and 40 kg ha^{-1} phosphorus rates in the Wagga experiment (Fig. 50.1) would, if they defined the usual broad plateau of near constant yields, reduce the R_a^2 values for models 1 and 2 relative to those for the other models, by introducing a predominance of relatively large deviations.

Similarly, more data between 0 and 5 kg ha^{-1} phosphorus would, if they defined a near linear or sigmoidal response form, have reduced the R_a^2 value of models 3 and 4 relative to the other models. Corresponding values favouring particular models could be similarly expected from a juggling of treatment rates for the Lake Grace experiment.

The simple inspection of the graphs and the deviations of the data points in each case show much more clearly the merits and weaknesses of the models relative to each set of data, and incidentally the irrelevancy of quasi-statistical statements such as 'an expected association between mean and variance' sometimes used to justify log Y models. The standard quadratic model 1 is clearly unsuitable for both sets of data because it imposes a symmetrical maximum that is quite unlike the usual broad plateau for a fertilizer– yield maximum, and similarly for model 3 in Fig. 50.2. The square root quadratic model 4 is correspondingly unsuitable for low fertilizer rates because it gives maximum curvature at the nil rate, though this weakness may be greatly reduced by using a quadratic of $(X + A)^{1/2}$ where A is a transposition parameter (Palmer *et al.* 1975). Model 2 is sometimes proposed as a compromise between the quadratic models 1 and 4, but because it has their combined weaknesses it gives too much curvature in the regions of maximum yield and nil fertilizer rate. The exponential model 5 gives a good representation except for the highest rate where it necessarily shows a continuing increase in yield even though the data suggets the opposite; under some circumstances yield depressions for high fertilizer rates can be quite severe (Colwell 1963*a*; Storrier 1965; Dann 1969). The log Y models 3 and 6 exhibit corresponding features to their counterpart models 4 and 5, but can show a pronounced bias in favour of the low-yield data due to the relative reduction of deviations by the logarithmic transformations for high values (particularly in Fig. 50.2). The 'broken stick' models 7, 8 and 9, clearly suit the type of data presented in Fig. 50.2 but not the diminishing form of Fig. 50.1. Model 9 may also be useful for sigmoidal types of response, as described for example by Steenbjerg and Jakobsen (1963).

Effect of model on fertilizer requirement

Optimal rates calculated by solving the equations $dY/dX = (C/V)\ (R+1)$ in Table 50.6 for the different models and listed in Table 50.7, for the examples, demonstrate the most worrying thing about choosing a model: *the model affects the estimate of the optimal rate.* Thus for the set of Wagga data, estimates of the optimal rate range from 6.8 to 21.7 kg ha^{-1} phosphorus (rates shown by **o** in Fig. 50.1), following the changes in curviness of the models, and similarly from 26.4 to 49.9 kg ha^{-1} phosphorus for the Lake Grace data: dramatic ranges in both cases for sets of exceptionally high quality data. Clearly the choice of a model as standard for a region

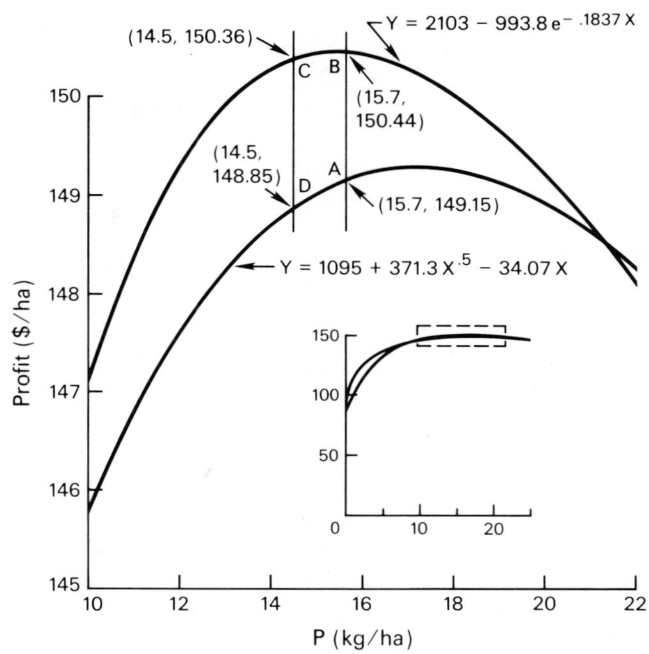

Fig. 50.3. Profits calculated from functions 4 and 5 for the Wagga Wagga data (Fig. 50.1) in the vicinity of the optimum rate. Insert is for the complete range of treatment rates.

can bias estimates, either up or down, and could lead to general losses of potential profits either through the use of too much or too little fertilizer. The importance of the effect of the model on estimates of fertilizer requirements is however easily exaggerated. The above ranges of optimal rates include models which should not be used for the respective types of data. Thus only models 3 to 6 warrant serious consideration for the Wagga data, and models 6 to 9 for the Lake Grace data, giving the much narrower ranges 14.0–17.0 and 28.9–35.8 respectively. Moreover the comparison of rates rather than of the effects of rates on profits is misleading because it does not allow for the basic *low elasticity* of yield functions near their respective optima. In fact the low elasticity of the functions makes the effects of such variations quite negligible. This is illustrated in Fig. 50.3 with a magnified portion of the calculated relationship between profit response to fertilizer and P rate, calculated using models 4 and 5 of Fig. 50.1 by $\pi = VY - CX$ where π = profit, V = value of wheat and C = cost of the fertilizer. Thus let the exponential model 5 be the appropriate choice rather than the square root quadratic model 4. If then the optimal rate of 15.7 kg ha^{-1} phosphorus is calculated incorrectly from the quadratic, the profit that will actually be obtained will be \$150.44 ha^{-1}, rather than the expected \$149.15 (points A,B). If however the rate of 14.5 kg ha^{-1} phosphorus had been correctly calculated from the exponential, then the profit that would have been obtained would have been \$150.36 ha^{-1}

(point C), that is \$0.08 ha^{-1} less because of the specification of $R = 0.2$. Hardly a worrying consequence. The comparison can be repeated vice versa with the quadratic now considered appropriate rather than the exponential. The use of the wrong model then gives a loss \$149.15–\$148.84 = \$0.30, (points D,A) again hardly a devastating effect when considered relative to the overall gain of about \$149 ha^{-1}, and the many other contingencies that can affect a farmer's income. Viewed on the full scale (insert in Fig. 50.3) the differences in profit in the vicinity of the optimum are in fact hardly perceptible.

It is possible then for relatively simple models to be good enough for most types of data, as with the square-root quadratic model for Fig. 50.1 type data, and the 'broken stick' model 8 for Fig. 50.2 type data.

Models for fertilizer application to pasture

The choice of a model for representing the effects of fertilizer on pasture production presents a particular set of problems, due to the complexity of the system controlling animal production. Yield in a grazing system is affected by the growth and composition of pasture throughout the year, supplementary feeding, frequency and intensity of grazing, weather conditions, etc., and the effects of all such factors are likely to interact with the effects of rate, time and frequency of fertilizer application. Moreover, production from a grazing system commonly involves more than one simply defined product e.g. animals of various ages, wool, hay etc. Such difficulties have been amply demonstrated in many countries (recently in Australia by Willoughby 1954; Reid 1970; Ayres 1977; Ayres *et al.* 1977 *a*, *b*; Winter *et al.* 1977 *a*, *b*; Cook *et al.* 1978 *a*, *b*; etc.) though there has not been much doubt, at least not since the 1950s (Gruen 1955), concerning ultimate economic value of fertilizing pastures. The difficulties are shown for example by Cook *et al.* (1978*a*, *b*) who conclude that pastures in the New England region of New South Wales require massive applications of 25–50 kg ha^{-1} phosphorus every year, with reduced grazing over the winter months when feed is in shortest supply, to maintain the quality of improved pastures; a conclusion that will hardly appeal to the farmers of the region.

The difficulty lies then in modelling a system with perhaps several products, and with many interacting factors affecting response to fertilizers on a continuous scale. Animal – pasture – fertilizer models have accordingly been developed (Morley 1975) though they necessarily suffer from a lack of sufficient definitive data. There are simply just not enough data to indicate appropriate mathematical forms for yield functions throughout a sequence of seasons. The problems of determining the fertilizer needs of pastures should nevertheless not be overstated. Some notable successes in this regard have been reported recently by Cannon (1972) and Curll (1977 *a*, *b*), and more carefully planned experiments for defining components of the system may provide the required information when results are combined.

Prediction of yield functions and fertilizer requirements

Fertilizer is applied before a crop has been grown so that optimal rates have to be predicted on the basis of expected yield responses. Yields and yield responses commonly vary, however, throughout agricultural regions in a more or less unpredictable fashion due to variations of soil, weather, farming method, etc., so that predicted rates are inevitably inaccurate most of the time, potential profits being lost from either over- or under-estimating fertilizer requirement. There is then a statistical problem in estimating optimal rates: to estimate fertilizer rates which will, in the long term, give an overall optimal return on the investment in fertilizer despite short term variations in yield response and the inevitable errors in individual estimates. Optimal rates are, however, not satisfactory variables for statistical estimation studies. Thus, although in theory, sets of optimal rate data can be obtained to represent a variable region for statistical studies, as by carrying out series of fertilizer experiments, in practice there are extrapolation problems due to the occurrence of high fertility sites where optimal rates may be in effect negative (yields would not be decreased by the removal of the nutrient from the soil) and low fertility sites where requirements exceed the highest treatment rates. It is more convenient therefore to investigate the problem of predicting the yield functions from which optimal rates are calculated rather than the actual optimal rates. One such procedure, illustrated here, involves studies on the parameters of orthogonal polynomial yield functions, corresponding to the polynomial yield functions of the preceding sections (Colwell 1967*b*, *c*, 1968). Another, described by Jardine (1969), is based on transformations of experiment data to principle components with corresponding studies on their variability and prediction. In each case studies are in effect made on sets of experiment data with the objective of predicting the results of similar experiments, if they were to be carried out, for other sites in the region represented by the data. The predicted results are then used to calculate the optimal fertilizer rates by the same procedure that are used for yield functions derived from actual experiment data.

Orthogonal polynomial trend coefficients

Given data from a series of n factorial-type fertilizer experiments, n polynomial yield functionsof the form

$$Y_i = b_{0i} + b_{1i}N^{1/2} + b_{2i}P^{1/2} + b_{3i}(NP)^{1/2} + b_{4i}N + b_{5i}P \quad \text{for} \quad i = 1, 2, \ldots, n \tag{7}$$

corresponding to the example model 3, can be estimated by least squares regression. Such a set of functions can then be regarded as samples from a sample space, defined in terms of the region or soils represented by the experiments. For statistical studies the polynomials can be transformed to orthogonal polynomials (for procedure, Bliss 1970; Colwell 1978),

$$Y_i = p_{0i}M + p_{1i}L_n + p_{2i}L_p + p_{3i}L_nL_p + p_{4i}Q_n + p_{5i}Q_p \quad i = 1, 2, \ldots, n \qquad (8)$$

where now the regressor variables are all orthogonal with respect to each other for the experiment treatments, M, L, Q representing zero, linear and quadratic trends for the nutrients indicated by the subscripts. The reason for the transformation is that the trend coefficients p_{0i}, p_{1i}, etc., provide direct measures of the amounts of independent (orthogonal) trends in the data for each experiment and can consequently be used as independent statistical variables for the calculation of means, variances, regressions etc. Thus the mean trends for the sample space are estimated by the sample means

$$\bar{p}_0 = \frac{1}{n}\Sigma p_{0i}, \quad \bar{p}_1 = \frac{1}{n}\Sigma p_{1i}, \ldots, \quad \bar{p}_n = \frac{1}{n}\Sigma p_{ni}$$

and their variability by the variances

$$\text{Var}\ (p_0) = \frac{1}{n-1}\ \Sigma\ (p_{0i} - \bar{p}_0)2,$$

$$\text{Var}\ (p_1) = \frac{1}{n-1}\ \Sigma\ (p_{1i} - \bar{p}_1)^2, \qquad \text{etc.}$$

and regression relationships are studied using the p_{ji} as dependent variables. The form of the trends and the mathematical procedure for the transformation is detailed elsewhere (Colwell 1978). Computer drawn illustrations of the trends in square-root quadratics corresponding to equations (7) and (8) may also be consulted (Colwell 1979).

The mean yield function and general fertilizer rates

Given a variable statistical population and a normal type frequency distribution, the best estimate, in the absence of more specific information, is provided by the sample mean. Applied to the trend coefficients p_{ji} this leads to a best estimate for the sample space represented by a series of yield functions such as equations (7) or (8) termed the mean yield function,

$$y = \bar{b}_0 + \bar{b}_1N^{1/2} + \bar{b}_2P^{1/2} + \bar{b}_3(NP)^{1/2} + \bar{b}_4N + \bar{b}_5P \qquad (9)$$

and

$$Y = \bar{p}_0M + \bar{p}_1L_n + \bar{p}_2L_p + \bar{p}_3L_nL_p + \bar{p}_4Q_n + \bar{p}_5Q_p \qquad (10)$$

where the $\bar{b}_j$ and $\bar{p}_j$ are means of the b_{ji} and p_{ji} *in* equations (7) and (8). If a sample space has been defined in terms of region, soil, crop, farming system etc., and the local weather, and if experiments have been distributed throughout the sample space at random, and repeated over several years to represent variations with time (season), then a first best estimate of the yield function for a field is provided by the mean yield function given by equation (9) or (10) – i.e. best in the sense that the only available information is that provided by the sample of n yield functions. Accordingly a first estimate of the optimal rate for a field within the sample space is that calculated from the mean function, and this is termed the *general* optimal fertilizer rate.

In practice, the general rate calculated from the mean function, serves as a first estimate, to be modified upwards or downwards for individual fields according to local assessments of their fertility levels. The main difficulty is in obtaining an adequate and representative sampling. Ideally large numbers of experiments to provide the yield function samples Y_i, will have been distributed at random throughout the sample space and repeated for several years but because personnel, objectives, crop varieties, priorities, etc., change with time, such a sampling is rare. Moreover, sampling bias can be easily and unknowingly introduced by the selection of better than average sites for the experiments, as emphasized by Davidson and Martin (1968). Examples of such difficulties are seen in estimates based on the National Soil Fertility Data Bank (Colwell 1977, 1979). Nevertheless, the general rates can often suggest real and important differences between regions and soils, especially when the differences are large. For example, general rates calculated from the above Data Bank show interesting contrasts in fertilizer requirements of different regions across Australia, e.g. high nitrogen and relatively low phosphorus requirements in parts of Western Australia, negligible nitrogen and high phosphorus requirements in parts of Victoria and New South Wales, and so on.

Regression estimates and soil test calibrations

The variation of polynomial yield functions within a region (sample space) is seen in corresponding variations in orthogonal trend coefficients, as in the p_{ji} in equation (8). If regression relationships can be established to explain some of this variation in terms of other site variables then these can be used to predict trend coefficient

Table 50.8. Trend coefficients for nitrogen-phosphorus fertilizer experiments with wheat estimated using regression.

*, $P < 0.05$; **, $P < 0.01$; ***, $P < 0.001$

Regression	R_a^2(%)
P_0 = 25228 – 88.39*** (sow date) + 1736 log($NaHCO_3$-P)	38.6***
P_1 = 1997 – 1193*** log(NO_3-M) + 13.34 (NO_3-L) + 1.113* (NO_3-L)^2 + 255.5* (pH)	62.3***
P_2 = 5430 – 1165* log($NaHCO_3$-P) – 49.78*** (NO_3-M) + 14.19 (rain – M)	47.3***
P_3 = 1309 – 397.9* log(NO_3-M) + 2.605 (NO_3-L)	26.3**
P_4 = 749.8 – 197.2* log($NaHCO_3$-P) + 7.598 (NO_3-M) – 87.82 (pH)	21.5*
P_5 = –574.2	–

values for individual sites, thus giving better and more specific estimates than the sample mean $\bar{p}_j$. The estimates can be combined in a corresponding manner to the assemblage of mean estimates given above by equation (10), and then transformed back to the non-orthogonal form to obtain an estimate of a polynomial that can be used in turn to estimate fertilizer rates for specific sites.

The general procedure is illustrated by the regressions in Table 50.8, estimated from the National Soil Fertility Data Bank for wheat in the southern wheat-belt region of New South Wales (Colwell 1979). Nitrogen–phosphorus fertilizer experiments were carried out at 35 sites in this region and these provided 35 orthogonal polynomial yield functions, corresponding in form to equation (8). The set of 35 values for each coefficient were then used as dependent variables for regression studies on other site variables, leading to the set of equations in Table 50.8. Thus for this example, 38.6% of the variance of the zero trend coefficient p_0, representing yield levels (p_0M = mean experiment yield) for the 35 sites, is explained by the regression for sowing date for the wheat and a phosphorus soil test ($NaHCO_3$ soluble P, Colwell 1963*b*), 62.3% of the variance of the linear (square root) trend for nitrogen fertilizer is explained by a regression for trends in water-soluble nitrate in the soil profile (NO_3-M = zero trend of NO_3, NO_3-L = linear trend, with depth) and soil pH (0–5 cm), and so on for the other trend coefficients in Table 50.8. The p_5 coefficient for the quadratic trend for the phosphorus treatment failed to give a significant regression and is consequently best estimated by its mean ($\bar{p}_5 = -574.2$). These equations can all be used to estimate trend coefficient values for particular sites in the region (sample space) represented by the data and these estimates combined to obtain an estimate of an orthogonal polynomial yield function corresponding to equation (8). The non-orthogonal polynomial form, equation (7), can then be obtained by algebraic transformation (as detailed, Colwell 1978) and used to calculate optimal rates for the particular site, as described in the preceding section.

The yield function estimated by this procedure using regression or mean estimates for the individual trends, can be shown to be mathematically identical with the estimate that would be obtained from the corresponding multiple regression using the same set of variables. Such a regression cannot be used, however, to test for the significance of the contribution by the regressor variables because its residual mean square, in an analysis of variance, is a combination of the very different residual mean squares for within and between experiments (as detailed, Colwell 1967*a, b*). In addition the component trends for a regression given by equation (8), and hence equation (7), are estimated with varying degrees of precision, and this must be allowed for in tests of significance. Contributions of variables are thus tested in the regressions for individual trends as in Table 50.8, and not in a grand multiple regression for all of the data. It may be noted, incidentally, that many published multiple regressions for crop yield as a function of fertilizer rates and site variables, are statistically invalid in this respect. It is unfortunate in this regard that the error results in an overestimation of the significance of site variables, many being rated significant when in fact they are not ($P < 0.05$).

The example demonstrates some of the problems encountered in predicting yield functions and hence fertilizer requirements, as by soil testing services. Some of the variables can serve as direct predictors of fertilizer requirements in that their values can be obtained before the fertilizer is applied, namely the sowing date, P soil test ($NaHCO_3$-P) and the soil pH for 0–5 cm soil samples. Values for other variables, as for the rainfall (rain – M), cannot be known however until after the crop has been produced and hence serve to explain rather than predict variations in fertilizer requirements. Values for such variables must be estimated, as from local climatic records, before the regressions can be used for predictions. Other values, such as the trends in water-soluble nitrate in the soil profiles, are also likely to be more useful for explaining rather than predicting, because there is not likely to be sufficient time for sampling and analysis by a routine testing service for the values to be used to predict fertilizer requirements before the fertilizer is actually applied. Clearly fertilizer requirements are determined by many factors, additional to the availability of the particular nutrient as measured by a simple soil test analysis, and estimates cannot therefore be expected to be very accurate. Soil test estimates will simply be a bit better than general estimates and will probably be of most value when used to supplement local experience based on previous practices. More detailed descriptions of such problems in predicting fertilizer requirements, and of the value of soil testing are given elsewhere (Colwell 1978, 1979).

The need for soil tests to assist in the estimation of fertilizer requirements, despite the problems involved, is seen in the extensive literature on soil testing procedures (e.g. Beckwith 1965; Colwell 1967*c*, 1976; Colwell and Esdaile 1968; Ozanne and Shaw 1967; Russell 1968*a, b, c*; Montgomery and Rubenis 1978; etc. and reviews by Skene *et al.* 1966; Colwell, 1967*a*; Smith 1971). Some soil tests work better for particular kinds of soil, but often there is little basis for choice if allowance is made for the effects of soil composition (Colwell 1971; Colwell and Donnelly 1971).

Fertilizer residues

Much of the nutrient applied to a crop in fertilizer may remain in the soil as a residue, improving the soil nutrient status and reducing fertilizer requirements for future crops. Such residue effects from the regular application of fertlizer may eventually eliminate the deficiency, or increase the available nutrient in the soil to a level for which only maintenance applications are required, to replace the nutrient removed during cropping in crop products, by leaching, etc. For some soils and agricultural systems the effects can be considerable, as for many pasture-grazing systems, where the effects of a single application of fertilizer can be seen for several years, or for the wheat-rice rotation system of many Australian irrigation regions, where phosphatic fertilizer is applied only to the wheat and nitrogenous fertilizer to the rice. In other regions residual effects seem to be negligible, possibly because of relatively low rates of application, as for much of the Australian wheat-belt where phosphatic fertilizer requirements seem to remain constant despite regular applications of phosphate fertilizers over many years. Effects are likely to be greater for relatively immobile fertilizer nutrients such as P rather than for mobile nutrients such as nitrogen (Engelstad 1970) and also to be greater for minimum tillage systems (e.g. Rowell *et al.* 1977; Rudd and Barrow 1973).

The form of residual effects

Models for the representation of the effects of fertilizer residues on crop yield, and the nature of experimental designs for their estimation are suggested by noting the analogy between the problem of estimating the effects of the *same* fertilizer nutrient applied at *different* times with that of estimating the effects of *different* nutrients applied at the *same* time. Thus standard multiple nutrient yield function models (for example model 3) suggest models such as

$$Y = a_0 + \sum_{i=0}^{n} b_i X_i^{0.5} + \sum_{i=0}^{n} \sum_{j=i+1}^{n} c_{ij} (X_i X_j)^{0.5} + \sum_{i=0}^{n} d_i X_i \qquad (11)$$

where now X_i is a nutrient applied in the year denoted by the subscript, X_0 = fresh application for the yield Y, X_1 = application for the previous year (crop), X_2 = application 2 years earlier, etc. The analogy also suggests the use of factorial type experiments with factorial combinations of the nutrient applied over the series of years (a serial design, John and Quenouille 1977). There is however, an important difference in the analogy. The availability of a nutrient to a crop several years after its application will be affected by many factors such as chemical reactions with the soil, incorporation into crop residues and soil organic matter, removal by farm products, leaching, erosion etc. and the magnitudes of the effects may vary with the rate of application. Consequently the effective experimental treatment rates vary with time, and X_i values for $i>0$ in models such as that given by equation (11) must be regarded as sources of the residues rather than the actual rates that determine the yield. The difficulty may be overcome by the substitution rate concept, described below, but there will nevertheless be practical difficulties in obtaining reliable data to estimate relationships for more than about three successive fertilizer applications i.e. for $n > 2$ in equation (11). Care is also required to prevent an overfitting of data by using too many regression variables and thereby introducing meaningless inflections to the fitted function.

An example yield function, evolved from the general form of equation (11), for applications of phosphatic fertilizer to successive crops of wheat is

$$Y = 882 + 406.9\, P_0^{0.5} + 24.24\, P_1 - 3.448\, P_0^{0.5} P_1 - 11.61\, P_0 \qquad (12)^*$$

$$(R_a^2 = 0.83,\ \text{probability} < 0.991)$$

where Y is the yield of wheat, P_0 is the fresh or current application rate and P_1 is the rate applied to the preceding wheat crop, 1 year previously. The function shows clear responses to both applications and residual effects from preceding application P_1 are indicated by decreases in requirement for a fresh application P_0 to obtain particular yield levels, as illustrated by the contour lines in Fig. 50.4. The function also indicates a highly significant ($P<0.01$) negative interaction such that there is a decrease in the yield response to residues from the preceding application (P_1) with increase in the fresh application rate (P_0).

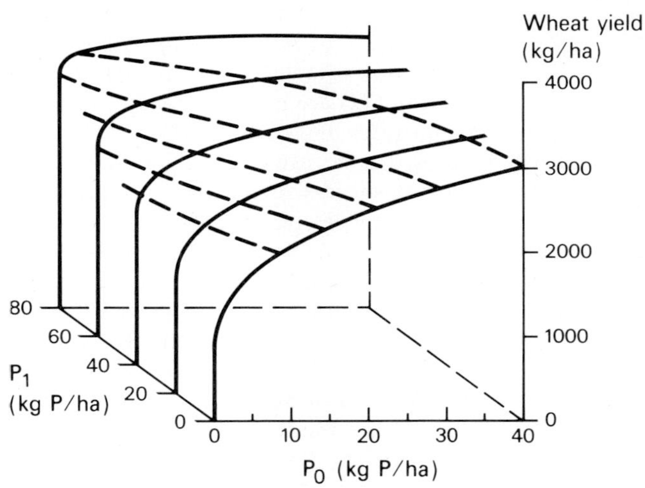

Fig. 50.4. Effects of residues from fertilizer applied to the preceding wheat crop (P_1), on response to the current application (P_0), as represented by equation (12). Contour lines for Y = 2000, 2250, 2500, 2750, 3000 kg wheat per ha show decrease in current P requirements with increase in previous application.

*Estimated from a two-year fertilizer experiment with wheat at a site near Temora, N.S.W. with a factorial combination of the treatment rates P_1 = 0, 10, 20, 40 kg ha^{-1} phosphorus in 1978 and P_1 = 0, 20, 40, 80 kg ha^{-1} phosphorus applied to a preceding wheat crop in 1977.

The analogy of fertilizer residue effects to multinutrient yield functions can also be extended to emphasize the essentially empirical nature of residue yield functions, as for the earlier discussion on yield functions. Thus although other models can be developed from theoretical considerations (e.g. Holford 1976; Holford and Mattingly 1976*a*, *b*; Russell 1977), caution is needed in assuming theoretical interpretations for their parameters.

Substitution rates

Although the effects of fertilizer residues can be quite obvious, as in the above example, there is no simple direct method for measuring the actual residues. Rather, residues are inferred from their effects and are usually represented in terms of equivalents of freshly applied fertilizer. It will be found for example, from calculations for the contour lines in Fig. 50.4, that to obtain a yield level of 3000 kg ha^{-1} wheat, 40.4 kg ha^{-1} P must be applied to the crop if none was applied to the preceding crop ($Y = 3000$ for $P_0 = 40.4$, $P_1 = 0$), whereas only 30 kg ha^{-1} need be applied if 44.4 kg ha^{-1} was applied previously ($Y = 3000$ for $P_0 = 30$, $P_1 = 44.4$). An application of 44.4 kg ha^{-1} to the previous crop has reduced the requirement for a fresh application by 40.4 − 30 = 10.4 for this yield level, or in other words the 1 year residue effect of 44.4 kg ha^{-1} is equivalent to 10.4 kg ha^{-1} of freshly applied fertilizer. More conveniently, the residue effect is expressed as a proportion of the original application, $s_1 = 10.4/44.4 = 0.23$ and termed the substitution rate i.e. the rate of substitution by the residue for fresh fertilizer is 23% of the previous application. Other substitution rates calculated by this same procedure using equation (12) are given in Table 50.9 for a range of application rates and yield levels.

Substitution rates can be estimated as in the above example for series of successive fertilizer applications, extending over several successive crops or years, and are conveniently denoted as a series, s_j with $j = 1, 2, \ldots, n$, the members of the series corresponding to the successive applications. Examples are the values $s_1 = 0.5$, $s_2 = 0.25$, $s_3 = 0.125$ suggested 'as a reasonably fair general compromise' for the residual effects of phosphatic fertilizer applied to cereals in Scotland 1, 2 and 3 years previously (Williams and Reith 1971), and the series $s_j = 1/(1+j)$ for P applied to cereals in Western Australia *j* years previously (DECIDE model of Bowden and Bennett 1976). Such simple series suggest that s_j values are constant for each year (value of *j*) irrespective of yield levels or application rates, and this seems unlikely. The values in Table 50.9 from equation (12) show, for example, a very wide range with fertilizer rates and yield levels. Nevertheless if s_j values can be specified, at least for restricted ranges of yield levels, and application rates, they provide simple means for estimating the effects of *previous* fertilizer applications on present levels of fertility and for estimating series of successive fertilizer applications for an optimal fertilizing program.

Table 50.9. Fertilizer substitution rates (s_1) for a range of application rates and yield levels.

	P_1 (kg ha^{-1})	P_0 (kg ha^{-1})	Reduction of P_0	s_1
Yield: 2000 kg ha^{-1}				
	0	9.0	—	0.25[A]
	16.1	5	4.0	0.25
	46.1	0	9.0	0.20
Yield: 2500 kg ha^{-1}				
	0	20.9	—	0.26[A]
	3.5	20	0.9	0.26
	19.9	15	5.9	0.30
	33.5	10	10.9	0.33
	46.3	5	15.9	0.35
	66.7	0	20.9	0.31
Yield: 3000 kg ha^{-1}				
	0	40.4	—	0.11
	3.7	40	0.4	0.12
	30.5	35	5.4	0.18
	44.4	30	10.4	0.23
	53.4	25	15.4	0.29
	60.1	20	20.4	0.34
	65.8	15	25.4	0.39
	71.0	10	30.4	0.43
	76.6	5	35.4	0.46
	87.4	0	40.4	0.46

[A] values of $\frac{dP_0}{dP_1}$ given for ($P_1 = 0$)

The first problem in the direct estimation of s_j values is that data are required from long-term and representative fertilizer experiments. Such experiments are expensive, and difficult to carry through to a successful conclusion, being particularly susceptible to the common hazards of weeds, disease, abnormal weather etc. because they must extend over several successive crops. Then there is a problem of choosing appropriate s_j values from estimated residue yield functions since values can be expected to vary both with yield levels and fertilizer application rates. Such variations are only to be expected because crops remove more of applied nutrients under favourable growing conditions than they do under poor conditions (e.g. during a drought). Greater proportions of fertilizer will remain as residues when application rates are high, and the effects of residues on yield will be less in the presence of low or nil fresh applications of fertilizer than of high. Such problems may be reduced by restricting the range of yield levels and application rates, or by developing residue function models to allow for nutrient losses during cropping etc. The effects of such factors on substitution rates, and appropriate procedures for predicting values for particular agricultural situations are however largely unknown, and represent a current research challenge. It is worth noting, incidentally, that problems in estimating substitution rates may be hidden by a naive choice of yield function models in which s_j values are mathematically constant, irrespective of any information to the contrary indicated by data. For example, the use of the model

$$Y = b_0 + b_1(X_0 + s_1X_1 + s_2X_2 \ldots)^{0.5}$$

$$+ b_1(X_0 - s_1X_1 - s_2X_2 \ldots) \qquad (13)$$

will ensure constant substitution rates, irrespective of variations with level Y or the X_i.

Unfortunately, few of the Australian publications on residue effects give sufficient data for studies on substitution rates, and most of these are only for 1 year residues i.e. for s_1. Barrow and Campbell (1972) used data from two successive crops in a glasshouse experiment but chose models to give constant s_1 values for restricted ranges of their data. Holford and Gleeson (1976) report similar data from field experiments. Arndt and McIntyre (1963) give data and substitution rates for several years, but their data do not allow for a check of an implicit assumption of nil interaction effects. High substitution rates are suggested in many instances by pronounced residue effects, especially with pastures (e.g. Teakle and Cariss 1943; Anderson and McLachlan 1951; Woodroffe and Williams 1953; Donald and Williams 1954; Tyson 1955; Piper and DeVries 1964; Jones 1968; Barrow 1973, 1975*a*, *b*, 1976; Tudsri and Whiteman 1977; Tupper 1978). Low values are similarly suggested by small effects for phosphorus in soils with high contents of iron and aluminium oxides, (e.g. Tyson 1955; Raupach and Piper 1959; Beckwith 1965). Some substitution rates may in fact be rather surprising, due for example to leaching in sandy soils (Ozanne *et al.* 1961), reduced availability after irrigation (Willett *et al.* 1978) or flooding (Williams and Simpson 1965), reactions involving soil organic matter (Williams 1954; Williams and Spencer 1954), release of previously unavailable nutrient from soils during plant growth (examples with potassium, Williams and Lipsett 1960; Spencer 1954), and when residues give larger yields of crops than can be obtained with fresh fertilizer (Cooke 1971). Increased rather than a decreased response to phosphorus in the presence of phosphorus residues on soils with extremely high phosphorus-sorbing capacity (Searle 1967) could also give unexpected values.

Fertilizer history and fertility level

If the substitution rate s_j indicates the residual effect of fertilizer applied j years (crops) previously, and values are available for n years, then the total effect of fertilizer applied over an n-year period can be expressed in terms of an equivalent fresh application by the sum

$$S = \Sigma\, s_jX_j \qquad (14)$$

For example Bowden and Bennett (1976) use the sum $S = \Sigma\,[P_j/(1+j)]$ to calculate the level of available phosphorus in soils (in place of more conventional tests) for estimating P fertilizer requirements in Western Australia, P_j being the fertilizer P applied j years previously. Here $s_j = 1/(1+j)$ as noted above.

Optimal rates with residue effects

The procedure for calculating optimal fertilizer rates described in a previous section is simplified in that it is based solely on considerations of profit from the crop to which fertilizer has been applied, ignoring the effects of residue on the yields and fertilizer requirements of following crops. Such effects constitute part of the return on the investment in fertilizer, and allowance should be made for them also in the calculations.

The extension of the procedure for calculating optimal rates to allow for residue effects depends essentially on calculating the expected additional profits that will accrue from the residue effects of a fertilizer application on future crops. These profits, discounted for time are then added to the profit from the crop to which the fertilizer was actually applied to obtain a total profit π_{TD}, from the investment I for the optimum rate calculation, by $d\pi_{TD}/dI = 0$.

The procedure is conveniently described for polynomial models and annual cropping systems in which returns are obtained from the sale of crops at about 1 year intervals following the investment in fertilizer, and can be extended to other models and systems. Given a basic yield function model, for an annual crop,

$$Y_1 = b_{01} + b_{11}X_0^{0.5} + b_{12}X_0 \qquad (15)$$

where X_0 = fresh application rate of a fertilizer and Y_1 = yield 1 year after the application, when the crop is sold, the optimal rate without residue effects is obtained by

$$\frac{d\pi_D}{dI} = 0, \qquad \frac{dY_D}{dX} = \frac{C}{V} \quad \text{or} \quad \frac{dY}{dX} = \frac{C}{V}(1+R)$$

as described previously, equation (3). Now if the crop Y_1 is followed by $(n-1)$ crops, at yearly intervals, and without any further application of fertilizer, they will show corresponding though decreasing responses to residues from the X_0 application which may be represented by similar functions

$$Y_i = b_{0i} + b_{li}X_0^{0.5} + b_{2i}X_0, \qquad i = 2,3,\ldots, n \qquad (16)$$

The total discounted yield from the n functions is then

$$Y_{TD} = \sum_{i=1}^{n} \frac{Y_i}{(1+R)^i} \qquad (17)$$

$$= \Sigma\, b_{0D} + \Sigma\, b_{1D}X_0^{0.5} + \Sigma\, b_{2D}X_0 \qquad (18)$$

where

$$\Sigma\, b_{jD} = \sum_{i=1}^{n} \frac{b_{ji}}{(1+R)^i}\,.$$

The optimal rate for X_0 can be then calculated for this total yield by $d\pi_{TD}/dI = 0$ etc., as for the simpler case when residues were ignored. Examples of the calculations and a detailed derivation are given by Pesek *et al.* (1961).

If fertilizer is also applied to each of the following crops the above procedure based on a total yield response to X_0 in equation (18) can be followed again, irrespective of responses to the other fertilizer treatments, only if the effects of the subsequent fertilizer applications are simply additive, without interaction effects. If there are interaction effects as in equation (12), the total discounted yield will depend on the following applications and allowance must be made for these in the otimization calculation. Thus given the series of yield functions, with provision for interactions with future fertilizer applications, for 3 successive crops,

$$Y_1 = b_{01} + b_{11}X_0^{0.5} + b_2X_0$$

$$Y_2 = b_{02} + b_{12}X_0^{0.5} + b_{22}X_0 + b_{32}X_1^{0.5} + b_{42}X_1 + c_{12}(X_0X_1)^{0.5}$$

$$Y_3 = b_{03} + b_{13}X_0^{0.5} + b_{23}X_0 + b_{33}X_1^{0.5} + b_{43}X_1 + b_{53}X_2^{0.5} + b_{63}X_2 + c_{13}(X_0X_1)^{0.5} + c_{23}(X_0X_2)^{0.5} + c_{33}(X_1X_2)^{0.5} \quad (19)$$

the total discounted yield response to the application X_0 is now obtained by

$$\Sigma\, Y_{TD} = \Sigma\, b_{0D} + \Sigma\, b_{1D}^{*}X_0^{0.5} + \Sigma\, b_{2D}X_0 \quad (20)$$

where $\Sigma\, b_{0D}$ and $\Sigma\, b_{2D}$ are as in equation (18), but $\Sigma\, b_{1D}^{*}$ differs from $\Sigma\, b_{1D}$ by the addition of interaction terms

$$\Sigma\, b_{1D}^{*} = \frac{b_{11}}{1+R} + \frac{b_{12} + c_{12}X_1^{0.5}}{(1+R)^2} + \frac{b_{13} + c_{13}X_1^{0.5} + c_{23}X_2^{0.5}}{(1+R)^3}$$

The calculation of an optimal rate for X_0 by $\mathrm{d}\pi_{TD}/\mathrm{d}I = 0$ now requires a specification of following rates and, ideally these rates will also be estimated with allowance for future interactive residue effects. In practice, however, interaction effects may be relatively small, especially for crops to be grown 2 or more years after the current crop so that rough estimates of future applications, based for example on substitution rates,

$$X_1 = X_0(1-s_1), \qquad X_2 = X_0\,[(1-s_1)^2 - s_2], \quad \text{etc.}$$

may be sufficient.

Experimental requirements

The need for a series of yield functions covering interaction effects, for the calculation of optimal rates suggests the need for a series of large, long-term fertilizer experiments which are likely to prove impracticable if extended over more than about three consecutive crops. Given, however, that yields are discounted with respect to time, and that residue effects after about two crops are likely to be very small, the actual need is likely to be for no more than about two or three crop experiments, to give estimates for Y_1, Y_2 and Y_3 functions corresponding to those in the series of equation (19). The effect of neglecting further residue effects will be to reduce estimated optimal rates, but given the many uncertainties affecting such estimates, a bias towards somewhat conservative estimates may not be very serious.

Repeated treatment experiments

A particular type of problem that is often not appreciated arises when residue effects are studied in experiments based on fertilizer treatments repeated on the same plots for successive crops over several years. Such experiments give no information on interaction effects and hence on the possible variability of s_j values. At best, data from such experiments will identify an optimal constant maintenance rate of application, because the effects of application for different years are completely confounded. Helyar and Godden (1977) suggest that data from such experiments can be used to derive 'maintenance requirement functions', representing the ultimate effects of continued application of a range of fixed rates. Though it may be argued that such input-output equilibria can be established for a given agricultural system, thus arriving at an optimal rate, it is difficult to see how such a rate would interest farmers given the pronounced disequilibrium states of severely deficient soils. Farmers usually simply wish to follow a fertilizing program, obtaining optimal returns each year on money spent on fertilizer with appropriate allowances for the values of residues to future crops i.e. with reductions each year towards maintenance rates as fertility is built up by the residues.

Data from some fixed-rate experiments can demonstrate very clearly another difficulty in studies of residue effects. Variations in seasonal conditions can have very disturbing effects, completely obscuring any consistent trend in response forms towards a 'maintenance requirement function', and corresponding difficulties can be anticipated in estimating substitution rates. Marked seasonal variations are shown, for example, with wheat and pasture experiments at Wagga Wagga, New South Wales, by Osborne *et al.* (1977) and Curll (1977*a*, *b*). Such seasonal effects can be hidden by converting yields to relative yields though then the smoothed trends may become rather unrealistic. For example, the relative yields of Rossiter (1964) as used by Helyar and Godden (1977) to illustrate their 'maintenance requirement function', show steady increases for nil and low fertilizer rates where the opposite would be expected due to continuous exploitative cropping for low rates.

Problems for research

In principle the determination of optimal rates of application of fertilizers is simple, particularly given modern computing facilities for carrying out the necessary calculations. The difficulties arise because little or none of the information required for the calculations, such as the effects that various rates will have on yield (yield functions), the value of fertilizer residues to future crops

(s_i values), the economic situation of the farmer (R value), and so on, is known with any accuracy at the time when the calculations have to be carried out i.e. well before the crop is sown. Errors, representing the accumulated effect of inaccurate estimates of all these things are thus inevitable, but nevertheless estimates of optimal rates must be made. The task for the scientist is to reduce the uncertainties and on this basis accordingly the present review suggests problems deserving some priority in research.

Models for fertilizer effects on crop yields

Clearly no single mathematical model can adequately describe the many kinds of functional relationship between yield and nutrient application rates for all crops in all situations. Although simple and convenient models, such as the square root quadratic, can represent perhaps most relationships, with due cognizance of their limitations over certain ranges, they may be quite inappropriate and misleading for some. At present there is little information on when or where particular models should or should not be used, and as a consequence models have to be chosen with a certain amount of faith and rationalization. Under many circumstances the chosen model will be good enough, as demonstrated by the comparison in Fig. 50.3, but this needs to be checked. A need clearly seems to exist for at least two types of model such as the square root quadratic for the type of data shown in Fig. 50.1 and intersecting straight lines for the type shown in Fig. 50.2. But for animal-pasture systems much more elaborate models are necessary and work on these has hardly started. There is thus a need for high quality experimental data covering many fertilizer rates for crops in contrasting environments, to provide bases for the choice of models. For pasture systems the data must also extend over a considerable time span and include animal production, suggesting a need for at least a few large-scale experiments with grazing animals, to be coupled perhaps with a series of smaller pasture experiments.

Estimation of yield functions

Given a satisfactory model to represent the effects of fertilizers on yield, yield functions then have to be estimated for individual sites for the calculation of optimal rates. The estimation is conveniently treated as a statistical problem, involving the calculation of sample means and regressions, but a problem immediately becomes apparent. There are very few consistent data available representing regions or soils. The National Soil Fertility Data Bank (Colwell 1977) provides by far the most extensive set of this type of data for Australia but inadequacies are obvious for the representation of regions and soils. A first basic need, then, is for series of fertilizer experiments to provide adequate samplings of important regions and soils with respect to temporal as well as locational variability.

Mean yield functions provide general first bases for estimating fertilizer requirements and more precise, site-specific estimates can be obtained using regressions on site variables, particularly on soil analyses or tests. The problem here concerns the validity of the regression estimates, again due to inadequate samplings, and the relative merits of alternative tests. There is at present much confusion on this matter with rather fruitless arguments on the relative merits of alternative predictors, including local farmer or adviser experience. There is thus a need to study the implications of the many inter-correlations between the wide range of potentially useful tests, with the ultimate objective of defining the most appropriate variables for regression estimates for particular situations. Apart from providing direct estimates, regressions on *a posteriori* site variables such as weather conditions can serve to provide better estimates of regional means, where extensive data are available (weather records). The need again is for statistical studies.

Bridging data

The statistical requirements of an adequate sampling of a well defined sample space are obvious enough, but involve a large amount of work for each space. Clearly such data requirements are impossible to satisfy for every soil, region, crop variety, climatic zone etc. and results from particular studies must be extended to other sample spaces. There is a need, then, for what may be called 'bridging experiments', designed to extend results from one sample space to another without the requirement of a complete repetition of experimental work.

Fertilizer residues

Allowances can be made for the value of fertilizer residues in calculating optimal rates of fertilizer application, if residue effects can be represented by yield functions extending over sequences of crops, but there are very few data available to estimate such functions. Data for such studies can be obtained from relatively simple fertilizer experiments with treatments extending over about 3 years. Methods have yet to be devised, however, for estimating appropriate residue functions for particular sites, as can be done for simple yield functions by soil testing.

Fertilizer use in cropping systems

Most studies on fertilizer use relate to particular crops though in practical farming systems fertilizers are applied to sequences of crops under some crop-pasture rotation system. In regions of intensive agriculture (e.g. India) the whole sequence of two, three or four crops may be compressed into the time span of 1 year whereas in Australia the sequence may cover several years, but the nature of the problem of optimising the distribution of fertilizer amongst the successive crops with appropriate allowances for residue effects remain the same. Little work on this problem seems to have been carried out anywhere in the world. Studies on the problem could develop logically out of studies on fertilizer residue effects, with similar types of experiments but different succeeding crops.

Avoidance devices

Fertilizers have to be used despite the existence of the above types of unsolved problems, and a general paucity of reliable information and data for the calculation of optimal rates. Accordingly, various devices have been developed, often out of the subconscious mind of research workers, to avoid the problems so that estimates can nevertheless be made. Two current examples are the use of relative yields, and mathematical models sometimes described as 'black boxes'.

Relative yields

Perhaps the commonest and most popular device for avoiding the difficulties of the variability of yield functions with seasonal weather conditions is to replace yields by relative quantities or abstractions, such as percentage of an asympototic maximum, $\% Y = 100\ (Y/A)$ where A is the maximum yield attainable with fertilizer, or percentage yield deficit, $\% D = 100[(A-Y)/A]$. The device is commonly associated with the Mitscherlich exponential yield function, possibly following from a suggestion by the mathematician Baule (1918) and promotion by Bray (1944, 1948). Thus it is asserted, quite ingenuously, that the asymptotic maximum yield A is known or can be realistically nominated from local experience, and that fertilizer should be applied to obtain 80%, 90%, 95% or such of this value, the percentage being varied to ensure reasonable estimates. Apart from the 'fudge factor' aspect of this procedure and the possibility of consequent uneconomic fertilizer use, there are statistical objections. Dividing yield data by the maximum, effectively weights error deviations very markedly as the inverse of the maximum, and this may result in serious bias in statistical estimates of relationships and tests of hypotheses. Moreover its use may entail the subjective selection of 'typical data' to avoid unmanageable negative yield responses or values >100, again resulting in statistical bias. It is hardly original to comment that a farmer derives his income from yields and not from percentages, and that the cost of fertilizers must be justified by increased profits and not by relative responses (e.g. Bondorff 1952).

Current Australian examples of the use of relative yields mostly relate to studies with pastures (e.g. Ozanne *et al.* 1967, 1976, 1978) for the plausible reason that pastures present formidable problems and that the alternatives are not very attractive. Although their use can thus be justified, they also indicate basic problems for research.

'Black boxes'

The complications of agricultural systems, and of the use of fertilizers, has led to the development of computer programs or models which rather miraculously seem to produce optimal rates despite a lack of essential information. The models have been developed on the basis of available information, with use of informed imagination of the scientist to specify mechanisms and produce the best estimates apparently possible under existing circumstances. The unfortunate and worrying feature about them is that it is virtually impossible for the outsider to test the validity of their output, that is without first obtaining the data and information, the absence of which led to the development of the models in the first place. Hence the term 'black box'.

Ways for tackling the problems raised by such 'black box' devices, which may in fact be well worthwhile, are suggested by some comments of the Holcomb Research Institute (1976) relating to computer models in the U.S. Thus 'only those models accompanied by evidence of adequate validation and documentation should be published . . .', 'the modeller should clearly explain the limitation of, and the degree of confidence warranted by the model,' and 'without a complete and consistent data base, model projections are of limited use'. Corresponding comments are given by Bennett and Bowden (1976) in relation to their Australian example of such a model (the DECIDE model, used for estimating fertilizer requirements in Western Australia). The problems raised by such models are considerable, and this may be their most valuable application, i.e. defining research needs for the future.

References

Anderson, A. J. and McLachlan, K. D. (1951). The residual effect of phosphorus on soil fertility and pasture development on acid soils. *Aust. J. Agric. Res.* **2**, 377-400.

Anderson, J. R. (1967). Economic interpretation of fertilizer response data. *Rev. Marketing Agric. Econ.* **35**, 43-57.

Anderson, R. L. (1957). Experimental Design for Fertilizer Experiments. *In* 'Economic and Technical Analysis of Fertilizer Innovations and Resource Use'. pp. 187-206. (Iowa State Univ. Press: Ames.)

Anderson, R. L. and Nelson, L. A. (1975). A family of models involving intersecting straight lines and concomitant experimental designs useful in evaluating response to fertilizer nutrients. *Biometrics* **31**, 303-18.

Arndt, W. and McIntyre, G. A. (1963). The initial and residual effects of superphosphate and rock phosphate for sorghum on a lateritic red earth. *Aust. J. Agric. Res.* **14**, 785-95.

Ayres, J. F. (1977). Thesis summary 'Superphosphate requirements of clover-ley farming'. *J. Aust. Inst. Agric. Sci.* **43**, 64.

Ayres, J. F., McFarlane, J. D., Gilmore, A. R. and McManus, W. R. (1977a). Superphosphate requirements of clover-ley farming. I. The effects of topdressing on productivity in the ley phase. *Aust. J. Agric. Res.* **28**, 269-85.

Ayres, J. F., McFarlane, J. D., Osbourne, G. J. and Corbin, E. J. (1977b) Superphosphate requirements of clover-ley farming. II. The residual effects of topdressing. *Aust. J. Agric. Res.* **28**, 287-99.

Balmukand, B. H. (1928). Studies in crop variation, V. The relation between yield and soil nutrients. *J. Agric. Sci.* **18**, 602-27.

Barrow, N. J. (1973). Relation between a soils ability to adsorb phosphate and the residual effectiveness of superphosphate. *Aust. J. Soil Res.* **11**, 57-63.

Barrow, N. J. (1975*a*). The response to phosphate of two annual pasture species. I. Effects of the soils ability to adsorb phosphate on the comparative phosphate requirement. *Aust. J. Agric. Res.* **26**, 137-43.

Barrow, N. J. (1975*b*). The response to phosphate of two annual pasture species. II. The specific rate of uptake of phosphate, its distribution and use for growth. *Aust. J. Agric. Res.* **26**, 145-56.

Barrow, N. J. (1976). The Adsorption and Desorption of Phosphate. *In* 'Reviews in Rural Science. III. Prospects for Improving the Efficiency of Phosphorus Utilization'. (Univ. of New England: Armidale, N. S. W.)

Barrow, N. J. and Campbell, N. A. (1972). Methods of measuring residual value of fertilizers. *Aust. J. Exp. Agric. Anim. Husb.* **12**, 502-10.

Baule, H. (1918). Zu Mitscherlichs Gesetz der biologischen Beziehugen Landwirtsch. *Jahrb. Schweiz.* **51**, 363-85.

Baum, E. L., Heady, E. O. and Blackmore, J. (1956). Methodological Procedures. *In* 'The Economic Analysis of Fertilizers Use Data'. (Iowa Univ. Press: Ames.)

Beckwith, R. S. (1965). Sorbed phosphate at standard supernatant concentration as an estimate of the phosphate needs of soils. *Aust. J. Exp. Agric. Anim. Husb.* **5**, 52-8.

Bennett, D. and Bowden, J. W. (1976). DECIDE — an Aid to Efficient Use of Phosphorus. *In* 'Reviews in Rural Science 3. Prospects for Improving the Efficiency of Phosphorus Utilization'. (Uni. of New England: Armidale, N. S. W.)

Blackman, F. F. (1905). Optima and limiting factors. *Ann. Botany* **19**, 281-95.

Blair, G. J. (Ed.) (1976). Reviews in Rural Science 3. Prospects for Improving Efficiency of Phosphorus Utilization'. (Uni. of New England: Armidale, N. S. W.)

Bliss, C. I. (1970). 'Statistics in Biology, Vol. 2'. (McGraw-Hill: New York.)

Bondorff, K. A. (1952). The evaluation of soil analysis. Trans. Internat. Soil Sci. Soc. Comm. II and IV, Dublin, pp. 290-5.

Bowden, J. W. and Bennett, D. (1976). Proc. of a symposium 'Phosphate in Agriculture', Nov. 1974). (Aust. Inst. Agric. Sci. : Melbourne.)

Boyd, D. A. (1970). Some recent ideas on fertilizer response curves. Proc. Congress Int. Potash Inst., Antibes. On the role of fertilization in the intensification of agricultural production. (Potash Inst.: Berne.)

Bray, R. H. (1944). Soil-plant relations. I. The quantitative relation of exchangeable potassium to crop yields and to crop response to potash additions. *Soil Sci.* **58**, 305-24.

Bray, R. H. (1948). *In* 'Diagnostic Techniques for Soil and Crops'. Chap. 2, pp. 53-85. (American Potash Inst. : Washington, D. C.)

Cady, F. B. and Laird, R. J. (1969). Bias error in yield functions as influenced by treatment design and postulated model. *Proc. Soil Sci. Soc. Am.* **33**, 282-6.

Campbell, N. A. and Keay, J. (1970). Flexible techniques in describing mathematically a range of response curves of pasture species. Proc. of the 11th Internat. Grasslands Congress. pp. 332-4. Surfers Paradise, Australia.

Cannon, D. J. (1972). The influence of rate of stock and application of superphosphate on the production and quality of wool, and the gross margin from merino wethers. *Aust. J. Exp. Agric. Anim. Husb.* **12**, 348-54.

Carmer, S. J. (1964). Exponential regression and a computer program for the estimation of parameters. *Agron. J.* **56**, 515-18.

Colwell, J. D. (1963*a*). The effect of fertilizers and season on the yield and composition of wheat in southern N. S. W. *Aust. J. Exp. Agric. Anim. Husb.* **3**, 51-61.

Colwell, J. D. (1963*b*). The estimation of phosphorus fertilizer requirements of wheat in southern N. S. W. by soil analysis. *Aust. J. Exp. Agric. Anim. Husb.* **3**, 190-7.

Colwell, J. D. (1967*a*). Soil tests for phosphorus fertilizer requirements. *J. Aust. Inst. Agric. Sci.* **33**, 289-95.

Colwell, J. D. (1967*b*). The calibration of soil tests. *J. Aust. Inst. Agric. Sci.* **33**, 321-30.

Colwell, J. D. (1967*c*). Calibration and assessment of soil tests. I. Statistical models and tests of significance. *Aust. J. Soil Res.* **5**, 275-93.

Colwell, J. D. (1968). Calibraton and assessment of soil tests. II. Fertilizer requirements and an evaluation of soil testing. *Aust. J. Soil Res.* **6**, 93-103.

Colwell, J. D. (1971). Effects of variations in soil composition on soil test values for phosphorus fertilizer requirements. Trans. Int. Symp. on Soil Fertility Evaluation, Int. Soil Sci. Soc., New Delhi, pp. 327-36.

Colwell, J. D. (1973*a*). Assessments of the relative values of nitrogen–phosphorus compound fertilizers for wheat production. *Aust. J. Agric. Econ.* **17**, 189-99.

Colwell, J. D. (1973*b*). The computation of optimal rates of application of fertilizers from quadratic response functions. CSIRO Aust. Div. Soils Tech. Pap. No. 21.

Colwell, J. D. (1975). The computation of optimal fertilizer rates for conditions of limited supply. Proc. of FAI-FAO Seminar on 'Optimising agricultural production under limited availability of nutrients, 1974'. (Fert. Assoc. of India: New Delhi.)

Colwell, J. D. (1976). Soil Testing and the Efficient Use of Fertilizers. *In* 'Reviews in Rural Science III. Prospects for Improving the Efficiency of Phosphorus Utilization'. (Univ. of New England: Armidale, N. S. W.)

Colwell, J. D. (1977). National Soil Fertility Project. Vol. 1. Objectives and Procedures. (CSIRO Aust. Div. Soils: Adelaide.)

Colwell, J. D. (1978). 'Computations for study of soil fertility and fertilizer requirements'. (Commonw. Agric. Bur: Slough/CSIRO: Australia.)

Colwell, J. D. (1979). 'National Soil Fertility Project, Vol. 2. Soil Fertility Relationships'. (CSIRO Aust., Div. Soils: Adelaide.)

Colwell, J. D. and Donnelly, J. D. (1971). Effects of soil composition on the relationships between soil test values for phosphorus fertilizer requirements. *Aust. J. Soil Res.* **9**, 43-54.

Colwell, J. D. and Esdaile, R. J. (1968). The calibration, interpretation and evaluation of tests for the phosphorus fertilizer requirements of wheat in northern N. S. W. *Aust. J. Soil Res.* **6**, 105-20.

Cook, S. J., Lazenby, A. and Blair, G. J. (1978*a*). Pasture degeneration. I. Effect on total and seasonal pasture production. *Aust. J. Agric. Res.* **29**, 9-18.

Cook, S. J., Blair, G. J. and Lazenby, A. (1978*b*). Pasture degeneration. II. The importance of superphosphate, nitrogen and grazing management. *Aust. J. Agric. Res.* **29**, 19-29.

Cooke, G. W. (1971). Value and valuation of fertilizer residues. G. B. Minist. Agric. Fish. Food Tech. Bull. No. 20, pp. 296-316.

Curll, M. L. (1977*a*). Superphosphate on perennial pastures. I. Effects of pasture response on sheep production. *Aust. J. Agric. Res.* **28**, 991-1005.

Curll, M. L. (1977*b*). Superphosphate on perennial pastures. II. Effects of a pasture response on steer beef production. *Aust. J. Agric. Res.* **28**, 1007-14.

Daniel, C. and Wood, F. S. (1971). 'Fitting Equations to Data'. (Wiley Interscience: New York.)

Dann, P. R. (1969). Response by wheat to phosphate and nitrogen, with particular reference to 'haying-off'. *Aust. J. Exp. Agric. Anim. Husb.* **9**, 625-9.

Davidson, B. R. and Martin, B. R. (1968). 'Experimental Research and Farm Production'. (Univ. of W. Aust. Press: Perth.)

Dillon, J. L. (1968). 'The Analysis of Response in Crop and Livestock Production'. (Pergamon Press: Oxford.)

Donald, C. M. and Williams, C. H. (1954). Fertility and productivity of a podzolic soil as influenced by subterranean clover (*Trifolium subterraneum L.)* and superphosphate. *Aust. J. Agric. Res.* **5**, 664-87.

Draper, N. R. and Smith, H. (1968). 'Applied Regression Analysis'. (Wiley: New York.)

Engelstad, O. P. (1970). (Ed.) 'Nutrient Mobility in Soils: Accumulation and Losses'. (Soil Sci. Soc. Am. : Madison.)

Fried, M. and Broeshart, H. (1967). 'The Soil-Plant System in Relation to Inorganic Nutrition'. (Academic Press: New York.)

Greenwood, D. J. and Cleaver, T. J. (1971). Nitrogen fertilizer requirements of vegetable crops. *J. Sci. Food Agric.* **22**, 435.

Griffiths, D. A. and Miller, A. J. (1973). Hyperbolic Regression – a model based on two-phase piecewise linear regression with a smooth transition between regimes. *Commun. Stat.* **2**, 561-06.

Gruen, F. H. (1955). Superphosphate use in N. S. W. *Rev. Marketing Agric. Econ.* **23**, 17-30.

Hagin, J. (1960). On the shape of the yield curve. *Plant Soil* **12**, 285-96.

Heady, E. O. and Dillon, J. L. (1961). 'Agricultural Production Functions'. (Iowa State Uni. Press: Ames.)

Heady, E. O. (Chairman), Albaugh, L. G., Engibous, J. L., French, B. L., Kehrberg, E. W., Mason, D. D. and Pesek, J. T. (1961). 'Status and methods of research in economic and agronomic aspects of fertilizer response and use'. Publ. 918. (Nat. Acad. Sci. N. R. C. : Washington, D. C.)

Heady, E. O., Pesek, J. T. and McCarthy, W. O. (1963). Production functions and methods of specifying optimum fertilizer use under various uncertainty conditions for hay. Agric. Exp. Sta. Res. Bull. No. 518. (Iowa State Uni. Press: Ames.)

Heady, E. O., Pesek, J. T. and Rao, V. Y. (1966). Fertilizer production functions from experimental data with associated supply and demand relationships. Agric. Home Econ. Exp. Sta. Res. Bull. No. 543. (Iowa State Uni. Press: Ames.)

Helyar, K. R. and Godden, D. P. (1977). The biology and modelling of fertilizer response. *J. Aust. Inst. Agric. Sci.* **43**, 22-30.

Holcomb Research Institute (1976). 'Environmental modelling and decision making; the United States experience. A report by the Holcomb Research Institute for the Scientific Committee on problems of the environment'. (Praeger: New York.)

Holford, I. C. R. (1976). Factors Affecting the Accumulation and Availability of Residual Fertilizer Phosphate. *In* 'Reviews in Rural Science. III. Prospects for Improving the Efficiency of Phosphorus Utilization'. (Uni. of New England, Armidale, N. S. W.)

Holford, I. C. R. and Gleeson, A. C. (1976). Residual effectiveness of phosphorus on white clover on granitic soils. *Aust. J. Agric. Res.* **27**, 509-18.

Holford, I. C. R. and Mattingly, G. E. G. (1976a). A model for the behaviour of labile phosphate in soil. *Plant Soil* **44**, 219-29.

Holford, I. C. R. and Mattingly, G. E. G. (1976b). Phosphate adsorption and plant availability of phosphate. *Plant Soil* **44**, 377-89.

Jackson, E. A. (1966). Phosphorus fertilizer research in Australia. CSIRO Aust. Bull. No. 284.

Jardine, R. (1969). The calibration of soil tests for fertilizer requirements. *Aust. J. Soil Res.* **7**, 223-8.

Jardine, R. (1975). Two cheers for optimality, Jardine comments further. *J. Aust. Inst. Agric. Sci.* **41**, 30-4, 192-201, 257.

John, J. A. and Quenouille, M. H. (1977). 'Experiments: Design and Analysis'. (Charles Griffin: London.)

Johnson, P. R. (1953). Alternative functions for analysing a fertilizer-yield relationship. *J. Farm Econ.* **35**, 519-21.

Jones, R. K. (1968). Initial and residual effects of superphosphate on a Townsville lucerne pasture in north-eastern Queensland. *Aust. J. Exp. Agric. Anim. Husb.* **8**, 521-7.

McLachlan, K. D. (1975). (Ed.) 'Sulphur in Australasian Agriculture'. (Sydney Uni. Press: Sydney.)

Mead, R. and Pike, D. J. (1975). A review of response surface methodology from a biometric point of view. *Biometrics* **31**, 803-51.

Miller, C. P. and Jones, R. K. (1977). Nutrient requirements of *Stylosanthes guianensis* pastures on a euchrozem in north Queensland. *Aust. J. Exp. Agric. Anim. Husb.* **17**, 607-13.

Mitscherlich, E. A. (1907). Eine chemische Bodenanalyse für pflanzenphysiologische Forschungen. *Landwirtsch. Jahrb. Schweiz.* **36**, 309.

Mitscherlich, E. A. (1909). Das Gesetz des minimums und das Gesetz des abnehmenden Bodenertrages. *Landwirtsch. Jahrb. Schweiz.* **38**, 537-52.

Montgomery, A. J. and Rubenis, G. (1978). Correlation of soil phosphorus tests with the response of irrigated perennial pasture to phosphorus fertilizer. *Aust. J. Exp. Agric. Anim. Husb.* **18**, 243-8.

Morley, R. H. W. (1975). Evaluation by annual production of increases in pasture growth using computer simulation. Proc. of the 12th Internat. Grasslands Congress Vol. 3, Moscow.

National Soil Test Work Group (1956). National soil and fertilizer research committee. North Carolina Agric. Exp. Sta. Tech. Bull. No. 121, pp. 1-36.

Nelder, J. A. (1966). Inverse polynomials, a useful group of multifactor response functions. *Biometrics* **22**, 128-41.

Nelson, L. A. and Anderson, R. L. (1977). Partitioning of soil test-crop response probability. Am. Soc. Agron. Spec. Publ. No. 29, pp. 19-38.

Osborne, G. T., Batten, G. D. and Kohn, G. D. (1977). Nitrogen and phosphorus requirements of wheat following subterranean clover pasture. *Aust. J. Exp. Agric. Anim. Husb.* **28**, 971-9.

Ozanne, P. G. and Shaw, T. C. (1967). Phosphate sorption by soils as a measure of the phosphate requirement for future growth. *Aust. J. Agric. Res.* **18**, 601-2.

Ozanne, P. G., Howes, K. M. W. and Petch, A. (1976). The comparative phosphate requirements of four annual pastures and two crops. *Aust. J. Agric. Res.* **27**, 479-88.

Ozanne, P. G., Kirton, D. J. and Shaw, T. C. (1961). The loss of phosphorus from sandy soils. *Aust. J. Agric. Res.* **12**, 409-23.

Ozanne, P. G and Petch, A. (1978). Effect of species and cultivation on the responses to phosphate of annual pasture and crops. *Aust. J. Agric. Res.* **29**, 225-33.

Palmer, B., McClelland, V. F. and Jardine, R. (1975). Soil measurements and wheat yield response to superphosphate in Victoria. *Aust. J. Exp. Agric. Anim. Husb.* **15**, 93-101.

Pesek, J. T., Heady, E. O. and Dumenil, L. C. (1961). Influence of residual fertilizer effects and discounting upon optimum fertilizer rates. Trans. Int. Soil Sci. Soc. Madison, Vol. 3, pp. 220-7. (Elsevier: Amsterdam.)

Piper, C. S. and DeVries, M. P. C. (1964). Residual value of superphosphate on a red-brown earth in South Australia. *Aust. J. Exp. Agric. Anim. Husb.* **15**, 234-72.

Ravpach, M. and Piper, C. S. (1959). Interactions of silicate and phosphate in a lateritic soil. *Aust. J. Agric. Res.* **10**, 818-31.

Reid, D. (1970). The effect of a wide range of nitrogen application rates on the yields from a perennial ryegrass sward with and without white clover. *J. Agric. Sci.* **74**, 227-40.

Ross, G. J. S. (1979). The efficient use of function minimisation in non-linear maximum – likelihood estimation. *Appl. Stat.* **19**, 205-21.

Rossiter, R. C. (1964). The effect of phosphate supply on the growth and botanical composition of annual-type pasture. *Aust. J. Agric. Res.* **15**, 61-75.

Rowell, D. L., Osbourne, G. J., Matthews, P. G., Stonebridge, W. C. and McNeill, A. A. (1977). The effects of a long term trial of minimum and reduced cultivation on wheat yields. *Aust. J. Exp. Agric. Anim. Husb.* **17**, 802-11.

Rudd, C. L. and Barrow, N. J. (1973). The effectiveness of several methods of applying superphosphate on yield response by wheat. *Aust. J. Exp. Agric. Anim. Husb.* **13**, 430-3.

Russell, E. J. (1937). 'Soil Conditions and Plant Growth'. 7th edn. pp. 133-43. (Longmans Green: London.)

Russell, J. S. (1957). Nitrogen fertilizer in Australian soils. *J. Aust. Inst. Agric. Sci.* **23**, 110-9.

Russell, J. S. (1968*a*). Nitrogen fertilizer and wheat in a semi-arid environment. 2. Climatic factors affecting response. *Aust. J. Exp. Agric. Anim. Husb.* **8**, 223-31.

Russell, J. S. (1968*b*). Nitrogen fertilizer and wheat in a semi-arid environment. 3. Soil and cultural factors affecting response. *Aust. J. Exp. Agric. Anim. Husb.* **8**, 340-8.

Russell, J. S. (1968*c*). Nitrogen fertilizer and wheat in a semi-arid environment. 4. Empirical yield response models and economic factors. *Aust. J. Exp. Agric. Anim. Husb.* **8**, 736-48.

Russell, J. S. (1972). A theoretical approach to plant nutrient response under conditions of variable maximum yield. *Soil Sci.* **114**, 387-94.

Russell, J. S. (1977). Evaluation of residual nutrient effects in soils. *Aust. J. Agric. Res.* **38**, 461-75.

Searle, P. G. E. (1967). The evaluation of phosphorus availability in a Queensland krasnozem. *J. Aust. Inst. Agric. Sci.* **33**, 43-4.

Skene, J. K. M., Hosking, W. J., Leverington, K. C., Russell, J. S. and Williams, C. H. (1966). A survey of soil and plant tissue testing activities in Australia. *J. Aust. Inst. Agric. Sci.* **32**, 159-72.

Smith, A. N. (1965). Aluminium and iron phosphates in soils. *J. Aust. Inst. Agric. Sci.* **31**, 110-26.

Smith, A. N. (1971). Soil testing as an aid in the diagnosis of mineral deficiencies in plants. *J. Aust. Inst. Agric. Sci.* **37**, 296-302.

Smith, A. N. (1977). Fertilization. *In* 'Soil Factors in Crop Production in a Semi-arid Environment. ' (Eds J. S. Russell and E. L. Greacen). pp. 262-78. (Univ. of Queensland Press: Brisbane.)

Spencer, K. (1954). Studies in soil fertility with special reference to organic manures. I. The field experiments. *Aust. J. Agric. Res.* **5**, 181-97.

Steenbjerg, F. and Jakobsen, S. T. (1963). Plant nutrition and yield curves. *Soil Sci.* **95**, 69-88.

Stephens, C. G. and Donald, C. M. (1958). Australian soils and their responses to fertilizers. *Adv. Agron.* **10**, 167-256.

Storrier, R. R. (1965). Excess soil nitrogen and the yield and uptake of nitrogen by wheat in southern N. S. W. *Aust. J. Exp. Agric. Anim. Husb.* **18**, 317-22.

Teakle, L. J. H. and Cariss, H. G. (1943). Superphosphate requirements for growing wheat in Western Australia. *J. Dep. Agric. West. Aust.* **20**, 1-28.

Tudsri, S. and Whiteman, P. C. (1977). Effects of initial and maintenance phosphorus levels on the establishment of four legumes oversown into *Setaria anceps* swards. *Aust. J. Exp. Agric. Anim. Husb.* **17**, 629-36.

Tupper, G. J. (1978). Effects of nitrogen and phosphorus fertilizers and gypsum on a *Danthonia caespitosa — Stipa variabilis* grassland. 2. Residual response. *Aust. J. Exp. Agric. Anim. Husb.* **18**, 262-7.

Tyson, A. G. (1955). Studies on the fertility of Seddon soil (Kangaroo Is. S. A.). I. The response of subterranean clover to various phosphatic fertilisers. *Aust. J. Agric. Res.* **6**, 398-422.

Willett, I. R., Muirhead, W. A. and Higgins, M. L. (1978). The effects of ricegrowing on soil phosphorus immobilization. *Aust. J. Exp. Agric. Anim. Husb.* **18**, 270-5.

Willey, R. W. and Heath, S. B. (1969). The quantitative relationships between plant population and crop yield. *Adv. Agron.* **21**, 281-321.

Williams, C. H. and Colwell, J. D. (1977). Inorganic Chemical Properties. *In* 'Soil Factors in Crop Production in a Semi-arid Environment. (Eds J. S. Russell and E. L. Greacen). pp. 105-26. (Univ. Queensland Press: Brisbane.)

Williams, C. H. and Lipsett, J. (1960). The buildup of available potassium under subterranean clover pastures on a podsolic soil. *Aust. J. Agric. Res.* **11**, 473-84.

Williams, C. H. and Simpson, J. R. (1965). Some effects of cultivation and waterlogging on the availability of phosphorus in pasture soils. *Aust. J. Agric. Res.* **16**, 413-27.

Williams, E. G. and Reith, J. W. S. (1971). Residual effects of phosphate and the relative effectiveness of annual and rotational dressings. G. B. Minist. Agric. Fish. Food. Tech. Bull. No. 20, pp. 16-33.

Williams, R. S. (1954). Studies in soil fertility with special reference to organic manures. II. Plant growth and nutrition in the field. *Aust. J. Agric. Res.* **5**, 198-223.

Williams, R. S. and Spencer, K. (1954). Studies in soil fertility with special reference to organic manures. III. Residual effects of the organic matter. *Aust. J. Agric. Res.* **5**, 224-34.

Willoughby, W. M. (1954). Some factors affecting grass-clover relationships. *Aust. J. Agric. Res.* **5**, 157-80.

Winter, W. H., Edye, L. A. and Williams, W. T. (1977*a*). Effects of fertilizer and stocking rate on pasture and beef production from sown pastures in northern Cape York Peninsula. I. Botanical and chemical composition of the pastures. *Aust. J. Exp. Agric. Anim. Husb.* **17**, 66-74.

Winter, W. H., Edye, L. A. and Williams, W. T. (1977*b*). Effects of fertilizer and stocking rate on pasture and beef production from sown pastures in northern Cape York Peninsula. II. Beef production and its relation to blood, faecal and pasture measurements. *Aust. J. Exp. Agric. Anim. Husb.* **17**, 187-96.

Woodroffe, K. and Williams, C. H. (1953). The residual effect of superphosphate in soils cultivated for wheat in South Australia. *Aust. J. Agric. Res.* **4**, 127-50.

Chapter 51
Production and uses of biological superphosphate

R. J. Swaby

In 1959 Swaby and Sperber (CSIRO Division of Soils 1959) gave their reasons for doubting whether organic acids from rhizosphere microorganisms dissolved sufficient applied rock phosphate to benefit plants growing in soils deficient in phosphorus. This led to reexamination of claims by Lipman and McLean (1916, 1917, 1918) and McLean (1918), that sulfuric acid from microbial oxidation of sulfur can dissolve admixed rock phosphate. Preliminary experiments showed that broadcasting compost with their 1:3 mixture of sulfur and rock phosphate benefits plants growing in light soils, but not in heavy ones. Since composting and 1:3 mixtures were uneconomic, compost was eliminated and wider ratios were tested, a 1:5 mixture being finally adopted. The sulfur and rock phosphate were pelleted to preserve close contact between sulfur and rock phosphate, and to decrease reactions between sulfuric acid and soil constituents. These pellets were called sulphaphos; they occasionally failed to benefit plants when tested in several soils, owing to the absence of indigenous thiobacilli. If pellets were inoculated with thiobacilli, their performance improved in many soils; this biological superphosphate was given the name biosuper.

After this work had been in progress for several years (CSIRO Division of Soils 1963), Kittams and Attoe (1965), Attoe and Olsen (1966) and Nimgade (1968) developed fusions of sulfur and rock phosphate. Compared with biosuper, these fusions were more expensive to manufacture, and were inferior in several comparative pot trials. Biosuper has been favourably compared with superphosphate on cattle pastures in Northern Australia by Fisher and Norman (1970), Gillman (1973), Swaby (1975) and Jones and Field (1976). Biosuper was ultimately superior to superphosphate as a non-leachable fertilizer for pine trees in sand in Western Australia (CSIRO Division of Soils 1972-3).

Production of pellets

Biosuper is produced by grinding 5 parts by weight of rock phosphate with 1 part of lump sulfur, moistened with an atomized mist of water to reduce the risk of explosion, until 70% passes through sieves with 100 mesh (<0.15 mm). The ingredients are mixed with 5% of biosuper inoculum (see below) containing active thiobacilli, and moistened with the minimum (8–10%) amount of water to promote granulation. Pellets are tipped on to 4 mm sieves, and then screened through 2 mm sieves, oversize and undersize material being returned to the pelletizer. The 2–4 mm diam. pellets are air-dried to approximately 4% moisture, then bagged or bulk stored in heaps until used. For laboratory and glasshouse use, gram quantities of biosuper are pelleted in a rotary drum, or kilogram amounts in a concrete mixer with the baffles removed and cast iron or stone balls present. For field experiments, Australian Mineral Development Laboratories, Adelaide, produced tonne lots continuously on a pelleting dish, 1 m diam. inclined at 44^0 and rotating at 19 rev. min^{-1}.

The poorer grades of rock phosphate, contaminated with Fe and Si minerals, that cannot be used in superphosphate manufacture, can be used in biosuper; and local rock containing as little as 18% P_2O_5 (7.9% P) may be used economically. Thiobacilli tolerate up to 5% F in apatite without being inhibited. Prior calcination greatly improves the yields obtained with phosphate rock containing insoluble crandallites.

Properties of pellets

The compression strength of biosuper pellets containing less than 4% moisture permits storage in heaps, without distortion of the pellets or caking of the mass. The density and porosity of pellets depend on percentage of water during granulation, and on particle size of ingredients; at 10% moisture the mixture flows together to give compact coherent pellets; the greater the percentage of fine particles (<0.15 mm) in the mixture, the denser the pellets and the poorer their aeration. Pellets made from coarser particles, or with lower percentages of water, were well aerated but are too fragile. Those made with 10% water and from ingredients which are 70% <0.15 mm, have, on being air-dried and capillary wetted, 10% of their pores free draining, and are therefore aerobic. Good quality biosuper pellets survived bumping by rail or road transport over 5000 km, and afterwards were distributed successfully by broadcaster, drill or plane, without shattering. Pellets on the surface of moist soil absorb moisture without slaking, but partly disintegrate under the impact of heavy rain. Pellets can be further stabilized, so that they survive heavy rain, by running them through hot air at 150°C for 3 s, to superficially sinter the sulfur with the rock phosphate.

Being neutral, biosuper does not corrode farm implements or manufacturing and bulk-handling machinery, and can be marketed in cheap paper bags. Since the sulfur content is 16.6%, it is well below the legal maximum of 22% considered a fire risk during aerial spreading. Superphosphate or double or triple super may be applied along with biosuper pellets, to provide some readily available P and S. Biosuper is compatible with most other fertilizers containing major or minor elements, and with most pesticides and herbicides. Since it contains no immediately water- or citrate-soluble P it does not qualify in Australia for a Government bounty, but it would in some other countries. In developing countries lacking superphosphate-manufacturing plants, biosuper can be manufactured from local crude rock with simple equipment and at a much lower price than imported superphosphate. The amounts of immediately available P and S can be increased for annual crops by incubating moist biosuper mixture for several months in heaps prior to granulating it, but this would increase production costs.

Preparation of inoculum

Almost half the Australian soils examined by Swaby and Vitolins (1968) contained insufficient thiobacilli to oxidize sulfur rapidly. The availability of P and S from biosuper is usually improved by inoculation with *Thiobacillus* species, particularly by *T. thioparus* which can decrease the pH from 7.0 to 4.5, and *T. thiooxidans* which drops it to pH 2.7. Unfortunately, air-drying kills over 99% of thiobacillus cells added to biosuper, so huge numbers need to be added to ensure survival of some thiobacilli after dehydration and storage. There is no improvement in the viability of thiobacilli air dried in the presence of added protective colloidal substances such as peat, clay, mucin, albumin, methyl cellulose, polyvinyl pyrrolidone and polyethylene glycol, or of smaller molecules, like inositol, $NH_4H_2PO_4$, glucosamine, dipicolinic acid and tetra-ethylammonium hydroxide. The addition of 0.1% of $(NH_4)_2SO_4$ to biosuper accelerates the development of thiobacilli, but no other major or minor elements seem beneficial; additional energy materials, such as sugars and amino acids, are inhibitory.

The procedure finally adopted for preparing the first inoculum for commercial use was as follows: several field soils with pH less than 6 were tested, by the methods used by Vitolins and Swaby (1969), for their ability to oxidize sulfur rapidly; the soils were enriched with 1% sulphur, and a soil was chosen whose pH dropped below 3.2 within a month. A quantity of moist biosuper mixture (1:5) containing 0.1% $(NH_4)_2$ SO_4 was added to this soil so that the pH was brought to 4.2, and the mixture was incubated at 25°C. When the pH fell to 3.2 additional biosuper was added to bring the pH up to 4.2, and to build up sufficient active bulk inoculum containing *T. thioparus* and *T. thiooxidans.* This bulk inoculum contains appreciable amounts of immediately available P and S and would be a suitable fertilizer for certain crops.

Laboratory experiments

Biosuper pellets were tested in the laboratory by mixing 2 g in 100 g of autoclaved sandy soil, moistening the mixture, incubating at 25°C, and at intervals subsampling 10 g, adding 200 ml water, shaking for 2 h, and determining soluble sulfate and phosphate. It was found that using particle sizes of rock phosphate or sulfur finer than 300 mesh (0.04 mm diam.) produces dense pellets, with all pores becoming waterlogged in moist soil; the sulfur oxidizes slowly on the surface, and only small amounts of phosphate are released. Particles coarser than 0.2 mm diam. give pellets which are aerobic, but too fragile. The most efficient pellets are those made from a mixture of particle sizes from 0.2 mm to 0.001 mm, giving pores of different sizes, some drainable and aerobic, but others non-drainable. It was found that the smaller the pellets the deeper they are aerated and the greater the amount of soluble P and S produced; pellets > 5 mm are poorly aerated at the centre, and sulfur there remains unoxidized.

Pure cultures of thiobacilli usually oxidize sulfur less rapidly than do mixtures of bacteria, particularly those including autotrophs such as *T. thioparus*, *T. thiooxidans* and *T. denitrificans*, facultative autotrophs like *T. intermedius*, and heterotrophs such as *Bacillus cereus* and *Pseudomonas* spp. The addition of 0.1% of $(NH_4)_2SO_4$ stimulates the oxidation of sulfur when biosuper is mixed with siliceous sand, but not when mixed with sandy loams, which seem able to supply sufficient nitrogen for the sulfur oxidizers.

Pellets can be prepared, without difficulty, with ratios of sulfur to rock phosphate ranging from 1:3 to 1:7; the narrower the ratio, the quicker sulfur is oxidized and phosphate released. Thus pellets with a ratio of 1:3 dropped to pH 3.5 in a week, oxidized 15% of S, and released 8% of P, whereas those with a ratio of 1:7 were better buffered, decreased pH to 6.0, oxidized 7% of S, and released < 1%P. Until the pH dropped below 4.0 very little water-soluble P was produced, indicating that very little monocalcium phosphate was formed. The longer biosuper is incubated, either in soil or in heaps, prior to planting crops, the greater the available P and S and the greater the response.

Glasshouse experiments

Sulfur was mixed in the ratios of 1:4 and 1:5 with phosphate rock from Christmas, Ocean, and Nauru Islands, from Duchess and Lady Annie mines in Queensland,

Table 51.1. Biosuper compared with other fertilizers

Involving a maximum of eleven phosphate rocks, two perennials, three annuals, four siliceous sands and three loams in an Adelaide glasshouse. Figures in brackets refer to the numbers of similar treatments averaged.

Fertilizer treatment	Year	Percentage increment relative to superphosphate		
		Herbage dry wt (%)	Phosphorus uptake (%)	Sulfur uptake (%)
Biosuper (1:5)	1	70 (69)	74 (67)	95 (61)
	1+2	81 (4)	72 (4)	135 (4)
Biosuper (1:4)	1	47 (4)	–	–
	1+2	–	–	–
Biosuper (1:5) + 10% Superphosphate	1	80 (6)	66 (2)	134 (2)
	1+2	150 (2)	82 (2)	122 (2)
Sulphaphos (1:5)	1	43 (10)	33 (9)	–
	1+2	26 (6)	32 (3)	–
Sulphaphos (1:4)	1	42 (18)	30 (5)	–
	1+2	26 (6)	31 (2)	–
Sulphaphos (1:5) + 10% superphosphate	1	43 (5)	–	–
Rockphosphate + gypsum (2:1)	1	47 (38)	17 (38)	70 (28)
	1+2	27 (11)	15 (7)	56 (4)
Rockphosphate	1	42 (20)	53 (20)	48 (20)
Rockphosphate + 10% superphosphate	1	47 (3)	–	–
Superphosphate	1	100 (25)	100 (25)	100 (16)
	1+2	100 (11)	100 (7)	100 (4)
Superphosphate increment per pot (150 mm diam.)	1	6.2 g (25)	6.9 mg (25)	9.8 mg (16)
	1+2	11.7 g (11)	15.4 mg (7)	15.8 mg (4)

and from Rum Jungle in the Northern Territory; the mixtures were then pelleted with and without thiobacilli inoculum to produce biosuper and sulphaphos respectively. These fertilizers were then compared with standard superphosphate and rock-phosphate-plus-gypsum at rates of P equivalent to a superphosphate dressing of 200 kg ha^{-1}, or sometimes at double this rate. Fertilizer was placed at a depth of 25 mm in 2.5 kg of soil in plastic pots (150 mm diam.), four replicates being used for each treatment.

In a heated glasshouse in Adelaide, three *siliceous sands* and three loams were sown with temperate plants, three being annuals, subterranean clover (*Trifolium subterraneum*) or burr medic (*Medicago denticulata*) or Wimmera rye-grass (*Lolium rigidum*) and two being perennials, lucerne (*Medicago sativa*) or couch-grass (*Cynodon dactylon*). The annuals were harvested at late flowering; the tops of the perennials were clipped at three-monthly intervals over a period of 2 years. The herbage was oven dried at 100^0C, weighed, and analysed for P and S by X-ray fluorescence methods. Increments were calculated by subtracting mean control values from means of dry weight, P-uptake, and S-uptake, from each treatment, then expressing them as percentages relative to the increment for superphosphate. Over several years many experiments were conducted; to reduce the volume of data the means of the percentage increments are presented in Table 51.1.

Data from all plants and all soils undergoing similar treatments were combined. Biosuper (1:5) averaged 70% of the increment obtained from superphosphate in year one and 81% in years one and two. For P uptake the corresponding values were 74% and 72%, while for S uptake they were 95% and 135% respectively. Thus biosuper (1:5) was inferior to superphosphate in producing herbage and supplying P, but equalled or exceeded it as a source of S. Biosuper (1:4) produced less herbage, than biosuper (1:5). Sulphaphos (1:5) and (1:4) were equally inferior to biosuper, and were no better than rock phosphate + gypsum (2:1), indicating that inoculation of biosuper with thiobacilli ensures quicker oxidation of sulfur and release of phosphate. When 10% superphosphate is added to biosuper or sulphaphos or rock phosphate, their performances are rarely improved, due to reversion. Further analysis of the data shows the effect of the fertilizers on annuals and perennials; biosuper (1:5) performed better on perennials than on annuals.

In a further series of experiments in pots, lucerne was grown for 15 months in the Adelaide glasshouse. Comparisons were made with biosuper of different pellet sizes, of several particle sizes of rock phosphate, of pellets broadcast versus buried, of different nitrogen sources for the thiobacilli, of various metallic sulfides as substitutes for sulfur, and of calcined versus non-calcined C-grade Christmas Island rock phosphate. Some results are summarized in Table 51.2.

Biosuper sometimes performed better when buried than broadcast, but differences were not clear cut. When

Table 51.2. Biosuper compared with superphosphate in glasshouse trials

Fertilizer treatment	Percentage increment relative to superphosphate		
	Herbage dry wt (%)	Phosphorus uptake (%)	Sulfur uptake (%)
	Adelaide glasshouse		
2–4 mm pellets	72	56	59
0.2–2 mm pellets	41	48	45
70% <100 mesh R.P. in biosuper	74	75	104
70% <300 mesh R.P. in biosuper	80	73	100
100% < 2μm R.P. in biosuper	71	74	89
Buried biosuper	80	83	105
Broadcast biosuper	71	71	96
Superphosphate increment	100	100	100
weight per 150 mm pot	4.8 g	6.7 mg	14.1 mg
	Townsville glasshouse		
2–4 mm pellets	97	42	111
0.2–2 mm pellets	63	38	113
Superphosphate increment	100	100	100
weight/150 mm pot	11.2 g	24.6 mg	46.0 mg
70% < 100 mesh R.P. in biosuper	51	25	55
70% < 300 mesh R.P. in biosuper	58	27	54
100% < 2 μm R.P. in biosuper	58	27	28
Superphosphate increment	100	100	100
weight per 150 mm pot	12.8 g	20.1 mg	46.7 mg

nitrogen compounds were added to biosuper, no improvement was noticed, indicating that thiobacilli obtained adequate nitrogen from the soils. None of the metal sulfides oxidized rapidly enough to be substituted for sulfur in biosuper, but sphalerite was superior to marcasite and pyrites. Biosuper prepared from calcined rock phosphate was greatly superior to pellets from non-calcined material which, in turn, was only slightly better than calcined rock phosphate alone; non-calcined material was almost useless.

Under temperate conditions, with low soil temperatures during the winter, biosuper was generally inferior to superphosphate in herbage production, P uptake, and S uptake. Experiments using *siliceous sand* or *solodic soil*, and applying similar treatments, were set up in a cooled glasshouse in Townsville with the tropical pasture plants Townsville stylo (*Stylosanthes humilus*), and occasionally fine-stemmed stylo (*S. guyanensis*), Centro (*Centrosema pubescens*), Glycine (*Glycine wightii*), Desmodium (*Desmodium intortum*), Setaria (*Setaria sphacelata*) or Guinea grass (*Panicum maximum*). Some results are summarized in Table 51.2. A thin, final coat of rock phosphate, to eliminate sulfur dust, slightly hindered sulfur oxidation in biosuper, and sintering, to enable biosuper to withstand tropical rain, did not interfere with its performance. Substituting metal sulfides for sulfur in biosuper gave inferior responses in both sand and solodic soil.

Once again, in the Townsville glasshouse, superphosphate mostly out-yielded biosuper, even under tropical conditions. Availability of P and S from biosuper as measured by plant uptake in the glasshouse, confirmed many of the findings based on extraction experiments in the laboratory.

Pasture Trials

Temperate Climatic Conditions

Biosuper, sulphaphos, or other fertilizers, were drilled at two rates of P, equivalent to superphosphate dressings at 200 and 400 kg ha^{-1}, into triplicate plots (0.001 ha in area) on two *lateritic podzolic* soils at Kangarilla and Kuitpo in South Australia, the former being sown to subterranean clover and Wimmera ryegrass, the latter to subterranean clover. At late flowering stage, rows 4.55 m in length were analysed for P; the percentage increments are presented in Table 51.3. Clearly, both biosuper and sulphaphos were inferior to superphosphate, but superior to rock phosphate + gypsum.

When the soil is moist enough for plants, in temperate climates the field temperatures are lower than the optimum of 28°C for thiobacilli. Hence it was decided to conduct further field trials in monsoonal Queensland and the Northern Territory.

Table 51.3. Biosuper and sulphaphos compared with other fertilizers

Sub. clover or Wimmera rye-grass grown in plots at two sites in South Australia. Figures in brackets show the number of similar treatments averaged.

Fertilizer treatment applied		Percentage increment relative to superphosphate			
		Herbage, dry wt		Phosphorus uptake	
		Kangarilla loam	Kuitpo loam	Kangarilla loam	Kuitpo loam
Biosuper (1:5)	single	–	43 (1)	–	65 (1)
	double	–	43 (1)	–	38 (1)
Biosuper (1:4)	single	–	13 (1)	–	14 (1)
	double	–	14 (1)	–	31 (1)
Sulphaphos (1:5)	single	6 (2)	41 (1)	8 (2)	36 (1)
	double	7 (2)	31 (1)	7 (2)	28 (1)
Sulphaphos (1:4)	single	18 (2)	25 (1)	19 (2)	36 (1)
	double	17 (2)	47 (1)	13 (2)	36 (1)
Sulphaphos (1:5) + 10% superphosphate	single	2 (2)	48 (1)	–	–
	double	4 (2)	20 (1)	–	–
Rock phosphate + gypsum	single	1 (2)	8 (1)	3 (2)	4 (1)
	double	2 (2)	8 (1)	2 (2)	6 (1)
Superphosphate	single	100 (2)	100 (1)	100 (2)	100 (1)
	double	100 (2)	100 (1)	100 (2)	100 (1)
Superphosphate increment per row of 4.55 m	single	211 g (2)	401 g (1)	210 mg (2)	177 mg (1)
	double	238 g (2)	504 g (1)	304 mg (2)	291 mg (1)

Table 51.4. Percentage increment from pasture strips under different rainfall at twelve Queensland sites over three years

Figures in brackets show the number of experimental sites averaged. Herbage figures are dry-weight values

Fertilizer treatment	Analysis	Increment (%) Rainfall: low	medium	high
Biosuper	Herbage	64	151	145
	P uptake	46	60	119
	S uptake	71	115	148
Rock phosphate + Gypsum	Herbage	27	59	–
	P uptake	23	29	–
	S uptake	33	55	–
Superphosphate	Herbage	100	100	100
	P uptake	100	100	100
	S uptake	100	100	100
Superphosphate increment (kg ha^{-1})	Herbage	1460	590	2050
	P uptake	1.23	0.98	4.57
	S uptake	1.89	1.07	2.75

Tropical Climatic Conditions

Queensland and New South Wales

At twelve sites in Queensland, fertilizer and Townsville stylo seed were applied to long strips (0.81 ha) by aeroplane or broadcaster. Superphosphate was applied at 251 kg ha^{-1} in low rainfall (460–635 mm) and medium rainfall (635–1015 mm) sites and at double the rate in high rainfall (>1015 mm) sites. Biosuper (1:5) and rock phosphate + gypsum (2:1) pellets (2–4 mm diam.) were prepared from Lady Annie rock phosphate (68% < 100 mesh). Equal amounts of P were added with all fertilizers in the initial year only, and residual effects were determined over a period of three years. The percentage increments are set out in Table 51.4.

Biosuper yielded more herbage, 151% and 145% respectively, compared with superphosphate in the medium and high rainfall districts, but less in the low rainfall areas (64%). Rock phosphate + gypsum was inferior to both superphosphate and biosuper. Swaby (1975) gives more details of these experiments.

Pasture trials are still in progress in the subtropics at Isis Junction in Queensland and at Grafton in New South Wales, where biosuper is equalling the performance of superphosphate.

Table 51.5. Effect of phosphate rock on percentage increment from pasture plots at six Northern Territory sites over five years

Trials conducted by Mr J. Austin, Northern Territory Administration. Herbage figures are dry weight values

Fertilizer treatment	Analysis	Increment (%)	
		Lady Annie Phosphate rock	Rum Jungle Phosphate rock
Biosuper	Herbage	95	66
	P uptake	73	42
	S uptake	113	109
Sulphaphos	Herbage	64	63
	P uptake	46	51
	S uptake	74	90
Rock phosphate	Herbage	39	27
	P uptake	32	23
	S uptake	12	19
Superphosphate	Herbage	100	100
	P uptake	100	100
	S uptake	100	100
Superphosphate increment (kg ha^{-1})	Herbage	1064	1350
	P uptake	1.00	1.29
	S uptake	1.81	2.04

Northern Territory

Four blocks of plots, each plot 0.0045 ha in area, were laid out at Katherine, Howard Springs, and Douglas-Daley River Experiment Farms, on several soil types, after cultivation to suppress spear grass (*Heteropogon contortus*). Biosuper and sulphaphos (1:5), prepared from Lady Annie and Rum Jungle rock phosphates, were drilled in with Townsville stylo seed and recommended trace elements, and compared with rock phosphate and superphosphate equivalent to 200 kg ha^{-1}. At the end of each growing season the ungrazed pastures were sampled by clipping quadrats one metre square from each plot. At Katherine the experiments ran for two years and at the other places for five years, to determine residual effects of a single application of fertilizer. The usual percentage increments were calculated, and are given in Table 51.5 with full acknowledgment to Mr J. Austin, Animal Industry Branch, N.T. Administration, who conducted the trials.

Biosuper from Lady Annie rock phosphate averaged 95% of the herbage achieved with superphosphate, 73% of the P uptake and 113% of S uptake. Biosuper performed better than sulphaphos, and much better than rock phosphate. Biosuper from Rum Jungle rock phosphate contained fluorapatite and crandalites; consequently it was not as beneficial as biosuper from Lady Annie fluorapatite, but nevertheless gave 66% of the herbage, 42% of P, and 109% of S provided by superphosphate, while sulphaphos was almost as good, but rock phosphate was poor.

Fiji

Here it is reported that biosuper equals the response of superphosphate on pastures (I. Partridge, personal communication). Biosuper may prove to be a useful substitute for superphosphate on pastures in the humid tropics but not in the drier tropics, or in temperate regions.

Forestry Trials

Western Australia, Northern Territory and Queensland

Trials are still in progress at Yanchep in Western Australia (CSIRO Division of Soils 1972–73) with maritime pine (*Pinus pinaster*), at Melville Is. in the Northern Territory and Cardwell in Queensland with Caribbean pine (*P. Caribaea*); at each place biosuper has promoted tree growth to the same extent or better than superphosphate, particularly on *siliceous sands*, where the more soluble superphosphate leaches, but not on gravelly *solodic soils*.

Crop Trials

Many field trials comparing biosuper with superphosphate have been conducted on wheat, maize, sorghum, sugarcane, and bananas, in Queensland, New South Wales, Northern Territory and Fiji. In all trials biosuper was inferior to superphosphate on crops, probably because it supplied phosphate at suboptimal rates for fast-growing plants. The water-soluble P in biosuper can be increased by incubating the mixture in aerobic heaps for several months prior to pelleting it.

Recently, several workers in Australia and New Zealand have initiated experiments with selected strains of acid-tolerant, vesicular-arbuscular mycorrhizal fungi to see if they were able to assist in nutrient uptake from biosuper.

Conclusions

Biosuper is a slow-release fertilizer for P and S; it is produced with simple equipment, by grinding phosphate rock with elemental sulfur, mixing with thiobacilli, pelleting with a little water, air drying, and storing in bags or heaps.

The most economical ratio of sulfur to rock phosphate is 1:5, optimal particle size is between 40 and 150 microns, and optimal pellet diameter is between 0.5 and 4.0 mm.

Pellets are neutral, non-corrosive, stable when transported, easily handled in bulk or in paper bags, compatible with most other fertilizers, and suitable for drilling, broadcasting or aerial topdressing.

Biosuper, which contains inoculum, performs better than sulphaphos, which requires colonization by soil thiobacilli, sometimes absent from or rare in soils.

Inoculum may be prepared with thiobacilli from active S-enriched soil, added to moist biosuper mixture containing 0.1% ammonium sulfate, incubating until the pH drops to 3.2, and adding moist biosuper mixture at intervals to build up sufficient bulk inoculum.

Metal sulfides oxidize too slowly to substitute for sulfur in biosuper.

Biosuper is suitable for most soils, from light sands to heavy clay loams, from siliceous phosphate-leaching types to sesquioxide phosphate-fixing types, and from acid to nearly neutral, but not calcareous or alkaline soils.

Biosuper performs better than superphosphate on pastures in the humid tropics, by producing more herbage; it provides more S, but less P. It is also useful as a maintenance fertilizer on pastures in the subtropics, but not in temperate regions.

In temperate and tropical forest plantations biosuper is equal or superior to superphosphate in the absence of applied nitrogen.

Under all climatic conditions biosuper supplies P and S too slowly for fast-growing annual and short-lived perennial crops. Incubated biosuper, containing large amounts of available P and S, would be more suitable for such crops.

In the developing tropical countries lacking facilities for the manufacture of superphosphate, biosuper could be made from local deposits of even low grades of rock phosphate and imported sulfur for approximately half the cost of imported superphosphate. Incubated biosuper would cost slightly more, but it would probably be economic for food and fibre crops.

References

Attoe, O. J. and Olson, R. A. (1966). Factors affecting the rate of oxidation in soils of elemental sulfur and that added in rock phosphate-sulfur fusions. *Soil Sci.* **101**, 317-25.

Audus, L. J. (1946). A new soil perfusion apparatus. *Nature* (London) **158**, 419.

CSIRO Division of Soils (1959). Solubilization of phosphate by microorganisms. CSIRO (Aust.) Div. Soils Annu. Rep. 1958-59. p.36.

CSIRO Division of Soils (1963). Sulphur-rock phosphate granules. CSIRO (Aust.) Div. Soils Annu. Rep. 1962-63, p.49.

CSIRO Division of Soils (1972-3). Biosuper trials. CSIRO (Aust.) Div. Soils Biennial Rep. p.113.

Fisher, M. J. and Norman, M.T. J. (1970). Tests of phosphates from Rum Jungle, Northern Territory. *Aust. J. Exp. Agric. Anim. Husb.* **10**, 597-8.

Gillman, G. P. (1973). Studies on some deep sandy soils in Cape York Peninsula, North Queensland. 3. Losses of applied phosphate and sulphur. *Aust. J. Exp. Agric. Anim. Husb.* **13**, 418-22.

Jones, R. K. and Field, J. B. (1976). A comparison of biosuper and superphosphate on a sandy soil in the monsoonal tropics of North Queensland. *Aust. J. Exp. Agric. Anim. Husb.* **16**, 99-102.

Kittams, H. A. and Attoe, O. J. (1965). Availability of phosphorus in rock phosphate-sulfur fusions. *Agron. J.* **57**, 331-4.

Lipman, J. G., and McLean, H. C. (1916). Sulfur oxidation in soils and its effect on availability of mineral phosphates. *Soil Sci.* **2**, 499-538.

Lipman, J. G. and McLean, H. C. (1917). Vegetation experiments on the availability of treated phosphates. *Soil Sci.* **4**, 337-42.

Lipman, J. G. and McLean, H. C. (1918). Experiments with sulfur – phosphate composts conducted under field conditions. *Soil Sci.* **5**, 243-50.

McLean, H. C. (1918). The oxidation of sulfur by microorganisms and its relation to the availability of phosphates. *Soil Sci.* **5**, 251-90.

Nimgade, N. M. (1968). Effects of mixing rock phosphate with molten sulfur on phosphorus availability. Trans. IX Congr. Int. Soil Sci. (Adelaide).Vol. 2, pp.765-74.

Swaby, R. J. (1975). Biosuper–Biological Superphosphate. *In* 'Sulphur in Australasian Agriculture'. (Ed. K. D. McLachlan) (Sydney University Press: Sydney.)

Swaby, R. J. and Vitolins, M. I. (1968). Sulphur oxidation in Australian Soils. Trans. IX Congr. Int. Soil Sci. (Adelaide) Vol. 4, pp.673-81.

Vitolins, M. I. and Swaby, R. J. (1969). Activity of sulphur-oxidizing microorganisms in some Australian soils. *Aust. J. Soil Res.* **7**, 171-83.

Chapter 52
Pesticide residues in soils

R. B. Jackson

PESTICIDES ARE BEING USED IN INCREASING AMOUNTS IN the production of food and fibre; some others are also important for the protection of human and animal health. Figures for the weight of pesticides used for crop protection in Australia are not readily accessible, but for formulated products the value at the factory door in 1974 was almost $A50 000 000 (Pesticides Branch, Dep. Primary Industry). Preliminary data indicate that this figure has been exceeded in subsequent years.

Although pesticides are often applied to the foliage of plants rather than directly to the soil, major proportions of all pesticides used for crop protection eventually reach the soil. Since pesticides are biologically active substances, it is important to understand their behaviour in soils so that harmful effects may be avoided and beneficial effects fully exploited. Pesticide residues in soils have the potential to enter water supplies, be absorbed by food crops, injure subsequent crops directly, and reduce crop yields indirectly by lowering soil fertility. The sale and use of agricultural chemicals in each Australian State is controlled by the State Government, with the Australian Government acting in a co-ordinating role to achieve uniformity in standards, registration, and control.

Sale of pesticides is not permitted unless registration has been granted by the appropriate State Authority. Although the registration authorities make considerable use of overseas data in reaching decisions, results from significant Australian research are a prerequisite for registration, because experience has shown that the behaviour of pesticides in Australia is often different from that in other countries. With regard to the behaviour of pesticides in soils, data are usually required on persistence, adsorption characteristics, movement in soils, and the soil types used in trials. In addition to data provided by the company seeking registration of a pesticide, the registration authority usually has access to results of experiments carried out by State Departments of Agriculture and certain other government departments.

From the legislation governing the registration and sale of pesticides, it can be inferred that a considerable amount of information about the behaviour of pesticides in Australian soils exists in the files of government departments and industrial companies. Unfortunately, much of this information remains unpublished and is not readily available for study.

Common and chemical names of pesticides mentioned in the text are listed in the Appendix on p. 839. Comprehensive lists of common and chemical names of pesticides have been published by the Standards Association of Australia (1975) and the Commonwealth Department of Health (1978). The Australian Department of Agriculture (1974) has published a list of agricultural and veterinary chemicals, including: withholding periods, residue limits, poison schedules, and summaries of situations of use. Swarbrick (1978) has compiled a comprehensive list of common and trade names of herbicides available in Australia, together with guidance for application and situations of use.

Persistence of pesticides in soils

The time that a pesticide remains intact and biologically active is important in determining its suitability for a particular purpose; persistence beyond the period necessary for control may lead to problems with residues. Ideally, a pesticide should remain active long enough to give the desired control, then decompose to harmless products before a further application is necessary. Estimation of persistence is influenced by the sensitivity of the analytical techniques. In bio-assays, loss of biological activity may be interpreted as lack of persistence, but adsorption by soil constituents or dilution below the activity threshold may mask the presence of intact pesticide. Changes in environmental conditions or cultural practices may result in release of adsorbed pesticide, or a greatly diluted pesticide may be concentrated by passage along a food chain and thus become hazardous. The persistence of a pesticide is determined mainly by its molecular structure, but environmental parameters and soil characteristics also are important. Kearney *et al.* (1969) reviewed about 80 sources concerned with pesticide persistence in American and European soils, and the values in Table 52.1 are derived from their data; these values resulted from normal rates of application and normal agricultural practice.

Table 52.1. Persistence of pesticides in soil

Data of Kearney *et al.* (1969) developed from a review of about 80 sources. The values shown resulted from normal rates of application and normal agricultural conditions.

Chlorinated hydrocarbon insecticides	*Period for 75–100% loss*
Aldrin	2 years
BHC	3 years
Chlordane	5 years
DDT	4 years
Dieldrin	3 years
Heptachlor	2 years
Organo-phosphate insecticides	
Diazinon	12 weeks
Disulfoton	4 weeks
Malathion	7 days
Parathion	7 days
Phorate	2 weeks
Herbicides	
Atrazine	10 months
Barban	2 weeks
Bensulide	10 months
CDAA	2 months
CDEC	6 weeks
Chloramben	3 months
Chlorpropham	2 months
Dicamba	2 months
Dichlobenil	4 months
Diphenamid	8 months
Diuron	8 months
2,4-D	4 weeks
2,2-DPA	2 months
EPTC	4 weeks
Fenuron	8 months
IPC	4 weeks
Linuron	4 months
MCPA	3 months
Monuron	10 months
Picloram	18 months
Prometryne	3 months
Propazine	18 months
TCA	3 months
Trifluralin	6 months
2,4,5-T	5 months
2,3,6-TBA	12 months

Insecticides persistence in soils

DDT

The use of DDT in Australia reached about 1000 t year^{-1} in 1966 and 1967, with a value then at the factory door of $A2 066 000 year^{-1} (Pesticides Branch, Dept of Primary Industry). Consumption remained static from 1966 until manufacture ceased in Australia in 1971. Australian production had high tariff protection, but since production ceased, supplies have been imported at half or less of the former domestic price. The value at the factory door reached $A2 730 000 year^{-1} in 1975, so there appears to have been a sharp increase in the use of DDT since 1971.

The major agricultural uses of DDT in 1972 were on vegetables, grain sorghum, and cotton (Australian Academy of Science 1972). The highest rates of annual application are usually on cotton (8 kg ha^{-1}), tobacco (5 kg ha^{-1}), and vegetables (4 kg ha^{-1}). Since most of this DDT eventually reaches the soil, it is surprising that so few data on residues in Australian soils have been published.

The greatest application of DDT in Australia is probably in the cotton-growing area of the Namoi Valley in New South Wales. Surveys made from 1969 to 1974 indicated little contamination of the Namoi River itself, but residues were found in the creeks and dams in the area (Gilbert and Singh 1976). Thirty-eight creek and river water samples contained levels of DDT ranging from less than 0.01 to 0.65 μg l^{-1} and 13 drain samples contained from 0.22 to 42.9 μg l^{-1}. Thirty-nine samples of creek and river sediments had residue levels ranging from less than 0.01 to 3.6 μg g^{-1} dry weight, and the DDT content of 14 samples of drain sediments ranged from 0.13 to 50.0 μg g^{-1} dry weight (Australian Academy of Science 1972). A further survey following severe flooding in 1974, revealed DDT contamination of the Namoi River as much as 300 km downstream from the cotton fields (Gilbert and Singh 1976); since DDT has a very low solubility in water, this was probably due to transport of contaminated soil from the DDT-treated fields. McDonald and Madsen (1973) determined DDT residues in soils from cotton fields in the Namoi Valley at the beginning of the growing season in December 1971; levels ranged from 1 to 1.5 ppm. Several applications of DDT were applied to the cotton crops during the summer, and the soils were sampled again late in February 1972; DDT levels ranged from 9 to 16 ppm, the high level being in an area which had received seven applications at a rate of 2.2 kg ha^{-1}.

Dieldrin and Heptachlor

Gilbert *et al.* (1976) reported that dieldrin or heptachlor applied to soil (1.1kg ha^{-1}) at the time of sowing pasture appeared in the herbage at levels exceeding 0.02 μg g^{-1}, and resulted in excessive residues in grazing animals. Data for soil residues were not given.

Persistence of herbicides in soil

Most assessments of the persistence of herbicides in Australian soils have been qualitative, being based on the extent of natural regeneration of plants or on the response of crop plants sown on treated areas. While such assessments may be useful for some practical purposes, they depend on the sensitivity of the indicator plants. Furthermore, they show only phytotoxicity in the root zone, giving no indication whether the herbicide has been decomposed or strongly adsorbed, or leached down the profile.

2,4-D

In common with experience in North America and Europe, 2,4-D apparently causes few problems as residues in Australian soils. The main difficulties in the use of 2,4-D seem to be the drift of spray on to sensitive crops in non-target areas, and injury to adjacent crops by the vapour of volatile esters of 2,4D.

R.B. Jackson (unpublished results) followed the disappearance of 2,4-D from an Urrbrae *red-brown earth* and a Monarto sandy *terra rossa* in South Australia. The applied dose was 22 $\mu g\ g^{-1}$, the soils were kept moist and incubated at 25°C; the concentration of 2,4-D in the Urrbrae soil fell to 6% of the applied dose after 16 days, and to 2% after 12 days in the Monarto soil.

2,4,5-T

In recent years the annual value at the factory door of formulated 2,4,5-T in Australia has been about half that of 2,4-D, and reached $A3 100 000 for 1975 (Pesticides Branch). In spite of this extensive usage, there are few published reports of its behaviour in Australian soils. Bachelard and Johnson (1969) found no effects on *Pinus radiata* seeded 2 months after application of the butyl ester of 2,4,5-T equivalent to 2,4,5-T acid at a rate of 5.6 kg ha^{-1}. Overseas studies, however, indicate persistence times of 5 to 7 months (Kearney *et al.* 1969; Alexander 1969).

The use of 2,4,5-T has given rise to concern in recent years because it is sometimes contaminated with a by-product, 2,3,7,8-tetrachlorodibenzo-*p*-dioxin (TCDD or dioxin); TCDD is one of the most toxic substances yet synthesized (Jenkins 1978). Sublethal doses of TCDD cause chloracne (a persistent and painful form of dermatitis), and induce birth defects.

The maximum level of TCDD permitted in 2,4,5-T in Australia is 0.1 ppm. There is a need, however, to determine whether dangerous amounts of TCDD can be formed during normal use of 2,4,5-T or during prolonged storage under adverse conditions. There appear to be no reports about the behaviour of TCDD in Australian soils. American studies indicate a half-life in soil of about 1 year (Casarett and Doull 1975). When applied to leaf surfaces, TCDD is rapidly broken down by sunlight, but on soil surfaces photolysis is much slower (Crosby and Wong 1977).

Atrazine

In a laboratory study, Bowmer (1973*a*) found that loss of atrazine from six contrasting Australian soils followed first-order kinetics, i.e. the rate of loss was independent of the initial concentration. This was in contrast to the results obtained with two English soils by Hance and McKone (1971) who found that the rate of atrazine loss increased as the initial concentration decreased. Persistence in the Australian soils varied with soil type, and decreased with increasing temperature; there was little relationship between rates of dissipation and clay or organic carbon contents. At 29.6°C, half-lives ranged from 6.5 weeks in Pelican Clay to 25 weeks in a grey medium clay from Carrathool, N.S.W.; at 15.2°C the half-lives obtained by extrapolation were respectively 35 and 167 weeks. Swain (pers. comm.) studied the dissipation in soil of atrazine applied to irrigated sorghum in southern New South Wales; like Bowmer, he found that atrazine loss followed first-order kinetics. Application rates up to 10 kg ha^{-1} left residues less than 0.05 ppm after 11 months; the depth of sampling was not stated. In laboratory experiments, atrazine disappeared from Birganbigil silty loam faster than from Willbriggie clay loam, Billabong clay, or Gundaline clay. The Birganbigil soil had the lowest clay content and cation exchange capacity of the soils used, but the highest organic carbon content (1.42%).

Simazine

Simazine is one of the more pesistent of the *s*-triazine herbicides. Residues may persist in soil for more than 12 months, so repeated annual applications could lead to an accumulation of residues. Bowmer (1972) determined residues in an apple orchard at Bathurst in New South Wales where simazine had been applied annually for 6 years; the soil was a light sandy clay loam A horizon, about 30 cm thick, on a medium clay B horizon. The annual rates of application were 6.7, 13.4 and 20.2 kg ha^{-1}, and a year after the sixth application the residues in the 0–5 cm layer were respectively 0.07, 0.42 and 0.73kg ha^{-1}. Residues deeper in the profile were measured in plots treated at the highest rate; the total amount of simazine in the profile, sampled to 30 cm, was 11% of the annual application, and more than 88% of the residue was in the top 15 cm.

Diuron

Bowmer (1972) also determined residues in plots which had been treated annually with diuron at the same rates used for simazine; the two lower rates were applied for 6 years, but the highest diuron treatment was omitted in the sixth year. The residues found in the seventh year in the 0–5 cm layer were respectively 2.4, 6.2 and 5.9 kg ha^{-1}. The total amount of diuron in the profile of the plots that received 13.4 kg ha^{-1} was 101% of the annual application.

2,2-DPA

Campbell (1976) found that to avoid deleterious effects following application of 2,2-DPA (8.7 kg ha^{-1}) to unploughed hill country near Turondale in New South Wales a period of 10 to 20 days should elapse before sowing pasture species.

Glyphosate

Glyphosate has been reported (Anon. 1973) to have no residual activity in soil. The low activity of glyphosate in soil appears to result from a combination of moderate adsorption, microbial degradation, and low intrinsic toxicity when applied to roots (Hance 1976; Nomura and Hilton 1977; Torstensson and Aamisepp 1977). Campbell (1974) studied the residual effects of glyphosate under glasshouse conditions and after application to unploughed hill country near Turondale, New South Wales (Campbell 1976); following application of 4.5 kg ha^{-1} to control resident species, the growth of several pasture species, especially legumes, was inhibited if sown less than 20–35 days later. The pasture species could have been inhibited by toxins produced during decay of the resident plant residues (Kimber 1973; Ellis and Lynch 1977), but this is unlikely because the extent of inhibition was related to herbicide dose, and the toxicity symptoms were typical of glyphosate.

Dichlobenil

Dichlobenil is a promising herbicide for control of *Elodea canadensis* in water supply and drainage channels in the irrigation systems of southern New South Wales and northern Victoria, but doubts have been expressed about residual effects and hazards to crops. Bowmer *et al.* (1976) found that 10 weeks after application at 22.7 kg ha^{-1} the residual dichlobenil in the soil was about 6% of the amount applied, most of it being in the top 12 cm. Loss of dichlobenil from soils in the field is largely by volatilization (Beynon and Wright 1972), particularly under wet conditions (Verloop 1972). Since the weather during the trials was exceptionally wet, the authors concluded that dissipation occurred mainly by volatilization; they pointed out that in a drier season, the persistence could be more prolonged.

2,3,6-TBA

Parsons *et al.* (pers. comm.) found that persistence of 2,3,6-TBA varied with soil type. On one soil type (not specified), wheat gave a normal yield when sown 12 months after application of 2,3,6-TBA at 22.4 kg ha^{-1}; on another soil type, normal yields were not obtained unless 24 months elapsed between herbicide application and sowing. On a *red-brown earth* the growth of sensitive species such as legumes was inhibited even 3 years after treatment.

Picloram

Welsh (1966) monitored the persistence of picloram in a sandy soil at Watchupga in the Victorian Mallee by measuring the growth of legumes sown at various intervals after spraying at rates of 0.14 to 0.56 kg ha^{-1}. Welsh concluded that the half-life of picloram was about 9 months under those conditions.

Wells (1972) reported that applying picloram at 35 g ha^{-1} to Kattyoong sand (Victorian Mallee) reduced the growth of *Medicago* species sown 12 months later; at a rate of 280 g ha^{-1}, toxic residues were still present 21 months after application. Bachelard and Johnson (1969) reported injury to *Pinus radiata* seedlings grown in a 1:1 mixture of silty loam and coarse sand that had been treated 6 months earlier with Tordon 50D (picloram + 2,4-D, 1:4) to add picloram at 0.56 kg ha^{-1}. This effect can be attributed to picloram, as 2,4-D is degraded quickly in soil.

Wheat yield was used by Welsh (pers. comm.) as an indicator of residual picloram in soils at Numurkah and Rutherglen in Victoria. Picloram was applied at various intervals before sowing; the length of this interval giving no reduction in yield ranged from less than 1 month for picloram at 0.14 kg ha^{-1} up to 12 months for 1.12 kg ha^{-1}. There was a positive correlation of yield with the rainfall between spraying and sowing. Marley (pers. comm.) studied the loss and movement of picloram in a heavy black clay soil near Warwick in Queensland. The surface soil contained 55% montmorillonitic clay, 1.8% organic carbon, and pH was 7.3. Picloram was applied at rates ranging from 0.023 to 1.89 kg ha^{-1}, and the plots were sampled down to 90 cm for bioassay 113, 225, and 510 days after application. At all sampling times, picloram was concentrated in the 0–15 cm horizon; smaller amounts were found in the 15–30 cm horizon, but only for the two highest rates of application. Thus leaching was minimal. Losses ranged from 80 to 85% after 113 days, 76 to 92% after 225 days, and 90 to 100% after 510 days. The losses reported for the first 113 days are unusally high and amount to 1.5 kg ha^{-1} for the highest rate of application. The first sampling period extended from late October to mid February, so temperatures would have been high, and, with 28 cm of rain, conditions would have been favourable for decomposition. Marley's data are of further interest because, on the whole, they fit first-order kinetics; that is, the percentage loss per unit time was independent of applied dose. By contrast, Hamaker *et al.* (1968) found that percentage loss per unit time increased with decreasing dose, and they considered that loss of picloram from 13 Californian and Texan soils was best described by half-order or Michaelis-Menten kinetics. The half-life of picloram calculated from Marley's data is about 6 weeks. In laboratory experiments of 12 months duration, R. B. Jackson and R. J. Swaby (unpublished results) followed the loss of picloram (applied at 0.24 ppm) from a mixture of four soils under non-leaching conditions at 25°C; they found the time for 50% loss to be about 3 months. The soil mixture contained equal parts of Monarto sandy *terra rossa*, Urrbrae *red-brown earth*, Athelstone black clay, and a well-rotted garden compost. When the black clay and compost were omitted from the mixture, the 50% disappearance time was about 22 months.

The very slow degradation in soils low in organic carbon suggests that leaching is the major factor in the removal of picloram from the root zone of plants in soils of this type. Following studies of more than 200 treatments at field locations in 18 states and provinces of the U.S. and Canada, Hamaker *et al.* (1967) found rainfall to be the most important variable in the disappearance of picloram from soils.

Many of the cereal-growing areas of southern Australia, where picloram would be useful for controlling some difficult weeds, have low and unreliable rainfall and their soils are low in organic carbon. The disadvantages of these factors, which retard dissipation, are emphasized by the great sensitivity of legumes to picloram, and the importance of cereal/legume rotations in these areas. Consequently, picloram is recommended only for spot spraying in these areas.

Most agricultural areas in Australia receive high levels of solar radiation. Picloram in aqueous solution is subject to photolysis by sunlight (Hall *et al.* 1968), but it is unlikely that photolysis is significant in the degradation of picloram that has reached the soil, because it is readily washed into the soil matrix and only the soil surface is irradiated by sunlight.

Persistence of fungicides in soil

There is little published information about the persistence in Australian soils of fungicides, in spite of their potential for adversely affecting microorganisms important to the maintenance of soil fertility.

Benzimidazoles are important commercially because they provide broad spectrum fungicidal control. Baker and Bertus (1976) compared the persistence in soil of three benzimidazole fungicides; persistence was in the order thiabendazole > benomyl = carbendazim. The equal persistence of benomyl and carbendazim is surprising in view of the ease of hydrolysis of benomyl to the relatively stable carbendazim, which is mainly responsible for the fungitoxic activity of benomyl.

Movement of pesticides in soil

The movement of a pesticide in soil is an important factor in determining its effectiveness and its potential for contaminating the environment. Pesticides can move through soil by diffusion and by mass flow. Diffusion is a result of random molecular motion, and leads to a net movement down a concentration gradient until the molecules become uniformly distributed throughout the system. Mass flow is caused by external forces. Several processes can result in mass flow of pesticides, the most important being mass flow of water in which pesticide is dissolved or suspended, and movement of soil particles on which the pesticide is absorbed or adsorbed. Such soil movement may occur by suspension in flowing water, by wind action, or cultivation, or the activities of various soil animals. Letey and Farmer (1974) have presented a concise review of the movement of pesticides in soils. A more detailed treatment of diffusion of organic chemicals in soil has been given by Hamaker (1972), and Letey and Oddson (1972) have given a detailed account of the principles involved in movement in water, air, and soil. Adsorption is one of the most important factors governing the distribution of organic compounds between soil particles and water; the subject has been reviewed in detail by Hamaker and Thompson (1972).

Both inorganic and organic components contribute to the adsorptive capacity of soil, but their intimate association makes it difficult to determine the contribution made by each. Water plays an important part in adsorption processes in soils. In recent years, characterized crystalline clays have been used in detailed studies of model systems of pesticide–clay–water; the topic has been reviewed by several authors (summarized by Green 1974). There is some evidence that the adsorptive surfaces of minerals are blocked in soils by organic matter. Thus the soil organic matter largely determines the pesticide–soil interactions; many authors have reported that the retention of pesticides by soils is more closely related to the organic matter content than to any other soil property. The recent literature on interactions between soil organic matter and organic pesticides has been reviewed by Weed and Weber (1974). With the advent of radioactive-labelled pesticides, it became apparent that in some cases residues are so firmly bound to soils that they cannot be removed by exhaustive extraction with a variety of solvents. The nature of these bound residues, and their significance with regard to accumulation, toxicity, and availability, have been the subject of a recent symposium (Kaufman *et al.* 1976). There are few published accounts of studies of pesticide movement in Australian soils.

Adsorption and pesticide movement

Bowmer (1971*a*) determined the Freundlich isotherms for atrazine adsorption by 26 Australian soils, mostly from the Riverina and Namoi areas. The soils had clay contents ranging from 17 to 73%. For 25 of the soils, the organic carbon contents ranged from 0.23 to 1.87%. In general, the adsorption of atrazine was related to the organic carbon content of the soils rather than to their clay content. Pillay and Tchan (1972) used an algal bioassay to study the effect of clay and organic matter content of soils on the adsorption of three phenyl-urea herbicides. Adsorption of these herbicides by Chullora clay and Mt Gambier peat decreased in the order neburon > diuron > monuron; the peat adsorbed 18 times more herbicide than the clay.

Leaching and pesticide movement

Donaldson (pers. comm.) studied the movement of 2,3,6-TBA (26 kg ha^{-1}) down the profile of a light sandy soil in the Victorian Mallee and a *red-brown earth* in the Goulburn Valley in Victoria. In the Mallee soil, the herbicide had moved out of the top metre within 12 months, but with the *red-brown earth* there was a high concentration in the surface layer and lower concentrations down to 1.5 m.

The movement of picloram down the pofile of a light sand in the Victorian Mallee was studied by Molnar (pers. comm.). After application of 2.2 kg ha^{-1} as the potassium salt, the picloram spread evenly down the profile to a depth of about 1.2 m. After 27 months the herbicide had disappeared from the profile except for a trace at 1.2 m.

Weiss (1971) compared the leaching of diuron and simazine in a Murray Valley sand (clay 10.6%, organic matter 0.9%). Diuron was leached more slowly than simazine, even though its solubility in water is about eight times that of simazine. This is consistent with the results of Hilton and Yuen (1963) for 10 Hawaiian sugar-cane soils.

By contrast, in a study at Bathurst in New South Wales, Bowmer (1972) found that after six annual applications diuron had moved further down the profile than simazine. This apparent contradiction in the relative mobilities of these two herbicides can be accounted for by the difference between their dissipation rates in Bowmer's long-term study; the total diuron residues in the profile were 101% of the annual application of 13.4 kg ha^{-1}, but the total simazine residue was only 11% of the annual application of 20.2 kg ha^{-1}. Thus even if diuron is less mobile than simazine, its greater persistence may allow it to reach greater depths than simazine.

In some situations, leaching may be necessary for a pesticide to be effective. In the Murrumbidgee Irrigation Areas, maize is usually grown in rows with furrow irrigation; furrow irrigation in an evaporative environment such as this tends to carry herbicides up out of the root zone of the target weeds into the surface of the raised soil in the row. Bowmer (1973*b*) reported that pre-emergence applications of a number of herbicides for control of barnyard grass (*Echinochloa crus-galli*) were ineffective when a period of dry weather followed treatment. Under these conditions, herbicide performance was improved by application and incorporation to a depth of about 7 cm before seeding. Bowmer (1971*b*) had previously concluded that leaching of the herbicides into the root zone by rain is necessary for effective weed control when incorporation is not practised.

Soil erosion and pesticide movement

There seem to have been no direct studies in Australia of the movement of pesticides by soil erosion. Contamination of the Namoi River with DDT has already been discussed; DDT has a very low solubility in water, and it seems likely that transport of soil particles carrying adsorbed DDT was the principal mechanism of contamination. Dust storms, which occur periodically in agricultural areas in southern Australia, have the potential for moving large quantities of pesticides over considerable distances. The movement of pesticides by wind erosion, however, does not seem to have been studied in Australia. In America, pesticides have been carried from western Texas and New Mexico to Cincinnati by transport on atmospheric dust (Cohen and Pinkerton 1966).

Volatilization and pesticide movement

Loss of pesticide by volatilization during and after application may be as much as 50%. Such losses may be important because of reduction in effective dose and also the possibility of contaminating nontarget areas. Pesticide volatilization from soil has been reviewed in detail by Spencer *et al.* (1973).

The factors controlling the volatilization of pesticides from soils include the physical and chemical properties of the pesticide, the adsorptive properties of the soil, the water content of the soil, the pesticide concentration, air movement, temperature, and rate of movement through soil; all of these factors are inter-related.

The molecular structure of a pesticide determines its vapour pressure, its solubility in water, and its adsorption behaviour. The rate of volatilization of a pesticide from soil is related to its vapour pressure in the soil, and this is lower than that of the pure compound because of adsorption. Soil water plays an important role in volatilization of pesticides from soil because it increases the vapour pressure of pesticides by displacing the chemical from adsorption sites. Water is involved also in the movement of pesticide to the soil surface for volatilization; as water evaporates from the soil surface, soil water moves towards the surface by capillary action and carries with it any pesticide in the soil solution.

Temperature affects a number of soil and pesticide properties, but the interactions are not completely understood. The effect of temperature on the rate of volatilization of pesticides from soil is largely due to its effect on the vapour pressure of the pesticide. Temperature also influences pesticide volatilization through its effect on movement of pesticide to the soil surface by diffusion and by mass flow of soil water to replace that lost by evaporation. Increasing the temperature of soil – pesticide systems usually increases the volatilization rate; in some circumstances, however, it can result in decreased volatilization rates. The rate of soil drying may increase with temperature, and if the water content of the soil falls below the level necessary for saturation of adsorption sites, the vapour density of the pesticide falls; the net effect on volatilization depends on the balance between this fall in vapour density and the increase caused by the direct effect of temperature on vapour density.

The concentration of a pesticide in soil has a direct effect on the volatilization rate through its effect on vapour density. Vapour density increases with increasing concentration until the soil air is saturated; this saturation vapour density is equal to the vapour density of the pesticide without soil.

Air movement directly affects pesticide volatilization by removing vapour from the soil surface, thus causing increased movement of pesticide vapour to the soil surface. There is also an indirect effect when the relative humidity of the air is less than 100%; in these circumstances, increased air flow hastens drying.

The evaporative loss of volatile pesticides applied to soil can be reduced by incorporating them into the soil as soon as possible after application; for example, the preemergent herbicide trifluralin, which is widely used in Australia, should be incorporated into soil within 4 hours of application. Molinate is used extensively in the irrigation areas of New South Wales to control barnyard grass (*Echinochloa* spp.) in rice. Molinate is readily lost from soil by volatilization when the soil surface is wet at the time of application (Weed Science Society of America 1974); when applied before seeding, such losses may be reduced by incorporating the molinate into the soil by harrowing. A different technique is necessary, however, when molinate is applied after seeding; in this case, the herbicide is sprayed onto a dry soil surface so that it may enter the pore spaces rapidly by capillarity (Hartley 1976). The soil is then flooded with water for at least 2 days; thus, evaporative loss from the soil – water system can occur only by diffusion from the pore spaces into the water, followed by volatilization from the floodwater solution (Swain 1974).

There appear to be no published measurements of the volatilization of pesticides from Australian soils. O'Neill *et al.* (1976) found that volatilization of dieldrin from aerated aqueous solution is proportional to the concentration remaining in solution. The vapour density of dieldrin in air is found to be a function of the solution concentration, and volatilization a function of temperature, with a Q_{10} of 2.0.

The relevance of such studies to the volatilization of pesticides from soils is doubtful, since Farmer *et al.* (1972) found that the loss of dieldrin from soil has a Q_{10} of 4, and furthermore, that as the concentration at the soil surface falls, the rate at which dieldrin moves to the surface becomes the controlling factor. With aerated solutions, rate of movement to the evaporative surface would not be limiting.

Degradation of Pesticides

Microbial degradation of pesticides

Microorganisms are one of the major factors involved in the degradation of pesticides in soils. Some pesticides are decomposed in soils within a few weeks, while others may persist for several years. The molecular characteristics that determine the biodegradability of organic compounds are not well understood. Compounds having a high chlorine content and/or low solubility in water (e.g. DDT) are usually resistant to attack, but many compounds freely soluble in water and having few or no chlorine substituents also resist microbial attack. The rate at which an organic compound is metabolized in soil is subject to a number of environmental factors. Conditions which favour microbial activity, such as adequate moisture, suitable pH, warm temperatures, and the presence of organic substrates, also promote pesticide decomposition. The mechanisms by which soil microbes develop the ability to degrade pesticides are not fully understood; the subject has been discussed recently by Hiltbold (1974) and by Kaufman and Kearney (1976).

Microbial degradation of pesticides in soils is of two kinds. In one, after a period of exposure to the pesticide during which little or no degradation occurs, some microorganisms adapt to use the added chemical as a source of carbon, energy, and sometimes nitrogen and sulfur; the population of the adapted cells increases at the expense of the pesticide, so the rate of degradation increases with time until the concentration falls to a low level.

The second type of degradation is catalyzed by enzymes already present in the existing microbial population, so decomposition begins as soon as the pesticide reaches the soil. This type of degradation takes place by co-metabolism; and co-metabolism is defined as metabolism of a compound that the cell cannot use as a source of energy or of an essential nutrient (Jensen 1963). The rate of co-metabolism depends on the size and biochemical activity of the existing population, which are largely determined by the nature and amount of the soil organic matter. It seems likely that both kinds of degradation occur in some cases, co-metabolism taking place during the adaptation phase. There is an extensive literature on the microbial degradation of pesticides, and it has been reviewed recently by Kaufman and Kearney (1976). There have been few Australian studies of microbial degradation of pesticides, and most have been concerned with pure cultures in nutrient solutions.

Microbial degradation of herbicides

Picloram and other pyridine compounds. Naik *et al.* (1972) studied the degradation of picloram, N-Serve and 36 other substituted pyridines in dilute soil suspensions in nutrient medium. There was a rough correlation between the number of substituents in the pyridine ring and the resistance to microbial degradation; in general, compounds with four or five substituents were more persistent than those with one or two. Decomposition was slower under anaerobic than aerobic conditions, except for pyridine, 2-picoline, and 2-hydroxypyridine. Chloro derivatives were usually resistant to microbial attack, but some nonhalogenated pyridines also were very resistant.

Allyl alcohol. Jackson (1971) isolated a *Trichoderma* species from soil previously treated with allyl alcohol; this fungus was able to use allyl alcohol as a carbon and energy source. In the presence of allyl alcohol, growing cultures and washed suspensions of the fungus accumulated considerable amounts of acrylate. This observation, together with studies of arsenite-inhibited suspensions, indicated that allyl alcohol is oxidized by a pathway in which acrylate, lactate (or their acyl-CoA esters), and pyruvate are intermediates (Jackson 1973).

Amitrole is degraded rapidly in soils. This had been attributed to chemical processes for several reasons including repeated failure to isolate amitrole-degrading microorganisms from soil (Kaufman *et al.* 1968). In these unsuccessful attempts to isolate amitrole-degrading organisms, the enrichment media contained amitrole as the sole carbon source. Conversely, Campacci *et al.* (1977), working at the University of Sydney, used amitrole as the sole source of nitrogen, and added sucrose as a source of carbon. They succeeded in isolating from soil 10 organisms able to degrade amitrole; these included *Bacillus* spp., *Corynebacterium* spp., and a species of *Pseudomonas*.

Microbial degradation of insecticides

Lindane is the gamma isomer of hexachlorocyclohexane. In general, it is less persistent than DDT and the chlorinated cyclodiene insecticides. MacRae *et al.* (1969) reported degradation of lindane by a species of *Clostridium* isolated from flooded soil. Later, at the University of Queensland, Heritage and MacRae (1977*a*) identified the organism as *Clostridium sphenoides*, and showed that anaerobic washed suspensions degraded lindane to γ-3,4,5,6-tetrachloro-1-cyclohexene; the α-isomer of 1,2,3,4,5,6-hexachlorocyclohexane was degraded to δ-3,4,5,6-tetrachloro-1-cyclohexene. Both of the tetrachlorocyclohexenes were degraded further, but the products were not identified. The dehydrohalogenation of lindane was also observed in cell-free preparations of *C. sphenoides* (Heritage and MacRae 1977*b*); the activity was associated with the cell membrane fraction, and reduced glutathione was an essential co-factor.

Organophosphate pesticides are in general lost from soils fairly quickly. Many processes may be involved, but probably the most important are chemical decomposition, microbial degradation, and volatilization. The major reactions in the microbial degradation of organophosphates are hydrolysis and oxidation; reduction also occurs, particularly under anaerobic conditions. There seem to be no published reports about microbial degradation of organophosphates in Australian soils. Wood and MacRae (1977) reported that washed suspensions of the nitrogen-fixing bacterium *Klebsiella pneumoniae* reduced fensulfothion to the sulphide.

Carbamates, because of their low toxicity for warm-blooded animals and their short persistence in the environment, are being used more and more in place of the organochlorine insecticides. In spite of this increasing importance, there is little information available about the behaviour of carbamate pesticides in Australian soils. Fung and Uren (1975) reported that microbial activity was largely responsible for dissipation of methomyl in perfused soils after a lag period of about 7 days; in previously enriched soils, loss of methomyl began without a lag phase.

Nonbiological degradation of pesticides

Nonbiological processes play an important part in the dissipation of many pesticides in soils. Crosby (1976*a*) has recently reviewed the subject with particular reference to herbicides. In general, hydrolytic reactions are the most important, but oxidation-reduction reactions are important for certain compounds. Many pesticides are degraded by photochemical and free radical induced reactions in soil-free systems, but the significance of these reactions in soils is difficult to assess.

Hydrolysis and pesticide degradation

The hydrolysis of many pesticides is faster in soils than in soil-free aqueous systems because of catalysis by soil constituents. Hydrolysis appears to be particularly important for two groups of pesticides, namely the chloro-*s*-triazine herbicides and the organophosphate insecticides.

Chloro-s-triazines. Evidence from various types of experiments indicates that non-biological degradation of certain *s*-triazines may be equally or more important than degradation by microorganisms. The subject has been reviewed by Jordan *et al.* (1970).

Organophosphate insecticides. The susceptibility of the organophosphate insecticides to degradation is a major factor in their increasing use in agriculture. Although many organophosphates can be degraded by microorganisms, their rate of hydrolysis in the presence of soil is usually much faster than in soil-free microbial cultures.

Oxidation and pesticide degradation

Oxidation is believed to be important in the degradation of aldrin and heptachlor. Aldrin is converted to its epoxide, dieldrin (Edwards 1966), and heptachlor to heptachlor epoxide (Wilkinson *et al.* 1964).

Photodecomposition and pesticide degradation

Photolytic reactions have been reported for a wide range of pesticides. The subject has been reviewed by Rosen (1971) and, with particular reference to herbicides, by Crosby (1976*b*). It is difficult to assess the significance of photodecomposition to pesticide degradation in soils. Many of the studies of photodecomposition of pesticides have involved the use of light of high intensity and of wavelength shorter than that which reaches the earth's surface. Photodegradation is of greatest significance when pesticides remain for a time on soil and plant surfaces exposed to sunlight; when they have been leached into soil or incorporated by cultivation, absorption of light by other soil components greatly reduces the amount reaching the pesticides.

There are few published reports concerning nonbiological degradation of pesticides in Australian soils. Kimber (1980) found that the experimental insecticide 1,1-bis(*p*-ethoxyphenyl)-2-nitropropane (Holan 1971*a*) was oxidized to 1,1-bis(*p*-ethoxyphenyl)-2-propanone in γ-irradiated soils, which presumably were sterile. The propanone was also produced during adsorption of Florisil, so the reaction was considered to be surface-activated.

Effects of pesticides on soil microorganisms

Pesticides are used in agriculture because they are toxic to specific groups of weeds, insects, or plant pathogens. A major concern about their use is the possibility of adverse effects on nontarget soil microorganisms and thus on soil fertility. In addition to the active pesticide, a commercial product usually contains surfactants and solvents and other such adjuvants that may have toxic effects.

Assessment of the effect of pesticides on the soil microflora is difficult because of the many interactions between pesticides and soil, and there is great diversity of types of microorganisms and their activities.

A great variety of methods have been used to assess the effects of pesticides on soil microbes, and many have been discussed in detail by Greaves *et al.* (1976). The methods fall into two main groups: measurement of changes in populations, and changes in the physiological activities of these populations. The literature on pesticide effects on microorganisms in soil has been reviewed by Helling *et al.* (1971), Parr (1974) and by Greaves *et al.* (1976).

Non-symbiotic nitrogen-fixation and pesticides

Australian studies of the effects of pesticides on soil microorganisms which fix nitrogen appear to have been confined to experiments with pure cultures in liquid media. Mackenzie and MacRae (1972) found that the insecticides DDT and lindane and the herbicides dalapon and 2,4,5-T had no effects on growth of, or acetylene reduction by, *A. vinelandii*. The effect of several organo-phosphorus insecticides on acetylene reduction by *A. vinelandii* was studied by Wood and MacRae (1974). Chlorpyrifos, Dowco 214, methidathion, diazinon and fensulfothion did not inhibit acetylene reduction, and the inhibitory effect of Nemacur was only transient. Dowco 217 inhibited acetylene reduction, but this was accounted for by its instability and consequent release of 2-hydroxy-3,5,6-trichloropyridine which inhibited acetylene reduction. 2-Hydroxy-3,5,6-trichloropyridine, which is also released by hydrolysis of both chlorpyrifos and Dowco 214, is phytotoxic and very resistant to attack by soil microbes (Naik *et al.* 1972). A large proportion of the studies of aerobic nitrogen-fixing bacteria has been concerned with *Azotobacter*; MacRae and Celo (1974) reported the effects of some organo-phosphorus pesticides on the respiration of washed suspensions of *A. vinelandii* with glucose as substrate. With a pesticide concentration of 2 ppm, the respiration rate was reduced by naled, coumaphos, dichlorvos, chlorpyrifos, Dowco 217, and Nemacur; at 100 ppm, respiration was inhibited by these pesticides and also by Terracur, malathion, and Dowco 214. Respiration was stimulated by malathion at 2 ppm and by diazinon at both levels. Endogenous respiration, also, was inhibited by Terracur, diazinon, Dowco 217, and Dowco 214, but Nemacur and dichlorvos stimulated it. Since nitrogen fixation by *Azotobacter* is dependent upon ATP generated by respiration,

$$N_2 + 6e^- + 6H^+ + 6n\ ATP \longrightarrow 2NH_3 + 6n\ ADP + 6n\ P_i,$$

it might be expected that these organo-phosphates would inhibit nitrogen fixation. Wood and MacRae (1974), however, reported that reduction of acetylene by *A. vinelandii* was not inhibited by concentrations up to 500 ppm, fensulfothion, diazinon, methidathion, chlorpyrifos, or Dowco 214. Nemacur and 2-hydroxy-3,5,6-trichloropyridine (the leaving-group of chloropyrifos, Dowco 214 and 217) had adverse effects on ethylene production when added to suspensions of *A. vinelandii* in the early logarithmic phase, but not if present from the time of inoculation; Wood and MacRae concluded that the microorganism develops a tolerance to these compounds with time; presumably, the amounts of reductant and ATP provided by the inhibited respiration were still adequate for reduction of acetylene by the nitrogenase system. It should be noted, however, that the reductant and ATP requirements for acetylene reduction are only one-third those for nitrogen reduction.

$$C_2H_2 + 2e^- + 2H^+ + 2n\ ATP \longrightarrow C_2H_4 + 2n\ ADP + 2n\ P_i$$

Photosynthetic bacteria belonging to *Rhodopseudomonas* spp. fix nitrogen under anaerobic conditions. Generally, these bacteria are not found in well-aerated soils, but sometimes they flourish in flooded soils and may fix significant amounts of nitrogen. Herbicides that inhibit photosynthesis might be expected to reduce nitrogen fixation by these organisms. Tchan and Gregory (1976) used the acetylene-reduction method to study the effects of amitrole (100 ppm), diuron (40 ppm), neburon (5 ppm), and simazine (4 ppm) on the nitrogenase system of *Rhodopseudomonas capsulatus*. The three photosynthesis-inhibiting herbicides, diuron, neburon and simazine, all inhibited acetylene reduction, but these concentrations are much higher than would be found in normal agricultural practice. At realistic concentrations (0.05–0.1 ppm) acetylene reduction was slightly inhibited by diuron, but was stimulated by neburon and simazine. Amitrole, photosynthesis, actually stimulated acetylene reduction when present at 100 ppm.

Symbiotic nitrogen fixation and pesticides

Symbiotic fixation of atmospheric nitrogen is the most important process by which nitrogen levels in soils are maintained or increased; it results from a symbiosis between leguminous plants and bacteria of the genus *Rhizobium*. When considering the effects of pesticides on symbiotic nitrogen fixation, their effect on the bacterium and also on the host plant must be taken into account. The susceptibilities of *Rhizobium* to herbicides and insecticides appear to vary from strain to strain, but in general the genus is relatively resistant to these biocides. The main danger to symbiotic nitrogen fixation from insecticides and herbicides seems to be the effects of herbicides on the host legumes.

In view of the importance of legume pastures in many agricultural systems in Australia, it is surprising that so few Australian studies of pesticide effects on symbiotic nitrogen fixation have been reported. Brockwell and Robinson (1976) studied the effects of four organophosphorus insecticides on nitrogen fixation in two symbiotic associations, lucerne/*R. meliloti* and subterranean clover/*R. trifolii*, growing in agar culture. The highest rates of malathion (700 ppm), methidathion (350 ppm), phosmet (350 ppm) and dimethoate (175 ppm) caused some leaf damage, but did not affect symbiotic nitrogen-fixation.

Effects of pesticides on the rhizosphere microflora

The kinds and numbers of microorganisms in the rhizosphere are profoundly influenced by the chemical composition of the plant-root exudates. There is increasing evidence that chemicals applied to foliage can cause changes in root exudates and thus alter the rhizosphere microflora. Although rhizosphere studies have figured prominently in Australian soil research in recent years, changes in the rhizosphere as a result of application of agricultual chemicals to foliage seem to have been ignored. Useful manipulation of the rhizosphere by chemical treatment of foliage would appear to be a promising field for research.

Effects of pesticides on nitrification and ammonification

Nitrification appears to be one of the most pesticide-sensitive processes in soils; consequently, it is probably the best single indicator of detrimental effects of pesticides on the microbiological equilibrium in soil. Furthermore, mineralization of nitrogen probably has more relevance to practical agriculture than most other processes used to monitor effects of pesticides on microbial activity in soils.

Herbicides and insecticides at normal application rates seem, from published reports, unlikely to affect the mineralization of organic nitrogen to ammonia. On the other hand, there are many reports of pesticides inhibiting nitrification. The subject has been reviewed on several occasions (Alexander 1969; Helling *et al.* 1971; Greaves *et al.* 1976; Grossbard 1976). In most studies the overall oxidation of ammonia to nitrate has been measured. When a distinction has been made between ammonium oxidation and subsequent nitrite oxidation, the first step has usually been the more susceptible to inhibition (Grossbard and Marsh 1974). In spite of its importance, there seem to be no reports about the effects of herbicides and insecticides on nitrogen mineralization in Australian soils.

Soil fumigants and fungicides, cause marked changes in the soil microflora because they are selected on the basis of their toxicity to microorganisms, and because the application rates are usually much higher than for herbicides and insecticides. The majority of fumigants and fungicides are not highly specific, and nontarget organisms are usually affected to a greater or lesser extent. The nitrifying bacteria are particularly sensitive and may not re-establish for periods of 3 months or more. The literature on the indirect effects of fumigants and fungicides on the soil microflora has been reviewed frequently, e.g. Alexander (1969), Martin (1972) and Parr (1974).

Rovira (1976) fumigated a *solonized brown soil* at Roseworthy in South Australia with methyl bromide or chloropicrin at 220 kg ha^{-1}. With methyl bromide, the ammonium concentrations indicated re-establishment of a nitrifying population 14 days after fumigation; with chloropicrin however, ammonium concentrations did not decline until 46 days after treatment, and 95 days elapsed before the amounts fell to those of control plots. The counts of 'total' aerobic bacteria were scarcely affected by methyl bromide (220 kg ha^{-1}), but chloropicrin at the same rate caused a 95% drop in numbers (Ridge 1976). Ten days later, 'total' numbers of aerobic bacteria in chloropicrin-treated plots were 10 times those in untreated plots, but the composition of the bacterial flora was markedly different; fluorescent pseudomonads constituted over 80% of the aerobes in chloropicrin-treated soil, but only 10% in the control soil.

Effects of pesticides on chemical properties of soils

Pesticides that reach the soil may persist for varying periods of time. Their presence represents changes in chemical properties of the soil which may be more or less important. Significant changes are most likely to occur with fungicides and fumigants because high rates are applied; for example, fumigation with bromo or chloro compounds may give rise to high concentrations of the respective halide ions. Decomposition of organisms killed by pesticides will also alter the chemical properties of soils, for example, the flush of ammonium following fumigation. Pesticides often cause increases in soluble Ca, Mg, K, P, S, Zn, Mn and other elements, the amounts sometimes being too great to have come only from organisms killed by the pesticide.

There are a few reports of changes in chemical properties of Australian soils following fumigation, but there seem to be none on the effects of fungicides. The increase in soil ammonium following fumigation with chloropicrin (Rovira 1976) has been mentioned previously in the section on nitrification; he also found that bicarbonate-extractable phosphate had doubled in 28 days after fumigation. Wheat plants grown in the fumigated soil had higher concentrations of Ca, K, and Cl than plants grown in untreated soil, but the concentrations of these elements in soil extracts were not determined. Ladd *et al.* (1976) determined the ninhydrin-reactive compounds in the soils fumigated with chloropicrin or methyl bromide by Rovira (1976). Both fumigants increased the amount of extractable ninhydrin-reactive material. Chloropicrin was much more effective than methyl bromide, and at the rate of 220 kg ha^{-1} there was a rapid release of ninhydrin-reactive material. A high concentration was then maintained for at least 75 days, and did not fall to that of the untreated soil until 117 days after fumigation. Ladd *et al.* (1976) did not determine the proportion of ammonium nitrogen in the ninhydrin-reactive fraction, but they concluded, from unpublished measurements of soils fumigated with chloroform: that immediately following fumigation the ammonium nitrogen would account for less than 50% of ninhydrin-reactive nitrogen, but that later (day 28 to day 75) the ammonium nitrogen would account for most of, if not all, the ninhydrin-reactive nitrogen. This, however, would be inconsistent with Rovira's (1976) finding that ammonium nitrogen reached a maximum at about day 28 and then fell rapidly.

The effects of fumigation on hydrolytic activity of soil towards a peptide derivative (benzyloxy-carbonyl-phenylalanylleucine), and benzoyl-arginineamide, and casein have been measured by Ladd *et al.* (1976). With the peptide and amide substrates, methylbromide (220 kg ha^{-1}) had little effect, but chloropicrin at the same rate depressed hydrolytic activity 40–50%; this depression persisted for the duration of the experiment (320 days). The effects of the fumigants on hydrolysis of casein contrasted with the effects on hydrolysis of the low molecular weight substrates; immediately following fumigation with methyl bromide or chloropicrin at 220 kg ha^{-1}, caseinase activity was depressed; activity then increased rapidly, so that by day 28 it exceeded that in the untreated controls. Subsequently, the caseinase activities in the fumigated and untreated soils fluctuated widely until they almost ceased as the soils dried with the approach of summer. No consistent relationship was found between changes in numbers of bacteria, in ninhydrin-reactive compounds, or in enzyme activities.

Effects of soils on biological activity of pesticides

The biological activity of pesticides is affected by interactions with soil constituents. Various interactions occur, depending on the properties of the pesticides and of the soil constituents, but in most cases adsorption is involved and usually leads to reduced biological activity.

Anionic pesticides are weakly adsorbed by clay minerals but more so by soil organic matter and metallic oxides; adsorption is usually greater under acidic conditions. Non-ionic pesticides are adsorbed primarily by soil organic matter. Basic pesticides are adsorbed by both clays and organic matter, adsorption being greater under acidic conditions. Cationic pesticides are ionically bound to clay minerals and soil organic matter; in the case of the herbicides diquat and paraquat, the binding is so strong that they are largely inactivated when applied to soils.

There are few published studies of the effects of Australian soils on the biological activity of pesticides. Bowmer (1971*a*) found that the effect of atrazine on barnyard grass (*Echinochloa* spp.) in soils of the Murrumbidgee Irrigation Area was inversely related to the extent of atrazine adsorption by the soils; adsorption was related to the organic carbon content rather than to the quantity and nature of the clay minerals. Similarly, the toxicity of diuron to oat seedlings is inversely related to the organic carbon content of the soil. Using 10 soils that ranged widely in organic carbon (0.69–4.8%) and clay (0.5–61.3%) contents, collected in New South Wales and South Australia, Pillay and Tchan (1971) found a high correlation ($r = 0.876$, significant at 0.1% level) between *ED*50 values (dose giving 50% inhibition) for diuron against oats, and organic carbon contents; the correlation with clay content was not significant. Donaldson (1973) determined the *ED*50 values for picloram against wheat and lucerne in four different soils, viz., a Mallee calcareous sandy loam, a Wimmera calcareous clay, a Wimmera red-brown loam and a Goulburn Valley red-brown loam. With each plant species the toxicity was greatest in the sandy soil and least in the clay soil, with intermediate values for the loams; the organic carbon contents of these soils were not reported. The *ED*50 values ranged from 0.22 to 0.6 ppm for wheat and 0.004 to 0.02 ppm for lucerne. The minimum concentrations which reduced growth ranged from 0.1 to 0.2 ppm for wheat and 0.001 to 0.003 ppm for lucerne.

Industrial companies, the State Departments of Agriculture, and some other State Government Departments carry out extensive field trials of pesticides in Australia. These trials probably encompass a wide variety of soil types, so the data would give much information about relationships between soil type and the biological activity of pesticides. Unfortunately, most of this information remains unpublished.

Attenuation of pesticide residues

It has already been pointed out that a large proportion of agricultural pesticide applications eventually reach the soil. The axiom that prevention is better than cure, is particularly pertinent to pesticide residues in soils. Some pesticides last only a few days, some for a few weeks or months, but many may persist for several years; the inorganic components of pesticides containing copper, mercury, lead, and arsenic are virtually permanent. There are several practices which help to keep pesticide residues to a minimum, namely the use of pesticides of short persistence; increasing the effectiveness of pesticides by improving formulations and techniques of application so that smaller amounts are required; rotation of crops and pesticides; promotion or introduction of biological control agents; and changes in fertilization and cultivation practices which reduce pest populations. A combination of two or more of these techniques in a pest control program is often more effective than any single method; this allows the use of smaller amounts of pesticides and minimizes any adverse side effects. This type of approach has been called 'integrated pest control'; its successful application requires an understanding of the interactions of the various components of the ecosystem.

When a pesticide residue problem is recognized, it is desirable to attenuate the residues as quickly as possible. Burnside (1974) has reviewed some of the possible methods for dissipating pesticide residues or reducing their availability to plants, and Blair *et al.* (1976) have reviewed information on protectants that render crops more tolerant of herbicides.

Altering the availability of pesticides by adsorption

Activated carbon adsorbs many pesticides, and may be used to decrease uptake of pesticides by plants; the most useful technique has been application to seeds and the roots of transplants. Overall soil application has been used to inactivate unintentional herbicide residues, but this is generally uneconomical.

Toth (1973) found reduced activity of a number of herbicides when applied to plots covered with ash of burnt vegetation and containing carbon equivalent to about 200 kg ha^{-1}; the substituted ureas and triazines are affected more than other herbicide groups. In a more detailed study with diuron, Toth and Milham (1975) found that the capacity of paspalum ash-carbon to adsorb diuron is about one-quarter that of commercial activated-carbon. Bio-assay experiments were carried out with diuron at rates of 8, 16, and 32 kg ha^{-1} and a carbon/diuron ratio of 5 to 1 (w/w); the carbon was spread evenly on the surface before spraying with diuron. Over the three diuron rates, the mean decrease in diuron availability to barley was 95% with activated-carbon and about 80% with ash-carbon. Toth and Milham suggested that in situations where the residue from a herbicide applied to one crop might interfere with growth of a following crop, the risk could be reduced by burning crop residues and working the ash-carbon into the seed bed. On the other hand, they pointed out that the burning of crop residues and the pre-harvest burning in sugarcane could cause an undesired reduction of herbicide activity in the following crop. Hagon (1977) found that activated carbon (200 kg ha^{-1}), applied as a band or a seed pellet, protected seedlings of four Australian native grasses from the toxic effects of chlorthal-dimethyl (6 and 12 kg ha^{-1}) and diuron (4 kg ha^{-1}) which had been used to reduce weed competition.

While addition of activated carbon to soils decreases the availability of pesticides to plants, it also decreases the availability to microorganisms and thus the rates of degradation (Helweg-Andersen 1969); Moyer *et al.* (1972) found, however, that decomposition rate was affected less than pesticide activity.

Antidote alteration of herbicide activity

In 1962, Hoffmann reported that certain compounds applied to the seed protect wheat against damage by barban. Two compounds, 1,8-naphthalic anhydride and *N,N*-diallyl-2,2-dichloracetamide, have been developed commercially in the United States as herbicide antidotes. The main crop/herbicide interactions with these two antidotes have been summarized by Blair *et al.* (1976).

These substances have received little attention in Australia. In pot experiments, Swain (pers. comm.) found that rice seed treated with 1,8-naphthalic anhydride is partly protected from the effects of molinate (5 kg ha^{-1}) incorporated in the soil but not from *EPTC* (5 kg ha^{-1}); when applied as a seed dressing with methyl cellulose as an adhesive, 1,8-naphthalic anhydride reduces germination and the vigour of emerging shoots and roots.

Leaching attenuation of pesticides

The effect of leaching on the behaviour of pesticides in soils has been discussed earlier. The use of controlled leaching to dissipate residues is restricted to areas that can be irrigated.

In the irrigation areas of southern New South Wales, empty channels are sprayed with diuron in the winter. To remove surplus residues in the spring, the first discharge of water is wasted to the drainage system before irrigation is begun. Bowmer and Adeney (1978) found that the success of this operation depends on the nature of the soil; about five times more diuron was removed from the heavier-textured Willbriggie clay loam than from the lighter-textured Wamoon sandy clay. This difference is attributed to greater leaching of diuron below the soil surface in the lighter soil.

Attenuation of residues by co-metabolism

The role of co-metabolism in decomposing pesticides in soils has been discussed in a previous section.

R. B. Jackson and R. J. Swaby (unpublished results) attempted to accelerate decomposition of picloram in some South Australian soils by promoting co-metabolism. Two types of substrates were added to soils: the polymeric substances wheat straw or cellulose or chitin, to stimulate broad groups of the soil microflora; and benzenecarboxylic acids or pyridine derivatives, to stimulate more specialized groups of organisms; small but statistically significant increases in the rate of picloram disappearance were observed when wheat straw or iso-phthalic acid had been added to soil.

The biocides chloropicrin and allyl alcohol were also used to promote the activity of certain groups of organisms. Following fumigation with chloropicrin, total numbers of aerobic bacteria drop, then rise rapidly to about 10 times the original numbers; fluorescent pseudomonads constitute about 80% of this post-fumigation population (Ridge 1976). Pseudomonads are noted for their ability to metabolize aromatic compounds (Clarke and Richmond 1975). When soil is treated with allyl alcohol, *Trichoderma* spp. become dominant members of the soil microflora (Garrett 1970, p. 173), and Rieck (1969) had reported that *T. viride* decomposes picloram. Jackson and Swaby (*loc. cit.*) found no increase in the rate of picloram degradation in some South Australian soils after treating them with chloropicrin or allyl alcohol.

Nutrition and crop tolerance to pesticide residues

Although not directly related to attenuation of residues, plant nutrition may have a bearing on responses of plants to pesticide residues. The literature contains very few references to interactions between plant nutrition and responses to pesticides. Drew (pers. comm.) found that trifluralin and alachlor damage cabbages and cauliflowers growing on *krasnozem* soils at Redlands in Queensland, the symptoms resembling those of molybdenum and boron deficiencies; a relationship between trifluralin damage and molybdenum deficiency in cauliflowers was demonstrated using plants in solution culture.

Design of non-persistent pesticides

Insecticides

Many theories have been proposed to explain the insecticidal activity of DDT. Mullins (1956) suggested that several DDT molecules fit into a pore of a nerve membrane and distort adjacent pores; this distortion was supposed to permit leakage of sodium ions, thus giving rise to disrupting trains of nerve impulses in insects. Holan (1969), working at the CSIRO Division of Applied Organic Chemistry, used Mullins' theory as a basis for the rational selection of compounds to be synthesized which would be insecticidal. Generally, the prediction of active and inactive compounds was good, but there were exceptions.

By comparing projections of steric atomic models of compounds with their insecticidal activity, Holan developed a three-dimensional model for DDT-type compounds. The shape obtained in this way resembled a wedge. For insecticidal activity, the wedge needed two characteristics: the projected van der Waals diameter of the apex, i.e. the trichloromethyl group in DDT, has to be 0.6 ± 0.05 nm, but the chemical composition is unimportant; and the two phenyl rings need to have electron-donating substituents, the size of this part of the wedge being much less critical than for the apex. Since DDT has high activity, the requirement for electron-donating substituents seems open to question. While chloro groups donate electrons to the phenyl rings through tautomeric effects, the inductive effects usually predominate to such an extent that chloro groups decrease electron density all over the ring (Badger 1954 pp. 207, 211).

One of the objectives of Holan's work was the discovery of useful insecticides that would be less persistent in the environment than DDT. A satisfactory balance between activity and persistence, however, has been elusive, and many of the new compounds are not persistent enough for commercial application (Holan 1971*a*, *b*). One of the new compounds selected for possible commercial development was 1,1-*bis*(p-ethoxyphenyl)-2-nitro-propane (ENP). Kimber (1980) has studied the persistence in soils of ENP and the homologous 2-nitrobutane (ENB). At concentrations ranging from 0.4 to 40 ppm, recoveries fell to 50% in 16–64 weeks in a sand, in 2–6 weeks in a clay loam, and in 2–10 weeks in a heavy clay. Adsorption, chemical breakdown, and microbial degradation all appeared to be involved in the disappearance. Unfortunately, no 'bench-mark' chemical (Goring 1972, pp. 824, 837) such as DDT was included in those studies for comparison.

Design of non-persistent herbicides.

The herbicide picloram has been studied extensively in Australia because it controls a number of difficult weed species. The practical use of picloram, however, has been restricted because of its persistence and high activity of residues. Naik *et al.* (1972) determined the phytotoxicity and microbial degradation of a number of substituted pyridines with a view to defining the structural requirements for phytotoxicity and for degradability. A chloro substituent in position 6 of picolinic acid greatly increased recalcitrance; phytotoxicity also was increased, but an even greater increase in toxicity occurred with an additional chloro group in position 3. Unfortunately, some key compounds, particularly 3-chloropicolinic acid, were not available for that study.

The most satisfactory theory about the structural requirements for auxin activity is that of Porter and Thimann (1965). This theory proposes that auxin activity depends on the presence of a fractional positive charge situated about 0.55 nm from the negative charge of a carboxyl group. The phytotoxicity of the chloropicolinic acids studied by Naik *et al.* (1972) can be accounted for by the Porter and Thimann theory. In picloram the three chlorine atoms strongly withdraw electrons from the ring, and the resonance with the 4-amino group gives the latter a strong partial positive charge; this charge is 0.55 nm from the negative charge of the carboxylate ion. In 6-chloropicolinate, the 5 position is about 0.55 nm from the carboxylate negative charge and suffers electron depletion through the electron-attracting effect of the 6-chloro group. In 3,6-dichloropicolinate the 3-chloro group has a purely inductive effect on position 5 and strongly reinforces the effect of the 6-chloro group, thus increasing the partial positive charge at position 5 and giving a more active compound.

R. B. Jackson (unpublished work) considered the possibility of replacing the chlorine atoms in chloropicolinates with substituents that would be more susceptible to microbial attack but would still induce the necessary partial positive charge at position 5. A survey of the literature indicated that aryl-O-CH_3 is more susceptible to microbial attack than aryl-Cl. Methoxyl substitution in position 3 of picolinate would induce a partial positive charge at position 5, but the net result at position 5 of the opposing inductive and tautomeric effects of a 6-methoxyl group could not be predicted. Since the Hammett substituent constants given by Badger (1954, p. 207) indicated that a *meta* ethoxyl group would induce a stronger partial charge than a *meta* methoxyl group, ethoxy compounds were selected for the preliminary studies. 6-Ethoxy- and 3-chloro-6-ethoxy-picolinic acids were readily prepared by treating the chloro analogues with sodium ethoxide, but attempts to prepare 3,6-diethoxy-picolinic acid were unsuccessful. In a lucerne seedling bio-assay, 6-ethoxypicolinate was less active than 6-chloropicolinate, but 3-chloro-6-ethoxypicolinate was slightly more active than 3,6-dichloropicolinate (Table 52.2).

Persistence was measured in moist soils at 25 °C; the percentages of the applied doses remaining after 98 days in a sandy *terra rossa* are shown in Table 52.2. Dicamba and 2,4-D were included in these studies as examples of compounds having medium and short disappearance times, respectively.

Table 52.2. Persistence in soil and toxicity to lucerne of some chloro- and ethoxypicolinates

The soil was classified as Um6.13 by Northcote *et al.* (1975). Physical composition; coarse sand 20%, fine sand 68%, silt 3%, clay 6%, pH, 6.2; organic carbon, 1.0%; nitrogen, 0.1%. Phytotoxicity: $-\log_{10}$ lethal and non-toxic molar concentrations in a soil-free system. ND, not determined

	% remaining at 98 days	Phytotoxicity range
6-Chloropicolinate	77	3–5
6-Ethoxypicolinate	12	2–5
3,6-Dichloropicolinate	87	4–8
3-Chloro-6-ethoxypicolinate	53	5–7
Dicamba	49	4–ND
2,4-D	[A]	4–ND

[A] 2% and 0% remained at 12 and 16 days.

Conclusion

There are comparatively few published papers dealing with the behaviour and levels of pesticides in Australian soils. The companies seeking registration of pesticides and the registration authorities rely heavily on data obtained in North America and Europe. This approach, however, is inadequate because of the peculiarities of much of the Australian agricultural environment.

Apart from the molecular structure of a pesticide, the two most important factors controlling its behaviour in soils are the level of soil organic matter and the climate. Australian soils are low in organic matter, and much Australian agriculture is practised in areas where rainfall is both low and unreliable. Consequently, it is likely that predictions based on North American and European data will often be unfulfilled. Furthermore, the maintenance of adequate levels of nitrogen in many Australian soils depends on leguminous plants which, in general, are particularly sensitive to residues of many commonly used herbicides.

For the above reasons, there is an increasing need for much more information about residue levels and the behaviour of pesticides in Australian soils.

Appendix

Chemical names of pesticides mentioned in this chapter

Names recommended in Australian Standard 1719 — 1975 have been used where possible. Other names, marked†, are well-known trade or common names listed in Chemical Abstracts.

Common or trade name*	Chemical name
alachlor	2-chloro-2′, 6′-diethyl-1-*N*-(methoxymethyl) acetanilide
aldrin	1,2,3,4,10,10-hexachloro-1,4,4a,5,8,8a-hexahydro-1,4-*endo-exo*-5,8-dimethano-naphthalene
allidochlor (CDAA)	*N,N*-diallyl-2-chloroacetamide
amitrole	3-amino-*s*-triazole
atrazine	2-chloro-4-ethylamino-6-isopropyl-amino-*s*-triazine

Appendix *continued*

Names recommended in Australian Standard 1719 — 1975 have been used where possible. Other names, marked[†], are well-known trade or common names listed in Chemical Abstracts.

Common or trade name*	Chemical name
barban	4-chlorobut-2-ynyl *N*-(3-chlorophenyl) carbamate
benomyl	methyl *N*-[1-(butylcarbamoyl) benzimidazol-2-y1] carbamate
bensulide	*O,O*-diisopropyl *S*-[2- (phenylsulphonamido) ethyl] phosphorodithioate
BHC	benzenehexachloride
carbendazim (†)	methyl (benzimidazol-2-yl) carbamate
chloramben	3-amino-2,5-dichlorobenzoic acid
chlordane	1,2,4,5,6,7,8,8-octachloro-3a,4,7,7a-tetrahydro-4,7-methanoindane
chloropicrin (†)	trichloronitromethane
chlorpropham	isopropyl *N*-(3-chlorophenyl)carbamate
chlorpyrifos	*O,O*-diethyl *O*-(3,5,6-trichloro-2-pyridyl) phosphorothioate
chlorthal-dimethyl	tetrachloroterephthalic acid, dimethyl ester
coumaphos	*O,O*-diethyl *O*-(3-chloro-4-methyl-2-oxo-2*H*-1-benzopyran-7-yl) phosphorothioate
dalapon (†)	2,2-dichloropropionic acid
DDT	1,1-*bis* (*p*-chlorophenyl)-2,2,2-trichloroethane
diazinon	*O,O*-diethyl *O*-(2-isopropyl-4-methyl-6-pyrimidyl) phosphorothioate
dicamba	3,6-dichloro-2-methoxybenzoic acid
dichlobenil	2,6-dichlorobenzonitrile
dichlorvos	2,2-dichlorovinyl dimethyl phosphate
dieldrin	1,2,3,4,10,10-hexachloro-6,7-epoxy-1,4,4a,5,6,7,8a-octa-hydro-1,4-*endo-exo*-5,8-di-methanonaphthalene
dimethoate	*O,O*-dimethyl *S*-(*N*-methylcarbamoyl-methyl)phosphorodithioate
diphenamid	*N,N*-dimethyl-2,2-diphenylacetamide
diquat	9,10-dihydro-8a,10a-diazoniaphenanthrene ion
disulfoton	*O,O*-diethyl *S*-[2-(ethylthio)ethyl] phosphorodithioate
diuron	3(3,4-dichlorophenyl)-1,1-dimethylurea
Dowco 214	*O,O*-dimethyl *O*-(3,5,6-trichloro-2-pyridyl) phosphorothioate
Dowco 217	3,5,6-trichloro-2-pyridyl dimethyl phosphate
2,4-D	2,4-dichlorophenoxyacetic acid
2,2-DPA	2,2-dichloropropionic acid
EPTC	*S*-ethyl-dipropylthiocarbamate
fensulfothion	*O,O*-diethyl *O*-[*p*-(methylsulphinyl) phenyl]phosphorothioate
fenuron	1,1-dimethyl-3-phenylurea
GH-74	1,1-*bis* (*p*-ethoxyphenyl)-2-nitropropane
GH-108	1,1-*bis* (*p*-ethoxyphenyl)-2-nitrobutane
glyphosate	*N*-(phosphonomethyl)glycine
heptachlor	1,4,5,6,7,8,8-heptachloro-3a,4,7,7a-tetrahydro-4,7-methanoindene
lindane	γ-1,2,3,4,5,6-hexachlorocyclohexane
linuron	3-(3,4-dichlorophenyl)-1-methoxy-1-methylurea
malathion (†)	*O,O*-dimethyl *S*-[1,2-di(ethoxycarbonyl)-ethyl] phosphorodithioate
MCPA	4-chloro-2-methylphenoxyacetic acid
methomyl	1-(methylthio)ethylidenimino *N*-methylcarbamate
molinate	*S*-ethyl hexahydro-1 *H*-azepine-1-carbothioate
monuron	3-(*p*-chlorophenyl)-1,1-dimethylurea
naled	1,2-dibromo-2,2-dichloroethyl dimethyl phosphate
neburon	1-butyl-3-(3,4-dichlorophenyl)-1-methylurea
Nemacur (†)	*O*-ethyl *O*-[3-methyl-4-(methylthio)phenyl]*N*-isopropyl phosphoramidate
N-Serve	2-chloro-6-trichloromethylpyridine
paraquat	1,1 -dimethyl-4,4 dipyridylium ion (as dichloride or bis methyl sulfate salts)
parathion	*O,O*-diethyl *O-p*-nitrophenyl phosphorothioate
phorate	*O,O*-diethyl *S*-[(ethylthio)methyl] phosphorodithioate
phosmet	*O,O*-dimethyl *S*-(phthalimido-methyl) phosphorodithioate
picloram	4-amino-3,5,6-trichloropicolinic acid
prometryn	2,4-*bis*(isopropylamino)-6-methylthio-*s*-triazine
propazine	2-chloro-4,6-*bis*(isopropylamino)-*s*-triazine
propham (IPC)	isopropyl N-phenylcarbamate
simazine	2-chloro-4,6-*bis*(ethylamino)-*s*-triazine
sulfallate (CDEC)	*S*-(2-chloroallyl)diethyldithiocarbamate
TCA	trichloroacetic acid
Terracur (†)	*O,O*-diethyl *O*-[*p*-(methyl-sulphinyl)phenyl]phosphorothioate
thiabendazole	2-(4'-thiazolyl)benzimidazole
trifluralin	*a,a,a*-trifluoro-2, 6-dinitro-*N,N*-dipropyl-*p*-toluidine
2,4,5-T	2,4,5-trichlorophenoxyacetic acid
2,3,6-TBA	2,3,6-trichlorobenzoic acid

References

Alexander, M. (1969). Microbial Degradation and Biological Effects of Pesticides in Soil. *In* 'Natural Resources Research. IX. Soil Biology. Reviews of Research'. pp.209-40. (UNESCO: Paris.)

Anon. (1973). Roundup (R) – Herbicide formulation of isopropylamine salt of glyphosphate (N-phosphonomethylglycine). Postemergent herbicide. Monsanto Agricultural Division, St. Louis, Missouri, Technical Bulletin Mon-0573-2-73.

Australian Academy of Science (1972). Report No. 14. The use of DDT in Australia. February, 1972.

Australian Department of Agriculture (1974). Agricultural and Veterinary Chemicals – withholding periods, maximum residue limits and poisons schedules. P.B. 158A.

Bachelard, E. P. and Johnson, M. E. (1969). A study of the persistence of herbicides in soil. *Aust. For.* **33**, 19-24.

Baker, H. J. and Bertus, A. L. (1976). Research Report for 1976 of Biological and Chemical Research Institute, New South Wales Department of Agriculture, p.47.

Badger, G. M. (1954). 'The Structures and Reactions of the Aromatic Compounds'. (Cambridge University Press: Cambridge.)

Beynon, K. I., and Wright, A. N. (1972). The fates of the herbicides chlorthiamid and dichlobenil in relation to residues in crops, soils and animals. *Residue Rev.* **43**, 23-53.

Blair, A. M., Parker, C., and Kasasian, L. (1976). Herbicide protectants and antidotes – a review. PANS **22**, 65-74.

Bowmer, K. H. (1971*a*). Atrazine adsorption and control of *Echinochloa* in Australian soils. *Proc. Weed Soc. N.S.W.* **4**, 12-20.

Bowmer, K. H. (1971*b*). Barnyard grass control in maize and sorghum. *Farmers' Newsletter, Large Area, Griffith.* No. 79: 12-7.

Bowmer, K. H. (1972). Measurement of residues of diuron and simazine in an orchard soil. *Aust. J. Exp. Agric. Anim. Husb.* **12**, 535-9.

Bowmer, K. H. (1973*a*). The effects of temperature and soil type on atrazine persistence. Proceedings of the Fourth Asian-Pacific Weed Science Society Conf., Rotorua, New Zealand, pp.242-7.

Bowmer, K. H. (1973*b*). Herbicides for the control of barnyard grass (*Echinochloa crus-galli*) in maize. Proc. Fourth Asian-Pacific Weed Science Society Conf., Rotorua, New Zealand, pp.378-83.

Bowmer, K. H. and Adeney, J. A. (1978). Residues of diuron and phytotoxic degradation products in aquatic situations. II. Diuron in irrigation water. *Pestic. Sci.* **9**, 354-364.

Bowmer, K. H., O'Loughlin, E. M., Shaw, K. and Sainty, G. R. (1976) Residues of dichlobenil in irrigation water. *J. Environ. Qual.* **5**, 315-9.

Brockwell, J., and Robinson, A. C. (1976). Effects of commercial organophosphorus insecticides on the growth in culture, viability in seed pellets, and symbiotic nitrogen fixation of *Rhizobium* spp. CSIRO (Aust.) Div. Plant Ind. Field Stn Rec. No. 15, pp.15-26.

Burnside, O. C. (1974). Prevention and Detoxification of Pesticide Residues in Soils. *In* 'Pesticides in Soil and Water'. (Ed. W. D. Guenzi). pp.387-412. (Soil Science Soc. America: Madison.)

Campacci, E. F., New, P. B. and Tchan, Y. T. (1977). Isolation of amitrole-degrading bacteria. *Nature (London)* **266**, 164-5.

Campbell, M. H. (1974). Effects of glyphosate on the germination and establishment of surface-sown pasture species. *Aust. J. Exp. Agric. Anim. Husb.* **14**, 557-60.

Campbell, M. H. (1976). Effect of timing of glyphosate and 2,2-DPA application on establishment of surface-sown pasture species. *Aust. J. Exp. Agric. Anim. Husb.* **16**, 491-9.

Casarett, L. D. and Doull, J. (Eds) (1975). 'Toxicology'. (Macmillan: New York.)

Clarke, P. H., and Richmond, M. H. (1975). 'Genetics and Biochemistry of *Pseudomonas.*' (Wiley – Interscience: London.)

Cohen, J. M., and Pinkerton, C. (1966). 'Organic Pesticides in the Environment'. (Ed. R. F. Gould). Advances in Chemistry Series, No. 60. pp.163-76. (Am. Chem. Soc. Publications: Washington.)

Commonwealth Department of Health (1978). Pesticides–synonyms and chemical names, fourth edition, (Australian Government Publishing Service: Canberra.)

Crosby, D. G. (1976*a*). Nonbiological Degradation of Herbicides in the Soil. *In* 'Herbicides – Physiology, Biochemistry, Ecology'. (Ed. L. J. Audus). 2nd edn, Vol. 2, pp.65-97. (Academic Press: London.)

Crosby, D. G. (1976*b*). Herbicide Photodecomposition. *In* 'Herbicides – Chemistry, Degradation and Mode of Action'. (Eds P. C. Kearney and D. D. Kaufman). 2nd edn, Vol. 2, pp.835-90. (Marcel Dekker: New York.)

Crosby, D. G. and Wong, A. S. (1977). Environmental degradation of 2,3,7,8-tetrachlorodibenzo-*p*-dioxin (TCDD). *Science* **195**, 1337-8.

Donaldson, T. W. (1973). Phytotoxicity of picloram to wheat and lucerne in soil and nutrient solution. Proc. Fourth Asian-Pacific Weed Science Society Conf., 239-41.

Edwards, C. A. (1966). Insecticide residues in soils. *Residue Rev.* **13**, 83-132.

Ellis, F. B. and Lynch, J. M. (1977). Problems associated with the persistence of plant residues in direct-drilled crops. Agricultural Research Council Letcombe Laboratory Annual Report 1976, pp.53-4.

Farmer, W. J., Igue, K., Spencer, W. F. and Martin, J. P. (1972). Volatility of organo-chlorine insecticides from soil. I. Effect of concentration, temperature, air flow rate, and vapour pressure. *Soil Sci. Soc. Am. Proc.* **36**, 443-7.

Fung, K. and Uren, N. C. (1975). Annual Report, Division of Agricultural Chemistry, Department of Agriculture Victoria, p.69.

Garrett, S. D. (1970). 'Pathogenic Root-infecting Fungi'. p.173. (Cambridge University Press: Cambridge.)

Gilbert, W. S. and Singh, G. (1976). Research Report for 1976 of Biological and Chemical Research Institute, New South Wales Department of Agriculture, p.46.

Gilbert, W. S., Ferris, I. G. and Lewis, C. E. (1976). Research Report for 1976 of Biological and Chemical Research Institute, New South Wales Department of Agriculture, p.46.

Goring, C. A . I. (1972). Agricultural Chemicals in the Environment: A Quantitative Viewpoint. *In* 'Organic Chemicals in the Soil Environment'. (Eds C. A. I. Goring and J. W. Hamaker). (Marcel Dekker: New York.)

Greaves, M. P., Davies, H. A., Marsh, J. A. P. and Wingfield, G. I. (1976). Herbicides and Soil Microorganisms. CRC Critical Reviews in Microbiology, Vol. 5, Issue 1 (Eds A. I. Laskin and H. Le Chevalier.)

Green, R. E. (1974). Pesticide–clay–water Interactions. *In* 'Pesticides in Soil and Water'. (Ed. W. D. Guenzi). pp.3-37. (Soil Science Soc. America: Madison.)

Grossbard, E. (1976). Effects on the Soil Microflora. *In* 'Herbicides – Physiology, Biochemistry, Ecology'. (Ed. L. J. Audus). 2nd edn, Vol. 2, p.99-147. (Academic Press: London.)

Grossbard, E. and Marsh, J. A. P. (1974). The effect of seven substituted urea herbicides on the soil microflora. *Pestic. Sci.* **5**, 609-23.

Hagon, M. W. (1977). Effects of competition, herbicides and activated carbon on establishment of Australian grasses. *Weed Res.* **17**, 297-301.

Hall, R. C., Giam, C. S. and Merkle, M. G. (1968). The photolytic degradation of picloram. *Weed Res.* **8**, 292-7.

Hamaker, J. W. (1972). Diffusion and Volatilization. *In* 'Organic Chemicals in the Soil Environment'. (Eds C. A. I. Goring and J. W. Hamaker). Vol. 1. pp.341-97. (Marcel Dekker: New York.)

Hamaker, J. W. and Thompson, J. M. (1972). Adsorption. *In* 'Organic Chemicals in the Soil Environment'.(Eds C. A. I. Goring and J. W. Hamaker.) Vol. 1. pp.49-143. (Marcel Dekker: New York.)

Hamaker, J. W., Youngson, C. R. and Goring, C. A. I. (1967). Prediction of the persistence and activity of Tordon herbicide in soils under field conditions. *Down to Earth* **23**, 30-36.

Hamaker, J. W. Youngson, C. R. and Goring, C. A. I. (1968). Rate of detoxification of 4-amino-3,5,6-trichloropicolinic acid in soil. *Weed Res.* **8**, 46-57.

Hance, R. J. (1976). Adsorption of glyphosate by soils. *Pestic. Sci.* **7**, 363-6.

Hance, R. J. and McKone, C. E. (1971). Effect of concentration on the decomposition rates in soil of atrazine, linuron and picloram. *Pestic. Sci.* **2**, 31-4.

Hartley, G. S. (1976). Physical Behaviour in the Soil. *In* 'Herbicides – Physiology, Biochemistry, Ecology'. (Ed. L. J. Audus). 2nd edn, Vol. 2, pp.1-28. (Academic Press: London.)

Helling, C. S., Kearney, P. C. and Alexander, M. (1971). Behaviour of pesticides in soils. *Advan. Agron.* **23**, 147-240.

Helweg-Andersen, A., (1969). Influence of charcoal-adsorbed herbicides on micro-organisms. *Weed Res.* **9**, 254-7.

Heritage, A. D., and MacRae, I. C. (1977*a*). Identification of intermediates formed during the degradation of hexachlorocyclohexanes by *Clostridium sphenoides. Appl. Environ. Microbiol.* **33**, 1295-7.

Heritage, A. D., and MacRae, I. C. (1977*b*). Degradation of lindane by cell-free preparations of *Clostridium sphenoides. Appl. Environ. Microbiol.* **34**, 222-4.

Hiltbold, A. E. (1974). Persistence of Pesticides in Soil. *In* 'Pesticides in Soil and Water'. (Ed. W. D. Guenzi). pp.203-22. (Soil Science Soc. America: Madison.)

Hilton, H. W. and Yuen, Q. H. (1963). Adsorption and leaching of herbicides in Hawaiian sugarcane soils. *J. Agric. Food Chem.* **14**, 86-90.

Hoffmann, O. L. (1962). Chemical seed treatments as herbicidal antidotes. *Weeds* **10**, 322-3.

Holan, G. (1969). New halocyclopropane insecticides and the mode of action of DDT. *Nature (London)* **221**, 1025-9.

Holan, G. (1971*a*). Rational design of degradable insecticides. *Nature (London)* **232**, 644-7.

Holan, G. (1971*b*). Rational design of insecticides. *Bull. W. H. O.* **44**, 355-62.

Jackson, R. B. (1971). Effect of composition of the medium on utilization of allyl alcohol by *Trichoderma viride. J. Gen. Appl. Microbiol.* **17**, 399-408.

Jackson, R. B. (1973). The metabolism of allyl alcohol in *Trichoderma viride. J. Gen. Appl. Microbiol.* **19**, 41-54.

Jenkins, I. D. (1978). The chemistry of 2,4,5-T. *Chem. Aust.* **45**, 455-6.

Jensen, H. L. (1963). Carbon nutrition of some micro-organisms decomposing halogen-substituted aliphatic acids. *Acta Agric. Scand.* **13**, 404-12.

Jordan, L. S., Farmer, W. J., Goodin, J. R. and Day, B. E. (1970). Volatilization and non-biological degradation of triazine herbicides *in vitro* and in soils. *Residue Rev.* **32**, 267-86.

Kaufman, D. D., Still, G. G., Paulson, G. D., Bandal, Suresh K. (Eds) (1976). American Chemical Society Symposium Series, No. 29 (1976). 'Bound and Conjugated Pesticide Residues'.

Kaufman, D. D., and Kearney, P. C. (1976). Microbial Transformations in the Soil. *In* 'Herbicides – Physiology, Biochemistry, Ecology'. (Ed. L. J. Audus). 2nd edn, Vol. 2, pp.29-64. (Academic Press: London.)

Kaufman, D. D, Plimmer, J. R., Kearney, P. C., Blake, J. and Guardia, F. S. (1968). Chemical versus microbial degradation of amitrole in soil. *Weed Sci.* **16**, 266-72.

Kearney, P. C., Nash, R. G. and Isensee, A. R. (1969). *In* 'Chemical Fallout: Current Research on Persistent Pesticides'. (Eds M. W. Miller and G. G. Berg). (Charles C. Thomas: Springfield, Ill.)

Kimber, R. W. L. (1973). Phytotoxicity from plant residues. III. The relative effect of toxins and nitrogen immobilization on the germination and growth of wheat. *Plant Soil* **38**, 543-55.

Kimber, R. W. L. (1980). An evaluation of the persistence in soil of two non-chlorinated insecticides analogous to DDT. *Pestic. Sci.* 11, 533-45.

Ladd, J. N., Brisbane, P. G., Butler, J. H. A. and Amato, M. (1976). Studies on soil fumigation – III. Effects on enzyme activities, bacterial numbers and extractable ninhydrin reactive compounds. *Soil Biol. Biochem.* **8**, 255-60.

Letey, J. and Farmer, W. J. (1974). Movement of Pesticides in Soil. *In* 'Pesticides in Soil and Water'. (Ed. W. D. Guenzi). pp.67-97. (Soil Science Soc. America: Madison.)

Letey, J. and Oddson, J. K. (1972). Mass Transfer. *In* 'Organic Chemicals in the Soil Environment'. (Eds C. A. I. Goring and J. W. Hamaker). Vol. 1. pp.399-440. (Marcel Dekker: New York.)

McDonald, F. J. D. and Madsen, J. F. (1973). DDT residues in Soil and water in the Narrabri district. *Residue* **2**,(4), 10-7. (Australian Government Analytical Laboratory: Melbourne.)

Mackenzie, K. A., and MacRae, I. C. (1972). Tolerance of the nitrogen-fixing system of *Azotobacter vinelandii* for four commonly used pesticides. *Antonie van Leeuwenhoek* **38**, 529-35.

MacRae, I. C., and Celo, J. S. (1974). The effects of organo-phosphorus pesticides on the respiration of *Azotobacter vinelandii. Soil Biol. Biochem.* **6**, 109-11.

MacRae, I. C., Raghu, K., and Bautista, E. M. (1969). Anaerobic degradation of the insecticide lindane by *Clostridium* sp. *Nature (London)* **221**, 859-60.

Martin, J. P. (1972). Side Effects of Organic Chemicals on Soil Properties and Plant Growth. *In* 'Organic Chemicals in the Soil Environment'. (Eds C. A. I. Goring and J. W. Hamaker). Vol. 2 pp.733-92. (Marcel Dekker: New York.)

Moyer, J. R., Hance, R. J. and McKone, C. E. (1972). The effect of adsorbents on the rate of degradation of herbicides incubated with soil. *Soil Biol. Biochem.* **4**, 307-11.

Mullins, L. J. (1956). *In* 'Molecular Structure and Functional Activity of Nerve Cells'. (Eds R. G. Grenfell and L. J. Mullins). p.123. (Am. Inst. for Biol. Sci.: Washington.)

Naik, M. N. Jackson, R. B., Stokes, J. and Swaby, R. J. (1972). Microbial degradation and phytotoxicity of picloram and other substituted pyridines. *Soil Biol. Biochem.* **4**, 313-23.

Nomura, N. S. and Hilton, H. W. (1977). The adsorption and degradation of glyphosate in five Hawaiian sugar cane soils. *Weed Res.* **17**, 113-21.

Northcote, K. H., Hubble, G. D., Isbell, R. F., Thompson, C. H. and Bettenay, E. (1975). 'A Description of Australian Soils'. (CSIRO: Melbourne.)

O'Neill, D. K., Barlow, E. W. R. and Ferris, I. G. (1976). Research Report, Biological and Chemical Research Institute, New South Wales Dept. Agriculture, p.46.

Parr, J. F. (1974). Effects of Pesticides on Micro-organisms in Soil and Water. *In* 'Pesticides in Soil and Water'. (Ed. W. D. Guenzi). (Soil Science Soc. America: Madison.)

Pesticides Branch, Dep. Primary Industry. P.B. 296. Agricultural and veterinary chemicals. Statistics of production, 1963-1975. (AGPS: Canberra.)

Pillay, A. R. and Tchan, Y. T. (1971). Phytotoxicity of diuron in some Australian soils. *Proe. Weed Soc. N.S.W.* **4**, 21-4.

Pillay, A. R. and Tchan, Y. T. (1972). Study of soil algae. VII. Adsorption of herbicides in soil and prediction of their rate of application by algal methods. *Plant Soil* **36**, 571-94.

Porter, W. L. and Thimann, K. V. (1965). Molecular requirements for auxin action I. Halogenated indoles and indoleacetic acid. *Phytochemistry.* **4**, 229-43.

Ridge, E. H. (1976). Studies on soil fumigation – II. Effects on bacteria. *Soil Biol. Biochem.* **8**, 249-53.

Rieck, C. E. (1969). Ph.D. Thesis, University of Nebraska, Lincoln, U.S.A.

Rosen, J. D. (1971). Photodecomposition of Organic Pesticides. *In* 'Organic Compounds in Aquatic Environments'. (Eds S. J. Faust and J. V. Hunter). pp.425-38. (Marcel Dekker: New York.)

Rovira, A. D. (1976). Studies on soil fumigation – I. Effects on ammonium, nitrate and phosphate in soil and on the growth, nutrition and yield of wheat. *Soil Biol. Biochem.* **8**, 241-7.

Spencer, W. F., Farmer, W. J. and Cliath, M. M. (1973). Pesticide volatilization. *Residue Rev.* **49**, 1-47.

Standards Association of Australia (1975). Recommended common names for pesticides, Australian Standard 1719 – 1975.

Swain, D. J. (1974). Molinate for the control of *Echinochloa* spp. in rice in New South Wales. *Weed Res.* **14**, 185-91.

Swarbrick, J. T. (1978). The Australian Weed Control Handbook 1979, Fourth Edition, (Plant Press: Toowoomba, Queensland.)

Tchan, Y. T. and Gregory, M. (1976). Influence of photosynthesis inhibiting herbicides on nitrogen fixation by photosynthetic bacteria. *Non-symbiotic Nitrogen Fixation Newsl.* **4**, 60-5.

Torstensson, N. T. L. and Aamisepp, A. (1977). Detoxification of glyphosate in soil. *Weed Res.* **17**, 209-12.

Toth, J. (1973). Burning reduces herbicide effect. *Agric. Gaz. N.S.W.* **84**, 62.

Toth, J. and Milham, P. J. (1975). Activated-carbon and ash-carbon effects on the adsorption and phytotoxicity of diuron. *Weed Res.* **15**, 171-6.

Verloop, A. (1972). Fate of the herbicide dichlobenil in plants and soil in relations to its biological activity. *Residue Rev.* **43**, 55-103.

Weed Science Society of America (1974). Herbicide Handbook, 3rd edn, pp.252-5.

Weed, S. B. and Weber, J. B. (1974). Pesticide–Organic Matter Interactions. *In* 'Pesticides in Soil and Water'. (Ed. W. D. Guenzi). pp.39-66. (Soil Sci. Soc. Am.: Madsion.)

Weiss, P. W. (1971). Leaching properties of simazine and diuron in a Murray Valley sand in relation to weed control in citrus. *Proc. Weed Soc. N.S.W.* **4**, 25-8.

Wells, G. J. (1972). Residual effect of some herbicides on *Medicago* species in the Victorian Mallee. *Aust. J. Exp. Agric. Anim. Husb.* **12**, 181-84.

Welsh, N. S. (1966). Legume establishment after use of picloram for weed control. *J. Aust. Inst. Agric. Sci.* **32**, 223-24.

Wilkinson, A. T. S., Finlayson, D. G. and Morley, H. V. (1964). Toxic residues in soil 9 years after treatment with aldrin and heptachlor. *Science* **143**, 681-82.

Wood, P. A. and MacRae, I. C. (1974). The effect of several organophosphorus insecticides upon the acetylene-reduction activity of *Azotobacter vinelandii. Bull. Environ. Contam. Toxicol.* **12**, 26-31.

Wood, P. A. and MacRae, I. C. (1977). Reduction of fensulfothion to fensulfothion sulphide by *Klebsiella pneumoniae. Appl. Environ. Microbiol.* **34**, 247-50.

Chapter 53
Management of salt-affected soils

J. Loveday and B. J. Bridge

In Australia, the management of salt-affected soils is a matter of considerable significance because of their wide and extensive distribution throughout the continent (Fig. 53.1). Northcote and Skene (1972) estimate that together saline and sodic soils cover about one third (2.4×10^6 km^2) of the total area of Australia. Many kinds of soil are involved, the major ones being *solonchaks, solonized brown soils, desert loams, grey-brown* and *red calcareous soils, grey, brown and red clays, red-brown earths, soloth, solodized-solonetz* and *solodic soils*. Reference to their morphology is made in Chapter 3, and discussion of their occurence and genesis can be found in Chapter 9.

The saline soils indicated on Fig. 53.1 occur mainly in the more arid interior, where extensive grazing of native vegetation is the main form of land use, and soil management *per se* is not generally feasible. On the other hand, sodic soils, many of which have saline subsoils, are especially common in south-western and southern and eastern Australia. For climatic and topographic reasons, these are areas in which a significant part of the nation's mixed farming (involving cereal cropping and grazing of native and improved pastures) and irrigation is practised. Considerable disturbance of the natural ecosystems has usually resulted, with consequent changes to hydrologic regimes; salt formerly distributed in deeper subsoil and substrate layers in the catchments and alluvial plains has become mobilized and concentrated in certain vulnerable parts of the landscape — the process of secondary salinization.

This chapter is concerned with the management and reclamation of soils suffering secondary salinization in both irrigated and dryland situations; it also deals with problems arising directly from surface and subsoil sodicity in cropped and grazed lands.

In general terms, these problems have counterparts elsewhere in the world, but the emphasis appears to be different, especially in relation to northern hemisphere soils. In the case of irrigated soils, this arises because much of Australia's irrigation water is applied to relatively impermeable clays and duplex (texture contrast) soils, rather than to the more permeable silty and sandy soils as commonly used in India, China, Central Asia, Eastern Europe, and the U.S. Irrigated clays of the Sudan Gezira (Greene 1928) apparently have features and problems in common with those of Australia. In the dry-land context, the widespread use of sodic duplex soils for farming operations in Australia has led to hillside seepage, valley salting, tunnel erosion, and scalding. Such degradation phenomena have received scant attention in the world literature, possibly because they are of less significance elsewhere.

Although this chapter deals with agricultural aspects of salinity and sodicity, it is worth noting that there are also important implications for engineering structures. The effect of electrolyte content on the swelling characteristics of clays, and hence on foundation stability, is dealt with in Chapter 54. Other effects include: increased rates of corrosion of steel structures embedded in saline soils, e.g. steel culverts (Crawford and Laimer 1968), and instability in embankments and excavations built in sodic soils (Ingles and Metcalf 1972).

Concepts and criteria for salinity and sodicity

Adequate diagnosis is necessary before appropriate management can be formulated. The brief discussion which follows mentions some of the concepts and criteria used in diagnosis, but not the methods. The latter can be found elsewhere, for example in Loveday (1974*a*).

The usual definition of the term 'salt-affected soil' is, 'that soil which contains sufficient soluble salt and/or exchangeable sodium to interfere with growth of most crop plants'. The interference may be, in the case of soluble salts, by way of osmotic effects on water uptake, or nutritional imbalance, or toxicity caused by specific ions; in the case of sodicity, it is particularly due to associated adverse physical properties: e.g. surface crusting, reduced permeability to water and air, and increased mechanical resistance to root penetration.

Plant species vary greatly in their sensitivity to saline conditions, and this offers opportunity for breeding and selection to increase their tolerance. Though this is an important avenue of research it is outside the scope of this chapter.

Besides species differences, plant responses to salt are conditioned by a range of soil and environmental

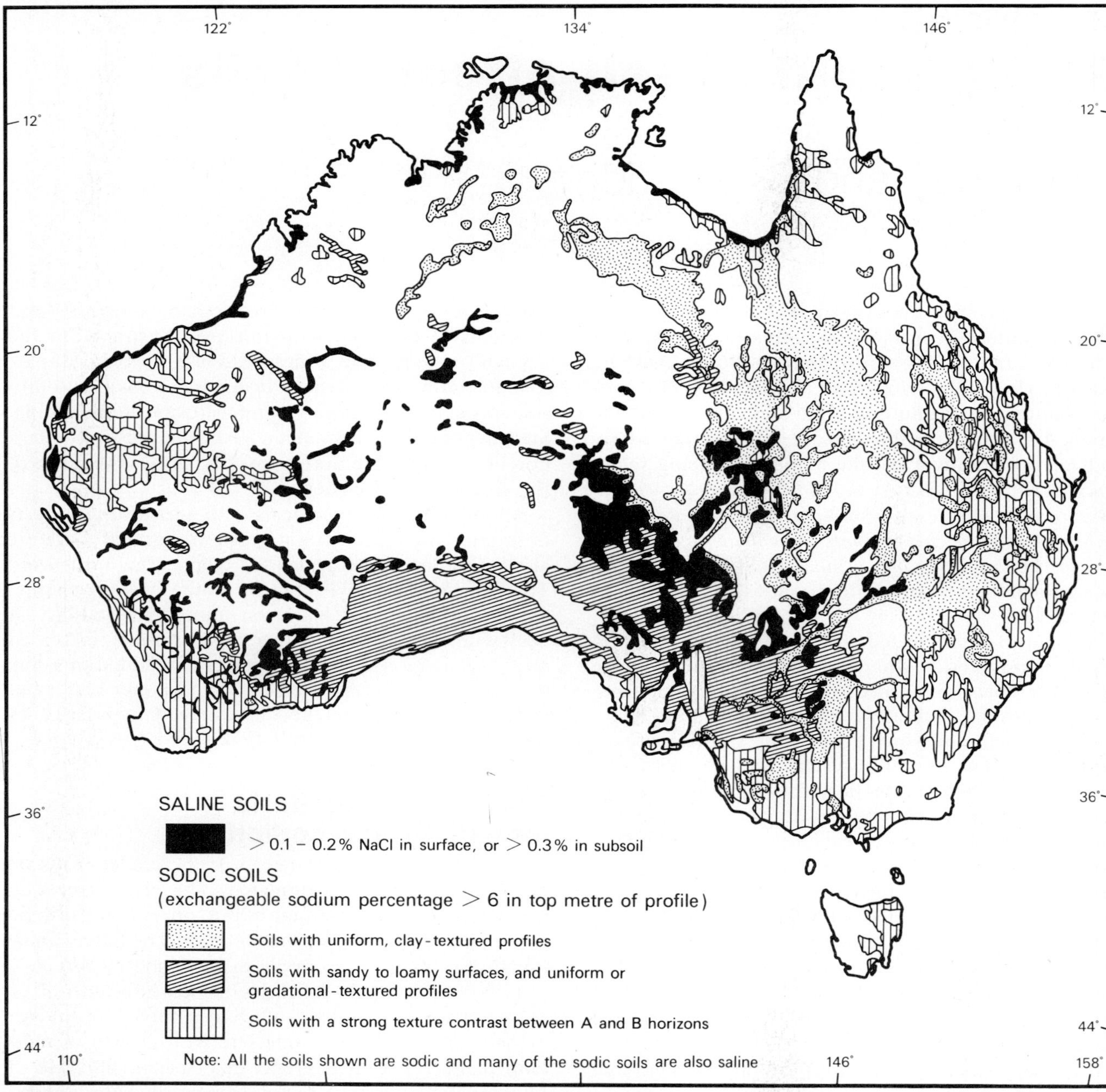

Fig. 53.1. The distribution of saline and sodic soils in Australia (after Northcote and Skene 1972).

factors: evapotranspiration demand, moisture regime of the soil, the composition of the salts, nutrient status, and mechanical composition of the soil profile. This means that ' there can be no universal definition of salt-affected soils in precise chemical levels of salinity, alkalinity and sodicity that satisfies all agricultural situations' (Northcote and Skene 1972).

Nevertheless, many attempts have been made to set criteria, either for individual crops or, in a more general sense, for delineating soils. Maas and Hoffman (1977) have recently reviewed the literature relating to the former, and have produced, for a large number of agricultural crops, 'best estimates' of threshold salinity levels and yield decrease per unit salinity increase beyond the threshold; salinity was expressed as the electrolytic conductivity of the saturation extract (EC_e) of the active root zone.

An example of a more general set of criteria is the one established by Northcote and Skene (1972) for distinguishing classes of saline and sodic soils in Australia.

Their three salinity classes are based on chloride levels both of surface and subsoil; the values they adopted for the surface soil limits (0.1% sodium chloride* for loams and coarser soils, and 0.2% in clay loams and clays) correspond roughly to an EC_e of 4 mS cm^{-1} at 25^{0}C, the value specified for saline soils by the United States Salinity Laboratory Staff (1954); the criterion for subsoil salinity is 0.3% sodium chloride or greater. For the three sodicity classes the criterion is the exchangeable sodium percentage (ESP), non sodic and sodic and strongly sodic classes being separated at ESP values of 6 and 15. The lower value of 6 was suggested by general Australian experience as being associated with the onset of the hardsetting characteristics of surface soils and reduced permeability. The value of 15, originally suggested by the United States Salinity Laboratory Staff (1954), is widely accepted as a value above which adverse physical properties are exhibited.

Management options

Before dealing with specific situations, the broad management options available for saline and sodic soils will be considered briefly. These fall into two categories; in the first are treatments aimed at directly changing the level of salinity or sodicity of the soil itself; in the second category are measures to change the hydrology of the soil or the landscape in which the salt-affected soil occurs, and so to indirectly manage salinity.

Modifying salinity and sodicity of the soil

When reclaiming saline soils, the objective is to leach salts from the profile and yet retain satisfactory hydraulic conductivity. This is achieved if there is a source of calcium sufficient to maintain adequate exchangeable calcium in the soil during the leaching process. In the absence of sufficient calcium, sodium occupies exchange sites, and hydrates as the electrolyte level is reduced; hydraulic conductivity declines until the soil freezes up. Even a thin layer of sodic soil at or near the surface can effectively prevent further leaching of salts from the profile.

In sodic soils that have attained their low hydraulic conductivity through normal soil forming processes, the problem is one of ameliorating existing undesirable physical properties. Affected surface soils are poorly structured, are difficult to prepare as seed beds, easily form crusts that cause poor emergence of seedlings, and have low infiltration and hence low water storage. In texture differentiated profiles (duplex soils), sodic subsoils of low permeability can cause waterlogging of surface horizons, restricted root development, and proneness to erosion. Permanent improvement requires replacement of exchangeable sodium by calcium, and removal of the sodium and any other soluble salts from the profile.

Thus for both saline and sodic soil reclamation, leaching is required, and the percolating water (at least in the upper soil horizons) must have a cation balance favouring and ensuring adequate calcium status of the exchange complex. Where sufficient soluble calcium is not present in the surface horizons, it must be supplied; the options include direct application of calcium compounds (e.g. gypsum, calcium carbonate, calcium chloride, calcium polysulphide), or deep tillage to bring subsoil horizons rich in gypsum or calcium carbonate to the surface. Where irrigation is practised, the calcium source may be dissolved in the normal irrigation water (Davidson and Quirk 1961) or in successive dilutions of more salty water (Reeve and Doering 1966). Generally, the more soluble the calcium source, the more efficient and rapid is the reclamation. The hydrolysis of calcium carbonate provides calcium only slowly (Bower and Goertzen 1958), unless the process is hastened by adding an acid amendment e.g. sulphur or sulphuric acid. Carbon dioxide produced by plant roots may play only a minor role in the hydrolysis of calcium carbonate (Goertzen and Bower 1958).

The leaching necessary for reclamation is not easily controlled under dryland conditions, but deliberate water harvesting, with ponding on areas to be reclaimed, can be used to increase the local effectiveness of rainfall.

With irrigation however, the control of leaching over extensive areas becomes possible. The amount of water required to reclaim a saline soil will depend on soil properties, especially hydraulic conductivity, and on methods and rates of water application. Rhoades (1974) states, that for continuously ponded leaching of highly saline soils, 30 cm of good quality water for each 30 cm depth of soil is usually adequate to reduce salinity by 70–80% and to permit crops to be grown. Intermittent application reduced by about one third the water required to achieve the same leaching (Miller *et al.* 1965); even more efficient leaching can be achieved by sprinkler irrigation (Nielsen *et al.* 1966). Natural or artificial drainage to convey water away from the root zone is a prerequisite for leaching to be successful. Where artificial drains are installed, their spacing and placement depth affect the degree of reclamation achieved (Talsma 1967).

Having reclaimed a site, the problem becomes one of maintaining a salt balance in the profile at a level which does not depress crop production. Water additional to

*Because chloride dominates the anions and sodium the cations in many Australian saline soils, it has been usual to analyze for chloride and to express it as sodium chloride. For convenience of the reader, the following conversion table is given:
% chloride x 10^4 = ppm
% chloride x 1.65 = % sodium chloride
% chloride ÷ 0.0355 = mequiv. $100g^{-1}$
% chloride x 10^4 x bulk density = $g/10^6$ cm^3
The units used in this chapter are those given by the particular author whose data are being quoted.

that required for evapotranspiration by the crop must be supplied to leach salts added by the irrigation water; this additional water is the leaching requirement (United States Salinity Laboratory Staff 1954). The desirability of keeping this requirement to a minimum, and thus also minimizing deep drainage and return of salt water to river systems, has prompted recent reappraisal of the procedures for estimating the leaching requirement. Rhoades (1974), in his review, has recommended a procedure that gives the desired depth of both irrigation and drainage water; the procedure takes into account the salinity of the irrigation water, the consumptive use by the crop, and its salt tolerance. The prospect of achieving nominated amounts of irrigation and drainage is greatly enhanced by use of sprinkler or trickle irrigations systems, which allow close control over water supply.

Modifying the hydrology of the soil and its landscape

The most obvious form of hydrologic modification is drainage to control groundwater levels. There is a critical water table depth above which capillary flow to the surface allows soil salinization to occur, and for soils of intermediate texture this depth is greater (in the vicinity of 2 m) than for either fine-textured soils of low permeability or coarse-textured sands of high permeability (Talsma 1963). Besides being dependent on soil physical properties, the critical depth is also a function of groundwater salinity, and climatic and crop factors influencing evapotranspiration and salt tolerance.

In the absence of effective drainage, the trend towards salinization is almost universal in irrigated soils. However, a great many drainage systems to suit all manner of conditions are in common use, including: open ditches, surface field drains, mole and tile drains, and tubewells (e.g. see Donnan and Schwab 1974).

Perhaps the greater problem associated with drainage, especially in large irrigation schemes such as those in the Murray Valley, is the need to provide for disposal of saline drainage water; re-use after dilution, evaporation in large holding basins, placement in deep aquifers, and outfall to the ocean, are some of the options usually considered. Of course the disposal problem is lessened to the extent that the volume of drainage water can be reduced by minimizing leaching, i.e. the leaching requirement consistent with minimal salinity damage to crops.

Generally, in dryland situations, major drainage works are not economically feasible. Sometimes a single ditch or tile drain can be installed to intercept foreign groundwater (Collis-George and Evans 1964); local surface drains may be used to protect saline areas from further additions of runoff water, and deep-rooting vegetation encouraged on the catchment areas (Cope 1958). By using more water where it falls, the local hydrology may be gradually changed, thus lowering water tables and allowing leaching and reclamation of salt patches.

Reclaiming irrigated saline soils

Reclaiming irrigated horticultural soils

Generally, the soils that have been used for irrigated horticulture are relatively permeable, either because of their coarse or medium textures (e.g. many *solonized brown soils* and aeolian *siliceous sands* associated with prior streams), or because of the subplastic nature of their clay subsoils (e.g. some *red-brown earths* (Talsma and Flint 1958)). Perhaps the most extensive areas of slowly permeable soils used for horticulture are the *red-brown earths* of the Goulburn Valley (Cockroft 1965).

Many of the more permeable soils, especially those in the lower rainfall zones along the Murray Valley, initially were saline at depth; irrigation has redistributed the salt, with upper slope sites bing leached, and mid to lower slopes often becoming salinized. Almost universally, slowly permeable layers occur at no great depth, and cause the development of water tables shallow enough for salt to be concentrated in the root zone of the overlying soils. The data of Table 53.1 (extracted from Marshall and Hooper 1935 and Marshall and Walkley 1937) illustrate these changes accompanying irrigation for two areas in the Murray Valley. In both areas, the soils sampled are *solonized brown soils*, saline in the subsoil before irrigation. The irrigated profiles from Berri in South Australia show the leaching that occurred in the absence of a water table; those collected from salt patches at Coomealla show accumulation of salt in the surface wherever shallow water tables occurred.

Salting problems in horticultural areas have long been in evidence, and their correction by tile drainage and leaching is well known (e.g. Thomas 1939; Lyon and Tisdall 1942). In some cases, the need for drainage appeared very soon after the start of irrigation, e.g. at Loxton in South Australia (Blackburn and Wright 1972); apparently these soils usually have sufficient soluble calcium so that leaching does not cause sodicity, and the attendant loss of permeability, to develop.

In the Goulburn Valley of Victoria, despite early prognostications about the improbability of salinity becoming a major problem (Skene and Poutsma 1962; Cockroft 1965), there has been a rapid spread of salt-affected areas in recent years, associated with widespread development of saline water tables at shallow depth (Bakker 1978). Tile drainage is not favoured on economic grounds (Trewhella and Webster 1978), though it seems likely also that many subsoils may be too impermeable for its satisfactory functioning. Instead, a grid of pumps is being installed, and effective control of the ground water is obtained where aquifers are within 20 m of the surface (Trewhella and Webster 1978). The extent to which the effluent can be re-used for irrigation

Table 53.1. Chloride (as % NaCl) in *solonized brown soils*

This table illustrates the leaching (Berri and surrounding areas) and surface salt accumulating effects (Coomealla) of irrigation. (From Marshall and Hooper 1935, Marshall and Walkley 1937)

Depth (cm)	Berri and surrounding areas Non-irrigated[A]	Irrigated[B]	Coomealla Non-irrigated	Irrigated
0–30	0.023	0.025	0.043 (37)[C]	0.128 (20)
30–60	0.120	0.013	0.133 (21)	0.038 (69)
60–90	0.140	0.013		
90–120	0.160	0.015	0.153 (92)	0.021 (34)
120–150	0.140	0.018		
150–180	0.160	0.020		
180–360	–	–	0.206 (15)	0.054 (9)

[A] 61 profiles were sampled from 7 soil types.
[B] 44 profiles were sampled from 7 soil types.
[C] Number of samples in brackets.

Table 53.2. Chloride (mequiv. per 100 g) in Riverina clay

Table shows leaching due to ponding for one rice crop (compare treatments A and B of April 1951), and some salt return after two subsequent years of intermittent flooding for annual pasture (compare treatment B April 1951 and April 1953). (Data from Barley and Hutton 1956)

Depth (cm)	April 1951		April 1953	
Treatment[A]:	A	B	A	B
0 – 7.5	0.4	0.2	0.7	0.5
7.5 – 30	1.5	0.2	1.0	0.6
30 – 52.5	3.1	1.0	3.0	1.6
52.5 – 75	5.2	2.3	6.8	3.2

[A] Treatment A, dry fallow in summer 1950-1; Treatment B, rice crop (ponded) in summer 1950-1. In May 1951 both treatments cultivated and sown to annual pasture, which was flood irrigated in October 1951 and in March, April, and October 1952.

without damage to soils and plantings is being investigated (Bakker 1978); aspects which need examining, according to Bakker, include: nutrient responses under saline conditions, methods of applying calcium to readjust the sodium adsorption ratio of recycled water, the improvement of subsoil permeability for leaching, and the monitoring of salinity trends in soils and groundwaters. A technique that has been used successfully on an experimental scale gives increased subsoil hydraulic conductivity by injecting gypsum into the B horizon, following ripping (Cockroft and Tisdall 1978).

Reclaiming irrigated crop and pasture soils

Large areas of fine-textured soils in northern Victoria, used mainly for irrigated pastures, have become salinized. For the most seriously affected region, Kerang, Gutteridge *et al.* (1970) estimated that 55 x 10^3 ha, or 17% of the total irrigated land, was salinized; the area is expected to increase in both Victoria and southern New South Wales as groundwater levels rise. In Queensland, similar problems may occur in the lower Burdekin delta (Smith, G.D., Qld Dept Prim. Ind., pers. comm.), and clay soils at Dalby are becoming salinized by irrigating with moderately saline groundwater (Cassidy 1971).

With surface drainage, and attention to irrigation layout, grading, frequency and amount of watering, and appropriate agronomic practices, reclamation in the Kerang district of Victoria has been possible even in the presence of high and saline water tables (Morgan and Garland 1954; Barkla *et al.* 1964). However, such reclamation is achieved at the expense of surrounding country, especially if it is not, or only lightly, irrigated. Only by sub-surface drainage to lower the water table can this be avoided.

Where shallow aquifers exist, pumping from them appears to be the most economically feasible form of subsurface drainage (Garland and Jones 1963; Pels 1967), but the possibility of using tile drainage is also being examined (Poulton 1978). To date, widespread implementation of subsurface drainage is prevented by unsolved problems associated with drainage effluent disposal.

The desalination of fine-textured soils, when achieved by aquifer pumping and leaching, frequently leaves a legacy of sodicity that affects surface soil structure. This can be countered by gypsum application as described later.

Data presented in the many soil survey reports relating to Riverine Plain soils show that inherently saline subsoils are widespread. The practice of growing rice in rotation with other crops and pastures has resulted in leaching of salt from the profile. This is shown in the data of Barley and Hutton (1956), here reproduced as Table 53.2. The data also show that there is some tendency for a salt return to the rootzone during subsequent years under annual pasture. Where rice has not been grown, the degree to which saline subsoils may be responsible for growth limitations seems worthy of investigation.

Reclaiming dryland saline soils

Seepage salting and valley waterlogging reclamation

In the undulating and rolling and hilly landscapes of southern and eastern and especially south-western Australia, there have been countless outbreaks of salinity associated with slope seepages and valley waterlogging (e.g. Fig. 53.2). Northcote and Skene (1972) estimated that not less than 197 000 ha have been affected in this way. A survey in Western Australia in 1962 suggested that the salinized area was increasing at 3400 ha yr^{-1} (Lightfoot *et al.* 1964).

In some cases, there has been an enlargement of naturally occurring saline areas, but, frequently, saline spots have appeared where none were previously apparent. The outbreaks vary considerably in size; though many are small, it is vital to control and reclaim them, because they act as focal points for erosion to develop, and spread, in the agriculturally more desirable lower-slope soils.

The soils affected by saline outbreaks are mainly the alkaline and non-alkaline sodic duplex soils (Northcote and Skene 1972); they belong to a wide range of soil groups, including: *solonetz*, *solodized-solonetz* and *solodic soils*, *solonized brown soils*, *red-brown earths*, and various *podzolic soils*. The distribution of the outbreaks has a degree of climatic control, probably resulting from the balance between infiltration and leaching; in Victoria, for example, they occur in the 300–760 mm rainfall zone (Cope 1958; Rowan and Downes 1963).

It has been held generally that this form of salinity has resulted from changed hydrologic regimes following clearing of deep-rooted native vegetation, especially trees. Overgrazing, bare fallowing, and failure to sow down and maintain improved pasture, are all contributing factors. Salts formerly dispersed in catchment soils and sediments have been moved and concentrated in areas where water evaporates, i.e. in lower slope and valley bottom positions where the groundwater either appears at, or approaches near to, the surface (Bettenay *et al.* 1964; Rowan 1971; Peck 1977).

This understanding has provided the rationale for a variety of control and reclamation practices; they are aimed at, firstly, reducing the flow of water, and therefore salt, through the upper catchment soils to the salt outbreaks lower down, and, secondly, re-establishing useful vegetation on the salted areas. The level of success of the latter operation clearly depends on that of the former.

In treating catchments, general replanting of trees at their former density is not economic, but the planting of stands in strategic positions, especially above and around saline outbreaks, is recommended. More significant increases in water use on the catchments may be obtained by establishing perennial species (e.g. phalaris, lucerne), and using contour furrowing to help retain the rain where it falls. For this purpose, the value of improved pastures based on annuals such as subterranean clover and rye grass is equivocal, and may be less than well-managed native perennial grasses (Dunin and Downes 1962). Perennial plants, either native or introduced, have several important advantages over annuals. Firstly, a surface cover is maintained throughout the year, restricting runoff and consequently reducing erosion; and secondly, with greater rooting depth and summer growth, perennials more fully exploit subsoil water (Fig. 53.3), so that a larger proportion of opening rains in autumn is used in soil water recharge.

Associated treatment of the saline outbreaks may include drainage ditches, carefully graded off the contour, along the upper edge of the salted area; the aims are to

Fig. 53.2. Saltpan on cropped interdune flat. Low sandhill in background almost devoid of original mallee vegetation. North-West Victoria. (Photo J. N. Rowan.)

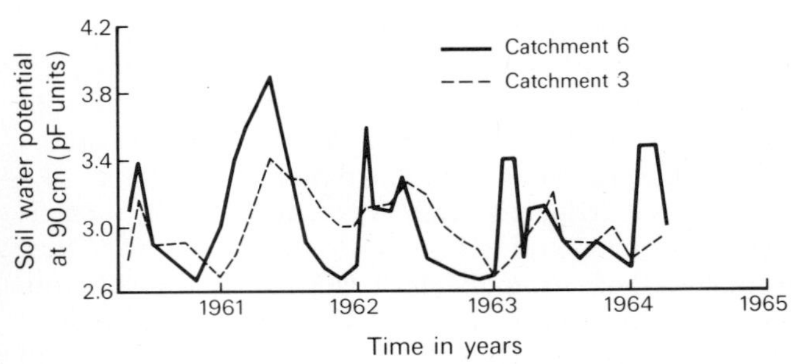

Fig. 53.3. Fluctuations in water potential of subsoil at 90 cm from contrasting conditions of natural grassland at Bacchus Marsh, Victoria. Catchment 6 contained a high proportion of perennial grasses (*Themeda australis* and *Poa caespitosa*), while catchment 3 was a degraded native grassland. (After Dunin 1970.)

divert potential run-on, and to intercept sub-surface water moving along the top of the B horizon. Also, salt tolerant species are established first, with the help of a straw mulch; then, as the watertable is lowered, rainfall gives the necessary leaching to allow the establishment of more productive species.

As might be expected, the desalination process is relatively slow, and has rarely been documented. Rowan (1971) gives some comparative data, from which Fig. 53.4 has been drawn. In this case, a salt pan (site 27 of Rowan) was fenced to exclude grazing stock, then sown to Wimmera rye grass and lucerne and barrel medic, following straw mulching; the adjacent catchment (a dune) was sown to lucerne. Ten years later, the treated area was sampled at the same time as other sites both visibly and not visibly salt affected. The chloride profiles show the reclaimed area to have been very similar to the site not visibly salt affected. Salinity at all sites remained high at depths below 50–60 cm.

If no attempt is made to use more water where it falls on the catchment, salted areas can be expected to remain or extend. Their treatment will be restricted to establishing salt-tolerant species, which at least initially must be protected from grazing (Lightfoot *et al.* 1964). At present rates of groundwater discharge from certain farmed catchments in south-western Australia, Peck and Hurle (1973) calculated that 30 to 400 years are needed before salt levels are reduced sufficiently to make stream water acceptable for drinking.

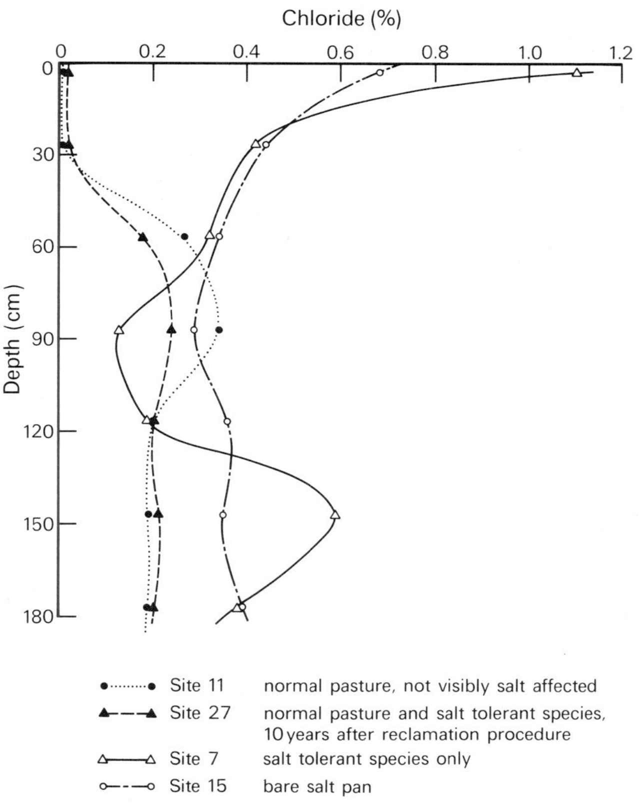

Fig. 53.4. Chloride profiles on interdune flats, north-west Victoria, sampled October 1965. (Drawn from data in Table 6 of Rowan 1971.)

Dryland salting and scald reclamation

Although the term 'dryland salting' is somewhat ambiguous, it is used here, as it has been in Western Australia by Smith (1961) and in Victoria by Rowan and Downes (1963), to describe situations where, following cultivation or grazing, salinity has developed despite the absence of seepage or high water tables. The problem occurs widely from Queensland to Western Australia on wind-eroded *red-brown earths* and *grey, brown and red clays* in semi-arid pastoral country and marginal wheat lands. Salt from the subsoil moves upwards to the surface, under the influence of evaporation from bare fallows or from surfaces laid bare by wind erosion. During periods of rainfall, the immediate surface is leached, and clay disperses to form a crust or scalded surface that retards further infiltration and inhibits plant growth.

Reclamation involves leaching the salt from a sufficient depth of soil to allow establishment of a vegetative cover, to increase transpiration and minimize evaporation. This process may take place naturally in those parts of scalded areas where water ponds during unusually wet years. (Jones 1966); it may also be aided by various forms of ploughing or banking to provide local ponding and roughened sites for seeds to lodge and germinate (Fig 53.5), and by gypsum applications to improve infiltration (Muirhead *et al.* 1974). The spreading of seed of suitable salt tolerant species may be worthwhile, and the control of subsequent grazing is essential (Lightfoot *et al.* 1964).

Fig. 53.5. A 'scalded' surface, bare except for salt-tolerant species establishing in plough furrows. Hay district, N.S.W. (Photo W. A. Muirhead.)

Gypsum amelioration in the management of sodic soils

Nature of the response to gypsum

Widespread responses to gypsum have been obtained under both dryland and irrigated agriculture on soils that have hardsetting or crusting surfaces due to clay dispersion. A severe case of dispersion, following a 5 cm flood irrigation by ordinary water (left), and its prevention by gypsum dissolved in the irrigation water (right), are shown in Fig. 53.6; the soil is a member of the *grey, brown and red clays*. More detail of the nature of the gypsum effect upon aggregate stability, total porosity, and interconnection of pores, can be seen in the vertical thin sections of dry surface aggregates illustrated in Fig. 53.7. The practical outcome is improved soil tilth, reduced crust strength, reduced runoff, increased water infiltration and storage, and increased drainage and leaching. Effects upon water storage, water use, and leaching of chloride, under irrigated cotton on a transitional *red-brown earth*, are shown in Fig. 53.8; rather more dramatic is the effect gypsum application has upon leaching of chloride under a rice crop on the same soil (Fig. 53.9). The flux of water (some 70–80 cm) beyond the root zone, which accompanied this leaching (Loveday *et al.* 1978), raises questions regarding the advisability of growing rice following amelioration by gypsum on soils such as this.

Although appropriate measurements have not been reported, it should be possible to evaluate, in terms of the tillage requirements for land grading and seed-bed preparation, the effect of improved surface soil aggregation. Reduced crust strength with greater seedling emergence should mean reduced seeding rates; and increased infiltration and water storage should mean reduced frequency of irrigation. Not so easy to evaluate, but nevertheless of great potential significance, is the improved timeliness of operations, especially cultivating and sowing, resulting from the wider moisture range over which gypsum-treated soil can be worked.

The expected outcome of these soil effects is improved crop and pasture yields, and frequently this outcome has been realized (see e.g. Sims and Rooney 1965; Myers and Squires 1970); however, rather variable results have sometimes been obtained. Loveday (1975) has suggested that interactions with seasonal conditions (especially likely for a treatment which affects soil water status) may be responsible; for example, for dryland wheat cropping in a season of inadequate rainfall in the latter part, greater plant density resulting from gypsum application could conceivably lead to premature haying-off and even yield reductions; on the other hand, for summer crops under irrigated conditions, the greatest yield increases due to gypsum may occur in drier seasons, when increased infiltration and water storage become effective.

In contrast to soils showing obvious responses to

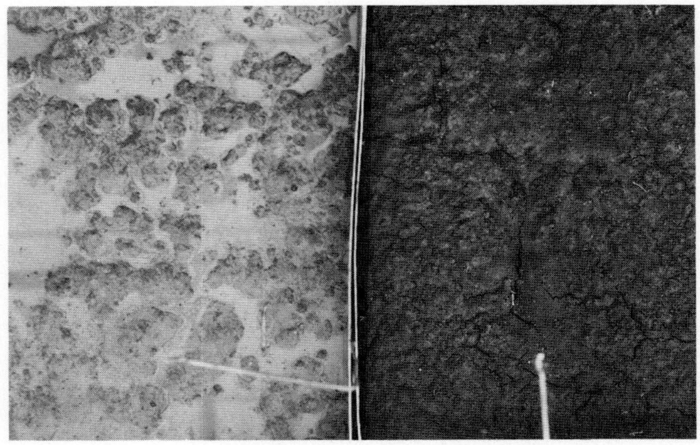

Fig. 53.6. Dispersion of clay at the surface of flood-irrigated soil (left) and its prevention by gypsum dissolved in the irrigation water (right). The soil is one of the *grey, brown* and *red clays*.

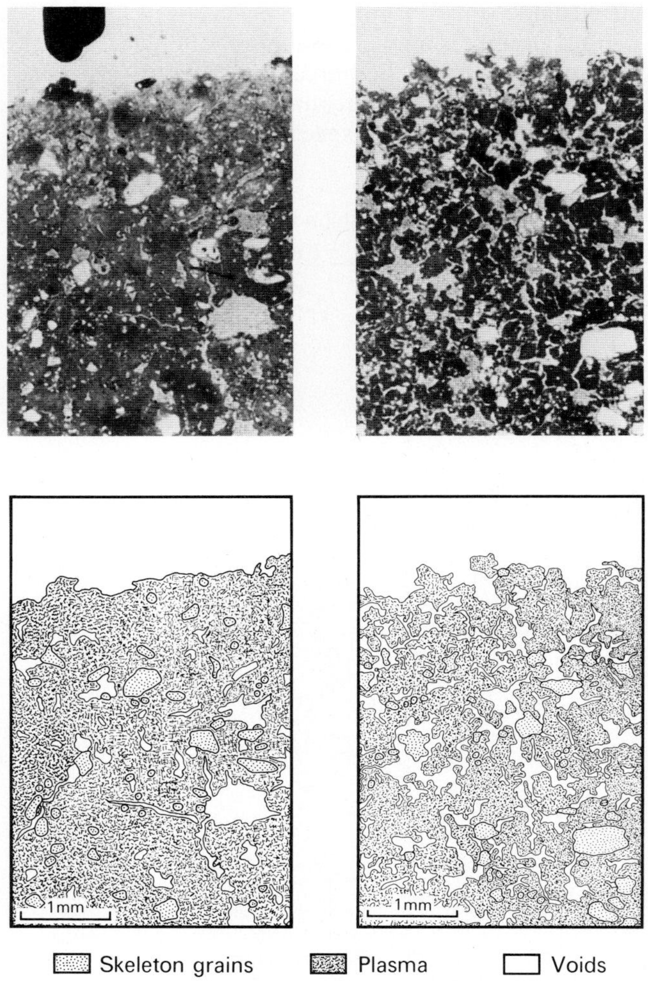

Fig. 53.7. Thin sections of surface aggregates irrigated with ordinary water (left) and with gypsum solution (right). The effect upon total porosity and interconnection of pores is apparent. (Reproduced with permission from Loveday and Scotter 1966.)

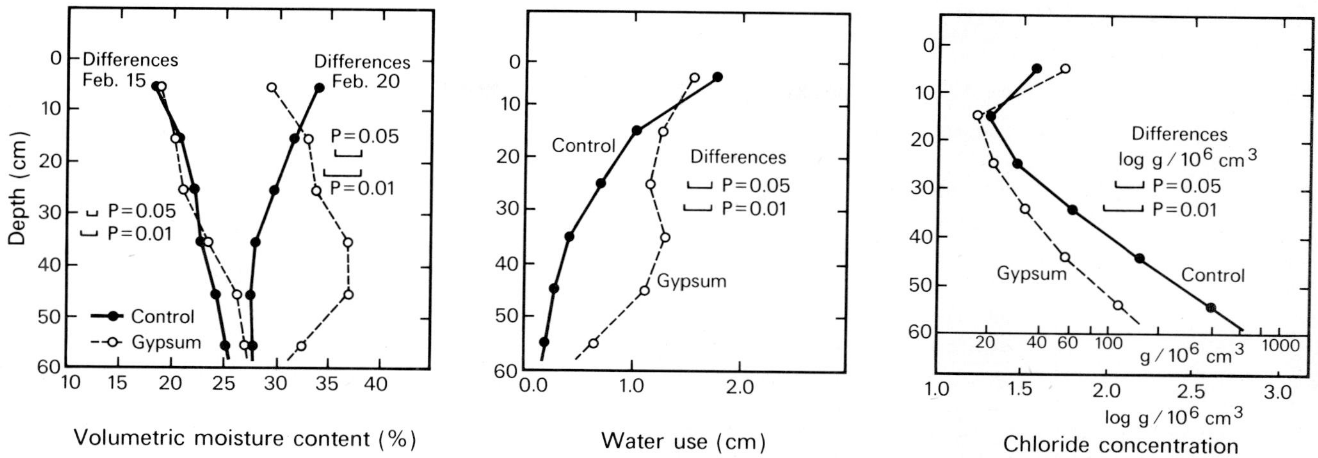

Fig. 53.8. The effect of gypsum (at 25 t ha^{-1}) applied 6 months previously to a transitional *red-brown earth* on: (left) soil water contents before and after the 7th irrigation of the season; (centre) water use by the cotton crop during a 10 day period between the 7th and 8th irrigations; and (right) chloride contents midway between the 5th and 6th irrigation following gypsum application. (After Loveday *et al.* 1970.)

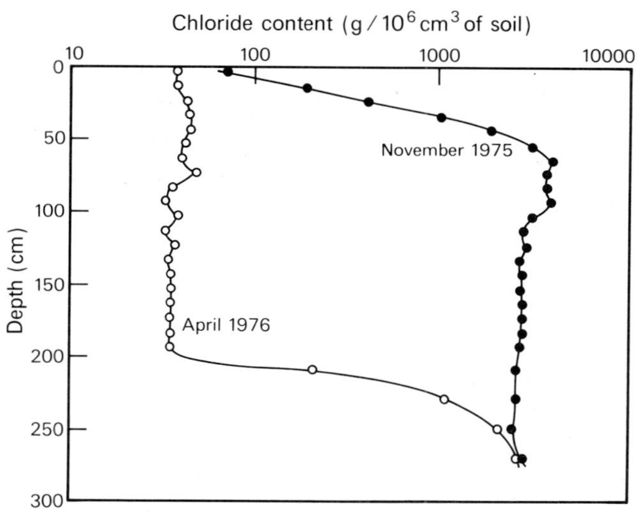

Fig. 53.9. Chloride profiles before (November 1975) and after (April 1976) growing a rice crop on a transitional *red-brown earth* treated with gypsum at 10 t ha^{-1}.

gypsum in the plough layer, are the duplex soils with non-sodic coarse-textured A horizons over impervious sodic subsoils. On one such soil (an irrigated *red-brown earth* with 12% clay in the sandy loam A horizon overlying a sandy clay B horizon with 35% clay), Bridge and Kleinig (1968) doubled the growth rate of white clover by broadcast additions of gypsum at 10 t ha^{-1}; the treatment effects upon hydraulic conductivity, especially of the B horizon (Table 53.3), with resulting increases in subsoil water storage and overall efficiency of water use, were considered to be responsible for the improved growth.

Duplex soils of this general kind, including, besides many *red-brown earths*, some *red* and *yellow podzolics* and *solodized-solonetz* and *solodic soils*, are very widespread in southern and eastern Australian areas of moderate to good rainfall. The impervious and dispersible nature of their subsoils has often led to development of tunnel erosion, which has attracted considerable attention and is dealt with separately later in this chapter. Far less consideration has been given to problems of growth restriction which, it is suggested, must accompany the seasonal waterlogging of surface horizons and the impenetrability of B horizons to both water and roots. Though it seems that gypsum may sometimes ameliorate these conditions, much remains to be known about their severity and about ways of overcoming them economically.

Recognizing gypsum-responsive soils

Both dryland and irrigated field experiments conducted by various agencies, and farmers' trials using observation of treated drill strips, have established the general categories of soils likely to respond to gypsum treatment; they include many *red-brown earths* with shallow hard-

Table 53.3. Effect of gypsum on exchangeable sodium percentage (ESP) and hydraulic conductivity of a *red-brown earth*

(From Bridge and Kleinig 1968)

Horizon	Depth (cm)	Gypsum treatment (t ha^{-1})	ESP	Hydraulic conductivity ($\times 10^{-6}$ cm s^{-1})
A1	0.5– 7.6	Nil	2.1*	41.4
		10	0.9	80.7
A2	7.6–15.2	Nil	3.6*	
		10	1.1	
B	15.2–22.9	Nil	6.8*	0.475***
		10	3.2	4.76
B	22.9–30.5	Nil	7.7	
		10	5.8	

* and *** Treatment means significantly different at $P < 0.05$ and 0.001.

Table 53.4. Regressions of gypsum-response measures on selected pairs of soil properties

(From Loveday 1974*b*)

Regression equation	R^2	Adjusted $(R^2)^A$	Confidence interval[B]
$Y_1^C = 2.30\text{–}0.06\ \mathrm{DI}_2 - 0.20\ \mathrm{pH}$	70	69	± 0.29
$Y_2 = -60 + 5.2\ \mathrm{DI}_2 + 6.2\ \mathrm{pH}$	73	72	± 22
$Y_3 = -5.9 + 0.16\ \mathrm{DI}_2 + 0.03\ \mathrm{DI}_2^2 + 0.84\ \mathrm{pH}$	77	76	± 1.8
$Y_1 = 2.04 - 0.05\ \mathrm{DI}_2 - 0.07\%\mathrm{Mg} + 0.0009\%\mathrm{Mg}^2$	81	79	± 0.24
$Y_2 = -26 + 4.5\ \mathrm{DI}_2 + 0.44\%\mathrm{Mg}$	73	72	± 22
$Y_3 = -2.76 + 0.34\ \mathrm{DI}_2 + 0.10\%\mathrm{Mg}$	82	81	± 1.6

[A] Adjusted $R^2 = 1 - \{(n-1)/(n-p-1)\}\,\{1-R^2\}$, where R^2 = coefficient of determination; n = number of observations; p = number of constants. This takes account of the additional predictors in multiple regressions and allows direct comparisons of predictive power.

[B] 'Confidence interval' indicates the interval on each side of the estimate of a new observation which will have 95% probability of including the true estimate. These values are the means of confidence intervals for the sample and so are approximate.

[C] Gypsum response measures: Y_1, resistance to penetrometer (gypsum/control); Y_2, percentage seedling emergence (gypsum minus control); Y_3, hydraulic conductivity $\{\log_e$ (gypsum/control)$\}$.

Table 53.5. Limits to gypsum-response categories in terms of three response measures

(From Loveday 1974*b*). G, gypsum; C, control.

Response measure[A]	Strong response (i.e. G > 20C)	Moderate response (i.e. 20C > G > 5C)	Little or no response (i.e. G < 5C)
Y_1	<0.3	0.3– 0.6	>0.6
Y_2	>30	10–30	<10
Y_3	>3.0	1.6–3.0	<1.6

[A] See Table 53.4.

setting loam surfaces over dense clay subsoils, hardsetting or crusting *grey, brown and red clays*, and some *solodized-solonetz* and *solodic soils* with fine sandy loam surfaces over dense clay subsoils. Several experiments on irrigated *solonized brown soils* (Thomas 1939; J.B. Robinson, priv. comm.) have failed to show any additional response to gypsum, over and above that due merely to drainage and leaching.

Although field trials doubtless will continue to be used to indicate the responsiveness of particular parcels of land, the less time- and labour-consuming method of predicting from soil analyses could receive more attention in the future. In an attempt to establish such a method, Loveday (1974*b*) measured gypsum response in terms of penetrometer resistance, seedling emergence, and hydraulic conductivity, and related these three reponse measures to a range of soil properties. Hydraulic conductivity proved to be the most sensitive response measure, and it is best predicted (R^2 = 86%) by three properties: a dispersion index (Loveday and Pyle 1973), ESP, and percentage exchangeable magnesium. The best pair of predictors is the dispersion index and percentage exchangeable magnesium (Table 53.4), but predictive ability is reduced only marginally by using the two easily determined properties, dispersion index and pH. Limits have been set for three categories of response, viz. strong, moderate or negligible, in terms of the three response measures used (Table 53.5). In practice, having determined the dispersion index and pH, the values for these properties can then be used in one or other of the first three regressions of Table 53.4; then, by reference to Table 53.5, the calculated value of the response measure (either penetrometer resistance, seedling emergence, or hydraulic conductivity) can be used to nominate the gypsum response category.

The soils for which this procedure is expected to give reasonably reliable estimates include *red-brown earths*, transitional *red-brown earths*, and *grey, brown and red clays*, i.e. medium- to fine-textured soils in which the ESP does not exceed 25 or so. At higher ESP levels, applied gypsum may not be able to maintain permeability (Quirk and Schofield 1955) because of its limited solubility; about 30 mequiv. L^{-1} is a saturated solution. In such instances, the use of high-salt waters may be necessary to achieve reclamation (Reeve and Doering 1966). Experimentally, dilute sea water has been used for reclaiming soils in the Burdekin delta of Queensland, with promising results (Smith, G.D., pers. comm.).

Gypsum application rates

As pointed out earlier, permanent improvement of

sodic soils requires replacement of most of the exchangeable sodium by calcium. The possibility of also having to replace exchangeable magnesium has received little or no attention, even though the enhancement of swelling and dispersion when sodium is associated with magnesium rather than calcium has been demonstrated (van der Merwe and Burger 1969; Bakker and Emerson 1973). In the field, a significant portion of the applied calcium entering into exchange displaces magnesium rather than sodium, as would be expected in an equilibrium reaction (Loveday 1976).

Both the standard methods (United States Salinity Laboratory Staff 1954) and more recent computer-based methods (e.g. Dutt *et al.* 1972) for determining 'gypsum requirement', i.e. the amount of gypsum to reduce the ESP to a desired level for a nominated depth of soil, are based on the residual effect of the exchange of calcium for sodium. Regardless of method, the amount of gypsum required for permanent reclamation is often found to be very considerable: for example the gypsum requirement of a 20 cm plough-depth of a typical irrigated clay from the Riverina of New South Wales is about 12.5 t ha^{-1} if calculated according to the United States Salinity Laboratory Staff (1954), and assuming a 50% replacement efficiency; and, for such a soil, Loveday (1976) estimated from field experiment data that 25 t ha^{-1} would be needed to reduce the ESP to 5 or less.

Fortunately, gypsum, besides having residual exchange effects, also acts as an electrolyte once dissolved by rain or irrigation water. Quirk and Schofield (1955), in a study of the effect of electrolyte concentration on permeability, proposed the 'threshold concentration' as that concentration of electrolyte which causes 10–15 % decrease in permeability. As mentioned previously, gypsum is sparingly soluble; nevertheless, the concentration achieved by its solution in percolating soil water is sufficient to maintain permeability in many soils, provided the ESP is no greater than 25 or so. Davidson and Quirk (1961) used this understanding to demonstrate for a 7.5 cm irrigation that irrigation water with as little gypsum as 10 mequiv. L^{-1} has a profound effect on the stability of surface soil structure and on the emergence of pasture seedlings.

Later, Loveday (1976) showed that the very significant initial effect an application of solid gypsum had on hydraulic conductivity could also be attributed to the electrolyte effect, whereas there was a much reduced effect attributable to residual exchange once the gypsum had dissolved. Thus the reason is apparent for the success of applications much lighter than the 'gypsum requirement'.

A commonly recommended gypsum application rate for wheat lands with hardsetting surface soils in southern Australia is 3–5 t ha^{-1} (Sims and Rooney 1965; Matheson 1969; Rudd 1974). It seems likely that this is the minimum needed to give a reasonable cover of the soil, in order to ensure that none of the surface soil is exposed to rainfall that has not had its electrolyte level raised by dissolving some gypsum. Such applications must be repeated at intervals, until permanent improvement has occurred by displacement of all or most of the exchangeable sodium. The interval between applications may be several years, depending on the amount of gypsum applied, its fineness, and the amount of rainfall or irrigation to dissolve it.

In the lower Burdekin delta of Queensland, where sodic duplex soils are used for sugar cane, gypsum at 10 t ha^{-1} is recommended to help stabilize porosity formed during deep ripping (Smith 1977). Using large machines with horizontal blades, the subsoil is disrupted into large blocks to a depth of 80–100 cm while preserving the general integrity of the profile: the coarser the blocks, the longer lasting the effect.

Control of tunnel erosion in sodic soils

Tunnel erosion manifests itself particularly in duplex soils with dispersible clay subsoils of low permeability, features which may be associated with sodicity and high exchangeable magnesium (Downes 1949; Charman 1969, 1970*a*). Critical soil analytical data have not usually been obtained.

Tunnelling has been reported from hilly or rolling country in north-eastern Victoria (Downes 1949), and, in New South Wales, from the Riverina (Newman and Phillips 1957), the Hunter Valley (Monteith 1954), and the Kempsey and Grafton and Taree areas (Charman 1969, 1970*a,b*). According to Charman and to Floyd (1974), it occurs especially on *solodic* and *solonetzic soils* (Dy 3.4), but also on *podzolics* and *red-brown earths*.

As with the seepage and valley salting, disturbance of the delicate hydrologic balance of these duplex soils sets the stage for erosion. Overgrazing and removal of vegetative cover on hardsetting surface soils produces runoff that concentrates and enters the soil through small depressions or cracks or stump holes (Downes 1946). The water moves along the top of the impermeable clay layer, dispersing it and bringing it to the surface lower down the slope, where the hydraulic head is sufficient to cause a break-out. Eventually, the surface soil bridge may collapse into the tunnel, and gully erosion develops. Tunnelling is more likely to result from heavy summer rains falling on dry soil with little vegetative cover than from winter rains (Floyd 1974).

Control measures aim to ensure that rain infiltrates evenly over the whole area. Contour ripping or chisel ploughing, and the establishment of a vigorous pasture, preferably with perennial grasses, are recommended. Banks and gully-control structures may be built, where stable soils exist above eroded areas; banks built with dispersible soil will themselves tunnel unless adequately compacted and treated with gypsum or hydrated lime (Rosewell 1970).

Conclusion

It is now generally recognised that in the short agricultural history of Australia the disturbance of natural ecosystems has wrought widespread changes in the hydrology of landscapes, with the consequent mobilization and redistribution of salt, leading to the exposure of dispersible sodic soils to erosion. Successful management to conserve the soil resource involves appropriate adjustments to the hydrologic cycle, to reduce runoff to a minimum, to prevent further mobilization of salt from deep subsoils, and to reclaim already salinized areas by leaching. For naturally occurring sodic soils, the addition of a soluble calcium source has frequently resulted in improvement of their adverse physical conditions. The provision of economic management techniques for overcoming plant growth restrictions caused by the impervious sodic subsoils of the extensive duplex soils is a challenge for the future.

References

Bakker, A. C. (1978). Disposal of drainage effluent in the Shepparton region. Aust. Soc. Soil Sci. Inc. (Riverina Branch) Symp. 'The Hydrogeology of the Riverine Plain of South East Australia'. (Eds R. R. Storrier and I. D. Kelly) pp.247-58.

Bakker, A. C. and Emerson, W. W. (1973). The comparative effects of exchangeable calcium, magnesium and sodium on some physical properties of red-brown earth subsoils. 111. The permeability of Shepparton soil and comparison of methods. *Aust. J. Soil Res.* **11**, 159-65.

Barley, K. P. and Hutton, J. T. (1956). Effects of a lowland rice crop and of gypsum on a saline alkali clay. *Aust. J. Agric. Res.* **1**, 110-25.

Barkla, F. J., Garland, K. R., and Jones, G. O. (1964). Reclamation trails at Swan Hill Research Farm. *J. Dep. Agric. Vic.* **62**, 291-6.

Bettenay, E., Blackmore, A. V. and Hingston, F. J. (1964). Aspects of the hydrologic cycle and related salinity in the Belka Valley, Western Australia. *Aust. J. Soil Res.* **2**, 187-210.

Blackburn, G. and Wright, D. A. (1972). Soil survey of the Loxton Irrigation Area, Sth. Australia. CSIRO Aust. Soil and Land Use Ser. No. 53.

Bower, C. A. and Goertzen, J. O. (1958). Replacement of adsorbed sodium by hydrolysis of calcium carbonate. *Proc. Soil Sci. Soc. Am.* **22**, 33-5.

Bridge, B. J. and Kleinig, C. R. (1968).The effect of gypsum on the water storage in a sandy loam soil under an irrigated perennial pasture. Trans. 9th Int. Congr. Soil Sci. Adelaide. Vol. 1, pp.313-23.

Cassidy, N. G. (1971). Effect of irrigation water on heavy clay soils at Dalby. Queensl. Dept. Primary Ind. Agric. Chem. Tech. Rept. No. 2.

Charman, P. E. V. (1969). The influence of sodium salts on soils with reference to tunnel erosion in coastal districts. Pt.1. Kempsey Area. *J. Soil Conserv. Serv. N.S.W.* **25**, 331-47.

Charman, P. E. V. (1970*a*). The influence of sodium salts on soils with reference to tunnel erosion in coastal districts. Pt. II. Grafton Area. *J. Soil Conserv. Serv. N.S.W.* **26**, 71-86.

Charman, P. E. V. (1970*b*). The influence of sodium salts on soils with reference to tunnel erosion in coastal districts. Pt. III. Taree area. *J. Soil Conserv. Serv. N.S.W.* **26**, 256-75.

Cockroft, B. (1965). Pedology of the Goulburn Valley area, Victoria. Dep. Agric. Victoria, Tech. Bull. No. 19.

Cockroft, B. and Tisdall, J. M. (1978). Soil Management, Soil Structure and Root Activity. *In* 'Modification of Soil Structure'. (Eds W. W. Emerson, R. D. Bond and A. R. Dexter). (John Wiley & Sons, New York.)

Collis-George, N. and Evans, G. N. (1964). A hydraulic investigation of salt affected soils in an alluvial plain of the Hawkesbury River, N.S.W. *Aust. J. Soil Res.* **2**, 20-8.

Cope, F. (1958). Catchment salting in Victoria. (Soil Conserv. Authority Vic.: Melbourne.)

Davidson, J. L. and Quirk, J. P. (1961). The influence of dissolved gypsum on pasture establishment on irrigated sodic clays. *Aust. J. Agric. Res.* **12**, 100-110.

Crawford, J. and Laimer, T. (1968). Corrosion of galvanised steel culverts in Queensland. Queensland Roads No. 11.

Donnan, W. W. and Schwab, G. O. (1974). Current Drainage Methods in the U.S.A. *In* 'Drainage for Agriculture'. (Ed. J. van Schilfgaarde). Agron. No. 17. (Am. Soc. Agron.: Madison.)

Downes, R. G. (1946). Tunnelling erosion in north-eastern Victoria. J.CSIR **19**, 283-92.

Downes, R. G. (1949). Soil, land-use, and erosion survey around Dookie, Victoria. CSIRO Aust. Bull. No. 243.

Dunin, F. X. (1970). Changes in water balance components with pasture management in south-eastern Australia. *J. Hydrol.* **10**, 90-102.

Dunin, F. X. and Downes, R. G. (1962). The effect of subterranean clover and wimmera rye grass in controlling surface run-off from four-acre catchments near Bacchus Marsh, Victoria. *Aust. J. Exp. Agric. Anim. Husb.* **2**, 148-52.

Dutt, G. R., Terkeltoub,R. W. and Rauschkolb, R. S. (1972). Prediction of gypsum and leaching requirements for sodium-affected soils. *Soil Sci.* **114**, 93-103.

Floyd, E. J. (1974). Tunnel erosion – a field study in the Riverina. *J. Soil Conserv. Serv. N.S.W.* **30**, 145-56.

Garland, K. R. and Jones, G. O. (1963). Reclamation of saline soils by ponding and aquifer pumping. *Aust. J. Exp. Agric. Anim. Husb.* **3**, 177-9.

Goertzen, J. O. and Bower, C. A. (1958). Carbon dioxide from plant roots as a factor in the replacement of adsorbed sodium in calcareous soils. *Proc. Soil Sci. Soc. Am.* **22**, 36-7.

Greene, H. (1928). Soil permeability in the Eastern Gezira. *J. Agric. Sci.* **18**, 531-43.

Gutteridge, Haskins and Davey (1970). 'Murray Valley Salinity Investigation'. A report to River Murray Commission, Canberra, Aust.

Ingles, O. G. and Metcalf, J. B. (1972). 'Soil Stabilization: Principles and Practice'. (Butterworths: Sydney.)

Jones, R. M. (1966). Scald reclamation in the Hay district, N.S.W. Pt.1. Natural reclamation of scalds. *J. Soil Conserv. Serv. N.S.W.* **22**, 147-60.

Lightfoot, L. C., Smith, S. T. and Malcolm, C. V. (1964). Salt land survey, 1962. *J. Agric. West. Aust.* (4) **5**, 396-410.

Loveday, J. (Ed.) (1974*a*). Methods for Analysis of Irrigated Soils. Commonw. Bur. Soils Tech. Commun. No. 54.

Loveday, J. (1974*b*). Recognition of gypsum-responsive soils. *Aust. J. Soil Res.* **12**, 87-96.

Loveday, J. (1975). The Use of Sulphur and its Compounds in Soil Amendment. *In* 'Sulphur in Australasian Agriculture'. (Ed. K. D. McLachlan). (Sydney Univ. Press: Sydney.)

Loveday, J. (1976). Relative significance of electrolyte and cation exchange effects when gypsum is applied to a sodic clay soil. *Aust. J. Soil Res.* **14**, 361-71.

Loveday, J. and Scotter, D. R. (1966). Emergence response of subterranean clover to dissolved gypsum in relation to soil properties and evaporative conditions. *Aust. J. Soil Res.* **4**, 55-68.

Loveday, J., Saunt, J. E., Fleming, P. M. and Muirhead, W. A. (1970). Soil and cotton responses to tillage and ameliorant treatments in a brown clay soil. 1. Soil responses and water use. *Aust. J. Exp. Agric. Anim. Husb.* **10**, 313-24.

Loveday, J. and Pyle, J. C. (1973). The Emerson dispersion test and its relationship to hydraulic conductivity. CSIRO Aust. Div. Soils Tech. Pap. No. 15.

Loveday, J., Watson, C. L. and McIntyre, D. S. (1978). Aspects of the hydrology of Riverine Plain soils. Aust. Soc. Soil Sci. Inc. (Riverina Branch). Symp. 'The Hydrogeology of the Riverine Plain of South East Australia'. (Eds R. R. Storrier and I. D. Kelly). pp.99-114.

Lyon, A. V. and Tisdall, A. L. (1942). Production of dried grapes in Murray Valley irrigation settlements. 11. Irrigation, drainage and reclamation. CSIR Aust. Bull. No. 149.

Maas, E. V. and Hoffman, G. J. (1977). Crop salt tolerance – current assessment. *J. Irrig. & Drainage, Am. Soc. Civil Eng.* **103**, 115-34.

Marshall, T. J. and Hooper, P. D. (1935). A soil survey of the Berri, Cobdogla, Kingston and Moorook Irrigation Areas, and Lyrup Village District, Sth. Australia. CSIR Aust. Bull. No. 86.

Marshall, T. J. and Walkley, A (1937). A soil survey of the Coomealla, Wentworth (Curlwaa), and Pomona Irrigation Settlements, N.S.W. CSIR Aust. Bull. No. 107.

Matheson, W. E. (1969). Gypsum for hardsetting soils. *J. Agric. S. Aust.* **72**, 402-6.

Miller, R. J., Nielsen, D. R. and Biggar, J. W. (1965). Chloride displacement in Panoche clay loam in relation to water movement and distribution. *J. Water Resour. Res.* **1**, 63-73.

Monteith, N. H. (1954). Problems of some Hunter Valley soils. *J. Soil Conserv. Serv. N.S.W.* **10**, 127-34.

Morgan, A. and Garland, K. R. (1954). Saltland reclamation at Swan Hill. *J. Dep. Agric. Vic.* **52**, 268-70.

Muirhead, W. A., Jones, R. M. and Williamson, D. R. (1974). Response by scalds to gypsum in the Riverina Plain, N.S.W. *J. Soil Conserv. Serv. N.S.W.* **30**, 112-9.

Myers, L. F. and Squires, V. R. (1970). Plant and animal production of an irrigated annual pasture following initial gypsum treatment of a heavy clay soil. *Aust. J. Exp. Agric. Anim. Husb.* **10**, 36-41.

Newman, J. C. and Phillips, J. R. H. (1957). Tunnel erosion in the Riverina. *J. Soil Conserv. Serv. N.S.W.* **13**, 159-69.

Nielsen, D. R., Biggar, J. W. and Luthin, J. N. (1966). Desalination of soils under controlled unsaturated flow conditions. 6th Congr. int. Comm. on Irrigation and Drainage (New Delhi, India). Q19, 15-24.

Northcote, K. H. and Skene, J. K. M. (1972). Australian soils with saline and sodic properties. CSIRO Aust. Div. Soils, Soil Publ. No. 27.

Peck, A. J. (1977). Development and Reclamation of Secondary Salinity. *In* 'Soil Factors in Crop Production in a Semi-Arid Environment'. (Eds J. S. Russell and E. L. Greacen). (Univ. Qld. Press: St. Lucia, Qld.)

Peck, A. J. and Hurle, D. H. (1973). Chloride balance of some farmed and forested catchments in southwestern Australia. *J. Water Resour. Res.* **9**, 648-57.

Pels, S. (1967). Soil salting and waterlogging problems in the Tullakool Irrigation Area and surrounding parts of Wakool Irrigation District, N.S.W. (unpublished, Water Conserv. & Irrig. Comm., N.S.W.).

Poulton, D. (1978). Operation and groundwater hydraulics of an experimental tile drainage system at Kerang. Aust. Soc. Soil Sci. Inc. (Riverina Branch). Symp. 'The Hydrogeology of the Riverine Plain of South East Australia'. (Eds R. R. Storrier and I. D. Kelly). pp.220-46.

Quirk, J. P. and Schofield, R. K. (1955). The effect of electrolyte concentration on soil permeability. *J. Soil Sci.* **6**, 163-78.

Reeve, R. C. and Doering, E. J. (1966). The high-salt water dilution method for reclaiming sodic soils. *Proc. Soil Sci. Soc. Am.* **30**, 498-504.

Rhoades, J. D. (1974). Drainage for Salinity Control. *In* 'Drainage for Agriculture'. (Ed. J. van Schilfgaarde). Agron. No. 17. (Am. Soc. Agron.: Madison.)

Rosewell, C. J. (1970). Investigations into the control of earth-work tunnelling. *J. Soil Conserv. Serv. N.S.W.* **26**, 188-203.

Rowan, J. M. (1971). Salting on dryland farms in north-western Victoria. (Soil Conserv. Authority Victoria: Melbourne.)

Rowan, J. M. and Downes, R. G. (1963). A study of the land in north-western Victoria. (Soil Conserv. Authority Victoria: Melbourne.)

Rudd, C. L. (1974). More about gypsum. *J. Agric. S. Aust.* **77**, 70-2.

Sims, H. J. and Rooney, D. R. (1965). Gypsum for difficult clay wheat-growing soils. *J. Dep. Agric. Vic.* **63**, 401-9.

Skene, J. K. M. and Poutsma, T. J. (1962). Soils and land-use in part of the Goulburn Valley, Victoria. Dep. Agric. Vic. Tech. Bull. No. 14.

Smith, G. D. (1977). Soil physical factors in land utilization in the lower Burdekin. Queensl. Dep. Prim. Ind. Agric. Chem. Branch Internal Report.

Smith, S. T. (1961). Soil salinity in Western Australia. *J. Agric. West. Aust.* (4) **2**, 757-60.

Talsma, T. (1963). The control of saline ground-water. *Meded. LandbHoogesh. Wageningen.* **63**, (10), 1-68.

Talsma, T. (1967). Leaching of tile-drained saline soils. *Aust. J. Soil Res.* **5**, 37-46.

Talsma, T. and Flint, S. E. (1958). Some factors determining the hydraulic conductivity of subsoils with special reference to tile drainage problems. *Soil Sci.* **85**, 198-206.

Thomas, J. E. (1939). An investigation of the problems of salt accumulation on a Mallee soil in the Murray Valley irrigation area. CSIR Aust. Bull. No. 128.

Trewhella, W. and Webster, A. (1978). Salinity control and drainage in the Shepparton and Kerang regions. Aust. Soc. Soil Sci. Inc. (Riverina Branch). Symp. 'The Hydrogeology of the Riverine plain of South East Australia'. (Eds R. R. Storrier and I. D. Kelly). pp.169-96.

United States Salinity Laboratory Staff (1954). Diagnosis and improvement of saline and alkali soils. U.S.D.A. Agric. Handb. No. 60.

van der Merwe, A. J. and Burger, R. DuT. (1969). The influence of exchangeable cations on certain physical properties of a saline-alkali soil. *Agrochemophysico* **1**, 63-6.

Chapter 54
Soil engineering in Australia

G. D. Aitchison

IN ANY DISCUSSION OF SOIL ENGINEERING IN A publication oriented toward pedology it is important that the difference between the engineering concept of 'soil' and the pedological concept of 'soil' should be fully appreciated.

The engineering concept of 'soil'

To an engineer, 'soil' is a material, usually within a complex array of such materials, upon which or within which or with which or adjacent to which engineering structures must be built, and, having been built, must retain an appropriate level of stability throughout the required period of utilization.

The engineer is concerned with the quantification of the forces and of the related displacements which may occur in the soil as a consequence both of engineering construction and of environmental influences, and with all forms of interaction between such forces and displacements in the soil and the associated forces and displacements in contiguous engineering structures. The range of engineering structures to be designed, each with proper attention to soil-structure interaction, is immense. Only a few examples can be given here: *Upon the soil*, there are road, airfield and railroad pavements; buildings (domestic, institutional and industrial) with shallow or deep footings; embankments and dams; *Within the soil*, there are tunnels and conduits (for transport or for location of services); open excavations (drainage ditches, open cut mines, etc.); basements and underground structures (car parks, etc.); *Construction with (or of) soil* relates to earth dams (for the retention of water or other fluids or semi-fluids); embankments (terrain modification for road or rail systems, etc.); fill (reclaimed areas etc.); *Construction problems from adjacent soils* relate to unstable landforms (natural or artificially created); overlap of soil displacements beyond the boundaries of a loaded site; etc.

The background to soil mechanics

Within the discipline of soil mechanics, on which the practices of soil engineering are based, soil is considered to be a multi-phase material, with each phase exhibiting a different stress-deformation response both in terms of magnitude and of rate. The stress-deformation behaviour of the soil as a whole is determined by the simultaneous interactions of the magnitudes and of the rates of response of all phases.

In the simplest mode, soil is treated as a two-phase material (Terzaghi 1943) in which a relatively incompressible fluid fills the pores of a relatively compressible 'skeleton' of mineral particles. Following the application of any load, volumetric deformation is initially limited to almost zero by the low compressibility of the pore fluid (water) and of the mineral particles themselves, with the load increment reflected in an increment of pressure in the pore fluid. If a drainage path is open to permit the dissipation of the increment of pore pressure, the soil will decrease in volume (consolidate) by an amount dependent on the stiffness of the mineral skeleton and at a rate dependent on the permeability of the soil. Sands will consolidate almost instantaneously while clays may require very long periods (decades or longer depending on the thickness of the clay layer and on the load increment) for consolidation to be complete. This simple model relates to saturated soils and is entirely adequate for those engineering purposes which apply to structurally stable soils in humid environments.

A further development of this model, intended to apply to unsaturated soils, involves three phases in which the soil pores contain both liquid and gas. This model is more relevant (Aitchison and Donald 1956) to most of the near surface soils of Australia. Due to the high compressibility of the gaseous phase within the pores such unsaturated soils exhibit relatively high values of initial volumetric deformation (compression) followed by lower values of consolidation. It is also significant that at equilibrium under any load the pore liquid pressures in unsaturated soils are less than the pressures in the gaseous phase (these normally tend towards ambient values approaching atmospheric pressure) and consequently drainage from such a soil can only occur to an external suction.

For a realistic comprehension of the stress-deformation behaviour of most soils it is necessary to consider a much more sophisticated model of the mineral 'skeleton', taking account of all of the recognizable fabric elements and of the factors determining their spatial distribution and orientation (Lafeber 1966). A 'soil mechanics model' involving some features of a complex fabric soil has been attempted (Aitchison 1960). Considerable attention has also been paid by engineers to the role of clay domains and to the influence of the constituents of the electrolyte on soil physical properties (Anon. 1960). However, while these topics represent a vital part of soil mechanics research, leading to the comprehension of soil engineering behaviour, they have not yet emerged as constituent components of soil engineering practice.

A brief comment on soil 'strength' may be appropriate. Soils, in common with other materials, will exhibit continually increasing deformations with increasing levels of stress, leading to ultimate rupture, usually in a plastic mode. In many stiff materials, such as steel or concrete, deformations at all levels significantly less than the failure load are quite tolerable and engineers are content to adopt the principle of a safe working load based upon the ultimate strength. Such is the case only in a few aspects of soil engineering, e.g. slope stability studies, while in most situations the soil engineer is concerned not with a safe working stress in the soil but with a limiting tolerable deformation in the soil, and in the structure interacting with the soil.

The depth of soil in an engineering sense

The depth of soil of concern to the engineer rarely coincides with the depth of soil of interest to the pedologist, although some exceptions are discussed later. To an engineer the significant depth of soil at any construction site is – *either* the whole depth of soil material to bedrock, or (if smaller) that depth of soil which is stressed significantly by the loads applied by the engineering structure. The larger the structure and the more heavily loaded the structure, the greater the depth of soil to be considered. As a simple guide to the 'significant' depth of soil it may be assumed that, beneath any footing or other areas of load application, a further depth amounting to $1\frac{1}{2}$ times the least dimension of the loaded area will be influenced by the load. Thus it is not uncommon for engineers to have to take cognisance of soil depths extending for many metres. In addition, it may well happen that environmentally induced deformations may occur at greater depths.

The range of engineering soils

The range of soils with which the engineer is concerned is not of itself particularly large. Indeed, in the disturbed or remoulded state, it is considered to be sufficient to recognize only 15 categories on the basis of textural and plasticity characteristics (Casagrande 1947). However, in the undisturbed state, other characteristics related to and determined by soil fabric (Lafeber 1964), stress-deformation history and the geochemical history are superimposed and are usually of dominant importance. Consequently, the number of effectively different categories of engineering soils may be very large indeed.

If cognisance is also taken of the normal lateral and vertical differentiations that occur in any soil mass, particularly over areas and throughout depths of concern to engineers, and if it is accepted that there is little or no logical relationship (in terms of engineering characteristics) between contiguous zones, horizons, layers or beds of soil occurrence, then it must be recognized that the number of possible combinations of categories and dimensions is virtually infinite. For this reason it is extremely rare, in engineering, for any combination of soil categories to be given identities (again, some exceptions are discussed later). Usual practice requires that each soil material should be described in quantitative physical terms and that the interaction between zones should be similarly quantified in stress-deformation terms.

Given this almost infinite array of potential combinations of soil materials, the question arises as to whether there is any simple basis for the comparison of the soils of separate continental land masses in engineering terms, and if so whether there are distinguishing features of Australian soils *vis-a-vis* the soils of other areas. Can soil engineering in Australia be seen as 'different' and if so, in what respects?

Engineering attributes of Australian soils

It is to be expected that the relatively unique geological and climatic history of Australia should have given rise to some relatively unique problems in soil engineering. However, the general practices of soil engineering in this country do not reflect a significant proportion of unusual patterns of soil behaviour and it has been found that the technology which originated in Europe and North America is widely adopted here.

Perhaps the reasons for the lack of emergence of an 'Australian' soil engineering technology can be stated: firstly, it is evident that a high degree of conservatism exists within the discipline and its practitioners and this tendency masks many features that may otherwise have been identified: secondly, there is no organized data bank or collated account of defined soil engineering experience throughout Australia: and thirdly, there is an enormous problem in the quantification and verification of the real engineering properties of any soil mass.

Terzaghi (1948), to whom the principal credit for the development of soil mechanics and soil engineering must be given, pointed out the obvious requirements for a proper definition of the parameters of soil engineering behaviour. These involved, in turn, a theoretical model for prediction of soil behaviour, a technique for measurement of each parameter within the model, the application of the model and the relevant measured parameters to a real situation, and the observation of the real performance of the structure. In the absence of the final stage of field observations no new development in soil engineering practice, however well supported by theory and measurement of soil parameters, can be considered to be fully acceptable.

Regrettably there are very few instances in Australia in which this desirable level of complete investigation has been approximated. Fraser (1976) has reported on some foundation studies on the interbedded sands of the Perth (W.A.) region: Aitchison *et al.* (1954) have reported studies of shallow foundations on expansive clays in the Adelaide region. Cox (1974) has reviewed geotechnical experience on the deeper clays of the Adelaide region.

The sum total of all of these limited studies is too meagre to permit the compilation of an overall pattern of engineering attributes of Australian soils based upon quantitative data. However, the fragmentary evidence that does exist is sufficient to permit an assessment of these attributes. Perhaps this assessment is best presented through comment on those manifestations of soil engineering behaviour that are of greatest concern world-wide, and on the frequency and severity of occurrence in Australia of these well-defined phenomena.

Soils of low strength and/or high deformability under modest load

Australia is fortunate in the fact that there are very few occurrences here of soils of the types that cause many of the world's most serious problems in foundation engineering.

'Quick clays', which are structurally unstable saturated soils of high void ratio which liquefy when strained and which are usually associated with recent glaciation, and which can contribute to catastrophic mud flows, are totally absent.

Underconsolidated *clays* which are the product of rapid deposition, and which are highly deformable have not been reported in Australia.

Clays of moderate to high *sensitivity*, which show a significant reduction in strength upon remoulding, occur to a limited extent only in the soft sediments of the drowned estuaries of the more sluggish rivers. Such soft sensitive clays are not a feature of dry-land soils.

Highly compressible volcanic ash deposits occur only in isolated pockets of no general significance.

Peats and highly organic soils occur only in extremely limited areas and are therefore of purely local significance.

Soils of moderate strength and low deformability

Most of the soils of Australia can be characterized as having moderate strength and relatively low deformability thus permitting the safe construction of most engineering structures following the adoption of rational design procedures.

Stiff clays

The vast areas of riverine and alluvial deposits that occur throughout Australia are composed principally of stiff clays. These stiff clays do not differ significantly in engineering properties from those occurring in other parts of the world. Cox (1970) has compared the London Clay of the U.K. with the 'Adelaide Clay' of South Australia and has demonstrated an essential similarity in response to applied loads (Fig. 54.1).

Soils of aeolian origin

Large areas of wind-deposited materials with widely varying particle composition occur in inland Australia.

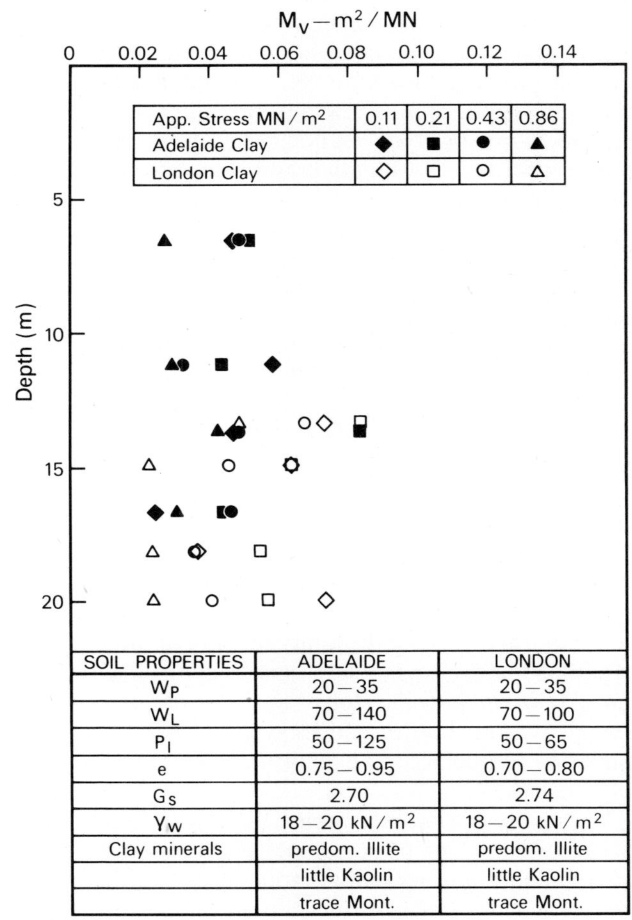

SOIL PROPERTIES	ADELAIDE	LONDON
W_P	20 – 35	20 – 35
W_L	70 – 140	70 – 100
P_I	50 – 125	50 – 65
e	0.75 – 0.95	0.70 – 0.80
G_s	2.70	2.74
γ_w	18 – 20 kN/m²	18 – 20 kN/m²
Clay minerals	predom. Illite	predom. Illite
	little Kaolin	little Kaolin
	trace Mont.	trace Mont.

Fig. 54.1. A comparison of volume compressibilities (M_v) and other general properties for the 'Adelaide' and London clays (after Cox 1970). In general the stiff fissured clays occurring widely in alluvial and riverine deposits in Australia are very similar in all normal stress-deformation properties to similarly described clays in the U.K. and U.S.

All such soils conform to the description of moderate strength and low deformability with the notable exception that most exhibit the phenomenon of 'collapse' when saturated while sustaining a critical load. The collapse behaviour of Australian soils has not been studied in any detail, although it can represent a serious engineering problem, which is, however, co fined to areas in which there is only a minor level of engineering activity. The limited data available on Australian collapsing soils (e.g. Parry 1960) suggest that overseas practices (Sultan 1969) would provide a conservative basis for engineering design.

Interbedded sands (with clay layers)

Coastal plains provide significant areas of interbedded sands and clays in which the sands are relatively loose and the clays firm to soft. McArthur (1970) has described the soils of the Swan coastal plain and Andrews (1971) and Fraser (1976) have defined the geotechnical characteristics of the interbedded sands of the Perth region within this plain. The overall engineering properties of these deposits conform to the general description of moderate strength and low deformability; and the engineering practices developed in Europe and North America for site investigation and foundation design on such soils are relevant.

Silcretes and ferricretes

There are widespread occurrences of indurated materials associated with past laterization. These silcretes and ferricretes have been used extensively in the construction of low-cost road pavements (Grant and Aitchison 1970). Although rarely required as foundation materials, these soils would exhibit high strength and low deformability.

Susceptibility of soils to environmental effects

Since most of the engineering soils of Australia possess the intrinsic characteristics of moderate strength and low deformability under applied loads, it might be assumed that the problems of soil engineering are limited, and are confined mainly to the few occurrences of softer and more deformable soils. However, this is not the situation for the reason that, over a large part of Australia, engineering soil behaviour is dominated by environmental influences.

With only a few exceptions, the Australian climatic environment is one of occasional aridity, in some places seasonal and in other places more or less continual. The impact of this climatic environment on soils is seen in two ways, desiccation and salinity.

Desiccation effects

Most of the soils of Australia are either permanently or occasionally in a state, of unsaturation (Aitchison 1956). Soil strength is higher in the unsaturated state than in the saturated state, while deformability is significantly lower. There are obvious engineering advantages in a sustained state of unsaturation in a loaded soil (e.g. most low cost roads depend for their continuing stability on such a sustained state in the subgrade) but there are concomitant engineering disadvantages when a pre-existing high level of unsaturation is reduced by wetting (e.g. many road pavement failures and many landslips in natural slopes attest to the dramatic influence of such a reduction in the state of unsaturation). The most widespread and obvious manifestation of a change in the state of unsaturation is the volume change (shrinkage or swelling) that occurs in clay soils. This poses a major engineering problem due to the fact that such volume changes may take place at any time in the post-construction period. Road pavements may distort seriously due to volume change in the subgrade, pipes and other conduits may fail due to diferential soil shrinking and swelling, and, of greatest apparent importance, buildings of all types may crack and distort due to soil movements beneath the footings.

Salinity effects in engineering soil behaviour

The widespread occurrences in Australia of saline soils have been reported by Northcote and Skene (1972). There are three principal directions in which the engineering behaviour of a soil is influenced by high levels of soluble salts or by high levels of electrolyte in the soil water. These are as follows.

The presence of soluble salts may cause an interaction with other additives in the processes of soil stabilization that is intended to form 'improved' materials for engineering construction purposes. High sulfate levels may be deleterious when Portland cement is used as an additive but there are many complex interactions which must be noted. Netterberg (1970) has discussed some of these interactions but very little evidence is available to demonstrate the presumed importance of such salinity effects in the stabilization of Australian soils.

Clay soils with high levels of sodium on the exchange complex (the sodic soils, p.845) pose a severe problem in the construction of water-retaining structures due to the potential for soil dispersion (in contact with water of low salinity) with consequential internal erosion or piping. This phenomenon has been observed to occur widely in most of the drier areas of Australia (Aitchison *et al.* 1963) leading to piping failures in a significant proportion of small earth dams (p.492). The engineering significance of this phenomenon lies in the need for adequate control of the compaction of such soils in earthen embankments, for it has been shown that there is a clear-cut interaction between the level of compaction and the consequent permeability and hence of the potential for post-construction deflocculation in such materials (Aitchison and Wood 1965).

A further feature of saline soils is reflected in the high

levels of solute suction that exist for considerable depths throughout the soil profile – with a consequential potential for serious swelling when the solute suction is reduced by interaction with another electrolyte. This point is discussed below in relation to the damage caused to buildings through heaves associated with this phenomenon.

Quantification of environmental influences

Environmental influences include water entry from rainfall or other causes; water removal by drainage, evaporation or evapotranspiration; increase of salinity by rise of watertable; decrease of salinity by leaching; changes of temperature, etc. All such influences must be expressed in the quantitative terminology of soil physics and soil mechanics.

It is obviously a matter of considerable importance for the engineer to comprehend the nature and magnitude of all environmental influences on soil stability and soil deformation, and to have available a technology to permit the quantification of all such influences and of all environmentally sensitive parameters of soil behaviour.

Such a technology is not yet fully established, although most of the procedures have been defined (Aitchison *et al.* 1973).

Soil suctions and engineering properties of soils

An attempt has been made to define the nature, extent and engineering significance of the circumstances of unsaturation in the soils of Australia (Aitchison 1956). A number of parameters may be involved in the quantification of a state of unsaturation in a soil but to an engineer the most meaningful is that of soil suction.

Soil suctions have been studied for many years within the discipline of soil physics (as reviewed by Marshall 1959) and the concepts of that discipline have been found to be applicable in engineering (Croney 1952; Aitchison 1956). A rather complex terminology has arisen to serve the many needs of soil physics, agriculture and soil engineering. For engineering purposes it is often necessary to quantify the two components of total suction separately, i.e. matrix suction and solute suction (Aitchison (ed.) 1965).

The pattern of desiccation in Australia has been expressed (Aitchison and Richards 1965) in pF terms (Schofield 1935), following a broad-scale study of the moisture status of soils beneath road pavements throughout the country. The range of equilibrium suction values recorded (from less than pF 2.5 to more than pF 5.5) is very wide and, to the engineer, indicates a wide range in values of soil strength and elastic modulus. Aitchison (1957) in field and laboratory studies, demonstrated a tenfold increase in the unconfined compressive strength of several clays through the range pF 2.5 to pF 5.0. Richards (1970), in reporting the results of studies of the effects of environment on road pavement design on a heavy clay subgrade in Queensland, demonstrated that the modulus of resilience of the subgrade increased by approximately 100% when the subgrade suction was increased from pF 3.9 to pF 4.4. The required pavement thickness (for a specific loading) decreased from 25 cm to 11.5 cm for the same increase in subgrade suction values.

Effective stresses and soil suctions

The most important single concept in soil mechanics is that of 'effective stress'. The effective stress in a soil is the difference between the applied stress (that due to load, either static or dynamic) and the pressure in the fluid within the pores in the soil. It is not appropriate in this text to discuss, in detail, the significance of 'effective stress' but it may suffice to state that not only is the shear strength of a soil determined directly by the effective stress operating in the soil but also any volume change in a compressible soil is determined by the relevant change in effective stress.

It has long been recognized on theoretical grounds

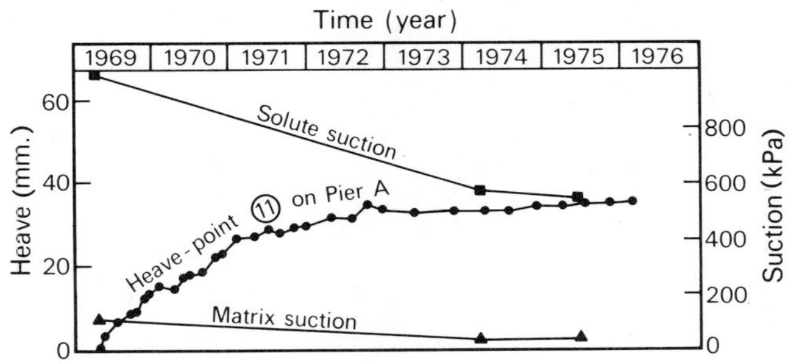

E.6. FIG. 2A.

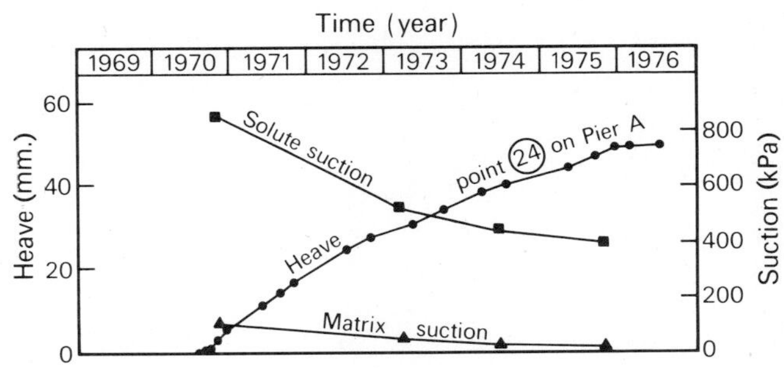

Fig. 54.2. Volume increases due to decreases of solute suction at considerable depth in the soil have been noted at a number of sites in Adelaide and have been shown to be the cause of serious distortions of buildings of all types. This problem could be widespread due to the (probable) general occurrence of high values of solute suction in many clay soils throughout Australia. (*a*, top). Time-movement of pier in relation to solute and matrix suction at Modbury High School. (*b*, bottom). Time-movement of a pier in relation to solute and matrix suction at Holden Hill Police Station.

(Terzaghi 1943) that a negative pore pressure, corresponding to a matrix suction, should contribute positively to effective stress and hence to strength and/or deformation; this point has been proved experimentally (Aitchison and Donald 1956; Skempton 1960).

More recently, it has been postulated (Aitchison 1965) that solute suction should also be a determinant of volume change in compressible soils. Experimental proof of this postulate has now been obtained. It has been clearly demonstrated (Aitchison *et al.* 1977) that a decrease in solute suction in a deep clay profile is associated with an increase in volume throughout the profile (see Fig. 54.2).

Despite the engineering significance of the two suction components of effective stress (i.e. matrix suction and solute suction) it is a regrettable fact that very little evidence is available anywhere in Australia (or elsewhere) to define these values.

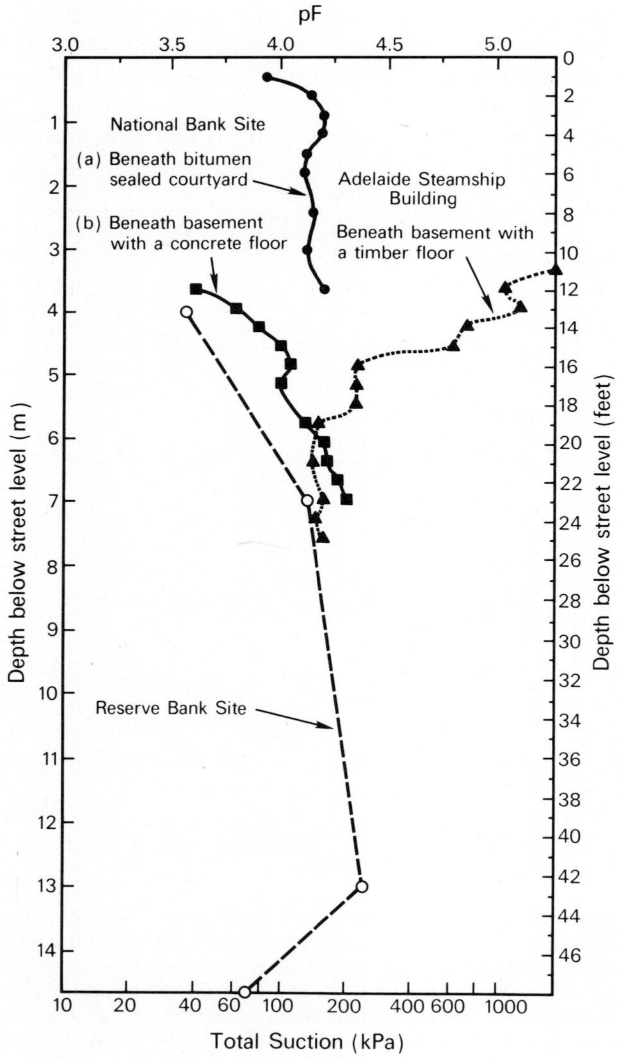

Fig. 54.3. Suction profiles in clay beneath major buildings in Adelaide, South Australia.

One of the few locations in which deep suction profiles are available is beneath the City of Adelaide (see Fig. 54.3).

Richards *et al.* (1970) have reported some data to depths of 4.5 m in clay soils of the Darling Downs, Queensland. Some further evidence was produced in a study of drought conditions in south-eastern Australia (Aitchison 1970).

All such data suggest that high values of total suction commonly exist to substantial depths in all soils of the semi arid to arid zones of Australia and also in many areas within the subhumid zones.

Such soil suctions may well be seen to be the major determinants of strength and of volume change in soils, and hence of most of the engineering attributes of Australian soils.

The lack of such suction data is to be regretted since not only do the few available figures suggest the need for engineering consideration but such high values may also be significant as constraints over the rate of moisture transfer in the soil-plant-atmosphere system.

The pragmatic approach

In the absence of accumulated data defining the values of solute and matrix suction and of the susceptibility of soils (in engineering terms) to changes in such suctions, it is rarely practicable to attempt to apply a complex technology involving the specific determination of such values for a specific engineering project.

Without diminishing the emphasis on the engineering desirability of a fund of quantitative data to permit the computation of the effect of a quantified environmental control on a quantified, intrinsic, environmentally sensitive soil property it is possible, and indeed unavoidable, for the engineer to adopt a pragmatic approach to the quantification of overall soil engineering behaviour. This involves simply the measurement of such environmentally influenced soil engineering properties as may be considered to be significant, and the establishment of a process of storage and retrieval of such information making use of any valid set of associated soil characteristics. An example of such a pragmatic approach to soil engineering was provided by the use of the pedological soil profile as the key to the engineering properties of the soils of the suburban areas of Adelaide, South Australia (Aitchison 1953).

The pedological soil profile in engineering

As mentioned above, it is relatively rare for engineering soil profiles to be given identities or for such profiles to correspond to pedological soil profiles. Some exceptions have been noted, particularly in those circumstances in which the form of engineering land use is such that the soil is stressed for only a shallow depth. Such is the case in the construction of road pavements and small buildings.

The relevance of the pedological soil profile as a key to the processes of engineering information storage and retrieval, pertaining specifically to road pavements, has been widely recognized in the U.S. (Anon. 1949, 1950, 1951) and later in Australia (Aitchison *et al.* 1962; Ingles and Metcalf 1972). The pedological soil profile also represents an important component of the PUCE system of terrain classifications (Grant 1973, 1974) which has wide application for all types of engineering land use.

Early attempts to interpret pedological soil surveys as the basis for foundation design for domestic buildings were largely intuitive and qualitative. Isaacs (1950) adopted a similar approach to the interpretation of Prescott's Soil Map of Australia (1944). Later work provided a quantitative base for the use of modified pedological soil classifications as an aid to foundation design for small buildings.

In the subsequent parts of this chapter, reference is made only to those studies in which an attempt has been made to fulfil the requirements of a conclusive investigation as emphasized by Terzaghi (see above).

Soil morphology and foundation engineering

In some parts of the urban and suburban region of the city of Adelaide (a total of approximately 900 km^2 situated in an uplifted alluvial plane between the Mt Lofty Range and St Vincent Gulf) domestic and institutional buildings, constructed traditionally with walls of brick or masonry, have had a history of serious foundation failures. It has been widely known for many years that buildings on clay soils have suffered unsightly cracking in walls (often exceeding 1 cm in width and occasionally reaching 5 cm) whereas on other soils similar structures have remained entirely sound.

Since so many structures were seriously affected by this phenomenon, a program of research was initiated in an attempt to define the specific causes of foundation failure, the types of soils contributing to such failure, the quantitative relationships (if any) between soil types and those soil properties affecting foundation performance, and finally the relationships (if any) between soil type and acceptable foundation practice. The detailed results of the study have been reported fully (Aitchison *et al.* 1954) and are the basis of the following discussion.

Foundation characteristics and the pedological soil profile

The cheapest and simplest foundation practice adopted (traditionally) within the study area was, and still is, the surface strip footing (see Fig. 54.4*). Such a footing, or equally a continuous concrete slab footing, on or slightly beneath the soil surface can be expected to reflect the whole of the physical response of the total soil profile to a change in environmental controls. Thus, if any layer tends to soften, due to wetting, and thus to yield under foundation loadings, the response of a building supported on a surface strip or slab should reflect the total of all such yields of all layers in the total soil profile. Similarly, if any layer should exhibit shrinkage or swelling behaviour, the response of a structure supported on a surface

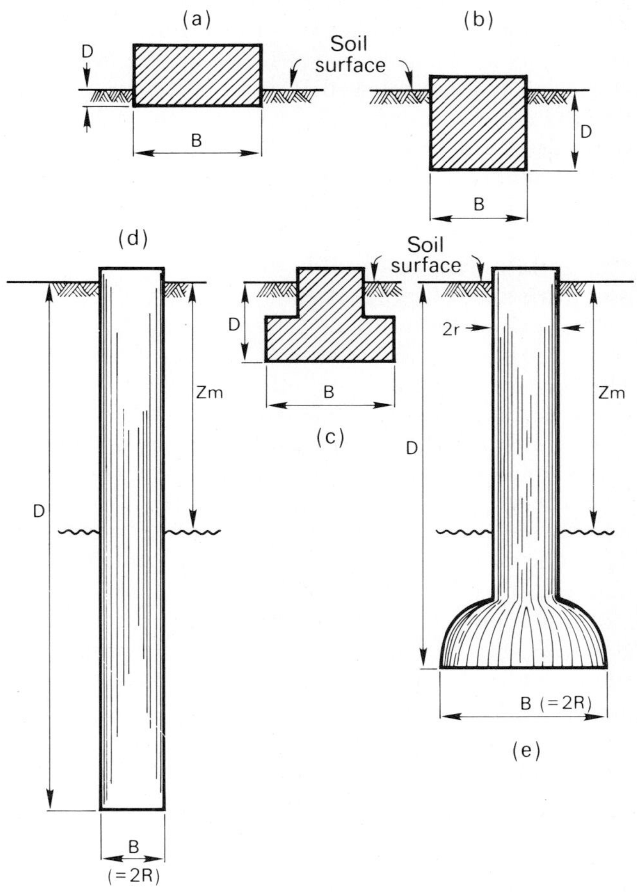

Fig. 54.4. Foundation practices on the soils of Adelaide and suburbs, (*a*), (*b*), and (*c*): variants of 'surface strip footings'; (*d*) bored pile; (*e*) under reamed pier. Types (*d*) and (*e*) support beams suspended clear of the soil surface. Z_m represents the depth of soil movement associated with a 'natural' soil moisture regime.

*Fig. 54.4 represents the foundation types which were current at the time of the investigation and against which observations of performance could be made. It is important to note that evidence of foundation performance can only be gained from records extending over many years. Hence, although new foundation designs may have been developed (e.g. Fargher and Steel 1969) and new techniques of foundation construction may have been postulated, the validity or otherwise of these new approaches has yet to be proven by long-term observations.

footing should reflect the total of all such shrinkage or swelling movements in all layers of the total soil profile.

Extensive observations of (pedological) soil profiles throughout the study area and of foundation performances throughout the same area showed that a high degree of correlation existed between foundation experience, as demonstrated by the severity of cracking in typical buildings supported upon surface footings, and the pedological soil profile (which was commonly taken to be not more than 2 m in effective depth).

By contrast, an alternative type of footing, the bored pile or pier-and-beam (see Fig.54.4), which penetrated the pedological soil profile, was observed to be physically insensitive to the various soils in which it was used. However, it was noted that certain design features, e.g. the minimum depth of the base of the pile or pier, were directly related to the soil profile or, in this case, to the depth of the lower boundary of the B (clay) horizon.

The pedological soil profile and intrinsic soil properties

The soils of the study area were widely divergent in characteristics and included sands, silts, marls and clays. The clays were dominantly illitic, but ranged from relatively inert types (which included significant proportions of kaolinitic clays) to highly expansive types (which included modest proportions of montmorillonite).

Soil surveys of the total area showed that a very large number of possible 'soil types' existed if a pedantic definition of pedological soil type was accepted. Quantitative evidence was then sought, in all profiles of apparent importance, to define the intrinsic volume change characteristics of each horizon and of the profile as a whole; and the intrinsic strength characteristics of each horizon and of the soil profile as a whole. Measurements of these intrinsic properties were made both in the field and in the laboratory. The values so obtained were used in part as the basis for justification of an engineering bias in the pedological classification which was subsequently adopted.

In principle, it was required that soils possessing significantly different intrinsic physical characteristics should be readily differentiated through the system of classification and identification to be adopted. The converse was not a demand on the system, although in the interest of efficiency in application it was clearly desirable that the number of 'types' with identical intrinsic properties should be a minimum.

In fact, it proved to be a simple matter to identify sensitive cases in which refined attention was required to separate one soil type from another. In all cases it was found that the usual morphological attributes of the soil profile were adequate to form the basis for separate identities.

The pedological soil profile and the environmental control of soil response

The actual physical response of a soil is, of course, the consequence of an interaction between the intrinsic properties of the soil and the controls exerted over these intrinsic properties by changes within the environment, principally associated with the seasonal moisture cycles.

It was logical to assume that since most, if not all, of the soils could be regarded as products of the current natural environment, the soil profile itself could be considered as an expression of that environment. Thus it could be postulated that significant changes in the natural environment should be clearly reflected in differences of soil profile, or conversely, and perhaps more importantly, that in the absence of any differences in soil profile, the environment could be considered to be unchanged.

This assumption was tested in a series of field experiments in which the soil moisture regimes (under a typical vegetative cover, and also under an adjacent building) were studied over a number of seasons. It was found that little or no change in soil moisture occurred at any depth below that at which the concentrations of soluble salts or of readily leached soil constituents suggested such absence of change. Not only did these experiments support the concept of an association between the natural seasonal moisture regime and the pedological profile, but it also appeared that those soils which differed significantly in intrinsic properties differed also in characteristic soil moisture regimes.

The concept of coupling

The conclusions from the general study mentioned above, *viz.* that different pedological soil types reflect different (and characteristic) foundation experiences; that different pedological soil types quantitatively reflect different intrinsic soil properties; that different pedological soil types quantitatively reflect different natural environmental controls over the intrinsic properties to create specific physical responses; and that the same differences of soil types serve in each case, all lead to one further conclusion, i.e. that the soil profile could serve as the basis for the quantitative extrapolation of knowledge basic to foundation design (which is based upon physical responses in the soil) or of foundation experience itself (see Aitchison 1953).

The fact that the intrinsic properties of the soil and the natural environmental controls over these properties are coupled together and both linked to the soil profile means that each separate soil profile identity holds a key to the transmission of foundation experience.

The soil identities

In seeking to use separate soil profile identities as the basis for communication of knowledge pertinent to foundation engineering, it is necessary to optimize the complete system of such communication. It is essential not only that cognisance be taken of the nature of the soil variations and of the level of sophistication required in the definition of soil properties, but also that proper attention be given to the process of communication itself. It would serve no purpose to present data and descriptions relating to the soils in terms meaningful only to pedologists, for the real requirement is to communicate with architects, engineers, geologists, builders, bankers and often the lay public.

These over-riding demands for simple communication led to the adoption of a codified classification reflecting certain Great Soil Group categories (Stephens 1956) with subdivisions based on soil-engineering characteristics.

The number of soil identities so defined was kept to the absolute minimum; 33 soil types were found to be adequate to cover all significant variations in this area, despite an extraordinary diversity of major profile forms.

Presentation, application and limitations

In the early stages of this study there was a clear-cut demand for a detailed map of soil occurrences. This supposed requirement followed naturally from the pragmatic interest of the users listed above. It is noteworthy, however, that this requirement was not met nor has there been any subsequent demand for such detail. The alternative, of course, lay in the adoption of a rational mapping unit coupled with a simplified procedure for recognition of soil identities.

The map involved the use of the concept of the soil association as the mapping unit. Within each soil association the number of soil types was limited so that the procedure for recognition would not be cumbersome.

Table 54.1. Example of soil type and foundation experience

Great Soil Group	Soil Type	Soil Morphology Principal Features of the Soil Profile	Foundation Experience on the Soil[A]		
			Incidence of Foundation Failure on the Soil	Soil Property Responsible for Failure of Foundation	Observed Satisfactory Foundation Practice
Red Brown Earths	RB 3	0–30cm A horizon: Brown silt loam. 30–85cm B horizon: Red brown clay with marked development of coarse prismatic structure. 85–300cm B-C horizon: Brown clay with lime. Concentration of lime diminishes with depth. Underlying soil is stiff fissured clay.	Failures are common. Cracks shown in about 50% of houses on this soil. Serious disfiguration occurs in less than 5% of normal dwellings.	Differential movements of the soil (as the result of seasonal shrinking and swelling of the clay horizons).	Pier-and-beam Piers must be effectively supported at considerable depth below soil surface. (Approx. 2.7–3.6 m is recent practice for bored piles.)
	RB 6	0–30cm A horizon: Brown loamy sand. 30–45cm B horizon: Reddish brown sandy clay with poorly developed coarse prismatic structure. 45–55cm B-C horizon: Light brown sandy clay with lime. 55–150cm Reddish brown sandy clay.	Failures are rare. Some minor cracks occur in about 5% of houses on this soil.	Differential settlement may occur if foundation loading is uneven. Water-table is shallow causing soft, saturated subsoils.	Continuous strip footing on the soil surface.
	RB 9	0–15cm Grey brown clay loam. 15–30cm Light grey mottled clay. 30–55cm Reddish brown clay of prismatic to columnar structure 55–60cm Light brown friable clay with moderate lime. 60–150cm Reddish brown friable clay.	Few failures have been observed since most houses on this soil are newly constructed. About 10% of older houses show minor to moderate cracking.	Some swelling and shrinking movements of small magnitude occur seasonally or as a result of changes of level of the water-table after underground drainage.	Pier-and-beam or rigid type of continuous strip footing on the soil surface. Short bored piles (up to 1.8 m long) have been used as piers.
Black Earths	BE 1	0–50cm Dark grey to black granular clay. 50–100cm Dark grey or brown or redbrown coarsely structured clay. 100–300cm Light brown clay with varying amounts of lime. Underlying soil is stiff fissured clay.	Failures are commonplace. Almost every house on this soil is cracked to some extent. About 10% of houses are noticeably disfigured.	The seasonal shrinking and swelling movements are large throughout the whole profile. In addition the soil exhibits wide lateral variation and differential movements are consequently accentuated.	Pier-and-beam. Short bored piles (1.2–2.4 m) commonly used but show some instability. Extended bored piles (3.0–4.5 m) or underreamed piers are suggested as superior alternatives.

[A] Foundation experience recorded relates to domestic type brick or masonry buildings with normal foundations — i.e. footings placed upon the soil surface or supported at shallow depth.

Eighteen associations were mapped involving a total of 33 soil types.

The information was presented to the user in the following manner:
the map with the accompanying text defined the soil association within which a limited number of soil types could occur;
each soil type was described by a coloured photograph of the complete profile and by further descriptions in the text;
the intrinsic properties and all relevant data were tabulated for each soil type; and
a summary was given of the profile characteristics of each soil type, the recorded foundation experience on the soil, and a suggested foundation practice to cope with the engineering properties and environmental controls (see Table 54.1).

The importance of two elementary aspects of this presentation cannot be over-emphasized. Firstly, there was a simply encoded descriptive term which could be easily memorized and used in discussion. Secondly, and most importantly, there was the coloured illustration of the profile presented with sufficient clarity for use by any member of the general public.

In use, a series of steps is required. From the map, the soil association and the possible range of soil identities are known. (This step may not be required). From an examination of a soil profile, preferably from an open pit but alternatively from an undisturbed continuous sequence of soil samples, the soil type can be recognized by matching with the coloured photograph, and occasionally by further reference to the type description. From the tabulated data (as in Table 54.1), the general nature of likely foundation problems with the soil can be understood and a recommendation obtained for a suitable foundation practice for a normal type of building.

The above steps can be taken, and indeed have been taken repeatedly and quite successfully over a period of 25 years by members of the general public, as well as by professionally qualified persons.

The concept developed above has been extended by further studies under the aegis of the S.A. Department of Mines with apparent success in application. (Taylor *et al.* 1974; Fargher and Steel 1969).

It is, nevertheless, an embarrassing necessity for this author, as the original proponent of the principle of coupling of the pedological identity to the environmental control as the basis for communication of foundation engineering experience, now to suggest the desirability of caution and, indeed, to indicate the importance of adopting, in certain circumstances, a somewhat different principle. The need for this new thinking has become obvious from experiences which have demonstrated that the environmental controls over intrinsic properties to produce physical responses in a totally urbanized environment are not always related, and, indeed, in exceptional circumstances may be totally unrelated, to those environmental controls which exist in a naturally vegetated area (and which are closely paralleled in an open-planned residential area with fully established gardens).

One early experience of this type involved the unsatisfactory performance of a block of flats erected on a black earth (designated BE1). All prior knowledge, as coupled to this soil type, suggested that foundation movements in a normal environment would not occur below a depth of 2.5 m. In actual fact the footings (or bored piles) were carried down to a greater depth (more than 4 m), but the structure, nevertheless, suffered serious distortions due to differential foundation movements of between 5 and 10 cm. As far as could be ascertained, these movements could be attributed to the modified moisture regime (at depths below 4 m) arising from increased water entry due to local concentrations of surface or drainage waters and to decreased water removal owing to the relative absence of vegetation.

A totally different type of inadequacy of the coupled soil-type foundation-practice system was observed to arise when excessive water entry – from broken or inadequate agricultural drains or from the exposure by surface grading of a permeable layer (and hence a potential aquifer) – caused leaching to such an extent that changes in the solute suction occurred. Volume increases due to solute suction decreases at zero matrix suctions represented a new form of response not accounted for in the coupled pedological soil-type foundation-experience system (see Aitchison *et al.* 1973). This point has been discussed above.

Intrinsic properties and environmental controls

All of the foundation-engineering implications of the various departures from the preceding rules of coupled behaviour can be quantified and handled in the language of soil mechanics, if the decision is taken to uncouple the intrinsic properties and the environmental controls. Both of these values, i.e. the intrinsic properties and the environmental controls, can be quantified (Aitchison *et al.* 1973) and the soil response can be computed e.g.

$$\Delta = \sum_{0}^{Z_m} I_p''.N''.dZ,$$

where Δ is the total vertical dimension change; I_p'' is the instability index (a measure of the stress-deformation response of the soil); N'' is the environmental control (a measure of the suction change in the soil; and Z_m is the depth of soil affected by the environmental control.

Values of the instability index I_p'' express the intrinsic properties of the soil *in situ*, and can be measured using appropriate laboratory techniques on undisturbed samples. Values of the environmental control N'' express the suction range to which the soil will be subjected and consequently it is through the varying values of N'', representing on the one hand the 'natural' moisture regime and on the other hand the 'artificial' or urbanized moisture regime, that different patterns of behaviour of the one soil can be defined. The principles involved in the prediction of the values of N'' follow those of basic soil physics. The estimation of the depth Z_m of soil influenced by a particular soil-moisture regime also accords with computations from basic soil-physics principles, but accurate input data for such computations are rarely available. In most localities it is not difficult to estimate a maximum value of Z_m, since factors such as the presence of a non-swelling layer ($I_p'' = 0$) or a permanent ground water table may be obvious determinants.

The whole of the technology required to implement the above approach has been described (Aitchison *et al.* 1973).

It is important, however, that note should be taken of the compatibility in practice of the coupled pedological soil-type/foundation-practice system and of the uncoupled specific computation method involving quantified values of instability indices and environmental controls.

As a first guide to foundation behaviour and desirable foundation practice, the coupled system following the pedological soil types and data as in Table 54.1 may be extremely valuable. A modified form of Table 54.1, expressing the susceptibility of the soil layers at greater depth, could be useful. Whenever there could be reason to anticipate a moisture regime other than the 'natural' one, the concept of coupling should be discarded and computations made as suggested above. Approximate values of instability indices may be associated with the various horizons of a soil type or alternatively the real values can be measured; each anticipated 'artificial' moisture regime may be introduced in terms firstly of the relevant 'environmental control' (N''), affecting the relevant horizon, and secondly the relevant depth of influence Z_m, and the consequential deformation Δ of the soil may be computed. It would then be the task of the building designer to match structural characteristics, and details of footings with the predicted pattern of soil behaviour.

Regrettably there is as yet no adequate documentation to assist in the process of combination of the advantages of the coupled and the uncoupled assessments of environmental controls of engineering behaviour. The coupled concept of soil morphology and foundation engineering continues to be applied at times in the Adelaide region while elsewhere in Australia the engineer lacks even this level of collated supporting data on soil properties.

Terrain evaluation for engineering

In any large country such as Australia the tasks of planning and of development for expanding urban areas and for major resource exploitation require knowledge of the earthen materials and of their properties over vast tracts of land. This knowledge should be in a form which permits early availability for feasibility studies but which also permits the later acquisition, storage and retrieval of all relevant geotechnical data. The procedures of engineering terrain evaluation have been developed for this purpose (Aitchison and Grant 1968) and within the space of 15 years have been widely adopted in Australia (see Fig. 54.5) for a multiplicity of purposes.

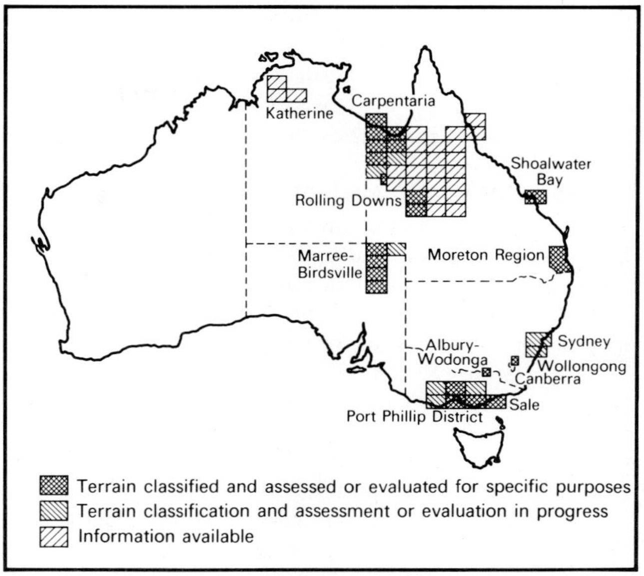

Fig. 54.5. Extent of terrain analysis for engineering and planning purposes in Australia and Papua New Guinea (from Grant and Finlayson 1978).

The earliest published works in the field of terrain evaluation were those of Christian and Stewart (1953) who developed a system of terrain evaluation, although it was not called so, as an aid to the development of Northern Australia; and Belcher (1943), who, at Purdue University, became interested in the spatial distribution of soils and other natural attributes that have a bearing on the engineering characteristics of terrain. Belcher (1942) was also an early advocate of the use of aerial photographs to determine these distributions.

A synthesis of these two approaches was achieved later, and independently, by Aitchison and Grant (1967) in Australia. They developed a system of terrain evaluation for general engineering and urban planning purposes. Beckett and Webster (1965) in England were interested in the extrapolation of terrain properties relevant to military use from known to unknown areas. Brink and Partridge (1967) in South Africa, considered aspects of terrain for road building and engineering geological purposes. Grabau (1963) in the U.S. attempted to make predictions of ground mobility for military purposes.

Many others, among whom are Dowling (1963) and

Clare and Beaven (1962) in England, Hittle (1949), Lueder (1951) and Miles (1962) in the U.S. and Kantey and Williams (1962) in South Africa, used some of the principles of terrain evaluation without, however, systematizing their approaches.

Principles of terrain evaluation

Terrain evaluation depends upon the ability to classify terrain and to evaluate for each member of each class the properties of the terrain that are significant for the desired purpose. Inherent in the process is the ability to predict terrain properties from known to unknown areas.

Many systems for classifying terrain for various purposes have been proposed. These range from landscape approaches, i.e. classifying the terrain in terms of its inherent characteristics, to parametric approaches in which the terrain is classified in terms of parameters significant for some given purpose (Grabau 1963).

The inherent difficulty in using the latter approach is that each purpose often requires parameters that are different, at least in some respect, from those required for any other purpose; hence each purpose requires not only its own terrain classification but also requires that classification to be based upon the parameters collected in a statistically significant number in a random manner. The predictive capability of such a system, based as it is on measured terrain properties, rather than naturalistic terrain attributes, is low.

On the other hand, the landscape approach assumes that parameters significant for any given purpose will be related to the inherent characteristics of the terrain; hence the terrain classification is completed at least in outline before the collection of parameters is attempted. The predictive capability of such a system is high. Experience has shown that the assumption, although difficult to prove, appears reasonable.

The various types of terrain evaluation systems available for use have been reviewed by Aitchison and Grant (1968).

The best documented and most widely used system of terrain evaluation is due to Grant (1973, 1974). This system, known as the PUCE (Pattern, Unit, Component, Evaluation) Programme for Terrain Evaluation, uses a combination of the landscape and parametric approaches for terrain classification. The parameters used are physical and relate only to the physiography of the terrain. The programme then considers methods of evaluation and makes provision for storing, processing and retrieving of information gathered both at the classification and evaluation stages.

Terrain classification

The PUCE programme divides all land into four basic classes.

1. Terrain component, defined rigidly in terms of its three-dimensional slope characteristics using two orthogonal slope axes; its soil, in which each layer can be classified into one class of the Unified Engineering Soil Classification (U.S. Dep. Interior 1963), and the whole profile into one class of its subdivided principal profile form (Northcote 1971); and its vegetation association. Terrain components can be mapped at a scale of 1:10 000 or greater.

2. Terrain units; physiographic features of the landscape (landforms) with consistent soil associations and vegetation formations. They consist of a limited repetitive association of a small number of terrain components. Terrain units are limited in number and can be listed. Terrain units can be mapped at a scale of 1:50 000 or greater.

3. Terrain Pattern; landscapes, which are usually coincident with a consistent pattern on an aerial photograph. Each terrain pattern has a consistent local relief amplitude and drainage pattern. They consist of a limited repetitive association of a small number of terrain units. Terrain patterns can be mapped at a scale of 1:250 000 or greater.

4. Provinces; consistent geology at the Group level (Anon. 1973). They consist of a limited repetitive association of a small number of terrain patterns.

The system of nomenclature used in the PUCE programme is essentially numerical and is entirely compatible with a digital computer for storing, processing and retrieving information.

Terrain components are allocated eight digits;

1, the types of slopes along the orthogonal axes, i.e. the combinations of convex, concave and planar slopes;

2,3, the maximum angles of slope along each of the orthogonal axes;

4,5, the soil (serial);

6, land use (in developed areas) surface cover (in undeveloped areas) (standardized by listing); and

7,8, the vegetation formation (serial).

Terrain units are allocated four units:

1,2, the landform (standardized by listing);

3, the soil (standardized by listing); and

4, the vegetation formation (standardized by listing).

Terrain patterns are allocated three digits:

1, the maximum local relief amplitude (classified using significant class intervals);

2, the stream density (classified using significant class intervals); and

3, serial.

Provinces are allocated five digits:

1, the geological era;

2, the geological period within each era; and

3,4,5, serial.

The terrain classification of any area is established under the PUCE programme by a process of aerial photograph interpretation using stereo pairs of photographs of suitable scale followed by a short period of field work to establish the soil and vegetation characteristics and to validate the interpretation of the members of each class and their boundaries. An example of terrain classification for the Melbourne area is given in Grant (1972).

Evaluation

The terrain evaluation system as specified in the programme is capable of working at three levels of detail consistent with the terrain classes, Terrain Pattern, Terrain Unit and Terrain Component. Thus, for initial broad-scale feasibility studies, provinces, terrain patterns and terrain units are recognized and evaluated; provinces and terrain patterns may be mapped. Over the more limited area defined by the feasibility study, terrain units may be mapped and evaluated in more detail for planning purposes. Finally, over a more limited area defined by the planning process, terrain components may be recognized, mapped, if necessary, and evaluated to the extent that construction can immediately proceed.

The technique of terrain evaluation is equally applicable to the planning of new developments in the open countryside or of re-development of city areas that are no longer viable in their present context within the city scene; each type of development, of course, imposes its own constraints on the purposes of and methods used for evaluation. In areas of new development, the number of degrees of freedom for location is generally large; in areas of re-development the number of degrees of freedom is mostly small.

The evaluation stage, in general, consists of specifying the properties of the terrain and its attributes that are critical for the desired purpose, and for those properties, the level of detail required to be measured or estimated for each phase in the feasibility, planning or construction process. The evaluation continues by measuring or estimating and recording the values of these necessary parameters in a sufficient number of terrain samples in terms of the level in the terrain classification appropriate to them, and then extrapolating to all occurrences of all members of all classes, the properties of each member of each class when these properties have been established by testing a significant number of terrain samples.

While the terrain classification stage in the process of terrain evaluation can be done in a standardized, ordered form, evaluation is always tied to a particular objective whether large or small, which, in general, requires the collection of a set of information different from the set required for any other objective. This is true whether broad-scale purposes or more detailed projects are under consideration. In the engineering sense, the required information can be considered as existing in three sets. One set of information is required for feasibility studies, another, more detailed set for planning and another, yet more detailed set for construction purposes. These sets of information may overlap and contain some items in common. For all purposes, there is virtually an unlimited number of sets of information required; such a situation defies any attempt at overall standardization of procedures, although the procedures for the collection of some items may be standardized.

Grant and Finlayson (1978) have described a number of the evaluations that have been published for some of the areas indicated in Fig. 54.5. Although these are, as yet, assessments of the real engineering behaviour of the Terrain Unit, still lacking the support of field observations to demonstrate the validity of the conclusions offered, they are based in many instances on the parameters often adopted in engineering design.

Handling terrain information

One of the advantages of using a terrain evaluation system for collecting information about the terrain is that, provided the system is properly constructed, the information obtained may be stored, processed and retrieved for future use. The actual system for information processing must necessarily be of an incremental type in which it is possible to add information continually as it is collected and retrieve data as they are required.

A key to a successful system for information processing is a method for encoding the information presented to the system. Encoding ensures that information is always presented to the system in a uniform manner. Because terminology is standardized, there is no duplication of information and no difficulites are experienced in interpreting the precise meaning of every item placed in the system.

The PUCE Programme is designed in such a way that it is compatible with a digital computer data processing medium. Other manual methods of data processing, however, may also be used.

Geomechanics in Australia

It would not be proper to imply that the state of the art of basic and applied soil and rock mechanics in Australia is adequately defined above. There is indeed a very comprehensive expertise in such subjects in this country, and necessarily so, for each major engineering project or resource development imposes a demand for investigation and design involving the best available knowledge worldwide. Such expertise therefore tends to be international in standard and character, and brief comment in this chapter would be futile.

The comments in the above paragraphs are merely intended as a commentary on those aspects of soil engineering in Australia that have an interface with other disciplines in soil science. Regrettably these contacts are all too few.

References

Aitchison, G. D. (1953). Soil morphology and foundation engineering. Proc. 3rd Int. Conf. Soil Mech. Found. Eng. Vol. 1, pp.3-7.

Aitchison, G. D. (1956). The circumstances of unsaturation in soils with particular reference to the Australian environment. Proc. 2nd Aust. N.Z. Conf. Soil Mech. Found. Eng. pp.173-92.

Aitchison, G. D. (1957).The strength of quasi saturated and unsaturated soils in relation to the pressure deficiency in the pore water. Proc. 4th Int. Conf. Soil Mech. Found. Eng. Vol. VI, pp.135-39.

Aitchison, G. D. (1960). Effective stresses in multi-phase systems. Proc. 3rd Aust.-N.Z. Conf. Soil Mech. Found. Eng. pp. 209-12.

Aitchison, G. D. (1965). Soil properties, shear strength and consolidation. Panel discussion to Session 2, Division 2. Proc. 6th Int. Conf. Soil Mech. Found. Eng. Vol. 3, pp.318-21.

Aitchison, G. D. (Ed.) (1965). 'Moisture Equilibria and Moisture Changes in Soils Beneath Covered Areas'. Symposium-in-print (Butterworths: Aust.) (In print).

Aitchison, G. D. (1970). Soil aridity during occasional drought in south eastern Australia. Proc. Symp. Soil Earth Structures in Arid Climates, pp.93-7.

Aitchison, G. D. and Donald, I. B. (1956). Effective stresses in unsaturated soils. Proc. 2nd Aust. N. Z. Conf. Soil Mech. Found. Eng. pp.192-99.

Aitchison, G. D. and Grant, K. (1967). The P.U.C.E. Programme of terrain description, evaluation and interpretation for engineering purposes. Proc. 4th Reg. Conf. Africa Soil Mech. Found. Eng. Vol. 1, pp.1-8.

Aitchison, G. D. and Grant, K. (1968). Terrain Evaluation for Engineering. *In* 'Land Evaluation'. (Ed. G. A. Stewart). (Macmillan: Melbourne.)

Aitchison, G. D. and Richards, B. G. (1965). 'A Broad Scale Study of Moisture Conditions in Pavement Subgrades Throughout Australia'. Symp.-in-print: Moisture Equilibria and Moisture Changes in Soils Beneath Covered Areas. pp.184-232. (Butterworths: Melbourne.)

Aitchison, G. D. and Wood, C. C. (1965). Some interactions of compaction, permeability and post-construction deflocculation affecting the probability of piping failure in small earth dams. Proc. 6th Int. Conf. Soil Mech. Found. Eng. Vol. 2, pp.442-46.

Aitchison, G. D., Sprigg, R. C. and Cochrane, G. W. (1954). The Soils and Geology of Adelaide and Suburbs. Dep. Mines S. Aust. Bull. No. 32.

Aitchison, G. D., Metcalf, J. B. and Richards, B. G. (1962). The pedological pattern of soil occurrences as the basis for an Australia-wide study of the engineering characteristics of soils in relation to pavement design and construction. Proc. 1st Conf. Aust. Road Res. Brd. Vol. 1, pp.864-92.

Aitchison, G. D., Ingles, O. G. and Wood, C. C. (1963). Post-construction deflocculation as a contributory factor in the failure of earth dams. Proc. 4th Aust. N.Z. Conf. Soil Mech. Found Eng. pp.275-79.

Aitchison, G. D., Peter, P. and Martin, R. (1973). Quantitative description of the stress-deformation behaviour of expansive soils. Proc. 3rd Int. Conf. Expansive Soils, Papers. pp.1-8.

Aitchison, G. D., Peter, P. and Power, R. F. (1977). Stress-deformation mechanisms in clays as determinants of design requirements for small and large structures — an evaluation through case histories in Adelaide of the relative significance of several mechanisms. Proc. IEA/Aust. Geomech. Soc. Symp. on How to Live with Geotechnical Risks, Sydney, pp.26-33.

Andrews, D. C. (1971). Soils of the Perth Area — The City Centre. CSIRO Aust. Div. Appl. Geomech. Tech. Rep. No. 13.

Anon. (1949). High. Res. Board, Bull. No. 22.

Anon. (1950). High. Res. Board, Bull. No. 28.

Anon. (1951). High. res. Board, Bull. No. 46.

Anon. (1960). Seminar on Interparticle Force in Clay-Water-Electrolyte Systems. CSIRO Soil Mech. Section and Civ. Eng. Dep. Univ. Melbourne 1v.

Beckett, P. H. T.and Webster, R. (1965). 'A classification system for terrain'. Military Eng. Exp. Establ., Christchurch, England, Report 872.

Belcher, D. J. (1942). 'Use of aerial photographs in war time engineering'. Rds. Str. 85, 7, 35.

Belcher, D. J. (1943). 'The engineering significance of soil patterns'. High. Res. Board, Proc. Annu. Meet. No. 23, pp.569.

Brink, A. B. A. and Partridge, T. C. (1967). Kyalami land system: An example of physiographic classification for the storage of terrain data. Proc. 4th Reg. Conf. Africa Soil Mech. Found. Eng. Vol. 1, pp.9.

Casagrande, A. (1947). Classification and Identification of Soils. *Proc. Am. Soc. Civ. Eng.* **113**, 901.

Christian, C. S. and Stewart, G. A. (1953). General report on survey of the Katherine-Darwin region, 1946. CSIRO Aust. Land Res. Ser. No. 1.

Clare, K. E. and Beavan, P. J. (1962). Roadmaking materials in British Borneo; 1. Geology, roadstone and soil classification. Lab. Note No. LN/86. 11. Properties of soils. Lab. Note No. LN/87. 111. Soil location and stabilization, general conclusions. Lab. Note No. LN/88. Road Research Laboratory, D.S.I.R., Harmondsworth.

Cox, J. B. (1970). A review of the geotechnical characteristics of the soils in the Adelaide city area. Proc. Symp. Soils and Earth Structures in Arid Climate, pp.72-86.

Croney, D. (1952). The movement and distribution of water in soils. *Geotechnique* **3**, 1-16.

Dowling, . W. F. (1963). The use of aerial photography in evaluating engineering soil conditions in a selected area of northern Nigeria. Lab. Note LN/379. Road Research Laboratory, D.S.I.R., Harmondsworth.

Fargher, P. and Steel, R. D. (1969). Footings for use on South Australian Soils. S. Aust. Dep. Mines Geol. Survey.

Fraser, R. A. (1976). Review of investigations on raft foundations for major buildings on deep soils, Perth Area. Proc. Symp. Raft Found. Perth, pp.23-40.

Grabau, W. E. (1963). A comparison of quantitative versus descriptive systems for mobility analysis. (U.S. Army Engineer Waterways Experiment Station: Vicksburg.)

Grant, K. (1972). Terrain classification for engineering purposes of the Melbourne area, Victoria. CSIRO Aust. Div. Appl. Geomech. Tech. Pap. No. 11.

Grant, K. (1973). The P.U.C.E. Programme for Terrain Evaluation — 1. Principles. CSIRO Aust. Div. Appl. Geomech. Tech. Pap. No. 15.

Grant, K. (1974). The P.U.C.E. Programme for Terrain Evaluation — 11. Procedures for Terrain Classification. CSIRO Aust. Div. Appl. Geomech. Tech. Pap. No. 19.

Grant, K. and Aitchison, G. D. (1970). Engineering Significance of Silcretes and Ferricretes in Australia. *Eng. Geol.* **4**, 93-120.

Grant, K. and Finlayson, A. A. (1978). The assessment and evaluation of geotechnical resources in urban or regional environments. *Eng. Geol.* **12**, 219-93.

Hittle, J. E. (1949). Airphoto interpretation of engineering sites and materials. *Photogramm. Eng.* **15**, 589.

Ingles, O. G. and Metcalf, J. B. (1972). 'Soil Stabilisation'. pp.44-51. (Butterworths: Melbourne.)

Isaacs, D. W. (1950). Foundation soils in Australia. D.D. 31 Com. Exp. Building Station, Sydney.

Kantey, B. A. and Williams, A. A. B. (1962). The use of soil engineering maps for road projects. *Civ. Engr S. Afr.* **4**, 149 and **5**, 41.

Lafeber, D. (1964). Soil fabric and soil mechanics *In* 'Soil Micromorphology'. (Ed. A. Jongerius). pp.351-60. (Elsevier: New York.)

Lafeber, D. (1966). Soil Structural Concepts. *Eng. Geol.* **1**, 261-69.

Lueder, D. R. (1951). The preparation of an engineering soil map of New Jersey. Symp. on Surface and Subsurface Reconnaissance. ASTM STP 122, 73.

McArthur, W. M. (1970). Soils of the Swan coastal plain. Proc. Symp. Soils and Earth Structures in Arid Climates, pp.107-110.

Marshall, T. J. (1959). Relations between water and soil. Comm. Bur. Soils. Tech. Comm. No. 50.

Miles, R. D. (1962). A concept of land forms, parent materials and soils in airphoto interpretation studies for engineering purposes. *In* 'Transactions of the Symposium on Photo Interpretation'. pp.462-76. (International Society for Photogrammetry: Delft.)
Netterberg, G. (1970). Occurrence and testing for deleterious salts in road construction materials with particular reference to calcretes. Proc. Symp. Soils and Earth Structures in Arid Climates, pp.87-92.
Northcote, K. H. (1971). 'A Factual Key for the Recognition of Australian Soils'. 3rd edn. (Rellim: Adelaide.)
Northcote, K. H. and Skene, J. K. M. (1972). Australian Soils with Saline and Sodic Properties. CSIRO Aust. Soil Pub. No. 27.
Parry, R. H. G. (1960). Consolidation and swelling properties of soils in the Victorian Mallee. Proc. 3rd Aust. New Zealand Conf. Soil Mech. Found. Eng. pp.73-80.
Prescott, J. A. (1944). A Soil Map of Australia. CSIR Aust. Bull. No. 17.
Richards, B. G. (1970). Effects of environment on pavement design. Proc. Symp. Soils and Earth Structures in Arid Climates, pp.20-24.
Richards, B. G., Murphy, J. W., Chan, C. Y. L. and Gordon, R. (1970). Preliminary observations on soil moisture and 'dry' compaction in pavement design on the Darling Downs, Queensland. Proc. 5th Aust. Rd Res. Brd Conf. Vol. 5(5), pp.116-46.
Russam, K. and Coleman, J. D. (1961). The effect of climatic factors on subgrade moisture conditions. *Geotechnique* **11**, 1-22.
Schofield, R. K. (1935). The pF of water in soil. Trans. 3rd Int. Congr. Soil Sci. Vol. 2, pp.37-48.
Skempton, A. W. (1960). Opening address: Conf. Pore Pressure and Suction in Soils. (Butterworth: Melbourne.)
Stephens, C. G. (1956). 'A Manual of Australian Soils'. 2nd edn. (CSIRO:Melbourne.)
Sultan, H. A. (1969). Collapsing Soils – State-of-the-Art Report 5th Speciality Session, VIIth Int. Conf. Soil Mech. Found. Eng., Mexico City.
Taylor, J. K., Thomson, B. P. and Shepherd, R. C. (1974). The Soils and Geology of the Adelaide Area. Geol. Survey S. Aust. Bull. 46.
Terzaghi, K. (1943). 'Theoretical Soil Mechanics'. (Wiley: London.)
Terzaghi, K. (1948). Presidential Address. 2nd Int. Conf. Soil Mech. Found. Eng. Rotterdam.
United States Department of the Interior, Bureau of Reclamation, 1963, 'Earth Manual'. 1st edn.

Epilogue

THE EARLY YEARS OF AUSTRALIAN SOIL RESEARCH WERE mainly characterized by scientific studies which sought to establish basic principles rather than tackle the detail of site-specific problems in agriculture, forestry and hydrology that daily distract the farmers and extension workers. This led to the view that Soil Science was somewhat isolated from the mainstream of agricultural science and its major Australian advances in plant breeding, trace element research and crop and pasture ecology. The isolation was more apparent than real. It should now be obvious from the present volume that, while the scope of Australian soil research has widened appreciably during its first 50 years, the search for basic principles has continued. This has not, however, prevented active collaboration with research workers interested in more practical objectives and it is to be hoped that the next half-century will see soil research fully integrated with other related natural sciences.

We also see from this book that a pedological and geomorphic framework or skeleton has been provided for our soil studies—still incomplete but with its outlines clearly manifest. Some flesh has been added by detailed field and laboratory research in soil chemistry, mineralogy, and to some extent in soil fertility and soil engineering. A great deal of new knowledge has been added in soil physics and biology, but so far without any particular reference to this pedological skeleton.

It is likely, by comparison with other developed countries, that we have suffered from too few workers and too much land, to the extent that our soil surveys are mostly large-scale, reconnaissance exercises, designed as initial soil and landscape appraisals that were not always followed up. Thus our attempts to devise an Australian system of soil classification, or to embrace the various overseas models, never attained generality because of the gaps and the approximations that, by necessity, we have had to live with. While soil classification is not an end in itself, it does provide an ordered system for the recognition of soils, to which other soil knowledge can be related. Australian opinions on this subject have been ambivalent and controversial, preventing the emergence of a unified system.

To be complete, statements on soils must have generality and the pedological framework is one important means of establishing generalizations, predictions and bases for planning; more importantly, of raising Soil Science to the disciplinary status accorded to it by its more vocal and enthusiastic practitioners. Possibly any pedological framework cannot usefully accommodate all the multi-disciplinary properties of soils, but in Australia at least this should be attempted. It is apparent from the contents of this volume that wide gaps still exist: soil-climate relations, i.e. the behaviour of particular soil classes under different climates; the provision of soil–water properties of our major soil types—essential in a country so dependent on rainfall; the problems raised by soil heterogeneity; and the influence of these, if any, on the ecological relationships of soil organisms and of economic plants.

Nevertheless, the Australian contribution to Soil Science has been substantial and in many ways pioneering. It is therefore appropriate to speculate, in the closing pages of this book, on the substantial problems that remain: the costs of research and development are escalating, Australia currently employs only about 400 research workers in Soil Science to cover its seven million square kilometres of land surface, and the resources required to halt the degradation of our landscapes completely appear unobtainable. For climatic reasons, there are few areas left in Australia for intensive agricultural development, and it is clear that new irrigation schemes, always a political temptation in arid regions, will produce their own special problems. To retain economic viability, Australia must upgrade its land management practices while ensuring landscape stability. We need, not only more research and extension, but better coordination of effort at all levels of government and private enterprise. In some instances, new legislation will be required in order to implement correct management and to protect our soils and water supplies.

However, there are encouraging signs that our predominantly urban community is becoming aware of the significance and fragility of our soil resource and there is increasing public concern with environmental issues such as salinity of water supplies, soil erosion and the impact of the resource boom on the quality of life. Soil and landscape studies are basic to these problems and, apart from its scientific record, we hope this account will provide new frames of reference for our users of land and convince our research planners of the vital importance of what lies under their feet.

A. E. Martin

Author index

Australian localities index

Subject index

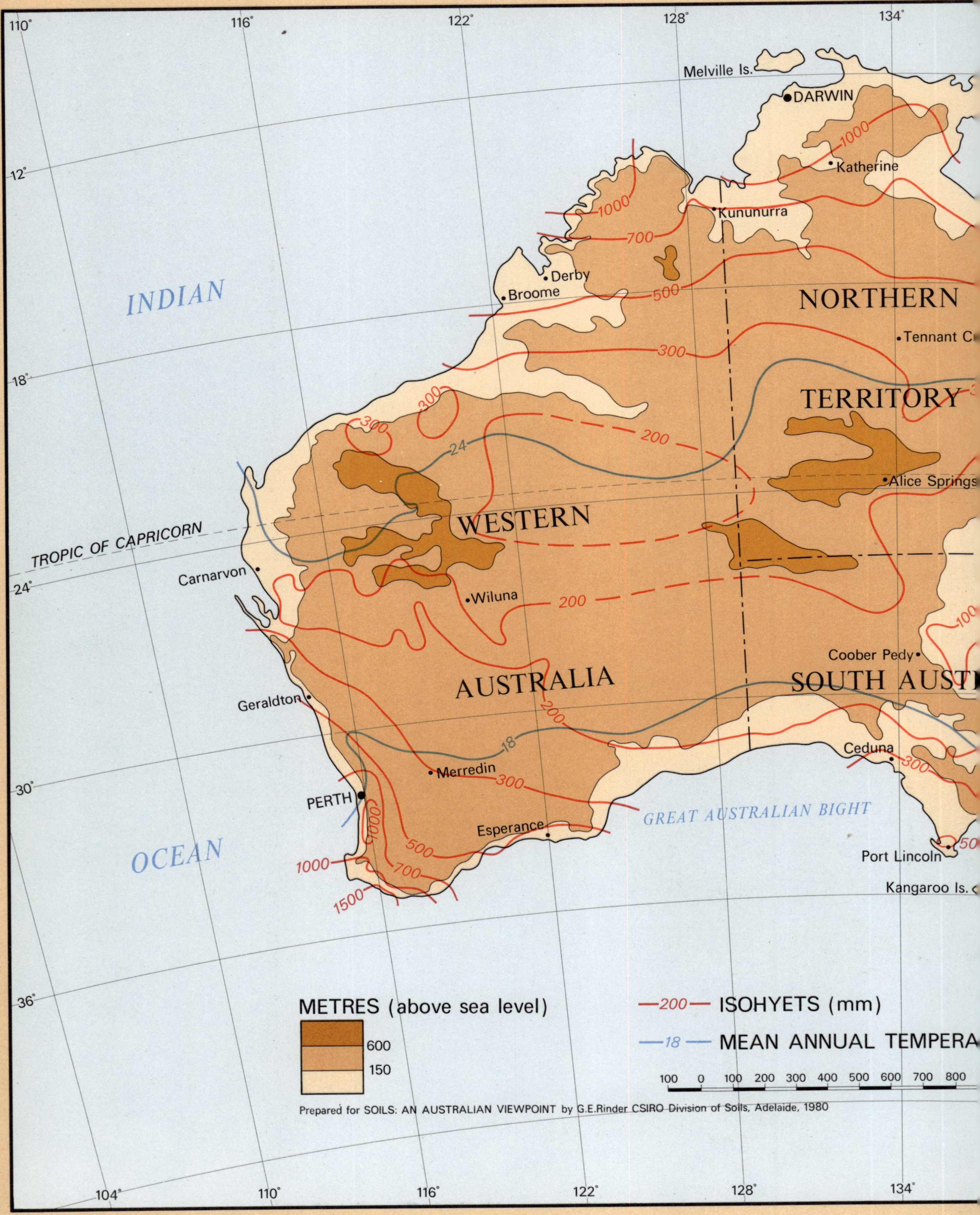

INDIAN
OCEAN
Melville Is.
DARWIN
Katherine
Kununurra
Derby
Broome
NORTHERN
TERRITORY
Tennant C
Alice Springs
WESTERN
AUSTRALIA
TROPIC OF CAPRICORN
Carnarvon
Wiluna
Coober Pedy
SOUTH AUST
Geraldton
Ceduna
Merredin
PERTH
Esperance
GREAT AUSTRALIAN BIGHT
Port Lincoln
Kangaroo Is.
METRES (above sea level)
600
150
200 ISOHYETS (mm)
18 MEAN ANNUAL TEMPERA
100 0 100 200 300 400 500 600 700 800
Prepared for SOILS: AN AUSTRALIAN VIEWPOINT by G.E.Rinder CSIRO Division of Soils, Adelaide, 1980